# Chiral Intermediates

# Chiral Intermediates

Edited by

Cynthia A. Challener

Ashgate

Published by
Ashgate Publishing Limited
Gower House
Croft Road
Aldershot
Hampshire GU11 3HR
England

Ashgate Publishing Company
131 Main Street
Burlington, VT 05401-5600 USA

British Library Cataloguing in Publication Data
Chiral intermediates
1. Chiral drugs
I. Challener, Cynthia A.
615.1'9
ISBN 0 566 08412 0

Library of Congress Control Number: 2001090547

Printed in Great Britain by MPG Books Ltd, Bodmin.

# CONTENTS

# PREFACE

Chiral molecules are ubiquitous in nature and have ever-increasing importance in the pharmaceutical, agrochemical, electronic, food, and flavor and fragrance industries. In recent years, several valuable texts have been published that cover in detail the various sources and methods of preparation for obtaining optically active materials.

There has been a need, however, for a guide for workers in the pharmaceutical and chemical industries seeking information on chiral molecules, processes, and commercially available chiral chemicals. The goal of this book is to present the chemical professional with a comprehensive listing of available chiral chemicals including specific data of interest for each entry in the listing.

Part I of the book, divided into four chapters, provides an introduction to topics relevant to the field of chiral chemistry and include a brief overview of chirality, a short discussion on the current market drivers in the area of chiral chemistry, and a basic presentation of the various sources and methods for obtaining chiral compounds.

This book will provide an introduction to the types of sources and methods currently in use for obtaining chiral molecules and will prove to be an invaluable resource for information on available chiral molecules. The reader is encouraged to investigate other sources for more detailed information on the technology and processes utilized for identifying, isolating, and preparing chiral molecules.

# FURTHER READING

Ager, D.J., (Ed.), *Handbook of Chiral Chemicals*, Marcel Dekker, Inc., New York, 1999.

Collins, A.N., Sheldrake, G.N., and Crosby, J., (Eds.), *Chirality in Industry*, John Wiley & Sons, New York, 1992.

Collins, A.N., Sheldrake, G.N., and Crosby, J., (Eds.), *Chirality in Industry II*, John Wiley & Sons, New York, 1997.

Sheldon, R.A., *Chirotechnology*, Marcel Dekker, Inc., New York, 1993.

# ACKNOWLEDGEMENTS

The Editor would like to acknowledge with the greatest appreciation the assistance and editorial expertise provided by Dr. Ellen Zeman. The Editor would also like to acknowledge the guidance provided by Dr. G.W.A. Milne through his multiple reviews of the text that appears in this book and also his work on collecting and creating many of the structures included in the listing. The efforts of Dr. Anthony B. Mauger and Dr. Marianne G. Patch in identifying and creating the structures are also appreciated. Lastly, the Editor would like to acknowledge Peter Nielsen for providing the opportunity to do this project.

# HOW TO USE THIS BOOK

*Chiral Intermediates* is divided into three parts. A brief description of each part is given below.

**PART I**

The four chapters in this part, entitled Chirality, provide an introduction to chiral chemistry. Chapter One, Overview of Chirality, introduces the reader to the definition of chirality, the importance of optical isomerism in chemistry and life science, issues involved in controlling chirality in synthesis, and methods for identifying the optical purity of a sample. Chapter Two, Drivers for the Chiral Market, describes key market issues, regulatory considerations, and recent technological developments in the field. Chapter Three, Sources of Chiral Compounds, discusses where the researcher can obtain chiral starting materials and derivatives as well as how to resolve racemic mixtures. Chapter Four, Methodologies for Obtaining Chiral Compounds, reviews methods for isolating optically active compounds as well as synthetic strategies.

**PART II**

The main entries in this part provide a comprehensive list of available chiral intermediates in alphabetical order. Each record is identical in structure, enabling the reader to select specific information efficiently. A unique record number has been assigned to every record. The three indexes in Part III allow quick cross-referencing according to the record number in Part II by CAS Number, EINECS number, or synonym. The Manufacturer and Supplier Directory in Part IV provides convenient access to information on where and how to obtain the chiral compound of interest.

## Record Structure

A typical record from the entries section of this book is shown below. The first line contains, in bold face, the record number (67) and the name of the material (N-Acetyl-β-D-glucosamine). The second line gives the Chemical Abstracts Service (CAS) Registry Number for the compound (7512-17-6), the corresponding *Merck Index* number [4466(11), where 11 represents the eleventh edition] and the European Inventory of Existing Commercial Chemical Substances (EINECS) number (231-368-2). These numbers always appear in the same position (left, center or right) enabling the reader to determine which source they belong to. Whenever CAS Registry Numbers are used in the text, they are always enclosed in brackets, for example [7512-17-6]. The molecular formula and structure of the compound are provided. A list of synonyms follows, including proprietary names and other trivial names.

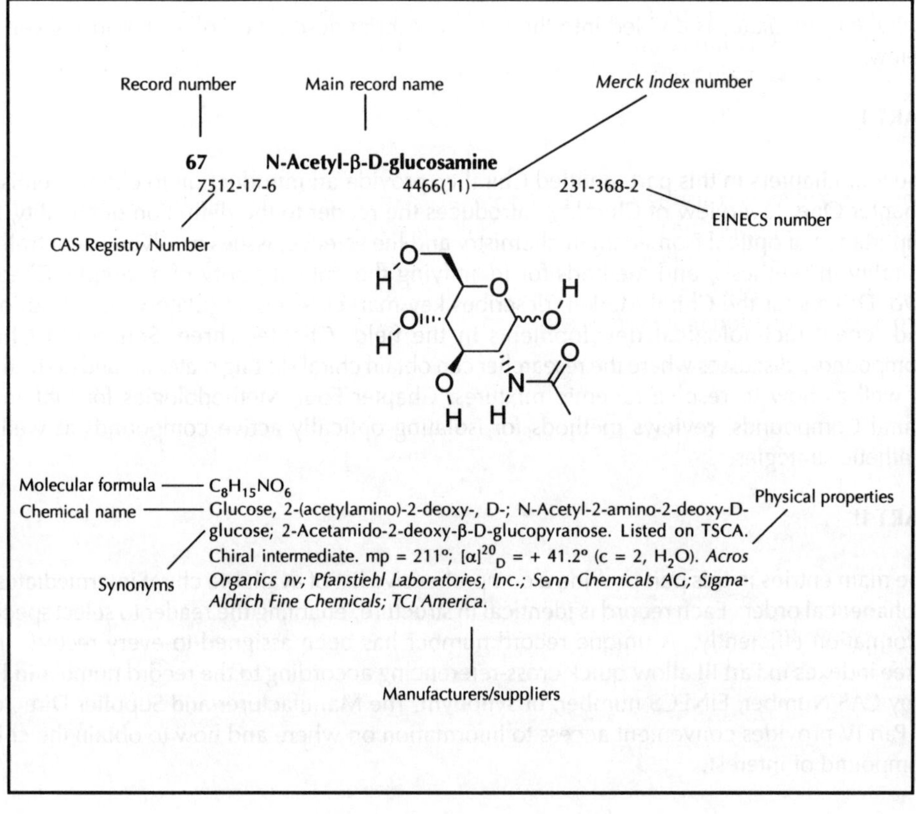

A description of the material and its known uses then follows. The record indicates whether the compound is listed under the Toxic Substances Control Act. Whenever possible, physical properties are presented. These include melting point, boiling point, and optical rotation, as well as density or specific gravity, uv absorption, solubility and acute toxicity, usually limited to oral dosage in rodents. Finally, the companies who supply the product are given.

**PART III**

This part contains three indexes. The purpose of each is described below:

- CAS Registry Number Index
  This index enables the reader to locate the record number and thereby find the main entry for a chiral compound based on its CAS Registry Number.

- EINECS Number Index
  This index enables the reader to locate the record number and thereby find the main entry for a chiral compound based on its EINECS number.

- Name and Synonym Index
  This is the master index containing all chemical and proprietary names found in Part II. It is the most convenient place for the reader to start if a name or synonym for a chiral molecule is known. This index enables the reader to locate the record number in Part II which relates to the main entry for that chemical.

**PART IV**

This part contains a listing of companies that provide contract manufacturing services or products that support the production of chiral compounds. Each listing includes the company name, address and contact information. In most cases, a brief description of the products and services is provided as well. Arranged alphabetically by company name, this directory provides information to help the reader to contact the organization directly.

# GLOSSARY OF UNITS

| Name | Description |
| --- | --- |
| Mass | Unless otherwise specified, mass is expressed in a multiple of grams (g), such as micrograms (μg; $10^{-6}$ g), milligrams (mg; $10^{-3}$ g), grams (g; $10^{0}$ g), kilograms (kg; $10^{+3}$ g), etc. |
| Volume | Volume is expressed in liters (l) or milliliters (ml) unless otherwise specified. |
| Temperature | When no units are cited, the temperature given is in degrees Celsius (°C). |
| Melting point | Melting points are cited in degrees Celsius (°C) unless otherwise specified. |
| Boiling point | When measured at atmospheric pressure, boiling points are cited with no pressure, e.g. bp = 167°. At other pressures, the pressure is also cited, e.g. $bp_{0.01}$ = 167°. |
| Density | The measurement temperature is given as a superscript; thus a density of 1.123 measured at 25° will appear as $d^{25}$ = 1.123. If the measurement was explicitly referenced to the density of water at 4°, the citation will carry both a superscript and a subscript, as in $d^{25}_{4}$ = 1.123. Specific gravities are denoted by the abbreviation 'sg'. |

**Optical rotation**

Denoted by the letter n, refractive indexes are usually determined at a temperature which is cited as a superscript, as in $n^{25} = 1.5432$. The wavelength of the light used in the measurement is cited as a subscript, as in $n^{25}_{546} = 1.5432$. Most commonly, the sodium D line (wavelength 549 nm) is used and in such cases, the subscript is a D, as in $n^{25}_{D} = 1.5432$.

**Refractive index**

As with refractive indexes, optical rotations (α) are cited with the measurement temperature superscripted, and the measurement wavelength (often the sodium D line) subscripted, as in $[\alpha]^{25}_{D} = 105°$. When mutarotation can occur, the rotation given is an equilibrium value, measured after some time interval, which is cited, as in $[\alpha]^{25}_{D} = 105°$(14 hr).

**UV absorption**

The ultraviolet absorption maxima given by the material are cited in nanometers (nm = $10^{-9}$ m) and the absorptivity (E, A, ε or log ε, all of which are unitless) may also be given.

**Acute toxicity**

Wherever possible the units of toxicity are $LD_{50}$, i.e. the dose which is lethal to 50% of the test animals. In most cases, acute toxicity is measured with the rat, orally administered, and the result is reported as $LD_{50}$ (rat orl) = 50 mg/kg. Other species (for example, mus = mouse; rbt = rabbit; pgn = pigeon; gpg = guinea pig; m = male; f = female) are occasionally cited as are other administration routes (sc = subcutaneous; ihl = inhalation; ip = intraperitoneal; iv = intravenous). Chronic toxicity data are not given.

# ABBREVIATIONS AND SYMBOLS

| | |
|---|---|
| abs config | absolute configuration |
| abs | absolute |
| Ac – | acetyl ($CH_3CO$ –) |
| ACE | angiotensin-converting enzyme |
| ACTH | adrenocorticotrophic hormone |
| AIDS | acquired immunodeficiency syndrome |
| alc | alcohol, alcoholic |
| amp.(s) | ampule(s) |
| AMP | adenosine 5′-monophosphate |
| aq | aqueous |
| atm | atmosphere, atmospheric |
| BINAP | 2,2′-bis(diphenylphosphino)-1,1′-binaphthalene ($C_{44}H_{16}P_2$) |
| BIPHEN | 1,2-bis(diphenylphosphino)ethane ($C_{26}H_{28}P_2$) |
| Bn- | benzyl ($C_7H_7$ –) |
| BOC | tert-butoxycarbonyl ($C_5H_9O_2$ –) |
| bp | boiling point |
| BPH | benign prostatic hypertrophy |
| Bu – | butyl ($C_2H_5$ –) |
| Bz – | benzoyl ($C_6H_5CO$ –) |
| c | concentration (g/100 ml), in rotations |
| C | Celsius (temperature scale) |
| cAMP | cyclic AMP |
| CBZ | carbobenzyloxy ($C_8H_7O_2$-) |
| $CH_3CN$ | acetonitrile |
| $C_5H_5N$ | pyridine |

| | |
|---|---|
| $C_6H_6$ | benzene |
| $C_7H_8$ | toluene |
| cc | cubic centimeters (milliters) |
| CCK | cholecystokinin |
| CCL | *Candida* cylindrical lipase |
| $CCl_4$ | carbon tetrachloride |
| CCK | cholecystokinin |
| $CH_2Cl_2$ | methylene chloride |
| $CHCl_3$ | chloroform |
| cm | centimeter |
| CNS | central nervous system |
| CoA | coenzyme A |
| COD | cyclooctadiene |
| COMT | catechol-O-methyltransferase |
| CPMA | chiral mobile phase additive |
| CPMP | Commission on Proprietary Medicinal Products |
| CSP | chiral stationary phase |
| d | dextro(rotatory) |
| d | density |
| dec | decompose, decomposition |
| DIPAMP | 1,2-bis(methylanisylphenylphsophino)ethane ($C_{28}H_{28}O_2P_2$) |
| DIPT | diisopropyltartrate |
| dl- | racemic |
| DL- | racemic |
| DMA | dimethylacetamide |
| DMF | dimethylformamide |
| DMSO | dimethylsulfoxide |
| DNA | deoxyribonucleic acid |
| DOPA | dihydroxyphenylalanine |
| (E)- | (entgegen) opposite |
| EC | Enzyme Commision |
| ee | enantiomeric equivalent |
| e.g. | for example |
| ED | effective dose |
| EDTA | ethylenediamine tetraacetic acid |
| EINECS | European Inventory of Existing Commercial Chemical Substances |
| endo- | stereochemical descriptor |
| Et- | ethyl ($C_2H_5$ –) |
| $Et_2O$ | diethyl ether |
| EtOAc | ethyl acetate |
| EtOH | ethanol |
| exo- | stereochemical descriptor |
| F | Fahrenheit (temperature scale) |
| FMOC | fluoromethoxycarbonyl ($C_2F_3O_2$-) |

| | |
|---|---|
| g | gram(s) |
| g/l | grams/liter |
| gal | gallon(s) |
| GI | gastrointestinal |
| GLC | gas liquid chromatography |
| gpg | guinea pig |
| $H_2O$ | water |
| $H_2SO_4$ | sulfuric acid |
| HCl | hydrochloric acid |
| HIV | human immunodeficiency virus |
| HKR | hydrolytic kinetic resolution |
| HMG-CoA | 3-hydroxy-3-methylglutaryl coenzyme A |
| hmtr | hamster |
| hr | hour |
| HT | hydroxytryptamine (serotonin) |
| ihl | inhalation |
| inj. | injection |
| im | intramuscular |
| ip | intraperitoneal |
| iPr – | isopropyl ($(CH_3)_2CH$ –) |
| IR | infrared |
| iv | intravenous |
| kcal | kilocalories |
| l | liter, levo(rotatory) |
| λ (lambda) | wavelength |
| LC | lethal concentration |
| $LC_{50}$ | median lethal concentration |
| LD | lethal dose |
| $LD_{50}$ | median lethal dose |
| log | common logarithm |
| LSR | lanthnide shift reagent |
| MAO | monoamine oxidase |
| max | maximum, maxima |
| Me – | methyl ($CH_3$ –) |
| $Me_2CO$ | acetone |
| MeOH | methanol |
| MEUF | micellar enhanced ultrafiltration |
| mg | milligram |
| min | minimum, minima, minute |
| MLD | minimum lethal dose |
| mp | melting point |
| µg | microgram |
| mµ | millimicron (nanometer) |
| Ms- | mesyl ($CH_3O_2S$-) |
| mus | mouse |

| | |
|---|---|
| N | normal, normality |
| NBD | norbornadiene |
| nm | nanometer ($10^{-9}$ m) |
| NMO | N-methylmorpholine N-oxide |
| NMR | nuclear magnetic resonance |
| NSAID | non-steroidal anti-inflammatory drug |
| NSC | National Service Center (of the National Cancer Institute) |
| NTP | normal temperature, pressure |
| o- | ortho |
| OD | optical density |
| orl | oral |
| p- | para |
| pgn | pigeon |
| Ph- | phenyl ($C_6H_5$-) |
| pH | acid-base scale (log of reciprocal hydrogen ion concentration) |
| pK | log of the reciprocal of the dissociation constant |
| PLE | pig liver esterase |
| PMA | Pharmaceutical Manufacturing Association |
| pOH | acid-base scale (log of reciprocal hydroxyl ion concentration) |
| ppb | parts-per-billion |
| PPL | porcine pancreatic lipase |
| ppm | parts-per-million |
| Pr- | propyl ($C_3H_7$ –) |
| (R) | rectus (stereochemical descriptor) |
| rbt | rabbit |
| Rh2(MEOX)4 | Doyle dirhodium catalyst |
| RNA | ribonucleic acid |
| (S) | sinister (stereochemical descriptor) |
| S- | symmetical |
| sc | subcutaneous |
| sec | second |
| sec- | secondary |
| SG, sg | specific gravity |
| SOM | site directed mutagenesis |
| spp. | species (plural) |
| STP | standard temperature, pressure |
| tabl. | tablet |
| TBHP | tert-butyl hydroperoxide |
| temp | temperature |
| tert- | tertiary |
| THF | tetrahydrofuran |
| THP | tetrahydopyranyl ($C_5H_7O$-) |
| Ts- | tosyl ($C_7H_7O_2S$-) |
| TSCA | Toxic Substances Control Act |

| | |
|---|---|
| UK | United Kingdom |
| USA | United States of America |
| USAN | United States Adopted Names |
| USP | United States Pharmacopeia |
| UV | ultraviolet |
| v/v | volume in volume |
| VIS | visible |
| viz. | namely |
| w/w | weight in weight |
| w/v | weight in volume |
| wt | weight |
| (Z)- | (zusammen) on the same side |
| > | greater than |
| < | less than |
| ~ | approximately |
| Å | Angstrom units ($10^{-8}$ cm) |

# PART I

# CHIRALITY

# CHAPTER 1

# OVERVIEW OF CHIRALITY

The existence of optical isomerism has been known since its discovery in 1815 by the French chemist Jean-Baptiste Biot.[1] However, it was only in the early twentieth century that Cushny[2] established the relevance of chirality to the pharmaceutical industry by showing that one enantiomer of hyoscyamine possessed greater pharmacological activity than the other.

Today, most new drugs and those under development consist of a single optically active isomer, and chirality is also becoming an issue for the agrochemical industry. Regulatory agencies throughout the world are currently reviewing the importance of chirality with regard to pharmaceutical and agrochemical products. New guidelines from agencies such as these have been key drivers for the focus on single enantiomer products in these industries. Optically pure compounds also have roles in electronics, foods, and flavor and fragrance products.

The "Chiral Pool" of readily available, relatively inexpensive chiral compounds has been expanding at a rapid rate as more and more products are produced in large quantities at economic prices. New developments in technologies for isolating, preparing, and purifying chiral materials have greatly increased the opportunities for utilizing optically pure compounds in commercial applications. Novel techniques for classical resolution, new methodologies for developing selective enzymes for biocatalysis, advances in the application of microorganisms for chemical production, and continued progress in the area of asymmetric synthesis have all contributed to the growth of this field.

## Definition of Chirality

### *Superimposability*

*Chiral molecules* are molecules whose mirror images are not superimposable upon one another. Conversely, *achiral compounds* have superimposable mirror images. The word chirality is derived from the Greek word for hand and means handedness, reflecting the left and right-handedness of molecules that are chiral in nature. In fact, the left and right hands are perfect examples for demonstrating the non-superimposable property of chiral objects. The left and right hands are mirror images of each other, and no matter how the two are arranged, one cannot be placed directly over the top of the other in the exact same orientation.

### *Stereoisomerism and enantiomers*

*Stereoisomers* are compounds that have the same atoms connected in the same order but differ from each other in the way the atoms are oriented in space. Chiral molecules that are structurally different from each other only in the left and right-handedness of their orientations are called *enantiomers* (Figure 1.1). Enantiomers have the same physical properties, but behave differently under certain conditions. They react at different rates with other chiral compounds and may react at different rates in the presence of chiral catalysts and optically active solvents.

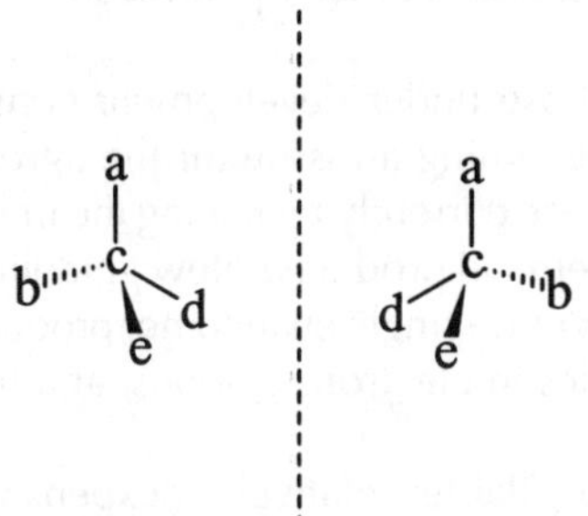


**Figure 1.1**

When polarized light is passed through a pure sample of each enantiomer, the plane of the polarized light is rotated in opposite directions in equal amounts by the two enantiomers. If rotation of the plane of the polarized light occurs, the material is considered to be optically active. An isomer that rotates light to the left is called *levo-* and indicated with the notation (-). An isomer that rotates light to the right is called *dextro-* and notated as (+).

A mixture of equal amounts of enantiomers will not rotate the plane of polarized light because the opposite rotations of the individual molecules cancel each other out. This *racemic mixture* is therefore optically inactive. The two enantiomers that comprise a racemic mixture are referred to as *racemates* (from the Latin racemus, meaning cluster of grapes). Any reaction that involves achiral or racemic starting materials, reagents, and solvents will result in a racemic mixture or optically inactive products. Desirable reactions produce an excess of one enantiomer over the other.

### *Absolute configuration*

The *absolute configuration* of a molecule indicates the actual arrangements of the substituents in the chiral compound. The direction of rotation of plane polarized light does not indicate the absolute configuration of a chiral molecule. In fact, it is possible for various unrelated compounds with the same absolute configuration to rotate light in opposite directions. In the early part of the century there was no method for determining absolute configuration empirically. Rosanoff chose glyceraldehyde to set the standard for configuration and assigned the isomer that rotates plane polarized light to the right as the D isomer and the isomer that rotates plane polarized light to the left as the L isomer.[3] The configuration of other compounds was in turn related back to that of glyceraldehydes using the Fischer convention.[4]

In 1951 Bijvoet used x-ray diffraction to examine the configuration of sodium rubidium tartrate and confirmed that Rosanoff's original assignment was indeed correct.[5] However, this old system has many flaws and has in general been replaced by the Cahn-Ingold-Prelog system for assigning absolute configuration to chiral molecules.[6] Only the amino acids and carbohydrates and their derivatives are still commonly assigned the D and L descriptors.

The Cahn-Ingold-Prelog system consists of a set of rules for prioritizing the substituents on a chiral carbon atom. If the substituents increase in priority going clockwise around the carbon atom, the configuration is assigned as R. Likewise, if the substituents increase in priority going counter-clockwise around the carbon atom, the configuration is assigned as S. Fischer projections[6] are often used to assist in the assignment of the absolute configuration of chiral molecule using the Cahn-Ingold-Prelog system.

Even with this newer system, absolute configurations must be determined either by x-ray crystallography or through relationships with compounds of known configuration. The relationship can be established through several means including: i) conversion to a known compound without loss of chirality; ii) conversion at the chiral center if the mechanism of the reaction is understood; iii) biochemical methods with enzymes specific for certain configurations, where the enzyme reacts selectively with either the (R) or (S) isomer and leaves the other enantiomer unconverted.

### *Diastereomers*

Many compounds have more than one chiral center in them. In the case of a compound with two chiral carbon atoms, there are a total of four possible stereoisomers. Two of the isomers will be mirror images of each other, and therefore enantiomers. The other two isomers are called *diastereomers* – they are stereoisomers that are not enantiomers (Figure 1.2).

CH3
H
Cl
Cl
H
CH2CH3
I
CH3
Cl
H
H
Cl
CH2CH3
II
CH3
H
Cl
H
Cl
CH2CH3
III
CH3
Cl
H
Cl
H
CH2CH3
IV
enantiomers
enantiomers
diastereomers

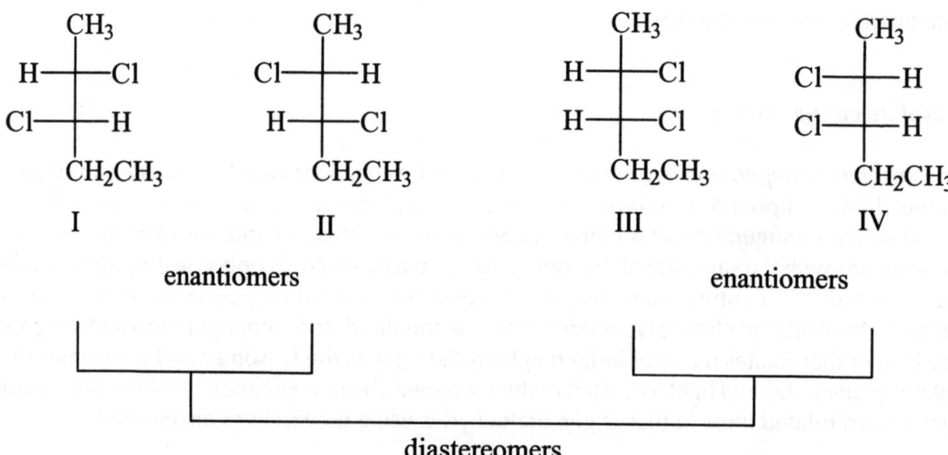

**Figure 1.2**

A diastereomer with two chiral centers where the four groups on each of the chiral carbon atoms contains a plane of symmetry within the molecule. As a result, this *meso compound* is not chiral and therefore is optically inactive (Figure 1.3).

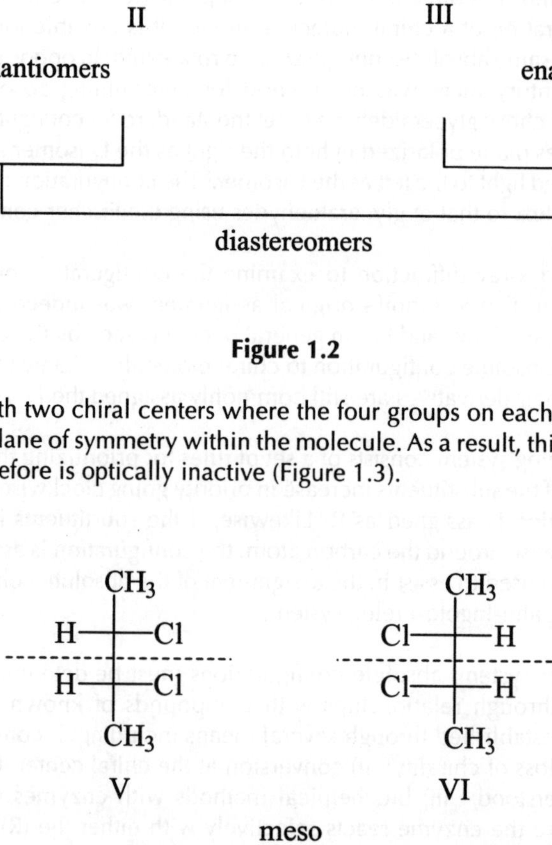

CH3
H
Cl
H
Cl
CH3
V
CH3
Cl
H
Cl
H
CH3
VI
meso

**Figure 1.3**

Unlike enantiomers, diastereomers have different physical and chemical properties, can be distinguished from one another using normal analytical techniques, and are often easy to separate from one another.

***Other chiral compounds***

Most chiral compounds are chiral because they contain a chiral carbon atom. There are, however, other types of molecules that exhibit chirality. Compounds with restricted rotation around an axis can display optical activity, as can those molecules with a helical shape. Other tetravalent atoms (like phosphorus) that have four different substituents on them will be chiral as well. In all of these examples, the molecules are asymmetric. As with chiral carbon atoms, the mirror images of these compounds are not superimposable upon each other.

A summary of terms introduced in this section is presented in Table 1.1.

**Table 1.1 Summary of Terms**

| Term | Definition |
|---|---|
| Chiral molecules | molecules whose mirror images are not superimposable upon each other |
| Achiral compounds | molecules whose mirror images are superimposable upon each other |
| Stereoisomers | compounds that have the same atoms connected in the same order but differ from each other in the way the atoms are oriented in space |
| Enantiomers | chiral molecules that are structurally different from each other only in the left and right-handedness of their orientations |
| Levo- | isomer that rotates the plane of polarized light to the left |
| Dextro- | isomer that rotates the plane of polarized light to the right |
| Racemic mixture | mixture of equal amounts of enantiomers |
| Racemates | the two enantiomers that comprise a racemic mixture |
| Absolute configuration | indicates the actual arrangements of the substituents in the chiral compound |
| Diastereomers | stereoisomers with multiple chiral centers that are not enantiomers |
| Meso compound | diastereomer with two or more chiral centers where the four groups on each of the chiral carbon atoms contains a plane of symmetry within the molecule |

## Significance of Chirality in Chemistry

After optical activity was discovered in 1815 by Biot, the next major advance in chiral chemistry was achieved by Louis Pasteur of the Ecole Normale in Paris.[7] In 1848, Pasteur prepared crystals of sodium ammonium tartrate and found that the mixture was comprised of two different crystals that were mirror images of each other. After separating the two types of crystals by hand, he further determined that each set was optically active, with the specific rotation being equal but in the opposite direction. In addition, the original mixture of crystals displayed no optical activity, and because the crystals maintained their optical activity in solution as well as in the crystal form, Pasteur concluded that the optical activity was a property of the molecules themselves and not of the crystals.

In 1874, two independent researchers – J.H. Van't Hoff[8] and J.A. LaBel[9] – proposed the tetrahedral nature of the carbon atom. Up to this point there were several possible geometric structures proposed for tetravalent carbon. Both Van't Hoff and LaBel used the known reactions of carbon compounds to form optically active products and showed that the only possible form for carbon was a tetrahedron.

There are several reasons chiral products have importance today. Frequently the biological activity of one isomer is much greater than that of the other, or the other isomer has no activity at all. In the worst case, the other isomer can have distinct, even undesirable biological activity. In some instances, the activity of the pure isomer is enhanced over that of the mixture. From another viewpoint, putting nonessential material into the environment is undesirable. Economically, companies have the opportunity to increase production by a factor of two by switching from generation of a racemic mixture to manufacture of the pure enantiomer. Overall there can be improved cost effectiveness and potential formulation advantages.

Some examples of pharmaceuticals for which one isomer has the desired effect and the other has harmful properties include thalidomide,[10] where the R-enantiomer is a sedative and the S-enantiomer is teratogenic; ethambutol,[11] where the S,S-enantiomer is tuberculostatic and the R,R-enantiomer causes blindness; and penicillamine,[12] where the S-enantiomer shows antiarthritic properties and the R-enantiomer is extremely toxic (Figure 1.4). Thalidomide is the best known of these drugs. When first discovered, the toxicity of the S-enantiomer was unknown, and Thalidomide was made available as a racemic mixture. Many pregnant women in Europe took the drug and as a result had babies with serious birth defects.

Taste and odor perception are determined by specific receptors on the tongue and in the nose. The interaction of molecules with these receptors is invariably enantioselective, and different enantiomers display different flavor and fragrance characteristics. Examples of compounds with enantiomers that have different properties include asparagine, where the S-enantiomer is bitter and the R-enantiomer is sweet, and carvone, where the S-enantiomer has the flavor of caraway and the R-enantiomer has the flavor of spearmint[13] (Figure 1.5).

Although the difference in pharmacological activity of enantiomers was known in the early 1900s, it was not until the late 1980s and early 1990s that chirality became a critical issue in the manufacture of drugs, agrochemicals, and other products. The interest of regulatory agencies and hence the pharmaceutical and agrochemical industries has resulted in increased investigations and new technological developments in all areas of chiral chemistry. Increased outsourcing of the critical development and testing phases of drug projects has also led to the formation of new businesses specializing in the production of chiral compounds, providing opportunities for even further advancements.

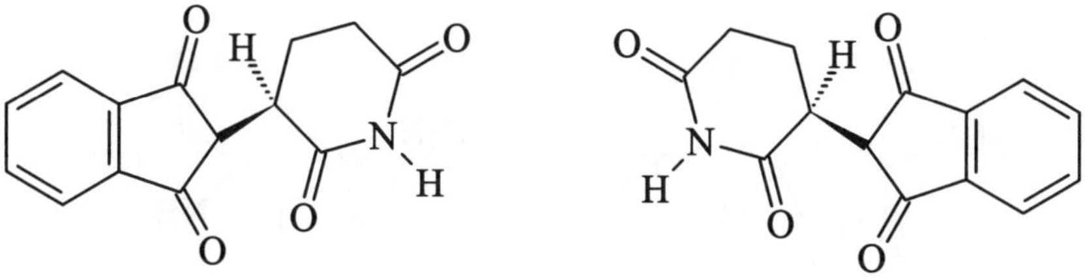

(S) - teratogen

(R) - sedative

Thalidomide

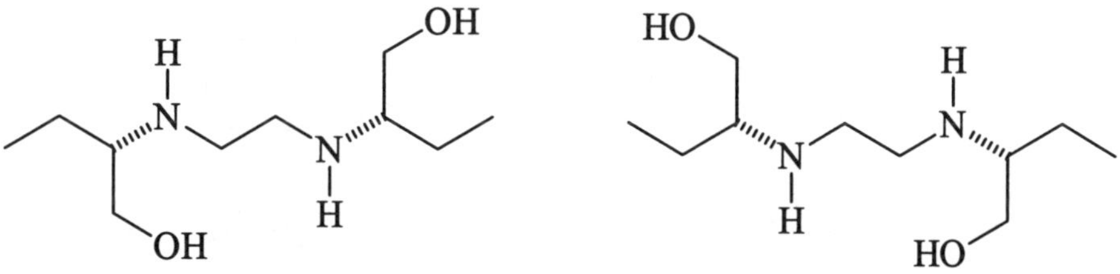

(S,S) - tuberculostatic

(R,R) - blindness

Ethambutol

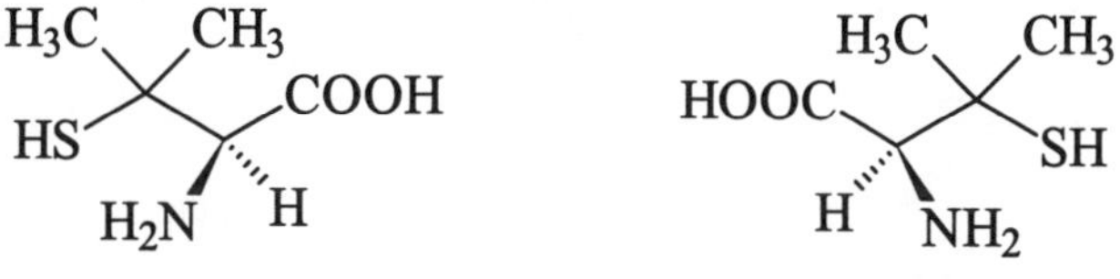

(S) - antiarthritic

(R) - mutagen

Penacillamine

**Figure 1.4**

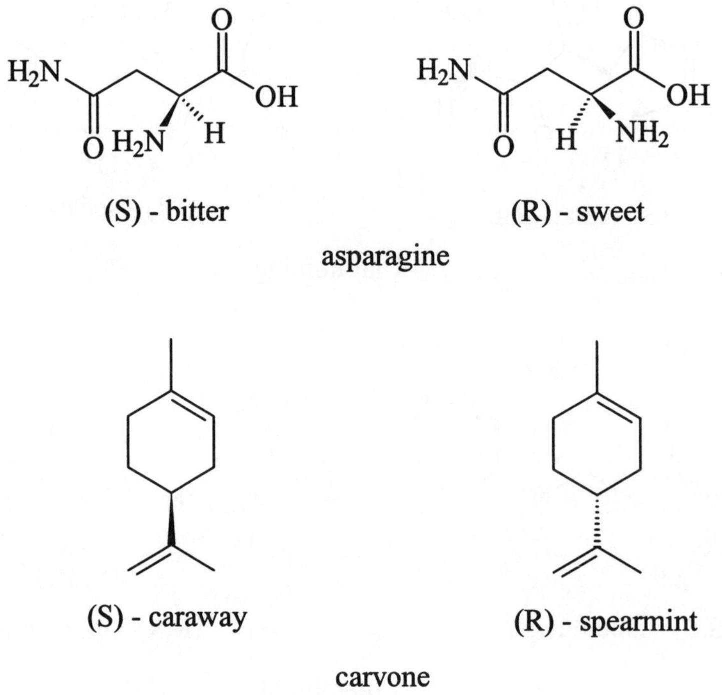
H2N
O
OH
O H2N
H
(S) - bitter
H2N
O
OH
O
H
NH2
(R) - sweet
asparagine
(S) - caraway
(R) - spearmint
carvone

**Figure 1.5**

### General Methods of Control

The synthesis of pharmaceuticals, agrochemicals, and other complex chiral chemicals often requires multiple steps to achieve the ultimate product that will serve as the active ingredient in a drug or agrochemical formulation. One factor in controlling the chirality of the final product lies in the choice of where in the production process to introduce the chiral center.

In general it is best to introduce the chiral center as early as possible in the synthetic route. In doing so the problem of carrying "ballast", or material in the form of the unwanted enantiomer, throughout the rest of the process is avoided.[14] Once the chiral center has been introduced, the synthesis must be designed to avoid loss of chirality through careful choice of reaction sequences and methodologies, thereby ensuring retention of configuration.

One way to introduce chirality early in a synthesic route is to use a chiral compound from the "Chiral Pool". These chemicals are widely available, relatively inexpensive, and well characterized. Alternatively, chirality may be introduced through a variety of reactions at an intermediate stage in the overall synthesis.

Chirality may be controlled by careful design of the synthesis methodology. In some cases, only one possible technique is available for preparing the desired molecule. In other instances, there are several possible routes for introducing the required chiral center, and usually the most economically viable route will be chosen.

If a racemic mixture is formed, separation techniques including classic resolution of diastereomers, kinetic resolution, and direct crystallization of enantiomers can be utilized to control the introduction of the chiral center. For some purposes, preparation of a single isomer through biocatalytic or chemocatalytic methods is the most appropriate means for controlling the chirality of the desired intermediate. A discussion and examples of these various methods for obtaining chiral compounds are presented in Chapters Three and Four.

## General Methods of Identification

Analysis of chiral compounds can be one of the more difficult aspects of working with optically active materials. In many cases, extremely high purities are required. Enantiomeric excess (ee) is the most widely accepted means of indicating the optical purity of a chiral compound. The ee can be calculated[15] as shown in Equation 1.1 (for [R] > [S]):

$$\text{enantiomeric excess} = \frac{[R] - [S]}{[R] + [S]} \times 100 = \%R - \%S$$

**Equation 1.1**

Methods for determining ee's for the isolation and preparation of chiral molecules must be very accurate. Enantiomeric purity of drugs and other commercial chiral products must be known precisely in order to correlate biological activity with the appropriate isomer. Certain products require much higher ee's than others, particularly those compounds prepared as pharmaceutical or agrochemical intermediates, and even minor differences in enantiomeric purity can change the activity of the compound.

### *Polarimetry*

One of the first methods available for analyzing chiral molecules was polarimetry.[16] As discussed above, chiral molecules display optical activity when one enantiomer exists in excess of the other. Pure enantiomers rotate the plane of polarized light in equal amounts but in opposite directions. A polarimeter is the instrument used to measure the amount of rotation, which is reported as the number of degrees from the original plane of polarization.

Several factors affect the amount that a molecule will rotate the plane of polarized light including the length of the sample vessel, the temperature, the solvent used, the concentration of the compound in the solvent, the pressure, and the wavelength of light used.[17] Two formulas are used to calculate the specific rotation [a] of a compound (Equation 1.2):

$$[a] = \frac{a}{lc} \text{ for solutions; and } [a] = \frac{a}{ld} \text{ for compounds;}$$

where:
a = observed rotation
l = sample vessel length
c = concentration in g/mL
d = density in g/mL

**Equation 1.2**

Polarimetry has one main drawback – samples of differing purities will give differing results. Lower specific rotations may therefore be obtained for compounds of higher enantiomeric purity due to contamination of the sample by chemical impurities. Because of this limitation, optical rotation measurement has been largely replaced with more reliable methods of determining enantiomeric excess (see below).

### *High Pressure Liquid Chromatography*

High pressure liquid chromatography (HPLC) is another valuable method for analyzing chiral compounds.[18] HPLC involves passing a liquid phase containing the compounds to be separated through a stationary phase under high pressures. One type of HPLC utilizes a chiral stationary phase (CSP) that interacts with the chiral molecules in the liquid phase.[19] The interaction is designed to be different with each of the isomers, thereby resulting in their separation. Once separated, the excess of one enantiomer over the other can be determined. CSP's have been categorized into five groups including chiral ligand exchange phases, which include affinity phases, helical polymers, cavity phases, and Pirkle type phases.[19a]

The use of chiral mobile phase additives (CMPAs) is an alternative HPLC method that involves the use of additives in the liquid phase that interact with the chiral•compounds to be separated.[20] In this case, less expensive achiral columns are utilized. Cyclodextrins, which provide a hydrophobic cavity for complexing the chiral compounds, have been used as CMPAs. Reversed phase ligand-exchange chromatography with $Cu^{2+}$ and chiral ligands in the mobile phase is another example.

### *Gas Liquid Chromatography*

Gas liquid chromatography (GLC) is another method for analyzing chiral molecules.[21] This method works well for compounds that are readily vaporized without decomposition. The compounds to be analyzed are injected onto a column packed with a chiral stationary phase. Upon injection the sample vaporizes and travels down the CSP. For this technique, CSPs can consist of ligand exchange phases, amino acid derivatives, or cyclodextrin derivatives coated on glass or fused silica.

### *Nuclear magnetic resonance spectroscopy*

Nuclear magnetic resonance (NMR) spectroscopy can also be utilized to evaluate the enantiomeric purity of a chiral compound. There are two basic methods that have been employed.[22] In the first case, diastereomers are prepared from the mixture of enantiomers. Different resonance peaks can be identified for each of the diastereomers, and a relative percentage of each can be determined. The second technique involves the addition of lanthanide shift reagents (LSRs) to the sample. These reagents form complexes with the different enantiomers and allow the resonances of the two isomers to be detected in a manner similar to that when diastereomers are actually prepared.

### *Summary*

In general, polarimetry provides information about optical rotation, but does not give quantitative results on the enantiomeric purity. The instrumentation for NMR experiments is very expensive, and the analysis may require the formation of diastereomers to be successful. HPLC and GLC have found the widest applications and continue to be the focus of new research investigations. These two techniques show the greatest promise for the future.

## References


1. Biot, J.B., *Bull. Soc. Philomath., Paris,* 190 (1815).
2. Cushny, A.R., *J. Phyiol.*, **30**, 193 (1904).
3. March, J., *Advanced Organic Chemistry, 3rd. ed.*, John Wiley and Sons, New York, 1985, pp. 94-95.
4. Fischer, E., *Ber. Dtsch. Chem. Ges.*, **524**, 129-138 (1919).
5. Bijvoet, J.M., Peerdeman, A.F., and Van Bommel, A.J., Nature, **168**, 271-272 (1951).
6. a) Cahn, R.S., Ingold, C.K., and Prelog, V., *Experientia*, **12**, 81 (1919). b) Cahn, R.S., Ingold, C.K., and Prelog, V., *Angew. Chem. Int. Ed. Engl.*, **5**, 385-415 (1966).
7. a) Pasteur, L.C.R., *Acad. Sci.*, **26**, 535-538 (1818). b) Pasteur, L.C.R., *Hebd. Seances Acad. Sci.*, **37**, 162 (1853).
8. Van't Hoff. J.H., *Arch. Neerl. Sci. Exacts Nat.*, **9**, 445-454 (1874).
9. Le Bel, J.A., *Bull. Soc. Chim. Fr.*, **22**, 337-347 (1874).
10. De Camp, W.H., *Chirality*, **1**, 2-6 (1989).
11. Hyneck, M., Dent, J., and Hook, J.B., in *Chirality in Drug Design and Synthesis*, Brown, C. (Ed.), Academic Press, New York, 1990, pp. 1-28, and references cited therein.
12. Ariëns, E.J., in *Chiral Separations by HPLC*, Krstulovic, A.M. (Ed.), Ellis Horwood, Chichester, 1989, pp. 31-68.
13. Crosby, J. in *Chirality in Industry*, Collins, A.N., Sheldrake, G.N., and Crosby, J., (Eds.), John Wiley & Sons, New York, 1992, p. 3.
14. Sheldon, R.A., *Chirotechnology*, Marcel Dekker, Inc., New York, 1993, p. 343.
15. March, J., *Advanced Organic Chemistry, 3rd. ed.*, John Wiley and Sons, New York, 1985, p. 107.
16. Morrison, R., T., and Boyd, R.N., *Organic Chemistry, 4th ed.*, Allyn and Bacon, Inc., Boston, 1983, pp. 126-127.
17. March, J., *Advanced Organic Chemistry, 3rd. ed.*, John Wiley and Sons, New York, 1985, pp. 83-84.
18. Sheldon, R.A., *Chirotechnology*, Marcel Dekker, Inc., New York, 1993, p. 27.
19. a) Däppen, R., Arm. H., and Meyer, V.R., *J. Chromatogr.*, **315**, 279 (1984). b) Davankov, V.A., Kurganov, A.A., and Bochkov, A.S., *Advan. Chromatogr.*, **22**, 71 (1983). c) Pirkle, W.H., and Pochapsky, T.C., *Advan. Chromatogr.*, **27**, 73 (1987). d) Pirkle, W.H., and Finn, J. in *Asymmetric Synthesis, Vol. 1*, Morrison, J.D. (Ed.), Academic Press, New York, 1983, pp. 87-124. e) Blaschke, G., *Angew. Chem. Intl. Ed. Engl.*, **19**, 13-24 (1980). f) Armstrong, D.W., *J. Liq. Chromatogr.*, **7**, (52), 353 (1984). g) Allenmark, S., J. Biochem. Biophys. Methods, **9**, 1 (1984). h) Wainer, I.W. in *Drug Stereochemistry: Analytical Methods and Pharmacology*, Wainer, I.W., and Drayer, D.E. (Eds.), Marcel Dekker, New York, 1988, pp. 147-173.
20. Sheldon, R.A., *Chirotechnology*, Marcel Dekker, Inc., New York, 1993, p. 31.
21. a) Schürig, V., and Nowotny, H.P., *Angew. Chem. Intl. Ed. Engl.*, **29**, 939-957 (1990); b) Schürig, V., in *Asymmetric Synthesis, Vol. 1.*, Morrison, J.D. (Ed.), Academic Press, New York, 1983, pp. 59-86.
22. Sheldon, R.A., *Chirotechnology*, Marcel Dekker, Inc., New York, 1993, p. 32.

# CHAPTER 2

# DRIVERS FOR THE CHIRAL MARKET

### Market Information

Today, approximately 80% of chiral intermediates and related products go into the pharmaceuticals market,[1] and it is expected that in the future, pharmaceuticals will remain the key driver for chiral chemistry development.[2] Optically active compounds also find applications in agrochemicals, flavor and fragrances, foodstuffs, liquid crystals and non-linear optical materials, electronics, biochemicals, polymers and other areas.

In 1998, worldwide sales of single enantiomer (optically active) drugs was $96.4 billion, an 11% increase over 1997 levels.[3] This growth is expected to continue, with the percentage of chiral drugs on the market projected to increase from 36% of all available drugs in 1998 to 40% in 2003.[4] In fact, a total of 950 chiral drugs were being evaluated in advanced clinical trials in 1998.[5] Of those compounds, 65% were made synthetically (using asymmetric synthesis, resolution of diastereomers, via enzymatic pathways, or as derivatives of chiral pool materials). The remaining 35% were obtained via isolation of natural products, through natural fermentation routes, or by other natural processes.

Excluding drugs, demand for chiral chemicals is growing at a healthy rate of about 10% per year and is anticipated to reach $9.4 billion in 2003.[4] This number includes chiral chemicals used as starting materials, reagents, or intermediates (compounds that are incorporated into the final product) for the preparation of optically active compounds. Sales of products used in chiral synthesis that do not get consumed in the process (catalysts, enzymes, chiral auxiliaries, solvents, etc.) are anticipated to grow at 5.8% per year to $135 million in 2000,[6] and revenues for the chiral separations sector (chiral HPLC columns, GLC columns, and other similar products) were determined to be $8 billion in 1999.[7]

The biopharmaceuticals market is another area of interest in chiral synthesis. This market includes mainly cell cultures that support synthesis and proteins that are primary products. Cell cultures are used in fermentation processes and enzymatic transformations that produce chiral compounds. In 1998, the biopharmaceuticals market was growing at 15% per year.[8]

In the agrochemicals area, chiral chemicals made as racemic mixtures were worth $7 billion in sales in 1997.[9] Overall, chiral molecules comprise 30.4% of the global pesticide market.[10] Out of approximately 650 pesticides in commercial use in 1997, 173 contained at least one chiral center with the potential to be prepared and marketed as single enantiomer products. At that time only 22 compounds in development stages were being made as enantiomerically pure products.

## Key Issues for Chiral Chemicals

As the market for chiral chemicals continues to grow, companies that participate in the industry will be wrestling with several issues. Because the majority of optically active compounds have applications in the pharmaceuticals sector, regulatory requirements for chiral pharmaceutical intermediates have a strong impact on the entire chiral chemical industry. Recent concerns over health care reforms and the cost effectiveness of drugs have also had repercussions for chiral chemical suppliers.

Competition from Indian and Chinese manufacturers has also increased.[11] Companies in these countries, using modern technologies, are now producing chiral intermediates and bulk active drugs that meet the regulatory requirements of the US Food and Drug Administration (FDA) and regulatory bodies in Europe and other key markets.

Developments in the technology used to produce chiral molecules continue at a tremendous pace. New reports on improved catalysts for asymmetric synthesis appear regularly in the literature. Progress is being made in separations methodology as well. New successes in the design and control of biochemical transformations are announced frequently. In the current regulatory environment, with the capabilities already available today and the continued rapid advancements in the area of chiral chemistry, there is little argument for selling biologically active products as racemic mixtures.

## Regulatory Issues

### *Pharmaceuticals*

Regulatory changes have significant effects on the production of chiral chemicals, particularly those prepared for use in formulated pharmaceutical and agrochemical products. Manufacturers of chiral drug intermediates and pesticides must pay close attention to rules and regulations issued by regulatory bodies throughout the world. Regulation of chiral chemical products has become an international issue as most regulatory agencies in the world look at issues related to chiral molecules in drugs and agrochemicals.

Currently the International Conference on Harmonization (ICH) is working to establish a set of guidelines for the analysis and application of chiral compounds in drugs.[12] The ICH is a unique project that brings together the regulatory authorities of Europe, Japan and the United States and experts from the pharmaceutical industry in the three regions to discuss scientific and technical aspects of product registration. In the area of optically active drugs, the goal of this group is to come to an agreement accepted by countries throughout the world on requirements for enantiomerically pure materials used in pharmaceuticals.

The US FDA published its first comments on the issue of chiral drug compounds in 1987. In the Drug Substance Guideline,[13] the FDA made some references to stereoisomers but placed little emphasis on the topic. The next step involved formation of a working group in 1989 to address the issue of regulating optically active drug compounds.[12] In response, the Pharmaceutical Manufacturing Association (PMA) set out its position on the development of drugs with chiral centers in a 1990 article in Pharmaceutical Technology.[14] The PMA recommended that its members evaluate the properties of each enantiomer during the drug development process in order to make an informed decision about marketing a new drug as a single enantiomer or racemic mixture, considering all available data and providing regulatory agencies with appropriate information to document the safety and effectiveness of the new products.

The Drug Information Association conference was held in Paris in 1992.[12] This group was assembled as an international forum to examine various topics related to the manufacture and sale of pharmaceuticals. The group spent time discussing the regulatory requirements for chiral drugs. A summary of these discussions was published in 1993 in the Drug Information Journal.[15]

Also in 1992, the FDA issued its proposed guidelines for regulating drugs containing stereogenic centers.[16] The proposal required manufacturers to research and characterize each enantiomer in all drugs that will be marketed as a racemic mixture. In the 1992 proposal, the development and marketing of chiral drugs as racemates was not prohibited by the FDA, but it was determined that final approval must be based on complete information, including the pharmacodynamics and pharmacokinetics of all individual isomers and the racemic mixture as well.

These guidelines were released as a proposal for public comment, and eight years later no revised document has yet been published. Even though the FDA did not mandate the manufacture of single enantiomer drugs, most pharmaceutical companies have taken the proposed guidelines as a signal and, unless the isomers are readily interconvertible within the body, are only making drugs as pure isomers.[17] In fact, in 1998 80% of all new drugs in development were being investigated as single enantiomer products.[18]

The European Commission on Proprietary Medicinal Products (CPMP) is the body that coordinates registration of applications for new drugs in Europe. The CPMP considers an isomer to be an impurity.[19] Guidelines for identification and qualification of impurities have been established and drug manufacturers must comply with these guidelines with respect to enantiomeric and diastereomeric impurities. FDA considers diastereomers as impurities but has no position on undesired enantiomers.[20]

Currently there is no single global standard for development of optically active pharmaceutical compounds. Manufacturers must develop chiral drug compounds according to the regulatory requirements of each individual country or region where they will be registering their products. The International Conference on Harmonization and other groups continue to work to develop a regulatory framework that can be implemented throughout the world.

In the United States, there is yet another mechanism for regulation of chiral drug chemicals and formulated final products. These additional regulatory requirements appear in the FDA's Chemistry, Manufacturing and Controls (CMC) section of the application for new drugs.[21] In this section the FDA gathers information on the identification, quality, purity, and strength of the drug substance (active ingredient) and product (formulated final drug). In the CMC the FDA requires stereochemically specific data for any drugs that contain one or more chiral centers.

The agency has published a decision tree that must be followed to determine whether a stereospecific identification test for configuration and/or a stereoselective assay for enantiomeric purity must be conducted. The decision tree applies to the active ingredient itself and not the formulated drug product. Different actions must be taken depending on whether the compound is a racemic mixture or a single isomer. Another determining characteristic is whether the chemical was obtained from a synthetic preparation or isolated from a natural source.

For the final drug product, the regulatory requirements focus mainly on stability issues. The formulated product must be tested under raised temperatures, high humidity, and increased light levels, and the tests involve monitoring of the isomeric purity to determine if interconversion has occurred. Long-term studies are required if any isomerization is observed.

Other issues in this section include control of starting materials and reagents, which involves conducting assays for compounds containing chiral centers. The FDA also addresses concerns that arise when introducing a chiral center into an intermediate. Once a stereogenic (chiral) center is present, the FDA requires monitoring of isomeric purity throughout the remainder of the synthesis. Any step with a potential to epimerize the chiral center must have controls to confirm that optical purity remains within established limits.

In 1997 the FDA made another proposal for public comment regarding optically active pharmaceutical products.[22] In this proposal, the FDA considered the possibility of granting a five year marketing exclusivity to developers of single enantiomer drugs of previously approved racemic mixtures. This idea has been termed a racemic switch. In the 1997 proposal, the FDA also introduced the possibility of requiring that racemate products be withdrawn if safer, single enantiomer versions of the products are brought to the market. No action has yet been taken on this proposal.

The idea of the racemic switch has aided pharmaceutical manufacturers in the last several years.[23] Single enantiomer versions of many drugs that were initially introduced as racemic mixtures have been subjected to clinical trials in order to get entirely new patents issued for a newly formulated product. With a racemic switch, a drug company can extend the life of a patent for a racemic mixture up to 20 years. Currently approximately $9 billion in sales are protected by selling single isomer versions of previously marketed optically inactive pharmaceuticals. Examples (Figure 2.1) include omeprazole (acid/peptic disorders, $4 billion), fluoxetine (depression and obsessive/compulsive disorders, \$3 billion), cisapride (acid/peptic disorders, \$1 billion), and levofloxacin (antibiotic, \$1 billion).

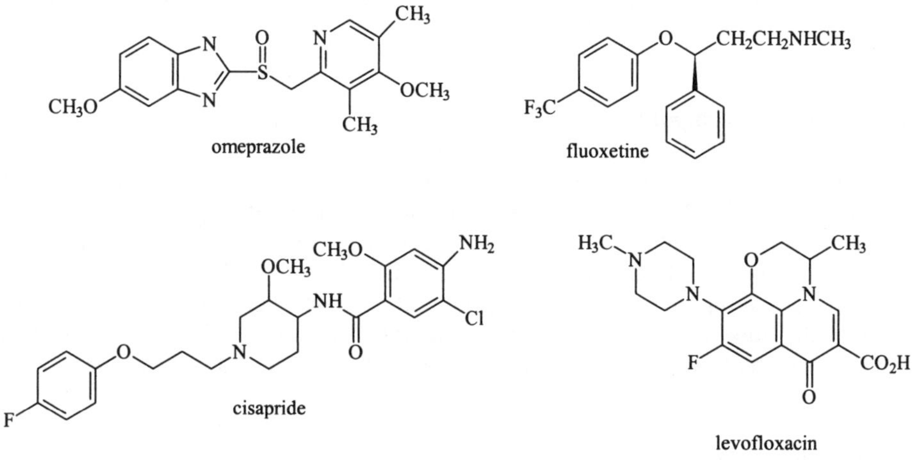

omeprazole

fluoxetine

cisapride

levofloxacin

**Figure 2.1**

Exclusivity of supply agreements can be confusing for chiral chemical manufacturers.[24] It can be very difficult to tell when exclusivity has expired on a patent, even though the FDA publishes this type of information. New rulings are made on a case-by-case basis, and generic drug companies may not have timely knowledge of these changes in exclusivity rights. Often suppliers of chiral starting materials and intermediates work with generic drug manufacturers to develop routes to products only to find out that the exclusivity period does not end when expected.

One example is the extension of a Hoffmann La Roche patent on the Rocephin brand of ceftriaxone, a third-generation cephalasporin antiobiotic. Roche licensed a patent from Hoechst that extended their original patent for Rocephin from 1999 to 2005. This type of extension is not included in the FDA publication known as the "Orange Book" that includes lists of approved drugs by active pharmaceutical ingredient, and patent and market exclusivity expiration. Therefore, generic firms might not be aware of the patent extension.

### *Agrochemicals*

Although at present there are relatively few agrochemicals sold as single enantiomer compounds, environmental pressures will most likely bring changes to this industry.[25] Compared to pharmaceuticals, which are consumed in small quantities, pesticides and other agrochemicals are applied in large amounts. Therefore, the environmental and economic impact of reduced application rates can be significant.

The use of single enantiomer products raises the possibility of reducing the application rate of an agchem product by a minimum of 50%. If the undesired isomer is an inhibitor of the active ingredient, then the application rate may be reduced by even greater than 50%. As a result, a significant quantity of chemicals would no longer need to be introduced into the environment. Both regulatory agencies and environmental activist groups have taken notice of this consequence of the use of chiral molecules.

### *Flavor and fragrance compounds*

In the food industry, there is an increasing demand for better quality products that are more healthful and natural.[26] One aspect of this change in consumer buying habits is the desire for foods with 'natural ingredients' as opposed to 'synthetic additives'. Natural products are those compounds that have been extracted from natural sources or made by natural processes (fermentation for example). For this industry, natural has often become synonymous with enantiomerically pure. The isomers of menthol are one example, with both having a minty aroma. However, the l-isomer is preferred because it is the naturally occurring enantiomer.

## Healthcare Reforms and Cost Effectiveness

Healthcare costs in the United States have been increasing rapidly and politicians in both Washington, D.C. and in many state governments have given serious attention to this issue. The managed care environment of today has been one result of the scrutiny of the healthcare system in this country.

A fundamental goal of the managed care business is to reduce the costs associated with healthcare. One focus of the industry has been on prescription medications. In order to survive in this environment, pharmaceutical companies must find a way to significantly reduce expenses. Some experts believe the pharmaceutical industry of the future will not have the high margins it has been used to over the past several decades.[27] In fact, the drug companies are moving away from a specialty business model towards one more like that for commodities.

Outsourcing of much of the drug development process has been one means for the industry to reduce costs and time to market for new drug products. A consequence of the increase in outsourcing has been the rapid rise in the number of chemical companies specializing in chiral technologies. Several companies have chosen to focus on small-scale process and product development, targeting the initial stages in the lifecycle of pharmaceutical compounds.

Companies created in the late 1980s and early 1990s as specialists in chiral chemistry have found that their businesses cannot survive on chirality alone,[28] and these firms are now redefining themselves as overall specialty chemical service providers for the pharmaceutical, agrochemical, and other industries interested in chiral starting materials and intermediates.

The consolidation of pharmaceutical companies has been occurring steadily over the past several years and this trend is expected to continue as another means of reducing costs.[29] The few, large companies that will be left will have much more power to exert over the specialty chemical companies providing them with outsourcing services. Many pharmaceutical firms will prefer to establish relationships with a limited number of suppliers. Chiral chemical manufacturers will have to develop the appropriate toolkits in order to attract the business of the surviving large pharmaceutical companies.

## Developments in Technology

New developments in the technology of chiral molecule preparation and isolation will be critical to the continued success of the industry. Advances are being made in all areas of the field of chiral chemistry, including classical resolution techniques, other separation methods, biological processes, biocatalysis, and asymmetric synthesis.

### *Classical resolution techniques*

Novel, relatively inexpensive resolving agents have been developed that do not depend on the extraction of chiral compounds from natural products. One example of these "designer resolving agents" is a group of chiral phosphoric acids that can be used for resolution of amines and underivatized amino acids.[30] These resolving agents were created based on years of experience working with classical resolution technology.

The use of computer aided molecular modeling for the rational design of resolving agents is an area that is under investigation. Studies have shown that crystal packing arrangements of diastereomeric salts are quite different from each other and correlate with solubility.[31] It has been found that insoluble diastereomers have very different crystal packing arrangements as compared to soluble isomers.

Other researchers have established that solubility and heat of fusion data can be related to lattice energy differences, allowing for the prediction of resolution efficiency.[32] Salts with lower solubilities have higher heats of fusion, which correspond to greater non-bonded interactions in the crystalline state. These results are encouraging, and researchers in the area hope that continued efforts will someday provide additional tools for the preparation of resolving agents designed for specific applications.

The use of dibasic acids such as tartaric acid allows for formation of both acidic and neutral salts that possess different thermal stabilities, which allows for separation of the two diastereomers.[33] A liquid racemic base was reacted with half an equivalent of L-tartaric acid. When heated under vacuum, some of the base was released, and both the distillate and the residue were enriched in the opposite isomers, up to 65% depending on the amine used. In another area, researchers have been investigating the use of polymers designed with specifically shaped cavities as a means of separating diastereomers.[34]

### *Other separations methods*

Discoveries have also been made in other types of separation technologies. One example is a new membrane method developed specifically for chiral separations.[35] The technology utilizes apoenzymes (enzymes that bind to a molecule but do not possess any catalytic activity) to bind to one enantiomer and allow transportation of that isomer across a membrane.

Recently, advances have been reported in the use of HPLC for the separation of stereoisomers. Daicel has licensed technology from Novartis to immobilize polysaccharides on chromatographic media for commercial scale separations of chiral chemicals.[36]

### *Biological processes and biocatalysis*

The use of enzymes in chiral chemicals synthesis is expected to grow rapidly. Biocatalysts are attractive because they offer a means of increasing the specificity of syntheses, which decreases costs.

Recombinant DNA technology (r-DNA) has allowed for the rational design of highly active microbial strains that possess selected catalytic properties.[37] These new strains produce proteins that the original microbes never did. Site-directed mutagenesis (SDM) is a process that involves replacement of one or more specific amino acid residues in an enzyme.[38] Thus, new enzymes can be designed with specific catalytic properties. Mapping the active sites of enzymes has provided yet another rationale for the design and use of enzymes as biocatalysts for chiral chemical synthesis.[39]

The availability of recently developed bioinformatics and genomics tools for use in the identification and design of enzymes for catalysis has also contributed to the development of novel enzymes.[40] These new enzymes serve as catalysts for the synthesis of chiral compounds and provide greater selectivities for the desired optically active products.

Extremophiles, or microbes that have been modified so they can function in extreme environments (high-pH, high temperature, high pressure) provide enzymes that have greater stability in chemical processes.[40] Gene shuffling is another new method for the rapid development of novel enzymes. This technology uses the natural variants of enzymes and introduces new variations through mutation and rearrangement of DNA.[40]

### *Asymmetric synthesis*

In the area of asymmetric organic synthesis, new ligands for chiral catalysts have resulted in processes that provide products of steadily increasing enantiomeric purities. Novel techniques for supporting chiral catalysts to aid in recovery and separation from the product are also being developed at a rapid pace. New discoveries in these and other areas are reported on a regular basis and are too numerous to cover in this book. Information can be found in industry trade magazines and organic chemistry journals. For a recent review of new chiral synthesis methodologies of interest to industry, see Reference 41. Many companies license technology directly from academic institutions in order to reap the benefits of the results in a timely manner.

### *Summary*

Significant advances have been made in all the technologies used to isolate and prepare chiral chemicals. The commercial scale preparation of optically active compounds has become much more efficient and economical over the past several years. The continued efforts to research and investigate new methodologies indicate great promise for the future of the chiral industry.

## References


1. Chirals to Grow, *Chemical Week*, Feb. 3, 1999, p. 33.
2. Chiral Demand Up, *Chemical Week*, April 10, 1996, p. 40.
3. Stinson, Stephen C., Chiral Drug Interactions, *Chemical & Engineering News*, October 11, 1999, p. 101.
4. U.S. Chiral Chemicals Demand Set to Grow, *Process Eng.*, **80**, 11,Nov. 16, 1999, p. 7.
5. Mirasol, F., Chirals Producers Expand to Meet Burgeoning Demand, *Chemical Market Reporter*, April 17, 1998.
6. Chirals to Grow, Chemical Week, Feb. 3, 1999, p.33.
7. Chirotechnology: Going from Strength to Strength, *Performance Chemicals International*, **14**, 13, May 1, 1999, pp. 15-26.
8. Scott, A., Biotech Synthesis Competition Toughens, *Chemical Week*, April 22, 1998.
9. Fine Chemicals: Mirroring Changes in Chiral Technology, *Chemical Week*, July 16, 1997, p. 32.
10. Chiral Pesticides, *Pesticide Outlook*, **8**, 6, Dec. 31, 1997, pp/ 15-19.
11. Stinson, Stephen C., Chiral Drug Interactions, *Chemical & Engineering News*, October 11, 1999, p. 106.
12. Blumenstein, J.J., in Chirality in Industry II, Collins, A.N., Sheldrake, G.N., and Crosby, J., (Eds.), John Wiley & Sons, New York, 1997, p. 11.
13. Center for Drugs and Biologics Food and Drug Administration Department of Health and Human Services, *Guideline for Submitting Supporting Documentation in Drug Applications for the Manufacture of Drug Substances*, FDA, Washington, D.C., February, 1987.
14. PMA Ad Hoc Committee on Racemic Mixtures, 'Comments on Enantiomerism in the Drug Development Process', Pharm. Technol., May, 46 (1990).
15. Gross, M., *et al.*, Regulatory Requirements for Chiral Drugs, *Drug Info. J.*, **27**, p. 453 (1993).
16. Food and Drug Administration, FDA's Policy Statement for the Development of New Stereoisomeric Drugs, FDA, Washington, D.C., May 1, 1992.
17. Richards, A., and McCague, R., The Impact of Chiral Technology on the Pharmaceutical Industry, *Chemistry and Industry*, June 2, 1997, pp. 422-425.
18. Chiral Competition Begins to Heat Up, *Performance Chemicals International*, **13**, 10, Feb. 2, 1998, p. 24.
19. CPMP Working Party on Quality of Medicinal Products, CPMP Working Party on Safety of Medicinal Products, CPMP Working Party on Efficacy of Medicinal Products, *Note for Guidance: Investigation of Chiral Active Substances*.
20. Blumenstein, J.J., in Chirality in Industry II, Collins, A.N., Sheldrake, G.N., and Crosby, J., (Eds.), John Wiley & Sons, New York, 1997, p. 16.
21. Blumenstein, J.J., in Chirality in Industry II, Collins, A.N., Sheldrake, G.N., and Crosby, J., (Eds.), John Wiley & Sons, New York, 1997, p. 14-17.
22. *Chem. Eng.*, **105**, 3, March 18. 1995, p. 61.
23. Stinson, Stephen C., Chiral Drug Interactions, *Chemical & Engineering News*, October 11, 1999, p. 102.
24. Stinson, Stephen C., Chiral Drug Interactions, *Chemical & Engineering News*,

October 11, 1999, p. 104.

25. Sheldon, R.A., Chirotechnology, Marcel Dekker, Inc., New York, 1993, p. 61.
26. Sheldon, R.A., Chirotechnology, Marcel Dekker, Inc., New York, 1993, pp. 67-69.
27. Breston, I., and Ward, M., Outsourcing Gets New Emphasis. Fine Chemicals Suppliers Key, *Chemical Week*, April 19, 1995, pp. 36-37.
28. Fine Chemicals: Mirroring Changes in Chiral Technology, *Chemical Week*, July 16, 1997, p. 32.
29. Custom Manufacturing, *Chemical Week*, May 19, 1997, p. 31.
30. Ten Hoeve, W., and Wynberg, H., *J. Org. Chem*, **50**, 4508-4514 (1985).
31. Sheldon, R.A., Hulshof, L.A., Bruggink, A., Leusen, F.S.S., Van der Haest, A.D., and Wynberg, H., *Chemistry Today*, **4** (Jul/Aug), 23-39 (1991).
32. a) Van der Haest, A.D., Wynberg, H., Leusen, F.J.J., and Bruggink, A., *Recl. Trav. Chim., Pays-Bas*, **109**, 523-528 (1990). b) see also: Leusen, F.J.J., Bruinslot, A.J., Noordik, J.H., Van der Haest, A.D., Wynberg, H., and Bruggink, A., *Recl. Trav. Chim. Pays-Bas*, **111**, 111-118 (1992).
33. Acs, M., Szili, T., and Fogassy, E., Tetrahedron Lett., 32, 7325-7328 (1991).
34. Wulff, G., in Bioorganic Chemistry in Healthcare and Technology, Pandit, U.K., and Alderweireld, F.C., (Eds.), Plenum Press, New York, 1991.
35. Enzymes: New Methods for Chiral Separations, *European Chemical News*, **68**, 1782, Sept. 29, 1997, p. 55.
36. Chiral Separations, Chemical Week, April 14, 1999, p. 35.
37. Sheldon, R.A., Chirotechnology, Marcel Dekker, Inc., New York, 1993, pp. 110-112.
38. Sheldon, R.A., Chirotechnology, Marcel Dekker, Inc., New York, 1993, p. 110, and references cited therein.
39. Sheldon, R.A., Chirotechnology, Marcel Dekker, Inc., New York, 1993, p. 389.
40. Scott, A., Harnessing Biocatalysts to Make Complex Drugs, *Chemical Week*, June 23, 1999, p.47.
41. Fertile Ferment in Chiral Chemistry, Chemical and Engineering News, May 8, 2000, p. 59.

# CHAPTER 3

# SOURCES OF CHIRAL COMPOUNDS

There are three primary sources of pure enantiomers:[1] the first technique involves isolation of naturally occurring molecules through extraction from plant materials; the second method is through *de novo* fermentation of inexpensive, available feedstocks; the third route is through synthesis using optically active compounds (obtained from one of the first two methods) or prochiral starting materials.

These three main sources can be characterized in more detail (see Figure 3.1).[1] Compounds included in the Chiral Pool (collection of chiral chemicals) are relatively inexpensive, readily available natural products. These optically active compounds often serve as building blocks for the synthesis of more complex chiral molecules.

In chemical processes, racemic mixtures are commonly prepared and the desired isomer must be separated from the unwanted enantiomer. Direct or preferential crystallization is the most attractive option if it is available (no derivatization is necessary). Separation of the diastereomeric salts of the two isomers is one of the most widely used methods for obtaining pure enantionmers. Crystallization, kinetic resolution, and preparative HPLC are three possible techniques for achieving this separation.

Biochemical routes using microbes and enzymes as catalysts have had increasing applications in the commercial preparation of chiral molecules. Catalysis with chiral chemical compounds has also received significant attention for the asymmetric synthesis of optically active entities.

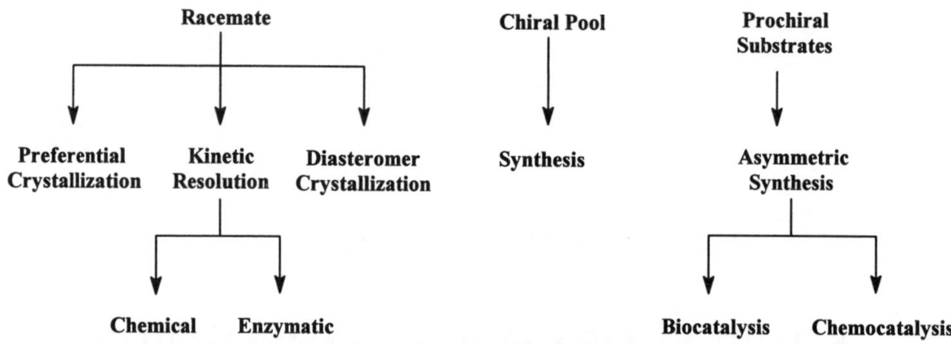


**Figure 3.1**[1]

## The "Chiral Pool"

As mentioned above, the Chiral Pool is a collection of relatively inexpensive, readily available, optically active natural products. These compounds are typically supplied in quantities ranging from 100 to 10 000 tonnes per year.[2] The main types of chemicals included in the Chiral Pool are carbohydrates, amino acids, hydroxy acids, terpenes, and alkaloids.

Recently the argument has been made that newly developed, synthetically derived chiral compounds that can serve as inexpensive starting materials and intermediates for established drugs should be considered as part of the Chiral Pool. These compounds could be intermediates or products themselves, or by-products formed in large-scale commercial processes.

*Carbohydrates* are the largest group of chiral compounds found in nature. Natural carbohydrates generally have the D configuration (see Chapter One, absolute configuration). The simplest sugars, the monosaccharides, have the most utility in commercial chiral processes.[3] Examples of carbohydrates included in the Chiral Pool are shown in Figure 3.2.

*Amino acids* are often considered the most important category of chemicals included in the Chiral Pool. Large quantities of natural α–amino acids (components of proteins) are prepared via fermentation and other biological processes. They can also be obtained from hydrolysis of plant and animal proteins. These compounds are attractive as starting materials because they are simple in structure, containing one or two chiral centers and functionality that can be readily modified.[4] Amino acids with both the L and D configurations are members of the Chiral Pool. Examples of available amino acids can be seen in Figure 3.3.

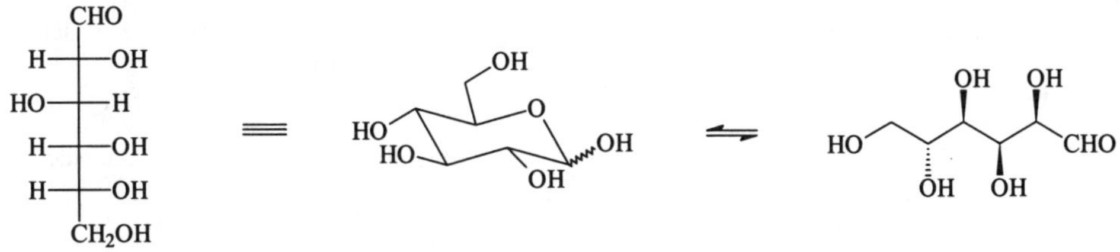

D-glucose

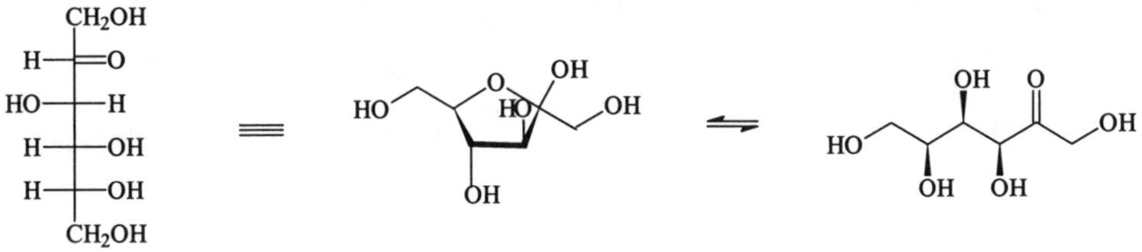

D-fructose

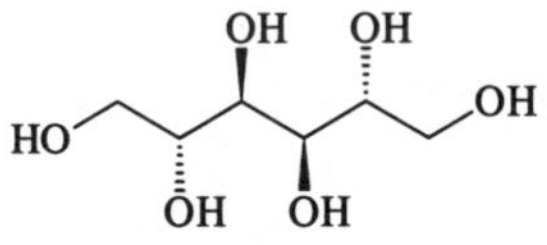

D-Sorbitol

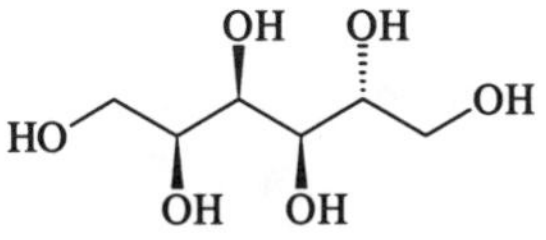

D-Mannitol

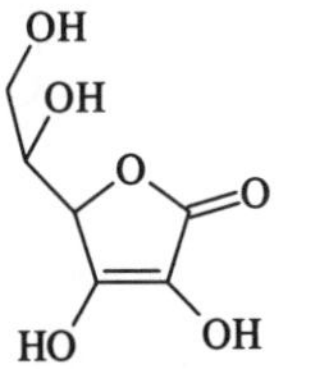

L-ascorbic acid

Others: D-sucrose, D-lactose, D-gluconic acid, D-isoascorbic acid

**Figure 3.2**

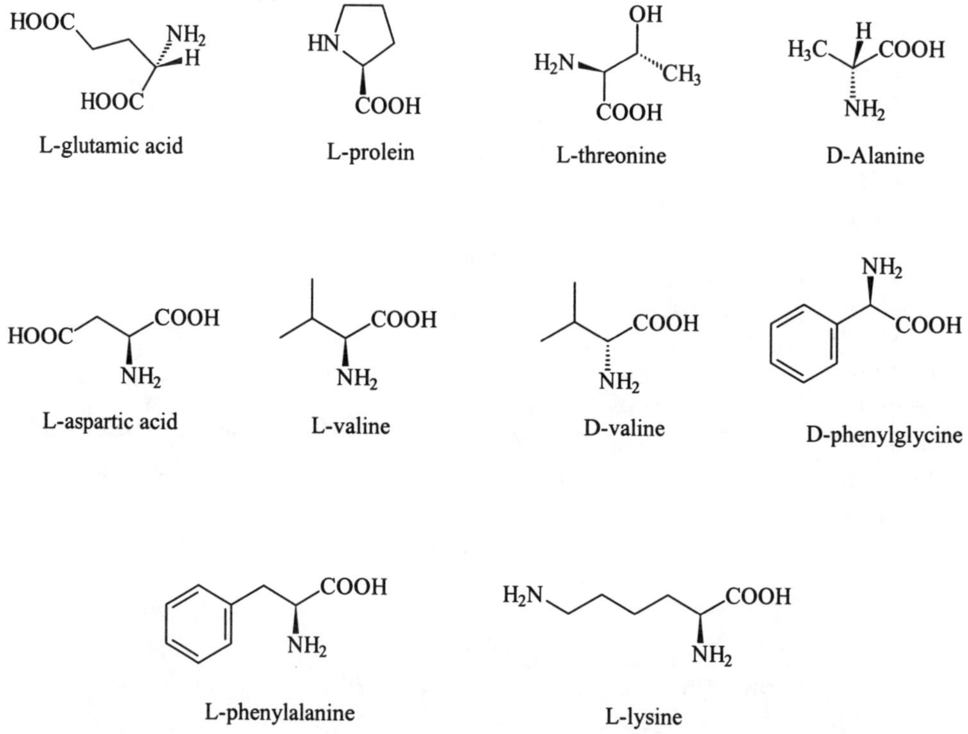
HOOC
NH2
H
HOOC
L-glutamic acid
HN
COOH
L-prolein
OH
H2N
CH3
COOH
L-threonine
H3C
H
COOH
NH2
D-Alanine
HOOC
COOH
NH2
L-aspartic acid
COOH
NH2
L-valine
COOH
NH2
D-valine
NH2
COOH
D-phenylglycine
COOH
NH2
L-phenylalanine
H2N
COOH
NH2
L-lysine

**Figure 3.3**

*Hydroxy acids* included in the Chiral Pool are generated either via fermentation of glucose or from selective microbial oxidation of aliphatic acids.[5] Chiral α–hydroxy acids have some direct commercial applications and are also finding use as chiral starting materials for the synthesis of optically active products.[6] β–Hydroxy acids have recently gained importance in the commercial preparation of pharmaceutical intermediates.[7] Figure 3.4 contains examples of chiral hydroxy acids included in the Chiral Pool.

*Terpenes* are available from several natural sources. The simplest compounds of the group, the monoterpenes, are used mostly for the synthesis of other terpenoid products.[8] They also find applications as starting materials for the preparation of resolving agents and chiral ligands for catalysts used in asymmetric synthesis.[9] The optical activity of the natural monoterpenes depends largely on the origin of the oil of turpentine, which is the main source for many of these compounds.[10] Examples of terpene members of the Chiral Pool are shown in Figure 3.5.

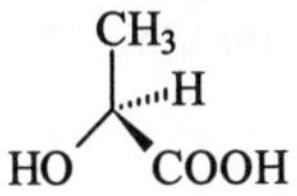

**Figure 3.4**

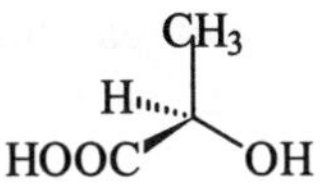

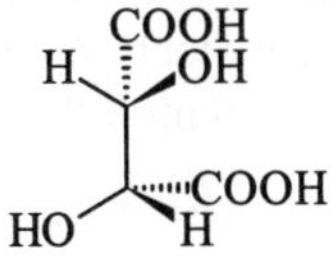

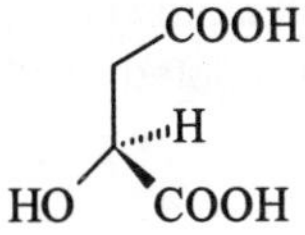

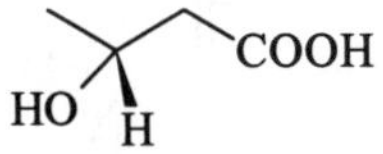

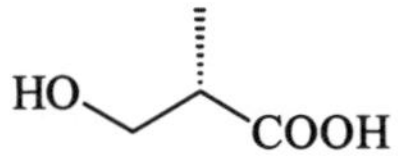

**Figure 3.5**

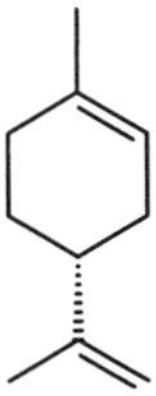

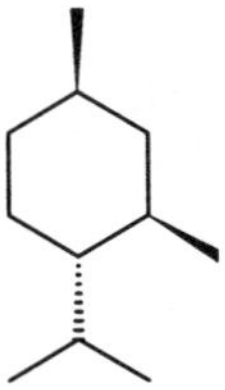

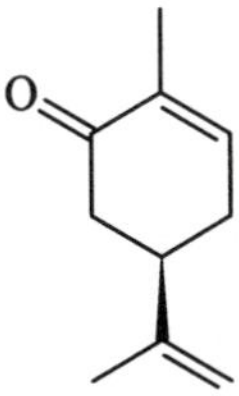

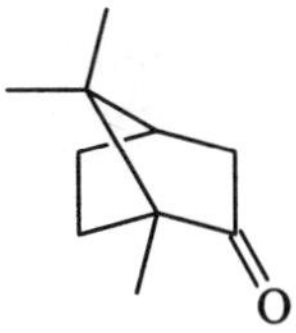

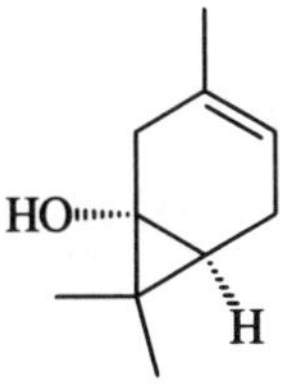

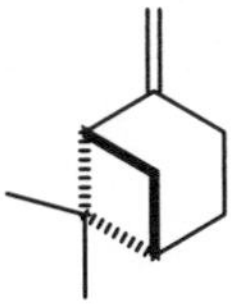

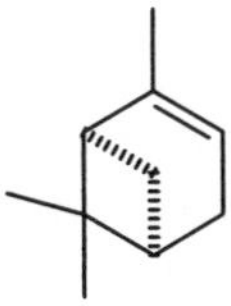

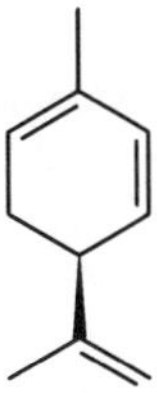

*Alkaloids*, which are usually isolated from plants, are the most expensive and complex members of the Chiral Pool.[11] Thus, these compounds find their main application as resolving agents in small-scale separations of racemic mixtures. One area of increasing use for the alkaloids is in asymmetric synthesis as chiral ligands for catalysts or as catalysts in base-promoted reactions.[12] Figure 3.6 includes examples of alkaloids included in the Chiral Pool.

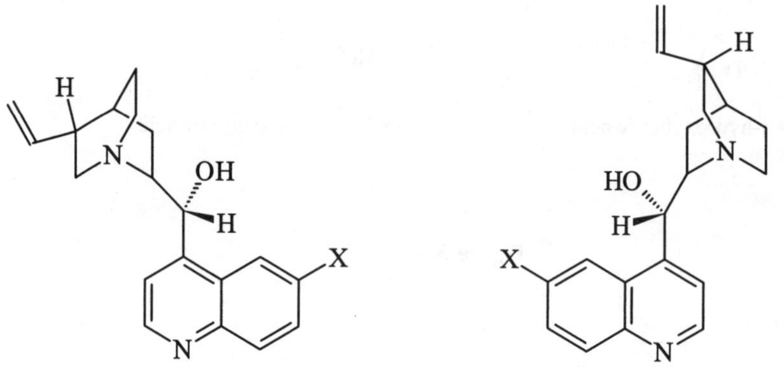

(8S, 9R)

X = OME Quinine
X = H Cinchonidine

(8R, 9S)

X = OME Quinidine
X = H Cinchonine

(1R, 2S)-ephedrine

(1S, 2S)-pseudoephedrine

**Figure 3.6**

*Steroids* are also a class of compounds that some consider members of the Chiral Pool. Basic steroid derivatives find use as starting materials for more complex products containing the steroid nucleus.

## Derivatives of Chiral Pool Materials

Many of the members of the Chiral Pool can be utilized in the synthesis of optically active compounds. Carbohydrates, amino acids, and hydroxy acids have all been utilized in commercial applications.

Monosaccharides undergo catalytic reduction to polyols and reductive amination[13] to aminopolyols. One example of the first step is the catalytic hydrogenation (yields ranging from 97-99%) of D-glucose to D-sorbitol using a Raney nickel catalyst.[14] D-sorbitol itself is used in the Reichstein-Grüssner process for manufacturing L-ascorbic acid (Vitamin C), which is a large-scale commercial route to this desired optically active chemical.[15] Oxidation of D-glucose (microbial, via fermentation, or noble metal catalyzed) affords a wide variety of useful chiral molecules such as D-gluconic acid, 5-keto-D-gluconic acid, D-glucuronic acid, D-glucaric acid, and others.[16]

The carbohydrates do pose difficulties for chemists seeking to utilize them as chiral building blocks.[17] The multiple numbers of hydroxyl groups with similar reactivity make it difficult to achieve any selectivity between them. Also, there are several chiral centers, and often procedures using carbohydrates become very complicated in addressing these two issues.

Recently interest in $C_3$-chirons (three carbon compounds containing an asymmetric center) as starting materials for chiral pharmaceuticals has resulted in a reevaluation of the utility of carbohydrates as commercial chiral raw materials. In particular, the $C_3$-chirons with the asymmetric atom at the central carbon position have been studied.[18] These compounds include protected glycerol, glyceraldehde, and glyceric acid derivatives, all of which can be prepared from protected or unprotected carbohydrates.

Two other commercial examples using carbohydrates as raw materials include the preparation of both enantiomers of lactic acid from glucose or lactose[19] and the synthesis of (S)-solketal[20] via cleavage of mannitol.

Amino acids find wide use in the synthesis of optically active commercial products. L-aspartic acid and L-phenylalanine are key materials in the preparation of the artificial sweetener aspartame.[21] The amino acids L-proline, L-lysine, and L-alanine are all used in the manufacture of ACE-inhibitor drugs[22] (lisinopril, enalapril, captopril for example). Several of the Chiral Pool amino acids have also been used in the synthesis of carbapenem antibiotics[23] (ampicillin is one).

Chiral hydroxy acids have been used for the synthesis of agrochemicals and pharmaceuticals. R-Lactic acid has been employed as an intermediate in the synthesis of optically active herbicides.[24] (R)-β–Hydroxybutyric acid serves as a key raw material for the preparation of carbapenam intermediates.[25] One synthesis of the ACE inhibitor captopril utilizes (R)-β–hydroxyisobutyric acid as a key chiral starting material.[26]
Figure 3.7 shows selected commercial products prepared from Chiral Pool starting materials.

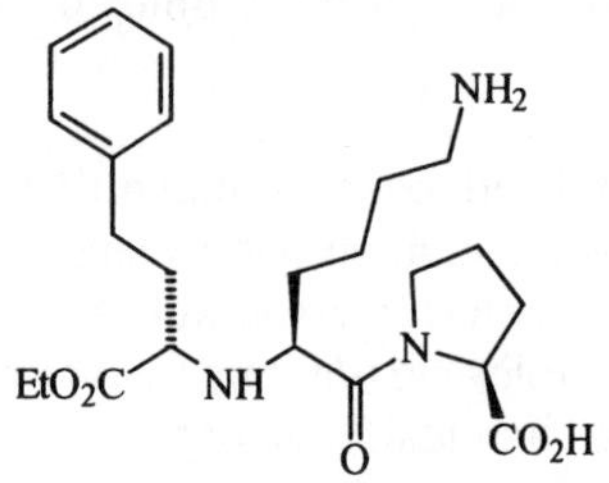

lisinopril

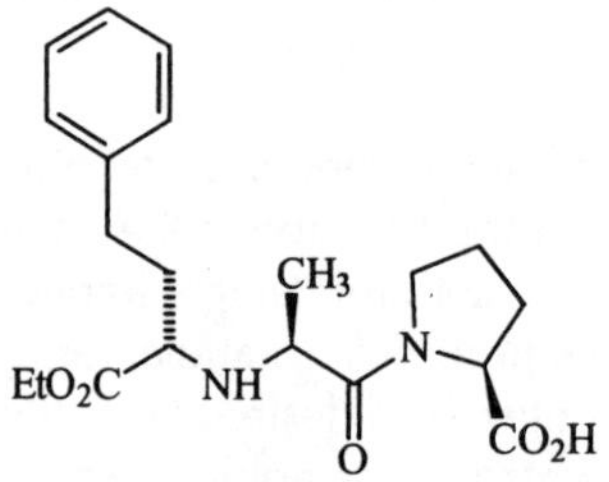

enalapril

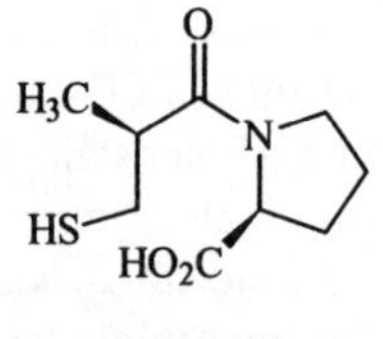

captopril

ACE (angiotensin-converting enzyme) inhibitors from L-proline[23a]

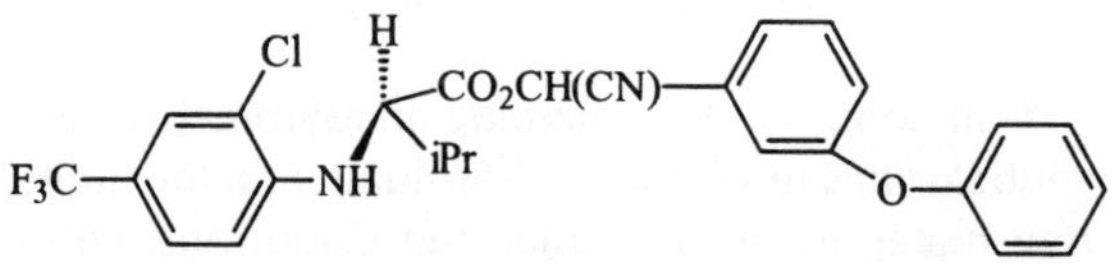

fluvalinate - insecticide from D-valine[27a]

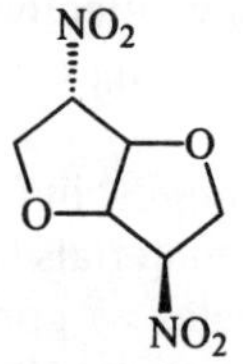

isosorbide dentrate - coronary vasodilator from D-sorbitol[27b]

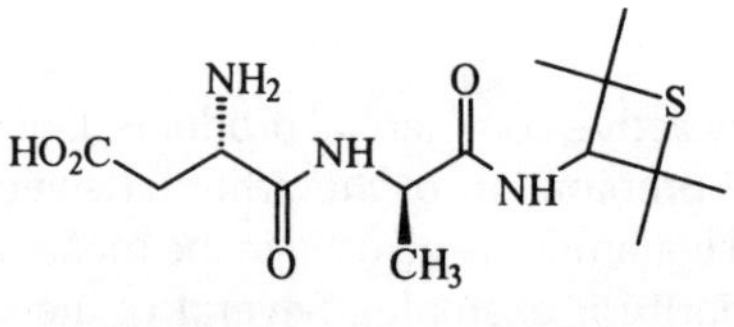

alitame - sweetener from D-alanine[27c]

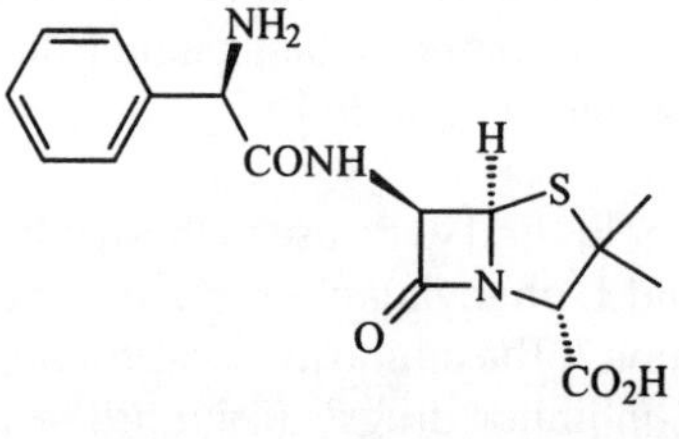

ampicillin - antibiotic from D-phenylglycine[27d]

**Figure 3.7**

## Resolution of Racemic Mixtures

Although most research efforts these days are focused on applying new asymmetric organic synthesis and enzymatic methodologies to commercial applications, classical resolution still remains a widely used technique for industrial scale separation of enantiomers. Issues associated with resolution of enantiomers include selection and recovery of the resolving agent (if required), racemization of the undesired isomer, and choice of biocatalysis or chemocatalysis (if applicable). There are three classes of resolutions – preferential crystallization, crystallization of diastereomeric salts, and kinetic resolution.

### *Preferential crystallization*

Separation of enantiomers via preferential, or direct crystallization is the more attractive option, for no derivatization or reaction of the starting material is required. Preferential crystallization, however, requires that the racemic mixture exists as a conglomerate.[28] In a conglomerate, the individual crystals form as pure enantiomers and as a result the crystalline racemic mixture is actually a physical combination of equal amounts of the two enantiomerically pure crystalline structures.

Approximately 10% of all compounds that exist as racemates can be classified as conglomerates, therefore preferential crystallization is not available as a general method for resolution of most mixtures of enantiomers. The other 90% of racemates are referred to as racemic compounds.[28] Crystals of these materials are composed of an ordered array of both enantiomers, with each crystal containing an equal amount of each isomer. Preferential crystallization will not work for these compounds.

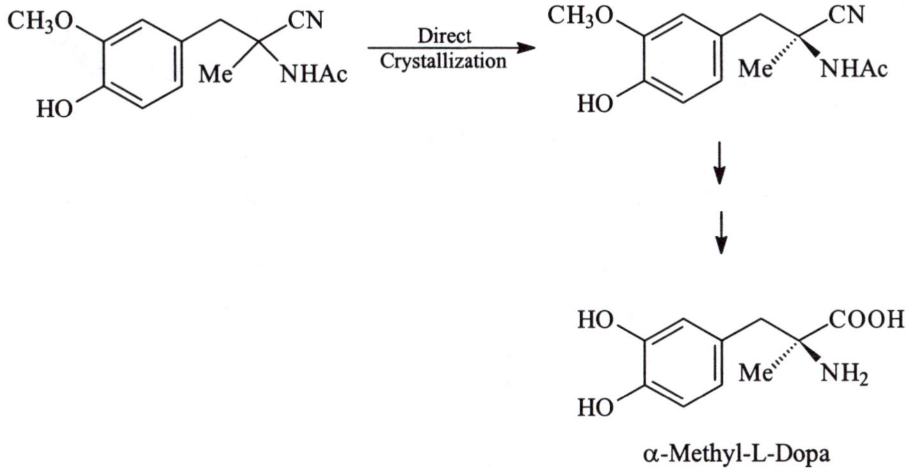

**Figure 3.8**[29]

One commercial process for preferential crystallization of conglomerates involves simultaneous crystallization of the two enantiomers in two separate chambers.[29] A supersaturated solution is passed through the chambers, which are seeded with crystals of each of the pure isomers. The material removed from these chambers through crystallization is replaced as the racemic mixture in a third chamber prior to re-cooling the solution and sending it through the process again. This method has been used in the preparation of α-methyl-L-dopa[30] (see Figure 3.8) and for the $C_3$-synthon glycidyl-3-nitrobenzenesulfonate.[31]

A second commercial process for direct crystallization of conglomerates involves the collection of alternating crops of each enantiomer in a single crystallization vessel.[32] This method is referred to as resolution by entrainment and has been used in the synthesis of chloramphenicol[33] and intermediates for benzothiazepines.[34] In this method, a supersaturated solution artificially enriched with one isomer is cooled and seeded, resulting in isolation of approximately twice the amount of the desired enantiomer originally contained in the solution. The removed material is replaced with the other enantiomer, and the solution cooled and seeded. The number of repetitions is limited only by the level of impurities that build up through each cycle.

If preferential crystallization occurs simultaneously with spontaneous *in-situ* racemization of the enantiomer that remains in excess in the solution, the process is termed crystallization induced asymmetric transformation or deracemization.[35] In this instance, a theoretical yield of 100% can be obtained. An example of the commercial application of this technique is the preferential crystallization of 1,4-benzodiazepene derivatives.[36] This crystallization occurs with spontaneous racemization in solution at room temperature, so that upon standing, one enantiomer crystallizes in 75% yield (see Figure 3.9).

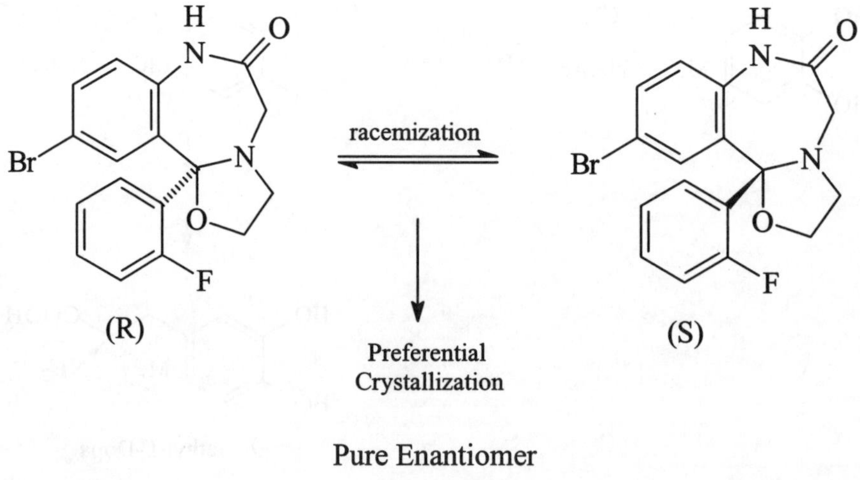

**Figure 3.9**[36]

### *Diastereomer crystallization*

For racemic compounds, which cannot be separated by preferential crystallization, crystallization of diastereomeric salts of the two enantiomers is often an attractive option. Diastereomeric salts are about three times as likely to exist as conglomerates when compared to general racemic mixtures.[37] In this case, the choice of resolving agent, solvent composition, the racemization of the unwanted isomer, and the recovery of the chiral auxiliary are all issues in determining the practical and economical viability of a process.

Advantages of diastereomer crystallization include the simplicity of the process, the ability to use standard production equipment, its flexibility, and that it is suited to intermittent batch production.[38] A limitation of this procedure is that only enantiomers with appropriate functionality can form diastereomeric salts of acids and bases. Thus simple hydrocarbons and other unfunctionalized compounds cannot be derivatized and then separated in this manner. Other negatives are associated with the costs of the production equipment, storage space for material that requires reworking, recovery of the resolving agent, and racemizing the unwanted isomer.[38]

### *Diastereomer crystallization – resolving agents*

Advancements in the use of computer aided molecular modeling have provided some assistance to professionals in the rational design and selection of resolving agents. A list of criteria for good resolving agents is included in Table 3.1.[39]

**Table 3.1: Criteria for Good Resolving Agents[39]**

| |
|---|
| Diastereomeric compound crystallizes well |
| Formation of diastereomeric salts occurs readily |
| Diastereomeric salts can be separated easily |
| Inexpensive |
| Available in optically active pure form |
| Chiral center is as close as possible to functional group responsible for salt formation |
| Available as both enantiomers |
| Have tight rigid structures |
| Strong acids/bases give best results |
| Molecular weight as low as possible |
| Easily recovered |

The Chiral pool is one key source for resolving agents. There are also several commercially available synthetic resolving agents. Figures 3.10[40] and 3.11[41, 40b] include examples of common chiral acids and bases used as resolving agents.

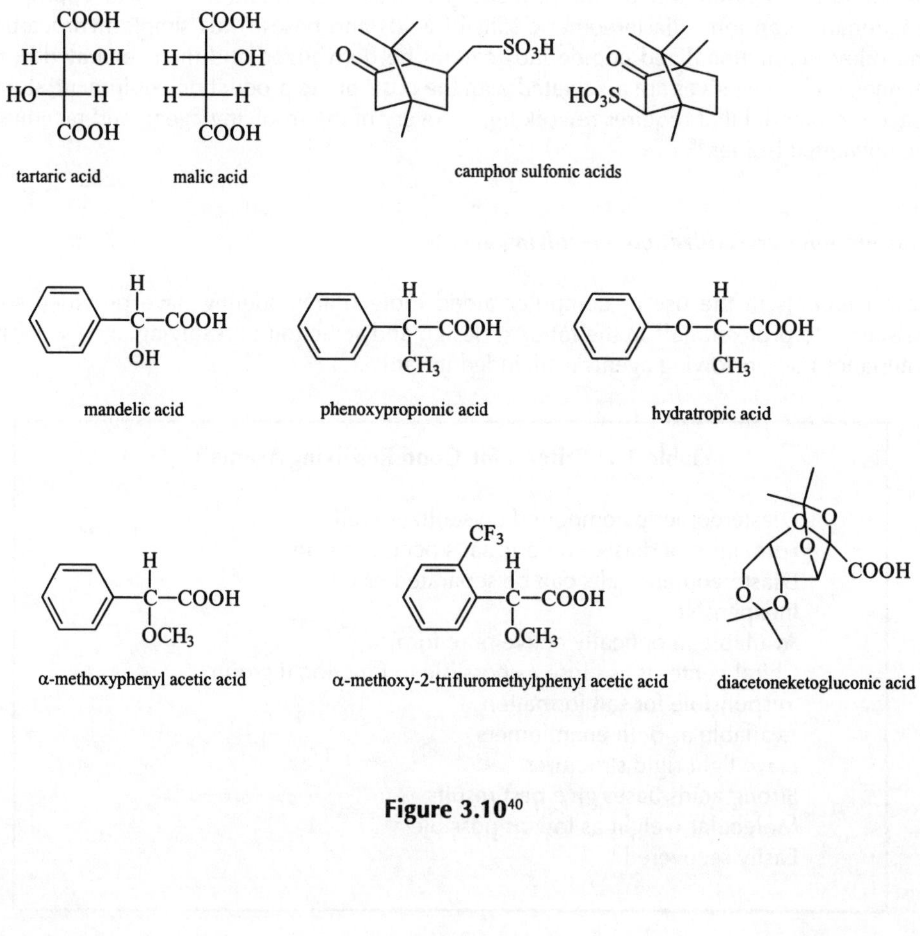

tartaric acid

malic acid

camphor sulfonic acids

mandelic acid

phenoxypropionic acid

hydratropic acid

α-methoxyphenyl acetic acid

α-methoxy-2-triflurومethylphenyl acetic acid

diacetoneketogluconic acid

**Figure 3.10**[40]

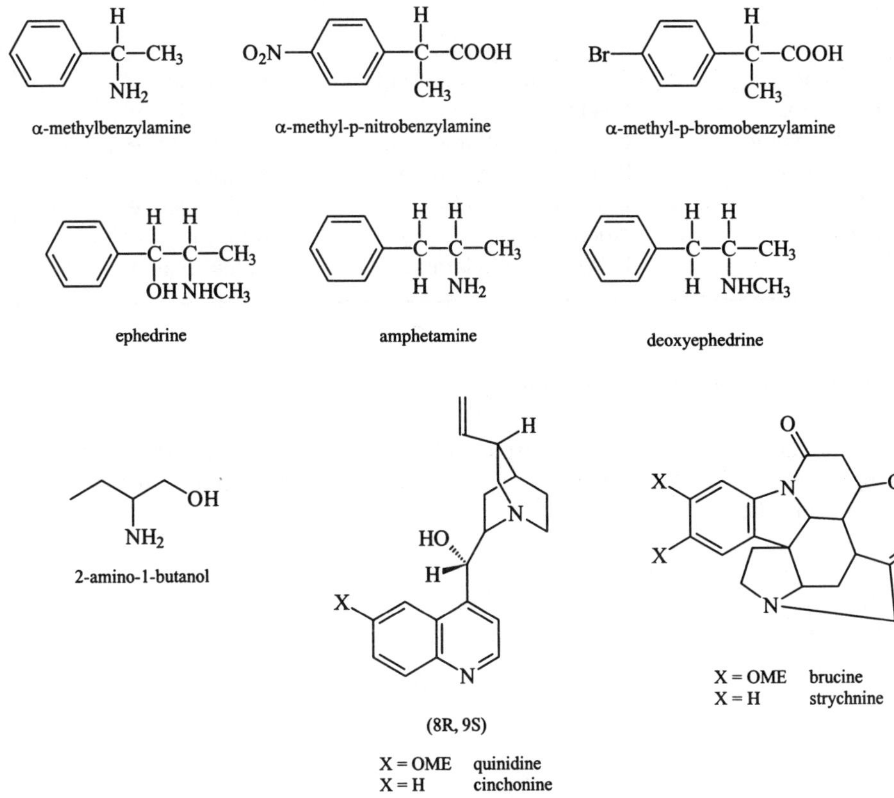

α-methylbenzylamine

α-methyl-p-nitrobenzylamine

α-methyl-p-bromobenzylamine

ephedrine

amphetamine

deoxyephedrine

2-amino-1-butanol

(8R, 9S)

X = OME quinidine
X = H cinchonine

X = OME brucine
X = H strychnine

**Figure 3.11**[41, 40b]

## *Diastereomer crystallization – commercial examples*

Many types of pharmaceuticals have been prepared utilizing crystallization of diastereomeric salts as a means of obtaining optically pure intermediates. Table 3.2 includes a listing of a few selected drugs and the resolving agents employed in the preparation of these products.[42]

**Table 3.2: Drugs Prepared via Classical Resolution**[42]

| Drug | Resolving Agent | Activity |
|---|---|---|
| Ampicillin | D-camphosulfonic Acid | Antibiotic |
| Ethambutol | L-(+)-tartartic acid | Tuberculostatic |
| Choramphenicol | D-camphosulfonic Acid | Anti-infective |
| Fosfomycin | R-(+)-phenethylamine | Antibiotic |
| Thiampenicol | D-(-)-tartartic acid | Anti-infective |
| Naproxen | cinchonidine | Anti-inflammatory |
| Diltiazem | R-(+)-phenethylamine | Calcium antagonist |

### *Diastereomer crystallization – designer resolving agents*

The new chiral phosphoric acid esters depicted in Figure 3.12[43] are an example of 'designer resolving agents,' or synthetic chemicals specifically prepared to serve as resolving agents for certain substrates.

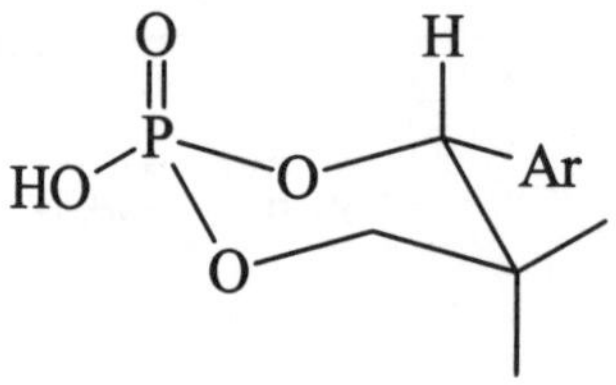

O H HO P O Ar O

Ar = Ph, o-ClPh, 2,4-$Cl_2$Ph, o-$CH_3$OPh

**Figure 3.12**[43]

### *Diastereomer crystallization – diastereomer interconversion*

Crystallization of diastereomeric salts can also occur in conjunction with racemization of the unwanted isomer. Asymmetric transformation of diastereomeric salts, also known as diastereomer interconversion, can yield a theoretical 100% of one isomer.

One example includes the conversion of L-proline to D-proline using (+)-L-tartaric acid (Figure 3.13).[44] In this reaction, a catalytic amount of aldehyde is used for the reversible formation of a Schiff's base. The L-proline forms a salt with the tartaric acid and precipitates out, driving the racemization. Yields are obtained in the range of 93-95% with 93-95% ee.

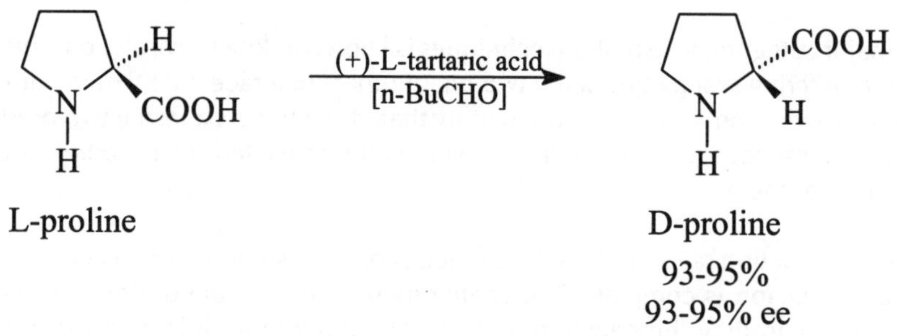

L-proline → (+)-L-tartaric acid [n-BuCHO] → D-proline
93-95%
93-95% ee

**Figure 3.13**[44]

Many derivatives of amino acids can be isolated as pure enantiomers using tartaric acid and a catalytic amount of a carbonyl compound (see Figure 3.14).[45]

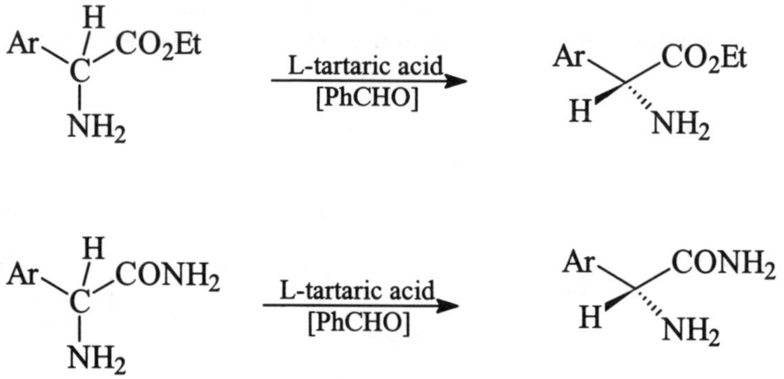

Ar–CH(NH$_2$)–$CO_2$Et → L-tartaric acid [PhCHO] → Ar–CH(NH$_2$)–$CO_2$Et

Ar–CH(NH$_2$)–CONH$_2$ → L-tartaric acid [PhCHO] → Ar–CH(NH$_2$)–CONH$_2$

**Figure 3.14**[45]

***Diastereomer crystallization – miscellaneous topics***

Since only one of the two diastereomers is desired, theoretically only ½ an equivalent of resolving agent should be needed. This 'method of half quantities' was first proposed by Marckwald.[46] It is most often more economical to use between ½ and one equivalent. Pope and Peachey developed a process that uses ½ an equivalent of chiral resolving agent with ½ equivalent of an achiral acid or base.[47] In this case, the slat formed from reaction with the chiral resolving agent precipitates while the salt formed with the achiral acid or base remains in solution. Another option is to react the diastereomeric salt (formed from an achiral acid or base) of a racemic mixture with ½ an equivalent of a chiral resolving agent.[48]

## *Kinetic resolution*

The third major category of resolution techniques is known as kinetic resolution. This process relies on the differing reactivities of the two enantiomers in a racemic mixture with a chiral reagent. One of the isomers reacts more readily than the other. As a result, two products are obtained from these reactions – one enantiomer remains unreacted and the other is converted into a different product.

The ratio of the reaction rates predicts the efficiency of the resolution. The reaction is stopped before the conversion is complete. The enantiomeric purity of the desired isomer can be controlled by controlling the degree of conversion. Some times it is worth sacrificing the yield of product to gain higher optical purities.

## *Kinetic resolution – chemical*

Kinetic resolutions can be achieved with chemical or enzymatic methods. One example of a stoichiometric chemical resolution is the use of the Sharpless epoxidation reaction to form chiral epoxides.[49] This reaction has also been developed as a catalytic process by Sharpless[50] (see Figure 3.15). The isomer of the allylic alcohol undergoes the epoxidation reaction while the other enantiomer does not react at all. Catalytic routes are preferred because they require minimal use of expensive reagents and allow for easier isolation of products.

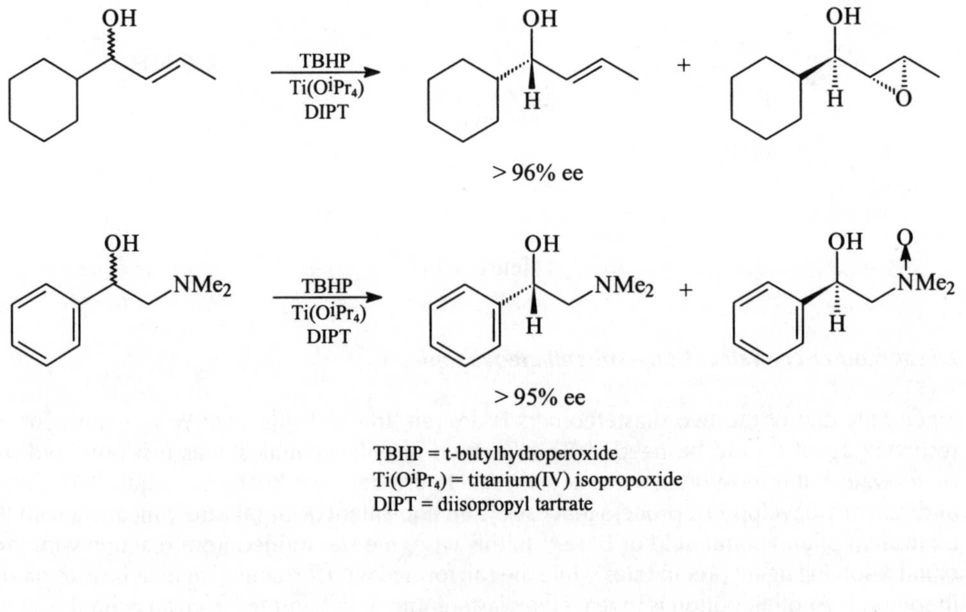

OH TBHP Ti(OiPr4) DIPT OH H + OH H O

> 96% ee

OH NMe2 TBHP Ti(OiPr4) DIPT OH NMe2 H + OH O NMe2 H

> 95% ee

TBHP = t-butylhydroperoxide
Ti(OiPr4) = titanium(IV) isopropoxide
DIPT = diisopropyl tartrate

**Figure 3.15**[50]

Hydrolytic kinetic resolution (HKR) of epoxides involves the enantioselective ring opening of terminal epoxides catalyzed by a chiral cobalt complex to give greater than 98% ee for each isomer.[51] This reaction is particularly attractive because it is carried out in water and the catalyst can be regenerated easily with acetic acid without loss of activity or selectivity.

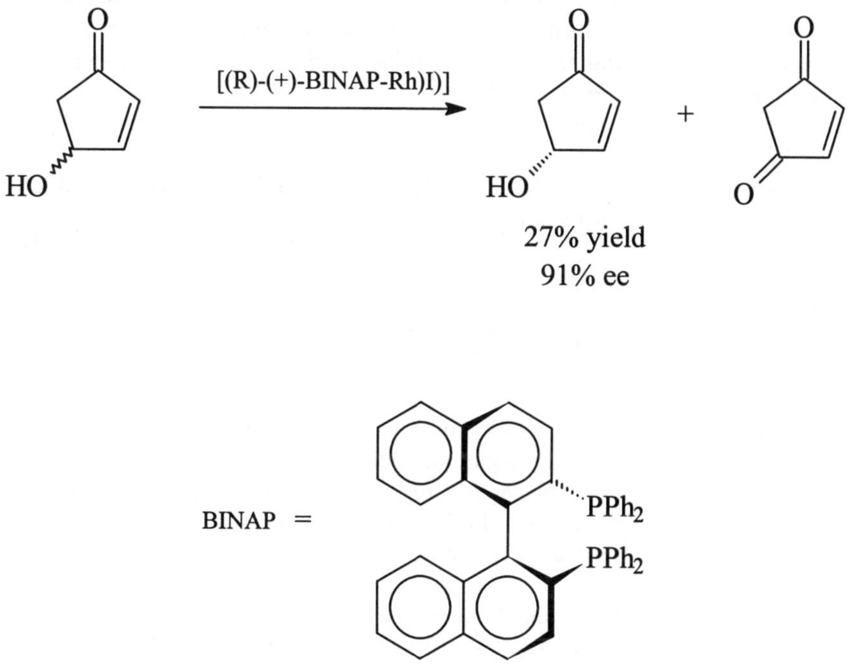

[(R)-(+)-BINAP-Rh)I)]

HO

HO

27% yield
91% ee

BINAP =

PPh$_2$
PPh$_2$

**Figure 3.16**[52]

A third example employs a Rhodium BINAP catalyst in a selective oxidation process.[52] In the example shown in Figure 3.16, one enantiomer remains as the hydroxy ketone while the other is oxidized to the diketone.

### *Kinetic resolution – enzymatic*

Enzymatic kinetic resolutions have received a great deal of attention over the last several years. New techniques for evaluating the structure and activity of enzymes and advances in the technology for preparing them have opened the door to many new opportunities. The use of enzymes for kinetic resolutions is attractive because the reactions are catalytic in nature, highly selective, environmentally benign, and can be conducted under mild reaction conditions.[53]

Kinetic resolution of alcohols, esters, amides, and acids has been achieved with the use of lipase and esterase enzymes. The reactions occur via enantioselective hydrolysis, esterification, and acylation reactions.

The *humida langinosa* lipase enzyme (lipase CE) has been used in the kinetic resolution of α,α-disubstituted α–amino acids via the hydrolysis of amino esters (see Figure 3.17).[54] These chemicals are found in natural peptide antibiotics, serve as chiral intermediates for pharmaceuticals and other biologically active agents, and act as inhibitors of acid decarboxylase enzymes. Routes to (S)-β–blocker intermediates have also utilized lipase based enzymatic kinetic resolution reactions.[55]

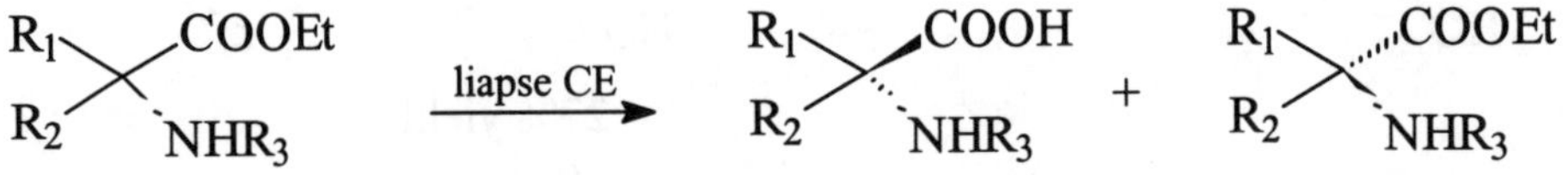

**Figure 3.17**[54]

Acylase, esterase, and aminopeptidase enzymes have been employed for the kinetic resolution of amino acids. As discussed previously, amino acids are major building blocks for a variety of pharmaceutical products. Aminoacylase immobilized on DEAE-Sephadex has been utilized to prepare various L-α–amino acids including L-methionine, L-valine, and L-phenylalanine from the corresponding α–ketocarboxylic acids.[56]

D-Amino acids specifically have found use as raw materials in the synthesis of angiotensin-converting enzyme (ACE) inhibitors and semi-synthetic β–lactam antibiotics. Another method for obtaining D-amino acids uses the D-hydantoinase enzymes.[57] Enantioselective ring cleavage of hydantoins has been achieved with this method of kinetic resolution. The products (D-carbamylamino acids) are then further hydrolyzed to the corresponding D-amino acids (see Chapter Four).

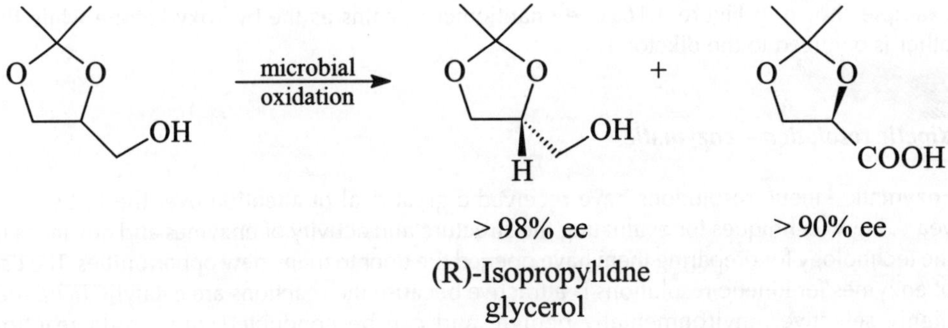

**Figure 3.18**[58]

Biological transformations have also been employed for the kinetic resolution of optically active compounds. Kinetic resolution of the chiral synthon R-isopropylidine glycerol has been achieved using microbial oxidation (see Figure 3.18), resulting in greater than 98% ee for the desired product and greater than 90% ee for the oxidized product.[58] Epichlorohydrin and epibromohydrin have both been obtained in high optical purities using selective microbial degradation of dihalo-substituted propanols.[59]

Dynamic kinetic resolutions are achieved when the unwanted isomer is racemized *in situ* under the reaction conditions. In this case, greater than 50% yield of the desired enantiomer can be obtained. ACE inhibitor intermediates have been prepared using this method.[60] The reaction with *candida humicola* results in greater than 50% yields of product with 98% ee (Figure 3.19).

CHO
$H_3C$ COOR
*candida humicola*
OH
$H_3C$ COOR
H
98% ee

**Figure 3.19**[60]

DL-α–amino-ε–caprolactam, an intermediate in the synthesis of L-lysine has also been hydrolyzed with simultaneous racemization of the D-isomer.[61] This dynamic kinetic resolution process utilizes two separate enzymes – L-aminocaprolactam hydrolase to form the desired product and D-amino caprolactam racemase to racemize the unwanted enantiomer.

Pyrethroid alkaloids have also been prepared from intermediates resolved via dynamic kinetic resolution using lipase enzymes in combination with basic substitution chemistry.[62]

There are other techniques available for the separation of racemic mixtures. Preparative HPLC,[63] simulated moving beds,[64] membranes[63] and supercritical fluids[65] have been investigated as viable technologies. Some successes have been achieved with large-scale HPLC and membrane separation technologies. The area of supercritical fluids presents many possibilities as well. Research in these areas is providing continued opportunities for further advances.

## Other Biological Processes

In recent years, the desire to develop more environmentally friendly approaches to the synthesis of commercial chemicals has brought attention to the use of biological methods for producing chiral molecules. Processes using microorganisms often can be carried out under mild conditions with minimal waste streams. Fermentation, microbial transformations (other than fermentation), and enzyme catalyzed reactions are the three main categories of biological processes available for preparing optically active molecules.

There are four important classes of microorganisms available for use in the synthesis of chiral chemicals: yeasts, molds, bacteria, and soil inhabiting actinomycetes.[66] Yeasts and molds are types of fungi, which are eukaryotic with a membrane enclosed nucleus and multiple chromosomes. Bacteria and soil-inhabiting actinomycetes are prokaryotic and are single cell organisms with no nuclear membrane and only one chromosome.

Because the structure of bacteria is much simpler than that of yeasts, these organisms have received most of the attention by investigators looking to enhance the activity and selectivity of microorganisms used in synthesis. Techniques such as recombinant DNA technology, site directed mutagenesis, and gene mapping have all contributed to the development of novel methods for the production of optically active products.

Careful control of process conditions must be accomplished in order to achieve successful syntheses with microorganisms.[67] Temperature and pH must be maintained within specific limits in order for the organisms to function properly. All necessary nutrients must also be made available in the correct quantities, and byproducts must be removed.

### *Fermentation*

Fermentation involves the reproduction of microorganisms in the presence of a source of carbon and energy and various nutrients.[68] Chiral Pool natural carbohydrates provide an excellent source of carbon and energy for the organisms that carry out the fermentation process. It is usually a several step process that is catalyzed by many different enzymes. In addition, the fermentation process cannot occur outside the living cell.

Fermentation is used in the production of antibiotics, amino acids, vitamins, enzymes, steroid hormones, and other biologically active pharmaceuticals and agrochemicals. The core structure of cefaclor, a cephalasporin antibiotic, has been prepared via fermentation.[69] The cholesterol-lowering drug lovastatin was also synthesized using fermentation as a method for establishing key chiral centers.[70] Two other coenzyme A (HmG-CoA) inhibiting drugs made using fermentation technology include prevastatin[71] and simvastatin[72] (see Figure 3.20).

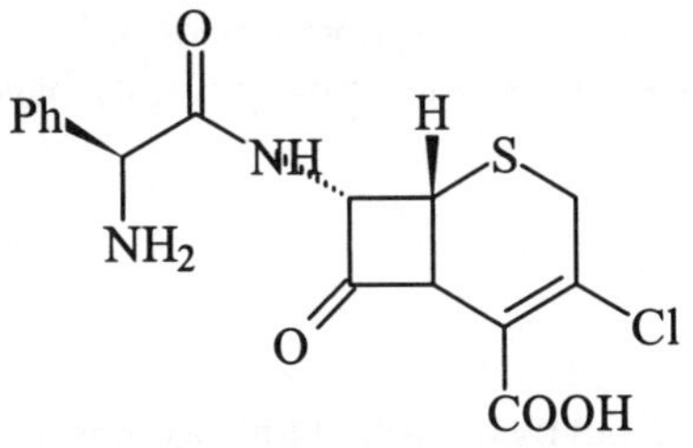

cefaclor

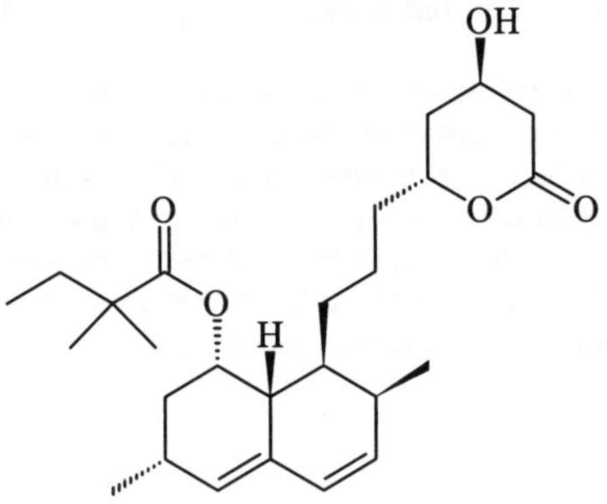

simvastatin

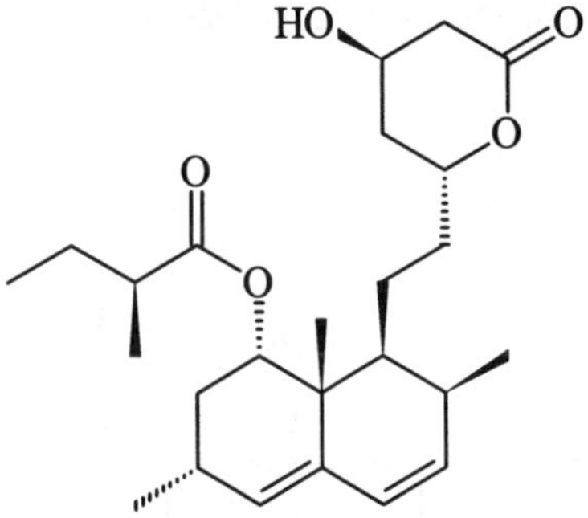

lovastatin

**Figure 3.20**

### *Precursor fermentation*

When microorganisms are used to convert foreign substrates into desired products, the process is referred to as precursor fermentation or microbial transformation. Unlike true fermentation, these reactions usually involve only one step with a single enzyme that often is cofactor dependent.[73]

Beyond the environmental considerations, there are two other distinct advantages in using microorganisms over other techniques for producing optically active entities.[74] Firstly, there is a tremendous diversity of cheap raw materials available for use in syntheses with microorganisms. Secondly, these processes provide very high regioselectivity and stereoselectivity in a single step.

There are disadvantages associated with the use of microorganism as well.[74] One problem is the low productivity (measured in $gl^{-1}hr^{-1}$) of many of these processes. Another is the low product concentrations (1-10% of total weight of organic material) obtained from the reactions, which leads to complex downstream processing (requiring extensive equipment) for isolation of the desired material. As a result of both of these issues, these processes generally have high fixed costs.

***Precursor fermentation – microbial oxidation***

Oxidations comprise the largest group of reactions carried out by microbial transformations. The oxidation of D-sorbitol to L-sorbose is one of the most well known processes in this class.[75] The microorganism *Rhizopus nigricans* has been employed in the regio- and stereoselective hydroxylation of the steroid nucleus for the synthesis of 11α–progesterone[76] (see Figure 3.21). Choice of the microorganism affects the site where oxidation occurs, allowing for formation of a total of four different products from the same starting material (via 11α–hydroxylation, 11β–hydroxylation, 16α–hydroxylation, and C(1)-dehydrogenation).

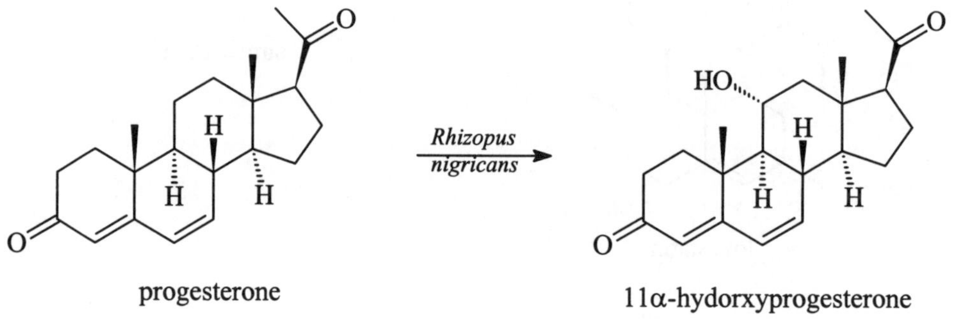

**Figure 3.21**

Enantioselective β–hydroxylation of aliphatic carboxylic acids[77] and the preparation of optically active dihydrocatechols via regioselective hydroxylation of aromatic compounds[78] have also been achieved with precursor fermentation.

Stereoselective epoxidation of prochiral olefins with *Pseudomonas oleovorans*,[79] *Corynebacterium equi*[80] or *Nocardia corallina*[81] has also found use in the synthesis of various optically active chemicals. The chiral epoxides have importance as chiral synthons for optically active alcohols, carboxylic acids, halohydrins, glycidyl ethers and other intermediates. *Nocardia corallina* in particular has been used in the preparation of ferroelectric liquid crystals.[82]

Stereoselective side chain oxidation with microorganisms has also found commercial applications. The drugs capotpril[83] and (S)-naproxen[84] as well as the herbicide fluazifop[85] have all been synthesized using *C. rugosa* for enantioselective oxidation of side chain substituents (Figure 3.21).

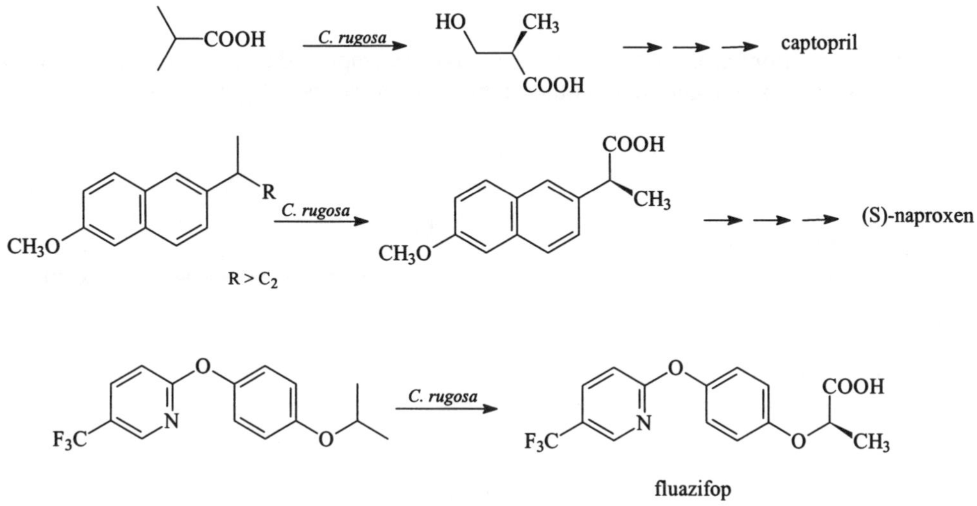

COOH
C. rugosa
HO
$CH_3$
COOH
captopril
R
C. rugosa
$CH_3O$
R > $C_2$
COOH
$CH_3$
$CH_3O$
(S)-naproxen
O
N
$F_3C$
O
C. rugosa
O
N
$F_3C$
O
COOH
$CH_3$
fluazifop

**Figure 3.22**

### *Precursor fermentation – microbial reduction*

Microbial reductions are another class of precursor fermentations. Most of these reactions are carried out with Baker's yeast (*Saccharomyces cerevisiae*).[86] They are attractive because the reagents are usually inexpensive, only simple equipment is required with no need for sterile conditions, and they are applicable to a wide range of substrates and provide high ee's.[86]

The disadvantages are related to the need for high dilution, causing difficult isolation of the product in addition to effluent problems. These reductions also occur with poor productivities, and therefore are only viable for production of Kg quantities of material.[87]

Examples of microbial reductions with Baker's yeast include reduction of β–ketoesters to give β–hydroxy carboxylic acids.[88] Figure 3.23 shows the preparation of optically active intermediates used for the synthesis of L-carnitine using this process.[89]

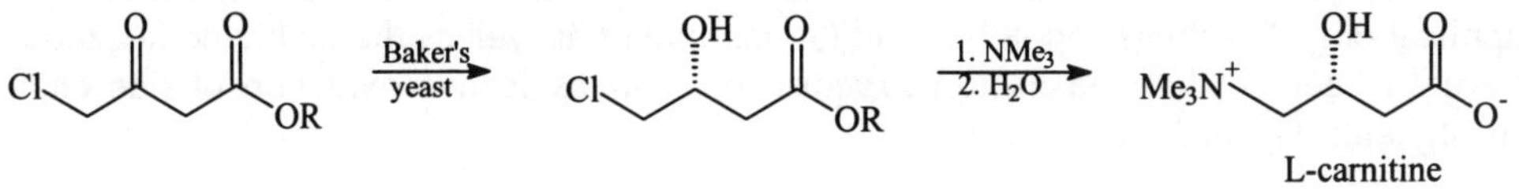

Baker's yeast

1. $NMe_3$ 2. $H_2O$

L-carnitine

**Figure 3.23**[89]

Reduction of oxoisophorone with bakers yeast has been employed in the synthesis of hydroxy carotenoids such as zeaxanthin and other terpenoid compounds (Figure 3.24).[90]

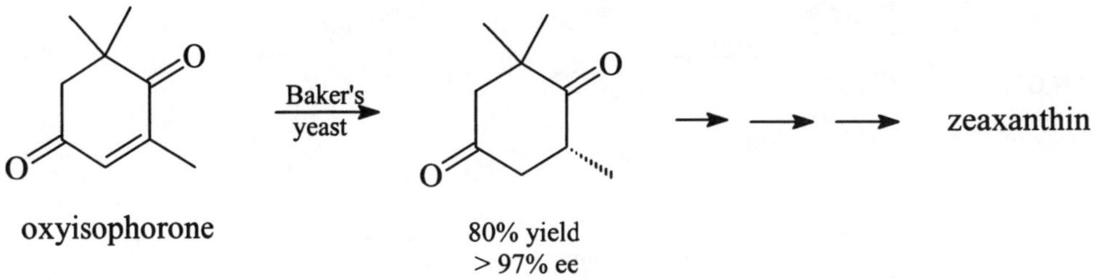

oxyisophorone

Baker's yeast

80% yield
> 97% ee

zeaxanthin

**Figure 3.24**[90]

Figure 3.25 shows the Baker's yeast reduction of 17-keto steroids to the 17-β–hydroxysteroids, which are used in the manufacture of steroid hormones.[91]

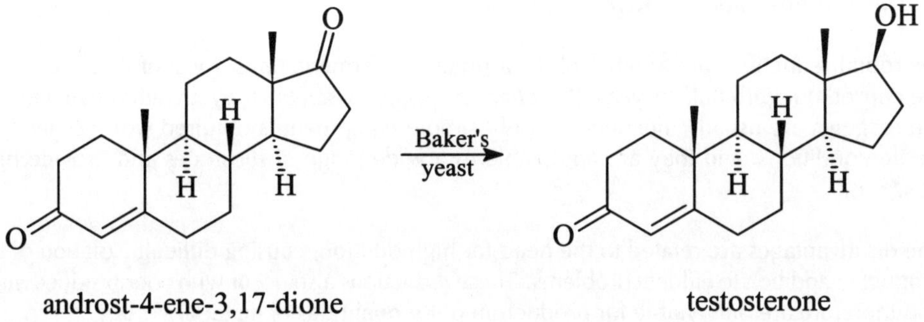

androst-4-ene-3,17-dione

Baker's yeast

testosterone

**Figure 3.25**[91]

*Agrobacterium sp* has also been used in microbial reductions. One example of this microbial transformation (Figure 3.26) is the preparation of D-pantoic acid,[92] which is an intermediate for the synthesis of panthothenic acid (vitamin $B_5$).

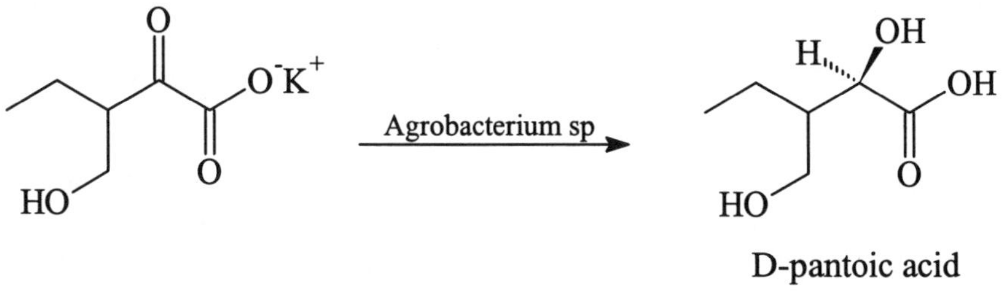
O
O⁻K⁺
O
HO
Agrobacterium sp
H
OH
OH
O
HO
D-pantoic acid

**Figure 3.26**[92]

***Precursor fermentation – microbial condensation reactions***

Microbial condensation reactions are a third group of precursor fermentations. The acyloin condensation of aromatic and α,β–unsaturated aldehydes with acetaldehyde mediated by Baker's yeast has been utilized in the preparation of the key intermediate for ephedrine and pseudoephedrine.[93] This reaction is shown in Figure 3.27.

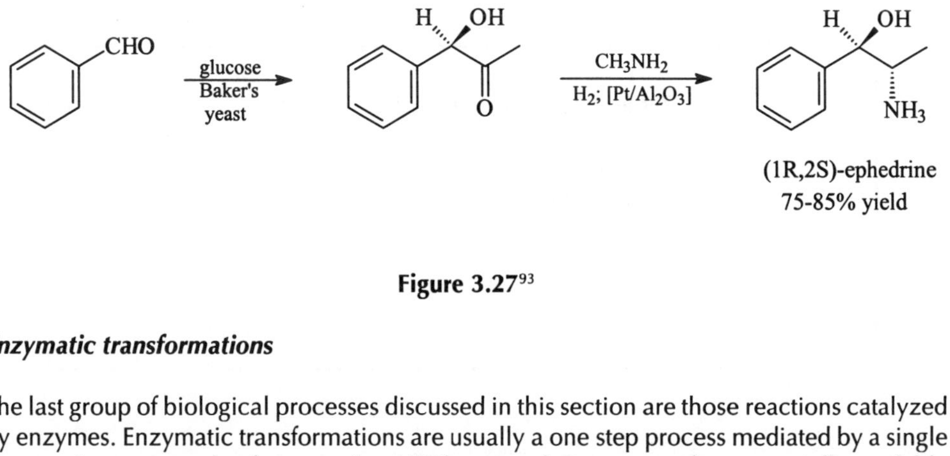
CHO
glucose
Baker's
yeast
H
OH
O
$CH_3NH_2$
$H_2$; [Pt/$Al_2O_3$]
H
OH
$NH_3$
(1R,2S)-ephedrine
75-85% yield

**Figure 3.27**[93]

***Enzymatic transformations***

The last group of biological processes discussed in this section are those reactions catalyzed by enzymes. Enzymatic transformations are usually a one step process mediated by a single enzyme (as compared to fermentations).[73] There are three groups of commercially available enzymes – those used for industrial purposes (detergents, dairy products [cheese making] and starch processing), those with therapeutic purposes, and those used in food and clinical diagnostic applications.[94]

There are several advantages to using enzymes for the asymmetric synthesis of compounds.[95] Like fermentations, they are very efficient, react under mild conditions, can often be used without any solvent, and provide complex molecules in fewer steps. Reactions catalyzed by enzymes proceed with a wide range of activities and high substrate, chemo-, regio-, and stereo-selectivity.

Disadvantages include the limited availability and higher price of specialty enzymes, instability, lack of flexibility, the possibility of substrate and/or product inhibition, and the need for cofactor regeneration for some.[95] Most enzymes have optimum temperature and pH ranges for peak performance (usually 40°C to 60°C and pH 5 –7). They are generally stable to high pressures, and supercritical fluids have been used in some enzymatic transformations.

Some enzymes have been found to provide enhanced reactivities in organic solvents.[96] The organic solvent can increase thermal stability and resistance to denaturization. The lack of hydrophilic groups in organic media often results in different substrate binding effects providing novel reactivities for traditional enzymes. Reactions in organic solvents are also more amenable to other organic synthesis techniques.

The stability of enzymes can also be increased by immobilizing them.[97] There are three possible methods for immobilizing enzymes – carrier binding, crosslinking, and entrapment. Membrane bioreactors are a specific example of the application of the entrapment method. This technique also aids in the recovery and recycling of these biological catalysts.

Sources of enzymes include the liver and pancreas of animals, plant extracts, and various microorganisms (yeast, bacteria, etc.).[98] Enzymes are classified by a four digit number called the Enzyme Commission Number (EC Number) and are assigned according to function.[99] The six major categories of enzymes are listed in Table 3.3. The category number is the first digit in the EC Number assigned to enzymes.

**Table 3.3: Major Categories of Enzymes**

Oxidoreductases
Transferases
Hydrolases
Lyases
Isomerases
Ligases

***Enzyme transformations – oxidoreductases***

Oxidoreductases include enzymes responsible for oxidation and reduction of various compounds. Dioxygenase has been demonstrated to catalyze the stereoselective cis-dihydroxylation of aromatic compounds,[100] which have recently been investigated as interesting intermediates for optically active products. Figure 3.28 shows one example of the use of this chemistry in the synthesis of antiviral drugs.[101]

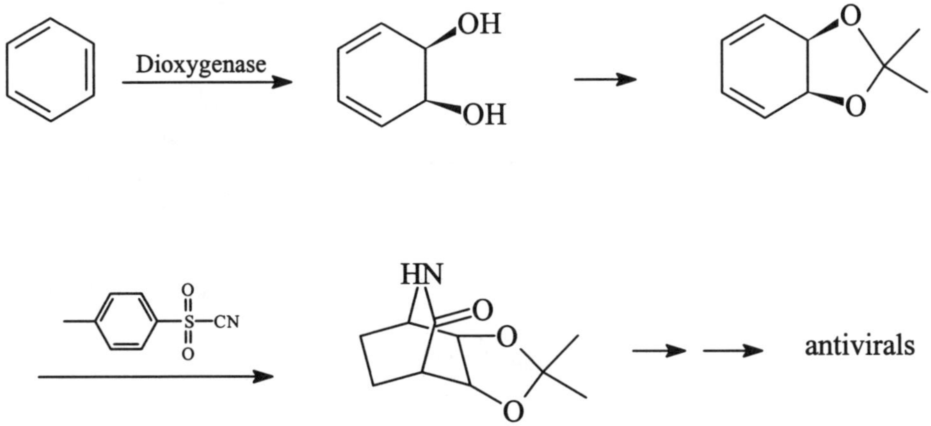

Dioxygenase
OH
OH
O
O
HN
O
O
O
S—CN
antivirals

**Figure 3.28**[101]

The major members of the oxidoreductase class of enzymes are the dehydrogenases, which catalyze oxidation and reduction processes, some of which require a cofactor to supply the hydride ion. Dehydrogenases are used for reduction of ketones[102] (lactate dehydrogenase) and reductive amination of a variety of α–keto acids to form amino acids.[103]

### *Enzyme transformations – transferases*

Transferases are used to transfer a substituent from one molecule to another. Threonine aldolase is used for the production of β–hydroxy α–amino acids. Natural amino acids can be prepared via reactions catalyzed by the aminotransferase enzyme.[104]

### *Enzyme transformations – hydrolases*

Hydrolases are the largest class of enzymes finding industrial applications (75%).[105] Most are used in the detergent and food industries.[106] Esterases, lipases, epoxidases, and proteases are all types of hydrolase enzymes.

Pig liver esterase (PLE) has found application in the synthesis of key optically active intermediates for carbapenem antiobiotics (Figure 3.29).[107] 2-Aminoglutaric acid, a symmetric diester is converted to the corresponding chiral amino acid using PLE and this intermediate is then further transformed into the pharmaceutically active product.

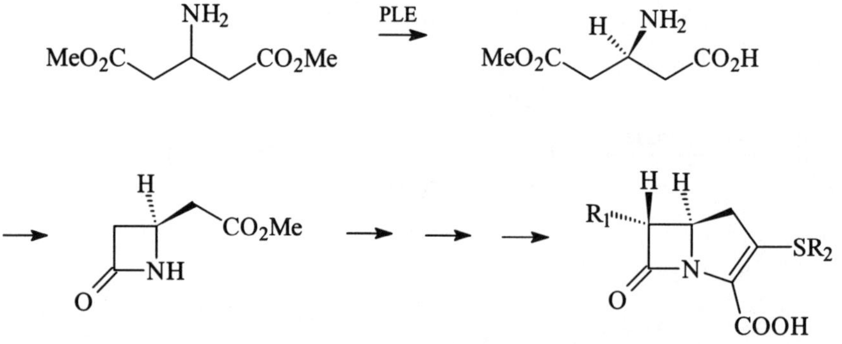


trans-carbapenem antibiotics

**Figure 3.29**[107]

Lipases, in addition to applications in kinetic resolution, are useful for the intramolecular esterification of hydroxycarbonyl compounds and esterification, transesterification, and interesterification reactions.[108] Porcine pancreatic lipase (PPL) has been used for the hydrolysis of meso-diacetals to prepare intermediates for pheromone syntheses.[109] PPL (or Chromobacterium viscosum lipase) has also been used in the synthesis of fusinopril (Figure 3.30).[110]

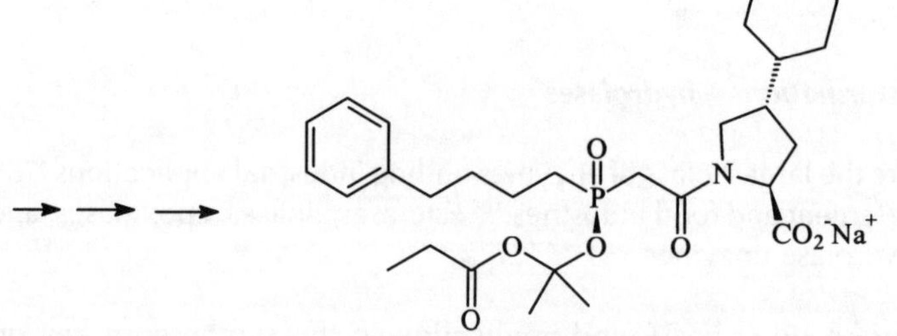


fusinopril

**Figure 3.30**[110]

Protease enzymes have been used in enantioselective and regioselective coupling reactions for the preparation of synthetic and semisynthetic peptides.[111] Another application for protease enzymes is shown in Figure 3.31, where they are used in the synthesis of the sweetener aspartame.[112]

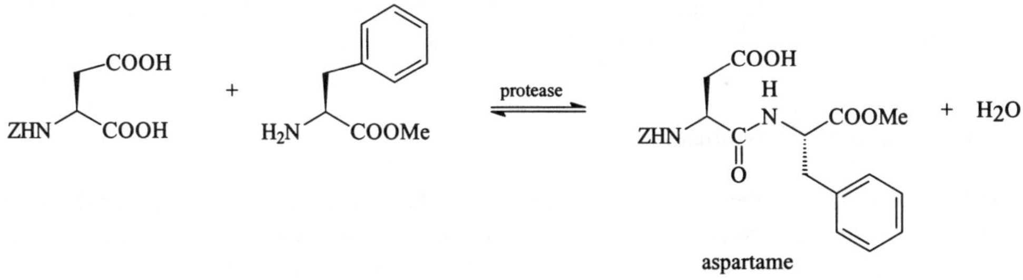

COOH ZHN COOH + $H_2N$ COOMe ⇌ (protease) ZHN C(=O) N(H) COOMe + $H_2O$

aspartame

**Figure 3.31**[112]

### *Enzyme transformations – lyases*

Lyases catalyze the cleavage of carbon-carbon, carbon-nitrogen, and carbon-oxygen bonds by means other than hydrolysis. Ammonia addition to fumaric acid has been employed for the synthesis of L-aspartic acid (Figure 3.32).[113] One application of the reverse reaction is in the synthesis of amino acids.[114] L-aspartate-β–decarboxylase has been used in whole cells for the synthesis of L-alanine on a multi-ton scale.[115]

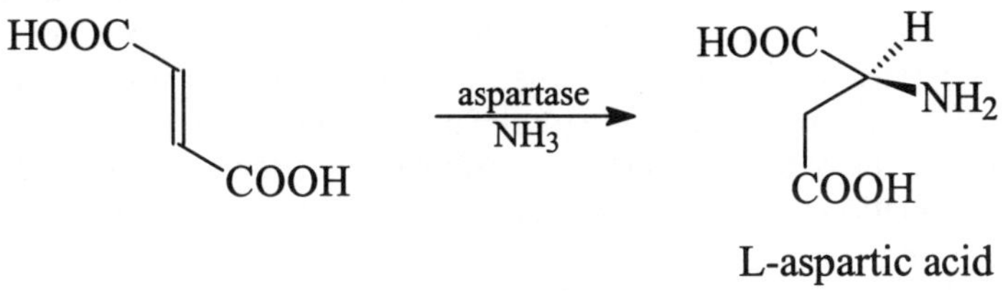

HOOC COOH → (aspartase, $NH_3$) HOOC H $NH_2$ COOH

L-aspartic acid

**Figure 3.32**[113]

### *Enzyme transformations – isomerases*

Isomerases include the racemases, epimerases, and mutases.[116] The racemases and epimerases all cause the stereochemistry of at least once chiral center to be changed. Mutases cause the transfer of a functional group to an adjacent carbon, resulting in the generation of a new asymmetric center. D-Xylose isomerase catalyzes the conversion of D-glucose to D-fructose, which is used in the production of high fructose corn syrup.[117] This enzyme is products in larger quantities than any other.

### *Enzyme transformations – ligases*

Ligases catalyze C-O, C-S, C-N, and C-C bonds. However, they have little application as biocatalysts for asymmetric synthesis of optically active molecules because reactions with these enzymes require the use of complex cofactors.[118]

## Asymmetric Organic Synthesis

Asymmetric organic synthesis has only become a field in its own right in the last two decades. The rapid advances in the area of organometallic chemistry have led to possibilities in the application of chiral transition metal catalysts to the synthesis of optically active compounds. Reactions with chiral acids and bases have also received attention. Substitution reactions also play a role in the asymmetric synthesis of chiral molecules.

### *Catalysts*

Catalytic asymmetric organic synthesis utilizes synthetic catalysts designed to promote formation of optically active products. Catalysts can be heterogeneous metals, homogeneous complexes, or soluble chiral acids and bases. Advantages of organometallic catalysts over enzymes include the broader range of reactivity available to the synthetic catalysts, the fact that these catalysts in general have greater stability to heat, oxidation, and pH, and the ability to optimize enantioselectivity and overall chirality by changing the chiral ligands incorporated into the catalysts.[119]

The success of asymmetric synthesis with chiral catalysts is due to the differing reactivities displayed by enantiomers when placed in a chiral environment. The optically active catalysts react at differing rates with each enantiomer. The two diastereomeric transition states that form possess different energies, and the more favored transition leads to product at a more rapid rate than the one of higher energy. The extent of the enantioselectivity depends on the degree of differentiation between the two diastereomeric transition states.

Transition metals are successful as catalysts for asymmetric synthesis for several reasons.[120] First, the availability of several oxidation states allows for the addition and elimination of substrates, reactants, and products. Secondly, transition metals possess the capability to coordinate to ligands in specific arrangements that establish stereo- and regioselectivities. Third, these catalysts preferentially bind small molecules for insertion into substrates. Lastly, transition metals are able to stabilize reactive intermediates, allowing for formation of products not available through traditional synthetic methods.

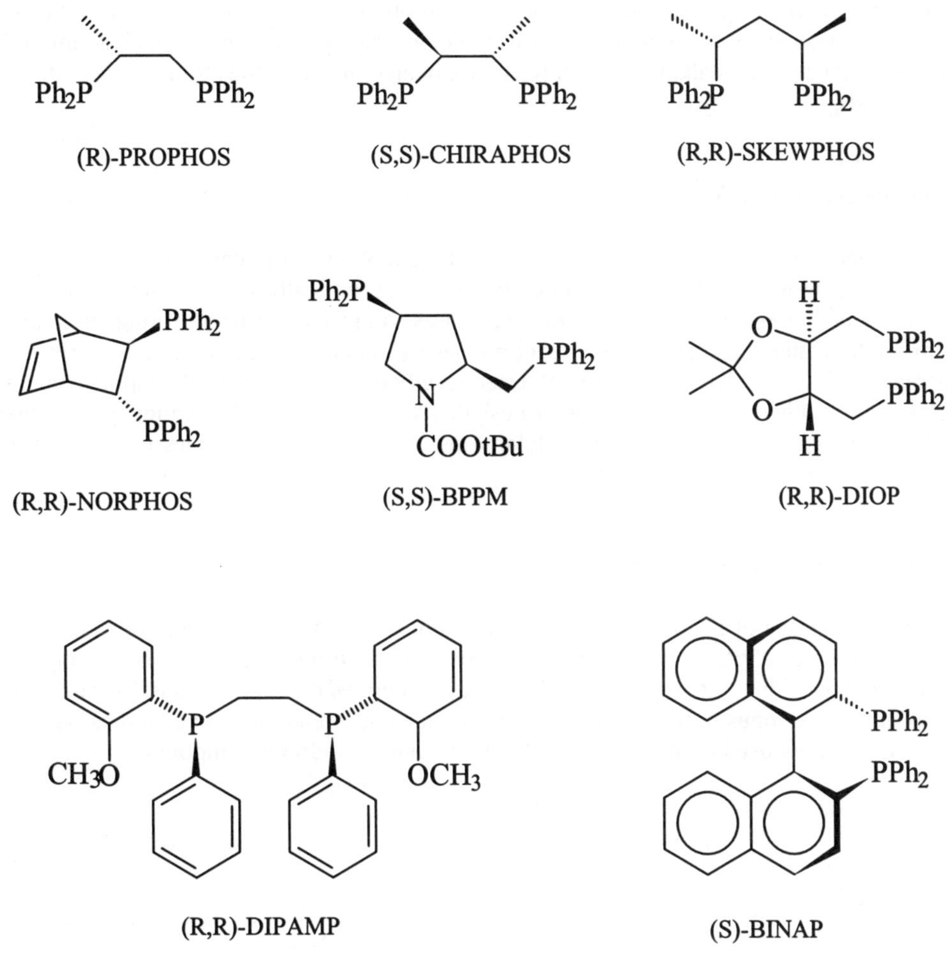

(R)-PROPHOS (S,S)-CHIRAPHOS (R,R)-SKEWPHOS

(R,R)-NORPHOS (S,S)-BPPM (R,R)-DIOP

(R,R)-DIPAMP (S)-BINAP

**Figure 3.33**[121]

The chiral ligands used to form these transition metal complexes are the key to establishing the regio and stereospecificity of catalytic asymmetric synthesis processes. Most are based on molecules found in the Chiral Pool. Almost all of the ligands successfully employed to prepare optically active compounds have been based on substituted (bis)phosphines. Some examples of these compounds are shown in Figure 3.33.[121]

### *Heterogeneous catalysis*

Heterogeneous catalysis involves reactions on a metal surface. Typically metals such as nickel, platinum and palladium are used for these processes. Hydrogenations are the key types of reactions done via heterogeneous asymmetric catalysis. For example, nickel-tartrate[122] and platinum-cinchona alkaloid[123] catalysts have been utilized for the reduction of carbonyl compounds.

### *Homogeneous catalysis*

Homogeneous catalysis has received most of the attention from academic and industrial chemists trying to develop new methods for preparing optically active compounds. One problem with homogeneous transition metal catalysts is the difficulty in recovering the catalyst from solution after completion of the reaction. Several different methods have been devised for overcoming this obstacle. Some catalysts have been placed on solid supports such as polymers.[124] Other catalysts have been designed with water soluble ligands (phosphines with sulfonate, carboxylate, or tetrabutylammonium groups).[125] Examples of some water soluble ligands are included in Figure 3.34.[126]

### *Homogeneous catalysis – hydrogenations*

The largest number of asymmetric reactions falls under the category of hydrogenations with homogeneous catalysts. A wide variety of substrates have been subjected to asymmetric hydrogenation with homogeneous catalysts. α,β–unsaturated carboxylic acids, β–keto esters, β–alkylamino ketones, allylic alcohols, z-enamines and z-enamides and imines have all been successfully used in the synthesis of commercially important compounds.

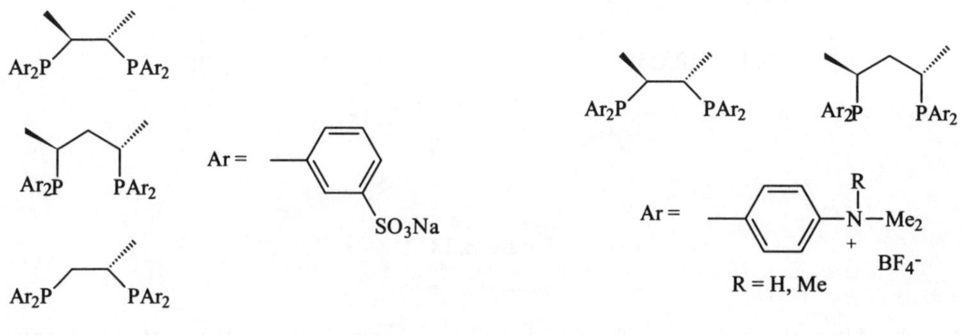

$Ar_2P$ $PAr_2$

$Ar_2P$ $PAr_2$

$Ar_2P$ $PAr_2$

Ar = (3-$SO_3Na$-phenyl)

$Ar_2P$ $PAr_2$ $Ar_2P$ $PAr_2$

Ar = (4-$N^+(R)Me_2$ $BF_4^-$ phenyl)

R = H, Me

**Figure 3.34**

One of the first industrial applications of asymmetric homogeneous hydrogenations was the Monsanto process for the preparation of L-Dopa.[127] The key step is shown in Figure 3.35.

Enantioselective hydrogenation of the unsaturated carboxylic acid intermediate was achieved using a Rhodium DIPAMP catalyst (see Figure 3.33 for DIPAMP structure). This catalyst was developed slowly over time as different ligands were tested to determine which would provided the highest selectivities.

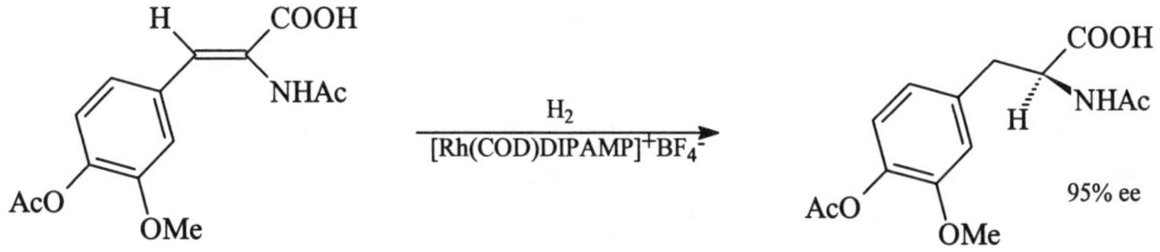

$$\xrightarrow[\text{[Rh(COD)DIPAMP]}^{+}\text{BF}_4^{-}]{H_2}$$

95% ee

**Figure 3.35**[127]

The hydrogenation of α,β–unsaturated carboxylic acids was also utilized in the synthesis of (S)-naproxen.[128] In this route, a ruthenium-BINAP catalyst provided high yields with high ee's (Figure 3.36).

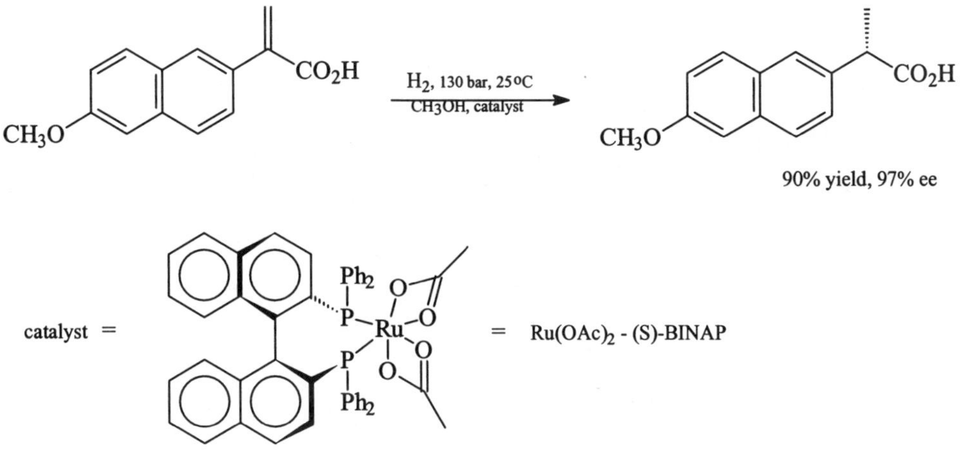

$$\xrightarrow[\text{CH}_3\text{OH, catalyst}]{H_2\text{, 130 bar, 25}^{o}\text{C}}$$

90% yield, 97% ee

catalyst = = $Ru(OAc)_2$ - (S)-BINAP

**Figure 3.36**[128]

β–keto esters have been hydrogenated in quantitative yields using $RuX_2$-(S)-BINAP as the chiral catalyst.[129] One example of this reaction is shown in Figure 3.37, where this method was utilized in the synthesis of an optically active intermediate to (R)-carnitine. This catalyst has also been found to provide successful asymmetric hydrogenations of 1,3-diketones, α–ketoesters, and γ–ketoesters.[130]

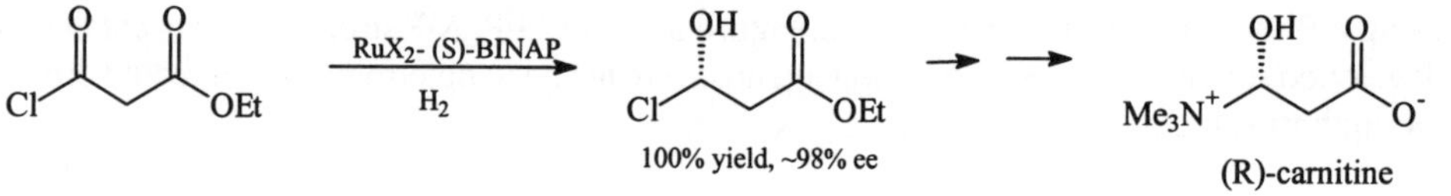

$RuX_2$- (S)-BINAP, $H_2$

100% yield, ~98% ee

(R)-carnitine

**Figure 3.37**

Enantioselective hydrogenation of β–alkylamino ketones has been used in the synthesis of intermediates for β–blockers[131] and the drug levamisol[132] (Figure 3.38). The catalyst for this asymmetric transformation is a rhodium complex of a sterically hindered bisphosphine.

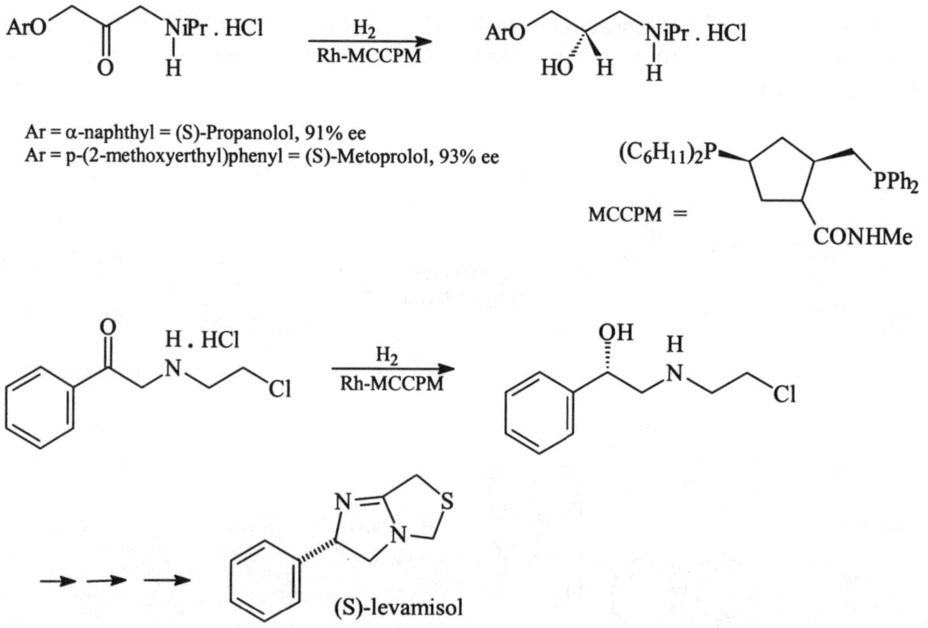

$H_2$, Rh-MCCPM

Ar = α-naphthyl = (S)-Propanolol, 91% ee
Ar = p-(2-methoxyerthyl)phenyl = (S)-Metoprolol, 93% ee

MCCPM = $(C_6H_{11})_2P$ … $PPh_2$ … CONHMe

$H_2$, Rh-MCCPM

(S)-levamisol

**Figure 3.38**[131, 132]

Hydrogenation of z-enamines and z-enamides has been accomplished with ee's as high as 95-100%.[133] The products resulting from these hydrogenation reactions are intermediates for alkaloids. Two important examples of these compounds that have been prepared using catalytic asymmetric hydrogenation of enamides and enamines are tetrahdyropapaverne and dextromethorphan.[134]

Ru-BINAP has also been applied to the preparation of key products in the flavor and fragrance industry. Figure 3.39 shows how the choice of substrate and chirality of the catalyst can control the optical activity of the product.[135]

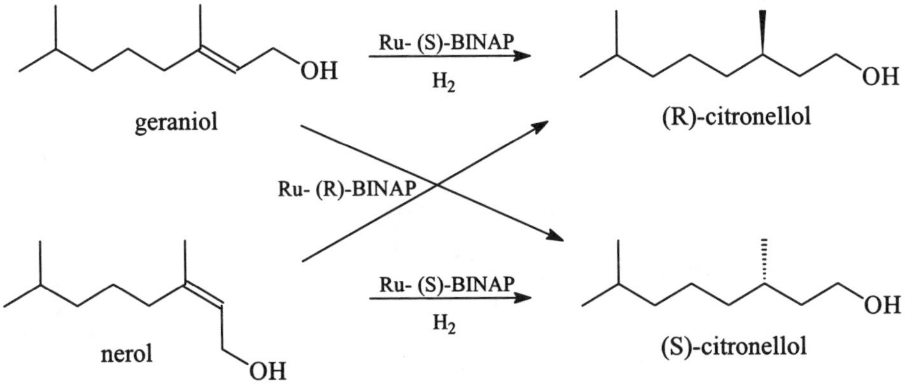

OH
geraniol
Ru- (S)-BINAP
$H_2$
OH
(R)-citronellol
Ru- (R)-BINAP
nerol
OH
Ru- (S)-BINAP
$H_2$
OH
(S)-citronellol

**Figure 3.39**

Chiral metallocene complexes have found increasing use in asymmetric synthesis over the last several years. These catalysts derive their chirality from the dissymmetric nature of the cylopentadienyl rings attached to the metal. An example of the use of such a catalyst is shown in Figure 3.40 for the preparation of the pesticide metolachlor.[136]

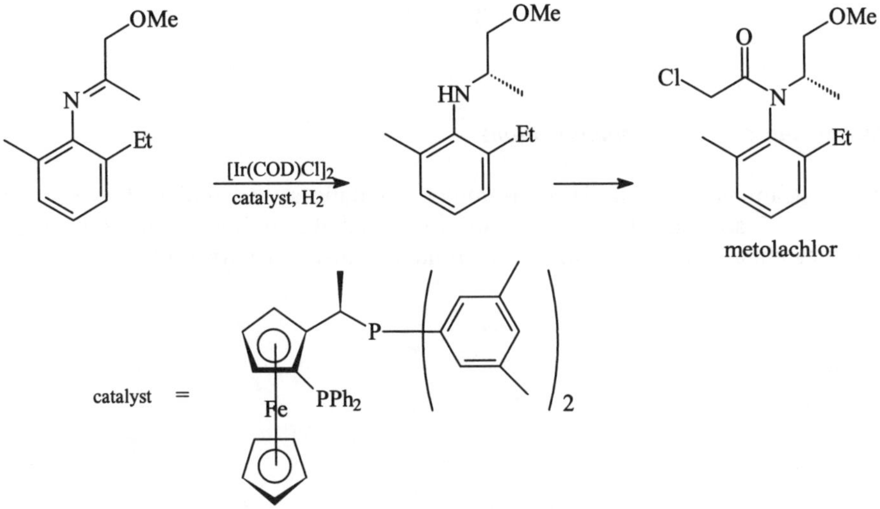

OMe
N
Et
[Ir(COD)Cl]$_2$
catalyst, $H_2$
OMe
HN
Et
OMe
O
Cl
N
Et
metolachlor
catalyst =
P
Fe
PPh$_2$
2

**Figure 3.40**[136]

Reduction of ketones with chiral boron complexes[137] has recently been used for the synthesis of pharmaceuticals and other optically active products. This chemistry involves the formation of a borane-tetrahydrofuran ($BH_3$:THF) complex in the presence of a catalytic amount of sterically hindered oxazaboraline. This chemistry has been utilized for the preparation of a key intermediate in the synthesis of setraline,[138] an antidepressent drug (Figure 3.41).

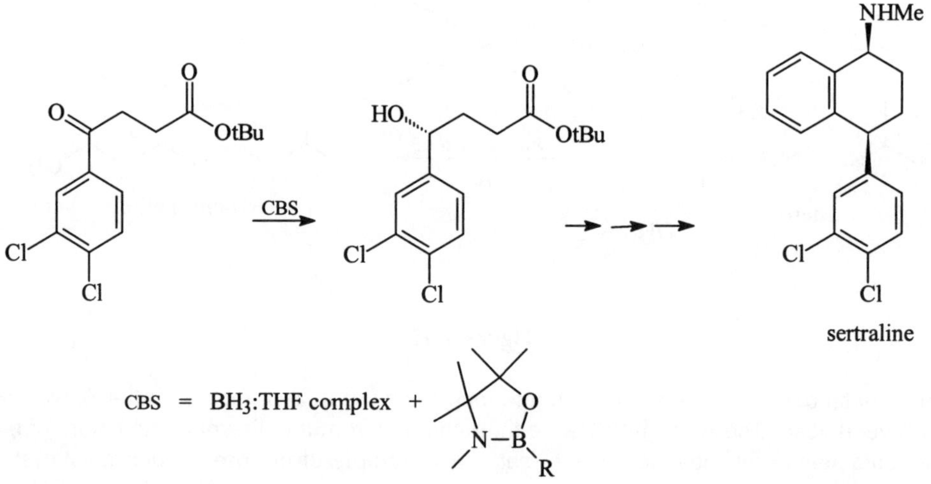

**Figure 3.41**[138]

### ***Homogeneous catalysis – isomerizations***

A truly elegant synthesis utilizing asymmetric catalysis was achieved by chemists at Takasago.[139] Enantioselective isomerization of an allylic amine was the critical step in establishing chirality for the synthesis. The route is shown in Figure 3.42.

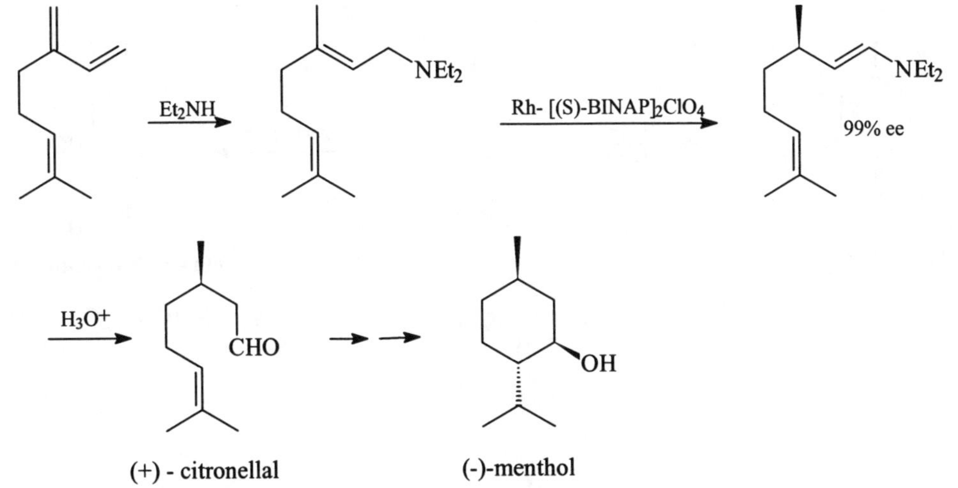

$Et_2NH$

$NEt_2$

Rh- $[(S)\text{-BINAP}]_2ClO_4$

$NEt_2$

99% ee

$H_3O^+$

CHO

OH

(+) - citronellal

(-)-menthol

**Figure 3.42**[139]

The citronellal derivative produced as an intermediate in this synthesis has been used for the preparation of several other important optically pure compounds.[140] Some of these are shown in Figure 3.43. Other terpenes prepared via asymmetric isomerization reactions are listed in Table 3.4.[141]

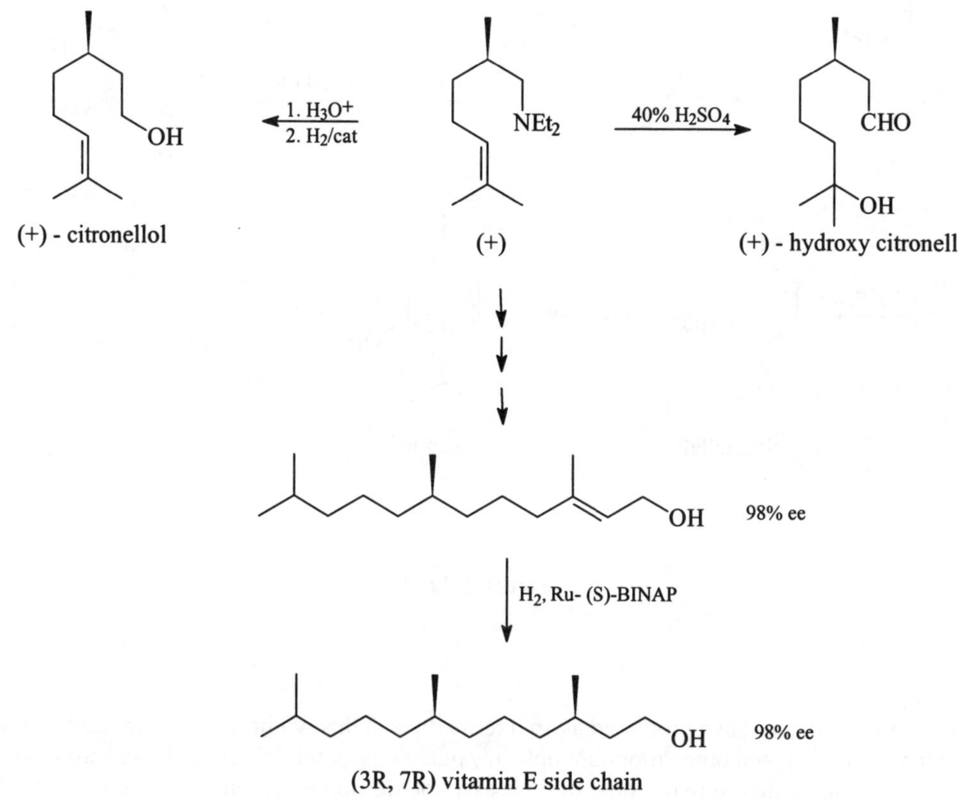

(+) - citronellol

1. $H_3O^+$
2. $H_2$/cat

$NEt_2$

(+)

40% $H_2SO_4$

(+) - hydroxy citronell

98% ee

$H_2$, Ru- (S)-BINAP

98% ee

(3R, 7R) vitamin E side chain

**Figure 3.43**[140]

| **Table 3.4: Other Terpenes Prepared via Catalytic Asymmetric Isomerization**[141] |
|---|
| (-)-citronellol |
| (-)-hydroxycitronellal |
| (-)-isopulegol |
| S-3,7-dimethoxy-1-octanal |
| S-7-methoxy-citronellal |

***Homogeneous catalysis – cyclopropanations***

Asymmetric cyclopropanation reactions[142] have been used for the synthesis of intermediates for pyrethroid insecticides. These reactions typically use a copper catalyst. Both cis/trans selectivity and enantiomeric ratios are fairly good. The synthesis of cypermethrin[142a] via an intermediate prepared via asymmetric cyclopropanation is shown in Figure 3.44.

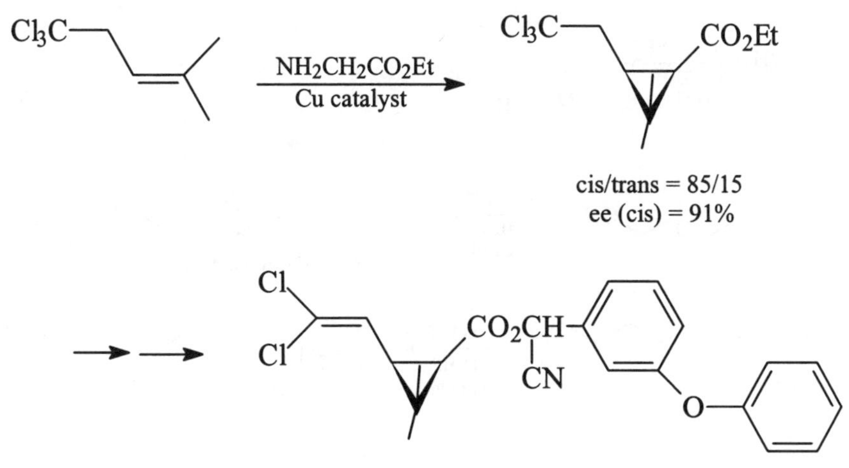

$Cl_3C$ — $NH_2CH_2CO_2Et$, Cu catalyst → $Cl_3C$ … $CO_2Et$

cis/trans = 85/15
ee (cis) = 91%

→ → $CO_2CH(CN)$ …

cypermethrin

**Figure 3.44**[142a]

### *Homogeneous catalysis – oxidations*

Asymmetric epoxidation of olefins has been successfully achieved via the Sharpless epoxidation of allylic alcohols.[143] The catalyst for this transformation form in situ from titanium ispropoxide and optically active diisopropyl tartrate (DIPT). The oxidant for the reaction is t-butyl hydroperoxide (TBHP). The choice of (+)- or (-)-DIPT allows control of the stereochemistry of the epoxidation. The chiral glycidols formed in the reaction are excellent chiral synthons for a host of optically active intermediates. A few examples of the asymmetric epoxidation of allylic alcohols are shown in Figure 3.45.[144]

The Sharpless epoxidation reaction is limited to allylic alcohols. Jacobsen has developed a method for epoxidizing other olefins using a Manganese(III) Salen complex[145] (see Chapter Four). In certain cases, reasonably good enantioselectivities can be obtained. Another oxidation process developed by Sharpless that has potential applications in chiral synthesis involves the asymmetric dihydroxylation of alkenes.[146] This reaction is catalyzed by osmium tetroxide in the presence of cinchona alkaloid derivatives and N-methylmorpholine N-oxide (NMO). Many groups have been exploring the possible applications for this chemistry in the synthesis of biologically active compounds.

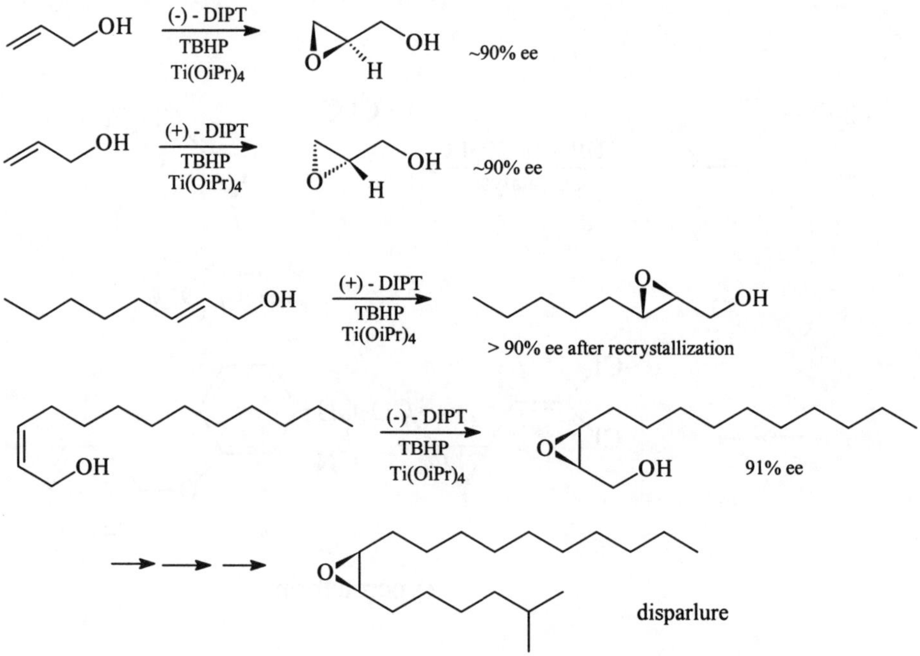

(-) - DIPT, TBHP, Ti(OiPr)4 ~90% ee

(+) - DIPT, TBHP, Ti(OiPr)4 ~90% ee

(+) - DIPT, TBHP, Ti(OiPr)4 > 90% ee after recrystallization

(-) - DIPT, TBHP, Ti(OiPr)4 91% ee

disparlure

**Figure 3.45**[144]

| **Table 3.35: Other Potentially Valuable Asymmetric Catalytic Processes** |
|---|
| Alkylation of aldehydes with dialkyl zinc compounds |
| Allylic alkylation |
| Carbonylation |
| Cycloaddition (diels-alder and ene reactions) |
| Dimerization and cyclodimerization of olefins |
| Grignard cross coupling |
| Hydrocyanation |
| Hydrodehalogenation |
| Hydroformylation |
| Hydrosylilation |
| Oxidation |
| Sulfoxidation |

There are several other catalytic asymmetric processes that may have potential applications in the synthesis of optically active compounds of commercial importance. Some of these are listed in Table 3.35. Many have been investigated in academic laboratories or have just recently been considered for large-scale production of chiral compounds. Others have limitations that must be surmounted before they can be viable processes for industrial synthesis.

### *Catalysis with soluble acids and bases*

Soluble chiral acids and bases have also been utilized in the synthesis of pharmaceutical and agrochemical products. The chiral acids and bases selected for these reactions typically originate from the Chiral Pool. Aldol reactions catalyzed by L-proline are one example of chiral-base catalyzed processes.[147] A bicyclic intermediate for the synthesis of 19-norsteroids has been prepared in this manner (see Figure 3.46).

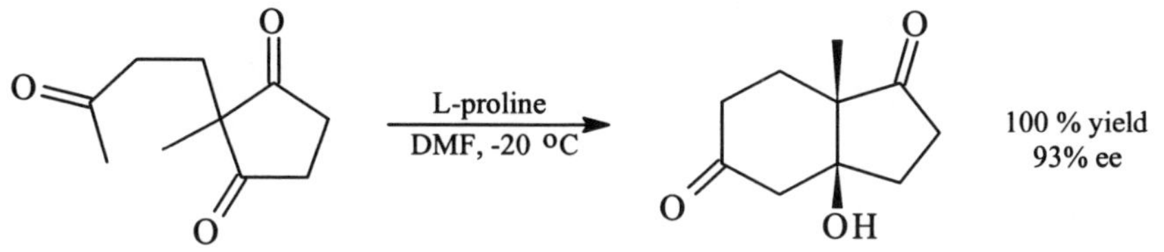

**Figure 3.46**

### *Phase transfer catalysis*

Phase transfer catalysis using chiral tetraalkylammonium salts has also been successfully applied to the preparation of chiral compounds. Figure 3.47 shows one example of this type of reaction. In this case, a precursor for the antihypertensive drug (S)-indacrinone[148] was the target.

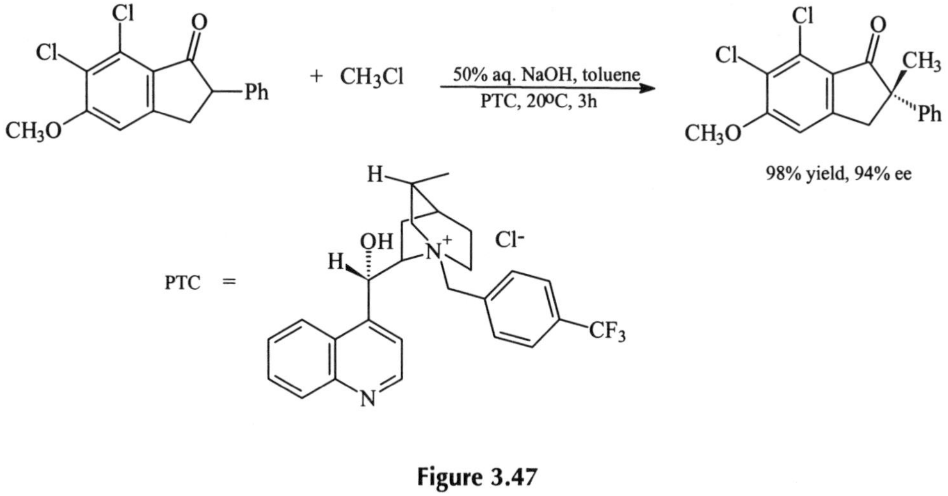

**Figure 3.47**

### ***Substitution reactions***

Substitution reactions have been an important tool for the synthetic chemist when preparing optically active materials. In order to be of use, the stereochemistry of the reaction must be understood. Often, inversion of the chiral center has been used to help identify the absolute configuration of chiral compounds. Examples of practical substitution reactions include the ring opening of epoxides, hydrolysis of halides with hydroxide, cyclic sulfate reactions, iodolactonization, and allylic substitutions.[149]

## References


1. Sheldon, R.A., *Chirotechnology*, Marcel Dekker, Inc., New York, 1993, pp. 73-75.
2. Crosby, J., in *Chirality in Industry*, Collins, A.N., Sheldrake, G.N., and Crosby, J. (Eds.), John Wiley & Sons, New York, 1992, p. 5.
3. a) Lichtenthaler, F.W. (Ed.), *Carbohydrates as Organic Raw Materials*, Verlag Chemie, Weinheim, 1991. b) Roper, H., in *Carbohydrates as Organic Raw Materials*, Lichtenthaler, F.W. (Ed.), Verlag Chemie, Weinheim, 1991, pp. 267-288. c) Röper, H., and Kod, H., *Starch/Stärke*, **40**, 453-464 (1988). d) Röper, H., and Kod, H., *Starch/Stärke*, **42**, 342-349 (1990).
4. Coppola, G.M., and Schuster, H.F., *Asymmetric Synthesis – Construction of Chiral Molecules Using Amino Acids*, Wiley, New York, 1987.
5. Sheldon, R.A., *Chirotechnology*, Marcel Dekker, Inc., New York, 1993, p. 155.
6. Sheldon, R.A., *Chirotechnology*, Marcel Dekker, Inc., New York, 1993, pp. 155-157, and references cited therein.
7. Sheldon, R.A., *Chirotechnology*, Marcel Dekker, Inc., New York, 1993, p. 159, and references cited therein.
8. Sheldon, R.A., *Chirotechnology*, Marcel Dekker, Inc., New York, 1993, pp. 163-164.
9. Crosby, J., in *Chirality in Industry*, Collins, A.N., Sheldrake, G.N., and Crosby, J. (Eds.), John Wiley & Sons, New York, 1992, p.15.
10. Erman, W.F., *Chemistry of the Monoterpenes, Parts A and B*, Marcel Dekker, New York, 1985.
11. Crosby, J., in *Chirality in Industry*, Collins, A.N., Sheldrake, G.N., and Crosby, J. (Eds.), John Wiley & Sons, New York, 1992, pp. 19-20.
12. Sheldon, R.A., *Chirotechnology*, Marcel Dekker, Inc., New York, 1993, p. 166.
13. Kelkenberg, H. *Tenside Surfactants Detergents*, **25**, 8-13 (1988).
14. Sheldon, R.A., *Chirotechnology*, Marcel Dekker, Inc., New York, 1993, p. 146.
15. Crawford, T.C., and Crawford, S.A., *Advan. Carbohydrate Chem.*, **37**, 79-155 (1980), and references cited therein.
16. Sheldon, R.A., *Chirotechnology*, Marcel Dekker, Inc., New York, 1993, pp. 147-149.
17. Crosby, J., in *Chirality in Industry*, Collins, A.N., Sheldrake, G.N., and Crosby, J., (Eds.), John Wiley & Sons, New York, 1992, p.12.
18. Sheldon, R.A., *Chirotechnology*, Marcel Dekker, Inc., New York, 1993, pp. 149-154, and references cited therein.
19. Crosby, J., in *Chirality in Industry*, Collins, A.N., Sheldrake, G.N., and Crosby, J. (Eds.), John Wiley & Sons, New York, 1992, pp.10-11.
20. Baer, E., *Biochem, Prep.*, **2**, 31 (1952).
21. Oyama, K., in *Chirality in Industry*, Collins, A.N., Sheldrake, G.N., and Crosby, J. (Eds.), John Wiley & Sons, New York, 1992, pp. 237-247.
22. Sheldon, R.A., Zeegers, H.J.M., Houbiers, J.P.M., and Hulshof, L.A., *Chimica Oggi (Chemistry Today)*, May, 1991, pp. 35-47.
23. a) Ohta, T., Kimura, T., Sato, N., and Nozoe, S., *Tetrahedron Lett.*, **29**, 4303-4305 (1988). b) Ohta, T, Sato, N., Kimura, T., Nozoe, S., and Izawa, K., *Tetrahedron*

*Lett.*, **29**, 4305-4308 (1988). c) Salzmann, T.N., Ratcliffe, R.W., Christensen, B.G., and Bouffard, F.A., *J. Am. Chem. Soc.*, **102**, 6161-6163 (1980). D) Shiozaki, M., Ishida, N., Hiraoka, T., and Yanagisawa, H., *Tetrahedron Lett.*, **22**, 5205-5208 (1981). e) Yanagisawa, H., Ando, A., Shiozaki, M., and Hisaoka, T., *Tetrahedron Lett.*, **24**, 1037-1040 (1983).

24. a) Eveleens, W., Spaans, J., and Wissink, H.G., *European Patent Appl.*, 9.285 (1980) to AKZO; CA: **93**, 95006p (1980). b) Gras, G., *German Patent* 3.024.265 (1981) to Rhone-Poulenc; CA: **94**, 191956q (1981). c) Crosby, J., *Tetrahedron*, **47**, 4789-4846 (1991).
25. a) Ohashi, T., and Hasegawa, J., *J. Synth. Org. Chem., Japan*, **45**, 331-345 (1987). b) Ohashi, T., *Proc. Chiral 90 Symp.*, Spring Innovations, Stockport, UK, 1990, pp. 65-71.
26. Shimazaki, M., Hasegawa, J., Kan K., Nomura, K., Nose, Y., Kundo, H., Ohashi, T., and Watanabe, K., *Chem Pharm. Bull.*, **30**, 3139-3146 (1982).
27. a) *Farm Chemicals Handbook, 72nd edn.*, Meister, Willoughby, 1986. b) Goldberg, L., *Acta Physiol. Scand.*, **15**, 173 (1948). c) Budavari, S. (Ed.), *The Merck Index, 11th edn.*, Merck, Rahway, New Jersey, 1989, entry 235, p. 42. d) Crosby, J., in *Chirality in Industry*, Collins, A.N., Sheldrake, G.N., and Crosby, J. (Eds.), John Wiley & Sons, New York, 1992, p. 7.
28. Liu, W., in *Handbook of Chiral Chemicals*, Ager, D.J. (Ed.), Marcel Dekker, Inc., New York, 1999, p. 116.
29. Crosby, J., in *Chirality in Industry*, Collins, A.N., Sheldrake, G.N., and Crosby, J., (Eds.), John Wiley & Sons, New York, 1992, pp. 24-25.
30. a) *Chem. Eng.*, 8 Nov., 247 (1965). b) Reinhold, D.F., Firestone, R.A., Gaines, W.A., Chemerda, J.M., and Sletzinger, M., *J. Org. Chem.*, **33**, 1209-1213 (1968).
31. Eur. Pat., 441471.
32. a) Gernez, D., *Compt. Rend.*, **63**, 843 (1866). b) Amiard, G., *Bull. Soc. Chim. Fr.*, 447 (1956).
33. Amiard, G., *Experientia*, **15**, 1-7, (1959).
34. Eur. Pat., 325965.
35. Sheldon, R.A., *Chirotechnology*, Marcel Dekker, Inc., New York, 1993, p. 180.
36. a) Okada, Y., Takebayashi, T., Hashimoto, M., Kasuoga, S., Sato, S., and Tamura, C., *J. Chem. Soc., Chem. Commun.*, 784-785 (1983). b) Okada, Y., and Takebayashi, T., *Chem. Pharm. Bull.*, **36**, 3787-3792 (1988).
37. Jacques, J., Collet, A., and Wilen, S., *Enantiomers, Racemates, and Resolutions*, Wiley, New York, 1981.
38. Bayley, C.R., and Vaidya, N.A. in *Chirality in Industry*, Collins, A.N., Sheldrake, G.N., and Crosby, J. (Eds.), John Wiley & Sons, New York, 1992, pp. 70-71.
39. a) Bayley, C.R., and Vaidya, N.A. in *Chirality in Industry*, Collins, A.N., Sheldrake, G.N., and Crosby, J. (Eds.), John Wiley & Sons, New York, 1992, pp. 72-75. b) Sheldon, R.A., *Chirotechnology*, Marcel Dekker, Inc., New York, 1993, pp. 191-192.
40. a) Sheldon, R.A., *Chirotechnology*, Marcel Dekker, Inc., New York, 1993, p. 188. b) .Bayley, C.R., and Vaidya, N.A. in *Chirality in Industry*, Collins, A.N., Sheldrake, G.N., and Crosby, J. (Eds.), John Wiley & Sons, New York, 1992, p. 73.
41. Sheldon, R.A., *Chirotechnology*, Marcel Dekker, Inc., New York, 1993, p. 189.
42. a) Bayley, C.R., and Vaidya, N.A. in *Chirality in Industry*, Collins, A.N., Sheldrake, G.N., and Crosby, J. (Eds.), John Wiley & Sons, New York, 1992, p. 71. b) Sheldon,

R.A., *Chirotechnology*, Marcel Dekker, Inc., New York, 1993, p. 174.
43. Ten Hoeve, W., and Wynberg, H., *J. Org. Chem.*, **50**, 4508-4514 (1985).
44. Shirawai, T., Shinjo, K., and Kurokawa, H., *Chem. Lett.*, 1413-1414 (1989).
45. a) Boesten, W.H.J., *Netherlands Patent*, 90.00386 and 90.00387 (1990) to Stamicarbon. b) Clark, C.J., Phillipps, G.H., and Steer, M.R., *J. Chem. Soc., Perkin Trans. I*, 475-481 (1976).
46. Marckwald, W., *Ber. Dtsch. Chem. Ges.*, **29**, 42-43 (1896).
47. Pope, W.J., and Peachey, S.J., *J. Chem. Soc.*, **75**, 1066 (1899).
48. Sheldon, R.A., *Chirotechnology*, Marcel Dekker, Inc., New York, 1993, p. 197.
49. a) Martin, V.S., Woodard, S.S., Katsuki, T., Yamada, Y., Ikeda, M., and Sharpless, K.B., *J. Am. Chem. Soc.*, **103**, 6237 (1981). b) Katsuki, T. and Sharpless, K.B., *J. Am. Chem. Soc.*, **102**, 5974 (1980). c) Myano, B., Lu, L.D.-l., Viti, S.M., and Sharpless, K.B., *J. Org. Chem.*, **48**, 3608 (1983).
50. Gao, Y., Hanson, R.M., Klundar, J.M., Ko, S.Y., Masamune, H., and Sharpless, K.B., *J. Am. Chem. Soc.*, **109**, 5765 (1987).
51. Liu, W. in *Handbook of Chiral Chemicals*, Ager, D.J. (Ed.), Marcel Dekker, Inc., New York, 1999, pp. 126-127.
52. Kitamura, M., Manabe, K., and Noyori, R., *Tetrahedron Lett.*, **28**, 4719 (1987).
53. Sheldon, R.A., *Chirotechnology*, Marcel Dekker, Inc., New York, 1993, p. 209.
54. Liu, W. in *Handbook of Chiral Chemicals*, Ager, D.J. (Ed.), Marcel Dekker, Inc., New York, 1999, pp. 128-134.
55. Crosby, J., in *Chirality in Industry*, Collins, A.N., Sheldrake, G.N., and Crosby, J. (Eds.), John Wiley & Sons, New York, 1992, pp. 30-34.
56. a) Chibata. I., in *Asymmetric Reactions and Processes in Chemistry* (ed. E.L. Eliel and S. Otsuka), ACS Symposium Series, Vol. 185, Amer. Chem. Soc., Washington, DC, 1982, p. 195.
57. a) Yamada, H., Takahashi, S., Kii, Y., and Kunegai, H., *J. Ferment. Technol.*, **56**, 484 (1978). b) Takahashi, T., Ohashi, T., Kii, Y., Kumagi, H., and Yamada, H., J. Ferment. Technol., 57, 328-332 (1979). c) Kanegafuchi Chemical Indsutries, *Jpn. Pat.*, 6225990, 1987. d) Yamada, H., Takahashi, S., Yoneda, K., and Amagasaki, H., *Ger. Pat.*, 2.757980, 1991. e) Olivieri, R., Fascetti, E., Angelini, L., and Degen, L., *Biotechnol. Bioeng.*, **23**, 2173 (1981).
58. a) *Eur. Pat.*, 244912. b) Kooreman, H.J., Van Nistelrooij, H., and Pryce, R.J. in Proceedings of the Chemical Syntehsis Symposium and Workshop; Manchester, UK, 18 April, 1989.
59. a) *Jpn. Pat.*, 86129797. b) *Eur. Pat.*, 207636.
60. a) Matzinger, P.K., and Leuenberger, H.G.W., *Appl. Microbiol. Biotechnol.*, **22**, 208 (1985). b) *Eur. Pat.*, 144832.
61. Fufkumura, T., *Agr. Biol. Chem.*, **40**, 1687 (1976).
62. Danda, H., Nagatomi, T., Machara, A., and Umemura, T., *Tetrahedron*, **47**, 8701 (1991).
63. Crosby, J., in *Chirality in Industry II*, Collins, A.N., Sheldrake, G.N., and Crosby, J., (Eds.), John Wiley & Sons, New York, 1997, p. 7.
64. Keurentjes, J.T.F., *PCT Int. Pat. Appl.*, WO 94/07814, 1994.
65. Crosby, J., in *Chirality in Industry II*, Collins, A.N., Sheldrake, G.N., and Crosby, J., (Eds.), John Wiley & Sons, New York, 1997, p. 8.
66. Sheldon, R.A., *Chirotechnology*, Marcel Dekker, Inc., New York, 1993, pp. 106-107.

67. Sheldon, R.A., *Chirotechnology*, Marcel Dekker, Inc., New York, 1993, p. 113.
68. Sheldon, R.A., *Chirotechnology*, Marcel Dekker, Inc., New York, 1993, p. 103.
69. Chauvette, R.R., Ger. Patent, 1974, 2408698.
70. Gbewonyo, K., Hunt, G., and Buckland, B., *Bioprocess Eng.*, **8**, 1 (1992).
71. Serizawa, N., *Biotechnol. Ann. Rev.*, **2**, 373 (1996).
72. Crosby, J., in *Chirality in Industry*, Collins, A.N., Sheldrake, G.N., and Crosby, J., (Eds.), John Wiley & Sons, New York, 1992, pp. 14-15.
73. Sheldon, R.A., *Chirotechnology*, Marcel Dekker, Inc., New York, 1993, pp. 127-128.
74. Sheldon, R.A., *Chirotechnology*, Marcel Dekker, Inc., New York, 1993, p. 115.
75. Sheldon, R.A., *Chirotechnology*, Marcel Dekker, Inc., New York, 1993, pp. 119, 128.
76. a) Peterson, D.H. and Murray, H.C., *J. Am. Chem. Soc.*, **73**, 5513 (1951). b) Peterson, D.H., *Steroids*, **45**, 1 (1985).
77. a) Hasegawa, J. Ogura, M., Hamaguchi, S., Shimazaki, M., Kawaharada, H., and Watanabe, K., *J. Ferment. Technol.*, **59**, 203-208 (1981). b) Hasegawa, J., Ogura, M., Kanema, H., Noda, N., Kawaharada, H., and Watanabe, K., **J. Ferment. Technol.**, **60**, 501-508 (1982). c) Ohashi, T., *Proc. Chiral Symp.*, Spring Innovations, Stockport, UK, 1990, pp. 65-71. d) Shimazaki, M., Hasegawa, J., Kan, K., Nomura, K., Nose, Y., Kondo, H., Ohashi, T., and Watanabe, K., *Chem. Pharm. Bull.*, **30**, 3139-3146 (1982).
78. Kulla, H.G., Chimia, 45, 81-85 (1991).
79. Furohashi, K., in *Chirality in Industry*, Collins, A.N., Sheldrake, G.N., and Crosby, J., (Eds.), John Wiley & Sons, New York, 1992, pp. 169-170.
80. Furohashi, K., in *Chirality in Industry*, Collins, A.N., Sheldrake, G.N., and Crosby, J., (Eds.), John Wiley & Sons, New York, 1992, p 170.
81. Furohashi, K., in *Chirality in Industry*, Collins, A.N., Sheldrake, G.N., and Crosby, J., (Eds.), John Wiley & Sons, New York, 1992, pp. 170-173.
82. a) Yoshizawa, A., Nishiyama, I., Fukumasa, M., Hirai, T., and Yamane, M., *Jpn. J. Appl. Phys.*, **28**, L1269 (1989). b) Shiratori, N., Yoshizawa, A., Nishiyama, I., Fukumasa, M., Yokoyama, A., Hirai, T., and Yamane, M., *Mol. Cryst. Liq. Cryst.*, **199**, 129 (1991).
83. a*) Jpn. Pat.*, 58 158188. b) *Ger. Pat.*, 3041224 (1984). c) Hasegawa, J., *Hakko Kogako Kaishi*, **62**, 341 (1984). d) *Chem. Abstr.*, **101**, 226524 (1984). e) *US Pat.*, 4981794.
84. Gist-Brocades/Shell Eur. Pat., EP 274146.
85. E.g. Shell, Eur. Pat., EP 319100.
86. a) Davies, H.G., Green, R.H., Kelly, D.R., and Roberts, S.M., *Biotransformations in Preparative Organic Chemistry*, Academic Press, London, 1989, p. 99. b) Servi, S., *Synthesis*, **1** (1990), and references cited therein. c) Gramatica, P., *Chim. Oggi*, **17** (1988), and references cited therein.
87. Crosby, J., in *Chirality in Industry*, Collins, A.N., Sheldrake, G.N., and Crosby, J., (Eds.), John Wiley & Sons, New York, 1992, pp. 49-55.
88. Sheldon, R.A., *Chirotechnology*, Marcel Dekker, Inc., New York, 1993, pp. 134-136 and references cited therein.
89. Zhou, B.-N., Gopalan, A.S., VanMiddlesworth, F., Shieh, W.-R., and Sih, C.J., *J. Am. Chem. Soc.*, **105**, 5925 (1983).
90. a) Leuenberger, H.G.W., Boguth, W., Widmer, E., and Zell, R., *Helv. Chim. Acta*,

**59**, 1832 (1976). b) Leuenberger, H.G.W., paper presented at Warwick Workshop, Enzymes as Reagents, Warwick University, April 1987.

91. Wipf, R., Kupfer, E., Bertozzi, R., and Leuenberger, H.G.W., *Helv. Chim. Acta*, **66**, 485 (1983).
92. Kataoka, M., Shimizu, S., and Yamada, H., *Agric. Biol. Chem.*, **54**, 177-182 (1990).
93. Rose, A.H., *Industrial Microbiology*, Butterworths, Washington, 1961, p. 264.
94. Gerhartz, W. (Ed.), *Enzymes in Industry, Production and Application*, VCH, Weinheim, 1990.
95. Sheldon, R.A., *Chirotechnology*, Marcel Dekker, Inc., New York, 1993, p. 209.
96. Sheldon, R.A., *Chirotechnology*, Marcel Dekker, Inc., New York, 1993, pp. 214-216 and references cited therein.
97. Sheldon, R.A., *Chirotechnology*, Marcel Dekker, Inc., New York, 1993, pp. 216-219.
98. Sheldon, R.A., *Chirotechnology*, Marcel Dekker, Inc., New York, 1993, p. 211.
99. Sheldon, R.A., *Chirotechnology*, Marcel Dekker, Inc., New York, 1993, p. 207.
100. Sheldrake, G.N., in *Chirality in Industry*, Collins, A.N., Sheldrake, G.N., and Crosby, J., (Eds.), John Wiley & Sons, New York, 1992, pp. 127-166.
101. Enzymatix, *PCT Int. Appl.*, WO 9012798, 1990.
102. Pantaleone, D. P., in *Handbook of Chiral Chemicals*, Ager, D.J. (Ed.), Marcel Dekker, Inc., New York, 1999, pp. 248-249, and references cited therein.
103. Pantaleone, D. P., in *Handbook of Chiral Chemicals*, Ager, D.J. (Ed.), Marcel Dekker, Inc., New York, 1999, pp. 249-250, and references cited therein.
104. a) Christen, P., and Metzler, D. E., in *Transaminases*; in series *Biochemistry, Vol. 2*, Meister, A. (Ed.), Wiley, New York, 1985. b) Crump, S.P., and Rozzell, J.D., in *Biocatalytic Production of Amino Acids and Derivatives*, Rozzell, J.D., and Wagner, F. (Eds.), Carl Hanser Verlag, Munich, 1992, p. 43.
105. Godfrey, T., and West, S.I., in *Industrial Enzymology*, Godfery, T., and West, S. I. (Eds.), Macmillan Press Ltd., London, 1996, p. 3.
106. Pantaleone, D.P. in *Handbook of Chiral Chemicals*, Ager, D.J. (Ed.), Marcel Dekker, Inc., New York, 1999, p. 259.
107. a) Ohno, M., in *Enzymes in Organic Synthesis*, Pitman, London, 1985, pp. 171-187. CIBA Foundations Symposium No. 111. b) Ohno, M., Kobayashi, S., and Adachi, K., in *Enzymes as Catalysts in Organic Synthesis*, Schneider, M.P. (Ed.), Rediel, Dordrecht, 1986, pp. 123-142.
108. Sheldon, R.A., *Chirotechnology*, Marcel Dekker, Inc., New York, 1993, p. 239.
109. Mori, K., *Synlett.* **11**, 1097 (1995).
110. a) Patel, R.N., *Adv. Appl. Microbiol.*, **43**, 91 (1997). b) Patel, R.N., Robinson, R.S., and Szarka, L.J., *Appl. Microb. Biotechnol.*, **34**, 10 (1990).
111. a) Fruton, J.S., *Adv. Enzymol. Relat. Areas Mol. Biol.*, **53**, 239 (1982). b) Chaiken, I.M., Komoriya, A., Ohno, M., and Widmer, F., *Appl. Biochem. Biotechnol.* **7**, 385 (1982). c) Jakubke, H.-D., Kuhl, P., and Könnecke, A., *Angew. Chem. Int. Ed. Engl.*, **24**, 85 (1985). d) Kullman, W. in *Enzymatic Peptide Synthesis*, CRC Press, Boca Raton, FL, 1987. e) Heiduschka, P., Dittrich, J., Barth, A., *Pharmazie*, **45**, 164 (1990). f) Schellenberger, V., and Jakubke, H.-D., *Angew. Chem. Int. Ed. Engl.*, **30**, 1437 (1991).
112. a) Oyama, K., in *Biocatalysis in Organic Media*, Laane, C., Tramper, J., and Lilly, M.D. (Eds)., in series *Studies in Organic Chemistry, Vol. 29*, Elsevier, Amsterdam, 1987, p. 209. b) Pantaleone, D.P., and Dikeman, R.N., in *Science for the Food*

*Industry of the 21st Century, Vol. 1,* Yalpani, M. (Ed.), ATL Press, Mt. Prospect, IL, 1993, p. 173. c) Ager, D.J., Pantaleone. D.P., Henderson, S.A., Katritzky, A.R., Prakash, L., and Walters, D.E., *Angew. Chem. Int., Ed. Engl.*, **37**, 1998.

113. a) Umemura, J., Takamatsu, S., Sato, T., Tosa, T., and Chibata, J., *Eur. J. Appl. Microbiol. Biotechnol.*, **20**, 291 (1984). b) Chibata, I., Tosa, T., and Sato, T., in *Biotechnology, Vol. 7a,* Rehm, H.J., and Reed, G., *Eds.), VCH, Weinheim, 1987, pp. 653-684.

114. a) Werf, M.J.Y.D., van den tweel, W.J.J., Kamphuis, J., Hartmans, S., and de Bont, J.A.M., *TIBETCH*, **12**, 95, (1994). b) Collins, A.N., Sheldrake, G.N., and Crosby, J., (Eds.), *Chirality in Industry*, John Wiley & Sons, New York, 1992. c) Gerhartz, W. (Ed.), in *Enzymes in Industry*, VCH, Weinheim, 1990.

115. a) 114b. b) Calton, G.J., in *Biocatalytic Production of Amino Acids and Derivatives*, Rozzell, J.D., and Wagner, F. (Eds)., Carl Hanser Verlag, Muich, 1992, p. 59.

116. Pantaleone, D.P. in *Handbook of Chiral Chemicals*, Ager, D.J. (Ed.), Marcel Dekker, Inc., New York, 1999, p. 270.

117. a) Rotheim, P. in *The Enzyme Industry: Specialty and Medical Applications*, Business Communications Company, Norwalk, CT, 1994. b) 114 c. c) Pedersen, S., *Bioprocess Technol.*, **16**, 185 (1993). d) Bhosale, S.H., Rao, M.B., and Deshpande, V.V., *Microbiol. Rev.*, **60**, 280 (1996).

118. Sheldon, R.A., *Chirotechnology*, Marcel Dekker, Inc., New York, 1993, p. 207.

119. Laneman, J.A., in *Handbook of Chiral Chemicals*, Ager, D.J. (Ed.), Marcel Dekker, Inc., New York, 1999, p. 144.

120. Sheldon, R.A., *Chirotechnology*, Marcel Dekker, Inc., New York, 1993, p. 280.

121. Sheldon, R.A., *Chirotechnology*, Marcel Dekker, Inc., New York, 1993, p. 275.

122. a) Fukawa, H., Izumi, Y., Komatsu, S., and Akabori, S., *Bull. Chem. Soc. Jpn,*, **35**, 1703 (1962). b) Izunmi, Y., *Advan. Catal.*, **32**, 215 (1983). c) Sachtler, W.M.H., in *Catalysis of Organic Reactions*, Augustine, R.L. (Ed.), Marcel Dekker, New York, 1985, pp. 189-206. d) Tai, A., and Harada, T., in *Tailored Metal Catalysts*, Iwasawa, Y. (Ed.), Rediel, Dordrecht, 1986, p. 265. e) Bakos, J., Toth, I., and Marko, L., *J. Org. Chem.*, **46**, 5427 (1981). f) Brünner, H., Muschiol, M., Wischert, T., Harada, T., and Tai, A., *Chem. Lett.*, 1267-1270 (1987).

123. Blaser, H.U., *Tetrahedron Asymm.*, **2**, 843-866.

124. Pittman, C.U., in *Comprehensive Organometallic Chemistry, Vol. 8,* Wilkinson, G., Stone, F.G.A., and Abel, E.W. (Eds.), Pergamon, Oxford, 1982, P. 553.

125. Kuntz, E.G., *Chemtech*, 570 (1987).

126. Sheldon, R.A., *Chirotechnology*, Marcel Dekker, Inc., New York, 1993, p. 286.

127. a) Knowles, W.S., and Sabacky, M.J., *J. Chem. Soc., Chem. Commun.*, 1445 (1968). b) Knowles, W.S., Sabacky, M.J., Vineyard,B.D., and Weinkauf, D.J., *J. Am. Chem. Soc.*, **97**, 2567 (1975). c) Knowles, W.S., *J. Chem. Educ.*, **63**, 222-225 (1986). d) Vineyard,B.D., Knowles, W.S., Sabacky, M.J., *J. Mol. Catal.*, **19**, 159-169 (1983). e) Knowles, W.S., *Acc, Chem. Res.*, **16**, 106-112 (1983).

128. Ohta, T., Takaya, H., Kitamura, M., Nagai, K., and Noyori, R., *J. Org, Chem.*, **52**, 3176-3178 (1987).

129. Noyori, R., Ohkuma, T., Kitamura, M., Takaya, H., Sayo, N.H., Kumobayashi, H., and Akutagawa, S., *J. Am. Chem. Soc.*, **109**, 5856-5858 (1987).

130. Sheldon, R.A., *Chirotechnology*, Marcel Dekker, Inc., New York, 1993, p. 293.

131. Takahashi, H., Sakuraba, S., Takeda, H., and Achiwa, K., *J. Am. Chem. Soc.*, **112**, 5876-5878 (1990).

132. Takeshi, H., Tachinami, T., Aburatani, M., Takahashi, H., Morimoto, T., and Achiwa, K., *Tetrahedron Lett.*, **30**, 363 (1989).
133. a) Kitamura, M., Hsiao, Y., Noyori, R., and Takaya, H., *Tetrahedron Lett.*, **28**, 4829-4832 (1987). B) Noyori, R., Kitamura, M., Takaya, H., Kumobayashi, H., and Akutagawa, S., *European Patent* 0.245.960 (1987) to Takasago Perfumery Co.
134. Sheldon, R.A., *Chirotechnology*, Marcel Dekker, Inc., New York, 1993, p. 292.
135. a) Takaya, H., Ohta, T., Sato, N., Kumobayahashi, H., Akutagawa, S., Inoue, S., Kasahara, I., and Noyori, R., *J. Am. Chem. Soc.*, **109**, 1596-1597 (1987). b) Takaya, H., Ohta, T., Sato, N., Kumobayahashi, H., Akutagawa, S., Kasahara, I., and Noyori, R., *J. Am. Chem. Soc.*, **109**, 4129-4130 (1987).
136. Stinson, S.C., *Chem. Eng. News*, **75**, 34 (1997).
137. Caille, J.-C., Bulliard, M., and Laboue, B., in *Chirality in Industry II*, Collins, A.N., Sheldrake, G.N., and Crosby, J., (Eds.), John Wiley & Sons, New York, 1997, pp. 391-401.
138. Quallich, G.J., and Woodall, T.M., *Tetrahedron*, **48**, 10239 (1992).
139. a) Akutagawa, S., and Tani, K., in *Catalytic Asymmetric Synthesis*, Ojima, I. (Ed.), VCH Publsihers, New York, 1993, P. 41. b) Otsuka, S., and Tani, K. in *Asymmetric Synthesis, Vol. 5*, Morrison, J.D. (Ed.), Academic Press, New York, 1985, p. 171. c) Otsuka, S., and Tani, K, Synthesis, 665-680 (1991). d) Tani, K., Yamagata, T., Tatsuno, Y., Yamagata, Y., Tomita, K., Akutagawa, S., Kumobayashi, H., and Otsuka, S., *Angew. Chem. Int. Ed. Engl.*, **24**, 217 (1985).
140. a) Sheldon, R.A., *Chirotechnology*, Marcel Dekker, Inc., New York, 1993, p. 304. b) 135a. c) Takabe, K., Uchiyama, Y., Okisaka, K., Yamada, T., Katagiri, T., Okazaki, T., Oketa, Y., Kumobayashi, H., and Akutagawara, S., *Tetrahderon Lett.*, **26**, 5153 (1985).
141. a) Akutagawa, S., and Tani, K., in *Catalytic Asymmetric Synthesis*, Ojima, I. (Ed.), VCH Publsihers, New York, 1993, P. 41.
142. a) Aratani, T., *Pure. Appl. Chem.*, **57**, 1839-1844 (1985) and references cited therein. b) Doyle, M.P., *Recl. Trav. Chim. Pays-Bas*, **110**, 305-316 (1991). c) Pfaltz, A., *Bull. Soc. Chim. Belg.*, **99**, 729-739 (1990). d) Pfaltz, A., in *Modern Synthetic Methods, Vol. 5*, Scheffold, R. (Ed.), Springer-Verlag, Heidelberg, 1989, pp. 199-248. e) Evans, D.A., Woerpel, K.A., and Hinman, M., *J. Am. Chem. Soc.*, **113**, 726-728 (1991). f) Lowenthal, R.E., Abiko, A., and Masamune, S., *Tetrahedron Lett.*, **31**, 6005 (1990).
143. a) Hanson, R.M., and Sharpless, K.B., *J. Org. Chem.*, **51**, 1922-1925 (1986). b) Gao, Y., Hanson, R.M., Klunder, J.M., Ko, S.Y., Masamune, H., and Sharpless, K.B., *J. Am. Chem. Soc.*, **109**, 5765-5780 (1987).
144. Sheldon, R.A., *Chirotechnology*, Marcel Dekker, Inc., New York, 1993, pp. 324-325.
145. a) Zhang, W., and Jacobsen, E.N., *J. Org. Chem.*, **56**, 2296-2298 (1991). b) Jacobsen, E.N., Zhang, W., and Güler, M.L., *J. Am. Chem. Soc.*, **113**, 6703-6704 (1991).
146. Sheldon, R.A., *Chirotechnology*, Marcel Dekker, Inc., New York, 1993, p. 320.
147. a) Jacobsen, E.N., Marko, I., Mungall, W.S., Schröder, G., and Sharpless, K.B., *J. Am. Chem. Soc.*, **110**, 1968-1970 (1988). b). Jacobsen, E.N., Marko, I., France, M.B., Svendsen, J.S.., and Sharpless, K.B., *J. Am. Chem. Soc.*, **111**, 737-739 (1989). c). Wai, J.S.M., Marko, I., Svendsen, J.S., Finn, M.G., Jacobsen, E.N., and Sharpless, K.B., *J. Am. Chem. Soc.*, **111**, 1123-1125 (1989). d) Marko, I.E., *Proc. Chiral 89 Symp.*, Spring Innovations, Stockport, UK, 1989, pp. 13-20. e) Kwong, H.L., Sorato, C., Ogino, Y., Chen, H., and Sharpless, K.B, *Tetrahedron Lett.*, **31**, 2999-3002

(1990). f) Ogino, Y., Chen, H., Kwong, H.L., and Sharpless, K.B, *Tetrahedron Lett.*, **32**, 3965-3968 (1991). g) Sharpless, K.B., Amberg, W., Beller, M., Chen, H., Hartung, J., Kawanami, Y., Lübben, D., Manoury, E., Ogino, Y., Shibata, T., and Ukita, T., *J. Org. Chem.*, **56**, 4585-4588 (1991). h) Kim, B.M., and Sharpless, K.B., *Tetrahedron Lett.*, **31**, 3003-3006 (1990).

148. a) Dolling, U.H., Davis, P., and Grabowski, E.J.J., *J. Am. Chem. Soc.*, **106**, 446-447 (1984). b) Hughes, D.L., Dolling, U.H., Ryan, K.M., Schoenwaldt, E.F., and Grabowski, E.J.J., *J. Org. Chem.*, **52**, 4745-4752 (1987). c) Dolling, U.H., Hughes, D.L., Battacharya, A., Ryan, K.M., Karady, S., Weinstock, L.M., Grenda, V.J., and Grabowski, E.J.J., in *Catalysis of Organic Reactions*, Rylander, P.N., Greefield, H., and Augustine, R.L. (Eds.), Marcel Dekker, New York, 1988, pp.65-85.

149. Ager, D.J., in *Handbook of Chiral Chemicals*, Ager, D.J. (Ed.), Marcel Dekker, Inc., New York, 1999, pp. 103-113.

# CHAPTER 4

# METHODOLOGIES FOR OBTAINING CHIRAL COMPOUNDS: SOME EXAMPLES

Synthesis of optically active chemicals poses additional problems for the chemist beyond the traditional challenges associated with preparation of racemic mixtures or achiral compounds. The two key issues associated with the selection of a route to a chiral compound are when and how to introduce the chirality. There are no set guidelines for where in a synthetic route to create the asymmetric center(s). Each target molecule must be evaluated individually. The same holds true when choosing a method of introduction, whether it will be biological or chemical, using resolution of a racemic mixture or preparation of a single enantiomer, etc. The most economically viable route will vary from one product to another.

In general it is best to introduce the chiral center as early on in the synthesis as possible in order to avoid carrying the unwanted isomer through the entire process. However, the chemist must rely on his or her own experience and knowledge of the physical properties of the compounds involved to make that decision. Solubility and crystallization are important characteristics that need to be evaluated. If the molecule exists as a conglomerate, crystallization techniques will most likely be utilized. If resolution is necessary, then the chiral center should be introduced where the unwanted isomer can be most efficiently recycled. If a second order asymmetric transformation is possible, there is no issue.

At whatever stage in the synthesis the chirality is introduced, there should be no steps beyond that point where racemization or epimerization of the asymmetric center can occur. A crystallization step downstream of the chiral center introduction can be beneficial as a means for increasing enantiomeric excess. In the end, the process economics, including material costs, capital costs, operating costs, and environmental costs will be the determining factor in the design of synthetic routes to chiral molecules.[1]

There are several factors that affect the cost or price of chiral products.[2] A key issue is the cost of the substrate used in generating the chiral center. The chemical and optical yields and overall productivity of the process must also be evaluated. If a resolution is performed, the cost of the resolving agent or biocatalyst will be a significant consideration. The ease of racemization of the unwanted isomer and position of the resolution step in the overall synthesis will affect the overall cost as well. The total number of steps in the overall synthesis must also be factored into a decision.

The following examples have been selected to illustrate the types of processes that have found application in the preparation and isolation of chiral molecules. These selected reactions are merely representative of the myriad of possible methodologies available to professionals in the pharmaceutical, agrochemical, and other specialty chemical industries. New developments in all of the areas covered in this chapter are reported on a regular basis and will continue to advance the field of chiral chemistry.

## Generic Isolation and Separation Techniques

### *Natural products*

Isolation of natural products can be achieved in several ways. Glutamic acid has been crystallized from gluten hydrolysate.[3] The very important carbohydrate sucrose (table sugar) is collected from sugar cane and beets.[4] Glucose, another key carbohydrate member of the Chiral Pool is obtained from the hydrolysis of starch.[5] Many terpenes are extracted from the various plants in which they are produced.

Steam distillation is often the technique used for extracting the essential oils from plant materials.[6] The monoterpenes are collected in this manner as a mixture of several compounds. The mixture is then further separated by repeated distillation. Dry distillation of the organic matter has also been utilized. The chemical and optical purities of the monoterpenes is dependent upon the specific plant source, where in the world it was grown, and the processing conditions. Table 4.1 lists different terpenes and the plant sources they are derived from.[7]

**Table 4.1: Monoterpenes and Their Plant Sources**

| Terpene | Source |
|---|---|
| α–Pinene | various pine trees |
| d-Limonene | orange, caraway, dill, grape, lemon oils |
| l-Limonene | turpentine oils |
| dl-Limonene | turpentine oils |
| l-Carvone | spearmint oil |
| d-Carvone | caraway oil |
| dl-Carvone | gingergrass oils |
| l-Menthone | Pinus palustris Mell |

Fermentation is another key route for preparing Chiral Pool materials. Many hydroxy acids, amino acids, and other more complex products are produced via fermentation processes. Some examples of molecules obtained through fermentation are presented in Chapter Three.

Because fermentation today seldom is the route of choice for preparation of optically active intermediates, no further discussion will be presented here.

## Preferential Crystallization

Preferential crystallization of conglomerates is one of the most attractive methods for separating optically active isomers. However, only a few percent of racemic mixtures exist as conglomerates, therefore this option has limited availability. Preferential crystallizations have been successfully implemented in two different ways. A schematic of the process for simultaneous crystallization[8] of the two enantiomers in racemic mixture is shown in Figure 4.1. The other method, known as resolution by entrainment involves alternation the isolation of each enantiomer from a single vessel.[9]

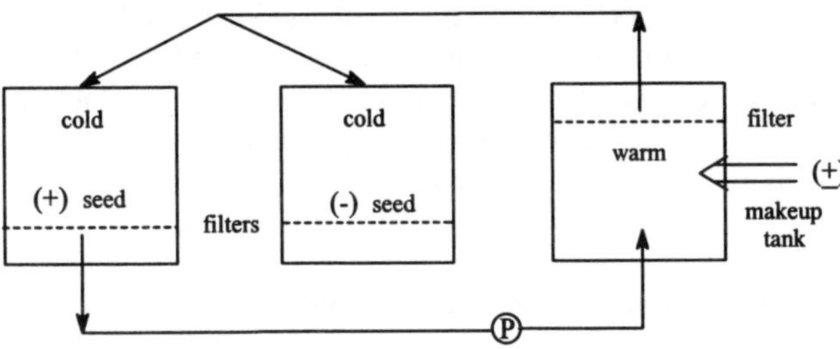


**Figure 4.1**[8]

If preferential crystallization occurs in conjunction with racemization of the unwanted isomer, then crystallization induced asymmetric transformation occurs. This deracemization process can provide a theoretical 100% yield of the desired enantiomer (see Figure 4.2).[10]

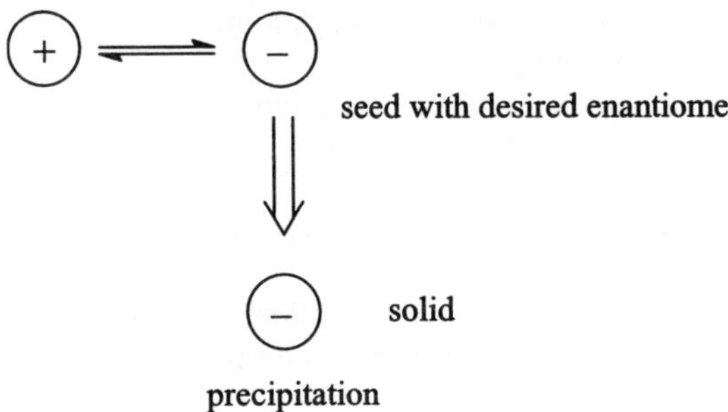


**Figure 4.2**[10]

## Classical Resolution

Classical resolution remains the most widely used technique for obtaining optically pure compounds for synthesis of pharmaceuticals, agrochemicals, and other biologically active products. A chiral acid or base resolving agent is reacted with a racemic mixture and two diastereomeric salts are formed. Because they are diastereomers, these salts have different physical properties and therefore can be separated much more easily than the enantiomers from which they are derived.

Separation can be done by several methods. Crystallization is preferred. Salts have two to three times the likelihood of existing as conglomerates, and therefore isolation of the desired compound by crystallization is often possible. One limitation to this method is the need for functionality in the compound to be separated. The diastereomeric salts are formed via reaction with chiral acids or bases.

Several issues are critical when selecting a resolving agent and establishing the appropriate system for crystallizing the diastereomers. Some important properties of resolving agents were presented in Chapter Three. In addition to these points, attention must be paid to the careful selection of the solvent for the crystallization.[11] Screening tests are conducted to determine which resolving agent will work best under what conditions. Table 4.2 lists a few of the common acid and base resolving agents in use today.[12]

**Table 4.2: Selected Common Resolving Agents[12]**

| Acids | Bases |
|---|---|
| 1-Camphor-10-sulfonic acid | *threo*-2-Amino-1-(p-nitrophenyl)-propane-1,3-diol |
| 2-Ketogluconic acid | 2-Aminobutane |
| Malic acid | Brucine |
| Mandelic acid | Cinchonidine |
| α–Methoxyphenylacetic acid | Cinchonine |
| α–Methoxy-α–trifluoromethyl-phenylacetic acid | Dehydroabietylamine |
| 2-Pyrrolidone-5-carboxylic acid | Ephedrine |
| Tartaric acid | α–Methylbenzylamine |
| | a -Methyl-p-bromobenzylamine |
| | N-Methylglucamine |
| | a -Methyl-p-nitrobenzylamine |
| | α–(1-Naphthyl)ethylamine |
| | Quinine |

Classical resolution via crystallization can also occur with racemization of the undesired enantiomer. Again, with this second order transformation, a theroetcial 100% yield of the desired product can be obtained in the crystallization.

## Kinetic Resolution

Kinetic resolution offers a means for obtaining optically pure chemicals when preferential crystallization or classical resolution methodologies are not available for the desired compound. This technology relies on the differing reactivities of enantiomers when placed in a chiral environmental. Typically one enantiomer remains unreactive and the other isomer forms a product with the chiral reagent. The two compounds can then be separated with relative ease and often with very high optical purities. Kinetic resolutions can be achieved with chemical or biological catalysts. Examples of both types of reactions are shown below.

*Hydrogenations via chemocatalysis*[13, 14]

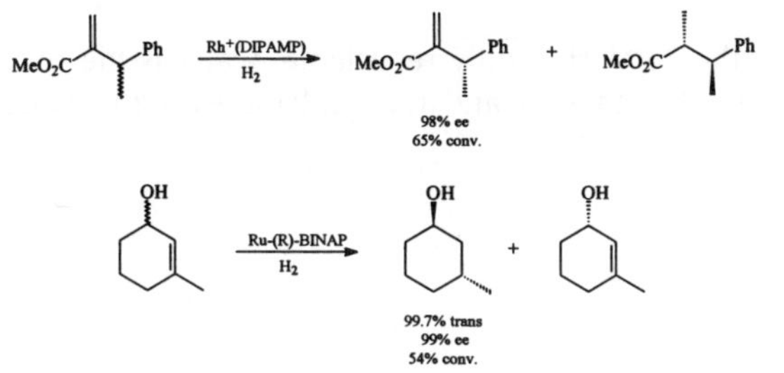

$MeO_2C$ Ph — $Rh^+(DIPAMP)$, $H_2$ → $MeO_2C$ Ph (98% ee, 65% conv.) + $MeO_2C$ Ph

OH — Ru-(R)-BINAP, $H_2$ → OH (99.7% trans, 99% ee, 54% conv.) + OH

**Figure 4.3**

Note: see Chapter Three for DIPAMP and BINAP structures.

*Epoxidation via chemocatalysis*[15]

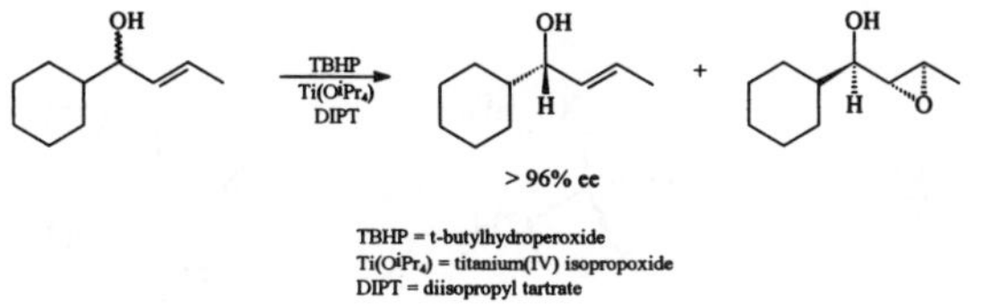

OH — TBHP, $Ti(O^iPr_4)$, DIPT → OH, H (> 96% ee) + OH, H, O

TBHP = t-butylhydroperoxide
$Ti(O^iPr_4)$ = titanium(IV) isopropoxide
DIPT = diisopropyl tartrate

**Figure 4.4**

*Hydrolytic kinetic resolution of epoxides via chemocatalysis*[16]

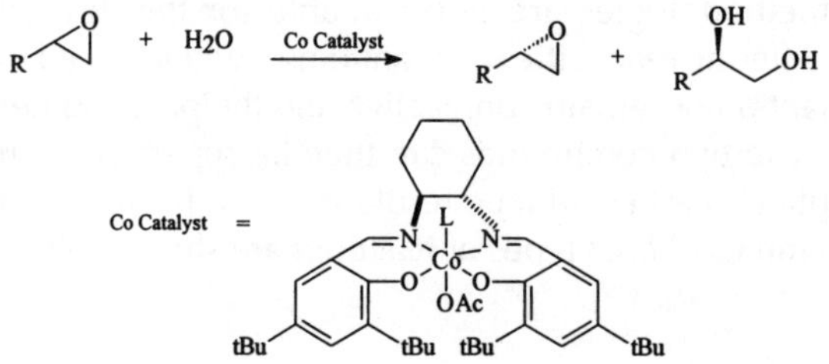


**Figure 4.5**

Note: this reaction is attractive because the catalyst is fully recyclable, water is the only reagent used, there is no solvent required for the reaction, and as a result there is virtually no waste stream.

*Acylation via biocatalysis*[17]

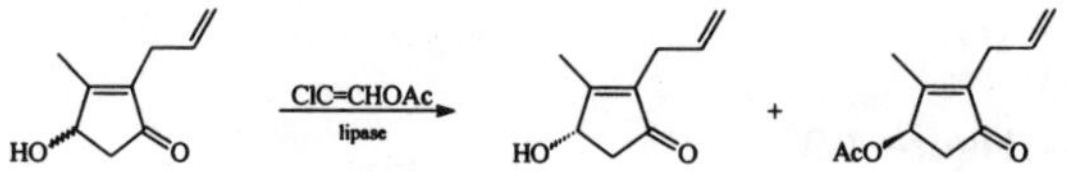


**Figure 4.6**

*Esterification*[18,19]

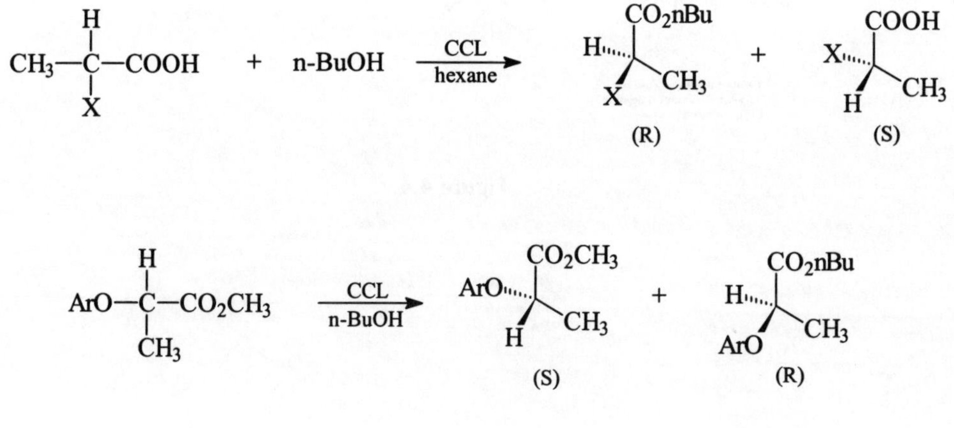

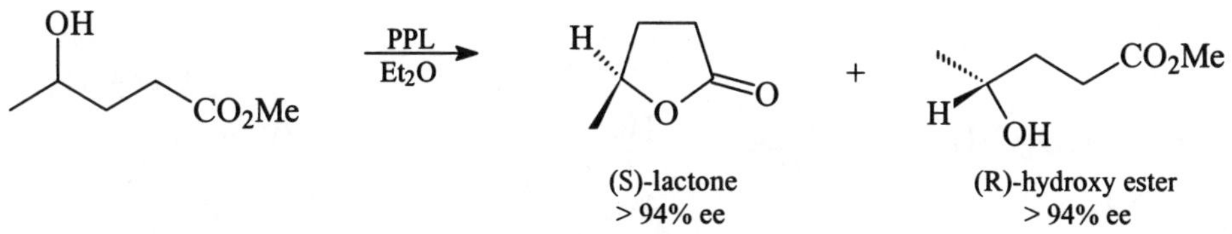

OH
CO2Me
PPL
Et2O
H
O
O
+
CO2Me
H
OH
(S)-lactone
> 94% ee
(R)-hydroxy ester
> 94% ee

**Figure 4.7**

Note: CCL = *Candida cylindrical* lipase; PPL = porcine pancreas lipase.

*Hydrolysis via biocatalysis*[20,21,22,17,22,23,24]

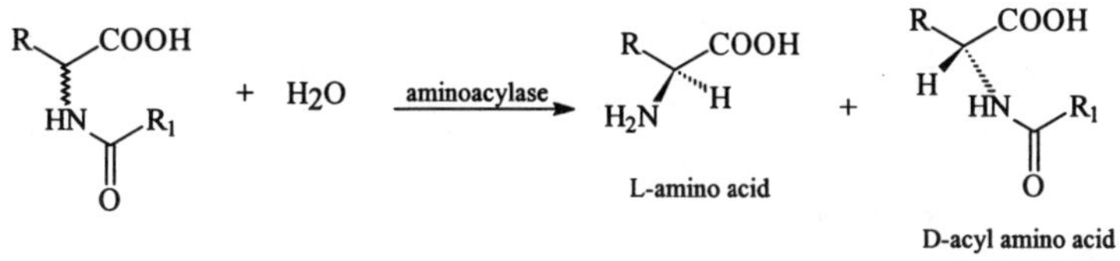

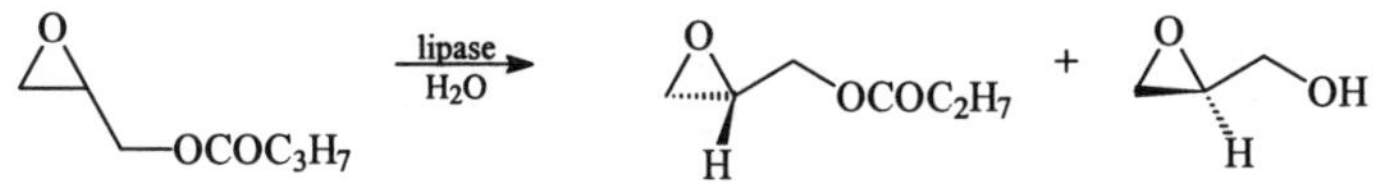

R
COOH
HN
R1
O
+ H2O
aminoacylase
R
COOH
H2N
H
+
R
COOH
H
HN
R1
O
L-amino acid
D-acyl amino acid

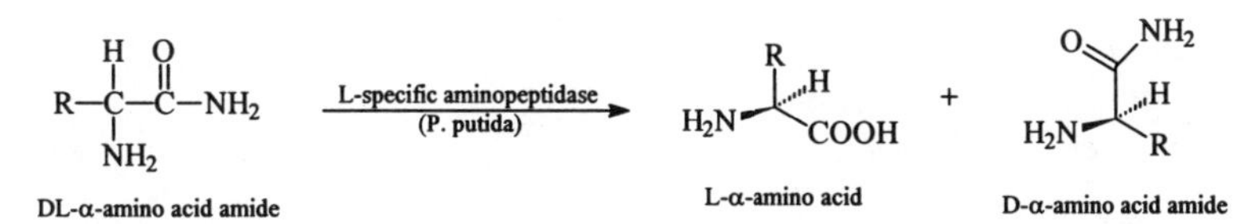

R—C—C—NH2
NH2
DL-α-amino acid amide
L-specific aminopeptidase
(P. putida)
H2N
COOH
L-α-amino acid
+
D-α-amino acid amide

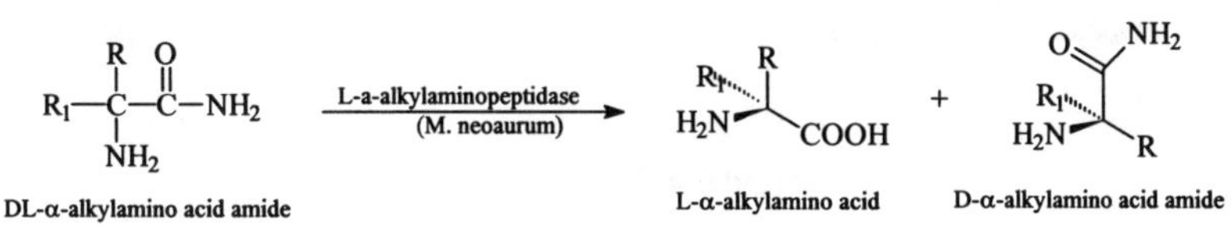

R1—C—C—NH2
NH2
DL-α-alkylamino acid amide
L-a-alkylaminopeptidase
(M. neoaurum)
H2N
COOH
L-α-alkylamino acid
+
D-α-alkylamino acid amide

AcO
O
lipase
AcO
O
+
HO
O

O
OCOC3H7
lipase
H2O
O
OCOC2H7
H
+
O
OH
H

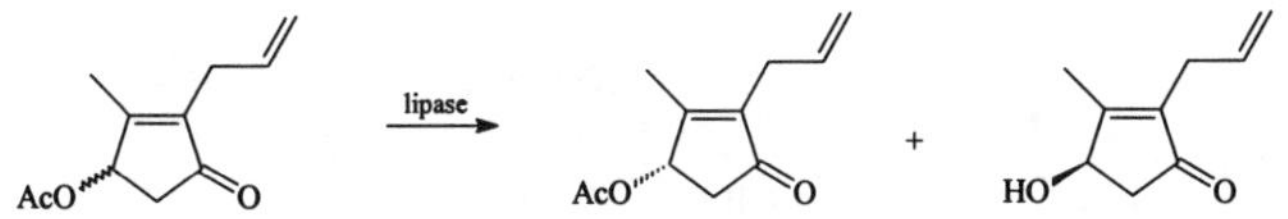

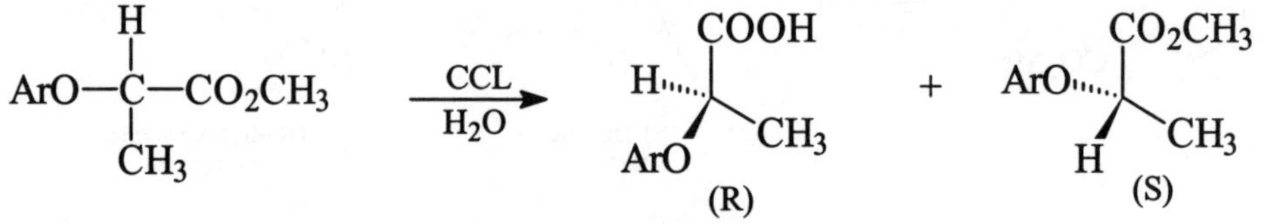

H
ArO—C—CO2CH3
CH3
CCL
H2O
COOH
H
ArO
CH3
(R)
+
CO2CH3
ArO
H
CH3
(S)

Figure 4.8

*Degradation via biocatalysis*[25]

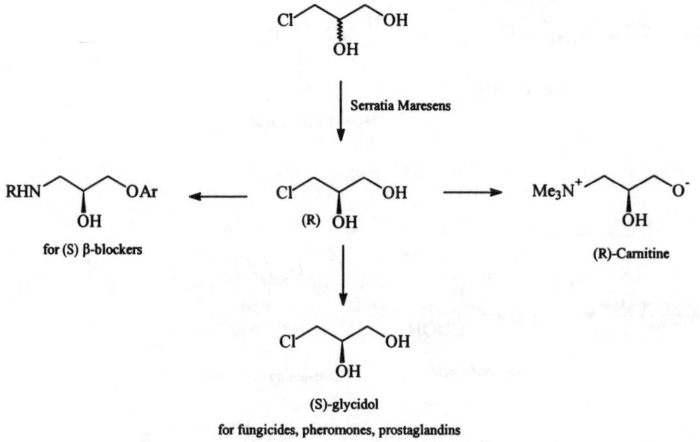

Cl OH
OH
Serratia Maresens
RHN OAr
OH
for (S) β-blockers
Cl OH
(R) OH
Me3N+ O-
OH
(R)-Carnitine
Cl OH
OH
(S)-glycidol
for fungicides, pheromones, prostaglandins

Figure 4.9

*Decarboxylation via biocatalysis*[26]

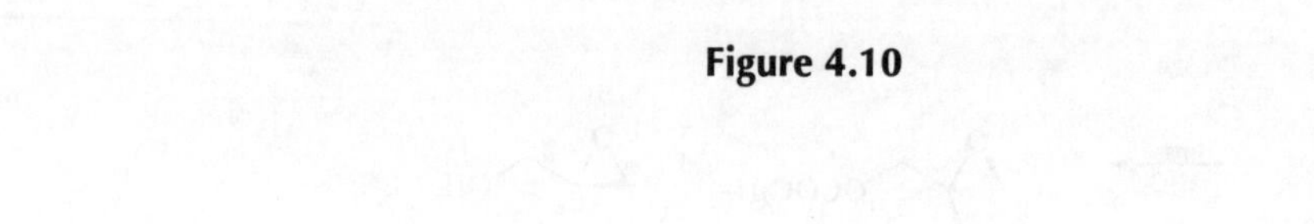

COOH
H3N COOH
E.C. 4.1.1.12
L-Asp-β-decarboxylase
H3N COOH
+
COOH
H3N COOH
+ CO2

Figure 4.10

*Dehalogenation vis biocatalysis*[27]

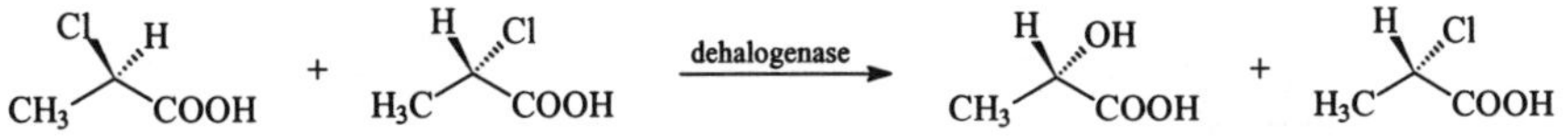

dehalogenase

Figure 4.11

*Oxidation via biocatalysis*[28]

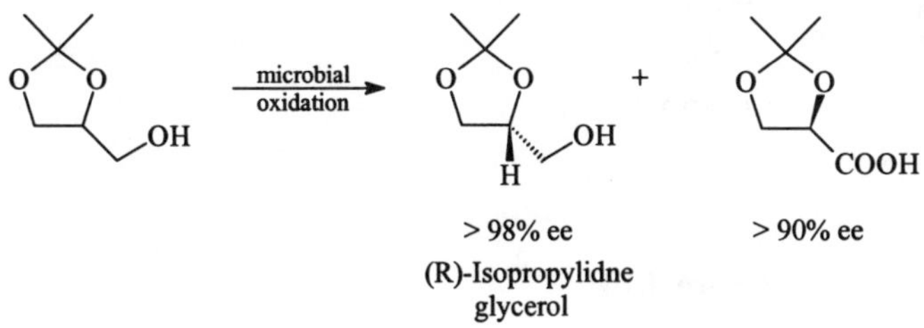

microbial oxidation

> 98% ee
(R)-Isopropylidne glycerol

> 90% ee

Figure 4.12

***Dynamic kinetic resolution***

Dynamic kinetic resolution is another example of a second order asymmetric transformation. In this case, the unreacted enantiomer is converted to the isomer that undergoes the desired transformation. As a result, a theoretical 100% yield can be obtained. Figure 4.13 shows an example of a hydrogenation reaction catalyzed by a Ruthenium BINAP complex (see Chapter Three for structure of BINAP ligand) where racemization occurs more quickly than the hydrogenation, resulting in production of a single isomer.[29]

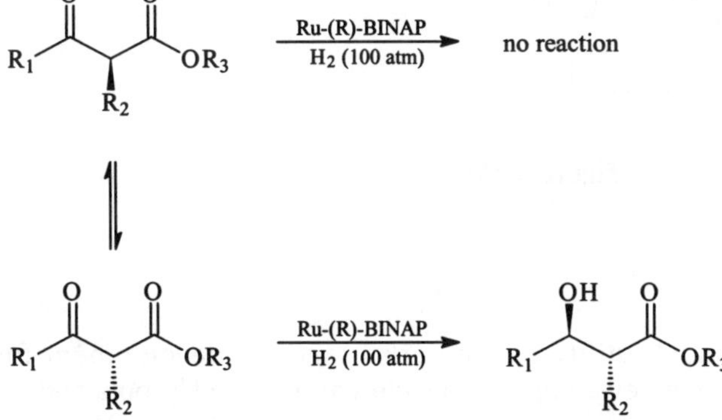

Ru-(R)-BINAP
$H_2$ (100 atm)

no reaction

Ru-(R)-BINAP
$H_2$ (100 atm)

Figure 4.13

In Figures 4.14 and 4.15, two different enzyme-mediated dynamic kinetic resolutions that have been successfully employed in the synthesis of biologically active products are presented.

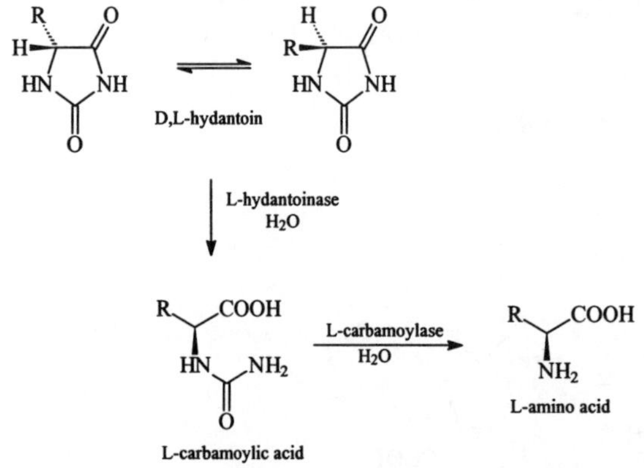

Figure 4.14[30]

Hydantoins have been found to racemize under basic conditions.[30] Thus, pure D and L amino acids can be prepared via the corresponding hydantoins through a two step process using hydantoinase and carbamylase enzymes (Figure 4.14).

In Figure 4.15, two microorganisms are used to achieve the dynamic kinetic resolution of a caprolactam.[31] *Cryptococcus laurentii* is used for the enantioselective hydrolysis of α–amino-ε–caprolactam to L-lysine. The microorganism *Achromobacter obae* (via a racemase enzyme) mediates the racemization of the unreacted D-α–amino-ε–caprolactam.

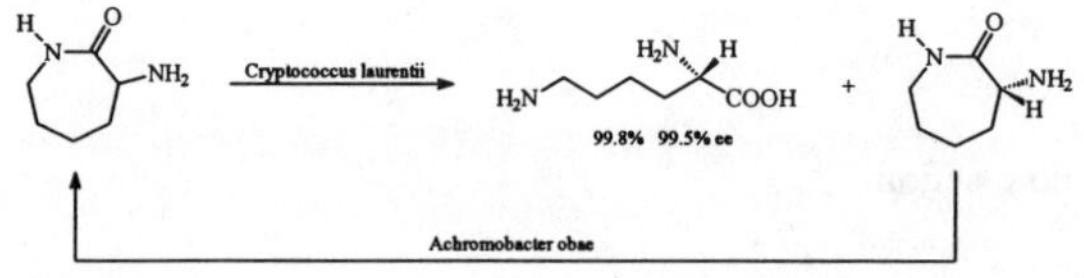

Figure 4.15[31]

### Preparative HPLC

Preparative HPLC has found much utility in the early stages of development for pharmaceuticals.[32] Both enantiomers of a chiral molecule can be quickly obtained by separating the racemic mixture using preparative HPLC and a chiral liquid phase. No asymmetric synthesis is required to obtain the pure isomers for testing and evaluation.

Recently Daicel has licensed technology from Novartis to immobilize polysaccharides on chromatographic media for commercial scale separations of chiral chemicals.[33] These and other advances in this area look promising for the future of preparative scale High Pressure Liquid Chromatography.

## Membrane Separations

Membrane separations have developed in two directions.[34] In the first approach, an enantioselective membrane is utilized to effect direct separation of a racemic mixture. In the second method, a non-selective membrane is employed to assist in an enantioselective process.

With enantioselective membranes, preferred transport of one isomer over the other occurs. The membrane is made of either a dense polymer or a liquid. The polymer is comprised of a porous, non-soluble support on which the membrane is supported. A pressure differential drives the transport across the membrane. Liquid membranes consist of a liquid film immobilized on part of a membrane by capillary forces. The membrane contains an enantioselective carrier that selectively forms a complex with one isomer and transports it across the membrane.

Examples of membrane assisted enantioselective processes include hollow-fiber membrane fractionation (liquid-liquid extraction), liquid-membrane fractionation, micellar enhanced ultra filtration (MEUF), and membrane assisted kinetic resolution.

Commercial examples include the Wandrey/Degussa membrane reactor[35] developed for amino acid separations. This technology uses ultrafiltration membranes. The Bend Research System[36] uses a selectively permeable membrane. Sepracor has developed a hollow-fiber system.[37]

## Supercritical Fluids

The use of supercritical fluids has recently received some attention as a potential aid in separating optically active isomers.[38] The higher pressures used to create the supercritical fluids results in lower viscosities and affects other pressure dependent physical properties. These changes may provide greater efficiencies in the separation of enantiomers.

## General Biological Transformations and Preparations

Precursor fermentation finds some applications in the synthesis of optically active molecules. The use of microorgansims to produce desired products from foreign substrates has received attention in the past several years as new technologies have been developed for manipulating the ability of the organisms to manufacture chiral compounds with high regio- and stereoselectivities. Some examples of precursor fermentation processes are presented below.

*Selective oxidation of carboxylic acids*[39]

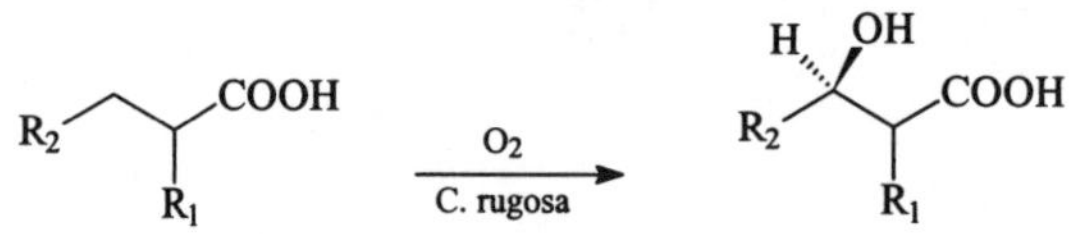

$$R_2CH_2CH(R_1)COOH \xrightarrow[\text{C. rugosa}]{O_2} R_2CH(OH)CH(R_1)COOH$$

**Figure 4.16**

***Selective microbial epoxidation***

Microbial epoxidation has been achieved with three different organisms – *P. oleovorans, C. equi,* and *Nocardia corallina*. Results of chiral epoxidation reactions catalyzed by *P. oleovorans* are shown in Table 4.3.[40] *N. corallina* has been found to epoxidize a much wider range of substrates.[41] Examples of some olefins that have been epoxidized using N. corallina are shown in Figure 4.17.

**Table 4.3 *P. oleovorans* epoxidations**[40]

| Epoxide | % ee |
|---|---|
| (R)-1,2-Epoxyoctane | 70 |
| (R)-1,2-Epoxydecane | 60 |
| (R)-7,8-Epoxyoct-1-ene | 80 |
| (R)-Butyl glycidyl ether | 85 |
| (R)-Allyl glycidyl ether | 81 |
| (R)-Benzyl glycidyl ether | 75 |
| (R)-Phenyl glycidyl ether | 92 |
| (R)-p-Methoxyphenyl glycidyl ether | 98 |
| (R)-p-Fluorophenyl glycidyl ether | 99 |
| (R)-p-Methoxyethylphenyl glycidyl ether | 100 |

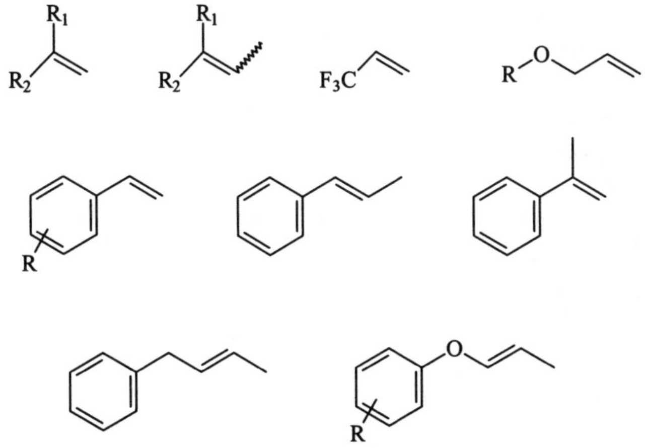

R = H, alkyl, alkenyl, for aliphatic substitutents

R= H, alkyl, alkoxy, halo for substiutents on aromatic ring

**Figure 4.17**[41]

### *Selective yeast reductions*

A wide variety of substrates can be reduced using yeasts. Baker's yeast is the most common type to be employed in the synthesis of optically active molecules, and enantioselectivities are generally high. Figure 4.18 shows several different substrates that have been accepted in yeast reductions along with the corresponding products.[42]

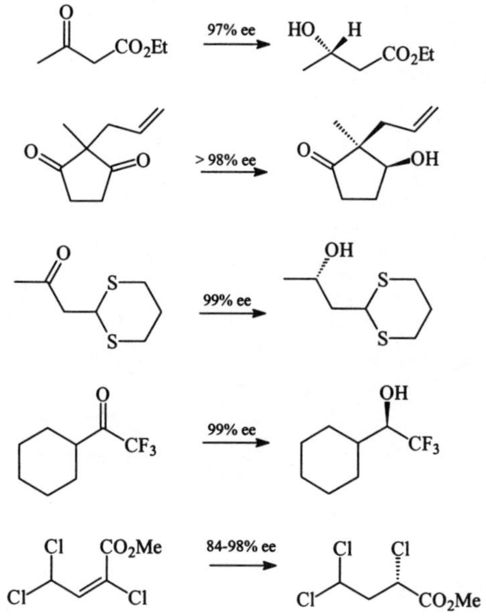

**Figure 4.18**

Enzymatic transformations (other than fermentations) have also seen increased applications in the synthesis of biologically active molecules. There are many types of enzymes available for a wide variety of organic transformations. Selected examples are presented below.

*Formation of chiral cyanohydrins*[43, 44, 45]

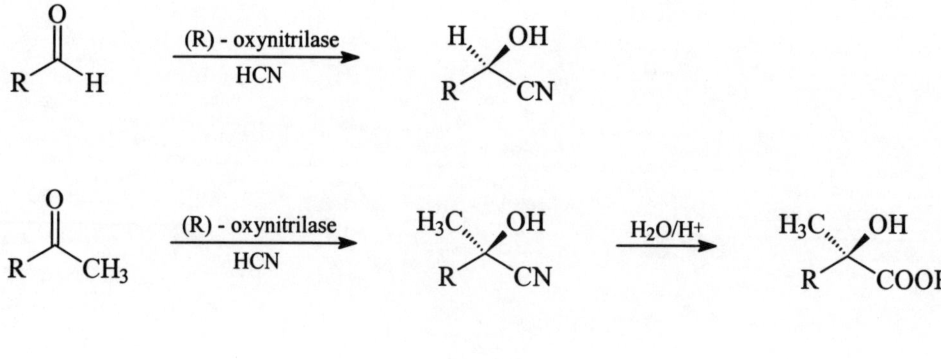

**Figure 4.19**

*Selective aldol condensations*[46]

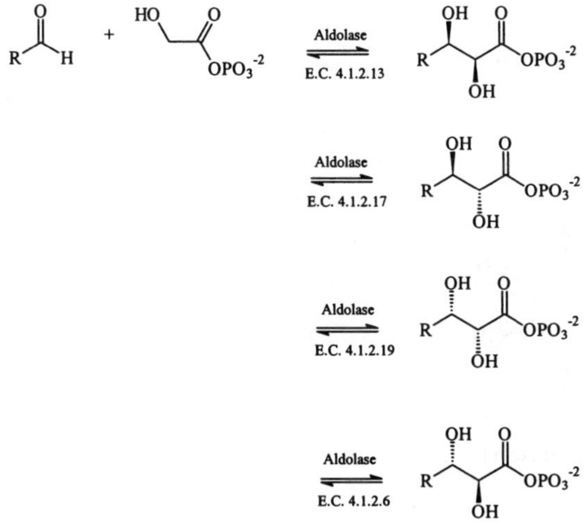

R–CHO + HO–CH$_2$–C(=O)–OPO$_3^{-2}$

Aldolase E.C. 4.1.2.13

Aldolase E.C. 4.1.2.17

Aldolase E.C. 4.1.2.19

Aldolase E.C. 4.1.2.6

Figure 4.20

*Selective addition of ammonia*[47]

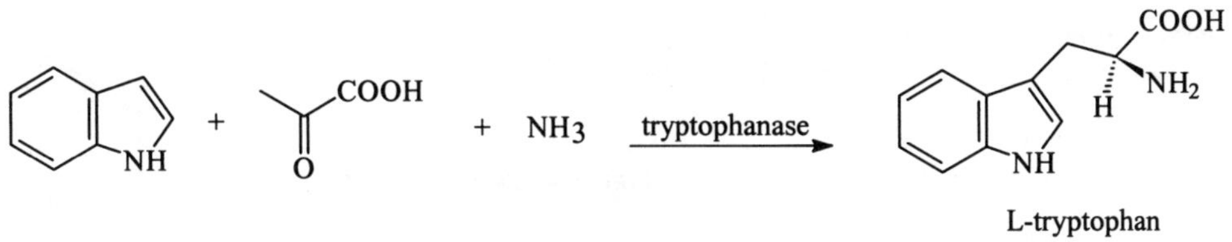

+ COOH + $NH_3$ tryptophanase NH NH COOH $NH_2$ H

L-tryptophan

Figure 4.21

*Selective hydration*[48]

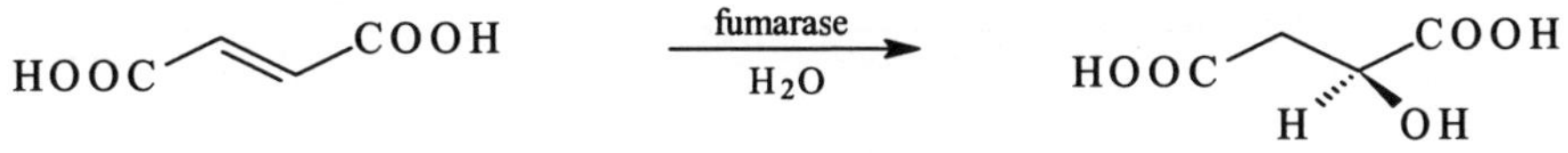

HOOC–CH=CH–COOH $\xrightarrow[H_2O]{\text{fumarase}}$ HOOC–CH$_2$–C(H)(OH)–COOH

Figure 4.22

*Asymmetric sulfoxidation*[49]

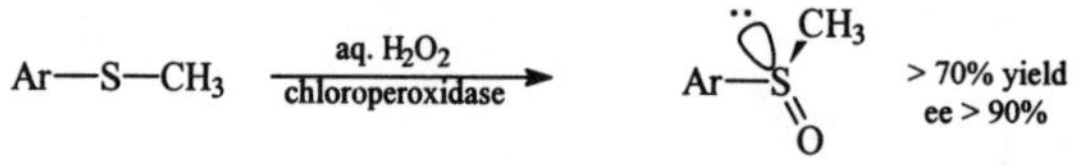

Ar—S—$CH_3$ —(aq. $H_2O_2$ / chloroperoxidase)→ Ar—S(=O)—$CH_3$   > 70% yield, ee > 90%

**Figure 4.23**

*Dihydroxylation of benzene derivatives*[50]

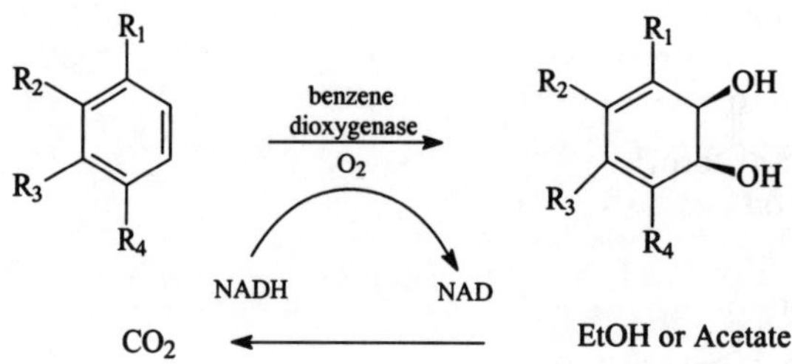


**Figure 4.24**

*Transamination*[51]

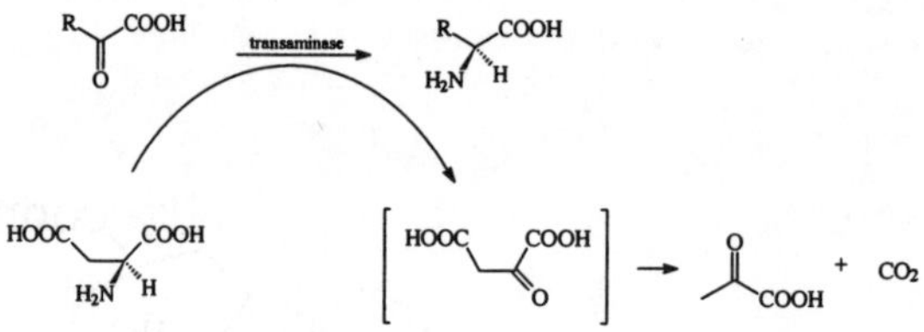

R–CO–COOH —(transaminase)→ R–CH($NH_2$)–COOH

HOOC–$CH_2$–CH($NH_2$)–COOH → [HOOC–$CH_2$–CO–COOH] → $CH_3$–CO–COOH + $CO_2$

**Figure 4.25**

*Reductive amination*[52]

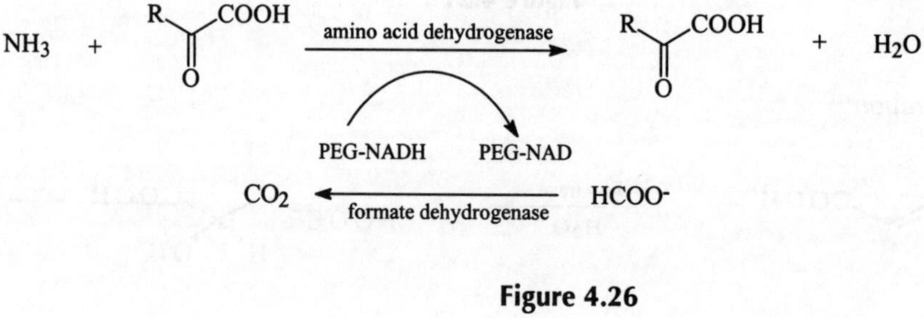

$NH_3$ + R–CO–COOH —(amino acid dehydrogenase)→ R–CH($NH_2$)–COOH + $H_2O$

PEG-NADH → PEG-NAD

$CO_2$ ←(formate dehydrogenase)— HCOO-

**Figure 4.26**

*Selective transfer of amino group*[53]

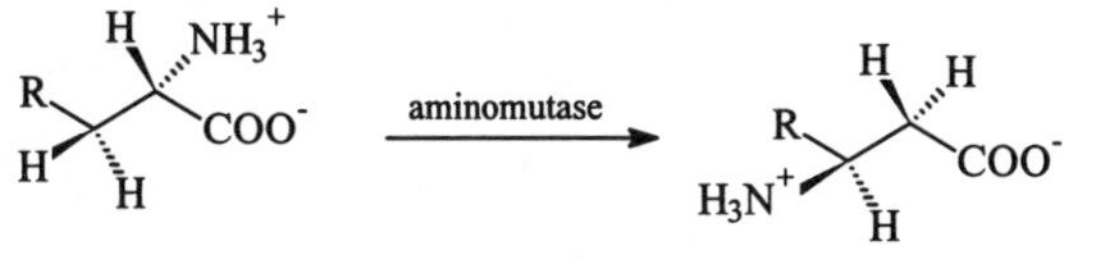
H
$NH_3^+$
R
$COO^-$
H
H
aminomutase
H
H
R
$COO^-$
$H_3N^+$
H

**Figure 4.27**

## Generic Asymmetric Synthesis Reactions

Many traditional catalytic organic synthetic methodologies have been adapted over the past two decades to provide optically pure compounds. Both heterogeneous and homogeneous catalysts have been utilized in these processes. Substitution reactions and transformations catalyzed by soluble chiral auxiliaries are also important means of carrying out asymmetric syntheses. Examples of selected reactions are shown below.

## Heterogeneous Catalysis

*Reduction of carbonyl compounds*[54, 55]

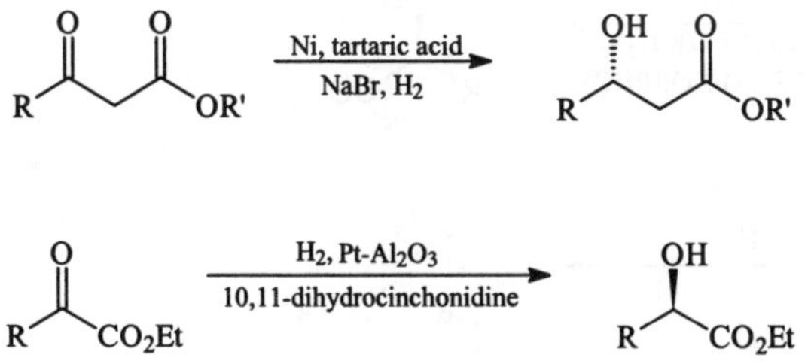
O
O
R
OR'
Ni, tartaric acid
NaBr, $H_2$
OH
O
R
OR'
O
R
$CO_2Et$
$H_2$, Pt-$Al_2O_3$
10,11-dihydrocinchonidine
OH
R
$CO_2Et$

**Figure 4.28**

*Hydrodehalogenation*[56]

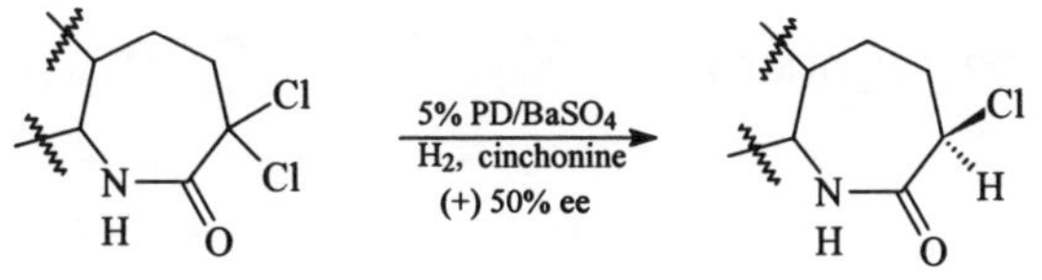
Cl
Cl
N
H
O
5% PD/$BaSO_4$
$H_2$, cinchonine
(+) 50% ee
Cl
H
N
H
O

**Figure 4.29**

## *Homogeneous catalysis*

*Hydrogenation*[57, 58, 59, 60, 61, 62, 63,]

H, R, COOH, NHAc → $H_2$, $[Rh(COD)DIPAMP]^+BF_4^-$ → R, COOH, NHAc, H

$CO_2Me$, NHAc, $R_1$, $R_2$ → $H_2$, [(R,R)-Et-DUPHOS-Rh] → $CO_2Me$, NHAc, $R_1$, $R_2$

$R_1$, $R_2$, $CO_2R$, NHAc → $H_2$, [(S,S)-Me-BPE-Rh] → $R_1$, $R_2$, $CO_2R$, NHAc

P, P

Me-BPE

O, R, $CO_2Et$ → $H_2$, $[Rh(NBD)Cl]_2$ - (2S,3S)-NORPHOS → OH, R, $CO_2Et$

Ar, COOH → $H_2$, $Ru(OAc)_2$ - (S)-BINAP → Ar, COOH

O, R, $CO_2CH_3$, X → $H_2$, Ru-BINAP → OH, R, $CO_2CH_3$, X

X = NHCOR, $CH_2NHCOR$, Cl

X = NHCOR, $CH_2NHCOR$, Cl

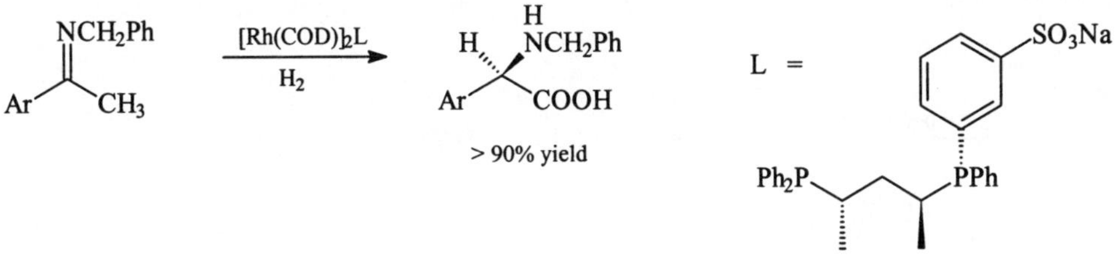

Figure 4.30

Note: COD = cyclooctadiene; NBD = norbornadiene; see Chapter Three for structures of DIPAMP, DUPHOS, NORPHOS, and BINAP ligands.

*Reduction with oxazaboralines*[64]

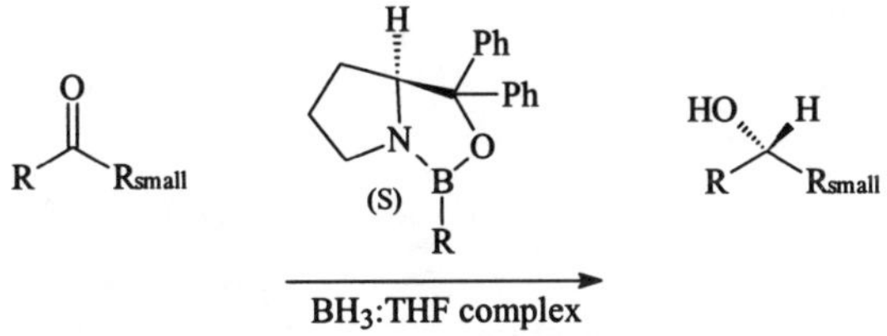

Figure 4.31

*Hydroformylation*[65]

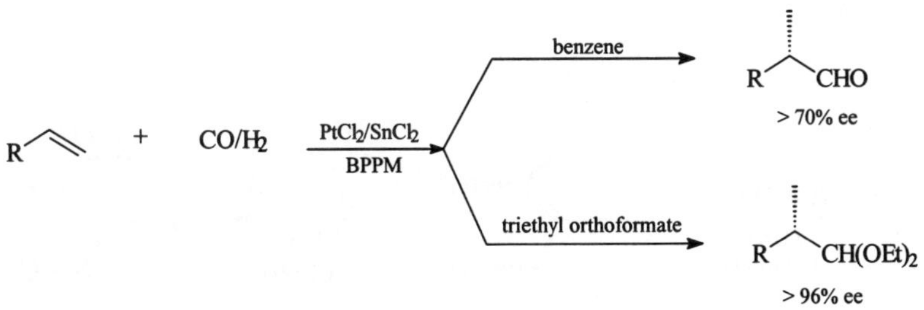

Figure 4.32

*Hydrosilylation*[66]

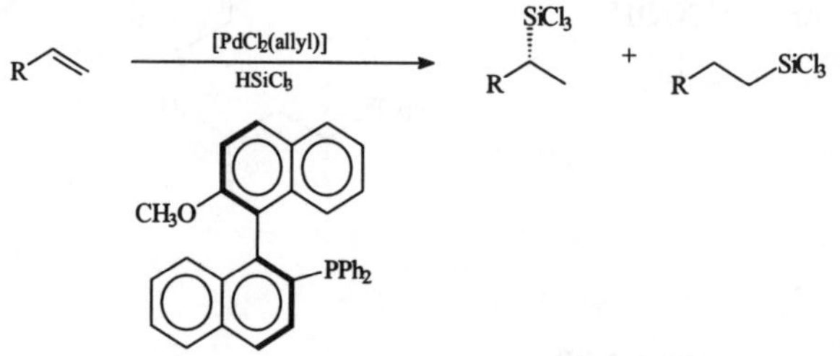

R⌄ —[$PdCl_2$(allyl)] / $HSiCl_3$→ $SiCl_3$ R + R⌄$SiCl_3$

$CH_3O$ … $PPh_2$

Figure 4.33

*Hydrogen transfer*[67, 68]

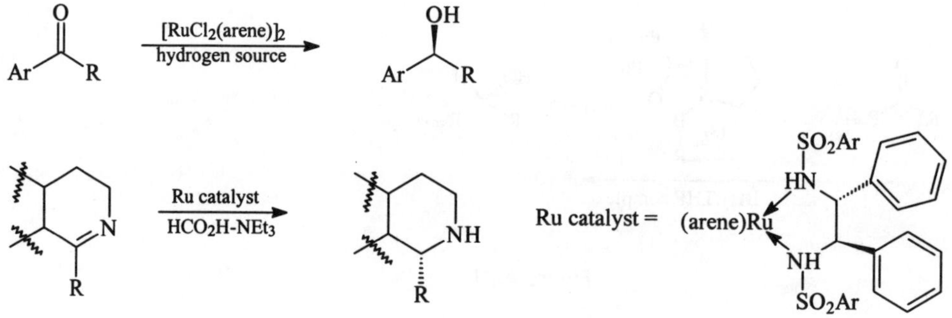

Ar(C=O)R —[$RuCl_2$(arene)]$_2$ / hydrogen source→ Ar(CHOH)R

—Ru catalyst / $HCO_2H$-$NEt_3$→ NH, R

Ru catalyst = (arene)Ru, HN, NH, $SO_2Ar$, $SO_2Ar$

Figure 4.34

*Sharpless epoxidation*[69]

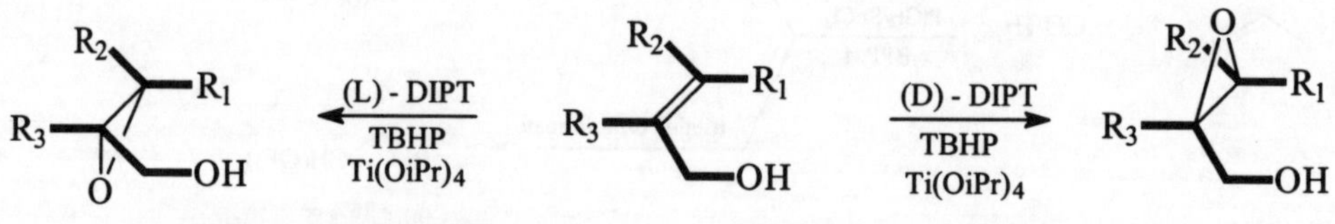

$R_2$, $R_1$, $R_3$, O, OH ←(L) - DIPT / TBHP / $Ti(OiPr)_4$— $R_2$, $R_1$, $R_3$, OH —(D) - DIPT / TBHP / $Ti(OiPr)_4$→ $R_2$, O, $R_1$, $R_3$, OH

DIPT = diispropyl tartrate
TBHP = tert-butyl hydroperoxide

Figure 4.35

*Jacobsen epoxidation*[70]

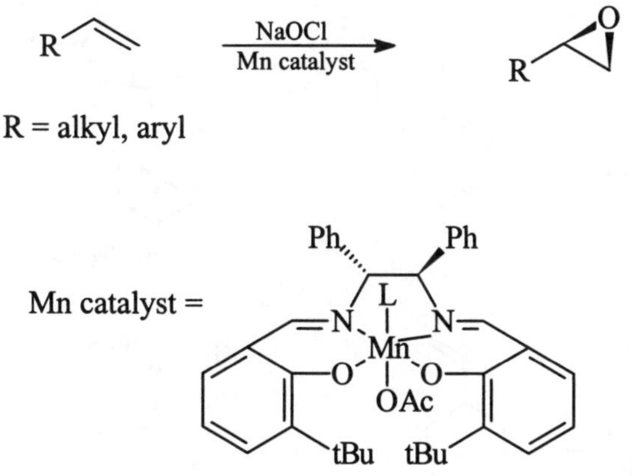
R
NaOCl
Mn catalyst
O
R
R = alkyl, aryl
Mn catalyst =
Ph
Ph
L
N
N
Mn
O
O
OAc
tBu
tBu

**Figure 4.36**

*Dihydroxylation*[71]

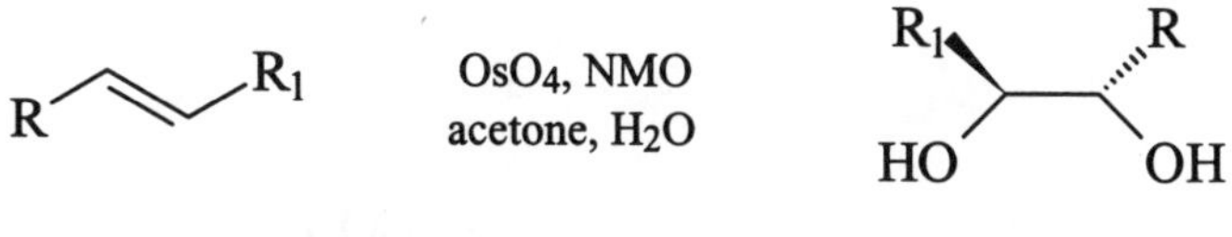
R
R1
OsO4, NMO
acetone, H2O
R1
R
HO
OH

NMO = N-methymorpholine N-oxide

**Figure 4.37**

*Sulfoxidation*[72]

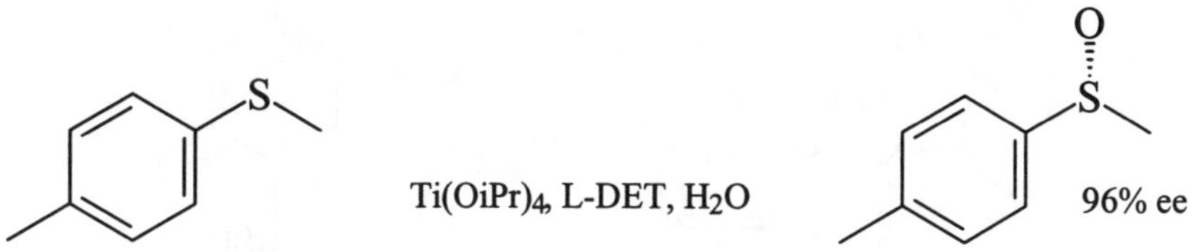
S
Ti(OiPr)4, L-DET, H2O
O
S
96% ee

DET = diethyl tartrate

**Figure 4.38**

*Codimerization*[73, 74]

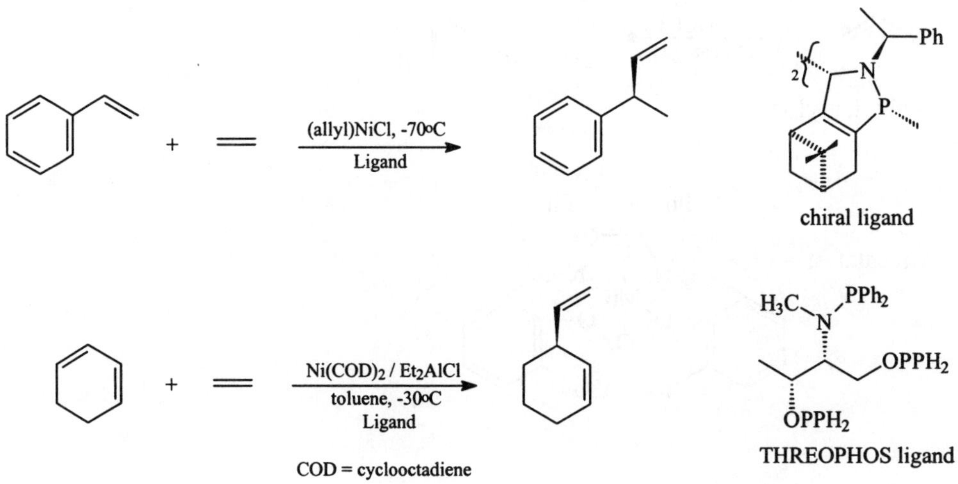

**Figure 4.39**

*Cyclopropanation*[75]

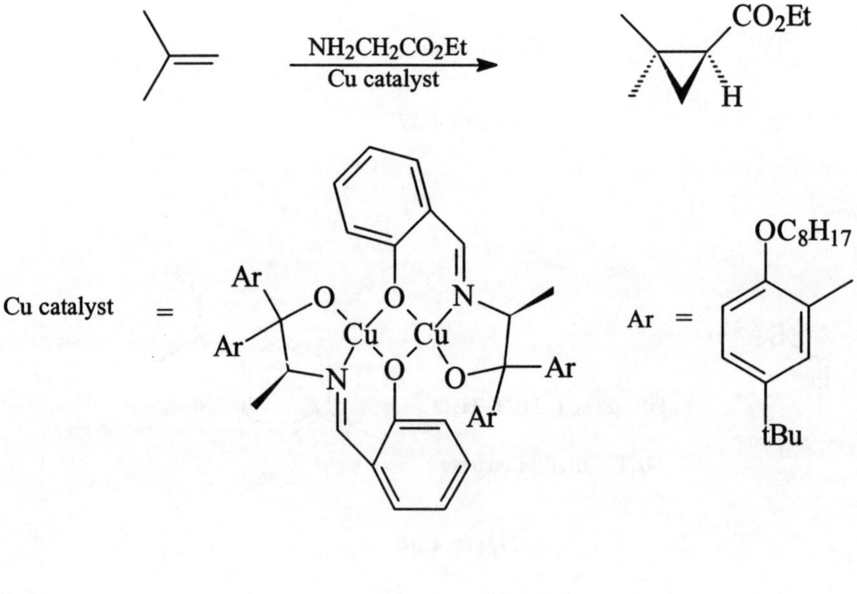

**Figure 4.40**

*Isomerization*[76]

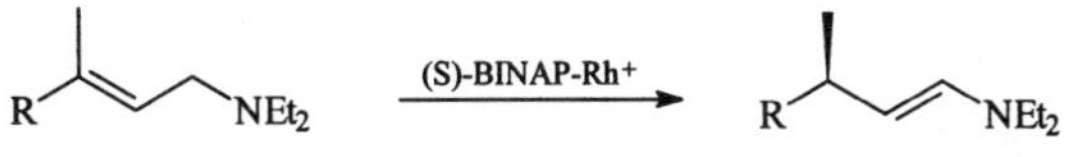

R–C(CH3)=CH–CH2–NEt2 —(S)-BINAP-Rh+→ R–CH(CH3)–CH=CH–NEt2

**Figure 4.41**

Note: see Chapter 3 for structure of BINAP ligand.

## *Other organometallic reactions*

*Reductive addition with dialkyl zinc*[77]

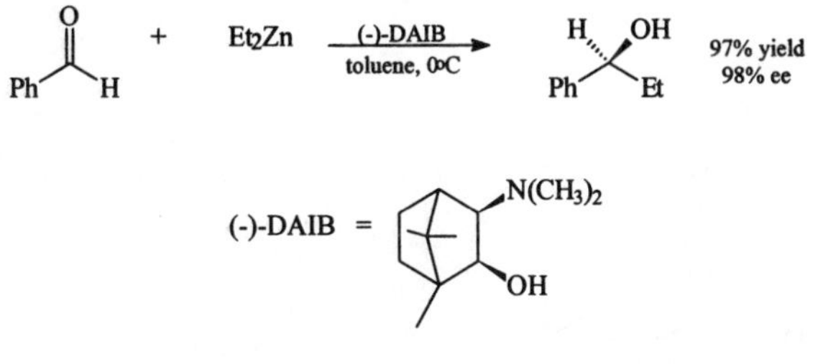

PhCHO + $Et_2Zn$ —(-)-DAIB, toluene, 0°C→ PhCH(OH)Et 97% yield 98% ee

(-)-DAIB = [bornane skeleton bearing N(CH3)2 and OH]

**Figure 4.42**

*Grignard cross-coupling*[78]

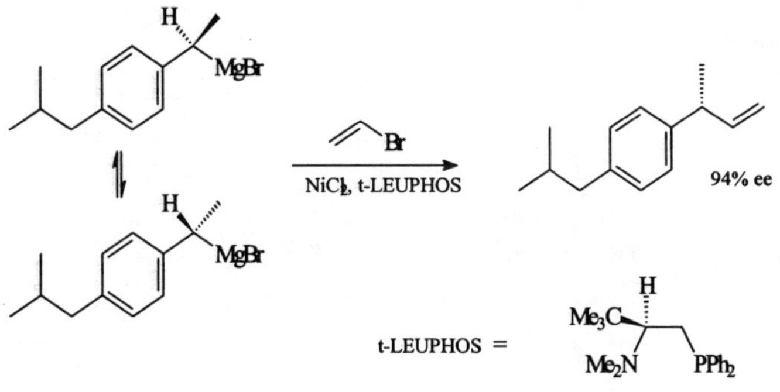

4-isobutylphenyl–CH(CH3)–MgBr (two epimers in equilibrium) —vinyl bromide, $NiCl_2$, t-LEUPHOS→ 4-isobutylphenyl–CH(CH3)–CH=CH2 94% ee

t-LEUPHOS = $Me_3C$–CH($NMe_2$)–$CH_2PPh_2$

**Figure 4.43**

*Palladium catalyzed coupling*[79]

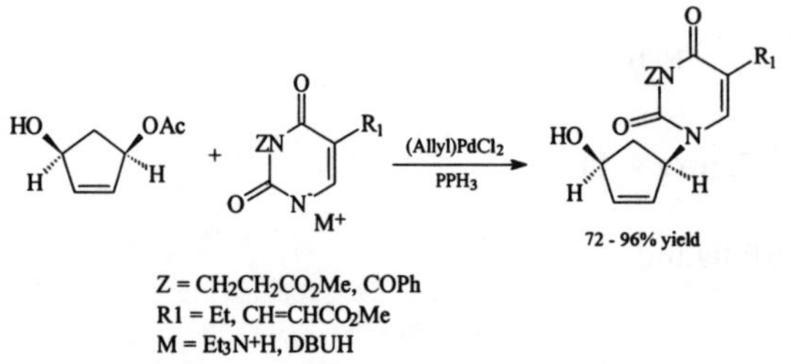
HO
OAc
H
H
+
O
ZN
R1
O
N
M+
(Allyl)PdCl2
PPH3
O
ZN
R1
O
N
HO
H
H
72 - 96% yield
Z = CH2CH2CO2Me, COPh
R1 = Et, CH=CHCO2Me
M = Et3N+H, DBUH

**Figure 4.44**

***Phase transfer catalysis***

*α–Alkylation of ketones*[80]

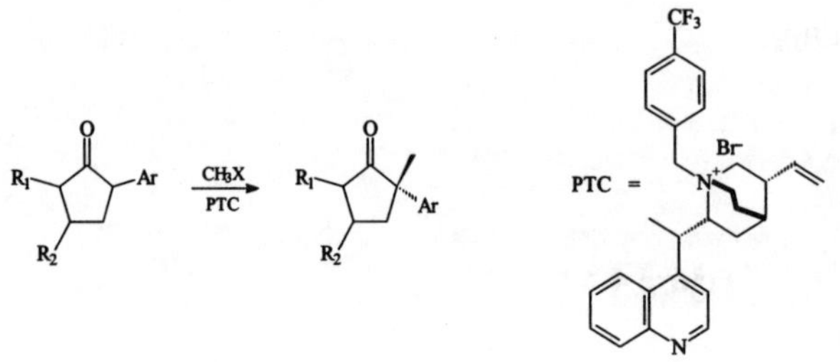
O
R1
Ar
R2
CH3X
PTC
O
R1
Ar
R2
CF3
Br-
PTC =
N+
N

**Figure 4.45**

*Alkylation of Schiff's Base esters*[81]

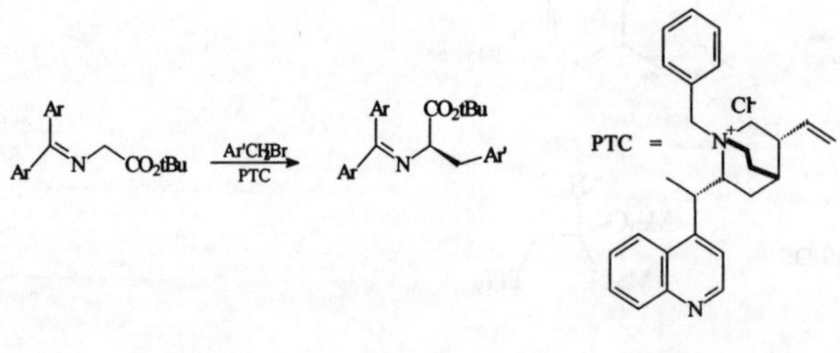
Ar
Ar
N
CO2tBu
Ar'CH2Br
PTC
Ar
CO2tBu
Ar
N
Ar'
Cl-
PTC =
N+
N

**Figure 4.46**

*Robinson Annulation*[82]

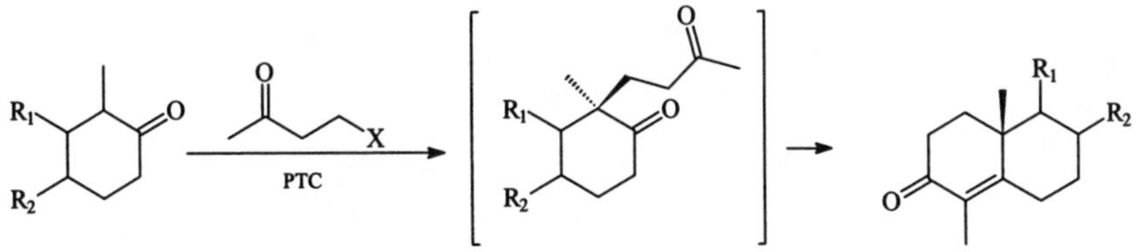

**Figure 4.47**

***Chiral auxiliaries***

*Examples of chiral auxiliaries*[83]

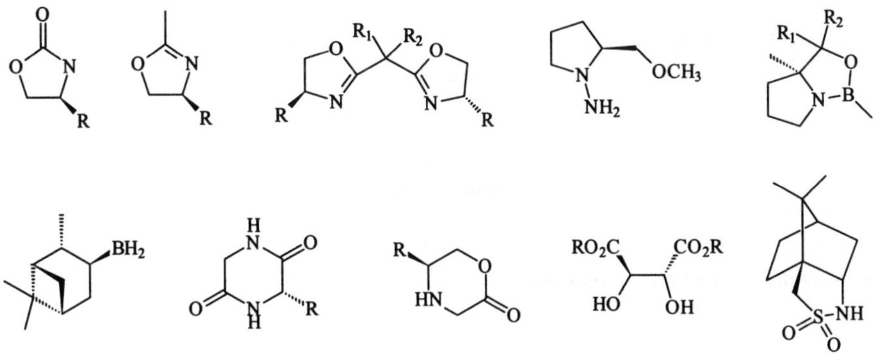

**Figure 4.48**

*Miscellaneous reactions with chiral auxiliaries*[84, 85]

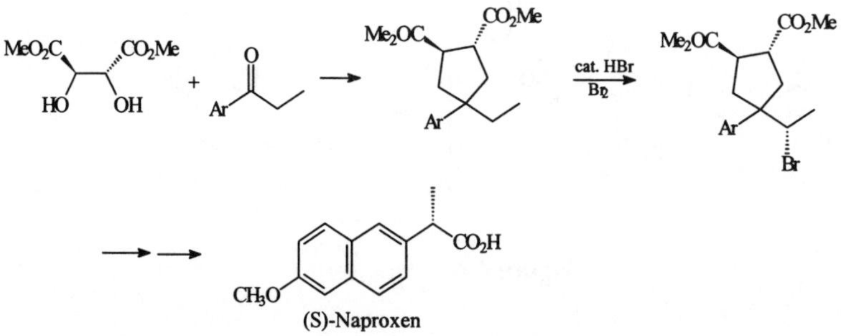

**Figure 4.49**

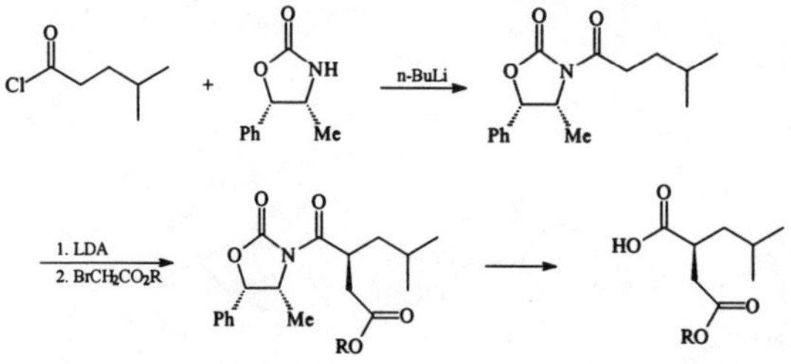

**Figure 4.50**

*Camphor derivatives as chiral auxiliaries*

Alkylations[86]

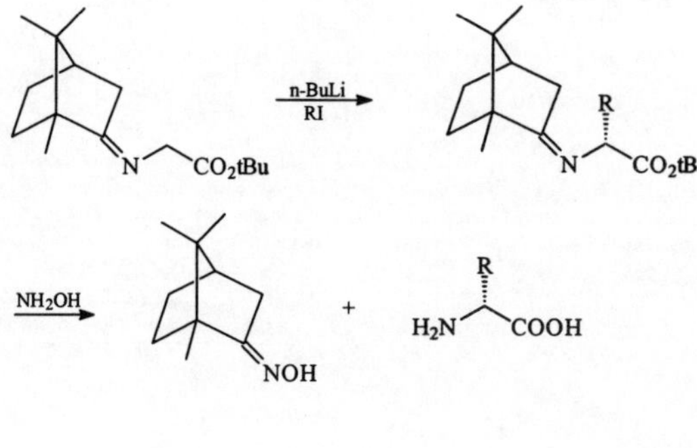

**Figure 4.51**

*Camphor derivatives as chiral auxiliaries (continued)*

Aldol condensations[87]

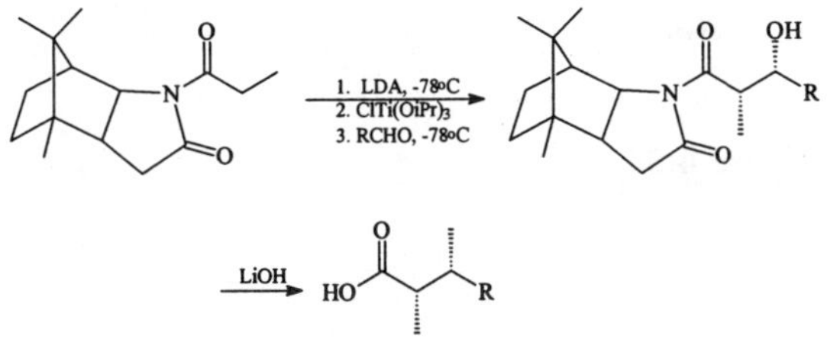

1. LDA, -78°C
2. ClTi(OiPr)$_3$
3. RCHO, -78°C

LiOH

**Figure 4.52**

Diels-Alder reactions[88]

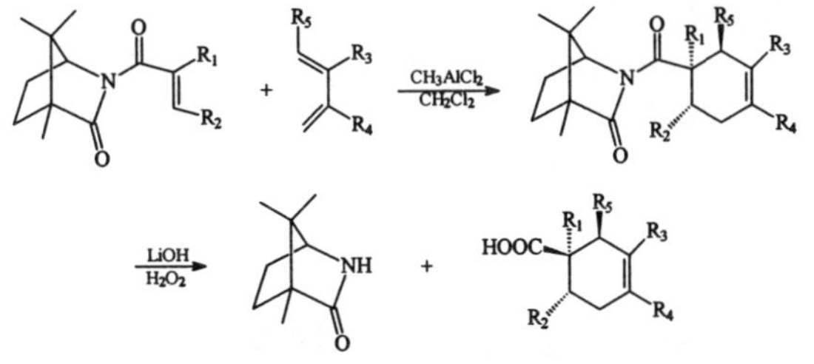

$CH_3AlCl_2$
$CH_2Cl_2$

LiOH
$H_2O_2$

**Figure 4.53**

***Substitution reactions***

*Reactions of alcohols*[89]

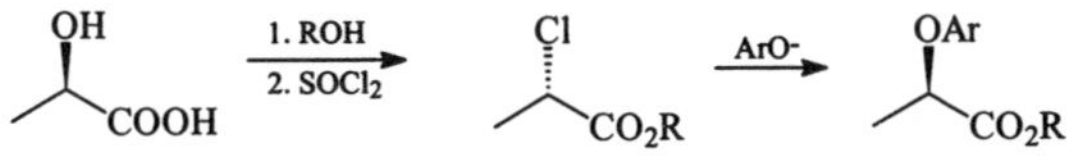

1. ROH
2. $SOCl_2$

ArO-

**Figure 4.53**

*Epoxide ring opening*[90]

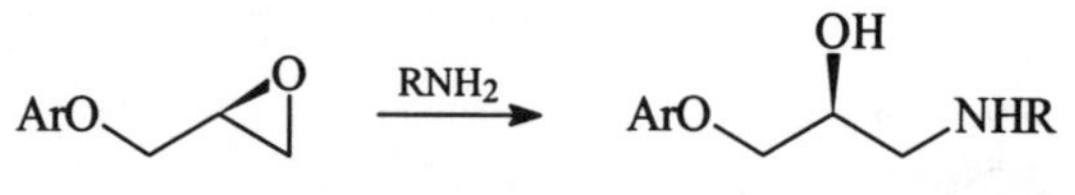

ArO–epoxide $\xrightarrow{RNH_2}$ ArO–CH$_2$–CH(OH)–CH$_2$–NHR

**Figure 4.54**

*Cyclic sulfate reactions*[91]

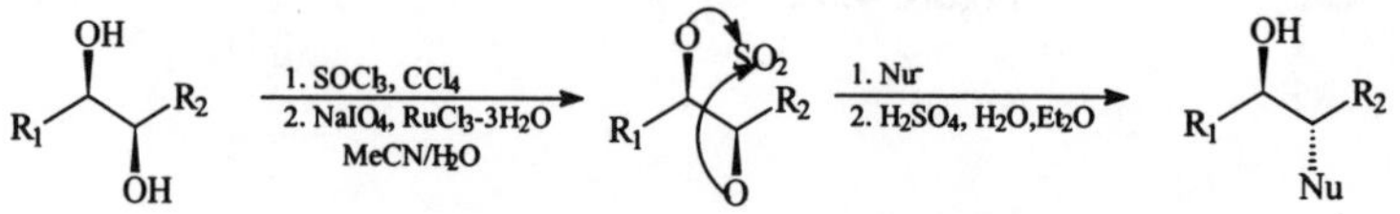

1. $SOCl_2$, $CCl_4$
2. $NaIO_4$, $RuCl_3$-$3H_2O$, $MeCN/H_2O$

1. $Nu^-$
2. $H_2SO_4$, $H_2O$, $Et_2O$

**Figure 4.55**

*Iodolactonizations*[92]

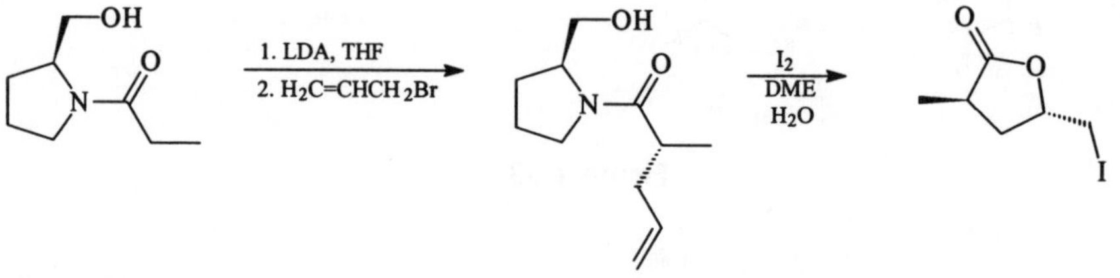

1. LDA, THF
2. $H_2C{=}CHCH_2Br$

$I_2$, DME, $H_2O$

**Figure 4.54**

*Allylic substitutions*[93]

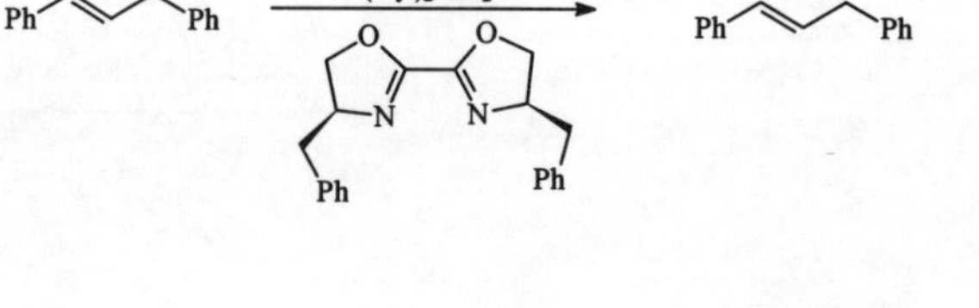

$NaCH(CO_2Me)_2$
$(allyl)_2PdCl_2$

**Figure 4.55**

## References


1. Crosby, J., in *Chirality in Industry*, Collins, A.N., Sheldrake, G.N., and Crosby, J. (Eds.), John Wiley & Sons, New York, 1992, p.55.
2. Sheldon, R.A., *Chirotechnology*, Marcel Dekker, Inc., New York, 1993, p. 89.
3. *J. Prakt. Chem.*, **99**, 6 (866).
4. Ager, D.J., in *Handbook of Chiral Chemicals*, Ager, D.J. (Ed.), Marcel Dekker, Inc., New York, 1999, p. 70.
5. Sheldon, R.A., *Chirotechnology*, Marcel Dekker, Inc., New York, 1993, p. 146
6. Liu, W., in *Handbook of Chiral Chemicals*, Ager, D.J. (Ed.), Marcel Dekker, Inc., New York, 1999, p. 84.
7. Liu, W., in *Handbook of Chiral Chemicals*, Ager, D.J. (Ed.), Marcel Dekker, Inc., New York, 1999, pp. 84-90 and references cited therein.
8. Crosby, J., in *Chirality in Industry*, Collins, A.N., Sheldrake, G.N., and Crosby, J. (Eds.), John Wiley & Sons, New York, 1992, p.24.
9. a) Gernez, D., *Compt. Rend.*, **63**, 843 (1866). b) Amiard, G., *Bull. Soc. Chim. Fr.*, **447** (1956).
10. Crosby, J., in *Chirality in Industry*, Collins, A.N., Sheldrake, G.N., and Crosby, J. (Eds.), John Wiley & Sons, New York, 1992, pp. 26-27.
11. Bayley, C.R., and Vaidya, N.A., in *Chirality in Industry*, Collins, A.N., Sheldrake, G.N., and Crosby, J. (Eds.), John Wiley & Sons, New York, 1992, pp. 75-76.
12. a) Sheldon, R.A., *Chirotechnology*, Marcel Dekker, Inc., New York, 1993, p. 188. b) Bayley, C.R., and Vaidya, N.A. in *Chirality in Industry*, Collins, A.N., Sheldrake, G.N., and Crosby, J. (Eds.), John Wiley & Sons, New York, 1992, p. 73. c) Sheldon, R.A., *Chirotechnology*, Marcel Dekker, Inc., New York, 1993, p. 189.
13. Kitamura, M., Kasahara, I., Manabe, K., and Noyori, R., *J. Org. Chem.*, **53**, 708 (1988).
14. Brown, J.M., *Angew. Chem. Int. Ed. Engl.*, **26**, 190 (1987) and references cisted therein.
15. a) Martin, V.S., Woodward, S.S., Katsuki, T., Yamada, Y., Ikeda, M., and Sharpless, K.B., *J. Am. Chem. Soc.*, **103**, 6237 (1981). b) Miyano, S., Lu, L.D.L., Viti, M., and Sharpless, K.B., *J. Org. Chem.*, **50**, 4350-4360 (1985).
16. Liu, W., in *Handbook of Chiral Chemicals*, Ager, D.J. (Ed.), Marcel Dekker, Inc., New York, 1999, pp. 126-127.
17. Sumitomo, *Jpn. Pat.*, 50013365, 1973.
18. a) Kirchner, G., Scollar, M.P., and Klibanov, A.M., *J. Am. Chem. Soc.*, **107**, 7072-7076 (1985). b) Cambou, B., and Klibanov, A.M., *Biotechnol. Bioeng.*, **26**, 144901454 (1984).
19. a) Yamada, H., Ohsawa, S., Sugai, T., Ohta, H., and Yoshikawa, S., *Chem. Lett.*, 1775-1776 (1989). B) Gutman, A.L., Zuobi, K., and Boltansky, A., *Tetrahedron Lett.*, **28**, 3861-3864 (1987).
20. Chibata, I., Tosa, T., and Shibatani, T., in *Chirality in Industry*, Collins, A.N., Sheldrake, G.N., and Crosby, J. (Eds.), John Wiley & Sons, New York, 1992, p. 356.
21. Kamphus, J., Boesten, W.H.J., Kaptein, B., Hermes, H.F.M., Sonke, T., Broxterman, Q.B., van den Tweel, W.J.J., and Schoemaker, H.E., in *Chirality in Industry*, Collins, A.N., Sheldrake, G.N., and Crosby, J. (Eds.), John Wiley & Sons, New York, 1992, pp. 189-192.

22. a) Kamphius, J., Hermes, H.F.M., van Balken, J.A.M., Scheomaker, H.E., Boesten, W.H.J., and Meijer, E.M., in *Amino Acids, Chemistry, Biology and Medicine*, Lubec, G., and Rosenthal, G.A. (Eds.), ESCOM Science Publ., Vienna, 1990, pp. 119-125. b) Kaptein, B., Boesten, W.H.J., Broxterman, Q.B., Meijer, E.M., Polinelli, S., Schoemaker, H.E., and Kamphuis, J., in *Proceedings of the Second International Symposium on Chiral Discrimination, Rome, 1991*, Misiti, D., (Ed.), 1991, p. 58.
23. Hoechst, *Eur. Pat.*, 321918, 1988.
24. a) Ladner, W.E., and Whitesides, G.M., *J. Am. Chem. Soc.*, **106**, 7250-7251 (1984). b) Kloosterman, M., Elfrink, V.H.M., Van Iersel, J., Roskam, J.H., Meijer, E.M., Hulshof, L.A., and Sheldon, R.A., *Trends Biotechnol.*, **6**, 251-256 (1988).
25. Ohashi, T., *Proc. Chiral 90 Symp.*, Spring Innovations, Stockport, UK, 1990, pp. 65-71.
26. Pantaleone, D.P., in *Handbook of Chiral Chemicals*, Ager, D.J. (Ed.), Marcel Dekker, Inc., New York, 1999, p. 270.
27. Taylor, S.C., paper presented at SCI Meeting, *Opportunities in Biotransformations*, Cambridge, UK, 3-5, April, 1990.
28. *Eur. Pat.*, 244912.
29. Kitamura, M., Ohkuna, T., Tokunga, M., and Noyori, R., *Tetrahedron Asymmetry*, **1**, 1-4 (1990).
30. a) Yamada, H., Takahashi, S., Kii, Y., and Kunegai, H., *J. Ferment. Technol.*, **56**, 484 (1978). b) Kanegafuchi Chemical Indsutries, *Jpn. Pat.*, 6225990, 1987. c) Yamada, H., Takahashi, S., Yoneda, K., and Amagasaki, H., *Ger. Pat.*, 2.757980, 1991. d) Olivieri, R., Fascetti, E., Angelini, L., an dDegen, L., *Biotechnol. Bioeng.*, **23**, 2173 (1981).
31. Fufkumura, T., *Agric. Biol. Chem.*, **40**, 1687 (1976).
32. Crosby, J., in *Chirality in Industry II*, Collins, A.N., Sheldrake, G.N., and Crosby, J. (Eds.), John Wiley & Sons, New York, 1997, p. 97.
33. Chiral Separations, Chemical Week, April 14, 1999, p. 35.
34. Keurentjes, J.T.F., and Voermans, F.J.M., in *Chirality in Industry II*, Collins, A.N., Sheldrake, G.N., and Crosby, J. (Eds.), John Wiley & Sons, New York, 1997, pp. 157-180.
35. Bommarius, A.S., Drauz, K., Groeger, U., and Wandrey, C. in *Chirality in Industry*, Collins, A.N., Sheldrake, G.N., and Crosby, J. (Eds.), John Wiley & Sons, New York, 1992, pp. 372-397.
36. van Eikeren, P., Brose, D.J., Muchmore, D.C., West, J.B., and Colton, R.H., in *Proceedings of Chiral '92 Symposium*, Manchester, 24-25 March 1992, Spring Innovations, 1992, p. 63.
37. See, for example, Young, J.W., in *Proceedings of the Chiral Synthesis Symposium and Workshop*, 18 April, 1989, Stockport, UK, Spring Innovations, 1989, p. 39.
38. Crosby, J., in *Chirality in Industry II*, Collins, A.N., Sheldrake, G.N., and Crosby, J. (Eds.), John Wiley & Sons, New York, 1997, p. 8.
39. a) Hasegawa, J., Ogura, M., Hamaguchi, S., Shimazaki, M., Kawaharanda, H., and Watanabe, K., *J. Ferment. Technol.*, **60**, 501-508 (1982). b) Hasegawa, J., Ogura, M., Kanema, H., Noda, N., Kawaharanda, H., and Watanabe, K., *J. Ferment. Technol.*, **59**, 203-208 (1981). c) Ohashi, T., *Proc. Chiral 90 Symp.*, Spring Innovations, Stockport, UK, 1990, pp. 65-71. d) Shimazaki, M., Hasegawa, J., Kan, K., Nomura, K., Nose, Y., Kondo, H., Ohashi, T., and Watanabe, K., *Chem. Pharm. Bull.*, **30**, 3139-3146 (1982).

40. Furuhashi, K., in *Chirality in Industry*, Collins, A.N., Sheldrake, G.N., and Crosby, J. (Eds.), John Wiley & Sons, New York, 1992, pp. 169-171.
41. Furuhashi, K., in *Chirality in Industry*, Collins, A.N., Sheldrake, G.N., and Crosby, J. (Eds.), John Wiley & Sons, New York, 1992, pp. 171-175.
42. Crosby, J., in *Chirality in Industry II*, Collins, A.N., Sheldrake, G.N., and Crosby, J. (Eds.), John Wiley & Sons, New York, 1997, p. 50 and references cited therein.
43. Sheldon, R.A., *Chirotechnology*, Marcel Dekker, Inc., New York, 1993, pp. 257-258 and references cited therein.
44. Effenberger, F., Hörsch, B., Weingart, F., Ziegler, T., and Kühner, S., *Tetrahedron Lett.*, **32**, 2605-2608 (1991).
45. Kruse, C.G., in *Chirality in Industry*, Collins, A.N., Sheldrake, G.N., and Crosby, J. (Eds.), John Wiley & Sons, New York, 1992, pp. 279-299.
46. Pantaleone, D.P., in *Handbook of Chiral Chemicals*, Ager, D.J. (Ed.), Marcel Dekker, Inc., New York, 1999, pp. 268-270.
47. Sheldon, R.A., *Chirotechnology*, Marcel Dekker, Inc., New York, 1993, pp. 230-232.
48. a) Umemura, J., Takamatsu, S., Sato, T., Tosa, T., and Chibata, J., *Eur. J. Appl. Microbiol. Biotechnol.*, **20**, 291 (1984). b) Chibata, I., Tosa, T., and Sato, T., in *Biotechnology, Vol. 7a*, Rehm, H.J., and Reed, G., *Eds.), VCH, Weinheim, 1987, pp. 653-684.
49. a) Colonna, S., Gaggero, N., Manfredi, A., Casella, L., Gullotti, M., Carrera, G., and Pasta, P., *Biochemistry*, **29**, 10465-10468 (1990). b) Colonna, S., Gaggero, N., Casella, L., Carrera, G., and Pasta, P., *Tetrahedron Asymm.*, **3**, 95-106 (1992).
50. Sheldrake, G.N., in *Chirality in Industry*, Collins, A.N., Sheldrake, G.N., and Crosby, J. (Eds.), John Wiley & Sons, New York, 1992, pp. 127-166.
51. Stirling, D.I., in *Chirality in Industry*, Collins, A.N., Sheldrake, G.N., and Crosby, J. (Eds.), John Wiley & Sons, New York, 1992, pp. 209-222.
52. Sheldon, R.A., *Chirotechnology*, Marcel Dekker, Inc., New York, 1993, pp. 228-229.
53. Pantaleone, D.P., in *Handbook of Chiral Chemicals*, Ager, D.J. (Ed.), Marcel Dekker, Inc., New York, 1999, p. 271 and references cited therein.
54. a) Fukawa, H., Izumi, Y., Komatsu, S., and Akabori, S., *Bull. Chem. Soc. Jpn,*, **35**, 1703 (1962). b) Izunmi, Y., *Advan. Catal.*, **32**, 215 (1983). c) Sachtler, W.M.H., in *Catalysis of Organic Reactions*, Augustine, R.L. (Ed.), Marcel Dekker, New York, 1985, pp. 189-206. d) Tai, A., and Harada, T., in *Tailored Metal Catalysts*, Iwasawa, Y. (Ed.), Rediel, Dordrecht, 1986, p. 265. e) Bakos, J., Toth, I., and Marko, L., *J. Org. Chem.*, **46**, 5427 (1981). f) Brünner, H., Muschiol, M., Wischert, T., Harada, T., and Tai, A., *Chem. Lett.*, 1267-1270 (1987).
55. Blaser, H.-U., *Tetrahedron Asymm.*, **2**, 843-866 (1991).
56. Blaser, H.-U., Boyer, S.K., and Pittelkow, U., *Tetrahedron: Asymmetry.*, **2**, 721 (1991).
57. a) Selke, R., and Pracejus, H., *J. Mol. Catal*, **37**, 213-225 (1986). b) Bosnich, B. (Ed.), *Asymmetric Catalysis*, NATO ASI Series, Martinus Nijhoff, Dordrecht, 1986. c) Arntz, D., and Schäfer, A., in *Metal Promoted Selectivity in Organic Synthesis*, Kluwer, Amsterdam, 1991, pp. 161-189.
58. Burk, M.J., in *Handbook of Chiral Chemicals*, Ager, D.J. (Ed.), Marcel Dekker, Inc., New York, 1999, pp. 346-347.
59. Burk, M.J., in *Handbook of Chiral Chemicals*, Ager, D.J. (Ed.), Marcel Dekker, Inc., New York, 1999, pp. 347-348.

60. a) Kumobayashi, H., *Proc. 2nd Int. Conf. Pharm. Ingredients and Intermediates*, Manufacturing Chemist, London, 1992, pp. 135-138. b) Ohta, T., Takaya, H., Kitamura, M., Nagai, K., and Noyori, R., *J. Org. Chem.*, **52**, 3176-3178 (1987).
61. Spindler, F., Pittelkow, U., and Blaser, H.-U., *Chirality*, **3**, 370 (1991).
62. a) Noyori, R., Ikeda, T., Ohkuma, T., Widhalm, M., Kitamira, M., Takaya, H., Akutagawa, S., Sayo, N., Saito, T., Taketomi, T., and Kumobayashi, H., *J. Am. Chem. Soc.*, **111**, 9134-9135 (1989). b) Kitamura, M., Ohkuma, T., Tokunga, M., and Noyori, R., *Tetrahedron Asymm.*, **1**, 1-4 (1990).
63. a) Bakos, J., Orosz, A., Heil, B., Laghmari, M., Lhoste, P., and Sinou, D., J. Chem. Soc., Chem. Commun., 1684-1685 (1991). b) Lensink, C., and De Vries, J.G., *Tetrahedropn Asymm.*, **3**, 235-238 (1992).
64. a) Caille, J.-C., Bulliard, M., and LaBoue, B., in Crosby, J., in *Chirality in Industry II*, Collins, A.N., Sheldrake, G.N., and Crosby, J. (Eds.), John Wiley & Sons, New York, 1997, pp. 391-401. b) Bulliard, M., in *Handbook of Chiral Chemicals*, Ager, D.J. (Ed.), Marcel Dekker, Inc., New York, 1999, pp. 211-225.
65. a) Parinello, G., and Stille, J.K., *J. Am. Chem. Soc.*, **109**, 7122-7127 (1987). b) See for example: Hayashi, T., Tanaka, M., and Ogata, I., *J. Mol. Catal.*, **26**, 17 (1984).
66. a) Uozumi, Y., and Hayashi, T., *J. Am. Chem. Soc.*, **113**, 9887 (1991). b) Uozumi, Y., Kitayama, K., Hayashi, T., Yanagi, K., and Fukuyo, E., *Bull. Chem. Soc. Jpn.*, **68**, 713 (1995).
67. a) Hashiguchi, S., Fujii, A., Takehara, J., Ikariya, T., and Noyori, R., *J. Am. Chem. Soc.*, **117**, 7562 (1995). b) Fujii, A., Hashiguchi, S., Uematsu, N., Ikariya, R., and Noyori, R., *J. Am. Chem. Soc.*, **118**, 2521 (1996). c) Takehara, J., Hashiguchi, S., Fujii, A., Inoue, S., Ikariya, T., and Noyori, R., *J. Chem. Soc. Chem. Commun.*, 233 (1996).
68. a) Lee, N.E., and Buchwald, S.L., *J. Am. Chem. Soc.*, **116**, 5985 (1994). b) Bakos, J., Toth, I.B.H., and Marko, L., *J. Organometal. Chem.*, **279**, 23 (1985). c) Becalski, A.G., Cullen, W.R., Fryzuk, M.D., James, B.R., Kang, G.-J., and Rettig, S.J., *Inorg. Chem.*, **30**, 5002, 1991. d) Bakos, J., Orosz, A., Heil, B., Laghmari, M., Lhoste, P., and Sinou, D., *J. Chem. Soc. Chem. Commun.*, 1684 (1991). e) Chan, Y., Ng, C., and Osborn, J.A., *J. Am. Chem. Soc.*, **112**, 9600 (1990). f) Morimoto, T., and Achiwa, K., *Tetrahedron: Asymmetry*, **6**, 2661 (1995). g) Hoveyda, A.H., and Morken, J.P., *Angew. Chem. Int. Ed. Engl.*, **35**, 1262 (1996). h) Uematsu, N., Fujii, A., Hashiguchi, S., Ikariya, T., and Noyori, R., *J. Am. Chem. Soc.*, **118**, 4916 (1996).
69. a) Hanson, R.M., and Sharpless, K.B., *J. Org. Chem.*, **51**, 1922-1925 (1986). B) Gao, Y., Hanson, R.M., Klunder, J.M., Ko, S.Y., Masamune, H., and Sharpless, K.B., *J. Am. Chem. Soc.*, **109**, 5765-5780 (1987).
70. Sheldon, R.A., *Chirotechnology*, Marcel Dekker, Inc., New York, 1993, pp. 324-325.
71. a) Zhang, W., and Jacobsen, E.N., *J. Org. Chem.*, **56**, 2296-2298 (1991). b) Jacobsen, E.N., Zhang, W., and Güler, M.L., *J. Am. Chem. Soc.*, **113**, 6703-6704 (1991).
72. Sheldon, R.A., *Chirotechnology*, Marcel Dekker, Inc., New York, 1993, pp. 328-330 and references cited therein.
73. Wilke, G., *Angew. Chem. Int. Ed. Engl.*, **27**, 185 (1988).
74. Buono, G., Siv, C., Peiffrer, G., Triantaphylides, C., Denis, P., Mortreux, A., and Petit, F., *J. Org. Chem.*, **50**, 1781 (1985).
75. Dolling, U.H., Davis, P., and Grabowski, E.J.J., *J. Am. Chem. Soc.*, **106**, 446 1984).

76. Sheldon, R.A., *Chirotechnology*, Marcel Dekker, Inc., New York, 1993, pp. 304-308 and references cited therein.
77. Sheldon, R.A., *Chirotechnology*, Marcel Dekker, Inc., New York, 1993, pp. 315-317.
78. Sheldon, R.A., *Chirotechnology*, Marcel Dekker, Inc., New York, 1993, pp. 314-315.
79. Bray, B.L., Goodyear, M.D., Partridge, J.S., and Tapolczay, D.J., in *Chirality in Industry II*, Collins, A.N., Sheldrake, G.N., and Crosby, J. (Eds.), John Wiley & Sons, New York, 1997, pp. 61-63.
80. a) Dolling, U.H., Davis, P., and Grabowski, E.J.J., *J. Am. Chem. Soc.*, **106**, 446-447 (1984). b) Hughes, D.L., Dolling, U.H., Ryan, K.M., Schoenewaldt, E.F., and Grabowski, E.J.J., *J. Org. Chem.*, **52**, 4745 – 4752 (1987). c) Dolling, U.H., Hughes, D.L., Battacharya, A., Ryan, K.M., Karady, S., Weinstock, L.M., Grenda, V.J., and Grabowski, E.J.J., in *Catalysis of Organic Reactions*, Rylander, P.N., Greenfield, H., and Augustine, R.L. (Eds.), Marcel Dekker, New York, 1988, pp. 65-85.
81. O'Donnel, M.J., Bennett, W.D., and Wu, S., J. Am. Chem. Soc., **111**, 2353 (1989).
82. Neriackx, W., and Vandewalle, M., Tetrahedron: Asymmetry, **1**, 265 (1990).
83. Schaod, D.R., in *Handbook of Chiral Chemicals*, Ager, D.J. (Ed.), Marcel Dekker, Inc., New York, 1999, p. 288.
84. a) Castaldi, G., and Giordano, C., *Synthesis*, 1039 (1987). b) Crosby, J., *Tetrahedron*, **47**, 4789 (1991). c) Giordano, C., Castaldi, G., Cavicchioli, S., *US Patent*, 4,888,433, 1989. d) Giordano, C., Castaldi, G., Cavicchioli, S., and Villa, M.G., *Tetrahedron*, **45**, 4243 (1989). e) Giordano, C., Castaldi, G., Uggeri, F., and Cavicchioli, S., *US Patent*, 4,855464, 1989.
85. Hoekstra, M.S., Sobieray, D.M., Schwindt, M.A., Mulhern, T.A., Grote, T.M., Huckabee, B.K., Hendrickson, V.S., Franklin, L.C., Granger, E.J., and Karrick, G.L., *Org. Process R. & D.*, **1**, 26 (1997).
86. a) McIntosh, J.M., and Mishra, P., *Can. J. Chem.*, **64**, 726 (1986). b) Schölllköpf, U., and Schütze, R., *Liebigs Ann. Chem.*, 45 (1987). c) Yaozhong, J., Guilan, L., Jinchu, L., and Changyou, Z., *Synth. Commun.*, **17**, 1545 (1987). d) Yaozhong, J., Guilan, L., and Jingen, D., *Synth. Commun.*, **18**, 1291 (1988). e) McIntosh, J.M., Cassidy, K.C., and Matassa, L., *Tetrahedron*, **45**, 5449 (1989). f) Yaozhong, J., Peng, G., and Guilan, L., *Synth. Commun.*, **20**, 15 (1990).
87. a) Oppolzer, W., Blagg, J., Rodriguez, I., and Walther, W., *J. Am. Chem. Soc.*, **112**, 2767 (1990). b) Boeckman, R.K., Johnson, A.T., and Musselman, R.A., *Tetrahedron Lett.*, **35** 8521 (1994). c) Kelly, T.R., and Arvanitis, A., *Tetrahedron Lett.*, **25**, 39 (1984). d) Bartlettm P.D., and Knox, L.H., *Org. Synth., Coll. Vol. 5*, 1973, p. 689. e) Anh, K.H., Lee, S., and Lim, A., *J. Org. Chem.*, **57**, 5065 (1992).
88. Terashima, S., Jew, S.-S., and Koga, K., *Chemistry Lett.*, 1109 (1977).
89. Gras, G., *Ger. Patent*, 3,024,265, 1979.
90. Sheldon, R.A., *Chirotechnology*, Marcel Dekker, Inc., New York, 1993.
91. Ager, D.J., in *Handbook of Chiral Chemicals*, Ager, D.J. (Ed.), Marcel Dekker, Inc., New York, 1999, p. 107.
92. Ager, D.J., in *Handbook of Chiral Chemicals*, Ager, D.J. (Ed.), Marcel Dekker, Inc., New York, 1999, p. 108 and references cited therein.
93. Ager, D.J., in *Handbook of Chiral Chemicals*, Ager, D.J. (Ed.), Marcel Dekker, Inc., New York, 1999, pp. 108-110 and references cited therein.

# PART II

# MAIN ENTRIES

1 **(S)-(+)-Abscisic acid**
21293-29-8

$C_{15}H_{20}O_8$
Chiral intermediate. *Toray International, Inc.*

2 **2-Acetamido-1-amino-1,2-dideoxy-β-D-glucopyranose**

$C_8H_{16}N_2O_5$
Chiral intermediate. *Senn Chemicals AG.*

3 **2-Acetamido-1-azido-1,2-dideoxy-β-D-glucopyranose**

$C_8H_{14}N_4O_5$
Chiral intermediate. *Senn Chemicals AG.*

4 **2-Acetamido-4,6-O-benzylidene-2-deoxy-D-glucopyranose**

$C_{15}H_{19}NO_6$
Chiral intermediate. *Pfanstiehl Laboratories, Inc.*

5 **2-Acetamido-2-deoxy-α-D-glucopyranose**
10036-64-3 233-115-1

$C_8H_{15}NO_6$
Glucopyranose, 2-(acetylamino)-2-deoxy-, α-D-. N-Acetyl-α-D-glucosamine. Chiral intermediate. mp = 211°; [α] = + 41.2° (c = 2, $H_2O$). *Acros Organics nv.*

6 **(+)-2-Acetamido-2-deoxyglucopyranosyl chloride 3,4,6-triacetate**
3068-34-6 221-325-6

$C_{14}H_{20}ClNO_8$
Chiral intermediate. mp = 214°; $[\alpha]_D^{20}$ = + 105° (c = 1, $CHCl_3$). *Sigma-Aldrich Fine Chemicals.*

7 **β-(1,4)-2-Acetamido-2-deoxy-D-glucose**
1398-61-4 215-744-3
$C_8H_{13}NO_5$
Chitin.
Poly-[1→4]-β-D-N-acetylglucosamine; Acetylated chitin; Clandosan; Poly-N-acetyl-D-glucosamine; poly-1-β-4GlcNAc; Chitan, N-acetyl-; Chitin Tc-L; Kimitsu Chitin; Regitex FA. Chiral intermediate. *Acros Organics nv.*

8 **2-Acetamido-2-deoxy-1,3,4,6-tetra-O-acetyl-β-D-glucopyranose**
7772-79-4 231-865-4

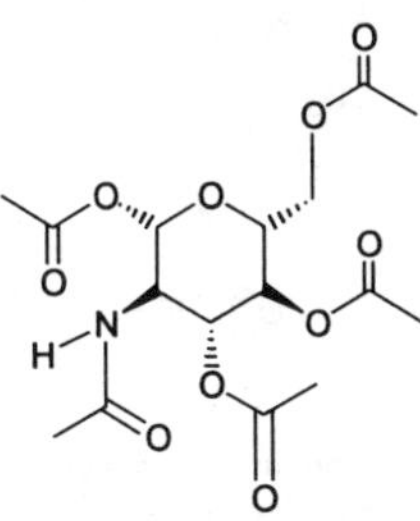

$C_{16}H_{23}NO_{10}$
Glucopyranose, 2-acetamido-2-deoxy-, 1,3,4,6-tetraacetate, β-D-. N-acetyl-β-D-glucosamine tetraacetate; 1,3,4,6-Tetra-O-acetyl-N-acetyl-β-D-glucosamine; 2-Acetamido-1,3,4,6-tetra-O-acetyl-2-deoxy-β-D-glucopyranose. Chiral intermediate. mp = 186-189°; $[\alpha]^{24}$ = + 3.9° (c = 10, $CHCl_3$). *Acros Organics nv; Sigma-Aldrich Fine Chemicals.*

9 **(3R)-(+)-3-Acetamidopyrrolidine**
$C_6H_{12}N_2O$
Chiral building block. *Sigma-Aldrich Fine Chemicals; TCI America.*

**10 (3S)-(-)-3-Acetamidopyrrolidine**
114636-31-6
$C_6H_{12}N_2O$
Chiral building block. mp = 56-58°; $[\alpha]_D^{20}$ = - 44.5 ± 2° (c = 1, EtOH). *Lancaster Synthesis Ltd.; Sigma-Aldrich Fine Chemicals; TCI America.*

**11 2-Acetamido-3,4,6-tri-O-acetyl-2-deoxy-β-D-glucopyranosyl azide**
6205-69-2

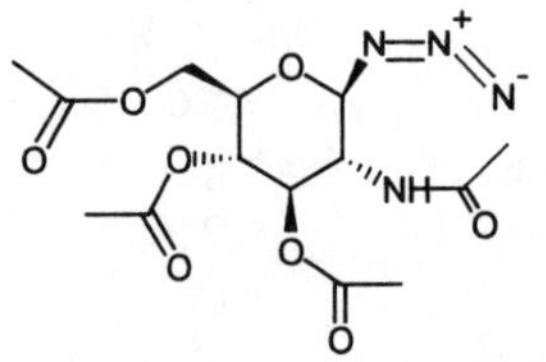

$C_{14}H_{20}N_4O_8$
Chiral intermediate. [α] = - 40° (c = 1.3, $CHCl_3$). *Acros Organics nv.*

**12 α-D-Acetobromogalactose**
3068-32-4 221-324-0
$C_{14}H_{19}BrO_9$
Galactopyranosyl bromide, tetraacetate, α-D. 2,3,4,6-Tetra-O-acetyl-α-D-galactopyranosyl bromide; Bromo 2,3,4,6-Tetra-O-acetyl-α-D-galactopyranoside; α-Bromotetraacetylgalactose. Chiral intermediate. *Pfanstiehl Laboratories, Inc.*

**13 α-D-Acetobromoglucose**
572-09-8 209-339-0

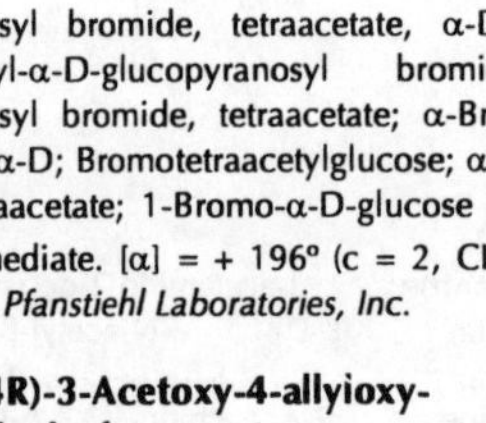

$C_{14}H_{19}BrO_9$
Glucopyranosyl bromide, tetraacetate, α-D-; 2,3,4,6-Tetra-O-acetyl-α-D-glucopyranosyl bromide; α-D-Glucopyranosyl bromide, tetraacetate; α-Bromoglucose tetraacetate; α-D; Bromotetraacetylglucose; α-D-Glucosyl bromide tetraacetate; 1-Bromo-α-D-glucose tetraacetate. Chiral intermediate. [α] = + 196° (c = 2, $CHCl_3$). *Acros Organics nv; Pfanstiehl Laboratories, Inc.*

**14 (3R,4R)-3-Acetoxy-4-allyioxy-tetrahydrofuran**

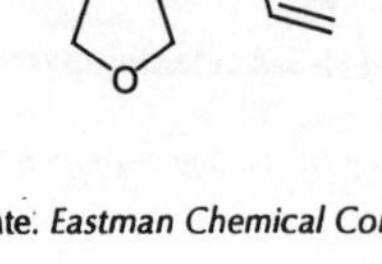

$C_7H_{12}O_3$
Chiral intermediate. *Eastman Chemical Company.*

**15 (R)-Acetoxy-4-bromobutyric acid**

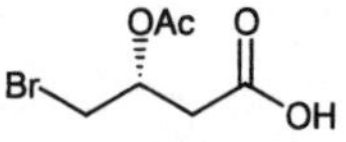

$C_6H_9BrO_4$
Butanoic acid, 3-(acetyloxy)-4-bromo-, (R)-. Chiral building block. *Synthon Chiragenics Corporation.*

**16 (S)-3-Acetoxy-4-bromobutyric acid**
191354-44-6

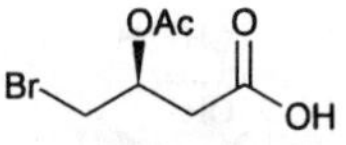

$C_6H_9BrO_4$
Butanoic acid, 3-(acetyloxy)-4-bromo-, (S)-. Chiral building block. *Synthon Chiragenics Corporation.*

**17 (R)-Acetoxy-4-bromobutyryl chloride**

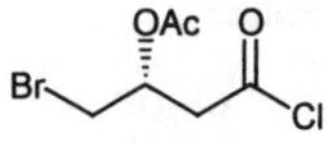

$C_6H_8BrClO_3$
Butanoyl chloride, 3-(acetyloxy)-4-bromo-, (R)-. Chiral building block. *Synthon Chiragenics Corporation.*

**18 (S)-3-Acetoxy-4-bromobutyryl chloride**
191354-46-8

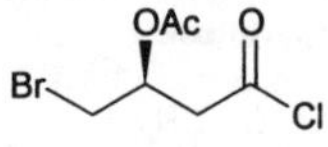

$C_6H_8BrClO_3$
Butanoyl chloride, 3-(acetyloxy)-4-bromo-, (S)-. Chiral building block. *Synthon Chiragenics Corporation.*

**19 (3R,4R)-4-Acetoxy-3-[(R)-(tert-butyl-dimethylsilyloxy)ethyl]-2-azetidinone**
76855-69-1

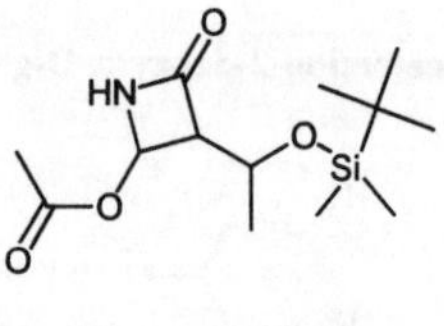

$C_{13}H_{25}NO_4Si$
Azetidin-2-one, 4-(acetyloxy)-3-[1-[[(1,1-dimethylethyl)-dimethylsilyl]oxy]ethyl]-, [3R-[3α(R*),4α]]-. Chiral intermediate. mp = 107-109°; $[\alpha]_D^{20}$ = + 51° (c = 1, $CHCl_3$). *Kaneka Corporation; Nippon Soda Co. Ltd.; Sigma-Aldrich Fine Chemicals; Takasago International Corporation; TCI America.*

**20 (R)-Acetoxy-γ-butyrolactone**
138666-02-1

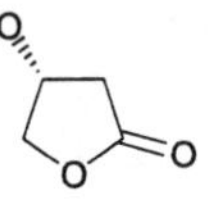

$C_6H_8O_4$
Furan-2(3H)-one, 4-(acetyloxy)dihydro-, (R)-; Chiral building block. *Synthon Chiragenics Corporation.*

**21 (S)-3-Acetoxy-γ-butyrolactone**
191403-65-3

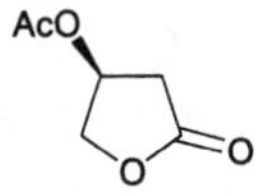

$C_6H_8O_4$
Furan-2(3H)-one, 4-(acetyloxy)dihydro-, (S)-; (S)-4-(Acetyloxy)-dihydro-2(3H)-furanone. Chiral building block. *Acros Organics nv; SK Energy and Chemical, Inc.; Synthon Chiragenics Corporation.*

**22 5-(R)-Acetoxy-1-chlorohexane**
154885-34-4
$C_8H_{15}ClO_2$
Chiral building block. bp = 220°; d = 1.02; n = 1.44. *Sigma-Aldrich Fine Chemicals.*

**23 (R)-Acetoxy-1,4-dibromobutane**

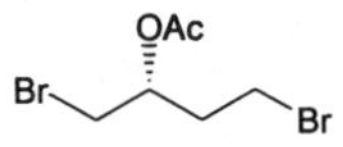

$C_6H_{10}Br_2O_2$
Butan-2-ol, 1,4-dibromo-, acetate, (R)-; Chiral building block. *Synthon Chiragenics Corporation.*

**24 (S)-2-Acetoxy-1,4-dibromobutane**
191354-48-0

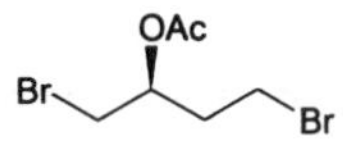

$C_6H_{10}Br_2O_2$
Butan-2-ol, 1,4-dibromo-, acetate, (S)-; Chiral building block. *Synthon Chiragenics Corporation.*

**25 (R)-(+)-2-Acetoxy-3,3-dimethylbutyronitrile**
126567-38-2

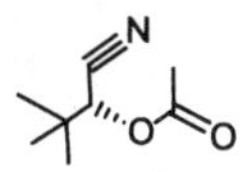

$C_8H_{13}NO_2$
Chiral building block. bp = 198°; d = 0.953; n = 1.416; $[\alpha]_D^{20}$ = + 106° (c = 5, $CHCl_3$). *Sigma-Aldrich Fine Chemicals.*

**26 (1S,2S,3R,6S)-3-Acetoxy-3-methyl-6-(1-methylethyl)-2-(3-oxobutyl)-cyclohexanenitrile**
131447-90-0

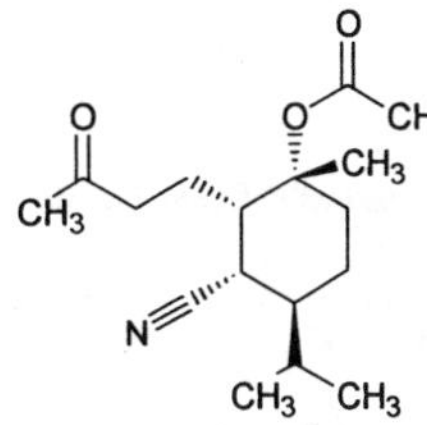

$C_{17}H_{27}NO_3$
Chiral intermediate. [α] = -5.6 (c = 1.5, $CH_2Cl_2$). *Acros Organics nv.*

**27 Acetoxyphenylacetonitrile**
11971-89-7

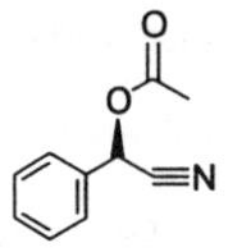

$C_{10}H_9NO_2$
Chiral intermediate. bp = 268°; d = 1.115; n = 1.506; $[\alpha]_D^{20}$ = + 8.0° (c = 10, $CHCl_3$). *Sigma-Aldrich Fine Chemicals.*

**28 (R)-(+)-2-Acetoxy-4-phenylbutyronitrile**
126641-88-1

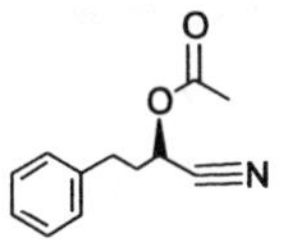

$C_{12}H_{13}NO_2$
Chiral building block. d = 1.072; n = 1.5; $[\alpha]_D^{20}$ = + 44° (c = 5, $CHCl_3$). *Sigma-Aldrich Fine Chemicals.*

**29 (S)-(-)-2-Acetoxypropionic acid**
6034-46-4

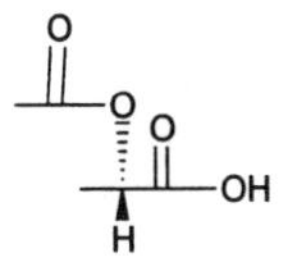

$C_5H_8O_4$
(S)-(-)-O-Acetyllactic acid. Chiral building block. $bp_{2.0}$ = 115-117°; d = 1.183; n = 1.0423; $[\alpha]_D^{20}$ = - 51° (c = 7, $CHCl_3$). *Sigma-Aldrich Fine Chemicals; TCI America.*

**30 (S)-(-)-2-Acetoxypropionyl chloride**
36394-75-9

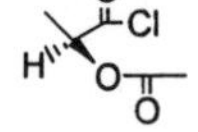

$C_5H_7ClO_3$
(S)-(-)-O-Acetyllactoyl chloride. Building block. $bp_5$ = 50°; d = 1.189; n = 1.423; $[\alpha]_D^{20}$ = - 31° (c = 4, $CHCl_3$). *Austin Chemical Company, Inc.; Saurefabrik Schweizer-hall; Sigma-Aldrich Fine Chemicals; TCI America.*

**31 (R)-(+)-2-Acetoxysuccinic anhydride**

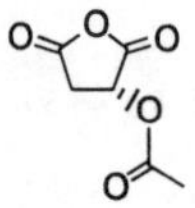

$C_6H_6O_5$
Chiral building block. mp = 56-58°; $[\alpha]_D^{20}$ = + 27° (c = 1, $CHCl_3$). *Sigma-Aldrich Fine Chemicals.*

**32 (S)-(-)-2-Acetoxysuccinic anhydride**

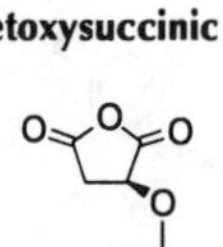

$C_6H_6O_5$
Chiral intermediate. mp = 56-58°; $[\alpha]_D^{20}$ = - 27° (c = 1, $CHCl_3$). *Sigma-Aldrich Fine Chemicals.*

**33 (R)-Acetoxy-2-thioxopyrrolidine**

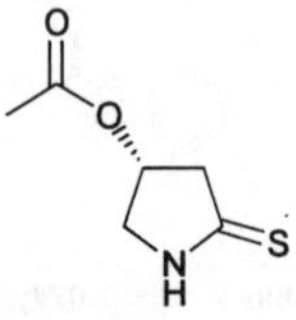

$C_6H_9NO_2S$
Pyrrolidine, 2-thioxo-4-acetoxy-, (R)-; Chiral building block. *Synthon Chiragenics Corporation.*

**34 (S)-4-Acetoxy-2-thioxopyrrolidine**

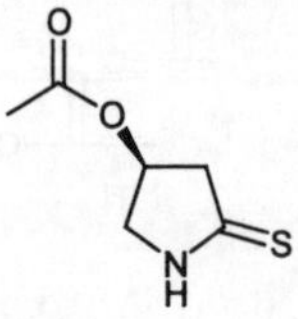

$C_6H_9NO_2S$
Pyrrolidine, 2-thioxo-4-acetoxy-, (S)-; Chiral building block. *Synthon Chiragenics Corporation.*

**35 Acetyl-5-acetoxy-L-tryptophan**

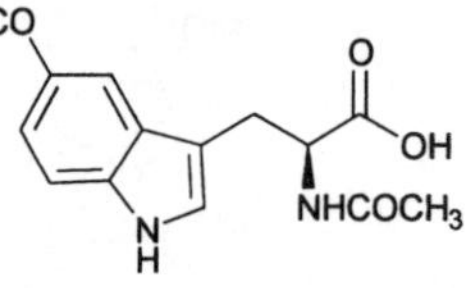

$C_{15}H_{19}N_2O_5$
Chiral intermediate. *Senn Chemicals AG.*

**36 Acetyl-L-alanine-1-naphthyl ester**

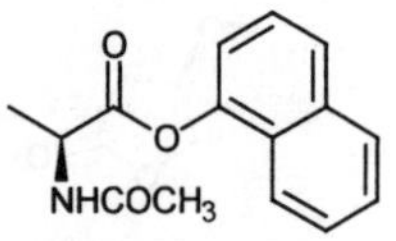

$C_{15}H_{15}NO_3$
Chiral intermediate. *Senn Chemicals AG.*

**37 N-Acetyl-L-alanyl-L-alanyl-L-alanine**
19245-85-3 242-912-3

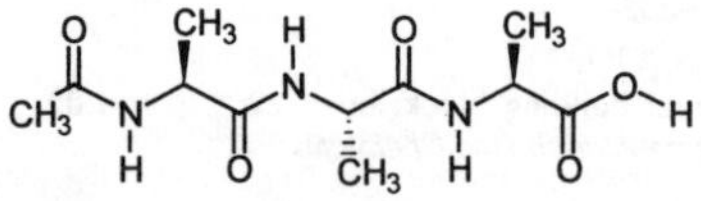

$C_{11}H_{19}N_3O_5$
Alanine, N-[N-(N-acetyl-L-alanyl)-L-alanyl]-, L-. Acetyltrialanine. Chiral intermediate. mp = 248°. *Acros Organics nv.*

**38 N-Acetyl-L-alanyl-L-glutamine**

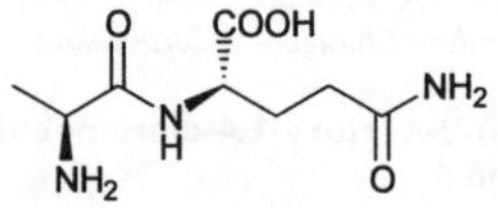

$C_8H_{15}N_3O_4$
Chiral intermediate. *Rexim S.A. Produits Chimiques.*

**39 N-Acetyl-L-asparagine**
4033-40-3 223-716-7

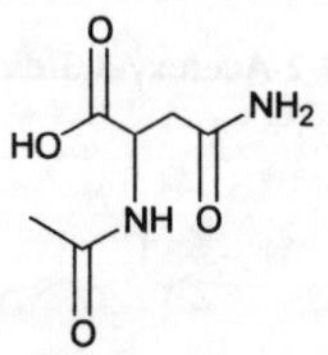

$C_6H_{10}N_2O_4$
Asparagine, $N^2$-acetyl-, L-. Chiral building block. mp = 168-170°. *Flamma s.p.a.; Sigma-Aldrich Fine Chemicals.*

**40 Acetyl-L-aspartic acid**
997-55-7 875(12) 213-643-9

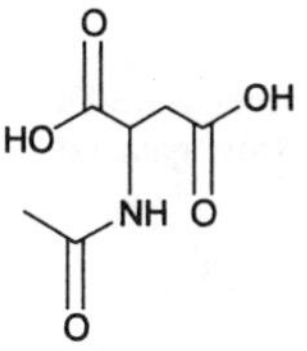

$C_6H_9NO_5$
Aspartic acid, N-acetyl-, L-; Chiral intermediate. mp = 137-140°; $[\alpha]^{25}$ = + 52° (c = 1, HOAc). *Austin Chemical Company, Inc.; Flamma s.p.a.; Rexim S.A. Produits Chimiques; Sigma-Aldrich Fine Chemicals.*

**41 N-Acetyl-S-benzyl-L-cysteine**
19542-77-9

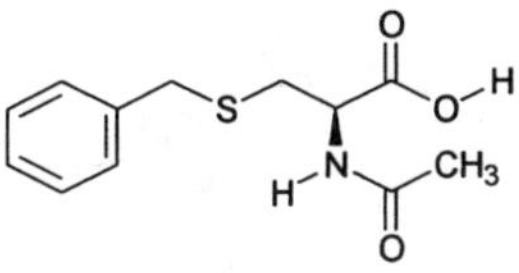

$C_{12}H_{15}NO_3S$
Chiral intermediate. mp = 147-149°. *Acros Organics nv.*

**42 N-Acetyl-S-benzyl-L-cysteine methyl ester**
77549-14-5

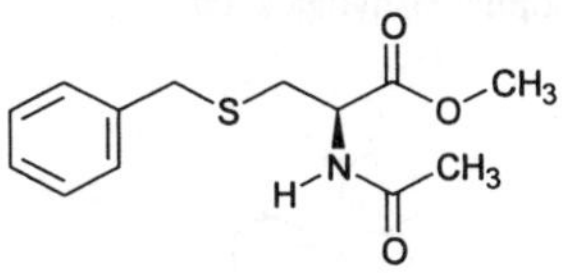

$C_{13}H_{17}NO_3S$
Chiral intermediate. *Acros Organics nv.*

**43 (-)-3-O-Acetyl-6-O-benzoyl-5-O-(methylsulfonyl)-1,2-O-isopropylidene-glucofuranose**

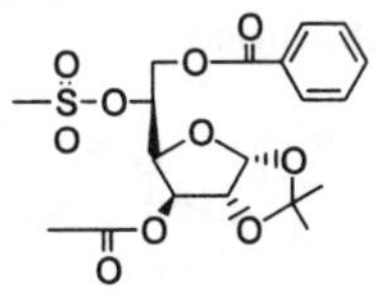

$C_{19}H_{24}O_{10}S$
Chiral intermediate. mp = 128-130°; $[\alpha]^{25}$ = - 9° (c = 1, $CHCl_3$). *Sigma-Aldrich Fine Chemicals.*

**44 (R)-(-)-3-Acetyl-4-benzyl-2-oxazolidinone**
184363-65-3

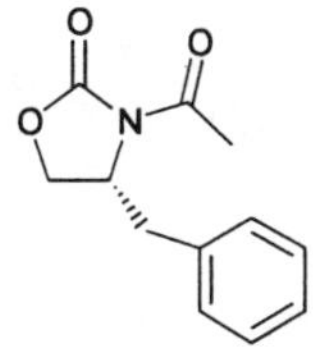

$C_{12}H_{13}NO_3$
Chiral intermediate. *Acros Organics nv; Oxford Asymmetry International plc; SK Energy and Chemical, Inc.*

**45 (S)-(+)-3-Acetyl-4-benzyl-2-oxazolidinone**
132836-66-9

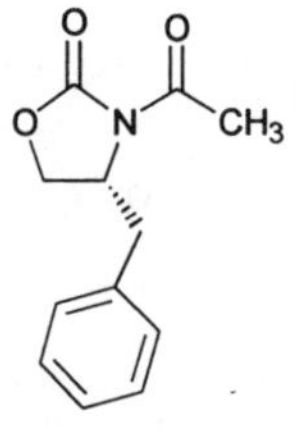

$C_{12}H_{13}NO_3$
Acetyl-(4S)-phenylmethyl-2-oxazolidinone, 3-. Chiral auxiliary. mp = 105-108°; $[\alpha]_D^{20}$ = + 110 ± 3° (c = 1, $CH_3OH$). *Fischer Chemicals AG; Great Lakes Fine Chemicals; Lancaster Synthesis Ltd.; Oxford Asymmetry International plc; Sigma-Aldrich Fine Chemicals; SK Energy and Chemical, Inc.*

**46 N-Acetyl-(2R)-bornane-10,2-sultam**

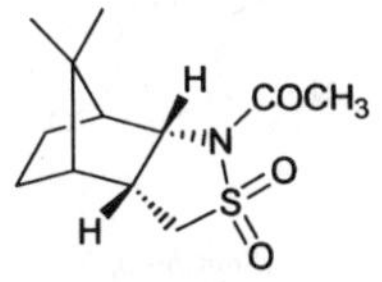

$C_{12}H_{19}NO_3S$
Resolving agent; Chiral auxiliary. *Acros Organics nv; Oxford Asymmetry International plc.*

**47 N-Acetyl-(2S)-bornane-10,2-sultam**

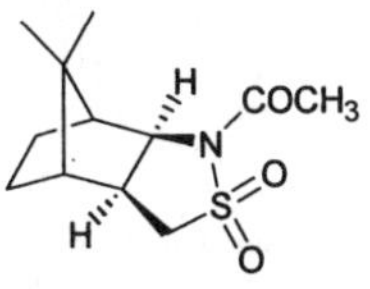

$C_{12}H_{19}NO_3S$
Resolving agent; Chiral auxiliary. *Oxford Asymmetry International plc.*

**48 Acetyl-L-carnitine**
3040-38-8

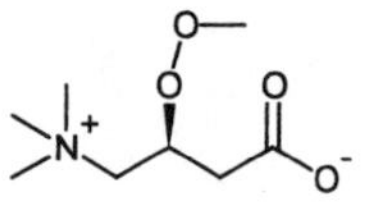

$C_9H_{17}NO_4$
Propan-1-aminium, 2-(acetyloxy)-3-carboxy-N,N,N-trimethyl-, hydroxide, inner salt, (R)-. ALCAR; Ammonium, - (3-carboxy-2-hydroxypropyl)-trimethyl-, hydroxide, inner salt, acetate, L-; L-Carnitine acetyl ester; Levocarnitine acetyl; Nicetile. Chiral building block. *Tanabe Seiyaku Co. Ltd.*

**49 Acetyl-L-carnitine hydrochloride**
5080-50-2 85(12)

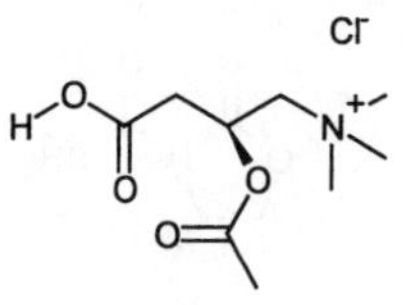

$C_9H_{18}ClNO_4$
Propan-1-aminium, 2-(acetyloxy)-3-carboxy-N,N,N-trimethyl-, chloride, (R)-; (R)-3-Acetoxy-4-(trimethylammonio)butyrate hydrochloride; Ammonium, (3-carboxy-2-hydroxypropyl)trimethyl-, chloride, acetate, (-)-. Chiral building block. mp = 194°; $[\alpha]^{25}$ = - 28° (c = 1, $H_2O$). *Acros Organics nv; Austin Chemical Company, Inc.; Flamma s.p.a.; KingChem, Inc.; Sigma-Aldrich Fine Chemicals; TCI America.*

**50 O-Acetyl-L-carnitine hydrochloride**

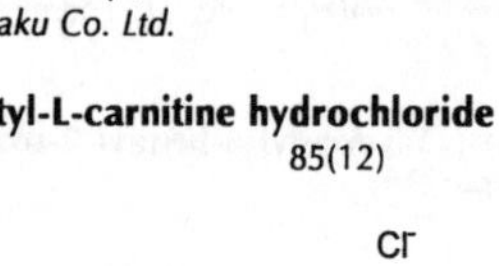

$C_9H_{18}ClNO_4$
Chiral intermediate. *KingChem, Inc.*

**51 N-Acetyl-L-cysteine**
616-91-1 89(12) 210-498-3

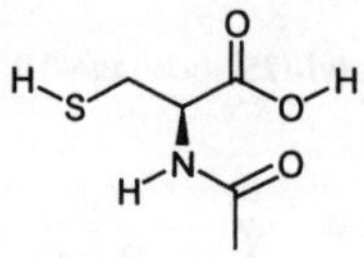

$C_5H_9NO_3S$
Cysteine, N-acetyl-, L-.
Airbron; Broncholysin; Fluibiotic; Fluimicil; Fluimucetin; Fluimucil; L-Acetylcysteine; Mercapturic acid, (R)-; Mucofilin; Mucolyticum-Lappe; Mucomyst; Mucosolvin; NAC; NSC 111180; Parvolex; Respaire. Listed on TSCA. Chiral building block. mp = 109-111°; $[\alpha]^{25}$ = + 4.5° (c = 2, $H_2O$). *Acros Organics nv; Austin Chemical Company, Inc.; Flamma s.p.a.; Interchem Corporation; KingChem, Inc.; Kyowa Hakko Kogyo Co., Ltd.; Sigma-Aldrich Fine Chemicals; Tanabe Seiyaku Co. Ltd.; Varsal Instruments, Inc.*

**52 (-)-4-O-Acetyl-3,6-Di-O-(tert-butyldimethylsilyl)galactal**

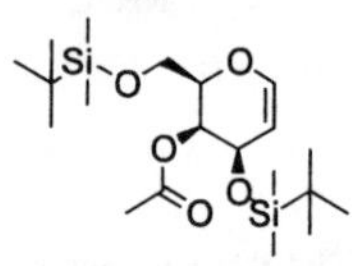

$C_{20}H_{40}O_5Si_2$
Chiral intermediate. mp = 56-59°; $[\alpha]^{22}$ = - 41° (c = 1, $CHCl_3$). *Sigma-Aldrich Fine Chemicals.*

**53 (-)-4-O-Acetyl-3,6-Di-O-(tert-butyldimethylsilyl)glucal**

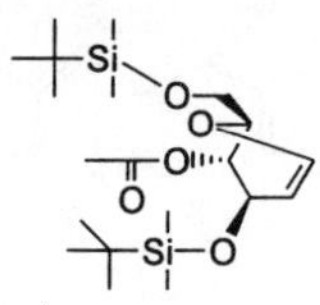

$C_{20}H_{40}O_5Si_2$
Chiral intermediate. bp = 324°; d = 0.968; n = 1.456; $[\alpha]_D^{20}$ = - 19° (c = 2.1, $CHCl_3$). *Sigma-Aldrich Fine Chemicals.*

**54 (+)-4-O-Acetyl-3,6-di-O-(tert-butyldiphenylsilyl)galactal**

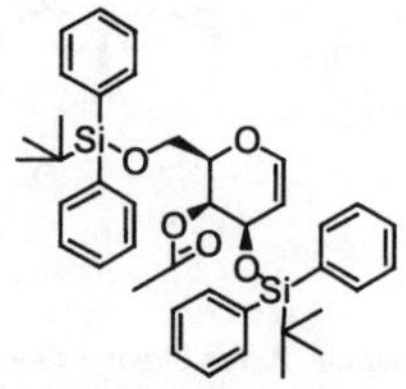

$C_{40}H_{48}O_5Si_2$
Chiral intermediate. mp = 108-112°. *Sigma-Aldrich Fine Chemicals.*

**55 (+)-4-O-Acetyl-3,6-di-O-(tert-butylphenylsilyl)glucal**

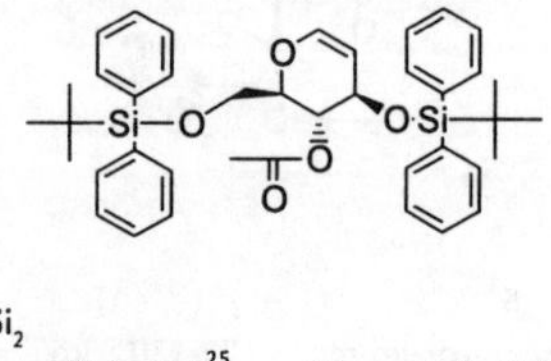

$C_{40}H_{48}O_5Si_2$
Chiral intermediate. $[\alpha]^{25}$ = + 7° (c = 1.5, $CHCl_3$). *Sigma-Aldrich Fine Chemicals.*

**56 (-)-4-O-Acetyl-3,6-di-O-(triisopropylsilyl)-galactal**

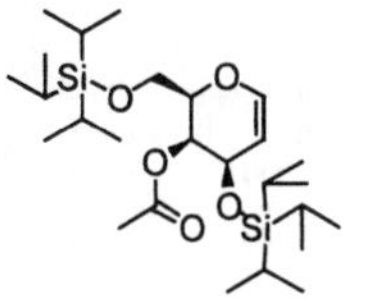

$C_{26}H_{52}O_5Si_2$
Chiral intermediate. n = 1.475; $[\alpha]^{23}$ = - 39° (c = 1, $CHCl_3$). *Sigma-Aldrich Fine Chemicals.*

**57 (R)-(-)-Acetylcarbonyl(+lh$^5$-2,4-cyclo-pentadien-1-yl)(triphenylphosphine)**
36548-61-5
$C_{26}H_{23}FeO_2P$
(R)-(-)-Acetyl-cyclopentadienyl-iron carbonyl triphenyl-phosphine complex (Iron acetyl chiral). Chiral catalyst. *Oxford Asymmetry International plc; TCI America.*

**58 (S)-(+)-Acetylcarbonyl(+lh$^5$-2,4-cyclo-pentadien-1-yl)(triphenylphosphine)Iron**
36548-60-4
$C_{26}H_{23}FeO_2P$
(S)-(+)-Acetylcyclopentadienyl-iron carbonyltriphenyl-phosphine complex; Iron acetyl complex (+)-(S)-enantiomer. Chiral intermediate. *Oxford Asymmetry International plc; TCI America.*

**59 N-Acetyl-3,5-diiodo-L-tyrosine**
1027-28-7 213-837-3
$C_{11}H_{11}I_2NO_4$
Tyrosine, N-acetyl-3,5-diiodo-, L-; L-Tyrosine, N-acetyl-3,5-diiodo-. Intermediate. *Pharmasyn, Inc.; TCI America.*

**60 N-Acetyl-3,5-dinitro-L-tyrosine**
20767-00-4

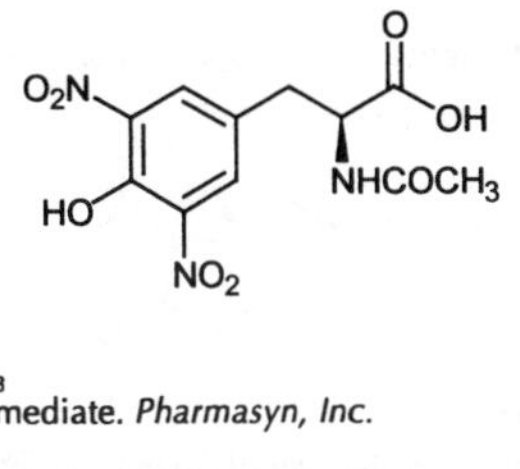

$C_{11}H_{11}N_3O_8$
Chiral intermediate. *Pharmasyn, Inc.*

**61 N-Acetyl-3,5-dinitro-L-tyrosine ethyl ester**
29358-99-4

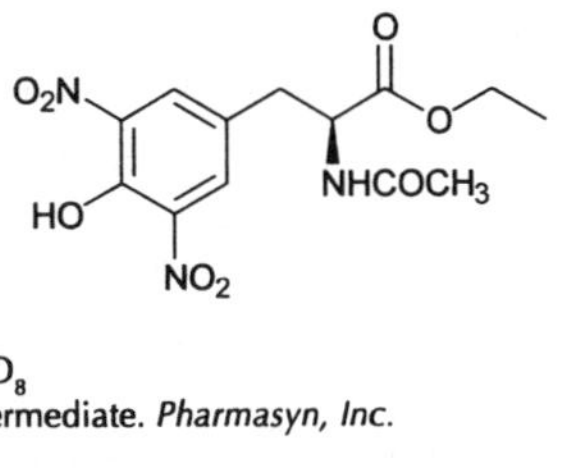

$C_{13}H_{15}N_3O_8$
Chiral intermediate. *Pharmasyn, Inc.*

**62 3-O-Acetyl-1,2:5,6-di-O-isopropylidene-α-D-glucofuranose**
16713-80-7

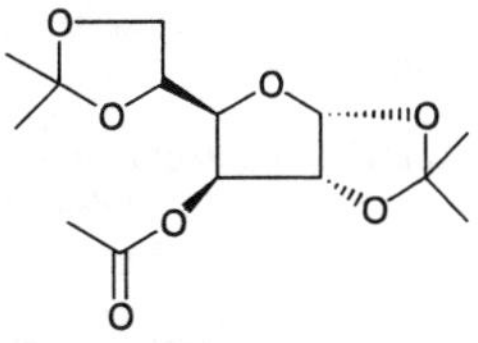

$C_{14}H_{22}O_7$
Chiral intermediate. mp = 60-62°; $[\alpha]^{25}$ = - 37° (c = 1, $CHCl_3$). *Acros Organics nv; Pfanstiehl Laboratories, Inc.; Sigma-Aldrich Fine Chemicals.*

**63 (S)-N-Acetyl-3-(3,4-dimethoxyphenyl)-alanine**

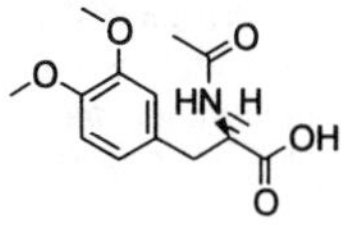

$C_{13}H_{17}NO_5$
Chiral intermediate. mp = 150-152°; $[\alpha]^{22}$ = - 26° (c = 4, 1N HCl). *Sigma-Aldrich Fine Chemicals.*

**64 (R)-N-Acetyl-3,4-dimethoxyphenylalanine**
33043-37-7

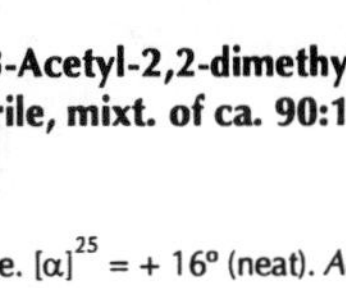

$C_{13}H_{17}NO_5$
Alanine, N-acetyl-3-(3,4-dimethoxyphenyl)-, D-; D-Tyrosine, N-acetyl-3-methoxy-O-methyl-; N-Acetyl-D-veratrylglycine. Chiral intermediate. *PCAS.*

**65 (1S,3S)-3-Acetyl-2,2-dimethylcyclobutane acetonitrile, mixt. of ca. 90:10 Z/E**
28353-00-6
$C_{10}H_{15}NO$
Chiral intermediate. $[\alpha]^{25}$ = + 16° (neat). *Acros Organics nv.*

**66 N-Acetyl-D-galactosamine**
14215-68-0

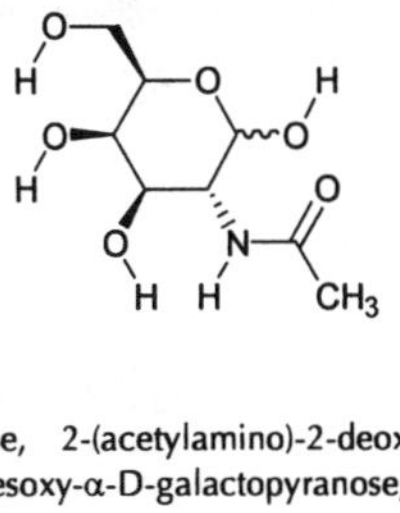

$C_8H_{15}NO_6$
Galactopyranose, 2-(acetylamino)-2-deoxy-, α-D-; 2-Acetamido-2-desoxy-α-D-galactopyranose; 2-Acetamido-

2-deoxy-α-D-galactopyranose; 2-Deoxy-2-acetamido-α-D-galactopyranose; 2-Acetamido-2-deoxy-α-D-galactose. Listed on TSCA. Chiral building block. mp = 160-161°; $[\alpha]_D^{20} = +86°$ (c = 2, $H_2O$). *Acros Organics nv; Pfanstiehl Laboratories, Inc.; Senn Chemicals AG; Sigma-Aldrich Fine Chemicals.*

**67 N-Acetyl-β-D-glucosamine**
7512-17-6 4466(12) 231-368-2

$C_8H_{15}NO_6$
Glucose, 2-(acetylamino)-2-deoxy-, D-; N-Acetyl-2-amino-2-deoxy-D-glucose; 2-Acetamido-2-deoxy-β-D-glucopyranose. Listed on TSCA. Chiral intermediate. mp = 211°; $[\alpha]_D^{20} = +41.2°$ (c = 2, $H_2O$). *Acros Organics nv; Pfanstiehl Laboratories, Inc.; Senn Chemicals AG; Sigma-Aldrich Fine Chemicals; TCI America.*

**68 N-Acetyl-L-glutamic acid**
1188-37-0 214-708-4

$C_7H_{11}NO_5$
Listed on TSCA. Chiral building block. mp = 199-201°; $[\alpha]^{22} = -15.6°$ (c = 1, $H_2O$). *Acros Organics nv; Austin Chemical Company, Inc.; Rexim S.A. Produits Chimiques; Sigma-Aldrich Fine Chemicals.*

**69 N-Acetyl-L-glutamine**
2490-97-3 219-647-7

$C_7H_{12}N_2O_4$
Glutamine, $N^2$-acetyl-, L-. Listed on TSCA. Chiral building block. *Flamma s.p.a.; Kyowa Hakko Kogyo Co., Ltd.*

**70 N-Acetyl-glycyl-L-glutamine**

$C_9H_{15}N_3O_5$
Chiral intermediate. *Rexim S.A. Produits Chimiques.*

**71 N-α-Acetyl-L-histidine monohydrate**
39145-52-3 219-678-6

$C_8H_{11}N_3O_3$
Chiral intermediate. mp = 157-159°; $[\alpha]^{24} = +46°$ (c = 1, $H_2O$). *Acros Organics nv; Sigma-Aldrich Fine Chemicals.*

**72 N-Acetyl-L-hydroxyproline**
33996-33-7 7035(12) 251-780-6

$C_7H_{11}NO_4$
Proline, L-, 1-acetyl-4-hydroxy-, (4R)-. Listed on TSCA. Chiral building block. mp = 132-133°; $[\alpha]_D^{20} = -119°$ (c = 4, $H_2O$). *Austin Chemical Company, Inc.; Kyowa Hakko Kogyo Co., Ltd.; Rexim S.A. Produits Chimiques; Sigma-Aldrich Fine Chemicals.*

**73 N-Acetyl-(4R)-isopropyl-2-oxazolidinone**

$C_8H_{13}NO_3$
Chiral intermediate. *Oxford Asymmetry International plc.*

**74 N-Acetyl-(4S)-isopropyl-2-oxazolidinone**

$C_8H_{13}NO_3$
Chiral intermediate. *Oxford Asymmetry International plc.*

**75 N-Acetyl-L-leucine**
1188-21-2 214-706-3

$C_8H_{15}NO_3$
Listed on TSCA. Chiral building block. mp = 187-190°;

$[\alpha]^{25}$ = - 23° (c = 2, EtOH). *Acros Organics nv; Austin Chemical Company, Inc.; Sigma-Aldrich Fine Chemicals; TCI America.*

**76 N-α-Acetyl-L-lysine**
1946-82-3 217-747-5

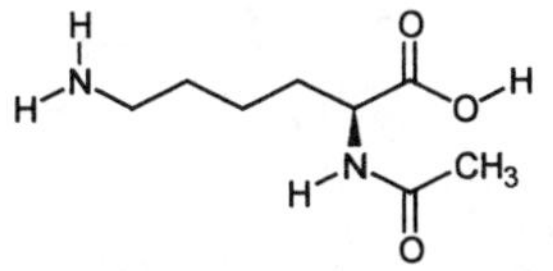

$C_8H_{16}N_2O_3$
$N^2$-acetyl-, L-; 6-Amino-L-2-acetamidohexanoic acid. Chiral building block. mp = 256-258°; $[\alpha]^{22}$ = + 5.7° (c = 1, $H_2O$). *Acros Organics nv; Sigma-Aldrich Fine Chemicals.*

**77 N-ε-Acetyl-L-lysine**
692-04-6 211-725-9

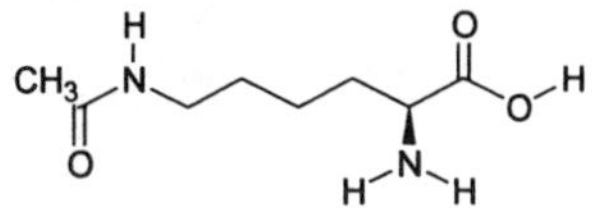

$C_8H_{16}N_2O_3$
$N^6$-acetyl-, L-. Chiral building block. mp = 250°; [α] = + 22.5° (c = 2, 5N HCl). *Acros Organics nv; Sigma-Aldrich Fine Chemicals.*

**78 N-α-Acetyl-L-lysine-N-methylamide monohydrate**
81013-00-5

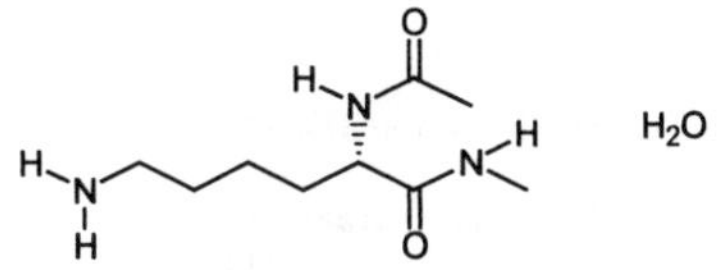

$C_9H_{19}N_3O_2$
Chiral intermediate. *Acros Organics nv.*

**79 N-α-Acetyl-L-lysine methyl ester hydrochloride**
20911-93-7 244-111-4

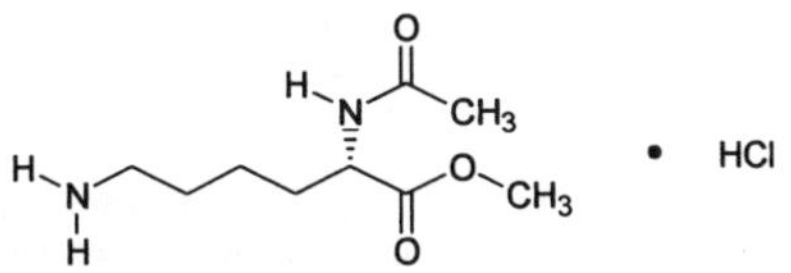

$C_9H_{18}N_2O_3$
$N^2$-acetyl-, methyl ester, monohydrochloride, L-. Methyl N-α-acetyl-L-lysinate hydrochloride. Chiral intermediate. mp = 108-114°; $[\alpha]^{22}$ = - 18° (c = 10 6N HCl). *Acros Organics nv; Sigma-Aldrich Fine Chemicals.*

**80 (S)-(-)-O-Acetylmalic anhydride**
59025-03-5
$C_6H_6O_5$
(S)-2-Acetoxysuccinic anhydride. Chiral building block. mp = 54-55°; $[\alpha]_D^{20}$ = - 24 ± 1° (c = 1, $CHCl_3$). *Lancaster Synthesis Ltd.; TCI America.*

**81 (R)-(-)-O-Acetylmandelic acid**
51019-43-3

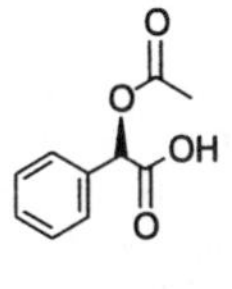

$C_{10}H_{10}O_4$
(R)-(-)-α-(Acetoxy)phenylacetic acid. Chiral building block; Resolving agent. mp = 98°; $[\alpha]_D^{20}$ = - 152.4° (c = 2, acetone). *Oxford Asymmetry International plc; Sigma-Aldrich Fine Chemicals; TCI America.*

**82 (S)-(+)-O-Acetylmandelic Acid**
7322-88-5

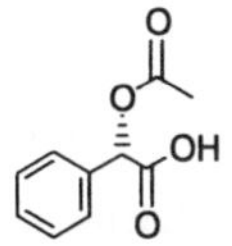

$C_{10}H_{10}O_4$
(S)-(+)-α-(Acetoxy)phenylacetic acid. Chiral building block. mp = 88°; $[\alpha]^{28}$ = + 153° (c = 2, acetone). *Oxford Asymmetry International plc; Sigma-Aldrich Fine Chemicals; TCI America.*

**83 N-Acetyl-D-mannosamine monohydrate**
14131-64-7
$C_8H_{15}O_6N$
2-Acetamido-2-deoxy-D-mannose monohydrate. Chiral building block. $[\alpha]_D^{20}$ = + 10.2° (c = 4, $H_2O$). *Pfanstiehl Laboratories, Inc.*

**84 L-Acetyl-3-mercapto-2-methylpropionic acid**
74431-52-0 277-868-4

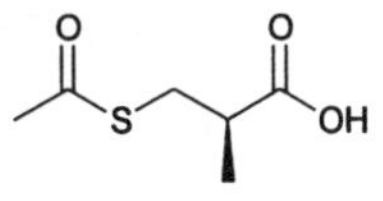

$C_6H_{10}O_3S$
Propanoic acid, 3-(acetylthio)-2-methyl-, (R)-. L-(+)-S-Acetyl-β-mercaptoisobutyric acid; (R)-3-(acetylthio)-2-methylpropionic acid. Chiral intermediate. *Mitsubishi Chemical Corporation.*

**85 N-Acetyl-L-methionine**
65-82-7 97(12) 200-617-7

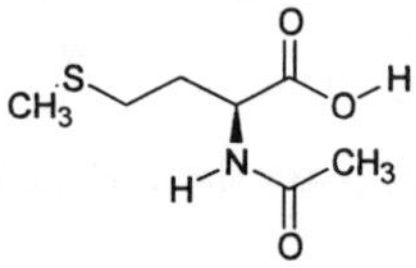

$C_7H_{13}NO_3S$
Methionine, N-acetyl-, L-; Methionamine; Thiomedon. Listed on TSCA. Chiral intermediate. mp = 104-107°; $[\alpha]_D^{20}$ = - 20° (c = 1, $H_2O$). *Acros Organics nv; Austin Chemical Company, Inc.; Rexim S.A. Produits Chimiques; Sigma-Aldrich Fine Chemicals; Varsal Instruments, Inc.*

**86 Acetyl-L-methionine 1-naphthyl ester**

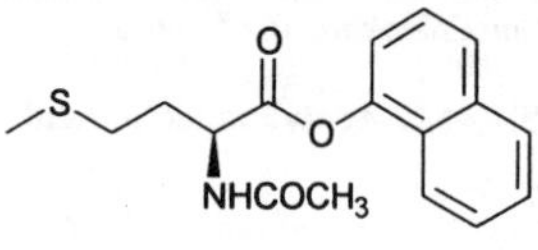

$C_{17}H_{19}NO_3S$
Chiral intermediate. *Senn Chemicals AG.*

**87 N-Acetyl-(4R,5S)-4-methyl-5-phenyl-2-oxazolidinone**

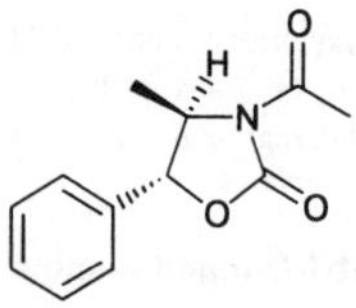

$C_{12}H_{13}NO_3$
Chiral intermediate. *Oxford Asymmetry International plc.*

**88 N-Acetyl-(4S,5R)-4-methyl-5-phenyl-2-oxazolidinone**

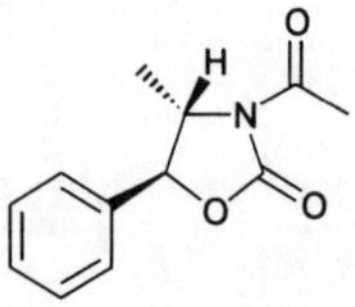

$C_{12}H_{13}NO_3$
Chiral intermediate. *Oxford Asymmetry International plc.*

**89 N-Acetyl-D-muramic acid**
10597-89-4 6384(12) 234-214-2

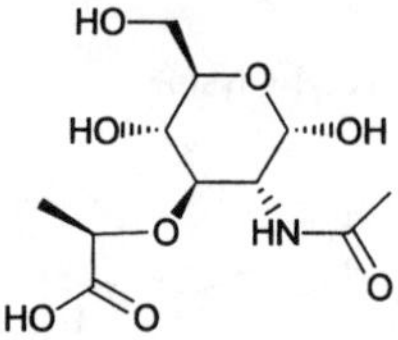

$C_{11}H_{19}NO_8$
D-Glucose, 2-(acetylamino)-3-O-(1-carboxyethyl)-2-deoxy-, (R)-; 2-Acetamido-2-deoxy-3-O-(D-1-carboxyethyl)-D-glucopyranose; N-acetyl-D-muramoate; (R)-2-acetamido-3-O-(1-carboxyethyl)-2-deoxy D-glucose. Chiral building block. mp = 125°; $[\alpha]_D^{20}$ = + 41.8° (c = 1.5, $H_2O$); $[\alpha]_D^{20}$ = + 65° (c = 0.1, $H_2O$). *Acros Organics nv; Pfanstiehl Laboratories, Inc.; Sigma-Aldrich Fine Chemicals.*

**90 (R)-N-Acetyl-2-naphthylalanine**

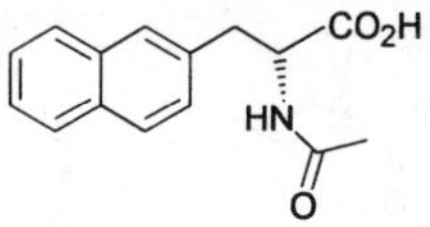

$C_{15}H_{13}NO_3$
Chiral building block. *ChiroTech Technology Ltd.*

**91 (S)-N-Acetyl-2-naphthylalanine**

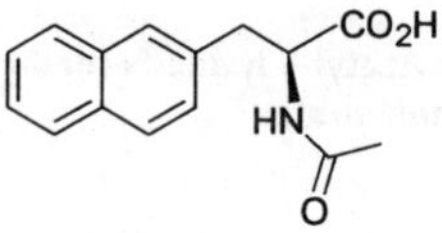

$C_{15}H_{13}NO_3$
Chiral building block. *ChiroTech Technology Ltd.*

**92 N-Acetyl-D-neuraminic acid**
131-48-6 8431(11) 205-023-1

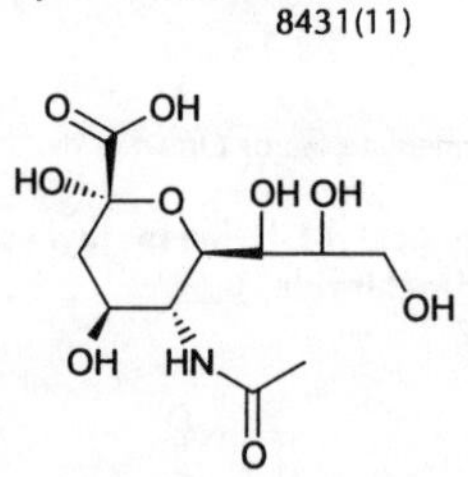

$C_{11}H_{19}NO_9$
Neuraminic acid, N-acetyl-; 5-Acetamido-3,5-dideoxy-D-glycero-D-galacto-nonulopyranosonic acid; o-Sialic acid; 5-Acetamido-3,5-dideoxy-D-glycero-D-galactonulosonic acid; Neu5Ac; NANA, Aceneuramic acid; N-Acetylsialic acid; Sialomucin; Lactaminic acid. Listed on TSCA. Chiral intermediate. mp = 186°; $[\alpha]_D^{20}$ = - 32° (c = 2, $H_2O$). *Acros Organics nv; Pfanstiehl Laboratories, Inc.; Sigma-Aldrich Fine Chemicals.*

**93 N-Acetyl-S-(4-nitrophenyl)-L-cysteine**

$C_{11}H_{12}N_2O_5S$

(S)-2-Acetamido-3-(4-nitrophenylthio)propionic acid. Chiral intermediate. mp = 158°. *TCI America.*

**94 N-Acetyl-3-nitro-L-tyrosine ethyl ester**

40642-95-3

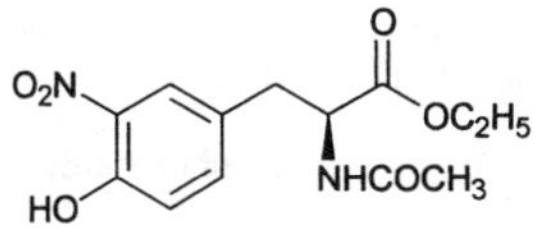

$C_{13}H_{16}N_2O_6$

Chiral intermediate. *Pharmasyn, Inc.*

**95 N-Acetyl-D-phenylalanine**

10172-89-1 233-447-7

$C_{11}H_{13}NO_3$

Alanine, N-acetyl-3-phenyl-, D-; (-)-2-Acetylamino-3-phenylpropanoic acid. Chiral intermediate. *TCI America; Varsal Instruments, Inc.*

**96 N-Acetyl-L-phenylalanine**

2018-61-3 217-959-8

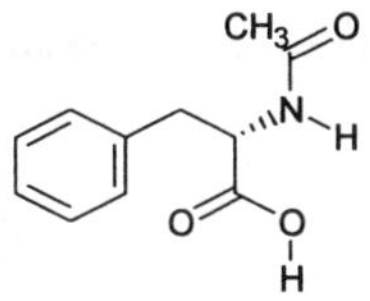

$C_{11}H_{13}NO_3$

Alanine, N-acetyl-3-phenyl-, L-; L-Phenylalanine, N-acetyl-; (S)-2-(Acetylamino)-3-phenylpropanoic acid. Listed on TSCA. Chiral intermediate. mp = 165-167°; [α] = + 40.7° (c = 1, EtOH). *Acros Organics nv; Schweizerhall Pharma; Sigma-Aldrich Fine Chemicals; Varsal Instruments, Inc.*

**97 N-Acetyl-L-phenylalanyl-3,5-diiodo-L-tyrosine**

3786-08-1 223-254-6

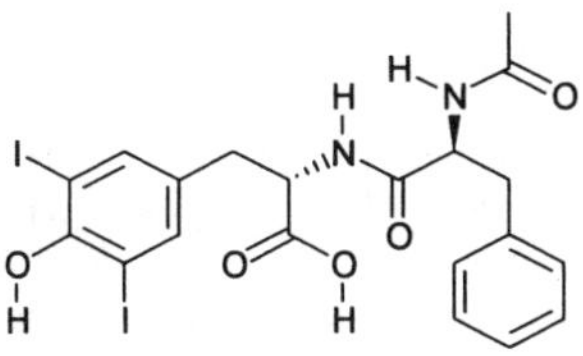

$C_{20}H_{20}I_2N_2O_5$

Tyrosine, N-(N-acetyl-3-phenyl-L-alanyl)-3,5-diiodo-, L-. Chiral intermediate. *Acros Organics nv.*

**98 N-Acetyl-L-proline**

68-95-1 200-698-9

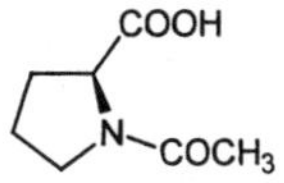

$C_7H_{11}NO_3$

Proline, 1-acetyl-, L-. Chiral building block. *Austin Chemical Company, Inc.; Rexim S.A. Produits Chimiques; Varsal Instruments, Inc.*

**99 N-Acetyl-L-thiazolidine-4-carboxylic acid**

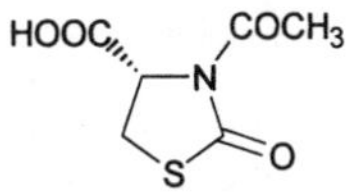

$C_6H_7NO_4$

Chiral intermediate. *Austin Chemical Company, Inc.; Flamma s.p.a.*

**100 (S)-(-)-3-(Acetylthio)isobutyric acid**

76497-39-7 278-480-8

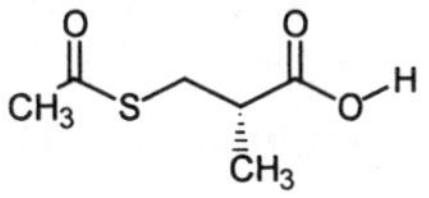

$C_6H_{10}O_3S$

Propanoic acid, 3-(acetylthio)-2-methyl-, (2S)-; (S)-(-)-3-(Acetylmercapto)isobutyric acid; (S)-3-(acetylthio)-2-methylpropionic acid; (S)-(-)-3-(Acetylthio)-2-methylpropanoic acid. Chiral building block. mp = 16°; $bp_2$ = 108°; d = 1.18; [α] = - 46° (c = 1, 96% EtOH). *Acros Organics nv; DSM Fine Chemcials Netherlands; Kaneka Corporation; Mitsubishi Rayon Co., Ltd.; Sumitomo Chemcial Co. Ltd.; TCI America.*

**101 (R)-(+)-3-(Acetylthio)isobutyric acid methyl ester**

86961-07-1

$C_7H_{12}O_3S$

(R)-(+)-3-(Acetylthio)-2-methylpropionic acid methyl ester. Chiral intermediate. *TCI America.*

**102 Acetylthio-2-methylpropanoyl chloride**

74345-73-6

$C_6H_9O_2ClS$

Ethanethioic acid, S-(3-chloro-2-methyl-3-oxopropyl) ester. Chiral intermediate. *Kaneka Corporation.*

**103 N-[3-(Acetylthio)-(2S)-methylpropyl]-L-proline hydrate**
64838-55-7 265-250-7

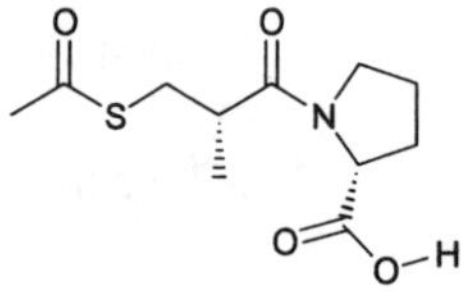

$C_{11}H_{17}NO_4S$
Proline, 1-[3-(acetylthio)-2-methyl-1-oxopropyl]-, (S)-. (S,S)-1-(D-3-Acetylthio-2-methylpropanoyl)-L-proline; DU 1163; S-Acetylcaptopril. Chiral intermediate. mp = 76-82°; $[\alpha]^{21}$ = - 158° (c = 1, $CH_3OH$). *Acros Organics nv; Sigma-Aldrich Fine Chemicals; Zhejiang Chemicals Import and Export Corporation.*

**104 (S)-2-Acetylthio-3-phenylpropionic acid dicyclohexylalanine salt**

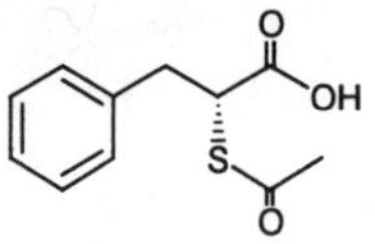

$C_{11}H_{12}O_3S$
Chiral intermediate. *Austin Chemical Company, Inc.*

**105 (R)-Acetylthio-2-pyrrolidinone**
142705-97-3
$C_6H_9NO_2S$
Pyrrolidin-2-one, 4-acetylthio-, (R)-. Chiral building block. *Synthon Chiragenics Corporation.*

**106 (S)-4-Acetylthio-2-pyrrolidinone**

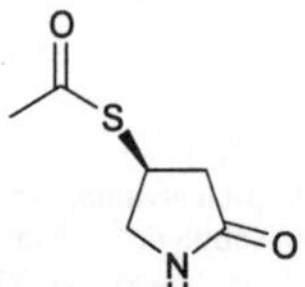

$C_6H_9NO_2S$
Pyrrolidin-2-one, 4-acetylthio-, (S)-. Chiral building block. *Synthon Chiragenics Corporation.*

**107 1-O-Acetyl-2,3,4-tri-O-benzyl-L-fucopyranose**
151909-89-6

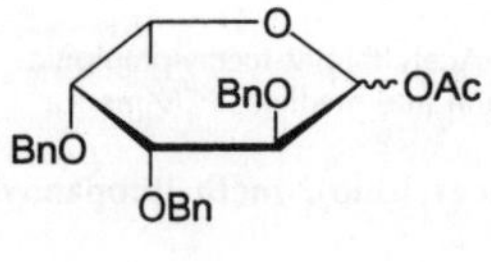

$C_{29}H_{32}O_6$
Chiral intermediate. *Pfanstiehl Laboratories, Inc.*

**108 N-Acetyl-D-tryptophan**
2280-01-5 218-912-4
$C_{13}H_{14}N_2O_3$
Tryptophan, N-acetyl-, D-. Chiral building block. *Austin Chemical Company, Inc.; TCI America.*

**109 N-Acetyl-L-tryptophan**
1218-34-4 214-935-9
$C_{13}H_{14}N_2O_3$
Tryptophan, N-Acetyl, (S)-; L-Tryptophan, N-acetyl-. Chiral intermediate. *TCI America.*

**110 N-Acetyl-L-tryptophanamide**
2382-79-8 219-189-8

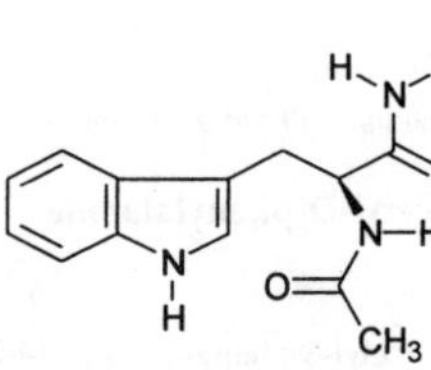

$C_{13}H_{15}N_3O_2$
Indole-3-propanamide, α-(acetylamino)-, 1H-, (S)-; (+)-2-Acetamido-3-(indol-3-yl)propionamide. Chiral intermediate. mp = 194-196°; $[\alpha]^{22}$ = + 17.5° (c = 2, $CH_3OH$). *Acros Organics nv; Sigma-Aldrich Fine Chemicals.*

**111 N-Acetyl-L-tryptophan ethyl ester**
2382-80-1 219-190-3

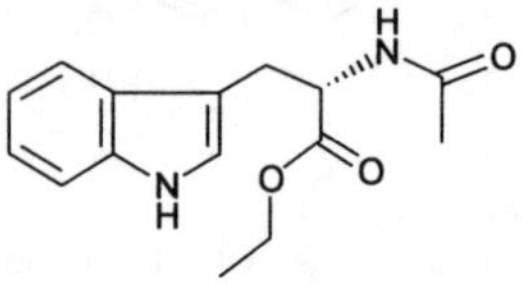

$C_{15}H_{18}N_2O_3$
Tryptophan, N-acetyl-, ethyl ester, L-; Ethyl N-acetyl-L-tryptophanate; Ac-trp-oet. Chiral building block. mp = 112-114°; $[\alpha]_D^{20}$ = + 45° (c = 0.5, $CHCl_3$). *Acros Organics nv; Sigma-Aldrich Fine Chemicals; TCI America.*

**112 N-Acetyl-L-tyrosinamide**
1948-71-6 217-756-4

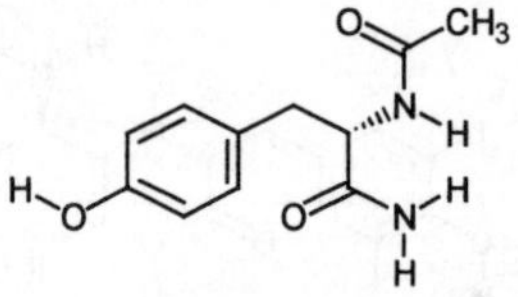

$C_{11}H_{14}N_2O_3$
Benzenepropanamide, α-(acetylamino)-4-hydroxy-, (αS)-; Hydrocinnamamide, α-acetamido-p-hydroxy-, L-; N-Acetyl-4-hydroxy-L-phenylalaninamide. Chiral intermediate. mp = 223-225°. *Acros Organics nv; Sigma-Aldrich Fine Chemicals.*

**113 N-Acetyl-L-tyrosine**
537-55-3 208-671-3

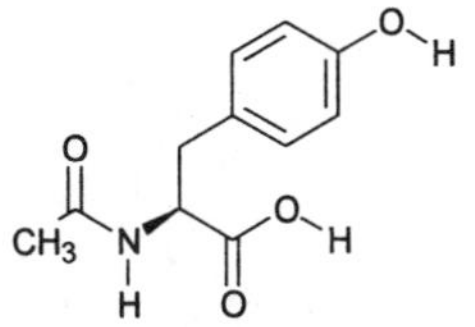

$C_{11}H_{13}NO_4$
Tyrosine, N-acetyl-, L-; (S)-2-Acetamido-3-(4-hydroxyphenyl) propanoic acid. Listed on TSCA. Chiral building block. mp = 145-150°; [α] = + 46.5 to + 49.0° (c = 4, $H_2O$). *Acros Organics nv; Austin Chemical Company, Inc.; Interchem Corporation; Kyowa Hakko Kogyo Co., Ltd.; Rexim S.A. Produits Chimiques; Sigma-Aldrich Fine Chemicals; Sochinaz SA; Tanabe Seiyaku Co. Ltd.*

**114 N-Acetyl-L-tyrosine ethyl ester**
840-97-1 212-663-5

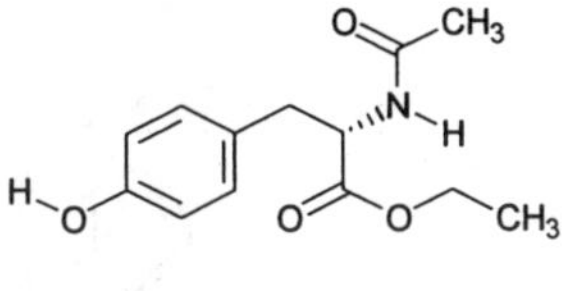

$C_{13}H_{17}NO_4$
Tyrosine, N-acetyl-, ethyl ester, L-; Ethyl N-acetyl-L-tyrosinate; ATEE. Listed on TSCA. Chiral building block. [α] = + 24.1° (c = 1, EtOH). *Acros Organics nv; TCI America.*

**115 N-Acetyl-L-tyrosine ethyl ester monohydrate**
36546-50-6

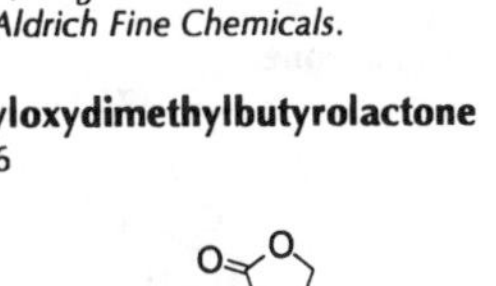

$C_{13}H_{18}NO_6$
ATEE monohydrate. Listed on TSCA. Chiral intermediate. mp = 80-81°; $[\alpha]_D^{20}$ = + 24.1° (c = 1, EtOH). *Pharmasyn, Inc.; Sigma-Aldrich Fine Chemicals.*

**116 Acryloxydimethylbutyrolactone**
102096-60-6

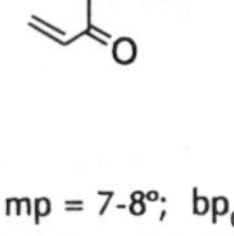

$C_9H_{12}O_4$
Chiral intermediate. mp = 7-8°; $bp_{0.1}$ = 84°; d = 1.125; n = 1.46; $[\alpha]_D^{20}$ = + 6.0° (c = 17, $CH_2Cl_2$). *Sigma-Aldrich Fine Chemicals.*

**117 Adenine-9-β-D-arabinofuranoside-5'-monophosphate**
29984-33-6 249-990-8

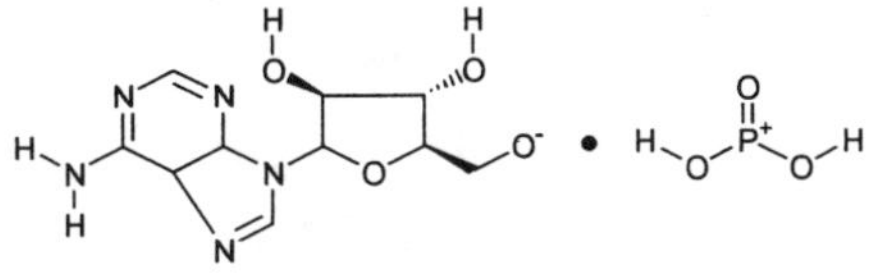

$C_{10}H_{16}N_5O_7P$
Purin-(9H)-6-amine, 9-(5-O-phosphono-β-D-arabinofuranosyl)-; 9-(5-O-Phosphono-β-D-arabinofuranosyl)-9H-purin-6-amine; 9-(5-O-Phosphono-β-D-arabinofuranosyl)adenine; 5'-Arabinosyladenine monophosphate; 9-β-D-Arabinofuranosyladenine 5'-monophosphate; Adenine arabinonucleoside 5'-phosphate; Adenine arabinoside 5. Chiral intermediate. *Acros Organics nv.*

**118 D-Adenosine**
58-61-7 152(12) 200-389-9

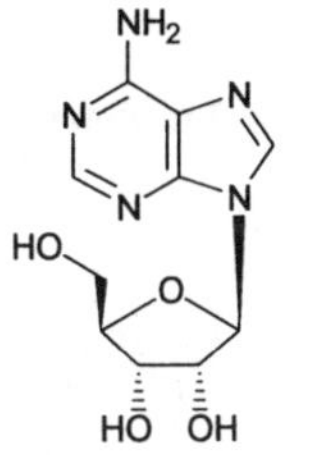

$C_{10}H_{13}N_5O_4$
Ribofuranose, 1-(6-amino-9H-purin-9-yl)-1-deoxy-, β-D-. 9H-Purin-6-amine, 9-β-D-ribofuranosyl-; 9-β-D-Ribofuranosyl-9H-purin-6-amine; 9-β-D-Ribofuranosyladenine; Adenine riboside; Adenocard; Adenoscan; Boniton; Myocol; Nucleocardyl; Riboadenosine; Sandesi. Listed on TSCA. Chiral intermediate. mp = 234-236°; [α] = - 60° (c = 0.7, $H_2O$). *Acros Organics nv; Sigma-Aldrich Fine Chemicals; TCI America.*

**119 Adenosine 3'-,5'-cyclic monophosphate**
362-74-3 206-649-8

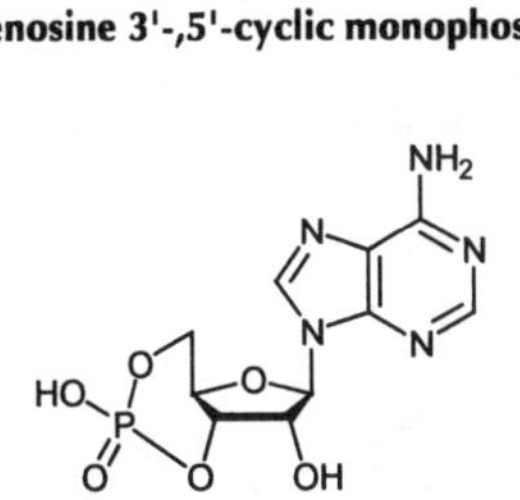

$C_{10}H_{12}N_5O_6P$
Butyramide, N-(9-β-D-ribofuranosyl-9H-purin-6-yl)-, 2'-butyrate, cyclic 3',5'-phosphate; Bucladesine; Adenosine, N-(1-oxobutyl)-, cyclic 3',5'-(hydrogen phosphate) 2'-butanoate; 3',5'-Cyclic AMP dibutyrate; Actosin; dbcAMP; Dibutyryl 3',5'-cyclic adenosine monophosphate; Dibutyryl 3',5'-cyclic AMP; Dibutyryl cAMP; Dibutyryl cyclic 3',5'-adeny. Chiral intermediate. [α] = - 53.7° (c = 0.7, $H_2O$). *Acros Organics nv.*

**120 Adenosine 3,5-cyclic monophosphate**
60-92-4 2777(12) 200-492-9

$C_{10}H_{12}N_5O_6P$
Cyclic AMP. Listed on TSCA. Chiral intermediate. mp = 260°; $[\alpha]^{22}$ = - 53.7° (c = 0.7, $H_2O$). *Sigma-Aldrich Fine Chemicals.*

**121 Adenosine 3-monophosphate hydrate**
147(11) 201-521-8

$C_{10}H_{14}N_5O_7P$
3'-Adenylic acid. Listed on TSCA. Chiral intermediate. mp = 210°; $[\alpha]^{22}$ = - 41.6° (c = 1, $H_2O$). *Sigma-Aldrich Fine Chemicals.*

**122 Adenosine 5-monophosphate monohydrate**
18422-05-4 148(11)

$C_{10}H_{14}N_5O_7P$
AMP; Adenosine-5'-phosphoric acid. Listed on TSCA. Chiral intermediate. mp = 200°; $[\alpha]^{22}$ = - 46.6° (c = 2, 2% NaOH). *Sigma-Aldrich Fine Chemicals.*

**123 (S)-Adenosyl-L-methionine disulfate tosylate**

$C_{22}H_{32}N_6O_{15}S_4$
SAMe. Chiral intermediate. *KingChem, Inc.*

**124 (S)-Adenosyl-L-methionine tosylate**

$C_{22}H_{28}N_6O_7S_2$
Chiral intermediate. *KingChem, Inc.*

**125 Adonitol**
488-81-3 207-685-7
$C_5H_{12}O_5$
Ribitol.
Adonit; Adonite. Listed on TSCA. Chiral building block. *TCI America.*

**126 Adrenosterone**
382-45-6 174(12) 206-843-2

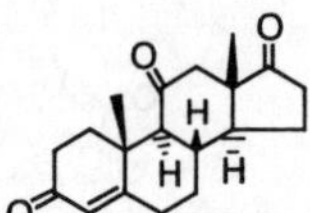

$C_{19}H_{24}O_3$
Androst-4-ene-3,11,17-trione; Reichstein's substance G. Pharmaceutical or derivative. mp = 219-222°; $[\alpha]_D^{20}$ = + 300° (c = 1, $CHCl_3$). *Sigma-Aldrich Fine Chemicals.*

**127 D-Alaninamide hydrochloride**

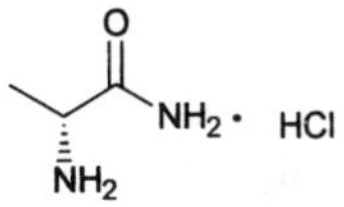

$C_3H_9ClN_2O$
Chiral building block. *Synthetech, Inc.*

**128 L-Alaninamide hydrochloride**
33208-99-0

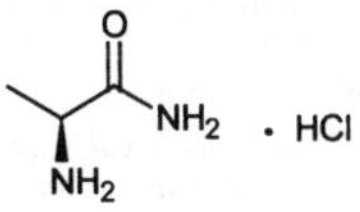

$C_3H_9ClN_2O$
Propanamide, 2-amino-, monohydrochloride, (S)-. Chiral building block. mp = 212-217°; $[\alpha]^{25}$ = + 11.0° (c = 1, $CH_3OH$). *Austin Chemical Company, Inc.; Sigma-Aldrich Fine Chemicals; Synthetech, Inc.*

**129 D-Alanine**
338-69-2 205(12) 206-418-1

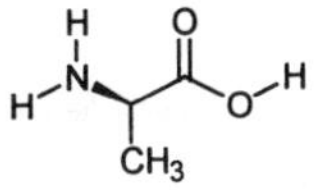

$C_3H_7NO_2$
Alanine, D-; D-2-Aminopropionic acid; Ba 2776. Listed on TSCA. Chiral building block. mp = 291°; $[\alpha]^{25}$ = - 14° (c = 10, 6N HCl). *Acros Organics nv; Ajinomoto Co. Inc.; Austin Chemical Company, Inc.; Kaneka Corporation; Kyowa Hakko Kogyo Co., Ltd.; Nippon Kayaku; Omega Chemical Company Inc.; Recordati S.p.A.; Rexim S.A. Produits Chimiques; Sigma-Aldrich Fine Chemicals; Tanabe Seiyaku Co. Ltd.; TCI America; Toray International, Inc.*

**130 L-Alanine**
56-41-7 205(12) 200-273-8

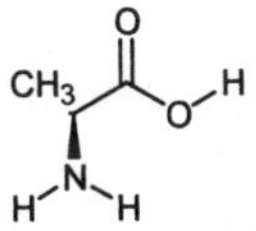

$C_3H_7NO_2$
Propanoic acid, 2-amino-, (S)-; L-2-Aminopropionic acid. Listed on TSCA. Chiral building block. mp = 29°; $[\alpha]_D^{20}$= + 14.5° (c = 10, 6N HCl). *Aceto Corporation; Acros Organics nv; Ajinomoto Co. Inc.; Austin Chemical Company, Inc.; KingChem, Inc.; Kyowa Hakko Kogyo Co., Ltd.; Nippon Kayaku; Rexim S.A. Produits Chimiques; Senn Chemicals AG; Shanghai DSL International Trading Company; Sigma-Aldrich Fine Chemicals; Tanabe Seiyaku Co. Ltd.; TCI America; Varsal Instruments, Inc.; Yoneyama Yakuhin Kogyo Co., Ltd.*

**131 β-Alanine benzyl ester p-toluenesulfonic acid salt**

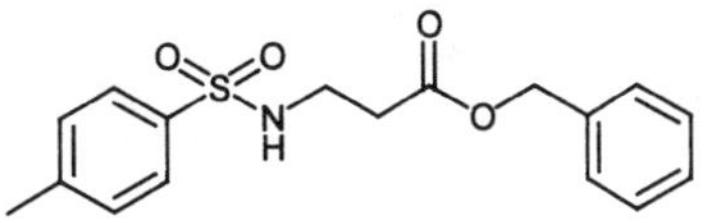

$C_{17}H_{19}NO_4S$
Chiral intermediate. *Synthetech, Inc.*

**132 D-Alanine benzyl ester p-toluenesulfonic acid salt**
41036-32-2 255-185-2

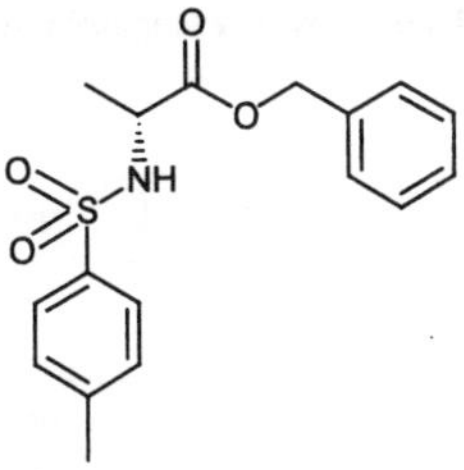

$C_{17}H_{21}NO_5S$
Alanine, phenylmethyl ester, 4-methylbenzenesulfonate, D-; O-Benzyl-D-alanine toluene-p-sulphonate; Benzyl D-alaninate p-toluenesulfonate. Chiral intermediate. *Austin Chemical Company, Inc.; Synthetech, Inc.*

**133 L-Alanine benzyl ester p-toluenesulfonic acid salt**
42854-62-6 255-969-4

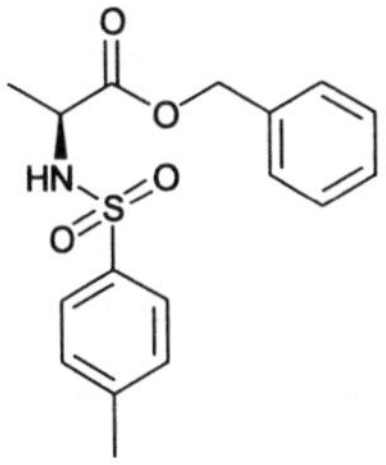

$C_{17}H_{21}NO_5S$
Alanine, phenylmethyl ester, 4-methylbenzenesulfonate, L-; O-benzyl-L-alanine toluene-p-sulphonate; Benzyl L-alaninate-p-toluenesulfonic acid (1:1). Chiral intermediate. *Flamma s.p.a.; Synthetech, Inc.*

**134 D-Alanine ethyl ester hydrochloride**

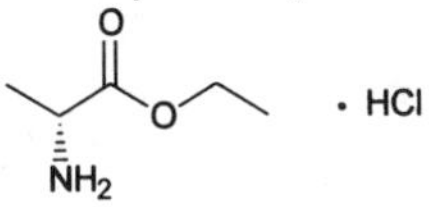

$C_5H_{12}ClNO_2$
Chiral building block. *Tanabe Seiyaku Co. Ltd.*

**135 L-Alanine ethyl ester hydrochloride**
1115-59-9 214-225-9

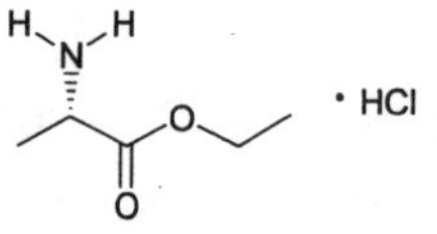

$C_5H_{12}ClNO_2$
Alanine, ethyl ester, hydrochloride, L-.
Ethyl alaninate hydrochloride. Chiral building block. mp = 69°; $[\alpha]^{20}$ = + 2.9° (c = 2.5, $H_2O$). *Acros Organics nv; Sigma-Aldrich Fine Chemicals.*

**136 L-Alanine hydroxysuccinimide ester**

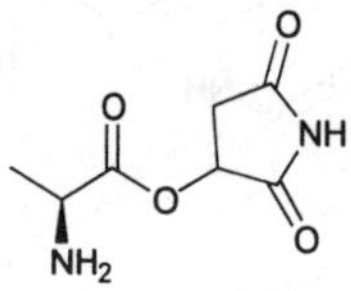

$C_7H_{10}N_2O_4$
Chiral intermediate. *Flamma s.p.a.*

**137 L-Alanine-4-methyl-7-coumarinylamide trifluoroacetate**
96594-10-4

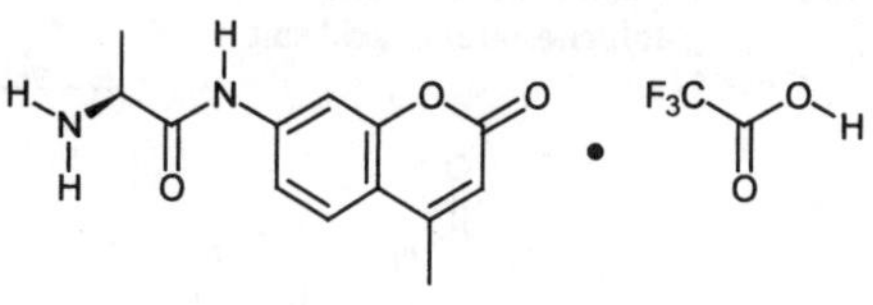

$C_{15}H_{15}F_3N_2O_5$
Chiral intermediate. *Acros Organics nv.*

**138 D-Alanine methyl ester hydrochloride**
14316-06-4 238-258-3

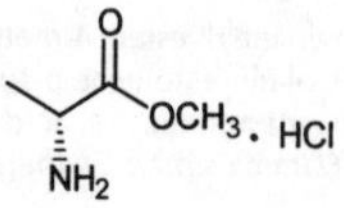

$C_4H_{10}ClNO_2$
Alanine, methyl ester, hydrochloride, D-; Methyl D-alaninate hydrochloride; (R)-2-Aminopropionic acid methyl ester hydrochloride. Chiral building block. mp = 108-110°. *Sigma-Aldrich Fine Chemicals; Tanabe Seiyaku Co. Ltd.*

**139 L-Alanine methyl ester hydrochloride**
2491-20-5 219-652-4

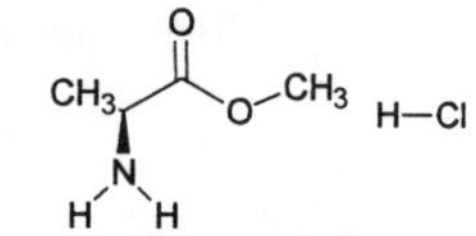

$C_4H_{10}ClNO_2$
Alanine, methyl ester, hydrochloride, L-; (S)-2-Aminopropionic acid methyl ester hydrochloride; Methyl L-alaninate hydrochloride. Chiral building block. mp = 109-111°; $[\alpha]^{25}$ = + 7.0° (c = 1.6, $CH_3OH$). *Acros Organics nv; Austin Chemical Company, Inc.; Flamma s.p.a.; Loba Feinchemie AG; Sigma-Aldrich Fine Chemicals; TCI America.*

**140 D-Alanine 1-naphthylester hydrochloride**

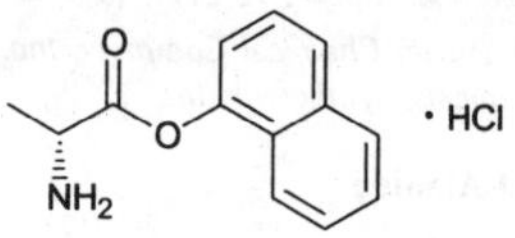

$C_{13}H_{14}ClNO_2$
Chiral building block. *Senn Chemicals AG.*

**141 D-Alaninol**
35320-23-1

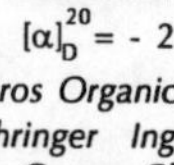

$C_3H_9NO$
(R)-β-Propanolamine; (R)-1-Hydroxy-2-aminopropane; (R)-1-Methyl-2-hydroxyethylamine;1-Propanol, 2-amino-, (R)-; (R)-2-Amino-1-hydroxypropane; (R)-2-Hydroxy-1-methylethylamine; (R)-2-Amino-2-methylethanol; (R)-Propanolamine. Chiral intermediate. bp = 173-176°; d = 0.9650; $n_D^{20}$ = 1.4493; $[\alpha]_D^{20}$ = - 22 ± 2° (c = 2, EtOH); $[\alpha]^{19}$ = - 18° (neat). *Acros Organics nv; Austin Chemical Company, Inc.; Boehringer Ingelheim Pharma KG; Lancaster Synthesis Ltd.; Omega Chemical Company Inc.; Oxford Asymmetry International plc; Sigma-Aldrich Fine Chemicals; Synthetech, Inc.*

**142 L-Alaninol**
2749-11-3 480(11) 220-388-7
$C_3H_9NO$
Propan-1-ol, 2-amino-, (2S)-; (S)-(+)-2-Amino-1-propanol. Chiral building block. $bp_{11}$ = 72-73°; d = 0.9650; $n_D^{20}$ = 1.4498; $[\alpha]_D^{20}$ = + 22 ± 2° (c = 2, EtOH); [α] = + 18 (neat). *Acros Organics nv; Boehringer Ingelheim Pharma KG; Lancaster Synthesis Ltd.; Loba Feinchemie AG; Omega Chemical Company Inc.; Oxford Asymmetry International plc; Senn Chemicals AG; Sigma-Aldrich Fine Chemicals; Synthetech, Inc.; TCI America; Zhejiang Chemicals Import and Export Corporation.*

**143 L-Alanyl-L-alanine**
1948-31-8 217-751-7

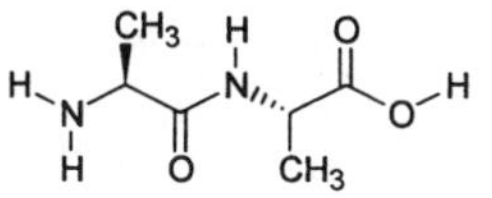

$C_6H_{12}N_2O_3$
Alanine, N-L-alanyl-, L-; Dialanine. Chiral intermediate. mp = 286-288°; $[\alpha]^{22}$ = - 35.9° (c = 2, 6N HCl). *Acros Organics nv; Sigma-Aldrich Fine Chemicals.*

**144 L-Alanyl-L-alanyl-L-alanine methyl ester acetate**
84794-58-1 284-163-5

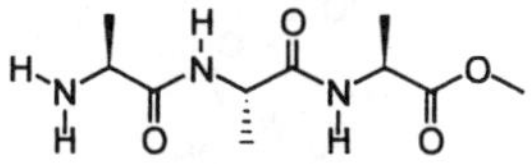

$C_{10}H_{19}N_3O_4$
Alanine, N-(N-L-alanyl-L-alanyl)-, methyl ester, monoacetate, L-; N-(N-L-alanyl-L-alanyl)-O-methyl-L-alanine monoacetate. Chiral intermediate. *Acros Organics nv.*

**145 L-Alanyl-L-glutamine**
16874-70-7
$C_8H_{15}N_3O_4$
Chiral intermediate. mp = 215°. *Kyowa Hakko Kogyo Co., Ltd.; Rexim S.A. Produits Chimiques; TCI America.*

**146 L-Alanyl-glycyl-glycine**
3146-40-5 221-559-9
$C_7H_{13}N_3O_4$
Glycine, L-alanylglycyl-; Ala-gly-gly. Chiral intermediate. *TCI America.*

**147 β-Alanyl-L-histidine**
305-84-0 1899(12) 206-169-9

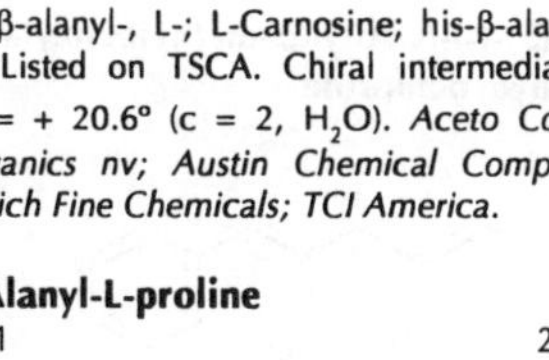

$C_9H_{14}N_4O_3$
Histidine, β-alanyl-, L-; L-Carnosine; his-β-ala; Ignotine; Karnozin. Listed on TSCA. Chiral intermediate. mp = 250°; [α] = + 20.6° (c = 2, $H_2O$). *Aceto Corporation; Acros Organics nv; Austin Chemical Company, Inc.; Sigma-Aldrich Fine Chemicals; TCI America.*

**148 L-Alanyl-L-proline**
13485-59-1 236-795-8
$C_8H_{14}N_2O_3$
Alanylproline, L-; L-Proline, L-alanyl-. Chiral intermediate. *Degussa-Huls AG.*

**149 L-Alanyl-L-tryptophan**
16305-75-2 240-392-2
$C_{14}H_{17}N_3O_3$
Tryptophan, N-L-alanyl-, L-; Chiral intermediate. *TCI America.*

**150 L-Alanyl-L-tyrosine**
730-08-5
$C_{12}H_{16}N_2O_4$
Chiral intermediate. *TCI America.*

**151 Albizziin**
1483-07-4 212(12) 216-046-1

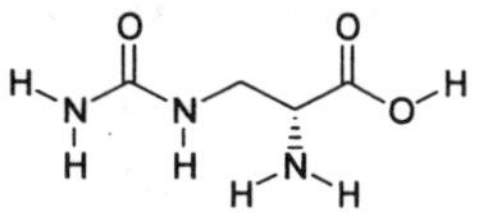

$C_4H_9N_3O_3$
Propionic acid, 2-amino-3-ureido-; L-2-Amino-3-ureidopropionic acid; L-β-Ureidoalanine; L-Alanine, 3-[(aminocarbonyl)amino]-. Chiral intermediate. mp = 220-223°; $[\alpha]^{23}$ = - 23.4° (c = 1, 1N HCl). *Acros Organics nv; Sigma-Aldrich Fine Chemicals.*

**152 D-(+)-Allose**
2595-97-3 288(12) 219-994-4

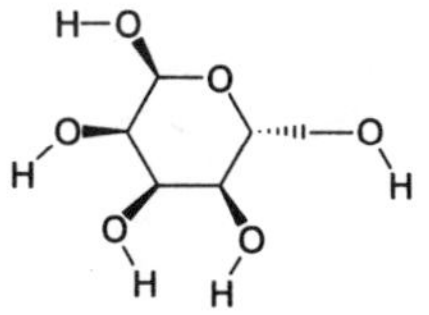

$C_6H_{12}O_6$
Allose, D-; β-D-Allopyranose. Chiral building block. mp = 148-150°; $[\alpha]^{19}$ = + 15° (c = 3.8, $H_2O$, 20h). *Acros Organics nv; Pfanstiehl Laboratories, Inc.; Senn Chemicals AG; Sigma-Aldrich Fine Chemicals.*

**153 L-(-)-Allose**
7635-11-2

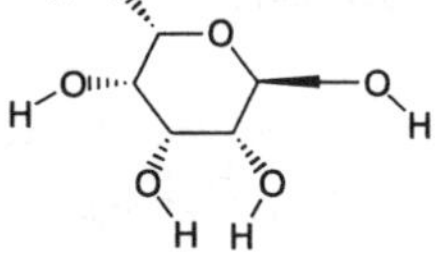

$C_6H_{12}O_6$
Chiral building block. [α] = - 15° (c = 5 $H_2O$). *Acros Organics nv; Senn Chemicals AG.*

**154 O-Allyl-N-(9-Anthracenylmethyl)-cinchonidinium bromide**

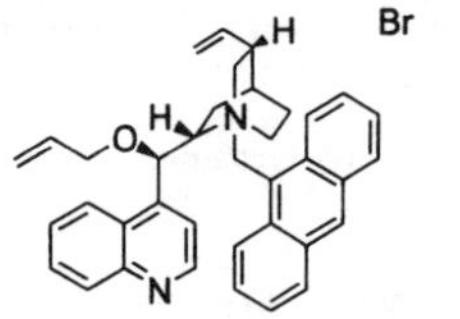

$C_{37}H_{37}BrN_2O$
Resolving agent; Chiral auxiliary. mp = 170°; $[\alpha]_D^{20}$ = - 340° (c = 0.45, $CHCl_3$). *Sigma-Aldrich Fine Chemicals.*

**155 O-Allyl-N-benzylcinchondinium bromide**

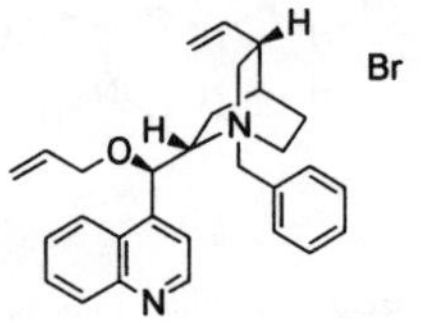

$C_{29}H_{33}BrN_2O$
Resolving agent; Chiral auxiliary. mp = 140-144°; $[\alpha]_D^{20}$= - 158° (c = 1, $CHCl_3$). *Sigma-Aldrich Fine Chemicals.*

**156 Allyl 7-[3(R)-tert-butylsilyloxy-5-oxo-cyclopent-1-enyl]heptanoate**

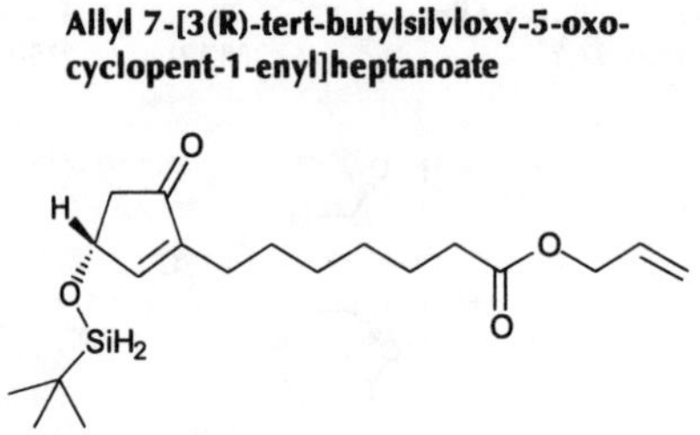

$C_{21}H_{36}O_4Si$
Chiral intermediate. *Nissan Chemical Industries, Ltd.*

**157 (S)-Allyl-L-cysteine**
21593-77-1
$C_6H_{11}NO_2S$
(S)-Allylthio-2-aminopropionic acid; L-Deoxyalliin. Chiral building block. *TCI America.*

**158 (+)-1-O-Allyl-2-deoxy-4,6-O-isopropylidene-2-(trifluoroacetamido)-glucopyranoside**
139629-59-7

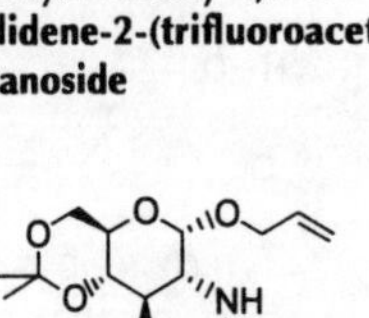

$C_{14}H_{20}F_3NO_6$
Chiral intermediate. mp = 94-96°; $[\alpha]_D^{20}$= + 95° (c = 1, $CHCl_3$). *Sigma-Aldrich Fine Chemicals.*

**159 (-)-7-Allyl-7,8-dihydro-8-oxoguanosine**

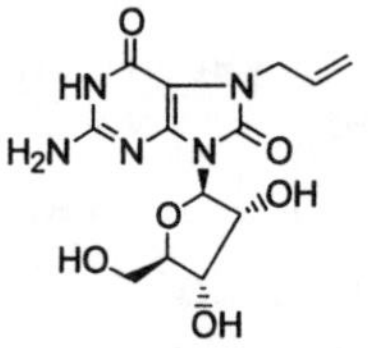

$C_{13}H_{17}N_5O_6$
Chiral intermediate. mp = 228-233°; $[\alpha]^{21}$ = - 32° (c = 1, $H_2O$). *Sigma-Aldrich Fine Chemicals.*

**160 (+)-Allylgalactopyranoside**

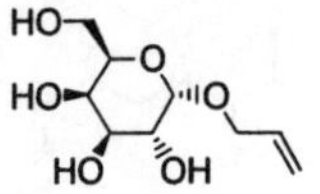

$C_9H_{16}O_6$
Chiral intermediate. mp = 141-145°; $[\alpha]_D^{20}$=+ 170° (c = 1.5, $H_2O$). *Sigma-Aldrich Fine Chemicals.*

**161 (S)-(-)-N-Allyl-α-methylbenzylamine**
115914-08-4

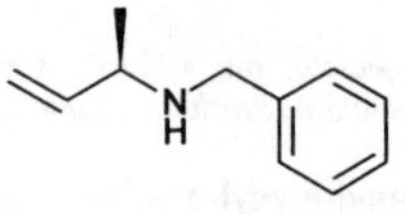

$C_{11}H_{15}N$
Chiral building block. $bp_1$ = 58-59°; d = 0.92; n = 1.51; $[\alpha]_D^{20}$ = - 56° (c = 1, $CH_2Cl_2$). *Sigma-Aldrich Fine Chemicals.*

**162 (R)-(+)-N-Allyl-α-methylbenzylamine hydrochloride**

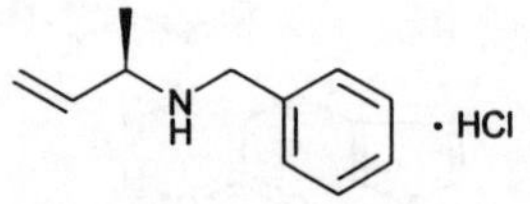

$C_{11}H_{16}ClN$
Chiral intermediate. *Oxford Asymmetry International plc.*

**163 (S)-(-)-N-Allyl-α-methylbenzylamine hydrochloride**

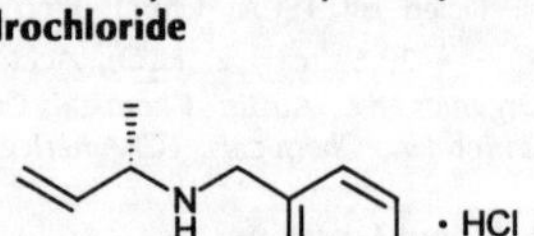

$C_{11}H_{16}ClN$
Chiral building block. *Oxford Asymmetry International plc.*

**164 (3S,4S)-3-Allyloxy-4-hydroxy-tetrahydrofuran**

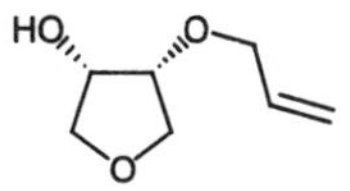

$C_7H_{12}O_3$
Chiral intermediate. *Eastman Chemical Company.*

**165 (2R)-3-(o-Allyloxyphenoxy)-1,2-epoxypropane**
66966-19-6

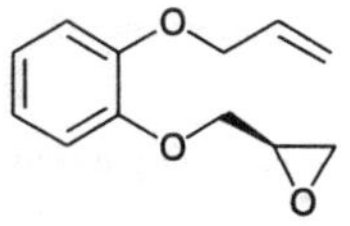

$C_{12}H_{14}O_3$
Chiral intermediate. *Omega Chemical Company Inc.*

**166 (2S)-3-(o-Allyloxyphenoxy)-1,2-epoxypropane**

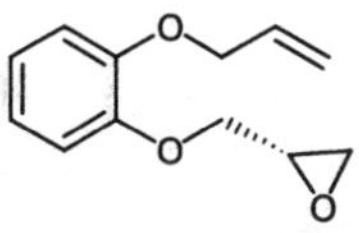

$C_{12}H_{14}O_3$
Chiral intermediate. *Omega Chemical Company Inc.*

**167 N-Allyl-1-phenylethylamine hydrochloride**
37696-17-6

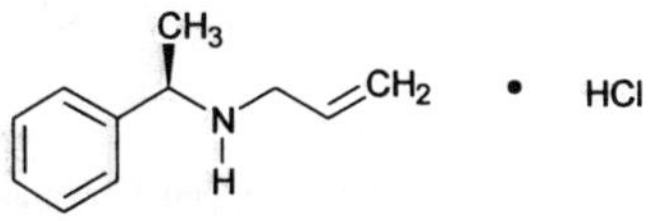

$C_{11}H_{16}ClN$
Chiral intermediate. *Acros Organics nv.*

**168 (R)-Alpine-boramine N,N-bis(monoisopino-campheylbornane)-N,N,N,N-tetramethyl-ethylenediamine**

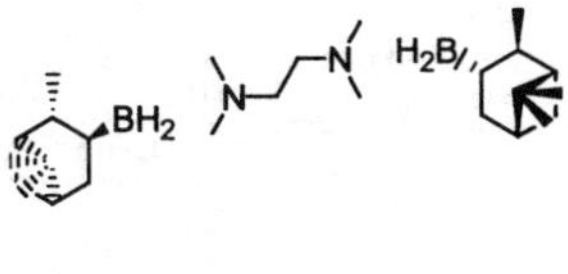

$C_{26}H_{54}B_2N_2$
N,N'-Bis(monoisopinocampheylborane)-N,N,N',N'-tetra-methylethylenediamine. Chiral auxiliary. mp = 144-146°. *Sigma-Aldrich Fine Chemicals.*

**169 (S)-Alpine-boramine N,N-bis(monoisopino-campheylbornane)-N,N,N,N-tetramethyl-ethylenediamine**
67826-92-0

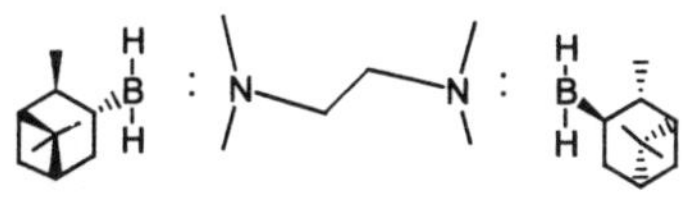

$C_{26}H_{54}B_2N_2$
(S)-N,N'-Bis(monoisopinocampheylborane)-N,N,N',N'-tetramethylethylenediamine. Chiral auxiliary. mp = 143-144°. *Sigma-Aldrich Fine Chemicals.*

**170 (S)-Alpine borane**
42371-63-1

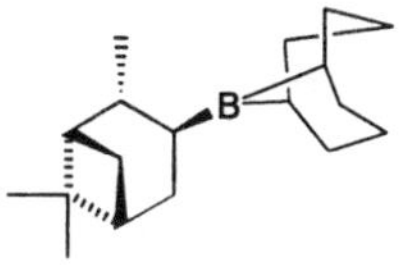

$C_{18}H_{31}B$
(+)-B-Isopinocampheyl-9-borabicyclo[3.3.1]nonane. Chiral auxiliary. bp = 55°; d = 0.947; $[\alpha]^{22}$ = + 20° (c = 12, THF). *Sigma-Aldrich Fine Chemicals.*

**171 (R)-Alpine borane**
73624-47-2

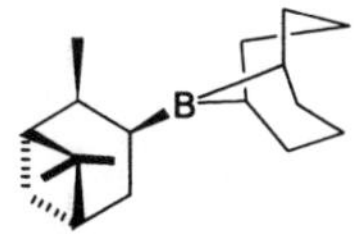

$C_{18}H_{31}B$
(-)-B-isopinocampheyl-9-borabicyclo[3.3.1]nonane. Chiral intermediate. bp = 55°; d = 0.947; $[\alpha]^{21}$ = - 22° (c = 12, THF). *Sigma-Aldrich Fine Chemicals.*

**172 (R)-Alpine hydride**
64081-12-5

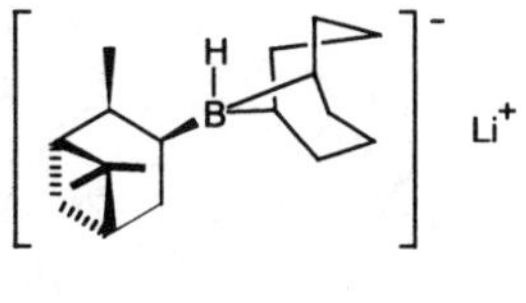

$C_{18}H_{32}BLi$
(-)-Lithium B-isopinocampheyl-9-borabicyclo[3.3.1]nonyl hydride (0.5M solution in tetrahydrofuran). Chiral auxiliary. d = 0.919; $[\alpha]_D^{20}$ = - 5.6° (neat). *Sigma-Aldrich Fine Chemicals.*

**173 (S)-Alpine hydride**
10013-07-8

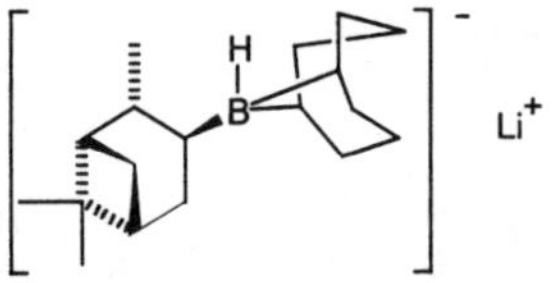

$C_{18}H_{32}BLi$
(+)-Lithium B-isopinocampheyl-9-borabicyclo[3.3.1]-nonyl hydride (0.5M solution in tetrahydrofuran). Chiral auxiliary. d = 0.92; $[\alpha]_D^{20}$ = + 5.5° (neat). *Sigma-Aldrich Fine Chemicals.*

**174 D-(+)-Altrose**
1990-29-0 217-870-4

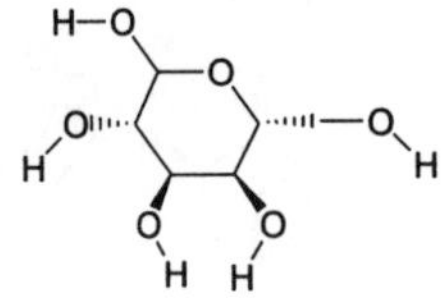

$C_6H_{12}O_6$
Altrose, D-. Chiral building block. mp = 106-108°; [α] = + 33° (c = 2 $H_2O$). *Acros Organics nv; Senn Chemicals AG.*

**175 Ambroxide**
229-861-2

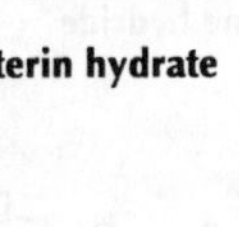

$C_{16}H_{28}O$
Listed on TSCA. Chiral intermediate. mp = 74-76°; $[\alpha]_D^{20}$= - 29° (c = 1, toluene). *Sigma-Aldrich Fine Chemicals.*

**176 D-Amethopterin hydrate**
257-482-2

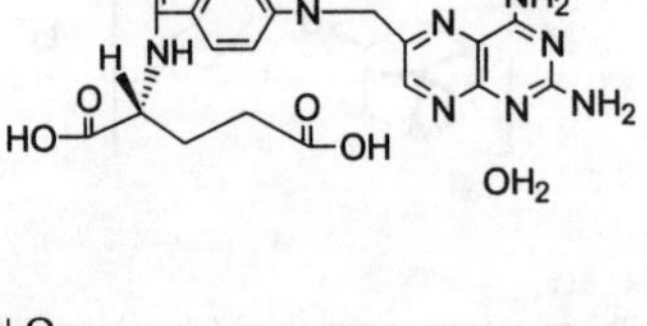

$C_{20}H_{22}N_8O_5$
D-4-Amino-$N^{10}$-methylpteroylglutamic acid. Chiral intermediate. mp = 195°; $[\alpha]_D^{20}$ = - 19.4° (c = 2, 0.1N NaOH). *Sigma-Aldrich Fine Chemicals.*

**177 L-Amethopterin hydrate**
133073-73-1 5908(11)

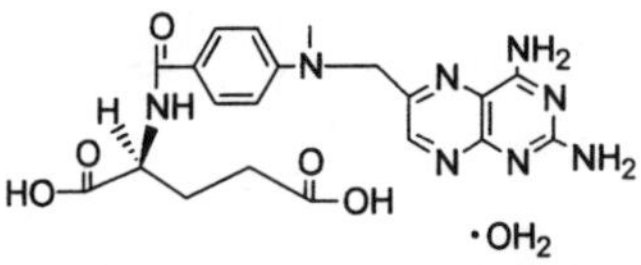

$C_{20}H_{22}N_8O_5$
L-4-Amino-$N^{10}$-methylpteroylglutamic acid; Methotrexate: MTX. Listed on TSCA. Chiral intermediate. $[\alpha]^{25}$ = + 17° (c = 1, 0.05M $Na_2CO_3$). *Sigma-Aldrich Fine Chemicals.*

**178 D-(-)-2-Aminoadipic acid**
7620-28-2 427(11)

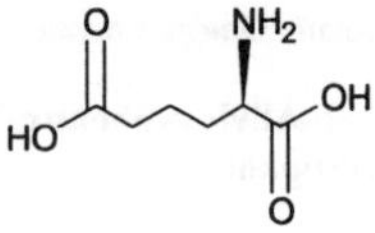

$C_6H_{11}NO_4$
Chiral building block. mp = 208-210°; $[\alpha]^{21}$ = - 24° (c = 1, 6N HCl). *Sigma-Aldrich Fine Chemicals.*

**179 L-(+)-2-Aminoadipic acid**
1118-90-7 436(12)

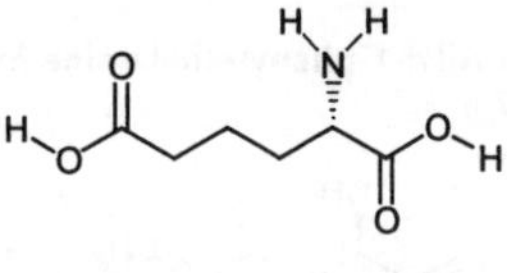

$C_6H_{11}NO_4$
Hexanedioic acid, 2-amino-, L-. (S)-2-Aminohexanedioic acid; L-Homoglutamic acid; L-2-Aminoadipate. Chiral building block. mp = 186°; $[\alpha]_D$ = + 23.5°; soluble in $H_2O$, aqueous NaOH, HCl. *Acros Organics nv; Lancaster Synthesis Ltd.; Mercian Corporation; Senn Chemicals AG; Sigma-Aldrich Fine Chemicals; TCI America.*

**180 (S)-(+)-2-Amino-3-(2-aminoethoxy)-propanoic acid monohydrochloride**

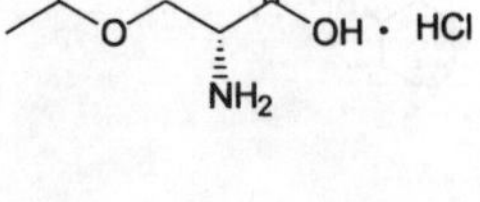

$C_5H_{13}ClN_2O_3$
Chiral building block. mp = 212-216°; $[\alpha]^{22}$ = + 4° (c = 1, $H_2O$). *Sigma-Aldrich Fine Chemicals.*

**181 4-(4-Aminobenzoyl)-L-glutamic acid**

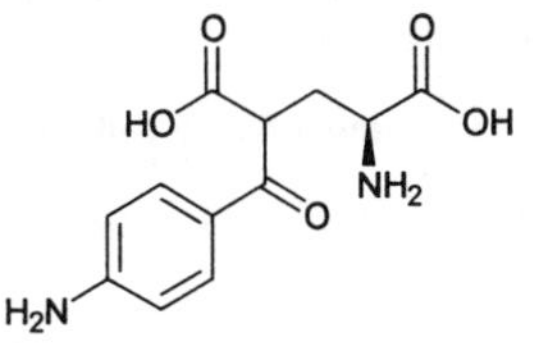

$C_{12}H_{14}N_2O_5$
Chiral intermediate. *KingChem, Inc.*

**182 N-(4-Aminobenzoyl)-L-glutamic acid diethyl ester**
13726-52-8 237-293-1

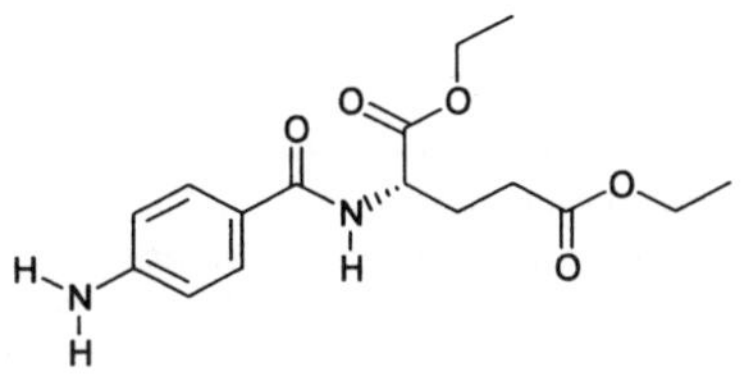

$C_{16}H_{22}N_2O_5$
Glutamic acid, N-(4-aminobenzoyl)-, diethyl ester, L-. Diethyl N-(4-aminobenzoyl)-L-glutamate. Chiral intermediate. mp = 139-142°; $[\alpha]^{21}$ = + 17.9° (c = 5, $CHCl_3$). *Acros Organics nv; Sigma-Aldrich Fine Chemicals.*

**183 (S)-2-Amino-3-(benzylamino)-propanoic acid**
119830-32-9

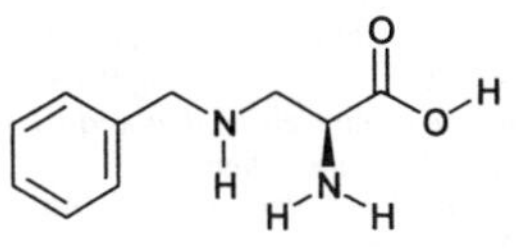

$C_{10}H_{14}N_2O_2$
Chiral building block. [α] = + 26° (c = 10, 6N HCl). *Acros Organics nv.*

**184 (R)-(-)-2-Amino-1-benzyloxybutane**
142559-11-3

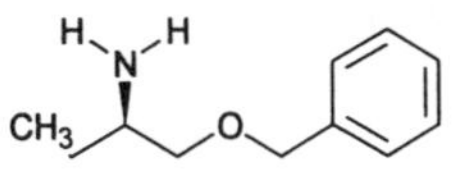

$C_{11}H_{17}NO$
(R)-(-)-1-(Benzyloxymethyl)propylamine. Chiral building block. $bp_{0.15}$ = 80°. *Acros Organics nv.*

**185 (R)-(+)-2-Amino-3-benzyloxy-1-propanol**

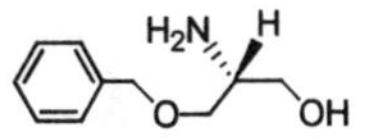

$C_{10}H_{15}NO_2$
Chiral building block. mp = 34-37°; bp = 307°; $[\alpha]_D^{20}$ = + 4° (c = 1, $CH_2Cl_2$). *Sigma-Aldrich Fine Chemicals.*

**186 (S)-(+)-2-Amino-4-bromobutyric acid hydrobromide**
15159-65-6
$C_4H_9Br_2NO_2$
Chiral building block. *Acros Organics nv; TCI America.*

**187 (R)-Amino-6-bromotetralin**

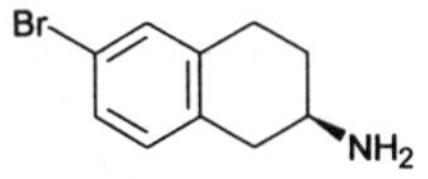

$C_{10}H_{12}BrN$
Chiral building block. *Chiragene, Inc.*

**188 (S)-2-Amino-6-bromotetralin**

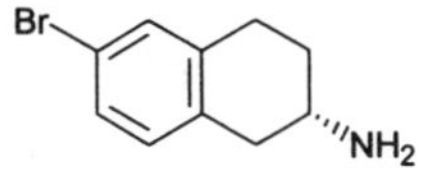

$C_{10}H_{12}BrN$
Chiral building block. *Chiragene, Inc.*

**189 (R)-(-)-2-Aminobutane**
13250-12-9 1544(11) 236-232-6

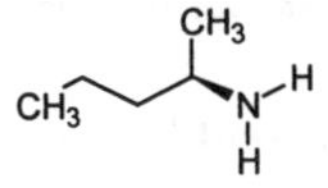

$C_4H_{11}N$
Butane, 2-amino-, (-)-. (R)-(-)-sec-Butylamine; (-)-2-Butylamine. Chiral building block. bp = 57-60°; d = 0.724; n = 0.3936; [α] = - 7.5° (c = neat). *Acros Organics nv; Norse Laboratories; Sigma-Aldrich Fine Chemicals.*

**190 (R)-Amino-1,3-butanediol**

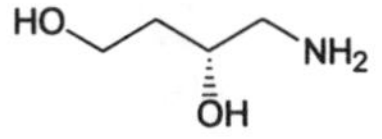

$C_4H_{11}NO_2$
Butane-1,3-diol, 4-amino, (R)-. Chiral building block. *Synthon Chiragenics Corporation.*

**191 (S)-4-Amino-1,3-butanediol**

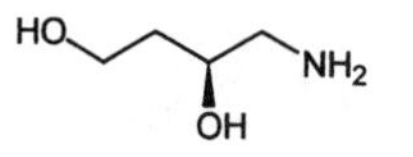

$C_4H_{11}NO_2$
Butane-1,3-diol, 4-amino, (S)-. Chiral building block. *Synthon Chiragenics Corporation.*

**192 (R)-(-)-2-Amino-1-butanol**
5856-63-3 4490(12) 227-476-4

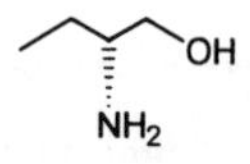

$C_4H_{11}NO$
Butan-1-ol, 2-amino-, (2R)-. L(-)-2-Amino-1-butanol. Listed on TSCA. Chiral auxiliary. bp = 172-174°; d = 0.9470; $n_D^{20}$= 1.4525; $[\alpha]_D^{20}$= - 8.5 ± 1° (c = 2, EtOH); $[\alpha]_D^{22}$= - 9.5° (neat). *Acros Organics nv; Arran Chemical Company Ltd.; Lancaster Synthesis Ltd.; Norse Laboratories; Sigma-Aldrich Fine Chemicals; TCI America; Themis Chemicals Ltd.*

**193 (S)-(+)-2-Amino-1-butanol**
5856-62-2 438(11) 227-475-9

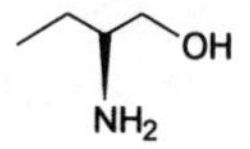

$C_4H_{11}NO$
Butan-1-ol, 2-amino-, (S)-. D(+)-2-Amino-1-butanol. Chiral auxiliary. bp = 172-174°; d = 0.9470; $n_D^{20}$= 1.4525; $[\alpha]_D^{20}$= + 12.5 ± 1° (c = 2, EtOH); $[\alpha]_D^{22}$= + 9.8° (neat). *Acros Organics nv; Arran Chemical Company Ltd.; Lancaster Synthesis Ltd.; Norse Laboratories; Sigma-Aldrich Fine Chemicals; TCI America; Themis Chemicals Ltd.*

**194 (R)-3-Amino-1-butanol**
61477-40-5

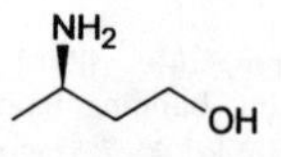

$C_4H_{11}NO$
Chiral building block. *IMI (TAMI) Institute for R&D.*

**195 (S)-3-Amino-1-butanol**
61477-39-2

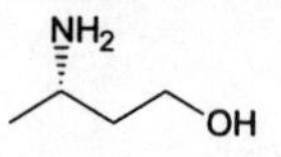

$C_4H_{11}NO$
Chiral building block. *IMI (TAMI) Institute for R&D.*

**196 (R)-Amino-3-buten-1-ol, benzoate salt**
$C_{11}H_{15}NO_3$
Chiral building block. *ChiroTech Technology Ltd.; Lancaster Synthesis Ltd.*

**197 (S)-2-Amino-3-buten-1-ol, benzoate salt**
$C_{11}H_{15}NO_3$
Chiral building block. *ChiroTech Technology Ltd.; Lancaster Synthesis Ltd.*

**198 (R)-(-)-2-Amino-n-butyric acid**
2623-91-8 220-084-4

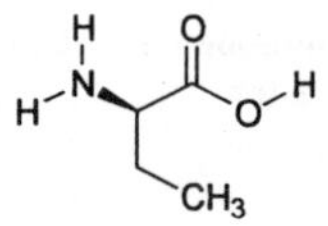

$C_4H_9NO_2$
Butanoic acid, 2-amino-, (2R)-. D-Butyrine; D-Ethylglycine. Chiral building block. mp > 300°; $[\alpha]_D^{20}$= - 7.94° (c = 4, $H_2O$). *Acros Organics nv; Recordati S.p.A.; Sigma-Aldrich Fine Chemicals; TCI America; Yoneyama Yakuhin Kogyo Co., Ltd.*

**199 (S)-(+)-2-Aminobutyric acid**
1492-24-6 439(11) 216-083-3

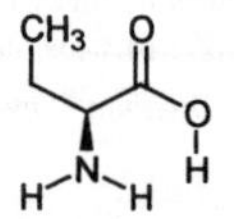

$C_4H_9NO_2$
Butanoic acid, 2-amino-, (2S)-. L-2-Aminobutyric acid; L-Butyrine; L-Ethylglycine; (S)-(+)-2-Amino-n-butyric acid. Listed on TSCA. Chiral intermediate. mp = 280°; [α] = + 20.4° (c = 2, 5N HCl). *Acros Organics nv; Degussa-Huls AG; DSM Fine Chemcials Netherlands; Fischer Chemicals AG; Great Lakes Fine Chemicals; Senn Chemicals AG; Sigma-Aldrich Fine Chemicals; TCI America; Yoneyama Yakuhin Kogyo Co., Ltd.*

**200 (-)-α-Amino-γ-butyrolactone hydrobromide**
15295-77-9

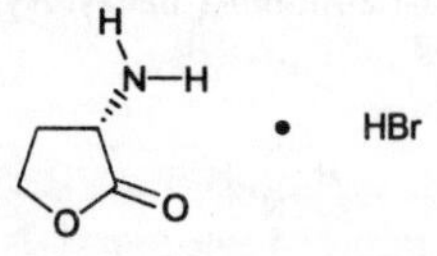

$C_4H_8BrNO_2$
Chiral intermediate. mp = 225°; [α] = - 21.6° (c = 4, $H_2O$). *Acros Organics nv; Sigma-Aldrich Fine Chemicals.*

**201 (R)-(+)-α-Amino-γ-butyrolactone hydrochloride**
104347-13-9

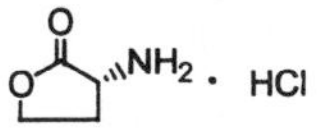

$C_4H_7NO_2$
Chiral building block. mp = 220°; $[\alpha]^{23}$ = + 28° (c = 1, $H_2O$). *Hamari Chemicals; Sigma-Aldrich Fine Chemicals.*

**202 (S)-(-)-α-Amino-γ-butyrolactone hydrochloride**
2185-03-7 218-571-1

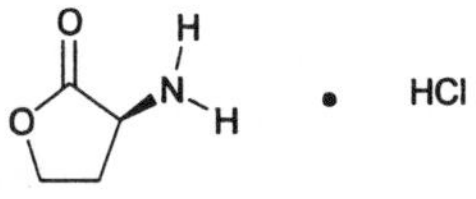

$C_4H_7NO_2$
Furan-2(3H)-one, 3-aminodihydro-, hydrochloride, (S)-. (S)-(-)-3-Aminotetrahydrofuran-2-one hydrochloride; (S)-[Dihydro-2-oxo-3-furyl]ammonium chloride; L-(-)-Homoserine lactone hydrochloride. Chiral intermediate. mp = 220°; -0.22; $[\alpha]_D^{20}$ = - 27.8° (c = 1, $H_2O$). *Acros Organics nv; Hamari Chemicals; Sigma-Aldrich Fine Chemicals; TCI America.*

**203 (-)-α-Amino-γ-butyrolactone hydroiodide**

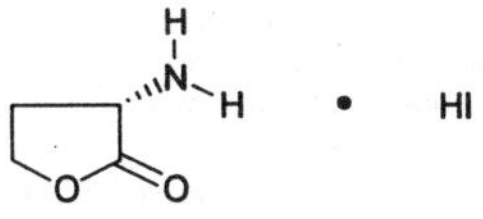

$C_4H_8INO_2$
Chiral intermediate. [α] = - 16° (c = 1, $H_2O$). *Acros Organics nv.*

**204 7-Aminocephalosporanic acid**
957-68-6 453(12) 213-485-0

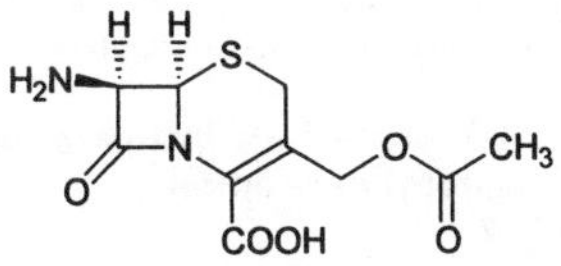

$C_{10}H_{12}N_2O_5S$
Azabicyclo[4.2.0]oct-2-ene-5-thia-2-carboxylic acid, 3-[(acetyloxy)methyl]-7-amino-8-oxo-, (6R,7R)-. 3-Acetoxymethylen-7-amino-8-oxo-5-thia-1-azabicyclo[4.2.0]oct-2-ene-2-carboxylic acid; 7β-Amino-3-acetoxymethylcephemcarboxylic acid. Pharmaceutical or derivative. mp > 300°; $[\alpha]^{19}$ = + 90° (c = 0.5, $KH_2PO_4$/trace NaOH). *Acros Organics nv; Sigma-Aldrich Fine Chemicals.*

**205 Amino-1-(3-chlorophenyl)ethanol**
121652-86-6

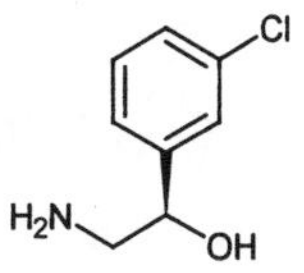

$C_8H_{10}ClNO$
Chiral building block. *Daicel Chemical Ind. Ltd.*

**206 (-)-2-Amino-6-chloropurine riboside**
2004-07-1 217-905-3

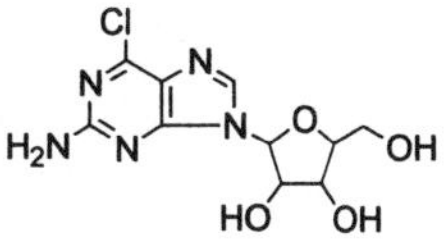

$C_{10}H_{12}ClN_5O_4$
Chiral intermediate. mp = 165-167°; $[\alpha]^{22}$ = - 36.3° (c = 0.6, $H_2O$). *Sigma-Aldrich Fine Chemicals.*

**207 (S)-6-Aminochroman-2-acetic acid**

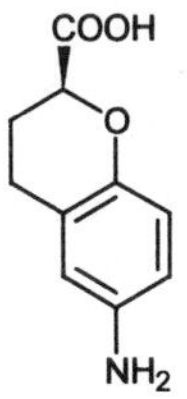

$C_{10}H_{11}NO_3$
Chiral intermediate. *Austin Chemical Company, Inc.*

**208 trans-4-Aminocyclohexanol**

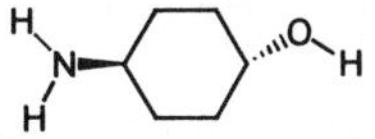

$C_6H_{13}NO$
Cyclohexanol, 4-amino-, trans-. 4-trans-Hydroxycyclohexylamine; trans-1-Amino-4-cyclohexanol. Chiral building block. *Acros Organics nv.*

**209 (S)-(+)-2-Amino-3-cyclohexyl-1-propanol hydrochloride**
117160-99-3

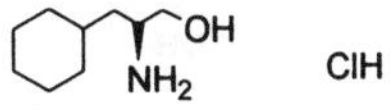

$C_9H_{20}ClNO$
Chiral building block. mp = 230°; $[\alpha]_D^{20}$ = + 2.6° (c = 1, $CH_3OH$). *Sigma-Aldrich Fine Chemicals.*

**210 (-)-(1R,2S)-2-Amino-1-cyclopentane-carboxylic acid**

$C_6H_{11}NO_2$
Chiral building block. *Acros Organics nv.*

**211 (+)-(1S,2R)-2-Amino-1-cyclopentane-carboxylic acid**
64191-14-6

$C_6H_{11}NO_2$
Chiral intermediate. *Acros Organics nv.*

**212 (1R,3S)-3-Aminocyclopentane-carboxylic acid**
$C_6H_{11}NO_2$
Chiral building block. *ChiroTech Technology Ltd.; Lancaster Synthesis Ltd.*

**213 (1S,3R)-3-Aminocyclopentane-carboxylic acid**
71830-07-4
$C_7H_8O_3$
Chiral intermediate. *ChiroTech Technology Ltd.; Lancaster Synthesis Ltd.*

**214 (1R,4S)-4-Amino-2-cyclopentene-1-carboxylic acid**
130468-01-8
$C_6H_9NO_2$
Chiral building block. *ChiroTech Technology Ltd.; Lancaster Synthesis Ltd.*

**215 (1S,4R)-4-Amino-2-cyclopentene-1-carboxylic acid**
102579-72-6
$C_6H_9NO_2$
Chiral building block. *ChiroTech Technology Ltd.; Lancaster Synthesis Ltd.*

**216 1-Amino-1-deoxy-β-D-glucose**

$C_6H_{13}NO_5$
β-D-Glucopyranosylamine. Chiral intermediate. *Senn Chemicals AG.*

**217 β-(1,4)-2-Amino-2-deoxy-D-glucose**
$C_{11}H_{23}NO_3$
Chiral intermediate. *Acros Organics nv.*

**218 2-Amino-2-deoxy-D-mannitol hydrochloride**

· HCl

$C_6H_{15}ClNO_5$
Chiral intermediate. *Senn Chemicals AG.*

**219 L-(+)-2-Amino-6-(O,O'-diethylphosphono)-hexanoic acid**

$C_{10}H_{22}NO_5P$
Chiral intermediate. [α] = + 17.1° (c = 0. 96, 2N HCl). *Acros Organics nv.*

**220 (R)-Amino-1,2-dihydroxypropane hydrochloride**

· HCl

$C_3H_{10}ClNO_2$
Propane-1,2-diol, 3-amino, hydrochloride, (R)-. Amino-1,2-propanediol, hydrochloride. Chiral building block. *Synthon Chiragenics Corporation.*

**221 (S)-3-Amino-1,2-dihydroxypropane hydrochloride**
209849-99-0

· HCl

$C_3H_{10}ClNO_2$
Propane-1,2-diol, 3-amino, hydrochloride, (S)-. (S)-3-Amino-1,2-propanediol hydrochloride. Chiral building block. *Synthon Chiragenics Corporation.*

**222 (S)-3-Amino-1,3-dihydroxypropane isopropylidene acetal**
82954-65-2

$C_6H_{13}NO_2$
Dioxolane-4-methanamine, 1,3-, 2,2-dimethyl-, (S)-. (S)-4-Aminomethyl-2,2-dimethyl-1,3-dioxolane. Chiral intermediate. *Synthon Chiragenics Corporation.*

**223 (S)-2-Amino-3-(dimethylamino)propanoic acid dihydrochloride monohydrate**
31697-39-9

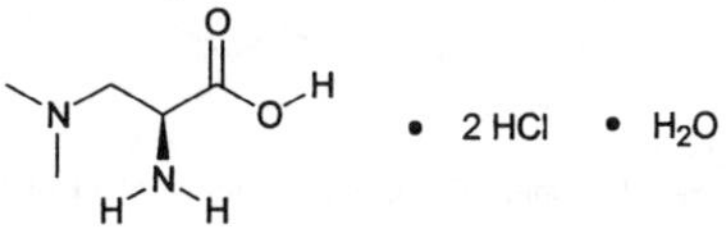

$C_5H_{16}Cl_2N_2O_3$
Chiral building block. [α] = + 16° (c = 1.5, 6N HCl). *Acros Organics nv.*

**224 Aminomethyl-2,2-dimethyl-1,3-dioxolane**
103883-30-3

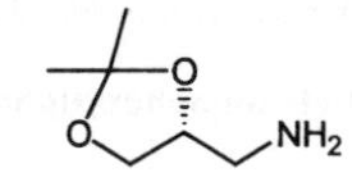

$C_6H_{13}NO_2$
Dioxolane-4-methanamine, 1,3-, 2,2-dimethyl-, (R)-. Chiral intermediate. *Synthon Chiragenics Corporation.*

**225 (1R,2R)-(+)-2-Amino-1,2-diphenylethanol**
$C_{14}H_{15}NO$
Chiral building block. mp = 143-147°; $[\alpha]_D^{20}$ = + 7° (c = 0.6, EtOH). *Sigma-Aldrich Fine Chemicals.*

**226 (1R,2S)-(-)-2-Amino-1,2-diphenylethanol**
23190-16-1 449(11)

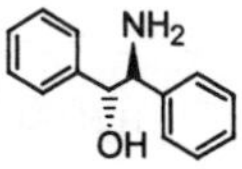

$C_{14}H_{15}NO$
Chiral building block. mp = 144°; $[\alpha]^{25}$ = - 7.0° (c = 0.6, EtOH). *Sigma-Aldrich Fine Chemicals; TCI America; Yamakawa Chemical Industry Co. Ltd.*

**227 (1S,2R)-(+)-2-Amino-1,2-diphenylethanol**
23364-44-5 449(11)

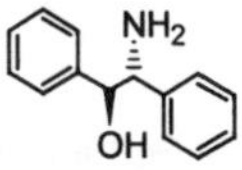

$C_{14}H_{15}NO$
Resolving agent; Chiral building block. mp = 144°; $[\alpha]^{25}$ = + 7.0° (c = 0.6, EtOH). *Austin Chemical Company, Inc.; SEAC; Sigma-Aldrich Fine Chemicals; TCI America; Yamakawa Chemical Industry Co. Ltd.*

**228 (1S,2S)-(-)-2-Amino-1,2-diphenylethanol**
$C_{14}H_{15}NO$
Resolving agent; Chiral building block. mp = 140-144°; $[\alpha]_D^{20}$ = - 7° (c = 0.6, EtOH). *Sigma-Aldrich Fine Chemicals.*

**229 (S)-(-)-2-Amino-1,1-diphenyl-1-propanol**
78603-91-5

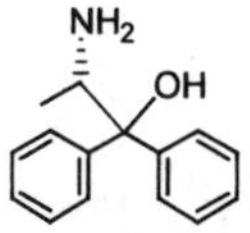

$C_{15}H_{17}NO$
Chiral building block. mp = 100-102°; $[\alpha]^{22}$ = - 85° (c = 1, $CHCl_3$). *Sigma-Aldrich Fine Chemicals.*

**230 (2S,3S)-2-Amino-3-ethoxybutanoic acid hydrochloride**

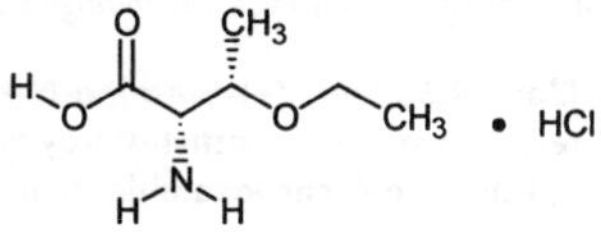

$C_6H_{13}NO_3$
Chiral building block. [α] = + 18° (c = 1.5, 6N HCl). *Acros Organics nv.*

**231 (S)-(2-Aminoethyl)-L-cysteine hydrochloride**
4099-35-8 223-862-1

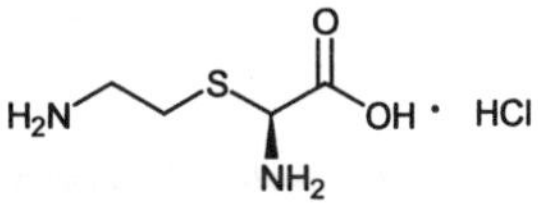

$C_5H_{13}ClN_2O_2S$
Alanine, 3-[(2-aminoethyl)thio]-, monohydrochloride, L-. L-Cysteine, S-(2-aminoethyl)-, monohydrochloride; Thialysine hydrochloride. Chiral intermediate. *Pharmasyn, Inc.*

**232 (R)-(-)-(1-Aminoethyl)phosphonic acid**
60687-36-7

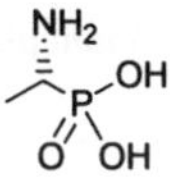

$C_2H_8NO_3P$
Resolving agent. mp = 290°; $[\alpha]_D^{20}$ = - 4.8° (c = 5, $H_2O$). *Sigma-Aldrich Fine Chemicals.*

**233 (S)-(+)-(1-Aminoethyl)phosphonic acid**
66068-76-6

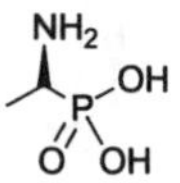

$C_2H_8NO_3P$
Resolving agent. mp = 290°; $[\alpha]_D^{20}$ = + 4.8° (c = 5, $H_2O$). *Sigma-Aldrich Fine Chemicals.*

**234 (3aR,4S,6R,6aS)-(-)-6-Amino-N-ethyl-tetrahydro-2,2-dimethyl-4H-cyclopenta-1,3-dioxole-4-carboxamide, benzoate salt**

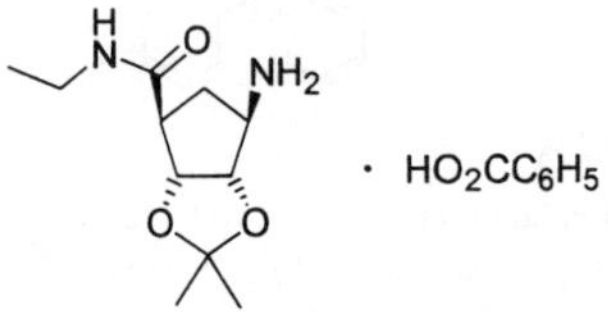

$C_{18}H_{26}N_2O_5$
Chiral intermediate. *ChiroTech Technology Ltd.*

**235 (3aS,4R,6S,6aR)-(+)-6-Amino-N-ethyl-tetrahydro-2,2-dimethyl-4H-cyclopenta-1,3-dioxole-4-carboxamide, benzoate salt**

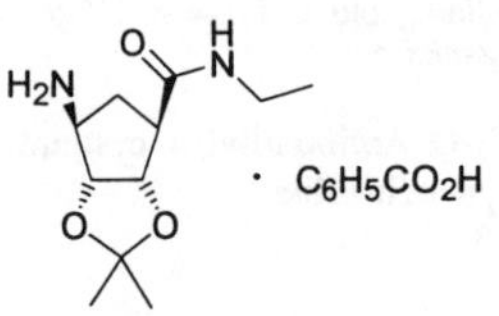

$C_{18}H_{26}N_2O_5$
Chiral intermediate. *ChiroTech Technology Ltd.*

**236 (R)-(-)-2-Aminoheptane**
6240-90-0

$C_7H_{17}N$
Heptan-2-amine, (2R)-. (-)-2-Heptylamine; Hexylamine, 1-methyl-, (R)-(-)-. Chiral building block. bp = 142-144°; d = 0.7660; $n_D^{20}$ = 1.4190; $[\alpha]_D^{20}$ = - 7.2 ± 0.5° (neat). *Kiralchem Ltd.; Lancaster Synthesis Ltd.; Norse Laboratories; Sigma-Aldrich Fine Chemicals.*

**237 (S)-(+)-2-Aminoheptane**
44745-29-1

$C_7H_{17}N$
Heptan-2-amine, (S)-. (+)-2-Heptylamine. Chiral building block. bp = 142-144°; d = 0.7660; $n_D^{20}$ = 1.4190; $[\alpha]_D^{20}$ = + 7.2 ± 0.5° (neat). *Kiralchem Ltd.; Lancaster Synthesis Ltd.; Norse Laboratories; Sigma-Aldrich Fine Chemicals.*

**238 (R)-(-)-2-Aminohexane**
70095-40-8

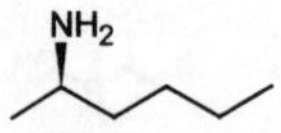

$C_6H_{15}N$
Hexan-2-amine, (R)-. Chiral building block. *Chiragene, Inc.*

**239 (S)-(-)-2-Aminohexane**
70492-67-0

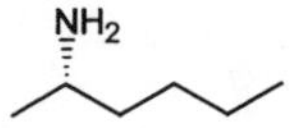

$C_6H_{15}N$
Hexan-2-amine, (S)-. Chiral building block. *Chiragene, Inc.*

**240 (R)-2-Amino-1-hexanol**

$C_6H_{15}NO$
Chiral building block. *Sigma-Aldrich Fine Chemicals.*

**241 (S)-2-Amino-1-hexanol**

$C_6H_{15}NO$
Chiral building block. *Sigma-Aldrich Fine Chemicals.*

**242 (1R)-(-)-(1-Aminohexyl)phosphonic acid**

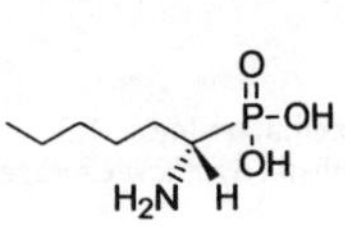

$C_6H_{16}NO_3P$
Chiral ligand. mp = 270-276°; $[\alpha]_D^{20}$ = - 25° (c = 5, 1N NaOH). *Sigma-Aldrich Fine Chemicals.*

**243 (1S)-(+)-(1-Aminohexyl)phosphonic acid**

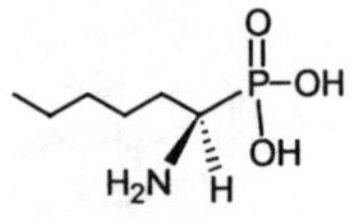

$C_6H_{16}NO_3P$
Resolving agent. mp = 270-276°; $[\alpha]_D^{20}$ = + 25° (c = 5, 1N NaOH). *Sigma-Aldrich Fine Chemicals.*

**244 (R)-Amino-3-hydroxybutanoic acid**
7013-07-2

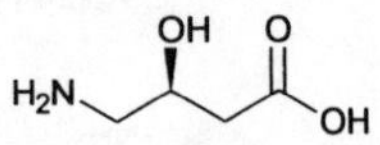

$C_4H_9NO_3$
Butanoic acid, 4-amino-3-hydroxy-, (R)-. (-)-β-Hydroxy-GABA; (R)-GABOB. Chiral building block. *Synthon Chiragenics Corporation.*

**245 (S)-4-Amino-3-hydroxybutanoic acid**
7013-05-0

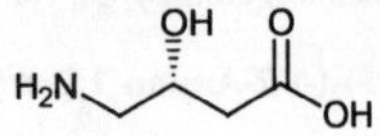

$C_4H_9NO_3$
Butanoic acid, 4-amino-3-hydroxy-, (S)-. (S)-(+)-3-

Hydroxy-GABA. Chiral building block. *Synthon Chiragenics Corporation.*

**246 (2S,3R)-2-Amino-3-hydroxybutanol**
515-93-5

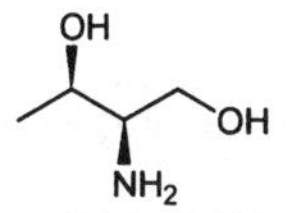

$C_4H_{11}NO_2$
L-Threoninol. Chiral building block. *Isochem; Senn Chemicals AG.*

**247 (S)-(-)-4-Amino-2-hydroxybutyric acid**
40371-51-5

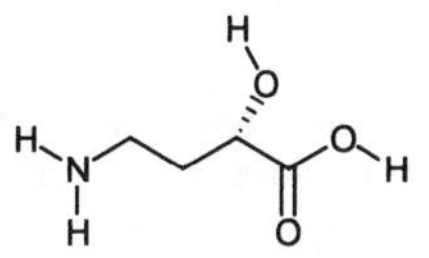

$C_4H_9NO_3$
Chiral building block. mp = 200-203°; $[\alpha]^{23}$ = - 30° (c = 1, $H_2O$). *Acros Organics nv; Sigma-Aldrich Fine Chemicals; TCI America.*

**248 (R)-(-)-2-Amino-3-hydroxy-3-methyl-butanoic acid**
2280-27-5

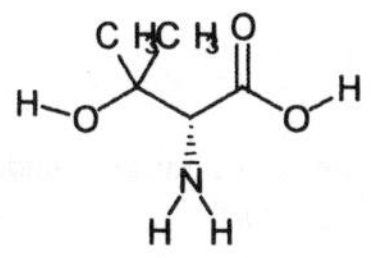

$C_5H_{11}NO_3$
Chiral building block. [α] = - 11.4° (c = 2, 6N HCl). *Acros Organics nv.*

**249 (S)-(+)-2-Amino-3-hydroxy-3-methyl-butanoic acid**

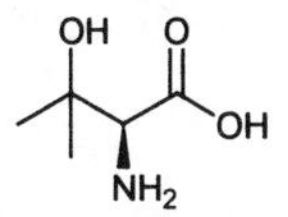

$C_5H_{11}NO_3$
Chiral building block. [α] = + 11° (c = 1.92, 6N HCl). *Acros Organics nv.*

**250 (3S,4S)-4-Amino-3-hydroxy-6-methylheptanoic acid**

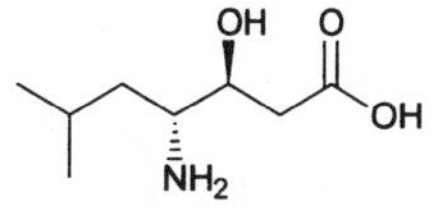

$C_8H_{17}NO_3$
Chiral intermediate. *Austin Chemical Company, Inc.; Senn Chemicals AG.*

**251 (R,S)-α-Amino-3-hydroxy-5-methyl-4-isoxazolepropionic acid**
74341-63-2

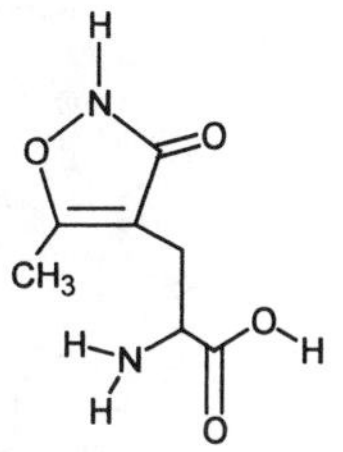

$C_7H_{10}N_2O_4$
Chiral intermediate. *Acros Organics nv.*

**252 (2R,3S)-(-)-2-Amino-3-hydroxy-4-methyl-pentanoic acid**
87421-23-6

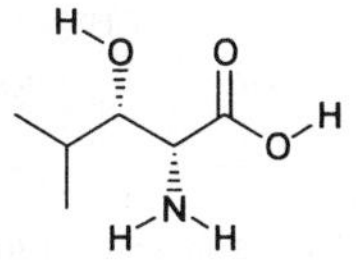

$C_6H_{13}NO_3$
Chiral intermediate. [α] = - 18.4° (c = 2.0, 6N HCl). *Acros Organics nv.*

**253 (2S,3R)-(+)-2-Amino-3-hydroxy-4-methyl-pentanoic acid**
10148-71-7

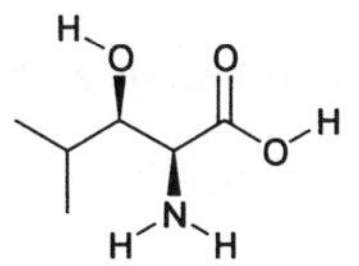

$C_6H_{13}NO_3$
threo-L-β-Hydroxyleucine. Chiral intermediate. [α] = + 18.5° (c = 1, 6N HCl). *Acros Organics nv.*

**254 (2R,3S)-3-Amino-2-hydroxy-4-phenyl-butanoic acid**
76647-67-1

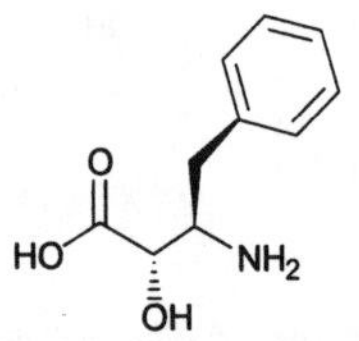

$C_{10}H_{13}NO_3$
Chiral building block. *Austin Chemical Company, Inc.; Kaneka Corporation; Schweizerhall Pharma.*

**255 N-(2S,3R)-3-Amino-2-hydroxy-4-phenyl-butyrylleucine**
9973(12) 261-529-2

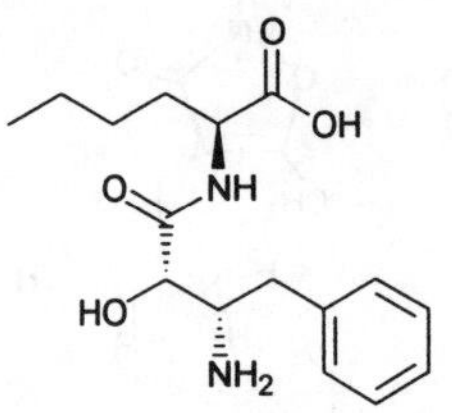

$C_{16}H_{24}N_2O_4$
Chiral intermediate. mp = 245°; $[\alpha]_D^{20}$ = - 11° (c = 1, 1N NaOH). *Acros Organics nv; Senn Chemicals AG; Sigma-Aldrich Fine Chemicals.*

**256 (R)-Amino-2-hydroxy-1-trityloxy-propane**

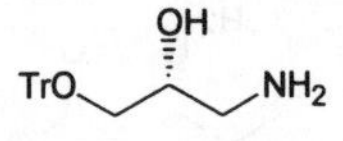

$C_{22}H_{23}NO_2$
Propane-1,3-diol, 3-amino-, 1-trityl-, (R)-. Chiral intermediate. *Synthon Chiragenics Corporation.*

**257 (S)-3-Amino-2-hydroxy-1-trityloxy-propane**

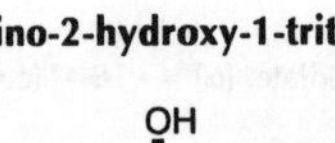

$C_{22}H_{23}NO_2$
Propane-1,3-diol, 3-amino-, 1-trityl-, (S)-. Chiral intermediate. *Synthon Chiragenics Corporation.*

**258 (R)-Amino-3-hydroxy-1-trityloxy-butane**

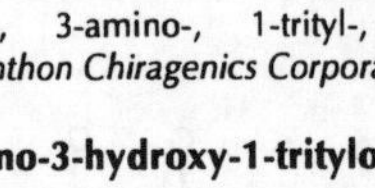

$C_{23}H_{25}NO_2$
Butane-1,3-diol, 4-amino-, 1-trityl-, (R)-. Chiral intermediate. *Synthon Chiragenics Corporation.*

**259 (S)-4-Amino-3-hydroxy-1-trityloxy-butane**

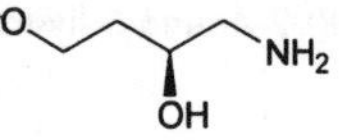

$C_{23}H_{25}NO_2$
Butane-1,3-diol, 4-amino-, 1-trityl-, (S)-. Chiral intermediate. *Synthon Chiragenics Corporation.*

**260 (R)-(-)-1-Aminoindan**
10277-74-4

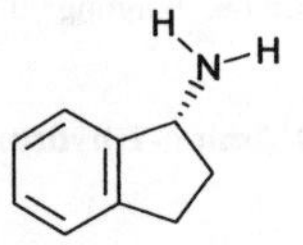

$C_9H_{11}N$
(R)-(-)-1-Indanamine. Chiral intermediate. mp = 15°; $bp_8$ = 96-97°; d = 1.038; $n_D^{20}$ = 1.5620; $[\alpha]_D^{20}$ = - 16.5 ± 1° (c = 1.5, $CH_3OH$). *Acros Organics nv; BASF Aktiengesellschaft; Chiragene, Inc.; Lancaster Synthesis Ltd.; Sigma-Aldrich Fine Chemicals.*

**261 (S)-(+)-1-Aminoindan**
61341-86-4

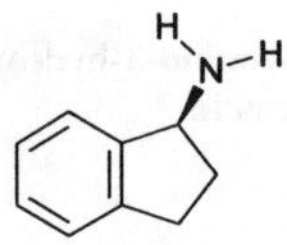

$C_9H_{11}N$
(S)-(+)-1-Indanamine. Chiral intermediate. $bp_{11}$ = 96-97°; d = 1.038; $n_D^{20}$ = 1.5620; $[\alpha]_D^{20}$ = + 16.5 ± 1° (c = 1.5, $CH_3OH$). *Acros Organics nv; Austin Chemical Company, Inc.; BASF Aktiengesellschaft; Chiragene, Inc.; FineTech, Ltd.; Lancaster Synthesis Ltd.; Sigma-Aldrich Fine Chemicals.*

**262 cis-(1R,2S)-Aminoindan-2-ol**
136030-00-7

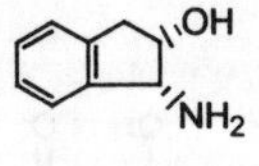

$C_9H_{11}NO$.
(1R,2S)-1-Amino-2-indanol. Chiral intermediate. mp = 117-121°; $[\alpha]_D$ = + 42.5° (c = 1, $CH_3OH$); $[\alpha]^{22}$ = + 63° (c = 0.2, $CHCl_3$). *Acros Organics nv; Sigma-Aldrich Fine Chemicals; Strem Chemicals, Inc.*

**263 cis-(1S,2R)-Aminoindan-2-ol**
126456-43-7

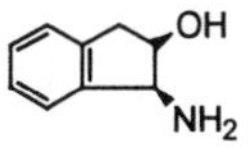

$C_9H_{11}NO$
Chiral building block. mp = 117-121°; $[\alpha]_D^{20}$ = - 61° (c = 0.5, $CHCl_3$). *Acros Organics nv; Sigma-Aldrich Fine Chemicals; Strem Chemicals, Inc.*

**264 2-Amino-6-mercaptopurine-9-D-riboside**
85-31-4 9474(12) 201-597-2

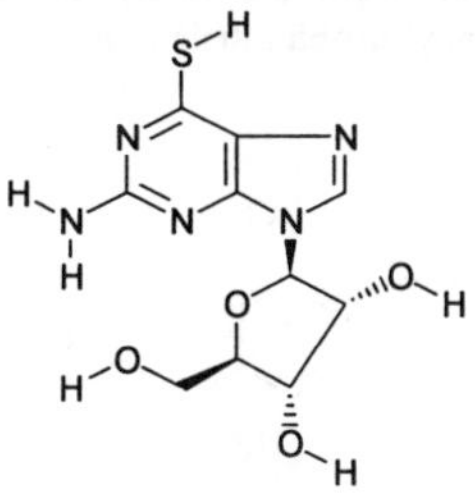

$C_{10}H_{13}N_5O_4S$
Purine-(6H)-6-thione, 2-amino-1,9-dihydro-9-β-D-ribofuranosyl-; 2-Amino-6-mercaptopurine ribonucleoside; 6-Mercaptoguanine riboside; 6-Mercaptoguanosine; 6-Thioguanine riboside; 6-Thioguanosine; 9-β-D-Ribofuranosylthioguanine; 9H-Purine-6(1H)-thione, 2-Amino-9-β-D-ribofuranosyl-; Guanosine, 6-thio-; NSC-29422. Chiral intermediate. mp = 230-231°; $[\alpha]^{22}$ = - 69.9° (c = 1.3, 0.1N NaOH). *Acros Organics nv; Sigma-Aldrich Fine Chemicals.*

**265 (2S,3S)-2-Amino-3-methoxybutanoic acid**
104195-80-4

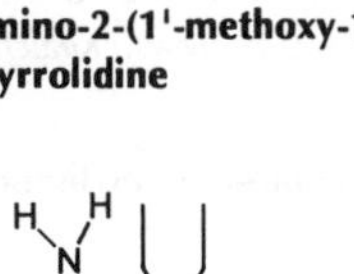

$C_5H_{11}NO_3$
Chiral building block. [α] = + 27.7° (c = 0.6, 6N HCl). *Acros Organics nv.*

**266 (S)-(-)-Amino-2-(1'-methoxy-1'-ethylpropyl)pyrrolidine**
118535-62-9

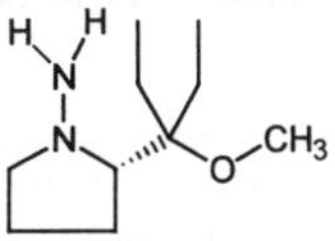

$C_{10}H_{22}N_2O$
SAEP. Chiral intermediate. *Acros Organics nv.*

**267 (S)-(-)-Amino-2-(1'-methoxy-1'-methylethyl)pyrrolidine**
118535-61-8

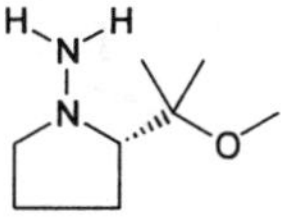

$C_8H_{18}N_2O$
SADP. Chiral intermediate. *Acros Organics nv.*

**268 (R)-(+)-1-Amino-2-(methoxymethyl)-pyrrolidine**
72748-99-3

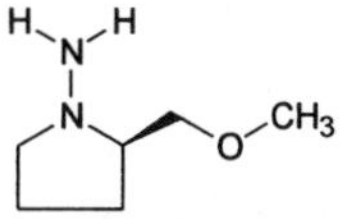

$C_6H_{14}N_2O$.
Pyrrolidin-1-amine, 2-(methoxymethyl)-, (2R)-. (R)-1-Amino-2-methoxymethylpyrrolidine; RAMP. Chiral auxiliary; Chiral intermediate. $bp_2$ = 45-47°; d = 0.970; $[\alpha]_D^{20}$ = + 83.0 ± 0.5° (neat). *Acros Organics nv; Kiralchem Ltd.; Oxford Asymmetry International plc; Sigma-Aldrich Fine Chemicals; TCI America.*

**269 (S)-(-)-1-Amino-2-(methoxymethyl)-pyrrolidine**
59983-39-0

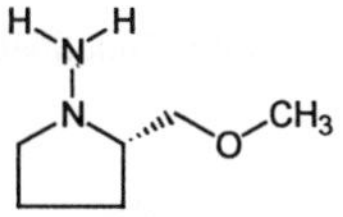

$C_6H_{14}N_2O$
(S)-(-)-2-(Methoxymethyl)-1-aminopyrrolidine; SAMP. Chiral auxiliary; Chiral intermediate. $bp_2$ = 45-47°; d = 0.970; $[\alpha]_D^{20}$ = - 83.0 ± 0.5° (neat). *Acros Organics nv; Kiralchem Ltd.; Oxford Asymmetry International plc; Sigma-Aldrich Fine Chemicals; TCI America.*

**270 (1S,2S)-(+)-2-Amino-3-methoxy-1-phenyl-1-propanol**
51594-34-4

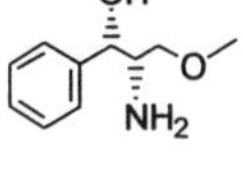

$C_{10}H_{15}NO_2$
Chiral building block. mp = 51-53°; $[\alpha]^{18}$ = + 25° (c = 10.6, $CHCl_3$). *Sigma-Aldrich Fine Chemicals.*

**271 (S)-2-Amino-3-methoxypropanoic acid**
32620-11-4

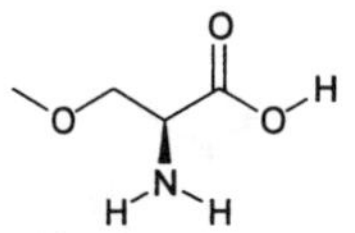

$C_4H_9NO_3$
Chiral building block. [α] = + 13° (c = 1, 6N HCl). *Acros Organics nv.*

**272 (S)-2-Amino-3-methoxypropanoic acid hydrochloride**

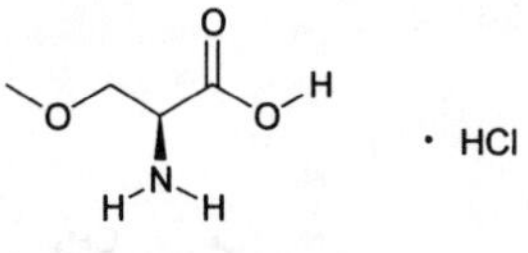

$C_4H_{10}ClNO_3$
Chiral building block. *Acros Organics nv.*

**273 (R)-(-)-2-Amino-2-methylbutanedioic acid**
14603-76-0

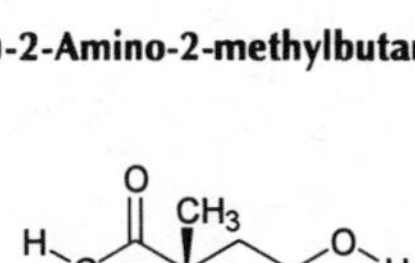

$C_5H_9NO_4$
Chiral building block. [α] = - 50.8° (c = 1, $H_2O$). *Acros Organics nv.*

**274 (S)-(+)-2-Amino-2-methylbutanedioic acid**
3227-17-6

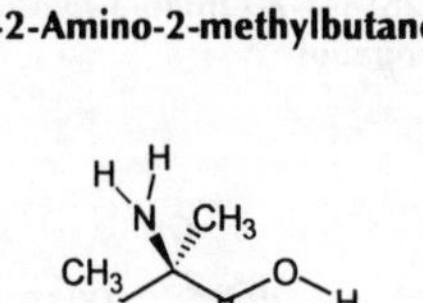

$C_5H_9NO_4$
Chiral building block. [α] = + 48.9° (c = 0.7,2N HCl). *Acros Organics nv.*

**275 (R)-(-)-2-Amino-2-methylbutanoic acid**
3059-97-0

$C_5H_{11}NO_2$
Chiral building block. [α] = - 10.4° (c = 5, $H_2O$). *Acros Organics nv.*

**276 (S)-(+)-2-Amino-2-methylbutanoic acid hydrate**
595-40-4

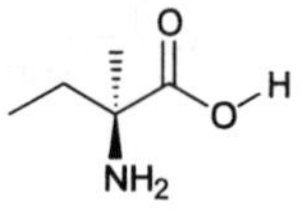

$C_5H_{11}NO_2$
Chiral building block. [α] = + 10.4° (c = 5, $H_2O$). *Acros Organics nv.*

**277 L-(+)-2-Amino-2-methyl-3-(3,4-dimethoxyphenyl)propionitrile hydrochloride**

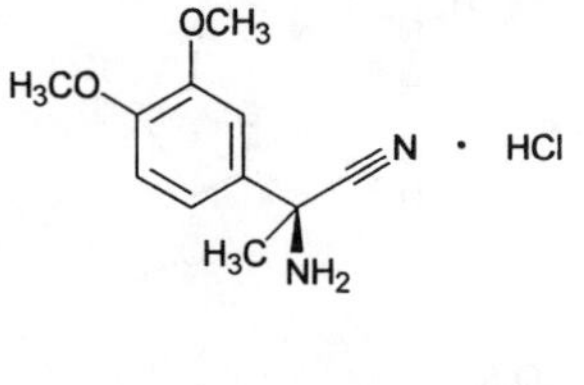

$C_{12}H_{17}ClN_2O_2$
Chiral intermediate. *Alfa Chemicals Italiana srl.*

**278 (R)-(+)-2-Aminomethyl-1-ethylpyrrolidine**
$C_7H_{16}N_2$
Pyrrolidine, 2-(aminomethyl)-1-ethyl-, (R)-. (R)-(+)-1-Ethyl-2-aminomethylpyrrolidine; (R)-2-(Aminomethyl)-N-ethylpyrrolidine. Chiral building block. *TCI America.*

**279 (S)-(-)-2-Aminomethyl-1-ethylpyrrolidine**
22795-99-9
$C_7H_{16}N_2$
(S)-(-)-1-Ethyl-2-aminomethylpyrrolidine. Chiral building block. *TCI America.*

**280 (R)-Amino-6-methylheptane**
70419-11-3 274-599-4
$C_8H_{19}N$
Heptan-2-amine, 6-methyl-, (R)-. 6-Methyl-(2R)-heptylamine; (R)-1,5-Dimethylhexylamine. Chiral building block. bp = 154-156°; d = 0.7670; $n_D^{20}$ = 1.4215; $[\alpha]_D^{20}$ = - 7.3 ± 0.2° (neat). *Kiralchem Ltd.; Lancaster Synthesis Ltd.*

**281 (S)-2-Amino-6-methylheptane**
70419-10-2 274-598-9
$C_8H_{19}N$
Heptan-2-amine, 6-methyl-, (S)-. 6-Methyl-(2S)-heptylamine; (S)-1,5-Dimethylhexylamine. Chiral building block. bp = 154-156°; d = 0.7670; $n_D^{20}$ = 1.4215; $[\alpha]_D^{20}$ = + 7.3 ± 0.2° (neat). *Kiralchem Ltd.; Lancaster Synthesis Ltd.*

**282 (R)-(-)-2-Amino-2-methyl-3-hydroxy-propanoic acid**
81132-44-7

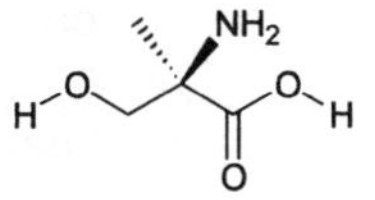

$C_4H_9NO_3$
Chiral building block. *Acros Organics nv.*

**283 (S)-(+)-2-Amino-2-methyl-3-hydroxy-propanoic acid**
16820-18-1

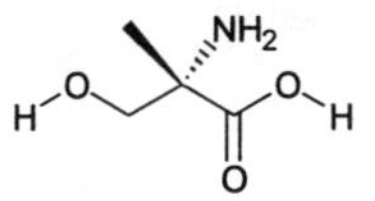

$C_4H_9NO_3$
Chiral building block. *Acros Organics nv.*

**284 D-(-)-threo-2-Amino-1-(4-methyl-mercaptophenyl)-1,3-propanediol**

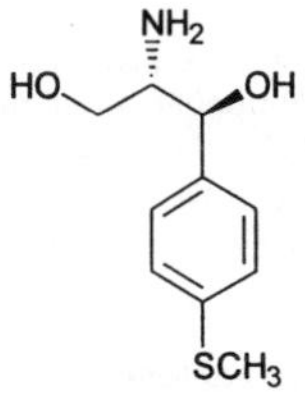

$C_{10}H_{15}NO_2$
Chiral intermediate. *Austin Chemical Company, Inc.*

**285 (2S,4R)-(+)-2-Amino-4-methyl-pentanedioic acid**

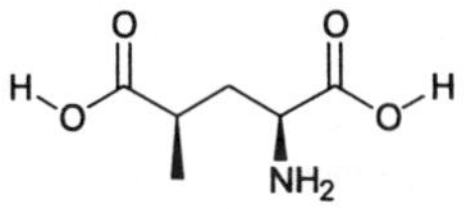

$C_6H_{11}NO_4$
Chiral building block. [α] = + 23.3° (c = 1, 6N HCl). *Acros Organics nv.*

**286 (R)-(-)-2-Amino-3-methyl-1-pentanol**

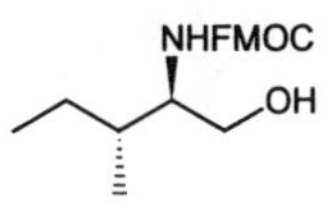

$C_6H_{15}NO$
Pentan-1-ol, 2-amino-3-methyl-, (2R,3R)-.
D-(-)-Isoleucinol. Chiral building block. *Omega Chemical Company Inc.*

**287 (R)-(-)-2-Amino-4-methyl-1-pentanol**
53448-09-2

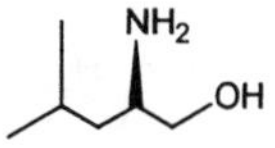

$C_6H_{15}NO$
D-(-)-Leucinol. Chiral building block. $bp_{768}$ = 198-200°; d = 0.917; n = 14496; $[\alpha]_D^{20}$ = - 4° (c = 9, EtOH). *Acros Organics nv; Omega Chemical Company Inc.; Sigma-Aldrich Fine Chemicals.*

**288 (S)-(-)-2-Amino-2-methyl-4-pentenoic acid monohydrate**
96886-55-4

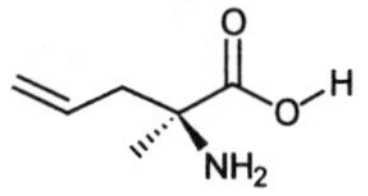

$C_6H_{11}NO_2$
Chiral intermediate. [α] = - 17.9° (c = 1, 6 N HCl). *Acros Organics nv.*

**289 (S)-(+)-2-Amino-2-methyl-3-phenyl-propanoic acid, monohydrate**
23239-35-2

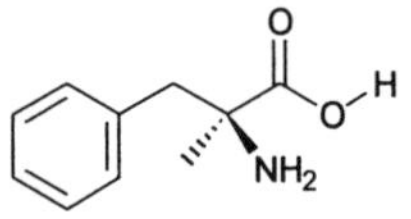

$C_{10}H_{13}NO_2$
Chiral building block. [α] = + 4.2° (c = 0.5 , 1N HCl). *Acros Organics nv.*

**290 (1R)-(+)-(1-Amino-2-methylpropyl)-phosphonic acid**
66254-56-6

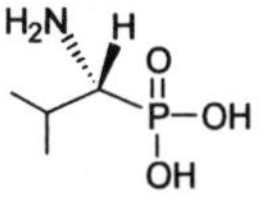

$C_4H_{12}NO_3P$
Chiral ligand. mp = 272-277°; $[\alpha]_D^{20}$ = + 1.0° (c = 1, 1N NaOH). *Sigma-Aldrich Fine Chemicals.*

**291 (1S)-(-)-(1-Amino-2-methylpropyl)-phosphonic acid**
66254-55-5

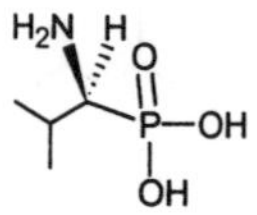

$C_4H_{12}NO_3P$
Resolving agent. mp = 272-277°; $[\alpha]_D^{20}$ = - 1.0° (c = 1, 1N NaOH). *Sigma-Aldrich Fine Chemicals.*

**292 (S)-(+)-2-(Aminomethyl)pyrrolidine**
69500-64-7

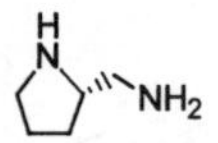

$C_5H_{12}N_2$
Chiral building block. $bp_{11}$ = 65°; d = 0.933; n = 1.482; $[\alpha]_D^{20}$ = + 20° (c = 1, $CHCl_3$). *Sigma-Aldrich Fine Chemicals.*

**293 (R)-(+)-2-Amino-methylthio-1-butanol**
87206-44-8

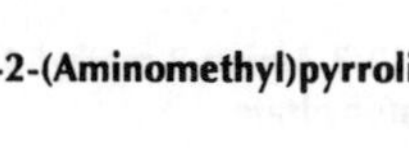

$C_5H_{13}NOS$
Butan-1-ol, 2-amino-4-(methylthio)-, (2R)-. D-Methioninol. Chiral building block. mp = 31-36°; d = 1.095; $[\alpha]^{23}$ = + 12° (neat). *Omega Chemical Company Inc.; Sigma-Aldrich Fine Chemicals.*

**294 [R-(R,R)]-2-Amino-1-[4-(methylthio)-phenyl]-1,3-propanediol**
23150-35-8 245-460-5

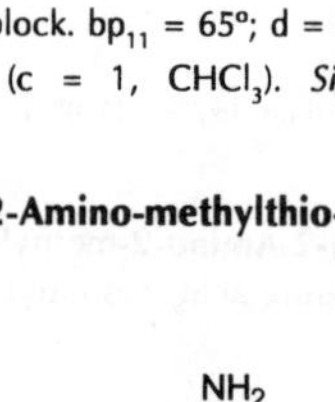

$C_{10}H_{15}NO_2S$
Propane-1,3-diol, 2-amino-1-[4-(methylthio)phenyl]-, [R-(R*,R*)]-. Chiral intermediate. *Zambon Group SpA.*

**295 [S-(R,R)]-2-Amino-1-[4-(methylthio)-phenyl]-1,3-propanediol**
16854-32-3 240-878-4

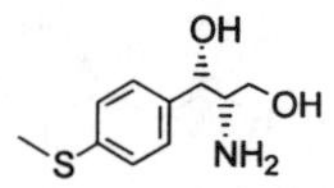

$C_{10}H_{15}NO_2S$
Propane-1,3-diol, 2-amino-1-[4-(methylthio)phenyl]-, (1S,2S)-. Thiomicamine. Chiral intermediate. mp = 151-154°; $[\alpha]^{25}$ = + 25° (c = 1, EtOH). *Sigma-Aldrich Fine Chemicals; Zambon Group SpA.*

**296 D-(-)-threo-2-Amino-1-(4-nitrophenyl)-1,3-propanediol**
716-61-0 211-938-7

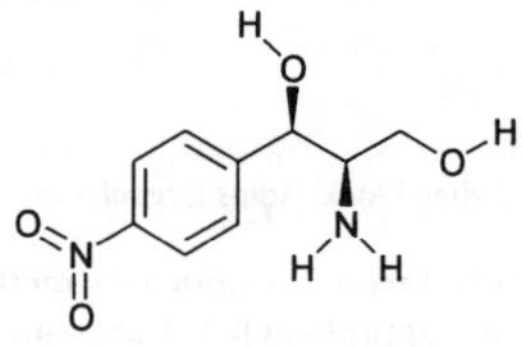

$C_9H_{12}N_2O_4$
Propane-1,3-diol, 2-amino-1-(4-nitrophenyl)-, (1R,2R)-. (-)-threo-1-(p-Nitrophenyl)-2-amino-1,3-propanediol; Chloramphenicol D base; Levoamine. Chiral intermediate. mp = 163°; $[\alpha]^{25}$ = - 30° (c = 1, 6N HCl). *Acros Organics nv; Sigma-Aldrich Fine Chemicals; TCI America.*

**297 (R)-Aminooctane**
34566-05-7
$C_8H_{19}N$
Octan-2-amine, (R)-. (R)-1-Methylheptylamine; (2R)-(-)-Octylamine. Chiral building block. bp = 163-165°; d = 0.7700; $n_D^{20}$ = 1.4235. *Lancaster Synthesis Ltd.*

**298 (S)-2-Aminooctane**
34566-04-6
$C_8H_{19}N$
Octan-2-amine, (S)-.
(S)-(+)-2-Octylamine; (S)-1-Methylheptylamine. Chiral building block. bp = 163-165°; d = 0.7700; $n_D^{20}$ = 1.4235. *Lancaster Synthesis Ltd.*

**299 6-Aminopenicillanic acid**
551-16-6 478(12) 208-993-4

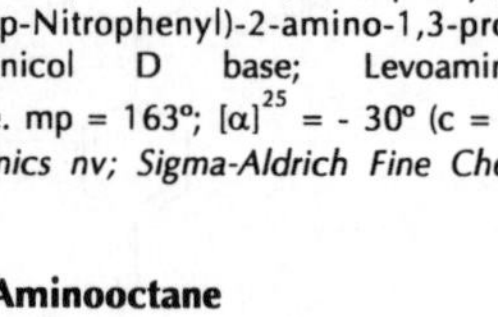

$C_8H_{12}N_2O_3S$
Azabicyclo[3.2.0]heptane-4-thia-2-carboxylic acid, 6-amino-3,3-dimethyl-7-oxo-, (2S,5R,6R)-; 2S-(2α,5α,6β)]-6-Amino-3,3-dimethyl-7-oxo-4-thia-1-azabicyclo-[3.2.0]-

heptane-2-carboxylic acid; 6-APA; 6-APS; Phenacyl 6-aminopenicillinate. Listed on TSCA. Pharmaceutical or derivative. mp = 198-200°; $[\alpha]^{22}$ = + 276.3° (c = 1.2, 0.1N HCl). *Acros Organics nv; Sigma-Aldrich Fine Chemicals.*

**300 (S)-(+)-2-Aminopentane**
545542-13-1

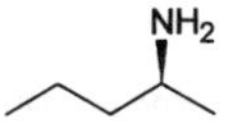

$C_5H_{13}N$
Chiral building block. *Norse Laboratories.*

**301 (R)-2-Amino-1-pentanol**
80696-30-6
$C_5H_{13}NO$
Chiral building block. *Sigma-Aldrich Fine Chemicals.*

**302 (S)-2-Amino-1-pentanol**
22724-81-8
$C_5H_{13}NO$
Chiral building block. *Sigma-Aldrich Fine Chemicals.*

**303 (S)-(-)-2-Amino-4-pentenoic acid**
16338-48-0

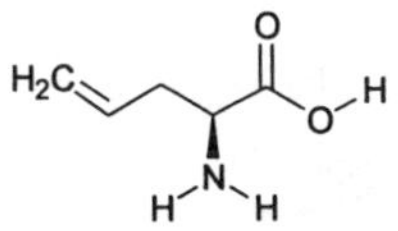

$C_5H_9NO_2$
L-α-Allylglycine. Chiral building block. mp = 283°; [α] = - 6.4° (c = 2.1, 6 N HCl); $[\alpha]_D^{20}$ = - 36° (c = 4, $H_2O$). *Acros Organics nv; Sigma-Aldrich Fine Chemicals.*

**304 (2R)-(-)-1-Amino-3-phenoxy-2-propanol**

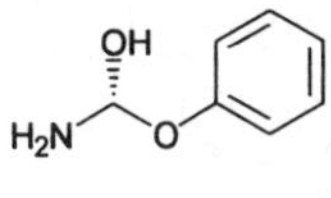

$C_7H_9NO_2$
Chiral building block. *Austin Chemical Company, Inc.*

**305 (2S)-(+)-1-Amino-3-phenoxy-2-propanol**

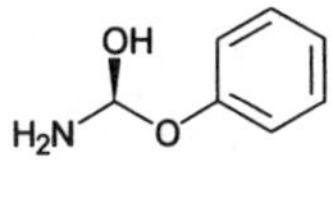

$C_7H_9NO_2$
Chiral building block. *Austin Chemical Company, Inc.*

**306 4-Aminophenyl-2-acetamido-2-deoxy-β-D-glucopyranoside**

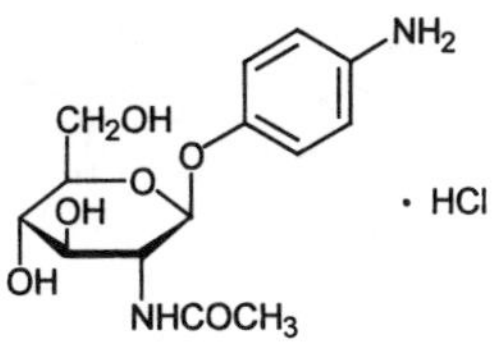

$C_{14}H_{21}ClN_2O_6$
Chiral intermediate. *Senn Chemicals AG.*

**307 L-(4-Aminophenyl)alanine**

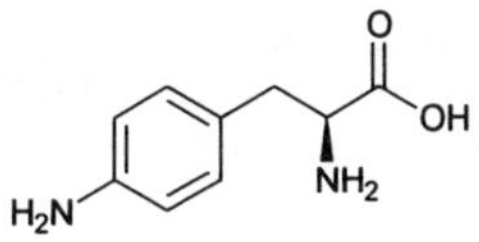

$C_9H_{12}N_2O_2$
Alanine, 3-(p-aminophenyl)-, L-. L-Phenylalanine, 4-amino-. Chiral building block. $[\alpha]_D^{20}$ = - 36° (c = 2, $H_2O$). *Fischer Chemicals AG; Great Lakes Fine Chemicals; Sigma-Aldrich Fine Chemicals.*

**308 (+)-4-Aminophenylalanine hydrate**
102281-45-8

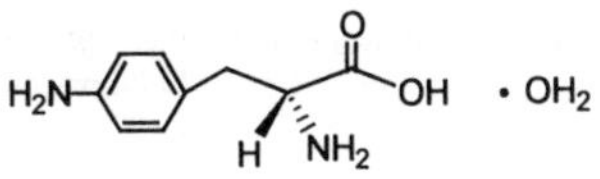

$C_9H_{12}N_2O_2$
Chiral building block. mp = 268-270°; $[\alpha]_D^{20}$ = + 37° (c = 2, $H_2O$). *Sigma-Aldrich Fine Chemicals.*

**309 4-Amino-L-phenylalanine hydrochloride hemihydrate**
24286-13-3 263-388-2

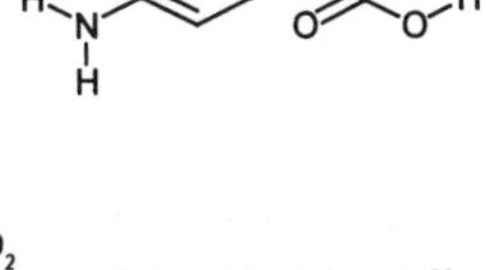

$C_9H_{13}ClN_2O_2$
Chiral intermediate. mp = 247-249°; $[\alpha]_D^{20}$ = - 8.5° (c = 2, $H_2O$). *Acros Organics nv; Sigma-Aldrich Fine Chemicals.*

**310 (3R)-Amino-1-phenylbutane**
937-52-0

$C_{10}H_{15}N$
(R)-1-Methyl-3-phenylpropylamine; 4-Phenyl-(2R)-butylamine. Chiral building block. bp = 220-222°; d = 0.9280; $n_D^{20}$ = 1.5140. *Acros Organics nv; Chiragene, Inc.; Dynamit Nobel GmbH; Lancaster Synthesis Ltd.; Toray International, Inc.*

**311 (3S)-Amino-1-phenylbutane**
4187-57-9
$C_{10}H_{15}N$
4-Phenyl-(2S)-butylamine; (S)-1-Methyl-3-phenylpropylamine. Chiral intermediate. bp = 220-222°; d = 0.9280; $n_D^{20}$ = 1.5140. *Chiragene, Inc.; Lancaster Synthesis Ltd.; Toray International, Inc.*

**312 (R)-(-)-2-Amino-2-phenylbutyric acid**
20318-28-9

$C_{10}H_{13}NO_2$
Benzeneacetic acid, α-amino-α-ethyl-, (R)-. Butyric acid, 2-amino-2-phenyl-, (R)-. Chiral building block. *PPG - Sipsy.*

**313 (S)-(+)-2-Amino-4-phenylbutyric acid ethyl ester**
46460-23-5
$C_{12}H_{17}NO_2$
Chiral building block. *Sigma-Aldrich Fine Chemicals.*

**314 (R)-Amino-1-phenylethanol**
2549-14-6

$C_8H_{11}NO$
Chiral building block. mp = 58-60°; $[\alpha]_D^{20}$ = - 41.0 ± 1.0° (c = 2, $CH_3OH$). *Kiralchem Ltd.; Oxford Asymmetry International plc; Sigma-Aldrich Fine Chemicals.*

**315 4-Aminophenyl-β-L-fucopyranoside**

$C_{12}H_{17}NO_6$
Chiral intermediate. *Senn Chemicals AG.*

**316 2-Aminophenyl-β-D-glucopyranoside hydrochloride**

$C_{12}H_{18}ClNO_6$
Chiral intermediate. *Senn Chemicals AG.*

**317 4-Aminophenyl-β-D-glucopyranoside hydrochloride**

$C_{12}H_{18}ClNO_6$
Chiral intermediate. *Senn Chemicals AG.*

**318 4-Aminophenyl-β-D-mannopyranoside hydrochloride**

$C_{12}H_{18}ClNO_6$
Chiral intermediate. *Senn Chemicals AG.*

**319 (R)-(-)-1-Amino-1-phenyl-2-methoxyethane**
64715-85-1

$C_9H_{13}NO$
Chiral building block. *Omega Chemical Company Inc.*

**320 (S)-(+)-1-Amino-1-phenyl-2-methoxyethane**

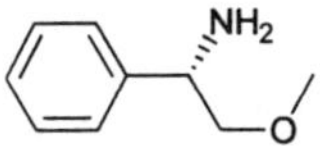

$C_9H_{13}NO$
Chiral building block. *Omega Chemical Company Inc.*

**321 (1R,2R)-2-Amino-1-phenylpropane-1,3-diol**
46032-98-8 256-250-8

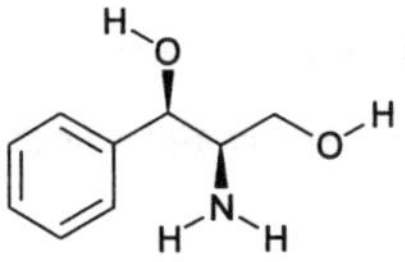

$C_9H_{13}NO_2$
Propane-1,3-diol, 2-amino-1-phenyl-, [R-(R*,R*)]-; D-threo-3-Phenyl-2-amino-1,3-propanediol. Chiral building block. mp = 112-118°; $[\alpha]^{23}$ = - 37° (c = 1, 1N HCl). *Acros Organics nv; Sigma-Aldrich Fine Chemicals.*

**322 (1S,2S)-(+)-2-Amino-1-phenyl-1,3-propanediol**
28143-91-1 248-867-6

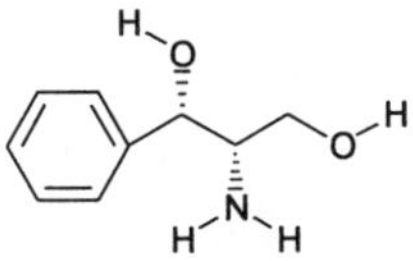

$C_9H_{13}NO_2$
Propane-1,3-diol, 2-amino-1-phenyl-, (1S,2S)-; (1S,2S)-(+)-1-Phenyl-2-amino-1,3-propanediol. Chiral building block. mp = 114°; [α] = + 25.7° (c = 1, $CH_3OH$); $[\alpha]^{25}$ = + 37° (c = 1, 1N HCl). *Acros Organics nv; KingChem, Inc.; Sigma-Aldrich Fine Chemicals; TCI America; Zhejiang Chemicals Import and Export Corporation.*

**323 (R)-Amino-1-phenyl-1-propanol**

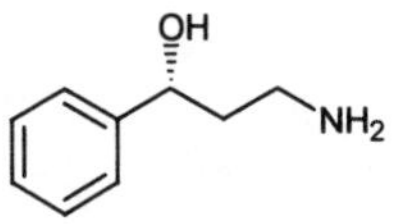

$C_9H_{13}NO$
Chiral building block. mp = 50-53°; $[\alpha]_D^{23}$ = + 43.5 ± 0.5° (c = 2, $CH_3OH$). *Kiralchem Ltd.*

**324 (S)-3-Amino-1-phenyl-1-propanol**

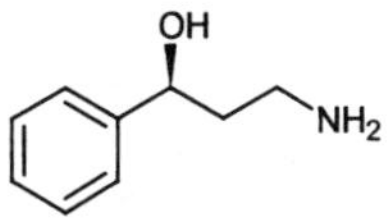

$C_9H_{13}NO$
Chiral building block. mp = 50-53°; $[\alpha]_D^{23}$ = - 43.5 ± 0.5° (c = 2, $CH_3OH$). *Kiralchem Ltd.*

**325 (R)-Amino-1-phenyl-2-propanol**

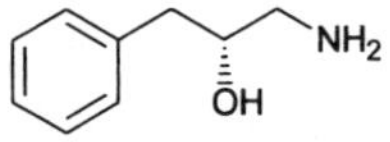

$C_9H_{13}NO$
Chiral building block. mp = 71-73°; $[\alpha]_D^{21}$ = + 15.0 ± 0.5° (c = 2, $CH_3OH$). *Kiralchem Ltd.*

**326 (S)-3-Amino-1-phenyl-2-propanol**

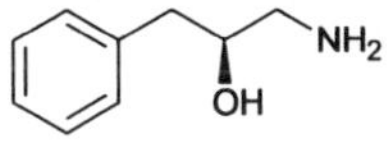

$C_9H_{13}NO$
Chiral building block. mp = 71-73°; $[\alpha]_D^{21}$ = - 15.0 ± 0.5° (c = 2, $CH_3OH$). *Kiralchem Ltd.*

**327 (R)-Amino-3-phenyl-1-propanol**
170564-98-4
$C_9H_{13}NO$
Chiral building block. *ChiroTech Technology Ltd.; Lancaster Synthesis Ltd.*

**328 (S)-3-Amino-3-phenyl-1-propanol**
82769-76-4
$C_9H_{13}NO$
(S)-1-Phenyl-3-propanolamine. Chiral building block. *Lancaster Synthesis Ltd.*

**329 3-Amino-(2R)-phenylsulfonylamino-propionic acid**

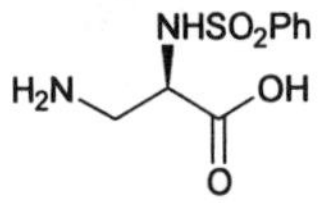

$C_9H_{12}N_2O_4S$
Resolving agent. *Omega Chemical Company Inc.*

**330 (2R,3S)-2-Amino-3-phenylthiobutanoic acid hydrochloride**

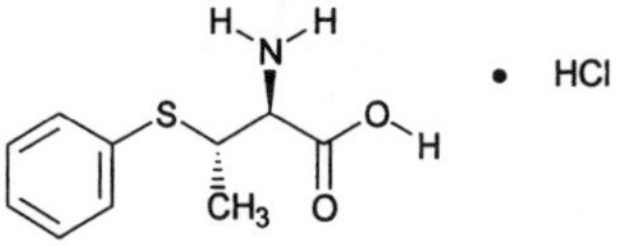

$C_{10}H_{14}ClNO_2S$
Chiral building block. [α] = + 4.1° (c = 1, 6N HCl). *Acros Organics nv.*

**331 4-Aminophenyl-1-thio-β-D-galactopyranoside**

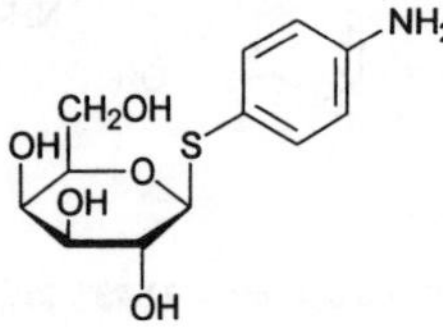

$C_{12}H_{17}NO_5S$
Chiral intermediate. *Senn Chemicals AG.*

**332 4-Aminophenyl-1-thio-β-D-glucopyranoside**

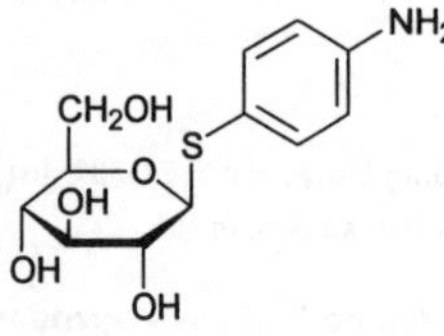

$C_{12}H_{17}NO_5S$
Chiral intermediate. *Senn Chemicals AG.*

**333 D-(+)-β-[(2-Aminophenyl)-thio]-α-hydroxy-4-methoxyphenylpropionic acid**
42399-48-4 255-798-5

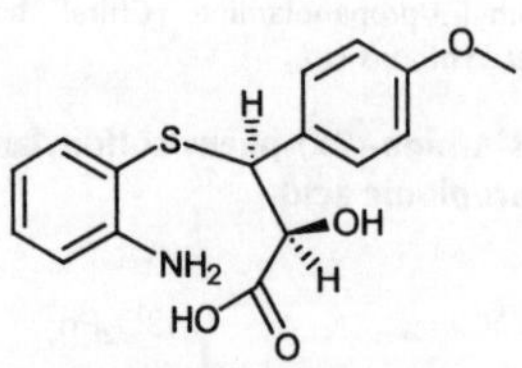

$C_{16}H_{17}NO_4S$
Benzenepropanoic acid, β-[(2-aminophenyl)thio]-α-hydroxy-4-methoxy-, [S-(R*,R*)]-; S-(R*,R*)]-3-[(o-Aminophenyl)thio]-3-(p-methoxyphenyl)lactic acid; δ-threo-2-Hydroxy-3-(2-aminophenylthio)-3-(4-methoxyphenyl)-propionic acid. Chiral intermediate. *D&O Group.*

**334 4-Aminophenyl-1-thio-β-D-mannopyranoside hydrochloride**

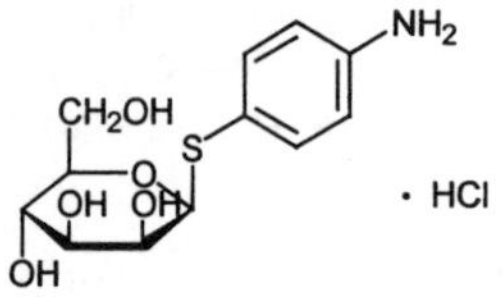

$C_{12}H_{18}ClNO_5S$
Chiral intermediate. *Senn Chemicals AG.*

**335 (S)-3-Aminopiperidine dihydrochloride**
$C_5H_{14}Cl_2N_2$
Chiral building block. *Sigma-Aldrich Fine Chemicals.*

**336 D-(-)-2-Amino-4-phosphonobutyric acid**
78739-01-2

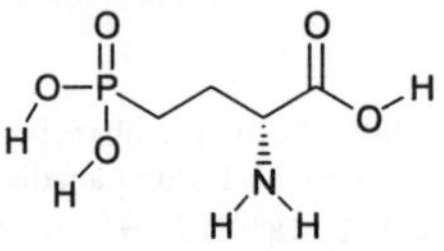

$C_4H_{10}NO_5P$
Chiral building block; Resolving agent. *Acros Organics nv.*

**337 L-(+)-2-Amino-4-phosphonobutyric acid**
23052-81-5

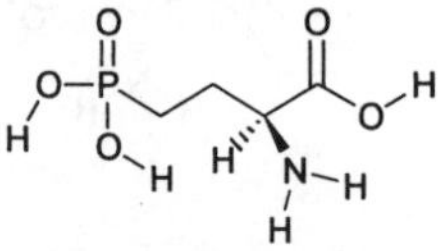

$C_4H_{10}NO_5P$
Chiral building block; Resolving agent. *Acros Organics nv.*

**338 (R)-(-)-2-Amino-6-phosphonohexanoic acid hydrate**
131177-53-2

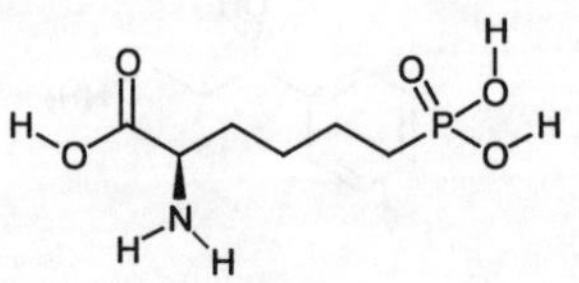

$C_6H_{14}NO_5P$
Chiral intermediate. *Acros Organics nv.*

**339 (S)-(+)-2-Amino-6-phosphonohexanoic acid**

$C_6H_{14}NO_5P$
Chiral intermediate. [α] = + 17.1° (c = 0.64, 2N HCl). *Acros Organics nv.*

**340 D-(-)-2-Amino-3-phosphonopropionic acid**
128241-72-5

$C_3H_8NO_5P$
Chiral building block; Resolving agent. *Acros Organics nv.*

**341 L-(+)-2-Amino-3-phosphonopropionic acid**
23052-80-4

$C_3H_8NO_5P$
Chiral building block; Resolving agent. *Acros Organics nv.*

**342 D-(-)-2-Amino-5-phosphonovaleric acid**
79055-68-8

$C_5H_{12}NO_5P$
Chiral building block; Resolving agent. [α] = - 20.8°. *Acros Organics nv.*

**343 L-(+)-2-Amino-5-phosphonovaleric acid**
79055-67-7

$C_5H_{12}NO_5P$
Chiral building block; Resolving agent. *Acros Organics nv.*

**344 (R)-Amino-1,2-propanediol**
66211-46-9

$C_3H_9NO_2$
Propane-1,2-diol, 3-amino, (R)-. Chiral building block. *Daiso Company, Ltd.; Synthon Chiragenics Corporation.*

**345 (S)-3-Amino-1,2-propanediol**
61278-21-5

$C_3H_9NO_2$
Propane-1,2-diol, 3-amino, (S)-. Chiral building block. $bp_{0.4}$ = 117-119°; d = 1.175; n = 1.483; $[\alpha]^{23}$ = - 12° (neat). *Daiso Company, Ltd.; Sigma-Aldrich Fine Chemicals; Synthon Chiragenics Corporation.*

**346 (R)-(-)-1-Amino-2-propanol**
2799-16-8 220-532-9

$C_3H_9NO$
Propan-2-ol, 1-amino-, (R)-; (R)-(-)-Isopropanolamine. Chiral auxiliary. mp = 24-26°; bp = 160°; d = 0.9600; $n_D^{20}$ = 1.4482; $[\alpha]_D^{20}$ = - 18 ± 1° (c = 1.8, $H_2O$); $[\alpha]_D^{20}$ = - 23.5° (c = 1, $CH_3OH$). *Acros Organics nv; Daiso Company, Ltd.; Lancaster Synthesis Ltd.; Omega Chemical Company Inc.; Sigma-Aldrich Fine Chemicals; TCI America.*

**347 (S)-(+)-1-Amino-2-propanol**
2799-17-9 220-533-4

$C_3H_9NO$
Propan-2-ol, 1-amino-, (S)-; (S)-(+)-Isopropanolamine. Chiral auxiliary. mp = 24-26°; bp = 160°; d = 0.9600; $n_D^{20}$ = 1.4437; $[\alpha]_D^{20}$ = + 18 ± 1° (c = 1.8, $H_2O$). *Acros Organics nv; Daiso Company, Ltd.; Lancaster Synthesis Ltd.; Omega Chemical Company Inc.; Sigma-Aldrich Fine Chemicals; TCI America.*

**348 (1R)-(-)-(1-Aminopropyl)phosphonic acid**
98049-00-4

$C_3H_{10}NO_3P$
Chiral ligand. mp = 265-269°; $[\alpha]_D^{20}$ = - 16.5° (c = 2, 1N NaOH). *Sigma-Aldrich Fine Chemicals.*

**349 (1S)-(+)-(1-Aminopropyl)phosphonic acid**
98048-99-8

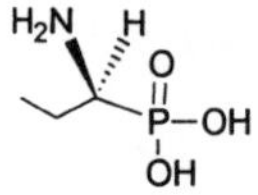

$C_3H_{10}NO_3P$
Resolving agent. mp = 265-269°; $[\alpha]_D^{20}$ = + 16.5° (c = 2, 1N NaOH). *Sigma-Aldrich Fine Chemicals.*

**350 (3R)-(+)-3-Aminopyrrolidine**
116183-82-5
$C_4H_{10}N_2$
Chiral building block. *Austin Chemical Company, Inc.; TCI America.*

**351 (3S)-(-)-3-Aminopyrrolidine**
128345-57-3
$C_4H_{10}N_2$
Chiral building block. *Austin Chemical Company, Inc.; TCI America.*

**352 (R)-(-)-3-Aminopyrrolidine dihydrochloride**
116183-81-4
$C_4H_{12}Cl_2N_2$
Chiral building block. $[\alpha]_D^{20}$ = - 1.2 ± 0.2° (c = 10, $H_2O$). *Lancaster Synthesis Ltd.; Sigma-Aldrich Fine Chemicals; TCI America.*

**353 (S)-(+)-3-Aminopyrrolidine dihydrochloride**
116183-83-6
$C_4H_{12}Cl_2N_2$
Chiral building block. $[\alpha]_D^{20}$ = + 1.2 ± 0.5° (c = 10, $H_2O$). *Lancaster Synthesis Ltd.; Sigma-Aldrich Fine Chemicals; TCI America.*

**354 (3R)-(+)-Aminoquinuclidine dihydrochloride**
123536-14-1

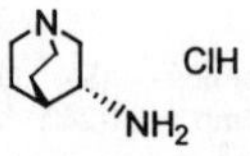

$C_7H_{16}Cl_2N_2$
Chiral intermediate. mp > 300°; $[\alpha]_D^{20}$ = + 24° (c = 1, $H_2O$). *Norchim SA; Sigma-Aldrich Fine Chemicals.*

**355 (S)-(-)-3-Aminoquinuclidine dihydrochoride**
119904-90-4

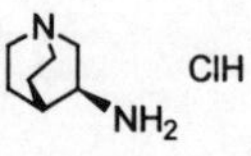

$C_7H_{16}Cl_2N_2$
Resolving agent; Chiral auxiliary. *Borregaard Synthesis; Sigma-Aldrich Fine Chemicals.*

**356 (R)-α-Aminosuberic acid**

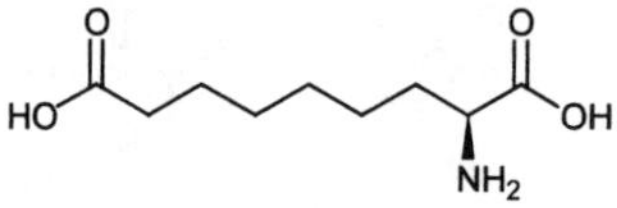

$C_8H_{15}NO_4$
Octanedioic acid, 2-amino-, (2R)-.; (R)-2-Aminooctanedioic acid. Chiral building block. *Austin Chemical Company, Inc.*

**357 (S)-α-Aminosuberic acid**
4254-88-0 224-231-3

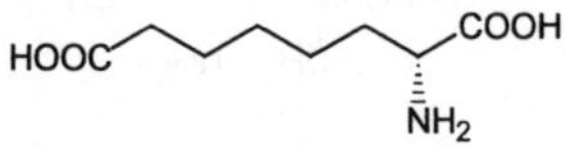

$C_8H_{15}NO_4$
Octanedioic acid, 2-amino-, (2S)-; (S)-2-Amino-octanedioic acid. Chiral building block. *Austin Chemical Company, Inc.*

**358 (R)-6-Amino-6,7,8,9-tetrahydro-5H-benzo-cycloheptene**
101692-74-4
$C_{11}H_{15}N$
Chiral intermediate. *Sigma-Aldrich Fine Chemicals.*

**359 (S)-6-Amino-6,7,8,9-tetrahydro-5H-benzo-cycloheptene**
$C_{11}H_{15}N$
Chiral intermediate. *Sigma-Aldrich Fine Chemicals.*

**360 (R)-2 Aminotetralin**

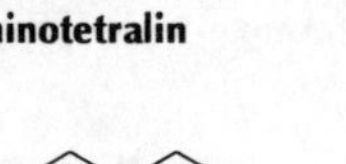

$C_{10}H_{13}N$
Naphthyl-2-amine, 1,2,3,4-tetrahydro-, (R)-; (R)-1,2,3,4-Tetrahydro-2-naphthylamine; (R)-2-Amino-1,2,3,4-tetrahydronaphthalene. Chiral building block. *Chiragene, Inc.*

**361 (S)-2-Aminotetralin**

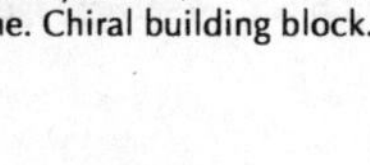

$C_{10}H_{13}N$
Naphthyl-2-amine, 1,2,3,4-tetrahydro-, (S)-; (S)-1,2,3,4-Tetrahydro-2-naphthylamine; (S)-2-Amino-1,2,3,4-tetrahydronaphthalene. Chiral building block. *Chiragene, Inc.*

**362 (+)-Aminotperin hydrate**
54-62-6 493(12) 200-209-9

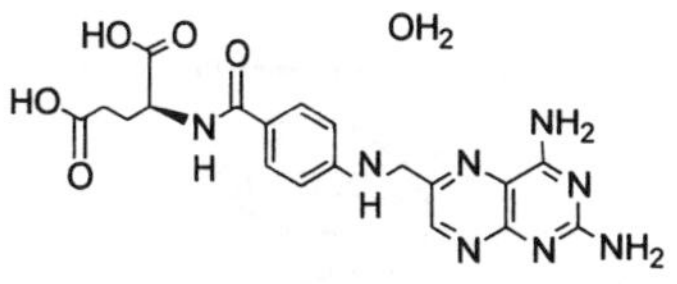

$C_{19}H_{20}N_8O_5$
4-Aminofolic acid; 4-Aminopteroylglutamic acid. Listed on TSCA. Chiral intermediate. mp = 225°; $[\alpha]^{24}$ = + 18° (c = 0.1, 0.1N NaOH). *Sigma-Aldrich Fine Chemicals.*

**363 3-Amino-L-tyrosine dihydrochloride monohydrate**
23279-22-3 245-552-5

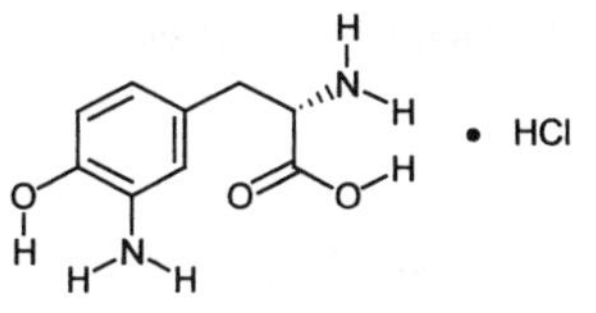

$C_9H_{16}Cl_2N_2O_4$
Tyrosine, 3-amino-, dihydrochloride, L-; Chiral intermediate. mp = 158-160°; $[\alpha]^{22}$ = - 1.6° (c = 2, $H_2O$). *Acros Organics nv; Senn Chemicals AG; Sigma-Aldrich Fine Chemicals.*

**364 L-2-Amino-3-ureido-propionic acid**

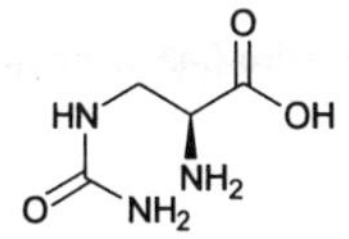

$C_4H_9N_3O_3$
Chiral intermediate. *Senn Chemicals AG.*

**365 Ammonium (-)-3-bromo-8-camphor-sulfonate**
55870-50-3 259-871-2

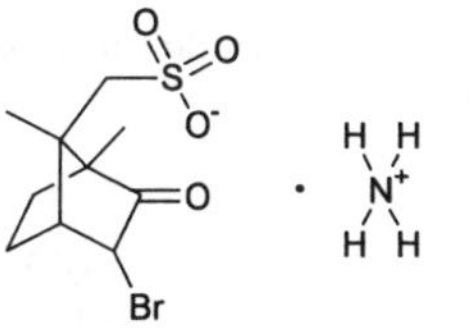

$C_{10}H_{18}BrNO_4S$
Bicyclo[2.2.1]heptane-7-methanesulfonic acid, 3-bromo-1,7-dimethyl-2-oxo-, ammonium salt, (1S,3R,4R,7S)-; L(-)-α-Bromocamphor-+lp-sulfonic acid, ammonium salt; ammonium L-5-bromo-6-oxo-9-bornanesulphonate. Resolving agent. mp = 289°; $[\alpha]^{25}$ = - 83° (c = 4, $H_2O$). *Acros Organics nv; Calaire Chimie s.a.; Sigma-Aldrich Fine Chemicals.*

**366 Ammonium D-xylonate**
5461-96-1

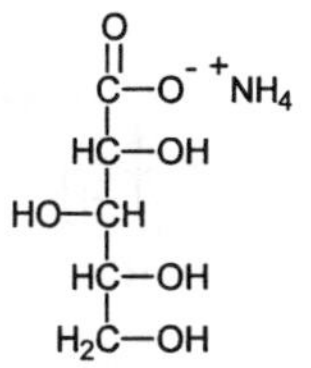

$C_5H_{13}NO_6$
Chiral building block. *Pfanstiehl Laboratories, Inc.*

**367 Anabasine**
494-52-0 661(12) 207-791-3
$C_{10}H_{14}N_2$
Pyridine, 3-(2-piperidinyl)-, (S)-; Neonicotine; 2-(3'-Pyridyl)piperidine; (S)-3-(2-Piperidinyl)pyridine. Pharmaceutical or derivative. mp = 9°; bp = 270-272°; d = 1.0460; $n_D^{20}$ = 1.5430; $[\alpha]_D^{20}$ = - 22 ± 2° (c = 1, $H_2O$). *Lancaster Synthesis Ltd.*

**368 L-Anabasine**
40774-73-0

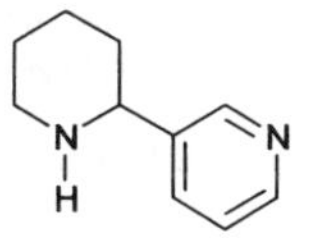

$C_{10}H_{14}N_2$
Chiral intermediate. *Acros Organics nv.*

**369 Andrographolide**
5508-58-7 673(12) 226-852-5

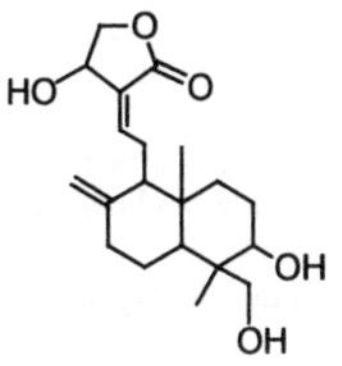

$C_{20}H_{30}O_5$
Chiral intermediate. mp = 231-234°; $[\alpha]_D^{20}$ = - 126° (c = 1.5, HOAc). *Sigma-Aldrich Fine Chemicals.*

**370 Androsta-1,4-diene-3,17-dione**
897-06-3 212-977-2

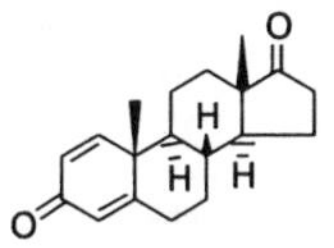

$C_{19}H_{24}O_2$
Pharmaceutical or derivative. mp = 138-139°; $[\alpha]^{22}$ = + 117.5° (c = 1, $CHCl_3$). *Sigma-Aldrich Fine Chemicals.*

**371 5-Androst-16-en-3-one**
18339-16-7 242-220-1

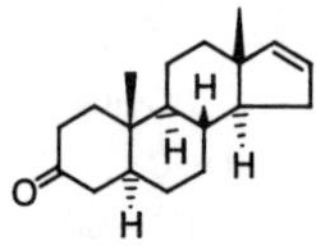

$C_{19}H_{28}O$
Pharmaceutical or derivative. mp = 140-144°; $[\alpha]^{17}$ = + 38° (c = 2, $CHCl_3$). *Sigma-Aldrich Fine Chemicals.*

**372 (+)-5-Androstrane-3,17-diol hydrate**
209-334-3

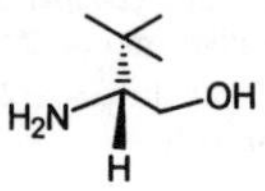

$C_{19}H_{32}O_2$
Chiral intermediate. mp = 165-167°; $[\alpha]^{25}$ = + 10.7° (c = 1, EtOH). *Sigma-Aldrich Fine Chemicals.*

**373 2,7-Anhydro-β-D-altroheptulopyranose**
469-90-9 207-424-7

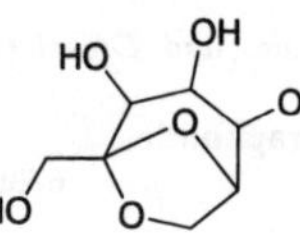

$C_7H_{12}O_6$
Heptulopyranose, 2,7-anhydro-, β-D-altro-; Sedoheptulosan. Chiral intermediate. *Senn Chemicals AG.*

**374 1,6-Anhydro-β-D-glucopyranose**
498-07-7 207-855-0

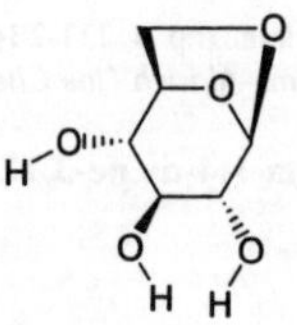

$C_6H_{10}O_5$
Glucopyranose, 1,6-anhydro-, β-D-; Laevoglucosan; Leucoglucosan; Levoglucosan. Chiral building block. mp = 170-180°; [α] = - 66° (c = 2, $H_2O$). *Acros Organics nv; Kaden Biochemicals GmbH; Pfanstiehl Laboratories, Inc.; Senn Chemicals AG.*

**375 1,6-Anhydro-β-D-glucose triacetate**
13242-55-2 236-222-1

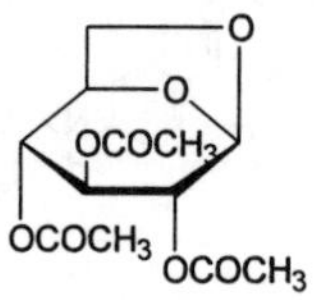

$C_{12}H_{16}O_8$
Glucopyranose, 1,6-anhydro-, triacetate, β-D-; 1,6-Anhydro-2,3,4-tri-O-acetyl-β-D-glucopyranose; 2,3,4-Tri-O-acetyl-1,6-anhydro-β-D-glucopyranose; Levoglucosan, triacetate; Triacetyllevoglucosan. Chiral intermediate. *Senn Chemicals AG.*

**376 2,5-Anhydro-D-mannitol**
41107-82-8 255-221-7

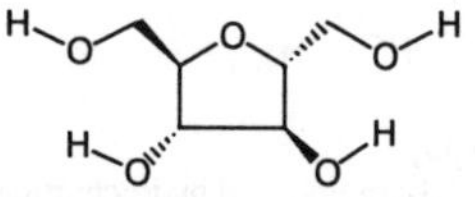

$C_6H_{12}O_5$
Mannitol, 2,5-anhydro-, D -. Chiral building block. mp = 101-103°; $[\alpha]^{19}$ = + 56.7° (c = 1, $H_2O$). *Acros Organics nv; Senn Chemicals AG; Sigma-Aldrich Fine Chemicals.*

**377 1,6-Anhydro-β-D-mannopyranoside**

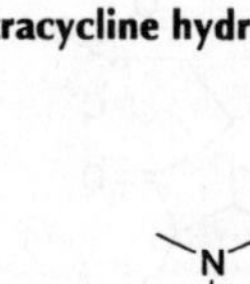

$C_6H_{10}O_5$
Chiral intermediate. *Senn Chemicals AG.*

**378 Anhydrotetracycline hydrochloride**
13803-65-1

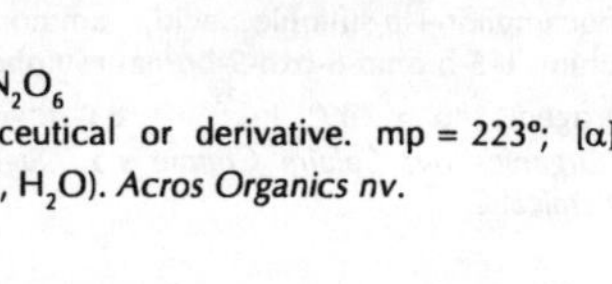

$C_{22}H_{22}N_2O_6$
Pharmaceutical or derivative. mp = 223°; [α] = - 264° (c = 0.5, $H_2O$). *Acros Organics nv.*

**379 1,5-Anhydro-3,4,6-tri-O-benzyl-2-deoxy-D-arabinohex-1-enitol**
55628-54-1

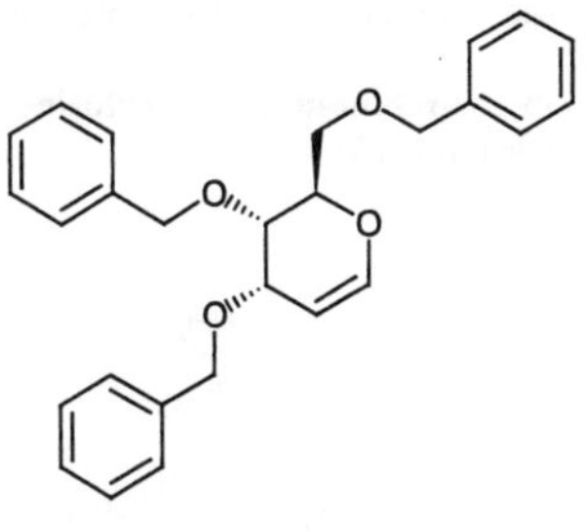

$C_{27}H_{28}O_4$
Chiral intermediate. mp = 57-58°; [α] = - 3 (c = 1, $CHCl_3$). *Acros Organics nv; Sigma-Aldrich Fine Chemicals.*

**380 (R)-(-)-2-(Anilinomethyl)pyrrolidine**
68295-45-4
$C_{11}H_{16}N_2$
(R)-2-(Phenylaminomethyl)pyrrolidine. Chiral building block. $bp_{0.55}$ = 111-112°; d = 1.05; n = 1.57; $[\alpha]_D^{20}$ = - 19° (c = 1, EtOH). *Kiralchem Ltd.; Sigma-Aldrich Fine Chemicals.*

**381 (S)-(+)-2-(Anilinomethyl)pyrrolidine**
64030-44-0

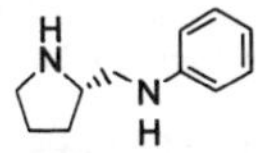

$C_{11}H_{16}N_2$
Pyrrolidine-2-methanamine, N-phenyl-, (2S)-; (S)-(+)-N-(2-Pyrrolidinomethyl)aniline; (S)-2-(Phenylaminomethyl)-pyrrolidine. Chiral building block. $bp_2$ = 45-47°; $bp_{0.55}$ = 111-112°; d = 1.05; n =0.574; $[\alpha]_D^{20}$ = + 48.0 ± 0.5° (c = 4, $C_6H_6$); $[\alpha]_D^{20}$ = + 19° (c = 1, EtOH). *Kiralchem Ltd.; Sigma-Aldrich Fine Chemicals; TCI America.*

**382 N-(9-Anthracenylmethyl)cinchonindinium chloride**

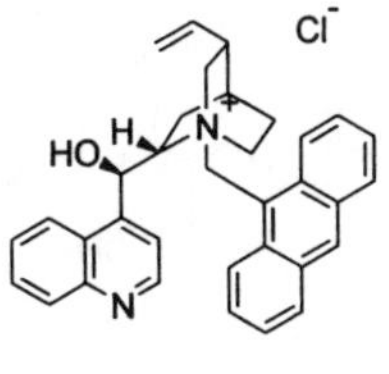

$C_{34}H_{33}ClN_2O$
Resolving agent; Chiral auxiliary. mp = 158-161°. *Sigma-Aldrich Fine Chemicals.*

**383 (R)-(-)-1-(9-Anthryl)-2,2,2-trifluoroethanol**
53531-34-3 258-611-5

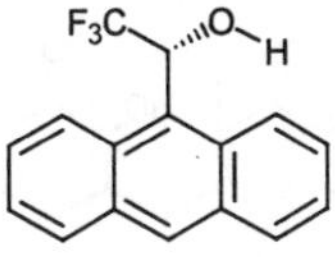

$C_{16}H_{11}F_3O$
Anthracene-9-methanol, a-(trifluoromethyl)-, (R)-; (R)-(-)-α-(Trifluoromethyl)anthracene-9-methanol; (-)-2,2,2-trifluoro-1-(9-anthryl)ethanol. Diagnostic reagent. mp = 132-135°; $[\alpha]_D^{20}$ = - 25 ± 2° (c = 6, $CHCl_3$). *Acros Organics nv; Austin Chemical Company, Inc.; FineTech, Ltd.; Lancaster Synthesis Ltd.; Oxford Asymmetry International plc; Sigma-Aldrich Fine Chemicals; TCI America.*

**384 (S)-(+)-1-(9-Anthryl)-2,2,2-trifluoroethanol**
60646-30-2 262-348-1

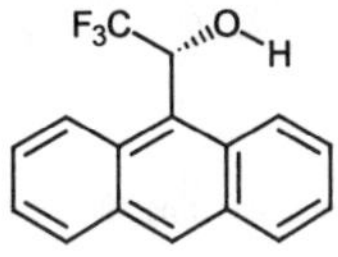

$C_{16}H_{11}F_3O$
Anthracene-9-methanol, α-(trifluoromethyl)-, (αS)-; (S)-(+)-2,2,2-Trifluoro-1-(9-anthryl)ethanol; (S)-α-(Trifluoromethyl)anthracene-9-methanol; Pirkle's alcohol; (S)-(+)-α-(Trifluoromethyl)anthracene-9-methanol. Chiral building block. mp = 132-135°; $[\alpha]_D^{20}$ = + 25.5 ± 1° (c = 6, $CHCl_3$). *Acros Organics nv; FineTech, Ltd.; Lancaster Synthesis Ltd.; Oxford Asymmetry International plc; Sigma-Aldrich Fine Chemicals; TCI America.*

**385 Apigenin-7-glucoside**
578-74-5 209-430-5
$C_{21}H_{20}O_{10}$
Benzo[4H]-pyran-4-one, 7-(β-D-glucopyranosyloxy)-5-hydroxy-2-(4-hydroxyphenyl)-; Apigetrin; 7-(β-D-glucopyranosyloxy)-5-hydroxy-2-(4-hydroxyphenyl)-4H-1-benzopyran-4-one; 4',5,7-Trihydroxyflavone 7-glucoside; Apigenin 7-monoglucoside; Apigenin, 7-β-D-glucopyranoside; Cosmetin; Cosmosiin; Cosmosiine; Cosmosioside. Pharmaceutical or derivative. *Kaden Biochemicals GmbH.*

**386 α-Apo-oxytetracycline**
18695-01-7

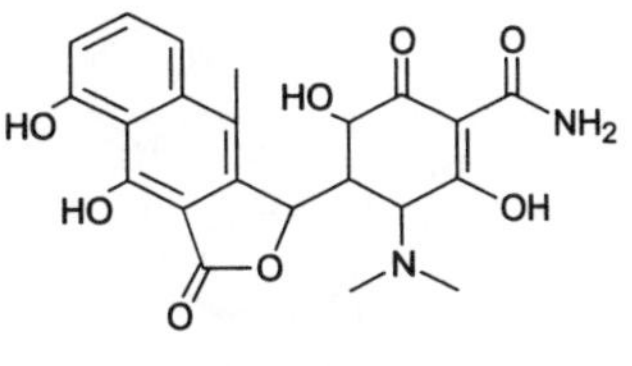

$C_{22}H_{22}N_2O_8$
Pharmaceutical or derivative. mp = 200-205°; [α = - 64° (c = 0.4, DMF). *Acros Organics nv.*

**387 β-Apo-oxytetracycline**
18751-99-0

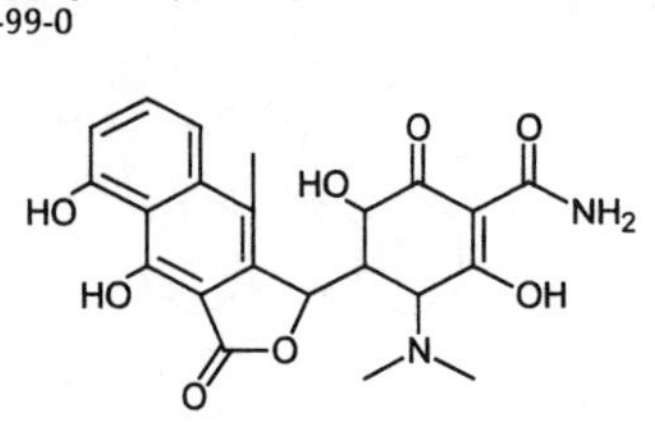

$C_{22}H_{22}N_2O_8$
Pharmaceutical or derivative. mp = 200-204°; [α] = - 207° (c = 0.5, DMF). *Acros Organics nv.*

**388 Arabic acid**
32609-14-6 251-128-0

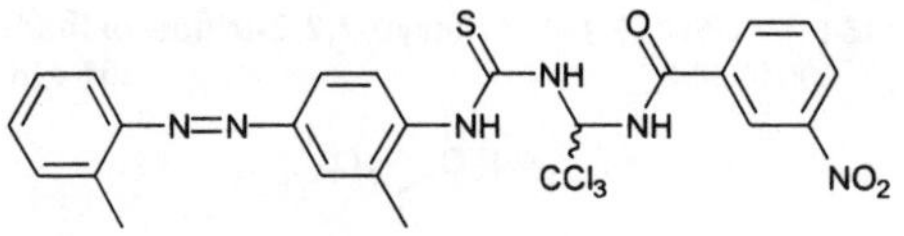

Unspecified
Arabic acid. Chiral building block. *Kaden Biochemicals GmbH.*

**389 9-β-D-Arabinofuranosyl-adenine hydrate**
5536-17-4 226-893-9

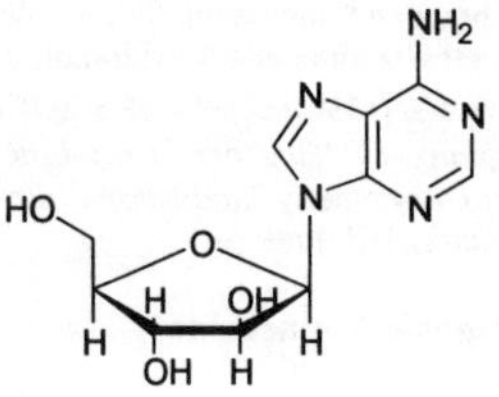

$C_{10}H_{13}N_5O_4$
Purin-(9H)-6-amine, 9-β-D-arabinofuranosyl-; Vidarabine; b-Ara-A; b-D-Arabinosyladenine; 6-Amino-9-β-D-arabinofuranosylpurine; Adenine arabinoside; Adenine, 9-β-D-arabinofuranosyl-; Ara-A; Araadenosine; CI-673; NSC 404241; Spongoadenosine. Listed on TSCA. Chiral intermediate. *Pfanstiehl Laboratories, Inc.*

**390 1-β-D-Arabinofuranosyl-cytosine**
147-94-4 205-705-9

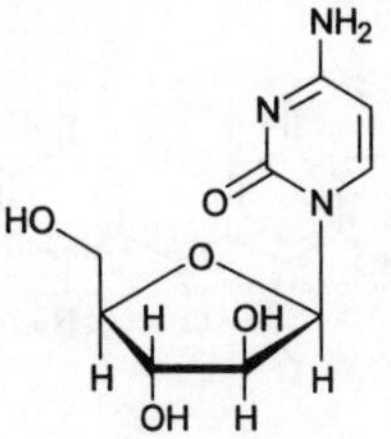

$C_9H_{13}N_3O_5$
Pyrimidin-2(1H)-one, 4-amino-1-β-D-arabinofuranosyl-; Arabinocytidine; 4-Amino-1-β-D-arabinofuranosyl-2(1H)-pyrimidinone; β-Cytosine arabinoside; Arabitin; Aracytidine; Cytosar; Erpalfa; Cytosar-U; Aracytine; Tarabine; Udicil; Ara-C; 4-Amino-1-arabinofuranosyl-2-oxo-1,2-dihydropyrimidine; 1-β-D-arabino. Chiral intermediate. mp = 212-213°. *Pfanstiehl Laboratories, Inc.*

**391 1-β-D-Arabinofuranosyl-cytosine hydrochloride**

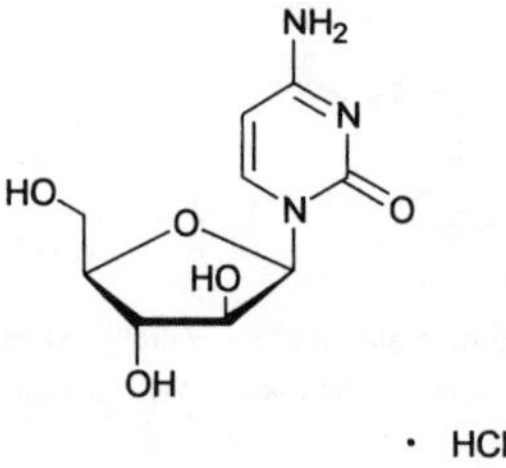

$C_9H_{14}N_3O_5Cl$
Chiral intermediate. mp = 212°; $[\alpha]_D^{20}$ = + 130° (c = 0.5, $H_2O$). *Pfanstiehl Laboratories, Inc.*

**392 9-β-D-Arabinofuranosyl-hypoxanthine**
7013-16-3 230-295-3
$C_{10}H_{12}N_4O_5$
Purin-(6H)-6-one, 9-β-D-arabinofuranosyl-1,9-dihydro-; 9-β-D-Arabinofuranosyl-1,9-dihydro-6H-purin-6-one; Arabinosylhypoxanthine; Arainosine; Hypoxanthine arabinoside; Hypoxanthine, 9-β-D-arabinofuranosyl-. Chiral intermediate. *Pfanstiehl Laboratories, Inc.*

**393 9-β-D-arabinofuranosyl-6-mercaptopurine**
90899-89-1
$C_{10}H_{12}N_4O_4S$
Chiral intermediate. *Pfanstiehl Laboratories, Inc.*

**394 1-β-D-Arabinfuranosyl-uracil**
3083-77-0
$C_8H_{12}N_2O_6$
Chiral intermediate. mp = 220°. *Pfanstiehl Laboratories, Inc.*

**395 L-(+)-Arabinogalactane**
9036-66-2 232-910-0

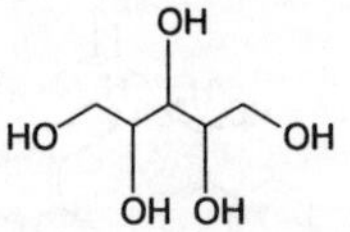

$[(C_5H_8O_4)(C_6H_{10}O_4)_6]_n$
Galacto-L-arabinan, D-. Listed on TSCA. Chiral building block. mp = 200°; $[\alpha]^{22}$ = + 10.3° (c = 1, $H_2O$). *Kaden Biochemicals GmbH; Sigma-Aldrich Fine Chemicals.*

**396 D-Arabino-1,4-lactone**
59990-13-5

$C_5H_8O_5$
Chiral building block. *Pfanstiehl Laboratories, Inc.*

**397 D-(-)-Arabinose**
28697-53-2

$C_5H_{10}O_5$
Arabinopyranose, D-; D-Arabinopyranose. Chiral building block. mp = 162-164°; [α] = - 104.3° (c = 3, $H_2O$, 20h). *Acros Organics nv; Sigma-Aldrich Fine Chemicals.*

**398 L-(+)-Arabinose**
10323-20-3 233-708-5
$C_5H_{10}O_5$
L-Arabinose; L-Arabinopyranose. Listed on TSCA. Chiral building block. mp = 150-160°; $[\alpha]_D^{20}$ = - 104.5° (c = 4, $H_2O$). *Austin Chemical Company, Inc.; Kaden Biochemicals GmbH; Pfanstiehl Laboratories, Inc.; Senn Chemicals AG; TCI America.*

**399 D-Arabitol**
488-82-4 789(11) 207-686-2

$C_5H_{12}O_5$
Arabinitol, D-. Listed on TSCA. Chiral building block. mp = 103°; $[\alpha]_D^{20}$ = + 130° (c = 0.4, excess acidified molybddate). *Acros Organics nv; Kaden Biochemicals GmbH; Pfanstiehl Laboratories, Inc.; Senn Chemicals AG; Sigma-Aldrich Fine Chemicals.*

**400 L-Arabitol**
7643-75-6 789(11) 231-582-6

$C_5H_{12}O_5$
Arabinitol, L-; L-Arabinitol. Listed on TSCA. Chiral building block. mp = 101-104°; $[\alpha]_D^{20}$ = - 130° (c = 0.4, excess acidified molybddate). *Acros Organics nv; Kaden Biochemicals GmbH; Pfanstiehl Laboratories, Inc.; Senn Chemicals AG; TCI America.*

**401 D-Araboascorbic acid**
89-65-6 5141(12) 201-928-0

$C_6H_8O_6$
Hexonic acid, 3-keto-, γ-lactone, D-erythro-; D-Erythorbic acid; γ-Lactone D-erythro-hex-2-enoic acid; 2,3-Didehydro-D-erythro-hexono-1,4-lactone; D-(-)-Iso-ascorbic acid; D-arabino-Ascorbic acid; E 315; Erycorbin; Glucosaccharonic acid; Isovitamin C; Mercate 5; Neo-Cebicure; Saccharosonic acid. Listed on TSCA. Chiral building block. mp = 167°; $[\alpha]^{25}$ = - 16.8° (c = 2, $H_2O$). *Acros Organics nv; Sigma-Aldrich Fine Chemicals; TCI America.*

**402 D-(-)-Arabonic acid potassium salt**
Chiral building block. *Senn Chemicals AG.*

**403 N-Arachidoyl-D-erythro-sphingosine**
7344-02-7

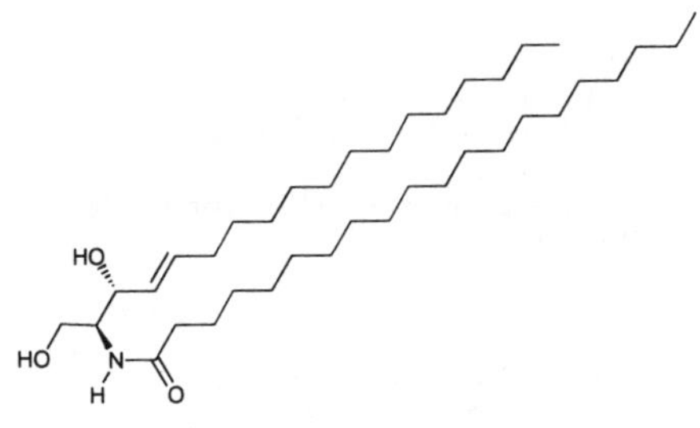


$C_{38}H_{75}NO_3$
Chiral intermediate. *Acros Organics nv.*

**404 D-(-)-Arginine**
157-06-2 205-866-5

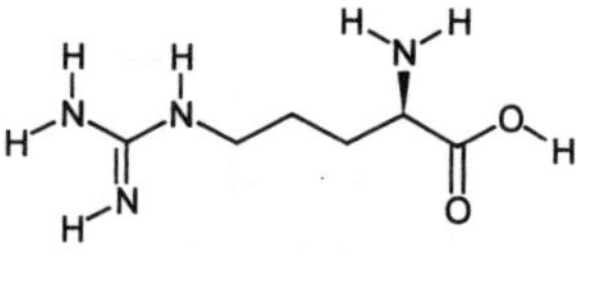

$C_6H_{14}N_4O_2$
D-Arginine; D-2-Amino-5-guanidinovaleric acid. Listed on TSCA. Chiral building block. mp = 221-224°; $[\alpha]^{23}$ = - 23° (c = 1.6, 6N HCl). *Acros Organics nv; Ajinomoto Co. Inc.; Rexim S.A. Produits Chimiques; Sigma-Aldrich Fine Chemicals; TCI America; Yoneyama Yakuhin Kogyo Co., Ltd.*

**405 L-(+)-Arginine**
74-79-3 817(12) 200-811-1

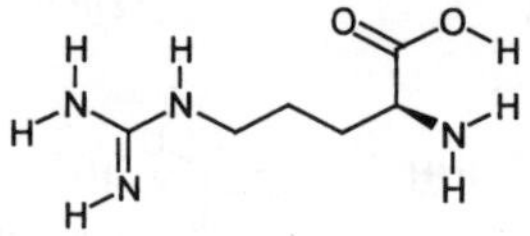

$C_6H_{14}N_4O_2$
Pentanoic acid, 2-amino-5-[(aminoiminomethyl)amino]-, (S)-; L-α-Amino-δ-guanidinovaleric acid; L-Norvaline, 5-[(aminoiminomethyl)amino]-; L-Ornithine, $N^5$-(aminoiminomethyl)-; (S)-2-Amino-5-[(aminoiminomethyl)-amino]pentanoic acid. Listed on TSCA. Chiral building block. mp = 244°; $[\alpha]^{24}$ = + 26.1° (c = 1.6, 6N HCl). *Aceto Corporation; Acros Organics nv; Ajinomoto Co. Inc.; Austin Chemical Company, Inc.; Flamma s.p.a.; KingChem, Inc.; Kyowa Hakko Kogyo Co., Ltd.; Rexim S.A. Produits Chimiques; Senn Chemicals AG; Sigma-Aldrich Fine Chemicals; Tanabe Seiyaku Co. Ltd.; TCI America; Yoneyama Yakuhin Kogyo Co., Ltd.*

**406 L-Arginine acetate**
71173-62-1 275-247-2

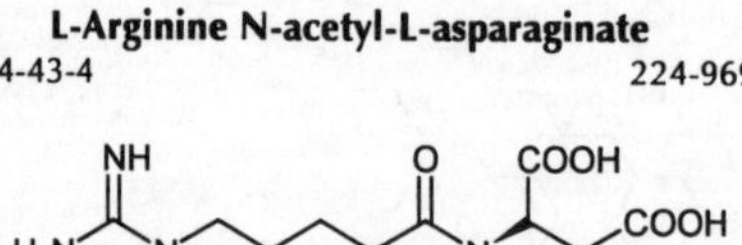

$C_8H_{18}N_4O_4$
Arginine, acetate, L-. Listed on TSCA. Chiral building block. *Rexim S.A. Produits Chimiques.*

**407 L-Arginine N-acetyl-L-asparaginate**
4584-43-4 224-969-6

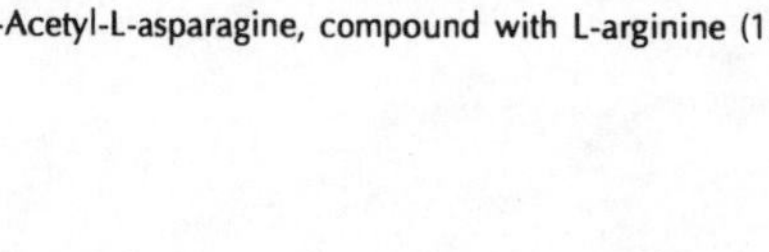

$C_{12}H_{24}N_6O_6$
Asparagine, $N^2$-acetyl-, L-, compd. with L-arginine (1:1); $N^2$-Acetyl-L-asparagine, compound with L-arginine (1:1). Chiral intermediate. *Austin Chemical Company, Inc.; Flamma s.p.a.*

**408 L-Arginine-L-aspartate**
7675-83-4 231-656-8

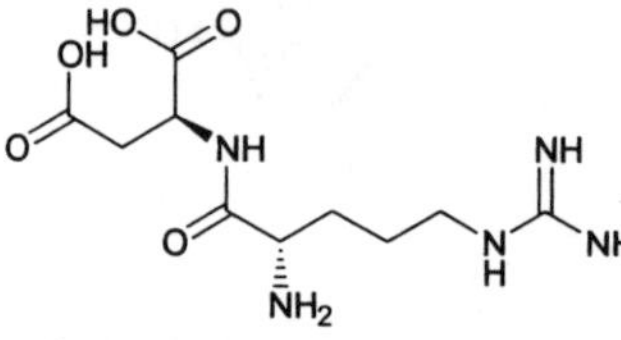

$C_{10}H_{21}N_5O_6$
Aspartic acid, L-, compd. with L-arginine (1:1). Listed on TSCA. Chiral intermediate. *Flamma s.p.a.; Rexim S.A. Produits Chimiques.*

**409 L-Arginine-L-glutamate**
4320-30-3 224-350-0

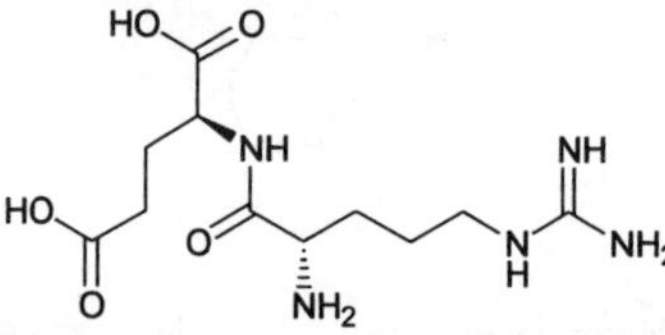

$C_{11}H_{23}N_5O_6$
Glutamic acid, L-, compd. with L-arginine (1:1); Argimate; Arginine glutamate; Ginamate; Glutargin; Glutepar; Modumate. Listed on TSCA. Chiral intermediate. *Flamma s.p.a.; Rexim S.A. Produits Chimiques.*

**410 D-(-)-Arginine hydrochloride**
627-75-8 211-010-1

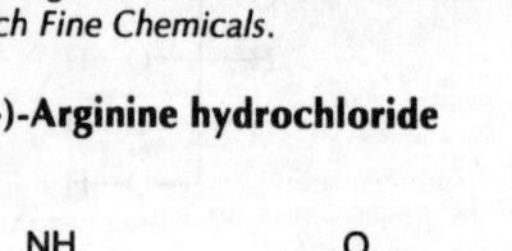

$C_6H_{15}ClN_4O_2$
Arginine, monohydrochloride, D-. Listed on TSCA. Chiral building block. mp = 216-218°; $[\alpha]^{23}$ = - 21° (c = 2, 6N HCl). *Acros Organics nv; Rexim S.A. Produits Chimiques; Sigma-Aldrich Fine Chemicals.*

**411 L-(+)-Arginine hydrochloride**
15595-35-4 239-674-8

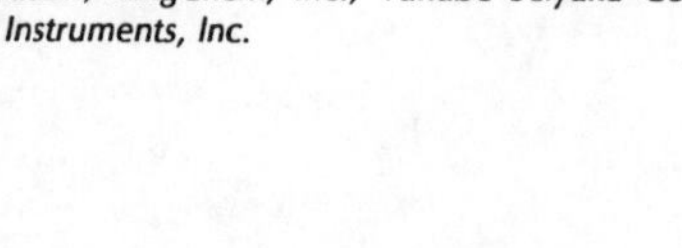

$C_6H_{14}N_4O_2.xClH$
Arginine, hydrochloride, L-; Arginine chlorhydrate; Arginine muriate. Chiral building block. *Aceto Corporation; KingChem, Inc.; Tanabe Seiyaku Co. Ltd.; Varsal Instruments, Inc.*

**412 L-Arginine-α-ketoglutarate (2:1)**
5256-76-8 226-059-4

$C_{17}H_{34}N_8O_9$
Glutaric acid, 2-oxo-, compd. with arginine (1:2), L-; Di-L-arginine 2-oxoglutarate; Di-argiceto; Eucol; L-Arginine, 2-oxopentanedioate (2:1). Chiral intermediate. *Rexim S.A. Produits Chimiques.*

**413 L-Arginine monohydrochloride**
1119-34-2 817(12) 214-275-1

$C_6H_{14}N_4O_2$
Arginine, monohydrochloride, L-; Argamine. Listed on TSCA. Chiral building block. mp = 235°; [α] = + 22° (c = 12, 10% HCl). *Acros Organics nv; Austin Chemical Company, Inc.; Kyowa Hakko Kogyo Co., Ltd.; Rexim S.A. Produits Chimiques; Shanghai DSL International Trading Company; Sigma-Aldrich Fine Chemicals; TCI America.*

**414 L-Arginine-L-pyroglutamate**

$C_{11}H_{19}N_5O_4$
Chiral intermediate. *Aceto Corporation; Rexim S.A. Produits Chimiques.*

**415 L-Arginine-thiazolidinecarboxylate**
Chiral intermediate. *Flamma s.p.a.*

**416 (-)-Asarinin**
133-05-1 1848(12)

$C_{20}H_{18}O_6$
Chiral intermediate. mp = 120-122°; $[\alpha]_D^{20}$ = - 114° (c = 1, $CHCl_3$). *Sigma-Aldrich Fine Chemicals.*

**417 L-Ascorbic acid**
50-81-7 867(12) 200-066-2

$C_6H_8O_6$
Hex-2-enonic acid, γ-lactone, L-threo-; Vitamin C; 3-keto-L-Gulofuranolactone; 3-Oxo-L-gulofuranolactone; Adenex; Allercorb; Antiscorbic vitamin; Antiscorbutic vitamin; Ascoltin; Ascorbajen; Ascorbutina; Ascorin; Ascorteal; Ascorvit; C-Quin; C-Vimin; Cantan; Cantaxin; Catavin C; Ce-Mi-Lin; Ce-V. Listed on TSCA. Chiral intermediate. mp = 192°; $[\alpha]^{25}$ = + 20.6° (c = 1, $H_2O$). *Aceto Corporation; Acros Organics nv; Sigma-Aldrich Fine Chemicals; TCI America.*

**418 (+)-Ascorbic acid 6-palmitate**
137-66-6 205-305-4

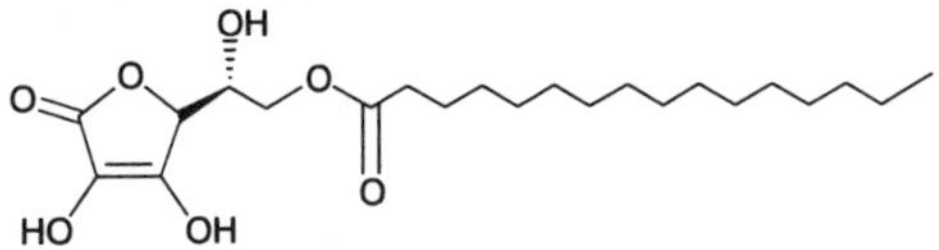

$C_{22}H_{38}O_7$
Listed on TSCA. Chiral intermediate. mp = 115-118°; $[\alpha]_D^{20}$ = + 22.9° (c = 2, $CH_3OH$). *Sigma-Aldrich Fine Chemicals.*

**419 L-Ascorbic acid sodium salt**
134-03-2 8723(12) 205-126-1

$C_6H_8O_6$
Ascorbic acid, monosodium salt, L-; Sodium L-ascorbate; Vitamin C sodium salt; Monosodium ascorbate; 3-Oxo-L-gulofuranolactone sodium; Ascorbicin; Ascorbin; Cebitate; Cenolate; Iskia-c; Natrascorb; Natri-c; Sodascorbate; E 301; HBL 508. Listed on TSCA. Chiral building block. mp = 218°; $[\alpha]^{25}$ = + 104° (c = 1, $H_2O$); soluble in $H_2O$ (> 10 g/100 ml, 17°). *Sigma-Aldrich Fine Chemicals; TCI America.*

**420 L-Ascorbyl 2,6-dibutyrate**
4337-04-6
$C_{14}H_{20}O_8$
Chiral intermediate. *TCI America.*

**421 L-Ascorbyl 2,6-dipalmitate**
4218-81-9
$C_{38}H_{68}O_8$
2,6-Dipalmitoyl-L-ascorbic acid. Chiral intermediate. mp = 113°. *TCI America.*

**422 L-Ascorbyl 6-stearate**
25395-66-8 246-944-9

$C_{24}H_{42}O_7$
Ascorbic acid, monooctadecanoate, L-; (Stearoyloxy)-L-ascorbic acid; E 304 (ii). Chiral intermediate. mp = 117°. *TCI America*.

**423 D-Asparagine**
5794-24-1 859(11) 218-163-3

$C_4H_8N_2O_3$
Chiral building block. mp = 275°; $[\alpha]_D^{20}$ = - 27° (c = 1, 1N HCl). *Sigma-Aldrich Fine Chemicals; TCI America; Varsal Instruments, Inc.; Yoneyama Yakuhin Kogyo Co., Ltd.*

**424 L-Asparagine**
70-47-3 872(12) 200-735-9

$C_4H_8N_2O_3$
Butanoic acid, 2,4-diamino-4-oxo-, (S)-; Aspartic acid β-amide; (S)-2,4-Diamino-4-oxobutanoic acid; α-Aminosuccinamic acid; Agedoite; Altheine; Asn; Asparagine acid; Asparamide; Aspartamic acid; Crystal VI; L-Aspartamine. Listed on TSCA. Chiral intermediate. mp = 235°; $[\alpha]_D^{20}$ = + 28.3° (c = 13.2, 1N HCl). *Fischer Chemicals AG; Kyowa Hakko Kogyo Co., Ltd.; Rexim S.A. Produits Chimiques; Senn Chemicals AG; Sigma-Aldrich Fine Chemicals; Varsal Instruments, Inc.*

**425 L-Asparagine hydroxysuccinimide ester**

$C_8H_{11}N_3O_5$
Chiral intermediate. *Flamma s.p.a.*

**426 L-Asparagine methyl ester**

$C_5H_{10}N_2O_3$
Chiral building block. *Flamma s.p.a.*

**427 D-(-)-Asparagine monohydrate**
2058-58-4 872(12) 218-163-3

$C_4H_8N_2O_3$
D-2-Aminosuccinamic acid monohydrate; D-2-Aminobutanedioic acid monohydrate. Chiral building block. mp = 275°; $[\alpha]_D^{20}$ = - 34° (c = 1, 6N HCl). *Acros Organics nv; Austin Chemical Company, Inc.; Sigma-Aldrich Fine Chemicals; Varsal Instruments, Inc.*

**428 L-(+)-Asparagine monohydrate**
5794-13-8 859(11)

$C_4H_{10}N_2O_4$
Asparagine, monohydrate, L-.
(S)-)(+)-2-Aminosuccinic acid; (S)-(+)-Asparagine; L-Ascorbyl 6-palmitate. Listed on TSCA. Chiral building block. mp = 233-235°; [α] = + 33.5° (c = 10, 1N HCl); $[\alpha]_D^{20}$ = + 22° (c = 1, 0.1N HCl). *Acros Organics nv; Alfa Chemicals Ltd.; Austin Chemical Company, Inc.; Great Lakes Fine Chemicals; Sigma-Aldrich Fine Chemicals; TCI America; Varsal Instruments, Inc.; Yoneyama Yakuhin Kogyo Co., Ltd.*

**429 L-Asparagine p-nitrophenyl ester**

$C_{10}H_{11}N_3O_5$
Chiral building block. *Flamma s.p.a.*

**430 L-Aspartate sodium**
3792-50-5 223-264-0

$C_4H_7NaNO_4$
Aspartic acid, monosodium salt, L-; Sodium hydrogen L-aspartate; Monosodium L-aspartate. Listed on TSCA. Chiral building block. *Flamma s.p.a.*

**431 D-Aspartic acid**
1783-96-6 875(12) 217-234-6

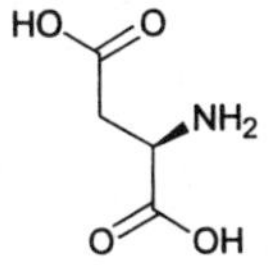

$C_4H_7NO_4$
D-Aminosuccinic acid; D-2-aminobutanedioic acid; D-Asparaginic acid. Chiral building block. mp > 300°; [α] = - 24° (c = 2.3, 6N HCl). *Acros Organics nv; Austin Chemical Company, Inc.; Rexim S.A. Produits Chimiques; Shanghai DSL International Trading Company; Sigma-Aldrich Fine Chemicals; Tanabe Seiyaku Co. Ltd.; TCI America; Toray International, Inc.; Yoneyama Yakuhin Kogyo Co., Ltd.*

**432 L-Aspartic acid**
56-84-8 875(12) 200-291-6

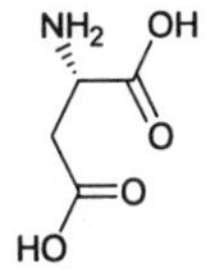

$C_4H_7NO_4$
Butanedioic acid, amino-, (S)-; L-Aminosuccinic acid; (+)-2-Aminobutanedioic acid; Asparagic acid; Asparaginic acid. Listed on TSCA. Chiral intermediate. mp > 300°; $[\alpha]_D^{20}$ = + 25° (c = 2, 5N HCl). *Aceto Corporation; Acros Organics nv; Ajinomoto Co. Inc.; Austin Chemical Company, Inc.; Daesang Corporation; DSM Fine Chemcials Netherlands; Fischer Chemicals AG; Flamma s.p.a.; Great Lakes Fine Chemicals; Kyowa Hakko Kogyo Co., Ltd.; Rexim S.A. Produits Chimiques; Senn Chemicals AG; Shanghai DSL International Trading Company; Sigma-Aldrich Fine Chemicals; Tanabe Seiyaku Co. Ltd.; TCI America; Varsal Instruments, Inc.; Yoneyama Yakuhin Kogyo Co., Ltd.*

**433 L-Aspartic acid α-(7-amido-4-methyl-coumarin) trifluoroacetic acid salt**

$C_{16}H_{15}F_3N_2O_7$
Chiral intermediate. *Acros Organics nv.*

**434 L-Aspartic acid 4-benzyl ester**
2177-63-1 218-541-8

$C_{11}H_{13}NO_4$
Aspartic acid, 4-benzyl ester, L-; Benzyl hydrogen β-L-aspartate; 4-Benzyl L-aspartate; L-Aspartic acid, 4-(phenylmethyl) ester. Chiral intermediate. *Austin Chemical Company, Inc.*

**435 D-Aspartic acid dibenzyl ester toluene-4-sulfonate**
4079-64-5

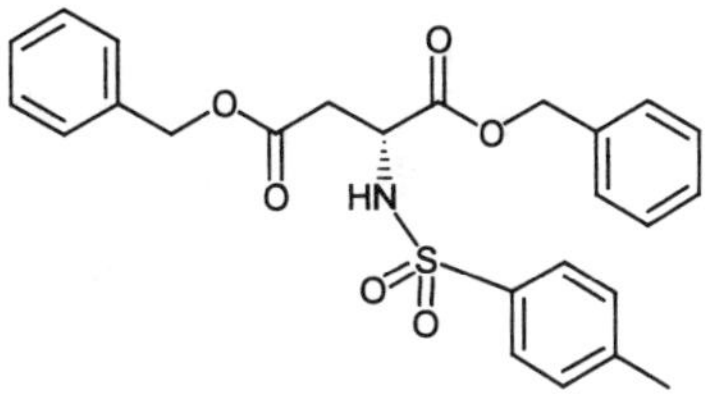

$C_{25}H_{27}NO_7S$
Aspartic acid, dibenzyl ester, p-toluenesulfonate, D-; D-Aspartic acid, bis(phenylmethyl) ester, 4-methylbenzenesulfonate. Chiral intermediate. *Austin Chemical Company, Inc.*

**436 (+)-Aspartic acid dibenyl ester p-toluenesulfonate**
2886-33-1 220-746-2

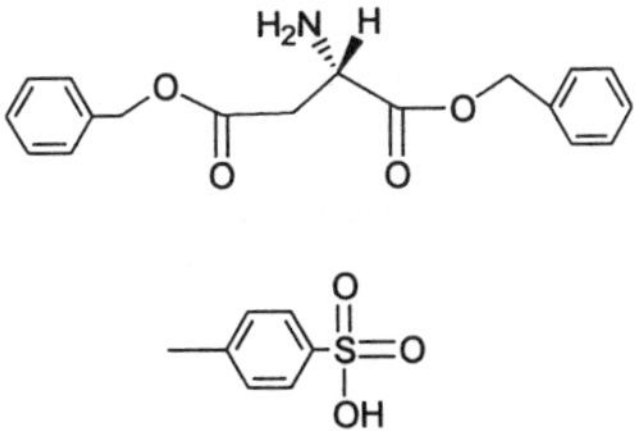

$C_{18}H_{19}NO_4$
Chiral intermediate. mp = 157-160°; $[\alpha]^{24}$ = + 9° (c = 1, $CHCl_3$). *Sigma-Aldrich Fine Chemicals.*

**437 L-Aspartic acid-1,4-dimethyl ester hydrochloride**
32213-95-9 250-957-7

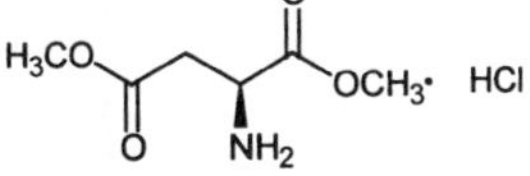

$C_6H_{12}ClNO_4$
Aspartic acid, dimethyl ester, hydrochloride, L-; Dimethyl L-aspartate hydrochloride. Chiral intermediate. mp = 115-117°. *Austin Chemical Company, Inc.; Fischer Chemicals AG; Sigma-Aldrich Fine Chemicals.*

**438 L-Aspartic acid 4-(4-methyl-7-coumarinylamide)**

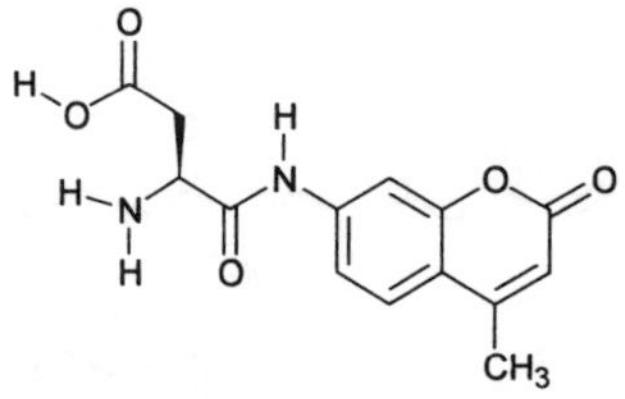

$C_{14}H_{14}N_2O_5$
Chiral intermediate. *Acros Organics nv.*

**439 L-Aspartic acid β-monomethyl ester hydrochloride**
16856-13-6 240-880-5

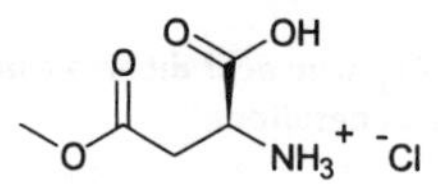

$C_5H_{10}ClNO_4$
Aspartic acid, 4-methyl ester, hydrochloride, L-; 4-Methyl hydrogen L-aspartate hydrochloride. Chiral intermediate. *Omega Chemical Company Inc.*

**440 L-Aspartic acid potassium salt**
1115-63-5 862(11) 214-226-4

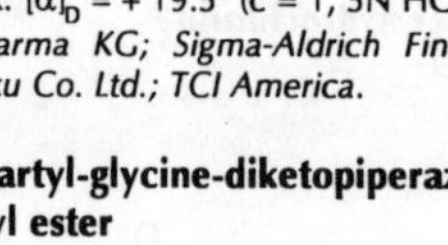

$C_4H_7NO_4$
Aspartic acid, monopotassium salt, L-; Potassium hydrogen aspartate; Potassium aspartate; Monopotassium aspartate; Potassium L-Aspartate. Listed on TSCA. Chiral building block. $[\alpha]_D^{20}$ = + 19.5° (c = 1, 5N HCl). *Boehringer Ingelheim Pharma KG; Sigma-Aldrich Fine Chemicals; Tanabe Seiyaku Co. Ltd.; TCI America.*

**441 L-Aspartyl-glycine-diketopiperazine methyl ester**

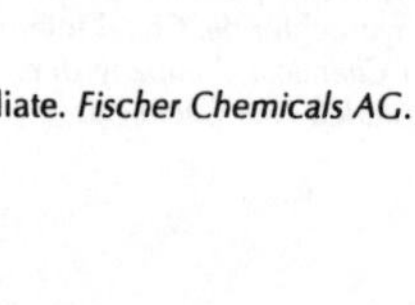

$C_7H_{10}N_2O_4$
Chiral intermediate. *Fischer Chemicals AG.*

**442 L-Aspartyl-L-phenylalanine**
13433-09-5 236-557-3

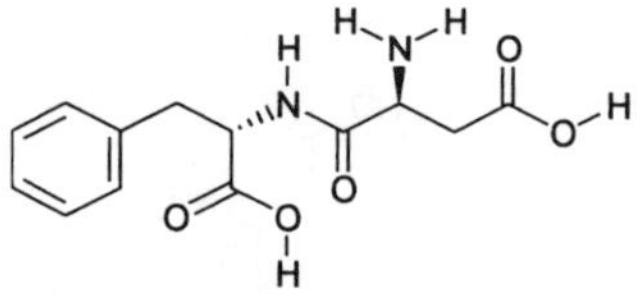

$C_{13}H_{16}N_2O_5$
Aspartylphenylalanine, α-; Demethylaspartame; L-Phenylalanine, L-α-aspartyl-. Chiral intermediate. mp = 236-239°; $[\alpha]_D^{20}$ = + 12.5° (c = 1, 0.5N HCl). *Acros Organics nv; Fischer Chemicals AG; Sigma-Aldrich Fine Chemicals.*

**443 L-Aspartyl-L-phenylalanine methyl ester**
22839-47-0 874(12) 245-261-3

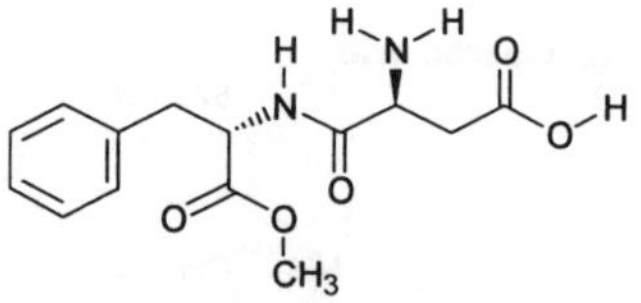

$C_{14}H_{18}N_2O_5$
Succinamic acid, 3-amino-N-(α-carboxyphenethyl)-, N-methyl ester, stereoisomer; Equal; Nutrasweet; L-Phenylalanine, N-L-α-aspartyl-, 1-methyl ester; Aspartame; Sweet dipeptide; Methyl aspartylphenylalanate; Canderel; Dipeptide sweetener; α-Sweet. Listed on TSCA. Chiral intermediate. mp = 248-250°; [α] = + 15° (c = 4, 15N $HCO_2H$). *Acros Organics nv; Fischer Chemicals AG; Sigma-Aldrich Fine Chemicals; Varsal Instruments, Inc.*

**444 β-Aspartyl-L-phenylalanine-1-methyl ester**
22839-61-8 245-262-9

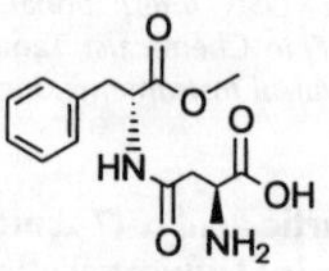

$C_{14}H_{18}N_2O_5$
Asparagine, N-(α-carboxyphenethyl)-, N-methyl ester, L-. L-Phenylalanine, L-β-aspartyl-, 2-methyl ester; β-Aspartame; 1-methyl N-L-β-aspartyl-3-phenyl-L-alaninate. Chiral intermediate. mp = 201-204°; $[\alpha]_D^{20}$ = + 6° (c = 1, $H_2O$). *Fischer Chemicals AG; Sigma-Aldrich Fine Chemicals.*

**445 (S)-(-)-Atenolol**
93379-54-5 879(11)

$C_{14}H_{22}N_2O_3$
(-)-4-[2-Hydroxy-3-[(1-methylethyl)amino]propoxy]benzeneacetamide. Chiral intermediate. mp = 148-152°; $[\alpha]^{25}$ = - 16° (c = 1, 1N HCl). *Sigma-Aldrich Fine Chemicals.*

**446 Atenolol**
56715-13-0 879(11)

$C_{14}H_{22}N_2O_3$
(+)-4-[2-Hydroxy-3-[(1-methylethyl)amino]propoxy]benzeneacetamide. Chiral intermediate. mp = 148-152°; $[\alpha]^{25}$ = + 16° (c = 1, 1N HCl). *Sigma-Aldrich Fine Chemicals.*

**447 [(3S)-4-Azabicyclo[4.4.0]dec-3-yl]-N-(tert-butyl)carboxamide**

$C_{14}H_{26}N_2O$
Chiral intermediate. [α] = - 70°. *Acros Organics nv.*

**448 (1R,4S)-2-Azabicyclo[2.2.1]heptan-3-one**
71830-08-5
$C_6H_9NO$
Chiral building block. *ChiroTech Technology Ltd.; Lancaster Synthesis Ltd.*

**449 (1S,4R)-2-Azabicyclo[2.2.1]heptan-3-one**
134003-03-5
$C_6H_{11}NO_2$
Chiral intermediate. *ChiroTech Technology Ltd.; Lancaster Synthesis Ltd.*

**450 (1R,4R)-(-)-2-Azabicyclo[2.2.1]hept-5-en-3-one**

$C_6H_7NO$
(1R,4S)-(-)-2-Azabicyclo[2.2.1]hept-5-en-3-one. Chiral intermediate. mp = 94-97°; $[\alpha]_D^{20}$ = - 565° (c = 1, $CHCl_3$). *Acros Organics nv; Celltech Chiroscience Ltd.; ChiroTech Technology Ltd.; Lancaster Synthesis Ltd.*

**451 (1S,4R)-(+)-2-Azabicyclo[2.2.1]hept-5-en-3-one**
130931-83-8

$C_6H_7NO$
Chiral intermediate. mp = 94-97°; $[\alpha]_D^{20}$ = + 565° (c = 1, $CHCl_3$). *Acros Organics nv; Celltech Chiroscience Ltd.; ChiroTech Technology Ltd.; Lancaster Synthesis Ltd.; Sigma-Aldrich Fine Chemicals.*

**452 (1S,4S)-(+)-2-Aza-5-oxabicyclo[2.2.1]-heptane hydrochloride**
31560-06-2

$C_5H_{10}ClNO$
Chiral intermediate. mp = 155-158°; $[\alpha]^{23}$ = + 84° (c = 1, $H_2O$). *Sigma-Aldrich Fine Chemicals.*

**453 (-)-6-Azauridine 2,3,5-triacetate 2-ribo-furanosyl-1,2,4-triazine-3,5(2H,4H)-dione 2,3,5-triacetate**
2169-64-4 937(12) 218-515-6

$C_{14}H_{17}N_3O_9$
2-β-D-ribofuranosyl-1,2,4-triazine-3,5(2H,4H)-dione 2',3',5'-triacetate. Chiral intermediate. mp = 99-101°; $[\alpha]^{22}$ = - 40.9° (c = 1, pyridine). *Sigma-Aldrich Fine Chemicals.*

**454 (2R,4R)-(+)-Azetidine-2,4-dicarboxylic acid**
161596-63-0

$C_5H_7NO_4$
Chiral intermediate. *Acros Organics nv.*

**455 (2S,4S)-(-)-Azetidine-2,4-dicarboxylic acid**
161596-62-9

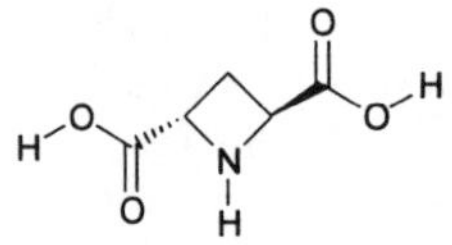

$C_5H_7NO_4$
tADA; trans-2,4-Azetidinedicarboxylic acid. Chiral intermediate. mp = 239-240°. *Acros Organics nv.*

**456 (R)-(+)-Azetidine-2-carboxylic acid**

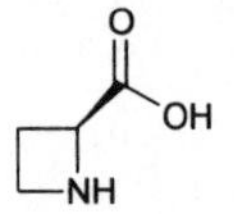

$C_4H_7NO_2$
Azetidine-2-carboxylic acid, (2S)-. Chiral building block. *FineTech, Ltd.*

**457 (S)-(-)-2-Azetidinecarboxylic acid**
2133-34-8 940(12) 218-362-5

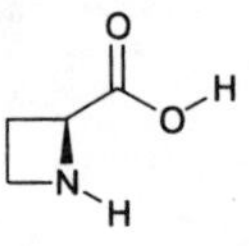

$C_4H_7NO_2$
Azetidine-2-carboxylic acid, (S)-; (S)-(-)-Azetidine-2-carboxylic acid; L-Trimethyleneimine-2-carboxylic acid. Chiral building block. mp = 217°; $[\alpha]_D^{20}$ = - 123 ± 2° (c = 4, $H_2O$). *Acros Organics nv; FineTech, Ltd.; Kaneka Corporation; Lancaster Synthesis Ltd.; Senn Chemicals AG; TCI America*; Aldrich.

**458 (+)-3αS-(3α,4,5,7α)-5-Azido-7-bromo-3α,4,5,7α-tetrahydro-2,2-dimethyl-1,3-benzodioxol-4-ol**

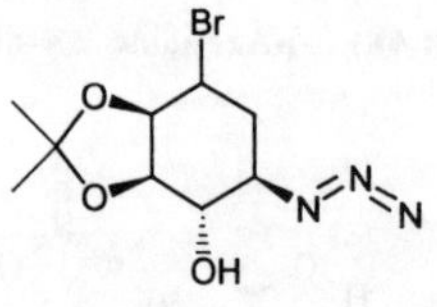

$C_9H_{12}BrN_3O_3$
Chiral intermediate. mp = 116-118°; $[\alpha]^{25}$ = + 24° (c = 1, $CHCl_3$). *Sigma-Aldrich Fine Chemicals.*

**459 (-)-1-Azido-1-deoxygalactopyranoside tetraacetate**

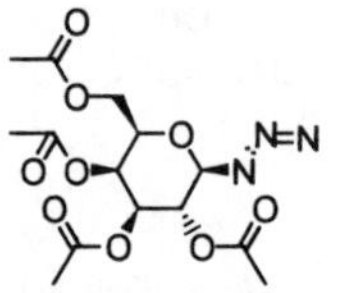

$C_{14}H_{19}N_3O_9$
Chiral intermediate. mp = 96-99°; $[\alpha]^{25}$ = -15° (c = 1, $CHCl_3$). *Sigma-Aldrich Fine Chemicals.*

**460 2-Azido-2-deoxy-1,3,4,6-tetra-O-acetyl-α-D-galactopyranose**

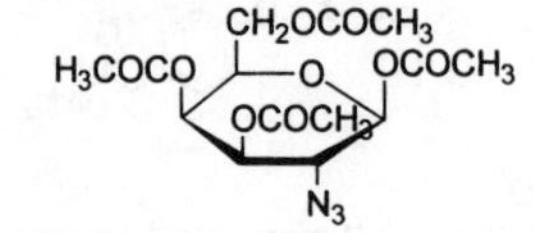

$C_{14}H_{19}N_3O_9$
Chiral intermediate. *Senn Chemicals AG.*

**461 2-Azido-2-deoxy-1,3,4,6-tetra-O-acetyl-α-D-glucopyranose**

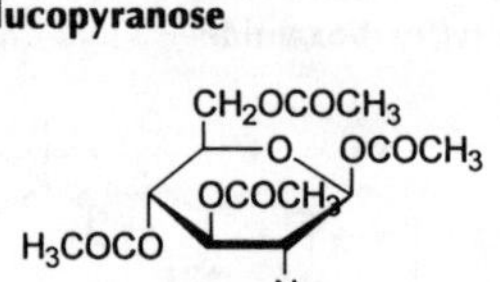

$C_{14}H_{19}N_3O_9$
Chiral intermediate. *Senn Chemicals AG.*

**462 2-Azido-2-deoxy-1,3,4,6-tetra-O-acetyl-α-D-mannopyranose**

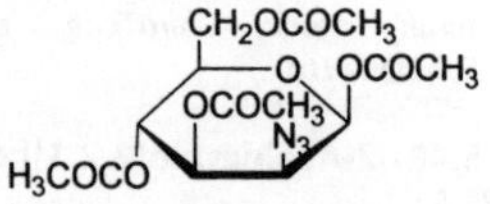

$C_{14}H_{19}N_3O_9$
Chiral intermediate. *Senn Chemicals AG.*

**463 (-)-3-Azido-2,3-dideoxy-1-O-(tert-butyl-dimethylsilyl)arabino-hexopyranose**

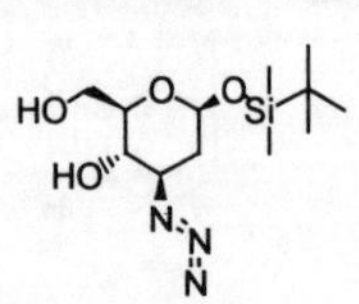

$C_{12}H_{25}N_3O_4Si$
Chiral intermediate. mp = 73-75°; $[\alpha]_D^{20}$ = - 32° (c = 1, $CHCl_3$). *Sigma-Aldrich Fine Chemicals.*

**464 D-(-)-α-Azidophenylacetic acid**
29125-25-5 249-455-9

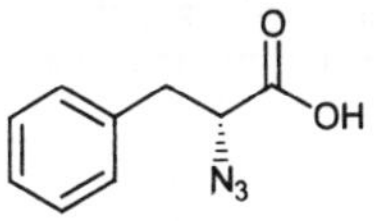

$C_8H_7N_3O_2$
Benzeneacetic acid, α-azido-, (R)-; Acetic acid, azidophenyl-, D-(-)-. Chiral intermediate. *Dynamit Nobel GmbH.*

**465 D-(-)-α-Azidophenylacetyl chloride**

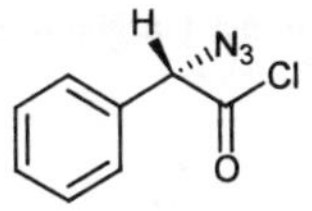

$C_8H_6ClN_3O$
Chiral intermediate. *Dynamit Nobel GmbH.*

**466 Baclofen**
967(12) 214-486-9

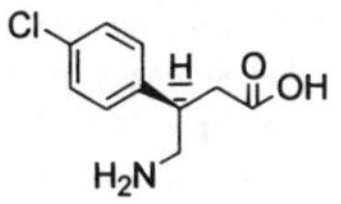

$C_{10}H_{12}ClNO_2$
β-(Aminomethyl)-4-chlorobenzenepropanoic acid. Chiral intermediate. mp = 189°; $[\alpha]_D^{20}$ = + 2° (c = 0.5, 1N HCl). *Sigma-Aldrich Fine Chemicals.*

**467 D-Baicalin hydrate**

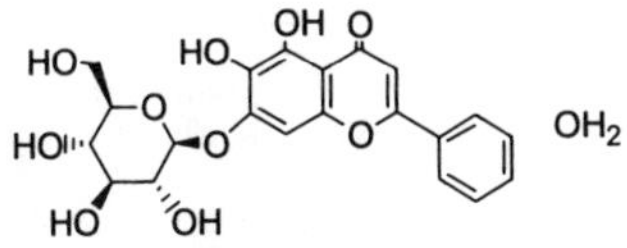

$C_{21}H_{18}O_{11}$
Baicalein 7-β-D-glucopyranosiduronate. Chiral intermediate. mp = 231-233°; $[\alpha]^{23}$ = - 120° (c = 1, 50% pyridine). *Sigma-Aldrich Fine Chemicals.*

**468 cis-(1R,2S)-(-)-2-Benzamidocyclohexane-carboxylic acid**
26693-55-0
$C_{14}H_{17}NO_3$
Chiral intermediate. mp = 205-206°; $[\alpha]_D^{20}$ = - 36.5 ± 1.5° (c = 0.5, EtOH). *Lancaster Synthesis Ltd.; Yamakawa Chemical Industry Co. Ltd.*

**469 cis-(1S,2R)-(+)-2-Benzamidocyclohexane-carboxylic acid**
26685-82-5
$C_{14}H_{17}NO_3$
Chiral intermediate. mp = 205-206°; $[\alpha]_D^{20}$ = + 36.5 ± 1.5° (c = 0.5, EtOH). *Lancaster Synthesis Ltd.; Yamakawa Chemical Industry Co. Ltd.*

**470 (S)-Benzenepropanoic acid, μ-hydrazino-3,4-dihydroxy-+1m-methyl-, monohydrate**
38821-49-7

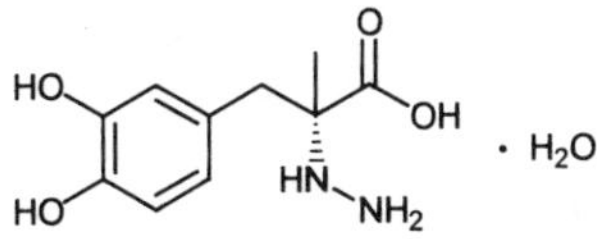

$C_{10}H_{16}N_2O_5$
Carbidopa monohydrate. Chiral intermediate. $[\alpha]^{25}$ = - 21.0 to - 23.5° (c = 1, $AlCl_3$ solution, pH 1.5). *Sochinaz SA.*

**471 (1R,2S,3R)-(+)-3-N-Benzenesulfonyl-N-(3,5-dimethylphenyl)amino-2-bornanol**
87420-26-6

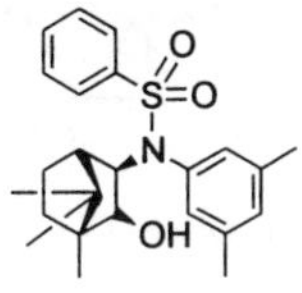

$C_{24}H_{31}NO_3S$
Chiral intermediate. mp = 125-128°; $[\alpha]_D^{20}$ = + 65° (c = 4, $CHCl_3$). *Sigma-Aldrich Fine Chemicals.*

**472 (R)-Benzodioxan-2-carboxylic acid**

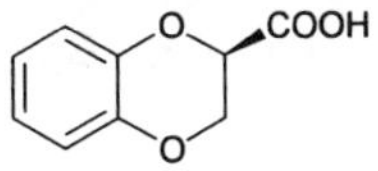

$C_9H_8O_4$
Chiral intermediate. *Austin Chemical Company, Inc.; FineTech, Ltd.*

**473 (S)-1,4-Benzodioxan-2-carboxylic acid**

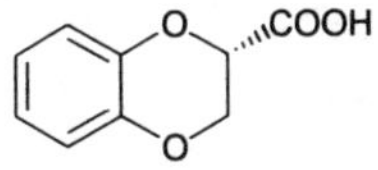

$C_9H_8O_4$
Chiral building block. *Austin Chemical Company, Inc.; FineTech, Ltd.*

**474 (R)-(-)-Benzoin**
5928-65-5

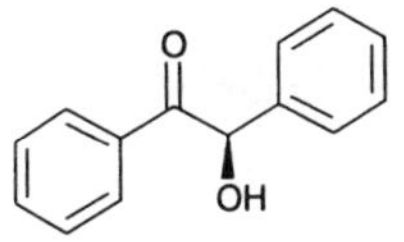

$C_{14}H_{12}O_2$
2-Hydroxy-2-phenylacetophenone. Chiral intermediate.

mp = 135-137°; $[\alpha]^{24}$ = - 115° (c = 1.5, acetone). *Sigma-Aldrich Fine Chemicals.*

**475 (S)-Benzoin**
5928-67-6 1103(11)

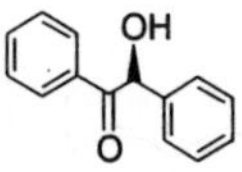

$C_{14}H_{12}O_2$
Resolving agent. mp = 135-137°; $[\alpha]^{19}$ = + 115° (c = 1.5, acetone). *Kankyo Kagaku Center Co., Ltd.; Sigma-Aldrich Fine Chemicals.*

**476 (3R,5S,7aS)-5-(Benzotriazol-1-yl)-3-phenyl[2,1-b]oxazolopyrrolidine**

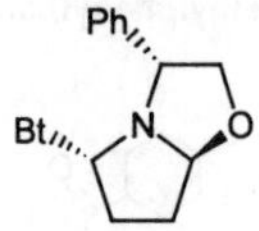

$C_{18}H_{18}N_4O$
Chiral intermediate. *Omega Chemical Company Inc.*

**477 (3S,5R,7aR)-5-(Benzotriazol-1-yl)-3-phenyl[2,1- b]oxazolopyrrolidine**
205442-89-3

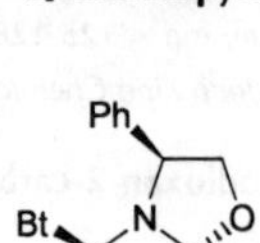

$C_{18}H_{18}N_4O$
Chiral intermediate. *Omega Chemical Company Inc.*

**478 N-Benzoyl-L-alanine methyl ester**
7244-67-9 230-649-7

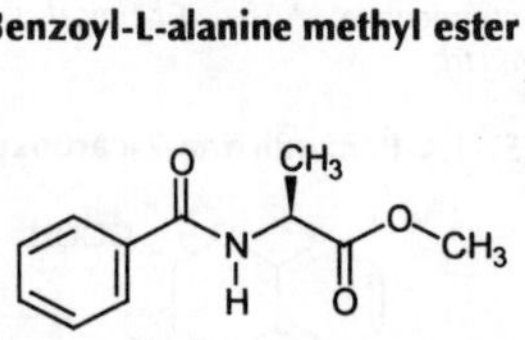

$C_{11}H_{13}NO_3$
Alanine, N-benzoyl-, methyl ester, L-; Methyl N-benzoyl-L-alaninate. Chiral intermediate. *Acros Organics nv.*

**479 N-α-Benzoyl-L-arginine**
154-92-7 205-837-7

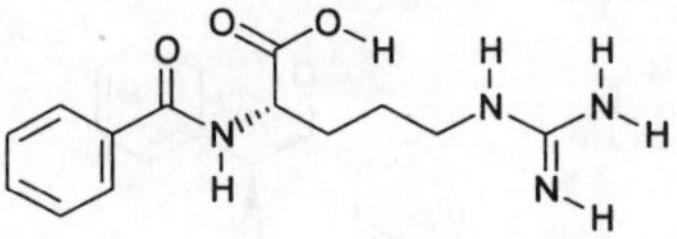

$C_{13}H_{18}N_4O_3$
$N^2$-benzoyl-, L-. Chiral intermediate. mp = 285°; $[\alpha]^{22}$ = - 9.4° (c = 1.2, 1N HCl). *Acros Organics nv; Sigma-Aldrich Fine Chemicals.*

**480 N-α-Benzoyl-L-arginine amide hydrochloride monohydrate**
965-03-7

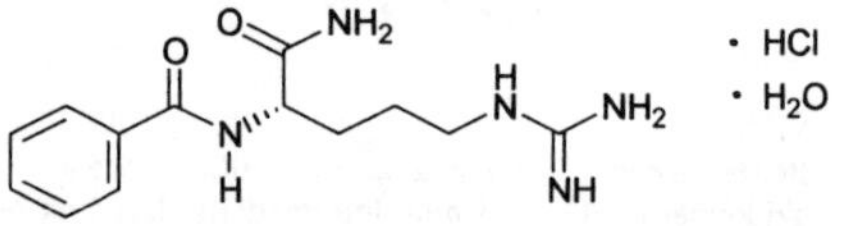

$C_{13}H_{20}ClN_5O_2$
Chiral intermediate. *Acros Organics nv.*

**481 N-α-Benzoyl-L-arginine ethyl ester hydrochloride**
2645-08-1 220-157-0

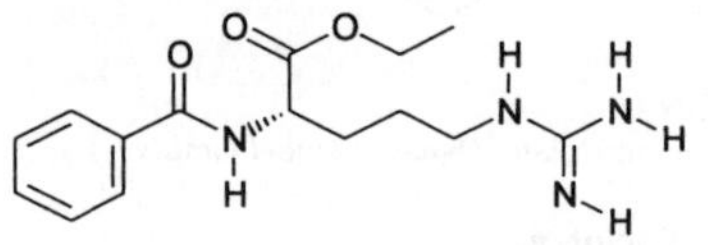

$C_{15}H_{23}ClN_4O_3$
$N^2$-benzoyl-, ethyl ester, monohydrochloride, L-; BAEE hydrochloride; Ethyl N-benzoyl-α-arginine hydrochloride. Chiral intermediate. mp = 127-131°; $[\alpha]^{22}$ = - 14.5° (c = 2, $H_2O$); $[\alpha]^{22}$ = - 15.5° (c = 2, 1N HCl). *Acros Organics nv; Sigma-Aldrich Fine Chemicals.*

**482 (S)-(+)-1-Benzoyl-2-tert-butyl-3-methyl-4-imidazolidnone**
101055-56-5

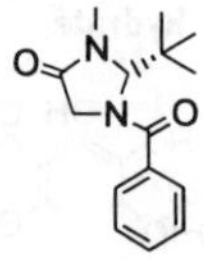

$C_{15}H_{20}N_2O_2$
Chiral intermediate. mp = 145-147°; $[\alpha]^{26}$ = + 125° (c = 1, $CH_2Cl_2$). *Sigma-Aldrich Fine Chemicals.*

**483 $N^4$-Benzoylcytidine**
13089-48-0

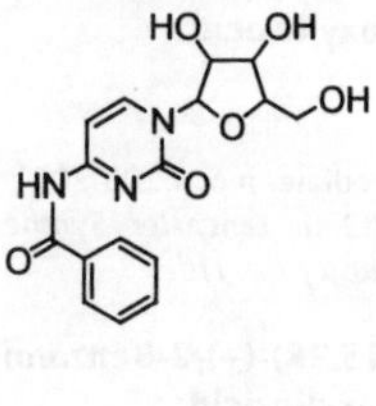

$C_{16}H_{17}N_3O_6$
Chiral intermediate. mp = 230-234°; $[\alpha]^{27}$ = + 55° (c = 0.1, $CH_3OH$). *Sigma-Aldrich Fine Chemicals.*

**484 (+)-6-O-Benzoylglucal**
58871-05-9

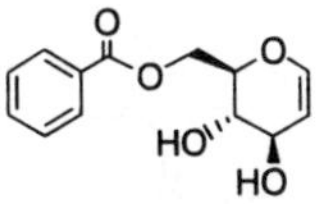

$C_{13}H_{14}O_5$
Chiral intermediate. mp = 106-111°; $[\alpha]^{25}$ = + 44° (c = 1.4, $CHCl_3$). *Sigma-Aldrich Fine Chemicals.*

**485 N-Benzoyl-D-Glucosamine**

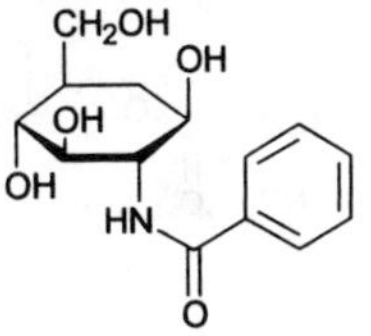

$C_{14}H_{19}NO_5$
Chiral intermediate. *Austin Chemical Company, Inc.*

**486 N-Benzoylglutamic acid**
58094-18-1

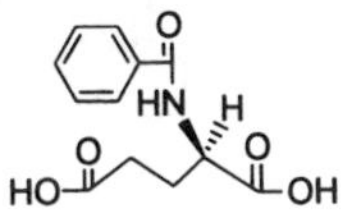

$C_{12}H_{13}NO_5$
Chiral intermediate. mp = 130-135°; $[\alpha]^{22}$ = + 12° (c = 5, $H_2O$). *Norse Laboratories; Sigma-Aldrich Fine Chemicals.*

**487 N-ε-Benzoyl-L-lysine**
98-09-1

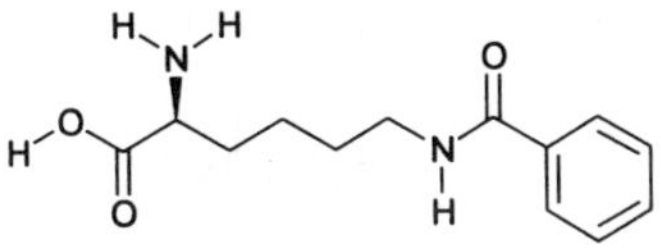

$C_{13}H_{18}N_2O_3$
Chiral building block. *Acros Organics nv.*

**488 O-Benzoylmalic acid**

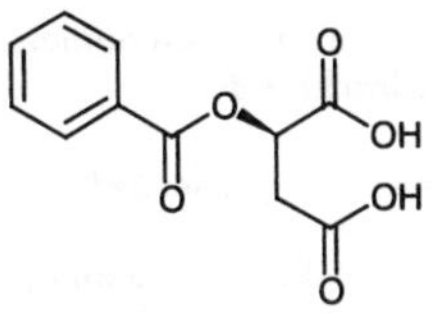

$C_{11}H_{10}O_6$
Chiral intermediate. *Norse Laboratories.*

**489 (6S)-(-)-4-Benzoyloxy-6-benzoyloxymethyl-2H-pyran-3(6H)-one**
75414-38-9
$C_{20}H_{16}O_6$
Chiral intermediate. mp = 101-103°; $[\alpha]_D^{20}$ = - 16 ± 2° (c = 1, $CHCl_3$). *Lancaster Synthesis Ltd.*

**490 (R)-(+)-1-Benzoyloxy-2-benzyloxy-3-tosyloxypropane**
109371-31-5
$C_{24}H_{24}O_6S$
(+)-BOBTOP. Chiral building block. mp = 97-99°; $[\alpha]_D^{20}$ = + 9.5 ± 0.5° (c = 2, $CHCl_3$). *Lancaster Synthesis Ltd.*

**491 (S)-(-)-1-Benzoyloxy-2-benzyloxy-3-tosyloxypropane**
109371-33-7
$C_{24}H_{24}O_6S$
(-)-BOBTOP. Chiral building block. mp = 97-99°; $[\alpha]_D^{20}$ = - 9.5 ± 0.5° (c = 2, $CHCl_3$). *Lancaster Synthesis Ltd.*

**492 N-Benzoyl-(2R,3S)-3-phenylisoserine**
132201-33-3

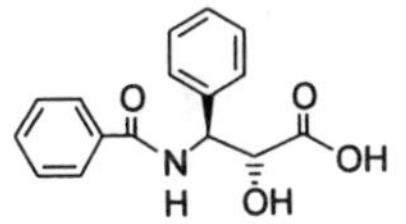

$C_{16}H_{15}NO_4$
Taxol side chain. Chiral intermediate. mp = 169-172°; $[\alpha]_D^{20}$ = - 40° (c = 1.0, EtOH). *Austin Chemical Company, Inc.; Sigma-Aldrich Fine Chemicals.*

**493 (+)-N-Benzoylthreonine**
27696-01-1

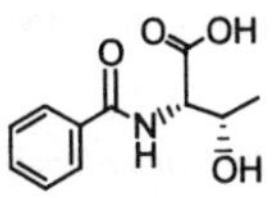

$C_{11}H_{13}NO_4$
Chiral intermediate. mp = 145-147°; $[\alpha]^{21}$ = + 25.6° (c = 1.5, $H_2O$). *Sigma-Aldrich Fine Chemicals.*

**494 (+)-N-Benzoylthreonine methyl ester**
79893-89-3

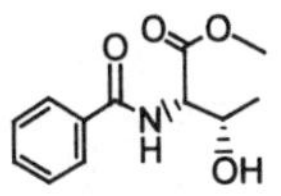

$C_{12}H_{15}NO_4$
Chiral intermediate. mp = 97-99°; $[\alpha]^{21}$ = + 23° (c = 5, EtOH). *Sigma-Aldrich Fine Chemicals.*

**495 N-Benzoyl-L-tyrosine ethyl ester**
3483-82-7 222-469-2

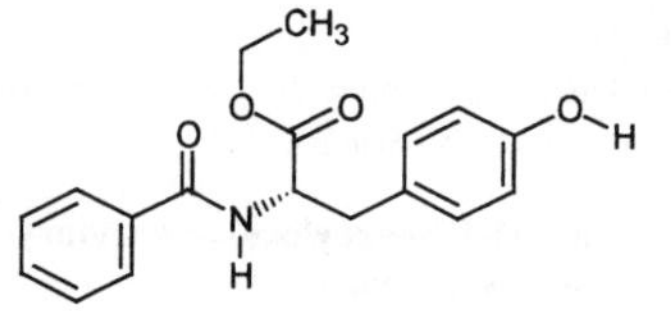

$C_{18}H_{19}NO_4$
Tyrosine, N-benzoyl-, ethyl ester, L-; Ethyl N-benzoyl-L-tyrosinate. Listed on TSCA. Chiral intermediate. mp = 118-121°; $[\alpha]^{22}$ = - 27.6° (c = 2, EtOH). *Acros Organics nv; Sigma-Aldrich Fine Chemicals.*

**496 Benzyl 2-acetamido-2-deoxy-6-O-triphenyl-methyl-α-D-glucopyranoside**
33493-71-9

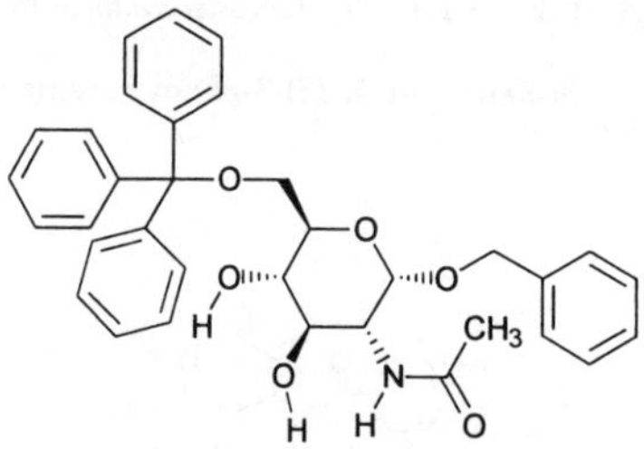

$C_{34}H_{35}NO_6$
Chiral intermediate. [α] = + 61.3° (c = 1.3, $CHCl_3$). *Acros Organics nv.*

**497 (R)-Benzyl-3-acetylaminopyrrolidine**
114636-33-8
$C_{13}H_{18}N_2O$
Chiral intermediate. *Sigma-Aldrich Fine Chemicals.*

**498 (S)-1-Benzyl-3-acetylaminopyrrolidine**
114636-30-5
$C_{13}H_{18}N_2O$
Chiral intermediate. *Sigma-Aldrich Fine Chemicals.*

**499 (R)-(-)-2-Benzylamino-1-butanol**
6257-49-4
$C_{11}H_{17}NO$
Chiral building block. mp = 73-75°; $[\alpha]_D^{20}$ = - 23 ± 1° (c = 2, $CHCl_3$). *Lancaster Synthesis Ltd.; Norse Laboratories.*

**500 (S)-(+)-2-Benzylamino-1-butanol**
26191-63-9

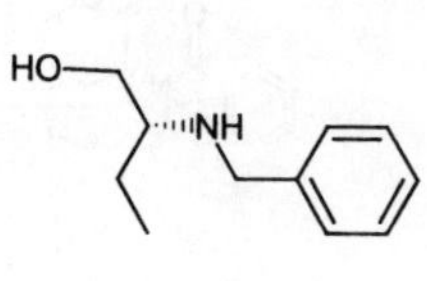

$C_{10}H_{15}NO$
Chiral building block. *Norse Laboratories.*

**501 Benzyl (R)-3-aminobutyrate sulfate salt**

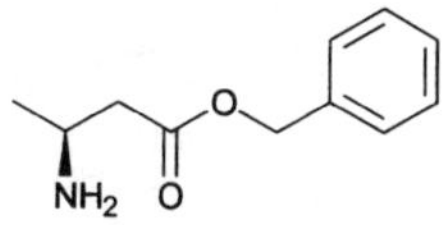

$C_{11}H_{15}NO_2$
Chiral building block. *Chiragene, Inc.*

**502 Benzyl (S)-3-aminobutyrate sulfate salt**

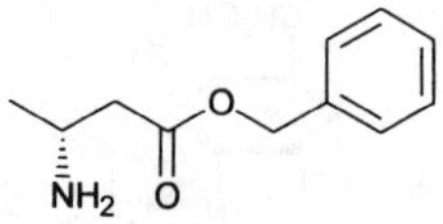

$C_{11}H_{15}NO_2$
Chiral building block. *Chiragene, Inc.*

**503 cis-(1R,2S)-(+)-2-Benzylamino-cyclohexanemethanol**
71581-92-5

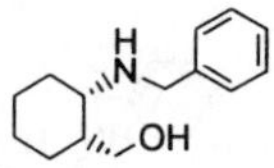

$C_{14}H_{21}NO$
(1S,2R)-(+)-cis-N-Benzyl-2-(hydroxymethyl)cyclohexyl-amine. Chiral building block; Resolving agent. mp = 67-68°; $[\alpha]_D^{20}$ = + 24 ± 1° (c = 1, $CH_3OH$). *Lancaster Synthesis Ltd.; Sigma-Aldrich Fine Chemicals; Yamakawa Chemical Industry Co. Ltd.*

**504 cis-(1S,2R)-(-)-2-Benzylamino-cyclohexanemethanol**
71581-93-6

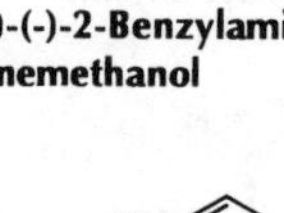

$C_{14}H_{21}NO$
(1R,2S)-(-)-cis-N-Benzyl-2-(hydroxymethyl)cyclohexyl-amine. Chiral intermediate. mp = 67-68°; $[\alpha]_D^{20}$ = -24 ± 1° (c = 1, $CH_3OH$). *Lancaster Synthesis Ltd.; Sigma-Aldrich Fine Chemicals; Yamakawa Chemical Industry Co. Ltd.*

**505 (S)-1-Benzyl-3-aminopiperidine dihydrochloride**
$C_{12}H_{20}Cl_2N_2$
Chiral intermediate. *Sigma-Aldrich Fine Chemicals.*

**506 (3R)-(-)-1-Benzyl-3-aminopyrrolidine**
114715-39-8
$C_{11}H_{16}N_2$
(3R)-(-)-3-Amino-1-benzylpyrrolidine; (R)-N-Benzyl-3-aminopyrrolidine. Chiral intermediate. d = 1.020; $n_D^{20}$ =

1.5430; $[\alpha]_D^{20}$ = - 9 ± 1° (c = 5, $H_2O$). *Austin Chemical Company, Inc.; Chiragene, Inc.; Lancaster Synthesis Ltd.; Sigma-Aldrich Fine Chemicals; TCI America.*

**507 (3S)-(+)-1-Benzyl-3-aminopyrrolidine**
114715-38-7

$C_{11}H_{16}N_2$
(3S)-(+)-3-Amino-1-benzylpyrrolidine; (S)-N-Benzyl-3-aminopyrrolidine. Chiral building block. $bp_3$ = 100-105°; d = 1.0200; $n_D^{20}$ = 1.5440; $[\alpha]_D^{20}$ = + 5 ± 0.5° (c = 10, EtOH). *Austin Chemical Company, Inc.; Chiragene, Inc.; Lancaster Synthesis Ltd.; Sigma-Aldrich Fine Chemicals; TCI America.*

**508 (S)-Benzyl 2-azetidinone-4-carboxylate**
72776-05-7

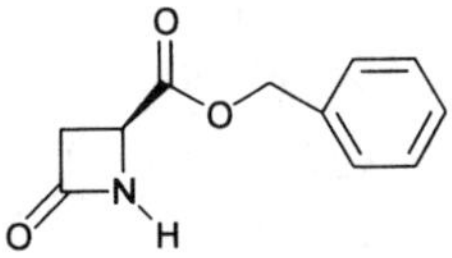

$C_{11}H_{11}NO_3$
Chiral intermediate. mp = 137-140°; $[\alpha]^{21}$ = - 44° (c = 3.3, $CHCl_3$). *Acros Organics nv; Sigma-Aldrich Fine Chemicals.*

**509 (R)-Benzylbenzylamine**

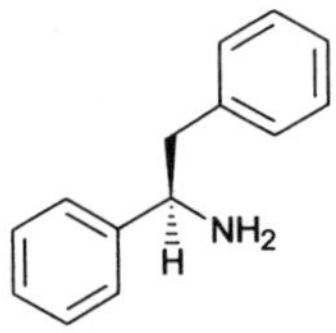

$C_{14}H_{15}N$
Benzeneethanamine, α-phenyl-, (R)-; (R)-1,2-Diphenylethylamine; Phenylphenethylamine; Phenylbenzeneethanamine; (R)-1-Amino-1,2-diphenylethane; Ethanamine, 1,2-diphenyl-, (R)-; Ethylamine, 1,2-diphenyl-, (R)-. Chiral building block. *Sumitomo Chemcial Co. Ltd.*

**510 (S)-α-Benzylbenzylamine**

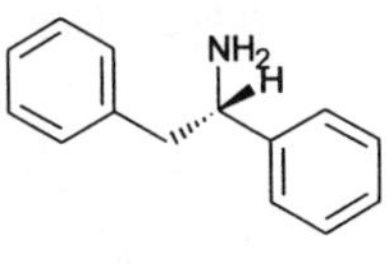

$C_{14}H_{15}N$
Benzeneethanamine, α-phenyl-, (S)-; (S)-1,2-Diphenylethylamine; (S)-α-Phenylphenethylamine; (S)-α-Phenylbenzeneethanamine; (S)-1-Amino-1,2-diphenylethane; Ethanamine, 1,2-diphenyl-, (S)-; Ethylamine, 1,2-diphenyl-, (S)-. Chiral building block. *Sumitomo Chemcial Co. Ltd.*

**511 (3R,4R)-(+)-1-Benzyl-3,4-bis(diphenylphosphino)pyrrolidine**
99135-95-2

$C_{35}H_{33}NP_2$
(R,R)-DEGUPHOS; (3R,4R)-(-)-Bis(diphenylphosphino)-benzylpyrrolidine. Chiral ligand. *Digital Specialty Chemicals, Inc.; Maxsyn.*

**512 (3S,4S)-(-)-1-Benzyl-3,4-bis(diphenylphosphino)pyrrolidine**

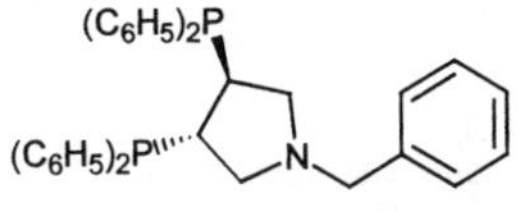

$C_{35}H_{33}NP_2$
(S,S)-DEGUPHOS; (3S,4S)-(-)-Bis(diphenylphosphino)-benzylpyrrolidine. Chiral ligand. *Digital Specialty Chemicals, Inc.; Maxsyn.*

**513 (3R)-(+)-1-Benzyl-3-(tert-butoxycarbonylamino)pyrrolidine**
131878-23-4

$C_{16}H_{24}N_2O_2$
Chiral intermediate. *TCI America.*

**514 (S)-(-)-N-α-Benzyl-N-β-(tert-butoxycarbonyl)-N-ε-carbobenzyloxy-D-hydrazinomethionine, RS-S-oxyde**

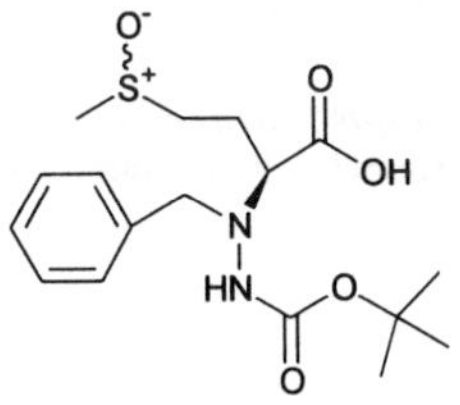

$C_{17}H_{26}N_2O_5S$
Chiral intermediate. *Acros Organics nv.*

**515 N-α-Benzyl-N-β-(tert-butoxycarbonyl)-N-ε-carbobenzyloxy-L-hydrazinomethionine RS-S-oxyde**

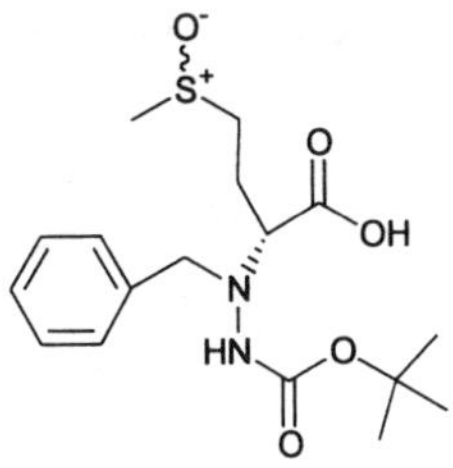

$C_{17}H_{26}N_2O_5S$
Chiral intermediate. *Acros Organics nv.*

**516 (R)-(-)-N-α-Benzyl-N-β-(tert-butoxycarbonyl)-D-hydrazinoalanine dicyclohexylamine salt**

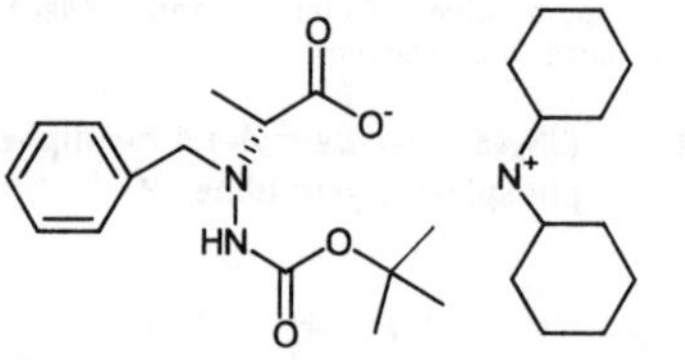

$C_{27}H_{45}N_3O_4$
Chiral intermediate. *Acros Organics nv.*

**517 (S)-(+)-N-α-Benzyl-N-β-(tert-butoxycarbonyl)-L-hydrazinoalanine dicyclohexylamine salt**

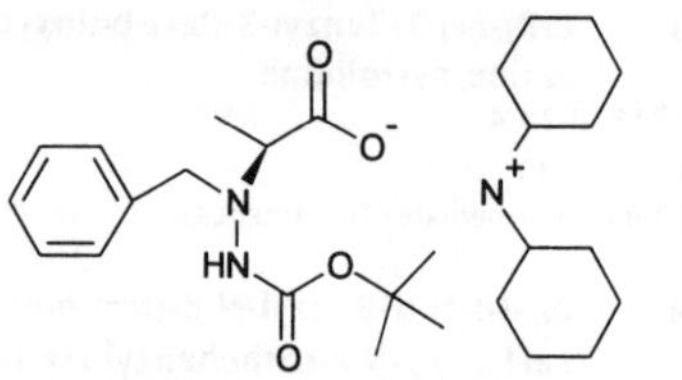

$C_{27}H_{45}N_3O_4$
Chiral intermediate. *Acros Organics nv.*

**518 (S)-(+)-N-α-Benzyl-N-β-(tert-butoxycarbonyl)-L-hydrazinoisoleucine**

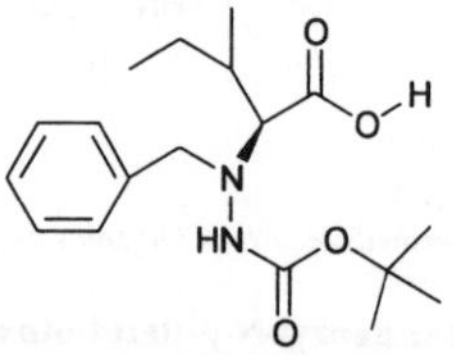

$C_{18}H_{28}N_2O_4$
Chiral intermediate. *Acros Organics nv.*

**519 (R)-(-)-N-α-Benzyl-N-β-(tert-butoxycarbonyl)-D-hydrazinoserine**

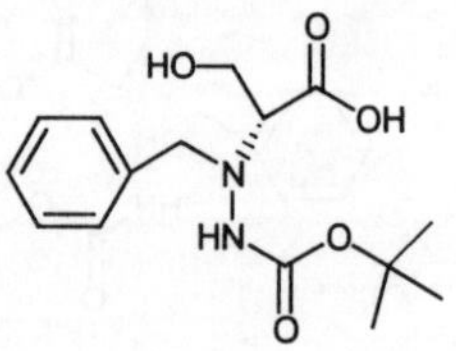

$C_{15}H_{22}N_2O_5$
Chiral intermediate. *Acros Organics nv.*

**520 (S)-(+)-N-α-Benzyl-N-β-(tert-butoxycarbonyl)-L-hydrazinoserine**

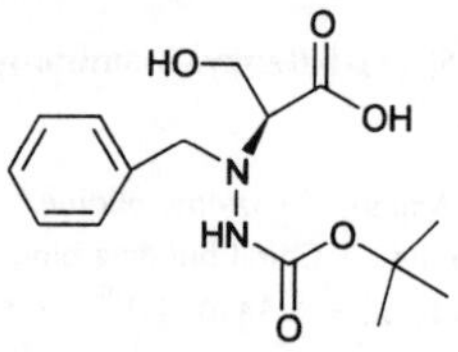

$C_{15}H_{22}N_2O_5$
Chiral intermediate. *Acros Organics nv.*

**521 (R)-(-)-N-α-Benzyl-N-β-(tert-butoxycarbonyl)-D-hydrazinotryptophane**

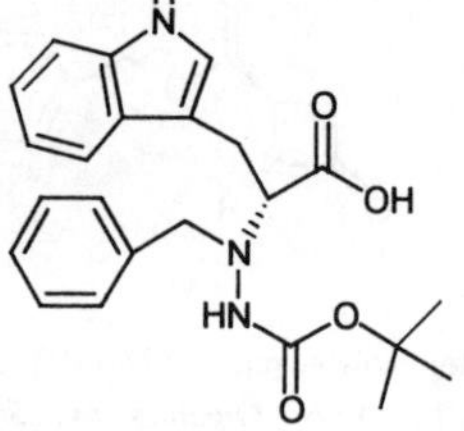

$C_{23}H_{27}N_3O_4$
Chiral intermediate. *Acros Organics nv.*

**522 (S)-(+)-N-α-Benzyl-N-β-(tert-butoxycarbonyl)-L-hydrazinotryptophane**

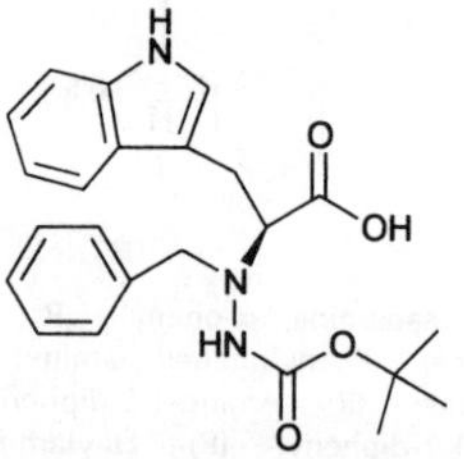

$C_{23}H_{27}N_3O_4$
Chiral intermediate. *Acros Organics nv.*

**523 (R)-(-)-N-α-Benzyl-N-β-(tert-butoxycarbonyl)-D-hydrazinovaline**

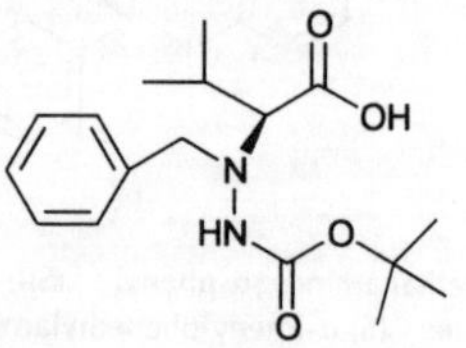

$C_{17}H_{26}N_2O_4$
Chiral intermediate. *Acros Organics nv.*

**524 (S)-(+)-N-α-Benzyl-N-β-(tert-butoxy-carbonyl)-L-hydrazinovaline**

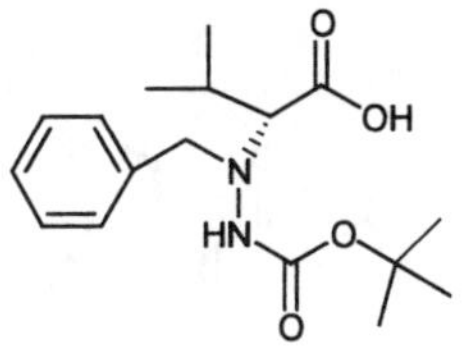

$C_{17}H_{26}N_2O_4$
Chiral intermediate. *Acros Organics nv.*

**525 (R)-(-)-N-α-Benzyl-N-β-(tert-butoxy-carbonyl)-O-benzyl-D-hydrazinotyrosine**

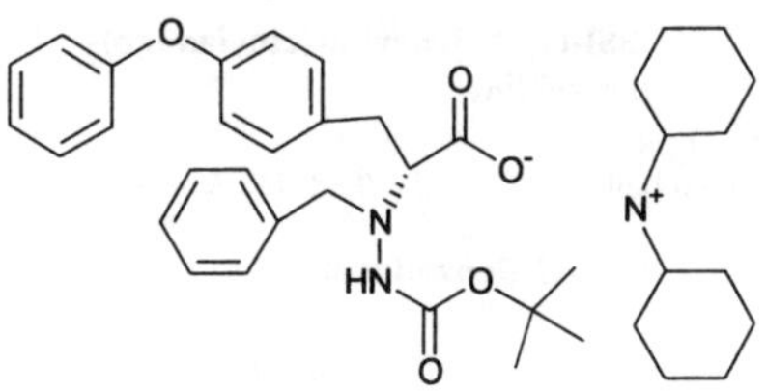

$C_{39}H_{53}N_3O_5$
Chiral intermediate. *Acros Organics nv.*

**526 (S)-(+)-N-α-Benzyl-N-β-(tert-butoxy-carbonyl)-O-benzyl-L-hydrazinotyrosine**

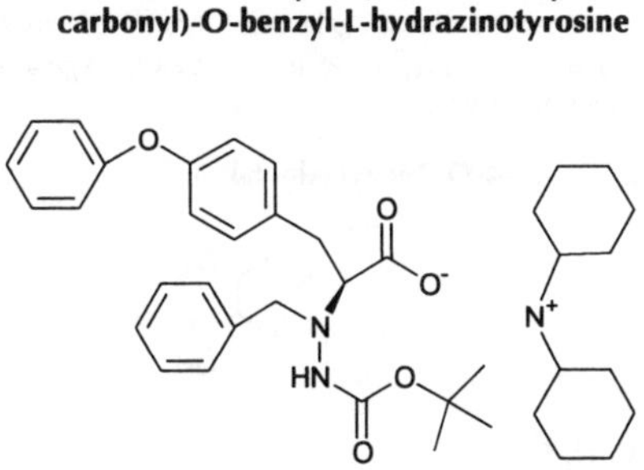

$C_{39}H_{53}N_3O_5$
Chiral intermediate. *Acros Organics nv.*

**527 N-Benzylcinchonidinium bromide**

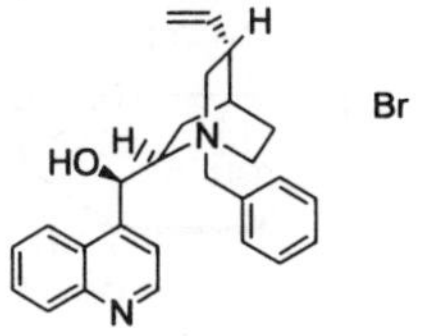

$C_{26}H_{29}BrN_2O$
Chiral auxiliary; Resolving agent. mp = 190°; $[\alpha]_D^{20}$ = - 138° (c = 1, $CHCl_3$). *Sigma-Aldrich Fine Chemicals.*

**528 (8S,9R)-(-)-N-Benzylcinchonidinium chloride**
69257-04-1 273-938-3

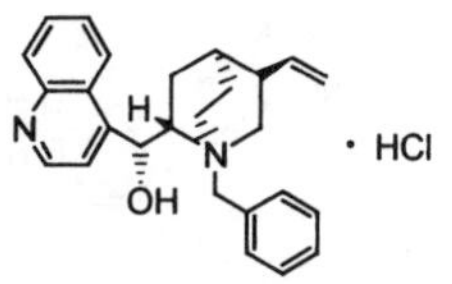

$C_{26}H_{29}ClN_2O$
Cinchonanium, 9-hydroxy-1-(phenylmethyl)-, chloride, (8α,9R)-. Chiral auxiliary; Resolving agent; Chiral ligand. mp = 210°; $[\alpha]_D^{20}$ = - 180° (c = 1.3, $H_2O$). *Omega Chemical Company Inc.; Sigma-Aldrich Fine Chemicals.*

**529 (S)-4-Benzyl-3-crotonyl-2-oxazolidinone**
133812-16-5
$C_{14}H_{15}NO_3$
N-Crotonyl-(4S)-benzyl-2-oxazolidinone. Chiral intermediate. *Sigma-Aldrich Fine Chemicals.*

**530 S-Benzyl-L-cysteine**
3054-01-1 221-273-4

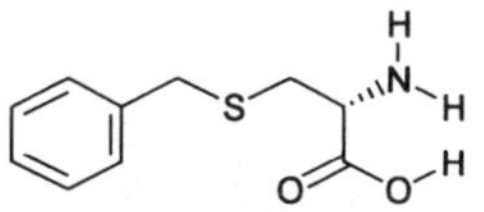

$C_{10}H_{13}NO_2S$
Alanine, 3-(benzylthio)-, L-; 3-Benzylthioalanine; L-Cysteine, S-(phenylmethyl)-. Listed on TSCA. Chiral intermediate. mp = 214°; [α] = + 26° (c = 0.4, 1N NaOH). *Acros Organics nv; Sigma-Aldrich Fine Chemicals.*

**531 (S)-Benzylcysteine-4-nitroanilide**
7436-62-6 231-090-1

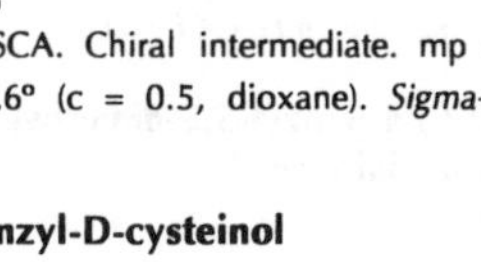

$C_{16}H_{17}N_3O_3S$
Listed on TSCA. Chiral intermediate. mp = 100-102°; $[\alpha]_D^{20}$ = - 68.6° (c = 0.5, dioxane). *Sigma-Aldrich Fine Chemicals.*

**532 S-Benzyl-D-cysteinol**
85803-43-6

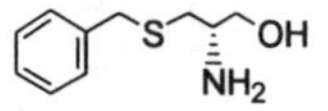

$C_{10}H_{15}NOS$
(R)-2-Amino-3-(benzylthio)-1-propanol. Chiral building block. mp = 42-46°; $[\alpha]_D^{20}$ = - 49° (c = 1.4, EtOH). *Sigma-Aldrich Fine Chemicals.*

**533 (1S,4S)-N-Benzy-2,5-diazabicyclo[2.2.1]-heptane dihydrobromide**
116258-17-4

$C_{12}H_{18}Br_2N_2$
Chiral intermediate. mp = 270°; $[\alpha]_D^{20}$= + 16.5° (c = 3, 2N NaOH). *Medinger & Sohn; Sigma-Aldrich Fine Chemicals.*

**534 (R)-N-Benzyl-3,4-dihydroxybutyramide**

$C_{11}H_{15}NO_3$
Butanamide, 3,4-dihydroxy-N-(phenylmethyl)-, (R)-. Chiral building block. *Synthon Chiragenics Corporation.*

**535 (S)-N-Benzyl-3,4-dihydroxybutyramide**
191354-49-1

$C_{11}H_{15}NO_3$
Butanamide, 3,4-dihydroxy-N-(phenylmethyl)-, (S)-. (S)-N-Phenylmethyl-3,4-dihydroxybutanamide. Chiral intermediate. *Synthon Chiragenics Corporation.*

**536 (R)-(+)-4-Benzyl-5,5-dimethyl-2-oxazolidinone**

$C_{12}H_{15}NO_2$
Chiral intermediate. mp = 60-62°; $[\alpha]_D^{20}$ = + 96° (c = 2, $CHCl_3$). *Sigma-Aldrich Fine Chemicals.*

**537 (S)-(-)-4-Benzyl-5,5-dimethyl-2-oxazolidinone**
168297-85-6

$C_{12}H_{15}NO_2$
Chiral intermediate. mp = 65-68°; $[\alpha]^{22}$ = - 98° (c = 2, $CHCl_3$). *Sigma-Aldrich Fine Chemicals.*

**538 [1(S)-Benzyl-2(S)-3-epoxypropyl]-carbamic acid tert-butyl ester**
98737-29-2

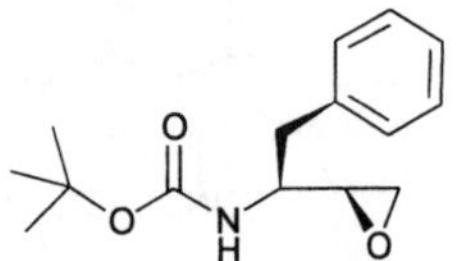


$C_{15}H_{21}NO_3$
Chiral intermediate. *Kaneka Corporation.*

**539 (3R)-(-)-1-Benzyl-3-(ethylamino)-pyrrolidine**
$C_{13}H_{20}N_2$
Chiral intermediate. d = 0.98. *TCI America.*

**540 (3S)-(+)-1-Benzyl-3-(ethylamino)-pyrrolidine**
$C_{13}H_{20}N_2$
Chiral building block. d = 0.98. *TCI America.*

**541 (-)-4-O-Benzylfucal**

$C_{13}H_{16}O_3$
4-O-Benzyl-6-deoxy-L-galactal. Chiral intermediate. mp = 72-77°; $[\alpha]_D^{20}$ = - 8° (c = 1, $CHCl_3$). *Sigma-Aldrich Fine Chemicals.*

**542 (-)-4-O-Benzylgalactal**

$C_{13}H_{16}O_4$
Chiral intermediate. mp = 101-103°; $[\alpha]_D^{20}$ = - 22° (c = 1, $CHCl_3$). *Sigma-Aldrich Fine Chemicals.*

**543 Benzyl-β-D-galactopyranoside**

$C_{13}H_{18}O_6$
Chiral intermediate. $[\alpha]_D^{20}$= - 38.6° (c = 1, $H_2O$). *Pfanstiehl Laboratories, Inc.*

**544 (+)-4-O-Benzylglucal**

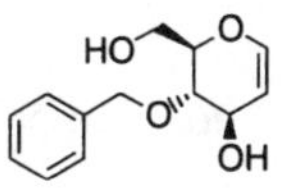

$C_{13}H_{16}O_4$
Chiral intermediate. mp = 99-103°; $[\alpha]_D^{20}$ = + 10° (c = 1, $CHCl_3$). *Sigma-Aldrich Fine Chemicals.*

**545 γ-Benzyl L-glutamate**
1676-73-9 216-826-1

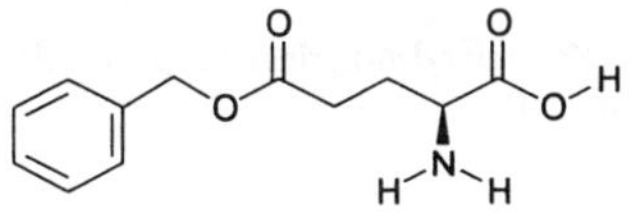

$C_{12}H_{15}NO_4$
Glutamic acid, 5-(phenylmethyl) ester; Benzyl hydrogen γ-L-glutamate; Glutamic acid, 5-benzyl ester, L-. Listed on TSCA. Chiral intermediate. mp = 160-163°; $[\alpha]^{22}$ = + 27.7° (c = 2.5, 1N HCl). *Acros Organics nv; Austin Chemical Company, Inc.; Flamma s.p.a.; Sigma-Aldrich Fine Chemicals.*

**546 (R)-(+)-1-Benzylglycerol**
56552-80-8

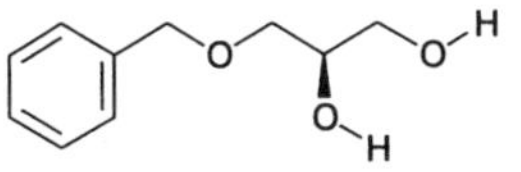

$C_{10}H_{14}O_3$
Chiral building block. mp = 25-29°; $[\alpha]$ = + 5.9° (neat); $[\alpha]_D^{20}$ = + 5.5° (c = 20, $CHCl_3$). *Acros Organics nv; Sigma-Aldrich Fine Chemicals.*

**547 (S)-(-)-1-Benzylglycerol**
17325-85-8

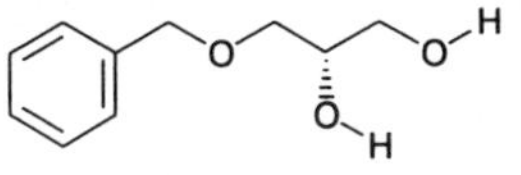

$C_{10}H_{14}O_3$
(S)-(-)-3-[(Phenylmethyl)oxy]-1,2-propanediol; (S)-3-benzyloxy-1,2-propanediol. Chiral building block. mp = 25-29°; bp = 261°; $bp_{0.5}$ = 130-133°; $[\alpha]$ = - 5.6° (neat); $[\alpha]_D^{20}$ = - 4.7° (c = 20, $CHCl_3$). *Acros Organics nv; Sigma-Aldrich Fine Chemicals.*

**548 Benzyl (R)-(-)-glycidyl ether**
14618-80-5

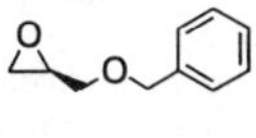

$C_{10}H_{12}O_2$
(R)-(-)-2-(Benzyloxymethyl)oxirane. Chiral building block. $bp_{0.1}$ = 130°; d = 1.0770; $n_D^{20}$ = 1.5170; $[\alpha]_D^{20}$ = - 5.1 ± 0.5° (c = 5, toluene). *Austin Chemical Company, Inc.; Daiso Company, Ltd.; Kaneka Corporation; Lancaster Synthesis Ltd.; Sigma-Aldrich Fine Chemicals.*

**549 Benzyl (S)-(+)-glycidyl ether**
16495-13-9

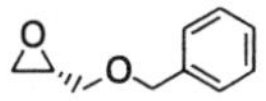

$C_{10}H_{12}O_2$
(S)-(+)-2-(Benzyloxymethyl)oxirane. Chiral building block. d = 1.0720; $n_D^{20}$ = 1.5170; $[\alpha]_D^{20}$ = + 5.1 ± 0.5° (c = 5, toluene). *Austin Chemical Company, Inc.; Daiso Company, Ltd.; Lancaster Synthesis Ltd.; Sigma-Aldrich Fine Chemicals.*

**550 L-5-Benzylhydantoin**

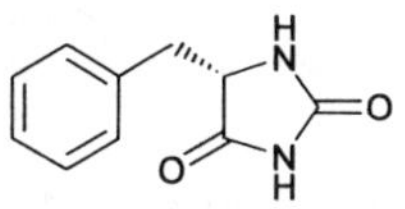

$C_{10}H_{10}N_2O_2$
Imidazolidine-2,4-dione, 5-(phenylmethyl)-, L-; L-5-(Phenylmethyl)imidazolidine-2,4-dione. Chiral intermediate. *Senn Chemicals AG.*

**551 (R)-Benzyl-4-hydroxymethyl-2-oxazolidinone**
157823-75-1

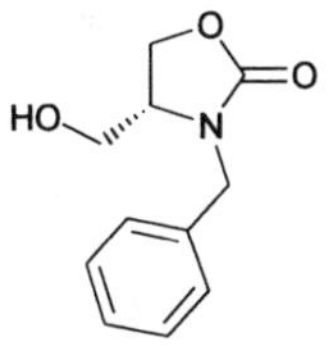

$C_{11}H_{13}NO_3$
Chiral intermediate. *Daiso Company, Ltd.*

**552 (S)-3-Benzyl-4-hydroxymethyl-2-oxazolidinone**
136015-39-9

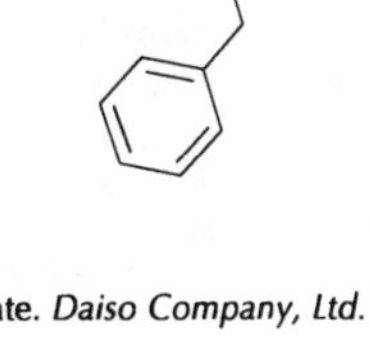

$C_{11}H_{13}NO_3$
Chiral intermediate. *Daiso Company, Ltd.*

**553 Benzyl (R)-(+)-2-hydroxy-3-phenylpropionate**
7622-22-2

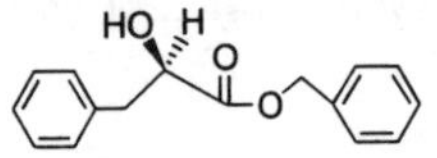

$C_{16}H_{16}O_3$
Chiral building block. mp = 20-30°; bp = 228°; d = 1.14; n = 1.559; $[\alpha]^{25}$ = + 55° (c = 1.8, $CH_2Cl_2$). *Sigma-Aldrich Fine Chemicals.*

**554 Benzyl (S)-(-)-2-hydroxy-3-phenylpropionate**
7622-21-1

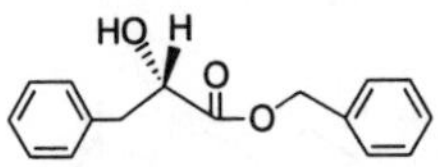

$C_{16}H_{16}O_3$
Chiral building block. bp = 288°; d = 1.142; n = 1.558; $[\alpha]^{24}$ = - 55° (c = 1.8, $CH_2Cl_2$). *Sigma-Aldrich Fine Chemicals.*

**555 (R)-(-)-1-Benzyl-3-hydroxypiperidine**
91599-81-4

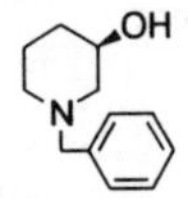

$C_{12}H_{17}NO$
(R)-1-Benzyl-3-piperidinol. Chiral building block. bp = 275°; d = 1.07; n = 1.547; $[\alpha]^{24}$ = - 12° (c = 1, $CH_3OH$). *Sigma-Aldrich Fine Chemicals.*

**556 (R)-N-Benzyl-3-hydroxypyrrolidine**
101930-07-8

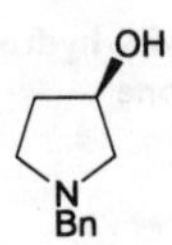

$C_{11}H_{15}NO$
Pyrrolidin-3-ol, 1-(phenylmethyl)-, (R)-; (R)-1-Benzyl-3-pyrrolidinol. Chiral building block. $bp_{0.9}$ = 116°; d = 1.07; n = 1.548; $[\alpha]^{23}$ = + 3.7° (c = 5, $CH_3OH$). *Austin Chemical Company, Inc.; Kaneka Corporation; Sigma-Aldrich Fine Chemicals; Synthon Chiragenics Corporation; TCI America; Toray International, Inc.*

**557 (S)-N-Benzyl-3-hydroxypyrrolidine**
101385-90-4

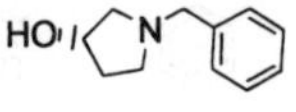

$C_{11}H_{15}NO$
Pyrrolidin-3-ol, 1-(phenylmethyl)-, (S)-; (S)-1-Benzyl-3-pyrrolidinol. Chiral building block. $bp_{0.8}$ = 115°; d = 1.07; n = 1.548; $[\alpha]^{24}$ = - 3.7° (c = 5, $CH_3OH$). *Austin Chemical Company, Inc.; Kaneka Corporation; Sigma-Aldrich Fine Chemicals; Synthon Chiragenics Corporation; TCI America; Toray International, Inc.*

**558 (R)-Benzyl-4-hydroxy-2-pyrrolidinone**
122089-39-8

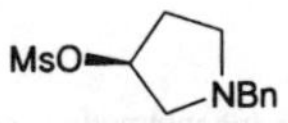

$C_{11}H_{13}NO_2$
Pyrrolidin-2-one, 4-hydroxy-1-(phenylmethyl)-, (R)-. Chiral intermediate. *Synthon Chiragenics Corporation.*

**559 (S)-1-Benzyl-4-hydroxypyrrolidnonone**
191403-66-4

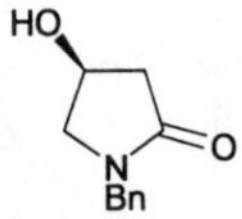

$C_{11}H_{13}NO_2$
Pyrrolidin-2-one, 4-hydroxy-1-(phenylmethyl)-, (S)-. Chiral intermediate. *Synthon Chiragenics Corporation.*

**560 1,3-O-Benzylidene-D-arabitol**
Chiral intermediate. *Senn Chemicals AG.*

**561 2,4-O-Benzylidene-D-glucitol**
32647-67-9 251-136-4

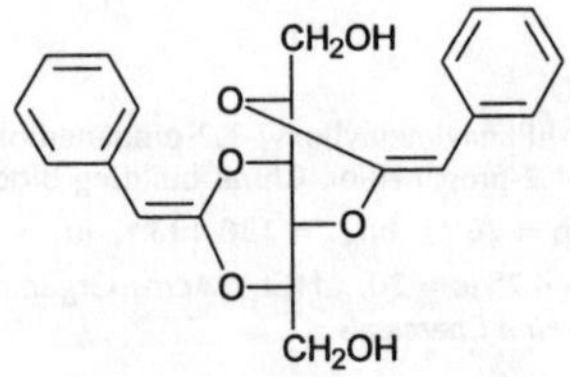

$C_{18}H_{22}O_6$
Glucitol, bis-O-(phenylmethylene)-, D-; Bis-O-(phenylmethylene)glucitol; D-Sorbitol, di-O-benzylidene-; Denon YK 2; Dibenzalsorbitol; Dibenzylidene sorbitol acetal; Dibenzylidene-D-sorbitol; EC Gel-D; EC-I-ME; Millithix 925; Sorbitol dibenzal. Listed on TSCA. Chiral intermediate. *Senn Chemicals AG.*

**562 (-)-(4,6-O-Benzylidene)phenyl-glucopyranoside**
75829-66-2

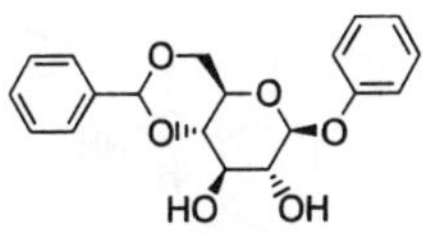

$C_{19}H_{20}O_6$
Chiral intermediate. mp = 194-196°; $[\alpha]^{21}$ = - 56° (c = 2, acetone). *Sigma-Aldrich Fine Chemicals.*

**563 (-)-3,4-O-Benzylideneribonic lactone**
20603-45-6

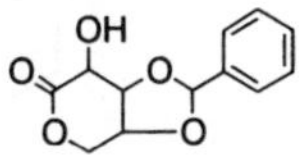

$C_{12}H_{12}O_5$
Chiral intermediate. mp = 234-236°; $[\alpha]_D^{20}$ = - 178° (c = 2.2, DMF). *Sigma-Aldrich Fine Chemicals.*

**564 (+)-2,3-O-Benzylidene-D-threitol**
58383-35-0

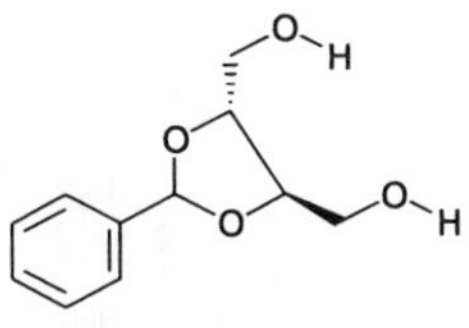

$C_{11}H_{14}O_4$
Chiral intermediate. mp = 72-74°; $[\alpha]^{20}$ = + 10° (c = 2, $CH_3OH$). *Acros Organics nv; Sigma-Aldrich Fine Chemicals.*

**565 (-)-2,3-O-Benzylidene-L-threitol**
35572-34-0

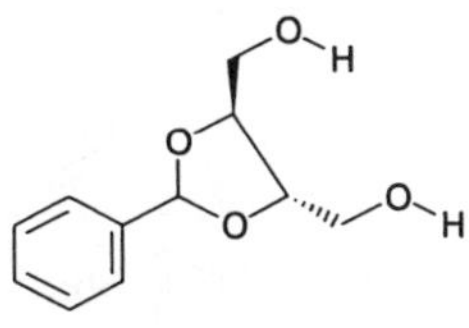

$C_{11}H_{14}O_4$
Chiral intermediate. mp = 72-74°; $[\alpha]^{20}$ = - 10° (c = 2, $CH_3OH$). *Acros Organics nv; Sigma-Aldrich Fine Chemicals.*

**566 2,4-O-Benzylidene-L-xylose**

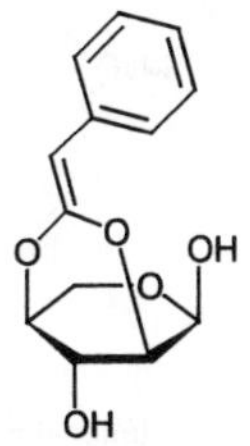

$C_{13}H_{14}O_5$
Chiral intermediate. *Senn Chemicals AG.*

**567 Benzyl L-lactate**
56777-24-3

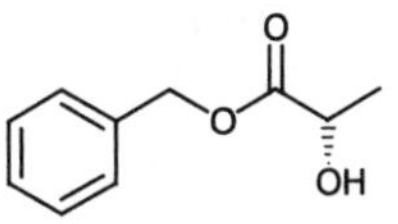

$C_{10}H_{12}O_3$
L-Lactic acid benzyl ester; Benzyl (S)-2-hydroxypropionate. Chiral building block. $bp_{0.1}$ = 105-109°; d = 1.1200; $n_D^{20}$ = 1.5130; $[\alpha]_D^{20}$ = - 12.9 ± 1° (c = 3.1, $CH_3OH$); $[\alpha]_D^{20}$ = - 19° (c = 1, acetone). *Lancaster Synthesis Ltd.; Sigma-Aldrich Fine Chemicals.*

**568 Benzyl (R)-(-)-mandelate**
97415-09-3

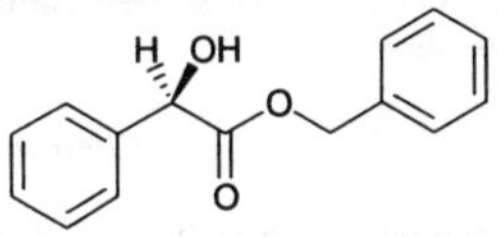

$C_{15}H_{14}O_3$
Chiral building block. mp = 104-107°; $[\alpha]^{25}$ = - 55° (c = 1, $CHCl_3$). *Sigma-Aldrich Fine Chemicals.*

**569 Benzyl (S)-(+)-mandelate**
62173-99-3

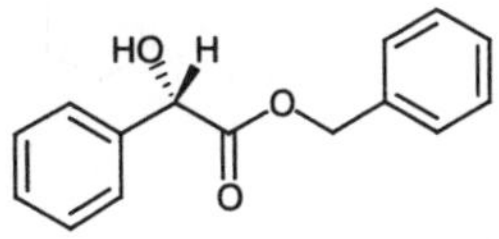

$C_{15}H_{14}O_3$
Chiral building block. mp = 104-107°; $[\alpha]^{25}$ = + 55° (c = 1, $CHCl_3$). *Sigma-Aldrich Fine Chemicals.*

**570 (R)-Benzyl-3-mesyloxy pyrrolidine**

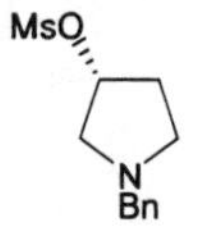

$C_{12}H_{17}NO_3S$
Pyrrolidin-3-ol, 1-(phenylmethyl)-methanesulfonate (ester), (R)-; Chiral intermediate. *Synthon Chiragenics Corporation.*

**571 (S)-1-Benzyl-3-mesyloxy pyrrolidine**
118354-71-5

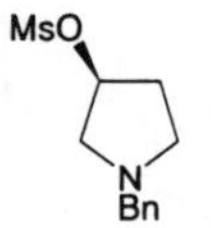

$C_{12}H_{17}NO_3S$
Pyrrolidin-3-ol, 1-(phenylmethyl)-methanesulfonate (ester), (S)-; Chiral intermediate. *Synthon Chiragenics Corporation.*

**572 (3R)-(-)-1-Benzyl-3-(methylamino)-pyrrolidine**

$C_{12}H_{18}N_2$
Chiral intermediate. d = 0.99. *TCI America.*

**573 (3S)-(+)-1-Benzyl-3-(methylamino)-pyrrolidine**

$C_{12}H_{18}N_2$
Chiral building block. d = 0.99. *TCI America.*

**574 1-Benzyl 2-methyl (S)-(-)-1,2-aziridine-dicarboxylate**

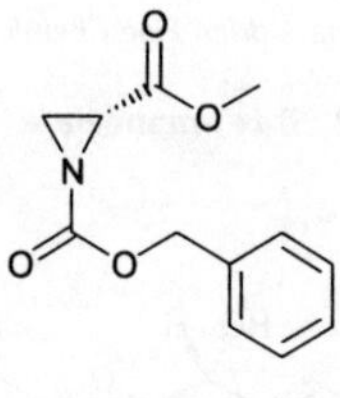

$C_{12}H_{13}NO_4$
Chiral intermediate. $bp_{2.0}$ = 160-165°; d = 1.19; n = 1.519; $[\alpha]_D^{20}$ = - 25° (c = 1, toluene). *Sigma-Aldrich Fine Chemicals.*

**575 (R)-Benzyl-2-methylbenzylamine**

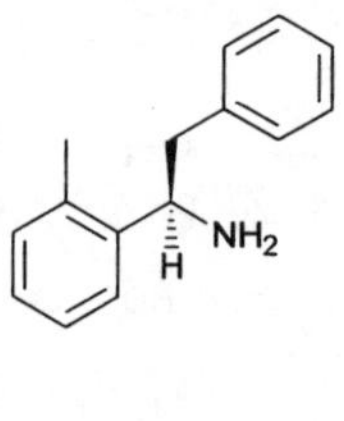

$C_{15}H_{17}N$
Chiral building block. *Sumitomo Chemcial Co. Ltd.*

**576 (S)-α-Benzyl-2-methylbenzylanine**

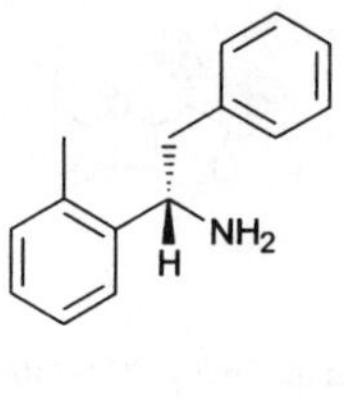

$C_{15}H_{17}N$
Chiral building block. *Sumitomo Chemcial Co. Ltd.*

**577 (R)-Benzyl-4-methylbenzylamine**

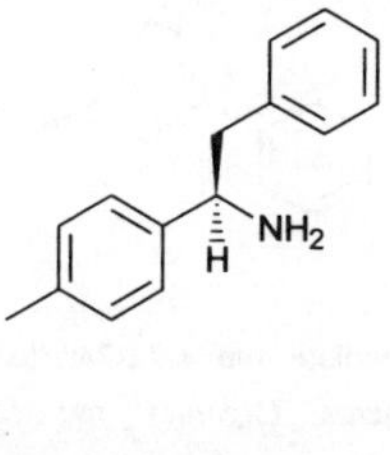

$C_{15}H_{17}N$
Chiral building block. *Sumitomo Chemcial Co. Ltd.*

**578 (S)-α-Benzyl-4-methylbenzylamine**

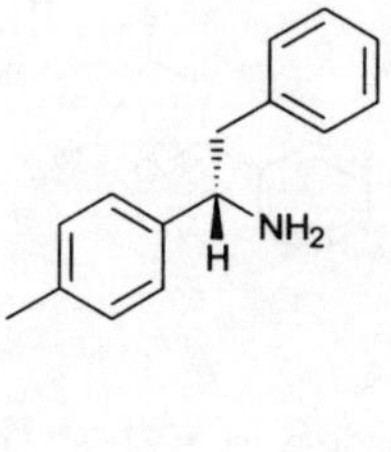

$C_{15}H_{17}N$
Chiral building block. *Sumitomo Chemcial Co. Ltd.*

**579 (R)-N-Benzyl-α-methylbenzylamine hydrochloride**
128593-66-8

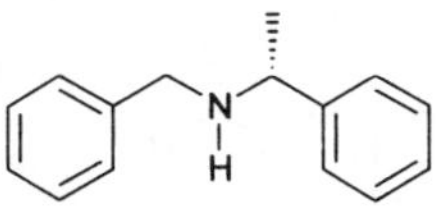

$C_{15}H_{17}ClN$
Chiral intermediate. *Acros Organics nv; Oxford Asymmetry International plc.*

**580 (S)-(-)-N-Benzyl-α-methylbenzylamine hydrochloride**

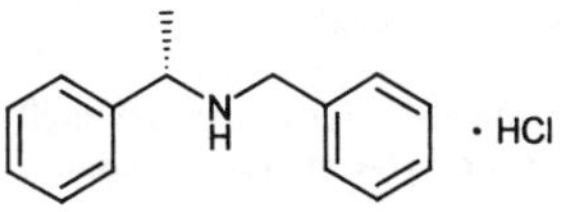

$C_{15}H_{17}ClN$
Chiral building block. *Oxford Asymmetry International plc.*

**581 N-Benzyl-N-methyl ephedrinium bromide**
58648-09-2 261-375-6

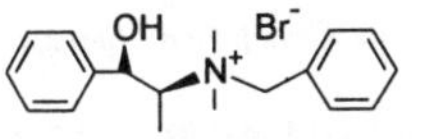

$C_{18}H_{24}BrNO$
Benzyl(2-hydroxy-1-methyl-2-phenylethyl)dimethylammonium bromide. mp = 209-211°; $[\alpha]_D^{20}$ = - 5.3° (c = 4.5, $CH_3OH$). *Sigma-Aldrich Fine Chemicals.*

**582 (S)-(-)-4-Benzyl-2-methyl-2-oxazoline**
75866-72-7

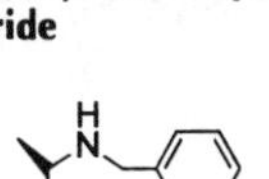

$C_{11}H_{13}NO$
Chiral intermediate. bp = 221°; d = 1.05; n = 1.525; $[\alpha]^{22}$ = - 50° (c = 2, $CHCl_3$). *Sigma-Aldrich Fine Chemicals.*

**583 (R)-(-)-N-Benzyl-1-(1-naphthyl)ethylamine hydrochloride**

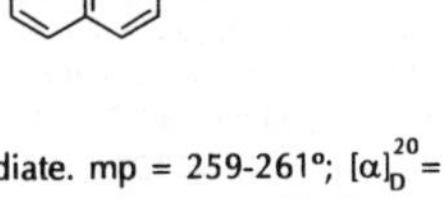

$C_{19}H_{20}ClN$
Chiral intermediate. mp = 259-261°; $[\alpha]_D^{20}$ = - 61° (c = 2, $CH_3OH$). *FineTech, Ltd.; Sigma-Aldrich Fine Chemicals.*

**584 (S)-(+)-N-Benzyl-1-(1-naphthyl)ethylamine hydrochloride**

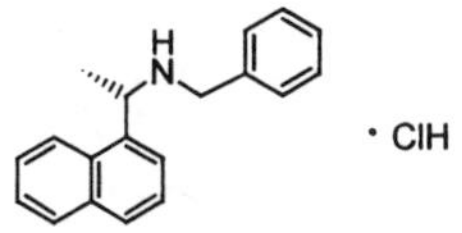

$C_{19}H_{20}ClN$
Chiral intermediate. mp = 259-261°; $[\alpha]_D^{20}$ = + 61° (c = 2, $CH_3OH$). 1553; *Sigma-Aldrich Fine Chemicals.*

**585 (R)-(+)-4-Benzyl-2-oxazolidinone**
102029-44-7

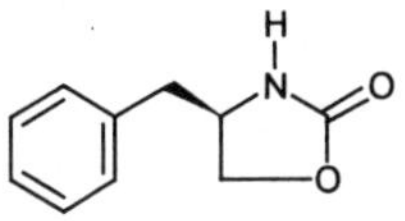

$C_{10}H_{11}NO_2$
Chiral auxiliary. mp = 86-88°; $[\alpha]_D^{20}$ = + 64 ± 2° (c = 1, $CHCl_3$). *Acros Organics nv; Boehringer Ingelheim Pharma KG; Fischer Chemicals AG; Great Lakes Fine Chemicals; Lancaster Synthesis Ltd.; Newport Synthesis Ireland Ltd.; Omega Chemical Company Inc.; Oxford Asymmetry International plc; PPG - Sipsy; Senn Chemicals AG; Sigma-Aldrich Fine Chemicals; SK Energy and Chemical, Inc.; TCI America; Urquima S.A.*

**586 (S)-(-)-4-Benzyl-2-oxazolidinone**
90719-32-7

$C_{10}H_{11}NO_2$
Chiral auxiliary. mp = 86-88°; $[\alpha]_D^{20}$ = - 64 ± 2° (c = 1, $CHCl_3$). *Acros Organics nv; Austin Chemical Company, Inc.; Boehringer Ingelheim Pharma KG; Davos Chemical Corporation; Degussa-Huls AG; Fischer Chemicals AG; Great Lakes Fine Chemicals; Lancaster Synthesis Ltd.; Newport Synthesis Ireland Ltd.; Omega Chemical Company Inc.; Oxford Asymmetry International plc; SEAC; Senn Chemicals AG; Sigma-Aldrich Fine Chemicals; SK Energy and Chemical, Inc.; TCI America; Urquima S.A.*

**587 Benzyl (2R,3S)-(-)-6-oxo-2,3-diphenyl-4-morpholinecarboxylate**
100516-54-9

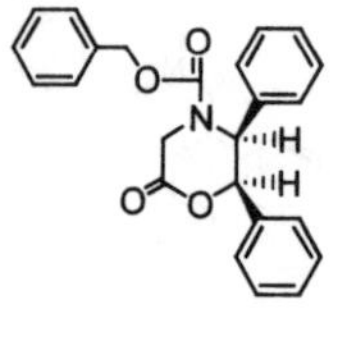

$C_{24}H_{21}NO_4$
Chiral intermediate. mp = 205-207°; $[\alpha]^{25}$ = - 66° (c = 5.5, $CH_2Cl_2$). *Sigma-Aldrich Fine Chemicals.*

**588 (R,R)-2-Benzyloxy-1-aminocyclohexane**
216394-06-8

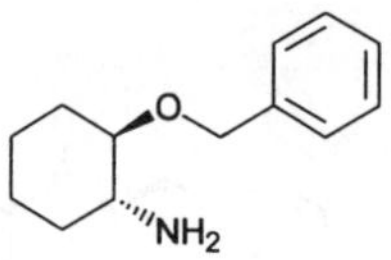

$C_{13}H_{19}NO$
Chiral intermediate. $bp_{0.1}$ = 79-80°. *BASF Aktiengesellschaft.*

**589 (S,S)-2-Benzyloxy-1-aminocyclohexane**
216394-07-9

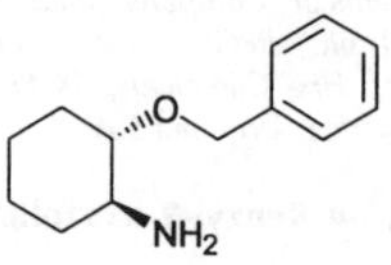

$C_{13}H_{19}NO$
Chiral intermediate. $bp_{0.1}$ = 79-80°. *BASF Aktiengesellschaft.*

**590 (R)-(+)-3-Benzyloxy-2-(tert-butoxycarbonylamino)-1-propanol**

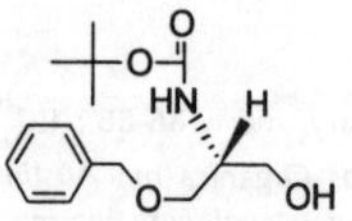

$C_{15}H_{23}NO_4$
Chiral intermediate. mp = 66-69°; $[\alpha]_D^{20}$ = + 14° (c = 1, $CHCl_3$). *Sigma-Aldrich Fine Chemicals.*

**591 (S)-4-Benzyloxycarbonylamino-2-hydroybutyric acid**
40371-50-4 254-892-3

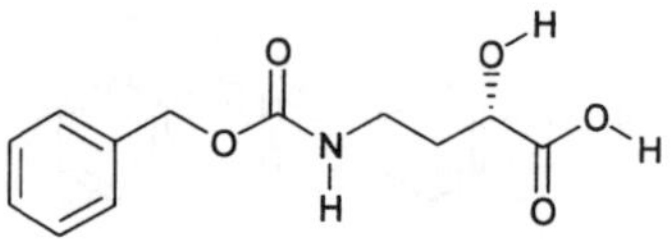

$C_{12}H_{15}NO_5$
Butanoic acid, 2-hydroxy-4-[[(phenylmethoxy)carbonyl]-amino]-, (S)-; CBZ-HABA; (S)-2-Hydroxy-4-[[(phenylmethoxy)carbonyl]amino]butyric acid; L-γ-Benzyloxycarbonylamino-α-hydroxybutyric acid. Chiral intermediate. mp = 75-78°; $[\alpha]_D^{20}$ = + 10° (c = 1, $CHCl_3$). *Acros Organics nv; D&O Group; KingChem, Inc.; Sigma-Aldrich Fine Chemicals.*

**592 (R)-(+)-1-(Benzyloxycarbonyl)-2-tert-butyl-3-methyl-4-imidazolidinone**

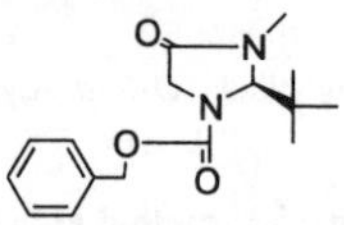

$C_{16}H_{22}N_2O_3$
Chiral intermediate. mp = 84-86°; $[\alpha]^{25}$ = + 11.5° (c = 1, $CH_2Cl_2$). *Sigma-Aldrich Fine Chemicals.*

**593 (S)-(-)-1-(Benzyloxycarbonyl)-2-tert-butyl-3-methyl-4-imidazolidinone**

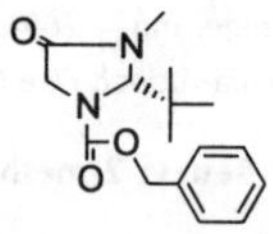

$C_{16}H_{22}N_2O_3$
Chiral intermediate. mp = 84-86°; $[\alpha]^{26}$ = - 11.5° (c = 1, $CH_2Cl_2$). *Sigma-Aldrich Fine Chemicals.*

**594 N-Benzyloxycarbonyl-L-glutamic acid**
1155-62-0 214-584-1

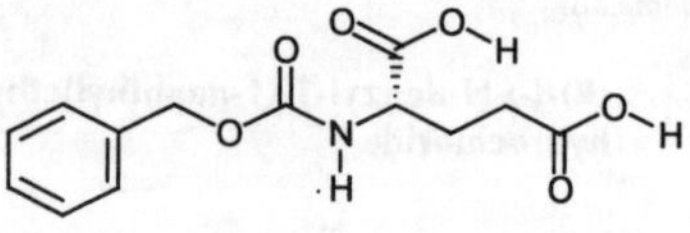

$C_{13}H_{15}NO_6$
Glutamic acid, N-[(phenylmethoxy)carbonyl]-, L-; Glutamic acid, N-carboxy-, N-benzyl ester, L-; N-[(Phenylmethoxycarbonyl]-L-glutamic acid; N-Carbobenzoxy-L-glutamic acid. Listed on TSCA. Chiral intermediate. mp = 118-120°; $[\alpha]^{22}$ = - 7.4° (c = 10, HOAc). *Acros Organics nv; Sigma-Aldrich Fine Chemicals.*

**595 (S)-(+)-3-(Benzyloxycarbonyl)-5-oxo-4-oxazolidineacetic acid**
23632-66-8

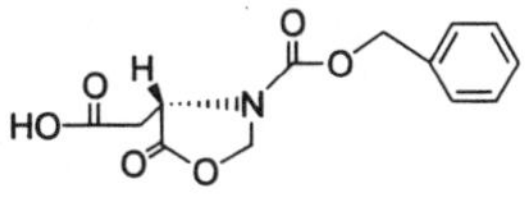

$C_{13}H_{13}NO_6$
Chiral intermediate. mp = 84-87°; $[\alpha]^{22}$ = + 137° (c = 3.5, $CH_3OH$). *Sigma-Aldrich Fine Chemicals.*

**596 (S)-(+)-3-(Benzyloxycarbonyl)-5-oxo-4-oxazolidinepropionic acid**
23632-67-9

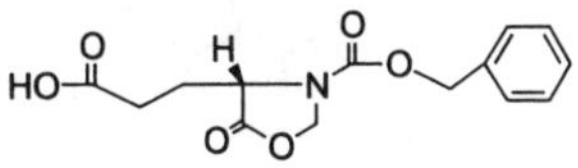

$C_{14}H_{15}NO_6$
Chiral intermediate. d = 1.05; n = 1.535; $[\alpha]^{22}$ = + 88° (c = 1, $CH_3OH$). *Sigma-Aldrich Fine Chemicals.*

**597 (3S)-2-[Benzyloxycarbonyl]-1,2,3,4-tetrahydroisoquinoline-3-carboxylic acid**
79261-58-8

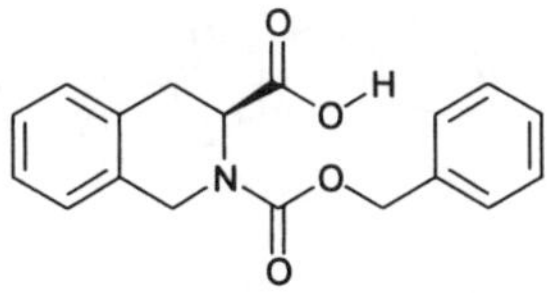

$C_{18}H_{17}NO_4$
Chiral intermediate. mp = 137-141°; $[\alpha]^{24}$ = + 22° (c = 1, $CH_3OH$). *Acros Organics nv; Sigma-Aldrich Fine Chemicals.*

**598 (1R,2R)-(+)-2-Benzyloxycyclohexylamine**
$C_{13}H_{19}NO$
(1R,2R)-(+)-1-Amino-2-benzyloxycyclohexane; (1R-trans)-(+)-2-(Phenylmethoxy)cyclohexanamine. Chiral building block; Chiral auxiliary. $bp_{0.08}$ = 79-80°; $n_D^{20}$ = 1.5275; $[\alpha]_D^{20}$ = + 85 ± 2° (neat). *Lancaster Synthesis Ltd.*

**599 (1S,2S)-(-)-2-Benzyloxycyclohexylamine**
$C_{13}H_{19}NO$
(1S-trans)-(-)-2-(Phenylmethoxy)cyclohexanamine; (1S,2S)-(-)-1-Amino-2-benzyloxycyclohexane. Chiral intermediate; Resolving agent. $bp_{0.08}$ = 79-80°; $n_D^{20}$ = 1.5275; $[\alpha]_D^{20}$ = - 85 ± 2° (neat). *Lancaster Synthesis Ltd.*

**600 (1R,2R)-2-Benzyloxycyclopentylamine**
35572-31-7
$C_{12}H_{17}NO$
(1R,2R)-1-Amino-2-benzyloxycyclopentane; (1R-trans)-2-(Phenylmethoxy)cyclopentanamine. Chiral building block; Chiral auxiliary. $bp_{0.23}$ = 90-93°; $n_D^{20}$ = 1.5305. *Lancaster Synthesis Ltd.*

**601 (1S,2S)-2-Benzyloxycyclopentylamine**
181657-57-8
$C_{12}H_{17}NO$
(1S,2S)-1-Amino-2-benzyloxycyclopentane; (1S-trans)-2-(Phenylmethoxy)cyclopentanamine. Chiral intermediate. $bp_{0.23}$ = 90-93°; $n_D^{20}$ = 1.5305. *BASF Aktiengesellschaft; Lancaster Synthesis Ltd.*

**602 (R)-Benzyloxy-3-hydroxybutanoic acid ethyl ester**

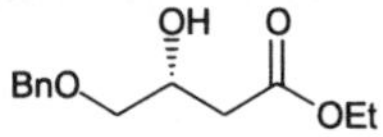

$C_{13}H_{18}O_4$
Chiral building block. *Syntai Chemicals and Pharmaceuticals.*

**603 (S)-4-Benzyloxy-3-hydroxybutanoic acid ethyl ester**

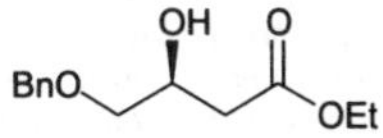

$C_{13}H_{18}O_4$
Chiral building block. *Syntai Chemicals and Pharmaceuticals.*

**604 (R)-(+)-2-Benzyloxypropionic acid**
100836-85-9
$C_{10}H_{12}O_3$
O-Benzyl-D-lactic acid. Chiral building block. mp = 51-53°; $[\alpha]_D^{20}$ = + 90 ± 2° (c = 1, IMS). *Lancaster Synthesis Ltd.*

**605 (R)-(-)-1-Benzyloxy-3-(p-tosyloxy)-2-propanol**
23214-66-6

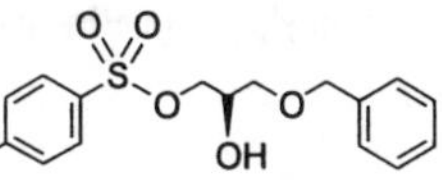

$C_{17}H_{20}O_5S$
Chiral intermediate. mp = 52-55°; $[\alpha]^{25}$ = - 7° (c = 10, toluene). *Sigma-Aldrich Fine Chemicals.*

**606 (R)-(+)-N-Benzyl-1-phenylethylamine**
38235-77-7

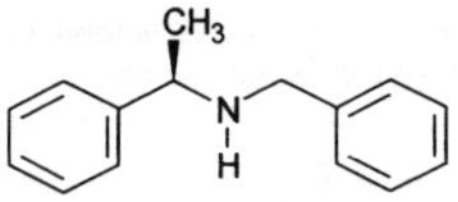

$C_{15}H_{17}N$
D-(+)-N-Benzyl-α-methylbenzylamine; N-Benzyl-α-methylbenzylamine. Chiral auxiliary; Chiral intermediate. $bp_{15}$ = 171°; $bp_3$ = 127-128°; d = 1.0100; $n_D^{20}$ = 1.5640; $[\alpha]_D^{20}$ = + 48.0 ± 0.2° (c = 6, cyclohexane); $[\alpha]_D^{20}$ = + 39 ± 1° (neat). *Acros Organics nv; Arran Chemical Company Ltd.; Austin Chemical Company, Inc.; FineTech, Ltd.; Kiralchem Ltd.; Lancaster Synthesis Ltd.; Norse Laboratories; Sigma-Aldrich Fine Chemicals; TCI America; Yamakawa Chemical Industry Co. Ltd.*

**607 (S)-(-)-N-Benzyl-1-phenylethylamine**
17480-69-2

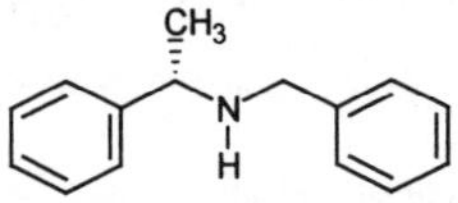

$C_{15}H_{17}N$
(S)-(-)-N-Benzyl-α-methylbenzylamine; L-(-)-N-Benzyl-α-methylbenzylamine. Chiral auxiliary; Chiral intermediate. $bp_{15}$ = 171°; $bp_3$ = 129°; d = 1.0100; $n_D^{20}$ = 1.5640; $[\alpha]_D^{20}$ = - 39 ± 1° (neat); $[\alpha]_D^{20}$ = - 48.0 ± 0.2° (c = 6, cyclohexane). *Acros Organics nv; Arran Chemical Company Ltd.; Austin Chemical Company, Inc.; FineTech, Ltd.; Kiralchem Ltd.; Lancaster Synthesis Ltd.; Norse Laboratories; Sigma-Aldrich Fine Chemicals; TCI America; Yamakawa Chemical Industry Co. Ltd.*

**608 (R)-(-)-N-Benzyl-2-phenylglycinol**
14231-57-3

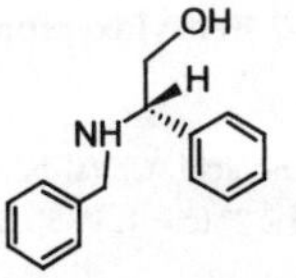

$C_{15}H_{17}NO$
Chiral intermediate. bp = 87-90°; $[\alpha]^{21}$ = - 80° (c = 0.85, EtOH). *Sigma-Aldrich Fine Chemicals.*

**609 (-)-N-Benzylproline ethyl ester**
955-40-8

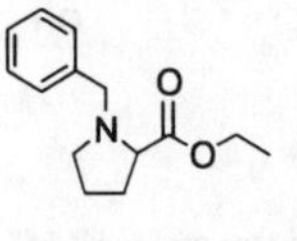

$C_{14}H_{19}NO_2$
Chiral intermediate. d = 1.048; n = 1.511; $[\alpha]_D^{20}$ = - 62° (neat). *Sigma-Aldrich Fine Chemicals.*

**610 (+)-N-Benzylproline ethyl ester**

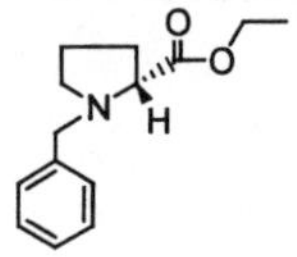

$C_{14}H_{19}NO_2$
Chiral intermediate. d = 1.046; n = 1.511; $[\alpha]_D^{20}$ = + 62° (neat). *Sigma-Aldrich Fine Chemicals.*

**611 (S)-(-)-2-(N-Benzylprolyl)amino-benzophenone**
96293-17-3

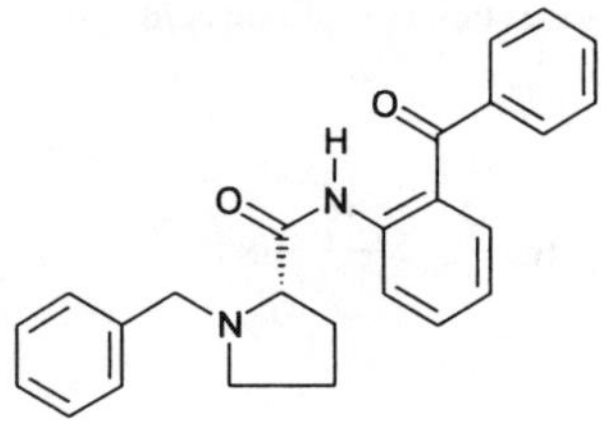

$C_{25}H_{24}N_2O_2$
Chiral auxiliary; Chiral intermediate. mp = 98-101°; $[\alpha]_D^{20}$ = - 135 ± 1° (c = 1, $CH_3OH$). *Acros Organics nv; Lancaster Synthesis Ltd.*

**612 (4R)-4-Benzyl-2-(2-pyridyl)-1,3-oxazol-2-ine**

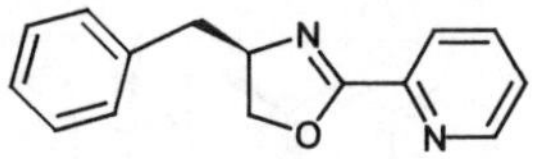

$C_{15}H_{14}N_2O$
Chiral intermediate. *FineTech, Ltd.*

**613 (4S)-4-Benzyl-2-(2-pyridyl)-1,3-oxazol-2-ine**

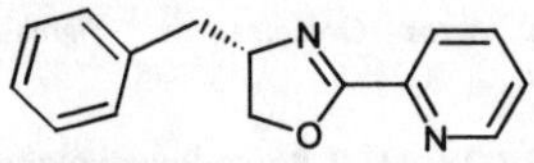

$C_{15}H_{14}N_2O$
Chiral intermediate. *FineTech, Ltd.*

**614 (S)-(-)-1-Benzyl-2-pyrrolidinemethanol**
53912-80-4

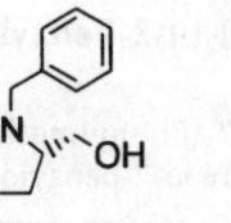

$C_{12}H_{17}NO$
Chiral intermediate. $bp_{0.5}$ = 115-120°; d = 1.08; n =

1.541; $[\alpha]_D^{20}$ = - 72.7° (neat). *Sigma-Aldrich Fine Chemicals.*

**615 (3R,4R)-(-)-1-Benzyl-3,4-pyrrolidinol**

$C_{11}H_{15}NO_2$
Chiral building block. *Digital Specialty Chemicals, Inc.*

**616 (3S,4S)-(+)-1-Benzyl-3,4-pyrrolidinol**

90365-74-5
$C_{11}H_{15}NO_2$
Chiral intermediate. *Digital Specialty Chemicals, Inc.*

**617 Benzyl (S)-(-)-2-(1-pyrrolidinylcarbonyl)-1-pyrrolinecarboxylate**

50888-84-1

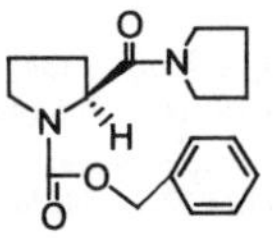

$C_{17}H_{22}N_2O_3$
Chiral intermediate. mp = 132-135°; $[\alpha]^{24}$ = - 14° (c = 1.6, $CH_3OH$). *Sigma-Aldrich Fine Chemicals.*

**618 (-)-N-Benzylquininium chloride**

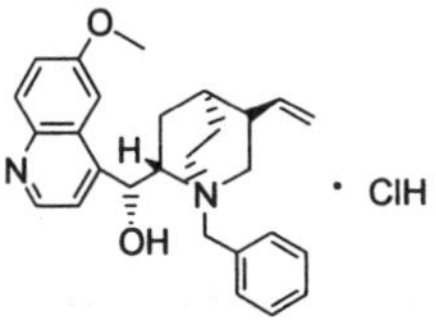

$C_{27}H_{31}ClN_2O_2$
Chiral intermediate. mp = 200-205°; $[\alpha]_D^{20}$ = - 235° (c = 1.5, $H_2O$). *Sigma-Aldrich Fine Chemicals.*

**619 (-)-4-O-Benzylrhamnal**

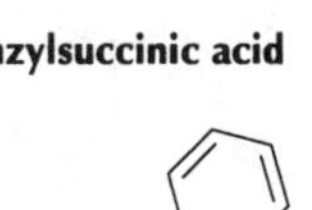

$C_{13}H_{16}O_3$
4-O-Benzyl-6-deoxy-L-glucal. Chiral intermediate. mp = 111-113°; $[\alpha]^{22}$ = - 9° (c = 1, $CHCl_3$). *Sigma-Aldrich Fine Chemicals.*

**620 (S)-2-Benzylsuccinic acid**

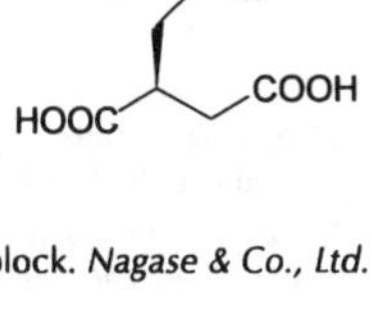

$C_{11}H_{12}O_4$
Chiral building block. *Nagase & Co., Ltd.*

**621 (R)-Benzylsuccinic acid 1-methyl ester**

119807-84-0
$C_{12}H_{14}O_4$
(R)-3-Methoxycarbonyl-4-phenylbutyric acid. Chiral intermediate. *ChiroTech Technology Ltd.; Lancaster Synthesis Ltd.*

**622 (S)-2-Benzylsuccinic acid 1-methyl ester**

182247-45-6
$C_{12}H_{14}O_4$
(S)-3-Methoxycarbonyl-4-phenylbutyric acid. Chiral intermediate. *ChiroTech Technology Ltd.; Lancaster Synthesis Ltd.*

**623 (R)-Benzyl superquat**

Chiral auxiliary. *Oxford Asymmetry International plc.*

**624 (S)-Benzyl superquat**

Chiral auxiliary. *Oxford Asymmetry International plc.*

**625 Benzyl 2,3,4,6-tetra-O-benzyl-1-thio-β-D-galactopyranoside**

210358-01-3
$C_{41}H_{42}O_5S$
Chiral intermediate. *Pfanstiehl Laboratories, Inc.*

**626 Benzyl (S)-1,2,3,4-tetrahydroisoquinoline-3-carboxylate p-toluenesulfonate**

77497-97-3

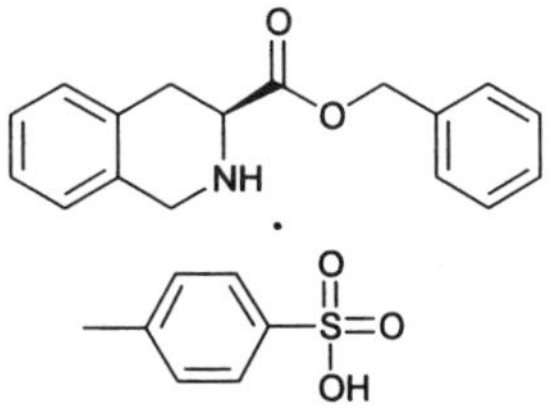

$C_{24}H_{25}NO_5S$
Isoquinoline-3-carboxylic acid, 1,2,3,4-tetrahydro-, phenylmethyl ester, (3S)-, 4-methylbenzenesulfonate; 1,2,3,4-Tetrahydroisoquinoline-(3S)-carboxylic acid benzyl ester p-toluenesulfonic acid salt. Chiral intermediate. mp = 145-150°. *Fine Organics Ltd.; Fischer Chemicals AG; Great Lakes Fine Chemicals; Sigma-Aldrich Fine Chemicals.*

**627 Benzyl (S)-(-)-tetrahydro-5-oxo-3-furanylcarbamate**

87219-29-2

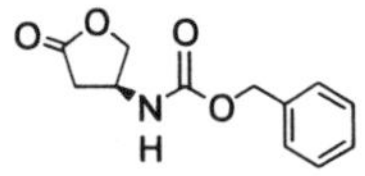

$C_{12}H_{13}NO_4$
Chiral intermediate. mp = 103-105°; $[\alpha]^{21}$ = - 49° (c = 1, acetone). *Sigma-Aldrich Fine Chemicals.*

**628 (S)-(+)-2-Benzyl-1-(p-tolylsulfonyl)aziridine**
62596-64-9

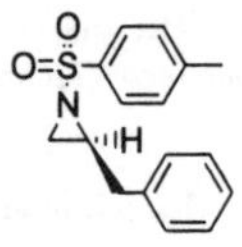

$C_{16}H_{17}NO_2S$
Chiral intermediate. mp = 92-94°; $[\alpha]_D^{20}$ = + 8.8° (c = 1.3, toluene). *Sigma-Aldrich Fine Chemicals.*

**629 Benzyl (R)-2-tosyloxypropionate**
117589-36-3

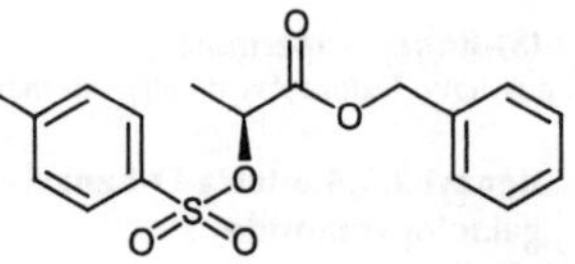

$C_{17}H_{18}O_5S$
Propanoic acid, 2-[[(4-methylphenyl)sulfonyl]oxy]-, phenylmethyl ester, (R)-; Benzyl 2-(R)-(p-toluenesulfonyloxy)propionate. Chiral intermediate. *Daicel Chemical Ind. Ltd.*

**630 (R)-Benzyl trolox**

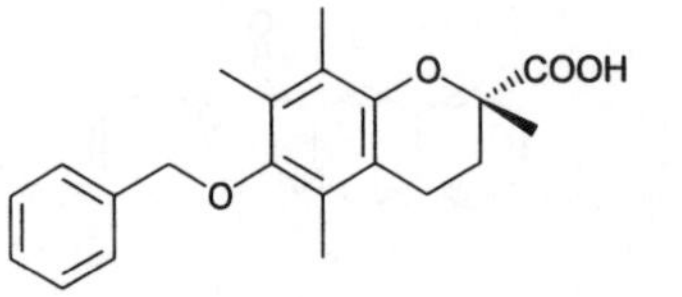

$C_{21}H_{24}O_4$
Chiral intermediate. *Expansia.*

**631 (S)-Benzyl trolox**

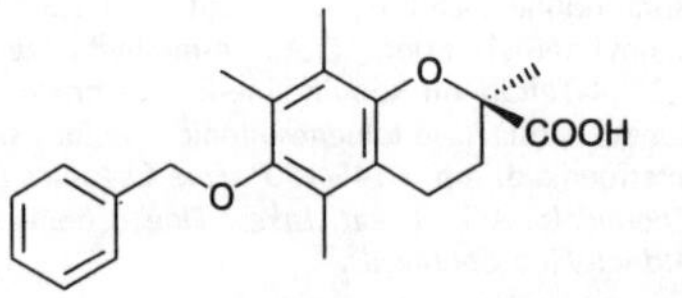

$C_{20}H_{14}O_4$
Chiral intermediate. *Expansia.*

**632 O-Benzyl-L-tyrosine**
16652-64-5 240-699-1

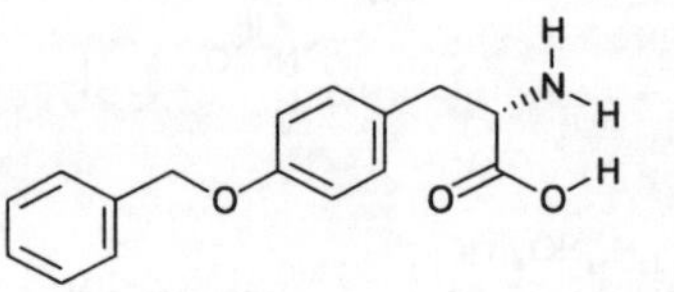

$C_{16}H_{17}NO_3$
Alanine, 3-[p-(benzyloxy)phenyl]-, L-; 4-Benzyloxy-L-phenylalanine; L-Tyrosine, O-(phenylmethyl)-; Tyrosine benzyl ether. Chiral intermediate. mp = 259°; $[\alpha]^{23}$ = - 8.9° (c = 1, 80% HOAc). *Acros Organics nv; Sigma-Aldrich Fine Chemicals.*

**633 (+)-Betulin**
473-98-3 1237(12) 207-475-5

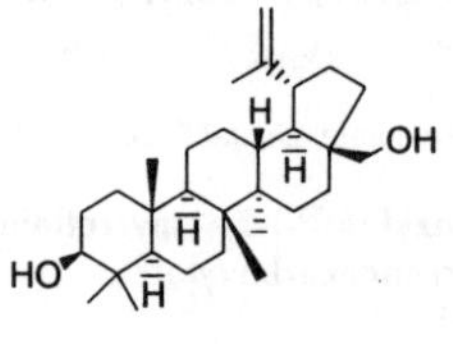

$C_{30}H_{50}O_2$
Lup-20(29)-ene-3,28-diol, (3β)-; Trochol; 3β,28-Dihydroxylup-20(29)-ene. Pharmaceutical or derivative. mp = 256-257°; $[\alpha]_D^{20}$ = + 19° (c = 2, pyridine). *Kaden Biochemicals GmbH; Sigma-Aldrich Fine Chemicals.*

**634 (R)-(+)-1,1'-Bi(2-naphthol)**
18531-94-7

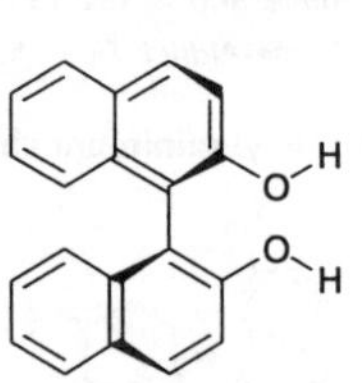

$C_{20}H_{14}O_2$
(R)-(+)-1,1'-Binaphthalene-2,2'-diol; R-(+)-2,2'-Dihydroxy-1,1'-dinaphthyl; (R)-(+)-BINOL; (R)-(+)-2,2'-Dihydroxy-1,1'-dinaphthyl. Chiral building block. mp = 208-211°; $[\alpha]_D^{20}$ = + 34.5 ± 1° (c = 1, THF). *Acros Organics nv; Austin Chemical Company, Inc.; Digital Specialty Chemicals, Inc.; FineTech, Ltd.; Kankyo Kagaku Center Co., Ltd.; Kiralchem Ltd.; Lancaster Synthesis Ltd.; Mitsubishi Chemical Corporation; Omega Chemical Company Inc.; Oxford Asymmetry International plc; RCA Reuter Chemischer Apparaebau KG; Shanghai DSL International Trading Company; Sigma-Aldrich Fine Chemicals; Syntai Chemicals and Pharmaceuticals; TCI America; Ultrafine.*

**635 (S)-(-)-1,1'-Bi(2-naphthol)**
18531-99-2

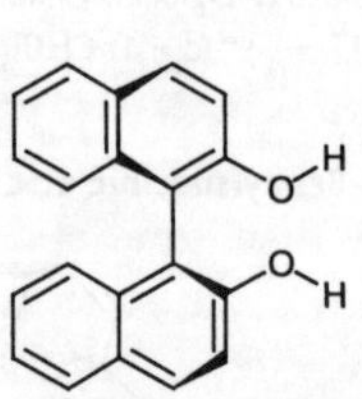

$C_{20}H_{14}O_2$
(S)-(-)-1,1'-Binaphthalene-2,2'-diol; (S)-(-)-BINOL; (S)-(+)-2,2'-Dihydroxy-1,1'-dinaphthyl. Chiral building block. mp = 207-210°; $[\alpha]_D^{20}$ = - 34.5 ± 1° (c = 1, THF). *Acros Organics nv; Austin Chemical Company, Inc.; Chattem*

*Chemicals; Digital Specialty Chemicals, Inc.; FineTech, Ltd.; Kankyo Kagaku Center Co., Ltd.; Kiralchem Ltd.; Lancaster Synthesis Ltd.; Mitsubishi Chemical Corporation; Omega Chemical Company Inc.; Oxford Asymmetry International plc; RCA Reuter Chemischer Apparaebau KG; Shanghai DSL International Trading Company; Sigma-Aldrich Fine Chemicals; Syntai Chemicals and Pharmaceuticals; TCI America; Ultrafine.*

**636 (R)-(-)-1,1'-Bi-2-naphthol bis(trifluoromethanesulfonate)**
126613-06-7

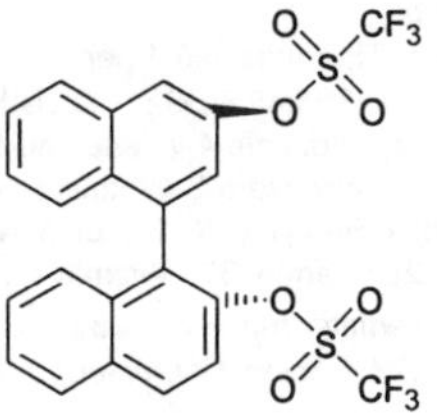

$C_{22}H_{12}F_6O_6S_2$
Chiral intermediate. mp = 83-85°; bp = 230-240°; $[\alpha]^{21}$ = - 148° (c = 1, $CHCl_3$). *Kiralchem Ltd.; Omega Chemical Company Inc.; Sigma-Aldrich Fine Chemicals.*

**637 (S)-(+)-1,1'-Bi-2-naphthol bis(trifluoromethanesulfonate)**
128544-05-8

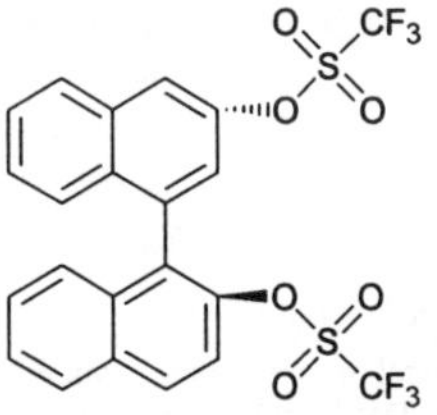

$C_{22}H_{12}F_6O_6S_2$
6,6-Dimethylbicyclo[3.1.1]hept-2-ene-2-ethanol; 6,6-Dimethyl-2-norpinene-2-ethanol. Chiral ligand. mp = 83-85°; bp = 230-240°; $[\alpha]^{21}$ = + 148° (c = 1, $CHCl_3$). *Kiralchem Ltd.; Omega Chemical Company Inc.; Sigma-Aldrich Fine Chemicals.*

**638 (+)-Bicuculline**
485-49-4

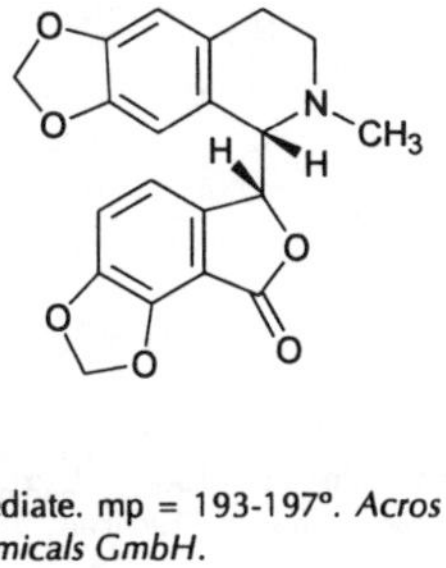

$C_{20}H_{17}NO_6$
Chiral intermediate. mp = 193-197°. *Acros Organics nv; Kaden Biochemicals GmbH.*

**639 (Bicyclo[2.2.1]hepta-2,5-diene)(2S,3S)-bis(diphenylphosphino)butanerhodium(I) perchlorate**
65012-74-0

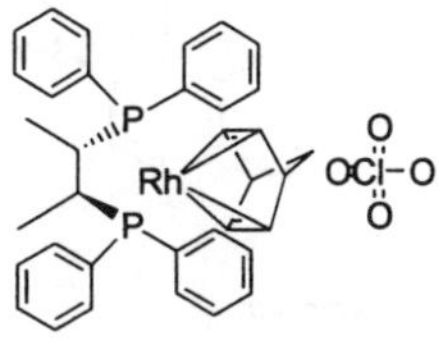

$C_{35}H_{36}ClO_4P_2Rh$
Chiral catalyst. mp = 210°. *Sigma-Aldrich Fine Chemicals.*

**640 (R)-(-)-1,1'-Binaphthyl-2,2'-dioxychlorophosphine**
155613-52-8

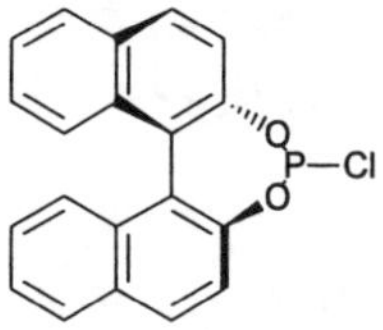

$C_{20}H_{12}ClO_2P$
Chiral ligand. *Digital Specialty Chemicals, Inc.*

**641 (S)-(+)-1,1'-Binaphthyl-2,2'-dioxychlorophosphine**
137156-22-0

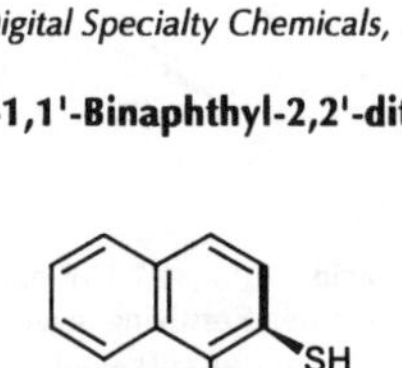

$C_{20}H_{12}ClO_2P$
Chiral ligand. *Digital Specialty Chemicals, Inc.*

**642 (R)-(+)-1,1'-Binaphthyl-2,2'-dithiol**

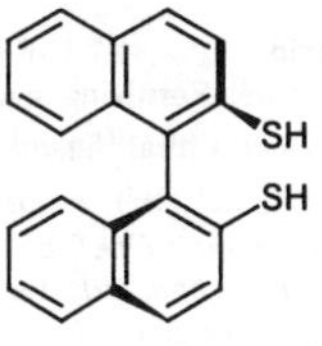

$C_{20}H_{12}ClO_2P$
Chiral building block; Chiral ligand. *Asymchem.*

**643 (S)-(+)-1,1'-Binaphthyl-2,2'-dithiol**

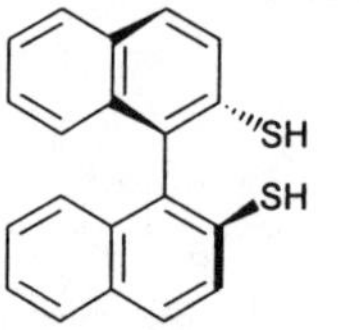

$C_{20}H_{12}ClO_2P$
Chiral ligand. *Asymchem.*

**644 (R)-(-)-1,1'-Binaphthyl-2,2'-diyl hydrogen phosphate**
39648-67-4

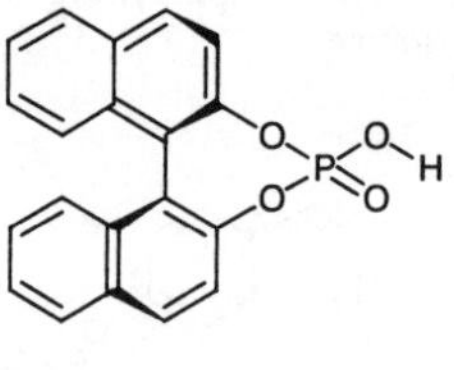

$C_{20}H_{13}O_4P$
(R)-(-)-BNP acid; (R)-(-)-1,1'-Binaphthalene-2,2'-diyl hydrogen phosphate; (R)-(-)-2,2'-Dihydroxy-1,1'-binaphthyl Hydrogenphosphate. Resolving agent; Chiral ligand; Chiral auxiliary. mp > 300°; $[\alpha]_D^{20}$ = - 605 ± 5° (c = 1, $CH_3OH$). *Acros Organics nv; Austin Chemical Company, Inc.; FineTech, Ltd.; Kankyo Kagaku Center Co., Ltd.; Kiralchem Ltd.; Lancaster Synthesis Ltd.; Omega Chemical Company Inc.; PCAS; RCA Reuter Chemischer Apparaebau KG; Sigma-Aldrich Fine Chemicals; TCI America.*

**645 (S)-(+)-1,1'-Binaphthyl-2,2'-diyl hydrogen phosphate**
35193-64-7

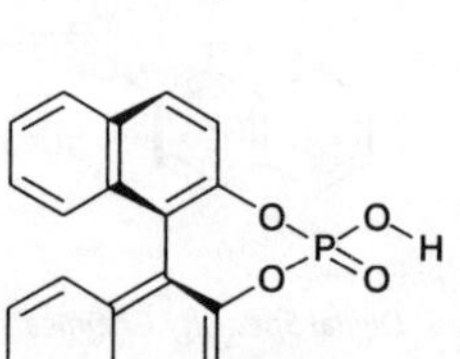

$C_{20}H_{13}O_4P$
(S)-(+)-BNP acid; (S)-(+)-1,1'-Binaphthalene-2,2'-diyl hydrogen phosphate. Resolving agent; Chiral auxiliary; Chiral intermediate; Chiral ligand. mp > 300°; $[\alpha]_D^{20}$ = + 605 ± 5° (c = 1, $CH_3OH$). *Acros Organics nv; Austin Chemical Company, Inc.; FineTech, Ltd.; Kankyo Kagaku Center Co., Ltd.; Kiralchem Ltd.; Lancaster Synthesis Ltd.; Omega Chemical Company Inc.; PCAS; RCA Reuter Chemischer Apparaebau KG; Sigma-Aldrich Fine Chemicals; TCI America.*

**646 D-(+)-Biotin**
58-85-5 1272(12) 200-399-3

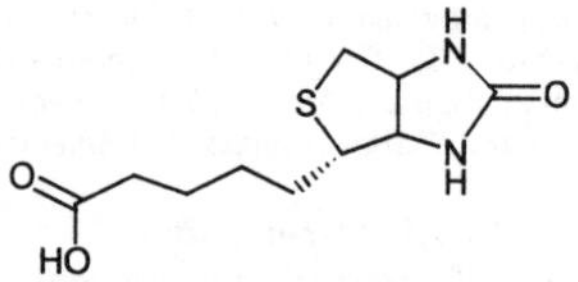

$C_{10}H_{16}N_2O_3S$
Thieno[3,4-d]-1H-imidazole-4-pentanoic acid, hexahydro-2-oxo-, (3αS,4S,6αR)-; (+)-cis-Hexahydro-2-oxo-1H-thieno[3,4]imidazole-4-valeric acid; Bioepiderm; Bios II; cis-(+)-Tetrahydro-2-oxothieno[3,4]imidazoline-4-valeric acid; Coenzyme R; Factor S (vitamin); Meribin; Rovimix H 2; Vitamin B7; Vitamin H. Listed on TSCA. Chiral intermediate. mp = 231-233°; $[\alpha]^{18}$ = + 91.1° (c = 1, 0.1N NaOH). *Aceto Corporation; Sigma-Aldrich Fine Chemicals.*

**647 (+)-Biotin 4-nitrophenyl ester**
33755-53-2

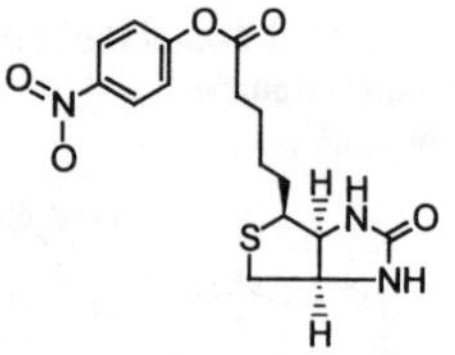

$C_{16}H_{19}N_3O_5S$
Pharmaceutical or derivative. mp = 166-167°; $[\alpha]^{25}$ = + 51° [c = 2, DMF/AcOH). *Sigma-Aldrich Fine Chemicals.*

**648 2,2-Bis(4S)-4-benzyl-2-oxazoline**
133463-88-4

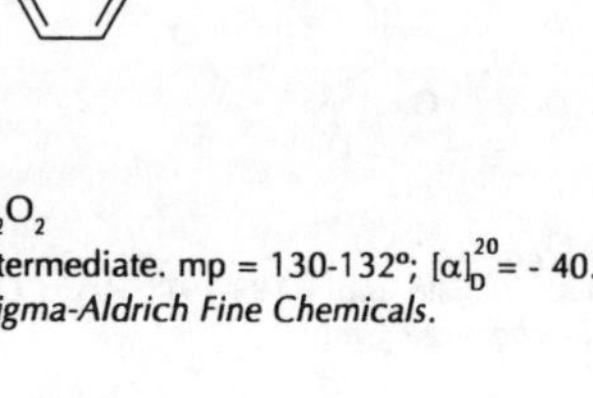

$C_{20}H_{20}N_2O_2$
Chiral intermediate. mp = 130-132°; $[\alpha]_D^{20}$ = - 40.6° (c = 1, EtOH). *Sigma-Aldrich Fine Chemicals.*

**649 (4R,5R)-Bis[bis(3',5'-dimethyl-4'-methoxyphenyl)phosphinomethyl]-2,2-dimethyl-1,3-dioxolane**

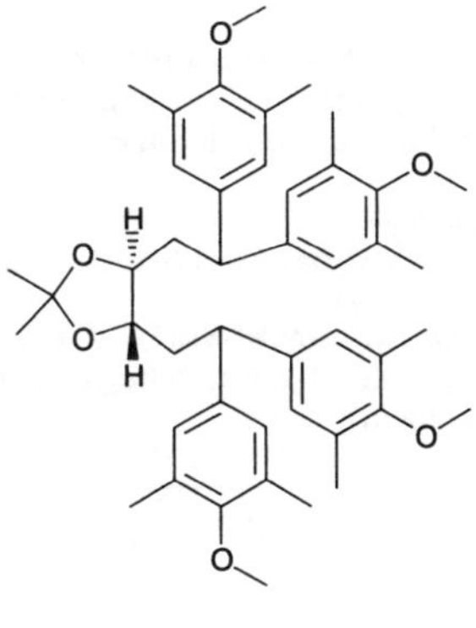

$C_{42}H_{56}O_6P_2$
(R,R)-MOD-DIOP. Chiral ligand. *Digital Specialty Chemicals, Inc.; Toyotama Perfumery Co., Ltd.*

**650 (4S,5S)-Bis[bis(3',5'-dimethyl-4'-methoxyphenyl)phosphinomethyl]-2,2-dimethyl-1,3-dioxolane**

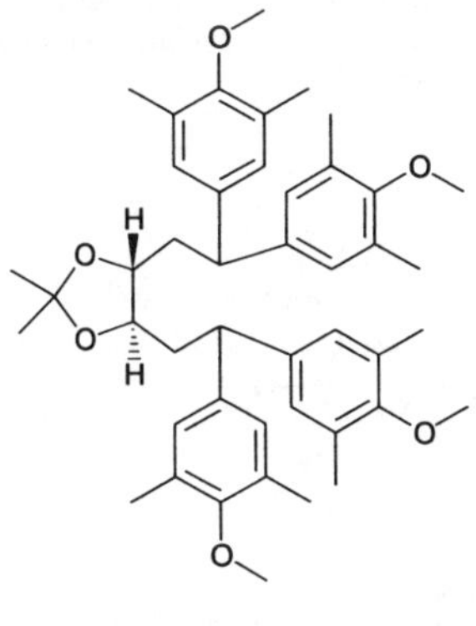

$C_{42}H_{56}O_6P_2$
(S,S)-MOD-DIOP. Chiral ligand. *Digital Specialty Chemicals, Inc.; Toyotama Perfumery Co., Ltd.*

**651 (R)-(+)-2,2'-Bis[bis(3,5-dimethylphenylphosphino)]-1,1'-binaphthyl**

$C_{46}H_{36}P_2$
Chiral ligand. *Digital Specialty Chemicals, Inc.*

**652 (S)-(+)-2,2'-Bis[bis(3,5-dimethylphenylphosphino)]-1,1'-binaphthyl**

$C_{46}H_{36}P_2$
Chiral ligand. *Digital Specialty Chemicals, Inc.*

**653 Bis(R)-(-)-2,2-bis(diphenylphosphino)-1,1-binaphthylrhodium(I) perchlorate**
95156-21-1

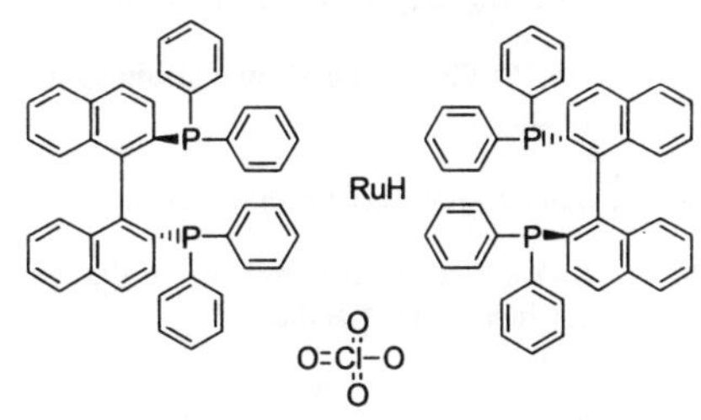


$C_{88}H_{64}ClO_4P_4Rh$
Chiral ligand. mp = 258-261°; $[\alpha]^{24}$ = - 109.7° (c = 0.2, $CH_2Cl_2$). *Sigma-Aldrich Fine Chemicals.*

**654 (R)-(+)-2,2'-Bis[bis(3,5-ditrifluoromethylphenyl)phosphino]-1,1'-binaphthyl**

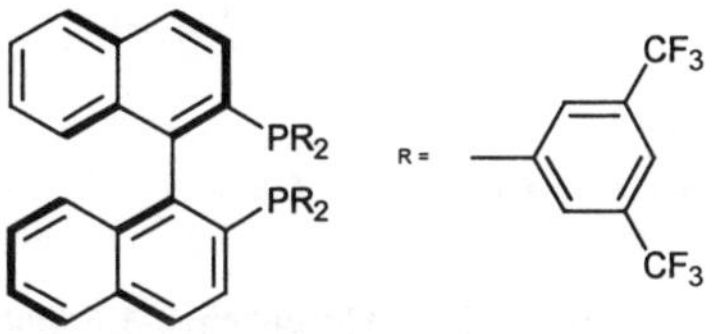

$C_{52}H_{24}F_{24}P_2$
Chiral ligand. *Digital Specialty Chemicals, Inc.*

**655 (S)-(-)-2,2'-Bis[bis(3,5-ditrifluoromethylphenyl)phosphino]-1,1'-binaphthyl**

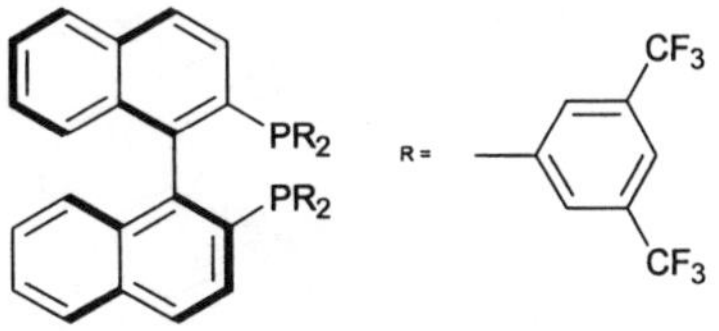

$C_{52}H_{24}F_{24}P_2$
Chiral ligand. *Digital Specialty Chemicals, Inc.*

**656 (1S,9S)-(-)-1,9-Bis(tert-butyldimethylsilyloxy)methyl-5-cyanosemicorrin**
105251-52-3

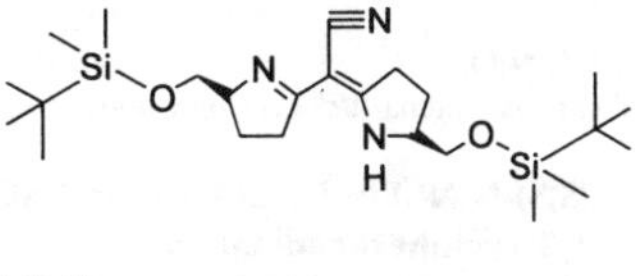

$C_{24}H_{45}N_3O_2Si_2$
Chiral intermediate. mp = 74-77°; $[\alpha]_D^{20}$ = - 64° (c = 1, $CHCl_3$). *Sigma-Aldrich Fine Chemicals.*

**657 (1R,2R)-Bis[tert-butyl(phenyl)phosphino]-ethane**

$C_{22}H_{32}P_2$
Chiral ligand. *Digital Specialty Chemicals, Inc.*

**658 (1S,2S)-Bis[tert-butyl(phenyl)phosphino]-ethane**

$C_{22}H_{32}P_2$
Chiral ligand. *Digital Specialty Chemicals, Inc.*

**659 (-)-1,6-Bis(2-chlorophenyl)-1,6-diphenyl-2,4-hexadiyne-1,6-diol**

86436-20-6

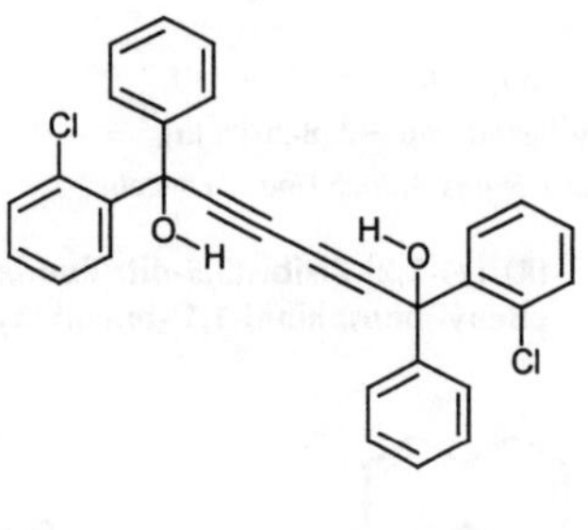

$C_{30}H_{20}Cl_2O_2$
Chiral intermediate. [α] = - 126° (c = 2, $CH_3OH$). *Acros Organics nv.*

**660 (+)-1,6-Bis(2-chlorophenyl)-1,6-diphenyl-2,4-hexadiyne-1,6-diol**

86436-19-3

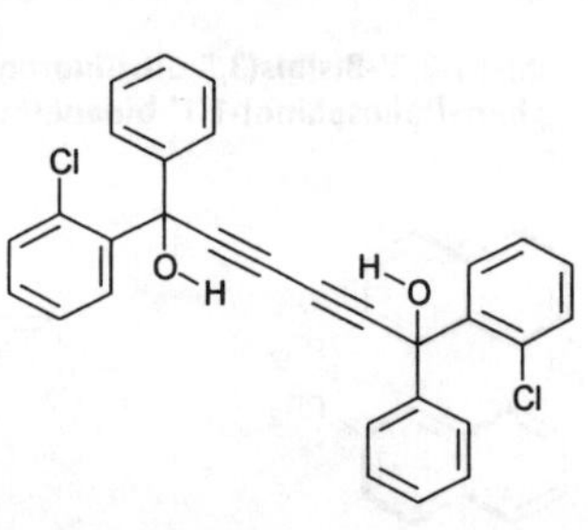

$C_{30}H_{20}Cl_2O_2$
Chiral intermediate. [α] = + 126° (c = 1 $CH_3OH$). *Acros Organics nv.*

**661 (R,R)-N,N'-Bis(3,5-di-tert-butyl-salicylidine)-1,2-cyclohexanediamino-aluminum(III) chloride**

250611-13-3

$C_{36}H_{52}AlClN_2O_2$
Chiral catalyst. *Sigma-Aldrich Fine Chemicals.*

**662 (S,S)-N,N'-Bis(3,5-di-tert-butylsalicylidine)-1,2-cyclohexanediamino-aluminum(III) chloride**

$C_{36}H_{52}AlClN_2O_2$
Chiral catalyst. *Sigma-Aldrich Fine Chemicals.*

**663 (R)-N,N'-Bis(3,5-Di-tert-butylsalicylidene)-1,2-cyclohexanediamino-chromium(III) chloride**

164931-83-3

$C_{36}H_{52}ClCrN_2O_2$
Chiral intermediate. *Sigma-Aldrich Fine Chemicals.*

**664 (S,S)-N,N'-Bis(3,5-Di-tert-butyl-salicylidene)-1,2-cyclohexanediamin-ochromium(III) chloride**

219143-92-7

$C_{36}H_{52}ClCrN_2O_2$
Chiral catalyst. *Sigma-Aldrich Fine Chemicals.*

**665 (1R,2R)-(-)-N,N'-Bis(3,5-di-t-butyl-salicydene)-1,2-cyclohexanediamino-cobalt(II)**

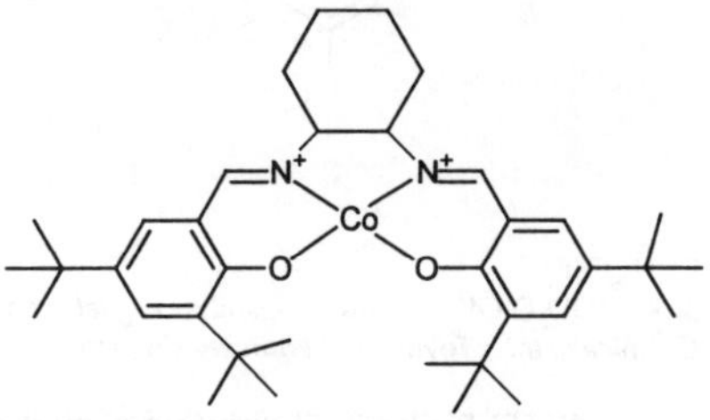

$C_{36}H_{52}CoN_2O_2$
Chiral catalyst. *Acros Organics nv.*

**666 (1S,2S)-(+)-N,N'-Bis(3,5-di-t-butyl-salicydene)-1,2-cyclohexanediamino-cobalt(II)**

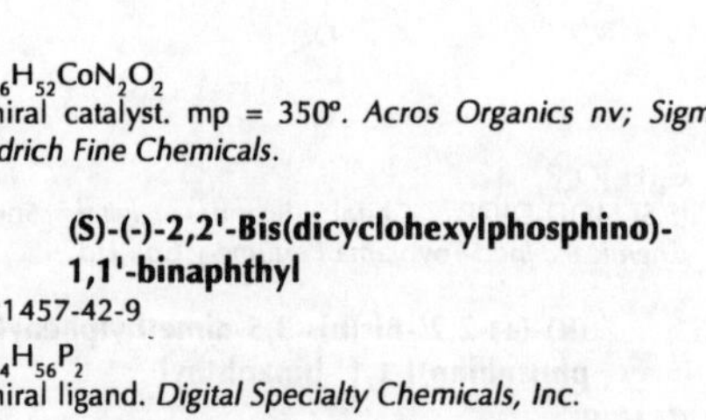

$C_{36}H_{52}CoN_2O_2$
Chiral catalyst. mp = 350°. *Acros Organics nv; Sigma-Aldrich Fine Chemicals.*

**667 (S)-(-)-2,2'-Bis(dicyclohexylphosphino)-1,1'-binaphthyl**

121457-42-9

$C_{44}H_{56}P_2$
Chiral ligand. *Digital Specialty Chemicals, Inc.*

**668 (4R,5R)-Bis[(dicyclohexylphosphino)-methyl]-2,2-dimethyl-1,3-dioxolane**

82239-68-7

$C_{31}H_{56}O_2P_2$
Chiral intermediate. *Digital Specialty Chemicals, Inc.; Maxsyn.*

**669 (4S,5S)-Bis[(dicyclohexylphosphino)-methyl]-2,2-dimethyl-1,3-dioxolane**

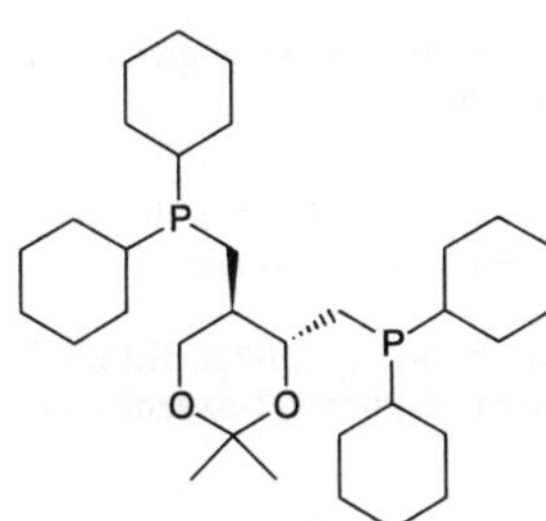

$C_{31}H_{56}O_2P_2$
Chiral ligand. *Digital Specialty Chemicals, Inc.*

**670 (-)-1,2-Bis((2R,5R)-2,5-diethylphosphalano)benzene**
136705-64-1

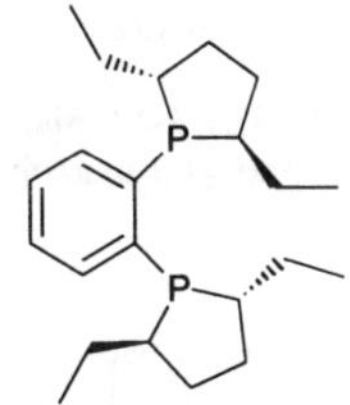

$C_{22}H_{36}P_2$
(R,R)-Et-DUPHOS. Chiral ligand. $[\alpha]_D$ = - 265° (c = 1, hexane). *Strem Chemicals, Inc.*

**671 (+)-1,2-Bis((2S,5S)-2.5-diethylphosphalano)benzene**
136779-28-7

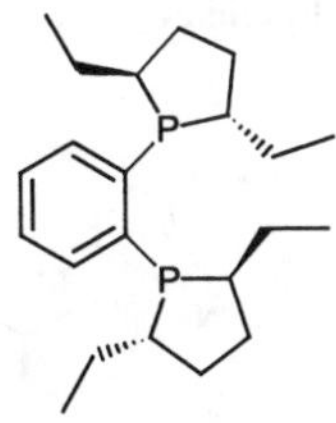

$C_{22}H_{36}P_2$
(S,S)-Et-DUPHOS. Chiral ligand. $[\alpha]_D$ = + 265° (c = 1, hexane). *Strem Chemicals, Inc.*

**672 (-)-1,2-Bis((2R,5R)-2,5-diethylphosphalano)benzene (cyclooctadiene)-rhodium(II) tetrafluorborate**

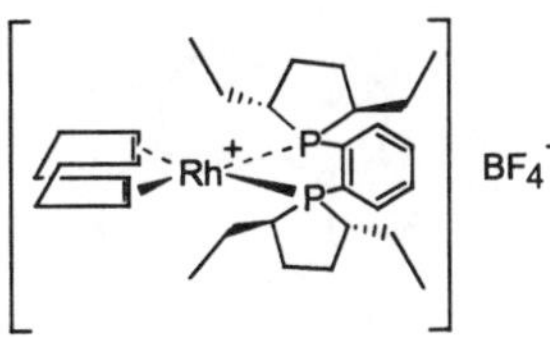


$C_{30}H_{48}BF_4P_2Rh$
Chiral catalyst. *Strem Chemicals, Inc.*

**673 (+)-1,2-Bis((2S,5S)-2,5-diethylphosphalano)benzene (cyclooctadiene)-rhodium(II) tetrafluorborate**

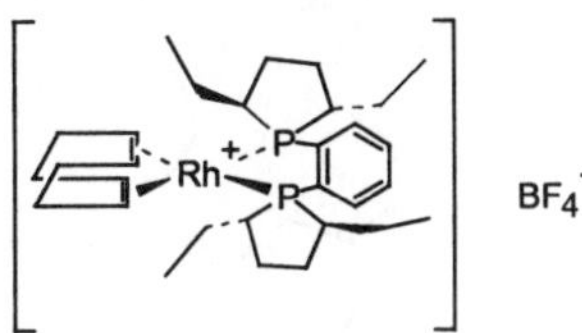


$C_{30}H_{48}BF_4P_2Rh$
Chiral catalyst. *Strem Chemicals, Inc.*

**674 (-)-1,2-Bis((2R,5R)-2,5-diethylphosphalano)benzene (cyclooctadiene)-rhodium(I) trifluoromethanesulfonate**
136705-77-6

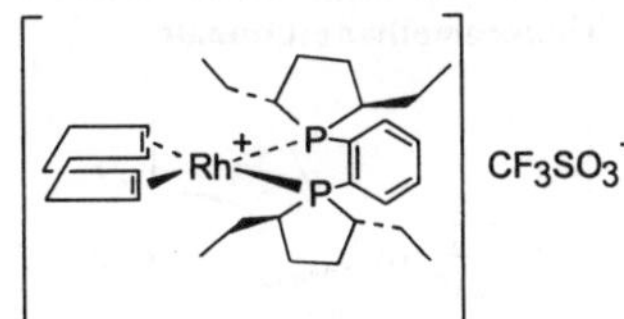


$C_{31}H_{48}F_3P_2RhS$
Chiral catalyst. *Strem Chemicals, Inc.*

**675 (+)-1,2-Bis((2S,5S)-2,5-diethylphosphalano)benzene (cyclooctadiene)-rhodium(I) trifluoromethanesulfonate**
142184-30-3

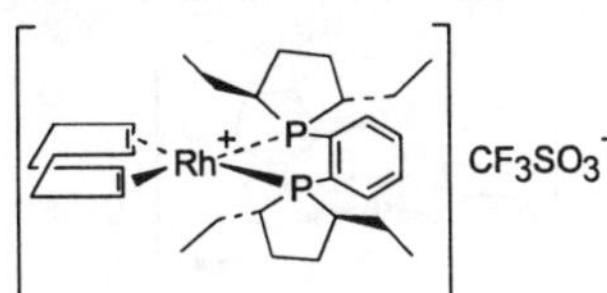


$C_{31}H_{48}F_3P_2RhS$
Chiral catalyst. *Strem Chemicals, Inc.*

**676 (+)-1,2-Bis((2R,5R)-2,5-diethylphospholano)ethane**
136705-62-9

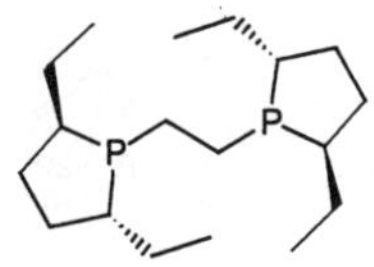

$C_{18}H_{36}P_2$
(R,R)-Et-BPE. Chiral ligand. $bp_{0.5}$ = 1 04-106°; $[\alpha]_D$ = + 320 ± 4° (c = 1, hexane). *Strem Chemicals, Inc.*

**677 (-)-1,2-Bis((2S,5S)-2,5-diethylphospholano)ethane**
136779-27-6

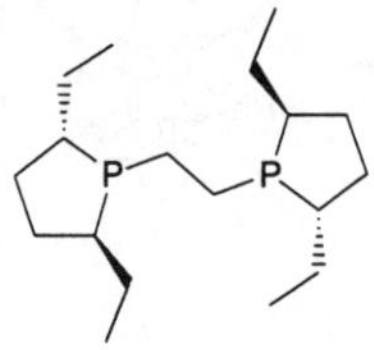

$C_{18}H_{36}P_2$
(S,S)-Et-BPE. Chiral ligand. $bp_{0.5}$ = 104-106°; $[\alpha]_D$ = - 320 ± 4° (c = 1, hexane). *Strem Chemicals, Inc.*

**678 (+)-1,2-Bis((2R,5R)-2,5-diethylphosphalano)ethane (cyclooctadiene)rhodium(I) trifluoromethanesulfonate**

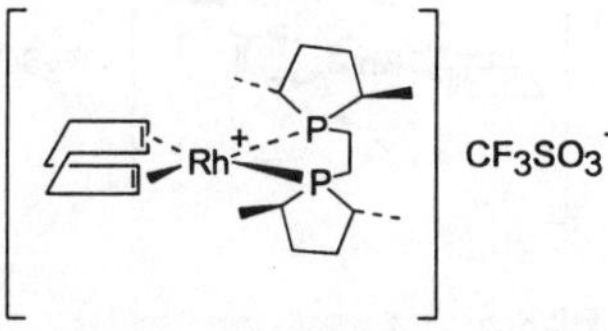

$CF_3SO_3^-$

$C_{23}H_{40}F_3O_3P_2RhS$
Chiral catalyst. *Strem Chemicals, Inc.*

**679 (-)-1,2-Bis((2S,5S)-2,5-diethylphosphalano)ethane (cyclooctadiene)rhodium(I) trifluoromethanesulfonate**

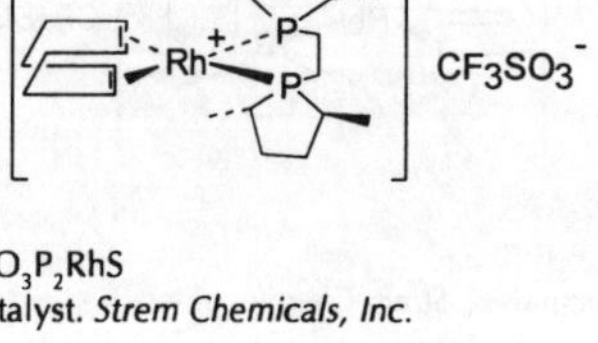

$CF_3SO_3^-$

$C_{23}H_{40}F_3O_3P_2RhS$
Chiral catalyst. *Strem Chemicals, Inc.*

**680 Bis(diethyl-D-tartrate glycolato)-diboron**

$C_{16}H_{24}B_2O_{12}$
Chiral auxiliary. *Sigma-Aldrich Fine Chemicals.*

**681 Bis(diethyl-L-tartrate glycolato)-diboron**

$C_{16}H_{24}B_2O_{12}$
Chiral auxiliary. d = 1.3; n = 1.46; $[\alpha]_D^{20}$ = - 20° (c = 1, $CHCl_3$). *Sigma-Aldrich Fine Chemicals.*

**682 (4R)-2(4R,5R),4,5-2,6-Bis(4,5-dihydro-4-methyl-5-phenyl-2-oxazolyl)pyridine**

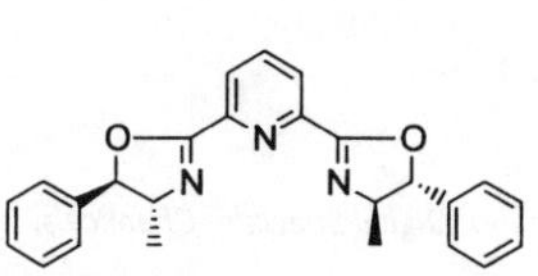

$C_{25}H_{23}N_3O_2$
Chiral intermediate. mp = 135-139°; $[\alpha]^{22}$ = + 32° (c = 1, $CHCl_3$). *Sigma-Aldrich Fine Chemicals.*

**683 (4S)-2(4R,5R),4,5-2,6-Bis(4,5-dihydro-4-methyl-5-phenyl-2-oxazolyl)pyridine**

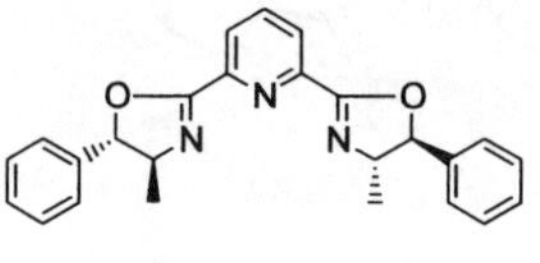

$C_{25}H_{23}N_3O_2$
Chiral intermediate. mp = 138-141°; $[\alpha]^{22}$ = - 132° (c = 1, $CHCl_3$). *Sigma-Aldrich Fine Chemicals.*

**684 (R)-(R,R)-2,6-Bis(4,5-dihydro-4-phenyl-2-oxazolyl)pyridine**

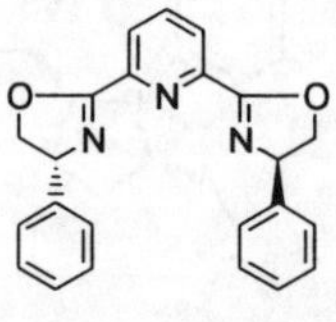

$C_{23}H_{19}N_3O_2$
Chiral intermediate. mp = 171-175°; $[\alpha]^{22}$ = + 228° (c = 1, $CHCl_3$). *Sigma-Aldrich Fine Chemicals.*

**685 (S)-(R,R)-2,6-Bis(4,5-dihydro-4-phenyl-2-oxazolyl)pyridine**

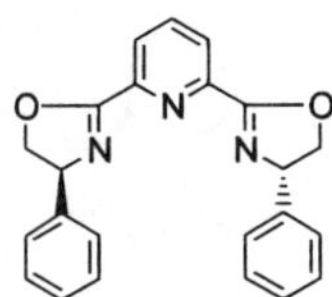


$C_{23}H_{19}N_3O_2$
Chiral intermediate. mp = 171-175°; $[\alpha]_D^{20}$ = - 224° (c = 1, $CHCl_3$). *Sigma-Aldrich Fine Chemicals.*

**686 (1R,2R)-N,N-Bis(3,3-dimethylbutyl)-1,2-cyclohexanediamine**

$C_{18}H_{38}N_2$
Chiral ligand. mp = 68-72°; $[\alpha]^{23}$ = - 88° (c = 2, $CHCl_3$). *Sigma-Aldrich Fine Chemicals.*

**687 (1S,2S)-N,N-Bis(3,3-dimethylbutyl)-1,2-cyclohexanediamine**

$C_{18}H_{38}N_2$
Resolving agent; Chiral building block. mp = 68-71°; $[\alpha]^{22}$ = + 87° (c = 2, $CHCl_3$). *Sigma-Aldrich Fine Chemicals.*

**688 (-)-1,2-Bis((2R,5R)-2,5-dimethylphosphalano)benzene**
147253-67-6

$C_{18}H_{28}P_2$
(R,R)-Me-DUPHOS. Chiral ligand. mp = 82-84°; $[\alpha]_D$ = - 476° (c = 1, hexane). *Strem Chemicals, Inc.*

**689 (+)-1,2-Bis((2S,5S)-2,5-dimethylphosphalano)benzene**

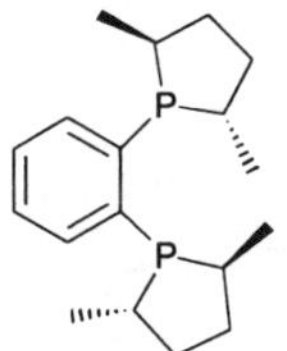


$C_{18}H_{28}P_2$
(S,S)-Me-DUPHOS. Chiral ligand. mp = 82-84°; $[\alpha]_D$ = + 476° (c = 1, hexane). *Strem Chemicals, Inc.*

**690 (+)-1,2-Bis((2R,5R)-2,5-dimethylphosphalano)benzene (cyclooctadiene)-rhodium(II) tetrafluorborate**
2150057-23-1
$C_{26}H_{40}BF_4P_2Rh$
Chiral catalyst. *Strem Chemicals, Inc.*

**691 (+)-1,2-Bis((2S,5S)-2,5-dimethylphosphalano)benzene (cyclooctadiene)-rhodium(II) tetrafluorborate**
205064-10-4

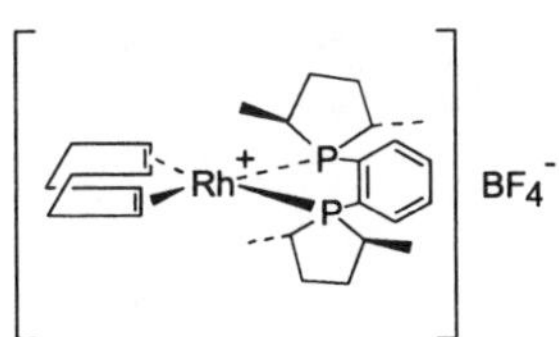


$C_{26}H_{40}BF_4P_2Rh$
Chiral catalyst. *Strem Chemicals, Inc.*

**692 (-)-1,2-Bis((2R,5R)-2,5-dimethylphosphalano)benzene (cyclooctadiene)-rhodium(I) trifluoromethanesulfonate**
187682-63-9

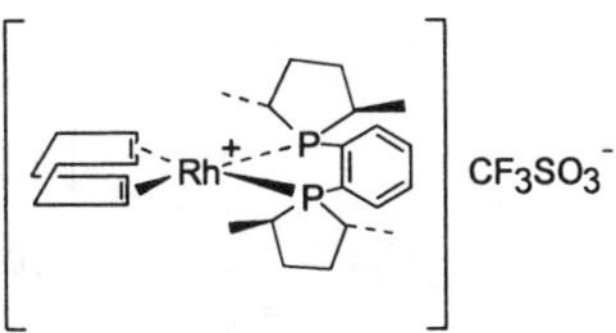


$C_{27}H_{40}F_3P_2RhS$
Chiral catalyst. *Strem Chemicals, Inc.*

**693 (+)-1,2-Bis((2S,5S)-2,5-dimethylphosphalano)benzene (cyclooctadiene)-rhodium(I) trifluoromethanesulfonate**
136705-75-4

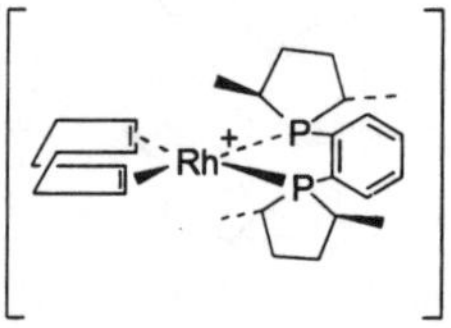

$CF_3SO_3^-$

$C_{27}H_{40}F_3P_2RhS$
Chiral catalyst. *Strem Chemicals, Inc.*

**694 (-)-1,2-Bis((2R,5R)-2,5-dimethylphosphalano)ethane**
129648-07-3

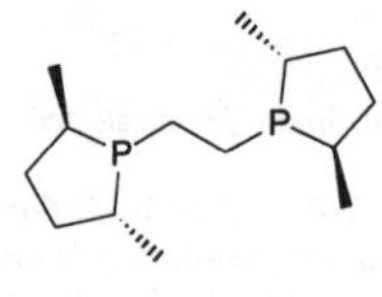

$C_{14}H_{28}P_2$
(R,R)-Me-BPE. Chiral ligand. $bp_{0.06}$ = 64-67°; $[\alpha]_D$ = + 263 ± 3° (c = 1, hexane). *Strem Chemicals, Inc.*

**695 (-)-1,2-Bis((2S,5S)-2,5-dimethylphosphalano)ethane**
136779-26-5

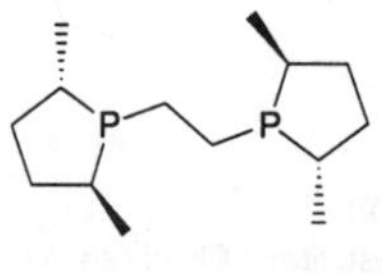

$C_{14}H_{28}P_2$
(S,S)-Me-BPE. Chiral ligand. $bp_{0.06}$ = 64-67°; $[\alpha]_D$ = - 263 ± 3° (c = 1, hexane). *Strem Chemicals, Inc.*

**696 (+)-1,2-Bis((2R,5R)-2,5-dimethylphosphalano)ethane (cyclooctadiene)-rhodium(II) tetrafluorborate**

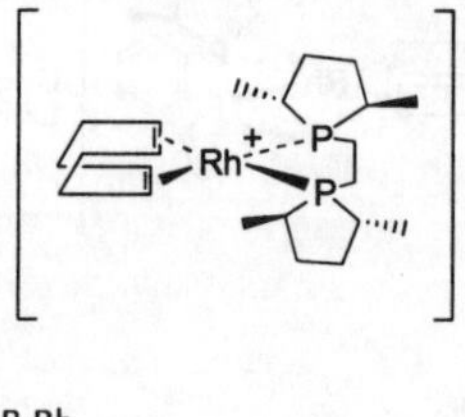

$BF_4^-$

$C_{22}H_{40}BF_4P_2Rh$
Chiral catalyst. *Strem Chemicals, Inc.*

**697 (R)-(+)-2,2'-Bis[(diphenylphosphino)-amino]-1,1'-binaphthyl**
74974-14-4

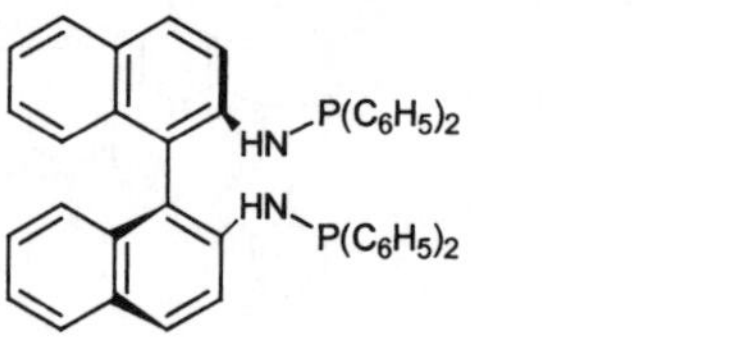

$C_{44}H_{34}N_2P_2$
Chiral ligand. *Digital Specialty Chemicals, Inc.*

**698 (S)-(-)-2,2'-Bis[(diphenylphosphino)-amino]-1,1'-binaphthyl**
74974-15-5

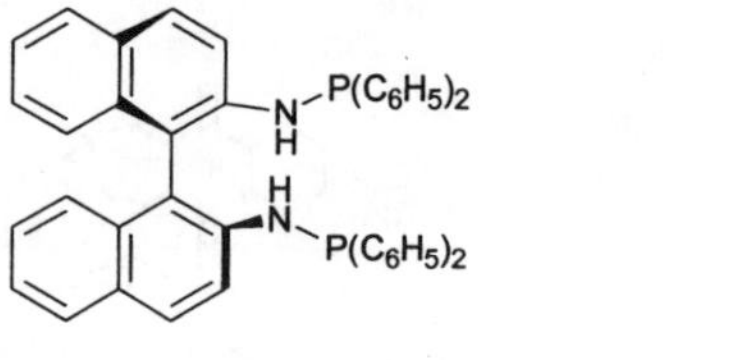

$C_{44}H_{34}N_2P_2$
Chiral ligand. *Digital Specialty Chemicals, Inc.*

**699 (2R,3R)-(+)-Bis[(diphenylphosphino)-amino]butane**

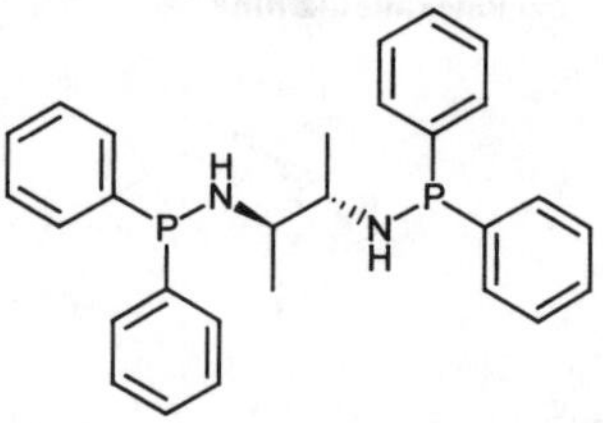

$C_{28}H_{30}N_2P_2$
Chiral ligand. *Digital Specialty Chemicals, Inc.*

**700 (2S,3S)-(-)-2,3-Bis[(diphenylphosphino)-amino]butane**
72090-84-7

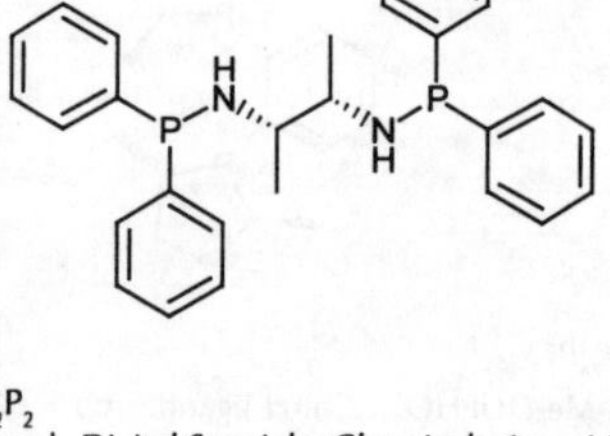

$C_{28}H_{30}N_2P_2$
Chiral ligand. *Digital Specialty Chemicals, Inc.; Maxsyn.*

**701 (1R,2R)-(-)-1,2-Bis[(N-diphenylphosphino)-amino]cyclohexane**
70708-37-1

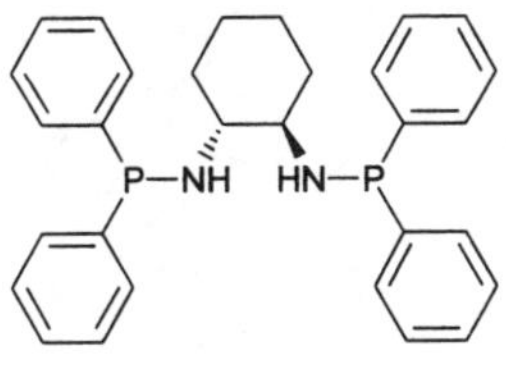

$C_{30}H_{32}N_2P_2$
Chiral ligand. *Digital Specialty Chemicals, Inc.*

**702 (1S,2S)-(+)-1,2-Bis[(N-diphenylphosphino)-amino]cyclohexane**
72090-83-6

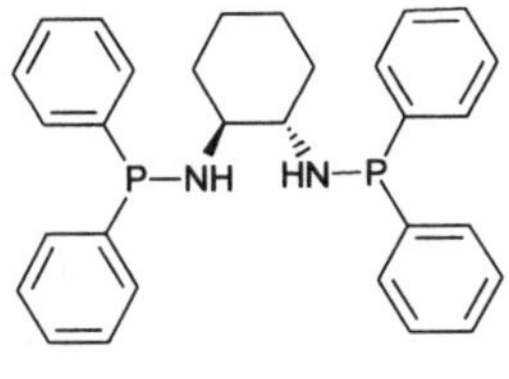

$C_{30}H_{32}N_2P_2$
Chiral ligand. *Digital Specialty Chemicals, Inc.*

**703 (2R,3R)-(-)-2,3-Bis(diphenylphosphino)-bicyclo[2.2.1]heptane**

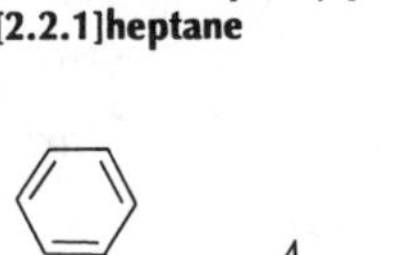

$C_{31}H_{30}P_2$
Chiral ligand. *Digital Specialty Chemicals, Inc.*

**704 (2S,3S)-(+)-2,3-Bis(diphenylphosphino)-bicyclo[2.2.1]heptane**
76740-45-9

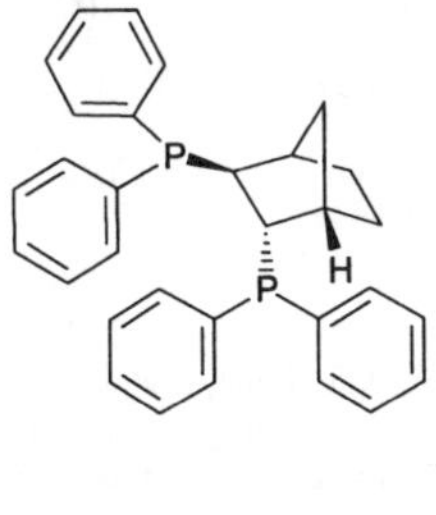

$C_{31}H_{30}P_2$
Chiral ligand. *Digital Specialty Chemicals, Inc.*

**705 (-)-(2R,3R)-2,3-Bis(diphenylphosphino)-bicyclo[2.2.1]hept-5-ene**
71042-54-1
$C_{31}H_{28}P_2$
(R,R)-NORPHOS. Chiral ligand. mp = 116-119°; $[\alpha]_D$ = - 47° (c = 1, $CHCl_3$). *Digital Specialty Chemicals, Inc.; Maxsyn; Strem Chemicals, Inc.*

**706 (+)-(2S,3S)-2,3-Bis(diphenylphosphino)-bicyclo[2.2.1]hept-5-ene**
71042-55-2
$C_{31}H_{28}P_2$
(S,S)-NORPHOS. Chiral ligand. mp = 116-119°; $[\alpha]_D$ = + 47° (c = 1, $CHCl_3$). *Digital Specialty Chemicals, Inc.; Maxsyn; Strem Chemicals, Inc.*

**707 (R)-(+)-2,2'-Bis(diphenylphosphino)-1,1'-binaphthyl**
76189-55-4

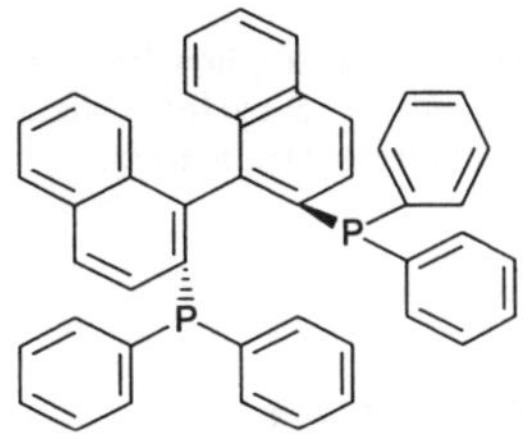

$C_{44}H_{36}P_2$
(R)-(+)-BINAP. Chiral ligand. mp = 239-241°; $[\alpha]_D^{20}$ = + 235 ± 3° (c = 0.3, toluene); $[\alpha]_D^{20}$ = + 224.5° (c = 0.7, $C_6H_6$). *Acros Organics nv; Digital Specialty Chemicals, Inc.; FineTech, Ltd.; Kiralchem Ltd.; Lancaster Synthesis Ltd.; Omega Chemical Company Inc.; Oxford Asymmetry International plc; Sigma-Aldrich Fine Chemicals; Strem Chemicals, Inc.; TCI America.*

**708 (S)-(-)-2,2'-Bis(diphenylphosphino)-1,1'-binaphthyl**
76189-56-5

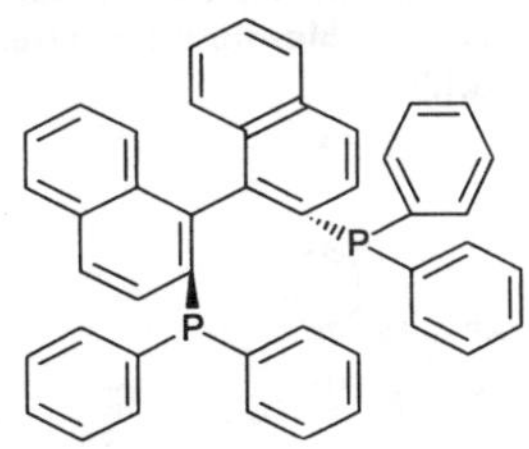

$C_{44}H_{34}P_2$
(S)-(-)-BINAP. Chiral ligand. mp = 238-240°; $[\alpha]_D^{20}$ = - 235 ± 3° (c = 0.3, toluene); $[\alpha]_D^{20}$ = - 228° (c = 0.68, $C_6H_6$). *Acros Organics nv; Digital Specialty Chemicals, Inc.; FineTech, Ltd.; Kiralchem Ltd.; Lancaster Synthesis Ltd.; Omega Chemical Company Inc.; Oxford Asymmetry International plc; Sigma-Aldrich Fine Chemicals; Strem Chemicals, Inc.; TCI America.*

**709 (R)-(+)-2,2-Bis(diphenylphosphino)-1,1-binaphthyl(1,5-cyclooctadiene)rhodium(I) perchlorate tetrahydrofuran complex (1:1)**

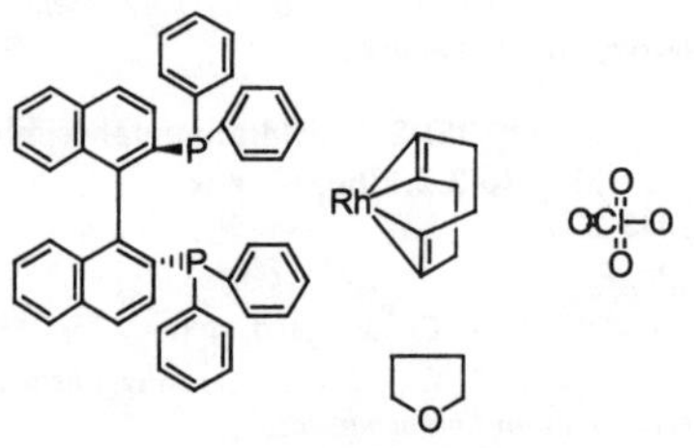

$C_{56}H_{52}ClO_5P_2Rh$
Chiral catalyst. mp = 211-214°. *Sigma-Aldrich Fine Chemicals.*

**710 (S)-(-)-2,2-Bis(diphenylphosphino)-1,1-binaphthyl(1,5-cyclooctadiene)rhodium(I) perchlorate tetrahydrofuran complex (1:1)**

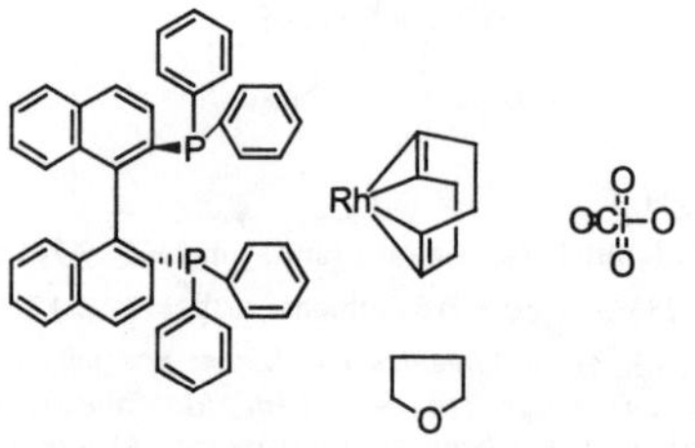

$C_{56}H_{52}ClO_5P_2Rh$
Chiral ligand. mp = 210-212°; $[\alpha]_D^{20}$ = - 180° (c = 0.05, $CH_2Cl_2$). *Sigma-Aldrich Fine Chemicals.*

**711 (R)-(+)-2,2-Bis(diphenylphosphino)-1,1-binaphthylchloro(p-cymene)ruthenium chloride**

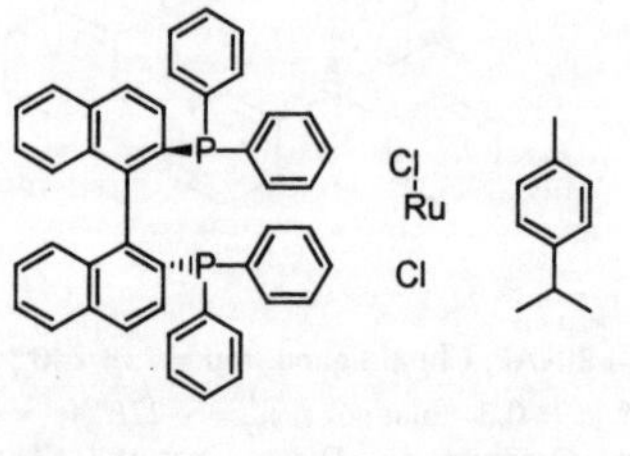

$C_{54}H_{46}Cl_2P_2Ru$
Chiral catalyst. $[\alpha]_D^{20}$ = + 57° (c = 0.1, $CHCl_3$). *Sigma-Aldrich Fine Chemicals.*

**712 (R)-(+)-2,2-Bis(diphenylphosphino)-1,1-binaphthylpalladium(II) chloride**
115826-95-4

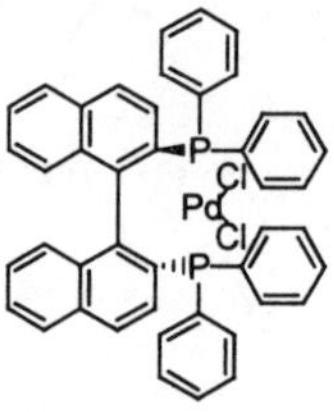

$C_{44}H_{32}Cl_2P_2Pd$
Chiral catalyst. $[\alpha]^{24}$ = + 682° (c = 0.5, $CHCl_3$). *Sigma-Aldrich Fine Chemicals.*

**713 (2R,3R)-(+)-2,3-Bis(diphenylphosphino)-butane**
74839-84-2

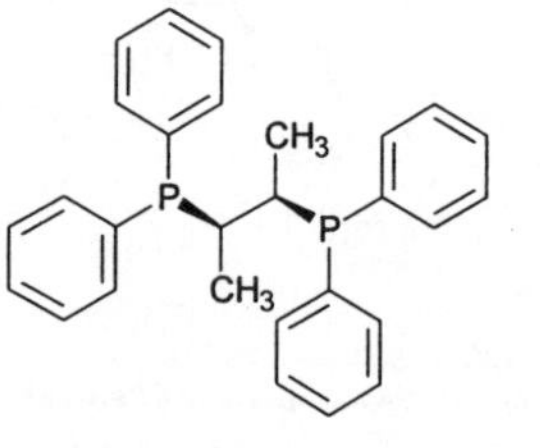

$C_{28}H_{28}P_2$
(R,R)-CHIRAPHOS. Chiral ligand. mp = 107-109°; [α] = + 195° (c = 1.5, $CHCl_3$). *Acros Organics nv; Digital Specialty Chemicals, Inc.; FineTech, Ltd.; Maxsyn; Sigma-Aldrich Fine Chemicals.*

**714 (2S,3S)-(-)-2,3-Bis(diphenylphosphino)-butane**
64896-28-2

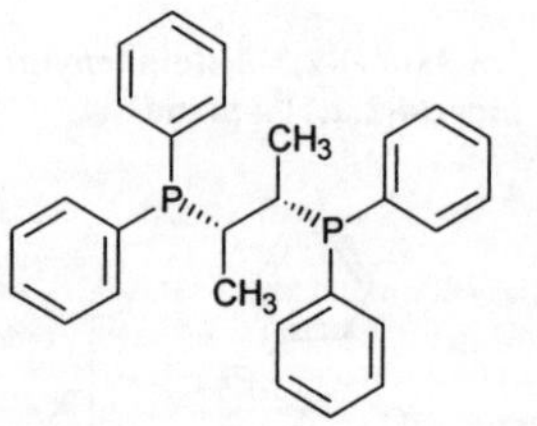

$C_{28}H_{28}P_2$
(S,S)-CHIRAPHOS. Chiral ligand. mp = 108-109°; $[\alpha]_D$ = - 190° (c = 1.5, $CHCl_3$). *Acros Organics nv; Digital Specialty Chemicals, Inc.; Maxsyn; Sigma-Aldrich Fine Chemicals; Strem Chemicals, Inc.*

**715 (2R,5R)-(+)-Bis(diphenylphosphino)hexane**

$C_{32}H_{32}P_2$
Chiral ligand. *Digital Specialty Chemicals, Inc.*

**716 (2S,5S)-(-)-Bis(diphenylphosphino)hexane**

$C_{32}H_{32}P_2$
Chiral ligand. *Digital Specialty Chemicals, Inc.*

**717 (2R,3R)-(-)-1,4-Bis(diphenylphosphino)-2,3-O-isopropylidene-2,3-butanediol**
32305-98-9 250-984-2

$C_{31}H_{32}O_2P_2$
Phosphine, [(2,2-dimethyl-1,3-dioxolane-4,5-diyl)bis-(methylene)]bis[diphenyl-, (4R-trans)-; (-)-DIOP; (-)-2,2-Dimethyl-4,5-((diphenylphosphino)dimethyl) dioxolane; (4R,5R)-(4,5-Bis(diphenylphosphinomethyl) 2,2-dimethyl-1,3-dioxolane); (4R,5R)-(-)-2,3-O-Isopropylidene-2,3-dihydroxy-1,4-bis(diphenylphosphino)butane. Chiral ligand. mp = 87-91°; $[\alpha]_D$ = - 12.7° (c = 9.5, $CHCl_3$); $[\alpha]^{19}$ = - 26° (c = 2.3, $CHCl_3$). *Acros Organics nv; Digital Specialty Chemicals, Inc.; FineTech, Ltd.; Loba Feinchemie AG; Maxsyn; Norse Laboratories; Sigma-Aldrich Fine Chemicals; Strem Chemicals, Inc.; TCI America.*

**718 (2S,3S)-(+)-1,4-Bis(diphenylphosphino)-2,3-O-isopropylidene-2,3-butanediol**
37002-48-5 253-307-9

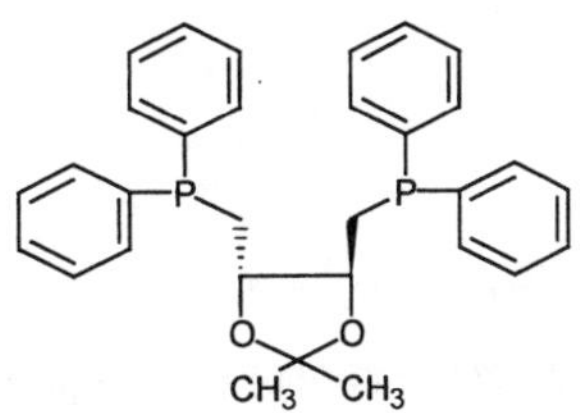

$C_{31}H_{32}O_2P_2$
Phosphine, [(2,2-dimethyl-1,3-dioxolane-4,5-diyl)bis-(methylene)]bis[diphenyl-, (4S-trans)-; (4S,5S)-(+)-2,3-O-Isopropylidene-2,3-dihydroxy-1,4 bis(diphenylphosphino)butane; (+)-DIOP; (+)-2,2-dimethyl-4,5-((diphenylphosphino)dimethyl)dioxolane. Chiral ligand. mp = 90°; $[\alpha]^{19}$ = + 26° (c = 2.3, $CHCl_3$). *Acros Organics nv; Loba Feinchemie AG; Sigma-Aldrich Fine Chemicals; Strem Chemicals, Inc.; TCI America.*

**719 (1R,2R)-(-)-1,2-Bis(diphenylphosphinomethyl)cyclohexane**
70774-28-6

$C_{32}H_{32}P_2$
Chiral ligand. *Digital Specialty Chemicals, Inc.*

**720 (1S,2S)-(+)-1,2-Bis(diphenylphosphinomethyl)cyclohexane**
70223-77-7

$C_{32}H_{32}P_2$
Chiral ligand. *Digital Specialty Chemicals, Inc.*

**721 (2R,3R)-(+)-Bis(diphenylphosphino)pentane**

$C_{29}H_{30}P_2$
Chiral ligand. *Digital Specialty Chemicals, Inc.*

**722 (2S,3S)-(-)-Bis(diphenylphosphino)pentane**

$C_{29}H_{30}P_2$
Chiral ligand. *Digital Specialty Chemicals, Inc.*

**723 (2R,4R)-(+)-2,4-Bis(diphenylphosphino)-pentane**
96183-46-9
$C_{29}H_3OP_2$
(R,R)-BDPP. Chiral ligand. mp = 78°; $[\alpha]_D$ = + 124° (c = 4.0, $CHCl_3$). *Digital Specialty Chemicals, Inc.; Strem Chemicals, Inc.*

**724 (2S,4S)-(-)-2,4-Bis(diphenylphosphino)-pentane**
77876-39-2
$C_{29}H_3OP_2$
(S,S)-BDPP. Chiral ligand. mp = 81°; $[\alpha]_D$ = - 124° (c = 3.0, $CHCl_3$). *Digital Specialty Chemicals, Inc.; Strem Chemicals, Inc.*

**725 (2S,4S)-(-)-2,4-Bis(diphenylphosphino)-pentane (norbornadiene)rhodium(I) hexaflurophosphate**

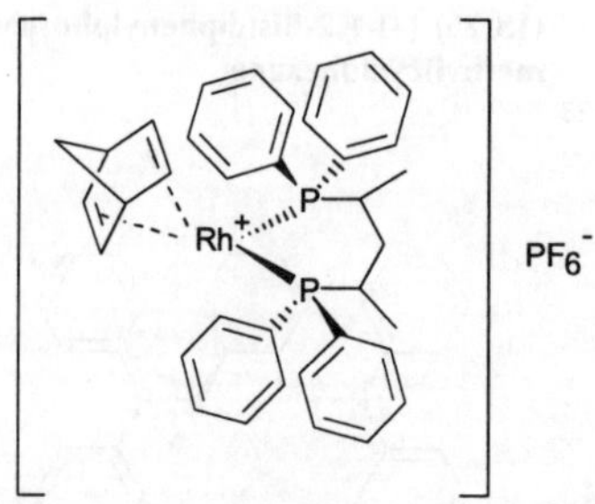


$C_{36}H_{38}F_6P_3Rh$
Chiral catalyst. *Strem Chemicals, Inc.*

**726 (R)-(+)-1,2-Bis(diphenylphosphino)propane**
67884-32-6

$C_{27}H_{26}P_2$
(R)-PROPHOS. Chiral ligand. mp = 69-71° (sealed tube); $[\alpha]_D$ = + 186° (c = 1.0, acetone). *Acros Organics nv; Digital Specialty Chemicals, Inc.; Maxsyn; Sigma-Aldrich Fine Chemicals; Strem Chemicals, Inc.*

**727 (S)-(-)-1,2-Bis(diphenylphosphino)propane**
67884-33-7

$C_{27}H_{26}P_2$
(S)-PROPHOS. Chiral ligand. *Digital Specialty Chemicals, Inc.*

**728 (3R,4R)-(+)-Bis(diphenylphosphino)-pyrrolidine**
99135-90-7
$C_{28}H_{27}NP_2$
(R,R)-PYROPHOS. Chiral ligand. *Digital Specialty Chemicals, Inc.; Maxsyn.*

**729 (3S,4S)-(-)-Bis(diphenylphosphino)-pyrrolidine**

$C_{28}H_{27}NP_2$
(S,S)-PYROPHOS. Chiral ligand. *Digital Specialty Chemicals, Inc.; Maxsyn.*

**730 (R)-(+)-2,2'-Bis(di-p-tolylphosphino)-1,1'-binaphthyl**
99646-28-3
$C_{48}H_{40}P_2$
(R)-Tol-BINAP. Chiral ligand. mp = 255-257; $[\alpha]_D$ = + 156° (c = 0.5, $C_6H_6$). *Digital Specialty Chemicals, Inc.; Strem Chemicals, Inc.*

**731 (S)-(-)-2,2'-Bis(di-p-tolylphosphino)-1,1'-binaphthyl**
100165-88-6
$C_{48}H_{40}P_2$
(S)-Tol-BINAP. Chiral ligand. mp = 255-257; $[\alpha]_D$ = - 160° (c = 0.5, $C_6H_6$). *Digital Specialty Chemicals, Inc.; Strem Chemicals, Inc.*

**732 Bis-(S)-1-(ethoxycarbonyl)ethyl fumarate**

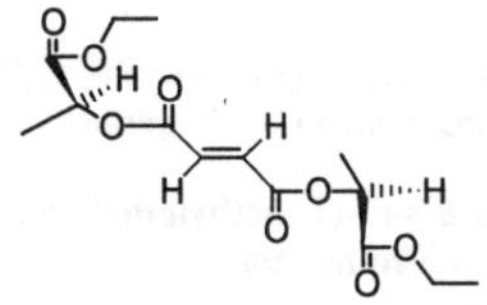

$C_{14}H_{20}O_8$
Chiral intermediate. $bp_{0.40}$ = 151°; d = 1.144; n = 1.45; $[\alpha]_D^{20}$ = - 8° (c = 0.5, $CHCl_3$). *Sigma-Aldrich Fine Chemicals.*

**733 (-)-4,5-Bis[hydroxy(diphenyl)methyl]-2,2-dimethyl-1,3-dioxolane**
93379-48-7

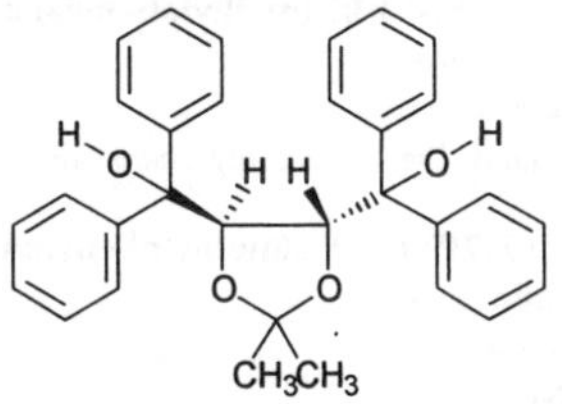

$C_{31}H_{30}O_4$
(-)-2,3-O-Isopropylidene-1,1,4,4-tetraphenyl L-threitol; (-)-trans-α,α'-(2,2-Dimethyl-1,3-dioxolane-4,5-diyl) bis-(di-phenylmethanol); (2R,3R)-1,1,4,4-Tetraphenyl-2,3-O-isopropylidene-1,2,3,4-butanetetrol; (-)-Taddol. Chiral intermediate. mp = 193-195°; $[\alpha]^{19}$ = - 62.6° (c = 1, $CHCl_3$). *Acros Organics nv; FineTech, Ltd.; Sigma-Aldrich Fine Chemicals; TCI America.*

**734 (+)-4,5-Bis[hydroxy(diphenyl)methyl]-2-methyl-2-phenyl-1,3-dioxolane**
109306-21-0
$C_{36}H_{32}O_4$
(2R,3R)-1,1,4,4-Tetraphenyl-2,3-O-(1-phenylethylidene)-1,2,3,4-butanetetrol; (2R,3R)-2,3-O-(1-Phenylethylidene)-1,1,4,4-tetraphenyl-1,2,3,4-butanetetrol. Chiral intermediate. *TCI America.*

**735 (S)-N,N'-Bis[2-hydroxy-1-(hydroxymethyl)-ethyl]-5-[(2-hydroxy-1-oxopropyl)amino]-2,4,6-triiodo-1,3-benzenedicarboxamide**
60166-93-0 262-093-6
$C_{17}H_{22}I_3N_3O_8$
Benzene-1,3-dicarboxamide, N,N'-bis[2-hydroxy-1-(hydroxymethyl)ethyl]-5-[(2-hydroxy-1-oxopropyl)amino]-2,4,6-triiodo-, (S)-; Lopamidol; (S)-N,N'-Bis[2-hydroxy-1-(hydroxymethyl)ethyl]-5-[(2-hydroxy-1-oxopropyl)amino]-2,4,6-triiodoisophthaldiamide; Lopamiro; Lopamiron; Isovue; Jopamiron; Niopam; Solutrast. Pharmaceutical or derivative. *Hovione SA.*

**736 trans-(4S,5S)-(-)-4,5-Bis(iodomethyl)-2,2-dimethyl-1,3-dioxolane**
60046-17-5
$C_7H_{12}I_2O_2$
Chiral intermediate. $n_D^{20}$ = 1.574; $[\alpha]_D^{20}$ = - 16 ± 1° (c = 3.5, $CH_3OH$). *Lancaster Synthesis Ltd.*

**737 2,6-Bis(4R)-(+)-isopropyl-2-oxazolin-2-ylpyridine**
131864-67-0

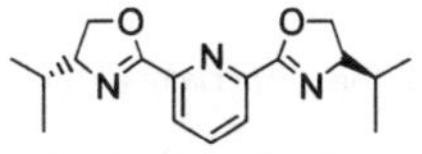

$C_{17}H_{23}N_3O_2$
Chiral intermediate. mp = 154-156°; $[\alpha]_D^{20}$ = + 118° (c = 0.7, $CH_2Cl_2$). *Sigma-Aldrich Fine Chemicals.*

**738 2,6-Bis(4S)-(-)-isopropyl-2-oxazolin-2-ylpyridine**
118949-61-4

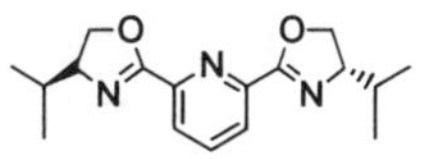

$C_{17}H_{23}N_3O_2$
Chiral intermediate. mp = 155-157°; $[\alpha]^{25}$ = - 118° (c = 0.7, $CH_2Cl_2$). *Sigma-Aldrich Fine Chemicals.*

**739 N,O-Bis(4-methoxybenzyloxycarbonyl)-(S)-2-amino-3-(4-hydroxyphenyl)propionitrile**
73148-72-8

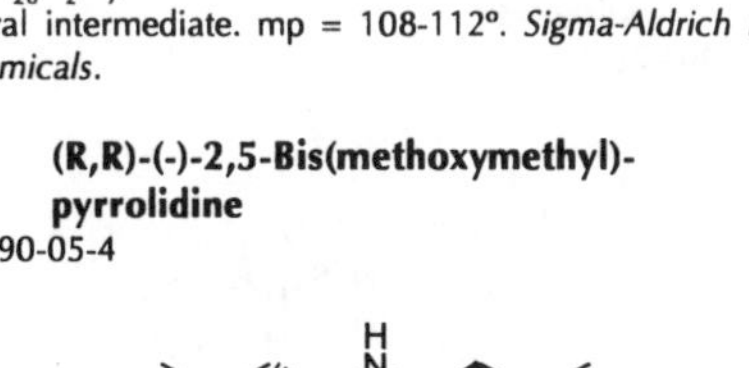

$C_{27}H_{26}N_2O_7$
Chiral intermediate. mp = 108-112°. *Sigma-Aldrich Fine Chemicals.*

**740 (R,R)-(-)-2,5-Bis(methoxymethyl)-pyrrolidine**
90290-05-4

$C_8H_{17}NO_2$
Chiral intermediate. $[\alpha]$ = - 7.5° (c = 3, EtOH). *Acros Organics nv; Oxford Asymmetry International plc.*

**741** **(S,S)-(+)-2,5-Bis(methoxymethyl)-pyrrolidine**
93621-94-4

$C_8H_{17}NO_2$
Chiral intermediate. [α] = + 7.5° (c = 3, EtOH). *Acros Organics nv; Oxford Asymmetry International plc.*

**742** **(1R,4R)-Bis[(2-methoxyphenyl)-phenylphosphino]butane**
$C_{30}H_{32}O_2P_2$
Chiral ligand. *Digital Specialty Chemicals, Inc.*

**743** **(1S,4S)-Bis[(2-methoxyphenyl)-phenylphosphino]butane**
$C_{30}H_{32}O_2P_2$
Chiral ligand. *Digital Specialty Chemicals, Inc.*

**744** **(R)-(R,R)-(+)-Bis(methylbenzyl)amine hydrochloride**

$C_{16}H_{20}ClN$
Chiral intermediate. mp = 260-263°; $[\alpha]^{24}$ = + 74° (c = 3, EtOH). *Sigma-Aldrich Fine Chemicals.*

**745** **(R,R)-(+)-N,N-Bis(methylbenzyl)sulfamide**
91410-68-3

$C_{16}H_{20}N_2O_2S$
Chiral intermediate. mp = 98-100°; $[\alpha]_D^{20}$ = + 80° (c = 2, EtOH). *Sigma-Aldrich Fine Chemicals.*

**746** **(S,S)-(-)-N,N-Bis(methylbenzyl)-sulfamide**
27304-75-2

$C_{16}H_{20}N_2O_2S$
Resolving agent. mp = 98-100°; $[\alpha]_D^{20}$ = - 80° (c = 2, EtOH). *Sigma-Aldrich Fine Chemicals.*

**747** **1,2-Bis-O-(1-methylethylidene)-D-chiroinositol**
40617-60-5
$C_{12}H_{20}O_6$
Chiral intermediate. mp = 153-156°; $[\alpha]_D^{20}$ = + 4° (c = 1.2, EtOH). *Sigma-Aldrich Fine Chemicals.*

**748** **1,2-Bis-O-(1-Methylethylidene)-L-chiroinositol**
65556-81-2
$C_{12}H_{20}O_6$
Chiral intermediate. mp = 153-156°; $[\alpha]_D^{20}$ = - 4° (c = 1.2, EtOH). *Sigma-Aldrich Fine Chemicals.*

**749** **1,2-Bis-O-(1-methylethylidene)-3-methyl-1D-chiroinositol**
57819-56-4
$C_{12}H_{20}O_6$
Chiral intermediate. mp = 105-108°; $[\alpha]_D^{20}$ = - 42.5° (c = 2, $CHCl_3$). *Sigma-Aldrich Fine Chemicals.*

**750** **(1R,2R)-(+)-Bis(methylphenylarsino)-benzene**
$C_{20}H_{20}As2$
Chiral ligand. *Digital Specialty Chemicals, Inc.*

**751** **(1S,2S)-(-)-Bis(methylphenylarsino)-benzene**
$C_{20}H_{20}As2$
Chiral ligand. *Digital Specialty Chemicals, Inc.*

**752** **(1R,2R)-(+)-Bis(methylphenylphosphino)-benzene**
72150-36-8
$C_{20}H_{20}NP_2$
Chiral ligand. *Digital Specialty Chemicals, Inc.*

**753** **(1S,2S)-(-)-Bis(methylphenylphosphino)-benzene**
72150-63-1
$C_{20}H_{20}NP_2$
Chiral ligand. *Digital Specialty Chemicals, Inc.*

**754** **Bis(R)-(-)-(1-naphthyl)ethylamine hydrochloride**

$C_{24}H_{24}ClN$
Resolving agent. mp = 280°; $[\alpha]^{23}$ = - 185° (c = 2, $CH_3OH$). *FineTech, Ltd.; Sigma-Aldrich Fine Chemicals.*

**755 Bis(S)-(+)-(1-naphthyl)ethylamine hydrochloride**

$C_{24}H_{24}ClN$
Resolving agent. mp = 280°; $[\alpha]^{23}$ = + 185° (c = 2, $CH_3OH$). *Sigma-Aldrich Fine Chemicals.*

**756 (1R,2R)-N,N'-Bis[3-oxo-2-(2,4,6-trimethylbenzoyl)butylidene]-1,2-diaminocobalt**
$C_{42}H_{42}CoN_2O_4$
(R)-MPAC. Chiral catalyst. *TCI America.*

**757 (1S,2S)-N,N'-Bis[3-oxo-2-(2,4,6-trimethylbenzoyl)butylidene]-1,2-diaminocobalt**
$C_{42}H_{42}CoN_2O_4$
(S)-MPAC. Chiral catalyst. *TCI America.*

**758 Bis[(R)-1-Phenylethyl]amine**

$C_{16}H_{19}N$
Chiral intermediate. *Austin Chemical Company, Inc.*

**759 (R,R')-Bis(1-phenylethyl)amine hydrochloride**
23294-41-9

$C_{16}H_{20}ClN$
Chiral ligand; Resolving agent. $bp_{0.05}$ = 86°; d = 0.985; n = 1.5523. *Arran Chemical Company Ltd.; Sigma-Aldrich Fine Chemicals.*

**760 (S,S')-Bis(1-phenylethyl)amine hydrochloride**
56210-72-1

$C_{16}H_{20}ClN$
Chiral auxiliary. $bp_{0.05}$ = 86°; d = 0.987; n = 1.5525; $[\alpha]_D^{20}$ = - 197.5° (neat). *Arran Chemical Company Ltd.; FineTech, Ltd.; Oxford Asymmetry International plc; Sigma-Aldrich Fine Chemicals.*

**761 Bis((R)-1-phenylethyl)amine hydrochloride**
82398-30-9

$C_{16}H_{20}ClN$
Chiral intermediate. *Acros Organics nv.*

**762 Bis((S)-1-phenylethyl)amine hydrochloride**
40648-92-8

$C_{16}H_{20}ClN$
Chiral intermediate. mp = 261-263°; $[\alpha]_D^{20}$ = -73.0° (c = 3, EtOH). *Acros Organics nv; Sigma-Aldrich Fine Chemicals.*

**763 N,N-Bis[(R)-1-phenylethyl)phthalamic acid**

$C_{24}H_{23}NO_3$
Chiral intermediate. mp = 110°; $[\alpha]^{22}$ = + 153° (c = 1, $CHCl_3$). *FineTech, Ltd.; Sigma-Aldrich Fine Chemicals.*

**764 N,N-Bis[(S)-1-phenylethyl)phthalamic acid**

$C_{24}H_{23}NO_3$
Chiral intermediate. mp = 100-110°; $[\alpha]^{22}$ = - 153° (c = 1, $CHCl_3$). *FineTech, Ltd.; Sigma-Aldrich Fine Chemicals.*

**765 Bis[(-)-Pinanediolato]diboron**
230299-05-5
$C_{20}H_{32}B_2O_4$
Bis[(1R,2R,3S,5R-pinanediolato)]diboron. Chiral auxiliary. mp = 147-151°; $[\alpha]_D^{20}$ = - 66° (c = 6.5,toluene). *Sigma-Aldrich Fine Chemicals.*

**766 Bis[(+)-Pinanediolato]diboron**
230299-17-9
$C_{20}H_{32}B_2O_4$
Bis[(1S,2S,3R,5S-pinanedolato)]diboron. Chiral auxiliary. mp = 147-151°; $[\alpha]_D^{20}$ = + 66° (c = 6.5, toluene). *Sigma-Aldrich Fine Chemicals.*

**767 Bis(N,N,N'-tetramethyl-L-tartaramide glycolato)diboron**
230299-42-0

$C_{16}H_{28}B_2N_4O_8$
Chiral auxiliary. *Sigma-Aldrich Fine Chemicals.*

**768 (R)-(+)-4,5-Bis(tosyloxymethyl)-2,2-dimethyl-1,3-dioxolane**
51064-65-4 256-943-5

$C_{21}H_{26}O_8S_2$
Dioxolane-4,5-dimethanol, 1,3-, 2,2-dimethyl-, bis(4-methylbenzenesulfonate), (4R,5R)-; (2R,3R)-1,4-Di-O-tosyl-2,3-O-isopropylidene-D-threitol; (4R-trans)-2,2-dimethyl-1,3-dioxolane-4,5-dimethyl bis(toluene-p-sulphonate); 2,3-Isopropylidene-D-threitol-1,4-ditosylate. Chiral building block. mp = 89-91°; $[\alpha]_D^{20}$ = + 12.3 ± 0.4° (c = 4, $CHCl_3$). *Acros Organics nv; Digital Specialty Chemicals, Inc.; Kiralchem Ltd.; Sigma-Aldrich Fine Chemicals; Strem Chemicals, Inc.*

**769 (S)-(-)-4,5-Bis(tosyloxymethyl)-2,2-dimethyl-1,3-dioxolane**
37002-45-2 253-306-6

$C_{21}H_{26}O_8S_2$
Dioxolane-4,5-dimethanol, 1,3-, 2,2-dimethyl-, bis(4-methylbenzenesulfonate), (4S,5S)-; (S,S)-(-)-1,4-Di-O-tosyl-2,3-O-isopropylidene-D-threitol; (4S-trans)-2,2-Dimethyl-1,3-dioxolane-4,5-dimethyl bis(toluene-p-sulphonate); (2S,3S)-(-)-2,3-O-Isopropylidene-1,4-di-O-tosyl-L-threitol. Chiral building block. mp = 92-93°; $[\alpha]^{23}$ = - 12.3° (c = 8.8, $CHCl_3$). *Acros Organics nv; Digital Specialty Chemicals, Inc.; FineTech, Ltd.; Kiralchem Ltd.; Sigma-Aldrich Fine Chemicals.*

**770 (R,R)-N,N'-Bis(trifluoromethanesulfonyl)-1,2-diphenylethylenediamine**
121788-73-6

$C_{16}H_{14}F_6N_2O_4S_2$
(R,R)-1,2-Bis(trifluoromethanesulfonamido)-1,2-diphenylethane. Chiral intermediate. *TCI America.*

**771 (S,S)-N,N'-Bis(trifluoromethanesulfonyl)-1,2-diphenylethylenediamine**
121788-77-0

$C_{16}H_{14}F_6N_2O_4S_2$
(S,S)-1,2-Bis(trifluoromethanesulfonamido)-1,2-diphenylethane. Chiral intermediate; ligand. *TCI America.*

**772 Boldine**
476-70-0 1355(12) 207-509-9

$C_{19}H_{21}NO_4$
Dibenzo[4H,de,g]quinoline-2,9-diol, 5,6,6α,7-tetrahydro-1,10-dimethoxy-6-methyl-, (S)-; Boldine-chloroform; 2,9-Dihydroxy-1,10-dimethoxyaporphine; 1,10-Dimethoxy-6aα-aporphine-2,9-diol; 6aα-Aporphine-2,9-diol, 1,10-dimethoxy-. Listed on TSCA. Pharmaceutical or derivative. mp = 162-165°; $[\alpha]^{22}$ = + 127° (c = 0.1, EtOH). *Kaden Biochemicals GmbH; Sigma-Aldrich Fine Chemicals.*

**773 (2R)-Bornane-10,2-sultam**

$C_{10}H_{17}NO_2S$
Chiral auxiliary; Chiral intermediate. mp = 179-181°; $[\alpha]_D^{20}$ = - 34.3 ± 0.3° (c = 2, 1,2-DCE). *Kiralchem Ltd.*

**774 (2S)-Bornane-10,2-sultam**

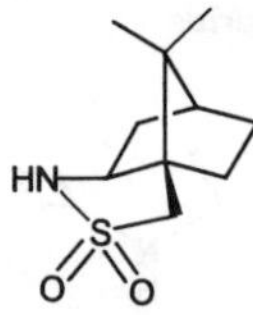

$C_{10}H_{17}NO_2S$
Chiral auxiliary; Chiral intermediate. mp = 179-181°; $[\alpha]_D^{20}$ = + 34.3 ± 0.3° (c = 2, 1,2-DCE). *Kiralchem Ltd.*

**775 (2S)-Bornane 10,2-sultam glycinate**
137819-66-0

$C_{15}H_{24}N_2O_3S_3$
N-(Bis(methytthio)methylene)glycine-(2S)-camphorsultam. Chiral intermediate. *Oxford Asymmetry International plc.*

**776 (2R)-Bornane 10,2-sultam glycinate**
127556-03-0

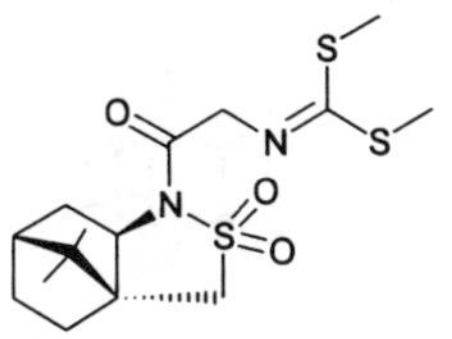

$C_{15}H_{24}N_2O_3S_3$
N-(Bis(methylthio)methylene)glycine-(2R)-camphor-sultam. Chiral intermediate. *Oxford Asymmetry International plc.*

**777 (-)-Borneol**
464-45-9 1366(12) 207-353-1

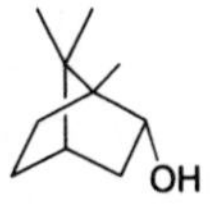

$C_{10}H_{18}O$
Bicyclo[2.2.1]heptan-2-ol, 1,7,7-trimethyl-, (1S,2R,4S)-; endo-(1S)-1,7,7-Trimethylbicyclo[2.2.1]heptan-2-ol; Linderol; Ngai camphor. Listed on TSCA. Chiral intermediate. mp = 205-208°; $[\alpha]_D^{20}$ = - 19 ± 3° (c = 5, EtOH). *Acros Organics nv; Lancaster Synthesis Ltd.; Sigma-Aldrich Fine Chemicals; Taiwan Tekho Camphor Co. Ltd.*

**778 (+)-Borneol**
464-43-7 207-352-6

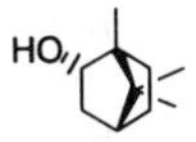

$C_{10}H_{18}O$
Chiral intermediate. mp = 207-208°; $[\alpha]_D^{20}$ = + 36° (c = 5, EtOH). *Sigma-Aldrich Fine Chemicals.*

**779 Bornyl acetate**
5655-61-8 1339(11) 227-101-4

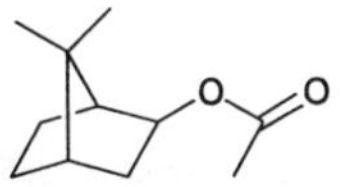

$C_{12}H_{20}O_2$
Bicyclo[2.2.1]heptan-2-ol, 1,7,7-trimethyl-, acetate, (1S,2R,4S)-; Borneol, acetate, (1S,2R,4S)-. Listed on TSCA. Chiral intermediate. bp = 223-224°; d = 0.982; n = 1.463; $[\alpha]_D^{20}$ = - 38.10° (neat). *Acros Organics nv; Sigma-Aldrich Fine Chemicals.*

**780 Bornylamine**
32511-345 251-076-9

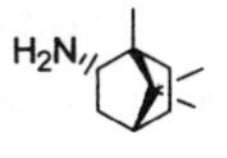

$C_{10}H_{19}N$
Chiral intermediate. mp = 160-163°; $[\alpha]_D^{20}$ = + 44° (c = 0.5, EtOH). *Sigma-Aldrich Fine Chemicals.*

**781 Brefeldin A**
20350-15-6

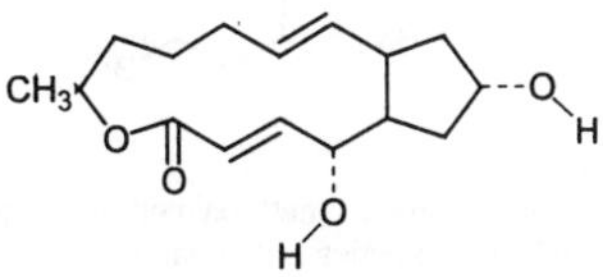

$C_{16}H_{24}O_4$
1,6,7,8,9,11a,12,13,14,14a-Decahydro-1,13-dihydroxy-6-methyl-4H-cyclopent[f]oxacyclotridecin-4-one; Ascotoxin; Cyanein. Pharmaceutical or derivative. mp = 203-204°; [α] = + 93° (c = 2, $CH_3OH$). *Acros Organics nv.*

**782 (-)-8-Bromoadenosine**
2946-39-6 220-959-0

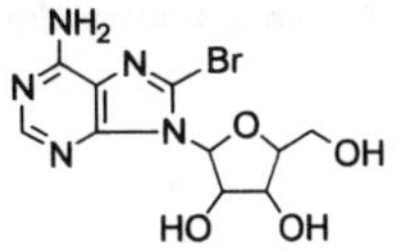

$C_{10}H_{12}BrN_5O_4$
Chiral intermediate. mp = 210-212°; $[\alpha]^{25}$ = - 54° (c = 2.5, DMSO). *Sigma-Aldrich Fine Chemicals.*

**783 (R)-6-Bromo-1,1'-bi-2-naphthol**
147053-25-6

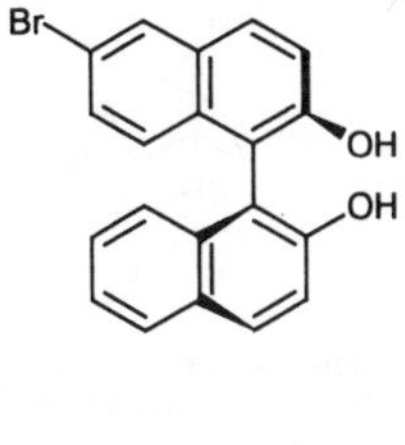

$C_{20}H_{13}BrO_2$
Chiral ligand. *Kankyo Kagaku Center Co., Ltd.*

**784 (1R)-3-Bromo-1-(bromomethyl)propyl-methoxymethyl ether**

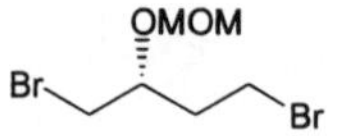

$C_6H_{12}Br_2O_2$
Butane, 1,4-dibromo-2-(methoxymethoxy)-, (R)-. Chiral intermediate. *Synthon Chiragenics Corporation.*

**785 (1S)-3-Bromo-1-(bromomethyl)propyl-methoxymethyl ether**
209806-87-1

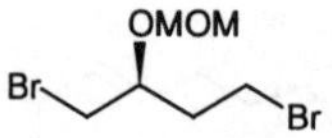

$C_6H_{12}Br_2O_2$
Butane, 1,4-dibromo-2-(methoxymethoxy)-, (S);. (2S)-1,4-Dibromo-2-(methoxymethoxy)butane. Chiral intermediate. *Synthon Chiragenics Corporation.*

**786 (R)-Bromo-1,3-butanediol**

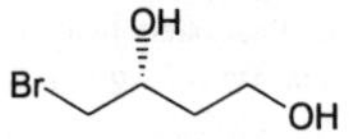

$C_4H_9BrO_2$
Butane-1,3-diol, 4-bromo, (R)-; Chiral building block. *Synthon Chiragenics Corporation.*

**787 (S)-4-Bromo-1,3-butanediol**
191354-42-4

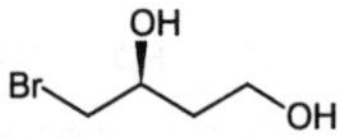

$C_4H_9BrO_2$
Butane-1,3-diol, 4-bromo, (S)-. Chiral building block. *Synthon Chiragenics Corporation.*

**788 (+)-3-Bromocamphor**
10293-06-8 1435(12) 233-652-1

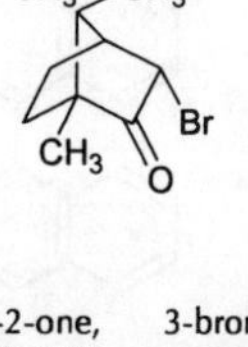

$C_{10}H_{15}BrO$
Bicyclo[2.2.1]heptan-2-one, 3-bromo-1,7,7-trimethyl-, (1R,3S,4S)-; (1R)-endo-3-Bromocamphor. Chiral intermediate. mp = 75-77°; $[\alpha]_D^{20}$ = + 133 ± 3° (c = 1, EtOH). *Acros Organics nv; Lancaster Synthesis Ltd.; Norse Laboratories; Sigma-Aldrich Fine Chemicals.*

**789 (1S)-endo-(-)-3-Bromocamphor**
64474-54-0

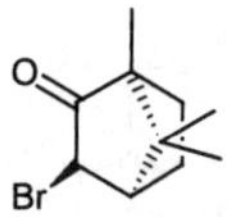

$C_{10}H_{15}BrO$
Resolving agent. mp = 75-77°; $[\alpha]_D^{20}$ = - 127° (c = 5, $CH_3OH$). *Sigma-Aldrich Fine Chemicals.*

**790 (1S)-(+)-3-endo-Bromo-10-camphorsulfonic acid**

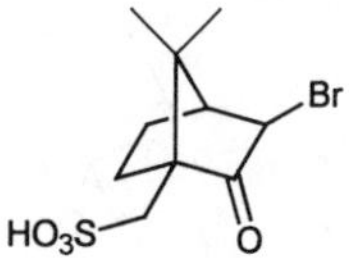

$C_{10}H_{15}BrO_4S$
Bicyclo[2.2.1]heptane-1-methanesulfonic acid, 3-bromo-7,7-dimethyl-2-oxo-, (S)-endo-; (1S)-(+)-3-endo-Bromo-2-oxobornane-10-sulphonic acid; 10-Bornanesulfonic acid, 3-bromo-2-oxo-, (S)-. Resolving agent. *Alfa Aesar; FineTech, Ltd.*

**791 (+)-3-Bromo-8-camphorsulfonic acid ammonium salt**
14575-84-9 238-616-9

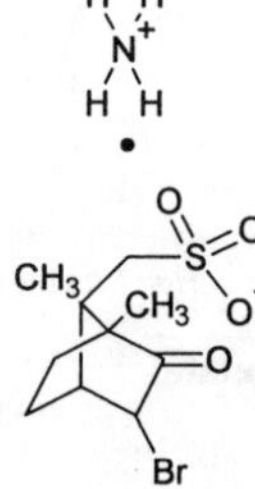

$C_{10}H_{18}BrNO_4S$
Bicyclo[2.2.1]heptane-7-methanesulfonic acid, 3-bromo-1,7-dimethyl-2-oxo-, ammonium salt, (1R,3S,4S,7R)-; D(+)-α-Bromocamphor-+lp-sulfonic acid ammonium salt; Ammonium (+)-3-bromo-8-camphorsulfonate; Ammonium D-5-bromo-6-oxo-9-bornanesulphonate; 8-Bornanesulfonic acid, 3-bromo-2-oxo-, ammonium salt, (+)-. Resolving agent; Chiral building block. mp = 284°; $[\alpha]^{25}$ = + 84.5° (c = 4, $H_2O$). *Acros Organics nv; Alfa Aesar; Calaire Chimie s.a.; Norse Laboratories; Sigma-Aldrich Fine Chemicals.*

**792 (1S)-(+)-3-Bromocamphor-10-sulfonic acid hydrate**

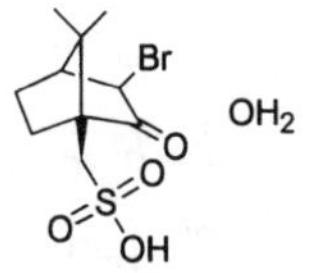

$C_{10}H_{15}BrO_4S$
Resolving agent. mp = 117-121°; $[\alpha]_D^{20}$ = + 94° (c = 1, $H_2O$). *Sigma-Aldrich Fine Chemicals.*

**793 2-Bromo-carbobenzyloxy-(tert-butoxy-carbonyl)-L-tyrosine**

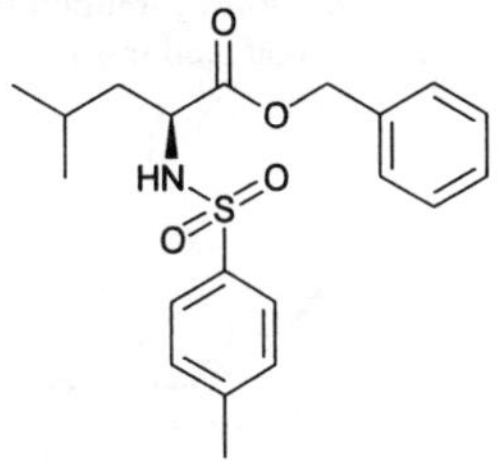

$C_{20}H_{25}NO_4S$
Chiral intermediate. *Synthetech, Inc.*

**794 (2R,3R)-4-Bromo-1-carboxy-2,3-dihydroxy-cyclohexa-4,6-diene**

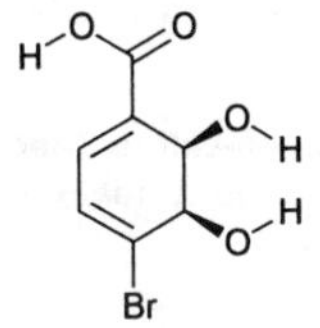

$C_7H_7BrO_4$
Chiral intermediate. *Acros Organics nv.*

**795 5-Bromo-4-chloro-3-indolyl-2-acetamido-2-deoxy-β-D-glucopyranoside**
4264-82-8
$C_{16}H_{18}BrClN_2O_6$
Chiral diagnostic reagent. Soluble in $H_2O$ (0.1%). *Diagnostic Chemicals Limited.*

**796 5-Bromo-4-chloro-3-indolyl-N-acetyl-β-D-glucosaminide**
Chiral intermediate. *Senn Chemicals AG.*

**797 5-Bromo-4-chloro-3-indolyl-β-D-cellobioside**

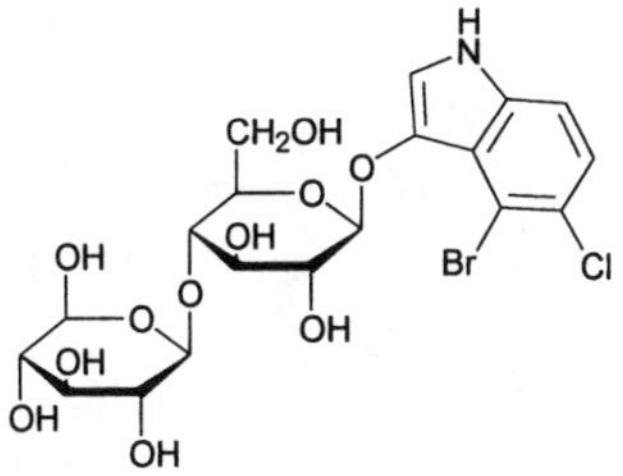

$C_{19}H_{23}BrClNO_{11}$
Chiral diagnostic reagent. *Diagnostic Chemicals Limited.*

**798 5-Bromo-4-chloro-3-indolyl-β-D-fucopyranoside**
17016-46-5
$C_{14}H_{15}BrClNO_5$
Galactopyranoside, 5-bromo-4-chloroindol-3-yl 6-deoxy-, β-D-; Chiral diagnostic reagent. Slightly soluble in $H_2O$. *Diagnostic Chemicals Limited; Senn Chemicals AG.*

**799 5-Bromo-4-chloro-3-indolyl-β-L-fucopyranoside**
125328-84-9

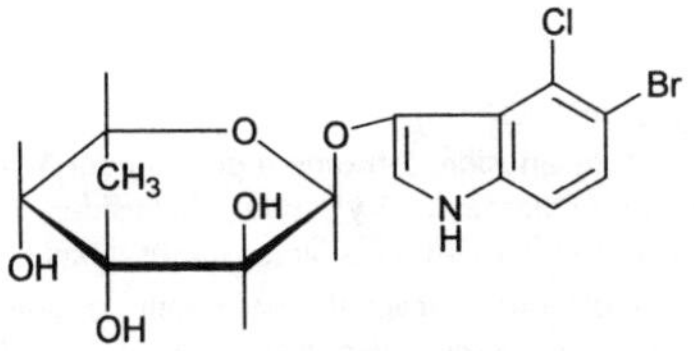

$C_{14}H_{15}BrClNO_5$
β-L-Galactopyranoside, 5-bromo-4-chloro-1H-indol-3-yl 6-deoxy-. Chiral diagnostic reagent. mp > 300°; slightly soluble in $H_2O$. *Diagnostic Chemicals Limited.*

**800 5-Bromo-4-chloro-3-indolyl-α-D-galactopyranoside**

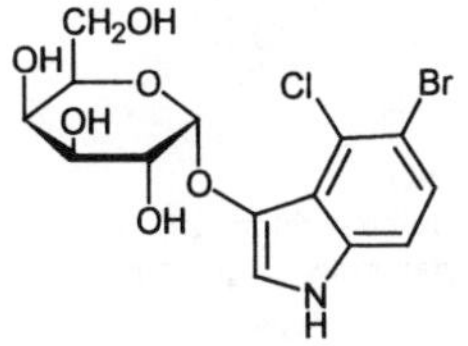

$C_{14}H_{15}BrClNO_6$
Chiral intermediate. *Senn Chemicals AG.*

**801 5-Bromo-6-chloro-3-indolyl-β-D-galactopyranoside**
93863-88-8

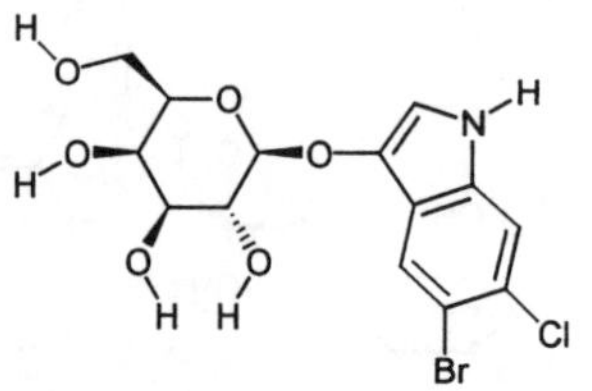

$C_{14}H_{15}BrClNO_6$
Chiral diagnostic reagent. insoluble in $H_2O$. *Acros Organics nv; Diagnostic Chemicals Limited; Senn Chemicals AG.*

**802 5-Bromo-4-chloro-3-indolyl-β-D-galactopyranoside**
7240-90-6 230-640-8

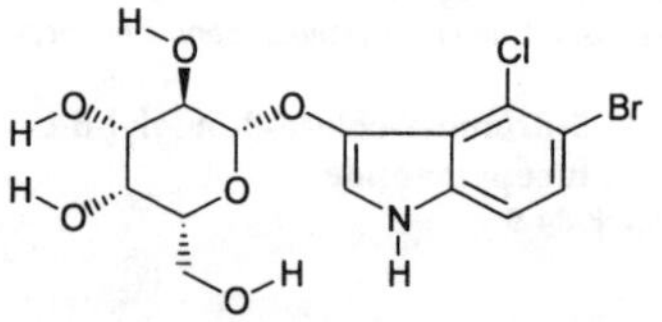

$C_{14}H_{15}BrClNO_6$
Galactopyranoside, 5-bromo-4-chloroindol-3-yl, β-D-; 5-Bromo-4-chloroindol-3-yl β-D-galactoside; Indole, 5-bromo-4-chloro-3-(β-D-galactopyranosyloxy)-; X-gal. Chiral diagnostic reagent. moderately soluble in $H_2O$. *Acros Organics nv; Diagnostic Chemicals Limited; Senn Chemicals AG.*

**803 5-Bromo-4-chloro-3-indolyl-β-D-glucopyranoside**
15548-60-4 239-603-0

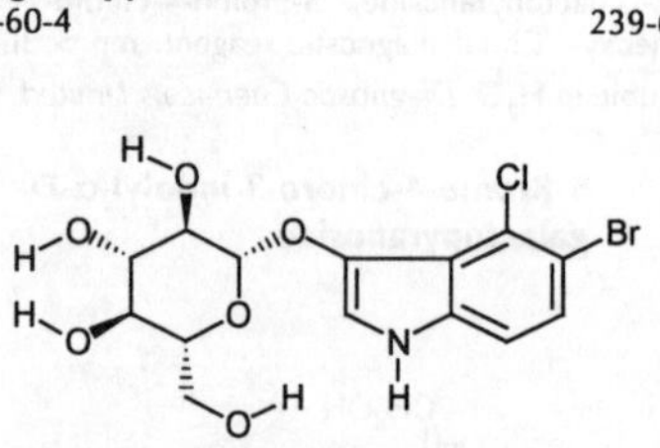

$C_{14}H_{15}BrClNO_6$
Indole, 5-bromo-4-chloro-3-(β-D-glucopyranosyloxy)-; β-D-Glucopyranoside, 5-bromo-4-chloro-1H-indol-3-yl. Chiral diagnostic reagent. mp = 249=251°; $[\alpha]_D$ = - 85 ± 10° (c = 1, 1:1 DMF/$H_2O$); soluble in $H_2O$. *Acros Organics nv; Diagnostic Chemicals Limited; Senn Chemicals AG.*

**804 5-Bromo-4-chloro-3-indolyl-β-D-glucuronic acid, cyclohexylammonium salt**
114162-64-0

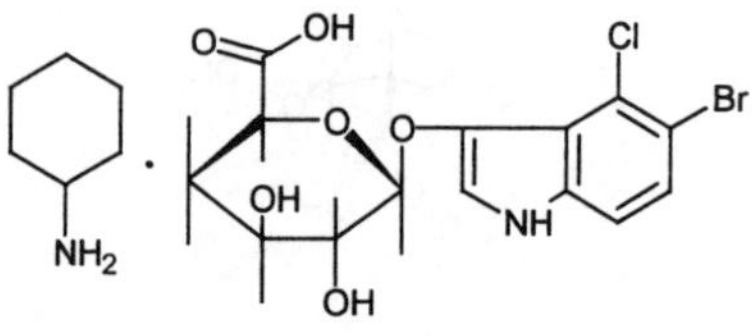

$C_{20}H_{26}BrClN_2O_7$
Glucopyranosiduronic acid, β-D-, 5-bromo-4-chloro-1H-indol-3-yl, compd. with cyclohexanamine (1:1). Chiral diagnostic reagent. *Acros Organics nv; Diagnostic Chemicals Limited; Senn Chemicals AG.*

**805 5-Bromo-4-chloro-3-indolyl-β-D-glucuronic acid, sodium salt**
129541-41-9

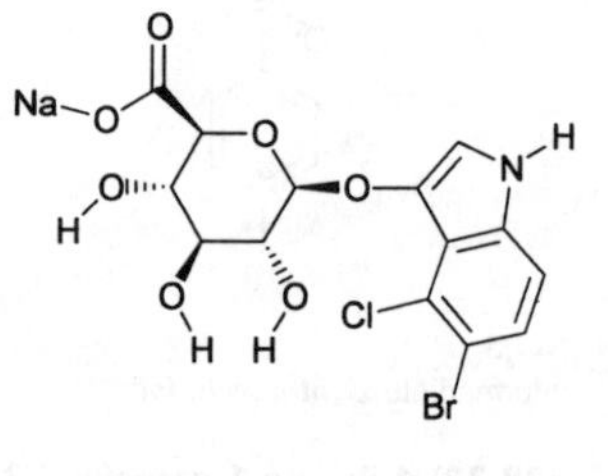

$C_{14}H_{12}BrClNO_7Na$
Chiral diagnostic reagent. Soluble in $H_2O$. *Acros Organics nv; Diagnostic Chemicals Limited; Senn Chemicals AG.*

**806 5-Bromo-4-chloro-3-indolyl-β-D-glucuronide cyclohexylammonium salt hydrate**
18656-96-7

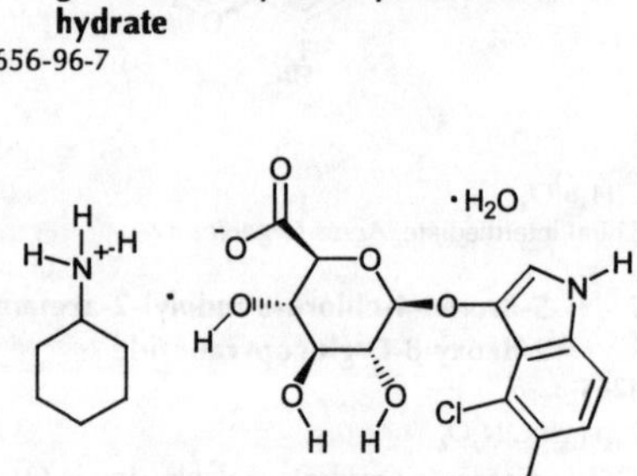

$C_{20}H_{26}BrClN_2O_7$
Chiral intermediate. *Acros Organics nv.*

**807 5-Bromo-6-chloro-3-indolyl-β-D-glucuronide cyclohexylammonium salt**
144110-43-0

$C_{20}H_{26}BrClN_2O_7$
Chiral intermediate. *Acros Organics nv.*

**808 (1S-cis)-3-Bromo-3,5-cyclohexadien-1,2-diol**

$C_6H_7BrO_2$
Chiral building block. mp = 90-94°; $[\alpha]^{22}$ = + 20° (c = 1, $CH_3OH$). *Sigma-Aldrich Fine Chemicals.*

**809 5-Bromo-5-deoxy-2,3-O-isopropylidene-D-ribonolactone**
94324-23-9

$C_8H_{11}BrO_4$
Chiral intermediate. [α] = - 44° (c = 1.5, $CHCl_3$). *Acros Organics nv.*

**810 (+)-5-Bromo-2'-deoxyuridine**
59-14-3 200-415-9

$C_9H_{11}BrN_2O_5$
Uridine, 5-bromo-2'-deoxy-; Broxuridine; 2'-Deoxy-5-bromouridine; 5-BDU; BUdR; NSC 38297. Listed on TSCA. Chiral intermediate. mp = 191-194°; [α] = + 23° (c = 1, $H_2O$). *Acros Organics nv; Sigma-Aldrich Fine Chemicals.*

**811 (R)-Bromo-2,3-dihydroxypropane**
14437-88-8

$C_3H_7BrO_2$
Propane-1,2-diol, 3-bromo-, (R)-; (R)-3-Bromo-1,2-propanediol. Chiral building block. *Synthon Chiragenics Corporation.*

**812 (S)-1-Bromo-2,3-dihydroxypropane**
137490-63-2
$C_3H_7BrO_2$
Propane-1,2-diol, 3-bromo-, (S)-; Chiral intermediate. *Synthon Chiragenics Corporation.*

**813 (1R)-(-)-B-Bromodiisopinocampheylborane**
104114-70-7

$C_{20}H_{34}BBr$
(-)-DIP-Bromide. Chiral auxiliary. *Sigma-Aldrich Fine Chemicals.*

**814 (1S)-(+)-B-Bromodiisopinocampheylborane**
112246-74-9

$C_{20}H_{34}BBr$
(+)-DIP-Bromide. Chiral auxiliary. mp = 62-72°. *Sigma-Aldrich Fine Chemicals.*

**815 (R)-Bromo-1,2-epoxybutane**
79413-93-7

$C_4H_7BrO$
Oxirane, (2-bromoethyl)-, (R)-. Chiral building block. *Synthon Chiragenics Corporation.*

**816 (S)-4-Bromo-1,2-epoxybutane**
61487-07-2
Chiral building block. *Synthon Chiragenics Corporation.*

**817 (2R,3R)-4-Bromo-cis-2,3-epoxybutyl-(1S)-10-camphorsulfonate**

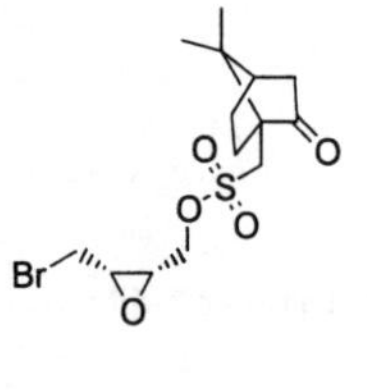

$C_{14}H_{21}BrO_5S$
Resolving agent. mp = 45-48°; $[\alpha]^{22}$ = + 80° (c = 1, $CHCl_3$). *Sigma-Aldrich Fine Chemicals.*

**818 (2S,3S)-4-Bromo-cis-2,3-epoxybutyl-(1S)-10-camphorsulfonate**

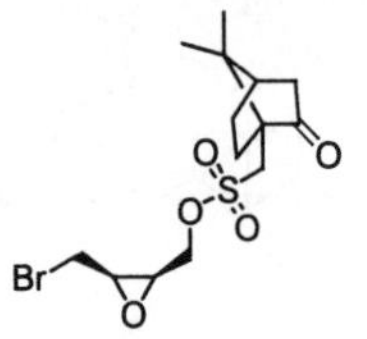

$C_{14}H_{21}BrO_5S$
Chiral intermediate. mp = 40-43°; $[\alpha]^{22}$ = - 6° (c = 1, $CHCl_3$). *Sigma-Aldrich Fine Chemicals.*

**819 (-)-8-Bromoguanosine hydrate**
223-677-6

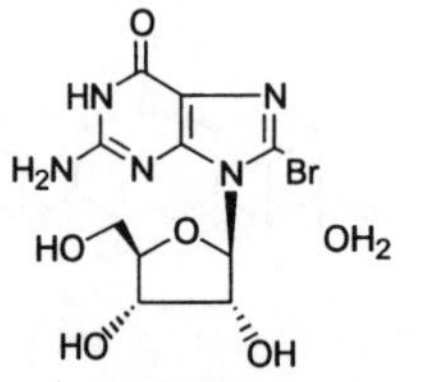

$C_{10}H_{12}BrN_5O_5$
Chiral intermediate. mp = 222°; $[\alpha]_D^{20}$ = - 28.8° (c = 2, DMSO/$H_2O$). *Sigma-Aldrich Fine Chemicals.*

**820 (R,S)-4-Bromo-homoibotenic acid**
71366-32-0

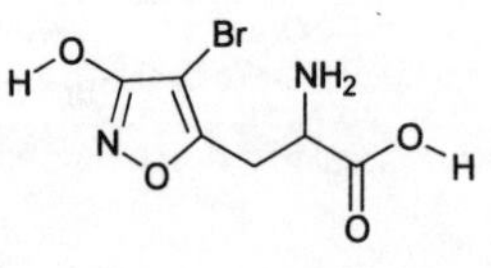

$C_6H_7BrN_2O_4$
Chiral intermediate. *Acros Organics nv.*

**821 5-Bromo-3-indolyl-β-D-galactopyranoside**
97753-82-7
$C_{14}H_{16}BrNO_6$
Galactopyranoside, 5-bromo-1H-indol-3-yl, β-D-. Indole, 5-bromo-3-(β-D-galactopyranosyloxy)-. Chiral diagnostic reagent. Sparingly soluble in $H_2O$. *Diagnostic Chemicals Limited; Senn Chemicals AG.*

**822 5-Bromo-3-indolyl-β-D-glucuronic acid, cyclohexylammonium salt**

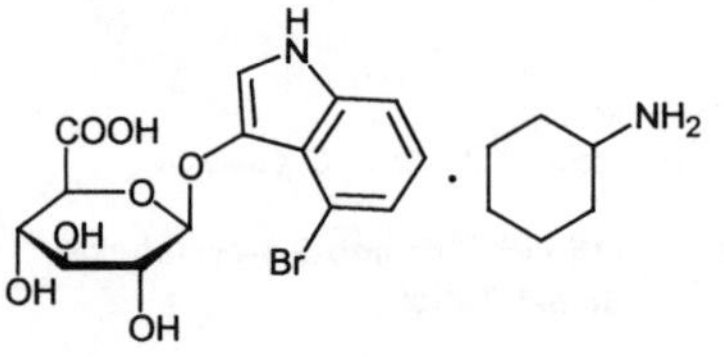

$C_{20}H_{26}BrN_2O_7$
Chiral diagnostic reagent. *Diagnostic Chemicals Limited.*

**823 (R)-Bromomandelic acid**

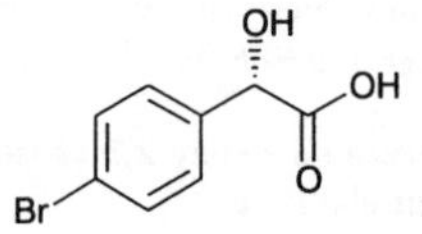

$C_8H_7BrO_3$
Benzeneacetic acid, 4-bromo-α-hydroxy-, (R)-; (R)-2-Hydroxy-2-(4-bromophenyl)acetic acid; Mandelic acid, p-bromo-, (R)-. Chiral building block. *Austin Chemical Company, Inc.*

**824 (S)-4-Bromomandelic acid**

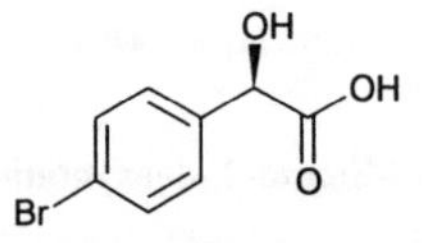

$C_8H_7BrO_3$
Benzeneacetic acid, 4-bromo-α-hydroxy-, (S)-; (S)-2-Hydroxy-2-(4-bromophenyl)acetic acid; Mandelic acid, p-bromo-, (S)-. Chiral building block. *Austin Chemical Company, Inc.*

**825 (R)-5-Bromo-4-(methoxymethoxy)-pentanenitrile**

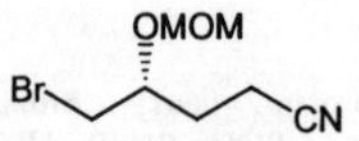

$C_7H_{12}BrNO_2$
Pentanenitrile, 5-bromo-4-(methoxymethoxy)-, (R)-. Chiral intermediate. *Synthon Chiragenics Corporation.*

**826 (S)-5-Bromo-4-(methoxymethoxy)-pentanenitrile**
209806-89-3

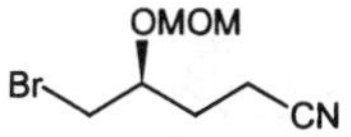

$C_7H_{12}BrNO_2$
Pentanenitrile, 5-bromo-4-(methoxymethoxy)-, (S)-. Chiral intermediate. *Synthon Chiragenics Corporation.*

**827 (R)-(+)-2-Bromomethylbenzyl alcohol**
76116-20-6 226-491-3

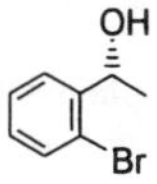

$C_8H_9BrO$
Chiral building block. mp = 54-56°; $[\alpha]^{22}$ = + 54° (c = 1, $CHCl_3$). *Sigma-Aldrich Fine Chemicals.*

**828 (S)-(-)-2-Bromomethylbenzyl alcohol**
114446-55-8

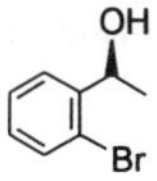

$C_8H_9BrO$
Chiral building block. mp = 56-58°; $bp_{15}$ = 128°; $[\alpha]^{30}$ = - 54° (c = 1, $CHCl_3$). *Sigma-Aldrich Fine Chemicals.*

**829 (S)-(+)-1-Bromo-2-methylbutane**
534-00-9 642(12) 208-583-5

$C_5H_{11}Br$
Butane, 1-bromo-2-methyl-, (2S)-; (+)-2-Methyl-1-bromobutane; (S)-2-Methylbutyl bromide; (S)-Amyl bromide. Listed on TSCA. Chiral building block. bp = 121-122°; d = 1.223; n = 1.4444; $[\alpha]^{21}$ = + 4.5° (c = 5, $CHCl_3$). *Norse Laboratories.*

**830 (R)-(+)-2-Bromo-3-methylbutyric acid**
76792-22-8

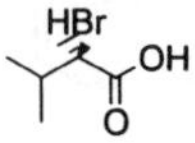

$C_5H_9BrO_2$
(R)-α-Bromoisopentanoic acid; (R)-2-Bromoisovaleric acid. Chiral building block. mp = 35-40°; $bp_{0.5}$ = 90-100°; $[\alpha]^{22}$ = + 21° (c = 37, $C_6H_6$). *Austin Chemical Company, Inc.; Sigma-Aldrich Fine Chemicals.*

**831 (S)-(-)-2-Bromo-3-methylbutyric acid**
26782-75-2

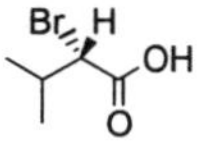

$C_5H_9BrO_2$
(S)-2-Bromoisovaleric acid. Chiral building block. mp = 39-44°; $bp_{0.5}$ = 90-100°; $[\alpha]^{22}$ = -21° (c = 37, $C_6H_6$). *Sigma-Aldrich Fine Chemicals.*

**832 (S)-(+)-6-Bromo-4-methylhexanal**

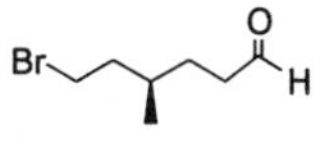

$C_7H_{13}BrO$
Chiral building block. bp = 147°; d = 1.284; n = 1.476; $[\alpha]^{22}$ = + 6° (neat). *Sigma-Aldrich Fine Chemicals.*

**833 (4R)-4-Bromomethyl-2-phenyl-1,3-dioxane**

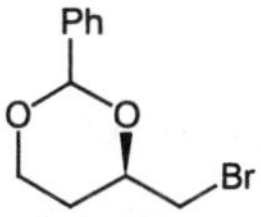

$C_{11}H_{13}BrO_2$
Dioxane, 1,3-, 4-(bromomethyl)-2-phenyl-, (4R)-. Chiral intermediate. *Synthon Chiragenics Corporation.*

**834 (4S)-4-Bromomethyl-2-phenyl-1,3-dioxane**
201743-52-4

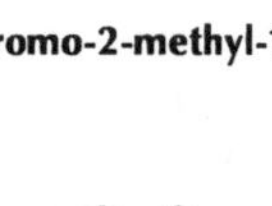

$C_{11}H_{13}BrO_2$
Dioxane, 1,3-, 4-(bromomethyl)-2-phenyl-, (4S)-. Chiral intermediate. *Acros Organics nv; Synthon Chiragenics Corporation.*

**835 (R)-(-)-3-Bromo-2-methyl-1-propanol**
93381-28-3

$C_4H_9BrO$
Chiral building block. $bp_9$ = 73-74°; d = 1.461; n = 1.484; $[\alpha]^{25}$ = - 6.6° (c = 2, $CHCl_3$). *Sigma-Aldrich Fine Chemicals.*

**836 (S)-(+)-3-Bromo-2-methyl-1-propanol**
98244-48-5

Br OH

$C_4H_9BrO$
Chiral building block. d = 1.461; n = 1.4836; $[\alpha]^{25}$ = + 7.3° (c = 2, $CHCl_3$). *Sigma-Aldrich Fine Chemicals.*

**837 6-Bromo-2-naphthyl-N-acetyl-β-D-glucosaminide**

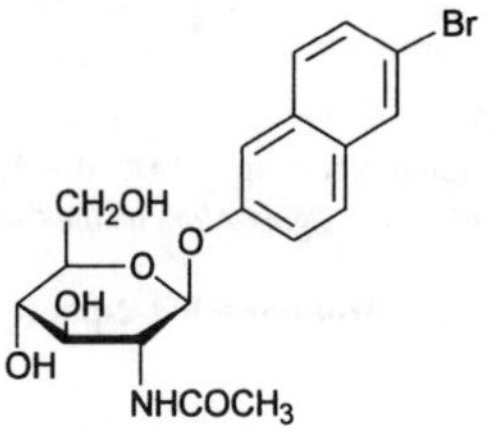

$C_{18}H_{20}BrNO_6$
Chiral intermediate. *Senn Chemicals AG.*

**838 6-Bromo-2-naphthyl-β-D-galactopyranoside**
15572-30-2 239-628-7

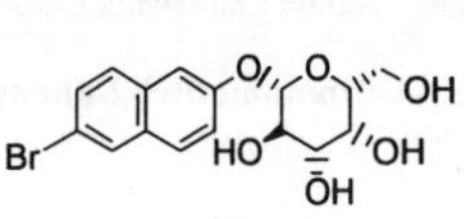

$C_{16}H_{17}BrO_6$
Galactopyranoside, 6-bromo-2-naphthyl, β-D-. Chiral intermediate. mp = 218-220°; [α] = - 44° (c = 1.2, pyridine). *Acros Organics nv; Senn Chemicals AG; Sigma-Aldrich Fine Chemicals.*

**839 6-Bromo-2-naphthyl-β-D-glucopyranoside**
15548-61-5 239-604-6

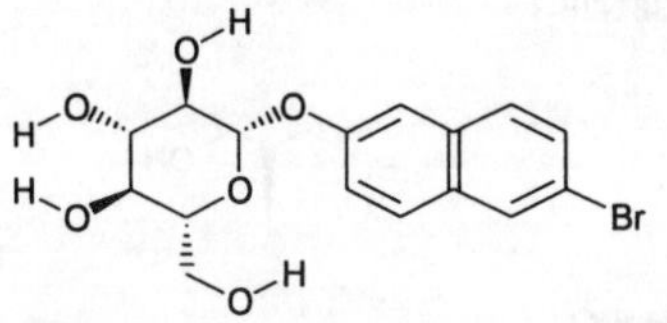

$C_{16}H_{17}BrO_6$
Glucopyranoside, 6-bromo-2-naphthyl, β-D-. Chiral intermediate. mp = 210-212°. *Acros Organics nv; Senn Chemicals AG.*

**840 6-Bromo-2-naphthyl-α-D-mannopyranoside**

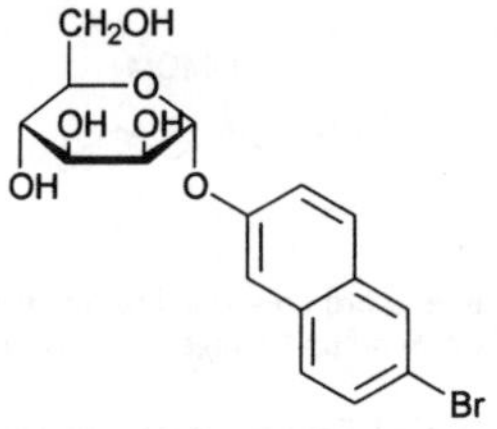

$C_{16}H_{17}BrO_6$
Chiral intermediate. *Senn Chemicals AG.*

**841 4-Bromo-D-phenylalanine**

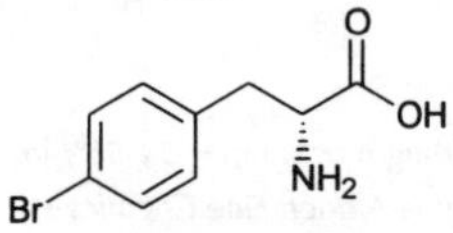

$C_9H_{12}BNO_4$
Chiral building block. *Digital Specialty Chemicals, Inc.*

**842 4-Bromo-L-phenylalanine**

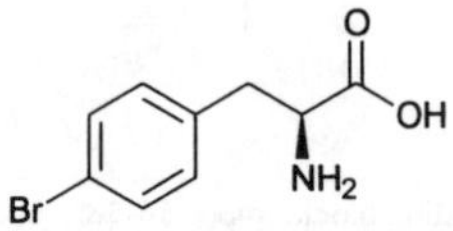

$C_9H_{12}BNO_4$
Chiral building block. *Digital Specialty Chemicals, Inc.*

**843 (R)-(+)-1-(4-Bromophenyl)ethylamine**
45791-36-4 256-245-0

$C_8H_{10}BrN$
Benzenemethanamine, 4-bromo-α-methyl-, (R)-; (R)-α-Methyl-p-bromobenzylamine. Chiral building block. $bp_{30}$ = 140-145°; $n_D^{20}$ = 1.5670; $[\alpha]_D^{20}$ = + 25 ± 2° (neat). *Arran Chemical Company Ltd.; FineTech, Ltd.; Lancaster Synthesis Ltd.; Norse Laboratories; Yamakawa Chemical Industry Co. Ltd.*

**844 (S)-(-)-1-(4-Bromophenyl)ethylamine**
27298-97-1

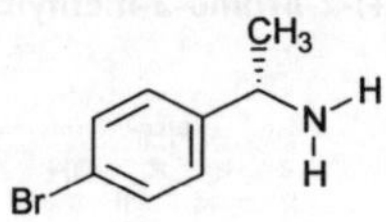

$C_8H_{10}BrN$
(S)-(-)-p-Bromo-α-methylbenzylamine. Chiral building block. $bp_{30}$ = 140-145°; d = 1.3710; $n_D^{20}$ = 1.5670; $[\alpha]_D^{20}$ = - 25 ± 2° (neat); $[\alpha]_D^{20}$ = - 19° (c = 2, $CH_3OH$). *Acros Organics nv; Arran Chemical Company Ltd.; Lancaster Synthesis Ltd.; Yamakawa Chemical Industry Co. Ltd.*

**845 (R)-(+)-1-(4-Bromophenyl)ethylamine hydrochloride**
64265-77-6

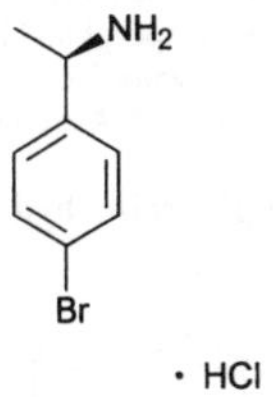

$C_8H_{11}BrClN$
Chiral building block. *Arran Chemical Company Ltd.*

**846 (R)-(+)-2-Bromopropanoic acid methyl ester**
20047-41-0

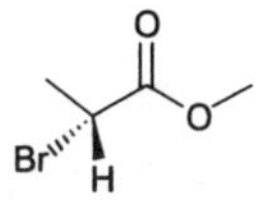

$C_4H_7BrO_2$
Chiral building block. $d_4^{20}$ = 1.498; $[\alpha]_D^{20}$ > +50° (neat). *Nitrokemia 2000 Rt.*

**847 (R)-(+)-2-Bromopropionic acid**
10009-70-8

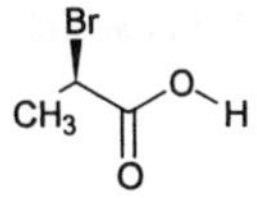

$C_3H_5BrO_2$
Chiral building block. mp = 25°; bp = 203°; d = 1.69; [α] = + 29.6° (c = 1, EtOH). *Acros Organics nv; Nitrokemia 2000 Rt.; TCI America.*

**848 (S)-(-)-2-Bromopropionic acid**
32644-15-8

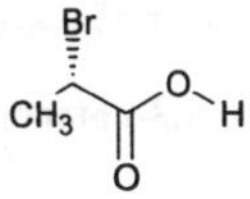

$C_3H_5BrO_2$
Chiral building block. mp = -35°; $bp_4$ = 78°; d = 1.696; n = 1.47; [α] = - 31.8° (c = 1, EtOH); $[\alpha]_D^{20}$ = - 25° (neat). *Acros Organics nv; Sigma-Aldrich Fine Chemicals; TCI America.*

**849 (3aS)-(3a,4,5,7a)-7-Bromo-3a,4,5,7a-tetrahydro-2,2-dimethyl-1,3-benzodioxol-4,5-diol**

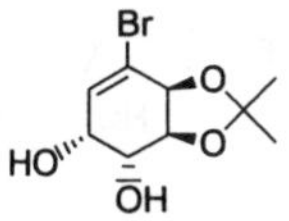

$C_9H_{13}BrO_4$
Chiral intermediate. mp = 121-124°; $[\alpha]^{22}$ = - 10° (c = 1, $CHCl_3$). *Sigma-Aldrich Fine Chemicals.*

**850 (+)-3aS-(3a,5a,6a,6b)-4-Bromo-3a,5a,6a,6b-tetrahydro-2,2-dimethyl-oxirene-1,3-benzodioxole**

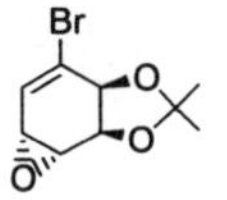

$C_9H_{11}BrO_3$
Chiral intermediate. mp = 76-79°; $[\alpha]^{25}$ = + 106° (c = 1, $CH_3OH$). *Sigma-Aldrich Fine Chemicals.*

**851 D-(+)-p-Bromotetramisole oxalate**
71461-24-0 275-471-0

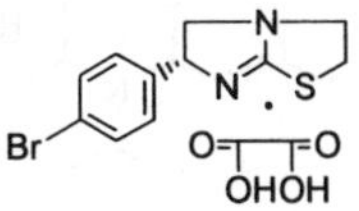

$C_{11}H_{11}BrN_2S$
Chiral intermediate. mp = 192°; $[\alpha]^{25}$ = + 105° (c = 0.5, $H_2O$). *Sigma-Aldrich Fine Chemicals.*

**852 L-(-)-p-Bromotetramisole oxalate**
62284-79-1 263-487-0

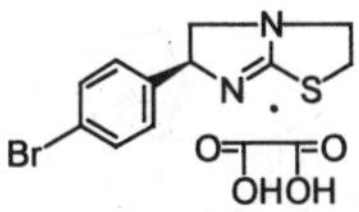

$C_{11}H_{11}BrN_2S$
Imidazo[2,1-b]thiazole, 6-(4-bromophenyl)-2,3,5,6-tetrahydro-, (S)-, ethanedioate (1:1); (S)-6-(p-Bromophenyl)-2,3,5,6-tetrahydroimidazo[2,1-b]thiazole oxalate. Chiral intermediate. mp = 192°; $[\alpha]^{25}$ = - 105° (c = 0.5, $H_2O$). *Acros Organics nv; Sigma-Aldrich Fine Chemicals.*

**853 (-)-5-Bromouridine**
957-75-5 213-486-6

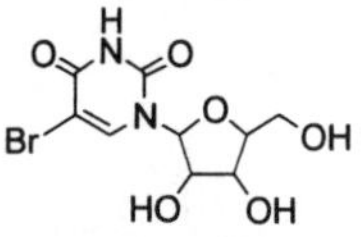

$C_9H_{11}BrN_2O_6$
Chiral intermediate. mp = 180-182°; $[\alpha]^{22}$ = - 24.1° (c = 2, $H_2O$). *Sigma-Aldrich Fine Chemicals.*

**854 Brucine**
1476(12) 020-664-7

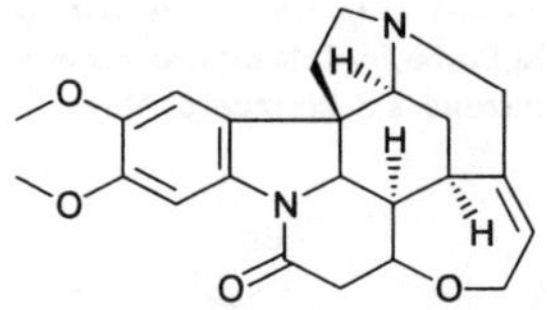

$C_{23}H_{26}N_2O_4$
Listed on TSCA. Resolving agent; Chiral auxiliary. mp = 175-178°; $[\alpha]_D^{20}$ = - 118° (c = 2.5, $CHCl_3$). *Sigma-Aldrich Fine Chemicals.*

**855 Brucine dihydrate**
145428-94-0 1443(11) 206-614-7

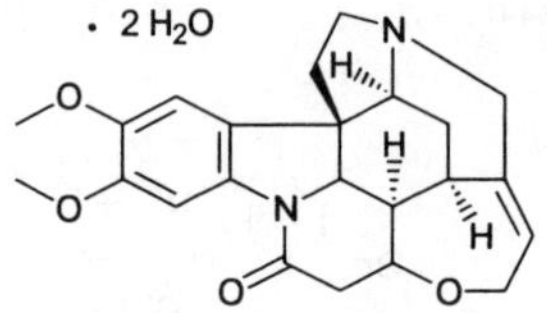

$C_{23}H_{26}N_2O_4$
Listed on TSCA. Chiral intermediate. mp = 177-180°; $[\alpha]_D^{20}$ = - 67° (c = 1.3, EtOH). *Acros Organics nv; Sigma-Aldrich Fine Chemicals.*

**856 Brucine sulfate hydrate**
38741-38-7 1443(11) 225-432-9

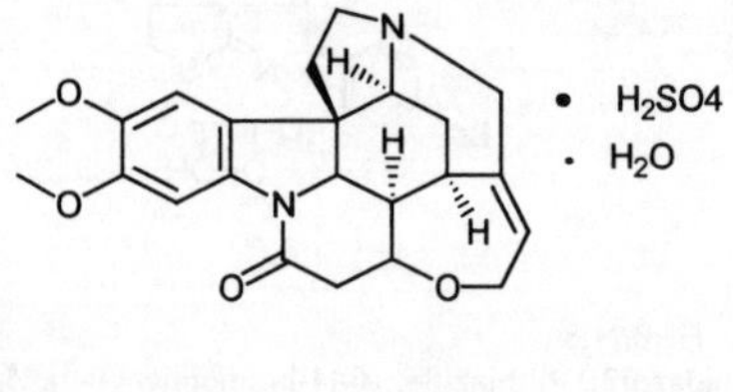

$C_{23}H_{26}N_2O_4$
Listed on TSCA. Chiral intermediate . mp = 180°; $[\alpha]^{22}$ = - 24° (c = 1, $H_2O$). *Acros Organics nv; Sigma-Aldrich Fine Chemicals.*

**857 Bulbocapnine hydrochloride**
632-47-3 211-178-6
$C_{19}H_{19}NO_4$
Benzo-[5H,g]-1,3-benzodioxolo[6,5,4-de]quinolin-12-ol, 6,7,7a,8-tetrahydro-11-methoxy-7-methyl-, hydrochloride, (S)-; 6aα-Aporphin-11-ol, 10-methoxy-1,2-(methylenedioxy)-, hydrochloride. Pharmaceutical or derivative. mp = 257-270°. *Kaden Biochemicals GmbH.*

**858 (S)-(-)-Bupivacaine hydrochloride**
27262-48-2

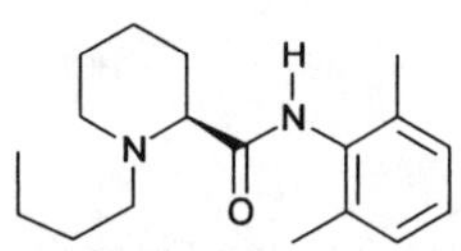

$C_{18}H_{22}N_2O$
Chiral intermediate. *Acros Organics nv.*

**859 (R)-(-)-1,3-Butanediol**
6290-03-5 228-532-0

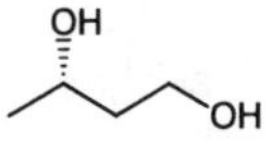

$C_4H_{10}O_2$
(R)-(-)-1,3-Butylene glycol; (R)-(-)-1,3-Dihydroxybutane. $bp_{23}$ = 108-109°; d = 1.0050; $n_D^{20}$ = 1.4400; $[\alpha]_D^{20}$ = - 29 ± 1° (neat). *Acros Organics nv; Daicel Chemical Ind. Ltd; Digital Specialty Chemicals, Inc.; FineTech, Ltd.; IMI (TAMI) Institute for R&D; Lancaster Synthesis Ltd.; Maxsyn; Sigma-Aldrich Fine Chemicals; Strem Chemicals, Inc.; Synthon Chiragenics Corporation; TCI America.*

**860 (S)-(+)-1,3-Butanediol**
24621-61-2 246-363-0

$C_4H_{10}O_2$
Butane-1,3-diol, (3S)-.
(S)-(+)-1,3-Butylene glycol; (S)-(+)-1,3-Dihydroxybutane. Chiral building block. $bp_{14}$ = 109°; d = 1.005; n = 1.44; $[\alpha]_D$ = - 31° (c = 1, EtOH). *Acros Organics nv; Digital Specialty Chemicals, Inc.; IMI (TAMI) Institute for R&D; Maxsyn; Rohner Ltd.; Sigma-Aldrich Fine Chemicals; Strem Chemicals, Inc.; TCI America.*

**861 (2R,3R)-(-)-2,3-Butanediol**
24347-58-8 1602(12) 246-186-9

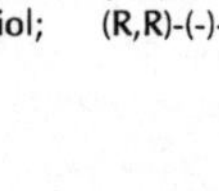

$C_4H_{10}O_2$
Butane-2,3-diol, (2R,3R)-(-)-; (R,R)-(-)-2,3-Butylene glycol; (R,R)-(-)-2,3-Butanediol; (R,R)-(-)-2,3-Dihydroxybutane.

Chiral building block. bp = 179-182°; $bp_{10}$ = 77-78°; d = 0.9880; $n_D^{20}$ = 1.4330; $[\alpha]_D^{20}$ = - 13 ± 1° (neat). *Acros Organics nv; Daicel Chemical Ind. Ltd; Digital Specialty Chemicals, Inc.; Lancaster Synthesis Ltd.; Maxsyn; Norse Laboratories; Sigma-Aldrich Fine Chemicals; Strem Chemicals, Inc.; TCI America; Urquima S.A.*

**862 (S,S)-(+)-2,3-Butanediol**
19132-06-0 1567(11)

$C_4H_{10}O_2$
(S,S)-(+)-2,3-Butylene Glycol; (S,S)-(+)-2,3-Dihydroxybutane. Chiral building block. bp = 179-182°; d = 0.987; n = 1.433; $[\alpha]_D$ = + 13° (neat). *Acros Organics nv; Digital Specialty Chemicals, Inc.; FineTech, Ltd.; Maxsyn; Norse Laboratories; Sigma-Aldrich Fine Chemicals; Strem Chemicals, Inc.; TCI America.*

**863 (2R,3R)-(+)-2,3-Butanediol di-p-tosylate**
(2R,3R)-(+)-Di-O-tosyl-2,3-butanediol. Chiral building block. mp = 65-67°; $[\alpha]_D^{20}$ = + 38 ± 2° (c = 2, $CHCl_3$). *Lancaster Synthesis Ltd.*

**864 (2S,3S)-(-)-2,3-Butanediol di-p-tosylate**
74839-83-1

$C_{18}H_{22}O_6S_2$
(2S,3S)-(-)-Di-O-tosyl-2,3-butanediol. Chiral building block. mp = 65-67°; $[\alpha]_D^{20}$ = - 38 ± 2° (c = 2, $CHCl_3$). *Lancaster Synthesis Ltd.; Sigma-Aldrich Fine Chemicals.*

**865 (R)-Butanediol-1-tosylate**
103745-07-9

$C_{11}H_{16}O_4S$
(R)-2-Hydroxybutyl p-tosylate. Chiral building block. *Acros Organics nv; Eastman Chemical Company.*

**866 (S)-1,2-Butanediol-1-tosylate**
143731-32-2

$C_{11}H_{16}O_4S$
(S)-2-Hydroxybutyl p-tosylate. Chiral building block. *Acros Organics nv; Eastman Chemical Company.*

**867 (R)-(+)-1,2,4-Butanetriol**
70005-88-8

$C_4H_{10}O_3$
Butane-1,2,4-triol, (R)-. Chiral building block. $bp_{0.04}$ = 150°; d = 1.19; n = 1.475; $[\alpha]_D^{20}$ = + 26° (c = 1, $CH_3OH$). *Sigma-Aldrich Fine Chemicals; Synthon Chiragenics Corporation.*

**868 (S)-(-)-1.2,4-Butanetriol**
42890-76-6

$C_4H_{10}O_3$
Butane-1,2,4-triol, (S)-. Chiral building block. $bp_{0.04}$ = 150°; d = 1.19; n = 1.475; $[\alpha]^{19}$ = - 27° (c = 1, $CH_3OH$). *Acros Organics nv; Boehringer Ingelheim Pharma KG; Daiso Company, Ltd.; Sigma-Aldrich Fine Chemicals; SK Energy and Chemical, Inc.; Synthon Chiragenics Corporation.*

**869 (R)-1,2,4-Butanetriol trimesylate**
203515-83-7

$C_7H_{16}O_9S_3$
Butane-1,2,4-triol, trimethanesulfonate, (R)-. Chiral intermediate. *Synthon Chiragenics Corporation.*

**870 (S)-1,2,4-Butanetriol trimesylate**
99520-81-7

$C_7H_{16}O_9S_3$
Butane-1,2,4-triol, trimethanesulfonate, (S)-; (S)-1,2,4-Butanetriol trimethanesulfonate; (S)-1,2,4-Tri-hydroxybutane trimesylate. Chiral intermediate. *Acros Organics*

*nv; Great Lakes Fine Chemicals; SK Energy and Chemical, Inc.; Synthon Chiragenics Corporation.*

**871 (R)-(-)-2-Butanol**
14898-79-4 1541(11) 238-967-8

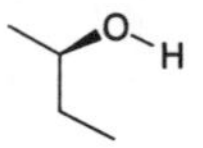

$C_4H_{10}O$
Butan-2-ol, (R)-; (R)-(-)-sec-Butyl alcohol; L-(-)-Ethylmethylcarbinol. Chiral building block. mp = - 114°; bp = 97-100°; d = 0.8060; $n_D^{20}$ = 1.3980; $[\alpha]_D^{20}$ = - 15 ± 1° (c = 10, $CH_3OH$); $[\alpha]_D^{20}$ = - 13° (neat). *Acros Organics nv; Lancaster Synthesis Ltd.; Norse Laboratories; Sigma-Aldrich Fine Chemicals; TCI America.*

**872 (S)-(+)-2-Butanol**
4221-99-2 1576(12) 224-168-1

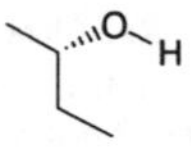

$C_4H_{10}O$
Butan-2-ol, (S)-; (S)-(+)-sec-Butyl alcohol. Chiral building block. bp = 97-100°; d = 0.8060; $n_D^{20}$ = 1.3980; $[\alpha]_D^{20}$ = + 13.5 ± 1° (c = 10, $CH_3OH$). *Acros Organics nv; Lancaster Synthesis Ltd.; Norse Laboratories; Sigma-Aldrich Fine Chemicals; TCI America.*

**873 N-Butanoyl-D-erythro sphingosine**
74713-58-9

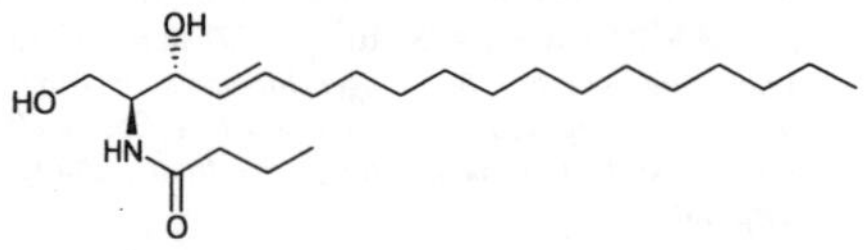

$C_{22}H_{43}NO_3$
Chiral intermediate. *Acros Organics nv.*

**874 (R)-3-Butene-1,2-diol**
86106-09-4

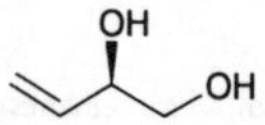

$C_4H_8O_2$
Chiral building block. *Eastman Chemical Company.*

**875 (S)-3-Butene-1,2-diol**
62214-39-5

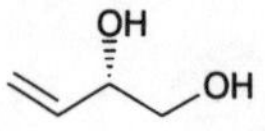

$C_4H_8O_2$
Chiral building block. *Eastman Chemical Company.*

**876 (R)-3-Buten-2-ol**
33447-72-2

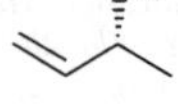

$C_4H_8O$
Chiral building block. *Boehringer Ingelheim Pharma KG; Celltech Chiroscience Ltd.*

**877 (S)-3-Buten-2-ol**
6118-13-4

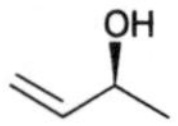

$C_4H_8O$
Chiral building block. *Boehringer Ingelheim Pharma KG; Celltech Chiroscience Ltd.*

**878 L-Buthionine-(S,R)-sulfoximine**
83730-53-4

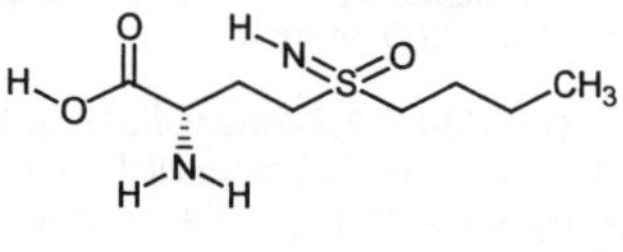

$C_8H_{18}N_2O_3S$
Chiral intermediate. mp = 224-226°; $[\alpha]_D^{20}$ = + 33° (c = 1, 1N HCl). *Acros Organics nv; Sigma-Aldrich Fine Chemicals.*

**879 (tert-Butoxycarbonyl)-D-alanine**
7764-95-6

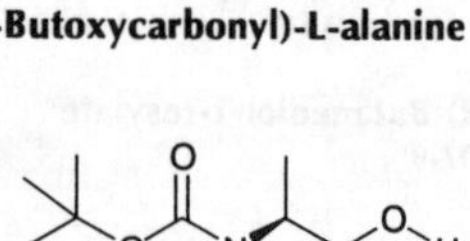

$C_8H_{15}NO_4$
Chiral intermediate. [α] = + 25.9° (c = 2, HOAc). *Acros Organics nv; Isochem; Synthetech, Inc.*

**880 (tert-Butoxycarbonyl)-L-alanine**
15761-38-3 239-847-8

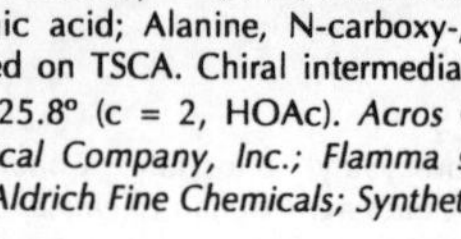

$C_8H_{15}NO_4$
$C_8H_{15}NO_4$Alanine, N-[(1,1-dimethylethoxy)carbonyl]-, L-; N-[(1,1-Dimethylethoxy)carbonyl]-L-alanine; N-(tert-Butoxycarbonyl)-L-alanine; (S)-2-(N-tert-Butoxycarbonyl)-aminopropionic acid; Alanine, N-carboxy-, N-tert-butyl ester, L-. Listed on TSCA. Chiral intermediate. mp = 79-82°; [α] = - 25.8° (c = 2, HOAc). *Acros Organics nv; Austin Chemical Company, Inc.; Flamma s.p.a.; PPG - Sipsy; Sigma-Aldrich Fine Chemicals; Synthetech, Inc.*

**881 (tert-Butoxycarbonyl)-L-alanine hydroxysuccinimide ester**

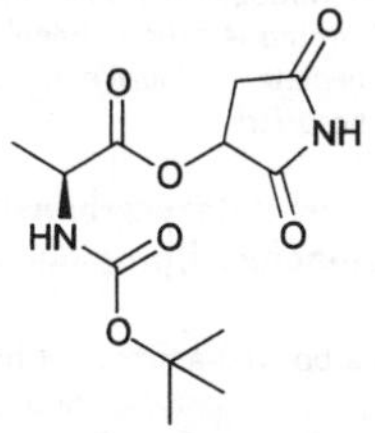

$C_{12}H_{18}N_2O_6$
Chiral intermediate. *Synthetech, Inc.*

**882 N-(tert-Butoxycarbonyl)alanine N-methoxy-N-methylamide**
87694-45-3

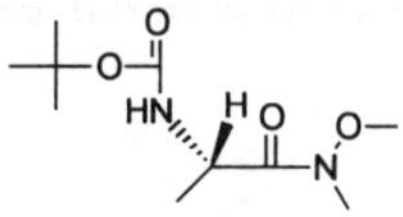

$C_{10}H_{20}N_2O_4$
Chiral intermediate. mp = 148-153°; $[\alpha]^{24}$ = - 26° (c = 1, $CH_3OH$). *Sigma-Aldrich Fine Chemicals.*

**883 (tert-Butoxycarbonyl)-D-alanine methyl ester**

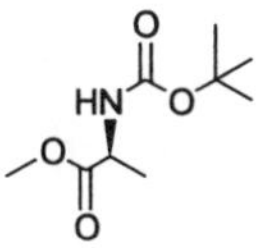

$C_9H_{17}NO_4$
Chiral intermediate. mp = 32-35°; d = 1.03; $[\alpha]^{21}$ = - 45° (c = 1, $CH_3OH$). *Sigma-Aldrich Fine Chemicals; Synthetech, Inc.*

**884 (tert-Butoxycarbonyl)-L-alanine methyl ester**
91103-47-8

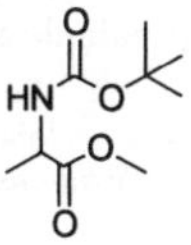

$C_9H_{17}NO_4$
Chiral intermediate. mp = 34-37°; d = 1.03; $[\alpha]_D^{20}$ = + 45° (c = 1, $CH_3OH$). *Sigma-Aldrich Fine Chemicals; Synthetech, Inc.*

**885 (tert-Butoxycarbonyl)-L-alanine β-naphthylamide**

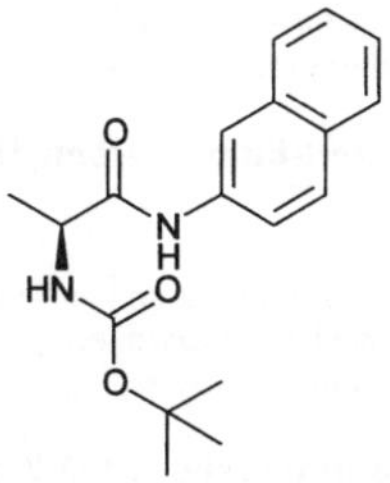

$C_{18}H_{22}N_2O_3$
Chiral intermediate. *Senn Chemicals AG.*

**886 (tert-Butoxycarbonyl)-D-alaninol**
106391-86-0

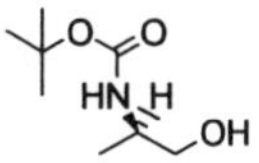

$C_8H_{17}NO_3$
N-t-BOC-D-alaninol. Chiral intermediate. mp = 60°; $[\alpha]_D^{20}$ = + 11° (c = 1, $CHCl_3$). *Daiso Company, Ltd.; Degussa-Huls AG; Omega Chemical Company Inc.; Sigma-Aldrich Fine Chemicals; Synthetech, Inc.*

**887 (tert-Butoxycarbonyl)-L-alaninol**
79069-13-9

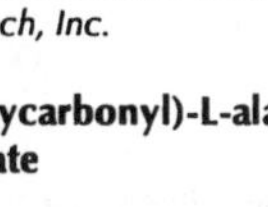

$C_8H_{17}NO_3$
Chiral intermediate. mp = 59-62°; $[\alpha]_D^{20}$ = - 11° (c = 1, $CHCl_3$). *Daiso Company, Ltd.; Omega Chemical Company Inc.; Senn Chemicals AG; Sigma-Aldrich Fine Chemicals; Synthetech, Inc.*

**888 (tert-Butoxycarbonyl)-L-alanyl-L-alanine monohydrate**

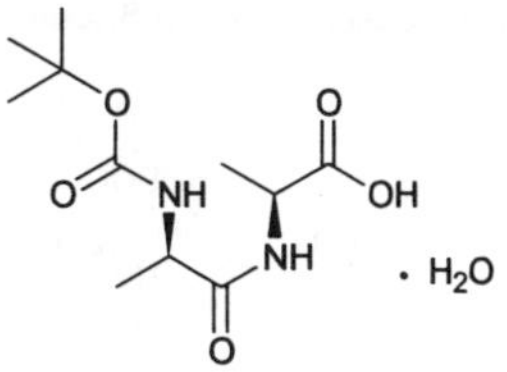

$C_{11}H_{22}N_2O_6$
Chiral intermediate. *Austin Chemical Company, Inc.; Synthetech, Inc.*

**889 (R)-N-(tert-Butoxycarbonyl)allylglycine**
90600-20-7
$C_{10}H_{17}NO_4$
N-tert-Butoxycarbonyl-(R)-2-amino-4-pentenoic acid. Chiral intermediate. *ChiroTech Technology Ltd.; Lancaster Synthesis Ltd.*

**890 (S)-N-(tert-Butoxycarbonyl)allylglycine**
170899-08-8
$C_{10}H_{17}NO_4$
N-tert-Butoxycarbonyl-(S)-2-amino-4-pentenoic acid. Chiral intermediate. *ChiroTech Technology Ltd.; Lancaster Synthesis Ltd.*

**891 (tert-Butoxycarbonyl)-D-2-aminoadipic acid**
110544-97-3

HOOC COOH HN O O

$C_{11}H_{19}NO_6$
Chiral intermediate. *Synthetech, Inc.*

**892 (tert-Butoxycarbonyl)-L-2-aminoadipic acid**
77302-72-8

HOOC COOH HN O O

$C_{11}H_{19}NO_6$
Chiral intermediate. *Synthetech, Inc.*

**893 (tert-Butoxycarbonyl)-β-amino-D-alanine**

NHBOC $H_2N$ OH O

$C_8H_{16}N_2O_4$
Chiral intermediate. *Omega Chemical Company Inc.*

**894 (tert-Butoxycarbonyl)-β-amino-L-alanine**
188016-53-7

NHBOC $H_2N$ OH O

$C_8H_{16}N_2O_4$
Chiral intermediate. mp = 216-220°; $[\alpha]_D^{20}$ = - 4° (c = 2, AcOH). *Omega Chemical Company Inc.; Sigma-Aldrich Fine Chemicals.*

**895 (R)-N-(tert-Butoxycarbonyl)-3-amino-3-(4-bromophenyl)propionic acid**
$C_{14}H_{18}BrNO_4$
(R)-N-Boc-β-(4-bromophenyl)-β-alanine; N-tert-Butoxycarbonyl-(R)-3-amino-(4-bromophenyl)propanoic acid. Chiral intermediate. *ChiroTech Technology Ltd.; Lancaster Synthesis Ltd.*

**896 (S)-N-(tert-Butoxycarbonyl)-3-amino-3-(4-bromophenyl)propionic acid**
$C_{14}H_{18}BrNO_4$
N-tert-Butoxycarbonyl-3-amino-3-(4-bromophenyl) propanoic acid; (S)-N-Boc-β-(4-bromophenyl)-β-alanine. Chiral intermediate. *ChiroTech Technology Ltd.; Lancaster Synthesis Ltd.*

**897 (S)-(+)-2-(tert-Butoxycarbonyl)amino-1,4-butanediol**
128427-10-1
$C_{10}H_{21}NO_4$
Chiral intermediate. *Sigma-Aldrich Fine Chemicals.*

**898 (tert-Butoxycarbonyl)-D-aminobutyric acid**
45121-22-0

COOH HN O O

$C_9H_{17}NO_4$
Chiral intermediate. *Synthetech, Inc.*

**899 (tert-Butoxycarbonyl)-L-2-aminobutyric acid**
34306-42-8

O OH HN O O

$C_9H_{17}NO_4$
Chiral intermediate. *Senn Chemicals AG; Synthetech, Inc.*

**900 (R)-N-(tert-Butoxycarbonyl)-2-amino-2-cyclohexylpropionic acid**
$C_{14}H_{25}NO_4$
N-tert-Butoxycarbonyl-α-cyclohexyl-D-alanine. Chiral intermediate. *Lancaster Synthesis Ltd.*

**901 (S)-N-(tert-Butoxycarbonyl)-2-amino-2-cyclohexylpropionic acid**
$C_{14}H_{25}NO_4$
N-tert-Butoxycarbonyl-α-cyclohexyl-L-alanine. Chiral intermediate. *Lancaster Synthesis Ltd.*

**902 (1R,3S)-N-(tert-Butoxycarbonyl)-1-aminocyclopentane-3-carboxylic acid**
151907-79-8
$C_{11}H_{19}NO_4$
N-tert-Butoxycarbonyl-(1R,3S)-1-aminocyclopentane-3-carboxylic acid. Chiral intermediate. *ChiroTech Technology Ltd.; Lancaster Synthesis Ltd.*

**903 (1S,3R)-N-(tert-Butoxycarbonyl)-1-aminocyclopentane-3-carboxylic acid**
161660-94-2
$C_6H_9NO$
(1S,3R)-N-tert-Butoxycarbonyl-1-aminocyclopentane-3-carboxylic acid. Chiral intermediate; Resolving agent. *ChiroTech Technology Ltd.; Lancaster Synthesis Ltd.*

**904 (1R,4S)-N-(tert-Butoxycarbonyl)-1-aminocyclopent-2-ene-4-carboxylic acid**
$C_{11}H_{17}NO_4$
N-tert-Butoxycarbonyl-(1R,4S)-1-amino-2-cyclopentene-4-carboxylic acid. Chiral intermediate. *ChiroTech Technology Ltd.; Lancaster Synthesis Ltd.*

**905 (1S,4R)-N-(tert-Butoxycarbonyl)-1-aminocyclopent-2-ene-4-carboxylic acid**
108999-93-5
$C_{11}H_{17}NO_4$
N-tert-Butoxycarbonyl-(1S,4R)-1-amino-2-cyclopentene-4-carboxylic acid. Chiral intermediate. *ChiroTech Technology Ltd.; Lancaster Synthesis Ltd.*

**906 (1R,3S,4R,6S)-N-(tert-Butoxycarbonyl)-6-amino-2,2-dimethyltetrahydrocyclopenta-1,3-dioxole-4-carboxylic acid**
129619-37-0
$C_{14}H_{23}NO_6$
(1R,3S,4R,6S)-6-(tert-Butoxycarbonylamino)-2,2-dimethyltetrahydrocyclopenta-1,3-dioxole-4-carboxylic acid. Chiral intermediate; Resolving agent. *ChiroTech Technology Ltd.; Lancaster Synthesis Ltd.*

**907 (1S,3R,4S,6R)-N-(tert-Butoxycarbonyl)-6-amino-2,2-dimethyltetrahydrocyclopenta-1,3-dioxole-4-carboxylic acid**
$C_{14}H_{23}NO_6$
(1S,3R,4S,6R)-6-(tert-Butoxycarbonylamino)-2,2-dimethyltetrahydrocyclopenta-1,3-dioxole-4-carboxylic acid. Chiral intermediate; Resolving agent. *ChiroTech Technology Ltd.; Lancaster Synthesis Ltd.*

**908 (2S,3S)-3-N-tert-Butoxycarbonylamino-2-hydroxy-4-phenylbutanoic acid**
116661-86-0

HO O O N H O OH

$C_{15}H_{21}NO_5$
Chiral intermediate. *Kaneka Corporation.*

**909 (S)-(+)-2-(tert-Butoxycarbonyl)amino-1,5-pentanediol**
162955-48-8
$C_{10}H_{21}NO_4$
Chiral intermediate. *Sigma-Aldrich Fine Chemicals.*

**910 (S)-(-)-2-(tert-Butoxycarbonylamino)-3-phenylpropanal**

O O HN H H O

$C_{14}H_{19}NO_3$
N-BOC-L-phenylalaninal. Chiral intermediate. mp = 86-88°; $[\alpha]_D^{20}$ = - 47° (c = 0.5, $CH_3OH$). *Sigma-Aldrich Fine Chemicals.*

**911 (R)-N-(tert-Butoxycarbonyl)-3-amino-3-phenylpropionic acid**
161024-80-2
$C_{14}H_{19}NO_4$
N-tert-Butoxycarbonyl-(R)-3-amino-3-phenylpropanoic acid. Chiral intermediate. *ChiroTech Technology Ltd.; Lancaster Synthesis Ltd.*

**912 (S)-N-(tert-Butoxycarbonyl)-3-amino-3-phenylpropionic acid**
103365-47-5
$C_{14}H_{19}NO_4$
N-tert-Butoxycarbonyl-(S)-3-amino-3-phenyl propanoic acid; (S)-N-Boc-β-phenyl-β-alanine. Chiral intermediate. *ChiroTech Technology Ltd.; Lancaster Synthesis Ltd.*

**913 (2R)-2-(tert-Butoxycarbonyl)-amino-3-phenylsulfonyl-1-(2-tetrahydropyranyloxy)-propane**

O S=O H N O O O O $CH_3$ $CH_3$ $CH_3$

$C_{19}H_{29}NO_6S$
Chiral intermediate. *Acros Organics nv.*

**914 (2S)-2-(tert-Butoxycarbonyl)-amino-3-phenylsulfonyl-1-(2-tetrahydropyranyloxy)-propane**

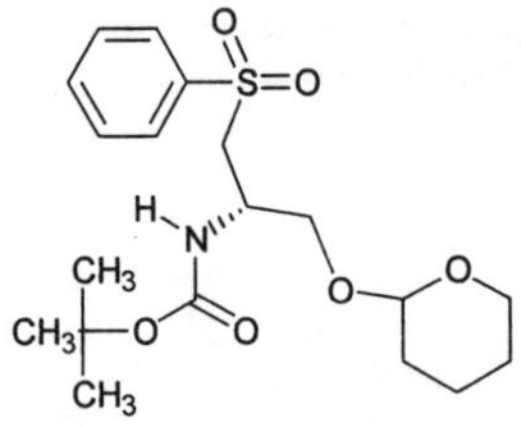

$C_{19}H_{29}NO_6S$
Chiral intermediate. [α] = + 7.2° (c = 1, $CH_3OH$, as R-OH). *Acros Organics nv.*

**915 (R)-N-(tert-Butoxycarbonyl)-4-amino-L-proline**
132622-69-6
$C_{10}H_{18}N_2O_4$
Chiral intermediate. *Sigma-Aldrich Fine Chemicals.*

**916 (S)-N-(tert-Butoxycarbonyl)-4-amino-L-proline**
132622-66-3
$C_{10}H_{18}N_2O_4$
Chiral intermediate. *Sigma-Aldrich Fine Chemicals.*

**917 (tert-Butoxycarbonyl)-(R)-1-amino-2-propanol**
119768-44-4

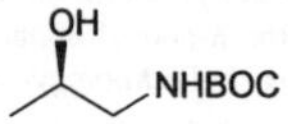

$C_8H_{17}NO_3$
Chiral intermediate. *Omega Chemical Company Inc.*

**918 (tert-Butoxycarbonyl)-(S)-1-amino-2-propanol**

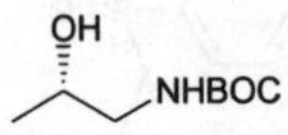

$C_8H_{17}NO_3$
Chiral intermediate. *Omega Chemical Company Inc.*

**919 (3R)-(+)-3-(tert-Butoxycarbonylamino)-pyrrolidine**
$C_9H_{18}N_2O_2$
(3R)-(+)-3-(BOC-amino)pyrrolidine; (R)-N-(tert-Butoxy-carbonyl)-3-aminopyrrolidine. Chiral building block. *Sigma-Aldrich Fine Chemicals; TCI America.*

**920 (3S)-(-)-3-(tert-Butoxycarbonylamino)-pyrrolidine**
$C_9H_{18}N_2O_2$
(S)-N-(tert-Butoxycarbonyl)-3-aminopyrrolidine. Chiral intermediate. *Sigma-Aldrich Fine Chemicals; TCI America.*

**921 (tert-Butoxycarbonyl)-D-2-aminosuberic acid**
75113-71-2

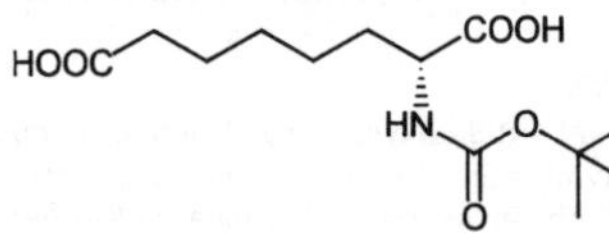

$C_{13}H_{23}NO_6$
Chiral intermediate. *Synthetech, Inc.*

**922 (tert-Butoxycarbonyl)-L-2-aminosuberic acid**
66713-87-9

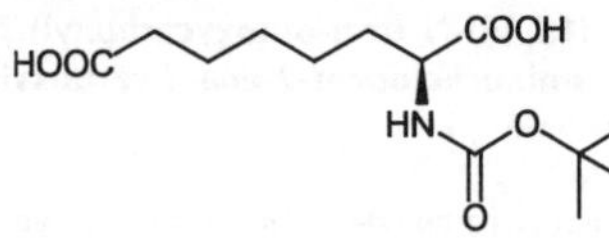

$C_{13}H_{23}NO_6$
Chiral intermediate. *Synthetech, Inc.*

**923 (S)-N-(tert-Butoxycarbonyl)-2-amino-3-(4-thiazolyl)propionic acid**

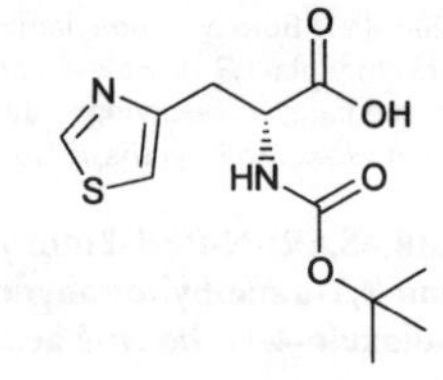

$C_{11}H_{16}N_2O_4S$
Chiral intermediate. *Celltech Chiroscience Ltd.*

**924 (tert-Butoxycarbonyl)-D-arginine**
78603-12-0

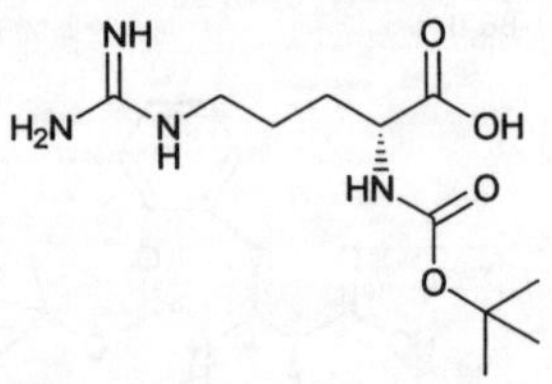

$C_{11}H_{22}N_4O_4$
Chiral intermediate. *Isochem.*

**925 (tert-Butoxycarbonyl)-L-arginine**
13726-76-6 237-295-2

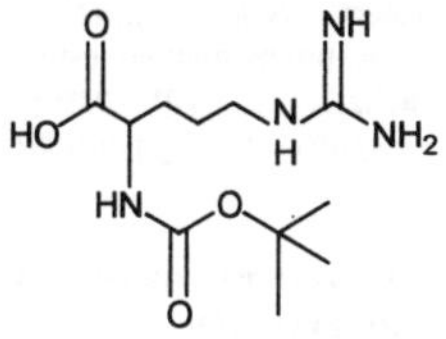

$C_{11}H_{22}N_4O_4$
Chiral intermediate. $[\alpha]_D^{20}$ = - 6.5° (c = 1, HOAc). *Davos Chemical Corporation; Sigma-Aldrich Fine Chemicals.*

**926 N-α-(tert-Butoxycarbonyl)-D-arginine hydrochloride hydrate**
114622-81-0

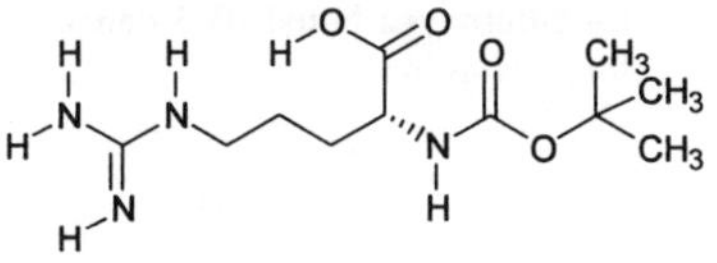

$C_{11}H_{23}ClN_4O_4$
N-(tert-Butoxycarbonyl)-D-arginine hydrochloride hydrate. Chiral intermediate. mp = 176-178°; [α] = 4 (c = 2, $H_2O$). *Acros Organics nv; Isochem; Synthetech, Inc.*

**927 (tert-Butoxycarbonyl)-L-arginine hydrochloride monohydrate**

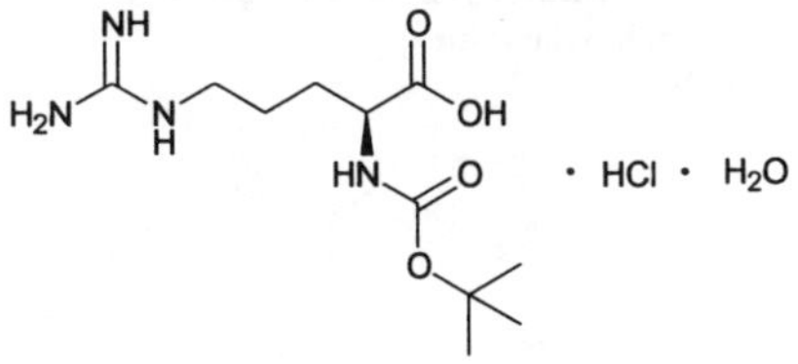

$C_{11}H_{25}ClN_4O_5$
Chiral intermediate. *Synthetech, Inc.*

**928 (tert-Butoxycarbonyl)-L-arginine nitro ester**
Chiral intermediate. *Synthetech, Inc.*

**929 (tert-Butoxycarbonyl)-D-arginine tosyl ester**

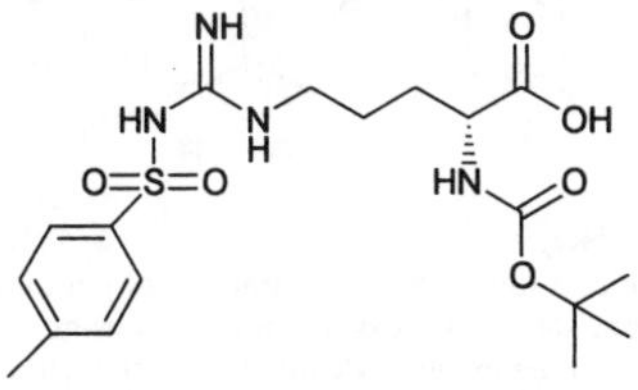

$C_{18}H_{26}N_4O_6S$
Chiral intermediate. *Synthetech, Inc.*

**930 (tert-Butoxycarbonyl)-L-arginine tosyl ester**

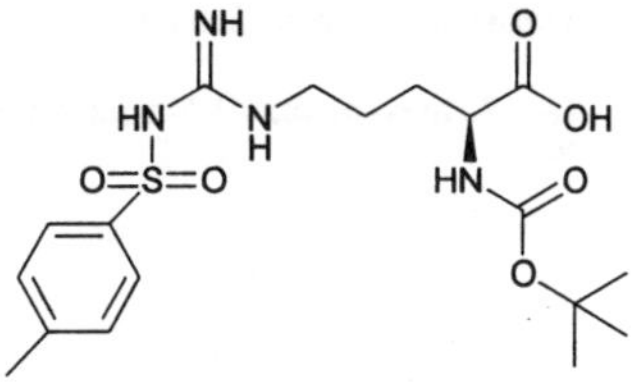

$C_{18}H_{26}N_4O_6S$
Chiral intermediate. *Synthetech, Inc.*

**931 (tert-Butoxycarbonyl)-D-asparagine**
75647-01-7

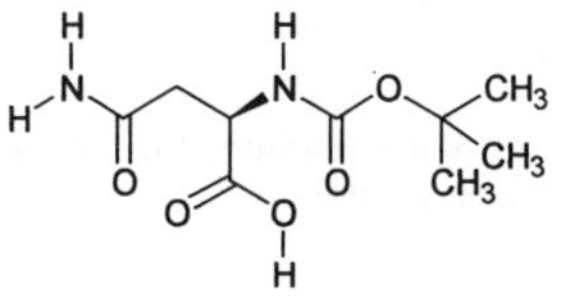

$C_9H_{16}N_2O_5$
Chiral intermediate. *Acros Organics nv; Isochem.*

**932 N-α-(tert-Butoxycarbonyl)-L-asparagine**
7536-55-2 231-405-2

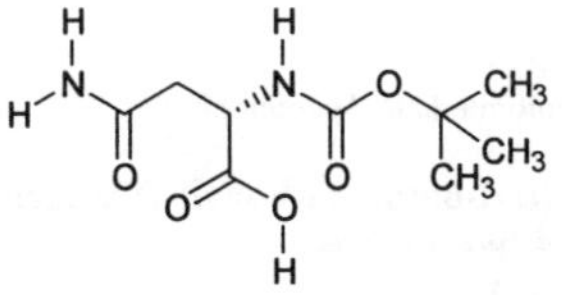

$C_9H_{16}N_2O_5$
$N^2$-[(1,1-Dimethylethoxy)carbonyl]-, L-; N-(tert-Butoxycarbonyl)-L-asparagine; Asparagine, $N^2$-carboxy-, $N^2$-tert-butyl ester, L-; $N^2$-[(1,1-Dimethylethoxy)carbonyl]-L-asparagine. Listed on TSCA. Chiral intermediate. mp = 175°; [α] = - 7.6° (c = 1, DMF). *Acros Organics nv; Flamma s.p.a.; Sigma-Aldrich Fine Chemicals; Synthetech, Inc.*

**933 (tert-Butoxycarbonyl)-L-asparagine trityl ester**

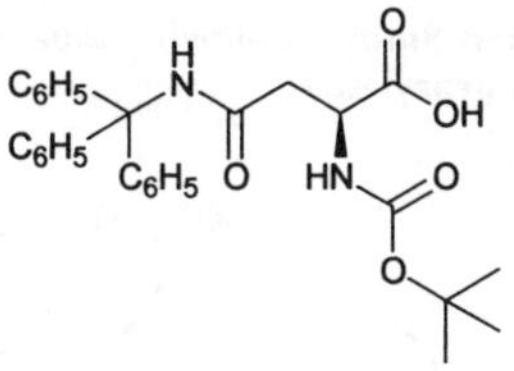

$C_{28}H_{130}N_2O_5$
Chiral intermediate. *Synthetech, Inc.*

**934 (tert-Butoxycarbonyl)-L-asparagine xanthyl ester**
Chiral intermediate. *Synthetech, Inc.*

**935 (tert-Butoxycarbonyl)-L-aspartic acid**
13726-67-5 237-294-7

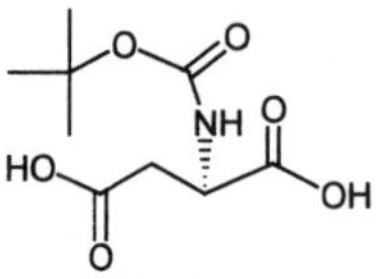

$C_9H_{15}NO_6$
Aspartic acid, N-carboxy-, N-tert-butyl ester; N-(tert-butoxycarbonyl)-L-aspartic acid. Chiral intermediate. mp = 116-118°; $[\alpha]_D^{20}$ = - 6.0° (c = 1, $CH_3OH$). *Acros Organics nv; PPG - Sipsy; Sigma-Aldrich Fine Chemicals; Synthetech, Inc.*

**936 (tert-Butoxycarbonyl)-L-aspartic acid 2-benzyl ester**

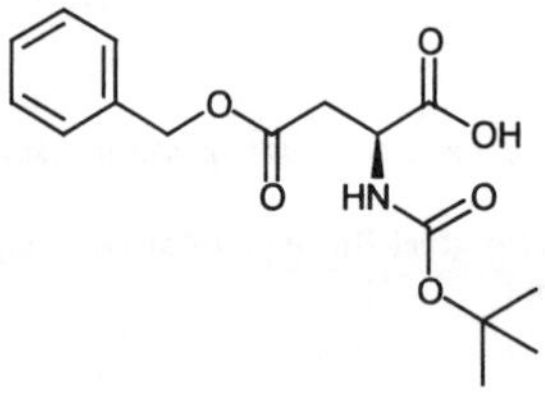

$C_{16}H_{21}NO_6$
Chiral intermediate. *Synthetech, Inc.*

**937 (tert-Butoxycarbonyl)-D-aspartic acid 4-benzyl ester**
51186-58-4

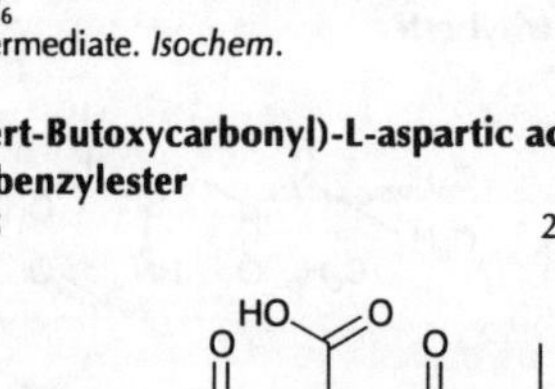

$C_{16}H_{21}NO_6$
Chiral intermediate. *Isochem.*

**938 (tert-Butoxycarbonyl)-L-aspartic acid 4-benzylester**
7536-58-5 231-406-8

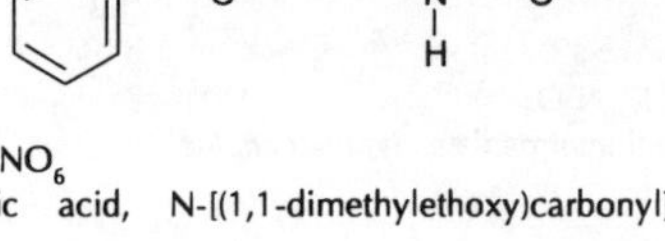

$C_{16}H_{21}NO_6$
Aspartic acid, N-[(1,1-dimethylethoxy)carbonyl]-, 4-(phenylmethyl) ester, L-; 4-Benzyl hydrogen N-(tert-butoxycarbonyl)-L-aspartate; β-Benzyl N-tert-butoxycarbonylaspartate; N-(tert-Butoxycarbonyl)-L-aspartic acid 4-benzyl ester; N-[(1,1-Dimethylethoxy)carbonyl]-L-aspartic acid, 4-(phenylmethyl) ester. Listed on TSCA. Chiral intermediate. mp = 101-103°; [α] = - 19.1° (c = 2, DMF). *Acros Organics nv; Sigma-Aldrich Fine Chemicals; Synthetech, Inc.*

**939 (tert-Butoxycarbonyl)-L-aspartic acid 4-cyclohexyl ester**
Chiral intermediate. *Synthetech, Inc.*

**940 (R)-N-(tert-Butoxycarbonyl)-azetidine-carboxylic acid**
203564-56-1
$C_9H_{15}NO_4$
1-tert-Butoxycarbonyl-(R)-azetidine-2-carboxylic acid. Chiral intermediate. *Lancaster Synthesis Ltd.*

**941 (tert-Butoxycarbonyl)-D-3-benzothienylalanine**
111082-76-9

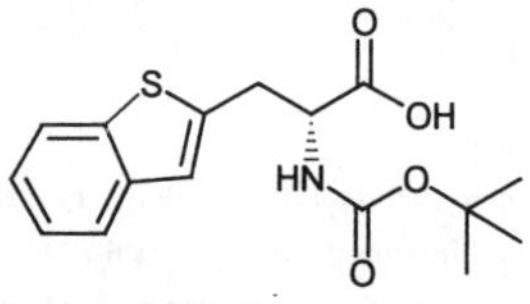

$C_{16}H_{19}N_{04}S$
Chiral intermediate. *Senn Chemicals AG; Synthetech, Inc.*

**942 (tert-Butoxycarbonyl)-L-3-benzothienylalanine**
154902-51-9

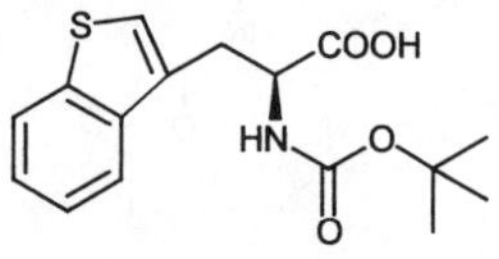

$C_{16}H_{19}N_{04}S$
Chiral intermediate. *Synthetech, Inc.*

**943 (tert-Butoxycarbonyl)-S-benzyl-L-cysteine**
5068-28-0 225-772-8

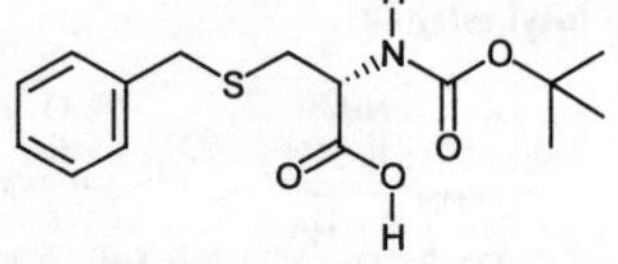

$C_{15}H_{21}NO_4S$
Alanine, 3-(benzylthio)-N-carboxy-, N-tert-butyl ester, L-; S-Benzyl-N-[tert-butoxycarbonyl]-L-cysteine; L-Cysteine, N-[(1,1-dimethylethoxy)carbonyl]-S-(phenylmethyl)-; N-[(1,1-Dimethylethoxy)carbonyl]-S-(phenylmethyl)-L-cysteine. Listed on TSCA. Chiral intermediate. [α] = - 44.5° (c = 1, HOAc). *Acros Organics nv; Isochem.*

**944 (tert-Butoxycarbonyl)-S-benzyl-L-cysteinyl-L-serine methyl ester**

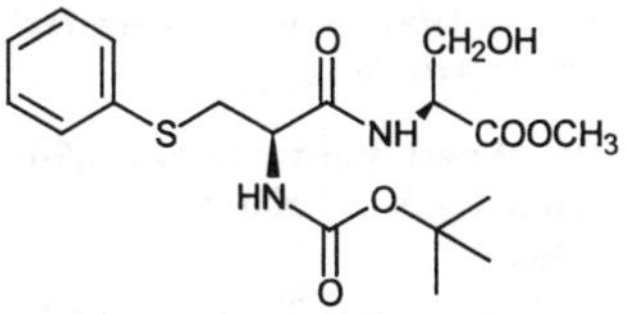

$C_{18}H_{26}N_2O_6S$
Chiral intermediate. *Austin Chemical Company, Inc.*

**945 (tert-Butoxycarbonyl)-O-benzyl-N-methyl-L-tyrosine**

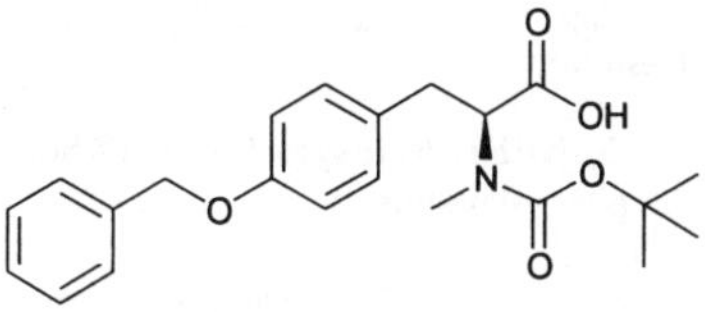

$C_{27}H_{22}NO_5$
Chiral intermediate. *Synthetech, Inc.*

**946 (tert-Butoxycarbonyl)-O-benzyl-D-serine**
47173-80-8

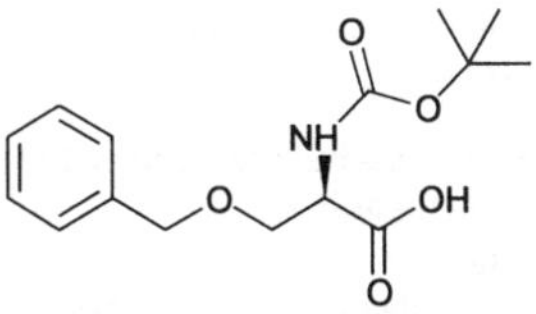

$C_{15}H_{21}NO_5$
Chiral intermediate. [α] = - 19°. *Acros Organics nv; Synthetech, Inc.*

**947 (tert-Butoxycarbonyl)-O-benzyl-L-serine**
23680-31-1 245-820-1

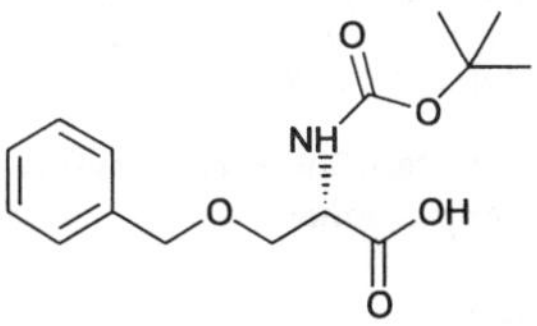

$C_{15}H_{21}NO_5$
Serine, N-[(1,1-dimethylethoxy)carbonyl]-O-(phenylmethyl)-, L-; N-[(1,1-Dimethylethoxy)carbonyl]-O-(phenylmethyl)-L-serine; Alanine, 3-(benzyloxy)-N-carboxy-, N-tert-butyl ester, L-; N-(tert-Butoxycarbonyl)serine benzyl ether; N-tert-Butoxycarbonyl-O-benzyl-L-serine; O-Benzyl-N-tert-butoxycarbonyl-L-serine. Listed on TSCA. Chiral intermediate. *Acros Organics nv.*

**948 N-(tert-Butoxycarbonyl)-O-benzyl-L-threonine**
15260-10-3 239-304-5

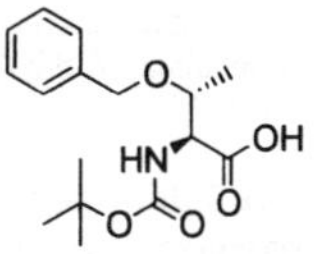

$C_{16}H_{23}NO_5$
Listed on TSCA. Chiral intermediate. mp = 116-120°; $[\alpha]^{22}$ = + 22° (c = 2, 95% EtOH). *Sigma-Aldrich Fine Chemicals; Synthetech, Inc.*

**949 (tert-Butoxycarbonyl)-O-benzyl-L-tyrosine**

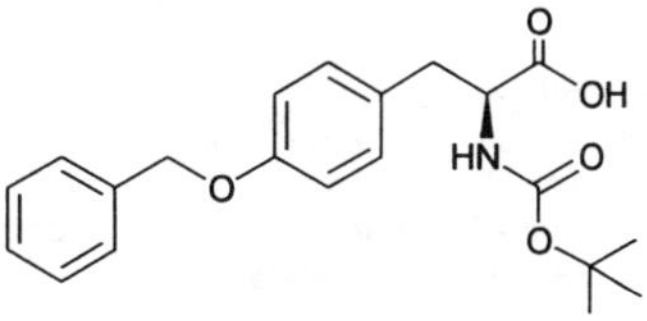

$C_{21}H_{25}NO_5$
Chiral intermediate. *Synthetech, Inc.*

**950 (tert-Butoxycarbonyl)-D-4,4'-biphenylalanine**
128779-47-5

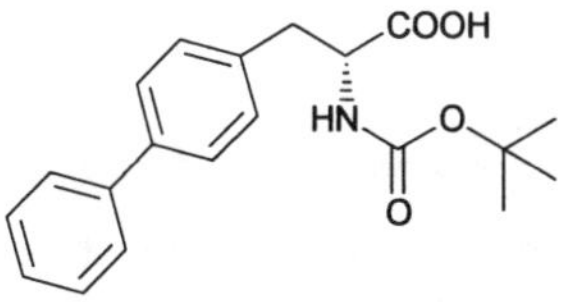

$C_{20}H_{23}NO_4$
Chiral intermediate. *Synthetech, Inc.*

**951 (tert-Butoxycarbonyl)-L-4,4'-biphenylalanine**
147923-08-8

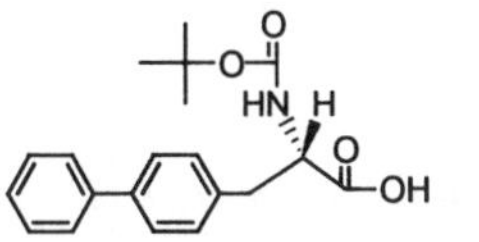

$C_{20}H_{23}NO_4$
Chiral intermediate. mp = 103-107°; $[\alpha]^{22}$ = + 21° (c = 2, $CH_3OH$). *Sigma-Aldrich Fine Chemicals; Synthetech, Inc.*

**952 (R)-N-(tert-Butoxycarbonyl)-(2-bromoallyl)glycine**
149930-92-7
$C_{10}H_{16}BrNO_4$
N-tert-Butoxycarbonyl-(R)-2-amino-4-bromo-4-pentenoic acid. Chiral intermediate. *ChiroTech Technology Ltd.; Lancaster Synthesis Ltd.*

**953 (S)-N-(tert-Butoxycarbonyl)-(2-bromoallyl)glycine**
151215-34-8
$C_{10}H_{16}BrNO_4$
N-tert-Butoxycarbonyl-(S)-2-amino-4-bromo-4-pentenoic acid. Chiral intermediate. *ChiroTech Technology Ltd.; Lancaster Synthesis Ltd.*

**954 (R)-N-(tert-Butoxycarbonyl)-5-bromo-2,4-dimethoxyphenylalanine**

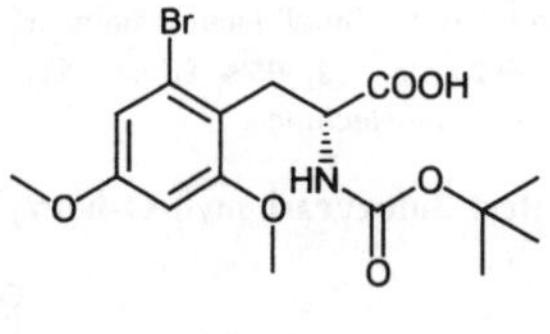

$C_{16}H_{22}BrNO_6$
Chiral intermediate. *ChiroTech Technology Ltd.*

**955 (S)-N-(tert-Butoxycarbonyl)-5-bromo-2,4-dimethoxyphenylalanine**

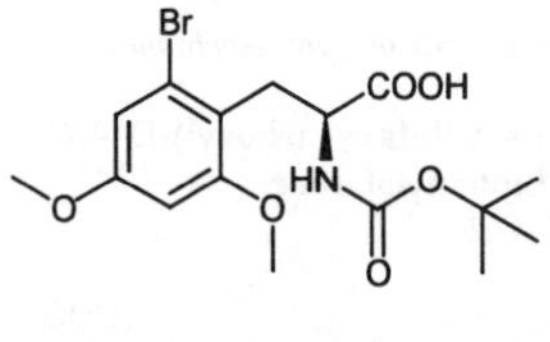

$C_{16}H_{22}BrNO_6$
Chiral intermediate. *ChiroTech Technology Ltd.*

**956 (R)-N-(tert-Butoxycarbonyl)-(3-bromo-4-methoxyphenyl)alanine**

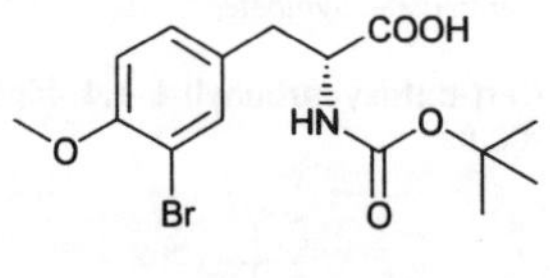

$C_{15}H_{20}BrNO_5$
Chiral intermediate. *ChiroTech Technology Ltd.*

**957 (S)-N-(tert-Butoxycarbonyl)-(3-bromo-4-methoxyphenyl)alanine**

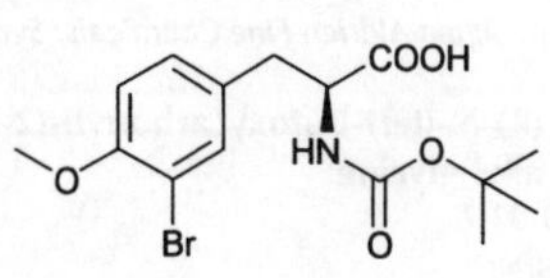

$C_{15}H_{20}BrNO_5$
Chiral intermediate. *ChiroTech Technology Ltd.*

**958 (R)-N-(tert-Butoxycarbonyl)-(5-bromo-2-methoxyphenyl)alanine**
$C_{15}H_{20}BrNO_5$
N-tert-Butoxycarbonyl-(5-bromo-2-methoxyphenyl)-D-alanine. Chiral intermediate. *ChiroTech Technology Ltd.; Lancaster Synthesis Ltd.*

**959 (S)-N-(tert-Butoxycarbonyl)-(5-bromo-2-methoxyphenyl)alanine**
$C_{15}H_{20}BrNO_5$
N-tert-Butoxycarbonyl-(5-bromo-2-methoxyphenyl)-L-alanine. Chiral intermediate. *ChiroTech Technology Ltd.; Lancaster Synthesis Ltd.*

**960 (R)-N-(tert-Butoxycarbonyl)-(2-bromophenyl)alanine**
$C_{14}H_{18}BrNO_4$
N-tert-Butoxycarbonyl-2-bromophenyl-D-alanine. Chiral intermediate. *ChiroTech Technology Ltd.; Lancaster Synthesis Ltd.*

**961 (S)-N-(tert-Butoxycarbonyl)-(2-bromophenyl)alanine**
$C_{14}H_{18}BrNO_4$
N-tert-Butoxycarbonyl-2-bromophenyl-L-alanine. Chiral intermediate. *ChiroTech Technology Ltd.; Lancaster Synthesis Ltd.*

**962 (R)-N-(tert-Butoxycarbonyl)-(3-bromophenyl)alanine**
$C_{14}H_{18}BrNO_4$
N-tert-Butoxycarbonyl-3-bromophenyl-D-alanine. Chiral intermediate. *ChiroTech Technology Ltd.; Lancaster Synthesis Ltd.*

**963 (S)-N-(tert-Butoxycarbonyl)-(3-bromophenyl)alanine**
$C_{14}H_{18}BrNO_4$
N-tert-Butoxycarbonyl-3-bromophenyl-L-alanine. Chiral intermediate. *ChiroTech Technology Ltd.; Lancaster Synthesis Ltd.*

**964 (R)-N-(tert-Butoxycarbonyl)-(4-bromophenyl)alanine**
79561-82-3
$C_{14}H_{18}BrNO_4$
N-tert-Butoxycarbonyl-4-bromophenyl-D-alanine; Boc-D-4-bromophenylalanine. Chiral intermediate. *ChiroTech Technology Ltd.; Lancaster Synthesis Ltd.; Synthetech, Inc.*

**965 (S)-N-(tert-Butoxycarbonyl)-(4-bromophenyl)alanine**
62129-39-9
$C_{14}H_{18}BrNO_4$
N-tert-Butoxycarbonyl-4-bromophenyl-L-alanine. Chiral intermediate. *ChiroTech Technology Ltd.; Lancaster Synthesis Ltd.; Synthetech, Inc.*

**966 (R)-N-(tert-Butoxycarbonyl)-(5-bromo-2-thienyl)alanine**
$C_{12}H_{16}BrNO_4S$
N-tert-Butoxycarbonyl-(5-bromo-2-thienyl)-D-alanine. Chiral intermediate. *ChiroTech Technology Ltd.; Lancaster Synthesis Ltd.*

**967 (S)-N-(tert-Butoxycarbonyl)-(5-bromo-2-thienyl)alanine**

190319-95-0

$C_{12}H_{16}BrNO_4S$

N-tert-Butoxycarbonyl-5-bromo-2-thienyl-L-alanine. Chiral intermediate. *ChiroTech Technology Ltd.; Lancaster Synthesis Ltd.*

**968 (R)-(+)-1-(tert-Butoxycarbonyl)-2-tert-butyl-3-methyl-4-imidazolidinone**

119838-44-7

$C_{13}H_{24}N_2O_3$

Chiral intermediate. mp = 68-70°; $[\alpha]^{25}$ = + 14° (c = 1, $CH_2Cl_2$). *Acros Organics nv; Senn Chemicals AG; Sigma-Aldrich Fine Chemicals.*

**969 (S)-(-)-1-(tert-Butoxycarbonyl)-2-tert-butyl-3-methyl-4-imidazolidinone**

119838-38-9

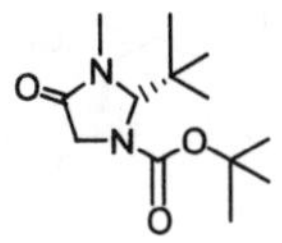

$C_{13}H_{24}N_2O_3$

Chiral intermediate. mp = 68-70°; $[\alpha]^{25}$ = - 14° (c = 1, $CH_2Cl_2$). *Sigma-Aldrich Fine Chemicals.*

**970 (tert-Butoxycarbonyl)-O-t-butyl-L-serine**

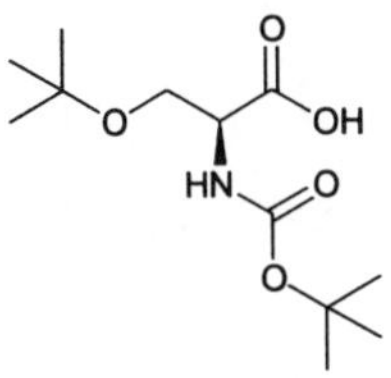

$C_{12}H_{23}NO_5$

Chiral intermediate. *Synthetech, Inc.*

**971 (tert-Butoxycarbonyl)-O-t-butyl-L-tyrosine**

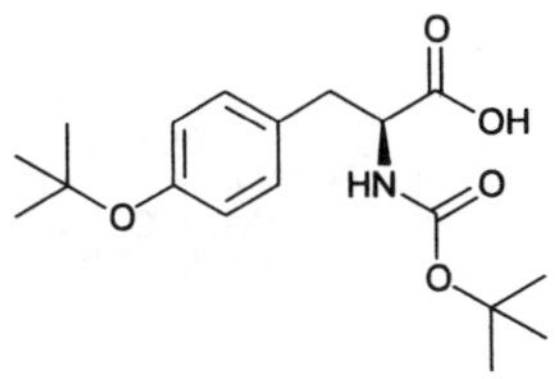

$C_{18}H_{27}NO_5$

Chiral intermediate. *Synthetech, Inc.*

**972 (tert-Butoxycarbonyl)-carbobenzyloxy-L-lysinamide**

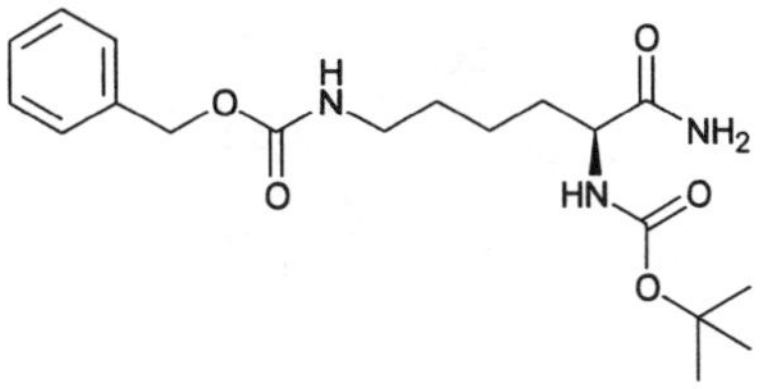

$C_{19}H_{29}N_3O_5$

Chiral intermediate. *Synthetech, Inc.*

**973 (tert-Butoxycarbonyl)-carbobenzyloxy-L-lysine**

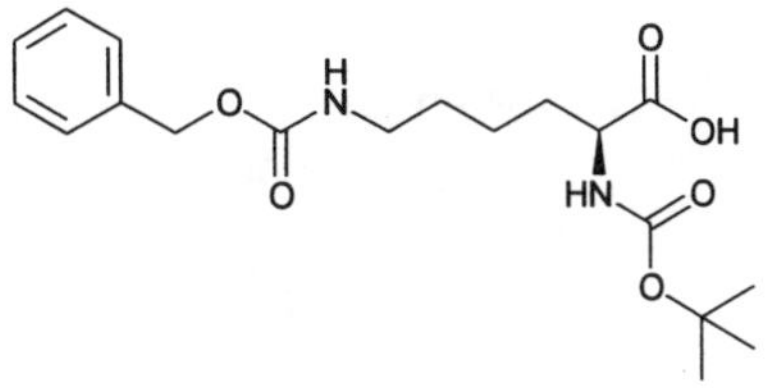

$C_{19}H_{28}N_2O_6$

Chiral intermediate. *Synthetech, Inc.*

**974 N-α-(tert-Butoxycarbonyl)-Nε-carbobenzyloxy-L-lysine dicyclohexylammonium salt**

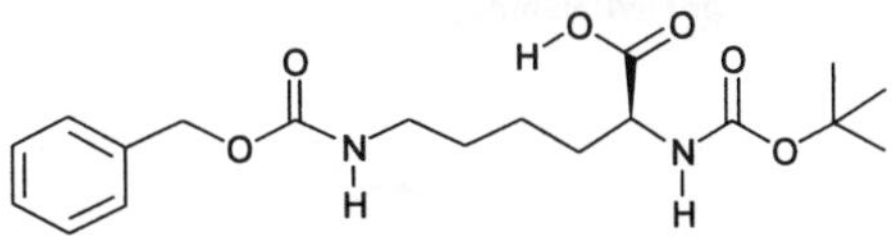

$C_{19}H_{28}N_2O_6$

Chiral intermediate. [α] = + 8.5° (c = 2, $CH_3OH$). *Acros Organics nv.*

**975 (tert-Butoxycarbonyl)-carbobenzyloxy-L-lysine hydroxysuccinimide ester**

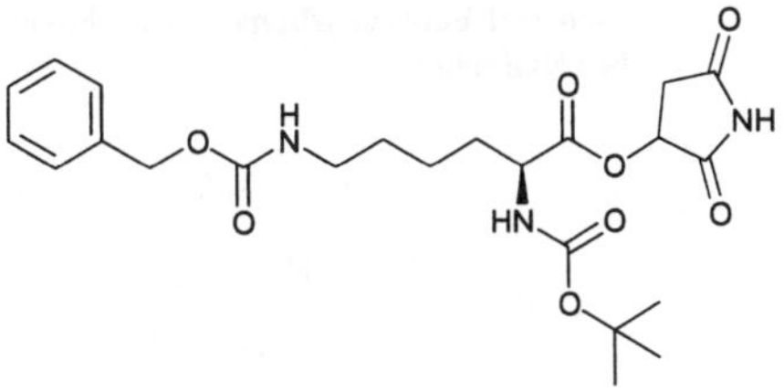

$C_{23}H_{31}N_3O_8$

Chiral intermediate. *Synthetech, Inc.*

**976 (R)-N-(tert-Butoxycarbonyl)-(2-carboxyphenyl)alanine**

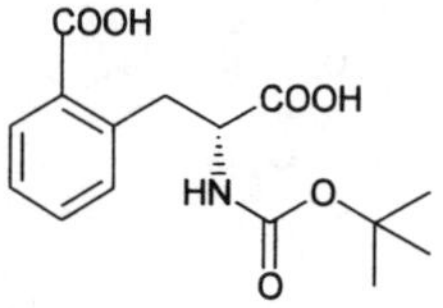

$C_{15}H_{19}NO_6$
Chiral intermediate. *ChiroTech Technology Ltd.*

**977 (S)-N-(tert-Butoxycarbonyl)-(2-carboxyphenyl)alanine**

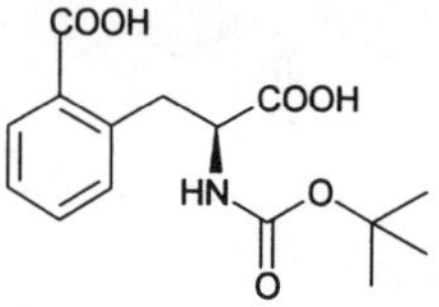

$C_{15}H_{19}NO_6$
Chiral intermediate. *ChiroTech Technology Ltd.*

**978 (R)-N-(tert-Butoxycarbonyl)-(4-carboxyphenyl)alanine**

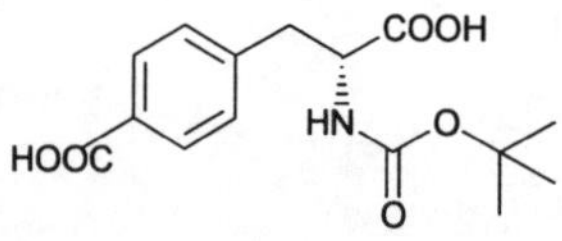

$C_{15}H_{19}NO_6$
Chiral intermediate. *ChiroTech Technology Ltd.*

**979 (S)-N-(tert-Butoxycarbonyl)-(4-carboxyphenyl)alanine**

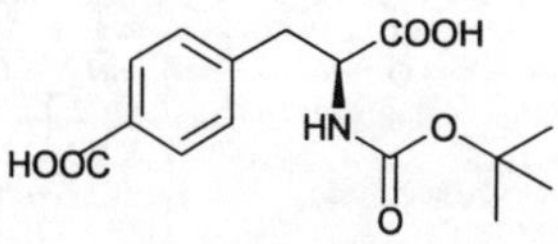

$C_{15}H_{19}NO_6$
Chiral intermediate. *ChiroTech Technology Ltd.*

**980 (R)-N-(tert-Butoxycarbonyl)-(2-chloro-5-fluorophenyl)alanine**

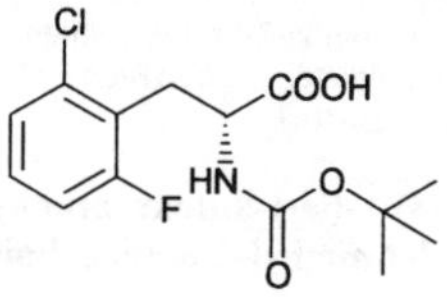

$C_{14}H_{17}ClFNO_4$
Chiral intermediate. *ChiroTech Technology Ltd.*

**981 (S)-N-(tert-Butoxycarbonyl)-(2-chloro-5-fluorophenyl)alanine**

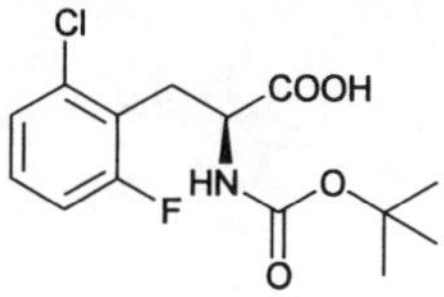

$C_{14}H_{17}ClFNO_4$
Chiral intermediate. *ChiroTech Technology Ltd.*

**982 (R)-N-(tert-Butoxycarbonyl)-(2-chloro-5-nitrophenyl)alanine**

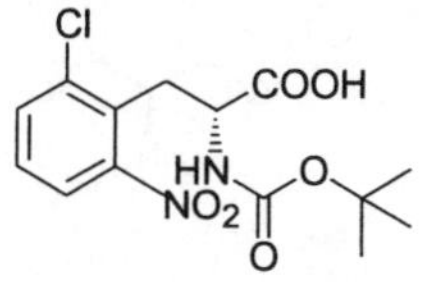

$C_{14}H_{17}ClN_2O_6$
Chiral intermediate. *ChiroTech Technology Ltd.*

**983 (S)-N-(tert-Butoxycarbonyl)-(2-chloro-5-nitrophenyl)alanine**

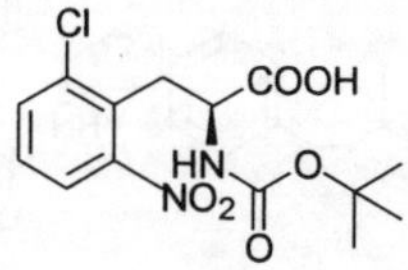

$C_{14}H_{17}ClN_2O_6$
Chiral intermediate. *ChiroTech Technology Ltd.*

**984 (R)-N-(tert-Butoxycarbonyl)-(4-chloro-3-nitrophenyl)alanine**

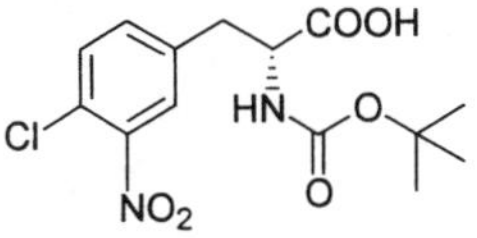

$C_{14}H_{17}ClN_2O_6$
Chiral intermediate. *ChiroTech Technology Ltd.*

**985 (S)-N-(tert-Butoxycarbonyl)-(4-chloro-3-nitrophenyl)alanine**

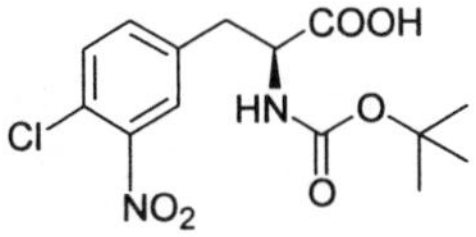

$C_{14}H_{17}ClN_2O_6$
Chiral intermediate. *ChiroTech Technology Ltd.*

**986 (R)-N-(tert-Butoxycarbonyl)-(2-chlorophenyl)alanine**
80102-23-4

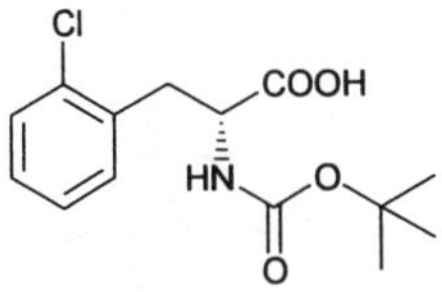

$C_{14}H_{18}ClNO_4$
Chiral intermediate. *ChiroTech Technology Ltd.; Synthetech, Inc.*

**987 (S)-N-(tert-Butoxycarbonyl)-(2-chlorophenyl)alanine**
114873-02-8

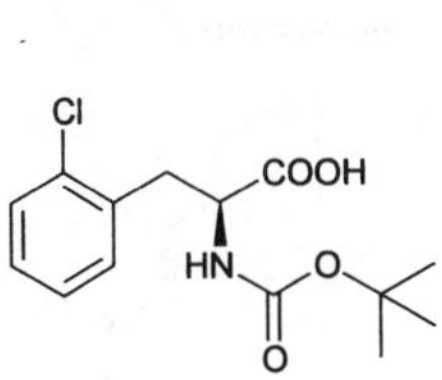

$C_{14}H_{18}ClNO_4$
Chiral intermediate. *ChiroTech Technology Ltd.; Synthetech, Inc.*

**988 (R)-N-(tert-Butoxycarbonyl)-(3-chlorophenyl)alanine**
80102-25-6

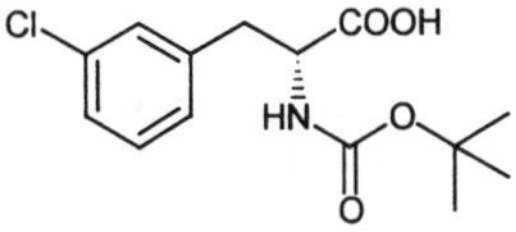

$C_{14}H_{18}ClNO_4$
Chiral intermediate. *ChiroTech Technology Ltd.; Synthetech, Inc.*

**989 (S)-N-(tert-Butoxycarbonyl)-(3-chlorophenyl)alanine**
114873-03-9

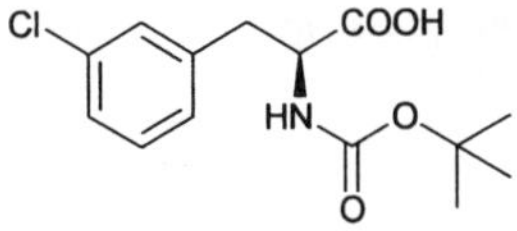

$C_{14}H_{18}ClNO_4$
Chiral intermediate. *ChiroTech Technology Ltd.; Synthetech, Inc.*

**990 (R)-N-(tert-Butoxycarbonyl)-(4-chlorophenyl)alanine**
57292-44-1
$C_{14}H_{18}ClNO_4$
N-tert-Butoxycarbonyl-4-chlorophenyl-D-alanine; Boc-D-4-chlorophenylalanine. Chiral intermediate. *ChiroTech Technology Ltd.; Lancaster Synthesis Ltd.; Synthetech, Inc.*

**991 (S)-N-(tert-Butoxycarbonyl)-(4-chlorophenyl)alanine**
68090-88-0
$C_{14}H_{18}ClNO_4$
N-tert-Butoxycarbonyl-4-chlorophenyl-L-alanine; Boc-L-4-chlorophenylalanine. Chiral intermediate. *Austin Chemical Company, Inc.; ChiroTech Technology Ltd.; Lancaster Synthesis Ltd.; Synthetech, Inc.*

**992 (R)-N-(tert-Butoxycarbonyl)-(2-cyanophenyl)alanine**

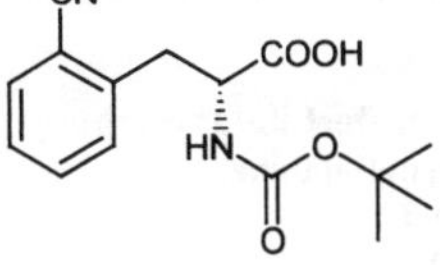

$C_{15}H_{18}N_2O_4$
Chiral intermediate. *ChiroTech Technology Ltd.*

**993 (S)-N-(tert-Butoxycarbonyl)-(2-cyanophenyl)alanine**

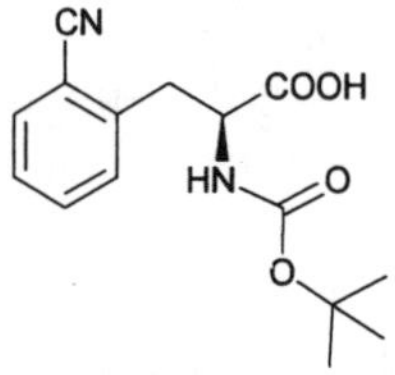

$C_{15}H_{18}N_2O_4$
Chiral intermediate. *ChiroTech Technology Ltd.*

**994 (R)-N-(tert-Butoxycarbonyl)-(3-cyanophenyl)alanine**
205445-56-3

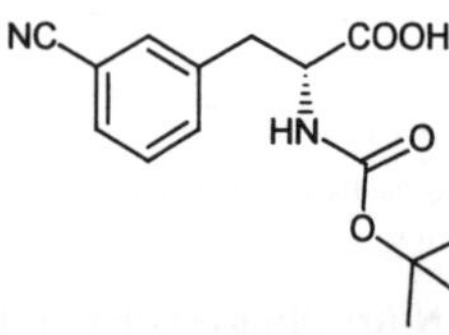

$C_{15}H_{18}N_2O_4$
Chiral intermediate. *ChiroTech Technology Ltd.; Synthetech, Inc.*

**995 (S)-N-(tert-Butoxycarbonyl)-(3-cyanophenyl)alanine**
131980-30-8
$C_{15}H_{18}N_2O_4$
N-tert-Butoxycarbonyl-3-cyanophenyl-L-alanine; Boc-L-3-cyanophenylalanine. Chiral intermediate. *ChiroTech Technology Ltd.; Lancaster Synthesis Ltd.*

**996 (R)-N-(tert-Butoxycarbonyl)-(4-cyanophenyl)alanine**
146727-62-0
$C_{15}H_{18}N_2O_4$
N-tert-Butoxycarbonyl-4-cyanophenyl-D-alanine; Boc-D-4-cyanophenylalanine. Chiral intermediate. *ChiroTech Technology Ltd.; Lancaster Synthesis Ltd.; Synthetech, Inc.*

**997 (S)-N-(tert-Butoxycarbonyl)-(4-cyanophenyl)alanine**
131724-45-3
$C_{15}H_{18}N_2O_4$
N-tert-Butoxycarbonyl-4-cyanophenyl-L-alanine; Boc-L-4-cyanophenylalanine. Chiral intermediate. *ChiroTech Technology Ltd.; Lancaster Synthesis Ltd.; Synthetech, Inc.*

**998 (tert-Butoxycarbonyl)-D-cyclohexylalanine**

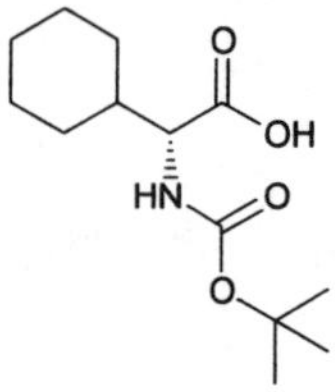

$C_{14}H_{25}NO_4$
Chiral intermediate. *Genzyme Pharmaceuticals; Synthetech, Inc.*

**999 (tert-Butoxycarbonyl)-L-cyclohexylalanine**
37736-82-6

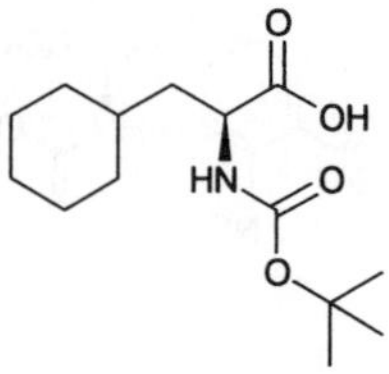

$C_{14}H_{25}NO_4$
Chiral intermediate. *Fischer Chemicals AG; Great Lakes Fine Chemicals; Senn Chemicals AG; Synthetech, Inc.*

**1000 (tert-Butoxycarbonyl)-L-cyclohexylalanine methyl ester**
98105-41-0

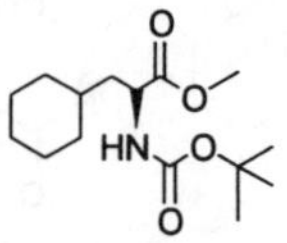

$C_{15}H_{27}NO_4$
Chiral intermediate. mp = 47-49°; $[\alpha]_D^{20}$ = - 20° (c = 1, $CH_3OH$). *Fischer Chemicals AG; Great Lakes Fine Chemicals; Sigma-Aldrich Fine Chemicals; Synthetech, Inc.*

**1001 (tert-Butoxycarbonyl)-D-cyclohexylalaninol**

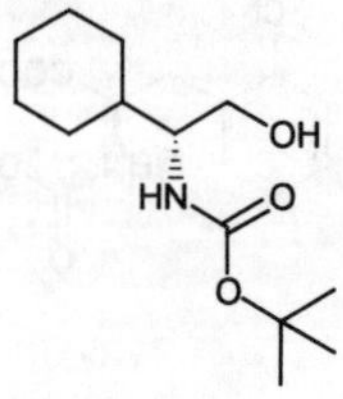

$C_{13}H_{25}NO_3$
Chiral intermediate. *Synthetech, Inc.*

**1002 (tert-Butoxycarbonyl)-L-cyclohexylalaninol**
103322-56-1

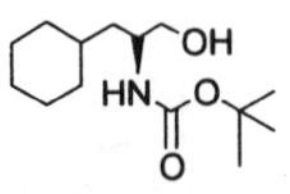

$C_{14}H_{27}NO_3$
Chiral intermediate. bp = 214°; d = 1.126; n = 0.464; $[\alpha]_D^{20}$ = - 25° (c = 1, EtOH). *Fischer Chemicals AG; Great Lakes Fine Chemicals; Sigma-Aldrich Fine Chemicals; Synthetech, Inc.*

**1003 (tert-Butoxycarbonyl)-D-cyclohexylglycine**
70491-05-3

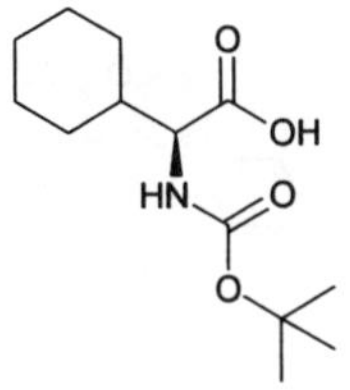

$C_{13}H_{23}NO_4$
Chiral intermediate. *Genzyme Pharmaceuticals; Isochem; Synthetech, Inc.*

**1004 (tert-Butoxycarbonyl)-L-cyclohexylglycine**

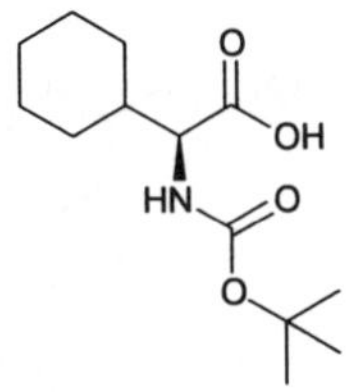

$C_{13}H_{23}NO_4$
Chiral intermediate. *Senn Chemicals AG; Synthetech, Inc.*

**1005 (R)-N-(tert-Butoxycarbonyl)cyclopropyl-β-alanine**

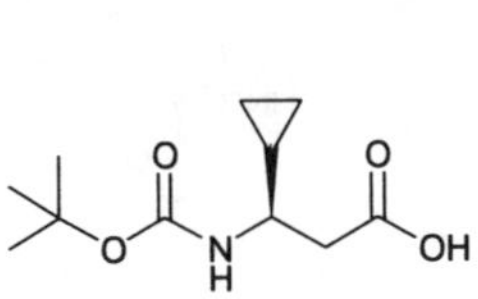

$C_{11}H_{19}NO_4$
Chiral intermediate. *Eastman Chemical Company.*

**1006 (S)-N-(tert-Butoxycarbonyl)cyclopropyl-β-alanine**

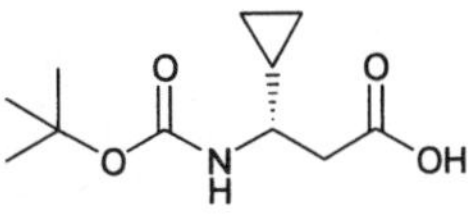

$C_{11}H_{19}NO_4$
Chiral intermediate. *Eastman Chemical Company.*

**1007 (R)-N-(tert-Butoxycarbonyl)cyclopropylglycine**

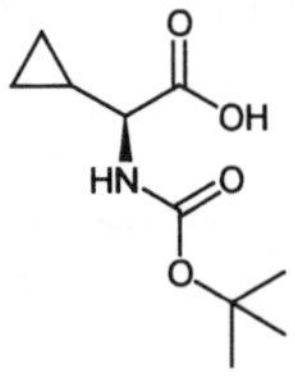

$C_{10}H_{17}NO_4$
Chiral intermediate. *Eastman Chemical Company.*

**1008 (S)-N-(tert-Butoxycarbonyl)cyclopropylglycine**
155976-13-9

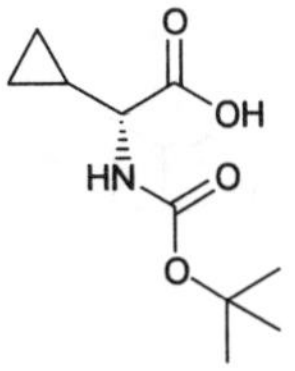

$C_{10}H_{17}NO_4$
Chiral intermediate. *Eastman Chemical Company.*

**1009 (tert-Butoxycarbonyl)-L-cysteine**

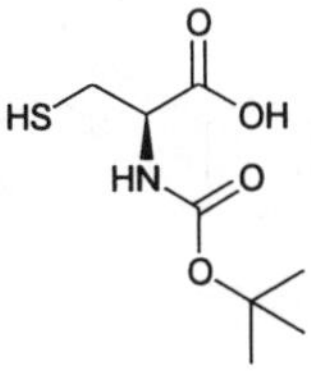

$C_8H_{15}NO_4S$
Chiral intermediate. *Austin Chemical Company, Inc.*

**1010 (tert-Butoxycarbonyl)-L-cysteine acetamidomethyl ester**

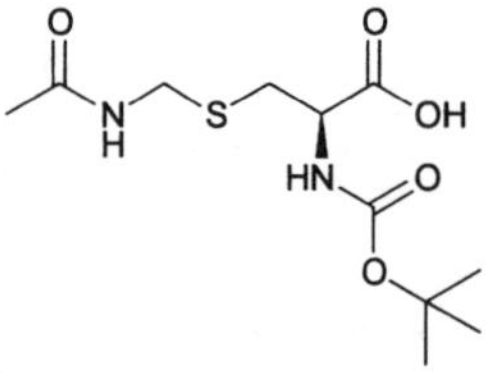

$C_{11}H_{20}N_2O_5S$
Chiral intermediate. *Synthetech, Inc.*

**1011 (tert-Butoxycarbonyl)-L-cysteine benzyl ester**

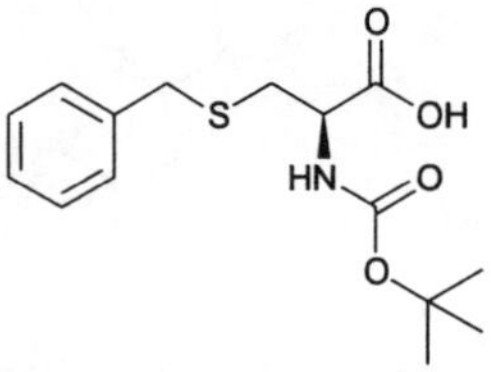

$C_{15}H_{21}NO_4S$
Chiral intermediate. *Synthetech, Inc.*

**1012 (tert-Butoxycarbonyl)-L-cysteine 4-methoxybenzyl ester**

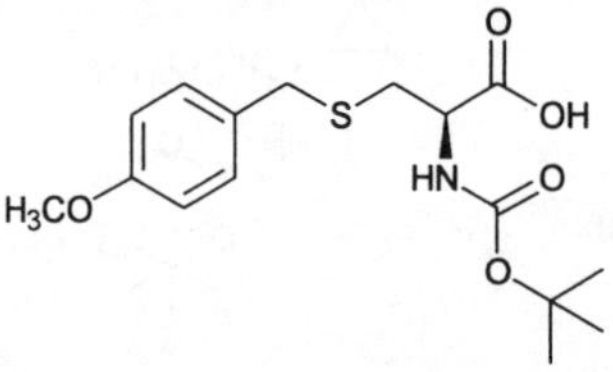

$C_{16}H_{23}NO_5S$
Chiral intermediate. *Synthetech, Inc.*

**1013 (tert-Butoxycarbonyl)-L-cysteine 4-methylbenzyl ester**

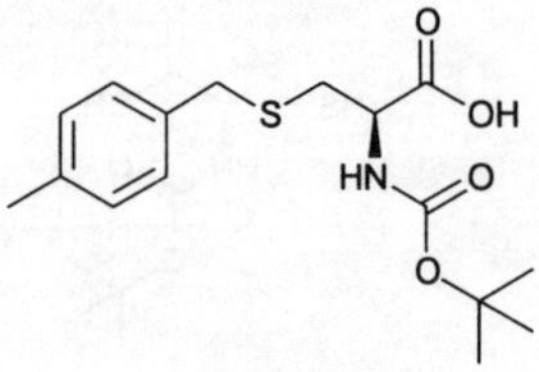

$C_{16}H_{23}NO_4S$
Chiral intermediate. *Synthetech, Inc.*

**1014 N-(tert-Butoxycarbonyl)-L-cysteine methyl ester**
55757-46-5

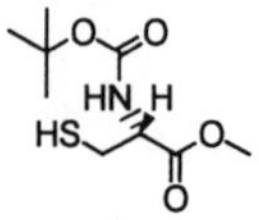

$C_9H_{17}NO_4S$
Chiral intermediate. bp = 214°; d = 1.143; n = 1.475; $[\alpha]^{22}$ = + 21° (c = 7.5, $CHCl_3$). *Sigma-Aldrich Fine Chemicals.*

**1015 (tert-Butoxycarbonyl)-L-cysteine trityl ester**
21947-98-8 244-674-6

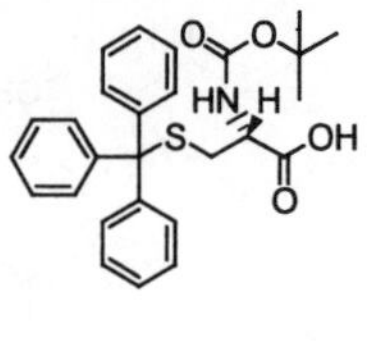

$C_{27}H_{29}NO_4S$
Chiral intermediate. mp = 143-146°; $[\alpha]_D^{20}$ = + 27.5° (c = 1, EtOH). *Sigma-Aldrich Fine Chemicals; Synthetech, Inc.*

**1016 (tert-Butoxycarbonyl)-L-2,4-diamino-butyric acid**

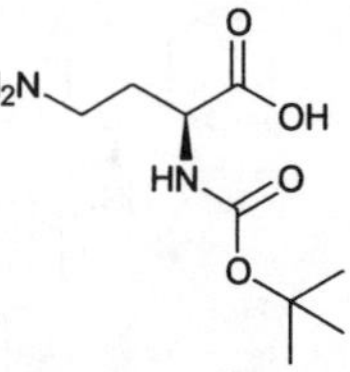

$C_9H_{18}N_2O_4$
Chiral intermediate. *Senn Chemicals AG.*

**1017 (1S,4S)-N-(tert-Butoxycarbonyl)-2,5-diaza-bicyclo[2.2.1]heptane dihydrobromide**
113451-59-5

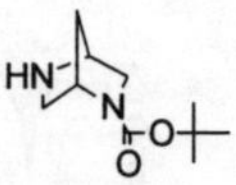

$C_{10}H_{20}Br_2N_2O_2$
Chiral intermediate. mp = 74-76°; $[\alpha]^{22}$ = - 44° (c = 1, $CHCl_3$). *Medinger & Sohn; Sigma-Aldrich Fine Chemicals.*

**1018 (R)-N-(tert-Butoxycarbonyl)-2,3-dichlorophenylalanine**

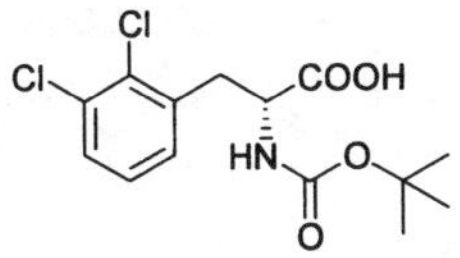

$C_{14}H_{17}Cl_2NO_4$
Chiral intermediate. *ChiroTech Technology Ltd.*

**1019 (S)-N-(tert-Butoxycarbonyl)-2,3-dichlorophenylalanine**

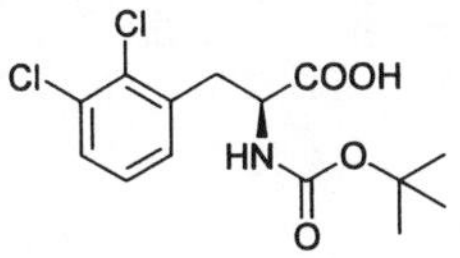

$C_{14}H_{17}Cl_2NO_4$
Chiral intermediate. *ChiroTech Technology Ltd.*

**1020 (R)-N-(tert-Butoxycarbonyl)-2,4-dichlorophenylalanine**

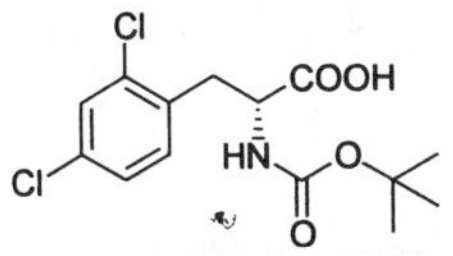

$C_{14}H_{17}Cl_2NO_4$
Chiral intermediate. *ChiroTech Technology Ltd.*

**1021 (S)-N-(tert-Butoxycarbonyl)-2,4-dichlorophenylalanine**

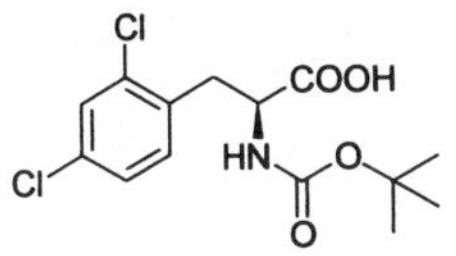

$C_{14}H_{17}Cl_2NO_4$
Chiral intermediate. *ChiroTech Technology Ltd.*

**1022 (R)-N-(tert-Butoxycarbonyl)-2,6-dichlorophenylalanine**

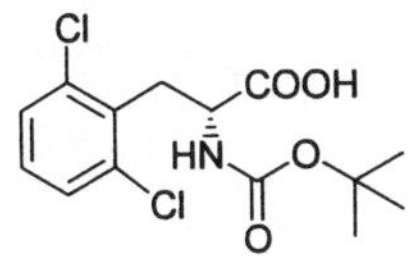

$C_{14}H_{17}Cl_2NO_4$
Chiral intermediate. *ChiroTech Technology Ltd.*

**1023 (S)-N-(tert-Butoxycarbonyl)-2,6-dichlorophenylalanine**

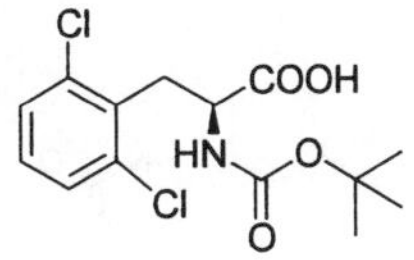

$C_{14}H_{17}Cl_2NO_4$
Chiral intermediate. *ChiroTech Technology Ltd.*

**1024 (R)-N-(tert-Butoxycarbonyl)-3,4-dichlorophenylalanine**
114873-13-1

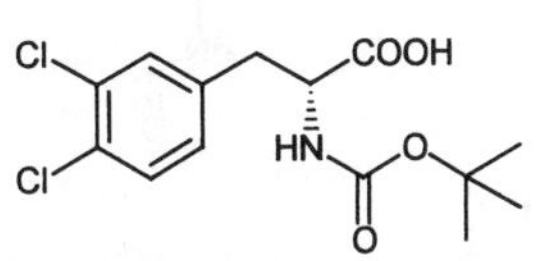

$C_{14}H_{17}Cl_2NO_4$
Chiral intermediate. *ChiroTech Technology Ltd.; Synthetech, Inc.*

**1025 (S)-N-(tert-Butoxycarbonyl)-3,4-dichlorophenylalanine**
80741-39-5

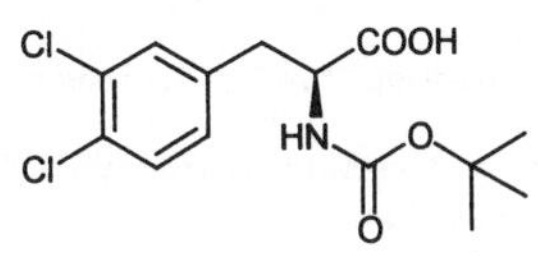

$C_{14}H_{17}Cl_2NO_4$
Chiral intermediate. *ChiroTech Technology Ltd.; Synthetech, Inc.*

**1026 (tert-Butoxycarbonyl)-O-2,6-dichlorobenzyl-L-tyrosine**

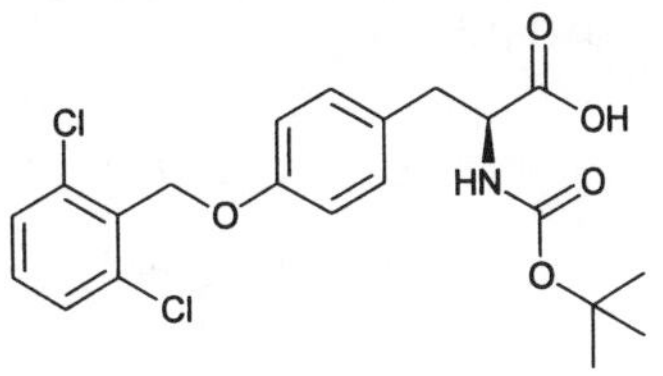

$C_{21}H_{25}Cl_2NO_5$
Chiral intermediate. *Synthetech, Inc.*

**1027 (R)-N-(tert-Butoxycarbonyl)-2,4-difluorophenylalanine**

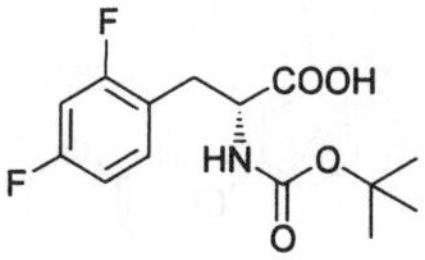

$C_{14}H_{17}F_2NO_4$
Chiral intermediate. *ChiroTech Technology Ltd.*

**1028 (S)-N-(tert-Butoxycarbonyl)-2,4-difluorophenylalanine**

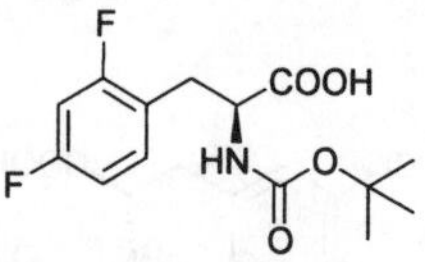

$C_{14}H_{17}F_2NO_4$
Chiral intermediate. *ChiroTech Technology Ltd.*

**1029 (R)-N-(tert-Butoxycarbonyl)-2,5-difluorophenylalanine**

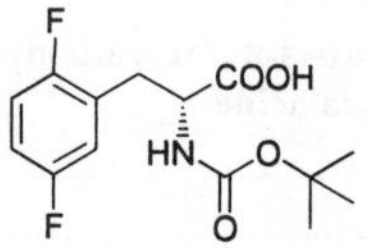

$C_{14}H_{17}F_2NO_4$
Chiral intermediate. *ChiroTech Technology Ltd.*

**1030 (S)-N-(tert-Butoxycarbonyl)-2,5-difluorophenylalanine**

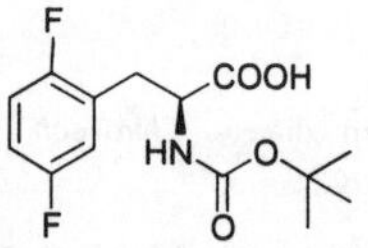

$C_{14}H_{17}F_2NO_4$
Chiral intermediate. *ChiroTech Technology Ltd.*

**1031 (tert-Butoxycarbonyl)-D-3,4-difluorophenylalanine**
205445-51-8

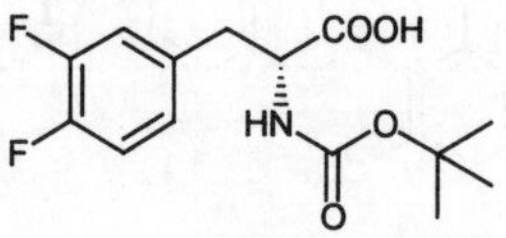

$C_{14}H_{17}F_2NO_4$
Chiral intermediate. *Synthetech, Inc.*

**1032 (tert-Butoxycarbonyl)-L-3,4-difluorophenylalanine**
198474-90-7

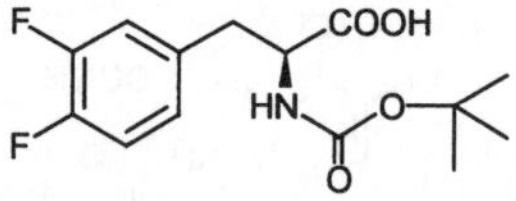

$C_{14}H_{17}F_2NO_4$
Chiral intermediate. *Synthetech, Inc.*

**1033 (tert-Butoxycarbonyl)-D-3,5-difluorophenylalanine**
205445-53-0

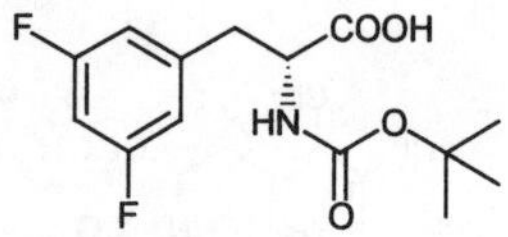

$C_{14}H_{17}F_2NO_4$
Chiral intermediate. *Synthetech, Inc.*

**1034 (tert-Butoxycarbonyl)-L-3,5-difluorophenylalanine**
205445-52-9

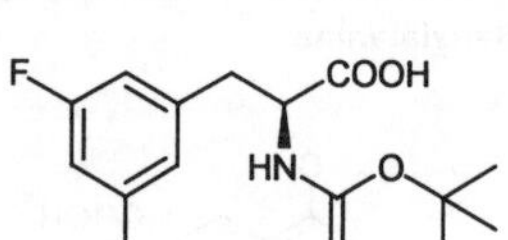

$C_{14}H_{17}F_2NO_4$
Chiral intermediate. *Synthetech, Inc.*

**1035 (R)-N-(tert-Butoxycarbonyl)-2,5-dimethoxyphenylalanine**

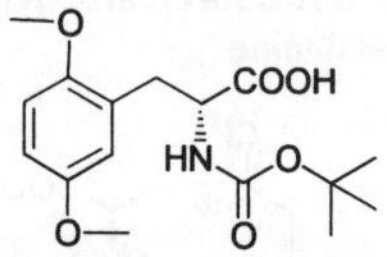

$C_{16}H_{23}NO_6$
Chiral intermediate. *ChiroTech Technology Ltd.*

**1036 (S)-N-(tert-Butoxycarbonyl)-2,5-dimethoxyphenylalanine**

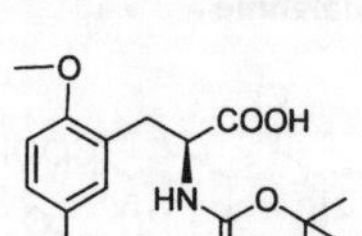

$C_{16}H_{23}NO_6$
Chiral intermediate. *ChiroTech Technology Ltd.*

**1037 (R)-N-(tert-Butoxycarbonyl)-3,4-dimethoxyphenylalanine**

218457-71-7

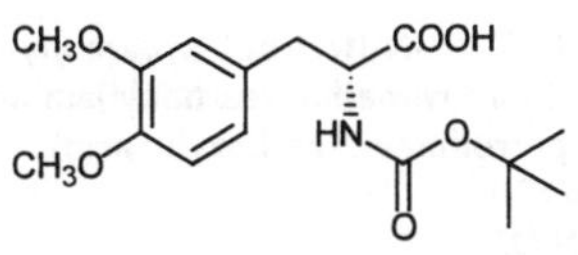

$C_{16}H_{23}NO_6$

Chiral intermediate. *ChiroTech Technology Ltd.; Synthetech, Inc.*

**1038 (S)-N-(tert-Butoxycarbonyl)-3,4-dimethoxyphenylalanine**

127095-97-0

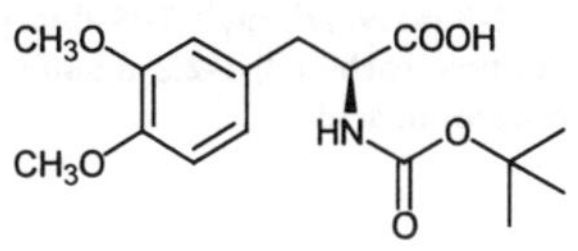

$C_{16}H_{23}NO_6$

Chiral intermediate. *ChiroTech Technology Ltd.; Synthetech, Inc.*

**1039 (R)-N-(tert-Butoxycarbonyl)-2,5-dimethylphenylalanine**

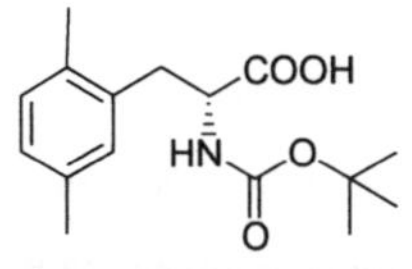

$C_{16}H_{23}NO_4$

Chiral intermediate. *ChiroTech Technology Ltd.*

**1040 (S)-N-(tert-Butoxycarbonyl)-2,5-dimethylphenylalanine**

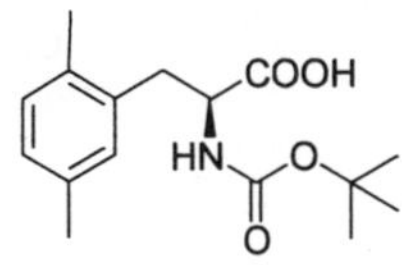

$C_{16}H_{23}NO_4$

Chiral intermediate. *ChiroTech Technology Ltd.*

**1041 (tert-Butoxycarbonyl)-D-3,3-diphenylalanine**

117027-46-0

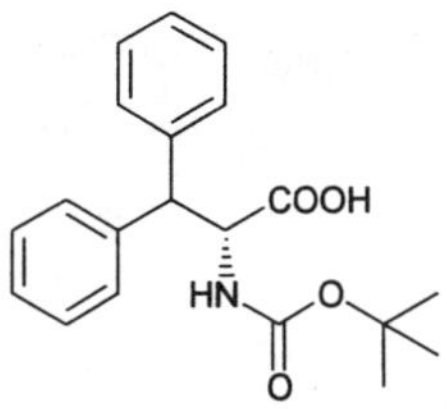

$C_{20}H_{23}NO_4$

Chiral intermediate. *Synthetech, Inc.*

**1042 (tert-Butoxycarbonyl)-L-3,3-diphenylalanine**

138662-63-2

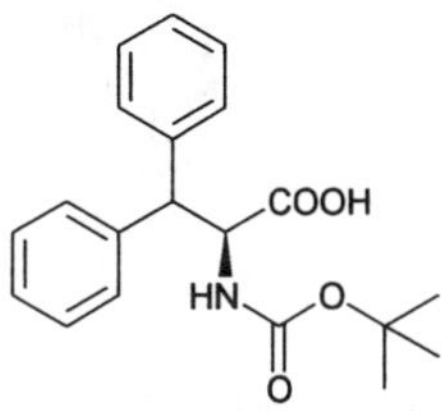

$C_{20}H_{23}NO_4$

Chiral intermediate. *Synthetech, Inc.*

**1043 (2R,4R)-N-Butoxycarbonyl-4-diphenylphosphino-2-diphenylphosphinomethylpyrrolidine**

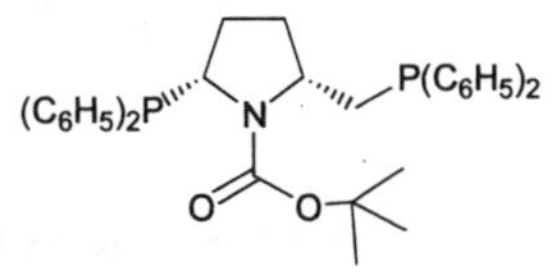

$C_{33}H_{37}NO_2P_2$

Chiral ligand. *Digital Specialty Chemicals, Inc.*

**1044 (2S,4S)-N-Butoxycarbonyl-4-diphenylphosphino-2-diphenylphosphinomethylpyrrolidine**

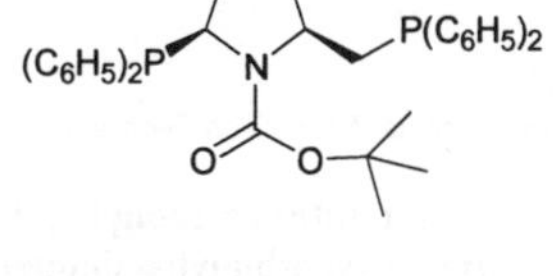

$C_{33}H_{37}NO_2P_2$

Chiral ligand. *Digital Specialty Chemicals, Inc.*

**1045 (tert-Butoxycarbonyl)-(S)-ethyl-L-cysteine**

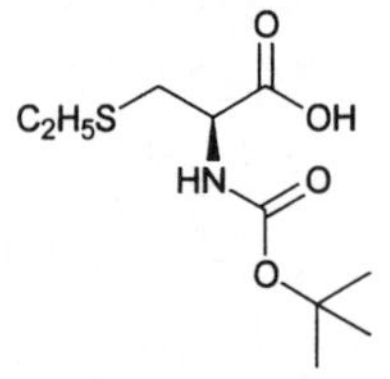

$C_{10}H_{19}NO_4S$
Chiral intermediate. *Austin Chemical Company, Inc.; Synthetech, Inc.*

**1046 (tert-Butoxycarbonyl)-N-ethylglycine**

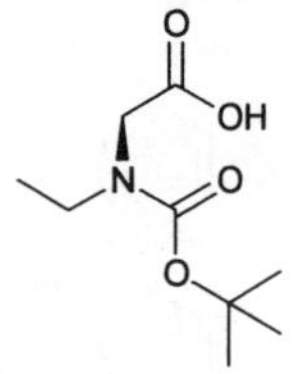

$C_9H_{17}NO_4S$
Chiral intermediate. *Synthetech, Inc.*

**1047 (R)-N-(tert-Butoxycarbonyl)-4-ethylphenylalanine**

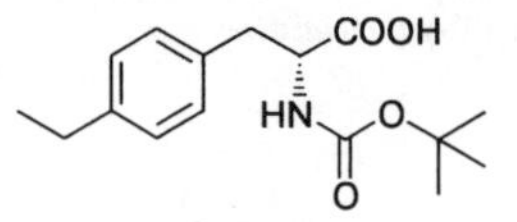

$C_{16}H_{23}NO_4$
Chiral intermediate. *ChiroTech Technology Ltd.*

**1048 (S)-N-(tert-Butoxycarbonyl)-4-ethylphenylalanine**

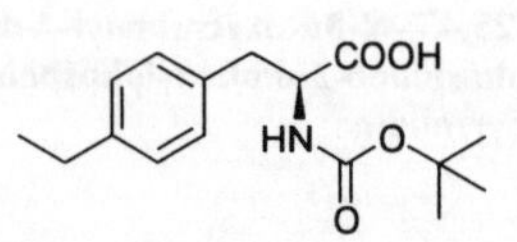

$C_{16}H_{23}NO_4$
Chiral intermediate. *ChiroTech Technology Ltd.*

**1049 (R)-1-(tert-Butoxycarbonyl)-3-N-(9-fluorenylmethoxycarbonyl)aminopyrrolidine**

$C_{24}H_{28}N_2O_4$
(R)-N-BOC-3-N-FMOC-aminopyrrolidine. Chiral intermediate. *Sigma-Aldrich Fine Chemicals.*

**1050 (S)-1-(tert-Butoxycarbonyl)-3-N-(9-fluorenylmethoxycarbonyl)aminopyrrolidine**

$C_{24}H_{28}N_2O_4$
(S)-N-BOC-3-N-FMOC-aminopyrrolidine. Chiral intermediate. *Sigma-Aldrich Fine Chemicals.*

**1051 ((2S,4R)-N-(tert-Butoxycarbonyl)-4-N-(9-fluorenylmethoxycarbonyl)amino-pyrrolidine-2-carboxylic acid)**

176486-63-8
$C_{25}H_{28}N_2O_6$
N-BOC-trans-4-N-FMOC-amino-L-proline. Chiral intermediate. *Sigma-Aldrich Fine Chemicals.*

**1052 ((2S,4S)-N-tert-Butoxycarbonyl-4-N-(9-fluorenylmethoxycarbonyl)amino-pyrrolidine-2-carboxylic acid)**

174148-03-9
$C_{25}H_{28}N_2O_6$
N-BOC-cis-4-N-FMOC-amino-L-proline. Chiral intermediate. *Sigma-Aldrich Fine Chemicals.*

**1053 (tert-Butoxycarbonyl)-3-(9-fluorenylmethoxycarbonyl)-L-2,3-diamino-propionic acid**

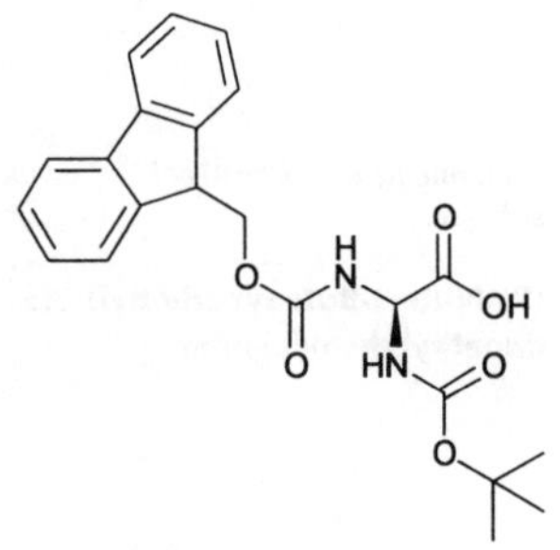

$C_{22}H_{24}N_2O_6$
Chiral intermediate. *Senn Chemicals AG.*

**1054 (tert-Butoxycarbonyl)-(9-fluorenylmethoxycarbonyl)-L-lysine**

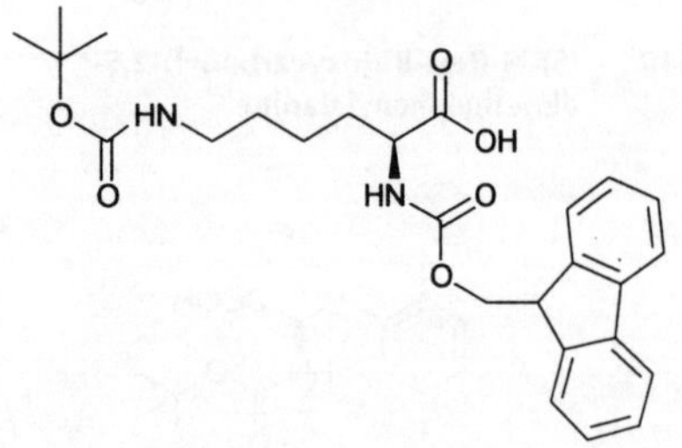

$C_{26}H_{32}N_2O_6$
Chiral intermediate. *Interchem Corporation; Synthetech, Inc.*

**1055 (R)-N-(tert-Butoxycarbonyl)-(2-fluorophenyl)alanine**
114873-10-8
$C_{14}H_{18}FNO_4$
N-tert-Butoxycarbonyl-2-fluorophenyl-D-alanine; Boc-D-2-fluorophenylalanine. Chiral intermediate. *ChiroTech Technology Ltd.; Lancaster Synthesis Ltd.; Synthetech, Inc.*

**1056 (S)-N-(tert-Butoxycarbonyl)-(2-fluorophenyl)alanine**
114873-00-6
$C_{14}H_{18}FNO_4$
N-tert-Butoxycarbonyl-2-fluorophenyl-L-alanine; Boc-L-2-fluorophenylalanine. Chiral intermediate. *ChiroTech Technology Ltd.; Lancaster Synthesis Ltd.; Synthetech, Inc.*

**1057 (R)-N-(tert-Butoxycarbonyl)-(3-fluorophenyl)alanine**
114873-11-9
$C_{14}H_{18}FNO_4$
N-tert-Butoxycarbonyl-3-fluorophenyl-D-alanine; Boc-D-3-fluorophenylalanine. Chiral intermediate. *ChiroTech Technology Ltd.; Lancaster Synthesis Ltd.; Synthetech, Inc.*

**1058 (S)-N-(tert-Butoxycarbonyl)-(3-fluorophenyl)alanine**
114873-01-7
$C_{14}H_{18}FNO_4$
N-tert-Butoxycarbonyl-3-fluorophenyl-L-alanine; Boc-L-3-fluorophenylalanine. Chiral intermediate. *ChiroTech Technology Ltd.; Lancaster Synthesis Ltd.; Synthetech, Inc.*

**1059 (R)-N-(tert-Butoxycarbonyl)-(4-fluorophenyl)alanine**
57292-45-2
$C_{14}H_{18}FNO_4$
N-tert-Butoxycarbonyl-4-fluorophenyl-D-alanine; Boc-D-4-fluorophenylalanine. Chiral intermediate. *ChiroTech Technology Ltd.; Lancaster Synthesis Ltd.; Synthetech, Inc.*

**1060 (S)-N-(tert-Butoxycarbonyl)-(4-fluorophenyl)alanine**
41153-30-4
$C_{14}H_{18}FNO_4$
N-tert-Butoxycarbonyl-4-fluorophenyl-L-alanine; Boc-L-4-fluorophenylalanine. Chiral intermediate. *ChiroTech Technology Ltd.; Lancaster Synthesis Ltd.; Senn Chemicals AG; Synthetech, Inc.*

**1061 (R)-N-(tert-Butoxycarbonyl)-(4-fluorophenyl)glycine**
196707-32-1
$C_{13}H_{16}FNO_4$
D-N-tert-Butoxycarbonyl-α-(4-fluorophenyl)acetic acid. Chiral intermediate. *ChiroTech Technology Ltd.; Lancaster Synthesis Ltd.*

**1062 (S)-N-(tert-Butoxycarbonyl)-(4-fluorophenyl)glycine**
142186-36-5
$C_{13}H_{16}FNO_4$
L-N-tert-Butoxycarbonyl-α-(4-fluorophenyl)acetic acid. Chiral intermediate. *ChiroTech Technology Ltd.; Lancaster Synthesis Ltd.*

**1063 (S)-(-)-3-tert-Butoxycarbonyl-4-formyl-2,2-dimethyl-1,3-oxazolidine**
102308-32-7

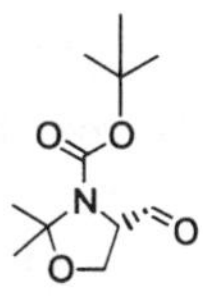

$C_{11}H_{19}NO_4$
(S)-(-)-3-BOC-4-formyl-2,2-dimethyl-1,3-oxazolidine; Garner's aldehyde. Chiral intermediate. $bp_1$ = 83-88°; d = 1.06; n = 1.445. *Sigma-Aldrich Fine Chemicals; TCI America.*

**1064 (tert-Butoxycarbonyl)-D-formyltryptophan**
64905-10-8

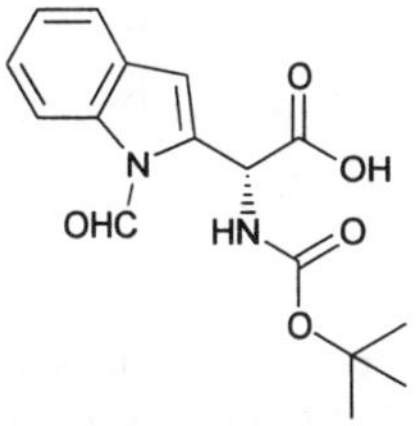

$C_{16}H_{18}N_2O_5$
Chiral intermediate. *Isochem.*

**1065 (R)-N-(tert-Butoxycarbonyl)-(2-furyl)-alanine**
$C_{12}H_{17}NO_5$
N-tert-Butoxycarbonyl-2-furyl-D-alanine. Chiral intermediate. *ChiroTech Technology Ltd.; Lancaster Synthesis Ltd.*

**1066 (S)-N-(tert-Butoxycarbonyl)-(2-furyl)-alanine**
145206-40-2
$C_{12}H_{17}NO_5$
N-tert-Butoxycarbonyl-2-furyl-L-alanine. Chiral intermediate. *ChiroTech Technology Ltd.; Lancaster Synthesis Ltd.*

**1067 (R)-N-(tert-Butoxycarbonyl)-3-furylalanine**

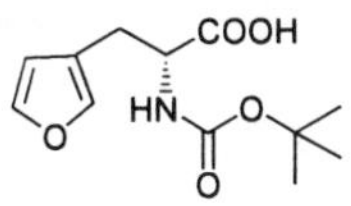

$C_{12}H_{17}NO_5$
Chiral intermediate. *ChiroTech Technology Ltd.*

**1068 (S)-N-(tert-Butoxycarbonyl)-3-furylalanine**

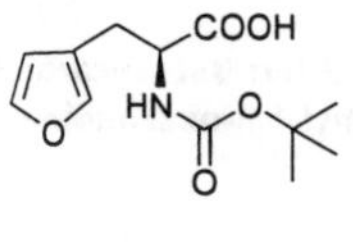

$C_{12}H_{17}NO_5$
Chiral intermediate. *ChiroTech Technology Ltd.*

**1069 (+)-N-(tert-Butoxycarbonyl)glucosamine**
75251-80-8

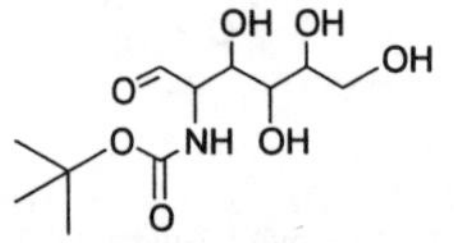

$C_{11}H_{21}NO_7$
Chiral intermediate. mp = 194°; $[\alpha]^{25}$ = + 64° (c = 1, $CH_3OH$). *Sigma-Aldrich Fine Chemicals.*

**1070 (tert-Butoxycarbonyl)-L-glutamic acid**
2419-94-5

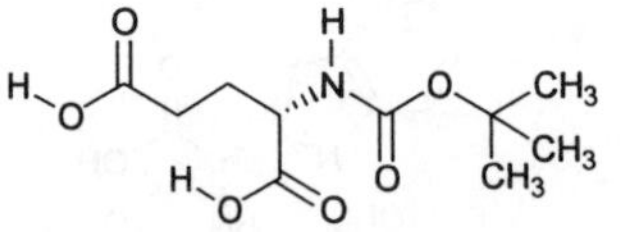

$C_{10}H_{17}NO_6$
N-(tert-BOC)-L-glutamic acid. Intermediate. $[\alpha]$ = - 14.9° (c = 1, $CH_3OH$). *Acros Organics nv; Synthetech, Inc.*

**1071 (tert-Butoxycarbonyl)-L-glutamic acid 2-benzyl ester**

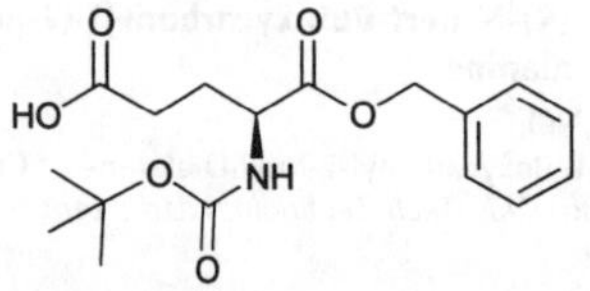

$C_{17}H_{23}NO_6$
Chiral intermediate. *Synthetech, Inc.*

**1072 (tert-Butoxycarbonyl)-L-glutamic acid 5-benzyl ester**
13574-13-5 237-007-5

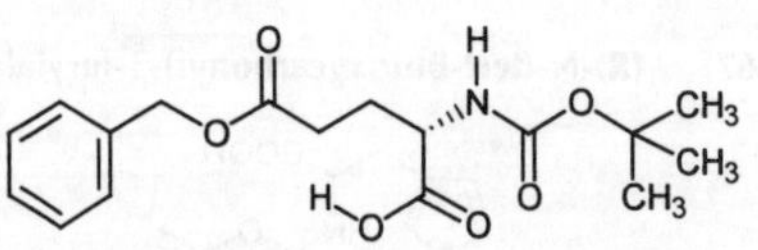

$C_{17}H_{23}NO_6$
Glutamic acid, N-[(1,1-dimethylethoxy)carbonyl]-, 5-(phenylmethyl) ester, L-; N-[(1,1-Dimethylethoxy)-carbonyl]-L-glutamic acid, 5-(phenylmethyl) ester; boc-glu(obzl); N-(tert-Butoxycarbonyl)-L-glutamic acid 5-benzylester; 5-benzyl N-[(1,1-dimethylethoxy)carbonyl]-L-glutamate. Chiral intermediate. mp = 66-68°; $[\alpha]$ = - 16.2° (c = 2, DMF). *Acros Organics nv; Synthetech, Inc.*

**1073 (tert-Butoxycarbonyl)-L-glutamic acid 5-benzylester dicyclohexylammonium salt**
13574-84-0 237-008-0

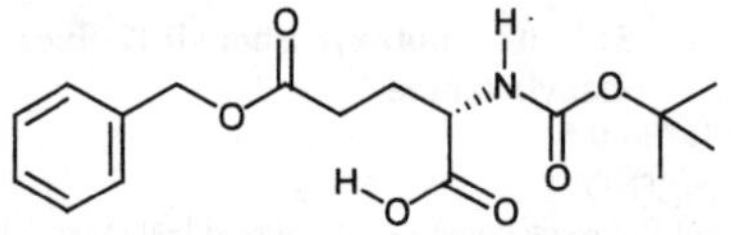

$C_{17}H_{23}NO_6$
Glutamic acid, N-carboxy-, 5-benzyl N-tert-butyl ester, compd. with dicyclohexylamine; N-(tert-Butoxycarbonyl)-L-glutamic acid 5-benzylester dicyclohexylamine salt; 5-Benzyl N-(tert-butoxycarbonyl)-L-2-aminoglutarate, compound with N-dicyclohexylamine (1:1); L-Glutamic acid, N-[(1,1-dimethylethoxy)carbonyl]-, 5-(phenylmethyl) ester compd. Chiral intermediate. mp = 141-143°; $[\alpha]$ = - 19.1° (c = 2, $CH_3OH$). *Acros Organics nv.*

**1074 N-α-(tert-Butoxycarbonyl)-D-glutamine**
61348-28-5

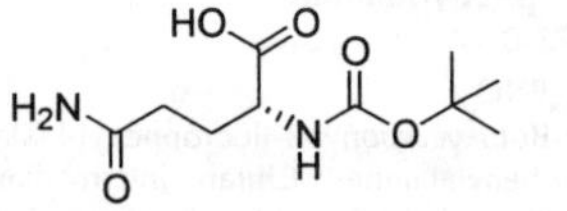

$C_{10}H_{18}N_2O_5$
Chiral intermediate. *Acros Organics nv.*

**1075 N-α-(tert-Butoxycarbonyl)-L-glutamine**
13726-85-7 237-296-8

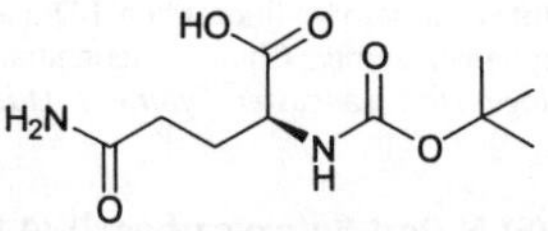

$C_{10}H_{18}N_2O_5$
$N^2$-[(1,1-dimethylethoxy)carbonyl]-, L-; $N^2$-[(1,1-Dimethylethoxy)carbonyl]-L-glutamine; N-(tert-Butoxycarbonyl)-L-glutamine; Glutamine, $N^2$-carboxy-, $N^2$-tert-butyl ester, L-. Listed on TSCA. Chiral intermediate. mp = 113-115°; $[\alpha]$ = - 3.5° (c = 2, EtOH). *Acros Organics nv; Sigma-Aldrich Fine Chemicals; Synthetech, Inc.*

**1076 (tert-Butoxycarbonyl)-L-glutamine trityl ester**

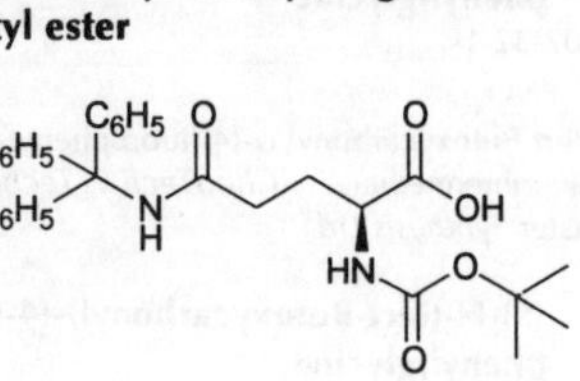

$C_{29}H_{32}N_2O_5$
Chiral intermediate. *Synthetech, Inc.*

**1077 (tert-Butoxycarbonyl)-L-glutamine xanthyl ester**

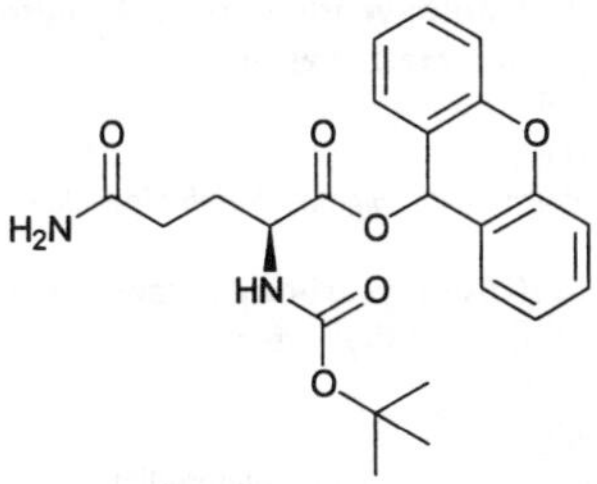

$C_{23}H_{26}N_2O_6$
Chiral intermediate. *Synthetech, Inc.*

**1078 N-α-(tert-Butoxycarbonyl)-D-histidine**
50654-94-9

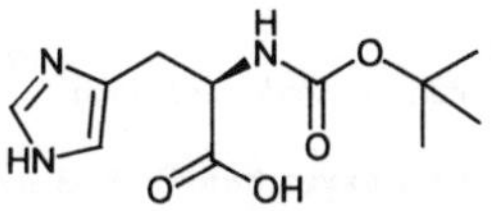

$C_{11}H_{17}N_3O_4$
Chiral intermediate. *Acros Organics nv.*

**1079 N-α-(tert-Butoxycarbonyl)-L-histidine**
17791-52-5 241-768-9

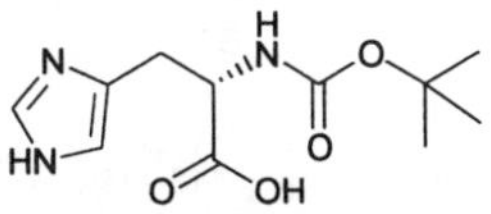

$C_{11}H_{17}N_3O_4$
Histidine, N-carboxy-, N-tert-butyl ester, L-; N-(tert-Butoxycarbonyl)-L-histidine; L-Histidine, N-[(1,1-dimethylethoxy)carbonyl]-. Chiral intermediate. mp = 196-198°; [α] = + 25.4° (c = 1, EtOH). *Acros Organics nv; Sigma-Aldrich Fine Chemicals; Synthetech, Inc.*

**1080 (tert-Butoxycarbonyl)-L-histidine benzyloxymethyl ester**
Chiral intermediate. *Synthetech, Inc.*

**1081 (tert-Butoxycarbonyl)-L-histidine hydrazide**

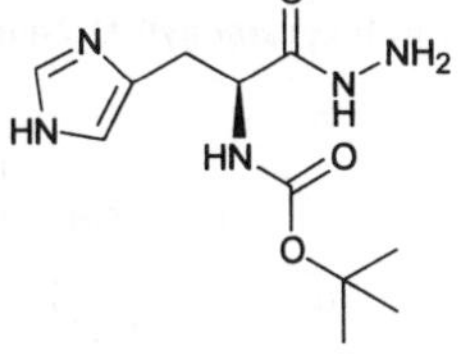

$C_{11}H_{19}N_5O_3$
Chiral intermediate. *Senn Chemicals AG.*

**1082 (tert-Butoxycarbonyl)-L-histidine tosyl ester dicyclohexylammonium salt**
Chiral intermediate. *Synthetech, Inc.*

**1083 (tert-Butoxycarbonyl)-D-homophenylalanine**
82732-07-8

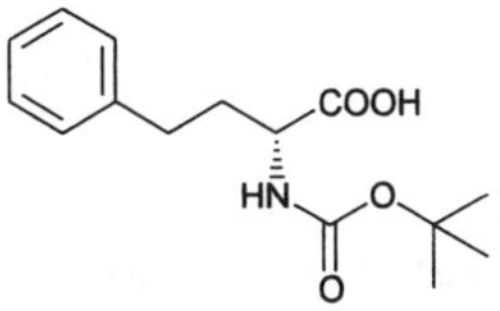

$C_{15}H_{21}NO_4$
Chiral intermediate. *Synthetech, Inc.*

**1084 (tert-Butoxycarbonyl)-L-homophenylalanine**
100564-78-1

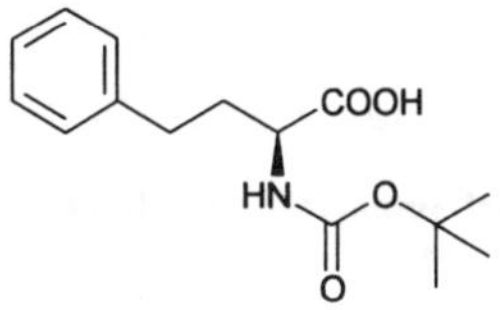

$C_{15}H_{21}NO_4$
Chiral intermediate. *Senn Chemicals AG; Synthetech, Inc.*

**1085 N-β-(tert-Butoxycarbonyl)-D-hydrazinoproline**

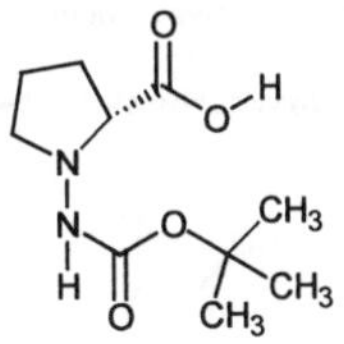

$C_{10}H_{18}N_2O_4$
Chiral intermediate. *Acros Organics nv.*

**1086 (S)-(-)-N-β-(tert-Butoxycarbonyl)-L-hydrazinoproline**

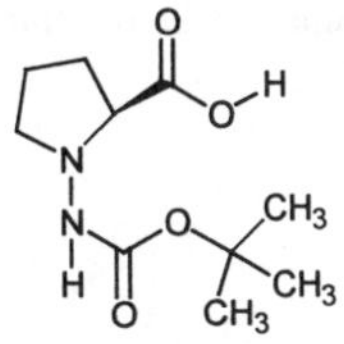

$C_{10}H_{18}N_2O_4$
Chiral intermediate. *Acros Organics nv.*

**1087 (tert-Butoxycarbonyl)-L-6-hydroxy-norleucine**

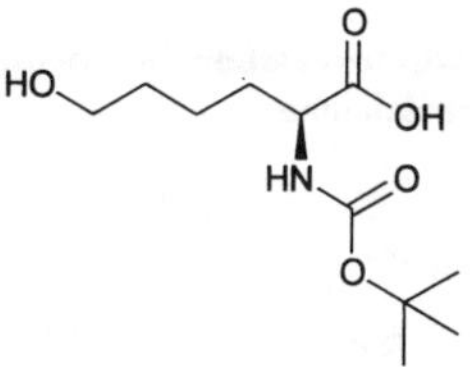

$C_{11}H_{21}NO_5$
Chiral intermediate. *Senn Chemicals AG.*

**1088 (tert-Butoxycarbonyl)-cis-4-hydroxy-D-proline**

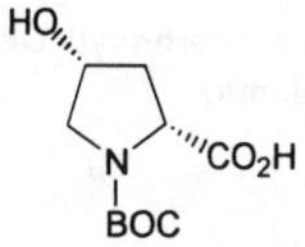

$C_{10}H_{17}NO_5$
Chiral intermediate. *Omega Chemical Company Inc.*

**1089 (tert-Butoxycarbonyl)-trans-4-hydroxy-D-proline**

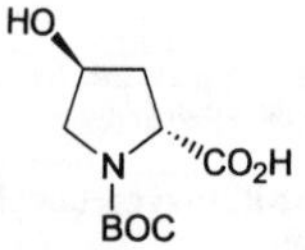

$C_{10}H_{17}NO_5$
Chiral intermediate. *Omega Chemical Company Inc.*

**1090 (tert-Butoxycarbonyl)-cis-4-hydroxy-L-proline**

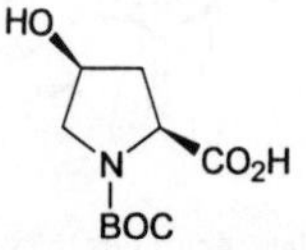

$C_{10}H_{17}NO_5$
Chiral intermediate. *Omega Chemical Company Inc.; Sigma-Aldrich Fine Chemicals.*

**1091 (tert-Butoxycarbonyl)-trans-4-hydroxy-L-proline**
13726-69-7

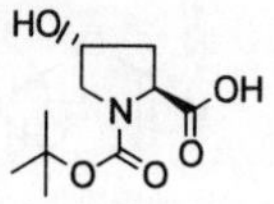

$C_{10}H_{17}NO_5$
Chiral intermediate. mp = 122°; $[\alpha]_D^{20}$ = - 78° (1, $H_2O$). *Omega Chemical Company Inc.; Sigma-Aldrich Fine Chemicals; Synthetech, Inc.*

**1092 (tert-Butoxycarbonyl)-cis-4-hydroxy-L-proline methyl ester**
114676-69-6
$C_{11}H_{19}NO_5$
Chiral intermediate. *Sigma-Aldrich Fine Chemicals.*

**1093 (tert-Butoxycarbonyl)-trans-4-hydroxy-L-proline methyl ester**
74844-91-0
$C_{11}H_{19}NO_5$
BOC Hyproester. Chiral intermediate. mp = 98-99°; $[\alpha]^{22}$ = - 104° (c = 0.9, $CH_2Cl_2$). *Austin Chemical Company, Inc.; Sigma-Aldrich Fine Chemicals; Synthetech, Inc.*

**1094 (S)-(-)-(tert-Butoxycarbonyl)-hydroxy-prolinol**
61748-26-0
$C_{10}H_{19}NO_4$
Chiral intermediate. *Medinger & Sohn.*

**1095 (tert-Butoxycarbonyl)-(R)-3-hydroxy-pyrrolidine**
109431-87-0

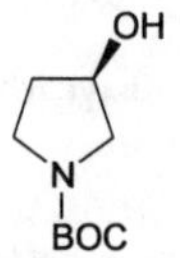

$C_9H_{17}NO_3$
Chiral intermediate. mp = 113-115°; $[\alpha]_D^{20}$ = - 7° (c = 1, $CH_3OH$). *Omega Chemical Company Inc.; Sigma-Aldrich Fine Chemicals.*

**1096 (tert-Butoxycarbonyl)-(S)-3-hydroxy-pyrrolidine**
101469-92-5

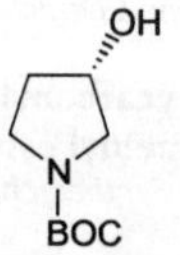

$C_9H_{17}NO_3$
Chiral intermediate. *Omega Chemical Company Inc.*

**1097 (tert-Butoxycarbonyl)-D-2-indanylglycine**
181227-48-5

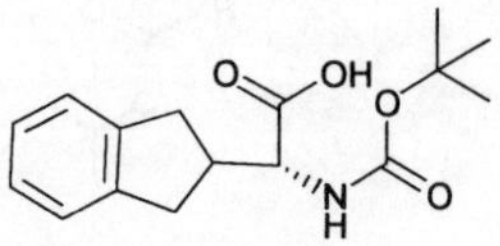

$C_{16}H_{21}NO_4$
Chiral intermediate. *Synthetech, Inc.*

**1098 (tert-Butoxycarbonyl)-L-2-indanylglycine**
181227-47-4

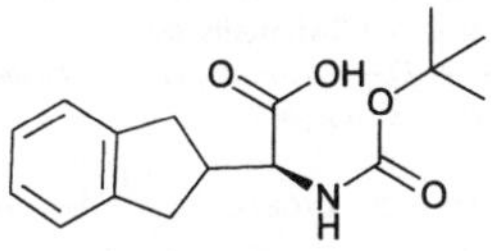

$C_{16}H_{21}NO_4$
Chiral intermediate. *Synthetech, Inc.*

**1099 (-)-N-(tert-Butoxycarbonyl)-3-iodoalanine methyl ester**
93267-04-0

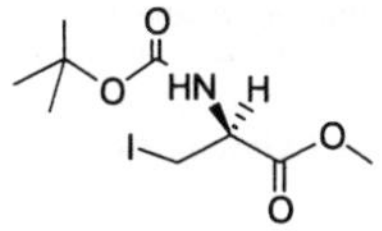

$C_9H_{16}INO_4$
Chiral intermediate. mp = 50-52°; $[\alpha]^{22}$ = - 4° (c = 2, $CH_3OH$). *Sigma-Aldrich Fine Chemicals.*

**1100 (+)-N-(tert-Butoxycarbonyl)-3-iodoalanine benzyl ester**
12594-29-2

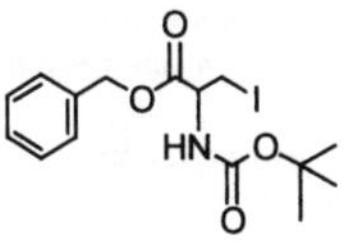

$C_{15}H_{20}INO_4$
Chiral intermediate. mp = 79-81°; $[\alpha]_D^{20}$ = + 18° (c = 1, EtOH). *Sigma-Aldrich Fine Chemicals.*

**1101 (tert-Butoxycarbonyl)-D-4-iodo-phenylalanine**
176199-35-2

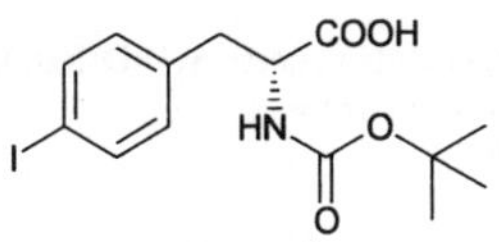

$C_{14}H_{18}INO_4$
Chiral intermediate. *Advanced Asymmetrics, Inc.; Synthetech, Inc.*

**1102 (tert-Butoxycarbonyl)-L-4-iodo-phenylalanine**
62129-44-6

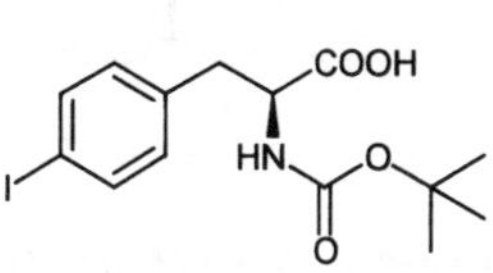

$C_{14}H_{18}INO_4$
Chiral intermediate. *Advanced Asymmetrics, Inc.; Synthetech, Inc.*

**1103 (tert-Butoxycarbonyl)-p-Iodo-D-phenyl-alanine methyl ester**

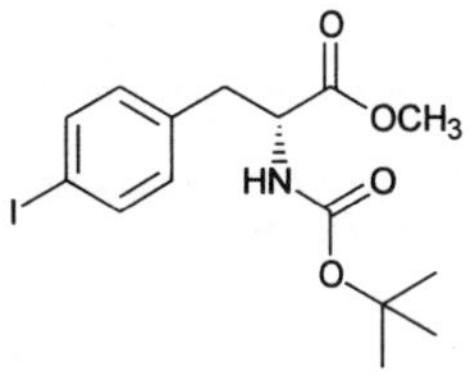

$C_{15}H_{20}INO_4$
Chiral intermediate. *Advanced Asymmetrics, Inc.*

**1104 (tert-Butoxycarbonyl)-p-Iodo-L-phenyl-alanine methyl ester**

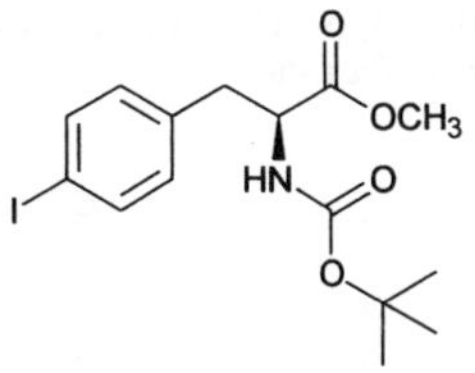

$C_{15}H_{20}INO_4$
Chiral intermediate. *Advanced Asymmetrics, Inc.*

**1105 (tert-Butoxycarbonyl)-L-isoleucine**
13139-16-7 236-074-8

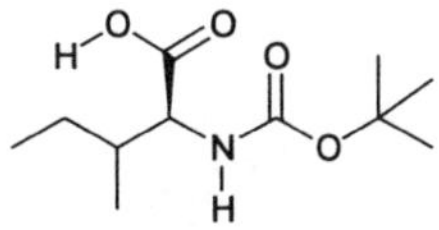

$C_{11}H_{21}NO_4$
Isoleucine, N-[(1,1-dimethylethoxy)carbonyl]-, L-; Boc-L-isoleucine hemihydrate; Isoleucine, N-carboxy-, N-tert-butyl ester, L-. Listed on TSCA. Chiral intermediate. mp = 58-61°; $[\alpha]$ = + 2.8° (c = 2, HOAc). *Acros Organics nv; PPG - Sipsy; Sigma-Aldrich Fine Chemicals; Synthetech, Inc.*

**1106 (tert-Butoxycarbonyl)-D-isoleucinol**

NHBOC OH

$C_{11}H_{23}NO_3$
Chiral intermediate. *Omega Chemical Company Inc.*

**1107 (tert-Butoxycarbonyl)-L-isoleucinol**
106946-74-1

NHBOC OH

$C_{11}H_{23}NO_3$
Chiral intermediate. *Omega Chemical Company Inc.*

**1108 (R)-N-(tert-Butoxycarbonyl)-(4-isopropylphenyl)alanine**

COOH HN O O

$C_{11}H_{25}NO_4$
Chiral intermediate. *ChiroTech Technology Ltd.*

**1109 (S)-N-(tert-Butoxycarbonyl)-(4-isopropylphenyl)alanine**

COOH HN O O

$C_{11}H_{25}NO_4$
Chiral intermediate. *ChiroTech Technology Ltd.*

**1110 (-)-N-(tert-Butoxycarbonyl)-D-leucine N-methoxy-N-methylamide**
87694-50-6

O HN H O N O

$C_{13}H_{26}N_2O_4$
Chiral intermediate. bp = 235°; d = 1.46; n = 1.454; $[\alpha]^{23}$ = - 25.5° (c = 2, $CH_3OH$). *Sigma-Aldrich Fine Chemicals.*

**1111 (tert-Butoxycarbonyl)-L-tert-leucine**
62965-35-9

$C_{11}H_{21}NO_4$
N-tert-Butoxycarbonyl-(S)-2-amino-3,3-butyric acid; (S)-N-Boc-2-amino-3,3-dimethylbutyric acid. Chiral intermediate. *Degussa-Huls AG; Lancaster Synthesis Ltd.; Senn Chemicals AG.*

**1112 (tert-Butoxycarbonyl)-L-leucine hydrate**
13139-15-6 236-073-2

O OH O N H O

$C_{11}H_{21}NO_4$
Leucine, N-[(1,1-dimethylethoxy)carbonyl]-, L-; N-[(1,1-Dimethylethoxy)carbonyl]-L-leucine; Boc-leu; BOC-L-leucine; N-tert-Butoxycarbonyl-L-leucine; Leucine, N-carboxy-, N-tert-butyl ester, L-. Chiral intermediate. mp = 82-85°; [α] = - 24.5 (c = 2, HOAc). *Acros Organics nv; Austin Chemical Company, Inc.; PPG - Sipsy; Synthetech, Inc.*

**1113 (tert-Butoxycarbonyl)-L-leucine hydroxysuccinimide ester**
3392-09-4 222-232-3

O O HN H O O N O O

$C_{15}H_{24}N_2O_6$
Chiral intermediate. mp = 110-113°; $[\alpha]^{25}$ = - 43° (c = 2, dioxane). *Sigma-Aldrich Fine Chemicals; Synthetech, Inc.*

**1114 N-(tert-Butoxycarbonyl)-D-leucine methyl ester**
63096-02-6

O O HN H O O

$C_{12}H_{23}NO_4$
Chiral intermediate. bp = 205°; d = 0.991; n = 1.44; $[\alpha]^{22}$ = - 18° (neat). *Sigma-Aldrich Fine Chemicals.*

**1115 (tert-Butoxycarbonyl)-D-leucine monohydrate**
16937-99-8

O O OH N H O

$C_{11}H_{21}NO_4$
Chiral intermediate. mp = 85-87°; $[\alpha]_D^{20}$ = - 24° (c = 1,

HOAc). *Acros Organics nv; Sigma-Aldrich Fine Chemicals; Synthetech, Inc.*

**1116 (tert-Butoxycarbonyl)-D-leucinol**
106930-51-2

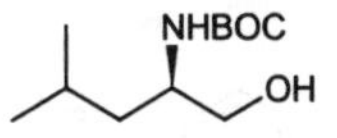

$C_{11}H_{23}NO_3$
Chiral intermediate. *Omega Chemical Company Inc.*

**1117 (tert-Butoxycarbonyl)-L-leucinol**
82010-31-9

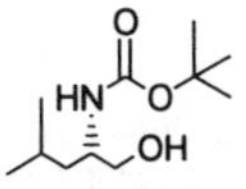

$C_{11}H_{23}NO_3$
Chiral intermediate. bp = 213°; d = 0.983; n = 1.449; $[\alpha]^{23}$ = - 27° (c = 2, $CH_3OH$). *Omega Chemical Company Inc.; Senn Chemicals AG; Sigma-Aldrich Fine Chemicals.*

**1118 (tert-Butoxycarbonyl)-D-lysine**

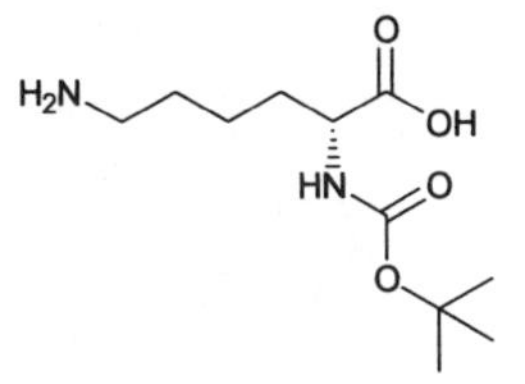

$C_{11}H_{22}N_2O_4$
Chiral intermediate. *Synthetech, Inc.*

**1119 N-α-(tert-Butoxycarbonyl)-L-lysine**
13734-28-6 237-303-4

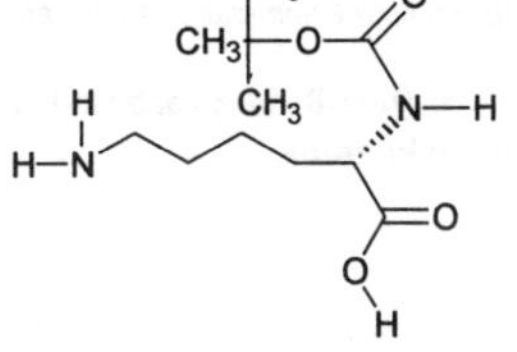

$C_{11}H_{22}N_2O_4$
$N^2$-[(1,1-dimethylethoxy)carbonyl]-, L-; $N^2$-[(1,1-Dimethylethoxy)carbonyl]-L-lysine; Lysine, $N^2$-carboxy-, $N^2$-tert-butyl ester, L-; α-tert-Butoxycarbonyl-L-lysine. Chiral intermediate. mp = 207°; $[\alpha]_D^{20}$ = + 22° (c = 2, $CH_3OH$). *Acros Organics nv; Sigma-Aldrich Fine Chemicals; Synthetech, Inc.*

**1120 N-ε-(tert-Butoxycarbonyl)-L-lysine**
2418-95-3

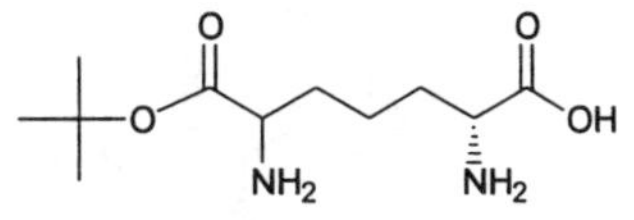

$C_{11}H_{22}N_2O_4$
Chiral intermediate. mp = 250°; $[\alpha]_D^{20}$ = + 18° (c = 1, HOAc). *Acros Organics nv; Sigma-Aldrich Fine Chemicals.*

**1121 (tert-Butoxycarbonyl)-D-methionine**
5241-66-7 226-043-7

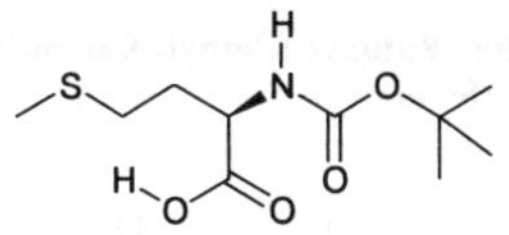

$C_{10}H_{19}NO_4S$
Methionine, N-[(1,1-dimethylethoxy)carbonyl]-, D-; N-tert-butoxycarbonyl-D-methionine, D-Methionine, N-carboxy-, N-tert-butyl ester. Chiral intermediate. *Acros Organics nv.*

**1122 (tert-Butoxycarbonyl)-L-methionine**
2488-15-5 219-639-3

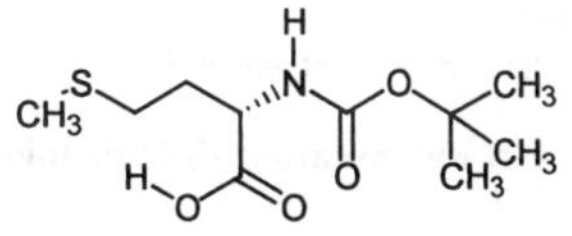

$C_{10}H_{19}NO_4S$
Methionine, N-[(1,1-dimethylethoxy)carbonyl]-, L-; N-(tert-Butoxycarbonyl)-L-methionine; Methionine, N-carboxy-, N-tert-butyl ester, L-. Chiral intermediate. mp = 49-52°; [α] = -22° (c = 1.3, $CH_3OH$). *Acros Organics nv; Fischer Chemicals AG; Sigma-Aldrich Fine Chemicals; Synthetech, Inc.*

**1123 N-(tert-Butoxycarbonyl)-methionine dicyclohexylammonium salt**
61315-59-1 262-707-2

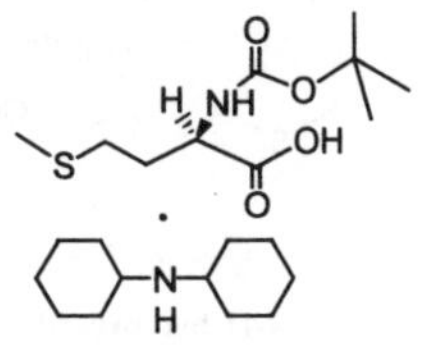

$C_{10}H_{19}NO_4S$
Chiral intermediate. mp = 141-143°; $[\alpha]_D^{20}$ = - 27° (c = 1, $CHCl_3$). *Sigma-Aldrich Fine Chemicals.*

**1124 (tert-Butoxycarbonyl)-L-methionine sulfone**

$C_{10}H_{19}NO_6S$
Chiral intermediate. *Synthetech, Inc.*

**1125 (tert-Butoxycarbonyl)-L-methionine sulfoxide**

$C_{10}H_{19}NO_5S$
Chiral intermediate. *Synthetech, Inc.*

**1126 (tert-Butoxycarbonyl)-D-methioninol**
91177-57-0

NHBOC
OH

$C_{10}H_{21}NO_3S$
Chiral intermediate. *Omega Chemical Company Inc.*

**1127 (tert-Butoxycarbonyl)-L-methioninol**
51372-93-1 200-098-7

NHBOC
OH

$C_{10}H_{21}NO_3S$
Benzene-1,2-diol, 4-[1-hydroxy-2-(methylamino)ethyl]-, (R)-; (-)-Adrenaline; Adnephrine; Adrenal; Adrin; Benzyl alcohol, 3,4-dihydroxy-α-[(methylamino)methyl]-, (-)-; Bosmin; Chelafrin; Epifrin; Epinefrina; Epirenan; Exadrin; Hemisine; Hemostasin; Hemostatin; Hypernephrin; Isoptoepinal; Levoepinephrine; Levorenin. Listed on TSCA. Chiral building block. *Omega Chemical Company Inc.; Sigma-Aldrich Fine Chemicals.*

**1128 (tert-Butoxycarbonyl)-S-(4-methoxy-benzyl)cysteine**
18942-46-6 242-695-5

$C_{16}H_{23}NO_5S$
Alanine, N-carboxy-3-[(p-methoxybenzyl)thio]-, N-tert-butyl ester, L-; N-(tert-Butoxycarbonyl)-S-(p-methoxy-benzyl)-L-cysteine; L-Cysteine, N-[(1,1-dimethyl-ethoxy)carbonyl]-S-[(4-methoxyphenyl)methyl]-; N-[(1,1-Dimethylethoxy)carbonyl]-S-[(4-methoxyphenyl)methyl]-L-cysteine. Listed on TSCA. Chiral intermediate. *Isochem.*

**1129 (S)-(-)-3-tert-Butoxycarbonyl-4-methoxy-carbonyl-2,2-dimethyl-1,3-oxazolidine**
108149-60-6

$C_{12}H_{21}NO_5$
(S)-(-)-3-BOC-4-methoxycarbonyl-2,2-dimethyl-1,3-oxazolidine; 2,2-Dimethyl oxazolidine-3,4-dicarboxylic acid 3-tert-butyl ester 4-methyl ester; Methyl (S)-3-(tert-butoxycarbonyl)-2,2-dimethyl-4-oxazolidinecarboxylate.
Chiral intermediate. $bp_2$ = 101-102°; d = 1.082; n = 1.443; $[\alpha]_D^{20}$ = - 55° (c = 1.3, $CHCl_3$). *Acros Organics nv; Sigma-Aldrich Fine Chemicals; TCI America.*

**1130 (R)-N-(tert-Butoxycarbonyl)-2-methoxy-phenylalanine**

COOH
HN

$C_{15}H_{21}NO_5$
Chiral intermediate. *ChiroTech Technology Ltd.*

**1131 (S)-N-(tert-Butoxycarbonyl)-2-methoxyphenylalanine**

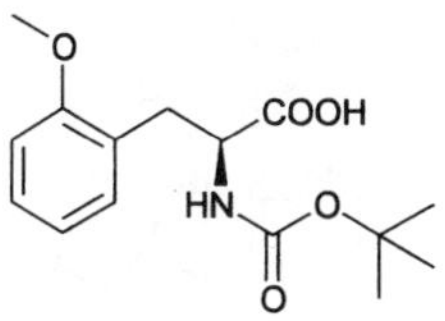

$C_{15}H_{21}NO_5$
Chiral intermediate. *ChiroTech Technology Ltd.*

**1132 (R)-N-(tert-Butoxycarbonyl)-3-methoxyphenylalanine**

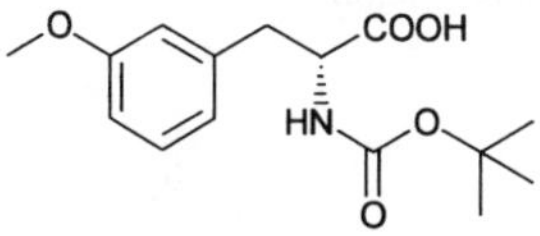

$C_{15}H_{21}NO_5$
Chiral intermediate. *ChiroTech Technology Ltd.*

**1133 (S)-N-(tert-Butoxycarbonyl)-3-methoxyphenylalanine**

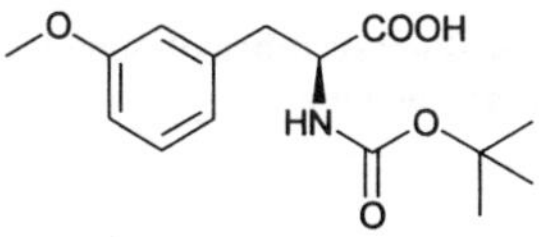

$C_{15}H_{21}NO_5$
Chiral intermediate. *ChiroTech Technology Ltd.*

**1134 (R)-N-(tert-Butoxycarbonyl)-4-methoxyphenylalanine**

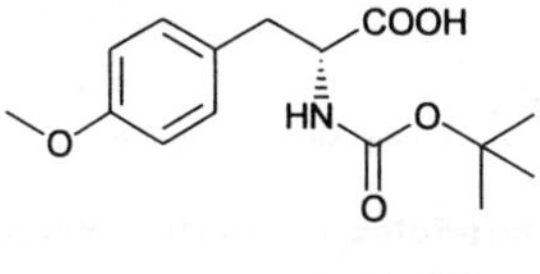

$C_{15}H_{21}NO_5$
Chiral intermediate. *ChiroTech Technology Ltd.*

**1135 (S)-N-(tert-Butoxycarbonyl)-4-methoxyphenylalanine**

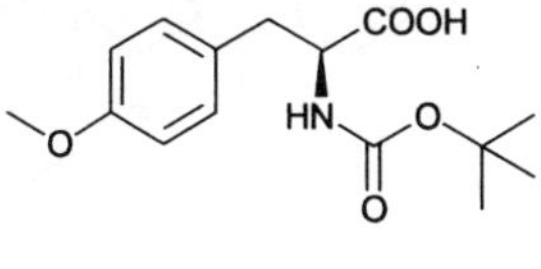

$C_{15}H_{21}NO_5$
Chiral intermediate. *ChiroTech Technology Ltd.*

**1136 (tert-Butoxycarbonyl)-(S)-(4-methylbenzyl)-cysteine**
61925-78-8

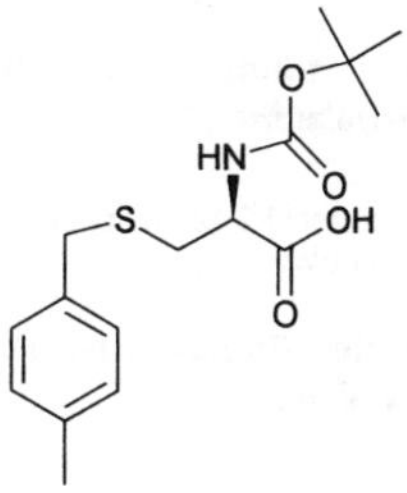

$C_{16}H_{23}NO_4S$
Chiral intermediate. *Isochem.*

**1137 (tert-Butoxycarbonyl)-N-methyl-L-isoleucine**

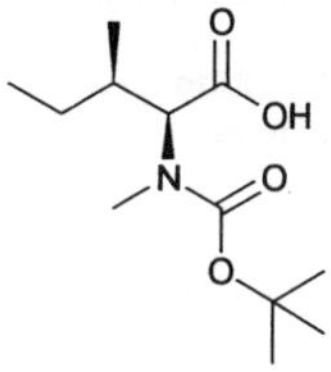

$C_{12}H_{23}NO_4$
Chiral intermediate. *Senn Chemicals AG.*

**1138 (tert-Butoxycarbonyl)-N-methyl-D-leucine**

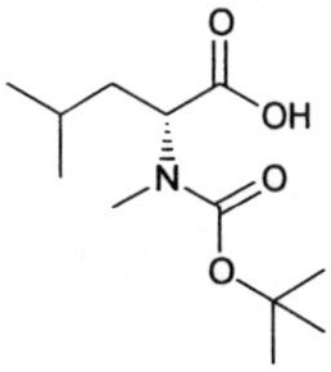

$C_{12}H_{23}NO_4$
Chiral intermediate. *Synthetech, Inc.*

**1139 (tert-Butoxycarbonyl)-N-methyl-L-leucine**

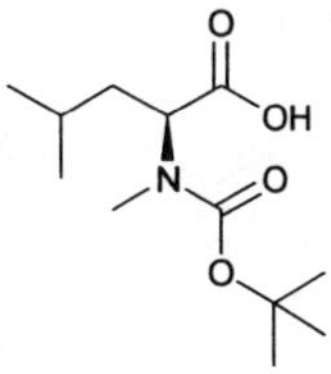

$C_{12}H_{23}NO_4$
Chiral intermediate. *Senn Chemicals AG.*

**1140 (R)-N-(tert-Butoxycarbonyl)-(2-methylphenyl)alanine**

$C_{15}H_{21}NO_4$
N-tert-Butoxycarbonyl-2-methylphenyl-D-alanine.
*ChiroTech Technology Ltd.; Lancaster Synthesis Ltd.*

**1141 (S)-N-(tert-Butoxycarbonyl)-(2-methylphenyl)alanine**

$C_{15}H_{21}NO_4$
N-tert-Butoxycarbonyl-2-methylphenyl-L-alanine.
*ChiroTech Technology Ltd.; Lancaster Synthesis Ltd.*

**1142 (R)-N-(tert-Butoxycarbonyl)-3-methylphenylalanine**

COOH
HN
O
O

$C_{15}H_{21}NO_4$
Chiral intermediate. *ChiroTech Technology Ltd.*

**1143 (S)-N-(tert-Butoxycarbonyl)-3-methylphenylalanine**

COOH
HN
O
O

$C_{15}H_{21}NO_4$
Chiral intermediate. *ChiroTech Technology Ltd.*

**1144 (R)-N-(tert-Butoxycarbonyl)-4-methylphenylalanine**

80102-27-8

COOH
HN
O
O

$C_{15}H_{21}NO_4$
Chiral intermediate. *ChiroTech Technology Ltd.; Synthetech, Inc.*

**1145 (S)-N-(tert-Butoxycarbonyl)-4-methylphenylalanine**

80102-26-7

COOH
HN
O
O

$C_{15}H_{21}NO_4$
Chiral intermediate. *ChiroTech Technology Ltd.; Synthetech, Inc.*

**1146 (tert-Butoxycarbonyl)-erythro-D-β-methylphenylalanine**

O
O
HN
O
H
O

$C_{15}H_{21}NO_4$
Chiral intermediate. *Acros Organics nv.*

**1147 (tert-Butoxycarbonyl)-erythro-L-β-methylphenylalanine**

O
HN
O
OH
O

$C_{15}H_{21}NO_4$
Chiral intermediate. *Acros Organics nv.*

**1148 (tert-Butoxycarbonyl)-N-methyl-D-phenylalanine**

O
OH
N
O
O

$C_{15}H_{21}NO_4$
Chiral intermediate. *Synthetech, Inc.*

**1149 (tert-Butoxycarbonyl)-N-methyl-L-phenylalanine**

O
OH
N
O
O

$C_{15}H_{21}NO_4$
Chiral intermediate. *Senn Chemicals AG.*

**1150 (R)-N(4)-t-Butoxycarbonyl-2-methylpiperazine**

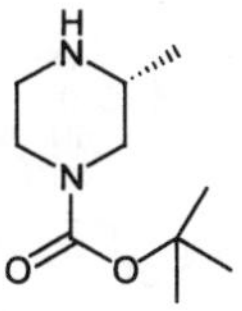

$C_{10}H_{20}N_2O_2$
Chiral building block. mp = 95-97°; $[\alpha]_D^{21}$ = - 38.5 ± -0.3° (c = 3, $C_6H_6$). *Kiralchem Ltd.*

**1151 (S)-N(4)-t-Butoxycarbonyl-2-methylpiperazine**

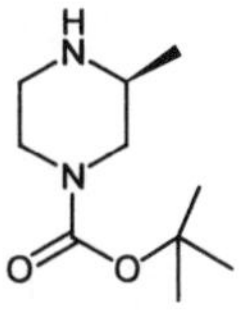

$C_{10}H_{20}N_2O_2$
Chiral building block. mp = 95-97°; $[\alpha]_D^{21}$ = + 38.5 ± -0.3° (c = 3, $C_6H_6$). *Kiralchem Ltd.*

**1152 (S)-(+)-N-tert-(Butoxycarbonyl)-2-methylpiperidine**

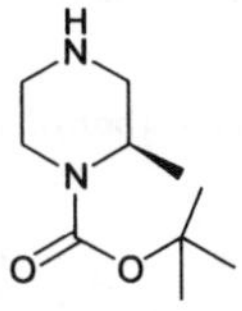

$C_{11}H_{21}NO_2$
Chiral intermediate. $bp_{0.05}$ = 54-56°; d = 0.937; n = 1.453; $[\alpha]_D^{20}$ = + 46° (c = 1, $CHCl_3$). *Sigma-Aldrich Fine Chemicals.*

**1153 (tert-Butoxycarbonyl)-N-methyl-L-serine**

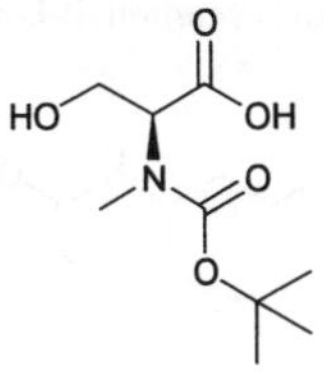

$C_9H_{17}NO_5$
Chiral intermediate. *Senn Chemicals AG.*

**1154 (tert-Butoxycarbonyl)-O-methyl-L-serine**

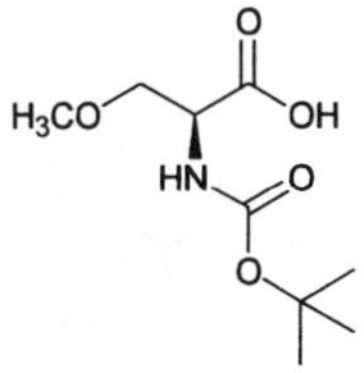

$C_9H_{17}NO_5$
Chiral intermediate. *Synthetech, Inc.*

**1155 (R)-N-(tert-Butoxycarbonyl)-4-methylthiophenylalanine**

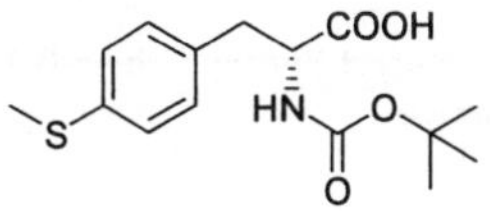

$C_{15}H_{21}SNO_4$
Chiral intermediate. *ChiroTech Technology Ltd.*

**1156 (S)-N-(tert-Butoxycarbonyl)-4-methylthiophenylalanine**

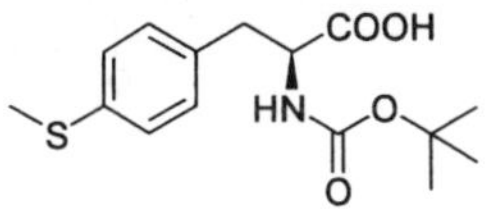

$C_{15}H_{21}SNO_4$
Chiral intermediate. *ChiroTech Technology Ltd.*

**1157 (tert-Butoxycarbonyl)-O-methyl-D-tyrosine**

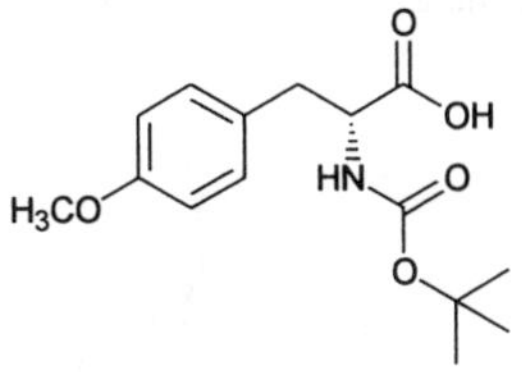

$C_{15}H_{21}NO_5$
Chiral intermediate. *Synthetech, Inc.*

**1158 (tert-Butoxycarbonyl)-O-methyl-L-tyrosine**

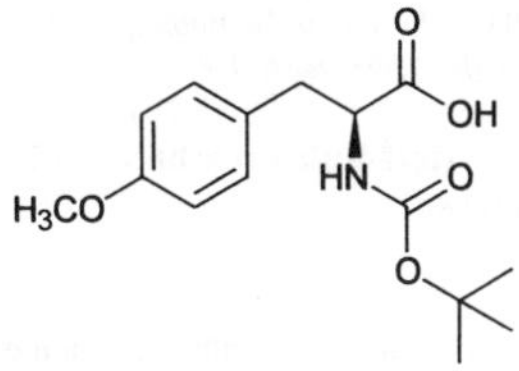

$C_{15}H_{21}NO_5$
Chiral intermediate. *Synthetech, Inc.*

**1159 (tert-Butoxycarbonyl)-N-methyl-L-valine**

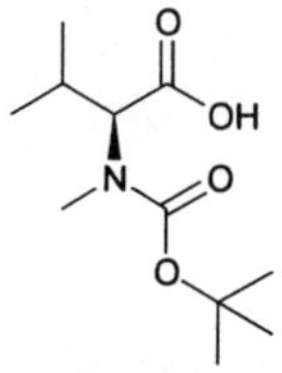

$C_{11}H_{21}NO_4$
Chiral intermediate. *Senn Chemicals AG; Synthetech, Inc.*

**1160 (R)-N-(tert-Butoxycarbonyl)-1-napthyl-alanine**
76932-48-4

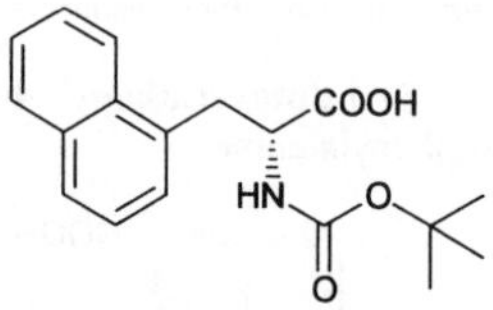

$C_{18}H_{21}NO_4$
Chiral intermediate. *ChiroTech Technology Ltd.; Synthetech, Inc.*

**1161 (S)-N-(tert-Butoxycarbonyl)-1-napthyl-alanine**
55447-00-2

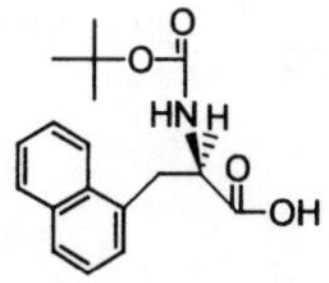

$C_{18}H_{21}NO_4$
Chiral intermediate. mp = 145-148°; $[\alpha]^{22}$ = + 53° (c = 2.5, $CH_3OH$). *ChiroTech Technology Ltd.; Sigma-Aldrich Fine Chemicals; Synthetech, Inc.*

**1162 (R)-N-(tert-Butoxycarbonyl)-(2-naphthyl)-alanine**
76985-10-9
$C_{18}H_{21}NO_4$
N-tert-Butoxycarbonyl-2-naphthyl-D-alanine; Boc-D-2-naphthylalanine. Chiral intermediate. *ChiroTech Technology Ltd.; Lancaster Synthesis Ltd.; Synthetech, Inc.*

**1163 (S)-N-(tert-Butoxycarbonyl)-(2-naphthyl)-alanine**
58438-04-3

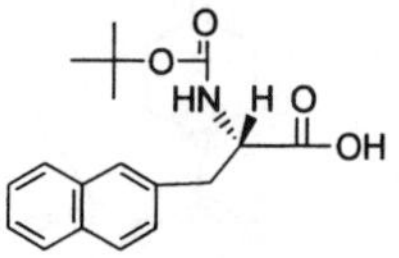

$C_{18}H_{21}NO_4$
N-tert-Butoxycarbonyl-2-naphthyl-L-alanine; N-BOC-L-3-(2-naphthyl)alanine. Chiral intermediate. mp = 92-95°; $[\alpha]^{22}$ = + 46° (c = 1, EtOH). *ChiroTech Technology Ltd.; Lancaster Synthesis Ltd.; PCAS; Senn Chemicals AG; Sigma-Aldrich Fine Chemicals; Synthetech, Inc.*

**1164 (tert-Butoxycarbonyl)-L-4-nitrophenyl-alanine**
33305-77-0 251-450-1

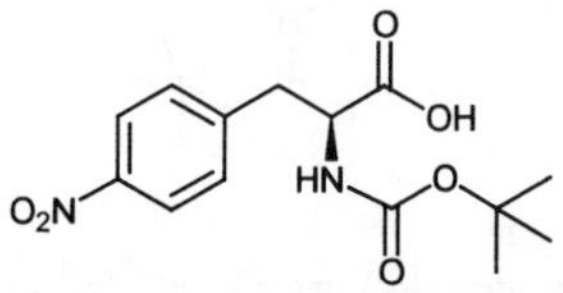

$C_{14}H_{18}N_2O_6$
Alanine, N-carboxy-3-(p-nitrophenyl)-, N-tert-butyl ester, L-; N-(tert-Butoxycarbonyl)-4-nitro-3-phenyl-L-alanine; N-[(1,1-dimethylethoxy)carbonyl]-4-nitro L-Phenylalanine. Chiral intermediate. *Advanced Asymmetrics, Inc.; Synthetech, Inc.*

**1165 (tert-Butoxycarbonyl)-L-norleucine**
6404-28-0

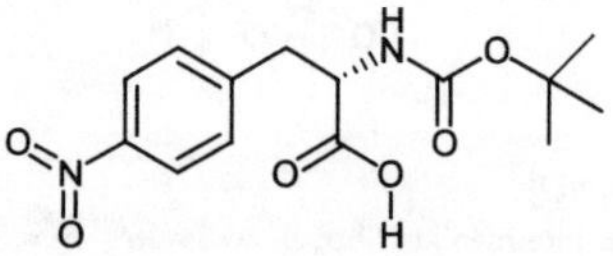

$C_{14}H_{18}N_2O_6$
Chiral intermediate. *Acros Organics nv; Senn Chemicals AG; Synthetech, Inc.*

**1166 (tert-Butoxycarbonyl)-L-norvaline**
53308-95-5

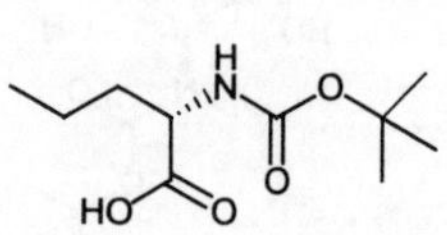

$C_{10}H_{19}NO_4$
Chiral intermediate. *Acros Organics nv; Senn Chemicals AG.*

**1167 (R)-N-(tert-Butoxycarbonyl)-octylglycine**

$C_{15}H_{29}NO_4$
N-tert-Butoxycarbonyl-(R)-2-aminodecanoic acid. Chiral intermediate. *ChiroTech Technology Ltd.; Lancaster Synthesis Ltd.*

**1168 (S)-N-(tert-Butoxycarbonyl)-octylglycine**
67862-03-7

$C_{15}H_{29}NO_4$
tert-Butoxycarbonyl-(S)-2-aminodecanoic acid. Chiral intermediate. *ChiroTech Technology Ltd.; Lancaster Synthesis Ltd.*

**1169 (tert-Butoxycarbonyl)-L-ornithine**

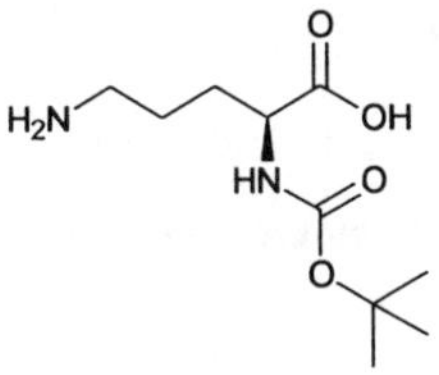

$C_{10}H_{20}N_2O_4$
Chiral intermediate. *Synthetech, Inc.*

**1170 (tert-Butoxycarbonyl)-D-pentafluoro-phenylalanine**
136207-26-6

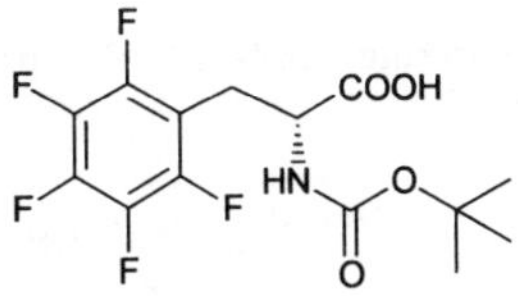

$C_{14}H_{14}F_5NO_4$
Chiral intermediate. *Synthetech, Inc.*

**1171 (tert-Butoxycarbonyl)-L-pentafluoro-phenylalanine**
34702-60-8

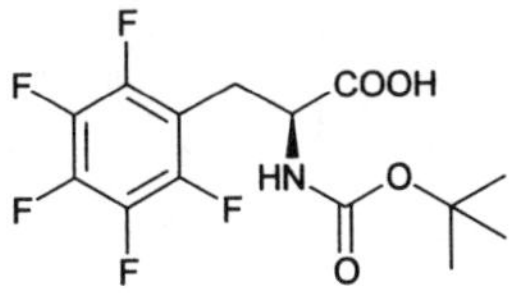

$C_{14}H_{14}F_5NO_4$
Chiral intermediate. *Synthetech, Inc.*

**1172 (tert-Butoxycarbonyl)-D-phenylalaninal**
77119-85-8

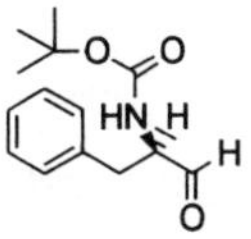

$C_{14}H_{19}NO_3$
N-t-BOC-D-phenylalaninal. Chiral intermediate. mp = 86-89°; $[\alpha]_D^{20}$ = + 45° (c = 0.5, $CH_3OH$). *Daiso Company, Ltd.; Omega Chemical Company Inc.; Sigma-Aldrich Fine Chemicals.*

**1173 (tert-Butoxycarbonyl)-L-phenylalaninal**
72155-45-4

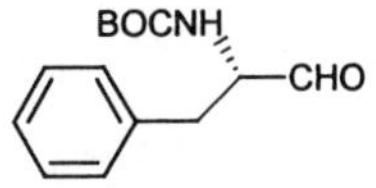

$C_{14}H_{19}NO_3$
N-t-BOC-L-phenylalaninal. Chiral intermediate. *Daiso Company, Ltd.; Omega Chemical Company Inc.*

**1174 (tert-Butoxycarbonyl)-D-phenylalanine**
18942-49-9

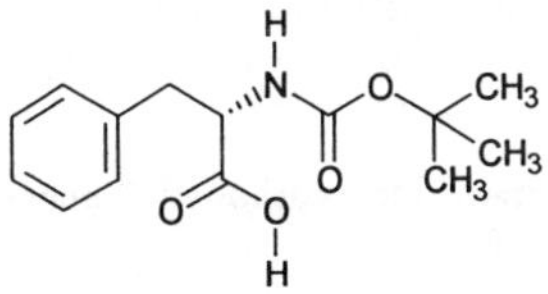

$C_{14}H_{19}NO_4$
Chiral intermediate. mp = 86-88°; [α] = - 24.8° (c = 2, EtOH). *Acros Organics nv; Davos Chemical Corporation; Isochem; Synthetech, Inc.*

**1175 (tert-Butoxycarbonyl)-L-phenylalanine**
13734-34-4 237-305-5

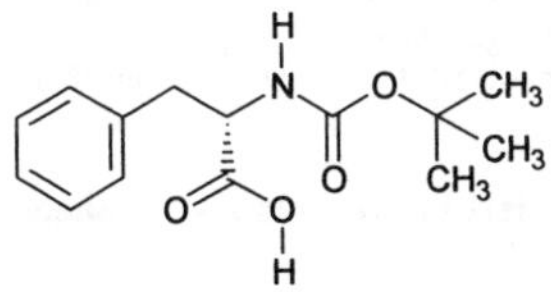

$C_{14}H_{19}NO_4$
Phenylalanine, N-[(1,1-dimethylethoxy)carbonyl]-, L-; N-[(1,1-Dimethylethoxy)carbonyl]-L-phenylalanine; Boc-phe; N-(tert-Butoxycarbonyl)-L-phenylalanine; Alanine, N-carboxy-3-phenyl-, N-tert-butyl ester, L-. Listed on TSCA. Chiral intermediate. mp = 86-88°; [α] = + 24.8° (c = 1.5, EtOH). *Acros Organics nv; Austin Chemical Company, Inc.; Fischer Chemicals AG; Great Lakes Fine Chemicals; PPG - Sipsy; Sigma-Aldrich Fine Chemicals; Synthetech, Inc.*

**1176 (tert-Butoxycarbonyl)-D-phenylalanine methyl ester**

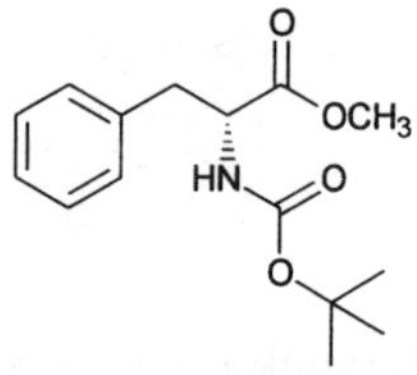

$C_{15}H_{21}NO_4$
Chiral intermediate. *Synthetech, Inc.*

**1177 (tert-Butoxycarbonyl)-L-phenylalanine methyl ester**
51987-73-6

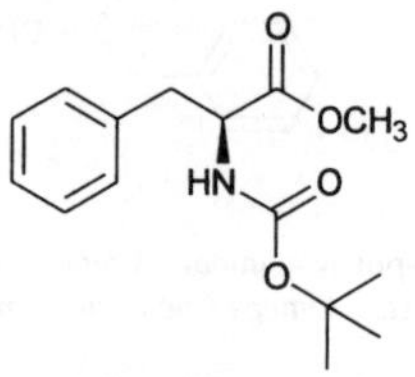

$C_{15}H_{21}NO_4$
Chiral intermediate. mp = 36-40°; $[\alpha]^{22}$ = - 4.0° (c = 2, $CH_3OH$). *Fischer Chemicals AG; Sigma-Aldrich Fine Chemicals; Synthetech, Inc.*

**1178 (tert-Butoxycarbonyl)-D-phenylalaninol**
106454-69-7

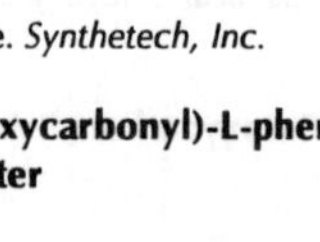

$C_{14}H_{21}NO_3$
(R)-(+)-2-(tert-Butoxycarbonylamino)-3-phenyl-1-propanol. Chiral intermediate. mp = 95-98°; $[\alpha]^{23}$ = + 24° (c = 1, $CHCl_3$). *Acros Organics nv; Newport Synthesis Ireland Ltd.; Omega Chemical Company Inc.; Sigma-Aldrich Fine Chemicals; SK Energy and Chemical, Inc.; Synthetech, Inc.*

**1179 (tert-Butoxycarbonyl)-L-phenylalaninol**
66605-57-0

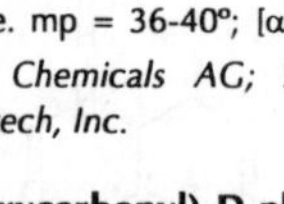

$C_{14}H_{21}NO_3$
1-(tert-Butoxycarbonyl)-L-phenylalaninol; (S)-2-(BOC-amino)-3-phenyl-1-propanol; (S)-(-)-BOC-phenylalaninol. Chiral intermediate. mp = 96-97°; $[\alpha]_D^{20}$ = - 27 ± 2° (c = 1, $CH_3OH$). *Acros Organics nv; Fischer Chemicals AG; Great Lakes Fine Chemicals; Lancaster Synthesis Ltd.; Newport Synthesis Ireland Ltd.; Omega Chemical Company Inc.; Senn Chemicals AG; Sigma-Aldrich Fine Chemicals; SK Energy and Chemical, Inc.; Synthetech, Inc.*

**1180 (tert-Butoxycarbonyl)-D-phenylglycine**

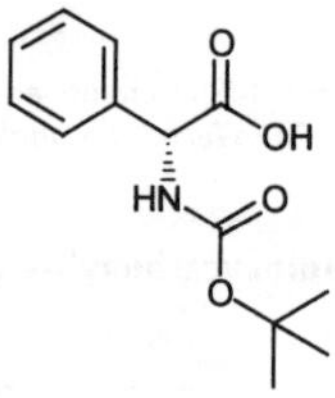

$C_{13}H_{17}NO_4$
Chiral intermediate. *Synthetech, Inc.*

**1181 (tert-Butoxycarbonyl)-L-phenylglycine**

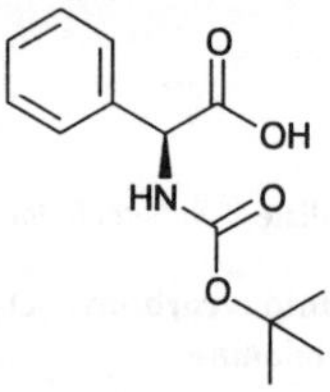

$C_{13}H_{17}NO_4$
Chiral intermediate. *Senn Chemicals AG; Synthetech, Inc.*

**1182 (tert-Butoxycarbonyl)-(R)-2-phenylglycinol**
102089-74-7

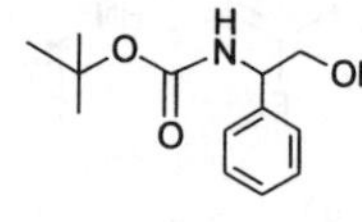

$C_{13}H_{19}NO_3$
Chiral intermediate. mp = 137-139°; $[\alpha]^{19}$ = - 38° (c = 1, $CHCl_3$). *Newport Synthesis Ireland Ltd.; Omega Chemical Company Inc.; Sigma-Aldrich Fine Chemicals; SK Energy and Chemical, Inc.*

**1183 (tert-Butoxycarbonyl)-(S)-2-phenylglycinol**
117049-14-6

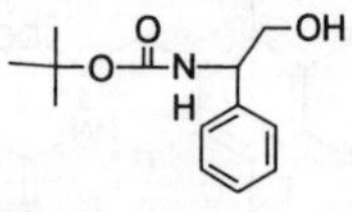

$C_{13}H_{19}NO_3$
Chiral intermediate. mp = 136-139°; $[\alpha]^{19}$ = + 37° (c = 1, $CHCl_3$). *Newport Synthesis Ireland Ltd.; Omega Chemical Company Inc.; Sigma-Aldrich Fine Chemicals; SK Energy and Chemical, Inc.*

**1184 (-)-N-(tert-Butoxycarbonyl)phenyl phenylalanine**

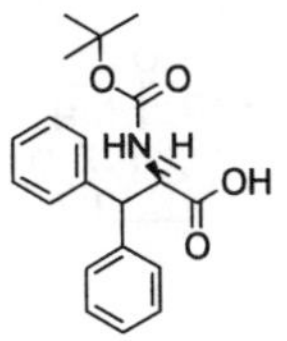

$C_{20}H_{23}NO_4$
Chiral intermediate. mp = 157°; $[\alpha]_D^{20}$ = - 29.5° (c = 1, EtOH). *Sigma-Aldrich Fine Chemicals.*

**1185 (+)-N-(tert-Butoxycarbonyl)phenyl phenylalanine**

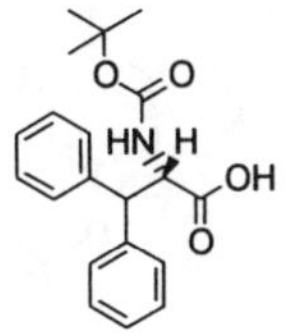

$C_{20}H_{23}NO_4$
Chiral intermediate. mp = 160°; $[\alpha]_D^{20}$ = + 29° (c = 1, $CHCl_3$). *Sigma-Aldrich Fine Chemicals.*

**1186 (-)-N-(tert-Butoxycarbonyl)phenyl phenylalaninol**

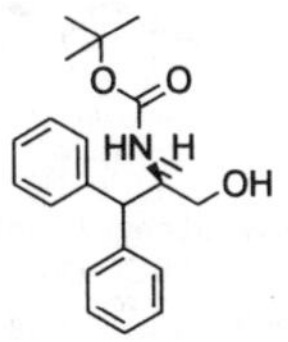

$C_{20}H_{25}NO_3$
Chiral intermediate. mp = 114-116°; $[\alpha]^{22}$ = - 21° (c = 1, $CH_3OH$). *Sigma-Aldrich Fine Chemicals.*

**1187 (+)-N-(tert-Butoxycarbonyl)phenyl phenylalaninol**

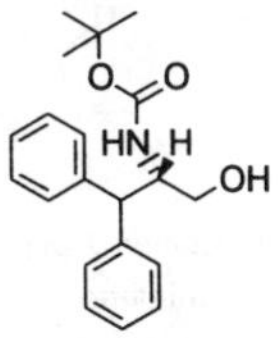

$C_{20}H_{25}NO_3$
Chiral intermediate. mp = 114-116°; $[\alpha]_D^{20}$ = + 25° (c = 1, $CHCl_3$). *Sigma-Aldrich Fine Chemicals.*

**1188 (tert-Butoxycarbonyl)-D-pipecolinic acid**

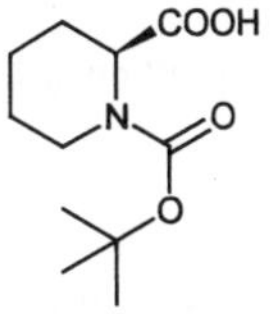

$C_{11}H_{19}NO_4$
Chiral intermediate. *Advanced Asymmetrics, Inc.*

**1189 (tert-Butoxycarbonyl)-L-pipecolinic acid**

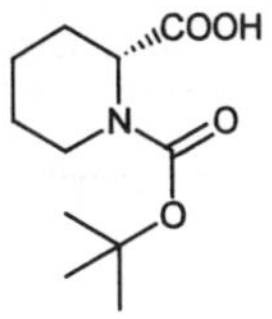

$C_{11}H_{19}NO_4$
Chiral intermediate. *Advanced Asymmetrics, Inc.*

**1190 (R)-(-)-1-(tert-Butoxycarbonyl)-2-piperidinecarboxylic acid**

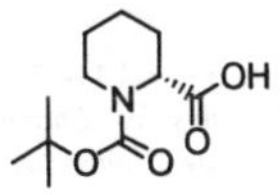

$C_{11}H_{19}NO_4$
N-t-BOC-(R)-(+)-pipecolinic acid. Chiral intermediate. mp = 116-119°; $[\alpha]^{23}$ = + 68° (c = 1, HOAc). *Sigma-Aldrich Fine Chemicals.*

**1191 (S)-(-)-1-(tert-Butoxycarbonyl)-2-piperidinecarboxylic acid**

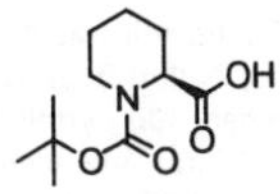

$C_{11}H_{19}NO_4$
N-t-BOC-(S)-(-)-pipecolinic acid. Chiral intermediate. mp = 122-126°; $[\alpha]^{23}$ = - 63.2° (c = 1, HOAc). *Sigma-Aldrich Fine Chemicals.*

**1192 (tert-Butoxycarbonyl)-D-prolinal**
73365-02-3

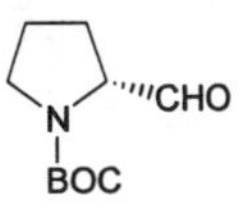

$C_{10}H_{17}NO_3$
Chiral intermediate. bp = 228°; d = 1.059; n = 1.461;

$[\alpha]^{23}$ = + 83° (neat). *Omega Chemical Company Inc.; Sigma-Aldrich Fine Chemicals.*

**1193 (tert-Butoxycarbonyl)-L-prolinal**
69610-41-9

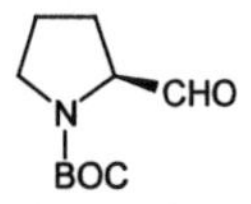

$C_{10}H_{17}NO_3$
Chiral intermediate. bp = 211°; d = 1.063; n = 1.462; $[\alpha]_D^{20}$ = - 101° (c = 0.66, $CHCl_3$). *Omega Chemical Company Inc.; Sigma-Aldrich Fine Chemicals.*

**1194 (tert-Butoxycarbonyl)-D-proline**
37784-17-1

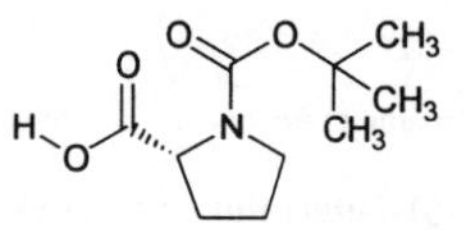

$C_{10}H_{17}NO_4$
Chiral intermediate. mp = 134-137°; $[\alpha]^{22}$ = + 60° (c = 2, HOAc). *Acros Organics nv; Omega Chemical Company Inc.; Sigma-Aldrich Fine Chemicals.*

**1195 (tert-Butoxycarbonyl)-L-proline**
15761-39-4 239-848-3

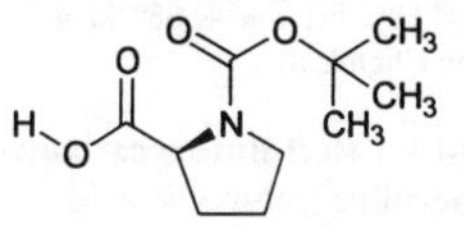

$C_{10}H_{17}NO_4$
Pyrrolidine-1,2-dicarboxylic acid, 1-(1,1-dimethylethyl) ester, (S)-; Boc-pro; N-(tert-Butoxycarbonyl)-L-proline; 1-(tert-Butyl) hydrogen (S)-pyrrolidine-1,2-dicarboxylate. Listed on TSCA. Chiral intermediate. mp = 133-136°; [α] = - 60.9° (c = 2, HOAc). *Acros Organics nv; Omega Chemical Company Inc.; Sigma-Aldrich Fine Chemicals; Synthetech, Inc.*

**1196 (tert-Butoxycarbonyl)-L-proline amide**

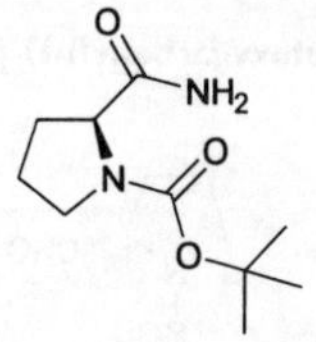

$C_{10}H_{18}N_2O_3$
Chiral intermediate. *Senn Chemicals AG.*

**1197 N-((tert-Butoxycarbonyl)-proline N-methoxy-N-methylamide**
115186-37-3

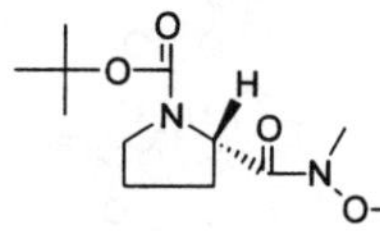

$C_{12}H_{22}N_2O_4$
Chiral intermediate. bp = 253°; d = 1.059; n = 1.471; $[\alpha]^{22}$ = - 38° (c = 1, $CH_3OH$). *Sigma-Aldrich Fine Chemicals.*

**1198 (tert-Butoxycarbonyl)-L-proline methyl ester**

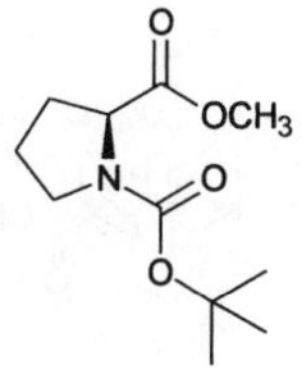

$C_{11}H_{19}NO_4$
Chiral intermediate. *Synthetech, Inc.*

**1199 (tert-Butoxycarbonyl)-D-prolinol**
83435-58-9

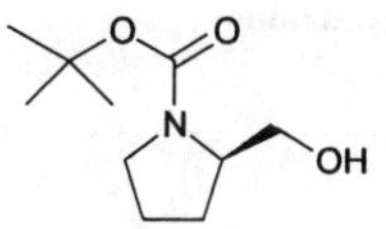

$C_{10}H_{19}NO_3$
(R)-(+)-1-tert-Butoxycarbonyl-2-pyrrolidinemethanol; (R)-(+)-BOC-Prolinol; N-t-BOC-D-prolinol. Chiral intermediate. mp = 59-62°; $[\alpha]_D^{20}$ = + 54 ± 2° (c = 5, $CH_3OH$). *Acros Organics nv; Daiso Company, Ltd.; Lancaster Synthesis Ltd.; Omega Chemical Company Inc.; Sigma-Aldrich Fine Chemicals.*

**1200 (tert-Butoxycarbonyl)-L-prolinol**
69610-40-8

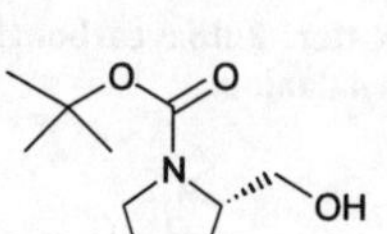

$C_{10}H_{19}NO_3$
(S)-(-)-1-tert-Butoxycarbonyl-2-pyrrolidinemethanol; N-t-BOC-L-prolinol. Chiral intermediate. mp = 59-62°; $[\alpha]_D^{20}$ = - 54 ± 2° (c = 5, $CH_3OH$); $[\alpha]^{21}$ = - 48° (c = 1.3, $CHCl_3$). *Acros Organics nv; Daiso Company, Ltd.; Lancaster Synthesis Ltd.; Omega Chemical Company Inc.; Senn Chemicals AG; Sigma-Aldrich Fine Chemicals; Synthetech, Inc.*

**1201 (R)-N-(tert-Butoxycarbonyl)-propargyl-glycine**
63039-46-3
$C_{10}H_{15}NO_4$
N-tert-Butoxycarbonyl-(R)-2-amino-4-pentynoic acid. Chiral intermediate. *ChiroTech Technology Ltd.; Lancaster Synthesis Ltd.*

**1202 (S)-N-(tert-Butoxycarbonyl)-propargyl-glycine**
63039-48-5
$C_{10}H_{15}NO_4$
N-tert-Butoxycarbonyl-(S)-2-amino-4-pentynoic acid. Chiral intermediate. *ChiroTech Technology Ltd.; Lancaster Synthesis Ltd.*

**1203 (R)-N-(tert-Butoxycarbonyl)-(2-pyridyl)-alanine**
98266-32-1
$C_{13}H_{18}N_2O_4$
N-tert-Butoxycarbonyl-2-pyridyl-D-alanine; Boc-D-2-pyridylalanine. Chiral intermediate. *ChiroTech Technology Ltd.; Lancaster Synthesis Ltd.; Synthetech, Inc.*

**1204 (S)-N-(tert-Butoxycarbonyl)-(2-pyridyl)-alanine**
71239-85-5
$C_{13}H_{18}N_2O_4$
N-tert-Butoxycarbonyl-2-pyridyl-L-alanine; Boc-L-2-pyridylalanine. Chiral intermediate. *ChiroTech Technology Ltd.; Lancaster Synthesis Ltd.; Synthetech, Inc.*

**1205 (R)-N-(tert-Butoxycarbonyl)-(3-pyridyl)-alanine**
98266-33-2
$C_{13}H_{18}N_2O_4$
N-tert-Butoxycarbonyl-3-pyridyl-D-alanine; Boc-D-3-pyridylalanine. Chiral intermediate. *ChiroTech Technology Ltd.; Lancaster Synthesis Ltd.; Synthetech, Inc.*

**1206 (S)-N-(tert-Butoxycarbonyl)-(3-pyridyl)-alanine**
117142-26-4
$C_{13}H_{18}N_2O_4$
N-tert-Butoxycarbonyl-3-pyridyl-L-alanine; Boc-L-3-pyridylalanine. Chiral intermediate. *ChiroTech Technology Ltd.; Lancaster Synthesis Ltd.; Synthetech, Inc.*

**1207 (R)-N-(tert-Butoxycarbonyl)-4-pyridyl-alanine**
37535-58-3

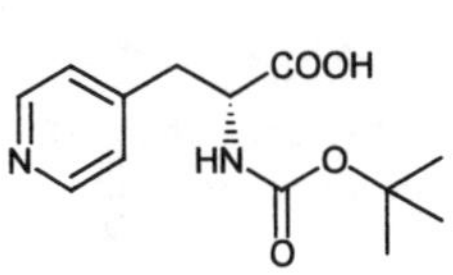

$C_{13}H_{18}N_2O_4$
Chiral intermediate. *ChiroTech Technology Ltd.; Synthetech, Inc.*

**1208 (S)-N-(tert-Butoxycarbonyl)-4-pyridyl-alanine**
37535-57-2

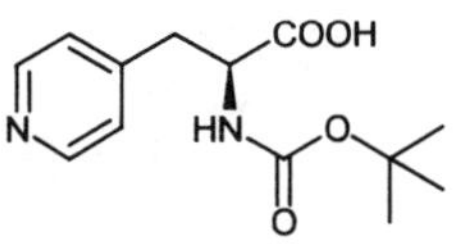

$C_{13}H_{18}N_2O_4$
Chiral intermediate. *ChiroTech Technology Ltd.; Synthetech, Inc.*

**1209 N-(tert-Butoxycarbonyl)-3,4-(1-pyrrolyl)-phenylalanine**

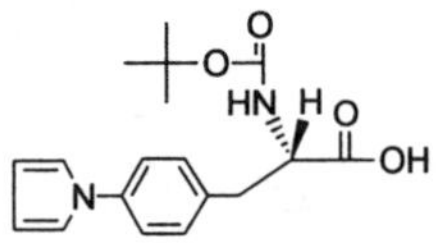

$C_{18}H_{22}N_2O_4$
Chiral intermediate. mp = 85°; $[\alpha]^{22}$ = + 20° (c = 1, $CH_3OH$). *Sigma-Aldrich Fine Chemicals.*

**1210 (tert-Butoxycarbonyl)-D-serine**
6368-20-3

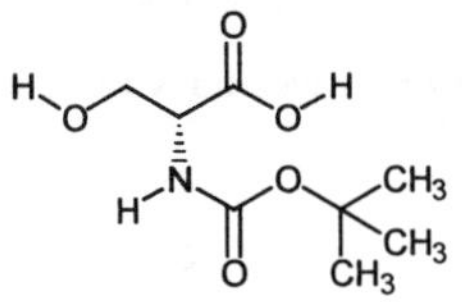

$C_8H_{15}NO_5$
Chiral intermediate. mp = 91-95°; $[\alpha]_D^{20}$ = + 3.5° (c = 2, HOAc). *Acros Organics nv; Sigma-Aldrich Fine Chemicals; Synthetech, Inc.*

**1211 (tert-Butoxycarbonyl)-D-serine benzyl ester**

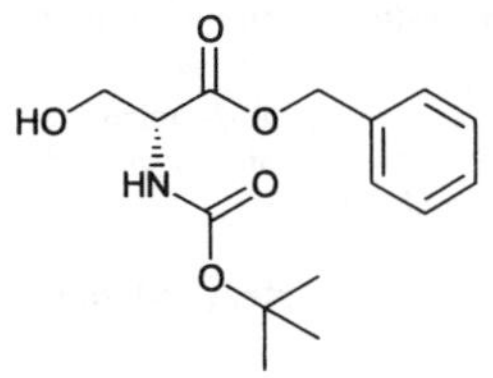

$C_{15}H_{21}NO_5$
Chiral intermediate. *Synthetech, Inc.*

**1212 (tert-Butoxycarbonyl)-L-serine hydrate**
3262-72-4 221-867-3

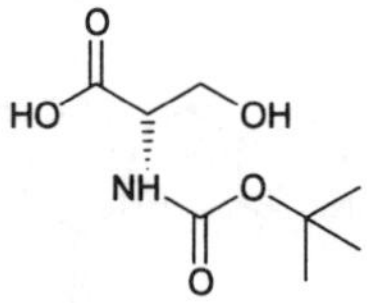

$C_8H_{15}NO_5$
Serine, N-carboxy-, N-tert-butyl ester, L-; N-[tert-Butoxycarbonyl]-L-serine. Chiral intermediate. mp = 91°; [α] = - 7.8° (c = 2.6, $H_2O$); $[\alpha]_D^{20}$ = - 3° (c = 2, HOAc). *Acros Organics nv; PPG - Sipsy; Sigma-Aldrich Fine Chemicals; Synthetech, Inc.*

**1213 (tert-Butoxycarbonyl)-D-serine methyl ester**

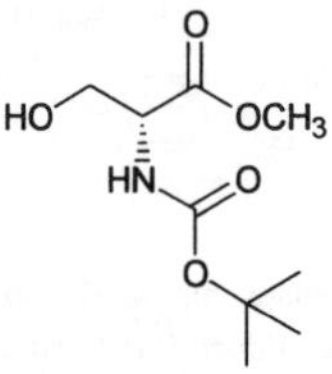

$C_9H_{17}NO_5$
Chiral intermediate. *Synthetech, Inc.*

**1214 (tert-Butoxycarbonyl)-L-serine methyl ester**
95715-85-8

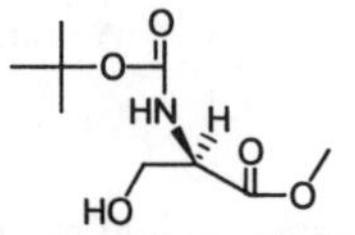

$C_9H_{17}NO_5$
Chiral intermediate. bp = 215°; d = 1.08; n = 1.453; $[\alpha]^{25}$ = + 18.8° (c = 5, $CH_3OH$). *Sigma-Aldrich Fine Chemicals; Synthetech, Inc.*

**1215 (R)-N-(tert-Butoxycarbonyl)-β-styrylalanine**
$C_{16}H_{21}NO_4$
N-tert-Butoxycarbonyl-(R)-2-amino-5-phenyl-4-pentenoic acid. Chiral intermediate. *ChiroTech Technology Ltd.; Lancaster Synthesis Ltd.*

**1216 (S)-N-(tert-Butoxycarbonyl)-β-styrylalanine**
$C_{16}H_{21}NO_4$
N-tert-Butoxycarbonyl-(S)-2-amino-5-phenyl-4-pentenoic acid. Chiral intermediate. *ChiroTech Technology Ltd.; Lancaster Synthesis Ltd.*

**1217 (tert-Butoxycarbonyl)-D-1,2,3,4-tetrahydro-3-carboline-3-carboxylic acid**

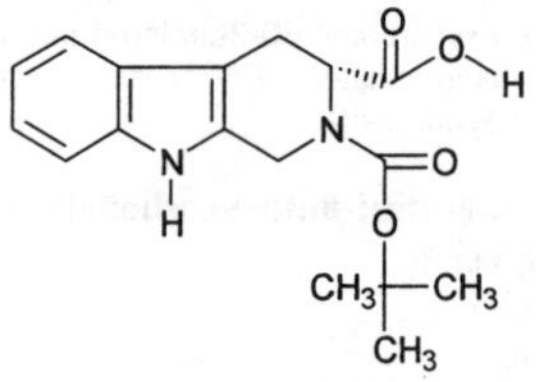

$C_{17}H_{20}N_2O_4$
Chiral intermediate. *Acros Organics nv.*

**1218 (tert-Butoxycarbonyl)-L-1,2,3,4-tetrahydro-β-carboline-3-carboxylic acid**

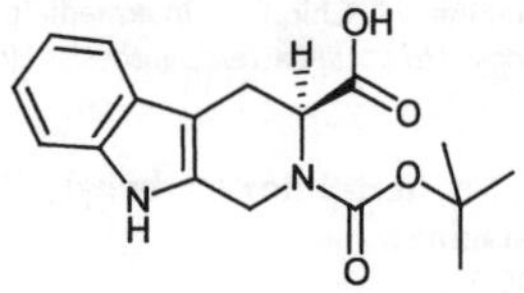

$C_{17}H_{20}N_2O_4$
Chiral intermediate. *Acros Organics nv.*

**1219 (tert-Butoxycarbonyl)-D-1,2,3,4-tetrahydroisoquinoline-3-carboxylic acid**
11592-35-1

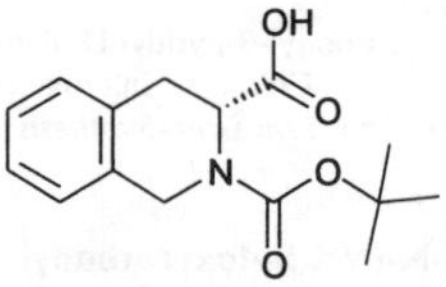

$C_{15}H_{19}NO_4$
Chiral intermediate. *Acros Organics nv; Senn Chemicals AG; Synthetech, Inc.*

**1220 (tert-Butoxycarbonyl)-L-1,2,3,4-tetrahydroisoquinoline-3-carboxylic acid**
78879-20-6

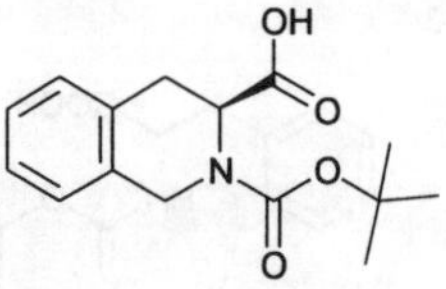

$C_{15}H_{19}NO_4$
Chiral intermediate. *Acros Organics nv; Synthetech, Inc.*

**1221 (tert-Butoxycarbonyl)-L-thiazolidine-4-carboxylic acid**

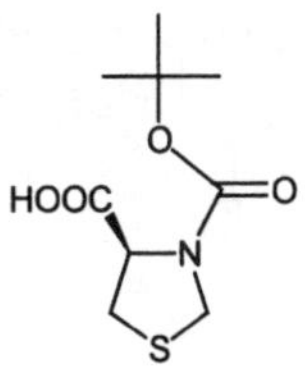

$C_9H_{15}NO_4S$
Chiral intermediate. *Senn Chemicals AG.*

**1222 (R)-N-(tert-Butoxycarbonyl)-(4-thiazolyl)-alanine**
134107-69-0
$C_{11}H_{16}N_2O_4S$
N-tert-Butoxycarbonyl-4-thiazolyl-D-alanine. Chiral intermediate. *ChiroTech Technology Ltd.; Lancaster Synthesis Ltd.; Synthetech, Inc.*

**1223 (S)-N-(tert-Butoxycarbonyl)-(4-thiazolyl)-alanine**
119434-75-2
$C_{11}H_{16}N_2O_4S$
N-tert-Butoxycarbonyl-4-thiazolyl-L-alanine. Chiral intermediate. *ChiroTech Technology Ltd.; Lancaster Synthesis Ltd.; Synthetech, Inc.*

**1224 (R)-N-(tert-Butoxycarbonyl)-(2-thienyl)-alanine**
75452-55-8
$C_{12}H_{17}NO_4S$
N-tert-Butoxycarbonyl-2-thienyl-D-alanine. Chiral intermediate. *ChiroTech Technology Ltd.; Lancaster Synthesis Ltd.; Synthetech, Inc.*

**1225 (S)-N-(tert-Butoxycarbonyl)-(2-thienyl)-alanine**
56675-37-7
$C_{12}H_{17}NO_4S$
N-tert-Butoxycarbonyl-2-thienyl-L-alanine. Chiral intermediate. *ChiroTech Technology Ltd.; Lancaster Synthesis Ltd.; Synthetech, Inc.*

**1226 (R)-N-(tert-Butoxycarbonyl)-(3-thienyl)-alanine**
$C_{12}H_{17}NO_4S$
N-tert-Butoxycarbonyl-3-thienyl-D-alanine. Chiral intermediate. *ChiroTech Technology Ltd.; Lancaster Synthesis Ltd.*

**1227 (S)-N-(tert-Butoxycarbonyl)-(3-thienyl)-alanine**
83825-42-7
$C_{12}H_{17}NO_4S$
N-tert-Butoxycarbonyl-3-thienyl-L-alanine. Chiral intermediate. *ChiroTech Technology Ltd.; Lancaster Synthesis Ltd.*

**1228 (tert-Butoxycarbonyl)-L-threonine**
2592-18-9 219-987-6

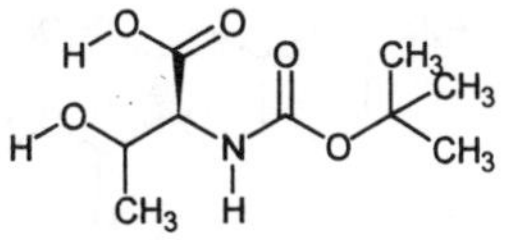

$C_9H_{17}NO_5$
Threonine, N-carboxy-, N-tert-butyl ester, L-; N-(tert-Butoxycarbonyl)-L-threonine; L-Threonine, N-[(1,1-dimethylethoxy)carbonyl]-. Chiral intermediate. mp = 80-82°; [α] = - 8.6° (c = 1, HOAc). *Acros Organics nv; Sigma-Aldrich Fine Chemicals; Synthetech, Inc.*

**1229 (tert-Butoxycarbonyl)-L-threonine hydrazide**

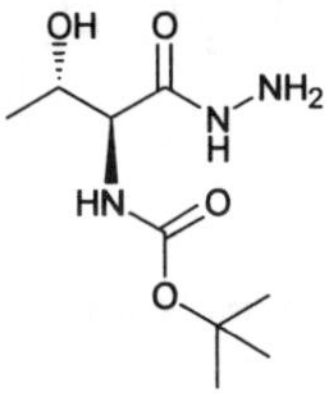

$C_9H_{19}N_3O_4$
Chiral intermediate. *Senn Chemicals AG.*

**1230 (R)-N-(tert-Butoxycarbonyl)-(2-trifluoromethylphenyl)alanine**

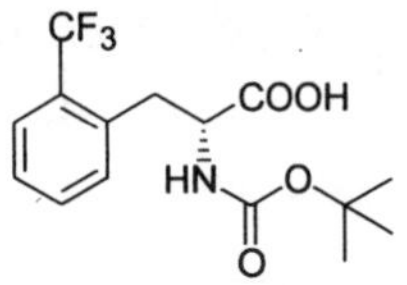

$C_{15}H_{18}F_3NO_4$
Chiral intermediate. *ChiroTech Technology Ltd.*

**1231 (S)-N-(tert-Butoxycarbonyl)-(2-trifluoromethylphenyl)alanine**

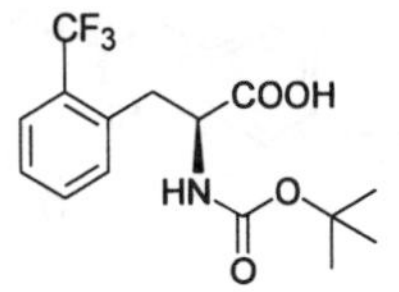

$C_{15}H_{18}F_3NO_4$
Chiral intermediate. *ChiroTech Technology Ltd.*

**1232 (R)-N-(tert-Butoxycarbonyl)-(3-trifluoromethylphenyl)alanine**
82317-82-6

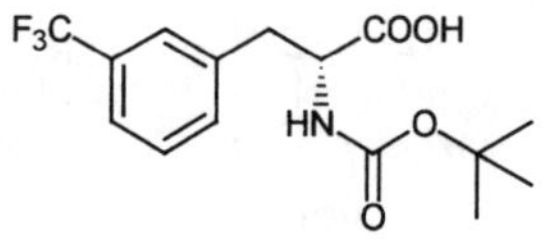

$C_{15}H_{18}F_3NO_4$
Chiral intermediate. *ChiroTech Technology Ltd.; Synthetech, Inc.*

**1233 (S)-N-(tert-Butoxycarbonyl)-(3-trifluoromethylphenyl)alanine**
142995-31-1

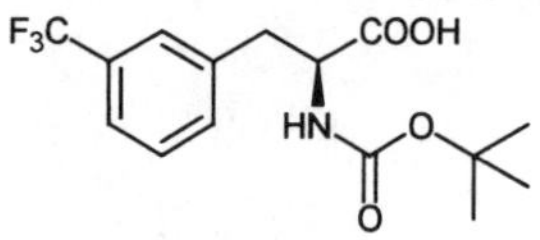

$C_{15}H_{18}F_3NO_4$
Chiral intermediate. *ChiroTech Technology Ltd.; Synthetech, Inc.*

**1234 (R)-N-(tert-Butoxycarbonyl)-(4-trifluoromethylphenyl)alanine**
82317-83-7

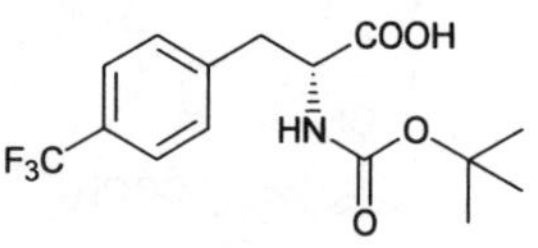

$C_{15}H_{18}F_3NO_4$
Chiral intermediate. *ChiroTech Technology Ltd.; Synthetech, Inc.*

**1235 (S)-N-(tert-Butoxycarbonyl)-(4-trifluoromethylphenyl)alanine**
114873-07-3

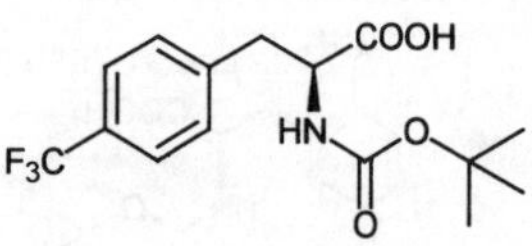

$C_{15}H_{18}F_3NO_4$
Chiral intermediate. *ChiroTech Technology Ltd.; Synthetech, Inc.*

**1236 (tert-Butoxycarbonyl)-D-3,4,5-trifluorophenylalanine**
205445-55-2

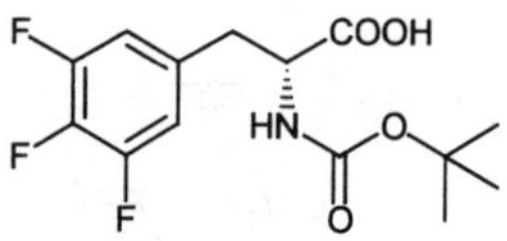

$C_{14}H_{16}F_3NO_4$
Chiral intermediate. *Synthetech, Inc.*

**1237 (tert-Butoxycarbonyl)-L-3,4,5-trifluorophenylalanine**
205445-54-1

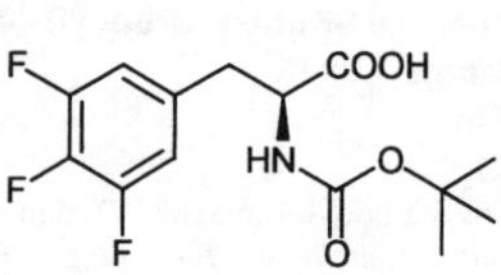

$C_{14}H_{16}F_3NO_4$
Chiral intermediate. *Synthetech, Inc.*

**1238 (tert-Butoxycarbonyl)-D-tryptophan**
5241-64-5 226-042-1

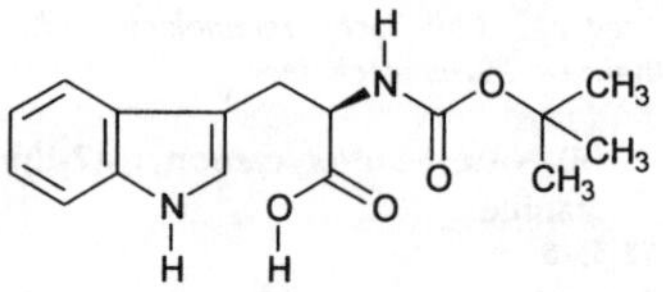

$C_{16}H_{20}N_2O_4$
Tryptophan, N-[(1,1-dimethylethoxy)carbonyl]-, D-; N-[(tert-Butoxy)carbonyl]-D-tryptophan; D-Tryptophan, N-carboxy-, N-tert-butyl ester. Chiral intermediate. [α] = + 20° (c = 2, HOAc). *Acros Organics nv; Isochem; Synthetech, Inc.*

**1239 N-α-(tert-Butoxycarbonyl)-L-tryptophan**
13139-14-5 236-072-7

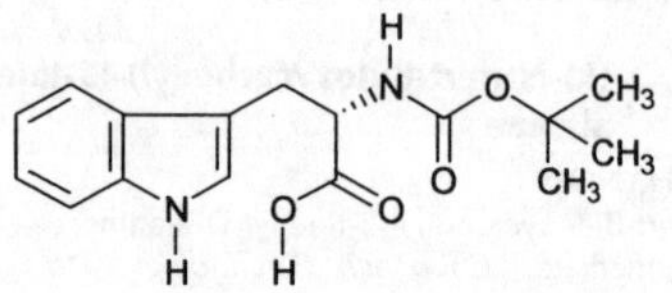

$C_{16}H_{20}N_2O_4$
Tryptophan, N-[(1,1-dimethylethoxy)carbonyl]-, L-; N-tert-Butoxycarbonyl-L-(-)-tryptophan; N-[(1,1-Dimethylethoxy)carbonyl]-L-tryptophan; boc-trp; BOC-L-tryptophane; Tryptophan, N-carboxy-, N-tert-butyl ester, L-. Listed on TSCA. Chiral intermediate. mp = 136°; [α] = - 19.8° (c = 2, HOAc). *Acros Organics nv; Sigma-Aldrich Fine Chemicals; Synthetech, Inc.*

**1240 (tert-Butoxycarbonyl)-L-tryptophan amide**

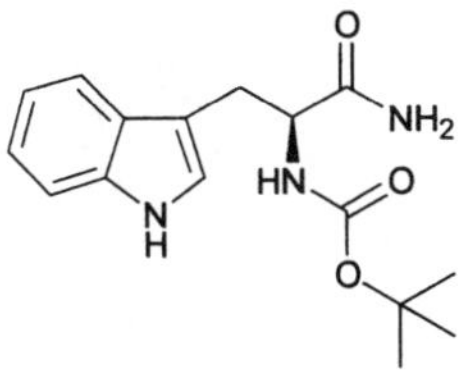

$C_{16}H_{21}N_3O_3$
Chiral intermediate. *Senn Chemicals AG.*

**1241 (tert-Butoxycarbonyl)-D-tryptophanol**

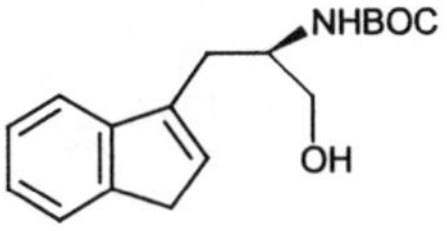

$C_{16}H_{22}N_2O_3$
Chiral intermediate. *Omega Chemical Company Inc.*

**1242 (tert-Butoxycarbonyl)-L-tryptophanol**
82689-19-8

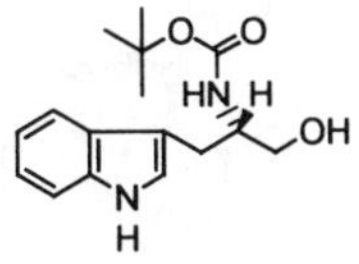

$C_{16}H_{22}N_2O_3$
Chiral intermediate. mp = 118-122°; $[\alpha]_D^{20}$ = - 51° (c = 2, HOAc). *Omega Chemical Company Inc.; Sigma-Aldrich Fine Chemicals.*

**1243 (tert-Butoxycarbonyl)-D-tyrosine**
70642-86-3

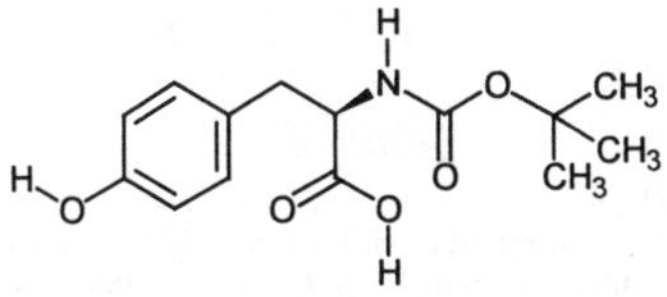

$C_{14}H_{19}NO_5$
Chiral intermediate. [α] = + 2.7° (c = 2, HOAc). *Acros Organics nv; Isochem; Synthetech, Inc.*

**1244 (tert-Butoxycarbonyl)-L-tyrosine**
3978-80-1 223-613-7

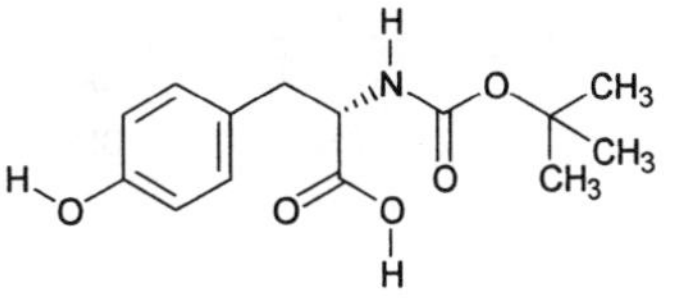

$C_{14}H_{19}NO_5$
Chiral intermediate. mp = 133-135°; [α] = + 2.5° (c = 2, HOAc). *Acros Organics nv; Sigma-Aldrich Fine Chemicals; Synthetech, Inc.*

**1245 (tert-Butoxycarbonyl)-D-tyrosine methyl ester**

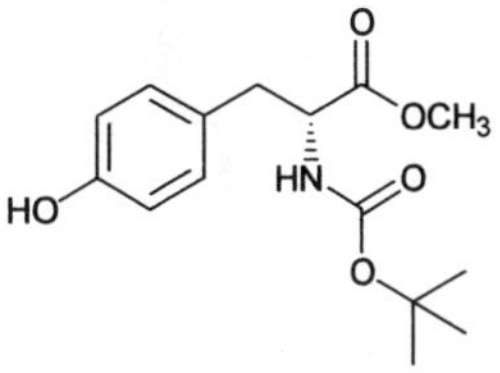

$C_{15}H_{21}NO_5$
Chiral intermediate. *Synthetech, Inc.*

**1246 (tert-Butoxycarbonyl)-L-tyrosine methyl ester**
4326-36-7

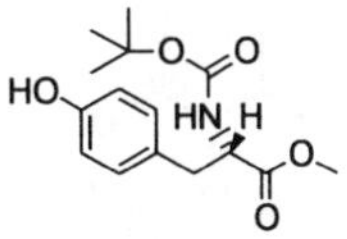

$C_{15}H_{21}NO_5$
Chiral intermediate. mp = 100-104°; $[\alpha]^{22}$ = + 51° (c = 1, $CHCl_3$). *Sigma-Aldrich Fine Chemicals; Synthetech, Inc.*

**1247 (tert-Butoxycarbonyl)-D-valine**
22838-58-0

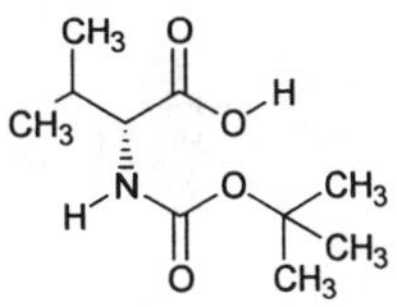

$C_{10}H_{19}NO_4$
Chiral intermediate. *Acros Organics nv; Synthetech, Inc.*

**1248 (tert-Butoxycarbonyl)-L-valine**
13734-41-3 237-307-6

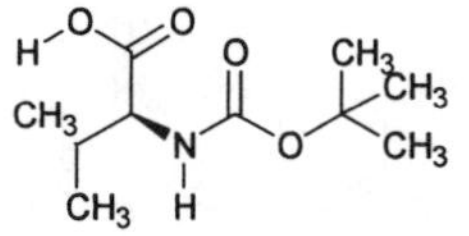

$C_{10}H_{19}NO_4$
Valine, N-[(1,1-dimethylethoxy)carbonyl]-, L-; N-[(1,1-Dimethylethoxy)carbonyl]-L-valine; N-(tert-Butoxy-carbonyl)-L-valine; Valine, N-carboxy-, N-tert-butyl ester, L-. Listed on TSCA. Chiral intermediate. mp = 78-80°; [α] = - 6.6° (c = 1, HOAc). *Acros Organics nv; Austin Chemical Company, Inc.; Flamma s.p.a.; Sigma-Aldrich Fine Chemicals; Synthetech, Inc.*

**1249 (tert-Butoxycarbonyl)-L-valine amide**

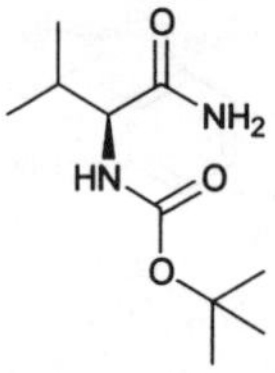

$C_{10}H_{20}N_2O_3$
Chiral intermediate. *Senn Chemicals AG.*

**1250 (tert-Butoxycarbonyl)-L-valine hydroxy-succinimide ester**

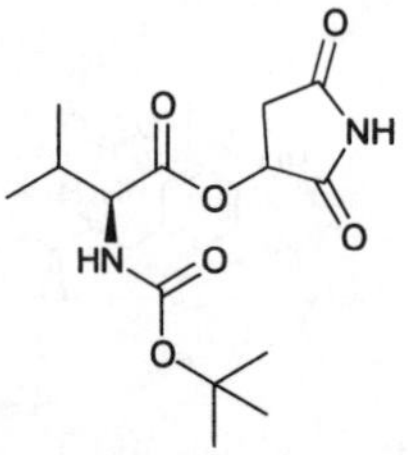

$C_{14}H_{22}N_2O_6$
Chiral intermediate. *Synthetech, Inc.*

**1251 N-(tert-Butoxycarbonyl)-valine N-methoxy-N-methylamide**
87694-52-8

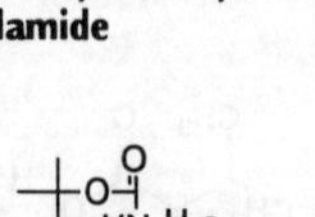

$C_{12}H_{24}N_2O_4$
Chiral intermediate. bp = 248°; d = 1.029; n = 1.456; $[\alpha]^{23}$ = - 18.5° (c = 1, $CH_3OH$). *Sigma-Aldrich Fine Chemicals.*

**1252 (tert-Butoxycarbonyl)-D-valine methyl ester**

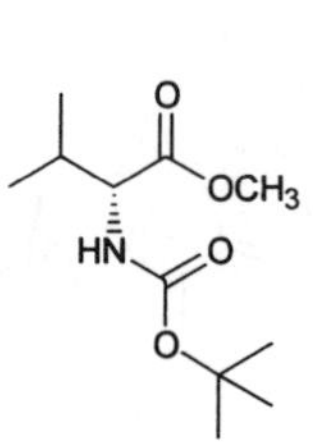

$C_{11}H_{21}NO_4$
Chiral intermediate. *Austin Chemical Company, Inc.; Synthetech, Inc.*

**1253 (tert-Butoxycarbonyl)-L-valine methyl ester**
58561-04-9

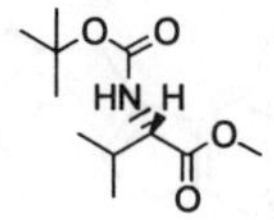

$C_{11}H_{21}NO_4$
Chiral intermediate. bp = 194°; d = 1.004; n = 1.44; $[\alpha]_D^{20}$ = - 22° (c = 1, $CH_3OH$). *Austin Chemical Company, Inc.; Sigma-Aldrich Fine Chemicals; Synthetech, Inc.*

**1254 (R)-(+)-(tert-Butoxycarbonyl)-valinol**
106391-87-1

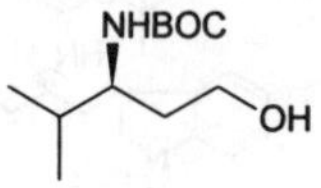

$C_{10}H_{21}NO_3$
Boc-D-valinol.
N-t-BOC-D-valinol. Chiral intermediate. *Daiso Company, Ltd.; Omega Chemical Company Inc.*

**1255 (tert-Butoxycarbonyl)-L-valinol**
79069-14-0

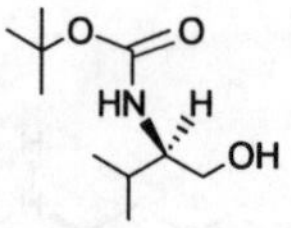

$C_{10}H_{21}NO_3$
N-(tert-Butoxycarbonyl)-L-valinol; N-tert-Butoxycarbonyl-(S)-2-amino-3-methyl-1-butanol; (S)-(-)-BOC-valinol. Chiral intermediate. bp = 208°; d = 0.9950; $n_D^{20}$ = 1.4510; $[\alpha]_D^{20}$ = - 21 ± 1° (c = 1, $CHCl_3$). *Daiso Company, Ltd.; Lancaster Synthesis Ltd.; Omega Chemical Company Inc.; Sigma-Aldrich Fine Chemicals; SK Energy and Chemical, Inc.; Synthetech, Inc.*

### 1256 (tert-Butoxycarbonyl)-L-valyl-L-proline

$C_{15}H_{26}N_2O_5$
Chiral intermediate. *Synthetech, Inc.*

### 1257 Nα-(tert-Butoxycarbonyl)-Nβ-xanthenyl-L-asparagine

65420-40-8
$C_{22}H_{24}N_2O_6$
Chiral intermediate. *Sigma-Aldrich Fine Chemicals.*

### 1258 (S)-(+)-2-Butylamine

513-49-5 1579(12) 208-164-7

$C_4H_{11}N$
Butan-2-amine, (2S)-; (+)-2-Aminobutane; (S)-1-Methylpropylamine. Listed on TSCA. Chiral building block. bp = 59-65°; d = 0.7220; $n_D^{20}$ = 1.3930; $[\alpha]_D^{20}$ = + 7.5 ± 0.5° (neat). *Acros Organics nv; Lancaster Synthesis Ltd.; Norse Laboratories; Sigma-Aldrich Fine Chemicals.*

### 1259 tert-Butyl-(R)-3-aminobutyrate

$C_8H_{17}NO_2$
Chiral building block. *Oxford Asymmetry International plc.*

### 1260 tert-Butyl-(3S)-3-aminobutanoate

161105-54-0

$C_8H_{17}NO_2$
Chiral intermediate. *Acros Organics nv; Oxford Asymmetry International plc.*

### 1261 tert-Butyl (3R)-3-amino-4-hexenoate

$C_{10}H_{19}NO_2$
Chiral building block. *Acros Organics nv; Oxford Asymmetry International plc.*

### 1262 tert-Butyl (3S)-3-amino-4-hexenoate

$C_{10}H_{19}NO_2$
Chiral building block. *Acros Organics nv; Oxford Asymmetry International plc.*

### 1263 tert-Butyl (2R,3R)-3-amino-2-hydroxy-butanoate

$C_8H_{17}NO_3$
Chiral building block. *Acros Organics nv; Oxford Asymmetry International plc.*

### 1264 tert-Butyl (2S,3S)-3-amino-2-hydroxy-butanoate

$C_8H_{17}NO_3$
Chiral building block. *Acros Organics nv; Oxford Asymmetry International plc.*

### 1265 tert-Butyl-(R)-3-amino-2-hydroxy-4-hexenoate

$C_{10}H_{19}NO_3$
Chiral building block. *Oxford Asymmetry International plc.*

**1266 tert-Butyl-(S)-3-amino-2-hydroxy-4-hexenoate**

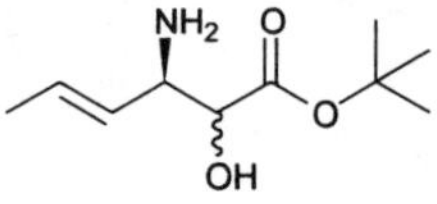

$C_{10}H_{19}NO_3$
Chiral building block. *Oxford Asymmetry International plc.*

**1267 tert-Butyl (2R,3S)-3-amino-2-hydroxy-4-phenylbutanoate**
119626-06-1

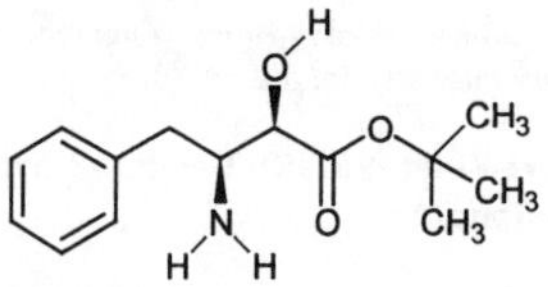

$C_{14}H_{21}NO_3$
Chiral intermediate. *Acros Organics nv; Oxford Asymmetry International plc.*

**1268 tert-Butyl-(S)-3-amino-2-hydroxy-4-phenylbutyrate**

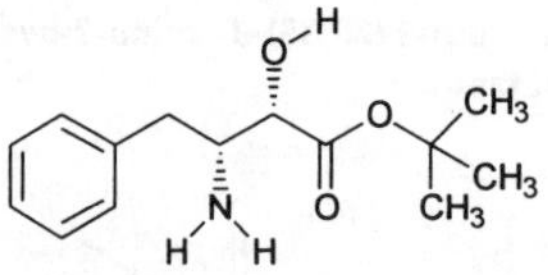

$C_{14}H_{21}NO_3$
Chiral building block. *Acros Organics nv; Oxford Asymmetry International plc.*

**1269 tert-Butyl-(R)-3-amino-2-hydroxy-3-phenylpropionate**

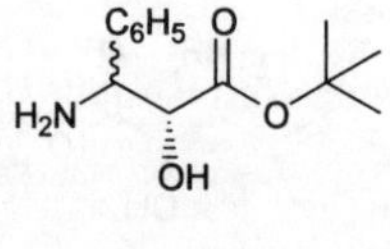

$C_{13}H_{19}NO_3$
Chiral building block. *Oxford Asymmetry International plc.*

**1270 tert-Butyl (2S,3S)-3-amino-2-hydroxy-3-phenylpropanoate**

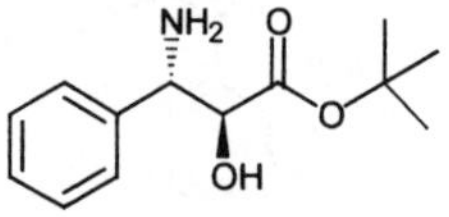

$C_{13}H_{19}NO_3$
Chiral intermediate. *Acros Organics nv; Oxford Asymmetry International plc.*

**1271 (4R,3R)-tert-Butyl-6-aminomethyl-2,2-dimethyl-1,3-dioxane-4-acetate**
$C_{13}H_{25}NO_4$
Chiral intermediate. *Sigma-Aldrich Fine Chemicals.*

**1272 tert-Butyl (3R)-3-amino-5-methylhexanoate**

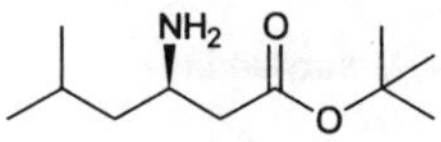

$C_{11}H_{23}NO_2$
Chiral building block. [α] = - 12°. *Acros Organics nv; Oxford Asymmetry International plc.*

**1273 tert-Butyl (3S)-3-amino-5-methylhexanoate**

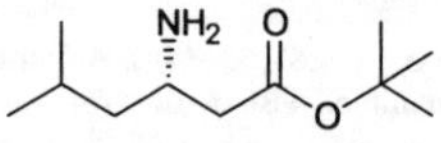

$C_{11}H_{23}NO_2$
Chiral building block. *Acros Organics nv; Oxford Asymmetry International plc.*

**1274 tert-Butyl (3R)-3-amino-4-phenylbutanoate**
166023-31-0

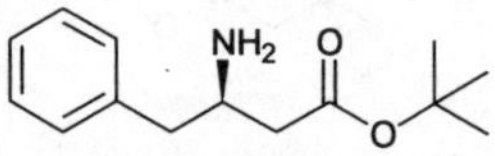

$C_{14}H_{21}NO_2$
Chiral intermediate. [α] = - 2.3°. *Acros Organics nv; Oxford Asymmetry International plc.*

**1275 tert-Butyl-(3S)-3-amino-4-phenylbutanoate**
120686-17-1

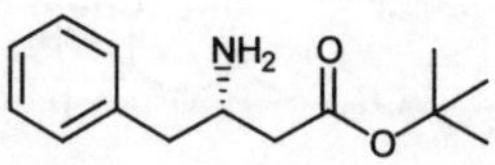

$C_{14}H_{21}NO_2$
Chiral intermediate. *Acros Organics nv; Oxford Asymmetry International plc.*

**1276 tert-Butyl-(3R)-3-amino-3-phenyl-propanoate**
161671-34-7

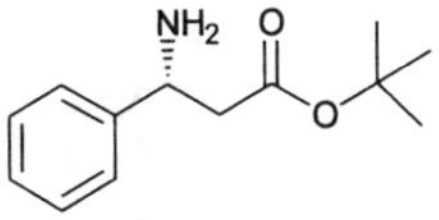

$C_{13}H_{19}NO_2$
Chiral intermediate. *Acros Organics nv; Oxford Asymmetry International plc.*

**1277 tert-Butyl (3S)-3-amino-3-phenyl-propanoate**
120686-18-2

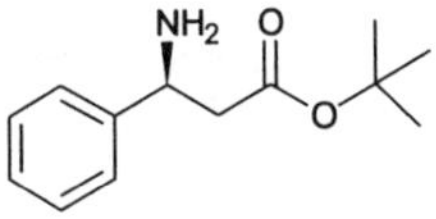

$C_{13}H_{19}NO_2$
Chiral intermediate. [α] = - 19°. *Acros Organics nv; Oxford Asymmetry International plc.*

**1278 (R)-(+)-3-tert-Butylamino-1,2-propanediol**

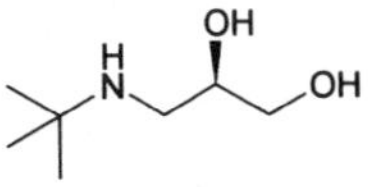

$C_7H_{17}NO_2$
Propane-1,2-diol, 3-[(1,1-dimethylethyl)amino]-, (R)-; 1,2-Propanediol, 3-(tert-butylamino)-, (R)-. Chiral building block. *Bachem AG.*

**1279 (S)-(-)-3-(3-tert-Butylamino)-1,2-propanediol**
30315-46-9 250-125-1

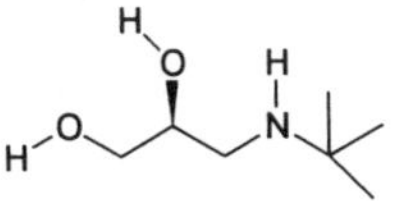

$C_7H_{17}NO_2$
Propane-1,2-diol, 3-[(1,1-dimethylethyl)amino]-, (S)-; 1,2-Propanediol, 3-(tert-butylamino)-, (S)-(-)-. Chiral intermediate. mp = 85-89°; $[\alpha]^{23}$ = - 30° (c = 2, 1N HCl). *Acros Organics nv; Bachem AG; D&O Group; DSM Fine Chemcials Netherlands; Shanghai DSL International Trading Company; Sigma-Aldrich Fine Chemicals; Vinchem; Zhejiang Chemicals Import and Export Corporation.*

**1280 tert-Butyl-(S)-(-)-5-benzyl-2-oxo-4-morpholinecarboxylate**

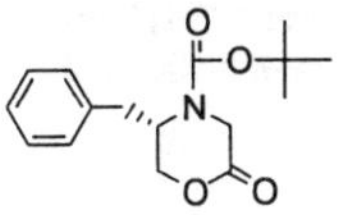

$C_{16}H_{21}NO_4$
Chiral intermediate. mp = 98-100°; $[\alpha]_D^{20}$ = - 14.5° (c = 2.45, $CHCl_3$). *Sigma-Aldrich Fine Chemicals.*

**1281 (R)-tert-Butylcarboxamide-4-tert-butoxy-carbonylpiperazine**
171866-36-7

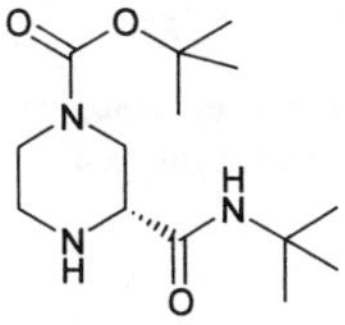

$C_{14}H_{27}N_3O_3$
Chiral intermediate. *Yamakawa Chemical Industry Co. Ltd.*

**1282 (S)-2-tert-Butylcarboxamide-4-tert-butoxy-carbonylpiperazine**
150323-35-6

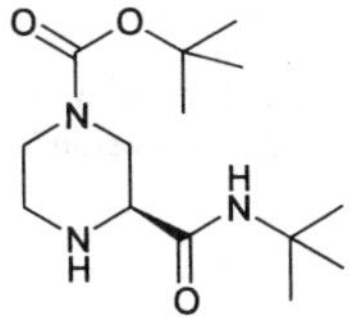

$C_{14}H_{27}N_3O_3$
Chiral intermediate. *Sigma-Aldrich Fine Chemicals; Yamakawa Chemical Industry Co. Ltd.*

**1283 (4R,3R)-tert-Butyl-6-cyanomethyl-2,2-dimethyl-1,3-dioxane-4-acetate**
125971-94-0
$C_{14}H_{23}NO_4$
Chiral intermediate. *Sigma-Aldrich Fine Chemicals.*

**1284 N-tert-Butyl-(3S)-decahydroisoquinoline-3-carboxamide**
136465-81-1

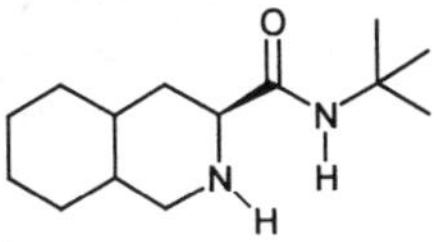

$C_{14}H_{26}N_2O$
(3S)-4-Azabicyclo[4.4.0]dec-3-yl]-N-(tert-butyl)carbox-

amide. Chiral intermediate. mp = 112-115°; [α] = - 70°. *Acros Organics nv; Fischer Chemicals AG; Great Lakes Fine Chemicals; Sigma-Aldrich Fine Chemicals.*

**1285 (4R)-N-(tert-Butyldimethylsilyl)azetidin-2-one-4-carboxylic acid**
162856-35-1

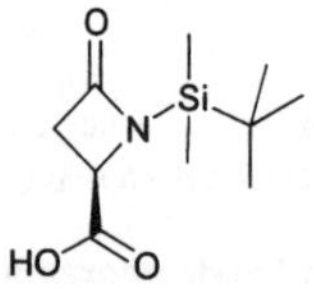

$C_{10}H_{19}NO_3Si$
Chiral intermediate. *Acros Organics nv; Oxford Asymmetry International plc.*

**1286 (4S)-N-(tert-Butyldimethylsilyl)azetidin-2-one-4-carboxylic acid**
82938-50-9

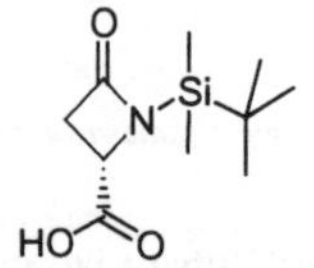

$C_{10}H_{19}NO_3Si$
Chiral intermediate. *Acros Organics nv; Oxford Asymmetry International plc.*

**1287 (-)-1-O-tert-Butyldimethylsilyl 2-azido-2-deoxyglucopyranoside 3,4,6-triacetate**

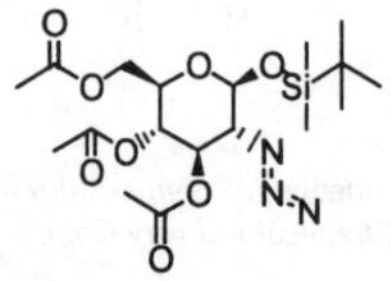

$C_{18}H_{31}N_3O_8Si$
Chiral intermediate. mp = 66-70°; $[\alpha]_D^{20}$ = - 3° (c = 1, $CHCl_3$). *Sigma-Aldrich Fine Chemicals.*

**1288 (+)-6-O-(tert-Butyldimethylsilyl)galactal**
124751-19-5

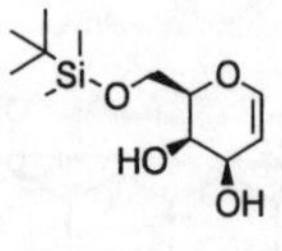

$C_{12}H_{24}O_4Si$
Chiral intermediate. n = 1.476; $[\alpha]^{23}$ = + 7° (c = 1.4, $CHCl_3$). *Sigma-Aldrich Fine Chemicals.*

**1289 (-)-6-O-(tert-Butyldimethylsilyl)galactal cyclic carbonate**
163228-38-4

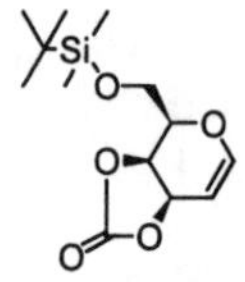

$C_{13}H_{22}O_5Si$
Chiral intermediate. mp = 45-49°; $[\alpha]_D^{20}$ = - 66° (c = 2, $CHCl_3$). *Sigma-Aldrich Fine Chemicals.*

**1290 tert-Butyldimethylsilyl (R)-(+)-glycidyl ether**

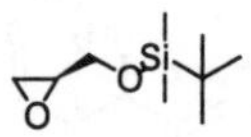

$C_9H_{20}O_2Si$
Chiral intermediate. bp = 197-199°; d = 0.89; n = 1.432; $[\alpha]_D^{20}$ = + 7° (neat). *Sigma-Aldrich Fine Chemicals.*

**1291 tert-Butyldimethylsilyl (S)-(-)-glycidyl ether**

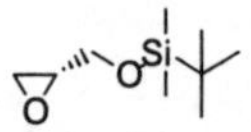

$C_9H_{20}O_2Si$
Chiral intermediate. bp = 195-199°; d = 0.87; n = 1.431. *Sigma-Aldrich Fine Chemicals.*

**1292 (-)-5-O-(tert-Butyldimethylsilyl)-2,3-O-isopropylideneribonic acid lactone**
75467-36-6

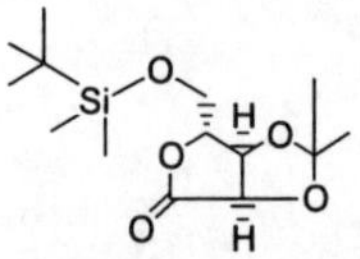

$C_{14}H_{26}O_5Si$
Chiral intermediate. mp = 72-74°; $[\alpha]_D^{20}$ = - 47° (c = 0.8, $CHCl_3$). *Sigma-Aldrich Fine Chemicals.*

**1293 (-)-3-O-(tert-Butyldimethylsilyl)-4,6-O-(4-methoxybenzilidene)glucal**
$C_{20}H_{30}O_5Si$
Chiral intermediate. mp = -30°; $[\alpha]^{25}$ = - 69° (c = 1.2, $CHCl_3$). *Sigma-Aldrich Fine Chemicals.*

**1294 (3S,4S)-3-[(R)-1-(tert-Butyldimethylsilyloxy)ethyl]-4-[(R)-1-carboxyethyl]-2-azetidinone**
90776-58-2
$C_{14}H_{27}NO_4Si$
(3S,4S)-3-Acetoxy-(R)-(tert-butyldimethylsilyloxy)ethyl)-4-((R)-carboxyethyl)-2-azetidinone. Chiral intermediate. *Kaneka Corporation; Takasago International Corporation; TCI America.*

**1295 (3αR,4S,5R,6αS)-(-)-(tert-Butyldimethylsilyloxy)hexahydro-5-(tetrahydro-2H-pyran-2-yloxy)-2hf**
65025-95-8

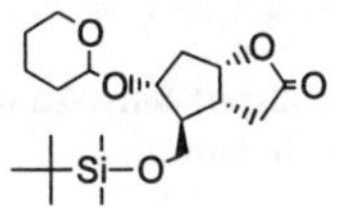

$C_{19}H_{34}O_5Si$
Chiral intermediate. mp = 40-53°; $[\alpha]^{22}$ = - 29° (c = 1, $CHCl_3$). *Sigma-Aldrich Fine Chemicals.*

**1296 (S)-4-(tert-Butyldimethylsilyloxy)-2-hydroxybutanoic acid**
$C_{11}H_{24}O_4Si$
Chiral intermediate. *Sigma-Aldrich Fine Chemicals.*

**1297 tert-Butyl-2-(2-diphenylphosphanyl phenyl)-4,5-dihydrooxazole**

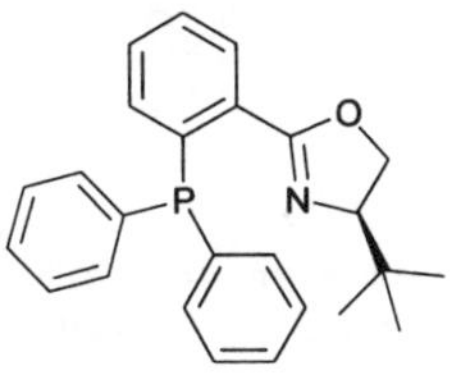

$C_{25}H_{26}NOP$
Chiral intermediate. *Acros Organics nv.*

**1298 (-)-6-O-(tert-Butyldiphenylsilyl)galactal cyclic carbonate**
151265-18-8

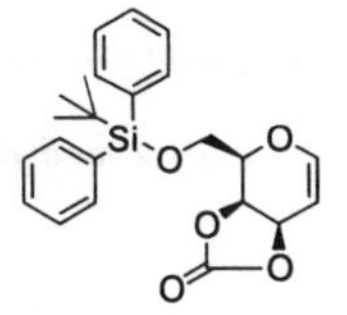

$C_{23}H_{26}O_5Si$
Chiral intermediate. mp = -3°; $[\alpha]_D^{20}$ = - 40.0° (c = 1, $CHCl_3$). *Sigma-Aldrich Fine Chemicals.*

**1299 (+)-6-O-(tert-Butyldiphenylsilyl)-galactal cyclic carbonate**
137893-35-7

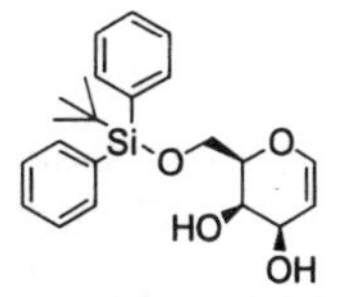

$C_{22}H_{28}O_4Si$
Chiral intermediate. $[\alpha]_D^{20}$ = + 4° (c = 1, $CHCl_3$). *Sigma-Aldrich Fine Chemicals.*

**1300 (+)-6-O-(tert-Butyldiphenylsilyl)glucal**
87316-22-1

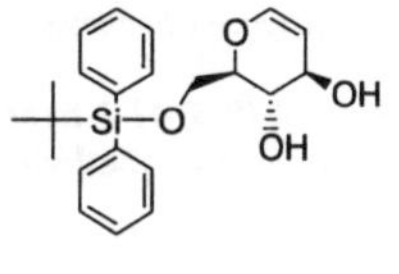

$C_{22}H_{28}O_4Si$
Chiral intermediate. mp = -9°; $[\alpha]^{25}$ = + 9° (c = 1.1, $CHCl_3$). *Sigma-Aldrich Fine Chemicals.*

**1301 n-Butyl (2R,3S)-(-)-2,3-epoxybutyrate**

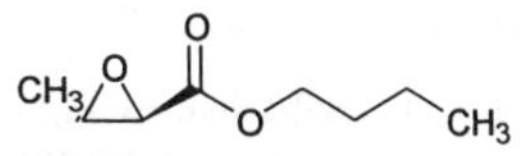

$C_8H_{14}O_3$
Chiral building block. [α] = - 17°. *Acros Organics nv; Mitsubishi Chemical Corporation.*

**1302 tert-Butyl (R)-(+)-4-formyl-2,2-dimethyl-3-oxazolidinecarboxylate**
95715-87-0

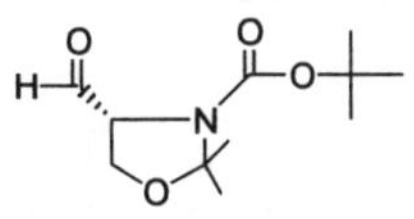

$C_{11}H_{19}NO_4$
Chiral intermediate. $bp_{0.3}$ = 67°; d = 1.06; n = 1.445; $[\alpha]^{23}$ = + 90° (c = 1, $CHCl_3$). *Sigma-Aldrich Fine Chemicals.*

**1303 tert-Butyl (R)-(-)-2-hydroxybutyrate**

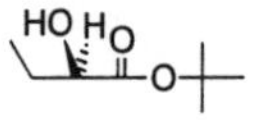

$C_8H_{16}O_3$
Chiral building block. mp = 52-54°; $[\alpha]_D^{20}$ = + 5.1° (c = 2, $CCl_4$). *Sigma-Aldrich Fine Chemicals.*

**1304 tert-Butyl (S)-(-)-2-hydroxybutyrate**
37787-90-9

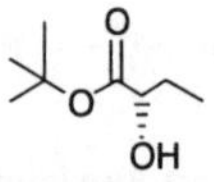

$C_8H_{16}O_3$
Chiral building block. mp = 52-54°; $[\alpha]_D^{20}$ = - 5.1° (c = 2, $CCl_4$). *Sigma-Aldrich Fine Chemicals.*

**1305 tert-Butyl R-(R,S)-N-2-hydroxy-2-(3-hydroxyphenyl)-1-methylethylcarbamate**
112113-57-2

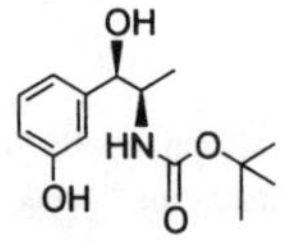

$C_{14}H_{21}NO_4$
[N-(tert-butoxycarbonyl)metaraminol. Chiral intermediate. mp = 60-65°; $[\alpha]_D^{20}$ = - 40° (c = 1, $CH_2Cl_2$). *Sigma-Aldrich Fine Chemicals.*

**1306 t-Butyl (3R,5S)-6-hydroxy-3,5-O-isopropylidene-3,5-dihyroxyhexanoate**
124655-09-0

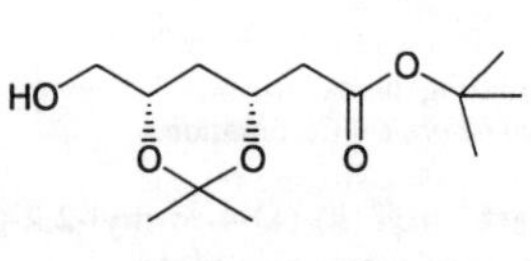

$C_{13}H_{24}O_5$
Chiral intermediate. *Kaneka Corporation.*

**1307 n-Butyl-L-lactate**
34451-19-9 252-036-3

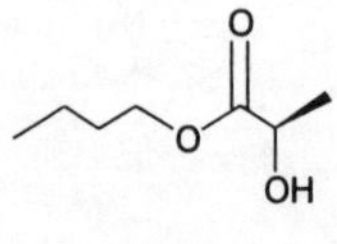

$C_7H_{14}O_3$
Propanoic acid, 2-hydroxy-, butyl ester, (S)-; n-Butyl (S)-(-)-2-hydroxypropionate; L-Lactic acid n-butyl ester; (S)-(-)-2-Hydroxypropanoic acid butyl ester. Chiral building block. mp = -28°; bp = 180-195°; d = 0.9800; $n_D^{20}$ = 1.4200; $[\alpha]_D^{20}$ = - 12 ± 1° (neat). *Lancaster Synthesis Ltd.; Nitrokemia 2000 Rt.; Sigma-Aldrich Fine Chemicals.*

**1308 tert-Butyl (R)-(+)-lactate**
68166-83-6

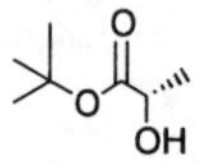

$C_7H_{14}O_3$
Chiral building block. mp = 38-42°; $[\alpha]_D^{20}$ = + 7.3° (c = 1.7, $CH_2Cl_2$). *Sigma-Aldrich Fine Chemicals.*

**1309 (3R,7aS)-3-tert-Butyl-7a-methyl bicyclic lactam**
$C_{11}H_{19}NO_2$
Chiral intermediate. *Oxford Asymmetry International plc.*

**1310 (3S,7aR)-3-tert-Butyl-7a-methyl bicyclic lactam**
$C_{11}H_{19}NO_2$
Chiral intermediate. *Oxford Asymmetry International plc.*

**1311 (2R,6R)-2-tert-Butyl-6-methyl-1,3-dioxan-4-one**
100017-18-3

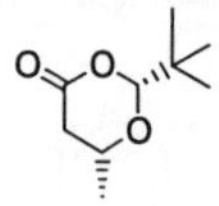

$C_9H_{16}O_3$
Chiral intermediate. mp = 82-84°; $[\alpha]_D^{20}$ = - 56° (c = 1, $CHCl_3$). *Sigma-Aldrich Fine Chemicals.*

**1312 (4R)-4-tert-Butyl-2-methyl-2-oxazoline**

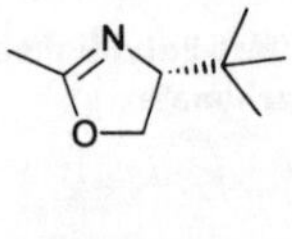

$C_8H_{15}NO$
Chiral intermediate. *Oxford Asymmetry International plc.*

**1313 (4S)-4-tert-Butyl-2-methyl-2-oxazoline**

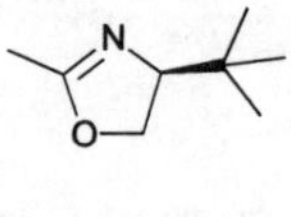

$C_8H_{15}NO$
Chiral intermediate. *Oxford Asymmetry International plc.*

**1314 (R)-(+)-4-tert-Butyl-2-oxazolidinone**

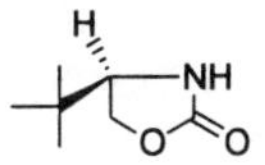

$C_7H_{13}NO_2$
Chiral intermediate. mp = 119-122°; $[\alpha]^{23}$ = + 17° (c = 6, EtOH). *Sigma-Aldrich Fine Chemicals.*

**1315 (S)-(-)-4-tert-Butyl-2-oxazolidinone**
54705-42-9

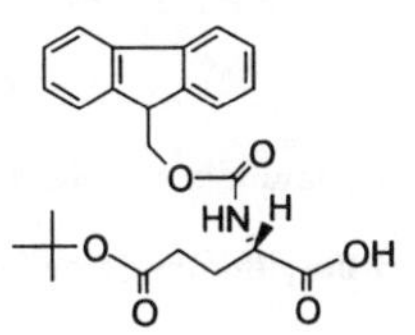

$C_7H_{13}NO_2$
Chiral intermediate. mp = 118-120°;$[\alpha]^{21}$ = - 16° (c = 6, EtOH). *Asymchem; Great Lakes Fine Chemicals; Sigma-Aldrich Fine Chemicals; Yoneyama Yakuhin Kogyo Co., Ltd.*

**1316 tert-Butyl-(S)-(R,R)-(-)-(1-oxiranyl-2-phenylethyl)carbamate**

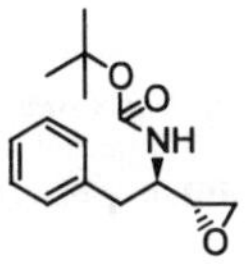

$C_{15}H_{21}NO_3$
Chiral intermediate. mp = 125-127°; $[\alpha]^{23}$ = - 7° (c = 0.6, $CH_3OH$). *Sigma-Aldrich Fine Chemicals.*

**1317 tert-Butyl (2R,3S)-(-)-6-oxo-2,3-diphenyl-4-morpholinecarboxylate**
112741-49-8

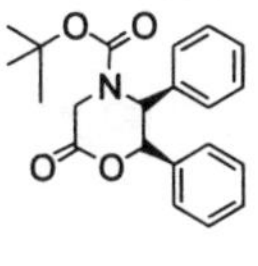

$C_{21}H_{23}NO_4$
Chiral intermediate. mp = 206°; $[\alpha]^{25}$ = - 87° (c = 5.5, $CH_2Cl_2$). *Sigma-Aldrich Fine Chemicals.*

**1318 tert-Butyl (2S,3R)-(+)-6-oxo-2,3-diphenyl-4-morpholinecarboxylate**
112741-50-1

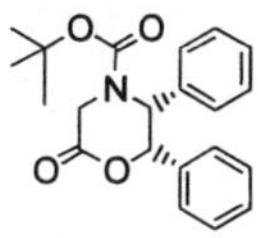

$C_{21}H_{23}NO_4$
Chiral intermediate. mp = 206°; $[\alpha]_D^{20}$ = + 86° (c = 5.5, $CH_2Cl_2$). *Sigma-Aldrich Fine Chemicals.*

**1319 tert-Butyl (3R,5S)-6-oxo-3,5-O-isopropylidene-3,5-dihyroxyhexanoate**
124752-23-4

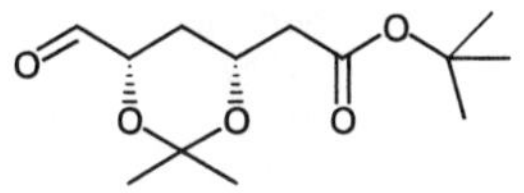

$C_{13}H_{22}O_5$
Chiral intermediate. *Kaneka Corporation.*

**1320 N-(4-tert-Butylphenylsulfonyl)proline**

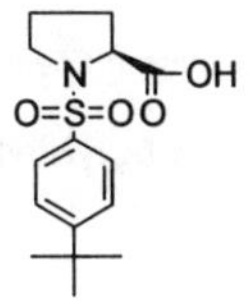

$C_{15}H_{21}NO_4S$
Chiral intermediate. mp = 158-160°; $[\alpha]^{21}$ = - 95° (c = 1, $CHCl_3$). *Sigma-Aldrich Fine Chemicals.*

**1321 (S)-3-tert-Butyl-2,5-piperazinedione**
65050-07-9

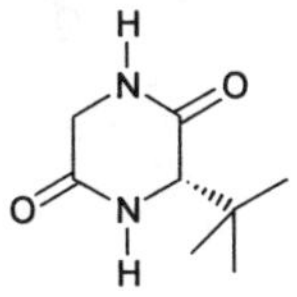

$C_8H_{14}N_2O_2$
Chiral building block. $[\alpha]$ = + 76.5° (c = 1, HOAc). *Acros Organics nv.*

**1322 Butyl (S)-(-)-2-pyrrolidone-5-carboxylate**
4931-68-4 225-568-9

$C_9H_{15}NO_3$
Butyl L-pyroglutamate. Chiral building block. $bp_{6.5}$ = 180°; d = 1.104; n = 1.403; $[\alpha]_D^{20}$ = - 13° (neat). *Sigma-Aldrich Fine Chemicals.*

**1323 L-threo-tert-Butylserine monohydrate**

$C_7H_{15}NO_3$
Chiral building block. *Acros Organics nv.*

**1324 (1E,3S)-3-tert-Butylsilyloxy-3-cyclohexyl-iodoprop-1-ene**
$C_{15}H_{29}IOSi$
Chiral intermediate. *Nissan Chemical Industries, Ltd.*

**1325 (1E,3S)-3-tert-Butylsilyloxy-1-iodooct-1-ene**
41138-67-4
$C_{14}H_{29}IOSi$
Chiral intermediate. *Nissan Chemical Industries, Ltd.*

**1326 (1E,3S,5R)-3-tert-Butylsilyloxy-5-methyl-1-iodonon-1-ene**

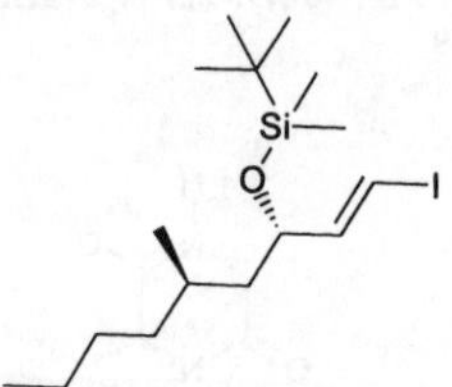


$C_{16}H_{33}IOSi$
Chiral intermediate. *Nissan Chemical Industries, Ltd.*

**1327 (R)-tert-Butylsulfinamide**

$C_4H_{11}NOS$
Chiral building block. *Advanced Asymmetrics, Inc.*

**1328 (S)-tert-Butylsulfinamide**

$C_4H_{11}NOS$
Resolving agent. *Advanced Asymmetrics, Inc.*

**1329 S-(tert-Butylthio)cysteine hydrate**
250-012-7

$C_7H_{15}NO_2S_2$
3-(tert-Butyldithio)-L-alanine. Chiral intermediate. mp = 177°; $[\alpha]^{21}$ = - 82° (c = 2, 1N HCl). *Sigma-Aldrich Fine Chemicals.*

**1330 O-tert-Butyl-L-tyrosine**
18822-59-8
$C_{13}H_{19}NO_3$
Chiral building block. *Sigma-Aldrich Fine Chemicals.*

**1331 (R)-(+)-3-Butyn-2-ol**
42969-65-3

$C_4H_6O$
Chiral building block. bp = 108-111°; d = 0.89; n = 1.426; $[\alpha]^{22}$ = + 45° (neat). *ChiroTech Technology Ltd.; DSM Fine Chemcials Netherlands; Sigma-Aldrich Fine Chemicals.*

**1332 (S)-(-)-3-Butyn-2-ol**
2914-69-4

$C_4H_6O$
Chiral building block. bp = 108-111°; d = 0.8940; $n_D^{20}$ = 1.4260; $[\alpha]_D^{20}$ = - 45° (neat). *ChiroTech Technology Ltd.; DSM Fine Chemcials Netherlands; Lancaster Synthesis Ltd.; Sigma-Aldrich Fine Chemicals.*

**1333 Caffeine chlorogenic acid complex (Potassium Salt)**

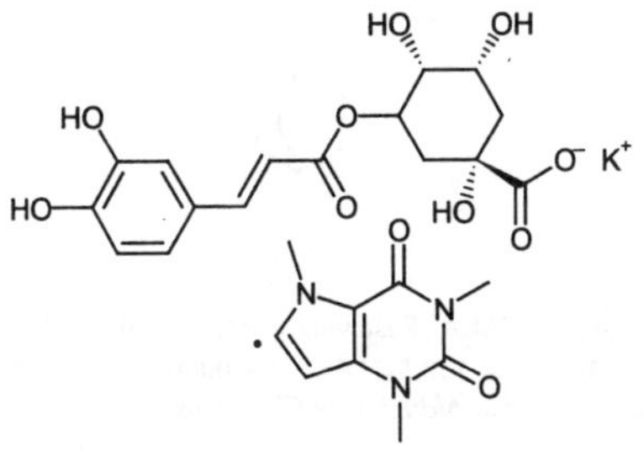

$C_{16}H_{17}KO_9 + C_8H_{10}N_4O_2$
Chiral intermediate. *Kaden Biochemicals GmbH.*

**1334 L-Calcium ascorbate dihydrate**
5743-28-2 1648(11)

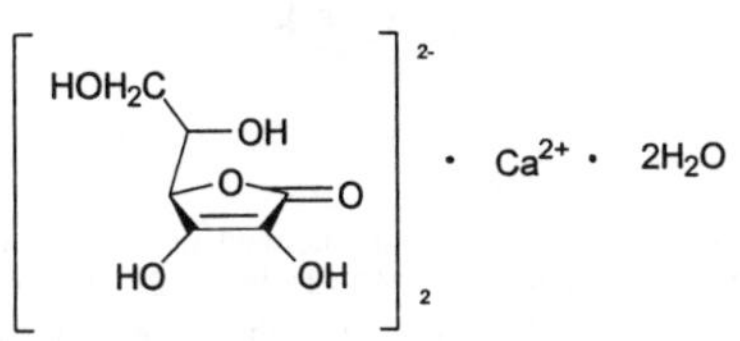

$C_6H_8O_6$
L-Ascorbic acid, calcium salt dihydrate. Chiral building block. mp = 166°; $[\alpha]_D^{20}$ = + 95° (c = 5, $H_2O$). *Sigma-Aldrich Fine Chemicals.*

**1335 Calcium L-aspartate**
21059-46-1 244-185-8

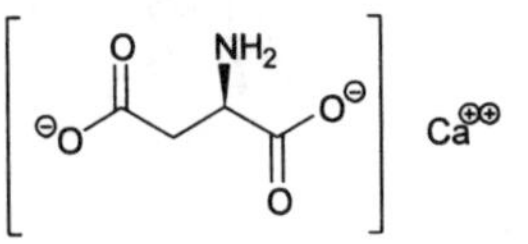

$C_4H_7NO_4.xCa$
Aspartic acid, calcium salt, L-. Chiral building block. *Varsal Instruments, Inc.*

**1336 Calcium (+)-10-camphorsulfonate**
1331-87-9 215-564-5

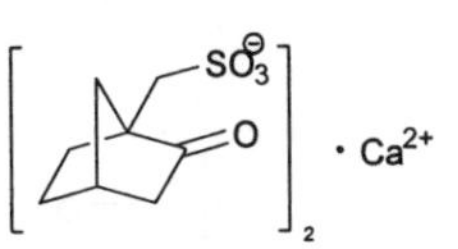

$C_{20}H_{30}CaO_8S_2$
Bornanesulfonic acid, 2-oxo-, calcium salt (2:1); Bicyclo[2.2.1]heptane-1-methanesulfonic acid, 7,7-dimethyl-2-oxo-, calcium salt, (1S)-; Calcium (1S)-[7,7-dimethyl-2-oxobicyclo[2.2.1]hept-1-yl]methanesulphonate. Chiral intermediate; Resolving agent. *Calaire Chimie s.a.*

**1337 Calcium-D-galactonate pentahydrate**
6622-52-2

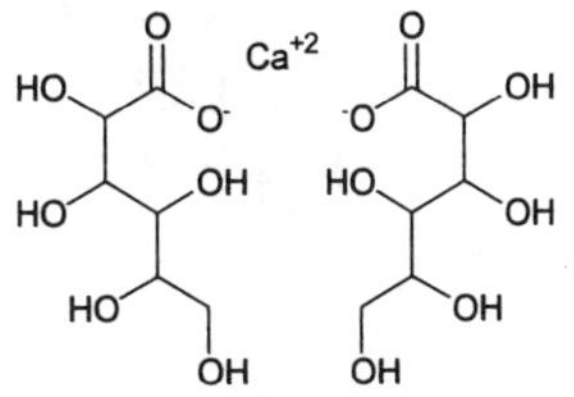

$C_{12}H_{22}CaO_{14}$
Chiral intermediate. *Pfanstiehl Laboratories, Inc.*

**1338 Calcium L-glutamate**
19238-49-4 242-905-5

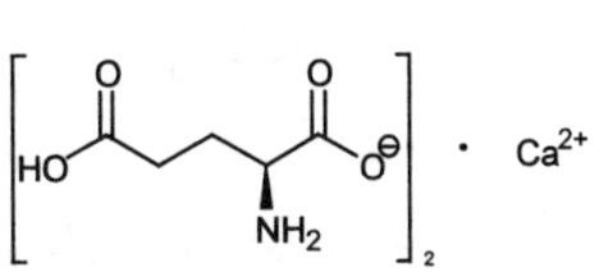

$C_5H_9NO_4.xCa$
Glutamic acid, calcium salt, L-. Listed on TSCA. Chiral building block. *Rexim S.A. Produits Chimiques.*

**1339 Calcium-D-saccharate**
5793-88-4 227-334-1

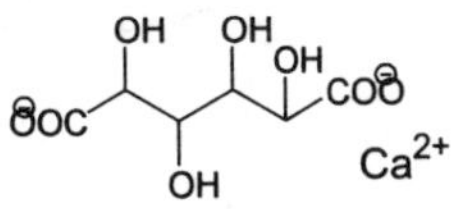

$C_6H_{10}CaO_8$
Glucaric acid, calcium salt (1:1), D-; Antacidin; Saccharated lime. Listed on TSCA. Chiral intermediate. *Aceto Corporation; KingChem, Inc.*

**1340 Calcium D-xylonate**

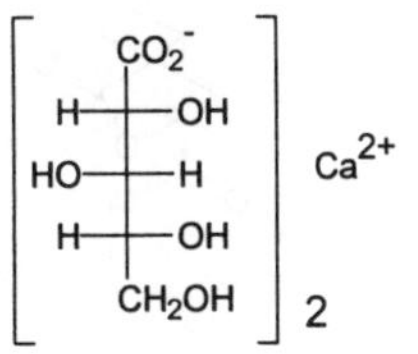

$C_{10}H_{18}CaO_{12}$
Chiral building block. $[\alpha]_D^{20}$ = + 11° (c = 5, $H_2O$). *Pfanstiehl Laboratories, Inc.*

**1341 Campesterol**
474-62-4 1776(12) 207-484-4

$C_{28}H_{48}O$
Pharmaceutical or derivative. mp = 158-160°; $[\alpha]^{25}$ = - 34° (c = 0.45, $CHCl_3$). *Sigma-Aldrich Fine Chemicals.*

**1342 (1R)-(+)-Camphanic acid**
67111-66-4

$C_{10}H_{14}O_4$
Resolving agent. mp = 200-202°; $[\alpha]_D^{20}$ = + 19° (c = 1, dioxane). *Sigma-Aldrich Fine Chemicals.*

**1343 (1S)-(-)-Camphanic acid**
13429-83-9 236-552-6

$C_{10}H_{14}O_4$
Camphanic acid, (-)-ω-; 4,7,7-Trimethyl-3-oxo-2-oxabicyclo[2.2.1]heptane-1-carboxylic acid. Chiral intermediate. mp = 201-203°; $[\alpha]^{22}$ = - 17.6° (c = 1, dioxane). *Acros Organics nv; FineTech, Ltd.; Sigma-Aldrich Fine Chemicals.*

**1344 (1S)-(-)-Camphanic chloride**
39637-74-6 4008(12) 254-552-4

$C_{10}H_{13}ClO_3$
Oxa-2-bicyclo[2.2.1]heptane-1-carbonyl chloride, 4,7,7-trimethyl-3-oxo-, (1S,4R)-; 3-Oxa-2-oxobornane-4-carbonyl chloride. Resolving agent; Chiral diagnostic reagent. mp = 67-70°; $[\alpha]_D^{20}$ = - 18 ± 1° (c = 2, $CCl_4$). *Acros Organics nv; FineTech, Ltd.; Lancaster Synthesis Ltd.; Sigma-Aldrich Fine Chemicals.*

**1345 (-)-Camphene**
5794-04-7 1736(11) 227-337-8

$C_{10}H_{16}$
Listed on TSCA. Resolving agent; Chiral auxiliary mp = 36-38°; bp = 157.5-158.5°; d = 0.84; $[\alpha]_D^{20}$ = - 48° (c = 4, EtOH). *Sigma-Aldrich Fine Chemicals.*

**1346 (+)-Camphene**
5794-03-6 1736(11) 227-336-2

$C_{10}H_{16}$
Listed on TSCA. Chiral intermediate; Chiral auxiliary; Resolving agent. mp = 40-43°; bp = 159-160°; d = 0.85; $[\alpha]^{21}$ = + 12.6° (c = 4, EtOH). *Sigma-Aldrich Fine Chemicals.*

**1347 D-(+)-Camphor**
464-49-3 1738(11) 207-355-2

$C_{10}H_{16}O$
Bicyclo[2.2.1]heptan-2-one, 1,7,7-trimethyl-, (1R,4R)-; (+)-Bornan-2-one; Alcanfor. Listed on TSCA. Chiral intermediate. mp = 179-181°; $[\alpha]^{25}$ = + 44.1° (c = 10, EtOH). *Acros Organics nv; Sigma-Aldrich Fine Chemicals.*

**1348 L-(-)-Camphor**
464-48-2 1738(11) 207-354-7

$C_{10}H_{16}O$
Bicyclo[2.2.1]heptan-2-one, 1,7,7-trimethyl-, (1S,4S)-; Bornan-2-one; (1S)-1,7,7-Trimethylbicyclo[2.2.1]heptan-2-one. Listed on TSCA. Chiral intermediate. mp = 178-180°; $[\alpha]$ = - 30.7° (c = 11.84, $CH_3OH$); $[\alpha]_D^{20}$ = - 43° (c = 10, EtOH). *Acros Organics nv; Loba Feinchemie AG.*

**1349 D-(+)-Camphor monobrominated**

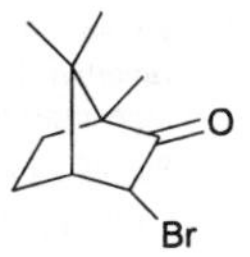

$C_{10}H_{15}BrO$
Resolving agent. mp = 74-77°; [α] = + 128-130°. *Taiwan Tekho Camphor Co. Ltd.*

**1350 (1R)-Camphor oxime**
2792-42-9 220-525-0

$C_{10}H_{17}NO$
Bicyclo[2.2.1]heptan-2-one, 1,7,7-trimethyl-, oxime, (1R,4R)-; (1R)-1,7,7-trimethylbicyclo[2.2.1]heptan-2-one oxime. Listed on TSCA. Chiral intermediate. mp = 119°. *TCI America.*

**1351 (1R)-(+)-Camphor p-tosylhydrazone**
4573-49-3

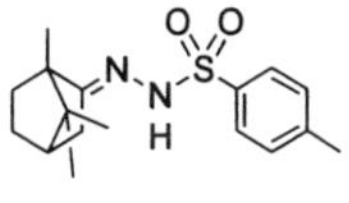

$C_{17}H_{24}N_2O_2S$
Resolving agent. mp = 162-165°; $[\alpha]_D^{20}$ = + 8.5° (c = 1, acetone). *Sigma-Aldrich Fine Chemicals.*

**1352 (1S)-(-)-Camphor p-tosylhydrazone**
123408-99-1

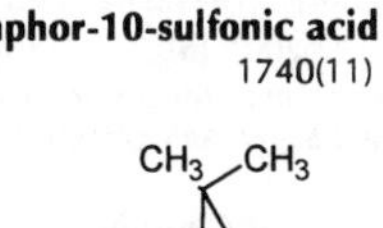

$C_{17}H_{24}N_2O_2S$
Resolving agent. mp = 164-167°; $[\alpha]_D^{20}$ = -8.5° (c = 1, acetone). *Sigma-Aldrich Fine Chemicals.*

**1353 (-)-Camphor-10-sulfonic acid**
35963-20-3 1740(11) 252-817-9

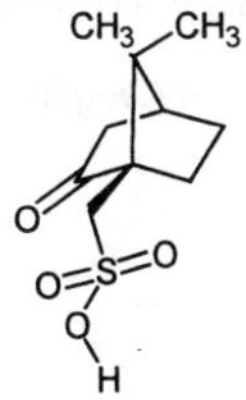

$C_{10}H_{16}O_4S$
Bicyclo[2.2.1]heptane-1-methanesulfonic acid, 7,7-dimethyl-2-oxo-, (1R)-; (1R)-[7,7-dimethyl-2-oxobicyclo[2.2.1]hept-1-yl]methanesulphonic acid. Resolving agent. mp = 198°; $[\alpha]_D^{20}$ = - 21° (c = 2, $H_2O$). *Acros Organics nv; Austin Chemical Company, Inc.; Calaire Chimie s.a.; Eastar Chemical Corporation; Interchem Corporation; Loba Feinchemie AG; Sigma-Aldrich Fine Chemicals; Taiwan Tekho Camphor Co. Ltd.; Yoneyama Yakuhin Kogyo Co., Ltd.*

**1354 (1S)-(+)-Camphor-10-sulfonic acid**
3144-16-9 1781(12) 221-554-1

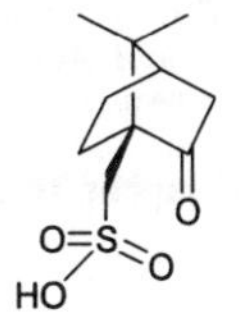

$C_{10}H_{16}O_4S$
Bicyclo[2.2.1]heptane-1-methanesulfonic acid, 7,7-dimethyl-2-oxo-, (1S,4R)-; (+)-CSA; D-2-Oxobornane-10-sulfonic acid; 10-Bornanesulfonic acid, 2-oxo-, (1S,4R)-(+)-; Reychler's acid; (+)-Camphor-10-sulfonic acid; D-(+)-Camphor-10-sulfonic acid. Resolving agent. mp = 194°; $[\alpha]_D^{20}$ = + 20 ± 2° (c = 2, $H_2O$). *Acros Organics nv; Austin Chemical Company, Inc.; Calaire Chimie s.a.; China Camphor Co. Ltd.; Interchem Corporation; Lancaster Synthesis Ltd.; Norse Laboratories; Senn Chemicals AG; Sigma-Aldrich Fine Chemicals; Taiwan Tekho Camphor Co. Ltd.; Yoneyama Yakuhin Kogyo Co., Ltd.*

**1355 L-10-Camphorsulfonic acid ammonium salt**
13867-85-1

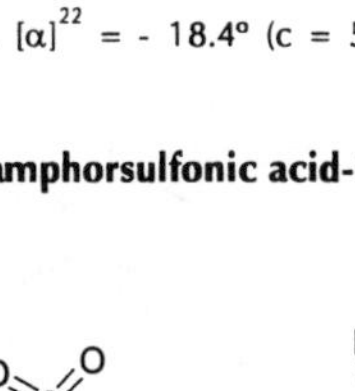

$C_{10}H_{19}NO_4S$
Resolving agent. $[\alpha]^{22}$ = - 18.4° (c = 5.3, $H_2O$). *Acros Organics nv.*

**1356 D-10-Camphorsulfonic acid-L-lysine**

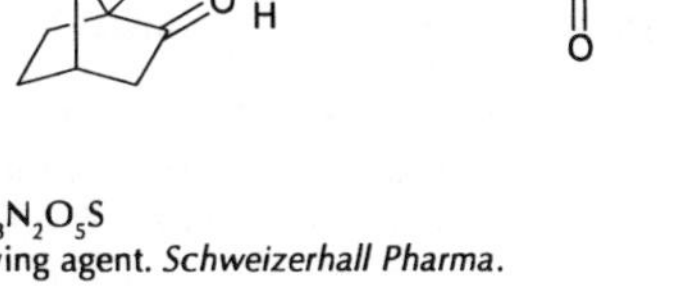

$C_{16}H_{28}N_2O_5S$
Resolving agent. *Schweizerhall Pharma.*

**1357 (-)-Camphor-10-sulfonyl chloride**
39262-22-1

$C_1OH_{15}ClO_3S$
Resolving agent; Chiral auxiliary. mp = 67-68°; $[\alpha]_D^{20}$ = - 32 ± 2° (c = 1, $CHCl_3$). *Lancaster Synthesis Ltd.; Loba Feinchemie AG; Oxford Asymmetry International plc; Sigma-Aldrich Fine Chemicals.*

**1358 (1S)-(+)-Camphor-10-sulfonyl chloride**
21286-54-4 244-314-8

$C_{10}H_{15}ClO_3S$
Bicyclo[2.2.1]heptane-1-methanesulfonyl chloride, 7,7-dimethyl-2-oxo-, (1S,4R)-; (+)-Camphor-10-sulfonyl chloride; (+)-2-oxobornane-10-sulphonyl chloride; 10-Bornanesulfonyl chloride, 2-oxo-, (+)-. Resolving agent. mp = 66-68°; $[\alpha]_D^{20}$ = + 32 ± 1° (c = 1, $CHCl_3$). *Acros Organics nv; Alfa Aesar; Lancaster Synthesis Ltd.; Loba Feinchemie AG; Oxford Asymmetry International plc; Sigma-Aldrich Fine Chemicals.*

**1359 (1S,3R)-(-)-Camphoric acid**
560-09-8

$C_{10}H_{16}O_4$
Resolving agent. mp = 188-190°; $[\alpha]_D^{20}$ = - 48° (c = 10, EtOH). *Sigma-Aldrich Fine Chemicals.*

**1360 (1R,3S)-(+)-Camphoric acid**
124-83-4 1780(12) 204-715-0

$C_{10}H_{16}O_4$
Camphoric acid, (+)-; cis-1,2,2-Trimethyl-1,3-cyclopentanedicarboxylic acid. Listed on TSCA. Resolving agent. mp = 186-189°; $[\alpha]_D^{20}$ = + 46 ± 2° (c = 1, $CH_3OH$). *Acros Organics nv; China Camphor Co. Ltd.; Lancaster Synthesis Ltd.; Sigma-Aldrich Fine Chemicals; Taiwan Tekho Camphor Co. Ltd.*

**1361 (1R,3S)-(-)-Camphoric anhydride**
595-29-9 209-860-3
$C_{10}H_{14}O_3$
Oxa-3-bicyclo[3.2.1]octane-2,4-dione, 1,8,8-trimethyl-, (1R,5S)-; (1R)-1,8,8-Trimethyl-3-oxabicyclo[3.2.1]octane-2,4-dione; 1,3-Cyclopentanedicarboxylic anhydride, 1,2,2-trimethyl-, (-)-. Resolving agent. mp = 223°; $[\alpha]_D^{20}$= - 7.2 ± 0.5° (c = 1, $CHCl_3$). *Lancaster Synthesis Ltd.; Norse Laboratories.*

**1362 (1S)-(+)-Camphorlactone sulphonyloxaziridine**
127411-75-0
$C_{10}H_{13}NO_5S$
Resolving agent. mp = 202-205°; $[\alpha]_D^{20}$= + 89 ± 2° (c = 1, acetone). *Lancaster Synthesis Ltd.*

**1363 D-Camphorquinone**
2767-84-2 220-446-1

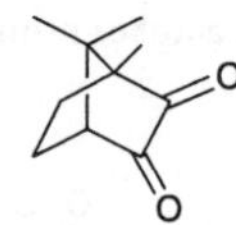

$C_{10}H_{14}O_2$
Bicyclo[2.2.1]heptane-2,3-dione, 1,7,7-trimethyl-, (1S,4R)-; (1R)-(-)-2,3-Bornanedione; (1R)-1,7,7-Trimethylbicyclo[2.2.1]heptane-2,3-dione. Listed on TSCA. Resolving agent. mp = 200-202°; $[\alpha]_D^{20}$= + 100 ± 2° (c = 2, EtOH). *Lancaster Synthesis Ltd.; Loba Feinchemie AG; Sigma-Aldrich Fine Chemicals; TCI America.*

**1364 L-Camphorquinone**
10334-26-6

$C_{10}H_{14}O_2$
Bicyclo[2.2.1]heptane-2,3-dione, 1,7,7-trimethyl-, (1R,4S)-; (1S)-(+)-2,3-Bornanedione; (1S)-1,7,7-Trimethylbicyclo[2.2.1]heptane-2,3-dione. Resolving agent. mp = 197-200°; $[\alpha]_D^{20}$= - 100 ± 2° (c = 2, EtOH). *Acros Organics nv; Lancaster Synthesis Ltd.; Loba Feinchemie AG; Sigma-Aldrich Fine Chemicals.*

**1365 anti-(1R)-(+)-Camphorquinone 3-oxime**
31571-14-9

$C_{10}H_{15}NO_2$
Resolving agent. mp = 154-156°; $[\alpha]^{25}$ = + 200° (c = 1, EtOH). *Sigma-Aldrich Fine Chemicals.*

**1366 anti-(1S)-(+)-Camphorquinone 3-oxime**

$C_{10}H_{15}NO_2$
Resolving agent. mp = 154-156°; $[\alpha]^{25}$ = - 200° (c = 1, EtOH). *Sigma-Aldrich Fine Chemicals.*

**1367 (1R)-10-Camphorsulfonamide**
72597-34-3

$C_{10}H_{17}NO_3S$
Resolving agent. mp = 133-135°; $[\alpha]_D^{20}$ = - 22° (c = 1, $CH_3OH$). *Sigma-Aldrich Fine Chemicals.*

**1368 (1S)-10-Camphosulfonamide**
60933-63-3

$C_{10}H_{17}NO_3S$
Resolving agent. mp = 129-132°; $[\alpha]_D^{20}$ = + 22° (c = 1, $CH_3OH$). *Sigma-Aldrich Fine Chemicals.*

**1369 (-)-10-Camphorsulfonimine**
60886-80-8

$C_{10}H_{15}NO_2S$
(3αS)-(-)-4,5,6,7-Tetrahydro-8,8-dimethyl-3H-3α,6-methano-2,1-benzisothiazole-2,2-dioxide; (7S)-(-)-10,10-Dimethyl-3-thia-4-azatricyclo[5.2.1.01''''5]dec-4-ene-3,3-dioxide. Resolving agent. mp = 228-230°; $[\alpha]^{19}$ = -34° (c = 1, $CHCl_3$). *Sigma-Aldrich Fine Chemicals; TCI America.*

**1370 (+)-10-Camphorsulfonimine**
107869-45-4

$C_{10}H_{15}NO_2S$
(3αR)-(+)-4,5,6,7-Tetrahydro-8,8-dimethyl-3H-3α,6-methano-2,1-benzisothiazole-2,2-dioxide; (7R)-(+)-10,10-Dimethyl-3-thia-4-azatricyclo[5.2.1.01''''5]dec-4-ene-3,3-dioxide. Resolving agent. mp = 229-230°; $[\alpha]_D^{20}$ = + 34° (c = 1, $CHCl_3$). *Sigma-Aldrich Fine Chemicals; TCI America.*

**1371 (1R)-(-)-(10-Camphorsulfonyl)oxaziridine**
104372-31-8

$C_{10}H_{15}NO_3S$
(-)-(Camphorsulfonyl)oxaziridine. Chiral building block. mp = 168-69°; $[\alpha]_D^{20}$ = - 45 ± 1 ° (c = 2, $CHCl_3$). *Acros Organics nv; Lancaster Synthesis Ltd.; Newport Synthesis Ireland Ltd.; Oxford Asymmetry International plc; Sigma-Aldrich Fine Chemicals; TCI America.*

**1372 (1S)-(+)-(10-Camphorsulfonyl)oxaziridine**
104322-63-6

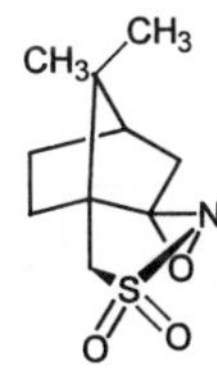


$C_{10}H_{15}NO_3S$
(+)-[(Camphoryl)sulphonyl]oxaziridine; (+)-(2R,8αS)-(Camphorylsulfonyl)oxaziridine. Chiral intermediate. mp = 167-169°; $[\alpha]_D^{20}$ = + 45 ± 1° (c = 2, $CHCl_3$). *Acros Organics nv; Lancaster Synthesis Ltd.; Newport Synthesis Ireland Ltd.; Oxford Asymmetry International plc; Sigma-Aldrich Fine Chemicals; TCI America.*

**1373 (1R,2S)-(+)-2,10-Camphorsultam**
108448-77-7

$C_{10}H_{17}NO_2S$
(2S)-Bornane-10,2-sultam; [3aR-(3aα,6α,7aβ)]-Hexahydro-8,8-dimethyl-3H-3a,6-methano-2,1-benzisothiazole 2,2-dioxide; (1R,5S)-10,10-Dimethyl-3-thia-4-azatricyclo[5.2.1.01''''5]decane 3,3-dioxide. Chiral

auxiliary. mp = 179-182°; $[\alpha]_D^{20}$ = + 33 ± 1° (c = 1, EtOH). *Acros Organics nv; Isochem; Lancaster Synthesis Ltd.; Newport Synthesis Ireland Ltd.; Oxford Asymmetry International plc; Senn Chemicals AG; Sigma-Aldrich Fine Chemicals; TCI America.*

**1374 (1S,2R)-(-)-2,10-Camphorsultam**
94594-90-8

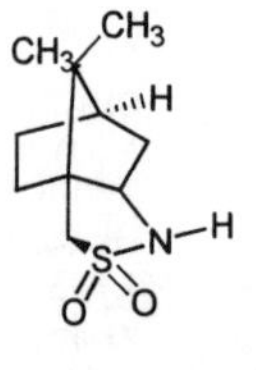

$C_{10}H_{17}NO_2S$
(2R)-Bornane-10,2-sultam; (1S,5R)-10,10-Dimethyl-3-thia-4-azatricyclo[5.2.1.01''''5]decane 3,3-dioxide; (-)-2,10-Camphorsultam. Resolving agent. mp = 182-184°; $[\alpha]_D^{20}$ = -33 ± 1° (c = 1, EtOH). *Acros Organics nv; Isochem; Lancaster Synthesis Ltd.; Newport Synthesis Ireland Ltd.; Oxford Asymmetry International plc; Senn Chemicals AG; Sigma-Aldrich Fine Chemicals; TCI America.*

**1375 (S)-(+)-Camptothecin**
7689-03-4 1783(12)

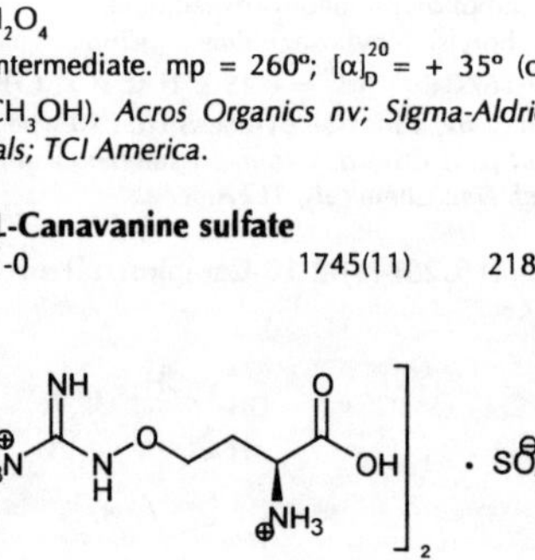

$C_{20}H_{16}N_2O_4$
Chiral intermediate. mp = 260°; $[\alpha]_D^{20}$ = + 35° (c = 0.4, $CHCl_3/CH_3OH$). *Acros Organics nv; Sigma-Aldrich Fine Chemicals; TCI America.*

**1376 L-Canavanine sulfate**
2219-31-0 1745(11) 218-728-4

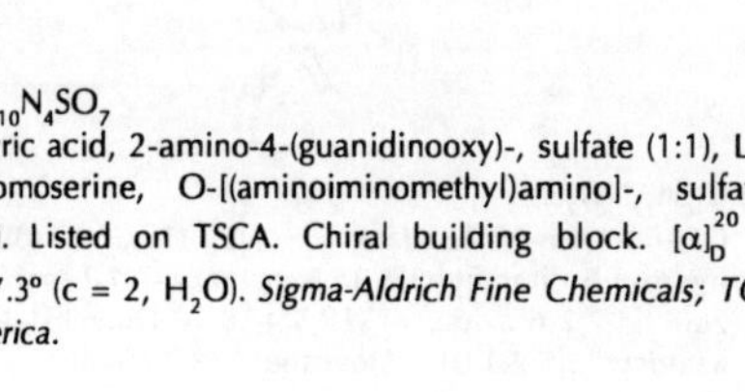

$C_5H_{10}N_4SO_7$
Butyric acid, 2-amino-4-(guanidinooxy)-, sulfate (1:1), L-; L-Homoserine, O-[(aminoiminomethyl)amino]-, sulfate (1:1). Listed on TSCA. Chiral building block. $[\alpha]_D^{20}$ = + 17.3° (c = 2, $H_2O$). *Sigma-Aldrich Fine Chemicals; TCI America.*

**1377 (S)-Carbamylcysteine**
2072-71-1 218-195-8

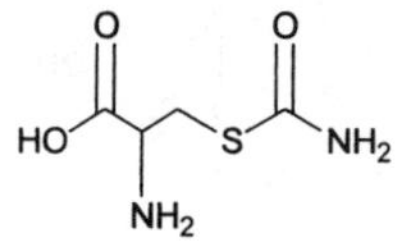

$C_4H_8N_2O_3S$
Chiral building block. mp = 164°; $[\alpha]^{19}$ = -45° (c = 1, 1N HCl). *Sigma-Aldrich Fine Chemicals.*

**1378 Carbobenzyloxy-D-alaninamide**

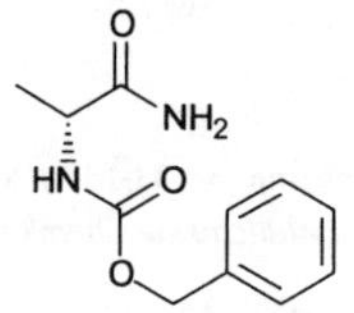

$C_{11}H_{14}N_2O_3$
Chiral intermediate. *Synthetech, Inc.*

**1379 Carbobenzyloxy-L-alaninamide**

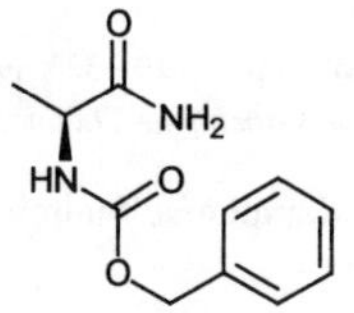

$C_{11}H_{14}N_2O_3$
Chiral intermediate. *Senn Chemicals AG; Synthetech, Inc.*

**1380 Carbobenzyloxy-D-alanine**

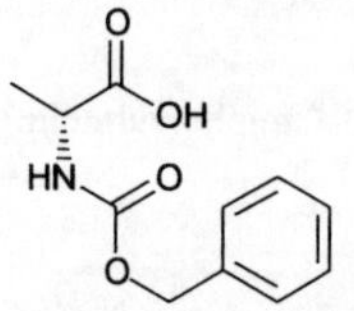

$C_{11}H_{14}NO_4$
Chiral intermediate. *Synthetech, Inc.*

**1381 Carbobenzyloxy-L-alanine**
1142-20-7 214-532-8

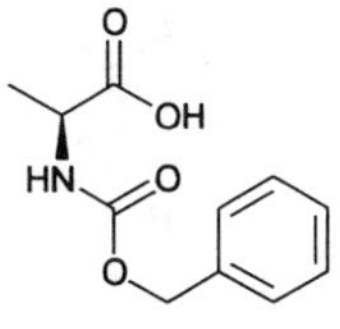

$C_{11}H_{13}NO_4$
Alanine, N-carboxy-, N-benzyl ester, L-; CBZ-L-alanine; N-Benzyloxycarbonyl-L-alanine. Chiral intermediate. mp = 82-84°; $[\alpha]^{23}$ = - 14.2° (c = 2, HOAc). *Acros Organics nv; Austin Chemical Company, Inc.; Fischer Chemicals AG; Flamma s.p.a.; PPG - Sipsy; Sigma-Aldrich Fine Chemicals; Synthetech, Inc.*

**1382 Carbobenzyloxy-D-alanine-β-(+)-fenchyl ester**

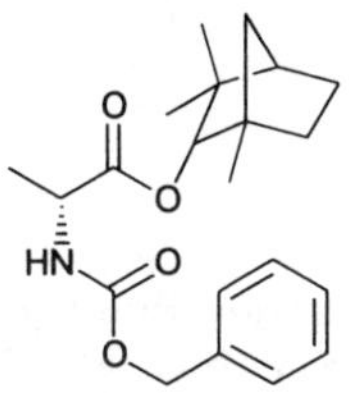

$C_{21}H_{29}NO_4$
Chiral intermediate. *Senn Chemicals AG.*

**1383 Carbobenzyloxy-L-alanine hydroxysuccinimide ester**

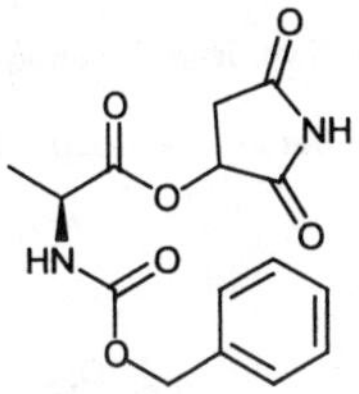

$C_{15}H_{16}N_2O_6$
Chiral intermediate. *Synthetech, Inc.*

**1384 Carbobenzyloxy-D-alanine methyl ester**

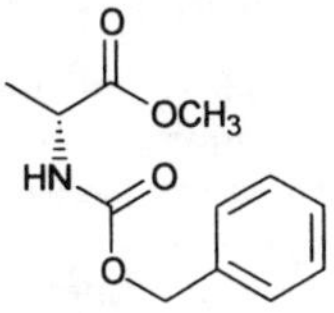

$C_{12}H_{15}NO_4$
Chiral intermediate. *Synthetech, Inc.*

**1385 Carbobenzyloxy-L-alanine methyl ester**

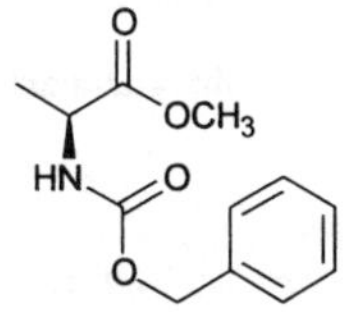

$C_{12}H_{15}NO_4$
Chiral intermediate. *Synthetech, Inc.*

**1386 Carbobenzyloxy-L-alanine-α-naphthylamide**

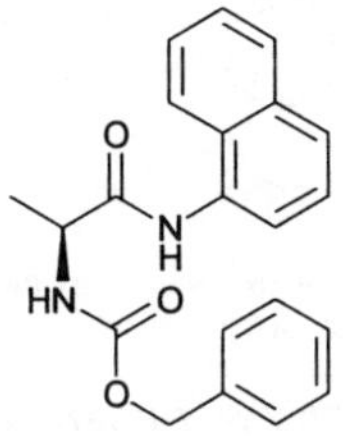

$C_{21}H_{22}N_2O_3$
Chiral intermediate. *Senn Chemicals AG.*

**1387 Carbobenzyloxy-L-alanine 1-naphthylester**

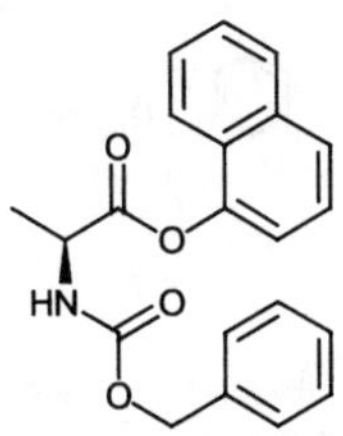

$C_{21}H_{29}NO_4$
Chiral intermediate. *Senn Chemicals AG.*

**1388 Carbobenzyloxy-L-alanine 2-naphthylester**

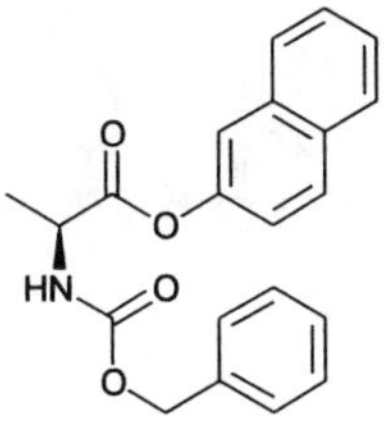

$C_{21}H_{29}NO_4$
Chiral intermediate. *Senn Chemicals AG.*

**1389 Carbobenzyloxy-L-alanine N-succinimide ester**
Chiral intermediate. *Austin Chemical Company, Inc.*

**1390 Carbobenzyloxy-D-alaninol**
61425-27-2
$C_{11}H_{15}NO_3$
Chiral intermediate. $[\alpha]_D^{20}$ = + 5.5° (c = 2, EtOH). *Sigma-Aldrich Fine Chemicals; Synthetech, Inc.*

**1391 Carbobenzyloxy-L-alaninol**
66674-16-6

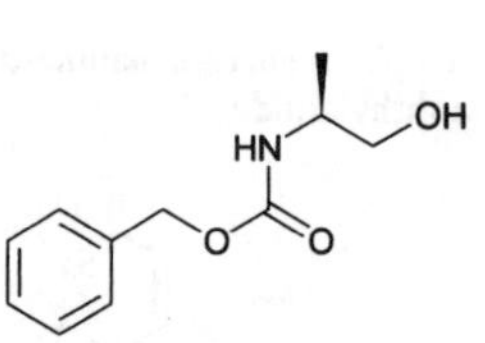

$C_{11}H_{15}NO_3$
Chiral intermediate. mp = 81-84°. *Degussa-Huls AG; Senn Chemicals AG.*

**1392 (3aS,7R,7aS)-7-(Carbobenzyloxyamino)-7,7a-dihydro-2,2-dimethyl-1,3-benzodioxal-4(3aH)-one**

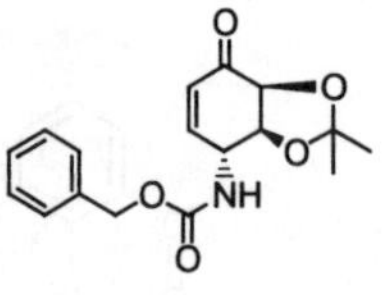

$C_{17}H_{19}NO_5$
Chiral intermediate. $[\alpha]^{25}$ = - 108° (c = 1, $CHCl_3$). *Sigma-Aldrich Fine Chemicals.*

**1393 (3aR,4S,7R,7aS)-7-(Carbobenzyloxyamino)-3a,4,7,7a-tetrahydro-2,2-diemthyl-1,3-benzodioxol-4-ol**

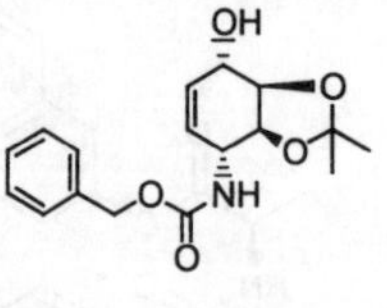

$C_{17}H_{21}NO_5$
Chiral intermediate. mp = 109-113°; $[\alpha]_D^{20}$ = - 44° (c = 1, $CHCl_3$). *Sigma-Aldrich Fine Chemicals.*

**1394 N-α-Carbobenzyloxy-D-arginine**
6382-93-0

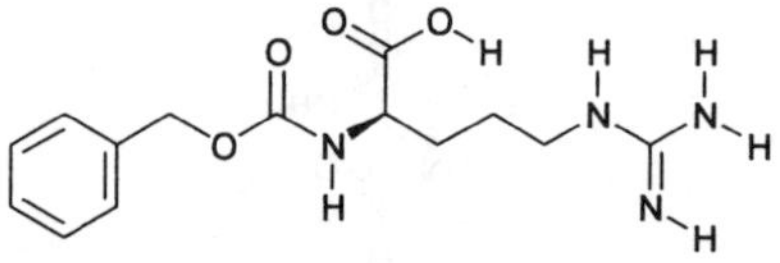

$C_{14}H_{20}N_4O_4$
Chiral intermediate. [α] = + 9.6° (c = 2, 1N HCl). *Acros Organics nv.*

**1395 N-α-Carbobenzyloxy-L-arginine**
1234-35-1 214-973-6

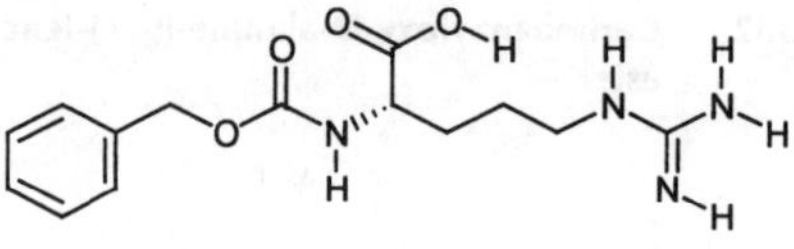

$C_{14}H_{20}N_4O_4$
$N^2$-carboxy-, N-benzyl ester; N-α-CBZ-L-arginine; N-α-Benzyloxycarbonyl-L-arginine; L-Arginine, $N^2$-[(phenylmethoxy)carbonyl]-. Listed on TSCA. Chiral intermediate. mp = 173-175°; $[\alpha]^{23}$ = - 9.3° (c = 5.3, 0.2N HCl). *Acros Organics nv; Sigma-Aldrich Fine Chemicals.*

**1396 Carbobenzyloxy-L-arginine-7-amido-4-methylcoumarin hydrochloride**
70375-22-3
Chiral intermediate. *Senn Chemicals AG.*

**1397 Carbobenzyloxy-L-asparagine**
2304-96-3 218-969-5

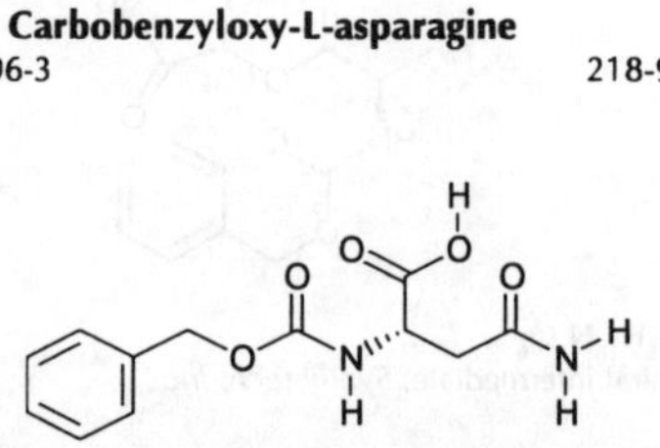

$C_{12}H_{14}N_2O_5$
Asparagine, $N^2$-carboxy-, $N^2$-benzyl ester, L-; N-α-Carbobenzyloxy-L-asparagine; Nα-CBZ-L-Asparagine; N-(Benzyloxycarbonyl)-L-asparagine; L-Asparagine, $N^2$-[(phenylmethoxy)carbonyl]-; $N^2$-[(Phenylmethoxy)-carbonyl]-L-asparagine. Listed on TSCA. Chiral intermediate. mp = 163-165°; $[\alpha]_D^{20}$ = + 6° (c = 1.6, HOAc). *Acros Organics nv; Fischer Chemicals AG; Flamma s.p.a.; Sigma-Aldrich Fine Chemicals.*

**1398 Carbobenzyloxy-L-asparagine 4-nitrobenzylester**

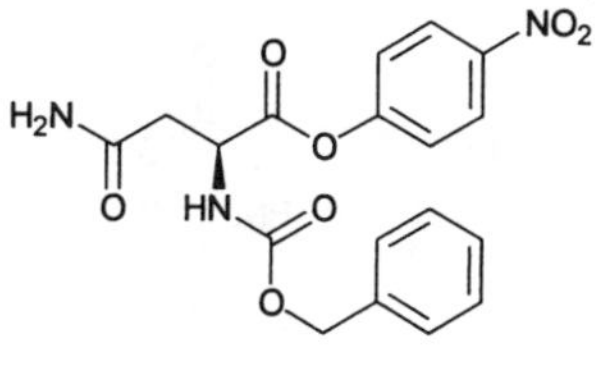

$C_{12}H_{17}N_3O_7$
Chiral intermediate. *Senn Chemicals AG.*

**1399 Carbobenzyloxy-D-aspartic acid**

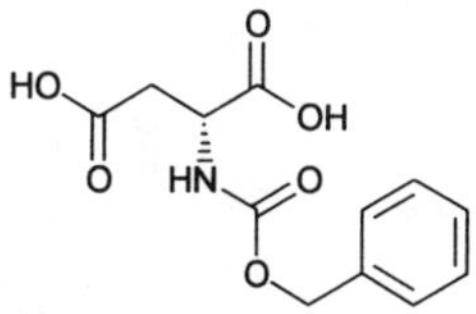

$C_{12}H_{13}NO_6$
Chiral intermediate. *Synthetech, Inc.*

**1400 Carbobenzyloxy-L-aspartic acid**
1152-61-0 214-568-4

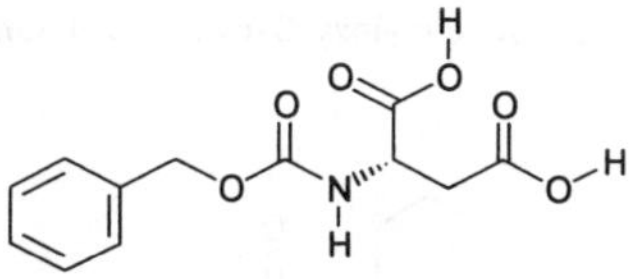

$C_{12}H_{13}NO_6$
Aspartic acid, N-[(phenylmethoxy)carbonyl]-, L-; N-Carbobenzyloxy-L-aspartic acid; N-CBZ-L-aspartic acid; N-(Benzyloxycarbonyl)-L-aspartic acid; Aspartic acid, N-carboxy-, N-benzyl ester, L-. Listed on TSCA. Chiral intermediate. mp = 117-119°; $[\alpha]^{23} = + 8.6°$ (c = 7, HOAc). *Acros Organics nv; Austin Chemical Company, Inc.; Fischer Chemicals AG; PPG - Sipsy; Sigma-Aldrich Fine Chemicals; Synthetech, Inc.*

**1401 Carbobenzyloxy-D-aspartic acid α-benzyl ester**

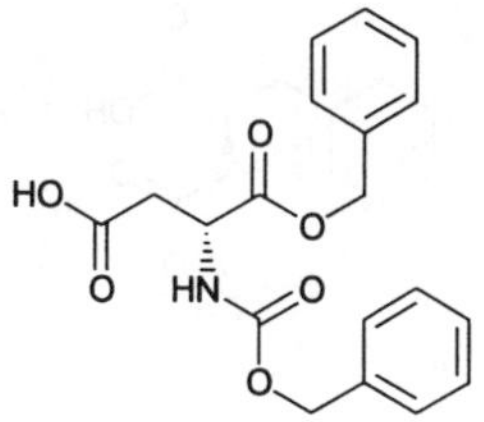

$C_{19}H_{19}NO_6$
Chiral intermediate. *Synthetech, Inc.*

**1402 Carbobenzyloxy-L-aspartic acid α-benzyl ester**

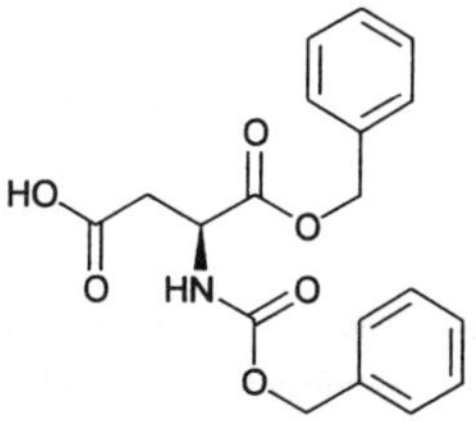

$C_{19}H_{19}NO_6$
Chiral intermediate. *Synthetech, Inc.*

**1403 Carbobenzyloxy-L-aspartic acid β-benzyl ester**

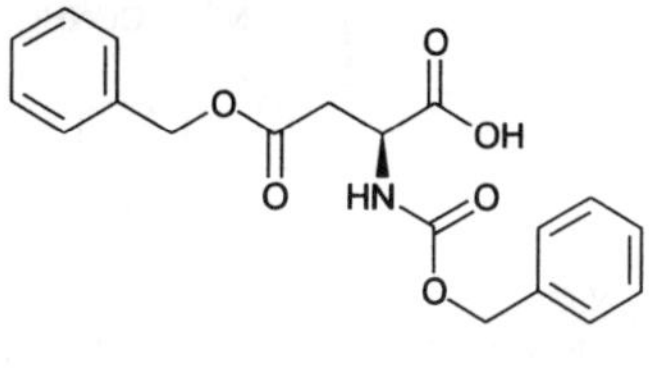

$C_{19}H_{19}NO_6$
Chiral intermediate. *Synthetech, Inc.*

**1404 Carbobenzyloxy-L-aspartic acid 4-benzyl ester**

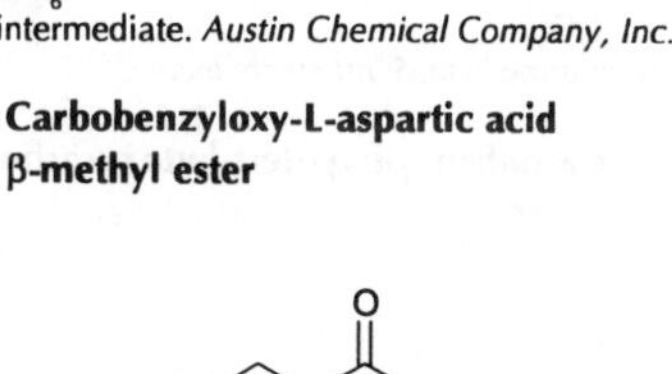

$C_{19}H_{19}NO_6$
Chiral intermediate. *Austin Chemical Company, Inc.*

**1405 Carbobenzyloxy-L-aspartic acid β-methyl ester**

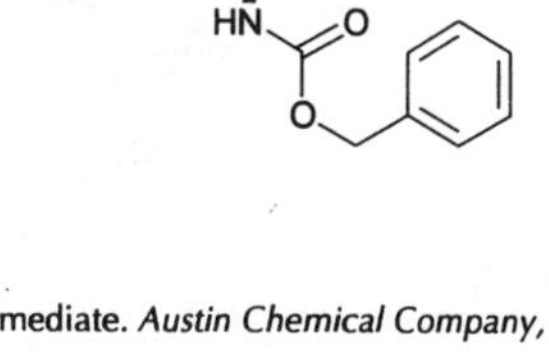

$C_{13}H_{15}NO_6$.
Chiral intermediate. *Austin Chemical Company, Inc.*

**1406 N-Carbobenzyloxy-aspartic anhydride**
4515-23-5 224-838-3

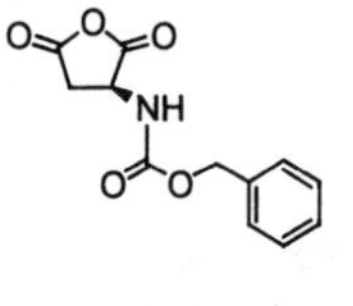

$C_{12}H_{11}NO_5$
Chiral intermediate. mp = 123-124°; $[\alpha]_D^{20}$ = - 25° (c = 1, $CH_3OH$). *Sigma-Aldrich Fine Chemicals.*

**1407 Carbobenzyloxy-L-aspartyl-L-phenylalanineamide**

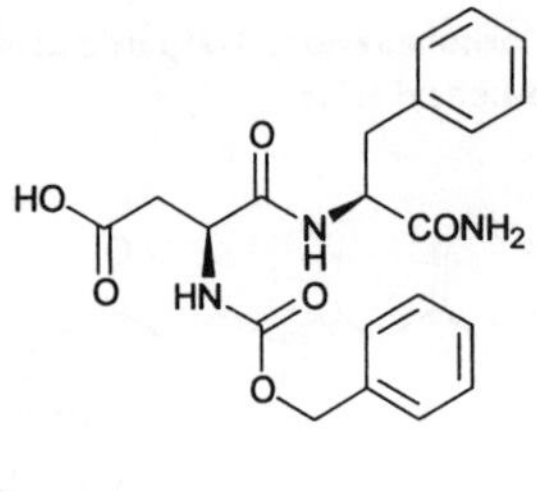

$C_{21}H_{23}N_3O_6$
Chiral intermediate. *Fischer Chemicals AG.*

**1408 Carbobenzyloxy-O-benzyl-L-threonine**
69863-36-1
$C_{19}H_{21}NO_5$
Chiral intermediate. *Sigma-Aldrich Fine Chemicals.*

**1409 Carbobenzyloxy-(tert-butoxycarbonyl)-D-lysine**

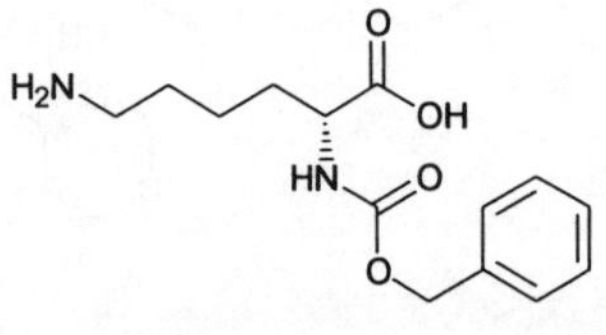

$C_{14}H_{20}N_2O_5$
Chiral intermediate. *Synthetech, Inc.*

**1410 Carbobenzyloxy-(tert-butoxycarbonyl)-L-lysine**

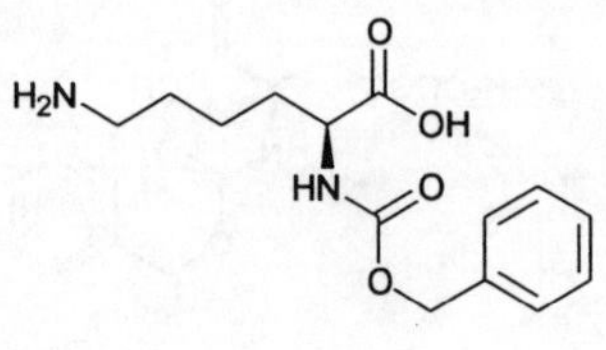

$C_{14}H_{20}N_2O_5$
Chiral intermediate. *Synthetech, Inc.*

**1411 Carbobenzyloxy-(tert-butoxycarbonyl)-D-ornithine**

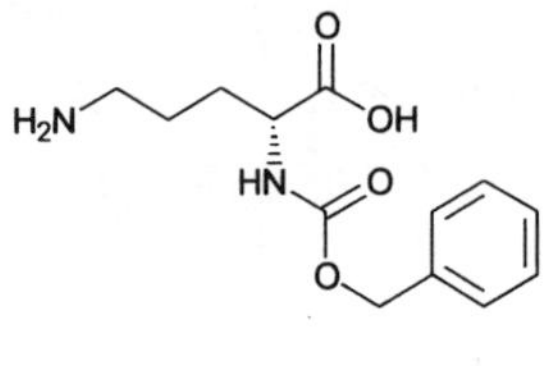

$C_{13}H_{18}N_2O_4$
Chiral intermediate. *Synthetech, Inc.*

**1412 Carbobenzyloxy-(tert-butoxycarbonyl)-L-ornithine**

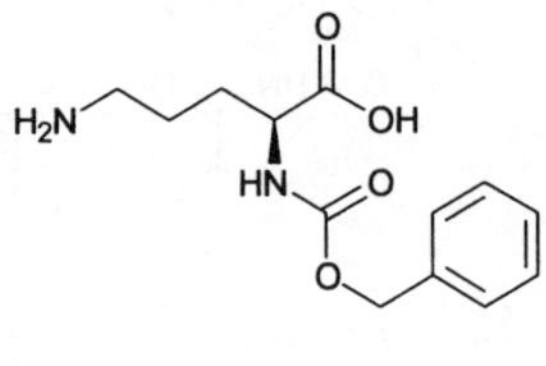

$C_{13}H_{18}N_2O_4$
Chiral intermediate. *Synthetech, Inc.*

**1413 Carbobenzyloxy-D-cyclohexylalanine**

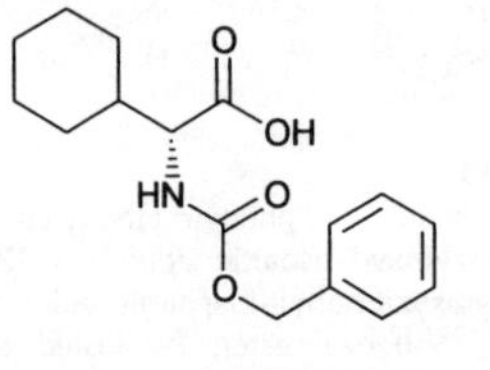

$C_{16}H_{21}NO_4$
Chiral intermediate. *Synthetech, Inc.*

**1414 Carbobenzyloxy-L-Cyclohexylalanine**
25341-42-8

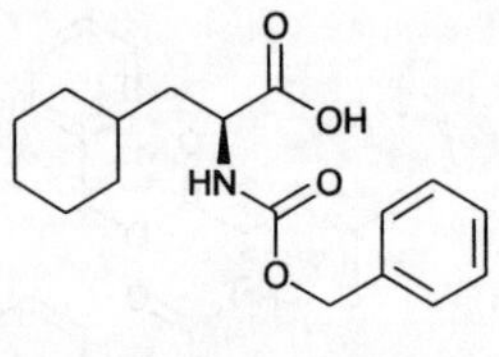

$C_{17}H_{23}NO_4$
Chiral intermediate. *Fischer Chemicals AG; Genzyme Pharmaceuticals; Great Lakes Fine Chemicals; Senn Chemicals AG.*

**1415 Carbobenzyloxy-D-cyclohexylalaninol**

$C_{16}H_{23}NO_3$
Chiral intermediate. *Synthetech, Inc.*

**1416 Carbobenzyloxy-L-cyclohexylalaninol**
113828-85-6

$C_{17}H_{25}NO_3$
Chiral intermediate. *Fischer Chemicals AG; Great Lakes Fine Chemicals; Synthetech, Inc.*

**1417 Carbobenzyloxy-L-cyclohexylglycine**

$C_{16}H_{21}NO_4$
Chiral intermediate. *Senn Chemicals AG.*

**1418 Carbobenzyloxy-L-2,4-diaminobutyric acid**

$C_{12}H_{16}N_2O_4$
Chiral intermediate. *Senn Chemicals AG.*

**1419 Carbobenzyloxy-L-2,3-diamino-propionic acid**

$C_{11}H_{24}N_2O_4$
Chiral intermediate. *Senn Chemicals AG.*

**1420 Carbobenzyloxy-D-galactosamine**

$C_{14}H_{19}NO_7$
Chiral intermediate. *Senn Chemicals AG.*

**1421 Carbobenzyloxy-D-glucosamine**

$C_{14}H_{19}NO_7$
Chiral intermediate. *Senn Chemicals AG.*

**1422 Carbobenzyloxy-L-glutamic acid**

$C_{13}H_{15}NO_6$
Chiral intermediate. *Synthetech, Inc.*

**1423 Carbobenzyloxy-L-glutamic acid 1-methylester**
5672-83-3

$C_{14}H_{17}NO_6$
Glutamic acid, N-carboxy-, N-benzyl 1-methyl ester, L-; α-Methyl N-benzyloxycarbonylgultamate. Chiral intermediate. mp = 68-70°; [α] = - 23°. *Acros Organics nv; Sigma-Aldrich Fine Chemicals.*

**1424 Carbobenzyloxy-L-glutamine**
2650-64-8 220-173-8

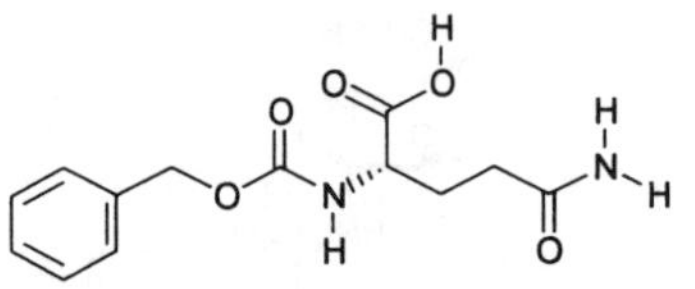

$C_{13}H_{16}N_2O_5$
$N^2$-carboxy-, N2-benzyl ester, L-; N-CBZ-L-Glutamine; N-Benzyloxycarbonyl-L-glutamine; $N^2$-Carbobenzoxy-L-glutamine; $N^2$-[(Phenylmethoxy)carbonyl]-L-glutamine. Listed on TSCA. Chiral intermediate. mp = 135-137°; $[\alpha]^{23}$ = - 7.3° (c = 2, EtOH). *Acros Organics nv; Sigma-Aldrich Fine Chemicals.*

**1425 N-Carbobenzyloxy-D-histidine**
14997-58-1 239-084-0

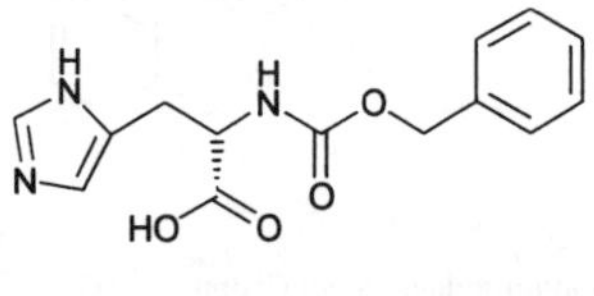

$C_{14}H_{15}N_3O_4$
Chiral intermediate. mp = 168°; $[\alpha]_D^{20}$= - 24° (c = 1, 6N HCl). *Sigma-Aldrich Fine Chemicals.*

**1426 Carbobenzyloxy-L-histidine hydrazide**

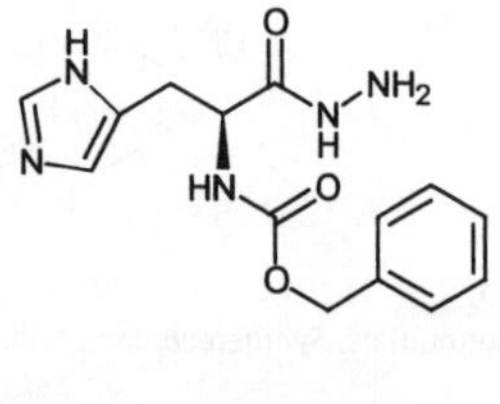

$C_{14}H_{17}N_5O_3$
Chiral intermediate. *Senn Chemicals AG.*

**1427 Carbobenzyloxy-L-hydroxyproline methyl ester**

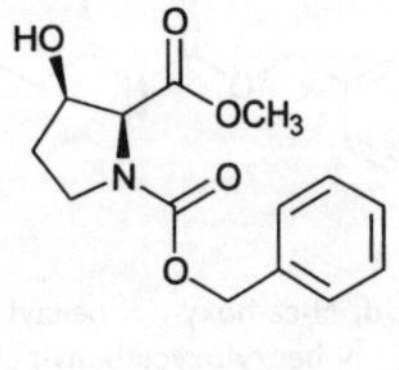

$C_{14}H_{17}NO_5$
Chiral intermediate. *Synthetech, Inc.*

**1428 Carbobenzyloxy-p-iodo-D-phenylalanine**

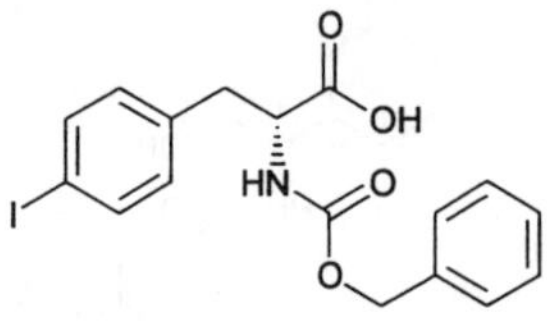

$C_{17}H_{16}INO_4$
Chiral intermediate. *Advanced Asymmetrics, Inc.*

**1429 Carbobenzyloxy-p-iodo-L-phenylalnine**

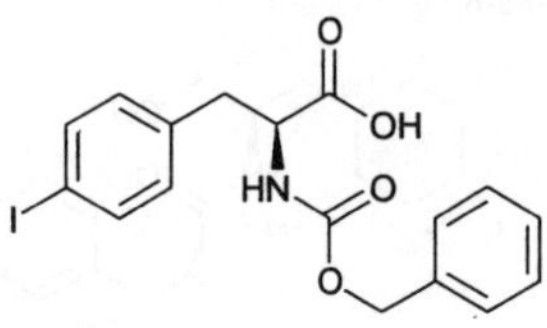

$C_{17}H_{16}INO_4$
Chiral intermediate. *Advanced Asymmetrics, Inc.*

**1430 Carbobenzyloxy-p-iodo-D-phenylalanine methyl ester**

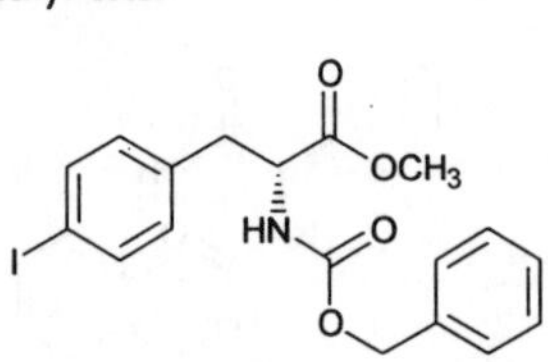

$C_{18}H_{18}INO_4$
Chiral intermediate. *Advanced Asymmetrics, Inc.*

**1431 Carbobenzyloxy-p-iodo-L-phenylalanine methyl ester**

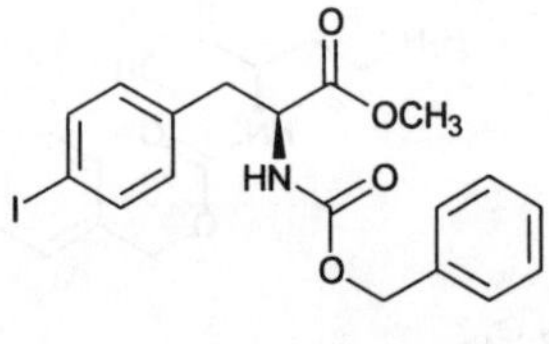

$C_{18}H_{18}INO_4$
Chiral intermediate. *Advanced Asymmetrics, Inc.*

**1432 Carbobenzyloxy-L-isoleucine**
3160-59-6 211-611-0

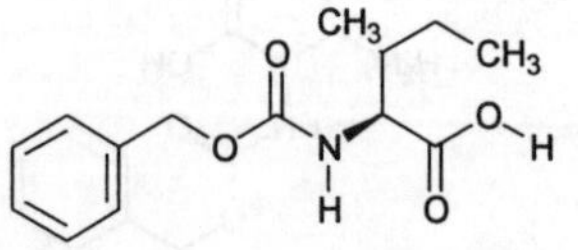

$C_{14}H_{19}NO_4$
Isoleucine, N-carboxy-, N-benzyl ester, L-; N-CBZ-L-isoleucine; N-Benzyloxycarbonyl-L-isoleucine; L-Iso-

leucine, N-[(Phenylmethoxy)carbonyl]-. Chiral intermediate. mp = 52-54°; [α] = + 6.7° (c = 6, EtOH). *Acros Organics nv; Sigma-Aldrich Fine Chemicals.*

**1433 Carbobenzyloxy-L-isoleucine dicyclohexylammonium salt**

$C_{18}H_{13}N_2O_2$
Chiral intermediate. *Synthetech, Inc.*

**1434 Carbobenzyloxy-L-isoleucine hydroxysuccinimide ester**

3391-99-9 222-228-1

$C_{18}H_{22}N_2O_6$
Carbamic acid, [1-[[(2,5-dioxo-1-pyrrolidinyl)oxy]carbonyl]-2-methylbutyl]-, phenylmethyl ester, [S-(R*,R*)]-; Benzyl [S-(R*,R*)]-[1-[[(2,5-dioxopyrrolidin-1-yl)oxy]carbonyl]-2-methylbutyl]carbamate; N-α-Carbobenzoxy-L-isoleucine-N-hydroxysuccinimide ester; Succinimide, N-[(N-carboxy-L-isoleucyl)oxy]-, benzyl ester. Chiral intermediate. *PPG - Sipsy.*

**1435 Carbobenzyloxy-L-isoleucine N-succinimide ester**

Chiral intermediate. *Austin Chemical Company, Inc.*

**1436 Carbobenzyloxy-D-leucine**

$C_{14}H_{19}NO_4$
Chiral intermediate. *Synthetech, Inc.*

**1437 Carbobenzyloxy-L-leucine**

2018-66-8 217-960-3

$C_{14}H_{19}NO_4$
Leucine, N-carboxy-, N-benzyl ester, L-. (Benzyloxycarbonyl)-L-leucine; CBZ-L-leucine; L-Leucine, N-[(phenylmethoxy)carbonyl]-. Chiral intermediate. d = 1; n = 1.512; $[\alpha]_D^{20}$ = - 17° (c = 2, EtOH). *Acros Organics nv; Sigma-Aldrich Fine Chemicals.*

**1438 Carbobenzyloxy-L-leucine amide**

$C_{14}H_{20}N_2O_3$
Chiral intermediate. *Senn Chemicals AG.*

**1439 Carbobenzyloxy-L-leucine dicyclohexylammonium salt**

$C_{18}H_{38}N_2O_2$
Chiral intermediate. *Synthetech, Inc.*

**1440 Carbobenzyloxy-L-leucine hydroxysuccinimde ester**

$C_{19}H_{24}N_2O_5$
Chiral intermediate. *Synthetech, Inc.*

**1441 Carbobenzyloxy-L-leucine β-naphthylamide monohydrate**

$C_{24}H_{26}N_2O_3$
Chiral intermediate. *Senn Chemicals AG.*

**1442 Carbobenzyloxy-L-leucine N-succinimide ester**
3397-35-1 222-256-4

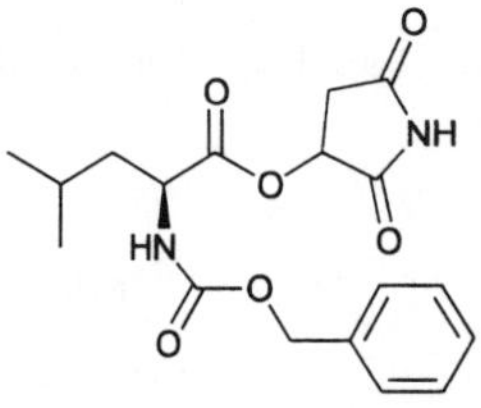

$C_{18}H_{22}N_2O_6$
Carbamic acid, [1-[[(2,5-dioxo-1-pyrrolidinyl)oxy]-carbonyl]-3-methylbutyl]-, phenylmethyl ester, (S)-; Benzyl (S)-[1-[[(2,5-dioxo-1-pyrrolidinyl)oxy]carbonyl]-3-methylbutyl]carbamate; N-(Benzyloxycarbonyl)-L-leucine hydroxysuccinimide ester; N-Carbobenzoxy-L-leucine succinimido ester; Succinimide, N-[(N-carboxy-L-leucyl)oxy]-, benzyl ester. Chiral intermediate. *Austin Chemical Company, Inc.; PPG - Sipsy.*

**1443 Carbobenzyloxy-D-lysine**
2212-75-1 218-662-6

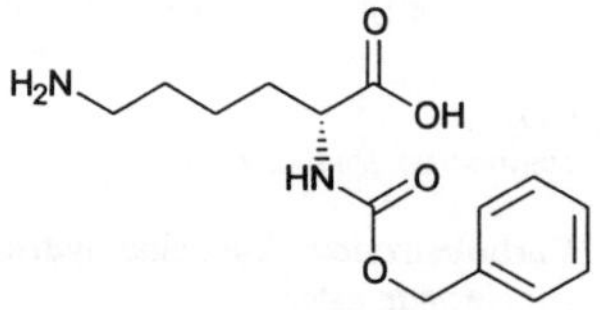

$C_{14}H_{20}N_2O_4$
Listed on TSCA. Chiral intermediate. mp = 226°; $[\alpha]_D^{20}$ = - 13° (c = 2, 0.2N HCl). *Sigma-Aldrich Fine Chemicals; Synthetech, Inc.*

**1444 Carbobenzyloxy-L-lysine**

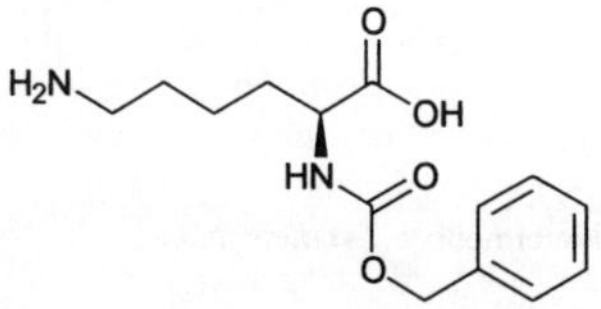

$C_{14}H_{20}N_2O_4$
Chiral intermediate. *Synthetech, Inc.*

**1445 N-ε-Carbobenzyloxy-L-lysine**
1155-64-2 214-585-7

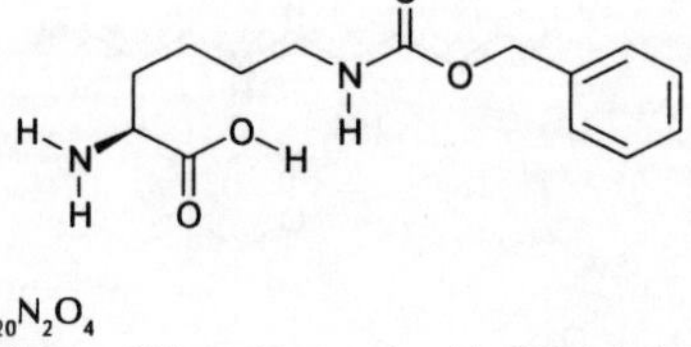

$C_{14}H_{20}N_2O_4$
$N^6$-carboxy-, N-benzyl ester, L-; Nε-CBZ-L-lysine; $N^6$-Benzyloxycarbonyl-L-lysine. Chiral intermediate. mp = 259°; $[\alpha]^{28}$ = + 16.5° (c = 1.5, 2N HCl). *Acros Organics nv; Sigma-Aldrich Fine Chemicals.*

**1446 Carbobenzyloxy-D-lysine benzyl ester hydrochloride**

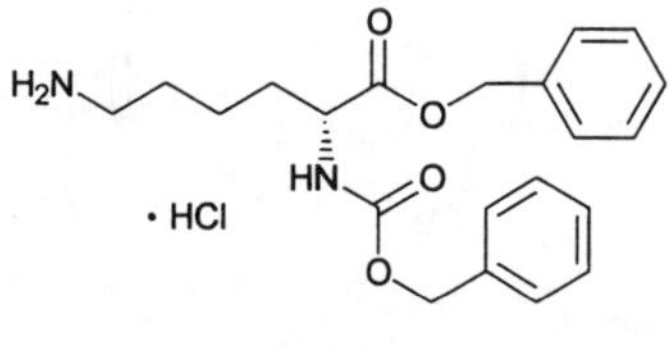

$C_{21}H_{27}ClN_2O_4$
Chiral intermediate. *Synthetech, Inc.*

**1447 Carbobenzyloxy-L-lysine-(tert-butoxycarbonyl)-dicyclohexylamine**

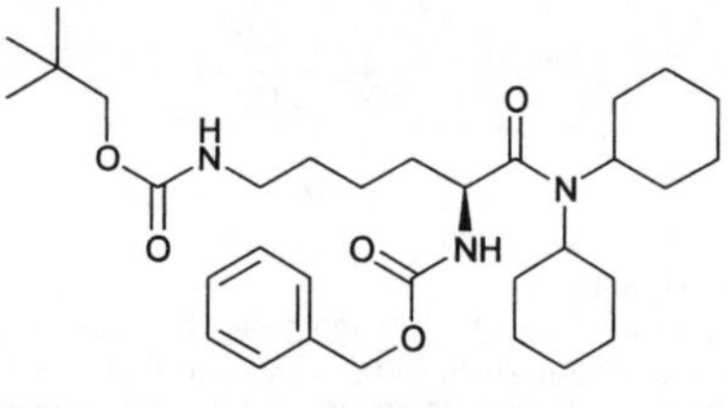

$C_{32}H_{51}N_3O_5$
Chiral intermediate. *Austin Chemical Company, Inc.*

**1448 Carbobenzyloxy-L-lysyl-(tert-butoxycarbonyl)-L-leucine methyl ester**

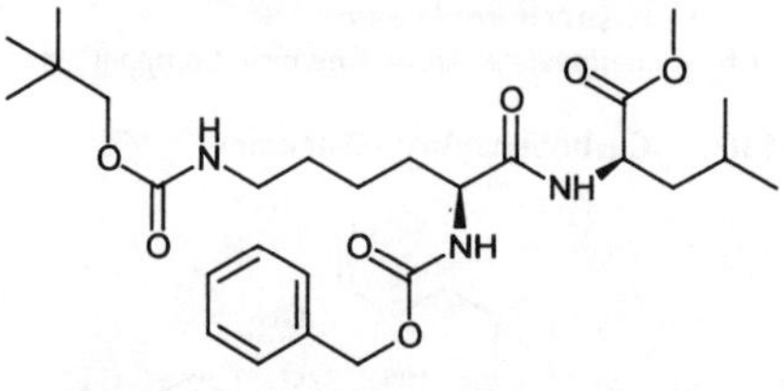

$C_{27}H_{43}N_3O_7$
Chiral intermediate. *Austin Chemical Company, Inc.*

**1449 Carbobenzyloxy-L-lysine-(tert-butoxycarbonyl)-N-succinimide ester**

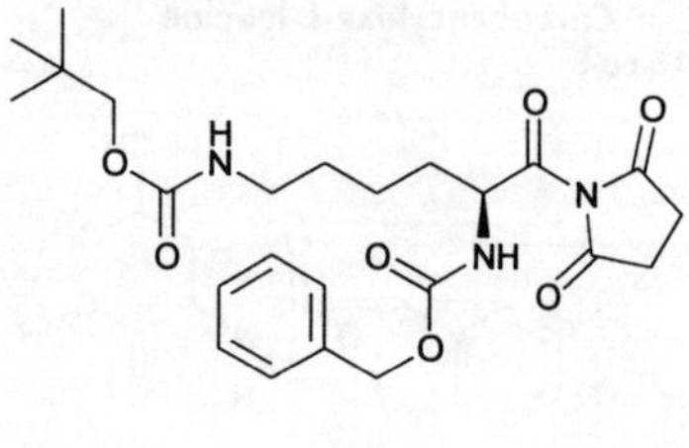

$C_{24}H_{33}N_5O_7$
Chiral intermediate. *Austin Chemical Company, Inc.*

**1450 Carbobenzyloxy-D-lysine methyl ester hydrochloride**

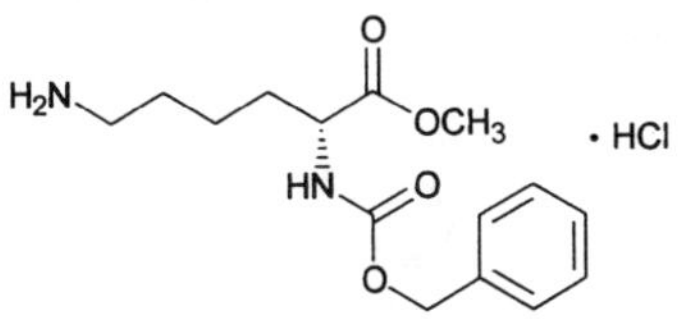

$C_{15}H_{23}ClN_2O_4$
Chiral intermediate. *Synthetech, Inc.*

**1451 N-α-Carbobenzyloxy-L-lysine 4-nitrophenylester hydrochloride**
4272-71-3 224-265-9

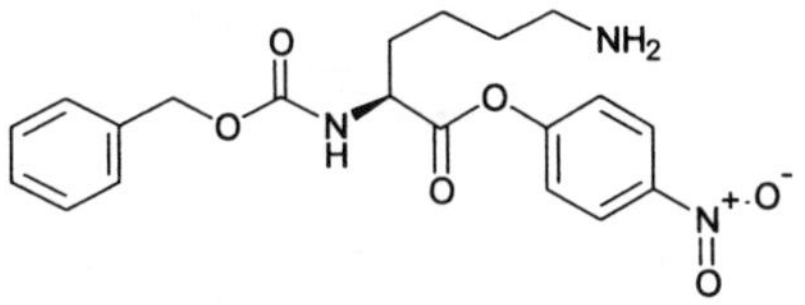

$C_{20}H_{23}N_3O_6$
$N^2$-benzyl p-nitrophenyl ester, L-; p-Nitrophenyl $N^2$-(benzyloxycarbonyl)-L-lysinate; Benzyloxycarbonyl-L-lysine 4-nitrophenyl ester; CBZ-L-lysine p-nitrophenyl ester. $[\alpha]$ = - 25° (c = 1, HOAc). *Acros Organics nv.*

**1452 Carbobenzyloxy-L-methionine**
1152-62-1 214-570-5

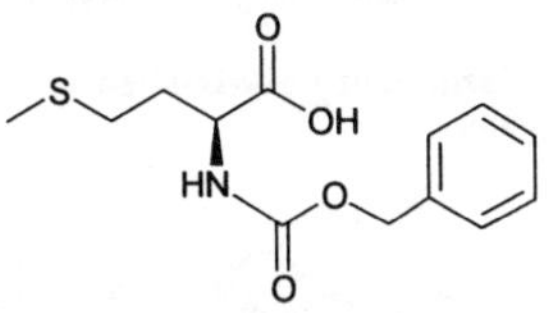

$C_{13}H_{17}NO_4S$
Methionine, N-carboxy-, N-benzyl ester, L-; N-(Benzyloxycarbonyl)-L-methionine; L-(Carbobenzyloxy)-methionine. Chiral intermediate. mp = 66-70°; $[\alpha]^{25}$ = - 18.5° (c = 2.4, 95% EtOH). *Sigma-Aldrich Fine Chemicals; Synthetech, Inc.*

**1453 Carbobenzyloxy-L-methionine methyl ester**

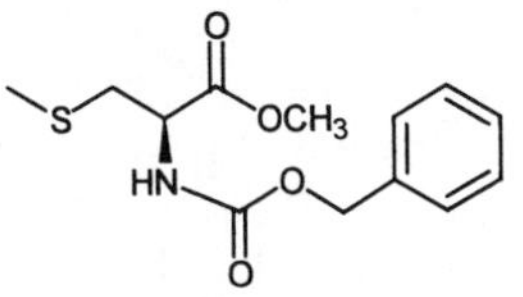

$C_{13}H_{17}NO_4S$
Chiral intermediate. *Synthetech, Inc.*

**1454 Carbobenzyloxy-N-methyl-D-phenylalanine**

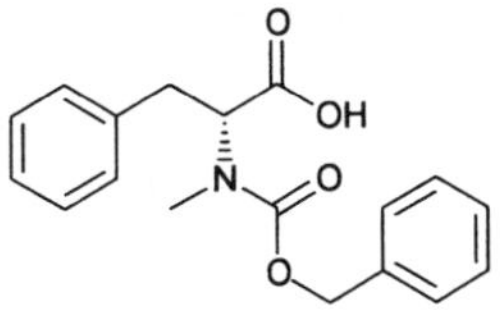

$C_{18}H_{19}NO_4$
Chiral intermediate. *Synthetech, Inc.*

**1455 Carbobenzyloxy-O-methyl-L-serine**

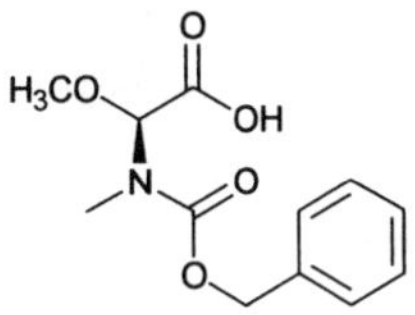

$C_{12}H_{15}NO_5$
Chiral intermediate. *Synthetech, Inc.*

**1456 Carbobenzyloxy-p-nitro-D-phenylalanine**

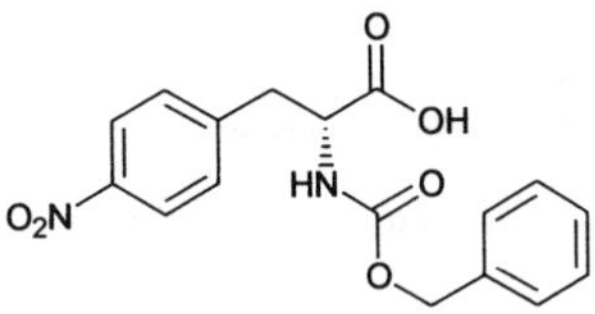

$C_{17}H_{16}N_2O_6$
Chiral intermediate. *Advanced Asymmetrics, Inc.*

**1457 Carbobenzyloxy-p-nitro-L-phenylalanine**

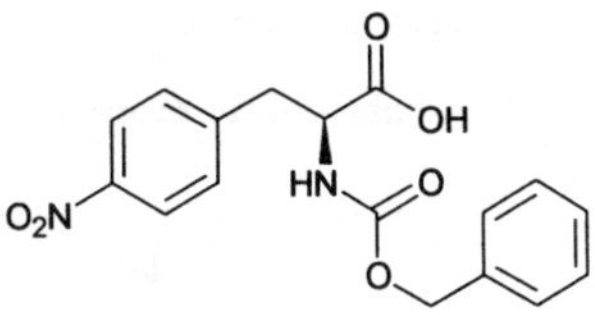

$C_{17}H_{16}N_2O_6$
Chiral intermediate. *Advanced Asymmetrics, Inc.*

**1458 Carbobenzyloxy-L-norleucine**

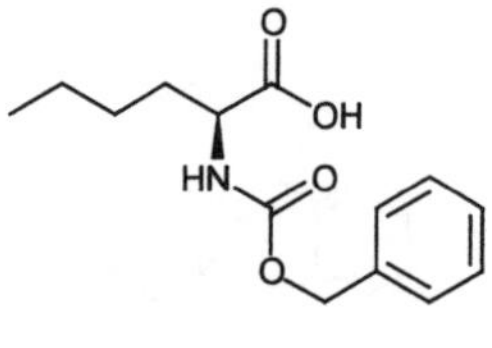

$C_{14}H_{19}NO_4$
Chiral intermediate. *Senn Chemicals AG.*

**1459 Carbobenzyloxy-L-norvaline**

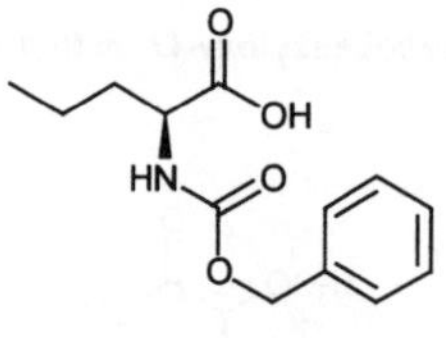

$C_{13}H_{17}NO_4$
Chiral intermediate. *Senn Chemicals AG.*

**1460 Carbobenzyloxy-D-ornithine**

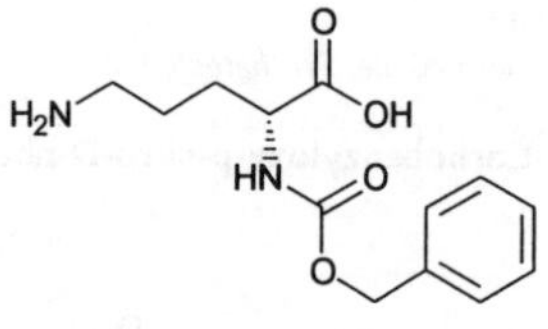

$C_{13}H_{18}N_2O_4$
Chiral intermediate. *Synthetech, Inc.*

**1461 Carbobenzyloxy-L-ornithine**

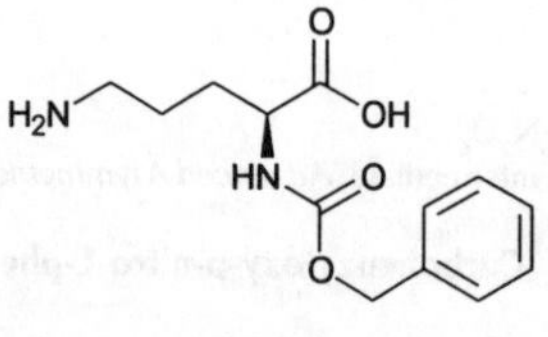

$C_{13}H_{18}N_2O_4$
Chiral intermediate. *Synthetech, Inc.*

**1462 (R)-Carbobenzyloxy-oxaproline**
97534-84-4

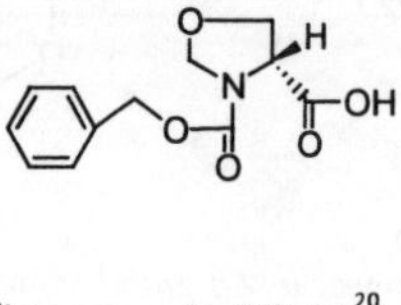

$C_{12}H_{13}NO_5$
Chiral intermediate. mp = 70-74°; $[\alpha]_D^{20}$ = + 92° (c = 1, $CHCl_3$). *Daiso Company, Ltd.; Omega Chemical Company Inc.; Sigma-Aldrich Fine Chemicals.*

**1463 (S)-Carbobenzyloxy-oxaproline**
97534-82-2

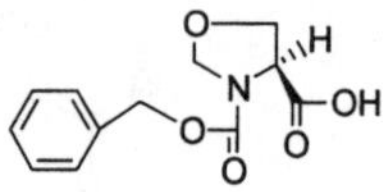

$C_{12}H_{13}NO_5$
Chiral intermediate. mp = 74-78°; $[\alpha]_D^{20}$ = - 92° (c = 1, $CHCl_3$). *Daiso Company, Ltd.; Omega Chemical Company Inc.; Sigma-Aldrich Fine Chemicals.*

**1464 Carbobenzyloxy-D-phenylalaninal**
63219-70-5

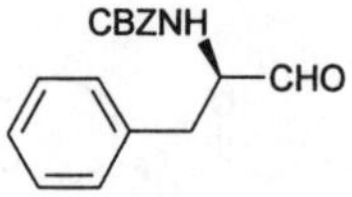

$C_{17}H_{17}NO_3$
Chiral intermediate. *Omega Chemical Company Inc.*

**1465 Carbobenzyloxy-L-phenylalaninal**
59830-60-3

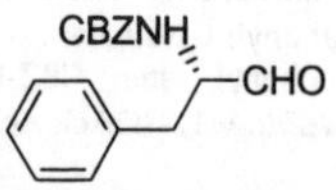

$C_{17}H_{17}NO_3$
Chiral intermediate. *Omega Chemical Company Inc.*

**1466 Carbobenzyloxy-D-phenylalanine**
28709-70-8

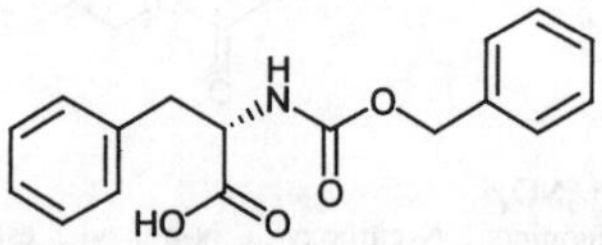

$C_{17}H_{17}NO_4$
N-CBZ-D-phenylalanine; N-Benzyloxycarbonyl-D-phenylalanine. Chiral intermediate. [α] = - 3.4° (c = 5, HOAc). *Acros Organics nv; Isochem; Synthetech, Inc.*

**1467 Carbobenzyloxy-L-phenylalanine**
1161-13-3 214-599-3

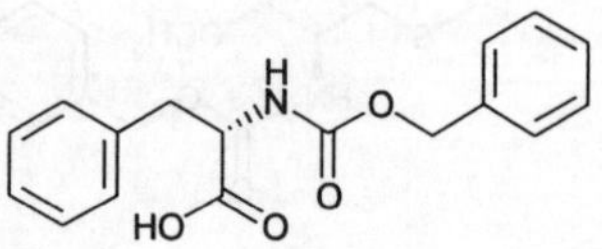

$C_{17}H_{17}NO_4$
Phenylalanine, N-[(phenylmethoxy)carbonyl]-, L-; N-(Carbobenzyloxy)-L-phenylalanine; N-Carboxy-3-phenyl-

L-alanine N-benzyl ester. Chiral intermediate. mp = 86-88°; [α] = + 4.2° (c = 5, HOAc). *Acros Organics nv; Austin Chemical Company, Inc.; Fischer Chemicals AG; PPG - Sipsy; Sigma-Aldrich Fine Chemicals; Synthetech, Inc.*

**1468 Carbobenzyloxy-L-phenylalanine amide**
Chiral intermediate. *Senn Chemicals AG.*

**1469 Carbobenzyloxy-D-phenylalanine hydroxysuccinimide ester**
Chiral intermediate. *Synthetech, Inc.*

**1470 Carbobenzyloxy-L-phenylalanine hydroxysuccinimide ester**
Chiral intermediate. *Austin Chemical Company, Inc.; Synthetech, Inc.*

**1471 Carbobenzyloxy-D-phenylalanine methyl ester**
Chiral intermediate. *Synthetech, Inc.*

**1472 Carbobenzyloxy-L-phenylalanine methyl ester**
35909-92-3
Chiral intermediate. *Fischer Chemicals AG; Great Lakes Fine Chemicals; Synthetech, Inc.*

**1473 Carbobenzyloxy-L-phenylalanine β-naphthylamide**

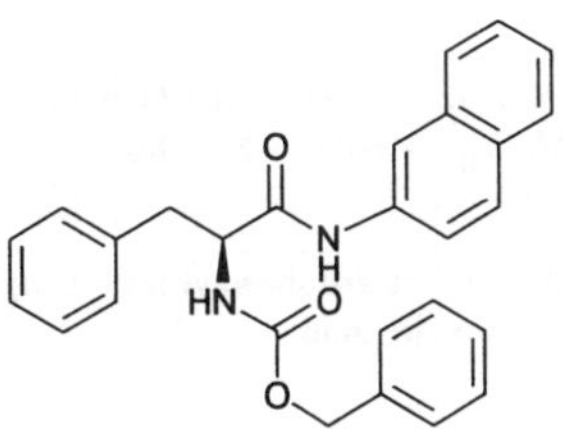

$C_{27}H_{23}N_2O_3$
Chiral intermediate. *Senn Chemicals AG.*

**1474 Carbobenzyloxy-L-phenylalanine N-succinimide ester**

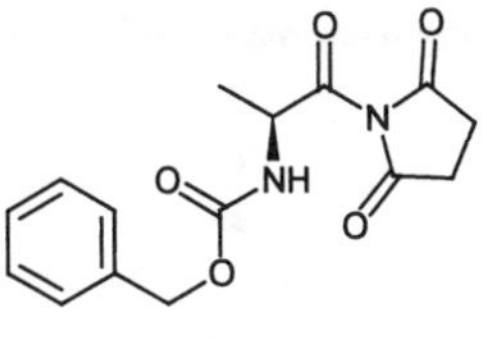

$C_{15}H_{16}N_2O_5$
Chiral intermediate. *Austin Chemical Company, Inc.*

**1475 Carbobenzyloxy-D-phenylalaninol**
58917-85-4

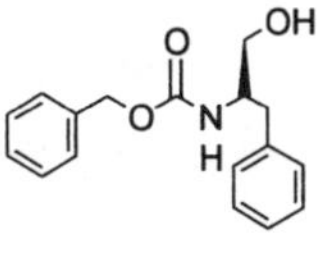

$C_{17}H_{19}NO_3$
(R)-(+)-Carbobenzyloxyamino-3-phenyl-1-propanol. Chiral intermediate. mp = 92-95°; $[\alpha]^{22}$ = + 30° (c = 1, $CHCl_3$). *Acros Organics nv; Newport Synthesis Ireland Ltd.; Sigma-Aldrich Fine Chemicals; SK Energy and Chemical, Inc.; Synthetech, Inc.*

**1476 Carbobenzyloxy-L-phenylalaninol**
6372-14-1

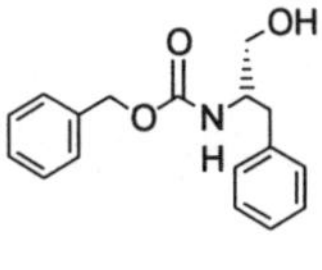

$C_{17}H_{19}NO_3$
N-Benzyloxycarbonyl-L-phenylalaninol; (S)-2-Carbobenzyloxyamino-3-phenyl-1-propanol. Chiral intermediate. mp = 92-95°; $[\alpha]_D^{20}$ = - 43 ± 1° (c = 2, $CH_3OH$); $[\alpha]_D^{20}$ = - 30° (c = 1, $CHCl_3$). *Acros Organics nv; Fischer Chemicals AG; Great Lakes Fine Chemicals; Lancaster Synthesis Ltd.; Newport Synthesis Ireland Ltd.; Nippon Rikagakuyakuhin; Senn Chemicals AG; Sigma-Aldrich Fine Chemicals; SK Energy and Chemical, Inc.; Synthetech, Inc.*

**1477 Carbobenzyloxy-L-phenylalanyl-chloromethyl ketone**
26049-94-5 247-432-8

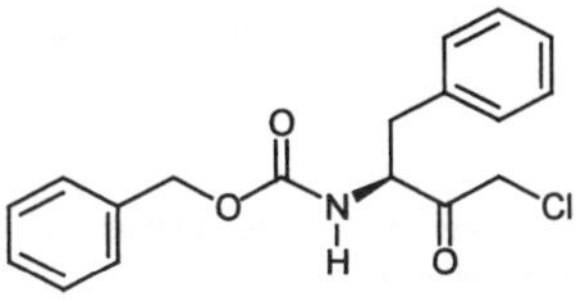

$C_{18}H_{18}ClNO_3$
Carbamic acid, [3-chloro-2-oxo-1-(phenylmethyl)propyl]-phenylmethyl ester, (S)-; N-CBZ-L-phenylalanyl-chloromethyl ketone; (S)-Benzyl [1-benzyl-3-chloro-2-oxopropyl]carbamate; Carbamic acid, [α-(chloroacetyl)phenethyl]-, benzyl ester, L-; L-[α-(Chloroacetyl)phenethyl]carbamic acid, benzyl ester; N-[(Benzyloxy)carbonyl]-L-phenylalan. Listed on TSCA. Chiral intermediate. mp = 107-108°; $[\alpha]^{23}$ = +30.1° (c = 1, $CHCl_3$). *Acros Organics nv; Sigma-Aldrich Fine Chemicals.*

**1478 Carbobenzyloxy-(S)-phenyl-L-cysteine**
159453-24-4

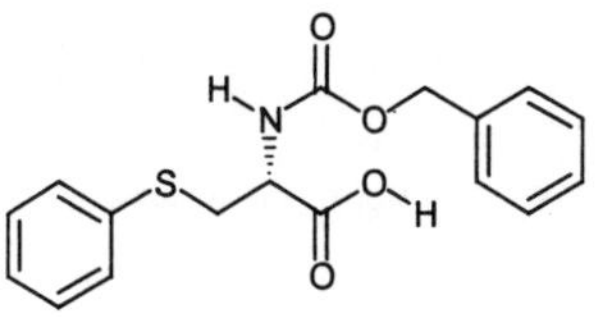

$C_{17}H_{17}NO_4S$
Chiral intermediate. mp = 94-97°; $[\alpha]_D^{20}$ = - 55° (c = 2, EtOH). *Acros Organics nv; Sigma-Aldrich Fine Chemicals; Synthetech, Inc.*

**1479 Carbobenzyloxy-(S)-phenyl-L-cysteine methyl ester**

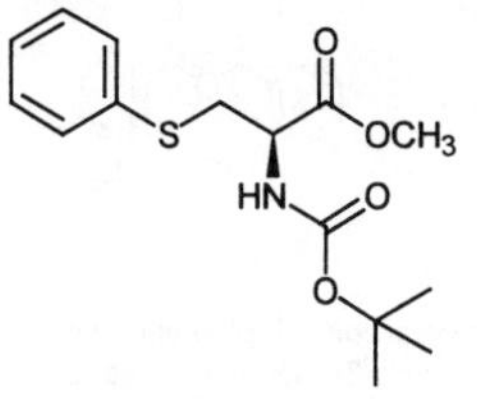

$C_{15}H_{21}NO_4S$
Chiral intermediate. *Synthetech, Inc.*

**1480 Carbobenzyloxy-(R)-2-phenylglycinol**

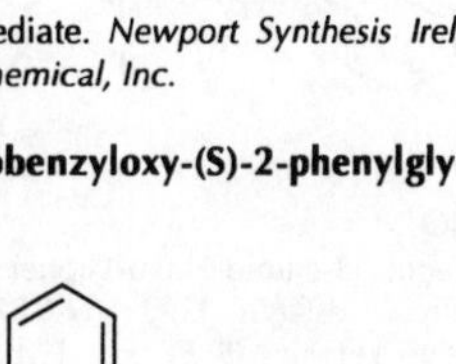

$C_{16}H_{17}NO_3$
Chiral intermediate. *Newport Synthesis Ireland Ltd.; SK Energy and Chemical, Inc.*

**1481 Carbobenzyloxy-(S)-2-phenylglycinol**

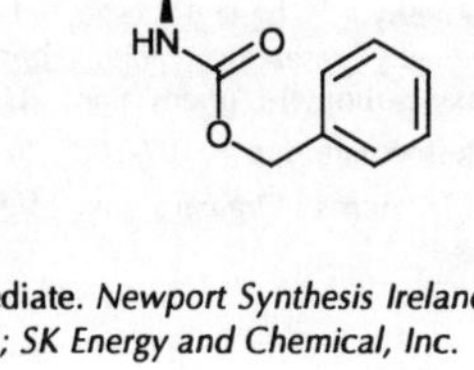

$C_{16}H_{17}NO_3$
Chiral intermediate. *Newport Synthesis Ireland Ltd.; Senn Chemicals AG; SK Energy and Chemical, Inc.*

**1482 Carbobenzyloxy-D-pipecolinic acid**

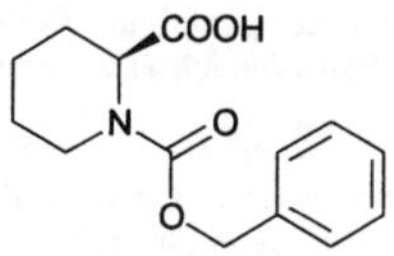

$C_{14}H_{17}NO_4$
Chiral intermediate. *Advanced Asymmetrics, Inc.*

**1483 Carbobenzyloxy-L-pipecolinic acid**

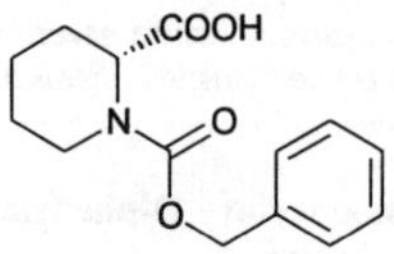

$C_{14}H_{17}NO_4$
Chiral intermediate. *Advanced Asymmetrics, Inc.*

**1484 (R)-(+)-1-(Carbobenzyloxy)-2-piperidine-carboxylic acid**

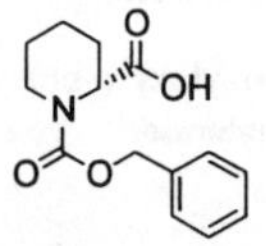

$C_{14}H_{17}NO_4$
N-CBZ-(R)-(+)-pipecolinic acid. Chiral intermediate. mp = 111-115°; $[\alpha]_D^{20}$ = + 60° (c = 2, HOAc). *Sigma-Aldrich Fine Chemicals.*

**1485 (S)-(-)-1-(Carbobenzyloxy)-2-piperidine-carboxylic acid**

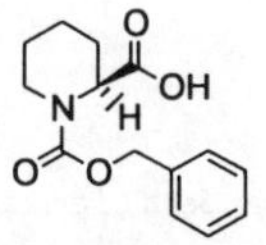

$C_{14}H_{17}NO_4$
Chiral intermediate. mp = 111-115°; $[\alpha]^{22}$ = - 59° (c = 2, HOAc). *Sigma-Aldrich Fine Chemicals.*

**1486 Carbobenzyloxy-L-prolinamide**

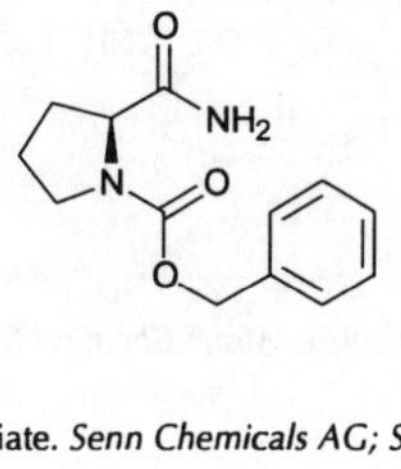

$C_{13}H_{13}N_2O_6$
Chiral intermediate. *Senn Chemicals AG; Synthetech, Inc.*

**1487 Carbobenzyloxy-D-proline**
6404-31-5 229-021-5

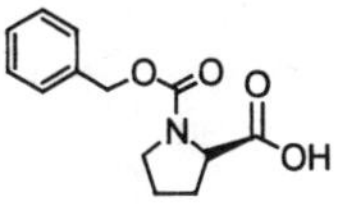

$C_{13}H_{15}NO_4$
Pyrrolidine-1,2-dicarboxylic acid, 1-(phenylmethyl) ester, (2R)-; N-(Benzyloxycarbonyl)-D-proline; Cbz-D-Pro-OH; 1-Benzyl hydrogen (R)-pyrrolidine-1,2-dicarboxylate. mp = 76-78°; $[\alpha]^{25}$ = + 40.2° (c = 2, EtOH). *Isochem; Sigma-Aldrich Fine Chemicals; Synthetech, Inc.*

**1488 Carbobenzyloxy-L-proline**
1148-11-4 214-557-4

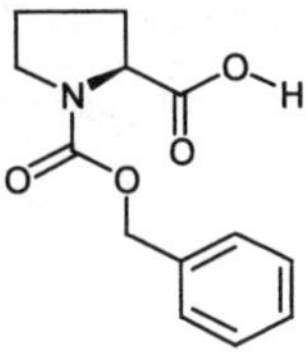

$C_{13}H_{15}NO_4$
Pyrrolidine-1,2-dicarboxylic acid, 1-(phenylmethyl) ester, (2S)-; N-CBZ-L-proline; N-Benzyloxycarbonyl-L-proline. Chiral intermediate. mp = 76-78°; [α] = - 53° (c = 0.55, 0.5N HCl); $[\alpha]_D^{20}$= - 42° (c = 2, EtOH). *Acros Organics nv; Austin Chemical Company, Inc.; Sigma-Aldrich Fine Chemicals; Synthetech, Inc.*

**1489 Carbobenzyloxy-L-proline hydroxy-succinimide ester**

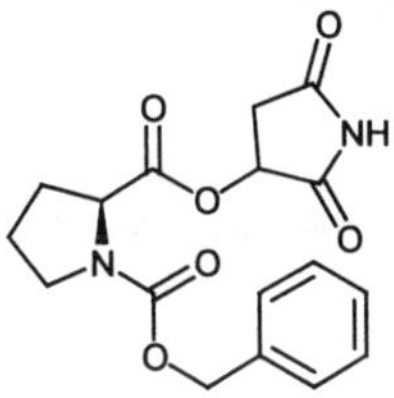

$C_{17}H_{18}N_2O_6$
Chiral intermediate. *Synthetech, Inc.*

**1490 N-(Carbobenzyloxy)-L-proline methyl ester**
5211-23-4

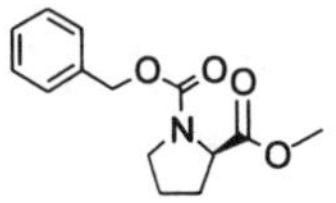

$C_{14}H_{17}NO_4$
Chiral intermediate. d = 1.175; n = 1.5217; $[\alpha]^{22}$ = - 36.8° (neat). *Sigma-Aldrich Fine Chemicals.*

**1491 Carbobenzyloxy-L-pyroglutamic acid**

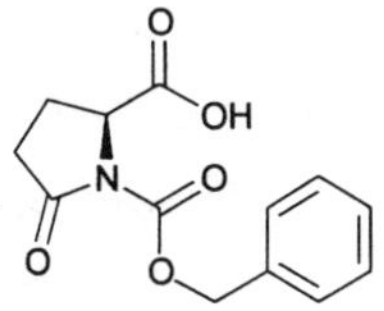

$C_{13}H_{13}NO_5$
Chiral intermediate. *Senn Chemicals AG; Synthetech, Inc.*

**1492 Carbobenzyloxy-L-serine**
1145-80-8 214-546-4
$C_{11}H_{13}NO_5$
Serine, N-carboxy-, N-benzyl ester, L-; N-CBZ-L-serine; N-Benzyloxycarbonyl-L-serine. Chiral intermediate. mp = 117-119°; [α] = + 5.8° (c = 2.7, HOAc). *Acros Organics nv; Austin Chemical Company, Inc.; PPG - Sipsy; Sigma-Aldrich Fine Chemicals; Synthetech, Inc.*

**1493 Carbobenzyloxy-L-serine amide**

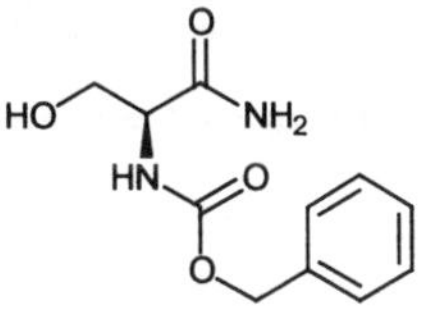

$C_{11}H_{14}N_2O_4$
Chiral intermediate. *Senn Chemicals AG.*

**1494 Carbobenzyloxy-L-serine benzyl ester**
21209-51-8
$C_{18}H_{19}NO_5$
Chiral intermediate. *Sigma-Aldrich Fine Chemicals.*

**1495 Carbobenzyloxy-L-serine cyclohexanonoximester**

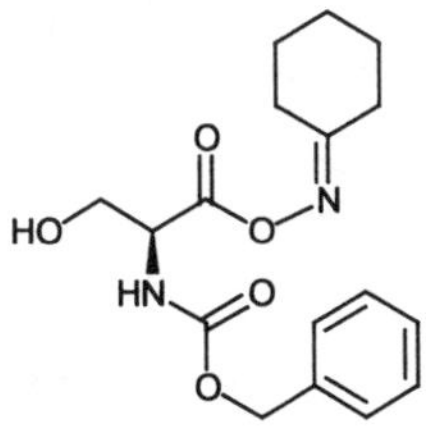

$C_{17}H_{22}N_2O_5$
Chiral intermediate. *Senn Chemicals AG.*

**1496 Carbobenzyloxy-L-serine hydrazide**

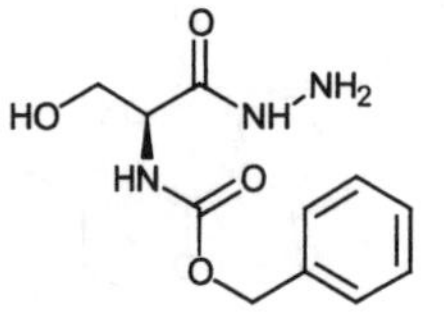

$C_{11}H_{15}N_3O_4$
Chiral intermediate. *Senn Chemicals AG.*

**1497 N-Carbobenzyloxyserine methyl ester**
93204-36-5

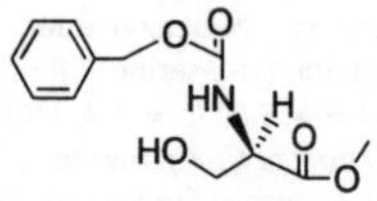

$C_{12}H_{15}NO_5$
Chiral intermediate. mp = 41-43°; $bp_{0.01}$ = 170°; $[\alpha]^{25}$ = + 15° (c = 1, $CH_3OH$). *Sigma-Aldrich Fine Chemicals.*

**1498 Carbobenzyloxy-L-serine methyl ester**

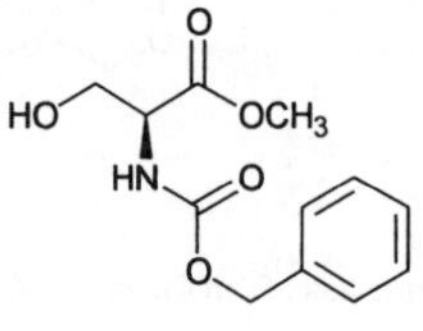

$C_{12}H_{15}NO_5$
Chiral intermediate. *Synthetech, Inc.*

**1499 Carbobenzyloxy-L-threonine**
19728-63-3 243-258-1

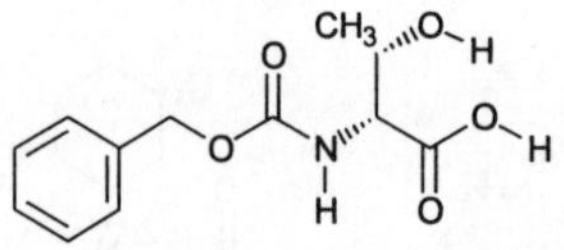

$C_{12}H_{15}NO_5$
Threonine, N-carboxy-, N-benzyl ester, L-; N-[(Benzyloxy)carbonyl]-L-threonine; L-Threonine, N-[(phenylmethoxy)carbonyl]-; Cbz-Thr-OH; N-[(Phenylmethoxy)carbonyl]-L-threonine; Z-Thr-OH. Chiral intermediate. mp = 100-103°; $[\alpha]_D^{20}$ = -4.4° (c = 4, HOAc). *Acros Organics nv; Sigma-Aldrich Fine Chemicals.*

**1500 Carbobenzyloxy-L-threonine amide**

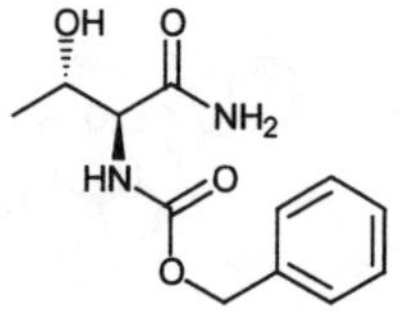

$C_{12}H_{16}N_2O_4$
Chiral intermediate. *Senn Chemicals AG.*

**1501 Carbobenzyloxy-L-threonine hydrazide**

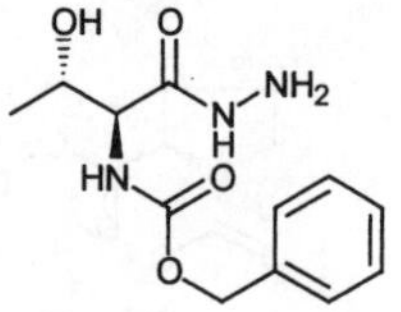

$C_{12}H_{17}N_3O_4$
Chiral intermediate. *Senn Chemicals AG.*

**1502 Carbobenzyloxy-L-threonine methyl ester**
57224-63-2

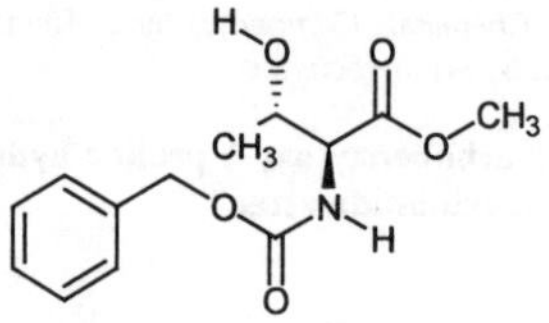

$C_{13}H_{17}NO_5$
Chiral intermediate. mp = 92-94°; $[\alpha]^{25}$ = - 18° (c = 1, $CH_3OH$). *Acros Organics nv; Sigma-Aldrich Fine Chemicals.*

**1503 Carbobenzyloxy-L-threonine-cyclohexanonoximester**

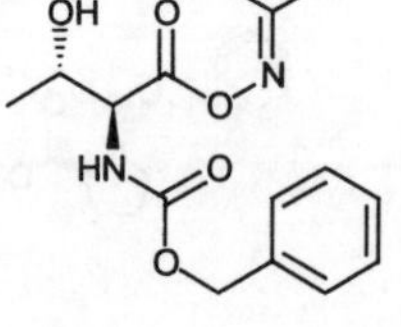

$C_{18}H_{24}N_2O_5$
Chiral intermediate. *Senn Chemicals AG.*

**1504 Carbobenzyloxy-L-threoninol**

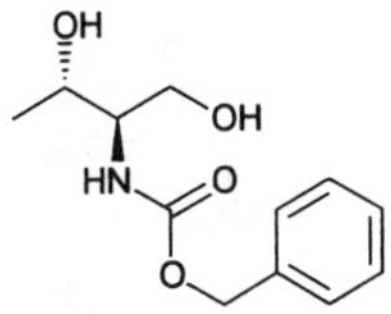

$C_{12}H_{17}NO_4$
Chiral intermediate. *Senn Chemicals AG.*

**1505 Carbobenzyloxy-D-tryptophan**

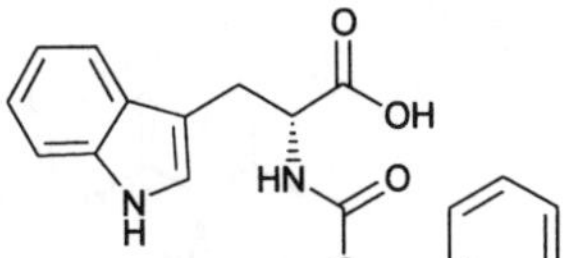

$C_{19}H_{18}N_2O_4$
Chiral intermediate. *Synthetech, Inc.*

**1506 N-α-Carbobenzyloxy-L-tryptophan**
7432-21-5 231-074-4

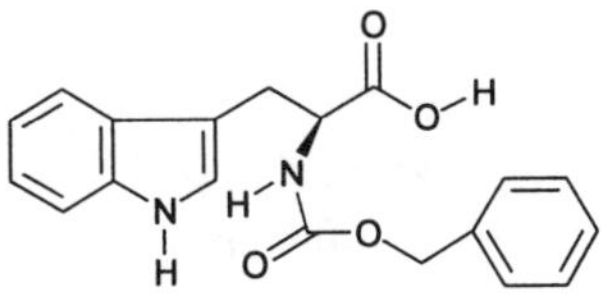

$C_{19}H_{18}N_2O_4$
Tryptophan, N-carboxy-, N-benzyl ester, L-; N-Benzyloxycarbonyl-L-tryptophan; N-Cbz-L-tryptophan. Chiral intermediate. mp = 125-127°; $[\alpha]_D^{20}$ = + 2.9° (c = 3, HOAc). *Acros Organics nv; Sigma-Aldrich Fine Chemicals; Synthetech, Inc.*

**1507 Carbobenzyloxy-L-tryptophan methyl amide**

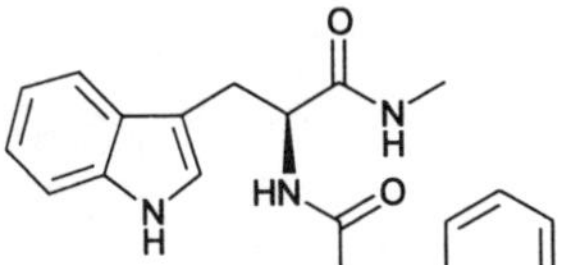

$C_{20}H_{21}N_3O_3$
Chiral intermediate. *Senn Chemicals AG.*

**1508 Carbobenzyloxy-D-tryrosine**

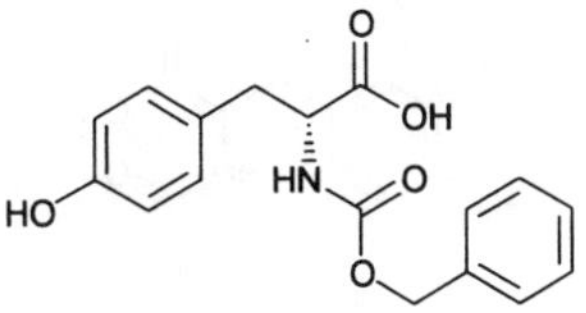

$C_{17}H_{17}NO_5$
Chiral intermediate. *Synthetech, Inc.*

**1509 Carbobenzyloxy-L-tyrosine**
1164-16-5 214-609-6

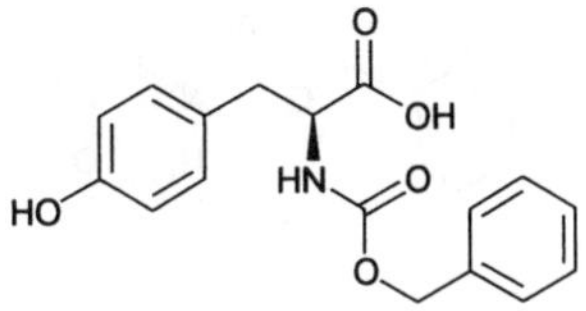

$C_{17}H_{17}NO_5$
Tyrosine, N-carboxy-, N-benzyl ester, L-; N-[(Benzyloxy)carbonyl]-L-tyrosine; Carbobenzoxy-L-tyrosine. Chiral intermediate. mp = 89-94°; $[\alpha]^{22}$ = + 11° (c = 1, HOAc). *Austin Chemical Company, Inc.; Sigma-Aldrich Fine Chemicals; Synthetech, Inc.*

**1510 Carbobenzyloxy-L-tyrosine amide**

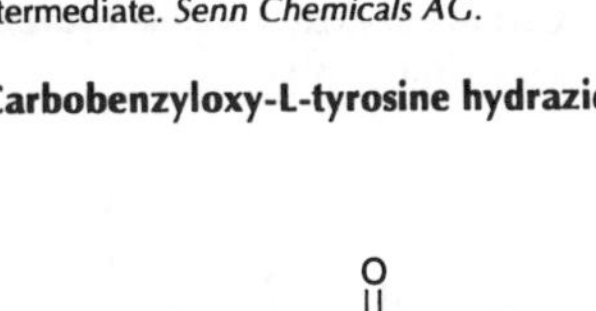

$C_{17}H_{18}N_2O_4$
Chiral intermediate. *Senn Chemicals AG.*

**1511 Carbobenzyloxy-L-tyrosine hydrazide**

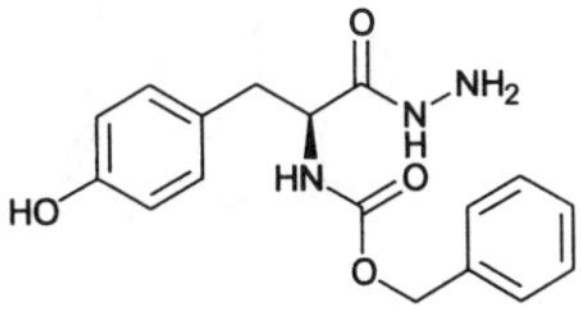

$C_{17}H_{19}N_3O_4$
Chiral intermediate. *Senn Chemicals AG.*

**1512 Carbobenzyloxy-L-tyrosine methyl ester**

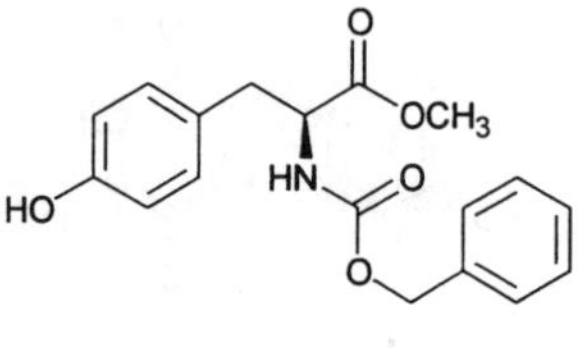

$C_{18}H_{19}NO_5$
Chiral intermediate. *Synthetech, Inc.*

**1513 Carbobenzyloxy-L-valine**
1149-26-4 214-562-1

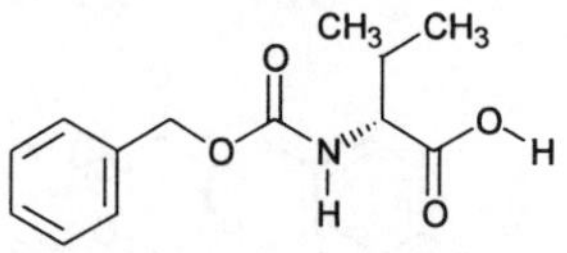

$C_{13}H_{17}NO_4$
Valine, N-[(phenylmethoxy)carbonyl]-, L-; N-CBZ-L-valine; N-Benzyloxycarbonyl-L-valine; L-valine, N-[(phenylmethoxy)carbonyl]-. Chiral intermediate. mp = 58-60°; [α] = - 4.4° (c = 2, HOAc). *Acros Organics nv; Austin Chemical Company, Inc.; Flamma s.p.a.; KingChem, Inc.; Sigma-Aldrich Fine Chemicals; Synthetech, Inc.*

**1514 Carbobenzyloxy-L-valine amide**

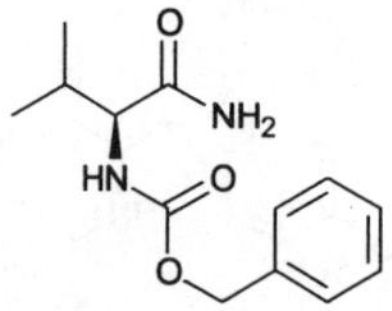

$C_{13}H_{18}N_2O_3$
Chiral intermediate. *Senn Chemicals AG.*

**1515 Carbobenzyloxy-L-valine hydroxysuccinimide ester**

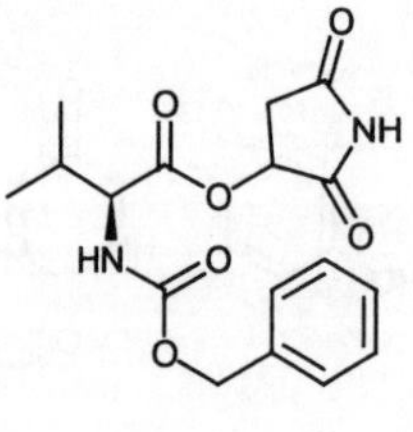

$C_{17}H_{20}N_2O_6$
Chiral intermediate. *Synthetech, Inc.*

**1516 Carbobenzyloxy-L-valine methyl ester**

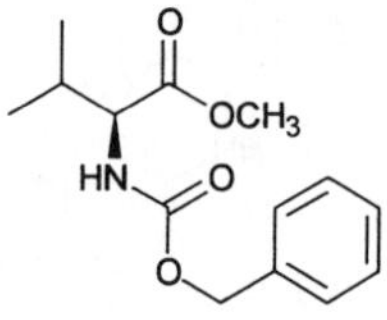

$C_{14}H_{19}NO_4$
Chiral intermediate. *Synthetech, Inc.*

**1517 Carbobenzyloxy-L-valine N-succinimide ester**

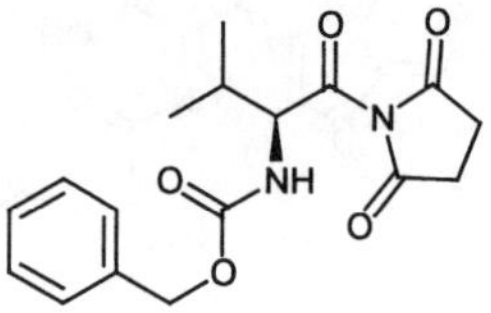

$C_{17}H_{20}N_2O_5$
Chiral intermediate. *Austin Chemical Company, Inc.*

**1518 Carbobenzyloxy-(R)-valinol**

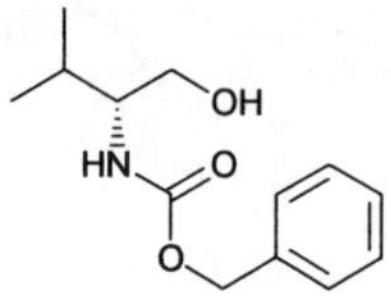

$C_{13}H_{19}NO_3$
Chiral intermediate. *SK Energy and Chemical, Inc.*

**1519 Carbobenzyloxy-(S)-valinol**

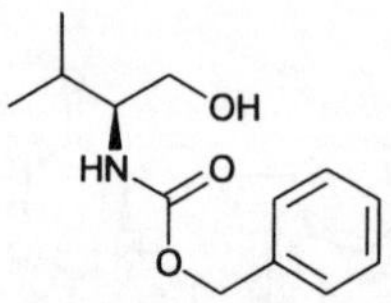

$C_{13}H_{19}NO_3$
Chiral intermediate. *SK Energy and Chemical, Inc.; Synthetech, Inc.*

**1520 Carboethoxy-L-phenylalanine**

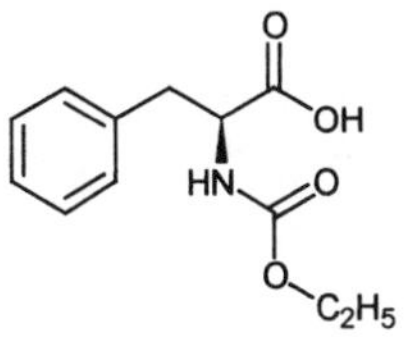

$C_{12}H_{15}NO_4$
Chiral intermediate. *Austin Chemical Company, Inc.*

**1521 (R)-(+)-2-(2'-Carbomethoxyethyl)-2-methylcyclohexanone**
94089-47-1

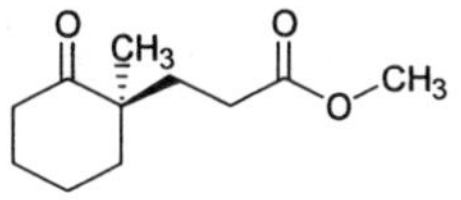

$C_{11}H_{18}O_3$
Chiral intermediate. $bp_{0.01}$ = 70°; d = 1.056; n = 1.469; $[\alpha]_D^{20}$ = + 37° (c = 3, EtOH). *Acros Organics nv; Sigma-Aldrich Fine Chemicals.*

**1522 (S)-(-)-2-(2'-Carbomethoxyethyl)-2-methylcyclohexanone**
112898-44-9

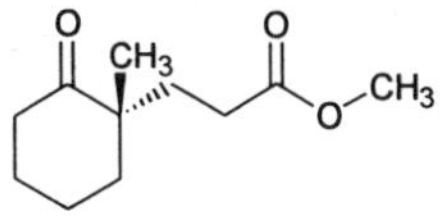

$C_{11}H_{18}O_3$
(S)-(-)-1-Methyl-2-oxo-cyclohexanepropanoic acid methylester; (-)-Methyl (S)-1-methyl-2-oxocyclohexane-propanoate. Chiral intermediate. bp = 280-283°; $bp_{0.01}$ = 70°; d = 1.055; n = 1.469; [α] = -37° (c = 3, EtOH). *Acros Organics nv; Sigma-Aldrich Fine Chemicals.*

**1523 5-O-Carbomethoxy-1,2-O-isopropylidene-3-O-(p-tolylsulfonyl)-α-D-xylofuranose**
74580-94-2
$C_{17}H_{22}O_9S$
Chiral intermediate. mp = 103°; $[\alpha]_D^{20}$ = - 12° (c = 2, $CH_3OH$). *Pfanstiehl Laboratories, Inc.*

**1524 5-O-Carbomethoxy-1,2-O-isopropylidene-α-D-xylofuranose**
5432-33-7

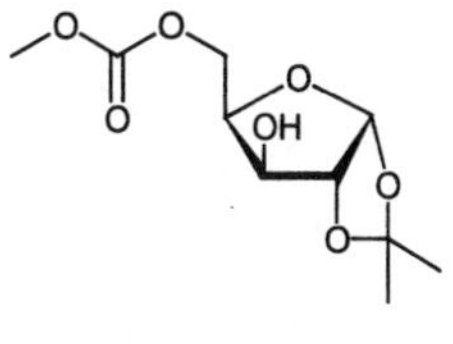

$C_{10}H_{16}O_7$
Chiral intermediate. mp = 103°; $[\alpha]_D^{20}$ = - 12° (c = 2, $CH_3OH$). *Pfanstiehl Laboratories, Inc.*

**1525 (2R,3S)-1-Carboxy-4-tert-butyl-2,3-dihydroxycyclohexa-4,6-diene potassium salt**

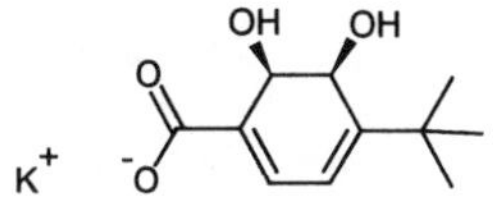

$C_{11}H_{15}KO_4$
Chiral intermediate. *Acros Organics nv.*

**1526 (R)-γ-Carboxy-γ-butyrolactone**
53558-93-3

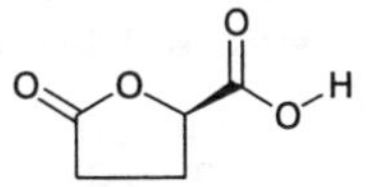

$C_5H_6O_4$
Furan-2-carboxylic acid, tetrahydro-5-oxo-, (2R)-; (R)-(-)-5-Oxo-2-tetrahydrofurancarboxylicacid; 5-Furancarboxylic acid, tetrahydro-2-oxo-, (R)-(-)-. Listed on TSCA. Chiral building block. mp = 71-73°; $bp_{0.3}$ = 165-167°; $[\alpha]_D^{20}$ = - 14° (c = 5, $CH_3OH$). *Acros Organics nv; Daicel Chemical Ind. Ltd; Oxford Asymmetry International plc; Sigma-Aldrich Fine Chemicals.*

**1527 (S)-γ-Carboxy-γ-butyrolactone**
21461-84-7

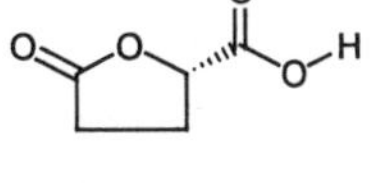

$C_5H_6O_4$
Furan-2-carboxylic acid, tetrahydro-5-oxo-, (S)-; (S)-(+)-5-Oxo-2-terahydrofurancarboxylic acid; 2-Furoic acid, tetrahydro-5-oxo-, (S)-(+)-; (S)-(+)-Tetrahydro-5-oxo-2-furanecarboxylic acid. Chiral intermediate. mp = 63-64°; $bp_2$ = 150-155°; $[\alpha]_D^{20}$ = + 14° (c = 5, $CH_3OH$). *Acros Organics nv; Daicel Chemical Ind. Ltd; Oxford Asymmetry International plc; Sigma-Aldrich Fine Chemicals.*

**1528 (2R,3R)-1-Carboxy-4-chloro-2,3-dihydroxycyclohexa-4,6-diene**
193338-31-7

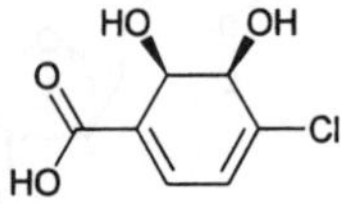

$C_7H_7ClO_4$
Chiral intermediate. *Acros Organics nv.*

**1529 (2R,3R)-1-Carboxy-4,5-dichloro-2,3-dihydroxycyclohexa-4,6-diene**

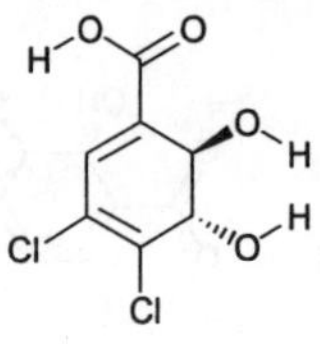

$C_7H_6Cl_2O_4$
Chiral intermediate. [α] = + 48.8° (c = 1, $CH_3OH$). *Acros Organics nv.*

**1530 (2R,3R)-1-Carboxy-2,3-dihydroxy-4-methylcyclohexa-3,5-diene**

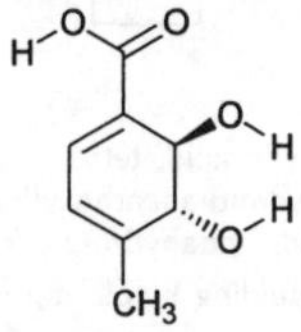

$C_8H_{10}O_4$
Chiral intermediate. *Acros Organics nv.*

**1531 (2R,3R)-1-Carboxy-2,3-dihydroxy-4-methylcyclohexa-4,6-diene**

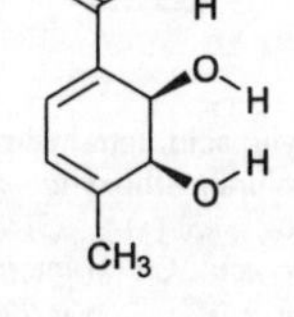

$C_8H_{10}O_4$
Chiral intermediate. *Acros Organics nv.*

**1532 (R)-1-(Carboxyethyl)-N-benzoylthiolglycinate**
150520-30-2

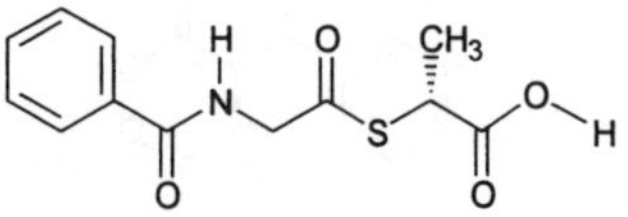

$C_{12}H_{13}NO_4S$
Chiral intermediate. *Acros Organics nv.*

**1533 (2R,3S)-1-Carboxy-4-ethyl-2,3-dihydroxycyclohexa-4,6-diene sodium salt**

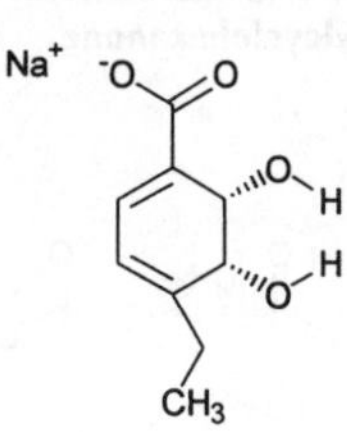

$C_9H_{11}NaO_4$
Chiral intermediate. *Acros Organics nv.*

**1534 (2R,3R)-1-Carboxy-4-iodo-2,3-dihydroxycyclohexa-4,6-diene**

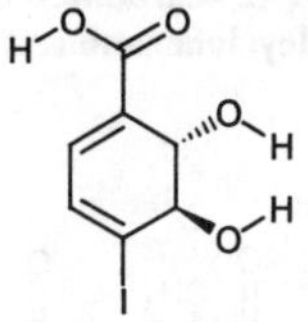

$C_7H_7IO_4$
Chiral intermediate. *Acros Organics nv.*

**1535 (2R,3S)-1-Carboxy-5-iodo-4-methyl-2,3-dihydroxycyclohexa-4,6-diene**

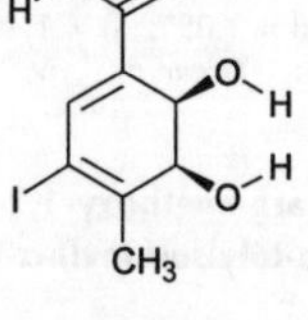

$C_8H_9IO_4$
Chiral intermediate. *Acros Organics nv.*

**1536 (2R,3S)-1-Carboxy-4-isopropyl-2,3-dihydroxycyclohexa-4,6-diene potassium salt**

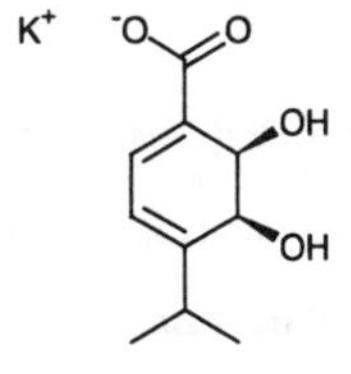

$C_{10}H_{13}KO_4$
Chiral intermediate. *Acros Organics nv.*

**1537 (2S,3S)-trans-3-(Carboxymethyl)-azetidine-2-acetic acid**

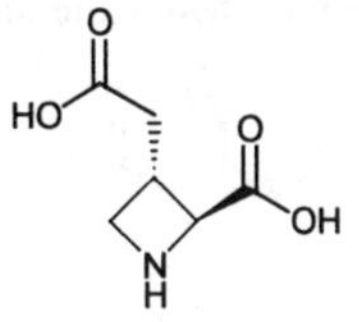

$C_6H_9NO_4$
Chiral intermediate. *Acros Organics nv.*

**1538 Carboxymethyl-N-benzoylthiolalaninate**
138079-74-0

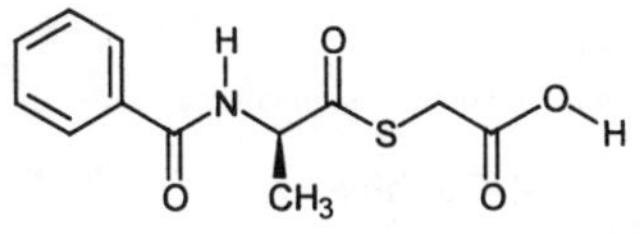

$C_{12}H_{13}NO_4S$
Chiral intermediate. *Acros Organics nv.*

**1539 (3S)-Carboxymethyl-(6S)-benzyl-2,5-diketopiperazine**
5262-10-2 226-075-1

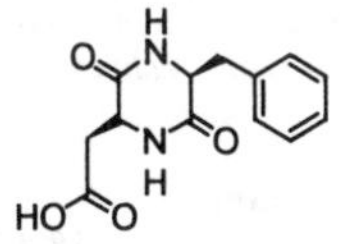

$C_{13}H_{14}N_2O_4$
Piperazine-2-acetic acid, 3,6-dioxo-5-(phenylmethyl)-, (2S,5S)-; (2S-cis)-5-Benzyl-3,6-dioxopiperazine-2-acetic acid; 2-Piperazineacetic acid, 5-benzyl-3,6-dioxo-. Chiral building block. mp = 270-272°; $[\alpha]_D^{20}$ = - 9.0° (c = 0.9, HOAc). *Fischer Chemicals AG; Great Lakes Fine Chemicals; Sigma-Aldrich Fine Chemicals.*

**1540 S-Carboxymethyl-L-cysteine**
638-23-3 211-327-5

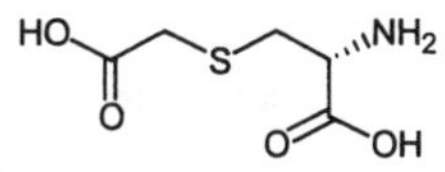

$C_5H_9NO_4S$
Alanine, 3-[(carboxymethyl)thio]-, L-; Thiodril; LJ 206; Mucodyne; Mucopront; Rhinathiol; Rinatiol; Carbocysteine; (L)-2-Amino-3-(carboxymethylthio)-propionic acid; 3-[(Carboxymethyl)thio]-L-alanine. Listed on TSCA. Chiral intermediate. [α] = - 34° (c = 10, 10% $NaHCO_3$). *Acros Organics nv; Austin Chemical Company, Inc.; Interchem Corporation; Kyowa Hakko Kogyo Co., Ltd.; Shanghai DSL International Trading Company; Varsal Instruments, Inc.*

**1541 (S)-3-Carboxymethyl-2,5-piperazinedione**

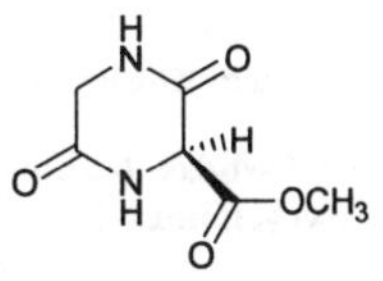

$C_7H_{10}N_2O_4$
Chiral intermediate. *Great Lakes Fine Chemicals.*

**1542 (2R,3S)-1-Carboxy-4-pentyl-2,3-dihydroxycyclohexa-4,6-diene potassium salt**

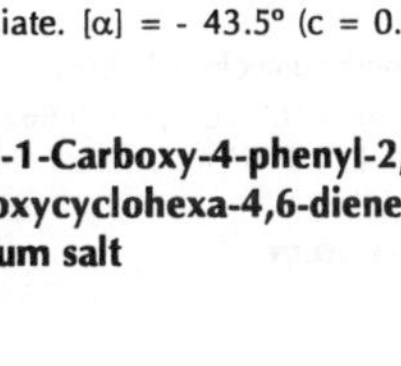

$C_{12}H_{17}KO_4$
Chiral intermediate. [α] = - 43.5° (c = 0.1, $H_2O$). *Acros Organics nv.*

**1543 (2R,3S)-1-Carboxy-4-phenyl-2,3-dihydroxycyclohexa-4,6-diene potassium salt**

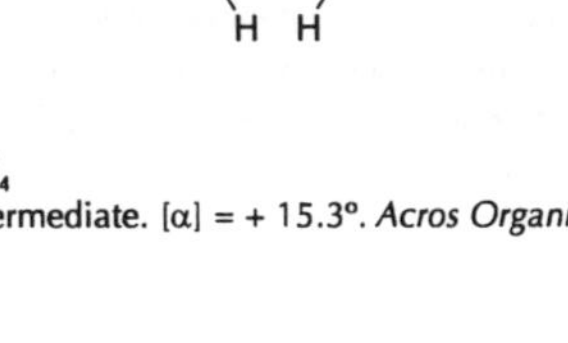

$C_{13}H_{11}KO_4$
Chiral intermediate. [α] = + 15.3°. *Acros Organics nv.*

**1544 (S)-(-)-1-[N-(1-Carboxy-3-phenylpropyl)-L-lysinyl]-L-proline dihydrate**
83915-83-7

$C_{21}H_{31}N_3O_5$
Lisinopril. Chiral intermediate. *Sigma-Aldrich Fine Chemicals.*

**1545 (2R,3S)-1-Carboxy-4-propyl-2,3-dihydroxy-cyclohexa-4,6-diene potassium salt**

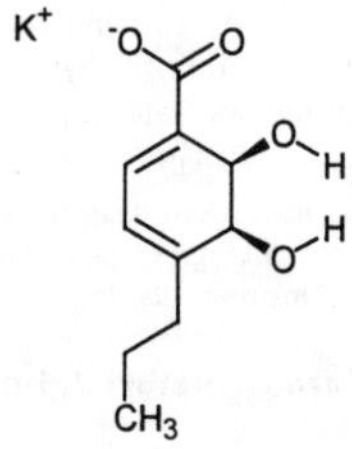

$C_{10}H_{13}KO_4$
Chiral intermediate. *Acros Organics nv.*

**1546 (2R,3S)-1-Carboxy-4-trifluoromethyl-2,3-dihydroxycyclohexa-4,6-diene**

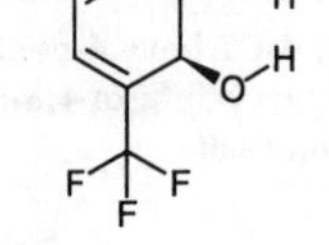

$C_8H_7F_3O_4$
Chiral intermediate. *Acros Organics nv.*

**1547 (+)-2-Carene**
554-61-0

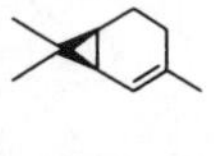

$C_{10}H_{16}$
(+)-3,7,7-Trimethylbicyclo[4.1.0]hept-2-ene. Chiral intermediate. bp = 167-168°; d = 0.862; n = 1.476; $[\alpha]_D^{20}$= + 90.0° (c = 6, EtOH). *Sigma-Aldrich Fine Chemicals.*

**1548 (+)-3-Carene**
13466-78-9 1885(12) 236-719-3

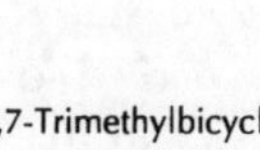

$C_{10}H_{16}$
δ-3-carene; 3,7,7-Trimethylbicyclo[4.1.0]hept-3-ene. Listed on TSCA. Chiral intermediate. $bp_{705}$ = 168-169°; d = 0.857; n = 1.474; $[\alpha]_D^{20}$= + 15° (neat). *Sigma-Aldrich Fine Chemicals.*

**1549 (1S)-(+)-3-Carene**
498-15-7 1843(11) 236-719-3

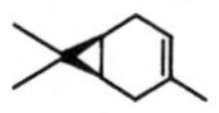

$C_{10}H_{16}$
(1S)-3,7,7-Trimethylbicyclo[4.1.0]hept-3-ene. Listed on TSCA. Chiral intermediate. bp = 170-172°; d = 0.865; n = 1.472; $[\alpha]_D^{20}$= + 17° (neat). *Sigma-Aldrich Fine Chemicals.*

**1550 L-Carnithine base**

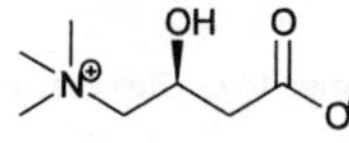

$C_7H_{15}NO_3$
Chiral building block. *Flamma s.p.a.*

**1551 L-Carnithine hydrochloride**

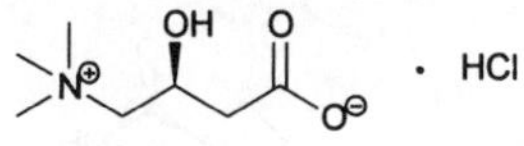

$C_7H_{16}ClNO_3$
Chiral building block. *Flamma s.p.a.*

**1552 L-Carnithine tartrate**

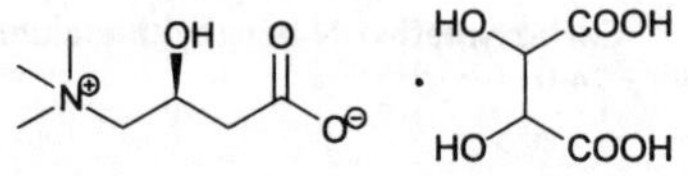

$C_{11}H_{21}NO_9$
Chiral building block. *Flamma s.p.a.*

**1553 D-(+)-Carnitine**
541-14-0

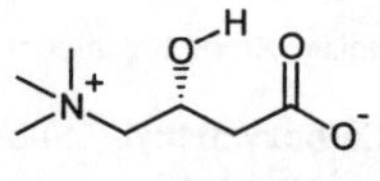

$C_7H_{15}NO_3$
Propan-1-aminium, 3-carboxy-2-hydroxy-N,N,N-tri-methyl-, inner salt, (S)-; Ammonium, (3-carboxy-2-hydroxypropyl)trimethyl-, hydroxide, inner salt, D-. Chiral building block. *Acros Organics nv; Yoneyama Yakuhin Kogyo Co., Ltd.*

**1554 L-(-)-Carnitine**
541-15-1 1898(12) 208-768-0

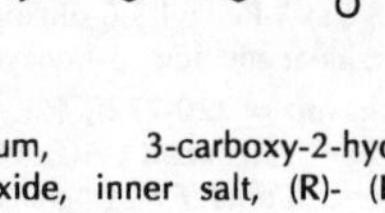

$C_7H_{16}NO_3$
Propan-1-aminium, 3-carboxy-2-hydroxy-N,N,N-tri-methyl-, hydroxide, inner salt, (R)- (R)-(3-Carboxy-2-hydroxypropyl)trimethylammonium hydroxide;

Ammonium, (3-carboxy-2-hydroxypropyl)trimethyl-, hydroxide, inner salt, L-; Levocarnitine; ST 198; Vitamin BT. Chiral building block. mp = 197-212°; $[\alpha]_D^{20}$ = - 31° (c = 10, $H_2O$). *Aceto Corporation; Acros Organics nv; Ajinomoto Co. Inc.; Flamma s.p.a.; Lonza Ltd; Senn Chemicals AG; Sigma-Aldrich Fine Chemicals; Tanabe Seiyaku Co. Ltd.; TCI America; Varsal Instruments, Inc.; Yoneyama Yakuhin Kogyo Co., Ltd.*

**1555 L-(-)-Carnitine hydrochloride**
6645-46-1 229-663-6

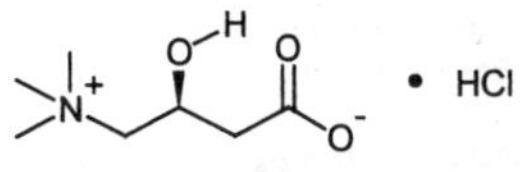

• HCl

$C_7H_{16}ClNO_3$
Propan-1-aminium, 3-carboxy-2-hydroxy-N,N,N-trimethyl-, chloride, (2R)-; (R)-(3-Carboxy-2-hydroxypropyl)trimethylammonium chloride; Ammonium, (3-carboxy-2-hydroxypropyl)trimethyl-, chloride, (-)-; LC 80; Levocarnitine chloride. Chiral intermediate. [α] = - 22° (c = 1, $H_2O$). *Aceto Corporation; Acros Organics nv; Tanabe Seiyaku Co. Ltd.*

**1556 L-Carnitine magnesium citrate**

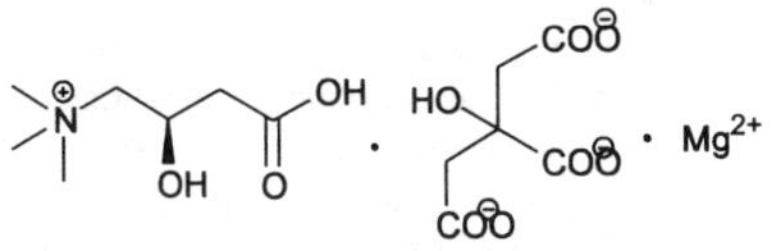

$C_{13}H_{21}MgNO_{10}$
Chiral building block. *Aceto Corporation.*

**1557 D-(+)-Carnitinenitrile chloride**

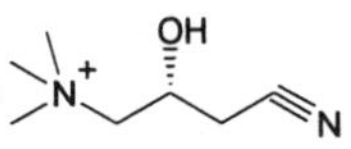

$C_7H_{15}ClN_2O$
Chiral intermediate. *Acros Organics nv.*

**1558 L-Carnitine tartrate**

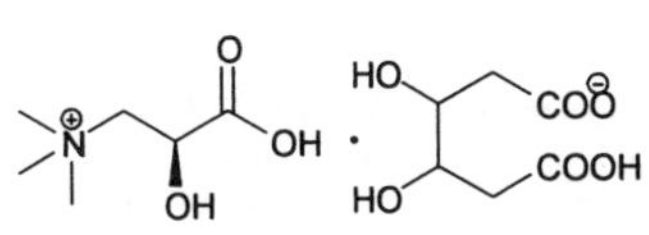

$C_{12}H_{23}NO_9$
Chiral building block. *Aceto Corporation.*

**1559 (-)-Carveol (mixture of cis and trans isomers)**
99-48-9 202-757-4

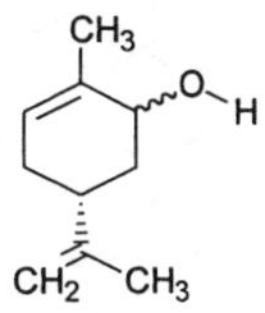

$C_{10}H_{16}O$
Cyclohex-2-en-1-ol, 2-methyl-5-(1-methylethenyl)-, (-); 2-Methyl-5-(1-methylvinyl)cyclohex-2-en-1-ol; p-Mentha-6,8-dien-2-ol, . Listed on TSCA. Chiral intermediate. $bp_{751}$ = 226-227°; d = 0.958; n = 1.496; $[\alpha]_D^{20}$ = - 112° (c = 1, $CHCl_3$). *Acros Organics nv; Sigma-Aldrich Fine Chemicals.*

**1560 (R)-(-)-Carvone**
6485-40-1 1925(12) 229-352-5

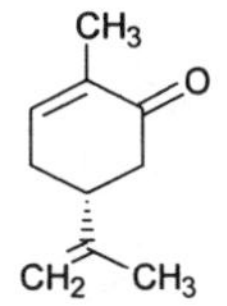

$C_{10}H_{14}O$
Cyclohex-2-en-1-one, 2-methyl-5-(1-methylethenyl)-, (5R)-; (R)-(-)-5-Isopropenyl-2-methyl-2-cyclohexenone; l-p-Mentha-1(6),8-dien-2-one; (R)-2-Methyl-5-(1-methylethenyl)-2-cyclohexen-1-one; (R)-(-)-p-Mentha-6,8-dien-2-one. Listed on TSCA. Chiral intermediate. bp = 228-230°; d = 0.9590; $n_D^{20}$ = 1.4990; $[\alpha]_D^{20}$ = - 60 ± 3° (neat). *Acros Organics nv; Lancaster Synthesis Ltd.; Sigma-Aldrich Fine Chemicals; TCI America.*

**1561 (S)-(+)-Carvone**
2244-16-8 1925(12) 218-827-2

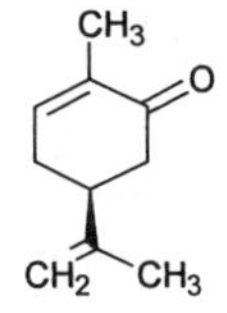

$C_{10}H_{14}O$
Cyclohex-2-en-1-one, 2-methyl-5-(1-methylethenyl)-, (5S)-. (S)-(+)-5-Isopropenyl-2-methyl-2-cyclohexenone; (S)-2-Methyl-5-(1-methylvinyl)cyclohex-2-en-1-one; p-Mentha-6,8-dien-2-one, (S)-(+)-. Talent. Listed on TSCA. Chiral intermediate. mp = 88.9°; bp = 228-230°; $bp_{10}$ = 98°; d = 0.9620; $n_D^{20}$ = 1.4970; $[\alpha]_D^{20}$ = + 61 ± 3° (neat); soluble in $H_2O$ (.5 0.1 g/100 ml, 25°). *Acros Organics nv; Lancaster Synthesis Ltd.; Sigma-Aldrich Fine Chemicals; TCI America.*

**1562 (-)-Carvyl acetate**
7053-79-4 202-580-2

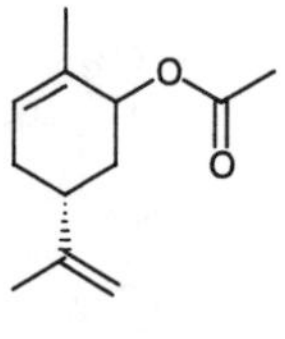

$C_{12}H_{18}O_2$
Cyclohex-2-en-1-ol, 2-methyl-5-(1-methylethenyl)-, acetate, (1S,5R)-; (1S-trans)-2-Methyl-5-(1-methylvinyl)-cyclohex-2-en-1-yl acetate; p-Mentha-6,8-dien-2-ol, acetate, (2S,4R)-. Listed on TSCA. Chiral intermediate. $bp_{0.1}$ = 77-79°; d = 0.976; n = 1.475; $[\alpha]^{23}$ = - 81.6° (neat). *Acros Organics nv; Sigma-Aldrich Fine Chemicals.*

**1563 (-)-Carvyl propionate**

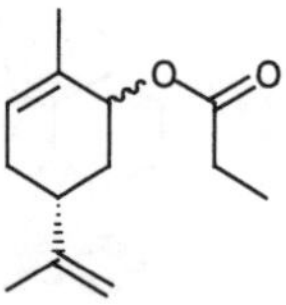

$C_{13}H_{20}O_2$
Cyclohex-2-en-1-ol, 2-methyl-5-(1-methylethenyl)-, propanoate, ; (1S-trans)-2-Methyl-5-(1-methylvinyl)cyclohex-2-en-1-yl propionate; p-Mentha-6,8-dien-2-ol, propionate, (2S,4R)-. (±) 97-45-0. Chiral intermediate. $bp_{0.2}$ = 78-80°; d = 0.952; n = 1.474; $[\alpha]^{22}$ = - 99° (neat). *Acros Organics nv.*

**1564 (-)-Caryophyllene oxide**
1139-30-6 214-519-7

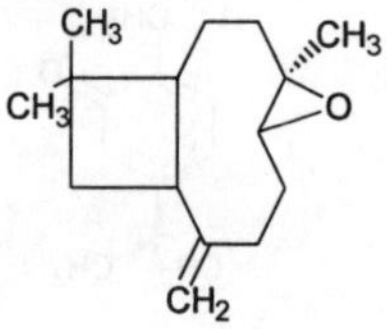

$C_{15}H_{24}O$
Oxa-5-tricyclo[8.2.0.04,6]dodecane, 4,12,12-trimethyl-9-methylene-, (1R,4R,6R,10S)-; (1R,4R,6R,10S)-4,12,12-Trimethyl-9-methylene-5-oxatricyclo[8.2.0.0(4,6)]dodecane; Caryophylene monoepoxide. Listed on TSCA. Chiral intermediate. mp = 61-62°; d = 0.966; $[\alpha]_D^{20}$ = - 71° (c = 2, $CHCl_3$). *Acros Organics nv; Sigma-Aldrich Fine Chemicals.*

**1565 (+)-Catechin hydrate**
1908(11)

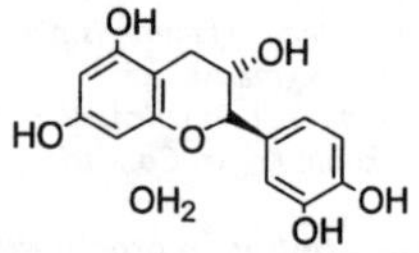

$C_{15}H_{14}O_6$
Chiral intermediate. $[\alpha]^{21}$ = + 16.0° (c = 1, $H_2O$). *Sigma-Aldrich Fine Chemicals.*

**1566 D-(+)-Cellobiose**
528-50-7 2008(12) 208-436-5

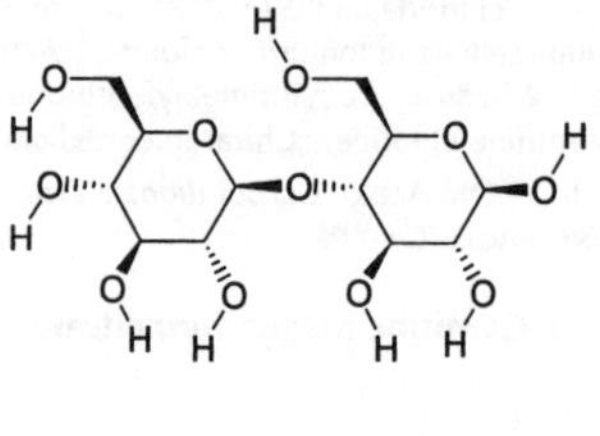

$C_{12}H_{22}O_{11}$
Glucose, 4-O-β-D-glucopyranosyl-, D-; 4-O-β-D-glucopyranosyl-D-glucopyranose; D-(+)-Cellose. Listed on TSCA. Chiral intermediate. mp = 239°; $[\alpha]_D^{20}$ = + 34.6° (c = 8, $H_2O$). *Acros Organics nv; Pfanstiehl Laboratories, Inc.; Senn Chemicals AG; Sigma-Aldrich Fine Chemicals; TCI America.*

**1567 α-D-Cellobiose octaacetate**
5346-90-7 1957(11) 226-304-5

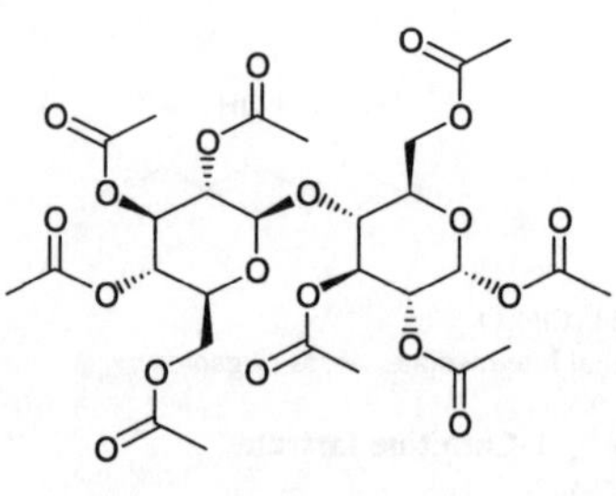

$C_{28}H_{38}O_{19}$
Glucopyranose, α-D-, 4-O-(2,3,4,6-tetra-O-acetyl-β-D-glucopyranosyl)-, tetraacetate; α-Cellobiopyranose octaacetate; Octa-O-acetyl-α-D-cellobiose. Chiral intermediate. mp = 224-226°; [α] = + 40° (c = 6, $CHCl_3$). *Acros Organics nv; Borregaard Synthesis; Sigma-Aldrich Fine Chemicals.*

**1568 D-(+)-Cellobiose octaacetate**
3616-19-1 222-799-7

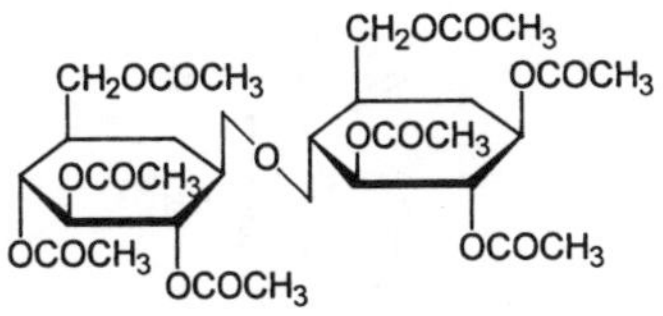

$C_{28}H_{38}O_{19}$
Glucopyranose, 4-O-(2,3,4,6-tetra-O-acetyl-β-D-glucopyranosyl)-, tetraacetate, D-; 4-O-β-D-Glucopyranosyl-D-glucose octaacetate; Octaacetyl-D-cellobiose. Listed on TSCA. Chiral intermediate. *Conti BPC NV; Senn Chemicals AG.*

**1569 (+)-Cembrene**
1898-13-1
$C_{20}H_{32}$
Chiral intermediate. mp = 59-61°; [α] = + 238° (c = 1, $CHCl_3$). *Acros Organics nv.*

**1570 Chelidonine**
476-32-4 207-504-1
$C_{20}H_{19}NO_5$
Benzo[1,3]dioxolo[5,6-c]-1,3-dioxolo[4,5-i]phenanthridin-6-ol, 5β,6,7,12b,13,14-hexahydro-13-methyl-, (5βR,6S,12βS)-; Stylophorine. Pharmaceutical or derivative. mp = 135-136°. *Kaden Biochemicals GmbH.*

**1571 (R)-Chiracamphox**

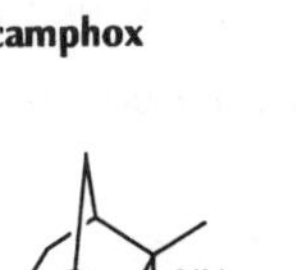

$C_{11}H_{17}NO_2$
Resolving agent; Chiral auxiliary. *Acros Organics nv.*

**1572 (S)-Chiracamphox**

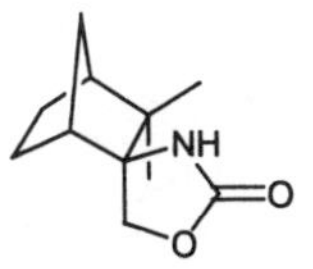

$C_{11}H_{17}NO_2$
Resolving agent. *Acros Organics nv.*

**1573 D-α-Chitobiose octaacetate**
7284-18-6 230-709-2
$C_{28}H_{40}N_2O_{17}$
Glucopyranose, 2-acetamido-4-O-(2-acetamido-2-deoxy-β-D-glucopyranosyl)-2-deoxy-, 1,3,3',4',6,6'-hexaacetate, α-D-; 2-Acetamido-4-O-(2-acetamido-2-deoxy-β-D-glucopyranosyl)-2-deoxy-α-D-glucose 1,3,3',4',6,6'-hexaacetate. Chiral intermediate. *Senn Chemicals AG.*

**1574 (-)-Chloralose**
2062(11) 240-429-2

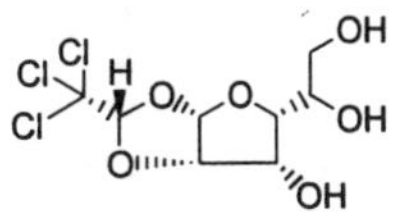

$C_8H_{11}Cl_3O_6$
Chiral intermediate. mp = 234-236°; $[\alpha]_D^{20}$ = - 17.2° (c = 2, pyridine). *Sigma-Aldrich Fine Chemicals.*

**1575 N-(Chloroacetyl)-(4R)-benzyl-2-oxazolidinone**

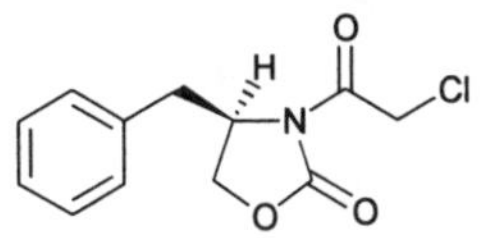

$C_{12}H_{12}ClNO_3$
Chiral intermediate. *Oxford Asymmetry International plc; Sigma-Aldrich Fine Chemicals.*

**1576 N-(Chloroacetyl)-(4S)-benzyl-2-oxazolidinone**
104324-16-5

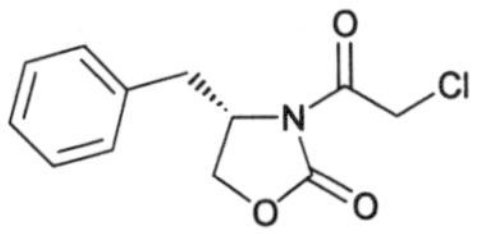

$C_{12}H_{12}ClNO_3$
N-Chloroacetyl-(4S)-benzyl-2-oxazolidinone. Chiral intermediate. *Oxford Asymmetry International plc; Sigma-Aldrich Fine Chemicals.*

**1577 N-(2-chloroacetyl)-L-tyrosine**
1145-56-8 214-544-3

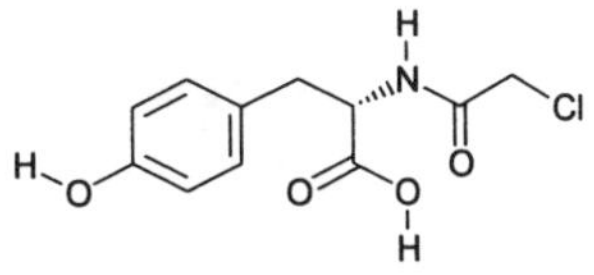

$C_{11}H_{12}ClNO_4$
Tyrosine, N-(chloroacetyl)-, L-; Chiral intermediate. mp = 153-154°; [α] = + 59° (c = 2, EtOH). *Acros Organics nv.*

**1578 3-Chloro-L-alanine**
2731-73-9

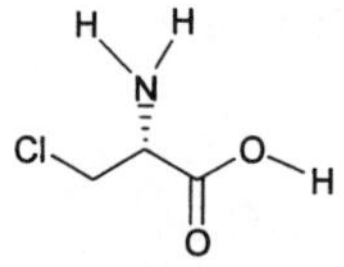

$C_3H_6NO_2Cl$
Chiral building block. *Acros Organics nv; Kaneka Corporation.*

**1579 (+)-3-Chloroalanine hydrochloride**
51887-89-9

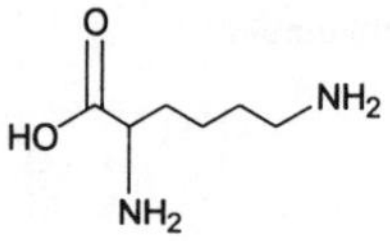

$C_3H_7Cl_2NO_2$
Chiral building block. mp = 205°; $[\alpha]_D^{20}$ = + 3.6° (c = 0.7, HCl). *Sigma-Aldrich Fine Chemicals.*

**1580 3-Chloro-L-alanine methylester hydrochloride**

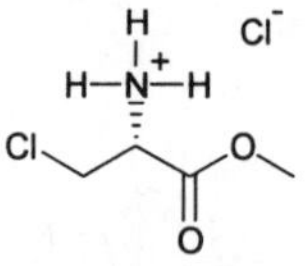

$C_4H_9Cl_2NO_2$
Chiral building block. *Acros Organics nv.*

**1581 (R)-α-(2-Chlorobenzyl)benzylamine**

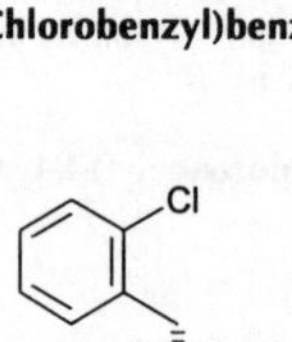

$C_{14}H_{14}ClN$
Chiral building block. *Sumitomo Chemcial Co. Ltd.*

**1582 (S)-α-(2-Chlorobenzyl)benzylamine**

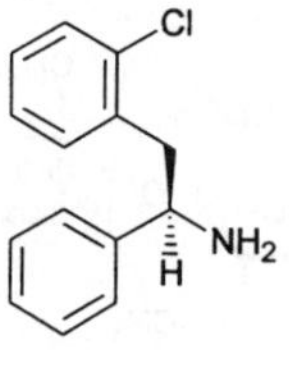

$C_{14}H_{14}ClN$
Chiral building block. *Sumitomo Chemcial Co. Ltd.*

**1583 (R)-α-(3-Chlorobenzyl)benzylamine**

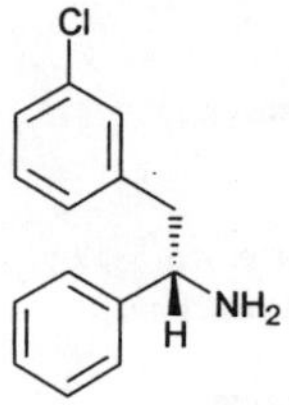

$C_{14}H_{14}ClN$
Chiral building block. *Sumitomo Chemcial Co. Ltd.*

**1584 (S)-α-(3-Chlorobenzyl)benzylamine**

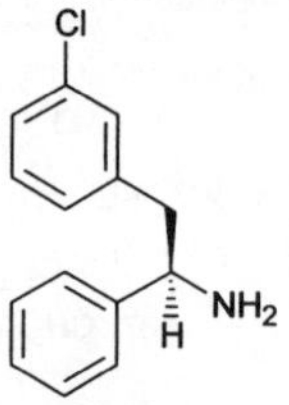

$C_{14}H_{14}ClN$
Chiral building block. *Sumitomo Chemcial Co. Ltd.*

**1585 (R)-α-(4-Chlorobenzyl)benzylamine**

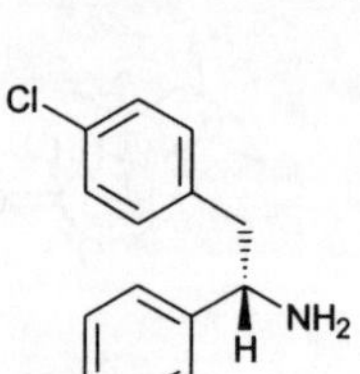

$C_{14}H_{14}ClN$
Chiral building block. *Sumitomo Chemcial Co. Ltd.*

**1586 (S)-α-(4-Chlorobenzyl)benzylamine**

$C_{14}H_{14}ClN$
Chiral building block. *Sumitomo Chemcial Co. Ltd.*

**1587 (R)-Chloro-1,3-butanediol**
125605-10-9
$C_4H_9C_1O_3$
Chiral building block. *Daiso Company, Ltd.; Rohner Ltd.*

**1588 (S)-4-Chloro-1,3-butanediol**
139013-68-6

$C_4H_9C_1O_3$
Chiral building block. *Daiso Company, Ltd.*

**1589 (S)-2-Chloro-n-butyric acid**
32653-32-0
$C_4H_7ClO_2$
Chiral building block. *TCI America.*

**1590 2-Chloro-carbobenzyloxy-(tert-butoxy-carbonyl)-L-lysine**

$C_{19}H_{27}ClN_2O_6$
Chiral intermediate. *Synthetech, Inc.*

**1591 (1S-cis)-3-Chloro-3,5-cyclohexadiene-1,2-diol**

$C_6H_7ClO_2$
Chiral building block. mp = 76-80°; $[\alpha]^{22}$ = + 54° (c = 1, $CHCl_3$). *Sigma-Aldrich Fine Chemicals.*

**1592 Chlorocyclopentadienyl-(4R,5R)-2,2-dimethyltetraphenyl-1,3-dioxolane-4,5-dimethanolatotitanium**
132068-98-5

$C_{36}H_{33}ClO_4Tl$
(R,R)-Duthaler-Hafner Reagent. Chiral catalyst. mp = 209-213°; $[\alpha]^{20}$ = - 246° (c = 1, $CHCl_3$). *Sigma-Aldrich Fine Chemicals.*

**1593 Chlorocyclopentadienyl-(4S,5S)-2,2-dimethyl-tetraphenyl-1,3-dioxolane-4,5-dimethanolatotitanium**
140462-73-3

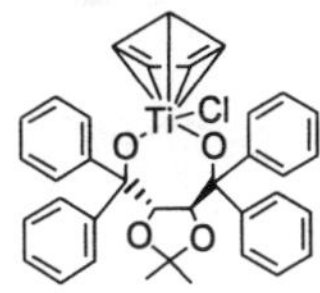

$C_{36}H_{33}ClO_4Tl$
(S,S)-Duthaler-Hafner Reagent. Chiral catalyst. mp = 209-213°; $[\alpha]^{20}$ = + 246° (c = 1, $CHCl_3$). *Sigma-Aldrich Fine Chemicals.*

**1594 (R)-Chloro-2-decanol**

$C_{10}H_{21}ClO$
Chiral building block. *Japan Energy Corporation.*

**1595 (S)-2-Chloro-1-decanol**

$C_{10}H_{21}ClO$
Chiral building block. *Japan Energy Corporation.*

**1596 (4S,5S)-(-)-2-Chloro-4,5-dimethyl-1,3,2-dioxaphospholane-2-oxide**
112966-13-9

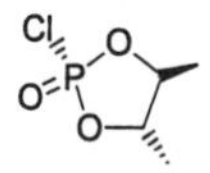

$C_4H_8ClO_3P$
Chiral intermediate. $bp_{0.2}$ = 84-86°; d = 1.3; n = 1.441; $[\alpha]_D^{20}$ = - 15° (neat). *Sigma-Aldrich Fine Chemicals.*

**1597 (2R,4R,5S)-(+)-2-Chloro-3,4-dimethyl-5-phenyl-1,3,2-oxazaphospholidine 2-sulfide**
28080-20-8

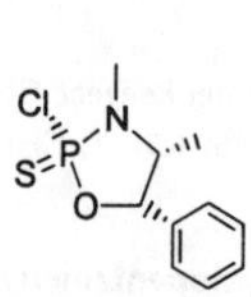

$C_{10}H_{13}ClNOPS$
Chiral ligand. mp = 128-130°; $[\alpha]_D^{20}$ = + 120° (c = 1, $CHCl_3$). *Sigma-Aldrich Fine Chemicals.*

**1598 (R)-Chloro-2-dodecanol**

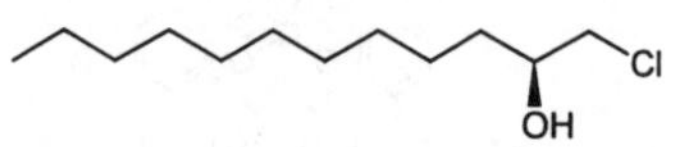

$C_{12}H_{25}ClO$
Chiral building block. *Japan Energy Corporation.*

**1599 (S)-2-Chloro-1-dodecanol**

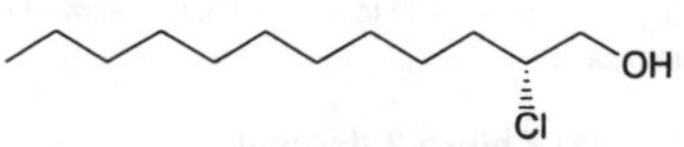

$C_{12}H_{25}ClO$
Chiral building block. *Japan Energy Corporation.*

**1600 2-Chloroethylfructopyranoside**
84543-36-2

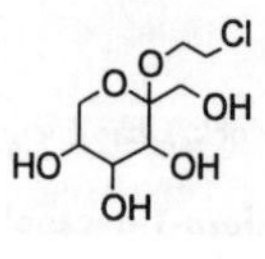

$C_8H_{15}ClO_6$
Chiral intermediate. mp = 139-141°; $[\alpha]^{23}$ = - 146° (c = 1, $H_2O$). *Sigma-Aldrich Fine Chemicals.*

**1601 (R)-Chloro-3-(4-fluorophenoxy)-2-propanol**
$C_9H_{10}ClFO_2$
Chiral building block. *Sigma-Aldrich Fine Chemicals.*

**1602 (S)-1-Chloro-3-(4-fluorophenoxy)-2-propanol**
$C_9H_{10}ClFO_2$
Chiral intermediate. *Sigma-Aldrich Fine Chemicals.*

**1603 (-)-Chlorogenic acid**
327-97-9

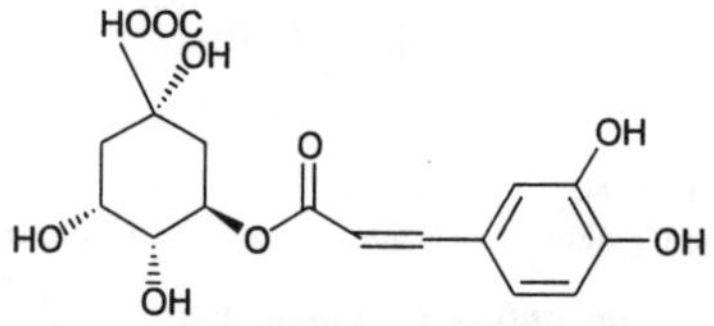

$C_{16}H_{18}O_9$
Chiral intermediate. mp = 208-210°; [α] = - 36° (c = 1, $H_2O$). *Acros Organics nv; Kaden Biochemicals GmbH.*

**1604 (S)-4-Chloro-3-hydroxybutanoic acid ethyl ester**

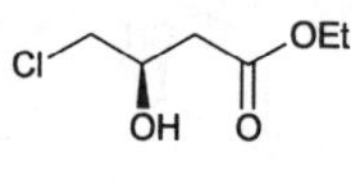

$C_6H_{11}ClO_3$
Chiral building block. *Austin Chemical Company, Inc.*

**1605 (R)-(+)-4-Chloro-3-hydroxybutyronitrile**
84367-31-7

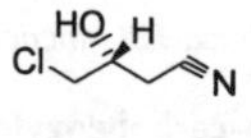

$C_4H_6ClNO$
Chiral building block. $bp_1$ = 110°; d = 1.3200; $n_D^{20}$ = 1.4740; $[\alpha]_D^{20}$ = + 11 ± 2° (neat). *Austin Chemical Company, Inc.; Daiso Company, Ltd.; Lancaster Synthesis Ltd.; Sigma-Aldrich Fine Chemicals.*

**1606 (S)-(-)-4-Chloro-3-hydroxybutyronitrile**
127913-44-4

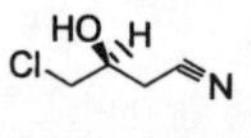

$C_4H_6ClNO$
Chiral building block. $bp_1$ = 110°; d = 1.3200; $n_D^{20}$ = 1.4740; $[\alpha]_D^{20}$ = - 8.0 ± 2° (neat). *Austin Chemical Company, Inc.; Daiso Company, Ltd.; Lancaster Synthesis Ltd.; Sigma-Aldrich Fine Chemicals.*

**1607 (R)-(+)-(3-Chloro-2-hydroxypropyl)-trimethylammonium chloride**
117604-42-9
$C_6H_{15}Cl_2NO$
Chiral building block. mp = 217°; $[\alpha]^{25}$ = + 29° (c = 1, $H_2O$). *Sigma-Aldrich Fine Chemicals.*

**1608 (S)-(-)-(3-Chloro-2-hydroxypropyl)-trimethylammonium chloride**
101396-91-2

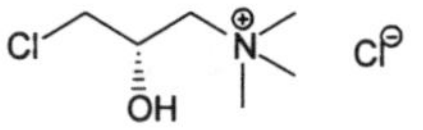

$C_6H_{15}Cl_2NO$
Chiral auxiliary. mp = 210°; $[\alpha]^{25}$ = - 29° (c = 1, $H_2O$). *Sigma-Aldrich Fine Chemicals.*

**1609 4-Chloro-3-indolyl-β-D-galactopyranoside**
135313-63-2

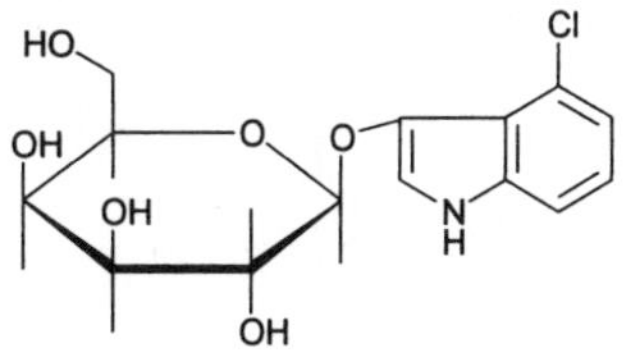

$C_{14}H_{16}ClNO_6$
Chiral diagnostic reagent. mp = 207-211°; moderately soluble in $H_2O$. *Diagnostic Chemicals Limited.*

**1610 6-Chloro-3-indolyl-β-D-galactopyranoside**

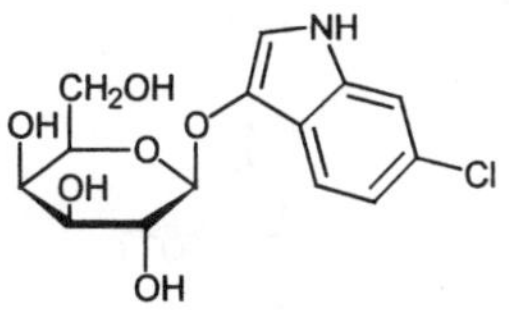

$C_{14}H_{16}ClNO_6$
Salmon Galactoside. Chiral intermediate. *Senn Chemicals AG.*

**1611 6-Chloro-3-indolyl-β-D-glucopyranoside**

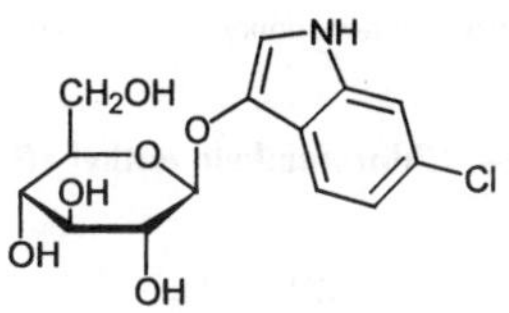

$C_{14}H_{16}ClNO_6$
Salmon Glucoside. Chiral intermediate. *Senn Chemicals AG.*

**1612 (R)-(-)-2-Chloromandelic acid**
52950-18-2

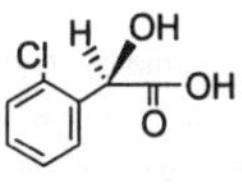

$C_8H_7ClO_3$
Chiral building block. mp = 119-121°; $[\alpha]_D^{23}$ = - 126° (c = 3, $H_2O$). *Acros Organics nv; Austin Chemical Company, Inc.; DSM Fine Chemcials Netherlands; Mitsubishi Rayon Co., Ltd.; Shanghai DSL International Trading Company; Sigma-Aldrich Fine Chemicals; Yamakawa Chemical Industry Co. Ltd.*

**1613 (S)-(+)-2-Chloromandelic acid**
52950-19-3

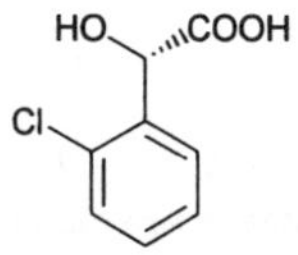

$C_8H_7ClO_3$
Benzeneacetic acid, 2-chloro-α-hydroxy-, (S)-; (S)-(2-Chlorophenyl)glycolic acid; (S)-2-Chloro-α-hydroxybenzeneacetic acid; Mandelic acid, o-chloro-, (S)-. Chiral building block. *Austin Chemical Company, Inc.; Yamakawa Chemical Industry Co. Ltd.*

**1614 (S)-3-Chloromandelic acid**

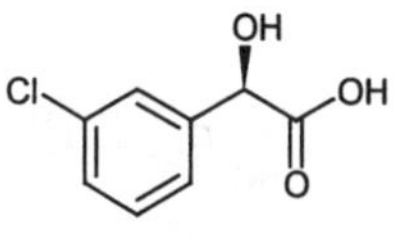

$C_8H_7ClO_3$
Benzeneacetic acid, 3-chloro-α-hydroxy-, (S)-; (S)-3-Chlorophenylglycolic acid; (S)-3-Chloro-α-hydroxybenzeneacetic acid; Mandelic acid, m-chloro-, (S)-. Chiral building block. *Austin Chemical Company, Inc.*

**1615 (S)-4-Chloromandelic acid**

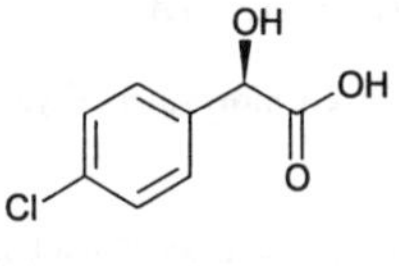

$C_8H_7ClO_3$
Benzeneacetic acid, 4-chloro-α-hydroxy-, (S)-. Chiral building block. *Austin Chemical Company, Inc.*

**1616 (R)-(3)-Chloromandelic acid**
61008-98-8

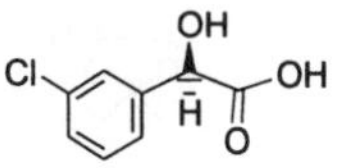

$C_8H_7ClO_3$
Chiral building block. mp = 100-104°; $[\alpha]^{30}$ = - 124° (c = 3, $H_2O$). *Acros Organics nv; Austin Chemical Company, Inc.; Mitsubishi Rayon Co., Ltd.; Sigma-Aldrich Fine Chemicals.*

**1617 (R)-(+)-4-Chloromandelonitrile**
97070-79-6

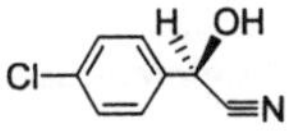

$C_8H_6ClNO$
Chiral building block. mp = 70-74°; $[\alpha]_D^{20}$ = + 39° (c = 1, $CHCl_3$). *Sigma-Aldrich Fine Chemicals.*

**1618 (-)-1-5-Chloro-2-(methylamino)phenyl-1,2,3,4-tetrahydroisoquinoline-(-)-tartrate**

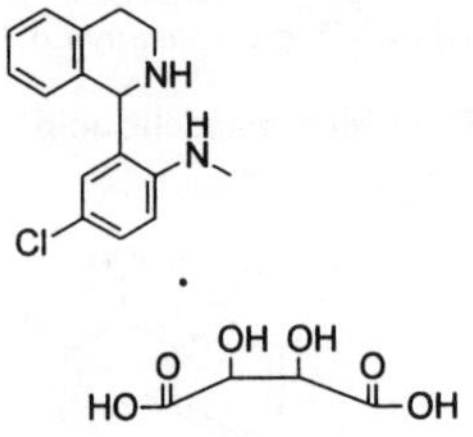

$C_{16}H_{17}ClN_2$
Chiral intermediate. mp = 211-213°; $[\alpha]^{22}$ = - 25° (c = 2, 6N HCl). *Sigma-Aldrich Fine Chemicals.*

**1619 (S)-(+)-1-Chloro-2-methylbutane**
40560-29-0
$C_5H_{11}Cl$
Chiral building block. *TCI America.*

**1620 (S)-(-)-2-Chloro-3-methylbutyric acid**
26782-74-1
$C_5H_9ClO_2$
(S)-2-Chloroisovaleric acid. Chiral building block. $bp_{16}$ = 98°; d = 1.1400; $[\alpha]_D^{20}$ = - 1.4 ± 0.2° (neat). *Lancaster Synthesis Ltd.; TCI America.*

**1621 (R)-(+)-4-(Chloromethyl)-2,2-dimethyl-1,3-dioxolane**
57044-24-3

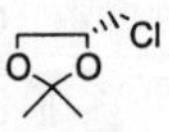

$C_6H_{11}ClO_2$
(R)-(-)-3-Chloro-1,2-propanediol acetonide. Chiral intermediate. $bp_{37}$ = 63°; d = 1.1030; $n_D^{20}$ = 1.4340; $[\alpha]_D^{20}$ = + 42.1 ± 3° (neat). *Daiso Company, Ltd.; Lancaster Synthesis Ltd.; Sigma-Aldrich Fine Chemicals.*

**1622 (S)-(-)-4-(Chloromethyl)-2,2-dimethyl-1,3-dioxolane**
60456-22-6

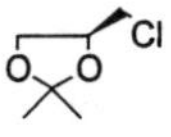

$C_6H_{11}ClO_2$
(S)-(+)-3-Chloro-1,2-propanediol acetonide. Chiral intermediate. $bp_{37}$ = 63°; d = 1.1030; $n_D^{20}$ = 1.4340; $[\alpha]_D^{20}$ = - 42.1 ± 3° (neat). *Daiso Company, Ltd.; Lancaster Synthesis Ltd.; Sigma-Aldrich Fine Chemicals.*

**1623 (+)-Chloromethyl isomenthyl ether**
144177-48-0

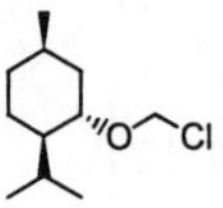

$C_{11}H_{21}ClO$
Chiral intermediate. bp = 206°; d = 0.988; n = 1.468; $[\alpha]_D^{20}$ = + 57° (neat). *Sigma-Aldrich Fine Chemicals.*

**1624 (-)-Chloromethylmenthyl ether**
26127-08-2

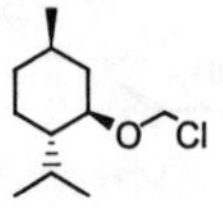

$C_{11}H_{21}ClO$
(1R,2S,4R)-2-(Chloromethyloxy)-1-isopropyl-4-methyl-cyclohexane. Chiral intermediate. $bp_{0.1}$ = 62°; d = 0.982; n = 1.467; $[\alpha]^{24}$ = - 177.0° (c = 1, $CH_2Cl_2$). *Oxford Asymmetry International plc; Sigma-Aldrich Fine Chemicals.*

**1625 (+)-Chloromethylmenthyl ether**
103128-76-3

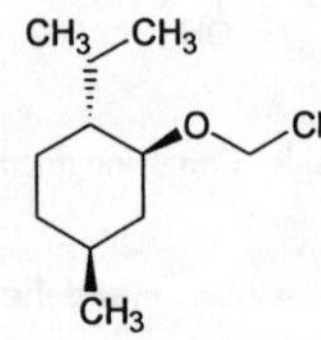

$C_{11}H_{21}ClO$
(1R,2S,4S)-2-(Chloromethyloxy)-1-isopropyl-4-methyl-cyclohexane. Chiral intermediate. $bp_{0.1}$ = 64°; d = 0.9876; n = 1.467; $[\alpha]$ = + 190.4 (c = 1, $CH_2Cl_2$). *Acros Organics nv; Oxford Asymmetry International plc; Sigma-Aldrich Fine Chemicals.*

**1626 (R)-(+)-2-(4-Chloro-2-methylphenoxy)-propionic acid**

16484-77-8 240-539-0

$C_{10}H_{11}ClO_3$
Propanoic acid, 2-(4-chloro-2-methylphenoxy)-, (2R)-. (+)-2-(4-Chloro-o-tolyloxy)propionic acid; (+)-MCPP; (+)-Mecoprop; 2M4XP; Duplosan. Chiral building block. *A.H. Marks & Co. Ltd.*

**1627 1-((S)-3-Chloro-2-methylpropanoyl)-L-proline**

80141-53-3

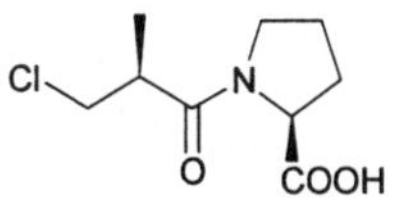

$C_9H_{14}NO_3Cl$
Proline, L-, 1-(3-chloro-2-methyl-1-oxopropyl)-, (S)-. Chiral intermediate. *Kaneka Corporation.*

**1628 (2S,3S)-2-Chloro-3-methyl-n-valeric acid**

32653-34-2

$C_6H_{11}ClO_2$
(2S,3S)-2-Chloro-3-methyl-n-pentanoic acid. Chiral building block. d = 1.12. *TCI America.*

**1629 (S)-2-Chloro-4-methyl-n-valeric acid**

28659-81-6

$C_6H_{11}ClO_2$
(S)-2-Chloro-4-methyl-n-pentanoic acid. Chiral building block. d = 1.09. *TCI America.*

**1630 (S)-2-Chloro-1-octanol**

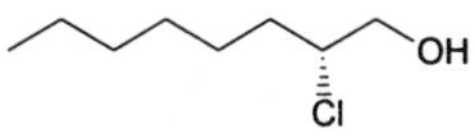

$C_8H_{17}ClO$
Chiral building block. *Japan Energy Corporation.*

**1631 (S)-(+)-Chloropheniramine maleate**

2438-32-6 2232(12) 219-450-6

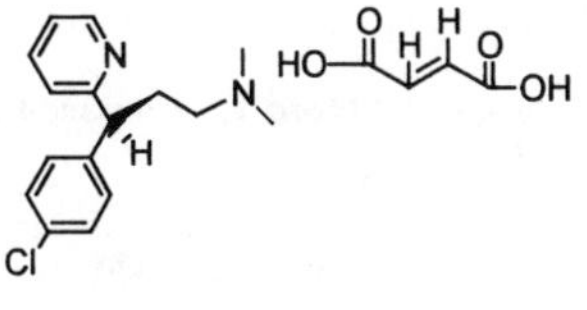

$C_{16}H_{19}ClN_2$
Chiral intermediate. mp = 112-115°; $[\alpha]_D^{20}$ = + 42° (c = 1, DMF). *Sigma-Aldrich Fine Chemicals.*

**1632 (-)-α-Chlorophenylacetyl chloride**

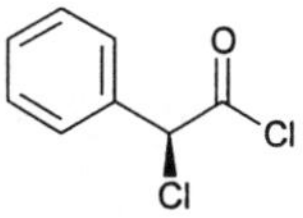

$C_8H_6Cl_2O$
Benzeneacetyl chloride, α-chloro-, (-)-; (-)-Chloro-(phenyl)acetyl chloride; Acetyl chloride, chlorophenyl-, (-)-; (-)-Phenyl(chloro)acetyl chloride. Chiral building block. *Norse Laboratories.*

**1633 D-3-(4-Chlorophenyl)alanine**

14091-08-8

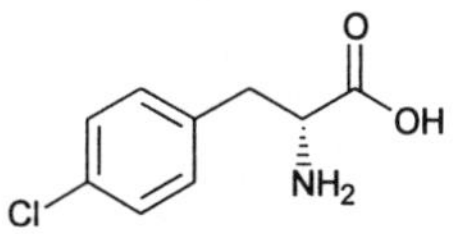

$C_9H_{10}ClNO_2$
Chiral building block. *Recordati S.p.A.*

**1634 L-3-(4-Chlorophenyl)alanine**

14173-39-8 238-023-5

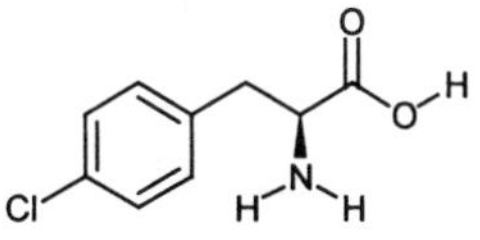

$C_9H_{10}ClNO_2$
Alanine, 3-(p-chlorophenyl)-, L-; L-4-Chloro-phenylalanine; L-Phenylalanine, 4-chloro-. Chiral building block. mp = 263°; $[\alpha]_D^{20}$ = -2 4° (c = 1, $H_2O$). *Acros Organics nv; Great Lakes Fine Chemicals; Sigma-Aldrich Fine Chemicals.*

**1635 (1S,4S)-(-)-2-(4-Chlorophenyl)-2,5-diaza-bicyclo[2.2.1]heptane hydrobromide**

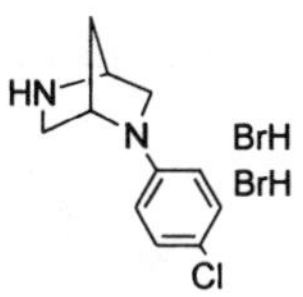

$C_{11}H_{14}BrClN_2$
Chiral intermediate. mp = 229°; $[\alpha]^{23}$ = - 94° (c = 1, $H_2O$). *Sigma-Aldrich Fine Chemicals.*

**1636 (R)-(+)-4-(2-Chlorophenyl)-5,5-dimethyl-2-hydroxy-1,3,2-dioxaphosphorinane-2-oxide**
98674-87-4

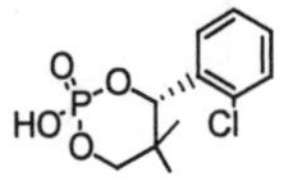

$C_{11}H_{14}ClO_4P$
Chiral ligand. mp = 228-234°; $[\alpha]^{22}$ = + 46° (c = 1, $CH_3OH$). *Sigma-Aldrich Fine Chemicals; Ultrafine.*

**1637 (S)-(-)-4-(2-Chlorophenyl)-5,5-dimethyl-2-hydroxy-1,3,2-dioxaphosphorinane-2-oxide**
98674-86-3

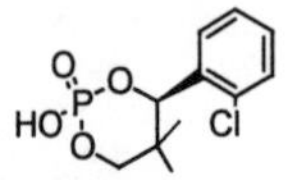

$C_{11}H_{14}ClO_4P$
Chiral auxiliary. mp = 228-234°; $[\alpha]^{22}$ = - 46° (c = 1, $CH_3OH$). *Sigma-Aldrich Fine Chemicals; Ultrafine.*

**1638 (R)-1-(3-Chlorophenyl)-1,2-ethanediol**

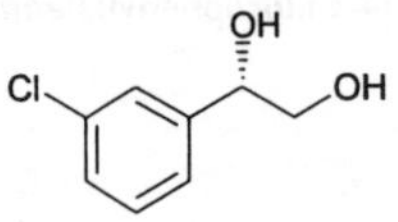

$C_8H_9ClO_2$
Chiral building block. *Austin Chemical Company, Inc.*

**1639 (1S)-1-(2-Chlorophenyl)ethane-1,2-diol**

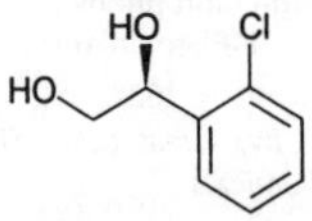

$C_8H_9ClO_2$
Chiral intermediate. bp = 293°; mp = 68-75°; $[\alpha]_D^{20}$ = + 70° (c = 0.2, $CHCl_3$). *Acros Organics nv; Sigma-Aldrich Fine Chemicals.*

**1640 (R)-(-)-2-Chloro-1-phenylethanol**
56751-12-3

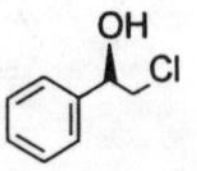

$C_8H_9ClO$
Chiral building block. d = 1.185; n = 1.552; $[\alpha]_D^{20}$ = - 48° (c = 2.8, cyclohexane). *Sigma-Aldrich Fine Chemicals.*

**1641 (S)-(+)-2-Chloro-1-phenylethanol**
70111-05-6

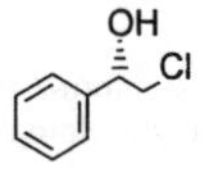

$C_8H_9ClO$
Chiral building block. d = 1.185; n = 1.552; $[\alpha]_D^{20}$ = + 43° (c = 2.8, cyclohexane). *Sigma-Aldrich Fine Chemicals.*

**1642 (R)-(+)-3-Chloro-1-phenyl-1-propanol**
100306-33-0

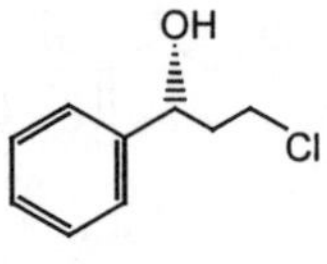

$C_9H_{11}ClO$
Chiral building block. mp = 58-60°; $[\alpha]^{24}$ = + 26° (c = 1, $CHCl_3$). *Sigma-Aldrich Fine Chemicals; Ultrafine.*

**1643 (S)-(-)-3-Chloro-1-phenyl-1-propanol**
100306-34-1

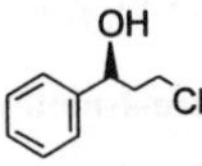

$C_9H_{11}ClO$
Chiral building block. mp = 58-60°; $[\alpha]^{19}$ = - 25° (c = 1, $CHCl_3$). *Sigma-Aldrich Fine Chemicals; Ultrafine.*

**1644 (R)-(-)-3-Chloro-1,2-propanediol**
57090-45-6

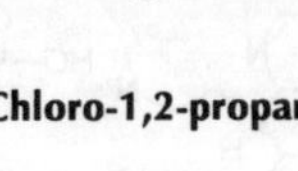

$C_3H_7ClO_2$
(R)-(-)-α-Chlorohydrin; (R)-(-)-α-Monochlorohydrin. Chiral building block. bp = 213°; d = 1.3220; $n_D^{20}$ = 1.4800; $[\alpha]_D^{20}$ = - 0.9° (neat). *Austin Chemical Company, Inc.; Chirex, Inc.; Daiso Company, Ltd.; Kaneka Corporation; Lancaster Synthesis Ltd.; Sigma-Aldrich Fine Chemicals; TCI America.*

**1645 (S)-(+)-3-Chloro-1,2-propanediol**
60827-45-4

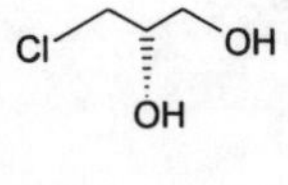

$C_3H_7ClO_2$
(S)-(+)-α-Chlorohydrin; (S)-(+)-α-Monochlorohydrin. Chiral building block. bp = 213°; d = 1.3220; $n_D^{20}$ = 1.4800; $[\alpha]_D^{20}$ = + 0.9° (neat). *Austin Chemical Company,*

*Inc.; Chirex, Inc.; Daiso Company, Ltd.; Kaneka Corporation; Lancaster Synthesis Ltd.; Sigma-Aldrich Fine Chemicals; TCI America.*

**1646 (R)-(+)-2-Chloropropanoic acid**
7474-05-7

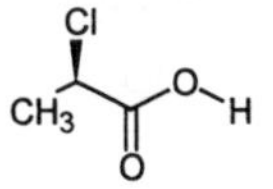

$C_3H_5ClO_2$
Propanoic acid, 2-chloro-, (R)-; Chiral building block. $bp_{10}$ = 77°; d = 1.249; n = 1.4345; $[\alpha]_D^{20}$ = + 15 ± 1° (neat). *Acros Organics nv; Nitrokemia 2000 Rt.; Norse Laboratories; Senn Chemicals AG; Sigma-Aldrich Fine Chemicals; Ultrafine.*

**1647 (S)-(-)-2-Chloropropionic acid**
29617-66-1

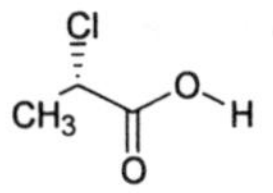

$C_3H_5ClO_2$
Propanoic acid, 2-chloro-, (S)-; Chiral intermediate. bp = 187°; $bp_{10}$ = 76-77°; d = 1.2490; $n_D^{20}$ = 1.4347; $[\alpha]_D^{20}$ = - 14 ± 0.5° (neat). *Acros Organics nv; Avecia Life Science Molecules; BASF Aktiengesellschaft; Lancaster Synthesis Ltd.; Norse Laboratories; Sigma-Aldrich Fine Chemicals; TCI America.*

**1648 (R)-(+)-2-Chloropropanoic acid butyl ester**
79398-16-6 279-148-5

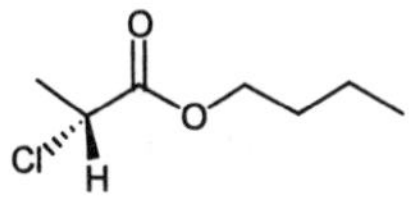

$C_7H_{13}ClO_2$
Propanoic acid, 2-chloro-, butyl ester, (R)-; Butyl (R)-2-chloropropionate. Chiral building block. $d_4^{20}$ = 1.0253; $[\alpha]_D^{20}$ = + 12 ± 0.5° (neat). *Nitrokemia 2000 Rt.*

**1649 (R)-(+)-2-Chloropropanoic acid ethyl ester**
42411-39-2

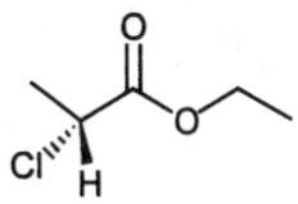

$C_5H_9ClO_2$
Chiral building block. $d_4^{20}$ = 1.0725; $[\alpha]_D^{20}$ = + 22 ± 0.5° (neat). *Nitrokemia 2000 Rt.*

**1650 (R)-(+)-2-Chloropropanoic acid methyl ester**
77287-29-7 278-658-5

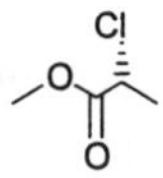

$C_4H_7ClO_2$
Propanoic acid, 2-chloro-, methyl ester, (2R)-; Methyl (R)-2-chloropropionate. Chiral intermediate. bp = 132-134°; $n_4^{20}$ = 1.138; $[\alpha]_D^{20}$ = + 30 ± 1° (neat). *Nitrokemia 2000 Rt.; Sigma-Aldrich Fine Chemicals.*

**1651 (R)-(-)-2-Chloropropan-1-ol**
37493-14-4

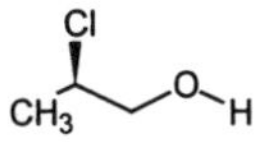

$C_3H_7ClO$
Chiral building block. $[\alpha]^{20}$ = - 17.5° (neat). *Acros Organics nv.*

**1652 (S)-(+)-2-Chloropropan-1-ol**
19210-21-0

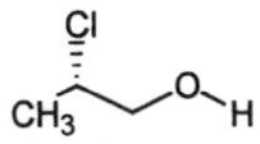

$C_3H_7ClO$
Propan-1-ol, 2-chloro-, (S)-; (S)-(+)-Propylene chlorohydrin. Chiral building block. $bp_{75}$ = 70.3°; d = 1.103; n = 1.438; [α]20 = + 17.5° (neat). *Acros Organics nv; Sigma-Aldrich Fine Chemicals.*

**1653 (R)-2-(4-Chlorophenyl)-3-phenylpropionic acid**

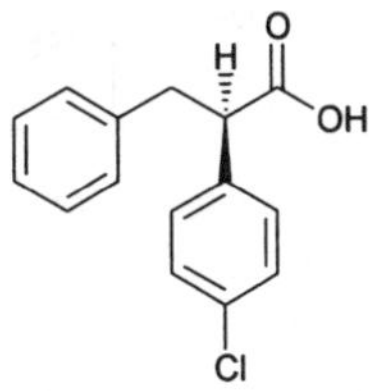

$C_{15}H_{13}ClO_2$
Chiral building block. *Sumitomo Chemcial Co. Ltd.*

**1654 (S)-2-(4-Chlorophenyl)-3-phenylpropionic acid**

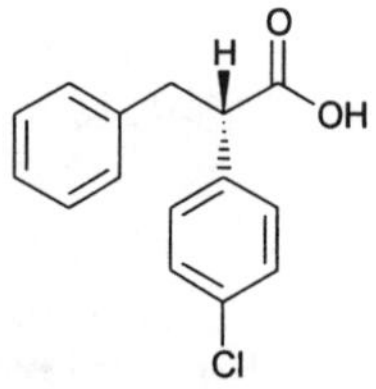

$C_{15}H_{13}ClO_2$
Chiral intermediate. *Sumitomo Chemcial Co. Ltd.*

**1655 (R)-4-Chlorostyrene oxide**
21019-51-2

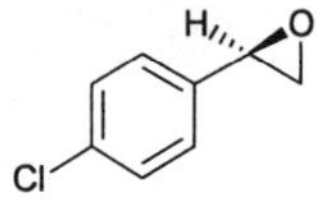

$C_8H_7ClO$
Chiral building block. *PPG - Sipsy.*

**1656 (S)-4-Chlorostyrene oxide**
97466-49-4

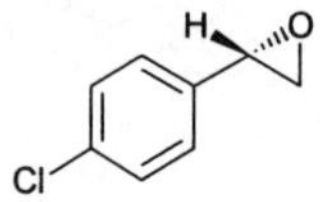

$C_8H_7ClO$
Chiral building block. *PPG - Sipsy.*

**1657 (R)-m-Chlorostyrene oxide**
62600-71-9

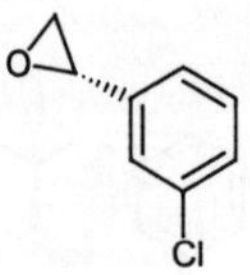

$C_8H_7OCl$
3-Chlorostyrene oxide. Chiral building block. $bp_1$ = 67-68°; d = 1.214; n = 1.551; $[\alpha]_D^{20}$ = + 21° (neat). *Austin Chemical Company, Inc.; Kaneka Corporation; Mitsubishi Rayon Co., Ltd.; Sigma-Aldrich Fine Chemicals.*

**1658 (S)-m-Chlorostyrene oxide**
115648-90-3

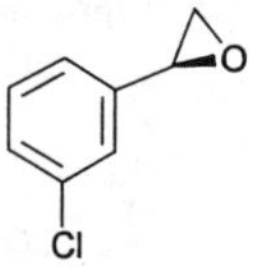

$C_8H_7OCl$
Chiral building block. *Kaneka Corporation.*

**1659 3-Chloro-L-tyrosine**
7423-93-0 231-050-3

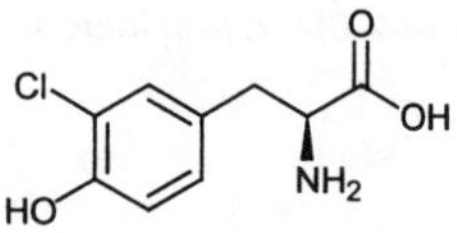

$C_9H_{10}ClNO_3$
Tyrosine, 3-chloro-, L-; Chiral building block. *Pharmasyn, Inc.*

**1660 3-Chloro-L-tyrosine hydrochloride**
35608-63-0

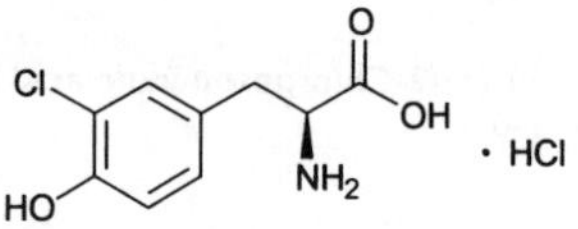

$C_9H_{11}Cl_2NO_3$
Chiral building block. *Pharmasyn, Inc.*

**1661 (+)-Cholesta-3,5-diene**
747-90-0 212-021-4

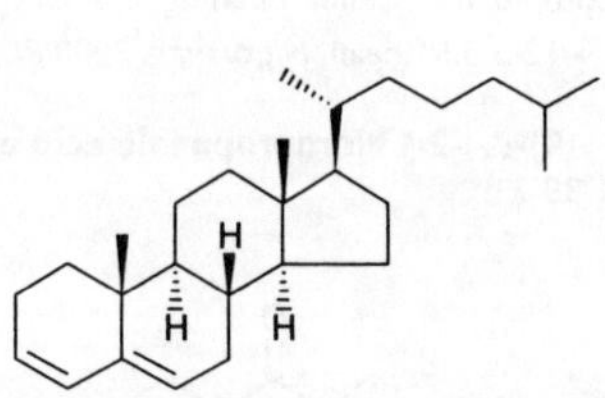

$C_{27}H_{44}$
Pharmaceutical or derivative. mp = 78-80°; $[\alpha]^{25}$ = - 119° (c = 1, $CHCl_3$). *Sigma-Aldrich Fine Chemicals.*

**1662 (+)-5-α-Cholestan-3-one**
15600-08-5 239-682-1

$C_{27}H_{46}O$
cholestan-3-one, 5α-; 3-Oxocholestane. Pharmaceutical or derivative. mp = 128-130°; [α] = + 42° (c = 1, $CHCl_3$). *Acros Organics nv.*

**1663 (+)-5-Cholestan-3-one**
566-88-1

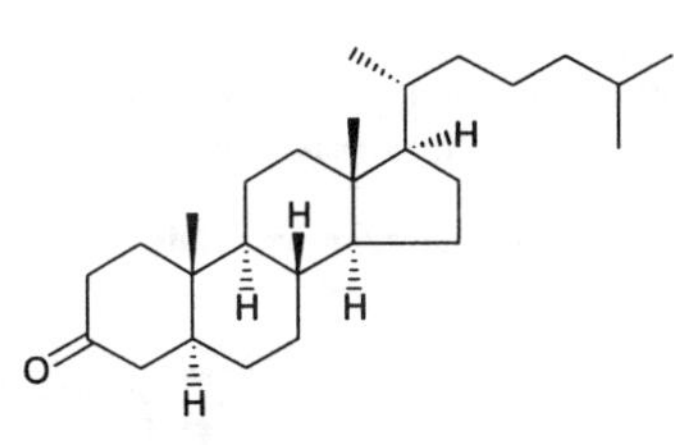

$C_{27}H_{46}O$
Pharmaceutical or derivative. mp = 128-130°; $[\alpha]_D^{20}$ = + 42° (c = 1, $CHCl_3$). *Sigma-Aldrich Fine Chemicals.*

**1664 (+)-5-α-Cholestane**
481-21-0 207-562-8

$C_{27}H_{48}$
Cholestane, (5α)-.
28,29,30-Trinorlanostane. Pharmaceutical or derivative. $[\alpha]^{19}$ = + 23° (c = 2.33, $CHCl_3$). *Acros Organics nv.*

**1665 (+)-4-Cholesten-3-one**
601-57-0 210-005-1

$C_{27}H_{44}O$
Listed on TSCA. Pharmaceutical or derivative. mp = 79-81°; $[\alpha]^{23}$ = + 91.0° (c = 2, $CHCl_3$). *Sigma-Aldrich Fine Chemicals.*

**1666 (-)-Cholesterol**
57-88-5 2256(12) 200-353-2

$C_{27}H_{46}O$
Cholest-5-en-3-ol (3β)-; 3β-Hydroxycholest-5-ene; Cholesteryl alcohol; Dythol; Lidinit; Lidinite; Provitamin D. Listed on TSCA. Pharmaceutical or derivative. mp = 147-149°; bp = 360°; d =1.067; $[\alpha]_D^{20}$ = - 39° (c = 2, $CHCl_3$). *Acros Organics nv; Sigma-Aldrich Fine Chemicals.*

**1667 (-)-Cholesteryl acetate**
604-35-3 210-066-4

$C_{28}H_{46}O_2$
Cholest-5-en-3-ol, acetate, (3β)-; 3β-Acetoxycholest-5-ene. Listed on TSCA. Pharmaceutical or derivative. *Acros Organics nv.*

**1668 (-)-Cholesteryl benzoate**
604-32-0 2204(11) 210-064-3

$C_{34}H_{50}O_2$
Cholest-5-en-3-ol (3β)-, benzoate; 3β-(Benzoyloxy)cholest-5-ene. Listed on TSCA. Pharmaceutical or derivative. mp = 149-151°; $[\alpha]_D^{20}$ = - 15.0° (c = 2, $CHCl_3$). *Acros Organics nv; Sigma-Aldrich Fine Chemicals.*

**1669 (-)-Cholesteryl caprylate**
1182-42-9 214-656-2

$C_{35}H_{60}O_2$
Cholest-5-en-3-ol (3β)-, octanoate; 3β-(Octanoyloxy)-cholest-5-ene. Pharmaceutical or derivative. mp = 110°; $[\alpha]^{24}$ = - 30° (c = 2, $CHCl_3$). *Acros Organics nv.*

**1670 (-)-Cholesteryl chloride**
910-31-6 213-004-4

$C_{27}H_{45}Cl$
Cholest-5-ene, 3-chloro-, (3β)-; 3β-Chlorocholest-5-ene. Listed on TSCA. Pharmaceutical or derivative. mp = 96-98°; $[\alpha]^{25}$ = - 24° (c = 1, $CHCl_3$). *Acros Organics nv; Sigma-Aldrich Fine Chemicals.*

**1671 (-)-Cholesteryl chloroformate**
7144-08-3

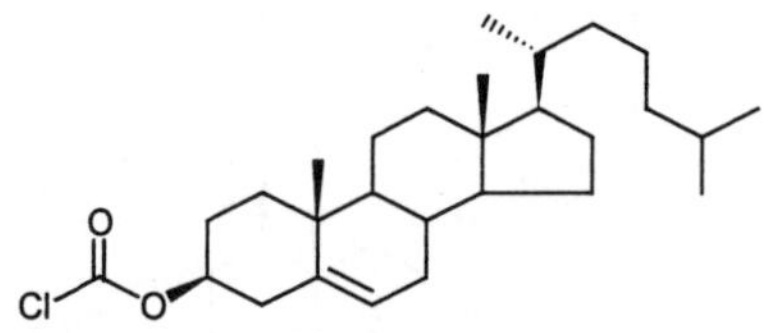


$C_{28}H_{45}ClO_2$
Pharmaceutical or derivative. mp = 115-117°; $[\alpha]^{27}$ = - 28° (c = 2, $CHCl_3$). *Acros Organics nv; Sigma-Aldrich Fine Chemicals.*

**1672 (-)-Cholesteryl hydrocinnamate**
14914-99-9 238-983-5

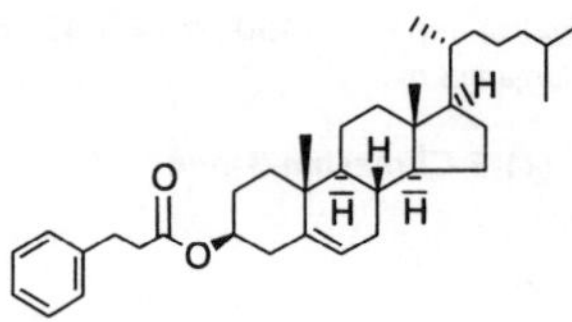


$C_{36}H_{54}O_2$
Pharmaceutical or derivative. mp = 109.5-110.5°; $[\alpha]^{27}$ = - 24° (c = 2, $CHCl_3$). *Sigma-Aldrich Fine Chemicals.*

**1673 (-)-Cholesteryl n-heptylate**
1182-07-6 214-564-1

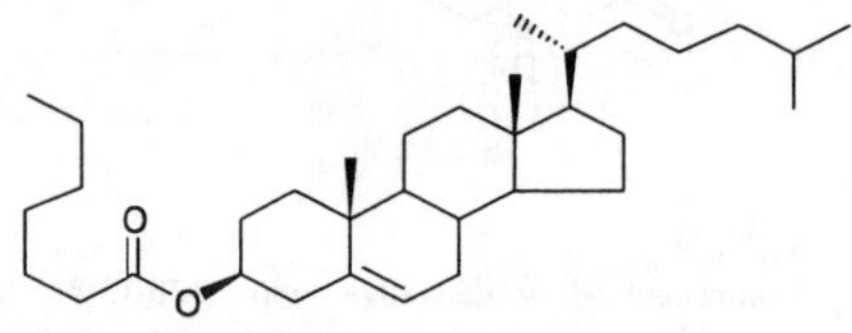


$C_{34}H_{58}O_2$
Cholest-5-en-3-ol (3β)-, heptanoate. Pharmaceutical or derivative. $[\alpha]^{21}$ = - 32° (c = 2, $CHCl_3$). *Acros Organics nv.*

**1674 (-)-Cholesteryl oleate**
303-43-5 201-142-1

$C_{45}H_{78}O_2$
Cholest-5-en-3-ol, 9-octadecenoate, (3β)-, (Z)-. Listed on TSCA. Pharmaceutical or derivative. mp = 44-47°; $[\alpha]_D^{20}$=

- 24° (c = 1, $CHCl_3$). *Acros Organics nv; Sigma-Aldrich Fine Chemicals.*

**1675 (-)-Cholesteryl oleyl carbonate**
17110-51-9 241-179-7

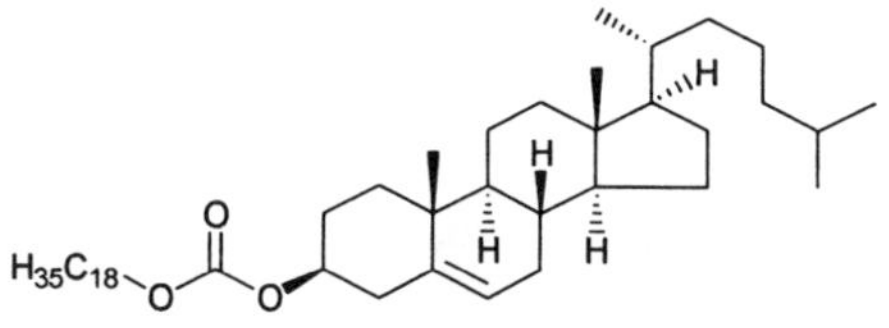

$C_{46}H_{80}O_3$
Cholest-5-en-3-ol (3β)-, (9Z)-9-octadecenyl carbonate; Carbonic acid, cholesteryl oleyl ester; Oleyl cholesteryl carbonate. Listed on TSCA. Pharmaceutical or derivative. mp = 20°; $[\alpha]^{27}$ = - 22° (c = 2, $CHCl_3$). *Acros Organics nv.*

**1676 Cholesteryl palmitate**
601-34-3 210-002-5

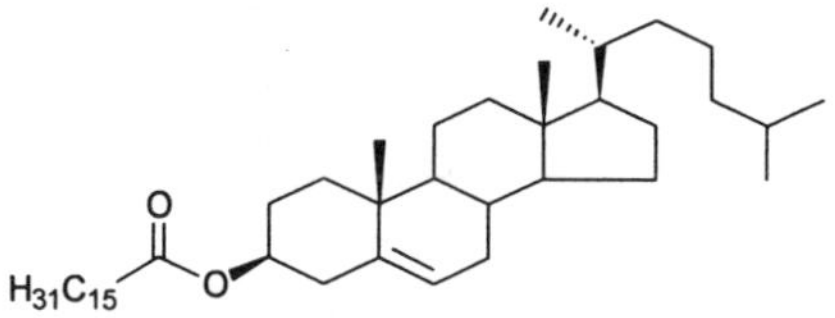

$C_{43}H_{76}O_2$
Cholest-5-en-3-ol (3β)-, hexadecanoate; Cholesteryl hexadecanoate. Listed on TSCA. Pharmaceutical or derivative. mp = 75-77°; $[\alpha]^{27}$ = - 24° (c = 2, $CHCl_3$). *Acros Organics nv; Sigma-Aldrich Fine Chemicals.*

**1677 (-)-Cholesteryl pelargonate**
1182-66-7 214-658-3

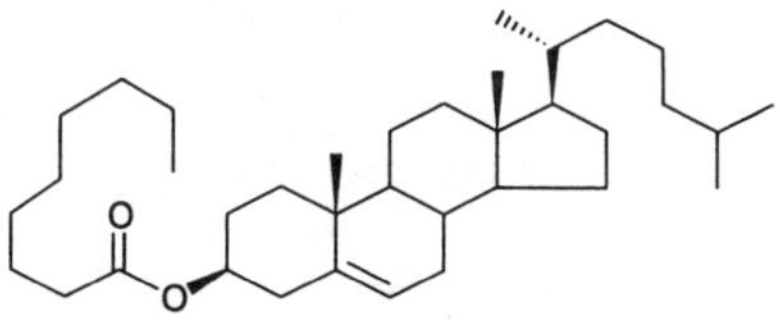

$C_{36}H_{62}O_2$
Cholest-5-en-3-ol (3β)-, nonanoate; Cholesteryl nonanoate. Pharmaceutical or derivative. mp = 74-77°; bp = 360°; $[\alpha]^{25}$ = - 30° (c = 2, $CHCl_3$). *Acros Organics nv; Sigma-Aldrich Fine Chemicals.*

**1678 (-)-Cholesteryl stearate**
35602-69-8 252-637-0

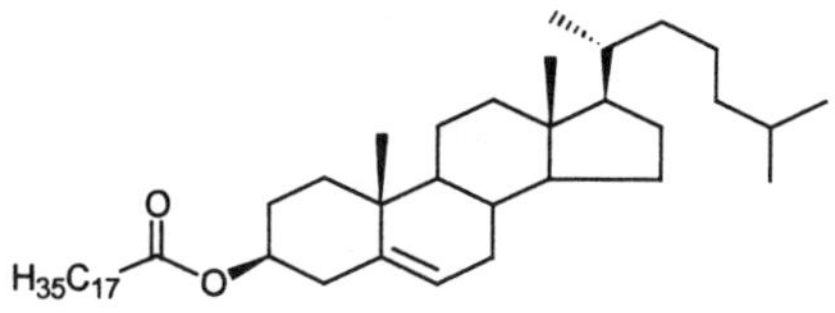

$C_{45}H_{80}O_2$
Cholest-5-en-3-ol, octadecanoate, (3β)-; Cholesteryl octadecanoate. Pharmaceutical or derivative. mp = 79-83°; $[\alpha]^{25}$ = - 21° (c = 2, $CHCl_3$). *Acros Organics nv; Sigma-Aldrich Fine Chemicals.*

**1679 (-)-Cholic acid sodium salt**
361-09-1 206-643-5

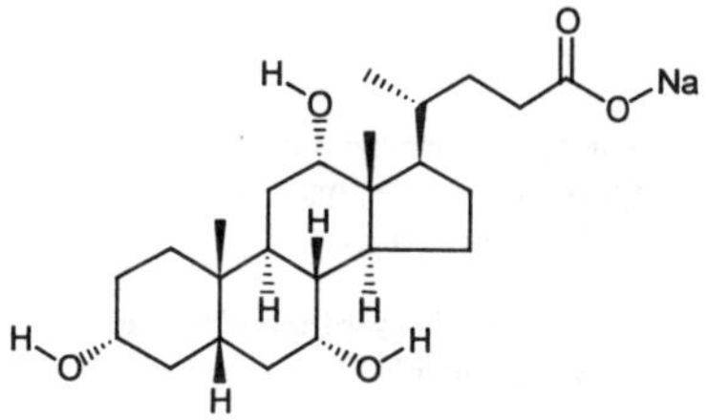

$C_{24}H_{39}NaO_5$
Cholan-24-oic acid, 3,7,12-trihydroxy-, monosodium salt, (3α,5β,7α,12α)-; (3α,5β,7α,12α)-3,7,12-Trihydroxy-cholan-24-oic acid monosodium salt; DS-Na; Sodium cholic acid. Listed on TSCA. Resolving agent. $[\alpha]^{20}$ = + 26°. *Acros Organics nv; Sigma-Aldrich Fine Chemicals.*

**1680 trans-(+)-Chrysanthemic acid**
4638-92-0 225-067-5

$C_{10}H_{16}O_2$
Cyclopropanecarboxylic acid, 2,2-dimethyl-3-(2-methyl-1-propenyl)-, (1R,3R)-; (1R-trans)-2,2-Dimethyl-3-(2-methylprop-1-enyl)cyclopropanecarboxylic acid. Listed on TSCA. Chiral intermediate; Resolving agent. $[\alpha]^{20}$ = + 14.2° (c = 1.5, EtOH). *Acros Organics nv.*

**1681 (1R)-Chrysanthemolactone**
14087-70-8

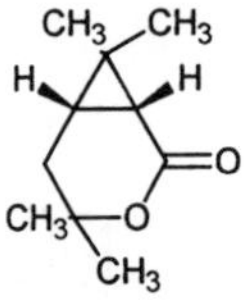

$C_{10}H_{16}O_2$
(1R,3S)-4,4,7,7-Trimethyl-3-oxabicyclo[4.1.0]heptan-2-one. Chiral intermediate. mp = 81-83°; [α] = + 74.9° (c = 1.5, $CHCl_3$). *Acros Organics nv.*

**1682 (1S)-Chrysanthemolactone**
14087-71-9

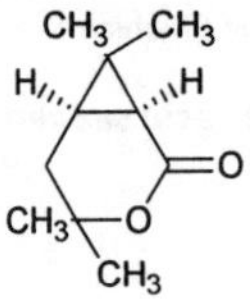

$C_{10}H_{16}O_2$
(1S,3R)-4,4,7,7-Trimethyl-3-oxabicyclo[4.1.0]heptan-2-one. Chiral intermediate. mp = 81-83°; [α] = - 74.2° (c = 1, $CHCl_3$). *Acros Organics nv.*

**1683 (1R)-cis-Chrysanthemylamine hydrochloride**

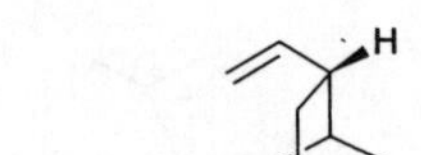

$C_{10}H_{20}ClN$
Chiral intermediate. [α] = + 86° (c = 3.12, $CHCl_3$). *Acros Organics nv.*

**1684 (-)-Cinchonidine**
485-71-2 2345(12) 207-622-3

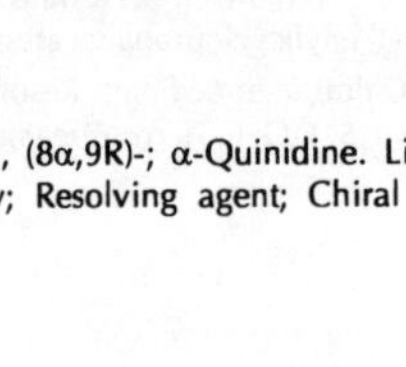

$C_{19}H_{22}N_2O$
Cinchonan-9-ol, (8α,9R)-; α-Quinidine. Listed on TSCA. Chiral auxiliary; Resolving agent; Chiral ligand. mp = 200-203°; $[\alpha]^{23}$ = - 105° (c = 1.5, EtOH). *Acros Organics nv; DSM Fine Chemcials Netherlands; Isochem; Sigma-Aldrich Fine Chemicals.*

**1685 (-)-Cinchonidine hydrochloride**
524-57-2 208-361-8

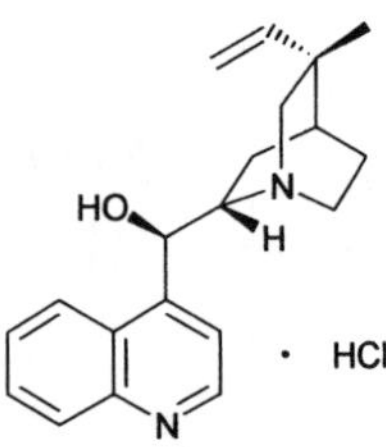

$C_{19}H_{23}ClN_2O$
(8a,9R)-9-Hydroxycinchonanium chloride. Resolving agent. *DSM Fine Chemcials Netherlands.*

**1686 (+)-Cinchonine**
118-10-5 2346(12) 204-234-6

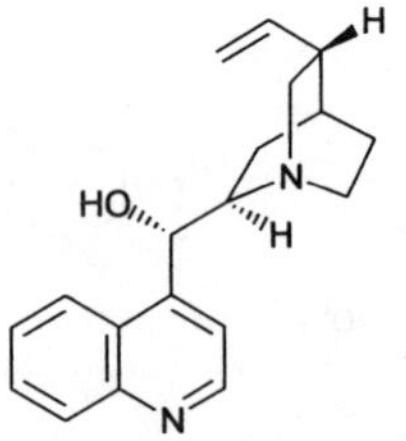

$C_{19}H_{22}N_2O$
Cinchonan-9-ol, (9S)-; (9S)-Cinchonan-9-ol. Listed on TSCA. Resolving agent. mp = 258-260; [α] = + 224° (c = 0.5, EtOH). *Acros Organics nv; DSM Fine Chemcials Netherlands; Isochem; Sigma-Aldrich Fine Chemicals.*

**1687 (+)-Cinchonine hydrochloride**
24302-67-8 246-139-2

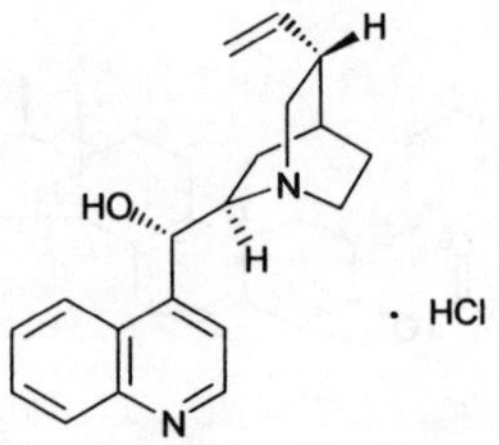

$C_{19}H_{22}N_2O.xClH$
Cinchonan-9-ol, hydrochloride, (9S)-. Resolving agent; Chiral auxiliary. *DSM Fine Chemcials Netherlands.*

**1688 (+)-Cinchonine monohydrochloride hydrate**
2289(11)

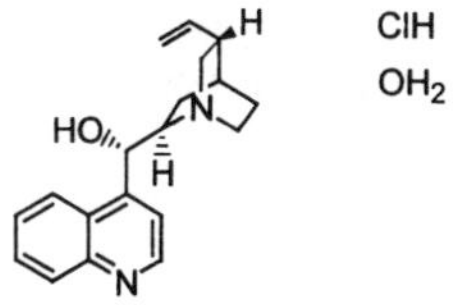

$C_{19}H_{23}ClN_2O$
Resolving agent; Chiral auxiliary. mp = 208-218°; $[\alpha]_D^{20}$ = + 101° (c = 1, $CHCl_3$). *Sigma-Aldrich Fine Chemicals.*

**1689 (-)-Cineole**
470-82-6

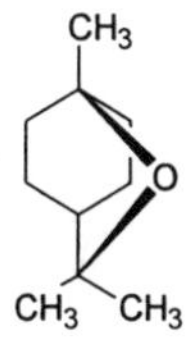

$C_{10}H_{18}O$
Chiral intermediate. [α] = - 0.3°. *Acros Organics nv.*

**1690 (R)-(-)-Citramalic acid**
6236-10-8 2325(11)

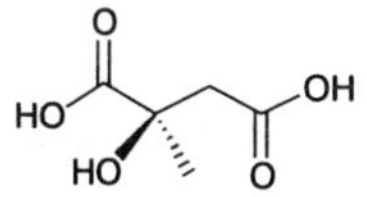

$C_5H_8O_5$
(R)-(-)-2-Methylmalic acid. Chiral building block. mp = 112°; -23 (c = 1, $H_2O$). *Acros Organics nv; Sigma-Aldrich Fine Chemicals; TCI America.*

**1691 (S)-(+)-Citramalic acid**
6236-09-5 2325(11) 228-350-1

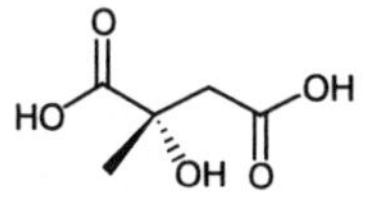

$C_5H_8O_5$
Butanedioic acid, 2-hydroxy-2-methyl-, (S)-; (S)-(+)-2-Methylmalic acid; (S)-2-Hydroxy-2-methylsuccinic acid; Malic acid, 2-methyl-, (S)-(+)-. Chiral building block. mp = 110°; [α] = + 23° (c = 1, $H_2O$). *Acros Organics nv; Sigma-Aldrich Fine Chemicals; TCI America.*

**1692 (-)-Citrinin**
518-75-2 208-257-2

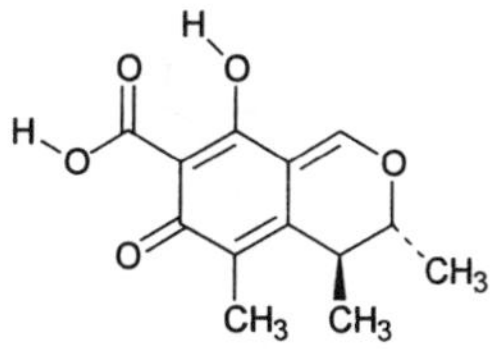

$C_{13}H_{14}O_5$
Benzo[3H-2]pyran-7-carboxylic acid, 4,6-dihydro-8-hydroxy-3,4,5-trimethyl-6-oxo-, (3R,4S)-; 3,4-Dihydro-8-hydroxy-3,4,5-trimethyl-6H-6-oxobenzo(c)pyran-7-carboxylic acid; 4,6-Dihydro-8-hydroxy-3,4,5-trimethyl-6-oxo-3H-2-benzopyran-7-carboxylic acid. Pharmaceutical or derivative. $[\alpha]^{23}$ = - 30.3° (c = 1, EtOH). *Acros Organics nv.*

**1693 (-)-Citronellal**
5949-05-3 227-707-9

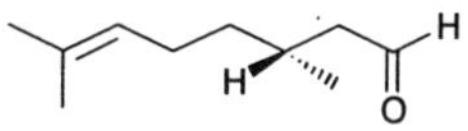

$C_{10}H_{18}O$
Oct-6-enal, 3,7-dimethyl-, (3S)-; (S)-3,7-Dimethyl-6-octenal; (S)-Citronellal. Listed on TSCA. Chiral intermediate. d = 0.851; n = 1.0446; [α] = - 13° (neat). *Acros Organics nv; Sigma-Aldrich Fine Chemicals; Takasago International Corporation; TCI America.*

**1694 (+)-Citronellal**
2385-77-5 2331(11) 219-194-5

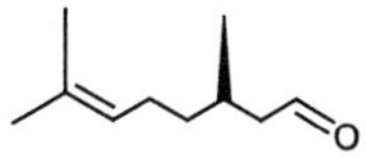

$C_{10}H_{18}O$
Oct-6-enal, 3,7-dimethyl-, (3R)-; (R)-3,7-Dimethyl-6-octenal. Listed on TSCA. Chiral intermediate. bp = 207°; d = 0.851; n = 1.448. *Sigma-Aldrich Fine Chemicals; TCI America.*

**1695 (S)-(-)-Citronellic acid**
2111-53-7

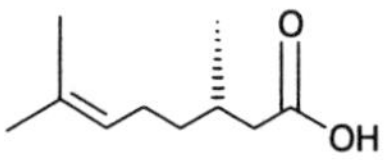

$C_{10}H_{18}O_2$
(S)-(-)-3,7-Dimethyl-6-octenoic acid. Chiral intermediate. $bp_{0.4}$ = 115-120°; d = 0.926; n = 1.453; $[\alpha]^{25}$ = - 8° (neat). *Sigma-Aldrich Fine Chemicals.*

**1696 (R)-(+)-Citronellic acid**
18951-85-4

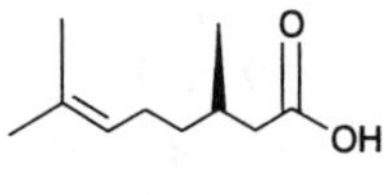

$C_{10}H_{18}O_2$
(+)-3,7-Dimethyl-6-octenoic acid. Chiral intermediate. $bp_{3.0}$ = 119°; d = 0.926; n = 1.4534; $[\alpha]^{25}$ = + 8° (neat). *Sigma-Aldrich Fine Chemicals.*

**1697 (-)-β-Citronellol**
7540-51-4 8186(11) 231-415-7

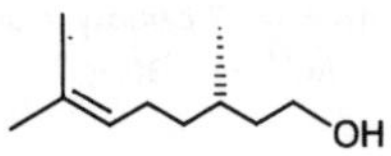

$C_{10}H_{20}O$
Oct-6-enol, 3,7-dimethyl-, (3S)-; (S)-(-)-3,7-Dimethyl-6-octen-1-ol; β-Rhodinol. Listed on TSCA. Chiral intermediate. bp = 225-226°; d = 0.856; n = 1.456; $[\alpha]_D^{20}$= - 5.3° (neat). *Sigma-Aldrich Fine Chemicals; TCI America.*

**1698 (+)-Citronellol**
1117-61-9 2332(11) 214-250-5

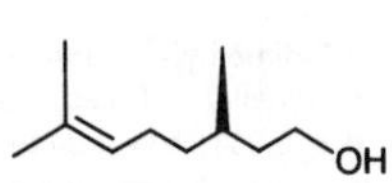

$C_{10}H_{20}O$
Listed on TSCA. Chiral intermediate. $bp_{12}$ = 112-113°; d = 0.857; n = 1.456; $[\alpha]^{19}$ = + 5.3° (neat). *Sigma-Aldrich Fine Chemicals.*

**1699 (R)-(-)-Citronellyl bromide**
10340-84-8

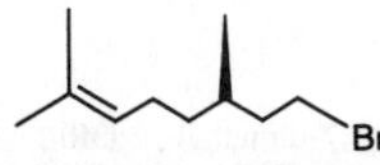

$C_{10}H_{19}Br$
(-)-8-Bromo-2,6-dimethyl-2-octene. Chiral intermediate. $bp_{12}$ = 111°; d = 1.11; n = 1.474; $[\alpha]_D^{20}$ = - 6.8° (neat). *Sigma-Aldrich Fine Chemicals.*

**1700 (S)-(+)-Citronellyl bromide**
143615-81-0

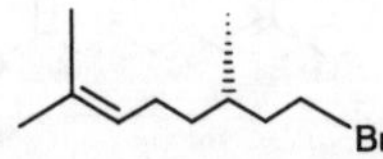

$C_{10}H_{19}Br$
(+)-8-Bromo-2,6-dimethyl-2-octene. Chiral intermediate. $bp_{12}$ = 111°; d = 1.11; n = 1.474; $[\alpha]_D^{20}$ = + 6.8° (neat). *Sigma-Aldrich Fine Chemicals.*

**1701 L-Citrulline**
372-75-8 206-759-6

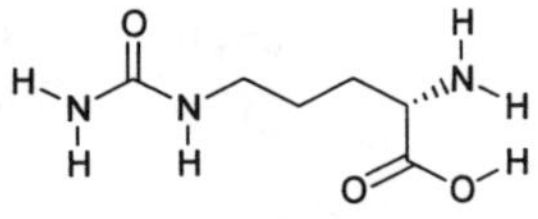

$C_6H_{13}N_3O_3$
$N^5$-(aminocarbonyl)-, L-; α-Amino-δ-ureidovaleric acid; δ-Ureidonorvaline; $N^5$-Carbamoyl-L-ornithine; Ornithine, $N^5$-carbamoyl-, L-. Listed on TSCA. Chiral building block. [α] = + 26.5° (c = 8, 6N HCl). *Acros Organics nv; Ajinomoto Co. Inc.; Senn Chemicals AG; TCI America.*

**1702 (-)-Colchicine**
64-86-8 2536(12) 200-598-5

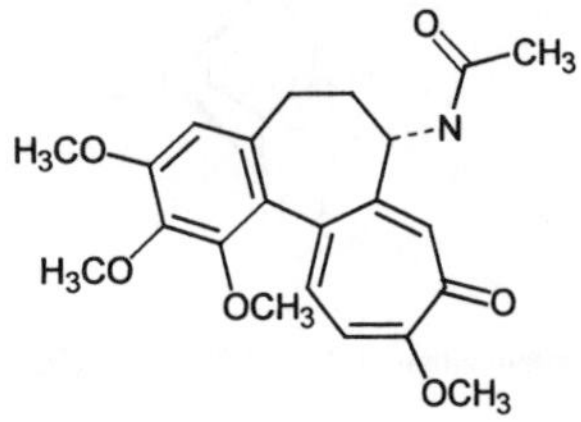

$C_{22}H_{25}NO_6$
Listed on TSCA. Resolving agent; Chiral auxiliary. mp = 155-157°; $[\alpha]^{24}$ = - 443° (c = 1.7, $H_2O$). *Sigma-Aldrich Fine Chemicals.*

**1703 (-)-Corey aldehyde benzoate**
39746-01-5

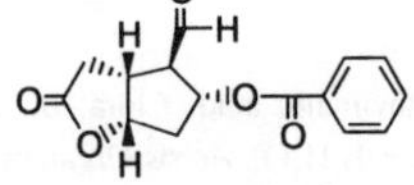

$C_{15}H_{14}O_5$
Chiral intermediate. mp = 128-138°; $[\alpha]^{23}$ = - 85° (c = 1, $CH_3OH$). 570; *Sigma-Aldrich Fine Chemicals.*

**1704 (-)-Corey lactone, 4-phenylbenzoate**
31752-99-5

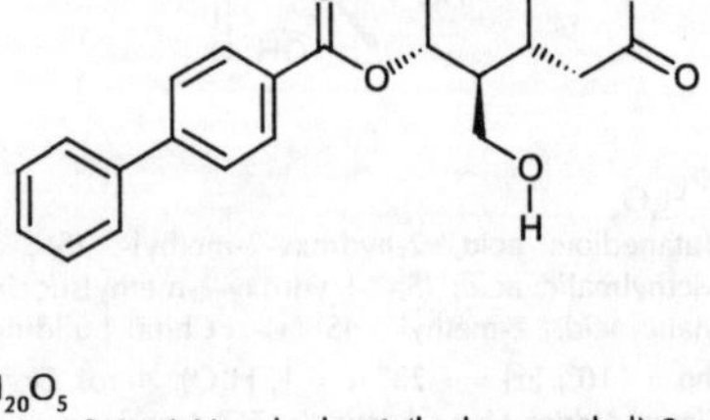

$C_{21}H_{20}O_5$
(3aα,4α,5β,6aα)-Hexahydro-4-(hydroxymethyl)-2-oxo-2H-cyclopenta[b]furan-5-yl 1,1'-biphenyl-4-carboxylate.

Chiral intermediate. mp = 135-137°; [α] = - 84° (c = 1; $CHCl_3$). *Acros Organics nv; Sigma-Aldrich Fine Chemicals.*

**1705 (+)-Corey lactone, 4-phenylbenzoate alcohol**
39265-57-1

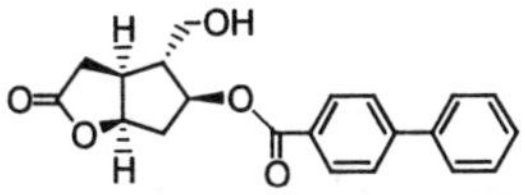

$C_{21}H_{20}O_5$
Chiral auxiliary. mp = 134-136°; $[\alpha]_D^{20}$ = + 87° (c = 1, $CHCl_3$). *Sigma-Aldrich Fine Chemicals.*

**1706 (-)-Corynanthine hydrochloride hydrate**
483-10-3 207-590-0

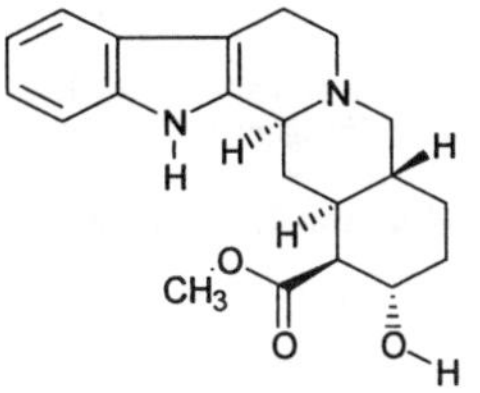

$C_{21}H_{26}N_2O_3$
Yohimban-16-carboxylic acid, 17-hydroxy-, methyl ester, (16β,17α)-; methyl (16b,17a)-17-hydroxyyohimban-16-carboxylate; Rauhimbine. Chiral intermediate. $[\alpha]^{21}$ = - 67° (c = 1, $H_2O$). *Acros Organics nv.*

**1707 (-)-Cotinine**
486-56-6 2619(12) 207-634-9

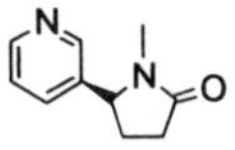

$C_{10}H_{12}N_2O$
Pyrrolidin-2-one, 1-methyl-5-(3-pyridinyl)-, (5S)-; (S)-1-Methyl-5-(3-pyridyl)-2-pyrrolidinone; NIH 10498. Chiral intermediate. $bp_{150}$ = 250°; $bp_3$ = 145-147°; $[\alpha]_D^{20}$ = - 20 ± 2° (c = 1, EtOH). *Lancaster Synthesis Ltd.; Sigma-Aldrich Fine Chemicals.*

**1708 Cresole-β-D-glucopyranoside**

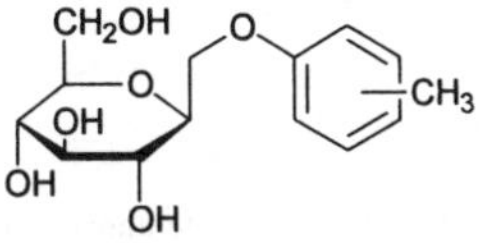

$C_{15}H_{24}O_6$
Chiral intermediate. *Senn Chemicals AG.*

**1709 N-Crotonyl-(4R)-benzyl-2-oxazolidinone**

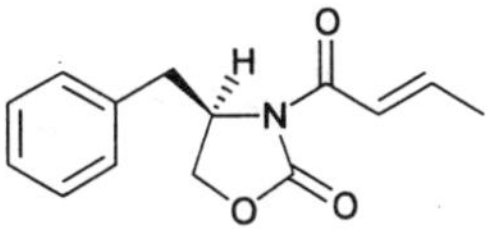

$C_{14}H_{15}NO_3$
Chiral intermediate. *Oxford Asymmetry International plc.*

**1710 N-Crotonyl-(4S)-benzyl-2-oxazolidinone**

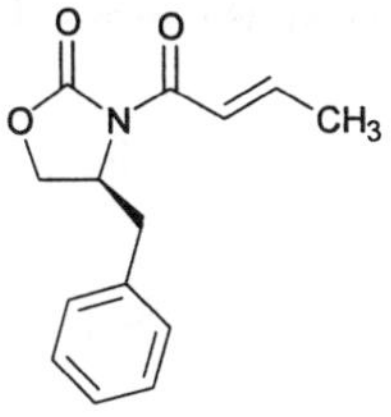

$C_{14}H_{15}NO_3$
Chiral intermediate. *Acros Organics nv; Oxford Asymmetry International plc.*

**1711 N-Crotonyl-(2R)-bornane-10,2-sultam**
94668-55-0

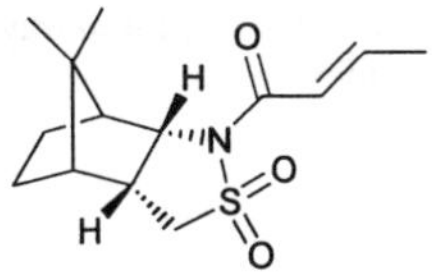

$C_{14}H_{21}NO_3S$
(R)-(2-Butenoyl)-2,10-camphorsultam. Chiral intermediate. *Oxford Asymmetry International plc; Sigma-Aldrich Fine Chemicals.*

**1712 N-Crotonyl-(2S)-bornane-10,2-sultam**
94594-81-7

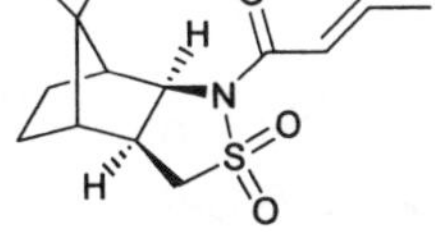

$C_{14}H_{21}NO_3S$
(S)-(2-Butenoyl)-2,10-camphorsultam. Chiral intermediate. *Oxford Asymmetry International plc; Sigma-Aldrich Fine Chemicals.*

**1713 N-Crotonyl-(4R)-isopropyl-2-oxazolidinone**
90719-30-5

$C_{14}H_{15}NO_3$
(R)-4-Benzyl-3-crotonyl-2-oxazolidinone. Chiral intermediate. *Acros Organics nv; Oxford Asymmetry International plc; Sigma-Aldrich Fine Chemicals.*

**1714 N-Crotonyl-(4S)-isopropyl-2-oxazolidinone**

$C_{10}H_{15}NO_3$
(S)-4-Benzyl-3-crotonyl-2-oxazolidinone. Chiral intermediate. *Oxford Asymmetry International plc.*

**1715 N-Crotonyl-(4R,5S)-4-methyl-5-phenyl-2-oxazolidinone**
90719-31-6
$C_{14}H_{15}NO_3$
Chiral intermediate. *Oxford Asymmetry International plc.*

**1716 N-Crotonyl-(4S,5R)-4-methyl-5-phenyl-2-oxazolidinone**
128440-43-7

$C_{14}H_{15}NO_3$
Chiral intermediate. *Oxford Asymmetry International plc.*

**1717 (+)-(18-Crown-6)-2,3,11,12-tetracarboxylic acid**
61696-54-6

$C_{16}H_{24}O_{14}$
Chiral auxiliary. mp = 210-212°; $[\alpha]_D^{20}$ = + 62° (c = 1, $CH_3OH$). *Sigma-Aldrich Fine Chemicals.*

**1718 (-)-Cubebene**
17699-14-8

$C_{15}H_{24}$
Chiral intermediate. bp = 245-246°; d = 0.889; n = 1.482; $[\alpha]_D^{20}$ = - 23° (neat). *Sigma-Aldrich Fine Chemicals.*

**1719 (R)-Cyano-1,2-epoxybutane**

$C_5H_7NO$
Propanenitrile, 2-(oxiran-2-yl)-, (R)-. Chiral building block. *Synthon Chiragenics Corporation.*

**1720 (S)-4-Cyano-1,2-epoxybutane**

$C_5H_7NO$
Propanenitrile, 2-(oxiran-2-yl)-, (S)-. Chiral building block. *Synthon Chiragenics Corporation.*

**1721 (-)-2-Cyano-6-phenyloxazolopiperidine**
88056-92-2

$C_{14}H_{16}N_2O$
Chiral intermediate. $[\alpha]$ = - 290° (c = 1, $CHCl_3$). *Acros Organics nv.*

**1722 (-)-Cyclocytidine hydrochloride**
10212-25-6 670(12) 233-515-6

$C_9H_{11}N_3O_4$
2,2'-Anhydro-1-β-D-arabinofuranosylcytosine hydrochloride. Chiral intermediate. mp = 269-270°; $[\alpha]^{22}$ = - 21.6° (c = 2, $H_2O$). *Sigma-Aldrich Fine Chemicals.*

**1723 (+)-α-Cyclodextrin hydrate**
10016-20-3 2724(11) 233-007-4

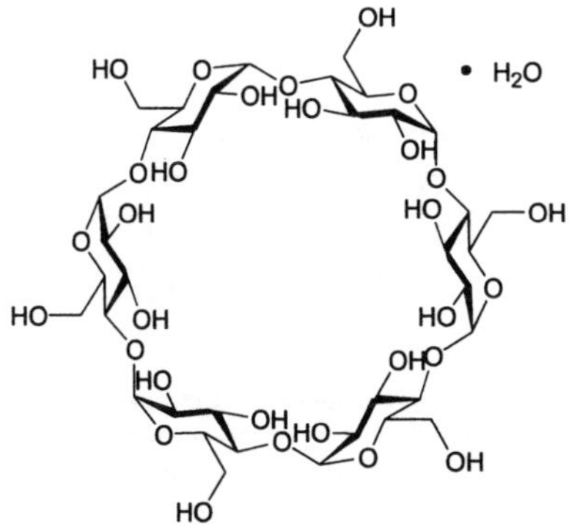

$C_{36}H_{60}O_{30}$
Cyclodextrin, α-; α-Cyclodextrin; α-Dextrin; α-Schardinger dextrin; α-Cycloamylose; Cyclomaltohexose; α-Cyclodextrin hydrate; Cyclohexapentylose; Alfadex; Celdex; Cyclohexaamylose; Cyclohexadextrin; Dextrin, α-cyclo; Dexy Pearl α-100; Ringdex A. Listed on TSCA. Chiral auxiliary. mp = 278°; $[\alpha]^{25} = +153°$ (c = 1, $H_2O$). *Acros Organics nv; Sigma-Aldrich Fine Chemicals.*

**1724 (+)-β-Cyclodextrin hydrate**
68168-23-0 2724(11)

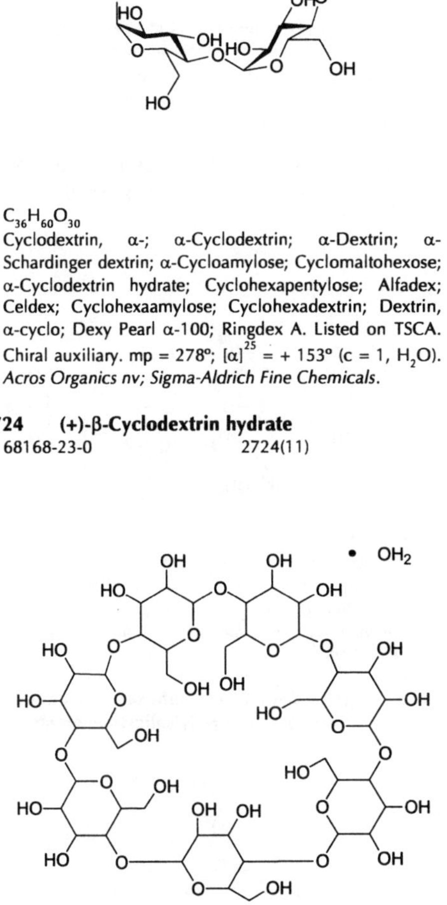

$C_{42}H_{70}O_{35}$
Cycloheptaamylose; β-Schardinger dextrin. Chiral auxiliary. mp = 260°; $[\alpha]_D^{20} = +142°$ (c = 1, $H_2O$, dry basis). *Acros Organics nv; Sigma-Aldrich Fine Chemicals.*

**1725 (+)-γ-Cyclodextrin hydrate**
17465-86-0 2724(11) 241-482-4

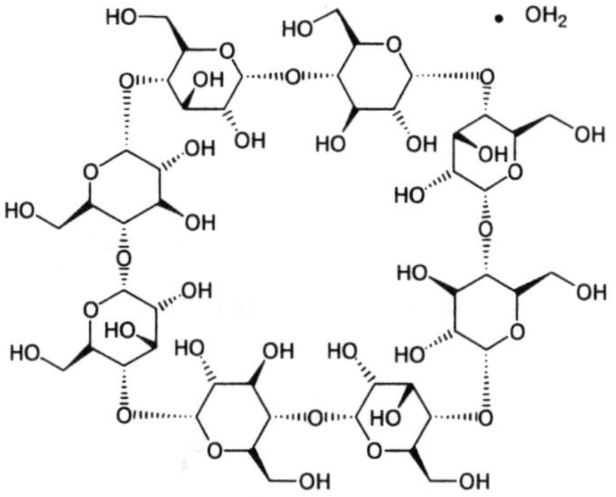

$C_{48}H_{82}O_{41}$
γ-Cyclodextrin; Cyclooctaamylose hydrate; γ-Schardinger dextrin hydrate; Dexy Pearl g-100; Ringdex C; cyclooctapentylose. Listed on TSCA. Chiral auxiliary. mp = 267°; $[\alpha]_D^{20} = +176.1°$ (c = 1, $H_2O$). *Acros Organics nv; Sigma-Aldrich Fine Chemicals.*

**1726 Cyclodextrin sulfated, sodium salt**
37191-69-8

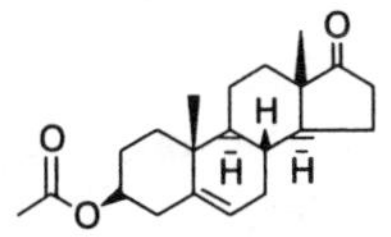

$C_{21}H_{30}O_3$
Chiral intermediate. mp = 190°; $[\alpha]^{23} = +78°$ (c = 1, $H_2O$). *Sigma-Aldrich Fine Chemicals.*

**1727 (R,R)-1,2-Cyclododecanediol**
118101-30-7

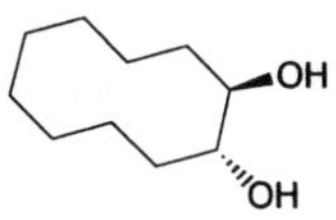

$C_{12}H_{24}O_2$
Chiral building block. *Kankyo Kagaku Center Co., Ltd.*

**1728 (S,S)-1,2-Cyclododecanediol**
118101-31-8

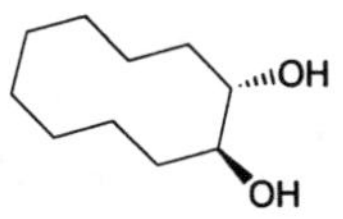

$C_{12}H_{24}O_2$
Chiral building block. *Kankyo Kagaku Center Co., Ltd.*

**1729 (R,R)-1,2-Cycloheptanediol**
108268-28-6

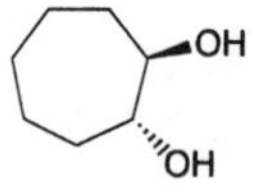

$C_7H_{14}O_2$
Chiral building block. *Kankyo Kagaku Center Co., Ltd.*

**1730 (S,S)-1,2-Cycloheptanediol**
108268-27-5

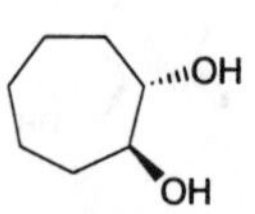

$C_7H_{14}O_2$
Chiral building block. *Kankyo Kagaku Center Co., Ltd.*

**1731 (1S,2S)-(-)-1,2-Cyclohexanediamine D-tartrate**
67333-70-4

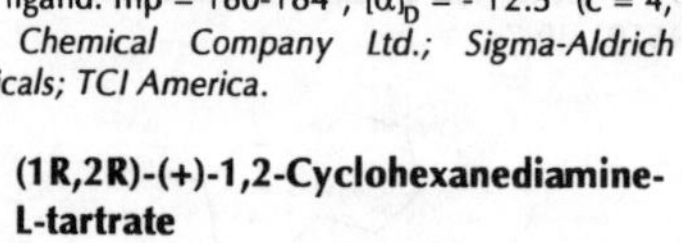

$C_6H_{14}N_2 \cdot C_4H_6O_7$
(1S)-trans-1,2-Diaminocyclohexane D-tartrate; (S,S)-1,2-Diaminocyclohexane tartrate. Chiral building block; Chiral ligand. mp = 180-184°; $[\alpha]_D^{20}$ = - 12.5° (c = 4, $H_2O$). *Arran Chemical Company Ltd.; Sigma-Aldrich Fine Chemicals; TCI America.*

**1732 (1R,2R)-(+)-1,2-Cyclohexanediamine-L-tartrate**
39961-95-0

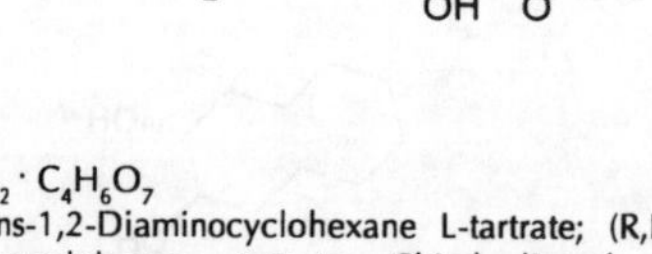

$C_6H_{14}N_2 \cdot C_4H_6O_7$
(1R)-trans-1,2-Diaminocyclohexane L-tartrate; (R,R)-1,2-Diaminocyclohexane tartrate. Chiral ligand; Chiral building block. mp = 273°; $[\alpha]_D^{20}$ = + 12.5° (c = 4, $H_2O$). *Arran Chemical Company Ltd.; Sigma-Aldrich Fine Chemicals; TCI America.*

**1733 (1R,2R)-(-)-1,2-Cyclohexanediamino-N,N'-bis-(3,5-di-tert-butylsalicylidene)**
135616-40-9

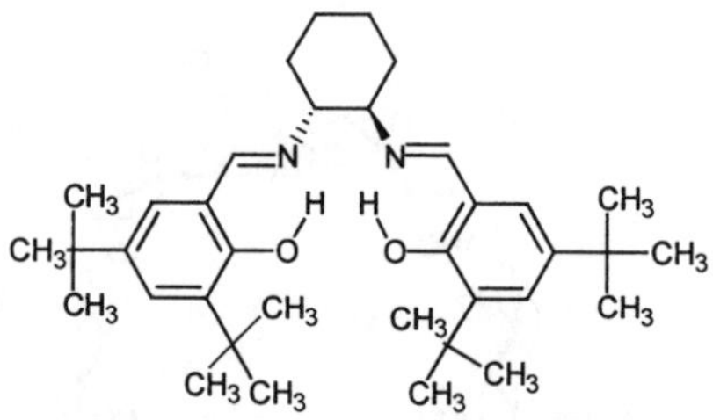

$C_{36}H_{54}N_2O_2$
(R,R)-Jacobsen Ligand; (R,R)-(-)-N,N[-Bis(3,5-di-tert-butylsalicylidene)-1,2-cyclohexanediamine. Chiral ligand. mp = 205-207°; $[\alpha]_D$ = - 315° (c = 1, $CH_2Cl_2$). *Acros Organics nv; Chirex, Inc.; Sigma-Aldrich Fine Chemicals; Strem Chemicals, Inc.*

**1734 (1S,2S)-(+)-1,2-Cyclohexanediamino-N,N'-bis-(3.5-di-tert-butylsalicylidene)**
135616-36-3

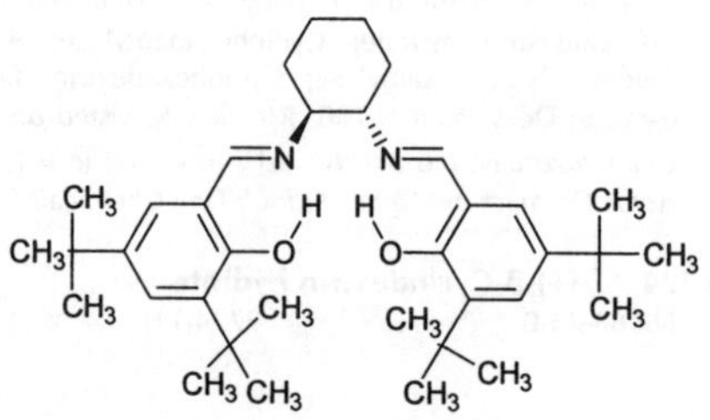

$C_{36}H_{54}N_2O_2$
(S,S)-Jacobsen Ligand; (S,S)-(+)-N,N[-Bis(3,5-di-tert-butylsalicylidene)-1,2-cyclohexanediamine. Chiral ligand. mp = 205-207°; $[\alpha]_D$ = + 305° (c = 1, $CH_2Cl_2$). *Acros Organics nv; Chirex, Inc.; Sigma-Aldrich Fine Chemicals; Strem Chemicals, Inc.*

**1735 (1R,2R)-(-)-1,2-Cyclohexanediamino-N,N'-bis-(3,5-di-tert-butylsalicylidene)cobalt(III)**

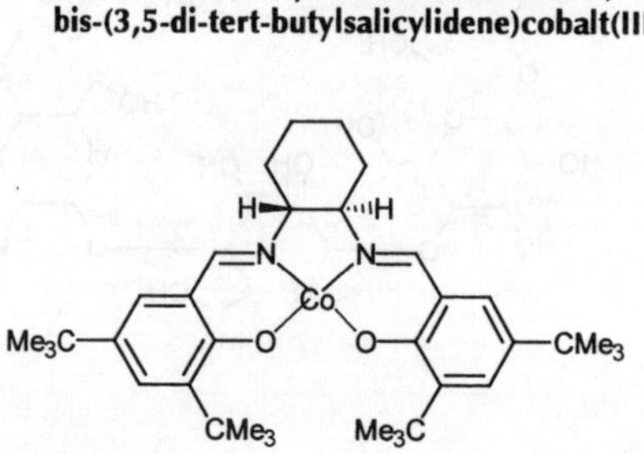

$C_{36}H_{52}CoN_2O_2$
Chiral catalyst. mp = 406-412°; $[\alpha]_D$ = - 1200° (c = 0.01, $CH_2Cl_2$). *Strem Chemicals, Inc.*

**1736 (1S,2S)-(-)-1,2-Cyclohexanediamino-N,N'-bis-(3,5-di-tert-butylsalicylidene)cobalt(III)**

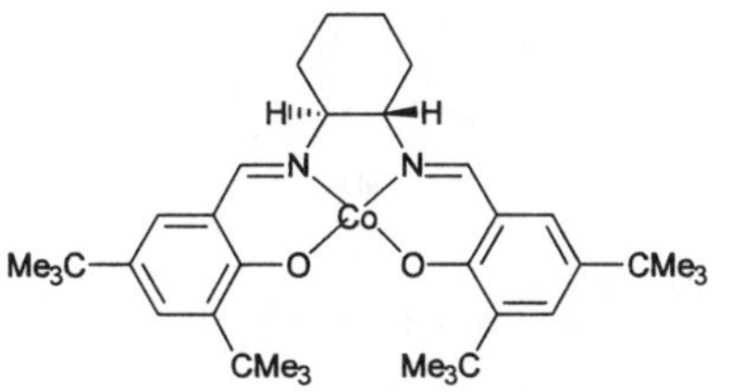

$C_{36}H_{52}CoN_2O_2$
Chiral catalyst. mp = 409-412°; $[\alpha]_D$ = + 1170° (c = 0.01, $CH_2Cl_2$). *Strem Chemicals, Inc.*

**1737 (1R,2R)-(-)-[1,2-Cyclohexanediamino-N,N'-bis-(3,5-di-tert-butylsalicylidene)]-manganese(III)chloride**
138124-32-0

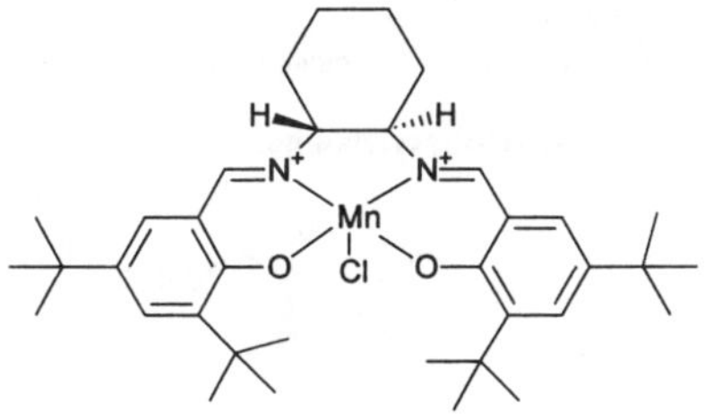

$C_{36}H_{52}N_2O_2ClMn$
(R,R)-Jacobsen Catalyst; (R,R)-(-)N,N'-Bis(3,5-di-tert-butylsalicylidene)-1,2-cyclohexanediamino manganese(III) chloride. Chiral catalyst. mp = 324-326°; $[\alpha]_D$ = - 1416° (c = 0.0072, $CH_2Cl_2$). *Acros Organics nv; Sigma-Aldrich Fine Chemicals; Strem Chemicals, Inc.*

**1738 (1S,2S)-(+)-(1,2-Cyclohexanediamino-N,N'-bis-(3,5-di-tert-butylsalicylidene)]-manganese(III)chloride**
135620-04-1

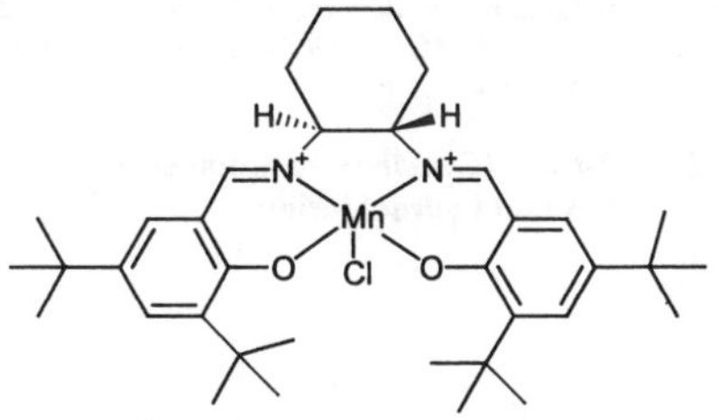

$C_{36}H_{52}N_2O_2ClMn$
(S,S)-Jacobsen Catalyst; S-(+)-Bis(3,5-di-tert-butylsalicylidene)-1,2-cyclohexane diaminomanganese(III) chloride. Chiral catalyst. mp = 325-326°. *Acros Organics nv; Sigma-Aldrich Fine Chemicals; Strem Chemicals, Inc.*

**1739 (1R,2R)-trans-1,2-Cyclohexanediol**
1072-86-2

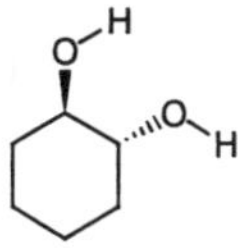

$C_6H_{12}O_2$
Chiral building block; Resolving agent. mp = 107-109°; $[\alpha]_D^{20}$ = - 39° (c = 1.6, $H_2O$). *Acros Organics nv; Digital Specialty Chemicals, Inc.; Oxford Asymmetry International plc; Sigma-Aldrich Fine Chemicals.*

**1740 (1S,2S)-trans-1,2-Cyclohexanediol**
57794-08-8

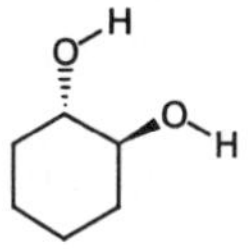

$C_6H_{12}O_2$
Chiral building block. mp = 107-109°; $[\alpha]_D^{20}$ = + 39° (c = 1.6, $H_2O$). *Acros Organics nv; Oxford Asymmetry International plc; Sigma-Aldrich Fine Chemicals.*

**1741 (1R)-trans-N,N-1,2-cyclohexanediyl-bis(1,1,1-trifluoromethanesulfonamide)**

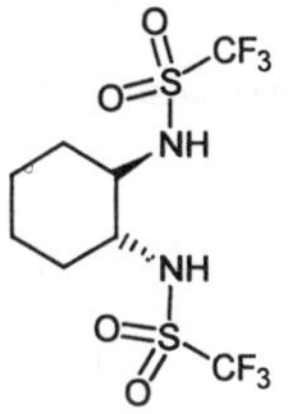

$C_8H_{12}F_6N_2O_4S_2$
Chiral intermediate. mp = 185-187°; $[\alpha]_D^{20}$ = -6.1° (c = 5, EtOH). *Sigma-Aldrich Fine Chemicals.*

**1742 (R)-Cyclohexen-1-amine**
153922-90-8

$C_6H_{11}N$
(R)-3-Aminocyclohexene; Cyclohex-2-enylamine. Chiral building block. *Lancaster Synthesis Ltd.*

**1743 (S)-2-Cyclohexen-1-amine**
153922-89-5

$C_6H_{11}N$
(S)-3-Aminocyclohexene; (S)-Cyclohex-2-enylamine. Chiral building block. *Lancaster Synthesis Ltd.*

**1744 D-Cyclohexylalanine**

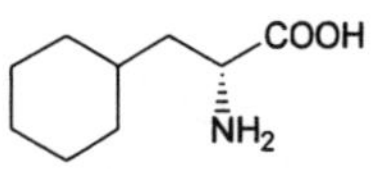

$C_9H_{17}NO_2$
Chiral building block. *SK Energy and Chemical, Inc.*

**1745 L-3-Cyclohexylalanine**
27527-05-5

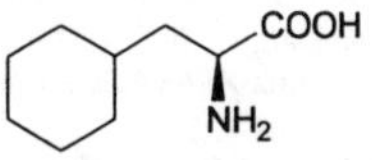

$C_9H_{17}NO_2$
Cyclohexylalanine, S-; (S)-(+)-Aminocyclohexanepropionic acid. Chiral building block. mp = 234-237°; $[\alpha]^{21}$ = + 12.0° (c = 1, 1N NaOH). *Degussa-Huls AG; Fischer Chemicals AG; Great Lakes Fine Chemicals; Senn Chemicals AG; Sigma-Aldrich Fine Chemicals; SK Energy and Chemical, Inc.*

**1746 D-Cyclohexylalanine amide**

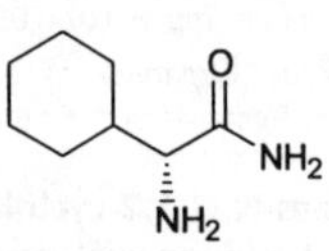

$C_8H_{16}N_2O$
Chiral building block. *Synthetech, Inc.*

**1747 L-Cyclohexylalanine amide**
145232-34-3

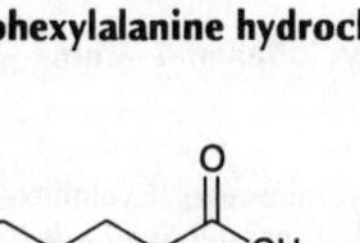

$C_8H_{16}N_2O$
Chiral building block. *Fischer Chemicals AG; Great Lakes Fine Chemicals; Synthetech, Inc.*

**1748 (S)-Cyclohexylalanine hydrochloride**
25528-71-6

$C_9H_{18}ClNO_2$
Chiral building block. *Fischer Chemicals AG.*

**1749 D-Cyclohexylalanine methyl ester hydrochloride**

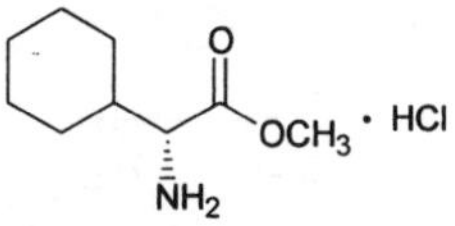

$C_9H_{18}ClNO_2$
Chiral building block. *Synthetech, Inc.*

**1750 L-Cyclohexylalanine methyl ester hydrochloride**

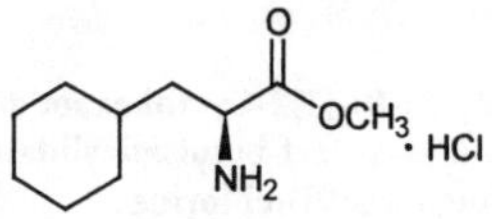

$C_{10}H_{20}ClNO_2$
Chiral building block. *Synthetech, Inc.*

**1751 (R)-Cyclohexylalaninol**

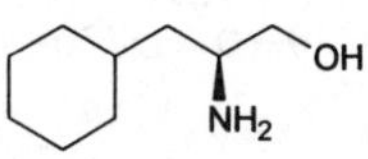

$C_9H_{19}NO$
Chiral building block. *SK Energy and Chemical, Inc.*

**1752 (S)-Cyclohexylalaninol**
131288-67-0

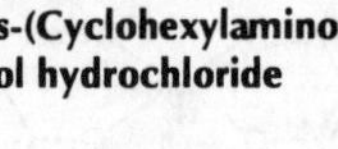

$C_9H_{19}NO$
Chiral building block. *Boehringer Ingelheim Pharma KG; Fischer Chemicals AG; Great Lakes Fine Chemicals; SK Energy and Chemical, Inc.*

**1753 2-trans-(Cyclohexylaminomethyl)-1-cyclohexanol hydrochloride**

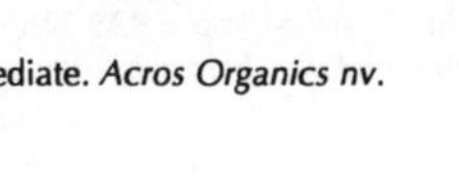

$C_{13}H_{26}ClNO$
Chiral intermediate. *Acros Organics nv.*

**1754 (R)-(-)-1-Cyclohexylethylamine**
5913-13-3 227-634-2

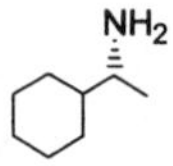

$C_8H_{17}N$
Cyclohexanemethylamine, α-methyl-, (R)-(-)-; (R)-α-Cyclohexanemethylamine. Chiral intermediate. d = 0.856; $[\alpha]_D^{20}$ = - 4° (neat). *BASF Aktiengesellschaft; Chemi SpA; Sigma-Aldrich Fine Chemicals; TCI America; Vinchem.*

**1755 (S)-(+)-1-Cyclohexylethylamine**
17430-98-7

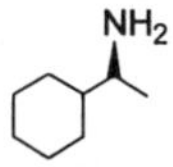

$C_8H_{17}N$
(S)-(+)-1-Aminoethylcyclohexane. Chiral intermediate. $bp_{12}$ = 60°; d = 0.856; n = 1.4614; $[\alpha]^{25}$ = + 3.5° (neat). *BASF Aktiengesellschaft; Chemi SpA; Sigma-Aldrich Fine Chemicals; TCI America.*

**1756 D-2-Cyclohexylglycine**
14328-52-0

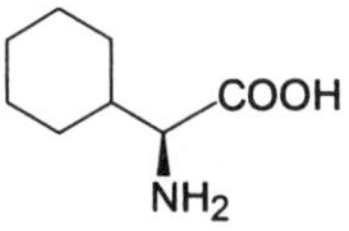

$C_8H_{15}NO_2$
Chiral building block. *Austin Chemical Company, Inc.; Degussa-Huls AG; Great Lakes Fine Chemicals.*

**1757 L-Cyclohexylglycine**

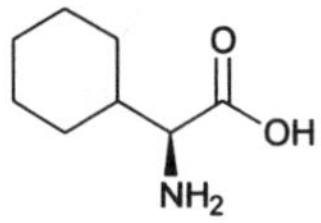

$C_8H_{15}NO_2$
Chiral building block. *Senn Chemicals AG.*

**1758 L-Cyclohexylglycine methyl ester hydrochloride**

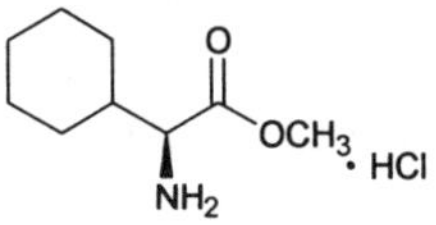

$C_9H_{18}ClNO_2$
Chiral building block. *Synthetech, Inc.*

**1759 (+)-5,6-O-Cyclohexylideneascorbic acid**
6614-52-4

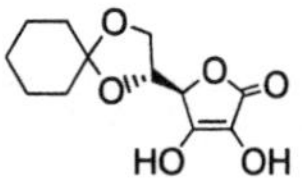

$C_{12}H_{16}O_6$
Chiral intermediate. mp = 180°. *Sigma-Aldrich Fine Chemicals.*

**1760 2,3-O-Cyclohexylidene-L-(-)-erythruronic acid**
85281-85-2
$C_{10}H_{14}O_5$
Chiral intermediate. *Sigma-Aldrich Fine Chemicals.*

**1761 1,2-O-Cyclohexylidene-α-D-glucofuranose**
16832-21-6

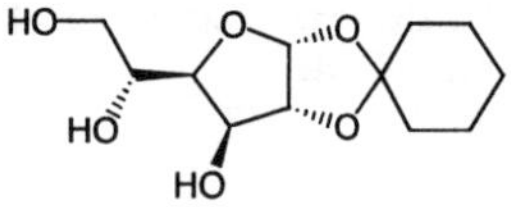

$C_{12}H_{20}O_6$
Chiral intermediate. mp = 152-155°. *Acros Organics nv.*

**1762 1,2-O-Cyclohexylidene-3-O-methyl-α-D-glucofuranose**
13322-87-7

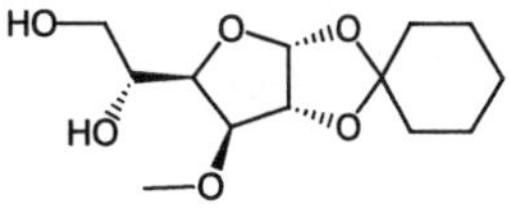

$C_{13}H_{22}O_6$
Chiral intermediate. mp = 98-101°; [α] = - 10° (c = 2.5, $CHCl_3$); $[\alpha]_D^{20}$ = - 24° (c = 3.3, $H_2O$). *Acros Organics nv; Sigma-Aldrich Fine Chemicals.*

**1763 2,3-Cyclohexylidene-D-riboinc acid γ-lactone**
27304-20-7
$C_{11}H_{16}O_5$
Chiral intermediate. *Sigma-Aldrich Fine Chemicals.*

**1764 1,2-O-Cyclohexylidene-α-D-xylopentodialdo-1,4-furanose dimer**
15356-27-1

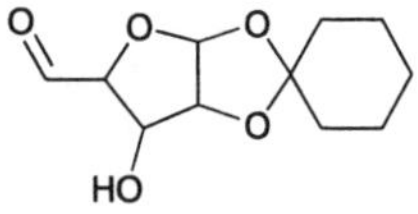

$C_{22}H_{32}O_{10}$
Chiral intermediate. mp = 182-184°; [α] = + 33° (c = 1.3, $CHCl_3$). *Acros Organics nv; Sigma-Aldrich Fine Chemicals.*

**1765 (R)-2-(Cyclohexylmethyl)succinic acid 1-methyl ester**
130165-88-7
$C_{12}H_{20}O_4$
(R)-4-Cyclohexyl-3-(methoxycarbonyl)butyric acid. Chiral building block. *ChiroTech Technology Ltd.; Lancaster Synthesis Ltd.*

**1766 (S)-2-(Cyclohexylmethyl)succinic acid 1-methyl ester**
$C_{12}H_{20}O_4$
(S)-4-Cyclohexyl-3-(methoxycarbonyl)butyric acid. Chiral intermediate. *ChiroTech Technology Ltd.; Lancaster Synthesis Ltd.*

**1767 (R)-Cyclohexylsuccinic acid 1-methyl ester**
$C_{11}H_{18}O_4$
(R)-3-Cyclohexyl-3-(methoxycarbonyl)propionic acid. Chiral intermediate. *ChiroTech Technology Ltd.; Lancaster Synthesis Ltd.*

**1768 (S)-2-Cyclohexylsuccinic acid 1-methyl ester**
213270-44-1
$C_{11}H_{18}O_4$
(S)-3-Cyclohexyl-3-(methoxycarbonyl)propionic acid. Chiral intermediate. *ChiroTech Technology Ltd.; Lancaster Synthesis Ltd.*

**1769 (R,R)-1,2-Cyclooctanediol**
108286-29-7

OH
OH

$C_8H_{16}O_2$
Chiral building block. *Kankyo Kagaku Center Co., Ltd.*

**1770 (S,S)-1,2-Cyclooctanediol**
20480-40-4

OH
OH

$C_8H_{16}O_2$
Chiral building block. *Kankyo Kagaku Center Co., Ltd.*

**1771 (R,R)-5-Cyclooctene-1,2-diol**
203319-31-7

OH
OH

$C_8H_{14}O_2$
Chiral building block. *Kankyo Kagaku Center Co., Ltd.*

**1772 (S,S)-5-Cyclooctene-1,2-diol**
69926-50-7

OH
OH

$C_8H_{14}O_2$
Chiral building block. *Kankyo Kagaku Center Co., Ltd.*

**1773 (1R,2R)-(-)-trans-1,2-Cyclopentanediol**
930-46-1

OH
OH

$C_5H_{10}O_2$
Chiral building block. mp = 47-51°; $[\alpha]_D^{20}$ = - 21° (c = 1, $CHCl_3$). *Sigma-Aldrich Fine Chemicals.*

**1774 (1S,2S)-(+)-trans-1,2-Cyclopentanediol**
63261-45-0

OH
OH

$C_5H_{10}O_2$
Chiral building block. mp = 47-51°; $bp_{21.5}$ = 136°; $[\alpha]_D^{20}$ = + 21.0° (c = 1, $CHCl_3$). *Sigma-Aldrich Fine Chemicals.*

**1775 (1R,3S)-(+)-4-Cyclopentene-1,3-diol 1-acetate**
60410-16-4

OH
O
O

$C_7H_{10}O_3$
Chiral intermediate. mp = 49-51°; $[\alpha]_D^{20}$ = + 68° (c = 2.3, $CHCl_3$). *Sigma-Aldrich Fine Chemicals.*

**1776 (R)-Cyclopropylglycine**
49606-99-7

O
OH
$NH_2$

$C_5H_9NO_2$
Chiral building block. *Eastman Chemical Company.*

**1777 (S)-Cyclopropylglycine**
15785-26-9

$C_5H_9NO_2$
Chiral building block. *Eastman Chemical Company.*

**1778 (+)-Cyclosativene**
22469-52-9

$C_{15}H_{24}$
Chiral intermediate. $bp_{10}$ = 87°; d = 0.929; n = 1.488; $[\alpha]_D^{20}$= + 78° (neat). *Sigma-Aldrich Fine Chemicals.*

**1779 D-(+)-Cycloserine**
68-41-7 2820(12) 200-688-4

$C_3H_6N_2O_2$
Isoxazolidin-3-one, 4-amino-, (R)-; 4-Amino-3-isoxazolidinone; Cyclorin; Farmiserina; Miroseryn; Novoserin; Orientomycin; Oxamycin; Tisomycin; Wasserina. Chiral intermediate. mp = 147°; $[\alpha]_D^{20}$= + 115 ± 3° (c = 2, $H_2O$); $[\alpha]_D$ = + 110 (c = 5, 1N NaOH). *Acros Organics nv; Lancaster Synthesis Ltd.; Sigma-Aldrich Fine Chemicals; TCI America.*

**1780 L-(-)-Cycloserine**
339-72-0 2758(11) 206-427-0

$C_3H_6N_2O_2$
Isoxazolidin-3-one, 4-amino-, (S)-; L-4-Amino-isoxazolidin-3-one; Levcycloserine. Chiral intermediate. mp = 146°; $[\alpha]^{22}$ = - 115° (c = 1, $H_2O$). *Acros Organics nv; Sigma-Aldrich Fine Chemicals.*

**1781 L-Cystamin hydrochloride**

$C_4H_{14}Cl_2N_2S_2$
Chiral building block. *Varsal Instruments, Inc.*

**1782 L-Cysteic acid**
498-40-8 207-861-3

$C_3H_7NO_5S$
Alanine, 3-sulfo-, L-; Cysteinesulfonic acid. Chiral building block. *Senn Chemicals AG; TCI America.*

**1783 L-Cysteic acid monohydrate**
23537-25-9 2786(11) 207-861-3

$C_3H_9NO_6S$
β-Sulfo-L-alanine monohydrate. Chiral building block. mp = 267°; $[\alpha]^{24}$ = + 8.2° (c = 7.5, $H_2O$). *Acros Organics nv; Sigma-Aldrich Fine Chemicals.*

**1784 D-Cysteine**
921-01-7 213-062-0

$C_3H_7NO_2S$
Cysteine, D-. Chiral building block. *Freedom Chemical Diamalt GmbH; Nippon Rikagakuyakuhin.*

**1785 L-Cysteine**
52-90-4 2850(12) 200-158-2

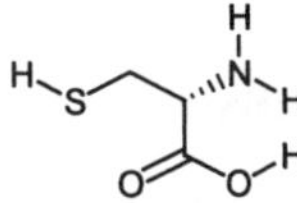

$C_3H_7NO_2S$
Propanoic acid, 2-amino-3-mercapto-, (R)-; Thioserine; L-Alanine, 3-mercapto-; NSC-8746; Half-cystine; (R)-2-Amino-3-mercaptopropanoic acid; β-Mercaptoalanine; α-Amino-β-thiolpropionic acid. Listed on TSCA. Chiral building block. mp = 220°; $[\alpha]^{20}$ = + 8.5° (c = 12, 2N HCl); $[\alpha]_D^{20}$= + 6.5° (c = 2, 5N HCl). *Aceto Corporation; Acros Organics nv; Ajinomoto Co. Inc.; Austin Chemical Company, Inc.; Flamma s.p.a.; Freedom Chemical Diamalt GmbH; KingChem, Inc.; Kyowa Hakko Kogyo Co., Ltd.; Shanghai DSL International Trading Company; Sigma-Aldrich Fine Chemicals; Sun-Orient Chemical Co. Ltd.; Tanabe Seiyaku Co. Ltd.; TCI America; Varsal Instruments, Inc.; Yoneyama Yakuhin Kogyo Co., Ltd.*

**1786 L-Cysteine n-amyl ester hydrochloride**

$C_8H_{18}ClNO_2S$
Chiral building block. *Senn Chemicals AG.*

**1787 L-Cysteine ethyl ester hydrochloride**
868-59-7 212-779-6

$C_5H_{12}ClNO_2S$
Cysteine, ethyl ester, hydrochloride, L-; Ethyl L-cysteinate hydrochloride. Listed on TSCA. Chiral intermediate. mp = 128°; [α] = - 12° (c = 1, 1N HCl). *Acros Organics nv; Austin Chemical Company, Inc.; Kyowa Hakko Kogyo Co., Ltd.; Sigma-Aldrich Fine Chemicals; TCI America.*

**1788 D-Cysteine hydrochloride**
32443-99-5 2787(11) 251-043-9

$C_3H_8ClNO_2S$
Cysteine, hydrochloride, D-; Chiral building block. $[\alpha]_D^{20}$ = - 5° (c = 2, 5N HCl). *Austin Chemical Company, Inc.; Sigma-Aldrich Fine Chemicals; TCI America.*

**1789 L-Cysteine hydrochloride**
52-89-1 2787(11) 200-157-7

$C_3H_8ClNO_2S$
Cysteine, hydrochloride, L-; Listed on TSCA. Chiral building block. $[\alpha]^{23}$ = + 5.2° (c = 2, 5N HCl). *Aceto Corporation; Acros Organics nv; KingChem, Inc.; Kyowa Hakko Kogyo Co., Ltd.; Shanghai DSL International Trading Company; Sigma-Aldrich Fine Chemicals; Sochinaz SA; Tanabe Seiyaku Co. Ltd.; Varsal Instruments, Inc.*

**1790 L-Cysteine hydrochloride monohydrate**
7048-04-6
$C_3H_{10}ClNO_3S$
Cysteine, hydrochloride, monohydrate, L-; (R)-2-Amino-3-mercaptopropionic acid monohydrochloride monohydrate. Chiral building block. $[\alpha]^{20}$ = + 6.0 to + 7.0° (c = 15, 1 N HCl). *Aceto Corporation; Austin Chemical Company, Inc.; KingChem, Inc.; Shanghai DSL International Trading Company; Sochinaz SA; Tanabe Seiyaku Co. Ltd.; TCI America; Varsal Instruments, Inc.*

**1791 L-Cysteine methyl ester hydrochloride**
18598-63-5 5828(12) 242-435-0

$C_4H_9NO_2S{\cdot}H_2O$
Cysteine, methyl ester, hydrochloride, L-; Mecysteine hydrochloride; Acdrile; Actiol; Methyl cysteine hydrochloride; Pectite; Visclair; Zeotin. Chiral building block. mp = 142°; $[\alpha]_D^{20}$ = - 1.8° (c = 10, $CH_3OH$). *Acros Organics nv; Austin Chemical Company, Inc.; Kyowa Hakko Kogyo Co., Ltd.; Sigma-Aldrich Fine Chemicals; TCI America.*

**1792 L-Cysteinesulfinic acid monohydrate**

$C_3H_7NO_4S$
3-Sulfino-L-alanine. Chiral intermediate. mp = 163°; $[\alpha]^{25}$ = + 24° (c = 1, 1N HCl). *Sigma-Aldrich Fine Chemicals.*

**1793 D-(+)-Cystine**
349-46-2 2851(12) 206-486-2

$C_6H_{12}N_2O_4S_2$
Cystine, D-; Chiral building block. mp = 265°; [α] = + 220° (c = 3, 1 N HCl). *Acros Organics nv; Austin Chemical Company, Inc.; Nippon Rikagakuyakuhin; Sigma-Aldrich Fine Chemicals; Varsal Instruments, Inc.*

**1794 L-(-)-Cystine**
56-89-3 2851(12) 200-296-3

$C_6H_{12}N_2O_4S_2$
Propanoic acid, 3,3'-dithiobis[2-amino-, [R-(R*,R*)]-; 3,3'-Dithiobis (2-aminopropionic acid); b,b'-Diamino-b,b'-dicarboxydiethyl disulfide; b,b'-Dithiodialanine; Cystine acid; Dicysteine; L-Alanine, 3,3'-dithiobis-; L-Cysteine disulfide; L-Cystin; Oxidized L-cysteine. Listed on TSCA. Chiral building block. mp = 261°; [α] = - 215° (c = 1, 1N HCl). *Aceto Corporation; Acros Organics nv; Ajinomoto Co. Inc.; Austin Chemical Company, Inc.; KingChem, Inc.; Kyowa Hakko Kogyo Co., Ltd.; Senn Chemicals AG; Shanghai DSL International Trading*

*Company; Sigma-Aldrich Fine Chemicals; Sinochem Ningbo; Sun-Orient Chemical Co. Ltd.; TCI America; Varsal Instruments, Inc.; Yoneyama Yakuhin Kogyo Co., Ltd.*

**1795 L-(-)-Cystine dihydrochloride**
30925-07-6 250-391-9

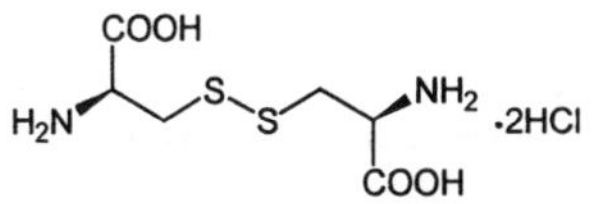

$C_6H_{14}Cl_2N_2O_4S_2$
Cystine, dihydrochloride, L-; Chiral building block. *Austin Chemical Company, Inc.; TCI America; Varsal Instruments, Inc.*

**1796 L-Cystine dimethyl ester dihydrochloride**
32854-09-4 251-261-4

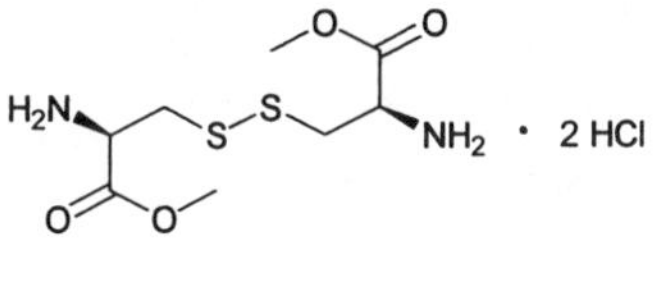

$C_8H_{16}N_2O_4S_2 \cdot 2HCl$
Chiral building block. mp = 182-183°; $[\alpha]^{22}$ = - 33.3° (c = 4, $CH_3OH$). *Acros Organics nv; Austin Chemical Company, Inc.; Sigma-Aldrich Fine Chemicals.*

**1797 (+)-Cytidine**
65-46-3 2855(12) 200-610-9

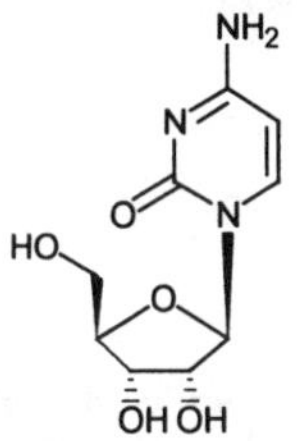

$C_9H_{13}N_3O_5$
Pyrimidin-2(1H)-one, 4-amino-1-β-D-ribofuranosyl-; 4-Amino-1-β-D-ribofuranosyl-2-(1H)-pyrimidione; 4-Amino-1-β-D-ribofuranosyl-2(1H)-pyrimidinone; Cytosine riboside; Cytosine, 1-β-D-ribosyl-; 1-β-D-Ribofuranosylcytosine. Listed on TSCA. Chiral intermediate. mp = 215-216°; $[\alpha]^{25}$ = + 31.5° (c = 0.6, $H_2O$). *Acros Organics nv; Sigma-Aldrich Fine Chemicals.*

**1798 (+)-Cytidine-5-monophosphoric acid hydrate**
200-556-6

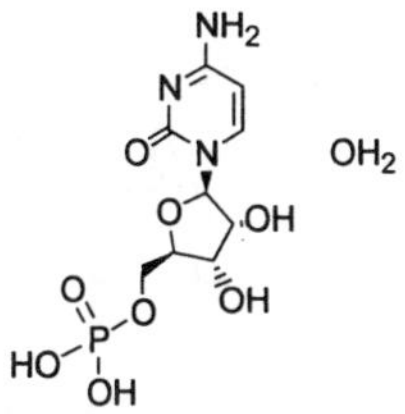

$C_9H_{14}N_3O_8P$
Listed on TSCA. Chiral intermediate. mp = 233°; $[\alpha]^{21}$ = + 21.8° (c = 0.5, $H_2O$). *Sigma-Aldrich Fine Chemicals.*

**1799 (-)-Cytisine**
485-35-8 2858(12) 207-616-0

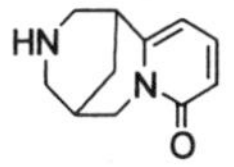

$C_{11}H_{14}N_2O$
Listed on TSCA. Chiral building block. mp = 154-156°; $bp_2$ = 218°; $[\alpha]^{25}$ = - 115° (c = 0.1, $H_2O$). *Sigma-Aldrich Fine Chemicals.*

**1800 (+)-Cytochalasin A**
14110-64-6 237-964-9

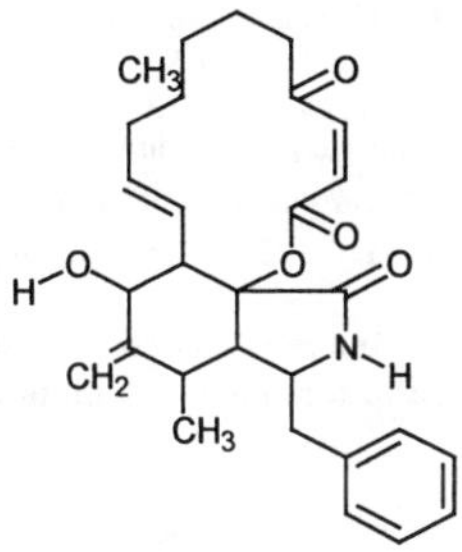

$C_{29}H_{35}NO_5$
24-Oxa[14]cytochalasa-6(12),13,21-triene-1,20,23-trione, 7-hydroxy-16-methyl-10-phenyl-, (7S,13E,16R,21E)-; 7(S)-Hydroxy-16(R)-methyl-10-phenyl-24-oxa[14]cytochalasa-6(12),13(E),21(E)-triene-1,20,23-trione; 2H-Oxacyclotetradecino[2,3-d]isoindole-2,5,18-trione, 16-benzyl-6,7,8,9,10,12a,13,14,15,15a,16,17-dodecahydro-13-hydroxy-9,15-dimethyl-14-methylene-; 5,5-Di. Pharmaceutical or derivative. mp = 193-195°; $[\alpha]^{21}$ = + 83.7° (c = 1, $CH_3OH$). *Acros Organics nv.*

**1801 (+)-Cytochalasin B**
14930-96-2 239-000-2

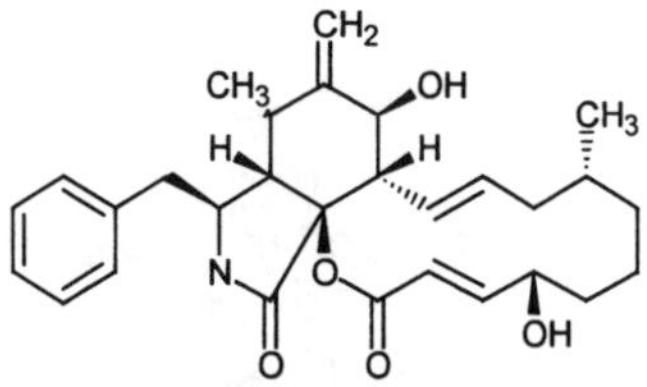

$C_{29}H_{37}NO_5$
24-Oxa[14]cytochalasa-6(12),13,21-triene-1,23-dione, 7,20-dihydroxy-16-methyl-10-phenyl-, (7S,13E,16R,20R,-21E)-; Phomin; 7(S),20(R)-Dihydroxy-16(R)-methyl-10-phenyl-24-oxa[14]cytochalasa-6(12),13(E),21(E)-triene-1,23-dione; 2H-Oxacyclotetradecino[2,3-d]isoindole-2,18(5H)-dione, 16-Benzyl-6,7,8,9,10,12α,13,14,15,-15α,16,17-dodecahydro-5,13-dihydroxy-9,15-dimethyl-. Pharmaceutical or derivative. mp = 221-223°; $[\alpha]^{21}$ = + 86.7° (c = 0.9, $CH_3OH$). *Acros Organics nv.*

**1802 (+)-Cytochalasin E**
36011-19-5 252-835-7

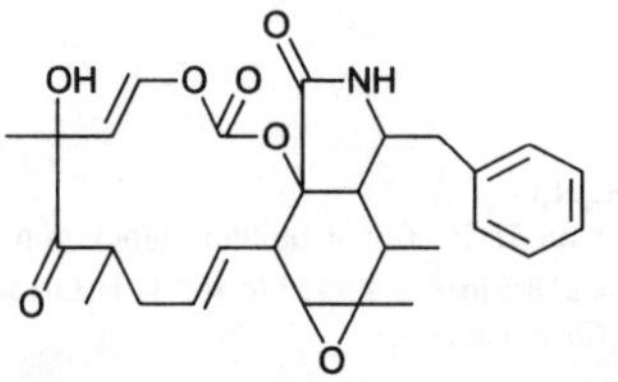

$C_{28}H_{33}NO_7$
[1,3]Dioxacyclotridecino[4,5-d]oxireno[f]isoindole-5,10,-12(4H,6H)-trione, 3,13,14,14α,15,15α,16α,16β-octa-hydro-6-hydroxy-4,6,15,15α-tetramethyl-14-(phenylmethyl)-, (1E,4S,6R,7E,11αS,14S,14αS,15S,15αR,16αS,-16βS)-; 21,23-Dioxa[13]cytochalasa-13,19-diene-1,17,22-trione, 6,7-epoxy-18-hydroxy-16,18-dimethyl-10-phenyl-, (7S,-13E,16S,18R,19E)-. Pharmaceutical or derivative. $[\alpha]^{21}$ = + 22.7° (c = 0.85, $CH_3OH$). *Acros Organics nv.*

**1803 (-)-10-Deacetylbaccatin III (ex taxus baccata) (contains 0.5 mole of methanol)**
32981-86-5

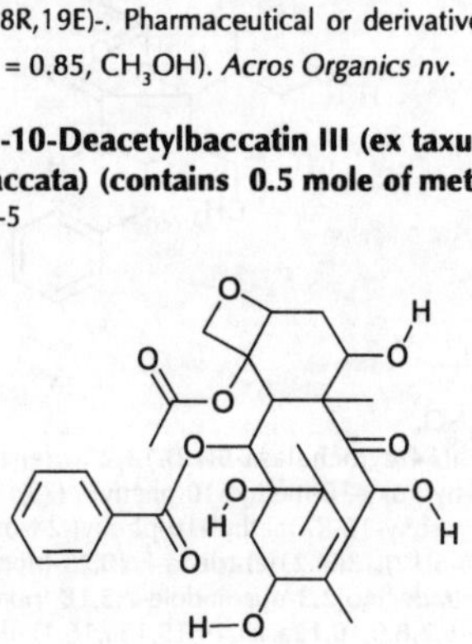

$C_{29}H_{36}O_{10}$
Chiral intermediate. mp = 234-236°; [α] = - 35° (c = 1, EtOH). *Acros Organics nv.*

**1804 Decahydroisoquinoline-(3S)-carboxylic acid**

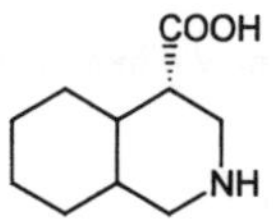

$C_{10}H_{17}NO_2$
Chiral intermediate. *Fischer Chemicals AG.*

**1805 (R)-Decanediol**
87827-60-9

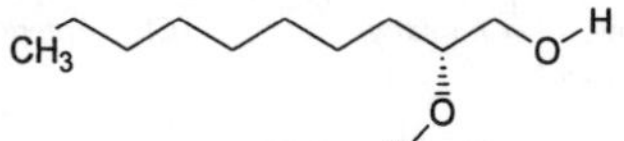

$C_{10}H_{22}O_2$
Chiral building block. mp = 60-62°; $[\alpha]^{23}$ = + 12° (c = 0.5, $CH_3OH$). *Acros Organics nv; Kankyo Kagaku Center Co., Ltd.; Sigma-Aldrich Fine Chemicals.*

**1806 (S)-1,2-Decanediol**
84276-14-2

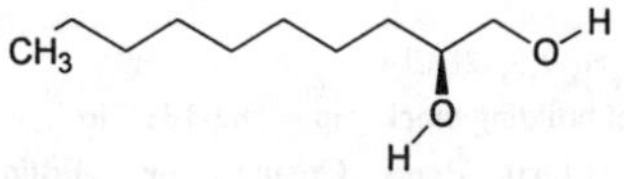

$C_{10}H_{22}O_2$
Chiral building block. mp = 59-61°; $[\alpha]^{24}$ = - 9° (c = 0.5, $CH_3OH$). *Acros Organics nv; Kankyo Kagaku Center Co., Ltd.; Sigma-Aldrich Fine Chemicals.*

**1807 (S)-2-Decanol**
33758-16-6

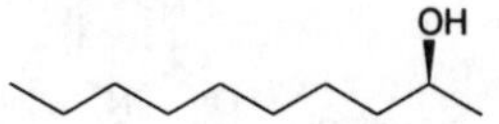

$C_{10}H_{22}O$
Decan-2-ol, (S)-; Chiral building block. bp = 211°; d = 0.827; n = 1.434. *Japan Energy Corporation; Sigma-Aldrich Fine Chemicals.*

**1808 (S)-4-Decanol**

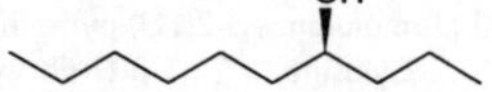

$C_{10}H_{22}O$
Decanol, (4S)-; (S)-1-Propylheptyl alcohol. Chiral building block. *Japan Energy Corporation.*

**1809 N-Decanoyl-D-erythro sphingosine**
111122-57-7

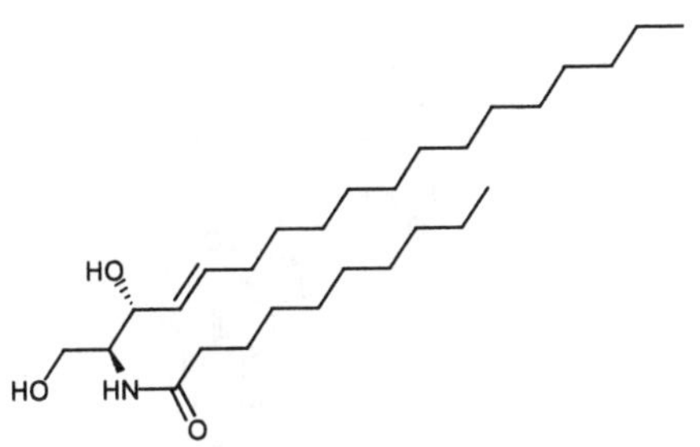


$C_{28}H_{55}NO_3$
Chiral intermediate. *Acros Organics nv.*

**1810 Dehydro-L-ascorbic acid dimer**
Chiral intermediate. *Senn Chemicals AG.*

**1811 (-)-7-Dehydrocholesterol**
434-16-2 2921(12) 207-100-5

$C_{27}H_{44}O$
Cholesta-5,7-dien-3-ol, (3β)-; 7,8-Didehydrocholesterol; Provitamin D3. Listed on TSCA. Pharmaceutical or derivative. $[\alpha]_D^{20}$ = - 118° (c = 1, $CHCl_3$). *Acros Organics nv; Sigma-Aldrich Fine Chemicals.*

**1812 3,4-Dehydro-D-proline**

$C_5H_7NO_2$
Pyrrole-2-carboxylic acid, 2,5-dihydro-, 1H-, (2R)-; (R)-2,5-Dihydro-1H-pyrrole-2-carboxylic acid; 3-Pyrroline-2-carboxylic acid, D-. Chiral building block. *Advanced Asymmetrics, Inc.*

**1813 3,4-Dehydro-L-proline**
4043-88-3 223-738-7

$C_5H_7NO_2$
Pyrrole-2-carboxylic acid, 2,5-dihydro-, 1H-, (2S)-; (S)-2,5-Dihydro-1H-pyrrole-2-carboxylic acid; 3-Pyrroline-2-carboxylic acid, L-. Chiral building block. mp = 248-250°; $[\alpha]_D^{20}$ = - 385° (c = 2, $H_2O$). *Advanced Asymmetrics, Inc.; Sigma-Aldrich Fine Chemicals.*

**1814 2-Deoxy-1,3,6-tri-O-acetyl-α-D-glucopyranose**

$C_{12}H_{18}O_8$
N-Acetyl neuraminic acid. Chiral intermediate. *Senn Chemicals AG.*

**1815 (2-Deoxyadenisinato)(pentamethylcyclopentadienyl)rhodium(III) triflate cyclic trimer octahydrate**

$C_{63}H_{81}F_9N_{15}O_{18}Rh_3S_3$
Chiral catalyst. mp = 150°. *Sigma-Aldrich Fine Chemicals.*

**1816 2'-Deoxyadenosine hydrate**
16373-93-6

$C_{10}H_{13}N_5O_3$
Chiral intermediate. *Acros Organics nv.*

**1817 (+)-Deoxycholic acid, sodium salt**
302-95-4 206-132-7

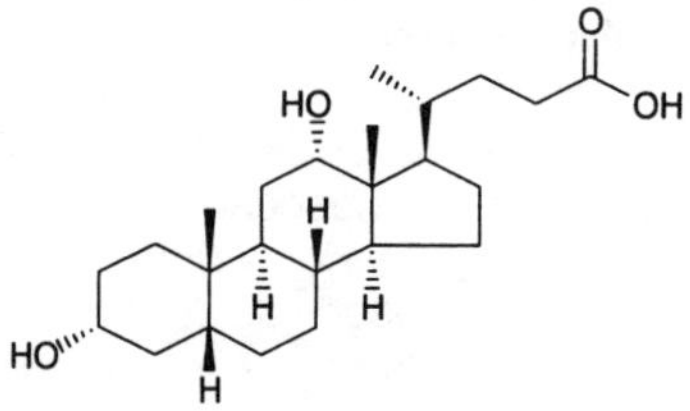

$C_{24}H_{40}O_4$
Cholan-24-oic acid, 3,12-dihydroxy-, monosodium salt, (3α,5β,12α)-; Sodium deoxycholate; 3α,12α-Dihydroxy-5β-cholan-24-oic acid, sodium salt; sodium 3-α,12-α-dihydroxy-5-β-cholan-24-oate; Desoxycholic acid sodium salt. Listed on TSCA. Chiral intermediate. [α] = + 42° (c = 1, EtOH). *Acros Organics nv.*

**1818 2-Deoxycytidine monohydrate**
213-454-1

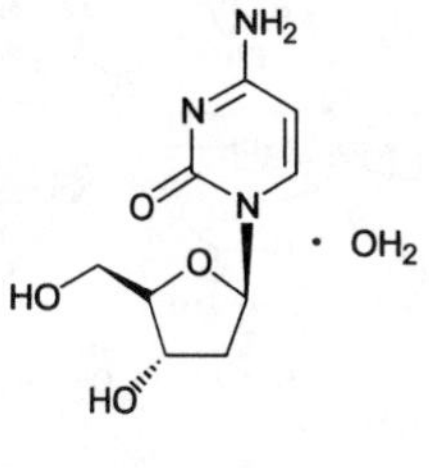

$C_9H_{13}N_3O_4 \cdot H_2O$
Listed on TSCA. Chiral intermediate. mp = 209-211°; $[\alpha]^{25}$ = + 54° (c = 2, $H_2O$). *Sigma-Aldrich Fine Chemicals.*

**1819 2-Deoxycytidine monohydrochoride**
3992-42-5 223-639-9

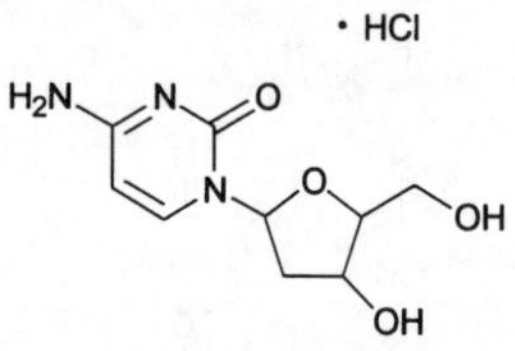

$C_9H_{14}ClN_3O_4$
Chiral intermediate. mp = 161-164°; $[\alpha]^{24}$ = + 54.3° (c = 6.4, $H_2O$). *Sigma-Aldrich Fine Chemicals.*

**1820 2-Deoxy-2-(4,5-dichlorophthalimdio)-glucopyranose 1,3,4,6-tetraacetate**

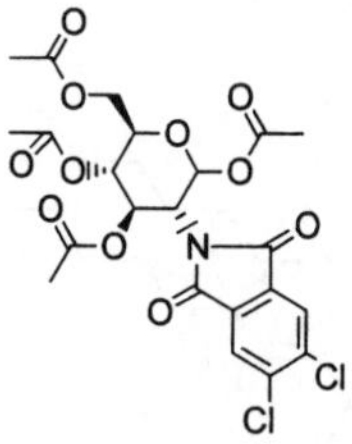

$C_{22}H_{21}Cl_2NO_{11}$
Chiral intermediate. mp = 171-174°; $[\alpha]_D^{20}$ = + 77° (c = 1, $CHCl_3$). *Sigma-Aldrich Fine Chemicals.*

**1821 2-Deoxy-2,2-difluoro-erythro-pentonic acid lactone 3,5-dibenzoate**
122116-01-7

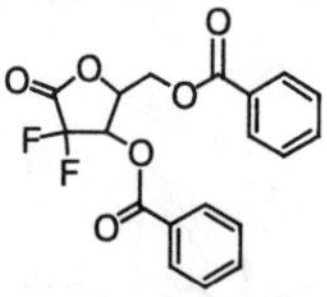

$C_{19}H_{14}F_2O_6$
Chiral intermediate. mp = 117-119°; $[\alpha]^{25}$ = + 62° (c = 1, EtOAc). *Sigma-Aldrich Fine Chemicals.*

**1822 (+)-3-Deoxy-3-fluorothymidine**
25526-93-6

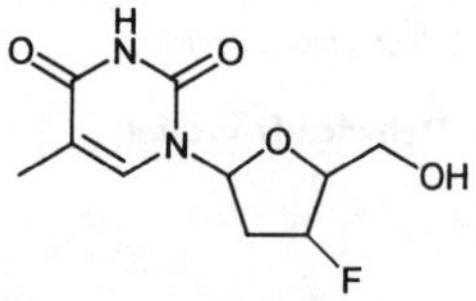

$C_{10}H_{13}FN_2O_4$
Chiral intermediate. mp = 176-178°; $[\alpha]^{35}$ = + 14° (c = 1, $CH_3OH$). *Sigma-Aldrich Fine Chemicals.*

**1823 2-Deoxy-2-fluoro-1,3,5-tri-O-benzyl-α-D-arabinofuranose**

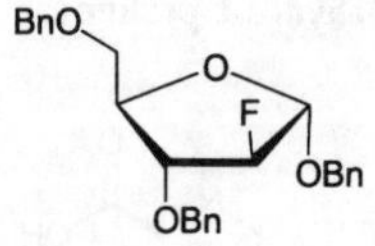

$C_{26}H_{27}FO_4$
Chiral intermediate. $[\alpha]_D^{20}$ = + 76.5° (c = 1, $CHCl_3$). *Pfanstiehl Laboratories, Inc.*

**1824 2-Deoxy-D-galactose**
1949-89-9 217-765-3

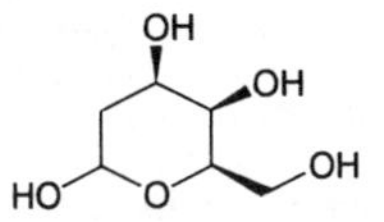

$C_6H_{12}O_5$
Galactose, 2-deoxy-; 2-Deoxy-D-lyxo-hexopyranose; D-lyxo-Hexose, 2-deoxy-. Chiral building block. mp = 109-111°; $[\alpha]_D^{20}$ =+ 60° (c = 4, $H_2O$). *Acros Organics nv; Pfanstiehl Laboratories, Inc.; Senn Chemicals AG; Sigma-Aldrich Fine Chemicals.*

**1825 (-)-1-Deoxy-gluco-hex-1-enopyranose tetrabenzoate**
14125-75-8

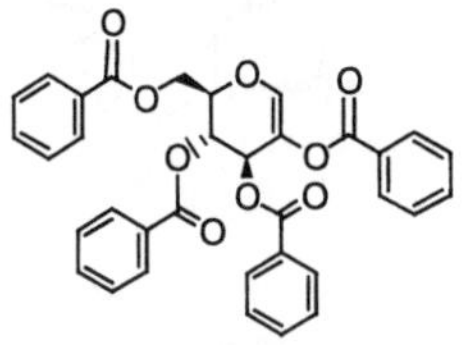

$C_{34}H_{26}O_9$
Chiral intermediate. mp = 122-124°; $[\alpha]^{18}$ = - 88° (c = 1, $CHCl_3$). *Sigma-Aldrich Fine Chemicals.*

**1826 2-Deoxy-D-glucose**
154-17-6 2951(12) 205-823-0

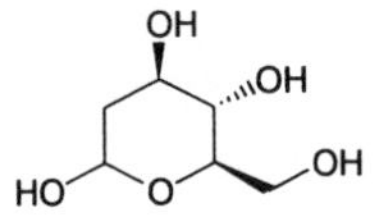

$C_6H_{12}O_5$
D-arabino-Hexose, 2-deoxy-; 2-Deoxy-D-arabino-hexopyranose; 2-Deoxy-D-arabino-hexose; D-Arabino-2-deoxyhexose; Ba 2758. Listed on TSCA. Chiral building block. mp = 146-147°; $[\alpha]_D^{20}$= + 46.6° (c = 2, $H_2O$). *Acros Organics nv; Austin Chemical Company, Inc.; Pfanstiehl Laboratories, Inc.; Sigma-Aldrich Fine Chemicals.*

**1827 (-)-2-Deoxyguanosine hydrate**
213-505-8

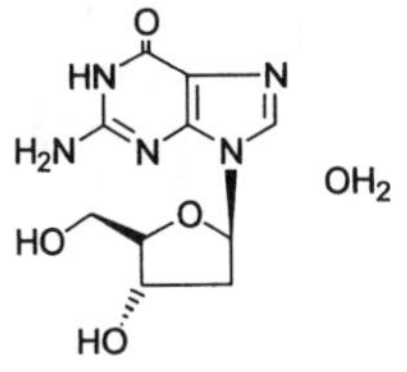

$C_{10}H_{13}N_5O_4$
Listed on TSCA. Chiral intermediate. mp > 300°; $[\alpha]_D^{20}$= - 30° (c = 0.2, $H_2O$). *Sigma-Aldrich Fine Chemicals.*

**1828 (+)-2'-Deoxyinosine**
890-38-0 212-964-1

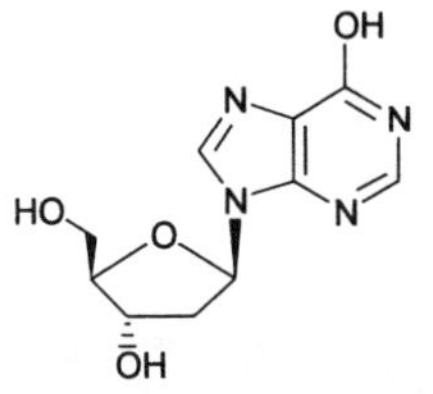

$C_{10}H_{12}N_4O_4$
Purin-(6H)-6-one, 9-(2-deoxy-β-D-erythro-pentofuranosyl)-1,9-dihydro-; Inosine, 2'-deoxy-; 9-(2-Deoxy-β-D-erythro-pentofuranosyl)-1,9-dihydro-6H-purin-6-one; Hypoxanthine deoxyriboside; Hypoxanthine, 9-(2-deoxy-β-D-erythro-pentofuranosyl)-. Chiral intermediate. $[\alpha]^{22}$ = + 21° (c = 1, $H_2O$). *Acros Organics nv.*

**1829 6-Deoxy-L-mannopyranose**
10030-85-0 8171(11)

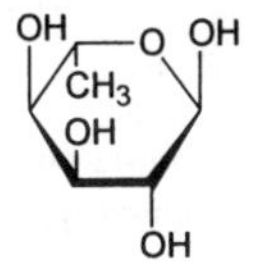

$C_6H_{14}O_6$
L-(+)-Rhamnose monohydrate. Listed on TSCA. Chiral building block. mp = 78-82°; $[\alpha]_D^{20}$= + 8.2° (c = 10, $H_2O$, 1hr). *Kaden Biochemicals GmbH; Sigma-Aldrich Fine Chemicals.*

**1830 (-)-1-Deoxy-1-(methylamino)galactitol**
7115-46-0

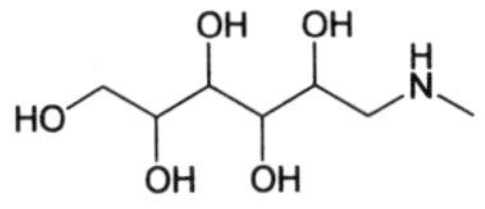

$C_7H_{17}NO_5$
Chiral intermediate. mp = 133-136°; $[\alpha]_D^{20}$= - 16° (c = 1, $CH_3OH$). *Sigma-Aldrich Fine Chemicals.*

**1831 (-)-1-Deoxy-1-nitroiditol hemihydrate**

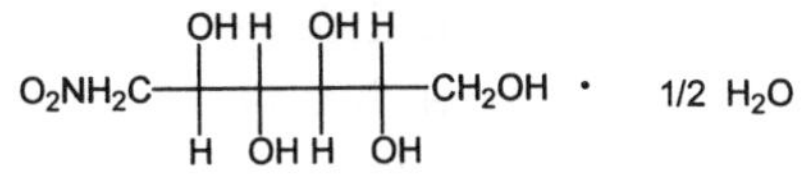

$C_6H_{13}NO_7$
Chiral intermediate. mp = 88°; $[\alpha]_D^{20}$= - 3.5° (c = 2, $H_2O$). *Sigma-Aldrich Fine Chemicals.*

**1832 (+)-1-Deoxy-1-nitroiditol hemihydrate**

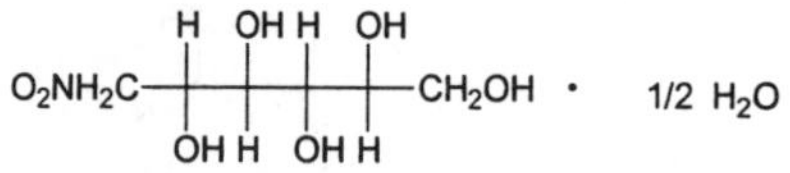

$C_6H_{13}NO_7$
Chiral intermediate. mp = 88°; $[\alpha]_D^{20}$ = + 4.0° (c = 2, $H_2O$). *Sigma-Aldrich Fine Chemicals.*

**1833 (-)-1-Deoxy-1-(octylamino)glucitol**
23323-37-7 245-582-9

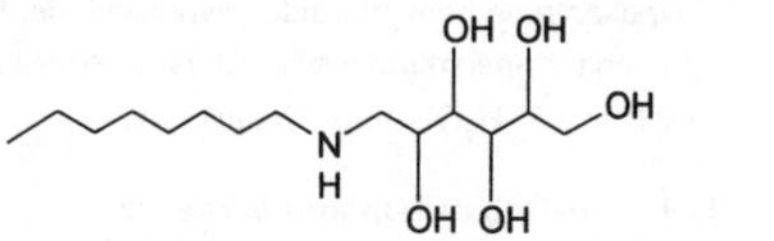

$C_{14}H_{31}NO_5$
Chiral intermediate. mp = 121-124°; $[\alpha]_D^{20}$ = - 15° (c = 1, $CH_3OH$). *Sigma-Aldrich Fine Chemicals.*

**1834 2-Deoxy-2-phthalimido-1,3,4,6-tetra-O-acetyl-β-D-glucopyranose**
10022-13-6
$C_{22}H_{23}NO_{11}$
Chiral intermediate. *Pfanstiehl Laboratories, Inc.*

**1835 (+)-2-Deoxypuridine**
951-78-0 2957(12) 213-455-7

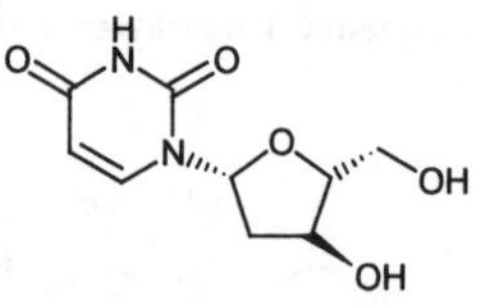

$C_9H_{12}N_2O_5$
Listed on TSCA. Chiral intermediate. mp = 167-169°; $[\alpha]^{26}$ = + 29.8° (c = 2, $H_2O$). *Sigma-Aldrich Fine Chemicals.*

**1836 2-Deoxy-D-ribose**
533-67-5 2955(12) 208-573-0

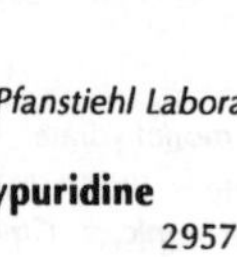

$C_5H_{10}O_4$
Pentose, 2-deoxy-, D-erythro-; 2-Deoxy-D-erythro-pentapyranose; 2-Deoxy-D-arabinose; 2-Deoxy-D-erythro-pentose; D-Ribose, 2-deoxy-; Thyminose; D-2-Deoxyribose. Listed on TSCA. Chiral building block. mp = 89-90°; $[\alpha]_D^{20}$ = - 57° (c = 1, $H_2O$). *Acros Organics nv; Austin Chemical Company, Inc.; Pfanstiehl Laboratories, Inc.; Senn Chemicals AG; Sigma-Aldrich Fine Chemicals; TCI America.*

**1837 O-Desmethyltramadol**
2-[(Dimethylamino)methyl]-1-(3-hydroxyphenyl)cyclohexanol; O-Demethyl tramadol. Pharmaceutical or derivative. mp = 132-134°. *Ultrafine.*

**1838 Diacetone-β-D-fructose**
20880-92-6

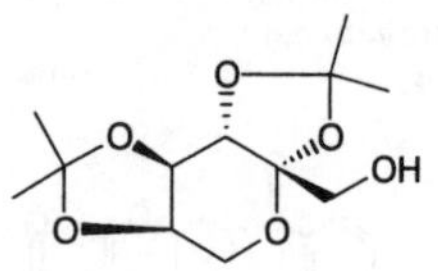

$C_{12}H_{20}O_6$
Chiral intermediate. *Boehringer Ingelheim Pharma KG.*

**1839 (S)-2,4-Diacetoxy-1-bromobutane**
95337-96-5

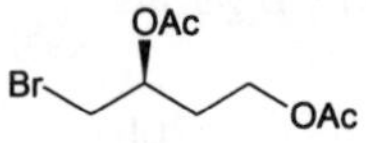

$C_8H_{13}BrO_4$
Butane-1,3-diol, 4-bromo, diacetate, (S)-; (2S)-1-Bromo-2,4-diacetoxybutane; (S)-4-bromo-1,3-butanediol, diacetate. Chiral building block. *Synthon Chiragenics Corporation.*

**1840 (-)-3,4-Di-O-acetyl-6-O-(tert-butyl-dimethylsilyl)galactal**

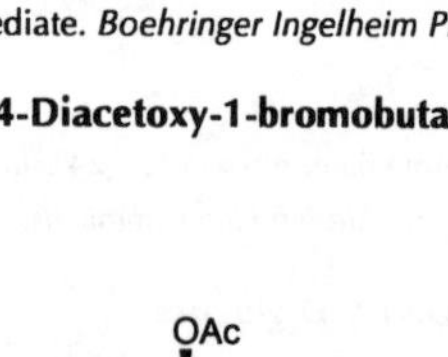

$C_{16}H_{28}O_6Si$
Chiral intermediate. bp = 278°; n = 1.458; $[\alpha]^{22}$ = - 26° (c = 1.1, $CHCl_3$). *Sigma-Aldrich Fine Chemicals.*

**1841 (-)-3,4-Di-O-acetyl-6-O-(tert-butyl-dimethylsilyl)glucal**

$C_{16}H_{28}O_6Si$
Chiral intermediate. d = 1.044; n = 1.456; $[\alpha]_D^{20}$ = - 7° (c = 1.4, $CHCl_3$). *Sigma-Aldrich Fine Chemicals.*

**1842 (-)-3,4-Di-O-acetyl-6-O-(tert-butyl-diphenylsilyl)galactal**

$C_{26}H_{32}O_6Si$
Chiral intermediate. n = 1.526; $[\alpha]^{22}$ = - 4° (c = 1.1, $CHCl_3$). *Sigma-Aldrich Fine Chemicals.*

**1843 (+)-3,4-Di-O-acetyl-6-O-(tert-butyl-diphenylsilyl)glucal**
151797-32-9

$C_{26}H_{32}O_6Si$
Chiral intermediate. n = 1.456; $[\alpha]_D^{20}$ = + 9° (c = 1.1, $CHCl_3$). *Sigma-Aldrich Fine Chemicals.*

**1844 N,S-Diacetyl-L-cysteine methyl ester**
19547-88-7 243-146-2

$C_8H_{13}NO_4S$
Cysteine, N-acetyl-, methyl ester, acetate (ester), L-; Methyl N,S-diacetyl-L-cysteinate; Mucothiol. Chiral intermediate. mp = 97-100°; $[\alpha]_D^{20}$ = - 36.0° (c = 1, $CH_3OH$). *Isochem; Sigma-Aldrich Fine Chemicals.*

**1845 3,4-Di-O-acetyl-6-deoxy-L-glucal**
34819-86-8 252-228-7

$C_{10}H_{14}O_5$
Arabino-Hex-1-enitol, 1,5-anhydro-2,6-dideoxy-, diacetate, L-; Di-O-acetyl-L-rhamnal; 1,5-dianhydro-2,6-dideoxy-L-arabino-hex-1-enitol diacetate; L-Rhamnal diacetate. Chiral intermediate. $bp_{0.06}$ = 68-69°; d = 1.116; n = 1.454; $[\alpha]_D^{20}$ = + 60° (c = 0.5, $CHCl_3$). *Pfanstiehl Laboratories, Inc.; Sigma-Aldrich Fine Chemicals.*

**1846 (-)-3,4-Di-O-acetyl-6-O-(triisopropylsilyl)-galactal**

$C_{19}H_{34}O_6Si$
Chiral intermediate. mp = 25-29°; $[\alpha]^{22}$ = - 24° (c = 3, $CHCl_3$). *Sigma-Aldrich Fine Chemicals.*

**1847 (-)-3,4-Di-O-acetyl-6-O-(triisopropylsilyl)-glucal**

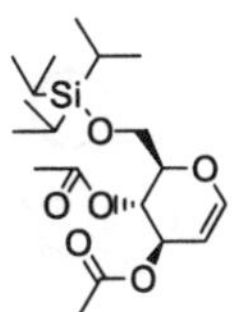

$C_{19}H_{34}O_6Si$
Chiral intermediate. d = 1.042; n = 1.468; $[\alpha]^{25}$ = - 5° (c = 1, $CHCl_3$). *Sigma-Aldrich Fine Chemicals.*

**1848 (+)-3,4-Di-O-acetylfucal**

$C_{10}H_{14}O_5$
Chiral intermediate. mp = 48-52°; $[\alpha]^{24}$ = + 8° (c = 1, acetone). *Sigma-Aldrich Fine Chemicals.*

**1849 Diacetyl-L-tartaric anhydride**
6283-74-5 228-502-7

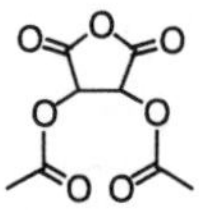

$C_8H_8O_7$
Furan-2,5-dione, 3,4-bis(acetyloxy)dihydro-, (3R,4R)-; (3R-trans)-Dihydro-2,5-dioxofuran-3,4-diyl diacetate; (R,R)-2,3-Diacetoxy-1,4-butanedioic anhydride; Tartaric anhydride, diacetate. Chiral building block; Chiral ligand. mp = 131-133°; $[\alpha]_D^{20}$ = + 59° (c = 6, acetone). *Acros Organics nv; KingChem, Inc.; Senn Chemicals AG; Sigma-Aldrich Fine Chemicals.*

**1850 L-(+)-N,N-Diallyl tartardiamide**
58477-85-3 261-277-3

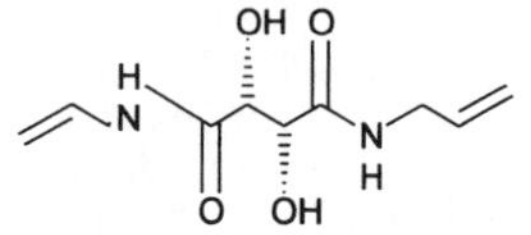

$C_{10}H_{16}N_2O_4$
(+)-N,N'-Diallyltartaramide. Listed on TSCA. Chiral building block; Chiral ligand. mp = 186-188°; $[\alpha]_D^{20}$= + 108° (c = 2.4, $H_2O$). *CU Chemie Uetikon GmbH; Sigma-Aldrich Fine Chemicals.*

**1851 (R)-(+)-2,2'-Diamino-1,1'-binaphthyl**
18741-85-0

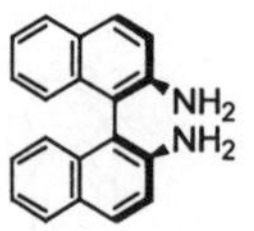

$C_{20}H_{16}N_2$
Chiral ligand. mp = 242-244°; $[\alpha]_D$ = + 157° (c = 1, pyridine). *Sigma-Aldrich Fine Chemicals; SK Energy and Chemical, Inc.; Strem Chemicals, Inc.*

**1852 (S)-(+)-2,2'-Diamino-1,1'-binaphthyl**
18531-95-8

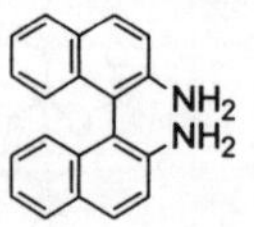

$C_{20}H_{16}N_2$
Chiral ligand. mp = 242-244°; $[\alpha]_D$ = - 157° (c = 1, pyridine). *Asymchem; Sigma-Aldrich Fine Chemicals; Strem Chemicals, Inc.*

**1853 (S,S)-2,3-Diaminobutane sulfate salt**

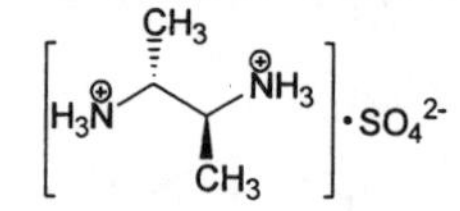

$C_4H_{14}N_2O_4S$
Chiral building block. *Austin Chemical Company, Inc.*

**1854 (S)-(+)-2,4-Diaminobutyric acid dihydrochloride**
1883-09-6 217-542-0

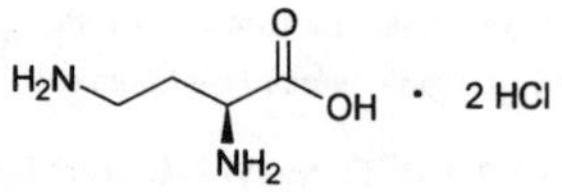

$C_4H_{10}N_2O_2 \cdot 2HCl$
Butanoic acid, 2,4-diamino-, dihydrochloride, (2S)-; Chiral building block. mp = 195°; $[\alpha]$ = + 18.2° (c = 2, 5N HCl); $[\alpha]_D^{20}$= + 12° (c = 1, $H_2O$). *Acros Organics nv; TCI America.*

**1855 (1R,2R)-(-)-1,2-Diaminocyclohexane**
20439-47-8

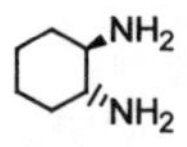

$C_6H_{14}N_2$
(1R)-trans-1,2-Cyclohexanediamine; (R,R)-DACH; (1R,2R)-(-)-1,2-Cyclohexanediamine; (1R)-(-)-trans-1,2-Diaminocyclohexane. Chiral building block; Resolving agent. mp = 43-45°; $bp_5$ = 65°; $bp_{40}$ = 104-110°; $[\alpha]_D^{20}$= - 25.5 ± 1° (c = 5, 1N HCl); $[\alpha]_D$ = - 40.4° (c = 7.13, $C_6H_6$). *Advanced Asymmetrics, Inc.; Arran Chemical Company Ltd.; Austin Chemical Company, Inc.; Kiralchem Ltd.; Lancaster Synthesis Ltd.; Oxford Asymmetry International plc; Sigma-Aldrich Fine Chemicals; Strem Chemicals, Inc.; TCI America; Toray International, Inc.*

**1856 (1S,2S)-(+)-1,2-Diaminocyclohexane**
21436-03-3

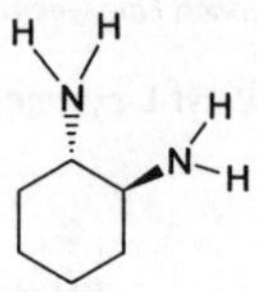

$C_6H_{14}N_2$
(1S)-trans-1,2-Cyclohexanediamine; (S,S)-DACH; (1S,2S)-(+)-1,2-Diphenylethylenediamine; (S,S)-1,2-Cyclohexanediamine. Chiral building block; Resolving agent. mp = 40-43°; $bp_{40}$ = 104-110°; $[\alpha]_D^{20}$= + 25.5 ± 1° (c = 5, 1N HCl); $[\alpha]_D^{20}$= + 43.0 ± 1.0° (c = 3, $CH_3OH$). *Acros Organics nv; Advanced Asymmetrics, Inc.; Arran Chemical Company Ltd.; Austin Chemical Company, Inc.; Kankyo Kagaku Center Co., Ltd.; Kiralchem Ltd.;*

*Lancaster Synthesis Ltd.; Oxford Asymmetry International plc; Sigma-Aldrich Fine Chemicals; Strem Chemicals, Inc.; TCI America; Toray International, Inc.*

**1857 (1R,2R)-(+)-1,2-Diaminocyclohexane-N,N'-bis(2'-diphenylphosphinobenzoyl)**
138517-61-0

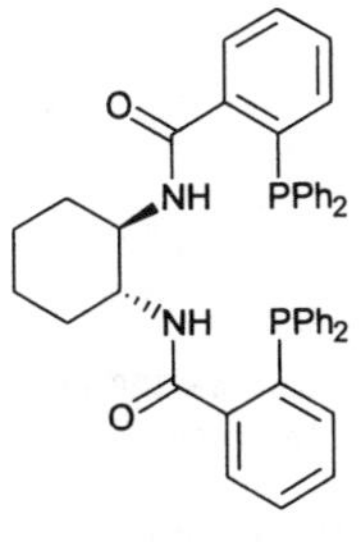

$C_{44}H_{40}N_2O_2P_2$
(+)-Trost Ligand. Chiral ligand. mp = 134-146; $[\alpha]_D$ = + 88° (c = 7, $CH_2Cl_2$). *Strem Chemicals, Inc.*

**1858 (1S,2S)-(-)-1,2-Diaminocyclohexane-N,N'-bis(2'-diphenylphosphinobenzoyl)**
169689-05-8

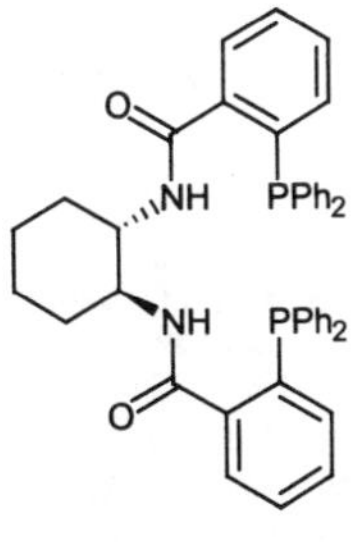

$C_{44}H_{40}N_2O_2P_2$
(-)-Trost Ligand. Chiral ligand. mp = 134-146; $[\alpha]_D$ = - 88° (c = 7, $CH_2Cl_2$). *Strem Chemicals, Inc.*

**1859 (S)-(+)-2,6-Diamino-1-hexanol**
110690-36-3
$C_8H_{16}N_2O$
L(+)-Lysinol. Chiral building block. *Boehringer Ingelheim Pharma KG.*

**1860 (4S,5S)-4,5-Di(aminomethyl)-2,2-dimethyldioxolane**
119322-88-2

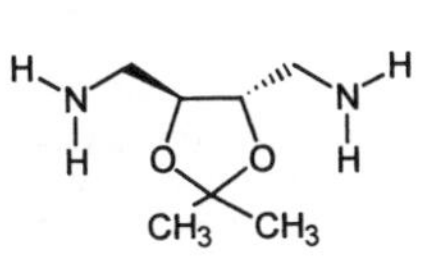

$C_7H_{16}N_2O_2$
Chiral intermediate. *Acros Organics nv.*

**1861 (1R,2R)-trans-Diaminomethylcyclohexane**

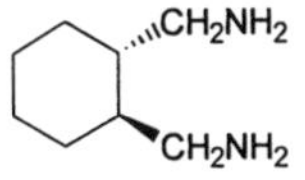

$C_3H_{18}N_2$
Chiral ligand; Chiral building block. *Oxford Asymmetry International plc.*

**1862 (1S,2S)-trans-Diaminomethylcyclohexane**

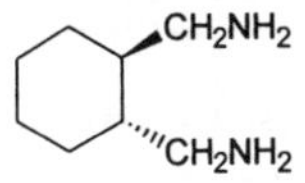

$C_3H_{18}N_2$
Resolving agent; Chiral building block. *Oxford Asymmetry International plc.*

**1863 (R)-1,2-Diaminopropane dihydrochloride**
19777-67-4

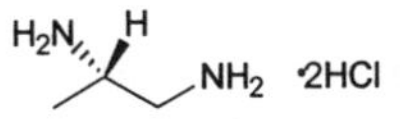

$C_3H_{10}N_2 \cdot 2HCl$
Chiral building block. mp = 241-244°; $[\alpha]_D^{20}$ = + 4.0° (c = 20, $H_2O$). *Austin Chemical Company, Inc.; Sigma-Aldrich Fine Chemicals; Toray International, Inc.*

**1864 (S)-1,2-Diaminopropane dihydrochloride**
19777-66-3

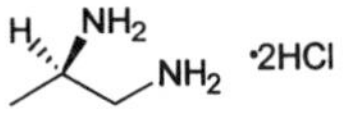

$C_3H_{10}N_2 \cdot 2HCl$
Chiral building block. mp = 227-229°; $[\alpha]^{22}$ = - 4° (c = 20, $H_2O$). *Sigma-Aldrich Fine Chemicals; Toray International, Inc.*

**1865 (R)-(-)-2,3-Diaminopropionic acid hydrochloride**

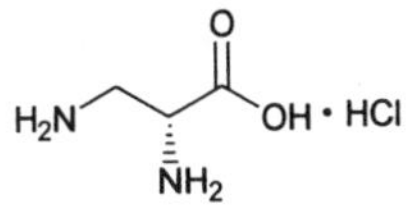

$C_3H_9ClN_2O_2$
Alanine, 3-amino-, monohydrochloride, D-; D-3-Aminoalanine hydrochloride. Chiral building block. *Acros Organics nv; TCI America.*

**1866 (S)-(+)-2,3-Diaminopropionic acid hydrochloride**
1482-97-9

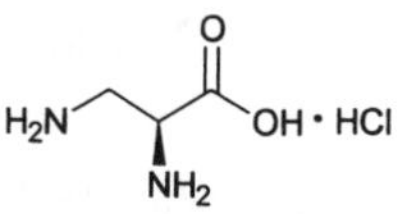

$C_3H_9ClN_2O_2$
Alanine, 3-amino-, monohydrochloride, L-; L-3-Aminoalanine hydrochloride. Chiral building block. *Acros Organics nv; Senn Chemicals AG; TCI America.*

**1867 (1S,4S)-(+)-2,5-Diazabicyclo[2.2.1]heptane dihydrobromide**
132747-20-7

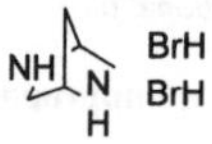

$C_5H_{12}Br_2N_2$
Chiral intermediate. mp > 300°; $[\alpha]^{23}$ = + 22° (c = 1, $H_2O$). *Sigma-Aldrich Fine Chemicals.*

**1868 (-)-2,3-Dibenzoyl-4-dimethyl-aminocarbonyl-1-butyric acid**

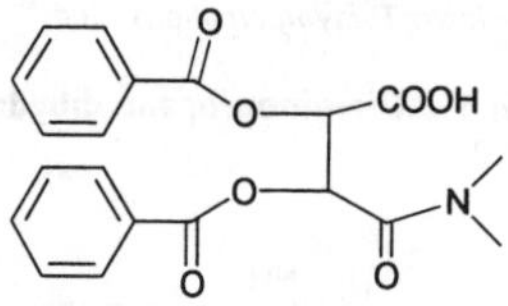

$C_{20}H_{19}O_7N$
Chiral intermediate. $[\alpha]_D^{20}$ = - 76 to -80° (c = 1, EtOH). *Elso Vegyi Industria Rt.*

**1869 (-)-3,6-Di-O-benzoylgalactal**

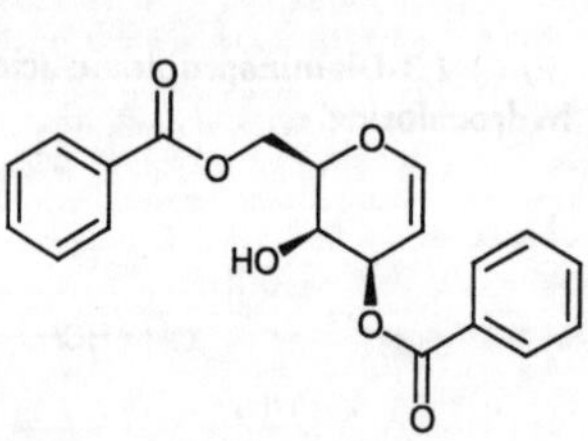

$C_{20}H_{18}O_6$
Chiral intermediate. mp = 122-124°; $[\alpha]^{22}$ = - 104° (c = 1.1, $CHCl_3$). *Sigma-Aldrich Fine Chemicals.*

**1870 (-)-3,6-Di-O-benzoylglucal**
58871-06-0

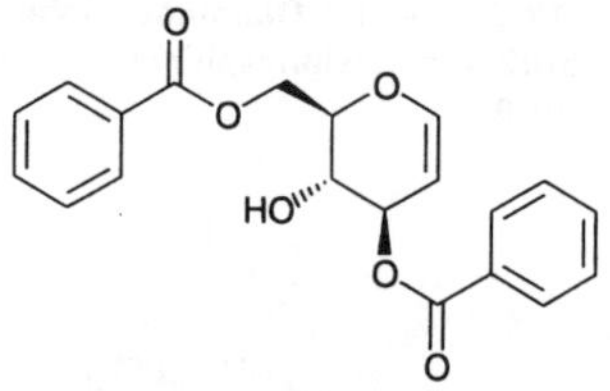

$C_{20}H_{18}O_6$
Chiral intermediate. mp = 129-134°; $[\alpha]^{25}$ = - 102° (c = 1, $CHCl_3$). *Sigma-Aldrich Fine Chemicals.*

**1871 Dibenzoyl-D-tartaric acid**
17026-42-5 241-097-1

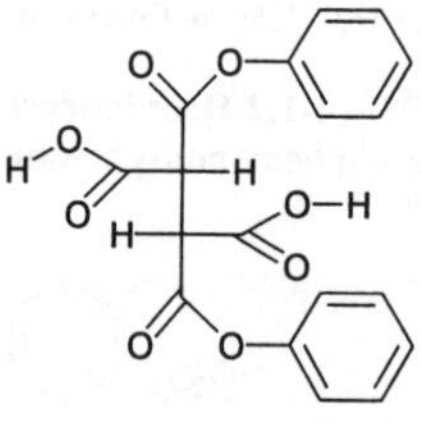

$C_{18}H_{14}O_8$
Butanedioic acid, 2,3-bis(benzoyloxy)-, (2S,3S)-; Tartaric acid, dibenzoate; [S-(R*,R*)]-2,3-bis(benzoyloxy)succinic acid. Listed on TSCA. Chiral building block; Chiral ligand. mp = 154-156°; $[\alpha]^{28}$ = + 116° (c = 9, EtOH). *Acros Organics nv; Elso Vegyi Industria Rt.; Norse Laboratories; Shanghai DSL International Trading Company; Sigma-Aldrich Fine Chemicals; Toray International, Inc.*

**1872 Dibenzoyl-L-(-)-tartaric acid**
2743-38-6 220-374-0

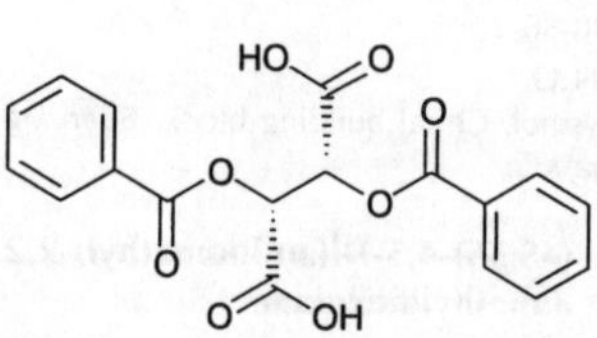

$C_{18}H_{14}O_8$
Butanedioic acid, 2,3-bis(benzoyloxy)-, (2R,3R)-; (R,R)-2,3-Dibenzoyloxysuccinic acid; Tartaric acid, dibenzoate. Listed on TSCA. Chiral building block; Chiral ligand. mp = 152-155°; $[\alpha]_D^{20}$ = - 116° (c = 9, EtOH). *Austin Chemical Company, Inc.; Elso Vegyi Industria Rt.; Knoll AG; Norse Laboratories; Sigma-Aldrich Fine Chemicals; Toray International, Inc.*

**1873 Dibenzoyl-D-tartartic acid anhydride**

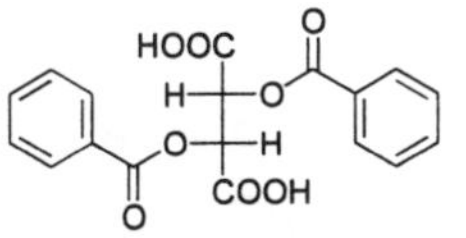

$C_{18}H_{14}O_8$
Furan-2,5-dione, 3,4-bis(benzoyloxy)dihydro-, (-); Tartaric anhydride dibenzoate. Chiral ligand; Chiral building block. mp = 189-196°. *Elso Vegyi Industria Rt.*

**1874 Dibenzoyl-L-tartaric acid anhydride**
17637-11-5 241-621-9

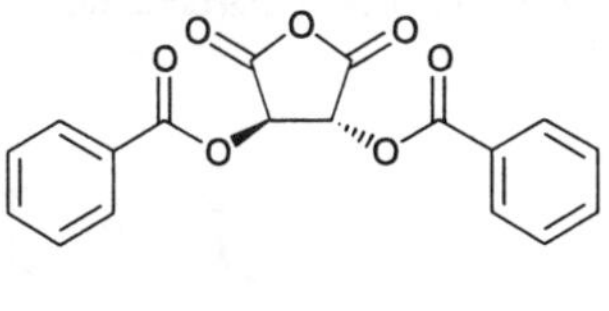

$C_{18}H_{12}O_7$
Furan-2,5-dione, 3,4-bis(benzoyloxy)dihydro-, (+)-; Tartaric anhydride dibenzoate. Chiral building block; Chiral ligand. *Acros Organics nv.*

**1875 L-(-)-Dibenzoyltartaric acid monohydrate**
62708-56-9

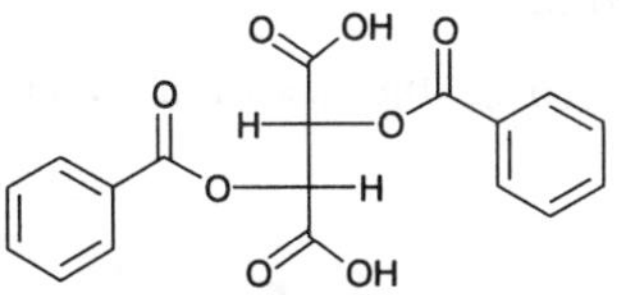

$C_{18}H_{14}O_8$
Butanedioic acid, 2,3-bis(benzoyloxy)-, (2R,3R)-, monohydrate. Chiral building block; Chiral ligand. $[\alpha]$ = - 112.5° (c = 1.5, EtOH). *Acros Organics nv.*

**1876 (S)-2-(N,N-Dibenzylamino)-3-methylbutanol**
111060-54-9
$C_{19}H_{25}NO$
Chiral intermediate. *Sigma-Aldrich Fine Chemicals.*

**1877 (S)-2-(N,N-Dibenzylamino)-4-methylpentanol**
$C_{20}H_{27}NO$
Chiral intermediate. *Sigma-Aldrich Fine Chemicals.*

**1878 (R)-(+)-2-(Dibenzylamino)-1-propanol**

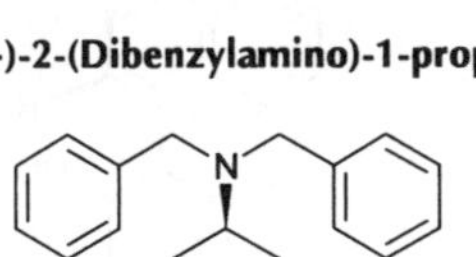

$C_{17}H_{21}NO$
Chiral building block. *Sumitomo Chemcial Co. Ltd.*

**1879 (S)-(+)-2-(Dibenzylamino)-1-propanol**
60479-65-4

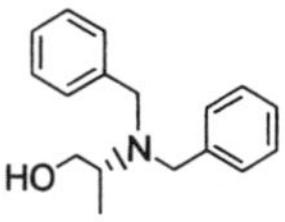

$C_{17}H_{21}NO$
Chiral building block. mp = 46-48°; $bp_{0.1}$ = 142-148°; $[\alpha]_D^{20}$ = + 90° (c = 1, $CHCl_3$). *Sigma-Aldrich Fine Chemicals; Sumitomo Chemcial Co. Ltd.*

**1880 (R)-(-)-2-(Dibenzylamino)-3-phenyl-1-propanol**
$C_{23}H_{25}NO$
Chiral building block. *Sigma-Aldrich Fine Chemicals.*

**1881 (S)-(+)-2-Dibenzylamino-3-phenyl-1-propanol**
111060-52-7
$C_{23}H_{25}NO$
N,N-Dibenzyl-L-phenylalaninol; Benzenepropanol, β-S-[(bis-phenyl)amino]-. Chiral intermediate. mp = 72-74°; $[\alpha]_D^{20}$ = + 8.5 ± 0.5° (c = 4, $CH_3OH$). *Fischer Chemicals AG; Great Lakes Fine Chemicals; Lancaster Synthesis Ltd.; Sigma-Aldrich Fine Chemicals; Synthetech, Inc.*

**1882 (-)-N,N-Dibenzyl-O-(tert-butyl-dimethylsilyl)-serine benzyl ester**

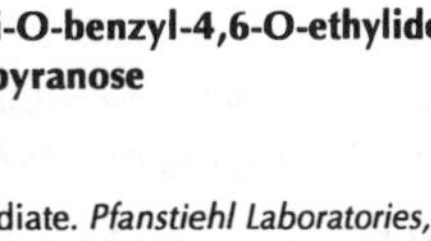

$C_{30}H_{39}NO_3Si$
Chiral intermediate. bp = 341-346°; d = 1.052; n = 1.536; $[\alpha]_D^{20}$ = - 55° (c = 1, $CH_3OH$). *Sigma-Aldrich Fine Chemicals.*

**1883 2,3-Di-O-benzyl-4,6-O-ethylidene-β-D-glucopyranose**
170078-65-6
$C_{22}H_{26}O_6$
Chiral intermediate. *Pfanstiehl Laboratories, Inc.*

**1884 2,3-Di-O-benzyl-D-glucopyranose**
18933-71-6
$C_{20}H_{24}O_6$
Chiral intermediate. *Pfanstiehl Laboratories, Inc.*

**1885 (S)-(-)-1,2-Di-O-benzylglycerol**
20196-71-8
$C_{17}H_{20}O_3$
Chiral intermediate. $bp_{0.4}$ = 153-156°; d = 1.08; n = 1.55; $[\alpha]_D^{20}$ = - 18° (c = 1, $CHCl_3$). *Sigma-Aldrich Fine Chemicals.*

**1886 (-)-2,4:3,5-Di-O-benzylidene aldehydo ribose hydrate**
32580-00-0

$C_{19}H_{18}O_5$
Chiral intermediate. mp = 147-149°; $[\alpha]^{23}$ = - 45° (c = 4, $CHCl_3$). *Sigma-Aldrich Fine Chemicals.*

**1887 1,3,4,6-Di-O-benzylidene-D-mannitol**

$C_{22}H_{22}O_6$
Chiral intermediate. *Senn Chemicals AG.*

**1888 (2R,3R)-(+)-1,4-Dibenzyloxy-2,3-butanediol**
91604-41-0

$C_{18}H_{22}O_4$
1,4-di-O-benzyl-D-threitol. Intermediate. mp = 55-57°; $[\alpha]_D^{20}$ = + 6° (c = 5, $CHCl_3$). *Sigma-Aldrich Fine Chemicals.*

**1889 (2S,3S)-(-)-1,4-Dibenzyloxy-2,3-butanediol**
17401-06-8

$C_{18}H_{22}O_4$
1,4-di-O-benzyl-L-threitol. Intermediate. mp = 55-57°; $[\alpha]_D^{20}$ = - 6° (c = 5, $CHCl_3$). *Sigma-Aldrich Fine Chemicals.*

**1890 N,N-Dibenzylserine benzyl ester**

$C_{24}H_{25}NO_3$
Chiral intermediate. bp = 290°; d = 1.06; n = 1.573; $[\alpha]_D^{20}$ = - 105° (c = 1, $CH_3OH$). *Sigma-Aldrich Fine Chemicals.*

**1891 (-)-N,N-Dibenzyltartaramide**
108321-43-3

$C_{18}H_{20}N_2O_4$
Chiral intermediate; Chiral ligand. mp = 198-200°; $[\alpha]_D^{20}$ = - 83° (c = 5.5, pyridine). *Sigma-Aldrich Fine Chemicals.*

**1892 (+)-N,N-Dibenzyltartaramide**
88393-56-0

$C_{18}H_{20}N_2O_4$
Chiral ligand; Chiral building block. mp = 198-200°; $[\alpha]_D^{20}$ = + 83° (c = 5.5, pyridine). *Sigma-Aldrich Fine Chemicals.*

**1893 (S)-3,3'-Dibromo-1,1'-bi-2-naphthol**

$C_{20}H_{12}Br_2O_2$
Chiral ligand. *Kankyo Kagaku Center Co., Ltd.*

**1894 (R)-3,3'-Dibromo-1,1'-bi-2-naphthol**

$C_{20}H_{12}Br_2O_2$
Chiral ligand. *Kankyo Kagaku Center Co., Ltd.*

**1895 (R)-(-)-6,6-Dibromo-1,1-bi-2-naphthol**
65283-60-5

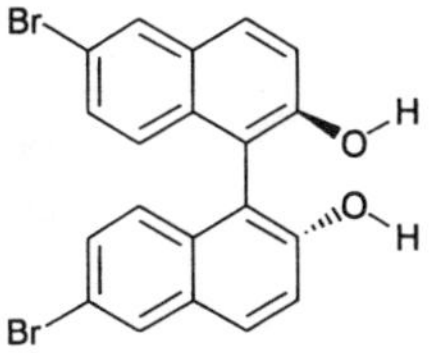

$C_{20}H_{12}Br_2O_2$
Chiral intermediate. mp = 195-199°; $[\alpha]_D^{20}$ = - 49° (c = 1.8, HOAc). *Acros Organics nv; Kankyo Kagaku Center Co., Ltd.; Sigma-Aldrich Fine Chemicals.*

**1896 (S)-(+)-6,6'-Dibromo-1,1'-bi-2-naphthol**
80655-81-8

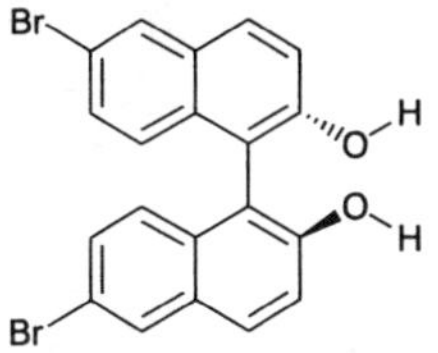

$C_{20}H_{12}Br_2O_2$
(S)-(+)-6,6'-Dibromo-1,1'-bi-2-naphthol. Chiral intermediate. mp = 195-199°; $[\alpha]_D^{20}$ = + 49° (c = 1.8, THF). *Acros Organics nv; Kankyo Kagaku Center Co., Ltd.; Sigma-Aldrich Fine Chemicals.*

**1897 (S)-1,4-Dibromo-2-butanol**
64028-90-6

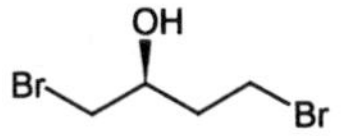

$C_4H_8Br_2O$
Butan-2-ol, 1,4-dibromo-, (S)-; Chiral building block. *Acros Organics nv; Synthon Chiragenics Corporation.*

**1898 3,9-Dibromo-(+)-camphor**
10293-10-4

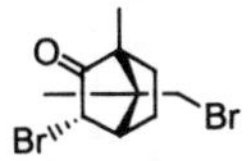

$C_{10}H_{14}Br_2O$
Resolving agent. mp = 156-159°; $[\alpha]^{19}$ = + 100° (c = 1, $CHCl_3$). *Sigma-Aldrich Fine Chemicals.*

**1899 (R)-6,6'-Dibromo-2,2'-dibutoxy-1,1'-binaphthyl**

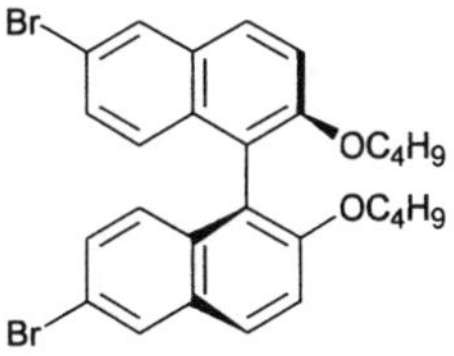

$C_{28}H_{28}Br_2O_2$
Chiral intermediate. *Kankyo Kagaku Center Co., Ltd.*

**1900 (S)-6,6'-Dibromo-2,2'-dibutoxy-1,1'-binaphthyl**

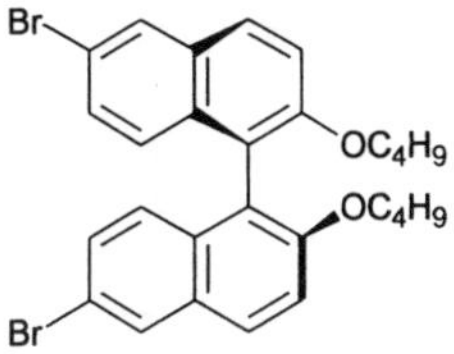

$C_{28}H_{28}Br_2O_2$
Chiral ligand. *Kankyo Kagaku Center Co., Ltd.*

**1901 (R)-3,3'-Dibromo-2,2'-dibutoxy-5,5',6,6',7,7',8,8'-octahydro-1,1'-binaphthyl**

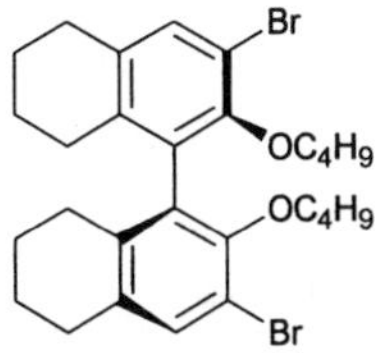

$C_{28}H_{36}Br_2O_2$
Chiral ligand. *Kankyo Kagaku Center Co., Ltd.*

**1902 (S)-3,3'-Dibromo-2,2'-dibutoxy-5,5',6,6',7,7',8,8'-octahydro-1,1'-binaphthyl**

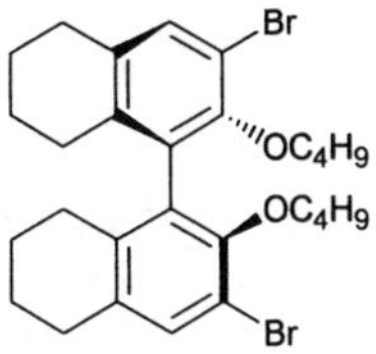

$C_{28}H_{36}Br_2O_2$
Chiral ligand. *Kankyo Kagaku Center Co., Ltd.*

**1903 (R)-6,6'-Dibromo-2,2'-diethoxy-1,1'-binaphthyl**

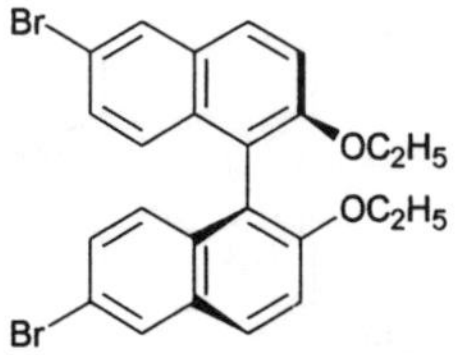

$C_{24}H_{20}Br_2O_2$
Chiral intermediate. *Kankyo Kagaku Center Co., Ltd.*

**1904 (S)-6,6'-Dibromo-2,2'-diethoxy-1,1'-binaphthyl**

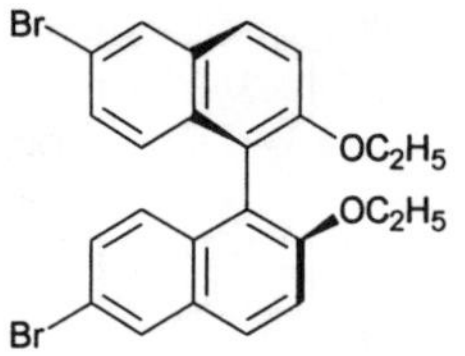

$C_{24}H_{20}Br_2O_2$
Chiral ligand. *Kankyo Kagaku Center Co., Ltd.*

**1905 (R)-3,3'-Dibromo-2,2'-diethoxy-5,5',6,6',7,7',8,8'-octahydro-1,1'-binaphthyl**

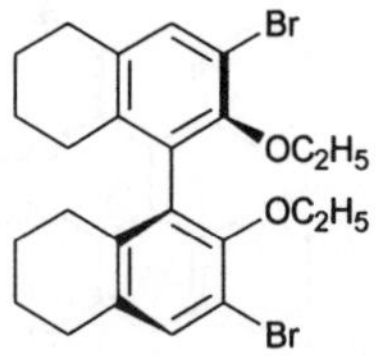

$C_{24}H_{28}Br_2O_2$
Chiral ligand. *Kankyo Kagaku Center Co., Ltd.*

**1906 (S)-3,3'-Dibromo-2,2'-diethoxy-5,5',6,6',7,7',8,8'-octahydro-1,1'-binaphthyl**

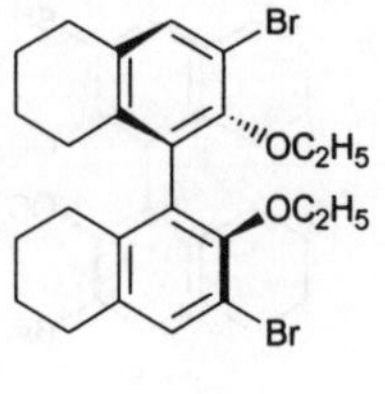

$C_{24}H_{28}Br_2O_2$
Chiral ligand. *Kankyo Kagaku Center Co., Ltd.*

**1907 (R)-6,6'-Dibromo-2,2'-diheptyloxy-1,1'-binaphthyl**

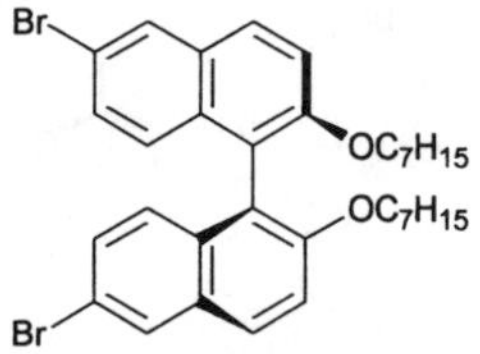

$C_{34}H_{40}Br_2O_2$
Chiral intermediate. *Kankyo Kagaku Center Co., Ltd.*

**1908 (S)-6,6'-Dibromo-2,2'-diheptyloxy-1,1'-binaphthyl**

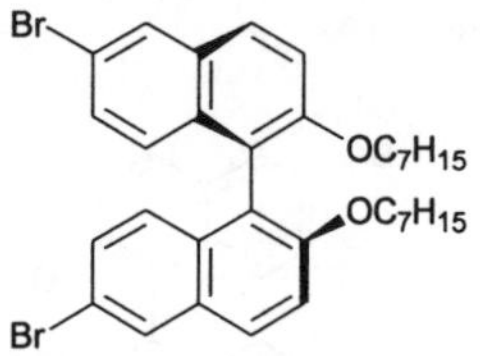

$C_{34}H_{40}Br_2O_2$
Chiral ligand. *Kankyo Kagaku Center Co., Ltd.*

**1909 (R)-3,3'-Dibromo-2,2'-diheptyloxy-5,5',6,6',7,7',8,8'-octahydro-1,1'-binaphthyl**

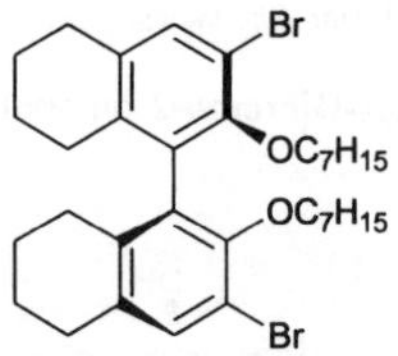

$C_{34}H_{48}Br_2O_2$
Chiral ligand. *Kankyo Kagaku Center Co., Ltd.*

**1910 (S)-3,3'-Dibromo-2,2'-diheptyloxy-5,5',6,6',7,7',8,8'-octahydro-1,1'-binaphthyl**

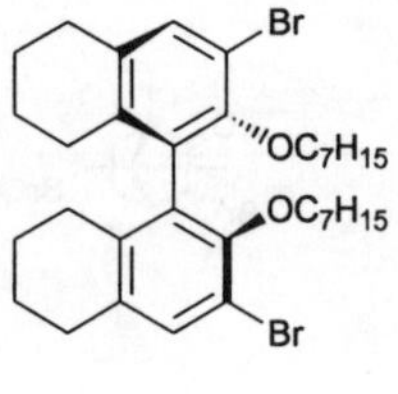

$C_{34}H_{48}Br_2O_2$
Chiral ligand. *Kankyo Kagaku Center Co., Ltd.*

**1911 (R)-6,6'-Dibromo-2,2'-dihexyloxy-1,1'-binaphthyl**

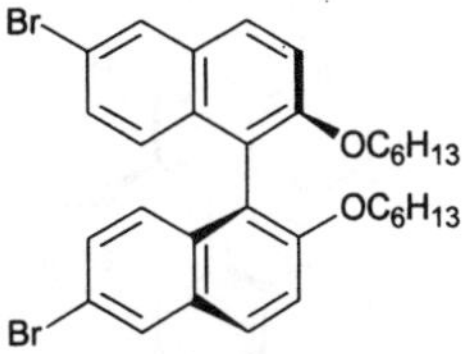

$C_{32}H_{36}Br_2O_2$
Chiral intermediate. *Kankyo Kagaku Center Co., Ltd.*

**1912 (S)-6,6'-Dibromo-2,2'-dihexyloxy-1,1'-binaphthyl**

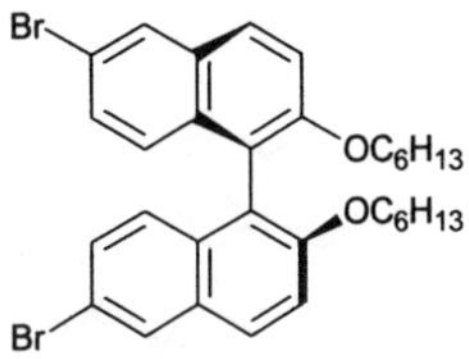

$C_{32}H_{36}Br_2O_2$
Chiral ligand. *Kankyo Kagaku Center Co., Ltd.*

**1913 (R)-3,3'-Dibromo-2,2'-dihexyloxy-5,5',6,6',7,7',8,8'-octahydro-1,1'-binaphthyl**

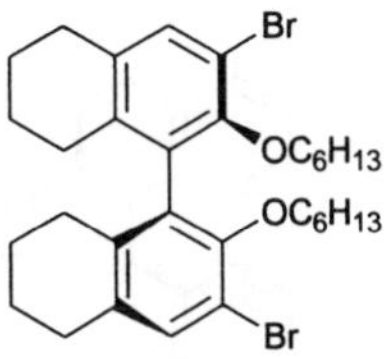

$C_{32}H_{44}Br_2O_2$
Chiral ligand. *Kankyo Kagaku Center Co., Ltd.*

**1914 (S)-3,3'-Dibromo-2,2'-dihexyloxy-5,5',6,6',7,7',8,8'-octahydro-1,1'-binaphthyl**

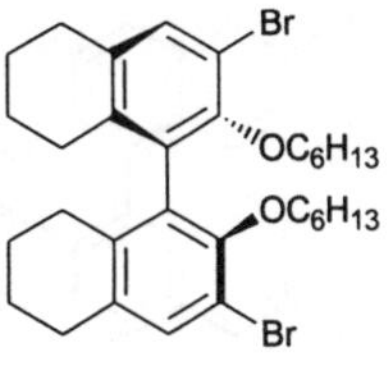

$C_{32}H_{44}Br_2O_2$
Chiral ligand. *Kankyo Kagaku Center Co., Ltd.*

**1915 (R)-3,3'-Dibromo-2,2'-dimethoxy-1,1'-binaphthyl**

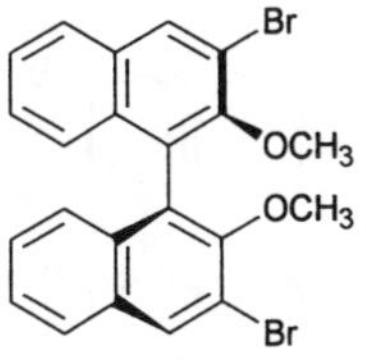

$C_{22}H_{16}Br_2O_2$
Chiral ligand. *Kankyo Kagaku Center Co., Ltd.*

**1916 (S)-3,3'-Dibromo-2,2'-dimethoxy-1,1'-binaphthyl**

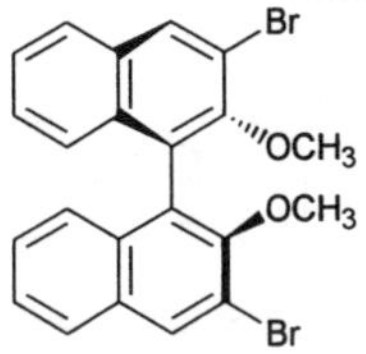

$C_{22}H_{16}Br_2O_2$
Chiral ligand. *Kankyo Kagaku Center Co., Ltd.*

**1917 (R)-6,6'-Dibromo-2,2'-dimethoxy-1,1'-binaphthyl**

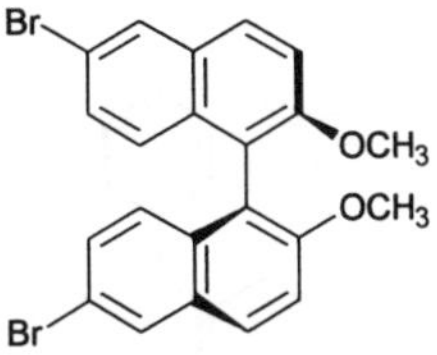

$C_{22}H_{16}Br_2O_2$
Chiral intermediate. *Kankyo Kagaku Center Co., Ltd.*

**1918 (S)-6,6'-Dibromo-2,2'-dimethoxy-1,1'-binaphthyl**

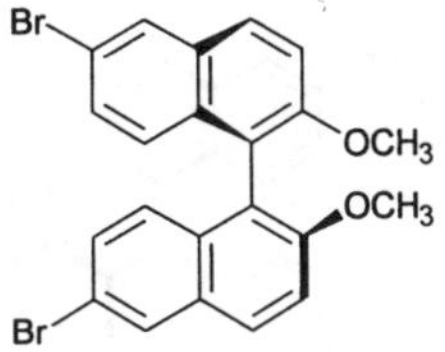

$C_{22}H_{16}Br_2O_2$
Chiral ligand. *Kankyo Kagaku Center Co., Ltd.*

**1919 (R)-3,3'-Dibromo-2,2'-dimethoxy-5,5',6,6',7,7',8,8'-octahydro-1,1'-binaphthyl**

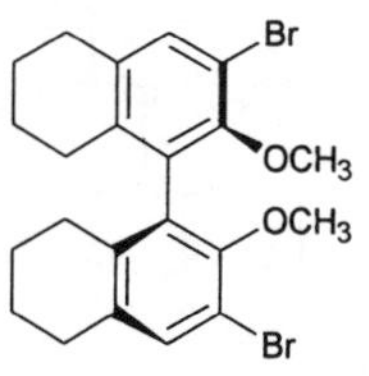

$C_{22}H_{24}Br_2O_2$
Chiral ligand. *Kankyo Kagaku Center Co., Ltd.*

**1920 (S)-3,3'-Dibromo-2,2'-dimethoxy-5,5',6,6',7,7',8,8'-octahydro-1,1'-binaphthyl**

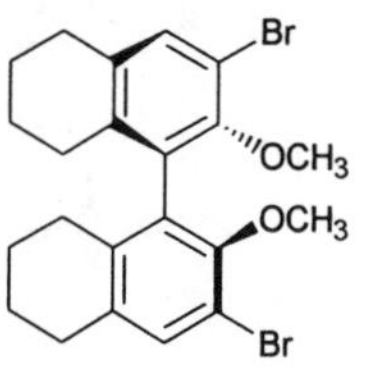

$C_{22}H_{24}Br_2O_2$
Chiral ligand. *Kankyo Kagaku Center Co., Ltd.*

**1921 (R)-6,6'-Dibromo-2,2'-dioctyloxy-1,1'-binaphthyl**

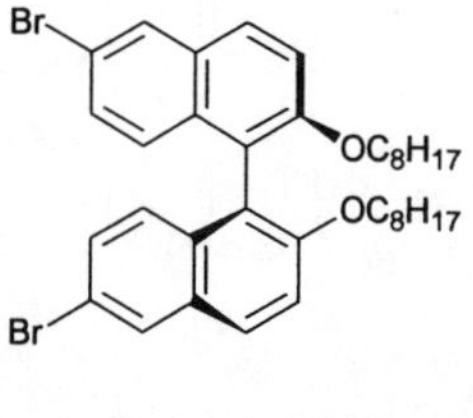

$C_{36}H_{44}Br_2O_2$
Chiral intermediate. *Kankyo Kagaku Center Co., Ltd.*

**1922 (S)-6,6'-Dibromo-2,2'-dioctyloxy-1,1'-binaphthyl**

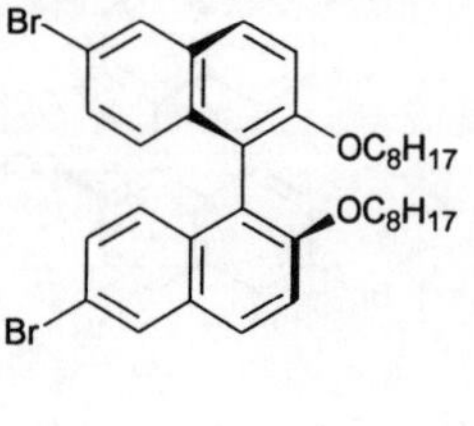

$C_{36}H_{44}Br_2O_2$
Chiral ligand. *Kankyo Kagaku Center Co., Ltd.*

**1923 (R)-3,3'-Dibromo-2,2'-dioctyloxy-5,5',6,6',7,7',8,8'-octahydro-1,1'-binaphthyl**

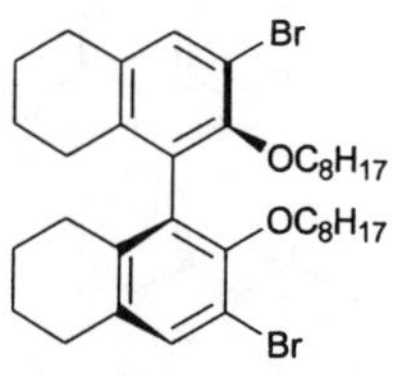

$C_{36}H_{52}Br_2O_2$
Chiral ligand. *Kankyo Kagaku Center Co., Ltd.*

**1924 (S)-3,3'-Dibromo-2,2'-dioctyloxy-5,5',6,6',7,7',8,8'-octahydro-1,1'-binaphthyl**

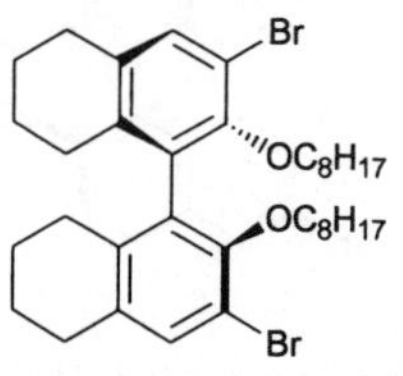

$C_{36}H_{52}Br_2O_2$
Chiral ligand. *Kankyo Kagaku Center Co., Ltd.*

**1925 (R)-6,6'-Dibromo-2,2'-dipentyloxy-1,1'-binaphthyl**

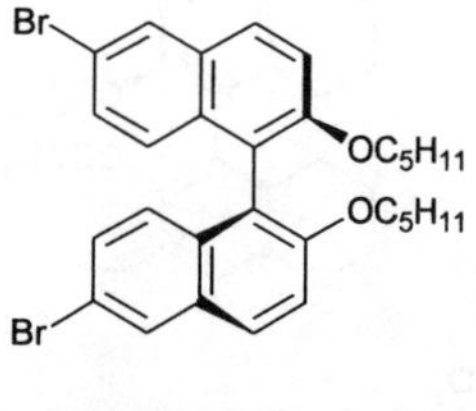

$C_{30}H_{32}Br_2O_2$
Chiral intermediate. *Kankyo Kagaku Center Co., Ltd.*

**1926 (S)-6,6'-Dibromo-2,2'-dipentyloxy-1,1'-binaphthyl**

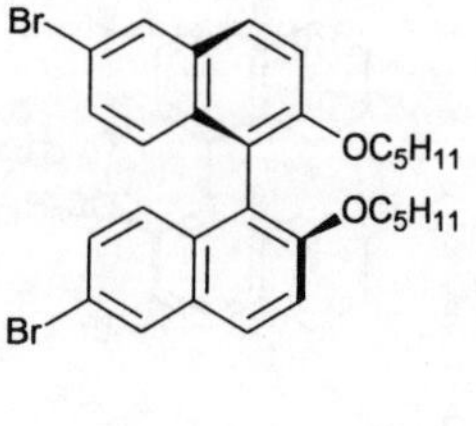

$C_{30}H_{32}Br_2O_2$
Chiral ligand. *Kankyo Kagaku Center Co., Ltd.*

**1927 (R)-3,3'-Dibromo-2,2'-dipentyloxy-5,5',6,6',7,7',8,8'-octahydro-1,1'-binaphthyl**

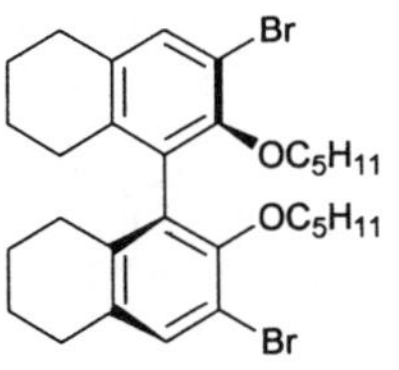

$C_{30}H_{40}Br_2O_2$
Chiral ligand. *Kankyo Kagaku Center Co., Ltd.*

**1928 (S)-3,3'-Dibromo-2,2'-dipentyloxy-5,5',6,6',7,7',8,8'-octahydro-1,1'-binaphthyl**

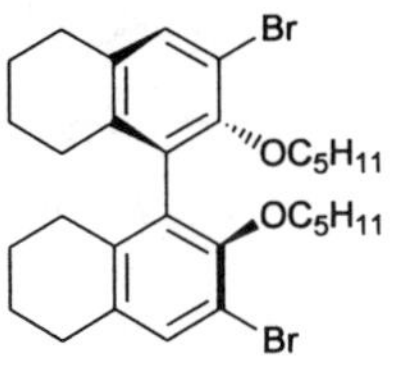

$C_{30}H_{40}Br_2O_2$
Chiral ligand. *Kankyo Kagaku Center Co., Ltd.*

**1929 (R)-6,6'-Dibromo-2,2'-dipropoxy-1,1'-binaphthyl**

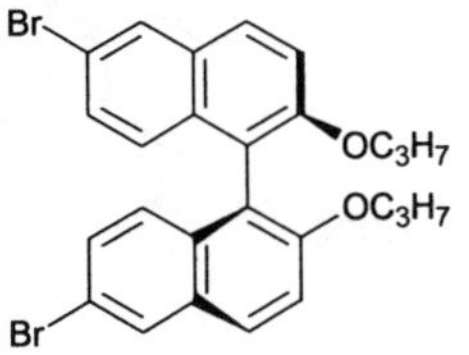

$C_{26}H_{24}Br_2O_2$
Chiral intermediate. *Kankyo Kagaku Center Co., Ltd.*

**1930 (S)-6,6'-Dibromo-2,2'-dipropoxy-1,1'-binaphthyl**

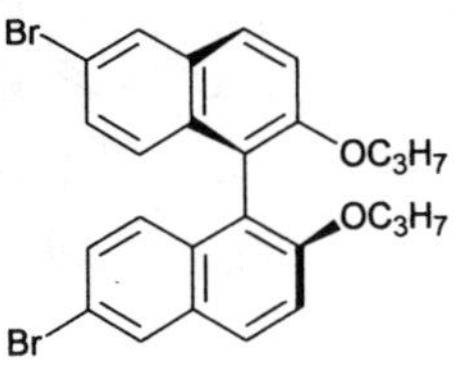

$C_{26}H_{24}Br_2O_2$
Chiral ligand. *Kankyo Kagaku Center Co., Ltd.*

**1931 (R)-3,3'-Dibromo-2,2'-dipropoxy-5,5',6,6',7,7',8,8'-octahydro-1,1'-binaphthyl**

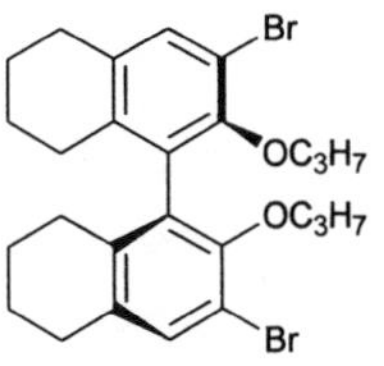

$C_{26}H_{32}Br_2O_2$
Chiral ligand. *Kankyo Kagaku Center Co., Ltd.*

**1932 (S)-3,3'-Dibromo-2,2'-dipropoxy-5,5',6,6',7,7',8,8'-octahydro-1,1'-binaphthyl**

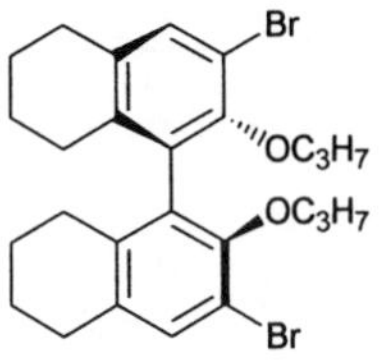

$C_{26}H_{32}Br_2O_2$
Chiral ligand. *Kankyo Kagaku Center Co., Ltd.*

**1933 (R)-6,6'-Dibromo-2,2'-methylenedioxy-1,1'-binaphthyl**

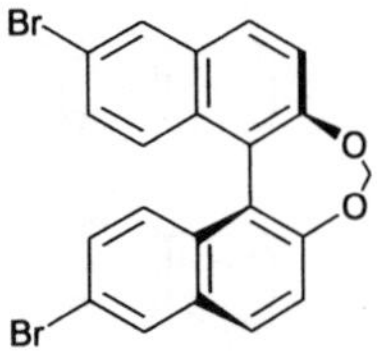

$C_{21}H_{12}Br_2O_2$
Chiral intermediate. *Kankyo Kagaku Center Co., Ltd.*

**1934 (S)-6,6'-Dibromo-2,2'-methylenedioxy-1,1'-binaphthyl**

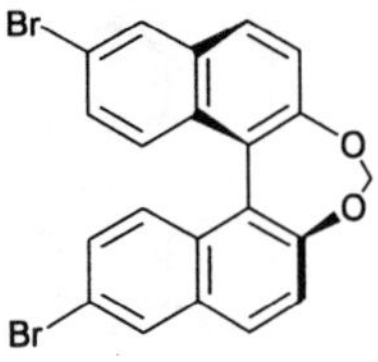

$C_{21}H_{12}Br_2O_2$
Chiral ligand. *Kankyo Kagaku Center Co., Ltd.*

**1935 (R)-3,3'-Dibromo-2,2'-methylenedioxy-5,5',6,6',7,7',8,8'-octahydro-1,1'-binaphthyl**

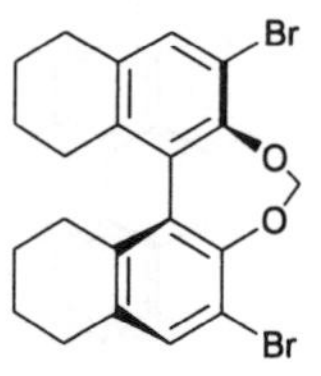

$C_{21}H_{20}Br_2O_2$
Chiral ligand. *Kankyo Kagaku Center Co., Ltd.*

**1936 (S)-3,3'-Dibromo-2,2'-methylenedioxy-5,5',6,6',7,7',8,8'-octahydro-1,1'-binaphthyl**

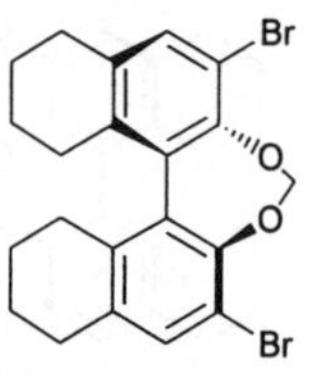

$C_{21}H_{20}Br_2O_2$
Chiral ligand. *Kankyo Kagaku Center Co., Ltd.*

**1937 (R)-3,3'-Dibromo-5,5',6,6',7,7',8,8'-octahydro-1,1'-bi-2-naphthol**
65355-08-0

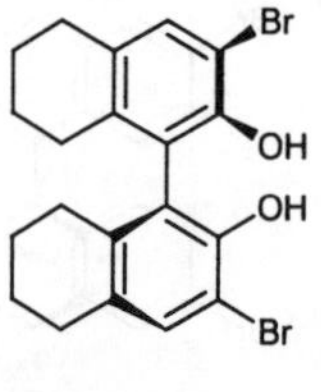

$C_{20}H_{20}Br_2O_2$
Chiral ligand. *Kankyo Kagaku Center Co., Ltd.*

**1938 (S)-3,3'-Dibromo-5,5',6,6',7,7',8,8'-octahydro-1,1'-bi-2-naphthol**

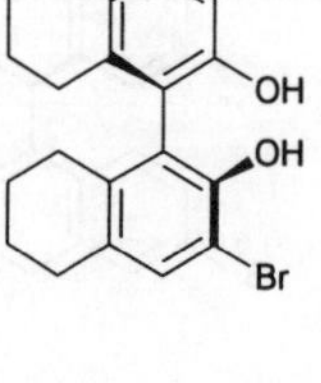

$C_{20}H_{20}Br_2O_2$
Chiral ligand. *Kankyo Kagaku Center Co., Ltd.*

**1939 (R)-2,2'-Dibutoxy-1,1'-binaphthalene-6,6'-dicarbaldehyde**

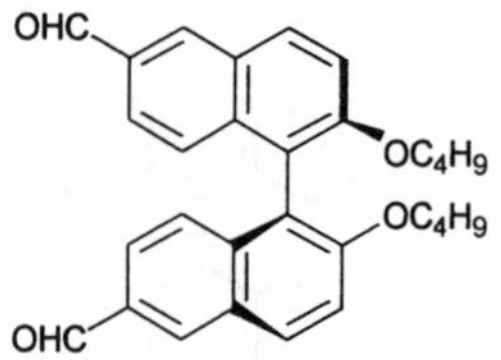

$C_{30}H_{30}O_4$
Chiral ligand. *Kankyo Kagaku Center Co., Ltd.*

**1940 (S)-2,2'-Dibutoxy-1,1'-binaphthalene-6,6'-dicarbaldehyde**

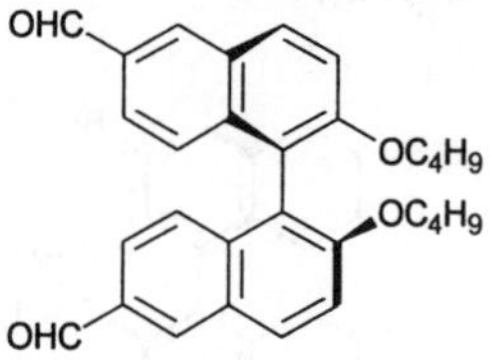

$C_{30}H_{30}O_4$
Chiral ligand. *Kankyo Kagaku Center Co., Ltd.*

**1941 (R)-2,2'-Dibutoxy-1,1'-binaphthalene-6,6'-dicarboxylic acid**

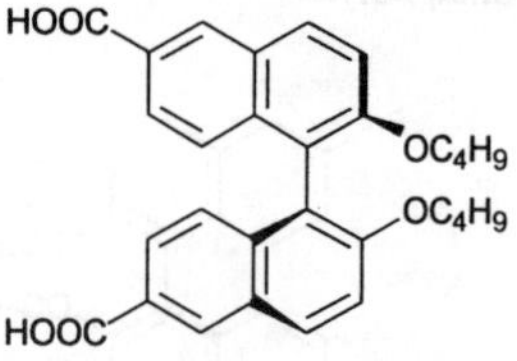

$C_{30}H_{30}O_6$
Chiral ligand. *Kankyo Kagaku Center Co., Ltd.*

**1942 (S)-2,2'-Dibutoxy-1,1'-binaphthalene-6,6'-dicarboxylic acid**

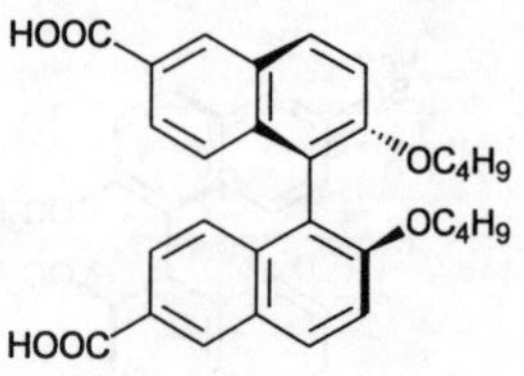

$C_{30}H_{30}O_6$
Chiral ligand. *Kankyo Kagaku Center Co., Ltd.*

**1943 (R)-2,2'-Dibutoxy-1,1'-binaphthyl**

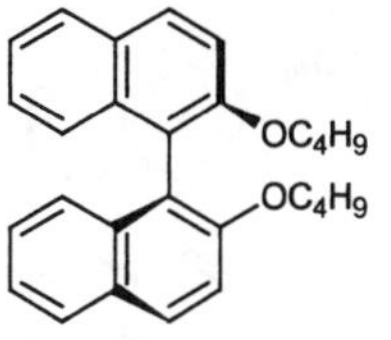

$C_{28}H_{30}O_2$
Chiral ligand. *Kankyo Kagaku Center Co., Ltd.*

**1944 (S)-2,2'-Dibutoxy-1,1'-binaphthyl**

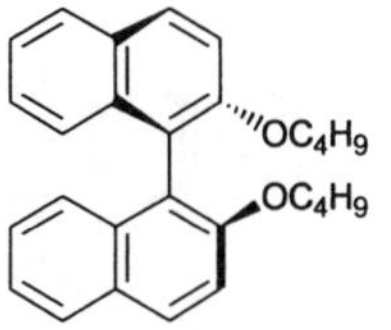

$C_{28}H_{30}O_2$
Chiral ligand. *Kankyo Kagaku Center Co., Ltd.*

**1945 (R)-2,2'-Dibutoxy-5,5',6,6',7,7',8,8'-octahydro-1,1'-binaphthyl**

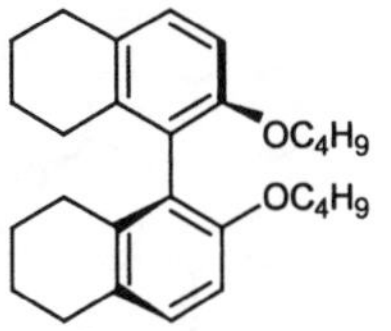

$C_{28}H_{38}O_2$
Chiral ligand. *Kankyo Kagaku Center Co., Ltd.*

**1946 (S)-2,2'-Dibutoxy-5,5',6,6',7,7',8,8'-octahydro-1,1'-binaphthyl**

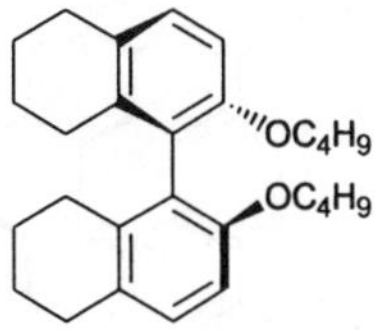

$C_{28}H_{38}O_2$
Chiral ligand. *Kankyo Kagaku Center Co., Ltd.*

**1947 (1R,2S)-2-Di-n-butylamino-1-phenyl-1-propanol**
115651-77-9

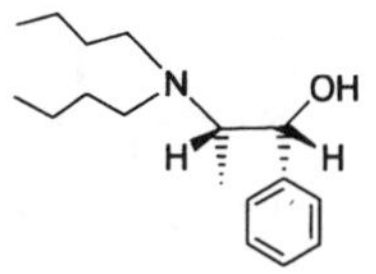

$C_{17}H_{29}NO$
(+)-α-[1-(N,N-Di-n-butylamino)ethyl]benzyl alcohol; (+)-N,N-Di-n-butylnorephedrin. Chiral intermediate. $bp_{0.10}$ = 121°; d = 0.94; n = 1.501; $[\alpha]_D^{20}$ = + 21° (c = 2, $CHCl_3$). *Sigma-Aldrich Fine Chemicals; TCI America.*

**1948 (1S,2R)-2-Di-n-butylamino-1-phenyl-1-propanol**
114389-70-7

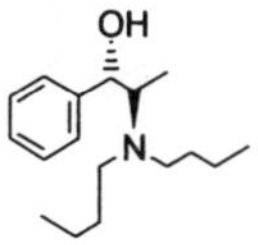

$C_{17}H_{29}NO$
(-)-α-[1-(N,N-Di-n-butylamino)ethyl]benzyl alcohol; (-)-N,N-Di-n-butylnorephedrin. Chiral intermediate. $bp_{2.0}$ = 170°; d = 0.948; n =1.502; $[\alpha]^{22}$ = - 21° (c = 2, $CHCl_3$). *Sigma-Aldrich Fine Chemicals; TCI America.*

**1949 (R)-3,3'-Dibutyl-1,1'-bi-2-naphthol**

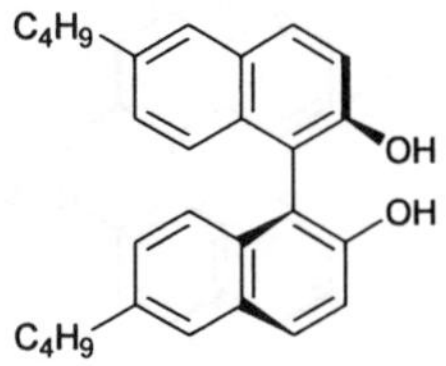

$C_{28}H_{30}O_2$
Chiral intermediate. *Kankyo Kagaku Center Co., Ltd.*

**1950 (S)-3,3'-Dibutyl-1,1'-bi-2-naphthol**

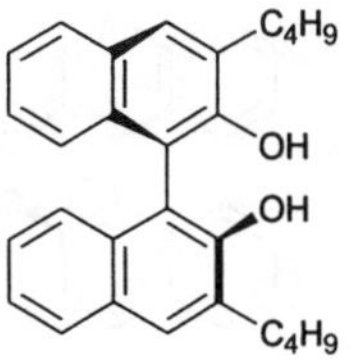

$C_{28}H_{30}O_2$
Chiral ligand. *Kankyo Kagaku Center Co., Ltd.*

**1951 (R)-6,6'-Dibutyl-1,1'-bi-2-naphthol**

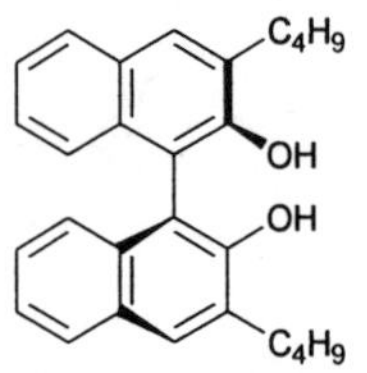

$C_{28}H_{30}O_2$
Chiral ligand. *Kankyo Kagaku Center Co., Ltd.*

**1952 (S)-6,6'-Dibutyl-1,1'-bi-2-naphthol**

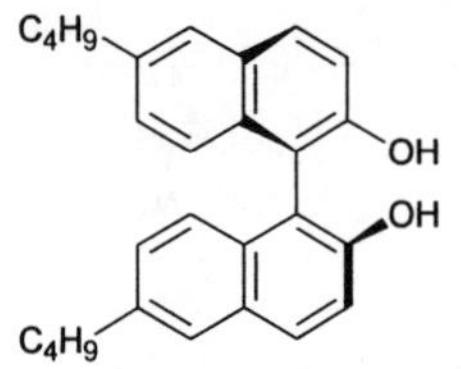

$C_{28}H_{30}O_2$
Chiral ligand. *Kankyo Kagaku Center Co., Ltd.*

**1953 (R)-2,2'-Dibutyl-1,1'-binaphthyl**

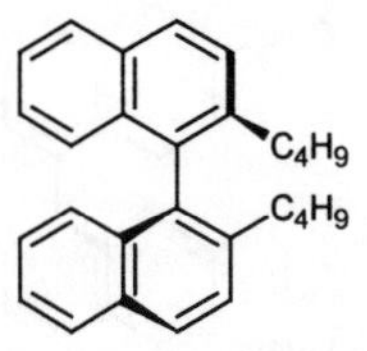

$C_{28}H_{30}$
Chiral ligand. *Kankyo Kagaku Center Co., Ltd.*

**1954 (S)-2,2'-Dibutyl-1,1'-binaphthyl**

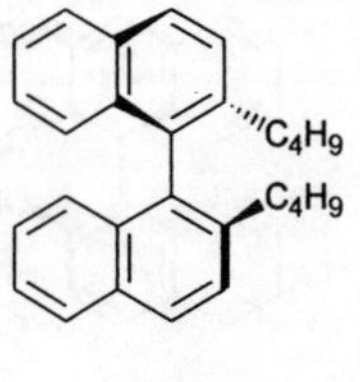

$C_{28}H_{30}$
Chiral ligand. *Kankyo Kagaku Center Co., Ltd.*

**1955 (R)-3,3'-Dibutyl-2,2'-dimethoxy-1,1'-binaphthyl**

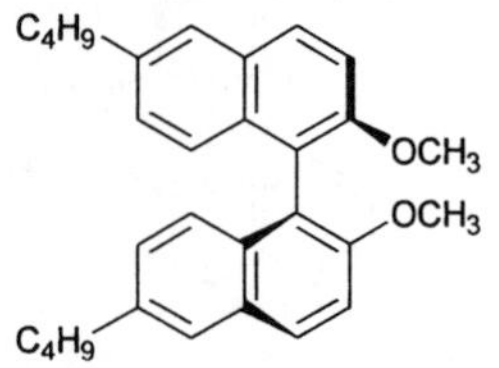

$C_{30}H_{34}O_2$
Chiral intermediate. *Kankyo Kagaku Center Co., Ltd.*

**1956 (S)-3,3'-Dibutyl-2,2'-dimethoxy-1,1'-binaphthyl**

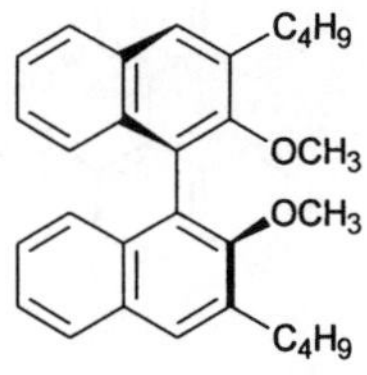

$C_{30}H_{34}O_2$
Chiral ligand. *Kankyo Kagaku Center Co., Ltd.*

**1957 (R)-6,6'-Dibutyl-2,2'-dimethoxy-1,1'-binaphthyl**

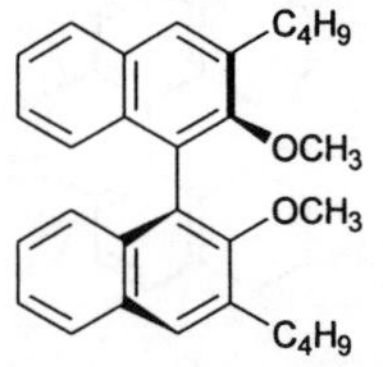

$C_{30}H_{34}O_2$
Chiral ligand. *Kankyo Kagaku Center Co., Ltd.*

**1958 (S)-6,6'-Dibutyl-2,2'-dimethoxy-1,1'-binaphthyl**

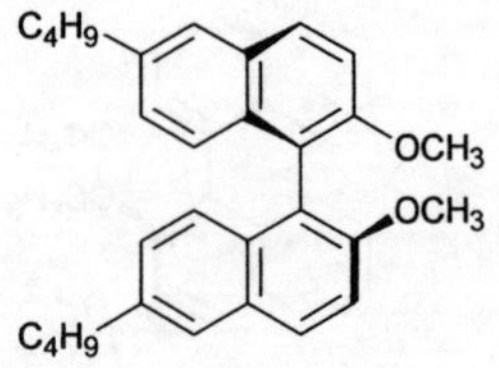

$C_{30}H_{34}O_2$
Chiral ligand. *Kankyo Kagaku Center Co., Ltd.*

**1959 (R)-3,3'-Dibutyl-2,2'-dimethoxy-5,5',6,6',7,7',8,8'-octahydro-1,1'-binaphthyl**

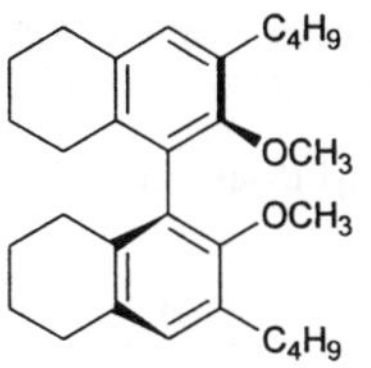

$C_{30}H_{42}O_2$
Chiral ligand. *Kankyo Kagaku Center Co., Ltd.*

**1960 (S)-3,3'-Dibutyl-2,2'-dimethoxy-5,5',6,6',7,7',8,8'-octahydro-1,1'-binaphthyl**

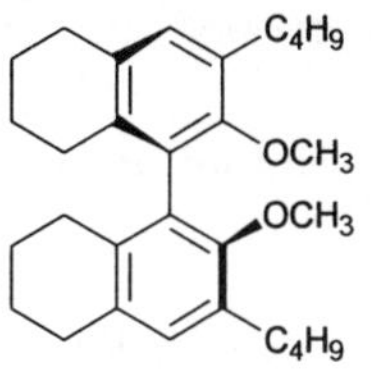

$C_{30}H_{42}O_2$
Chiral ligand. *Kankyo Kagaku Center Co., Ltd.*

**1961 (-)-3,6-Di-O-(tert-Butyldimethylsilyl)-galactal**

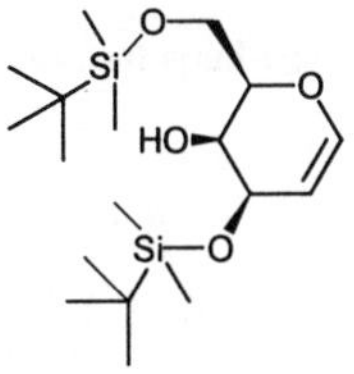

$C_{18}H_{38}O_4Si_2$
Chiral intermediate. bp = 235°; n = 1.458; $[\alpha]^{22}$ = - 40° (c = 1, $CHCl_3$). *Sigma-Aldrich Fine Chemicals.*

**1962 (-)-3,6-Di-O-(tert-butyldimethylsilyl)glucal**
111830-53-6

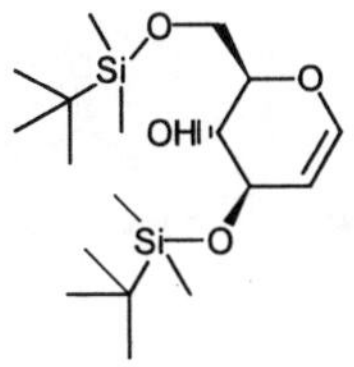

$C_{18}H_{38}O_4Si_2$
Chiral intermediate. mp = 43-45°; $[\alpha]_D^{20}$ = - 44° (c = 0.9, $CHCl_3$). *Sigma-Aldrich Fine Chemicals.*

**1963 (+)-6,6-Di-O-(tert-Butyldimethylsilyl)lactal**
142800-37-1

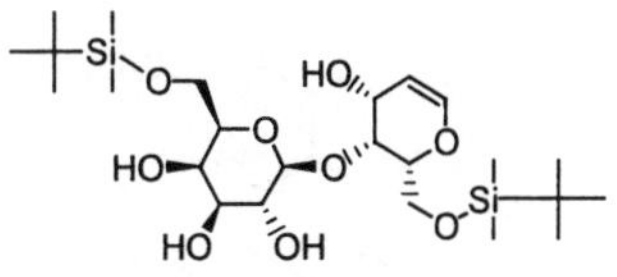

$C_{24}H_{48}O_9Si_2$
Chiral intermediate. mp = 56-62°; $[\alpha]_D^{20}$ = + 16.0° (c = 1.1, $CHCl_3$). *Sigma-Aldrich Fine Chemicals.*

**1964 (-)-3,6-Di-O-(tert-butyldiphenylsilyl)-galactal**

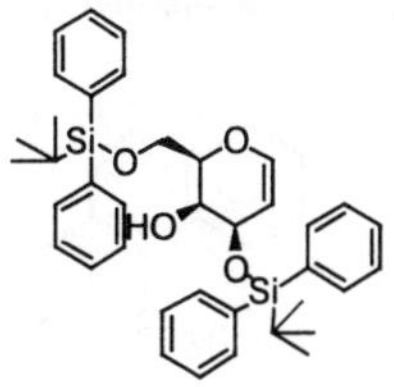

$C_{38}H_{46}O_4Si_2$
Chiral intermediate. n = 1.559; $[\alpha]^{22}$ = - 10° (c = 1, $CHCl_3$). *Sigma-Aldrich Fine Chemicals.*

**1965 (+)-6,6-Di-O-(tert-butyldiphenylsilyl)lactal**

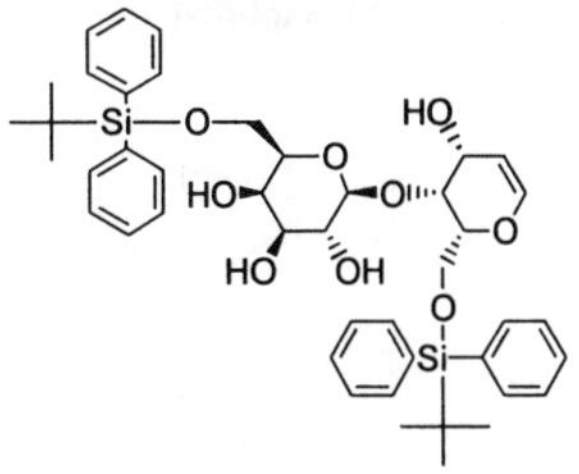

$C_{44}H_{56}O_9Si_2$
Chiral intermediate. mp = 85-92°; $[\alpha]_D^{20}$ = + 10° (c = 1, $CHCl_3$). *Sigma-Aldrich Fine Chemicals.*

**1966 (R)-3,3'-Dibutyl-5,5',6,6',7,7',8,8'-octahydro-1,1'-bi-2-naphthol**

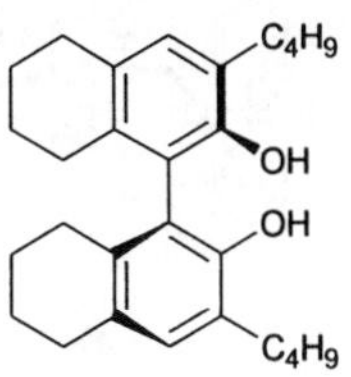

$C_{28}H_{38}O_2$
Chiral ligand. *Kankyo Kagaku Center Co., Ltd.*

**1967 (S)-3,3'-Dibutyl-5,5',6,6',7,7',8,8'-octahydro-1,1'-bi-2-naphthol**

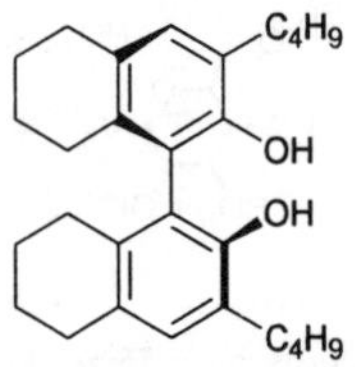

$C_{28}H_{38}O_2$
Chiral ligand. *Kankyo Kagaku Center Co., Ltd.*

**1968 (R)-2,2'-Dibutyl-5,5',6,6',7,7',8,8'-octahydro-1,1'-binaphthyl**

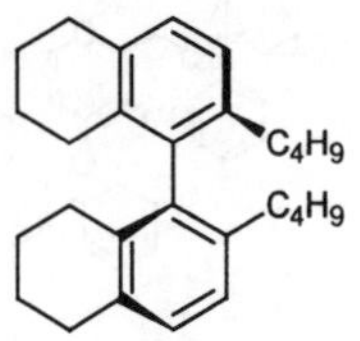

$C_{28}H_{38}$
Chiral ligand. *Kankyo Kagaku Center Co., Ltd.*

**1969 (S)-2,2'-Dibutyl-5,5',6,6',7,7',8,8'-octahydro-1,1'-binaphthyl**

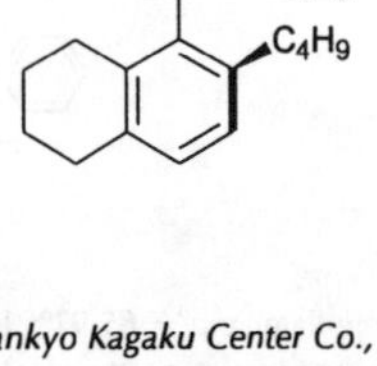

$C_{28}H_{38}$
Chiral ligand. *Kankyo Kagaku Center Co., Ltd.*

**1970 (+)-3,6-Di-O-(tert-butylphenylsilyl)glucal**

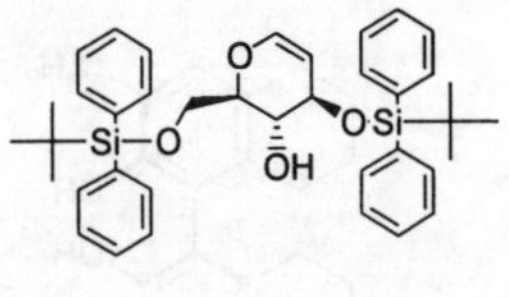

$C_{38}H_{46}O_4Si_2$
Chiral intermediate. $[\alpha]^{25}$ = + 6° (c = 1, $CHCl_3$). *Sigma-Aldrich Fine Chemicals.*

**1971 (S)-(-)-2-(3,5-Di-tert-butylsalicylideneamino)-3,3-dimethyl-1-butanol**
174022-08-3
$C_{21}H_{35}NO_2$
Chiral intermediate. mp = 88-92°; $[\alpha]_D^{20}$ = - 43° (c = 1,$CHCl_3$). *Sigma-Aldrich Fine Chemicals.*

**1972 Dibutyl-D-tartrate**

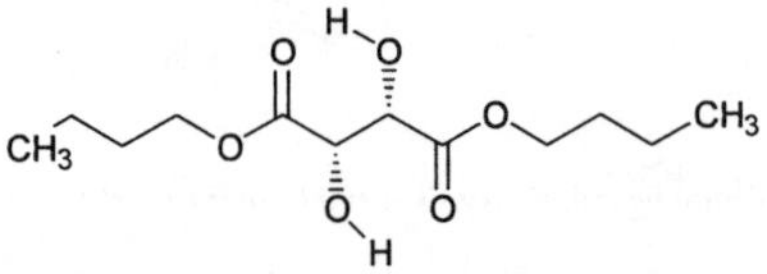

$C_{12}H_{22}O_6$
Butanedioic acid, 2,3-dihydroxy- (2S,3S)-, dibutyl ester.; D-Tartaric acid dibutyl ester. Chiral ligand; Chiral building block. *Acros Organics nv.*

**1973 (-)-Di-tert-butyl tartrate**
117384-46-0

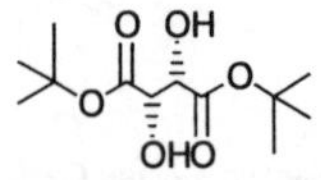

$C_{12}H_{22}O_6$
Chiral ligand; Chiral building block. mp = 90-92°; $[\alpha]_D^{20}$= - 11° (c = 1, acetone). *Sigma-Aldrich Fine Chemicals.*

**1974 (+)-Di-tert-butyl tartrate**
117384-45-9

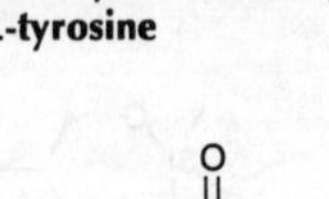

$C_{12}H_{22}O_6$
Chiral ligand; Chiral building block. mp = 90-92°; $[\alpha]_D^{20}$= + 11° (c = 1, acetone). *Sigma-Aldrich Fine Chemicals.*

**1975 Dichlorobenzyl-(tert-Butoxycarbonyl)-N-methyl-L-tyrosine**

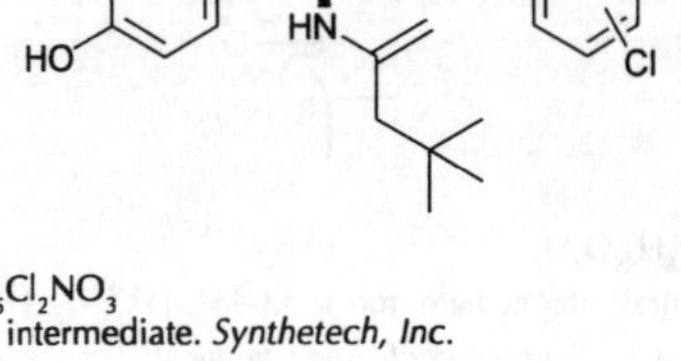

$C_{25}H_{35}Cl_2NO_3$
Chiral intermediate. *Synthetech, Inc.*

**1976 (+)-Di-μ-Chlorobis[2-[1-(dimethylamino)-ethyl]phenyl-C,N]-dipalladium**
34424-15-2

$C_{20}H_{28}Cl_2N_2Pd_2$
Chiral catalyst. mp = 190°; $[\alpha]_D^{20}$ = + 69° (c = 0.4, toluene). *Sigma-Aldrich Fine Chemicals.*

**1977 Dichloro[(R)-(+)-2,2'-bis(diphenyl-phosphino)-1,1'-binaphthyl]ruthenium(III)**
132071-87-5

$[C_{44}H_{32}Cl_2P_2Ru]x$
Chiral catalyst. *Strem Chemicals, Inc.*

**1978 Dichloro[(S)-(+)-2,2'-bis(diphenyl-phosphino)-1,1'-binaphthyl]ruthenium(III)**
134524-84-8

$[C_{44}H_{32}Cl_2P_2Ru]x$
Chiral catalyst. *Strem Chemicals, Inc.*

**1979 (-)-((8,8-Dichlorocamphoryl)sulfonyl)-oxaziridine**
139628-16-3

$C_{10}H_{13}Cl_2NO_3S$
Chiral intermediate. *Acros Organics nv; Oxford Asymmetry International plc.*

**1980 (+)-((8,8-Dichlorocamphoryl)sulfonyl)-oxaziridine**
127184-05-8

$C_{10}H_{13}Cl_2NO_3S$
Resolving agent. *Acros Organics nv; Oxford Asymmetry International plc.*

**1981 Dichloro-(S)-N,N-dimethyl-1-(R)-2-(diphenylphosphino)ferrocenylethyl-aminepalladium(II)**
79767-72-9

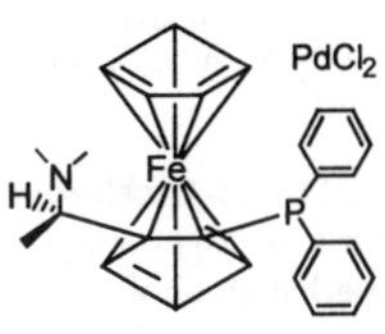


$C_{26}H_{28}Cl_2FeNPPd$
Chiral catalyst. mp = 170°; $[\alpha]_D^{20}$ = - 170° (c = 0.1, $CHCl_3$). *Sigma-Aldrich Fine Chemicals.*

**1982 Dichloro-(R,R)-ethylenebis(4,5,6,7-tetra-hydro-1-indenyl)titanium(IV)**

$C_{20}H_{24}Cl_2Ti$
Chiral catalyst. mp = 175-180°; $[\alpha]^{21}$ = + 3489° (c = 0.07, $CH_2Cl_2$, Hg lamp, 436nm). *Sigma-Aldrich Fine Chemicals.*

**1983 Dichloro-(S,S)-ethylenebis(4,5,6,7-tetra-hydro-1-indenyl)titanium(IV)**
$C_{20}H_{24}Cl_2Ti$
Chiral catalyst. mp = 175-180°; $[\alpha]^{21}$ = - 3486° (c = 0.07, $CH_2Cl_2$, Hg lamp, 436nm). *Sigma-Aldrich Fine Chemicals.*

**1984 Dichloro[(R,R)-ethylenebis(4,5,6,7-tetra-hydro-1-indenyl)]zirconium(IV)**
150131-28-5
$C_{20}H_{24}Cl_2Zr$
Chiral catalyst. *Sigma-Aldrich Fine Chemicals.*

**1985 Dichloro[(S,S)-ethylenebis(4,5,6,7-tetra-hydro-1-indenyl)]zirconium(IV)**
109429-79-0
$C_{20}H_{24}Cl_2Zr$
Chiral catalyst. *Sigma-Aldrich Fine Chemicals.*

**1986 (R)-(+)-2-(2,4-Dichlorophenoxy)-propionic acid**
15165-67-0 403-980-1

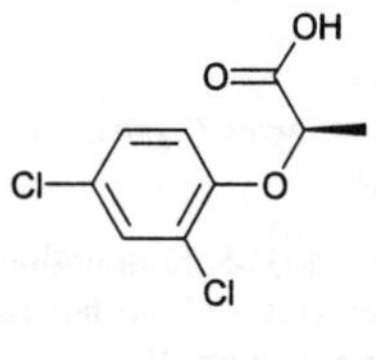

$C_9H_8Cl_2O_3$
Propanoic acid, 2-(2,4-dichlorophenoxy)-, (2R)-. (+)-2,4-DP; (+)-Dichlorprop; Duplosan DP. Chiral intermediate. mp = 117.5-118.1°. *A.H. Marks & Co. Ltd.; BASF Aktiengesellschaft.*

**1987 (R)-(cis-[2-(2,4-Dichlorophenyl)-2-1H-1,2,4-triazol-1-yl methyl-1,3-dioxolan-4-yl]methyl alcohol**

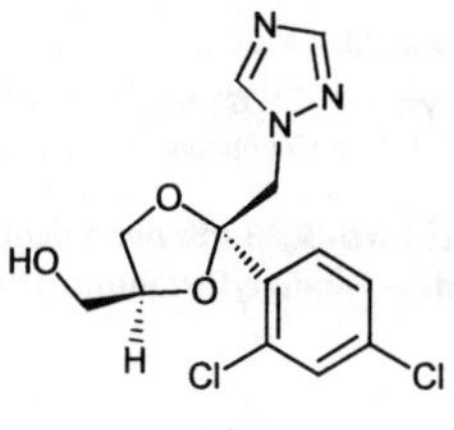

$C_{14}H_{15}Cl_2N_3O_3$
cis-DTMDM-OH. Chiral intermediate. *Syntai Chemicals and Pharmaceuticals.*

**1988 (S)-(cis-[2-(2,4-Dichlorophenyl)-2-1H-1,2,4-triazol-1-yl methyl-1,3-dioxolan-4-yl]methyl methane sulfonate**

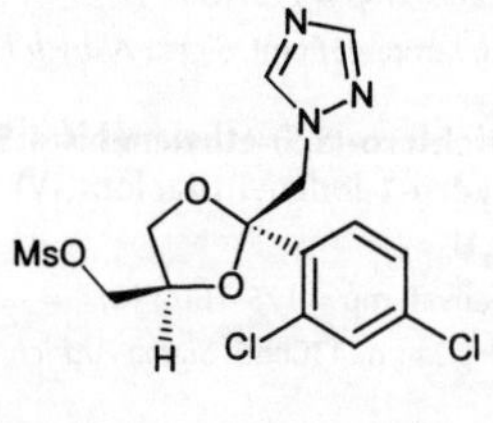

$C_{54}H_{17}Cl_2N_3O_5S$
cis-DTMDM-Oms. Chiral intermediate. *Syntai Chemicals and Pharmaceuticals.*

**1989 (R)-3,3'-Dicyclohexyl-1,1'-bi-2-naphthol**

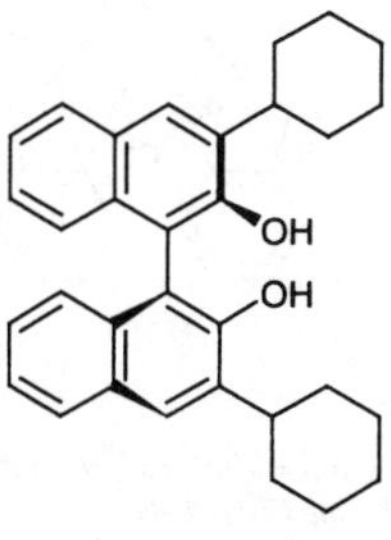

$C_{32}H_{34}O_2$
Chiral ligand. *Kankyo Kagaku Center Co., Ltd.*

**1990 (S)-3,3'-Dicyclohexyl-1,1'-bi-2-naphthol**

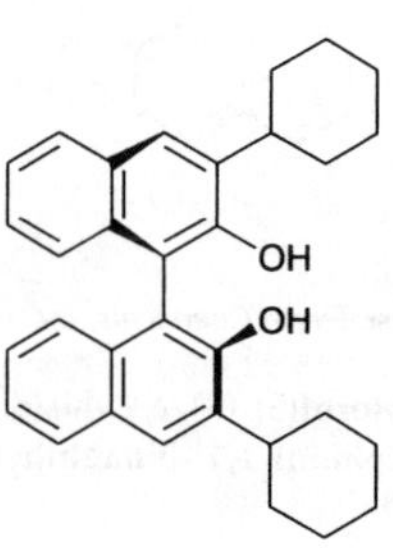

$C_{32}H_{34}O_2$
Chiral ligand. *Kankyo Kagaku Center Co., Ltd.*

**1991 (R)-6,6'-Dicyclohexyl-1,1'-bi-2-naphthol**

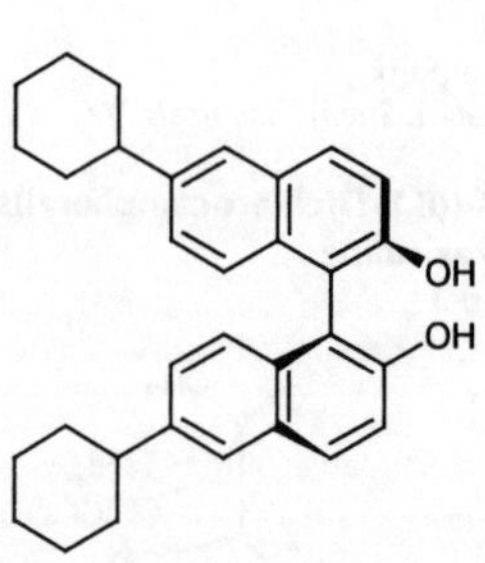

$C_{32}H_{34}O_2$
Chiral intermediate. *Kankyo Kagaku Center Co., Ltd.*

**1992 (S)-6,6'-Dicyclohexyl-1,1'-bi-2-naphthol**

OH
OH

$C_{32}H_{34}O_2$
Chiral ligand. *Kankyo Kagaku Center Co., Ltd.*

**1993 (R)-3,3'-Dicyclohexyl-2,2'-dimethoxy-1,1'-binaphthyl**

$OCH_3$
$OCH_3$

$C_{34}H_{38}O_2$
Chiral ligand. *Kankyo Kagaku Center Co., Ltd.*

**1994 (S)-3,3'-Dicyclohexyl-2,2'-dimethoxy-1,1'-binaphthyl**

$OCH_3$
$OCH_3$

$C_{34}H_{38}O_2$
Chiral ligand. *Kankyo Kagaku Center Co., Ltd.*

**1995 (R)-6,6'-Dicyclohexyl-2,2'-dimethoxy-1,1'-binaphthyl**

$OCH_3$
$OCH_3$

$C_{34}H_{38}O_2$
Chiral intermediate. *Kankyo Kagaku Center Co., Ltd.*

**1996 (S)-6,6'-Dicyclohexyl-2,2'-dimethoxy-1,1'-binaphthyl**

$OCH_3$
$OCH_3$

$C_{34}H_{38}O_2$
Chiral ligand. *Kankyo Kagaku Center Co., Ltd.*

**1997 (R)-3,3'-Dicyclohexyl-2,2'-dimethoxy-5,5',6,6',7,7',8,8'-octahydro-1,1'-binaphthyl**

$OCH_3$
$OCH_3$

$C_{34}H_{46}O_2$
Chiral ligand. *Kankyo Kagaku Center Co., Ltd.*

**1998 (S)-3,3'-Dicyclohexyl-2,2'-dimethoxy-5,5',6,6',7,7',8,8'-octahydro-1,1'-binaphthyl**

$C_{34}H_{46}O_2$
Chiral ligand. *Kankyo Kagaku Center Co., Ltd.*

**1999 (R,R)-(-)-1,2-Dicyclohexyl-1,2-ethanediol**
120850-92-2

$C_{14}H_{26}O_2$
Chiral building block. mp = 135-137°; [α] = - 2.6° (c = 0.78, $CHCl_3$). *Acros Organics nv; Sigma-Aldrich Fine Chemicals.*

**2000 (S,S)-(+)-1,2-Dicyclohexyl-1,2-ethanediol**
120850-91-1

$C_{14}H_{26}O_2$
Chiral building block. *Acros Organics nv.*

**2001 (R)-3,3'-Dicyclohexyl-5,5',6,6',7,7',8,8'-octahydro-1,1'-bi-2-naphthol**

$C_{32}H_{42}O_2$
Chiral ligand. *Kankyo Kagaku Center Co., Ltd.*

**2002 (S)-3,3'-Dicyclohexyl-5,5',6,6',7,7',8,8'-octahydro-1,1'-bi-2-naphthol**

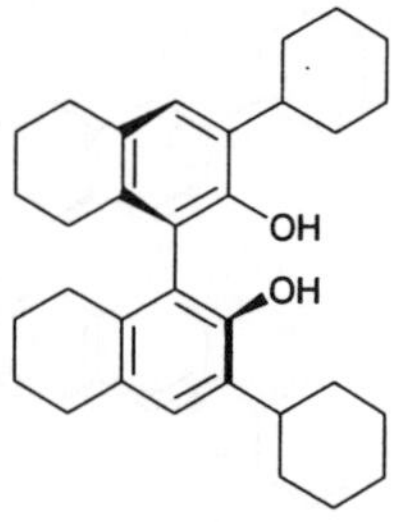


$C_{32}H_{42}O_2$
Chiral ligand. *Kankyo Kagaku Center Co., Ltd.*

**2003 (R)-(-)-1-[(S)-2-(Dicyclohexylphosphino)-ferrocenyl)ethyldicyclohexylphosphine**
167416-28-6

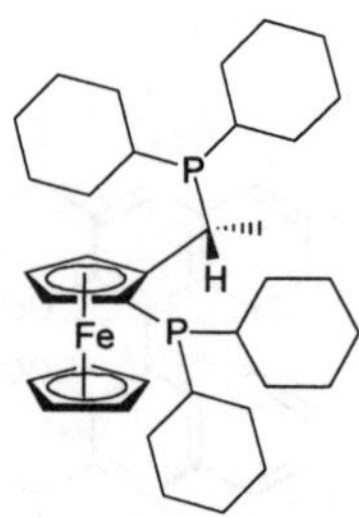


$C_{36}H_{56}FeP_2$
Chiral catalyst. $[\alpha]_D$ = - 135 ± 4° (c = 0.4, $CHCl_3$). *Strem Chemicals, Inc.*

**2004 (S)-(+)-1-[(R)-2-(Dicyclohexylphosphino)-ferrocenylethyldicyclohexylphosphine**
158923-07-0

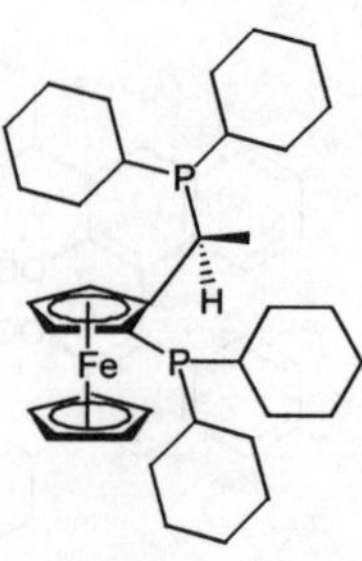


$C_{36}H_{56}FeP_2$
Chiral catalyst. $[\alpha]_D$ = + 135 ± 4° (c = 0.4, $CHCl_3$). *Strem Chemicals, Inc.*

**2005 (R)-(-)-1-[(S)-2-(Dicyclohexylphosphino)-ferrocenyl]ethyldiphenylphosphine**

158923-09-2

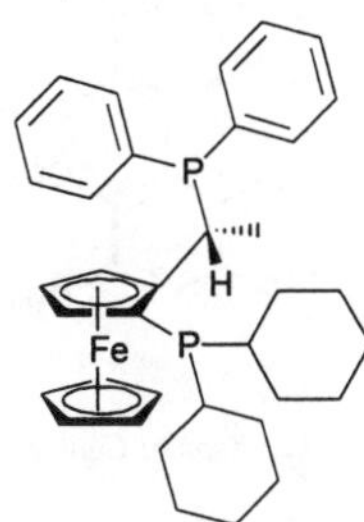


$C_{36}H_{44}FeP_2$

Chiral catalyst. $[\alpha]_D$ = - 102 ± 4° (c = 0.4, $CHCl_3$). *Strem Chemicals, Inc.*

**2006 (S)-(+)-1-[(R)-2-(Dicyclohexylphosphino)-ferrocenyl]ethyldiphenylphosphine**

162291-01-2

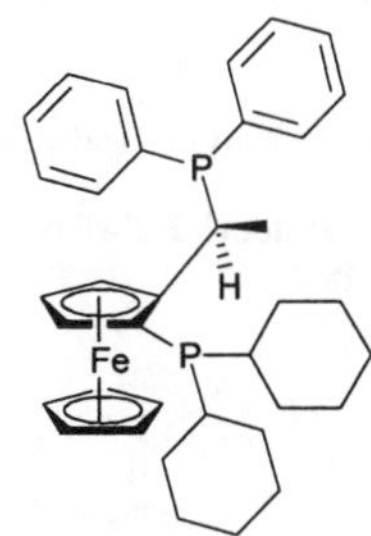


$C_{36}H_{44}FeP_2$

Chiral catalyst. $[\alpha]_D$ = + 102 ± 4° (c = 0.4, $CHCl_3$). *Strem Chemicals, Inc.*

**2007 (-)-10-Dicyclohexylsulfamoylisoborneol**

99295-72-4

$C_{22}H_{39}NO_3S$

Chiral intermediate. mp = 163-164°; $[\alpha]_D^{20}$ = - 25° (c = 0.76, EtOH). *Sigma-Aldrich Fine Chemicals.*

**2008 (+)-10-Dicyclohexylsulfamoylisoborneol**

96303-88-7

$C_{22}H_{39}NO_3S$

Chiral intermediate. mp = 161-164°; $[\alpha]_D^{20}$ = + 25° (c = 0.8, EtOH). *Sigma-Aldrich Fine Chemicals.*

**2009 (R)-3,3'-Didecyl-1,1'-bi-2-naphthol**

$C_{40}H_{54}O_2$

Chiral ligand. *Kankyo Kagaku Center Co., Ltd.*

**2010 (S)-3,3'-Didecyl-1,1'-bi-2-naphthol**

$C_{40}H_{54}O_2$

Chiral ligand. *Kankyo Kagaku Center Co., Ltd.*

**2011 (R)-6,6'-Didecyl-1,1'-bi-2-naphthol**

$C_{40}H_{54}O_2$

Chiral intermediate. *Kankyo Kagaku Center Co., Ltd.*

**2012 (S)-6,6'-Didecyl-1,1'-bi-2-naphthol**

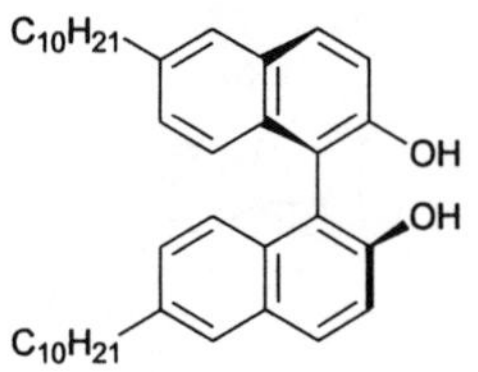

$C_{40}H_{54}O_2$
Chiral ligand. *Kankyo Kagaku Center Co., Ltd.*

**2013 (R)-2,2'-Didecyl-1,1'-binaphthyl**

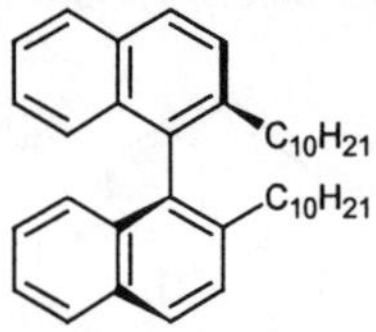

$C_{40}H_{54}$
Chiral ligand. *Kankyo Kagaku Center Co., Ltd.*

**2014 (S)-2,2'-Didecyl-1,1'-binaphthyl**

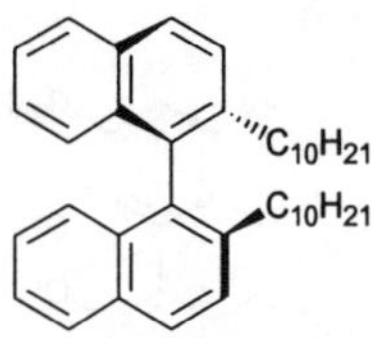

$C_{40}H_{54}$
Chiral ligand. *Kankyo Kagaku Center Co., Ltd.*

**2015 (R)-3,3'-Didecyl-2,2'-dimethoxy-1,1'-binaphthyl**

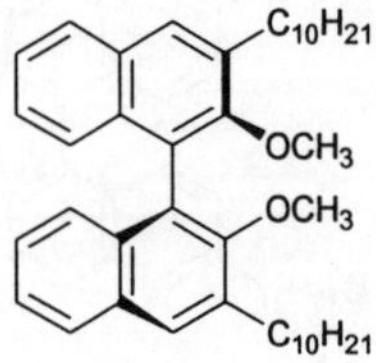

$C_{42}H_{58}O_2$
Chiral ligand. *Kankyo Kagaku Center Co., Ltd.*

**2016 (S)-3,3'-Didecyl-2,2'-dimethoxy-1,1'-binaphthyl**

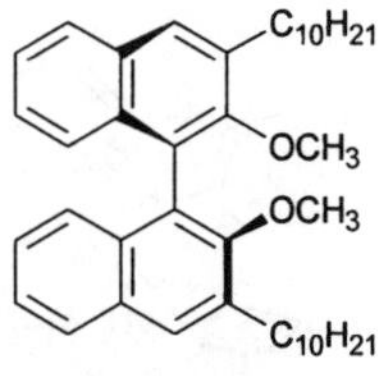

$C_{42}H_{58}O_2$
Chiral ligand. *Kankyo Kagaku Center Co., Ltd.*

**2017 (R)-6,6'-Didecyl-2,2'-dimethoxy-1,1'-binaphthyl**

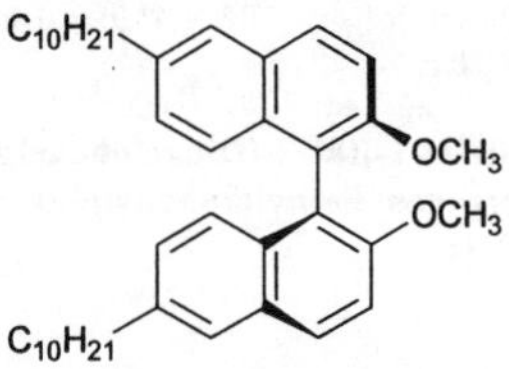

$C_{42}H_{58}O_2$
Chiral intermediate. *Kankyo Kagaku Center Co., Ltd.*

**2018 (S)-6,6'-Didecyl-2,2'-dimethoxy-1,1'-binaphthyl**

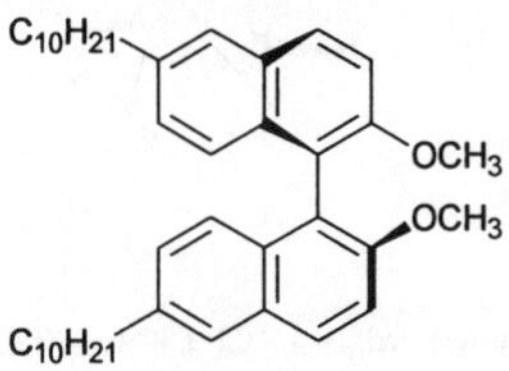

$C_{42}H_{58}O_2$
Chiral ligand. *Kankyo Kagaku Center Co., Ltd.*

**2019 (R)-3,3'-Didecyl-2,2'-dimethoxy-5,5',6,6',7,7',8,8'-octahydro-1,1'-binaphthyl**

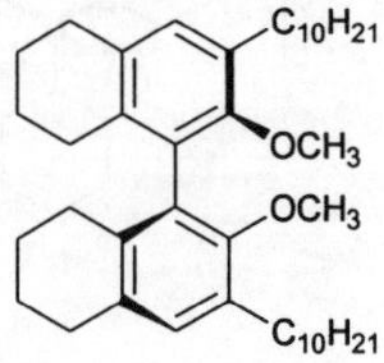

$C_{42}H_{66}O_2$
Chiral ligand. *Kankyo Kagaku Center Co., Ltd.*

**2020** **(S)-3,3'-Didecyl-2,2'-dimethoxy-5,5',6,6',7,7',8,8'-octahydro-1,1'-binaphthyl**

$C_{10}H_{21}$
$OCH_3$
$OCH_3$
$C_{10}H_{21}$

$C_{42}H_{66}O_2$
Chiral ligand. *Kankyo Kagaku Center Co., Ltd.*

**2021** **(R)-3,3'-Didecyl-5,5',6,6',7,7',8,8'-octahydro-1,1'-bi-2-naphthol**

$C_{10}H_{21}$
OH
OH
$C_{10}H_{21}$

$C_{40}H_{62}O_2$
Chiral ligand. *Kankyo Kagaku Center Co., Ltd.*

**2022** **(S)-3,3'-Didecyl-5,5',6,6',7,7',8,8'-octahydro-1,1'-bi-2-naphthol**

$C_{10}H_{21}$
OH
OH
$C_{10}H_{21}$

$C_{40}H_{62}O_2$
Chiral ligand. *Kankyo Kagaku Center Co., Ltd.*

**2023** **(R)-2,2'-Didecyl-5,5',6,6',7,7',8,8'-octahydro-1,1'-binaphthyl**

$C_{10}H_{21}$
$C_{10}H_{21}$

$C_{40}H_{62}$
Chiral ligand. *Kankyo Kagaku Center Co., Ltd.*

**2024** **(S)-2,2'-Didecyl-5,5',6,6',7,7',8,8'-octahydro-1,1'-binaphthyl**

$C_{10}H_{21}$
$C_{10}H_{21}$

$C_{40}H_{62}$
Chiral ligand. *Kankyo Kagaku Center Co., Ltd.*

**2025** **(R)-2,2'-Didecyloxy-1,1'-binaphthalene-6,6'-dicarbaldehyde**

OHC
$OC_{10}H_{21}$
$OC_{10}H_{21}$
OHC

$C_{42}H_{54}O_4$
Chiral ligand. *Kankyo Kagaku Center Co., Ltd.*

**2026** **(S)-2,2'-Didecyloxy-1,1'-binaphthalene-6,6'-dicarbaldehyde**

OHC
$OC_{10}H_{21}$
$OC_{10}H_{21}$
OHC

$C_{42}H_{54}O_4$
Chiral ligand. *Kankyo Kagaku Center Co., Ltd.*

**2027** **(R)-2,2'-Didecyloxy-1,1'-binaphthalene-6,6'-dicarboxylic acid**

HOOC
$OC_{10}H_{21}$
$OC_{10}H_{21}$
HOOC

$C_{42}H_{54}O_6$
Chiral ligand. *Kankyo Kagaku Center Co., Ltd.*

**2028 (S)-2,2'-Didecyloxy-1,1'-binaphthalene-6,6'-dicarboxylic acid**

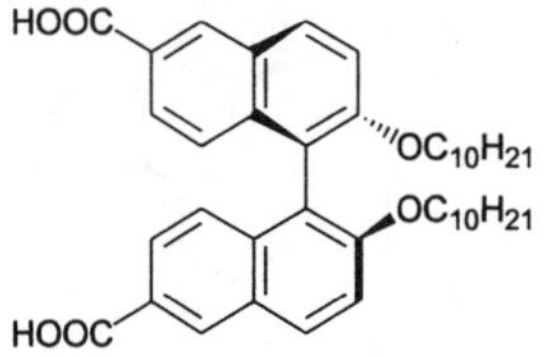

$C_{42}H_{54}O_6$
Chiral ligand. *Kankyo Kagaku Center Co., Ltd.*

**2029 (-)-2,3-Dideoxyadenosine**
4097-22-7 3151(12) 223-853-2

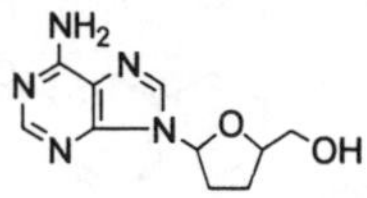

$C_{10}H_{13}N_5O_2$
Chiral intermediate. mp = 181-184°; $[\alpha]_D^{20}$ = - 27° (c = 1, $H_2O$). *Sigma-Aldrich Fine Chemicals.*

**2030 (-)-2,3-Dideoxyinosine**
69655-05-6 3148(12)

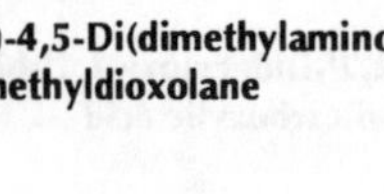

$C_{10}H_{12}N_4O_3$
Chiral intermediate. $[\alpha]_D^{20}$ = - 27° (c = 1, $H_2O$). *Sigma-Aldrich Fine Chemicals.*

**2031 (4R,5R)-4,5-Di(dimethylaminocarbonyl)-2,2-dimethyldioxolane**

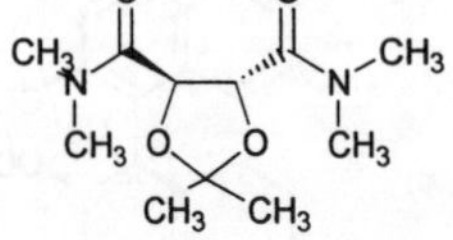

$C_{11}H_{20}N_2O_4$
Chiral intermediate. *Acros Organics nv.*

**2032 (R)-3,3'-Didodecyl-1,1'-bi-2-naphthol**

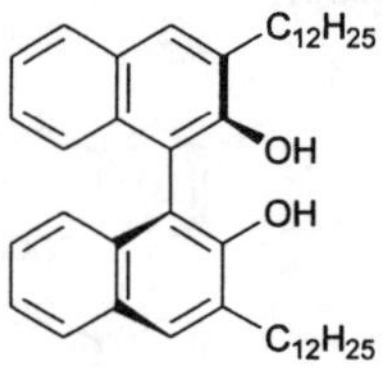

$C_{44}H_{62}O_2$
Chiral ligand. *Kankyo Kagaku Center Co., Ltd.*

**2033 (S)-3,3'-Didodecyl-1,1'-bi-2-naphthol**

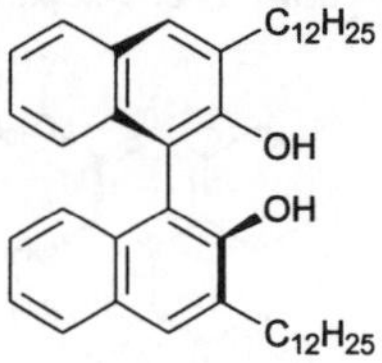

$C_{44}H_{62}O_2$
Chiral ligand. *Kankyo Kagaku Center Co., Ltd.*

**2034 (R)-6,6'-Didodecyl-1,1'-bi-2-naphthol**

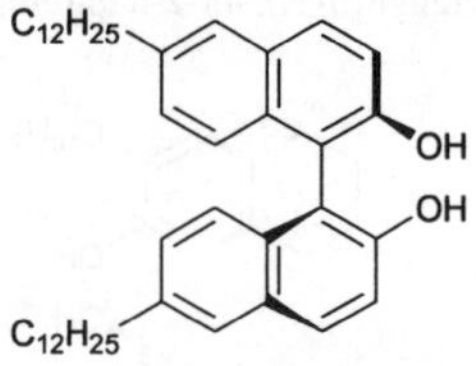

$C_{44}H_{62}O_2$
Chiral intermediate. *Kankyo Kagaku Center Co., Ltd.*

**2035 (S)-6,6'-Didodecyl-1,1'-bi-2-naphthol**

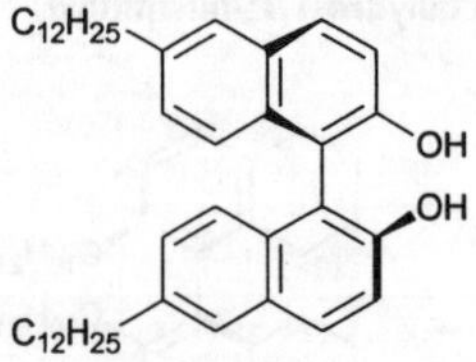

$C_{44}H_{62}O_2$
Chiral ligand. *Kankyo Kagaku Center Co., Ltd.*

**2036 (R)-2,2'-Didodecyl-1,1'-binaphthyl**

$C_{12}H_{25}$
$C_{12}H_{25}$

$C_{44}H_{62}$
Chiral ligand. *Kankyo Kagaku Center Co., Ltd.*

**2037 (S)-2,2'-Didodecyl-1,1'-binaphthyl**

$OC_{10}H_{21}$
$OC_{10}H_{21}$

$C_{44}H_{62}$
Chiral ligand. *Kankyo Kagaku Center Co., Ltd.*

**2038 (R)-3,3'-Didodecyl-2,2'-dimethoxy-1,1'-binaphthyl**

$C_{12}H_{25}$
$OCH_3$
$OCH_3$
$C_{12}H_{25}$

$C_{46}H_{66}O_2$
Chiral intermediate. *Kankyo Kagaku Center Co., Ltd.*

**2039 (S)-3,3'-Didodecyl-2,2'-dimethoxy-1,1'-binaphthyl**

$C_{12}H_{25}$
$OCH_3$
$OCH_3$
$C_{12}H_{25}$

$C_{46}H_{66}O_2$
Chiral ligand. *Kankyo Kagaku Center Co., Ltd.*

**2040 (R)-6,6'-Didodecyl-2,2'-dimethoxy-1,1'-binaphthyl**

$C_{12}H_{25}$
$OCH_3$
$OCH_3$
$C_{12}H_{25}$

$C_{46}H_{66}O_2$
Chiral ligand. *Kankyo Kagaku Center Co., Ltd.*

**2041 (S)-6,6'-Didodecyl-2,2'-dimethoxy-1,1'-binaphthyl**

$C_{12}H_{25}$
$OCH_3$
$OCH_3$
$C_{12}H_{25}$

$C_{46}H_{66}O_2$
Chiral ligand. *Kankyo Kagaku Center Co., Ltd.*

**2042 (R)-3,3'-Didodecyl-2,2'-dimethoxy-5,5',6,6',7,7',8,8'-octahydro-1,1'-binaphthyl**

$C_{12}H_{25}$
$OCH_3$
$OCH_3$
$C_{12}H_{25}$

$C_{46}H_{74}O_2$
Chiral ligand. *Kankyo Kagaku Center Co., Ltd.*

**2043 (S)-3,3'-Didodecyl-2,2'-dimethoxy-5,5',6,6',7,7',8,8'-octahydro-1,1'-binaphthyl**

$C_{12}H_{25}$
$OCH_3$
$OCH_3$
$C_{12}H_{25}$

$C_{46}H_{74}O_2$
Chiral ligand. *Kankyo Kagaku Center Co., Ltd.*

**2044 (R)-3,3'-Didodecyl-5,5',6,6',7,7',8,8'-octahydro-1,1'-bi-2-naphthol**

$C_{44}H_{70}O_2$
Chiral ligand. *Kankyo Kagaku Center Co., Ltd.*

**2045 (S)-3,3'-Didodecyl-5,5',6,6',7,7',8,8'-octahydro-1,1'-bi-2-naphthol**

$C_{44}H_{70}O_2$
Chiral ligand. *Kankyo Kagaku Center Co., Ltd.*

**2046 (R)-2,2'-Didodecyl-5,5',6,6',7,7',8,8'-octahydro-1,1'-binaphthyl**

$C_{44}H_{70}$
Chiral ligand. *Kankyo Kagaku Center Co., Ltd.*

**2047 (S)-2,2'-Didodecyl-5,5',6,6',7,7',8,8'-octahydro-1,1'-binaphthyl**

$C_{44}H_{70}$
Chiral ligand. *Kankyo Kagaku Center Co., Ltd.*

**2048 (R)-2,2'-Didodecyloxy-1,1'-binaphthalene-6,6'-dicarbaldehyde**

$C_{46}H_{62}O_4$
Chiral ligand. *Kankyo Kagaku Center Co., Ltd.*

**2049 (S)-2,2'-Didodecyloxy-1,1'-binaphthalene-6,6'-dicarbaldehyde**

$C_{46}H_{62}O_4$
Chiral ligand. *Kankyo Kagaku Center Co., Ltd.*

**2050 (R)-2,2'-Didodecyloxy-1,1'-binaphthalene-6,6'-dicarboxylic acid**

$C_{46}H_{62}O_6$
Chiral ligand. *Kankyo Kagaku Center Co., Ltd.*

**2051 (S)-2,2'-Didodecyloxy-1,1'-binaphthalene-6,6'-dicarboxylic acid**

$C_{46}H_{62}O_6$
Chiral ligand. *Kankyo Kagaku Center Co., Ltd.*

**2052 (R,R)-1,2,9,10-Diepoxydecane**
144741-95-7

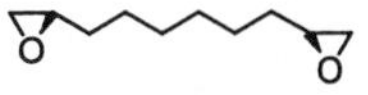

$C_{10}H_{18}O_2$
Oxirane, 2,2'-(1,6-hexanediyl)bis-, (R,R)-; (R,R)-2,2'-Hexane-1,6-diylbisoxirane; (R)-(Hexamethylene)dioxirane; (R)-1,4-Bis(2,3-epoxypropyl)butane. Chiral building block. $bp_{1.7}$ = 95.5°; d = 0.96; $[\alpha]^{25}$ = + 21° (neat). *Japan Energy Corporation; Sigma-Aldrich Fine Chemicals.*

**2053 Diethanolamine (3R)-(+)-tetrahydrofuranylboronate**
100858-40-0

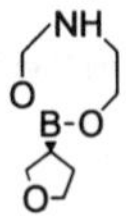

$C_8H_{16}BNO_3$
Chiral auxiliary. mp = 192-193°; $[\alpha]_D^{20}$ = + 18.0° (c = 1, $CHCl_3$). *Sigma-Aldrich Fine Chemicals.*

**2054 (R)-2,2'-Diethoxy-1,1'-binaphthalene-6,6'-dicarbaldehyde**

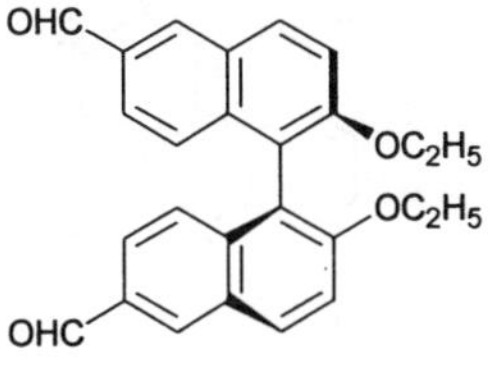

$C_{26}H_{22}O_4$
Chiral ligand. *Kankyo Kagaku Center Co., Ltd.*

**2055 (S)-2,2'-Diethoxy-1,1'-binaphthalene-6,6'-dicarbaldehyde**

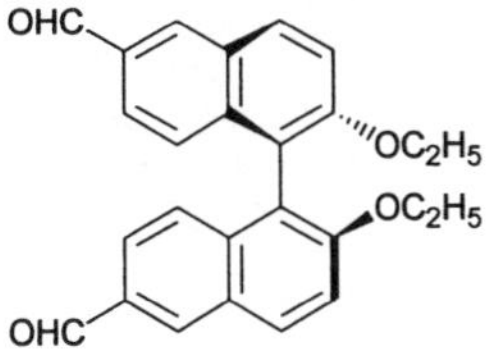

$C_{26}H_{22}O_4$
Chiral ligand. *Kankyo Kagaku Center Co., Ltd.*

**2056 (R)-2,2'-Diethoxy-1,1'-binaphthalene-6,6'-dicarboxylic acid**

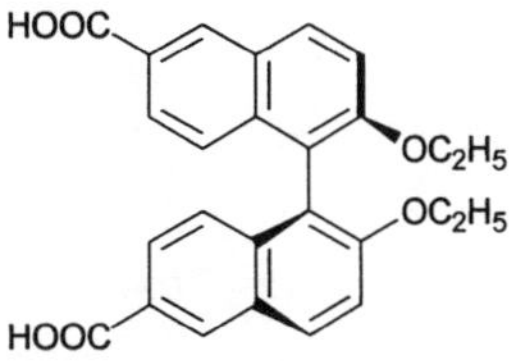

$C_{26}H_{22}O_6$
Chiral ligand. *Kankyo Kagaku Center Co., Ltd.*

**2057 (S)-2,2'-Diethoxy-1,1'-binaphthalene-6,6'-dicarboxylic acid**

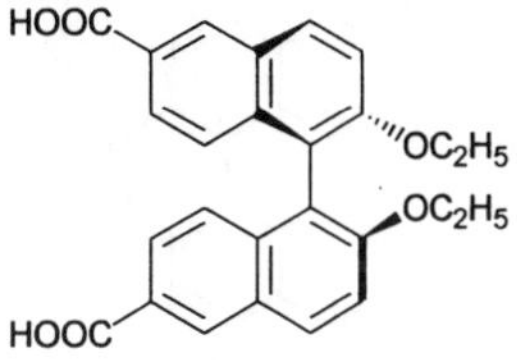

$C_{26}H_{22}O_6$
Chiral ligand. *Kankyo Kagaku Center Co., Ltd.*

**2058 (R)-2,2'-Diethoxy-1,1'-binaphthyl**

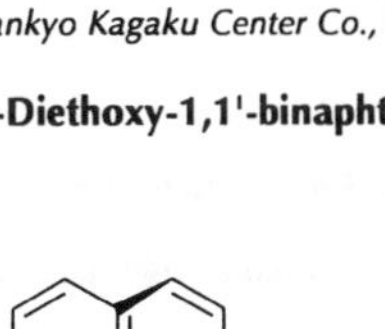

$C_{24}H_{22}O_2$
Chiral ligand. *Kankyo Kagaku Center Co., Ltd.*

**2059 (S)-2,2'-Diethoxy-1,1'-binaphthyl**

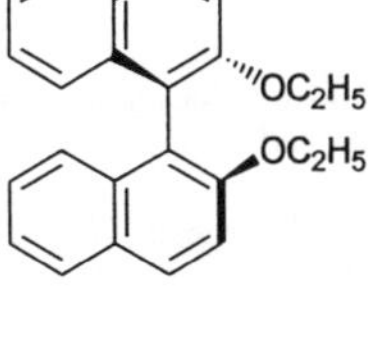

$C_{24}H_{22}O_2$
Chiral ligand. *Kankyo Kagaku Center Co., Ltd.*

**2060 (4R,5R)-4,5-Diethoxycarbonyl-2,2-dimethyldioxolane**
59779-75-8

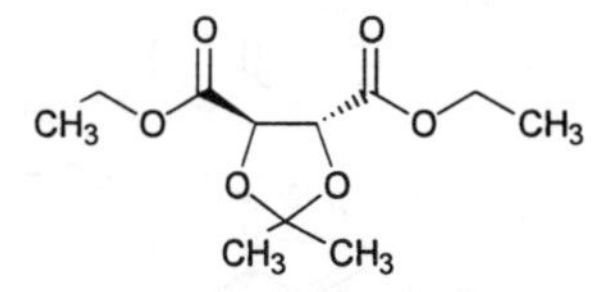

$C_{11}H_{18}O_6$
Chiral intermediate. *Acros Organics nv.*

**2061 (R)-2,2'-Diethoxy-5,5',6,6',7,7',8,8'-octahydro-1,1'-binaphthyl**

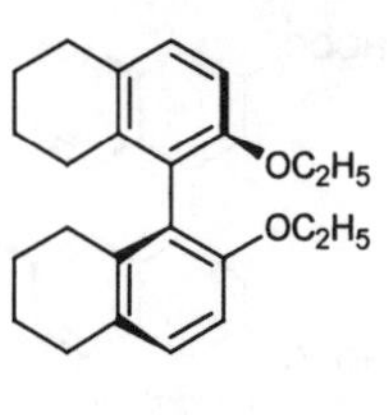

$C_{24}H_{30}O_2$
Chiral ligand. *Kankyo Kagaku Center Co., Ltd.*

**2062 (S)-2,2'-Diethoxy-5,5',6,6',7,7',8,8'-octahydro-1,1'-binaphthyl**

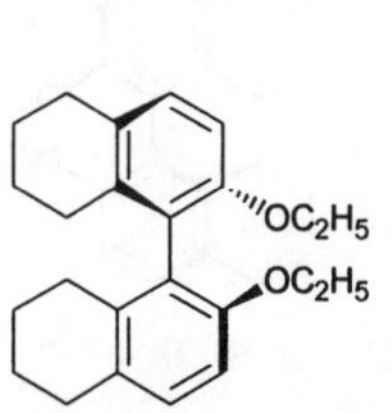

$C_{24}H_{30}O_2$
Chiral ligand. *Kankyo Kagaku Center Co., Ltd.*

**2063 1-Diethylamino-(R)-4-aminopentane**
67459-50-1
$C_9H_{22}N_2$
N,N-Diethyl-(4R)-1,4-pentanediamine. Chiral intermediate. d = 0.8200. *Lancaster Synthesis Ltd.*

**2064 1-Diethylamino-(S)-4-aminopentane**
67459-52-3
$C_9H_{22}N_2$
N,N-Diethyl-(4S)-1,4-pentanediamine; (S)-2-Amino-5-(diethylamino)pentane. Chiral intermediate. d = 0.8200. *Lancaster Synthesis Ltd.*

**2065 2-(Diethylamino)methyl-(4R)-tert-butylsilyloxycyclopent-2-enone**
117254-07-6

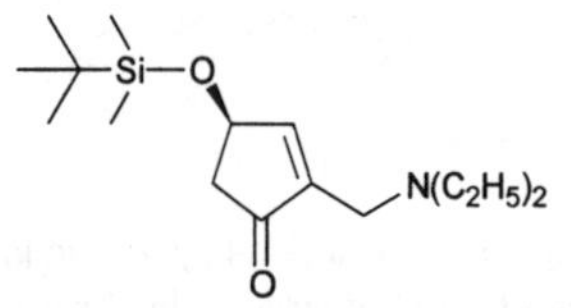

$C_{16}H_{31}NO_2Si$
Chiral intermediate. *Nissan Chemical Industries, Ltd.*

**2066 (2R,3R)-(-)-Diethyl 2,3-O-benzylidene-tartrate**
(2R,3R)-(-)-2,3-O-Benzylidenetartaric acid diethyl ester. Chiral building block; Chiral ligand. mp = 45-48°; $[\alpha]_D^{20}$ = - 37 ± 1° (c = 5, toluene). *Lancaster Synthesis Ltd.*

**2067 (2S,3S)-(+)-Diethyl 2,3-O-benzylidene-tartrate**
141042-56-0
$C_{15}H_{18}O_6$
(2S,3S)-(+)-2,3-O-Benzylidenetartaric acid diethyl ester. Chiral building block; Chiral ligand. mp = 45-48°; $[\alpha]_D^{20}$ = + 33 ± 1° (c = 5, toluene). *Lancaster Synthesis Ltd.*

**2068 (R)-3,3'-Diethyl-1,1'-bi-2-naphthol**

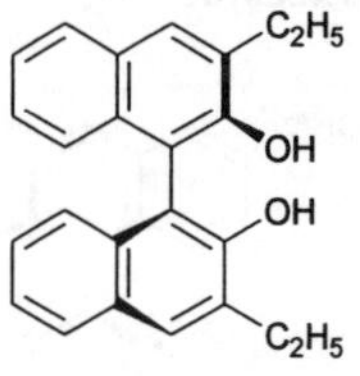

$C_{24}H_{22}O_2$
Chiral ligand. *Kankyo Kagaku Center Co., Ltd.*

**2069 (S)-3,3'-Diethyl-1,1'-bi-2-naphthol**

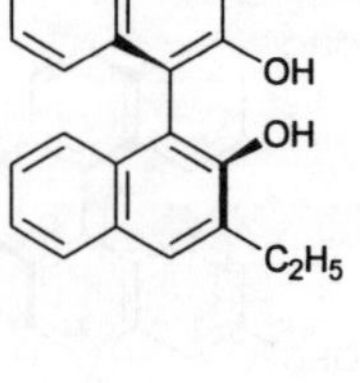

$C_{24}H_{22}O_2$
Chiral ligand. *Kankyo Kagaku Center Co., Ltd.*

**2070 (R)-6,6'-Diethyl-1,1'-bi-2-naphthol**

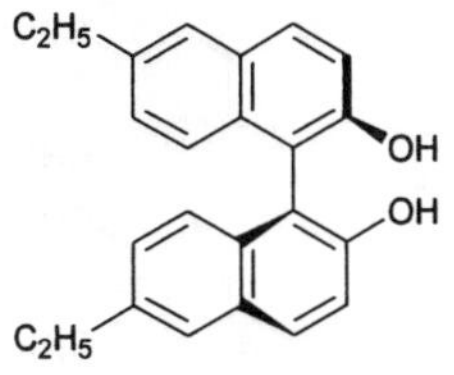

$C_{24}H_{22}O_2$
Chiral intermediate. *Kankyo Kagaku Center Co., Ltd.*

**2071 (S)-6,6'-Diethyl-1,1'-bi-2-naphthol**

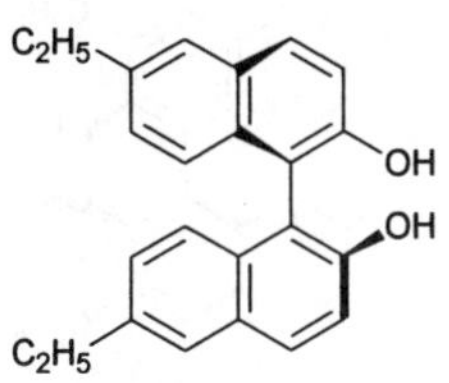

$C_{24}H_{22}O_2$
Chiral ligand. *Kankyo Kagaku Center Co., Ltd.*

**2072 (R)-2,2'-Diethyl-1,1'-binaphthyl**

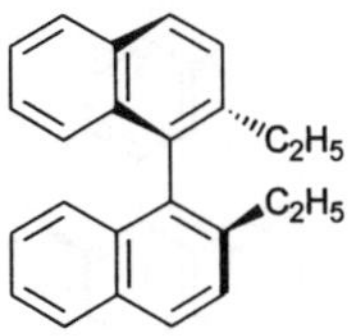

$C_{24}H_{22}$
Chiral ligand. *Kankyo Kagaku Center Co., Ltd.*

**2073 (S)-2,2'-Diethyl-1,1'-binaphthyl**

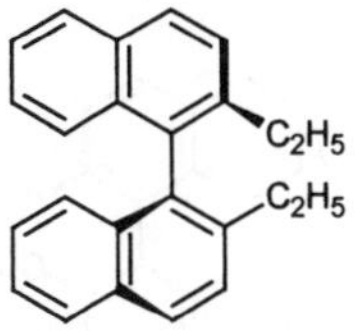

$C_{24}H_{22}$
Chiral ligand. *Kankyo Kagaku Center Co., Ltd.*

**2074 (R)-3,3'-Diethyl-2,2'-dimethoxy-1,1'-binaphthyl**

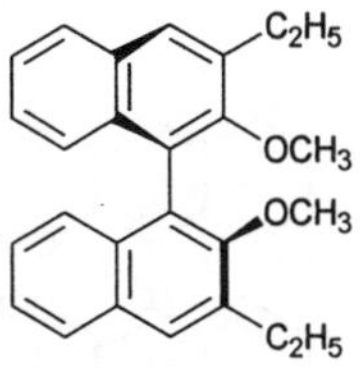

$C_{26}H_{26}O_2$
Chiral ligand. *Kankyo Kagaku Center Co., Ltd.*

**2075 (S)-3,3'-Diethyl-2,2'-dimethoxy-1,1'-binaphthyl**

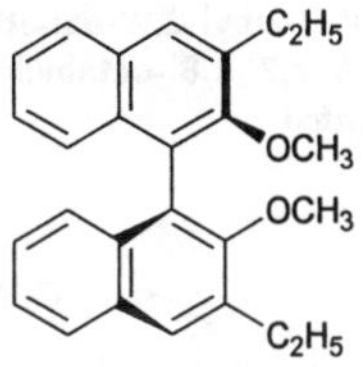

$C_{26}H_{26}O_2$
Chiral ligand. *Kankyo Kagaku Center Co., Ltd.*

**2076 (R)-6,6'-Diethyl-2,2'-dimethoxy-1,1'-binaphthyl**

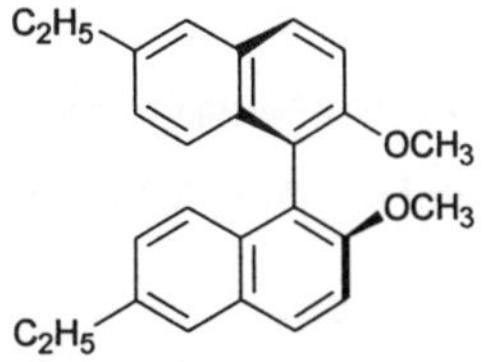

$C_{26}H_{26}O_2$
Chiral ligand. *Kankyo Kagaku Center Co., Ltd.*

**2077 (S)-6,6'-Diethyl-2,2'-dimethoxy-1,1'-binaphthyl**

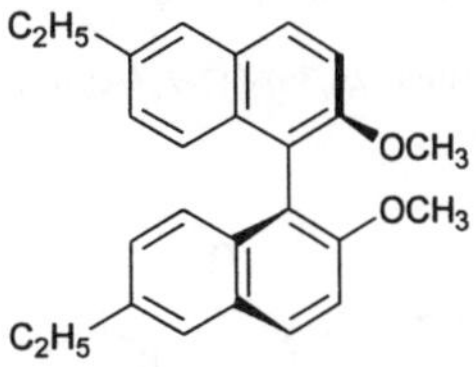

$C_{26}H_{26}O_2$
Chiral intermediate. *Kankyo Kagaku Center Co., Ltd.*

**2078 (R)-3,3'-Diethyl-2,2'-dimethoxy-5,5',6,6',7,7',8,8'-octahydro-1,1'-binaphthyl**

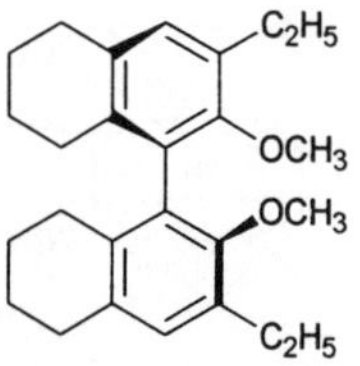

$C_{26}H_{34}O_2$
Chiral ligand. *Kankyo Kagaku Center Co., Ltd.*

**2079 (S)-3,3'-Diethyl-2,2'-dimethoxy-5,5',6,6',7,7',8,8'-octahydro-1,1'-binaphthyl**

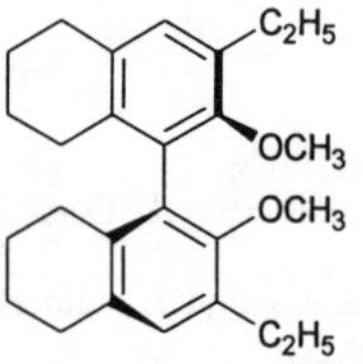

$C_{26}H_{34}O_2$
Chiral ligand. *Kankyo Kagaku Center Co., Ltd.*

**2080 Diethyl (2R,3R)-(-)-2,3-epoxysuccinate**
74243-85-9

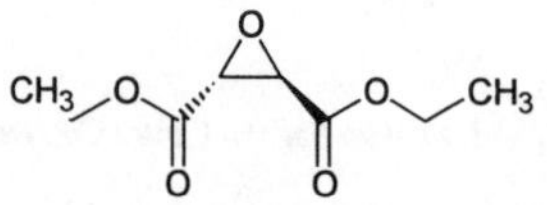

$C_8H_{12}O_5$
Chiral building block. [α] = - 109°. *Acros Organics nv.*

**2081 Diethyl (2S,3S)-(+)-2,3-epoxysuccinate**
73890-18-3

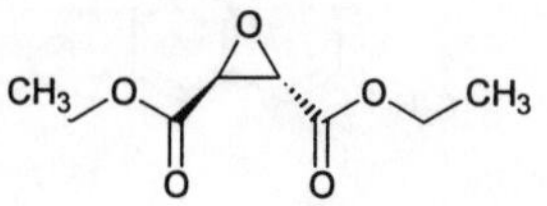

$C_8H_{12}O_5$
Chiral building block. *Acros Organics nv.*

**2082 (R)-3,3'-Diethyl-5,5',6,6',7,7',8,8'-octahydro-1,1'-bi-2-naphthol**

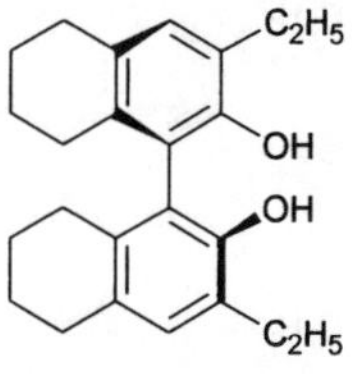

$C_{24}H_{30}O_2$
Chiral ligand. *Kankyo Kagaku Center Co., Ltd.*

**2083 (S)-3,3'-Diethyl-5,5',6,6',7,7',8,8'-octahydro-1,1'-bi-2-naphthol**

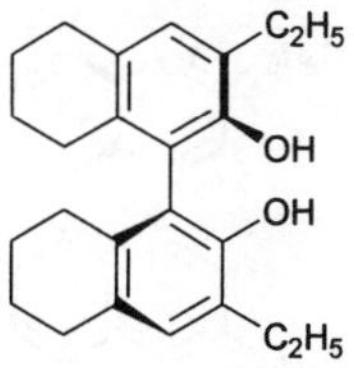

$C_{24}H_{30}O_2$
Chiral ligand. *Kankyo Kagaku Center Co., Ltd.*

**2084 (R)-2,2'-Diethyl-5,5',6,6',7,7',8,8'-octahydro-1,1'-binaphthyl**

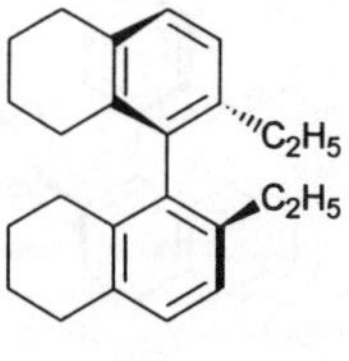

$C_{24}H_{30}$
Chiral ligand. *Kankyo Kagaku Center Co., Ltd.*

**2085 (S)-2,2'-Diethyl-5,5',6,6',7,7',8,8'-octahydro-1,1'-binaphthyl**

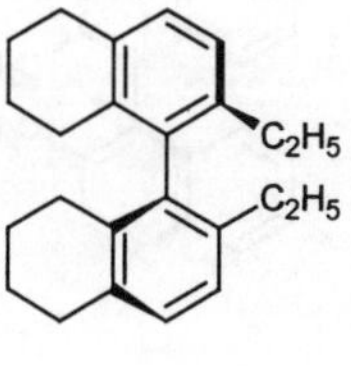

$C_{24}H_{30}$
Chiral ligand. *Kankyo Kagaku Center Co., Ltd.*

**2086 N-(Diethylphosphonoacetyl)-(4R)-benzyl-2-oxazolidinone**

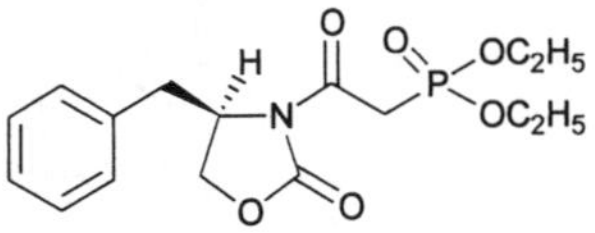

$C_{16}H_{22}NO_6P$
Chiral intermediate. *Oxford Asymmetry International plc.*

**2087 N-(Diethylphosphonoacetyl)-(4S)-benzyl-2-oxazolidinone**

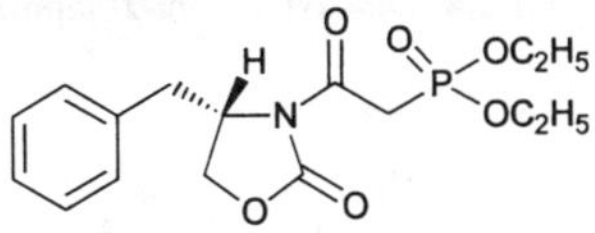

$C_{16}H_{22}NO_6P$
Chiral intermediate. *Oxford Asymmetry International plc.*

**2088 5,7-Difluoro-β-(S)-aminotetralin**

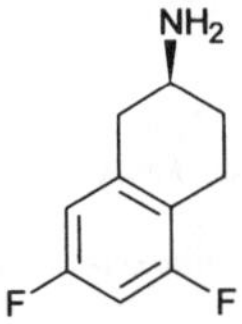

$C_{10}H_{11}F_2N$
Chiral intermediate. *Austin Chemical Company, Inc.*

**2089 (R)-N-(2,2-Difluoroethylidene)-1-phenylethylamine**
160797-29-5
$C_{10}H_{11}F_2N$
(R)-N-(2,2-Difluoroethylidene)-α-methylbenzylamine. Chiral intermediate. *TCI America.*

**2090 (S)-N-(2,2-Difluoroethylidene)-1-phenylethylamine**
161754-60-5
$C_{10}H_{11}F_2N$
Resolving agent. *TCI America.*

**2091 (R)-(+)-3,3-Difluoro-1,2-heptanediol**
158358-96-4

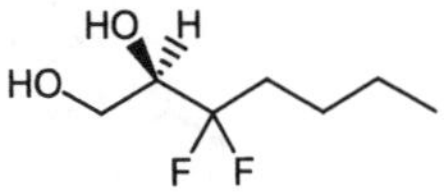

$C_7H_{14}F_2O_2$
Chiral building block. $bp_{30}$ = 130-132°; d = 0.783; n = 1.419; $[\alpha]^{22}$ = + 13° (c = 1.3, $CHCl_3$). *Sigma-Aldrich Fine Chemicals.*

**2092 D-(+)-Digitoxose**
527-52-6 208-416-6

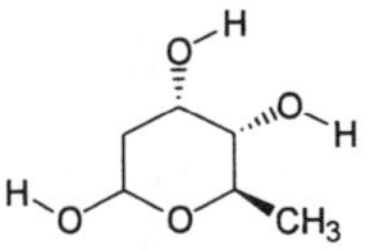

$C_6H_{12}O_4$
Ribohexose, 2,6-dideoxy-, D-; 2,6-Dideoxy-D-altrose; 2,6-Dideoxy-D-ribo-hexose. Pharmaceutical or derivative. mp = 112°; [α] = + 50° (c = 1.3, pyridine, 24h). *Acros Organics nv; TCI America.*

**2093 D-(+)-Digitoxose 2-deoxy-D-altro-methylose**
Pharmaceutical or derivative. *Senn Chemicals AG.*

**2094 (+)-Digoxigenin hydrate**
3149(11) 216-806-2

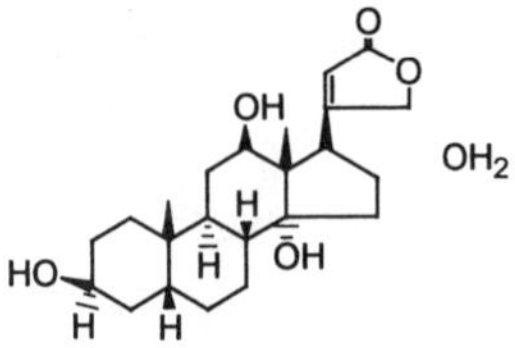

$C_{28}H_{48}O_2$
Pharmaceutical or derivative. mp = 205-210°; $[\alpha]^{23}$ = + 21.9° (c = 2, $CH_3OH$). *Sigma-Aldrich Fine Chemicals.*

**2095 (R)-2,2'-Diheptyloxy-1,1'-binaphthyl**

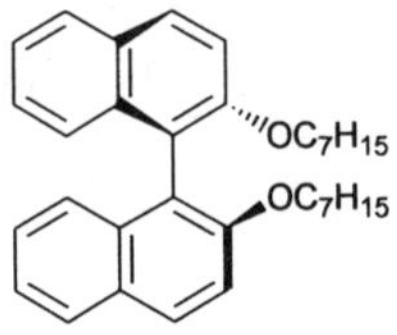

$C_{34}H_{42}O_2$
Chiral ligand. *Kankyo Kagaku Center Co., Ltd.*

**2096 (S)-2,2'-Diheptyloxy-1,1'-binaphthyl**

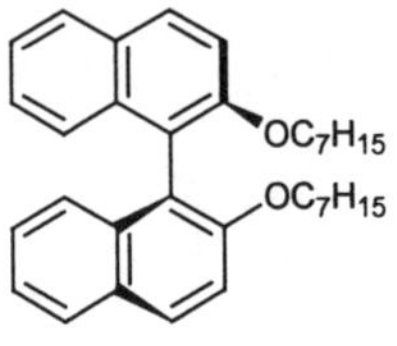

$C_{34}H_{42}O_2$
Chiral ligand. *Kankyo Kagaku Center Co., Ltd.*

**2097 (R)-2,2'-Diheptyloxy-5,5',6,6',7,7',8,8'-octahydro-1,1'-binaphthyl**

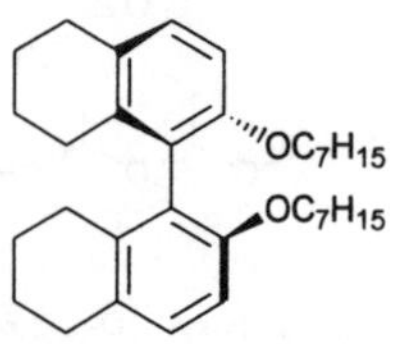

$C_{34}H_{50}O_2$
Chiral ligand. *Kankyo Kagaku Center Co., Ltd.*

**2098 (S)-2,2'-Diheptyloxy-5,5',6,6',7,7',8,8'-octahydro-1,1'-binaphthyl**

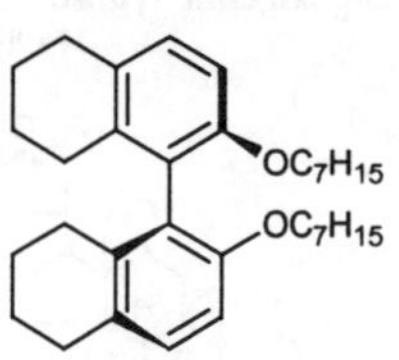

$C_{34}H_{50}O_2$
Chiral ligand. *Kankyo Kagaku Center Co., Ltd.*

**2099 (R)-3,3'-Dihexyl-1,1'-bi-2-naphthol**

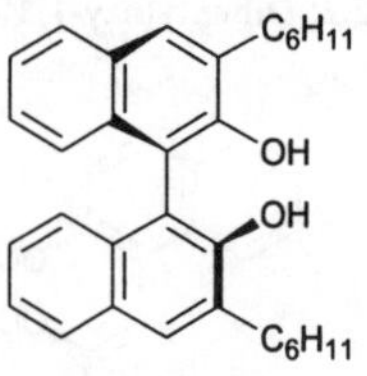

$C_{32}H_{38}O_2$
Chiral ligand. *Kankyo Kagaku Center Co., Ltd.*

**2100 (S)-3,3'-Dihexyl-1,1'-bi-2-naphthol**

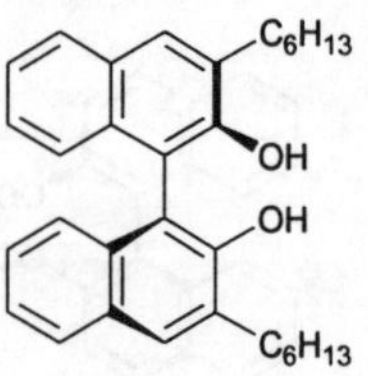

$C_{32}H_{38}O_2$
Chiral ligand. *Kankyo Kagaku Center Co., Ltd.*

**2101 (R)-6,6'-Dihexyl-1,1'-bi-2-naphthol**

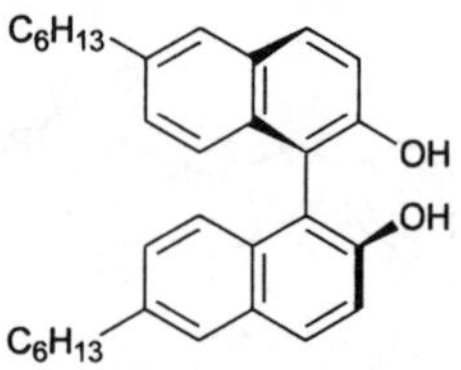

$C_{32}H_{38}O_2$
Chiral ligand. *Kankyo Kagaku Center Co., Ltd.*

**2102 (S)-6,6'-Dihexyl-1,1'-bi-2-naphthol**

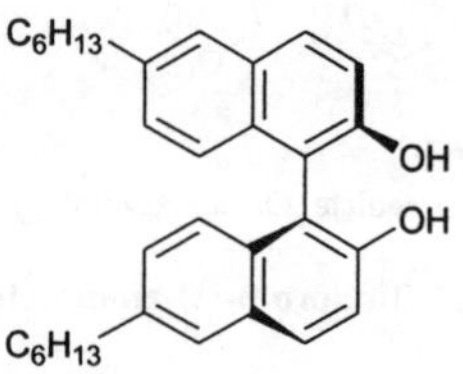

$C_{32}H_{38}O_2$
Chiral intermediate. *Kankyo Kagaku Center Co., Ltd.*

**2103 (R)-2,2'-Dihexyl-1,1'-binaphthyl**

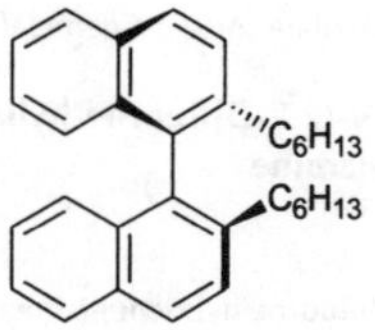

$C_{32}H_{38}$
Chiral ligand. *Kankyo Kagaku Center Co., Ltd.*

**2104 (S)-2,2'-Dihexyl-1,1'-binaphthyl**

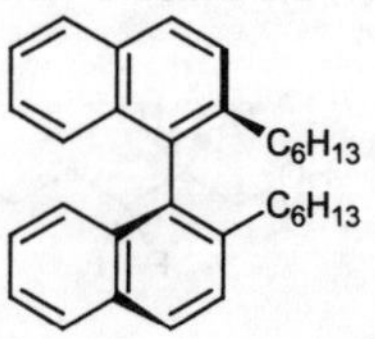

$C_{32}H_{38}$
Chiral ligand. *Kankyo Kagaku Center Co., Ltd.*

**2105 (R)-3,3'-Dihexyl-2,2'-dimethoxy-1,1'-binaphthyl**

$C_6H_{11}$
$OCH_3$
$OCH_3$
$C_6H_{11}$

$C_{34}H_{42}O_2$
Chiral ligand. *Kankyo Kagaku Center Co., Ltd.*

**2106 (S)-3,3'-Dihexyl-2,2'-dimethoxy-1,1'-binaphthyl**

$C_6H_{13}$
$OCH_3$
$OCH_3$
$C_6H_{13}$

$C_{34}H_{42}O_2$
Chiral ligand. *Kankyo Kagaku Center Co., Ltd.*

**2107 (R)-6,6'-Dihexyl-2,2'-dimethoxy-1,1'-binaphthyl**

$C_6H_{13}$
$OCH_3$
$OCH_3$
$C_6H_{13}$

$C_{34}H_{42}O_2$
Chiral ligand. *Kankyo Kagaku Center Co., Ltd.*

**2108 (S)-6,6'-Dihexyl-2,2'-dimethoxy-1,1'-binaphthyl**

$C_6H_{13}$
$OCH_3$
$OCH_3$
$C_6H_{13}$

$C_{34}H_{42}O_2$
Chiral intermediate. *Kankyo Kagaku Center Co., Ltd.*

**2109 (R)-3,3'-Dihexyl-2,2'-dimethoxy-5,5',6,6',7,7',8,8'-octahydro-1,1'-binaphthyl**

$C_6H_{11}$
$OCH_3$
$OCH_3$
$C_6H_{11}$

$C_{34}H_{50}O_2$
Chiral ligand. *Kankyo Kagaku Center Co., Ltd.*

**2110 (S)-3,3'-Dihexyl-2,2'-dimethoxy-5,5',6,6',7,7',8,8'-octahydro-1,1'-binaphthyl**

$C_6H_{13}$
$OCH_3$
$OCH_3$
$C_6H_{13}$

$C_{34}H_{50}O_2$
Chiral ligand. *Kankyo Kagaku Center Co., Ltd.*

**2111 (R)-3,3'-Dihexyl-5,5',6,6',7,7',8,8'-octahydro-1,1'-bi-2-naphthol**

$C_6H_{11}$
OH
OH
$C_6H_{11}$

$C_{32}H_{46}O_2$
Chiral ligand. *Kankyo Kagaku Center Co., Ltd.*

**2112 (S)-3,3'-Dihexyl-5,5',6,6',7,7',8,8'-octahydro-1,1'-bi-2-naphthol**

$C_6H_{13}$
OH
OH
$C_6H_{13}$

$C_{32}H_{46}O_2$
Chiral ligand. *Kankyo Kagaku Center Co., Ltd.*

2113 **(R)-2,2'-Dihexyl-5,5',6,6',7,7',8,8'-octahydro-1,1'-binaphthyl**

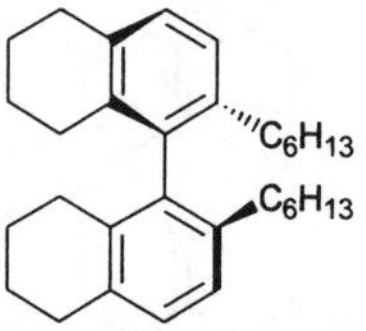

$C_{32}H_{46}$
Chiral ligand. *Kankyo Kagaku Center Co., Ltd.*

2114 **(S)-2,2'-Dihexyl-5,5',6,6',7,7',8,8'-octahydro-1,1'-binaphthyl**

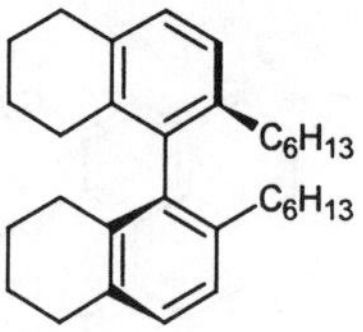

$C_{32}H_{46}$
Chiral ligand. *Kankyo Kagaku Center Co., Ltd.*

2115 **(R)-2,2'-Dihexyloxy-1,1'-binaphthalene-6,6'-dicarbaldehyde**

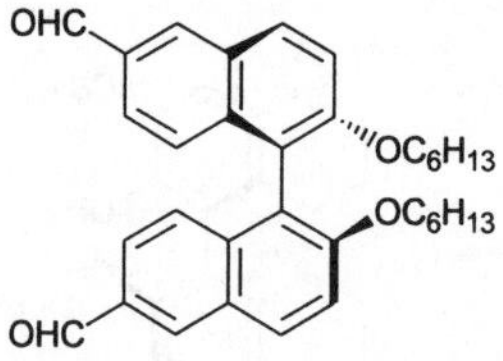

$C_{34}H_{38}O_4$
Chiral ligand. *Kankyo Kagaku Center Co., Ltd.*

2116 **(S)-2,2'-Dihexyloxy-1,1'-binaphthalene-6,6'-dicarbaldehyde**

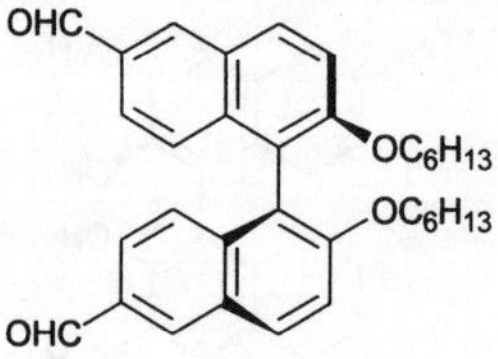

$C_{34}H_{38}O_4$
Chiral ligand. *Kankyo Kagaku Center Co., Ltd.*

2117 **(R)-2,2'-Dihexyloxy-1,1'-binaphthalene-6,6'-dicarboxylic acid**

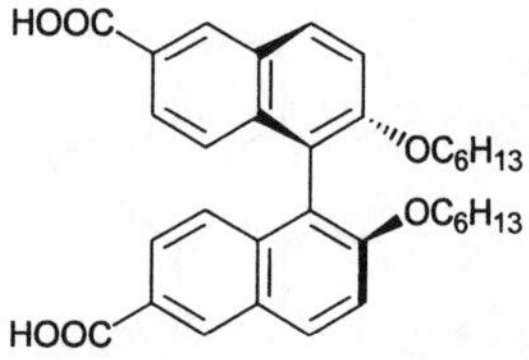

$C_{34}H_{38}O_6$
Chiral ligand. *Kankyo Kagaku Center Co., Ltd.*

2118 **(S)-2,2'-Dihexyloxy-1,1'-binaphthalene-6,6'-dicarboxylic acid**

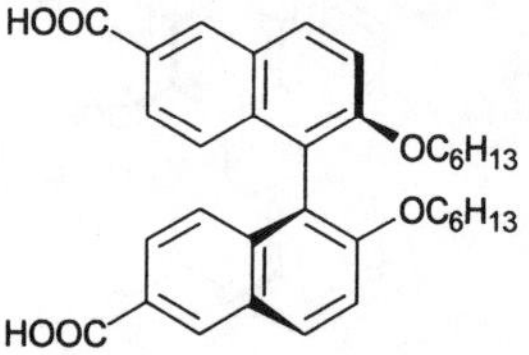

$C_{34}H_{38}O_6$
Chiral ligand. *Kankyo Kagaku Center Co., Ltd.*

2119 **(R)-2,2'-Dihexyloxy-1,1'-binaphthyl**

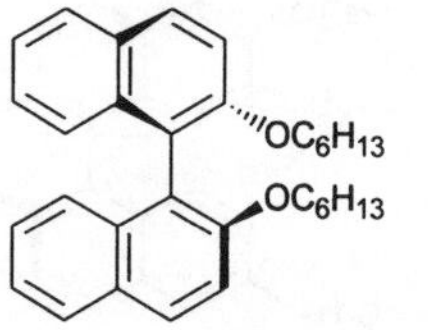

$C_{32}H_{38}O_2$
Chiral ligand. *Kankyo Kagaku Center Co., Ltd.*

2120 **(S)-2,2'-Dihexyloxy-1,1'-binaphthyl**

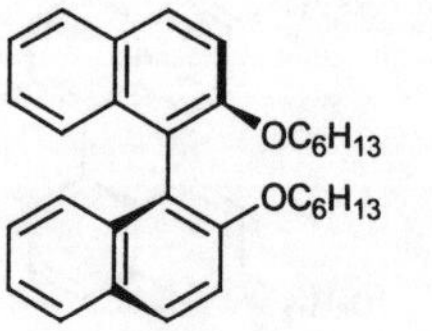

$C_{32}H_{38}O_2$
Chiral ligand. *Kankyo Kagaku Center Co., Ltd.*

**2121 (R)-2,2'-Dihexyloxy-5,5',6,6',7,7',8,8'-octahydro-1,1'-binaphthyl**

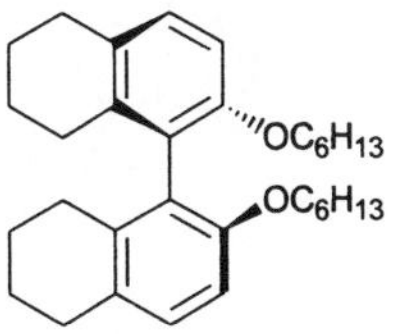

$C_{32}H_{46}O_2$
Chiral ligand. *Kankyo Kagaku Center Co., Ltd.*

**2122 (S)-2,2'-Dihexyloxy-5,5',6,6',7,7',8,8'-octahydro-1,1'-binaphthyl**

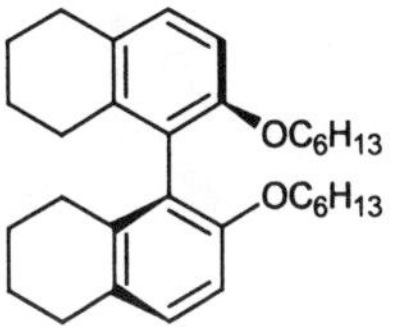

$C_{32}H_{46}O_2$
Chiral ligand. *Kankyo Kagaku Center Co., Ltd.*

**2123 (+)-Dihydrocarvone (mixture of isomers)**
5524-05-0 226-872-4

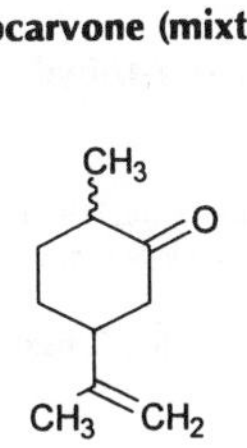

$C_{10}H_{16}O$
Cyclohexanone, 2-methyl-5-(1-methylethenyl)-, (2R,5R)-; (2R-trans)-2-Methyl-5-(1-methylvinyl)cyclohexan-1-one; p-Menth-8-en-2-one, . Listed on TSCA. Chiral intermediate. $bp_6$ = 87-88°; d = 0.929; n = 1.471; $[\alpha]^{22}$ = + 17° (neat). *Acros Organics nv; Sigma-Aldrich Fine Chemicals.*

**2124 (+)-Dihydrocarvyl acetate (mixture of isomers)**
20777-49-5 244-029-9

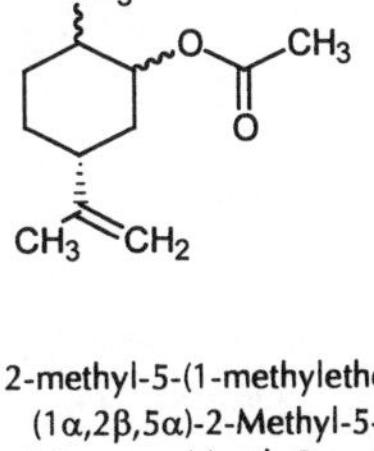

$C_{12}H_{20}O_2$
Cyclohexanol, 2-methyl-5-(1-methylethenyl)-, acetate, (1R,2R,5R)-; (1α,2β,5α)-2-Methyl-5-(1-methylvinyl)-cyclohexyl acetate; p-Menth-8-en-2-ol, acetate; Carhydrine; 8-p-Menthen-2-yl acetate; Turberyl acetate. Listed on TSCA. Chiral intermediate. bp = 232-234°; d = 0.947; n = 1.459; $[\alpha]^{25}$ = -13.5° (neat). *Acros Organics nv; Sigma-Aldrich Fine Chemicals.*

**2125 cis-(1S,2R)-1,2-Dihydro-3-chlorocatechol**

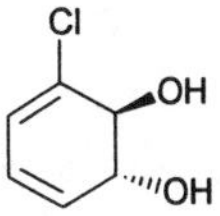

$C_6H_7ClO_2$
Chiral building block. *Genencor International.*

**2126 (+)-Dihydrocholesterol**
80-97-7 2255(12) 201-315-8

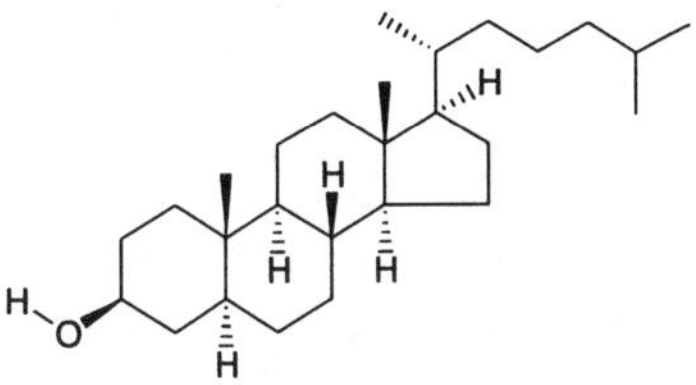

$C_{27}H_{48}O$
Cholestan-3-ol, (3β,5α)-; β-Cholestanol; 3β-Hydroxy-5α-cholestane; Zymostanol. Listed on TSCA. Pharmaceutical or derivative. mp = 140-142°; $[\alpha]^{21}$ = + 23.8° (c = 1.3, $CHCl_3$). *Acros Organics nv; Sigma-Aldrich Fine Chemicals.*

**2127 cis-(1S,2R)-1,2-Dihydro-3-cyanocatechol**

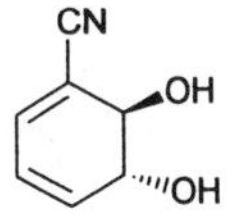

$C_7H_7NO_2$
Chiral building block. *Genencor International.*

**2128 cis-(1S,2S)-1,2-Dihydro-3-fluorocatechol**
131101-27-4

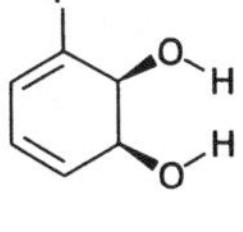

$C_6H_7FO_2$
Chiral building block. *Acros Organics nv; Lancaster Synthesis Ltd.*

**2129 (3αS,7R,7αS)-7,7α-Dihydro-7-hydroxy-2,2-dimethyl-1,3-benzodioxol-4(3aH)-one**

$C_9H_{12}O_4$
Chiral intermediate. mp = 65-69°; $[\alpha]_D^{20}$ = - 52° (c = 0.5, $CHCl_3$). *Sigma-Aldrich Fine Chemicals.*

**2130 (2S,3S)-(+)-2,3-Dihydro-3-hydroxy-2-(4-methoxyphenyl)-1,5-benzothiazepin-4(5H)-one**
42399-49-5 255-799-0

$C_{16}H_{15}NO_3S$
Benzo-1,5-thiazepin-4(5H)-one, 2,3-dihydro-3-hydroxy-2-(4-methoxyphenyl)-, (2S,3S)-; (2S,3S)-3-Hydroxy-2-(4-methoxyphenyl)-2,3-dihydro-1,5-benzothiazepin-4(5H)-one. Chiral intermediate. mp = 203-206°; $[\alpha]_D^{20}$ = + 112° (c = 0.5, $CH_3OH$). *Acros Organics nv; Delmar Chemicals; Sigma-Aldrich Fine Chemicals; Zambon Group SpA.*

**2131 cis-(1S,2R)-1,2-Dihydro-3-iodocatechol**

$C_6H_7IO_2$
Chiral building block. *Genencor International.*

**2132 cis-(1S,2R)-1,2-Dihydro-1,2-isopropylidere-3-cyanocatechol**

$C_{10}H_{11}NO_2$
Chiral intermediate. *Genencor International.*

**2133 (3R-cis)-(-)-2,3-Dihydro-3-isopropyl-7a-methylpyrrolo-[2,1-b]oxazol-5(7aH)-one**

$C_{10}H_{15}NO_2$
Chiral intermediate. mp = 47-50°; $[\alpha]^{22}$ = - 37° (c = 1, $CHCl_3$). *Acros Organics nv; Sigma-Aldrich Fine Chemicals.*

**2134 (3S-cis)-(+)-2,3-Dihydro-3-isopropyl-7a-methylpyrrolo-[2,1-b]oxazol-5(7aH)-one**
116910-11-3

$C_{10}H_{15}NO_2$
Chiral intermediate. mp = 46-49°; $[\alpha]_D^{20}$ = + 41.0° (c = 1, $CHCl_3$). *Acros Organics nv; Sigma-Aldrich Fine Chemicals.*

**2135 cis(1S,2R)-1,2-Dihydro-3-methylcatechol**
41977-20-2
$C_7H_{10}O_2$
cis-(1S,2R)-3-Methyl-3,5-cyclohexadiene-1,2-diol. Chiral intermediate. *Lancaster Synthesis Ltd.*

**2136 (S)-4,5-Dihydro-2-methyl-phenylmethyl-oxazole**

$C_{11}H_{13}NO$
Chiral intermediate. *Fischer Chemicals AG; Great Lakes Fine Chemicals.*

**2137 (3S-cis)-2,3-Dihydro-7a-methyl-3-phenyl-pyrrolo[2,1-b]oxazol-5(6H)-one**
143140-06-1

$C_{13}H_{13}NO_2$
Chiral intermediate. mp = 90-92°; $[\alpha]_D^{20}$ = + 124° (c = 1, $CHCl_3$). *Acros Organics nv; Sigma-Aldrich Fine Chemicals.*

**2138 (3R-cis)-2,3-Dihydro-7a-methyl-3-phenyl-pyrrolo[2,1-b]oxazol-5(7aH)-one**

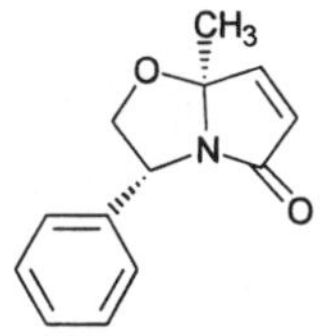

$C_{13}H_{13}NO_2$
Chiral intermediate. mp = 90-92°; $[\alpha]_D^{20}$ = - 124° (c = 1, $CHCl_3$). *Acros Organics nv; Sigma-Aldrich Fine Chemicals.*

**2139 (1R-cis)-1,2-Dihydro-1,2-naphthalenediol**

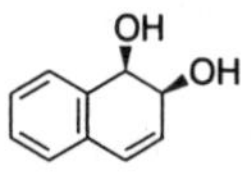

$C_{10}H_{10}O_2$
Chiral building block. mp = 121-127°; $[\alpha]^{22}$ = + 229° (c = 1, $CH_3OH$). *Sigma-Aldrich Fine Chemicals.*

**2140 D-(-)-2-(2,5-Dihydrophenyl)glycine**
26774-88-9 247-999-1

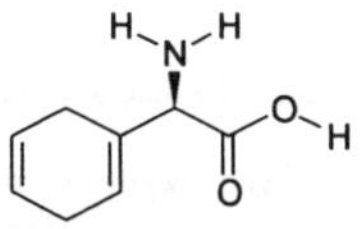

$C_8H_{11}NO_2$
Cyclohexa-1,4-diene-1-acetic acid, α-amino-, (αR)-. (R)-(-)-2-Amino-2-(cyclohexa-1,4-dienyl)acetic acid; D-2-(1,4-Cyclohexadienyl)glycine. Chiral building block. mp = 280°; [α] = - 160° (c = 1, 1N HCl). *Acros Organics nv; DSM Fine Chemcials Netherlands; Omega Chemical Company Inc.; Senn Chemicals AG; Sigma-Aldrich Fine Chemicals; TCI America.*

**2141 (S)-(+)-2,3-Dihydro-1H-pyrrolo-2,1-C1,4-benzodiazepine-5,11(10H,11αH)dione**
18877-34-4

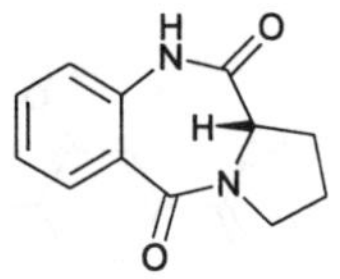

$C_{12}H_{12}N_2O_2$
Chiral intermediate. mp = 218-220°; $[\alpha]^{23}$ = + 525° (c = 0.6, $CH_3OH$). *Sigma-Aldrich Fine Chemicals.*

**2142 (R)-3,4-Dihydroxybutyramide**

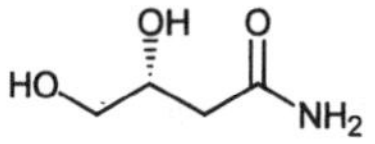

$C_4H_9NO_3$
Butanamide, 3,4-dihydroxy-, (R)-. Chiral building block. *Synthon Chiragenics Corporation.*

**2143 (S)-3,4-Dihydroxybutyramide**
126495-84-9

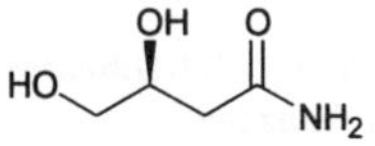

$C_4H_9NO_3$
Butanamide, 3,4-dihydroxy-, (S)-. Chiral building block. *Synthon Chiragenics Corporation.*

**2144 cis-1(R,2S)-1,2-Dihydroxy-1,2,3,4-tetra-hydronaphthalene**
57495-92-8

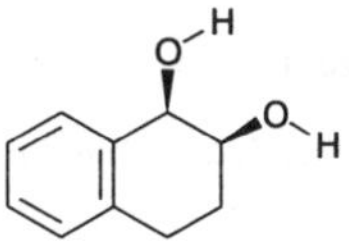

$C_{10}H_{12}O_2$
cis-(1R)-1,2,3,4-Tetrahydro-1,2-naphthalenediol. Chiral intermediate. *Acros Organics nv; Genencor International.*

**2145 cis-(2R,3S)-2,3-Dihydroxy-2,3-dihydro-benzonitrile**
138769-96-7

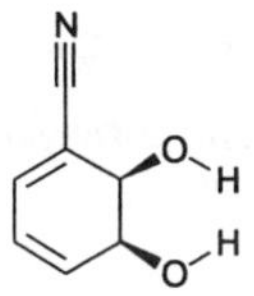

$C_7H_7NO_2$
Chiral intermediate. *Acros Organics nv.*

**2146 cis-2(R),3(S)-2,3-Dihydroxy-2,3-dihydro-benzonitrile acetonide**
150767-96-7

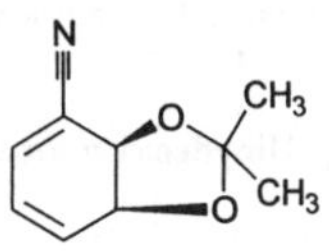

$C_{10}H_{11}NO_2$
Chiral intermediate. *Acros Organics nv.*

**2147 cis-(2S,3S)-2,3-Dihydroxy-2,3-dihydrobromobenzene**
130792-45-9

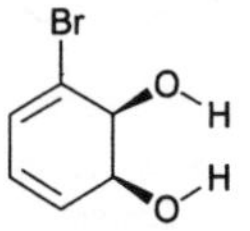

$C_6H_7BrO_2$
Chiral intermediate. *Acros Organics nv; Genencor International.*

**2148 cis-2(S),3(S)-2,3-Dihydroxy-2,3-dihydrochlorobenzene**
65986-73-4

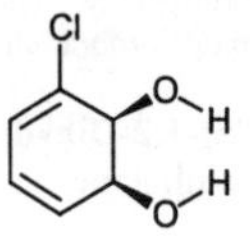

$C_6H_7ClO_2$
Chiral intermediate. *Acros Organics nv.*

**2149 cis-2(S),3(S)-2,3-Dihydroxy-2,3-dihydroiodobenzene**
138769-92-3

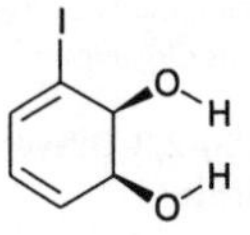

$C_6H_7IO_2$
Chiral intermediate. mp = 238-242°. *Acros Organics nv.*

**2150 cis-1(R,2S)-1,2-Dihydroxy-1,2-dihydronaphtalene**
51268-88-3

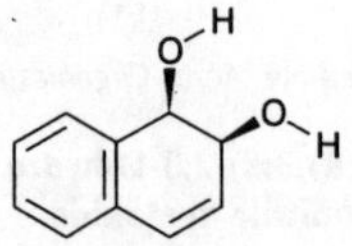

$C_{10}H_{10}O_2$
cis-(1R)-1,2-Dihydro-1,2-naphthalenediol. Chiral intermediate. *Acros Organics nv; Genencor International.*

**2151 2β,17β-Dihydroxyandrost-4-en-3-one**
10390-14-4
$C_{19}H_{28}O_3$
4-Androstene-2β,17β-diol-3-one; 2α-Hydroxytestosterone. Pharmaceutical or derivative. mp = 162-163°. *Ultrafine.*

**2152 (1R,2S,3R,4R)-2,3-Dihydroxy-4-(hydroxymethyl)-1-aminocyclopentane hydrochloride**
79200-56-9

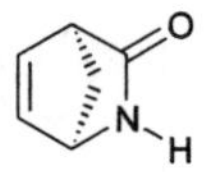

$C_6H_{14}ClNO_3$
Chiral intermediate. mp = 94-97°; [α] = - 560° (c = 1, $CH_2Cl_2$). *Acros Organics nv; ChiroTech Technology Ltd.; Lancaster Synthesis Ltd.; Sigma-Aldrich Fine Chemicals.*

**2153 (1S,2R,3S,4S)-2,3-Dihydroxy-4-(hydroxymethyl)-1-aminocyclopentane hydrochloride**
77841-56-6
$C_6H_{14}ClNO_3$
Chiral intermediate. *ChiroTech Technology Ltd.; Lancaster Synthesis Ltd.*

**2154 3α,5α-Dihydroxy-2β-(3S-hydroxy-trans-1-octenyl)-1α-cyclopentaneacetic acid**
56188-04-6

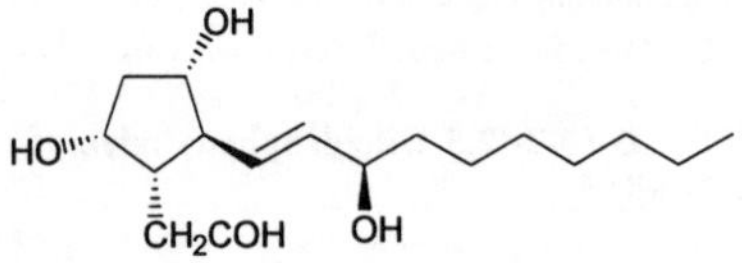

$C_{15}H_{25}O_5$
Chiral intermediate. *Pharmacia & Upjohn Inc.*

**2155 (-)-3,4-Dihydroxynorephedrine**
829-74-3 6785(12)

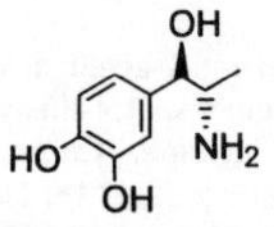

$C_9H_{13}NO_3$
α-(1-Aminoethyl)-3,4-dihydroxybenzyl alcohol. Pharmaceutical or derivative. $[\alpha]_D^{20}$ = - 18° (c = 0.5, 0.1N HCl). *Sigma-Aldrich Fine Chemicals.*

**2156 (4(S)-trans)-4,5-Dihydroxy-3-oxo-cyclohexene-1-carboxylic acid**
2922-42-1

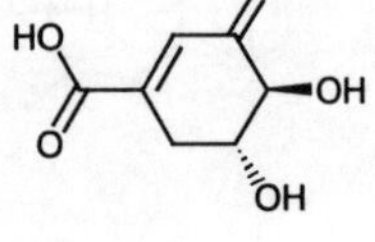

$C_7H_8O_5$
Chiral intermediate. *Acros Organics nv.*

**2157 D-(3,4-Dihydroxy)-α-phenylglycine**
56144-53-7

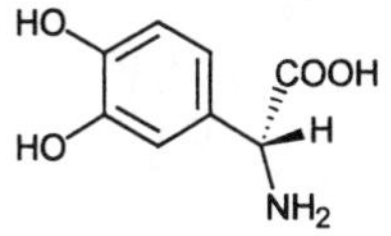

$C_8H_9NO_4$
Chiral building block. *Kaneka Corporation.*

**2158 L-3-(3,4-Dihydroxyphenyl)-2-methylalanine**
555-30-6 5974(11) 209-089-2

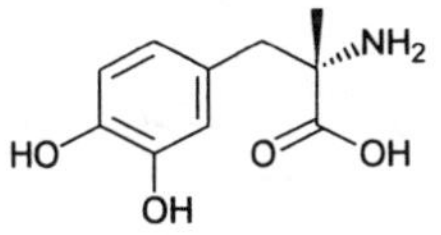

$C_{10}H_{13}NO_4$
Alanine, 3-(3,4-dihydroxyphenyl)-2-methyl-, L-; α-Methyl-L-DOPA; 3-Hydroxy-α-methyl-L-tyrosine; α-Methyl-L-3,4-dihydroxyphenylalanine; Aldomet; Aldometil; Aldomin; α-medopa; α-methyldopa; AMD; Bayer 1440L; Baypresol; Dopamet; Dopatec; Dopegyt; Medomet; Medopren; Methoplain; MK 351l; Nr.C 2294. Listed on TSCA. Chiral intermediate. mp > 300°; $[\alpha]^{25}$ = - 13° (c = 1, $H_2O$). *Acros Organics nv; Sigma-Aldrich Fine Chemicals; TCI America.*

**2159 3,5-Diiodo-D-thyronine**
5563-89-3 226-937-7

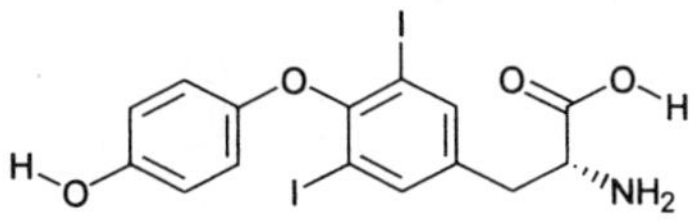

$C_{15}H_{13}I_2NO_4$
D-Alanine, 3-[4-(p-hydroxyphenoxy)-3,5-diiodophenyl]. O-(4-hydroxyphenyl)-3,5-diiodo-D-tyrosine; D-Tyrosine, O-(4-hydroxyphenyl)-3,5-diiodo-. Chiral intermediate. [α] = - 27° (c = 1, 1N HCl). *Acros Organics nv.*

**2160 3,5-Diiodo-L-tyrosine dihydrate**
300-39-0 3177(11) 206-092-0

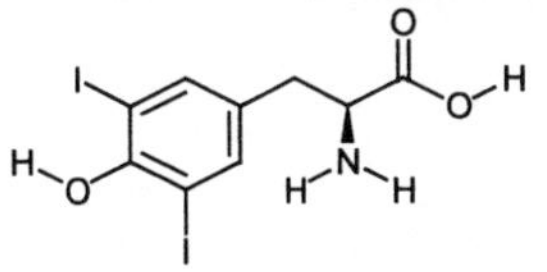

$C_9H_9I_2NO_3$
Tyrosine, 3,5-diiodo-, L-; 3,5-Diiodo-β-(p-hydroxyphenyl)alanine. Listed on TSCA. Chiral intermediate. mp = 200°; $[\alpha]_D^{20}$ = + 1.5° (c = 4, 1N HCl). *Acros Organics nv; Pfanstiehl Laboratories, Inc.; Sigma-Aldrich Fine Chemicals.*

**2161 (-)-Diisopinocampheylchloroborane**
85116-37-6

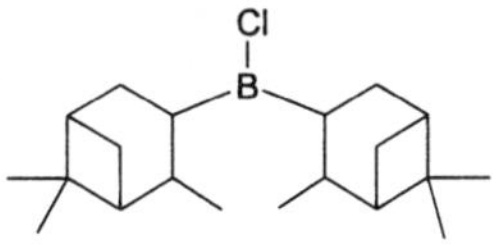

$C_{20}H_{34}BCl$
(-)-DPC. Chiral auxiliary. mp = 52-56°. *Callery Chemical Company; Sigma-Aldrich Fine Chemicals.*

**2162 (+)-Diisopinocampheylchloroborane**
112246-73-8

$C_{20}H_{34}BCl$
(+)-DPC. Chiral auxiliary. mp = 53-55°. *Callery Chemical Company; Sigma-Aldrich Fine Chemicals.*

**2163 (-)-Diisopropyl O,O-bis(trimethylsilyl)-tartrate**

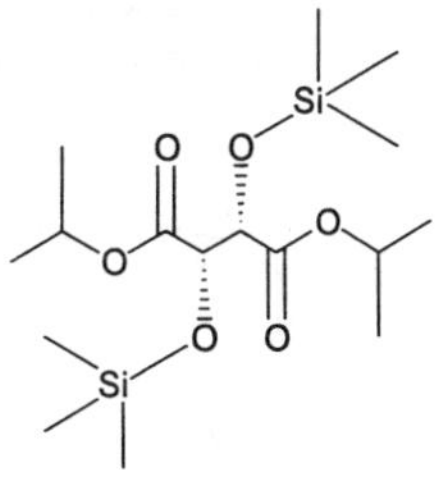

$C_{16}H_{34}O_6Si_2$
Chiral ligand; Chiral building block. $bp_{1.0}$ = 105-107°; d = 0.972; n = 1.428; $[\alpha]_D^{20}$ = - 57° (c = 1, $CHCl_3$). *Sigma-Aldrich Fine Chemicals.*

**2164 (+)-Diisopropyl O,O-bis(trimethylsilyl)-tartrate**

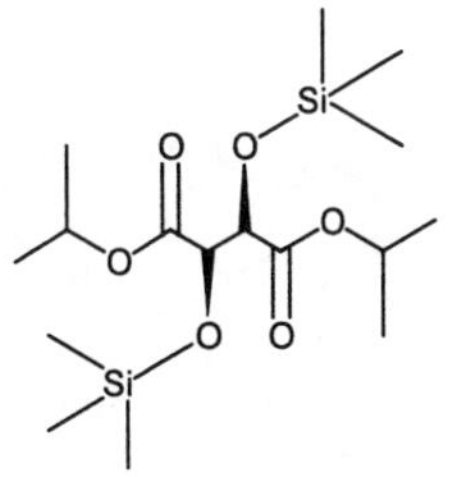

$C_{16}H_{34}O_6Si_2$
Chiral ligand; Chiral building block. $bp_{1.0}$ = 105-107°; d = 0.972; n = 1.428; $[\alpha]_D^{20}$ = + 57° (c = 1, $CHCl_3$). *Sigma-Aldrich Fine Chemicals.*

**2165 Diisopropyl (R)-(+)-malate**
83540-97-0

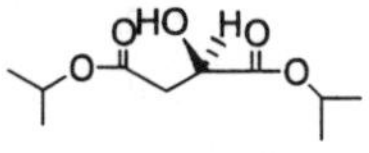

$C_{10}H_{18}O_5$
Chiral building block. bp = 237°; d = 1.055; n = 1.43; $[\alpha]^{25}$ = + 12.8° (neat). *Sigma-Aldrich Fine Chemicals.*

**2166 Diisopropyl (S)-(-)-malate**
83541-68-8

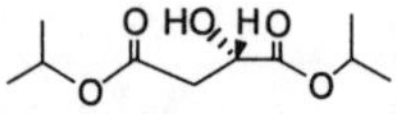

$C_{10}H_{18}O_5$
Chiral building block. bp = 237°; d = 1.055; n = 1.43; $[\alpha]_D^{20}$ = - 13° (neat). *Sigma-Aldrich Fine Chemicals.*

**2167 2,6-Diisopropylphenylimidoneophylidene [(R)-(+)-BIPHEN]molybdenum(VI)**

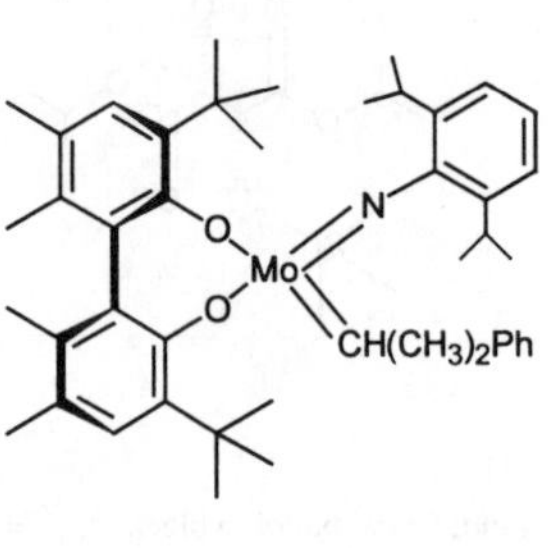

$C_{46}H_{61}MoNO_2$
(R)- Schrock-Hoveyda catalyst. Chiral catalyst. *Strem Chemicals, Inc.*

**2168 2,6-Diisopropylphenylimidoneophylidene [(S)-(-)-BIPHEN]molybdenum(VI)**

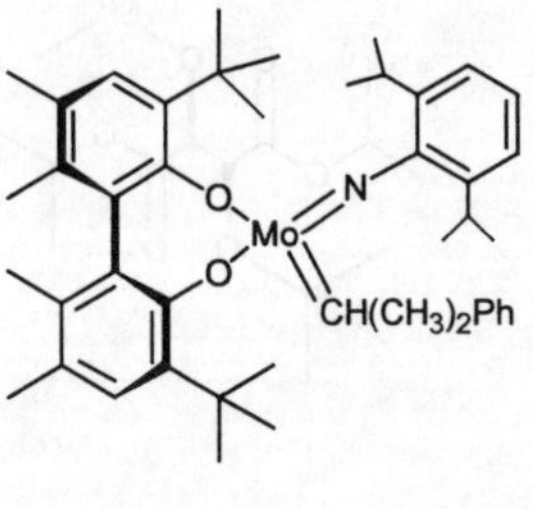

$C_{46}H_{61}MoNO_2$
(S)- Schrock-Hoveyda Catalyst. Chiral catalyst. *Strem Chemicals, Inc.*

**2169 (-)-Dimenthyl fumarate**
34675-24-6

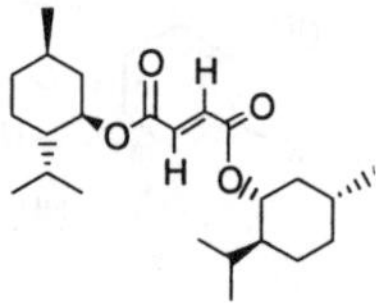

$C_{24}H_{40}O_4$
Chiral intermediate. mp = 59-61°; $bp_5$ = 211°; $[\alpha]_D^{20}$ = - 102° (c = 1, $CHCl_3$). *Sigma-Aldrich Fine Chemicals.*

**2170 (1R)-(-)-Dimenthyl succinate**

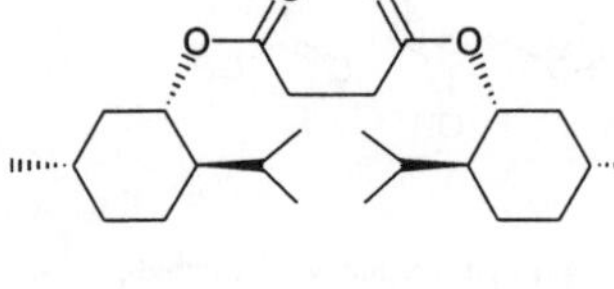

$C_{24}H_{42}O_4$
Chiral intermediate. mp = 62-64°; $bp_2$ = 200-205°; $[\alpha]_D^{20}$ = - 89° (c = 5, $CHCl_3$). *Sigma-Aldrich Fine Chemicals.*

**2171 (1S)-(+)-Dimenthyl succinate**
$C_{24}H_{42}O_4$
Chiral intermediate. mp = 62-64°; $bp_2$ = 200-205°; $[\alpha]_D^{20}$ = + 89° (c = 5, $CHCl_3$). *Sigma-Aldrich Fine Chemicals.*

**2172 (S)-(-)-(3,4-Dimethoxy)benzyl-1-phenylethylamine**

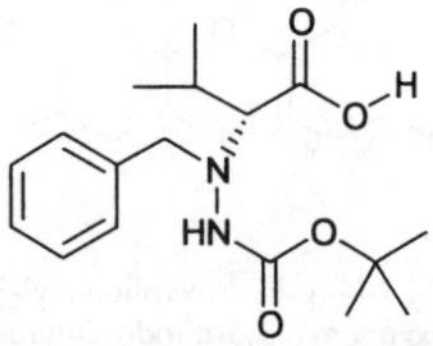

$C_{17}H_{21}NO_2$
Resolving agent. [α] = - 44.4°. *Acros Organics nv.*

**2173 (R)-(+)-N-(3,4-Dimethoxybenzyl)-1-phenylethylamine hydrochloride**

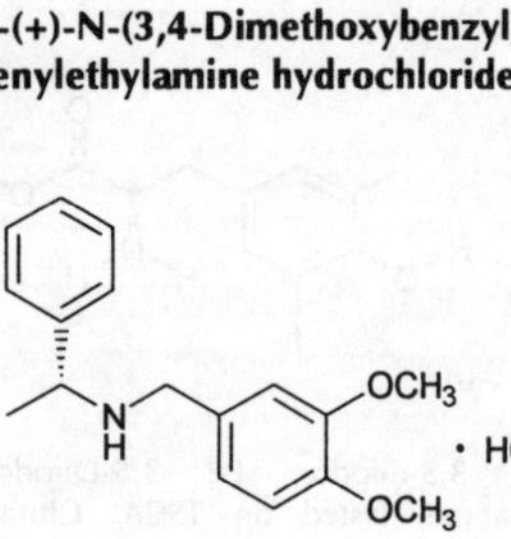

$C_{17}H_{22}ClNO_2$
Chiral intermediate. *Oxford Asymmetry International plc.*

**2174 (S)-(-)-N-(3,4-Dimethoxybenzyl)-1-phenylethylamine hydrochloride**
134430-93-6

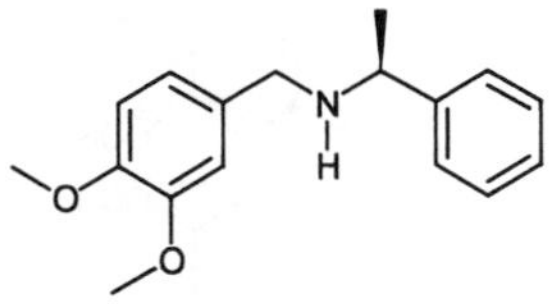

$C_{17}H_{21}NO_2$
Chiral intermediate. [α] = - 44.4°. *Acros Organics nv; Oxford Asymmetry International plc.*

**2175 (R)-2,2'-Dimethoxy-1,1'-binaphthalene-6,6'-dicarbaldehyde**

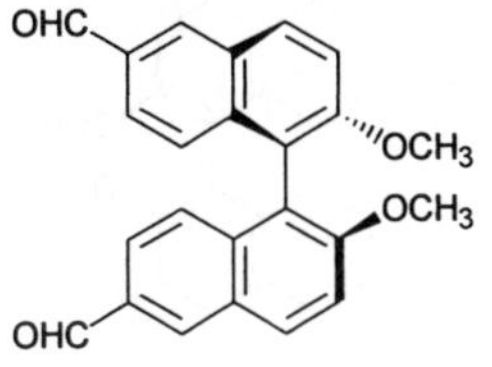

$C_{24}H_{18}O_4$
Chiral ligand. *Kankyo Kagaku Center Co., Ltd.*

**2176 (S)-2,2'-Dimethoxy-1,1'-binaphthalene-6,6'-dicarbaldehyde**

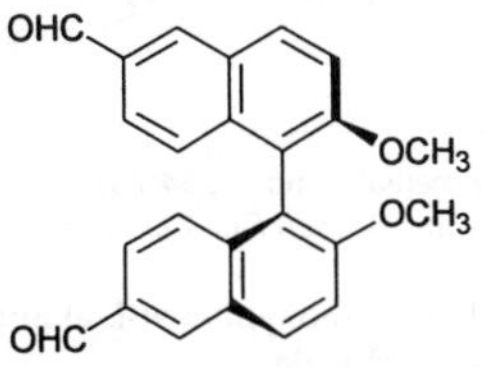

$C_{24}H_{18}O_4$
Chiral ligand. *Kankyo Kagaku Center Co., Ltd.*

**2177 (R)-2,2'-Dimethoxy-1,1'-binaphthalene-6,6'-dicarboxylic acid**

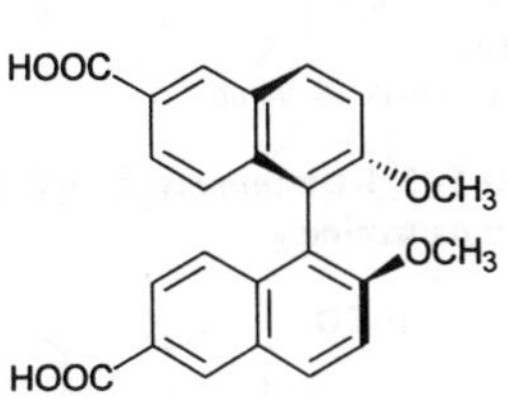

$C_{24}H_{18}O_6$
Chiral ligand. *Kankyo Kagaku Center Co., Ltd.*

**2178 (S)-2,2'-Dimethoxy-1,1'-binaphthalene-6,6'-dicarboxylic acid**

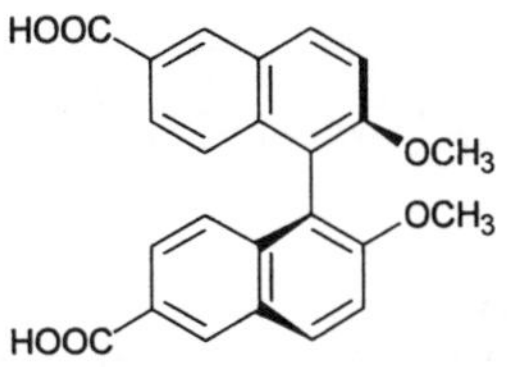

$C_{24}H_{18}O_6$
Chiral ligand. *Kankyo Kagaku Center Co., Ltd.*

**2179 (R)-2,2'-Dimethoxy-1,1'-binaphthyl**

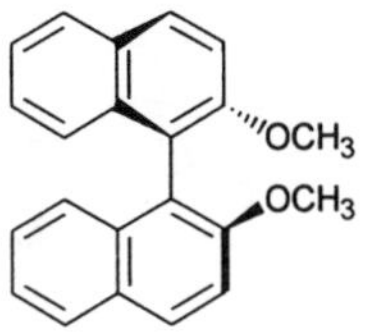

$C_{22}H_{18}O_2$
Chiral ligand. *Kankyo Kagaku Center Co., Ltd.*

**2180 (S)-2,2'-Dimethoxy-1,1'-binaphthyl**

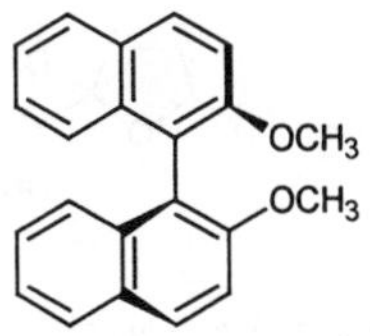

$C_{22}H_{18}O_2$
Chiral ligand. *Kankyo Kagaku Center Co., Ltd.*

**2181 (R,R)-(-)-2,3-Dimethoxy-1,4-bis(dimethylamino)butane**
26549-22-4 247-795-2

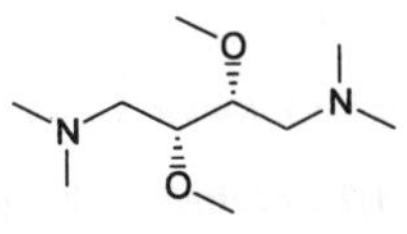

$C_{10}H_{24}N_2O_2$
Butane-1,4-diamine, 2,3-dimethoxy-N,N,N',N'-tetramethyl-, (R,R)-(-)-; (R,R)-(-)-2,3-Dimethoxy-N,N,N',N'-tetramethyl-1,4-butanediamine; (R,R)-1,4-Bis(dimethylamino)-2,3-dimethoxybutane; (R,R)-(-)-DDB. Chiral building block. [α] = - 14.4° (neat). *Acros Organics nv; TCI America.*

**2182 (S,S)-(+)-2,3-Dimethoxy-1,4-bis(dimethylamino)butane**
26549-21-3 247-794-7

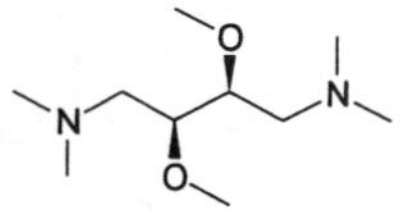

$C_{10}H_{24}N_2O_2$
Butane-1,4-diamine, 2,3-dimethoxy-N,N,N',N'-tetramethyl-, (S,S)-(+)-; (S,S)-(+)-2,3-Dimethoxy-N,N,N',N'-tetramethyl-1,4-butanediamine; (S,S)-1,4-Bis(dimethylamino)-2,3-dimethoxybutane; (S,S)-(+)-DDB. Chiral intermediate; Chiral ligand. *TCI America.*

**2183 (1R)-(2,5-Dimethoxy-2,5-dihydrofuran-2-yl)ethanol**

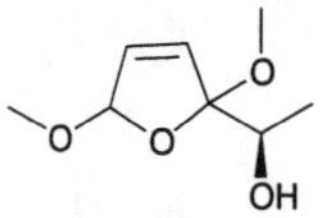

$C_8H_{14}O_4$
Chiral intermediate. *Acros Organics nv.*

**2184 (1S)-(2,5-Dimethoxy-2,5-dihydrofuran-2-yl)ethanol**

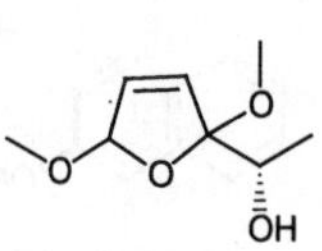

$C_8H_{14}O_4$
Chiral intermediate. *Acros Organics nv.*

**2185 (R,R)-1,2-Dimethoxy-1,2-diphenylethane**

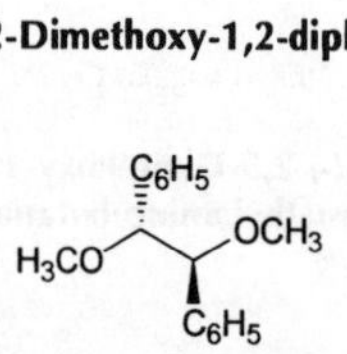

$C_{16}H_{18}O_2$
Chiral building block. *Kankyo Kagaku Center Co., Ltd.*

**2186 (S,S)-1,2-Dimethoxy-1,2-diphenylethane**

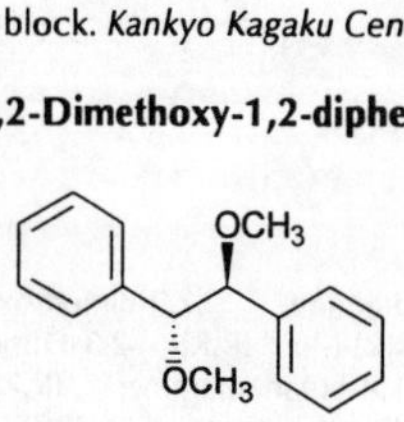

$C_{16}H_{18}O_2$
Chiral building block. *Kankyo Kagaku Center Co., Ltd.*

**2187 (R)-2,2'-Dimethoxy-5,5',6,6',7,7',8,8'-octahydro-1,1'-binaphthyl**

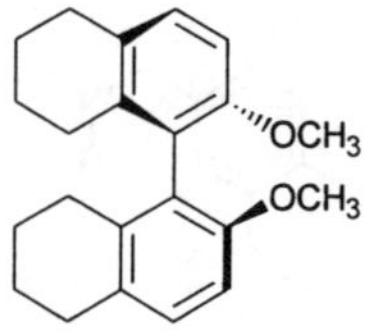

$C_{22}H_{26}O_2$
Chiral ligand. *Kankyo Kagaku Center Co., Ltd.*

**2188 (S)-2,2'-Dimethoxy-5,5',6,6',7,7',8,8'-octahydro-1,1'-binaphthyl**

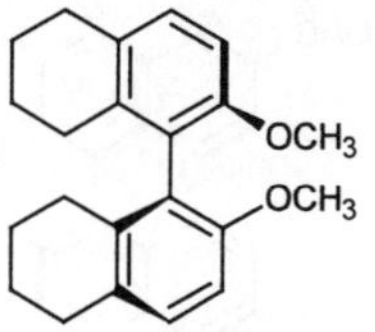

$C_{22}H_{26}O_2$
Chiral ligand. *Kankyo Kagaku Center Co., Ltd.*

**2189 (-)-3-(3,4-Dimethoxyphenyl)alanine**
32161-30-1

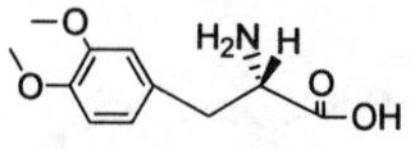

$C_{11}H_{15}NO_4$
Chiral intermediate. mp = 254-257°; $[\alpha]^{22}$ = - 5° (c = 4, 1N HCl). *Sigma-Aldrich Fine Chemicals.*

**2190 3,4-Dimethoxyphenyl-L-alanine hydrochloride**
69274-24-4

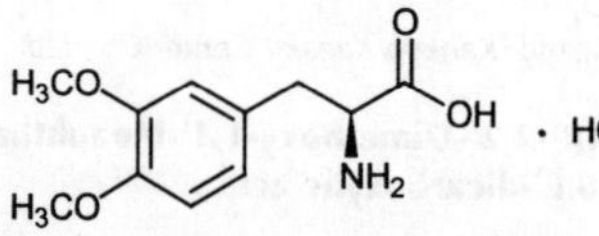

$C_{11}H_{16}ClNO_4$
Chiral intermediate. *Norchim SA.*

**2191 (R)-1-(3,4-Dimethoxyphenyl)-2-propanamine**

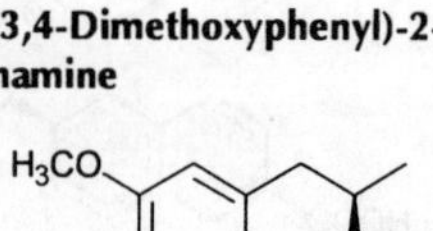

$C_{11}H_{17}NO_2$
Chiral building block. *Austin Chemical Company, Inc.*

**2192 (S)-1-(3,4-Dimethoxyphenyl)-2-propanamine**

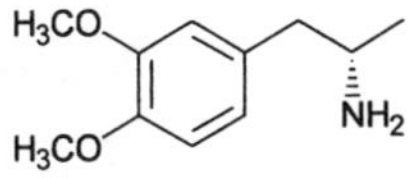

$C_{11}H_{17}NO_2$
Chiral building block. *Austin Chemical Company, Inc.*

**2193 (S)-1-(3,4-Dimethoxyphenyl)-2-propanol**
161121-02-4

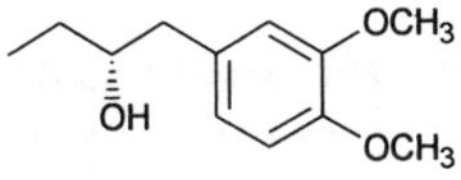

$C_{11}H_{16}O_3$
Chiral building block. *Daicel Chemical Ind. Ltd.*

**2194 (2R,3S)-(-)-4-Dimethylamino-1,2-diphenyl-3-methyl-2-butanol**
72541-03-8
$C_{19}H_{25}NO$
Chiral intermediate. mp = 54-55°. *Sigma-Aldrich Fine Chemicals.*

**2195 (2S,3R)-(+)-4-Dimethylamino-1,2-diphenyl-3-methyl-2-butanol**
38345-66-3 253-893-6

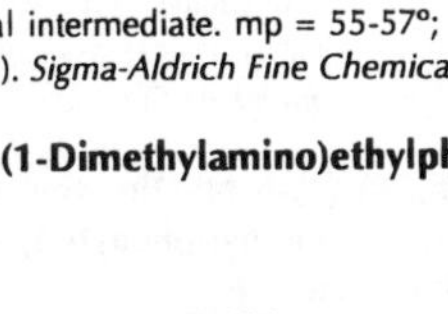

$C_{19}H_{25}NO$
Chirald. Chiral intermediate. mp = 55-57°; $[\alpha]^{21}$ = + 8.2° (c = 10, EtOH). *Sigma-Aldrich Fine Chemicals.*

**2196 (S)-3-(1-Dimethylamino)ethylphenol**
139306-10-8

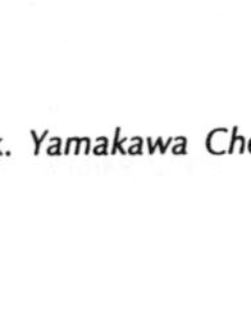

$C_{10}H_{15}NO$
Chiral building block. *Yamakawa Chemical Industry Co. Ltd.*

**2197 (R)-2,2-Dimethyl-4-aminomethyl-1,3-dioxane**

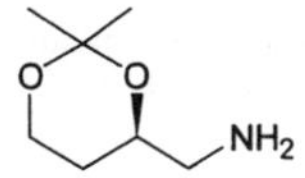

$C_7H_{15}NO_2$
Dioxane-4-methanamine, 1,3-, 2,2-dimethyl-, (R)-. Chiral intermediate. *Synthon Chiragenics Corporation.*

**2198 (S)-2,2-Dimethyl-4-aminomethyl-1,3-dioxane**
191354-52-6

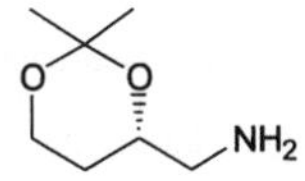

$C_7H_{15}NO_2$
Dioxane-4-methanamine, 1,3-, 2,2-dimethyl-, (S)-. Chiral intermediate. *Synthon Chiragenics Corporation.*

**2199 (R)-(-)-1-Dimethylamino-2-propanol**
15636-15-0

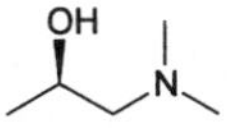

$C_5H_{13}NO$
Chiral building block. *Norse Laboratories.*

**2200 (S)-(+)-1-Dimethylamino-2-propanol**
53636-17-2

$C_5H_{13}NO$
Chiral building block. bp = 121-127°; d = 0.837; n = 1.419; $[\alpha]^{19}$ = + 27° (neat). *Norse Laboratories; Sigma-Aldrich Fine Chemicals.*

**2201 (3R)-(+)-3-(Dimethylamino)pyrrolidine**
132958-72-6
$C_6H_{14}N_2$
Chiral building block. *TCI America.*

**2202 (3S)-(-)-3-(Dimethylamino)pyrrolidine**
132883-44-4
$C_6H_{14}N_2$
Chiral building block. d = 0.9. *TCI America.*

**2203 (R)-(-)-4-(N,N-Dimethylaminosulfonyl)-7-(3-aminopyrrolidin-1-yl)-2,1,3-benzoxadiazole**
$C_{12}H_{17}N_5O_3S$
Chiral intermediate. *TCI America.*

**2204 (S)-(+)-4-(N,N-Dimethylaminosulfonyl)-7-(3-aminopyrrolidin-1-yl)-2,1,3-benzoxadiazole**

$C_{12}H_{17}N_5O_3S$
(S)-(+)-4-(N,N-Dimethylaminosulfonyl)-7-(3-aminopyrrolidin-1-yl) benzofurazan. Chiral intermediate. *TCI America.*

**2205 (R)-(+)-4-(N,N-Dimethylaminosulfonyl)-7-(2-chloroformylpyrrolidin-1-yl)-2,1,3-benzoxadiazole**

$C_{13}H_{15}ClN_4O_4S$
Chiral intermediate. *TCI America.*

**2206 (S)-(-)-4-(N,N-Dimethylaminosulfonyl)-7-(2-chloroformylpyrrolidin-1-yl)-2,1,3-benzoxadiazole**

$C_{13}H_{15}ClN_4O_4S$
(S)-(-)-4-(N,N-Dimethylaminosulfonyl)-7-(2-chloroformylpyrrolidin-1-yl); (S)-(-)-4-(N,N-Dimethylaminosulfonyl)-7-(2-chloroformylpyrrolidin-1-yl)- . Chiral intermediate. *TCI America.*

**2207 (R)-4-(N,N-Dimethylaminosulfonyl)-7-(3-isothiocyanatopyrrolidino)-2,1,3-benzoxadiazole**

163927-31-9
$C_{13}H_{15}N_5O_3S_2$
Chiral intermediate. *TCI America.*

**2208 (S)-4-(N,N-Dimethylaminosulfonyl)-7-(3-isothiocyanatopyrrolidino)-2,1,3-benzoxadiazole**

163927-32-0
$C_{13}H_{15}N_5O_3S_2$
(S)-4-(N,N-Dimethylaminosulfonyl)-7-(3-isothiocyanatopyrrolidino)benzofurazane; (S)-4-(N,N-Dimethylaminosulfonyl)-7-(3-isothiocyanatopyrrolidino)-. Chiral intermediate. *TCI America.*

**2209 (R,R)-(-)-Dimethyl-2,3-O-benzylidene-L-tartrate**

38270-72-3

$C_{13}H_{14}O_6$
mp = 74-76°; $[\alpha]_D^{20}$ = - 41.0° (c = 2, EtOH). *FineTech, Ltd.; Senn Chemicals AG; Sigma-Aldrich Fine Chemicals.*

**2210 (+)-Dimethyl 2,3-O-benzylidenetartrate**

38270-70-1

$C_{13}H_{14}O_6$
(4S,5S)-2-Phenyl-1,3-dioxolane-4,5-dicarboxylic acid, dimethyl ester. Chiral building block; Chiral ligand. mp = 74-76°; $[\alpha]_D^{20}$ = + 41.9° (c = 2, EtOH). *Sigma-Aldrich Fine Chemicals.*

**2211 (R)-N,N-Dimethyl-1-[(S)-1',2-bis(diphenylphosphino)ferrocenyl]ethylamine**

74311-56-1

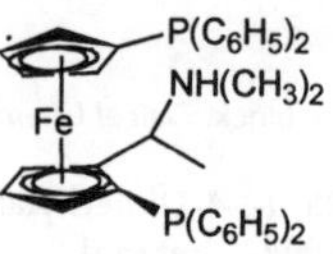


$C_{38}H_{37}FeNP_2$
(R)-(S)-BPPFA. Chiral catalyst. mp = 143-144°; $[\alpha]_D^{20}$ = - 350° (c = 0.5, $CHCl_3$). *Digital Specialty Chemicals, Inc.; Maxsyn; Sigma-Aldrich Fine Chemicals; TCI America.*

**2212 (S)-N,N-Dimethyl-1-[(R)-1',2-bis(diphenylphosphino)ferrocenyl]ethylamine**

55650-59-4

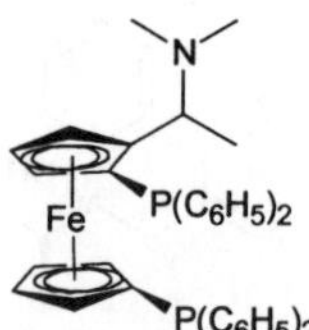


$C_{38}H_{37}FeNP_2$
(S)-(R)-BPPFA. Chiral catalyst. mp = 138-142°; $[\alpha]_D^{20}$ = + 350° (c = 0.5, $CHCl_3$). *Digital Specialty Chemicals, Inc.; Maxsyn; Sigma-Aldrich Fine Chemicals; TCI America.*

**2213 (1R,2R)-(+)-N,N'-Dimethyl-1,2-bis[3-(trifluoromethyl)phenyl]-1,2-ethanediamine**

$C_{18}H_{18}F_6N_2$
Chiral ligand. *Acros Organics nv.*

**2214 (1S,2S)-(-)-N,N-Dimethyl-1,2-bis[(3-trifluoromethyl)phenyl]-1,2-ethanediamine**

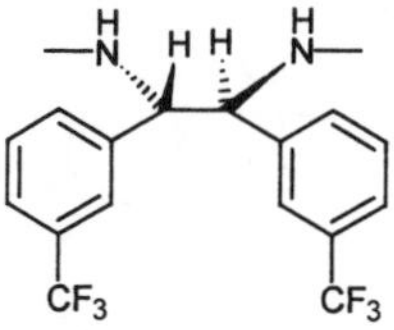

$C_{18}H_{18}F_6N_2$
Resolving agent; Chiral building block. mp = 113-115°; $[\alpha]_D^{20}$ = - 15° (c = 0.5, $CHCl_3$). *Acros Organics nv; Sigma-Aldrich Fine Chemicals.*

**2215 (R)-3,3-Dimethyl-2-butylamine**
66228-31-7
$C_6H_{15}N$
Butan-2-amine, 3,3-dimethyl-, (R)-; (R)-2-Amino-3,3-dimethylbutane; (-)-2,2-Dimethyl-3-aminobutane; (R)-1,2,2-Trimethylpropylamine; (R)-3,3-Dimethyl-2-aminobutane. Chiral intermediate. bp = 103°; $n_D^{20}$ = 1.4127. *Austin Chemical Company, Inc.; BASF Aktiengesellschaft; Chiragene, Inc.; Lancaster Synthesis Ltd.*

**2216 (S)-3,3-Dimethyl-2-butylamine**
22526-47-2
$C_6H_{15}N$
Butan-2-amine, 3,3-dimethyl-, (S)-; (S)-2-Amino-3,3-dimethylbutane; (S)-1,2,2-Trimethylpropanamine; Propylamine, 1,2,2-trimethyl-, (S)-. Chiral intermediate. bp = 103°; $n_D^{20}$ = 1.4126. *BASF Aktiengesellschaft; Chiragene, Inc.; Lancaster Synthesis Ltd.*

**2217 (R)-2,2-Dimethyl-4-[(tert-butylamino)-methyl]-1,3-dioxolane**

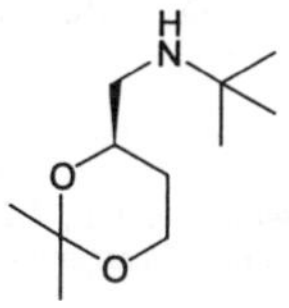

$C_{10}H_{21}NO_2$
Chiral intermediate. *Bachem AG.*

**2218 (S)-2,2-Dimethyl-4-[(tert-butylamino)-methyl]-1,3-dioxolane**

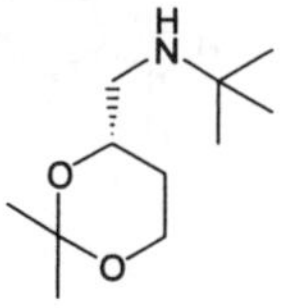

$C_{10}H_{21}NO_2$
Chiral intermediate. *Bachem AG.*

**2219 Dimethyl (1S,9S)-5-cyanosemicorrin-1,9-dicarboxylate**
105251-49-8

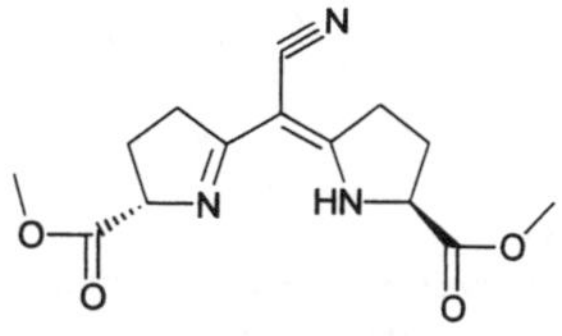

$C_{14}H_{17}N_3O_4$
Chiral intermediate. mp = 77-79°; $[\alpha]_D^{20}$ = - 144° (c = 1, $CHCl_3$). *Sigma-Aldrich Fine Chemicals.*

**2220 (+)-Dimethyl-β-cyclodextrin**
51166-71-3

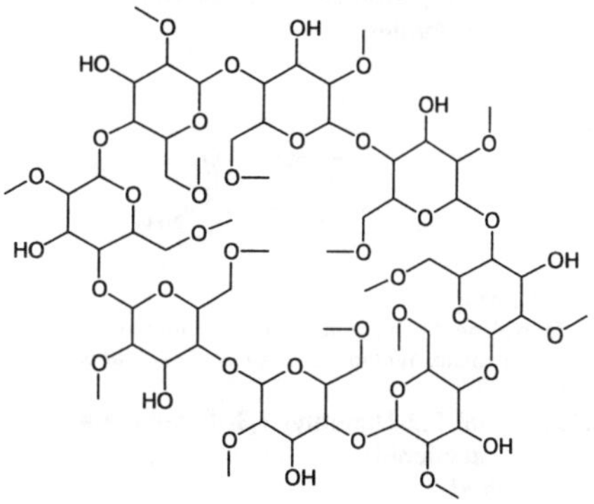

$C_{64}H_{112}O_{40}$
Chiral auxiliary. $[\alpha]^{20}$ = + 163° (c = 1, $H_2O$); $[\alpha]_D^{20}$ = + 126° (c = 1, $CHCl_3$). *Acros Organics nv; Sigma-Aldrich Fine Chemicals.*

**2221 (1R,2R)-(-)-N,N'-Dimethylcyclohexane-1,2-diamine**
68737-65-5
$C_8H_{18}N_2$
(1R,2R)-(-)-1,2-Bis(methylamino)cyclohexane. Chiral building block; Resolving agent. *Acros Organics nv; TCI America.*

**2222 (1S,2S)-(+)-N,N'-Dimethylcyclohexane-1,2-diamine**
87583-89-9
$C_8H_{18}N_2$
(1S,2S)-(+)-1,2-Bis(methylamino)cyclohexane. Chiral building block; Resolving agent. *TCI America.*

**2223 (+)-2,5-Dimethyl-1,4-cyclohexanediol**
$C_8H_{16}O_2$
Chiral building block. *Digital Specialty Chemicals, Inc.*

**2224 (S)-(+)-2,2-Dimethylcyclopropane-carboxamide**
75885-58-4 278-334-3

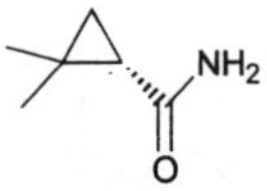

$C_6H_{11}NO$
Chiral intermediate. mp = 135-137°; $[\alpha]_D^{20}$ = + 82° (c = 1, $CH_3OH$). *Sigma-Aldrich Fine Chemicals.*

**2225 (1R,2S,6R,7S)-4,4-Dimethyl-3,5-dioxa-8-azatricyclo[5.2.1]decan-9-one**
79200-57-0
$C_9H_{13}NO_3$
(+)-exo-cis-5,6-Dimethylmethylenedioxy-2-azabicyclo-[2.2.1]heptan-3-one. Chiral intermediate. mp = 150-152°. *ChiroTech Technology Ltd.; Lancaster Synthesis Ltd.*

**2226 (R)-2,2-Dimethyl-1,3-dioxolane-4-acetamide**

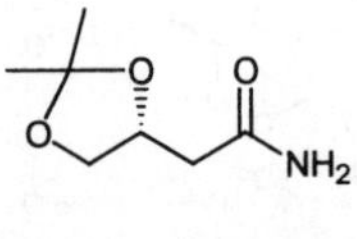

$C_7H_{13}NO_3$
Dioxolane-4-acetamide, 1,3-, 2,2-dimethyl-, (R)-. Chiral intermediate. *Synthon Chiragenics Corporation.*

**2227 (S)-2,2-Dimethyl-1,3-dioxolane-4-acetamide**
185996-33-2

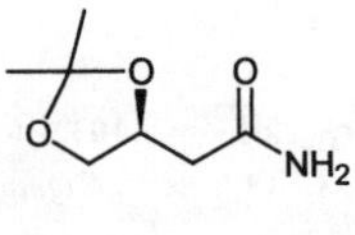

$C_7H_{13}NO_3$
Dioxolane-4-acetamide, 1,3-, 2,2-dimethyl-, (S)-. Chiral intermediate. *Acros Organics nv; Synthon Chiragenics Corporation.*

**2228 (R)-2,2-Dimethyl-1,3-dioxolane-4-acetic acid, methyl ester**
112031-10-4

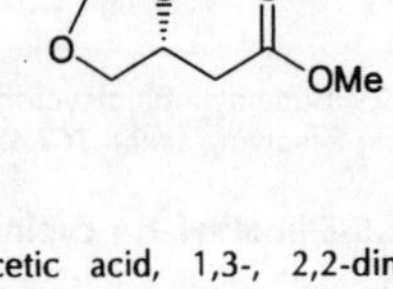

$C_8H_{14}O_4$
Dioxolane-4-acetic acid, 1,3-, 2,2-dimethyl-, methyl ester, (R)-. Chiral intermediate. *Synthon Chiragenics Corporation.*

**2229 (S)-2,2-Dimethyl-1,3-dioxolane-4-acetic acid, methyl ester**
95422-24-5

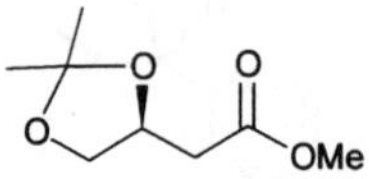

$C_8H_{14}O_4$
Dioxolane-4-acetic acid, 1,3-, 2,2-dimethyl-, methyl ester, (S); Methyl (S)-2,2-dimethyl-1,3-dioxolane-4-acetate. Chiral intermediate. bp = 192-194°; d = 1.07; n = 1.615; $[\alpha]_D^{20}$ = + 17° (neat). *Sigma-Aldrich Fine Chemicals; Synthon Chiragenics Corporation.*

**2230 (R)-2,2-Dimethyl-1,3-dioxolane-4-acetonitrile**
74923-97-0

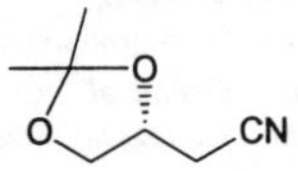

$C_7H_{11}NO_2$
Dioxolane-4-acetonitrile, 1,3-, 2,2-dimethyl-, (R)-. Chiral intermediate. *Synthon Chiragenics Corporation.*

**2231 (S)-2,2-Dimethyl-1,3-dioxolane-4-acetonitrile**
131724-43-1

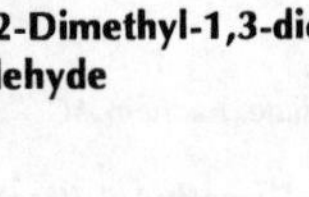

$C_7H_{11}NO_2$
Dioxolane-4-acetonitrile, 1,3-, 2,2-dimethyl-, (S)-. Chiral intermediate. *Synthon Chiragenics Corporation.*

**2232 (R)-(+)-2,2-Dimethyl-1,3-dioxolane-4-carboxaldehyde**

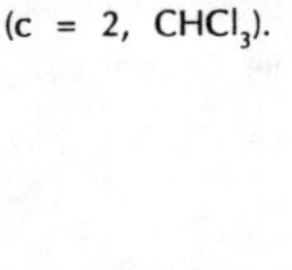

$C_6H_{10}O_3$
Chiral intermediate. bp = 139°; d = 1.045; n = 1.454; $[\alpha]^{22}$ = + 53.8° (c = 2, $CHCl_3$). *Sigma-Aldrich Fine Chemicals.*

**2233 (R)-(+)-2,2-Dimethyl-1,3-dioxolane-4-carboxylic acid methyl ester**
52373-72-5

$C_7H_{12}O_4$
(+)-2,3-O-Isopropylidene-D-glyceric acid methyl ester; (+)-Methyl 2,3-O-Isopropylidene-D-glycerate; (R)-(+)-Methyl 2,2-Dimethyl-1,3-dioxolane-4-carboxylate. Chiral intermediate. $bp_{10}$ = 70-75°; d = 1.106; n = 1.426; $[\alpha]_D^{20}$ = + 8.5° (c = 1.5, acetone). *Sigma-Aldrich Fine Chemicals; TCI America.*

**2234 (4R,5R)-2,2-Dimethyl-1,3-dioxolane-4,5-dicarboxylic acid dimethyl ester**
37031-29-1

$C_9H_{14}O_6$
(-)-2,3-O-Isopropylidene-L-tartaric acid dimethyl ester; (-)-Dimethyl 2,3-O-Isopropylidene-L-tartrate; Dimethyl (4R,5R)-2,2-Dimethyl-1,3-dioxolane-4,5-dicarboxylate; (R,R)-(-)-2,3-O-Dimethylisopropylidene-L-tartrate. Chiral intermediate. $bp_{19}$ = 150°; d = 1.188; n = 1.439; [α] = - 55° (c = 10, abs. EtOH). *Acros Organics nv; FineTech, Ltd.; Sigma-Aldrich Fine Chemicals; TCI America.*

**2235 (4S,5S)-2,2-Dimethyl-1,3-dioxolane-4,5-dicarboxylic acid dimethyl ester**
37031-30-4

$C_9H_{14}O_6$
(+)-2,3-O-Isopropylidene-D-tartaric acid dimethyl ester; (+)-Dimethyl 2,3-O-isopropylidene-D-tartrate; Dimethyl (4S,5S)-2,2-dimethyl-1,3-dioxolane-4,5-dicarboxylate. Chiral intermediate. $bp_{0.1}$ = 80-82°; d = 1.19; n = 1.439; [α] = + 55° (c = 10, abs EtOH). *Acros Organics nv; Sigma-Aldrich Fine Chemicals; TCI America.*

**2236 trans-(2,2-Dimethyl-1,3-dioxolane-4,5-diyl)bis(diphenylmethanol)**
93379-49-8

$C_{31}H_{30}O_4$
Chiral intermediate. mp = 194-196°; $[\alpha]^{19}$ = + 67° (c = 1, $CHCl_3$). *Sigma-Aldrich Fine Chemicals.*

**2237 (R)-2,2-Dimethyl-1,3-dioxolane-4-propanol**
133008-08-9

$C_8H_{16}O_3$
Dioxolane-4-propanol, 1,3-, 2,2-dimethyl-, (R)-. Chiral intermediate. *Synthon Chiragenics Corporation.*

**2238 (R)-(-)-2,2-Dimethyl-1,3-dioxolan-4-yl methyl mesylate**
83461-40-9

$C_7H_{14}O_5S$
Chiral intermediate. *Chemi SpA.*

**2239 (S)-(+)-2,2-Dimethyl-1,3-dioxolan-4-yl methyl mesylate**
90129-42-3

$C_7H_{14}O_5S$
Chiral intermediate. *Chemi SpA.*

**2240 (R)-(-)-2,2-Dimethyl-1,3-dioxolane-4-yl methyl p-toluenesulfonate**
23788-74-1

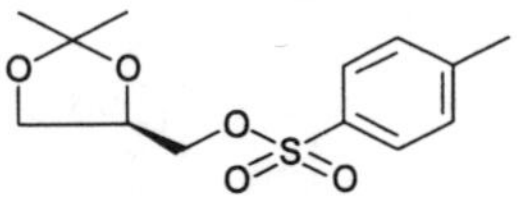

$C_{13}H_{18}O_5S$
(R)-Isopropylideneglycerol tosylate; (R)-(-)-2,2-Dimethyl-4-(hydroxymethyl)-1,3-dioxolane p-toluenesulfonate. Chiral intermediate. mp = 41-43°; d = 1.208; n = 1.506; [α] = - 3.7° (c = 10, abs.EtOH); $[\alpha]_D^{20}$ = - 7.0° (neat). *Acros Organics nv; Bachem AG; Chemi SpA; Interchem Corporation; Sigma-Aldrich Fine Chemicals; TCI America; Vinchem.*

**2241 (S)-(+)-2,2-Dimethyl-1,3-dioxolane-4-yl-methyl p-toluenesulfonate**
23735-43-5

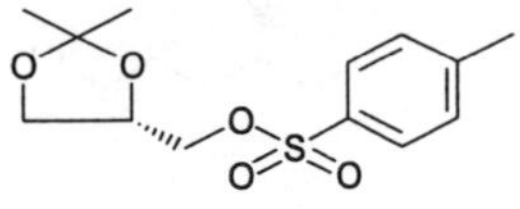

$C_{13}H_{18}O_5S$
(S)-Isopropylideneglycerol tosylate; (S)-(-)-2,2-Dimethyl-4-(hydroxymethyl)-1,3-dioxolane p-toluenesulfonate. Chiral intermediate. d = 1.208; n = 1.506; [α] = + 4.1° (c = 10, abs.EtOH); $[\alpha]^{26}$ = + 6.0° (neat). *Acros Organics nv; Bachem AG; Chemi SpA; Interchem Corporation; Sigma-Aldrich Fine Chemicals; TCI America; Vinchem.*

**2242 3-((4S)-2,2-Dimethyl-1,3-dioxolan-4-yl)-propanol**
51268-87-2

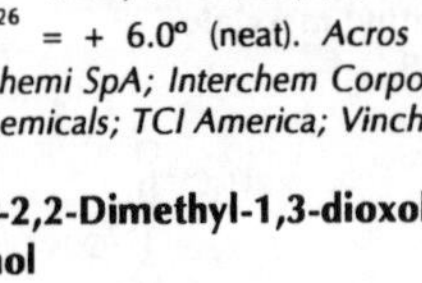

$C_8H_{16}O_3$
Dioxolane-4-propanol, 1,3-, 2,2-dimethyl-, (S)-; (S)-2,2-Dimethyl-1,3-dioxolane-4-propanol. Chiral intermediate. *Acros Organics nv; Synthon Chiragenics Corporation.*

**2243 Dimethyl (4R,5R)-1,3,2-dioxathiolane-4,5-dicarboxylate-2,2-dioxide**
117470-90-3

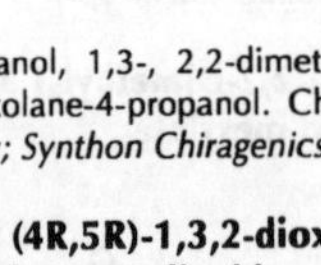

$C_6H_8O_8S$
Chiral intermediate. mp = 68-71°; $[\alpha]_D^{20}$ = - 73° (c = 2, $CHCl_3$). *Oxford Asymmetry International plc; Sigma-Aldrich Fine Chemicals.*

**2244 Dimethyl (4S,5S)-1,3,2-dioxathiolane-4,5-dicarboxylate-2,2-dioxide**
127854-46-0

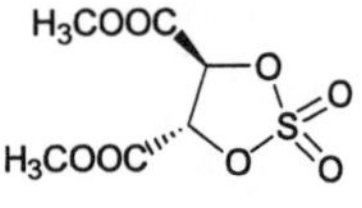

$C_6H_8O_8S$
Chiral intermediate. *Oxford Asymmetry International plc.*

**2245 (1R,2R)-(+)-N,N'-Dimethyl-1,2-diphenyl-ethylenediamine**
118628-68-5

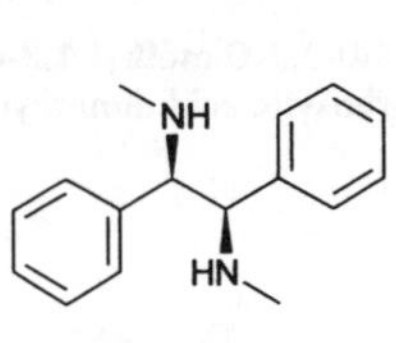

$C_{16}H_{20}H_2$
Chiral ligand. mp = 49-51°; [α] = + 20° (c = 0.15, $CHCl_3$). *Acros Organics nv; Advanced Asymmetrics, Inc.*

**2246 (1S,2S)-(-)-N,N'-Dimethyl-1,2-diphenyl-ethylenediamine**
70749-06-3

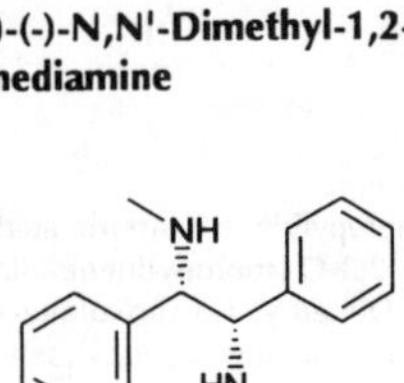

$C_{16}H_{20}N_2$
Resolving agent; Chiral building block. [α] = - 20° (c = 0.15, $CHCl_3$). *Acros Organics nv; Advanced Asymmetrics, Inc.*

**2247 3-(1,3-Dimethyl-(4S,5S)-diphenyl-imidazolidin-2-yl)pyridine**

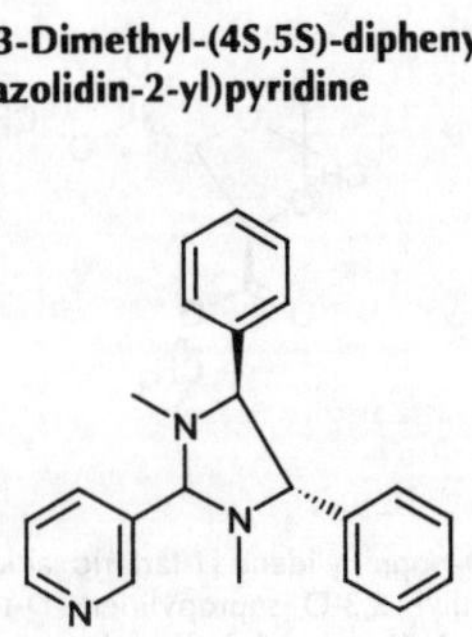

$C_{22}H_{23}N_3$
Chiral intermediate. *Acros Organics nv.*

**2248 (R)-N,N-Dimethyl-1-[(R)-2-(diphenylphosphino)ferrocenyl]ethylamine**

74311-54-9

$C_{26}H_{28}FeNP$

Chiral intermediate. *Digital Specialty Chemicals, Inc.*

**2249 (R)-N,N-Dimethyl-1-[(S)-2-(diphenylphosphino)ferrocenyl]ethylamine**

55700-44-2

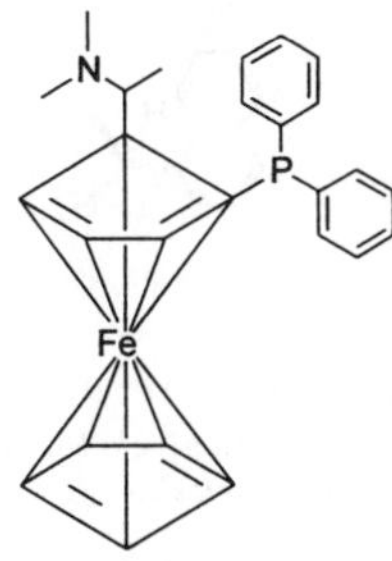


$C_{26}H_{28}FeNP$

(R)-(S)-PPFA. Chiral catalyst. mp = 141-143°; $[\alpha]_D^{20}$ = - 355° (c = 0.6, EtOH). *Digital Specialty Chemicals, Inc.; Maxsyn; Sigma-Aldrich Fine Chemicals; TCI America.*

**2250 (S)-N,N-Dimethyl-1-[(R)-2-(diphenylphosphino)ferrocenyl]ethylamine**

55650-58-3

$C_{26}H_{28}FeNP$

(S)-(R)-PPFA. Chiral catalyst. mp = 140-142°; $[\alpha]_D^{20}$ = + 360° (c = 0.6, EtOH). *Digital Specialty Chemicals, Inc.; Maxsyn; Sigma-Aldrich Fine Chemicals; TCI America.*

**2251 (R)-(+)-N,N-Dimethyl-1-ferrocenylethylamine**

31886-58-5

$C_{14}H_{19}FeN$

[1-(Dimethylamino)ethyl]ferrocene. Chiral catalyst. $bp_{0.70}$ = 120-121°; d = 0; n = 1.589; $[\alpha]_D^{20}$ = + 13° (c = 1, EtOH, 5h). *Sigma-Aldrich Fine Chemicals; TCI America.*

**2252 (S)-(-)-N,N-Dimethyl-1-ferrocenylethylamine**

31886-57-4

$C_{14}H_{19}FeN$

(S)-(-)-[1-(Dimethylamino)ethyl]ferrocene. Chiral catalyst. $bp_{0.70}$ = 120-121°; $[\alpha]_D^{20}$ = - 13° (c = 1, EtOH, 5h). *Sigma-Aldrich Fine Chemicals; TCI America.*

**2253 (2R)-(+)-3,3-Dimethylglycidyl 4-nitrobenzoate**

$C_{12}H_{13}NO_5$

Chiral intermediate. mp = 110-113°; $[\alpha]^{19}$ = + 36° (c = 3, $CHCl_3$). *Sigma-Aldrich Fine Chemicals.*

**2254 (R)-2,2-Dimethyl-4-[(isopropylamino)methyl]-1,3-dioxolane**

$C_9H_{19}NO_2$

Chiral intermediate. *Bachem AG.*

**2255 (S)-2,2-Dimethyl-4-[(isopropylamino)methyl]-1,3-dioxolane**

$C_{11}H_{21}NO_2$

Chiral intermediate. *Bachem AG.*

**2256 Dimethyl (R)-(+)-malate**
70681-41-3

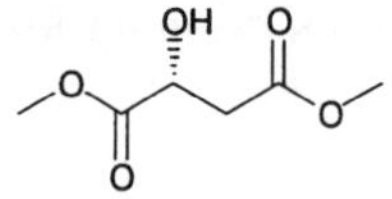

$C_6H_{10}O_5$
Chiral building block. $bp_{1.0}$ = 104-108°; d = 1.232; n = 1.44; $[\alpha]_D^{20}$ = + 6.6° (neat). *Sigma-Aldrich Fine Chemicals.*

**2257 Dimethyl (S)-(-)-malate**
617-55-0

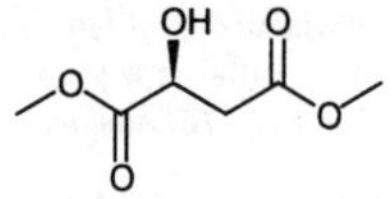

$C_6H_{10}O_5$
Chiral building block. d = 1.223; n = 1.435; $[\alpha]_D^{20}$ = - 6.5° (neat). *Sigma-Aldrich Fine Chemicals.*

**2258 (5R,11R)-(+)-2,8-Dimethyl-6H,12H-5,11-methanodibenzo[b,f][1,5]diazocin**
21451-74-1

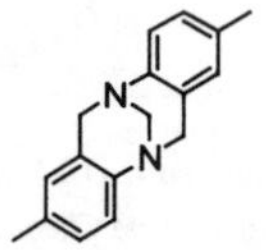

$C_{17}H_{18}N_2$
(+)-Troger's base. Chiral auxiliary; Resolving agent. mp = 129-131°; $[\alpha]_D^{20}$ = + 280° (c = 0.5, hexane). *RCA Reuter Chemischer Apparaebau KG; Sigma-Aldrich Fine Chemicals.*

**2259 (5S,11S)-(-)-2,8-Dimethyl-6H,12H-5,11-methanodibenzo[b,f][1,5]diazocin**
14645-24-0
$C_{17}H_{18}N_2$
(-)-Troger's base. Chiral auxiliary; Resolving agent. *RCA Reuter Chemischer Apparaebau KG.*

**2260 (S,S)-(+)-2,2-Dimethyl-5-methylamino-4-phenyl-1,3-dioxane**

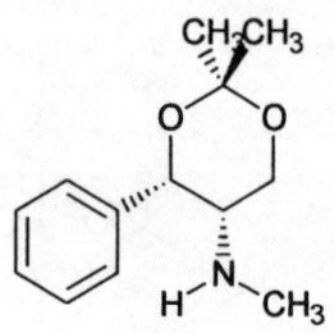

$C_{13}H_{19}NO_2$
Chiral intermediate. *Acros Organics nv.*

**2261 exo-cis-5,6-Dimethylmethylenedioxy-2-azabicyclo[2.2.1]heptan-3-one**
130468-02-9
$C_9H_{13}NO_3$
Chiral intermediate. mp = 150-152°. *ChiroTech Technology Ltd.; Lancaster Synthesis Ltd.*

**2262 (R)-(+)-N,N-Dimethyl-α-(1-naphthyl)-ethylamine**
119392-95-9

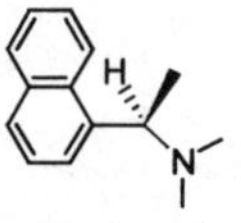

$C_{14}H_{17}N$
Chiral intermediate. bp = 218°; d = 1; n = 1.591; $[\alpha]^{23}$ = + 43° (c = 2, $CH_3OH$). *FineTech, Ltd.; Norse Laboratories; Sigma-Aldrich Fine Chemicals.*

**2263 (S)-(-)-N,N-Dimethyl-+1-(1-naphthyl)-ethylamine**
121045-73-6

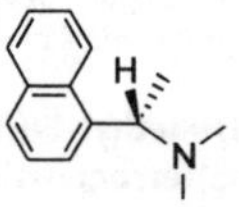

$C_{14}H_{17}N$
Chiral building block. bp = 212°; d = 1.006; n = 1.591; $[\alpha]^{23}$ = - 43° (c = 2, $CH_3OH$). *FineTech, Ltd.; Norse Laboratories; Sigma-Aldrich Fine Chemicals.*

**2264 2,7-Dimethyl-(3R,6R)octanediol**
151765-22-9
$C_{10}H_{22}O_2$
(3R,6R)-2,7-Dimethyloctane-3,6-diol. Chiral intermediate. *ChiroTech Technology Ltd.; Lancaster Synthesis Ltd.*

**2265 2,7-Dimethyl-(3S,6S)octanediol**
129705-30-2
$C_{10}H_{18}O_2$
(3S,6S)-2,7-Dimethyloctane-3,6-diol. Chiral intermediate. *ChiroTech Technology Ltd.; Lancaster Synthesis Ltd.*

**2266 2,7-Dimethyl-(3R,6R)oct-4-ynediol**
$C_{10}H_{18}O_2$
(3R,6R)-2,7-Dimethyloct-4-yne-3,6-diol. Chiral intermediate. *ChiroTech Technology Ltd.; Lancaster Synthesis Ltd.*

**2267 2,7-Dimethyl-(3S,6S)oct-4-ynediol**
$C_{10}H_{18}O_2$
(3S,6S)-2,7-Dimethyloct-4-yne-3,6-diol. Chiral intermediate. *ChiroTech Technology Ltd.; Lancaster Synthesis Ltd.*

**2268 (2R,3S)-1,1-Dimethyl-2-(3-oxobutyl)-3-(3-cyano-1-methylenepropyl) cyclobutane**
$C_{15}H_{23}NO$
Chiral intermediate. [α] = + 79° (neat). *Acros Organics nv.*

**2269 2-[(4S)-2,2-Dimethyl-5-oxo-1,3-dioxolan-4-yl]acetic acid**
73991-95-4

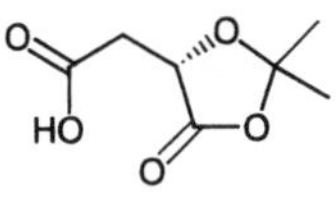

$C_7H_{10}O_5$
Chiral intermediate. mp = 105-112°; $[\alpha]^{23}$ = + 3.7° (c = 1, $CHCl_3$). *Acros Organics nv; Sigma-Aldrich Fine Chemicals.*

**2270 (R)-(-)-2,2-Dimethyl-5-oxo-1,3-dioxolane-4-acetic acid**
113278-68-5

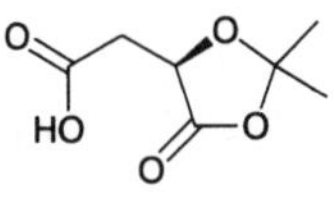

$C_7H_{10}O_5$
Chiral intermediate. mp = 108-112°; $[\alpha]_D^{20}$ = - 3.5° (c = 1, $CHCl_3$). *Sigma-Aldrich Fine Chemicals.*

**2271 (1R,3S)-2,2-Dimethyl-3-(2-oxopropyl)-cyclopropaneacetonitrile**
110847-02-4

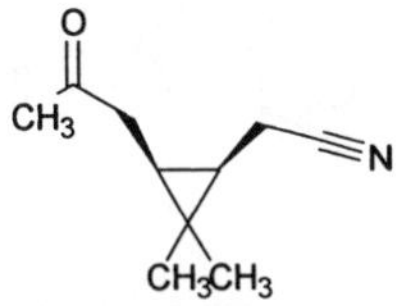

$C_{10}H_{15}NO$
Chiral intermediate. $bp_7$ = 132-133 °; $[\alpha]^{25}$ = - 13° (neat). *Acros Organics nv.*

**2272 N,N-Dimethyl-L-phenylalanine**
17469-89-5

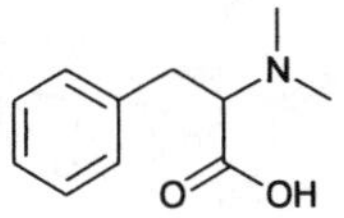

$C_{11}H_{15}NO_2$
Chiral building block. mp = 225-227°; $[\alpha]_D^{20}$ = + 77° (c = 1.3, $H_2O$). *Senn Chemicals AG; Sigma-Aldrich Fine Chemicals.*

**2273 (R)-N,N-Dimethyl-1-phenylethylamine**
19342-01-9 242-976-2

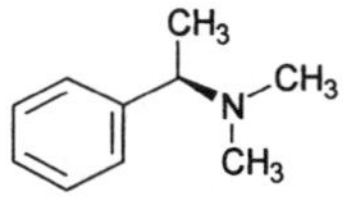

$C_{10}H_{15}N$
Benzenemethanamine, N,N,α-trimethyl-, (R)-; α-Methyl-N,N-dimethylbenzylamine; (R)-N,N,α-trimethylbenzylamine; N,N-Dimethyl-α-methylbenzylamine. Chiral building block. bp = 195°; bpsi16 = 81°; d = 0.9; n = 1.503; $[\alpha]_D^{20}$ = + 50.2 (c = 1, $CH_3OH$); $[\alpha]_D^{20}$ = + 72.6 ± 0.3° (neat). *Acros Organics nv; FineTech, Ltd.; Kiralchem Ltd.; Norse Laboratories; Sigma-Aldrich Fine Chemicals; TCI America.*

**2274 (S)-N,N-Dimethyl-1-phenylethylamine**
17279-31-1 241-310-8

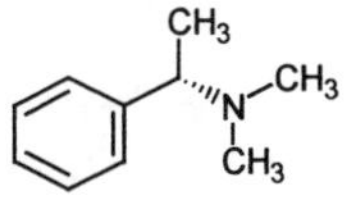

$C_{10}H_{15}N$
Benzenemethanamine, N,N,α-trimethyl-, (S)-; (S)-N,N,α-Trimethylbenzylamine; (S)-α-Methylbenzyldimethylamine; Benzylamine, N,N,α-trimethyl-, D-(-)-. Resolving agent. bp = 195°; bpsi16 = 81°; d = 0.899; n =1.503; $[\alpha]_D^{20}$ = -49.2° (c = 1, $CH_3OH$); $[\alpha]_D^{20}$ = - 72.6 ± 0.3° (neat). *Acros Organics nv; FineTech, Ltd.; Kiralchem Ltd.; Norse Laboratories; Sigma-Aldrich Fine Chemicals; TCI America.*

**2275 (4S,5R)-(+)-1,5-Dimethyl-4-phenyl-2-imidazolidinone**
112791-04-5

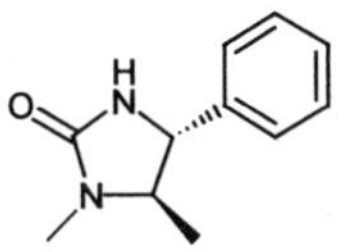

$C_{11}H_{14}N_2O$
Chiral intermediate. mp = 175-177°; $[\alpha]_D^{20}$ = + 45° (c = 1, $CH_3OH$). *Sigma-Aldrich Fine Chemicals.*

**2276 (4R,5S)-(-)-1,5-Dimethyl-4-phenyl-2-imidazolidninone**
92841-65-1

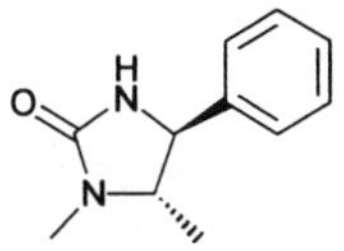

$C_{11}H_{14}N_2O$
Chiral intermediate. mp = 171-174°; $[\alpha]_D^{20}$ = - 42° (c = 1, $CH_3OH$). *Sigma-Aldrich Fine Chemicals.*

**2277 (R)-(-)-5,5-Dimethyl-4-phenyl-2-oxazolidinone**
170918-42-0

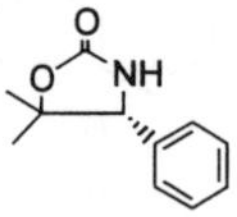

$C_{11}H_{13}NO_2$
Chiral intermediate. mp = 151-156°; $[\alpha]^{25}$ = - 71° (c = 2, $CHCl_3$). *Sigma-Aldrich Fine Chemicals.*

**2278 (S)-(+)-5,5-Dimethyl-4-phenyl-2-oxazolidinone**

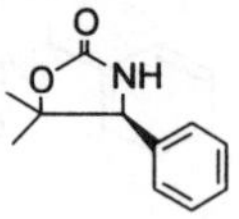

$C_{11}H_{13}NO_2$
Chiral intermediate. mp = 151-156°; $[\alpha]^{25}$ = + 71° (c = 2, $CHCl_3$). *Sigma-Aldrich Fine Chemicals.*

**2279 (R)-(-)-N,S-Dimethyl-(S)-phenylsulphoximine**
20414-85-1

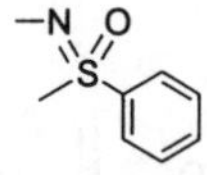

$C_8H_{11}NOS$
Resolving agent; Chiral auxiliary; Chiral intermediate. $bp_{0.4}$ = 110°; d = 1.1600; $n_D^{20}$ = 1.5658; $[\alpha]_D^{20}$ = - 179 ± 3° (c = 3, acetone); $[\alpha]^{24}$ = - 140° (c = 4.3, $CH_3OH$). *Kiralchem Ltd.; Lancaster Synthesis Ltd.; Sigma-Aldrich Fine Chemicals.*

**2280 (S)-(+)-N,S-Dimethyl-(S)-phenylsulphoximine**
33993-53-2

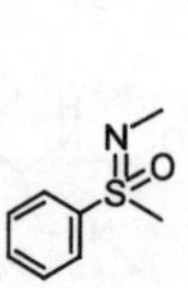

$C_8H_{11}NOS$
Resolving agent; Chiral auxiliary; Chiral intermediate. $bp_{0.4}$ = 110°; d = 1.1600; $n_D^{20}$ = 1.5658; $[\alpha]_D^{20}$ = + 179 ± 3° (c = 3, acetone); $[\alpha]^{24}$ = + 140° (c = 4.4, $CH_3OH$). *Kiralchem Ltd.; Lancaster Synthesis Ltd.; Sigma-Aldrich Fine Chemicals.*

**2281 (2R,5R)-2,5-Dimethylpyrrolidine**
62617-70-3

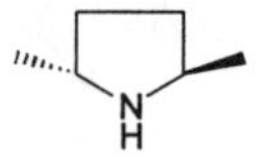

$C_6H_{13}N$
Chiral building block. *Acros Organics nv.*

**2282 (2S,5S)-2,5-Dimethylpyrrolidine**
117968-50-0

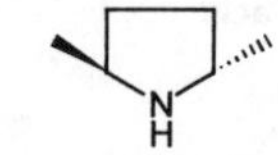

$C_6H_{13}N$
Chiral building block. *Acros Organics nv.*

**2283 (-)-(4R-trans)-2,2-Dimethyl-tetra(1-naphthyl)-1,3-dioxolane-4,5-dimethanol**
137536-94-8

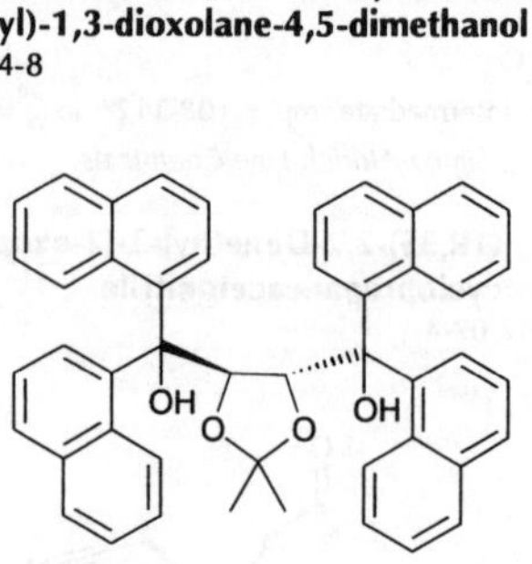

$C_{47}H_{38}O_4$
Chiral intermediate. mp = 200°; $[\alpha]_D^{20}$ = - 290° (c = 1, EtOAc). *Sigma-Aldrich Fine Chemicals.*

**2284 (+)-(4R-trans)-2,2-Dimethyl-tetra(1-naphthyl)-1,3-dioxolane-4,5-dimethanol**

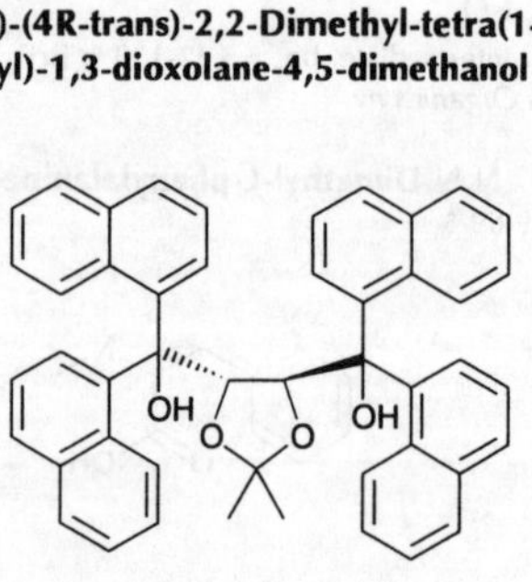

$C_{47}H_{38}O_4$
Chiral intermediate. mp = 200°; $[\alpha]_D^{20}$ = + 280° (c = 1, EtOAc). *Sigma-Aldrich Fine Chemicals.*

**2285 (4S-trans)-2,2-Dimethyl-tetra(2-naphthyl)-1,3-dioxolane-4,5-dimethanol**
137365-16-3

$C_{47}H_{38}O_4$
Chiral intermediate. mp = 214-216°; $[\alpha]_D^{20}$ = + 116° (c = 1, EtOAc). *Sigma-Aldrich Fine Chemicals.*

**2286 L-5,5-Dimethylthiazolidine-4-carboxylic acid**
72778-00-8

$C_6H_{11}NO_2S$
Chiral intermediate. *Pharmasyn, Inc.*

**2287 D-2,2-Dimethyl-4-thiazolidine-carboxylic acid hydrochloride**
67089-92-3

$C_6H_{12}ClNO_2S$
Chiral intermediate. *Ultrafine.*

**2288 L-2,2-Dimethylthiazolidine-4-carboxylic acid hydrochloride**

$C_6H_{12}ClNO_2S$
Thiazolidine-4-carboxylic acid, 2,2-dimethyl-, hydrochloride, (L)-. Chiral intermediate. *Senn Chemicals AG.*

**2289 (R,R)-(+)-1,2-Di(1-naphthyl)-1,2-ethanediol**

$C_{22}H_{18}O_2$
Chiral intermediate. mp = 149-151°; $[\alpha]^{23}$ = + 25° (c = 1, $CHCl_3$). *Sigma-Aldrich Fine Chemicals.*

**2290 (S,S)-(-)-1,2-Di(1-naphthyl)-1,2-ethanediol**

$C_{22}H_{18}O_2$
Chiral intermediate. mp = 149-151°; $[\alpha]^{23}$ = - 25° (c = 1, $CHCl_3$). *Sigma-Aldrich Fine Chemicals.*

**2291 (R)-Dinaphthylprolinol**
127986-84-9

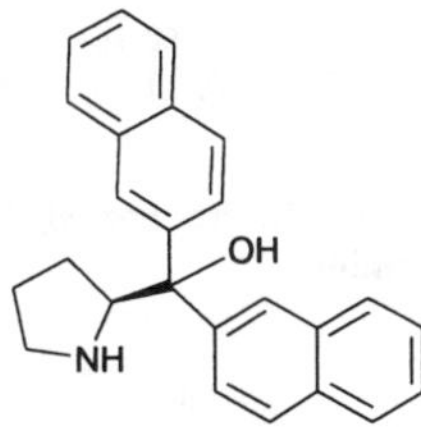

$C_{25}H_{23}NO$
Di-2-naphthylprolinol. Chiral building block. mp = 135-138°; $[\alpha]_D^{20}$ = - 101° (c = 0.7, $CH_3OH$). *Acros Organics nv; Oxford Asymmetry International plc; Sigma-Aldrich Fine Chemicals.*

**2292 (S)-Dinaphthylprolinol**
130798-48-0

$C_{25}H_{23}NO$
(S)-Di-2-naphthylprolinol. Chiral intermediate. mp = 135-138°; $[\alpha]_D^{20}$ = + 101° (c = 0.7, $CH_3OH$). *Acros Organics nv; Oxford Asymmetry International plc; Sigma-Aldrich Fine Chemicals.*

**2293 (-)-N-(3,5-Dinitrobenzoyl)leucine**
7495-01-4

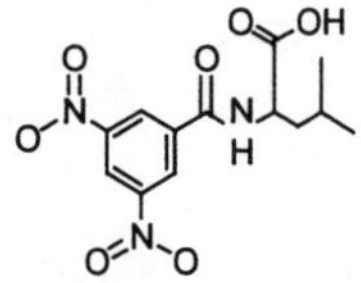

$C_{13}H_{15}N_3O_7$
Chiral intermediate. mp = 186-188°; $[\alpha]_D^{20}$ = - 14.3° (c = 3, EtOH). *Sigma-Aldrich Fine Chemicals.*

**2294 (S)-(+)-N-(3,5-Dinitrobenzoyl)-α-rnethylbenzylamine**
69632-31-1

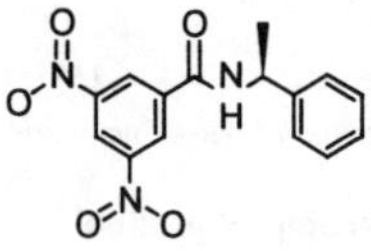

$C_{15}H_{13}N_3O_5$
Chiral intermediate. mp = 160-161°; $[\alpha]_D^{20}$ = + 46.2° (c = 0.9, acetone). *FineTech, Ltd.; Sigma-Aldrich Fine Chemicals.*

**2295 (R)-N-(3,5-Dinitrobenzoyl)-1-(1-naphthyl)-ethylamine**

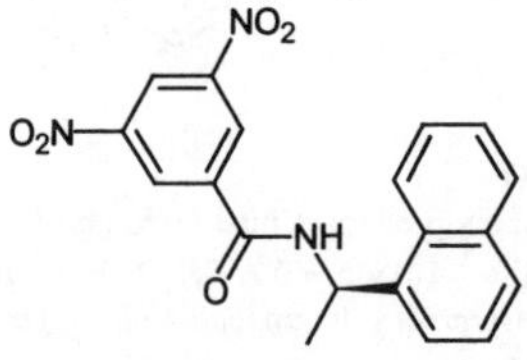

$C_{19}H_{15}N_3O_5$
Chiral intermediate. *FineTech, Ltd.*

**2296 (S)-N-(3,5-Dinitrobenzoyl)-1-(1-naphthyl)-ethylamine**

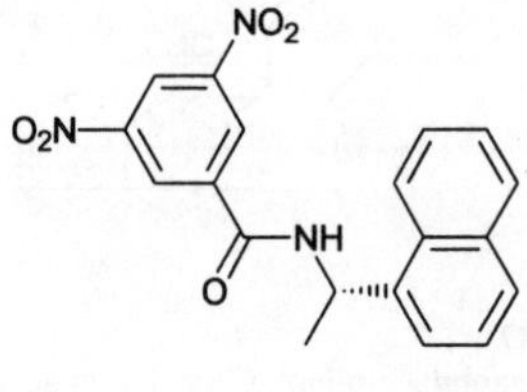

$C_{19}H_{15}N_3O_5$
Resolving agent. *FineTech, Ltd.*

**2297 (R)-(-)-N-(3,5-Dinitrobenzoyl)-α-phenylethylamine**
69632-32-2

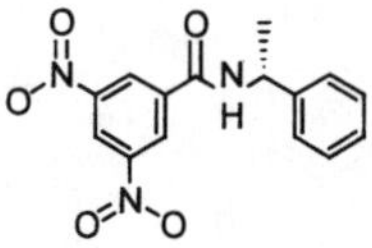

$C_{15}H_{13}N_3O_5$
(R)-(-)-N-(3,5-Dinitrobenzoyl)-α-methylbenzylamine. mp = 160-161°; $[\alpha]^{18}$ = - 46° (c = 0.9, acetone). *FineTech, Ltd.; Sigma-Aldrich Fine Chemicals; TCI America.*

**2298 (R)-(-)-N-(3,5-Dinitrobenzoyl)-α-phenylglycine**
74927-72-3

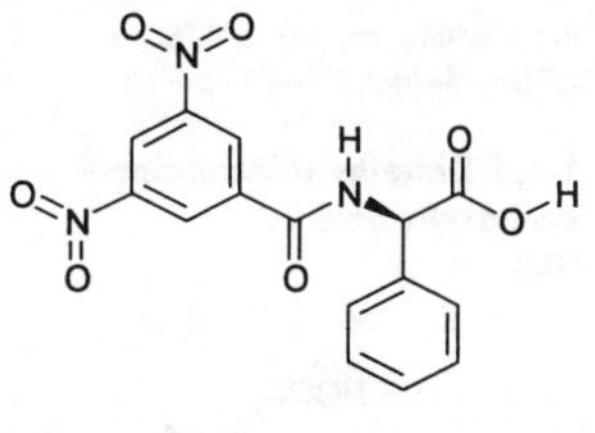

$C_{15}H_{11}N_3O_7$
Chiral intermediate. mp = 216-217°; [α = - 91.8° (c = 0.9, THF). *Acros Organics nv; Sigma-Aldrich Fine Chemicals; TCI America.*

**2299 (+)-N-(2,4-Dinitro-5-fluorophenyl)-alaninamide**
95713-52-3

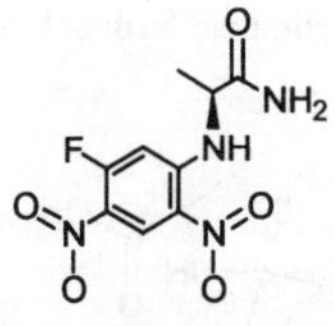

$C_9H_9FN_4O_5$
Marfey's Reagent. Intermediate. mp = 227-228°; $[\alpha]_D^{20}$ = + 56° (c = 1, acetone). *Sigma-Aldrich Fine Chemicals.*

**2300 (+)-N-(2,4-Dinitro-5-fluorophenyl)-valinamide**

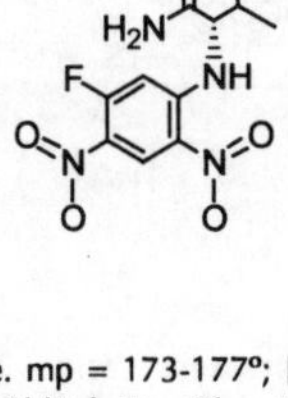

$C_{11}H_{13}FN_4O_5$
Chiral intermediate. mp = 173-177°; $[\alpha]_D^{20}$ = + 14.0° (c = 2, acetone). *Sigma-Aldrich Fine Chemicals.*

**2301 (S)-(+)-N-(3,5-Dinitrobenzoyl)-phenylglycine**
90761-62-9

$C_{15}H_{11}N_3O_7$
Chiral intermediate. mp = 218-220°; $[\alpha]_D^{20}$ = + 102° (c = 1, THF). *Sigma-Aldrich Fine Chemicals.*

**2302 3,5-Dinitro-L-tyrosine monohydrate**
71876-88-5

$C_9H_{11}N_3O_8$
Chiral intermediate. mp = 220°; $[\alpha]_D^{20}$ = + 12° (c = 1, 1N HCl). *Pharmasyn, Inc.; Sigma-Aldrich Fine Chemicals.*

**2303 (R)-3,3'-Dioctyl-1,1'-bi-2-naphthol**

$C_{36}H_{46}O_2$
Chiral ligand. *Kankyo Kagaku Center Co., Ltd.*

**2304 (S)-3,3'-Dioctyl-1,1'-bi-2-naphthol**

$C_{36}H_{46}O_2$
Chiral ligand. *Kankyo Kagaku Center Co., Ltd.*

**2305 (R)-6,6'-Dioctyl-1,1'-bi-2-naphthol**

$C_{36}H_{46}O_2$
Chiral ligand. *Kankyo Kagaku Center Co., Ltd.*

**2306 (S)-6,6'-Dioctyl-1,1'-bi-2-naphthol**

$C_{36}H_{46}O_2$
Chiral intermediate. *Kankyo Kagaku Center Co., Ltd.*

**2307 (R)-2,2'-Dioctyl-1,1'-binaphthyl**

$C_{36}H_{46}$
Chiral ligand. *Kankyo Kagaku Center Co., Ltd.*

**2308 (S)-2,2'-Dioctyl-1,1'-binaphthyl**

$C_{36}H_{46}$
Chiral ligand. *Kankyo Kagaku Center Co., Ltd.*

**2309 (R)-3,3'-Dioctyl-2,2'-dimethoxy-1,1'-binaphthyl**

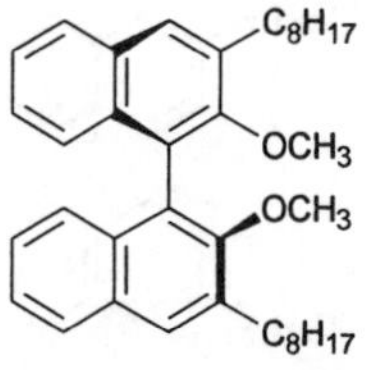

$C_{38}H_{50}O_2$
Chiral ligand. *Kankyo Kagaku Center Co., Ltd.*

**2310 (S)-3,3'-Dioctyl-2,2'-dimethoxy-1,1'-binaphthyl**

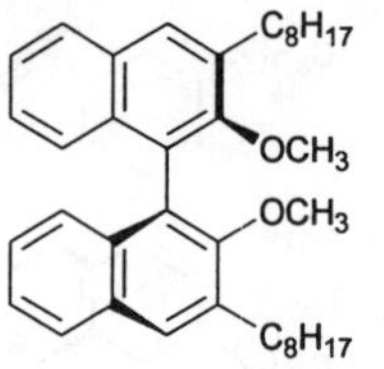

$C_{38}H_{50}O_2$
Chiral ligand. *Kankyo Kagaku Center Co., Ltd.*

**2311 (R)-6,6'-Dioctyl-2,2'-dimethoxy-1,1'-binaphthyl**

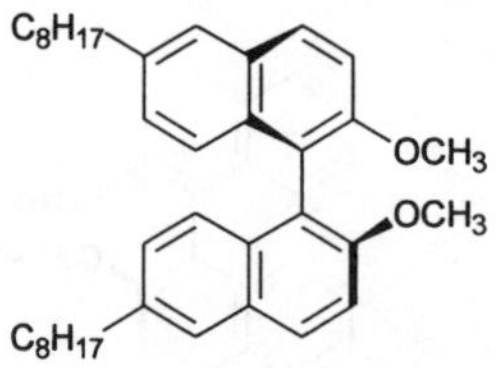

$C_{38}H_{50}O_2$
Chiral ligand. *Kankyo Kagaku Center Co., Ltd.*

**2312 (S)-6,6'-Dioctyl-2,2'-dimethoxy-1,1'-binaphthyl**

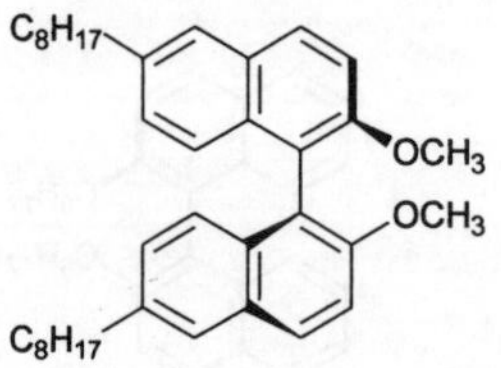

$C_{38}H_{50}O_2$
Chiral intermediate. *Kankyo Kagaku Center Co., Ltd.*

**2313 (R)-3,3'-Dioctyl-2,2'-dimethoxy-5,5',6,6',7,7',8,8'-octahydro-1,1'-binaphthyl**

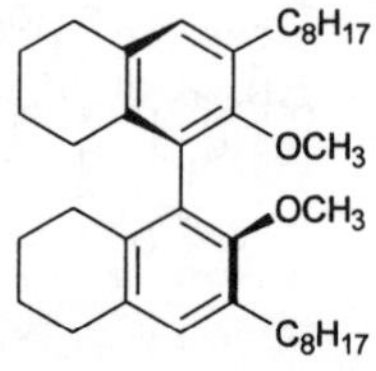

$C_{38}H_{58}O_2$
Chiral ligand. *Kankyo Kagaku Center Co., Ltd.*

**2314 (S)-3,3'-Dioctyl-2,2'-dimethoxy-5,5',6,6',7,7',8,8'-octahydro-1,1'-binaphthyl**

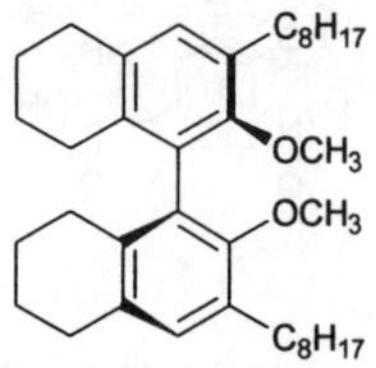

$C_{38}H_{58}O_2$
Chiral ligand. *Kankyo Kagaku Center Co., Ltd.*

**2315 (R)-3,3'-Dioctyl-5,5',6,6',7,7',8,8'-octahydro-1,1'-bi-2-naphthol**

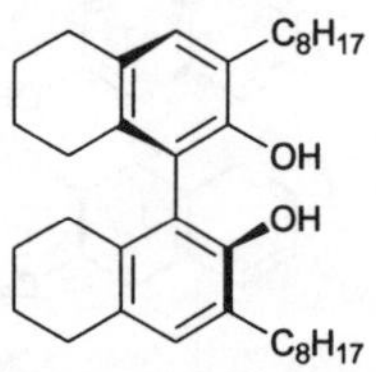

$C_{36}H_{54}O_2$
Chiral ligand. *Kankyo Kagaku Center Co., Ltd.*

**2316 (S)-3,3'-Dioctyl-5,5',6,6',7,7',8,8'-octahydro-1,1'-bi-2-naphthol**

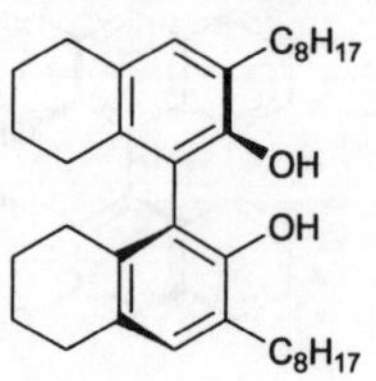

$C_{36}H_{54}O_2$
Chiral ligand. *Kankyo Kagaku Center Co., Ltd.*

**2317 (R)-2,2'-Dioctyl-5,5',6,6',7,7',8,8'-octahydro-1,1'-binaphthyl**

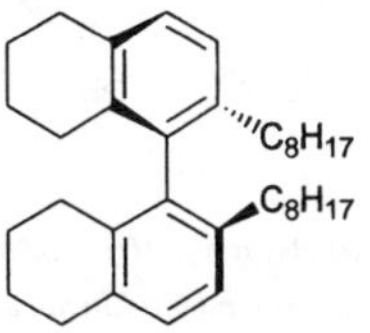

$C_{36}H_{54}$
Chiral ligand. *Kankyo Kagaku Center Co., Ltd.*

**2318 (S)-2,2'-Dioctyl-5,5',6,6',7,7',8,8'-octahydro-1,1'-binaphthyl**

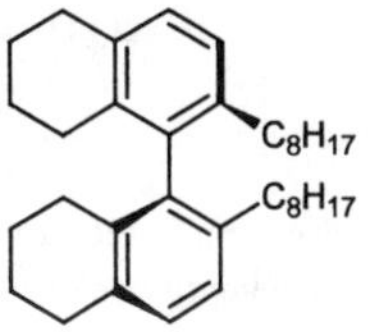

$C_{36}H_{54}$
Chiral ligand. *Kankyo Kagaku Center Co., Ltd.*

**2319 (R)-2,2'-Dioctyloxy-1,1'-binaphthalene-6,6'-dicarbaldehyde**

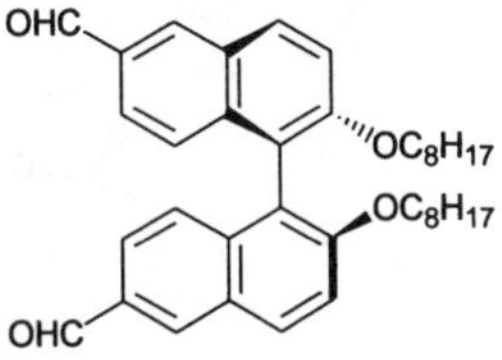

$C_{38}H_{46}O_4$
Chiral ligand. *Kankyo Kagaku Center Co., Ltd.*

**2320 (S)-2,2'-Dioctyloxy-1,1'-binaphthalene-6,6'-dicarbaldehyde**

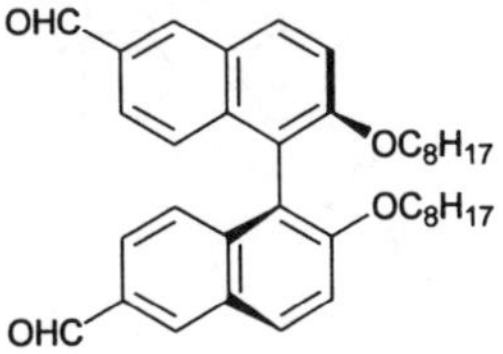

$C_{38}H_{46}O_4$
Chiral ligand. *Kankyo Kagaku Center Co., Ltd.*

**2321 (R)-2,2'-Dioctyloxy-1,1'-binaphthalene-6,6'-dicarboxylic acid**

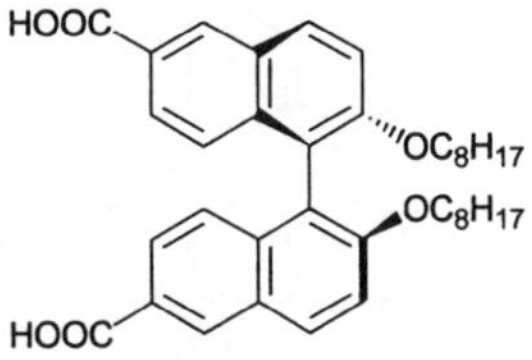

$C_{38}H_{46}O_6$
Chiral ligand. *Kankyo Kagaku Center Co., Ltd.*

**2322 (S)-2,2'-Dioctyloxy-1,1'-binaphthalene-6,6'-dicarboxylic acid**

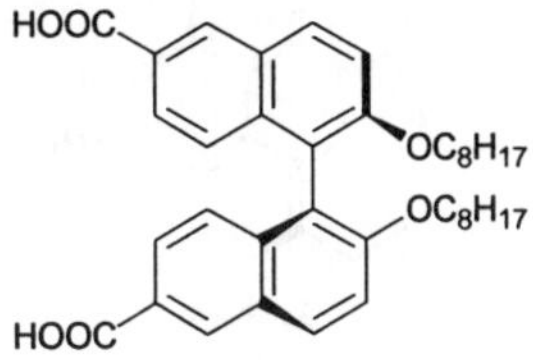

$C_{38}H_{46}O_6$
Chiral ligand. *Kankyo Kagaku Center Co., Ltd.*

**2323 (R)-2,2'-Dioctyloxy-1,1'-binaphthyl**

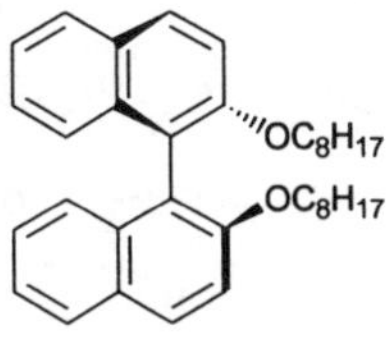

$C_{36}H_{46}O_2$
Chiral ligand. *Kankyo Kagaku Center Co., Ltd.*

**2324 (S)-2,2'-Dioctyloxy-1,1'-binaphthyl**

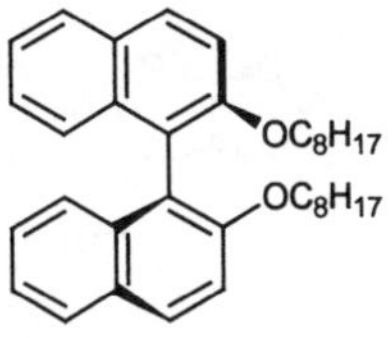

$C_{36}H_{46}O_2$
Chiral ligand. *Kankyo Kagaku Center Co., Ltd.*

**2325 (R)-2,2'-Dioctyloxy-5,5',6,6',7,7',8,8'-octahydro-1,1'-binaphthyl**

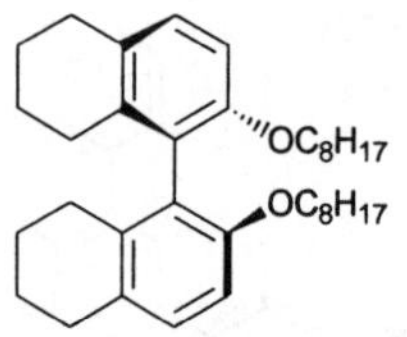

$C_{36}H_{54}O_2$
Chiral ligand. *Kankyo Kagaku Center Co., Ltd.*

**2326 (S)-2,2'-Dioctyloxy-5,5',6,6',7,7',8,8'-octahydro-1,1'-binaphthyl**

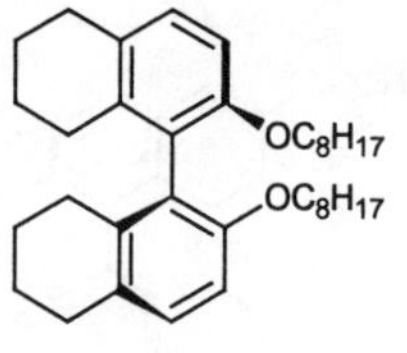

$C_{36}H_{54}O_2$
Chiral ligand. *Kankyo Kagaku Center Co., Ltd.*

**2327 1,2:5,6-Di-O-cyclohexylidene-3-cyano-α-D-allo-furanose**
62293-19-0

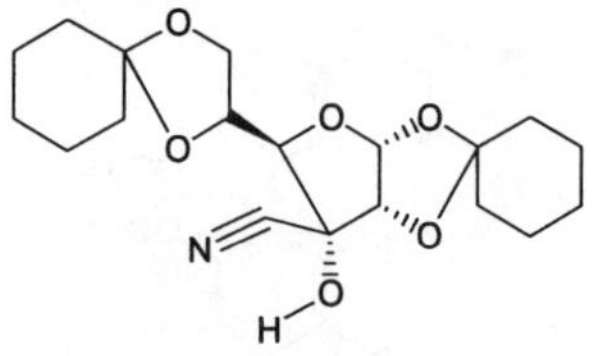

$C_{19}H_{27}NO_6$
Chiral intermediate. [α] = + 50° (c = 1, $CHCl_3$). *Acros Organics nv.*

**2328 1,2:5,6-Di-O-cyclohexylidene-α-D-glucofuranose**
23397-76-4

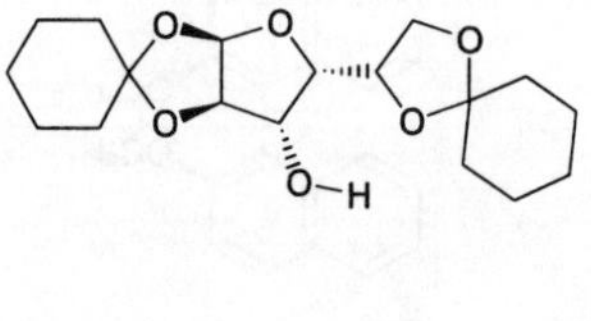

$C_{18}H_{28}O_6$
Chiral intermediate. mp = 132-135°; [α] =+ 1.6° (c = 2.5, $CHCl_3$). *Acros Organics nv; Sigma-Aldrich Fine Chemicals.*

**2329 (+)-1,2:5,6-Di-O-cyclohexylidinemannitol**
76779-67-4

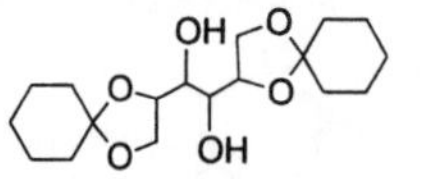

$C_{18}H_{30}O_6$
Chiral intermediate. mp = 103-105°; $[\alpha]_D^{20}$ = + 5.8° (c = 3, $CH_2Cl_2$). *Sigma-Aldrich Fine Chemicals.*

**2330 (+)-1,2:5,6-Di-O-isopropylidene mannitol cyclic sulfate**
106571-12-4

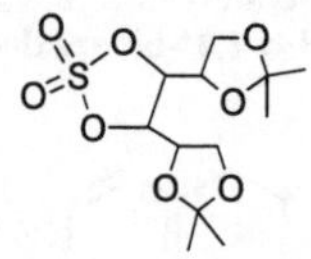

$C_{12}H_{20}O_8S$
Chiral intermediate. mp = 126-128°; $[\alpha]_D^{20}$ = + 27° (c = 2.9, $CHCl_3$). *Sigma-Aldrich Fine Chemicals.*

**2331 2,3:5,6-Di-O-cyclohexylidene-α-D-mannofuranose**
61489-23-4

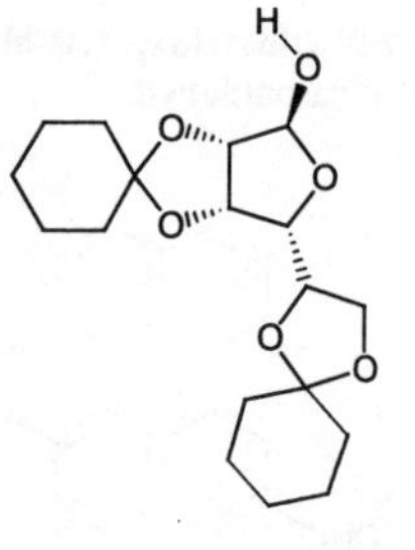

$C_{18}H_{28}O_6$
Chiral intermediate. [α] = + 13.9° (c = 1.19, $CHCl_3$). *Acros Organics nv.*

**2332 2,3:5,6-Di-O-cyclohexylidene-D-mannolactone**

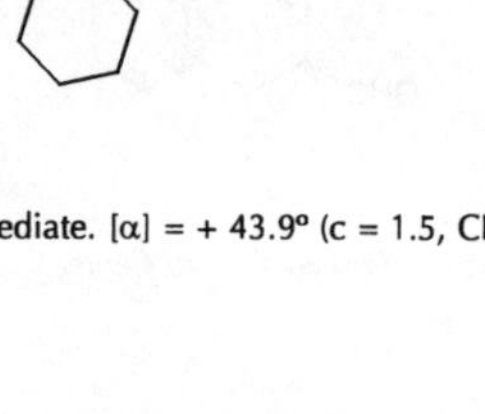

$C_{18}H_{26}O_6$
Chiral intermediate. [α] = + 43.9° (c = 1.5, $CHCl_3$). *Acros Organics nv.*

**2333 1,2-Di-O-cyclohexylidene-5-O-methyl-L-chiroinositol**
6848-53-9
$C_{19}H_{30}O_6$
Chiral intermediate. mp = 121-124°; $[\alpha]_D^{20}$ = - 18° (c = 1, $CHCl_3$). *Sigma-Aldrich Fine Chemicals.*

**2334 1,2:5,6-Di-O-cyclohexylidene-3-O-methyl-α-D-glucofuranose**
13440-19-2

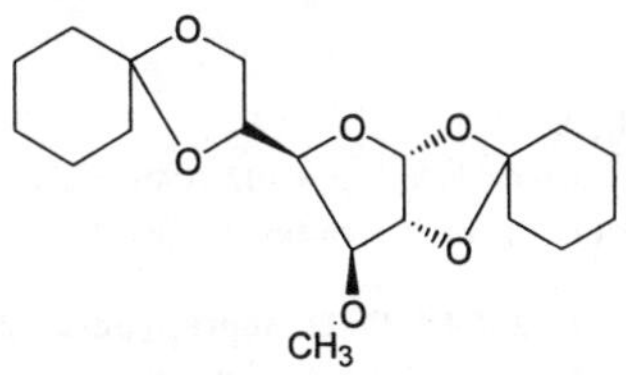

$C_{19}H_{30}O_6$
Chiral intermediate. mp = 35-37°; [α] = - 21.2° (c = 1, $CHCl_3$). *Acros Organics nv.*

**2335 1,2:3,5-Di-O-isopropylidene-α-D-apiose**
94943-41-6
$C_{11}H_{19}O_5$
Chiral intermediate. mp = 82°; $[\alpha]_D^{20}$ = + 56° (c = 2, EtOH). *Pfanstiehl Laboratories, Inc.; Senn Chemicals AG.*

**2336 1,2,3,4-Di-O-isopropylidene-L-arabinopyranose**

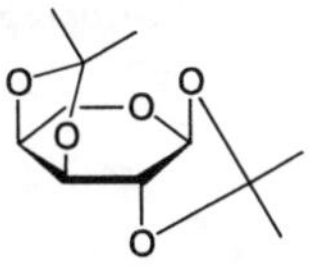

$C_{11}H_{19}O_5$
Chiral intermediate. *Senn Chemicals AG.*

**2337 O-Isopropylidene-2,3-dihydroxy-1,4-bis[bis(3,5-difluoromethylphenyl)-phosphino]butane**

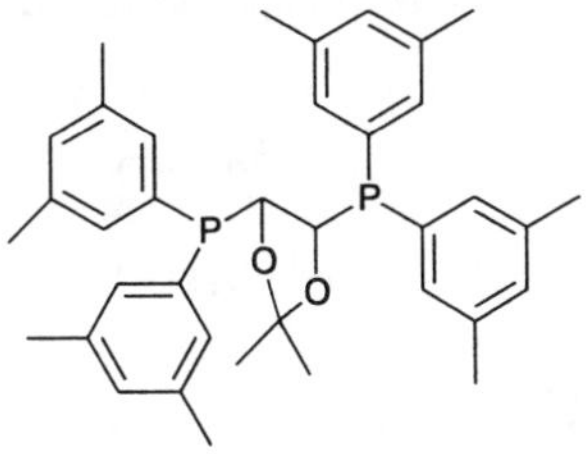

$C_{39}H_{24}F_{24}O_2P_2$
Chiral catalyst. *Digital Specialty Chemicals, Inc.*

**2338 O-Isopropylidene-2,3-dihydroxy-1,4-bis[bis(3,5-dimethylphenyl)-phosphino]butane**

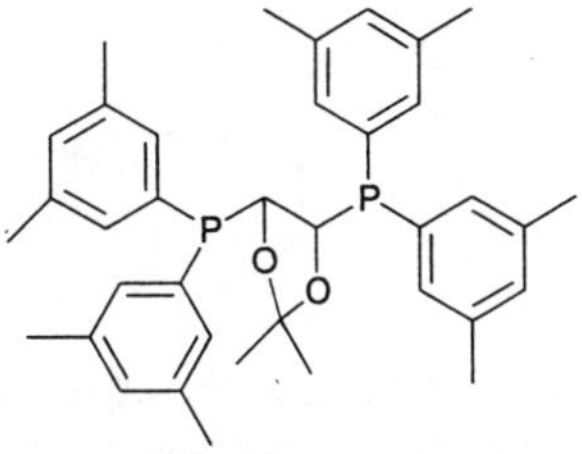

$C_{39}H_{48}O_2P_2$
Chiral ligand. *Digital Specialty Chemicals, Inc.*

**2339 1,2,4,5-Di-O-isopropylidene-β-D-fructopyranose**

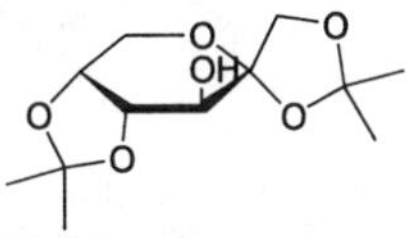

$C_{12}H_{20}O_6$
Chiral intermediate. *Senn Chemicals AG.*

**2340 2,3,4,5-Di-O-isopropylidene-β-D-fructopyranose**

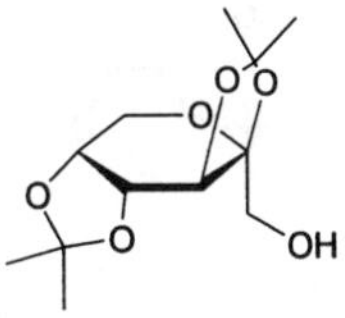

$C_{12}H_{20}O_6$
Chiral intermediate. *Senn Chemicals AG.*

**2341 1,2,3,4-Di-O-isopropylidene-α-D-fucose**

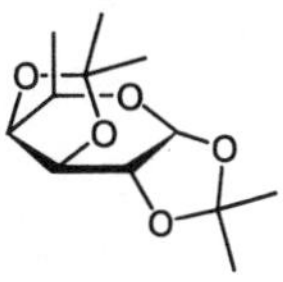

$C_{12}H_{20}O_5$
Chiral intermediate. *Senn Chemicals AG.*

**2342 1,2:5,6-Di-O-isopropylidene-α-D-allofuranose**
2595-05-3

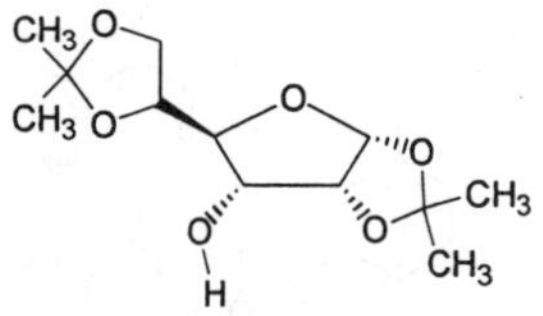

$C_{12}H_{20}O_6$
Chiral intermediate. mp = 75°; $[\alpha]_D^{20}$ = + 37.5° (c = 1, $CHCl_3$). *Acros Organics nv; Pfanstiehl Laboratories, Inc.; Senn Chemicals AG; Sigma-Aldrich Fine Chemicals.*

**2343 1,2:3,4-Di-O-isopropylidene-α-D-galactopyranose**
4064-06-6 223-771-7

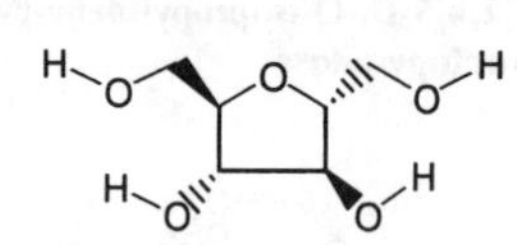

$C_{12}H_{20}O_6$
Galactopyranose, 1,2:3,4-bis-O-(1-methylethylidene)-, α-D-; 1,2:3,4-Di-O-isopropylidene-D-galactose. Chiral intermediate. $bp_{0.02}$ = 117°; d = 1.143; n = 1.466; [α] = - 52.6° (c = 3, $CHCl_3$). *Acros Organics nv; Pfanstiehl Laboratories, Inc.; Sigma-Aldrich Fine Chemicals.*

**2344 1,2:5,6-Di-O-isopropylidene-α-D-glucofuranose**
582-52-5 3009(12) 209-486-0

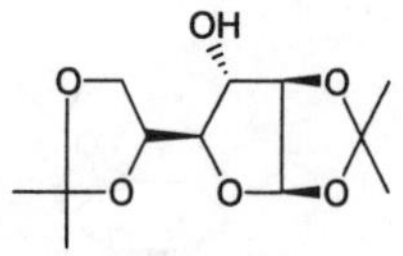

$C_{12}H_{20}O_6$
Glucofuranose, 1,2:5,6-di-O-isopropylidene-, α-D-; Di-isopropylidene-α-D-glucose; (-)-1,2:5,6-Di-O-isopropylidene-α-D-glucofuranose. Listed on TSCA. Chiral intermediate. mp = 108°; $[\alpha]_D^{20}$ = - 18.5° (c = 2, $H_2O$). *Acros Organics nv; Austin Chemical Company, Inc.; Pfanstiehl Laboratories, Inc.; Senn Chemicals AG; Sigma-Aldrich Fine Chemicals.*

**2345 1,2:5,6-Di-O-isopropylidene-α-L-glucofuranose**

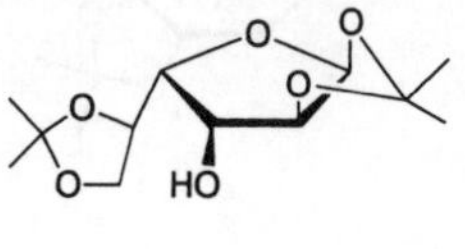

$C_{12}H_{20}O_6$
Glucofuranose, 1,2:5,6-di-O-isopropylidene-, α-L-; Diacetone-L-glucose. Chiral intermediate. $[\alpha]_D^{20}$ = + 18.5° (c = 2, $H_2O$). *Pfanstiehl Laboratories, Inc.; Senn Chemicals AG.*

**2346 (-)-1,2:4,5-Di-O-isopropylidene-erythro-2,3-hexodiulo-2,6-pyranose**

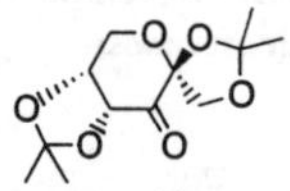

$C_{12}H_{18}O_6$
Chiral intermediate. mp = 102-104°; $[\alpha]_D^{20}$ = - 120.9° (c = 1, $CHCl_3$). *Sigma-Aldrich Fine Chemicals.*

**2347 (-)-2,3:4,6-Di-O-isopropylidene-2-keto-gulonic acid monohydrate**
68539-16-2 3184(11)

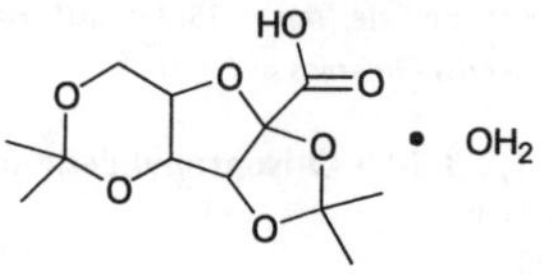

$C_{12}H_{18}O_7 \cdot H_2O$
2,3-Di-O-isopropylidene-α-L-xylo-2-hexulofuranosonic acid. Listed on TSCA. Chiral intermediate. mp = 88-90°; $[\alpha]^{21}$ = - 20° (c = 2, $CH_3OH$). *Sigma-Aldrich Fine Chemicals.*

**2348 (-)-1,2:4,5-Di-O-isopropylidenemannitol**
3969-61-7

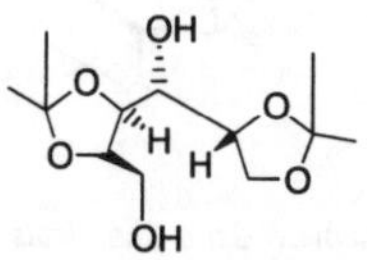

$C_{12}H_{22}O_6$
Chiral intermediate. mp = 49-52°; $[\alpha]_D^{20}$ = -9.2° (c = 1, $CHCl_3$). *Sigma-Aldrich Fine Chemicals.*

**2349 1,2,5,6-Di-O-isopropylidene-D-mannitol**
1707-77-3 216-954-8

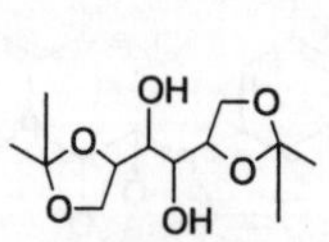

$C_{12}H_{22}O_6$
Mannitol, 1,2:5,6-di-O-isopropylidene-, D-; Diisopropylidene mannitol; Mannitol diacetonide. Chiral intermediate. mp = 120-122°; $[\alpha]_D^{20}$ = + 1.9° (c = 2, $CH_3OH$). *Oxford Asymmetry International plc; Senn Chemicals AG; Sigma-Aldrich Fine Chemicals.*

**2350 2,3:5,6-Di-O-isopropylidene-α-D-mannofuranose**

14131-84-1

$C_{12}H_{20}O_6$

Diacetone D-mannose. Chiral intermediate. mp = 120-122°; $[\alpha]_D^{20}$ = + 23° (c = 1, acetone). *Acros Organics nv; Senn Chemicals AG; Sigma-Aldrich Fine Chemicals.*

**2351 2,3:5,6-Di-O-isopropylidene-L-mannono-1,4-lactone**

49561-21-9

$C_{12}H_{18}O_6$

Chiral intermediate. *Pfanstiehl Laboratories, Inc.*

**2352 1,2:5,6-Di-O-isopropylidene-3-O-methylsulfonyl-α-D-glucofuranose**

5450-26-0 226-675-3

$C_{13}H_{22}O_8S$

Glucofuranose, 1,2:5,6-bis-O-(1-methylethylidene)-, α-D-, methanesulfonate; 3-O-Methylsulphonyl-1,2,5,6-di-O-isopropylidene-α-D-glucofuranose. Chiral intermediate. mp = 79-81°; $[\alpha]^{25}$ = - 30° (c = 2, $CH_3OH$). *Acros Organics nv; Pfanstiehl Laboratories, Inc.; Sigma-Aldrich Fine Chemicals.*

**2353 1,2:5,6-Di-O-isopropylidene-α-D-ribo-3-hexulofuranose**

2847-00-9 220-646-9

$C_{12}H_{18}O_6$

D-ribo-Hexofuranos-3-ulose, 1,2:5,6-di-O-isopropylidene-, α-. Chiral intermediate. [α] = + 36° (c = 1 $H_2O$). *Acros Organics nv.*

**2354 1,2:4,6-Di-O-isopropylidene-α-L-sorbofuranose**

32717-65-0 251-169-4

$C_{12}H_{20}O_6$

Sorbose, di-O-isopropylidene-, L-; L-Sorbose, bis-O-(1-methylethylidene)-; Diacetone sorbose. Listed on TSCA. Chiral intermediate. *Pfanstiehl Laboratories, Inc.; Senn Chemicals AG.*

**2355 2,3,4,6-Di-O-isopropylidene-α-L-sorbofuranose**

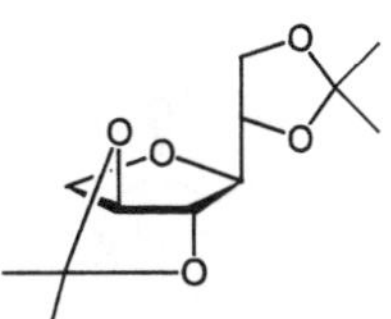

$C_{12}H_{20}O_5$

Chiral intermediate. *Senn Chemicals AG.*

**2356 1,2:3,4-Di-O-isopropylidene-6-O-p-tolylsulfonyl-α-D-galactose**

4478-43-7

$C_{18}H_{24}O_8S$

Chiral intermediate. *Pfanstiehl Laboratories, Inc.*

**2357 1,2:3,5-Di-O-isopropylidene-α-D-xylofuranose**

20881-04-3

$C_{11}H_{18}O_5$

Chiral intermediate. mp = 44°; $bp_2$ = 102-104°; $[\alpha]_D^{20}$ = + 19.5° (c = 4, acetone). *Acros Organics nv; Pfanstiehl Laboratories, Inc.; Senn Chemicals AG.*

**2358 1,2:3,5-Di-O-isopropylidene-α-L-xylofuranose**

$C_{11}H_{18}O_5$

Chiral intermediate. *Acros Organics nv.*

**2359 (3R)-1,3-Dioxane-2-methyl-4-carboxylic acid**

$C_6H_{10}O_4$

Dioxane-4-carboxylic acid, 1,3-, 2-methyl, (3R)-. Chiral intermediate. *Synthon Chiragenics Corporation.*

**2360 (3S)-1,3-Dioxane-2-methyl-4-carboxylic acid**

158817-45-9

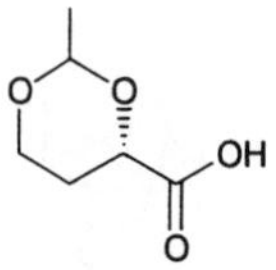

$C_6H_{10}O_4$

Dioxane-4-carboxylic acid, 1,3-, 2-methyl-, (3S)-. Chiral intermediate. *Synthon Chiragenics Corporation.*

**2361 (+)-1,4-Dioxaspiro[4.5]deacane-2-methanol**

95335-91-4

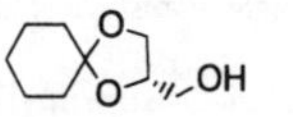

$C_9H_{16}O_3$

Chiral intermediate. $bp_{5.0}$ = 118-120°; d = 1.096; n = 1.477; $[\alpha]_D^{20}$ = + 9.5° (neat). *Sigma-Aldrich Fine Chemicals.*

**2362 L-α-Dipalmitoyl phosphatidylcholine**

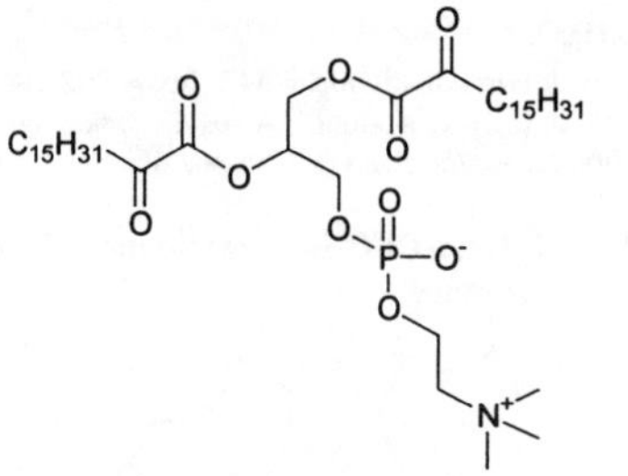

$C_{42}H_{80}NO_{10}P$

Chiral intermediate. [α] = + 7° (c = 2.0, $CHCl_3$). *Acros Organics nv.*

**2363 (R)-3,3'-Dipentyl-1,1'-bi-2-naphthol**

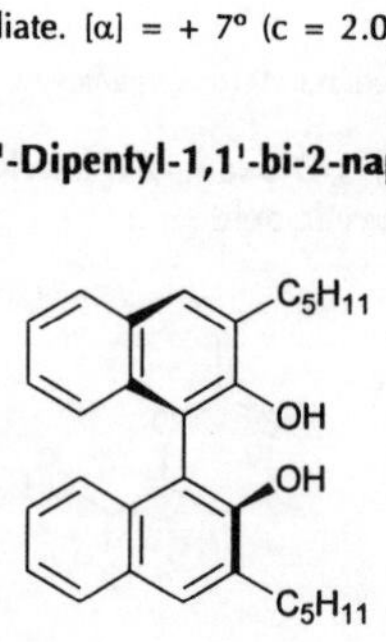

$C_{30}H_{34}O_2$

Chiral ligand. *Kankyo Kagaku Center Co., Ltd.*

**2364 (S)-3,3'-Dipentyl-1,1'-bi-2-naphthol**

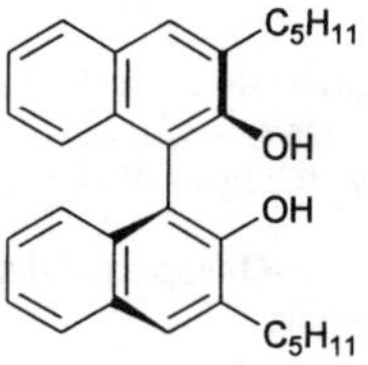

$C_{30}H_{34}O_2$

Chiral ligand. *Kankyo Kagaku Center Co., Ltd.*

**2365 (S)-6,6'-Dipentyl-1,1'-bi-2-naphthol**

$C_{30}H_{34}O_2$

Chiral ligand. *Kankyo Kagaku Center Co., Ltd.*

**2366 (R)-6,6'-Dipentyl-1,1'-bi-2-naphthol**

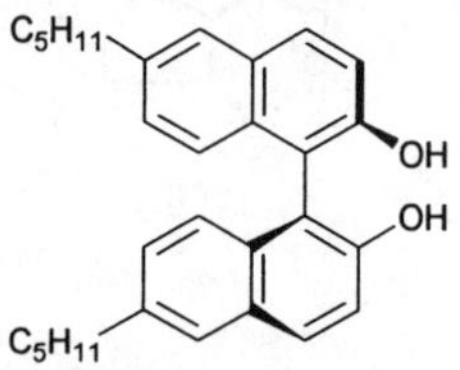

$C_{30}H_{34}O_2$

Chiral intermediate. *Kankyo Kagaku Center Co., Ltd.*

**2367 (R)-3,3'-Dipentyl-2,2'-dimethoxy-1,1'-binaphthyl**

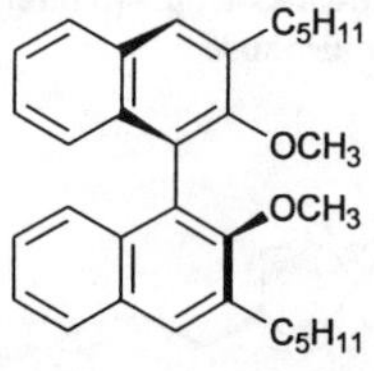

$C_{32}H_{38}O_2$

Chiral ligand. *Kankyo Kagaku Center Co., Ltd.*

**2368 (S)-3,3'-Dipentyl-2,2'-dimethoxy-1,1'-binaphthyl**

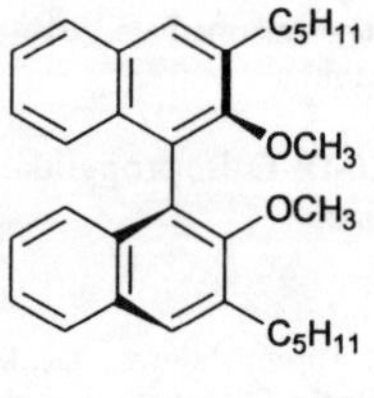

$C_{32}H_{38}O_2$

Chiral ligand. *Kankyo Kagaku Center Co., Ltd.*

2369 **(R)-6,6'-Dipentyl-2,2'-dimethoxy-1,1'-binaphthyl**

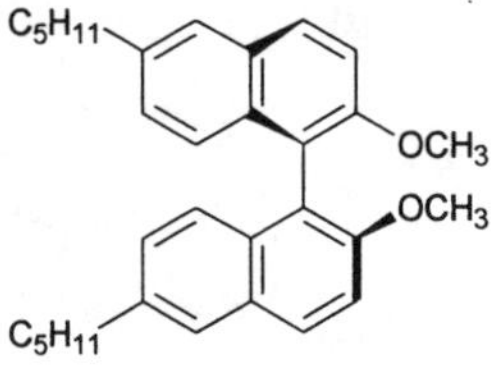

$C_{32}H_{38}O_2$
Chiral ligand. *Kankyo Kagaku Center Co., Ltd.*

2370 **(S)-6,6'-Dipentyl-2,2'-dimethoxy-1,1'-binaphthyl**

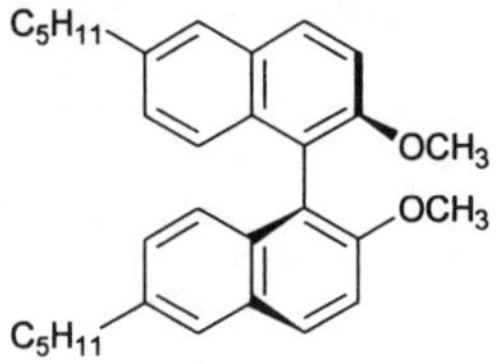

$C_{32}H_{38}O_2$
Chiral intermediate. *Kankyo Kagaku Center Co., Ltd.*

2371 **(R)-3,3'-Dipentyl-2,2'-dimethoxy-5,5',6,6',7,7',8,8'-octahydro-1,1'-binaphthyl**

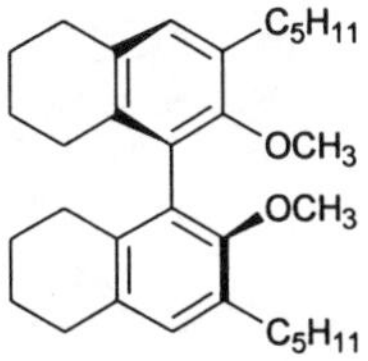

$C_{32}H_{46}O_2$
Chiral ligand. *Kankyo Kagaku Center Co., Ltd.*

2372 **(S)-3,3'-Dipentyl-2,2'-dimethoxy-5,5',6,6',7,7',8,8'-octahydro-1,1'-binaphthyl**

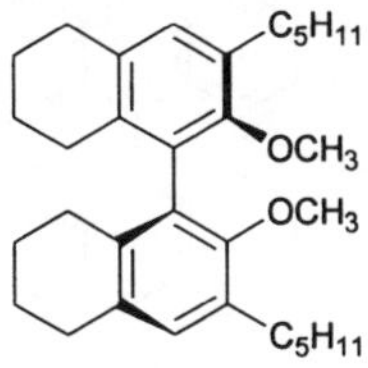

$C_{32}H_{46}O_2$
Chiral ligand. *Kankyo Kagaku Center Co., Ltd.*

2373 **(R)-3,3'-Dipentyl-5,5',6,6',7,7',8,8'-octahydro-1,1'-bi-2-naphthol**

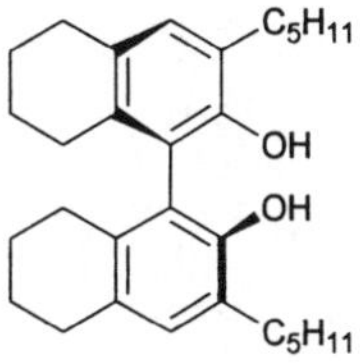

$C_{30}H_{42}O_2$
Chiral ligand. *Kankyo Kagaku Center Co., Ltd.*

2374 **(S)-3,3'-Dipentyl-5,5',6,6',7,7',8,8'-octahydro-1,1'-bi-2-naphthol**

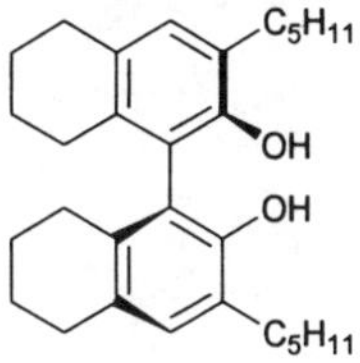

$C_{30}H_{42}O_2$
Chiral ligand. *Kankyo Kagaku Center Co., Ltd.*

2375 **(R)-2,2'-Dipentyloxy-1,1'-binaphthyl**

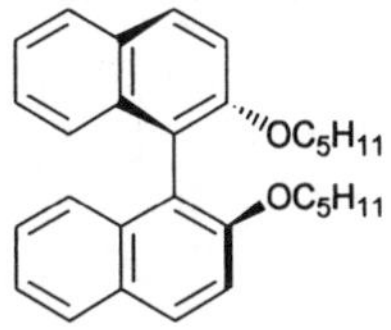

$C_{30}H_{34}O_2$
Chiral ligand. *Kankyo Kagaku Center Co., Ltd.*

2376 **(S)-2,2'-Dipentyloxy-1,1'-binaphthyl**

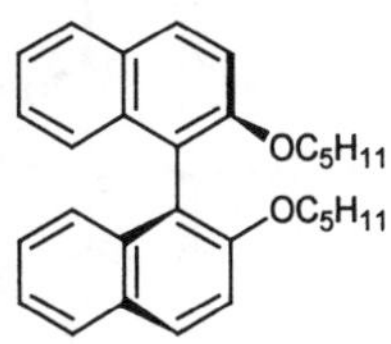

$C_{30}H_{34}O_2$
Chiral ligand. *Kankyo Kagaku Center Co., Ltd.*

**2377 (R)-2,2'-Dipentyloxy-5,5',6,6',7,7',8,8'-octahydro-1,1'-binaphthyl**

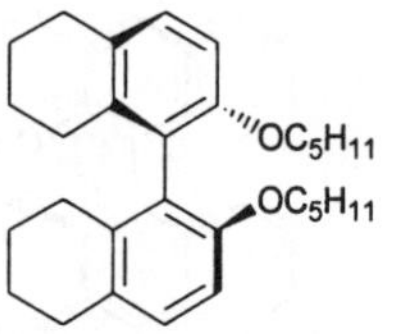

$C_{30}H_{42}O_2$
Chiral ligand. *Kankyo Kagaku Center Co., Ltd.*

**2378 (S)-2,2'-Dipentyloxy-5,5',6,6',7,7',8,8'-octahydro-1,1'-binaphthyl**

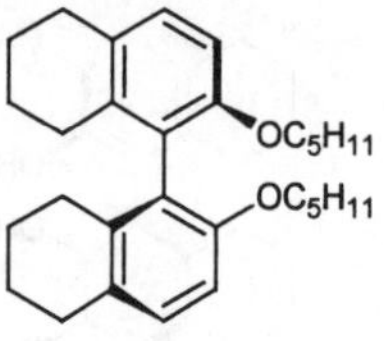

$C_{30}H_{42}O_2$
Chiral ligand. *Kankyo Kagaku Center Co., Ltd.*

**2379 (R)-3,3'-Diphenyl-1,1'-bi-2-naphthol**

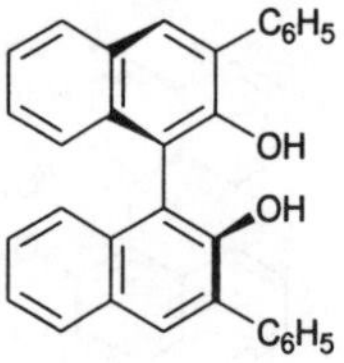

$C_{32}H_{22}O_2$
Chiral ligand. *Kankyo Kagaku Center Co., Ltd.*

**2380 (S)-3,3'-Diphenyl-1,1'-bi-2-naphthol**

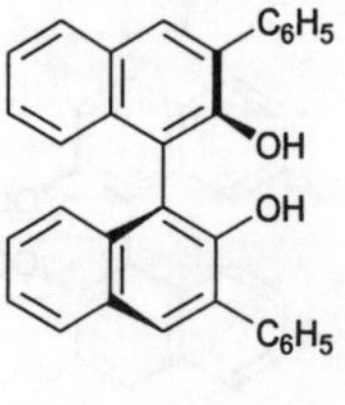

$C_{32}H_{22}O_2$
Chiral ligand. *Kankyo Kagaku Center Co., Ltd.*

**2381 (R)-6,6'-Diphenyl-1,1'-bi-2-naphthol**

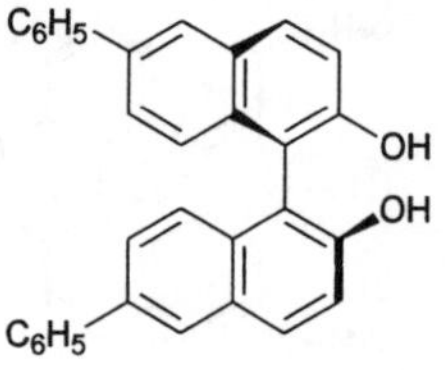

$C_{32}H_{22}O_2$
Chiral ligand. *Kankyo Kagaku Center Co., Ltd.*

**2382 (S)-6,6'-Diphenyl-1,1'-bi-2-naphthol**

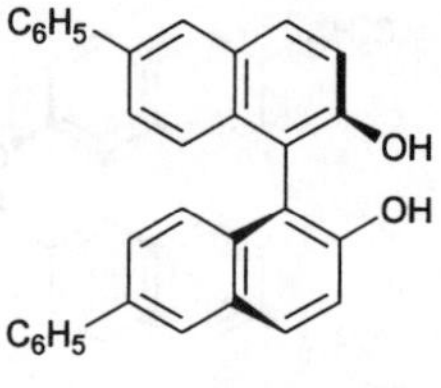

$C_{32}H_{22}O_2$
Chiral intermediate. *Kankyo Kagaku Center Co., Ltd.*

**2383 (R)-3,3'-Diphenyl-2,2'-dimethoxy-1,1'-binaphthyl**

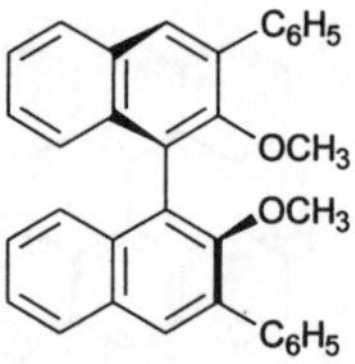

$C_{34}H_{26}O_2$
Chiral ligand. *Kankyo Kagaku Center Co., Ltd.*

**2384 (S)-3,3'-Diphenyl-2,2'-dimethoxy-1,1'-binaphthyl**

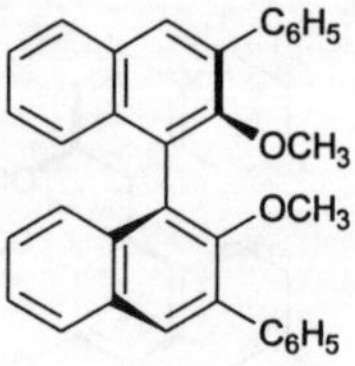

$C_{34}H_{26}O_2$
Chiral ligand. *Kankyo Kagaku Center Co., Ltd.*

**2385 (R)-6,6'-Diphenyl-2,2'-dimethoxy-1,1'-binaphthyl**

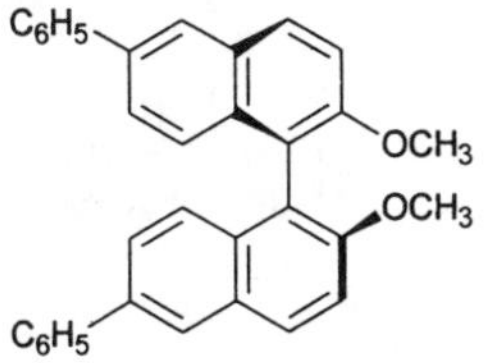

$C_{34}H_{26}O_2$
Chiral ligand. *Kankyo Kagaku Center Co., Ltd.*

**2386 (S)-6,6'-Diphenyl-2,2'-dimethoxy-1,1'-binaphthyl**

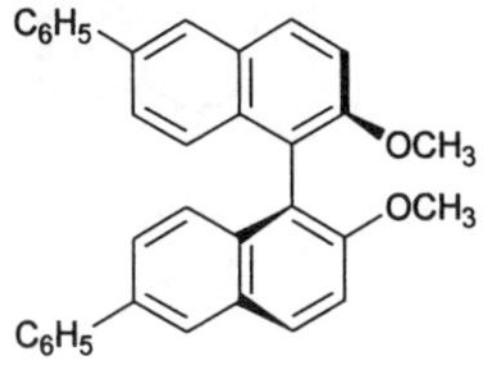

$C_{34}H_{26}O_2$
Chiral intermediate. *Kankyo Kagaku Center Co., Ltd.*

**2387 (R)-3,3'-Diphenyl-2,2'-dimethoxy-5,5',6,6',7,7',8,8'-octahydro-1,1'-binaphthyl**

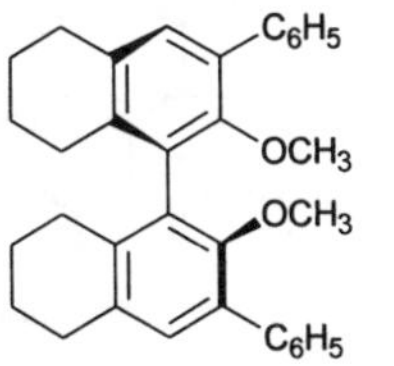

$C_{34}H_{34}O_2$
Chiral ligand. *Kankyo Kagaku Center Co., Ltd.*

**2388 (S)-3,3'-Diphenyl-2,2'-dimethoxy-5,5',6,6',7,7',8,8'-octahydro-1,1'-binaphthyl**

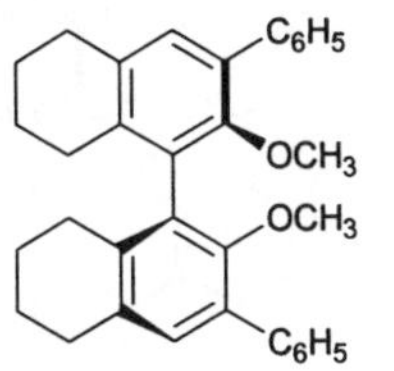

$C_{34}H_{34}O_2$
Chiral ligand. *Kankyo Kagaku Center Co., Ltd.*

**2389 (1R,2R)-(-)-1,2-Diphenyl-1,2-ethanediamine**
29841-69-8

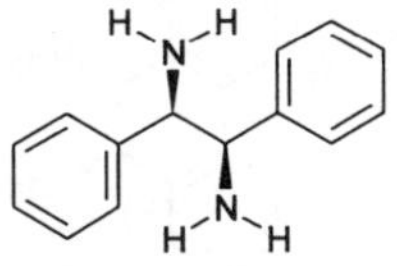

$C_{14}H_{16}N_2$
(1R,2R)-(-)-1,2-Diamino-1,2-diphenylethane; (1R,2R)-(+)-1,2-Diphenylethylenediamine. Chiral auxiliary. mp = 79-81°; $[\alpha]_D^{20}$ = - 100 ± 5° (c = 1, $CH_3OH$). *Arran Chemical Company Ltd.; FineTech, Ltd.; Kankyo Kagaku Center Co., Ltd.; Lancaster Synthesis Ltd.; Oxford Asymmetry International plc; Sigma-Aldrich Fine Chemicals; Strem Chemicals, Inc.*

**2390 (1S,2S)-(+)-1,2-Diphenyl-1,2-ethanediamine**
35132-20-8

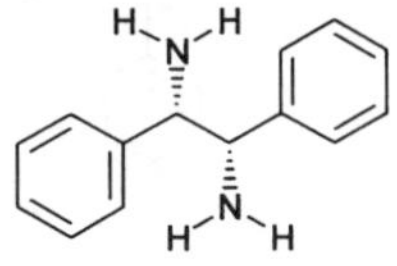

$C_{14}H_{16}N_2$
(S,S)-1,2-Diphenylethylenediamine. Chiral diagnostic reagent. mp = 79-81°; $[\alpha]_D^{20}$ = - 105 ± 2° (c = 1, $CH_3OH$). *Acros Organics nv; Advanced Asymmetrics, Inc.; Arran Chemical Company Ltd.; FineTech, Ltd.; Kankyo Kagaku Center Co., Ltd.; Lancaster Synthesis Ltd.; Oxford Asymmetry International plc; Sigma-Aldrich Fine Chemicals; Strem Chemicals, Inc.; Yoneyama Yakuhin Kogyo Co., Ltd.*

**2391 (S,S)-(-)-1,2-Diphenyl-1,2-ethanediol**
2325-10-2 4700(11)

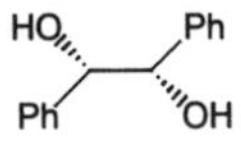

$C_{14}H_{14}O_2$
Chiral intermediate. mp = 148-150°; $[\alpha]^{24}$ = - 94° (c = 2.5, EtOH). *Acros Organics nv; Sigma-Aldrich Fine Chemicals.*

**2392 (R,R)-1,2-Diphenylethylenediamine**

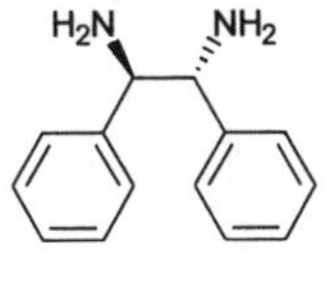

$C_{14}H_{16}N_2$
Chiral building block; Chiral auxiliary. mp = 79-81°; $[\alpha]_D^{20}$ = + 104.0 ± 1.0° (c = 1, $CH_3OH$). *Kiralchem Ltd.*

**2393 (S,S)-1,2-Diphenylethylenediamine**

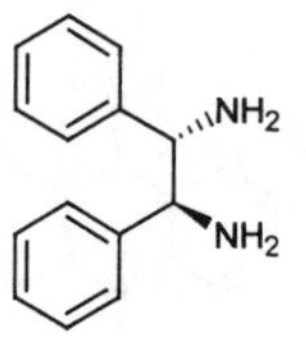

$C_{14}H_{16}N_2$
Chiral building block; Chiral auxiliary. mp = 79-81°; $[\alpha]_D^{20}$ = - 104.0 ± 1.0° (c = 1, $CH_3OH$). *Kiralchem Ltd.*

**2394 (R)-2,2-Diphenyl-2-hydroxy-1-methyl-ethylamine**

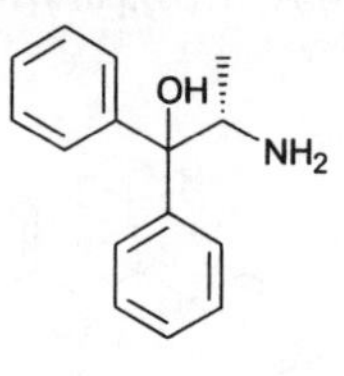

$C_{15}H_{17}NO$
Chiral intermediate. *Sumitomo Chemcial Co. Ltd.*

**2395 (S)-2,2-Diphenyl-2-hydroxy-1-methyl-ethylamine**

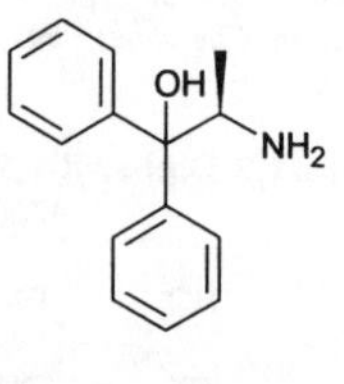

$C_{15}H_{17}NO$
Chiral intermediate. *Sumitomo Chemcial Co. Ltd.*

**2396 (R)-(+)-4-(Diphenylmethyl)-2-oxazolidinone**

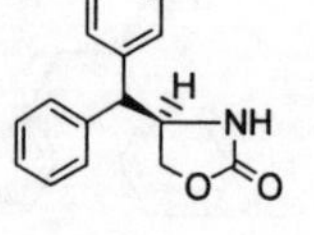

$C_{16}H_{15}NO_2$
Chiral intermediate. mp = 145-148°; $[\alpha]_D^{20}$ = + 50° (c = 1, $CHCl_3$). *Sigma-Aldrich Fine Chemicals.*

**2397 (S)-(-)-4-(Diphenylmethyl)-2-oxazolidinone**

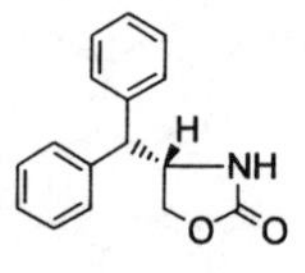

$C_{16}H_{15}NO_2$
Chiral intermediate. mp = 144-147°; $[\alpha]_D^{20}$ = - 49° (c = 2, $CHCl_3$). *Sigma-Aldrich Fine Chemicals.*

**2398 (S)-(-)-2-(Diphenylmethyl)pyrrolidine**
119237-64-8

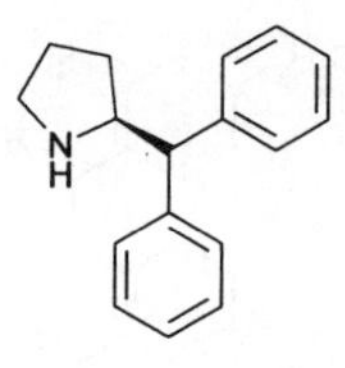

$C_{17}H_{19}N$
Chiral intermediate. $bp_{0.01}$ = 135°; d = 1.062; n = 1.587; $[\alpha]^{25}$ = - 3° (c = 1, $CHCl_3$). *Sigma-Aldrich Fine Chemicals.*

**2399 (R)-3,3'-Diphenyl-5,5',6,6',7,7',8,8'-octahydro-1,1'-bi-2-naphthol**

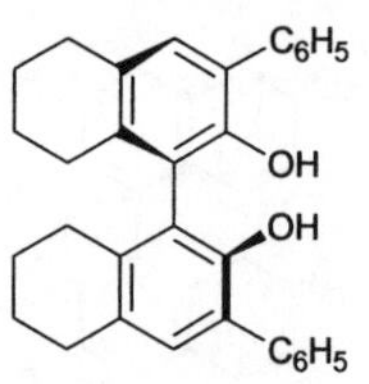

$C_{32}H_{32}O_2$
Chiral ligand. *Kankyo Kagaku Center Co., Ltd.*

**2400 (S)-3,3'-Diphenyl-5,5',6,6',7,7',8,8'-octahydro-1,1'-bi-2-naphthol**

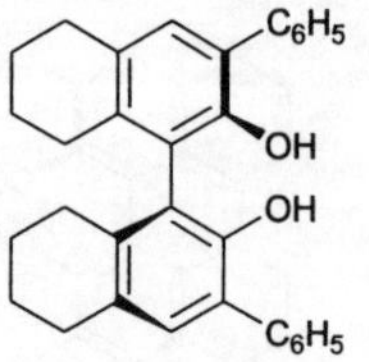

$C_{32}H_{32}O_2$
Chiral ligand. *Kankyo Kagaku Center Co., Ltd.*

**2401 (4R,5S)-(+)-cis-4,5-Diphenyl-2-oxazolidninone**
86286-50-2

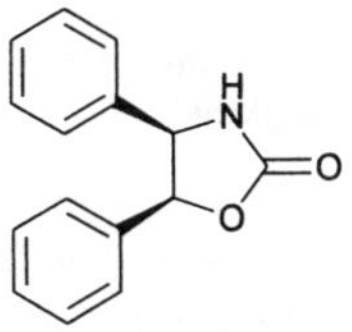

$C_{15}H_{13}NO_2$
Chiral intermediate. mp = 227-230°; $[\alpha]^{22}$ = + 56° (c = 2, $CHCl_3$). *Sigma-Aldrich Fine Chemicals.*

**2402 (4S,5R)-(-)-cis-4,5-Diphenyl-2-oxazolidninone**
23204-70-8

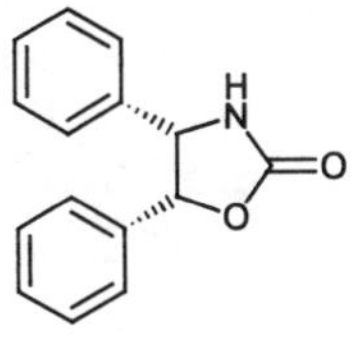

$C_{15}H_{13}NO_2$
Chiral intermediate. mp = 227-232°; $[\alpha]^{23}$ = - 56° (c = 2, $CHCl_3$). *Sigma-Aldrich Fine Chemicals.*

**2403 (2R,4R)-4-Diphenylphosphino-2-diphenyl-phosphinomethyl-pyrrolidine**
155830-69-6

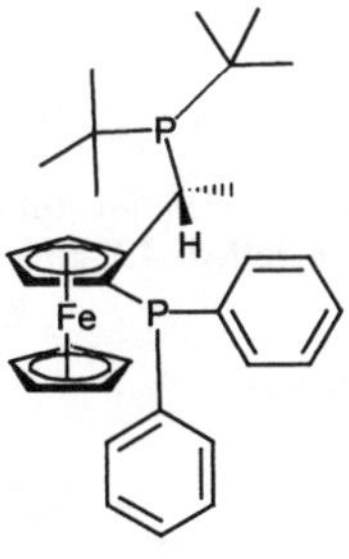

$C_{32}H_{40}FeP_2$
Chiral catalyst. $[\alpha]_D$ = - 412 ± 5° (c = 0.4, $CHCl_3$). *Strem Chemicals, Inc.*

**2404 (2S,4S)-4-Diphenylphosphino-2-diphenylphosphinomethyl-pyrrolidine**

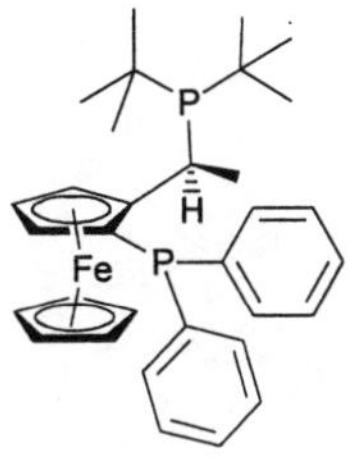

$C_{32}H_{40}FeP_2$
Chiral catalyst. $[\alpha]_D$ = + 412 ± 5° (c = 0.4, $CHCl_3$). *Strem Chemicals, Inc.*

**2405 (R)-(-)-1-[(S)-2-(Diphenylphosphino)-ferrocenyl]ethyldicyclohexylphosphine**
155806-35-2

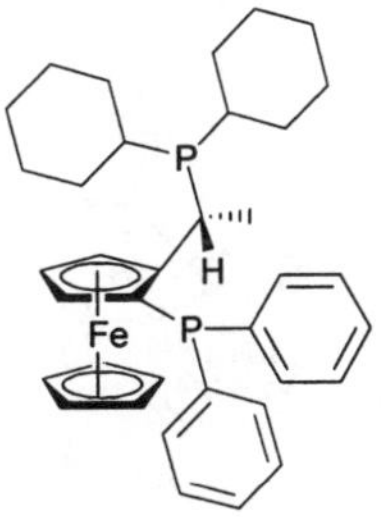

$C_{36}H_{44}FeP_2$
(R)-(S)-JOSIPHOS. Chiral catalyst. $[\alpha]_D$ = - 356 ± 6° (c = 0.4, $CHCl_3$). *Strem Chemicals, Inc.*

**2406 (S)-(+)-1-[(R)-2-(Diphenylphosphino)-ferrocenyl]ethyldicyclohexylphosphine**
162291-02-3

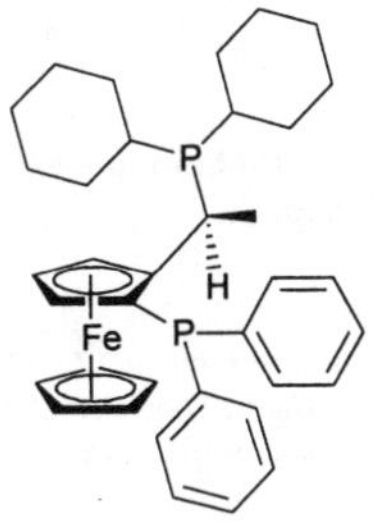

$C_{36}H_{44}FeP_2$
(S)-(R)-JOSIPHOS. Chiral catalyst. $[\alpha]_D$ = + 356 ± 6° (c = 0.4, $CHCl_3$). *Strem Chemicals, Inc.*

**2407 (R)-(-)-1-[(S)-2-(Diphenylphosphino)-ferrocenyl]ethyldi-t-butylphosphine**
82863-72-7

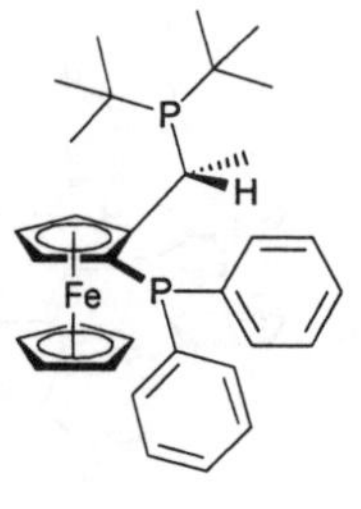

$C_{34}H_{44}FeP_2$
Chiral intermediate. mp = 127-128°; $[\alpha]_D^{20}$ = - 337° (c = 1, $CHCl_3$). *Sigma-Aldrich Fine Chemicals.*

**2408 (S)-(+)-1-[(R)-2-(Diphenylphosphino)-ferrocenyl]ethyldi-t-butylphosphine**
74286-11-6

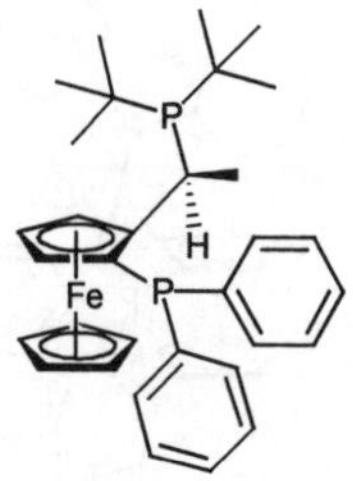

$C_{34}H_{44}FeP_2$
Chiral catalyst. mp = 127-128°; $[\alpha]_D^{20}$ = + 337° (c = 1, $CHCl_3$). *Sigma-Aldrich Fine Chemicals.*

**2409 (R)-(+)-2-(Diphenylphosphino)-2'-methoxy-1,1'-binaphthyl**
145964-33-6
$C_{33}H_{25}OP$
(R)-MOP. Chiral ligand. mp = 177-179°; $[\alpha]_D$ = + 67 ± 2° (c = 1, toluene). *Oxford Asymmetry International plc; Strem Chemicals, Inc.*

**2410 (S)-(+)-2-(Diphenylphosphino)-2'-methoxy-1,1'-binaphthyl**
134484-36-9
$C_{33}H_{25}OP$
(S)-MOP. Chiral ligand. mp = 175-179°; $[\alpha]_D$ = - 67 ± 2° (c = 1, toluene). *Digital Specialty Chemicals, Inc.; Oxford Asymmetry International plc; Strem Chemicals, Inc.*

**2411 (R)-Diphenylphosphino-2'-hydroxy-1,1'binaphthyl**
149917-88-4
$C_{32}H_{23}OP$
Chiral intermediate. *Digital Specialty Chemicals, Inc.*

**2412 (S)-(-)-Diphenylphosphino-2'-hydroxy-1,1'binaphthyl**
144868-15-5
$C_{32}H_{23}OP$
Chiral ligand. *Digital Specialty Chemicals, Inc.*

**2413 (R)-(+)-2-(Diphenylphosphino)-2'-methoxy-1,1'-binaphthyl**
149341-34-4
$C_{31}H_{22}NP$
(R)-QUINAP. Chiral intermediate. mp = 226-230°; $[\alpha]_D$ = + 153° (c = 21, $CHCl_3$). *Strem Chemicals, Inc.*

**2414 (S)-(+)-2-(Diphenylphosphino)-2'-methoxy-1,1'-binaphthyl**
149341-33-3
$C_{31}H_{22}NP$
S-QUINAP. Chiral intermediate. mp = 226-230°; $[\alpha]_D$ = - 153° (c = 21, $CHCl_3$). *Strem Chemicals, Inc.*

**2415 (R)-(+)-1-(2-Diphenylphosphino-1-naphthyl)isoquinoline**
164858-78-0

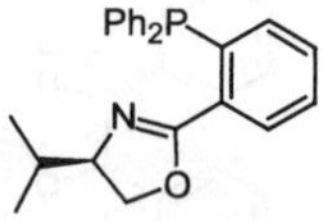

$C_{24}H_{24}NOP$
Chiral ligand. *Strem Chemicals, Inc.*

**2416 (S)-(+)-1-(2-Diphenylphosphino-1-naphthyl)isoquinoline**
148461-14-7

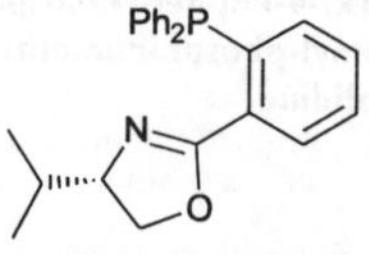

$C_{24}H_{24}NOP$
Chiral ligand. *Strem Chemicals, Inc.*

**2417 (R)-(+)-2-[2-(Diphenylphosphino)phenyl]-4-(1-methylethyl)-4,5-dihydrooxazole**

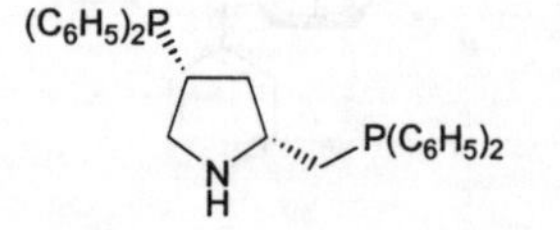

$C_{29}H_{29}NP_2$
Pyrrolidine, 4-(diphenylphosphino)-2-[(diphenylphosphino)methyl]-, (2R-cis)-. Chiral ligand. *Digital Specialty Chemicals, Inc.*

**2418 (S)-(+)-2-[2-(Diphenylphosphino)phenyl]-4-(1-methylethyl)-4,5-dihydrooxazole**
61478-29-3

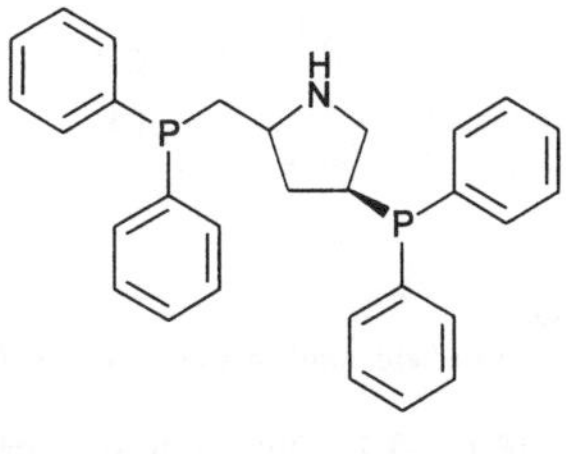

$C_{29}H_{29}NP_2$
Pyrrolidine, 4-(diphenylphosphino)-2-[(diphenylphosphino)methyl]-, (2S-cis)-. Chiral ligand. mp = 102-104°; $[\alpha]^{19}$ = - 16° (c = 1, toluene). *Digital Specialty Chemicals, Inc.; Sigma-Aldrich Fine Chemicals.*

**2419 (R)-(+)-2-Diphenylphosphino-2'-methoxy-1,1'-binaphthyl**
134484-37-0

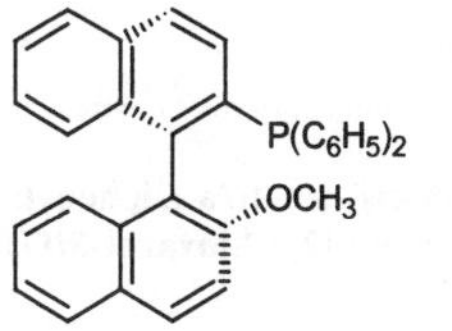

$C_{33}H_{25}OP_2$
Chiral ligand. *Digital Specialty Chemicals, Inc.; Oxford Asymmetry International plc.*

**2420 (2S,4S)-(-)-2-Diphenylphosphinomethyl-4-diphenylphosphino-1-t-butoxy-carbonyl-pyrrolidine**

$C_{34}H_{37}NO_2P_2$
(2S,4S)-(-)-Diphenylphosphonomethyl-4-diphenylphosphino-1-BOC-pyrrolidine. Chiral ligand. mp = 97-107°; $[\alpha]_D$ = - 36 to - 41° (c = 1, $C_6H_6$). *Advanced Asymmetrics, Inc.; Strem Chemicals, Inc.*

**2421 (2R,4R)-(+)-Diphenylphosphinomethyl-4-diphenylphosphino-1-(tert-butoxy-carbonyl)pyrrolidine**

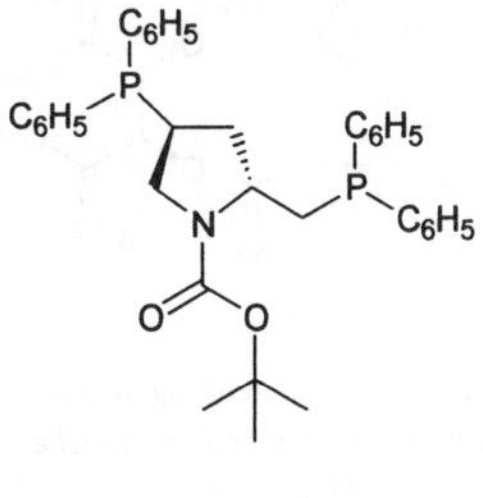

$C_{34}H_{37}NO_2P_2$
Chiral ligand. *Advanced Asymmetrics, Inc.*

**2422 (R)-(+)-Diphenylprolinol**
22348-32-9

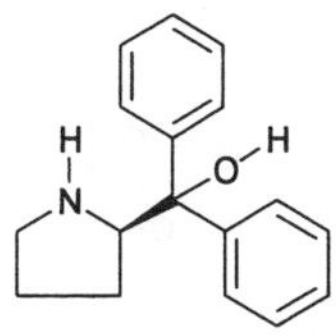

$C_{17}H_{19}NO$
α,α-Diphenyl-2-pyrrolidinemethanol; α,α-Diphenylprolinol. Chiral building block. mp = 76-78°; $[\alpha]_D^{20}$ = + 57.5 ± 2° (c = 3, $CH_3OH$). *Acros Organics nv; Boehringer Ingelheim Pharma KG; Lancaster Synthesis Ltd.; Orsynetics (Tradining) Ltd.; Oxford Asymmetry International plc; PPG - Sipsy; Sigma-Aldrich Fine Chemicals; TCI America.*

**2423 (S)-(-)-α,α-Diphenylprolinol**
112068-01-6

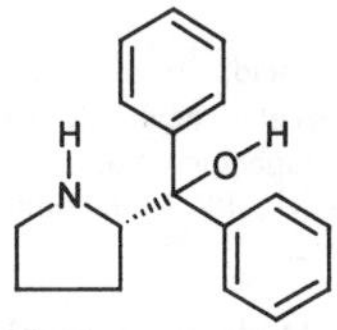

$C_{17}H_{19}NO$
(S)-(-)-α,α-Diphenyl-2-pyrrolidinemethanol; (S)-(-)-2-(Diphenylhydroxymethyl)pyrrolidine; (S)-(-)-Diphenyl-2-pyrrolidinemethanol; (S)-(-)-Diphenylprolinol. Chiral intermediate; Chiral auxiliary. mp = 76-78°; $[\alpha]_D^{20}$ = - 57.5 ± 2° (c = 3, $CH_3OH$); $[\alpha]_D^{20}$ = - 67° (c = 3, $CHCl_3$). *Acros Organics nv; Austin Chemical Company, Inc.; Boehringer Ingelheim Pharma KG; Lancaster Synthesis Ltd.; Sigma-Aldrich Fine Chemicals.*

**2424 (S)-(-)-1,1-Diphenyl-1,2-propanediol**
46755-94-6

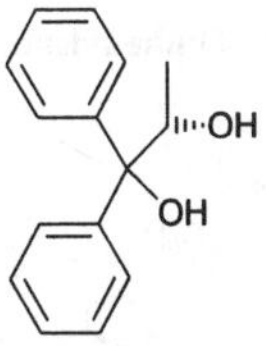

$C_{15}H_{16}O_2$
Chiral building block. mp = 89-92°; $[\alpha]^{25}$ = - 100° (c = 1, $CH_3OH$). *Sigma-Aldrich Fine Chemicals.*

**2425 (S)-(-)-1,1-Diphenyl-1,2-propanediol cyclic sulfite**
51226-54-1

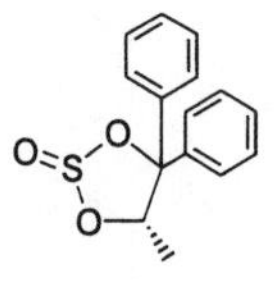

$C_{15}H_{14}O_3S$
Chiral intermediate. mp = 110-113°; $[\alpha]_D^{20}$ = - 245° (c = 1, $CHCl_3$). *Sigma-Aldrich Fine Chemicals.*

**2426 (R)-2,3-Diphenylpropionic acid**

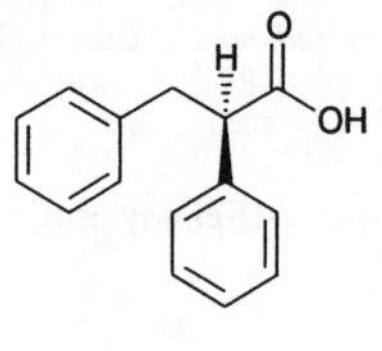

$C_{15}H_{14}O_2$
Benzeneacetic acid, α-(phenylmethyl)-, (R)-; (R)-α-Benzylbenzeneacetic acid; (R)-α-Phenylhydrocinnamic acid; Benzenepropanoic acid, α-phenyl-, (R)-; Propanoic acid, 2,3-diphenyl-, (R)-. Chiral building block. *Sumitomo Chemcial Co. Ltd.*

**2427 (S)-2,3-Diphenylpropionic acid**

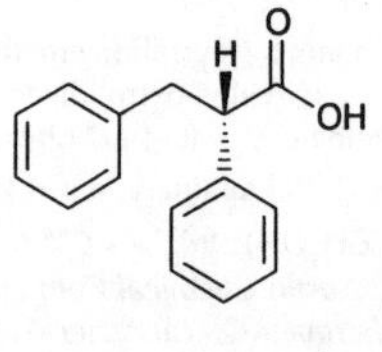

$C_{15}H_{14}O_2$
Benzeneacetic acid, α-(phenylmethyl)-, (S)-; (S)-α-Benzylbenzeneacetic acid; (S)-α-Phenylhydrocinnamic acid; Benzenepropanoic acid, α-phenyl-, (S)-; Propanoic acid, 2,3-diphenyl-, (S)-. Chiral building block. *Sumitomo Chemcial Co. Ltd.*

**2428 (R,R)-2,3-Diphenylsuccinic acid**
21037-34-3

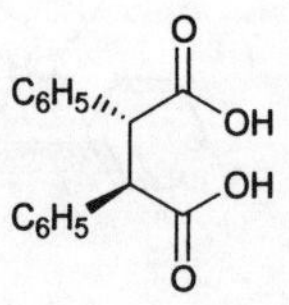

$C_{16}H_{14}O_4$
Chiral intermediate. *Kankyo Kagaku Center Co., Ltd.*

**2429 (S,S)-2,3-Diphenylsuccinic acid**
74431-38-2

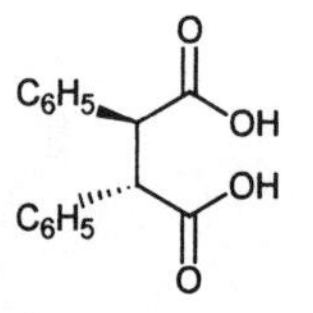

$C_{16}H_{14}O_4$
Chiral intermediate. *Kankyo Kagaku Center Co., Ltd.*

**2430 (3R-cis)-3,7a-Diphenyltetrahydropyrrolo-[2,1-b]oxazol-5(6H)-one**
132959-39-8

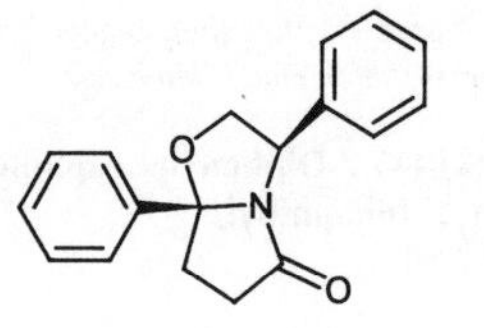

$C_{18}H_{17}NO_2$
Chiral intermediate. *Acros Organics nv.*

**2431 (3S-cis)-(+)-3,7a-Diphenyltetrahydro-pyrrolo-[2,1-b]oxazol-5(6H)-one**
161970-71-4

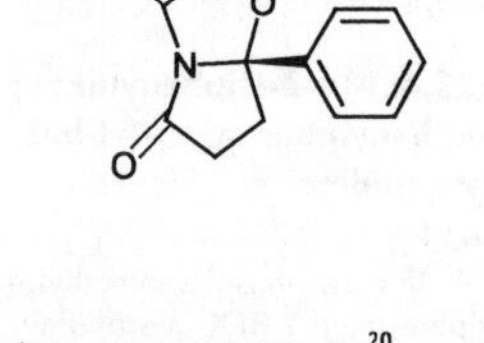

$C_{18}H_{17}NO_2$
Chiral intermediate. mp = 82-85°; $[\alpha]_D^{20}$ = + 113° (c = 1, $CHCl_3$). *Acros Organics nv; Sigma-Aldrich Fine Chemicals.*

**2432 Dipivaloyl-D-tartaric acid**
76769-55-6

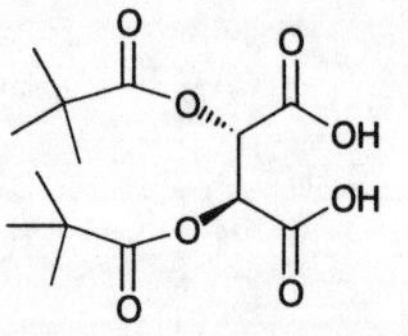

$C_{14}H_{22}O_8$
Chiral ligand; Chiral building block. mp = 127-130°; $[\alpha]_D^{20}$ = + 24° (c = 1, dioxane). *Senn Chemicals AG; Sigma-Aldrich Fine Chemicals.*

**2433 Dipivaloyl-L-tartaric acid**

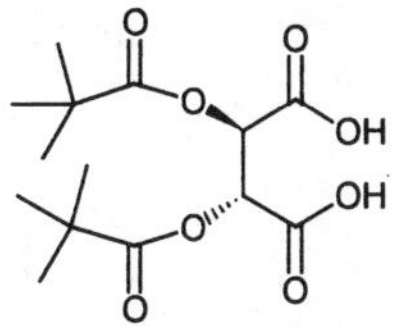

$C_{14}H_{22}O_8$
Chiral ligand; Chiral building block. mp = 130-132°; $[\alpha]^{25}$ = - 24° (c = 1.7, dioxane). *Senn Chemicals AG; Sigma-Aldrich Fine Chemicals.*

**2434 (R)-2,2'-Dipropoxy-1,1'-binaphthalene-6,6'-dicarbaldehyde**

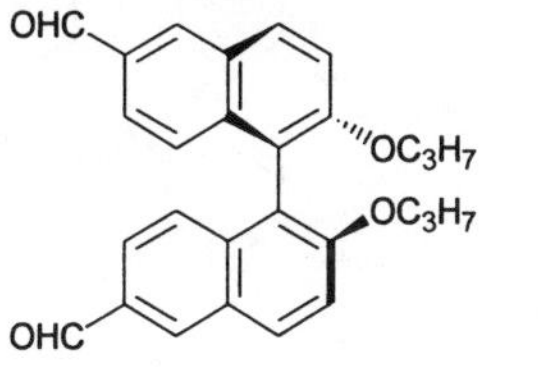

$C_{28}H_{26}O_4$
Chiral ligand. *Kankyo Kagaku Center Co., Ltd.*

**2435 (S)-2,2'-Dipropoxy-1,1'-binaphthalene-6,6'-dicarbaldehyde**

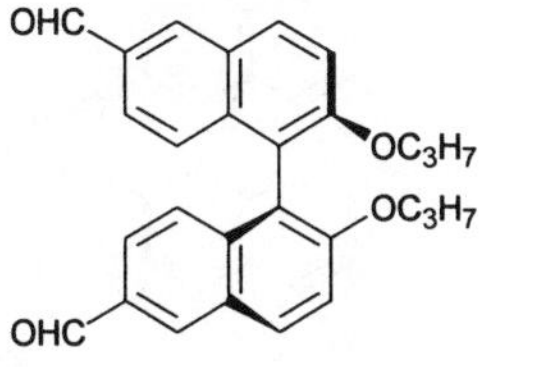

$C_{28}H_{26}O_4$
Chiral ligand. *Kankyo Kagaku Center Co., Ltd.*

**2436 (R)-2,2'-Dipropoxy-1,1'-binaphthalene-6,6'-dicarboxylic acid**

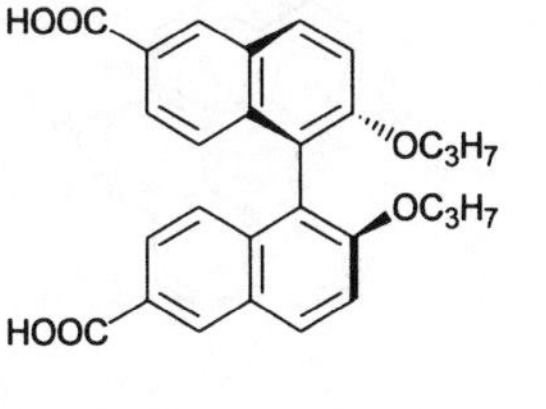

$C_{28}H_{26}O_6$
Chiral ligand. *Kankyo Kagaku Center Co., Ltd.*

**2437 (S)-2,2'-Dipropoxy-1,1'-binaphthalene-6,6'-dicarboxylic acid**

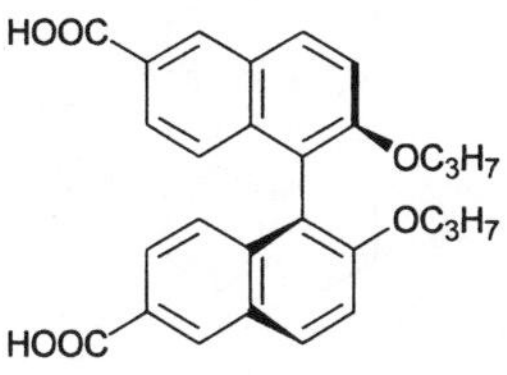

$C_{28}H_{26}O_6$
Chiral ligand. *Kankyo Kagaku Center Co., Ltd.*

**2438 (R)-2,2'-Dipropoxy-1,1'-binaphthyl**

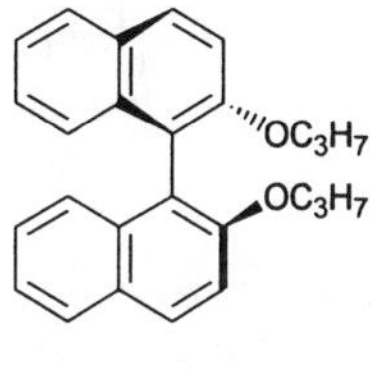

$C_{26}H_{26}O_2$
Chiral ligand. *Kankyo Kagaku Center Co., Ltd.*

**2439 (S)-2,2'-Dipropoxy-1,1'-binaphthyl**

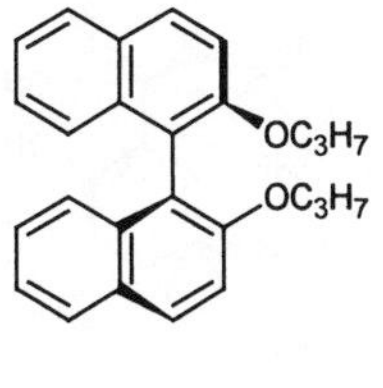

$C_{26}H_{26}O_2$
Chiral ligand. *Kankyo Kagaku Center Co., Ltd.*

**2440 (R)-2,2'-Dipropoxy-5,5',6,6',7,7',8,8'-octahydro-1,1'-binaphthyl**

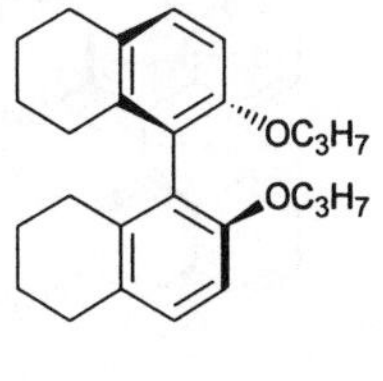

$C_{26}H_{34}O_2$
Chiral ligand. *Kankyo Kagaku Center Co., Ltd.*

**2441 (S)-2,2'-Dipropoxy-5,5',6,6',7,7',8,8'-octahydro-1,1'-binaphthyl**

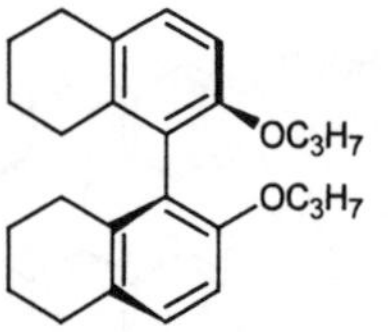

$C_{26}H_{34}O_2$
Chiral ligand. *Kankyo Kagaku Center Co., Ltd.*

**2442 (S)-N,N-Dipropylalanine**

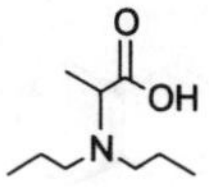

$C_9H_{19}NO_2$
(S)-(+)-N,N-Dipropylalanine. Chiral building block. mp = 94-95°; $[\alpha]_D^{20}$ = + 15.6° (c = 2, $CH_3OH$). *Sigma-Aldrich Fine Chemicals.*

**2443 (R)-3,3'-Dipropyl-1,1'-bi-2-naphthol**

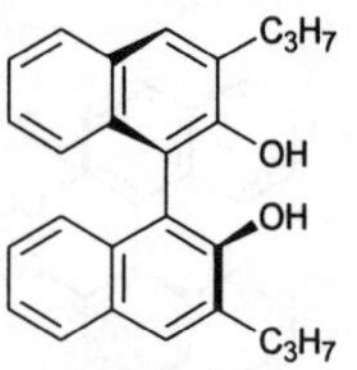

$C_{26}H_{26}O_2$
Chiral ligand. *Kankyo Kagaku Center Co., Ltd.*

**2444 (S)-3,3'-Dipropyl-1,1'-bi-2-naphthol**

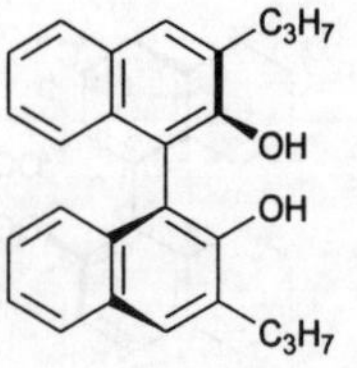

$C_{26}H_{26}O_2$
Chiral ligand. *Kankyo Kagaku Center Co., Ltd.*

**2445 (R)-6,6'-Dipropyl-1,1'-bi-2-naphthol**

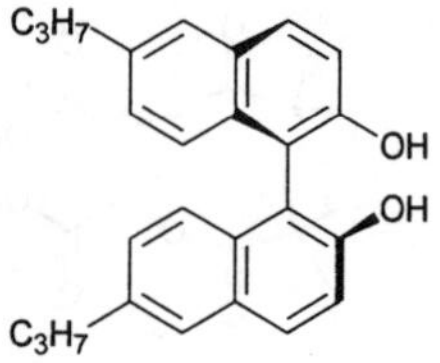

$C_{26}H_{26}O_2$
Chiral ligand. *Kankyo Kagaku Center Co., Ltd.*

**2446 (S)-6,6'-Dipropyl-1,1'-bi-2-naphthol**

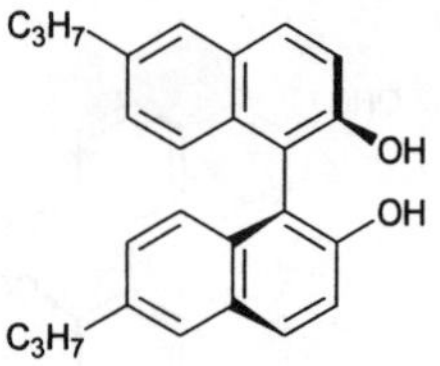

$C_{26}H_{26}O_2$
Chiral intermediate. *Kankyo Kagaku Center Co., Ltd.*

**2447 (R)-2,2'-Dipropyl-1,1'-binaphthyl**

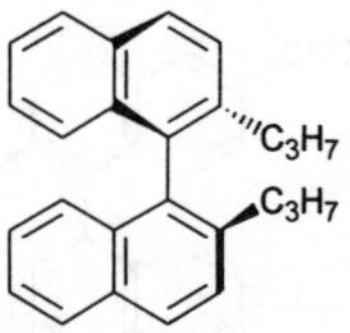

$C_{26}H_{26}$
Chiral ligand. *Kankyo Kagaku Center Co., Ltd.*

**2448 (S)-2,2'-Dipropyl-1,1'-binaphthyl**

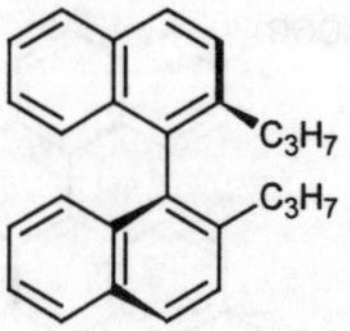

$C_{26}H_{26}$
Chiral ligand. *Kankyo Kagaku Center Co., Ltd.*

**2449 (R)-3,3'-Dipropyl-2,2'-dimethoxy-1,1'-binaphthyl**

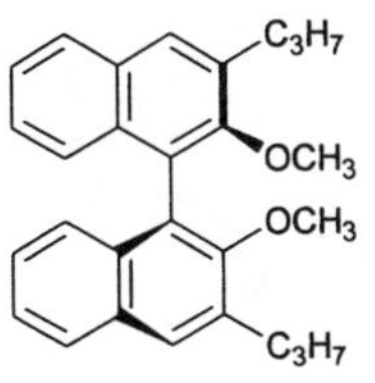

$C_{28}H_{30}O_2$
Chiral ligand. *Kankyo Kagaku Center Co., Ltd.*

**2450 (S)-3,3'-Dipropyl-2,2'-dimethoxy-1,1'-binaphthyl**

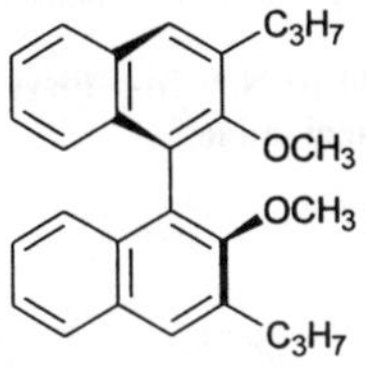

$C_{28}H_{30}O_2$
Chiral ligand. *Kankyo Kagaku Center Co., Ltd.*

**2451 (R)-6,6'-Dipropyl-2,2'-dimethoxy-1,1'-binaphthyl**

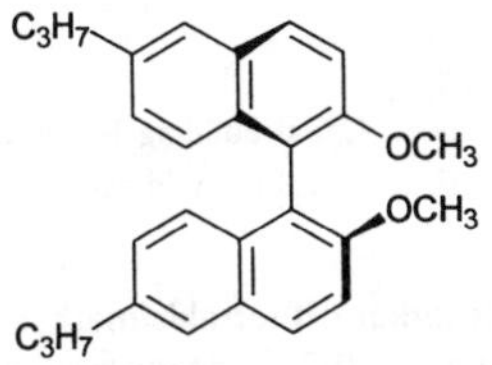

$C_{28}H_{30}O_2$
Chiral ligand. *Kankyo Kagaku Center Co., Ltd.*

**2452 (S)-6,6'-Dipropyl-2,2'-dimethoxy-1,1'-binaphthyl**

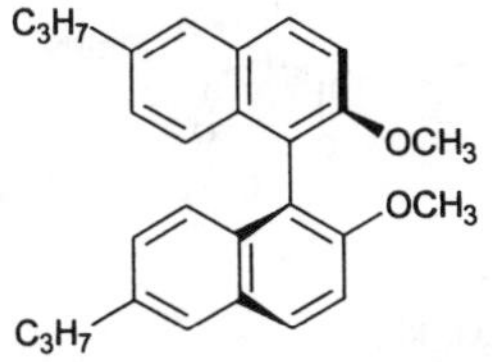

$C_{28}H_{30}O_2$
Chiral intermediate. *Kankyo Kagaku Center Co., Ltd.*

**2453 (R)-3,3'-Dipropyl-2,2'-dimethoxy-5,5',6,6',7,7',8,8'-octahydro-1,1'-binaphthyl**

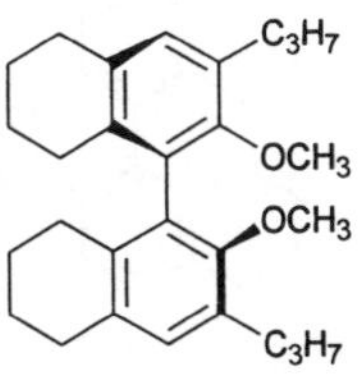

$C_{28}H_{38}O_2$
Chiral ligand. *Kankyo Kagaku Center Co., Ltd.*

**2454 (S)-3,3'-Dipropyl-2,2'-dimethoxy-5,5',6,6',7,7',8,8'-octahydro-1,1'-binaphthyl**

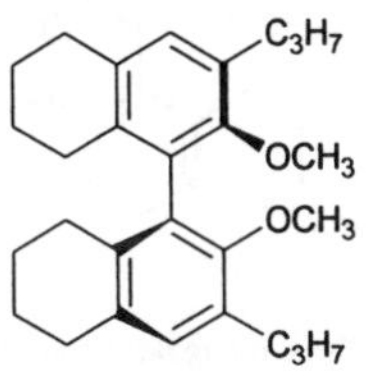

$C_{28}H_{38}O_2$
Chiral ligand. *Kankyo Kagaku Center Co., Ltd.*

**2455 (R)-3,3'-Dipropyl-5,5',6,6',7,7',8,8'-octahydro-1,1'-bi-2-naphthol**

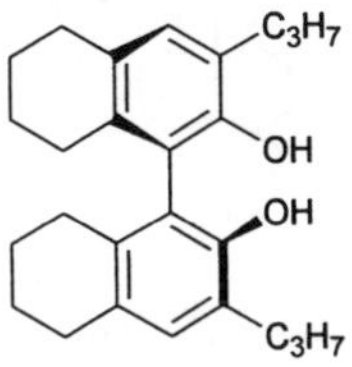

$C_{26}H_{34}O_2$
Chiral ligand. *Kankyo Kagaku Center Co., Ltd.*

**2456 (S)-3,3'-Dipropyl-5,5',6,6',7,7',8,8'-octahydro-1,1'-bi-2-naphthol**

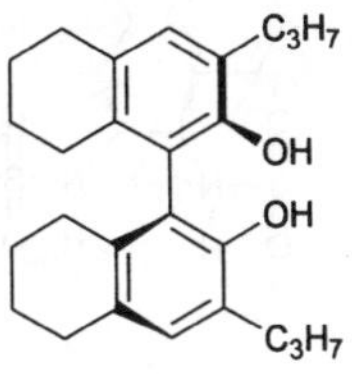

$C_{26}H_{34}O_2$
Chiral ligand. *Kankyo Kagaku Center Co., Ltd.*

**2457 (R)-2,2'-Dipropyl-5,5',6,6',7,7',8,8'-octahydro-1,1'-binaphthyl**

$C_{26}H_{34}$
Chiral ligand. *Kankyo Kagaku Center Co., Ltd.*

**2458 (S)-2,2'-Dipropyl-5,5',6,6',7,7',8,8'-octahydro-1,1'-binaphthyl**

$C_{26}H_{34}$
Chiral ligand. *Kankyo Kagaku Center Co., Ltd.*

**2459 (+)-Dipterocarpol**
471-69-2

$C_{30}H_{50}O_2$
Chiral intermediate. mp = 133-135°; $[\alpha]_D^{20}$ = + 67° (c = 1, $CHCl_3$). *Sigma-Aldrich Fine Chemicals.*

**2460 (1S,2S)-N,N'-Di-p-toluenesulfonyl-1,2-diphenyl-1,2-ethylenediamine**
170709-41-8

$C_{28}H_{28}N_2O_4S_2$
Chiral building block; Resolving agent. *Acros Organics nv.*

**2461 (1R,2R)-(+)-N,N-Di-p-tosyl-1,2-cyclohexanediamine**

$C_{20}H_{26}N_2O_4S_2$
Chiral ligand. mp = 170-173°; $[\alpha]^{23}$ = + 3° (c = 2.33, pyridine). *Sigma-Aldrich Fine Chemicals.*

**2462 (1S,2S)-(-)-N,N-Di-p-tosyl-1,2-cyclohexanediamine**

$C_{20}H_{26}N_2O_4S_2$
Resolving agent; Chiral building block. mp = 170-173°; $[\alpha]^{22}$ = - 3° (c = 2.33, pyridine). *Sigma-Aldrich Fine Chemicals.*

**2463 Dirhodium(II)tetrakis[methyl 1-(3-phenylpropanoyl)-2-imidazolidone-4(S)-carboxylate, acetonitrile complex (1:2)**
171230-55-0

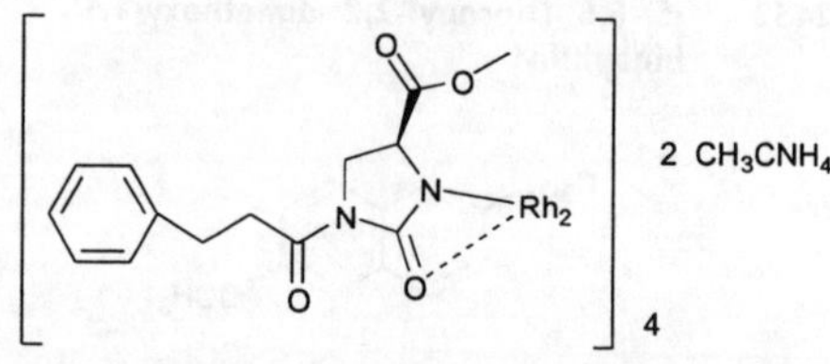


$C_{60}H_{66}N_{10}O_{16}Rh_2$
Doyle dirhodium catalyst - Rh2(4S-MPPIM)4. Chiral catalyst. mp = 201-205°; $[\alpha]^{23}$ = - 316° (c = 0.3, CH3CN). *Acros Organics nv; Sigma-Aldrich Fine Chemicals.*

**2464 Dirhodium(II)tetrakis[methyl 2-pyrrolidone-5(R)-carboxylate], acetonitrile/2-propanol complex**
131769-58-2

$C_{24}H_{32}N_4O_{12}Rh_2$
Doyle dirhodium catalyst - RH2(5R-MEPY)4. Chiral catalyst. *Acros Organics nv; Sigma-Aldrich Fine Chemicals.*

**2465 Dirhodium(II)tetrakis[methyl 2-pyrrolidone-5(S)-carboxylate], acetonitrile/2-propanol complex**
132435-65-5

X $CH_3CNH_4$
Y $CH_3CH_2OHCH_3$

$C_9H_{18}NO_3$
Doyle dirhodium catalyst - RH2(5S-MEPY)4. Chiral catalyst. *Acros Organics nv; Sigma-Aldrich Fine Chemicals.*

**2466 (+)-Disparlure**
259-390-8

$C_{19}H_{38}O$
(2S-cis)-2-Decyl-3-(5-methylhexyl)oxirane. Chiral intermediate. $bp_{0.25}$ = 146-148°; d = 0.828; n = 1.445; $[\alpha]_D^{20}$ = + 0.8° (c = 1, $CHCl_3$). *Sigma-Aldrich Fine Chemicals.*

**2467 Dithio-D-erythritol**
6892-68-8 229-998-8
$C_4H_{10}O_2S_2$
Butane-2,3-diol, 1,4-dimercapto-, (2R,3S)-; DTE; Erythritol, 1,4-dithio-; erythro-1,4,-Dimercapto-2,3-butanediol. Listed on TSCA. Chiral diagnostic reagent. mp > 81°; soluble in $H_2O$( 5% w/v). *Diagnostic Chemicals Limited.*

**2468 1,4-Dithio-L-threitol**
16096-97-2 240-263-0

$C_4H_{10}O_2S_2$
Butane-2,3-diol, 1,4-dimercapto-, [R-(R*,R*)]-; Cleland's Reagent chiral; [R-(R*,R*)]-1,4-dimercaptobutane-2,3-diol; Threitol, 1,4-dithio-, L-; (2R,3R)-1,4-Dimercapto-2,3-butanediol; L-DTT; (2R,3R)-(-)-2,3-Dihydroxy-1,4-butanedithiol. Chiral intermediate. mp = 49-51°; $[\alpha]_D^{20}$ = - 14 ± 1° (c = 2.4, $CHCl_3$). *Acros Organics nv; FineTech, Ltd.; Lancaster Synthesis Ltd.; TCI America.*

**2469 (-)-Dititonin**
11024-24-1 3204(12) 234-255-6

$C_{56}H_{92}O_{29}$
Listed on TSCA. Pharmaceutical or derivative. mp = 230-240°; $[\alpha]_D^{20}$ = - 50.1° (c = 1, 75% HOAc). *Acros Organics nv; Sigma-Aldrich Fine Chemicals.*

**2470 (R)-1,4-Ditosyl-2-butanol**
99520-82-8

$C_{18}H_{22}O_7S_2$
Butane-1,2,4-triol, 1,4-bis(4-methylbenzenesulfonate), (S)-; (S)-1,4-Bis(4-methylbenzenesulfonate)-1,2,4-butanetriol. Chiral building block. *Synthon Chiragenics Corporation.*

**2471 (S)-1,4-Ditosyl-2-butanol**
170305-50-7

$C_{18}H_{22}O_7S_2$
Butane-1,2,4-triol, 1,4-bis(4-methylbenzenesulfonate),

(R)-. Chiral building block. *Synthon Chiragenics Corporation.*

**2472 (1R,2R)-N,N'-Ditosyl-1,2-diphenyl-1,2-ethanediamine**

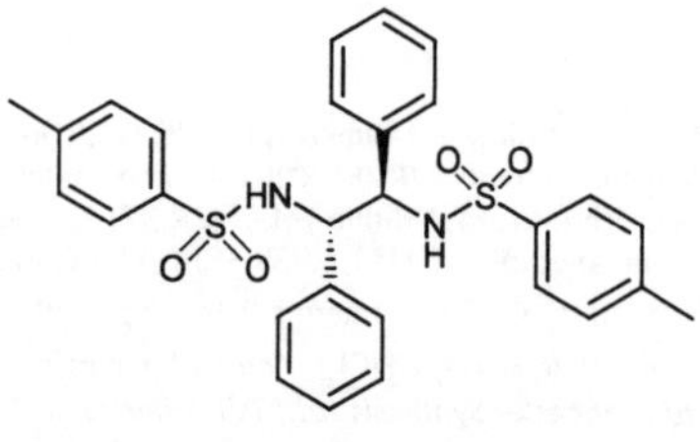

$C_{28}H_{28}N_2$
Chiral ligand. *Oxford Asymmetry International plc.*

**2473 (1S,2S)-N,N'-Ditosyl-1,2-diphenyl-1,2-ethanediamine**

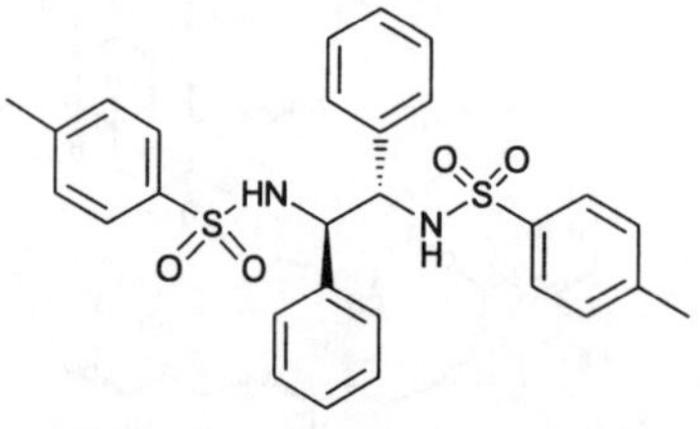

$C_{28}H_{28}N_2$
(1R,2R)-N,N'-Di-p-toluenesulfonyl-1,2-diphenyl-1,2-ethylenediamine. Chiral building block; Chiral auxiliary. *Acros Organics nv; Oxford Asymmetry International plc.*

**2474 (-)-3,6-Di-O-(triisopropylsilyl)galactal**

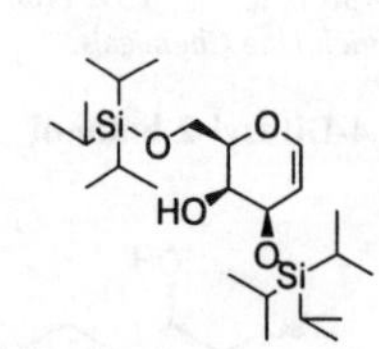

$C_{24}H_{50}O_4Si_2$
Chiral intermediate. n = 1.478; $[\alpha]^{22}$ = - 35° (c = 2.5, $CHCl_3$). *Sigma-Aldrich Fine Chemicals.*

**2475 (-)-3,6-Di-O-(triisopropylsilyl)glucal**

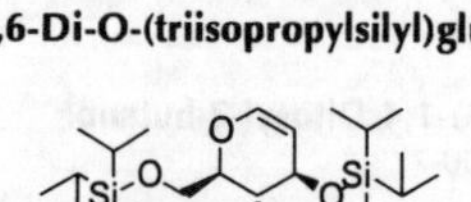

$C_{24}H_{50}O_4Si_2$
Chiral intermediate. d = 0.965; n = 1.48; $[\alpha]^{25}$ = - 22° (c = 1, $CHCl_3$). *Sigma-Aldrich Fine Chemicals.*

**2476 (R)-3,3'-Divinyl-2,2'-dimethoxy-1,1'-binaphthyl**

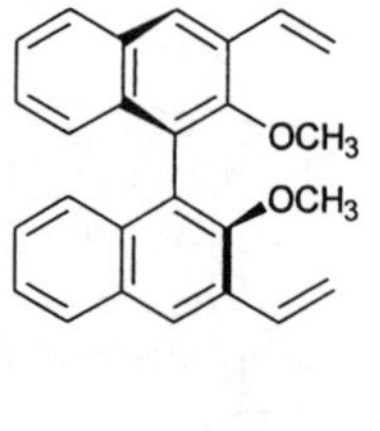

$C_{26}H_{22}O_2$
Chiral ligand. *Kankyo Kagaku Center Co., Ltd.*

**2477 (S)-3,3'-Divinyl-2,2'-dimethoxy-1,1'-binaphthyl**

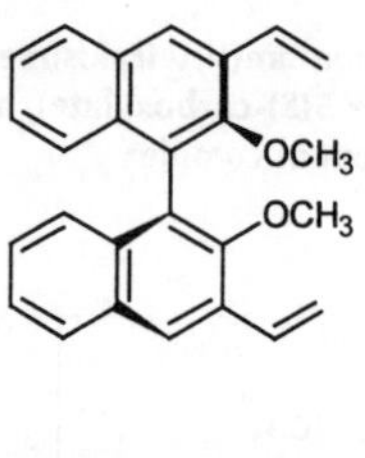

$C_{26}H_{22}O_2$
Chiral ligand. *Kankyo Kagaku Center Co., Ltd.*

**2478 (R)-6,6'-Divinyl-2,2'-dimethoxy-1,1'-binaphthyl**

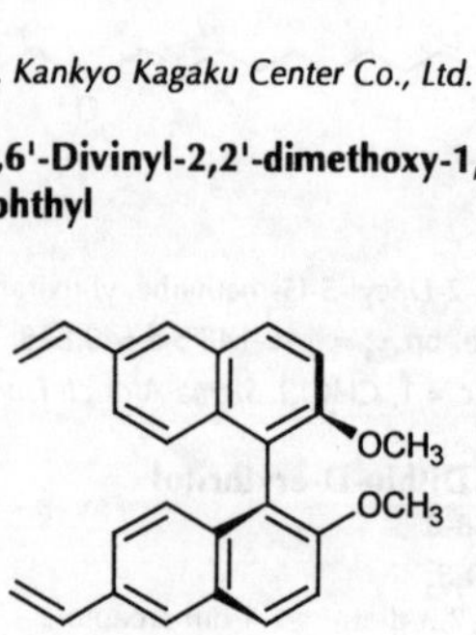

$C_{26}H_{22}O_2$
Chiral ligand. *Kankyo Kagaku Center Co., Ltd.*

**2479 (S)-6,6'-Divinyl-2,2'-dimethoxy-1,1'-binaphthyl**

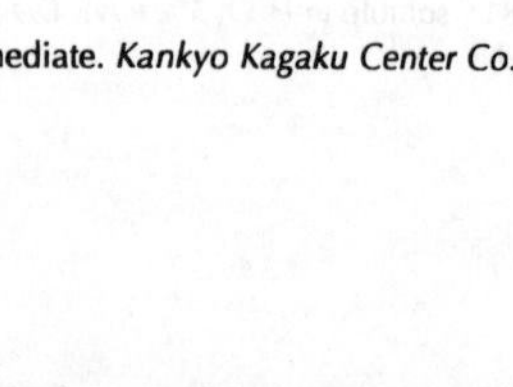

$C_{26}H_{22}O_2$
Chiral intermediate. *Kankyo Kagaku Center Co., Ltd.*

**2480 (R)-3,3'-Divinyl-2,2'-dimethoxy-5,5',6,6',7,7',8,8'-octahydro-1,1'-binaphthyl**



$C_{26}H_{30}O_2$
Chiral ligand. *Kankyo Kagaku Center Co., Ltd.*

**2481 (S)-3,3'-Divinyl-2,2'-dimethoxy-5,5',6,6',7,7',8,8'-octahydro-1,1'-binaphthyl**



$C_{26}H_{30}O_2$
Chiral ligand. *Kankyo Kagaku Center Co., Ltd.*

**2482 (R)-1,2-Dodecanediol**
85514-85-8



$C_{12}H_{26}O_2$
Chiral building block. mp = 69-72°; $[\alpha]^{22}$ = - 13° (c = 2.5, EtOH). *Acros Organics nv; Kankyo Kagaku Center Co., Ltd.; Sigma-Aldrich Fine Chemicals.*

**2483 (S)-1,2-Dodecanediol**
85514-84-7



$C_{12}H_{26}O_2$
Chiral building block. mp = 69-72°; $[\alpha]^{23}$ = + 13° (c = 2.5, EtOH). *Acros Organics nv; Kankyo Kagaku Center Co., Ltd.; Sigma-Aldrich Fine Chemicals.*

**2484 (S)-2-Dodecanol**
91681-57-1



$C_{12}H_{26}O$
Dodecan-2-ol, (S)-. Chiral building block. *Japan Energy Corporation.*

**2485 (S)-4-Dodecanol**



$C_{12}H_{26}O$
Dodecan-4-ol, (S)-. (S)-4-Hydroxydodecane. Chiral building block. *Japan Energy Corporation.*

**2486 n-Dodecyl-β-D-maltoside**
69227-93-6



$C_{26}H_{50}O_{11}$
Chiral intermediate. mp = 224-226°; $[\alpha]^{25}$ = + 47° (c = 1, $CH_3OH$). *Acros Organics nv; Senn Chemicals AG; Sigma-Aldrich Fine Chemicals.*

**2487 N-Dodecyl-N-methylephedrinium bromide**
31351-20-9 250-579-0



$C_{23}H_{42}BrNO$
Dodecyl(2-hydroxy-1-methyl-2-phenylethyl)dimethylammonium bromide. Chiral intermediate. mp = 80-82°; $[\alpha]_D^{20}$ = - 12.5° (c = 1, $CHCl_3$). *Sigma-Aldrich Fine Chemicals.*

**2488 3-Doxyl-5α-cholestane**
18353-76-9

$C_{31}H_{54}NO_2$
4',4'-Dimethylspiro[5α-cholestane-3,2'-oxazolidin]-3'-yloxy. Pharmaceutical or derivative. mp = 177-179°. *Acros Organics nv.*

**2489 Doyle dirhodium catalyst - RH2(4S-MEOX)4**
167693-36-9

$C_7H_{12}NO_4$
Chiral catalyst. mp = 293°; $[\alpha]^{25}$ = - 213.5° (c = 0.3, CH3CN). *Acros Organics nv; Sigma-Aldrich Fine Chemicals.*

**2490 (-)-Eburnamonine**
4880-88-0 3537(12) 225-490-5

$C_{19}H_{22}N_2O$
Chiral intermediate. mp = 174-177°; $[\alpha]^{23}$ = - 93.7° (c = 0.4, $CHCl_3$). *Sigma-Aldrich Fine Chemicals.*

**2491 (+)-Emetine dihydrochloride hydrate**
3517(11) 206-259-8

$C_{29}H_{42}Cl_2N_2O_4$
Listed on TSCA. Chiral intermediate. mp = 235°; $[\alpha]^{25}$ = + 9.0° (c = 1, $H_2O$). *Sigma-Aldrich Fine Chemicals.*

**2492 L-(-)-Ephedrine sulfate**
134-72-5
$C_{10}H_{15}NO$
L-α-(1-Methylaminoethyl)benzyl alcohol sulfate. Chiral intermediate. *Sigma-Aldrich Fine Chemicals.*

**2493 (+)-Epiandrosterone**
481-29-8 207-563-3

$C_{19}H_{30}O_2$
Androstan-17-one, 3-hydroxy-, (3β,5α)-; 3β-Hydroxy-5α-androstan-17-one; 3β-Hydroxyetioallocholan-17-one. Pharmaceutical or derivative. $[\alpha]^{25}$ = + 92° (c = 1, $CH_3OH$). *Acros Organics nv; Sigma-Aldrich Fine Chemicals.*

**2494 (-)-4-Epianhydrotetracycline hydrochloride**
4465-65-0

$C_{22}H_{22}N_2O_7$
Pharmaceutical or derivative. [α] = - 279° (c = 0.5, $H_2O$). *Acros Organics nv.*

**2495 L-(-)-Epicatechin**
490-46-0 207-710-1

$C_{15}H_{14}O_6$
Benzo-(2H)-1-pyran-3,5,7-triol, 2-(3,4-dihydroxyphenyl)-3,4-dihydro-, (2R,3R)-; 2-(3,4-Dihydroxyphenyl)-2,3,4-trihydro-3,5,7-trihydroxychromene; (-)-Epicatechol; epi-Catechin; epi-Catechol; l-Acacatechin. Chiral intermediate. mp = 240°; $[\alpha]_D^{20}$ = - 54° (c = 1, acetone/$H_2O$). *Acros Organics nv; Senn Chemicals AG; Sigma-Aldrich Fine Chemicals.*

**2496 (R)-(-)-Epichlorohydrin**
51594-55-9 3563(11)

$C_3H_5ClO$
(R)-(-)-1-Chloro-2,3-epoxypropane. Chiral building block. mp = - 48°; bp = 113-115°; d = 1.1800; $n_D^{20}$ = 1.4380; $[\alpha]_D^{20}$ = - 35 ± 2° (c = 1, $CH_3OH$). *Austin Chemical Company, Inc.; Chirex, Inc.; Daiso Company, Ltd.; Kaneka Corporation; Lancaster Synthesis Ltd.; Mitsubishi Chemical Corporation; Omega Chemical Company Inc.; Sigma-Aldrich Fine Chemicals.*

**2497 (S)-(+)-Epichlorohydrin**
67843-74-7 3563(11)

$C_3H_5ClO$
(S)-(+)-1-Chloro-2,3-epoxypropane. Chiral building block. mp = - 48°; $bp_{360}$ = 92-93°; d = 1.1800; $n_D^{20}$ = 1.4380; $[\alpha]_D^{20}$ = + 35 ± 2° (c = 1, $CH_3OH$). *Austin Chemical Company, Inc.; Chirex, Inc.; Daiso Company, Ltd.; Lancaster Synthesis Ltd.; Mitsubishi Chemical Corporation; Omega Chemical Company Inc.; Sigma-Aldrich Fine Chemicals.*

**2498 D-Epilactose monohydrate**

$C_{12}H_{24}O_{12}$
4-O-β-D-Galactopyranosyl-α-D-mannopyranoside monohydrate. Chiral intermediate. *Senn Chemicals AG.*

**2499 (-)-4-Epioxytetracycline**
35259-39-3

$C_{22}H_{24}N_2O_9$
Pharmaceutical or derivative. [α] = - 247° (c = 1.0, 0.03N HCl). *Acros Organics nv.*

**2500 (-)-4-Epitetracycline hydrochloride**
23313-80-6

$C_{22}H_{24}N_2O_9$
Pharmaceutical or derivative. [α] = - 322° (c = 0.5, $H_2O$). *Acros Organics nv.*

**2501 (S)-1,2-Epithiodecane**

$C_{10}H_{20}S$
(S)-(-)-Octylthiirane. Chiral intermediate. $bp_{0.8}$ = 66-67°; d = 0.87; n = 1.472; $[\alpha]^{25}$ = - 53° (neat). *Japan Energy Corporation; Sigma-Aldrich Fine Chemicals.*

**2502 (S)-1,2-Epithiododecane**

$C_{12}H_{24}S$
(S)-(-)-Decylthiirane. Chiral intermediate. $bp_{4.0}$ = 119-120°; d = 0.86; n = 1.472; $[\alpha]^{25}$ = - 44° (neat). *Japan Energy Corporation; Sigma-Aldrich Fine Chemicals.*

**2503 (S)-(-)-1,2-Epithiooctane**

$C_8H_{16}S$
(S)-(-)-Hexylthiirane. Chiral intermediate. $bp_5$ = 83°; d = 0.88; n = 1.47; $[\alpha]^{25}$ = - 66° (neat). *Japan Energy Corporation; Sigma-Aldrich Fine Chemicals.*

**2504 (+)-Epoxy-9-decene**
137310-67-9

$C_{10}H_{18}O$
Chiral building block. $bp_{0.3}$ = 39.5-40.5°; d = 0.86; $[\alpha]^{25}$ = + 13° (neat). *Japan Energy Corporation; Sigma-Aldrich Fine Chemicals.*

**2505 (2S,3S)-2,3-Epoxybutane**

$C_4H_8O$
Oxirane, 2,3-dimethyl-, (2S,3S)-; (2S,3S)-2,3-Dimethyloxirane; (2S,3S)-2,3-Butylene oxide. Chiral building block. *Acros Organics nv.*

**2506 (R)-(+)-1,2-Epoxybutane**
3760-95-0
$C_4H_8O$
(2R)-Ethyloxirane. Chiral building block. bp = 63°; d = 0.84; n = 1.39; $[\alpha]_D^{20}$ = - 10° (neat). *Sigma-Aldrich Fine Chemicals.*

**2507 (S)-(-)-1,2-Epoxybutane**
30608-62-9
$C_4H_8O$
(2S)-Ethyloxirane. Chiral building block. bp = 63°; d = 0.84; n = 1.39; $[\alpha]_D^{20}$ = - 10° (neat). *Sigma-Aldrich Fine Chemicals.*

**2508 (1S,3S,5R,6R,7S,8R)-6,7-Epoxy-8-butyl-3-[(S)-tropoyloxy]tropanulum bromide**

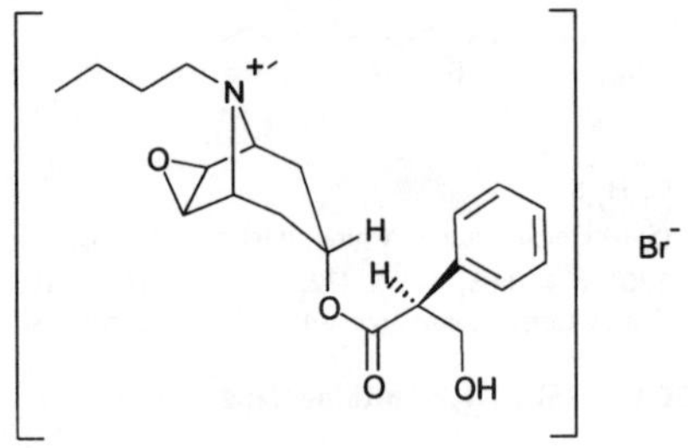

$C_{21}H_{30}BrNO_4$
[7(S)-(1α,β,4β,5α,7β)]-9-Butyl-7-(3-hydroxy-1-oxo-2-phenylpropoxy)-9-methyl-3-oxa-9-azoniatricyclo[3.3.1.-0.2.4]nonane bromide; Hycoscine-N-butylbromide; Scopalamine-N-butylbromide; N-Butyl scopalammonium bromide. Chiral intermediate. *Linnea S.A.*

**2509 (+)-Epoxydecane**
67210-36-0

$C_{10}H_{20}O$
Oxirane, octyl-, (R)-; (R)-Octyloxirane; Decene epoxide; Decylene oxide. Chiral building block. $bp_{15}$ = 94°; d = 0.84; n = 1.429; $[\alpha]^{25}$ = + 12° (neat). *Japan Energy Corporation; Sigma-Aldrich Fine Chemicals.*

**2510 (+)-Epoxydodecane**
109856-85-1

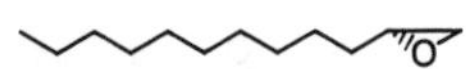

$C_{12}H_{24}O$
Oxirane, decyl-, (R)-; (R)-Decyloxirane; (R)-α-Dodecene oxide; (R)-1-Dodecene oxide; Dodecane, 1,2-epoxy-, (R)-. Chiral building block. $bp_{15}$ = 124-125°; d = 0.844; n = 1.436; $[\alpha]^{25}$ = + 10° (neat). *Japan Energy Corporation; Sigma-Aldrich Fine Chemicals.*

**2511 (+)-Epoxyheptadecane**

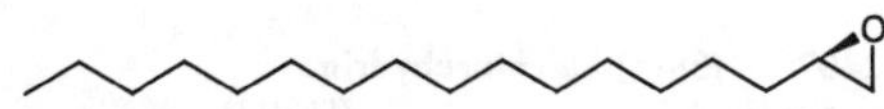

$C_{17}H_{34}O$
Oxirane, pentadecyl-, (R)-; (R)-Pentadecyloxirane; Heptadecane oxide; Heptadecane, 1,2-epoxy-, (R)-. Chiral building block. *Japan Energy Corporation.*

**2512 (+)-Epoxyheptane**
110549-07-0

$C_7H_{14}O$
Oxirane, pentyl-, (R)-; (R)-1-Heptene oxide; (R)-2-Pentyloxirane; Heptane, 1,2-epoxy-, (R)-. Chiral building block. $bp_{30}$ = 56°; d = 0.836; n = 1.415; $[\alpha]^{25}$ = + 15° (neat). *Japan Energy Corporation; Sigma-Aldrich Fine Chemicals.*

**2513 (+)-Epoxyhexadecane**

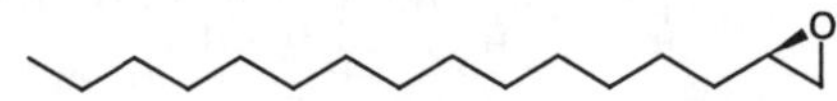

$C_{16}H_{32}O$
Oxirane, tetradecyl-, (R)-; (R)-Tetradecyloxirane; Hexadecane oxide; Hexadecene epoxide; Hexadecylene oxide; Hexadecane, 1,2-epoxy-, (R)-. Chiral building block. *Japan Energy Corporation.*

**2514 (+)-Epoxyhexane**
104898-06-8

$C_6H_{12}O$
Oxirane, butyl-, (R)-; (R)-Butyloxirane; (R)-1-Hexene epoxide; (R)-2-Butyloxirane; Hexane, 1,2-epoxy-, (R)-. Chiral building block. bp = 118-120°; d = 0.831; n = 1.406; $[\alpha]^{25}$ = + 13° (neat). *Japan Energy Corporation; Sigma-Aldrich Fine Chemicals.*

**2515 (-)-4-(S,S)-2,3-Epoxyhexyloxyphenyl-4-(deecyloxy)benzoate**
107133-34-6

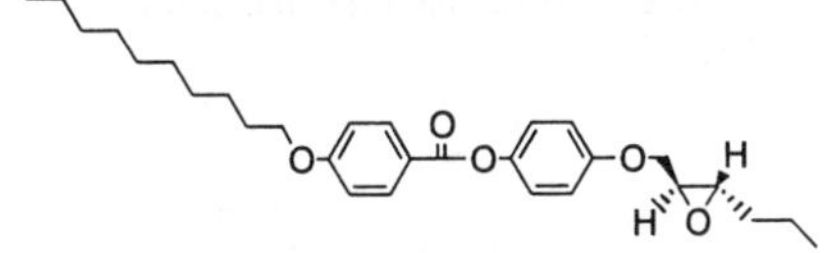

$C_{29}H_{40}O_5$
Chiral intermediate. $[\alpha]_D^{20}$ = - 12° (c = 1, $CDCl_3$). *Sigma-Aldrich Fine Chemicals.*

**2516 (+)-1-(2,3-Epoxylyxofuranosyl)uracil**
14042-38-7

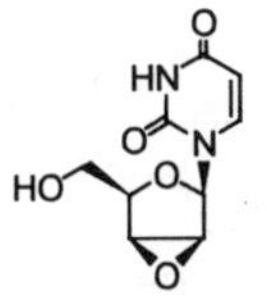

$C_9H_{10}N_2O_5$
Chiral intermediate. mp = 128-130°; $[\alpha]^{24}$ = + 34.2° (c = 1, $CH_3OH$). *Sigma-Aldrich Fine Chemicals.*

**2517 (+)-Epoxynonane**
130466-96-5

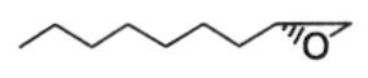

$C_9H_{18}O$
Oxirane, heptyl-, (R)-; Nonene oxide; (R)-2-Heptyloxirane; Nonane, 1,2-epoxy-, (R)-. Chiral building block. $bp_6$ = 59°; d = 0.837; n = 1.425; $[\alpha]^{25}$ = + 13° (neat). *Japan Energy Corporation; Sigma-Aldrich Fine Chemicals.*

**2518 (+)-Epoxyoctadecane**

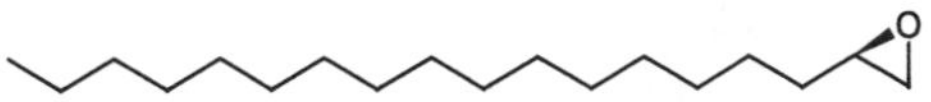

$C_{18}H_{36}O$
Oxirane, hexadecyl-, (R)- (R)-Hexadecyloxirane; Octadecene epoxide; Octadecylene oxide; Octadecane, 1,2-epoxy-, (R)-. Chiral building block. *Japan Energy Corporation.*

**2519 (R)-(+)-1,2-Epoxyoctane**
77495-66-0

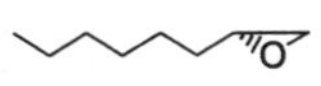

$C_8H_{16}O$
Chiral building block. $bp_{15}$ = 60-62°; d = 0.839; n = 1.42; $[\alpha]^{25}$ = + 14° (neat). *Japan Energy Corporation; Sigma-Aldrich Fine Chemicals.*

**2520 (+)-Epoxypentadecane**
96938-06-6

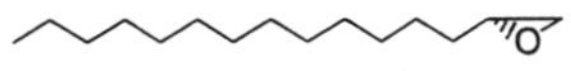

$C_{15}H_{30}O$
Oxirane, tridecyl-, (R)-; (R)-Tridecyloxirane; Pentadecane oxide; (R)-1-Pentadecene oxide; Pentadecane, 1,2-epoxy-, (R)-. Chiral building block. $bp_{2.5}$ = 128°; d = 0.846; $[\alpha]^{25}$ = + 8° (neat). *Japan Energy Corporation; Sigma-Aldrich Fine Chemicals.*

**2521 (+)-Epoxytetradecane**
116619-64-8

$C_{14}H_{28}O$
Oxirane, dodecyl-, (R)-; (R)-Dodecyloxirane; Tetradecene oxide; Tetradecylene oxide; Tetradecane, 1,2-epoxy-, (R)-. Chiral building block. $bp_{0.4}$ = 95-96°; d = 0.845; n = 1.441; $[\alpha]^{25}$ = + 9° (neat). *Japan Energy Corporation; Sigma-Aldrich Fine Chemicals.*

**2522 (+)-Epoxytridecane**
59829-81-1

$C_{13}H_{26}O$
Oxirane, undecyl-, (R)-; (R)-Undecyloxirane, tridecane, 1,2-epoxy-, (R)-. Chiral building block. $bp_{15}$ = 138°; d = 0.841; n = 1.438; $[\alpha]^{25}$ = + 10° (neat). *Japan Energy Corporation; Sigma-Aldrich Fine Chemicals.*

**2523 (+)-Epoxyundecane**
123493-71-0

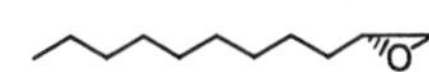

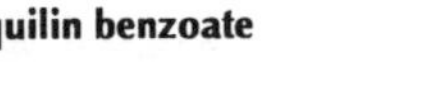

$C_{11}H_{22}O$
Oxirane, nonyl-, (R)-; Undecylene oxide; (R)-1-Undecene oxide; (R)-Nonyloxirane; Undecane, 1,2-epoxy-, (R)-. Chiral building block. $bp_{10}$ = 103°; d = 0.842; $[\alpha]^{25}$ = + 11° (neat). *Japan Energy Corporation; Sigma-Aldrich Fine Chemicals.*

**2524 (-)-Equilin benzoate**
6030-80-4

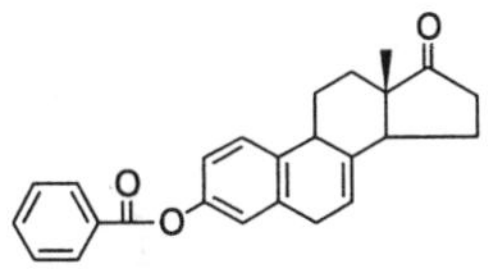

$C_{25}H_{24}O_3$
Chiral intermediate. mp = 193-196°; $[\alpha]^{22}$ = + 225° (c = 1, $CHCl_3$). *Sigma-Aldrich Fine Chemicals.*

**2525 Erbium tris3-(heptafluoropropylhydroxymethylene)-(-)-camphorate**

$C_{42}H_{42}ErF_{21}O_6$
(-)-Er(hfc)3. Chiral catalyst. mp = 140°; $[\alpha]^{26}$ = - 118° (c = 1.2, $CH_2Cl_2$). *Sigma-Aldrich Fine Chemicals.*

**2526 Erbium tris3-(heptafluoropropylhydroxymethylene)-(+)-camphorate**

$C_{42}H_{42}ErF_{21}O_6$
(+)-Er(hfc)3. Chiral catalyst. mp = 142°; $[\alpha]^{26}$ = + 118° (c = 1.2, $CH_2Cl_2$). *Sigma-Aldrich Fine Chemicals.*

**2527 D-Erythronolactone**
15667-21-7

$C_4H_6O_4$
(3R,4R)-Dihydroxydihydro-2(3H)-furanone. Chiral intermediate. mp = 99-102°; [α] = - 73° (c = 0.5, $H_2O$). *Acros Organics nv; Pfanstiehl Laboratories, Inc.; Sigma-Aldrich Fine Chemicals; TCI America.*

**2528 D-(-)-Erythrose**
583-50-6 3731(12) 209-505-2

$C_4H_8O_4$
Butanal, 2,3,4-trihydroxy-, (2R,3R)-. Chiral building block. n = 1.498; $[\alpha]^{24}$ = - 9.3° (c = 10.5, $H_2O$, 6 days). *Acros Organics nv; Sigma-Aldrich Fine Chemicals.*

**2529 D-Erythruronolactone acetonide**
85254-46-2

$C_7H_{10}O_5$
Chiral intermediate. *Acros Organics nv.*

**2530 (-)-Esculin monohydrate**
531-75-9 3739(12) 208-517-5

$C_{15}H_{16}O_9$
Listed on TSCA. Chiral intermediate. mp = 195-198°; $[\alpha]_D^{20}$ = - 78.4° (c = 2.2, 50% dioxane). *Sigma-Aldrich Fine Chemicals.*

**2531 N-Ethanoyl-D-erythro-sphingosine**
3102-57-6

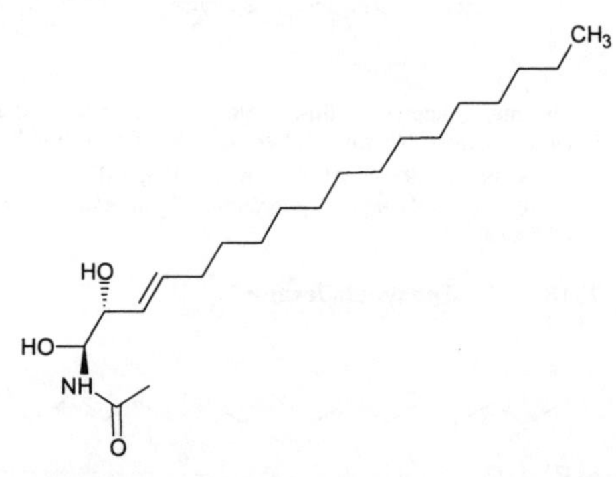


$C_{20}H_{39}NO_3$
Chiral intermediate. *Acros Organics nv.*

**2532 (R)-4-Ethenyl-1,3-dioxolan-2-one**
205673-79-6

$C_5H_6O_3$
Chiral intermediate. *Acros Organics nv; Eastman Chemical Company.*

**2533 (S)-4-Ethenyl-1,3-dioxolan-2-one**
140238-53-5

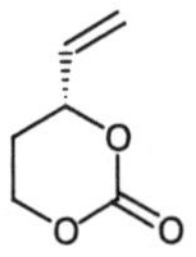

$C_5H_{16}O_3$
Chiral building block. *Eastman Chemical Company.*

**2534 D-Ethionine**
535-32-0 3693(11) 208-612-1

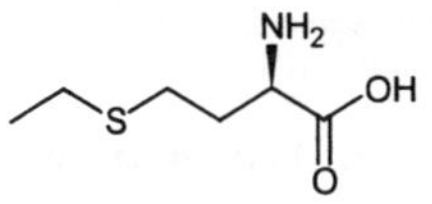

$C_6H_{13}NO_2S$
Butyric acid, 2-amino-4-(ethylthio)-, D-; D-2-Amino-4-(ethylthio)butyric acid; D-Homocysteine, S-ethyl-. Chiral intermediate. mp = 278°; $[\alpha]^{22}$ = - 21° (c = 1, 1N HCl). *Acros Organics nv; Sigma-Aldrich Fine Chemicals.*

**2535 L-Ethionine**
13073-35-3 3784(12) 235-966-4

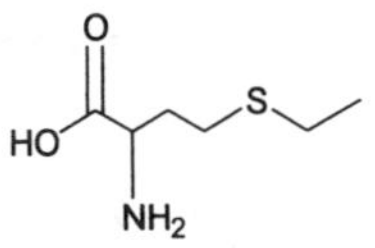

$C_6H_{13}NO_2S$
Butanoic acid, 2-amino-4-(ethylthio)-, (S)-; L-Homocysteine, S-ethyl; S-Ethyl-L-homocysteine. Listed on TSCA. Chiral building block. mp = 280°; $[\alpha]^{22}$ = + 21.7° (c = 1, 1N HCl). *Sigma-Aldrich Fine Chemicals; TCI America.*

**2536 (R)-(-)-Ethoxycarbonyl-butyrolactone**
33019-03-3

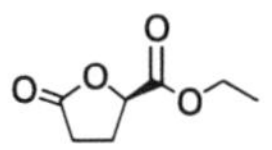

$C_7H_{10}O_4$
Chiral building block. $bp_{12}$ = 151-155°; d = 1.163; n = 1.448; $[\alpha]_D^{20}$ = - 12° (c = 0.4, EtOH). *Sigma-Aldrich Fine Chemicals.*

**2537 (S)-(+)-Ethoxycarbonyl-butyrolactone**
55094-96-7

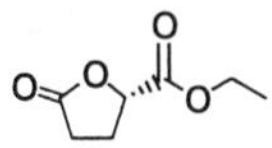

$C_7H_{10}O_4$
Chiral intermediate. $bp_{10}$ = 135-140°; d = 1.163; n = 1.447; $[\alpha]^{25}$ = + 7.5° (neat). *Sigma-Aldrich Fine Chemicals.*

**2538 Ethoxycarbonyl-L-phenylalanine**

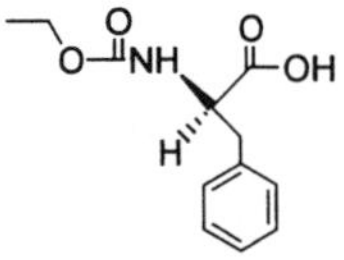

$C_{12}H_{15}NO_4$
Chiral intermediate. mp = 78-82°; $[\alpha]_D^{20}$ = + 25° (c = 2, EtOH). *Fischer Chemicals AG; Sigma-Aldrich Fine Chemicals.*

**2539 N-1-(S)-Ethoxycarbonyl-3-phenylpropyl)-L-alanine**
82717-96-2

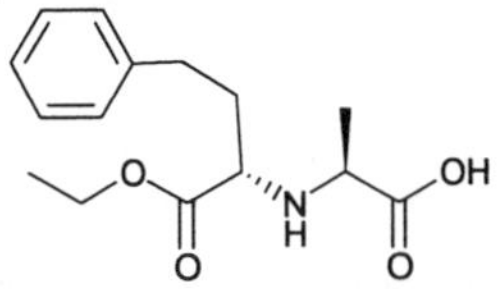

$C_{15}H_{21}NO_4$
Benzenebutanoic acid, α-[[(1S)-1-carboxyethyl]amino]-, monoethyl ester, (αS)-; ECPPA. Chiral intermediate. mp = 150-152°; $[\alpha]_D^{20}$ = + 28° (c = 1, $CH_3OH$). *Acros Organics nv; D&O Group; Kaneka Corporation; Sigma-Aldrich Fine Chemicals; Syntai Chemicals and Pharmaceuticals; Synthetech, Inc.*

**2540 N-(1-(S)-Ethoxycarbonyl-3-phenylpropyl)-L-alanyl-N-carboxyanhydride**
84793-24-8
$C_{16}H_{19}NO_5$
Chiral intermediate. *Kaneka Corporation.*

**2541 (S)-(-)-1-[N-(1-Ethoxycarbonyl-3-phenylpropyl)-L-alanyl]-L-proline**
75847-73-3
$C_{20}H_{28}N_2O_5$
Chiral intermediate. *Sigma-Aldrich Fine Chemicals.*

**2542 (S)-(-)-1-[N-(1-Ethoxycarbonyl-3-phenylpropyl)-N-trifluoroacetyl]-L-lysine**
130414-30-1
$C_{20}H_{27}F_3N_2O_5$
Chiral intermediate. *Sigma-Aldrich Fine Chemicals.*

**2543 (R)-2-(1-Ethoxyethoxy)-1,4-butanediol**

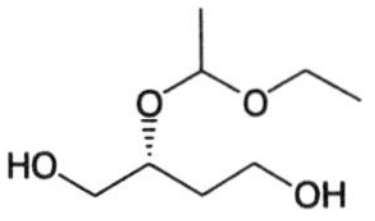

$C_8H_{18}O_4$
Butane-1,4-diol, 2-(1-ethoxyethoxy)-, (R)-; Chiral building

block and intermediate. *Synthon Chiragenics Corporation.*

**2544 (S)-2-(1-Ethoxyethoxy)-1,4-butanediol**
188790-85-4

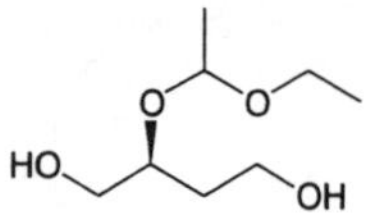

$C_8H_{18}O_4$
Butane-1,4-diol, 2-(1-ethoxyethoxy)-, (S)-; Chiral intermediate. *Synthon Chiragenics Corporation.*

**2545 (R)-3-(1-Ethoxyethoxy)-γ-butyrolactone**

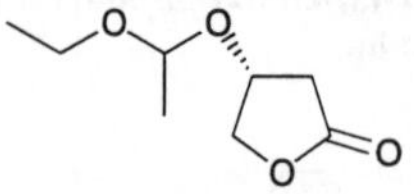

$C_8H_{14}O_3$
Furan-2(3H)-one, 4-(1-ethoxyethoxy)dihydro-, (R)-. Chiral building block. *Synthon Chiragenics Corporation.*

**2546 (S)-3-(1-Ethoxyethoxy)-γ-butyrolactone**
263164-11-0

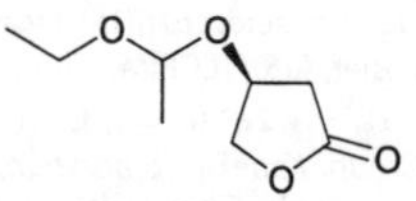

$C_8H_{14}O_3$
Furan-2(3H)-one, 4-(1-ethoxyethoxy)dihydro-, (S)-. Chiral building block. *Synthon Chiragenics Corporation.*

**2547 (-)-1,2-O-(1-Ethoxyethylidene)mannopyranose triacetate**

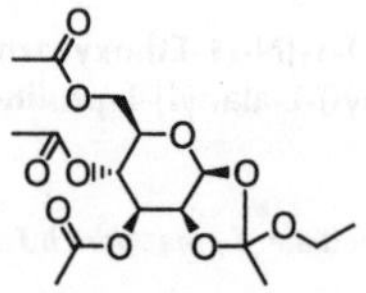

$C_{16}H_{24}O_{10}$
Chiral intermediate. mp = 101-103°; $[\alpha]_D^{20}$ = - 14° (c = 0.2, $CHCl_3$). *Sigma-Aldrich Fine Chemicals.*

**2548 L-4-Ethoxyphenylalanine**

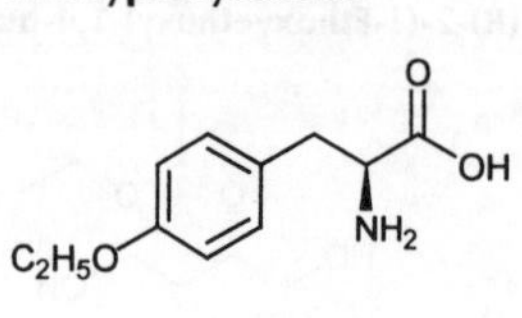

$C_{11}H_{15}NO_3$
Chiral building block. *Synthetech, Inc.*

**2549 Ethyl 2-acetamido-2-deoxy-D-glucopyranoside**
3055-47-8

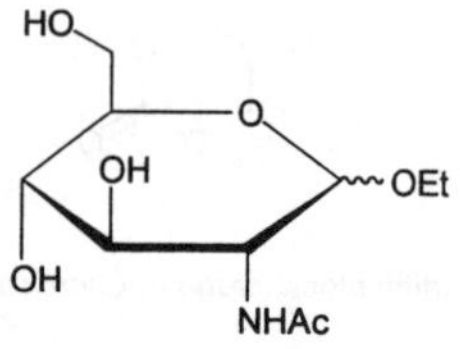

$C_{10}H_{19}NO_6$
Glucopyranoside, ethyl 2-acetamido-2-deoxy-, D-; Ethyl-N-acetyl-glucosamine. Chiral intermediate. *Pfanstiehl Laboratories, Inc.*

**2550 (S)-Ethyl 3-aminobutyrate**
115880-49-4

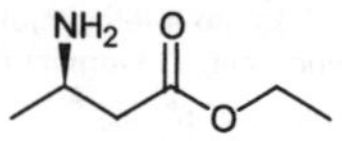

$C_6H_{13}NO_2$
Chiral building block. *IMI (TAMI) Institute for R&D.*

**2551 (R)-Ethyl 3-aminobutyrate**
22657-48-3

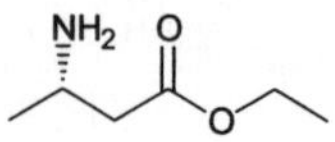

$C_6H_{13}NO_2$
Chiral building block. *IMI (TAMI) Institute for R&D.*

**2552 Ethyl (R)-3-aminobutyrate hydrochloride**

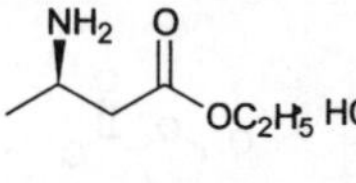

$C_6H_{14}ClNO_2$
Chiral building block. *Oxford Asymmetry International plc.*

**2553 Ethyl (S)-3-aminobutyrate hydrochloride**

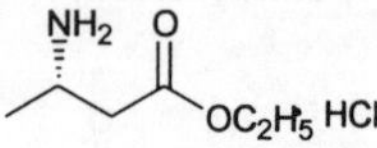

$C_6H_{14}ClNO_2$
Chiral building block. *Oxford Asymmetry International plc.*

**2554 (S)-(-)-1-Ethyl-2-aminomethyl pyrrolidine**

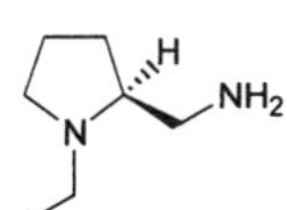


$C_7H_{16}N_2$
Chiral intermediate. *Shanghai DSL International Trading Company.*

**2555 (3R)-(+)-3-(Ethylamino)pyrrolidine**
$C_6H_{14}N_2$
Chiral building block. *TCI America.*

**2556 (3S)-(-)-3-(Ethylamino)pyrrolidine**
$C_6H_{14}N_2$
Chiral building block. bp = 176°; d = 0.92. *TCI America.*

**2557 Ethyl 2-O-benzoyl-4,6-O-benzylidene-β-D-galactopyranoside**
161765-88-4
$C_{29}H_{34}O_4S$
Chiral intermediate. *Pfanstiehl Laboratories, Inc.*

**2558 (R)-(-)-Ethylbenzylamine**

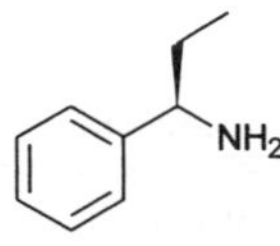


$C_7H_{13}N$
Chiral building block. *Daicel Chemical Ind. Ltd.*

**2559 (S)-(-)-Ethylbenzylamine**

$C_7H_{13}N$
Chiral building block. *Daicel Chemical Ind. Ltd.*

**2560 Ethyl 4,6-O-benzylidene-β-D-galactopyranoside**
101833-22-1
$C_{15}H_{20}O_6$
Chiral intermediate. *Pfanstiehl Laboratories, Inc.*

**2561 Ethyl (5S)-benzoyloxy-O-oxohexanoate**
82493-58-1
$C_{15}H_{18}O_5$
Chiral intermediate. *Ultrafine.*

**2562 Ethyl (R)-4-benzyloxy-3-hydroxybutyrate**
106058-91-7

$C_{13}H_{18}O_4$
Chiral building block. *Acros Organics nv.*

**2563 Ethyl (S)-4-benzyloxy-3-hydroxybutyrate**
112348-04-6

$C_{13}H_{18}O_4$
Chiral building block. *Acros Organics nv.*

**2564 Ethyl (R)-4-bromo-3-hydroxybutanoate**
95310-94-4

$C_6H_{11}BrO_3$
Butanoic acid, 4-bromo-3-hdroxy-, ethyl ester, (R)-. Chiral building block. *Oxford Asymmetry International plc; Synthon Chiragenics Corporation.*

**2565 (S)-(-)-Ethyl-4-bromo-3-hydroxybutyrate**
95537-36-3

$C_6H_{11}BrO_3$
Butanoic acid, 4-bromo-3-hydroxy-, ethyl ester, (S)-; Ethyl (S)-4-bromo-3-hydroxybutanoate. Chiral building block. $bp_2$ = 94-96°; d = 1.468; n = 1.474; $[\alpha]^{22}$ = - 11° (neat). *Acros Organics nv; Borregaard Synthesis; Sigma-Aldrich Fine Chemicals; SK Energy and Chemical, Inc.; Synthon Chiragenics Corporation.*

**2566 Ethyl (S)-(-)-2-(tert-butyldimethylsilyloxy)-propionate**
106513-42-2

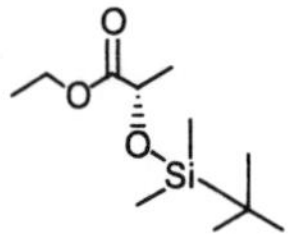

$C_{11}H_{24}O_3Si$
Chiral intermediate. bp = 187°; d = 0.875; n = 1.425; $[\alpha]_D^{20}$ = - 30° (c = 1.26, $CHCl_3$). *Sigma-Aldrich Fine Chemicals.*

**2567 (S)-Ethylcarbamoyl-L-cysteine**

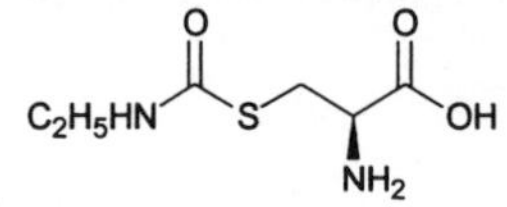

$C_6H_{12}NO_3S$
Chiral intermediate. *Senn Chemicals AG.*

**2568 Ethyl (R)-(+)-4-chloro-3-hydroxybutyrate**
90866-33-4

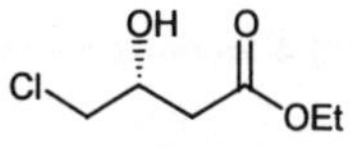

$C_6H_{11}ClO_3$
Butanoic acid, 4-chloro-3-hydroxy-, ethyl ester, (R)-; (R)-(+)-4-Chloro-3-hydroxybutyric acid ethyl ester. Chiral building block. $bp_5$ = 93-95°; d = 1.1900; $n_D^{20}$ = 1.4530; $[\alpha]_D^{20}$ = + 14 ± 2° (neat). *Daicel Chemical Ind. Ltd; Daiso Company, Ltd.; Lancaster Synthesis Ltd.; Oxford Asymmetry International plc; Rohner Ltd.; Sigma-Aldrich Fine Chemicals; Synthon Chiragenics Corporation.*

**2569 Ethyl (S)-(-)-4-chloro-3-hydroxybutyrate**
86728-85-0

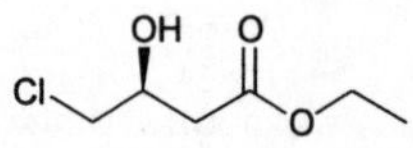

$C_6H_{11}ClO_3$
Butanoic acid, 4-chloro-3-hydroxy-, ethyl ester, (S)-; (R)-(-)-4-Chloro-3-hydroxybutyric acid ethyl ester; Ethyl (S)-4-chloro-3-hydroxybutanoate. Chiral building block. $bp_5$ = 93-95°; d = 1.1900; $n_D^{20}$ = 1.4530; $[\alpha]_D^{20}$ = - 14 ± 2° (neat). *Daicel Chemical Ind. Ltd; Daiso Company, Ltd.; Lancaster Synthesis Ltd.; Oxford Asymmetry International plc; Sigma-Aldrich Fine Chemicals; Synthon Chiragenics Corporation.*

**2570 Ethyl (S)-(-)-2-chloropropionate**
74497-15-7

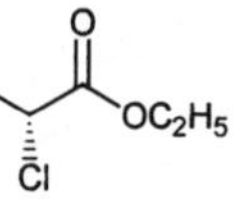

$C_5H_9ClO_2$
Propanoic acid, 2-chloro-, ethyl ester, (S)-; (S)-2-Chloropropionic acid ethyl ester. Chiral building block. *Daicel Chemical Ind. Ltd.*

**2571 Ethyl (S)-4-cyano-3-hydroxybutanamide**
158008-69-6

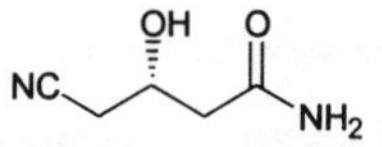

$C_5H_8N_2O_2$
Butanamide, 4-cyano-3-hydroxy, (S)-. Chiral building block. *Synthon Chiragenics Corporation.*

**2572 Ethyl (R)-4-cyano-3-hydroxybutanoate**
141942-85-0

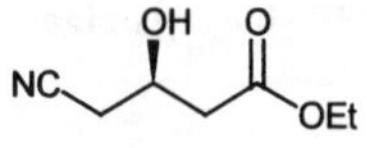

$C_7H_{11}NO_3$
Butanoic acid, 4-cyano-3-hydroxy-, ethyl ester, (R)-. Chiral building block. bp = 270°; d = 1.114; n = 1.448; $[\alpha]^{25}$ = - 5° (neat). *Acros Organics nv; Sigma-Aldrich Fine Chemicals; SK Energy and Chemical, Inc.; Synthon Chiragenics Corporation.*

**2573 Ethyl (S)-4-cyano-3-hydroxybutanoate**

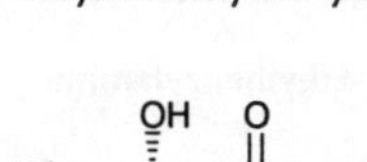

$C_7H_{11}NO_3$
Butanoic acid, 4-cyano-3-hydroxy-, ethyl ester, (S)-. Chiral building block. *Synthon Chiragenics Corporation.*

**2574 S-Ethyl-L-cysteine**

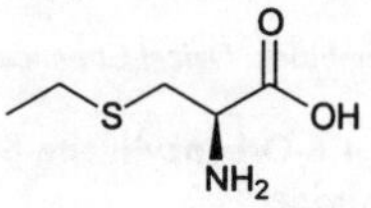

$C_5H_{11}NO_2S$
Chiral building block. *Austin Chemical Company, Inc.*

**2575 (S)-2-Ethyldecanoic acid**

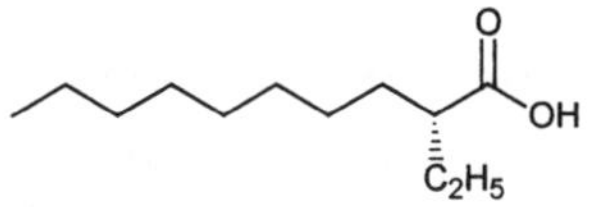

$C_{12}H_{24}O_2$
Decanoic acid, 2-ethyl-, (S)-; (S)-2-Ethylcapric acid. Chiral building block. *Japan Energy Corporation.*

**2576 (S)-2-Ethyldecanol**

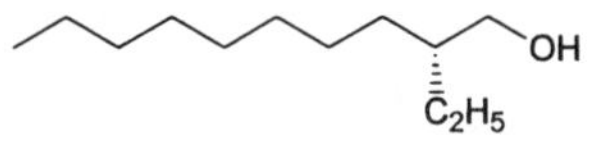

$C_{12}H_{26}O$
Decan-1-ol, 2-ethyl-, (S)-. Chiral building block. *Japan Energy Corporation.*

**2577 Ethyl-6-Deoxy-1-thio-2,3,4-tri-O-acetyl-α-L-mannopyranoside**
125520-01-6
$C_{14}H_{22}O_7S$
Chiral intermediate. *Pfanstiehl Laboratories, Inc.*

**2578 Ethyl 4,6-Di-O-acetyl-2,3-Dideoxy-erythro-hex-2-enopyranoside**
3323-72-6

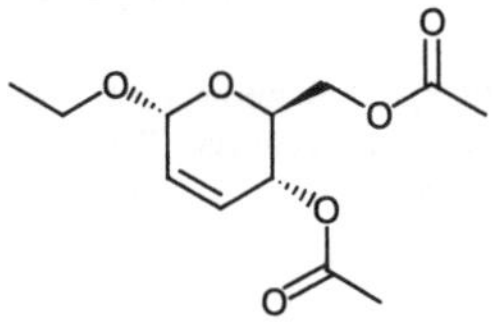

$C_{12}H_{18}O_6$
Chiral intermediate. mp = 77-79°; $[\alpha]^{26}$ = + 122° (c = 1, $CH_2Cl_2$). *Sigma-Aldrich Fine Chemicals.*

**2579 Ethyl (1R,3S)-2,2-dimethyl-3-(2-chloro-2-propenyl)cyclopropanecarboxylate (cis/trans)**

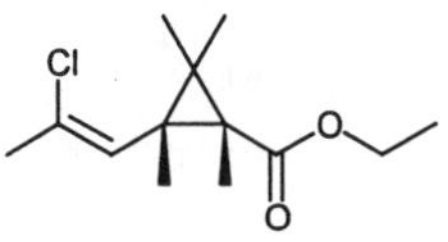

$C_{13}H_{21}ClO_2$
Chiral intermediate. *Acros Organics nv.*

**2580 Ethyl (R)-(-)-3-(2,2-dimethyl-1,3-dioxolan-4-yl)-trans-2-propenoate**
104321-62-2

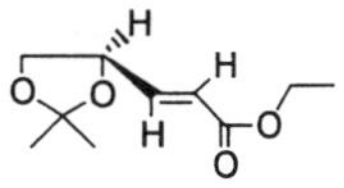

$C_{10}H_{16}O_4$
Chiral intermediate. d = 1.05; n = 1.452; $[\alpha]_D^{20}$ = - 41° (c = 1, $CHCl_3$). *Sigma-Aldrich Fine Chemicals.*

**2581 Ethyl (S)-(+)-3-(2,2-dimethyl-1,3-dioxolan-4-yl)-cis-2-propenoate**
91926-90-8

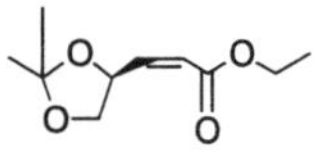

$C_{10}H_{16}O_4$
Chiral intermediate. d = 1.038; rf =0.447; $[\alpha]_D^{20}$ = + 123° (c = 1, $CHCl_3$). *Sigma-Aldrich Fine Chemicals.*

**2582 (2R)-2-[(4-Ethyl-2,3-dioxopiperazinyl)-carbonylamino]-2-(4-hydroxyphenyl)-acetic acid**
62893-24-7

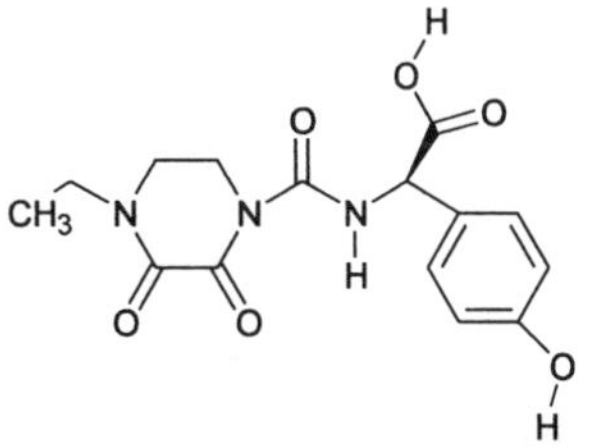

$C_{15}H_{17}N_3O_6$
Benzeneacetic acid, α-[[(4-ethyl-2,3-dioxo-1-piperazinyl)carbonyl]amino]-4-hydroxy-, (R)-; D-(-)-α-(4-Ethyl-2,3-dioxo-1-piperazine carboxamide)-p-hydroxy-phenylglycine. Chiral intermediate. mp = 205°; $[\alpha]_D^{20}$ = -0.85° (c = 1, $CH_3OH$). *Acros Organics nv; Sigma-Aldrich Fine Chemicals.*

**2583 (2R)-2-[(4-Ethyl-2,3-dioxopiperazinyl)-carbonylamino]-2-phenylacetic acid**
63422-71-9 264-133-8

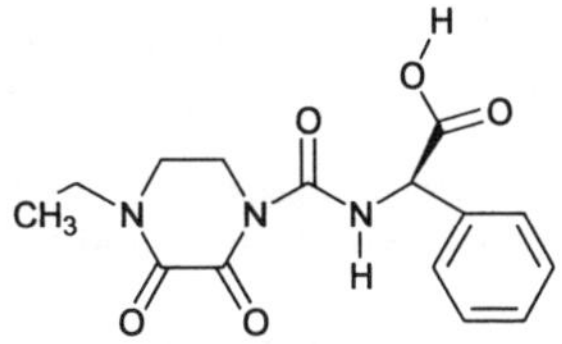

$C_{15}H_{17}N_3O_5$
Benzeneacetic acid, α-[[(4-ethyl-2,3-dioxo-1-piper-

azinyl)carbonyl]amino]-, (R)-. Chiral intermediate. mp = 171°; $[\alpha]_D^{20}$ = - 63° (c = 1, $CH_3OH$). *Acros Organics nv; Sigma-Aldrich Fine Chemicals.*

**2584 (S)-2-Ethyldodecanoic acid**

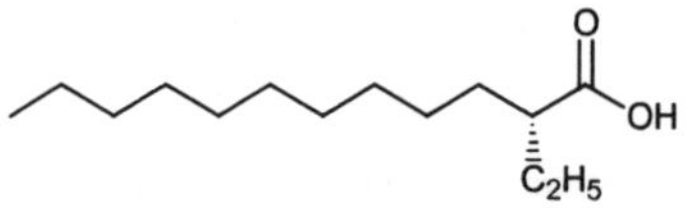

$C_{14}H_{28}O_2$
Dodecanoic acid, 2-ethyl-, (R)-; (R)-α-Ethyllauric acid. Chiral building block. *Japan Energy Corporation.*

**2585 (S)-2-Ethyl-1-dodecanol**

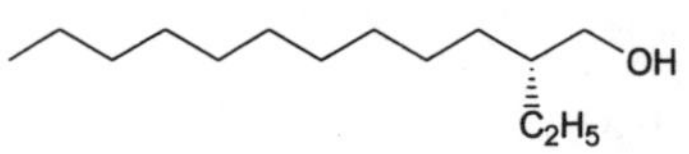

$C_{14}H_{30}O$
Chiral building block. *Japan Energy Corporation.*

**2586 (R,R)-Ethylenebis(4,5,6,7-tetrahydro-1-indenyl)difluorotitanium(IV)**

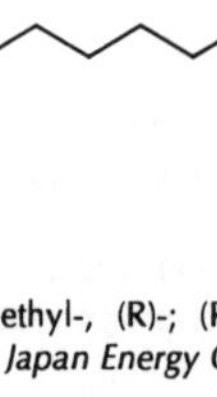

$C_{20}H_{24}F_2Ti$
Chiral catalyst. mp = 230°; $[\alpha]^{21}$ = + 1468° (c = 0.07, $CH_2Cl_2$, Hg lamp, 436nm). *Sigma-Aldrich Fine Chemicals.*

**2587 (S,S)-Ethylenebis(4,5,6,7-tetrahydro-1-indenyl)difluorotitanium(IV)**

$C_{20}H_{24}F_2Tl$
Chiral catalyst. mp = 230°; $[\alpha]^{21}$ = - 1161° (c = 0.07, $CH_2Cl_2$, Hg lamp, 436nm). *Sigma-Aldrich Fine Chemicals.*

**2588 (R,R)-Ethylenebis(4,5,6,7-tetrahydro-1-indenyl)dimethyltitanium(IV)**

$C_{22}H_{30}Ti$
Chiral catalyst. mp = 76-80°; $[\alpha]^{21}$ = - 3006° (c = 0.07, $CH_2Cl_2$, Hg lamp, 436nm). *Sigma-Aldrich Fine Chemicals.*

**2589 (S,S)-Ethylenebis(4,5,6,7-tetrahydro-1-indenyl)dimethyltitanium(IV)**

$C_{22}H_{30}Ti$
Chiral catalyst. mp = 76-80°; $[\alpha]^{21}$ = + 2837° (c = 0.07, $CH_2Cl_2$, Hg lamp, 436nm). *Sigma-Aldrich Fine Chemicals.*

**2590 [(R,R)-Ethylenebis(4,5,6,7-tetrahydro-1-indenyl)]dimethylzirconium(IV)**
110173-61-0

$C_{22}H_{30}Zr$
Chiral catalyst. *Sigma-Aldrich Fine Chemicals.*

**2591 [(S,S)-Ethylenebis(4,5,6,7-tetrahydro-1-indenyl)]dimethylzirconium(IV)**
132881-67-5

$C_{22}H_{30}Zr$
Chiral catalyst. *Sigma-Aldrich Fine Chemicals.*

**2592 (R,R)-Ethylenebis(4,5,6,7-tetrahydro-1-indenyl)titanium(IV) (R)-1,1-binaphthyl-2,2-diolate**

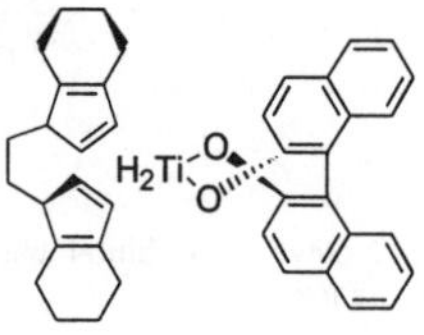

$C_{40}H_{38}O_2Ti$
Chiral catalyst. mp = 250°; $[\alpha]_D^{20}$ = - 750° (c = 0.07, $CH_2Cl_2$, Hg lamp, 436nm). *Sigma-Aldrich Fine Chemicals.*

**2593 (S,S)-Ethylenebis(4,5,6,7-tetrahydro-1-indenyl)titanium(IV) (S)-1,1-binaphthyl-2,2-diolate**

$C_{40}H_{38}O_2Ti$
Chiral catalyst. mp = 245°; $[\alpha]_D^{20}$ = + 700° (c = 0.07, $CH_2Cl_2$, Hg lamp, 436nm). *Sigma-Aldrich Fine Chemicals.*

**2594 [(R,R)-Ethylenebis(4,5,6,7-tetrahydro-1-indenyl)]zirconium(IV)-(R)-1,1'-binaphth-2-olate**
123236-85-1

$C_{40}H_{36}O_2Zr$
Chiral catalyst. *Sigma-Aldrich Fine Chemicals.*

**2595 [(S,S)-Ethylenebis(4,5,6,7-tetrahydro-1-indenyl)]zirconium(IV)-(R)-1,1'-binaphth-2-olate**
132881-66-4

$C_{40}H_{36}O_2Zr$
Chiral catalyst. *Sigma-Aldrich Fine Chemicals.*

**2596 4,6-O-Ethylidene-D-glucopyranose**
13224-99-2 236-198-2

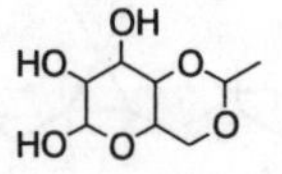

$C_8H_{14}O_6$
Glucopyranose, 4,6-O-ethylidene-, α-D-. Chiral intermediate. mp = 168-170°. *Pfanstiehl Laboratories, Inc.; Senn Chemicals AG; Sigma-Aldrich Fine Chemicals.*

**2597 L-N-Ethylephedrine**
48141-64-6 256-359-0

$C_{12}H_{19}NO$
Benzenemethanol, α-[1-(ethylmethylamino)ethyl]-, [R-(R*,S*)]-; R-(R*,S*)]-α-[1-(Ethylmethylamino)ethyl]benzyl alcohol; (-)-Etafedrine; l-α-[1-(Ethylmethylamino)ethyl]-benzyl alcohol. Pharmaceutical or derivative. *Knoll AG.*

**2598 Ethyl (R)-3,4-epoxybutanoate**
112083-64-4

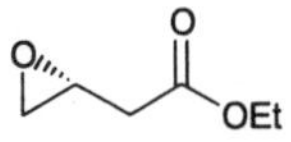

$C_6H_{10}O_3$
Oxiraneacetic acid, ethyl ester, (R)-. Chiral building block. *Synthon Chiragenics Corporation.*

**2599 Ethyl (S)-3,4-epoxybutanoate**
112083-63-3

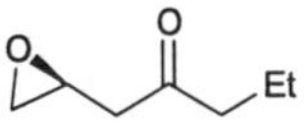

$C_6H_{10}O_3$
Oxiraneacetic acid, ethyl ester, (S)-; Ethyl (S)-oxirane acetate; (S)-3,4-Oxabutyric acid ethyl ester. Chiral building block. *Synthon Chiragenics Corporation.*

**2600 Ethyl (2R,3R)-2,3-epoxy-3-methyl-propanoate**
19780-35-9

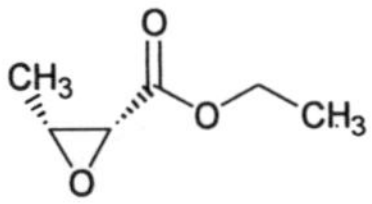

$C_6H_{10}O_3$
Chiral building block. $bp_{20}$ = 77°; [α] = -1.4° (c = 5.0, $CH_3OH$). *Acros Organics nv.*

**2601 Ethyl (2S,3S)-2,3-epoxy-3-methyl-propanoate**
110508-08-2

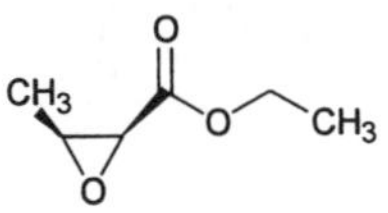

$C_6H_{10}O_3$
Ethyl (2S,3S)-3-methyl-2-oxiranecarboxylate. Chiral building block. $bp_{20}$ = 77; [α] = - 0.5° (neat). *Acros Organics nv.*

**2602 Ethyl (2R)-2,3-epoxypropanoate**
111058-33-4

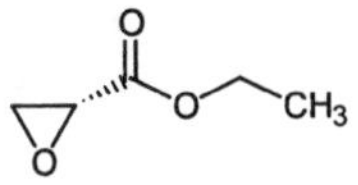

$C_5H_8O_3$
(R)-(+)-Ethyl glycidate; (+)-Ethyl (2R)-oxiranecarboxylate. Chiral intermediate. $bp_{15}$ = 68-69°. *Acros Organics nv.*

**2603 Ethyl (2S)-2,3-epoxypropanoate**
111058-34-5

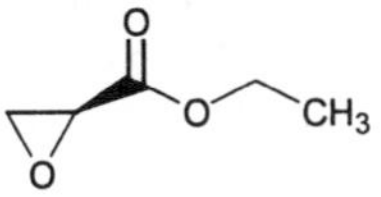

$C_5H_8O_3$
(-)-Ethyl (2S)-oxiranecarboxylate; (S)-(-)-Ethyl glycidate. Chiral building block. $bp_{15}$ = 68-69°; [α] = - 12.8 (c = 4.8, $CH_3OH$). *Acros Organics nv.*

**2604 Ethyl β-D-galactopyranoside**
18997-88-1
Chiral intermediate. *Pfanstiehl Laboratories, Inc.*

**2605 (R)-(-)-2-Ethylhexylamine**

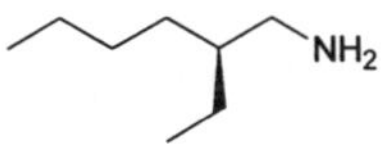

$C_8H_{19}N$
Hexan-1-amine, 2-ethyl-, (R)-; 1-Amino-2-ethylhexane, (R)-; Hexylamine, 2-ethyl-, (R)-. Chiral building block. *Norse Laboratories.*

**2606 Ethyl-(1R,2S)-cis-2-hydroxycyclopentane-carboxylate**
61586-79-6

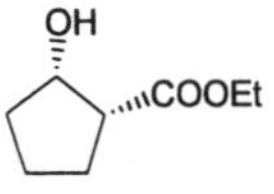

$C_8H_{14}O_3$
Chiral building block. *Rohner Ltd.*

**2607 Ethyl (R)-3-hydroxy-decanoate**
130322-00-8

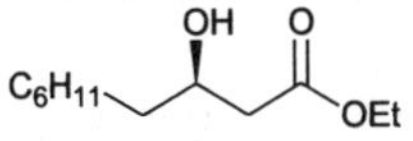

$C_{12}H_{24}O_3$
Decanoic acid, 3-hydroxy-, ethyl ester, (R)-. Chiral building block. *Synthon Chiragenics Corporation.*

**2608 Ethyl (S)-3-hydroxy-decanoate**

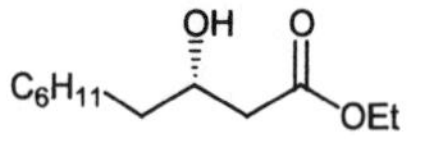

$C_{12}H_{24}O_3$
Decanoic acid, 3-hydroxy-, ethyl ester, (S)-. Chiral building block. *Synthon Chiragenics Corporation.*

**2609 Ethyl (S)-3-hydroxy-3-phenylpropanoate**
33401-74-0

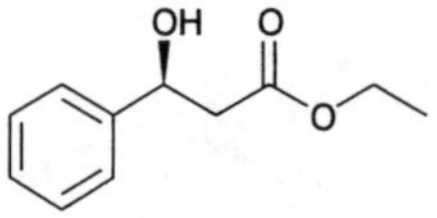

$C_{11}H_{14}O_3$
Chiral building block. *Kaneka Corporation.*

**2610 Ethyl (R)-3-hydroxy-tetradecanoate**
151763-75-6

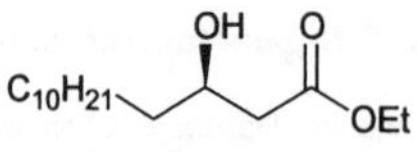

$C_{16}H_{32}O_3$
Tetradecanoic acid, 3-hydroxy-, ethyl ester, (R)-. Chiral building block. *Synthon Chiragenics Corporation.*

**2611 Ethyl (S)-3-hydroxy-tetradecanoate**
214193-71-2

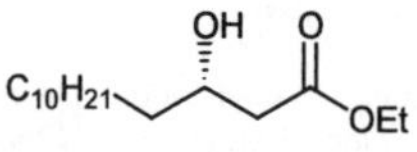

$C_{16}H_{32}O_3$
Tetradecanoic acid, 3-hydroxy-, ethyl ester, (S)-. Chiral building block. *Synthon Chiragenics Corporation.*

**2612 Ethyl (R)-4-iodo-3-hydroxybutanoate**
11283-27-9

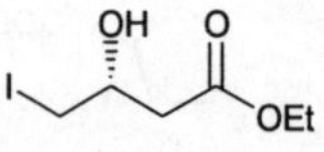

$C_6H_{11}IO_3$
Butanoic acid, 3-hydroxy-4-iodo-, ethyl ester, (R)-. Chiral building block. *Oxford Asymmetry International plc; Synthon Chiragenics Corporation.*

**2613 Ethyl (S)-4-iodo-3-hydroxybutanoate**
112100-39-7

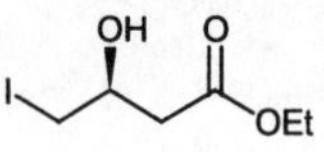

$C_6H_{11}IO_3$
Butanoic acid, 3-hydroxy-4-iodo-, ethyl ester, (S)-. Chiral building block. *Oxford Asymmetry International plc; SK Energy and Chemical, Inc.; Synthon Chiragenics Corporation.*

**2614 Ethyl (S)-(E)-4,5-O-isopropylidene-4,5-dihydroxy-2-pentenoate**
64520-58-7

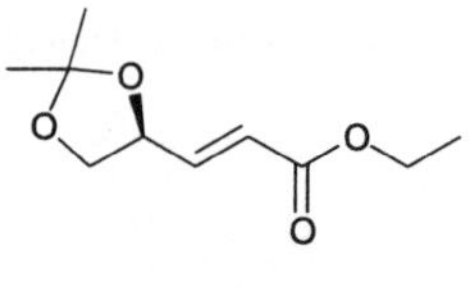

$C_{10}H_{16}O_4$
Ethyl (S)-3-(2,2-dimethyldioxolan-4-yl)-(E)-acrylate. Chiral building block. $bp_{12}$ = 115-117°; d = 1.0490; $n_D^{20}$ = 1.4520; $[\alpha]_D^{20}$ = + 41 ± 2° (c = 1CHCl$_3$). *Acros Organics nv; Lancaster Synthesis Ltd.; Oxford Asymmetry International plc; Sigma-Aldrich Fine Chemicals.*

**2615 (R)-Ethyl cis-2-isothiocyanato-1-cyclohexanecarboxylate**

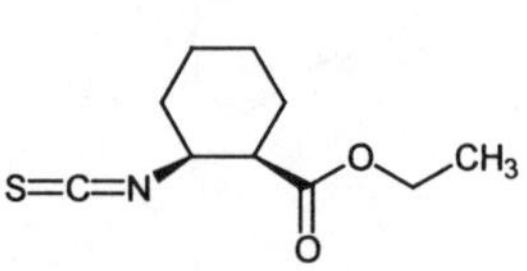

$C_{10}H_{15}NO_2S$
Chiral intermediate. *Acros Organics nv.*

**2616 (S)-Ethyl cis-2-isothiocyanato-1-cyclopentanecarboxylate**

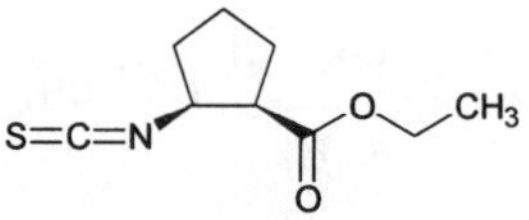

$C_9H_{13}NO_2S$
Chiral intermediate. *Acros Organics nv.*

**2617 Ethyl (R)-(-)-mandelate**
10606-72-1

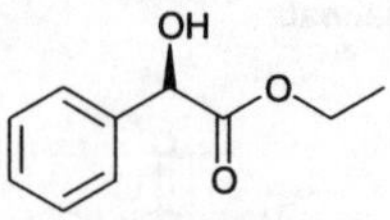

$C_{10}H_{12}O_3$
Benzeneacetic acid, α-hydroxy-, ethyl ester, (R)-; (R)-(-)-α-Hydroxyphenylacetic acid ethyl ester; D-(-)-Mandelic acid ethyl ester. Chiral building block. mp = 33-34°; $bp_2$ = 103-105°; $[\alpha]_D^{20}$ = - 134 ± 3° (c = 3, CHCl$_3$). *Lancaster Synthesis Ltd.; Norse Laboratories; Sigma-Aldrich Fine Chemicals; Yamakawa Chemical Industry Co. Ltd.*

**2618 Ethyl (S)-(+)-mandelate**
13704-09-1

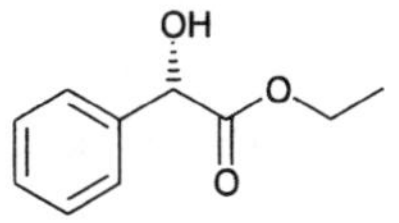

$C_{10}H_{12}O_3$
L-(+)-Mandelic acid ethyl ester; (S)-(+)-α-Hydroxyphenylacetic acid ethyl ester. Chiral building block. mp = 32-34°; $bp_4$ = 106-107°; d = 1.1200; $[\alpha]_D^{20}$ = + 134 ± 3° (c = 3, $CHCl_3$). *Lancaster Synthesis Ltd.; Sigma-Aldrich Fine Chemicals; TCI America; Yamakawa Chemical Industry Co. Ltd.*

**2619 Ethyl (R)-(-)-2-methoxypropionate**
40105-20-2

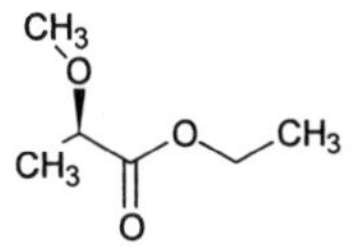

$C_6H_{12}O_3$
Propanoic acid, 2-methoxy-, ethyl ester, (R)-. Chiral building block. *Acros Organics nv.*

**2620 Ethyl (S)-(+)-2-methoxypropionate**

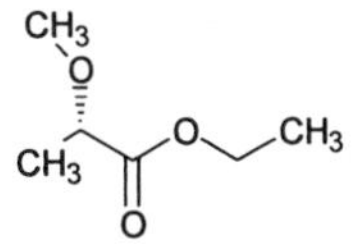

$C_6H_{12}O_3$
Chiral building block. *Acros Organics nv.*

**2621 (R)-Ethylnipecotate**
25137-01-3

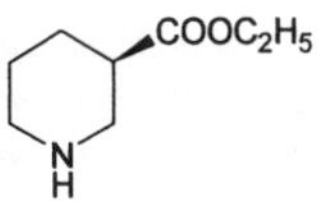

$C_8H_{15}NO_2$
Piperidine-3-carboxylic acid, ethyl ester, (R-); Ethyl piperidine-3-carboxylate; 3-Piperidinecarboxylic acid ethyl ester; Nipecotic acid, ethyl ester. Chiral building block. *Chemi SpA; Sigma-Aldrich Fine Chemicals; Vinchem.*

**2622 (S)-Ethyl nipecotate**
37675-18-6
$C_8H_{15}NO_2$
(S)-Nipecotic acid ethyl ester. Chiral building block. *Austin Chemical Company, Inc.; Chemi SpA; Sigma-Aldrich Fine Chemicals; Vinchem.*

**2623 Ethyl (R)-nipecotate L-tartarate**
167392-57-6

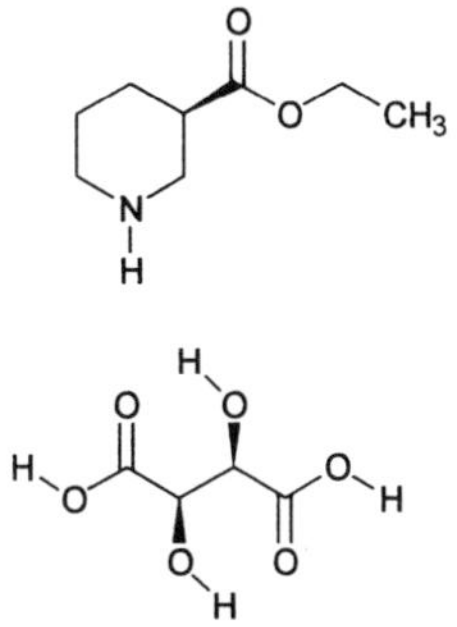

$C_{12}H_{21}NO_8$
Chiral building block. mp = 157-159°; $[\alpha]_D^{20}$ = + 10° (c = 5, $H_2O$). *Acros Organics nv; Sigma-Aldrich Fine Chemicals; Zeeland Chemicals, Inc.*

**2624 (S)-2-Ethyloctanoic acid**

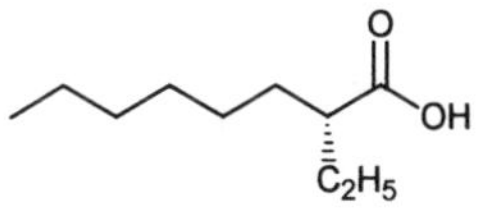

$C_{10}H_{20}O_2$
Octanoic acid, 2-ethyl-, (S)-; (S)-α-Ethylcaprylic acid. Chiral building block. *Japan Energy Corporation.*

**2625 (S)-2-Ethyl-1-octanol**

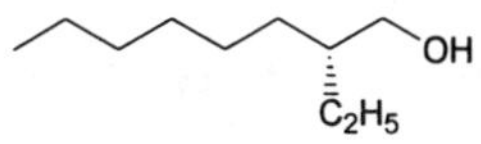

$C_{10}H_{22}O$
Octan-1-ol, 2-ethyl-, (S)-. Chiral building block. *Japan Energy Corporation.*

**2626 Ethyl (R)-(-)-2-Oxo-4-thiazolidine-carboxylate**
98155-24-9

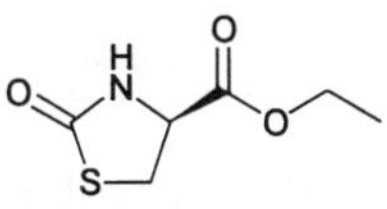

$C_6H_9NO_3S$
Chiral intermediate. d = 1.323; n = 1.525; $[\alpha]_D^{20}$ = - 42° (c = 2, $CHCl_3$). *Sigma-Aldrich Fine Chemicals.*

**2627 Ethyl (3R)-3-(N-((1R)-phenylethyl)-N-benzylamino)butyrate**
52673-22-8

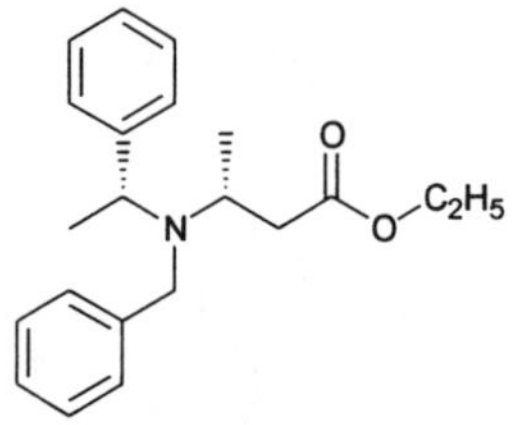

$C_{21}H_{25}NO_3$
Chiral intermediate. *Oxford Asymmetry International plc.*

**2628 Ethyl (3S)-3-(N-((1S)-phenylethyl)-N-benzylamino)butyrate**
152673-22-8

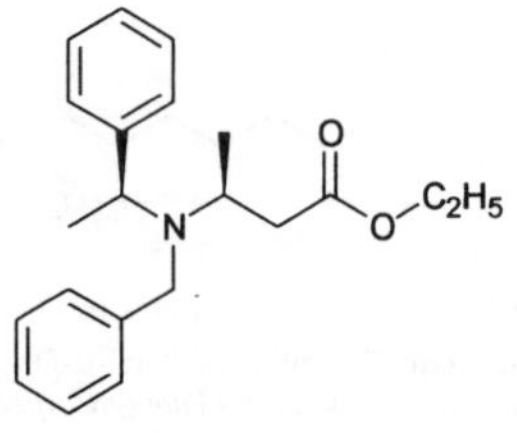

$C_{21}H_{25}NO_2$
Chiral intermediate. *Oxford Asymmetry International plc.*

**2629 (R)-(+)-1-Ethyl-2-pyrrolidinecarboxamide**
$C_7H_{14}N_2O$
Chiral building block. mp = 111°. *TCI America.*

**2630 (S)-(-)-1-Ethyl-2-pyrrolidinecarboxamide**
55446-83-8
$C_7H_{14}N_2O$
Chiral intermediate. *TCI America.*

**2631 (R)-(+)-1-Ethyl-2-pyrrolidinecarboxylic acid ethyl ester**
$C_9H_{17}NO_2$
Ethyl (R)-(+)-1-Ethyl-2-pyrrolidinecarboxylate. Chiral intermediate. d = 0.97. *TCI America.*

**2632 (S)-(-)-1-Ethyl-2-pyrrolidinecarboxylic acid ethyl ester**

$C_9H_{17}NO_2$
(S)-(-)-1-Ethylproline ethyl ester; Ethyl (S)-(-)-1-ethyl-2-pyrrolidinecarboxylate. Chiral intermediate. *TCI America.*

**2633 Ethyl (R)-(-)-2-pyrrolidone-5-carboxylate**
68766-96-1

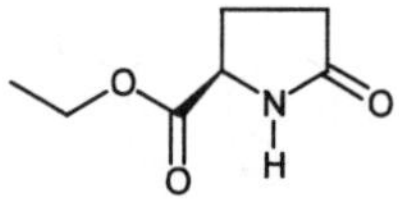

$C_7H_{11}NO_3$
Proline, 5-oxo-, ethyl ester, D-; L-Ethyl 5-oxo-L-prolinate; Ethyl (R)-pyroglutamate; Ethyl D-2-oxo-5-pyrrolidinecarboxylate; D-2-Ethoxycarbonyl-5-pyrrolidone; D-Pyroglutamic acid ethyl ester. Chiral building block. mp = 49-53°; $bp_{12}$ = 176°; n = 1.478; $[\alpha]_D^{20}$ = - 3.3° (c = 10, EtOH). *Acros Organics nv; Sigma-Aldrich Fine Chemicals.*

**2634 Ethyl (S)-(+)-2-pyrrolidone-5-carboxylate**
7149-65-7 230-480-9

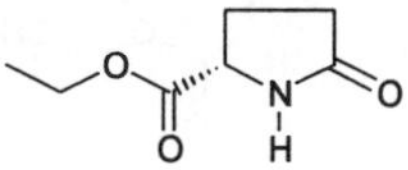

$C_7H_{11}NO_3$
Chiral building block. mp = 47-51°; $bp_{12}$ = 176°; $[\alpha]^{19}$ = + 3.3° (c = 10, EtOH). *Sigma-Aldrich Fine Chemicals.*

**2635 Ethyl (S)-(-)-2-(trifluoromethylsulfonyl)-oxypropionate**
84028-88-6

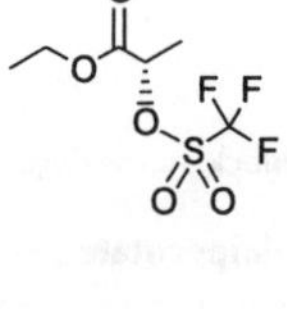

$C_6H_9F_3O_5S$
Ethyl L-lactate trifluoromethanesulfonate. Chiral building block. $bp_{0.01}$ = 34° / 0.01 mm; d = 1.342; n = 1.374; $[\alpha]_D^{20}$ = - 44° (neat). *Sigma-Aldrich Fine Chemicals.*

**2636 (4S,2RS)-2-Ethylthiazolidine-4-carboxylic acid**

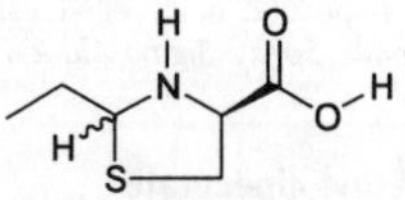

$C_6H_{11}NO_2S$
Chiral intermediate. $[\alpha]$ = - 186° (c = 0.2, 2% $Na_2CO_3$). *Acros Organics nv.*

**2637 Europium tris (D,D-dicampholylmethanate)**
52351-64-1

$C_{63}H_{105}EuO_6$
Chiral catalyst. *Acros Organics nv.*

**2638 Europium tris3-(heptafluoropropylhydroxy-methylene)-(-)-camphorate**

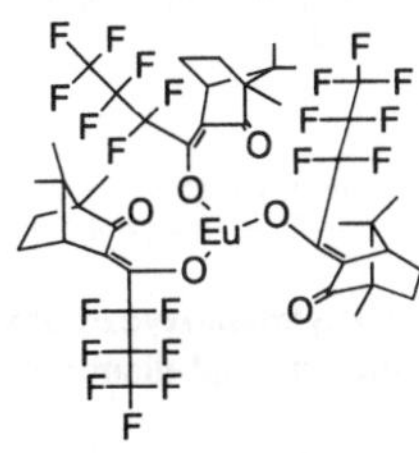

$C_{42}H_{42}EuF_{21}O_6$
(-)-Eu(hfc)3. Chiral catalyst. mp = 152-157°; $[\alpha]_D^{20}$ = - 140° (c = 1, $CH_2Cl_2$). *Sigma-Aldrich Fine Chemicals.*

**2639 Europium tris3-(heptafluoropropylhydroxy-methylene)-(+)-camphorate**
34788-82-4 252-214-0

$C_{42}H_{42}EuF_{21}O_6$
(+)-Eu(hfc)3. Chiral catalyst. mp = 156-158°; $[\alpha]_D^{20}$ = + 158.0° (c = 1, $CHCl_3$). *Sigma-Aldrich Fine Chemicals.*

**2640 (-)-Fenchone**
7787-20-4 232-107-5

$C_{10}H_{16}O$
Norbornan-2-one, 1,3,3-trimethyl-, (1R,4S)-; 1,3,3-Trimethylnorcamphor; (1R)-1,3,3-Trimethylbicyclo-[3.3.1]heptan-2-one; 1,3,3-Trimethylnorbornan-2-one. Listed on TSCA. Chiral intermediate. bp = 191-195°; d = 0.9450; $n_D^{20}$ = 1.4630; $[\alpha]_D^{20}$ = - 52 ± 2° (neat). *Acros Organics nv; Lancaster Synthesis Ltd.; Sigma-Aldrich Fine Chemicals.*

**2641 (+)-Fenchone**
4695-62-9 255-160-0
$C_{10}H_{16}O$
Bicyclo[2.2.1]heptan-2-one, 1,3,3-trimethyl-, (1S,4R)-. 1,3,3-Trimethylnorcamphor; 2-Norbornanone, 1,3,3-trimethyl-, (1S,4R)-; (1S)-1,3,3-Trimethylbicyclo-[2.2.1]heptan-2-one. Listed on TSCA. Chiral intermediate. mp = 5-7°; $bp_{13}$ = 63-65°; d = 0.9450; $n_D^{20}$ = 1.4630; $[\alpha]_D^{20}$ = + 60 ± 1° (neat). *Lancaster Synthesis Ltd.*

**2642 (1R)-(+)-Fenchyl acetate**
99341-77-2 237-588-5
$C_{12}H_{20}O_2$
1,3,3-Trimethyl-2-norbornyl acetate. Chiral intermediate. $n_D^{20}$ = 1.4555; $[\alpha]_D^{20}$ = + 48 ± 2° (c = 10, EtOH). *Lancaster Synthesis Ltd.*

**2643 (1R)-endo-(+)-Fenchyl alcohol**
2217-02-9 216-639-5

$C_{10}H_{18}O$
1,3,3-Trimethyl-2-norbornanol; (+)-Fenchol. Chiral intermediate. mp = 39-43°; bp = 201-202°; $[\alpha]_D^{20}$ = +0.5±1° (c = 10, EtOH). *Acros Organics nv; Lancaster Synthesis Ltd.; Sigma-Aldrich Fine Chemicals.*

**2644 (+)-Fluocinolone acetonide**
67-73-2 4185(12) 200-668-5

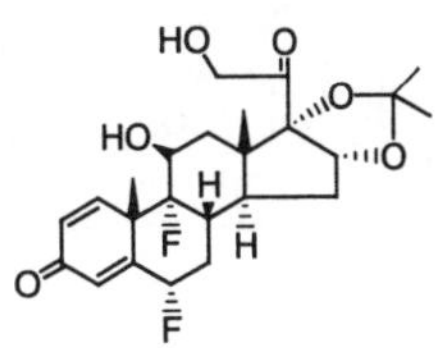

$C_{24}H_{30}F_2O_6$
Chiral intermediate. mp = 267-269°; $[\alpha]^{25}$ = + 98° (c = 1, $CHCl_3$). *Sigma-Aldrich Fine Chemicals.*

**2645 (+)-1-(9-Fluorenyl)ethyl chloroformate**
105764-39-4

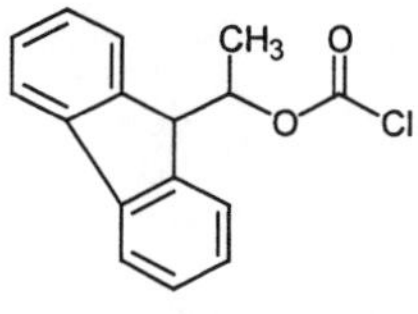

$C_{16}H_{13}ClO_2$
(+)-Flec. Chiral building block. mp = - 94°; bp = 56°; d = 0.79; [α] = + 70° (c = 1, $CH_2Cl_2$). *Acros Organics nv.*

**2646 (-)-1-(9-Fluorenyl)ethyl chloroformate (0.5 w/v in acetone, 18 millimolar)**
107474-79-3

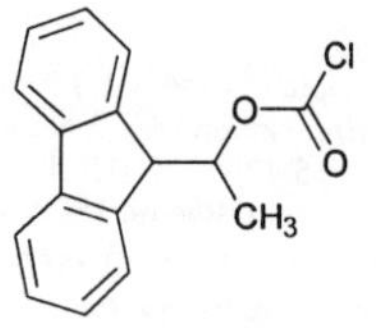

$C_{16}H_{13}ClO_2$
(-)-Flec. Chiral diagnostic reagent. mp = - 94°; bp = 56°; d = 0.79; n = 1.3602; [α] = - 70° (c = 1, $CH_2Cl_2$). *Acros Organics nv; Sigma-Aldrich Fine Chemicals.*

**2647 (9-Fluorenylmethoxycarbonyl)-D-alanine**

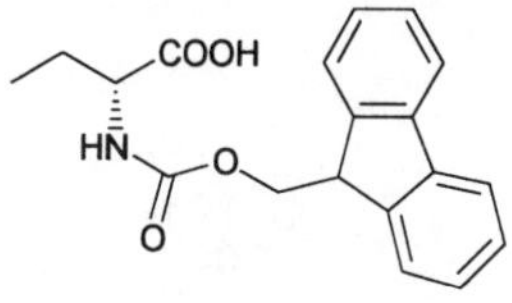

$C_{18}H_{17}NO_4$
Alanine, N-[(9H-fluoren-9-ylmethoxy)carbonyl]-, D-. Chiral intermediate. *Synthetech, Inc.*

**2648 (9-Fluorenylmethoxycarbonyl)-L-α-alanine**
35661-39-3 252-660-6

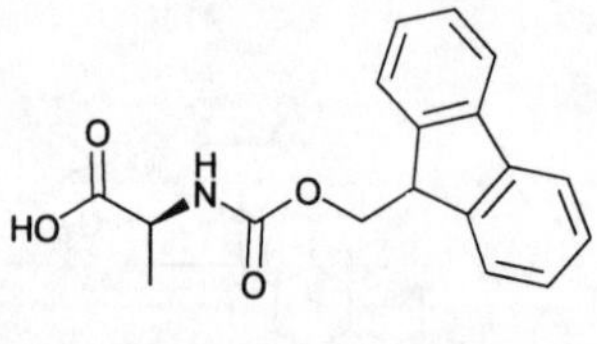

$C_{18}H_{17}NO_4$
Alanine, N-[(9H-fluoren-9-ylmethoxy)carbonyl]-, L-; FMOC-L-alanine; NPC 14688. Chiral intermediate. mp = 155-158°; $[\alpha]_D^{20}$ = - 18° (c = 1, DMF). *Acros Organics nv; Interchem Corporation; Sigma-Aldrich Fine Chemicals; Synthetech, Inc.*

**2649 (9-Fluorenylmethoxycarbonyl)-D-alaninol**

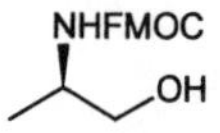

$C_{18}H_{19}NO_3$
Chiral intermediate. *Omega Chemical Company Inc.*

**2650 (9-Fluorenylmethoxycarbonyl)-L-alaninol**

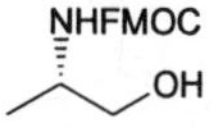

$C_{18}H_{19}NO_3$
Chiral intermediate. *Omega Chemical Company Inc.; Senn Chemicals AG.*

**2651 (R)-N-(9-Fluorenylmethoxycarbonyl)-allylglycine**
170642-28-1
$C_{20}H_{19}NO_4$
N-(9-Fluorenylmethoxycarbonyl)-(R)-2-amino-4-pentenoic acid. Chiral intermediate. *Lancaster Synthesis Ltd.*

**2652 (S)-N-(9-Fluorenylmethoxycarbonyl)-allylglycine**
146549-21-5
$C_{20}H_{19}NO_4$
N-(9-Fluorenylmethoxycarbonyl)-(S)-2-amino-4-pentenoic acid. Chiral intermediate. *Lancaster Synthesis Ltd.*

**2653 (9-Fluorenylmethoxycarbonyl)-4-allyloxycarbonyl-L-2,4-diaminobutyric acid**

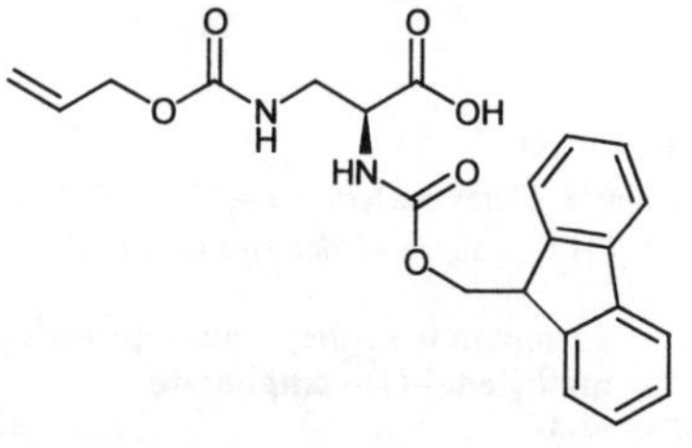

$C_{22}H_{22}N_2O_6$
Chiral intermediate. *Senn Chemicals AG.*

**2654 (9-Fluorenylmethoxycarbonyl)-D-2-aminoadipic acid**
218457-73-9

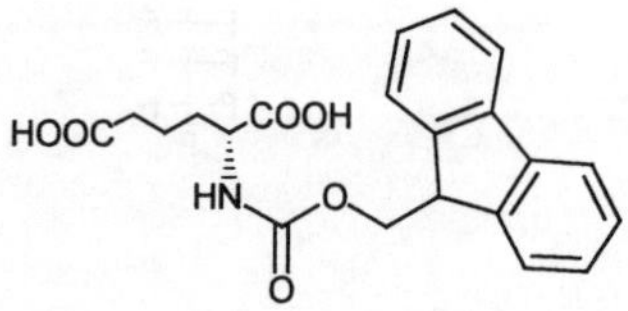

$C_{21}H_{21}NO_6$
Chiral intermediate. *Synthetech, Inc.*

**2655 (9-Fluorenylmethoxycarbonyl)-L-2-amino-adipic acid**

205384-77-1

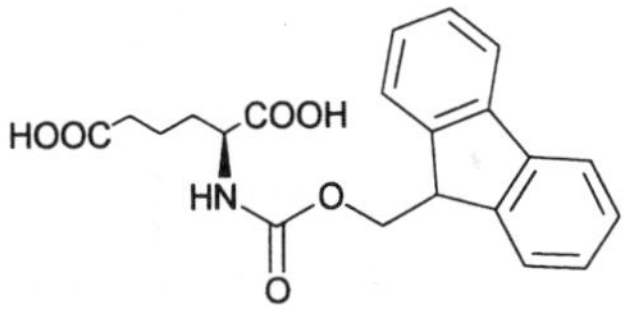

$C_{21}H_{21}NO_6$
Chiral intermediate. *Synthetech, Inc.*

**2656 (R)-N-(9-Fluorenylmethoxycarbonyl)-3-amino-3-(4-bromophenyl)propionic acid**

$C_{24}H_{20}BrNO_4$
(R)-N-Fmoc-β-(4-bromophenyl)-β-alanine. Chiral intermediate. *Lancaster Synthesis Ltd.*

**2657 (S)-N-(9-Fluorenylmethoxycarbonyl)-3-amino-3-(4-bromophenyl)propionic acid**

$C_{24}H_{20}BrNO_4$
(S)-N-Fmoc-β-(4-bromophenyl)-β-alanine. Chiral intermediate. *Lancaster Synthesis Ltd.*

**2658 (9-Fluorenylmethoxycarbonyl)-D-2-amino-butyric acid**

170642-27-0

$C_{19}H_{19}NO_4$
Chiral intermediate. *Synthetech, Inc.*

**2659 (9-Fluorenylmethoxycarbonyl)-L-2-amino-butyric acid**

135112-27-5

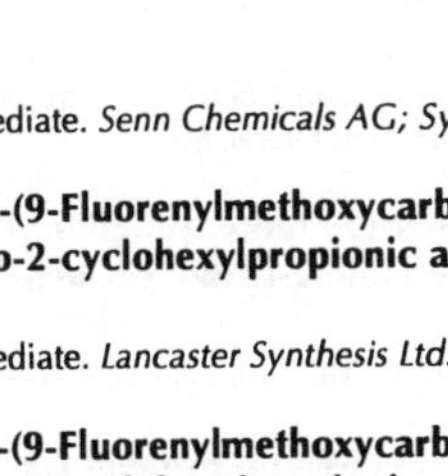

$C_{19}H_{19}NO_4$
Chiral intermediate. *Senn Chemicals AG; Synthetech, Inc.*

**2660 (R)-N-(9-Fluorenylmethoxycarbonyl)-2-amino-2-cyclohexylpropionic acid**

$C_{24}H_{27}NO_4$
Chiral intermediate. *Lancaster Synthesis Ltd.*

**2661 (S)-N-(9-Fluorenylmethoxycarbonyl)-2-amino-2-cyclohexylpropionic acid**

$C_{24}H_{27}NO_4$
N-(9-Fluorenylmethoxycarbonyl)-α-cyclohexyl-L-alanine. Chiral intermediate. *Lancaster Synthesis Ltd.*

**2662 (1R,3S)-N-(9-Fluorenylmethoxycarbonyl)-1-aminocyclopentane-3-carboxylic acid**

$C_{21}H_{21}NO_4$
Chiral intermediate. *Lancaster Synthesis Ltd.*

**2663 (1S,3R)-N-(9-Fluorenylmethoxycarbonyl)-1-aminocyclopentane-3-carboxylic acid**

$C_{21}H_{21}NO_4$
Chiral intermediate. *Lancaster Synthesis Ltd.*

**2664 (1R,4S)-N-(9-Fluorenylmethoxycarbonyl)-1-aminocyclopent-2-ene-4-carboxylic acid**

$C_{21}H_{19}NO_4$
Chiral intermediate. *Lancaster Synthesis Ltd.*

**2665 (1S,4R)-N-(9-Fluorenylmethoxycarbonyl)-1-aminocyclopent-2-ene-4-carboxylic acid**

$C_{21}H_{19}NO_4$
Chiral intermediate. *Lancaster Synthesis Ltd.*

**2666 (R)-N-(9-Fluorenylmethoxycarbonyl)-3-amino-3-(3-nitrophenyl)propionic acid**

$C_{24}H_{20}N_2O_6$
(R)-N-Fmoc-β-(3-nitrophenyl)-β-alanine. Chiral intermediate. *Lancaster Synthesis Ltd.*

**2667 (S)-N-(9-Fluorenylmethoxycarbonyl)-3-amino-3-(3-nitrophenyl)propionic acid**

$C_{24}H_{20}N_2O_6$
(S)-N-Fmoc-β-(3-nitrophenyl)-β-alanine. Chiral intermediate. *Lancaster Synthesis Ltd.*

**2668 (R)-N-(9-Fluorenylmethoxycarbonyl)-3-amino-3-phenylpropionic acid**

$C_{24}H_{21}NO_4$
(R)-N-Fmoc-β-phenyl-β-alanine. Chiral intermediate. *Lancaster Synthesis Ltd.*

**2669 (S)-N-(9-Fluorenylmethoxycarbonyl)-3-amino-3-phenylpropionic acid**

209252-15-3

$C_{24}H_{21}NO_4$
(S)-N-Fmoc-β-phenyl-β-alanine. Chiral intermediate. *Lancaster Synthesis Ltd.*

**2670 (9-Fluorenylmethoxycarbonyl)-D-2-amino-suberic acid**

218457-78-4

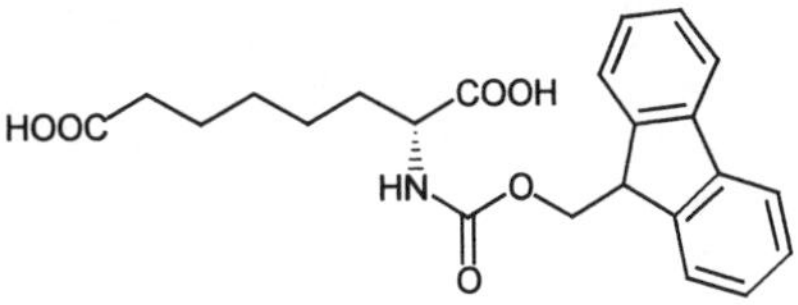

$C_{23}H_{25}NO_6$
Chiral intermediate. *Synthetech, Inc.*

**2671 (9-Fluorenylmethoxycarbonyl)-L-2-aminosuberic acid**
218457-76-2

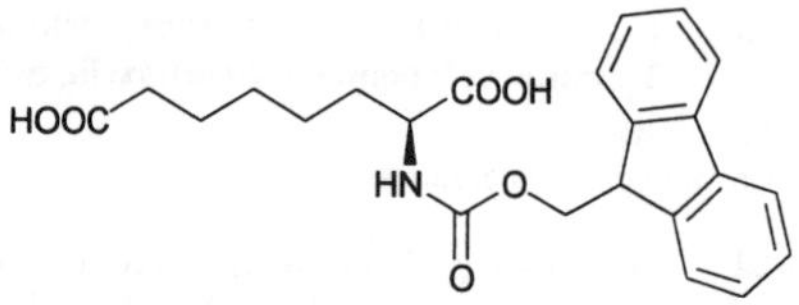

$C_{23}H_{25}NO_6$
Chiral intermediate. *Synthetech, Inc.*

**2672 N-α-(9-Fluorenylmethoxycarbonyl)-L-arginine**
91000-69-0

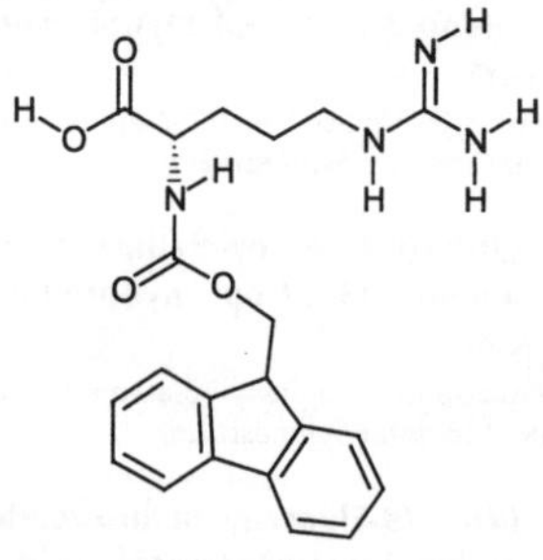

$C_{21}H_{24}N_4O_4$
Chiral intermediate. mp = 145-150°; [α] = + 2° (c = 0.5, 80% HOAc); $[\alpha]_D^{20}$ = + 8° (c = 1, DMF). *Acros Organics nv; Sigma-Aldrich Fine Chemicals.*

**2673 N-α-(9-Fluorenylmethoxycarbonyl)-L-asparagine**
71989-16-7 276-252-2

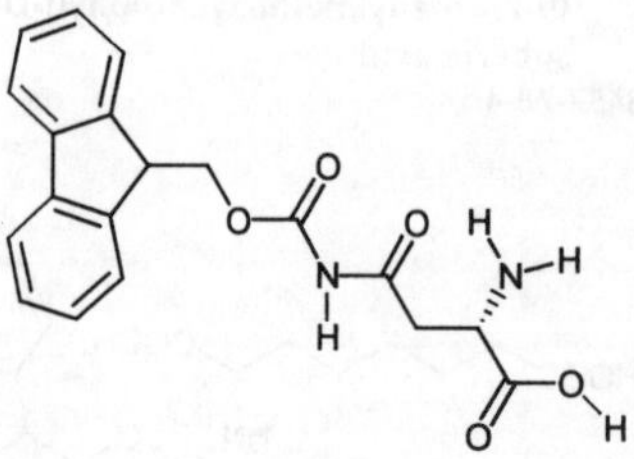

$C_{19}H_{18}N_2O_5$
$N^2$-[(9H-fluoren-9-ylmethoxy)carbonyl]-, L-. Chiral intermediate. mp = 190°; $[\alpha]_D^{20}$ = - 13° (c = 1, DMF). *Acros Organics nv; Interchem Corporation; Sigma-Aldrich Fine Chemicals; Synthetech, Inc.*

**2674 (9-Fluorenylmethoxycarbonyl)-L-asparagine trityl ester**
132388-59-1

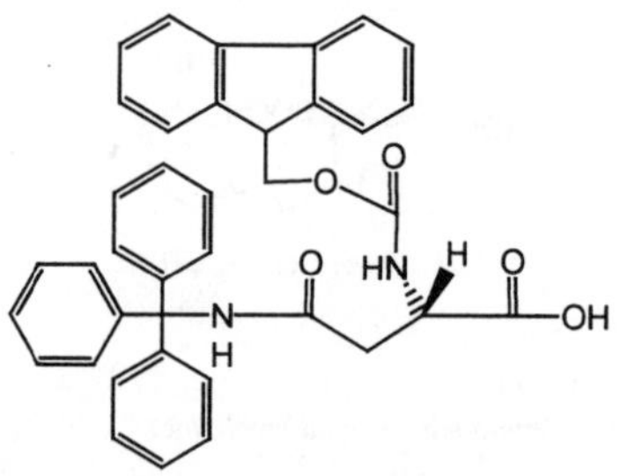

$C_{38}H_{32}N_2O_5$
Chiral intermediate. mp = 201-204°; $[\alpha]^{25}$ = - 14° (c = 1, $CH_3OH$). *Interchem Corporation; Sigma-Aldrich Fine Chemicals; Synthetech, Inc.*

**2675 (9-Fluorenylmethoxycarbonyl)-L-aspartic acid β-tert-butyl ester**
71989-14-5 276-251-7

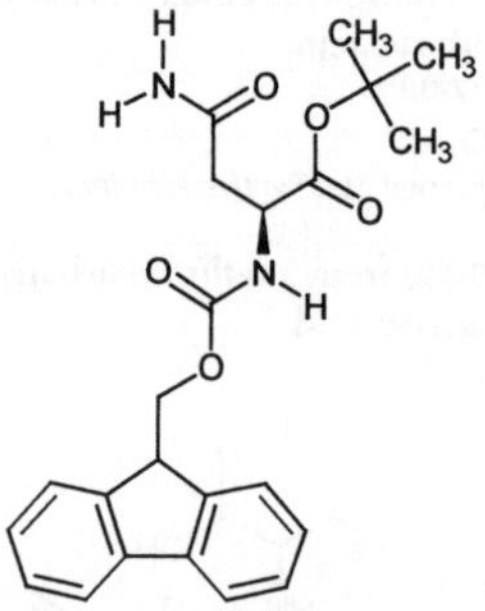

$C_{23}H_{26}N_2O_5$
Aspartic acid, N-[(9H-fluoren-9-ylmethoxy)carbonyl]-, 4-(1,1-dimethylethyl) ester, L-; 4-tert-Butyl hydrogen N-[(9H-fluoren-9-ylmethoxy)carbonyl]-L-aspartate; N-(9-Fluorenylmethoxycarbonyl)aspartic acid 4-tert-butyl ester. Chiral intermediate. $[\alpha]^{24}$ = - 25° (c = 1, DMF). *Acros Organics nv; Austin Chemical Company, Inc.; Interchem Corporation; Synthetech, Inc.*

**2676 (R)-N-(9-Fluorenylmethoxycarbonyl)-azetidine-2-carboxylic acid**

$C_{19}H_{17}NO_4$
Chiral intermediate. *Lancaster Synthesis Ltd.*

**2677 (S)-N-(9-Fluorenylmethoxycarbonyl)-azetidine-2-carboxylic acid**
136552-06-2

$C_{19}H_{17}NO_4$
Chiral intermediate. *Lancaster Synthesis Ltd.*

**2678 (9-Fluorenylmethoxycarbonyl)-p-azido-L-phenylalanine**

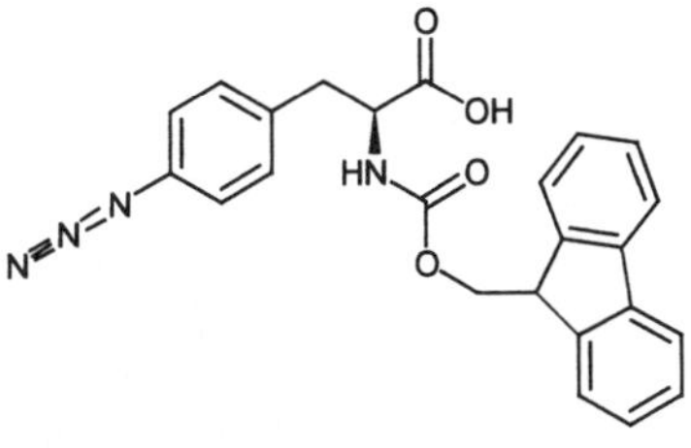

$C_{24}H_{20}N_4O_4$
Chiral intermediate. *Senn Chemicals AG.*

**2679 (9-Fluorenylmethoxycarbonyl)-D-3-benzothienylalanine**
177966-61-9

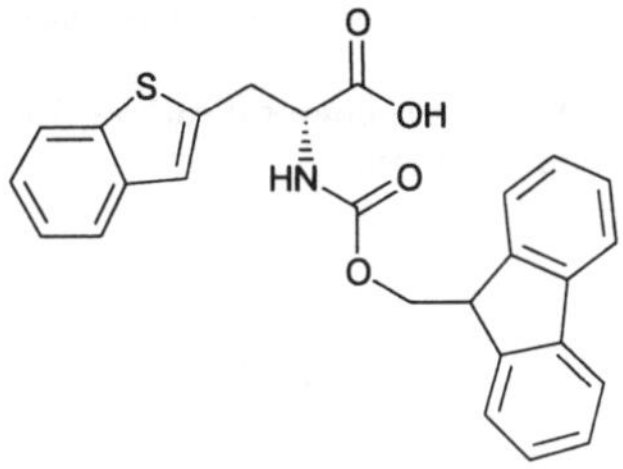

$C_{26}H_{21}NO_4S$
Chiral intermediate. *Senn Chemicals AG; Synthetech, Inc.*

**2680 (9-Fluorenylmethoxycarbonyl)-L-3-benzothienylalanine**
177966-60-8

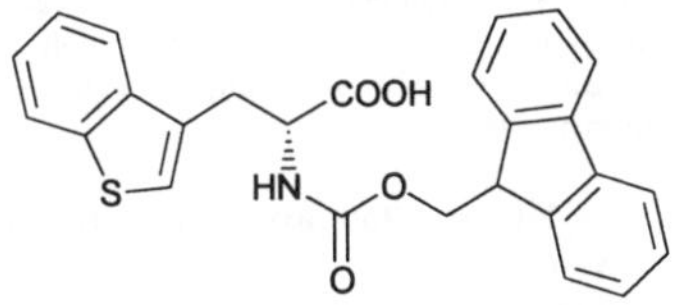

$C_{26}H_{21}NO_4S$
Chiral intermediate. *Synthetech, Inc.*

**2681 (9-Fluorenylmethoxycarbonyl)-O-benzyl-L-tyrosine**

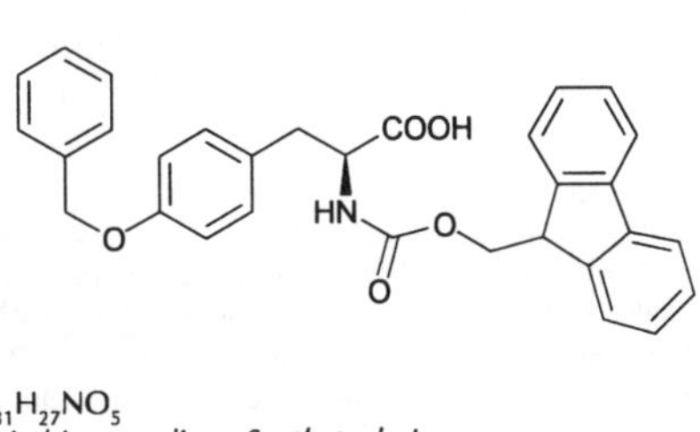

$C_{31}H_{27}NO_5$
Chiral intermediate. *Synthetech, Inc.*

**2682 (9-Fluorenylmethoxycarbonyl)-D-4,4'-biphenylalanine**
205526-38-1

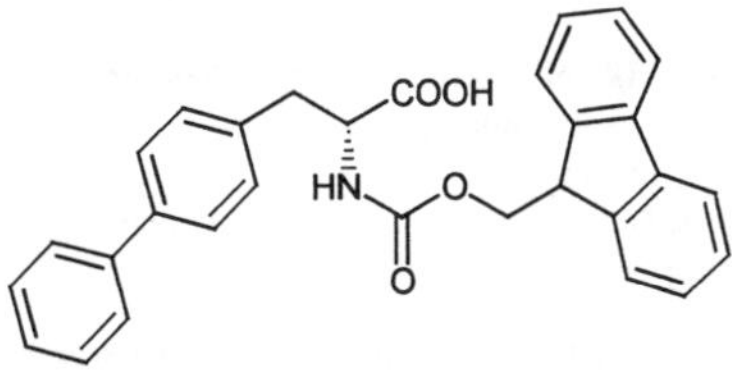

$C_{30}H_{25}NO_4$
Chiral intermediate. *Synthetech, Inc.*

**2683 (9-Fluorenylmethoxycarbonyl)-L-4,4'-biphenylalanine**
199110-64-0

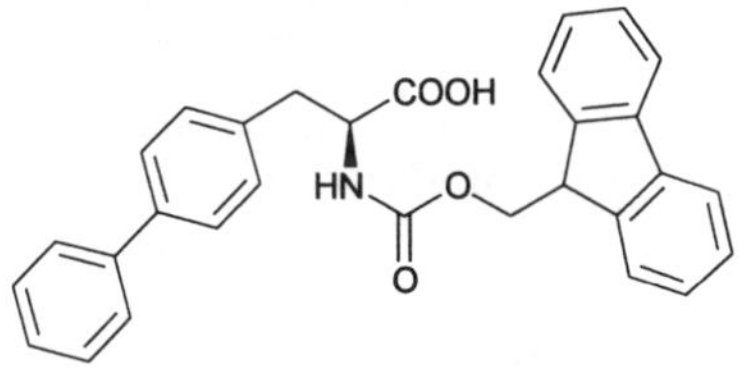

$C_{30}H_{25}NO_4$
Chiral intermediate. *Synthetech, Inc.*

**2684 (R)-N-(9-Fluorenylmethoxycarbonyl)-(2-bromoallyl)glycine**

$C_{20}H_{18}BrNO_4$
Chiral intermediate. *Lancaster Synthesis Ltd.*

**2685 (S)-N-(9-Fluorenylmethoxycarbonyl)-(2-bromoallyl)glycine**

$C_{20}H_{18}BrNO_4$
N-(9-Fluorenylmethoxycarbonyl)-(S)-2-amino-4-bromo-4-pentenoic acid. Chiral intermediate. *Lancaster Synthesis Ltd.*

**2686 (R)-N-(9-Fluorenylmethoxycarbonyl)-(5-bromo-2-methoxyphenyl)alanine**

$C_{25}H_{22}BrNO_5$
Chiral intermediate. *Lancaster Synthesis Ltd.*

**2687 (S)-N-(9-Fluorenylmethoxycarbonyl)-(5-bromo-2-methoxyphenyl)alanine**

$C_{25}H_{22}BrNO_5$
Chiral intermediate. *Lancaster Synthesis Ltd.*

**2688 (R)-N-(9-Fluorenylmethoxycarbonyl)-(2-bromophenyl)alanine**
220497-79-0

$C_{24}H_{20}BrNO_4$
Chiral intermediate. *Lancaster Synthesis Ltd.*

**2689 (S)-N-(9-Fluorenylmethoxycarbonyl)-(2-bromophenyl)alanine**
220497-47-2
$C_{24}H_{20}BrNO_4$
Chiral intermediate. *Lancaster Synthesis Ltd.*

**2690 (R)-N-(9-Fluorenylmethoxycarbonyl)-(3-bromophenyl)alanine**
$C_{24}H_{20}BrNO_4$
Fmoc-D-3-Bromophenylalanine. Chiral intermediate. *Lancaster Synthesis Ltd.*

**2691 (S)-N-(9-Fluorenylmethoxycarbonyl)-(3-bromophenyl)alanine**
$C_{24}H_{20}BrNO_4$
Fmoc-L-3-Bromophenylalanine. Chiral intermediate. *Lancaster Synthesis Ltd.*

**2692 (R)-N-(9-Fluorenylmethoxycarbonyl)-(4-bromophenyl)alanine**
198545-76-5
$C_{24}H_{20}BrNO_4$
Fmoc-D-4-Bromophenylalanine. Chiral intermediate. *Lancaster Synthesis Ltd.; Synthetech, Inc.*

**2693 (S)-N-(9-Fluorenylmethoxycarbonyl)-(4-bromophenyl)alanine**
198561-04-5
$C_{24}H_{20}BrNO_4$
Fmoc-L-4-Bromophenylalanine. Chiral intermediate. *Lancaster Synthesis Ltd.; Synthetech, Inc.*

**2694 (R)-N-(9-Fluorenylmethoxycarbonyl)-(5-bromo-2-thienyl)alanine**
$C_{22}H_{18}BrNO_4S$
Fmoc-D-5-Bromothienylalanine. Chiral intermediate. *Lancaster Synthesis Ltd.*

**2695 (S)-N-(9-Fluorenylmethoxycarbonyl)-(5-bromo-2-thienyl)alanine**
$C_{22}H_{18}BrNO_4S$
Fmoc-L-5-Bromothienylalanine. Chiral intermediate. *Lancaster Synthesis Ltd.*

**2696 (9-Fluorenylmethoxycarbonyl)-4-(tert-butoxycarbonyl)-L-2,4-diamino-butyric acid**

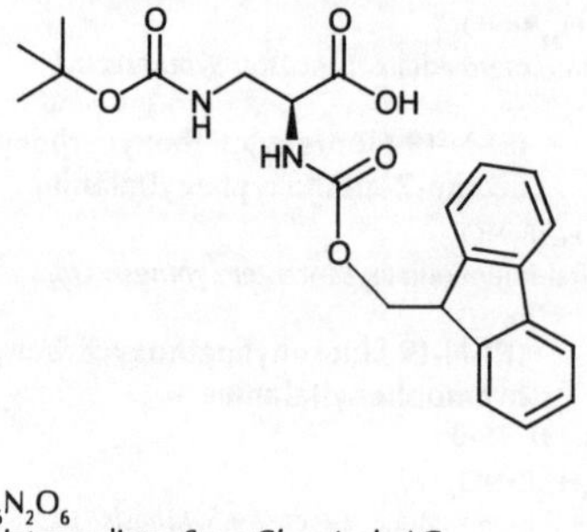

$C_{23}H_{26}N_2O_6$
Chiral intermediate. *Senn Chemicals AG.*

**2697 N-α-(9-Fluorenylmethoxycarbonyl)-N-ε-(tert-butoxycarbonyl)-L-lysine**
71989-26-9 276-256-4

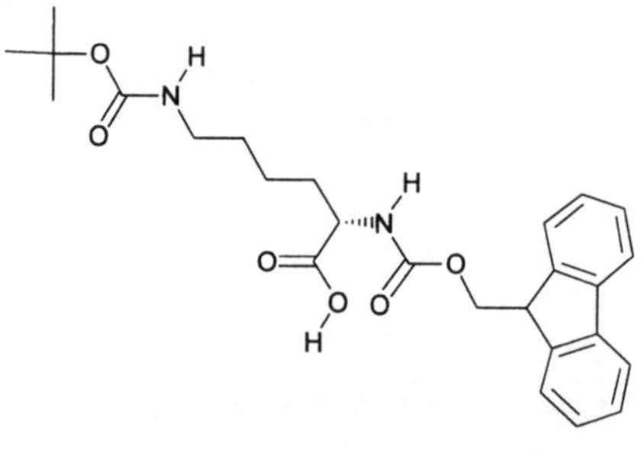

$C_{26}H_{32}N_2O_6$
$N^6$-(tert-Butoxycarbonyl)-$N^2$-[(9H-fluoren-9-ylmethoxy)-carbonyl]-L-lysine. Chiral intermediate. $[\alpha]^{24}$ = - 12.4° (c = 2, DMF). *Acros Organics nv; Synthetech, Inc.*

**2698 (9-Fluorenylmethoxycarbonyl)-(S-tert-butyl)cysteine**
73724-43-3 277-574-6

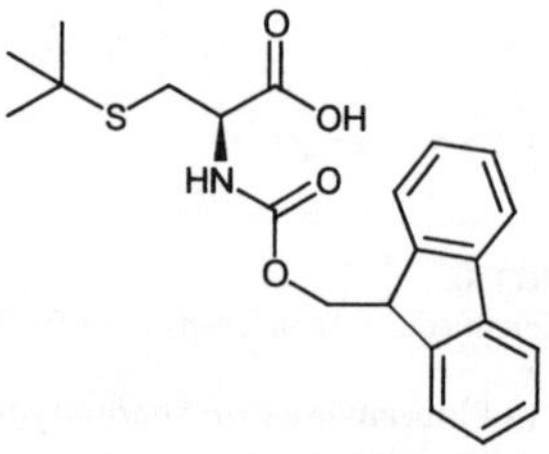

$C_{22}H_{25}NO_4S_2$
Alanine, 3-[(1,1-dimethylethyl)dithio]-N-[(9H-fluoren-9-ylmethoxy)carbonyl]-, L-; 3-[(1,1-dimethylethyl)dithio]-N-[(9H-fluoren-9-ylmethoxy)carbonyl]-L-alanine. Chiral intermediate. *Isochem.*

**2699 (9-Fluorenylmethoxycarbonyl)-O-t-butyl-L-serine**
71989-33-8 276-260-6

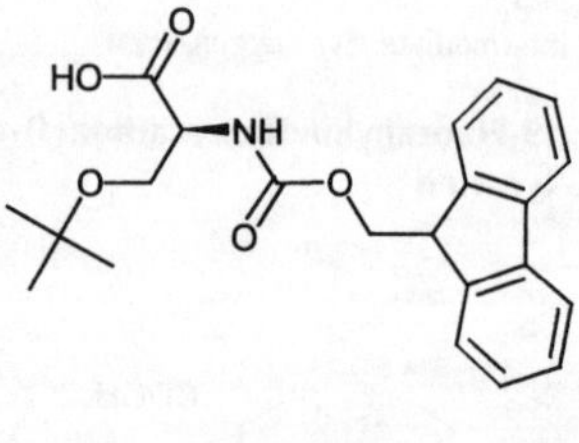

$C_{22}H_{25}NO_5$
Serine, O-(1,1-dimethylethyl)-N-[(9H-fluoren-9-ylmeth-oxy)carbonyl]-, L-; O-(tert-butyl)-N-[(9H-fluoren-9-ylmeth-oxy)carbonyl]-L-serine; N-(9-Fluorenylmethoxy-carbonyl)-O-tert-butylserine. Chiral intermediate. $[\alpha]^{24}$ = + 25° (c = 1, EtOAc). *Acros Organics nv; Interchem Corporation; Sigma-Aldrich Fine Chemicals; Synthetech, Inc.*

**2700 (9-Fluorenylmethoxycarbonyl)-O-tert-butyl-D-threonine**

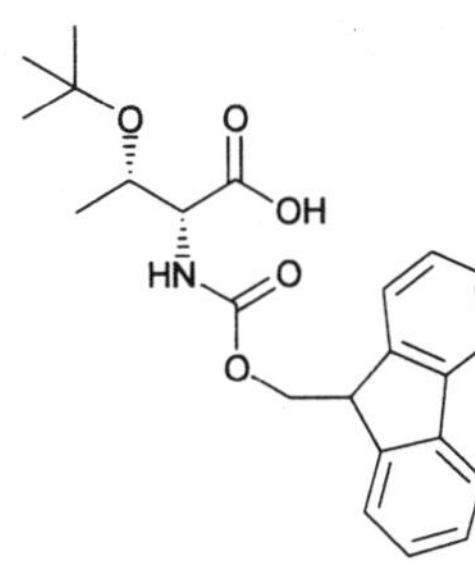

$C_{23}H_{27}NO_5$
Chiral intermediate. *Genzyme Pharmaceuticals.*

**2701 (9-Fluorenylmethoxycarbonyl)-O-tert-butyl-L-threonine**
71989-35-0 276-261-1

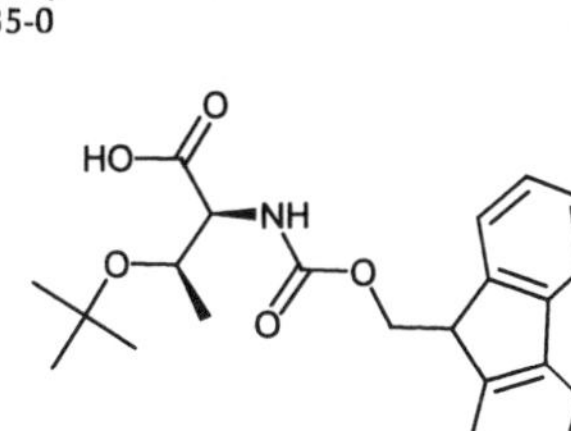

$C_{23}H_{27}NO_5$
Threonine, O-(1,1-dimethylethyl)-N-[(9H-fluoren-9-ylmethoxy)carbonyl]-, L-; Chiral intermediate. $[\alpha]^{24}$ = + 16° (c = 1, EtOAc). *Acros Organics nv; Sigma-Aldrich Fine Chemicals; Synthetech, Inc.*

**2702 N-ε-(9-Fluorenylmethoxycarbonyl)-N-α-carbobenzyloxy-L-lysine**
105751-18-6

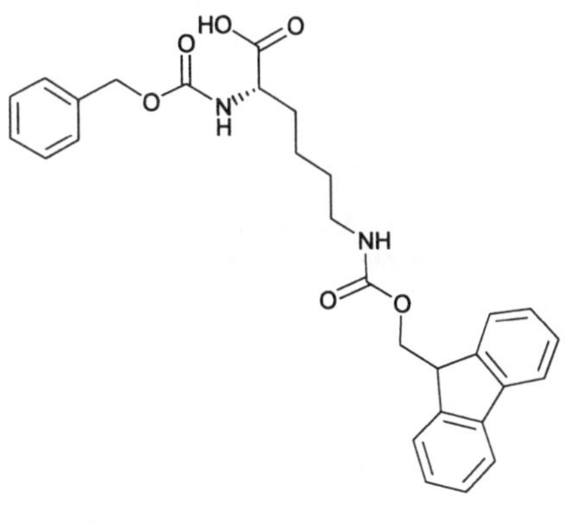

$C_{29}H_{30}N_2O_6$
FMOC-Nε-CBZ-L-Lysine; N-(9-Fluorenylmethoxycarbonyl)-Nε-CBZ-L-lysine. Chiral intermediate. mp = 101-105°; $[\alpha]^{24}$ = -11° (c = 1, DMF). *Acros Organics nv.*

**2703 (9-Fluorenylmethoxycarbonyl)-L-4-carboxyglutamic acid di-tert-butyl ester**

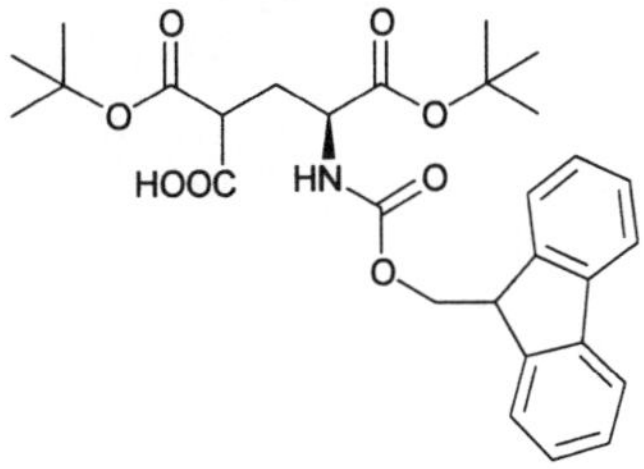

$C_{29}H_{35}NO_8$
Chiral intermediate. *Senn Chemicals AG.*

**2704 (9-Fluorenylmethoxycarbonyl)-(3-chloro)-L-alanine**

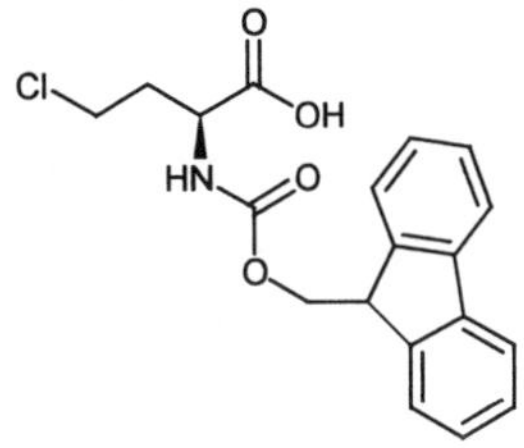

$C_{19}H_{18}ClNO_4$
Chiral intermediate. *Senn Chemicals AG.*

**2705 (9-Fluorenylmethoxycarbonyl)-D-2-chlorophenylalanine**
205526-22-3

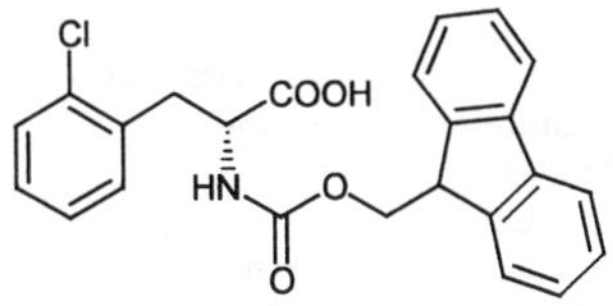

$C_{24}H_{20}ClNO_4$
Chiral intermediate. *Synthetech, Inc.*

**2706 (9-Fluorenylmethoxycarbonyl)-L-2-chlorophenylalanine**
198560-41-7

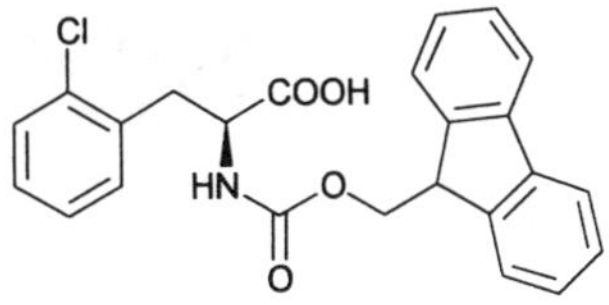

$C_{24}H_{20}ClNO_4$
Chiral intermediate. *Synthetech, Inc.*

**2707 (9-Fluorenylmethoxycarbonyl)-D-3-chlorophenylalanine**
205526-23-4

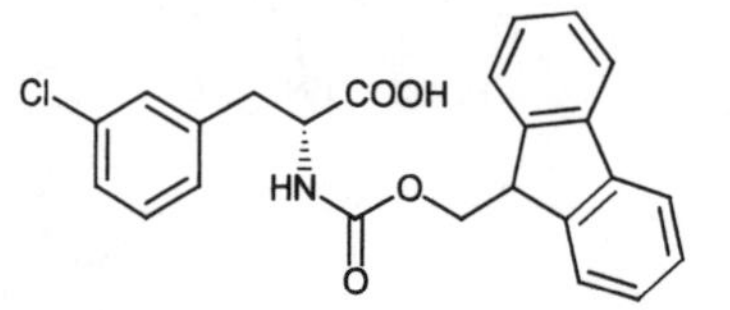

$C_{24}H_{20}ClNO_4$
Chiral intermediate. *Synthetech, Inc.*

**2708 (9-Fluorenylmethoxycarbonyl)-L-3-chlorophenylalanine**
198560-44-0

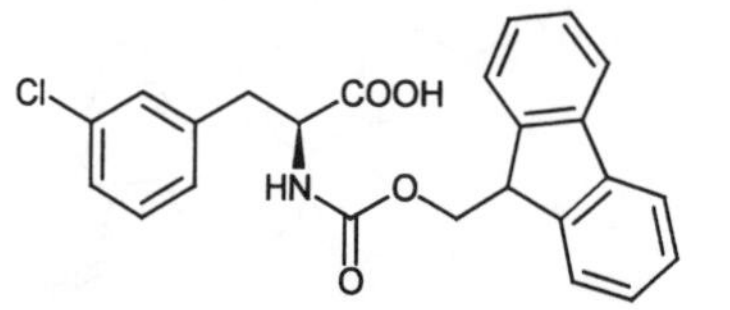

$C_{24}H_{20}ClNO_4$
Chiral intermediate. *Synthetech, Inc.*

**2709 (R)-N-(9-Fluorenylmethoxycarbonyl)-(4-chlorophenyl)alanine**
142994-19-2
$C_{24}H_{20}ClNO_4$
Fmoc-D-4-Chlorophenylalanine. Chiral intermediate. *Lancaster Synthesis Ltd.; Synthetech, Inc.*

**2710 (S)-N-(9-Fluorenylmethoxycarbonyl)-(4-chlorophenyl)alanine**
175453-08-4
$C_{24}H_{20}ClNO_4$
Fmoc-L-4-Chlorophenylalanine. Chiral intermediate. *Lancaster Synthesis Ltd.; Synthetech, Inc.*

**2711 (9-Fluorenylmethoxycarbonyl)-L-citrulline**

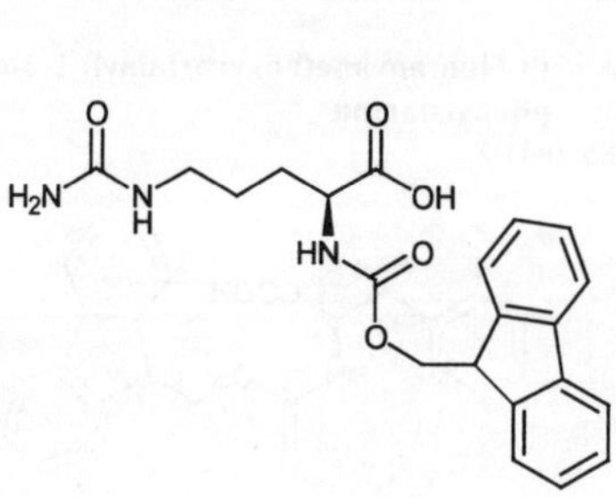

$C_{21}H_{23}N_3O_5$
Chiral intermediate. *Senn Chemicals AG.*

**2712 (9-Fluorenylmethoxycarbonyl)-(2R,3S)-(cis)-3-cyanomethyl proline**

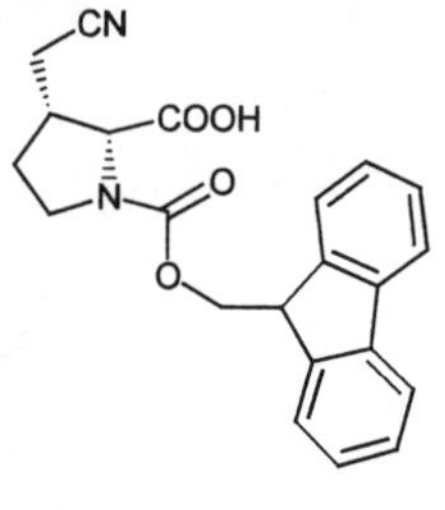

$C_{22}H_{20}N_2O_4$
Chiral intermediate. *Senn Chemicals AG.*

**2713 (9-Fluorenylmethoxycarbonyl)-(2S,3R)-(cis)-3-cyanomethyl proline**

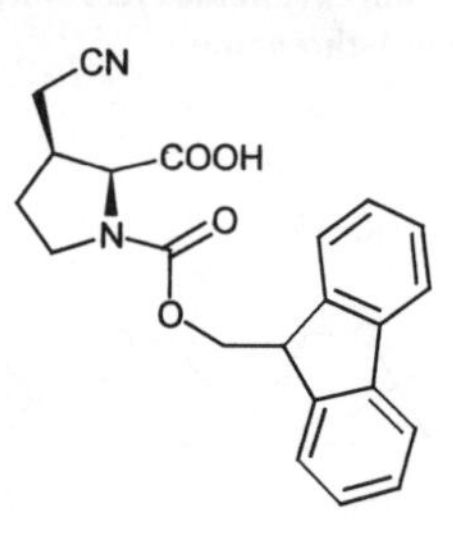

$C_{22}H_{20}N_2O_4$
Chiral intermediate. *Senn Chemicals AG.*

**2714 (9-Fluorenylmethoxycarbonyl)-D-3-cyanophenylalanine**
205526-37-0

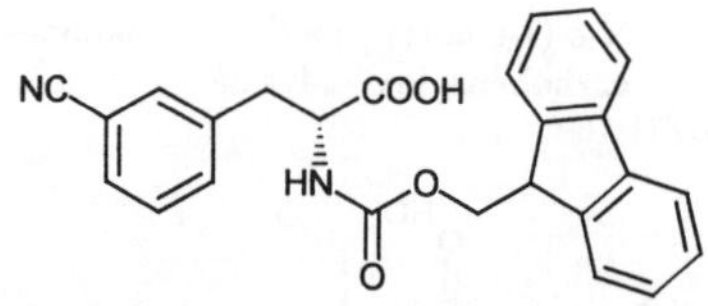

$C_{25}H_{20}N_2O_4$
Chiral intermediate. *Synthetech, Inc.*

**2715 (S)-N-(9-Fluorenylmethoxycarbonyl)-(3-cyanophenyl)alanine**
205526-36-9

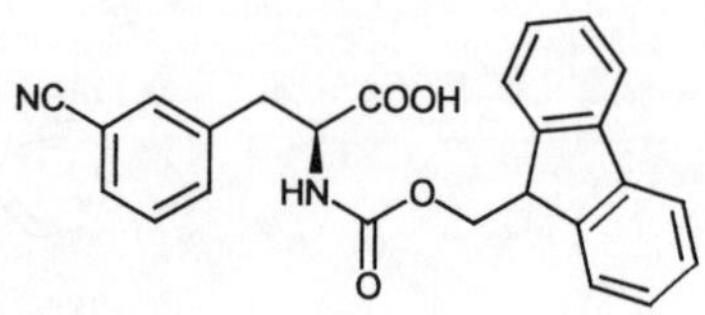

$C_{25}H_{20}N_2O_4$
Chiral intermediate. *Lancaster Synthesis Ltd.; Synthetech, Inc.*

**2716 (R)-N-(9-Fluorenylmethoxycarbonyl)-(4-cyanophenyl)alanine**
205526-34-7

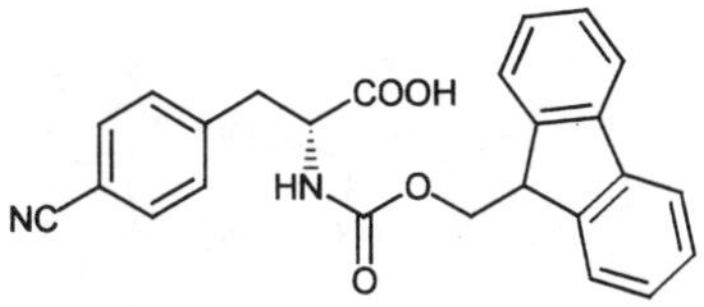

$C_{25}H_{20}N_2O_4$
Chiral intermediate. *Lancaster Synthesis Ltd.; Synthetech, Inc.*

**2717 (S)-N-(9-Fluorenylmethoxycarbonyl)-(4-cyanophenyl)alanine**
173963-93-4

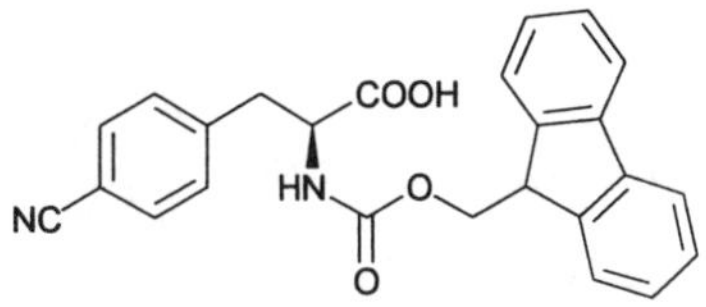

$C_{25}H_{20}N_2O_4$
Fmoc-L-4-Cyanophenylalanine. Chiral intermediate. *Lancaster Synthesis Ltd.; Synthetech, Inc.*

**2718 (9-Fluorenylmethoxycarbonyl)-L-cyclohexylalanine**

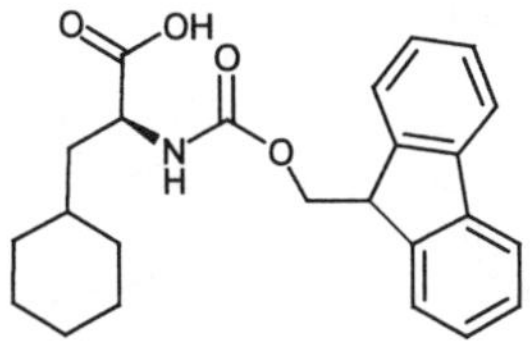

$C_{24}H_{27}NO_4$
Chiral intermediate. *Senn Chemicals AG.*

**2719 (9-Fluorenylmethoxycarbonyl)-L-cyclohexylglycine**

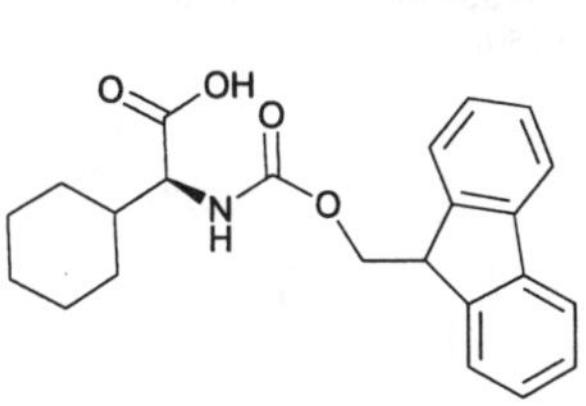

$C_{24}H_{27}NO_4$
Chiral intermediate. *Senn Chemicals AG.*

**2720 (R)-N-(9-Fluorenylmethoxycarbonyl)-cyclopropylglycine**

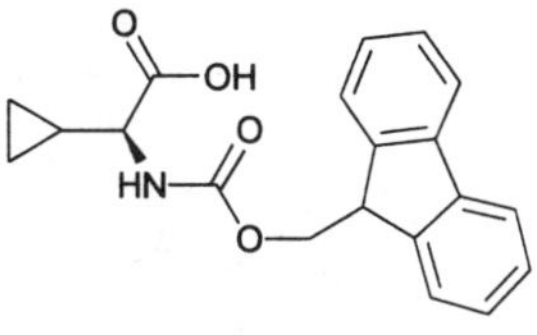

$C_{20}H_{19}NO_4$
Chiral intermediate. *Eastman Chemical Company.*

**2721 (S)-N-(9-Fluorenylmethoxycarbonyl)-cyclopropylglycine**

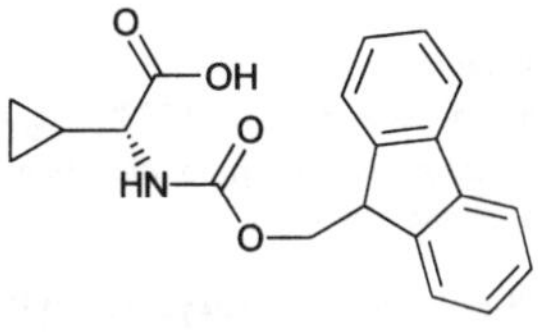

$C_{20}H_{19}NO_4$
Chiral intermediate. *Eastman Chemical Company.*

**2722 (9-Fluorenylmethoxycarbonyl)-L-cysteine acetamidomethyl ester**

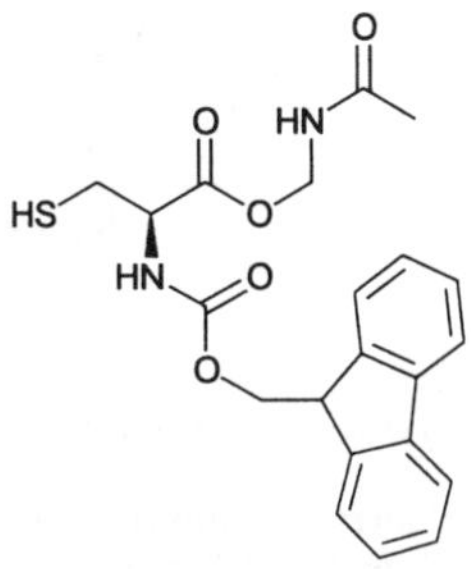

$C_{21}H_{22}N_2O_3S$
Chiral intermediate. *Interchem Corporation; Synthetech, Inc.*

**2723 (9-Fluorenylmethoxycarbonyl)-L-cysteine phenylacetate monohydrate**

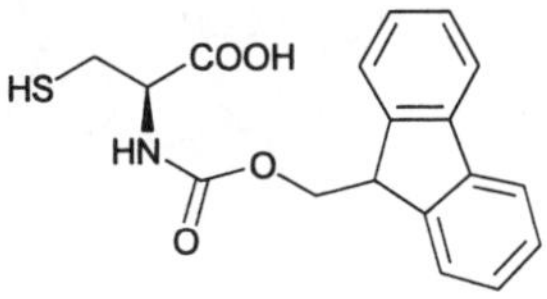

$C_{18}H_{17}NO_4S$
Chiral intermediate. *Synthetech, Inc.*

**2724 (9-Fluorenylmethoxycarbonyl)-L-cysteine trityl ester**
103213-32-7

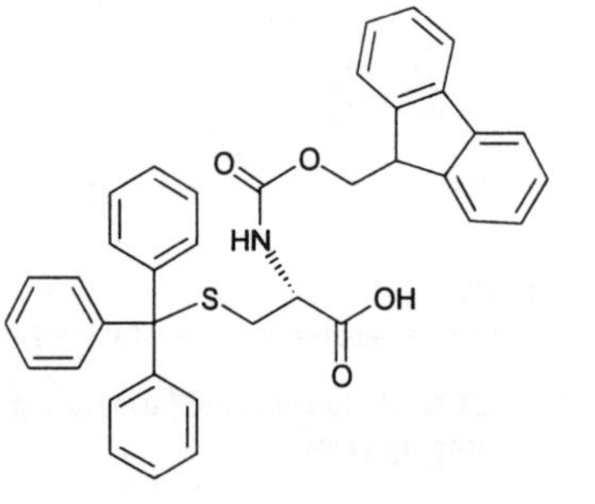

$C_{37}H_{31}NO_4S$
Chiral intermediate. mp = 170-173°; $[\alpha]_D^{20}$ = + 24° (c = 1, $CH_3OH$). *Interchem Corporation; Sigma-Aldrich Fine Chemicals; Synthetech, Inc.*

**2725 (9-Fluorenylmethoxycarbonyl)-L-2,3-diaminopropionic acid hydrochloride**

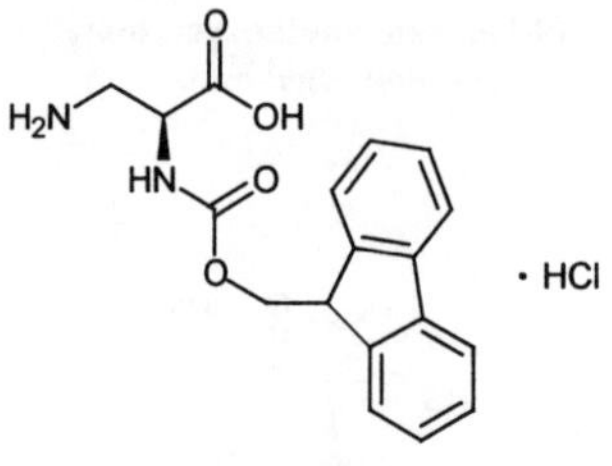

$C_{18}H_{19}ClN_2O_4$
Chiral intermediate. *Senn Chemicals AG.*

**2726 (9-Fluorenylmethoxycarbonyl)-D-3,4-dichlorophenylalanine**
177966-58-4

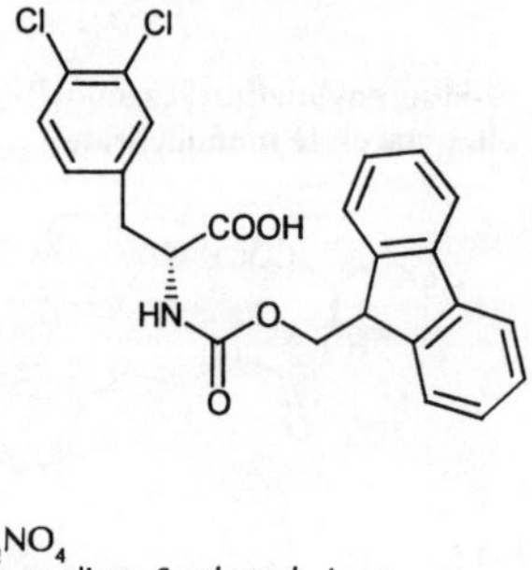

$C_{24}H_{19}Cl_2NO_4$
Chiral intermediate. *Synthetech, Inc.*

**2727 (9-Fluorenylmethoxycarbonyl)-L-3,4-dichlorophenylalanine**
17766-59-5

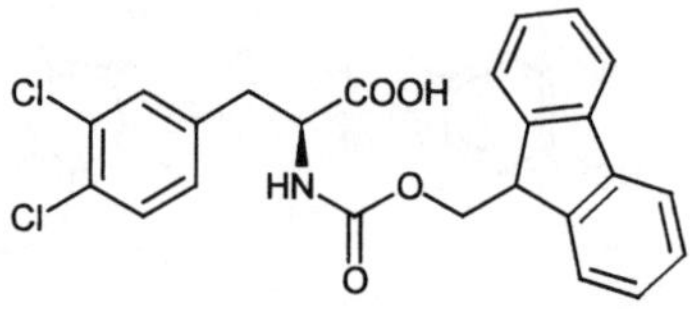

$C_{24}H_{19}Cl_2NO_4$
Chiral intermediate. *Synthetech, Inc.*

**2728 (9-Fluorenylmethoxycarbonyl)-D-3,4-difluorophenylalanine**
198545-59-4

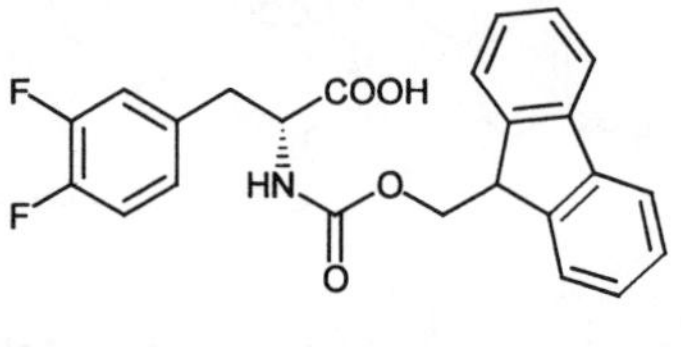

$C_{24}H_{19}F_2NO_4$
Chiral intermediate. *Synthetech, Inc.*

**2729 (9-Fluorenylmethoxycarbonyl)-L-3,4-difluorophenylalanine**
198560-43-9

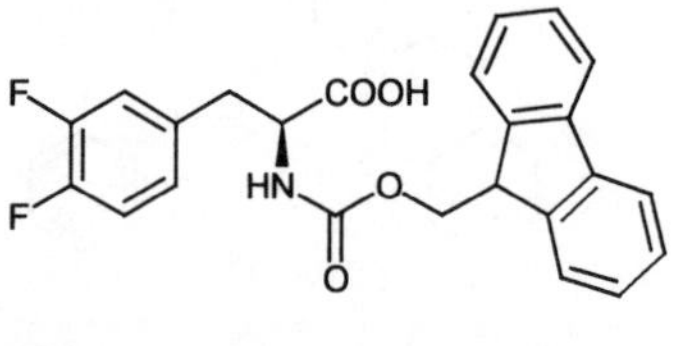

$C_{24}H_{19}F_2NO_4$
Chiral intermediate. *Synthetech, Inc.*

**2730 (9-Fluorenylmethoxycarbonyl)-D-3,5-difluorophenylalanine**
205526-25-6

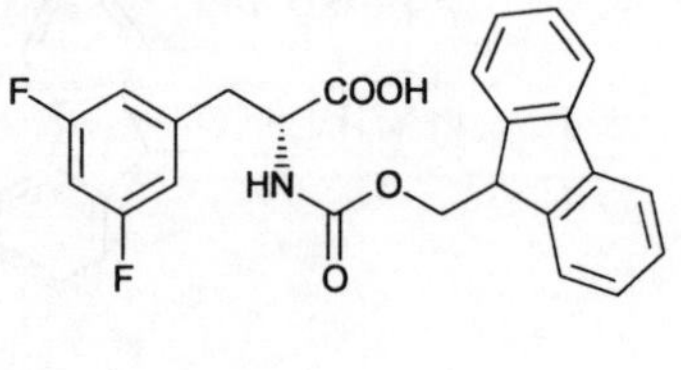

$C_{24}H_{19}F_2NO_4$
Chiral intermediate. *Synthetech, Inc.*

**2731 (9-Fluorenylmethoxycarbonyl)-L-3,5-difluorophenylalanine**
205526-24-5

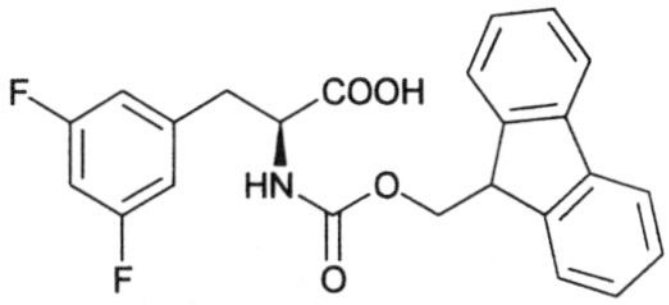

$C_{24}H_{19}F_2NO_4$
Chiral intermediate. *Synthetech, Inc.*

**2732 (9-Fluorenylmethoxycarbonyl)-D-3,4-dimethoxyphenylalanine**
218457-81-9

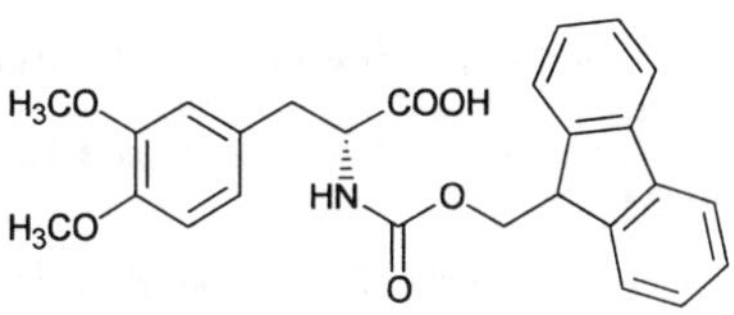

$C_{26}H_{25}NO_6$
Chiral intermediate. *Synthetech, Inc.*

**2733 (9-Fluorenylmethoxycarbonyl)-L-3,4-dimethoxyphenylalanine**
184962-88-7

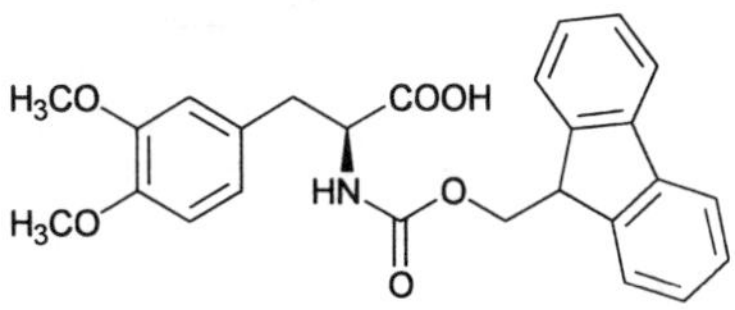

$C_{26}H_{25}NO_6$
Chiral intermediate. *Synthetech, Inc.*

**2734 (9-Fluorenylmethoxycarbonyl)-D-3,3-diphenylalanine**
189937-46-0

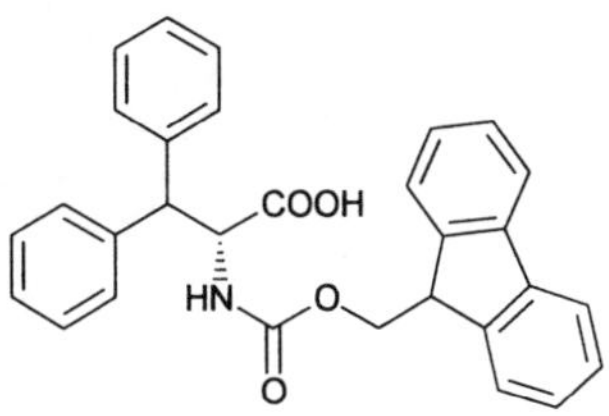

$C_{30}H_{25}NO_4$
Chiral intermediate. *Synthetech, Inc.*

**2735 (9-Fluorenylmethoxycarbonyl)-L-3,3-diphenylalanine**
201484-50-6

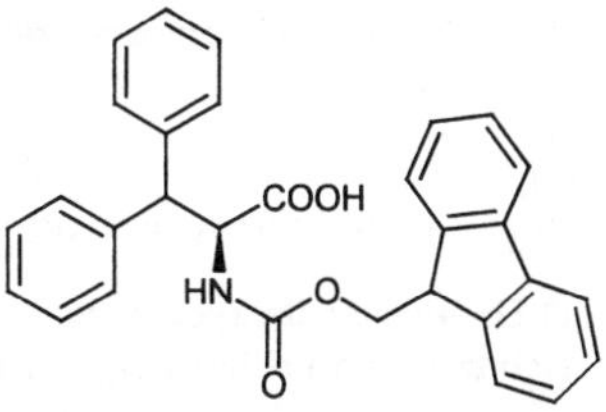

$C_{30}H_{25}NO_4$
Chiral intermediate. *Synthetech, Inc.*

**2736 (R)-N-(9-Fluorenylmethoxycarbonyl)-(2-fluorophenyl)alanine**
198545-46-9
$C_{24}H_{20}FNO_4$
Fmoc-D-2-fluorophenylalanine. Chiral intermediate. *Lancaster Synthesis Ltd.; Synthetech, Inc.*

**2737 (S)-N-(9-Fluorenylmethoxycarbonyl)-(2-fluorophenyl)alanine**
205526-26-7
$C_{24}H_{20}FNO_4$
Chiral intermediate. *Lancaster Synthesis Ltd.; Synthetech, Inc.*

**2738 (R)-N-(9-Fluorenylmethoxycarbonyl)-(3-fluorophenyl)alanine**
198545-72-1
$C_{24}H_{20}FNO_4$
Chiral intermediate. *Lancaster Synthesis Ltd.; Synthetech, Inc.*

**2739 (S)-N-(9-Fluorenylmethoxycarbonyl)-(3-fluorophenyl)alanine**
198560-68-8
$C_{24}H_{20}FNO_4$
Chiral intermediate. *Lancaster Synthesis Ltd.*

**2740 (R)-N-(9-Fluorenylmethoxycarbonyl)-(4-fluorophenyl)alanine**
177966-64-2
$C_{24}H_{20}FNO_4$
Fmoc-D-4-Fluorophenylalanine. Chiral intermediate. *Lancaster Synthesis Ltd.; Synthetech, Inc.*

**2741 (S)-N-(9-Fluorenylmethoxycarbonyl)-(4-fluorophenyl)alanine**
169243-86-1
$C_{24}H_{20}FNO_4$
Fmoc-L-4-Fluorophenylalanine. Chiral intermediate. *Lancaster Synthesis Ltd.; Synthetech, Inc.*

**2742 (R)-N-(9-Fluorenylmethoxycarbonyl)-(4-fluorophenyl)glycine**
$C_{23}H_{18}FNO_4$
Chiral intermediate. *Lancaster Synthesis Ltd.*

**2743 (R)-N-(9-Fluorenylmethoxycarbonyl)-(2-furyl)alanine**

$C_{22}H_{19}NO_5$
Chiral intermediate. *Lancaster Synthesis Ltd.*

**2744 (S)-N-(9-Fluorenylmethoxycarbonyl)-(2-furyl)alanine**

159611-02-6
$C_{22}H_{19}NO_5$
Chiral intermediate. *Lancaster Synthesis Ltd.*

**2745 (9-Fluorenylmethoxycarbonyl)-L-glutamic acid-α-tert-butyl ester**

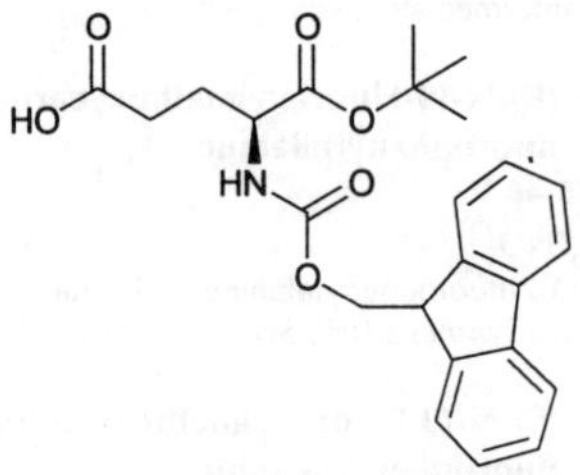

$C_{24}H_{27}NO_6$
Chiral intermediate. *Interchem Corporation; Synthetech, Inc.*

**2746 (9-Fluorenylmethoxycarbonyl)-L-glutamic acid-O-tert-butyl ester monohydrate**

71989-18-9 276-253-8

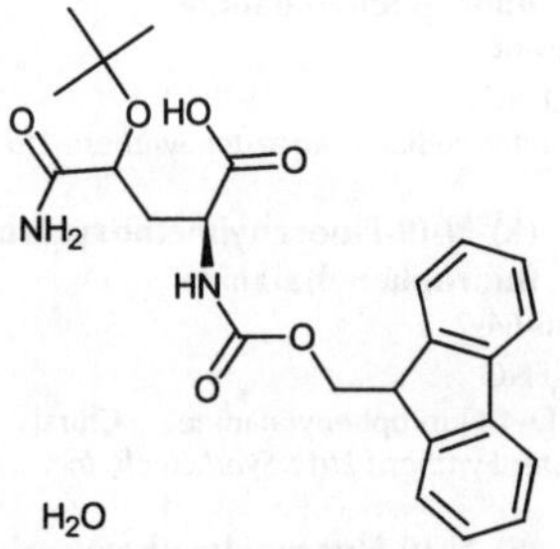

$H_2O$

$C_{24}H_{28}N_2O_6 \cdot H_2O$
Glutamic acid, N-[(9H-fluoren-9-ylmethoxy)carbonyl]-, 5-(1,1-dimethylethyl) ester, L-; 5-tert-butyl N-[(9H-fluoren-9-ylmethoxy)carbonyl]-2-aminoglutarate; N-(9-Fluorenylmethoxycarbonyl)-L-glutamic acid γ-tert-butyl ester. Chiral intermediate. mp = 86-89°; $[\alpha]_D^{20}$ = -17.5° (c = 1, DMF). *Acros Organics nv; Interchem Corporation; Sigma-Aldrich Fine Chemicals.*

**2747 N-α-(9-Fluorenylmethoxycarbonyl)-L-glutamine**

71989-20-3 276-254-3

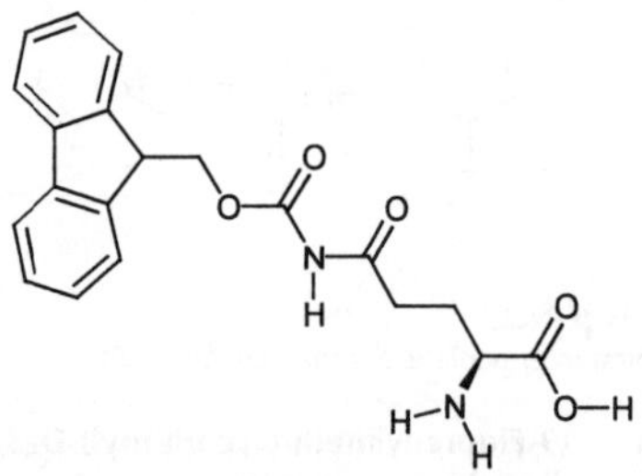

$C_{20}H_{20}N_2O_5$
$N^2$-[(9H-fluoren-9-ylmethoxy)carbonyl]-, L-. Chiral intermediate. mp = 213-217°; [α] = - 18° (c = 1, DMF). *Acros Organics nv; Interchem Corporation; Sigma-Aldrich Fine Chemicals; Synthetech, Inc.*

**2748 (9-Fluorenylmethoxycarbonyl)-L-glutamine trityl ester**

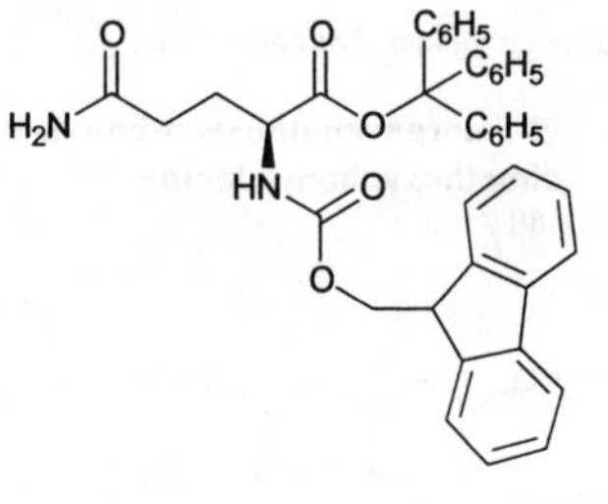

$C_{39}H_{24}N_2O_5$
Chiral intermediate. *Interchem Corporation; Synthetech, Inc.*

**2749 (9-Fluorenylmethoxycarbonyl)-L-histidine trityl ester**

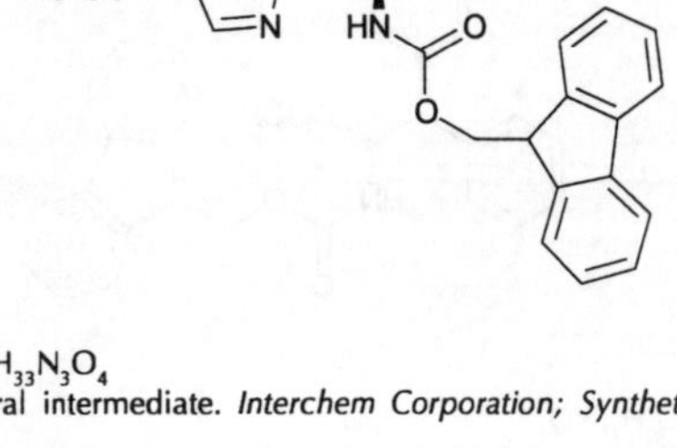

$C_{40}H_{33}N_3O_4$
Chiral intermediate. *Interchem Corporation; Synthetech, Inc.*

**2750 (9-Fluorenylmethoxycarbonyl)-L-homoleucine**

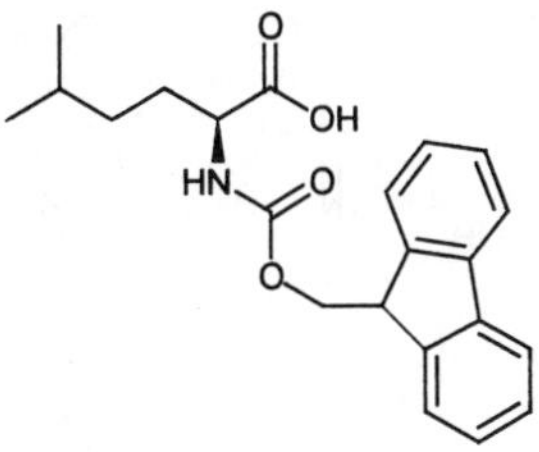

$C_{22}H_{25}NO_4$
Chiral intermediate. *Senn Chemicals AG.*

**2751 (9-Fluorenylmethoxycarbonyl)-D-homophenylalanine**
135944-09-1

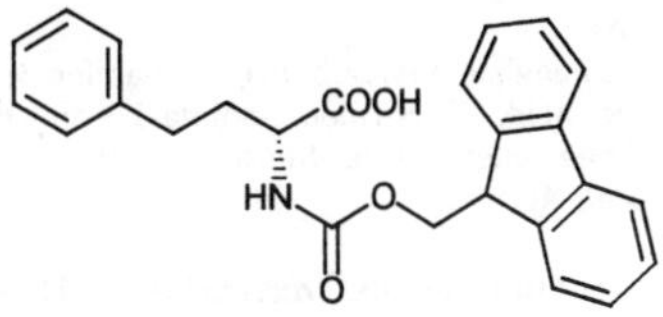

$C_{25}H_{23}NO_4$
Chiral intermediate. *Synthetech, Inc.*

**2752 (9-Fluorenylmethoxycarbonyl)-L-homophenylalanine**
132684-59-4

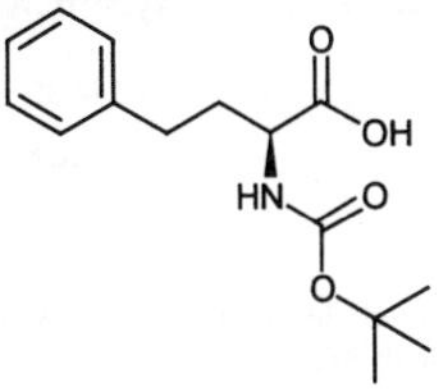

$C_{25}H_{23}NO_4$
Chiral intermediate. *Senn Chemicals AG; Synthetech, Inc.*

**2753 (9-Fluorenylmethoxycarbonyl)-(R)-3-hydroxypyrrolidine**

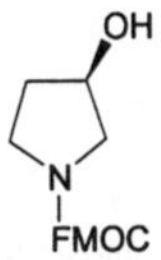

$C_{19}H_{19}NO_3$
Chiral intermediate. *Omega Chemical Company Inc.*

**2754 (9-Fluorenylmethoxycarbonyl)-(S)-3-hydroxypyrrolidine**

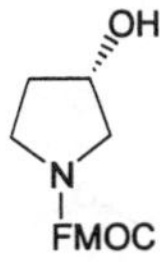

$C_{19}H_{19}NO_3$
Chiral intermediate. *Omega Chemical Company Inc.*

**2755 (9-Fluorenylmethoxycarbonyl)-D-2-indanylglycine**
205526-40-5

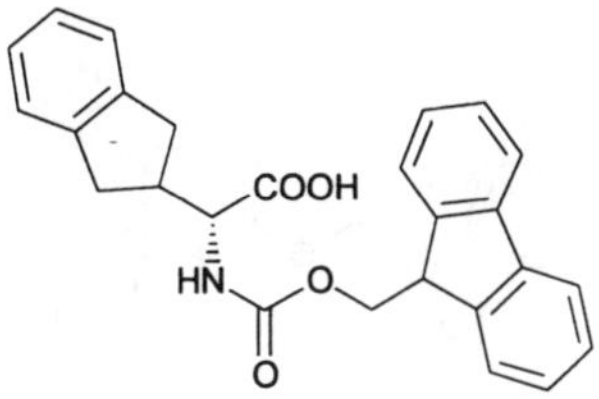

$C_{26}H_{23}NO_4$
Chiral intermediate. *Synthetech, Inc.*

**2756 (9-Fluorenylmethoxycarbonyl)-L-2-indanylglycine**
205526-39-2

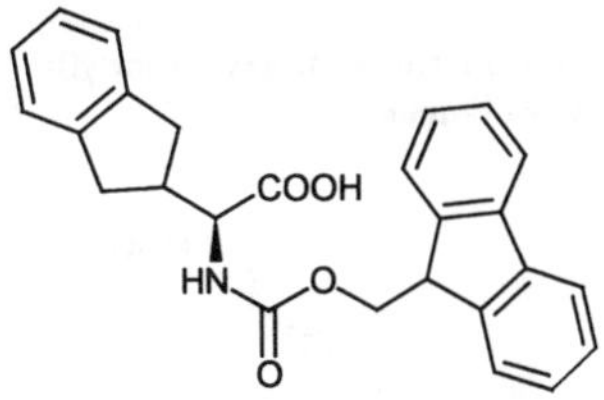

$C_{26}H_{23}NO_4$
Chiral intermediate. *Synthetech, Inc.*

**2757 (9-Fluorenylmethoxycarbonyl)-D-4-iodophenylalanine**
205526-29-0

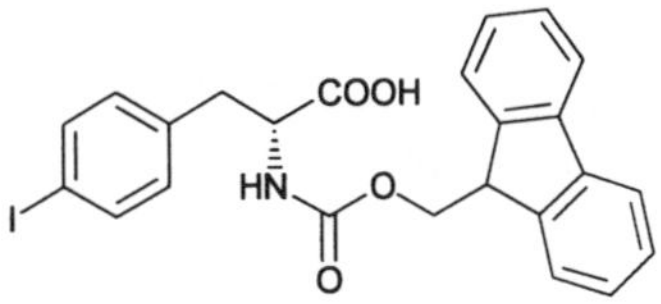

$C_{24}H_{20}INO_4$
Chiral intermediate. *Advanced Asymmetrics, Inc.; Synthetech, Inc.*

**2758 (9-Fluorenylmethoxycarbonyl)-L-4-iodo-phenylalanine**
82565-68-2

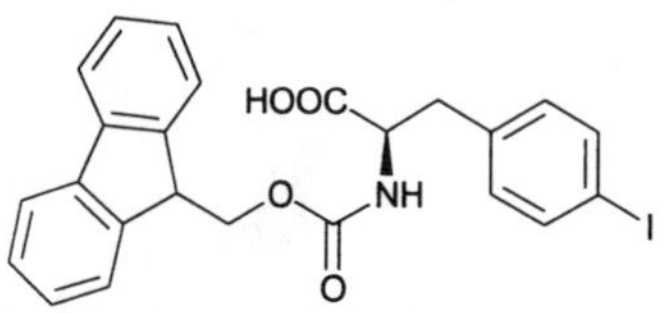

$C_{24}H_{20}INO_4$
Chiral intermediate. *Advanced Asymmetrics, Inc.; Synthetech, Inc.*

**2759 (9-Fluorenylmethoxycarbonyl)-L-isoleucine**
71989-23-6 276-255-9

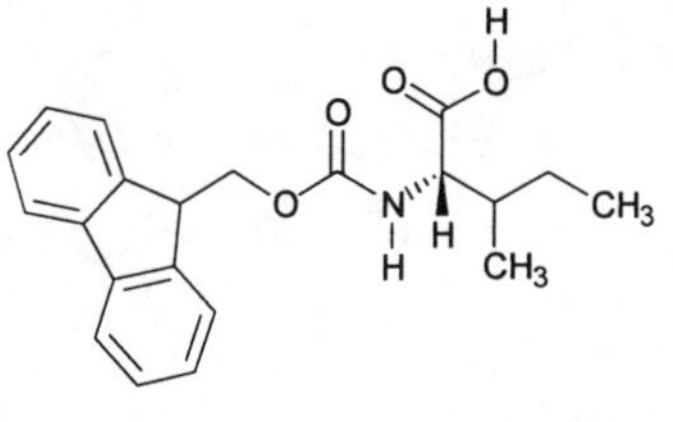

$C_{21}H_{23}NO_4$
Isoleucine, N-[(9H-fluoren-9-ylmethoxy)carbonyl]-, L-. Chiral intermediate. mp = 145-147°; $[\alpha]_D^{20}$ = - 12° (c = 1, DMF). *Acros Organics nv; Interchem Corporation; Sigma-Aldrich Fine Chemicals; Synthetech, Inc.*

**2760 (9-Fluorenylmethoxycarbonyl)-D-isoleucinol**

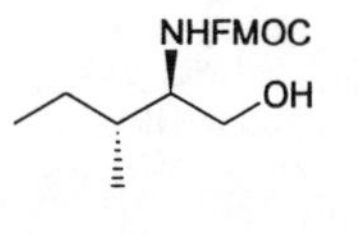

$C_{21}H_{25}NO_3$
Chiral intermediate. *Omega Chemical Company Inc.*

**2761 (9-Fluorenylmethoxycarbonyl)-L-isoleucinol**

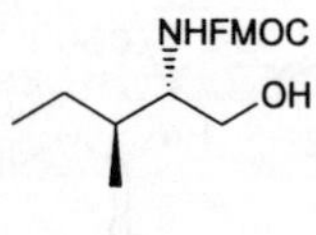

$C_{21}H_{25}NO_3$
Chiral intermediate. *Omega Chemical Company Inc.; Senn Chemicals AG.*

**2762 (9-Fluorenylmethoxycarbonyl)-L-leucine**
35661-60-0 252-662-7

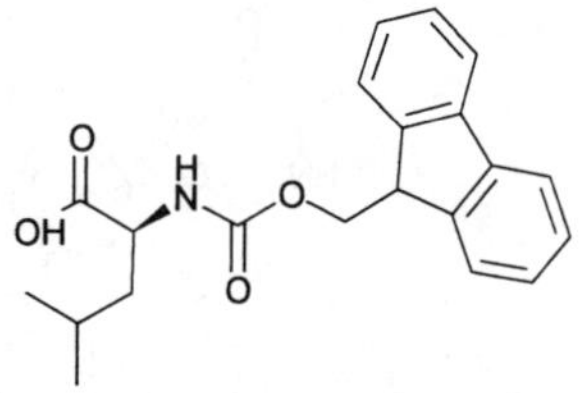

$C_{21}H_{23}NO_4$
Leucine, N-[(9H-fluoren-9-ylmethoxy)carbonyl]-, L-; NPC 15199. Chiral intermediate. mp = 156-158°; $[\alpha]^{24}$ = - 26° (c = 1, DMF). *Acros Organics nv; Interchem Corporation; Sigma-Aldrich Fine Chemicals; Synthetech, Inc.*

**2763 (9-Fluorenylmethoxycarbonyl)-L-tert-leucine**
132684-60-7
$C_{21}H_{23}NO_4$
N-(9-Fluorenylmethoxycarbonyl)-(S)-2-amino-3,3-dimethylbutyric acid; (S)-N-Fmoc-2-amino-3,3-dimethylbutyric acid. Chiral intermediate. *Lancaster Synthesis Ltd.; Senn Chemicals AG.*

**2764 (9-Fluorenylmethoxycarbonyl)-D-leucinol**

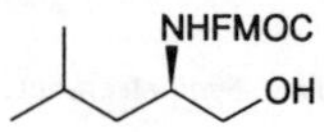

$C_{21}H_{25}NO_3$
Chiral intermediate. *Omega Chemical Company Inc.*

**2765 (9-Fluorenylmethoxycarbonyl)-L-leucinol**

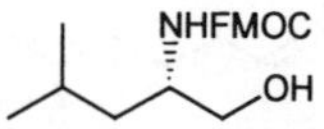

$C_{21}H_{25}NO_3$
Chiral intermediate. *Omega Chemical Company Inc.; Senn Chemicals AG.*

**2766 (9-Fluorenylmethoxycarbonyl)-L-methionine**
71989-28-1 276-258-5

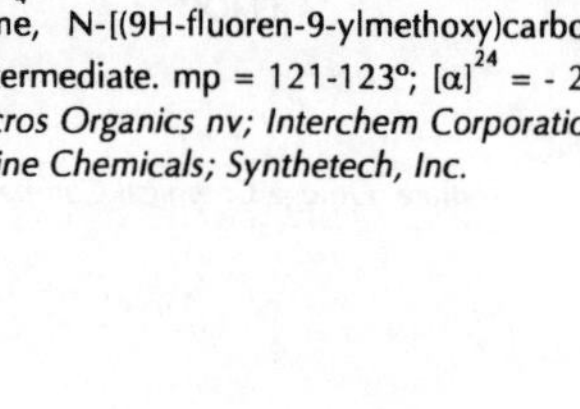

$C_{20}H_{21}NO_4S$
Methionine, N-[(9H-fluoren-9-ylmethoxy)carbonyl]-, L-. Chiral intermediate. mp = 121-123°; $[\alpha]^{24}$ = - 29° (c = 1, DMF). *Acros Organics nv; Interchem Corporation; Sigma-Aldrich Fine Chemicals; Synthetech, Inc.*

**2767 (9-Fluorenylmethoxycarbonyl)-L-methionine sulfone**

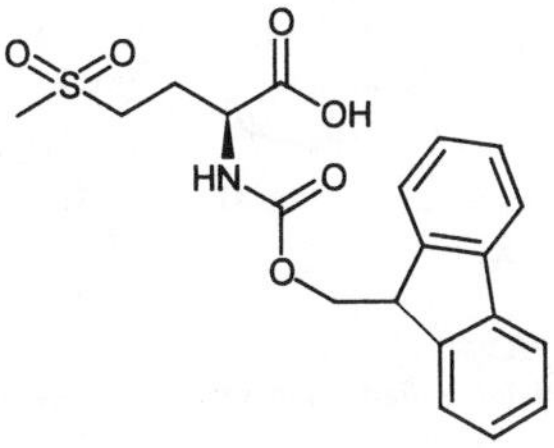

$C_{20}H_{21}NO_6S$
Chiral intermediate. *Senn Chemicals AG.*

**2768 (9-Fluorenylmethoxycarbonyl)-D-methioninol**

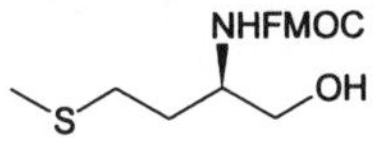

$C_{20}H_{23}NO_3S$
Chiral intermediate. *Omega Chemical Company Inc.*

**2769 (9-Fluorenylmethoxycarbonyl)-L-methioninol**

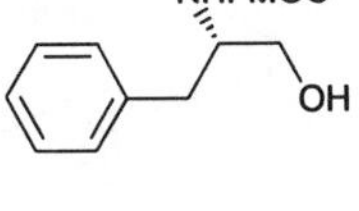

$C_{20}H_{23}NO_3S$
Chiral intermediate. *Omega Chemical Company Inc.; Senn Chemicals AG.*

**2770 (9-Fluorenylmethoxycarbonyl)-N-methyl-L-alanine**

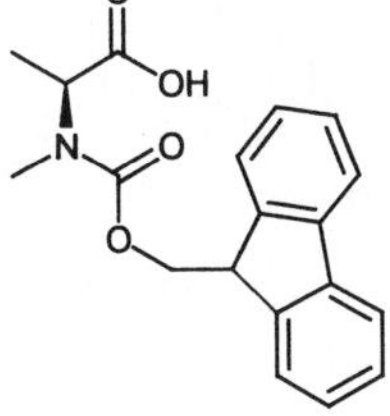

$C_{19}H_{19}NO_4$
Chiral intermediate. *Senn Chemicals AG.*

**2771 (9-Fluorenylmethoxycarbonyl)-O-methyl-L-homoserine**

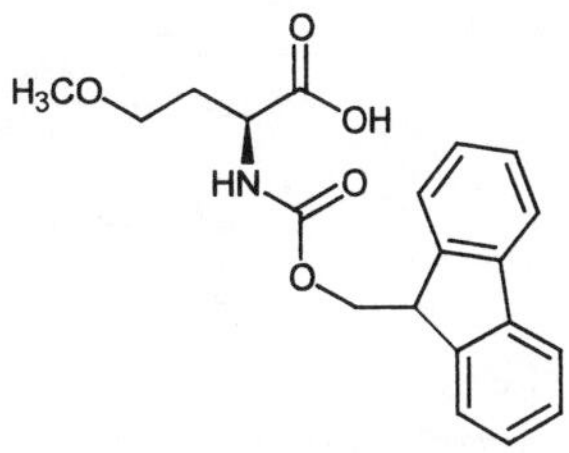

$C_{20}H_{21}NO_5$
Chiral intermediate. *Senn Chemicals AG.*

**2772 (9-Fluorenylmethoxycarbonyl)-N-methyl-L-isoleucine**

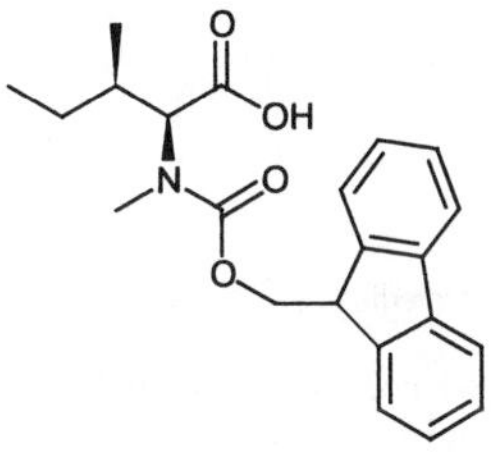

$C_{22}H_{25}NO_4$
Chiral intermediate. *Senn Chemicals AG.*

**2773 (9-Fluorenylmethoxycarbonyl)-N-methyl-L-leucine**

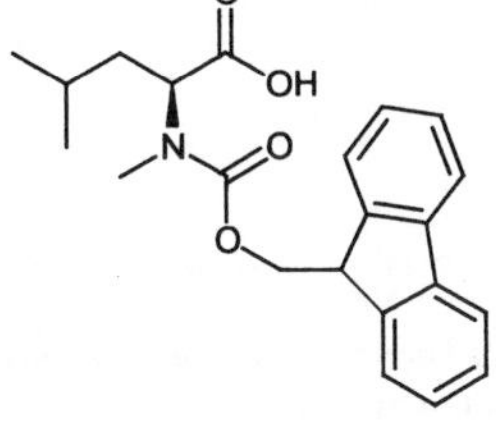

$C_{22}H_{25}NO_4$
Chiral intermediate. *Senn Chemicals AG.*

**2774 (R)-N-(9-Fluorenylmethoxycarbonyl)-(2-methylphenyl)alanine**

$C_{25}H_{23}NO_4$
Chiral intermediate. *Lancaster Synthesis Ltd.*

**2775 (S)-N-(9-Fluorenylmethoxycarbonyl)-(2-methylphenyl)alanine**

$C_{25}H_{23}NO_4$
Chiral intermediate. *Lancaster Synthesis Ltd.*

**2776** **(9-Fluorenylmethoxycarbonyl)-D-4-methylphenylalanine**
204260-38-8

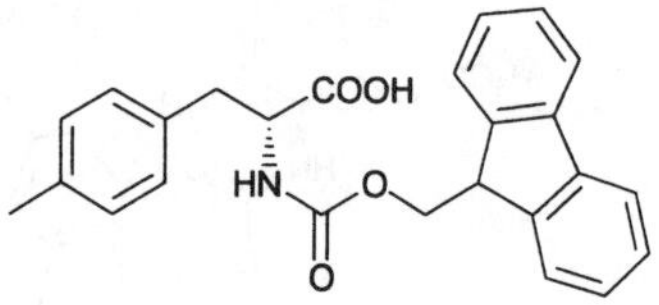

$C_{25}H_{23}NO_4$
Chiral intermediate. *Synthetech, Inc.*

**2777** **(9-Fluorenylmethoxycarbonyl)-L-4-methylphenylalanine**
199006-54-7

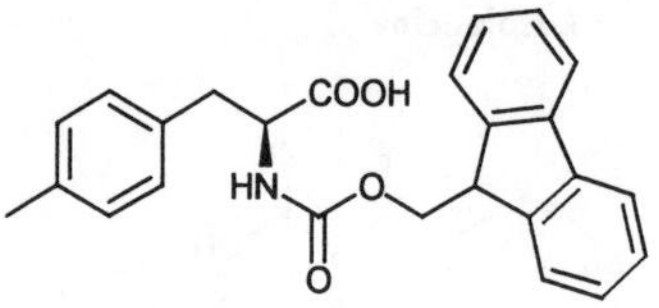

$C_{25}H_{23}NO_4$
Chiral intermediate. *Synthetech, Inc.*

**2778** **(9-Fluorenylmethoxycarbonyl)-N-methyl-D-phenylalanine**

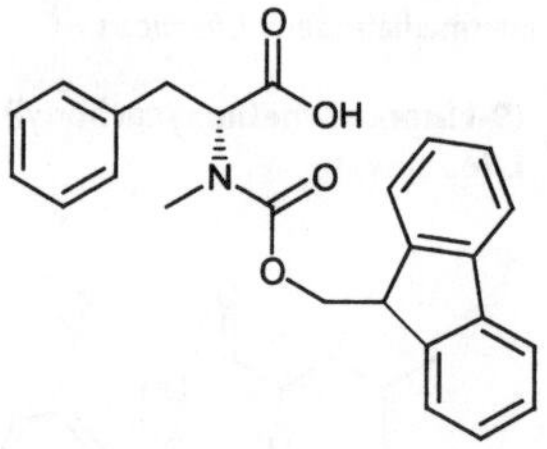

$C_{25}H_{23}NO_4$
Chiral intermediate. *Senn Chemicals AG.*

**2779** **(9-Fluorenylmethoxycarbonyl)-N-methyl-L-phenylalanine**

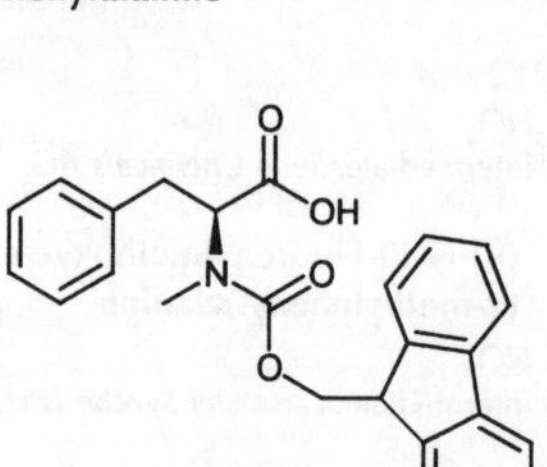

$C_{25}H_{23}NO_4$
Chiral intermediate. *Senn Chemicals AG.*

**2780** **(9-Fluorenylmethoxycarbonyl)-N-methyl-D-valine**

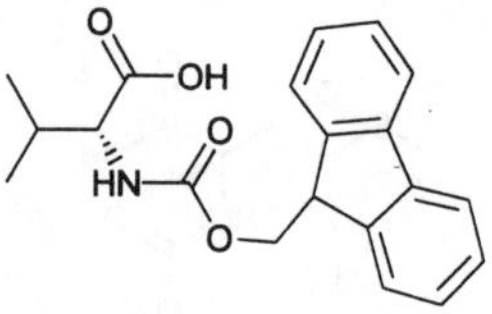

$C_{21}H_{23}NO_4$
Chiral intermediate. *Genzyme Pharmaceuticals.*

**2781** **(9-Fluorenylmethoxycarbonyl)-N-methyl-L-valine**

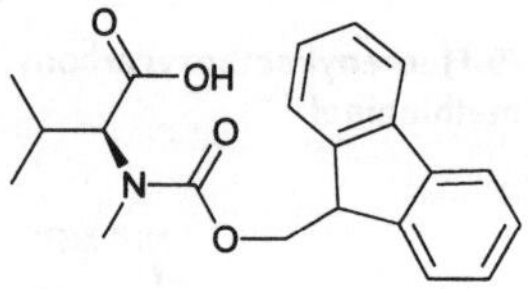

$C_{21}H_{23}NO_4$
Chiral intermediate. *Senn Chemicals AG.*

**2782** **(9-Fluorenylmethoxycarbonyl)-D-1-naphthylalanine**
138774-93-3

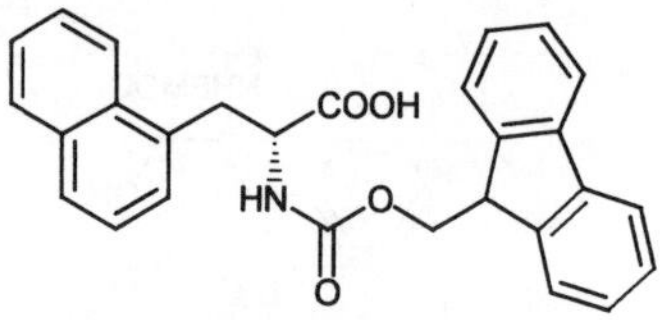

$C_{28}H_{23}NO_4$
Chiral intermediate. *Synthetech, Inc.*

**2783** **(9-Fluorenylmethoxycarbonyl)-L-1-naphthylalanine**
96402-49-2

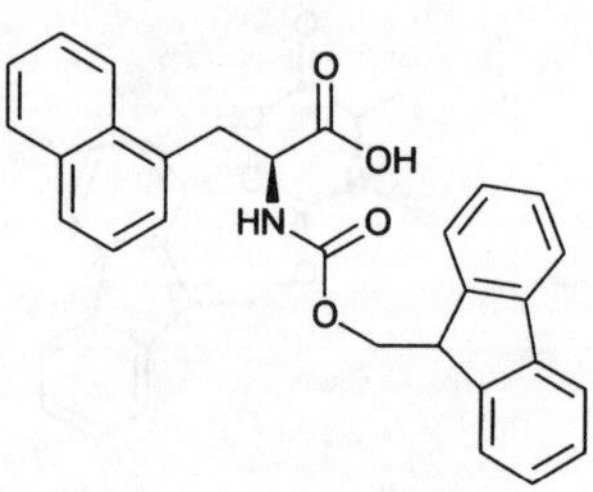

$C_{28}H_{23}NO_4$
Chiral intermediate. *Senn Chemicals AG; Synthetech, Inc.*

**2784 (R)-N-(9-Fluorenylmethoxycarbonyl)-(2-naphthyl)alanine**
138774-94-4
$C_{28}H_{23}NO_4$
Fmoc-D-2-naphthylalanine. Chiral intermediate. *Isochem; Lancaster Synthesis Ltd.; Synthetech, Inc.*

**2785 (S)-N-(9-Fluorenylmethoxycarbonyl)-(2-naphthyl)alanine**
112883-43-9
$C_{28}H_{23}NO_4$
Fmoc-L-2-naphthylalanine. Chiral intermediate. *Lancaster Synthesis Ltd.; Senn Chemicals AG; Synthetech, Inc.*

**2786 (9-Fluorenylmethoxycarbonyl)-D-4-nitrophenylalanine**
177966-63-1

COOH
HN
$O_2N$
O
O

$C_{24}H_{20}N_2O_4$
N-FMOC-p-Nitro-D-phenylalanine. Chiral intermediate. *Advanced Asymmetrics, Inc.; Synthetech, Inc.*

**2787 (9-Fluorenylmethoxycarbonyl)-L-4-nitrophenylalanine**
95753-55-2

COOH
HN
$O_2N$
O
O

$C_{24}H_{20}N_2O_4$
N-FMOC-p-Nitro-L-phenylalanine. Chiral intermediate. *Advanced Asymmetrics, Inc.; Senn Chemicals AG; Synthetech, Inc.*

**2788 (9-Fluorenylmethoxycarbonyl)-L-norleucine**
77284-32-3

O
OH
HN
O
O

$C_{21}H_{23}NO_4$
Chiral intermediate. mp = 142-144°; $[\alpha]_D^{20}$ = - 18.5° (c = 1, DMF). *Interchem Corporation; Sigma-Aldrich Fine Chemicals.*

**2789 (9-Fluorenylmethoxycarbonyl)-L-norvaline**

O
OH
HN
O
O

$C_{20}H_{21}NO_4$
Chiral intermediate. *Senn Chemicals AG.*

**2790 (R)-N-(9-Fluorenylmethoxycarbonyl)-octylglycine**
$C_{25}H_{31}NO_4$
N-(9-Fluorenylmethoxycarbonyl)-(R)-2-aminodecanoic acid. Chiral intermediate. *Lancaster Synthesis Ltd.*

**2791 (S)-N-(9-Fluorenylmethoxycarbonyl)-octylglycine**
193885-59-5
$C_{25}H_{31}NO_4$
N-(9-Fluorenylmethoxycarbonyl)-(S)-2-aminodecanoic acid. Chiral intermediate. *Lancaster Synthesis Ltd.*

**2792 (9-Fluorenylmethoxycarbonyl)-D-pentafluorophenylalanine**
198545-85-6

F
F
COOH
HN
F
F
O
F
O

$C_{24}H_{16}F_5NO_4$
Chiral intermediate. *Synthetech, Inc.*

**2793 (9-Fluorenylmethoxycarbonyl)-L-pentafluorophenylalanine**
205526-32-5

F
F
COOH
HN
F
F
O
F
O

$C_{24}H_{16}F_5NO_4$
Chiral intermediate. *Synthetech, Inc.*

**2794 (9-Fluorenylmethoxycarbonyl)-D-phenylalanine**

86123-10-6

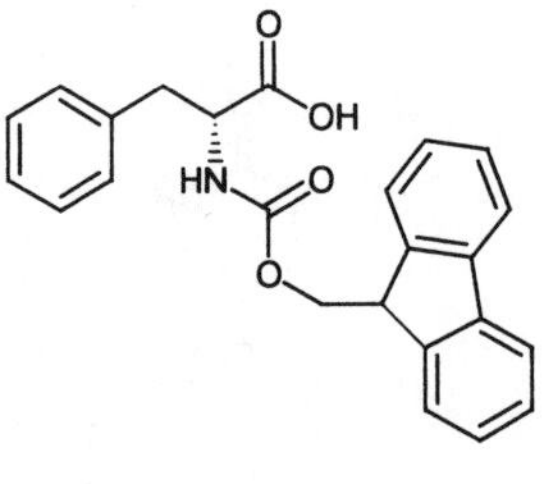

$C_{24}H_{21}NO_4$

Chiral intermediate. *Genzyme Pharmaceuticals; Isochem.*

**2795 (9-Fluorenylmethoxycarbonyl)-L-phenylalanine**

35661-40-6 252-661-1

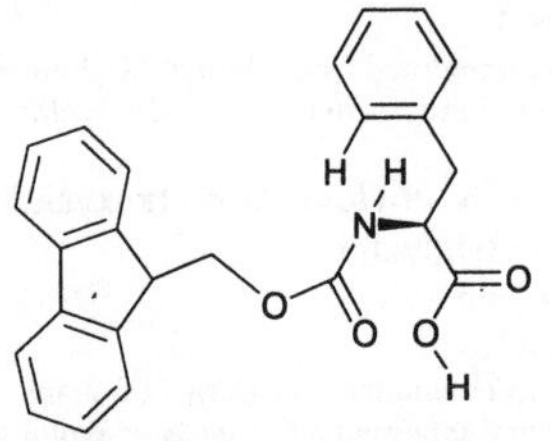

$C_{24}H_{21}NO_4$

Phenylalanine, N-[(9H-fluoren-9-ylmethoxy)carbonyl]-, L-.

Chiral intermediate. mp = 183-186°; $[\alpha]_D^{20}$ = - 37° (c = 1, DMF). *Acros Organics nv; Interchem Corporation; Sigma-Aldrich Fine Chemicals; Synthetech, Inc.*

**2796 (9-Fluorenylmethoxycarbonyl)-L-phenylalanine pentafluorophenyl ester**

86060-92-6

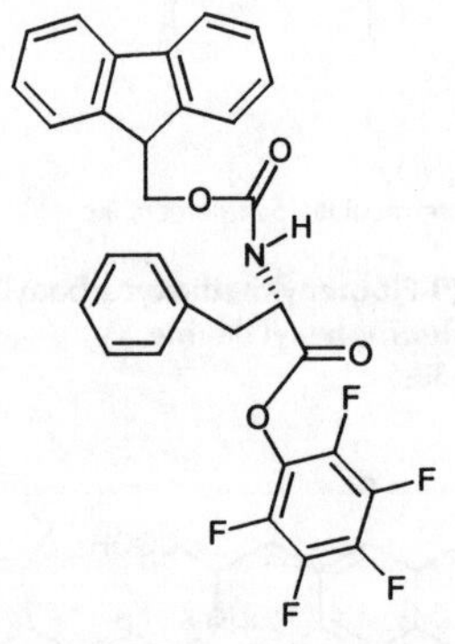

$C_{30}H_{20}F_5NO_4$

Chiral intermediate. mp = 149-151°; $[\alpha]_D^{20}$ = - 20° (c = 1, $CHCl_3$). *Acros Organics nv; Sigma-Aldrich Fine Chemicals.*

**2797 (9-Fluorenylmethoxycarbonyl)-D-phenylalaninol**

130406-30-3

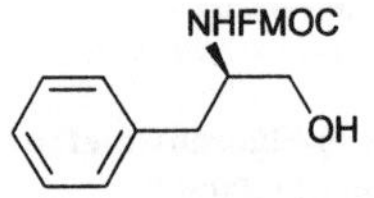

$C_{24}H_{23}NO_3$

Chiral intermediate. *Omega Chemical Company Inc.; SK Energy and Chemical, Inc.*

**2798 (9-Fluorenylmethoxycarbonyl)-L-phenylalaninol**

129397-83-7

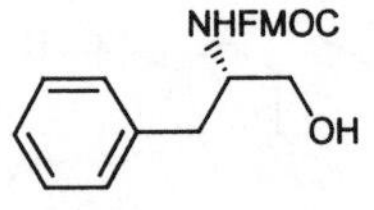

$C_{24}H_{23}NO_3$

Chiral intermediate. *Omega Chemical Company Inc.; Senn Chemicals AG; SK Energy and Chemical, Inc.*

**2799 (9-Fluorenylmethoxycarbonyl)-L-phenylglycine**

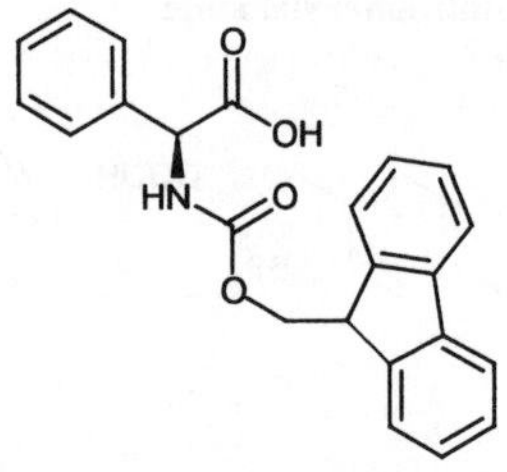

$C_{23}H_{19}NO_4$

Chiral intermediate. *Senn Chemicals AG.*

**2800 (9-Fluorenylmethoxycarbonyl)-(R)-2-phenylglycinol**

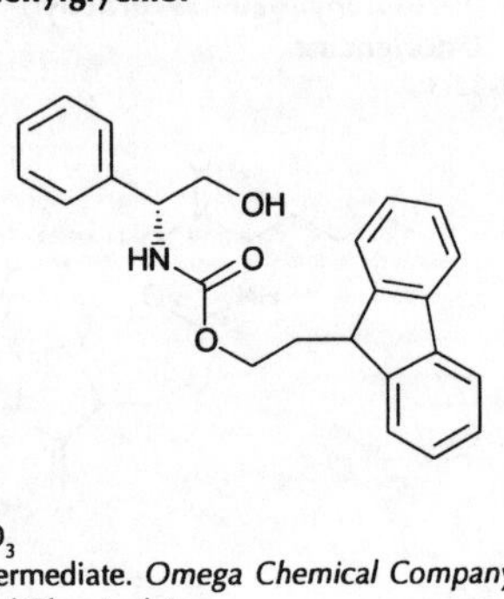

$C_{23}H_{21}NO_3$

Chiral intermediate. *Omega Chemical Company Inc.; SK Energy and Chemical, Inc.*

**2801 (9-Fluorenylmethoxycarbonyl)-(S)-2-phenylglycinol**

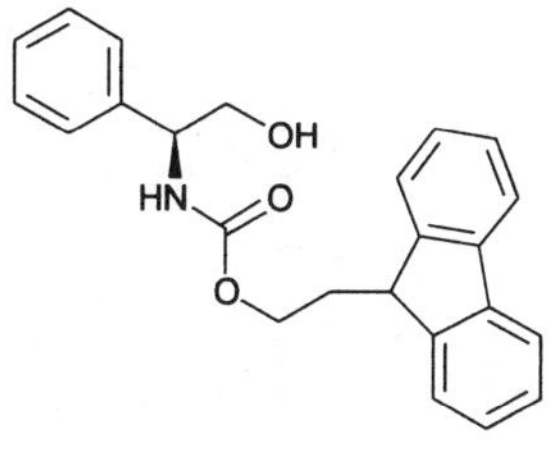

$C_{23}H_{21}NO_3$
Chiral intermediate. *Omega Chemical Company Inc.; SK Energy and Chemical, Inc.*

**2802 (-)-N-(9-Fluorenylmethoxycarbonyl)phenyl phenylalanine**

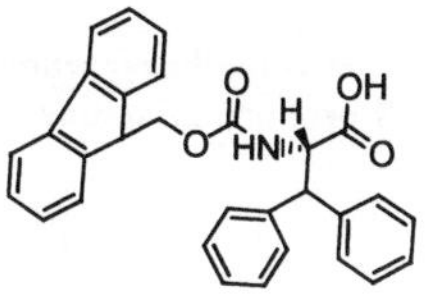

$C_{30}H_{25}NO_4$
Chiral intermediate. mp = 125-129°; $[\alpha]_D^{20}$ = - 12.0° (c = 1, $CHCl_3$). *Sigma-Aldrich Fine Chemicals.*

**2803 (+)-N-(9-Fluorenylmethoxycarbonyl)phenyl phenylalanine**

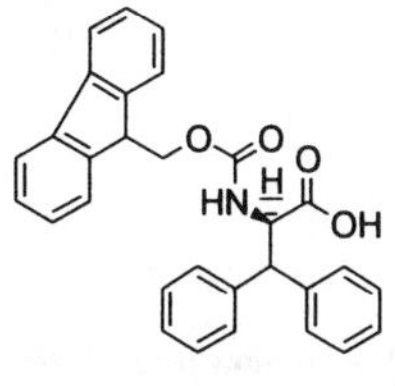

$C_{30}H_{25}NO_4$
Chiral intermediate. mp = 116.5-119.5°; $[\alpha]^{26}$ = + 12.0° (c = 1, $CHCl_3$). *Sigma-Aldrich Fine Chemicals.*

**2804 (-)-N-(9-Fluorenylmethoxycarbonyl)phenyl phenylalaninol**

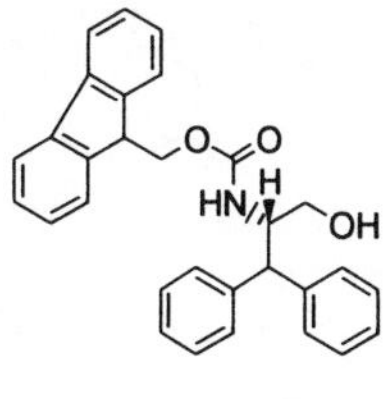

$C_{30}H_{27}NO_3$
Chiral intermediate. mp = 114-119°; $[\alpha]_D^{20}$ = - 7.0° (c = 1, $CHCl_3$). *Sigma-Aldrich Fine Chemicals.*

**2805 (+)-N-(9-Fluorenylmethoxycarbonyl)phenyl phenylalaninol**

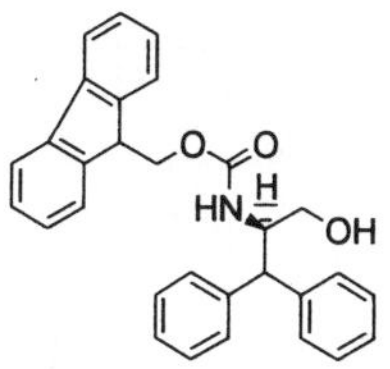

$C_{30}H_{27}NO_3$
Chiral intermediate. mp = 115-120°; $[\alpha]_D^{20}$ = + 8.0° (c = 1, $CHCl_3$). *Sigma-Aldrich Fine Chemicals.*

**2806 (9-Fluorenylmethoxycarbonyl)-D-pipecolinic acid**

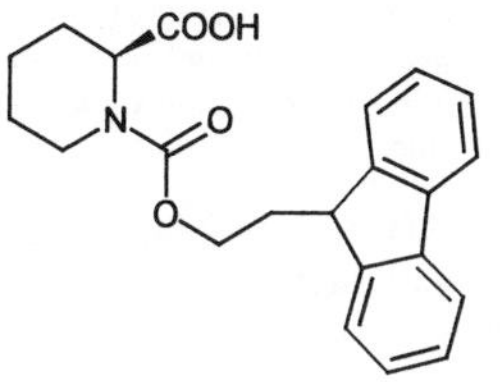

$C_{22}H_{23}NO_4$
Chiral intermediate. *Advanced Asymmetrics, Inc.*

**2807 (9-Fluorenylmethoxycarbonyl)-L-pipecolinic acid**

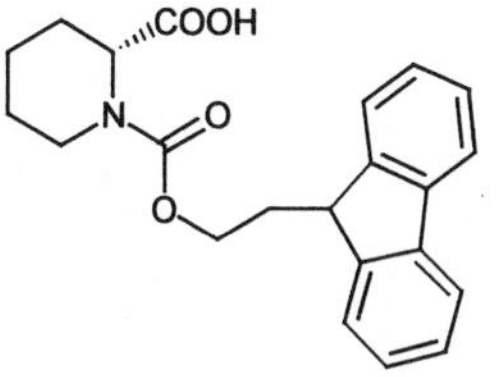

$C_{22}H_{23}NO_4$
Chiral intermediate. *Advanced Asymmetrics, Inc.*

**2808 (9-Fluorenylmethoxycarbonyl)-D-proline**
101555-62-8

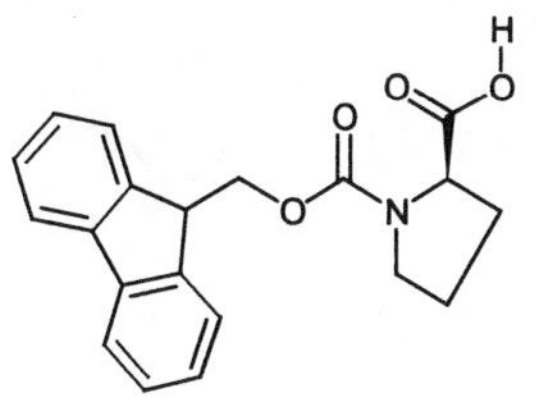

$C_{20}H_{19}NO_4$
Chiral intermediate. *Omega Chemical Company Inc.*

**2809 (9-Fluorenylmethoxycarbonyl)-L-proline**
71989-31-6 276-259-0

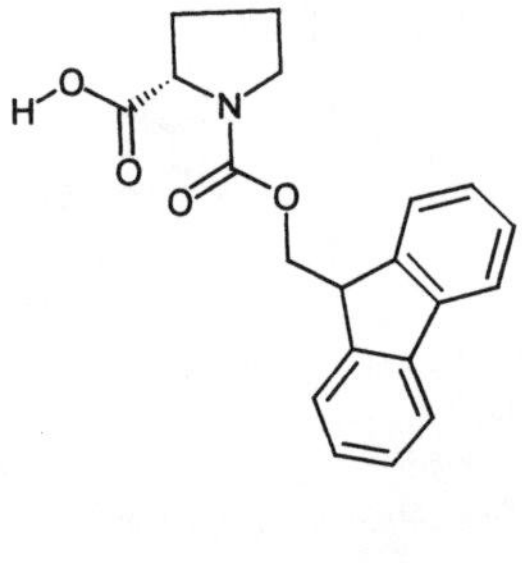

$C_{20}H_{19}NO_4$
Pyrrolidine-1,2-dicarboxylic acid, 1-(9H-fluoren-9-ylmethyl) ester, (2S)-; 1-(9H-Fluoren-9-ylmethyl) hydrogen (S)-pyrrolidine-1,2-dicarboxylate. Chiral intermediate. mp = 117-118°; $[\alpha]_D^{20}$ = - 32° (c = 1, DMF). *Acros Organics nv; Interchem Corporation; Omega Chemical Company Inc.; Sigma-Aldrich Fine Chemicals; Synthetech, Inc.*

**2810 N-(9-Fluorenylmethoxycarbonyl)proline pentaflurophenyl ester**
86060-90-4

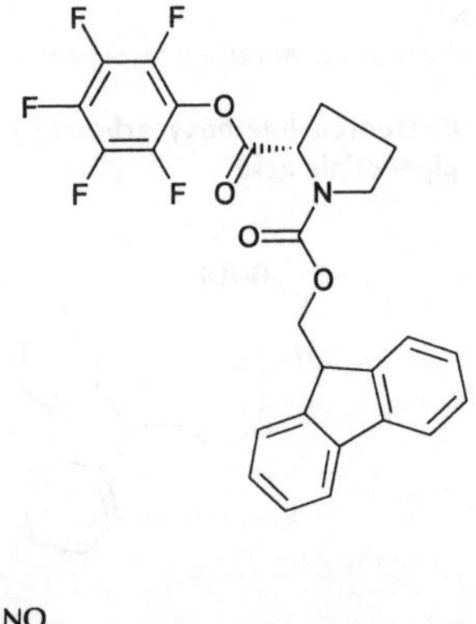

$C_{26}H_{18}F_5NO_4$
Chiral intermediate. mp = 127-129°; $[\alpha]^{25}$ = - 59° (c = 1, $CHCl_3$). *Sigma-Aldrich Fine Chemicals.*

**2811 (9-Fluorenylmethoxycarbonyl)-L-prolinol**
148625-77-8

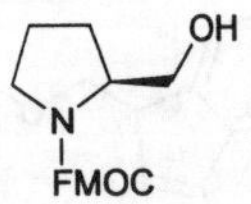

$C_{20}H_{21}NO_3$
Fmoc-(S)-2-(hydroxymethyl)pyrrolidine. Chiral intermediate. *Omega Chemical Company Inc.*

**2812 (9-Fluorenylmethoxycarbonyl)-(2R,3S)-(cis)-3-prolinomethionine**

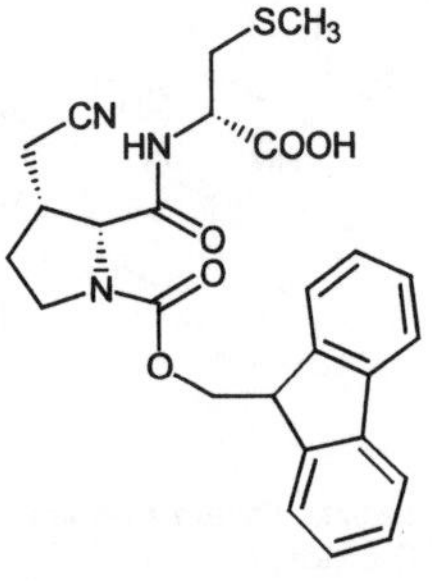

$C_{26}H_{27}N_3O_5S$
Chiral intermediate. *Senn Chemicals AG.*

**2813 (9-Fluorenylmethoxycarbonyl)-(2S,3R)-(cis)-3-prolinomethionine**

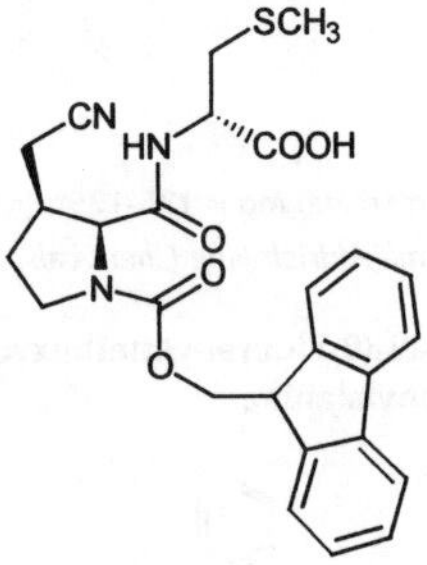

$C_{26}H_{27}N_3O_5S$
Chiral intermediate. *Senn Chemicals AG.*

**2814 (9-Fluorenylmethoxycarbonyl)-(2R,3S)-(cis)-3-prolinovaline**

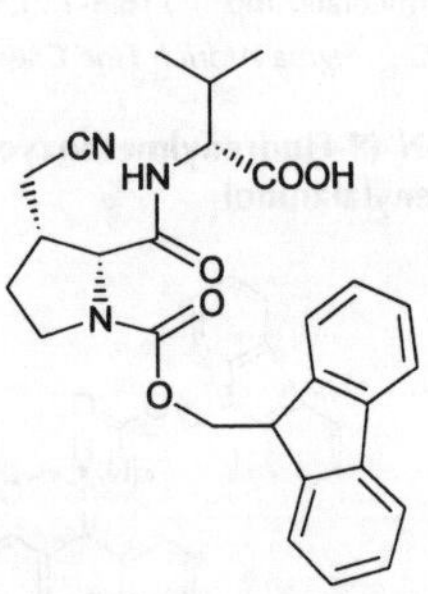

$C_{27}H_{29}N_3O_5$
Chiral intermediate. *Senn Chemicals AG.*

**2815 (9-Fluorenylmethoxycarbonyl)-(2S,3R)-(cis)-3-prolinovaline**

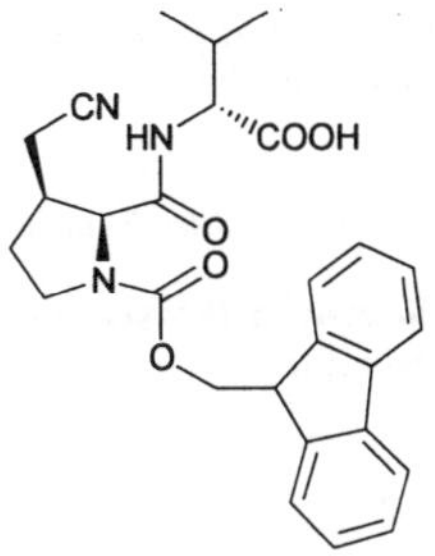

$C_{27}H_{29}N_3O_5$
Chiral intermediate. *Senn Chemicals AG.*

**2816 (R)-N-(9-Fluorenylmethoxycarbonyl)-propargylglycine**

$C_{20}H_{17}NO_4$
N-(9-Fluorenylmethoxycarbonyl)-(R)-2-amino-4-pentynoic acid. Chiral intermediate. *Lancaster Synthesis Ltd.*

**2817 (S)-N-(9-Fluorenylmethoxycarbonyl)-propargylglycine**

198561-07-8
$C_{20}H_{17}NO_4$
N-(9-Fluorenylmethoxycarbonyl)-(S)-2-amino-4-pentynoic acid. Chiral intermediate. *Lancaster Synthesis Ltd.*

**2818 (R)-N-(9-Fluorenylmethoxycarbonyl)-(2-pyridyl)alanine**

185379-39-9
$C_{23}H_{20}N_2O_4$
Fmoc-D-2-pyridylalanine. Chiral intermediate. *Lancaster Synthesis Ltd.; Synthetech, Inc.*

**2819 (S)-N-(9-Fluorenylmethoxycarbonyl)-(2-pyridyl)alanine**

185379-40-2
$C_{23}H_{20}N_2O_4$
Chiral intermediate. *Lancaster Synthesis Ltd.; Synthetech, Inc.*

**2820 (R)-N-(9-Fluorenylmethoxycarbonyl)-(3-pyridyl)alanine**

142994-45-4
$C_{23}H_{20}N_2O_4$
Fmoc-D-3-pyridylalanine. Chiral intermediate. *Lancaster Synthesis Ltd.; Synthetech, Inc.*

**2821 (S)-N-(9-Fluorenylmethoxycarbonyl)-(3-pyridyl)alanine**

175453-07-3
$C_{23}H_{20}N_2O_4$
Fmoc-L-3-pyridylalanine. Chiral intermediate. *Lancaster Synthesis Ltd.; Senn Chemicals AG; Synthetech, Inc.*

**2822 (9-Fluorenylmethoxycarbonyl)-D-4-pyridylalanine**

205528-30-9

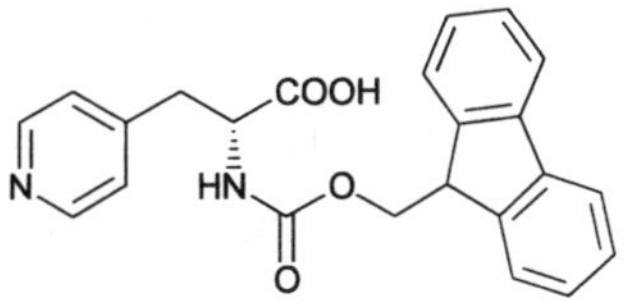

$C_{23}H_{20}N_2O_4$
Chiral intermediate. *Synthetech, Inc.*

**2823 (9-Fluorenylmethoxycarbonyl)-L-4-pyridylalanine**

169555-95-7

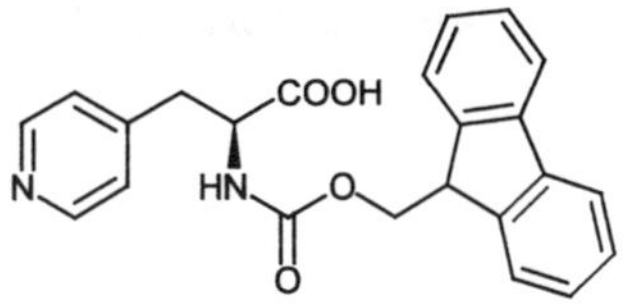

$C_{23}H_{20}N_2O_4$
Chiral intermediate. *Synthetech, Inc.*

**2824 (R)-N-(9-Fluorenylmethoxycarbonyl)-β-styrylalanine**

$C_{26}H_{23}NO_4$
Chiral intermediate. *Lancaster Synthesis Ltd.*

**2825 (S)-N-(9-Fluorenylmethoxycarbonyl)-β-styrylalanine**

159610-82-9
$C_{26}H_{23}NO_4$
N-(9-Fluorenylmethoxycarbonyl)-(S)-2-amino-5-phenyl-4-pentenoic acid. Chiral intermediate. *Lancaster Synthesis Ltd.*

**2826 (9-Fluorenylmethoxycarbonyl)-D-tetrahydroisoquinoline-3-carboxylic acid**

130309-33-0

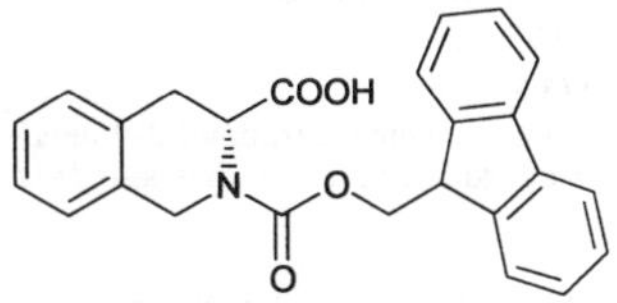

$C_{25}H_{21}NO_4$
Chiral intermediate. *Acros Organics nv; Synthetech, Inc.*

**2827 (9-Fluorenylmethoxycarbonyl)-L-tetrahydroisoquinoline-3-carboxylic acid**
136030-33-6

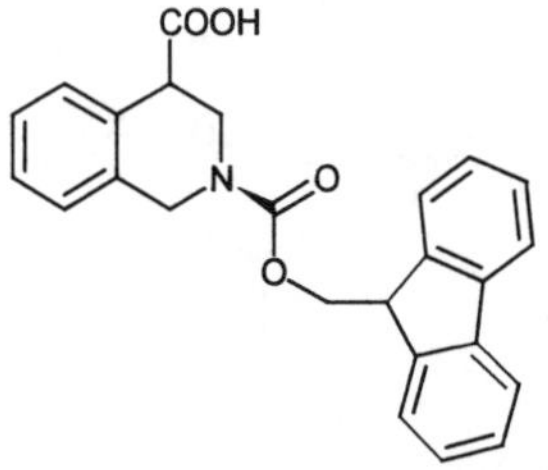

$C_{25}H_{21}NO_4$
Fmoc (3S)-1,2,3,4-tetrahydroisoquinoline-3-carboxylic acid. Chiral intermediate. *Acros Organics nv; Senn Chemicals AG; Synthetech, Inc.*

**2828 (9-Fluorenylmethoxycarbonyl)-L-thiazolidine-4-carboxylic acid**

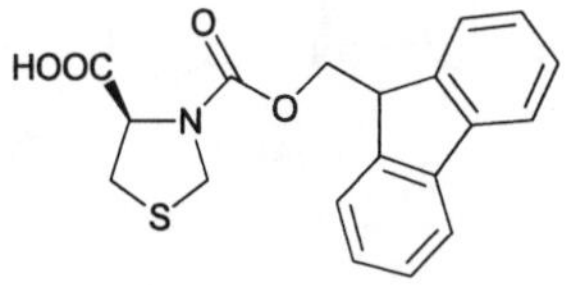

$C_{19}H_{17}NO_4S$
Chiral intermediate. *Senn Chemicals AG.*

**2829 (R)-N-(9-Fluorenylmethoxycarbonyl)-(4-thiazolyl)alanine**
205528-33-2
$C_{21}H_{18}N_2O_4S$
Fmoc-D-4-Thiazolylalanine. Chiral intermediate. *Lancaster Synthesis Ltd.; Synthetech, Inc.*

**2830 (S)-N-(9-Fluorenylmethoxycarbonyl)-(4-thiazolyl)alanine**
205528-32-1
$C_{21}H_{18}N_2O_4S$
Fmoc-L-4-Thiazolylalanine. Chiral intermediate. *Lancaster Synthesis Ltd.; Synthetech, Inc.*

**2831 (R)-N-(9-Fluorenylmethoxycarbonyl)-(2-thienyl)alanine**
201532-42-5
$C_{22}H_{19}NO_4S$
N-(9-Fluorenylmethoxycarbonyl)-2-thienyl-D-alanine. Chiral intermediate. *Lancaster Synthesis Ltd.; Synthetech, Inc.*

**2832 (S)-N-(9-Fluorenylmethoxycarbonyl)-(2-thienyl)alanine**
130309-35-2
$C_{22}H_{19}NO_4S$
Fmoc-L-2-Thienylalanine. Chiral intermediate. *Lancaster Synthesis Ltd.; Senn Chemicals AG; Synthetech, Inc.*

**2833 (R)-N-(9-Fluorenylmethoxycarbonyl)-(3-thienyl)alanine**
$C_{22}H_{19}NO_4S$
Chiral intermediate. *Lancaster Synthesis Ltd.*

**2834 (S)-N-(9-Fluorenylmethoxycarbonyl)-(3-thienyl)alanine**
186320-06-9
$C_{22}H_{19}NO_4S$
Chiral intermediate. *Lancaster Synthesis Ltd.*

**2835 (9-Fluorenylmethoxycarbonyl)-L-threonine t-butyl ester**

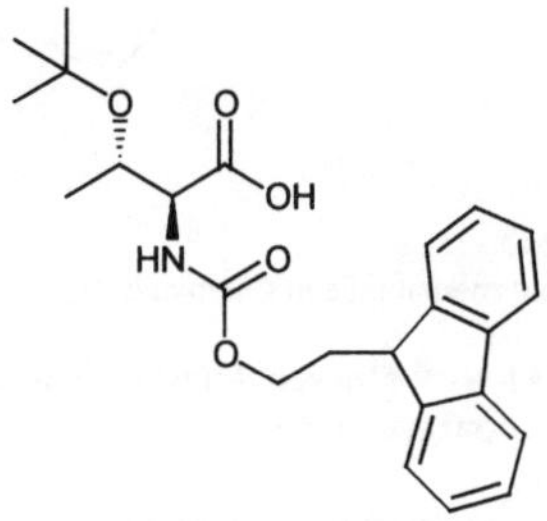

$C_{25}H_{29}NO_5$
Chiral intermediate. *Interchem Corporation.*

**2836 (9-Fluorenylmethoxycarbonyl)-L-threoninol**

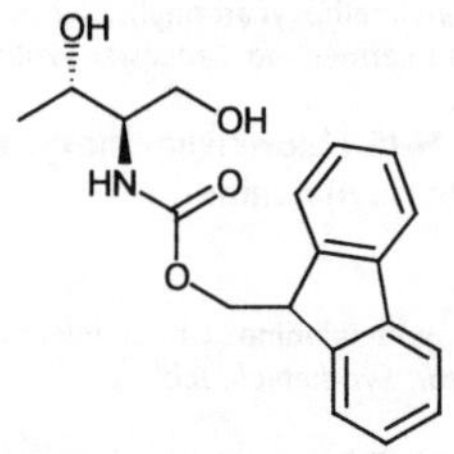

$C_{19}H_{21}NO_4$
Chiral intermediate. *Senn Chemicals AG.*

**2837 (9-Fluorenylmethoxycarbonyl)-D-3-trifluoromethylphenylalanine**
205526-28-9

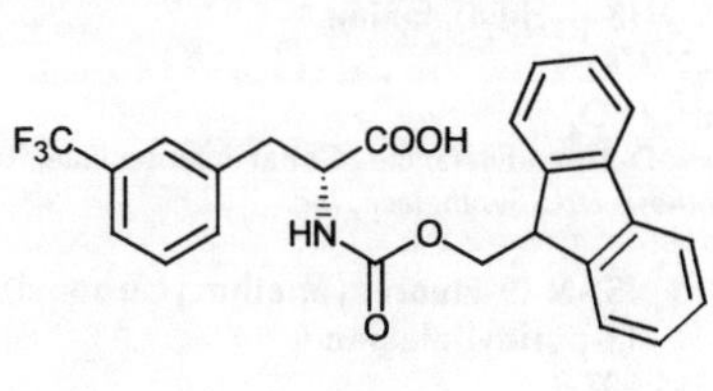

$C_{25}H_{20}F_3NO_4$
Chiral intermediate. *Synthetech, Inc.*

**2838 (9-Fluorenylmethoxycarbonyl)-L-3-trifluoromethylphenylalanine**

205526-27-8

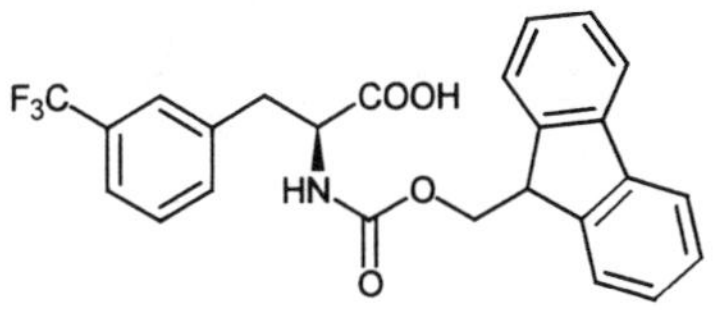

$C_{25}H_{20}F_3NO_4$
Chiral intermediate. *Synthetech, Inc.*

**2839 (9-Fluorenylmethoxycarbonyl)-D-4-trifluoromethylphenylalanine**

238742-88-6

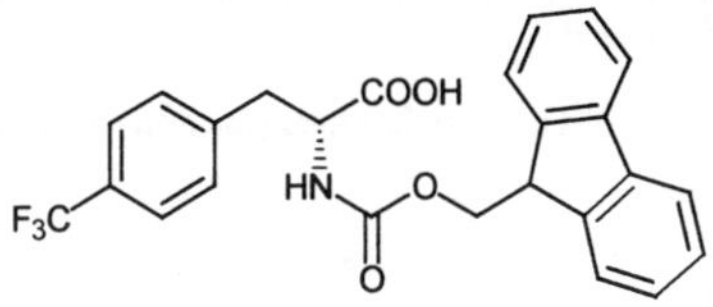

$C_{25}H_{20}F_3NO_4$
Chiral intermediate. *Synthetech, Inc.*

**2840 (9-Fluorenylmethoxycarbonyl)-L-4-trifluoromethylphenylalanine**

247113-86-6

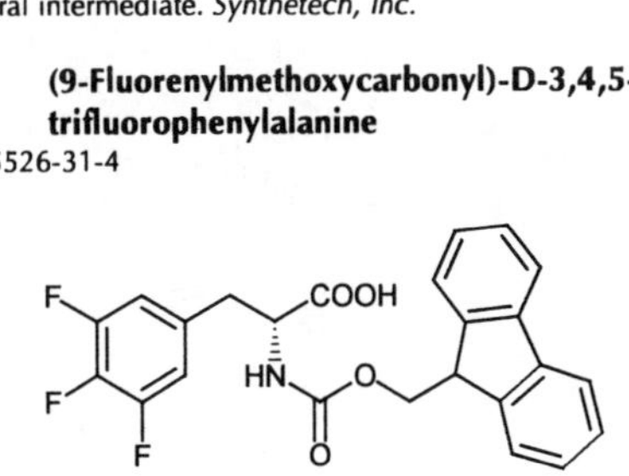

$C_{25}H_{20}F_3NO_4$
Chiral intermediate. *Synthetech, Inc.*

**2841 (9-Fluorenylmethoxycarbonyl)-D-3,4,5-trifluorophenylalanine**

205526-31-4

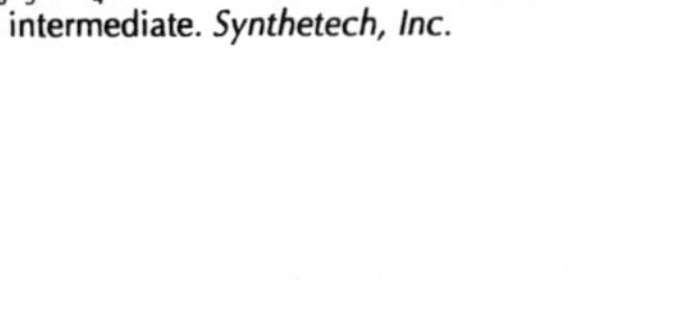

$C_{24}H_{18}F_3NO_4$
Chiral intermediate. *Synthetech, Inc.*

**2842 (9-Fluorenylmethoxycarbonyl)-L-3,4,5-trifluorophenylalanine**

205526-30-3

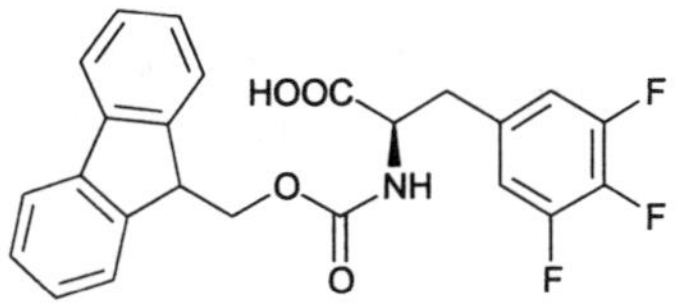

$C_{24}H_{18}F_3NO_4$
Chiral intermediate. *Synthetech, Inc.*

**2843 (9-Fluorenylmethoxycarbonyl)-D-tryptophan**

86123-11-7

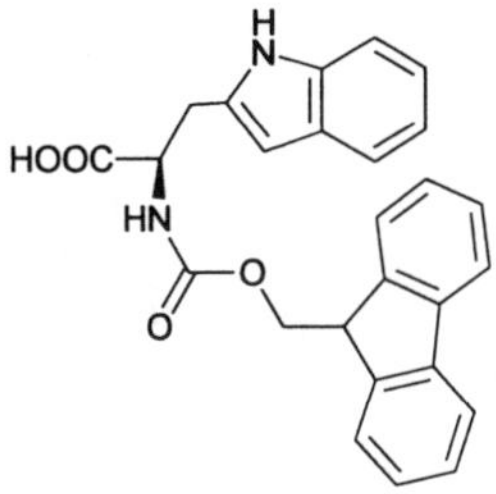

$C_{26}H_{22}N_2O_4$
Chiral intermediate. mp = 182-185°; $[\alpha]_D^{20}$ = + 28° (c = 1, DMF). *Isochem; Sigma-Aldrich Fine Chemicals; Synthetech, Inc.*

**2844 N-α-(9-Fluorenylmethoxycarbonyl)-L-tryptophan**

35737-15-6 252-706-5

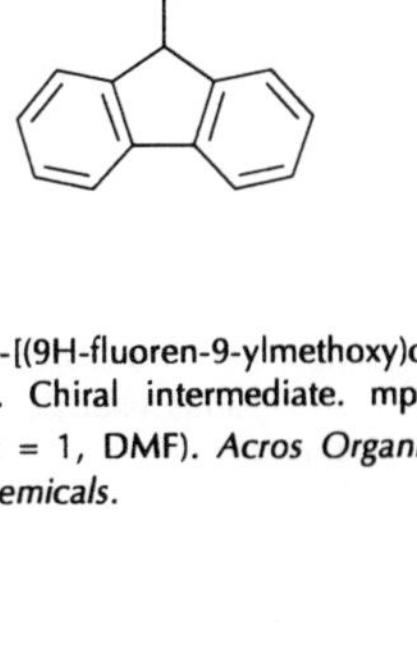

$C_{26}H_{22}N_2O_4$
Tryptophan, N-[(9H-fluoren-9-ylmethoxy)carbonyl]-, L-. FMOC-Trp-OH. Chiral intermediate. mp = 182-185°; $[\alpha]_D^{20}$ = - 28° (c = 1, DMF). *Acros Organics nv; Sigma-Aldrich Fine Chemicals.*

**2845 (9-Fluorenylmethoxycarbonyl)-O-trityl-L-homoserine**

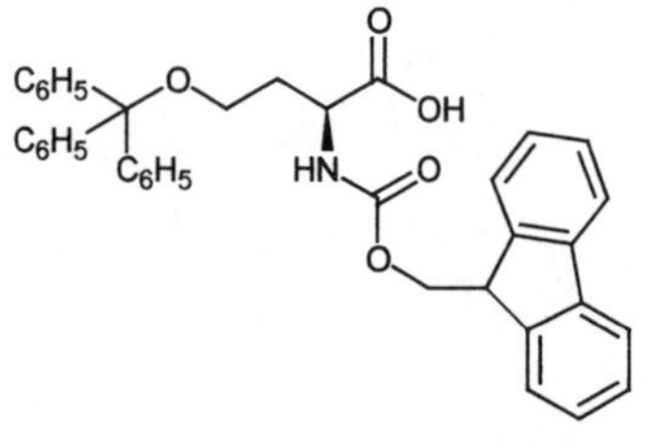

$C_{38}H_{33}NO_5$
Chiral intermediate. *Senn Chemicals AG.*

**2846 (9-Fluorenylmethoxycarbonyl)-L-tyrosine t-butyl ester**
71989-38-3 276-262-7

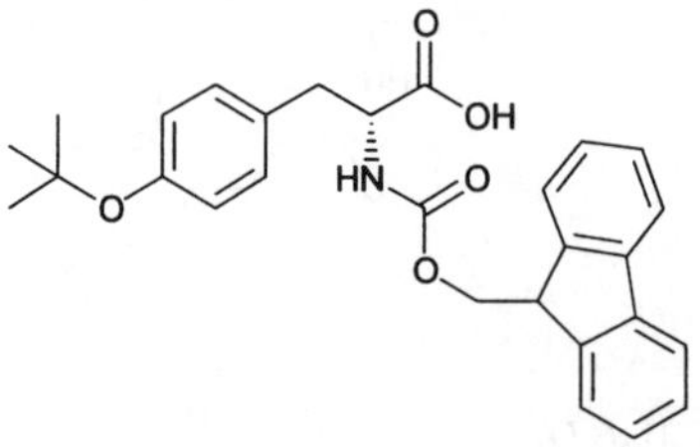

$C_{28}H_{29}NO_5$
Tyrosine, O-(1,1-dimethylethyl)-N-[(9H-fluoren-9-ylmethoxy)carbonyl]-, L-.; O-(tert-butyl)-N-[(9H-fluoren-9-ylmethoxy)carbonyl]-L-tyrosine. Chiral intermediate. mp = 153-156°; $[\alpha]_D^{20}$ = - 280° (c = 1, DMF). *Interchem Corporation; Sigma-Aldrich Fine Chemicals; Synthetech, Inc.*

**2847 (9-Fluorenylmethoxycarbonyl)-D-valine**
84624-17-9

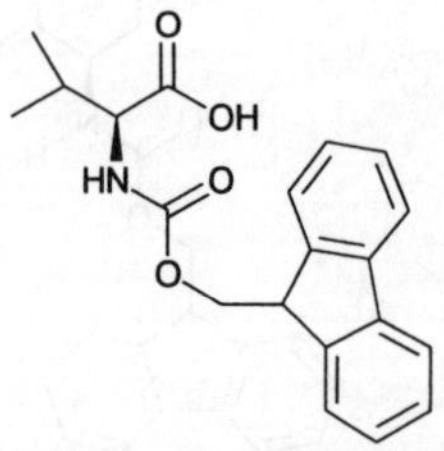

$C_{20}H_{21}NO_4$
Chiral intermediate. mp = 143-144°; $[\alpha]_D^{20}$ = + 17° (c = 1, DMF). *Genzyme Pharmaceuticals; Sigma-Aldrich Fine Chemicals; Synthetech, Inc.*

**2848 (9-Fluorenylmethoxycarbonyl)-L-valine**
68858-20-8 272-515-0

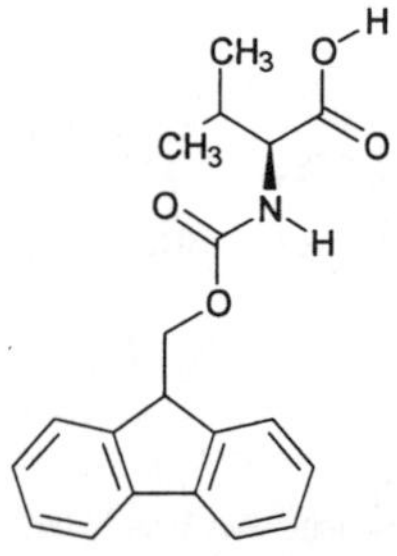

$C_{20}H_{21}NO_4$
Valine, N-[(9H-fluoren-9-ylmethoxy)carbonyl]-, L-. Chiral intermediate. mp = 143-145°; $[\alpha]_D^{20}$ = - 17° (c = 1, DMF). *Acros Organics nv; Sigma-Aldrich Fine Chemicals; Synthetech, Inc.*

**2849 (9-Fluorenylmethoxycarbonyl)-(R)-valinol**

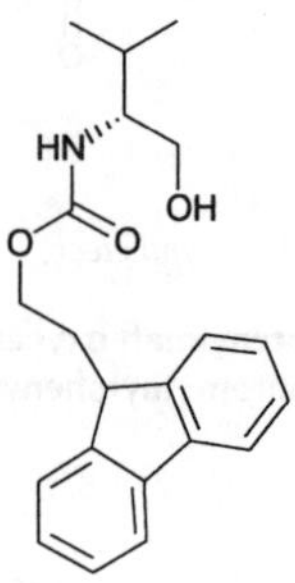

$C_{20}H_{23}NO_3$
Chiral intermediate. *Omega Chemical Company Inc.; SK Energy and Chemical, Inc.*

**2850 (9-Fluorenylmethoxycarbonyl)-(S)-valinol**

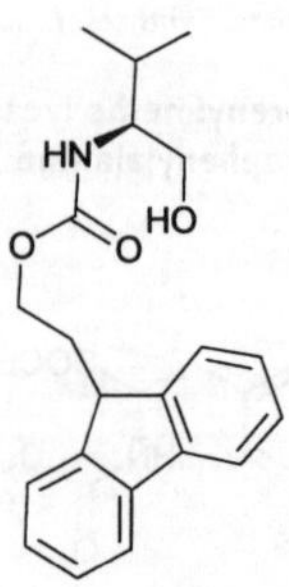

$C_{21}H_{25}NO_3$
Chiral intermediate. *SK Energy and Chemical, Inc.*

**2851 (9-Fluorenylmethoxycarbonyl)-L-valine pentafluorophenyl ester**
86060-87-9

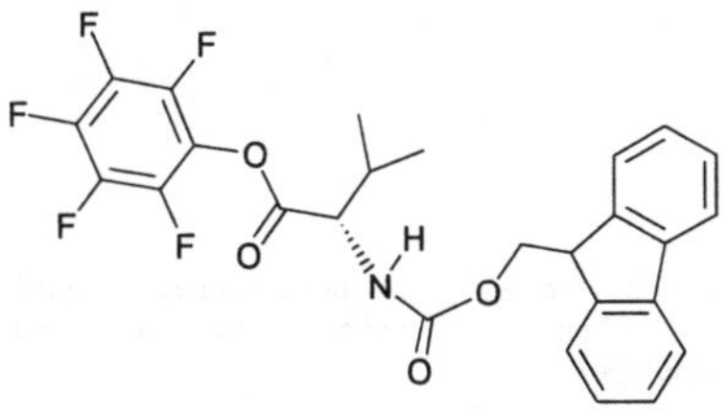

$C_{26}H_{20}F_5NO_4$
Chiral intermediate. mp = 121-122°; $[\alpha]_D^{20}$ = - 22° (c = 1, $CHCl_3$). *Acros Organics nv; Sigma-Aldrich Fine Chemicals.*

**2852 (9-Fluorenylmethoxycarbonyl)-L-valinol**
160885-98-3

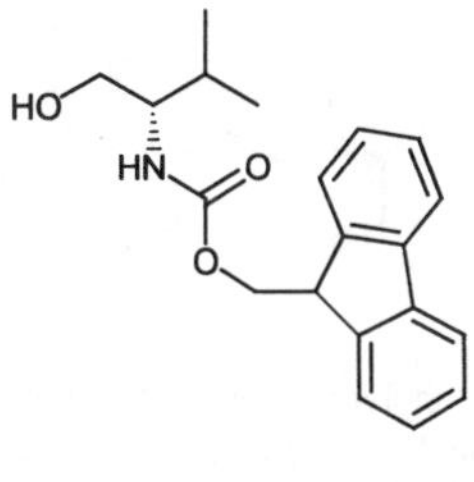

$C_{20}H_{23}NO_3$
Chiral intermediate. mp = 129-133°; $[\alpha]_D^{20}$ = - 20° (c = 1, $CHCl_3$). *Omega Chemical Company Inc.; Senn Chemicals AG; Sigma-Aldrich Fine Chemicals.*

**2853 (+)-3-Fluoro-3-deoxy-glucose**
14049-03-7

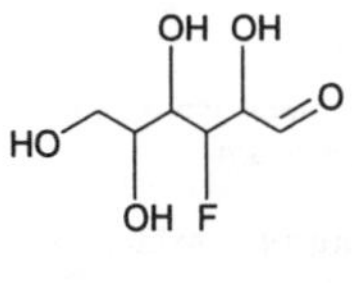

$C_6H_{11}FO_5$
3-Deoxy-3-fluoro-D-glucose. Chiral intermediate. mp = 115-117°; $[\alpha]^{25}$ = + 61° (c = 0.7, $H_2O$). *Sigma-Aldrich Fine Chemicals.*

**2854 (+)-5-Fluoro-5'-deoxyuridine**
3094-09-5
$C_9H_{11}FN_2O_5$
Chiral intermediate. *Sigma-Aldrich Fine Chemicals.*

**2855 (+)-5-Fluorodeoxyuridine**
50-91-9 4148(12) 200-072-5

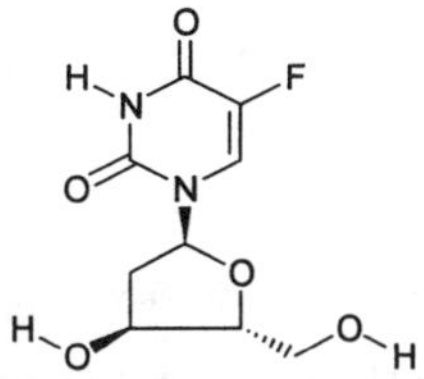

$C_9H_{11}FN_2O_5$
2'-Deoxy-5-fluorouridine; FUDR. Chiral intermediate. mp = 148°; $[\alpha]^{22}$ = + 37.2° (c = 1, $H_2O$). *Sigma-Aldrich Fine Chemicals.*

**2856 2-Fluoro-α-D-glucosylfluoride**
75414-45-8

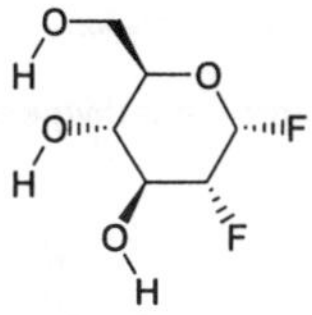

$C_6H_{10}F_2O_4$
Chiral intermediate. [α] = + 87.9° (c = 2, $CH_3OH$). *Acros Organics nv.*

**2857 (S)-(+)-2-Fluoromethyl-4-biphenyl-acetic acid**
4234(12) 257-263-1

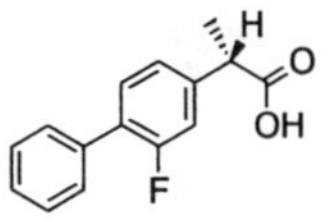

$C_{15}H_{13}FO_2$
(S)-(+)-Flurbiprofen. Chiral intermediate. mp = 109-110°; $[\alpha]^{24}$ = + 43° (c = 1, $CHCl_3$). *Sigma-Aldrich Fine Chemicals.*

**2858 (R)-4-(4-Fluorophenoxy)methyl butyrolactone**

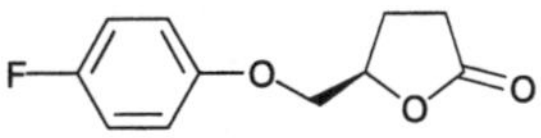

$C_{11}H_{11}FO_3$
Furan-2(3H)-one, 5-[(4-fluorophenoxy)methyl]dihydro-, (R)-. Chiral intermediate. *Synthon Chiragenics Corporation.*

**2859 (S)-2-Fluoro-2-methylheptanoic acid**

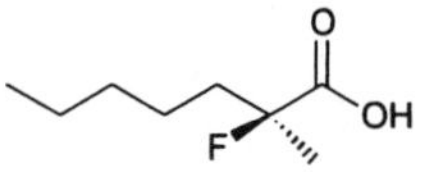

$C_8H_{15}FO_2$
Chiral building block. *Japan Energy Corporation.*

**2860 (S)-2-Fluoro-2-methylhexanoic acid**

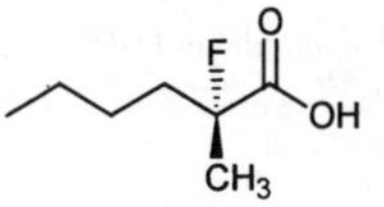

$C_7H_{13}FO_2$
Hexanoic acid, 2-fluoro-2-methyl-, (S)-. Chiral building block. *Japan Energy Corporation.*

**2861 (S)-2-Fluoro-2-methylpentanoic acid**

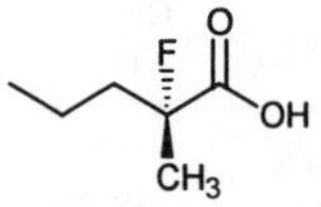

$C_6H_{11}FO_2$
Chiral building block. *Japan Energy Corporation.*

**2862 (+)-Fluoro-2-octanol**
110270-42-3

$C_8H_{17}FO$
Chiral building block. $bp_{22}$ = 96-97°; d = 1.254; $[\alpha]^{25}$ = + 10° (neat). *Japan Energy Corporation; Sigma-Aldrich Fine Chemicals.*

**2863 (S)-4-Fluoro-3-phenoxybenzladehyde cyanohydrin**

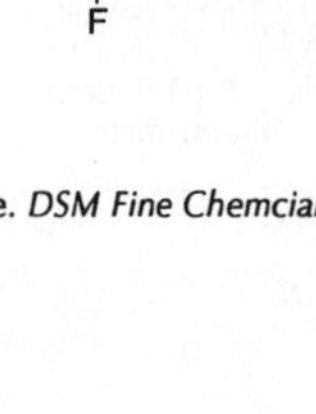

$C_{14}H_{10}FNO_2$
Chiral intermediate. *DSM Fine Chemcials Netherlands.*

**2864 (S)-4-(4-Flurophenoxy)methyl butyrolactone**
175212-40-5

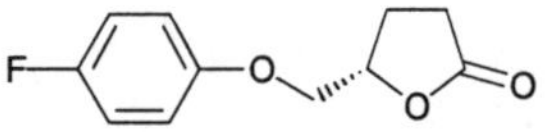

$C_{11}H_{11}FO_3$
Furan-2(3H)-one, 5-[(4-fluorophenoxy)methyl]dihydro-, (S)-. Chiral intermediate. *Synthon Chiragenics Corporation.*

**2865 (R)-1-(4-Fluorophenoxy)-2-propanol**
$C_9H_{11}FO_2$
Chiral building block. *Sigma-Aldrich Fine Chemicals.*

**2866 (S)-1-(4-Fluorophenoxy)-2-propanol**
206125-75-9
$C_9H_{11}FO_2$
Chiral building block. *Sigma-Aldrich Fine Chemicals.*

**2867 L-2-Fluorophenylalanine**

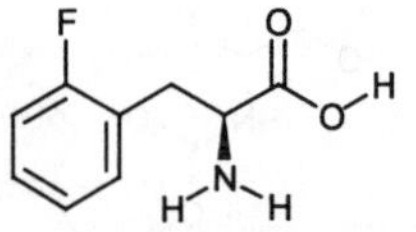

$C_9H_{10}FNO_2$
Alanine, 3-(o-fluorophenyl)-, L-; L-Phenylalanine, 2-fluoro-. Chiral building block. [α] = - 13.9° (c = 1, $H_2O$). *Acros Organics nv.*

**2868 D-4-Fluorophenylalanine**
18125-46-7

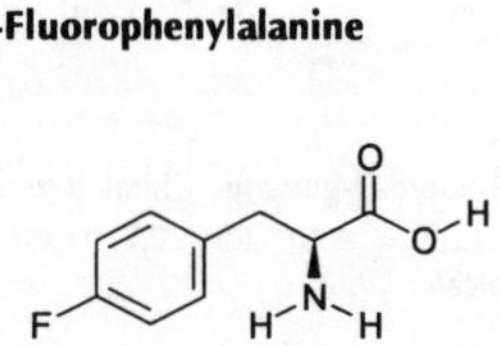

$C_9H_{10}FNO_2$
Chiral building block. mp = 240°; [α] = + 3.1° (c = 2,1N HCl). *Acros Organics nv.*

**2869 L-4-Fluorophenylalanine**
1132-68-9

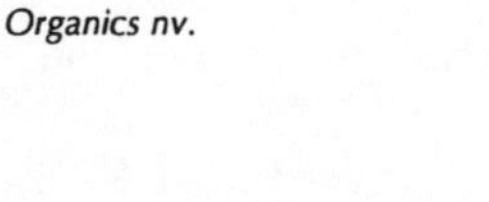

$C_9H_{10}FNO_2$
Chiral building block. mp = 253°; [α] = - 26.9° (c = 0.03, $H_2O$). *Acros Organics nv.*

**2870 D-4-Fluorphenylalanine hydrochloride**
122839-52-5

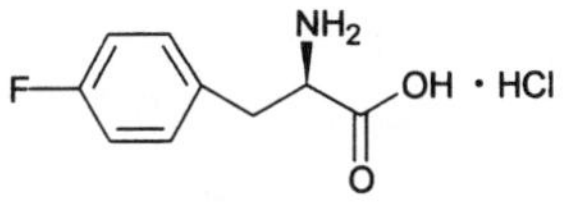

$C_9H_{11}ClFNO_2$
Chiral building block. mp = 245°; $[\alpha]^{23}$ = + 11° (c = 1, $H_2O$). *Sigma-Aldrich Fine Chemicals.*

**2871 L-4-Fluorphenylalanine hydrochloride**

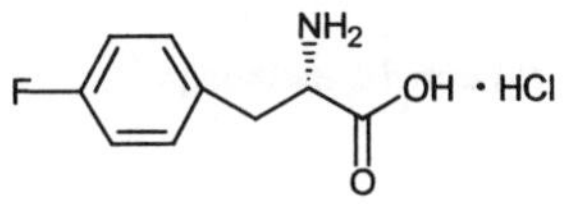

$C_9H_{11}ClFNO_2$
Chiral building block. mp = 245-251°; $[\alpha]^{24}$ = - 8° (c = 1, $H_2O$). *Sigma-Aldrich Fine Chemicals.*

**2872 (1S,4S)-(-)-2-(3-Fluorophenyl)-2,5-diazabicyclo[2.2.1]heptane hydrobromide**

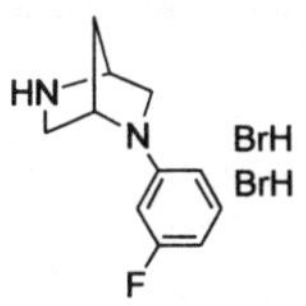

$C_{11}H_{14}BrFN_2$
Chiral intermediate. mp = 176-180°; $[\alpha]^{23}$ = - 98° (c = 1, $H_2O$). *Sigma-Aldrich Fine Chemicals.*

**2873 (1S,4S)-(-)-2-(4-Fluorphenyl)-2,5-diazabicyclo[2.2.1]heptane hydrobromide**

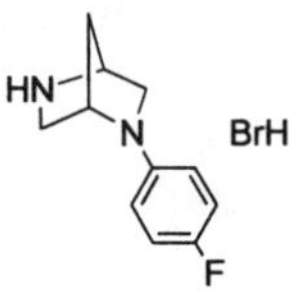

$C_{11}H_{14}BrFN_2$
Chiral intermediate. mp = 225-230°; $[\alpha]^{22}$ = - 68° (c = 1, $H_2O$). *Sigma-Aldrich Fine Chemicals.*

**2874 (S)-2-(4-Fluorophenyl)-3-methylbutyric acid**
110311-45-0

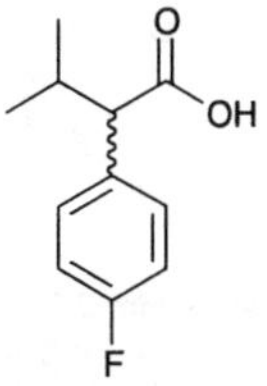

$C_{11}H_{12}FO_2$
Chiral intermediate. *Toray International, Inc.*

**2875 (+)-5-Fluorouridine**
316-46-1 206-260-3

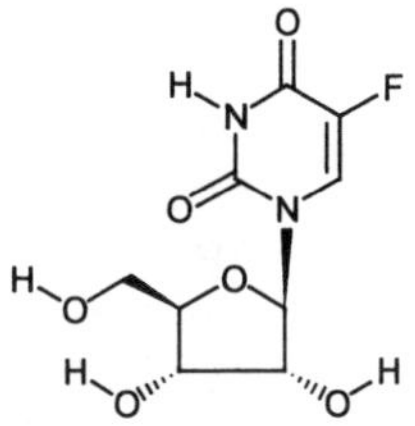

$C_9H_{11}FN_2O_6$
Uridine, 5-fluoro-; 5-Fluorouracil-1β-D-ribofuranoside. Chiral intermediate. $[\alpha]^{25}$ = + 17° (c = 2, $H_2O$). *Acros Organics nv; Sigma-Aldrich Fine Chemicals.*

**2876 Folic acid dihydrate**
75708-92-8 4140(11) 200-419-0

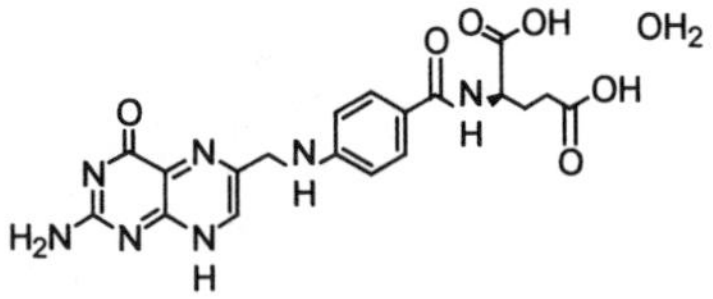

$C_{19}H_{19}N_7O_6$
PGA, Pteroylglutamic acid. Listed on TSCA. Chiral intermediate. $[\alpha]_D^{20}$= + 18° (c = 0.5, 0.1N NaOH). *Sigma-Aldrich Fine Chemicals.*

**2877 N-Formyl-L-alanine**
10512-86-4 234-045-4

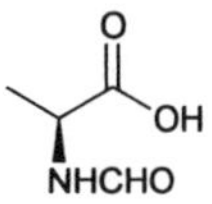

$C_4H_7NO_3$
Alanine, N-formyl-, L-. Chiral intermediate. *Flamma s.p.a.*

**2878 N-L-Formylaspartic acid anhydride**
33605-73-1 4254(12) 251-589-8

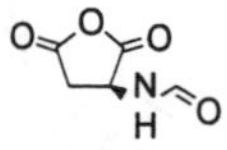

$C_5H_5NO_4$
Formamide, N-(tetrahydro-2,5-dioxo-3-furanyl)-, (S)-; (S)-N-(Tetrahydro-2,5-dioxo-3-furyl)formamide; (S)-(-)-2-Formamidosuccinic anhydride. Chiral intermediate. mp = 240-250°; $[\alpha]_D^{20}$ = + 14.3° (c = 3.4, $H_2O$). *Fischer Chemicals AG; Sigma-Aldrich Fine Chemicals.*

**2879 3-Fomyl-(6S)-tert-butylsilyloxymethyl-(7R)-tetrahydropyranyloxy-(1S,6S)-bicyclo[3,3,0]oct-2(E)-ene**

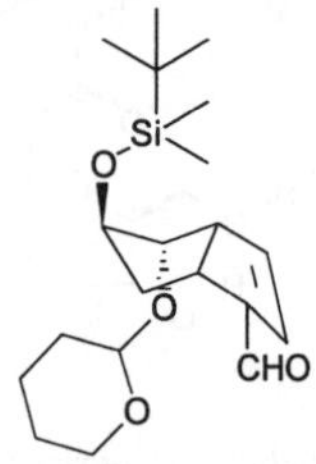

$C_{21}H_{36}O_4Si$
Chiral intermediate. *Nissan Chemical Industries, Ltd.*

**2880 Formyl-2,2-dimethyl-4-thiazolidine-carboxylic acid**
55878-44-9

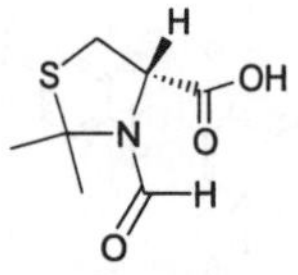

$C_7H_{11}NO_3S$
Chiral intermediate. *Ultrafine.*

**2881 (R)-(-)-Formylmandeloyl chloride**
29169-64-0 249-478-4

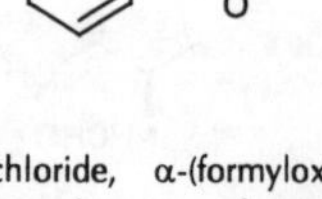

$C_9H_7ClO_3$
Benzeneacetyl chloride, α-(formyloxy)-, (R)-; (R)-α-(Chlorocarbonyl)benzyl formate; (R)-α-(Formyloxy)benzeneacetyl chloride; Mandeloyl chloride, formate, D-. Chiral intermediate. bp = 221°; d = 1.275; n = 1.523; $[\alpha]^{23}$ = - 228° (neat). *Austin Chemical Company, Inc.; DSM Fine Chemcials Netherlands; Sigma-Aldrich Fine Chemicals; Toray International, Inc.*

**2882 (S)-(+)-O-Formyl-L-mandeloyl chloride**

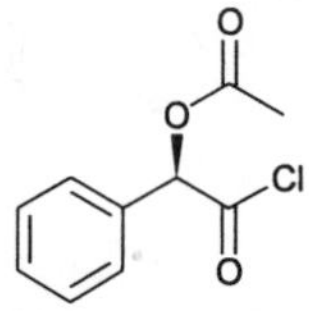

$C_9H_7ClO_3$
Benzeneacetyl chloride, α-(formyloxy)-, (S)-; (S)-α-(Formyloxy)benzeneacetyl chloride; (S)-α-(Chlorocarbonyl)benzyl formate; L-Phenyl(formyloxy)acetyl chloride. Chiral intermediate. *Austin Chemical Company, Inc.*

**2883 N-Formyl-L-methonine**

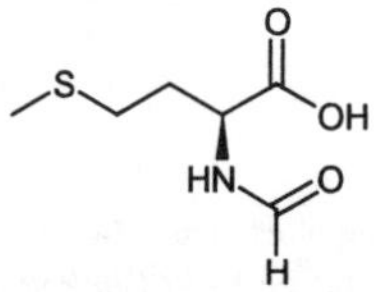

$C_6H_{11}NO_3S$
Methionine, N-formyl-, L-. Chiral intermediate. *Austin Chemical Company, Inc.*

**2884 N-Formyl-L-phenylalanine**
13200-85-6 236-166-8

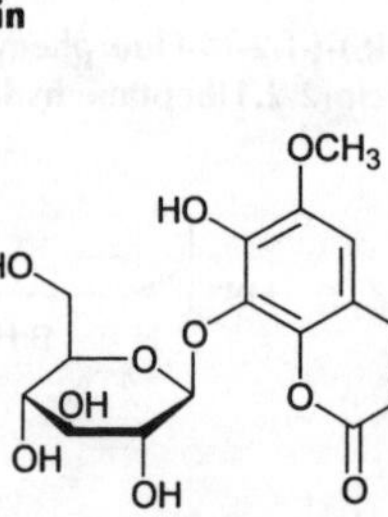

$C_{10}H_{11}NO_3$
Alanine, N-formyl-3-phenyl-, L-; L-Phenylalanine, N-formyl-. Chiral intermediate. *Fischer Chemicals AG.*

**2885 D-Fraxin**
524-30-1 208-355-5

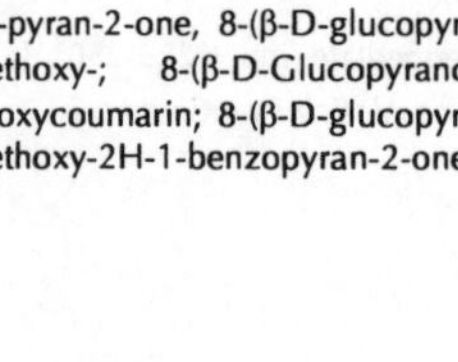

$C_{16}H_{18}O_{10}$
Benzo-(2H)-1-pyran-2-one, 8-(β-D-glucopyranosyloxy)-7-hydroxy-6-methoxy-; 8-(β-D-Glucopyranosyloxy)-7-hydroxy-6-methoxycoumarin; 8-(β-D-glucopyranosyloxy)-7-hydroxy-6-methoxy-2H-1-benzopyran-2-one; 7,8-Di-

hydroxy-6-methoxycoumarin 8-β-D-glucopyranoside; Fraxetin 8-β-D-glucopyranoside; Fraxetol 8-glucoside; Fraxoside. Chiral intermediate. mp = 205°. *Kaden Biochemicals GmbH.*

**2886 Friedelin**
559-74-0 4294(12) 209-205-1

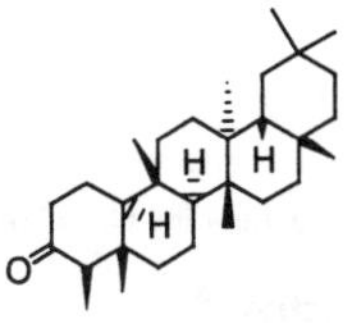

$C_{30}H_{50}O$
DA-Friedooleanan-3-one. Chiral intermediate. mp = 262-265°. *Sigma-Aldrich Fine Chemicals.*

**2887 D-(-)-Fructose**
57-48-7 4295(12) 200-333-3

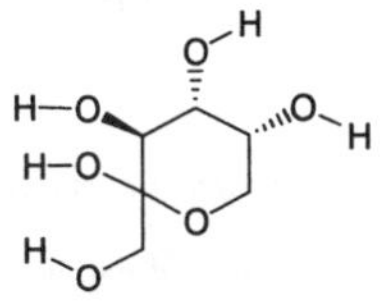

$C_6H_{12}O_6$
Fructose, D-; D-(-)-Levulose; arabino-Hexulose; Fruit sugar; Furucton; Hi-Fructo 970; Krystar 300; Nevulose. Listed on TSCA. Chiral building block. mp = 119-122°; $[\alpha]_D^{20}$ = - 92° (c = 10, $H_2O$). *Acros Organics nv; Pfanstiehl Laboratories, Inc.; Senn Chemicals AG; Sigma-Aldrich Fine Chemicals; TCI America.*

**2888 L-(+)-Fructose**

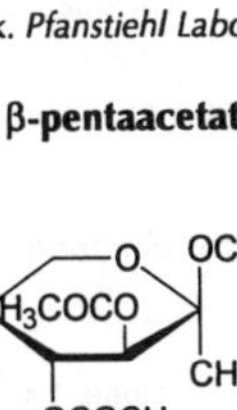

$C_6H_{12}O_6$
Chiral building block. *Pfanstiehl Laboratories, Inc.*

**2889 D-Fructose β-pentaacetate**

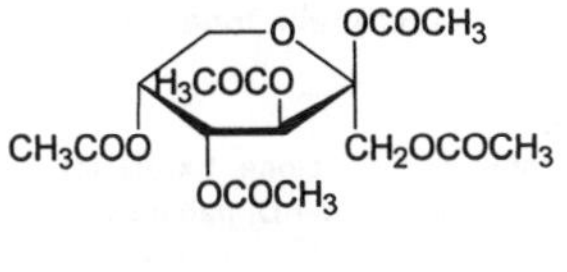

$C_6H_{22}O_{11}$
Chiral intermediate. *Senn Chemicals AG.*

**2890 Fructose 2,6-diphosphate sodium salt hydrate**
123333-75-5
$C_6H_{14}O_{12}P_2$
Chiral intermediate. *Sigma-Aldrich Fine Chemicals.*

**2891 D-Fructose-1,6-diphosphate, disodium salt**
26177-85-5 247-504-9

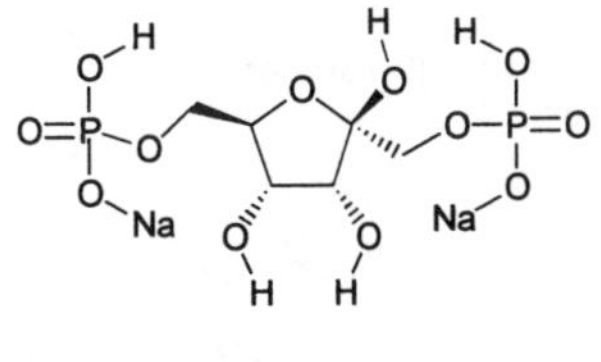

$C_6H_{12}Na_2O_{12}P_2$
Fructose, 1,6-bis(dihydrogen phosphate), disodium salt, D-; FDP; Disodium fructose 1,6-diphosphate. Listed on TSCA. Chiral diagnostic reagent. *Acros Organics nv; Senn Chemicals AG.*

**2892 D-(+)-Fucose**
3615-37-0 4301(12) 222-792-9

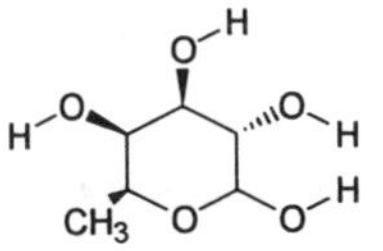

$C_6H_{12}O_5$
Fucose, D-; 6-Deoxy-D-galactose; D-Galactose, 6-deoxy-; Rhodeose; D-(+)-Rhodeose; D-Fucopyranose. Listed on TSCA. Chiral building block. mp = 137°; $[\alpha]_D^{20}$ = + 75.6° (c = 4, $H_2O$). *Acros Organics nv; Pfanstiehl Laboratories, Inc.; Senn Chemicals AG; Sigma-Aldrich Fine Chemicals; TCI America.*

**2893 L-(-)-Fucose**
2438-80-4 4193(11) 219-452-7

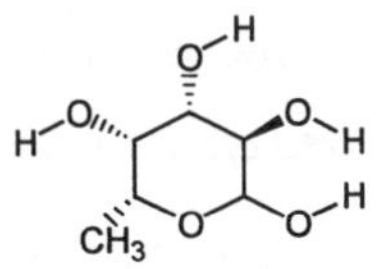

$C_6H_{12}O_5$
Galactose, 6-deoxy-, L-; L-Fucopyranose; L-Fuculose; 6-Deoxy-L-galactose; L-(-)-Rhodeose. Listed on TSCA. Chiral building block. mp = 140°; $[\alpha]_D^{20}$ = - 75.9° (c = 4, $H_2O$). *Acros Organics nv; Kaden Biochemicals GmbH; Pfanstiehl Laboratories, Inc.; Senn Chemicals AG; Sigma-Aldrich Fine Chemicals; TCI America.*

**2894 (R)-(+)-1-(2-Furyl)ethanol**
27948-61-4

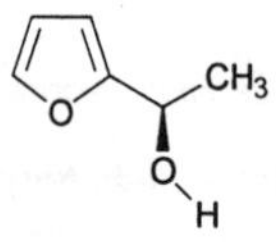

$C_6H_8O_2$
Chiral building block. *Acros Organics nv.*

**2895 (S)-(-)-1-(2-Furyl)ethanol**
85828-09-7

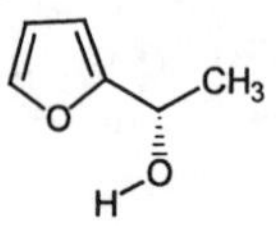

$C_6H_8O_2$
Chiral building block. [α] = - 24.4°. *Acros Organics nv.*

**2896 (R)-(-)-1-(2-Furyl)ethyl acetate**

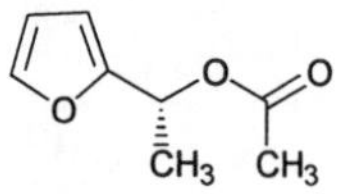

$C_8H_{10}O_3$
Chiral building block. [α] = - 3.4°. *Acros Organics nv.*

**2897 D-Galactal**
21193-75-9 244-264-7

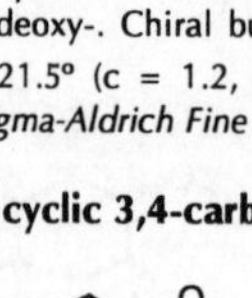

$C_6H_{10}O_4$
Hex-5-enitol, 2,6-anhydro-5-deoxy-, D-arabino-; 1,5-Anhydro-2-deoxy-D-lyxo-hex-1-enitol; D-lyxo-Hex-1-enopyranose, 1,2-dideoxy-. Chiral building block. mp = 99-103°; $[\alpha]^{22}$ = - 21.5° (c = 1.2, $CH_3OH$). *Pfanstiehl Laboratories, Inc.; Sigma-Aldrich Fine Chemicals.*

**2898 D-Galactal cyclic 3,4-carbonate**

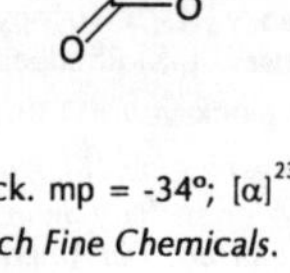

$C_7H_8O_5$
Chiral building block. mp = -34°; $[\alpha]^{23}$ = - 51° (c = 1.2, $CHCl_3$). *Sigma-Aldrich Fine Chemicals.*

**2899 D-Galactaric acid**

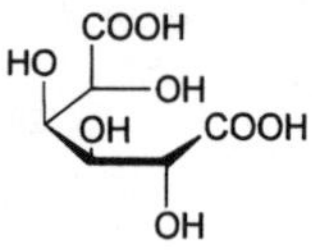

$C_6H_{10}O_8$
Mucic acid. Chiral building block. *Senn Chemicals AG.*

**2900 D-Galactitol**

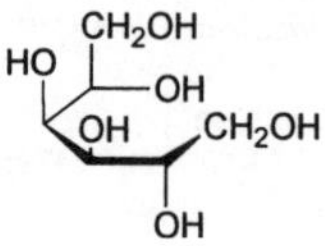

$C_6H_{14}O_6$
Chiral building block. *Senn Chemicals AG.*

**2901 D-Galactonic acid calcium salt pentahydrate**

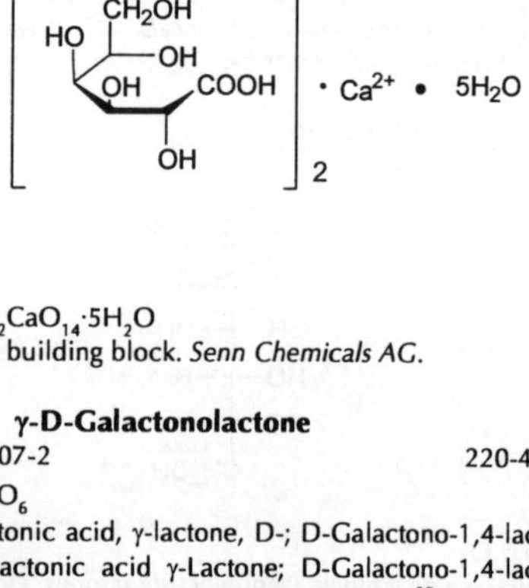

$C_{12}H_{22}CaO_{14}\cdot 5H_2O$
Chiral building block. *Senn Chemicals AG.*

**2902 γ-D-Galactonolactone**
2782-07-2 220-484-9
$C_6H_{10}O_6$
Galactonic acid, γ-lactone, D-; D-Galactono-1,4-lactone; D-Galactonic acid γ-Lactone; D-Galactono-1,4-lactone. Listed on TSCA. Chiral building block. $[\alpha]_D^{20}$ = - 77° (c = 4, $H_2O$). *Pfanstiehl Laboratories, Inc.; Senn Chemicals AG; TCI America.*

**2903 γ-L-Galactonolactone**
1668-08-2
$C_6H_{10}O_6$
L-Galactonic acid γ-Lactone; L-Galactono-1,4-lactone; 3-O-β-D-Galactopyranosyl-D-arabinose. Chiral building block. mp = 137°; $[\alpha]_D^{20}$ = - 62° (c = 1, $H_2O$). *Pfanstiehl Laboratories, Inc.; Senn Chemicals AG; TCI America.*

**2904 6-O-α-D-Galactopyranosyl-D-glucopyranose monohydrate**
66009-10-7

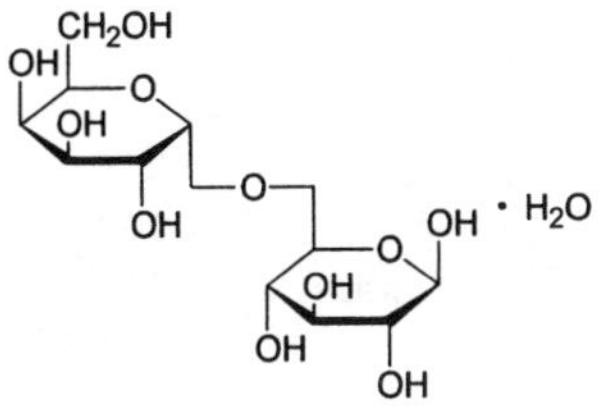

$C_{12}H_{24}O_{12}$
Glucose, 6-O-α-D-galactopyranosyl-, monohydrate, D-; α,β-Melibiose monohydrate. Chiral intermediate. mp = 179-181°. *Kaden Biochemicals GmbH.*

**2905 D-Galactopyranuronic acid monohydrate**
685-73-4 4242(11) 211-682-6

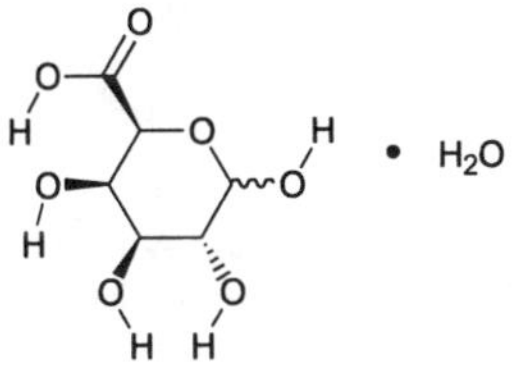

$C_6H_{12}O_8$
Galacturonic acid, D-; D-Galactopyranuronic acid. Listed on TSCA. Chiral intermediate. $[\alpha]_D^{20}$ = + 51.9° (c = 4, $H_2O$). *Acros Organics nv; Pfanstiehl Laboratories, Inc.; Sigma-Aldrich Fine Chemicals.*

**2906 D-(+)-Galactosamine hydrochloride**
1772-03-8 4240(11) 217-198-1

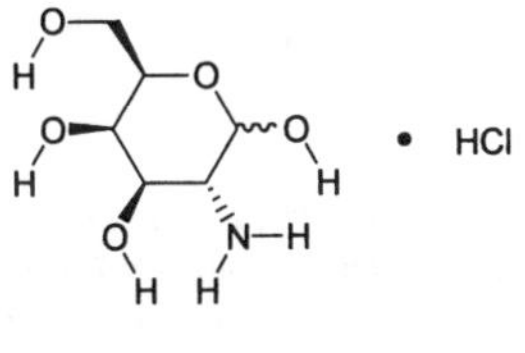

$C_6H_{14}ClNO_5$
D-Galactose, 2-amino-2-deoxy-, hydrochloride; Chondrosamine hydrochloride; Galactose, 2-amino-2-deoxy-, hydrochloride, D-; 2-Amino-2-deoxy-D-galactose hydrochloride; 2-Amino-2-deoxy-D-galactopyranose hydrochloride. Listed on TSCA. Chiral intermediate. mp = 182-185°; $[\alpha]_D^{20}$ = + 95° (c = 2, $H_2O$). *Acros Organics nv; Austin Chemical Company, Inc.; Pfanstiehl Laboratories, Inc.; Senn Chemicals AG; Sigma-Aldrich Fine Chemicals; TCI America.*

**2907 D-Galactosamine pentaacetate**

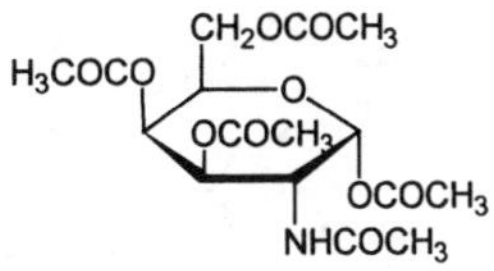

$C_{16}H_{23}NO_{10}$
Chiral building block. *Senn Chemicals AG.*

**2908 D-(+)-Galactose**
59-23-4 4353(12) 200-416-4

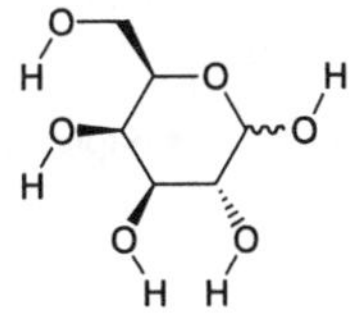

$C_6H_{12}O_6$
D-Galactose; D-Galactopyranose. Listed on TSCA. Chiral building block. mp = 165-168°; $[\alpha]_D^{20}$ = + 80.2° (c = 4, $H_2O$). *Acros Organics nv; Alfa Chemicals Ltd.; Kaden Biochemicals GmbH; Pfanstiehl Laboratories, Inc.; Senn Chemicals AG; Sigma-Aldrich Fine Chemicals; TCI America.*

**2909 L-(-)-Galactose**
15572-79-9 239-630-8

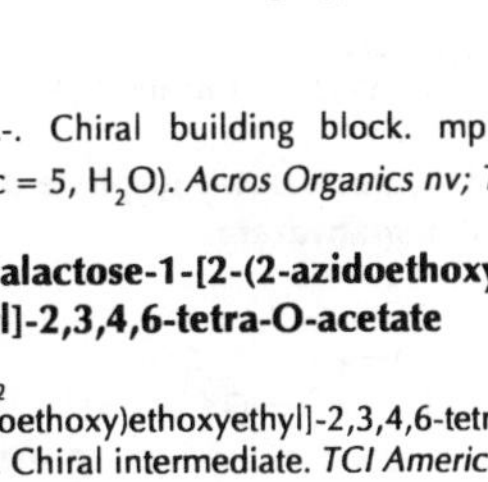

$C_6H_{12}O_6$
Galactose, L-. Chiral building block. mp = 163-165°; [α] = - 79° (c = 5, $H_2O$). *Acros Organics nv; TCI America.*

**2910 D-Galactose-1-[2-(2-azidoethoxy)ethoxyethyl]-2,3,4,6-tetra-O-acetate**

$C_{20}H_{31}N_3O_{12}$
1-[2-(2-Azidoethoxy)ethoxyethyl]-2,3,4,6-tetra-O-acetyl-D-galactose. Chiral intermediate. *TCI America.*

**2911 α,β-D-Galactose pentaacetate**
25878-60-8

$C_{16}H_{22}O_{11}$
Chiral intermediate. *Pfanstiehl Laboratories, Inc.*

**2912 β-D-Galactose pentaacetate**
4163-60-4 224-008-0

$C_{16}H_{22}O_{11}$
Galactopyranose, pentaacetate, β-D-; 1,2,3,4,6-Penta-O-acetyl-β-D-galactopyranoside. Listed on TSCA. Chiral intermediate. mp = 139-142°; $[\alpha]^{24}$ = + 23° (c = 1, $CHCl_3$). *Acros Organics nv; Interchem Corporation; Kaden Biochemicals GmbH; Senn Chemicals AG; Sigma-Aldrich Fine Chemicals.*

**2913 (+)-Galactose pentapivalate**
118711-42-5

$C_{31}H_{52}O_{11}$
Chiral intermediate. $[\alpha]_D^{20}$ = + 2.0° (c = 1, ether). *Sigma-Aldrich Fine Chemicals.*

**2914 α-D-(+)-Galactose-1-phosphate barium salt trihydrate**

$C_{12}H_{26}BaO_{18}P_2 \cdot 3H_2O$
Chiral intermediate. *Senn Chemicals AG.*

**2915 α-D-Galactose 1-phosphate dipotassium salt pentahydrate**
19046-60-7 242-783-3

$C_6H_{11}K_2O_{10}P$
Galactopyranose, 1-(dihydrogen phosphate), dipotassium salt, α-D-. Chiral intermediate. *Acros Organics nv; Senn Chemicals AG.*

**2916 β-D-Gentiobiose octaacetate**

$C_{28}H_{38}O_{19}$
Chiral intermediate. *Senn Chemicals AG.*

**2917 D-Gentobiose**
554-91-6 209-074-0

$C_{12}H_{22}O_{11}$
Glucose, 6-O-β-D-glucopyranosyl-, D-; Amygdalose; 6-O-β-D-glucopyranosyl-D-glucopyranose. Chiral intermediate. $[\alpha]^{24}$ = + 11° (c = 4, $H_2O$). *Acros Organics nv; Pfanstiehl Laboratories, Inc.; Senn Chemicals AG.*

**2918 (+)-Gitoxigenin**
545-26-6 208-886-2

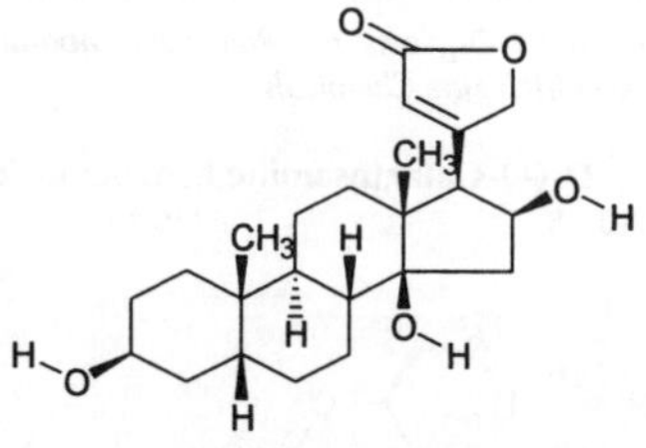

$C_{23}H_{34}O_5$
Card-20(22)-enolide, 3,14,16-trihydroxy-, (3β,5β,16β)-; 3β,14,16β-Trihydroxy-5β,14β-card-20(22)-enolide; 16β-Hydroxydigitoxigenin. Pharmaceutical or derivative. $[\alpha]^{22}$ = + 30.6° (c = 0.8, $CH_3OH$, 22 °). *Acros Organics nv.*

**2919 (S)-Glaucine**
475-81-0 207-501-5
$C_{21}H_{25}NO_4$
Dibenzo[4H,de,g]quinoline, 5,6,6α,7-tetrahydro-1,2,9,10-tetramethoxy-6-methyl-, (S)-; (S)-5,6,6a,7-Tetrahydro-1,2,9,10-tetramethoxy-6-methyl-4H-dibenzo[de,g]-quinoline; Glauvent; N,O,O'-Trimethyllaurelliptine; O,O-Dimethylboldine; 6aα-Aporphine, 1,2,9,10-tetramethoxy-; Boldine dimethyl ether; Bromcholitin. Pharmaceutical or derivative. mp = 116-120°. *Acros Organics nv.*

**2920 D-Glucal**
13265-84-4 236-259-3

$C_6H_{10}O_4$
(1,5-Anhydro-2-deoxy-D-arabino-hex-1-enitol). Chiral building block. mp = 58-60°; $[\alpha]^{21}$ = - 8° (c = 1.9, $H_2O$). *Sigma-Aldrich Fine Chemicals.*

**2921 D-Glucamine**
488-43-7 207-677-3
$C_6H_{15}NO_5$
Glucitol, 1-amino-1-deoxy-, D-; 1-Amino-1-deoxy-D-sorbitol; 1-amino-1-deoxy-D-glucitol; Sorbitol, 1-amino-1-deoxy-. Chiral building block. *TCI America.*

**2922 D-Glucaric acid potassium salt**

$C_6H_9KO_8$
Potassium hydrogen D-glucarate. Chiral building block. *Senn Chemicals AG.*

**2923 α-D-Glucoheptonic acid γ-lactone**
60046-25-5 4349(11) 262-037-0

$C_7H_{12}O_7$
Gluco-heptonic acid, γ-lactone, (2ε)-, D-. Chiral intermediate. mp = 152-155°; $[\alpha]_D^{20}$ = - 51° (c = 4.3, $H_2O$). *Acros Organics nv; Pfanstiehl Laboratories, Inc.; Sigma-Aldrich Fine Chemicals.*

**2924 D-Glucoheptose**
5349-37-1 226-310-8
$C_7H_{14}O_7$
Gluco-2-heptulose, D-. Chiral building block. $[\alpha]_D^{20}$ = - 20.2° (c = 4, $H_2O$). *Pfanstiehl Laboratories, Inc.*

**2925 Gluconic acid (45-50 +wt. % solution in $H_2O$)**
526-95-4 4464(12) 208-401-4
$C_6H_{12}O_7$
Listed on TSCA. Chiral building block. d = 1.234; n = 1.4161; $[\alpha]_D^{20}$ = + 12.5° (neat). *Sigma-Aldrich Fine Chemicals.*

**2926 D-Gluconic acid, calcium salt**
299-28-5 1712(12) 206-075-8

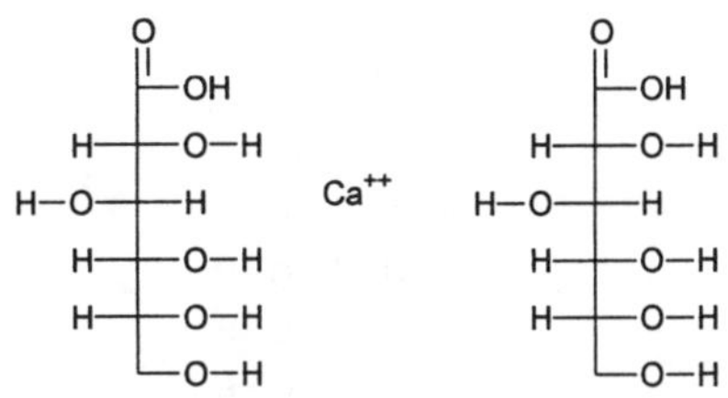

$C_{12}H_{22}CaO_{14}$
Gluconic acid, calcium salt (2:1), D-; Glucobiogen; Calciofon; Neocalglucon; Calcicol; Calcipur; Calcium D-gluconate; Calcium hexagluconate; Calglucol; Calglucon; Dragocal; Ebucin; Glucal; Kalpren; Novoca. Listed on TSCA. Chiral building block. $[\alpha]^{22}$ = + 10.2° (c = 1, $H_2O$). *Acros Organics nv; Sigma-Aldrich Fine Chemicals.*

**2927 D-Gluconic acid, copper(II) salt**
13005-35-1 235-844-0

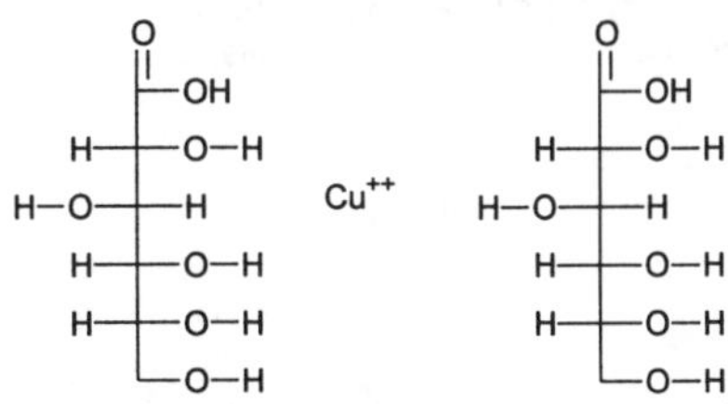

$C_{12}H_{24}CuO_{14}$
Copper gluconate, D-; Copper(II) D-gluconate. Resolving agent. *Acros Organics nv.*

**2928 D-Gluconic acid, potassium salt**
299-27-4 7796(12) 206-074-2

$C_6H_{11}KO_7$
Gluconic acid, monopotassium salt, D-; Potassium Gluconate; D-Gluconic acid, monopotassium salt; Gluconsan K; K-IAO; Kalium-β; Katorin; Potalium; Potasoral; Potassium D-gluconate; Potassuril; Sirokal. Listed on TSCA. Chiral building block. mp = 183°; $[\alpha]^{21}$ = + 14.1° (c = 1.2, $H_2O$). *Acros Organics nv; Sigma-Aldrich Fine Chemicals.*

**2929 D-Gluconic acid, sodium salt**
527-07-1 8766(12) 208-407-7

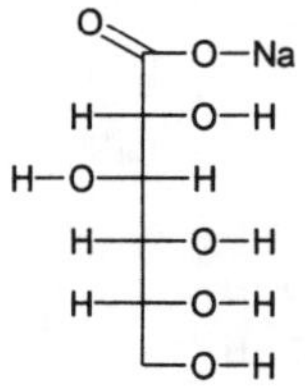

$C_6H_{11}NaO_7$
Gluconic acid, D-, monosodium salt; Clewat GL; D-Glulonic acid, monosodium salt; Despalite DV; Glonsen; Monosodium gluconate; Pasexon 100T; PMP Sodium Gluconate; Recover; Sodium 2,3,4,5,6-pentahydroxy-1-hexanoate; Sodium D-gluconate; Sunmorl N 60S. Listed on TSCA. Resolving agent. mp = 206°; $[\alpha]^{23}$ = + 12.1° (c = 1.75, $H_2O$). *Acros Organics nv; Sigma-Aldrich Fine Chemicals.*

**2930 D-(+)-Glucono-1,5-lactone**
90-80-2 4465(12) 202-016-5

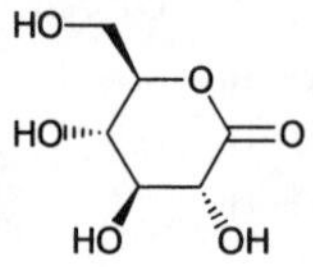

$C_6H_{10}O_6$
Gluconic acid, δ-lactone, D-; Gluconic acid δ-lactone; Gluconodeltalactone; Gluconic acid, anhydride. Listed on TSCA. Chiral building block. mp = 153°; $[\alpha]_D^{20}$ = + 66°, initial (c = 5, $H_2O$). *Aceto Corporation; Acros Organics nv; Pfanstiehl Laboratories, Inc.; Sigma-Aldrich Fine Chemicals; TCI America.*

**2931 L-Glucono-1,5-lactone**
52153-09-0

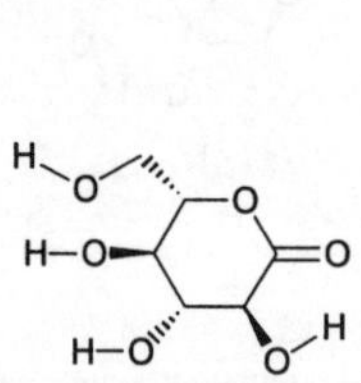

$C_6H_{10}O_6$
Chiral building block. mp = 142-144°; $[\alpha]_D^{20}$ = - 66° initial (c = 5, $H_2O$). *Acros Organics nv; Pfanstiehl Laboratories, Inc.; Sigma-Aldrich Fine Chemicals.*

**2932 D-Glucooctanoic lactone**
6968-62-3 230-185-5

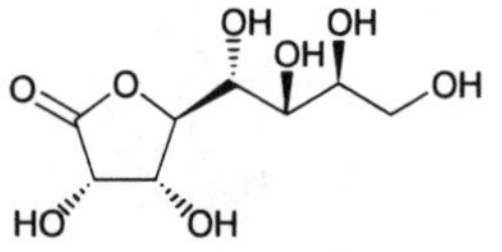

$C_8H_{14}O_8$
Chiral intermediate. mp = 188°; $[\alpha]^{24}$ = + 26° (c = 3, $H_2O$). *Sigma-Aldrich Fine Chemicals.*

**2933 (+)-Glucopyranose pentabenzoate**
22415-91-4

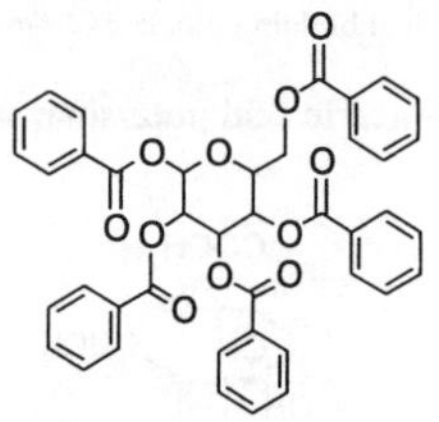

$C_{41}H_{32}O_{11}$
Chiral intermediate. mp = 185-188°; $[\alpha]_D^{20}$= + 127° (c = 1, $CHCl_3$). *Sigma-Aldrich Fine Chemicals.*

**2934 O-α-Glucopyranosyl (1,3)-β-D-fructofuranosyl-α-D-glucopyranoside**
597-12-6 209-894-9

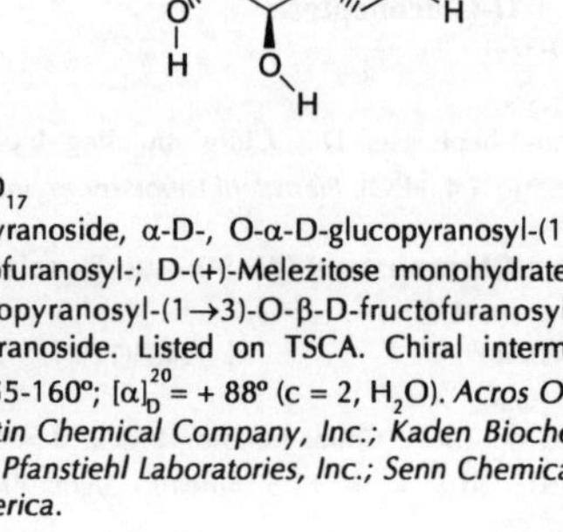

$C_{18}H_{34}O_{17}$
Glucopyranoside, α-D-, O-α-D-glucopyranosyl-(1→3)-β-D-fructofuranosyl-; D-(+)-Melezitose monohydrate; O-α-D-Glucopyranosyl-(1→3)-O-β-D-fructofuranosyl-α-D-glucopyranoside. Listed on TSCA. Chiral intermediate. mp = 155-160°; $[\alpha]_D^{20}$= + 88° (c = 2, $H_2O$). *Acros Organics nv; Austin Chemical Company, Inc.; Kaden Biochemicals GmbH; Pfanstiehl Laboratories, Inc.; Senn Chemicals AG; TCI America.*

**2935 (+)-Glucopyranosyl bromide tetrabenzoate**
14218-11-2

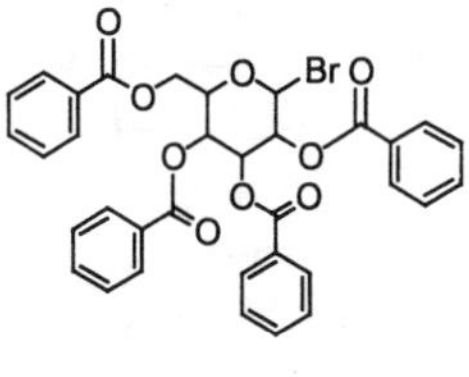

$C_{34}H_{27}BrO_9$
Chiral intermediate. mp = 128-131°; $[\alpha]_D^{20}$ = + 125° (c = 1, $CHCl_3$). *Sigma-Aldrich Fine Chemicals.*

**2936 (+)-Glucopyranosyl fluoride tetraacetate**

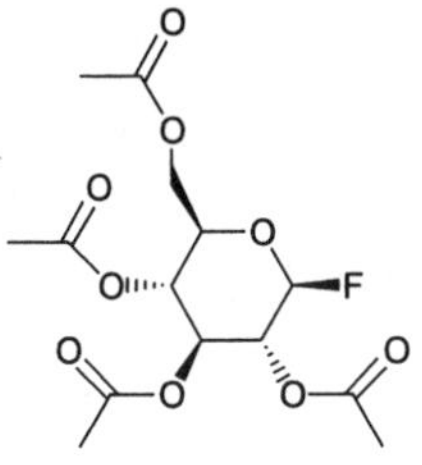

$C_{14}H_{19}FO_9$
Chiral intermediate. mp = 80-85°; $[\alpha]^{23}$ = + 21° (c = 1, $CHCl_3$). *Sigma-Aldrich Fine Chemicals.*

**2937 α-D-Glucopyranosyl-α-D-glucopyranoside dihydrate**
6138-23-4 9496(11)

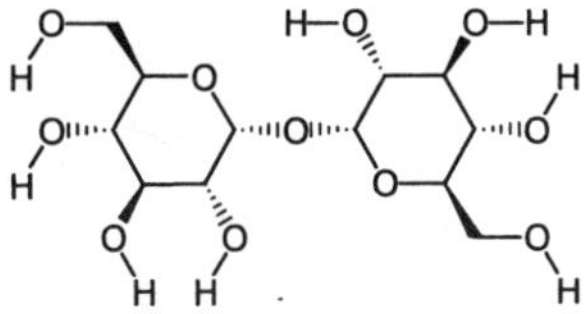

$C_{12}H_{26}O_{13}$
Glucopyranoside, α-D-, glucopyranosyl, α-D-, dihydrate; D-Trehalose dihydrate. Chiral intermediate. mp = 96-98°; $[\alpha]^{22}$ = + 179.9° (c = 7.5, $H_2O$). *Acros Organics nv; Kaden Biochemicals GmbH; Sigma-Aldrich Fine Chemicals; TCI America.*

**2938 4-O-α-D-Glucopyranosyl-D-glucopyranose**
69-79-4 200-716-5
$C_{12}H_{24}O_{12}$
Glucose, D-, 4-O-α-D-glucopyranosyl-; D-(+)-Maltose; Finetose; Malt sugar; Maltobiose; Maltodiose; Sunmalt; D-Maltose. Listed on TSCA. Chiral intermediate. mp = 117-121°. *Kaden Biochemicals GmbH; TCI America.*

**2939 7-[(β-D-Glucopyranosyloxy)-methyl]-4-methoxy-5H-furo{3,2-γ][1]benzopyran-5-one**
17226-75-4 241-263-3

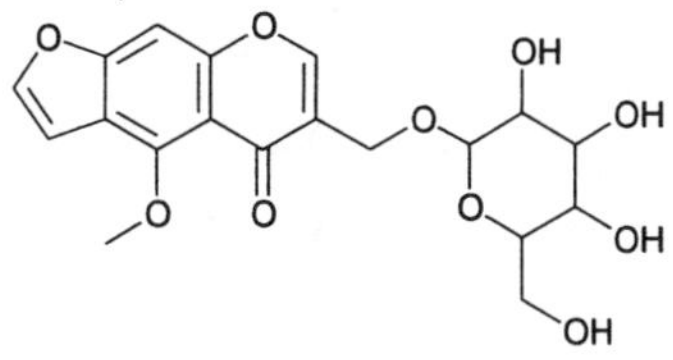

$C_{19}H_{20}O_{10}$
Glucopyranoside, (4-methoxy-5-oxo-5H-furo[3,2-γ][1]-benzopyran-7-yl)methyl, β-D-; Khellol glucoside; Khelloside; 5H-Furo[3,2-γ][1]benzopyran-5-one, 7-[(β-D-glucopyranosyloxy)methyl]-4-methoxy-; Khellinin; Khellol β-D-glucopyranoside. Chiral intermediate. mp = 179°. *Kaden Biochemicals GmbH.*

**2940 D-Glucosamic acid**
3646-68-2 222-879-1
$C_6H_{13}NO_6$
Gluconic acid, 2-amino-2-deoxy-, D-; D-Glucosaminic acid; 2-Amino-2-deoxy-D-gluconic acid. Listed on TSCA. Chiral building block. *Senn Chemicals AG; TCI America.*

**2941 D-Glucosamine-oxime hydrochloride**
54947-34-1
$C_6H_{15}ClN_2O_5$
Chiral building block. *TCI America.*

**2942 D-Glucosamine pentaacetate**

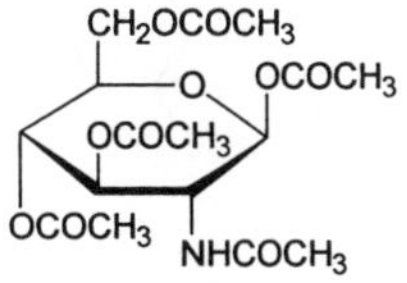

$C_{16}H_{23}NO_{10}$
2-Acetamido-2-deoxy-1,3,4,6-tetra-O-acetyl-D-glucopyranose. Chiral intermediate. *Senn Chemicals AG.*

**2943 D-(+)-Glucosamine tetraacetate hydrochloride**

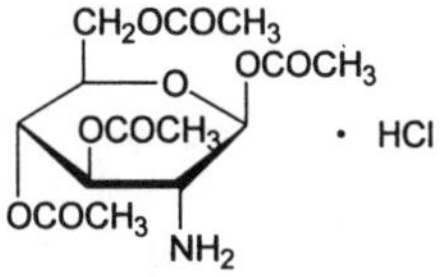

$C_{14}H_{22}ClNO_9$
2-Amino-2-deoxy-1,3,4,6-tetra-O-acetyl-α-D-glucopyranose hydrochloride. Chiral intermediate. *Senn Chemicals AG.*

**2944 α-D-Glucose**
50-99-7 200-075-1

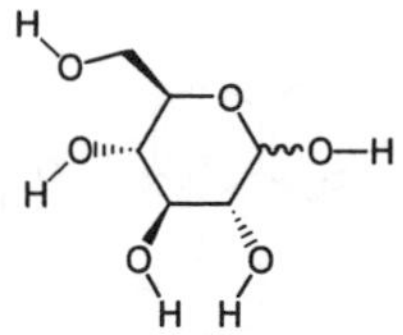

$C_6H_{12}O_6$
Glucose, α-D-; Cartose; Cerelose; Corn sugar; Dextropur; Dextrose; Dextrosol; Glucolin; Glucosteril; Grape sugar; Staleydex; Sugar, grape; Tabfine; 097(HS); Vadex; D-Glucopyranose; D-(+)-Dextrose; D-(+)-Glucose. Listed on TSCA. Chiral building block. mp = 146°; $[\alpha]_D^{20}$ = + 52.5° (c = 10, $H_2O$). *Acros Organics nv; Pfanstiehl Laboratories, Inc.; TCI America.*

**2945 β-D-Glucose**
492-61-5 207-756-2
$C_6H_{12}O_6$
Glucopyranose, β-D-; β-Dextrose; β-D-Glucopyranose. Listed on TSCA. Chiral building block. $[\alpha]_D^{20}$ = + 19°,*itial (c = 10, $H_2O$). Austin Chemical Company, Inc.; Pfanstiehl Laboratories, Inc.; Senn Chemicals AG.*

**2946 L-(-)-Glucose**
921-60-8 213-068-3

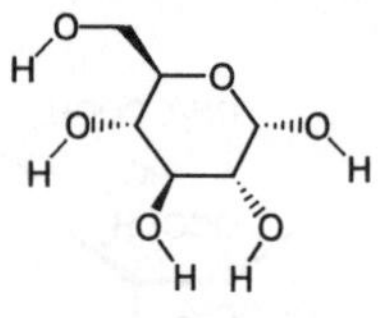

$C_6H_{12}O_6$
Glucose, L-(-)-. Chiral building block. mp = 153-156°; $[\alpha]_D^{20}$ = - 52.5° (c = 4, $H_2O$). *Acros Organics nv; Pfanstiehl Laboratories, Inc.; Senn Chemicals AG; Sigma-Aldrich Fine Chemicals; TCI America.*

**2947 D-Glucose-L-cysteine**

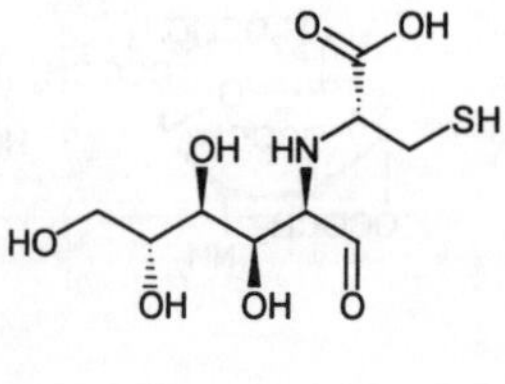

$C_9H_{17}NO_7S$
Chiral intermediate. *Austin Chemical Company, Inc.*

**2948 D-Glucose diethylmercaptal**
1941-52-2 217-728-1

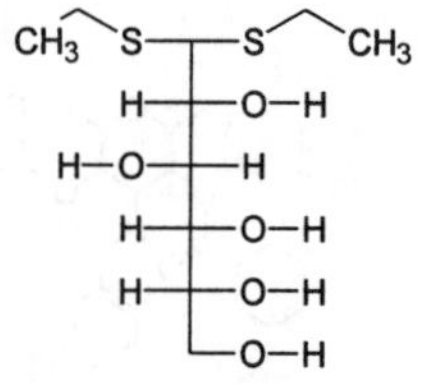

$C_{10}H_{22}O_5S_2$
D-Glucose, diethyl dithioacetal. Chiral intermediate. mp = 128°; $[\alpha]^{50}$ = - 31° (c = 5, $H_2O$). *Acros Organics nv; TCI America.*

**2949 α-D-Glucose pentaacetate**
604-68-2 210-073-2

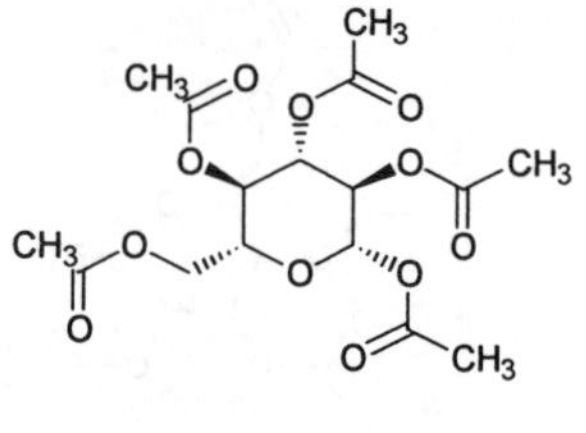

$C_{16}H_{22}O_{11}$
Glucopyranose, pentaacetate, α-D-; 1,2,3,4,6-Penta-O-acetyl-α-D-glucopyranose. Listed on TSCA. Chiral intermediate. mp = 111°; $[\alpha]_D^{20}$ = + 101.6° (c = 5, $CHCl_3$). *Acros Organics nv; Pfanstiehl Laboratories, Inc.; Sigma-Aldrich Fine Chemicals.*

**2950 β-D-Glucose pentaacetate**
604-69-3 210-074-8

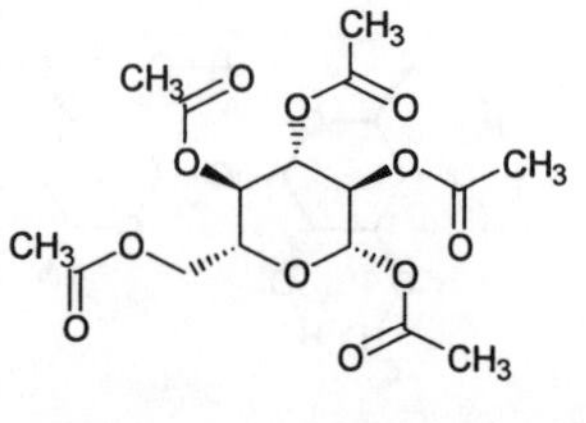

$C_{16}H_{22}O_{11}$
Glucopyranose, pentaacetate, β-D-; 1,2,3,4,6-Penta-O-acetyl-β-D-glucopyranose; 2,3,4,6-Tetra-O-acetyl-β-D-glucopyranosyl acetate. Chiral intermediate. mp = 130-132°; $[\alpha]_D^{23}$ = + 4° (c = 4, $CHCl_3$); $[\alpha]_D$ = + 2.3 (c = 1, $CH_3OH$). *Acros Organics nv; Kaden Biochemicals GmbH; Pfanstiehl Laboratories, Inc.; Senn Chemicals AG; Sigma-Aldrich Fine Chemicals.*

**2951 α,β-D-Glucose pentaacetate (High β)**
3891-59-6 223-439-1
$C_{16}H_{22}O_{11}$
Glucose, 2,3,4,5,6-pentaacetate. D-; 2,3,4,5,6-Penta-O-

acetyl-D-glucose. Listed on TSCA. Chiral intermediate. *Pfanstiehl Laboratories, Inc.*

**2952 D-Glucose-6-phosphate barium salt**
5996-16-7 227-836-0

$C_6H_{13}O_9P$
Glucose, 6-(dihydrogen phosphate), barium salt (1:1), D-; Barium glucose 6-phosphate. Listed on TSCA. Chiral intermediate. *TCI America.*

**2953 D-Glucose-6-phosphate dipotassium salt**
5996-17-8 227-837-6

$C_6H_{13}O_9P$
Glucose, 6-(dihydrogen phosphate), dipotassium salt, D-; Glucose 6-(dipotassium phosphate); Dipotassium D-glucose-6-phosphate. Listed on TSCA. Chiral building block. *TCI America.*

**2954 α-D-Glucose-1-phosphate, dipotassium salt dihydrate**
5996-14-5 4470(12) 229-788-6

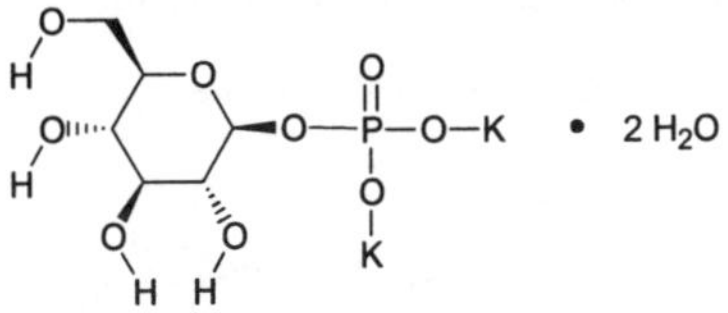

$C_6H_{11}K_2O_9P{\cdot}2H_2O$
Cori ester, dipotassium salt. Listed on TSCA. Chiral intermediate. mp = 209°; $[\alpha]^{23}$ = + 78.2° (c = 4, $H_2O$). *Acros Organics nv; Sigma-Aldrich Fine Chemicals.*

**2955 (+)-Glucose-1-phosphate disodium salt hydrate**

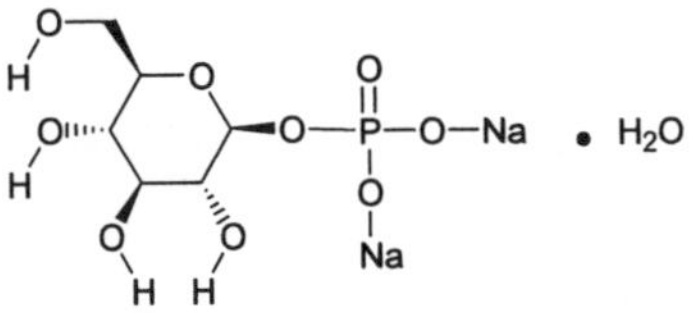

$C_6H_{13}Na_2O_9P{\cdot}H_2O$
Chiral intermediate. $[\alpha]^{21}$ = + 79.9° (c = 4.1, $H_2O$). *Sigma-Aldrich Fine Chemicals.*

**2956 Glucose-6-phosphate monosodium salt**
54010-71-8 4355(11) 258-921-0

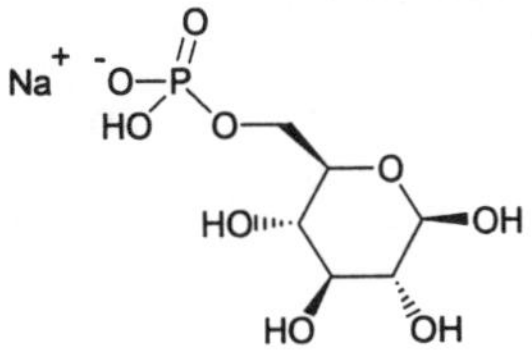

$C_6H_{13}NaO_9P$
Chiral intermediate. mp = 204°; $[\alpha]^{25}$ = + 33° (c = 10, $H_2O$). *Sigma-Aldrich Fine Chemicals.*

**2957 D-Glucuronamide**
3789-97-7

$C_6H_{11}NO_6$
Chiral building block. mp = 167°. *TCI America.*

**2958 D-Glucuronic acid**
6556-12-3 4474(12) 229-486-4

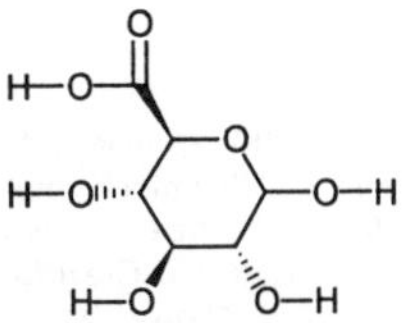

$C_6H_{10}O_7$
Listed on TSCA. Chiral building block. mp = 159-161°; $[\alpha]_D^{20}$ = + 36° (c = 3, $H_2O$). *Acros Organics nv; Sigma-Aldrich Fine Chemicals.*

**2959 D-Glucuronic acid, sodium salt, monohydrate**
14984-34-0 4360(11) 239-065-7

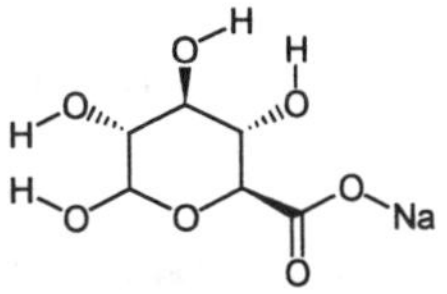

$C_6H_9NaO_7$
Glucuronic acid, monosodium salt, D-; Sodium D-glucuronate. Chiral building block. mp = 138°; $[\alpha]_D^{20}$ = + 21.5° (c = 2, $H_2O$). *Acros Organics nv; Senn Chemicals AG; Sigma-Aldrich Fine Chemicals.*

**2960 D-Glucurono-6,3-lactone**
32449-92-6 4476(12) 251-05303

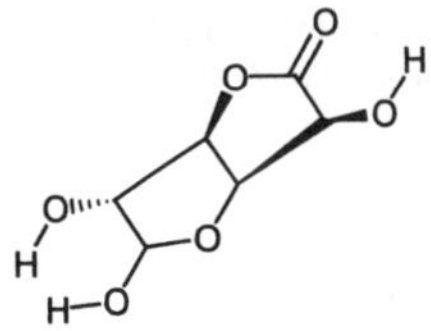

$C_6H_8O_6$
Glucuronic acid, γ-lactone, D-; D-Glucurone; Anhydroglucuronate; Dicurone; Glucoxy; Glucuronic acid lactone; Glucurolactone; Reulatt S.S.; Glucuronosan; Glycurone; Guronsan; D-Glucuronolactone. Listed on TSCA. Chiral building block. mp = 178°; $[\alpha]_D^{20}$ = + 18.5° (c = 8, $H_2O$). *Acros Organics nv; Austin Chemical Company, Inc.; Pfanstiehl Laboratories, Inc.; Senn Chemicals AG; Sigma-Aldrich Fine Chemicals; TCI America.*

**2961 D-Glutamic acid**
6893-26-1 4477(12) 230-000-8

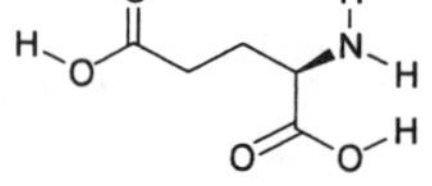

$C_5H_9NO_4$
Glutamic acid, D-; (2R)-2-Aminopentanedioic acid. Listed on TSCA. Chiral building block. mp = 200-202°; [α] = - 29° (c = 1, 6N HCl). *Acros Organics nv; Ajinomoto Co. Inc.; Fischer Chemicals AG; Great Lakes Fine Chemicals; Kaneka Corporation; Kyowa Hakko Kogyo Co., Ltd.; Norse Laboratories; Rexim S.A. Produits Chimiques; Shanghai DSL International Trading Company; Sigma-Aldrich Fine Chemicals; Sochinaz SA; TCI America; Varsal Instruments, Inc.; Yoneyama Yakuhin Kogyo Co., Ltd.*

**2962 L-Glutamic acid**
56-86-0 4477(12) 200-293-7

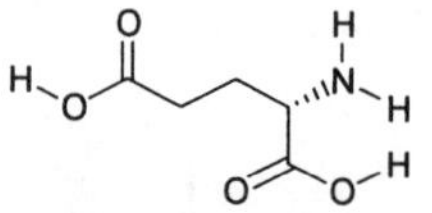

$C_5H_9NO_4$
Pentanedioic acid, 2-amino-, (S)-; 1-Aminopropane-1,3-dicarboxylic acid; (2S)-2-Aminopentanedioic acid; Aciglut; Glusate; Glutacid; Glutamicol; Glutamidex; Glutaminol; Glutaton; L-α-Aminoglutaric acid; L-Glutaminic acid; L-Gutamic acid. Listed on TSCA. Chiral building block. mp = 225°; [α] = + 29° (c = 1, 6N HCl). *Aceto Corporation; Acros Organics nv; Ajinomoto Co. Inc.; Flamma s.p.a.; Rexim S.A. Produits Chimiques; Senn Chemicals AG; Shanghai DSL International Trading Company; Sigma-Aldrich Fine Chemicals; Tanabe Seiyaku Co. Ltd.; TCI America; Varsal Instruments, Inc.; Yoneyama Yakuhin Kogyo Co., Ltd.*

**2963 γ-L-Glutamic acid 7-amido-4-methylcoumarin hydrate**
72669-53-5

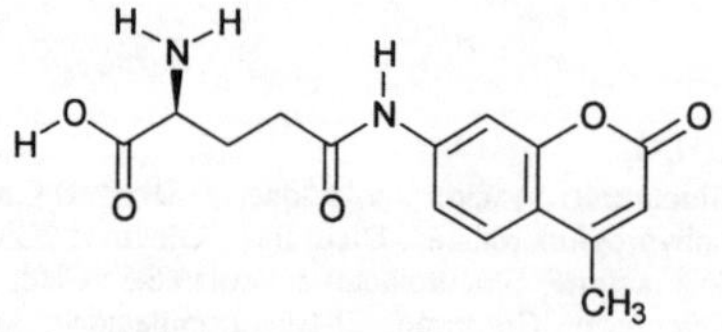

$C_{15}H_{16}N_2O_5$
Chiral intermediate. *Acros Organics nv.*

**2964 L-Glutamic acid ammonium salt**
7558-63-6 231-447-1

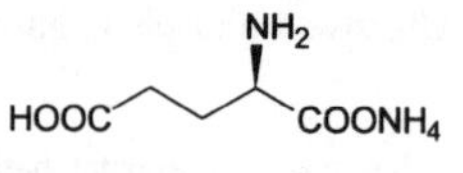

$C_5H_{12}N_2O_4$
Glutamic acid, monoammonium salt, L-; Ammonium glutamate; Monoammonium L-glutamate. Listed on TSCA. Chiral building block. *Aceto Corporation.*

**2965 L-Glutamic acid-α-anilide**

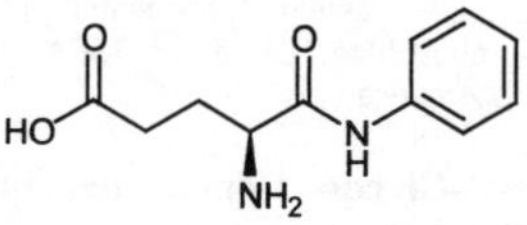

$C_{11}H_{14}N_2O_3$
Chiral building block. *Senn Chemicals AG.*

**2966 L-Glutamic acid-γ-anilide**

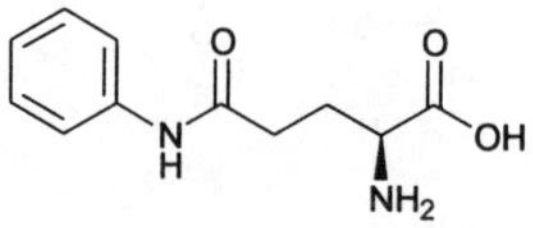

$C_{11}H_{14}N_2O_3$
Chiral building block. *Senn Chemicals AG.*

**2967 L-Glutamic acid diethyl ester hydrochloride**
1118-89-4 214-270-4

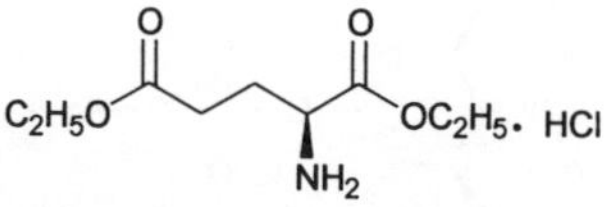

$C_9H_{17}NO_4$
Glutamic acid, diethyl ester, hydrochloride, L-; Diethyl L-glutamate hydrochloride. Listed on TSCA. Chiral intermediate. mp = 108-110°; $[\alpha]^{18}$ = + 22° (c = 1, $H_2O$). *Austin Chemical Company, Inc.; Flamma s.p.a.; Sigma-Aldrich Fine Chemicals; TCI America.*

**2968 L-Glutamic acid dimethyl ester hydrochloride**
23150-65-4 245-461-0

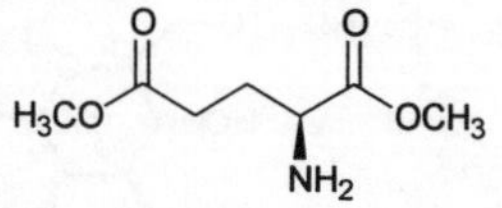

$C_7H_{14}ClNO_4$
Glutamic acid, dimethyl ester, hydrochloride, L-; Dimethyl L-2-aminoglutarate hydrochloride. Chiral intermediate. *Austin Chemical Company, Inc.*

**2969 L-(+)-Glutamic acid hydrochloride**
138-15-8 4477(12) 205-315-9

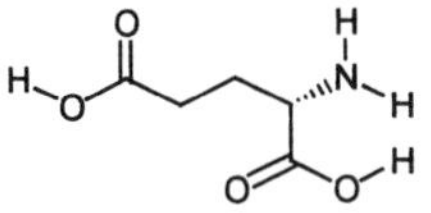

$C_5H_{10}ClNO_4$
Glutamic acid, hydrochloride, L-; Feracid; Acidulin; Acidoride; Antalka; Aciglumin; Pepsdol; Glutamidin; Acidogen; Aclor; Gastuloric; Glutan-HC1; Hydrionic; Muriamic; Glutasin; 2-Aminopentanedioic acid hydrochloride; Acidalin; Acidothyn; Acridogen; Acridoride; Glusatin; Glutan hydrochlori. Listed on TSCA. Chiral building block. mp = 214°; $[\alpha]^{19}$ = + 24.91° (c = 5.54, $H_2O$). *Aceto Corporation; Acros Organics nv; Rexim S.A. Produits Chimiques; Sigma-Aldrich Fine Chemicals; TCI America.*

**2970 L-Glutamic acid 5-methyl ester**
1499-55-4 216-110-9

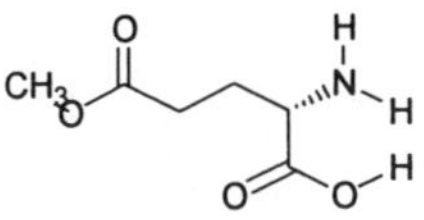

$C_6H_{11}NO_4$
Glutamic acid, 5-methyl ester, L-; (5)-Methyl L-hydrogen glutamate; (4S)-Carboxy-4-aminobutanoic acid methyl ester; γ-Methyl (S)-glutamate. Chiral intermediate. mp = 182°; [α] = + 29° (c = 2, 6N HCl). *Acros Organics nv; Sigma-Aldrich Fine Chemicals.*

**2971 L-(+)-Glutamic acid monosodium salt monohydrate**
6106-04-3 6165(11)

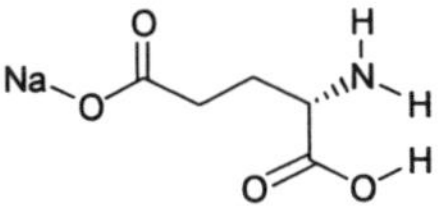

$C_5H_8NNaO_4$
Glutamic acid, monosodium salt, monohydrate, L-; Monosodium L-glutamate monohydrate; MSG monohydrate; Sodium glutamate monohydrate. Chiral building block. mp = 232°; [α] = + 25° (c = 5, 1N HCl). *Acros Organics nv; Kyowa Hakko Kogyo Co., Ltd.; Sigma-Aldrich Fine Chemicals.*

**2972 L-Glutamic acid sodium salt**
142-47-2 205-538-1
$C_5H_9NO_4$
Glutamic acid, monosodium salt, L-; Sodium L-Glutamate; Monosodium-L-glutamate; Sodium hydrogen glutamate; Accent; Ajinomoto; Ancoma; Chinese seasoning; Glutacyl; Glutavene; MSG; RL 50; Vetsin; Zest. Listed on TSCA. Chiral building block. mp = 232°; soluble in $H_2O$ (.g 10 g/100 ml, 20°). *Aceto Corporation; TCI America.*

**2973 D-Glutamine**
5959-95-5

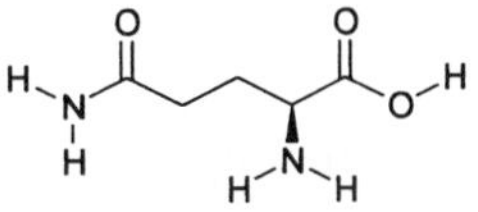

$C_5H_{10}N_2O_3$
Chiral building block. [α] = - 6.15° (c = 1, $H_2O$). *Acros Organics nv; Yoneyama Yakuhin Kogyo Co., Ltd.*

**2974 L-Glutamine**
56-85-9 4479(12) 200-292-1

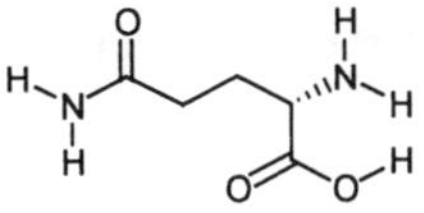

$C_5H_{10}N_2O_3$
Pentanoic acid, 2,5-diamino-5-oxo-, (S)-; Levoglutamide; (S)-2,5-Diamino-5-oxopentanoic acid; Cebrogen; Glumin; L-2-Aminoglutaramidic acid; L-Glutamic acid γ-amide; Stimulina. Listed on TSCA. Chiral building block. mp = 186°; $[\alpha]^{25}$ = + 31° (c = 2, 1N HCl). *Aceto Corporation; Acros Organics nv; Flamma s.p.a.; KingChem, Inc.; Kyowa Hakko Kogyo Co., Ltd.; Rexim S.A. Produits Chimiques; Senn Chemicals AG; Sigma-Aldrich Fine Chemicals; Tanabe Seiyaku Co. Ltd.; TCI America; Varsal Instruments, Inc.; Yoneyama Yakuhin Kogyo Co., Ltd.; Zhejiang Chemicals Import and Export Corporation.*

**2975 γ-L-Glutamyl-L-alanine**
5875-41-2

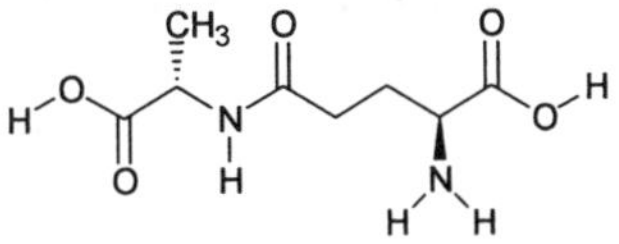

$C_8H_{14}N_2O_5$
Chiral intermediate. mp = 188-192°; $[\alpha]_D^{20}$ = - 26.5° (c = 5, $H_2O$). *Acros Organics nv; Sigma-Aldrich Fine Chemicals.*

**2976 γ-L-Glutamyl-L-glutamic acid**
1116-22-9 214-233-2

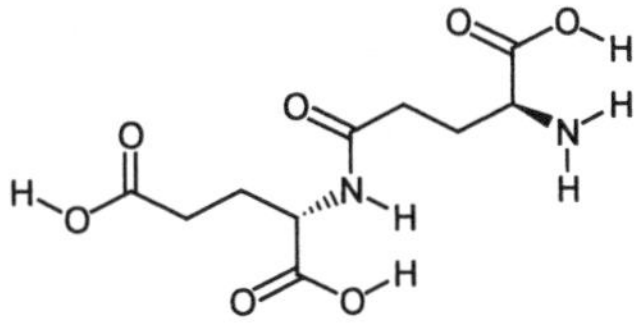

$C_{10}H_{16}N_2O_7$
Glutamic acid, N-L-γ-glutamyl-, L-. Chiral intermediate. mp = 179-182°. *Acros Organics nv.*

**2977 N-(γ-L-Glutamyl)-α-naphthylamide monohydrate**
81012-91-1

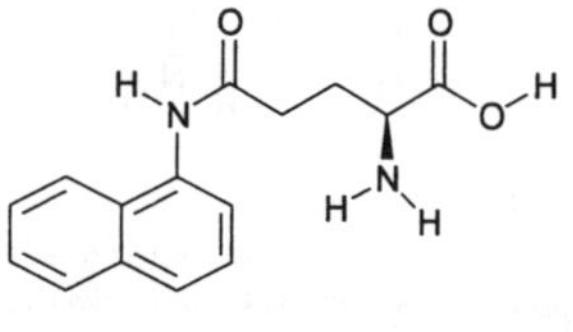

$C_{15}H_{16}N_2O_3$
Chiral intermediate. *Acros Organics nv.*

**2978 L-γ-5-Glutamyl-p-nitroanilide monohydrate**
7300-59-6 230-748-5

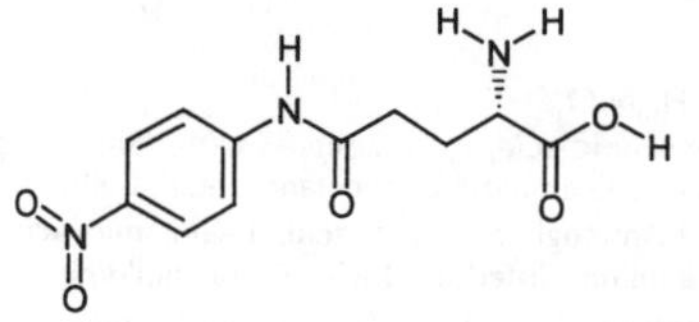

$C_{11}H_{15}N_3O_6$
Glutaranilic acid, 2-amino-4'-nitro-, L-; L-Glutamine, N-(4-nitrophenyl)-; N-(4-Nitrophenyl)-L-glutamine; (S)-2-Amino-5-[(4-nitrophenyl)amino]-5-oxopentanoic acid; L-2-Amino-4'-nitroglutaranilic acid. Listed on TSCA. Chiral diagnostic reagent. Soluble in $H_2O$. *Acros Organics nv; Diagnostic Chemicals Limited; TCI America.*

**2979 L-(-)-Glutathione hydrate, oxidized**
121-24-4

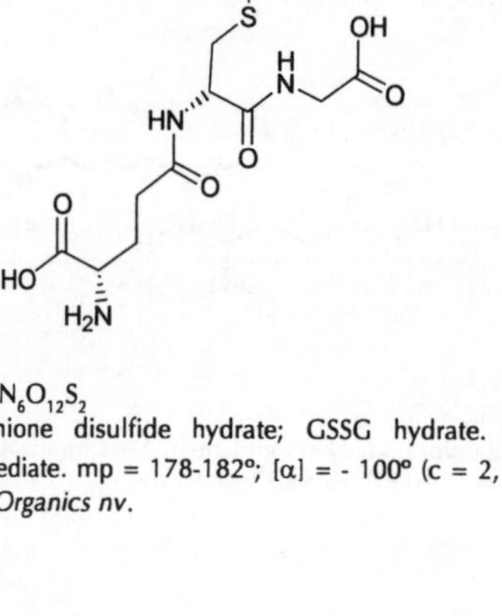

$C_{20}H_{32}N_6O_{12}S_2$
Glutathione disulfide hydrate; GSSG hydrate. Chiral intermediate. mp = 178-182°; [α] = - 100° (c = 2, $H_2O$). *Acros Organics nv.*

**2980 L-(-)-Glutathione, oxidized**
27025-41-8 4483(12) 248-170-7

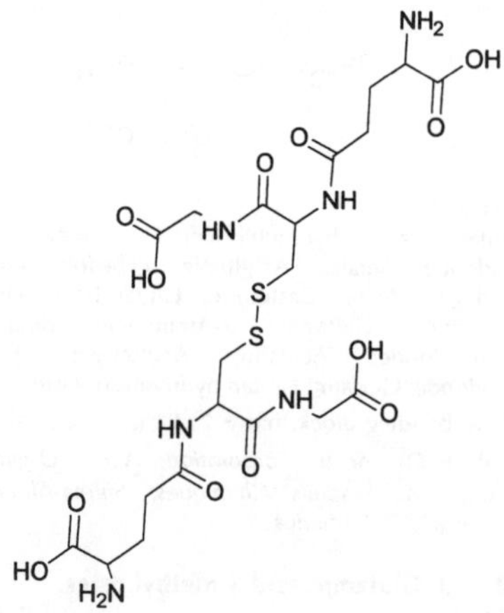

$C_{20}H_{32}N_6O_{12}S_2$
Glycine, L-γ-glutamyl-L-cysteinyl-, bimol. (2-(R)-2')-disulfide; Bi(glutathion-S-yl); Oxiglutatione; Sleep-promoting factor B; GSSG; Glutamine, N,N'-[dithiobis[1-[(carboxymethyl)carbamoyl]ethylene]]di-, L-; Glutathione-SSG; Glutathone disulfide; Glutathione-S-S-glutathione. Chiral building block. mp = 178°; $[\alpha]^{22}$ = - 98° (c = 1, $H_2O$). *Acros Organics nv; Kyowa Hakko Kogyo Co., Ltd.; Sigma-Aldrich Fine Chemicals.*

**2981 D-(+)-Glyceraldehyde**
453-17-8 4376(11) 207-217-1

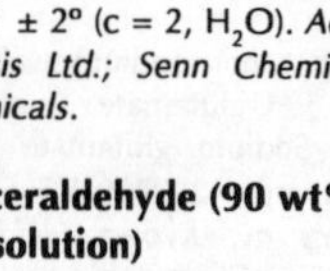

$C_3H_6O_3$
Propanal, 2,3-dihydroxy-, (2R)-; (R)-(+)-2,3-Dihydroxypropanal; Triose. Chiral building block. $n_D^{20}$ = 1.494; $[\alpha]_D^{20}$ = + 11 ± 2° (c = 2, $H_2O$). *Acros Organics nv; Lancaster Synthesis Ltd.; Senn Chemicals AG; Sigma-Aldrich Fine Chemicals.*

**2982 L-(-)-Glyceraldehyde (90 wt% aqueous solution)**
497-09-6 207-836-7

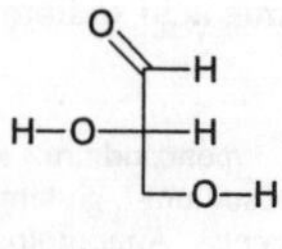

$C_3H_6O_3$
Propanal, 2,3-dihydroxy-, (2S)-; (S)-(-)-2,3-Dihydroxypropanal. Chiral building block. [α] = - 14° (c = 2, $H_2O$). *Acros Organics nv.*

**2983 (-)-Glyceric acid calcium salt dihydrate**
6057-35-8

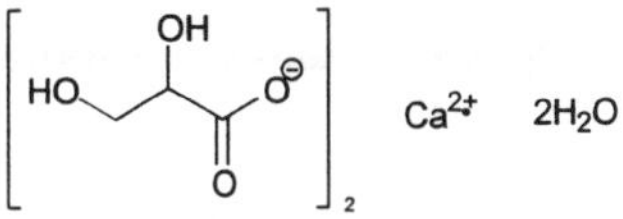

$C_3H_6O_4$
Chiral building block. mp = 134°; $[\alpha]_D^{20}$ = - 13° (c = 5, $H_2O$). *Sigma-Aldrich Fine Chemicals.*

**2984 (+)-Glyceric acid calcium salt hydrate**
14028-62-7 4378(11)
$C_3H_6O_4$
Listed on TSCA. Chiral intermediate. mp = 134°; $[\alpha]_D^{20}$= + 13° (c = 5, $H_2O$). *Sigma-Aldrich Fine Chemicals.*

**2985 (-)-Glycero-gulo-heptose**
3146-50-7 221-560-4

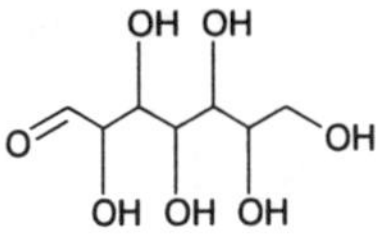

$C_7H_{14}O_7$
Chiral building block. mp = 176-178°; $[\alpha]^{24}$ = - 18° (c = 4, $H_2O$). *Sigma-Aldrich Fine Chemicals.*

**2986 (R)-(-)-Glycerol-1-tosylate**
41274-09-3

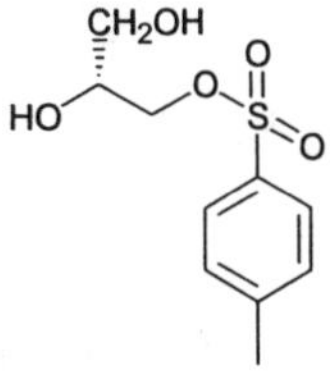

$C_{10}H_{14}O_5S$
Chiral building block. *Chemi SpA.*

**2987 (S)-(+)-Glycerol-1-tosylate**
50765-70-3

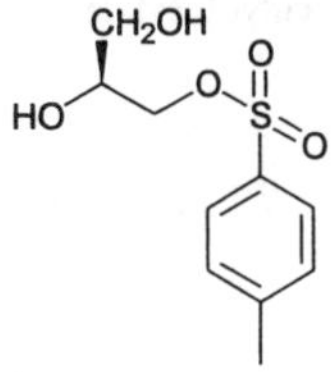

$C_{10}H_{14}O_5S$
Chiral building block. *Chemi SpA.*

**2988 (R)-Glycidol**
57044-25-4 4499(12)

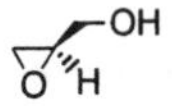

$C_3H_6O_2$
Chiral building block. $bp_{19}$ = 66-67°; d = 1.117; n = 1.43; $[\alpha]^{23}$ = + 15° (neat). *Sigma-Aldrich Fine Chemicals.*

**2989 (S)-Glycidol**
60456-23-7 4499(12)

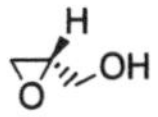

$C_3H_6O$
Oxiranemethanol, (S)-; (S)-2,3-Epoxypropan-1-ol; (S)-1,2-Epoxy-3-hydroxypropane; (S)-1-Hydroxy-2,3-epoxy-propane; 1-Propanol, 2,3-epoxy-, (S)-; (S)-2-(Hydroxy-methyl)oxirane; (S)-3-Hydroxypropylene oxide; (S)-Allyl alcohol oxide; (S)-Epihydrin alcohol. Chiral building block. $bp_{19}$ = 66-67°; d = 1.117; n = 1.433; $[\alpha]_D^{20}$ = - 15° (neat). *Orsynetics (Tradining) Ltd.; Sigma-Aldrich Fine Chemicals.*

**2990 (R)-(-)-Glycidyl butyrate**
60456-26-0

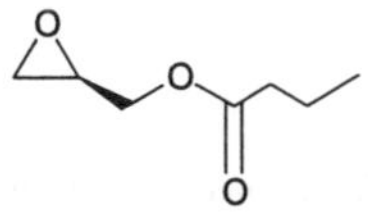

$C_7H_{12}O_3$
Chiral building block. $bp_{19}$ = 90°; d = 1.018; n = 1.428; $[\alpha]^{25}$ = -31° (neat). *Acros Organics nv; Daiso Company, Ltd.; DSM Fine Chemcials Netherlands; Sigma-Aldrich Fine Chemicals.*

**2991 (S)-(+)-Glycidyl butyrate**
65031-96-1

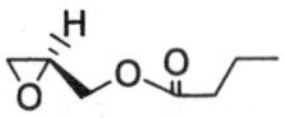

$C_7H_{12}O_3$
Chiral building block. $bp_{19}$ = 90°; d = 1.0180; $n_D^{20}$ = 1.4280; $[\alpha]_D^{20}$ = + 31 ± 2° (neat). *Daiso Company, Ltd.; Lancaster Synthesis Ltd.; Sigma-Aldrich Fine Chemicals.*

**2992 (S)-(+)-Glycidyl methyl ether**
64491-68-5

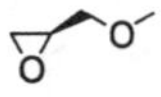

$C_4H_8O_2$
(S)-(+)-2-(Methoxymethyl)oxirane; (S)-(+)-Methyl glycidyl ether. Chiral building block. bp = 110-111°; d = 0.9820;

$n_D^{20}$ = 1.4060; $[\alpha]_D^{20}$ = + 15.6 ± 2° (c = 5, toluene). *Daiso Company, Ltd.; Lancaster Synthesis Ltd.; Sigma-Aldrich Fine Chemicals.*

**2993 (2R)-(-)-Glycidyl 4-nitrobenzoate**
106268-95-5

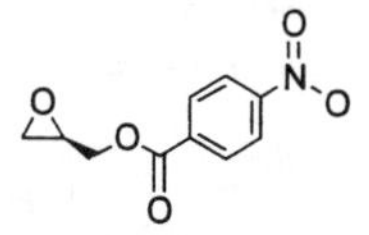

$C_{10}H_9NO_5$
Chiral intermediate. mp = 60-62°; $[\alpha]^{19}$ = - 37° (c = 1.08, $CHCl_3$). *Sigma-Aldrich Fine Chemicals.*

**2994 (2S)-(+)-Glycidyl 4-nitrobenzoate**
115459-65-9

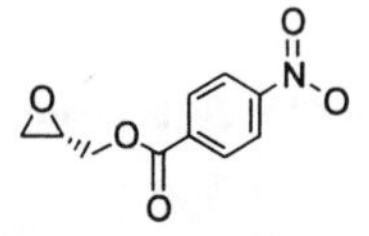

$C_{10}H_9NO_5$
Chiral intermediate. mp = 60-62°; $[\alpha]^{19}$ = + 37° (c = 1.08, $CHCl_3$). *Sigma-Aldrich Fine Chemicals.*

**2995 (R)-(-)-Glycidyl-3-nosylate**
115314-17-5

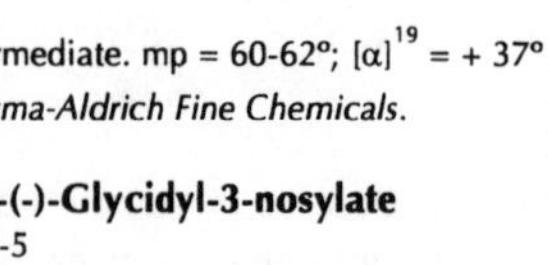

$C_9H_9NO_6S$
(2R)-(-)-Glycidyl-3-nitrobenzenesulfonate. Chiral building block. mp = 62-64°; $[\alpha]^{21}$ = - 22° (c = 2, $CHCl_3$). *Acros Organics nv; Chirex, Inc.; Daiso Company, Ltd.; Nagase & Co., Ltd.; Omega Chemical Company Inc.; Sigma-Aldrich Fine Chemicals.*

**2996 (S)-(+)-Glycidyl-3-nosylate**
115314-14-2

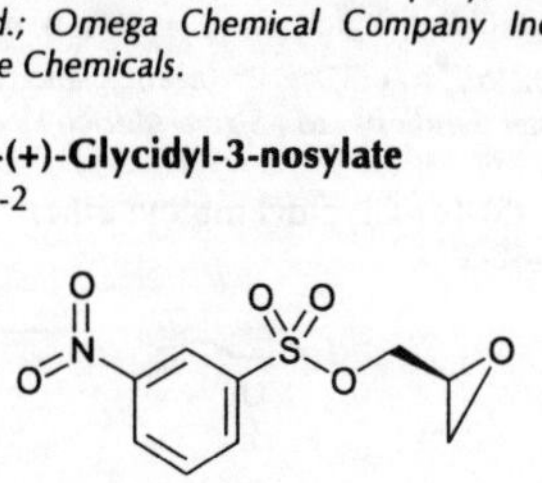

$C_9H_9NO_6S$
(2S)-(-)-Glycidyl-3-nitrobenzenesulfonate. Chiral building block. mp = 62-64°; $[\alpha]^{26}$ = + 22° (c = 2, $CHCl_3$). *Acros Organics nv; Chirex, Inc.; Daiso Company, Ltd.; Nagase & Co., Ltd.; Omega Chemical Company Inc.; Orsynetics (Tradining) Ltd.; Sigma-Aldrich Fine Chemicals.*

**2997 Glycidyl nosylate - 4 nitro derivative**
123750-60-7

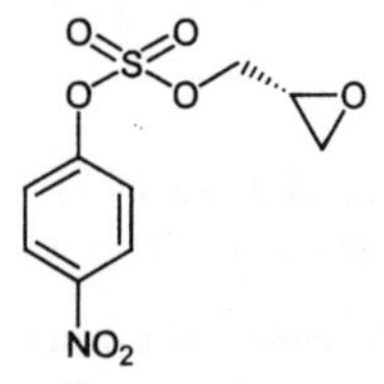

$C_9H_9NO_7S$
Chiral building block. *Chirex, Inc.*

**2998 (S)-Glycidyl nosylate -4-nitro derivative**
118712-60-0

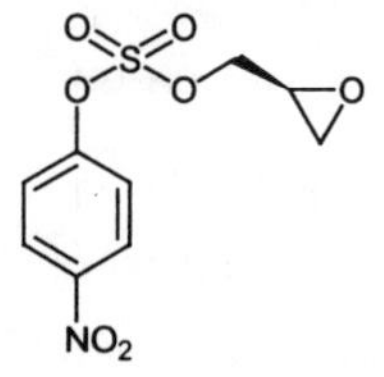

$C_9H_9NO_7S$
Chiral building block. *Chirex, Inc.*

**2999 (2R)-(-)-Glycidyl tosylate**
113826-06-5

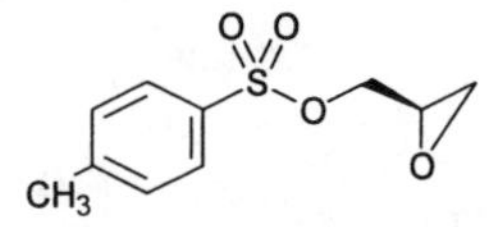

$C_{10}H_{12}O_4S$
p-Toluenesulfonic acid (2R)-(-)-glycidyl ester; (S)-Oxiranemethanol 4-methylbenzenesulfonate; (2R)-(-)-Glycidyl p-toluenesulfonate. Chiral building block. mp = 48°; $[\alpha]^{19}$ = - 17° (c = 2.75, $CHCl_3$). *Acros Organics nv; Chirex, Inc.; Daiso Company, Ltd.; DSM Fine Chemcials Netherlands; Omega Chemical Company Inc.; Sigma-Aldrich Fine Chemicals; TCI America.*

**3000 (S)-Glycidyl tosylate**

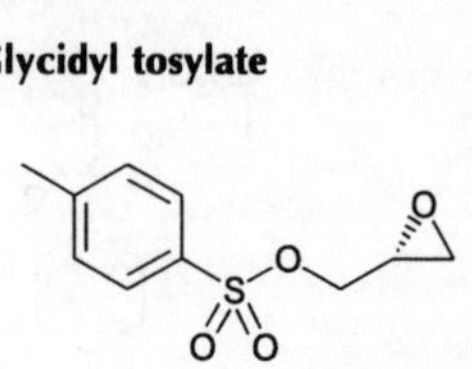

$C_{10}H_{12}O_4S$
Chiral building block. *Austin Chemical Company, Inc.; Interchem Corporation; Nagase & Co., Ltd.; Orsynetics (Tradining) Ltd.; Vinchem.*

**3001 (R)-(+)-Glycidyl triphenylmethyl ether**
65291-30-7

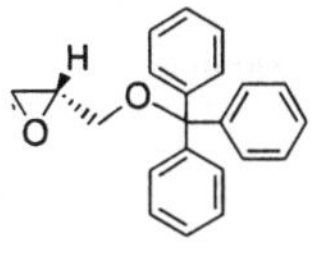

$C_{22}H_{20}O_2$
(R)-(+)-Glycidyl trityl ether; (R)-Tritylglycidol. Chiral intermediate. mp = 99-102°; $[\alpha]_D^{20}$ = + 10.5 ± 2° (c = 1, $CHCl_3$). *Daiso Company, Ltd.; Lancaster Synthesis Ltd.; Sigma-Aldrich Fine Chemicals.*

**3002 (S)-(-)-Glycidyl triphenylmethyl ether**
129940-50-7

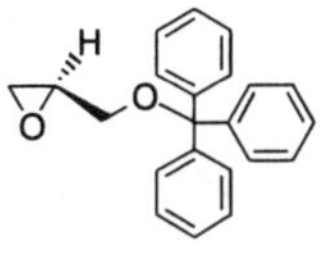

$C_{22}H_{20}O_2$
(S)-(-)-Glycidyl trityl ether; S-Trityl glycidyl ether; (S)-Tritylglycidol. Chiral intermediate. mp = 99-102°; $[\alpha]_D^{20}$ = - 10.5 ± 2° (c = 1, $CHCl_3$). *Daiso Company, Ltd.; Lancaster Synthesis Ltd.; Sigma-Aldrich Fine Chemicals.*

**3003 Glycyl-(-)-camphor sultam, hydrochloride**

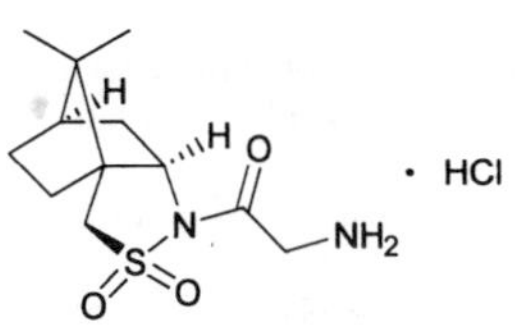

$C_{12}H_{21}ClN_2O_3S$
Resolving agent. *Isochem.*

**3004 Glycyl-(+)-camphor sultam, hydrochloride**

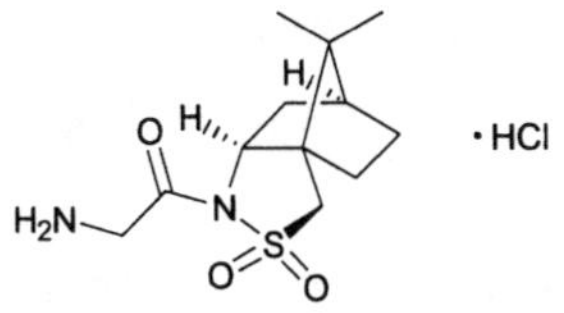

$C_{12}H_{21}ClN_2O_3S$
Chiral intermediate. *Isochem.*

**3005 Glycyl-L-glutamic acid**
7412-78-4 231-019-4

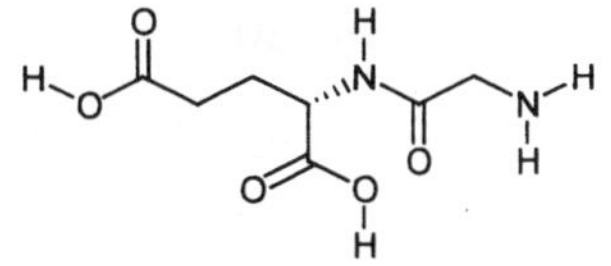

$C_7H_{12}N_2O_5$
Glutamic acid, N-glycyl-, L-.
Chiral intermediate. mp = 155-158°; $[\alpha]^{22}$ = - 4.7° (c = 5, $H_2O$). *Acros Organics nv; Sigma-Aldrich Fine Chemicals.*

**3006 Glycyl-L-glutamine monohydrate**
13115-71-4

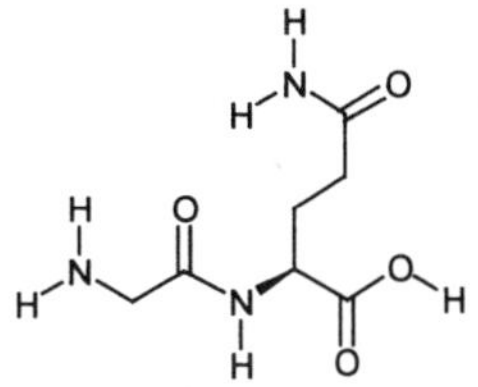

$C_7H_{13}N_3O_4$
(S)-Atrolactic acid. Chiral intermediate. $[\alpha]$ = - 1.7° (c = 2, $H_2O$). *Acros Organics nv; Degussa-Huls AG; Kyowa Hakko Kogyo Co., Ltd.; Rexim S.A. Produits Chimiques.*

**3007 Glycyl-L-histidyl-L-lysine acetate**
62024-09-3

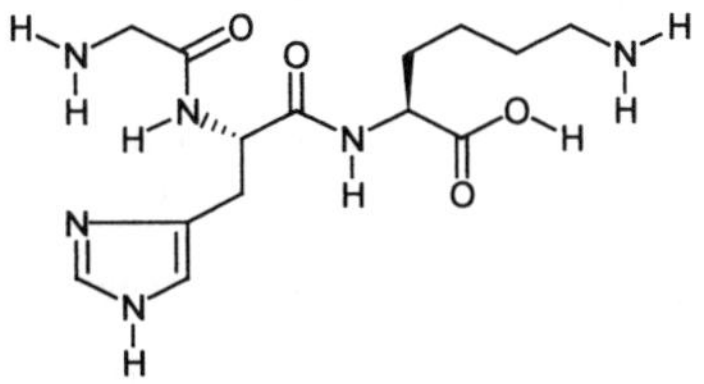

$C_{14}H_{24}N_6O_4$
Chiral intermediate. *Acros Organics nv.*

**3008 N-Glycyl-L-isoleucine**
19461-38-2 243-085-1

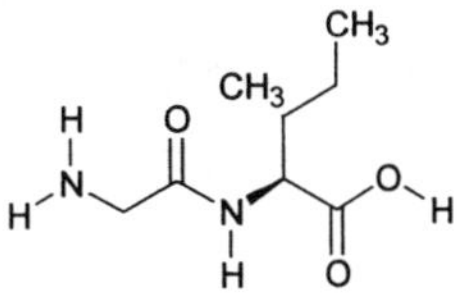

$C_8H_{16}N_2O_3$
Isoleucine, N-glycyl-, L-. Chiral intermediate. $[\alpha]$ = - 14.5° (c = 2 $H_2O$). *Acros Organics nv.*

**3009 N-Glycyl-L-leucine**
869-19-2 212-785-9

$C_8H_{16}N_2O_3$
Leucine, N-glycyl-, L-. Chiral intermediate. mp = 233-235°; $[\alpha]^{22}$ = - 34.5° (c = 2, $H_2O$). *Acros Organics nv; Sigma-Aldrich Fine Chemicals.*

**3010 N-Glycyl-L-tyrosine hydrate**
658-79-7 211-525-1

$C_{11}H_{14}N_2O_4$
Tyrosine, N-glycyl-, L-. Chiral intermediate. [α] = + 45.6° (c = 1, $H_2O$). *Acros Organics nv; Rexim S.A. Produits Chimiques; Sigma-Aldrich Fine Chemicals.*

**3011 N-Glycyl-L-valine**
1963-21-9 217-806-5

$C_7H_{14}N_2O_3$
Valine, N-glycyl-, L-. Chiral intermediate. mp = 248°; [α = - 21° (c = 2, $H_2O$). *Acros Organics nv.*

**3012 (+)-Glycyrrhizic acid monoammonium salt trihydrate**
53956-04-0 4401(11) 258-887-7

$C_{42}H_{71}NO_{19}$
Glucopyranosiduronic acid, (3b,20b)-20-carboxy-11-oxo-30-norolean-12-en-3-yl 2-O-β-D-glucopyranuronosyl-, α-D-, monoammonium salt. Listed on TSCA. Chiral intermediate. mp = 209°; $[\alpha]^{25}$ = + 47.8° (c = 1.6, 40% EtOH). *Acros Organics nv; Sigma-Aldrich Fine Chemicals.*

**3013 (+)-Griseofulvin**
126-07-8 4571(12) 204-767-4

$C_{17}H_{17}ClO_6$
Chiral intermediate. mp = 218-220°; $[\alpha]^{17}$ = + 370° (c = 1, $CHCl_3$). *Sigma-Aldrich Fine Chemicals.*

**3014 (-)-Guaiol**
489-86-1 4583(12) 207-702-8

$C_{15}H_{26}O$
Listed on TSCA. Chiral intermediate. mp = 91-93°; $[\alpha]_D^{20}$ = - 28° (c = 4, EtOH). *Sigma-Aldrich Fine Chemicals.*

**3015 (S)-(-)-2-Guanidinoglutaric acid**
73477-53-9
$C_6H_{11}N_3O_4$
Chiral intermediate. *TCI America.*

**3016 (S)-(-)-2-Guanidinoglutaric acid (S)-methylisothiourea salt**
$C_8H_{17}N_5O_4S$
Chiral intermediate. *TCI America.*

**3017 (-)-Guanosine**
118-00-3 204-227-8

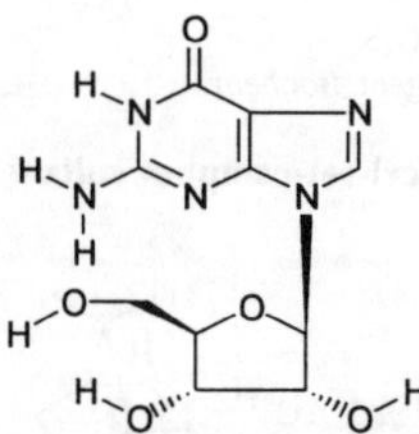

$C_{10}H_{13}N_5O_5$
Guanosine; 2-Amino-9-β-D-ribofuranosyl-9H-purine-6-(1H)-one; 6H-Purin-6-one, 2-amino-1,9-dihydro-9-β-D-ribofuranosyl-; Guanine, 9-β-D-ribofuranosyl-; Inosine, 2-amino-; Vernin; β-D-Ribofuranoside, guanine-9. Chiral intermediate. mp = 250°; [α] = - 62° (c = 3, 0.1N NaOH). *Acros Organics nv.*

**3018 D-Gulono-1,4-lactone**
6322-07-2 228-678-5

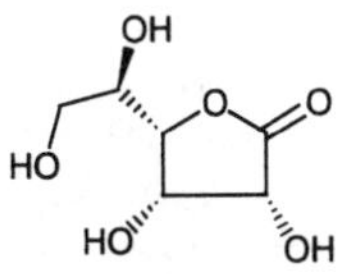

$C_6H_{10}O_6$
Gulonic acid, γ-lactone, D-; D-Glucono-γ-lactone. Chiral building block. mp = 182-188°; $[\alpha]_D^{20}$ = - 56.2° initial (c = 4, $H_2O$). *Acros Organics nv; Mitsubishi Rayon Co., Ltd.; Pfanstiehl Laboratories, Inc.; Senn Chemicals AG; Sigma-Aldrich Fine Chemicals.*

**3019 L-Gulonolactone**
1128-23-0

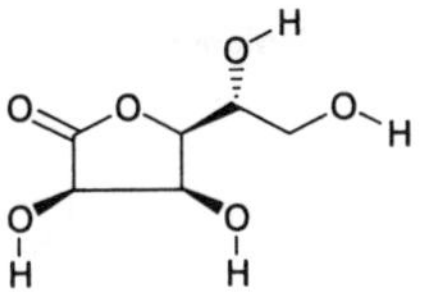

$C_6H_{10}O_6$
L-Gulono-1,4-lactone; L-(+)-Gulonic acid γ-lactone. Chiral building block. mp = 188°; [α] = + 55° (c = 4, $H_2O$). *Acros Organics nv; Senn Chemicals AG; Sigma-Aldrich Fine Chemicals; TCI America.*

**3020 D-(-)-Gulose**
4205-23-6 224-118-9

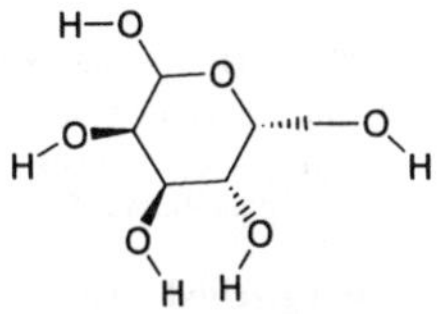

$C_6H_{12}O_6$
Gulose, D-. Chiral building block. [α] = - 22° (c = 5 $H_2O$). *Acros Organics nv.*

**3021 L-(+)-Gulose**
6027-89-0 227-897-3

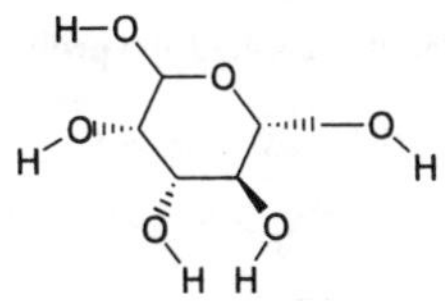

$C_6H_{12}O_6$
Gulose, L-; Chiral building block. *Acros Organics nv; TCI America.*

**3022 (S)-H8-BINAP**
139139-93-8
$C_{44}H_{40}P_2$
Chiral ligand. *Digital Specialty Chemicals, Inc.*

**3023 (R)-H8-BINAP**
139139-86-9
$C_{44}H_{40}P_2$
Chiral ligand. *Digital Specialty Chemicals, Inc.*

**3024 (-)-Hecogenin acetate**
915-35-5 213-021-7

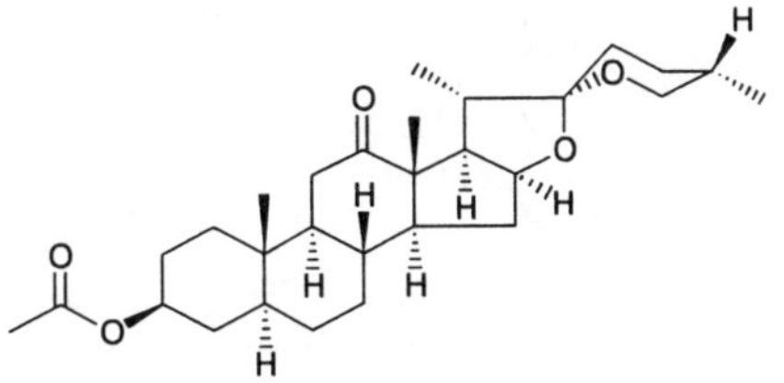

$C_{29}H_{44}O_5$
Spirostan-12-one, 3-(acetyloxy)-, (3β,5α,25R)-; 12-oxo-5-α-spirostan-3-β-yl acetate. Chiral intermediate. mp = 246-248°; $[\alpha]_D^{20}$ = - 6.0° (c = 1, $CHCl_3$). *Acros Organics nv; Sigma-Aldrich Fine Chemicals.*

**3025 (+)-Hederin hydrate**
4541(11)

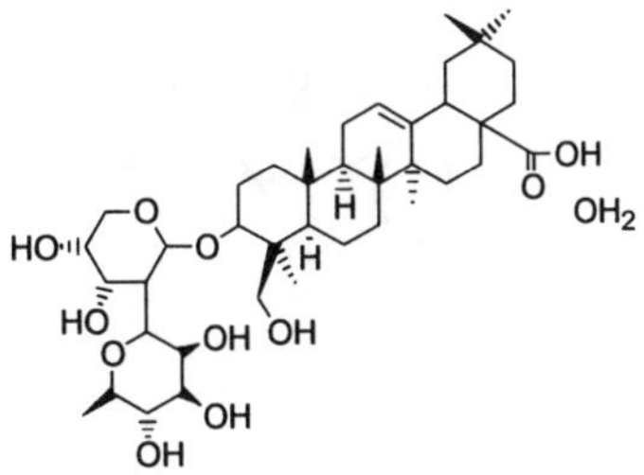

$C_{41}H_{66}O_{12}$
Chiral intermediate. mp = 261°; $[\alpha]^{19}$ = + 18.2° (c = 1, $CH_3OH$). *Sigma-Aldrich Fine Chemicals.*

**3026 (-)-Helicin**
618-65-5 4658(12) 210-558-9

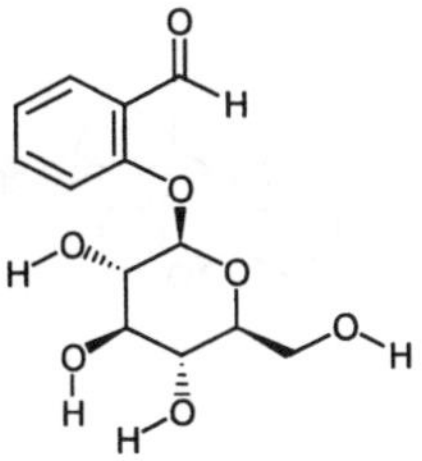

$C_{13}H_{16}O_7$
Benzaldehyde, 2-(β-D-glucopyranosyloxy)-; 2-(β-D-glu-

copyranosyloxy)benzaldehyde; Salicylaldehyde 3-O-acetyl-β-D-glucopyranoside. Chiral intermediate. mp = 178°; $[\alpha]^{25}$ = - 57.1° (c = 1.5, $H_2O$). *Acros Organics nv; Senn Chemicals AG; Sigma-Aldrich Fine Chemicals.*

**3027 (-)-Hepta-O-acetyl-cellobiosyl-β-azide**
33012-50-9

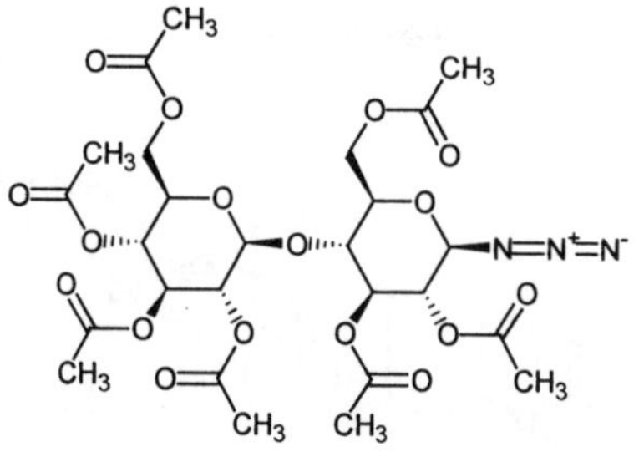

$C_{26}H_{35}N_3O_{17}$
Chiral intermediate. mp = 174-176°; [α] = - 31° (c = 0.44, $CHCl_3$). *Acros Organics nv.*

**3028 3-Heptafluorobutyryl-(+)-camphor**
51800-99-8 257-430-9

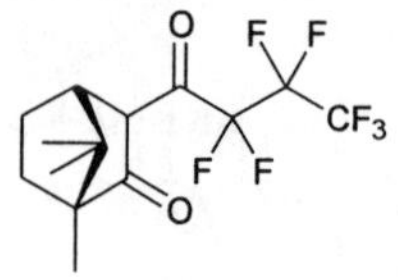

$C_{14}H_{15}F_7O_2$
Resolving agent. $bp_{0.2}$ = 60-70°; d = 1.218; n = 1.421; $[\alpha]_D^{20}$ = + 125° (c = 2.6, $CHCl_3$). *Sigma-Aldrich Fine Chemicals.*

**3029 3-Heptaflurobutyryl-(-)-camphor**
115224-00-5

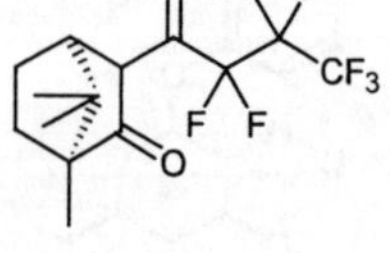

$C_{14}H_{15}F_7O_2$
Resolving agent. d = 1.218; n = 1.422; $[\alpha]^{19}$ = - 123° (c = 2.6, $CHCl_3$). *Sigma-Aldrich Fine Chemicals.*

**3030 (+)-Heptakis(2,3,6-tri-O-methyl)-cyclodextrin**
55216-11-0

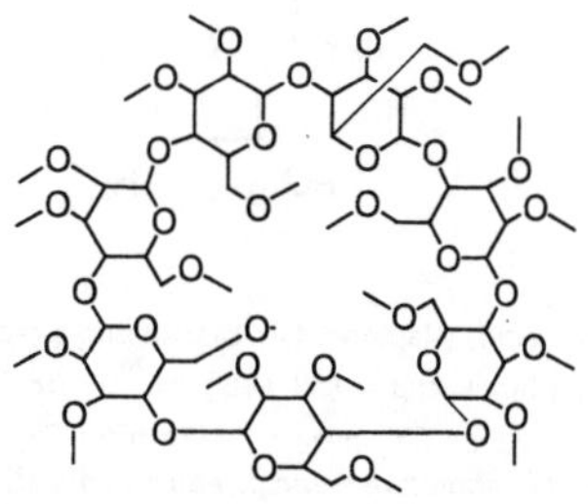

$C_{63}H_{112}O_{35}$
Chiral auxiliary. mp = 170-178°; $[\alpha]_D^{20}$ = + 156° (c = 1, $H_2O$). *Sigma-Aldrich Fine Chemicals.*

**3031 (R)-(-)-2-Heptanol**
6033-24-5

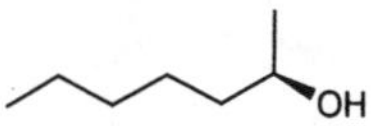

$C_7H_{16}O$
Heptan-2-ol, (R)-; l-2-Hydroxyheptane; l-Amyl-methyl-carbinol. Chiral building block. *Norse Laboratories.*

**3032 (S)-(+)-2-Heptanol**
6033-23-4 4583(11)

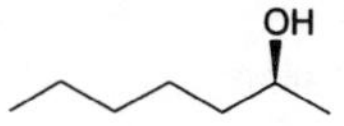

$C_7H_{16}O$
Heptan-2-ol, (S)-; D-2-Hydroxyheptane; D-Amyl-methylcarbinol. Chiral building block. $bp_{23}$ = 74-75°; d = 0.818; n = 1.419; [α] = + 10.2° (neat). *Acros Organics nv; Norse Laboratories; Sigma-Aldrich Fine Chemicals.*

**3033 (+)-Heptyl glycidyl ether**
121906-44-3

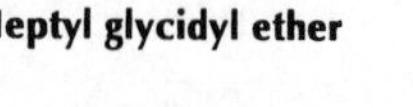

$C_{10}H_{20}O_2$
Chiral building block. $bp_2$ = 70-71°; d = 0.895; n = 1.432; $[\alpha]_D^{20}$ = + 8.6° (neat). *Japan Energy Corporation; Sigma-Aldrich Fine Chemicals.*

**3034 N-Heptyl-L-4-hydroxyproline monohydrate**

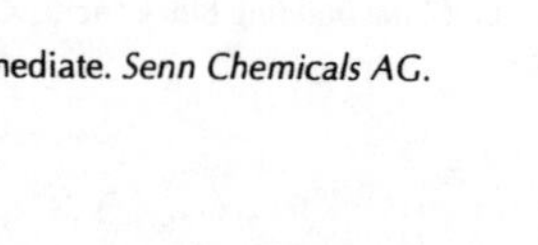

$C_{12}H_{25}NO_4$
Chiral intermediate. *Senn Chemicals AG.*

**3035 (S)-1-Heptyloxy-2-propanol**

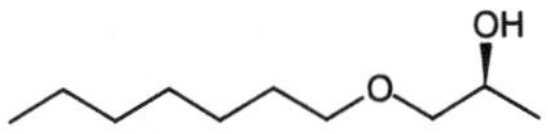

$C_{10}H_{22}O_2$
Chiral building block. *Japan Energy Corporation.*

**3036 n-Heptyl-β-D-thioglycoside**

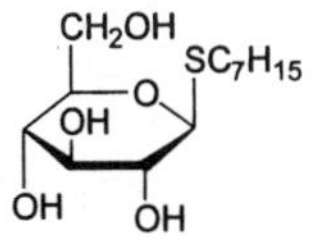

$C_{13}H_{26}O_5S$
Chiral intermediate. *Senn Chemicals AG.*

**3037 (S)-Hesperetin**
520-33-2 208-290-2
$C_{16}H_{14}O_6$
Benzo[4H]-pyran-4-one, 2,3-dihydro-5,7-dihydroxy-2-(3-hydroxy-4-methoxyphenyl)-, (2S)-; 3',5,7-Trihydroxy-4'-methoxyflavon; (S)-2,3-Dihydro-5,7-dihydroxy-2-(3-hydroxy-4-methoxyphenyl)-4-benzopyrone; Eriodictyol 4'-monomethyl ether; Flavanone, 3',5,7-trihydroxy-4'-methoxy-. Pharmaceutical or derivative. mp = 226-228°. *Kaden Biochemicals GmbH.*

**3038 L-Hesperidin**
520-26-3 4705(12) 208-288-1

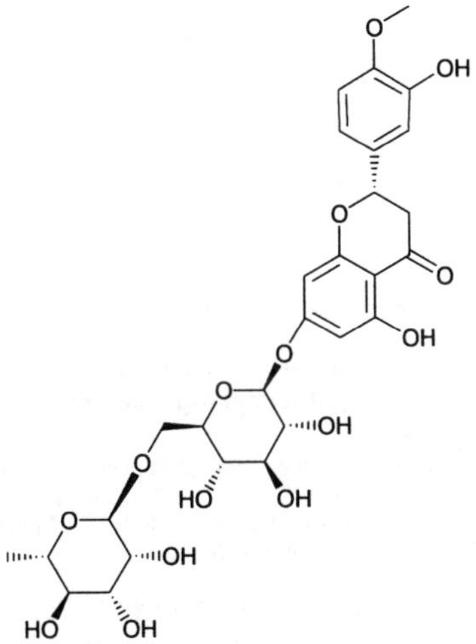

$C_{28}H_{34}O_{15}$
Benzo[4H]-pyran-4-one, 7-[[6-O-(6-deoxy-α-L-mannopyranosyl)-β-D-glucopyranosyl]oxy]-2,3-dihydro-5-hydroxy-2-(3-hydroxy-4-methoxyphenyl)-, (2S)-; 7-[[6-O-(6-deoxy-α-L-mannopyranosyl)-β-D-glucopyranosyl]oxy]-2,3-dihydro-5-hydroxy-2-(3-hydroxy-4-methoxyphenyl)-4H-1-benzopyran-4-one; Cirantin; Flavanone, 3',5,7-trihydroxy-4'-methoxy-, 7-(6-O-α-L-rhamnosyl-D-glucoside); Hesperetin 7-rhamnoglucosi. Listed on TSCA. Pharmaceutical or derivative. mp = 257-260°; $[\alpha]_D^{20}$ = - 64.8° (c = 2, pyridine). *Acros Organics nv; Kaden Biochemicals GmbH; Sigma-Aldrich Fine Chemicals.*

**3039 (R)-Hexadecanediol**
61490-71-9

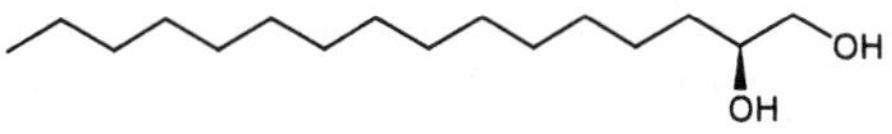

$C_{16}H_{34}O_2$
Chiral building block. *Kankyo Kagaku Center Co., Ltd.*

**3040 (S)-1,2-Hexadecanediol**
61490-70-8

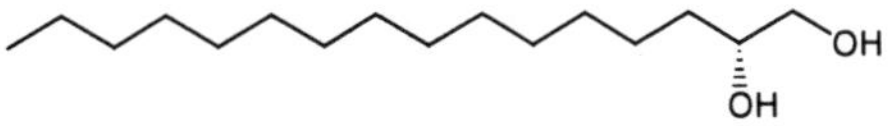

$C_{16}H_{34}O_2$
Chiral building block. *Kankyo Kagaku Center Co., Ltd.*

**3041 (3αS,4R,5S,6αR)-(+)-Hexahydro-5-hydroxy-4-(hydroxymethyl)-2H-cyclopentafuran-2-one**
76704-05-7

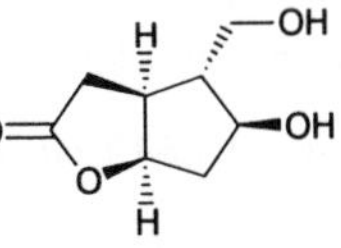

$C_8H_{12}O_4$
(+)-Corey lactone. Chiral intermediate. mp = 114-118°; $[\alpha]_D^{20}$ = + 44° (c = 2.3, $CH_3OH$). *Sigma-Aldrich Fine Chemicals.*

**3042 (+)-Hexahydro-3a-hydroxy-7a-methyl-1H-inden-1,5(6H)-dione**
33879-04-8

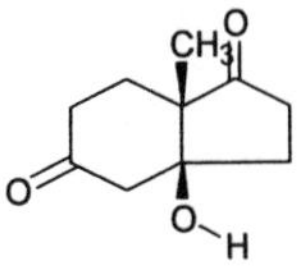

$C_{10}H_{14}O_3$
Chiral intermediate. mp = 116-118°; $[\alpha]^{19}$ = + 55° (c = 1, $CHCl_3$). *Acros Organics nv; Sigma-Aldrich Fine Chemicals.*

**3043 (R)-(-)-Hexahydromandelic acid**
53585-93-6

$C_8H_{14}O_3$
(-)-α-Hydroxycyclohexaneacetic acid. Chiral intermediate. mp = 127-129°; $[\alpha]^{18}$ = - 23° (c = 1, HOAc). *Sigma-Aldrich Fine Chemicals.*

**3044 (S)-(+)-Hexahydromandelic acid**
61475-31-8

$C_8H_{14}O_3$
(+)-α-Hydroxycyclohexaneacetic acid. Chiral intermediate. mp = 129-131°; $[\alpha]^{18}$ = + 23° (c = 1, HOAc). *Sigma-Aldrich Fine Chemicals.*

**3045 N,N-Hexamethylenebis(tributyl-ammonium-lactate) dihydrate**

$C_{36}H_{76}N_2O_6$
Chiral intermediate. d = 1.011; n = 1.481; $[\alpha]^{22}$ = - 4° (c = 4, $H_2O$). *Sigma-Aldrich Fine Chemicals.*

**3046 (2R,5R)-(-)-2,5-Hexanediol**

$C_6H_{14}O_2$
Hexane-2,5-diol, (2R,5R)-; (2R,5R)-Dihydroxyhexane; (2R,5R)-Diisopropanol. Chiral building block. mp = 50-53°; bp = 212-215°; $[\alpha]_D^{20}$ = - 35 ± 2° (c = 9, $CHCl_3$). *ChiroTech Technology Ltd.; Digital Specialty Chemicals, Inc.; Lancaster Synthesis Ltd.; Strem Chemicals, Inc.*

**3047 (2S,5S)-Hexanediol**
34338-96-0

$C_8H_{18}O_2$
(2S,5S)-Dihydroxyhexane. Chiral building block. mp = 50-53°; bp = 212-215°; $[\alpha]_D^{20}$ = + 35 ± 2° (c = 9, $CHCl_3$); $[\alpha]_D$ = + 39.4° (c = 1, $CHCl_3$). *Acros Organics nv; ChiroTech Technology Ltd.; Digital Specialty Chemicals, Inc.; Lancaster Synthesis Ltd.; Rohner Ltd.; Sigma-Aldrich Fine Chemicals; Strem Chemicals, Inc.*

**3048 (R)-(-)-2-Hexanol**
26549-24-6

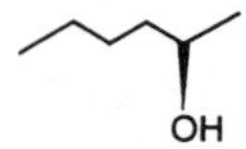

$C_6H_{14}O$
Hexan-2-ol, (R)-. Chiral building block. bp = 137-138°; d = 0.818; n = 1.414; $[\alpha]^{24}$ = - 11° (neat). *Norse Laboratories; Sigma-Aldrich Fine Chemicals.*

**3049 (S)-(+)-2-Hexanol**
52019-78-0

$C_6H_{14}O$
Hexan-2-ol, (S)-. Chiral building block. bp = 137-138°; d = 0.8180; $n_D^{20}$ = 1.4140; $[\alpha]_D^{20}$ = + 10.5 ± 0.5° (neat). *Acros Organics nv; Lancaster Synthesis Ltd.; Norse Laboratories; Sigma-Aldrich Fine Chemicals.*

**3050 N-Hexanoyl-D-erythro sphingosine**
124753-97-5

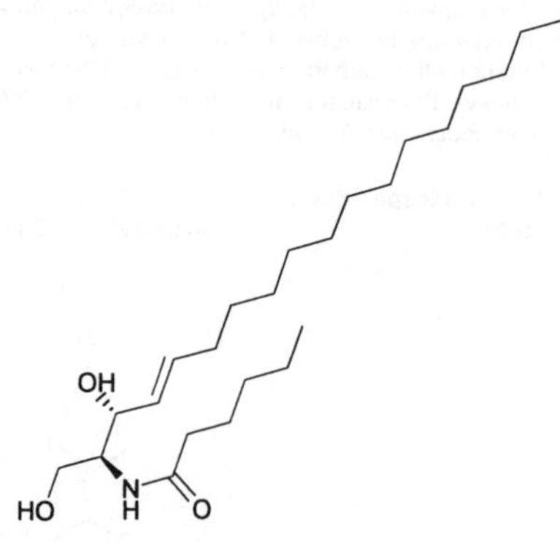

$C_{24}H_{47}NO_3$
Chiral intermediate. *Acros Organics nv.*

**3051 (4R,7R)-(-)-1,1,4,7,10,10-Hexaphenyl-1,4,7,10-tetraphosphadecane**

$C_{42}H_{42}P_4$
Phosphine, ethylenebis[[2-(diphenylphosphino)ethyl]-phenyl-, (4R,7R)-; 1,4,7,10-Tetraphosphadecane, 1,1,4,7,10,10-hexaphenyl-, (4R,7R)-. Chiral ligand. *Digital Specialty Chemicals, Inc.*

**3052 (4S,7S)-(+)-1,1,4,7,10,10-Hexaphenyl-1,4,7,10-tetraphosphadecane**

$C_{42}H_{42}P_4$
Phosphine, ethylenebis[[2-(diphenylphosphino)ethyl]-phenyl-, (4S,7S)-; 1,4,7,10-Tetraphosphadecane, 1,1,4,7,10,10-hexaphenyl-, (4S,7S)-. Chiral ligand. *Digital Specialty Chemicals, Inc.*

**3053 D-Hexose diphosphate calcium salt**
34378-77-3 251-977-7
$C_6H_{14}O_{12}P_2$
Fructose, 1,6-bis(dihydrogen phosphate), calcium salt, D-. Calcium fructose 1,6-diphosphate. Chiral intermediate. *TCI America.*

**3054 Hexyl-β-D-glucopyranoside**

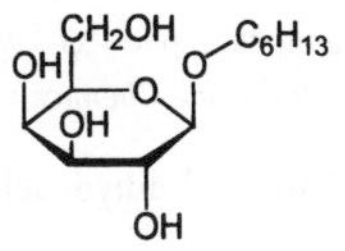

$C_{12}H_{24}O_6$
Chiral intermediate. *Senn Chemicals AG.*

**3055 D-S-2-3-(Hexyloxy)benzoylvinylglutathione**

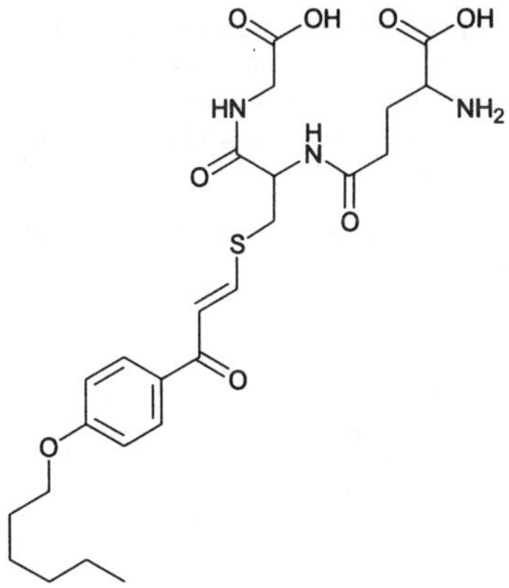

$C_{25}H_{35}N_3O_8S$
Chiral intermediate. mp = 195°. *Sigma-Aldrich Fine Chemicals.*

**3056 (S)-1-Hexyloxy-2-propanol**

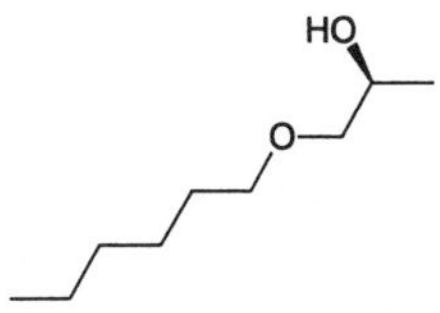

$C_9H_{20}O_2$
Propan-2-ol, 1-(hexyloxy)-, (S)-. Chiral building block. *Japan Energy Corporation.*

**3057 Hexyl-β-D-thioglucopyranoside**

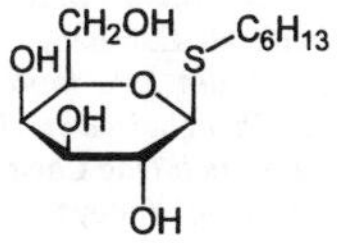

$C_{12}H_{24}O_5S$
Chiral intermediate. *Senn Chemicals AG.*

**3058 Hippuryl-L-histidyl-L-leucine, tetrahydrate**
31373-65-6 250-597-9

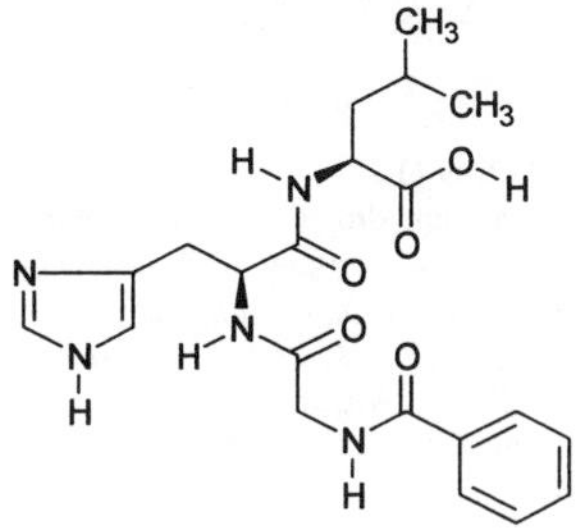

$C_{21}H_{27}N_5O_5$
Leucine, N-(N-hippuroyl-L-histidyl)-, L-. N-[N-(N-Benzoylglycyl)-L-histidyl]-L-leucine. Chiral intermediate. mp = 108-110°; [α] = + 6.8° (c = 2, $CH_3OH$). *Acros Organics nv; Sigma-Aldrich Fine Chemicals.*

**3059 D-Histidine**
351-50-8 206-513-8

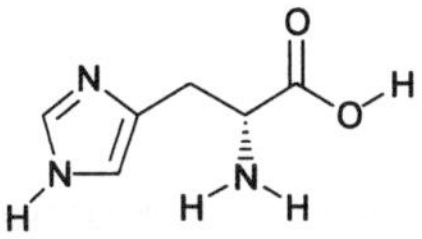

$C_6H_9N_3O_2$
Histidine, D-. Listed on TSCA. Chiral building block. [α] = - 10° (c = 2, 2.5N HCl). *Acros Organics nv; Austin Chemical Company, Inc.; Kyowa Hakko Kogyo Co., Ltd.; Rexim S.A. Produits Chimiques.*

**3060 L-Histidine**
71-00-1 4758(12) 200-745-3

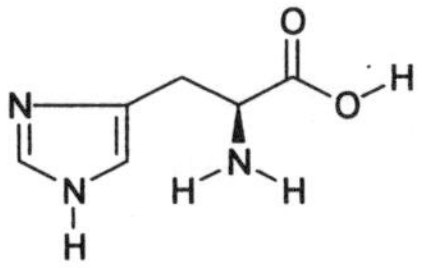

$C_6H_9N_3O_2$
Imidazole-4-propanoic acid, α-amino-, 1H-, (S)-; (S)-α-Amino-1H-imidazole-4-propanoic acid; (S)-4-(2-Amino-2-carboxyethyl)imidazole; 1H-Imidazole-4-alanine, (S)-;

Glyoxaline-5-alanine; L-Alanine, 3-(1H-imidazol-4-yl)-. Listed on TSCA. Chiral building block. mp = 277°; $[\alpha]^{23}$ = + 38.4° (c = 3, $H_2O$). *Aceto Corporation; Acros Organics nv; Austin Chemical Company, Inc.; Kyowa Hakko Kogyo Co., Ltd.; Rexim S.A. Produits Chimiques; Senn Chemicals AG; Shanghai DSL International Trading Company; Sigma-Aldrich Fine Chemicals; Tanabe Seiyaku Co. Ltd.; TCI America; Yoneyama Yakuhin Kogyo Co., Ltd.*

### 3061 L-Histidine hydrochloride monohydrate

5934-29-2 4642(11)

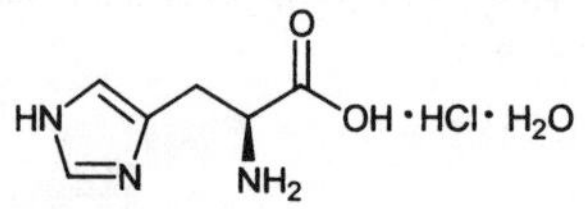

$C_6H_{10}ClN_3O_2 \cdot H_2O$

Histidine, monohydrochloride, monohydrate, L-. Chiral building block. mp = 254°; $[\alpha]_D^{20}$ = + 9.5° (c = 1, 6N HCl). *Aceto Corporation; Acros Organics nv; Kyowa Hakko Kogyo Co., Ltd.; Rexim S.A. Produits Chimiques; Shanghai DSL International Trading Company; Sigma-Aldrich Fine Chemicals; Tanabe Seiyaku Co. Ltd.; TCI America.*

### 3062 L-(+)-Histidine methyl ester dihydrochloride

7389-87-9 230-973-9

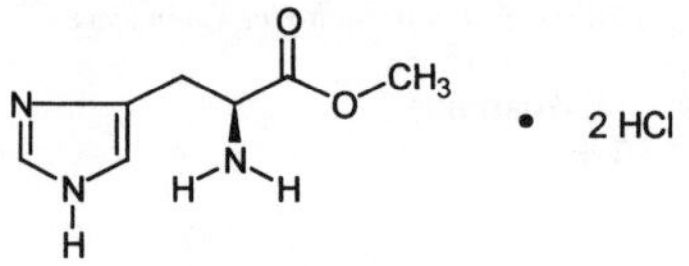

$C_7H_{13}Cl_2N_3O_2$

Histidine, methyl ester, dihydrochloride, L-; Methyl L-histidinate dihydrochloride. Chiral building block. mp = 207°; $[\alpha]_D^{20}$ = + 9.0° (c = 2, $H_2O$). *Acros Organics nv; Austin Chemical Company, Inc.; Sigma-Aldrich Fine Chemicals.*

### 3063 D-Histidine monohydrochloride monohydrate

6341-24-8 4642(11) 228-733-3

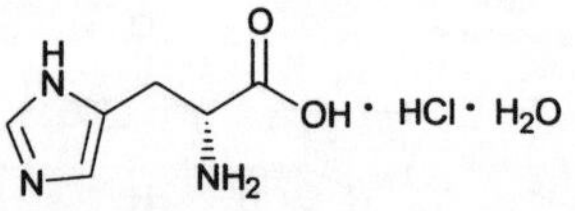

$C_6H_{10}ClN_3O_2$

(R)-(-)-Histidine monohydrochloride. Listed on TSCA. Chiral building block. mp = 254°; $[\alpha]^{25}$ = - 10.0° (c = 2, 5N HCl). *Sigma-Aldrich Fine Chemicals.*

### 3064 D-(+)-Histidinol dihydrochloride

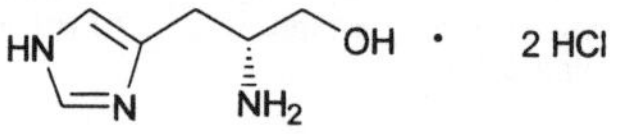

$C_6H_{13}Cl_2N_3O$

Chiral building block. mp = 193-195°; $[\alpha]_D^{20}$ = + 3° (c = 1, $H_2O$). *Sigma-Aldrich Fine Chemicals.*

### 3065 L-(-)-Histidinol dihydrochloride

1596-64-1 216-482-2

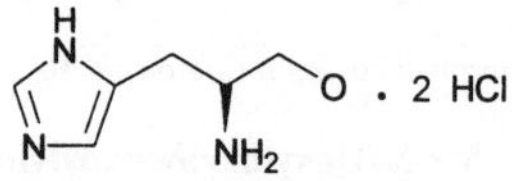

$C_6H_{13}Cl_2N_3O$

Imidazole-4-propanol, β-amino-, dihydrochloride, 1H-, (S)-; L-β-Amino-1H-imidazole-4-propanol dihydrochloride. Chiral building block. mp = 198-201°; $[\alpha]^{24}$ = - 2.5° (c = 2, $H_2O$). *Acros Organics nv; Sigma-Aldrich Fine Chemicals.*

### 3066 L-(+)-Homoarginine hydrochloride

1483-01-8 216-045-6

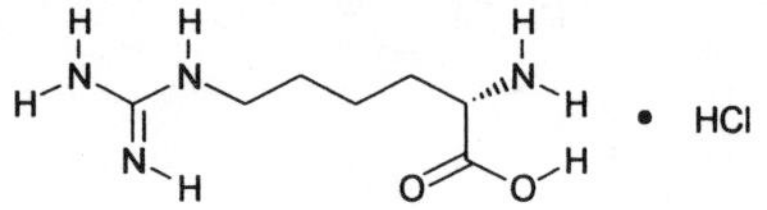

$C_7H_{17}ClN_4O_2$

$N^6$-amidino-, monohydrochloride, L-; L-Lysine, $N^6$-(aminoiminomethyl)-, monohydrochloride. Chiral building block. mp = 213-215°; $[\alpha]^{26}$ = + 19° (c = 1, 1N HCl). *Acros Organics nv; Senn Chemicals AG; Sigma-Aldrich Fine Chemicals.*

### 3067 L-Homocitrulline

1190-49-4 214-722-0

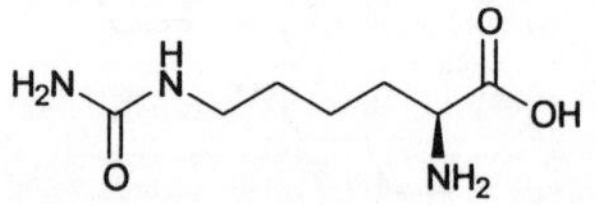

$C_7H_{15}N_3O_3$

$N^6$-(aminocarbonyl)-, L-; N-ε-Carbamyl-L-lysine. Chiral building block. *Advanced Asymmetrics, Inc.*

**3068 L-Homocysteine**
6027-13-0 227-891-0

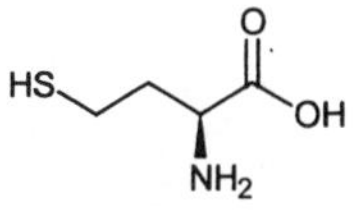

$C_4H_9NO_2S$
Butanoic acid, 2-amino-4-mercapto-, (S)-. (S)-2-Amino-4-mercaptobutanoic acid. Chiral building block. *DSM Fine Chemcials Netherlands; Tanabe Seiyaku Co. Ltd.*

**3069 L-Homocystine methyl ester hydrochloride**

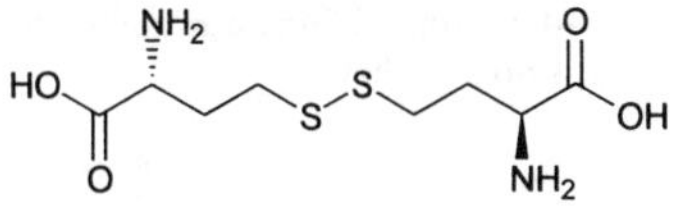

$C_8H_{16}N_4O_4S_2$
Chiral building block. *Senn Chemicals AG.*

**3070 D-(+)-Homophenylalanine**
82795-51-5

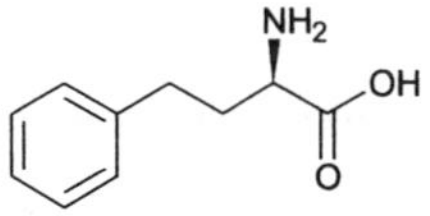

$C_{10}H_{13}NO_2$
Chiral building block. mp > 300°; $[\alpha]^{19}$ = - 45° (c = 1, 3N HCl). *Acros Organics nv; Kaneka Corporation; Recordati S.p.A.; Sigma-Aldrich Fine Chemicals.*

**3071 L-(-)-Homophenylalanine**
943-73-7 213-403-3

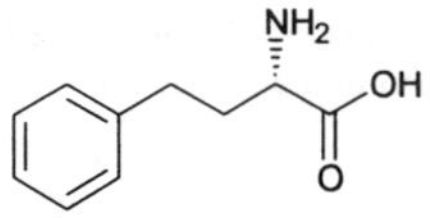

$C_{10}H_{13}NO_2$
Benzenebutanoic acid, α-amino-, (S)-; (+)-2-Amino-4-phenylbutyric acid; Butyric acid, 2-amino-4-phenyl-, L-; L-γ-Phenylbutyrine; L-Benzylalanine. Chiral building block. mp > 300°; $[\alpha]^{19}$ = + 45° (c = 1, 3N HCl). *Acros Organics nv; Daicel Chemical Ind. Ltd; Fischer Chemicals AG; Great Lakes Fine Chemicals; Kaneka Corporation; Senn Chemicals AG; Sigma-Aldrich Fine Chemicals; Tanabe Seiyaku Co. Ltd.*

**3072 D-Homoserine**
6027-21-0

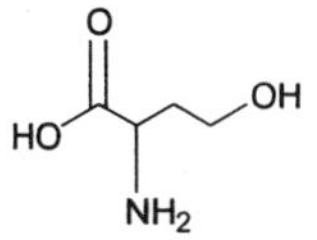

$C_4H_9NO_3$
Butyric acid, 2-amino-4-hydroxy-, D-. Chiral building block. mp = 205°; $[\alpha]_D^{20}$ = + 9° (c = 5, $H_2O$). *Hamari Chemicals; Sigma-Aldrich Fine Chemicals.*

**3073 L-Homoserine**
672-15-1 4777(12) 211-590-6

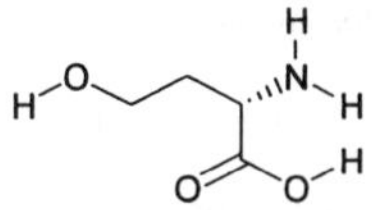

$C_4H_9NO_3$
Butanoic acid, 2-amino-4-hydroxy-, (S)-. (S)-2-Amino-4-hydroxybutanoic acid. Chiral building block. mp = 205°; $[\alpha]^{26}$ = - 8.0° (c = 5, $H_2O$). *Acros Organics nv; Degussa-Huls AG; Hamari Chemicals; Richman Chemical Inc.; Sigma-Aldrich Fine Chemicals; Tanabe Seiyaku Co. Ltd.; Yoneyama Yakuhin Kogyo Co., Ltd.*

**3074 (1S,2S)-(-)-Hydrobenzoin**
2352-10-2
$C_{14}H_{14}O_2$
(-)-1,2-Diphenylethanediol. Chiral building block; Chiral auxiliary. mp = 148-150°; $[\alpha]_D^{20}$ = - 94.0 ± 0.5° (c = 1, $CH_3OH$). *Kankyo Kagaku Center Co., Ltd.; Kiralchem Ltd.; Omega Chemical Company Inc.; TCI America.*

**3075 (1R,2R)-(+)-Hydrobenzoin**
52340-78-0 4700(11)

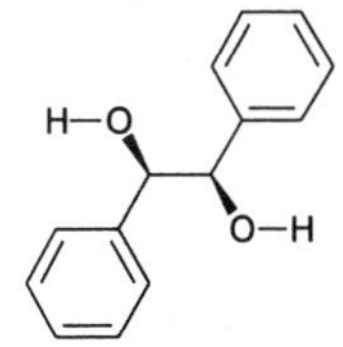

$C_{14}H_{14}O_2$
(+)-1,2-Diphenylethanediol. Chiral building block; Chiral auxiliary. mp = 148-150°; $[\alpha]_D^{20}$ = + 94.0 ± 0.5° (c = 1, $CH_3OH$). *Acros Organics nv; Kankyo Kagaku Center Co., Ltd.; Kiralchem Ltd.; Omega Chemical Company Inc.; Sigma-Aldrich Fine Chemicals; TCI America.*

**3076 Hydrocinchonine**
485-65-4 4824(12) 207-621-8

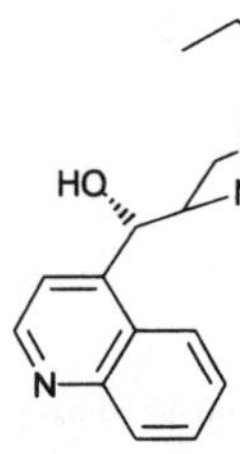

$C_{19}H_{24}N_2O$
Chiral auxiliary; Resolving agent. mp = 269-272°; $[\alpha]_D^{20}$ = + 194° (c = 0.5, EtOH). *Sigma-Aldrich Fine Chemicals.*

**3077 Hydromorphone-3-glucuronide**
$C_{19}H_{28}O_5S$
Pharmaceutical or derivative. mp > 230. *Ultrafine.*

**3078 (R)-(-)-Hydroorotic acid**

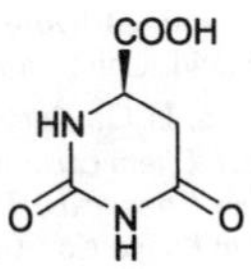

$C_5H_6N_2O_4$
Chiral building block. mp = 264-266°; $[\alpha]_D^{20}$ = - 33.0 ± 1.0° (c = 2, 1% $NaHCO_3$). *Kiralchem Ltd.*

**3079 (S)-(+)-Hydroorotic acid**
5988-19-2 4849(12)

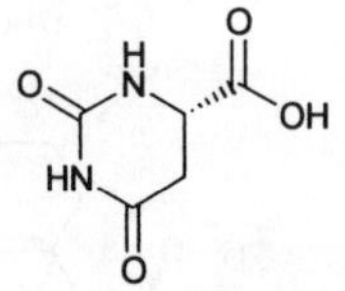

$C_5H_6N_2O_4$
L-4,5-Dihydroorotic acid; (S)-(+)-2,6-Dioxohexahydro-4-pyrimidinecarboxylic acid. Chiral building block. mp = 264-266°; $[\alpha]_D^{20}$ = + 33 ± 1° (c = 2, 1% $NaHCO_3$). *Kiralchem Ltd.; Lancaster Synthesis Ltd.; Sigma-Aldrich Fine Chemicals.*

**3080 D-Hydroorotic acid**
5988-53-4
$C_5H_6NO_4$
Chiral building block. *Sigma-Aldrich Fine Chemicals.*

**3081 (+)-Hydroquinidine**
1435-55-8 4851(12) 215-862-5

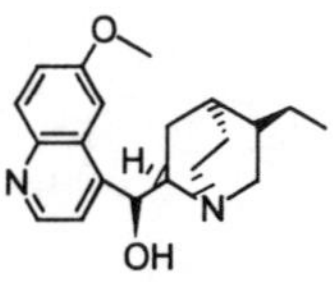

$C_{20}H_{26}N_2O_2$
Cinchonan-9-ol, 10,11-dihydro-6'-methoxy-, (9S)-. Hydroconquinine; (9S)-10,11-Dihydro-6'-methoxy-cinchonan-9-ol; 10,11-Dihydroquinidine. Resolving agent. mp = 169-170°; $[\alpha]^{25}$ = + 226° (c = 2, EtOH). *Isochem; Sigma-Aldrich Fine Chemicals.*

**3082 (-)-Hydroquinidine anthraquinone-1,4-diyl diether**

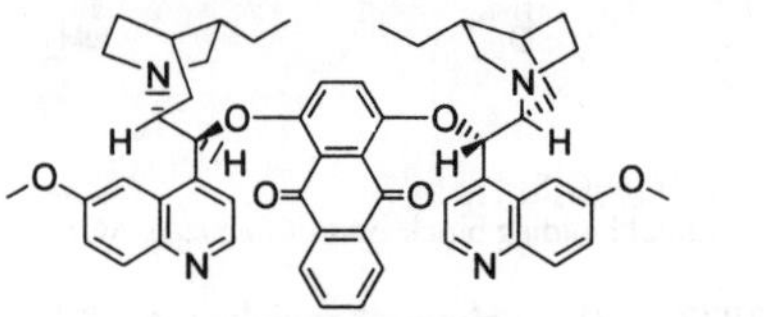

$C_{54}H_{56}N_4O_6$
(DHQD)2AQN. Intermediate. mp = 231-234°; $[\alpha]_D^{20}$ = - 468° (c = 1, $CHCl_3$). *Sigma-Aldrich Fine Chemicals.*

**3083 (-)-Hydroquinidine 4-chlorobenzoate**

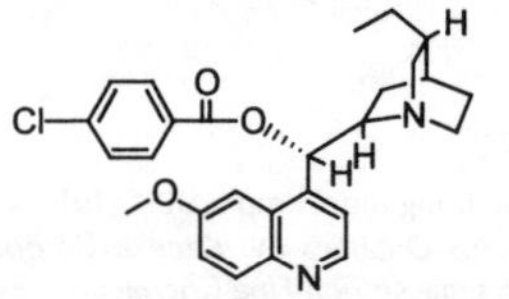

$C_{27}H_{29}ClN_2O_3$
Resolving agent; Chiral auxiliary. mp = 102-105°; $[\alpha]^{26}$ = - 73° (c = 1, EtOH). *Sigma-Aldrich Fine Chemicals.*

**3084 (-)-Hydroquinidine 2,5-diphenyl-4,6-pyrimidinediyl diether**

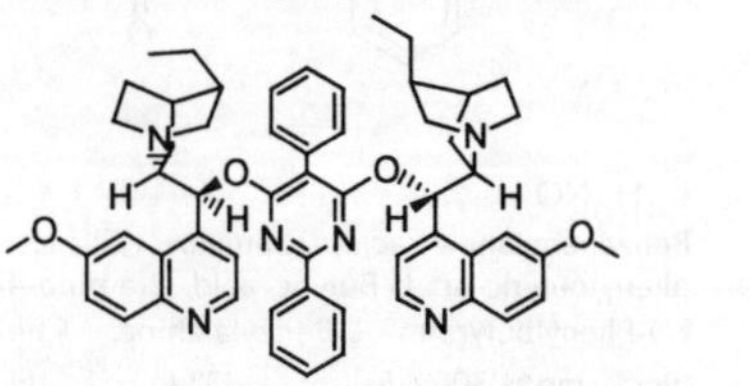

$C_{56}H_{60}N_6O_4$
(DHQD)2PYR. Intermediate. mp = 253-254°; $[\alpha]_D^{20}$ = - 390° (c = 1.2, $CH_3OH$). *Sigma-Aldrich Fine Chemicals.*

**3085 (-)-Hydroquinidine gluconate**
18253-58-2

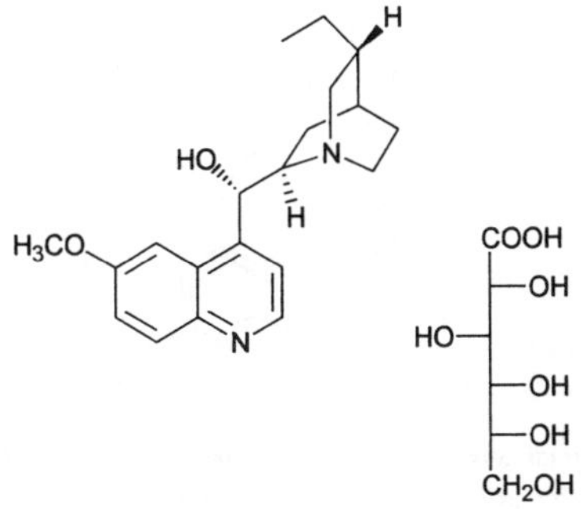

$C_{26}H_{38}N_2O_9$
Chiral auxiliary; Resolving agent. *Isochem.*

**3086 (-)-Hydroquinidine hydrochloride**
1476-98-8 216-024-1

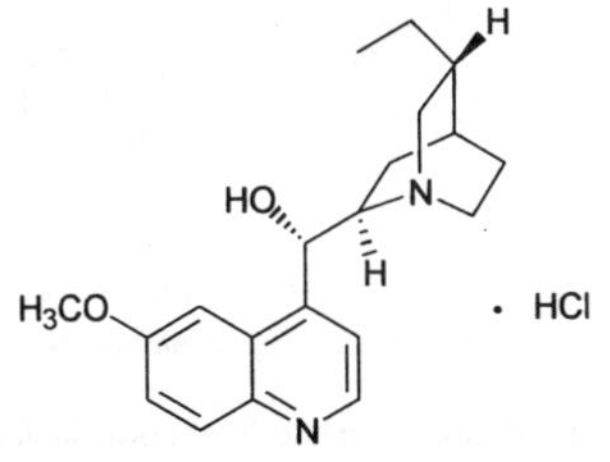

$C_{20}H_{27}ClN_2O_2$
Cinchonan-9-ol, 10,11-dihydro-6'-methoxy-, monohydrochloride, (9S)-; Hydroquinidine monohydrochloride; DHQ hydrochloride. Resolving agent. *DSM Fine Chemcials Netherlands; Isochem.*

**3087 (-)-Hydroquinidine 4-methyl-2-quinolyl ether**

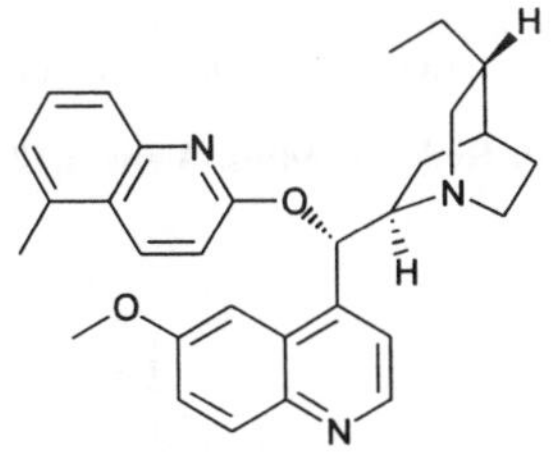

$C_{30}H_{33}N_3O_2$
Resolving agent; Chiral auxiliary. mp = 151-153°; $[\alpha]^{25}$ = - 168° (c = 1, EtOH). *Sigma-Aldrich Fine Chemicals.*

**3088 (-)-Hydroquinidine 1,4-phthalazinediyl diether**
140853-10-7

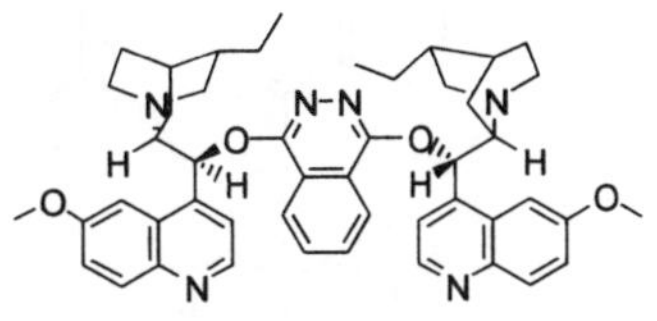

$C_{48}H_{54}N_6O_4$
(DHQD)2PHAL. Chiral intermediate. mp = 160°; $[\alpha]^{22}$ = - 262° (c = 1.2, $CH_3OH$). *Sigma-Aldrich Fine Chemicals.*

**3089 (-)-Hydroquinine**
522-66-7 4852(12) 208-334-0

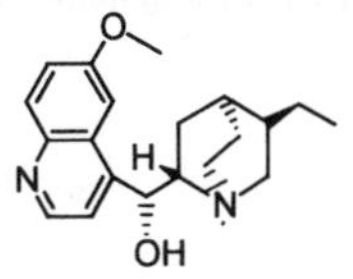

$C_{20}H_{26}N_2O_2$
Resolving agent. mp = 173-175°; $[\alpha]^{25}$ = - 148° (c = 1, EtOH). *Sigma-Aldrich Fine Chemicals.*

**3090 (+)-Hydroquinine anthraquinone-1,4-diyl diether**

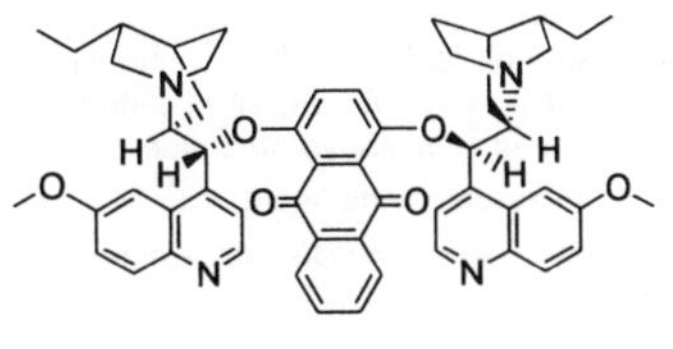

$C_{54}H_{56}N_4O_6$
(DHQ)2AQN. Chiral intermediate. mp = 175-180°; $[\alpha]_D^{20}$ = + 495° (c = 1, $CHCl_3$). *Sigma-Aldrich Fine Chemicals.*

**3091 (+)-Hydroquinine 2,5-diphenyl-4,6-pyrimidinediyl diether**

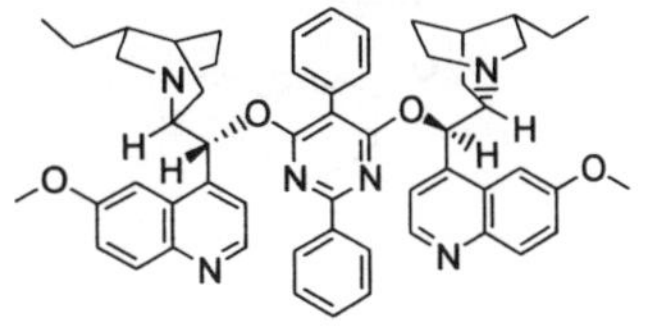

$C_{56}H_{60}N_6O_4$
(DHQ)2PYR. Chiral intermediate. mp = 247-249°; $[\alpha]_D^{20}$ = + 455° (c = 1.2, $CH_3OH$). *Sigma-Aldrich Fine Chemicals.*

**3092 (+)-Hydroquinine 4-chlorobenzoate**

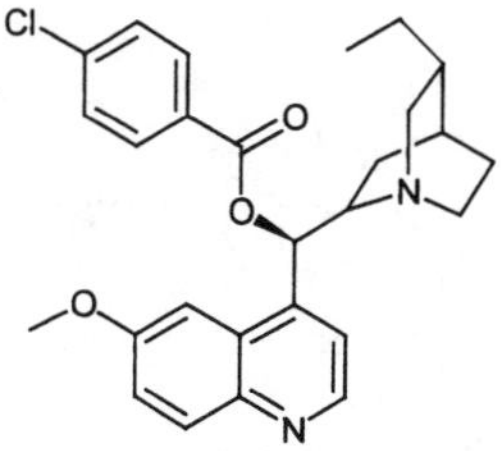

$C_{27}H_{29}ClN_2O_3$
Chiral auxiliary; Resolving agent; Chiral ligand. mp = 130-133°; $[\alpha]^{26}$ = + 150° (c = 1, EtOH). *Sigma-Aldrich Fine Chemicals.*

**3093 (+)-Hydroquinine hydrobromide**
85153-19-1 285-797-5

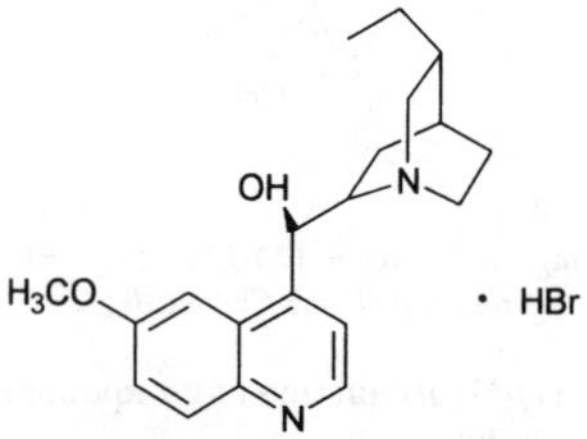

$C_{20}H_{27}BrN_2O_2$
Cinchonan-9-ol, 10,11-dihydro-6'-methoxy-, monohydrobromide, (8α,9R)-; (8α,9R)-10,11-Dihydro-6'-methoxycinchonan-9-ol monohydrobromide; Dihydroquinine hydrobromide. Resolving agent. $[\alpha]^{25}$ = + 100° (c = 1.9, $H_2O$). *DSM Fine Chemcials Netherlands; Sigma-Aldrich Fine Chemicals.*

**3094 (+)-Hydroquinine 4-methyl-2-quinolyl ether**

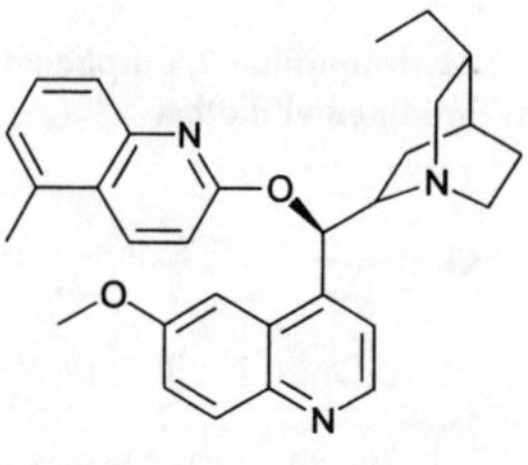

$C_{30}H_{33}N_3O_2$
Chiral auxiliary; Resolving agent; Chiral ligand. mp = 137-140°; $[\alpha]_D^{20}$ = + 260° (c = 1, EtOH). *Sigma-Aldrich Fine Chemicals.*

**3095 (+)-Hydroquinine 9-phenanthryl ether**

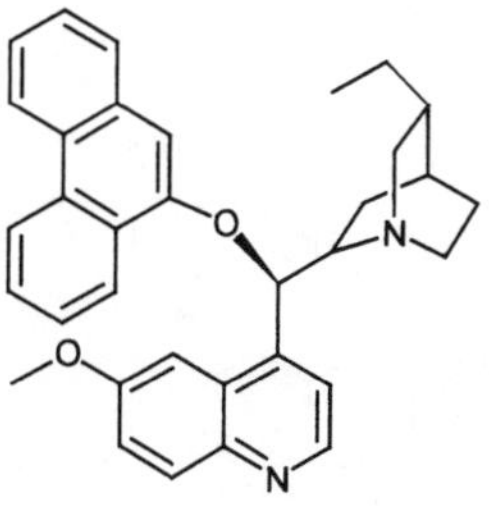

$C_{34}H_{34}N_2O_2$
Chiral auxiliary; Resolving agent; Chiral ligand. mp = 120°; $[\alpha]_D^{20}$ = + 420° (c = 1, EtOH). *Sigma-Aldrich Fine Chemicals.*

**3096 (+)-Hydroquinine 1,4-phthalazinediyl diether**
140924-50-1

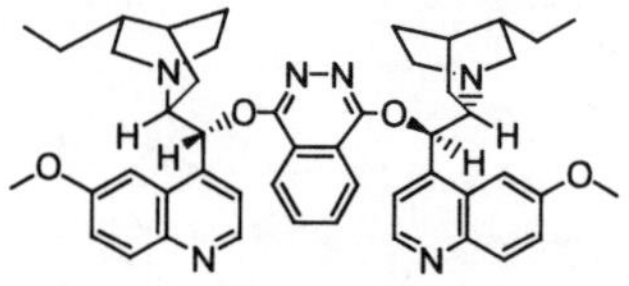

$C_{48}H_{54}N_6O_4$
(DHQ)2PHAL. Chiral intermediate. mp = 178°; $[\alpha]^{22}$ = + 336° (c = 1.2, $CH_3OH$). *Sigma-Aldrich Fine Chemicals.*

**3097 (-)-Hydroquinone 9-phenanthryl ether**

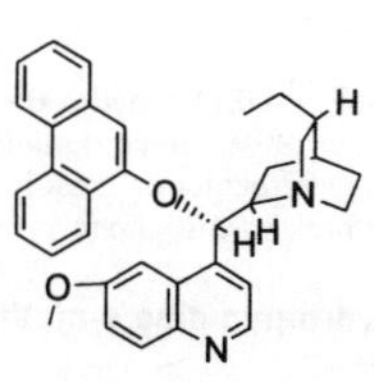

$C_{34}H_{34}N_2O_2$
Resolving agent; Chiral auxiliary. mp = 112°; $[\alpha]_D^{20}$ = - 348° (c = 1, EtOH). *Sigma-Aldrich Fine Chemicals.*

**3098 19-Hydroxyandrost-4-ene-3,17-dione**
510-64-5 208-116-5

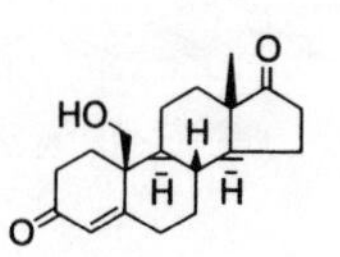

$C_{19}H_{26}O_3$
Pharmaceutical or derivative. mp = 168-169°. *Sigma-Aldrich Fine Chemicals.*

**3099 (R)-Hydroxybutanoic acid ethyl ester**
24915-95-5

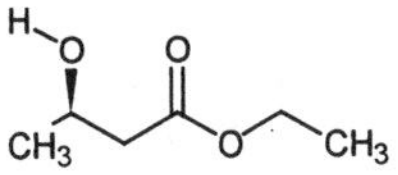

$C_6H_{12}O_3$
Butanoic acid, 3-hydroxy-, ethyl ester, (3R)-. (R)-(-)-Ethyl 3-hydroxybutyrate. Chiral building block. $bp_{12}$ = 75-76°; d = 1.017; n = 1.42; $[\alpha]_D^{20}$ = - 46° (c = 1, $CHCl_3$). *Acros Organics nv; IMI (TAMI) Institute for R&D; Sigma-Aldrich Fine Chemicals; Syntai Chemicals and Pharmaceuticals.*

**3100 (S)-3-Hydroxybutanoic acid ethyl ester**
56816-01-4 260-393-1

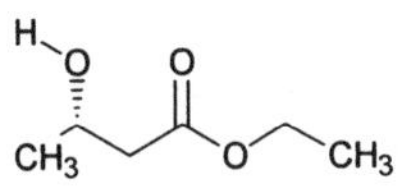

$C_6H_{12}O_3$
Butanoic acid, 3-hydroxy-, ethyl ester, (3S)-; (S)-(-)-Ethyl 3-hydroxybutyrate. Chiral building block. bp = 180-182°; d = 1.012; n = 1.421; $[\alpha]_D^{20}$ = + 43° (c = 1, $CHCl_3$). *Acros Organics nv; IMI (TAMI) Institute for R&D; Rohner Ltd.; Sigma-Aldrich Fine Chemicals; Syntai Chemicals and Pharmaceuticals.*

**3101 (R)-Hydroxybutanoic acid methyl ester**

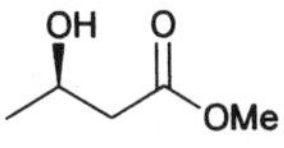

$C_5H_{10}O_3$
Butanoic acid, 3-hydroxy-, methyl ester, (3R)-; Methyl hydroxybutyrate. Chiral building block. *Syntai Chemicals and Pharmaceuticals.*

**3102 (S)-3-Hydroxybutanoic acid methyl ester**
53562-86-0 258-628-8

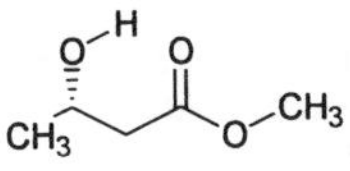

$C_5H_{10}O_3$
Butanoic acid, 3-hydroxy-, methyl ester, (3S)-; Methyl (S)-3-hydroxybutyrate. Chiral building block. $bp_{10}$ = 63°; d = 1.071; n = 1.421; $[\alpha]_D^{20}$ = + 19.8° (neat). *Acros Organics nv; Great Lakes Fine Chemicals; Kaneka Corporation; Sigma-Aldrich Fine Chemicals; Syntai Chemicals and Pharmaceuticals; TCI America.*

**3103 (R)-2-Hydroxy-3-buten-1-yl tosylate**
138249-07-7

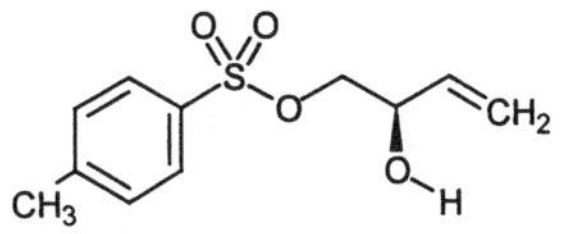

$C_{11}H_{14}O_4S$
Chiral building block. *Acros Organics nv; Eastman Chemical Company.*

**3104 (S)-2-Hydroxy-3-buten-1-yl tosylate**
133095-74-6

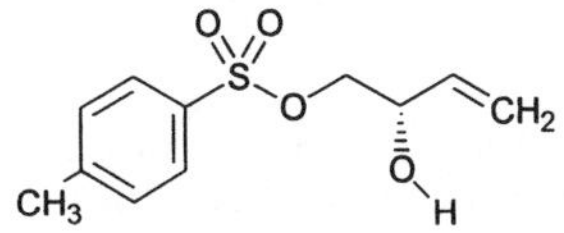

$C_{11}H_{14}O_4S$
Chiral building block. *Acros Organics nv; Eastman Chemical Company; Synthon Chiragenics Corporation.*

**3105 (3R)-(-)-3-Hydroxybutyric acid sodium salt**
13613-65-5

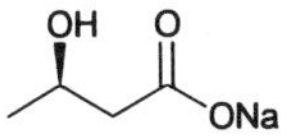

$C_4H_7NaO_3$
Chiral building block. mp = 149-155°; $[\alpha]^{22}$ = - 14° (c = 10, $H_2O$). *FineTech, Ltd.; Sigma-Aldrich Fine Chemicals.*

**3106 (3S)-(+)-3-Hydroxybutyric acid sodium salt**
127604-16-4 228-209-4
$C_4H_7NaO_3$
Chiral building block. mp = 150-155°; $[\alpha]_D^{20}$ = + 13.5° (c = 5, $H_2O$). *Sigma-Aldrich Fine Chemicals.*

**3107 (+)-Hydroxy-γ-butyrolactone**
56881-90-4

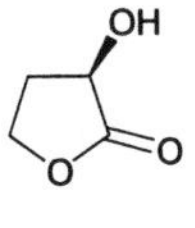

$C_4H_6O_3$
Furan-2(3H)-one, dihydro-3-hydroxy-, (R)-. Chiral building block. $bp_{10}$ = 133°; d = 1.309; n = 1.467; $[\alpha]^{23}$ = + 66° (c = 1.15, $CHCl_3$). *Sigma-Aldrich Fine Chemicals; Synthon Chiragenics Corporation.*

**3108 (-)-Hydroxy-γ-butyrolactone**
58081-05-3

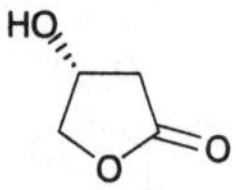

$C_4H_6O_3$
Furan-2(3H)-one, dihydro-4-hydroxy-, (R)-. Chiral building block. *Synthon Chiragenics Corporation.*

**3109 (S)-2-Hydroxy-γ-butyrolactone**
52079-23-9

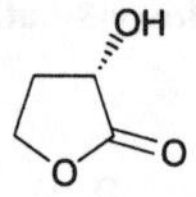

$C_4H_6O_4$
Furan-2(3H)-one, dihydro-3-hydroxy-, (S)-. Chiral building block. $bp_{10}$ = 133°; d = 1.309; n = 1.467; $[\alpha]^{23}$ = - 68° (c = 1.15, $CHCl_3$). *Sigma-Aldrich Fine Chemicals; Synthon Chiragenics Corporation.*

**3110 (S)-3-Hydroxybutyrolactone**
7331-52-4

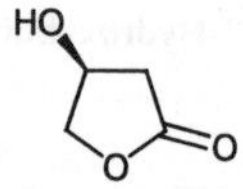

$C_4H_6O_3$
Furan-2(3H)-one, dihydro-4-hydroxy-, (S)-; (S)-3-Hydroxy-γ-butyrolactone. Chiral building block. $bp_{0.3}$ = 98-100°; d = 1.241; n = 1.464; $[\alpha]^{21}$ = - 81° (c = 2, EtOH). *Acros Organics nv; Kaneka Corporation; Sigma-Aldrich Fine Chemicals; SK Energy and Chemical, Inc.; Synthon Chiragenics Corporation.*

**3111 (1S,2S)-(-)-3-exo-Hydroxy-2,10-camphorsultam**
130748-66-2
$C_{10}H_{17}NO_3S$
(2S)-3-exo-Hydroxybornane-10,2-sultam. Chiral ligand. mp = 185-188°; $[\alpha]_D^{20}$ = - 4.5 ± 0.5° (c = 2, EtOH). *Lancaster Synthesis Ltd.*

**3112 2-Hydroxy-4(4-chlorophenyl)]butylimidazole**

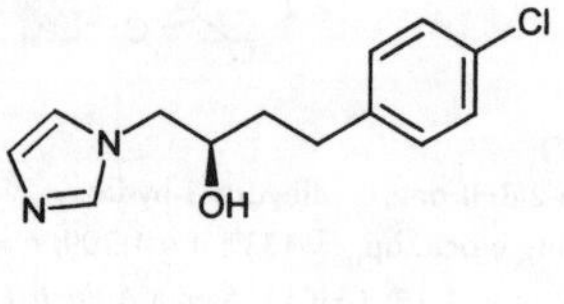

$C_{13}H_{15}ClN_2O$
Chiral intermediate. *Austin Chemical Company, Inc.*

**3113 (+)-Hydroxyethylcyclodextrin**
98513-20-3

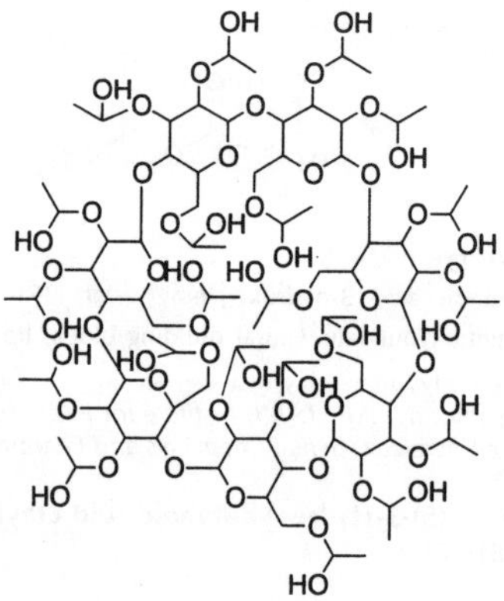

$C_{84}H_{154}O_{56}$
Chiral auxiliary. mp = 260°; $[\alpha]_D^{20}$ = + 132° (c = 1, $H_2O$). *Sigma-Aldrich Fine Chemicals.*

**3114 (1R, 2S)-cis-2-Hydroxycyclohexanecarboxylic acid ethyl ester**
61586-78-5

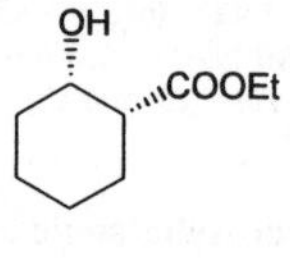

$C_9H_{16}O_3$
Chiral building block. *Rohner Ltd.*

**3115 (1S,3R)-3-Hydroxy-cyclopentanecarboxylic acid**

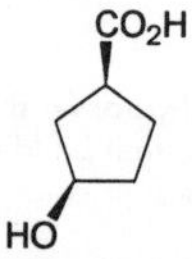

$C_6H_{10}O_3$
Cyclopentanecarboxylic acid, 3-hydroxy-, (1S-trans)-. Chiral intermediate. *Synthon Chiragenics Corporation.*

**3116 (1S,3S)-3-Hydroxy-cyclopentanecarboxylic acid**
107983-78-8

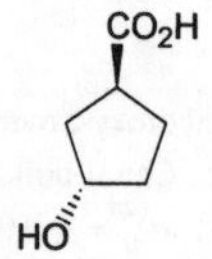

$C_6H_{10}O_3$
Cyclopentanecarboxylic acid, 3-hydroxy-, (1S-cis)-. Chiral intermediate. *Synthon Chiragenics Corporation.*

**3117 (R)-3-Hydroxy-1-cyclopentene-1-carboxylic acid**

$C_6H_8O_3$
Cyclopent-1-ene-1-carboxylic acid, 3-hydroxy-, (R)-. Chiral building block. *Synthon Chiragenics Corporation.*

**3118 (S)-3-Hydroxy-1-cyclopentene-1-carboxylic acid**

$C_6H_8O_3$
Cyclopent-1-ene-1-carboxylic acid, 3-hydroxy-, (S)-. Chiral intermediate. *Synthon Chiragenics Corporation.*

**3119 (4R)-Hydroxy-2-cyclopentenone**

$C_5H_6O_2$
Chiral building block. *Austin Chemical Company, Inc.*

**3120 (4S)-Hydroxy-2-cyclopentenone**

$C_5H_6O_2$
Chiral building block. *Austin Chemical Company, Inc.*

**3121 (S)-(-)-2-Hydroxy-3,3-dimethylbutyric acid**

$C_6H_{12}O_3$
Chiral building block. mp = 48-50°; $[\alpha]_D^{20}$ = - 62° (c = 1, $H_2O$/molybdate). *Sigma-Aldrich Fine Chemicals.*

**3122 (-)-7-Hydroxy-3,7-dimethyloctanal**
107-75-5 203-518-7

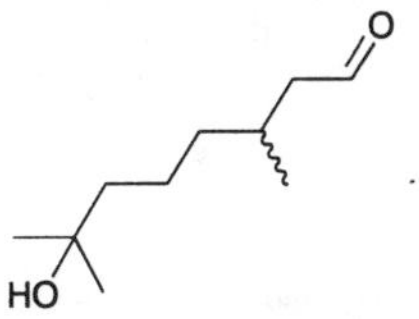


$C_{10}H_{20}O_2$
Octanal, 7-hydroxy-3,7-dimethyl-; 3,7-Dimethyl-7-hydroxyoctan-1-al; Cyclalia; Cyclosia; Laurine; Phixia; Hydroxycitronellal. Chiral intermediate. bp = 241°; d = 0.92; n = 1.447. *Takasago International Corporation.*

**3123 (R)-(-)-2-[Hydroxy(diphenyl)methyl]-1-methylpyrrolidine**
144119-12-0
$C_{18}H_{21}NO$
(R)-(-)-Diphenyl(1-methylpyrrolidin-2-yl)methanol. Chiral intermediate. *TCI America.*

**3124 (S)-(+)-2-[Hydroxy(diphenyl)methyl]-1-methylpyrrolidine**
110529-22-1
$C_{18}H_{21}NO$
(S)-(+)-Diphenyl(1-methylpyrrolidin-2-yl)methanol. Chiral intermediate. *TCI America.*

**3125 (4R)-4-(2-Hydroxyethyl)-2-phenyl-1,3-dioxolane**
187102-96-1

$C_{11}H_{14}O_3$
Dioxolane-4-ethanol, 1,3-, 2-phenyl-, (4R)-. Chiral intermediate. *Synthon Chiragenics Corporation.*

**3126 (R)-N-(2-Hydroxyethyl)-α-phenylethylamine**

$C_{10}H_{15}NO$
Chiral intermediate. *Acros Organics nv; Syntai Chemicals and Pharmaceuticals.*

**3127 (S)-N-(2-Hydroxyethyl)-α-phenylethylamine**

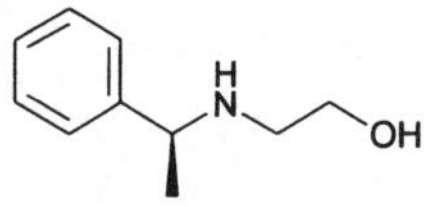

$C_{10}H_{15}NO$
Chiral building block. *Acros Organics nv; Syntai Chemicals and Pharmaceuticals.*

**3128 (R)-(-)-3-Hydroxyisobutyric acid methyl ester**
72657-23-9

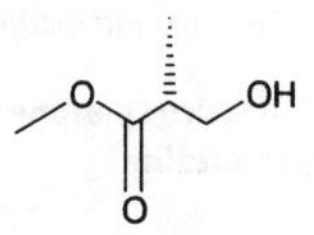

$C_5H_{10}O_3$
(R)-(-)-3-Hydroxy-2-methylpropionic acid methyl ester; Methyl (R)-(-)-3-hydroxy-2-methylpropionate; Methyl (R)-(-)-3-hydroxyisobutyrate. Chiral building block. $bp_{12}$ = 76-77°; d = 1.066; n = 1.425; $[\alpha]^{19}$ = - 26° (c = 4, $CH_3OH$). *Kaneka Corporation; Sigma-Aldrich Fine Chemicals; TCI America.*

**3129 (S)-(+)-3-Hydroxyisobutyric acid methyl ester**
80657-57-4

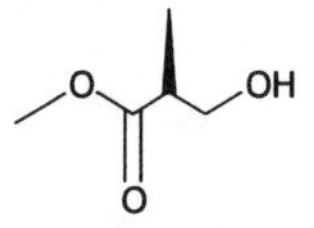

$C_5H_{10}O_3$
(S)-(+)-3-Hydroxy-2-methylpropionic acid methyl ester; (S)-(+)-3-Hydroxy-isobutyric acid methyl ester; Methyl (S)-(+)-3-Hydroxy-2-methylpropionate. Chiral building block. $bp_{10}$ = 74°; d = 1.071; n = 1.425; $[\alpha]^{19}$ = + 26° (c = 4, $CH_3OH$). *Daiso Company, Ltd.; Kaneka Corporation; Mitsubishi Rayon Co., Ltd.; Omega Chemical Company Inc.; Sigma-Aldrich Fine Chemicals; TCI America.*

**3130 (+)-3-Hydroxylamino isoborneol hydrochloride**

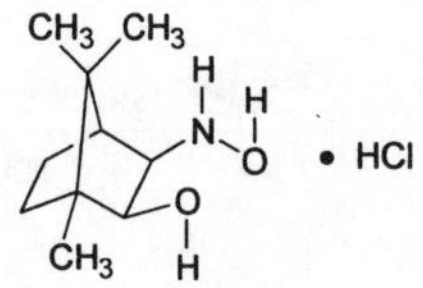

$C_{10}H_{20}ClNO_2$
Chiral intermediate. *Acros Organics nv.*

**3131 L-5-Hydroxylysine dihydrochloride monohydrate**

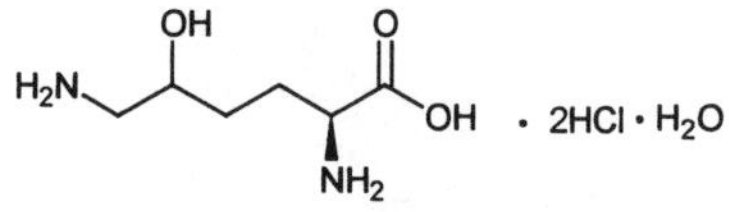

$C_6H_{18}Cl_2N_2O_4$
Chiral building block. *Senn Chemicals AG.*

**3132 (R)-Hydroxymandelic acid**
13244-78-5
$C_8H_8O_4$
Benzeneacetic acid, α,4-dihydroxy-, (R)-; (αR)-α,4-Dihydroxybenzeneacetic acid; Mandelic acid, p-hydroxy-, D-; Pisolithin B. Chiral building block. *Austin Chemical Company, Inc.*

**3133 (S)-4-Hydroxymandelic acid**
13244-75-2
$C_8H_8O_4$
Benzeneacetic acid, α,4-dihydroxy-, (S)-; Mandelic acid, p-hydroxy-, L-. Chiral building block. *Austin Chemical Company, Inc.*

**3134 (R)-(+)-2-Hydroxy-4-(2-methoxyphenyl)-5,5-dimethyl-1,3,2-dioxaphosphorinane 2-oxide**
98674-82-9

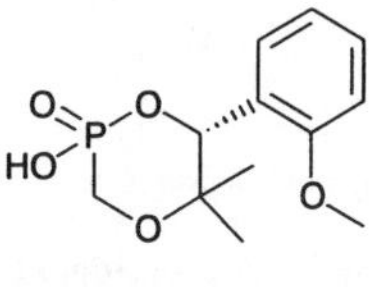

$C_{12}H_{17}O_5P$
Chiral ligand. mp = 197-200°; $[\alpha]^{23}$ = + 58° (c = 1, $CH_3OH$). *Sigma-Aldrich Fine Chemicals.*

**3135 (S)-(-)-2-Hydroxy-4-(2-methoxyphenyl)-5,5-dimethyl-1,3,2-dioxaphosphorinane 2-oxide**
98674-83-0

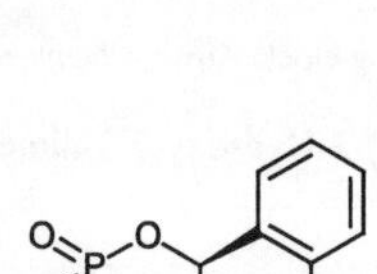

$C_{12}H_{17}O_5P$
Chiral intermediate. mp = 194-196°; $[\alpha]^{22}$ = - 59° (c = 1, $CH_3OH$). *Sigma-Aldrich Fine Chemicals.*

**3136 (+)-4-Hydroxy-3-methoxyphenylglycol, piperazine salt**

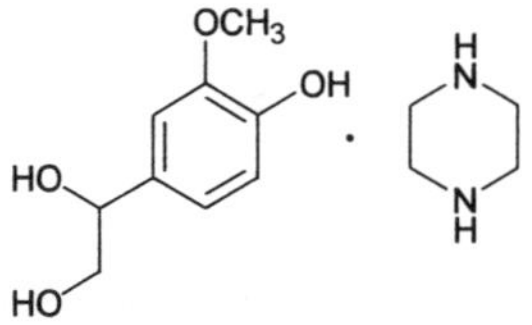

$C_{13}H_{22}N_2O_4$
(±) 67423-45-4. Chiral intermediate. *Pharmasyn, Inc.*

**3137 Hydroxymethyl butyrolactone**
52813-63-5

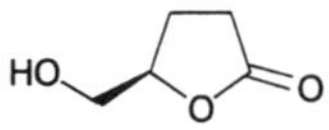

$C_5H_8O_3$
Furan-2(3H)-one, dihydro-5-(hydroxymethyl)-, (R)-. Chiral building block. $bp_{0.05}$ = 101-102°; d = 1.237; n = 1.47; $[\alpha]_D^{20}$ = - 56° (c = 3, $CHCl_3$). *Oxford Asymmetry International plc; Sigma-Aldrich Fine Chemicals; Synthon Chiragenics Corporation.*

**3138 (1S,5R,6S,7R)-6β-hydroxymethyl-7α-benzoyloxy-2-oxabicyclo[3.3.0]octane-3-one**
39746-00-4 254-615-6

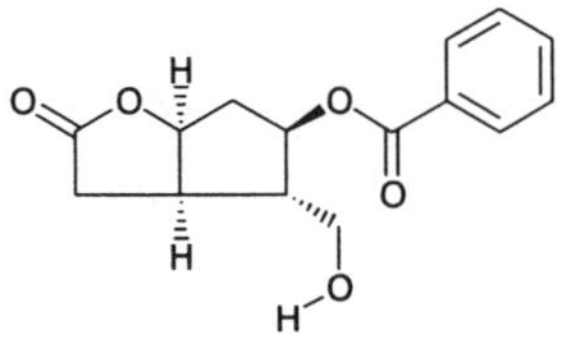

$C_{15}H_{16}O_5$
(3αR,4S,5R,6αS)-(-)-5-(Benzyloxy)hexahydro-4-(hydroxymethyl)-2H-cyclopenta[b]furan-2-one. Chiral intermediate. mp = 119-123°; -80 (c = 0.8, $CHCl_3$). *Acros Organics nv; Interchem Corporation; Sigma-Aldrich Fine Chemicals.*

**3139 (R)-Hydroxy-3-methylbutanoic acid**

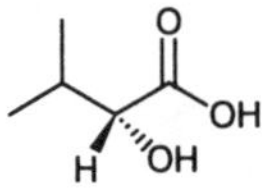

$C_5H_{10}O_3$
Butanoic acid, 2-hydroxy-3-methyl-, (R)- ; Hydroxysopentanoic acid; Hydroxyisovaleric acid. Chiral building block. *Great Lakes Fine Chemicals.*

**3140 (S)-(+)-2-Hydroxy-3-methylbutanoic acid**
17407-55-5

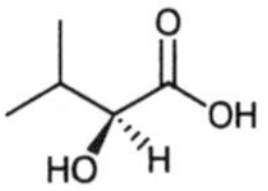

$C_5H_{10}O_3$
Chiral building block. mp = 68-70°; $bp_{13}$ = 124-125°; $[\alpha]_D^{20}$ = + 19° (c = 1, $CHCl_3$). *Acros Organics nv; Great Lakes Fine Chemicals; Sigma-Aldrich Fine Chemicals.*

**3141 (S)-(+)-5-Hydroxymethyl-γ-butyrolactone**
32780-06-6

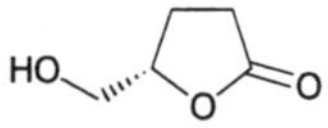

$C_5H_8O_3$
Furan-2(3H)-one, dihydro-5-(hydroxymethyl)-, (S)-; (S)-(+)-4,5-Dihydro-5-hydroxymethyl-2(3H)-furanone; (S)-(+)-2',3'-Dideoxyribonolactone. Chiral intermediate. $bp_{0.2}$ = 110-115°; d = 1.2370; $n_D^{20}$ = 1.4710; $\alpha]_D^{20}$ = + 34 ± 1° (c = 3. EtOH); $[\alpha]_D^{20}$ = + 56° (c = 3, $CHCl_3$). *Sigma-Aldrich Fine Chemicals; Synthon Chiragenics Corporation.*

**3142 (1R,4S)-4-(Hydroxymethyl)-2-cyclopenten-1-ol**

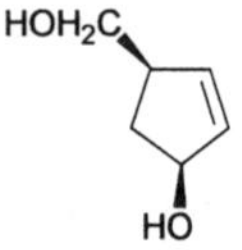

$C_6H_{10}O_2$
Chiral building block. *Oxford Asymmetry International plc.*

**3143 (1S,4R)-4-(Hydroxymethyl)-2-cyclopenten-1-ol**

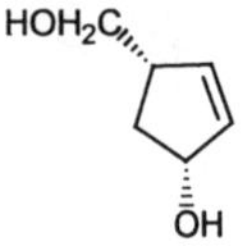

$C_6H_{10}O_2$
Chiral building block. *Oxford Asymmetry International plc.*

**3144 (S)-(+)-5-(1-Hydroxy-1-methylethyl)-2-methyl-2-cyclohexen-1-one**
60593-11-5

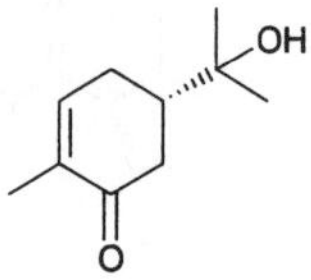

$C_{10}H_{16}O_2$
(S)-(+)-Carvone hydrate. Chiral intermediate. mp = 40-42°; $bp_{14}$ = 157-158°; d = 1.043; n = 1.507; $[\alpha]^{19}$ = + 41° (c = 10, EtOH). *Sigma-Aldrich Fine Chemicals.*

**3145 (R)-(+)-Hydroxymethyl-2(5H)-furanone**
112837-17-9

$C_5H_6O_3$
Chiral building block. [α] = + 147° (c = 1.4, $H_2O$). *Acros Organics nv; Daicel Chemical Ind. Ltd; Oxford Asymmetry International plc; Sigma-Aldrich Fine Chemicals.*

**3146 (S)-(-)-5-Hydroxymethyl-2(5H)-furanone**
78508-96-0

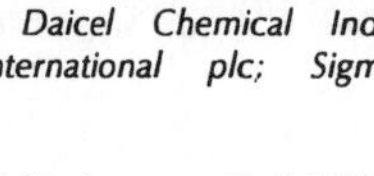

$C_5H_6O_3$
Chiral intermediate. mp = 41-43°; $[\alpha]_D^{20}$ = - 148 ± 5° (c = 1.4, $H_2O$). *Acros Organics nv; Lancaster Synthesis Ltd.; Oxford Asymmetry International plc; Sigma-Aldrich Fine Chemicals.*

**3147 L-5-Hydroxymethylhydantoin**

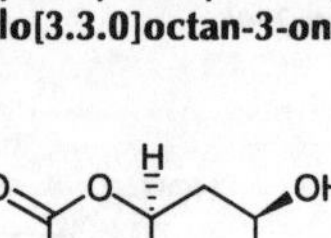

$C_4H_9N_2O_3$
Chiral building block. *Senn Chemicals AG.*

**3148 (-)-6-β-Hydroxymethyl-7α-hydroxy-cis-2-oxabicyclo[3.3.0]octan-3-one**
32233-40-2

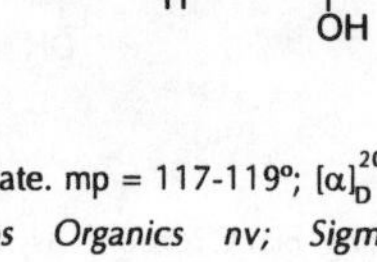

$C_8H_{12}O_4$
Chiral intermediate. mp = 117-119°; $[\alpha]_D^{20}$= -44° (c = 1.4, $CH_3OH$). *Acros Organics nv; Sigma-Aldrich Fine Chemicals.*

**3149 (4S,5S)-(-)-4-Hydroxymethyl-2-methyl-5-phenyl-2-oxazoline**
53732-41-5 258-730-2

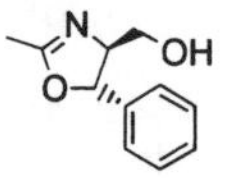

$C_{11}H_{13}NO_2$
(4S,5S)-(-)-2-Methyl-4-hydroxymethyl-5-phenyl-2-oxazoline; (4S,5S)-(-)-2-Methyl-5-phenyl-2-oxazoline-4-methanol. Chiral intermediate. mp = 61-65°; $[\alpha]_D^{20}$ = - 172° (c = 10.8, $CHCl_3$). *Sigma-Aldrich Fine Chemicals; TCI America.*

**3150 (R)-(+)-4-(Hydroxymethyl)-2-oxazolidinone benzoate**

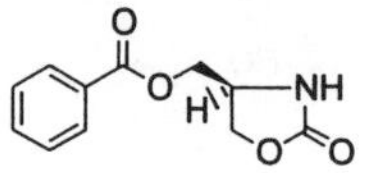

$C_{11}H_{11}NO_4$
Chiral intermediate. mp = 114-117°; $[\alpha]_D^{20}$ = + 30° (c = 1, $CHCl_3$). *Sigma-Aldrich Fine Chemicals.*

**3151 (1S,5R,6S,7R)-6β-Hydroxymethyl-7α-p-phenylbenzoyloxy-2-oxabicyclo[3.3.0]-octane-3-one**

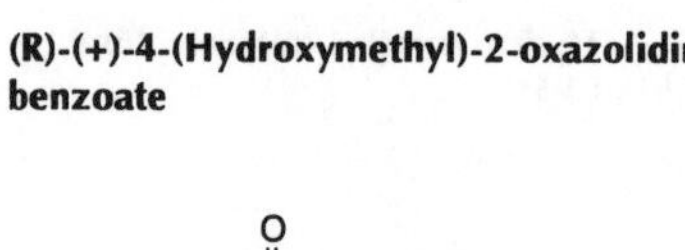

$C_{21}H_{20}O_5$
Chiral intermediate. *Interchem Corporation.*

**3152 (S)-(-)-N-1-(Hydroxymethyl)-2-phenylethyl-4-methylbenzenesulfonamide**
82495-70-3

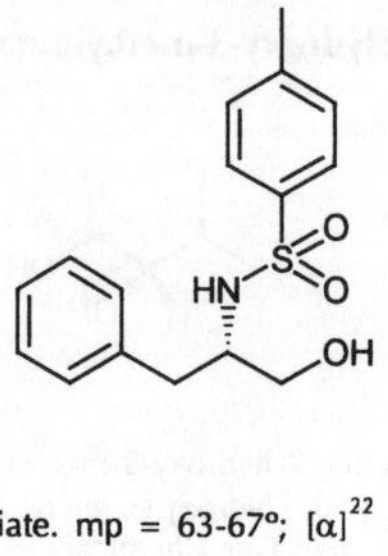

$C_{16}H_{19}NO_3S$
Chiral intermediate. mp = 63-67°; $[\alpha]^{22}$ = - 19° (c = 1, $CHCl_3$). *Sigma-Aldrich Fine Chemicals.*

**3153 (+)-20-(Hydroxymethyl)pregna-1,4-dien-3-one**
35525-27-0

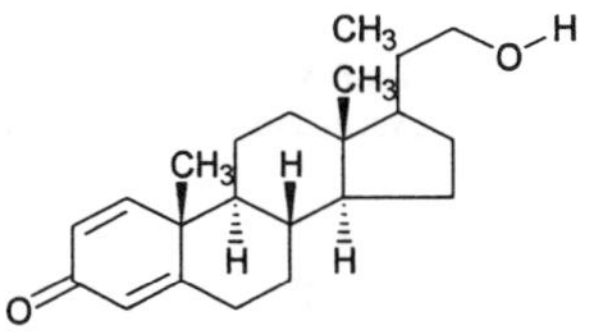

$C_{22}H_{32}O_2$
Chiral intermediate. [α] = + 29° (c = 1, $CHCl_3$). *Acros Organics nv.*

**3154 (R)-(-)-5-(Hydroxymethyl)-2-pyrrolidinone**
66673-40-3

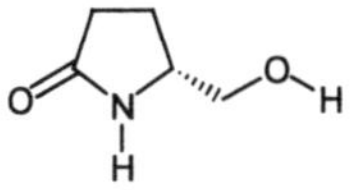

$C_5H_9NO_2$
D-Pyroglutaminol. Chiral building block. mp = 83-85°; $bp_{0.06}$ = 147-149°; $[\alpha]_D^{20}$ = - 35 ± 1° (c = 5, EtOH). *Acros Organics nv; Lancaster Synthesis Ltd.; Senn Chemicals AG; Sigma-Aldrich Fine Chemicals.*

**3155 (S)-(+)-5-(Hydroxymethyl)-2-pyrrolidinone**
17342-08-4

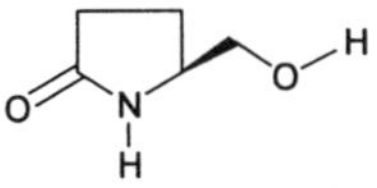

$C_5H_9NO_2$
L-Pyroglutaminol. Chiral building block. mp = 83-85°; $bp_{0.06}$ = 147-149°; $[\alpha]_D^{20}$ = + 35 ± 1° (c = 5, EtOH). *Acros Organics nv; Lancaster Synthesis Ltd.; Senn Chemicals AG; Sigma-Aldrich Fine Chemicals; TCI America.*

**3156 (S)-(+)-5-(Hydroxymethyl)-2-pyrrolidinone p-toluenesulfonate**
51693-17-5

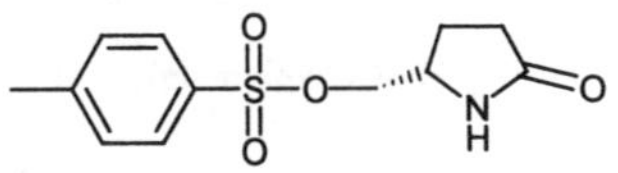

$C_{12}H_{15}NO_4S$
Chiral intermediate. mp = 124-128°; $[\alpha]_D^{20}$ = + 8.5° (c = 1, EtOH). *Sigma-Aldrich Fine Chemicals.*

**3157 (S)-(-)-2-Hydroxy-N-methylsuccinimide**
104612-35-3

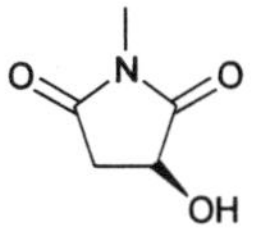

$C_5H_7NO_3$
Chiral building block. mp = 82-87°; $[\alpha]_D^{20}$ = - 75° (c = 3.5, EtOH). *Sigma-Aldrich Fine Chemicals.*

**3158 (R)-(-)-3-Hydroxy myristic acid**
28715-21-1

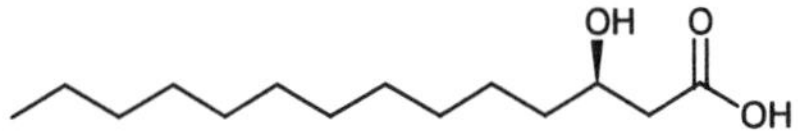

$C_{14}H_{28}O_3$
Chiral building block. *Hamari Chemicals.*

**3159 (S)-3-Hydroxymyristic acid**
35683-15-9

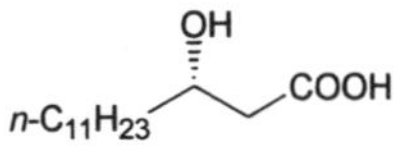

$C_{14}H_{28}O_3$
Tetradecanoic acid, 3-hydroxy-, (S)-; (S)-3-Hydroxytetradecanoic acid. Chiral building block. *Hamari Chemicals; Nagase & Co., Ltd.*

**3160 (R)-(-)-3-Hydroxynonanol**
$C_{11}H_{24}O_3$
Chiral building block. *Sigma-Aldrich Fine Chemicals.*

**3161 L-6-Hydroxynorleucine**
6033-32-5
$C_6H_{13}NO_3$
(S)-(+)-2-Amino-6-hydroxyhexanoic acid. Chiral building block. *Richman Chemical Inc.; TCI America.*

**3162 (1R,4S,5R)-(-)-4-Hydroxy-2-oxabicyclo-[3.3.0]oct-7-en-3-one**
$C_7H_8O_3$
3,3a,4,6a-Tetrahydro-3-hydroxy-(3S,3aR,6aR)-cyclopenta[b]furan-2-one. Chiral intermediate. *ChiroTech Technology Ltd.; Lancaster Synthesis Ltd.*

**3163 (1S,4R,5S)-(+)-4-Hydroxy-2-oxabicyclo-[3.3.0]oct-7-en-3-one**
$C_7H_8O_3$
3,3a,4,6a-Tetrahydro-3-hydroxy-(3R,3aS,6aS)-2H-cyclopenta[b]furan-2-one. Chiral intermediate. *ChiroTech Technology Ltd.; Lancaster Synthesis Ltd.*

**3164 (R)-(+)-3-Hydroxy-5-oxo-1-cyclopentene-1-heptanoic acid**
54996-33-7

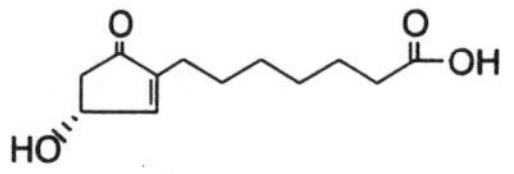

$C_{12}H_{18}O_4$
Chiral building block. mp = 40-43°; $[\alpha]^{22}$ = + 17° (c = 1, $CHCl_3$). *Sigma-Aldrich Fine Chemicals.*

**3165 (R)-(+)-2-(4-Hydroxyphenoxy)-propionic acid**
94050-90-5 407-960-3

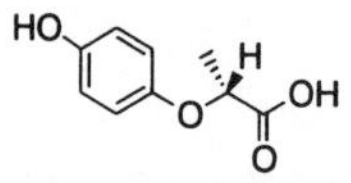

$C_9H_{10}O_4$
Propanoic acid, 2-(4-hydroxyphenoxy)-, (2R)-; D-HPPA. Chiral building block. mp = 145-148°. *Sigma-Aldrich Fine Chemicals; TCI America.*

**3166 (R)-Hydroxy-4-phenylbutyric acid**
29678-81-7

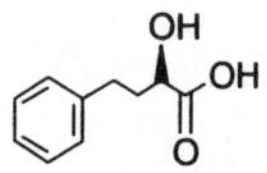

$C_{10}H_{12}O_3$
Chiral building block. mp = 114-117°; $[\alpha]_D^{20}$ = - 9.5° (c = 2.8, EtOH). *Kaneka Corporation; Sigma-Aldrich Fine Chemicals.*

**3167 (S)-2-Hydroxy-4-phenylbutyric acid**
115016-95-0

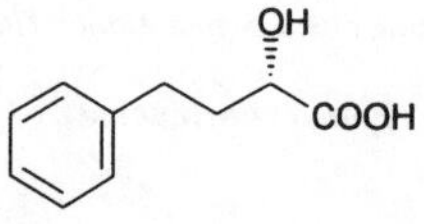

$C_{10}H_{12}O_3$
Chiral building block. *Yamakawa Chemical Industry Co. Ltd.*

**3168 (-)-Hydroxy-4-phenylbutyric acid ethyl ester**
90315-82-5

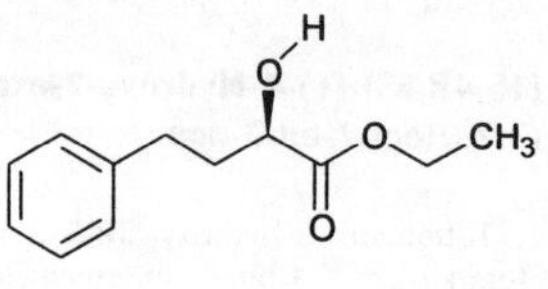

$C_{12}H_{16}O_3$
Ethyl hydroxy-4-phenylbutyrate. Chiral intermediate. bp = 212°; d = 1.075; n = 1.504; $[\alpha]^{21}$ = - 10° (neat). *Acros Organics nv; Daicel Chemical Ind. Ltd; Kaneka Corporation; Recordati S.p.A.; Sigma-Aldrich Fine Chemicals; Syntai Chemicals and Pharmaceuticals; TCI America.*

**3169 (-)-1-(2-Hydroxy-1-phenylethyl)-1,5-dihydropyrrol-2-one**
158271-95-5

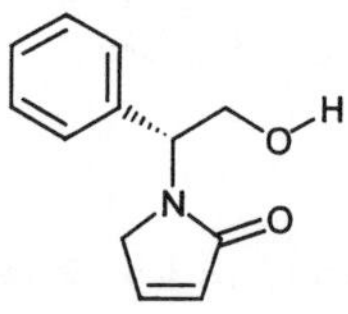

$C_{12}H_{13}NO_2$
Chiral intermediate. $[\alpha]$ = - 21°. *Acros Organics nv.*

**3170 N-[(1S)-2-Hydroxy-1-phenethyl)]ethoxy-carboxamide**

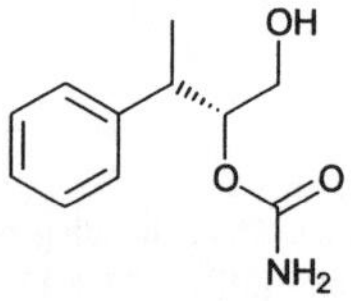

$C_{11}H_{15}NO_3$
Chiral intermediate. *Acros Organics nv.*

**3171 D-(-)-2-(4-Hydroxyphenyl)glycine**
22818-40-2 245-247-7

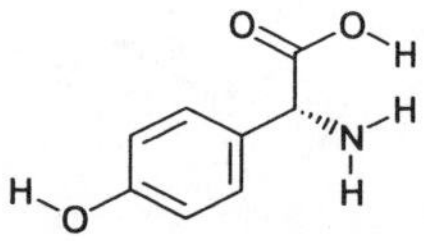

$C_8H_9NO_3$
Benzeneacetic acid, α-amino-4-hydroxy-, (αR)-; Amino-(4-hydroxyphenyl)acetic acid; Glycine, 2-(p-hydroxyphenyl)-, D-; D-(-)-p-Hydroxyphenylglycine. Listed on TSCA. Chiral building block. mp = 240°; $[\alpha]^{23}$ = - 156° (c = 1, 1N HCl). *Acros Organics nv; Alfa Chemicals Italiana srl; Austin Chemical Company, Inc.; Kaneka Corporation; Nippon Kayaku; Omega Chemical Company Inc.; Recordati S.p.A.; Senn Chemicals AG; Sigma-Aldrich Fine Chemicals; TCI America; Varsal Instruments, Inc.; Yoneyama Yakuhin Kogyo Co., Ltd.*

**3172 L-(+)-p-Hydroxyphenylglycine**

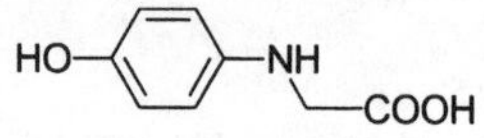

$C_8H_9NO_3$
Benzeneacetic acid, α-amino-4-hydroxy-, (L)-; (L-

Amino(p-hydroxyphenyl)acetic acid; Glycine, 2-(p-hydroxyphenyl)-, L-. Chiral intermediate. *D&O Group.*

**3173 D-(-)-p-Hydroxyphenylglycine chloride hydrochloride**

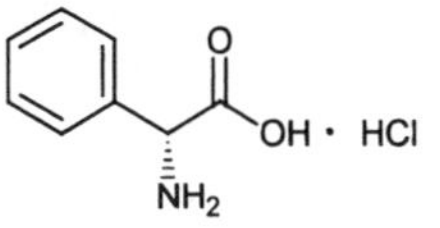

$C_8H_9Cl_2NO_2$
Chiral building block. *Nippon Kayaku; Sanofi Chimie.*

**3174 D-(-)-p-Hydroxyphenylglycine Dane salt (potassium, ethyl)**

$C_{14}H_{16}KNO_5$
Chiral building block. *Nippon Kayaku.*

**3175 D-(-)-p-Hydroxyphenylglycine Dane salt (sodium, methyl)**

$C_{13}H_{14}NNaO_5$
Chiral intermediate. *Nippon Kayaku.*

**3176 D-(-)-p-Hydroxyphenylglycine methyl ester hydrochloride**

57591-61-4
$C_8H_{10}ClNO_3$
(R)-Amino-(4-hydroxyphenyl)acetic acid methyl ester hydrochloride. Chiral intermediate. *Oxford Asymmetry International plc; Sigma-Aldrich Fine Chemicals.*

**3177 L-(+)-p-Hydroxyphenylglycine methyl ester hydrochloride**

127369-30-6
$C_9H_{12}ClNO_3$
(S)-Amino-(4-hydroxyphenyl)acetic acid methyl ester hydrochloride. Chiral intermediate. *Oxford Asymmetry International plc; Sigma-Aldrich Fine Chemicals.*

**3178 D-p-Hydroxyphenylglycine potassium salt**

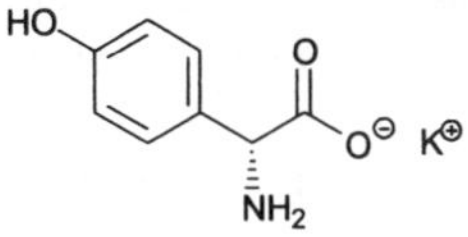

$C_8H_8KNO_3$
Chiral building block. *Varsal Instruments, Inc.*

**3179 (R)-(-)-2-Hydroxy-2-phenylpropionic acid**

3966-30-1
$C_7H_9NOS$
(R)-(-)-Atrolactic acid; (R)-(-)-2-Phenyllactic acid. Chiral building block. mp = 115-117°; $[\alpha]_D^{20}$ = - 49 ± 1° (c = 2, $H_2O$). *Acros Organics nv; Kiralchem Ltd.; Lancaster Synthesis Ltd.*

**3180 (S)-(+)-2-Hydroxy-2-phenylpropionic acid**

13113-71-8
$C_5H_{11}NO_2$
(S)-(+)-Atrolactic acid; (S)-(+)-2-Phenyllactic acid. Chiral building block. mp = 115-117°; $[\alpha]_D^{20}$ = + 49 ± 1° (c = 2, $H_2O$). *Acros Organics nv; Kiralchem Ltd.; Lancaster Synthesis Ltd.*

**3181 (R)-Hydroxy-3-phenylpropionic acid**

2768-42-5 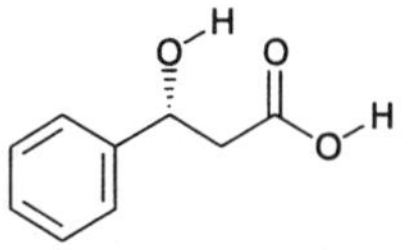
$C_9H_{10}O_3$
Chiral building block. mp = 117-119°; $[\alpha]_D^{20}$ = + 22.4 ± 0.2° (c = 4, $CH_3OH$). *Acros Organics nv; Eastman Chemical Company; Kiralchem Ltd.*

**3182 (S)-3-Hydroxy-3-phenylpropionic acid**

36567-72-3 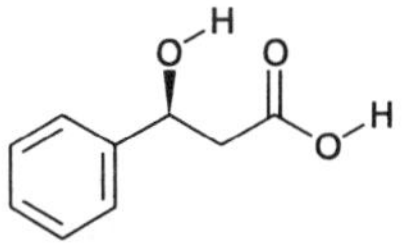
$C_9H_{10}O_3$
Chiral building block. mp = 117-119°; $[\alpha]_D^{20}$ = - 22.4 ± 0.2° (c = 4, $CH_3OH$). *Acros Organics nv; Eastman Chemical Company; Kiralchem Ltd.*

**3183 (R)-Hydroxy-3-phenylpropionitrile**

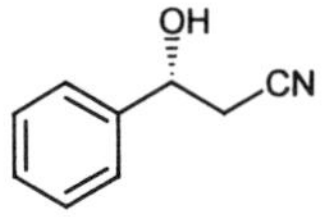

$C_9H_9NO$
Chiral building block. mp = 26-28°; $[\alpha]_D^{19}$ = + 59.4 ± 0.4° (c = 2, $CH_3OH$). *Kiralchem Ltd.*

**3184 (S)-3-Hydroxy-3-phenylpropionitrile**

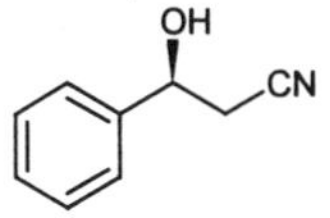

$C_9H_9NO$
Chiral building block. mp = 26-28°; $[\alpha]_D^{19}$ = - 59.4 ± 0.4° (c = 2, $CH_3OH$). *Kiralchem Ltd.*

**3185 (S)-(+)-2-Hydroxy-4-phthalimido-butyric acid**
48172-10-7

$C_{12}H_{11}NO_5$
(2S)-4-(1,3-Dioxoisoindolin-2-yl)-2-hydroxybutanoic acid. Chiral intermediate. mp = 157-159°; $[\alpha]_D^{20}$ = + 6.5° (c = 1, EtOH). *Acros Organics nv; Sigma-Aldrich Fine Chemicals; TCI America.*

**3186 (1R,2R,5R)-(+)-2-Hydroxy-3-pinanone**
24047-72-1

$C_{10}H_{16}O_2$
(1R,2R,5R)-(+)-2-Hydroxy-2,6,6-trimethylbicyclo[3.1.1]-heptan-3-one. Chiral intermediate. mp = 37-39°; bp = 245°; d = 1.059; $[\alpha]_D^{20}$ = + 39° (c = 0.5, $CHCl_3$). *Sigma-Aldrich Fine Chemicals; TCI America.*

**3187 (1S,2S,5S)-(-)-2-Hydroxy-3-pinanone**
1845-25-6

$C_{10}H_{16}O_2$
(1S,2S,5S)-(-)-2-Hydroxy-2,6,6-trimethylbicyclo[3.1.1]-heptan-3-one. Chiral intermediate. mp = 36-38°; bp = 245°; d = 1.059; $[\alpha]_D^{20}$ = - 39° (c = 0.5, $CHCl_3$). *Expansia; Sigma-Aldrich Fine Chemicals; TCI America.*

**3188 (R)-(+)-3-Hydroxypiperidine hydrochloride**

$C_5H_{11}NO$
Chiral building block. mp = 195-196°; $[\alpha]_D^{20}$ = + 8.6° (c = 2, $CH_3OH$). *Sigma-Aldrich Fine Chemicals.*

**3189 trans-3-Hydroxy-L-proline**
4298-08-2

$C_5H_9NO_3$
Chiral intermediate. [α] = - 17.2°. *Acros Organics nv; Senn Chemicals AG.*

**3190 cis-4-Hydroxy-D-proline**
2584-71-6 4775(11) 219-963-5

$C_5H_9NO_3$
D-Proline, 4-hydroxy-, (4R)-; (R)-allo-Hydroxyproline. Chiral building block. mp = 243°; $[\alpha]_D^{20}$ = + 58° (c = 2, $H_2O$). *Advanced Asymmetrics, Inc.; Rexim S.A. Produits Chimiques; Sigma-Aldrich Fine Chemicals.*

**3191 cis-4-Hydroxy-L-proline**
618-27-9 4887(12) 210-542-1

$C_5H_9NO_3$
Proline, 4-allo-hydroxy-, L-; L-allo-4-Hydroxyproline; (4S)-Hydroxy-(2S)-pyrrolidinecarboxylic acid. Listed on TSCA. Chiral building block. mp = 257°; [α] = - 55° (c = 2, $H_2O$). *Acros Organics nv; Advanced Asymmetrics, Inc.; Sigma-Aldrich Fine Chemicals.*

**3192 trans-4-Hydroxy-D-proline**

Proline, D-, 4-hydroxy-, (4S)-; (+)-4-Hydroxy-2-pyrrolidinecarboxylic acid. Chiral building block. *Advanced Asymmetrics, Inc.*

**3193 trans-4-Hydroxy-L-proline**
51-35-4 4887(12) 200-091-9

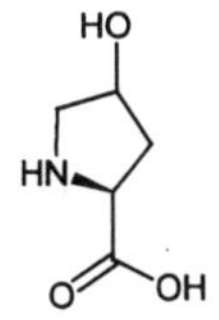

$C_5H_9NO_3$
Proline, 4-hydroxy-, L-; (2S,4R)-(-)-4-Hydroxypyrrolidine-2-carboxylic acid; L-Hypro; L-4-trans-Hydroxyproline. Listed on TSCA. Chiral building block. mp = 273°; $[\alpha]_D^{20}$ = - 76 ± 2° (c = 4, $H_2O$). *Acros Organics nv; Advanced Asymmetrics, Inc.; Austin Chemical Company, Inc.; Degussa-Huls AG; Kyowa Hakko Kogyo Co., Ltd.; Lancaster Synthesis Ltd.; Omega Chemical Company Inc.; Rexim S.A. Produits Chimiques; Senn Chemicals AG; Shanghai DSL International Trading Company; Sigma-Aldrich Fine Chemicals; TCI America; Varsal Instruments, Inc.; Yoneyama Yakuhin Kogyo Co., Ltd.*

**3194 D-cis-4-Hydroxyproline hydrochloride**
77449-94-6

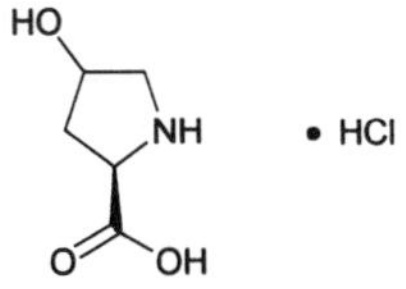

$C_5H_{10}ClNO_3$
Chiral building block. *Medinger & Sohn.*

**3195 L-(+)-Hydroxyprolinol**
104587-51-1

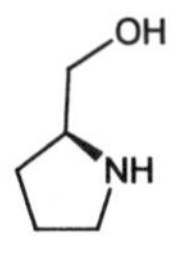

$C_5H_{11}NO_2$
Chiral building block. *Boehringer Ingelheim Pharma KG.*

**3196 (-)-Hydroxypropylcyclodextrin**
99241-24-4
Resolving agent; Chiral auxiliary. mp = 245°; $[\alpha]_D^{20}$ = - 125° (c = 1, $H_2O$). *Sigma-Aldrich Fine Chemicals.*

**3197 (+)-Hydroxypropylcyclodextrin**
94035-02-6

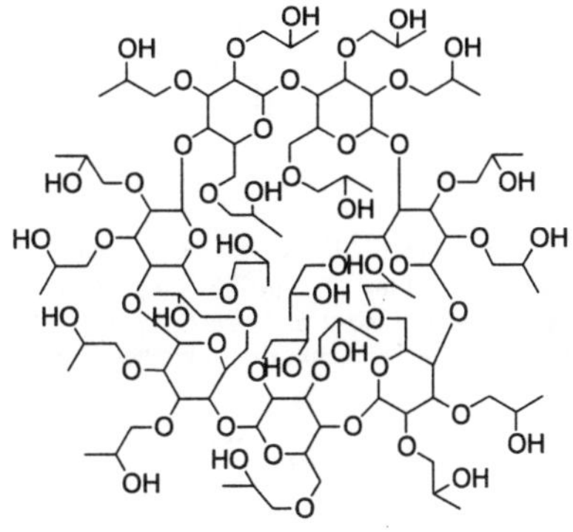

$C_{105}H_{196}O_{56}$
Chiral auxiliary. mp = 278°; $[\alpha]^{26}$ = + 127 to + 139° (c = 1, $H_2O$). *Sigma-Aldrich Fine Chemicals.*

**3198 (R)-3-Hydroxypyrrolidine**
2799-21-5

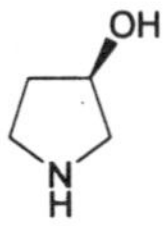

$C_4H_9NO$
Pyrrolidin-3-ol, (R)-. Chiral building block. $bp_8$ = 108-110°; d = 1.048; n = 1.488; $[\alpha]_D^{20}$ = + 6.5° (c = 3.5, $CH_3OH$). *Acros Organics nv; Austin Chemical Company, Inc.; Daiso Company, Ltd.; Omega Chemical Company Inc.; Sigma-Aldrich Fine Chemicals; Synthon Chiragenics Corporation; Toray International, Inc.*

**3199 (S)-3-Hydroxypyrrolidine**
100243-39-8

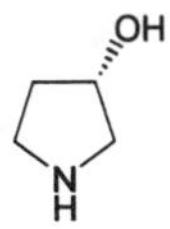

$C_4H_9NO$
Pyrrolidin-3-ol, (S)-; (S)-3-Pyrrolidinol. Chiral building block. *Acros Organics nv; Austin Chemical Company, Inc.; Daiso Company, Ltd.; Omega Chemical Company Inc.; Synthon Chiragenics Corporation; Toray International, Inc.*

**3200 (+)-Hydroxy-2-pyrrolidone**
22677-21-0

$C_4H_7NO_2$
Pyrrolidin-2-one, 4-hydroxy-, (R)-. Chiral building block. mp = 156-159°; $[\alpha]^{23}$ = + 43° (c = 1, EtOH). *Daicel Chemical Ind. Ltd; Daiso Company, Ltd.; Sigma-Aldrich Fine Chemicals; Synthon Chiragenics Corporation.*

**3201 (R)-3-Hydroxypyrrolidone**

$C_4H_7NO_2$
Chiral building block. *Daiso Company, Ltd.*

**3202 (S)-3-Hydroxypyrrolidone**

$C_4H_7NO_2$
Chiral building block. *Daiso Company, Ltd.*

**3203 (S)-4-Hydroxypyrolidone**
68108-18-9

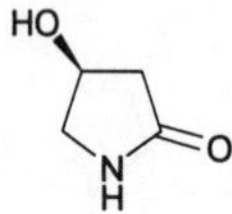


$C_4H_7NO_2$
Pyrrolidin-2-one, 4-hydroxy-, (S)-; (S)-4-Hydroxy-2-pyrrolidone. Chiral building block. mp = 156-159°; $[\alpha]^{23}$ = - 43° (c = 1, EtOH). *Daicel Chemical Ind. Ltd; Daiso Company, Ltd.; Sigma-Aldrich Fine Chemicals; SK Energy and Chemical, Inc.; Synthon Chiragenics Corporation.*

**3204 (3S)-3-Hydroxyquinidine**

$C_{20}H_{24}N_2O_3$
Resolving agent; Chiral auxiliary; Chiral ligand. mp = 211-214°. *Ultrafine.*

**3205 16α-Hydroxytestosterone**
63-01-4
$C_{19}H_{28}O_3$
16α,17β-Dihydroxyandrost-4-en-3-one. Pharmaceutical or derivative. mp = 192-194°. *Ultrafine.*

**3206 6α-Hydroxytestosterone**
2944-87-8
$C_{19}H_{28}O_3$
6α,17β-Dihydroxyandrost-4-en-3-one; 4-Androstene-6α,17β-diol-3-one. Pharmaceutical or derivative. *Ultrafine.*

**3207 (R)-3-Hydroxytetrahydrofuran**
86087-24-3

$C_4H_8O_2$
Furan-3-ol, tetrahydro-, (R). Chiral building block. bp = 181°; d = 1.097; n = 1.45; $[\alpha]_D^{20}$ = - 18° (c = 2.4, $CH_3OH$). *Daiso Company, Ltd.; Nagase & Co., Ltd.; Omega Chemical Company Inc.; Oxford Asymmetry International plc; Sigma-Aldrich Fine Chemicals; Synthon Chiragenics Corporation.*

**3208 (S)-3-Hydroxytetrahydrofuran**
86087-23-2

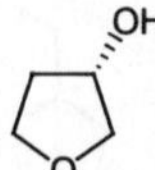


$C_4H_8O_2$
Furan-3-ol, tetrahydro-, (S); (S)-Tetrahydrofuran-3-ol. Chiral building block. $bp_{15}$ = 80°; d = 1.103; n = 1.45; $[\alpha]_D^{20}$ = + 17.5° (c = 2.4, $CH_3OH$). *Acros Organics nv; Austin Chemical Company, Inc.; Daiso Company, Ltd.; Kaneka Corporation; Omega Chemical Company Inc.; Oxford Asymmetry International plc; Sigma-Aldrich Fine Chemicals; SK Energy and Chemical, Inc.; Synthon Chiragenics Corporation; TCI America.*

**3209 L-7-Hydroxy-1,2,3,4-tetrahydroisoquinoline-3-carboxylic acid**
128502-56-7

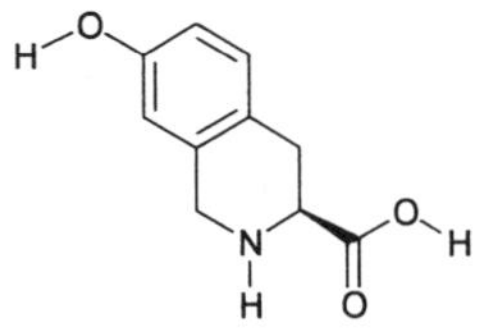

$C_{10}H_{11}NO_3$
Chiral intermediate. *Acros Organics nv.*

**3210 (R)-Hydroxytetrahydrothiophene**
100937-75-5

$C_4H_8OS$
Chiral building block. *Daiso Company, Ltd.*

**3211 (S)-3-Hydroxytetrahydrothiophene**
79107-75-8

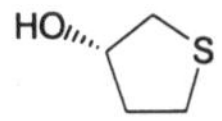

$C_4H_8OS$
Chiral building block. *Daiso Company, Ltd.*

**3212 (R)-(+)-6-Hydroxy-2,5,7,8-tetramethylchroman-2-carboxylic acid**
53101-49-8

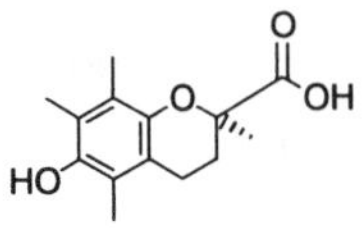

$C_{14}H_{18}O_4$
Chiral intermediate. mp = 162°; $[\alpha]^{27}$ = + 65° (c = 1.2, EtOH). *Sigma-Aldrich Fine Chemicals.*

**3213 (S)-(-)-6-Hydroxy-2,5,7,8-tetramethylchroman-2-carboxylic acid**
53174-06-4

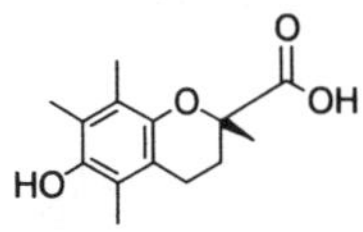

$C_{14}H_{18}O_4$
Chiral intermediate. mp = 160°; $[\alpha]^{26}$ = - 65° (c = 1.2, EtOH). *Sigma-Aldrich Fine Chemicals.*

**3214 (R)-(-)-3-Hydroxy-4,4,4-trichlorobutyric lactone**
16493-62-2

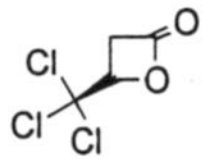

$C_4H_3Cl_3O_2$
(-)-4-(Trichloromethyl)-2-oxetanone, (-)-β-trichloromethyl-β-propiolactone. Chiral intermediate. mp = 52-54°; $bp_{13}$ = 108°; $[\alpha]_D^{20}$ = - 15° (c = 1, cyclohexane). *Sigma-Aldrich Fine Chemicals; TCI America.*

**3215 (S)-(+)-3-Hydroxy-4,4,4-trichlorobutyric lactone**
16493-63-3

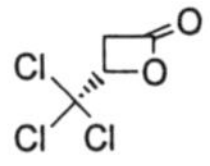

$C_4H_3Cl_3O_2$
(+)-4-(Trichloromethyl)-2-oxetanone, (+)-β-Trichloromethyl-β-propiolactone. Chiral intermediate. mp = 52-54°; $bp_{12}$ = 103°; $[\alpha]_D^{20}$ = + 15° (c = 1, cyclohexane). *Sigma-Aldrich Fine Chemicals; TCI America.*

**3216 (1R)-(2-endo,3-exo)-3-Hydroxy-4,7,7-trimethylbicyclo2.2.1]heptane-2-acetic acid**

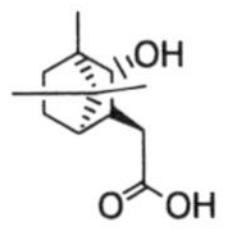

$C_{12}H_{20}O_3$
(-)-Isoborneolactic acid. Chiral intermediate. mp = 107-111°; $[\alpha]_D^{20}$ = - 18° (c = 1, $CHCl_3$). *Sigma-Aldrich Fine Chemicals.*

**3217 L-5-Hydroxytryptophan**
4350-09-8 4784(11) 224-411-1

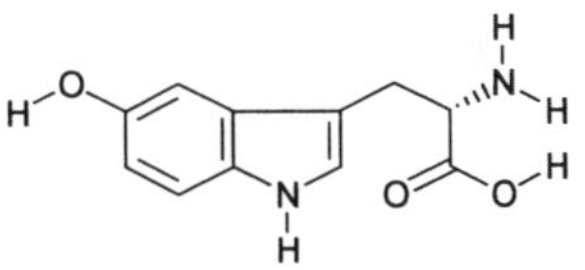

$C_{11}H_{12}N_2O_3$
Tryptophan, 5-hydroxy-, L-; Oxitriptan; Pretonine; Quietim. Listed on TSCA. Chiral intermediate. mp = 275°; $[\alpha]_D^{20}$ = - 34.5° (c = 1, $H_2O$). *Acros Organics nv; Austin Chemical Company, Inc.; Kaden Biochemicals GmbH; KingChem, Inc.; Linnea S.A.; MultiFerm A.B.; Norchim SA; Omega Chemical Company Inc.; Senn Chemicals AG; Sigma-Aldrich Fine Chemicals; TCI America.*

**3218 L-5-Hydroxytryptophan methyl ester hydrochloride**
Chiral building block. *Pfanstiehl Laboratories, Inc.; Senn Chemicals AG.*

**3219 3-Hydroxy-L-tyrosine**
59-92-7 5490(12) 200-445-2

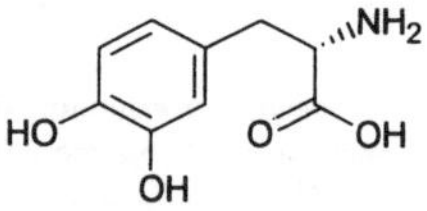

$C_9H_{11}NO_4$
L-DOPA. Listed on TSCA. Chiral building block. mp = 295°; $[\alpha]^{25}$ = - 11° (c = 1, 1N HCl). *Austin Chemical Company, Inc.; Sigma-Aldrich Fine Chemicals.*

**3220 (4S)-3-(2-Hydroythyl)-2-phenyl-1,3-dioxolane**
191354-62-8

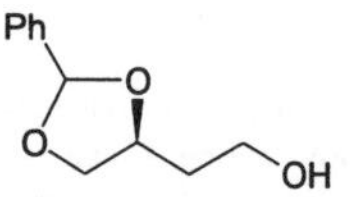

$C_{11}H_{14}O_3$
Dioxolane-4-ethanol, 1,3-, 2-phenyl-, (4S)-. Chiral intermediate. *Synthon Chiragenics Corporation.*

**3221 (S)-Ibuprofen**

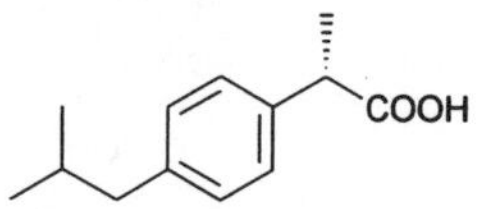

$C_{13}H_{18}O_2$
Benzeneacetic acid, α-methyl-4-(2-methylpropyl)-, (S)-; (S)-p-(2-Methylpropyl)-α-methylphenylacetic acid; (S)-2-(4-Isobutylphenyl)propionic acid; (S)-(4-Isobutylphenyl)-α-methylacetic acid; (S)-4-Isobutyl-α-methylphenylacetic acid; (S)-4-Isobutylhydratropic acid; Hydratropic acid, p-isobutyl-, (S)-; (S)-. Chiral intermediate. *Albemarle Corporation; Nagase & Co., Ltd.*

**3222 L-Iditol**
488-45-9

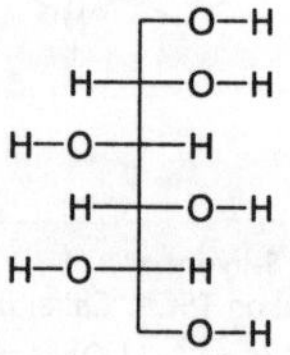

$C_6H_{14}O_6$
Chiral building block. mp = 78-80°; $[\alpha]^{19}$ = - 3.1° (c = 1, $H_2O$). *Acros Organics nv; Sigma-Aldrich Fine Chemicals.*

**3223 D-(+)-Idose**
5978-95-0 227-780-7

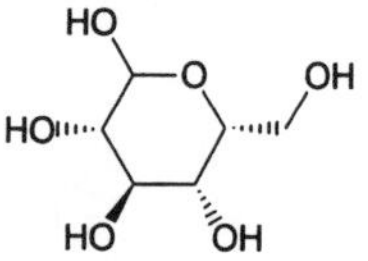

$C_6H_{12}O_6$
Idose, D-. Chiral building block. *Acros Organics nv; Senn Chemicals AG.*

**3224 2-(1'-Imidazoylsulfonyl)-1,3,5-tri-O-benzoyl-α-D-ribofuranose**
97614-42-1
$C_{23}H_{25}N_2O_7S$
Chiral intermediate. *Pfanstiehl Laboratories, Inc.*

**3225 L-N$^6$-(1-Iminoethyl)-lysine dihydrochloride**
150403-89-7

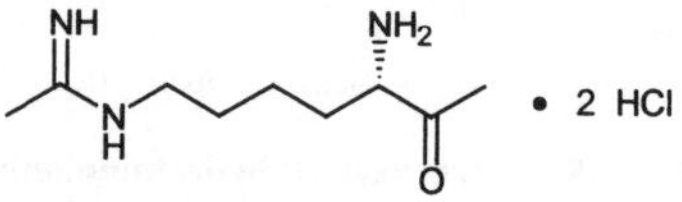

$C_8H_{19}Cl_2N_3O_2$
Chiral intermediate. *Acros Organics nv.*

**3226 L-N$^5$-(1-Iminoethyl)-ornithine dihydrochloride**
36889-13-1

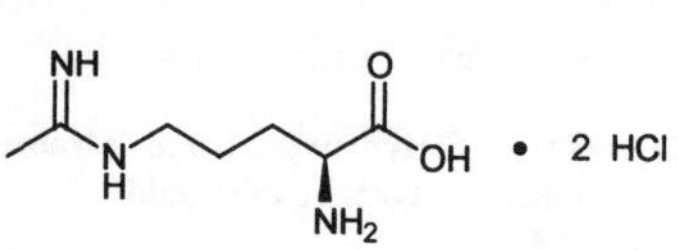

$C_7H_{17}Cl_2N_3O_2$
Chiral intermediate. *Acros Organics nv.*

**3227 (R)-(-)-1-Indanol**
697-64-3

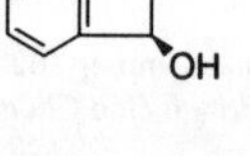

$C_9H_{10}O$
Chiral building block. mp = 72-73°; $[\alpha]^{30}$ = - 29° (c = 2, $CHCl_3$). *Sigma-Aldrich Fine Chemicals.*

**3228 (S)-(+)-1-Indanol**
25501-32-0

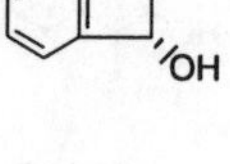

$C_9H_{10}O$
Chiral building block. mp = 73-75°; $[\alpha]_D^{20}$ = + 30° (c = 2, $CHCl_3$). *Sigma-Aldrich Fine Chemicals.*

**3229 Indole-3-acetyl-R-alanine**
57105-39-2

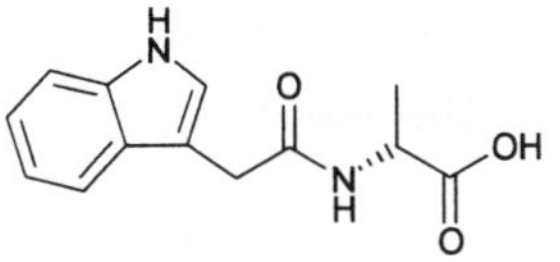

$C_{13}H_{14}N_2O_3$
Chiral intermediate. mp = 138-140°; $[\alpha]^{24}$ = - 21° (c = 1, $CH_3OH$). *Austin Chemical Company, Inc.; Sigma-Aldrich Fine Chemicals.*

**3230 Indole-3-acetyl-L-isoleucine**
57105-45-0

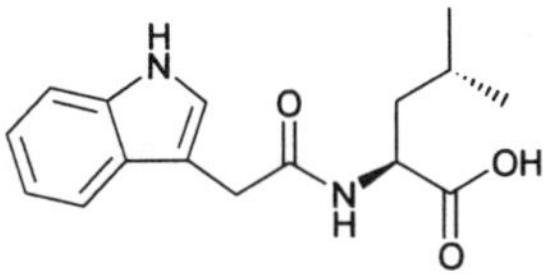

$C_{16}H_{20}N_2O_3$
Chiral intermediate. mp = 182-183°; $[\alpha]_D^{20}$ = + 6° (c = 3, $CH_3OH$). *Austin Chemical Company, Inc.; Sigma-Aldrich Fine Chemicals.*

**3231 Indole-3-acetyl-L-phenylalanine**
57105-50-7

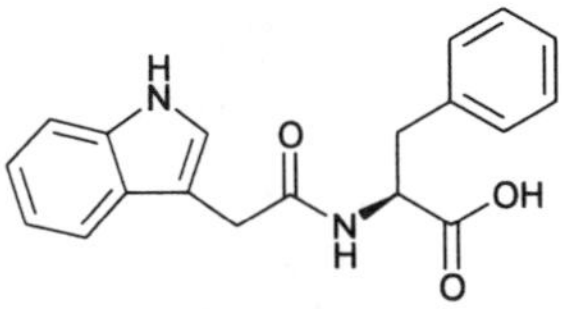

$C_{19}H_{18}N_2O_3$
Chiral intermediate. mp = 154-156°; $[\alpha]_D^{20}$ = + 36° (c = 3, EtOH). *Austin Chemical Company, Inc.; Sigma-Aldrich Fine Chemicals.*

**3232 Indole-3-acetyl-L-valine**
57105-42-7

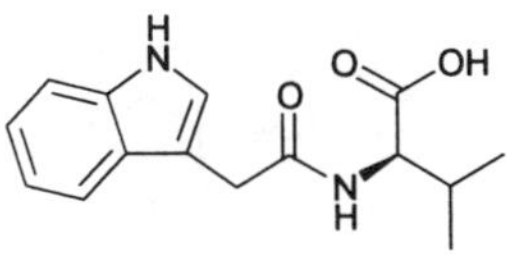

$C_{15}H_{18}N_2O_3$
Chiral intermediate. mp = 196-199°; $[\alpha]_D^{20}$ = + 9.2° (c = 3, EtOH). *Austin Chemical Company, Inc.; Sigma-Aldrich Fine Chemicals.*

**3233 (S)-(-)-Indoline-2-carboxylic acid**
79815-20-6

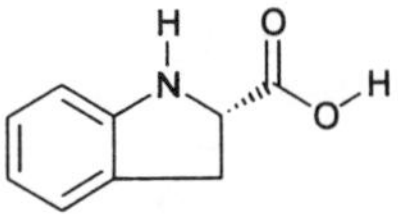

$C_9H_9NO_2$
Chiral intermediate. mp = 177°; $[\alpha]_D^{20}$ = - 114° (c = 1, 1N HCl). *Acros Organics nv; DSM Fine Chemcials Netherlands; Kawaken Fine Chemicals Co. Ltd.; Omega Chemical Company Inc.; Sigma-Aldrich Fine Chemicals; TCI America.*

**3234 (-)-N-(3-Indolyl)leucine**
36838-63-8

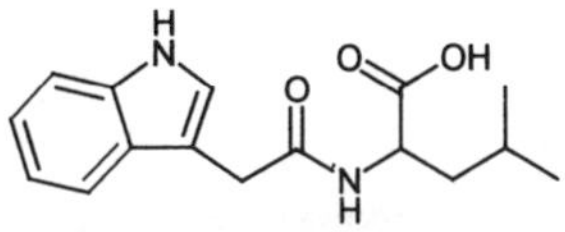

$C_{16}H_{20}N_2O_3$
Chiral intermediate. mp = 198-200°; $[\alpha]_D^{20}$ = - 17° (c = 3, EtOH). *Sigma-Aldrich Fine Chemicals.*

**3235 (R)-(-)-4-(1H-Indol-3-ylmethyl)-2-oxazolidinone**

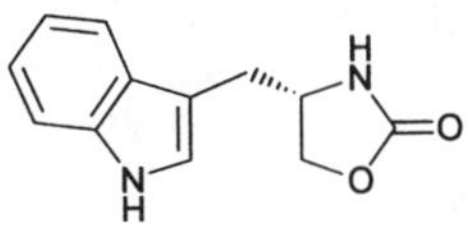

$C_{12}H_{12}N_2O_2$
Chiral intermediate. mp = 157-160°; $[\alpha]^{21}$ = - 12° (c = 1, $CH_3OH$). *Sigma-Aldrich Fine Chemicals.*

**3236 (S)-(+)-4-(1H-Indol-3-ylmethyl)-2-oxazolidinone**

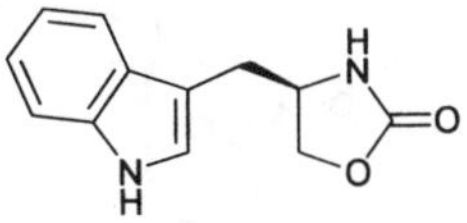

$C_{12}H_{12}N_2O_2$
Chiral intermediate. mp = 157-160°; $[\alpha]^{21}$ = + 9° (c = 1, $CH_3OH$). *Senn Chemicals AG; Sigma-Aldrich Fine Chemicals.*

**3237 Indoxyl-β-D-galactopyranoside**
126787-65-3

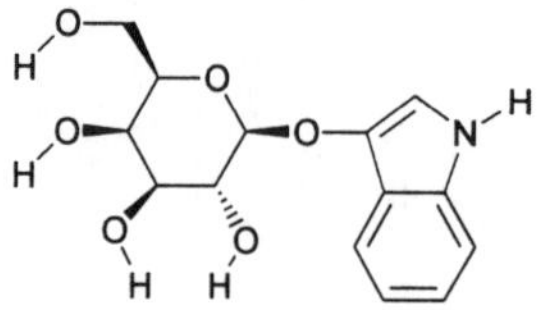

$C_{14}H_{17}NO_6$
Chiral intermediate. *Acros Organics nv.*

**3238 3-Indoxyl-β-D-glucopyranoside trihydrate**
487-60-5

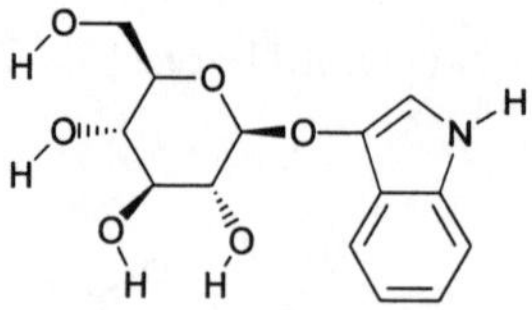

$C_{14}H_{17}NO_6$
Pharmaceutical or derivative. *Acros Organics nv.*

**3239 Indoxyl-β-D-glucuronic acid, cyclohexylammonium salt**
35804-66-1

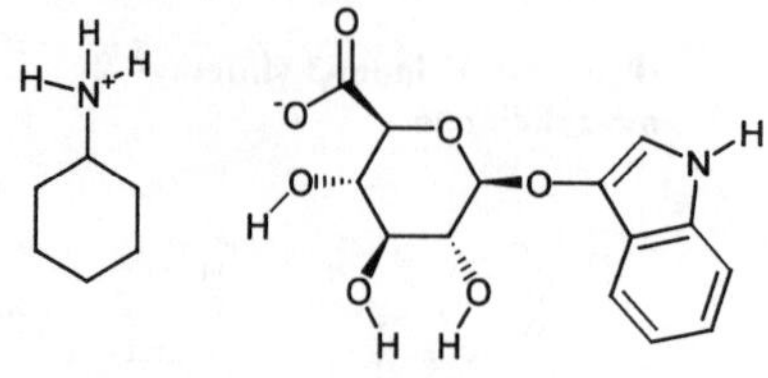

$C_{20}H_{28}N_2O_7$
Chiral diagnostic reagent. Soluble in $H_2O$. *Acros Organics nv; Diagnostic Chemicals Limited.*

**3240 D-Inosine**
58-63-9 5005(12) 200-390-4

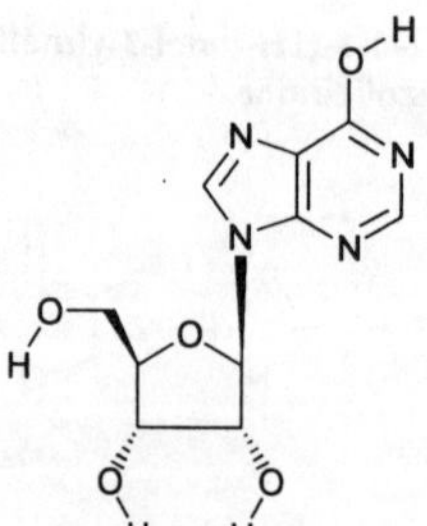

$C_{10}H_{12}N_4O_5$
Purin-(6H)-6-one, 1,9-dihydro-9-β-D-ribofuranosyl-; 9-β-D-Ribofuranosylhypoxanthine; 1,9-Dihydro-9-β-D-ribofuranosyl-6H-purin-6-one' Atorel; HXR; Hypoxanthine 9-β-D-ribofuranoside; Hypoxanthine riboside; Hypoxanthosine; Inosie; Oxiamin; Panholic-L; Ribonosine; Selfer; Trophicardyl. Listed on TSCA. Chiral intermediate. mp = 222-226°; $[\alpha]^{23}$ = -5 0.4° (c = 1, $H_2O$). *Acros Organics nv; Sigma-Aldrich Fine Chemicals.*

**3241 (-)-chiro-Inositol**
551-72-4 209-000-7

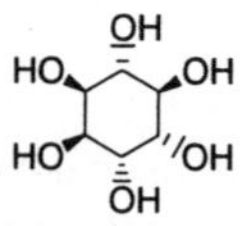

$C_6H_{12}O_6$
Chiral intermediate. mp = 230°; $[\alpha]_D^{20}$ = - 60° (c = 1.3, $H_2O$). *Sigma-Aldrich Fine Chemicals.*

**3242 (+)-chiro-Inositol**
643-12-9 211-394-0

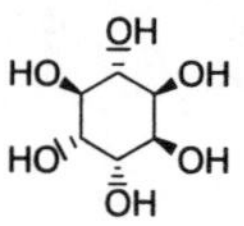

$C_6H_{12}O_6$
Chiral intermediate. mp = 230°; $[\alpha]^{23}$ = + 55° (c = 1.2, $H_2O$). *Sigma-Aldrich Fine Chemicals.*

**3243 (+)-5-Iodo-2-deoxypuridine**
54-42-2 4934(12) 200-207-8

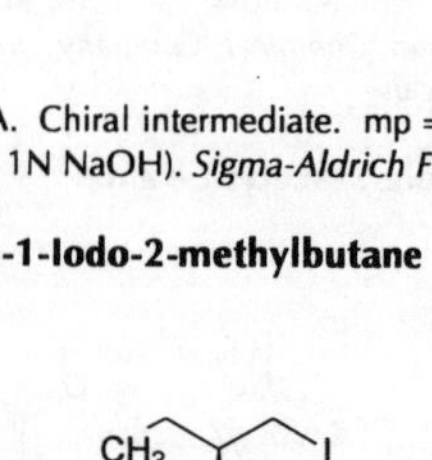

$C_9H_{11}IN_2O_5$
Listed on TSCA. Chiral intermediate. mp = 194°; $[\alpha]^{21}$ = + 29.9° (c = 1, 1N NaOH). *Sigma-Aldrich Fine Chemicals.*

**3244 (S)-(+)-1-Iodo-2-methylbutane**
29394-58-9

$C_5H_{11}I$
Chiral building block. bp = 148°; d = 1.525; n = 1.497; $[\alpha]_D^{20}$ = + 5.7° (neat). *Acros Organics nv; Sigma-Aldrich Fine Chemicals.*

**3245 (R)-Iodo-2-methyl-2-heptanol**

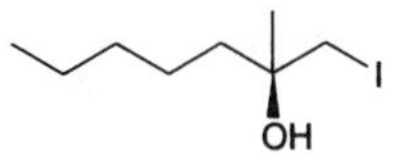

$C_8H_{17}IO$
Chiral building block. *Japan Energy Corporation.*

**3246 (S)-Iodo-2-methyl-2-pentanol**

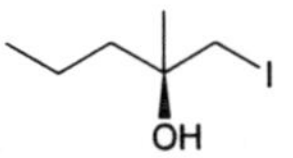

$C_6H_{13}IO$
Chiral building block. *Japan Energy Corporation.*

**3247 4-Iodo-D-phenylalanine**
62561-75-5

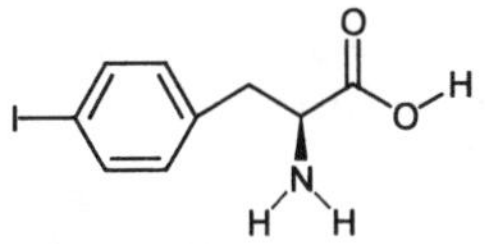

$C_9H_{10}INO_2$
p-Iodo-D-phenylalanine. Chiral intermediate. *Acros Organics nv; Advanced Asymmetrics, Inc.; Omega Chemical Company Inc.*

**3248 4-Iodo-L-phenylalanine**
24250-85-9

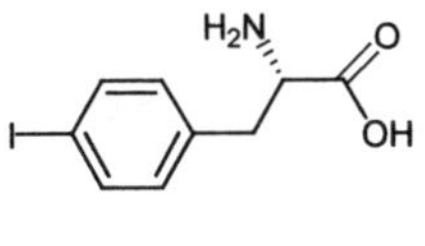

$C_9H_{10}INO_2$
Chiral building block. *Advanced Asymmetrics, Inc.; Omega Chemical Company Inc.*

**3249 p-Iodo-D-phenylalanine methyl ester**

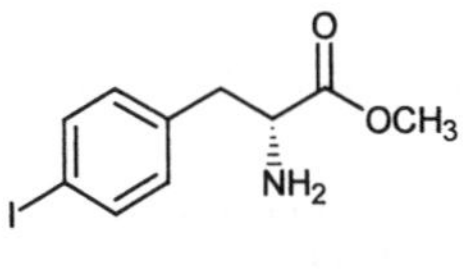

$C_{10}H_{12}INO_2$
Chiral building block. *Advanced Asymmetrics, Inc.*

**3250 p-Iodo-L-phenylalanine methyl ester**

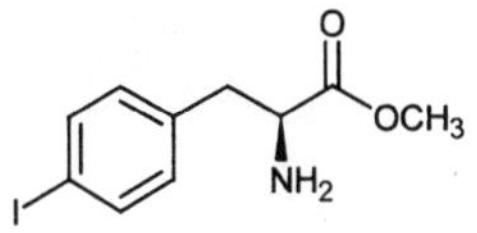

$C_{10}H_{12}INO_2$
Chiral building block. *Advanced Asymmetrics, Inc.*

**3251 (S)-2-(4-Iodophenyl) propanoic acid**

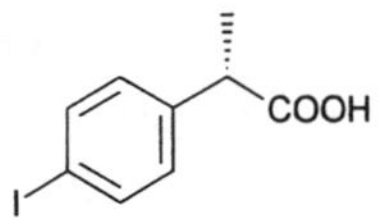

$C_9H_9IO_2$
Chiral intermediate. *Nagase & Co., Ltd.*

**3252 3-Iodo-L-tyrosine**
70-78-0 5065(12) 200-744-8

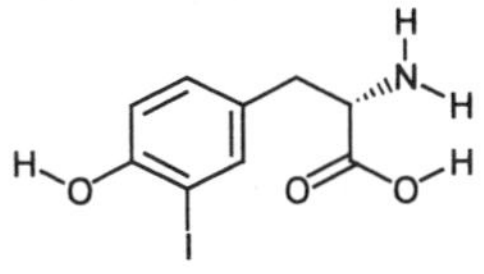

$C_9H_{10}INO_3$
Tyrosine, 3-iodo-, L-; 3-Iodo-4-hydroxyphenylalanine; MIT. Chiral intermediate. mp = 210°; $[\alpha]_D^{20}$ = - 4° (c = 5, 1N HCl). *Acros Organics nv; Sigma-Aldrich Fine Chemicals.*

**3253 D-(+)-Isoascorbic acid, sodium salt**
6381-77-7 228-973-9

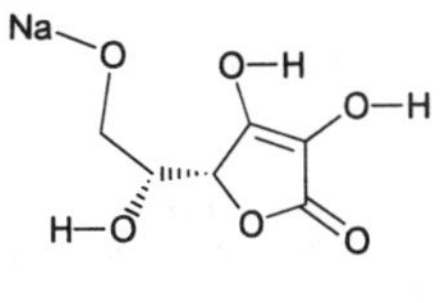

$C_6H_8NaO_6$
Araboascorbic acid, monosodium salt, D-; E 316; Eribate N; Erythorbic acid sodium salt; Isona; Mercate 20; Neocebitate; Ozoban; Sodium D-isoascorbate; Sodium erythorbate; D-erythro-Hex-2-enoic acid, γ-lactone, monosodium salt; 2,3-didehydro-3-O-sodio-D-erythro-hexono-1,4-lactone. Listed on TSCA. Chiral building block. $[\alpha]$ = + 95.5° (c = 10, $H_2O$). *Acros Organics nv; TCI America.*

**3254 Isobutyl (S)-(-)-2-chloropropionate**
83261-15-8 280-349-5

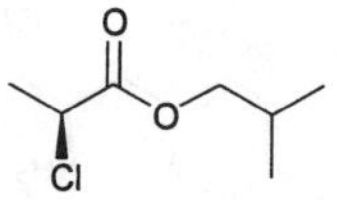

$C_7H_{13}ClO_2$
Propanoic acid, 2-chloro-, 2-methylpropyl ester, (2S)-; (S)-2-Chloropropionic acid 2-methylpropyl ester. Chiral intermediate. bp = 175-177°. *BASF Aktiengesellschaft.*

**3255 Isobutyl (R)-lactate**
61597-96-4 407-770-0

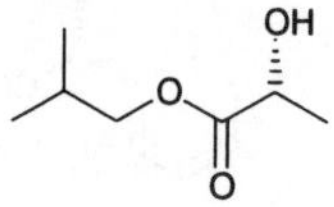

$C_7H_{14}O_3$
Propanoic acid, 2-hydroxy-, 2-methylpropyl ester, (2R)-; (R)-(+)-Lactic acid isobutyl ester; Isobutyl (R)-2-hydroxypropionate. Chiral intermediate. $bp_{13}$ = 73°; d = 0.971; n = 1.419; $[\alpha]^{19}$ = + 15° (neat). *BASF Aktiengesellschaft; Sigma-Aldrich Fine Chemicals.*

**3256 (S)-(+)-4-Isobutylmethlphenylacetic acid**
51146-56-6 4812(11)

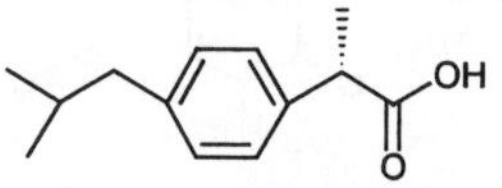

$C_{13}H_{18}O_2$
(S)-(+)-Ibuprofen. Chiral intermediate. mp = 51-53°; $[\alpha]_D^{20}$ = + 59° (c = 2, EtOH). *Sigma-Aldrich Fine Chemicals.*

**3257 (R)-2-Isobutylsuccinic acid 1-methyl ester**
130165-76-3
$C_9H_{16}O_4$
(R)-3-Methoxycarbonyl-5-methylhexanoic acid. Chiral intermediate. *ChiroTech Technology Ltd.; Lancaster Synthesis Ltd.*

**3258 (S)-2-Isobutylsuccinic acid 1-methyl ester**
213270-36-1
$C_9H_{16}O_4$
(S)-3-Methoxycarbonyl-5-methylhexanoic acid. Chiral intermediate. *ChiroTech Technology Ltd.; Lancaster Synthesis Ltd.*

**3259 (-)-Isocaryophyllene**
118-65-0 1926(12) 204-267-6

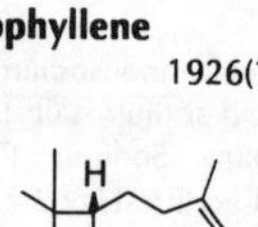

$C_{15}H_{24}$
Listed on TSCA. Chiral intermediate. bp = 271-273°; d = 0.894; n = 1.497; $[\alpha]_D^{20}$ = - 31.0° (neat). *Sigma-Aldrich Fine Chemicals.*

**3260 (+)-Isocordyne hydrochloride**
13552-72-2 5046(11)

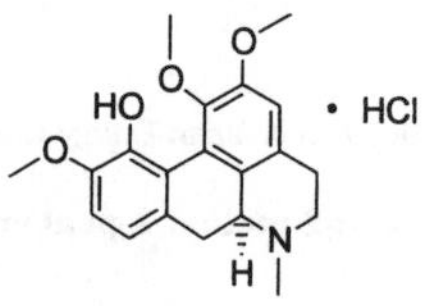

$C_{20}H_{24}ClNO_4$
Chiral intermediate. mp = 215-218°; $[\alpha]_D^{20}$ = + 170° (c = 0.5, $CHCl_3$). *Sigma-Aldrich Fine Chemicals.*

**3261 (S)-(+)-2-Isocyanato-3-tert-butoxypropionic acid methyl ester**
$C_9H_{15}NO_4$
Methyl (S)-(+)-2-isocyanato-3-tert-butoxypropionate. Chiral intermediate. *TCI America.*

**3262 (S)-(-)-2-Isocyanatoglutaric acid diethyl ester**
$C_{10}H_{15}NO_5$
(S)-(-)-2-Isocyanatopentanedioic acid diethyl ester; Diethyl (S)-(-)-2-isocyanatopentanedioate. Chiral intermediate. *TCI America.*

**3263 (S)-(-)-2-Isocyanato-3-methylbutyric acid methyl ester**
30293-86-8

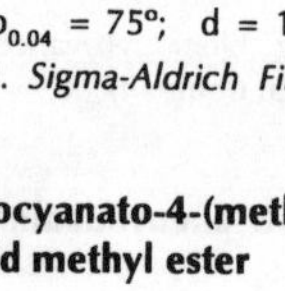

$C_7H_{11}NO_3$
Methyl (S)-(-)-2-isocyanato-3-methylbutyrate. Chiral building block. $bp_{0.04}$ = 75°; d = 1.053; n = 1.428; $[\alpha]_D^{20}$ = - 22° (neat). *Sigma-Aldrich Fine Chemicals; TCI America.*

**3264 (S)-(-)-2-Isocyanato-4-(methylthio)-butyric acid methyl ester**
93778-88-2
$C_7H_{11}NO_3S$
Methyl (S)-(-)-2-isocyanato-4-(methylthio)butyrate. Chiral intermediate. *TCI America.*

**3265 (2S,3S)-2-Isocyanato-3-methylvaleric acid methyl ester**
120219-17-2
$C_8H_{13}NO_3$
Methyl (2S,3S)-2-Isocyanato-3-methylvalerate. Chiral intermediate. *TCI America.*

**3266 (S)-(-)-2-Isocyanato-4-methylvaleric acid methyl ester**
39570-63-3

$C_8H_{13}NO_3$
Methyl (S)-(-)-2-isocyanato-4-methylvalerate. Chiral intermediate. *TCI America.*

**3267 (S)-2-Isocyanato-3-phenylpropionic acid methyl ester**
40203-94-9

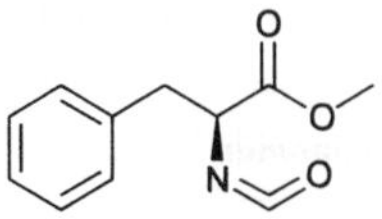

$C_{11}H_{11}NO_3$
Methyl (S)-2-isocyanato-3-phenylpropionate. Chiral intermediate. bp = 244°; d = 1.13; n = 1.513; $[\alpha]_D^{20}$ = - 80.0° (neat). *Sigma-Aldrich Fine Chemicals; TCI America.*

**3268 (S)-(-)-2-Isocyanatopropionic acid methyl ester**
$C_5H_7NO_3$
Methyl (S)-(-)-2-isocyanatopropionate. Chiral intermediate. *TCI America.*

**3269 Isocyanic acid (S)-(+)-1-(1-naphthyl)-ethyl ester**
73671-79-1

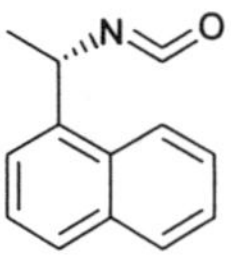

$C_{13}H_{11}NO$
(S)-(+)-1-(1-Naphthyl)ethyl isocyanate. Chiral intermediate. $bp_{0.16}$ = 106-108°; d = 1.128; n = 1.6045; $[\alpha]_D^{20}$ = + 45° (c = 3.5, toluene). *Norse Laboratories; Sigma-Aldrich Fine Chemicals; TCI America.*

**3270 (R)-(-)-2-(2-Isoindolinyl)butan-1-ol**
135711-18-1

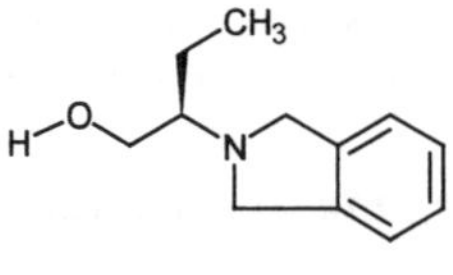

$C_{12}H_{17}NO$
(R)-(-)-2-(1,3-Dihydro-isoindol-2-yl)butan-1-ol. Chiral intermediate. mp = 57-58°. *Acros Organics nv.*

**3271 D-Isoleucine**
319-78-8 206-269-2

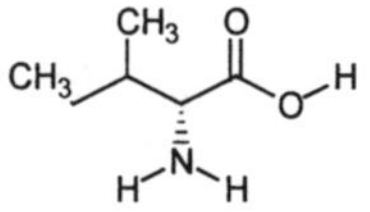

$C_6H_{13}NO_2$
Isoleucine, D-. Chiral building block. $[\alpha]^{25}$ = - 39° (c = 1, 5M HCl). *Acros Organics nv; Sigma-Aldrich Fine Chemicals.*

**3272 L-Isoleucine**
73-32-5 5196(12) 200-798-2

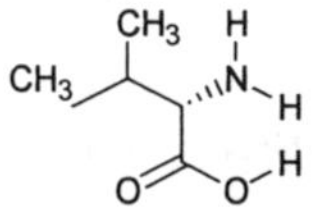

$C_6H_{13}NO_2$
Pentanoic acid, 2-amino-3-methyl-, [S-(R*,R*)]-; (2S,3S)-2-Amino-3-methylpentanoic acid; (2S,3S)-α-Amino-β-methylvaleric acid; L-Norvaline, 3-methyl-, erythro-. Listed on TSCA. Chiral building block. mp = 288°; $[\alpha]_D^{20}$ = + 40 ± 2° (c = 5, 6N HCl). *Aceto Corporation; Acros Organics nv; Ajinomoto Co. Inc.; Flamma s.p.a.; KingChem, Inc.; Kyowa Hakko Kogyo Co., Ltd.; Lancaster Synthesis Ltd.; Nippon Kayaku; Rexim S.A. Produits Chimiques; Senn Chemicals AG; Shanghai DSL International Trading Company; Sigma-Aldrich Fine Chemicals; Tanabe Seiyaku Co. Ltd.; TCI America; Varsal Instruments, Inc.; Yoneyama Yakuhin Kogyo Co., Ltd.*

**3273 D-allo-Isoleucine**
1509-35-9 216-143-9

$C_6H_{13}NO_2$
Alloisoleucine, D-; (2R,3S)-(-)-2-Amino-3-methylpentanoic acid. Chiral intermediate. *Great Lakes Fine Chemicals; Senn Chemicals AG; Sigma-Aldrich Fine Chemicals; TCI America; Yamakawa Chemical Industry Co. Ltd.*

**3274 L-allo-Isoleucine**
1509-34-8 216-142-3

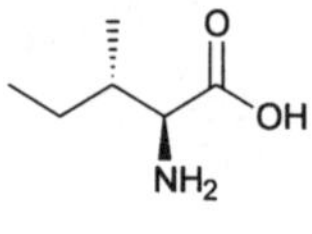

$C_6H_{13}NO_2$
Pentanoic acid, 2-amino-3-methyl-, [S-(R*,S*)]-; L-Norvaline, 3-methyl-, threo-. Chiral building block. *Senn Chemicals AG.*

**3275 L-Isoleucine amide hydrochloride**

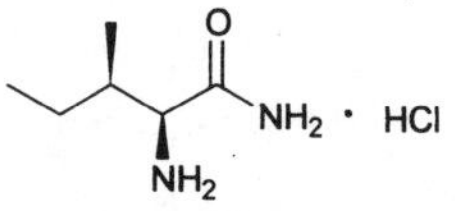

$C_6H_{15}ClN_2O$
Chiral building block. *Senn Chemicals AG.*

**3276 L-Isoleucine hydrazide**

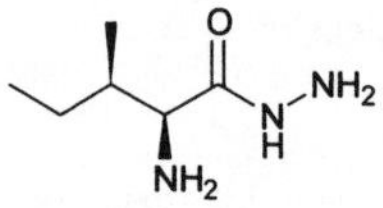

$C_6H_{15}N_3O$
Chiral building block. *Senn Chemicals AG.*

**3277 L-Isoleucine methyl ester hydrochloride**
18598-74-8 242-437-1
$C_7H_{16}ClNO_2$
Isoleucine, methyl ester, hydrochloride, L-; Methyl L-isoleucinate hydrochloride. Chiral building block. *TCI America.*

**3278 L-Isoleucinol**
24629-25-2 246-371-4

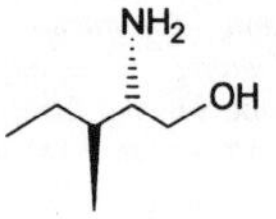

$C_6H_{15}NO$
Pentan-1-ol, 2-amino-3-methyl-, (2S,3S)-; (S)-(+)-2-Amino-3-methyl-1-pentanol. Chiral building block. mp = 30°; $bp_{14}$ = 97°; $[\alpha]^{22}$ = + 5.4° (c = 1.6, EtOH). *Acros Organics nv; Boehringer Ingelheim Pharma KG; Omega Chemical Company Inc.; Senn Chemicals AG; Sigma-Aldrich Fine Chemicals; TCI America.*

**3279 D-Isomaltitol**
534-73-6 208-605-3

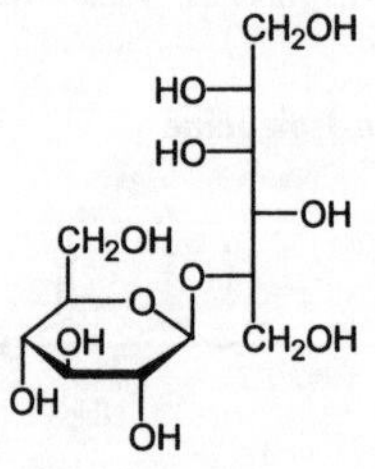

$C_{12}H_{24}O_{11}$
Glucitol, 6-O-α-D-glucopyranosyl-, D-; 6-O-α-D-Glucopyranosyl-D-glucitol. Chiral intermediate. *Senn Chemicals AG.*

**3280 D-Isomaltose**
499-40-1 207-879-1

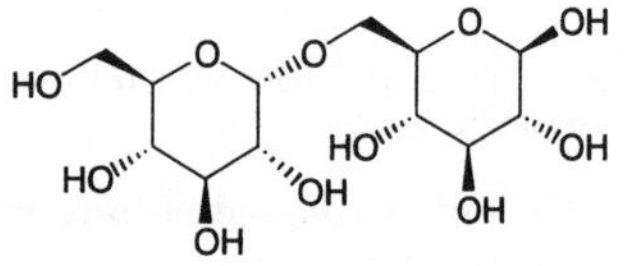

$C_{12}H_{22}O_{11}$
Glucose, 6-O-α-D-glucopyranosyl-, D-; 6-O-α-D-Glucopyranosyl-D-glucose. Chiral intermediate. *Senn Chemicals AG.*

**3281 D-Isomannide**
641-74-7 211-374-1

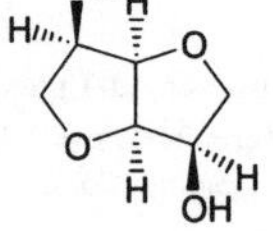

$C_6H_{10}O_4$
1,4-Dianhydro-D-mannitol. Chiral building block. mp = 80-85°; $[\alpha]^{25}$ = + 89° (c = 3, $H_2O$). *Sigma-Aldrich Fine Chemicals.*

**3282 (1R)-(+)-Isomenthol**
23283-97-8 245-554-6

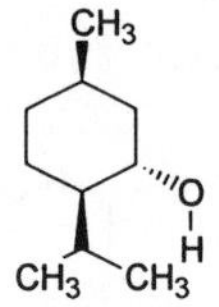

$C_{10}H_{20}O$
Cyclohexanol, 5-methyl-2-(1-methylethyl)-, (1S,2R,5R)-; [1S-(1α,2β,5β)]-5-Methyl-2-(1-methylethyl)cyclohexanol; (1S-(1α,2β,5β))-5-Methyl-(1-isopropyl)cyclohexan-2-ol. Listed on TSCA. Resolving agent; Chiral auxiliary; Chiral intermediate. mp = 77-83°; bp = 218-219°; $[\alpha]^{21}$ = + 24.0° (c = 4.2, EtOH). *Acros Organics nv; Sigma-Aldrich Fine Chemicals.*

**3283 (+)-Isopilocarpine nitrate**
5984-94-1 227-800-4
$C_{11}H_{17}N_3O_5$
Fura2(3H)one, 3-ethyldihydro-4-[(1-methyl-1H-imidazol-5-yl)methyl]-, (3R-trans)-, mononitrate. Pharmaceutical or derivative. $[\alpha]^{22}$ = + 35° (c = 10, $H_2O$). *Acros Organics nv.*

**3284 (1R,2R,3R,5S)-(-)-Isopinocampheol**
25465-65-0 247-011-9
$C_{10}H_{18}O$
(-)-3-Pinanol. Chiral intermediate. mp = 51-53°; bp = 219°; $[\alpha]^{22}$ = - 34° (c = 20, EtOH). *Sigma-Aldrich Fine Chemicals.*

**3285 (1S,2S,3S,5R)-(+)-Isopinocampheol**
27779-29-9 248-657-4

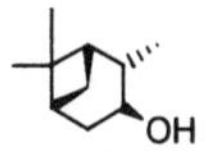

$C_{10}H_{18}O$
(+)-3-Pinanol. Chiral intermediate. mp = 51-53°; bp = 219°; $[\alpha]_D^{20}$ = + 35.1° (c = 20, EtOH). *Sigma-Aldrich Fine Chemicals.*

**3286 (1R,2R,3R,5S)-(-)-Isopinocampheylamine**
69460-11-3

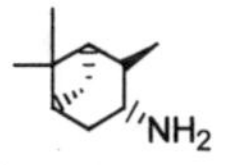

$C_{10}H_{19}N$
Resolving agent. $bp_{18}$ = 90°; d = 0.909; n = 1.48; $[\alpha]^{21}$ = - 42° (neat). *Sigma-Aldrich Fine Chemicals.*

**3287 (1S,2S,3S,5R)-(+)-Isopinocampheylamine**

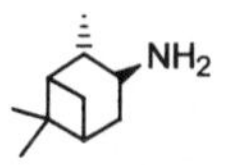

$C_{10}H_{19}N$
Resolving agent. $bp_{18}$ = 90°; d = 0.909; n = 1.481; $[\alpha]^{22}$ = + 44° (neat). *Sigma-Aldrich Fine Chemicals.*

**3288 Isopropoxycarbonyl-L-valine**

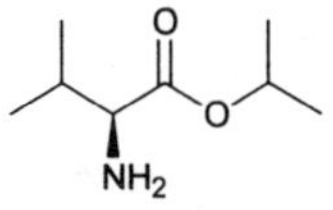

$C_8H_{17}NO_2$
Chiral intermediate. *Synthetech, Inc.*

**3289 (S)-2-Isopropylamino-3-methyl-1-butanol**
$C_8H_{19}NO$
(S)-2-iso-Propylamino-3-methyl-1-butanol; (S)-N-Iso-propyl-2-(3-methyl-1-butanol)amine. Chiral building block. *TCI America.*

**3290 (S)-1-Isopropylaminopropanediol**

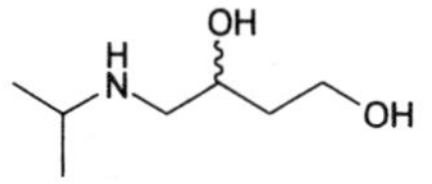

$C_7H_{17}NO_2$
Chiral building block. *Vinchem.*

**3291 (R)-(+)-3-Isopropylamino-1,2-propanediol**

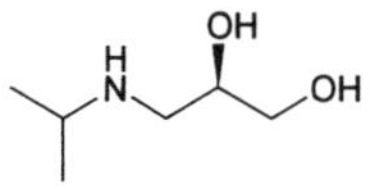

$C_6H_{15}NO_2$
Propane-1,2-diol, 3-[(1-methylethyl)amino]-, (R)-; (R)-1,2-Dihydroxy-3-isopropylaminopropane. Chiral building block. *Bachem AG.*

**3292 (S)-(-)-3-Isopropylamino-1,2-propanediol**

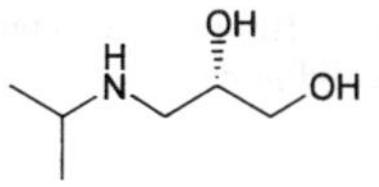

$C_6H_{15}NO_2$
Propane-1,2-diol, 3-[(1-methylethyl)amino]-, (S)-; (S)-1,2-Dihydroxy-3-isopropylaminopropane. Chiral building block. *Bachem AG.*

**3293 (3R)-3-Isopropyl bicyclic lactam**
$C_8H_{15}NO_2$
Chiral intermediate. *Oxford Asymmetry International plc.*

**3294 (3S)-3-Isopropyl-bicyclic lactam**
122383-35-1
$C_8H_{15}NO_2$
Chiral intermediate. *Oxford Asymmetry International plc.*

**3295 Isopropyl-(tert-Butoxycarbonyl)-L-lysine**

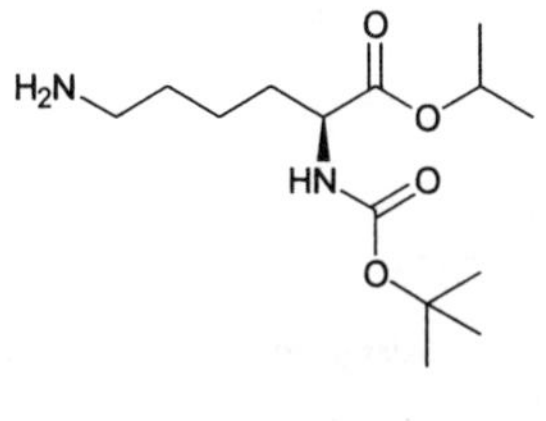

$C_{14}H_{28}N_2O_4$
Chiral intermediate. *Synthetech, Inc.*

**3296 Isopropyl-carbobenzyloxy-(tert-Butoxy-carbonyl)-L-lysine**

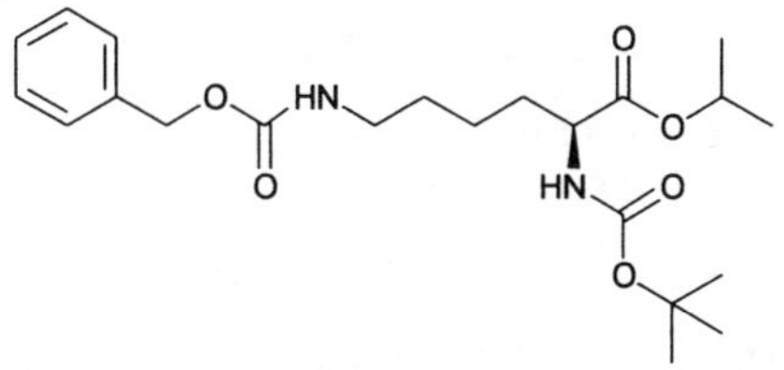

$C_{22}H_{34}N_2O_6$
Chiral intermediate. *Synthetech, Inc.*

**3297 (R)-(+)-4-Isopropyl-5,5-dimethyl-2-oxazolidinone**

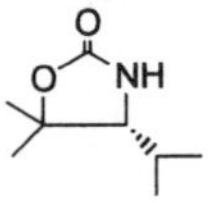

$C_8H_{15}NO_2$
Chiral intermediate. mp = 86-87°; $[\alpha]_D^{20}$ = + 47° (c = 1, $H_2O$). *Sigma-Aldrich Fine Chemicals.*

**3298 (S)-(-)-4-Isopropyl-5,5-dimethyl-2-oxazolidinone**

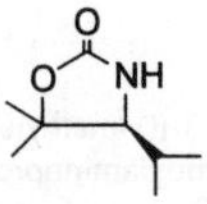

$C_8H_{15}NO_2$
Chiral intermediate. mp = 86-87°; $[\alpha]^{22}$ = - 47° (c = 1, $H_2O$). *Sigma-Aldrich Fine Chemicals.*

**3299 1,2-Isopropylidene-6-amino-6-deoxy-α-D-glucopyranoside hydrochloride**

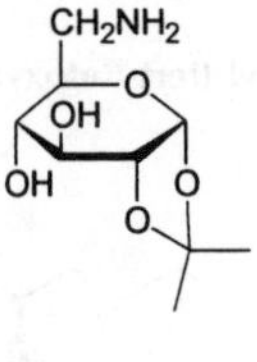

$C_9H_{17}NO_5$
Chiral intermediate. *Senn Chemicals AG.*

**3300 (+)-5,6-Isopropylideneascorbic acid**
15042-01-0

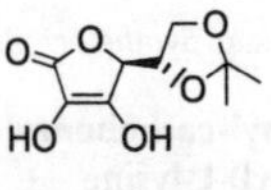

$C_9H_{12}O_6$
Chiral building block. mp = 210°; $[\alpha]^{19}$ = + 25.3° (c = 1, $H_2O$). *Sigma-Aldrich Fine Chemicals.*

**3301 (3R,8aS)-3-Isopropyl-7-dimethyl-8a-methyl bicyclic lactam**
$C_{13}H_{23}NO_2$
Chiral intermediate. *Oxford Asymmetry International plc.*

**3302 (3S,8aR)-3-Isopropyl-7-dimethyl-8a-methyl bicyclic lactam**
$C_{13}H_{23}NO_2$
Chiral intermediate. *Oxford Asymmetry International plc.*

**3303 (R)-Isopropyl-5,5-dimethyloxazolidin-2-one**

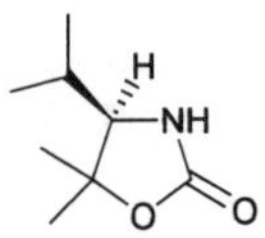

$C_8H_{15}NO_2$
Chiral intermediate. *Oxford Asymmetry International plc.*

**3304 (S)-Isopropyl-5,5-dimethyloxazolidin-2-one**

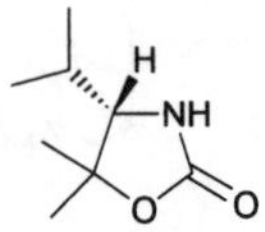

$C_8H_{15}NO_2$
Chiral intermediate. *Oxford Asymmetry International plc.*

**3305 (R)-(+)-2,2'-Isopropylidenebis(4-benzyl-2-oxazoline)**

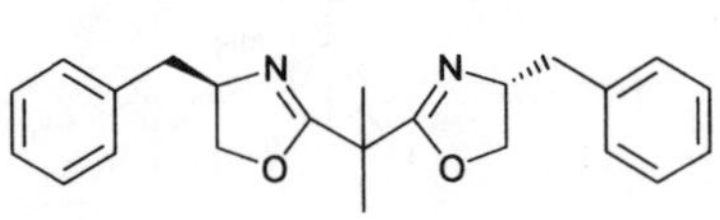

$C_{23}H_{26}N_2O_2$
Chiral intermediate. mp = 54-60°; $[\alpha]^{22}$ = + 65° (c = 1, EtOH). *Great Lakes Fine Chemicals; Sigma-Aldrich Fine Chemicals.*

**3306 (S)-(-)-2,2'-Isopropylidenebis(4-benzyl-2-oxazoline)**

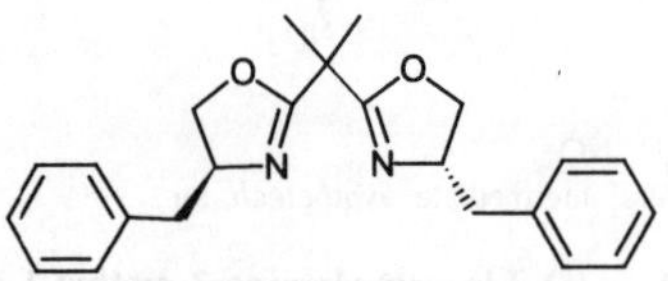

$C_{23}H_{26}N_2O_2$
Chiral ligand. *Great Lakes Fine Chemicals.*

**3307 2,2'-Isopropylidenebis((4R)-4-tert-butyl-2-oxazoline)**

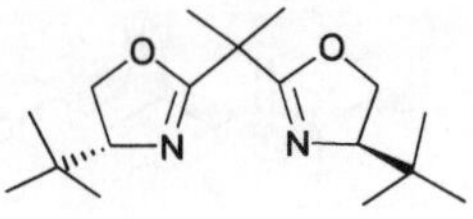

$C_{17}H_{30}N_2O_2$
Chiral intermediate. *Great Lakes Fine Chemicals.*

**3308 2,2'-Isopropylidenebis(4S)-4-tert-butyl-2-oxazoline**

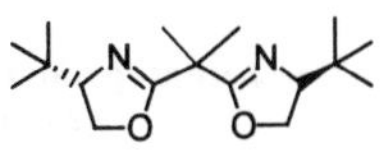

$C_{17}H_{30}N_2O_2$
Chiral intermediate. mp = 89-91°; $[\alpha]_D^{20}$ = - 120° (c = 5, $CHCl_3$). *Sigma-Aldrich Fine Chemicals.*

**3309 (R)-(+)-2,2'-Isopropylidenebis(4-phenyl-2-oxazoline)**

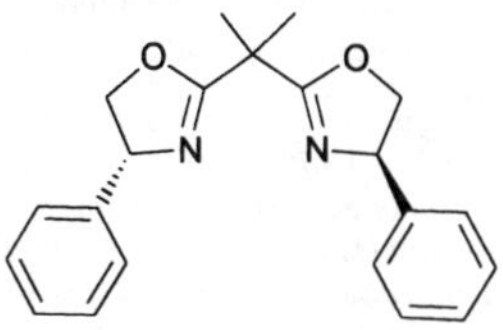

$C_{21}H_{22}N_2O_2$
Chiral intermediate. mp = 56-58°; $[\alpha]_D^{20}$ = + 160° (c = 1, EtOH). *Great Lakes Fine Chemicals; Sigma-Aldrich Fine Chemicals.*

**3310 (S)-(-)-2,2'-Isopropylidenebis(4-phenyl-2-oxazoline)**
131457-46-0

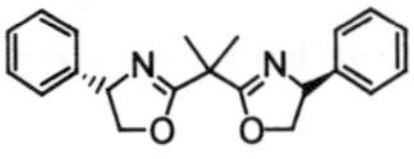

$C_{21}H_{22}N_2O_2$
Chiral ligand. mp = 37-41°; $bp_{0.03}$ = 193°; d = 1; $[\alpha]_D^{20}$ = - 160° (c = 1, EtOH). *Great Lakes Fine Chemicals; Sigma-Aldrich Fine Chemicals.*

**3311 O-Isopropylidene-2,3-dihydroxy-1,4-bis-[bis(3,5-dimethylphenyl)phosphino]butane**

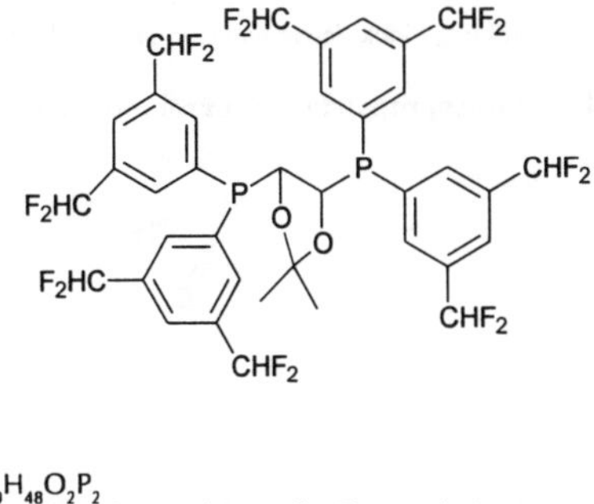

$C_{39}H_{48}O_2P_2$
Chiral ligand. *Digital Specialty Chemicals, Inc.*

**3312 O-Isopropylidene-2,3-dihydroxy-1,4-bis-[[3,5-di(trifluoromethyl)phenyl]phosphino]-butane**

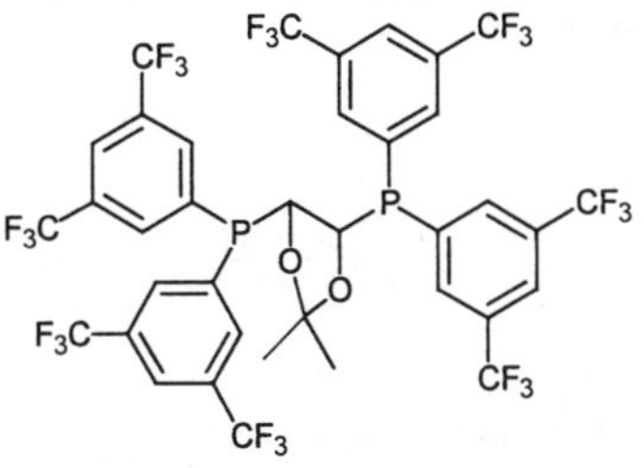

$C_{39}H_{24}F_{24}O_2P_2$
Chiral ligand. *Digital Specialty Chemicals, Inc.*

**3313 (6S,2E)-6,7-Isopropylidenedioxy-3,7-dimethyl-2-octen-1-ol**

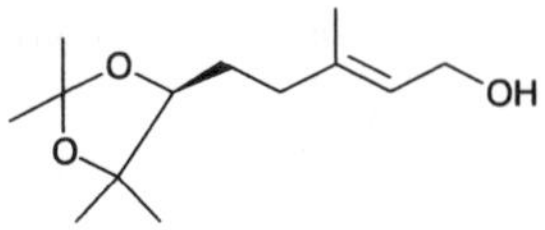

$C_{13}H_{24}O_3$
Chiral intermediate. [α] = + 6° (c = 1.12, $CHCl_3$). *Acros Organics nv.*

**3314 (6S,2Z)-6,7-Isopropylidenedioxy-3,7-dimethyl-2-octen-1-ol**
61262-96-2

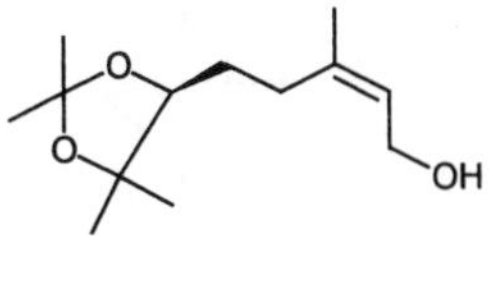

$C_{13}H_{24}O_3$
Chiral intermediate. *Acros Organics nv.*

**3315 (5S)-5,6-Isopropylidenedioxy-6-methyl-heptan-2-one**
61262-94-0

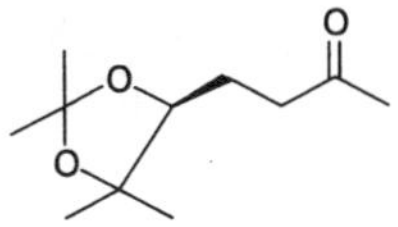

$C_{11}H_{20}O_3$
Chiral intermediate. [α] = - 12.1° (c = 1, $CHCl_3$). *Acros Organics nv.*

**3316 2,3-O-Isopropylidene-D-erythronolactol**
51607-16-0
$C_7H_{12}O_4$
Chiral intermediate. *Pfanstiehl Laboratories, Inc.*

**3317 2,3-O-Isopropylidene-D-erythronolactone**
25581-41-3

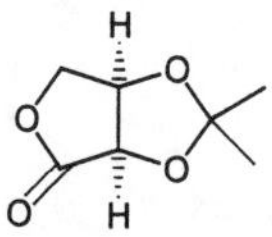

$C_7H_{10}O_4$
Chiral intermediate. mp = 67-69°; $[\alpha]_D^{20}$ = - 118° (c = 1, $H_2O$). *Pfanstiehl Laboratories, Inc.; Sigma-Aldrich Fine Chemicals.*

**3318 1,2-O-Isopropylidene-D-glucofuranose**
18549-40-1 242-420-9

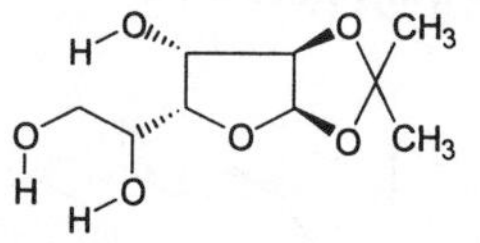

$C_9H_{16}O_6$
Glucofuranose, 1,2-O-(1-methylethylidene)-, α-D-; 1,2-Mono-O-isopropylidene-α-D-glucofuranose; 1,2-O-(1-Methylethylidene)-α-D-glucofuranose. Listed on TSCA. Chiral intermediate. mp = 160°; $[\alpha]_D^{20}$ = - 11.8° (c = 2, $H_2O$). *Acros Organics nv; Pfanstiehl Laboratories, Inc.; Senn Chemicals AG; Sigma-Aldrich Fine Chemicals.*

**3319 2,3-O-Isopropylidene-(R)-glyceraldehyde**
15186-48-8

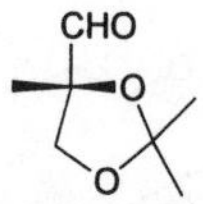

$C_6H_{10}O_3$
Chiral intermediate. *Oxford Asymmetry International plc.*

**3320 (S)-(+)-1,2-O-Isopropylideneglycerol**
22323-82-6 5232(12) 244-910-8

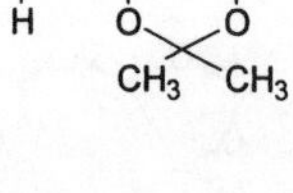

$C_6H_{12}O_3$
Dioxolane-4-methanol, 1,3-, 2,2-dimethyl-, (4S)-; (S)-(+)-2,2-Dimethyl-1,3-dioxolane-4-methanol; (+)-2,3-O-Isopropylidene-sn-glycerol; (S)-(+)-Solketal; S-Glycerol acetonide. Chiral building block. $bp_{14}$ = 82-83°; d = 1.0700; $n_D^{20}$ = 1.4340; $[\alpha]_D^{20}$ = + 11.5 ± 1° (c = 5, $CH_3OH$); $[\alpha]_D^{25}$ = + 15.2° (neat). *Acros Organics nv; Bachem AG; Chemi SpA; Daiso Company, Ltd.; Inalco S.p.A.; Interchem Corporation; Lancaster Synthesis Ltd.; Loba Feinchemie AG; TCI America; Vinchem.*

**3321 (R)-(-)-2,3-O-Isopropylideneglycerol**
14347-78-5 5232(12)

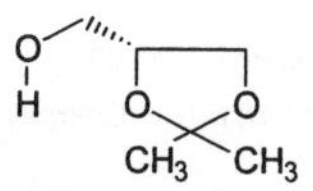

$C_6H_{12}O_3$
(R)-(-)-2,2-Dimethyl-1,3-dioxolane-4-methanol; (R)-(-)-Solketal; R-Glycerol acetonide; L-(-)-1,2-Isopropylideneglycerol. Chiral building block. $bp_8$ = 72-73°; d = 1.0620; $n_D^{20}$ = 1.4340; $[\alpha]_D^{20}$ = - 11.5 ± 1° (c = 5, $CH_3OH$); $[\alpha]_D^{20}$ = - 13.7° (neat). *Acros Organics nv; Austin Chemical Company, Inc.; Bachem AG; Chemi SpA; Daiso Company, Ltd.; Interchem Corporation; Lancaster Synthesis Ltd.; Loba Feinchemie AG; Senn Chemicals AG; Sigma-Aldrich Fine Chemicals; TCI America; Vinchem.*

**3322 (S)-(+)-2,3-O-Isopropylideneglycerol**

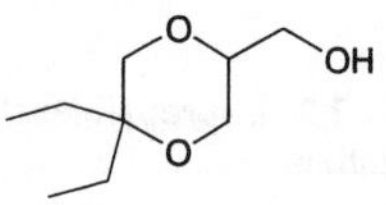

$C_8H_{17}O_3$
L-Skolketal. Chiral building block. *Pfanstiehl Laboratories, Inc.*

**3323 (R)-Isopropylideneglycerol mesylate**

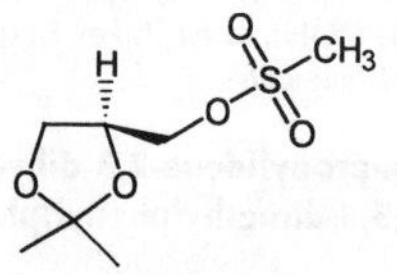

$C_7H_{14}O_5S$
Chiral building block. *Interchem Corporation; Vinchem.*

**3324 (S)-Isopropylideneglycerol mesylate**

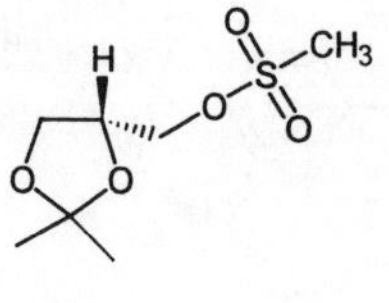

$C_7H_{14}O_5S$
Chiral building block. *Interchem Corporation; Vinchem.*

**3325 (+)-1,2-O-Iisopropylidenemannitol**
4306-35-8

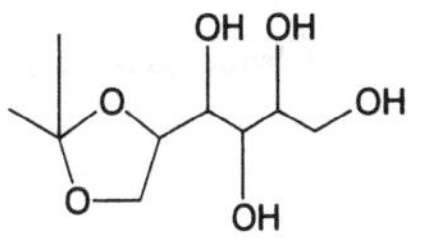

$C_9H_{18}O_6$
Chiral intermediate. mp = 158-161°; $[\alpha]_D^{20}$ = + 4° (c = 4, $H_2O$). *Sigma-Aldrich Fine Chemicals.*

**3326 3,4-O-Isopropylidene-D-mannitol**
3969-84-4

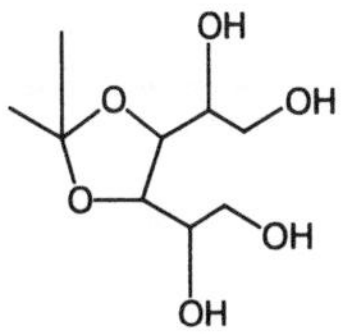

$C_9H_{18}O_6$
Chiral intermediate. mp = 85-88°; $[\alpha]^{28}$ = + 29° (c = 3, $H_2O$). *Senn Chemicals AG; Sigma-Aldrich Fine Chemicals.*

**3327 9-(2',3'-O-Isopropylidene-β-D-ribofuranosyl) adenine**
362-75-4 206-650-3

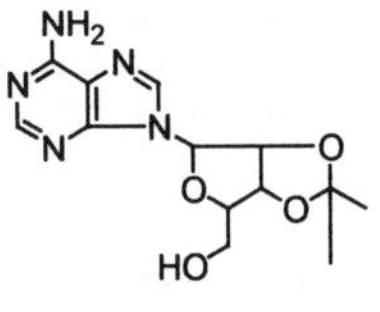

$C_{13}H_{17}N_5O_4$
Adenosine, 2',3'-O-(1-methylethylidene)-; 2',3'-Isopropylideneadenosine. Chiral intermediate. mp = 219°; $[\alpha]_D^{20}$ = - 98.5° (c = 1, dioxane). *Pfanstiehl Laboratories, Inc.; Sigma-Aldrich Fine Chemicals.*

**3328 2,3-O-Isopropylidene-D-ribono-1,4-lactone**
30725-00-9

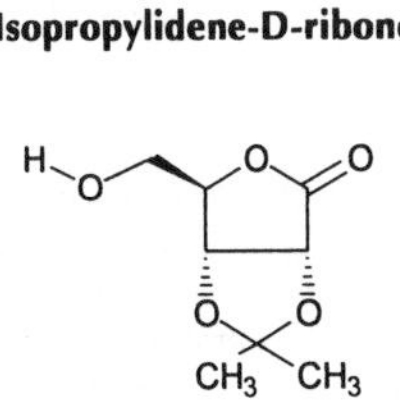

$C_8H_{12}O_5$
2,3-O-Isopropylidene-D-ribonic γ-lactone. Chiral intermediate. mp = 135-138°; $[\alpha]_D^{20}$ = - 69° (c = 2, pyridine). *Acros Organics nv; Pfanstiehl Laboratories, Inc.; Sigma-Aldrich Fine Chemicals.*

**3329 (R,R)-(-)-2,3-O-Isopropylidene-1,1,4,4-tetra(2-naphthyl)-L-threitol**
137365-09-4

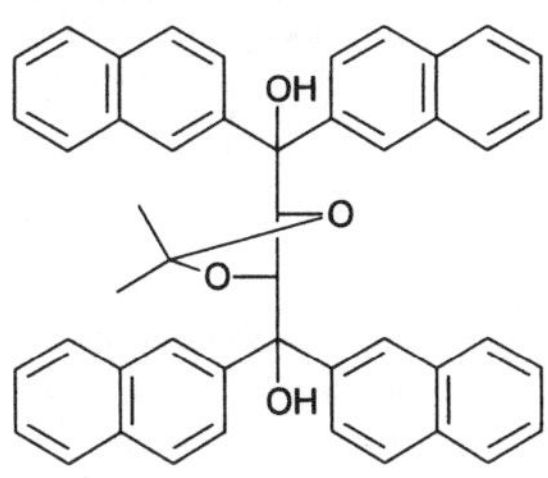

$C_{47}H_{38}O_4$
Chiral intermediate. *FineTech, Ltd.*

**3330 (-)-2,3-O-Isopropylidene-D-threitol**
73346-74-4 277-391-1

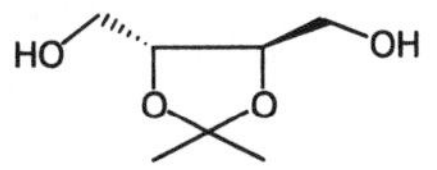

$C_7H_{14}O_4$
Dioxolane-4,5-dimethanol, 1,3-, 2,2-dimethyl-, (4R,5R)-; (4R,5R)-2,2-Dimethyl 1,3-dioxolane-4,5-dimethanol; (4R,5R)-(-)-4,5-Bis(hydroxymethyl) 2,2-dimethyl-1,3-dioxolane. Chiral intermediate. mp = 48-51°; $bp_{0.05}$ = 88-90°; $[\alpha]_D^{20}$ = - 4.5 ± 0.5° (c = 5, $CHCl_3$). *Acros Organics nv; Lancaster Synthesis Ltd.; Loba Feinchemie AG; Sigma-Aldrich Fine Chemicals; TCI America.*

**3331 (+)-2,3-O-Isopropylidene-L-threitol**
50622-09-8 256-658-6

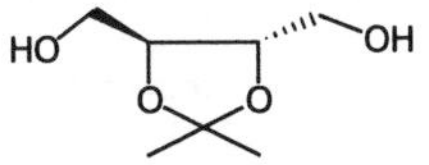

$C_7H_{14}O_4$
Dioxolane-4,5-dimethanol, 1,3-, 2,2-dimethyl-, (4S,5S)-; (4S,5S)-2,2-Dimethyl 1,3-dioxolane-4,5-dimethanol; (4S,5S)-(+)-4,5-Bis(hydroxymethyl) 2,2-dimethyl-1,3-dioxolane. Chiral intermediate. mp = 48-51°; $bp_{0.1}$ = 92-94°; $[\alpha]_D^{20}$ = + 3.0 ± 0.5° (c = 5, EtOH). *Acros Organics nv; FineTech, Ltd.; Lancaster Synthesis Ltd.; Loba Feinchemie AG; Sigma-Aldrich Fine Chemicals; TCI America.*

**3332 (+)-3,4-O-Isopropylidenethreonic acid calcium salt**
98733-24-5

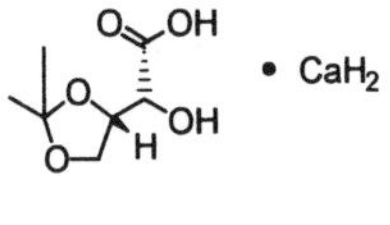

$C_7H_{12}O_5$
Chiral building block. mp > 300°; $[\alpha]_D^{20}$ = + 24° (c = 1, $H_2O$). *Sigma-Aldrich Fine Chemicals.*

**3333 1,2-O-Isopropylidene-α-D-xylofuranose**
20031-21-4

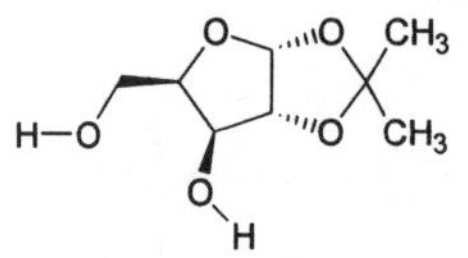

$C_8H_{14}O_5$
Xylofuranose, 1,2-O-isopropylidene-, α-D-. Listed on TSCA. Chiral intermediate. mp = 68-70°; $bp_{0.06}$ = 112-114; $[\alpha]_D^{20}$ = - 18° (c = 4, $H_2O$). *Acros Organics nv; Pfanstiehl Laboratories, Inc.; Senn Chemicals AG; Sigma-Aldrich Fine Chemicals.*

**3334 Isopropyl (S)-(-)-lactate**
63697-00-7 264-417-1

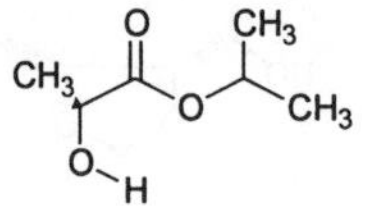

$C_6H_{12}O_3$
Propanoic acid, 2-hydroxy-, 1-methylethyl ester, (2S)-; Isopropyl (S)-2-hydroxypropionate. Listed on TSCA. Chiral intermediate. bp = 166-168°; d = 0.988; n = 1.41; $[\alpha]^{25}$ = - 10.3° (neat). *Acros Organics nv; Sigma-Aldrich Fine Chemicals.*

**3335 (3R,7aS)-3-isopropyl-7a-methyl bicyclic lactam**
12308-97-9
$C_{10}H_{17}NO_2$
Chiral intermediate. *Oxford Asymmetry International plc.*

**3336 (3R-cis)-(-)-3-Isopropyl-7a-methyltetra-hydropyrrolo[2,1-b]oxazol-5(6H)-one**
123808-97-9
$C_{10}H_{17}NO_2$
Chiral intermediate. $bp_{0.4}$ = 70°; d = 1.0270; $n_D^{20}$ = 1.4730; $[\alpha]_D^{20}$ = - 90 ± 3° (c = 2.8, EtOH). *Lancaster Synthesis Ltd.*

**3337 (3S-cis)-(+)-3-Isopropyl-7a-methyltetra-hydropyrrolo[2,1-b]oxazol-5(6H)-one**
98203-44-2

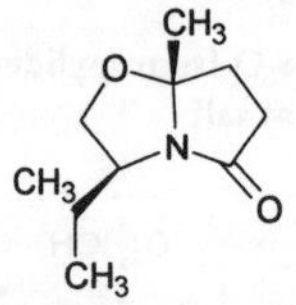

$C_{10}H_{17}NO_2$
(3S-cis)-Tetrahydro-7a-methyl-3-(1-methylethyl)pyrrolo-[2,1-b]oxazol-5(6H)-one; (3S,7aR)-3-Isopropyl-7a-methyl bicyclic lactam. Chiral intermediate. $bp_{0.4}$ = 70°; d = 1.0270; $n_D^{20}$ = 1.4730; $[\alpha]_D^{20}$ = - 90 ± 3° (c = 2.8, EtOH). *Acros Organics nv; Lancaster Synthesis Ltd.; Oxford Asymmetry International plc; Sigma-Aldrich Fine Chemicals.*

**3338 (4S)-(-)-4-Isopropyl-1,3-oxazolidine-2-thione**
84272-19-5

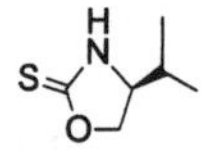

$C_6H_{11}NOS$
Chiral intermediate. mp = 45-46°; $[\alpha]_D^{20}$ = - 25 ± 1° (c = 1, $CHCl_3$). *Lancaster Synthesis Ltd.; Sigma-Aldrich Fine Chemicals.*

**3339 (4S)-(-)-4-Isopropyl-2-oxazolidinone**
17016-83-0

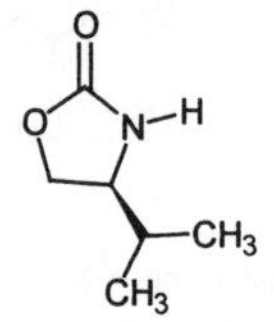

$C_6H_{11}NO_2$
(S)-(-)-4-Isopropyl-2-oxazolidinone. Chiral auxiliary; Chiral intermediate. mp = 70-71°; $[\alpha]_D^{20}$ = - 14 ± 0.5° (c = 7, $CHCl_3$); $[\alpha]_D^{20}$ = - 22.5 ± 0.4° (c = 4, $CH_3OH$). *Acros Organics nv; Austin Chemical Company, Inc.; Boehringer Ingelheim Pharma KG; Daiso Company, Ltd.; Fischer Chemicals AG; Great Lakes Fine Chemicals; Kiralchem Ltd.; Lancaster Synthesis Ltd.; Omega Chemical Company Inc.; Oxford Asymmetry International plc; Senn Chemicals AG; Sigma-Aldrich Fine Chemicals; TCI America.*

**3340 (R)-(+)-4-Isopropyl-2-oxazolidinone**
95530-58-8

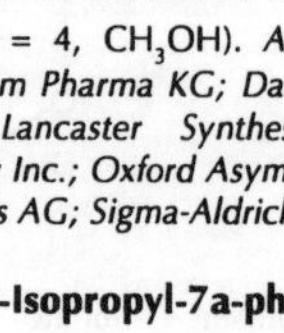

$C_6H_{11}NO_2$
Chiral auxiliary; Chiral intermediate. mp = 71-72°; $[\alpha]_D^{20}$ = + 22.5 ± 0.4° (c = 4, $CH_3OH$). *Acros Organics nv; Boehringer Ingelheim Pharma KG; Daiso Company, Ltd.; Kiralchem Ltd.; Lancaster Synthesis Ltd.; Omega Chemical Company Inc.; Oxford Asymmetry International plc; Senn Chemicals AG; Sigma-Aldrich Fine Chemicals.*

**3341 (3R,7aS)-3-Isopropyl-7a-phenyl bicyclic lactam**
$C_{15}H_{19}NO_2$
Chiral intermediate. *Oxford Asymmetry International plc.*

**3342 (3S,7aR)-3-Isopropyl-7a-phenyl bicyclic lactam**
88670-16-0

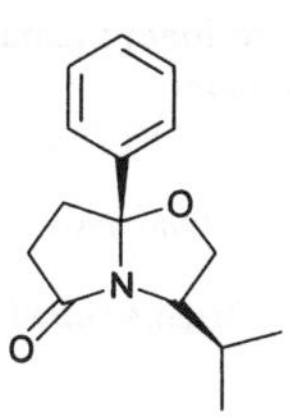

$C_{15}H_{19}NO_2$
Chiral intermediate. mp = 56-58°; $[\alpha]^{25}$ = + 70° (c = 1.5, $CHCl_3$). *Oxford Asymmetry International plc; Sigma-Aldrich Fine Chemicals.*

**3343 (3R)-3-Isopropyl-α-phenyl unsaturated bicyclic lactam**
$C_{15}H_{19}NO_2$
Chiral intermediate. *Oxford Asymmetry International plc.*

**3344 (3S)-3-Isopropyl-α-phenyl unsaturated bicyclic lactam**
$C_{15}H_{19}NO_2$
Chiral intermediate. *Oxford Asymmetry International plc.*

**3345 (R)-3-Isopropyl-2,5-piperazinedione**
143673-66-9

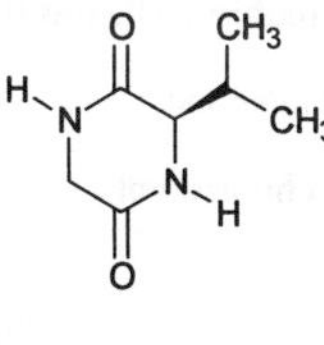

$C_7H_{12}N_2O_2$
Chiral intermediate. mp = 258-262°; $[\alpha]_D^{20}$ = - 31.5° (c = 1, $H_2O$). *Acros Organics nv; Sigma-Aldrich Fine Chemicals.*

**3346 (S)-3-Isopropyl-2,5-piperazinedione**
16944-60-8

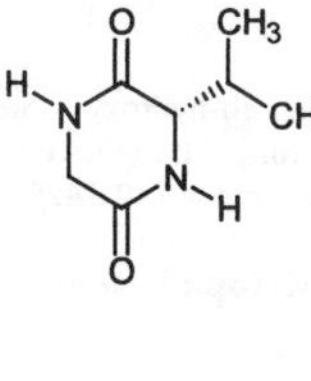

$C_7H_{12}N_2O_2$
Cyclo-(glycyl-L-valine). Chiral intermediate. mp = 250-260°. *Acros Organics nv.*

**3347 (R)-Isopropylsuccinic acid 1-methyl ester**
$C_8H_{14}O_4$
(R)-3-Methoxycarbonyl-4-methylpentanoic acid. Chiral intermediate. *ChiroTech Technology Ltd.; Lancaster Synthesis Ltd.*

**3348 (S)-2-Isopropylsuccinic acid 1-methyl ester**
208113-95-5
$C_8H_{14}O_4$
(S)-3-Methoxycarbonyl-4-methylpentanoic acid. Chiral intermediate. *ChiroTech Technology Ltd.; Lancaster Synthesis Ltd.*

**3349 Isopropyl-β-D-thiogalactopyranoside**
367-93-1 206-703-0

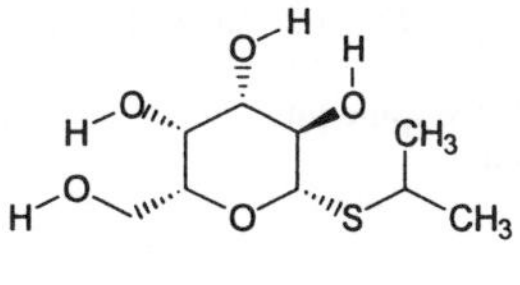

$C_9H_{18}O_5S$
Galactopyranoside, 1-methylethyl 1-thio-, β-D-. Listed on TSCA. Chiral diagnostic reagent. mp = 108-114°; $[\alpha]_D^{20}$= - 29° (c = 1, $H_2O$); soluble in $H_2O$. *Acros Organics nv; Diagnostic Chemicals Limited; Senn Chemicals AG; Sigma-Aldrich Fine Chemicals.*

**3350 (R)-(-)-Isoproterenol**
51-31-0 5105(11)

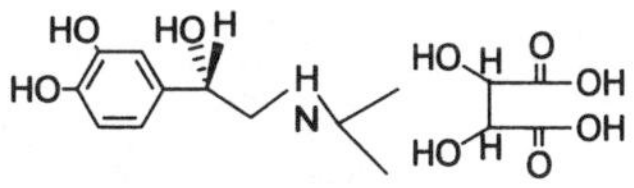

$C_{11}H_{17}NO_3$
Chiral building block. mp = 164-165°; $[\alpha]_D^{20}$= - 42° (c = 2, 2N HCl). *Sigma-Aldrich Fine Chemicals.*

**3351 Isoproterenol D-bitartrate**
14638-70-1 238-683-4

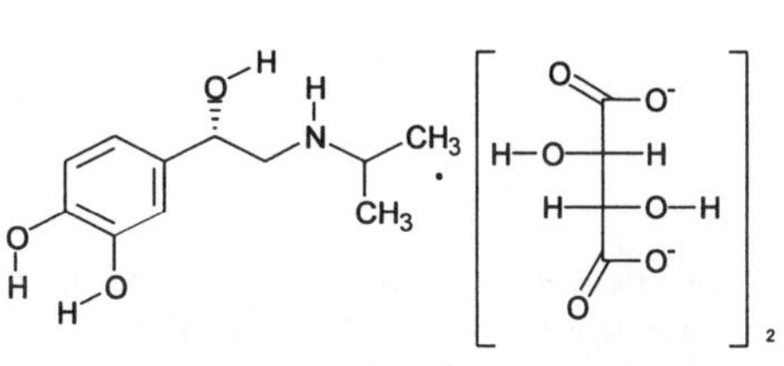

$C_{19}H_{25}NO_{15}$
Benzyl alcohol, 3,4-dihydroxy-α-[(isopropylamino)methyl]-, (+)-, (+)-tartrate (1:1) (salt); 1,2-Benzenediol, 4-[(1S)-1-hydroxy-2-[(1-methylethyl)amino]-ethyl]-, (2R,3R)-2,3-dihydroxybutanedioate (1:1) (salt); (S)-(isopropyl)(β,3,4-trihydroxyphenethyl)ammonium [R-(R*,R*)]-hydrogen tartrate. Chiral building block; Chiral ligand. mp = 165-170°; $[\alpha]^{22}$ = + 30.5° (c = 1.1, $H_2O$). *Acros Organics nv; Sigma-Aldrich Fine Chemicals.*

**3352 (-)-Isopulegol**
89-79-2 201-940-6

$C_{10}H_{18}O$
(1R,2S,5R)-2-Isopropenyl-5-methylcyclohexanol. Listed on TSCA. Chiral intermediate. $bp_{12}$ = 91°; d = 0.912; n = 1.471; $[\alpha]_D^{20}$= - 22° (neat). *Sigma-Aldrich Fine Chemicals.*

**3353 (+)-Isopulegol**
104870-56-6

$C_{10}H_{18}O$
(1S,2R,5S)-2-Isopropenyl-5-methylcyclohexanol. Chiral intermediate. $bp_{12}$ = 91°; d = 0.912; n = 1.471; $[\alpha]_D^{20}$= + 22° (neat). *Sigma-Aldrich Fine Chemicals.*

**3354 (-)-Isoreserpine**

$C_{33}H_{40}N_2O_9$
Chiral intermediate. mp = 148-155°; $[\alpha]^{25}$ = - 160° (c = 1, $CHCl_3$). *Sigma-Aldrich Fine Chemicals.*

**3355 D-Isosorbide**
652-57-5 5244(12) 211-492-3

$C_6H_{10}O_4$
1,4-Dianhydro-D-sorbitol. Listed on TSCA. Chiral building block. mp = 61-63°; $[\alpha]_D^{20}$= + 45.2° (c = 3, $H_2O$). *Sigma-Aldrich Fine Chemicals.*

**3356 D-Isosorbide-5-mononitrate**
16051-77-7 240-197-2

$C_6H_9NO_6$
Glucitol, 1,4:3,6-dianhydro-, 5-nitrate, D-; D-Glucitol, 1,4:3,6-dianhydro-, 5-nitrate; 1,4:3,6-dianhydro-D-glucitol 5-nitrate; Imdur; IS 5MN; Monizid. Chiral building block. mp = 88-91°; [α] = + 170° (c = 1, EtOH). *Acros Organics nv; SIFA Chemicals AG.*

**3357 (4S)-3-(Isothiocyanatoacetyl)-4-benzyl-2-oxazolidinone**
104324-18-7
$C_{13}H_{12}NO_3S$
Chiral intermediate. *Oxford Asymmetry International plc.*

**3358 (R,R)-Jacobsen's cobalt(II) catalyst**
176763-62-5

$C_{36}H_{54}CoN_2O_2$
Chiral catalyst. mp = 350°. *Chirex, Inc.; Sigma-Aldrich Fine Chemicals.*

**3359 (S,S)-Jacobsen's cobalt(II) catalyst**
188264-84-8
Chiral catalyst. *Chirex, Inc.*

**3360 (R,R)-Jacobsen's ligand (E/Z unknown)**
151433-25-9
Chiral ligand. *Chirex, Inc.*

**3361 (S,S)-Jacobsen's ligand (E/Z unknown)**
151380-44-8
Chiral ligand. *Chirex, Inc.*

**3362 6-Ketocholestanol**
1175-06-0 214-640-5

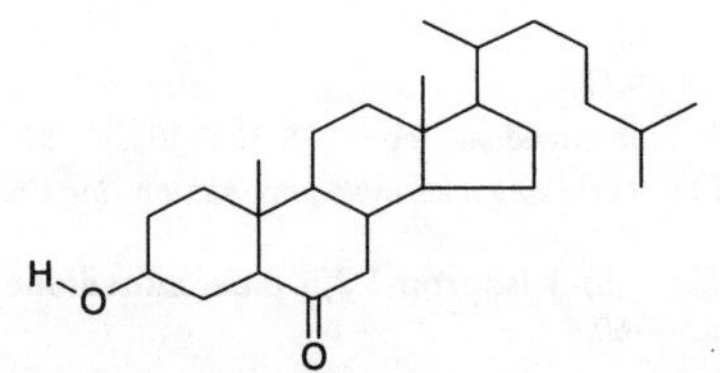


$C_{27}H_{46}O_2$
Cholestan-6-one, 3b-hydroxy-, 5α-; 6-Oxocholestanol; 5α-Cholestan-6-one, 3β-hydroxy-; YS 164. Pharmaceutical or derivative. mp = 140-142°. *Acros Organics nv.*

**3363 (S)-(+)-Ketopinic acid**
40724-67-2

$C_{10}H_{14}O_3$
Chiral intermediate. mp = 235°; $[\alpha]^{23}$ = + 58° (c = 1, $CHCl_3$). *Sigma-Aldrich Fine Chemicals; TCI America.*

**3364 (S)-(+)-Ketoprofen**

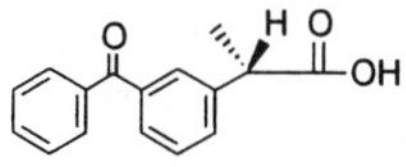

$C_{16}H_{14}O_3$
(S)-(+)-3-Benzoyl-α-methylbenzeneacetic acid. Chiral intermediate. mp = 75-78°; $[\alpha]^{22}$ = + 49° (c = 1, $CH_3OH$). *Sigma-Aldrich Fine Chemicals.*

**3365 L-Kynurenine**
2922-83-0
$C_{10}H_{12}N_2O_3$
Chiral intermediate. *TCI America.*

**3366 L-Kynurenine sulfate**
17268-44-9
$C_{10}H_{12}N_2O_7S$
Chiral intermediate. *TCI America.*

**3367 (+)-Lactal**
65207-55-8

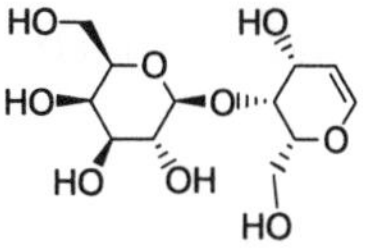

$C_{12}H_{20}O_9$
Chiral intermediate. mp = 196-199°; $[\alpha]_D^{20}$ = + 26.5° (c = 1, $H_2O$). *Sigma-Aldrich Fine Chemicals.*

**3368 (S)-(-)-Lactamide**
89673-71-2

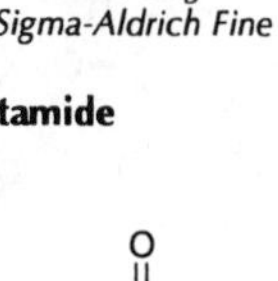

$C_3H_7NO_2$
(S)-α-Hydroxypropionamide; (S)-Lactic acid amide. Chiral building block. mp = 53-55°; $[\alpha]_D^{20}$ = - 20.5° (c = 10, $H_2O$). *Acros Organics nv; Sigma-Aldrich Fine Chemicals.*

**3369 (R)-(+)-Lactamide**
598-81-2

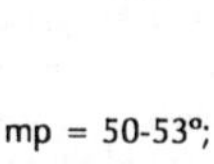

$C_3H_7NO_2$
Chiral building block. mp = 50-53°; $[\alpha]_D^{20}$ = + 21.0° (c = 10, $H_2O$). *Sigma-Aldrich Fine Chemicals.*

**3370 L-Lactic acid**
79-33-4 5350(12) 201-196-2

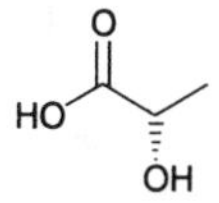

$C_3H_6O_3$
Propanoic acid, 2-hydroxy-, (2S)-; (S)-2-Hydroxypropionic acid. Listed on TSCA. Chiral building block. mp = 52-54°; $bp_{12}$ = 119°; d = 1.206; n = 1.427; $[\alpha]_D^{20}$ = - 13.5° (c = 2.5, 1.5N NaOH). *Acros Organics nv; Austin Chemical Company, Inc.; Lancaster Synthesis Ltd.; Pfanstiehl Laboratories, Inc.; Senn Chemicals AG; Sigma-Aldrich Fine Chemicals.*

**3371 L-(-)-Lactic acid ethyl ester**
687-47-8 3773(11) 211-694-1

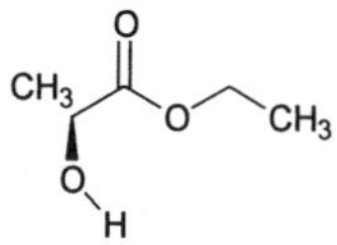

$C_5H_{10}O_3$
Propanoic acid, 2-hydroxy-, ethyl ester, (S)-; (S)-(-)-2-Hydroxypropanoic acid ethyl ester; Ethyl (S)-2-hydroxypropionate; (-)-Ethyl lactate. Chiral building block. bp = 148-153°; $d_4^{20}$ = 1.034; $[\alpha]_D^{20}$ = - 11 ± 0.5° (neat). *Acros Organics nv; Nitrokemia 2000 Rt.; Sigma-Aldrich Fine Chemicals.*

**3372 D-Lactic acid methyl ester**
17392-83-5 6007(11) 241-420-6

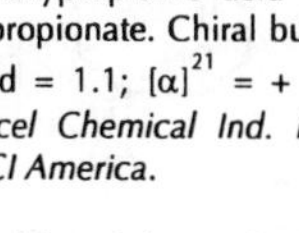

$C_4H_8O_3$
Propanoic acid, 2-hydroxy-, methyl ester, (R)-; (R)-Methyl lactate; (+)-2-Hydroxypropionic acid methyl ester; (+)-Methyl 2-hydroxypropionate. Chiral building block. bp = 144°; $bp_8$ = 33°; d = 1.1; $[\alpha]^{21}$ = + 8.4° (neat). *Acros Organics nv; Daicel Chemical Ind. Ltd; Sigma-Aldrich Fine Chemicals; TCI America.*

**3373 L-Lactic acid, calcium salt, hydrated**
28305-25-1
$C_6H_{10}CaO_6$
Propanoic acid, 2-hydroxy-, calcium salt (2:1), (S)-; Calcium (S)-lactate; calcium (S)-2-hydroxypropionate. Chiral building block. *Pfanstiehl Laboratories, Inc.*

**3374 D-(-)-Lactic acid, lithium salt**
27848-81-3 248-693-0

$C_3H_5LiO_3$
Propanoic acid, 2-hydroxy-, monolithium salt, (2R)-; Lithium D-(-)-lactate; (R)-(+)-2-Hydroxypropanoic acid, monolithium salt. Chiral building block. $[\alpha]_D^{20}$ = + 14.3° (c = 1, $H_2O$). *Acros Organics nv; Sigma-Aldrich Fine Chemicals.*

**3375 L-Lactic acid, zinc salt, hydrated**
6155-68-6 228-175-0

$C_6H_{14}O_7Zn$
Propanoic acid, 2-hydroxy-, zinc salt (1:1), (2S)-; Zinc (S)-3-oxidopropionate. Listed on TSCA. Chiral building block. $[\alpha]_D^{20}$ = - 8.0° (c = 4, $H_2O$). *Pfanstiehl Laboratories, Inc.*

**3376 L-(-)-Lactide**
4511-42-6 224-832-0

$C_6H_8O_4$
Dioxane-2,5-dione, 3,6-dimethyl-, 1,4-, (3S,6S)-; (3S,6S)-(-)-3,6-Dimethyl-1,4-dioxane-2,5-dione; L,L-Dilactide. Listed on TSCA. Chiral building block. mp = 97°; $[\alpha]_D^{20}$ = - 285° (c = 1, toluene). *Sigma-Aldrich Fine Chemicals; TCI America.*

**3377 D-Lactitol**
585-86-4 209-566-5

$C_{12}H_{24}O_{11}$
Glucitol, 4-O-β-D-galactopyranosyl-, D-; 4-O-β-D-Galactopyranosyl-D-glucitol; Finlac DC; Lactit; Lactite; Lactositol; Miruhen. Chiral intermediate. *Senn Chemicals AG.*

**3378 D-Lactitol monohydrate**
81025-04-9 5352(12)

$C_{12}H_{24}O_{11}$
4-O-β-D-Galactopyranosyl-D-glucitol monohydrate. Chiral intermediate. mp = 95-98°; $[\alpha]_D^{20}$ = + 13° (c = 1, $H_2O$). *Sigma-Aldrich Fine Chemicals.*

**3379 D-Lactobionic acid**
96-82-2 5354(12) 202-538-3

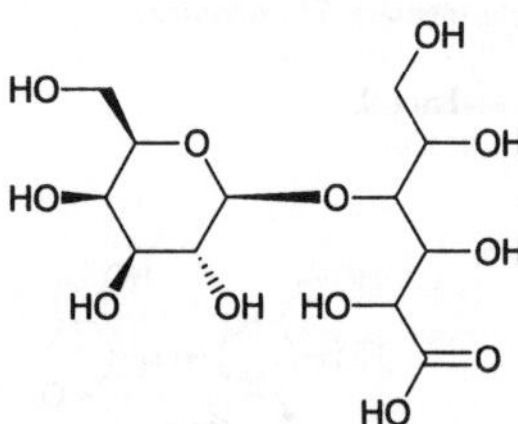

$C_{12}H_{22}O_{12}$
Gluconic acid, 4-O-β-D-galactopyranosyl-, D-; 4-O-β-D-Galactopyranosyl-D-gluconic acid. Chiral building block. mp = 113-118°; [α] = + 23.3° (c = 10.04, $H_2O$, 24h). *Acros Organics nv; Sigma-Aldrich Fine Chemicals.*

**3380 D-(+)-Lactose**
14641-93-1 238-691-8

$C_{12}H_{22}O_{11}$
Glucopyranose, 4-O-β-D-galactopyranosyl-, α-D-; 4-O-β-D-Galactopyranosyl-α-D-glucopyranose. Listed on TSCA. Chiral intermediate. *TCI America.*

**3381 β-D-Lactose (80% β, 20% α)**
5965-66-2 5221(11) 227-751-9

$C_{12}H_{22}O_{11}$
Glucopyranose, 4-O-β-D-galactopyranosyl-, β-D-; 4-O-β-D-Galactopyranosyl-β-D-glucopyranose. Listed on TSCA. Chiral building block. $[\alpha]^{25}$ = + 54.2° (c = 8, $H_2O$). *Acros Organics nv; Sigma-Aldrich Fine Chemicals.*

**3382 (+)-α-Lactose monohydrate**
5989-81-1 5221(11)

$C_{12}H_{22}O_{11}$
4-O-β-D-Galactopyranosyl-α-D-glucopyranose monohydrate; GranuLac 200; Pharmatose; Zeparox. Chiral building block. mp = 219°; $[\alpha]^{22}$ = + 52.4° (c = 4.5, $H_2O$). *Acros Organics nv; Sigma-Aldrich Fine Chemicals.*

**3383 D-Lactose monohydrate**

$C_{12}H_{24}O_{12}$
Lactobiose; 4-O-β-D-galactopyranosyl-D-glucopyranose. Chiral building block. $[\alpha]_D^{20}$ = + 52.6° (c = 8, $H_2O$). *Pfanstiehl Laboratories, Inc.*

**3384 D-Lactose octaacetate**

$C_{30}H_{42}O_{19}$
Chiral intermediate. *Senn Chemicals AG.*

**3385 D-Lactulose**
4618-18-2 225-027-7

$C_{12}H_{22}O_{11}$
Fructofuranose, 4-O-β-D-galactopyranosyl-, D-; 4-O-β-D-Galactopyranosyl-α-D-fructose; Isolactose; Laevolac; Bifiteral. Chiral intermediate. [α] = - 48° (c = 4, $H_2O$, 12h). *Acros Organics nv; Pfanstiehl Laboratories, Inc.; Senn Chemicals AG.*

**3386 D-Lactulose octaacetate**

$C_{30}H_{42}O_{19}$
4-O-β-D-Galactopyranosyl-D-fructose-octaacetate. Chiral intermediate. *Senn Chemicals AG.*

**3387 Lasalocid sodium salt**
25999-20-6 5243(11) 247-400-3
$C_{34}H_{54}O_8$
Pharmaceutical or derivative. mp = 180°; $[\alpha]_D^{20}$ = - 28° (c = 1, $CH_3OH$). *Sigma-Aldrich Fine Chemicals.*

**3388 Lauroyl-L-carnitine chloride**
7023-03-2

$C_{19}H_{57}NO_4$
Chiral intermediate. *Pharmasyn, Inc.*

**3389 N-Lauroyl-D-erythro sphingosine**
74713-60-3

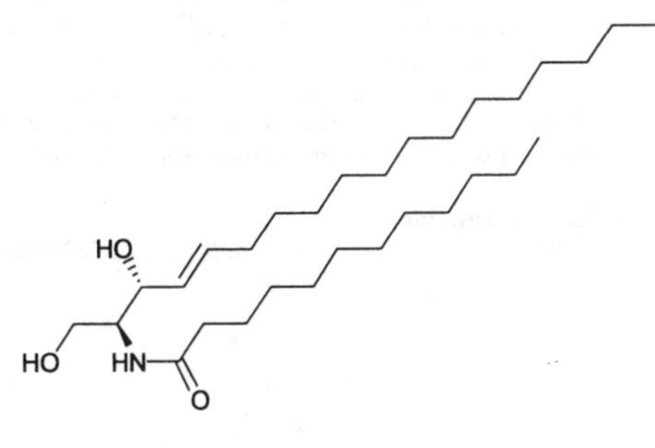


$C_{30}H_{59}NO_3$
Chiral intermediate. *Acros Organics nv.*

**3390 L-Leucic acid**
13748-90-8 237-329-6

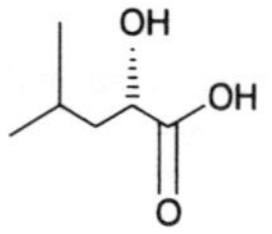

$C_6H_{12}O_3$
Pentanoic acid, 2-hydroxy-4-methyl-, (2S)-; L-2-Hydroxyisocaproic acid; L-2-Hydroxy-4-methylvaleric acid; Valeric acid, 2-hydroxy-4-methyl-, L-(+)-. Listed on TSCA. Chiral building block. mp = 81°; $[\alpha]^{22}$ = - 26.3° (c = 1, 1N NaOH). *Acros Organics nv; Austin Chemical Company, Inc.; Sigma-Aldrich Fine Chemicals; TCI America.*

**3391 L-Leucinamide hydrochloride**
10466-61-2 233-952-2

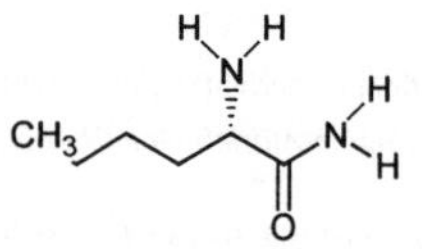

$C_6H_{15}ClN_2O$
Pentanamide, 2-amino-4-methyl-, monohydrochloride, (2S)-; (S)-2-Amino-4-methylvaleramide monohydrochloride. Chiral building block. mp = 254-256°; $[\alpha]^{25}$ = + 10° (c = 5, $H_2O$). *Acros Organics nv; Sigma-Aldrich Fine Chemicals; TCI America.*

**3392 D-Leucine**
328-38-1 206-327-7

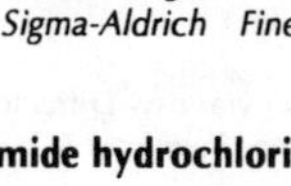

$C_9H_{10}O_3$
(R)-2-Amino-4-methylpentanoic acid. Listed on TSCA. Chiral building block. mp > 300°; $[\alpha]_D^{20}$ = - 15 ± 1° (c = 5, 5N HCl). *Acros Organics nv; Austin Chemical Company, Inc.; Lancaster Synthesis Ltd.; Sigma-Aldrich Fine Chemicals; Tanabe Seiyaku Co. Ltd.; TCI America; Varsal Instruments, Inc.; Yoneyama Yakuhin Kogyo Co., Ltd.*

**3393 L-Leucine**
61-90-5 5475(12) 200-522-0

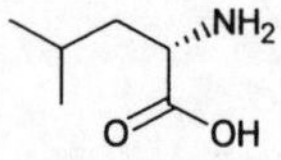

$C_6H_{13}NO_2$
Pentanoic acid, 2-amino-4-methyl-, (S)-; (S)-2-Amino-4-methylpentanoic acid; L-α-Aminoisocaproic acid; L-Norvaline, 4-methyl-; (S)-2-Amino-4-methylvaleric acid. Listed on TSCA. Chiral building block. mp > 300°; $[\alpha]_D^{20}$ = + 15.5 ± 1° (c = 2, 5N HCl). *Aceto Corporation; Acros Organics nv; Ajinomoto Co. Inc.; Austin Chemical Company, Inc.; Flamma s.p.a.; KingChem, Inc.; Kyowa Hakko Kogyo Co., Ltd.; Lancaster Synthesis Ltd.; Senn Chemicals AG; Shanghai DSL International Trading Company; Sigma-Aldrich Fine Chemicals; Tanabe Seiyaku Co. Ltd.; TCI America; Varsal Instruments, Inc.; Yoneyama Yakuhin Kogyo Co., Ltd.*

**3394 D-tert-Leucine**
26782-71-8

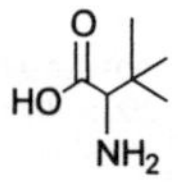

$C_6H_{13}NO_2$
(R)-2-Amino-3,3-dimethylbutyric acid; D-2-tert-Butylglycine. Chiral building block. mp > 300°; $[\alpha]_D^{20}$ = - 7 ± 1° (c = 5, 5N HCl). *Austin Chemical Company, Inc.; DSM Fine Chemcials Netherlands; Lancaster Synthesis Ltd.; Sigma-Aldrich Fine Chemicals; TCI America.*

**3395 L-tert-Leucine**
20859-02-3

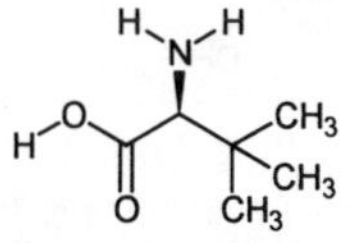

$C_6H_{13}NO_2$
(S)-2-Amino-3,3-dimethylbutyric acid; L-tert-Butylglycine. Chiral intermediate. mp > 300°; $[\alpha]_D^{20}$ = + 7 ± 1° (c = 5, 5N HCl). *Acros Organics nv; Austin Chemical Company, Inc.; Celltech Chiroscience Ltd.; Degussa-Huls AG; DSM Fine Chemcials Netherlands; Fischer Chemicals AG; Great Lakes Fine Chemicals; Lancaster Synthesis Ltd.; Mitsubishi Rayon Co., Ltd.; Senn Chemicals AG; Sigma-Aldrich Fine Chemicals; TCI America.*

**3396 L-Leucine allyl ester p-toluenesulfonic acid salt**

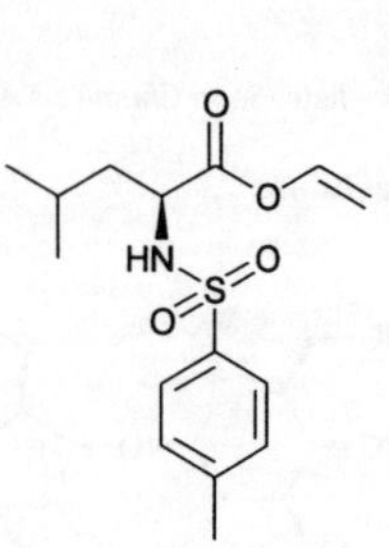

$C_{15}H_{21}NO_4S$
Chiral intermediate. *Synthetech, Inc.*

**3397 L-Leucine benzyl ester p-toluene-sulfonic acid salt**
1738-77-8 217-095-1

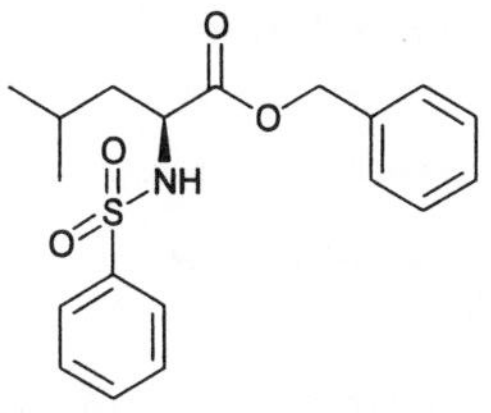

$C_{20}H_{27}NO_5S$
Leucine, benzyl ester, p-toluenesulfonate, L-; O-benzyl-L-leucine toluene-p-sulphonate; L-Leucine, phenylmethyl ester, 4-methylbenzenesulfonate. Chiral intermediate. *Synthetech, Inc.*

**3398 L-Leucine ethyl ester hydrochloride**
2743-40-0 220-375-6
$C_8H_{18}ClNO_2$
Leucine, ethyl ester, hydrochloride, L-; Ethyl L-leucinate hydrochloride. Building block. mp = 136°. *TCI America.*

**3399 D-Leucine methyl ester hydrochloride**
5845-53-4

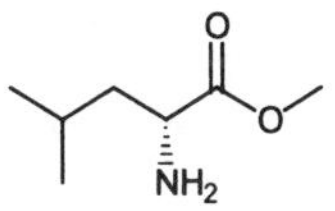

$C_7H_{16}ClNO_2$
Chiral building block. *Acros Organics nv; Tanabe Seiyaku Co. Ltd.*

**3400 L-Leucine methyl ester hydrochloride**
7517-19-3 231-375-0

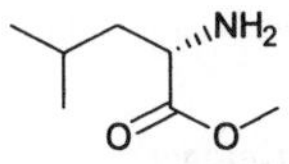

$C_7H_{16}ClNO_2$
Leucine, methyl ester, hydrochloride, L-; Methyl L-leucinate hydrochloride. Chiral building block. mp = 148-150°; $[\alpha]^{22}$ = + 13° (c = 2, $H_2O$). *Acros Organics nv; Austin Chemical Company, Inc.; Sigma-Aldrich Fine Chemicals; TCI America.*

**3401 L-tert-Leucine methylamide**
89226-12-0

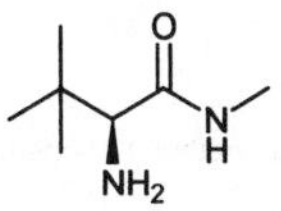

$C_7H_{16}N_2O$
Chiral building block. *Degussa-Huls AG; Fischer Chemicals AG; Great Lakes Fine Chemicals.*

**3402 L-Leucine-(4-methyl-7-coumarinylamide) hydrochloride**
62480-44-8

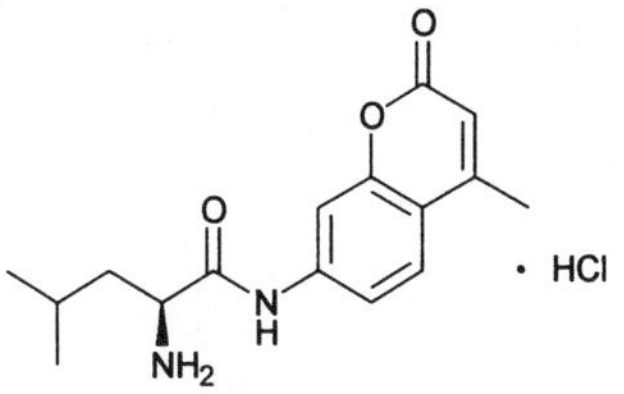

$C_{16}H_{21}ClN_2O_3$
Chiral intermediate. mp = 283° $[\alpha]_D^{20}$ = + 54° (c = 1, EtOH). *Acros Organics nv; Sigma-Aldrich Fine Chemicals.*

**3403 L-Leucine-4-nitroanilide**
4178-93-2 224-047-3

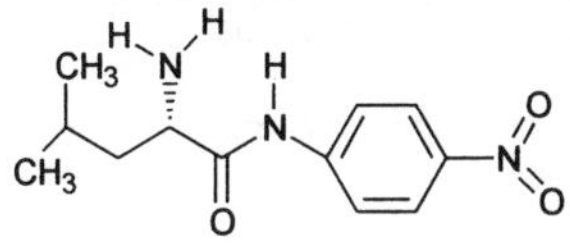

$C_{12}H_{17}N_3O_3$
Pentanamide, 2-amino-4-methyl-N-(4-nitrophenyl)-, (2S)-; (S)-2-Amino-4-methyl-N-(4-nitrophenyl)valeramide; Valeranilide, 2-amino-4-methyl-4'-nitro-, L-. Listed on TSCA. Chiral intermediate. mp = 88-90°; $[\alpha]_D^{20}$ = + 96° (c = 1, 1N HCl). *Acros Organics nv; Sigma-Aldrich Fine Chemicals.*

**3404 D-(-)-Leucinol**

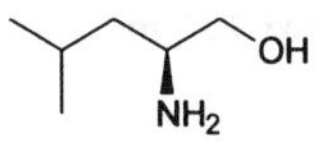

$C_6H_{15}NO$
Pentan-1-ol, 2-amino-4-methyl-, (2R)-; 2-Amino-4-methylpentan-1-ol. Chiral building block. *Oxford Asymmetry International plc.*

**3405 L-(+)-Leucinol**
7533-40-6 231-400-5

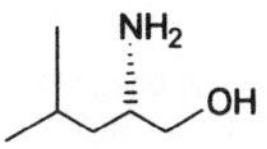

$C_6H_{15}NO$
Pentan-1-ol, 2-amino-4-methyl-, (2S)-; (S)-2-Amino-4-methylpentan-1-ol. Chiral building block. $bp_{768}$ = 198-200°; d = 0.917; n = 1.4511; $[\alpha]_D^{20}$ = + 4° (c = 9, EtOH). *Acros Organics nv; Boehringer Ingelheim Pharma KG; Fischer Chemicals AG; Great Lakes Fine Chemicals; Loba Feinchemie AG; Omega Chemical Company Inc.; Oxford Asymmetry International plc; Senn Chemicals AG; Sigma-Aldrich Fine Chemicals; TCI America.*

**3406 (R)-(-)-tert-Leucinol**
$C_6H_{15}NO$
((R)-2-amino-3,3-dimethyl-1-butanol). Chiral building block. mp = 30-33°; $bp_{0.4}$ = 70°; d = 0.9; $[\alpha]_D^{20}$ = - 37° (c = 1.5, EtOH). *Sigma-Aldrich Fine Chemicals.*

**3407 (S)-(+)-tert-Leucinol**
112245-13-3

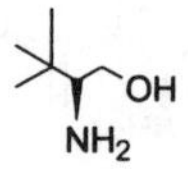

$C_6H_{15}NO$
2-Amino-3,3-dimethyl-1-butanol. Chiral building block. mp = 33-35°; d = 0.9; $[\alpha]^{26}$ = + 37° (c = 1.5, EtOH). *Sigma-Aldrich Fine Chemicals.*

**3408 D-Leucrose**
5-O-α-D-Glucopyranosyl-β-D-fructose. Chiral intermediate. *Senn Chemicals AG.*

**3409 L-Leucyl-L-alanine**
7298-84-2 230-737-5

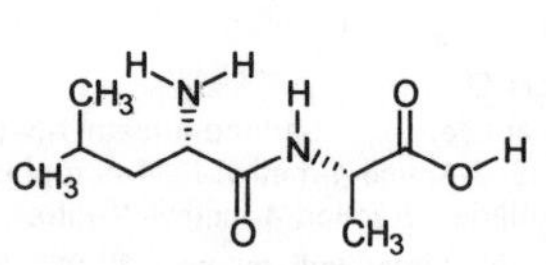

$C_9H_{18}N_2O_3$
Alanine, N-L-leucyl-, L-. Listed on TSCA. Chiral intermediate. [α] = + 18.8° (c = 5, $CH_3OH$). *Acros Organics nv; TCI America.*

**3410 L-Leucyl-glycine**
686-50-0 211-688-9
$C_8H_{16}N_2O_3$
Leucylglycine, L-. Chiral intermediate. *TCI America.*

**3411 D-Leucyl-glycyl-glycine**
18625-22-4 242-457-0
$C_{10}H_{19}N_3O_4$
Glycine, N-(N-D-leucylglycyl)-. Chiral intermediate. *TCI America.*

**3412 L-Leucyl-glycyl-glycine**
2576-67-2 219-925-8
$C_{10}H_{19}N_3O_4$
Glycine, N-(N-glycyl-L-leucyl)-; Gly-leu-gly; N-(N-glycyl-L-leucyl)glycine. Chiral intermediate. *TCI America.*

**3413 D-Leucyl-D-leucine**
38689-30-4
$C_{12}H_{24}N_2O_3$
Chiral intermediate. *TCI America.*

**3414 L-Leucyl-D-leucine**
17665-02-0
$C_{12}H_{24}N_2O_3$
Chiral intermediate. *TCI America.*

**3415 L-Leucyl-2-naphthylamide hydrochloride**
$C_{16}H_{21}ClN_2O$
Chiral intermediate. *TCI America.*

**3416 D-Leucyl-L-tyrosine**
3303-29-5
$C_{15}H_{22}N_2O_4$
Chiral intermediate. *TCI America.*

**3417 L-Leucyl-L-tyrosine**
968-21-8 213-527-8
$C_{15}H_{22}N_2O_4$
Tyrosine, N-L-leucyl-, L-. Chiral intermediate. *TCI America.*

**3418 (S)-Levoglucosenone**
37112-31-5
$C_6H_6O_3$
(1S)-6,8-Dioxabicyclo[3.2.1]oct-2-en-4-one. Chiral intermediate. d = 1.31. *TCI America.*

**3419 (R)-(+)-Limonene**
5989-27-5 5518(12) 227-813-5

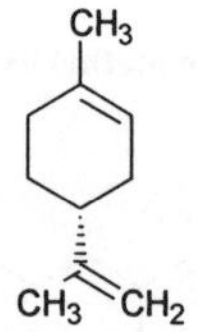

$C_{10}H_{16}$
Cyclohexene, 1-methyl-4-(1-methylethenyl)-, (4R)-; 4-Isopropenyl-1-methylcyclohexene; (+)-p-Mentha-1,8-diene. Listed on TSCA. Chiral building block. mp = -75 to -73°; bp = 175-176°; d = 0.8410; $n_D^{20}$ = 1.4730; $[\alpha]^{20}$ = + 115 ± 5° (c = 10, EtOH). *Acros Organics nv; Lancaster Synthesis Ltd.; Sigma-Aldrich Fine Chemicals.*

**3420 (S)-(-)-Limonene**
5989-54-8 5518(12) 227-815-6

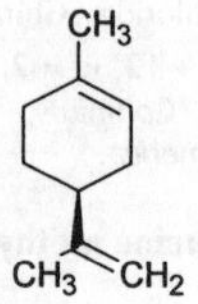

$C_{10}H_{16}$
Cyclohexene, 1-methyl-4-(1-methylethenyl)-, (4S)-; (-)-p-Mentha-1,8-diene; (S)-(-)-4-Isopropenyl-1-methylcyclohexene. Listed on TSCA. Chiral intermediate. bp = 175-177°; d = 0.8440; $n_D^{20}$ = 1.4720; $[\alpha]_D^{20}$ = -90 ± 5° (c = 10, EtOH). *Acros Organics nv; Lancaster Synthesis Ltd.; Sigma-Aldrich Fine Chemicals.*

**3421 (-)-Limonene oxide (mixture of cis and trans)**

$C_{10}H_{16}O$
Oxa-7-bicyclo[4.1.0]heptane, 1-methyl-4-(1-methylethenyl)-; Limonene 1,2-epoxide; 1-Methyl-4-(1-methylvinyl)-7-oxabicyclo[4.1.0]heptane; p-Menth-8-ene, 1,2-epoxy, . Chiral intermediate. $bp_{50}$ = 113-114°; d = 0.929; n = 1.4664; $[\alpha]^{21}$ = - 69° (neat). *Acros Organics nv; Sigma-Aldrich Fine Chemicals.*

**3422 (+)-Limonene oxide (mixture of cis and trans)**
1195-92-2 214-805-1

$C_{10}H_{16}O$
Limonene 1,2-epoxide; 1,2-Epoxy-p-menth-8-ene; 7-Oxabicyclo[4.1.0]heptane, 1-methyl-4-(1-methylethenyl)-; 1-Methyl-4-(1-methylvinyl) 7-oxabicyclo[4.1.0]heptane. Chiral intermediate. $bp_{50}$ = 113-114°; d = 0.929; $[\alpha]^{22}$ = + 69° (neat). *Acros Organics nv; Sigma-Aldrich Fine Chemicals.*

**3423 Lithium L-lactate**
27848-80-2 248-692-5
$C_3H_5LiO_3$
Propanoic acid, 2-hydroxy-, monolithium salt, (2S)-; L-Lactic acid lithium salt; (S)-(+)-2-Hydroxypropanoic acid, monolithium salt. Chiral building block. mp > 300°; $[\alpha]_D^{20}$ = - 13.5 ± 1° (c = 5, $H_2O$). *Acros Organics nv; Lancaster Synthesis Ltd.; Pfanstiehl Laboratories, Inc.; TCI America.*

**3424 (+)-Lithocholic acid**
434-13-9 207-099-1

$C_{24}H_{40}O_3$
Cholan-24-oic acid, 3-hydroxy-, (3α,5β)-; 17β-(1-Methyl-3-carboxypropyl)etiocholan-3α-ol; 3-Hydroxycholan-24-oic acid. Chiral intermediate. [α] = + 33.7° (c = 1.5, EtOH). *Acros Organics nv.*

**3425 (+)-Longifoline**
475-20-7 5597(12) 207-491-2

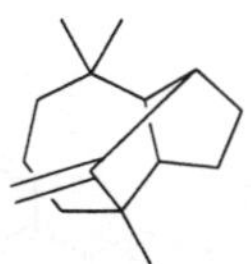

$C_{15}H_{24}$
[1S-(1α,3aβ,4α,8aβ)]-decahydro-4,8,8-trimethyl-9-methylene-1,4-methanoazulene. Listed on TSCA. Chiral intermediate. $bp_{706}$ = 254°; d = 0.928; n = 1.504; $[\alpha]^{22}$ = + 46° (neat). *Sigma-Aldrich Fine Chemicals.*

**3426 D-Luciferin**
2591-17-5 219-981-3

$C_{11}H_8N_2O_3S_2$
Thiazole-4-carboxylic acid, 4,5-dihydro-2-(6-hydroxy-2-benzothiazolyl)-, (S)-; (S)-4,5-Dihydro-2-(6-hydroxybenzothiazol-2-yl)thiazole-4-carboxylic acid; 2-Thiazoline-4-carboxylic acid, 2-(6-hydroxy-2-benzothiazolyl)-, (-)-; Firefly luciferin. Chiral intermediate. *Acros Organics nv; Pharmasyn, Inc.*

**3427 D-Luciferin potassium salt**
115144-35-9
Chiral intermediate. *Pharmasyn, Inc.*

**3428 D-(-)-Luciferin sodium salt monohydrate**
103404-75-7

$C_{11}H_7N_2NaO_3S_2$
Chiral building block.

**3429 (R)-Lupinine**
134003-02-4 5637(12) 207-638-0
$C_{10}H_{19}NO$
(1R-trans)-Octahydro-2H-quinolizine-1-methanol. Chiral intermediate. mp = 66-68°; $bp_4$ = 160-164°; $[\alpha]_D^{20}$ = - 26 ± 2° (c = 3, $H_2O$). *Lancaster Synthesis Ltd.*

**3430 D-Luteolin-7-O-glucoside**
5373-11-5 226-365-8
$C_{21}H_{20}O_{11}$
Benzopyran-(4H)-1-4-one, 2-(3,4-dihydroxyphenyl)-7-(β-D-glucopyranosyloxy)-5-hydroxy-; 2-(3,4-Dihydroxyphenyl)-7-(β-D-glucopyranosyloxy)-5-hydroxy-4H-1-benzopyran-4-one; 7-β-D-Glucosylluteolin; 7-Glucoluteolin;

Cinaroside; Cynaroside; Luteolin 7-β-monoglucoside; Luteolin 7-O-β-glucopyranoside; Luteoloside; Nephrocizin; Nephrocizine. Chiral intermediate. *Kaden Biochemicals GmbH.*

**3431 D-Lysergic acid hydrate**
82-58-6 201-431-9

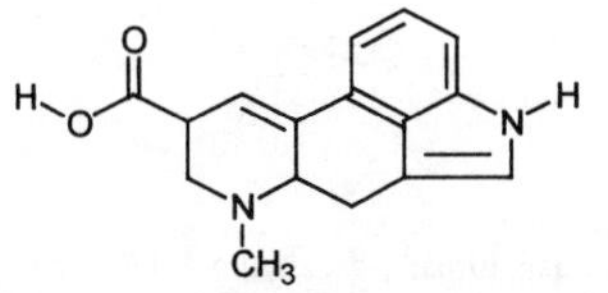

$C_{16}H_{16}N_2O_2$
Ergoline-8-carboxylic acid, 9,10-didehydro-6-methyl-, (8β)-. Chiral building block. *Acros Organics nv.*

**3432 (+)-Lysergol**
602-85-7 210-024-5

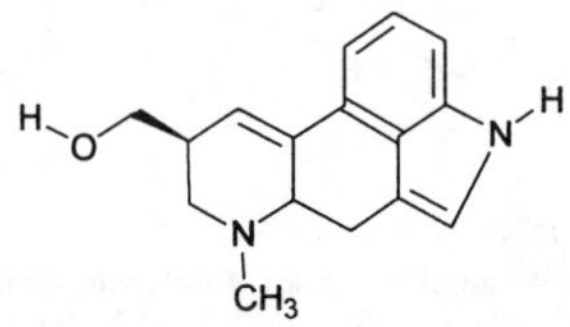

$C_{16}H_{18}N_2O$
Ergoline-8-methanol, 9,10-didehydro-6-methyl-, (8β)-; 9,10-didehydro-6-methylergoline-8β-methanol. Pharmaceutical or derivative. [α] = + 52.2° (c = 0.5, pyridine). *Acros Organics nv.*

**3433 D-(-)-Lysine**
213-091-9

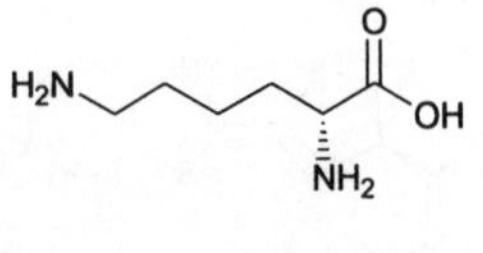

$C_6H_{14}N_2O_2$
(R)-(-)-Lysine. Chiral building block. mp = 218°; $[\alpha]_D^{20}$ = - 14° (c = 1, $H_2O$). *Sigma-Aldrich Fine Chemicals.*

**3434 L-(+)-Lysine**
56-87-1 5667(12) 200-294-2

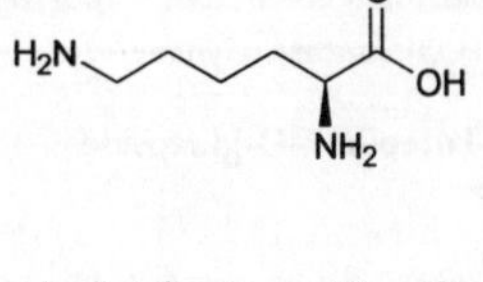

$C_6H_{14}N_2O_2$
Hexanoic acid, 2,6-diamino-, (S)-; (S)-α,ε-Diaminocaproic acid; (S)-2,6-Diaminohexanoic acid; L-Norleucine, 6-amino-. Listed on TSCA. Chiral building block. mp = 212°; $[\alpha]_D^{20}$ = + 25.2° (c = 2, 6N HCl). *Aceto Corporation; Rexim S.A. Produits Chimiques; Schweizerhall Pharma; Sigma-Aldrich Fine Chemicals; TCI America.*

**3435 L-Lysine acetate**
57282-49-2 260-664-4

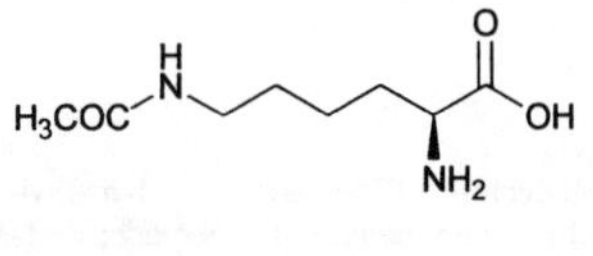

$C_8H_{18}N_2O_4$
Lysine, monoacetate, L-. Listed on TSCA. Chiral intermediate. *Kyowa Hakko Kogyo Co., Ltd.; Rexim S.A. Produits Chimiques; Shanghai DSL International Trading Company; Tanabe Seiyaku Co. Ltd.; Varsal Instruments, Inc.*

**3436 L-Lysine L-aspartate**
27348-32-9 248-423-1

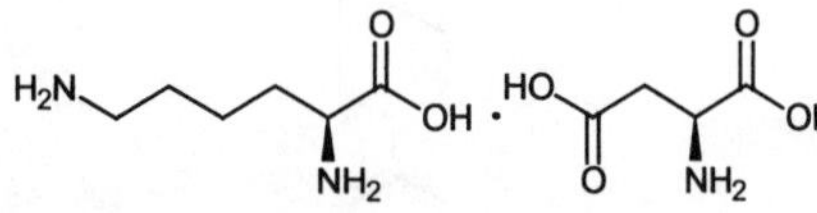

$C_{10}H_{21}N_3O_6$
Aspartic acid, L-, compd. with L-lysine (1:1). Chiral intermediate. *Rexim S.A. Produits Chimiques.*

**3437 L-(+)-Lysine dihydrochloride**
657-26-1 5509(11) 211-518-3

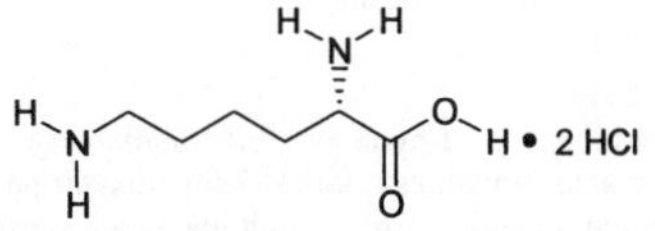

$C_6H_{16}Cl_2N_2O_2$
Lysine, dihydrochloride, L-. Listed on TSCA. Chiral building block. mp = 200-206°; $[\alpha]_D^{20}$ = + 17° (c = 2, 6N HCl). *Acros Organics nv; Sigma-Aldrich Fine Chemicals; TCI America.*

**3438 L-Lysine ethyl ester dihydrochloride**
3844-53-9 223-340-3

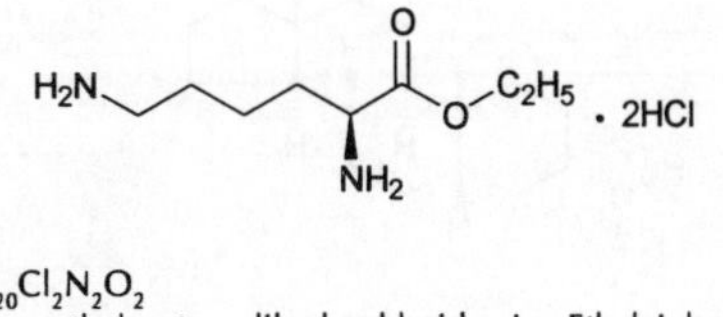

$C_8H_{20}Cl_2N_2O_2$
Lysine, ethyl ester, dihydrochloride, L-; Ethyl L-lysinate dihydrochloride. Chiral building block. *Austin Chemical Company, Inc.*

**3439 L-Lysine-L-glutamate**
5408-52-6 226-474-0

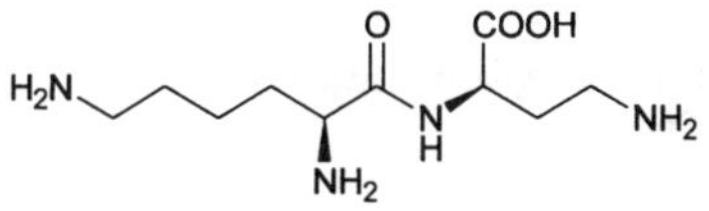

$C_{11}H_{23}N_3O_6$
Glutamic acid, L-, compd. with L-lysine (1:1). Listed on TSCA. Chiral intermediate. *Rexim S.A. Produits Chimiques.*

**3440 L-Lysine L-malate**

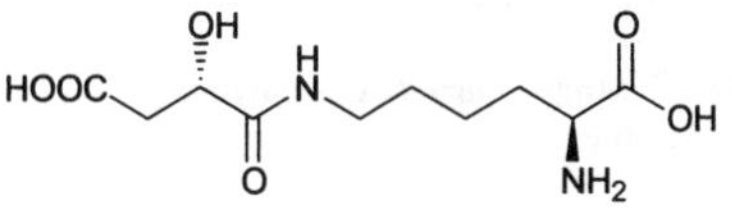

$C_{10}H_{18}N_2O_6$
Chiral intermediate. *Rexim S.A. Produits Chimiques.*

**3441 D-Lysine hexadecyl ester**

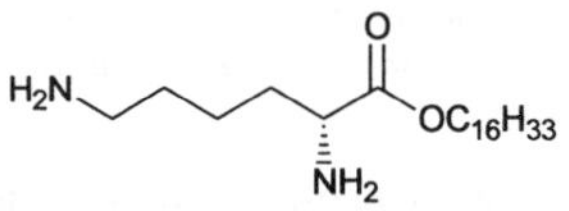

$C_{22}H_{46}N_2O_2$
Chiral intermediate. *Senn Chemicals AG.*

**3442 L-Lysine hexadecyl ester**

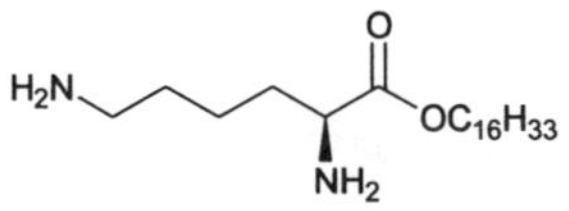

$C_{22}H_{46}N_2O_2$
Chiral intermediate. *Senn Chemicals AG.*

**3443 L-Lysine methyl ester dihydrochloride**
26348-70-9 247-625-7
$C_7H_{18}Cl_2N_2O_2$
Lysine, methyl ester, dihydrochloride, L-; (S)-2,6-Diaminohexanoic acid methyl ester dihydrochloride; Methyl L-lysinate dihydrochloride. Chiral building block. mp = 213-215°. *Lancaster Synthesis Ltd.*

**3444 D-Lysine monohydrate**

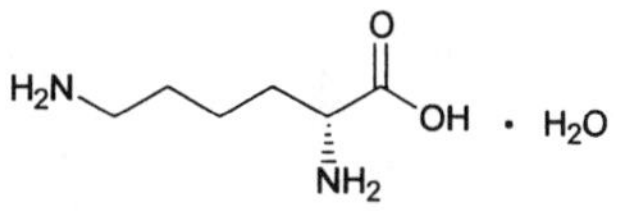

$C_6H_{16}N_2O_3$
Chiral building block. *Varsal Instruments, Inc.*

**3445 L-Lysine monohydrate**
39665-12-8 5667(12) 200-294-2

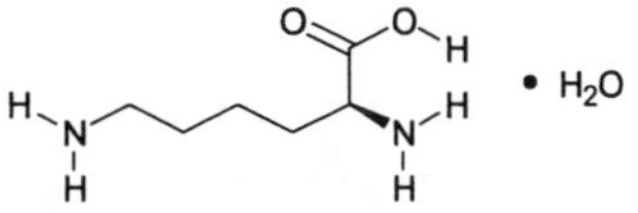

$C_6H_{14}N_2O_2$
(S)-2,6-Diaminohexanoic acid monohydrate. Chiral building block. mp = 215°; $[\alpha]_D^{20} = +25 \pm 1°$ (c = 6, 1N HCl). *Acros Organics nv; Lancaster Synthesis Ltd.; Rexim S.A. Produits Chimiques; Sigma-Aldrich Fine Chemicals.*

**3446 D-(-)-Lysine monohydrochloride**
7274-88-6 230-691-6

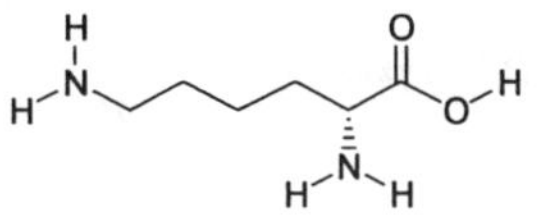

$C_6H_{15}ClN_2O_2$
Lysine, monohydrochloride, D-; (R)-2,6-Diaminohexanoic acid monohydrochloride. Listed on TSCA. Chiral building block. mp = 263-264°; $[\alpha]_D^{20} = -20.5 \pm 1°$ (c = 5, 5N HCl). *Acros Organics nv; Kaneka Corporation; Lancaster Synthesis Ltd.; Rexim S.A. Produits Chimiques; Sigma-Aldrich Fine Chemicals; TCI America; Yoneyama Yakuhin Kogyo Co., Ltd.*

**3447 L(+)-Lysine monohydrochloride**
657-27-2 5667(12) 211-519-9

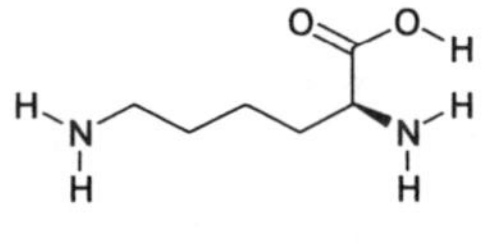

$C_6H_{15}ClN_2O_2$
Lysine, monohydrochloride, L-; Darvyl; L-Gen; Lyamine; Lysion; (S)-2,6-Diaminohexanoic acid monohydrochloride. Listed on TSCA. Chiral intermediate. mp = 263°; $[\alpha]_D^{20} = +20.5 \pm 1°$ (c = 5, 5N HCl). *Aceto Corporation; Acros Organics nv; Ajinomoto Co. Inc.; Austin Chemical Company, Inc.; BASF Aktiengesellschaft; KingChem, Inc.; Kyowa Hakko Kogyo Co., Ltd.; Lancaster Synthesis Ltd.; Rexim S.A. Produits Chimiques; Sigma-Aldrich Fine Chemicals; TCI America; Varsal Instruments, Inc.; Yoneyama Yakuhin Kogyo Co., Ltd.*

**3448 L-Lysine L-pyroglutamate**

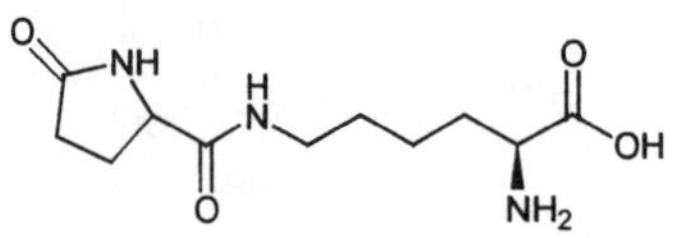

$C_{11}H_{19}N_3O_4$
Chiral intermediate. *Rexim S.A. Produits Chimiques.*

**3449 D-Lysine-D-serine-D-tryptophan**

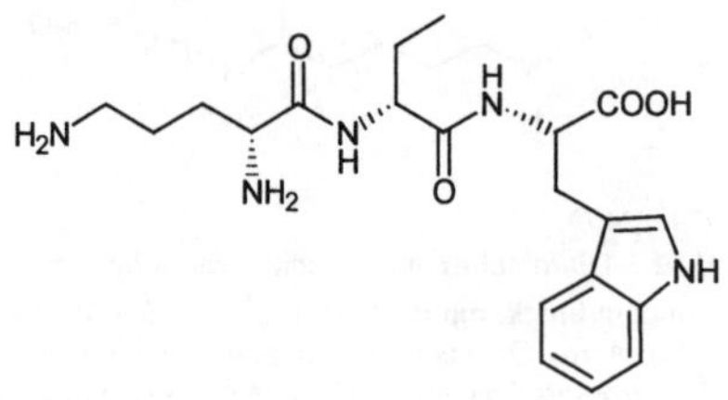

$C_{20}H_{29}N_5O_4$
Chiral intermediate. *Kyowa Hakko Kogyo Co., Ltd.*

**3450 D-(-)-Lyxose**
1114-34-7 5673(12) 214-212-8

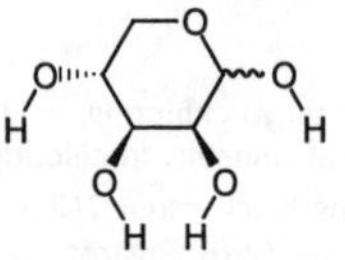

$C_5H_{10}O_5$
D-Lyxose; D-Lyxopyranose. Listed on TSCA. Chiral building block. mp = 108-112°; $[\alpha]_D^{20}$ = - 13.8° (c = 4, $H_2O$). *Acros Organics nv; Pfanstiehl Laboratories, Inc.; Senn Chemicals AG; Sigma-Aldrich Fine Chemicals; TCI America.*

**3451 L-(+)-Lyxose**
1949-78-6 217-763-2

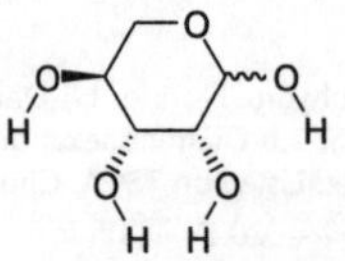

$C_5H_{10}O_5$
Lyxose, L-; . Listed on TSCA. Chiral building block. mp = 116-121°; $[\alpha]_D^{20}$ = + 13.8° (c = 4, $H_2O$). *Acros Organics nv; Pfanstiehl Laboratories, Inc.; Senn Chemicals AG; Sigma-Aldrich Fine Chemicals; TCI America.*

**3452 D-Lyxosylamine**
39840-37-4
$C_5H_{11}NO_4$
D-Lyxosimine. Chiral building block. mp = 138°. *Pfanstiehl Laboratories, Inc.*

**3453 L-Magnesium aspartate**
18962-61-3 242-703-7

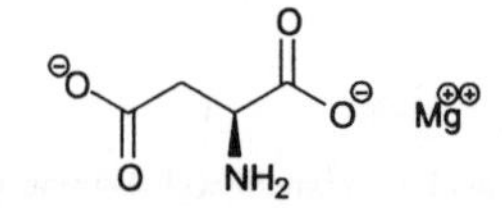

$C_8H_{12}MgN_2O_8$
Aspartic acid, magnesium salt, L-; Magnesate(2-), bis[L-aspartato(2-)-N,O1]-, dihydrogen, (T-4)-; dihydrogen bis[L-aspartato(2-)-N,O1]magnesate(2-). Listed on TSCA. Chiral building block. *Rexim S.A. Produits Chimiques; Varsal Instruments, Inc.*

**3454 Magnesium di-L-aspartate**
2068-80-6 218-191-6

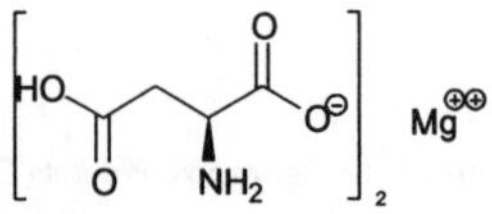

$C_8H_{14}MgN_2O_8$
Aspartic acid, magnesium salt (2:1), L-; Magnesium dihydrogen di-L-aspartate; Magnesium hydrogen aspartate. Listed on TSCA. Chiral building block. *Austin Chemical Company, Inc.; Flamma s.p.a.; Tanabe Seiyaku Co. Ltd.*

**3455 Magnesium α-D-glucoheptonate**
68475-44-5 270-642-6
$C_{14}H_{26}O_{16}Mg$
Magnesium D-glycero-D-guloheptonate; Bis(D-glycero-D-ido-heptonato)magnesium; D-Glycero-D-ido-heptonic acid, magnesium salt (2:1). Listed on TSCA. Chiral building block. *Pfanstiehl Laboratories, Inc.*

**3456 Magnesium D-gluconate hydrate**
4350(11)

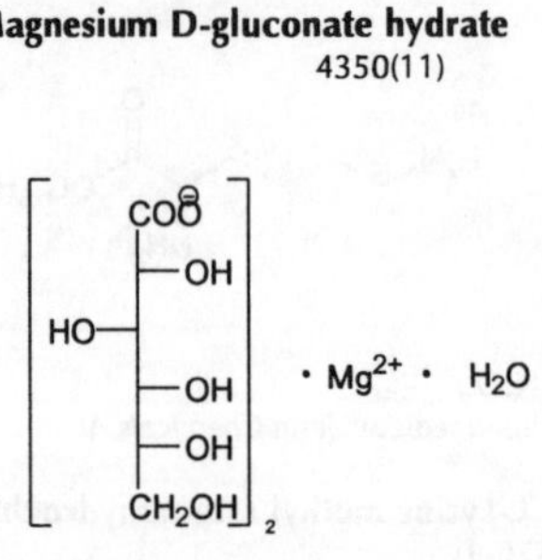

$C_6H_{12}O_7$
D-Gluconic acid, magnesium salt hydrate. Chiral building block. mp = 200°; $[\alpha]_D^{20}$ = + 11° (c = 1.8, $H_2O$). *Sigma-Aldrich Fine Chemicals.*

**3457 Magnesium L-glutamate**
64407-99-4 264-873-1

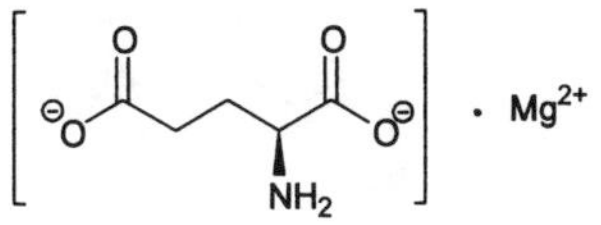

$C_5H_9MgNO_4$
Glutamic acid, magnesium salt (1:1), L-; Monomagnesium glutamate. Chiral building block. *Rexim S.A. Produits Chimiques.*

**3458 Magnesium-L-hydrogen aspartate dihydrate**

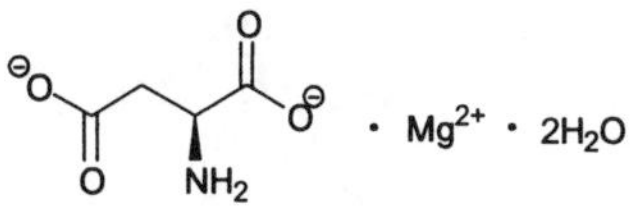

$C_4H_9MgNO_6$
Chiral intermediate. *Boehringer Ingelheim Pharma KG.*

**3459 D-(+)-Malic acid**
636-61-3 5747(12) 211-262-2

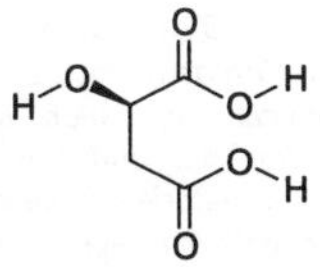

$C_4H_6O_5$
Butanedioic acid, hydroxy-, (R)-; D-(+)-2-Hydroxysuccinic acid. Chiral building block; Resolving agent. mp = 98-102°; $[\alpha]_D^{20}$ = + 27° (c = 5.5, pyridine). *Acros Organics nv; Sigma-Aldrich Fine Chemicals.*

**3460 L-(-)-Malic acid**
97-67-6 5747(12) 202-601-5

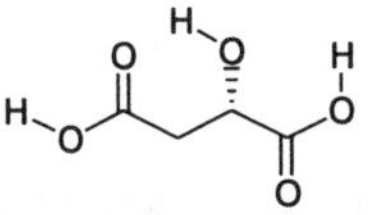

$C_4H_6O_5$
Butanedioic acid, hydroxy-, (2S)-; (S)-(-)-Hydroxysuccinic acid; Apple acid; S-2-Hydroxybutanedioic acid; (S)-(-)-Malic acid. Listed on TSCA. Resolving agent. mp = 101-103°; $[\alpha]_D^{20}$ = - 26 ± 2 ° (c = 5.5, pyridine). *Acros Organics nv; Austin Chemical Company, Inc.; Kyowa Hakko Kogyo Co., Ltd.; Lancaster Synthesis Ltd.; Sigma-Aldrich Fine Chemicals; TCI America; Varsal Instruments, Inc.*

**3461 L-(-)-Malic acid, disodium salt monohydrate**
64887-73-6

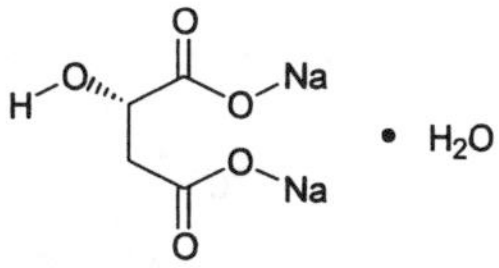

$C_4H_4Na_2O_5 \cdot H_2O$
Chiral building block. $[\alpha]^{21}$ = - 7° (c = 15, $H_2O$). *Acros Organics nv.*

**3462 (+)-Maltitol**
585-88-6 209-567-0

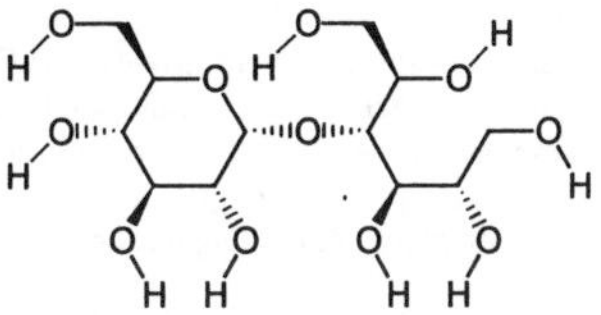

$C_{12}H_{24}O_{11}$
Glucitol, 4-O-α-D-glucopyranosyl-, D-; Amalty; Mabit; Maltidex 100; Maltisorb; Maltit. Listed on TSCA. Chiral building block. mp = 149-152°; $[\alpha]^{21}$ = + 107.0° (c = 10, $H_2O$). *Acros Organics nv; Senn Chemicals AG; Sigma-Aldrich Fine Chemicals.*

**3463 (+)-Maltopentaose hydrate**
123333-77-7

$C_{30}H_{52}O_{26}$
Chiral intermediate. $[\alpha]^{25}$ = + 179° (c = 1, $H_2O$). *Sigma-Aldrich Fine Chemicals.*

**3464 D-(+)-Maltose monohydrate**
6363-53-7 200-716-5

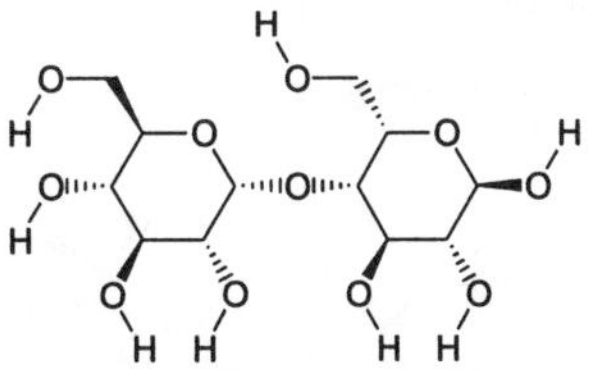

$C_{12}H_{22}O_{11}$
Glucose, 4-O-α-D-glucopyranosyl-, monohydrate, D-; Maltobiose monohydrate; 4-O-α-D-glucopyranosyl-D-glucopyranose monohydrate; D-Maltose monohydrate. Chiral building block. mp = 119-121°; $[\alpha]_D^{20}$ = + 130.4° (c = 4, $H_2O$). *Acros Organics nv; Pfanstiehl Laboratories, Inc.; Senn Chemicals AG; Sigma-Aldrich Fine Chemicals.*

**3465 D-Maltotetraose**
34612-38-9 252-111-0

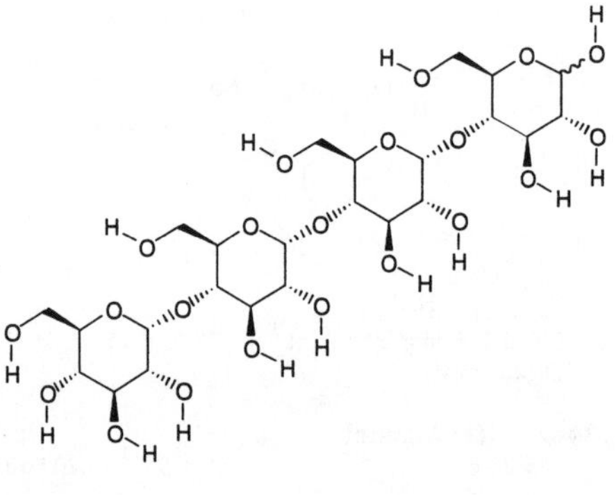

$C_{24}H_{42}O_{21}$
D-Glucose, O-α-D-glucopyranosyl-(1→4)-O-α-D-glucopyranosyl-(1→4)-O-α-D-glucopyranosyl-(1→4)-; O-α-D-glucopyranosyl-(1→4)-O-α-D-glucopyranosyl-(1→4)-O-α-D-glucopyranosyl-(1→4)-D-glucose; Amylotetraose. Chiral building block. *Acros Organics nv.*

**3466 D-(+)-Maltotriose**
1109-28-0 214-174-2

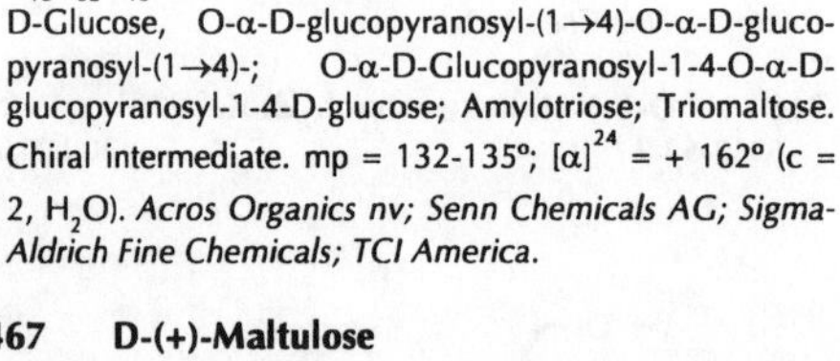

$C_{18}H_{32}O_{16}$
D-Glucose, O-α-D-glucopyranosyl-(1→4)-O-α-D-glucopyranosyl-(1→4)-; O-α-D-Glucopyranosyl-1-4-O-α-D-glucopyranosyl-1-4-D-glucose; Amylotriose; Triomaltose. Chiral intermediate. mp = 132-135°; $[\alpha]^{24}$ = + 162° (c = 2, $H_2O$). *Acros Organics nv; Senn Chemicals AG; Sigma-Aldrich Fine Chemicals; TCI America.*

**3467 D-(+)-Maltulose**
17606-72-3 241-578-6

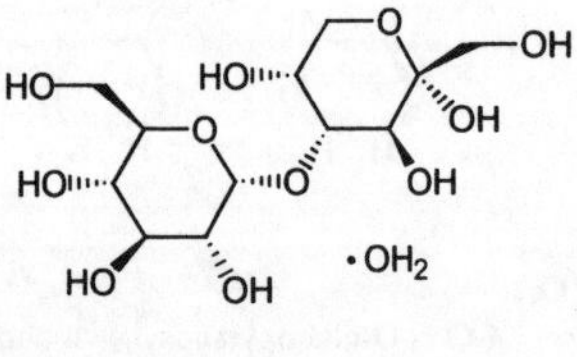

$C_{12}H_{22}O_{11}$
Fructose, 4-O-α-D-glucopyranosyl-, D-; 4-O-α-D-Glucopyranosyl-D-fructose monohydrate. Chiral intermediate. mp = 137°; $[\alpha]^{22}$ = + 57° (c = 2, $H_2O$). *Senn Chemicals AG; Sigma-Aldrich Fine Chemicals.*

**3468 (S)-Mandelamide**
24008-63-7

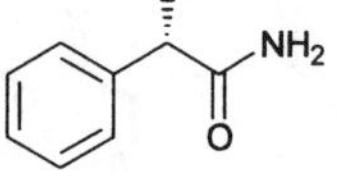

$C_8H_9NO_3$
Chiral building block. mp = 123-124°. *Mitsubishi Rayon Co., Ltd.*

**3469 D-(-)-Mandelic acid**
611-71-2 210-276-6

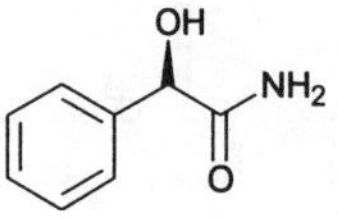

$C_8H_8O_3$
Benzeneacetic acid, α-hydroxy-, (R)-; (R)-(-)-α-Hydroxyphenylacetic acid; (R)-α-Hydroxybenzeneacetic acid; D-2-Phenylglycolic acid; (R)-(-)-Mandelic acid; (R)-(-)-Amygdalic acid. Listed on TSCA. Used as both a chiral building block and resolving agent. mp = 129-134°; $[\alpha]_D^{20}$ = - 155 ± 5° (c = 5, $H_2O$). *Acros Organics nv; Austin Chemical Company, Inc.; Lancaster Synthesis Ltd.; Mitsubishi Chemical Corporation; Mitsubishi Rayon Co., Ltd.; Nippon Chemical Industrial Co., Ltd.; Nippon Kayaku; Norse Laboratories; Omega Chemical Company Inc.; Senn Chemicals AG; Sigma-Aldrich Fine Chemicals; Tanabe Seiyaku Co. Ltd.; TCI America; Yamakawa Chemical Industry Co. Ltd.; Yoneyama Yakuhin Kogyo Co., Ltd.; Zeeland Chemicals, Inc.*

**3470 (L)-(+)-Mandelic acid**
17199-29-0 241-240-8

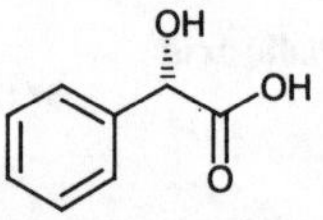

$C_8H_8O_3$
Benzeneacetic acid, α-hydroxy-, (S)-; (S)-(+)-α-Hydroxyphenylacetic acid; L-(+)-Mandelic acid. Listed on TSCA. Chiral intermediate; Resolving agent. mp = 131-134°; $[\alpha]_D^{20}$ = + 155 ± 5° (c = 5, $H_2O$). *Acros Organics nv; Austin Chemical Company, Inc.; Lancaster Synthesis Ltd.; Nippon Kayaku; Norse Laboratories; Omega Chemical Company Inc.; Senn Chemicals AG; Sigma-Aldrich Fine Chemicals; Tanabe Seiyaku Co. Ltd.; TCI America; Yamakawa Chemical Industry Co. Ltd.; Yoneyama Yakuhin Kogyo Co., Ltd.; Zeeland Chemicals, Inc.*

### 3471 Mandelonitrile
10020-96-9 5759(12)

$C_8H_7NO$
α-Hydroxybenzeneacetonitrile. Chiral building block. mp = 28-30°; bp = 170°; d = 1.117; n = 1.532; $[\alpha]_D^{20}$ = + 42° (c = 1, $CHCl_3$). *Sigma-Aldrich Fine Chemicals.*

### 3472 D-Mandelonitrile-β-D-glucosido-6-β-D-glucoside
29883-15-6 638(12) 249-925-3

$C_{20}H_{27}NO_{11}$
Benzeneacetonitrile, α-[(6-O-β-D-glucopyranosyl-β-D-glucopyranosyl)oxy]-, (R)- ; D-Amygdalin; Amygdaloside; Mandelonitrile-β-gentiobioside; NSC 15780. Pharmaceutical or derivative. mp = 223-226°; $[\alpha]^{22}$ = - 38° (c = 1.2, $H_2O$). *Acros Organics nv; Kaden Biochemicals GmbH; Senn Chemicals AG; Sigma-Aldrich Fine Chemicals.*

### 3473 D-Mannitol
69-65-8 5788(12) 200-711-8

$C_6H_{14}O_6$
Mannitol, D-; D-Mannite; Cordycepic acid; Mannitolum; Osmitrol; Osmosal; Diosmol; Isotol; Maniton S; Manna sugar; Mannidex; Mannigen; Mannistol. Listed on TSCA. Chiral building block. mp = 166°; $[\alpha]^{22}$ = + 141.0° (c = 0.4, acidified ammonium molybdate). *Acros Organics nv; Pfanstiehl Laboratories, Inc.; Sigma-Aldrich Fine Chemicals.*

### 3474 L-Mannitol
643-01-6

$C_6H_{14}O_6$
Chiral building block. *TCI America.*

### 3475 D-Mannitol hexaacetate

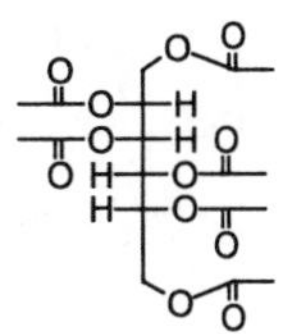

$C_{18}H_{26}O_{12}$
Chiral intermediate. mp = 120-127°; $[\alpha]^{23}$ = + 25° (c = 1, $CHCl_3$). *Sigma-Aldrich Fine Chemicals.*

### 3476 D-Mannoheptulose
3615-44-9 222-795-5

$C_7H_{14}O_7$
Mannoheptulose, D-. Listed on TSCA. Chiral building block. $[\alpha]_D^{20}$ = + 29° (c = 2, $H_2O$). *Pfanstiehl Laboratories, Inc.*

### 3477 D-Mannono-1,4-lactone
26301-79-1 247-596-0

$C_6H_{10}O_6$
Mannonic acid, γ-lactone, D-. Chiral building block. *TCI America.*

### 3478 L-Mannono-1,4-lactone
22430-23-5

$C_6H_{10}O_6$
L-Mannono-γ-lactone. Chiral building block. mp = 153-155°; $[\alpha]_D^{20}$ = - 51.5° (c = 4, $H_2O$). *Pfanstiehl Laboratories, Inc.; Senn Chemicals AG; Sigma-Aldrich Fine Chemicals.*

### 3479 D-Mannosamine hydrochloride
5505-63-5 226-847-8

$C_6H_{14}ClNO_5$
Mannose, 2-amino-2-deoxy-, hydrochloride, D-; 2-Amino-2-deoxy-D-mannose hydrochloride. Listed on TSCA. Chiral building block. mp = 168°; $[\alpha]_D^{20}$ = - 3.7° (c = 8, $H_2O$); $[\alpha]_D^{20}$ = - 4.7° (c = 10, 5% HCl). *Pfanstiehl Laboratories, Inc.; Senn Chemicals AG; Sigma-Aldrich Fine Chemicals.*

**3480 D-(+)-Mannose**
3458-28-4 222-392-4

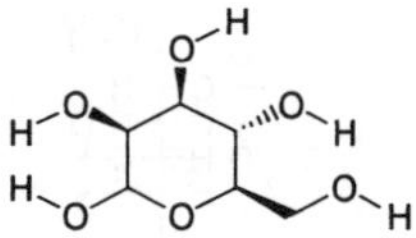

$C_6H_{12}O_6$
Mannose, D-; D-Mannopyranose; Carubinose; Seminose. Listed on TSCA. Chiral building block. mp = 133-140°; $[\alpha]_D^{20}$ = + 14.2° (c = 4, $H_2O$). *Acros Organics nv; Austin Chemical Company, Inc.; Kaden Biochemicals GmbH; Pfanstiehl Laboratories, Inc.; Senn Chemicals AG; TCI America.*

**3481 L-Mannose**
10030-80-5 233-080-2

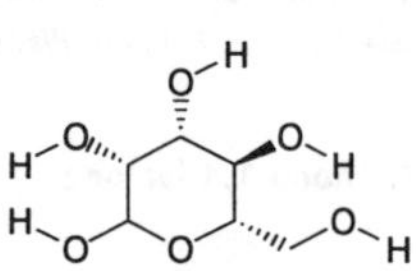

$C_6H_{12}O_6$
Mannose, 6-deoxy-, monohydrate, L-; Listed on TSCA. Chiral building block. mp = 129-131°; $[\alpha]_D^{20}$ = - 14.2° (c = 4, $H_2O$). *Acros Organics nv; Pfanstiehl Laboratories, Inc.; Senn Chemicals AG; Sigma-Aldrich Fine Chemicals.*

**3482 α-D-Mannose pentaacetate**

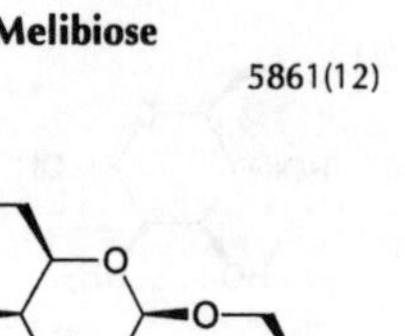

$C_{16}H_{22}O_{11}$
Chiral intermediate. *Senn Chemicals AG.*

**3483 D-(+)-Melibiose**
585-99-9 5861(12) 209-568-6

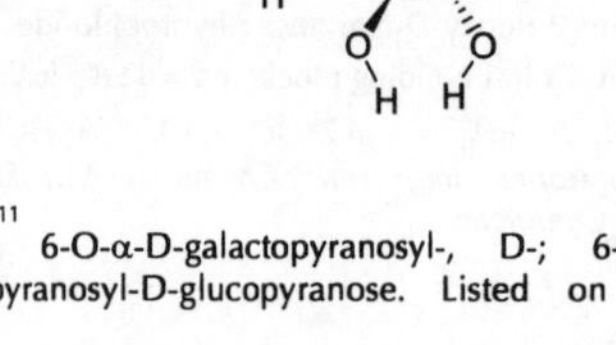

$C_{12}H_{22}O_{11}$
Glucose, 6-O-α-D-galactopyranosyl-, D-; 6-O-α-D-Galactopyranosyl-D-glucopyranose. Listed on TSCA. Chiral intermediate. mp = 182°; $[\alpha]_D^{20}$ = + 135.2° (c = 4, $H_2O$). *Acros Organics nv; Pfanstiehl Laboratories, Inc.; Senn Chemicals AG; Sigma-Aldrich Fine Chemicals; TCI America.*

**3484 (+)-4-Menth-1-en-9-ol**
13835-75-1 237-548-7

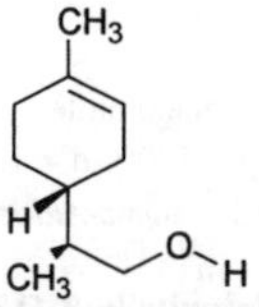

$C_{10}H_{18}O$
Cyclohex-3-ene-1-ethanol, β,4-dimethyl-, (βS,1R)-. Chiral intermediate. $bp_{10}$ = 115-116; $[\alpha]^{22}$ = + 94.5° (c = 4.3, $C_6H_6$). *Acros Organics nv.*

**3485 (+)-p-Menth-1-en-9-ol (mixture of isomers)**
18479-68-0 242-366-6

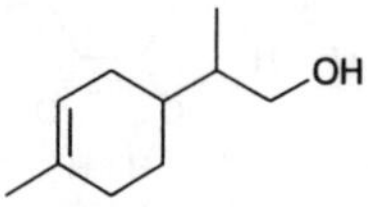

$C_{10}H_{18}O$
Listed on TSCA. Chiral intermediate. $bp_{10}$ = 115-116°; d = 0.941; n = 1.486; $[\alpha]_D^{20}$ = + 99° (c = 4.2, toluene). *Sigma-Aldrich Fine Chemicals.*

**3486 D-Menthol**
15356-60-2 239-387-8

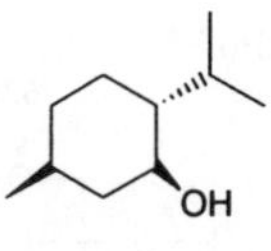

$C_{10}H_{20}O$
Cyclohexanol, 5-methyl-2-(1-methylethyl)-, (1S,2R,5S)-; (1S,2R,5S)-2-Isopropyl-5-methylcyclohexanol; (+)-Menthol. Listed on TSCA. Chiral intermediate. mp = 43-44°; $bp_{10}$ = 104-105°; $[\alpha]_D^{20}$ = + 49 ± 1° (c = 10, EtOH). *Lancaster Synthesis Ltd.; Sigma-Aldrich Fine Chemicals.*

**3487 L-Menthol**
2216-51-5 5882(12) 218-690-9

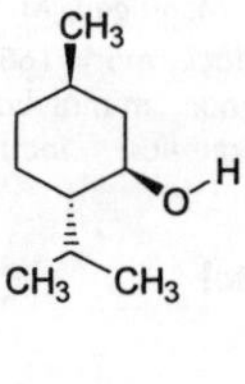

$C_{10}H_{20}O$
Cyclohexanol, 5-methyl-2-(1-methylethyl)-, [1R-(1α,2β,5α)]-; (1R,2S,5R)-2-Isopropyl-5-methylcyclo-

hexanol; L-Hexahydrothymol; 5-Methyl-2-isopropyl hexahydrophenol; 1-Methyl-4-isopropyl cyclohexan-3-ol; 1-isopropyl-4-methyl cyclohexan-2-ol; Levomenthol; (-)-Menthol. Chiral intermediate. mp = 43-45°; bp = 216°; d = 0.89; $[\alpha]_D^{20}$ = - 50 ± 1° (c = 10, EtOH). *Acros Organics nv; Haarmann & Reimer GmbH; Lancaster Synthesis Ltd.; Sigma-Aldrich Fine Chemicals; Takasago International Corporation.*

**3488 L-Menthone**
14073-97-3 5883(12) 237-926-1

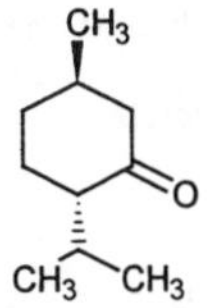

$C_{10}H_{18}O$
Cyclohexanone, 5-methyl-2-(1-methylethyl)-, (2S,5R)-; (-)-5-Methyl-2-(1-methylethyl)cyclohexanone; L-Menthan-3-one. Listed on TSCA. Chiral intermediate. bp = 207°; d = 0.893; n = 1.45; [α] = - 15° (neat). *Acros Organics nv; Sigma-Aldrich Fine Chemicals.*

**3489 (-)-Menthoxyacetyl chloride**
15356-62-4

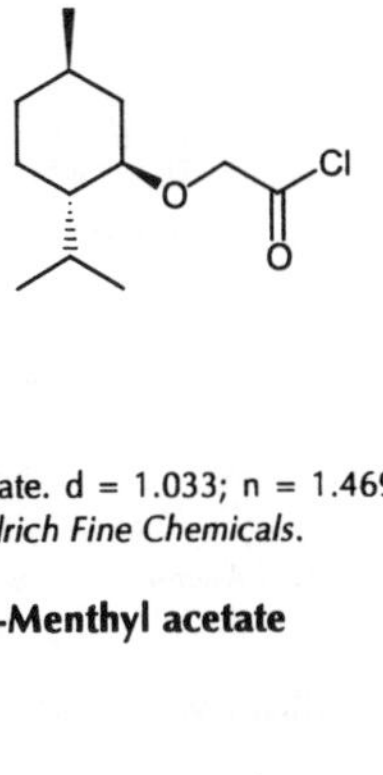

$C_{12}H_{21}ClO_2$
Chiral intermediate. d = 1.033; n = 1.469; $[\alpha]^{25}$ = - 10° (neat). *Sigma-Aldrich Fine Chemicals.*

**3490 (1S)-(+)-Menthyl acetate**
5157-89-1

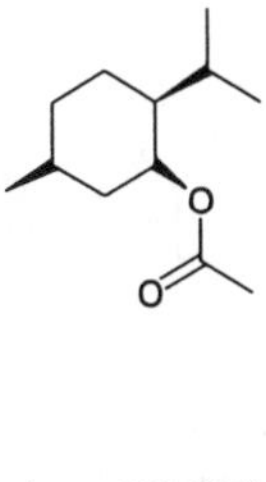

$C_{12}H_{22}O_2$
Chiral intermediate. bp = 229-230°; d = 0.9250; $n_D^{20}$ = 1.4470; $[\alpha]_D^{20}$ = + 80.5 ± 1° (c = 8, $C_6H_6$). *Lancaster Synthesis Ltd.; Oxford Asymmetry International plc; Sigma-Aldrich Fine Chemicals.*

**3491 (-)-Menthyl acetate**
2623-23-6 5884(12) 220-076-0

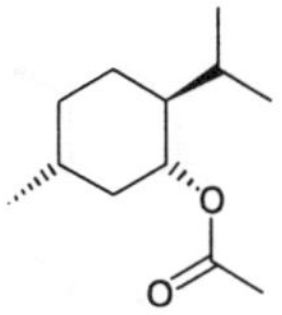

$C_{12}H_{22}O_2$
Cyclohexanol, 5-methyl-2-(1-methylethyl)-, acetate, (1R,2S,5R)-; (1R-[1α, 2β, 5α]-5-Methyl-2-[1-methylethyl]cyclohexylacetate; Menthol, acetate, (1R,3R,4S)-(-)-. Listed on TSCA. Chiral intermediate. bp = 229-231°; d = 0.9250; $n_D^{20}$ = 1.4470; $[\alpha]_D^{20}$ = - 80.5 ± 1° (c = 8, $C_6H_6$). *Haarmann & Reimer GmbH; Lancaster Synthesis Ltd.; Oxford Asymmetry International plc; Sigma-Aldrich Fine Chemicals; Takasago International Corporation.*

**3492 (-)-Menthyl chloride**
16052-42-9 240-200-7

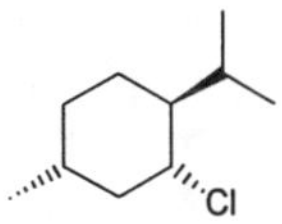

$C_{10}H_{19}Cl$
Cyclohexane, 2-chloro-4-methyl-1-(1-methylethyl)-, (1S,2R,4R)-; L-3-Chloro-p-menthane; [1S-(1α,2β,4β)]-2-chloro-1-isopropyl-4-methylcyclohexane. Chiral intermediate. mp = -16.5°; $bp_{21}$ = 101-101.5°; d = 0.936; n = 1.4634; $[\alpha]_D^{20}$ = - 52.4° (neat). *Norse Laboratories; Sigma-Aldrich Fine Chemicals; TCI America.*

**3493 (-)-Menthylamine**
2216-54-8 218-693-5

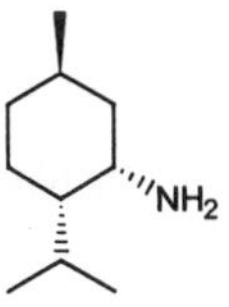

$C_{10}H_{21}N$
Cyclohexanamine, 5-methyl-2-(1-methylethyl)-, (1R,2S,5R)-; p-Menthan-3-amine, (-); [1R-(1α,2β,5α)]-2-(isopropyl)-5-methylcyclohexylamine. Chiral intermediate. *Norse Laboratories.*

**3494 (1R)-(-)-Menthyl chloroformate**
14602-86-9

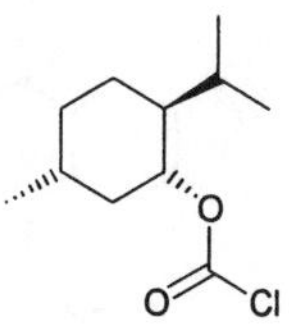

$C_{11}H_{19}ClO_2$
Chloroformic acid (1R)-menthyl ester; L-Menthyl chloroformate. Chiral diagnostic reagent. $bp_{11}$ = 108-109°; d = 1.0250; $n_D^{20}$ = 1.4580; $[\alpha]_D^{20}$ = - 82 ± 2° (c = 1, $CHCl_3$). *Lancaster Synthesis Ltd.; Oxford Asymmetry International plc; Sigma-Aldrich Fine Chemicals.*

**3495 (1S)-(+)-Menthyl chloroformate**
7635-54-3

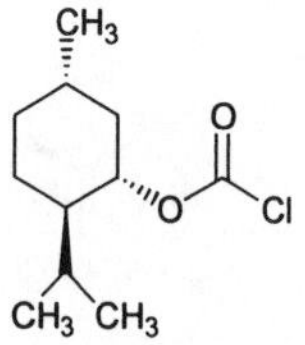

$C_{11}H_{19}ClO_2$
Chloroformic acid (1S)-menthyl ester; D-Menthyl chloroformate. Chiral diagnostic reagent. d = 1.0310; n = 1.4590; $[\alpha]_D^{20}$ = + 82 ± 2° (c = 1, $CHCl_3$). *Acros Organics nv; Lancaster Synthesis Ltd.; Oxford Asymmetry International plc; Sigma-Aldrich Fine Chemicals.*

**3496 (1R,2S,5R)-(-)-Menthyldiphenylarsine**
$C_{22}H_{29}As$
Chiral ligand. *Digital Specialty Chemicals, Inc.*

**3497 (1S,2R,5S)-(-)-Menthyldiphenylarsine**
$C_{22}H_{29}As$
Chiral ligand. *Digital Specialty Chemicals, Inc.*

**3498 (1R,2S,5R)-(-)-Menthyldiphenylphosphine**
43077-31-2
$C_{22}H_{29}P$
(R)-NMDPP. Chiral intermediate. *Digital Specialty Chemicals, Inc.*

**3499 (1S,2R,5S)-(+)-Menthyldiphenylphosphine**
91796-57-5
$C_{22}H_{29}P$
Chiral ligand. *Digital Specialty Chemicals, Inc.*

**3500 L-Menthyl glyoxylate hydrate**

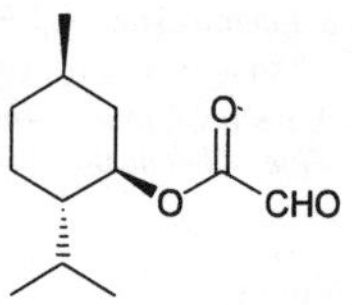

$C_{12}H_{22}O_4$
Chiral building block. *DSM Fine Chemcials Netherlands.*

**3501 Menthyl lactate**
59259-38-0

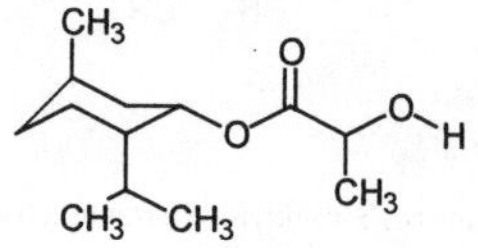

$C_{13}H_{24}O_3$
Propanaoic acid, 2-hydroxy-, 5-methyl-2-(1-methylethyl)cyclohexyl ester, [1R-[1α(R*),2β,5α]]-. Chiral intermediate. $[\alpha]^{20}$ = - 74° (c = 10, EtOH). *Acros Organics nv.*

**3502 (-)-Menthyloxyacetic acid**
40248-63-3 254-857-2

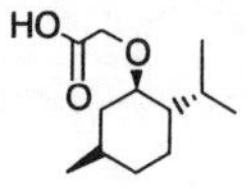

$C_{22}H_{22}O_3$
Acetic acid, [[5-methyl-2-(1-methylethyl)cyclohexyl]-oxy]-, [1R-(1α,2β,5α)]-; L-p-Menth-3-yloxyacetic acid. Chiral intermediate. mp = 52-55°; $bp_{10}$ = 163-164°; d = 1.02; n = 1.4672; $[\alpha]^{25}$ = - 92.5° (c = 4, $CH_3OH$). *Norse Laboratories; Oxford Asymmetry International plc; Sigma-Aldrich Fine Chemicals.*

**3503 (+)-Menthyloxyacetic acid**
94133-41-2 302-700-4

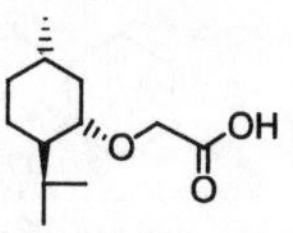

$C_{12}H_{22}O_3$
Acetic acid, [[5-methyl-2-(1-methylethyl)cyclohexyl]-oxy]-, [1S-(1α,2β,5α)]-; [1S-(1α,2β,5α)]-[[5-methyl-2-(1-methylethyl)cyclohexyl]oxy]acetic acid. Chiral intermediate. $bp_{10}$ = 163-164°; d = 1.02; n = 1.465; $[\alpha]_D^{20}$ = + 92.5° (c = 10, EtOH). *Oxford Asymmetry International plc; Sigma-Aldrich Fine Chemicals.*

**3504 (R)-5-(1R)-Menthyloxy-2(5H)-furanone**
77934-87-3

$C_{14}H_{22}O_3$
Chiral intermediate. $bp_{0.01}$ = 120-123°; mp = 78-80°; $[\alpha]_D^{20}$ = - 137° (c = 1, $CHCl_3$). *Sigma-Aldrich Fine Chemicals.*

**3505 (1R,2S,5R)-(-)-Menthyl (S)-p-toluene-sulfinate**
1517-82-4

$C_{17}H_{26}O_2S$
Chiral intermediate. mp = 102-104°; $[\alpha]_D^{20}$= - 195° (c = 2, acetone). *Advanced Asymmetrics, Inc.; Sigma-Aldrich Fine Chemicals; TCI America.*

**3506 (1S,2R,5S)-(+)-Menthyl (R)-p-toluene-sulfinate**
$C_{17}H_{26}O_2S$
Chiral intermediate. mp = 104-106°; $[\alpha]_D^{20}$= + 200° (c = 2, acetone). *Advanced Asymmetrics, Inc.; Sigma-Aldrich Fine Chemicals; TCI America.*

**3507 (S)-1-Mercaptoglycerol**
120785-95-7

$C_3H_8O_2S$
Chiral building block. *Kaneka Corporation.*

**3508 (1S)-(-)-10-Mercaptoisoborneol**
71242-58-5
$C_{10}H_{18}OS$
(1S)-(-)-10-Mercapto-isoborneol. Chiral intermediate. mp = 72°. *TCI America.*

**3509 6-Mercaptopurine-9-D-riboside**
574-25-4 209-371-5

$C_{10}H_{12}N_4O_4S$
Purine-(6H)-6-thione, 1,9-dihydro-9-β-D-ribofuranosyl-; 6-Thioinosine; 6-Mercaptoinosine; 6-MPR; 6-Thiopurine ribonucleoside; 9-β-D-Ribofuranosyl-9H-purine-6-thione; 9H-Purine-6-thiol, 9-β-D-ribofuranosyl-; Inosine, 6-thio-; NSC 4911; Ribosyl-6-thiopurine. Chiral intermediate. $[\alpha]^{25}$ = - 73° (c = 1, 0.1N NaOH). *Acros Organics nv.*

**3510 (R)-4-Mercapto-2-pyrrolidinone**
157429-42-0

$C_4H_7NOS$
Pyrrolidin-2-one, 4-mercapto-, (R)-; Hydroxy-2-pyrrolidinethione; Pyrrolidine-2-thione, 4-hydroxy, (R)-. Chiral building block. *Synthon Chiragenics Corporation.*

**3511 (S)-4-Mercapto-2-pyrrolidinone**
184759-58-8

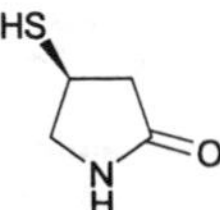


$C_4H_7NOS$
Pyrrolidin-2-one, 4-mercapto-, (S)-; (S)-4-Hydroxy-2-pyrrolidinethione; Pyrrolidine-2-thione, 4-hydroxy, (S)-. Chiral building block. *Synthon Chiragenics Corporation.*

**3512 L-(-)-Methanesulfonylethyllactate**
63696-99-1

$C_6H_{12}O_5S$
Chiral intermediate. $[\alpha]^{22}$ = - 52° (c = 10, EtOH). *Acros Organics nv.*

**3513 D-Methionine**
348-67-4 6053(12) 206-483-6

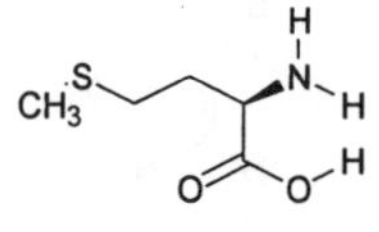

$C_5H_{11}NO_2S$
Methionine, D-; (R)-2-Amino-4-(methylthio)butyric acid. Listed on TSCA. Chiral building block. mp = 273°; $[\alpha]_D^{20}$ = - 23 ± 1.5° (c = 2, 2N HCl). *Acros Organics nv; Austin Chemical Company, Inc.; Kaneka Corporation; Lancaster Synthesis Ltd.; Omega Chemical Company Inc.; Rexim S.A. Produits Chimiques; Sigma-Aldrich Fine Chemicals; Tanabe Seiyaku Co. Ltd.; TCI America; Yoneyama Yakuhin Kogyo Co., Ltd.*

**3514 L-Methionine**
63-68-3 6053(12) 200-562-9

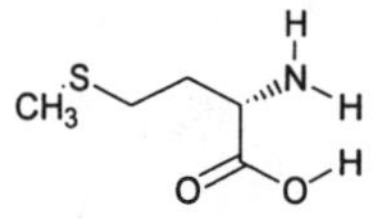

$C_5H_{11}NO_2S$
Butanoic acid, 2-amino-4-(methylthio)-, (S)-; (S)-2-Amino-4-(methylthio)butyric acid; α-Amino-γ-methylmercaptobutyric acid; Cymethion; L-Homocysteine, S-methyl-. Listed on TSCA. Chiral building block. mp = 84°; $[\alpha]_D^{20}$ = + 23 ± 1.5° (c = 2, 2N HCl). *Acros Organics nv; Ajinomoto Co. Inc.; Austin Chemical Company, Inc.; KingChem, Inc.; Kyowa Hakko Kogyo Co., Ltd.; Lancaster Synthesis Ltd.; Nippon Kayaku; Rexim S.A. Produits Chimiques; Senn Chemicals AG; Shanghai DSL International Trading Company; Sigma-Aldrich Fine Chemicals; Tanabe Seiyaku Co. Ltd.; TCI America; Yoneyama Yakuhin Kogyo Co., Ltd.*

**3515 L-Methionine amide hydrochloride**

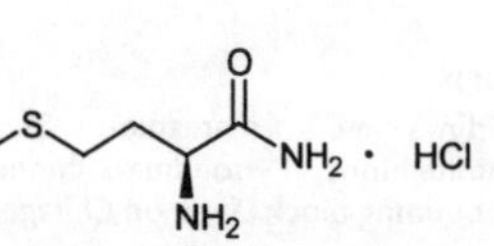

$C_5H_{13}ClO_2OS$
Chiral building block. *Senn Chemicals AG.*

**3516 L-Methionine ethyl ester hydrochloride**
2899-36-7 220-787-6

$C_7H_{16}ClNO_2S$
Methionine, ethyl ester, hydrochloride, L-; Ethyl L-methionate hydrochloride. Chiral building block. mp = 90-92°; $[\alpha]^{21}$ = + 17.7° (c = 2, EtOH). *Acros Organics nv; Sigma-Aldrich Fine Chemicals.*

**3517 L-Methionine methyl ester hydrochloride**
2491-18-1 219-651-9

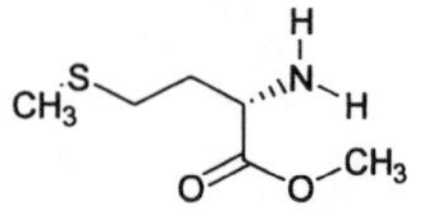

$C_6H_{14}ClNO_2S$
Methionine, methyl ester, hydrochloride, L-; Methyl L-methionate hydrochloride. Chiral building block. mp = 151°; $[\alpha]^{22}$ = + 25.9° (c = 1, $H_2O$). *Acros Organics nv; Sigma-Aldrich Fine Chemicals; TCI America.*

**3518 D-Methionine methylsulfonium bromide**

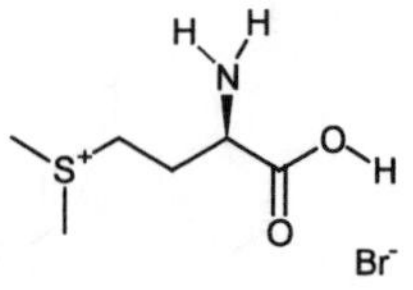

$C_6H_{14}BrNO_2S$
Chiral building block. *Acros Organics nv.*

**3519 L-Methionine methylsulfonium bromide**

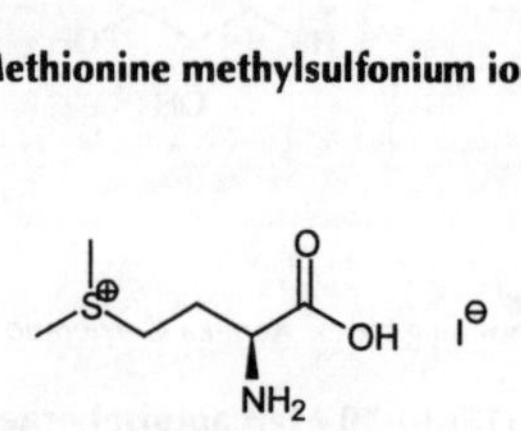

$C_6H_{14}BrNO_2S$
Chiral building block. [α] = + 20° (c = 6, $H_2O$). *Acros Organics nv.*

**3520 L-Methionine methylsulfonium iodide**

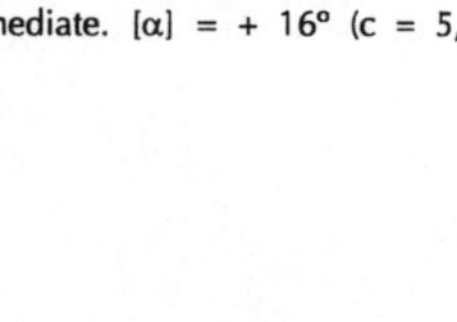

$C_6H_{14}INO_2S$
Chiral intermediate. [α] = + 16° (c = 5, $H_2O$). *Acros Organics nv.*

**3521 L-Methionine sulfoxide**
3226-65-1 221-758-0

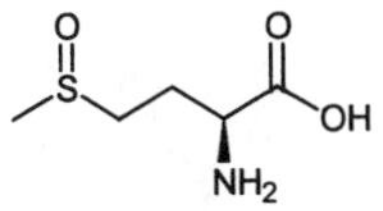

$C_5H_{11}NO_3S$
Butanoic acid, 2-amino-4-(methylsulfinyl)-, (2S)-. Chiral building block. *Senn Chemicals AG.*

**3522 L-Methionine sulfoximine**
15985-39-4

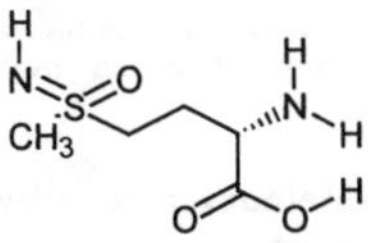

$C_5H_{12}N_2O_3S$
Chiral intermediate. mp = 210°; $[\alpha]^{25}$ = + 12° (c = 1, $H_2O$). *Acros Organics nv; Sigma-Aldrich Fine Chemicals.*

**3523 L-Methioninol**
2899-37-8 220-788-1

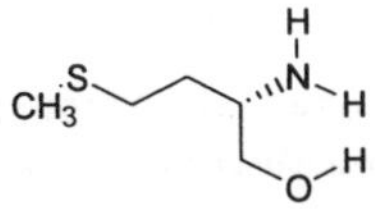

$C_{10}H_{13}ClN_2O_4$
Butan-1-ol, 2-amino-4-(methylthio)-, (2S)-; (S)-(-)-2-Amino-4-methylthio-1-butanol. Chiral building block. mp = 31-33°; n = 1.5216; $[\alpha]_D^{20}$ = - 21 ± 1° (c = 5, $H_2O$); $[\alpha]_D^{21}$ = - 12.7° (c = 1.4, EtOH). *Acros Organics nv; Lancaster Synthesis Ltd.; Loba Feinchemie AG; Omega Chemical Company Inc.; Senn Chemicals AG; Sigma-Aldrich Fine Chemicals; TCI America.*

**3524 L-Methionylglycine**
14486-03-4 238-487-9
$C_7H_{14}N_2O_3S$
Glycine, L-methionyl-.
Chiral building block. *TCI America.*

**3525 (S)-5-Methoxy-2-aminotetralin**

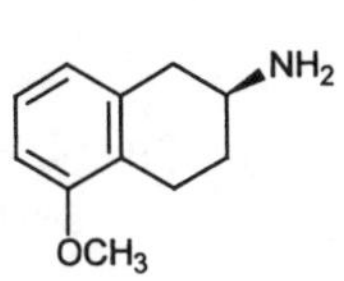

$C_{10}H_{15}NO$
Chiral building block. *Chiragene, Inc.*

**3526 (R)-6-Methoxy-2-aminotetralin**

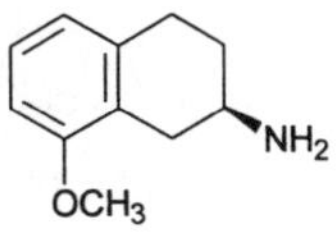

$C_{10}H_{15}NO$
Chiral intermediate. *Chiragene, Inc.*

**3527 (S)-6-Methoxy-2-aminotetralin**

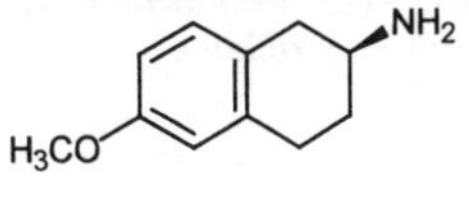

$C_{10}H_{15}NO$
Chiral building block. *Chiragene, Inc.*

**3528 (R)-7-Methoxy-2-aminotetralin**

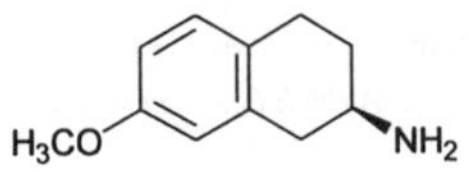

$C_{10}H_{15}NO$
Chiral intermediate. *Chiragene, Inc.*

**3529 (S)-7-Methoxy-2-aminotetralin**

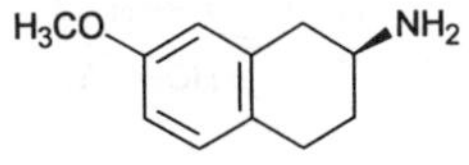

$C_{10}H_{15}NO$
Chiral building block. *Chiragene, Inc.*

**3530 (R)-8-Methoxy-2-aminotetralin**

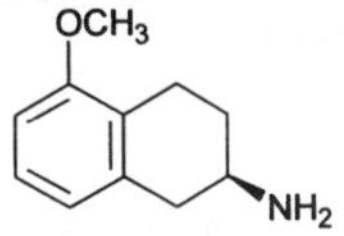

$C_{10}H_{15}NO$
Chiral building block. *Chiragene, Inc.*

**3531 (S)-8-Methoxy-2-aminotetralin**

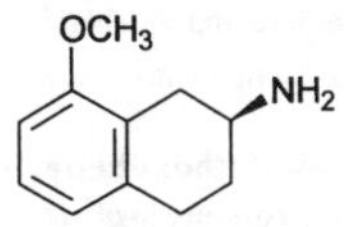

$C_{10}H_{15}NO$
Chiral building block. *Chiragene, Inc.*

**3532 (-)-4,6-O-(4-Methoxybenzilidene)glucal**

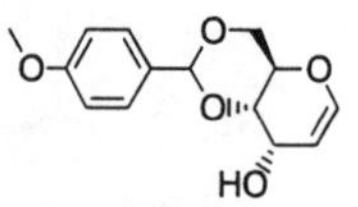

$C_{14}H_{16}O_5$
Chiral intermediate. mp = 120-138°; bp = 276-289°; $[\alpha]^{22}$ = - 18° (c = 1, $CHCl_3$). *Sigma-Aldrich Fine Chemicals.*

**3533 (S)-(-)-1-(2-Methoxybenzoyl)-2-(methoxymethyl)pyrrolidine**
102069-84-1

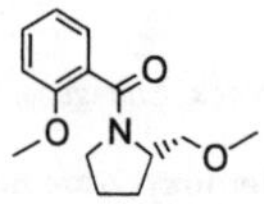

$C_{14}H_{19}NO_3$
Chiral intermediate. n = 1.54; $[\alpha]_D^{20}$ = - 122° (c = 1, $CHCl_3$). *Sigma-Aldrich Fine Chemicals.*

**3534 (R)-(+)-1-(2-Methoxybenzoyl)-2-pyrrolidinemethanol**

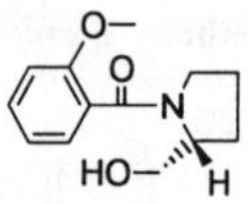

$C_{13}H_{17}NO_3$
Chiral intermediate. mp = 104-106°; $[\alpha]_D^{20}$ = + 91° (c = 1.2, $CHCl_3$). *Sigma-Aldrich Fine Chemicals.*

**3535 (S)-(-)-1-(2-Methoxybenzoyl)-2-pyrrolidinemethanol**
102069-83-0

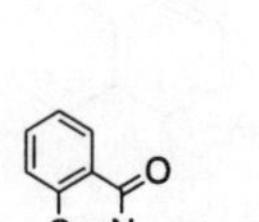

$C_{13}H_{17}NO_3$
Chiral intermediate. mp = 104-107°; $[\alpha]_D^{20}$ = - 92° (c = 1.2, $CHCl_3$). *Sigma-Aldrich Fine Chemicals.*

**3536 (R)-1-(4-Methoxybenzyl)-1,2,3,4,5,6,7,8-octahydroisoquinoline**
30356-08-2
$C_{17}H_{23}NO$
Chiral intermediate. *Sigma-Aldrich Fine Chemicals.*

**3537 2-[[(1E)-2-(Methoxycarbonyl)-1-methylvinyl]amino]-(2R)-2-cyclohexa-1,4-dienylacetic acid potassium salt**
85896-06-6 288-783-7

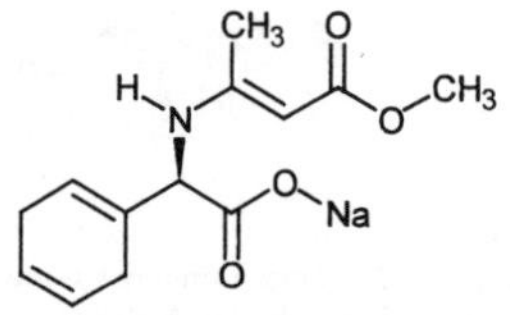

$C_{14}H_{19}KNO_4$
Cyclohexα-1,4-diene-1-acetic acid, α-[(3-ethoxy-1-methyl-3-oxo-1-propenyl)amino]-, monopotassium salt, (R)-; Potassium (R)-α-[(3-ethoxy-1-methyl-3-oxo-1-propenyl)amino]cyclohexa-1,4-diene-1-acetate. Chiral intermediate. *Acros Organics nv; Sigma-Aldrich Fine Chemicals.*

**3538 (-)-N-Methoxycarbonyltryptophan methyl ester**
58635-46-4

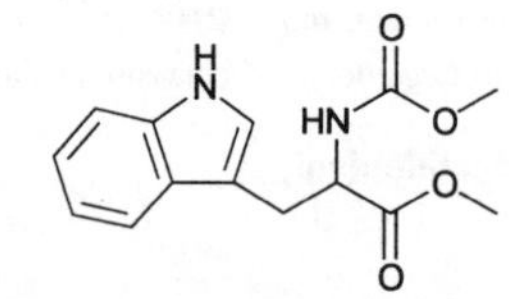

$C_{14}H_{16}N_2O_4$
Chiral intermediate. mp = 99-101°; $[\alpha]_D^{20}$ = - 1.4° (c = 1, $CH_3OH$). *Sigma-Aldrich Fine Chemicals.*

**3539 N-Methoxycarboxy-L-phenylalanine methyl ester**

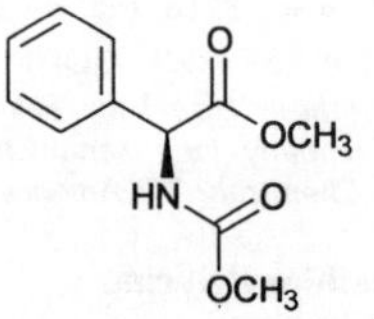

$C_{11}H_{13}NO_4$
Chiral intermediate. *Fischer Chemicals AG.*

**3540 (-)-β-Methoxydiisopinocampheylborane**
99438-28-5

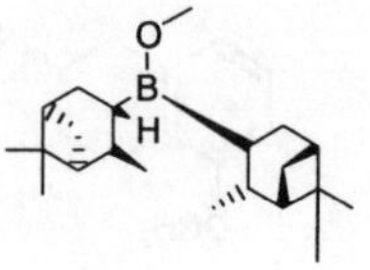

$C_{21}H_{37}BO$
Chiral intermediate. *Sigma-Aldrich Fine Chemicals.*

**3541 (+)-β-Methoxydiisopinocampheylborane**
85134-98-1

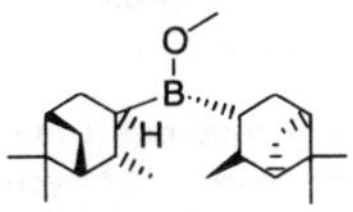

$C_{21}H_{37}BO$
Chiral auxiliary. *Sigma-Aldrich Fine Chemicals.*

**3542 L-(+)-p-Methoxylhomophenylalanine**
82310-97-2

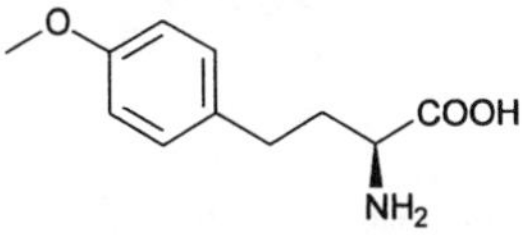

$C_{11}H_{15}NO_3$
Chiral building block. *Kaneka Corporation.*

**3543 (R)-4-Methoxymandelic acid**

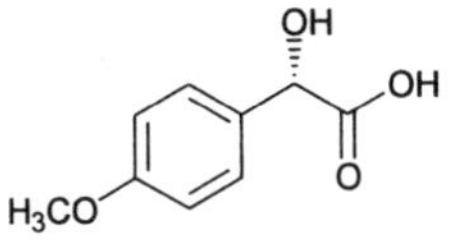

$C_9H_{10}O_4$
Benzeneacetic acid, α-hydroxy-4-methoxy-, (R)-; Methoxyphenylglycolic acid; (R)-2-Hydroxy-2-(4-methoxyphenyl)acetic acid; Mandelic acid, p-methoxy-, (R)-. Chiral building block. *Austin Chemical Company, Inc.*

**3544 (S)-4-Methoxymandelic acid**

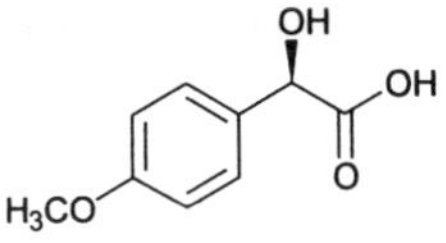

$C_9H_{10}O_4$
Benzeneacetic acid, α-hydroxy-4-methoxy-, (S)-. Chiral building block. *Austin Chemical Company, Inc.*

**3545 (R)-(+)-4-Methoxymandelonitrile**
97070-73-0

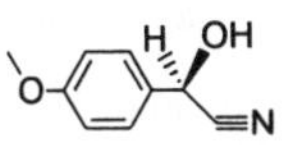

$C_9H_9NO_2$
Chiral building block. mp = 84-87°; $[\alpha]_D^{20}$ = + 47.5° (c = 1, $CHCl_3$). *Sigma-Aldrich Fine Chemicals.*

**3546 Methoxy-α-methylbenzylamine**

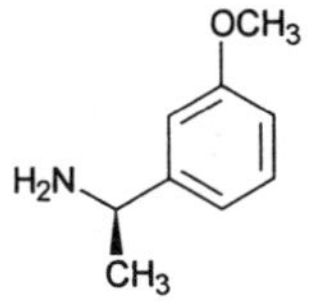

$C_9H_{13}NO$
Chiral building block. *Austin Chemical Company, Inc.*

**3547 (R)-(+)-4-(Methoxymethyl)-1,3-dioxolan-2-one**

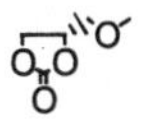

$C_5H_8O_4$
(R)-3-Methoxy-1,2-propanediol cyclic carbonate. Chiral intermediate. $bp_{0.9}$ = 105-106°; d = 1.24; n = 1.437; $[\alpha]_D^{20}$ = + 44° (neat). *Sigma-Aldrich Fine Chemicals.*

**3548 (S)-(-)-4-(Methoxymethyl)-1,3-dioxolan-2-one**
135682-18-7

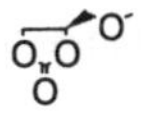

$C_5H_8O_4$
(S)-3-Methoxy-1,2-propanediol cyclic carbonate. Chiral intermediate. $bp_{0.9}$ = 105-106°; d = 1.24; n = 1.437; $[\alpha]_D^{20}$ = - 44° (neat). *Sigma-Aldrich Fine Chemicals.*

**3549 (R)-2-(1-Methoxy-1-methylethoxy)-butanediol**

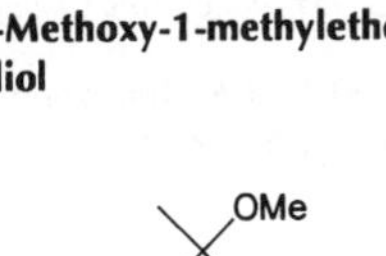

$C_8H_{18}O_4$
Butane-1,4-diol, 2-(1-methoxy-1-methylethoxy)-, (R)-. Chiral building block. *Synthon Chiragenics Corporation.*

**3550 (S)-2-(1-Methoxy-1-methylethoxy)-butanediol**
66348-33-2

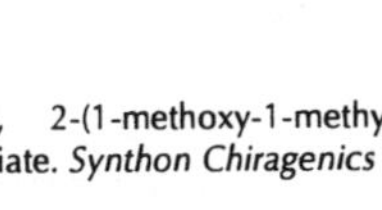

$C_8H_{18}O_4$
Butane-1,4-diol, 2-(1-methoxy-1-methylethoxy)-, (S)-. Chiral intermediate. *Synthon Chiragenics Corporation.*

**3551 (R)-3-(1-Methoxy-1-methylethoxy)-γ-butyolactone**

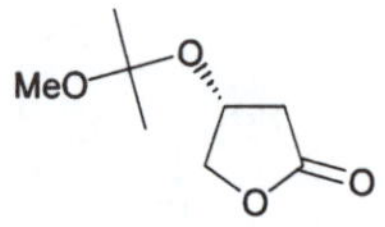

$C_8H_{14}O_4$
Furan-2(3H)-one, dihydro-3-(1-methoxy-1-methylethoxy)-, (R)-; Chiral building block. *Synthon Chiragenics Corporation.*

**3552 (S)-3-(1-Methoxy-1-methylethoxy)-γ-butyolactone**

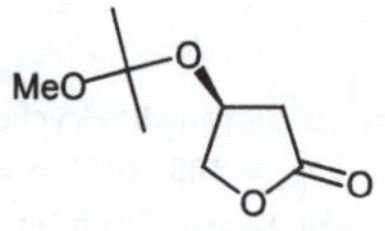

$C_8H_{14}O_4$
Furan-2(3H)-one, dihydro-3-(1-methoxy-1-methylethoxy)-, (S)-. Chiral intermediate. *Synthon Chiragenics Corporation.*

**3553 (S)-2-(1-Methoxy-1-methylethyl)pyrrolidine**
118971-00-9

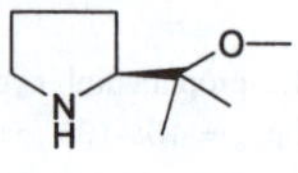

$C_8H_{17}NO$
Chiral intermediate. *Acros Organics nv.*

**3554 (S)-(+)-(Methoxymethyl)phenethylamine hydrochloride**
64715-81-7

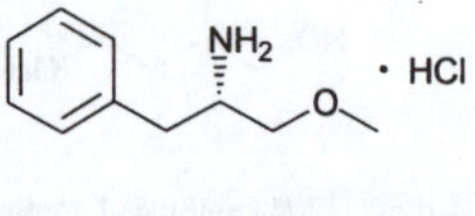

$C_{10}H_{16}ClNO$
Chiral building block. mp = 149-151°; $[\alpha]_D^{20}$ = + 20° (c = 2.5, EtOH). *Sigma-Aldrich Fine Chemicals.*

**3555 (4S,5S)-(-)-4-Methoxymethyl-2-methyl-5-phenyl-2-oxazoline**
52075-14-6 257-641-6

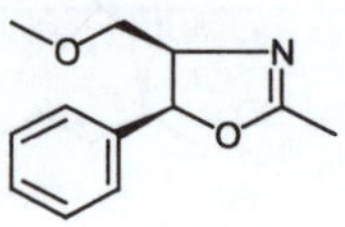

$C_{12}H_{15}NO_2$
Oxazole, 4,5-dihydro-4-(methoxymethyl)-2-methyl-5-phenyl-, (4S,5S)-; Meyer's oxazoline. Chiral intermediate. $bp_{0.05}$ = 79-82°; d = 1.07; n = 1.516; $[\alpha]^{22}$ = - 113.2° (c = 10.5, $CHCl_3$). *Acros Organics nv; Sigma-Aldrich Fine Chemicals; TCI America.*

**3556 (R)-(+)-(3-Methoxy-1-methyl-3-oxo-1-propenyl)amino-1,4-cyclohexadiene-1-acetic acid sodium salt**

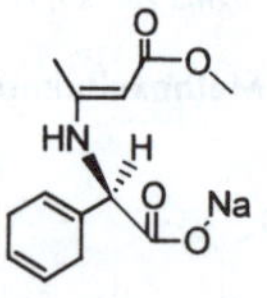

$C_{13}H_{16}NaNO_4$
Chiral intermediate. mp = 240°; $[\alpha]_D^{20}$ = + 83° (c = 1, $H_2O$). *Sigma-Aldrich Fine Chemicals.*

**3557 (S)-(-)-3-Methoxy-2-methyl-3-oxopropylzinc bromide (0.5 molar solution in tetrahydrofuran)**

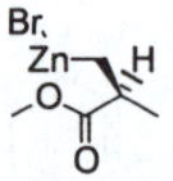

$C_5H_9BrO_2ZN$
Chiral catalyst. d = 0.965. *Sigma-Aldrich Fine Chemicals.*

**3558 (R)-(-)-2-Methoxymethylpyrrolidine**
84025-81-0
$C_6H_{13}NO$
Chiral auxiliary; Chiral intermediate. bp = 153-155°; $[\alpha]_D^{20}$ = - 2.0 ± 0.2° (c = 2, $C_6H_6$). *Kiralchem Ltd.; Omega Chemical Company Inc.; Oxford Asymmetry International plc; Senn Chemicals AG; TCI America.*

**3559 (S)-(+)-2-Methoxymethylpyrrolidine**
63126-47-6

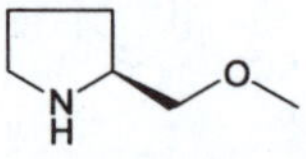

$C_6H_{13}NO$
O-Methyl-L-prolinol. Chiral building block. $bp_{40}$ = 61-62°; d = 0.9330; $n_D^{20}$ = 1.4457; $[\alpha]_D^{20}$ = + 2.2 ± 0.3° (c = 2, $C_6H_6$). *Acros Organics nv; Kiralchem Ltd.; Lancaster Synthesis Ltd.; Omega Chemical Company Inc.; Oxford Asymmetry International plc; Senn Chemicals AG; Sigma-Aldrich Fine Chemicals; TCI America.*

**3560 (R)-(+)-2-(Methoxymethyl)-1-pyrrolidinecarboxaldehyde**
121817-71-8
$C_7H_{13}NO_2$
(R)-(+)-1-Formyl-2-(methoxymethyl)pyrrolidine. Chiral intermediate. *TCI America.*

**3561 (S)-(-)-2-(Methoxymethyl)-1-pyrrolidine-carboxaldehyde**
63126-45-4

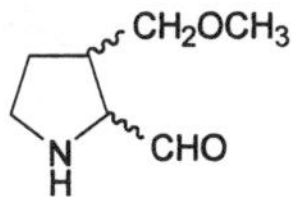

$C_7H_{13}NO_2$
(S)-(-)-1-Formyl-2-(methoxymethyl)pyrrolidine. Chiral intermediate. $bp_{0.25}$ = 67°; d = 1.057; n = 1.4755; $[\alpha]_D^{20}$ = - 34° (c = 2, toluene). *Sigma-Aldrich Fine Chemicals; TCI America.*

**3562 (S)-(+)-2-(6-Methoxy-2-naphthyl)-propionic acid**
22204-53-1 6504(12) 244-838-7

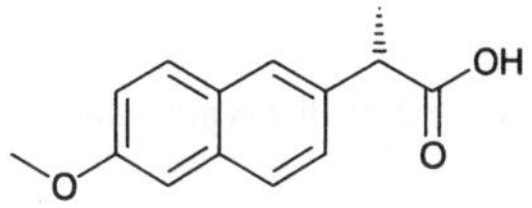

$C_{14}H_{14}O_3$
Naphthalene-2-acetic acid, 6-methoxy-α-methyl-, (S)-; (+)-6-Methoxy-α-methyl-2-naphthaleneacetic acid; Naproxen; Naprosyn; (+)-2-(6-Methoxy-2-naphthyl)-propionic acid; 2-naphthaleneacetic acid, 6-methoxy-α-methyl-,(+). Chiral intermediate. mp = 154-155°; $[\alpha]_D^{20}$ = + 66 ± 2° (c = 1, $CHCl_3$). *Lancaster Synthesis Ltd.; Sigma-Aldrich Fine Chemicals; TCI America.*

**3563 (R)-(-)-α-Methoxyphenylacetic acid**
3966-32-3 223-580-9

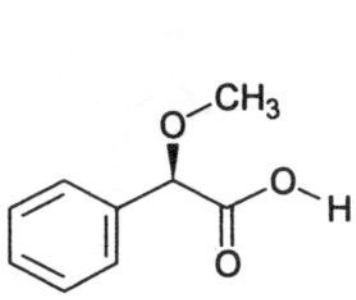

$C_9H_{10}O_3$
Benzeneacetic acid, α-methoxy-, (R)-; O-Methyl-D-mandelic acid; (R)-2-Phenyl-2-methoxyacetic acid; Acetic acid, methoxyphenyl-, (R)-(-)-. Chiral diagnostic reagent; Chiral building block; Resolving agent. mp = 62-65°; $[\alpha]_D^{20}$ = - 160.0 ± 1.0° (c = 1, $CH_3OH$); $[\alpha]_D^{20}$ = - 137 ± 2° (c = 6.4, $CCl_4$); $[\alpha]_D^{20}$ = - 146 ± 3° (c = 0.5, EtOH). *Acros Organics nv; Kiralchem Ltd.; Lancaster Synthesis Ltd.; Norse Laboratories; Omega Chemical Company Inc.; Sigma-Aldrich Fine Chemicals; TCI America; Yamakawa Chemical Industry Co. Ltd.*

**3564 (S)-(+)-α-Methoxyphenylacetic acid**
26164-26-1 247-492-5

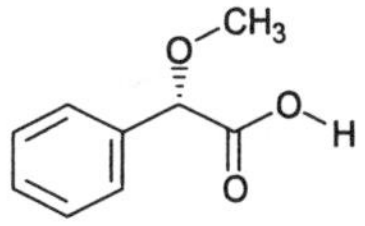

$C_9H_{10}O_3$
Acetic acid, methoxyphenyl-, (S)-; (2S)-O-Methyl-mandelic acid; (S)-α-Methoxybenzeneacetic acid; Benzeneacetic acid, α-methoxy-, (αS)-. Chiral diagnostic reagent; Chiral building block; Resolving agent. mp = 62-65°; $[\alpha]_D^{20}$ = + 146 ± 3° (c = 0.5, EtOH). *Kiralchem Ltd.; Lancaster Synthesis Ltd.; Norse Laboratories; Omega Chemical Company Inc.; Sigma-Aldrich Fine Chemicals; TCI America; Yamakawa Chemical Industry Co. Ltd.*

**3565 (R)-(-)-2-Methoxy-2-phenylethanol**
17628-72-7

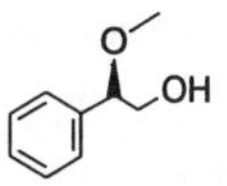

$C_9H_{12}O_2$
(R)-(-)-β-Methoxyphenethyl alcohol. Chiral building block. $bp_{0.1}$ = 65°; d = 1.0500; $n_D^{20}$ = 1.5200; $[\alpha]_D^{20}$ = - 133 ± 2° (c = 1, acetone). *Lancaster Synthesis Ltd.; Sigma-Aldrich Fine Chemicals.*

**3566 (S)-(+)-2-Methoxy-2-phenylethanol**
66051-01-2

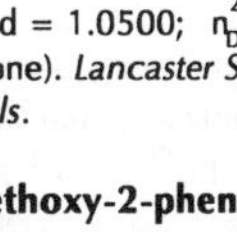

$C_9H_{12}O_2$
(S)-(+)-β-Methoxyphenethyl alcohol. Chiral building block. $bp_{0.1}$ = 90-92°; d = 1.0540; $n_D^{20}$ = 1.5200; $[\alpha]_D^{20}$ = + 133 ± 3° (c = 1, acetone). *Lancaster Synthesis Ltd.; Sigma-Aldrich Fine Chemicals.*

**3567 (R)-1-(2-Methoxyphenyl)ethylamine**
68285-23-4

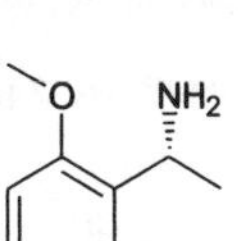

$C_9H_{13}NO$
Chiral intermediate. $bp_{115}$ = 25°. *BASF Aktiengesellschaft.*

**3568 (S)-1-(2-Methoxyphenyl)ethylamine**
68285-24-5

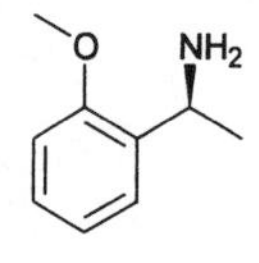

$C_9H_{13}NO$
Chiral intermediate. $bp_{115}$ = 25°. *BASF Aktiengesellschaft.*

**3569 (R)-1-(3-Methoxyphenyl)ethylamine**
88196-70-7
$C_9H_{13}NO$
(R)-m-Methoxy-α-methylbenzylamine. Chiral intermediate. $n_D^{20}$ = 1.5340. *Arran Chemical Company Ltd.; BASF Aktiengesellschaft; Lancaster Synthesis Ltd.; Yamakawa Chemical Industry Co. Ltd.*

**3570 (S)-1-(3-Methoxyphenyl)ethylamine**
82796-69-8
$C_9H_{13}NO$
(S)-m-Methoxy-α-methylbenzylamine. Chiral intermediate. $bp_{0.38}$ = 66°; $n_D^{20}$ = 1.5330. *Arran Chemical Company Ltd.; BASF Aktiengesellschaft; Lancaster Synthesis Ltd.; Yamakawa Chemical Industry Co. Ltd.*

**3571 (R)-(+)-1-(4-Methoxyphenyl)ethylamine**
22038-86-4
$C_9H_{13}NO$
(R)-(+)-p-Methoxy-α-methylbenzylamine; (R)-1-(4-Methoxyphenyl)ethylamine. Chiral intermediate. $bp_{0.38}$ = 65°; $n_D^{20}$ = 1.5328; $[\alpha]_D^{20}$ = + 34 ± 1° (neat). *BASF Aktiengesellschaft; Lancaster Synthesis Ltd.*

**3572 (S)-(-)-1-(4-Methoxyphenyl)ethylamine**
41851-59-6
$C_9H_{13}NO$
(S)-p-Methoxy-α-methylbenzylamine. Chiral intermediate. bp = 65°/0.38mm; $n_D^{20}$ = 1.5330; $[\alpha]_D^{20}$ = -34 ± 1°(neat). *BASF Aktiengesellschaft; Lancaster Synthesis Ltd.*

**3573 (R)-(-)-1-Methoxy-2-propanol**
4984-22-9

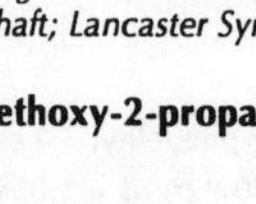

$C_4H_{10}O_2$
Chiral building block. *Acros Organics nv.*

**3574 (S)-(+)-1-Methoxy-2-propanol**
26550-55-0

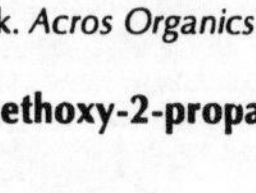

$C_4H_{10}O_2$
Chiral building block. *Acros Organics nv.*

**3575 (R)-(-)-2-Methoxypropanol**
6131-59-5

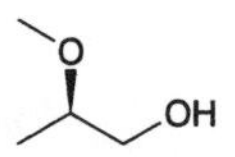

$C_4H_{10}O_2$
Chiral building block. *Acros Organics nv.*

**3576 (S)-(+)-2-Methoxypropanol**

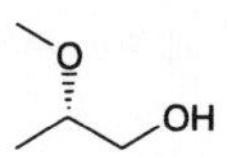

$C_4H_{10}O_2$
Propan-1-ol, 2-methoxy-, (S)-; (S)-2-Methoxy-1-hydroxypropane. Chiral building block. *Acros Organics nv.*

**3577 (R)-(-)-2-Methoxypropionamide**

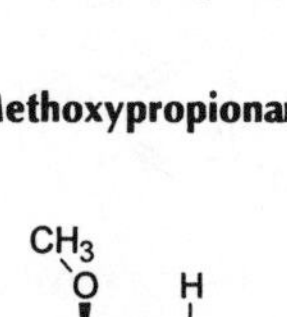

$C_4H_9NO_2$
Chiral building block. *Acros Organics nv.*

**3578 (S)-(+)-2-Methoxypropionamide**

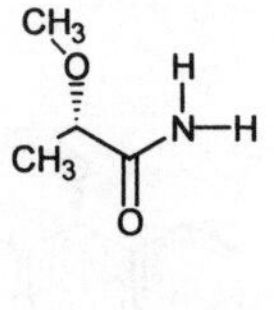

$C_4H_9NO_2$
Chiral building block. *Acros Organics nv.*

**3579 (R)-(-)-2-Methoxypropionic acid**
23943-96-6

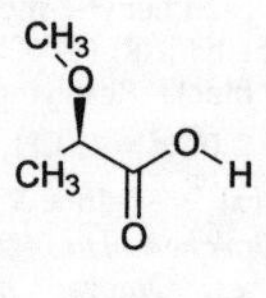

$C_4H_8O_3$
Chiral building block. *Acros Organics nv.*

**3580 (S)-(+)-2-Methoxypropionic acid**
23953-00-6

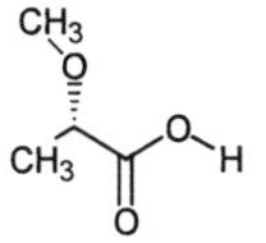

$C_4H_8O_3$
Chiral building block. *Acros Organics nv.*

**3581 (R)-(-)-2-Methoxypropionitrile**

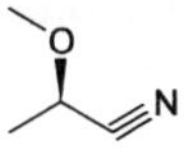

$C_4H_7NO$
Chiral building block. *Acros Organics nv.*

**3582 (S)-(+)-2-Methoxypropionitrile**

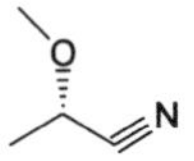

$C_4H_7NO$
Chiral building block. *Acros Organics nv.*

**3583 (S)-1-Methoxy-2-propylamine**
99636-32-5
$C_4H_{11}NO$
Propan-2-amine, 1-methoxy-, (2S)-; (S)-2-Amino-1-methoxypropane. Listed on TSCA. Chiral building block. mp = - 95°; $bp_{68}$ = 92-94°; $n_D^{20}$ = 1.4044. *Lancaster Synthesis Ltd.*

**3584 (R)-(+)-α-Methoxy-α-trifluoromethyl-phenylacetic acid**
20445-31-2 243-829-5

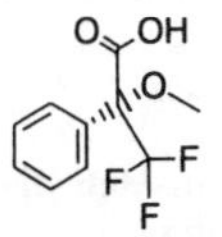

$C_{10}H_9F_3O_3$
Benzeneacetic acid, α-methoxy-α-(trifluoromethyl)-, (R)-; (+)-Mosher's acid; (+)-3,3,3-Trifluoro-2-methoxy-2-phenylpropionic acid; (+)-MTPA; Hydratropic acid, β,β,β-trifluoro-α-methoxy-, (+)-; (R)-(+)-α-Methoxy-α-(trifluoromethyl)phenylacetic acid. Chiral intermediate; Resolving agent. mp = 43-45°; $bp_1$ = 105-107°; $[\alpha]_D^{20}$= + 72 ± 2° (c = 2, $CH_3OH$). *Lancaster Synthesis Ltd.; Norse Laboratories; Oxford Asymmetry International plc; Sigma-Aldrich Fine Chemicals.*

**3585 (S)-(-)-α-Methoxy-α-trifluoromethyl-phenylacetic acid**
17257-71-5 241-292-1

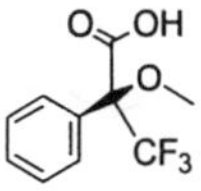

$C_{10}H_9F_3O_3$
Benzeneacetic acid, α-methoxy-α-(trifluoromethyl)-, (S)-; (-)-Mosher's acid; (S)-MTPA; (-)-MTPA. Chiral diagnostic reagent. mp = 43-45°; $bp_{0.1}$ = 90°; $[\alpha]_D^{20}$= - 72 ± 2° (c = 2, $CH_3OH$). *Lancaster Synthesis Ltd.; Norse Laboratories; Oxford Asymmetry International plc; Sigma-Aldrich Fine Chemicals.*

**3586 (R)-(-)-α-Methoxy-α-trifluoromethyl-phenylacetyl chloride**
39637-99-5

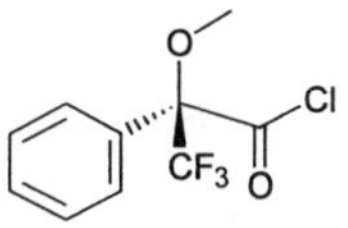

$C_{10}H_8ClF_3O_2$
(R)-(-)-Mosher's acid chloride; (R)-(-)-MTPA-Cl. Diagnostic reagent. bp = 213-214°; d = 1.3530; $n_D^{20}$= 1.4690; $[\alpha]_D^{20}$= - 137 ± 2° (c = 6.4, $CCl_4$). *Acros Organics nv; Lancaster Synthesis Ltd.; Sigma-Aldrich Fine Chemicals.*

**3587 (S)-(+)-α-Methoxy-α-trifluoromethyl-phenylacetyl chloride**
20445-33-4

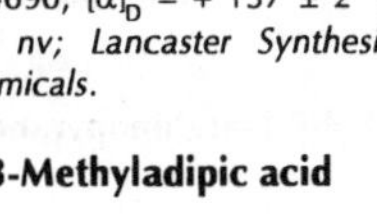

$C_{10}H_8ClF_3O_2$
(S)-(+)-Mosher's acid chloride; (S)-(+)-MTPA-Cl. Chiral diagnostic reagent. bp = 213-214°; $bp_{1.4}$ = 68-72°; d = 1.3530; $n_D^{20}$= 1.4690; $[\alpha]_D^{20}$= + 137 ± 2° (c = 6.4, $CCl_4$). *Acros Organics nv; Lancaster Synthesis Ltd.; Sigma-Aldrich Fine Chemicals.*

**3588 (R)-(+)-3-Methyladipic acid**
623-82-5 210-816-0

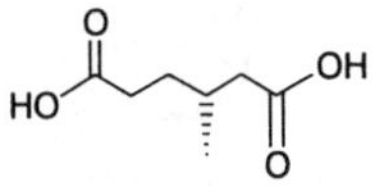

$C_7H_{12}O_4$
(+)-3-Methylhexanedioic acid. Chiral building block. mp = 81-84°; $bp_{30}$ = 230°; $[\alpha]^{21}$ = + 7.2° (c = 5, $CH_3OH$). *Sigma-Aldrich Fine Chemicals.*

**3589 (R)-2-(Methylamino)-1-phenylethanol**

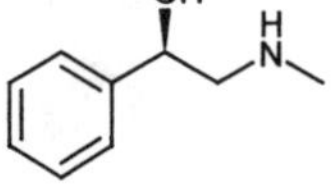

$C_9H_{13}NO$
Benzenemethanol, α-[(methylamino)methyl]-, (R)-; (R)-α-[(methylamino)methyl]benzyl alcohol; (R)-Halostachine; (R)-N-Methyl-β-hydroxyphenylethylamine. Chiral building block. mp = 44-46°; $[\alpha]_D^{20}$ = - 44.5 ± 0.5° (c = 2, $CH_3OH$). *Kiralchem Ltd.*

**3590 (S)-2-(Methylamino)-1-phenylethanol**

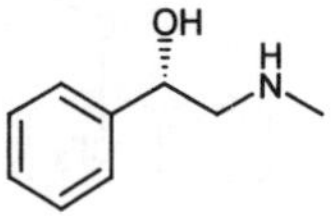

$C_9H_{13}NO$
Benzenemethanol, α-[(methylamino)methyl]-, (S)-; (S)-α-[(Methylamino)methyl]benzyl alcohol; (S)-Halostachine; (S)-N-Methyl-β-hydroxyphenylethylamine. Chiral building block. mp = 44-46°; $[\alpha]_D^{20}$ = + 44.5 ± 0.5° (c = 2, $CH_3OH$). *Kiralchem Ltd.*

**3591 Methyl-β-D-arabinopyranoside**
5328-63-2 226-215-1

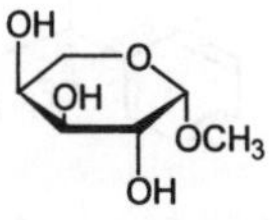

$C_6H_{12}O_5$
Arabinopyranoside, methyl, β-D-. Chiral intermediate. *Senn Chemicals AG.*

**3592 Methyl-β-L-arabinopyranoside**
1825-00-9 217-362-2

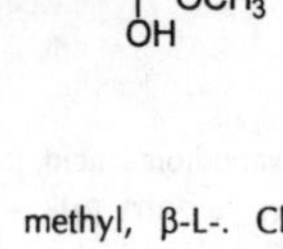

$C_6H_{12}O_5$
Arabinopyranoside, methyl, β-L-. Chiral intermediate. *Senn Chemicals AG.*

**3593 N-Methyl-D-aspartic acid**
6384-92-5

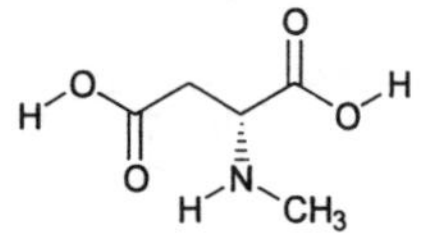

$C_5H_9NO_4$
Chiral building block. *Acros Organics nv; Senn Chemicals AG.*

**3594 N-Methyl-L-aspartic acid**
4226-18-0

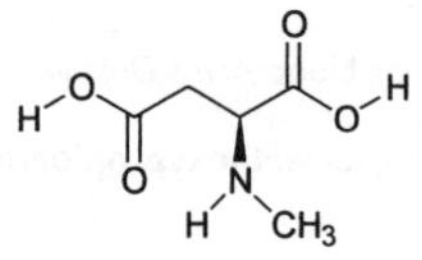

$C_5H_9NO_4$
Chiral intermediate. *Acros Organics nv.*

**3595 (+)-Methylbenzylamine**
3886-69-9 223-423-4

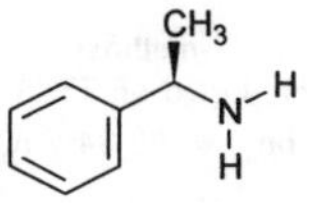

$C_8H_{11}N$
1-Phenylethylamine. Listed on TSCA. Chiral building block; Chiral auxiliary; Resolving agent. $bp_{765}$ = 180-181°; d = 0.94; n = 1.526; $[\alpha]^{23}$ = + 38° (neat). *Sigma-Aldrich Fine Chemicals.*

**3596 (R)-(+)-2-(α-Methylbenzylamino)-5-nitropyridine**
64138-65-4
$C_{13}H_{13}N_3O_2$
R-(+)-5-Nitro-2-(α-methylbenzylamino)pyridine. Chiral intermediate. mp = 83°. *TCI America.*

**3597 (S)-(-)-2-(α-Methylbenzylamino)-5-nitropyridine**
84249-39-8
$C_{13}H_{13}N_3O_2$
(S)-(-)-5-Nitro-2-(α-methylbenzylamino)pyridine. Chiral intermediate. mp = 84°. *TCI America.*

**3598 (R)-α-(4-Methylbenzyl)benzylamine**

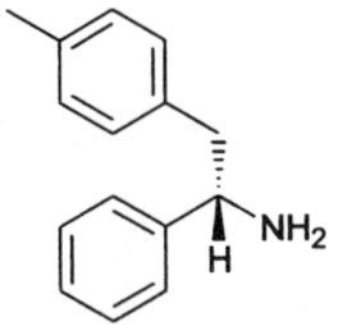

$C_{15}H_{17}N$
Chiral building block. *Sumitomo Chemcial Co. Ltd.*

**3599 (S)-α-(4-Methylbenzyl)benzylamine**

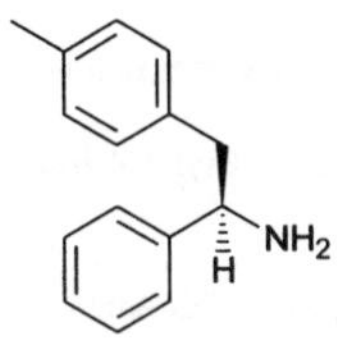

$C_{15}H_{17}N$
Chiral building block. *Sumitomo Chemcial Co. Ltd.*

**3600 (R)-α-(4-Methylbenzyl)-N,N-dimethylbenzylamine**

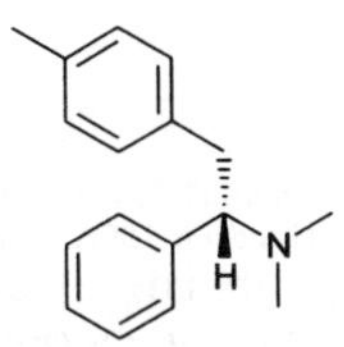

$C_{17}H_{21}N$
Chiral building block. *Sumitomo Chemcial Co. Ltd.*

**3601 (S)-α-(4-Methylbenzyl)-N,N-dimethylbenzylamine**

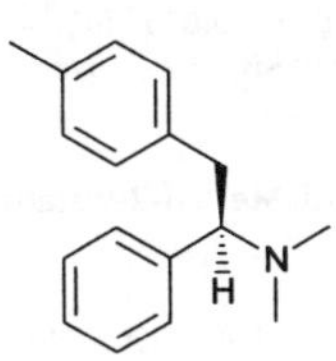

$C_{17}H_{21}N$
Chiral building block. *Sumitomo Chemcial Co. Ltd.*

**3602 (R)-α-(4-Methylbenzyl)-N-hydroxyethylbenzylamine**

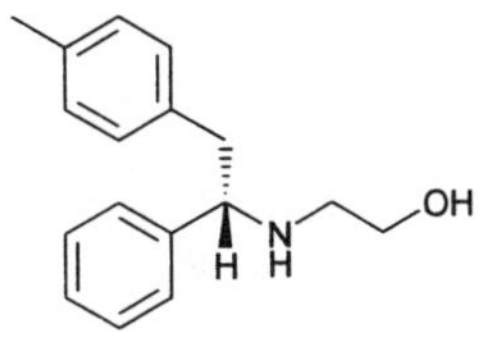

$C_{17}H_{20}NO$
Chiral building block. *Sumitomo Chemcial Co. Ltd.*

**3603 (S)-α-(4-Methylbenzyl)-N-hydroxyethylbenzylamine**

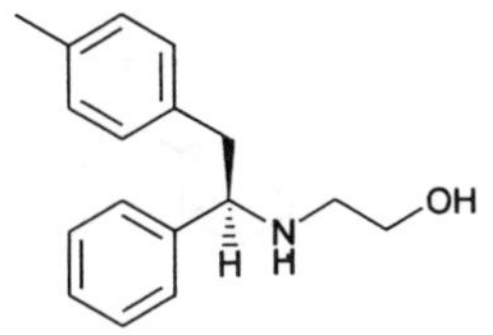

$C_{17}H_{2014}O$
Chiral building block. *Sumitomo Chemcial Co. Ltd.*

**3604 Methyl 4,6-O-Benzylidene-α-D-glucopyranoside**
3162-96-7 221-615-2

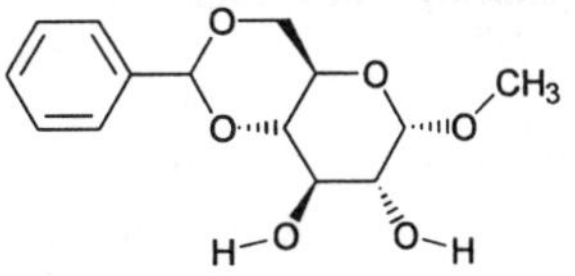

$C_{14}H_{18}O_6$
Glucopyranoside, methyl 4,6-O-benzylidene-, α-D-; (+)-(4,6-O-Benzylidene)methyl-α-D-glucopyranoside. Chiral intermediate. mp = 165°; $[\alpha]_D^{20}$ = + 109° (c = 2, $CHCl_3$). *Acros Organics nv; Pfanstiehl Laboratories, Inc.; Senn Chemicals AG; Sigma-Aldrich Fine Chemicals.*

**3605 (R)-(+)α-Methylbenzylisocyanide**
21872-33-3
$C_9H_9N$
Chiral building block. *Sigma-Aldrich Fine Chemicals.*

**3606 (R)-α-Methyl-3-bromobenzylamine**

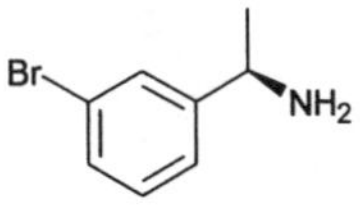

$C_8H_{10}BrN$
Benzenemethanamine, 3-bromo-α-methyl-, (R)-; (R)-3-

Bromo-α-methylbenzylamine; (+)-1-(3-Bromophenyl)-ethylamine. Chiral building block. *Chiragene, Inc.*

**3607 (S)-α-Methyl-3-bromobenzylamine**

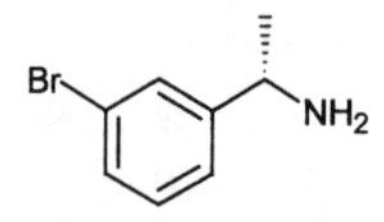

$C_8H_{10}BrN$
Benzenemethanamine, 3-bromo-α-methyl-, (S)-; (S)-3-Bromo-α-methylbenzylamine; (-)-1-(3-Bromophenyl)-ethylamine. Chiral building block. *Chiragene, Inc.*

**3608 (S)-(-)-α-Methyl-4-bromobenzylamine**

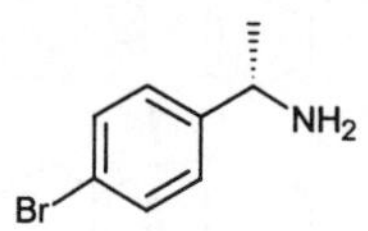

$C_8H_{10}BrN$
Benzenemethanamine, 4-bromo-α-methyl-, (S)-; (S)-4-Bromo-α-methylbenzylamine; (-)-1-(4-Bromophenyl)-ethylamine. Chiral building block. *FineTech, Ltd.; Norse Laboratories.*

**3609 Methyl (S)-(-)-4-bromo-3-tert-butyl-dimethylsilyloxybutanoate**
101703-35-9

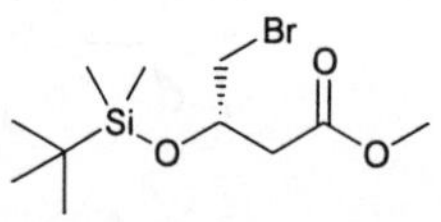

$C_{11}H_{23}BrO_3Si$
Chiral intermediate. *Acros Organics nv.*

**3610 Methyl (R)-(+)-3-bromo-2-methyl-propionate**
110556-33-7

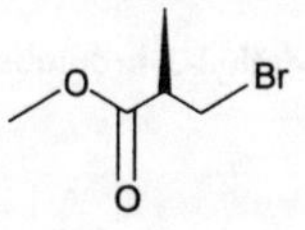

$C_5H_9BrO_2$
Chiral building block. $bp_{37}$ = 85°; d = 1.422; n = 1.455; $[\alpha]_D^{20}$ = + 16° (neat). *Sigma-Aldrich Fine Chemicals.*

**3611 Methyl (S)-(-)-3-bromo-2-methylpropionate**
98190-85-3

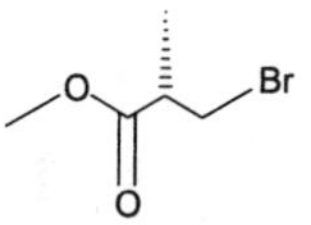

$C_5H_9BrO_2$
Chiral building block. $bp_{62}$ = 97.5-98.5°; d = 1.422; n = 1.455; $[\alpha]_D^{20}$ = - 16° (neat). *Sigma-Aldrich Fine Chemicals.*

**3612 (R)-(+)-2-Methyl-1,4-butanediol**
22644-28-6
$C_5H_{12}O_2$
Chiral building block. *TCI America.*

**3613 (S)-(-)-2-Methyl-1,4-butanediol**
70423-38-0
$C_5H_{12}O_2$
Chiral building block. *TCI America.*

**3614 (S)-(-)-2-Methyl-1-butanol**
1565-80-6 5952(11) 216-366-1

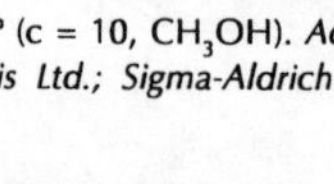

$C_5H_{12}O$
Butan-1-ol, 2-methyl-, (2S)-; active-Amyl alcohol; (-)-sec-Butylcarbinol. Listed on TSCA. Chiral building block. mp = - 70°; bp = 129-131°; d = 0.8190; $n_D^{20}$ = 1.4110; $[\alpha]_D^{20}$ = - 6.3 ± 0.3° (c = 10, $CH_3OH$). *Acros Organics nv; Lancaster Synthesis Ltd.; Sigma-Aldrich Fine Chemicals; TCI America.*

**3615 (R)-(-)-3-Methyl-2-butanol**
1572-93-6
$C_5H_{12}O$
Butan-2-ol, 3-methyl-, (R)-(-)-; (R)-(-)-Isopropyl methyl carbinol. Chiral building block. bp = 114-115°; d = 0.8100; $n_D^{20}$ = 1.4075; $[\alpha]_D^{20}$ = - 3.9 ± 0.5° (neat). *Lancaster Synthesis Ltd.; Norse Laboratories; TCI America.*

**3616 (S)-(+)-3-Methyl-2-butanol**
1517-66-4
$C_3H_5ClO_2$
Butan-2-ol, 3-methyl-, (S)-; (S)-(+)-Isopropyl methyl carbinol; (S)-1-Isopropylethanol. Chiral building block. bp = 111-112°; d = 0.8100; $n_D^{20}$ = 1.4090; $[\alpha]_D^{20}$ = + 5 ± 0.5° (neat). *Lancaster Synthesis Ltd.; Norse Laboratories; TCI America.*

**3617 (S)-(+)-4-(2-Methylbutoxy)-4-biphenyl-carbonitrile**

261-360-4

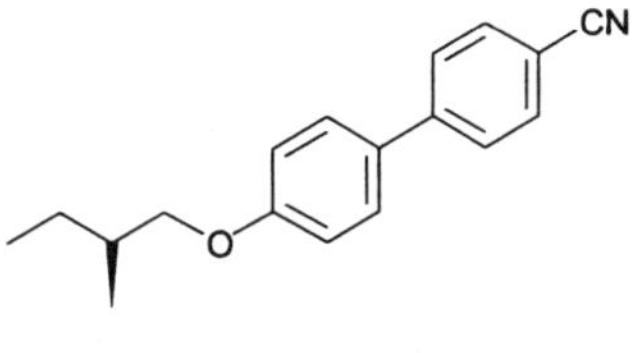

$C_{18}H_{19}NO$

Listed on TSCA. Chiral intermediate. mp = 53-55°; $[\alpha]_D^{20}$ = + 12° (c = 1, $CH_3OH$). *Sigma-Aldrich Fine Chemicals.*

**3618 Methyl (R)-(+)-3-(tert-butoxycarbonyl)-2,2-dimethyl-4-oxazolidinecarboxylate**

95715-86-9

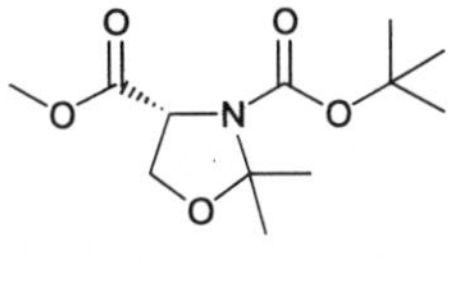

$C_{12}H_{21}NO_5$

Chiral intermediate. $bp_{0.8}$ = 89°; d = 1.082; n = 1.444; $[\alpha]_D^{20}$ = + 54° (c = 1.3, $CHCl_3$). *Sigma-Aldrich Fine Chemicals.*

**3619 (S)-(-)-2-Methylbutylamine**

20626-52-2

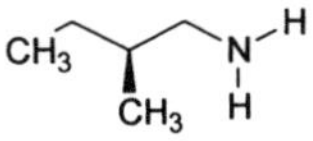

$C_5H_{13}N$

(S)-(-)-1-Amino-2-methylbutane. Chiral building block. bp = 96-97°; $bp_{12}$ = 40-45°; d = 0.7560; $n_D^{20}$ = 1.4120; $[\alpha]_D^{20}$ = - 5.9 ± 0.5° (neat); $[\alpha]_D^{20}$ = - 3.4 (c = 1, $H_2O$). *Acros Organics nv; Kiralchem Ltd.; Lancaster Synthesis Ltd.; Norse Laboratories; Omega Chemical Company Inc.; Sigma-Aldrich Fine Chemicals.*

**3620 (R)-(-)-3-Methyl-2-butylamine**

34701-33-2

$C_5H_{13}N$

(R)-(-)-2-Amino-3-methylbutane. Chiral building block. bp = 85-87°; $n_D^{20}$ = 1.4060; $[\alpha]_D^{20}$ = - 5.2 ± 0.5 ° (neat). *BASF Aktiengesellschaft; Lancaster Synthesis Ltd.*

**3621 (S)-(+)-3-Methyl-2-butylamine**

22526-46-1

$C_5H_{13}N$

(S)-(+)-2-Amino-3-methylbutane. Chiral intermediate. bp = 85-87°; $n_D^{20}$ = 1.4060; $[\alpha]_D^{20}$ = + 5.2 ± 0.5° (neat). *BASF Aktiengesellschaft; Lancaster Synthesis Ltd.*

**3622 (+)-2-Methylbutyl-p-aminocinnamate**

62742-50-1

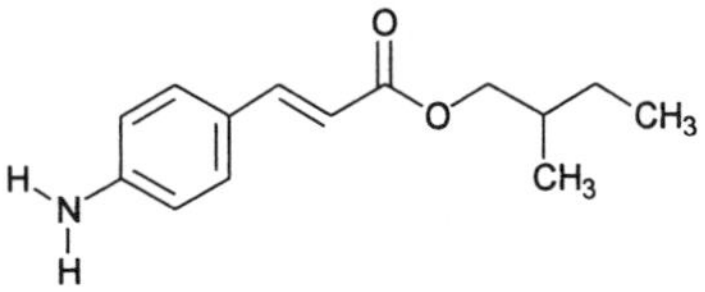

$C_{14}H_{19}NO_2$

Prop-2-enoic acid, 3-(4-aminophenyl)-, (2R)-2-methylbutyl ester, (2E)-. Listed on TSCA. Chiral intermediate. *Acros Organics nv.*

**3623 (S)-(+)-4-(2-Methylbutyl)-4-biphenyl-carbonitrile**

264-464-8

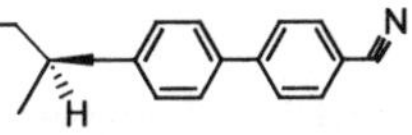

$C_{18}H_{19}N$

Listed on TSCA. Chiral intermediate. $bp_{0.5}$ = 140-150°; d = 1.01; n = 1.594; $[\alpha]_D^{20}$ = + 12° (c = 10, $CHCl_3$). *Sigma-Aldrich Fine Chemicals.*

**3624 Methyl (R)-(+)-3-(tert-butyldimethyl-silyloxy)-5-oxo-1-cyclopenten-1-heptanoate**

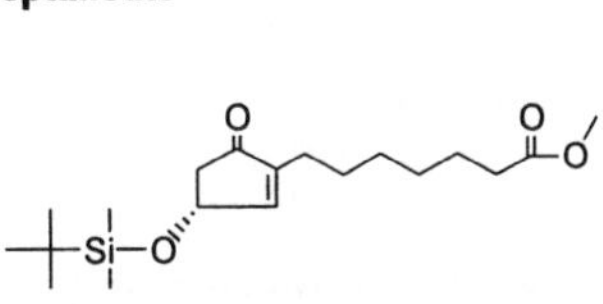

$C_{19}H_{34}O_4Si$

Chiral intermediate. bp = 210°; d = 0.977; n = 1.467; $[\alpha]^{24}$ = + 14° (c = 1, $CHCl_3$). *Sigma-Aldrich Fine Chemicals.*

**3625 Methyl (6R)-[(tert-butyl)diphenylsilyloxy]-(2E,4E,8Z)-tetradecatrienoate**

116279-81-3

$C_{31}H_{42}O_3Si$

Chiral intermediate. *Ultrafine.*

**3626 (S)-(+)-2-Methylbutyl methanesulfonate**

104418-40-8

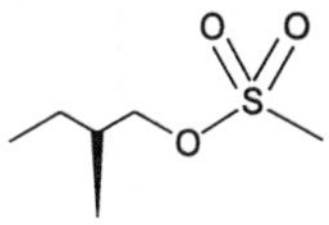

$C_6H_{14}O_3S$

Chiral building block. d = 1.08; n = 1.432; $[\alpha]_D^{20}$ = + 2.7° (neat). *Sigma-Aldrich Fine Chemicals.*

**3627 (+)-2-Methylbutyl p-[(p-methoxybenzylidene)amino]cinnamate**
24140-30-5

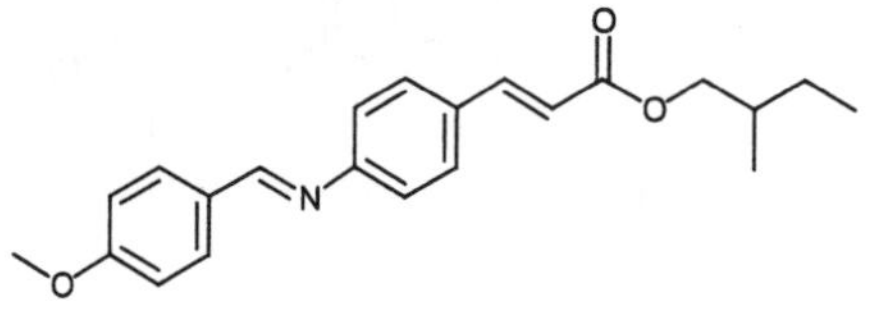

$C_{22}H_{25}NO_3$
Prop-2-enoic acid, 3-[4-[(E)-[(4-methoxyphenyl)methylene]amino]phenyl]-, (2S)-2-methylbutyl ester, (2E)-; 3-(4-(((4-Methoxyphenyl)methylene)amino)phenyl) 2-propenoic acid, 2-methylbutyl ester; act-Amyl p-anisalaminocinnamate; Cinnamic acid, p-[(p-methoxybenzylidene)amino]-, 2-methylbutyl ester, (S)-(+)-. Listed on TSCA. Chiral intermediate. *Acros Organics nv.*

**3628 Methyl 7-[3(R)-tert-butylsilyloxy-5-oxo-cyclopent-1-enyl]heptanoate**
41138-69-6

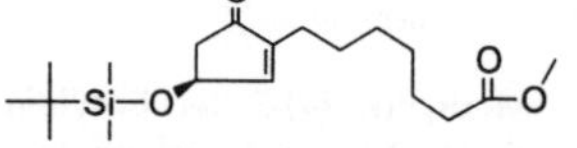

$C_{19}H_{34}O_4Si$
Chiral intermediate. bp = 210°; d = 0.977; n = 1.467; $[\alpha]_D^{20}$= - 14° (c = 1, $CHCl_3$). *Nissan Chemical Industries, Ltd.; Sigma-Aldrich Fine Chemicals.*

**3629 (R)-(-)-2-Methylbutyric acid**

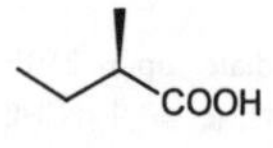

COOH

$C_5H_{10}O_2$
Butanoic acid, 2-methyl-, (R)-; (R)-Methylethylacetic acid; (R)-Ethylmethylacetic acid. Chiral building block. *Nagase & Co., Ltd.*

**3630 (S)-(+)-2-Methylbutyric acid**
1730-91-2

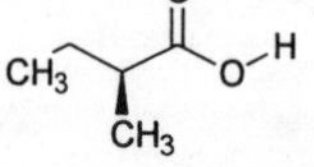

$C_5H_{10}O_2$
Chiral building block. $bp_{15}$ = 78-80°; d = 0.9380; $n_D^{20}$ = 1.4060; $[\alpha]_D^{20}$= + 19.5 ± 0.5° (neat). *Acros Organics nv; Lancaster Synthesis Ltd.; Norse Laboratories; Sigma-Aldrich Fine Chemicals.*

**3631 (S)-(+)-2-Methylbutyric anhydride**
84131-91-9

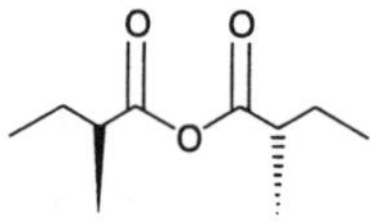

$C_{10}H_{18}O_3$
Chiral building block. $bp_{0.2}$ = 60°; d = 0.934; n = 1.418; $[\alpha]_D^{20}$= + 35° (neat). *Sigma-Aldrich Fine Chemicals.*

**3632 (S)-(+)-2-Methylbutyronitrile**
25570-03-0

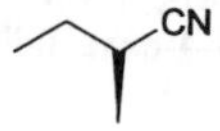

$C_5H_9N$
Chiral building block. bp = 125-126°; d = 0.786; n = 1.39; $[\alpha]_D^{20}$= + 31° (neat). *Sigma-Aldrich Fine Chemicals.*

**3633 (S)-α-methyl-4-carboxyphenylglycine**

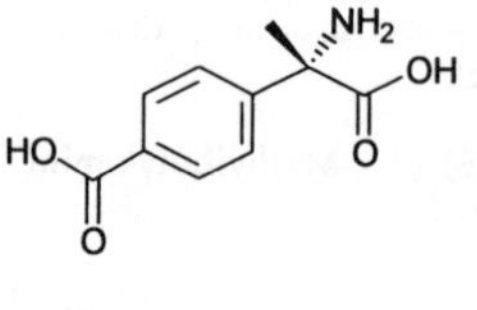

$C_{10}H_{11}NO_4$
Chiral building block. *Acros Organics nv.*

**3634 (R)-Methyl-CBS-oxazaborolidine**
112022-83-0

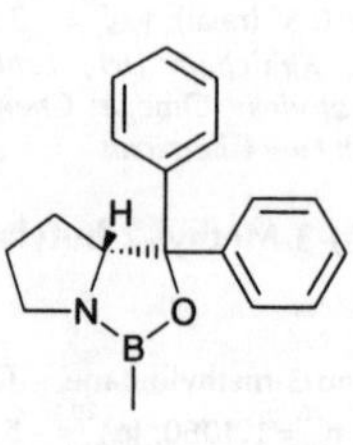

$C_{18}H_{20}BNO$
(R)-3,3-Diphenyl-1-methylpyrrolidino[1,2-c] 1,3,2-oxazaborole; (R)-MeCBS; (R)-Corey Catalyst; (R)-5,5-Diphenyl-2-methyl-3,4-propano-1,3,2-oxazaborolidine. Chiral catalyst. bp = 111°; d = 0.9250. *Callery Chemical Company; Lancaster Synthesis Ltd.; Sigma-Aldrich Fine Chemicals; Strem Chemicals, Inc.; TCI America.*

**3635 (S)-2-Methyl-CBS-oxazaborolidine**
112022-81-8

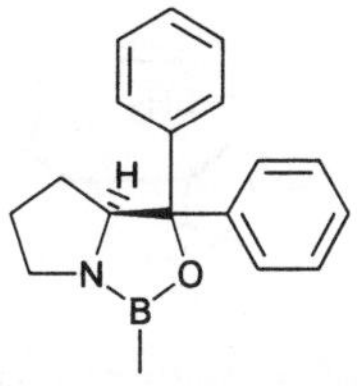

$C_{18}H_{20}BNO$
(S)-3,3-Diphenyl-1-methylpyrrolidino[1,2,c] 1,3,2-oxazaborole; (S)-5,5-Diphenyl-2-methyl-3,4-propano-1,3,2-oxazaborolidine; (S)-MeCBS; (S)-Corey Catalyst. Chiral catalyst. bp = 111°; d = 0.925. *Callery Chemical Company; Lancaster Synthesis Ltd.; Sigma-Aldrich Fine Chemicals; Strem Chemicals, Inc.; TCI America.*

**3636 (R)-α-Methyl-4-chlorobenzylamine**

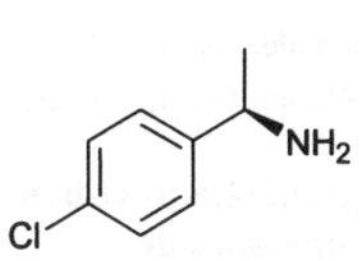

$C_8H_{10}ClN$
Benzenemethanamine, 4-chloro-α-methyl-, (R)-; (R)-4-Chloro-α-methylbenzylamine; (+)-1-(4-Chlorophenyl)-ethylamine. Chiral building block. *Chiragene, Inc.*

**3637 (S)-α-Methyl-4-chlorobenzylamine**
$C_8H_{10}ClN$
Benzenemethanamine, 4-chloro-α-methyl-, (S)-; (S)-4-Chloro-α-methylbenzylamine; (-)-1-(4-Chlorophenyl)-ethylamine. Chiral intermediate. bp = 232°. *BASF Aktiengesellschaft; Chiragene, Inc.*

**3638 (S)-Methyl-2-chloro-3-hydroxypropionate**

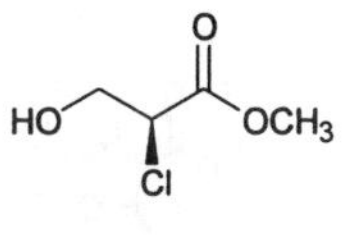

$C_4H_6ClO_3$
Chiral building block. *Mitsubishi Chemical Corporation.*

**3639 (5R)-Methyl-1-(chloromethyl)-2-pyrrolidinone**

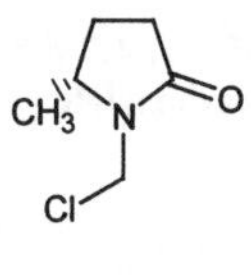

$C_6H_{10}ClNO$
Chiral building block. *Acros Organics nv.*

**3640 Methyl (S)-(-)-2-chloropropionate**
73246-45-4

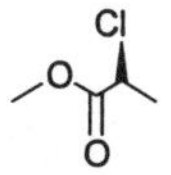

$C_4H_7ClO_2$
Propanoic acid, 2-chloro-, methyl ester, (2S)-; (S)-(-)-2-Chloropropionic acid methyl ester. Listed on TSCA. Chiral building block. $bp_{110}$ = 80-82°; d = 1.143; n = 1.417; $[\alpha]_D^{20}$ = - 26° (neat). *Daicel Chemical Ind. Ltd; Sigma-Aldrich Fine Chemicals.*

**3641 (+)-Methylcyclodextrin**

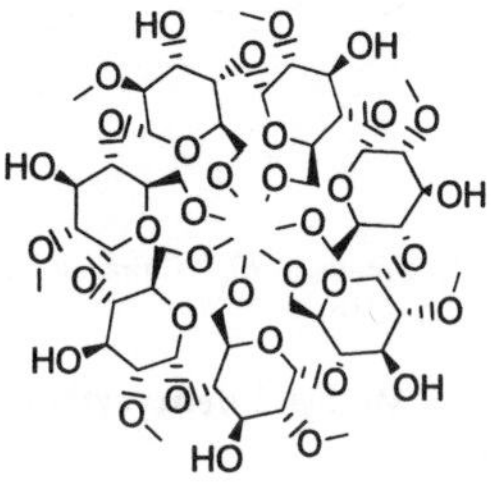

$C_{56}H_{98}O_{35}$
Chiral intermediate. mp = 180-182°; $[\alpha]^{25}$ = + 159° (c = 1, $H_2O$). *Sigma-Aldrich Fine Chemicals.*

**3642 (R)-(+)-3-Methylcyclohexanone**
13368-65-5 236-438-6
$C_7H_{12}O$
Chiral building block. bp = 168-169°; d = 0.916; n = 1.446; $[\alpha]^{24}$ = + 13.5° (neat). *Sigma-Aldrich Fine Chemicals.*

**3643 (R)-(+)-3-Methylcyclopentanone**
6672-30-6 229-711-6

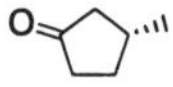

$C_6H_{10}O$
Chiral building block. bp = 143-144°; d = 0.914; n = 1.434; $[\alpha]^{23}$ = + 148° (c = 4.5, $CH_3OH$). *Sigma-Aldrich Fine Chemicals.*

**3644 S-Methyl-L-cysteine**
1187-84-4 214-701-6

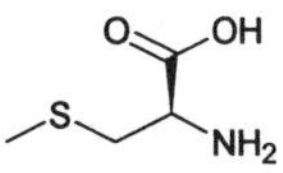

$C_4H_9NO_2S$
Chiral building block. $[\alpha]_D^{20}$ = - 30° (c = 1, $H_2O$). *Acros Organics nv; Austin Chemical Company, Inc.; Sigma-Aldrich Fine Chemicals; Synthetech, Inc.*

**3645 (S)-2-Methyldecanoic acid**
74036-83-2

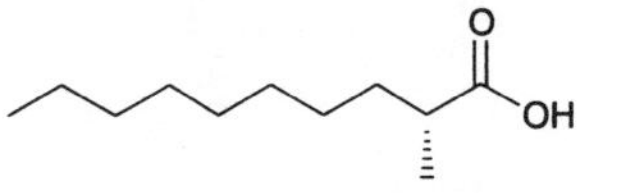

$C_{11}H_{22}O_2$
Decanoic acid, 2-methyl-, (S)-. Chiral building block. *Japan Energy Corporation.*

**3646 (S)-2-Methyl-1-decanol**

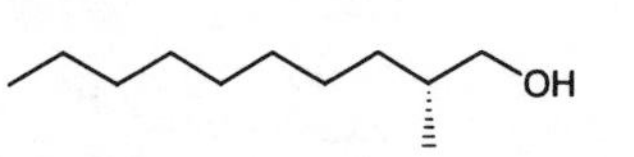

$C_{11}H_{24}O$
Decan-1-ol, 2-methyl-, (S)-. Chiral building block. *Japan Energy Corporation.*

**3647 (+)-5-Methyl-2-deoxycytidine**
838-07-3 212-655-1

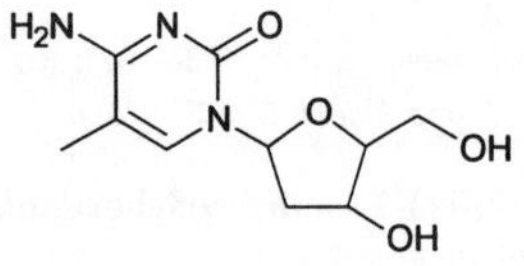

$C_{10}H_{15}N_3O_4$
Chiral intermediate. mp = 217-219°; $[\alpha]^{25}$ = + 63° (c = 1, 1N NaOH). *Sigma-Aldrich Fine Chemicals.*

**3648 Methyl 2-deoxy-3,5-di-O-p-nitrobenzoyl-D-ribofuranoside**
62279-73-6

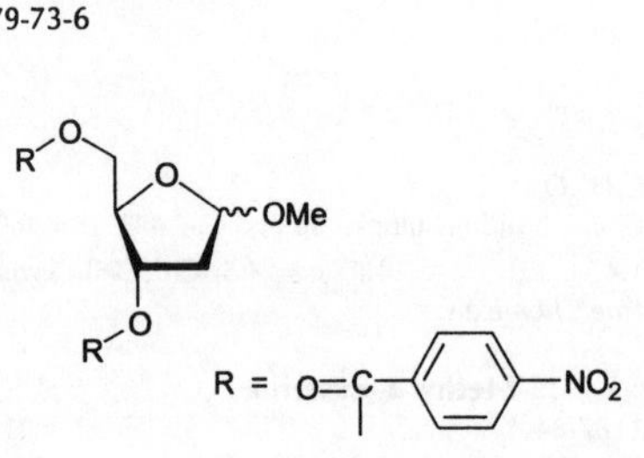

$C_{20}H_{18}N_2O_{10}$
Chiral intermediate. *Pfanstiehl Laboratories, Inc.*

**3649 Methyl 2-deoxy-2-phthalimido-1-thio-3,4,6-tri-O-acetyl-β-D-glucopyranoside**
79528-48-6
$C_{21}H_{23}NO_9S$
Chiral intermediate. *Pfanstiehl Laboratories, Inc.*

**3650 Methyl 6-deoxy-1-thio-2,3,4-tri-O-acetyl-β-L-galactopyranoside**

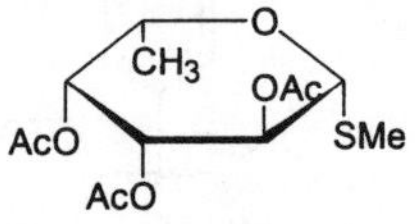

$C_{10}H_{20}O_7S$
Chiral intermediate. *Pfanstiehl Laboratories, Inc.*

**3651 (1S,4S)-(+)-2-Methyl-2,5-diazabicyclo-[2.2.1]heptane dihydrobromide**

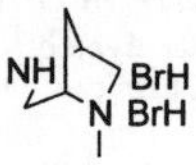

$C_6H_{14}Br_2N_2$
Chiral intermediate. mp = 248°; $[\alpha]^{23}$ = + 12° (c = 1, $H_2O$). *Sigma-Aldrich Fine Chemicals.*

**3652 Methyl 2,3:4,6-di-O-benzylidene-α-D-mannopyranoside**
4148-71-4

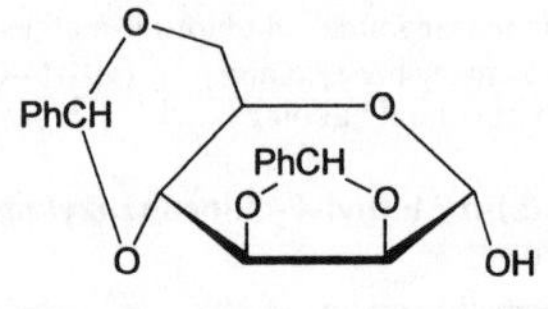

$C_{20}H_{20}O_6$
Chiral intermediate. *Pfanstiehl Laboratories, Inc.*

**3653 Methyl (4S,5S)-dihydro-5-methyl-2-phenyl-4-oxazolecarboxylate**
82659-84-5

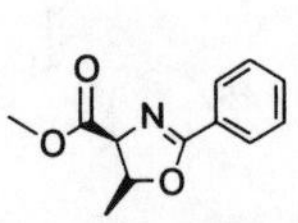

$C_{12}H_{13}NO_3$
Chiral intermediate. mp = 74-76°; $bp_{0.01}$ = 110-115°; $[\alpha]^{19}$ = + 66° (c = 8, EtOH). *Sigma-Aldrich Fine Chemicals.*

**3654 Methyl (S)-(+)-4,5-dihydro-2-phenyl-4-oxazolecarboxylate**
78715-83-0
$C_{11}H_{11}NO_3$
Chiral intermediate. bp = 131-132°; d = 1.2; n = 1.55; $[\alpha]_D^{20}$ = + 118° (c = 1, EtOH). *Sigma-Aldrich Fine Chemicals.*

**3655 Methyl (R)-4,5-dihydroxyisopropylidene-pentanoate**

$C_9H_{16}O_4$
Dioxolane-4-propanoic acid, 1,3-, 2,2-dimethyl-, methyl ester, (R)-. Chiral intermediate. *Synthon Chiragenics Corporation.*

**3656 Methyl (2R,3S)-(+)-2,3-dihydroxy-3-phenyl-propionate**

$C_{10}H_{12}O_4$
Chiral building block. mp = 84-88°; $[\alpha]^{23}$ = + 31° (c = 1, $H_2O$). *Sigma-Aldrich Fine Chemicals.*

**3657 Methyl (2S,3R)-(-)-2,3-dihydroxy-3-phenyl-propionate**

$C_{10}H_{12}O_4$
Chiral building block. mp = 85-88°; $[\alpha]^{21}$ = - 10° (c = 1, $CHCl_3$). *Sigma-Aldrich Fine Chemicals.*

**3658 Methyl (S)-(-)-2,2-dimethyl-1,3-dioxolane-4-carboxylate**
60456-21-5

$C_7H_{12}O_4$
Methyl α,β-isopropylidene-L-glycerate. Chiral intermediate. $bp_{15}$ = 84-86°; d = 1.106; n = 1.425; $[\alpha]_D^{20}$ = - 8.5° (c = 1.5, acetone). *Sigma-Aldrich Fine Chemicals.*

**3659 Methyl (S)-(+)-3-(2,2-dimethyl-1,3-dioxolan-4-yl)-cis-2-propenoate**
81703-94-8

$C_9H_{14}O_4$
Chiral intermediate. $bp_{0.2}$ = 60-62°; d = 1.067; n = 1.449; $[\alpha]_D^{20}$ = + 118° (c = 18, $CHCl_3$). *Sigma-Aldrich Fine Chemicals.*

**3660 Methyl (S)-(+)-3-(2,2-dimethyl-1,3-dioxolan-4-yl)-trans-2-propenoate**
81703-93-7

$C_9H_{14}O_4$
Chiral intermediate. d = 1.079; n = 1.454; $[\alpha]_D^{20}$ = + 44° (c = 18.5, $CHCl_3$). *Sigma-Aldrich Fine Chemicals.*

**3661 (-)-Methyl-3-((4S)-2,2-dimethyl-1,3-dioxolan-4-yl)propionate**
90472-93-8

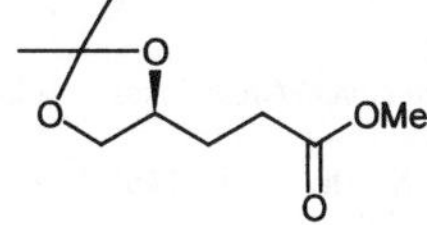


$C_9H_{16}O_4$
Dioxolane-4-propanoic acid, 1,3-, 2,2-dimethyl-, methyl ester, (S)-; Methyl (S)-4,5-dihydroxyisopropylidenepentanoate. Chiral intermediate. *Synthon Chiragenics Corporation.*

**3662 Methyl (1R,3S)-2,2-dimethyl-3-(2-oxo-propyl)-cyclopropaneacetate**
54878-01-2

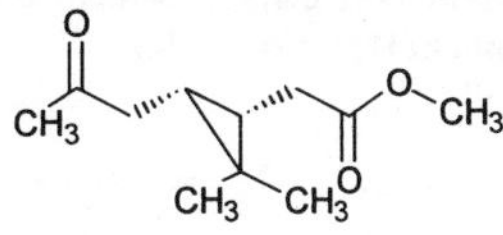


$C_{11}H_{18}O_3$
Chiral intermediate. [α] = - 26.6°. *Acros Organics nv.*

**3663 (S)-2-Methyldodecanoic acid**

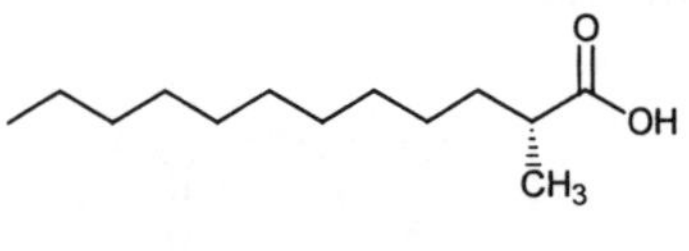

$C_{13}H_{26}O_2$
Dodecanoic acid, 2-methyl-, (S)-; (S)-α-Methyllauric acid. Chiral building block. *Japan Energy Corporation.*

**3664 (S)-2-Methyl-1-dodecanol**

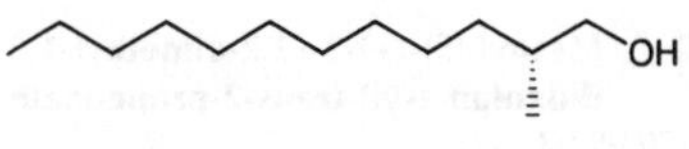

$C_{13}H_{28}O$
Dodecan-1-ol, 2-methyl-, (S)-; Chiral building block. *Japan Energy Corporation.*

**3665 2,2'-Methylenebis((4S)-4-benzyl-2-oxazoline)**

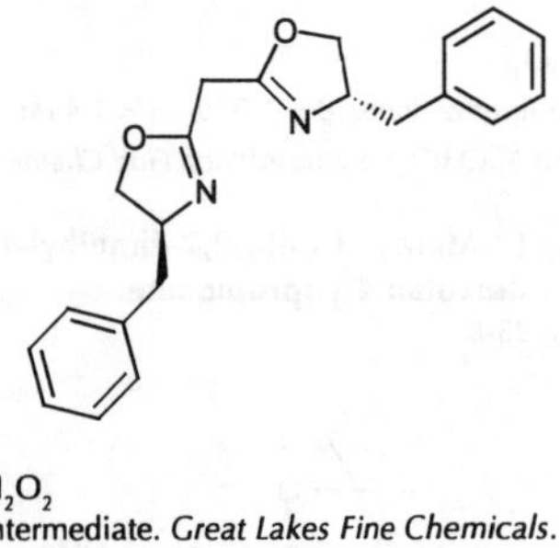

$C_{20}H_{20}N_2O_2$
Chiral intermediate. *Great Lakes Fine Chemicals.*

**3666 2,2-Methylenebis(4S)-4-tert-butyl-2-oxazoline**
132098-54-5

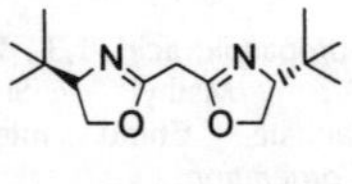

$C_{15}H_{26}N_2O_2$
Chiral intermediate. mp = 51-53°; $[\alpha]_D^{20}$ = - 118° (c = 0.5, $CHCl_3$). *Sigma-Aldrich Fine Chemicals.*

**3667 [3aS-[2(3a'R*,8a'S*),3a'α,8'aα]]-(-)-2,2'-Methylene-bis[3a,8a-dihydro-8H-indeno[1,2-d]oxazole]**
175166-49-1

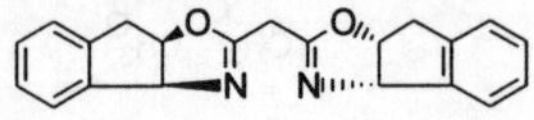

$C_{21}H_{18}N_2O_2$
Chiral intermediate. mp = 215-219°; $[\alpha]^{22}$ = - 355° (c = 3.7, $CHCl_3$). *Sigma-Aldrich Fine Chemicals.*

**3668 [3aS-[2(3a'R*,8a'S*),3a'β,8'aβ]]-(-)-2,2'-Methylene-bis[3a,8a-dihydro-8H-indeno[1,2-d]oxazole]**

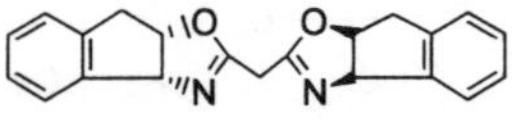

$C_{21}H_{18}N_2O_2$
Chiral intermediate. mp = 204-206°; $[\alpha]^{22}$ = + 353° (c = 3.7, $CHCl_3$). *Sigma-Aldrich Fine Chemicals.*

**3669 2,2-Methylenebis(4R,5S)-4,5-diphenyl-2-oxazoline**
139021-82-2

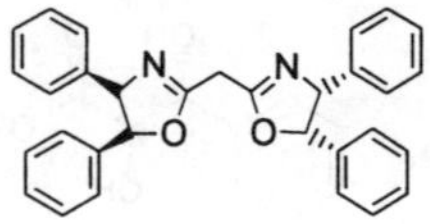

$C_{31}H_{26}N_2O_2$
Chiral intermediate. mp = 205-208°; $[\alpha]^{24}$ = + 160° (c = 0.5, $CHCl_3$). *Sigma-Aldrich Fine Chemicals.*

**3670 2,2'-Methylenebis((4R)-4-phenyl-2-oxazoline)**

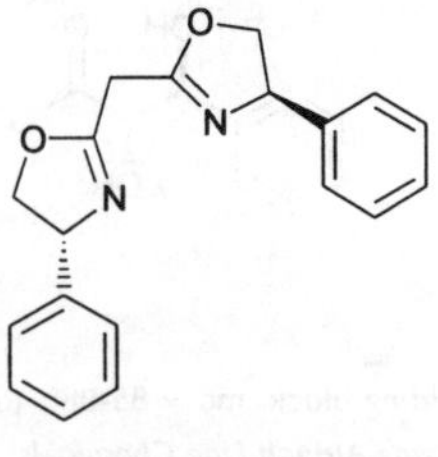

$C_{18}H_{16}N_2O_2$
Chiral intermediate. *Great Lakes Fine Chemicals.*

**3671 2,2-Methylenebis(4S)-4-phenyl-2-oxazoline**
132098-59-0

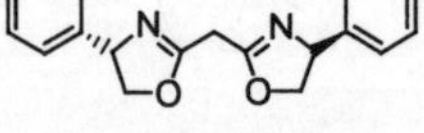

$C_{19}H_{18}N_2O_2$
Chiral intermediate. $bp_{0.01}$ = 131-134°; d = 1.28; n = 1.586; $[\alpha]_D^{20}$ = - 90° (c = 1, EtOH). *Great Lakes Fine Chemicals; Sigma-Aldrich Fine Chemicals.*

3672 **(R)-2,2'-Methylenedioxy-1,1'-binaphthyl**

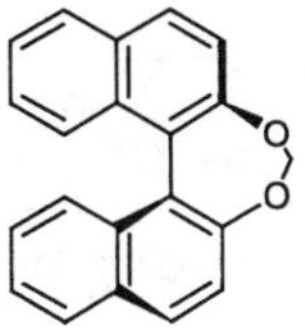

$C_{21}H_{14}O_2$
Chiral ligand. *Kankyo Kagaku Center Co., Ltd.*

3673 **(S)-2,2'-Methylenedioxy-1,1'-binaphthyl**

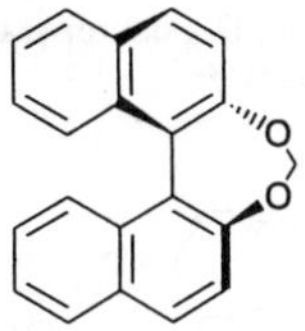

$C_{21}H_{14}O_2$
Chiral ligand. *Kankyo Kagaku Center Co., Ltd.*

3674 **(R)-2,2'-Methylenedioxy-5,5',6,6',7,7',8,8'-octahydro-1,1'-binaphthyl**

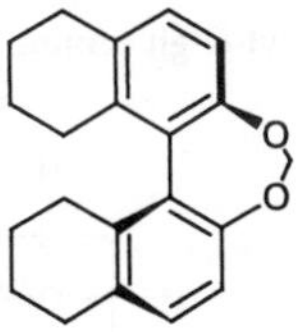

$C_{21}H_{22}O_2$
Chiral ligand. *Kankyo Kagaku Center Co., Ltd.*

3675 **(S)-2,2'-Methylenedioxy-5,5',6,6',7,7',8,8'-octahydro-1,1'-binaphthyl**

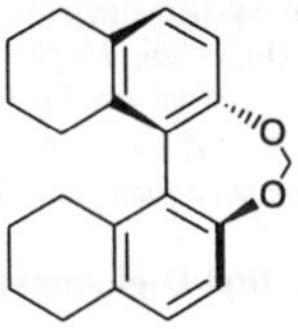

$C_{21}H_{22}O_2$
Chiral ligand. *Kankyo Kagaku Center Co., Ltd.*

3676 **(S)-Methyl-1,2-epoxydecane**

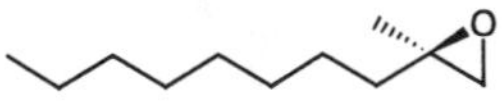

$C_{11}H_{22}O_2$
Chiral building block. *Japan Energy Corporation.*

3677 **Methyl-(5S,6S)-epoxy-(7E,9E,11Z,14Z)-eicosatetraenoate**
73466-12-3

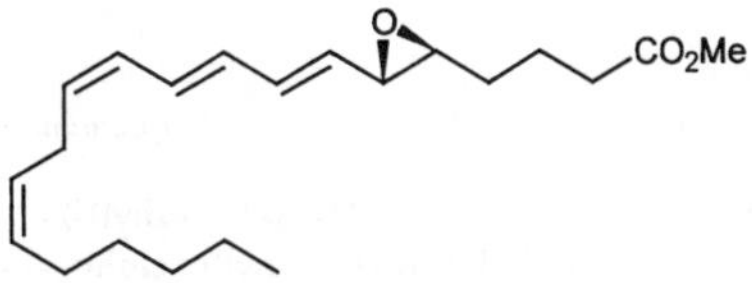

CO2Me

$C_{21}H_{32}O_3$
Chiral intermediate. *Ultrafine.*

3678 **(S)-Methyl-1,2-epoxyheptane**

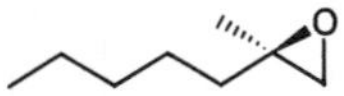

$C_8H_{16}O$
Chiral building block. *Japan Energy Corporation.*

3679 **(S)-Methyl-1,2-epoxyhexane**

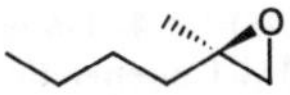

$C_7H_{14}O$
Chiral building block. *Japan Energy Corporation.*

3680 **Methyl-(-)-(5S,6S)-epoxy-7-hydroxy-heptanoate**
73957-99-0
$C_8H_{14}O_4$
Chiral intermediate. *Ultrafine.*

3681 **(S)-Methyl-1,2-epoxynonane**

$C_{10}H_{20}O$
Chiral building block. *Japan Energy Corporation.*

3682 **(S)-Methyl-1,2-epoxyoctane**

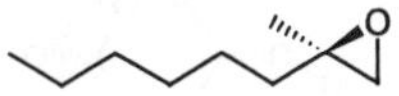

$C_9H_{18}O$
Chiral building block. *Japan Energy Corporation.*

3683 **Methyl-(5S,6S)-epoxy-11-oxo(7E,9E)-undecadienoate**
73958-00-6
$C_{12}H_{16}O_4$
Chiral intermediate. *Ultrafine.*

**3684 (S)-Methyl-1,2-epoxypentane**

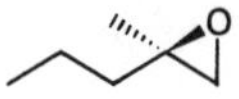

$C_6H_{12}O$
Chiral building block. *Japan Energy Corporation.*

**3685 (1R,2S,4R)-(-)-4-(1-Methylethyl)-2-(1-methyl-1-phenylethyl)cyclohexanol**

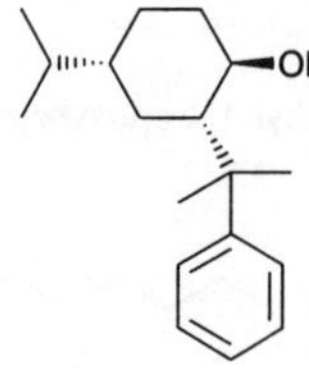

$C_{18}H_{28}O$
Chiral intermediate. mp = 68-71°; $[\alpha]_D^{20}$ = - 35° (c = 1, EtOH). *Sigma-Aldrich Fine Chemicals.*

**3686 (1S,2R,4S)-(+)-4-(1-Methylethyl)-2-(1-methyl-1-phenylethyl)cyclohexanol**

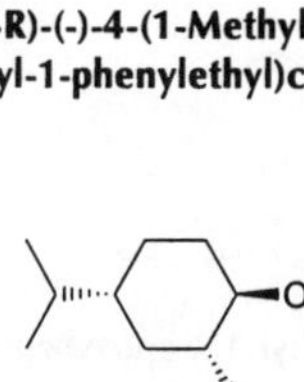

$C_{18}H_{28}O$
Chiral intermediate. mp = 68-71°; $[\alpha]_D^{20}$ = + 35° (c = 1, EtOH). *Sigma-Aldrich Fine Chemicals.*

**3687 Methyl α-D-galactopyranoside**
3396-99-4 222-251-7

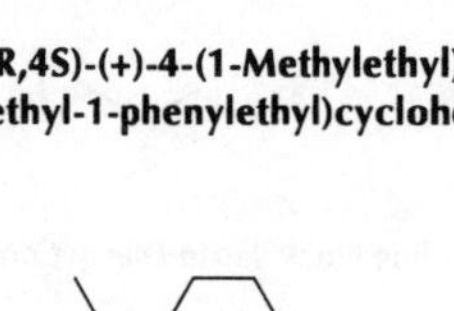

$C_7H_{14}O_6$
Galactopyranoside, methyl, α-D-. Chiral intermediate. $[\alpha]_D^{20}$ = + 179° (c = 4, $H_2O$). *Acros Organics nv; Pfanstiehl Laboratories, Inc.; Senn Chemicals AG.*

**3688 Methyl β-D-galactopyranoside**
1824-94-8 217-361-7

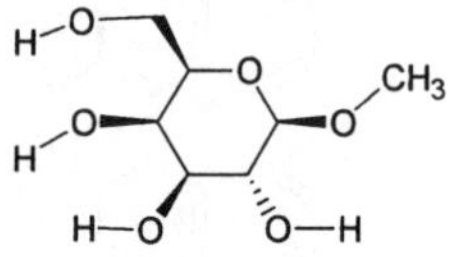

$C_7H_{14}O_6$
Galactopyranoside, methyl, β-D-. Chiral intermediate. mp = 176-179°; $[\alpha]_D^{20}$ = 0° (c = 4, $H_2O$). *Acros Organics nv; Pfanstiehl Laboratories, Inc.; Senn Chemicals AG; Sigma-Aldrich Fine Chemicals.*

**3689 Methyl-α-D-galactopyranoside monohydrate**
34004-14-3

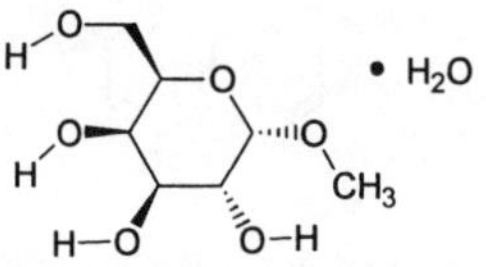

$C_7H_{14}O_6$
Chiral intermediate. [α] = + 181° (c = 1 $H_2O$). *Acros Organics nv.*

**3690 N-Methyl-D-glucamine**
6284-40-8 6154(12) 228-506-9

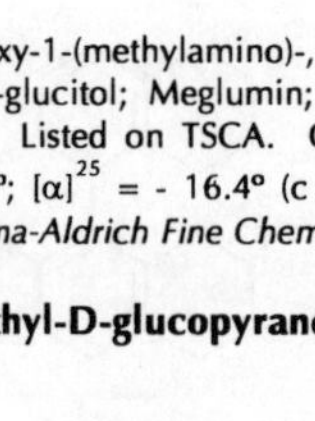

$C_7H_{17}NO_5$
Glucitol, 1-deoxy-1-(methylamino)-, D-; 1-Deoxy-1-(methylamino)-D-glucitol; Meglumin; Sorbitol, 1-deoxy-1-methylamino-. Listed on TSCA. Chiral intermediate. mp = 129-131.5°; $[\alpha]^{25}$ = - 16.4° (c = 10, $H_2O$). *Acros Organics nv; Sigma-Aldrich Fine Chemicals.*

**3691 3-O-Methyl-D-glucopyranose**
3370-81-8

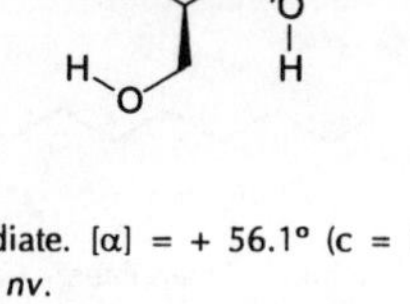

$C_7H_{14}O_6$
Chiral intermediate. [α] = + 56.1° (c = 1, $H_2O$, 24 h). *Acros Organics nv.*

**3692 Methyl α-D-glucopyranoside**
97-30-3 6156(12) 202-571-3

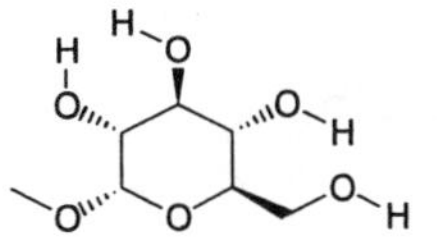

$C_7H_{14}O_6$
Glucopyranoside, methyl, α-D-; α-Methylglucoside; α-Methyl D-glucose ether. Listed on TSCA. Chiral intermediate. mp = 169-171°; $[\alpha]_D^{20}$ = + 158.9° (c = 10, $H_2O$). *Acros Organics nv; Pfanstiehl Laboratories, Inc.; Senn Chemicals AG; Sigma-Aldrich Fine Chemicals.*

**3693 Methyl β-D-glucopyranoside**
709-50-2 211-909-9

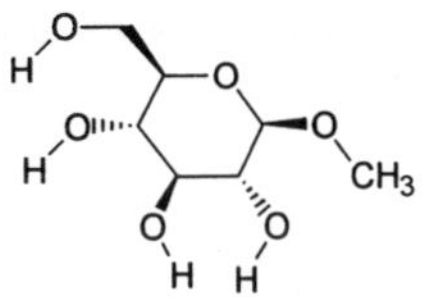

$C_7H_{14}O_6$
Glucopyranoside, methyl, β-D-; β-Methylglucoside. Listed on TSCA. Chiral intermediate. $[\alpha]_D^{20}$ = - 34.2° (c = 10, $H_2O$). *Acros Organics nv; Austin Chemical Company, Inc.; Pfanstiehl Laboratories, Inc.; Senn Chemicals AG.*

**3694 Methyl α-D-glucopyranoside-2,6-dibenzoate-3,4-di(bis(3,5-dimethylphenyl)-phosphite)**
158214-06-3
$C_{53}H_{56}O_8P_2$
Carbophos. Chiral intermediate. mp = 162-164°. *Strem Chemicals, Inc.*

**3695 (-)-Methylglucopyranoside hemihydrate**
7000-27-3 221-581-9

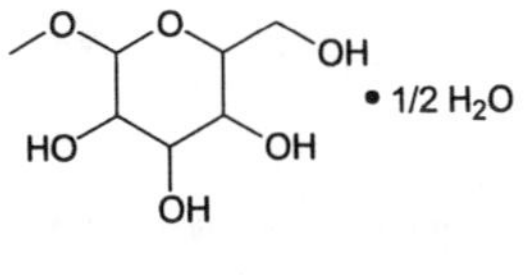

$C_7H_{14}O_6 \cdot 1/2H_2O$
Listed on TSCA. Chiral intermediate. mp = 111-113°; $[\alpha]_D^{20}$ = - 32.6° (c = 1, $H_2O$). *Sigma-Aldrich Fine Chemicals.*

**3696 3-O-Methyl-D-glucose**
13224-94-7 236-197-7
$C_7H_{14}O_6$
Glucopyranose, 3-O-methyl-, α-D-; 3-O-Methyl-α-D-glucopyranose. Chiral building block. mp = 167-169°; $[\alpha]_D^{20}$ = + 56 ° (c = 1, $H_2O$). *Pfanstiehl Laboratories, Inc.; Senn Chemicals AG.*

**3697 (2R,4R)-4-Methylglutamic acid hydrochloride**

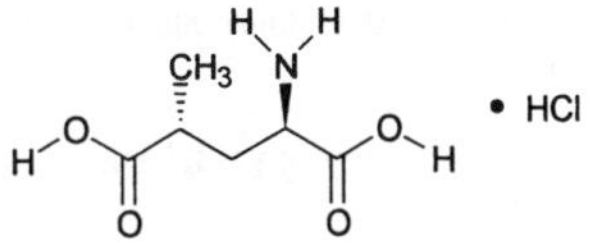

$C_6H_{12}ClNO_4$
Chiral building block. *Acros Organics nv.*

**3698 (2S,4S)-4-Methylglutamic acid hydrochloride**

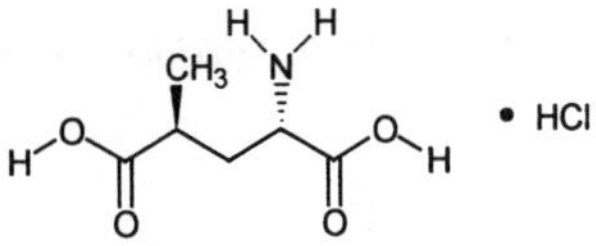

$C_6H_{12}ClNO_4$
Chiral building block. *Acros Organics nv.*

**3699 (R)-(+)-2-Methylglutaric acid**
1115-81-7
$C_6H_{10}O_4$
Chiral building block. *TCI America.*

**3700 (S)-(-)-2-Methylglutaric acid**
1115-82-8
$C_6H_{10}O_4$
Chiral building block. *TCI America.*

**3701 (S)-(+)-2-Methylglutaric acid dimethyl ester**
10171-92-3

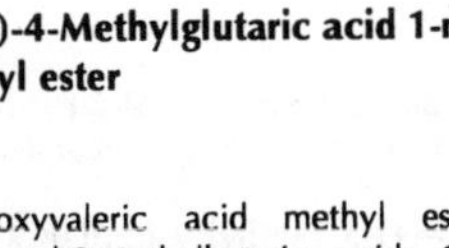

$C_8H_{14}O_4$
Pentanedioic acid, 2-methyl-, dimethyl ester, (S)-; Dimethyl (S)-(+)-2-methylglutarate. Chiral intermediate. *TCI America.*

**3702 (R)-(-)-4-Methylglutaric acid 1-monomethyl ester**
80986-17-0
$C_7H_{12}O_4$
(R)-(-)-4-Carboxyvaleric acid methyl ester; (R)-(-)-4-Methoxycarbonyl-2-methylbutyric acid; 1-Monomethyl (R)-(-)-4-methylglutarate. Chiral intermediate. *TCI America.*

**3703 (R)-Methyl glycidate**
111058-32-3
$C_4H_6O_3$
Methyl 2,3-epoxypropionate; R-methyl oxiranecarboxylate. Chiral building block. *Chirex, Inc.; Daiso Company, Ltd.*

**3704 (S)-Methyl glycidate**
118712-39-3
$C_4H_6O_3$
(S)-Methyl 2,3-epoxypropionate; S-Methyl oxiranecarboxylate. Chiral building block. *Chirex, Inc.; Daiso Company, Ltd.*

**3705 (R)-(-)-Methyl glycidyl ether**
64491-70-9

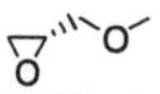

$C_4H_8O_2$
Chiral building block. bp = 110-111°; d = 0.982; n = 1.406; $[\alpha]_D^{20}$ = - 15° (c = 5, toluene). *Daiso Company, Ltd.; Sigma-Aldrich Fine Chemicals.*

**3706 (2R)-(-)-2-Methylglycidyl 4-nitrobenzoate**
106268-96-6

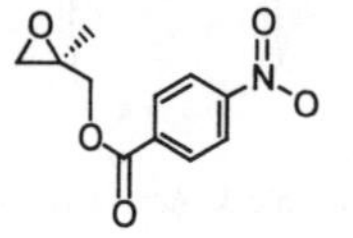

$C_{11}H_{11}NO_5$
Chiral intermediate. mp = 87-89°; $[\alpha]^{19}$ = - 5.9° (c = 2.4, $CHCl_3$). *Sigma-Aldrich Fine Chemicals.*

**3707 (2S)-(+)-2-Methylglycidyl 4-nitrobenzoate**

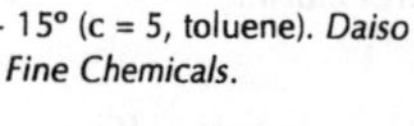

$C_{11}H_{11}NO_5$
Chiral intermediate. mp = 87-89°; $[\alpha]^{19}$ = + 5.9° (c = 2.4, $CHCl_3$). *Sigma-Aldrich Fine Chemicals.*

**3708 (S)-3-Methyl-3-heptanol**

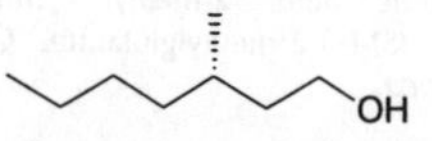

$C_8H_{18}O$
Heptan-3-ol, 3-methyl, (S)-; (S)-2-Ethyl-2-hexanol. Chiral building block. *Japan Energy Corporation.*

**3709 (S)-(+)-5-Methyl-1-heptanol**
57803-73-3
$C_8H_{18}O$
Chiral building block. d = 0.83. *TCI America.*

**3710 Methyl-6-O-(N-heptylcarbamoyl)-α-D-glucopyranoside**
115457-83-5
$C_{15}H_{29}NO_7$
Chiral intermediate. *Acros Organics nv.*

**3711 (R)-4-(1-Methylheptyloxy)carbonylphenyl 4-octyloxy-4-biphenylcarboxylate**
123286-51-1

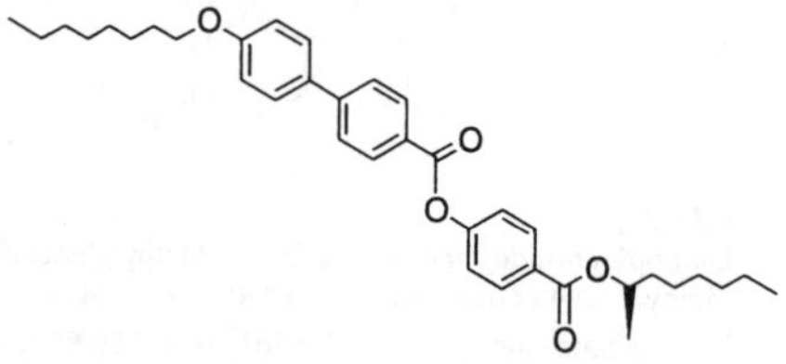

$C_{36}H_{46}O_5$
Chiral intermediate. $[\alpha]_D^{20}$ = - 25° (c = 5, acetone). *Sigma-Aldrich Fine Chemicals.*

**3712 (S)-4-(1-Methylheptyloxy)carbonylphenyl 4-octyloxy-4-biphenylcarboxylate**
112901-67-4

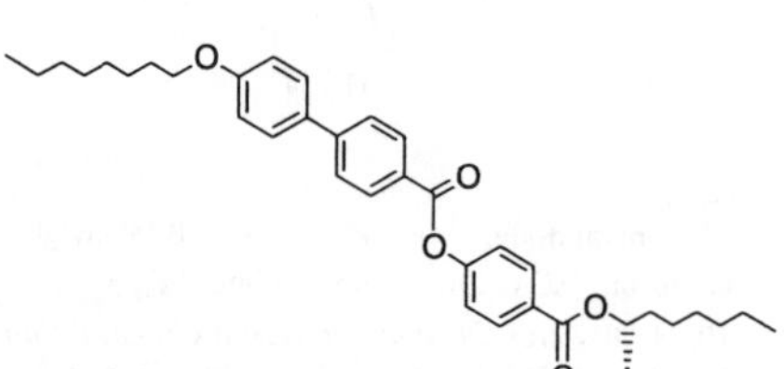

$C_{36}H_{46}O_5$
Chiral intermediate. $[\alpha]_D^{20}$ = + 27° (c = 5, acetone). *Sigma-Aldrich Fine Chemicals.*

**3713 O-Methyl-D-hexahydrotyrosine**

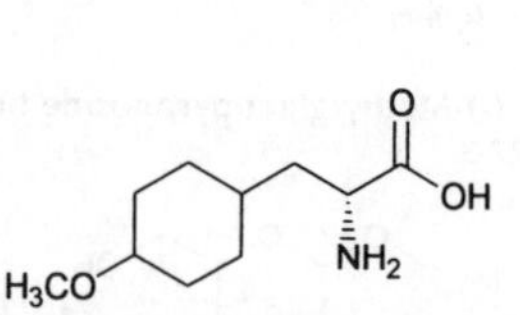

$C_{10}H_{19}NO_3$
Chiral intermediate. *Senn Chemicals AG.*

**3714 O-Methyl-L-hexahydrotyrosine**

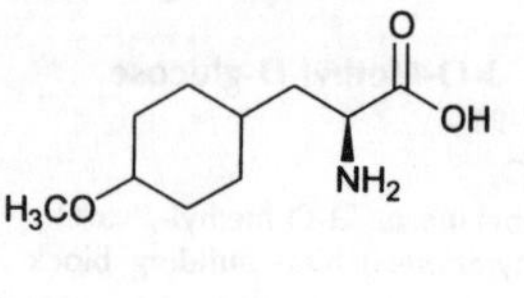

$C_{10}H_{19}NO_3$
Chiral intermediate. *Senn Chemicals AG.*

**3715 (R)-Methylhexanoic acid**
51703-97-0

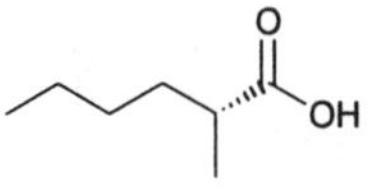

$C_8H_{16}O_2$
Hexanoic acid, 2-methyl-, (R)-. Chiral building block. *Chiragene, Inc.*

**3716 (S)-2-Methylhexanoic acid**
49642-51-5

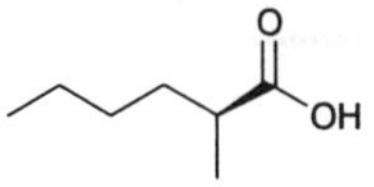

$C_8H_{16}O_2$
Hexanoic acid, 2-methyl-, (S)-. Chiral building block. *Chiragene, Inc.*

**3717 (S)-3-Methyl-3-hexanol**

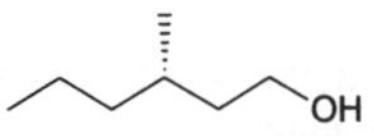

$C_7H_{16}O$
Hexan-3-ol, 3-methyl-, (S)-; (S)-2-Ethyl-2-pentanol. Chiral building block. *Japan Energy Corporation.*

**3718 (S)-(+)-4-Methyl-1-hexanol**
1767-46-0
$C_7H_{16}O$
Hexan-1-ol, 4-methyl-, (S)-. Chiral building block. *TCI America.*

**3719 1-Methyl-L-histidine**
15507-76-3
(S)-1-Methylimidazole-4-alanine. Chiral building block. *Advanced Asymmetrics, Inc.*

**3720 3-Methyl-L-histidine**
368-16-1 206-704-6

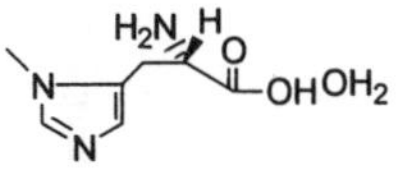

$C_7H_{11}N_3O_2$
Histidine, 3-methyl-, L-; (S)-1-Methylimidazole-5-alanine. Chiral building block. mp = 254°; $[\alpha]_D^{20}$ = + 9° (c = 1, 6N HCl). *Advanced Asymmetrics, Inc.; Sigma-Aldrich Fine Chemicals.*

**3721 Methyl-(2R)-hydroxy-(3S)-N-benzoyl-amino-3-phenylpropionate**

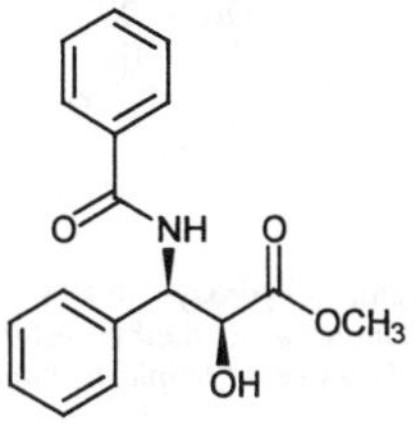

$C_{17}H_{17}NO_4$
Chiral intermediate. *Austin Chemical Company, Inc.*

**3722 Methyl (R)-(-)-3-hydroxybutyrate**
3976-69-0 223-610-0

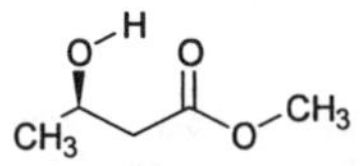

$C_5H_{10}O_3$
Butanoic acid, 3-hydroxy-, methyl ester, (R)-; (R)-(-)-3-Hydroxybutyric acid methyl ester; (R)-(-)-Methyl 3-hydroxybutyrate. Chiral intermediate. $bp_{11}$ = 56-58°; d = 1.0550; $n_D^{20}$ = 1.4220; $[\alpha]_D^{20}$ = - 48.5 ± 0.5° (c = 1.3, $CH_3OH$); $[\alpha]_D^{20}$ = - 19.5° (neat). *Acros Organics nv; Celltech Chiroscience Ltd.; Great Lakes Fine Chemicals; Kaneka Corporation; Lancaster Synthesis Ltd.; Sigma-Aldrich Fine Chemicals; TCI America.*

**3723 Methyl (R)-(+)-3-hydroxy-5-oxo-1-cyclo-pentene-1-heptanoate**
41138-61-8

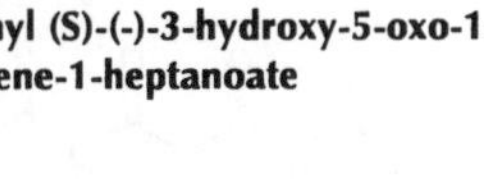

$C_{13}H_{20}O_4$
Chiral intermediate. mp = 59-63°; $[\alpha]^{23}$ = + 10° (c = 1, $CHCl_3$). *Sigma-Aldrich Fine Chemicals.*

**3724 Methyl (S)-(-)-3-hydroxy-5-oxo-1-cyclo-pentene-1-heptanoate**
42038-75-5

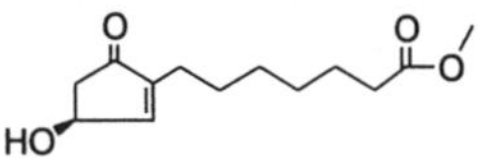

$C_{13}H_{20}O_4$
Chiral intermediate. mp = 59-62°; $[\alpha]_D^{20}$ = - 10.4° (c = 1, $CHCl_3$). *Sigma-Aldrich Fine Chemicals.*

**3725 Methyl (R)-β-hydroxypentanoate**
60793-22-8

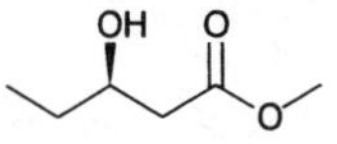

$C_6H_{12}O_3$
Pentanoic acid, 3-hydroxy-, methyl ester, (R)-; (R)-3-Hydroxypentanoic acid methyl ester. Chiral building block. *Great Lakes Fine Chemicals; Kaneka Corporation.*

**3726 Methyl (S)-β-hydroxypentanoate**
42558-50-9

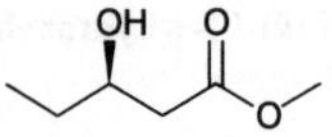

$C_6H_{12}O_3$
Pentanoic acid, 3-hydroxy-, methyl ester, (S)-; Chiral building block. *Great Lakes Fine Chemicals; Kaneka Corporation.*

**3727 Methyl (R)-(+)-2-(4-hydroxyphenoxy)-propanoate**
96562-58-2

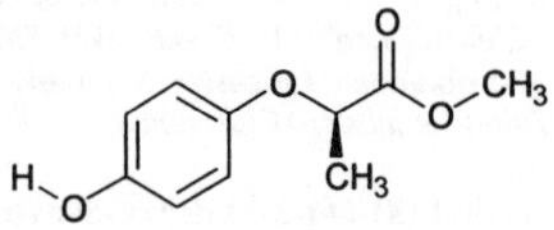

$C_{10}H_{12}O_4$
Chiral intermediate. mp = 64-67°; $[\alpha]^{22}$ = + 43° (c = 1, $CHCl_3$). *Acros Organics nv; Sigma-Aldrich Fine Chemicals.*

**3728 Methyl (R)-3-hydroxy-3-phenylpropanoate**
58692-70-9

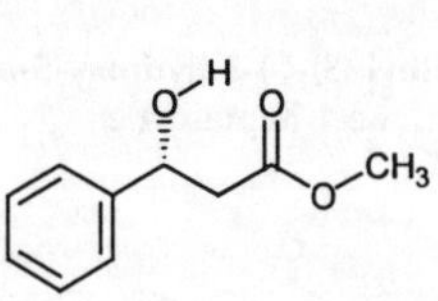

$C_{10}H_{112}O_3$
Chiral building block. $bp_1$ = 120°; $[\alpha]_D^{21}$ = + 44.2 ± 0.2° (c = 3, 1,2-DCE). *Acros Organics nv; Eastman Chemical Company; Kiralchem Ltd.*

**3729 (S)-Methyl 3-hydroxy-3-phenyl propanoate**
36615-45-9

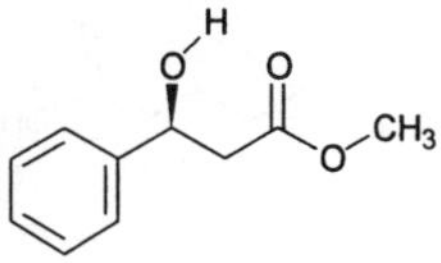

$C_{10}H_{12}O_3$
Chiral building block. $bp_1$ = 120°; $[\alpha]_D^{21}$ = - 44.2 ± 0.2° (c = 3, 1,2-DCE). *Acros Organics nv; Eastman Chemical Company; Kiralchem Ltd.*

**3730 (R)-1-Methyl-3-(4-hydroxyphenyl)-propylamine**

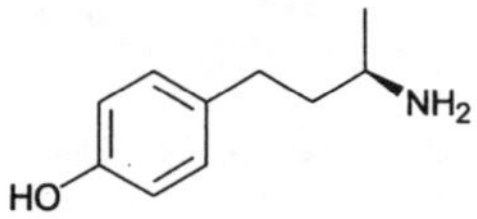

$C_{10}H_{15}NO$
Chiral building block. *Chiragene, Inc.*

**3731 (S)-1-Methyl-3-(4-hydroxyphenyl)-propylamine**

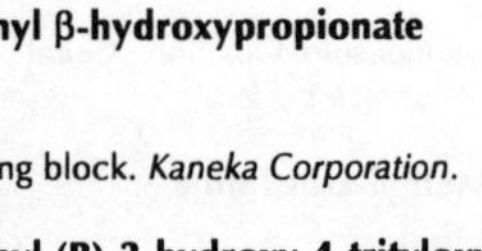

$C_{10}H_{15}NO$
Chiral intermediate. *Chiragene, Inc.*

**3732 Methyl β-hydroxypropionate**
6149-41-3
$C_4H_8O_3$
Chiral building block. *Kaneka Corporation.*

**3733 Methyl (R)-3-hydroxy-4-trityloxy-butanoate**
144754-24-5

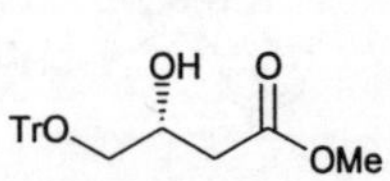

$C_{24}H_{24}O_4$
Butanoic acid, 3-hydroxy-4-(triphenylmethoxy)-, methyl ester, (R)-. Chiral intermediate. *Synthon Chiragenics Corporation.*

**3734 Methyl (S)-3-hydroxy-4-trityloxy-butanoate**
113240-53-2

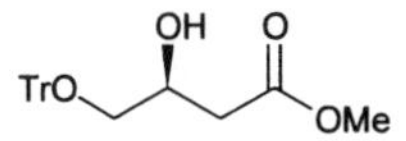

$C_{24}H_{24}O_5$
Butanoic acid, 3-hydroxy-4-(triphenylmethoxy)-, methyl ester, (S)-. Chiral intermediate. *Synthon Chiragenics Corporation.*

**3735 2-Methylidene-(3R)-[(3S)-tert-butylsilyloxy-(5S)-methylnon-1(E)-ene-1-yl-(4R)-tert-butylsilyloxycyclopentanone**

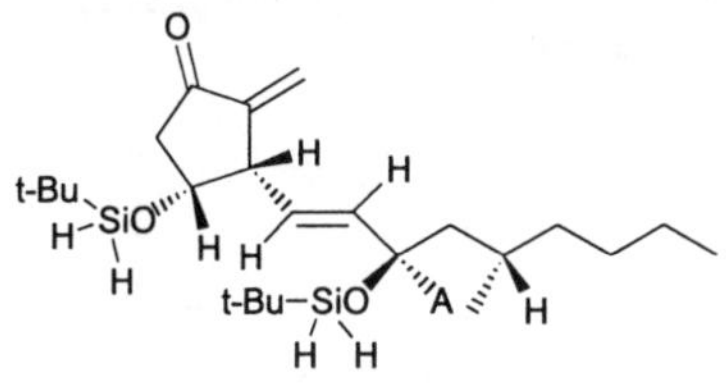

$C_{28}H_{50}O_3Si_2$
Chiral intermediate. *Nissan Chemical Industries, Ltd.*

**3736 2-Methylidene-(3R)-[(3S)-tert-butylsilyloxy-oct-1(E)-ene-1-yl]-(4R)-tert-butylsilyl-oxycyclopentanone**
83058-84-8
$C_{26}H_{50}O_3Si_2$
Chiral intermediate. *Nissan Chemical Industries, Ltd.*

**3737 N-Methyl-3-indolyl-β-D-galactopyranoside**

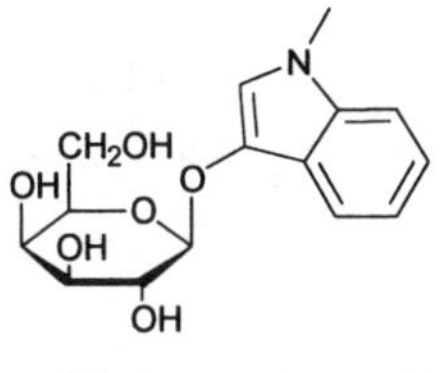

$C_{15}H_{19}NO_6$
Chiral intermediate. *Senn Chemicals AG.*

**3738 Methylisoborneol [(1R-exo)-1,2,7,7-tetra-methylbicyclo[2.2.1]heptan-2-ol]**
$C_{11}H_{20}O$
Chiral intermediate. *Ultrafine.*

**3739 Methyl 2,3-O-Isopropylidene-D-ribofuranoside**
4099-85-8 223-865-8
$C_9H_{16}O_5$
Ribofuranoside, methyl 2,3-O-isopropylidene-, β-D-. Chiral intermediate. *Pfanstiehl Laboratories, Inc.*

**3740 (+)-Methyl 3,4-O-isopropylidenethreonate**
92973-40-5

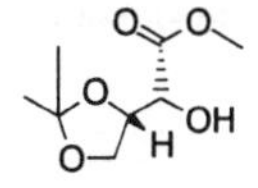

$C_8H_{14}O_5$
Chiral intermediate. $bp_{12}$ = 125°; d = 1.175; n = 1.447; $[\alpha]_D^{20}$ = + 18.5° (c = 2, acetone). *Sigma-Aldrich Fine Chemicals.*

**3741 Methyl 2,3-O-Isopropylidene-5-O-(p-tolyl-sulfonyl)-β-D-ribofuranoside**
4137-56-8
$C_{16}H_{22}O_7S$
Chiral intermediate. mp = 83°; $[\alpha]_D^{20}$ = - 36.5° (c = 1, EtOH). *Pfanstiehl Laboratories, Inc.*

**3742 Methyl (S)-(-)-lactate**
27871-49-4 248-704-9

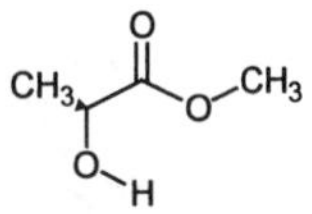

$C_4H_8O_3$
Propanoic acid, 2-hydroxy-, methyl ester, (S)-; (-)-2-Hydroxypropionic acid methyl ester; L-Lactic acid methyl ester. Chiral building block. bp = 135-138°; d = 1.0920; $n_D^{20}$ = 1.4140; $[\alpha]_D^{20}$ = -8.2 ± 0.5° (neat). *Acros Organics nv; Lancaster Synthesis Ltd.; Nitrokemia 2000 Rt.; Sigma-Aldrich Fine Chemicals.*

**3743 N-ε-methyl-L-lysine hydrochloride**
7622-29-9 231-539-1

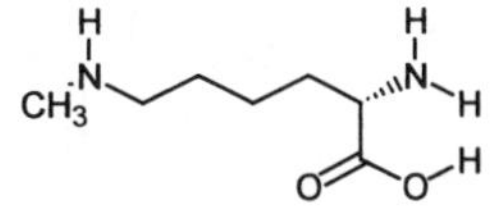

$C_7H_{17}ClN_2O_2$
$N^6$-methyl-, monohydrochloride, L-. Chiral building block. *Acros Organics nv.*

**3744 Methyl (R)-(-)-mandelate**
20698-91-3

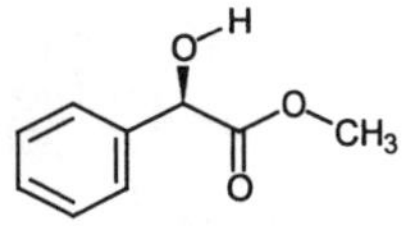

$C_9H_{10}O_3$
D-(-)-Mandelic acid methyl ester; (R)-(-)-α-Hydroxyphenylacetic acid methyl ester. Chiral diagnostic reagent. mp = 56-58°; $[\alpha]_D^{20}$ = - 144 ± 2° (c = 1, $CH_3OH$). *Acros Organics nv; Lancaster Synthesis Ltd.; Omega*

*Chemical Company Inc.; Sigma-Aldrich Fine Chemicals; Yamakawa Chemical Industry Co. Ltd.*

**3745 Methyl (S)-(+)-mandelate**
21210-43-5

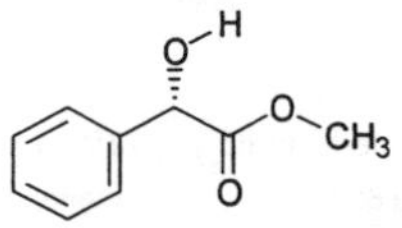

$C_9H_{10}O_3$
L-(+)-Mandelic acid methyl ester; (S)-(+)-α-Hydroxyphenylacetic acid methyl ester. Chiral diagnostic reagent. mp = 56-58°; $[\alpha]_D^{20}$ = + 144° (c = 1, $CH_3OH$). *Acros Organics nv; Lancaster Synthesis Ltd.; Omega Chemical Company Inc.; Sigma-Aldrich Fine Chemicals; Yamakawa Chemical Industry Co. Ltd.*

**3746 (R)-Methylmandelic acid**

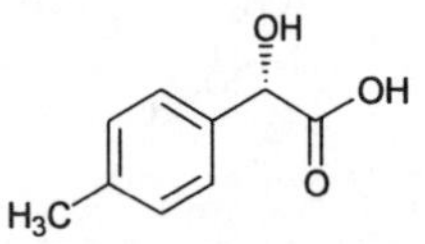

$C_9H_{10}O_3$
Chiral building block. *Austin Chemical Company, Inc.*

**3747 (S)-4-Methylmandelic acid**

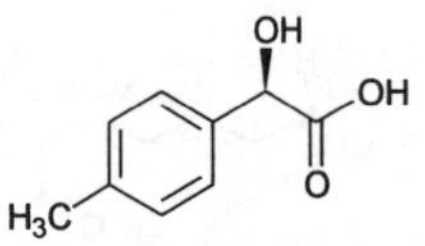

$C_9H_{10}O_3$
Chiral building block. *Austin Chemical Company, Inc.*

**3748 (R)-(+)-4-Methylmandelonitrile**
10017-04-6

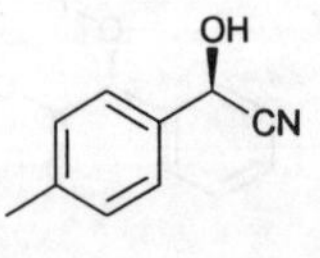

$C_9H_9NO$
Chiral building block. mp = 52-58°; $[\alpha]^{18}$ = + 39° (c = 1, $CHCl_3$). *Sigma-Aldrich Fine Chemicals.*

**3749 Methyl α-D-mannopyranoside**
617-04-9 210-502-3

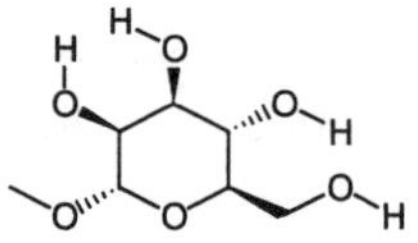

$C_7H_{14}O_6$
Mannopyranoside, methyl, α-D-. Chiral intermediate. mp = 187-193°; $[\alpha]_D^{20}$ = + 79.2° (c = 4, $H_2O$). *Acros Organics nv; CU Chemie Uetikon GmbH; Kaden Biochemicals GmbH; Pfanstiehl Laboratories, Inc.; Senn Chemicals AG; Sigma-Aldrich Fine Chemicals.*

**3750 (+)-Methyl mannopyranoside 2,3,4,6-tetra-acetate**
225-703-1

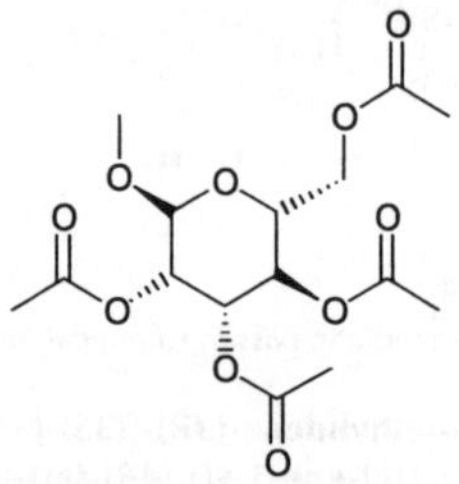

$C_{15}H_{22}O_{10}$
Chiral intermediate. mp = 64-66°; $[\alpha]_D^{20}$ = + 50° (c = 0.4, $CH_3OH$). *Sigma-Aldrich Fine Chemicals.*

**3751 (R)-α-Methyl-2-methoxybenzylamine**

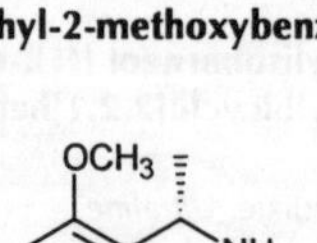

$C_9H_{13}NO$
Chiral building block. *Chiragene, Inc.; Sumitomo Chemcial Co. Ltd.*

**3752 (S)-α-Methyl-2-methoxybenzylamine**

$C_9H_{13}NO$
Chiral building block. *Chiragene, Inc.; Sumitomo Chemcial Co. Ltd.*

**3753 (R)-α-Methyl-3-methoxybenzylamine**

$C_9H_{13}NO$
Chiral building block. *Chiragene, Inc.; Sumitomo Chemcial Co. Ltd.*

**3754 (S)-α-Methyl-3-methoxybenzylamine**

$C_9H_{13}NO$
Chiral building block. *Chiragene, Inc.; Sumitomo Chemcial Co. Ltd.*

**3755 (R)-α-Methyl-4-methoxybenzylamine**

$C_9H_{13}NO$
Chiral building block. *Chiragene, Inc.; Sumitomo Chemcial Co. Ltd.*

**3756 (S)-α-Methyl-4-methoxybenzylamine**

$C_9H_{13}NO$
Chiral building block. *Chiragene, Inc.; Sumitomo Chemcial Co. Ltd.*

**3757 (R)-1-Methyl-3-(4-methoxyphenyl)-propylamine**

$C_{11}H_{17}NO$
Propylamine, 3-(p-methoxyphenyl)-1-methyl-, (R)-; p-Methoxyphenyl-3-butylamine; (R)-3-(4-Methoxyphenyl)-1-methylpropylamine; (R)-3-Amino-1-(4-methoxyphenyl)-butane; Benzenepropanamine, 4-methoxy-α-methyl-, (R)-. Chiral building block. *Chiragene, Inc.*

**3758 (S)-1-Methyl-3-(4-methoxyphenyl)-propylamine**

$C_{11}H_{17}NO$
Propylamine, 3-(p-methoxyphenyl)-1-methyl-, (S)-; (S)-1-p-Methoxyphenyl-3-butylamine; (S)-3-(4-Methoxy-phenyl)-1-methylpropylamine; (S)-3-Amino-1-(4-meth-oxyphenyl)butane; Benzenepropanamine, 4-methoxy-α-methyl-, (S)-. Chiral building block. *Chiragene, Inc.*

**3759 (R)-(+)-N-Methyl-α-4-methylbenzylamine**
5933-40-4

$C_9H_{14}N$
(R)-α-Methyl-4-methylbenzylamine; N-Methyl-1-phenyl-ethylamine. Chiral building block. $bp_{11}$ = 74-76°; d = 0.93; n = 1.0511; $[\alpha]_D^{20}$ = + 70° (c = 2, $CHCl_3$). *Arran Chemical Company Ltd.; Chiragene, Inc.; FineTech, Ltd.; Norse Laboratories; Sigma-Aldrich Fine Chemicals; Zeeland Chemicals, Inc.*

**3760 (S)-(-)-N-Methyl-α-methylbenzylamine**
19131-99-8

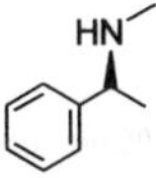


$C_9H_{13}N$
(S)-α-Methyl-4-methylbenzylamine; (S)-(-)-N-Methyl-1-phenylethylamine. Chiral building block. $bp_{11}$ = 74-76°; d = 0.93; n = 1.512; $[\alpha]_D^{20}$ = - 72° (c = 2, $CHCl_3$). *Arran Chemical Company Ltd.; Chiragene, Inc.; FineTech, Ltd.; Norse Laboratories; Sigma-Aldrich Fine Chemicals; Zeeland Chemicals, Inc.*

**3761 Methyl (R)-(+)-3-methylglutarate**
63473-60-9

$C_7H_{12}O_4$
Chiral building block. $bp_{0.4}$ = 103-104°; d = 1.125; n = 1.438; $[\alpha]_D^{20}$ = + 0.45° (neat). *Sigma-Aldrich Fine Chemicals.*

**3762 (S)-Methyl-2-methyl-3-hydroxypropionate**

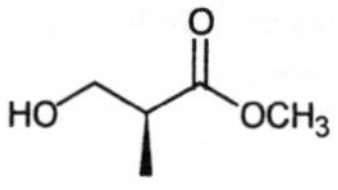

$C_5H_9O_3$
Chiral building block. *Mitsubishi Chemical Corporation.*

**3763 (1R,2S,5R)-(+)-5-Methyl-2-(1-methyl-1-phenylethyl)cycohexyl chloroacetate**
71804-27-8

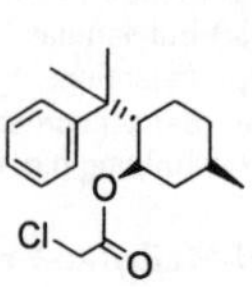

$C_{18}H_{25}ClO_2$
(+)-8-Phenylmenthyl chloroacetate. Chiral intermediate. mp = 82-84°; $[\alpha]^{25}$ = + 22° (c = 1.2, $CCl_4$). *Sigma-Aldrich Fine Chemicals.*

**3764 Methyl (methyl 1-thio-2,3,4-tri-O-acetyl-β-D-galactopyranoside)uronate**

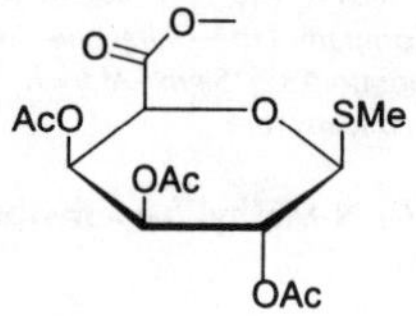

$C_{14}H_{20}O_9S$
Chiral intermediate. *Pfanstiehl Laboratories, Inc.*

**3765 Methyl (methyl 1-thio-2,3,4-tri-O-acetyl-β-D-glucopyranoside)uronate**
29587-10-8
$C_{14}H_{20}O_9S$
Chiral intermediate. *Pfanstiehl Laboratories, Inc.*

**3766 (R)-(+)-Methyl-2-naphthalenemethanol**
52193-85-8

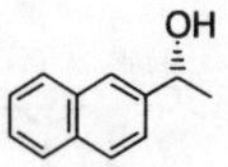

$C_{12}H_{12}O$
Chiral building block. mp = 68-70°; $[\alpha]_D^{20}$ = + 38° (c = 5, EtOH). *Sigma-Aldrich Fine Chemicals.*

**3767 (S)-(-)-Methyl-2-naphthalenemethanol**
27544-18-9

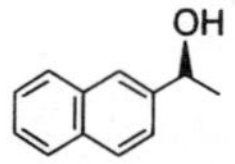

$C_{12}H_{12}O$
Chiral building block. mp = 70-71°; $[\alpha]_D^{20}$ = - 40° (c = 5, EtOH). *Sigma-Aldrich Fine Chemicals.*

**3768 (S)-(-)-1-Methyl-2-(1-naphthylamino-methyl)pyrrolidine**
82160-07-4
$C_{16}H_{20}N_2$
(S)-(-)-N-(1-Methyl-2-pyrrolidinylmethyl)-1-naphthylamine. Chiral intermediate. mp = 82°. *TCI America.*

**3769 (R)-(+)-N-Methyl-α-(1-naphthyl)ethylamine**
15297-33-3

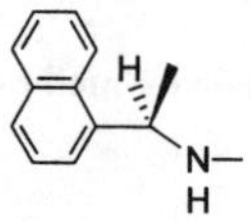

$C_{13}H_{15}N$
Chiral building block. bp = 261°; d = 1.041; n = 1.605; $[\alpha]^{24}$ = + 89° (c = 4, EtOH). *FineTech, Ltd.; Norse Laboratories; Sigma-Aldrich Fine Chemicals.*

**3770 (S)-(-)-N-Methyl-α-(1-naphthyl)ethylamine**
20218-55-7

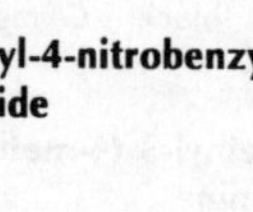

$C_{13}H_{15}N$
Chiral building block. bp = 203°; d = 1.04; n = 1.605; $[\alpha]^{23}$ = - 89° (c = 4, EtOH). *FineTech, Ltd.; Norse Laboratories; Sigma-Aldrich Fine Chemicals.*

**3771 (R)-α-Methyl-4-nitrobenzylamine hydrochloride**
57233-86-0

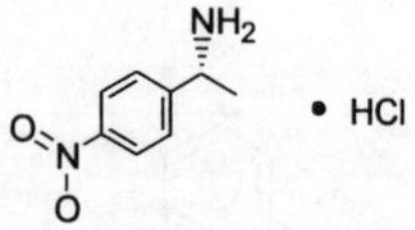

$C_8H_{11}ClN_2O_2$
Chiral building block. mp = 250°; $[\alpha]^{25}$ = + 6.0° (c = 1, 0.05N NaOH). *Sigma-Aldrich Fine Chemicals.*

**3772 (S)-α-Methyl-4-nitrobenzylamine hydrochloride**
132873-57-5

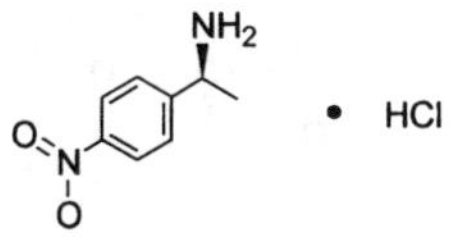

$C_8H_{11}ClN_2O_2$
(S)-1-(4-Nitrophenyl)ethylamine hydrochloride; (S)-Nitresolve. Chiral building block. mp = 250°; $[\alpha]^{25}$ = - 6.5° (c = 1, 0.05N NaOH). *EMS-Dottikon AG; FineTech, Ltd.; Sigma-Aldrich Fine Chemicals; TCI America; Yamakawa Chemical Industry Co. Ltd.*

**3773 (1R)-(1,2,3)-(+)-3-Methyl-2-(nitromethyl)-5-oxocyclopentaneacetic acid**
86023-17-8

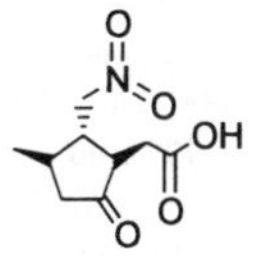

$C_9H_{13}NO_5$
Chiral intermediate. mp = 58-60°; $[\alpha]_D^{20}$= + 110° (c = 1, $CHCl_3$). *Sigma-Aldrich Fine Chemicals.*

**3774 (R)-(-)-10-Methyl-1(9)-octal-2-one**
63975-59-7

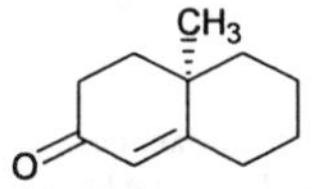

$C_{11}H_{16}O$
Chiral intermediate. $bp_{17}$ = 155°; d = 1.013; n = 1.525; $[\alpha]^{19}$ = - 208.3° (c = 1, EtOH). *Acros Organics nv; Oxford Asymmetry International plc; Sigma-Aldrich Fine Chemicals.*

**3775 (S)-(+)-10-Methyl-1(9)-octal-2-one**
4087-39-2

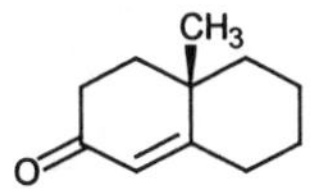

$C_{11}H_{16}O$
Chiral intermediate. [α] = + 207° (c = 1, EtOH). *Acros Organics nv; Oxford Asymmetry International plc.*

**3776 (S)-2-Methyloctanoic acid**

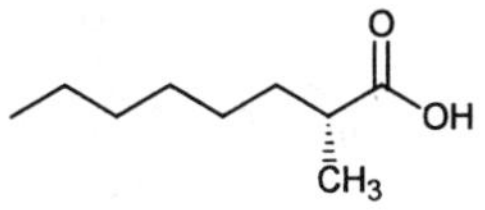

$C_9H_{18}O_2$
Octanoic acid, 2-methyl-, (S)-; (S)-α-Methylcaprylic acid. Chiral building block. *Japan Energy Corporation.*

**3777 (S)-2-Methyl-1-octanol**

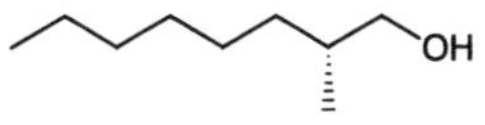

$C_9H_{20}O$
Octan-1-ol, 2-methyl-, (S)-; (S)-1-Hydroxy-2-methyloctane. Chiral building block. *Japan Energy Corporation.*

**3778 (S)-3-Methyl-3-octanol**

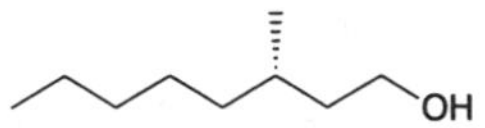

$C_9H_{20}O$
Octan-3-ol, 3-methyl-, (S)-; (S)-Amylethylmethylcarbinol. Chiral building block. *Japan Energy Corporation.*

**3779 (S)-(+)-6-Methyl-1-octanol**
110453-78-6
$C_9H_{20}O$
Chiral building block. d = 0.83. *TCI America.*

**3780 (R)-(-)-2-Methyl-2,4-pentanediol**
99210-90-9

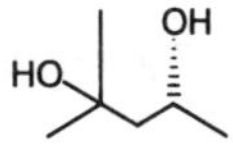

$C_6H_{14}O_2$
(R)-(-)-Hexylene glycol. Chiral building block. $bp_2$ = 197-198°; d = 0.9200; $[\alpha]_D^{20}$= - 21 ± 1° (neat). *Kiralchem Ltd.; Lancaster Synthesis Ltd.; Sigma-Aldrich Fine Chemicals.*

**3781 (S)-(+)-3-Methyl-1-pentanol**
42072-39-9
$C_6H_{14}O$
Pentan-1-ol, 3-methyl-, (S)-. Chiral building block. *TCI America.*

**3782 N-Methyl-D-phenylalanine-N-methylamide**

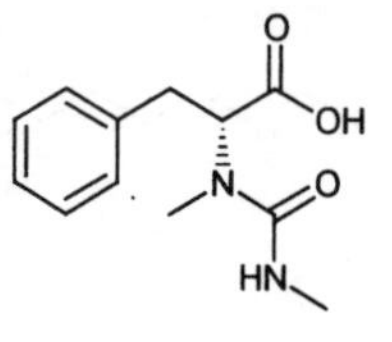

$C_{12}H_{16}N_2O_3$
Chiral intermediate. *Synthetech, Inc.*

**3783 (1S,4S)-(-)-2-(2-Methylphenyl)-2,5-diazabicyclo[2.2.1]heptane dihydrobromide**

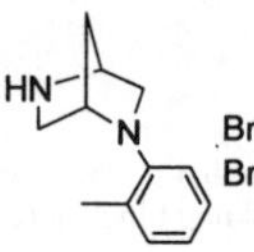

$C_{12}H_{18}Br_2N_2$
Chiral intermediate. mp = 180-184°; $[\alpha]^{23}$ = - 23° (c = 1, $H_2O$). *Sigma-Aldrich Fine Chemicals.*

**3784 (1S,4S)-(-)-2-(4-Methylphenyl)-2,5-diazabicyclo[2.2.1]heptane maleate salt**

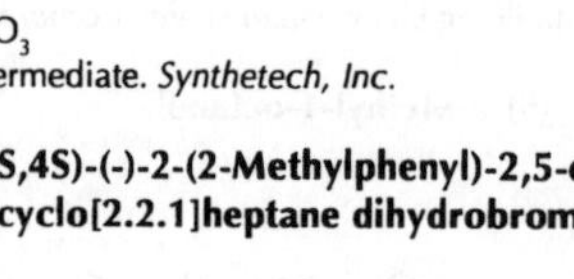

$C_{12}H_{16}N_2$
Chiral intermediate. mp = 174-178°; $[\alpha]_D^{20}$ = - 72° (c = 1, $H_2O$). *Sigma-Aldrich Fine Chemicals.*

**3785 Methyl (R)-2-phenyl-1,3-dioxolane-4-acetate**

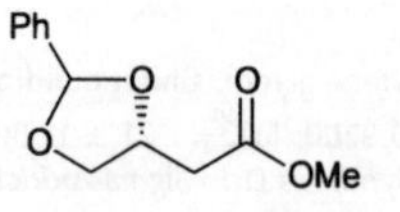

$C_{12}H_{14}O_4$
Dioxolane-4-acetic acid, 1,3-, 2-phenyl-, methyl ester, (R)-; Chiral intermediate. *Synthon Chiragenics Corporation.*

**3786 Methyl (4S)-2-phenyl-1,3-dioxolane-4-acetate**
191354-59-3

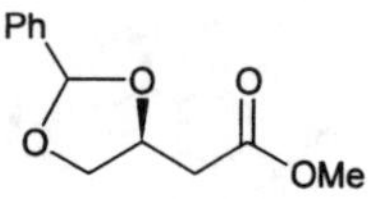

$C_{12}H_{14}O_4$
Dioxolane-4-acetic acid, 1,3-, 2-phenyl-, methyl ester, (S)-. Chiral intermediate. *Synthon Chiragenics Corporation.*

**3787 (R)-1-(4-Methylphenyl)ethylamine**
4187-38-6

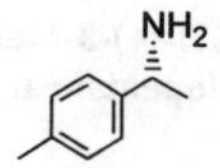

$C_9H_{13}N$
(R)-α,p-Dimethylbenzylamine; (R)-(+)-4-(1-Aminoethyl)toluene; (R)-(+)-1-(p-Tolyl)ethylamine. Chiral intermediate. bp = 205°; $bp_{27}$ = 105°; $bp_{11.4}$ = 77°; d = 0.9190; $n_D^{20}$ = 1.5215; $[\alpha]_D^{20}$ = + 37° (neat). *BASF Aktiengesellschaft; Lancaster Synthesis Ltd.; Sigma-Aldrich Fine Chemicals; TCI America; Yamakawa Chemical Industry Co. Ltd.*

**3788 (S)-1-(4-Methylphenyl)ethylamine**
27298-98-2

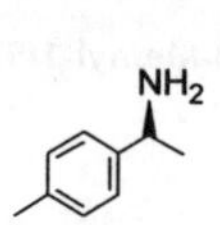

$C_9H_{13}N$
(S)-α,p-Dimethylbenzylamine; (S)-(-)-1-(p-Tolyl)ethylamine; (S)-(-)-4-(1-Aminoethyl)toluene. Chiral building block. bp = 205°; $bp_{27}$ = 105°; $bp_{11.4}$ = 77°; d = 0.9190; $n_D^{20}$ = 1.5220; $[\alpha]_D^{20}$ = - 37° (neat). *BASF Aktiengesellschaft; Lancaster Synthesis Ltd.; Sigma-Aldrich Fine Chemicals; TCI America; Yamakawa Chemical Industry Co. Ltd.*

**3789 (1R,2S)-(-)-trans-2-(1-Methyl-1-phenylethyl)cyclohexanol**
109527-43-7

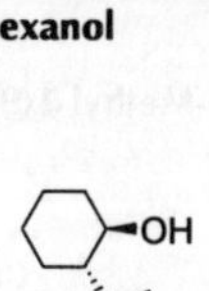

$C_{15}H_{22}O$
Chiral intermediate. bp = 300°; d = 1.026; n = 1.541; $[\alpha]^{24}$ = - 29° (c = 1.7, $CH_3OH$). *Sigma-Aldrich Fine Chemicals.*

**3790 (1S,2R)-(+)-trans-2-(1-Methyl-1-phenylethyl)cyclohexanol**
109527-45-9

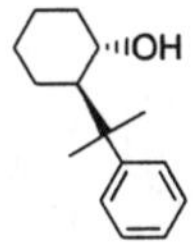

$C_{15}H_{22}O$
Chiral building block. bp = 300°; d = 1.008; n = 1.541; $[\alpha]^{24}$ = + 29° (c = 1.7, $CH_3OH$). *Sigma-Aldrich Fine Chemicals.*

**3791 (2S,3S)-(-)-2-Methyl-3-phenylglycidol**
107033-44-3

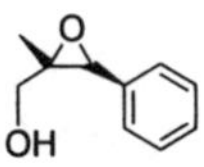

$C_{10}H_{12}O_2$
Chiral building block. $[\alpha]_D^{20}$ = - 15° (c = 2, $CHCl_3$). *Sigma-Aldrich Fine Chemicals.*

**3792 L-α-Methylphenylglycine**

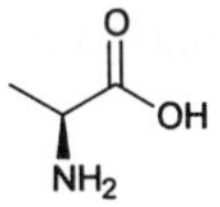

$C_3H_7NO_2$
Chiral building block. *DSM Fine Chemcials Netherlands.*

**3793 (S,E)-(-)-Methyl-4-(phenylmethoxy)-pent-2-enoate**

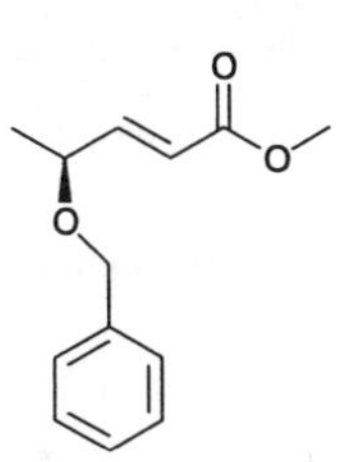

$C_{13}H_{16}O_3$
Chiral intermediate. *Acros Organics nv.*

**3794 (4R,5S)-(+)-4-Methyl-5-phenyl-1,3-oxazolidine-2-thione**
91794-28-4
$C_{10}H_{11}NOS$
Resolving agent. mp = 90°; $[\alpha]_D^{20}$ = + 220 ± 4° (c = 0.4, $CHCl_3$). *Kiralchem Ltd.; Lancaster Synthesis Ltd.*

**3795 (4S,5R)-(-)-4-Methyl-5-phenyl-1,3-oxazolidine-2-thione**
$C_{10}H_{11}NOS$
Chiral building block. mp = 81-82°; $[\alpha]_D^{20}$ = - 223.0 ± 2.0° (c = 1, $CHCl_3$). *Kiralchem Ltd.*

**3796 (4R,5S)-(+)-4-Methyl-5-phenyl-2-oxazolidinone**
77943-39-6

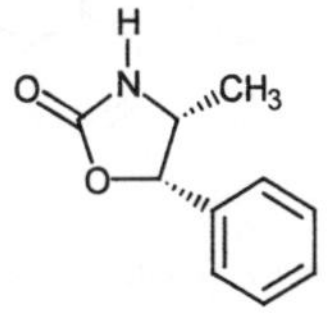

$C_{10}H_{11}NO_2$
Chiral auxiliary. mp = 121-123°; $[\alpha]_D^{20}$ = + 160 ± 3° (c = 1, $CHCl_3$). *Acros Organics nv; Lancaster Synthesis Ltd.; Oxford Asymmetry International plc; Sigma-Aldrich Fine Chemicals.*

**3797 (4S,5R)-(-)-4-Methyl-5-phenyl-2-oxazolidinone**
16251-45-9

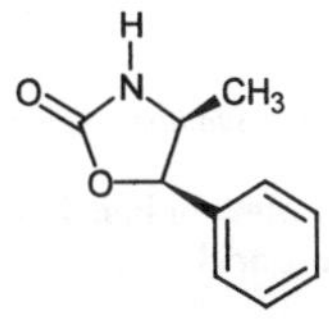

$C_{10}H_{11}NO_2$
Chiral intermediate. mp = 106-109°; $[\alpha]^{25}$ = - 168° (c = 2, $CHCl_3$). *Acros Organics nv; Lancaster Synthesis Ltd.; Oxford Asymmetry International plc; Sigma-Aldrich Fine Chemicals.*

**3798 (R)-(+)-2-Methyl-1-phenyl-1-propanol**
14898-86-3

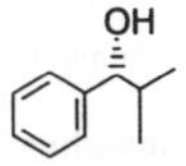

$C_{10}H_{14}O$
Benzenemethanol, α-(1-methylethyl)-, (R)-; (+)-α-Isopropylbenzyl alcohol; (R)-α-(1-Methylethyl)-benzenemethanol. Chiral building block. $bp_{15}$ = 124-125°; $bp_{10}$ = 110-11°; d = 0.9640; $n_D^{20}$ = 1.5130; $[\alpha]_D^{20}$ = + 21 ± 2° (neat). *Kiralchem Ltd.; Lancaster Synthesis Ltd.; Sigma-Aldrich Fine Chemicals.*

**3799 (S)-(-)-2-Methyl-1-phenyl-1-propanol**
34857-28-8
$C_{10}H_{14}O$
Benzenemethanol, α-(1-methylethyl)-, (S)-. (S)-(-)-Isopropylphenylcarbinol. Chiral building block. $bp_{10}$ =

103-105°; d = 0.9640; $n_D^{20}$ = 1.5130; $[\alpha]_D^{20}$ = - 21 ± 2° (neat). *Lancaster Synthesis Ltd.*

**3800 (R)-2-(2-Methylphenyl)-3-phenylpropionic acid**

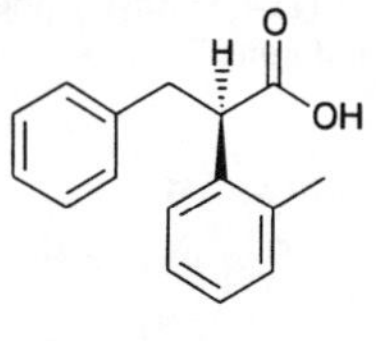

$C_{16}H_{16}O_2$
Chiral building block. *Sumitomo Chemcial Co. Ltd.*

**3801 (S)-2-(2-Methylphenyl)-3-phenylpropionic acid**

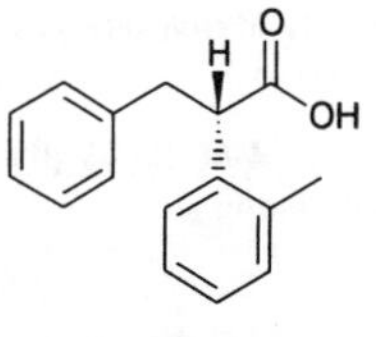

$C_{16}H_{16}O_2$
Chiral intermediate. *Sumitomo Chemcial Co. Ltd.*

**3802 (R)-2-(3-Methylphenyl)-3-phenylpropionic acid**

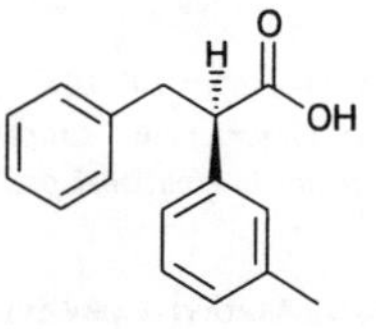

$C_{16}H_{16}O_2$
Chiral building block. *Sumitomo Chemcial Co. Ltd.*

**3803 (S)-2-(3-Methylphenyl)-3-phenylpropionic acid**

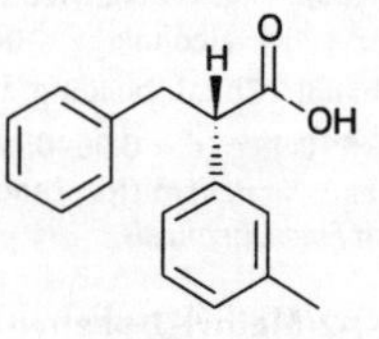

$C_{16}H_{16}O_2$
Chiral intermediate. *Sumitomo Chemcial Co. Ltd.*

**3804 (R)-2-(4-Methylphenyl)-3-phenylpropionic acid**

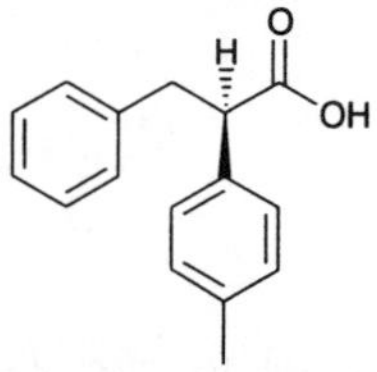

$C_{16}H_{16}O_2$
Chiral building block. *Sumitomo Chemcial Co. Ltd.*

**3805 (S)-2-(4-Methylphenyl)-3-phenylpropionic acid**

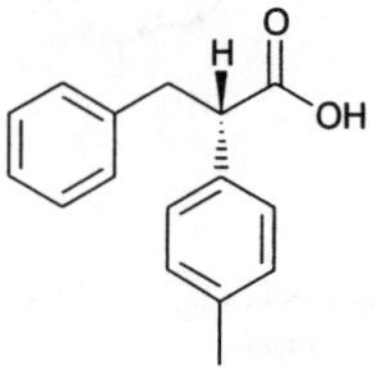

$C_{16}H_{16}O_2$
Chiral intermediate. *Sumitomo Chemcial Co. Ltd.*

**3806 (R)-(-)-1-Methyl-3-phenylpropylamine**

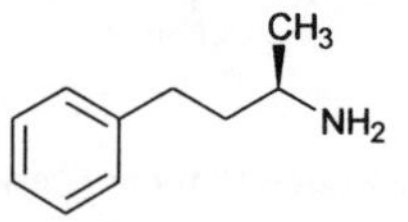

$C_{10}H_{15}N$
Benzenepropanamine, α-methyl-, (R)-; (R)-α-Methylbenzenepropanamine; (R)-1-Phenyl-3-aminobutane; (R)-2-Amino-4-phenylbutane; (R)-3-Amino-1-phenylbutane; (R)-4-Phenyl-2-aminobutane; Propylamine, 1-methyl-3-phenyl-, (R)-. Chiral building block. *Austin Chemical Company, Inc.*

**3807 (S)-(+)-1-Methyl-3-phenylpropylamine**

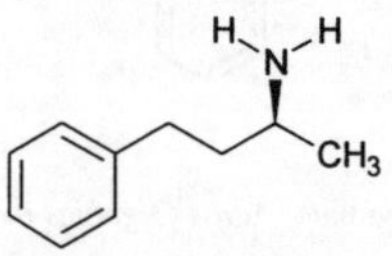

$C_{10}H_{15}N$
Benzenepropanamine, α-methyl-, (S)-; (S)-α-Methylbenzenepropanamine; (S)-1-Phenyl-3-aminobutane; (S)-2-Amino-4-phenylbutane; (S)-3-Amino-1-phenylbutane; (S)-4-Phenyl-2-aminobutane; Propylamine, 1-methyl-3-

phenyl-, (S)-. Chiral building block. *Acros Organics nv; Austin Chemical Company, Inc.*

**3808 (R)-N-Methyl-1-phenyl-2-(1-pyrrolidino)-ethylamine**
136329-39-0

$C_{13}H_{20}N_2$
(R)-N-(2-Propyl)-1-phenyl-2-(1-pyrrolidino)ethylamine; (R)-N-Isopropyl-1-phenyl-2-(1-pyrrolidino)ethylamine. Chiral intermediate. *Lancaster Synthesis Ltd.*

**3809 (R)-(-)-S-Methyl-S-phenylsulphoximine**
60933-65-5

$C_8H_{13}NOS$
Chiral building block. mp = 29-31°; $[\alpha]_D^{20}$ = - 36 ± 2° (c = 1, acetone). *Kiralchem Ltd.; Lancaster Synthesis Ltd.*

**3810 (S)-(+)-S-Methyl-S-phenylsulphoximine**
33903-50-3

$C_7H_{11}NOS$
Chiral building block. mp = 30-32°; $[\alpha]_D^{20}$ = + 36 ± 1.5° (c = 1, acetone). *Kiralchem Ltd.; Lancaster Synthesis Ltd.*

**3811 (3R-cis)-7a-Methyl-3-phenyltetrahydro-pyrrolo[2,1-b]oxazol-5(6H)-one**
137869-70-6

$C_{15}H_{18}O_6$
(3R-cis)-Tetrahydro-7a-methyl-3-phenylpyrrolo[2,1-b]-oxazol-5(6H)-one. Chiral intermediate. mp = 133-135°; $[\alpha]_D^{20}$ = - 188 ± 5° (c = 1, $CCl_4$). *Acros Organics nv; Lancaster Synthesis Ltd.*

**3812 (3S-cis)-7a-Methyl-3-phenyltetrahydro-pyrrolo[2,1-b]oxazol-5(6H)-one**
153745-22-3

$C_{13}H_{15}NO_2$
(3S-cis)-Tetrahydro-7a-methyl-3-phenylpyrrolo[2,1-b]-oxazol-5(6H)-one. Chiral intermediate. mp = 133-135°; $[\alpha]_D^{20}$ = + 188 ± 5° (c = 1, $CCl_4$). *Acros Organics nv; Lancaster Synthesis Ltd.*

**3813 (R)-(-)-2-Methylpiperazine**
75336-86-6

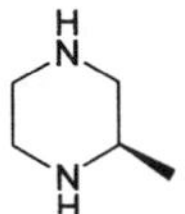

$C_5H_{12}N_2$
Chiral building block. mp = 89-91°; bp = 155-156°; $[\alpha]_D^{20}$ = - 7 ± 0.5° (c = 1, EtOH); $[\alpha]_D^{20}$ = - 16.8 ± 0.2° (c = 5, $C_6H_6$). *Kiralchem Ltd.; Lancaster Synthesis Ltd.; Sigma-Aldrich Fine Chemicals; Toray International, Inc.*

**3814 (S)-(+)-2-Methylpiperazine**
74879-18-8

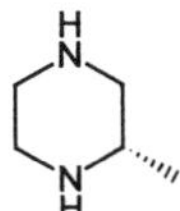

$C_5H_{12}N_2$
Chiral building block. mp = 89-91°; bp = 155-156°; $[\alpha]_D^{20}$ = + 7 ± 0.5° (c = 1, EtOH); $[\alpha]_D^{20}$ = + 16.8 ± 0.2° (c = 5, $C_6H_6$). *Austin Chemical Company, Inc.; Daicel Chemical Ind. Ltd; Kiralchem Ltd.; Lancaster Synthesis Ltd.; Sigma-Aldrich Fine Chemicals; Toray International, Inc.*

**3815 (R)-(+)-Methylpiperidine**

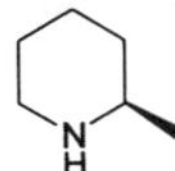

$C_6H_{13}N$
Piperidine, 2-methyl-, (R)-; (R)-α-Pipecoline. Chiral building block. bp = 118-119°; d = 0.844; n = 1.446; $[\alpha]_D^{20}$ = + 35.0 ± 0.2° (neat). *Austin Chemical Company, Inc.; Kiralchem Ltd.; Sigma-Aldrich Fine Chemicals.*

**3816 (S)-(-)-2-Methylpiperidine**

$C_6H_{13}N$
Piperidine, 2-methyl-, (S)-; (S)-α-Pipecoline. Chiral building block. bp = 118-119°; d = 0.844; $[\alpha]_D^{20}$ = - 35.0 ± 0.2° (neat). *Kiralchem Ltd.*

**3817 (S)-(-)-1-Methyl-2-(1-piperidinomethyl)-pyrrolidine**
84466-85-3

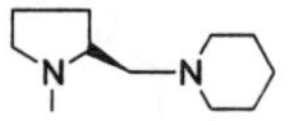

$C_{11}H_{22}N_2$
(S)-(-)-1-[(1-Methyl-2-pyrrolidinyl)methyl]piperidine. Chiral intermediate. $bp_{22}$ = 121-123°; d = 0.909; n = 1.478; $[\alpha]_D^{20}$ = - 69° (c = 0.5, EtOH). *Sigma-Aldrich Fine Chemicals; TCI America.*

**3818 (R)-(+)-2-Methyl-2-propanesulfinamide**

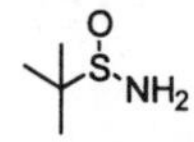

$C_4H_{11}NOS$
Chiral building block. mp = 103-107°; $[\alpha]_D^{20}$ = + 4° (c = 1.0242, $CHCl_3$ + amylenes). *Sigma-Aldrich Fine Chemicals.*

**3819 (S)-(-)-2-Methyl-2-propanesulfinamide**

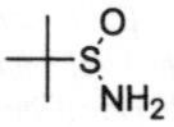

$C_4H_{11}NOS$
Chiral building block. mp = 97-101°; $[\alpha]_D^{20}$ = - 4.5° (c = 1, $CHCl_3$). *Sigma-Aldrich Fine Chemicals.*

**3820 (R)-1-Methylpropyl 4-methylbenzyl-sulfonate**
61530-30-1
$C_{11}H_{16}O_3S$
Chiral intermediate. *Sigma-Aldrich Fine Chemicals.*

**3821 (S)-1-Methylpropyl 4-methylbenzene-sulfonate**
50896-54-3
$C_{11}H_{16}O_3S$
Chiral intermediate. *Sigma-Aldrich Fine Chemicals.*

**3822 (R)-(+)-Methyl-4-pyridinemethanol**
27854-88-2

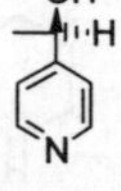

$C_7H_9NO$
1-(4-Pyridyl)ethanol. Chiral building block. mp = 67-69°; $[\alpha]_D^{20}$ = + 55° (c = 1, $CHCl_3$). *Sigma-Aldrich Fine Chemicals.*

**3823 (S)-(-)-Methyl-4-pyridinemethanol**
54656-96-1

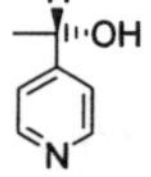

$C_7H_9NO$
(S)-(-)-1-(4-Pyridyl)ethanol. Chiral building block. mp = 67-69°; $[\alpha]_D^{20}$ = - 58° (c = 1, $CHCl_3$). *Sigma-Aldrich Fine Chemicals.*

**3824 (2S,3S)-3-Methylpyrrolidine-2-carboxylic acid**

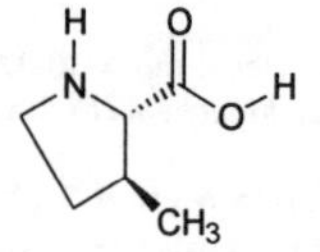

$C_6H_{11}NO_2$
Chiral building block. $[\alpha]$ = - 29.5° (c = 0.5, $H_2O$). *Acros Organics nv.*

**3825 (2S)-5-Methylpyrrolidine-2-carboxylic acid, mixture of cis/trans, ca. 1:2**
ID1039
$C_6H_{11}NO_2$
Chiral intermediate. $[\alpha]$ = - 73.5° (c = 0.6, $CH_3OH$). *Acros Organics nv.*

**3826 Methyl (S)-(+)-2-pyrrolidine-5-carboxylate**
4931-66-2 225-567-3

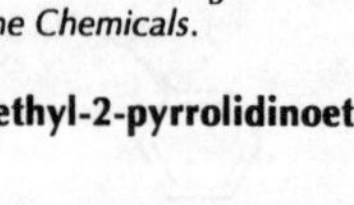

$C_6H_9NO_3$
Methyl L-pyroglutamate. Chiral building block. $bp_{0.3}$ = 90°; d = 1.226; n = 1.486; $[\alpha]_D^{20}$ = + 10.5° (c = 1, EtOH). *Sigma-Aldrich Fine Chemicals.*

**3827 (+)-1-Methyl-2-pyrrolidinoethanol**

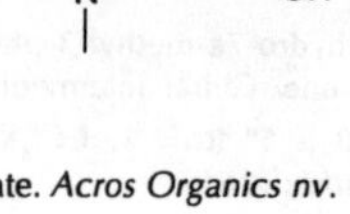

$C_7H_{15}NO$
Chiral intermediate. *Acros Organics nv.*

**3828 (S)-(-)-1-Methyl-2-pyrrolidinemethanol**
34381-71-0 251-981-9

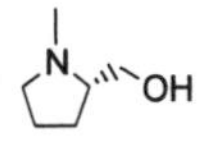

$C_6H_{13}NO$
Chiral intermediate. $bp_{12}$ = 67-69°; d = 0.968; n = 1.469; $[\alpha]^{19}$ = - 49.5° (c = 5, $CH_3OH$). *Sigma-Aldrich Fine Chemicals.*

**3829 (+)-1-Methyl-2-pyrrolidinoethanol**

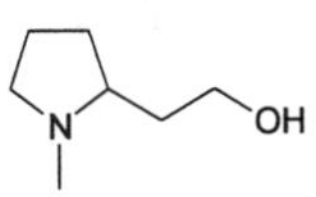

$C_7H_{15}NO$
Chiral intermediate. *Acros Organics nv.*

**3830 (R)-(+)-Methylsuccinic acid**
3641-51-8 207-857-1

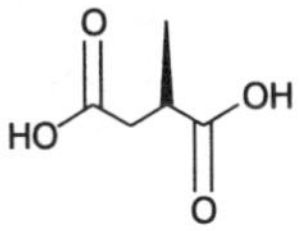

$C_5H_8O_4$
(R)-1,2-Propanedicarboxylic acid; (R)-2-Methylbutanedioic acid; (R)-Pyrotartaric acid; Succinic acid, methyl-, (R)-. Listed on TSCA. Chiral building block. mp = 114-117°; $[\alpha]^{25}$ = + 8.0° (c = 5, $H_2O$). *Mitsubishi Chemical Corporation; Sigma-Aldrich Fine Chemicals; TCI America; Toyotama Perfumery Co., Ltd.*

**3831 Methylsuccinic acid dimethyl ester**
22644-27-5

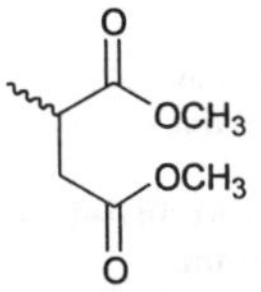

$C_7H_{12}O_4$
Pyrotartaric acid dimethyl ester; Dimethyl methylsuccinate. Chiral intermediate. $bp_{12}$ = 80-81°; d = 1.076; n = 1.418; $[\alpha]_D^{20}$ = + 4.8° (c = 2.9, $CHCl_3$). *Sigma-Aldrich Fine Chemicals; TCI America.*

**3832 (S)-(-)-Methylsuccinic acid dimethyl ester**
63163-08-6
$C_7H_{12}O_4$
(S)-(-)-Pyrotartaric acid dimethyl ester; Dimethyl (S)-(-)-methylsuccinate; Dimethyl (S)-(-)-pyrotartarate. Chiral building block. *TCI America.*

**3833 (R)-(+)-3-Methylsuccinic acid 1-monomethyl ester**
81025-83-4
$C_6H_{10}O_4$
(R)-(+)-3-Carboxy-n-butyric acid methyl ester; (R)-(+)-3-Methoxycarbonyl-2-methylpropionic acid; 1-Monomethyl (R)-(+)-3-methylsuccinate. Chiral intermediate. *TCI America.*

**3834 Methyl 2,3,4,6-tetra-O-acetyl-1-thio-β-D-galactopyranoside**
55722-48-0

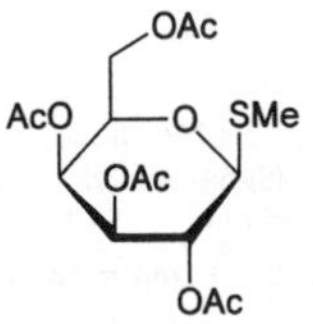

$C_{15}H_{22}O_9S$
Chiral intermediate. *Pfanstiehl Laboratories, Inc.*

**3835 Methyl 2,3,4,6-tetra-O-benzyl-1-thio-β-D-galactopyranoside**
97205-08-8
$C_{35}H_{38}O_5S$
Chiral intermediate. *Pfanstiehl Laboratories, Inc.*

**3836 Methyl 2,3,4,6-tetra-O-acetyl-1-thio-β-D-glucopyranoside**
13350-45-3
$C_{15}H_{22}O_9S$
Chiral intermediate. *Pfanstiehl Laboratories, Inc.*

**3837 Methyl 2,3,4,6-tetra-O-acetyl-1-thio-α-D-mannopyranoside**
64550-71-6

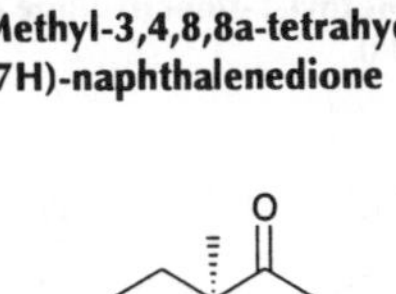

$C_{15}H_{22}O_9S$
Chiral intermediate. *Pfanstiehl Laboratories, Inc.*

**3838 (R)-8a-Methyl-3,4,8,8a-tetrahydro-1,6(2H,7H)-naphthalenedione**

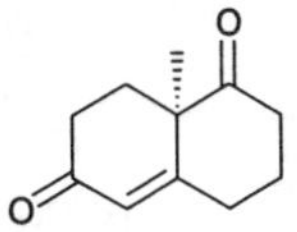

$C_{11}H_{14}O_2$
Naphthalene-1,6(2H,7H)-dione, 3,4,8,8a-tetrahydro-8a-methyl, (R)-; (R)-8a-Methyl-1,2,3,4,6,7,8,8a-octahydro-1,6-naphthalenedione, (R)-9-Methyl-D5,10-octalin-1,6-

dione. Chiral building block. mp = 48-50°; $[\alpha]_D^{20}$ = - 99.0 ± 0.5° (c = 1, $C_6H_6$). *Kiralchem Ltd.*

**3839 (S)-8a-Methyl-3,4,8,8a-tetrahydro-1,6(2H,7H)-naphthalenedione**

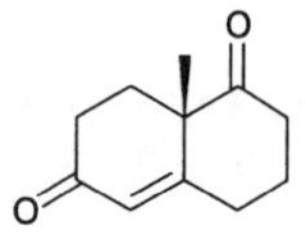

$C_{11}H_{14}O_2$
Naphthalene-1,6(2H,7H)-dione, 3,4,8,8a-tetrahydro-8a-methyl, (S)-; (S)-8a-Methyl-1,2,3,4,6,7,8,8a-octahydro-1,6-naphthalenedione, (S)-9-Methyl-D5,10-octalin-1,6-dione. Building block. mp = 48-50°; $[\alpha]_D^{20}$ = + 99.0 ± 0.5° (c = 1, $C_6H_6$). *Kiralchem Ltd.*

**3840 (1S,2R)-1-Methyl cis-1,2,3,6-tetrahydro-phthalate**
88335-93-7

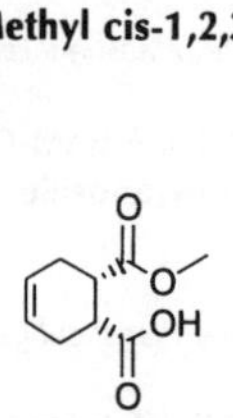

$C_9H_{12}O_4$
Chiral intermediate. mp = 65-67°; $[\alpha]_D^{20}$ = + 13° (c = 1, acetone). *Mitsubishi Chemical Corporation; Sigma-Aldrich Fine Chemicals.*

**3841 (4S,2RS)-2-Methylthiazolidine-4-carboxylic acid**

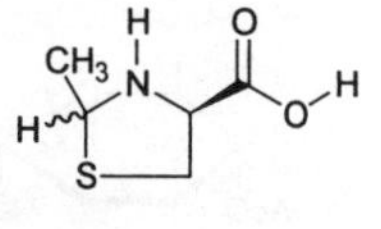

$C_5H_9NO_2S$
Chiral intermediate. *Acros Organics nv.*

**3842 S-Methyl-L-thiocitrulline dihydrochloride**
156719-39-0

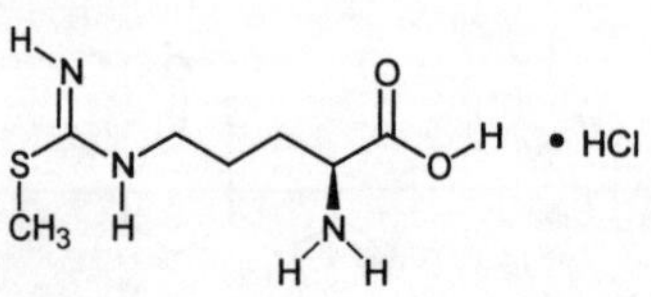

$C_7H_{17}ClN_3O_2S$
$N^5$-(Aminothioxomethyl)-L-ornithine; 2-Thioureido-L-norvaline. Chiral intermediate. *Acros Organics nv.*

**3843 (R)-4-(2-Methylthioethyl)-2-oxazolidinone**

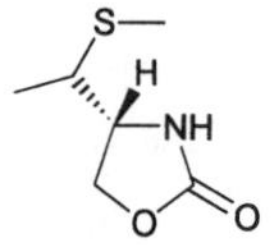

$C_6H_{11}NO_2$
Chiral intermediate. *Senn Chemicals AG.*

**3844 (S)-4-(2-Methylthioethyl)-2-oxazolidinone**

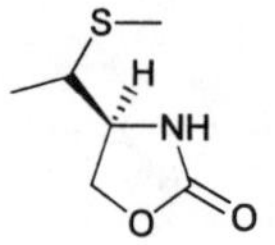

$C_6H_{11}NO_2$
Chiral intermediate. *Senn Chemicals AG.*

**3845 (R)-(+)-Methyl p-tolyl sulfoxide**
1519-39-7

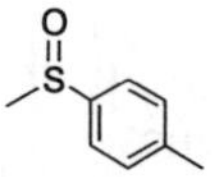

$C_8H_{10}OS$
Chiral building block. mp = 73-75°; bp = 74-76°; $[\alpha]_D^{20}$ = + 145° (c = 2, acetone). *Sigma-Aldrich Fine Chemicals.*

**3846 (S)-(-)-Methyl p-tolyl sulfoxide**
5056-07-5

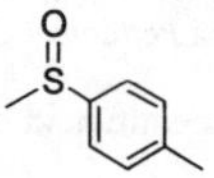

$C_8H_{10}OS$
Resolving agent. mp = 75-77°; $[\alpha]_D^{20}$ = - 145° (c = 2, acetone). *Sigma-Aldrich Fine Chemicals.*

**3847 (R)-(+)-4-Methyl-4-(trichloromethyl)-2-oxetanone**
93239-42-0

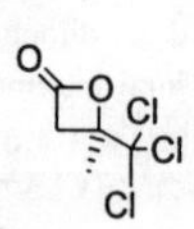

$C_5H_5Cl_3O_2$
(R)-(+)-4-(Trichloromethyl)-4-methyl-2-oxetanone; (R)-(+)-β-Trichloromethyl-β-butyrolactone. Chiral intermediate. mp = 42°; $bp_{0.1}$ = 120°; $[\alpha]^{26}$ = + 6.0° (c = 2, EtOH). *Sigma-Aldrich Fine Chemicals; TCI America.*

**3848 (S)-(-)-4-Methyl-4-(trichloromethyl)-2-oxetanone**
93206-60-1

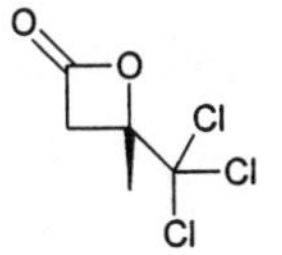

$C_5H_5Cl_3O_2$
(S)-(-)-4-(Trichloromethyl)-4-methyl-2-oxetanone; (S)-(-)-β-Trichloromethyl-β-butyrolactone. Chiral intermediate. mp = 42°; $bp_{0.1}$ = 120°; $[\alpha]^{27}$ = - 6.0° (c = 2, EtOH). *Sigma-Aldrich Fine Chemicals; TCI America.*

**3849 Methyl (4S)-trans-2,2,5-trimethyl-1,3-dioxolane-4-carboxylate**
38410-80-9

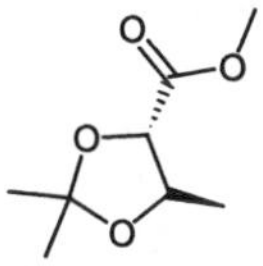

$C_8H_{14}O_4$
Chiral intermediate. $bp_1$ = 70-75°; d = 1.056; n = 1.422; $[\alpha]^{21}$ = - 19° (c = 1, $CHCl_3$). *Sigma-Aldrich Fine Chemicals.*

**3850 (-)-1-Methyltryptophan**
21339-55-9

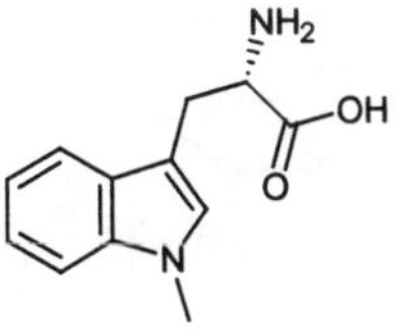

$C_{12}H_{14}N_2O_2$
Chiral intermediate. mp = 242-245°; $[\alpha]^{24}$ = - 8.0° (c = 2, HOAc). *Sigma-Aldrich Fine Chemicals.*

**3851 (+)-1-Methyltryptophan**
110117-83-4

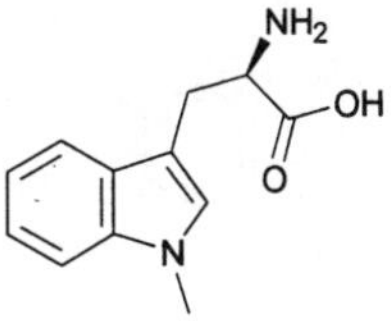

$C_{12}H_{14}N_2O_2$
Chiral building block. mp = 242-245°; $[\alpha]^{22}$ = +12.4° (c = 2, HOAc). *Sigma-Aldrich Fine Chemicals.*

**3852 N-Methyl-L-tryptophan**
526-31-8 6(12) 208-388-5

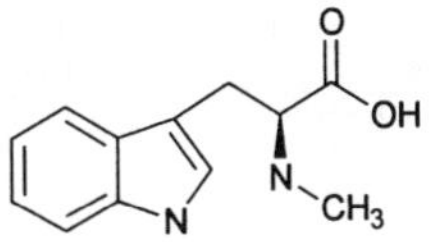

$C_{12}H_{14}N_2O_2$
Tryptophan, N-methyl-, L-; L-Abrine. Chiral building block. mp > 300°; [α] = + 44.6° (c = 2.8, 0.5N HCl); $[\alpha]_D^{20}$ = + 65° (c = 1, 0.5N NaOH). *Acros Organics nv; Sigma-Aldrich Fine Chemicals.*

**3853 Methyl (S)-(-)-1-trityl-2-aziridine-carboxylate**

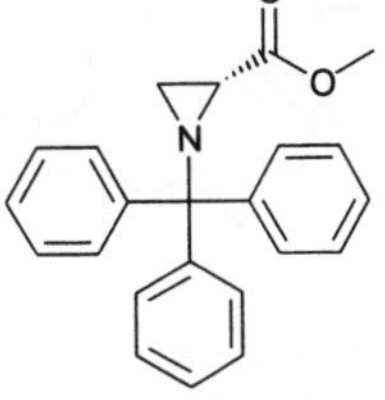

$C_{23}H_{21}NO_2$
Chiral intermediate. mp = 124-128°; $[\alpha]^{23}$ = - 88.6° (c = 1, $CHCl_3$). *Sigma-Aldrich Fine Chemicals.*

**3854 L-α-Methyltyrosine**
672-87-7 6248(12) 211-599-5

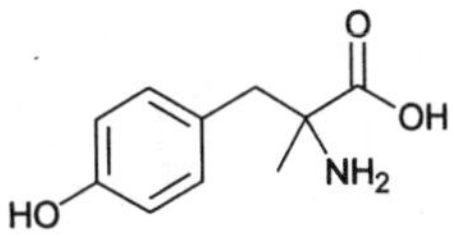

$C_{10}H_{13}NO_3$
Tyrosine, α-methyl-, L-; Metirosine. Chiral building block. $[\alpha]_D^{20}$ = - 7° (c = 1, 0.1N HCl). *Sigma-Aldrich Fine Chemicals.*

**3855 O-Methyl-D-tyrosine**

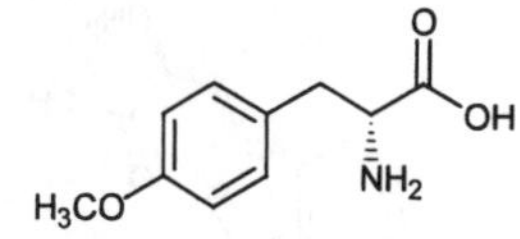

$C_{10}H_{13}NO_3$
Alanine, 3-(p-methoxyphenyl)-, D-; p-Methoxy-D-phenylalanine. Chiral intermediate. *Synthetech, Inc.*

**3856 O-Methyl-L-tyrosine**
6230-11-1 228-333-9

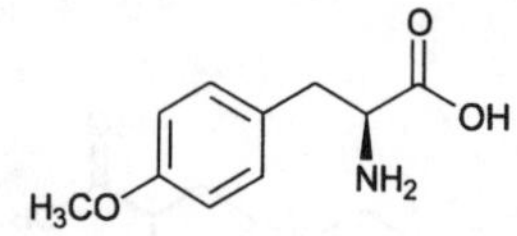

$C_{10}H_{13}NO_3$
Alanine, 3-(p-methoxyphenyl)-, L-; p-Methoxy-L-phenylalanine. Chiral building block. mp = 259-261°; [α] = - 9.5° (c = 2, 1N HCl). *Acros Organics nv; Sigma-Aldrich Fine Chemicals; Synthetech, Inc.*

**3857 4-Methylumbelliferyl-2-acetamide-2-deoxy-β-D-galactopyranoside**
36476-29-6 253-052-3

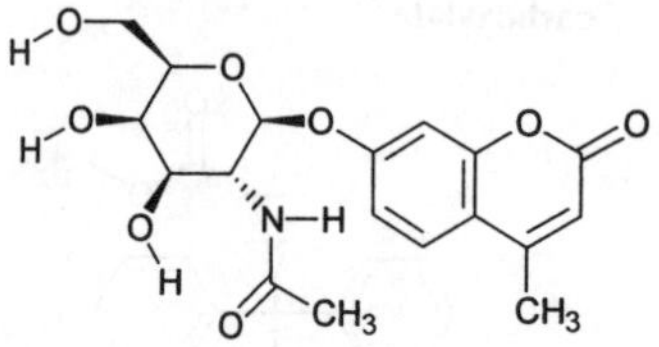

$C_{18}H_{21}NO_8$
Benzo-(2H)-1-pyran-2-one, 7-[[2-(acetylamino)-2-deoxy-β-D-galactopyranosyl]oxy]-4-methyl-; 7-[[2-Acetamido-2-deoxy-β-D-galactopyranosyl]oxy]-4-methyl-2H-1-benzo-pyran-2-one; 4-Methylumbelliferyl β-D-acetyl-galactosaminide. Chiral diagnostic reagent. *Acros Organics nv; Diagnostic Chemicals Limited; Senn Chemicals AG.*

**3858 4-Methylumbelliferyl-2-acetamide-2-deoxy-β-D-glucopyranoside**
37067-30-4 243-333-0

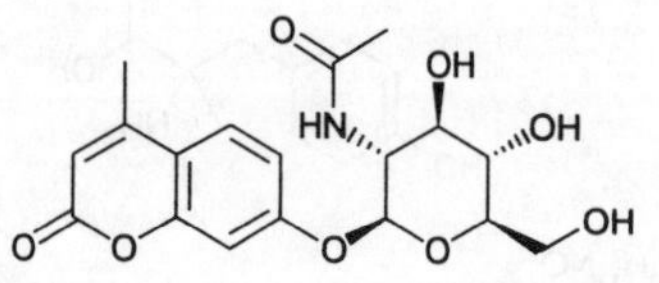

$C_{18}H_{21}NO_8$
Benzo-(2H)-1-pyran-2-one, 7-[[2-(acetylamino)-2-deoxy-β-D-glucopyranosyl]oxy]-4-methyl-; 4-Methylumbelliferyl-N-acetyl-β-D-glucosaminide; 7-[[2-acetamido-2-deoxy-β-D-glucopyranosyl]oxy] 4-methyl-2H-1-benzo-pyran-2-one; Coumarin, 7-[(2-acetamido-2-deoxy-β-D-glucopyranosyl)oxy]-4-methyl-. Listed on TSCA. Chiral diagnostic reagent. *Acros Organics nv; Diagnostic Chemicals Limited; Senn Chemicals AG.*

**3859 4-Methylumbelliferyl-α-D-arabinoside**
69414-26-2 273-991-2

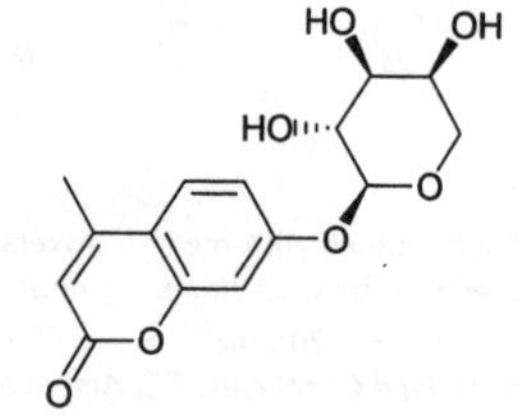

$C_{15}H_{16}O_7$
Benzo-2H-1-pyran-2-one, 7-(α-L-arabinopyranosyloxy)-4-methyl-; 4-Methylumbelliferyl α-L-arabinopyranoside. Chiral intermediate. *Senn Chemicals AG.*

**3860 4-Methylumbelliferyl-α-L-arabinoside**

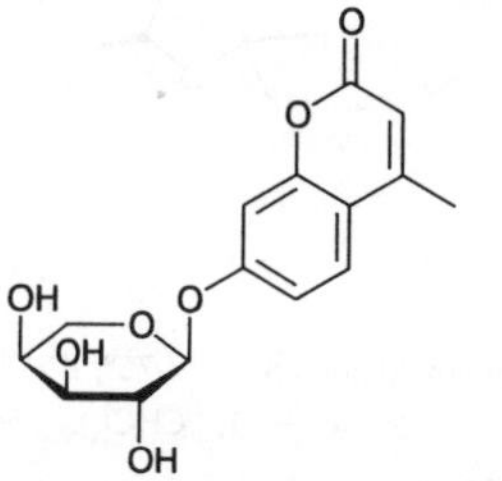

$C_{15}H_{16}O_7$
Chiral intermediate. *Senn Chemicals AG.*

**3861 4-Methylumbelliferyl-β-L-arabinoside**

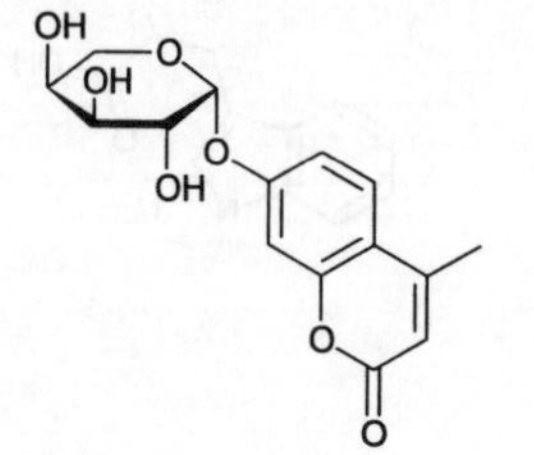

$C_{15}H_{16}O_7$
Chiral intermediate. *Senn Chemicals AG.*

**3862 4-Methylumbelliferyl-β-D-cellobioside**
72626-61-0 276-744-7

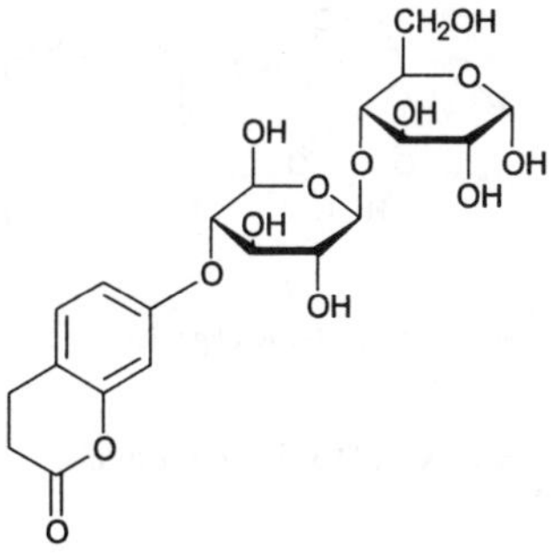

$C_{22}H_{28}O_{13}$
Benzo-(2H)-1-pyran-2-one, 7-[(4-O-β-D-glucopyranosyl-β-D-glucopyranosyl)oxy]-4-methyl-; 7-[(4-O-β-D-Glucopyranosyl-β-D-glucopyranosyl)oxy]-4-methyl-2H-1-benzopyran-2-one. Chiral diagnostic reagent. Soluble in $H_2O$. *Diagnostic Chemicals Limited.*

**3863 4-Methylumbelliferyl-α-L-fucopyranoside**
54322-38-2 259-098-0

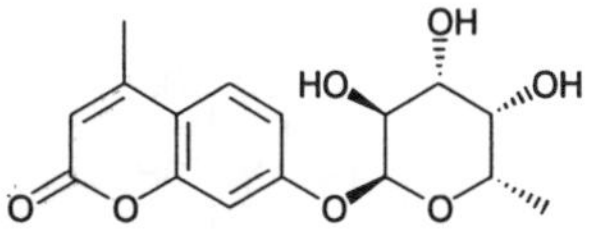

$C_{16}H_{18}O_7$
Benzo-(2H)-1-pyran-2-one, 7-[(6-deoxy-α-L-galactopyranosyl)oxy]-4-methyl-; 7-[(6-Deoxy-α-L-galactopyranosyl)oxy]-4-methyl-2H-1-benzopyran-2-one; 4-Methylumbelliferyl α-L-fucoside. Chiral diagnostic reagent. Soluble in $H_2O$. *Diagnostic Chemicals Limited.*

**3864 4-Methylumbelliferyl-β-L-fucopyranoside**
72601-82-2 276-730-0

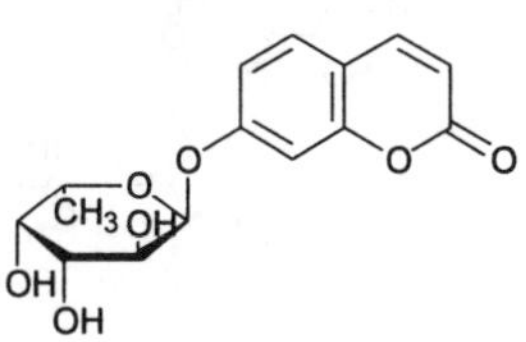

$C_{16}H_{18}O_7$
Benzo-(2H)-1-pyran-2-one, 7-[(6-deoxy-β-L-galactopyranosyl)oxy]-4-methyl-; 7-[(6-Deoxy-β-L-galactopyranosyl)oxy]-4-methyl-2H-1-benzopyran-2-one. Chiral diagnostic reagent. *Diagnostic Chemicals Limited.*

**3865 4-Methylumbelliferyl-α-D-galactopyranoside**
38597-12-5 254-031-1

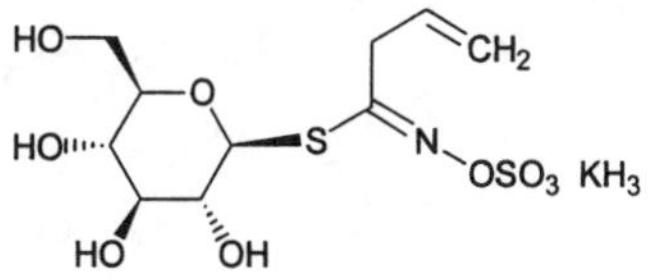

$C_{16}H_{18}O_8$
Benzo-(2H)-1-pyran, 7-(α-D-galactopyranosyloxy)-4-methyl-; 7-(α-D-Galactopyranosyloxy)-4-methyl-2H-1-benzopyran-2-one. Chiral diagnostic reagent. *Acros Organics nv; Diagnostic Chemicals Limited.*

**3866 4-Methylumbelliferyl-β-D-galactopyranoside**
6160-78-7 228-185-5

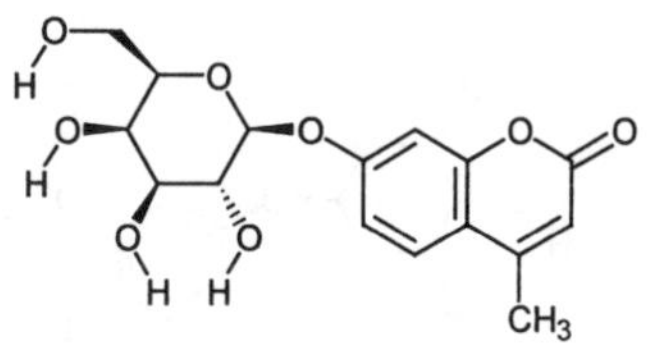

$C_{16}H_{18}O_8$
Benzo-(2H)-1-pyran-2-one, 7-(β-D-galactopyranosyloxy)-4-methyl-; 7-(β-D-Galactopyranosyloxy)-4-methyl-2H-1-benzopyran-2-one; 4-Methylumbelliferone β-D-galactoside; Coumarin, 7-(β-D-galactopyranosyloxy)-4-methyl-. Chiral diagnostic reagent. Soluble in $H_2O$. *Acros Organics nv; Diagnostic Chemicals Limited; Senn Chemicals AG.*

**3867 4-Methylumbelliferyl-α-D-glucopyranoside**
17833-43-1 241-794-0

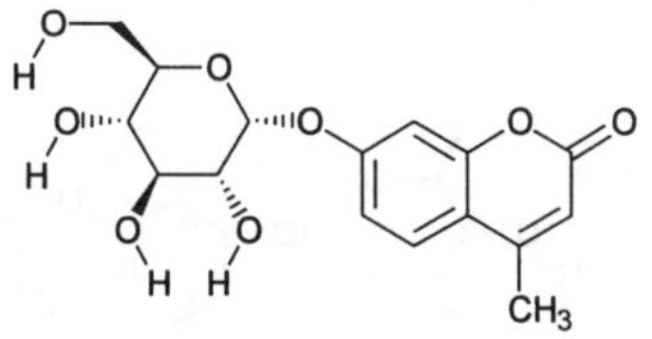

$C_{16}H_{18}O_8$
Benzo-(2H)-1-pyran, 7-(α-D-glucopyranosyloxy)-4-methyl-; 7-(α-D-Glucopyranosyloxy)-4-methyl-2H-1-benzopyran-2-one; Coumarin, 7-(α-D-glucopyranosyloxy)-4-methyl-. Chiral diagnostic reagent. Soluble in $H_2O$. *Acros Organics nv; Diagnostic Chemicals Limited.*

**3868 4-Methylumbelliferyl-β-D-glucopyranoside**
18997-57-4 242-736-7

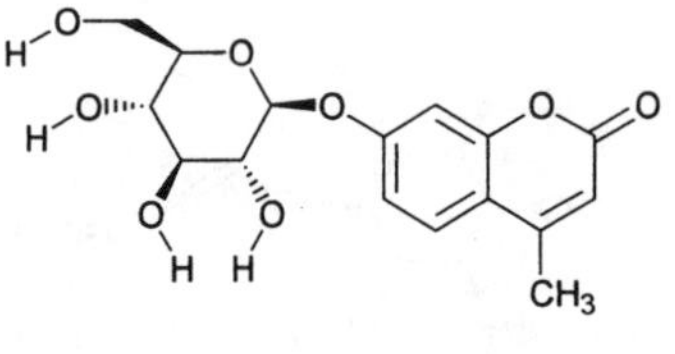

$C_{16}H_{18}O_8$
Benzo-(2H)-1-pyran-2-one, 7-(β-D-glucopyranosyloxy)-4-methyl-; Coumarin, 7-(β-D-glucopyranosyloxy)-4-methyl-; 7-(β-D-Glucopyranosyloxy) 4-methyl-2H-1-benzopyran-2-one. Chiral diagnostic reagent. $[\alpha]_D^{20} = -95 \pm 10°$ (c = 0.5, $H_2O$); very soluble in $H_2O$. *Acros Organics nv; Diagnostic Chemicals Limited; Senn Chemicals AG.*

**3869 4-Methylumbelliferyl-β-D-glucuronic acid**
6160-80-1 228-186-0

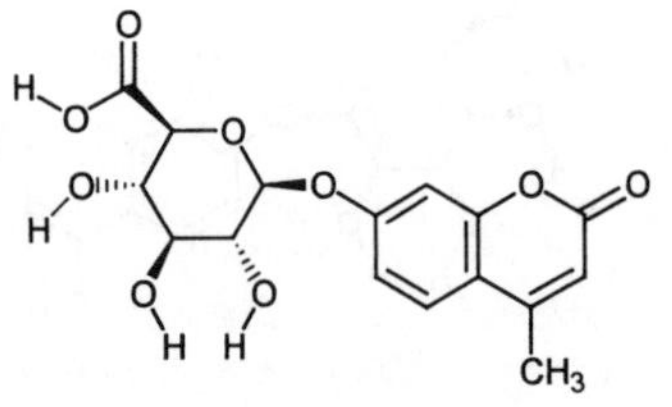

$C_{16}H_{16}O_9$
Glucopyranosiduronic acid, 4-methyl-2-oxo-2H-1-benzopyran-7-yl, β-D-; 4-Methyl-2-oxo-2H-1-benzopyran-7-yl β-D-glucopyranosiduronic acid; 4-Methylumbelliferone β-D-glucuronide. Listed on TSCA. Chiral diagnostic reagent. mp = 143-144°; slightly soluble in $H_2O$. *Acros Organics nv; Diagnostic Chemicals Limited; Senn Chemicals AG.*

**3870 4-Methylumbelliferyl-α-D-mannopyranosidehydrate**
28541-83-5 249-073-2

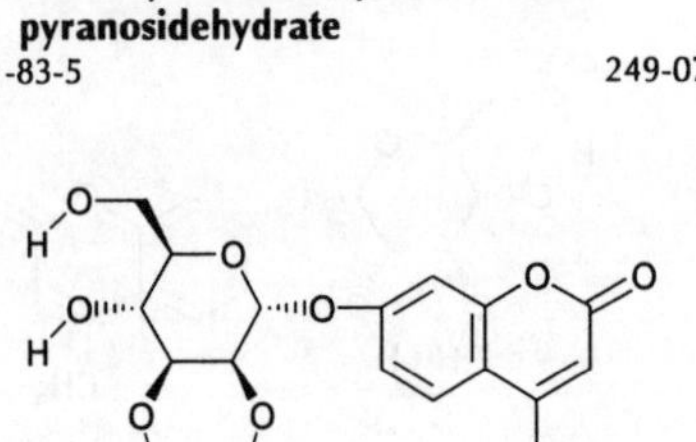

$C_{16}H_{18}O_8$
Benzo-(2H)-1-pyran-2-one, 7-(α-D-mannopyranosyloxy)-4-methyl-; 7-(α-D-Mannopyranosyloxy)-4-methyl-2H-1-benzopyran-2-one; Mannopyranoside, 4-methyl-2-oxo-2H-1-benzopyran-7-yl, α-D-. Chiral intermediate. *Acros Organics nv.*

**3871 4-Methylumbelliferyl-β-D-xylopyranoside**
6734-33-4

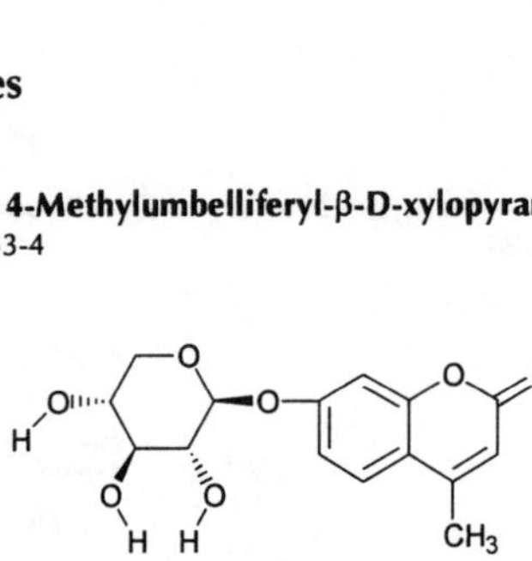

$C_{15}H_{16}O_7$
Chiral intermediate. *Acros Organics nv; Senn Chemicals AG.*

**3872 Methyl α-D-xylopyranoside**
91-09-8 202-040-6

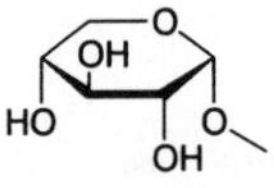

$C_6H_{12}O_5$
Xylopyranoside, methyl, α-D-; α-Methylxyloside. Listed on TSCA. Chiral intermediate. $[\alpha]_D^{20} = +153.9°$ (c = 10, $H_2O$). *Pfanstiehl Laboratories, Inc.; Senn Chemicals AG.*

**3873 Methyl β-D-xylopyranoside**
612-05-5 210-289-7

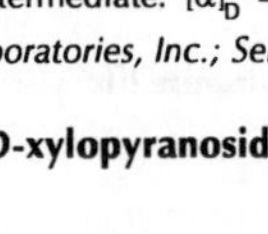

$C_6H_{12}O_5$
Xylopyranoside, methyl, β-D-; β-Methylxyloside. Listed on TSCA. Chiral intermediate. mp = 155-158°; $[\alpha]_D^{20} = -65.5°$ (c = 10, $H_2O$). *Pfanstiehl Laboratories, Inc.; Senn Chemicals AG; Sigma-Aldrich Fine Chemicals.*

**3874 D-Metrizamide**
31112-62-6 6240(12) 250-475-5

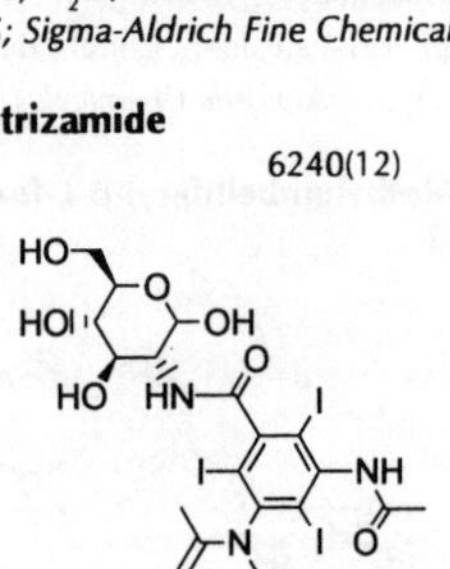

$C_{18}H_{22}I_3N_3O_8$
D-Glucose, 2-[[3-(acetylamino)-5-(acetylmethylamino)-2,4,6-triiodobenzoyl]amino]-2-deoxy-; 2-[3-Acetamido-5-(N-methylacetamido)-2,4,6-triiodobenzamido]-2-deoxy-D-glucose; Amipaque; Telebrix 300; D-Glucose, 2-[3-acetamido-2,4,6-triiodo-5-(N-methylacetamido)benzamido]-2-deoxy-. Chiral intermediate. mp = 222-224°. *Acros Organics nv; Sigma-Aldrich Fine Chemicals.*

**3875 (S)-Mimosine**
500-44-7 6286(12) 207-905-1

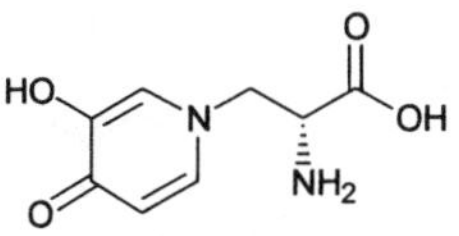

$C_8H_{10}N_2O_4$
Pyridine-1(4H)-propanoic acid, α-amino-3-hydroxy-4-oxo-, (S)-; (S)-α-Amino-3-hydroxy-4-oxo-1(4H)-pyridylpropionic acid; Leucaenine; Leucaenol; Leucenine; Leucenol. Chiral intermediate. mp = 228°; $[\alpha]^{22}$ = - 21° ($H_2O$). *Acros Organics nv; Sigma-Aldrich Fine Chemicals.*

**3876 Mono-(1R)-(-)-menthyl phthalate**
33744-74-0

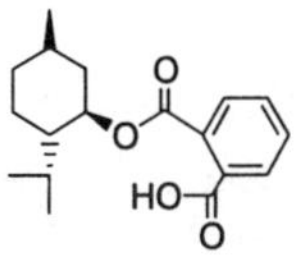

$C_{18}H_{24}O_4$
Chiral intermediate. mp = 103-108°; $[\alpha]_D^{20}$= - 95° (c = 5, $CHCl_3$). *Sigma-Aldrich Fine Chemicals.*

**3877 Mono-(1S)-(+)-menthyl phthalate**
53623-42-0

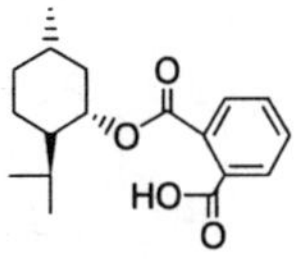

$C_{18}H_{24}O_4$
Chiral intermediate. mp = 111-114°; $[\alpha]_D^{20}$= + 93° (c = 5, $CHCl_3$). *Sigma-Aldrich Fine Chemicals.*

**3878 N-Monomethyl-D-arginine monoacetate**
137694-75-8

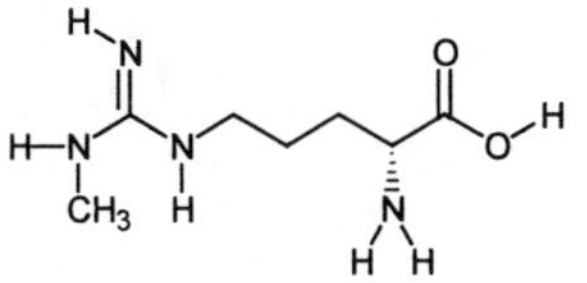

$C_9H_{19}N_4O_4$
Chiral intermediate. *Acros Organics nv.*

**3879 N-Monomethyl-L-arginine monoacetate**
53308-83-1

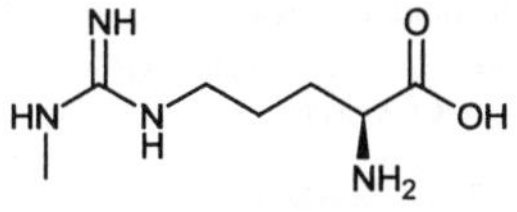

$C_7H_{19}N_4O_2$
Chiral intermediate. *Acros Organics nv.*

**3880 (+)-Muramic acid hydrate**
1114-41-6 6384(12) 214-214-9

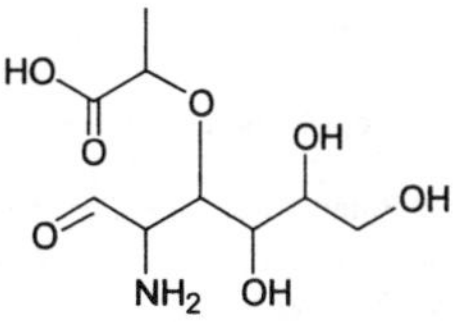

$C_9H_{17}NO_7$
Chiral intermediate. mp = 148°; $[\alpha]_D^{20}$= + 115.5° (c = 2, $H_2O$). *Sigma-Aldrich Fine Chemicals.*

**3881 N-Myristoyl-D-erythro sphingosine**
34227-72-0

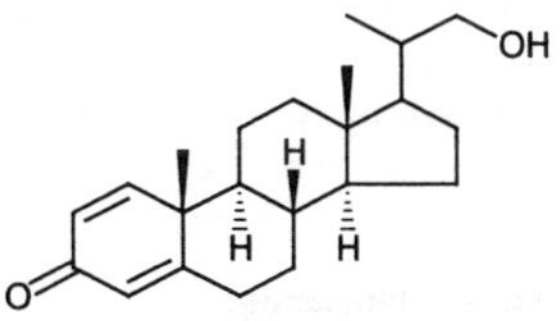

$C_{32}H_{63}NO_3$
Chiral intermediate. *Acros Organics nv.*

**3882 (1S,2S,5S)-(-)-Myrtanol**
53369-17-8 258-499-8

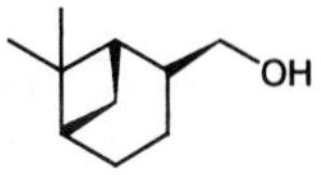

$C_{10}H_{18}O$
Chiral ligand. $bp_{15}$ = 117-119°; d = 0.966; n = 1.489; $[\alpha]_D^{20}$= - 28° (c = 4, $CHCl_3$). *Sigma-Aldrich Fine Chemicals.*

**3883 (1R)-(-)-Myrtenal**
564-94-3 209-274-8

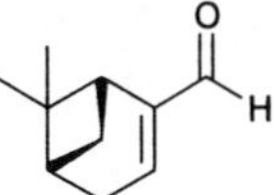


$C_{10}H_{14}O$
Bicyclo[3.1.1]hept-2-ene-2-carboxaldehyde, 6,6-dimeth-

yl-, (-); Pin-2-ene-1-carbaldehyde; 2-Norpinene-2-carboxaldehyde, 6,6-dimethyl-, 6,6-Dimethylbicyclo[3.1.1]-hept-2-ene-2-carboxaldehyde. Listed on TSCA. Chiral intermediate. bp = 220-221°; d = 0.987; n = 1.5039; $[\alpha]^{22}$ = - 15° (neat). *Acros Organics nv; Sigma-Aldrich Fine Chemicals.*

**3884 (1R)-(-)-Myrtenol**
19894-97-4 243-409-1

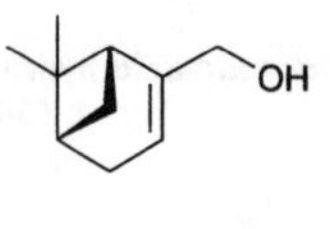

$C_{10}H_{16}O$
Bicyclo[3.1.1]hept-2-ene-2-methanol, 6,6-dimethyl-, (1R,5S)-; (1R)-6,6-Dimethylbicyclo[3.1.1]hept-2-ene-2-methanol; 2-Pinen-10-ol, (1R,5S)-. Chiral intermediate. bp = 221-222°; d = 0.954; n = 1.496; $[\alpha]^{22}$ = - 47.5° (neat). *Acros Organics nv; Sigma-Aldrich Fine Chemicals.*

**3885 (1S)-(+)-Myrtenol**
6712-78-3

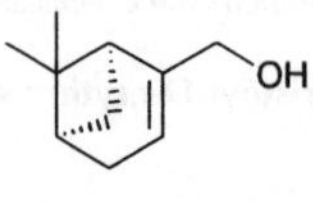

$C_{10}H_{16}O$
Chiral intermediate. $bp_{14}$ = 106-107°; d = 0.981; n = 1.496; $[\alpha]^{23}$ = + 51° (neat). *Sigma-Aldrich Fine Chemicals.*

**3886 cis-Myrtanylamine**
38235-68-6

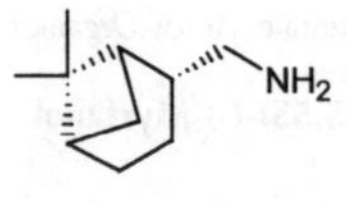

$C_{10}H_{19}N$
Chiral intermediate. $bp_{27}$ = 94-99°; d = 0.915; n = 1.4877; $[\alpha]^{22}$ = - 30.5° (neat). *Acros Organics nv; Sigma-Aldrich Fine Chemicals.*

**3887 Naloxone hydrochloride dihydrate**
51481-60-8 6449(12)

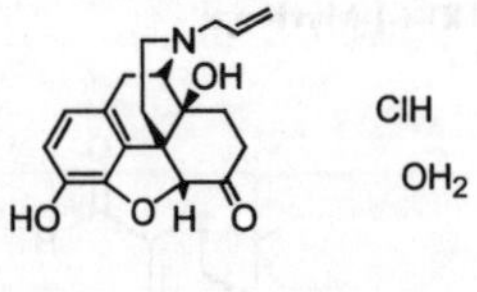

$C_{19}H_{22}ClNO_4$
Chiral intermediate. mp = 182°; $[\alpha]_D^{20}$ = - 160° (c = 1, $H_2O$). *Sigma-Aldrich Fine Chemicals.*

**3888 L-3-(1-Naphthyl)alanine**
55516-54-6

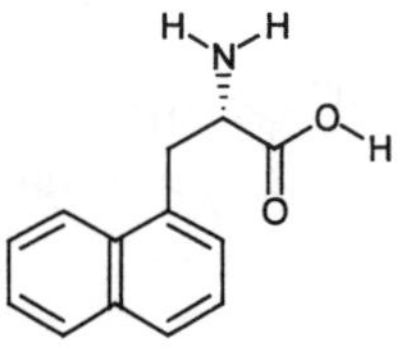

$C_{13}H_{13}NO_2$
(S)-(-)-2-Amino-3-(1-naphthyl)propanoic acid. Chiral building block. $[\alpha]$ = - 14.7° (c = 0.8, 0.3N HCl). *Acros Organics nv; Fischer Chemicals AG; Great Lakes Fine Chemicals; Senn Chemicals AG.*

**3889 D-3-(2-Naphthyl)alanine**
76985-09-6

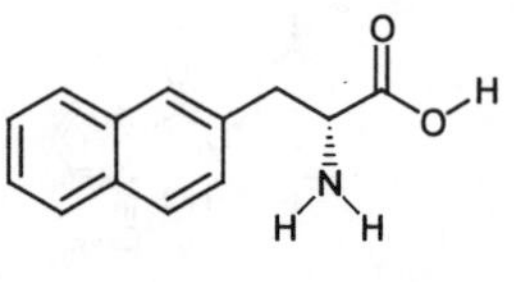

$C_{13}H_{13}NO_2$
3-(2-Naphthyl)-D-alanine. Chiral building block. *Acros Organics nv; Boehringer Ingelheim Pharma KG; Great Lakes Fine Chemicals.*

**3890 L-3-(2-Naphthyl)alanine**
58438-03-2

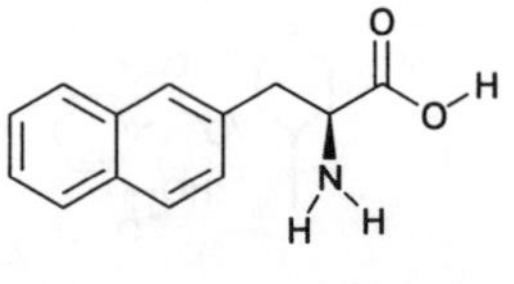

$C_{13}H_{13}NO_2$
Chiral building block. *Acros Organics nv; Boehringer Ingelheim Pharma KG; Degussa-Huls AG; Fischer Chemicals AG; Great Lakes Fine Chemicals; Senn Chemicals AG.*

**3891 (R)-(-)-1-(2-Naphthyl)-1,2-ethanediol**

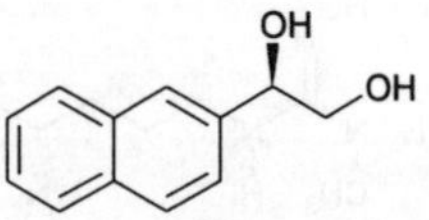

$C_{12}H_{12}O_2$
Chiral building block. mp = 131-135°; $[\alpha]^{23}$ = - 31° (c = 1, EtOH). *Sigma-Aldrich Fine Chemicals.*

**3892 (S)-(+)-1-(2-Naphthyl)-1,2-ethanediol**

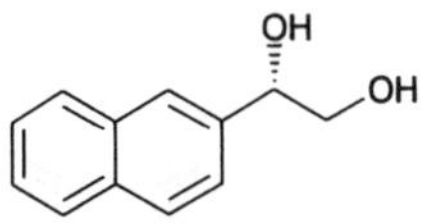

$C_{12}H_{12}O_2$
Chiral building block. mp = 131-135°; $[\alpha]^{23}$ = + 34° (c = 1, EtOH). *Sigma-Aldrich Fine Chemicals.*

**3893 (R)-(+)-1-(1-Naphthyl)ethanol**
42177-25-3

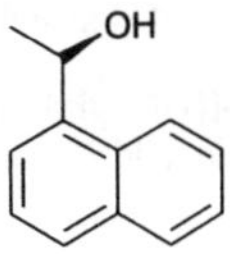

$C_{12}H_{12}O$
Chiral building block. mp = 42-45°; $[\alpha]_D^{23}$ = + 94.0 ± 0.5° (c = 5, $C_6H_6$); $[\alpha]_D^{20}$ = + 78° (c = 1, $CH_3OH$). *FineTech, Ltd.; Kiralchem Ltd.; Sigma-Aldrich Fine Chemicals.*

**3894 (S)-(-)-1-(1-Naphthyl)ethanol**
15914-84-8

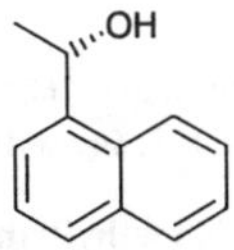

$C_{12}H_{12}O$
Chiral building block. mp = 42-45°; $[\alpha]_D^{23}$ = - 94.0 ± 0.5° (c = 5, $C_6H_6$); $[\alpha]_D^{20}$ = - 77° (c = 1, $CH_3OH$). *FineTech, Ltd.; Kiralchem Ltd.; Sigma-Aldrich Fine Chemicals.*

**3895 (R)-(+)-1-(1-Naphthyl)ethylamine**
3886-70-2 223-425-5

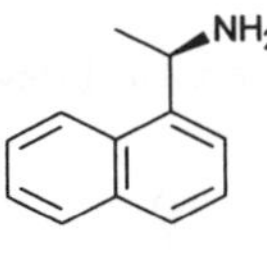

$C_{12}H_{13}N$
Naphthalene-1-methanamine, α-methyl-, (αR)-; (R)-(+)-α-(1-Aminoethyl)naphthalene; (R)-α-Methyl-1-naphthalene-methanamine. Listed on TSCA. Chiral auxiliary; Resolving agent; Chiral intermediate. $bp_{11}$ = 152-153°; $bp_{0.4}$ = 93°; d = 1.0600; $n_D^{20}$ = 1.6230; $[\alpha]_D^{20}$ = + 55 ± 2° (c = 2, EtOH). *Acros Organics nv; Arran Chemical Company Ltd.; Austin Chemical Company, Inc.; BASF Aktiengesellschaft; Lancaster Synthesis Ltd.; Norse Laboratories; SEAC; Sigma-Aldrich Fine Chemicals; TCI America; Yamakawa Chemical Industry Co. Ltd.*

**3896 (S)-(-)-1-(1-Naphthyl)ethylamine**
10420-89-0

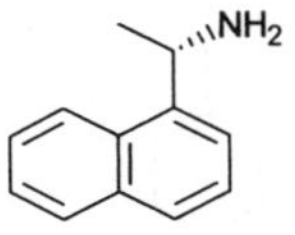

$C_{12}H_{13}N$
Naphthalene-1-methanamine, α-methyl-, (αS)-; (S)-α-Methyl-1-naphthalenemethanamine. Listed on TSCA. Chiral auxiliary; Chiral resolving agent; Chiral intermediate. $bp_{16}$ = 190-191°; $bp_{11}$ = 153°; d = 1.0609; $n_D^{20}$ = 1.6230; $[\alpha]_D^{20}$ = - 59 ± 2° (c = 5, $CH_3OH$). *Acros Organics nv; Arran Chemical Company Ltd.; Austin Chemical Company, Inc.; BASF Aktiengesellschaft; FineTech, Ltd.; Lancaster Synthesis Ltd.; Norse Laboratories; Sigma-Aldrich Fine Chemicals; TCI America; Yamakawa Chemical Industry Co. Ltd.*

**3897 (R)-(+)-1-(2-Naphthyl)ethylamine**
3906-16-9
$C_{12}H_{13}N$
Naphthalene-2-methanamine, α-methyl-, (R)-; (+)-α-Methyl-2-naphthalenemethylamine; (R)-(+)-α-(2-Naphthyl)ethylamine. Chiral building block. mp = 52°; $bp_{0.15}$ = 90°. *BASF Aktiengesellschaft; FineTech, Ltd.; Lancaster Synthesis Ltd.; Norse Laboratories.*

**3898 (S)-(-)-1-(2-Naphthyl)ethylamine**
3082-62-0
$C_{12}H_{13}N$
Naphthalene-2-methanamine, α-methyl-, (S)-; (S)-(-)-α-(2-Naphthyl)ethylamine. Chiral intermediate. mp = 52°; $bp_{0.15}$ = 90°. *BASF Aktiengesellschaft; FineTech, Ltd.; Lancaster Synthesis Ltd.; Norse Laboratories.*

**3899 (R)-(-)-N-1-(1-Naphthyl)ethyl-3,5-dinitrobenzamide**
85922-30-1

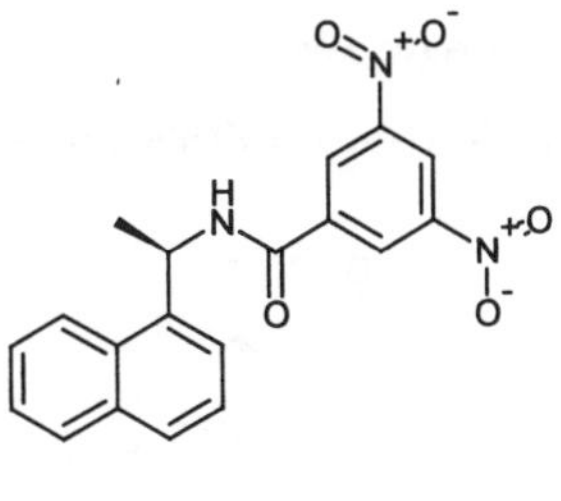

$C_{19}H_{15}N_3O_5$
Chiral intermediate. mp = 224-226°; $[\alpha]^{22}$ = - 112° (c = 2, acetone). *Sigma-Aldrich Fine Chemicals.*

**3900 (S)-(+)-N-1-(1-Naphthyl)ethyl-3,5-dinitrobenzamide**
85922-31-2

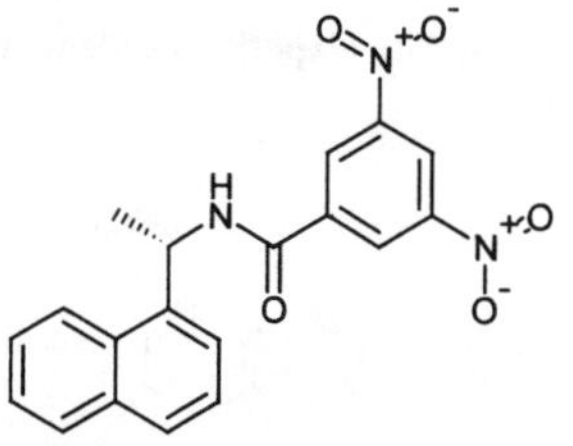

$C_{19}H_{15}N_3O_5$
Chiral intermediate. mp = 224-226°; $[\alpha]^{22}$ = + 112° (c = 2, acetone). *Sigma-Aldrich Fine Chemicals.*

**3901 (R)-(-)-1-(1-Naphthyl)ethyl isocyanate**
42340-98-7 255-759-2

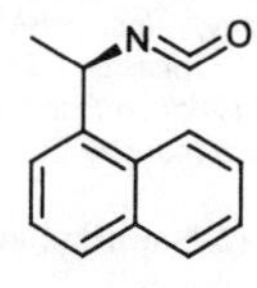

$C_{13}H_{11}NO$
Naphthalene, 1-(1-isocyanatoethyl)-, (R)-; Isocyanic acid (R)-(-)-1-(1-Naphthyl)ethyl ester; (R)-(-)-1-(1-Isocyanatoethyl)naphthalene. Chiral diagnostic reagent. $bp_{0.16}$ = 106-108°; d = 1.1180; $n_D^{20}$ = 1.6045; $[\alpha]_D^{20}$ = - 47 ± 2° (c = 3.5, toluene). *Lancaster Synthesis Ltd.; Norse Laboratories; Sigma-Aldrich Fine Chemicals; TCI America.*

**3902 (R)-N-[(1-Naphthyl)ethyl]maleimide**

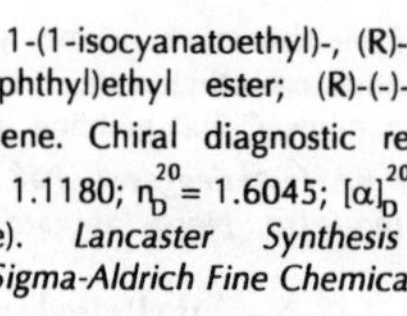

$C_{16}H_{13}NO_2$
Chiral intermediate. *FineTech, Ltd.*

**3903 (S)-N-[(1-Naphthyl)ethyl]maleimide**

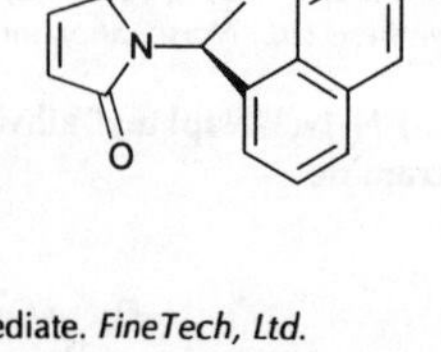

$C_{16}H_{13}NO_2$
Chiral intermediate. *FineTech, Ltd.*

**3904 (R)-(-)-N-[(1-Naphthyl)ethyl]-phthalamic acid**

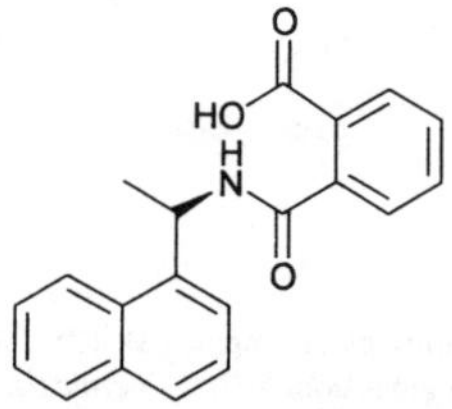

$C_{20}H_{17}NO_3$
Chiral intermediate. mp = 166°; $[\alpha]^{26}$ = - 48° (c = 1, $CH_3OH$). *FineTech, Ltd.; Sigma-Aldrich Fine Chemicals.*

**3905 (S)-(+)-N-[(1-Naphthyl)ethyl]-phthalamic acid**

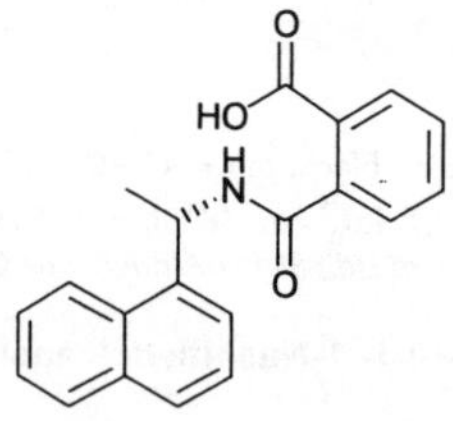

$C_{20}H_{17}NO_3$
Chiral intermediate. mp = 167°; $[\alpha]^{25}$ = + 48° (c = 1, $CH_3OH$). *FineTech, Ltd.; Sigma-Aldrich Fine Chemicals.*

**3906 (R)-N-[(1-Naphthyl)ethyl]phthalimide**

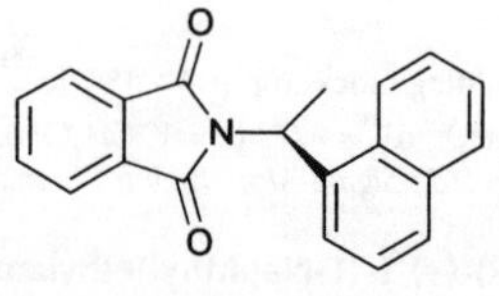

$C_{20}H_{15}NO_2$
Chiral intermediate. *FineTech, Ltd.*

**3907 (S)-N-[(1-Naphthyl)ethyl]phthalimide**

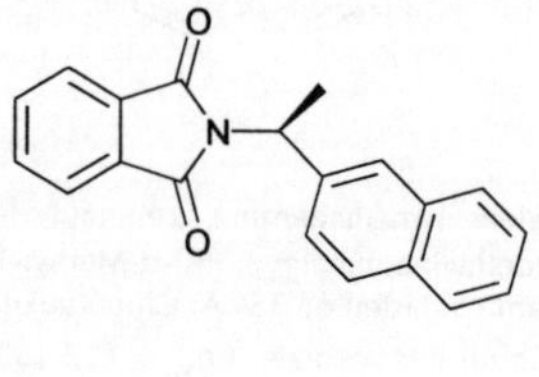

$C_{20}H_{15}NO_2$
Chiral intermediate. *FineTech, Ltd.*

**3908 (R)-N-[(1-Naphthyl)ethyl]succinamic acid**
78681-09-1

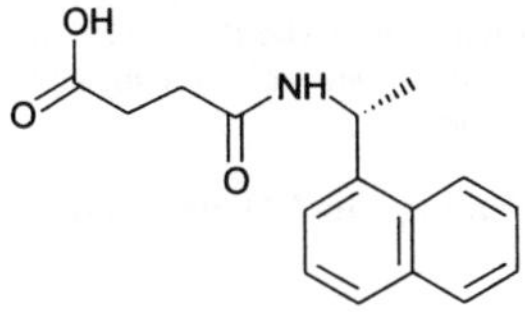

$C_{16}H_{17}NO_3$
Chiral intermediate. mp = 157-159°; $[\alpha]^{24}$ = + 45° (c = 2, EtOH). *FineTech, Ltd.; Sigma-Aldrich Fine Chemicals.*

**3909 (S)-N-[(1-Naphthyl)ethyl]succinamic acid**

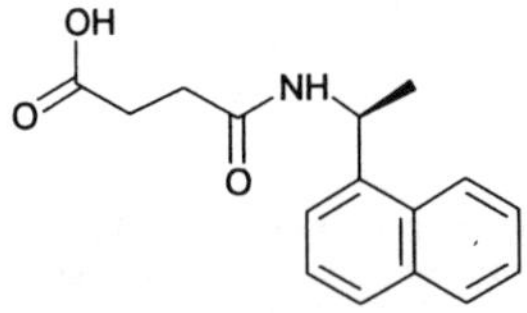

$C_{16}H_{17}NO_3$
Chiral intermediate. mp = 157-159°; $[\alpha]^{24}$ = - 45° (c = 2, EtOH). *FineTech, Ltd.; Sigma-Aldrich Fine Chemicals.*

**3910 2-Naphthyl-β-D-galactopyranoside**
33993-25-8 251-778-5

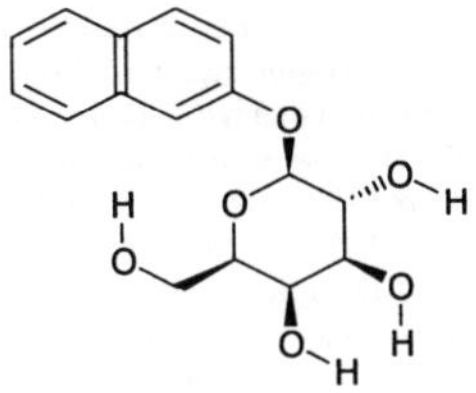

$C_{16}H_{18}O_6$
Galactopyranoside, 2-naphthyl, β-D-; Chiral diagnostic reagent. $[\alpha]_D^{20}$ = - 55° (c = 1, EtOH). *Acros Organics nv; Diagnostic Chemicals Limited; Sigma-Aldrich Fine Chemicals.*

**3911 Naphthyl-α-D-glucopyranoside**
2532-79-0

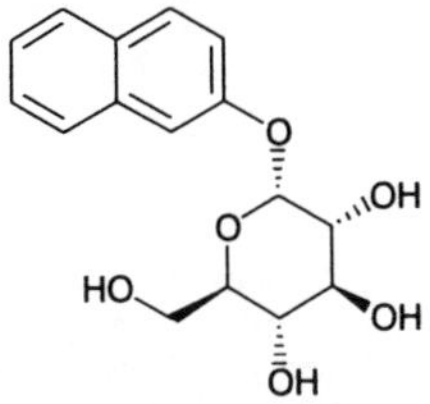

$C_{16}H_{18}O_6$
Chiral diagnostic reagent. *Diagnostic Chemicals Limited.*

**3912 2-Naphthyl-β-D-glucopyranoside**
6044-30-0 227-933-8

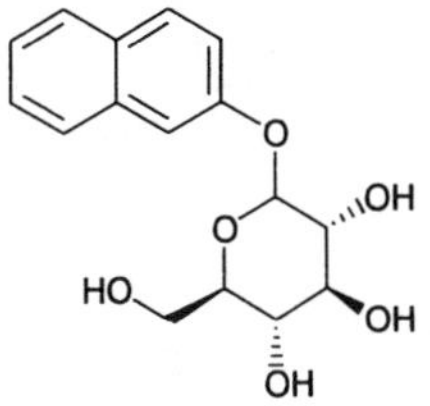

$C_{16}H_{18}O_6$
Glucopyranoside, 2-naphthyl, β-D-. Chiral intermediate. *Senn Chemicals AG.*

**3913 (R)-2-(2-Naphthyl)glycolic acid**
43210-73-7

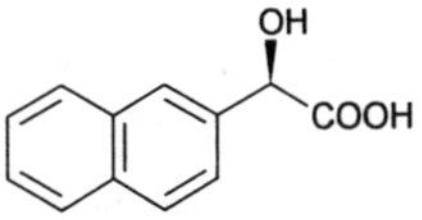

$C_{12}H_{10}O_3$
Chiral building block. *Yamakawa Chemical Industry Co. Ltd.*

**3914 (S)-2-(2-Naphthyl)glycolic acid**
144371-23-3

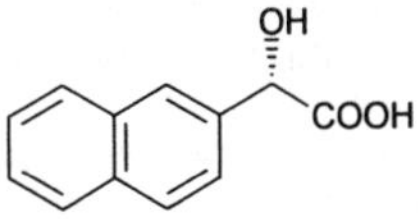

$C_{12}H_{10}O_3$
Chiral building block. *Yamakawa Chemical Industry Co. Ltd.*

**3915 1-Naphthyl-β-D-mannopyranoside**

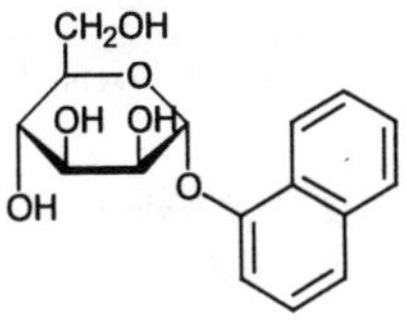

$C_{16}H_{18}O_6$
Chiral intermediate. *Senn Chemicals AG.*

**3916 2-Naphthyl-β-D-mannopyranoside**

$C_{16}H_{18}O_6$
Chiral intermediate. *Senn Chemicals AG.*

**3917 (R,R)-(+)-1-(1-Naphthyl)-2-(2-naphthyl)-1,2-ethanediol**

$C_{22}H_{18}O_2$
Chiral intermediate. mp = 100-105°; $[\alpha]^{24}$ = + 147° (c = 1, $CHCl_3$). *Sigma-Aldrich Fine Chemicals.*

**3918 (S,S)-(-)-1-(1-Naphthyl)-2-(2-naphthyl)-1,2-ethanediol**

$C_{22}H_{18}O_2$
Chiral intermediate. mp = 100-105°; $[\alpha]^{24}$ = - 147° (c = 1, $CHCl_3$). *Sigma-Aldrich Fine Chemicals.*

**3919 (R)-2-(1-Naphthylmethyl)succinic acid 1-methyl ester**
119807-82-8
$C_{16}H_{16}O_4$
(R)-3-Methoxycarbonyl-4-(1-naphthyl)butyric acid. Chiral intermediate. *ChiroTech Technology Ltd.; Lancaster Synthesis Ltd.*

**3920 (S)-2-(1-Naphthylmethyl)succinic acid 1-methyl ester**
130693-96-8
$C_{16}H_{16}O_4$
(S)-3-Methoxycarbonyl-4-(1-naphthyl)butyric acid. Chiral intermediate. *ChiroTech Technology Ltd.; Lancaster Synthesis Ltd.*

**3921 (R)-2-(2-Naphthylmethyl)succinic acid 1-methyl ester**
213270-42-9
$C_{16}H_{16}O_4$
(R)-3-Methoxycarbonyl-4-(2-naphthyl)butyric acid. Chiral intermediate. *ChiroTech Technology Ltd.; Lancaster Synthesis Ltd.*

**3922 (S)-2-(2-Naphthylmethyl)succinic acid 1-methyl ester**
$C_{16}H_{16}O_4$
(S)-3-Methoxycarbonyl-4-(2-naphthyl)butyric acid. Chiral intermediate. *ChiroTech Technology Ltd.; Lancaster Synthesis Ltd.*

**3923 Naringenin-7-D-glucoside**
529-55-5 208-464-8

$C_{21}H_{22}O_{10}$
Benzopyran-(4H)-1-4-one, 7-(β-D-glucopyranosyloxy)-2,3-dihydro-5-hydroxy-2-(4-hydroxyphenyl)-, (2S)-; Prunin; (S)-7-(b-D-glucopyranosyloxy)-2,3-dihydro-5-hydroxy-2-(4-hydroxyphenyl)-4H-1-benzopyran-4-one; 4',5,7-Trihydroxyflavanone 7-O-β-D-glucopyranoside; Naringenin 7-O-β-D-glucopyranoside. Chiral intermediate. mp = 220-223°. *Kaden Biochemicals GmbH.*

**3924 D-Naringin**
10236-47-2 233-566-4
$C_{27}H_{32}O_{14}$
Benzopyran-(4H)-1-4-one, 7-[[2-O-(6-deoxy-α-L-mannopyranosyl)-β D-glucopyranosyl]oxy]-2,3-dihydro-5-hydroxy-2-(4-hydroxyphenyl)-, (2S)-; 4H-1-Benzopyran-4-one, 7-[[2-O-(6-deoxy-α-L-mannopyranosyl)-β-D-glucopyranosyl]oxy]-2,3-dihydro-5-hydroxy-2-(4-hydroxyphenyl)-, (S)-; Naringenin-7-α-neo-hesperidoside (dihydrate); 7-(2-O-(6-Desoxy-α-L-mannopyranosyl)-β-D-glucopyranosyloxy)-2,3-d. Chiral intermediate. mp = 166°; [α = - 91° (c = 1, EtOH). *Acros Organics nv.*

**3925 (+)-Neomenthol**
2216-52-6 218-691-4

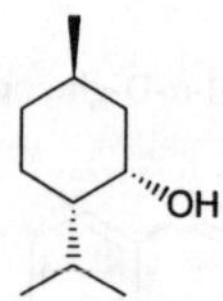


$C_{10}H_{20}O$
Cyclohexanol, 5-methyl-2-(1-methylethyl)-, (1R,2S,5R)-; (1S,2S,5R)-2-Isopropyl-5-methylcyclohexanol; [1S-(1α,2α,5β)]-5-methyl-2-(1-methylethyl)-cyclohexanol; Menthol, (1R,3S,4S)-. Chiral intermediate. mp = - 22°; $bp_{12}$ = 95°; d = 0.899; n = 1.461; $[\alpha]^{22}$ = + 17.3° (neat). *Sigma-Aldrich Fine Chemicals; TCI America.*

**3926 (1R)-(-)-Neomenthyl acetate**

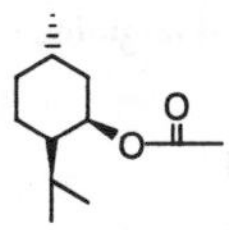

$C_{12}H_{22}O_2$
Chiral intermediate. mp = 36-38°; $bp_3$ = 81°; d = 0.912; $[\alpha]_D^{20}$ = - 48° (c = 8, $C_6H_6$). *Sigma-Aldrich Fine Chemicals.*

**3927 (1S)-(+)-Neomenthyl acetate**
2552-91-2 219-856-3

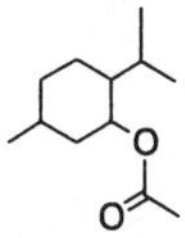

$C_{12}H_{22}O_2$
Listed on TSCA. Chiral intermediate. mp = 36-38°; $bp_3$ = 81°; d = 0.912; $[\alpha]_D^{20}$ = + 48° (c = 8, $C_6H_6$). *Sigma-Aldrich Fine Chemicals.*

**3928 (1R,2R,5S)-(-)-Neomenthyldiphenylarsine**
$C_{22}H_{29}As$
Chiral ligand. *Digital Specialty Chemicals, Inc.*

**3929 (1S,2S,5R)-(+)-Neomenthyldiphenylarsine**
$C_{22}H_{29}As$
Chiral ligand. *Digital Specialty Chemicals, Inc.*

**3930 (1R,2R,5S)-(-)-Neomenthyldiphenylphosphine**
$C_{22}H_{29}P$
Chiral ligand. *Digital Specialty Chemicals, Inc.*

**3931 (S)-(+)-Neomenthyldiphenylphosphine**
43077-29-8

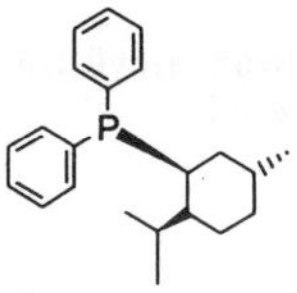

$C_{22}H_{29}P$
(S)-NMDPP. Chiral ligand. mp = 90-92°; $[\alpha]_D$ = + 95.5° (c = 1.26, $CH_2Cl_2$). *Digital Specialty Chemicals, Inc.; Sigma-Aldrich Fine Chemicals; Strem Chemicals, Inc.*

**3932 (R)-N-Neopentyl-1-phenyl-2-(1-piperidino)-ethylamine**
153837-28-6
$C_{18}H_{30}N_2$
(R)-N-(2,2-Dimethylpropyl)-1-phenyl-1-(1-piperidino)-ethylamine. Chiral intermediate. *Lancaster Synthesis Ltd.*

**3933 (R)-2-Neopentylsuccinic acid 1-methyl ester**
213270-41-8
$C_{10}H_{18}O_4$
(R)-5,5-Dimethyl-3-(methoxycarbonyl)hexanoic acid. Chiral intermediate. *ChiroTech Technology Ltd.; Lancaster Synthesis Ltd.*

**3934 (S)-2-Neopentylsuccinic acid 1-methyl ester**
$C_{10}H_{18}O_4$
(S)-5,5-Dimethyl-3-(methoxycarbonyl)hexanoic acid. Chiral intermediate. *ChiroTech Technology Ltd.; Lancaster Synthesis Ltd.*

**3935 D-(+)-Neopterin**
2009-64-5 217-924-7

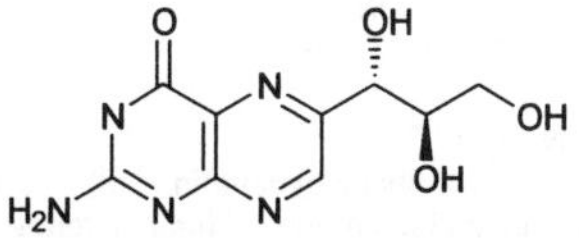

$C_9H_{11}N_5O_4$
Propane-1,2,3-triol, 1-(2-amino-4-hydroxy-6-pteridinyl)-, D-erythro-; 6-(D-erythro-1,2,3-Trihydroxypropyl)pterin; 4(1H)-Pteridinone, 2-amino-6-[(1S,2R)-1,2,3-trihydroxy-propyl]-; [S-(R*,S*)]-2-amino-6-(1,2,3-trihydroxypropyl)-1H-pteridin-4-one. Chiral intermediate. $[\alpha]_D^{20}$ = + 83° (c = 1, phosphate buffer). *Acros Organics nv; Sigma-Aldrich Fine Chemicals.*

**3936 (S)-(-)-Nicotine**
54-11-5 6611(12) 200-193-3

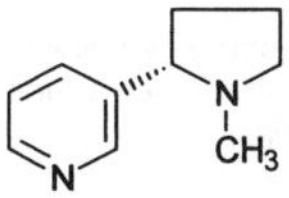

$C_{10}H_{14}N_2$
Pyridine, 3-(1-methyl-2-pyrrolidinyl)-, (S)-; (S)-3-(1-Methyl-2-pyrrolidinyl)pyridine; Flux Maag; Nicoderm; Niconil; Nicotin; Nicotinell; XL All Insecticide. Listed on TSCA. Chiral intermediate. mp = - 79°; bp = 243-248°; d = 1.0170; $n_D^{20}$ = 1.5265; $[\alpha]_D^{20}$ = - 168 ± 5° (c = 5, $H_2O$). *Acros Organics nv; Austin Chemical Company, Inc.; Lancaster Synthesis Ltd.; Sigma-Aldrich Fine Chemicals.*

**3937 Nicotinoyl-(tert-Butoxycarbonyl)-D-lysine**

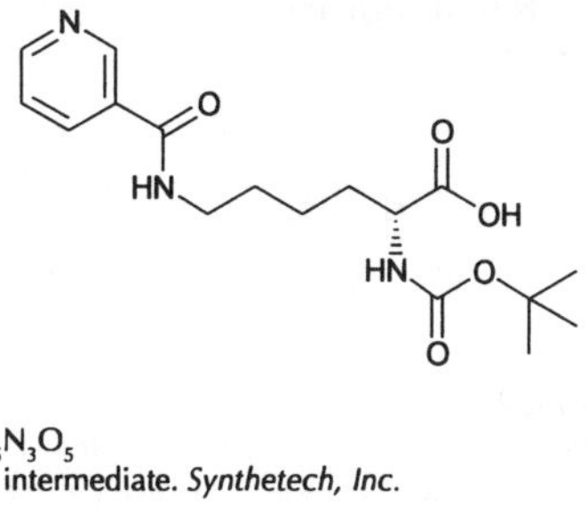

$C_{17}H_{25}N_3O_5$
Chiral intermediate. *Synthetech, Inc.*

**3938 (R)-Nipecotic acid**
25137-00-2
$C_6H_{11}NO_2$
Chiral building block. *Sigma-Aldrich Fine Chemicals.*

**3939 (S)-Nipecotic acid**
59045-82-8
$C_6H_{11}NO_2$
Chiral building block. *Sigma-Aldrich Fine Chemicals.*

**3940 (R)-(-)-Nirvanol**
65567-32-01

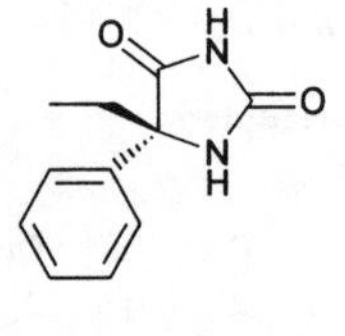

$C_{11}H_{12}N_2O_2$
(R)-(-)-5-Ethyl-5-phenylhydantoin; (R)-(-)-5-Ethyl-5-phenyl-2,4-imidazolidinedione. Chiral intermediate. mp = 232-237°. *Ultrafine.*

**3941 (S)-(+)-Nirvanol**
65567-34-2

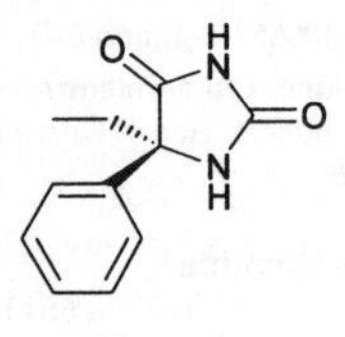

$C_{11}H_{12}N_2O_2$
(S)-(+)-5-Ethyl-5-phenylhydantoin; (S)-(+)-5-Ethyl-5-phenyl-2,4-imidazolidinedione. Chiral intermediate. mp = 232-237°. *Ultrafine.*

**3942 (R)-(-)-4-Nitro-7-(3-aminopyrrolidin-1-yl)-2,1,3-benzoxadiazole**
$C_{10}H_{11}N_5O_3$
Chiral intermediate. *TCI America.*

**3943 (S)-(+)-4-Nitro-7-(3-aminopyrrolidin-1-yl)-2,1,3-benzoxadiazole**
$C_{10}H_{11}N_5O_3$
(S)-(+)-4-Nitro-7-(3-aminopyrrolidin-1-yl)benzofurazan. Chiral intermediate. *TCI America.*

**3944 Nitro-L-arginine**
2149-70-4 218-418-9

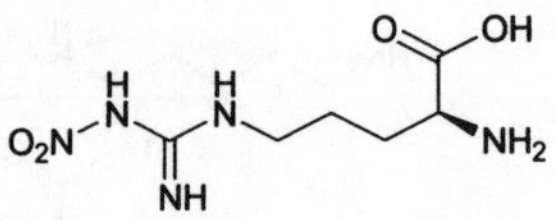

$C_6H_{13}N_5O_4$
$N^5$-(nitroamidino)-, L-; L-Arginine, ω-nitro-; L-Ornithine, $N^5$-[imino(nitroamino)methyl]-. Listed on TSCA. Chiral intermediate. mp = 257°; $[\alpha]^{23}$ = + 23° (c = 2, 2N HCl). *Acros Organics nv; Sigma-Aldrich Fine Chemicals.*

**3945 N-γ-Nitro-L-arginine methyl ester**

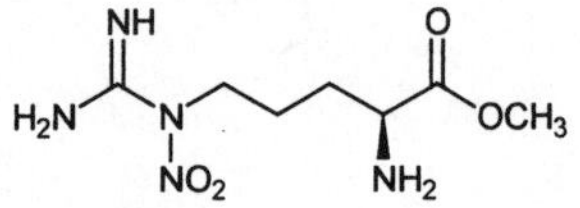

$C_7H_{15}N_5O_4$
Chiral intermediate. *Austin Chemical Company, Inc.*

**3946 N-Nitro-L-arginine-methyl ester hydrochloride**
51298-62-5 257-116-1

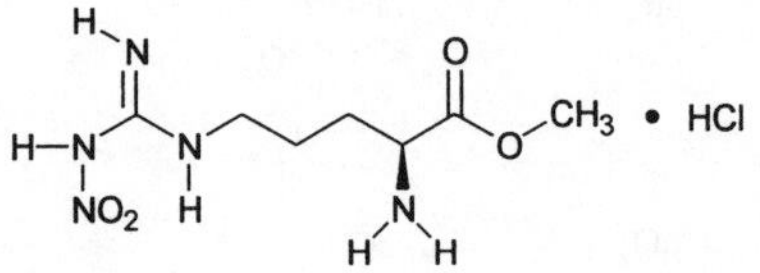

$C_7H_{16}ClN_5O_4$
$N^5$-[Imino(nitroamino)methyl]-, methyl ester, monohydrochloride, L-; Methyl $N^5$-[imino(nitroamino)methyl]-L-ornithine monohydrochloride. Chiral intermediate. *Acros Organics nv.*

**3947 N-Nitro-D-arginine methyl ester hydrochloride**
50912-92-0

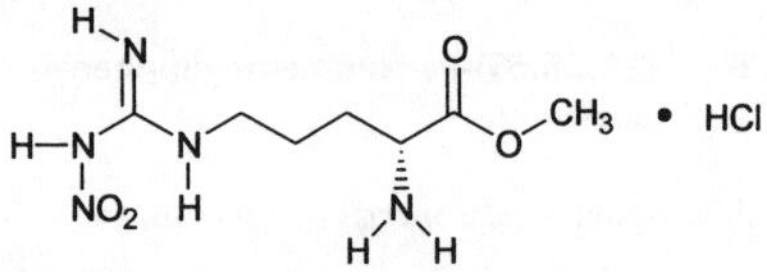

$C_7H_{16}ClN_5O_4$
Chiral intermediate. *Acros Organics nv.*

**3948 N-(4-Nitrobenzoyl)-L-glutamic acid diethyl ester**
7148-24-5 230-464-1

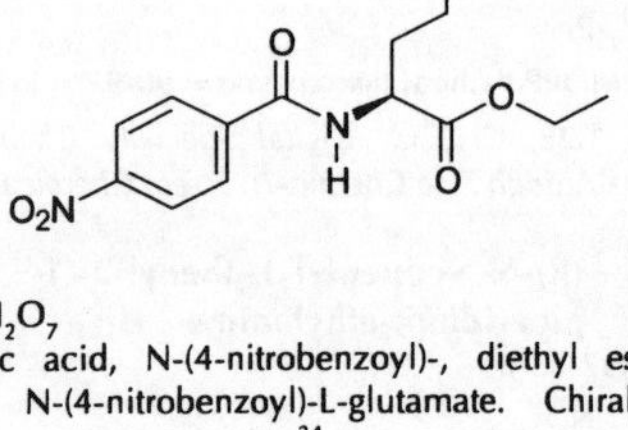

$C_{16}H_{20}N_2O_7$
Glutamic acid, N-(4-nitrobenzoyl)-, diethyl ester, L-; Diethyl N-(4-nitrobenzoyl)-L-glutamate. Chiral intermediate. mp = 92-94°; $[\alpha]^{24}$ = + 20.5° (c = 2, $CHCl_3$). *Acros Organics nv; Sigma-Aldrich Fine Chemicals.*

**3949 N-(4-Nitrobenzoyl)-L-glutamic acid hemihydrate**
6758-40-3 229-820-9

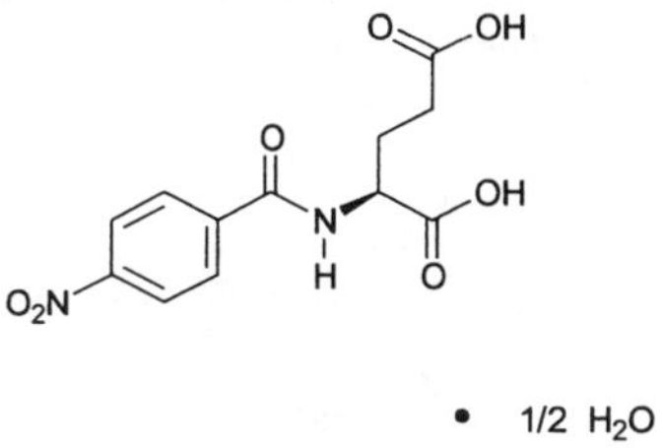

• 1/2 $H_2O$

$C_{12}H_{12}N_2O_7$
Glutamic acid, N-(p-nitrobenzoyl)-, L-; Chiral intermediate. mp = 117-118°; $[\alpha]^{24}$ = + 15.1° (c = 2, 1N NaOH). *Acros Organics nv; Sigma-Aldrich Fine Chemicals.*

**3950 p-Nitrobenzyl-(2S)-carboxy-trans-(4R)-hydroxy-L-proline**

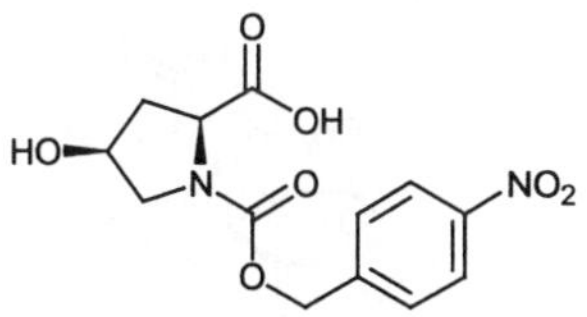

$C_{13}H_{14}N_2O_7$
Chiral intermediate. *Austin Chemical Company, Inc.*

**3951 p-Nitrobenzyl (1R,5R,6S)-6-[(1R)-1-hydroxyethyl]-2-[(diphenylphosphino)oxy]-1-methylcarbapen-2-em-3-carboxylate**
90776-59-3

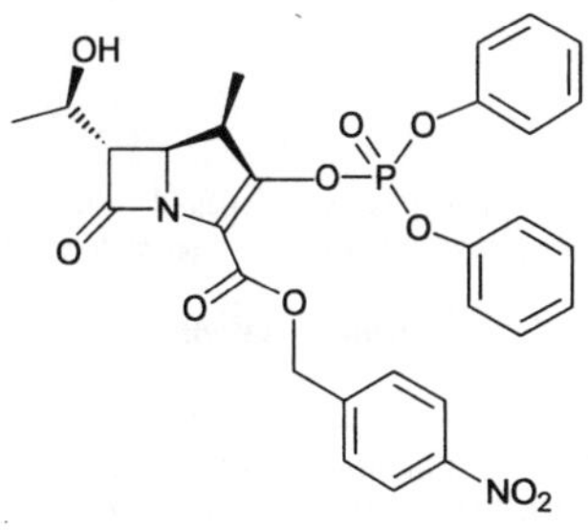

$C_{29}H_{27}N_2O_{10}P$
Chiral intermediate. *Kaneka Corporation.*

**3952 D-S-(4-Nitrobenzyl)-6-thioguanosine**
13153-27-0

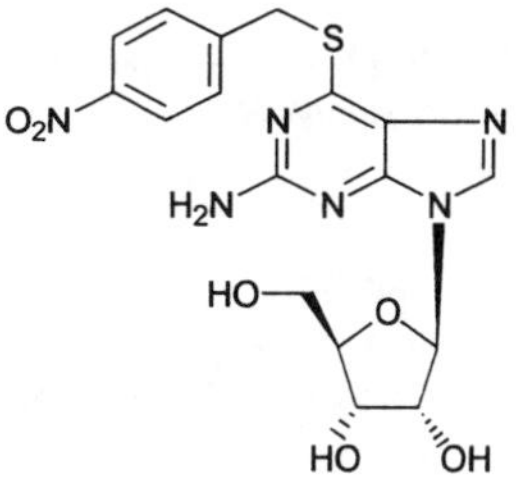

$C_{17}H_{18}N_6O_6S$
Amino-6-[(4-nitrobenzyl)thio]-9-β-D-ribofuranosylpurine; NBTGR. Chiral intermediate. mp = 203-205°. *Acros Organics nv.*

**3953 D-S-(4-Nitrobenzyl)-6-thioinosine**
38048-32-7 253-753-4

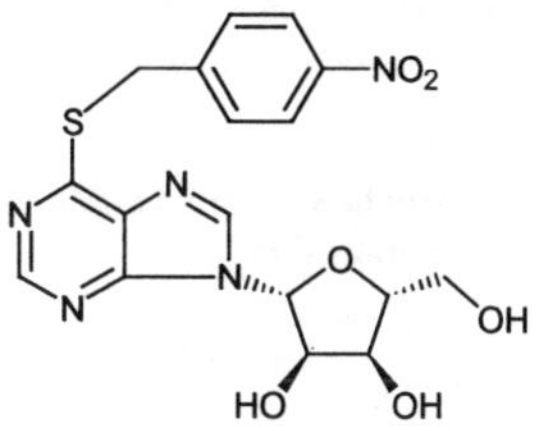

$C_{17}H_{17}N_5O_6S$
Inosine, 6-S-[(4-nitrophenyl)methyl]-6-thio-; 6-((4-nitrobenzyl)thio)-9-β-D-ribofuranosylpurine. Chiral intermediate. *Acros Organics nv.*

**3954 (R)-(+)-4-Nitro-7-(2-chloroformyl-pyrrolidin-1-yl)-2,1,3-benzoxadiazole**
$C_{11}H_9ClN_4O_4$
Chiral intermediate. *TCI America.*

**3955 (S)-(-)-4-Nitro-7-(2-chloroformylpyrrolidin-1-yl)-2,1,3-benzoxadiazole**
$C_{11}H_9ClN_4O_4$
(S)-(-)-4-Nitro-7-(2-chloroformylpyrrolidin-1-yl)benzofurazane. Chiral intermediate. *TCI America.*

**3956 p-Nitrophenyl-2-acetamide-2-deoxy-β-D-galactopyranoside**
14948-96-0 239-024-3

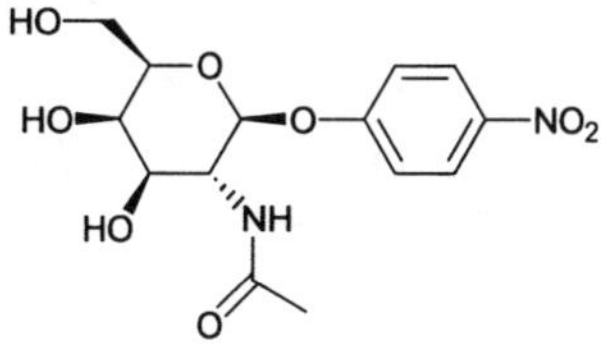

$C_{14}H_{18}N_2O_8$
Galactopyranoside, p-nitrophenyl 2-acetamido-2-deoxy-,

β-D-; 4-Nitrophenyl N-acetyl-β-D-galactosaminide monohydrate. Chiral diagnostic reagent. *Acros Organics nv; Diagnostic Chemicals Limited.*

**3957 2'-Nitrophenyl-2-acetamido-2-deoxy-α-D-glucopyranoside**
10139-01-2 233-391-3

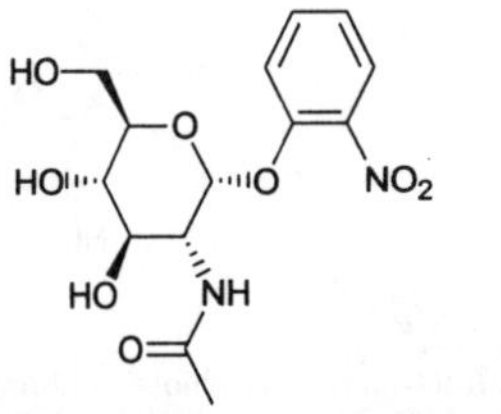

$C_{14}H_{18}N_2O_8$
Glucopyranoside, 2-nitrophenyl 2-(acetylamino)-2-deoxy-, α-D-; (2-Nitrophenyl)-N-acetyl-α-D-glucosaminide; α-D-Glucopyranoside, 2-nitrophenyl 2-(acetylamino)-2-deoxy-. Chiral intermediate. mp = 208-210°; $[\alpha]^{22}$ = + 217.1° (c = 0.47, $H_2O$). *Acros Organics nv; Sigma-Aldrich Fine Chemicals.*

**3958 4'-Nitrophenyl-2-acetamido-2-deoxy-α-D-glucopyranoside**
10139-02-3 233-392-9

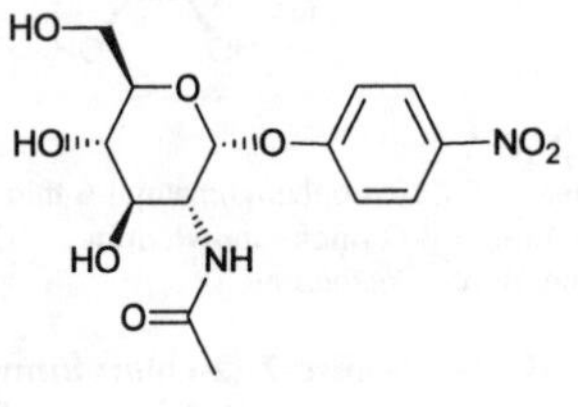

$C_{14}H_{18}N_2O_8$
Glucopyranoside, 4-nitrophenyl 2-(acetylamino)-2-deoxy-, α-D-; (4-Nitrophenyl)-N-acetyl-α-D-glucosaminide; p-Nitrophenyl 2-Acetamido-2-deoxy-α-D-glucopyranoside; GlcNAc1-α-PNP; α-D-Glucopyranoside, 4-nitrophenyl 2-(acetylamino)-2-deoxy-. Chiral intermediate. mp = 264-266°; $[\alpha]^{22}$ = + 280.3° (c = 0.5, $CH_3OH$). *Acros Organics nv.*

**3959 o-Nitrophenyl-2-acetamide-2-deoxy-β-D-glucopyranoside**
13264-92-1 236-258-8

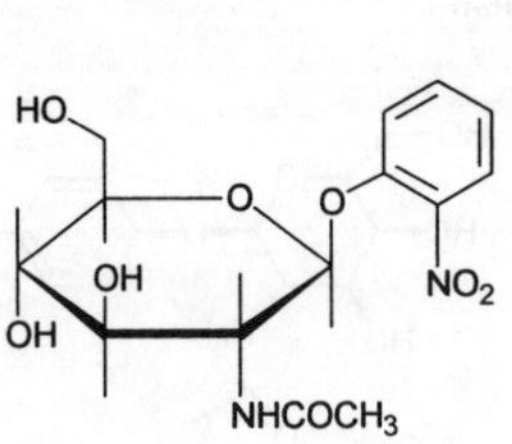

$C_{14}H_{18}N_2O_8$
Glucopyranoside, 2-nitrophenyl 2-(acetylamino)-2-deoxy-, β-D-; Glucopyranoside, o-nitrophenyl 2-acetamido-2-deoxy-, β-D-. Chiral diagnostic reagent. Soluble in $H_2O$ (0.1%). *Diagnostic Chemicals Limited.*

**3960 p-Nitrophenyl-2-acetamide-2-deoxy-β-D-glucopyranoside**
3459-18-5 222-398-7

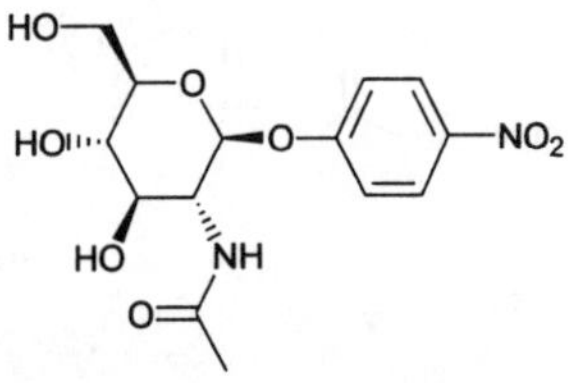

$C_{14}H_{18}N_2O_8$
Glucopyranoside, p-nitrophenyl 2-acetamido-2-deoxy-, β-D-; β-D-Glucoside, p-nitrophenyl 2-acetamido-2-deoxy-. Chiral diagnostic reagent. mp = 210-212°; $[\alpha]^{22}$ = - 16.0° (c = 0.5, $H_2O$); soluble in $H_2O$ (0.5%). *Acros Organics nv; Diagnostic Chemicals Limited; Senn Chemicals AG; Sigma-Aldrich Fine Chemicals.*

**3961 2-Nitrophenyl-N-acetyl-β-D-glucosaminide**

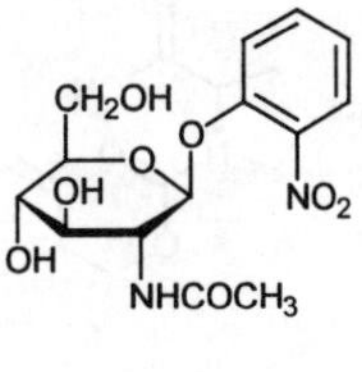

$C_{14}H_{18}N_2O_8$
Chiral intermediate. *Senn Chemicals AG.*

**3962 D-3-(4-Nitrophenyl)alanine**
56613-61-7

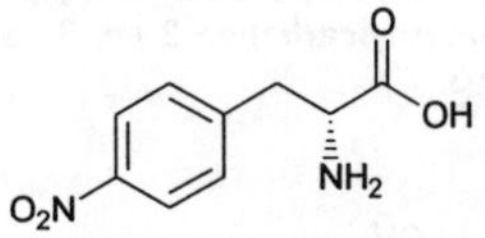

$C_9H_{10}N_2O_4$
p-Nitro-D-phenylalanine. Chiral building block. *Advanced Asymmetrics, Inc.; Ajinomoto Co. Inc.*

**3963 (S)-4-Nitrophenylalanine**
949-99-5 213-446-8

$C_9H_{10}N_2O_4$
Alanine, 3-(p-nitrophenyl)-, L-; L-3-(4-Nitrophenyl)alanine; L-Phenylalanine, 4-nitro-; p-Nitro-L-phenylalanine. Chiral building block. mp = 230°; $[\alpha]^{25}$ = + 6.8° (c = 1.3, 3N HCl). *Advanced Asymmetrics, Inc.; Ajinomoto Co. Inc.; Austin Chemical Company, Inc.; Fischer Chemicals AG; Great Lakes Fine Chemicals; Lancaster Synthesis Ltd.; Sigma-Aldrich Fine Chemicals.*

**3964 p-Nitro-L-phenylalanine ethyl ester**

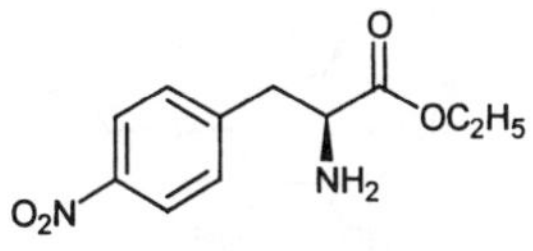

$C_{11}H_{14}N_2O_4$
Chiral building block. *Advanced Asymmetrics, Inc.*

**3965 p-Nitro-L-phenylalanine methyl ester**

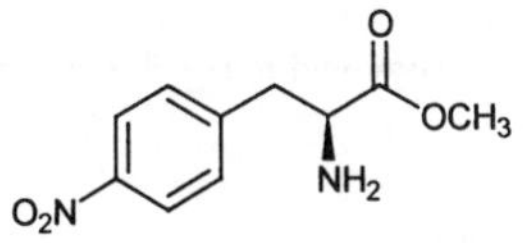

$C_{11}H_{14}N_2O_4$
Chiral building block. *Advanced Asymmetrics, Inc.*

**3966 (S)-4-Nitrophenylalanine methyl ester hydrochloride**

$C_{18}H_{21}BClNO$
Chiral building block. mp = 232°. *Lancaster Synthesis Ltd.*

**3967 4-Nitrophenyl-β-L-arabinofuranoside**

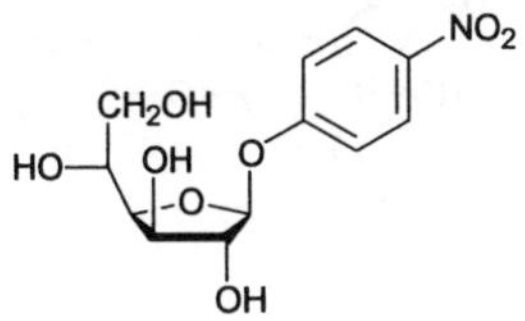

$C_{12}H_{15}NO_8$
Chiral intermediate. *Senn Chemicals AG.*

**3968 4-Nitrophenyl-α-L-arabinopyranoside**

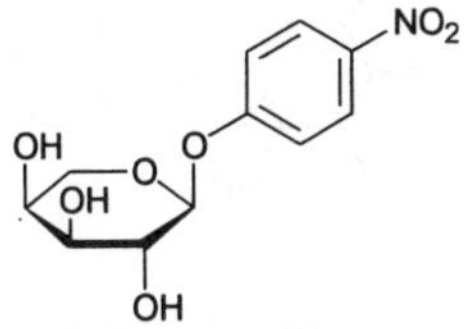

$C_{11}H_{13}NO_7$
Chiral intermediate. *Senn Chemicals AG.*

**3969 4-Nitrophenyl-β-D-cellobioside**
3482-57-3

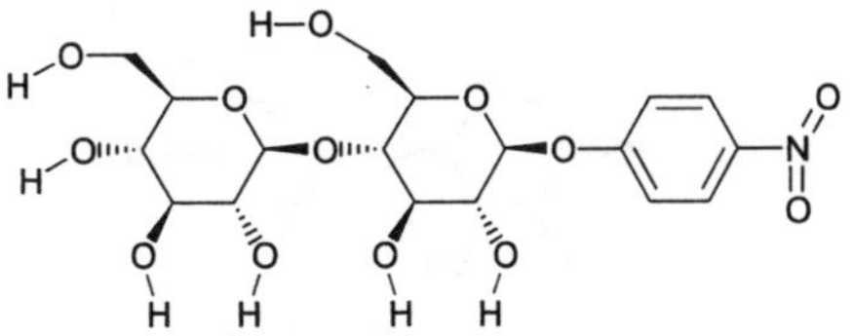

$C_{18}H_{25}NO_{13}$
Chiral diagnostic reagent. Soluble in $H_2O$. *Acros Organics nv; Diagnostic Chemicals Limited.*

**3970 (R)-4-(2-Nitrophenyl)-1,4-dihydro-2,6-dimethyl-3,5-pyridine dicarboxylic acid mono-2-acetamidoethyl ester**

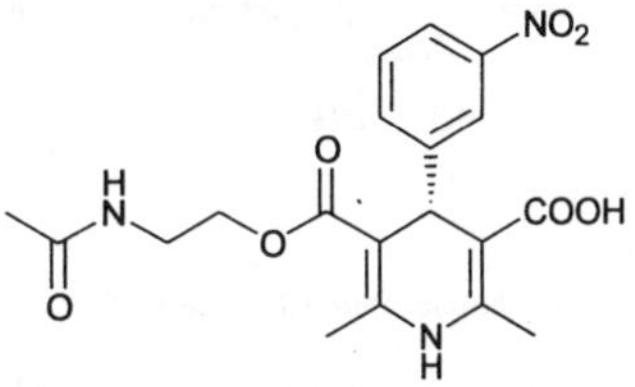

$C_{19}H_{21}N_3O_7$
Chiral intermediate. $[\alpha]_D$ = - 73.0°. *Mercian Corporation.*

**3971 (R)-4-(2-Nitrophenyl)-1,4-dihydro-2,6-dimethyl-3,5-pyridine dicarboxylic acid mono-methyl ester**

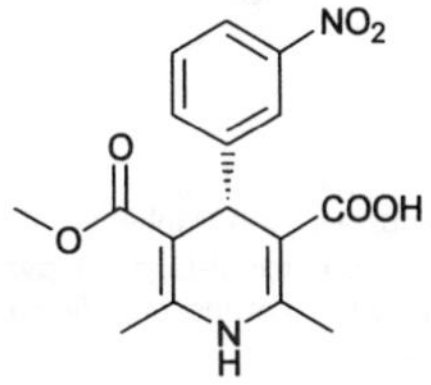

$C_{16}H_{16}N_2O_6$
Chiral intermediate. $[\alpha]_D$ = - 22.6°. *Mercian Corporation.*

**3972 (R)-1-(4-Nitrophenyl)ethylamine hydrochloride**

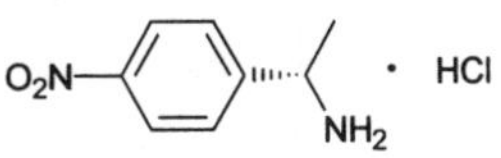

$C_8H_{11}ClN_2O_2$
Chiral intermediate. *FineTech, Ltd.; Yamakawa Chemical Industry Co. Ltd.*

**3973 4-Nitrophenyl-β-D-fucopyranoside**
1226-39-7

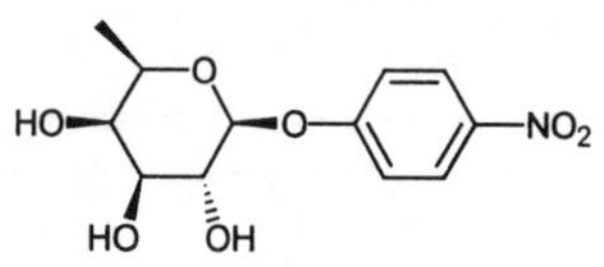

$C_{12}H_{15}NO_7$
Chiral intermediate. *Acros Organics nv.*

**3974 4-Nitrophenyl-α-L-fucopyranoside**
10231-84-2

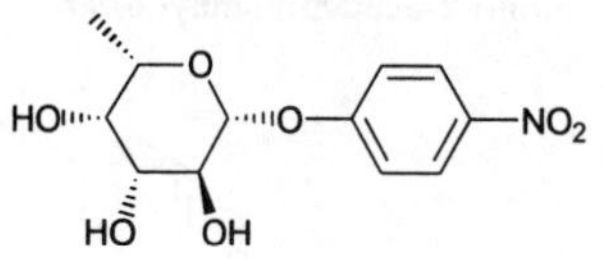

$C_{12}H_{15}NO_7$
Chiral intermediate. *Senn Chemicals AG.*

**3975 4-Nitrophenyl-β-L-fucopyranoside**
22153-71-5 244-808-3

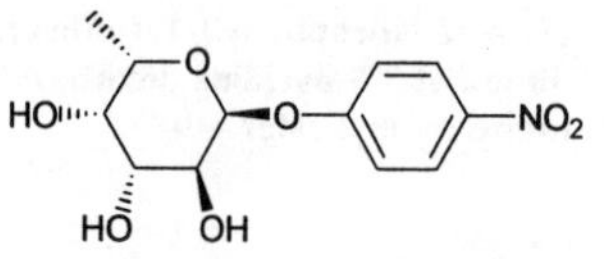

$C_{12}H_{15}NO_7$
Galactopyranoside, p-nitrophenyl 6-deoxy-, β-L-; p-Nitrophenyl 6-deoxy-β-L-galactopyranoside. Chiral diagnostic reagent. *Diagnostic Chemicals Limited; Senn Chemicals AG.*

**3976 2-Nitrophenyl-α-D-galactopyranoside**

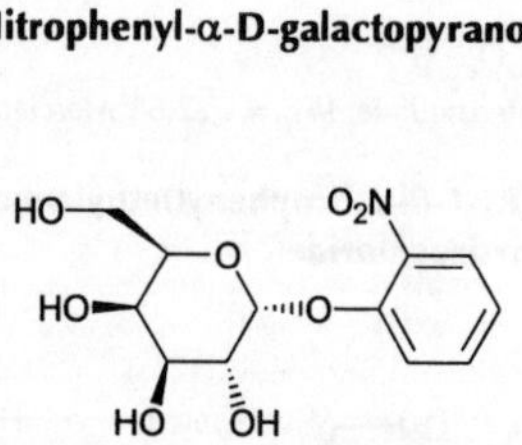

$C_{12}H_{15}NO_8$
Chiral intermediate. *Senn Chemicals AG.*

**3977 2-Nitrophenyl-β-D-galactopyranoside**
369-07-3 206-716-1

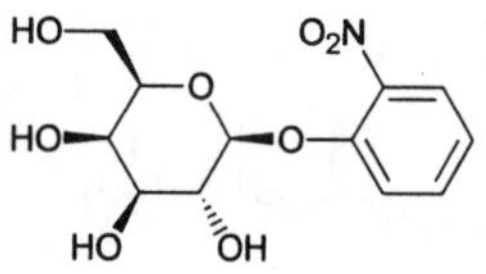

$C_{12}H_{15}NO_8$
Galactopyranoside, o-nitrophenyl, β-D-; Chiral diagnostic reagent. mp = 196-200°; $[\alpha]_D$ = - 66 ± 7° (c = 1, $H_2O$); soluble in $H_2O$. *Acros Organics nv; Diagnostic Chemicals Limited; Senn Chemicals AG; Sigma-Aldrich Fine Chemicals.*

**3978 3-Nitrophenyl-α-D-galactopyranoside**

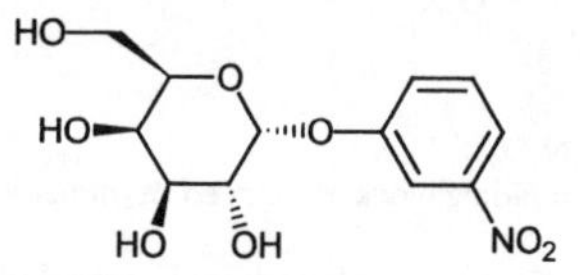

$C_{12}H_{15}NO_8$
Chiral intermediate. *Senn Chemicals AG.*

**3979 3-Nitrophenyl-β-D-galactopyranoside**
3150-25-2 221-585-0

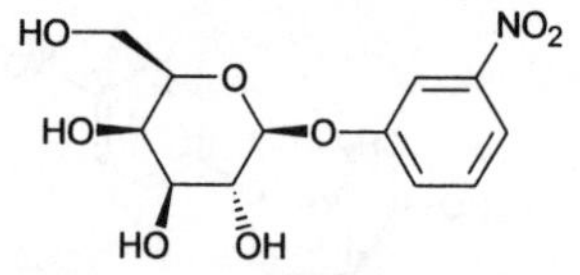

$C_{12}H_{15}NO_8$
Galactopyranoside, 3-nitrophenyl, β-D-; Chiral intermediate. *Senn Chemicals AG.*

**3980 4-Nitrophenyl-α-D-galactopyranoside**
7493-95-0

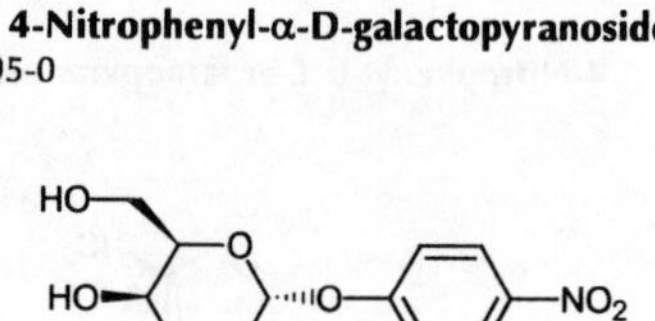

$C_{12}H_{15}NO_8$
4-Nitrophenyl-α-D-galactopyranoside. Chiral diagnostic reagent. $[\alpha]_D^{20}$ = + 240 ± 20°; slightly soluble in $H_2O$. *Acros Organics nv; Diagnostic Chemicals Limited; Senn Chemicals AG.*

**3981 4-Nitrophenyl-β-D-galactopyranoside**
3150-24-1 221-584-5

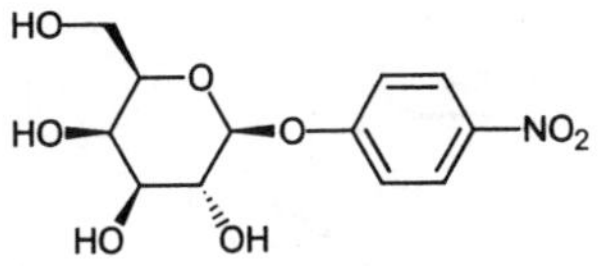

$C_{12}H_{15}NO_8$
Galactopyranoside, p-nitrophenyl, β-D-; p-Nitrophenyl β-D-galactoside. Chiral diagnostic reagent. mp = 178-184°; $[\alpha]_D^{20}$ = - 77 to - 87° (c = 0.3, $H_2O$); soluble in $H_2O$. *Acros Organics nv; Diagnostic Chemicals Limited; Sigma-Aldrich Fine Chemicals.*

**3982 2-Nitrophenyl-β-D-glucopyranoside**
2816-24-2 220-568-5

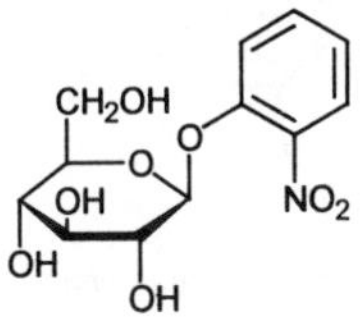

$C_{12}H_{15}NO_8$
Glucopyranoside, o-nitrophenyl, β-D-; o-Nitrophenyl-β-glucoside. Chiral diagnostic reagent. mp = 164-170°; soluble in $H_2O$ (2.4%). *Diagnostic Chemicals Limited; Senn Chemicals AG.*

**3983 3-Nitrophenyl-β-D-glucopyranoside**
20838-44-2 244-074-4

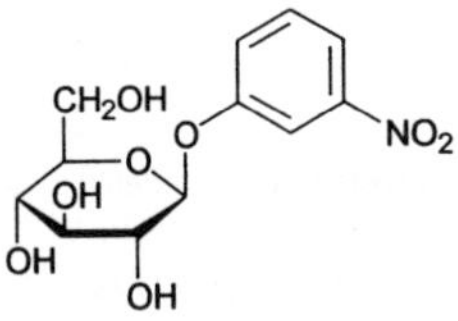

$C_{12}H_{15}NO_8$
Glucopyranoside, 3-nitrophenyl, β-D-; Chiral intermediate. *Senn Chemicals AG.*

**3984 4-Nitrophenyl-α-D-glucopyranoside**
3767-28-0 223-189-3

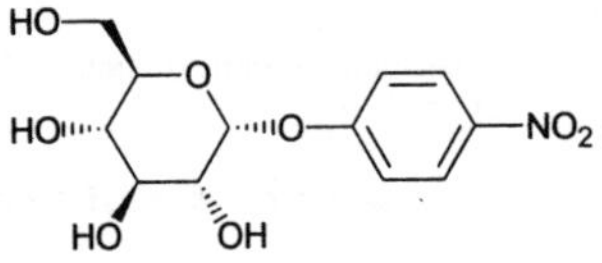

$C_{12}H_{15}NO_8$
Glucopyranoside, p-nitrophenyl, α-D-; 4-Nitrophenyl-α-D-glucopyranoside. Chiral diagnostic reagent. mp = 209-213°; $[\alpha]_D^{20}$ = +215° (c = 1, $H_2O$); soluble in $H_2O$. *Acros Organics nv; Diagnostic Chemicals Limited; Senn Chemicals AG; Sigma-Aldrich Fine Chemicals.*

**3985 4-Nitrophenyl-β-D-glucopyranoside**
2492-87-7 219-661-3

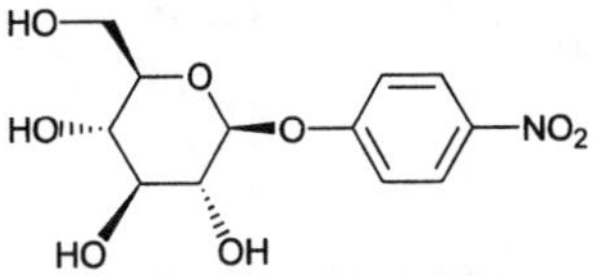

$C_{12}H_{15}NO_8$
Glucopyranoside, p-nitrophenyl, β-D-; 4-Nitrophenyl-β-D-glucopyranoside. Listed on TSCA. Chiral diagnostic reagent. mp = 165-168°. *Acros Organics nv; Diagnostic Chemicals Limited; Senn Chemicals AG; Sigma-Aldrich Fine Chemicals.*

**3986 4-Nitrophenyl-β-D-glucuronic acid**
10344-94-2 233-753-0

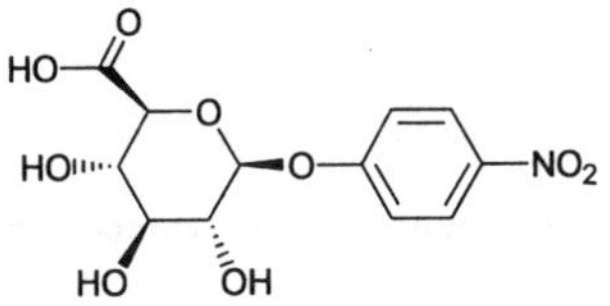

$C_{12}H_{13}NO_9$
Glucopyranosiduronic acid, 4-nitrophenyl, β-D-; 4-Nitrophenyl-β-D-glucuronide; Glucopyranosiduronic acid, p-nitrophenyl; p-Nitrophenyl β-D-glucosiduronic acid. Chiral diagnostic reagent. *Acros Organics nv; Diagnostic Chemicals Limited.*

**3987 S-Nitroso-L-glutathione**
57564-91-7
$C_{10}H_{16}N_4O_7S$
Chiral intermediate. *Acros Organics nv.*

**3988 4-Nitrophenyl-β-D-lactopyranoside**

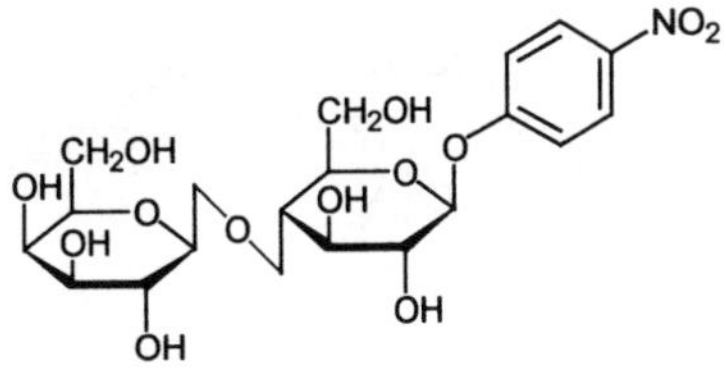

$C_{20}H_{29}NO_{13}$
Chiral intermediate. *Senn Chemicals AG.*

**3989 4-Nitrophenyl-β-D-maltoside**

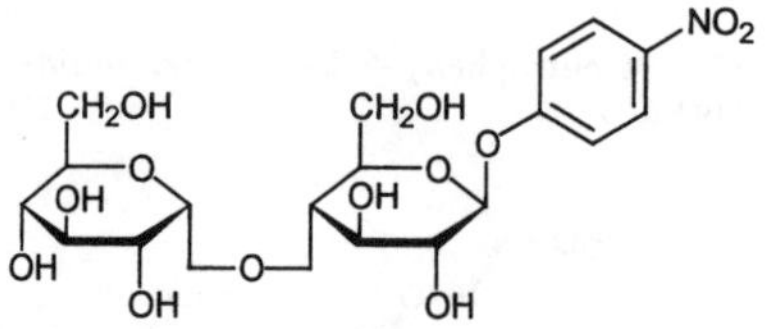

$C_{20}H_{29}NO_{13}$
Chiral intermediate. *Senn Chemicals AG.*

**3990 4-Nitrophenyl-α-D-maltoside hydrate**
17400-77-0

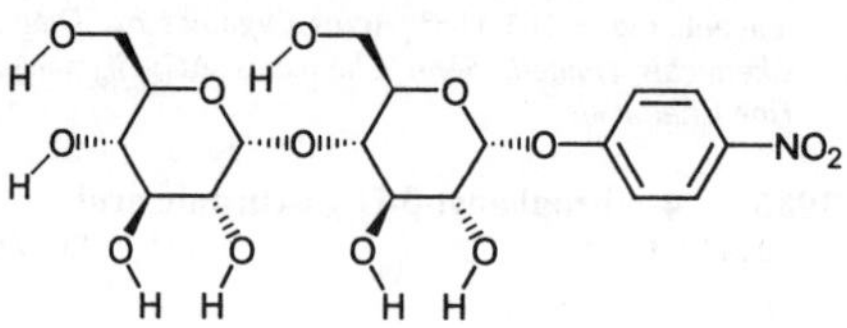

$C_{18}H_{25}NO_{13}$
Chiral intermediate. *Acros Organics nv.*

**3991 4-Nitrophenyl-β-D-mannopyranoside**

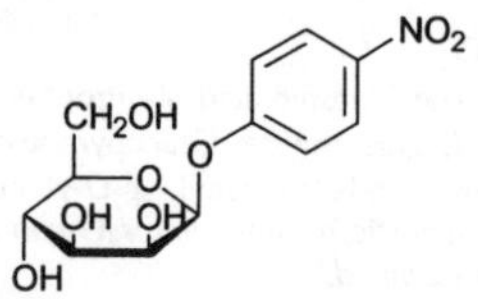

$C_{12}H_{15}NO_8$
Chiral intermediate. *Senn Chemicals AG.*

**3992 4-Nitrophenyl-α-D-mannopyranoside hydrate**
10357-27-4 233-776-6

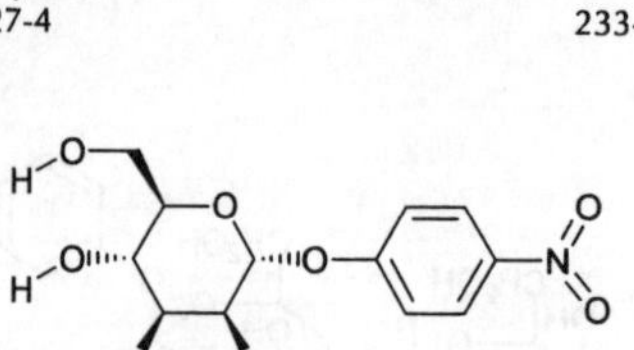

$C_{12}H_{15}NO_8$
Mannopyranoside, 4-nitrophenyl, α-D-; Mannopyranoside, p-nitrophenyl. Chiral intermediate. *Acros Organics nv; Senn Chemicals AG.*

**3993 4-Nitrophenyl-β-D-melibioside**

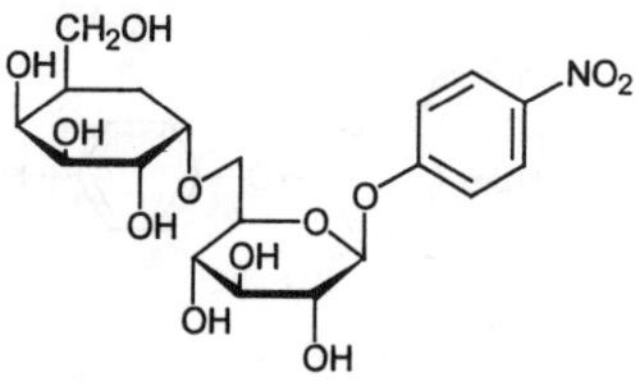

$C_{19}H_{27}NO_{12}$
Chiral intermediate. *Senn Chemicals AG.*

**3994 4-Nitrophenyl-α-L-rhamnoside**

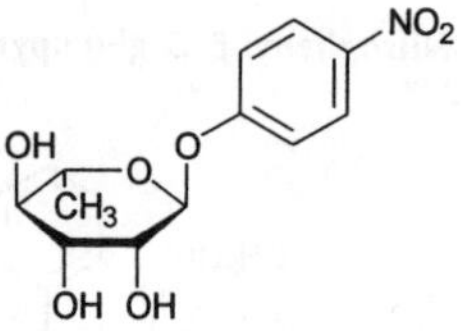

$C_{12}H_{15}NO_7$
Chiral intermediate. *Senn Chemicals AG.*

**3995 2-Nitrophenyl-β-D-thiogalactopyranoside**
1158-17-4 214-593-0

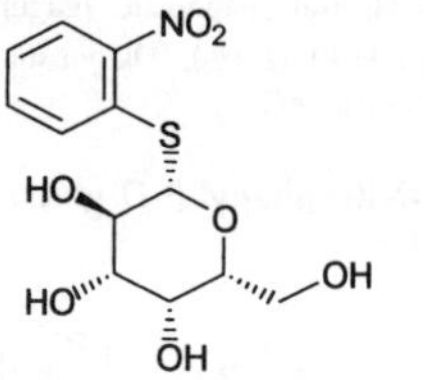

$C_{12}H_{15}NO_7S$
Galactopyranoside, o-nitrophenyl 1-thio-, β-D-; Chiral intermediate. mp = 201-202°; $[\alpha]^{25}$ = - 229.7° (c = 0.71, $CH_3OH$). *Acros Organics nv; Sigma-Aldrich Fine Chemicals.*

**3996 (S)-(-)-N-(5-Nitro-2-pyridyl)alaninol**
115416-52-9
$C_8H_{11}N_3O_3$
Chiral intermediate. *TCI America.*

**3997 (S)-N-(5-Nitro-2-pyridyl)phenylalaninol**
115416-53-0
$C_{14}H_{15}N_3O_3$
2-(N-L-Phenylalaninol)-5-nitropyridine. Chiral intermediate. *TCI America.*

**3998 (S)-(-)-N-(5-Nitro-2-pyridyl)prolinol**
88374-37-2
$C_{10}H_{13}N_3O_3$
(S)-(-)-1-(5-Nitro-2-pyridinyl)-2-pyrrolidinemethanol; 2-(N-Prolinol)-5-nitropyridine. Chiral intermediate. mp = 83°. *TCI America.*

**3999 3-Nitro-L-tyrosine**
621-44-3 210-688-6

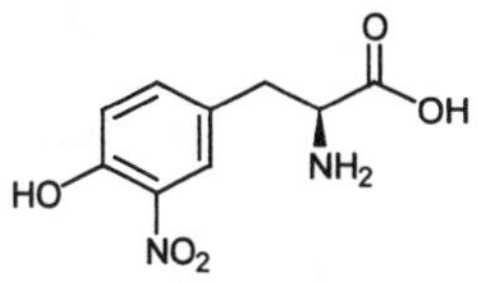

$C_9H_{10}N_2O_5$
Tyrosine, 3-nitro-, L-; Chiral intermediate. mp = 233-235°; $[\alpha]^{22}$ = + 3.8° (c = 1, 1N HCl). *Acros Organics nv; Sigma-Aldrich Fine Chemicals.*

**4000 2-Nitrophenyl-β-D-xylopyranoside**
10238-27-4 233-571-1

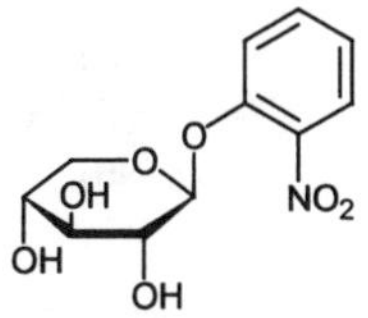

$C_{11}H_{13}NO_7$
Xylopyranoside, 2-nitropheny, β-D-; β-D-Xylopyranoside, 2-nitrophenyl. Chiral diagnostic reagent. *Diagnostic Chemicals Limited; Senn Chemicals AG.*

**4001 4-Nitrophenyl-α-D-xylopyranoside**
10238-28-5 233-572-7

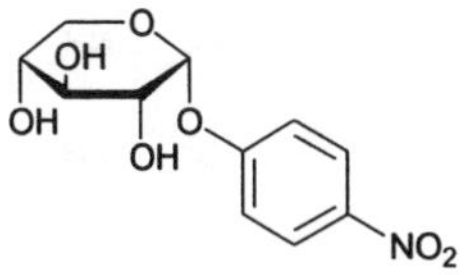

$C_{11}H_{13}NO_7$
Xylopyranoside, 4-nitrophenyl, α-D-; α-D-Xylopyranoside, 4-nitrophenyl. Chiral diagnostic reagent. *Diagnostic Chemicals Limited.*

**4002 4-Nitrophenyl-β-D-xylopyranoside**
2001-96-9 217-897-1

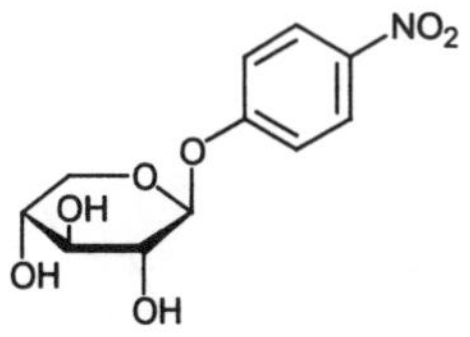

$C_{11}H_{13}NO_7$
Xylopyranoside, 4-nitrophenyl, β-D-; Chiral intermediate. *Senn Chemicals AG.*

**4003 (-)-Noe's reagent**
108031-79-4
$C_{24}H_{38}O_3$
MBF-OH dimer; Bis[(2S,3αR,4S,7αR)-octahydro-7,8,8-trimethyl-4,7-methanobenzo[b]furan-2-yl] ether. Chiral intermediate. mp = 150-152°; $[\alpha]_D^{20}$ = - 202 ± 2° (c = 2.5, THF). *Lancaster Synthesis Ltd.; Sigma-Aldrich Fine Chemicals.*

**4004 (+)-Noe's Reagent**
87248-50-8

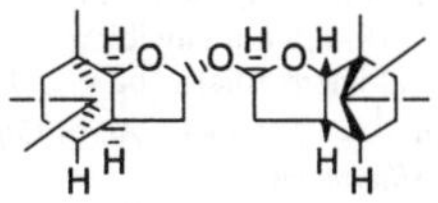

$C_{24}H_{38}O_3$
Bis[(2R,3αS,4R,7αS)-octahydro-7,8,8-trimethyl-4,7-methanobenzo[b]furan-2-yl] ether; MBF-OH dimer. Resolving agent. mp = 148-150°; $[\alpha]_D^{20}$ = + 202 ± 2° (c = 2.5, THF). *Lancaster Synthesis Ltd.; Sigma-Aldrich Fine Chemicals.*

**4005 (R)-(-)-1,3-Nonanediol**
23433-07-0
$C_9H_{20}O_2$
Chiral building block. *Sigma-Aldrich Fine Chemicals.*

**4006 (S)-3-Nonanol**

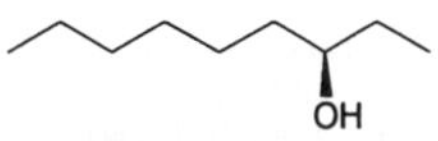

$C_9H_{20}O$
Nonan-3-ol, (3S)-; Chiral building block. *Japan Energy Corporation.*

**4007 (R)-Nonyl glycidyl ether**

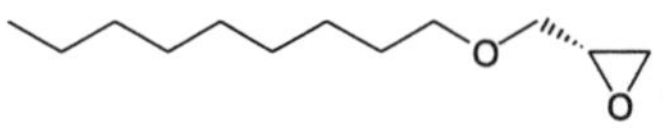

$C_{12}H_{24}O_2$
Oxirane, [(nonyloxy)methyl]-, (R)-; (R)-[(Nonyloxy)methyl]oxirane; Propane, 1,2-epoxy-3-(nonyloxy)-, (R)-. Chiral building block. *Japan Energy Corporation.*

**4008 (1R)-(+)-Nopinone**
38651-65-9

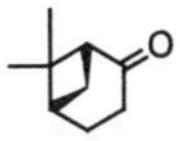

$C_9H_{14}O$
6,6-Dimethylbicyclo[3.1.1]heptan-2-one. Chiral intermediate. bp = 209°; d = 0.981; n = 1.479; $[\alpha]_D^{20}$ = + 16° (neat). *Sigma-Aldrich Fine Chemicals.*

**4009 (-)-Nopol**
35836-73-8 252-744-2

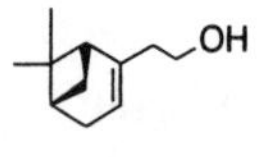

$C_{11}H_{18}O$
Bicyclo[3.1.1]hept-2-ene-2-ethanol, 6,6-dimethyl-, (1R,5S)-; (1R)-6,6-dimethylbicyclo[3.1.1]hept-2-en-2-ethanol. Chiral intermediate. bp = 230-240°; d = 0.973; n = 1.493; $[\alpha]^{24}$ = - 37° (neat). *Acros Organics nv; Sigma-Aldrich Fine Chemicals.*

**4010 (-)-Nopol benzyl ether**

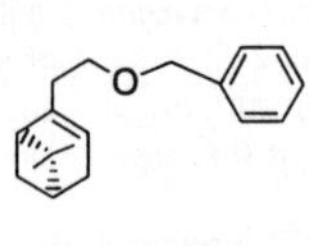

$C_{18}H_{24}O$
Chiral intermediate. $bp_{0.02}$ = 112-114°; d = 0.982; n = 1.5204; $[\alpha]_D^{20}$= - 26° (c = 10, $CHCl_3$). *Sigma-Aldrich Fine Chemicals.*

**4011 (1R,2R,4R)-(+)-endo-Norbornenol**
36779-79-0
$C_7H_{10}O$
Bicyclo[2.2 1]hept-5-en-2-ol, (1R,2R,4R)-; Chiral intermediate. *Austin Chemical Company, Inc.; Celltech Chiroscience Ltd.; ChiroTech Technology Ltd.; Lancaster Synthesis Ltd.*

**4012 (1S,2R,4R)-(-)-endo-Norborneol**
61277-90-5

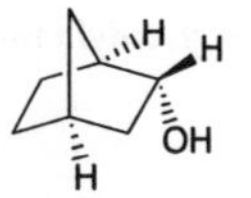

$C_7H_{12}O$
(1S,2R,4R)-Bicyclo[2.2.1]heptan-2-ol. Chiral intermediate. *Celltech Chiroscience Ltd.*

**4013 D-Norleucine**
327-56-0 6624(11) 206-320-9

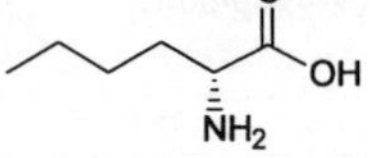

$C_6H_{13}NO_2$
Norleucine, D-; D-2-Amino-n-caproic acid; (R)-2-Aminohexanoic acid. Chiral building block. mp > 300°; $[\alpha]^{22}$ = - 21.2° (c = 4.7, 6N HCl). *Acros Organics nv; Austin Chemical Company, Inc.; Sigma-Aldrich Fine Chemicals; TCI America; Yoneyama Yakuhin Kogyo Co., Ltd.*

**4014 L-Norleucine**
327-57-1 6800(12) 206-321-4

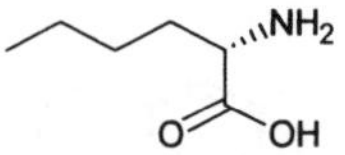

$C_6H_{13}NO_2$
Norleucine, L-; L-2-Amino-n-caproic Acid; (S)-2-Aminohexanoic acid. Listed on TSCA. Chiral building block. mp > 300°; $[\alpha]^{24}$ = + 23.3° (c = 4.2, 5N HCl). *Acros Organics nv; Austin Chemical Company, Inc.; Senn Chemicals AG; Sigma-Aldrich Fine Chemicals; TCI America; Yoneyama Yakuhin Kogyo Co., Ltd.*

**4015 D-Norvaline**
2013-12-9 6813(12) 217-936-2

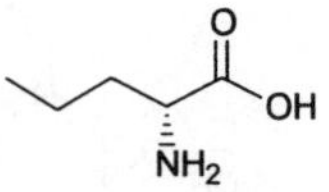

$C_5H_{11}NO_2$
Norvaline, D-; (R)-2-Aminopentanoic acid; D-α-Aminovaleric acid. Chiral building block. mp > 300°; $[\alpha]_D^{20}$= - 25 ± 1° (c = 10, 20% HCl). *Acros Organics nv; Austin Chemical Company, Inc.; Lancaster Synthesis Ltd.; Sigma-Aldrich Fine Chemicals; Yoneyama Yakuhin Kogyo Co., Ltd.*

**4016 L-Norvaline**
6600-40-4 6813(12) 229-543-3

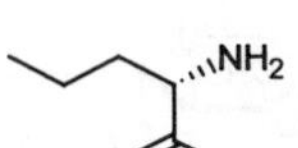

$C_5H_{11}NO_2$
Pentanoic acid, 2-amino-, (S)-; (S)-2-Aminopentanoic acid; (S)-2-Aminovaleric acid. Building block. mp > 300°; $[\alpha]_D^{20}$= + 25 ± 2° (c = 10, 20% HCl). *Acros Organics nv; Ajinomoto Co. Inc.; Austin Chemical Company, Inc.; Lancaster Synthesis Ltd.; Omega Chemical Company Inc.; Senn Chemicals AG; Sigma-Aldrich Fine Chemicals; Yoneyama Yakuhin Kogyo Co., Ltd.*

**4017 (R)-1,2-Octadecanediol**
61468-71-1

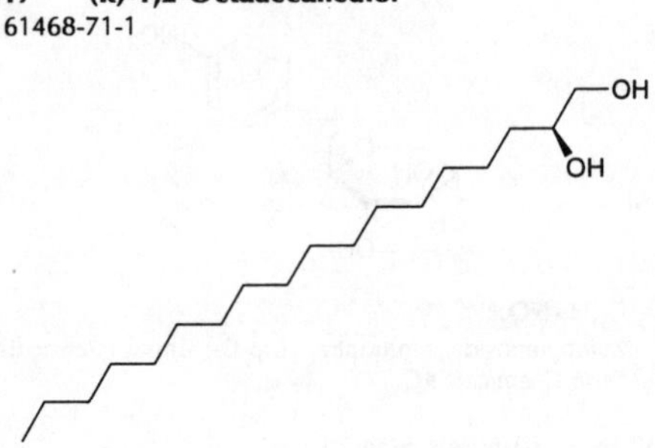

$C_{18}H_{38}O_2$
Chiral building block. *Kankyo Kagaku Center Co., Ltd.*

**4018 (S)-1,2-Octadecanediol**
125555-78-4

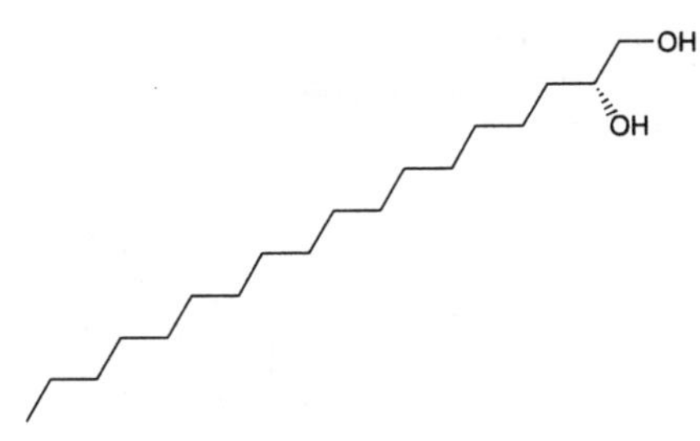

$C_{18}H_{38}O_2$
Zinc α-glucoheptonate; Chiral building block. *Kankyo Kagaku Center Co., Ltd.*

**4019 (R)-5,5',6,6',7,7',8,8'-octahydro-1,1'-bi-2-naphthol**
65355-14-8

$C_{20}H_{22}O_2$
Chiral intermediate. *Kankyo Kagaku Center Co., Ltd.*

**4020 (S)-5,5',6,6',7,7',8,8'-octahydro-1,1'-bi-2-naphthol**
65355-00-2

$C_{20}H_{22}O_2$
Chiral ligand. *Kankyo Kagaku Center Co., Ltd.*

**4021 (S)-Octahydroindole-2-carboxylic acid**

$C_9H_{15}NO_2$
Chiral intermediate. *Fischer Chemicals AG; Great Lakes Fine Chemicals; Kawaken Fine Chemicals Co. Ltd.*

**4022 (R)-1,2-Octanediol**
87720-90-9

$C_8H_{18}O_2$
Chiral building block. *Kankyo Kagaku Center Co., Ltd.*

**4023 (S)-1,2-Octanediol**
87720-91-0

$C_8H_{18}O_2$
Chiral building block. *Kankyo Kagaku Center Co., Ltd.*

**4024 (3R,6R)-Octanediol**
$C_8H_{18}O_2$
(3R,6R)-Dihydroxyoctane. Chiral building block. *ChiroTech Technology Ltd.; Lancaster Synthesis Ltd.*

**4025 (3S,6S)-Octanediol**
136705-66-3
$C_{11}H_{19}NO_4$
(3S,6S)-Dihydroxyoctane. Chiral building block. *Lancaster Synthesis Ltd.*

**4026 (R)-(-)-2-Octanol**
5978-70-1 6850(12) 227-777-0

$C_8H_{18}O$
Octan-2-ol, (2R)-; (R)-(-)-n-Hexylmethylcarbinol; (R)-1-Methylheptanol. Listed on TSCA. Chiral building block. bp = 174-175°; d = 0.8380; $n_D^{20}$ = 1.4260; $[\alpha]_D^{20}$ = - 10 ± 1° (c = 1, EtOH). *Acros Organics nv; Digital Specialty Chemicals, Inc.; Lancaster Synthesis Ltd.; Norse Laboratories; Sigma-Aldrich Fine Chemicals; TCI America.*

**4027 (S)-(+)-2-Octanol**
6169-06-8 6850(12) 228-213-6

$C_8H_{18}O$
Octan-2-ol, (2S)-; (S)-(+)-n-Hexylmethylcarbinol; (S)-1-Methylheptyl alcohol. Listed on TSCA. Chiral building block. bp = 174-175°; d = 0.8190; $n_D^{20}$ = 1.4260; $[\alpha]_D^{20}$ = + 10 ± 1° (c = 1, EtOH); $[\alpha]^{17}$ = + 9 (neat). *Acros Organics nv; Digital Specialty Chemicals, Inc.; Japan Energy Corporation; Lancaster Synthesis Ltd.; Norse Laboratories; Sigma-Aldrich Fine Chemicals; TCI America.*

**4028 N-Octanoyl-D-erythro sphingosine**
74713-59-0

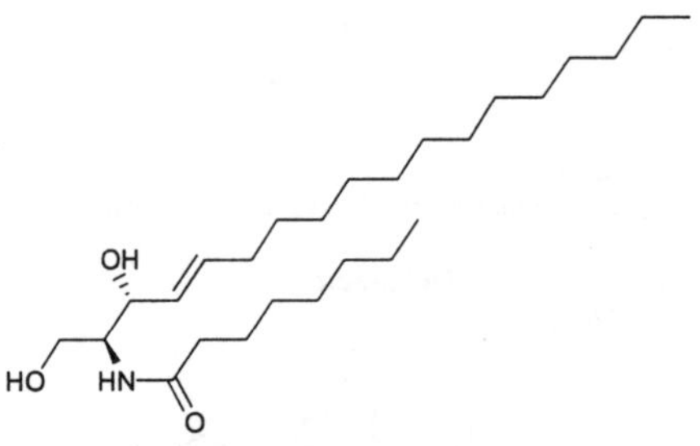

$C_{26}H_{51}NO_3$
Chiral intermediate. *Acros Organics nv.*

**4029 (R)-(-)-1-Octen-3-ol**
3687-48-7

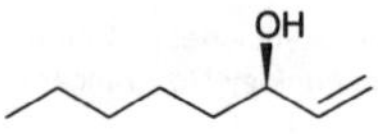

$C_8H_{16}O$
Galactopyranoside, isopropyl 1-thio-, β-D-; IPTG; Thioisopropyl β-D-galactopyranoside. Chiral building block. [α] = - 21°. *Acros Organics nv.*

**4030 (+)-Octopine**
34522-32-2

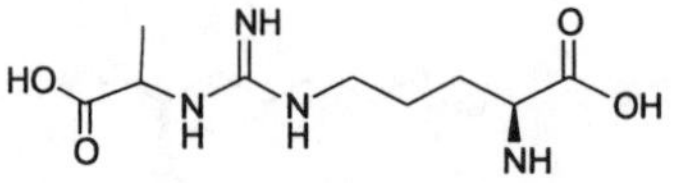

$C_9H_{18}N_4O_4$
$N^2$-(1-carboxyethyl)-, L-; $N^2$-(1-Carboxyethyl)-L-arginine. Chiral intermediate. mp = 283-284°; $[\alpha]^{25}$ = + 19.5° (c = 2, $H_2O$). *Acros Organics nv; Sigma-Aldrich Fine Chemicals.*

**4031 Octyl α-D-glucopyranoside**
29781-80-4
$C_{14}H_{28}O_6$
Chiral intermediate. *Pfanstiehl Laboratories, Inc.*

**4032 Octyl β-D-glucopyranoside**
29836-26-8 249-887-8

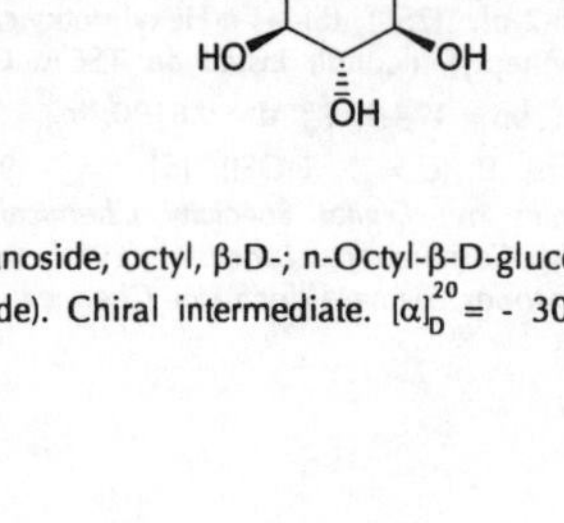

$C_{14}H_{28}O_6$
Glucopyranoside, octyl, β-D-; n-Octyl-β-D-glucoside; AG 8 (glycoside). Chiral intermediate. $[\alpha]_D^{20}$ = - 30° (c = 1, $CH_3OH$). *Acros Organics nv; Pfanstiehl Laboratories, Inc.; Senn Chemicals AG; Sigma-Aldrich Fine Chemicals.*

**4033 (-)-1-O-Octylglucopyranoside 2,3,4,6-tetraacetate**
38954-67-5

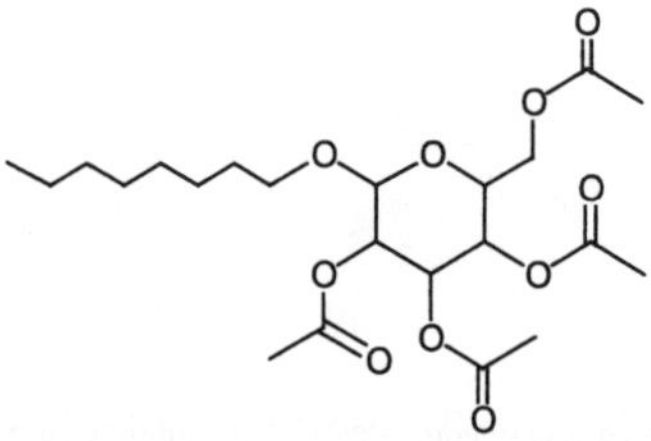

$C_{22}H_{36}O_{10}$
Chiral intermediate. mp = 50-52°; $[\alpha]_D^{20}$ = - 16° (c = 1.8, $CHCl_3$). *Sigma-Aldrich Fine Chemicals.*

**4034 (R)-Octyl glycidyl ether**

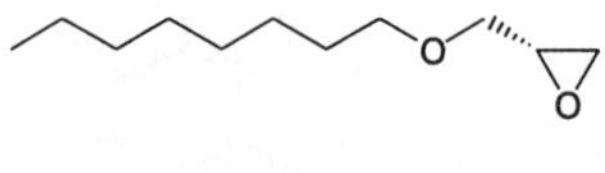

$C_{11}H_{22}O_2$
Oxirane, [(octyloxy)methyl]-, (R)-; (R)-[(Octyloxy)-methyl]oxirane; (R)-1,2-Epoxy-3-(octyloxy)propane; (R)-Glycidyl 1-octyl ether; Propane, 1,2-epoxy-3-(octyloxy)-, (R)-. Chiral building block. *Japan Energy Corporation.*

**4035 (S)-1-Octyloxy-2-propanol**

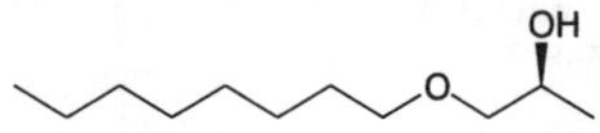

$C_{11}H_{24}O_2$
Chiral building block. *Japan Energy Corporation.*

**4036 n-Octyl-(S)-2-pyrrolidinone-5-carboxylate**
4931-70-8 225-569-4
$C_{13}H_{23}NO_3$
Proline, 5-oxo-, octyl ester, -; n-Octyl L-pyroglutamate; Octyl 5-oxo-L-prolinate; L-Pyroglutamic acid n-octyl ester; (S)-2-Pyrrolidinone-5-carboxylic acid octyl ester. Chiral intermediate. $n_D^{20}$ = 1.472. *Lancaster Synthesis Ltd.*

**4037 Octyl-β-D-thioglucopyranolide**

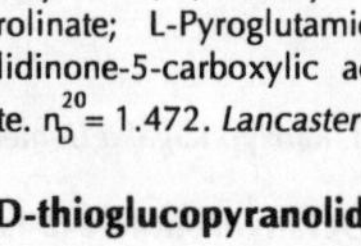

$C_{14}H_{28}O_5S$
Chiral intermediate. *Austin Chemical Company, Inc.*

**4038 S-1-Octyl-β-D-thioglucopyranoside**
85618-21-9

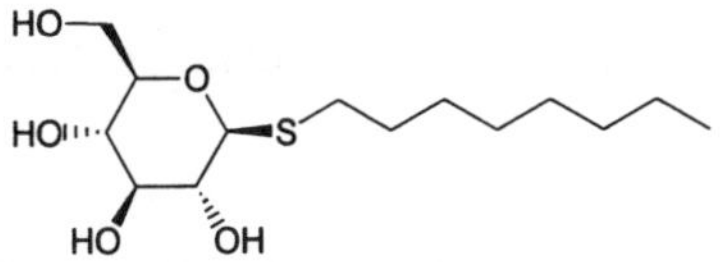

$C_{14}H_{28}O_5S$
Chiral intermediate. $[\alpha]_D^{20} = -52°$ (c = 1, $CH_3OH$). *Acros Organics nv; Pfanstiehl Laboratories, Inc.; Sigma-Aldrich Fine Chemicals.*

**4039 (R)-(+)-1-Octyn-3-ol**
32556-70-0

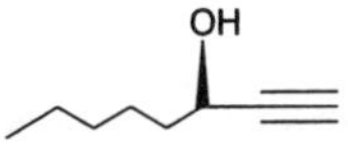

$C_8H_{14}O$
Chiral building block. $bp_{20} = 100°$; d = 0.864; n = 1.442; $[\alpha]_D^{20} = +6.5°$ (c = 2, $CH_2Cl_2$). *Sigma-Aldrich Fine Chemicals.*

**4040 (S)-(-)-1-Octyn-3-ol**
32556-71-1

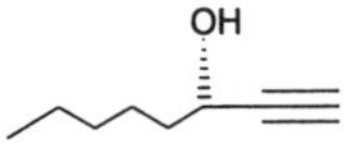

$C_8H_{14}O$
Chiral building block. $bp_{20} = 100°$; d = 0.864; n = 1.442; $[\alpha]_D^{20} = -6.5°$ (c = 2, $CH_2Cl_2$). *Sigma-Aldrich Fine Chemicals; TCI America.*

**4041 Oleanolic acid**
508-02-1 6964(12) 208-081-6

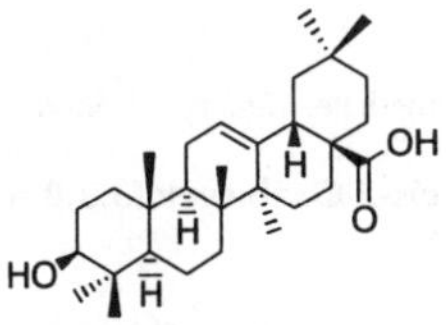

$C_{30}H_{48}O_3$
Olean-12-en-28-oic acid, 3-hydroxy-, (3β)-; 3β-hydroxyolean-12-en-28-oic acid; Astrantiagenin C; Caryophyllin; Giganteumgenin C; Virgaureagenin B. Listed on TSCA. Pharmaceutical or derivative. mp > 300°; $[\alpha]^{19} = +75.2°$ (c = 1, $CHCl_3$). *Kaden Biochemicals GmbH; Sigma-Aldrich Fine Chemicals.*

**4042 L-Ornithine**
70-26-8 200-731-7

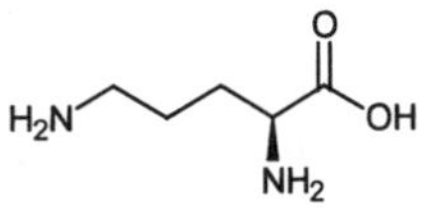

$C_5H_{12}N_2O_2$
Pentanoic acid, 2,5-diamino-, (S)-; (S)-2,5-Diaminopentanoic acid; L-Norvaline, 5-amino-. Listed on TSCA. Chiral building block. *Varsal Instruments, Inc.*

**4043 L-Ornithine-L-aspartate**
3230-94-2 221-772-7

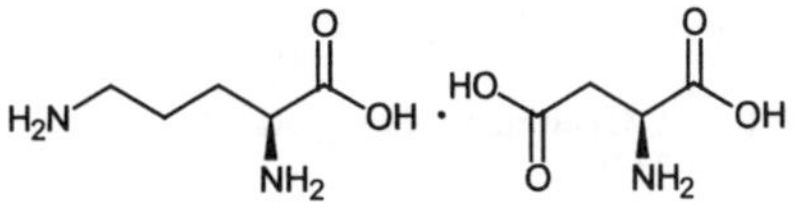

$C_9H_{19}N_3O_6$
Aspartic acid, L-, compd. with L-ornithine (1:1); L-Aspartic acid, compd. with L-ornithine (1:1); Hepamerz; Ormeta; Orparan. Listed on TSCA. Chiral intermediate. *Flamma s.p.a.; Kyowa Hakko Kogyo Co., Ltd.; Rexim S.A. Produits Chimiques.*

**4044 L-Ornithine dihydrochloride**
6211-16-1 228-277-5
$C_5H_{14}Cl_2N_2O_2$
Ornithine, dihydrochloride, L-; Listed on TSCA. Chiral building block. *TCI America.*

**4045 L-Ornithine ethyl ester hydrochloride**
94231-37-5 303-854-5

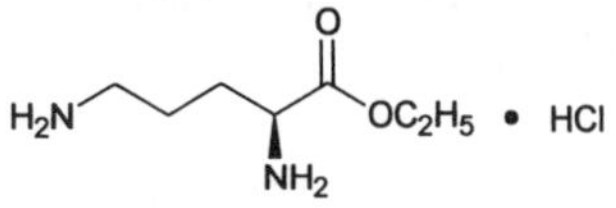

$C_7H_{17}ClN_2O_2$
Ornithine, ethyl ester, monohydrochloride, L-; Ethyl L-ornithine monohydrochloride. Chiral building block. *Austin Chemical Company, Inc.*

**4046 D-Ornithine hydrochloride**
16682-12-5 6826(11) 240-729-3
$C_5H_{13}ClN_2O_2$
Ornithine, monohydrochloride, D-; (R)-(-)-2,5-Diaminopentanoic acid hydrochloride. Chiral building block. mp = 238°; $[\alpha]_D^{20} = -23 \pm 1°$ (c = 5, 5N HCl). *Acros Organics nv; Kaneka Corporation; Lancaster Synthesis Ltd.; Rexim S.A. Produits Chimiques; Sigma-Aldrich Fine Chemicals; Yoneyama Yakuhin Kogyo Co., Ltd.*

**4047 L-Ornithine hydrochloride**
3184-13-2 7002(12) 221-678-6

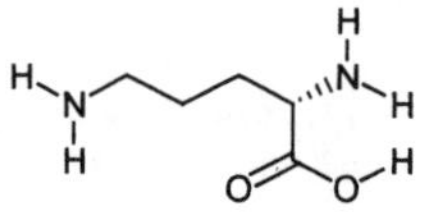

$C_5H_{13}ClN_2O_2$
Ornithine, monohydrochloride, L-; (S)-(-)-2,5-Diaminopentanoic acid hydrochloride; L-2,5-Diaminovaleric acid monohydrochloride. Listed on TSCA. Chiral building block. mp = 245°; $[\alpha]_D^{20}$ = + 23 ± 1° (c = 5, 5N HCl). *Aceto Corporation; Acros Organics nv; Ajinomoto Co. Inc.; Kyowa Hakko Kogyo Co., Ltd.; Lancaster Synthesis Ltd.; Rexim S.A. Produits Chimiques; Senn Chemicals AG; Sigma-Aldrich Fine Chemicals; Tanabe Seiyaku Co. Ltd.; TCI America; Yoneyama Yakuhin Kogyo Co., Ltd.*

**4048 L-Ornithine α-ketoglutarate (1:1)**
34414-83-0 252-003-3

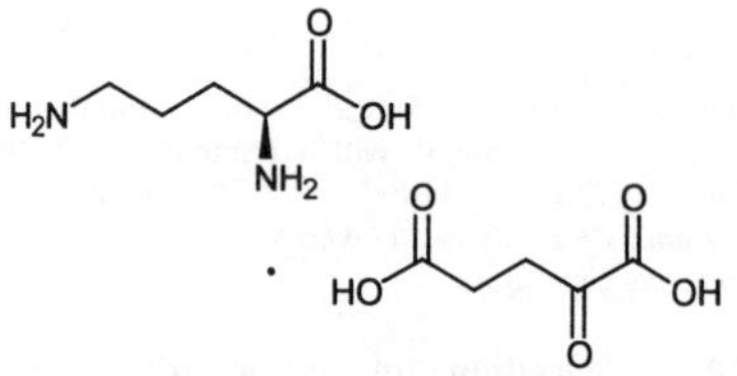

$C_{11}H_{18}N_2O_7$
Glutaric acid, 2-oxo-, compd. with L-ornithine; L-Ornithine 2-oxoglutarate; α-Ketoglutaric acid compd. with L-ornithine; 2-Oxoglutaric acid compd. with L-ornithine; Ornicetil. Chiral intermediate. *Rexim S.A. Produits Chimiques.*

**4049 L-Ornithine α-ketoglutarate (2:1)**
5144-42-3 225-914-9

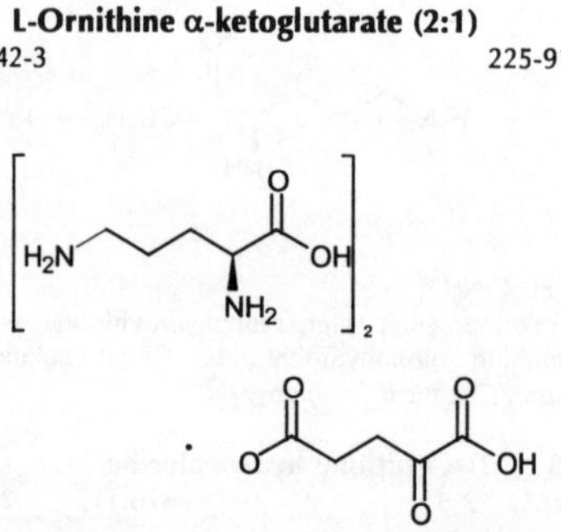

$C_{15}H_{30}N_2O_9$
Glutaric acid, 2-oxo-, compd. with L-ornithine (1:2); L-Ornithine 2-oxoglutarate (2:1); Di-L(+)-ornithine-α-oxoglutarate; L-Ornithine, 2-oxopentanedioate (2:1). Chiral intermediate. *Rexim S.A. Produits Chimiques.*

**4050 L-Ornithine methyl ester dihydrochloride**
40216-82-8 254-841-5

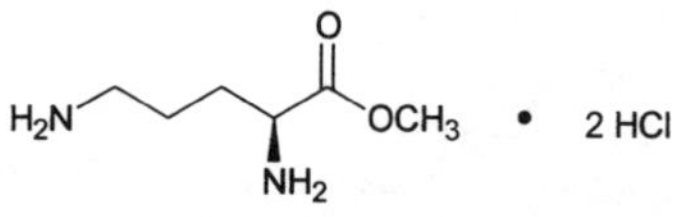

$C_6H_{16}Cl_2N_2O_2$
Ornithine, methyl ester, dihydrochloride, L-; Methyl L-ornithine dihydrochloride. Chiral building block. *Sigma-Aldrich Fine Chemicals.*

**4051 L-Ornithine methyl ester hydrochloride**
60080-69-5 262-049-6

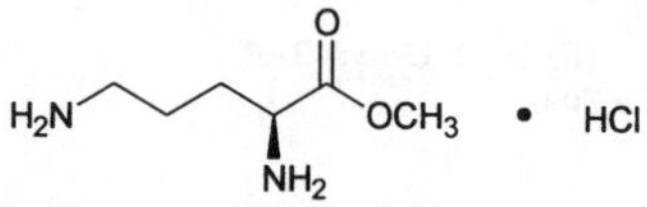

$C_6H_{15}ClN_2O_2$
Ornithine, methyl ester, monohydrochloride, L-; Methyl L-ornithine monohydrochloride. Chiral building block. *Austin Chemical Company, Inc.*

**4052 L-Ornithine sulfate (2:1) monohydrate**

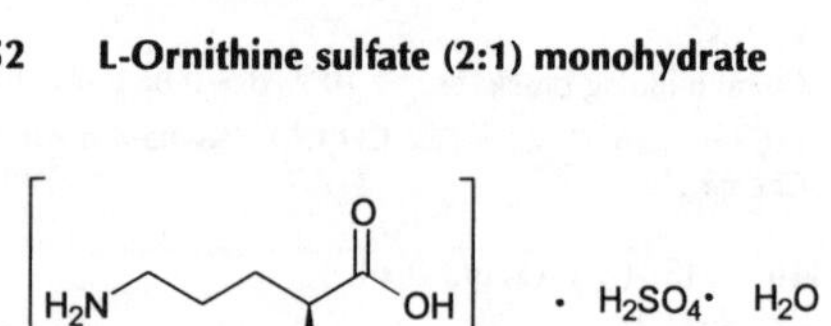

$C_{10}H_{28}N_2O_7S$
Chiral building block. *Rexim S.A. Produits Chimiques.*

**4053 (1S,4S)-2-Oxa-5-azabicyclo[2.2.1]heptane hydrochloride**

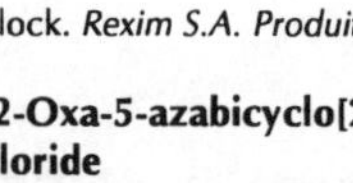

$C_5H_{10}ClNO$
Chiral intermediate. *Medinger & Sohn.*

**4054 (-)-cis-2-Oxabicyclo[3.3.0]oct-6-en-3-one**
43119-28-4

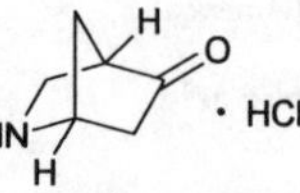

$C_7H_8O_2$
Chiral building block. mp = 45-48°; $[\alpha]^{22}$ = - 103° (c = 1, $CH_3OH$). *Acros Organics nv; Sigma-Aldrich Fine Chemicals.*

**4055 (+)-cis-2-Oxabicyclo[3.3.0]oct-6-en-3-one**
54483-22-6

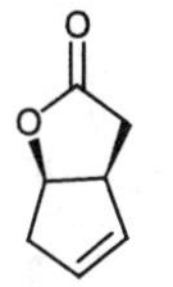

$C_7H_8O_2$
Chiral intermediate. mp = 45-48°; [α] = + 104° (c = 0.7, $CH_3OH$). *Acros Organics nv; Sigma-Aldrich Fine Chemicals.*

**4056 (S)-(-)-4-Oxo-2-azetidinecarboxylic acid**
16404-94-7

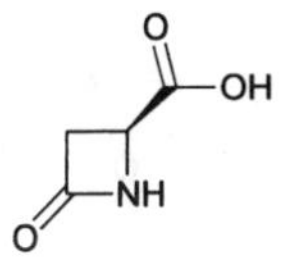

$C_4H_5NO_3$
Chiral intermediate. mp = 99-102°; $[\alpha]_D^{20}$ = - 46° (c = 1, $CH_3OH$). *Sigma-Aldrich Fine Chemicals.*

**4057 (-)-3-Oxo-6-β-t-butyldimethylsilyloxy-methyl-7-α-benzoyloxy-2-oxa-bicyclo[3.3.0]octane**

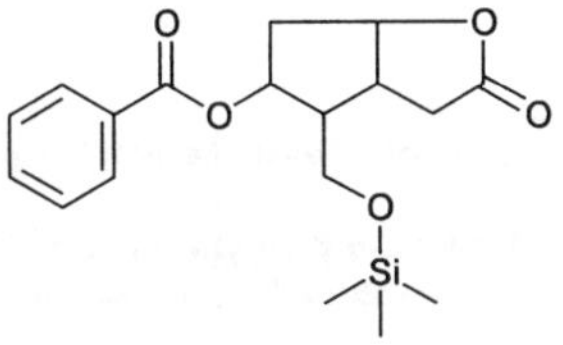

$C_{18}H_{24}O_5Si$
Chiral intermediate. *Acros Organics nv.*

**4058 (1S)-(-)-3-Oxocamphorsulphonylimine**
119106-38-6
$C_{10}H_{13}NO_3S$
Chiral auxiliary. mp = 195-198°; $[\alpha]_D^{20}$ = - 191 ± 2° (c = 2, acetone). *Lancaster Synthesis Ltd.*

**4059 Oxocyclopentanecarboxylic acid methyl ester**
132076-27-8
$C_7H_{10}O_3$
Chiral intermediate. *Boehringer Ingelheim Pharma KG.*

**4060 1-Oxo-dodecyl-α-D-glucopyranoside**
64395-91-1

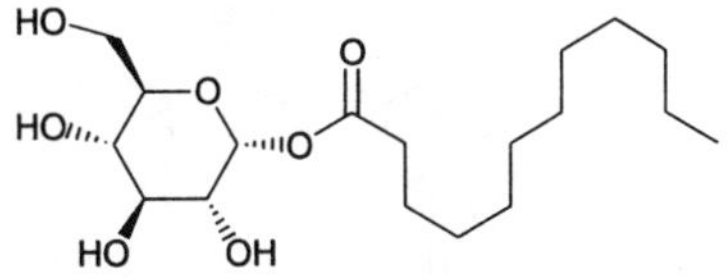

$C_{18}H_{34}O_7$
Chiral intermediate. *Acros Organics nv.*

**4061 1-Oxo-dodecyl-β-D-glucopyranoside**
60415-67-0

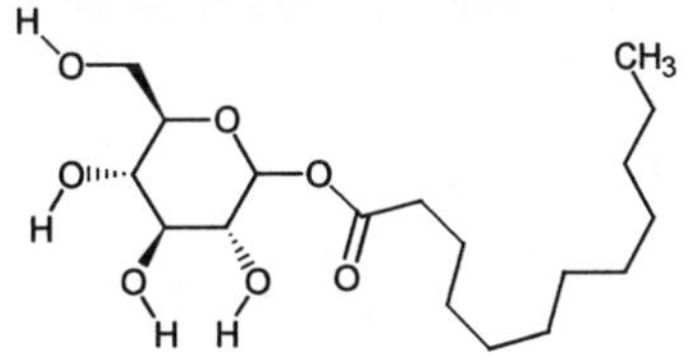

$C_{18}H_{34}O_7$
Chiral intermediate. *Acros Organics nv.*

**4062 1-Oxo-dodecyl-β-D-maltoside**

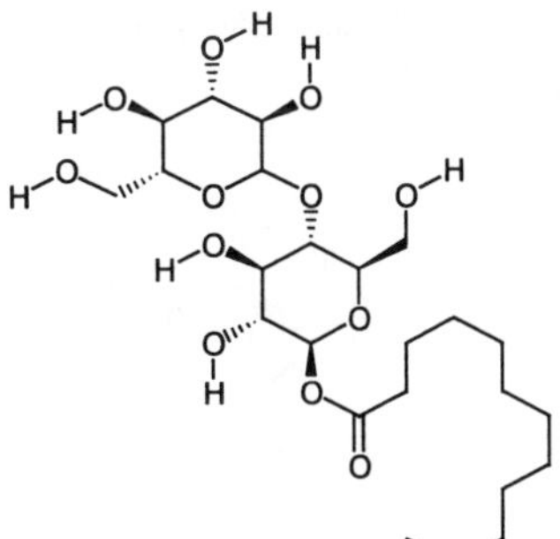

$C_{24}H_{44}O_{12}$
Chiral intermediate. *Acros Organics nv.*

**4063 (S)-(-)-2-Oxo-1,5-imidazolidinedicarboxylic acid 1-benzyl ester**
59760-01-9

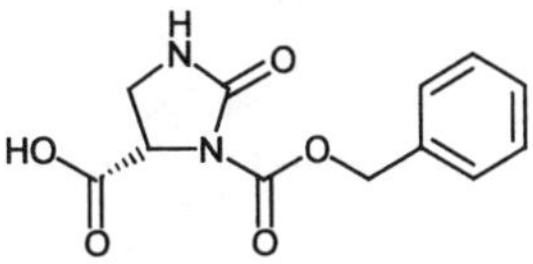

$C_{12}H_{12}N_2O_5$
Chiral intermediate. mp = 189°; $[\alpha]^{24}$ = - 63.0° (c = 2.5, $CH_3OH$). *Sigma-Aldrich Fine Chemicals.*

**4064 (1S)-(+)-4-Oxomyrtenyl pivalate**
113473-32-8

$C_{15}H_{22}O_3$
Chiral intermediate. mp = 40-44°; $[\alpha]_D^{20}$ = + 176° (c = 1, $CHCl_3$). *Sigma-Aldrich Fine Chemicals.*

**4065 Oxo-N-(R)-1-phenylethylphenylacetamide**
10549-15-2

$C_{16}H_{15}NO_2$
Chiral intermediate. mp = 110-112°; $[\alpha]_D^{20}$ = + 105° (c = 2, EtOH). *Sigma-Aldrich Fine Chemicals.*

**4066 (S)-(+)-2-Oxo-4-phenyl-3-oxazolidine-acetic acid**
99333-54-7

$C_{11}H_{11}NO_4$
Chiral intermediate. mp = 103-105°; $[\alpha]^{28}$ = + 165° (c = 2, $CHCl_3$). *Sigma-Aldrich Fine Chemicals.*

**4067 (R)-(-)-3-(1-Oxopropyl)-4-phenyl-2-oxazolidinone**
160695-26-1

$C_{12}H_{13}NO_3$
Chiral intermediate. *SK Energy and Chemical, Inc.*

**4068 (S)-(+)-3-(1-Oxopropyl)-4-phenyl-2-oxazolidinone**
184363-66-4

$C_{12}H_{13}NO_3$
Chiral intermediate. *SK Energy and Chemical, Inc.*

**4069 (S)-(+)-5-Oxo-2-tetrahydrofuran-carboxylic acid**

COOH

$C_6H_8O_3$
Chiral building block. *Austin Chemical Company, Inc.*

**4070 L-2-Oxothiazolidine-4-carboxylic acid ethyl ester**

$C_2H_5OOC$ NH S O

$C_6H_9NO_3S$
Chiral intermediate. *Austin Chemical Company, Inc.*

**4071 (-)-3-Oxo-6-β-trityloxymethyl-7-α-benzoyl-oxy-2-oxabicyclo[3.3.0]octane**

$C_{34}H_{30}O_5$
Chiral intermediate. [α] = - 59° (c = 1.0, $CHCl_3$). *Acros Organics nv.*

**4072 (2S)-(-)-2,2-Oxybis(octahydro-7,8,8-trimethyl-4,7-methanobenzofuran)**
108031-80-7

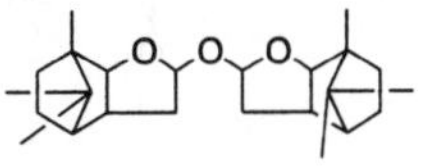

$C_{24}H_{38}O_3$
Chiral intermediate. mp = 149-151°; $[\alpha]_D^{20}$ = - 172° (c = 0.8, $CH_2Cl_2$). *Sigma-Aldrich Fine Chemicals.*

**4073 D-Palatinose**
13718-94-0 237-282-1

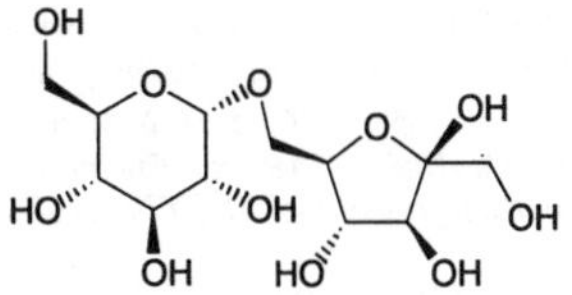

$C_{12}H_{22}O_{11}$
Fructose, 6-O-α-D-glucopyranosyl-, D-; 6-O-α-D-Glucopyranosyl-D-fructofuranose; Isomaltulose. Chiral intermediate. mp = 122-124°; $[\alpha]^{25}$ = + 97.3° (c = 2, $H_2O$). *Acros Organics nv; Senn Chemicals AG; Sigma-Aldrich Fine Chemicals.*

**4074 Palmitoyl-carnitine hydrochloride**
18877-64-0 242-642-6
$C_{23}H_{46}ClNO_4$
Chiral intermediate. mp = 168-172°; $[\alpha]_D^{20}$ = - 8.0° (c = 1, $H_2O$). *Sigma-Aldrich Fine Chemicals.*

**4075 N-Palmitoyl-D-erythro-sphingosine**
24696-26-2
$C_{34}H_{67}NO_3$
Chiral intermediate. *Acros Organics nv.*

**4076 D-Panose**
33401-87-5 251-500-2

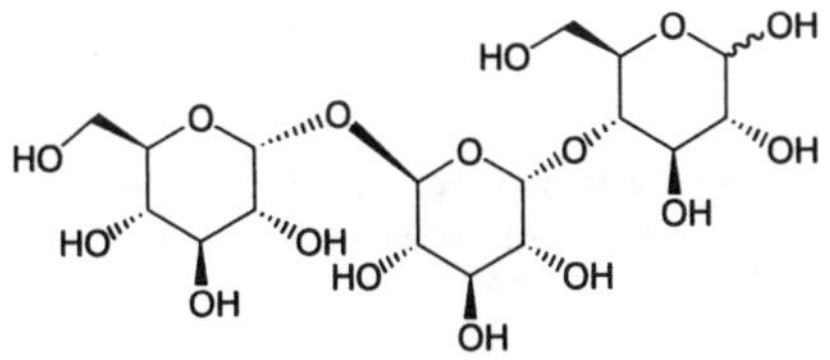

$C_{18}H_{32}O_{16}$
Glucose, O-α-D-glucopyranosyl-(1→6)-O-α-D-glucopyranosyl-(1→4), D-; O-α-D-glucopyranosyl-(1→6)-O-α-D-glucopyranosyl-(1→4)-D-glucose; 4-α-Isomaltosyl-glucose; Ethanone, 1-[4-(β-D-glucopyranosyloxy)phenyl]-; p-Hydroxyacetophenone-D-glucoside; Piceoside; Salicinerein; Salinigrin. Chiral intermediate. *Senn Chemicals AG.*

**4077 D-(-)-Pantoyl lactone**
599-04-2 7145(12) 209-963-3

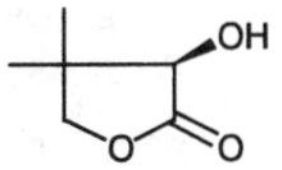

$C_6H_{10}O_3$
Fura2(3H)one, dihydro-3-hydroxy-4,4-dimethyl-, (3R)-. D-(-)-Pantoic acid lactone; α-Hydroxy-β,β-dimethyl-γ-butyrolactone; D-Pantolactone. Listed on TSCA. Pharmaceutical or derivative. mp = 91°; $bp_{15}$ = 120-122°; $[\alpha]^{25}$ = - 49.8° (c = 2, $H_2O$). *Acros Organics nv; Richman Chemical Inc.; Sigma-Aldrich Fine Chemicals; TCI America.*

**4078 (L)-(+)-Pantolactone**
5405-40-3

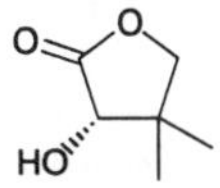

$C_6H_{10}O_3$
L-Pantolactone; (S)-Dihydro-3-hydroxy-4,4-dimethyl-2(3H)-furanone. Pharmaceutical or derivative. mp = 90-91°; $bp_{15}$ = 120-122°; $[\alpha]_D^{20}$ = + 53° (c = 2.4, $H_2O$). *Sigma-Aldrich Fine Chemicals.*

**4079 (-)-Parthenolide**
20554-84-1 7180(12)

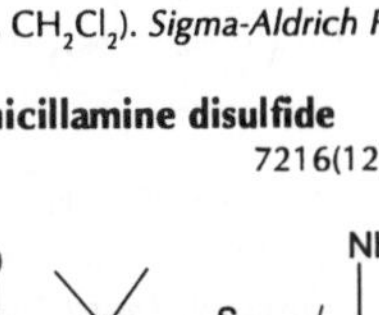

$C_{15}H_{20}O_3$
Pharmaceutical or derivative. mp = 115-116°; $[\alpha]^{25}$ = - 84° (c = 0.25, $CH_2Cl_2$). *Sigma-Aldrich Fine Chemicals.*

**4080 (-)-Penicillamine disulfide**
20902-45-8 7216(12) 244-107-2

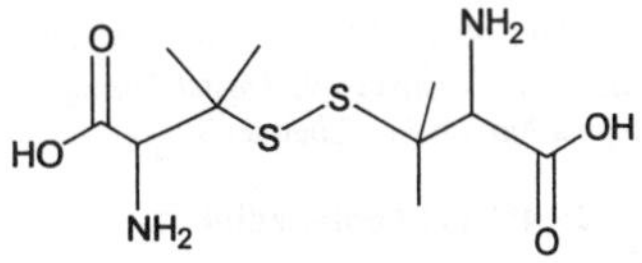

$C_{10}H_{20}N_2O_4S_2$
[3,3'-Dithiobis(2-amino-3-methylbutyric acid). Pharmaceutical or derivative. mp = 204°; $[\alpha]^{25}$ = - 76° (c = 1, 1N NaOH). *Sigma-Aldrich Fine Chemicals.*

**4081 (S)-(-)-1,1,1,2,2-Pentafluorododecan-3-ol**
$C_{12}H_{21}F_5O$
Chiral building block. *Sigma-Aldrich Fine Chemicals.*

**4082 (R)-(+)-1-(Pentafluorophenyl)ethanol**

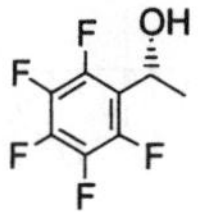

$C_8H_5F_5O$
Chiral building block. mp = 41-42°; $[\alpha]_D^{20}$ = + 7.5° (c = 1, pentane). *Sigma-Aldrich Fine Chemicals.*

**4083 (S)-(-)-1-(Pentafluorophenyl)ethanol**

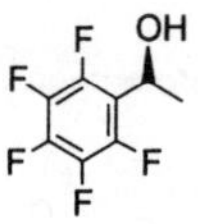

$C_8H_5F_5O$
Chiral building block. mp = 41-42°; $[\alpha]_D^{20}$ = - 7.5° (c = 1, pentane). *Sigma-Aldrich Fine Chemicals.*

**4084 (R)-Pentafluorostyrene oxide**

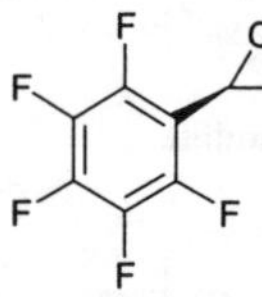

$C_8H_3F_5O$
Chiral building block. *Japan Energy Corporation.*

**4085 (2R,4R)-(-)-Pentanediol**
42075-32-1

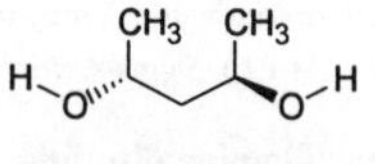

$C_5H_{12}O_2$
([R-(R*,R*)]-(-)-2,4-Pentanediol). Chiral building block. mp = 48-50°; $bp_{19}$ = 111-113°; $[\alpha]^{21}$ = - 40.4° (c = 10, $CHCl_3$). *Acros Organics nv; Digital Specialty Chemicals, Inc.; Sigma-Aldrich Fine Chemicals.*

**4086 (2S,4S)-(+)-Pentanediol**
72345-23-4

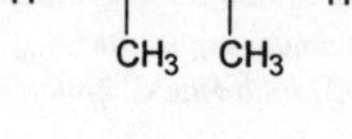

$C_5H_{12}O_2$
Chiral building block. mp = 45-48°; [α] = + 41.8° (c = 10, $CHCl_3$). *Acros Organics nv; Digital Specialty Chemicals, Inc.; Sigma-Aldrich Fine Chemicals.*

**4087 (S)-1,2-Pentanediol**
29117-54-2

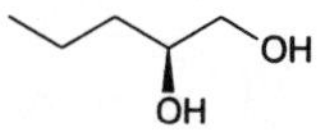

$C_5H_{12}O_2$
Chiral building block. *Kaneka Corporation.*

**4088 (R)-(-)-2-Pentanol**
31087-44-2 7075(11)

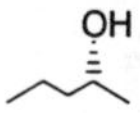

$C_5H_{12}O$
Pentan-2-ol, (R)-; (R)-(-)-Methyl-n-propylcarbinol; (R)-1-Methylbutyl alcohol; (R)-sec-Amyl alcohol. Chiral building block. bp = 119°; d = 0.814; n = 1.406; $[\alpha]^{25}$ = - 13° (neat). *Digital Specialty Chemicals, Inc.; Norse Laboratories; Sigma-Aldrich Fine Chemicals; TCI America.*

**4089 (S)-(+)-2-Pentanol**
26184-62-3 7075(11)

$C_5H_{12}O$
Pentan-2-ol, (S)-; (S)-(+)-Methyl-n-propylcarbinol. Chiral building block. bp = 118-119°; d = 0.810; $n_D^{20}$ = 1.4060; $[\alpha]_D^{20}$ = + 13.7 ± 0.5° (neat). *Acros Organics nv; Digital Specialty Chemicals, Inc.; Lancaster Synthesis Ltd.; Norse Laboratories; Sigma-Aldrich Fine Chemicals; TCI America.*

**4090 1,2,3,5,6-Penta-O-propanoyl-D-glucofuranose**

$C_{21}H_{32}O_{11}$
Chiral intermediate. mp = 75-76°; $[\alpha]_D^{20}$ = - 31° (c = 5, $CHCl_3$). *Sigma-Aldrich Fine Chemicals.*

**4091 (R)-Pentyl glycidyl ether**
121906-42-1

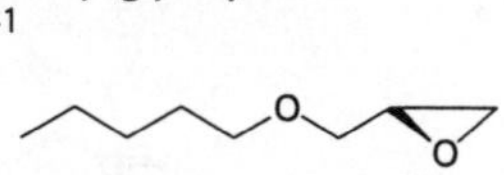

$C_8H_{16}O_2$
Chiral building block. $bp_8$ = 65-66°; d = 0.91; n = 1.425; $[\alpha]_D^{20}$ = + 8.9° (neat). *Japan Energy Corporation; Sigma-Aldrich Fine Chemicals.*

**4092 (S)-1-Pentyloxy-2-propanol**

$C_8H_{18}O_2$
Chiral building block. *Japan Energy Corporation.*

**4093 (R)-N-(3-Pentyl)-1-phenylethylamine hydrochloride**

$C_{13}H_{22}ClN$
(R)-N-(1-Ethylpropyl)-1-phenylethylamine hydrochloride. Chiral intermediate. *Lancaster Synthesis Ltd.*

**4094 (S)-N-(3-Pentyl)-1-phenylethylamine hydrochloride**

$C_{13}H_{22}ClN$
(S)-N-(1-Ethylpropyl)-1-phenylethylamine hydrochloride. Chiral intermediate. *Lancaster Synthesis Ltd.*

**4095 (R)-Pentylsuccinic acid 1-methyl ester**
213270-38-3

$C_{10}H_{18}O_4$
(R)-(3-Methoxycarbonyl)octanoic acid. Chiral intermediate. *ChiroTech Technology Ltd.; Lancaster Synthesis Ltd.*

**4096 (S)-2-Pentylsuccinic acid 1-methyl ester**

$C_{10}H_{18}O_4$
(S)-(3-Methoxycarbonyl)octanoic acid. Chiral intermediate. *ChiroTech Technology Ltd.; Lancaster Synthesis Ltd.*

**4097 (-)-Pepstatin A**
26305-03-3 7290(12) 247-600-0

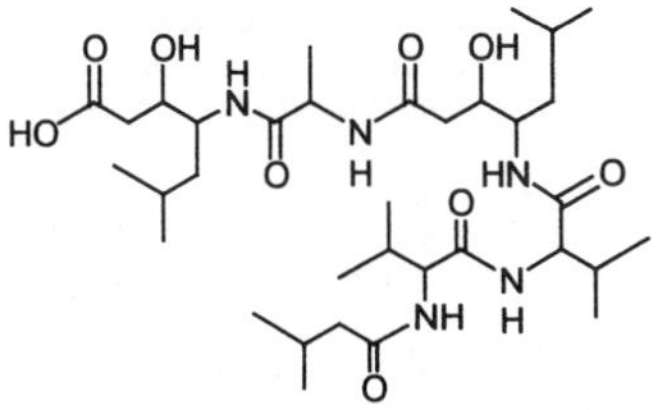

$C_{34}H_{63}N_5O_9$
Pharmaceutical or derivative. mp = 233°; $[\alpha]_D^{20}$ = - 90° (c = 0.3, $CH_3OH$). *Sigma-Aldrich Fine Chemicals.*

**4098 L-(-)-Perillaldehyde**
18031-40-8 7119(11)

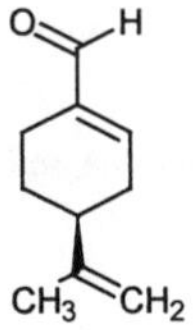

$C_{10}H_{14}O$
Cyclohex-1-ene-1-carboxaldehyde, 4-(1-methylethenyl)-, (S)-.; (4S)-p-Mentha-1,8-dien-7-al; L-4-(1-Methylethenyl)-1-cyclohexene-1-carboxaldehyde; 4-Isopropenyl-1-cyclohexene-1-carboxaldehyde. Chiral building block. $bp_{10}$ = 104-105°; d = 0.965; n = 1.5072; $[\alpha]$ = - 106° (c = 10, EtOH). *Acros Organics nv; Richman Chemical Inc.; Sigma-Aldrich Fine Chemicals.*

**4099 L-(-)-Perillic acid**
7694-45-3

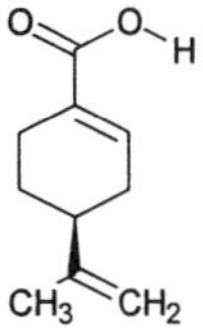

$C_{10}H_{14}O_2$
Chiral building block. $[\alpha]^{21}$ = - 102° (c = 2, $CH_3OH$). *Acros Organics nv.*

**4100 D-(+)-Perillyl alcohol**
57717-97-2

$C_{10}H_{16}O$
Chiral building block. $bp_{11}$ = 119-121°; d = 0.958; n = 1.501; $[\alpha]_D^{20}$ = + 109° (neat). *Sigma-Aldrich Fine Chemicals.*

**4101 L-(-)-Perillyl alcohol**
536-59-4 208-639-9

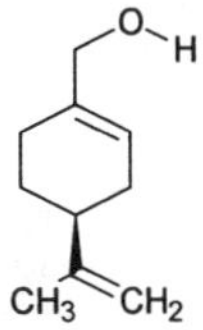

$C_{10}H_{16}O$
Cyclohex-1-ene-1-methanol, 4-(1-methylethenyl)-, L-. L-4-Isopropenylcyclohex-1-en-1-ylmethanol, L-p-Mentha-1,8-dien-7-ol. Listed on TSCA. Chiral building block. $bp_{11}$ = 119-121°; d = 0.96; n = 1.501; $[\alpha]^{22}$ = - 88° (c = 1, $CH_3OH$). *Acros Organics nv; Sigma-Aldrich Fine Chemicals.*

**4102 Perseitol**
524-06-0

Mannoheptitol. Pharmaceutical or derivative. $[\alpha]_D^{20}$= + 145° (c = 0.4, acidified molybdate). *Pfanstiehl Laboratories, Inc.*

**4103 (S)-3-Phenoxybenzaldehyde cyanohydrin**
61826-76-4

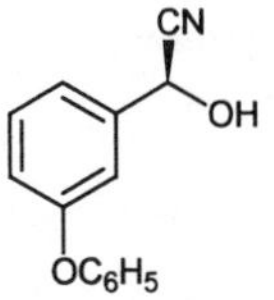

$C_{14}H_{11}NO_2$
Benzeneacetonitrile, α-hydroxy-3-phenoxy-, (S)-; (S)-(-)-α-Cyano-3-phenoxybenzyl alcohol; α-Hydroxy-3-

phenoxybenzeneacetonitrile-, (S)-. Chiral intermediate. *DSM Fine Chemcials Netherlands.*

**4104 (R)-1-Phenoxy-2-propylamine**

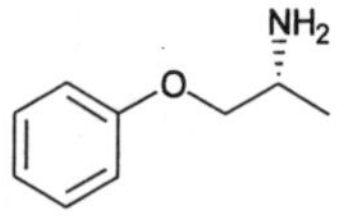

$C_9H_{13}NO$
Ethylamine, 1-methyl-2-phenoxy-, (R)-; Methyl-2-phenoxyethylamine; (R)-α-Methyl-2-phenoxyethanamine; (R)-2-Phenoxy-1-methylethylamine; 2-Propanamine, 1-phenoxy-, (R)-; (R)-Phenoxyisopropylamine. Chiral building block. *Chiragene, Inc.*

**4105 (S)-1-Phenoxy-2-propylamine**

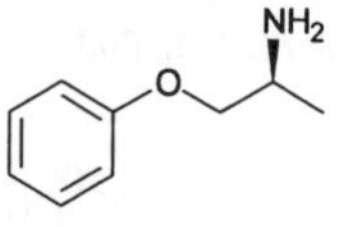

$C_9H_{13}NO$
Ethylamine, 1-methyl-2-phenoxy-, (S)-; (S)-1-Methyl-2-phenoxyethylamine; (S)-2-Phenoxy-1-methylethylamine; 2-Propanamine, 1-phenoxy-, (S)-; (S)-Phenoxyisopropylamine. Chiral building block. *Chiragene, Inc.*

**4106 Phenyl-N-acetyl-β-D-glucosaminide**
5574-80-1 226-946-6

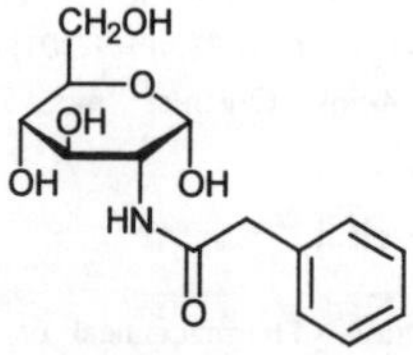

$C_{14}H_{19}NO_6$
Glucopyranoside, phenyl 2-acetamido-2-deoxy-, β-D-; Phenyl 2-acetamido-2-deoxy-β-D-glucopyranoside. Chiral intermediate. *Senn Chemicals AG.*

**4107 D-Phenylalanine**
673-06-3 7425(12) 211-603-5

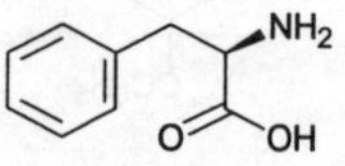

$C_9H_{11}NO_2$
Alanine, phenyl-, D-; (R)-2-Amino-3-phenylpropionic acid. Listed on TSCA. Chiral building block. mp = 273°; $[\alpha]_D^{20}$ = + 35 ± 2° (c = 2, $H_2O$). *Acros Organics nv; Ajinomoto Co. Inc.; Austin Chemical Company, Inc.; DSM Fine Chemcials Netherlands; Fischer Chemicals AG; Great Lakes Fine Chemicals; Kaneka Corporation; Lancaster Synthesis Ltd.; Nippon Kayaku; Omega Chemical Company Inc.; Recordati S.p.A.; Rexim S.A. Produits Chimiques; Shanghai DSL International Trading Company; Sigma-Aldrich Fine Chemicals; TCI America; Varsal Instruments, Inc.; Yoneyama Yakuhin Kogyo Co., Ltd.; Zhejiang Chemicals Import and Export Corporation.*

**4108 L-Phenylalanine**
63-91-2 7425(12) 200-568-1

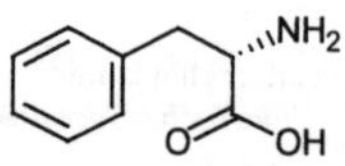

$C_9H_{11}NO_2$
Benzenepropanoic acid, α-amino-, (S)-; (S)-2-Amino-3-phenylpropionic acid; (S)-α-Aminobenzenepropanoic acid; (S)-α-Aminohydrocinnamic acid; Antibiotic FN 1636; L-Alanine, 3-phenyl-. Listed on TSCA. Chiral building block, Resolving agent. mp = 270-275°; $[\alpha]_D^{20}$ = - 34 ± 2° (c = 2, $H_2O$). *Aceto Corporation; Acros Organics nv; Ajinomoto Co. Inc.; Austin Chemical Company, Inc.; D&O Group; Daesang Corporation; DSM Fine Chemcials Netherlands; Fischer Chemicals AG; Great Lakes Fine Chemicals; Kyowa Hakko Kogyo Co., Ltd.; Lancaster Synthesis Ltd.; Nippon Kayaku; Rexim S.A. Produits Chimiques; Senn Chemicals AG; Shanghai DSL International Trading Company; Sigma-Aldrich Fine Chemicals; Tanabe Seiyaku Co. Ltd.; TCI America; Varsal Instruments, Inc.; Yoneyama Yakuhin Kogyo Co., Ltd.*

**4109 L-Phenylalanine amide**
5241-58-7

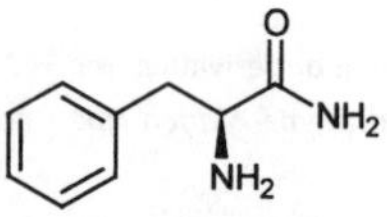

$C_9H_{12}N_2O$
Chiral building block. *Austin Chemical Company, Inc.; Fischer Chemicals AG.*

**4110 L-Phenylalanine amide hydrochloride**

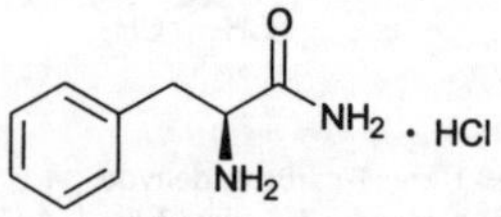

$C_9H_{13}ClN_2O$
Chiral building block. *Senn Chemicals AG; Synthetech, Inc.*

**4111 L-Phenylalanine benzyl ester hydrochloride**
2462-32-0 219-558-3

$C_{16}H_{18}ClNO_2$
Benzyl 3-phenyl-L-alaninate hydrochloride. Chiral intermediate. *Fischer Chemicals AG; Synthetech, Inc.*

**4112 L-Phenylalanine ethyl ester hydrochloride**
3182-93-2 221-673-9

$C_{11}H_{16}ClNO_2$
Alanine, phenyl-, ethyl ester, hydrochloride, L-(-); Ethyl L-phenylalaninate hydrochloride; (S)-Phenylalanine ethyl ester hydrochloride. Chiral building block. mp = 155-156°; $[\alpha]^{21}$ = + 33.2° (c = 5, EtOH). *Acros Organics nv; Fischer Chemicals AG; Sigma-Aldrich Fine Chemicals.*

**4113 L-Phenylalanine hydrazide**

$C_9H_{13}N_3O$
Chiral building block. *Senn Chemicals AG.*

**4114 L-Phenylalanine N-methylamide**

$C_{11}H_{14}N_2O_3$
Chiral building block. *Fischer Chemicals AG.*

**4115 L-Phenylalanine methyl ester**
2577-90-4 219-934-7

$C_{10}H_{19}N_3O_4$
Alanine, phenyl-, methyl ester, L-; Methyl 3-phenyl-L-alaninate. Chiral intermediate. *Fischer Chemicals AG.*

**4116 L-Phenylalanine methyl ester hydrochloride**
7524-50-7 231-383-4

$C_{10}H_{14}ClNO_2$
Alanine, phenyl-, methyl ester, hydrochloride, L-; Methyl L-phenylalaninate hydrochloride. Chiral building block. mp = 158-162°; [α] = + 37° (c = 2, EtOH). *Acros Organics nv; Austin Chemical Company, Inc.; Fischer Chemicals AG; Great Lakes Fine Chemicals; Sigma-Aldrich Fine Chemicals; TCI America.*

**4117 L-Phenylalanine-β-naphthylamide**
740-57-8

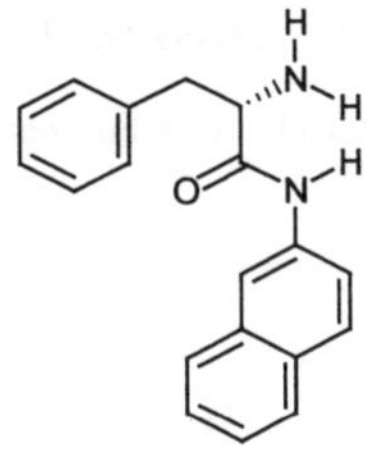

$C_{19}H_{18}N_2O$
Benzenepropanamide, α-amino-N-2-naphthalenyl-, (S)-; (S)-2-Amino-N-(2-naphthyl)-3-phenylpropionamide; Hydrocinnamamide, α-amino-N-2-naphthyl-, L-. Chiral intermediate. mp = 129-131°; $[\alpha]_D^{20}$ = + 83° (c = 1, $CH_3OH$). *Acros Organics nv; Sigma-Aldrich Fine Chemicals.*

**4118 D-Phenylalaninol**
5267-64-1 226-086-1

$C_9H_{13}NO$
Propan-1-ol, 2-amino-3-phenyl-, (R)-(+)-; (R)-(+)-2-Amino-3-phenyl-1-propanol; (R)-(+)-Phenylalaninol; Benzenepropanol, β-amino-, (R)-. Chiral auxiliary. mp = 93-95°; $[\alpha]_D^{20}$ = + 23 ± 2° (c = 5, EtOH). *Acros Organics nv; Boehringer Ingelheim Pharma KG; Lancaster Synthesis Ltd.; Newport Synthesis Ireland Ltd.; Omega Chemical Company Inc.; Oxford Asymmetry International plc; Richman Chemical Inc.; Sigma-Aldrich Fine Chemicals; SK Energy and Chemical, Inc.; Synthetech, Inc.; TCI America.*

**4119 L-Phenylalaninol**
3182-95-4 221-674-4
$C_9H_{13}NO$
Propan-1-ol, 2-amino-3-phenyl-, L-; (S)-(-)-2-Amino-3-phenyl-1-propanol; (S)-2-Benzylethanolamine; Benzenepropanol, β-amino-, (S)-; (S)-(-)-Phenylalaninol. Chiral

building block. mp = 92-94°; $[\alpha]_D^{20}$ = - 23 ± 2° (c = 5, EtOH). *Acros Organics nv; Boehringer Ingelheim Pharma KG; Degussa-Huls AG; Fischer Chemicals AG; Great Lakes Fine Chemicals; Lancaster Synthesis Ltd.; Newport Synthesis Ireland Ltd.; Omega Chemical Company Inc.; Oxford Asymmetry International plc; SEAC; Sigma-Aldrich Fine Chemicals; SK Energy and Chemical, Inc.; Synthetech, Inc.; TCI America.*

**4120 (R)-(+)-1-Phenyl-1-butanol**
22144-60-1

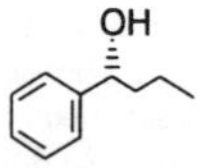

$C_{10}H_{14}O$
Chiral building block. mp = 44-46°; $bp_{14}$ = 115°; $[\alpha]_D^{20}$ = + 55° (c = 5, $CHCl_3$). *Sigma-Aldrich Fine Chemicals.*

**4121 (S)-(-)-1-Phenyl-1-butanol**
22135-49-5

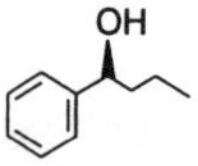

$C_{10}H_{14}O$
Chiral building block. mp = 45-47°; $bp_{0.05}$ = 120°; $[\alpha]^{21}$ = - 48.6° (c = 5, $CHCl_3$). *Sigma-Aldrich Fine Chemicals.*

**4122 (R)-(-)-2-Phenylbutyric acid**
938-79-4

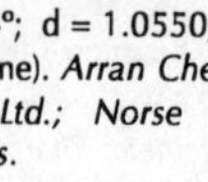

$C_{10}H_{12}O_2$
Benzeneacetic acid, α-ethyl-, (R)-; (R)-(-)-α-Ethylphenylacetic acid; (R)-α-Ethylbenzeneacetic acid; Butyric acid, 2-phenyl-, (R)-(-)-. Chiral building block. bp = 270-272°; $bp_{0.4}$ = 102-104°; d = 1.0550; $n_D^{20}$ = 1.5160; $[\alpha]_D^{20}$ = - 95 ± 2° (c = 5, toluene). *Arran Chemical Company Ltd.; Lancaster Synthesis Ltd.; Norse Laboratories; Sigma-Aldrich Fine Chemicals.*

**4123 (S)-(+)-2-Phenylbutyric acid**
4286-15-1

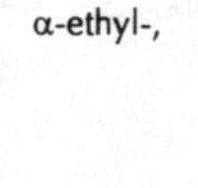

$C_{10}H_{12}O_2$
Benzeneacetic acid, α-ethyl-, (S)-; (S)-(+)-α-Ethylphenylacetic acid; (S)-α-Ethylbenzeneacetic acid. Chiral building block. bp = 270-272°; $bp_{0.3}$ = 98-100°; d = 1.0550; $n_D^{20}$ = 1.5160; $[\alpha]_D^{20}$ = + 95 ± 2° (c = 5, toluene). *Arran Chemical Company Ltd.; Lancaster Synthesis Ltd.; Norse Laboratories; Sigma-Aldrich Fine Chemicals.*

**4124 (R)-(+)-2-(Phenylcarbamoyloxy)-propionic acid**
145987-00-4
$C_{10}H_{11}NO_4$
(R)-(+)-Carbamalactic acid. Resolving agent. mp = 148-150°; $[\alpha]_D^{20}$ = +12.1 ± 1° (c = 1, EtOH); $[\alpha]_D^{20}$ = + 20.6 ± 0.3° (c = 4, $CH_3OH$). *Kiralchem Ltd.; Lancaster Synthesis Ltd.*

**4125 (S)-(-)-2-(Phenylcarbamoyloxy)-propionic acid**
102936-05-0

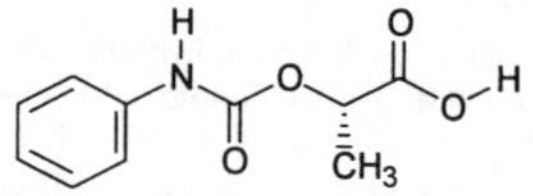

$C_{10}H_{11}NO_4$
(S)-(-)-Brown's reagent; S-Pacopa; (S)-(-)-Carbamalactic acid. Resolving agent. mp = 145-150°; $[\alpha]_D^{20}$ = - 12 ± 1° (c = 1, EtOH). *Acros Organics nv; Kiralchem Ltd.; Lancaster Synthesis Ltd.; Sigma-Aldrich Fine Chemicals.*

**4126 (R)-2-Phenyl-3-(2-chlorophenyl)-propionic acid**

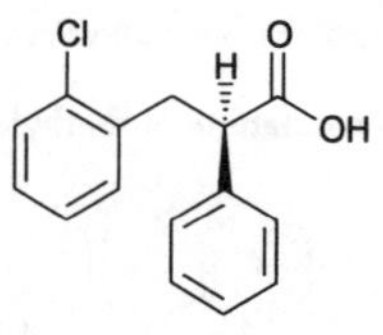

$C_{15}H_{13}ClO_2$
Chiral building block. *Sumitomo Chemcial Co. Ltd.*

**4127 (S)-2-Phenyl-3-(2-chlorophenyl)-propionic acid**

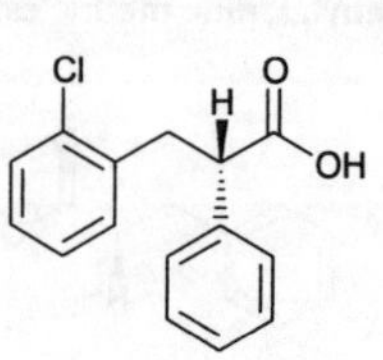

$C_{15}H_{13}ClO_2$
Chiral intermediate. *Sumitomo Chemcial Co. Ltd.*

**4128 (R)-2-Phenyl-3-(3-chlorophenyl)-propionic acid**

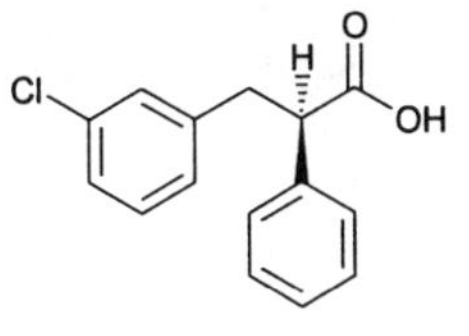

$C_{15}H_{13}ClO_2$
Chiral building block. *Sumitomo Chemcial Co. Ltd.*

**4129 (S)-2-Phenyl-3-(3-chlorophenyl)-propionic acid**

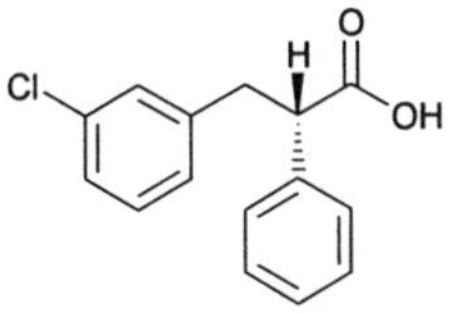

$C_{15}H_{13}ClO_2$
Chiral intermediate. *Sumitomo Chemcial Co. Ltd.*

**4130 (R)-2-Phenyl-3-(4-chlorophenyl)-propionic acid**

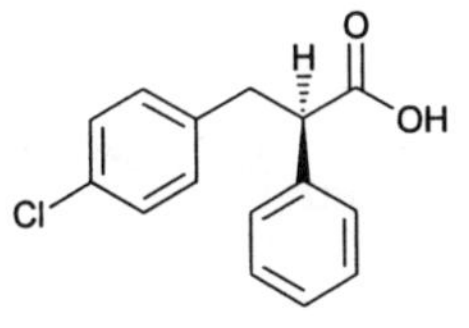

$C_{15}H_{13}ClO_2$
Chiral building block. *Sumitomo Chemcial Co. Ltd.*

**4131 (S)-2-Phenyl-3-(4-chlorophenyl)-propionic acid**

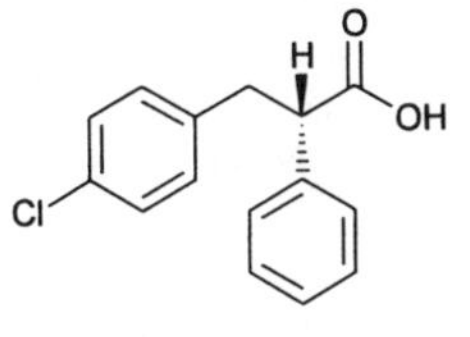

$C_{15}H_{13}ClO_2$
Chiral intermediate. *Sumitomo Chemcial Co. Ltd.*

**4132 (+)-(1S-cis)-3-Phenyl-3,5-cyclohexadiene-1,2-diol**

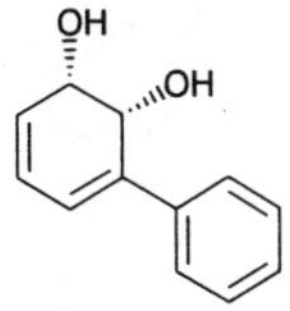

$C_{12}H_{12}O_2$
Chiral building block. mp = 79-84°; $[\alpha]_D^{20}$ = + 220° (c = 1.064, $CH_3OH$). *Sigma-Aldrich Fine Chemicals.*

**4133 4-trans-Phenylcyclohexane-(1R,2-cis)-dicarboxylic acid**

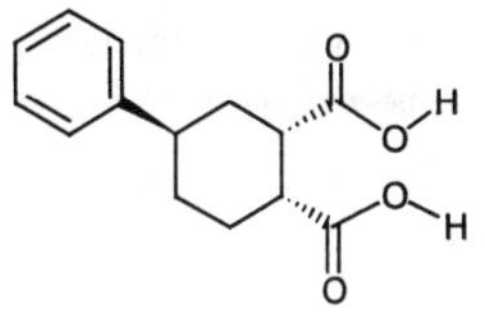

$C_{14}H_{16}O_4$
Chiral intermediate. *Acros Organics nv.*

**4134 4-trans-Phenylcyclohexane-(1R,2-cis)-dicarboxylic anhydride**

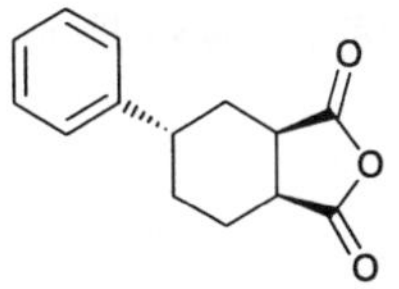

$C_{14}H_{14}O_3$
Chiral intermediate. *Acros Organics nv.*

**4135 4-trans-Phenylcyclohexane-(1R,2-cis)-dicarboxylic imide**

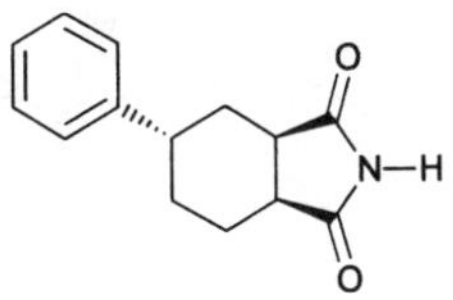

$C_{14}H_{15}NO_2$
Chiral intermediate. *Acros Organics nv.*

**4136 (R,R)-(-)-1-Phenylcyclohexane-cis-1,2-diol**

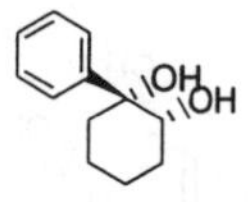

$C_{12}H_{16}O_2$
Chiral building block. mp = 121-123°; $[\alpha]^{23}$ = - 6° (c = 1, $CH_3OH$). *Sigma-Aldrich Fine Chemicals.*

**4137 (S,S)-(+)-1-Phenylcyclohexane-cis-1,2-diol**

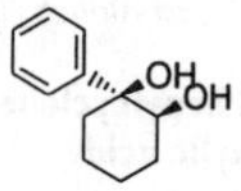

$C_{12}H_{16}O_2$
Chiral building block. mp = 121-123°; $[\alpha]^{23}$ = + 8° (c = 1, $CH_3OH$). *Sigma-Aldrich Fine Chemicals.*

**4138 (1R,2S)-(-)-trans-2-Phenyl-1-cyclohexanol**
98919-68-7

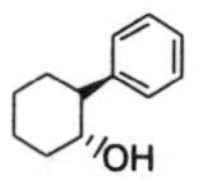

$C_{12}H_{16}O$
Chiral intermediate. mp = 64-66°; bp = 276-281°; $[\alpha]_D^{20}$= - 58° (c = 10, $CH_3OH$). *Sigma-Aldrich Fine Chemicals.*

**4139 (1S,2R)-(+)-trans-2-Phenyl-1-cyclohexanol**
34281-92-0

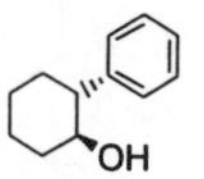

$C_{12}H_{16}O$
Chiral building block. mp = 65°; bp = 276-281°; $[\alpha]_D^{20}$= + 58° (c = 7, $CH_3OH$). *Sigma-Aldrich Fine Chemicals; TCI America.*

**4140 S-Phenyl-L-cysteine**
34317-61-8

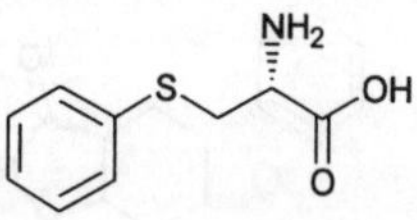

$C_9H_{11}NO_2S$
Chiral building block. mp = 200°; $[\alpha]_D^{20}$= + 10° (c = 1.5, 1N NaOH). *Acros Organics nv; Austin Chemical Company, Inc.; Mitsui Chemicals, Inc.; Sigma-Aldrich Fine Chemicals.*

**4141 (S)-(-)-1-Phenyl-1-decanol**
112419-76-8

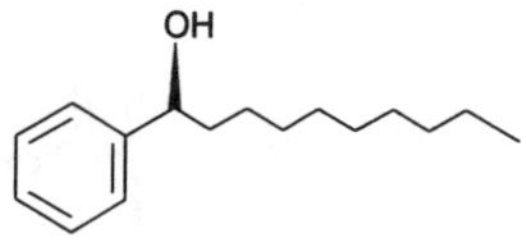

$C_{16}H_{26}O$
Chiral building block. $bp_{0.35}$ = 128°; d = 0.922; n = 1.497; $[\alpha]_D^{20}$ = - 14.4° (neat). *Sigma-Aldrich Fine Chemicals.*

**4142 (R)-Phenyl-5,5-dimethyl-2-hydroxy-1,3,2-dioxaphosphorinane-2-oxide**
98674-80-7

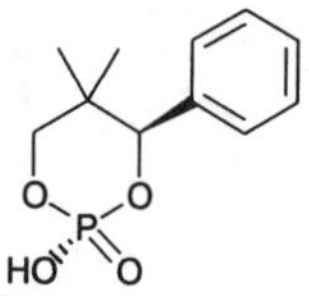

$C_{11}H_{15}O_4P$
Chiral intermediate. mp = 224-227°; $[\alpha]_D^{20}$= - 62° (c = 0.5, $CH_3OH$). *Sigma-Aldrich Fine Chemicals; Ultrafine.*

**4143 (S)-4-Phenyl-5,5-dimethyl-2-hydroxy-1,3,2-dioxaphosphorinane-2-oxide**
98674-81-8

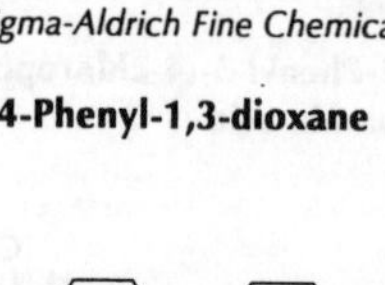

$C_{11}H_{15}O_4P$
Chiral intermediate. mp = 223-227°; $[\alpha]^{22}$ = + 62° (c = 0.5, $CH_3OH$). *Sigma-Aldrich Fine Chemicals; Ultrafine.*

**4144 (R)-(+)-4-Phenyl-1,3-dioxane**
107796-29-2

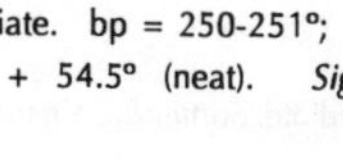

$C_{10}H_{12}O_2$
Chiral intermediate. bp = 250-251°; d = 1.111; n = 1.53; $[\alpha]_D^{20}$ = + 54.5° (neat). *Sigma-Aldrich Fine Chemicals.*

**4145 (S)-(-)-4-Phenyl-1,3-dioxane**
107796-30-5

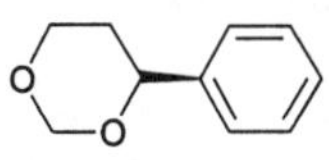

$C_{10}H_{12}O_2$

Chiral building block. bp = 250-251°; d = 1.111; n = 1.53; $[\alpha]_D^{20}$ = - 54.5° (neat). *Sigma-Aldrich Fine Chemicals.*

**4146 (-)-Phenylephrine hydrochloride**
61-76-7 7440(12) 200-517-3

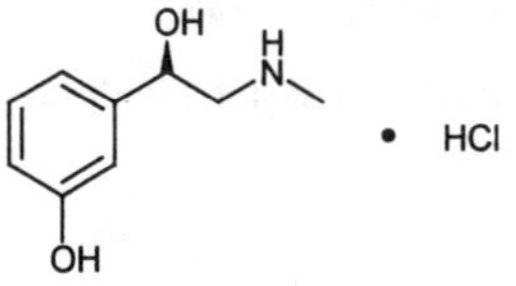

$C_9H_{14}ClNO_2$
Benzenemethanol, 3-hydroxy-α-[(methylamino)methyl]-, hydrochloride, (R)-; (R)-3-Hydroxy-[α-(methylamino)methyl]benzene methanol hydrochloride; Adrianol; Almefrin; Benzyl alcohol, m-hydroxy-α-[(methylamino)methyl]-, hydrochloride; Isophrin hydrochloride; l-m-Hydroxy-α-[(methylamino)methyl]-benzyl alcohol hydrochloride; Levo. Listed on TSCA. Chiral intermediate. mp = 143-145°; $[\alpha]^{25}$ = - 43.5° (c = 5, $H_2O$). *Boehringer Ingelheim Pharma KG; Sigma-Aldrich Fine Chemicals.*

**4147 (R)-(-)-Phenyl-1,2-ethanediol**
16355-00-3 8831(11)

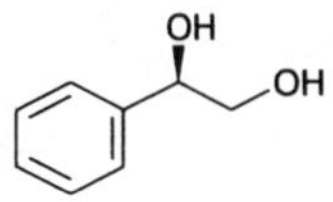

$C_8H_{10}O_2$
(R)-(-)-Styrene glycol; (R)-(-)-α,β-Dihydroxyethylbenzene; (R)-(-)-Phenylethane-1,2-diol. Chiral building block. mp = 67-69°; $bp_{755}$ = 272-274°; $[\alpha]_D^{20}$ = - 39 ± 1° (c = 3, EtOH); $[\alpha]_D^{20}$ = - 68° (c = 1, $CHCl_3$). *Acros Organics nv; Boehringer Ingelheim Pharma KG; Kiralchem Ltd.; Lancaster Synthesis Ltd.; Sigma-Aldrich Fine Chemicals; TCI America.*

**4148 (S)-(+)-Phenyl-1,2-ethanediol**
25779-13-9 8831(11)

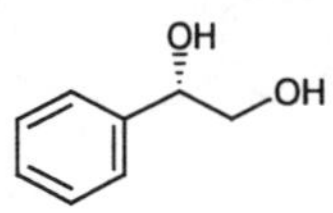

$C_8H_{10}O_2$
(S)-(+)-Styreneglycol; (S)-(+)-α,β-Dihydroxyethylbenzene; (S)-(+)-1-Phenylethane-1,2-diol. Chiral building block. mp = 64-66°; $[\alpha]_D^{20}$ = + 39 ± 1° (c = 3, EtOH); $[\alpha]_D^{20}$ = + 67° (c = 1, $CHCl_3$). *Acros Organics nv; Austin Chemical Company, Inc.; Boehringer Ingelheim Pharma KG; Kaneka Corporation; Kankyo Kagaku Center Co., Ltd.; Kiralchem Ltd.; Lancaster Synthesis Ltd.; Sigma-Aldrich Fine Chemicals; TCI America.*

**4149 (R)-(-)-1-Phenyl-1,2-ethanediol 2-tosylate**
40434-87-5 254-918-3
$C_{15}H_{16}O_4S$
Ethane-1,2-diol, 1-phenyl-, 2-(4-methylbenzenesulfonate), (R)-; 2-Hydroxy-2-phenylethyl [R-(-)]-p-toluenesulphonate. Chiral building block. mp = 79-81°; $[\alpha]_D^{20}$ = - 40 ± 1° (c = 2, 1,2-dichloroethane). *Lancaster Synthesis Ltd.*

**4150 (S)-(+)-1-Phenyl-1,2-ethanediol-2-tosylate**
40435-14-1

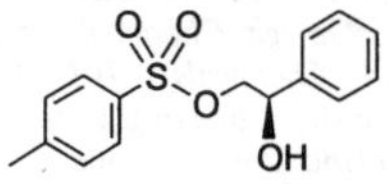

$C_{15}H_{16}O_4S$
Chiral building block. mp = 73-75°; $[\alpha]_D^{20}$ = + 31.5 ± 1° (c = 2, $CH_3OH$). *Lancaster Synthesis Ltd.; Sigma-Aldrich Fine Chemicals.*

**4151 (R)-(+)-1-Phenylethanol**
1517-69-7

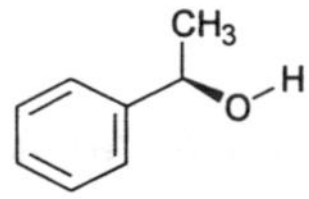

$C_8H_{10}O$
Benzenemethanol, α-methyl-, (R)-; (R)-(+)-α-Methylbenzyl alcohol; (R)-(+)-α-Hydroxyethylbenzene; D-(+)-Methylphenylcarbinol; (R)-(+)-sec-Phenethyl alcohol. Chiral diagnostic reagent; Chiral auxiliary; Chiral building block. mp = 9-11°; $bp_{27}$ = 98°; $bp_{10}$ = 88-89°; d = 1.0120; $n_D^{20}$ = 1.5270; $[\alpha]_D^{20}$ = + 45 ± 2° (c = 5, $CH_3OH$); $[\alpha]_D^{22}$ = + 39.5° (neat). *Acros Organics nv; BASF Aktiengesellschaft; FineTech, Ltd.; Kiralchem Ltd.; Lancaster Synthesis Ltd.; Norse Laboratories; TCI America.*

**4152 (S)-(-)-1-Phenylethanol**
1445-91-6

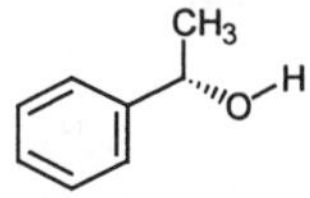

$C_8H_{10}O$
Benzenemethanol, α-methyl-, (S)-; (S)-(-)-sec-Phenethyl alcohol; (S)-(-)-α-Methylbenzyl alcohol. Chiral building block. mp = 9-11°; $bp_{20}$ = 98°; d = 1.0120; $n_D^{20}$ = 1.5280; $[\alpha]_D^{20}$ = - 45 ± 1° (c = 5, $CH_3OH$). *Acros Organics nv; BASF Aktiengesellschaft; FineTech, Ltd.; Kiralchem Ltd.; Lancaster Synthesis Ltd.; Norse Laboratories; TCI America.*

**4153 (S)-(-)-1-Phenylethylamine**
2627-86-3 6106(12) 220-098-0

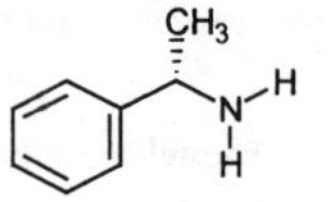

$C_8H_{11}N$
Benzenemethanamine, α-methyl-, (αS)-; L-(-)-α-Methylenzylamine; (S)-α-methyl-benzenemethanamine; (-)-1-Amino-1-phenylethane; Benzylamine, α-methyl-, (S)-(-)-. Listed on TSCA. Resolving agent; Chiral intermediate. bp = 187°; $bp_{18}$ = 80-81°; d = 0.9480; $n_D^{20}$ = 1.5260; $[\alpha]_D^{20}$ = - 30 ± 2° (c = 10, EtOH); $[\alpha]_D^{20}$ = - 40° (neat). *Acros Organics nv; BASF Aktiengesellschaft; Dynamit Nobel GmbH; FineTech, Ltd.; Lancaster Synthesis Ltd.; Norse Laboratories; Omega Chemical Company Inc.; Sigma-Aldrich Fine Chemicals; TCI America; Yamakawa Chemical Industry Co. Ltd.; Yoneyama Yakuhin Kogyo Co., Ltd.; Zeeland Chemicals, Inc.*

**4154 (R)-2-[(1-Phenylethyl)-amino]ethanol**

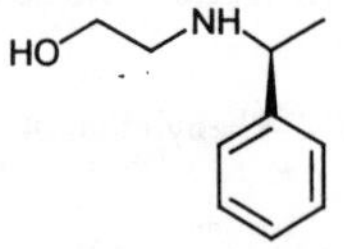

$C_{10}H_{15}NO$
Chiral building block. *FineTech, Ltd.*

**4155 (S)-2-[(1-Phenylethyl)-amino]ethanol**

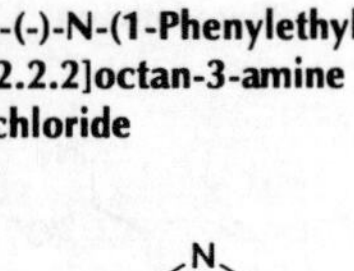

$C_{10}H_{15}NO$
Chiral building block. *FineTech, Ltd.*

**4156 (R)-(R,S)-(-)-N-(1-Phenylethyl)-1-azabicyclo[2.2.2]octan-3-amine dihydrochloride**

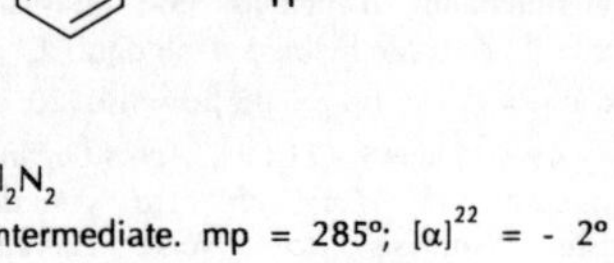

• 2 HCl

$C_{15}H_{24}Cl_2N_2$
Chiral intermediate. mp = 285°; $[\alpha]^{22}$ = - 2° (c = 2, $CHCl_3$). *Sigma-Aldrich Fine Chemicals.*

**4157 (S)-(R,S)-(+)-N-(1-Phenylethyl)-1-azabicyclo[2.2.2]octan-3-amine dihydrochloride**

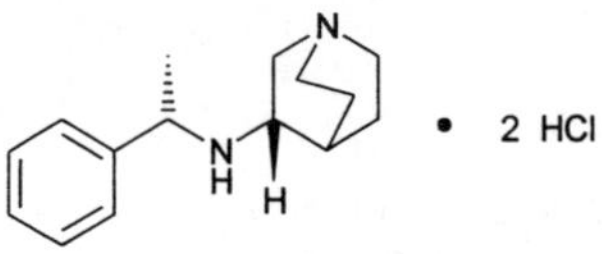

• 2 HCl

$C_{15}H_{24}Cl_2N_2$
Chiral intermediate. mp = 285°; $[\alpha]_D^{20}$ = + 2° (c = 2, $H_2O$). *Sigma-Aldrich Fine Chemicals.*

**4158 (R)-Phenylethyl-3,5-dinitrobenzoate**
3205-33-2

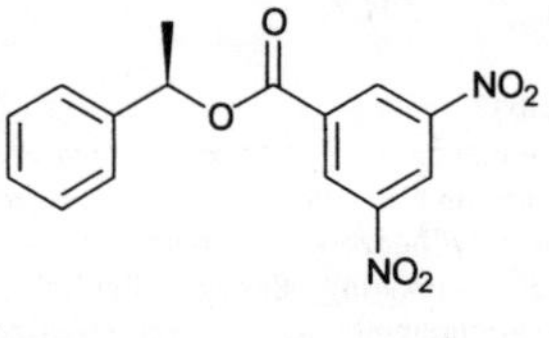

$C_{15}H_{12}N_2O_6$
Chiral intermediate. mp = 122-125°; $[\alpha]^{22}$ = - 47° (c = 2, acetone). *FineTech, Ltd.; Sigma-Aldrich Fine Chemicals.*

**4159 (S)-1-Phenylethyl-3,5-dinitrobenzoate**
3205-18-3

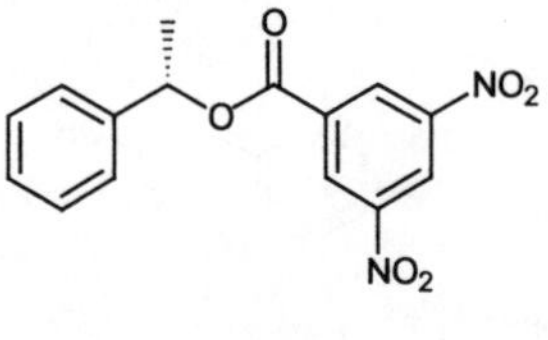

$C_{15}H_{12}N_2O_6$
Chiral intermediate. mp = 122-125°; $[\alpha]^{22}$ = + 47° (c = 2, acetone). *FineTech, Ltd.; Sigma-Aldrich Fine Chemicals.*

**4160 (2R,3R)-2,3-O-(1-Phenylethylidene)-L-tartaric acid dimethyl ester**
104333-83-7

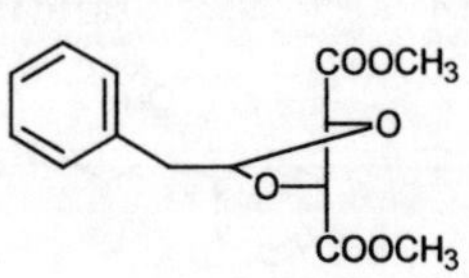

$C_{14}H_{16}O_6$
Dimethyl (2R,3R)-2,3-O-(1-phenylethylidene)-L-tartrate. Chiral intermediate. *TCI America.*

**4161 (R)-(+)-1-Phenylethyl isocyanate**
33375-06-3 251-485-2

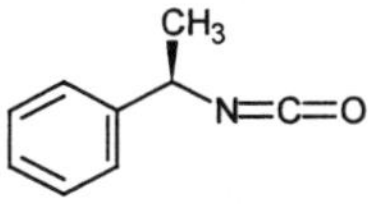

$C_9H_9NO$
Benzene, (1-isocyanatoethyl)-, (R)-; (R)-(+)-α-Methyl-benzyl isocyanate; Isocyanic acid (R)-(+)-α-Methylbenzyl ester. Chiral diagnostic reagent; Chiral building block. bp = 210-211°; $bp_{2.5}$ = 55-56°; d = 1.0450; $n_D^{20}$ = 1.5145; $[\alpha]_D^{20}$ = + 10.5 ± 1° (neat). *Acros Organics nv; BASF Aktiengesellschaft; Lancaster Synthesis Ltd.; Norse Laboratories; Sigma-Aldrich Fine Chemicals; TCI America.*

**4162 (S)-(-)-1-Phenylethyl isocyanate**
14649-03-7 238-698-6

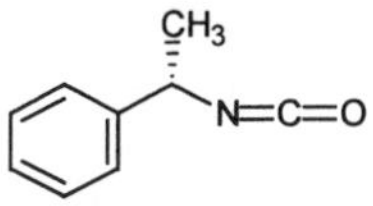

$C_9H_9NO$
Isocyanic acid, α-methylbenzyl ester, (-)-; (S)-(-)-α-Methylbenzyl isocyanate; Benzene, [(1S)-1-isocyanatoethyl]-; Isocyanic acid (S)-(-)-α-Methylbenzyl ester. Chiral building block. $bp_{2.5}$ = 55-56°; d = 1.0450; $n_D^{20}$ = 1.5145; $[\alpha]_D^{20}$ = - 10.5 ± 1° (neat). *Acros Organics nv; Lancaster Synthesis Ltd.; Norse Laboratories; Sigma-Aldrich Fine Chemicals; TCI America.*

**4163 (R)-N-(1-Phenylethyl)maleimide**
6129-15-3

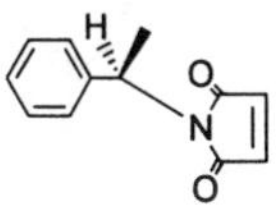

$C_{12}H_{11}NO_2$
Chiral building block. bp = 110-115°; d = 1.138; n = 1.556; $[\alpha]^{24}$ = + 61° (c = 2, EtOH). *FineTech, Ltd.; Sigma-Aldrich Fine Chemicals.*

**4164 (S)-N-(1-Phenylethyl)maleimide**
60925-76-0

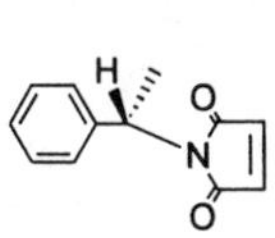

$C_{12}H_{11}NO_2$
Chiral building block. mp = 33-35°; $[\alpha]^{24}$ = - 61° (c = 2, EtOH). *FineTech, Ltd.; Sigma-Aldrich Fine Chemicals.*

**4165 (1'R,3R)-1-(1'-Phenylethyl)-5-oxo-3-pyrrolidinecarboxylic acid**
99735-43-0

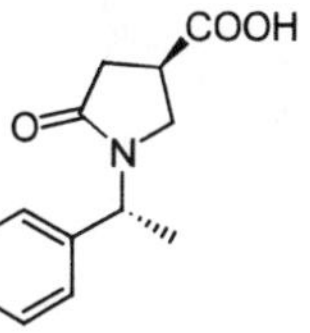

$C_{13}H_{15}NO_3$
Chiral intermediate. *Sigma-Aldrich Fine Chemicals; Yamakawa Chemical Industry Co. Ltd.*

**4166 (1'S,3S)-1-(1'-Phenylethyl)-5-oxo-3-pyrrolidinecarboxylic acid**
99735-44-1

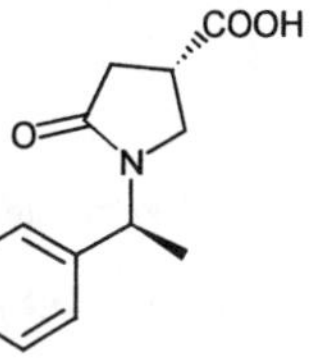

$C_{13}H_{15}NO_3$
Chiral intermediate. *Sigma-Aldrich Fine Chemicals; Yamakawa Chemical Industry Co. Ltd.*

**4167 (R)-N-(1-Phenylethyl)phthalamic acid**
21752-35-2 244-570-0

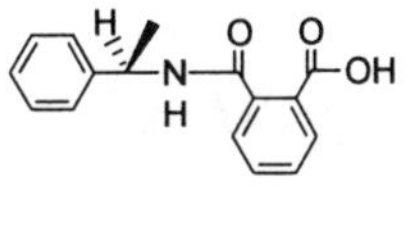

$C_{16}H_{15}NO_3$
Phthalamic acid, N-(α-methylbenzyl)-. Chiral intermediate. mp = 130-135°; $[\alpha]^{24}$ = + 46° (c = 2, EtOH). *FineTech, Ltd.; Norse Laboratories; Sigma-Aldrich Fine Chemicals; Yamakawa Chemical Industry Co. Ltd.*

**4168 (S)-(-)-N-(1-Phenylethyl)phthalamic acid**
21752-36-3 244-571-6

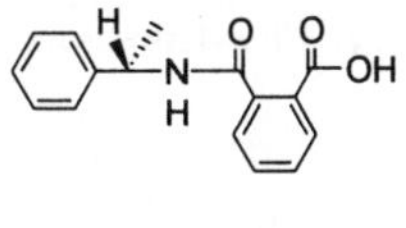

$C_{16}H_{15}NO_3$
Benzoic acid, 2-[[(1-phenylethyl)amino]carbonyl]-, (S)-; (S)-o-[[(1-Phenylethyl)amino]carbonyl]benzoic acid; (-)-N-(α-Methylbenzyl)phthalic acid monoamide; Phthalamic acid, N-(α-methylbenzyl)-, (S)-. Chiral intermediate. mp = 132°; $[\alpha]^{24}$ = - 46° (c = 2, EtOH). *FineTech, Ltd.; Norse Laboratories; Sigma-Aldrich Fine Chemicals; TCI America; Yamakawa Chemical Industry Co. Ltd.*

**4169 (R)-N-(1-Phenylethyl)phthalimide**

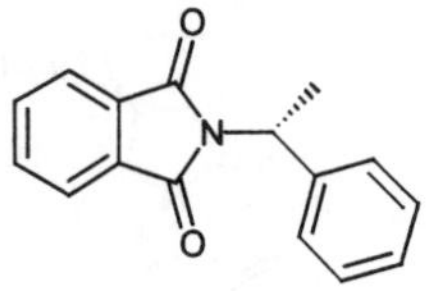

$C_{16}H_{13}NO_2$
Chiral intermediate. *FineTech, Ltd.; Norse Laboratories.*

**4170 (S)-(+)-N-(1-Phenylethyl)phthalimide**
3976-26-9

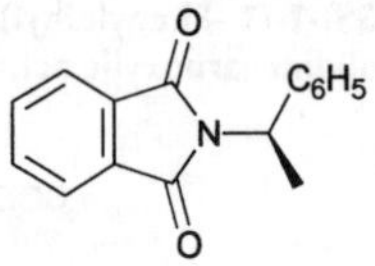

$C_{16}H_{13}NO_2$
Chiral intermediate. *FineTech, Ltd.; Norse Laboratories.*

**4171 (R)-(+)-5-(α-Phenylethyl)semioxamizide**
$C_{10}H_{13}N_3O_2$
Chiral intermediate. *Norse Laboratories.*

**4172 (S)-(-)-5-(α-Phenylethyl)semioxamizide**
$C_{10}H_{13}N_3O_2$
Chiral intermediate. *Norse Laboratories.*

**4173 (R)-N-(1-Phenylethyl)succinamic acid**
21752-33-0 244-567-4

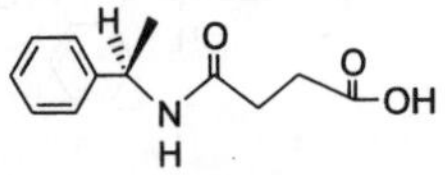

$C_{12}H_{15}NO_3$
Butanoic acid, 4-oxo-4-[(1-phenylethyl)amino]-, (R)-; (R)-4-Oxo-4-[(1-phenylethyl)amino]butyric acid; Succinamic acid, N-(α-methylbenzyl)-, (R)-(+). Chiral intermediate. mp = 100-104°; $[\alpha]^{24}$ = + 111° (c = 2, EtOH). *FineTech, Ltd.; Sigma-Aldrich Fine Chemicals.*

**4174 (S)-N-(1-Phenylethyl)succinamic acid**
21752-34-1 244-569-5

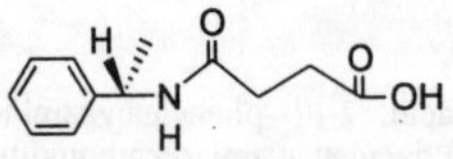

$C_{12}H_{15}NO_3$
Butanoic acid, 4-oxo-4-[(1-phenylethyl)amino]-, (S)-; (S)-4-Oxo-4-[(1-phenylethyl)amino]butyric acid; Succinamic acid, N-(α-methylbenzyl)-, (S)-. Chiral intermediate. mp = 100-104°; $[\alpha]^{24}$ = - 111° (c = 2, EtOH). *FineTech, Ltd.; Sigma-Aldrich Fine Chemicals.*

**4175 (R)-(1-Phenylethyl)succinate**
107832-33-7

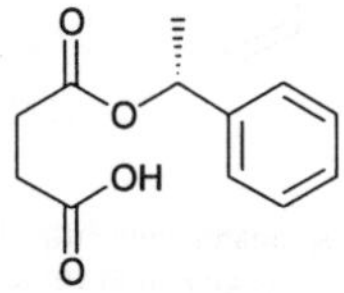

$C_{12}H_{14}O_4$
Chiral building block. *FineTech, Ltd.*

**4176 (S)-(1-Phenylethyl)succinate**

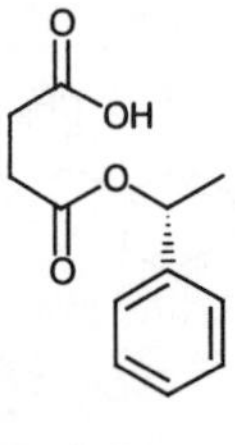

$C_{12}H_{14}O_4$
Chiral intermediate. *FineTech, Ltd.*

**4177 (R)-(+)-α-Phenylethylsulfamic acid**

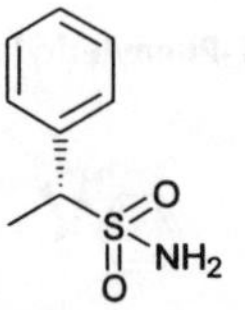

$C_8H_{11}NO_3S$
Chiral building block. *Norse Laboratories.*

**4178 (S)-(-)-α-Phenylethylsulfamic acid**

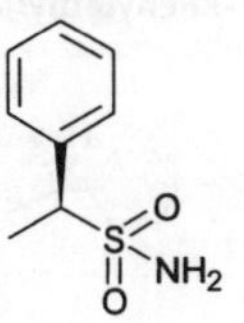

$C_8H_{11}NO_3S$
Resolving agent. *Norse Laboratories.*

**4179 Phenylethyl-β-D-thiogalactopyranoside**
63407-54-5

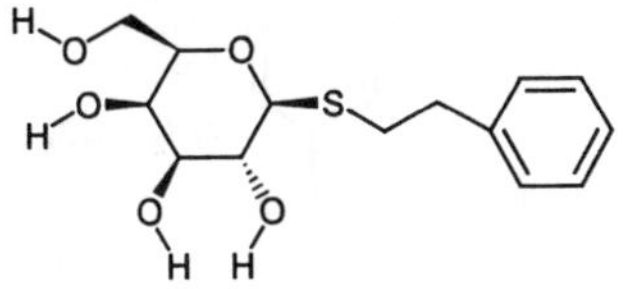

$C_{14}H_{20}O_5S$
2-Phenylethyl-β-D-thiogalactoside. Chiral diagnostic reagent. $[\alpha]^{25}$ = - 32 ± 12° (c = 1, $CH_3OH$); slightly soluble in $H_2O$, soluble in $CH_3OH$. *Acros Organics nv; Diagnostic Chemicals Limited.*

**4180 (R)-(+)-1-Phenylethylurea**
16849-91-5
$C_9H_{12}N_2O$
Chiral building block. mp = 124-125°. *Lancaster Synthesis Ltd.; Norse Laboratories.*

**4181 (S)-(-)-α-Phenylethylurethane**
33290-12-9

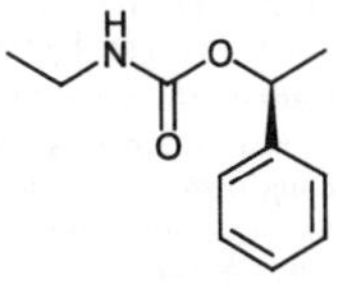

$C_{11}H_{15}NO_2$
Chiral intermediate. *Norse Laboratories.*

**4182 Phenyl-β-D-galactopyranoside**
2818-58-8 220-573-2

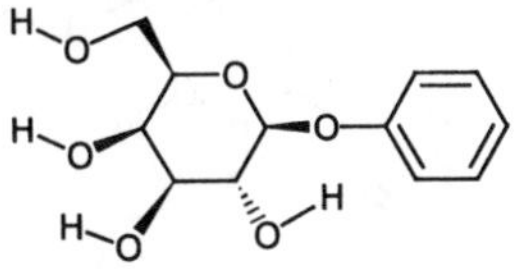

$C_{12}H_{16}O_6$
Galactopyranoside, phenyl, β-D-; Phenyl β-D-galactoside. Chiral diagnostic reagent. mp = 152-156°; $[\alpha]^{25}$ = - 40° (c = 2.3, $H_2O$); soluble in $H_2O$. *Acros Organics nv; Diagnostic Chemicals Limited; Senn Chemicals AG; Sigma-Aldrich Fine Chemicals.*

**4183 Phenyl-α-D-glucopyranoside**
4630-62-0

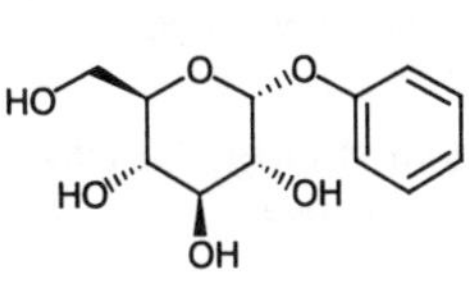

$C_{12}H_{16}O_6$
Chiral intermediate. *Senn Chemicals AG.*

**4184 Phenyl β-D-glucopyranoside**
1464-44-4 215-978-6

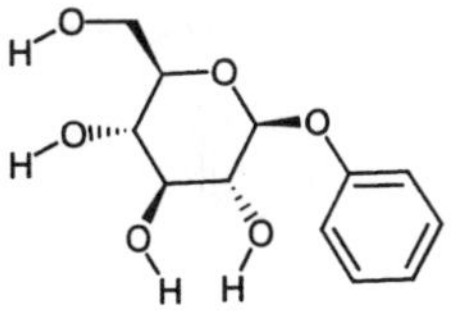

$C_{12}H_{16}O_6$
Glucopyranoside, phenyl, β-D-; β-Phenylglucoside. Listed on TSCA. Chiral intermediate. mp = 172-174°; $[\alpha]_D^{20}$ = - 72.5° (c = 2, $H_2O$). *Acros Organics nv; Pfanstiehl Laboratories, Inc.; Senn Chemicals AG; Sigma-Aldrich Fine Chemicals.*

**4185 Phenyl-β-D-glucuronide hydrate**
17685-05-1 241-666-4

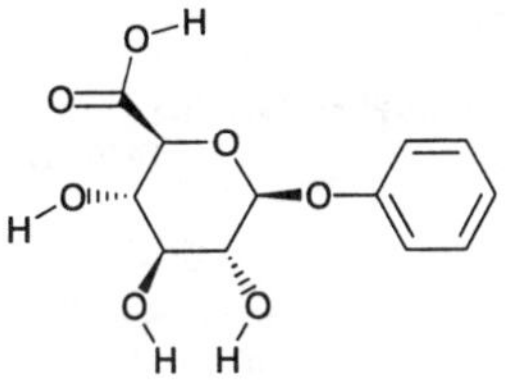

$C_{12}H_{14}O_7$
Glucopyranosiduronic acid, phenyl, β-D-; Phenyl β-D-glucopyranosiduronic acid. Chiral intermediate. *Acros Organics nv.*

**4186 (2R,3R)-3-Phenyl glycidol**
98819-68-2

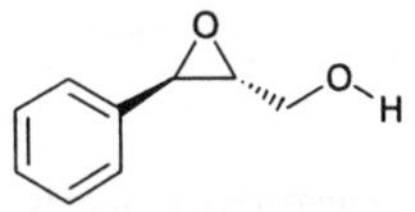

$C_9H_{10}O_2$
Chiral building block. mp = 51-53°; $[\alpha]_D^{20}$ = + 49° (c = 2, $CHCl_3$). *Acros Organics nv; Orsynetics (Tradining) Ltd.; Sigma-Aldrich Fine Chemicals.*

**4187 (2S,3S)-3-Phenyl glycidol**
104196-23-8

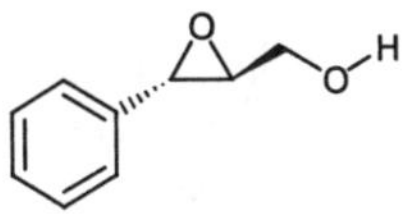

$C_9H_{10}O_2$
(2S,3S)-2,3-Epoxy-3-phenyl-1-propanol. Chiral building block. mp = 51-53°; $[\alpha]_D^{20}$ = - 49° (c = 2, $CHCl_3$). *Acros Organics nv; Sigma-Aldrich Fine Chemicals.*

**4188 D-(-)-α-Phenylglycine**
875-74-1 7446(12) 212-876-3

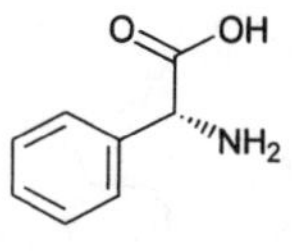

$C_8H_9NO_2$
Benzeneacetic acid, α-amino-, (R)-; (R)-(-)-α-Aminophenylacetic acid; (R)-α-Aminobenzeneacetic acid; Glycine, 2-phenyl-, D-; D-2-Phenylglycine. Listed on TSCA. Chiral building block. mp = 302°; $[\alpha]_D^{20}$= - 154 ± 2° (c = 1, 1N HCl). *Acros Organics nv; Alfa Chemicals Italiana srl; Austin Chemical Company, Inc.; Kaneka Corporation; KingChem, Inc.; Lancaster Synthesis Ltd.; Nippon Kayaku; Omega Chemical Company Inc.; Recordati S.p.A.; Richman Chemical Inc.; Sigma-Aldrich Fine Chemicals; Sinochem Ningbo; Siris Limited; TCI America; Varsal Instruments, Inc.; Yoneyama Yakuhin Kogyo Co., Ltd.*

**4189 L-(+)-α-Phenylglycine**
2935-35-5 7446(12) 220-909-8

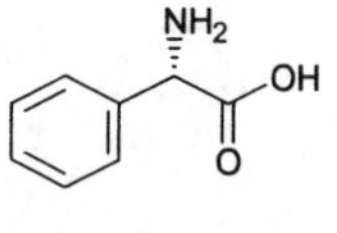

$C_8H_9NO_2$
Benzeneacetic acid, α-amino-, (S)-; (S)-(+)-α-Aminophenylacetic acid; (S)-(+)-2-Phenylglycine. Chiral building block. mp > 300°; $[\alpha]_D^{20}$= + 156 ± 2° (c = 1, 1N HCl). *Acros Organics nv; Austin Chemical Company, Inc.; DSM Fine Chemcials Netherlands; Lancaster Synthesis Ltd.; Nippon Kayaku; Omega Chemical Company Inc.; Richman Chemical Inc.; Senn Chemicals AG; Shanghai DSL International Trading Company; Sigma-Aldrich Fine Chemicals; TCI America; Varsal Instruments, Inc.; Yoneyama Yakuhin Kogyo Co., Ltd.*

**4190 D-(-)-Phenylglycine amide**
6485-67-2
$C_8H_{10}N_2O$
Benzeneacetamide, α-amino-, (R)-; D-α-Amino-α-phenylacetoamide; Acetamide, 2-amino-2-phenyl-, D-. Chiral building block. *Acros Organics nv; Shanghai DSL International Trading Company.*

**4191 D-(-)-Phenylglycine amide hydrochloride**
63291-39-4

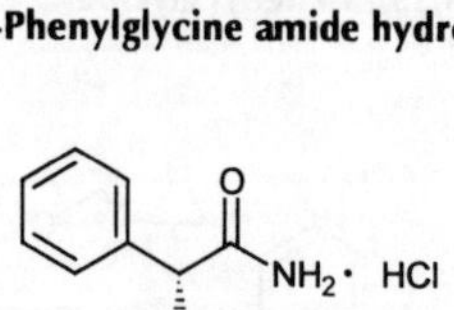

$C_8H_{11}ClN_2O$
(R)-(+)-2-Amino-2-phenylacetamide hydrochloride. Chiral intermediate. *Acros Organics nv; Sigma-Aldrich Fine Chemicals.*

**4192 L-Phenylglycine amide hydrochloride**

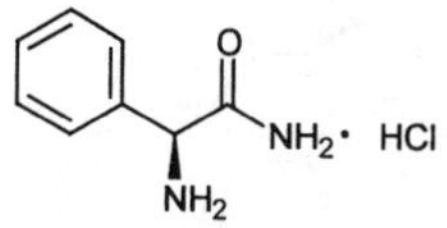

$C_8H_{11}ClN_2O$
Chiral building block. *Senn Chemicals AG.*

**4193 D-(-)-α-Phenylglycine chloride hydrochloride**
39878-87-0 254-668-5

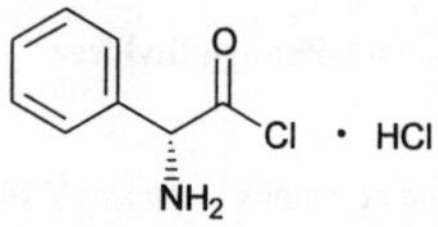

$C_8H_9Cl_2NO$
Benzeneacetyl chloride, α-amino-, hydrochloride, (αR)-; (-)-α-(Chloroformyl)benzylammonium chloride; (R)-α-Aminobenzeneacetyl chloride hydrochloride; D-(-)-α-Aminophenylacetyl chloride hydrochloride. Listed on TSCA. Chiral intermediate. mp = 117°; $[\alpha]^{29}$ = - 116° (c = 6, 1N HCl). *Alfa Chemicals Italiana srl; Kaneka Corporation; KingChem, Inc.; Nippon Kayaku; Sanofi Chimie; Shanghai DSL International Trading Company; Sigma-Aldrich Fine Chemicals.*

**4194 D-(-)-α-Phenylglycine Dane salt (potassium, ethyl)**
961-69-3 213-510-5

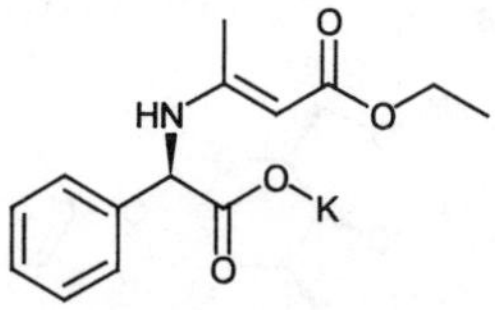

$C_{14}H_{16}NO_4$
Benzeneacetic acid, α-[(3-ethoxy-1-methyl-3-oxo-1-propenyl)amino]-, monopotassium salt, (R)-; Crotonic acid, 3-[(α-carboxybenzyl)amino]-, 1-ethyl ester, monopotassium salt, D-(-)-; Potassium D-(1-ethoxycarbonylpropen-2-yl)-α-aminophenyl acetate; Potassium (R)-[(3-ethoxy-1-methyl-3-oxoprop-1-enyl)-amino]phenylacetate. Listed on TSCA. Chiral intermediate. mp = 230-234°; $[\alpha]^{22}$ = - 78° (c = 1, 1N HCl). *Daesang Corporation; Sigma-Aldrich Fine Chemicals; Sinochem Ningbo; Siris Limited; Varsal Instruments, Inc.*

**4195 D-(-)-α-Phenylglycine ethyl ester hydrochloride**
17609-48-2 241-581-2
$C_{10}H_{14}ClNO_2$
Benzeneacetic acid, α-amino-, ethyl ester, hydrochloride, (R)-; (R)-(-)-α-Aminophenylacetic acid ethyl ester

hydrochloride; Ethyl D-phenylglycinate hydrochloride Glycine, 2-phenyl-, ethyl ester, monohydrochloride, D-(-)-. Chiral intermediate. mp = 189-191°; $[\alpha]_D^{20}$ = - 118 ± 5° (c = 1, $H_2O$). *Lancaster Synthesis Ltd.*

**4196 D-(-)-α-Phenylglycine methyl ester hydrochloride**
19883-41-1 243-399-9

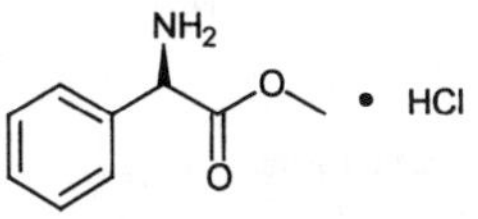

$C_9H_{12}ClNO_2$
Benzeneacetic acid, α-amino-, methyl ester, hydrochloride, (R)-; (R)-(-)-α-Aminophenylacetic acid methyl ester hydrochloride; Methyl (R)-aminophenylacetate hydrochloride; Glycine, 2-phenyl-, methyl ester, hydrochloride, D-; Methyl (R)-phenylglycinate hydrochloride. Chiral intermediate. mp = 189-191°; $[\alpha]_D^{20}$ = - 118° (c = 1, $H_2O$). *Lancaster Synthesis Ltd.; Oxford Asymmetry International plc; Sigma-Aldrich Fine Chemicals.*

**4197 L-(+)-2-Phenylglycine methyl ester hydrochloride**
15028-39-4

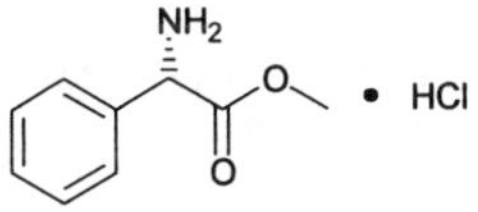

$C_9H_{11}NO_2$
Chiral building block. mp = 200°; $[\alpha]_D^{20}$ = + 120° (c = 1, $H_2O$). *Oxford Asymmetry International plc; Sigma-Aldrich Fine Chemicals.*

**4198 D-Phenylglycine potassium salt**
37661-32-8 253-582-5

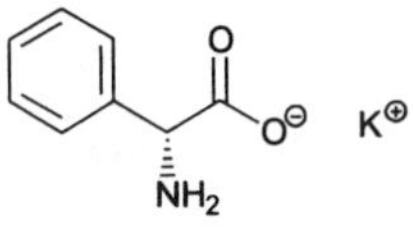

$C_8H_9KNO_2$
Benzeneacetic acid, α-amino-, monopotassium salt, (R)-; Monopotassium (R)-aminophenylacetate. Chiral building block. *Varsal Instruments, Inc.*

**4199 (R)-(-)-2-Phenylglycinol**
56613-80-0 260-287-5

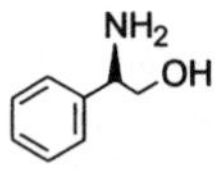

$C_8H_{11}NO$
Benzeneethanol, β-amino-, (R)- ;(R)-(-)-2-Amino-2-phenylethanol; (-)-(R)-β-Aminobenzeneethanol; (R)-(-)-α-Phenylglycinol; D-Phenylglycinol. Chiral building block. mp = 76-78°; $[\alpha]_D^{20}$ = - 30 ± 2° (c = 0.75, 1N HCl). *Acros Organics nv; Boehringer Ingelheim Pharma KG; Degussa-Huls AG; DSM Fine Chemcials Netherlands; Fischer Chemicals AG; Great Lakes Fine Chemicals; Lancaster Synthesis Ltd.; Newport Synthesis Ireland Ltd.; Omega Chemical Company Inc.; Oxford Asymmetry International plc; Sigma-Aldrich Fine Chemicals; SK Energy and Chemical, Inc.; Synthetech, Inc.; TCI America.*

**4200 (S)-(+)-2-Phenylglycinol**
20989-17-7

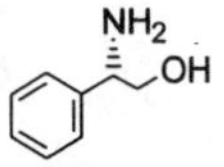

$C_8H_{11}NO$
(S)-(+)-2-Amino-2-phenylethanol. Resolving agent. mp = 76-78°; $[\alpha]_D^{20}$ = + 30 ± 1° (c = 0.75, 1N HCl). *Acros Organics nv; DSM Fine Chemcials Netherlands; Fischer Chemicals AG; Great Lakes Fine Chemicals; Kankyo Kagaku Center Co., Ltd.; Lancaster Synthesis Ltd.; Newport Synthesis Ireland Ltd.; Omega Chemical Company Inc.; Oxford Asymmetry International plc; Sigma-Aldrich Fine Chemicals; SK Energy and Chemical, Inc.; TCI America.*

**4201 L-Phenylglycinol**
56613-81-1

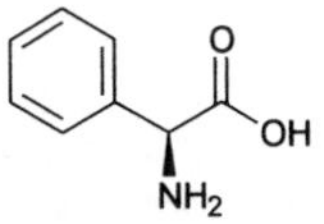

$C_8H_{11}NO$
Benzenemethanol, α-(aminomethyl)-, L-; (S)-1-Phenylethanolamine; (S)-2-Amino-1-phenylethanol; L-α-(Aminomethyl)benzenemethanol; L-β-Hydroxy-β-phenylethylamine; (S)-2-Phenyl-2-hydroxyethylamine; Benzeneethanamine, β-hydroxy-, L-; Benzyl alcohol, α-(aminomethyl)-, (S)-. Chiral building block. mp = 58-60°; $[\alpha]_D^{20}$ = + 41.0 ± 1.0° (c = 2, $CH_3OH$).

**4202 (2S,3S,8aR)-2-Phenyl-3-(hydroxymethyl)-8a-methyl bicyclic lactam**
116950-01-7

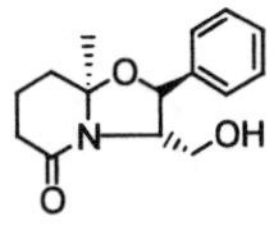

$C_{15}H_{19}NO_3$
Chiral intermediate. mp = 96-100°; $[\alpha]^{23}$ = + 14° (c = 1, EtOH). *Oxford Asymmetry International plc; Sigma-Aldrich Fine Chemicals.*

**4203 (2R,3S)-3-Phenylisoserine**
$C_9H_{11}NO_3$
Chiral building block. *Chiragene, Inc.*

**4204 (2S,3R)-3-Phenylisoserine**
$C_9H_{11}NO_3$
Chiral building block. *Chiragene, Inc.*

**4205 D-(+)-3-Phenyllactic acid**
7326-19-4 230-803-3

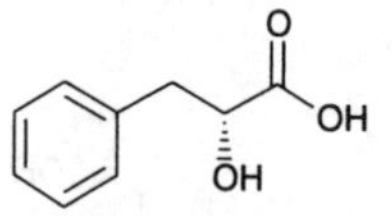

$C_9H_{10}O_3$
Benzenepropanoic acid, α-hydroxy-, (R)-; +)-2-Hydroxy-3-phenylpropionic acid; (R)-3-Phenyl-2-hydroxypropanoic acid; Lactic acid, 3-phenyl-, (R)-. Chiral building block. mp = 122-124°; $[\alpha]_D^{20}$ = + 19° (c = 1, EtOH). *Acros Organics nv; Fischer Chemicals AG; Great Lakes Fine Chemicals; Omega Chemical Company Inc.; Sigma-Aldrich Fine Chemicals.*

**4206 L-(-)-3-Phenyllactic acid**
20312-36-1 243-726-5

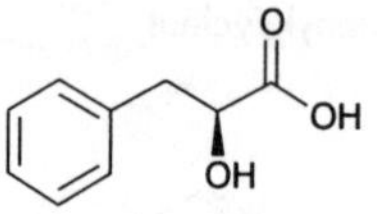

$C_9H_{10}O_3$
Benzenepropanoic acid, α-hydroxy-, (S)-; (S)-(-)-2-Hydroxy-3-phenylpropionic acid; Lactic acid, 3-phenyl-, (S)-; (S)-(-)-3-Phenyllactic acid. Chiral building block. mp = 122-124°; $[\alpha]_D^{20}$ = - 22 ± 2 ° (c = 1, $H_2O$). *Acros Organics nv; Fischer Chemicals AG; Great Lakes Fine Chemicals; Lancaster Synthesis Ltd.; Norse Laboratories; Omega Chemical Company Inc.; Sigma-Aldrich Fine Chemicals; TCI America.*

**4207 (-)-8-Phenylmenthol**
65253-04-5

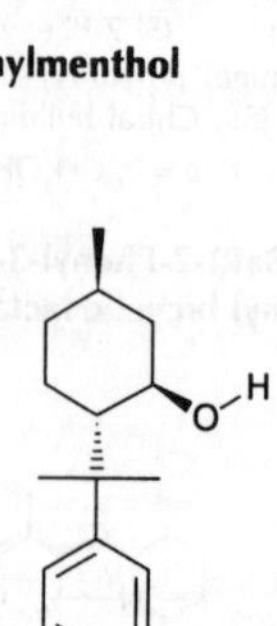

$C_{15}H_{24}O$
Chiral intermediate. d = 0.999; n = 1.53; $[\alpha]_D^{20}$ = - 26° (c = 2, EtOH). *Acros Organics nv; Oxford Asymmetry International plc; Sigma-Aldrich Fine Chemicals.*

**4208 (1R-trans)-2-(Phenylmethoxy)cyclopentaneamine**
181657-56-7

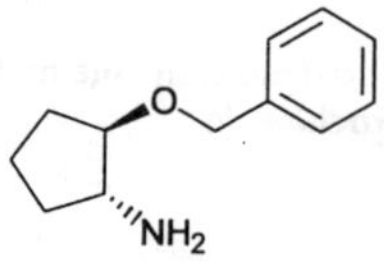

$C_{12}H_{17}NO$
Chiral intermediate. $bp_{0.3}$ = 90-93°. *BASF Aktiengesellschaft.*

**4209 (R)-2-(Phenyl)-3-(4-methylphenyl)-propionic acid**

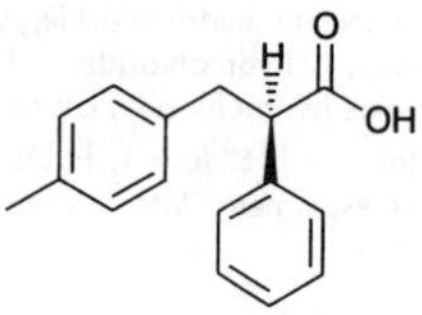

$C_{16}H_{13}O_2$
Chiral building block. *Sumitomo Chemcial Co. Ltd.*

**4210 (S)-2-(Phenyl)-3-(4-methylphenyl)-propionic acid**

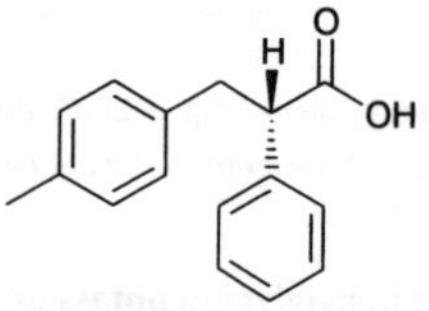

$C_{16}H_{13}O_2$
Chiral intermediate. *Sumitomo Chemcial Co. Ltd.*

**4211 (R)-(-)-4-Phenyl-2-oxazolidinone**
90319-52-1

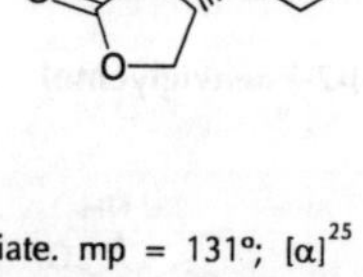

$C_9H_9NO_2$
Chiral intermediate. mp = 131°; $[\alpha]^{25}$ = - 48° (c = 2, $CHCl_3$). *Acros Organics nv; Asymchem; Great Lakes Fine Chemicals; Omega Chemical Company Inc.; Oxford Asymmetry International plc; PPG - Sipsy; Senn Chemicals AG; Sigma-Aldrich Fine Chemicals; SK Energy and Chemical, Inc.; TCI America; Urquima S.A.*

**4212 (S)-(+)-4-Phenyl-2-oxazolidinone**
99395-88-7

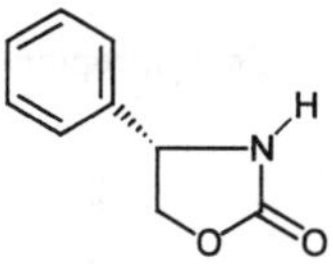

$C_9H_9NO_2$
Chiral intermediate. mp = 131°; $[\alpha]_D^{20}$ = + 48° (c = 2, $CHCl_3$). *Acros Organics nv; Asymchem; D&O Group; Davos Chemical Corporation; Degussa-Huls AG; Fischer Chemicals AG; Great Lakes Fine Chemicals; Omega Chemical Company Inc.; Oxford Asymmetry International plc; Senn Chemicals AG; Sigma-Aldrich Fine Chemicals; SK Energy and Chemical, Inc.; TCI America; Urquima S.A.*

**4213 (4S)-(+)-Phenyl-(4S)-phenyloxazolidin-2-ylidene-2-oxazoline-2-acetonitrile**
150639-33-1

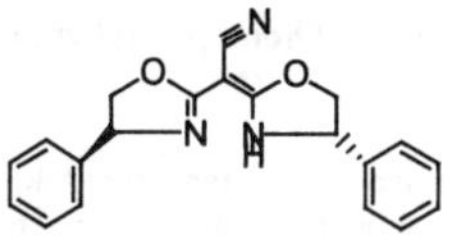

$C_{20}H_{17}N_3O_2$
Chiral intermediate. mp = 146-148°; $[\alpha]_D^{20}$ = + 364° (c = 1, $CHCl_3$). *Sigma-Aldrich Fine Chemicals.*

**4214 (R)-(+)-1-Phenyl-1,3-propanediol**
103548-16-9

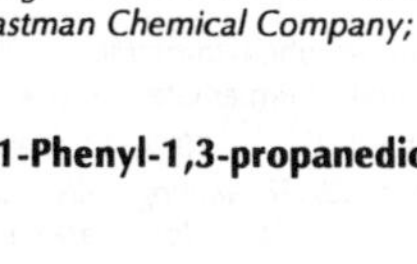

$C_9H_{12}O_2$
(R)-1-Phenyl-1,3-propanediol. Chiral building block. mp = 61-63°; $[\alpha]_D^{22}$ = + 54.5 ± 0.5° (c = 1, 1,2-DCE). *Acros Organics nv; Eastman Chemical Company; Kiralchem Ltd.*

**4215 (S)-(-)-1-Phenyl-1,3-propanediol**
96854-34-1

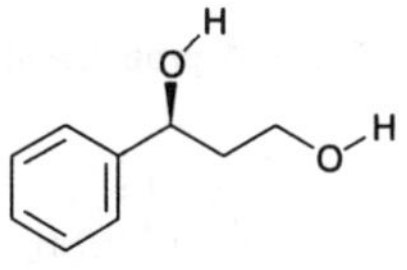

$C_9H_{12}O_2$
Chiral building block. mp = 61-63°; $[\alpha]_D^{22}$ = - 54.5 ± 0.5° (c = 1, 1,2-DCE). *Acros Organics nv; Eastman Chemical Company; Kaneka Corporation; Kiralchem Ltd.*

**4216 (R)-3-Phenyl-1,2-propanediol**

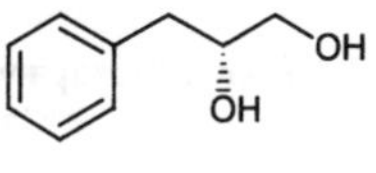

$C_9H_{12}O_2$
Chiral building block. mp = 42-45°; $[\alpha]_D^{20}$ = + 26.0 ± 1.0° (c = 3, 1,2-DCE). *Kiralchem Ltd.*

**4217 (S)-3-Phenyl-1,2-propanediol**

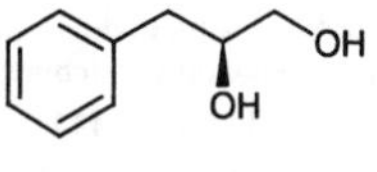

$C_9H_{12}O_2$
Chiral building block. mp = 42-45°; $[\alpha]_D^{20}$ = - 26.0 ± 1.0° (c = 3, 1,2-DCE). *Kiralchem Ltd.*

**4218 (R)-(+)-1-Phenyl-1-propanol**
1565-74-8 3727(11)

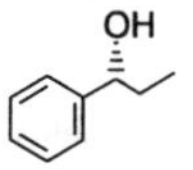

$C_9H_{12}O$
Benzenemethanol, α-ethyl-, (R)-; (R)-(+)-α-Ethylbenzyl alcohol; (R)-α-Hydroxypropylbenzene. Chiral building block. $bp_{14}$ = 102-103°; d = 0.9930; $n_D^{20}$ = 1.5200; $[\alpha]_D^{20}$ = + 47 ± 1° (c = 2.2, hexane); $[\alpha]_D^{20}$ = + 28.2 ± 0.2° (neat). *Kiralchem Ltd.; Lancaster Synthesis Ltd.; Sigma-Aldrich Fine Chemicals.*

**4219 (S)-(-)-1-Phenyl-1-propanol**
613-87-6

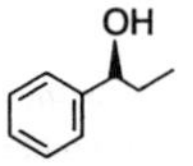

$C_9H_{12}O$
Benzenemethanol, α-ethyl-, (S)-; (S)-(-)-α-Ethylbenzyl alcohol; (S)-α-Ethylbenzenemethanol; (S)-α-Hydroxypropylbenzene. Chiral building block. bp = 218-220°; $bp_{12}$ = 105-106°; d = 0.9930; $n_D^{20}$ = 1.52; $[\alpha]_D^{20}$ = - 47 ± 1° (c = 2.2, hexane). *Kiralchem Ltd.; Lancaster Synthesis Ltd.; Sigma-Aldrich Fine Chemicals.*

**4220 (R)-(+)-1-Phenyl-2-propanol**
1572-95-8

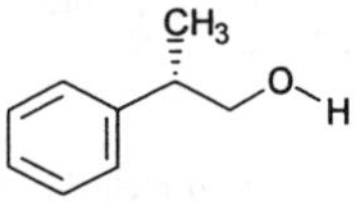

$C_9H_{12}O$
Benzeneethanol, α-methyl-, (R)-; (-)-α-Methyl-benzene-ethanol; (-)-α-Methylphenethyl alcohol; Phenethyl

alcohol, α-methyl-, (-)-. Chiral building block. [α] = + 18° (neat). *Acros Organics nv.*

**4221 (S)-(-)-1-Phenyl-2-propanol**
1517-68-6 1576(12)

$C_9H_{12}O$
Benzeneethanol, α-methyl-, (S)-; (+)-α-Methylbenzeneethanol; Phenethyl alcohol, α-methyl-, (S)-(+)-. Chiral building block. $bp_{15}$ = 111-112°; d = 0.9750; $n_D^{20}$ = 1.5260; $[\alpha]_D^{20}$ = - 18 ± 1° (neat). *Acros Organics nv; Lancaster Synthesis Ltd.*

**4222 (R)-(+)-2-Phenyl-1-propanol**
19141-40-3 6110(12)

$C_9H_{12}O$
Benzeneethanol, β-methyl-, (R)-; (+)-β-Methylbenzeneethanol; Phenethyl alcohol, β-methyl-, (R)-(+)-. Chiral building block. $bp_2$ = 80-82°; d = 0.975; $n_D^{20}$ = 1.5260; $[\alpha]_D^{18}$ = + 18.0 ± 0.2° (neat). *FineTech, Ltd.; Kiralchem Ltd.; Lancaster Synthesis Ltd.; Sigma-Aldrich Fine Chemicals.*

**4223 (S)-(-)-2-Phenyl-1-propanol**
37778-99-7

$C_9H_{12}O$
Benzeneethanol, β-methyl-, (S)-; (-)-β-Methylbenzeneethanol. Chiral building block. bp = 214°; $bp_2$ = 80-82°; d = 0.975; $[\alpha]_D^{18}$ = - 18.0 ± 0.2° (neat). *FineTech, Ltd.; Kiralchem Ltd.; Sigma-Aldrich Fine Chemicals.*

**4224 (R)-(-)-2-Phenylpropionic acid**
7782-26-5

$C_9H_{10}O_2$
Benzeneacetic acid, α-methyl-, (R)-; (R)-(-)-α-Methylphenylacetic acid; (-)-Hydratropic acid; (R)-α-Methylbenzeneacetic acid; l-PPA. Chiral diagnostic reagent; Chiral building block; Resolving agent. mp = 29-30°; $bp_1$ = 115°; d = 1.1; $n_D^{20}$ = 1.5230; $[\alpha]_D^{20}$ = - 73 ± 1° (c = 1.6, $CHCl_3$). *Acros Organics nv; Austin Chemical Company, Inc.; FineTech, Ltd.; Kiralchem Ltd.; Lancaster Synthesis Ltd.; Norse Laboratories; Sigma-Aldrich Fine Chemicals; TCI America.*

**4225 (S)-(+)-2-Phenylpropionic acid**
7782-24-3

$C_9H_{10}O_2$
Benzeneacetic acid, α-methyl-, (S)-; (S)-(+)-α-Methylphenylacetic acid; (+)-(S)-Hydratropic acid; (S)-α-Methylbenzeneacetic acid; d-PPA; (S)-(+)-Hydratropic acid. Chiral diagnostic reagent. mp = 29-30°; $bp_1$ = 115°; d = 1.1; $n_D^{20}$ = 1.5220; $[\alpha]_D^{20}$ = + 73 ± 1° (c = 1.6, $CHCl_3$). *Acros Organics nv; Austin Chemical Company, Inc.; FineTech, Ltd.; Kiralchem Ltd.; Lancaster Synthesis Ltd.; Norse Laboratories; Sigma-Aldrich Fine Chemicals; TCI America.*

**4226 (R)-(+)-1-Phenylpropylamine**
3082-64-2
$C_9H_{13}N$
(R)-(+)-1-Amino-1-phenylpropane; (R)-(+)-α-Ethylbenzylamine. Chiral building block. mp = - 69°; bp = 205°; d = 1.0510; $n_D^{20}$ = 1.5200; $[\alpha]_D^{20}$ = + 20 ± 1° (neat). *Arran Chemical Company Ltd.; BASF Aktiengesellschaft; Chiragene, Inc.; Lancaster Synthesis Ltd.; Norse Laboratories.*

**4227 (+)-Phenyl-1-propylamine**
28163-64-6

$C_9H_{13}N$
Chiral building block. bp = 197°; d = 0.945; n = 1.525; $[\alpha]^{22}$ = + 35° (c = 1, EtOH). *FineTech, Ltd.; Sigma-Aldrich Fine Chemicals.*

**4228 (S)-(-)-1-Phenylpropylamine**
3789-59-1
$C_9H_{13}N$
(S)-(-)-1-Amino-1-phenylpropane; (S)-(-)-α-Ethylbenzylamine. Chiral intermediate. mp = - 69°; bp = 205°; $n_D^{20}$ = 1.5200; $[\alpha]_D^{20}$ = - 20 ± 1° (neat). *Arran Chemical Company Ltd.; BASF Aktiengesellschaft; Chiragene, Inc.; Lancaster Synthesis Ltd.; Norse Laboratories.*

**4229 (S)-2-Phenyl-1-propylamine**
17596-79-1

$C_9H_{13}N$
Chiral building block. mp = 207°; d = 0.945; n = 1.525; $[\alpha]^{22}$ = - 35° (c = 1, EtOH). *FineTech, Ltd.; Sigma-Aldrich Fine Chemicals.*

**4230 (1R,2R)-(+)-1-Phenylpropyleneoxide**
14212-54-5

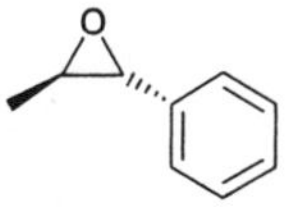

$C_9H_{10}O$
(2R,3R)-(+)-2-Methyl-3-phenyloxirane. Chiral building block. bp = 201-207°; d = 1.006; n = 1.518; $[\alpha]_D^{20}$ = + 110° (neat). *Sigma-Aldrich Fine Chemicals.*

**4231 (1S,2S)-(-)-1-Phenylpropyleneoxide**
4518-66-5

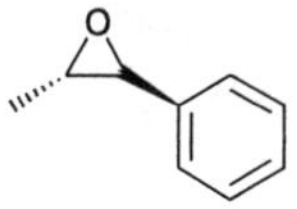

$C_9H_{10}O$
(2S,3S)-(-)-2-Methyl-3-phenyloxirane. Chiral building block. d = 1.003; n = 1.517; $[\alpha]^{19}$ = - 110° (neat). *Sigma-Aldrich Fine Chemicals.*

**4232 (R)-2-(Phenylpropyl)succinic acid 1-methyl ester**
156109-61-4
$C_{14}H_{18}O_4$
(R)-3-Methoxycarbonyl-6-phenylhexanoic acid. Chiral intermediate. *ChiroTech Technology Ltd.; Lancaster Synthesis Ltd.*

**4233 (S)-2-(Phenylpropyl)succinic acid 1-methyl ester**
$C_{14}H_{18}O_4$
(S)-3-Methoxycarbonyl-6-phenylhexanoic acid. Chiral intermediate. *ChiroTech Technology Ltd.; Lancaster Synthesis Ltd.*

**4234 (2S,3R)-3-Phenylpyrrolidine-2-carboxylic acid**

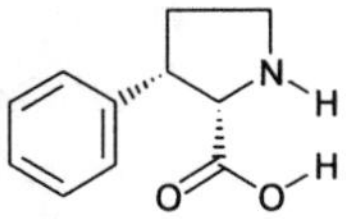

$C_{11}H_{13}NO_2$
Chiral building block. $[\alpha]$ = + 81.5° (c = 1, 6N HCl). *Acros Organics nv.*

**4235 (2R,3S)-3-Phenylserine**

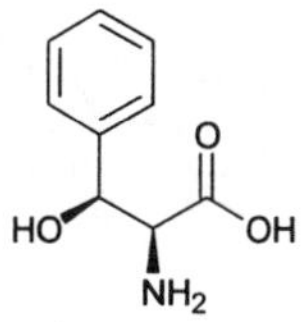

$C_9H_{11}NO_3$
Phenylalanine, β-D-hydroxy-, (βS)-; β-hydroxy-3-phenyl-D-alanine; (2R,3S)-2-Amino-3-hydroxy-3-phenylpropionic acid. Chiral building block. *Chiragene, Inc.*

**4236 (2S,3R)-3-Phenylserine**
6254-48-4 228-383-1

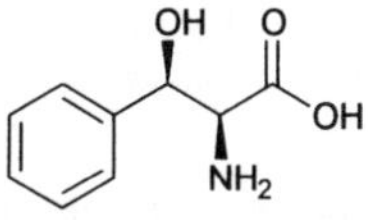

$C_9H_{11}NO_3$
Phenylalanine, β-L-hydroxy-, (βR)-; threo-β-Hydroxy-L-phenylalanine. Chiral building block. *Chiragene, Inc.*

**4237 (R)-3-Phenylserinol**

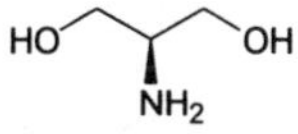

$C_3H_9NO_2$
Chiral building block. *Chiragene, Inc.*

**4238 (S)-3-Phenylserinol**

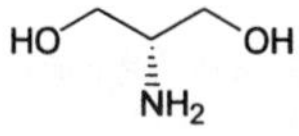

$C_3H_9NO_2$
Chiral building block. *Chiragene, Inc.*

**4239 (R)-(-)-Phenylsuccinic acid**
46292-93-7

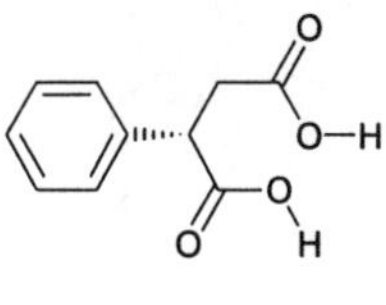

$C_{10}H_{10}O_4$
Chiral building block. $[\alpha]$ = - 180° (c = 1, acetone). *Acros Organics nv; Ultrafine.*

**4240 (S)-(+)-Phenylsuccinic acid**
4036-30-0 223-719-3

$C_{10}H_{10}O_4$
Butanedioic acid, phenyl-, (S)-; Chiral building block. mp = 172-174°; $[\alpha]_D^{20}$ = + 18° (c = 1, acetone). *Acros Organics nv; Lancaster Synthesis Ltd.; Norse Laboratories; TCI America; Ultrafine.*

**4241 (R)-(-)-N-(Phenylsulphonyl)glutamic acid**
20531-36-6
$C_{11}H_{13}NO_6S$
(R)-(-)-N-(Benzenesulphonyl)glutamic acid. Chiral intermediate. mp = 137-138°; $[\alpha]_D^{20}$ = - 14.2 ± 0.5° (c = 5, $H_2O$). *Lancaster Synthesis Ltd.; Sigma-Aldrich Fine Chemicals.*

**4242 (S)-(+)-N-(Phenylsulphonyl)glutamic acid**
20531-37-7
$C_{11}H_{13}NO_6S$
(S)-(+)-N-(Benzenesulphonyl)glutamic acid. Chiral intermediate. mp = 138-139°; $[\alpha]_D^{20}$ = + 14.2 ± 0.5° (c = 5, $H_2O$). *Lancaster Synthesis Ltd.; Norse Laboratories.*

**4243 (R)-Phenyl superquat**

$C_{11}H_{13}NO_2$
Chiral auxiliary. *Acros Organics nv; Oxford Asymmetry International plc.*

**4244 (S)-Phenyl superquat**

$C_{11}H_{13}NO_2$
Chiral auxiliary. *Acros Organics nv; Oxford Asymmetry International plc.*

**4245 (3R-cis)-(-)-3-Phenyltetrahydropyrrolo-[2,1-b]oxazol-5(6H)-one**
133007-27-9

$C_{12}H_{13}NO_2$
Chiral intermediate. *Acros Organics nv.*

**4246 (3S-cis)-(-)-3-Phenyltetrahydropyrrolo-[2,1-b]oxazol-5(6H)-one**
122383-34-0

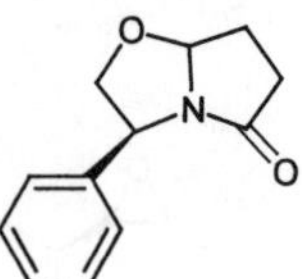


$C_{12}H_{13}NO_2$
Chiral intermediate. *Acros Organics nv.*

**4247 (4S,2RS)-2-Phenylthiazolidine-4-carboxylic acid**

$C_{10}H_{11}NO_2S$
Chiral intermediate. [α] = - 134.7° (c = 0.2, 2% $Na_2CO_3$). *Acros Organics nv.*

**4248 (S)-1-Phenyl-2-(p-tolyl)ethylamine**
30339-30-1
$C_{15}H_{17}N$
Benzeneethanamine, 4-methyl-α-phenyl-, (S)-; (S)-2-(p-Tolyl)-1-phenylethylamine; Phenethylamine, p-methyl-α-phenyl-, (+)-. Building block. d = 1.02. *TCI America.*

**4249 (S)-(+)-2-Phenyl-N-(trifluoroacetyl)glycine**
155894-96-5

$C_{10}H_8F_3NO_3$
Chiral intermediate. mp = 156-159°; $[\alpha]_D^{20}$ = + 163° (c = 1, $CH_3OH$). *Sigma-Aldrich Fine Chemicals.*

**4250 (S)-(+)-2-Phenyl-N-(trifluoroacetyl)glycine methyl ester**

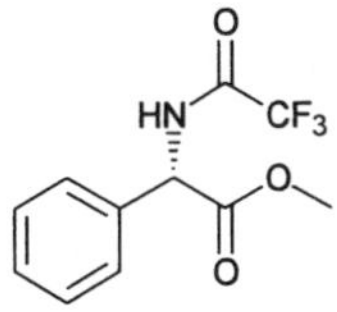

$C_{11}H_{10}F_3NO_3$
Chiral intermediate. mp = 79-80°; $[\alpha]_D^{20}$ = + 196° (c = 1, $CHCl_3$). *Sigma-Aldrich Fine Chemicals.*

**4251 (R)-1-Phenyl-2,2,2-trifluoroethanol**
10531-50-7 234-094-1

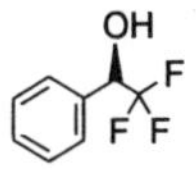

$C_8H_7F_3O$
Benzenemethanol, α-(trifluoromethyl)-, (αR)-; (R)-α-(Trifluoromethyl)benzyl alcohol; (-)-2,2,2-Trifluoro-1-phenylethanol. Chiral intermediate. mp = 20°; $bp_9$ = 73-76°; d = 1.292; n = 1.459; $[\alpha]^{21}$ = - 29° (c = 2.6, $CHCl_3$). *FineTech, Ltd.; Sigma-Aldrich Fine Chemicals.*

**4252 (S)-1-Phenyl-2,2,2-trifluoroethanol**
340-06-7 206-430-7

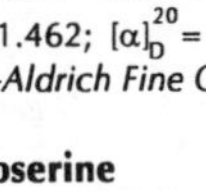

$C_8H_7F_3O$
Benzenemethanol, α-(trifluoromethyl)-, (αS)-; (S)-α-(Trifluoromethyl)benzyl alcohol; Benzyl alcohol, α-(trifluoromethyl)-, (S)-(+)-. Chiral building block. $bp_9$ = 73-76°; d = 1.3; n= 1.462; $[\alpha]_D^{20}$ = + 31.3 ± 0.5° (neat). *FineTech, Ltd.; Sigma-Aldrich Fine Chemicals.*

**4253 L-O-Phosphoserine**
407-41-0 7520(12) 206-986-0

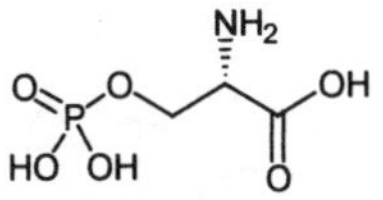

$C_3H_8NO_6P$
Serine, dihydrogen phosphate (ester), L-.
Dexfosfoserine; Fosforina; L-O-Serine phosphate; Phospho-L-serine. Chiral building block. mp = 190°; $[\alpha]_D^{20}$ = + 16.0° (c = 1, 2N HCl). *Flamma s.p.a.; Sigma-Aldrich Fine Chemicals; TCI America.*

**4254 L-Phosphothreonine**
1114-81-4 214-217-5

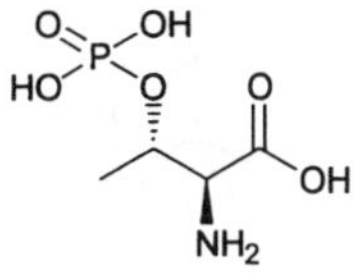

$C_4H_{10}NO_6P$
Threonine, dihydrogen phosphate (ester), L-; Threoninium dihydrogen phosphate. Chiral building block. *Austin Chemical Company, Inc.; Flamma s.p.a.*

**4255 (R)-Phthalimido-3-buten-1-ol**
$C_{12}H_{11}NO_3$
(R)-N-Phthaloyl-2-aminobut-3-en-1-ol. Chiral intermediate. *Lancaster Synthesis Ltd.*

**4256 (S)-2-Phthalimido-3-buten-1-ol**
174810-05-0
$C_{12}H_{11}NO_3$
(S)-N-Phthaloyl-2-aminobut-3-en-1-ol. Chiral intermediate. *Lancaster Synthesis Ltd.*

**4257 (R)-N-Phthaloyl-2-bromovaleric acid**
179090-36-9
$C_{13}H_{12}BrNrO_4$
Chiral intermediate. *Sigma-Aldrich Fine Chemicals.*

**4258 (S)-5-N-Phthaloyl-2-bromovaleric acid**
179090-37-0
$C_{13}H_{12}BrNrO_4$
Chiral intermediate. *Sigma-Aldrich Fine Chemicals.*

**4259 N-Phthaloyl-L-glutamic acid**
340-90-9

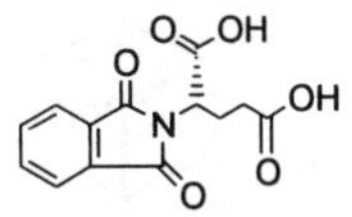

Chiral intermediate. mp = 160-162°. *Pharmasyn, Inc.; Sigma-Aldrich Fine Chemicals; Varsal Instruments, Inc.*

**4260 D-Picein**
530-14-3 208-473-7

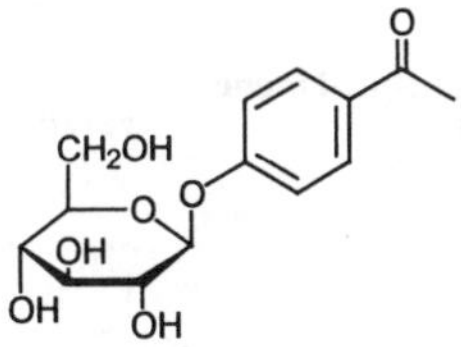

$C_{14}H_{18}O_7$
Ethanone, 1-[4-(β-D-glucopyranosyloxy)phenyl]-; 4-Hydroxyacetophenone-β-D-glucoside; 1-[4-(β-D-glucopyranosyloxy)phenyl]ethan-1-one; Ameliaroside. Chiral intermediate. *Senn Chemicals AG.*

**4261 (1R,2R,3S,5R)-(-)-2,3-Pinanediol**
22422-34-0

$C_{10}H_{18}O_2$
(1R-(1α,2β,3β,5α))-2,6,6-Trimethylbicyclo[3.1.1]heptane-2,3-diol. Chiral intermediate. mp = 57-59°; $bp_1$ = 101-102; [α] = - 8.6° (c = 6.5, toluene). *Acros Organics nv; Sigma-Aldrich Fine Chemicals.*

**4262 (1S,2S,3R,5S)-2,3-Pinanediol**
18680-27-8

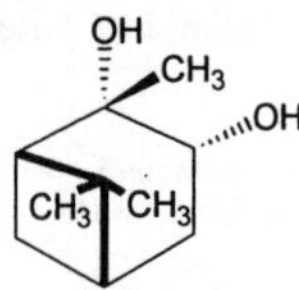

$C_{10}H_{18}O_2$
(1S,2S,3R,5S)-2,6,6-Trimethylbicyclo[3.1.1]heptane-2,3-diol; (+)-2-Hydroxyisopinocampheol. Chiral intermediate. mp = 56-58°; $[\alpha]_D^{20}$ = + 8.7 ± 0.5° (c = 6.5, toluene). *Acros Organics nv; Lancaster Synthesis Ltd.; Sigma-Aldrich Fine Chemicals.*

**4263 (1R)-(+)-α-Pinene**
7785-70-8 7414(11) 232-087-8

$C_{10}H_{16}$
Bicyclo[3.1.1]hept-2-ene, 2,6,6-trimethyl-, (1R,5R)-; (1R)-2,6,6-Trimethylbicyclo[3.1.1]hept-2-ene. Listed on TSCA. Chiral intermediate. bp = 156°; d = 0.86; $[\alpha]^{22}$ = + 50.7° (neat). *Acros Organics nv; Sigma-Aldrich Fine Chemicals; TCI America.*

**4264 (1S)-(-)-α-Pinene**
7785-26-4 7414(11) 232-077-3

$C_{10}H_{16}$
Bicyclo[3.1.1]hept-2-ene, 2,6,6-trimethyl-, (1S,5S)-; (1S)-2,6,6-Trimethylbicyclo[3.1.1]hept-2-ene. Listed on TSCA. Chiral intermediate. mp = -64°; bp = 155°; d = 0.86; n = 1.465; $[\alpha]^{22}$ = - 50.7° (neat). *Acros Organics nv; Sigma-Aldrich Fine Chemicals; TCI America.*

**4265 (1R)-(+)-β-Pinene**
19902-08-0

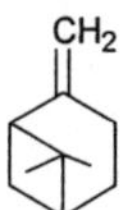

$C_{10}H_{16}$
Chiral intermediate. bp = 164-165°; d = 0.872; n = 1.478; $[\alpha]_D^{20}$ = + 21° (neat). *Sigma-Aldrich Fine Chemicals.*

**4266 (1S)-(-)-β-Pinene**
18172-67-3 7415(11) 242-060-2

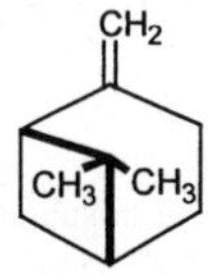

$C_{10}H_{16}$
Bicyclo[3.1.1]heptane, 6,6-dimethyl-2-methylene-, (1S,5S)-; (1S)-6,6-Dimethyl-2-methylenebicyclo[3.1.1]-heptane; (-)-Nopinene. Listed on TSCA. Chiral intermediate. mp = -61°; bp = 165-167°; d = 0.859; n = 1.476; $[\alpha]^{25}$ = - 22° (neat). *Acros Organics nv; Sigma-Aldrich Fine Chemicals.*

**4267 (-)-α-Pinene oxide**
72936-74-4

$C_{10}H_{16}O$
Chiral intermediate. *Acros Organics nv.*

**4268 (+)-β-Pinene oxide**
6931-54-0 230-055-8

$C_{10}H_{16}O$
Spiro[bicyclo[3.1.1]heptane-2,2'-oxirane], 6,6-dimethyl-; 2,10-Epoxypinane; β-Pinene epoxide; Pinane, 2,10-epoxy-. Listed on TSCA. Chiral intermediate. [α] = + 8.4° (neat). *Acros Organics nv.*

**4269 Pinitol**
10284-63-6

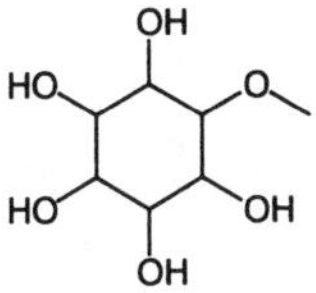

$C_7H_{14}O_6$
3-O-Methyl-D-chiro-inositol. Chiral building block. mp = 179-185°; $[\alpha]_D^{20}$ = + 60° (c = 1, $H_2O$). *Sigma-Aldrich Fine Chem.*

**4270 D-Pipecolinic acid**
1723-00-8 7425(11) 217-024-4

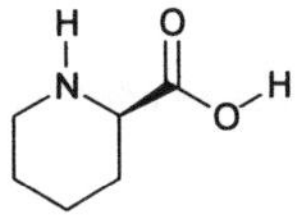

$C_6H_{11}NO_2$
Piperidine-2-carboxylic acid, (R)-; (R)-(+)-2-Piperidine-carboxylic acid; D-Homoproline. Chiral building block. mp = 277°; [α] = + 26.5° (c = 5, $H_2O$). *Acros Organics NV; Advanced Asymmetrics, Inc.; Austin Chem. Co.; RCA Reuter Chem.; Sigma-Aldrich Fine Chem.*

**4271 L-Pipecolinic acid**
3105-95-1 7425(11) 221-462-1

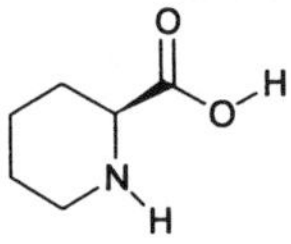

$C_6H_{11}NO_2$
Piperidine-2-carboxylic acid, (2S)-; (S)-2-Piperidine-carboxylic acid. Chiral building block. mp = 278°; $[\alpha]_D$ = - 31.2°; $[\alpha]_D^{25}$ = - 26.4° (c = 1, $H_2O$); soluble in $CH_3OH$, EtOH/ $H_2O$ solutions. *Acros Organics NV; Advanced Asymmetrics, Inc.; Austin Chem. Co.; Mercian Corp.; RCA Reuter Chem.; See Chem. AG; Sigma-Aldrich Fine Chem.*

**4272 (R)-Piperazine-2-carboxylic acid**

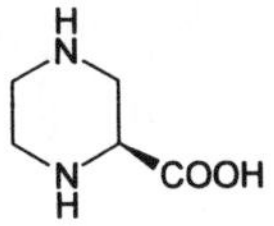

$C_5H_{10}N_2O_2$
Chiral building block. *Austin Chem. Co.; SK Energy & Chem.*

**4273 (S)-Piperazine-2-carboxylic acid**

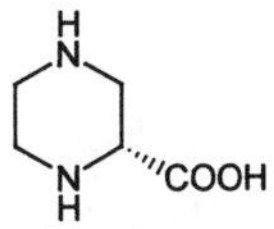

$C_5H_{10}N_2O_2$
Chiral intermediate. *Austin Chem. Co.; SK Energy & Chem.*

**4274 (R)-Piperonylsuccinic acid monomethyl ester**

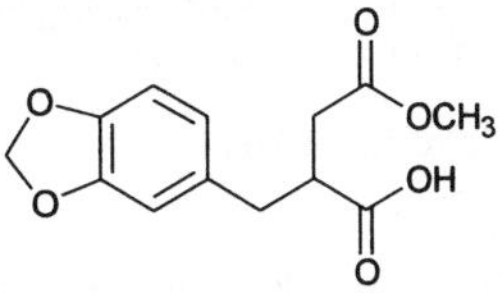

$C_{13}H_{14}O_6$
Chiral building block. *Mitsubishi Corp.; Toyotama Perfumery Co. Ltd.*

**4275 (S)-Piperonylsuccinic acid monomethyl ester**

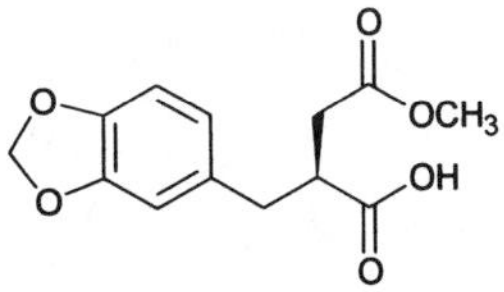

$C_{13}H_{14}O_6$
Chiral intermediate. *Mitsubishi Corp.; Toyotama Perfumery Co. Ltd.*

**4276 Potassium D-erythronate**
88759-55-1
$C_4H_7KO_5$
Chiral building block. *Pfanstiehl Labs Inc.*

**4277 Potassium L-glutamate monohydrate**
24595-14-0 246-336-3

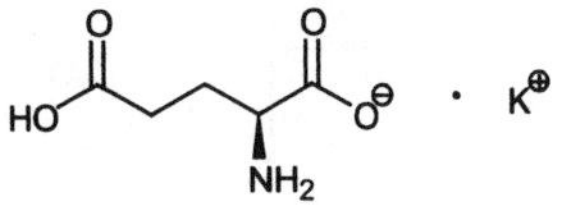

$C_5H_9NO_4.xK$
Glutamic acid, potassium salt, L-. Chiral building block. *Rexim/Degussa Europe.*

**4278 L-(+)-Potassium hydrogen tartrate**
868-14-4 212-769-1

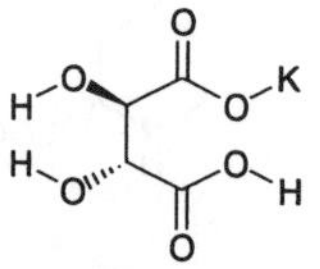

$C_4H_5KO_6$
Butanedioic acid, 2,3-dihydroxy- (2R,3R)-, monopotassium salt; 2,3-Dihydroxy-[R-(R*,R*)]-butanedioic acid monopotassium salt; Acid potassium tartrate; Cream of tartar; E 336 (i); Faccla; Faccula; Faecla; Faecula; Monopotassium tartrate; Potassium acid tartrate; Potassium bitartrate; Potassium hydrotartrate. Listed on TSCA. Chiral ligand; Chiral building block. [α] = + 32° (c = 10, 1N NaOH). *Acros Organics NV.*

**4279 Potassium L-lactate**
85895-78-9 288-752-8
$C_3H_6KO_3$
Propanoic acid, 2-hydroxy-, monopotassium salt, (S)-. L-Lactic acid potassium salt. Chiral building block. *Pfanstiehl Labs Inc.*

**4280 Potassium D-(-)-N-(1-methoxycarbonyl-propen-2-yl)-α-amino-p-hydroxy-phenylacetate**
69416-61-1 273-992-8

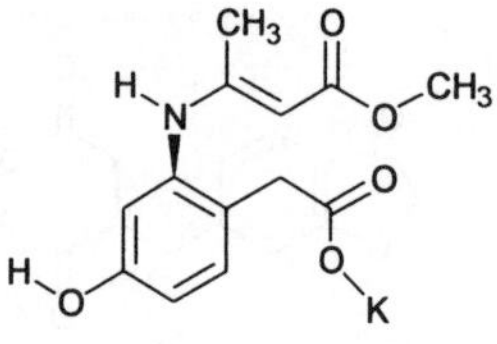

$C_{13}H_{14}NO_5K$
Benzeneacetic acid, 4-hydroxy-α-[(3-methoxy-1-methyl-3-oxo-1-propenyl)amino]-, monopotassium salt, (R)-; D-(-)-p-Hydroxyphenylglycine Dane salt (potassium, methyl); potassium (R)-(4-hydroxyphenyl)[(3-methoxy-1-methyl-3-oxoprop-1-enyl)amino]acetate. Chiral intermediate. *Acros Organics NV; Alfa Chem. Italiana S.r.L.; Kaneka Corp.; Recordati S.p.A.*

**4281 Potassium L-tartrate hydrate**
7669(11)

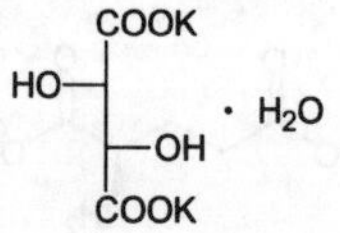

$C_4H_6O_6$
L-Tartaric acid, dipotassium salt hydrate. Listed on TSCA. Chiral building block; Chiral ligand. $[\alpha]_D^{20}$ = + 26° (c = 1, $H_2O$). *Sigma-Aldrich Fine Chem.*

**4282 (+)-5-β-Pregnane-3-α,20-α-diol**
80-92-2 201-313-7

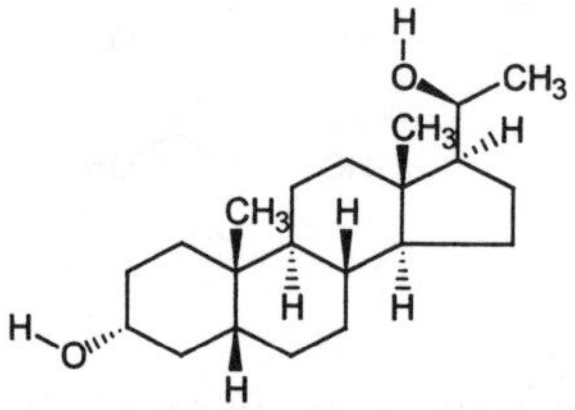

$C_{21}H_{36}O_2$
Pregnane-3,20-diol, (3α,5β,20S)-; 3α,20α-Dihydroxy-5β-pregnane; Diol; Pregnanediol. Pharmaceutical or derivative. $[\alpha]^{25}$ = + 26.2° (c = 0.7, $CH_3OH$). *Acros Organics NV.*

**4283 L-Procysteine ethyl ester**

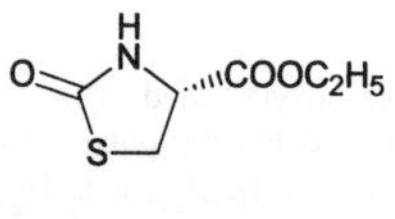

$C_6H_9NO_3S$
L-Oxothizaolidine-4-carboxylic acid. Chiral building block. *Austin Chem. Co.*

**4284 D-Prolinamide**

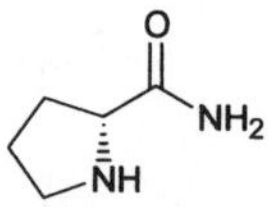

$C_5H_{10}N_2O$
Pyrrolidine-2-carboxamide, (2R)-; (R)-Proline amide. Chiral building block. *Synthetech.*

**4285 L-Prolinamide**
7531-52-4 231-397-0

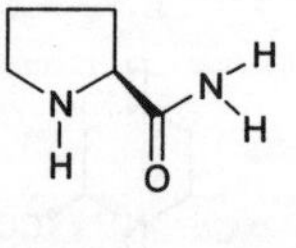

$C_5H_{10}N_2O$
Pyrrolidine-2-carboxamide, (2S)-; (S)-Pyrrolidine-2-carboxamide. Chiral building block. mp = 95-97°; $[\alpha]_D^{20}$ = - 106° (c = 2, EtOH). *Acros Organics NV; Sigma-Aldrich Fine Chem.; Synthetech.*

**4286 L-Proline-7-amido-4-methylcoumarin hydrobromide**
115388-93-7

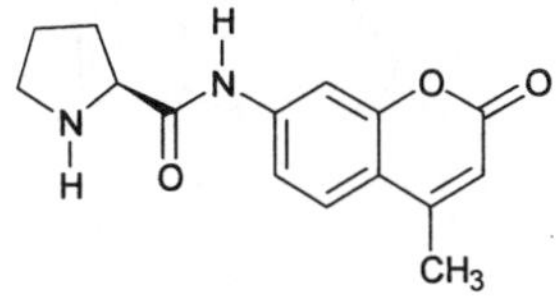

$C_{15}H_{17}BrN_2O_3$
Chiral intermediate. *Acros Organics NV.*

**4287 D-Proline**
344-25-2 206-452-7

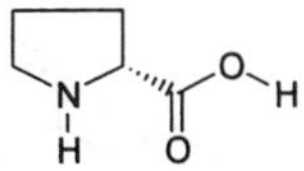

$C_5H_9NO_2$
Proline, D-; (R)-Pyrrolidine-2-carboxylic acid. Listed on TSCA. Chiral building block. mp = 223°; $[\alpha]_D^{20}$ = + 84 ± 3° (c = 1, $H_2O$). *Acros Organics NV; Ajinomoto Co. Inc.; DSM Fine Chem.; Kyowa Hakko Kogyo Co. Ltd.; Lancaster Synthesis Ltd.; Rexim/Degussa Europe; Sigma-Aldrich Fine Chem.; TCI America.*

**4288 L-Proline**
147-85-3 7963(12) 205-702-2

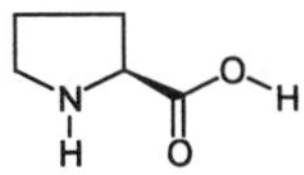

$C_5H_9NO_2$
Proline, L-; (S)-Pyrrolidine-2-carboxylic acid. Listed on TSCA. Chiral catalyst; Chiral auxiliary; Chiral intemediate. mp = 228°; $[\alpha]_D^{20}$ = - 84 ± 3° (c = 1, $H_2O$). *Acros Organics NV; Ajinomoto Co. Inc.; KingChem Inc.; Kyowa Hakko Kogyo Co. Ltd.; Lancaster Synthesis Ltd.; Rexim/Degussa Europe; See Chem. AG; Shanghai DSL Intl. trading Co.; Sigma-Aldrich Fine Chem.; Tanabe Seiyaku Co. Ltd.; TCI America; Varsal Instruments, Inc.; Yoneyama Yakuhin Kogyo Co. Ltd.*

**4289 L-Proline benzyl ester hydrochloride**
16652-71-4 240-700-5

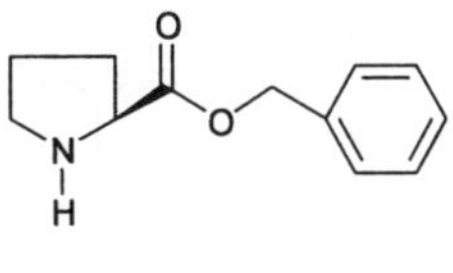

$C_{12}H_{16}ClNO_2$
Proline, benzyl ester, hydrochloride, L-; Benzyl L-prolinate hydrochloride. Chiral intermediate. mp = 148-151°; $[\alpha]_D^{20}$ = - 48° (c = 1, $H_2O$). *Acros Organics NV; Sigma-Aldrich Fine Chem.; Synthetech.*

**4290 L-Proline methyl ester hydrochloride**
2133-40-6 218-363-0

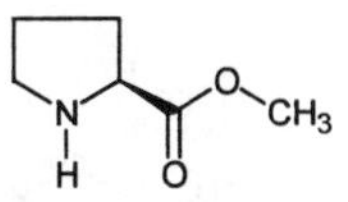

$C_6H_{12}ClNO_2$
Proline, methyl ester, hydrochloride, L-; Methyl L-prolinate hydrochloride. Chiral building block. mp = 69-71°; $[\alpha]_D^{20}$ = - 31° (c = 0.5, $H_2O$). *Acros Organics NV; Austin Chem. Co.; Sigma-Aldrich Fine Chem.; TCI America.*

**4291 (D)-(-)-Prolinol**
68832-13-3

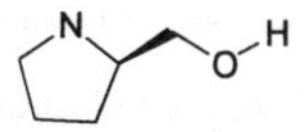

$C_5H_{11}NO$
(R)-(-)-2-Pyrrolidinemethanol; (R)-(-)-2-(Hydroxymethyl)-pyrrolidine; D-Prolinol. Chiral building block. $bp_2$ = 74-76°; d = 1.0250; $n_D^{20}$ = 1.4849; $[\alpha]_D^{20}$ = - 31 ± 2° (c = 1, toluene). *Acros Organics NV; Lancaster Synthesis Ltd.; Omega Chem. Co. Inc.; Sigma-Aldrich Fine Chem.; TCI America.*

**4292 L-(+)-Prolinol**
23356-96-9 245-605-2

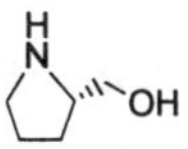

$C_5H_{11}NO$
Pyrrolidine-2-methanol, (S)-; (S)-(+)-2-Pyrrolidine-methanol; (S)-(+)-2-Hydroxymethylpyrrolidine; (S)-(+)-Prolinol. Chiral building block. $bp_2$ = 74-76°; d = 1.0250; $n_D^{20}$ = 1.4853; $[\alpha]_D^{20}$ = + 31 ± 2° (c = 1, toluene). *Alfa Chem. Ltd.; Boehringer Ingelheim Pharma KG; Lancaster Synthesis Ltd.; Loba Feinchemie AG; Omega Chem. Co. Inc.; Pfanstiehl Labs Inc.; See Chem. AG; Sigma-Aldrich Fine Chem.; TCI America.*

**4293 (R)-(-)-1,2-Propanediol**
4254-14-2 8040(12)

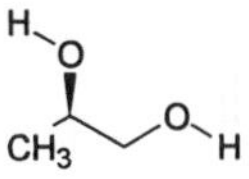

$C_3H_8O_2$
Propane-1,2-diol, (R)-; (R)-(-)-Propylene glycol; (R)-1,2-Dihydroxypropane; (R)-Methylethyl glycol. Chiral building block. bp = 186-188°; d = 1.0400; $n_D^{20}$ = 1.4320; $[\alpha]_D^{20}$ = -16.5 ± 0.5° (neat). *Acros Organics NV; Boehringer Ingelheim Pharma KG; Chirex Inc.; Daicel Chem. Ind. Ltd; Daiso Co. Ltd.; Digital Specialty Chem.; Lancaster Synthesis Ltd.; Loba Feinchemie AG; Maxsyn; PPG - Sipsy; Sigma-Aldrich Fine Chem.; TCI America.*

**4294 (S)-(+)-1,2-Propanediol**
4254-15-3 8040(12)

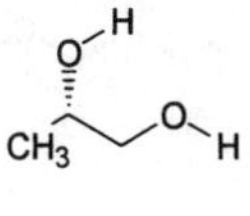

$C_3H_8O_2$
Propane-1,2-diol, (S)-; (S)-(-)-Propylene glycol; (S)-1,2-Dihydroxypropane; (S)-Methylethyl glycol. Chiral building block. bp = 186-188°; $bp_9$ = 77°; d = 1.0400; $n_D^{20}$ = 1.4320; $[\alpha]_D$ = + 16.7° (neat). *Acros Organics NV; Boehringer Ingelheim Pharma KG; Chirex Inc.; Daiso Co. Ltd.; Digital Specialty Chem.; Lancaster Synthesis Ltd.; Loba Feinchemie AG; Maxsyn; Sigma-Aldrich Fine Chem.; Strem Chem. Inc; TCI America.*

**4295 (S)-1,2-Propanediol carbonate**
51260-39-0

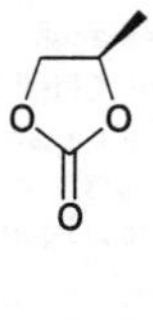

$C_4H_6O_3$
(S)-1,2-Propylene glycol carbonate. Chiral building block. bp = 240°; d = 1.189; n = 1.422; $[\alpha]_D^{20}$ = - 2.0° (neat). *Daiso Co. Ltd.; Sigma-Aldrich Fine Chem.*

**4296 (R)-(+)-1,2-Propanediol di-p-tosylate**
40299-67-0
$C_{18}H_{22}O_6S_2$
(R)-(+)-1,2-Propanediol ditosylate. Chiral building block. mp = 66-68°; $[\alpha]_D^{20}$ = + 17.6 ± 0.5° (c = 2, $CH_3OH$). *Digital Specialty Chem.; Kiralchem Ltd.; Lancaster Synthesis Ltd.; Maxsyn.*

**4297 (S)-(-)-1,2-Propanediol di-p-tosylate**
60434-71-1

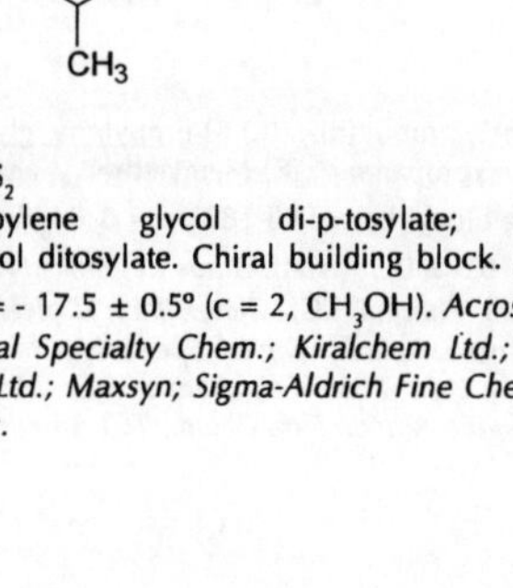

$C_{17}H_{20}O_6S_2$
(S)-(-)-Propylene glycol di-p-tosylate; (S)-(-)-1,2-Propanediol ditosylate. Chiral building block. mp = 68-70°; $[\alpha]_D^{20}$ = - 17.5 ± 0.5° (c = 2, $CH_3OH$). *Acros Organics NV; Digital Specialty Chem.; Kiralchem Ltd.; Lancaster Synthesis Ltd.; Maxsyn; Sigma-Aldrich Fine Chem.; Strem Chem. Inc.*

**4298 (R)-(-)-1,2-Propanediol-1-tosylate**

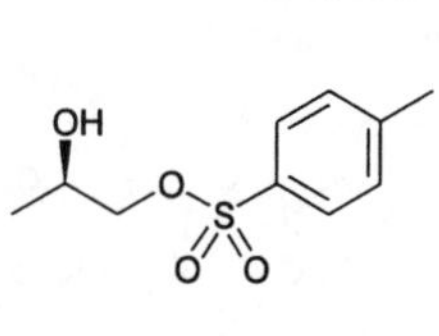

$C_{10}H_{14}O_4S$
Chiral building block. mp = 32-34°; $[\alpha]_D^{20}$ = - 9.8 ± 0.2° (c = 4, 1,2-DCE). *Kiralchem Ltd.*

**4299 (S)-(+)-1,2-Propanediol-1-tosylate**

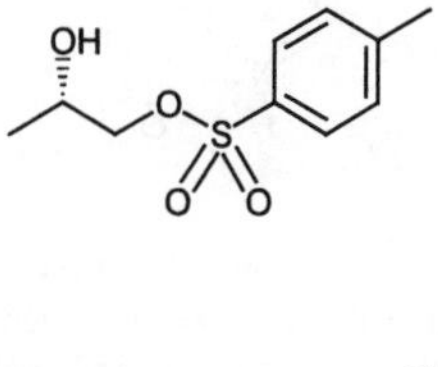

$C_{10}H_{14}O_4S$
Chiral building block. mp = 32-34°; $[\alpha]_D^{20}$ = + 9.8 ± 0.2° (c = 4, 1,2-DCE). *Kiralchem Ltd.*

**4300 (R)-(+)-Propanolol hydrochloride**
13071-11-9 7852(11) 235-961-7

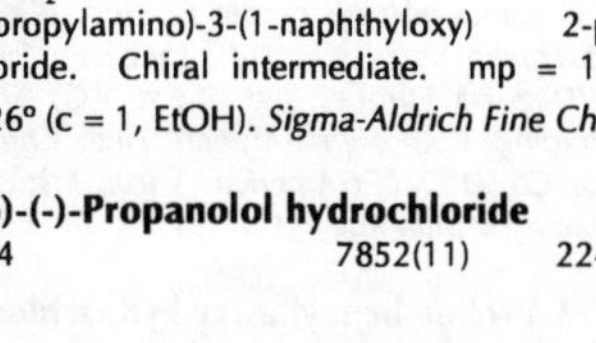

$C_{16}H_{22}ClNO_2$
(R)-1-(Isopropylamino)-3-(1-naphthyloxy) 2-propanol hydrochloride. Chiral intermediate. mp = 196-198°; $[\alpha]_D^{20}$ = + 26° (c = 1, EtOH). *Sigma-Aldrich Fine Chem.*

**4301 (S)-(-)-Propanolol hydrochloride**
4199-10-4 7852(11) 224-096-0

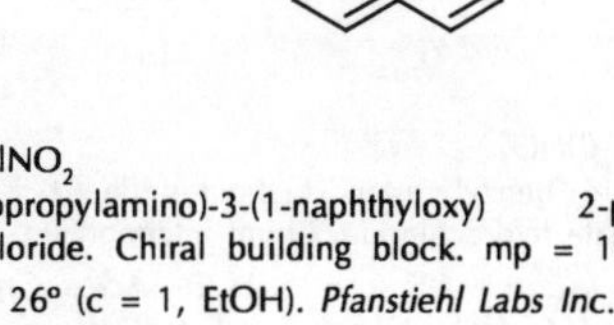

$C_{16}H_{22}ClNO_2$
(S)-1-(Isopropylamino)-3-(1-naphthyloxy) 2-propanol hydrochloride. Chiral building block. mp = 196-198°; $[\alpha]_D^{20}$ = - 26° (c = 1, EtOH). *Pfanstiehl Labs Inc.; Sigma-Aldrich Fine Chem.*

**4302 (4R)-N-(Propionyl)-4-benzyl-2-oxazolidinone**
131685-53-5

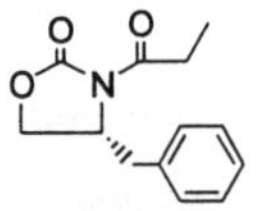

$C_{13}H_{15}NO_3$
Chiral intermediate. mp = 44-46°; $[\alpha]_D^{20}$ = - 102° (c = 1, EtOH). *Oxford Asymmetry Ltd.; Sigma-Aldrich Fine Chem.*

**4303 (4S)-N-(Propionyl)-4-benzyl-2-ozazolidinone**
101711-78-8

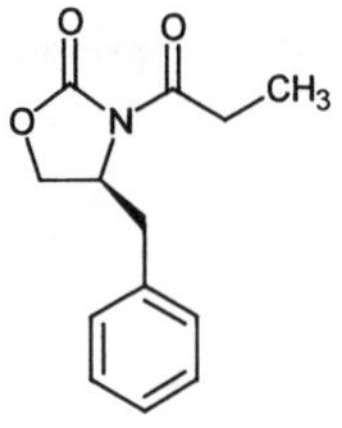

$C_{13}H_{15}NO_3$
Chiral intermediate. mp = 44-46°; $[\alpha]_D^{20}$ = + 97° (c = 1, EtOH). *Acros Organics NV; Fischer Chem. AG; Great Lakes Fine Chem.; Oxford Asymmetry Ltd.; Sigma-Aldrich Fine Chem.*

**4304 N-Propionyl-(2R)-bornane-10,2-sultam**
125664-95-1

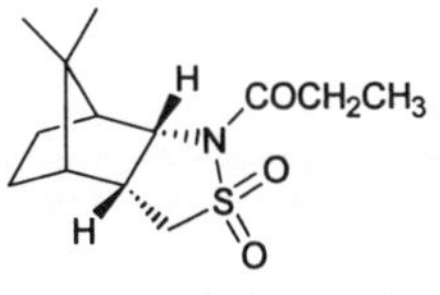

$C_{13}H_{21}NO_3S$
Resolving agent; Chiral auxiliary. *Oxford Asymmetry Ltd.*

**4305 N-Propionyl-(2S)-bornane-10,2-sultam**
128947-19-3

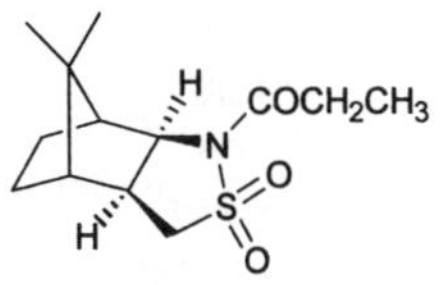

$C_{13}H_{21}NO_3S$
Resolving agent; Chiral auxiliary. *Oxford Asymmetry Ltd.*

**4306 Propionyl-L-carnithine hydrochloride**

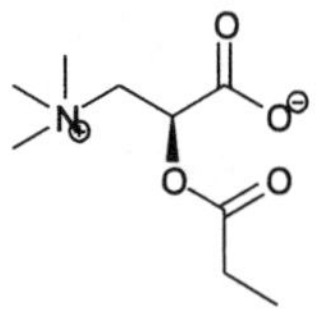

$C_9H_{17}NO_4$
Chiral intermediate. *Flamma s.p.a.*

**4307 N-Propionyl-(4R)-isopropyl-2-oxazolidinone**
89028-40-0

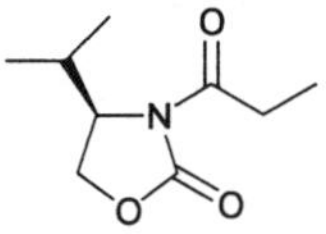

$C_9Hl_5NO_3$
Chiral intermediate. *Newport Synthesis Ireland Ltd.; Oxford Asymmetry Ltd.*

**4308 N-Propionyl-(4S)-isopropyl-2-oxazolidinone**
77877-19-1

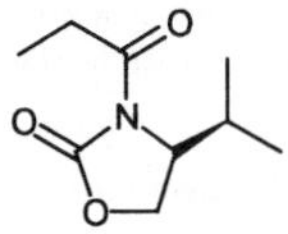

$C_9H_{15}NO_3$
Chiral intermediate. $bp_{0.75}$ = 102-106°; d = 1.094; n = 1.464; $[\alpha]^{25}$ = + 93° (c = 8.7, $CH_2Cl_2$). *Newport Synthesis Ireland Ltd.; Oxford Asymmetry Ltd.; Sigma-Aldrich Fine Chem.*

**4309 N-Propionyl-(4R,5S)-4-methyl-5-phenyl-2-oxazolidinone**

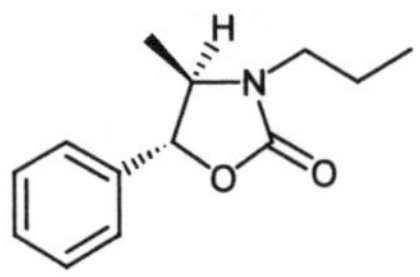

$C_{13}H_{15}NO_3$
Chiral intermediate. *Oxford Asymmetry Ltd.*

**4310 N-Propionyl-(4S,5R)-4-methyl-5-phenyl-2-oxazolidinone**

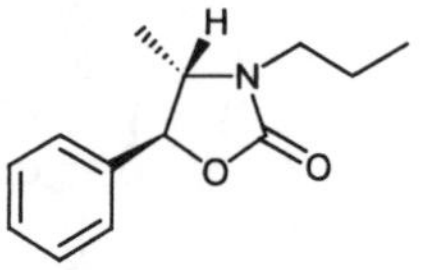

$C_{13}H_{15}NO_3$
Chiral intermediate. *Oxford Asymmetry Ltd.*

**4311 (-)-(3R,4R)-3-(Propylamino)-6-methoxy-1-benzopyran-4-ol**

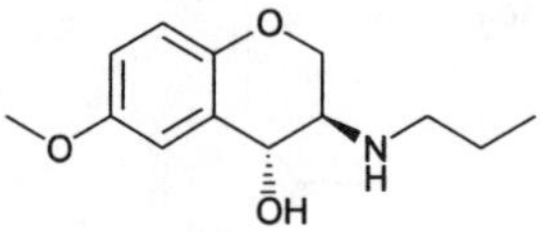

$C_{13}H_{19}NO_3$
Chiral intermediate. *Ultrafine.*

**4312 (R)-Propylene carbonate**
16606-55-6

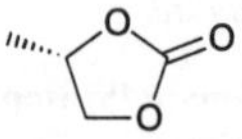

$C_4H_6O_3$
(R)-1,2-Propanediol carbonate. Chiral building block. bp = 240°; d = 1.189; n = 1.422; $[\alpha]_D^{20}$ = + 2.0° (neat). *Chirex Inc.; Daiso Co. Ltd.; PPG - Sipsy; Sigma-Aldrich Fine Chem.*

**4313 (R)-(+)-Propyleneoxide**
15448-47-2

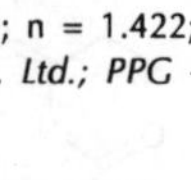

$C_3H_6O$
Oxirane, methyl-, (R)-; (+)-Methyloxirane. Chiral building block. bp = 35°; d = 0.829; n = 1.366; $[\alpha]^{24}$ = + 14° (neat). *Chirex Inc.; Omega Chem. Co. Inc.; Sigma-Aldrich Fine Chem.*

**4314 (S)-(-)-Propylene oxide**
16088-62-3 240-241-0

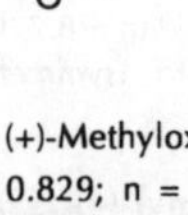

$C_3H_6O$
Oxirane, methyl-, (S)-; (S)-(-)-1,2-Epoxypropane; (S)-Methyloxirane. Chiral building block. bp = 35°; d = 0.829; n = 1.366; $[\alpha]_D^{20}$ = - 7.2° (c = 1, $CHCl_3$). *Acros Organics NV; Chirex Inc.; Omega Chem. Co. Inc.; Sigma-Aldrich Fine Chem.; TCI America.*

**4315 3-Propyl-5-hydroxy-5-D-arabinotetrahydroxybutyl-3-thiazolidine-2-thione**
63348-12-9
$C_{10}H_{19}NO_5S_2$
Chiral intermediate. *Sigma-Aldrich Fine Chem.*

**4316 (2R,3R)-(+)-3-Propyloxiranemethanol**
92418-71-8

$C_6H_{12}O_2$
Chiral building block. mp = 19°; $bp_{0.35}$ = 31-33°; d = 0.96; n = 1.434; $[\alpha]_D^{20}$ = + 46° (c = 1, $CHCl_3$). *Sigma-Aldrich Fine Chem.*

**4317 (2S,3S)-(-)-3-Propyloxiranemethanol**
89321-71-1

$C_6H_{12}O_2$
Chiral building block. mp = 19°; $bp_{0.35}$ = 31-33°; d = 0.96; n = 1.434; $[\alpha]_D^{20}$ = - 47° (c = 1, $CHCl_3$). *Sigma-Aldrich Fine Chem.*

**4318 (R)-N-(2-Propyl)-1-phenylethylamine hydrochloride**
128593-72-6
$C_{11}H_{18}ClN$
(R)-N-Isopropyl-1-phenylethylamine hydrochloride. Chiral intermediate. *Lancaster Synthesis Ltd.*

**4319 (S)-N-(2-Propyl)-1-phenylethylamine hydrochloride**
116297-12-2
$C_{11}H_{18}ClN$
(S)-N-Isopropyl-1-phenylethylamine hydrochloride. Chiral intermediate. *Lancaster Synthesis Ltd.*

**4320 (R)-N-(2-Propyl)-1-phenyl-2-(1-piperidino)-ethylamine**
129157-10-4
$C_{16}H_{26}N_2$
(R)-N-Isopropyl-1-phenyl-2-(1-piperidino)ethylamine. Chiral intermediate. *Lancaster Synthesis Ltd.*

**4321 D-Psicose**
551-68-8 208-999-7

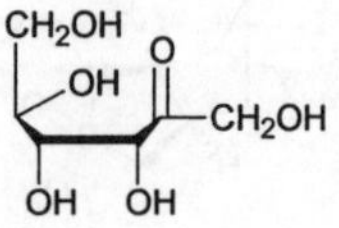

$C_6H_{12}O_6$
Psicose, D-; D-Allulose. Chiral building block. *See Chem. AG.*

**4322 (1S,2S)-Psuedoephedrine-α-fluoro-acetamide**
204323-36-4
$C_{12}H_{16}FNO_2$
Chiral ligand. mp = 85-89°; $[\alpha]_D^{20}$ = + 122° (c = 1, $CHCl_3$). *Sigma-Aldrich Fine Chem.*

**4323 (R)-5-Pthalimido-2-bromovaleric acid**

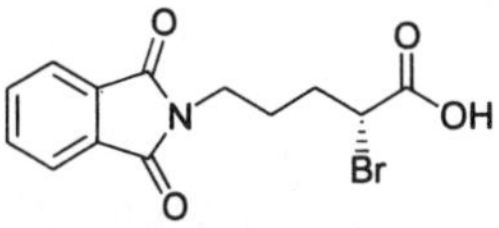

$C_{13}H_{12}BrNO_4$
Chiral intermediate. *ChiroTech Technologies.*

**4324 (S)-5-Pthalimido-2-bromovaleric acid**

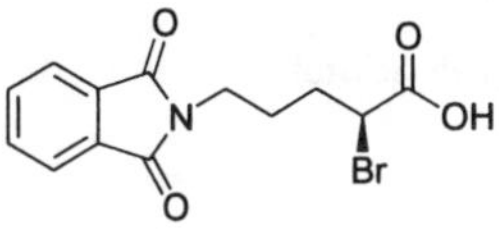

$C_{13}H_{12}BrNO_4$
Chiral intermediate. *ChiroTech Technologies.*

**4325 (S)-(-)-Pulegone**
3391-90-0 7955(11)

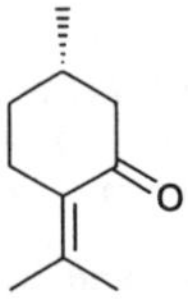

$C_{10}H_{16}O$
Chiral intermediate. bp = 223-224°; d = 0.937; n = 1.488; $[\alpha]_D^{20}$ = - 22° (neat). *Sigma-Aldrich Fine Chem.*

**4326 (R)-(+)-Pulegone**
89-82-7 8124(12) 201-943-2

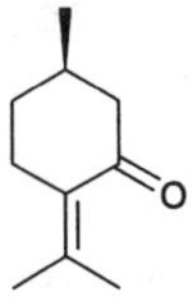

$C_{10}H_{16}O$
p-Menth-4(8)-en-3-one. Chiral intermediate. bp = 224°; d = 0.937; n = 1.486; $[\alpha]_D^{20}$ = + 22° (neat). *Sigma-Aldrich Fine Chem.*

**4327 (R)-5-(3-Pyridyl)-2-pentylamine**

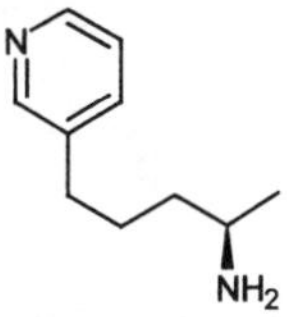

$C_{10}H_{16}N_2$
Chiral intermediate. *Chiragene.*

**4328 (S)-5-(3-Pyridyl)-2-pentylamine**

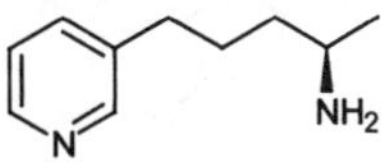

$C_{10}H_{16}N_2$
Chiral intermediate. *Chiragene.*

**4329 L-Pyroglutamic acid 4-methyl-7-coumarin-ylamide hydrate**
66642-36-2

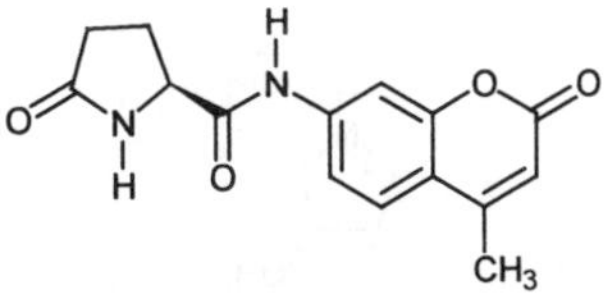

$C_{15}H_{14}N_2O_4$
Chiral intermediate. *Acros Organics NV.*

**4330 L-Pyroglutamic acid amide**

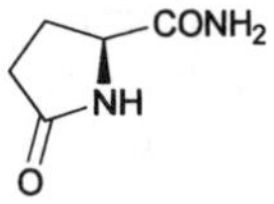

$C_5H_8N_2O_2$
Chiral building block. *See Chem. AG.*

**4331 (R)-(3-Pyrrolidineoxy)tetrahydro-2H-pyran**

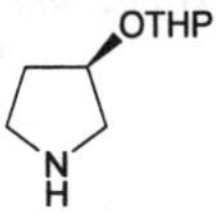

$C_9H_{17}NO_2$
Chiral intermediate. *Omega Chem. Co. Inc.*

**4332 (S)-(3-Pyrrolidineoxy)tetrahydro-2H-pyran**

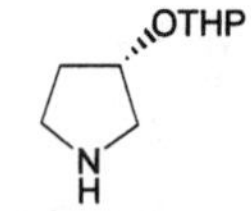

$C_9H_{17}NO_2$
Chiral intermediate. *Omega Chem. Co. Inc.*

**4333 (R)-(-)-3-Pyrrolidinol hydrochloride**
104706-47-0

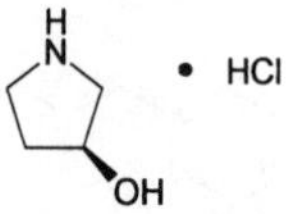

$C_4H_{10}ClNO$
(R)-(-)-3-Hydroxypyrrolidine hydrochloride. Chiral building block. mp = 109°; $[\alpha]_D^{20}$ = - 7.6° (c = 3.5, $CH_3OH$). *Kaneka Corp.; Sigma-Aldrich Fine Chem.; TCI America.*

**4334 (S)-3-Pyrrolidinol hydrochloride**
122536-94-1

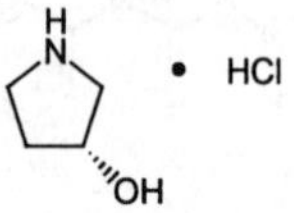

$C_4H_{10}ClNO$
Chiral building block. *Kaneka Corp.*

**4335 (R)-(+)-2-Pyrrolidinone-5-carboxylic acid**
4042-36-8 223-735-0

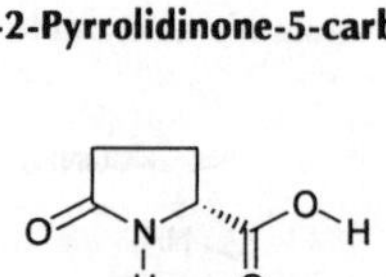

$C_5H_7NO_3$
Pyrrolidin-2-one-5-carboxylic acid, D-; 5-Oxo-D-proline; D-Pyroglutamic acid; D-Proline, 5-oxo-. Chiral building block. mp = 158-160°; $[\alpha]_D^{20}$ = + 10.5 ± 1 ° (c = 5, $H_2O$). *Acros Organics NV; Lancaster Synthesis Ltd.; See Chem. AG; Sigma-Aldrich Fine Chem.; Varsal Instruments, Inc.*

**4336 (S)-(-)-2-Pyrrolidinone-5-carboxylic acid**
98-79-3 8185(12) 202-700-3

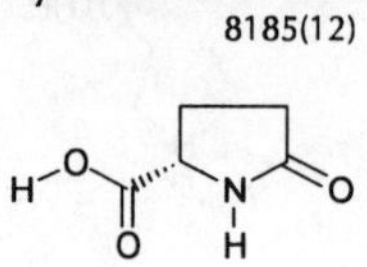

$C_5H_7NO_3$
Pyrrolidone-5-carboxylic acid, L-; L-Pyroglutamic acid; Proline, 5-oxo-, L-; L-5-Carboxy-2-pyrrolidinone; L-5-Oxo-2-pyrrolidinecarboxylic acid; L-5-Oxoproline; L-Glutamic acid, γ-lactam; L-Glutimic acid; L-Glutiminic acid; Ajidew A 100. Listed on TSCA. Resolving agent. mp = 160-163°; $[\alpha]_D^{20}$ = - 10.5 ± 1° (c = 5, $H_2O$). *Aceto Corp.; Acros Organics NV; Austin Chem. Co.; KingChem Inc.; Lancaster Synthesis Ltd.; See Chem. AG; Shanghai DSL Intl. trading Co.; Sigma-Aldrich Fine Chem.; TCI America; Varsal Instruments, Inc.*

**4337 (S)-(+)-1-(2-Pyrrolidinylmethyl)pyrrolidine**
51207-66-0

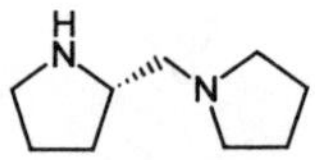

$C_9H_{18}N_2$
Chiral intermediate. $bp_2$ = 99-101°; d = 0.946; n = 1.4871; $[\alpha]^{19}$ = + 7.0° (c = 2.4, EtOH). *Sigma-Aldrich Fine Chem.; TCI America.*

**4338 Quebrachitol**
642-38-6

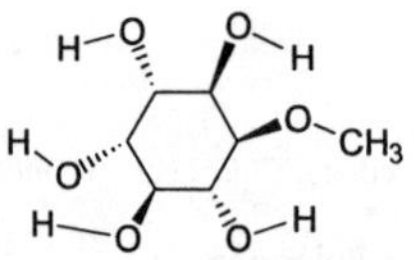

$C_7H_{14}O_6$
2-O-Methyl-L-inositol. Chiral intermediate. mp = 189-192°; $[\alpha]_D^{20}$ = - 79° (c = 1.2, $H_2O$). *Acros Organics NV; Sigma-Aldrich Fine Chem.*

**4339 D-(-)-Quinic acid**
77-95-2 8243(12) 201-072-8

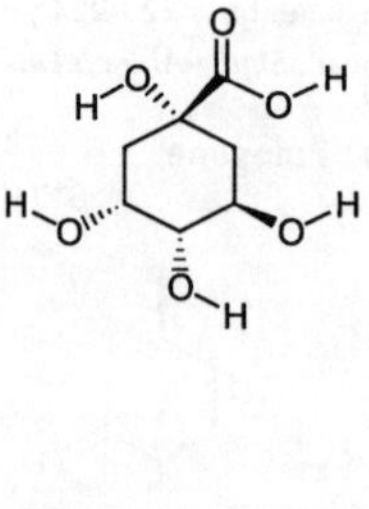

$C_7H_{12}O_6$
Cyclohexanecarboxylic acid, 1,3,4,5-tetrahydroxy-, (1α,3R,4α,5R)-; D-(-)-China acid; 1,3,4,5-Tetrahydroxycyclohexanecarboxylic acid; D-(-)-Quina Acid; D-(-)-Kina Acid; Hexahydro-1,3,4,5-tetrahydroxybenzoic acid. Chiral building block. mp = 166°; $[\alpha]_D^{20}$ = - 43.9° (c = 11.2, $H_2O$). *Acros Organics NV; Sigma-Aldrich Fine Chem.; TCI America.*

**4340 Quinidine**
56-54-2 8244(12) 200-279-0

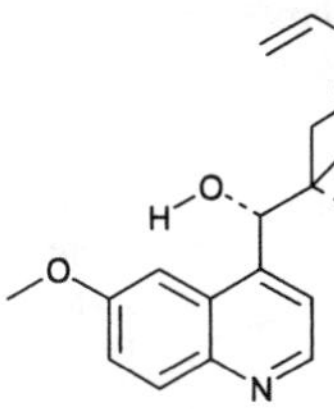

$C_{20}H_{24}N_2O_2$
Cinchonan-9-ol, 6'-methoxy-, (9S)-; (S)-6-Methoxy-α-(5-vinyl-2-quinuclidinyl)-4-quinolinemethanol; Chinidin; Conchinin; Conquinine; Pitayine; (S)-6'-Methoxy-cinchonan-9-ol. Listed on TSCA. Chiral auxiliary; Resolving agent; Chiral ligand. mp = 173-175°; [α] = + 256° (c = 1, EtOH). *Acros Organics NV; DSM Fine Chem.; Isochem; Sigma-Aldrich Fine Chem.; Ultrafine.*

**4341 Quinidine bisulphate**
747-45-5 212-020-9

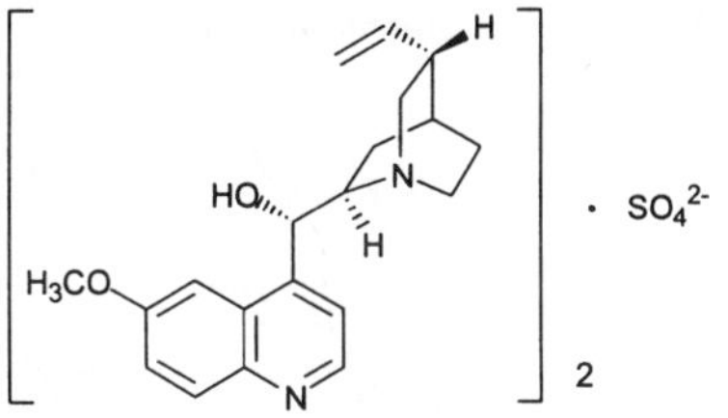

$C_{20}H_{26}N_2O_6S$
Cinchonan-9-ol, 6'-methoxy-, (9S)-, sulfate (1:1) (salt); Biquin Durules; Chinidin Duriles; Chinidin VUFB; Kinidin Duretter; Kinidin Durules; Kinilentin; Optochinidi; Quinidine hydrogen sulfate; Quinidine, sulfate (1:1) (salt). Listed on TSCA. Resolving agent; Chiral ligand; Chiral auxiliary. *DSM Fine Chem.; Isochem.*

**4342 Quinidine galacturonate**
7681-28-9 231-663-6
$C_{26}H_{34}N_2O_9$
Galacturonic acid, compd. with (9S)-6'-methoxy-cinchonan-9-ol (1:1); Galacturonic acid, compd. with quinidine. Resolving agent. *Isochem.*

**4343 Quinidine gluconate**
7054-25-3 230-333-9

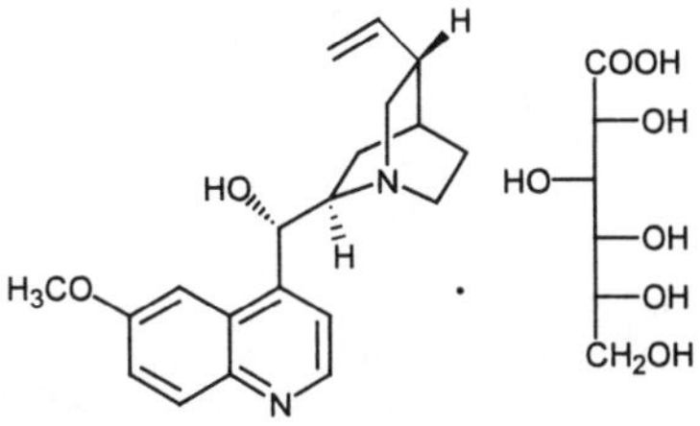

$C_{26}H_{36}N_2O_9$
Gluconic acid, D-, compd. with quinidine (1:1); Gluquinate; Quinaglute; Quinate; Dura-Tab; D-Gluconic acid, compd. with (9S)-6'-methoxycinchonan-9-ol (1:1). Resolving agent. *DSM Fine Chem.; Isochem.*

**4344 Quinidine hydrochloride**
1668-99-1 216-792-8

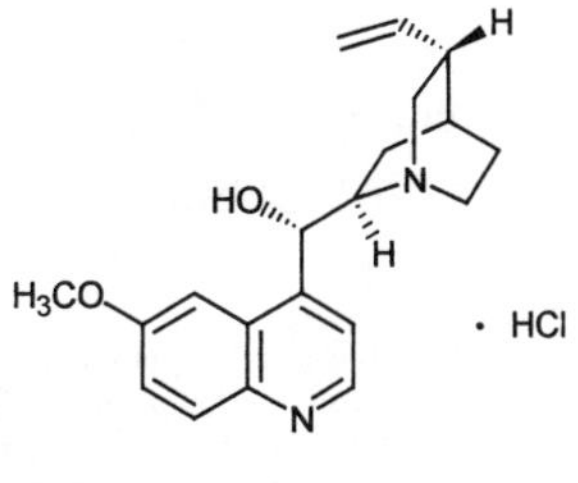

$C_{20}H_{25}ClN_2O_2$
Cinchonan-9-ol, 6'-methoxy-, monohydrochloride, (9S)-; (9S)-6'-Methoxycinchonan-9-ol monohydrochloride. Resolving agent. *DSM Fine Chem.*

**4345 Quinidine sulfate**
50-54-4 200-046-3

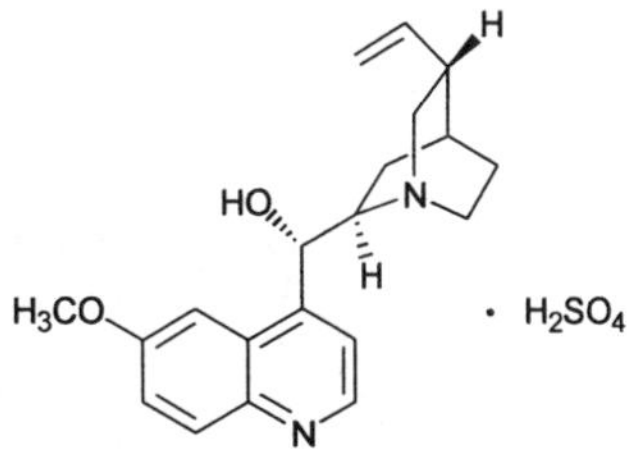

$C_{40}H_{50}N_4O_8S$
Cinchonan-9-ol, 6'-methoxy-, (9S)-, sulfate (2:1) (salt); Quinora; Systodin; Chinidine sulfate; Quinicardine; Quinidate; Quinidex. Listed on TSCA. Chiral ligand; Chiral building block; Chiral auxiliary. *Isochem.*

**4346 Quinidine sulphate hydrate**
6591-63-5 8244(12)

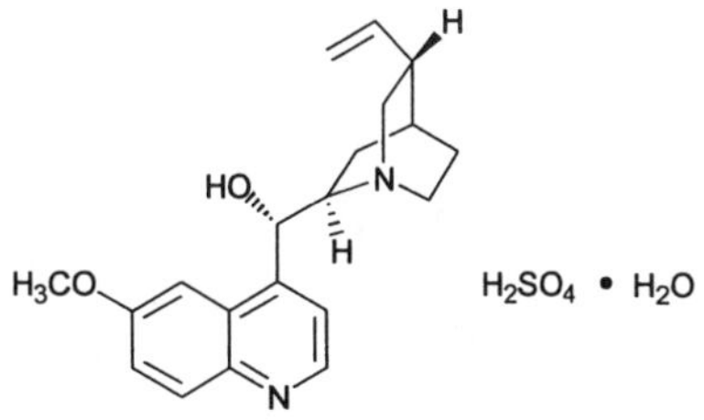

Cinchonan-9-ol, 6'-methoxy-, (9S)-, sulfate (2:1) (salt), monohydrate; Resolving agent; Chiral auxiliary. mp = 212-214°; $[\alpha]^{25}$ = + 220° (c = 1.1, EtOH). *DSM Fine Chem.; Sigma-Aldrich Fine Chem.*

## 4347 Quinine

130-95-0 8245(12) 205-003-2

$C_{20}H_{24}N_2O_2$

Quinine, (-)-; Legatrin; Quin-260; Quin-amino; Quinamm; Quindan; Quiphile; Q-VEL; 6'-Methoxycinchonan-9-ol; Novoquinine; Strema; 6-Methoxy-α-(5-vinyl-2-quinuclidinyl)-4-quinolinemethanol; (R)-6'-Methoxycinchonidine; Chinine; Cinchonan-9-ol, 6'-methoxy-, (8α,9R)- . Listed on TSCA. Resolving agent; Chiral auxiliary; Chiral ligand. mp = 177°; $[\alpha] = -172°$ ($c = 1$, 0.2 N $H_2SO_4$); $[\alpha]^{25} = -165°$ ($c = 2$, EtOH). *Acros Organics NV; DSM Fine Chem.; Isochem; Sigma-Aldrich Fine Chem.*

## 4348 Quinine ascorbate

146-40-7 205-671-5

$C_{32}H_{40}N_2O_{14}$

Ascorbic acid, compd. with quinine (2:1), L-; Combatin; L-Ascorbic acid, compd. with (8α,9R)-6'-methoxycinchonan-9-ol (2:1); Quinine diascorbate; Quinine biascorbate. Chiral auxiliary; Chiral ligand; Resolving agent. *Isochem.*

## 4349 Quinine benzoate

750-88-9 212-026-1

$C_{27}H_{30}N_2O_4$

Cinchonan-9-ol, 6'-methoxy-, (8α,9R)-, monobenzoate (salt); Quinine, monobenzoate (salt). Chiral ligand; Chiral building block; Chiral auxiliary. *DSM Fine Chem.; Isochem.*

## 4350 Quinine bisulphate

549-56-4 208-970-9

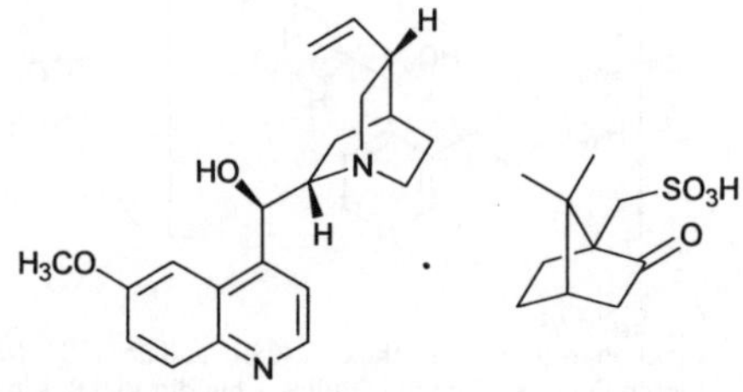


$C_{20}H_{26}N_2O_6S$

Cinchonan-9-ol, 6'-methoxy-, (8α,9R)-, sulfate (1:1) (salt); Quinine hydrogen sulfate; 8α,9R-6'-Methoxycinchonan-9-ol, sulfate (1:1) salt. Listed on TSCA. Chiral ligand; Chiral building block; Chiral auxiliary. *DSM Fine Chem.; Isochem.*

## 4351 Quinine camphosulfonate

84848-19-1 284-305-6

$C_{30}H_{40}N_2O_6S$

Cinchonan-9-ol, 6'-methoxy-, (8α,9R)-, mono[(1S)-7,7-dimethyl-2-oxobicyclo[2.2.1]heptane 1-methanesulfonate] (salt); (8a)-6'-Methoxycinchonan-(9R)-ol mono[(1S)-7,7-dimethyl-2-oxobicyclo[2.2.1]heptane 1-methanesulphonate]. Resolving agent. *Isochem.*

## 4352 Quinine dihydrochloride

60-93-5 200-493-4

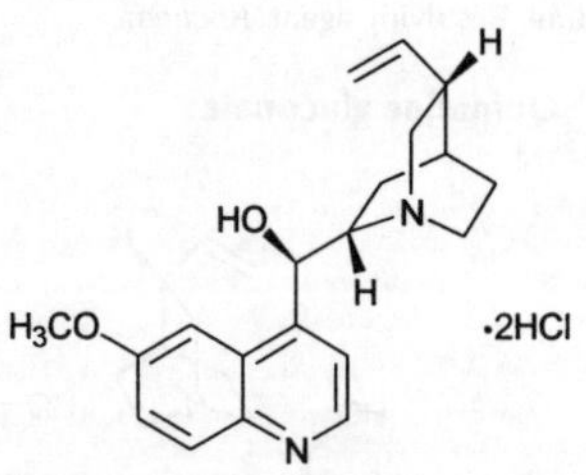


$C_{20}H_{26}Cl_2N_2O_2$

Cinchonan-9-ol, 6'-methoxy-, dihydrochloride, (8α,9R)-; Acid quinine hydrochloride; Quinine bimuriate. Listed on TSCA. Resolving agent; Chiral ligand; Chiral auxiliary. *Isochem.*

**4353 Quinine ethylcarbonate**
83-75-0 201-500-3

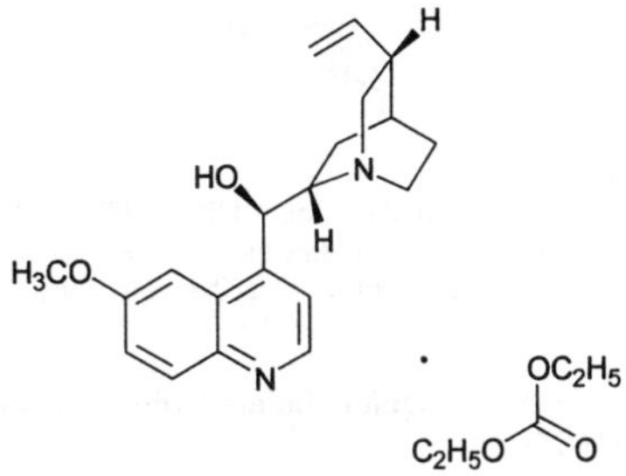

$C_{23}H_{28}N_2O_4$
Cinchonan-9-ol, 6'-methoxy-, ethyl carbonate (ester), (8α,9R)-; Ethyl quinine carbonate; Euquinin. Resolving agent. *Isochem.*

**4354 Quinine formate**
130-90-5 205-002-7

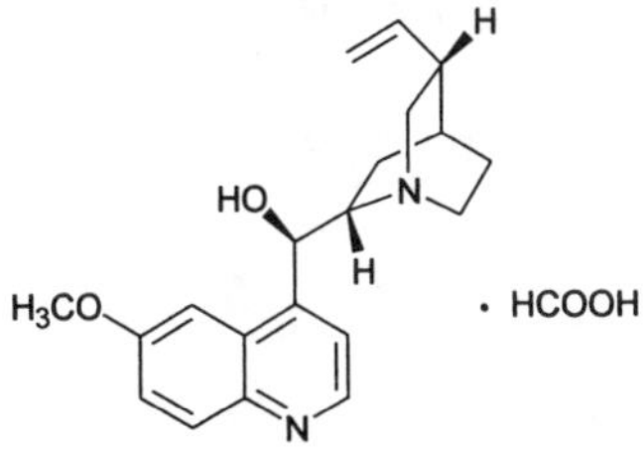

$C_{21}H_{26}N_2O_4$
Cinchonan-9-ol, 6'-methoxy-, (8α,9R)-, monoformate (salt); Quinoform; Quinaform. Chiral ligand; Chiral building block; Chiral auxiliary. *Isochem.*

**4355 Quinine gluconate**
4325-25-1 224-358-4

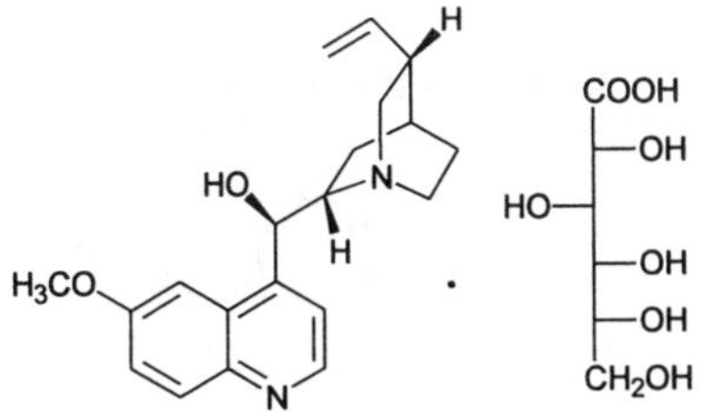

$C_{26}H_{36}N_2O_9$
Gluconic acid, compd. with (8α,9R)-6'-methoxycinchonan-9-ol (1:1); Gluconic acid, compd. with quinine (1:1). Chiral intermediate. *DSM Fine Chem.; Isochem.*

**4356 Quinine hydrochloride dihydrate**
6119-47-7 8245(12)

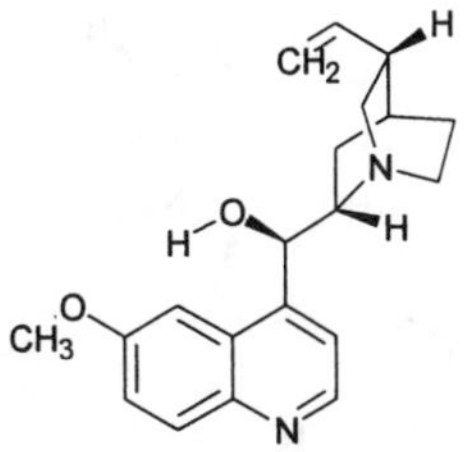

$C_{20}H_{25}ClN_2O_2$
Cinchonan-9-ol, 6'-methoxy-, monohydrochloride, dihydrate, (8α,9R)-; Resolving agent; Chiral auxiliary. mp = 115-116°; $[\alpha]_D^{20}$ = - 144° (c = 1, $H_2O$). *Acros Organics NV; DSM Fine Chem.; Sigma-Aldrich Fine Chem.*

**4357 Quinine monohydrobromide**
549-49-5 208-967-2

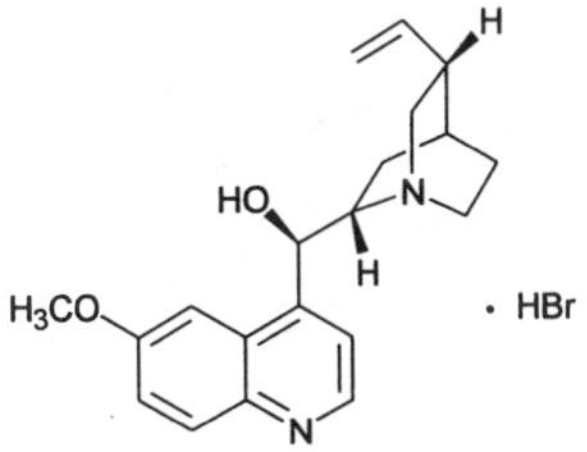

$C_{20}H_{25}BrN_2O_2$
Cinchonan-9-ol, 6'-methoxy-, monohydrobromide, (8α,9R)-; Quinine hydrobromide. Chiral ligand; Chiral building block; Chiral auxiliary. *Isochem.*

**4358 Quinine monohydrochloride**
130-89-2 205-001-1

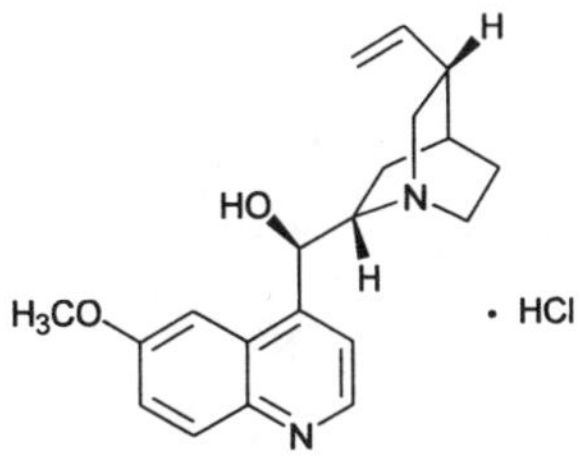

$C_{20}H_{25}ClN_2O_2$
Cinchonan-9-ol, 6'-methoxy-, monohydrochloride, (8α,9R)-; (8a,9R)-6'-Methoxycinchonan-9-ol monohydrochloride; Chinimetten; Quinine muriate. Resolving agent. *Isochem.*

**4359 Quinine salicylate**
750-90-3 212-027-7

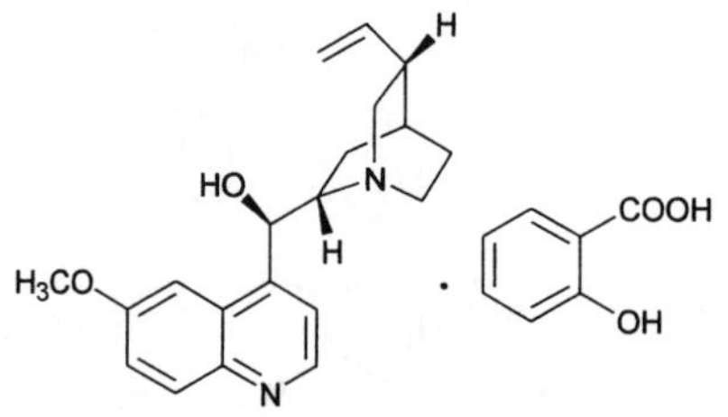

$C_{27}H_{30}N_2O_5$
Cinchonan-9-ol, 6'-methoxy-, (8α,9R)-, mono(2-hydroxybenzoate) (salt); Quinine, monosalicylate. Chiral ligand; Chiral building block; Chiral auxiliary. *Isochem.*

**4360 Quinine sulfate dihydrate**
6119-70-6

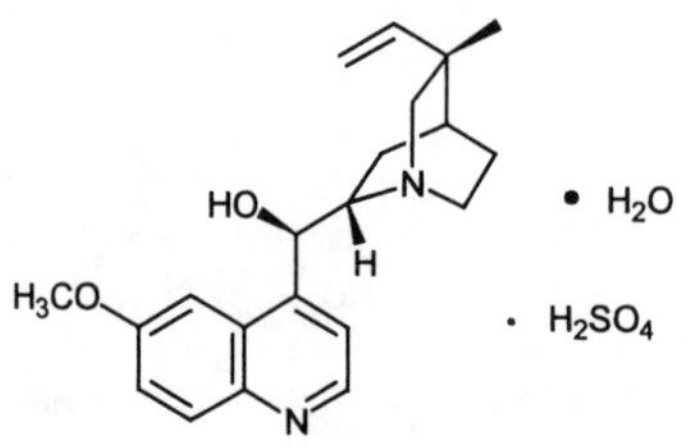

$C_{20}H_{28}N_2O_8S$
Cinchonan-9-ol, 6'-methoxy-, (8α,9R)-, sulfate(2:1) (salt), dihydrate. Chiral ligand; Resolving agent; Chiral auxiliary. [α] = - 245° (c = 2, 0, 1N HCl). *Acros Organics NV.*

**4361 Quinine sulphate**
804-63-7 8089(11) 212-359-2

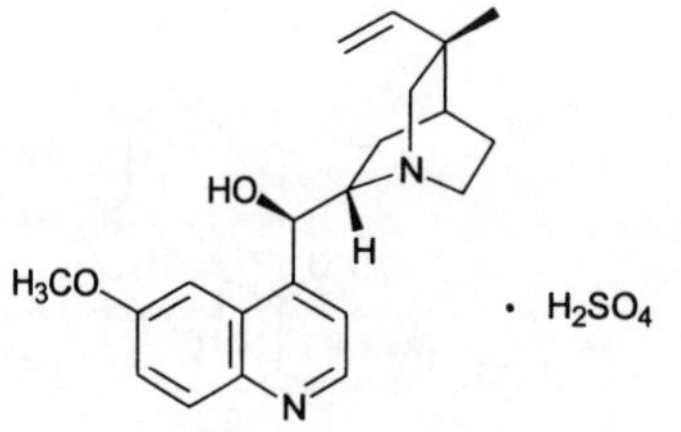

$C_{40}H_{50}N_4O_8S$
Cinchonan-9-ol, 6'-methoxy-, (8α,9R)-, sulfate (2:1) (salt); Aflukin; Coco-Quinine; Quinine hemisulfate. Listed on TSCA. Resolving agent; Chiral ligand; Chiral auxiliary. mp = 233-235°; $[\alpha]^{25}$ = - 229° (c = 2, 0.1M HCl). *DSM Fine Chem.; Isochem; Sigma-Aldrich Fine Chem.*

**4362 (R)-(-)-3-Quinuclidinol**
25333-42-0 246-857-6

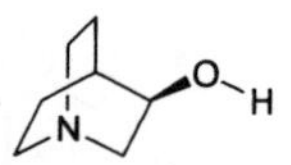

$C_7H_{13}NO$
Azabicyclo[2.2.2]octan-3-ol, (3R)-; (R)-(-)-3-Hydroxyquinuclidine. Chiral building block. [α] = - 45° (c = 3, 1N HCl). *Acros Organics NV; KingChem Inc.; Nagase & Co. Ltd.*

**4363 (R)-(-)-3-Quinuclidinol hydrochloride**
42437-96-7

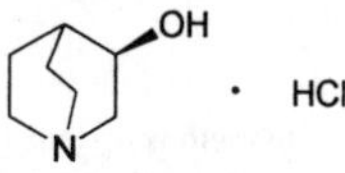

$C_7H_{14}ClNO$
Resolving agent; Chiral auxiliary; Chiral ligand. mp = 344-350°; $[\alpha]_D^{20}$ = - 35° (c = 1, $H_2O$). *Borregaard Synth.; Sigma-Aldrich Fine Chem.*

**4364 L-Quisqualic acid**
52809-07-1

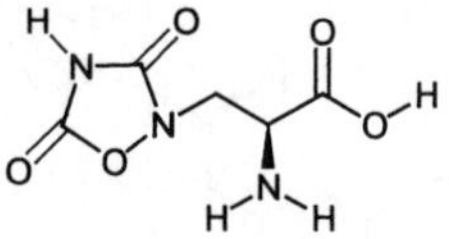

$C_5H_7N_3O_5$
Chiral intermediate. *Acros Organics NV.*

**4365 D-Raffinose**
512-69-6 208-146-9

$C_{18}H_{32}O_{16}$
Glucopyranoside, β-D-fructofuranosyl O-α-D-galactopyranosyl-(1→6)-, α-D-; Melitose; Gossypose; Melitriose; Gossypose pentahydrate; α-Galactosylsucrose pentahydrate. Listed on TSCA. Chiral building block. $[\alpha]_D^{20}$ = + 105.2° (c = 4, $H_2O$). *Pfanstiehl Labs Inc.; See Chem. AG.*

**4366 D-Raffinose pentahydrate**
17629-30-0 8120(11)

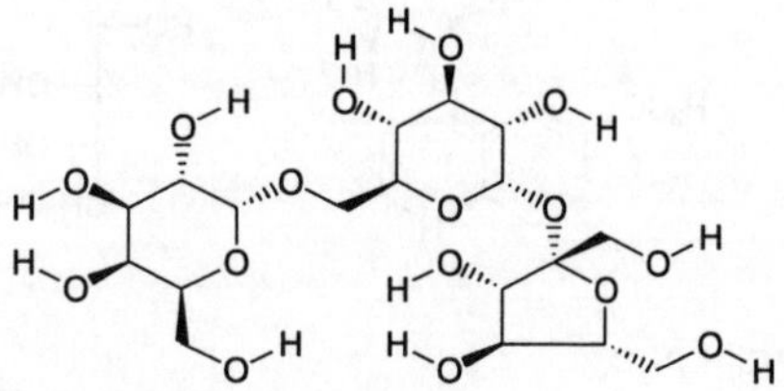

$C_{18}H_{42}O_{21}$
Glucopyranoside, α-D-, β-D-fructofuranosyl O-α-D-galactopyranosyl-(1→6)-, pentahydrate; Melitose; O-α-D-galactopyranosyl-[1-6]-α-D-glucopyranosyl β-D-fructo-

furanoside; O-α-D-Galactopyranosyl-(1,6)-α-D-glucopyranosyl-β-D-fructofuranoside. Chiral building block. mp = 80°; $[\alpha]^{24}$ = + 130° (c = 4, $H_2O$). *Acros Organics NV; Kaden BioChem. GmbH; Sigma-Aldrich Fine Chem.; TCI America.*

### 4367 D-(+)-Raffinose undecaacetate

6424-12-0 229-181-6

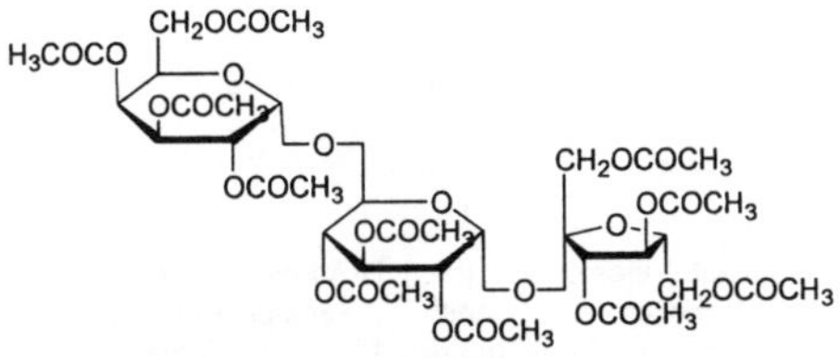

$C_{40}H_{54}O_{27}$

Glucopyranoside, α-D-, 1,3,4,6-tetra-O-acetyl-β-D-fructofuranosyl O-2,3,4,6-tetra-O-acetyl-α-D-galactopyranosyl-(1→6)-, triacetate; 1,3,4,6-tetra-O-acetyl-β-D-fructofuranosyl-O-2,3,4,6-tetra-O-acetyl-α-D-galactopyranosyl-(1→6)-α-D-glucopyranoside triacetate; Raffinose hendecaacetate. Chiral intermediate. *See Chem. AG.*

### 4368 Reduced L-glutathione

70-18-8 4483(12) 200-725-4

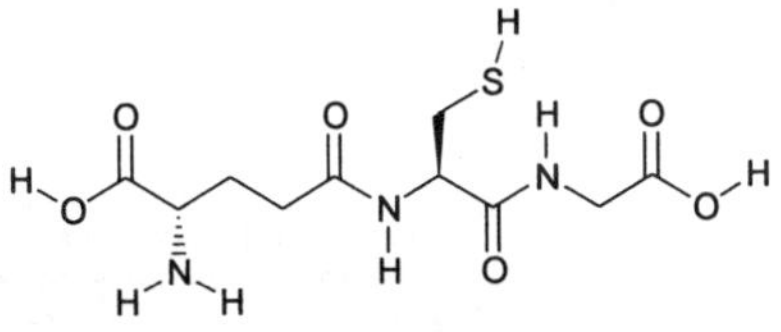

$C_{10}H_{17}N_3O_6S$

Glycine, L-γ-glutamyl-L-cysteinyl-; Glutathione, reduced; Tathion; Tathione; Triptide; Agifutol S; Copren; Deltathione; Glutathione (GSH); Glutathione-SH; Glutide; Glutinal; GSH; Isethion; L-Glutathione; Neuthion. Listed on TSCA. Chiral building block. mp = 192-195°; $[\alpha]^{25}$ = - 17.0° (c = 2, $H_2O$). *Aceto Corp.; Acros Organics NV; Austin Chem. Co.; Kaneka Corp.; Kyowa Hakko Kogyo Co. Ltd.; Sigma-Aldrich Fine Chem.*

### 4369 (S)-Reichstein's substance

152-58-9 205-805-2

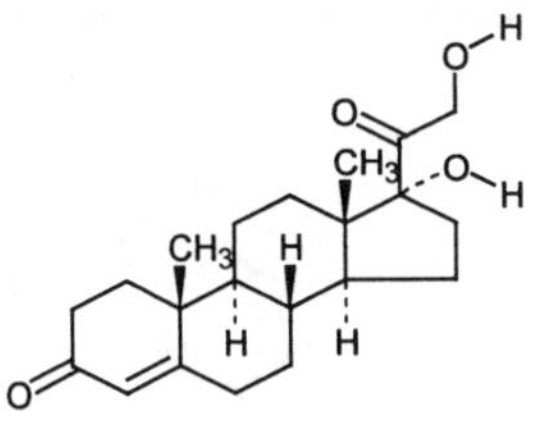

$C_{32}H_{30}O_4$

Chiral intermediate. mp = 208°; [α] = + 120°. *Acros Organics NV; KingChem Inc.*

### 4370 (-)-Retrorsine

480-54-6 8335(12)

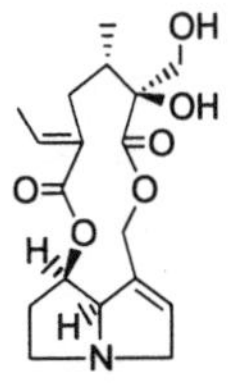

$C_{18}H_{25}NO_6$

Chiral intermediate. mp = 208-211°; $[\alpha]^{28}$ = - 18° (c = 2, EtOH). *Sigma-Aldrich Fine Chem.*

### 4371 L-(+)-Rhamnose

3615-41-6 222-793-4

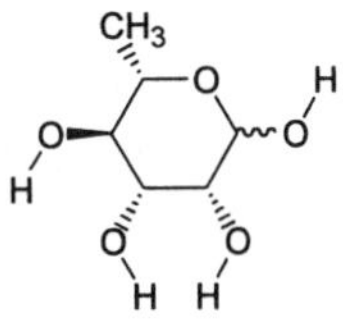

$C_6H_{12}O_5$

Mannose, 6-deoxy-, L-; 6-Deoxy-L-mannose; Isodulcit; Isodulcitol; L-Mannomethylose. Listed on TSCA. Chiral building block. $[\alpha]_D^{20}$ = + 8.2° (c = 4, $H_2O$). *Acros Organics NV; Austin Chem. Co.; Pfanstiehl Labs Inc.; See Chem. AG; TCI America.*

### 4372 D-Ribo-2-ketose

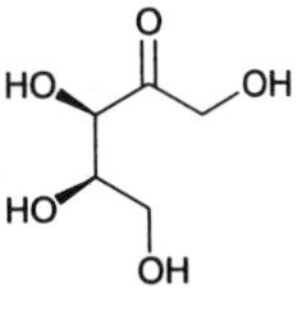

$C_5H_{10}O_5$

Chiral intermediate. *See Chem. AG.*

### 4373 D-Ribono-1,4-lactone

5336-08-3 226-256-5

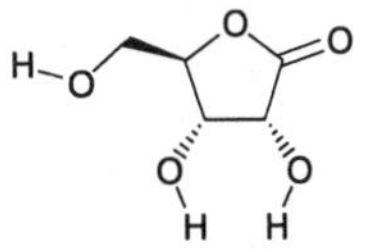

$C_5H_8O_5$

Ribonic acid, γ-lactone, D-; D-(+)-Ribonic acid lactone; D-Ribopentono-1,4-lactone. Chiral intermediate. mp = 85-87°; $[\alpha]_D^{20}$ = + 17.5° (c = 5, $H_2O$). *Acros Organics NV; Pfanstiehl Labs Inc.; See Chem. AG; Sigma-Aldrich Fine Chem.*

**4374 (-)-Ribopyranose 1,2,3,4-teraacetate**
4049-34-7 223-750-2

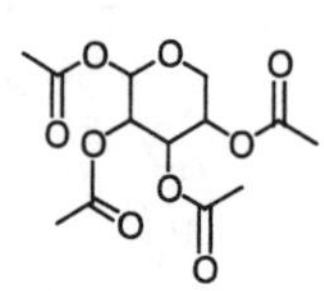

$C_{13}H_{18}O_9$
Chiral intermediate. mp = 110-112°; $[\alpha]^{22}$ = - 55.4° (c = 3, $CH_3OH$). *Sigma-Aldrich Fine Chem.*

**4375 D-Ribose**
50-69-1 8371(12) 200-059-4

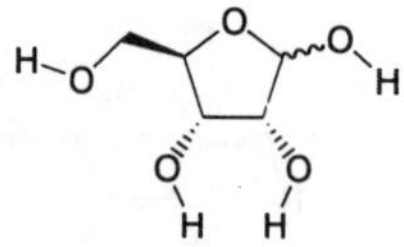

$C_5H_{10}O_5$
Ribose, D-; D-Ribopyranose. Listed on TSCA. Chiral building block. mp = 88-92°; $[\alpha]_D^{20}$ = - 20.4° (c = 4, $H_2O$). *Aceto Corp.; Acros Organics NV; Austin Chem. Co.; KingChem Inc.; Pfanstiehl Labs Inc.; See Chem. AG; Shanghai DSL Intl. trading Co.; Sigma-Aldrich Fine Chem.; TCI America.*

**4376 L-Ribose**
24259-59-4 246-110-4

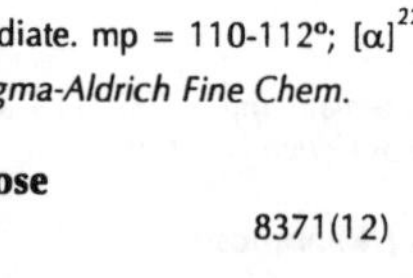

$C_5H_{10}O_5$
Ribose, L-; Chiral building block. mp = 81-82°; $[\alpha]_D^{20}$ = + 19° (c = 2, $H_2O$). *Acros Organics NV; Austin Chem. Co.; Omega Chem. Co. Inc.; Sigma-Aldrich Fine Chem.; TCI America.*

**4377 D-Ribose-5-phosphate disodium salt hydrate**
18265-46-8 242-140-7

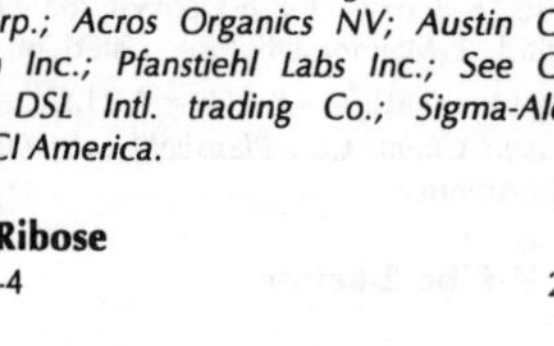

$C_5H_9Na_2O_8P$
Ribose, 5-(dihydrogen phosphate), disodium salt, D-; Sodium ribose 5-phosphate; D-Ribose, 5-(dihydrogen phosphate), disodium salt; Disodium D-ribose 5-phosphate. Chiral intermediate. $[\alpha]_D^{20}$ = + 13.9° (c = 4, $H_2O$). *Acros Organics NV; Sigma-Aldrich Fine Chem.*

**4378 D-(-)-Ribulose**
488-84-6 207-687-8

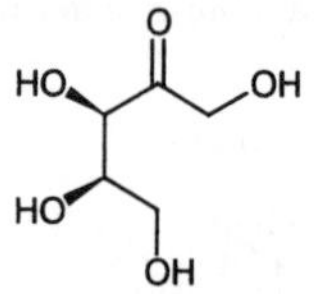

$C_5H_{10}O_5$
Pent-2-ulose, D-erythro-; D-Adonose; D-Arabinulose; D-Araboketose; D-erythro-2-Keptopentose; D-erythro-2-Pentulose. Chiral building block. *See Chem. AG.*

**4379 (-)-Rotenone**
83-79-4 8427(12) 201-501-9

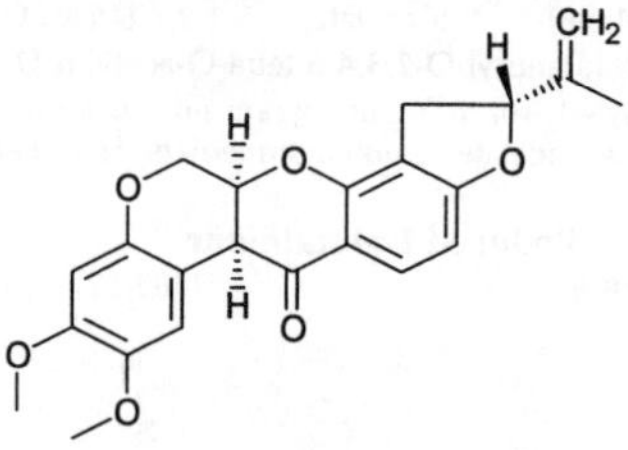

$C_{23}H_{22}O_6$
Benzo[1]pyrano[3,4-b]furo[2,3-h][1]benzopyran-6(6aH)-one, 1,2,12,12a-tetrahydro-8,9-dimethoxy-2-(1-methylethenyl)-, [2R-(2a,6aα,12aα)]-; (2R,6αS,12aS)-1,2,6,6a,12,12a-hexahydro-2-isopropenyl-8,9-dimethoxychromeno[3,4-b]furo[2,3-h]chromen-6-one. Chiral intermediate. mp = 159-164°; $bp_{0.5}$ = 210-220°; $[\alpha]_D^{20}$ = - 115° (c = 1.4, $CHCl_3$); $[\alpha]^{24}$ = - 213° (c = 2, toluene). *Acros Organics NV; Sigma-Aldrich Fine Chem.*

**4380 D-Saccharic acid monopotassium salt**
576-42-1 209-402-2
$C_6H_{10}O_8$
D-Glucaric acid, monopotassium salt; D-glucosaccharic acid, monopotassium salt. Chiral building block. mp = 188°; $[\alpha]^{21}$ = + 5° (c = 1, $H_2O$). *Sigma-Aldrich Fine Chem.*

**4381 Saccharic 1,4-lactone monohydrate**
389-36-6 4344(11) 206-865-2

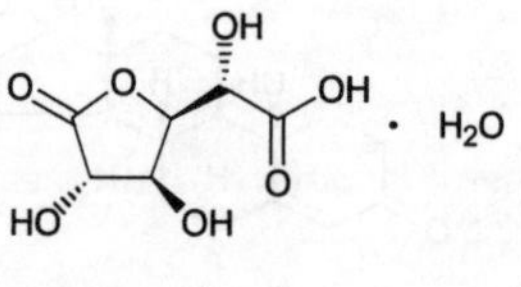

$C_6H_8O_7 \cdot H_2O$
Chiral building block. mp = 91-93°; $[\alpha]^{25}$ = + 32° (c = 1.2, $H_2O$). *Sigma-Aldrich Fine Chem.*

**4382 D-(+)-Saccharose**
57-50-1 9051(12) 200-334-9

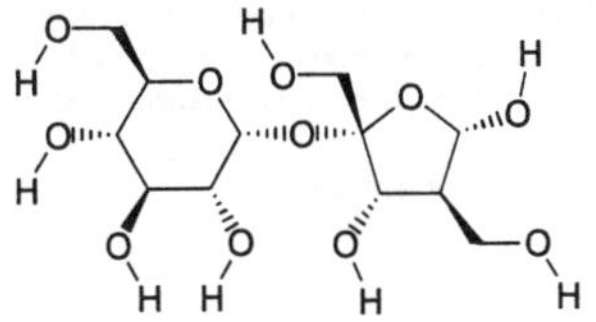

$C_{12}H_{22}O_{11}$
Glucopyranoside, β-D-fructofuranosyl, α-D-; D-(+)-Sucrose; β-D-Fructofuranosyl-α-D-glucopyranoside; Amerfond; Beet sugar; Cane sugar; Confectioner's sugar; Granulated sugar; Manalox AS; Microse; Rock candy; Saccharum; Sucralox; Sugar; White sugar; 1-O-α-D-Glucopyranosyl-β-D-fructofuranoside. Listed on TSCA. Chiral intermediate. mp = 189-191°; $[\alpha]_D^{20}$ = + 66.5° (c = 10, $H_2O$). *Acros Organics NV; Pfanstiehl Labs Inc.; Sigma-Aldrich Fine Chem.; TCI America.*

**4383 D-(+)-Saccharose octaacetate**
126-14-7 9052(12) 204-772-1

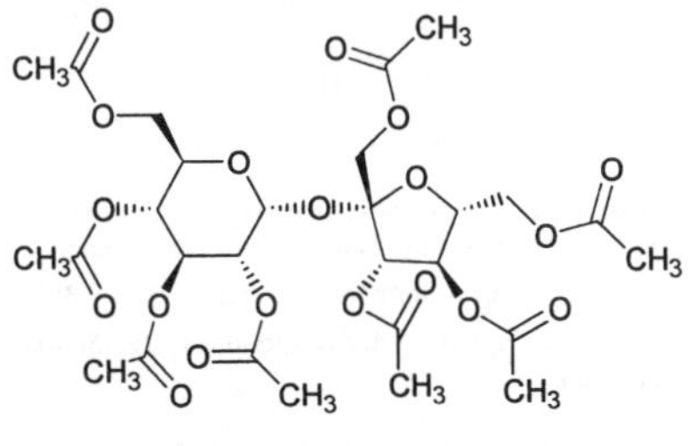

$C_{28}H_{38}O_{19}$
Sucrose, 2,3,4,6,1',3',4',6'-Octa-O-acetyl; D-(+)-Sucrose octaacetate; α-D-Glucopyranoside, 1,3,4,6-tetra-O-acetyl-β-D-fructofuranosyl, tetraacetate; Octaacetyl D-(+)-Sucrose. Listed on TSCA. Chiral intermediate. mp = 86°; d = 1.28; $[\alpha]^{22}$ = + 58.8° (c = 2.5, EtOH). *Acros Organics NV; See Chem. AG; Sigma-Aldrich Fine Chem.; TCI America.*

**4384 D-(-)-Salicin**
138-52-3 205-331-6

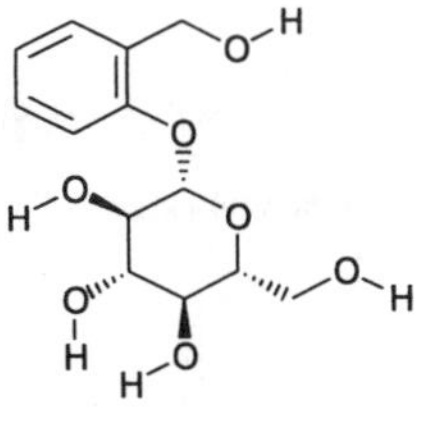

$C_{13}H_{18}O_7$
Glucopyranoside, 2-(hydroxymethyl)phenyl, β-D-; 2-(Hydroxymethyl)phenyl-β-D-glucopyranoside; Salicoside; 1-β-D-Glucosyloxy-2-hydroxymethylbenzene; Salicyl alcohol glucoside; Saligenin β-D-glucopyranoside. Chiral intermediate. mp = 197-200°; $[\alpha]_D^{20}$ = - 62.3° (c = 5, $H_2O$). *Acros Organics NV; CU Chemie Uetikon; Pfanstiehl Labs Inc.*

**4385 Salicyl D-glucuronide**

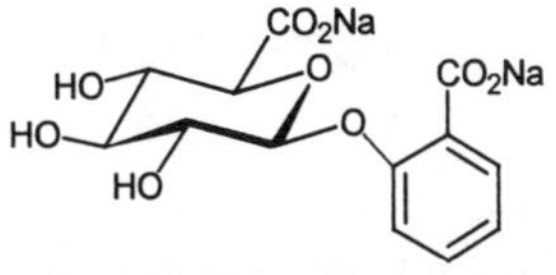

$C_{13}H_{12}Na_2O_9$
2-Carboxyphenyl β-D-glucopyranosiduronic acid disodium salt. Chiral intermediate. *Ultrafine.*

**4386 (-)-Sclareol**
515-03-7 208-194-0

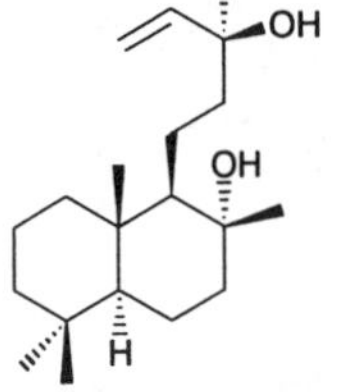

$C_{20}H_{36}O_2$
Listed on TSCA. Pharmaceutical or derivative. mp = 96-98°; $bp_{19}$ = 218-220°; $[\alpha]^{25}$ = - 13° (c = 4, $CCl_4$). *Sigma-Aldrich Fine Chem.*

**4387 (3αR)-(+)-Sclareolide**
564-20-5 209-269-0

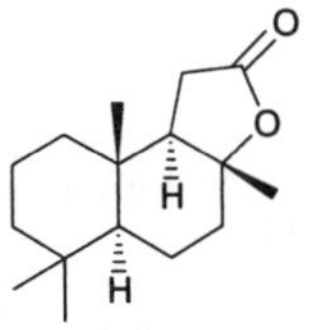

$C_{16}H_{26}O_2$
Listed on TSCA. Chiral intermediate. mp = 124-126°; $[\alpha]_D^{20}$ = + 47° (c = 2, $CHCl_3$). *Sigma-Aldrich Fine Chem.*

**4388 L-Selenocystine**
29621-88-3 221-001-4

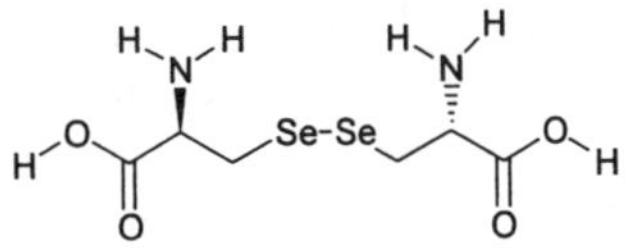

$C_6H_{12}N_2O_4Se_2$
Chiral intermediate. mp = 163-166°. *Acros Organics NV; Sigma-Aldrich Fine Chem.*

**4389 L-(+)-Selenomethionine**
3211-76-5 8585(12)

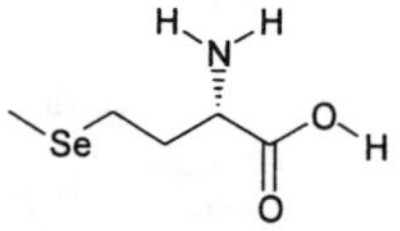

$C_5H_{11}NO_2Se$
(S)-(+)-2-Amino-4-(methylseleno)butanoic acid. Chiral auxiliary. mp = 265-267°; $[\alpha]_D^{20}$ = + 18° (c = 2, 2N HCl). *Acros Organics NV; Sigma-Aldrich Fine Chem.*

**4390 L-Sepiapterin**
17094-01-8

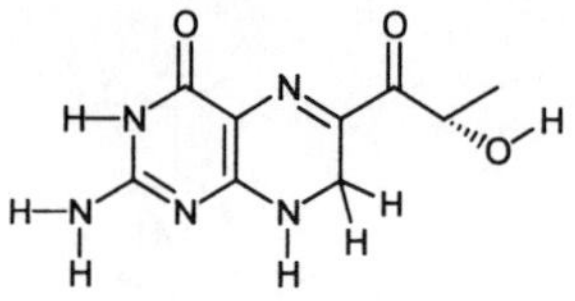

$C_9H_{11}N_5O_3$
Chiral intermediate. *Acros Organics NV.*

**4391 D-Serine**
312-84-5 8604(12) 206-229-4

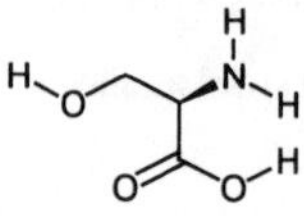

$C_3H_7NO_3$
Serine, D-; (R)-2-Amino-3-hydroxypropionic acid. Listed on TSCA. Chiral building block. mp = 220°; $[\alpha]_D^{20}$ = - 13 ± 1° (c = 5, 5N HCl). *Acros Organics NV; Austin Chem. Co.; Lancaster Synthesis Ltd.; Omega Chem. Co. Inc.; Recordati S.p.A.; Rexim/Degussa Europe; Sigma-Aldrich Fine Chem.; Tanabe Seiyaku Co. Ltd.; TCI America; Yamakawa Chemical Indsutry Co. Ltd.; Yoneyama Yakuhin Kogyo Co. Ltd.*

**4392 L-Serine**
56-45-1 8604(12) 200-274-3

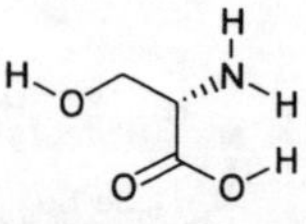

$C_3H_7NO_3$
Propanoic acid, 2-amino-3-hydroxy-, (S)-; (S)-2-Amino-3-hydroxypropionic acid; β-Hydroxy-L-alanine; L-3-Hydroxy-2-aminopropionic acid; L-Alanine, 3-hydroxy-. Listed on TSCA. Chiral building block. mp = 222°; $[\alpha]_D^{20}$= + 14.5 ± 1° (c = 9, 1N HCl). *Acros Organics NV; Ajinomoto Co. Inc.; Austin Chem. Co.; Flamma s.p.a.; Kyowa Hakko Kogyo Co. Ltd.; Lancaster Synthesis Ltd.; Mitsui Pharm. Inc.; Rexim/Degussa Europe; See Chem. AG; Shanghai DSL Intl. trading Co.; Sigma-Aldrich Fine Chem.; Tanabe Seiyaku Co. Ltd.; TCI America; Varsal Instruments, Inc.; Yoneyama Yakuhin Kogyo Co. Ltd.*

**4393 L-Serine amide**
6791-49-7 229-862-8

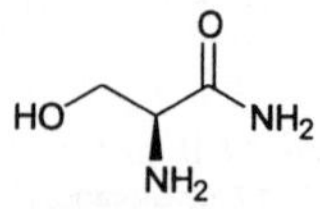

$C_3H_8N_2O_2$
Propanamide, 2-amino-3-hydroxy-, (2S)-; (S)-2-amino-3-hydroxypropionamide. Chiral building block. *See Chem. AG.*

**4394 L-Serine amide hydrochloride**
65414-74-6 265-753-1

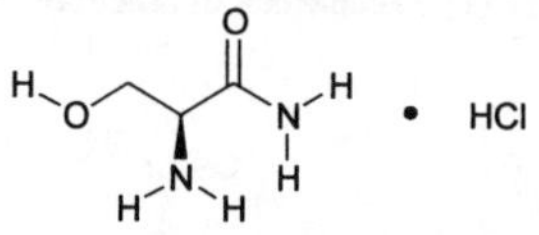

$C_3H_9ClN_2O_2$
Propanamide, 2-amino-3-hydroxy-, monohydrochloride, (2S)-; (S)-2-Amino-3-hydroxypropionamide hydrochloride. Chiral intermediate. mp = 185-188°; $[\alpha]_D^{20}$ = + 14° (c = 1, $H_2O$). *Acros Organics NV; Sigma-Aldrich Fine Chem.*

**4395 L-Serine-7-amido-4-methylcoumarin hydrochloride hydrate**
115918-60-0

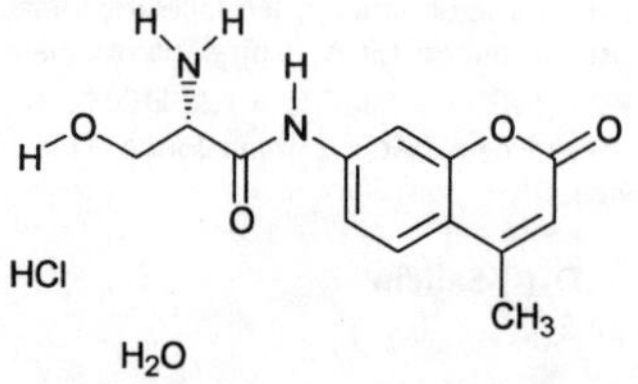

$C_{13}H_{14}N_2O_4 \cdot HCl \cdot H_2O$
Chiral intermediate. *Acros Organics NV.*

**4396 L-Serine benzyl ester**

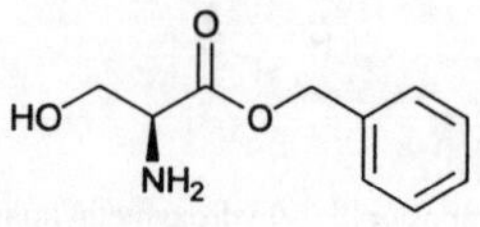

$C_{10}H_{13}NO_3$
Chiral building block. *Schweizerhall Pharma.*

**4397 L-Serine benzyl ester monohydrochloride**

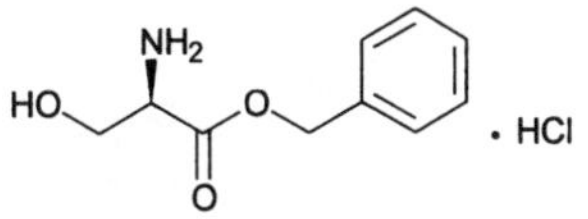

$C_{10}H_{14}ClNO_3$
Chiral building block. *Synthetech.*

**4398 L-Serine benzyl ester p-toluene-sulfonic acid salt**

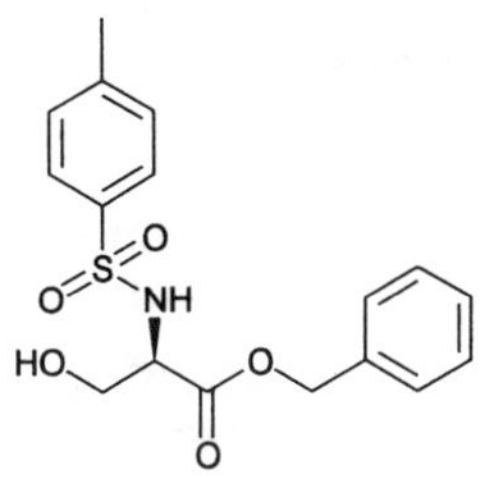

$C_{17}H_{19}NO_5S$
Chiral building block. *Synthetech.*

**4399 L-Serine ethyl ester hydrochloride**
26348-61-8 247-624-1

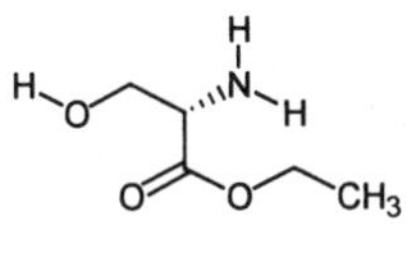

$C_5H_{12}ClNO_3$
Serine, ethyl ester, hydrochloride, L-; Ethyl L-serinate hydrochloride. Chiral building block. mp = 130-132°; $[\alpha]_D^{20}$ = - 4.4° (c = 2, $H_2O$). *Acros Organics NV; Sigma-Aldrich Fine Chem.*

**4400 L-Serine methyl ester hydrochloride**
5680-80-8 227-140-7

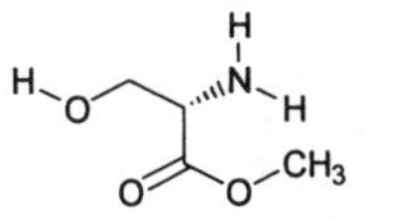

$C_4H_{10}ClNO_3$
Serine, methyl ester, hydrochloride, L-; Methyl L-serinate hydrochloride. Chiral building block. mp = 163°; $[\alpha]^{21}$ = + 3.4° (c = 4, MEtOH). *Acros Organics NV; Austin Chem. Co.; Degussa-Hüls AG; Sigma-Aldrich Fine Chem.; Synthetech; TCI America.*

**4401 D-Serine methyl ester hydrochloride**
5874-57-7

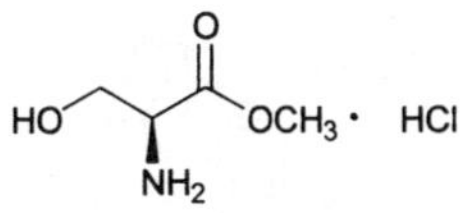

$C_4H_{10}ClNO_3$
Chiral building block. mp = 163-166°; $[\alpha]_D^{20}$ = - 4° (c = 4, EtOH). *Sigma-Aldrich Fine Chem.; Synthetech.*

**4402 L-Serine-N-pentylamide oxalate**

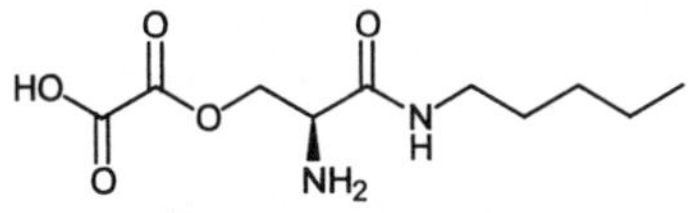

$C_{10}H_{18}N_2O_5$
Chiral building block. *Austin Chem. Co.*

**4403 Shikimic acid**
138-59-0 8625(12) 205-334-2

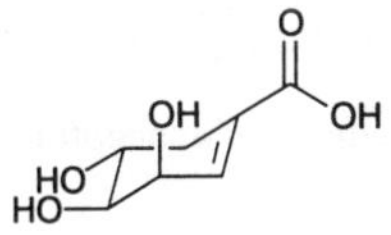

$C_7H_{10}O_5$
Cyclohex-1-ene-1-carboxylic acid, 3,4,5-trihydroxy-, (3R,4S,5R)-; 3,4,5-Trihydroxy-1-cyclohexene-1-carboxylic acid. Chiral building block; Resolving agent. mp = 185-187°; $[\alpha]^{25}$ = - 180° (c = 4, $H_2O$). *Acros Organics NV; Sigma-Aldrich Fine Chem.; Synthon Corp.*

**4404 Sinigrin monohydrate**
3952-98-5 8691(12) 223-545-8

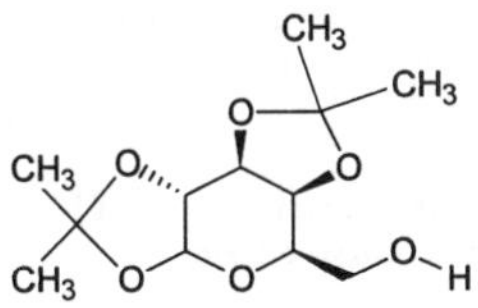

$C_{10}H_{16}KNO_9S_2$
Glucopyranose, 1-thio-, 1-(3-butenohydroximate) NO-(hydrogen sulfate), monopotassium salt, β-D-; Potassium 1-(β-D-glucopyranosylthio)but-3-enylideneaminooxy-sulphonate; 2-Propenyl glucosinolate; Allyl glucosinolate. Pharmaceutical or derivative. mp = 128°; $[\alpha]^{25}$ = - 17.0° (c = 2, $H_2O$). *Acros Organics NV; Sigma-Aldrich Fine Chem.*

**4405 Sinomenine**
115-53-7 8692(12) 204-094-6

$C_{19}H_{23}NO_4$
Pharmaceutical or derivative. mp = 180°. *Sigma-Aldrich Fine Chem.*

**4406 Sodium (+)-10-camphorsulfonate**
21791-94-6

$C_{10}H_{16}O_4S.Na$
Bicyclo[2.2.1]heptane-1-methanesulfonic acid, 7,7-dimethyl-2-oxo-, sodium salt, (1S,4R)-; D-Camphorsulfonic acid sodium salt; 10-Bornanesulfonic acid, 2-oxo-, sodium salt, (+)-; Sodium 2-oxobornane-10-sulphonate. Resolving agent. *Calaire Chimie S.A.*

**4407 Sodium α-D-glucoheptonate dihydrate**
13007-85-7 235-849-8
$C_7H_{1N}aO_{10}$
Gluceptate sodium; Sodium D-glycero-D-gulo-heptonate dihydrate; D-Glycero-D-gulo-heptonic acid, monosodium salt; Seqlene 190; Sodium α-glucoheptonate. Chiral intermediate. *Pfanstiehl Labs Inc.*

**4408 Sodium L-aspartate**
17090-93-6 241-155-6

$C_4H_7NO_4.xNa$
Aspartic acid, sodium salt, L-; Chiral building block. *Kyowa Hakko Kogyo Co. Ltd.*

**4409 Sodium L-lactate**
867-56-1 8781(12) 212-762-3
$C_3H_5NaO_3$
Propanoic acid, 2-hydroxy-, monosodium salt, (2S)-. L-Lactic acid sodium salt; (+)-Lactate sodium; (S)-2-Hydroxypropionic acid sodium salt. Listed on TSCA. Chiral building block. mp = 163-165°; $[\alpha]_D^{20}$ = - 12 ± 0.5° (c = 5, $H_2O$). *Acros Organics NV; Lancaster Synthesis Ltd.; Pfanstiehl Labs Inc.; Sigma-Aldrich Fine Chem.*

**4410 Sodium-L-malate**
138-09-0 205-313-8

$C_4H_4Na_2O_5$
Butanedioic acid, hydroxy-, disodium salt, (S)-; L-Hydroxysuccinic acid disodium salt. Chiral building block. $[\alpha]^{21}$ = - 7.0° (c = 15, $H_2O$). *Loba Feinchemie AG; Sigma-Aldrich Fine Chem.*

**4411 α-Sophorose**
20880-64-2 244-095-9

$C_{12}H_{22}O_{11}$
Glucopyranose, 2-O-β-D-glucopyranosyl-, α-D-; 2-O-β-D-Glucopyranosyl-D-glucose. Chiral intermediate. *See Chem. AG.*

**4412 D-Sorbitol**
50-70-4 8873(12) 200-061-5

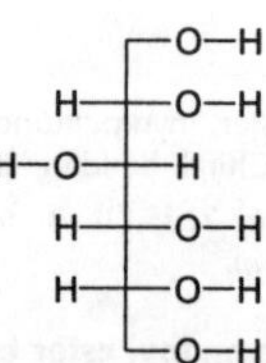

$C_6H_{14}O_6$
Sorbitol, D-; D-Glucitol; Cholaxine; Diakarmon; Esasorb; Foodol D 70; Glucarine; Karion; Multitol; Neosorb; Nivitin; Sionit; Sionon; Siosan; Sorbex; Sorbilande; Sorbit; Sorbo; Sorbol; Sorbostyl. Listed on TSCA. Chiral building block. mp = 98-100°; $[\alpha]_D^{20}$ = + 104° (c = 0.4, acidified ammonium molybdate). *Acros Organics NV; Pfanstiehl Labs Inc.; Sigma-Aldrich Fine Chem.*

**4413 L-Sorbitol**
$C_6H_{14}O_6$
Chiral building block. *TCI America.*

**4414 D-Sorbitol hexaacetate**
7208-47-1 230-588-6

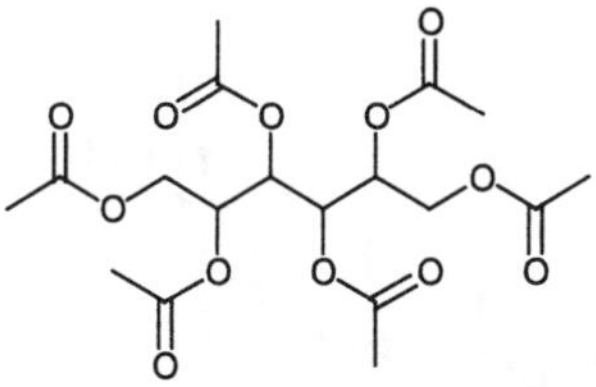

$C_{18}H_{26}O_{12}$
Glucitol, hexaacetate, D-; 1,2,3,4,5,6-Hexa-O-acetyl-D-glucitol. Listed on TSCA. Chiral intermediate. mp = 100-104°; $[\alpha]_D^{20}$ = + 9.5° (c = 1, acetone). *See Chem. AG; Sigma-Aldrich Fine Chem.*

**4415 D-Sorbose**
3615-56-3 222-796-0

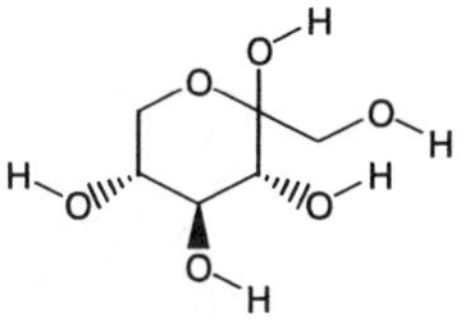

$C_6H_{12}O_6$
Sorbose, D-; Sorbinose. Chiral intermediate. mp = 163-165°; $[\alpha]^{25}$ = + 42° (c = 1, $H_2O$). *Acros Organics NV; See Chem. AG; Sigma-Aldrich Fine Chem.; TCI America.*

**4416 L-Sorbose**
87-79-6 8874(12) 201-771-8

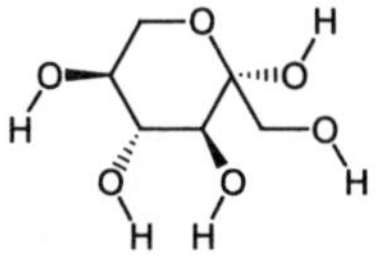

$C_6H_{12}O_6$
Sorbose, L-; L-Sorbopyranose; Esorben; L-1,3,4,5,6-Pentahydroxyhexan-2-one; L-Sorbinose; L-xylo-2-Hexulose. Listed on TSCA. Chiral building block. mp = 186°; $[\alpha]_D^{20}$ = - 43.3° (c = 12, $H_2O$). *Acros Organics NV; Pfanstiehl Labs Inc.; See Chem. AG; Sigma-Aldrich Fine Chem.; TCI America.*

**4417 L-Sorbose-1-phosphate potassium salt**

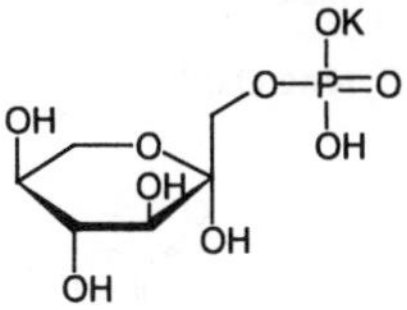

$C_6H_{12}KO_9P$
Chiral building block. *See Chem. AG.*

**4418 (-)-Sparteine free base**
90-39-1 8887(12) 201-988-8

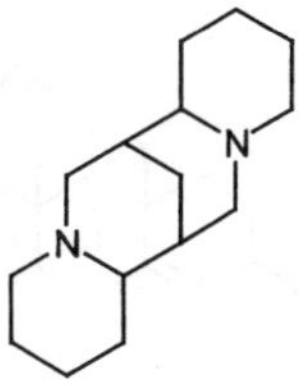

$C_{15}H_{26}N_2$
Methano-2H,6H-7,14-dipyrido[1,2-a:1',2'-e][1,5]diazocine, dodecahydro-, (7S,7αR,14S,14αS)-; 6β,7α,9α,11α-Pachycarpine; Lupinidine. Chiral intermediate. $bp_{1.0}$ = 137-138°; d = 1.02; n = 1.528; $[\alpha]_D^{20}$ = - 16.5° (c = 10, EtOH). *Austin Chem. Co.; Omega Chem. Co. Inc.; Sigma-Aldrich Fine Chem.*

**4419 (-)-Sparteine sulfate pentahydrate**
6160-12-9 8887(12) 206-078-4

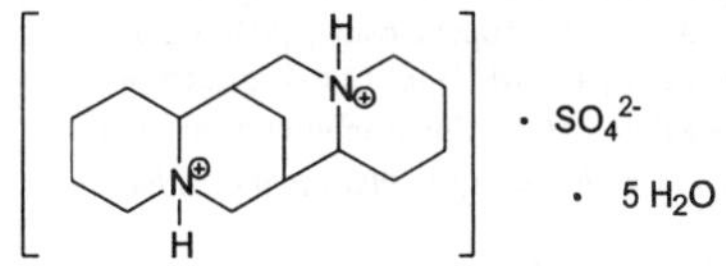

$C_{15}H_{26}N_2$
Chiral intermediate. mp = 133-140°; $[\alpha]^{25}$ = - 12.4° (c = 1, EtOH). *Acros Organics NV; Sigma-Aldrich Fine Chem.*

**4420 D-erythro-Spinghosine**
123-78-4

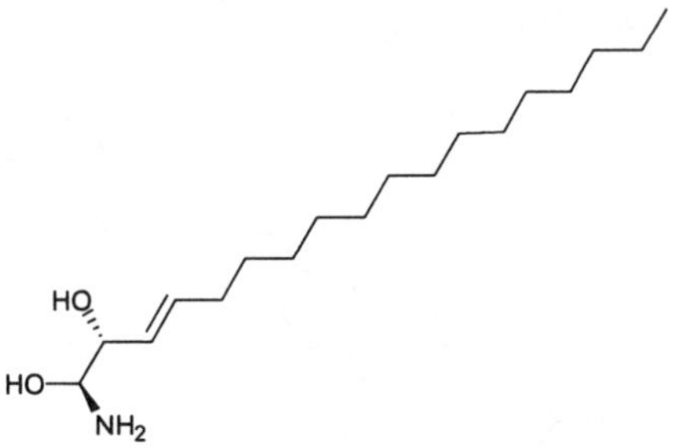

$C_{18}H_{37}NO_2$
Chiral intermediate. *Acros Organics NV.*

**4421 (-)-Spironolactone**
52-01-7 8917(12) 200-133-6

$C_{24}H_{32}O_4S$
Listed on TSCA. Chiral intermediate. mp = 207-208°; $[\alpha]_D^{20}$ = - 34° (c = 1, $CHCl_3$). *Sigma-Aldrich Fine Chem.*

**4422 (25R)-Spirost-5-ene-1-β-3-β-diol**
472-11-7 207-447-2
$C_{27}H_{42}O_4$
Spirost-5-ene-1,3-diol, (1β,3β,25R)-; Ruscogenin; (25R)-spirost-5-ene-1β,3β-diol. Pharmaceutical or derivative. mp = 197-199°. *Kaden BioChem. GmbH.*

**4423 D-Stachyose tetrahydrate**
470-55-3 207-427-3
$C_{24}H_{42}O_{22}$
Glucopyranoside, β-D-fructofuranosyl O-α-D-galactopyranosyl-(1→6)-O-α-D-galactopyranosyl-(1→6)-, α-D-; α-D-Glucopyranoside, β-D-fructofuranosyl O-α-D-galactopyranosyl-(1→6)-O-α-D-galactopyranosyl-(1→6)-. Listed on TSCA. Chiral intermediate. mp = 110°; $[\alpha]_D^{20}$ = + 131.3° (c = 4, $H_2O$). *Pfanstiehl Labs Inc.*

**4424 (3S,4S)-(-)-Statine**
49642-07-1

$C_8H_{17}NO_3$
Chiral intermediate. *Acros Organics NV.*

**4425 N-Stearoyl-D-erythro-spinghosine**
2304-81-6

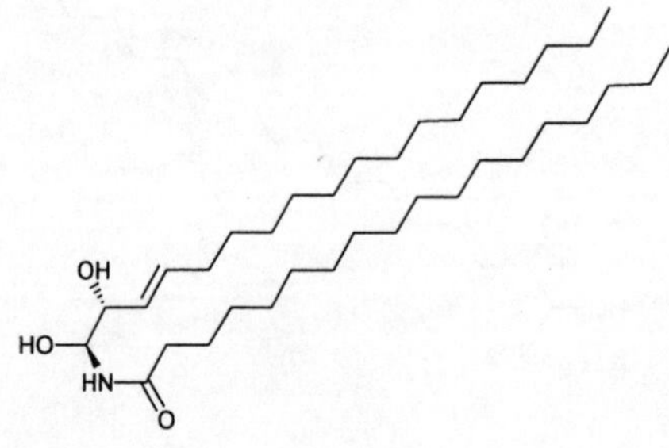


$C_{36}H_{71}NO_3$
Chiral intermediate. *Acros Organics NV.*

**4426 Stigmasterol**
83-48-7 8969(12) 201-482-7

$C_{29}H_{48}O$
Stigmasta-5,22-dien-3-ol, (3β,22E)-; Stigmasterin; 24-Ethyl-5,22-cholestadien-3β-ol; (24S)-5,22-Stigmastadien-3β-ol. Pharmaceutical or derivative. mp = 165-167°; $[\alpha]^{22}$ = - 51° (c = 2, $CHCl_3$). *Acros Organics NV; Sigma-Aldrich Fine Chem.*

**4427 l-Strychnine**
57-24-9 200-319-7

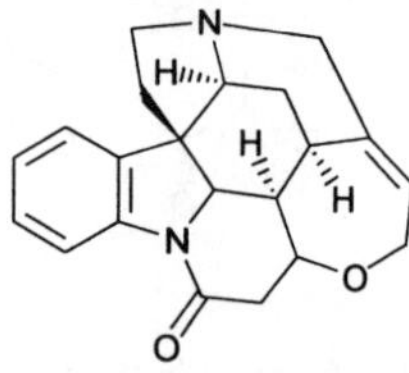


$C_{21}H_{22}N_2O_2$
Strychnidin-10-one; Listed on TSCA. Chiral intermediate. *Acros Organics NV.*

**4428 (R)-Styrene oxide**
20780-53-4

$C_8H_8O$
Oxirane, phenyl-, (R)-; (+)-Phenyloxirane; Phenyl-epoxyethane; (R)-Phenylethylene oxide; Benzene, (epoxyethyl)-, (R)-; Epoxystyrene. Chiral building block. mp = -37°; bp = 194°; $bp_{23}$ = 89-90°; d = 1.0510; $n_D^{20}$ = 1.5350; $[\alpha]_D^{20}$ = + 32 ± 1° (neat). *Acros Organics NV; Austin Chem. Co.; Chirex Inc.; Kaneka Corp.; Lancaster Synthesis Ltd.; Orsynetics (Tradining) Ltd.; PPG - Sipsy; Sigma-Aldrich Fine Chem.*

**4429 (S)-Styrene Oxide**
20780-54-5

$C_8H_8O$
Oxirane, phenyl-, (S)-; (-)-Phenyloxirane; (S)-(-)-

Phenylepoxyethane; (S)-Epoxystyrene; (S)-Phenylethylene oxide; Benzene, (epoxyethyl)-, (S)-. Chiral building block. bp = 192-194°; d = 1.051; n = 1.535; $[\alpha]_D^{20}$ = - 33° (neat). *Chirex Inc.; Kaneka Corp.; Orsynetics (Tradining) Ltd.; PPG - Sipsy; Sigma-Aldrich Fine Chem.*

**4430 (R)-N-Succinimidyl-2-methoxy-2-phenylacetic acid**

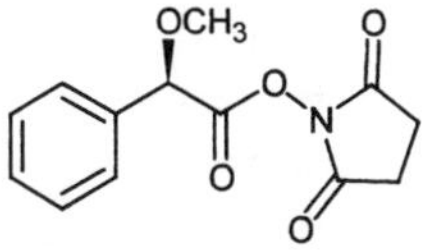

$C_{13}H_{13}NO_5$
Resolving agent. mp = 95-97°; $[\alpha]_D^{20}$ = + 48.0 ± -0.5° (c = 2, 1,2-DCE). *Kiralchem Ltd.*

**4431 (S)-N-Succinimidyl-2-methoxy-2-phenylacetic acid**

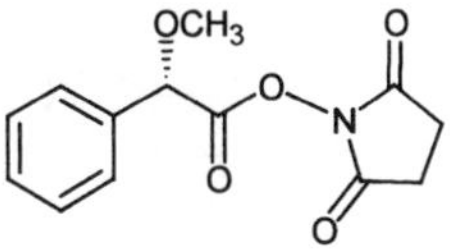

$C_{13}H_{13}NO_5$
Resolving agent. mp = 95-97°; $[\alpha]_D^{20}$ = - 48.0 ± -0.5° (c = 2, 1,2-DCE). *Kiralchem Ltd.*

**4432 D-Sucrose diacetate hexaisobutyrate**
126-13-6 204-771-6

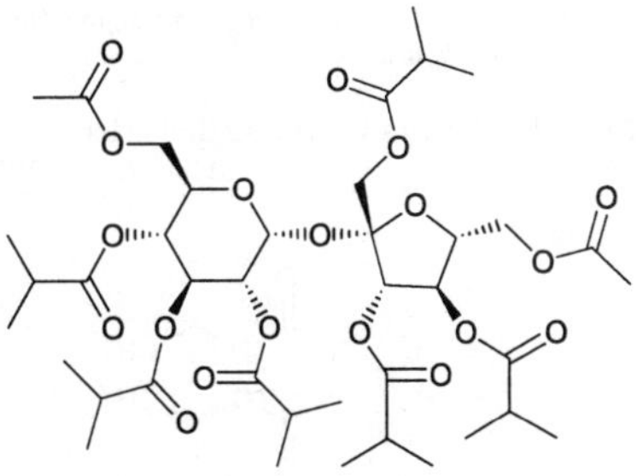

$C_{40}H_{62}O_{19}$
Saccharose acetate isobutyrate; SAIB; Sucrose acetate isobutyrate; α-D-Glucopyranoside, 6-O-Acetyl-1,3,4-tris-O-(2-methyl-1-oxopropyl)-β-D-fructofuranosyl, 6-acetate 2,3,4-tris(2-methylpropanoate); 6-O-Acetyl-1,3,4-tris-O-(2-methyl-1-oxopropyl)-β-D-fructofuranosyl α-D-glucopyran. Chiral intermediate. d = 1.15. *Acros Organics NV.*

**4433 (S)-(-)-Sulpuride**
23672-07-3 9163(12)

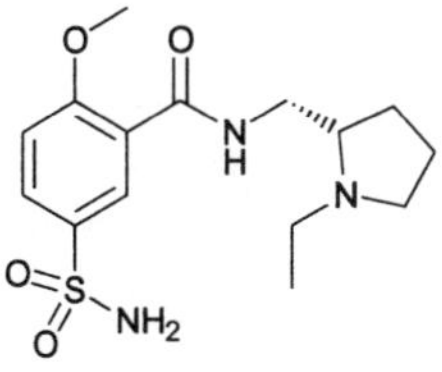

$C_{15}H_{23}N_3O_4S$
(S)-(-)-5-(Aminosulfonyl)-N-[(1-ethyl-2-pyrrolidinyl)-methyl]-2-methoxybenzamide. Chiral intermediate. mp = 183-186°; $[\alpha]_D^{20}$ = - 66° (c = 2, DMF). *Sigma-Aldrich Fine Chem.*

**4434 D-Tagatose**
87-81-0 9202(12) 201-772-3

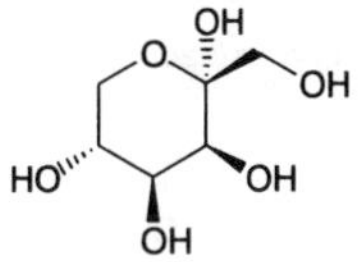

$C_6H_{12}O_6$
Tagatose, D-; Chiral building block. mp = 130-133°; $[\alpha]_D^{20}$ = - 5° (c = 1, $H_2O$). *Acros Organics NV; See Chem. AG; Sigma-Aldrich Fine Chem.; TCI America.*

**4435 D-Talitol**
643-03-8
$C_6H_{14}O_6$
Chiral building block. mp = 88°. *TCI America.*

**4436 D-(+)-Talose**
2595-98-4 219-996-5

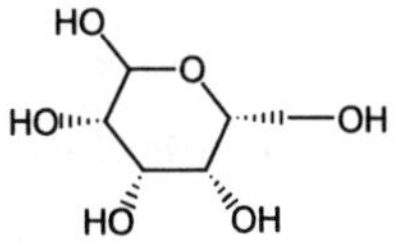

$C_6H_{12}O_6$
Chiral building block. mp = 133-135°; $[\alpha]_D^{20}$ = + 19° (c = 1, $H_2O$). *Sigma-Aldrich Fine Chem.*

**4437 L-(-)-Talose**
23567-25-1 245-744-9

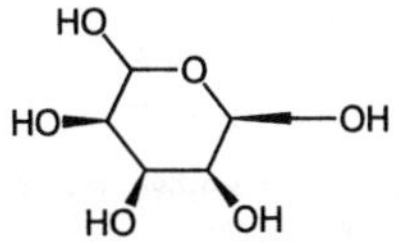

$C_6H_{12}O_6$
Talose, L-; Chiral building block. [α] = - 21° (c = 2 $H_2O$). *Acros Organics NV.*

**4438 D-(-)-Tartaric acid**
147-71-7 9235(12) 205-696-6

$C_4H_6O_6$
Butanedioic acid, 2,3-dihydroxy-, (2S,3S)-; D-(-)-Dihydroxysuccinic acid; [S-(R*,R*)]-2,3-Dihydroxybutanedioic acid. Listed on TSCA. Chiral building block. mp = 172-174°; [α] = - 11.6° (c = 20, $H_2O$). *Acros Organics NV; Austin Chem. Co.; CU Chemie Uetikon; Norse Labs; Shanghai DSL Intl. trading Co.; Sigma-Aldrich Fine Chem.; Sinochem Ningbo; TCI America; Toray International, Inc.; Yoneyama Yakuhin Kogyo Co. Ltd.*

**4439 L-(+)-Tartaric acid**
87-69-4 9237(12) 201-766-0

$C_4H_6O_6$
Butanedioic acid, 2,3-dihydroxy- (2R,3R)-; D-α,β-Dihydroxysuccinic acid; Dextrotartaric acid; E 334; Threaric acid; (+)-2,3-Dihydroxybutanedioic acid; (+)-1,2-Dihydroxyethane-1,2-dicarboxylic acid. Listed on TSCA. Chiral building block. mp = 171-174°; $[\alpha]_D^{20}$ = + 12° (c = 20, $H_2O$). *Aceto Corp.; Acros Organics NV; Richman Chemical Inc.; Shanghai DSL Intl. trading Co.; Sigma-Aldrich Fine Chem.; TCI America.*

**4440 L-(+)-Tartaric acid diammonium salt**
3164-29-2 238-245-2

$C_4H_{12}N_2O_6$
Butanedioic acid, 2,3-dihydroxy- (2R,3R)-, ammonium salt; Ammonium tartrate; 2,3-Dihydroxybutanedioic acid, Diammonium salt; Diammonium tartrate. Listed on TSCA. Chiral ligand; Chiral building block. d = 1.601; $[\alpha]^{15}$ = + 32.4° (c = 1.84, $H_2O$). *Acros Organics NV; Sigma-Aldrich Fine Chem.*

**4441 L-(-)-Tartaric acid dibenzyl ester**
622-00-4
$C_{18}H_{18}O_6$
L-(-)-Tartaric acid-O,O'-dibenzoyl ester. *Austin Chem. Co.; CU Chemie Uetikon; TCI America.*

**4442 L-(+)-Tartaric acid di-n-butyl ester**
87-92-3 201-784-9

$C_{12}H_{22}O_6$
Butanedioic acid, 2,3-dihydroxy- (2R,3R)-, dibutyl ester; (+)-Tartaric acid dibutyl ester; (2R,3R)-Di-n-butyl tartrate; ENT 396. Listed on TSCA. Chiral building block; Chiral ligand. d = 1.09. *Acros Organics NV; TCI America.*

**4443 D-(-)-Tartaric acid diethyl ester**
13811-71-7 237-458-8

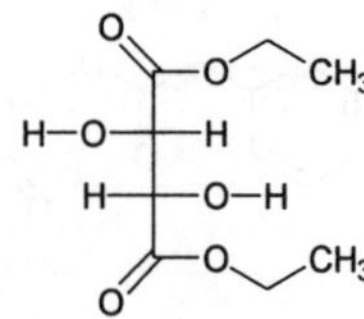

$C_8H_{14}O_6$
Butanedioic acid, 2,3-dihydroxy-, diethyl ester, [S-(R*,R*)]-; Diethyl D-(-)-tartrate; (2S,3S)-(-)-Dihydroxybutane-1,4-dioic acid diethyl ester; (-)-Diethyl-2,3-dihydroxysuccinate. Chiral ligand; Chiral building block. mp = 17°; bp = 280°; $bp_{19}$ = 162°; d = 1.21; n = 1.446; [α] = - 8.5° (neat). *Acros Organics NV; KingChem Inc.; Shanghai DSL Intl. trading Co.; Sigma-Aldrich Fine Chem.; TCI America.*

**4444 L-(+)-Tartaric acid diethyl ester**
87-91-2 3900(12) 201-783-3

$C_8H_{14}O_6$
Butanedioic acid, 2,3-dihydroxy- (2R,3R)-, diethyl ester; Diethyl L-(+)-tartrate; Diethyl 1,2-dihydroxy-1,2-ethanedicarboxylate. Listed on TSCA. Chiral ligand; Chiral building block. bp = 280°; d = 1.204; n = 1.446; [α] = + 9° (neat). *Acros Organics NV; CU Chemie Uetikon; Norse Labs; Sigma-Aldrich Fine Chem.; TCI America.*

**4445 D-(-)-Tartaric acid diisopropyl ester**
62961-64-2 263-771-4

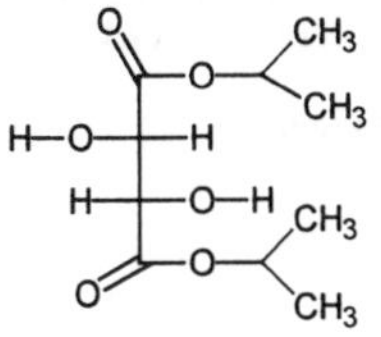

$C_{10}H_{18}O_6$
Butanedioic acid, 2,3-dihydroxy-, bis(1-methylethyl) ester, [S-(R*,R*)]-; [S-(R*,R*)]-2,3-Dihydroxybutanedioic acid, bis(1-methylethyl) ester; Di-isopropyl D-(-)-tartrate; Diisopropyl D-(-)-Tartrate. Listed on TSCA. Chiral ligand; Chiral building block. d = 1.12; $[\alpha]^{23}$ = -16.8° (neat). *Acros Organics NV; Austin Chem. Co.; CU Chemie Uetikon; Sigma-Aldrich Fine Chem.; TCI America; Toray International, Inc.*

**4446 L-(+)-Tartaric acid diisopropyl ester**
2217-15-4 218-709-0

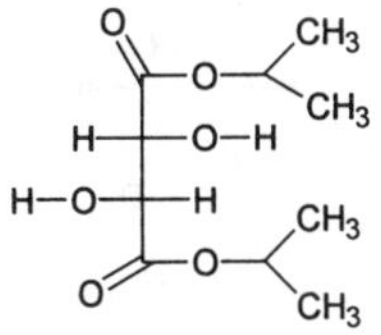

$C_{10}H_{18}O_6$
Butanedioic acid, 2,3-dihydroxy- (2R,3R)-, bis(1-methylethyl) ester; Diisopropyl L-(+)-tartrate; Bis(1-methylethyl) [R-(R*,R*)]-2,3-dihydroxybutanedioate. Chiral ligand; Chiral building block. bp = 275°; $bp_{12}$ = 152°; d = 1.12; n = 1.439; [α] = + 17° (neat). *Acros Organics NV; CU Chemie Uetikon; Sigma-Aldrich Fine Chem.; TCI America.*

**4447 D-(-)-Tartaric acid dimethyl ester**
13171-64-7 236-118-6

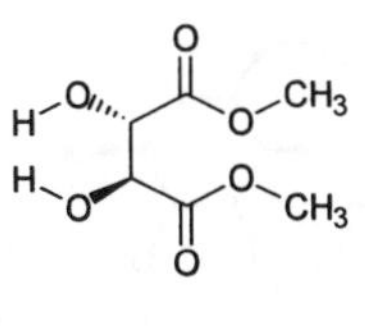

$C_6H_{10}O_6$
Dimethyl tartrate, (S,S)-; Dimethyl D-(-)-tartrate; Butanedioic acid, 2,3-dihydroxy-, dimethyl ester, (2S,3S)-. Chiral ligand; Chiral building block. mp = 60°; $bp_{15}$ = 158; [α] = - 21° (c = 2.5, $H_2O$). *Acros Organics NV; Austin Chem. Co.; Norse Labs; Shanghai DSL Intl. trading Co.; Sigma-Aldrich Fine Chem.; TCI America.*

**4448 L-(+)-Tartaric acid dimethyl ester**
608-68-4 210-166-8

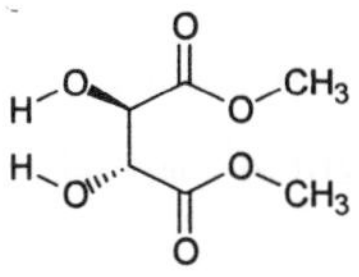

$C_6H_{10}O_6$
Butanedioic acid, 2,3-dihydroxy- (2R,3R)-, dimethyl ester; Dimethyl L-(+)-tartrate. Listed on TSCA. Chiral ligand; Chiral building block. mp = 62°; $bp_{23}$ = 163°; [α] = + 21° (c = 2.5, $H_2O$). *Acros Organics NV; Austin Chem. Co.; CU Chemie Uetikon; Shanghai DSL Intl. trading Co.; Sigma-Aldrich Fine Chem.; TCI America.*

**4449 L-(+)-Tartaric acid disodium salt dihydrate**
6106-24-7

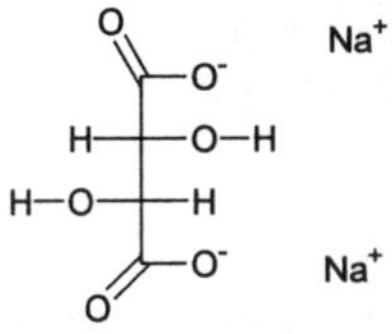

$C_4H_4Na_2O_6$
Butanedioic acid, 2,3-dihydroxy- (2R,3R)-, disodium salt, dihydrate; Disodium tartrate dihydrate; Sodium tartrate dihydrate. Chiral building block; Chiral ligand. $[\alpha]_D^{20}$ = + 26° (c = 1, $H_2O$). *Acros Organics NV; Sigma-Aldrich Fine Chem.*

**4450 D-(+)-Tartaric acid-O,O'-ditoluoyl ester**

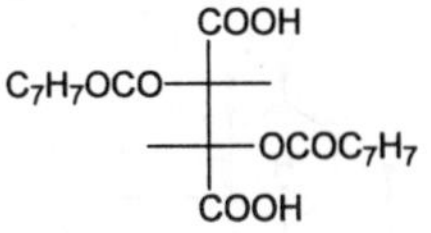

$C_{20}H_{18}O_8$
(+)-Di-p-toluoyl-D-tartaric acid; (+)-Di-O,O'-ditoluoyl-D-tartaric acid; Tartaric acid, di-p-toluate, (+)-. Chiral ligand; Chiral building block. *Austin Chem. Co.; CU Chemie Uetikon.*

**4451 L-(-)-Tartaric acid-O,O'-ditoluoyl ester**

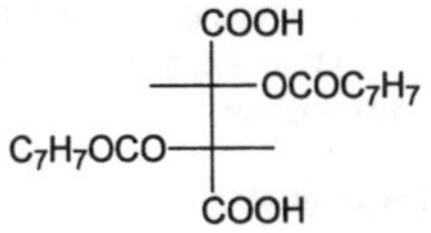

$C_{20}H_{18}O_8$
(-)-Di-p-toluoyl-L-tartaric acid; Tartaric acid, di-p-toluate, (-)-; (-)-Di-O,O'-ditoluoyl-L-tartaric acid. Chiral ligand; Chiral building block. *Acros Organics NV; Austin Chem.*

*Co.; CU Chemie Uetikon; D & O Group; Shanghai DSL Intl. trading Co.; Toray International, Inc.*

**4452 Tartranil**

$C_{16}H_{14}N_2O_4$
Chiral intermediate. *Norse Labs.*

**4453 (2R,3R)-Tartranilic acid**

3019-58-7
$C_{10}H_{11}NO_5$
(2R,3R)-2,3-Dihydroxy-3-(phenylcarbamoyl)propionic acid. Chiral intermediate; Resolving agent. *TCI America.*

**4454 Taurocholic acid, sodium salt hydrate**

145-42-6 9044(11) 205-653-7

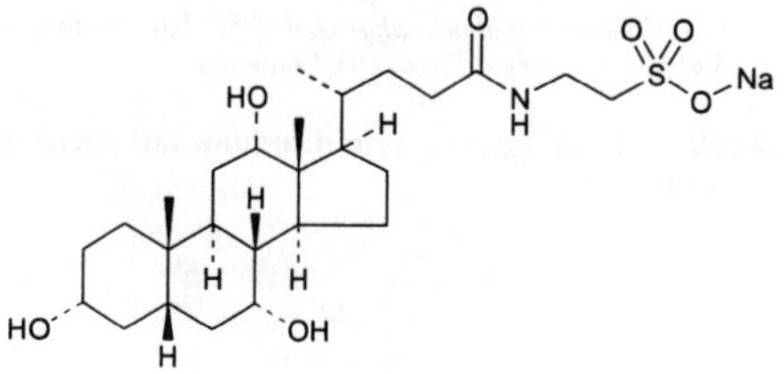

$C_{26}H_{44}NNaO_7S$
Ethanesulfonic acid, 2-[[(3α,5β,7α,12α)-3,7,12-trihydroxy-24-oxocholan-24-yl]amino]-, monosodium salt; Sodium taurocholate; Sodium N-choloyltaurinate; Taurine, N-choloyl-, monosodium salt. Listed on TSCA. Chiral building block. $[\alpha]_D^{20} = +22°$ (c = 3, $H_2O$). *Acros Organics NV; Sigma-Aldrich Fine Chem.*

**4455 Taurodeoxycholic acid sodium salt hydrate**

1180-95-6

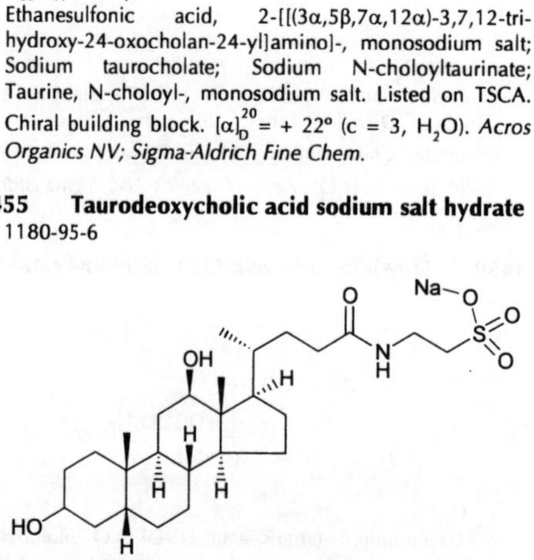

$C_{26}H_{45}NO_6S$
Sodium taurodeoxycholate; Sodium 3α,12α-dihydroxy-5β-cholanoyltaurate; STDC; Taurine, N-(3α,12α-dihydroxy-5β-cholan-24-oyl)-, monosodium salt; sodium 2-[[(3α,5β,12α)-3,12-dihydroxy-24-oxocholan-24-yl]-amino]ethane-1-sulphonate. Chiral intermediate. mp = 168°; $[\alpha]^{25} = +33°$ (c = 2.5, $H_2O$). *Acros Organics NV; Sigma-Aldrich Fine Chem.*

**4456 (-)-Terpinen-4-ol**

20126-76-5
$C_{10}H_{18}O$
Cyclohex-3-en-1-ol, 4-methyl-1-(1-methylethyl)-, (R)-; p-Menth-1-en-4-ol, (R)-. Chiral intermediate. $[\alpha] = -32°$ (neat). *Acros Organics NV.*

**4457 (+)-Terpinen-4-ol**

$C_{10}H_{18}O$
Cyclohex-3-en-1-ol, 4-methyl-1-(1-methylethyl)-, (S)-; p-Menth-1-en-4-ol, (S)-. (±) 562-74-3. Chiral intermediate. $[\alpha] = +29°$. *Acros Organics NV.*

**4458 L-Terpineol**

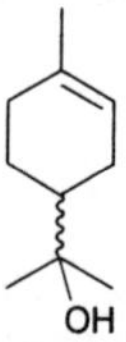

$C_{10}H_{18}O$
Chiral intermediate. *Taiwan Tekho Camphor Co. Ltd.*

**4459 (-)-Tetra-O-acetyl-6,6-di-O-(tert-butyl-dimethylsilyl)lactal**

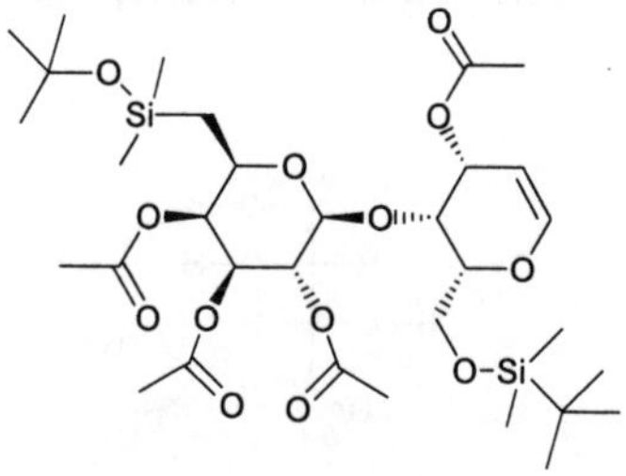

$C_{32}H_{56}O_{13}Si_2$
Chiral intermediate. $[\alpha]_D^{20} = -7°$ (c = 1.1, $CHCl_3$). *Sigma-Aldrich Fine Chem.*

**4460 (-)-Tetra-O-acetyl-6,6-di-O-(tert-butyl-diphenylsilyl)lactal**

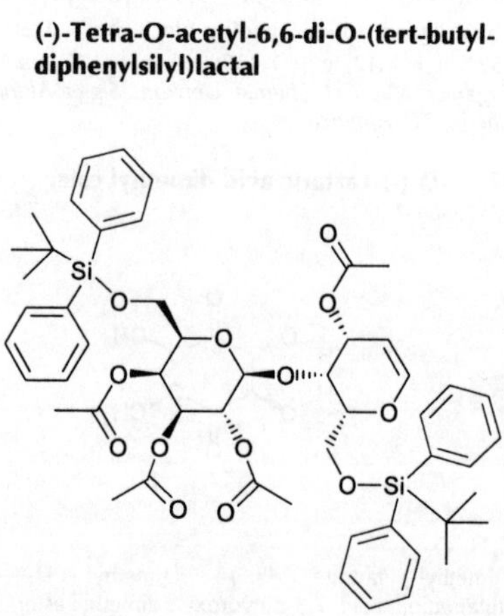

$C_{52}H_{64}O_{13}Si_2$
Chiral intermediate. mp = 67-70°; $[\alpha]_D^{20} = -20°$ (c = 1, $CHCl_3$). *Sigma-Aldrich Fine Chem.*

**4461 2,3,4,6-Tetra-O-acetyl-β-D-galacto-pyranosyl(1→4)-2-azido**
Chiral intermediate. *See Chem. AG.*

**4462 (+)-1,2,3,4-Tetra-O-acetylglucopyranose**
13100-46-4

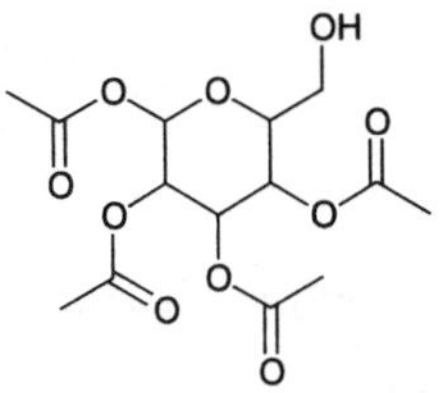

$C_{14}H_{20}O_{10}$
Chiral intermediate. mp = 126-128°; $[\alpha]_D^{20}$ = + 11° (c = 6, $CHCl_3$). *See Chem. AG; Sigma-Aldrich Fine Chem.*

**4463 2,3,4,6-Tetra-O-acetyl-β-D-glucopyranose**

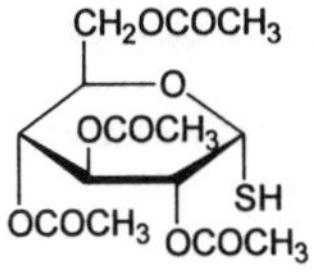

$C_{14}H_{20}O_9S$
Chiral intermediate. *See Chem. AG.*

**4464 2,3,4,6-Tetra-O-acetyl-α-D-gluco-pyranosylazide**
20369-61-3

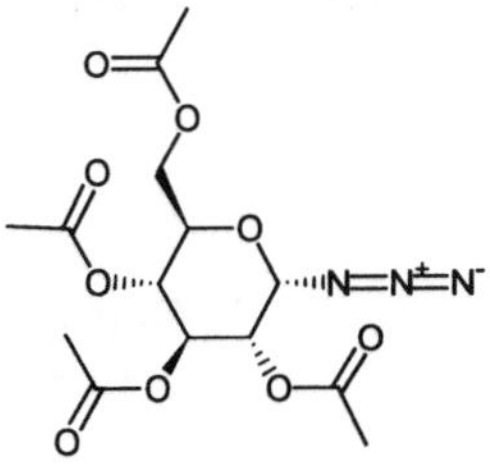

$C_{14}H_{19}N_3O_9$
Chiral intermediate. [α] = + 156° (c = 1, $CHCl_3$). *Acros Organics NV.*

**4465 2,3,4,6-Tetra-O-acetyl-β-D-gluco-pyranosylazide**
13992-25-1

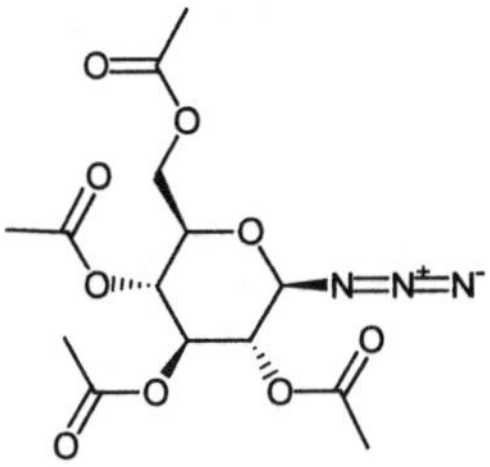

$C_{14}H_{19}N_3O_9$
Chiral intermediate. mp = 127-129°. *Acros Organics NV.*

**4466 2,3,4,6-Tetra-O-acetylglucopyranosyl isothiocyanate**
14152-97-7

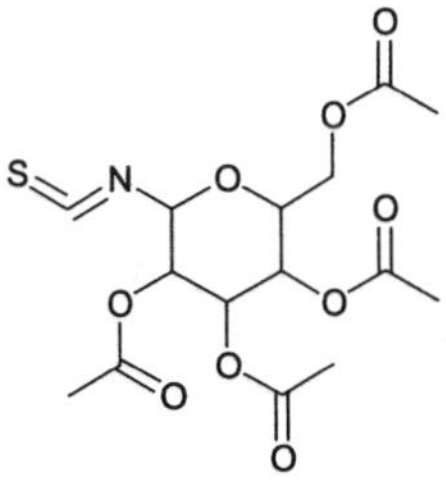

$C_{15}H_{19}NO_9S$
Chiral intermediate. mp = 114-116°. *Sigma-Aldrich Fine Chem.*

**4467 2-(2,3,4,6-Tetra-O-acetyl-β-D-gluco-pyranosyl)-2-thiopseudourea hydrobromide**
40591-65-9 254-989-0
$C_{15}H_{23}BrN_2O_9S$
Glucopyranose, 1-thio-, 2,3,4,6-tetraacetate 1-carbamimidate, monohydrobromide, β-D-; 2,3,4,6-Tetra-O-acetyl-1-carbamimidoyl-1-thio-β D-glucopyranose monohydrobromide; 2,3,4,6-tetra-O-acetyl-β-D-gluco-pyranosyl carbamimidothioate hydrobromide; Pseudourea 2-β-D-glucopyranosyl-2-thio-, tetraacetate, hydrobromide; S-(2,3,4,6-Tetra-O-acet. Chiral intermediate. *Pfanstiehl Labs Inc.*

**4468 1,3,4,6-Tetra-O-acetyl-β-D-glucosamine hydrochloride**
10034-20-5
$C_{14}H_{22}NO_9Cl$
Chiral intermediate. *Pfanstiehl Labs Inc.*

**4469 (-)-1,3,4,6-Tetra-O-acetylmannopyrannose**
18968-05-3

$C_{14}H_{20}O_{10}$
Chiral intermediate. mp = 160-161°; $[\alpha]^{19}$ = - 68° (c = 0.7, pyridine). *Sigma-Aldrich Fine Chem.*

**4470 1,3,4,6-Tetra-O-acetyl-2-(p-methoxybenzyl)imino-2-deoxy-β-D-glucopyranose**
7597-81-1
$C_{20}H_{26}NO_9$
Chiral intermediate. *Pfanstiehl Labs Inc.*

**4471 Tetra-O-acetyl-β-D-ribofuranose**
13035-61-5 235-898-5

$C_{13}H_{18}O_9$
β-D-Ribofuranose 1,2,3,5-tetraacetate; 1,2,3,5-Tetra-O-acetyl-D-ribofuranose; 1,2,3,5-tetraacetyl-β-D-ribose. Chiral intermediate. mp = 82°; $[\alpha]_D^{26}$ = - 14.6° (c = 5, $CH_3OH$); $[\alpha]_D^{20}$ = - 11.4° (c = 10, $CHCl_3$). *Interchem Corp.; Pfanstiehl Labs Inc.; See Chem. AG; Sigma-Aldrich Fine Chem.*

**4472 Tetraacetyl-D-ribose**
28708-32-9 249-174-1

$C_{13}H_{18}O_9$
Ribofuranose, tetraacetate, D-; tetra-O-Acetyl-D-ribofuranose. Chiral intermediate. *Boehringer Ingelheim Pharma KG.*

**4473 Tetra-O-acetyl-6-O-tosyl-β-D-glucopyranose**
6619-10-9

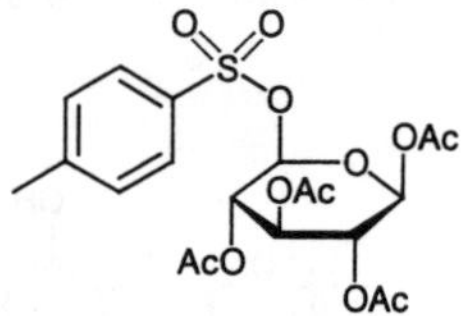

$C_{21}H_{26}O_{12}S$
Chiral intermediate. *Pfanstiehl Labs Inc.*

**4474 2,3,4,6-Tetra-O-acetyl-β-D-thiogalactopyranose**

$C_{14}H_{20}O_9S$
Chiral intermediate. *See Chem. AG.*

**4475 2,3,4,6-Tetra-O-acetyl-β-D-thioglucopyranose**

$C_{14}H_{20}O_9S$
Chiral intermediate. *See Chem. AG.*

**4476 (-)-1,3,4,6-Tetra-O-acetyl-2-O-trifluoromethanesulfonylmannopyranose**
92051-23-5

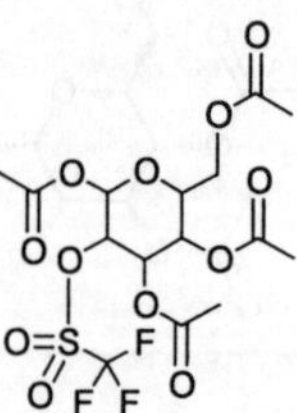

$C_{15}H_{19}F_3O_{12}S$
Chiral intermediate. $[\alpha]_D^{20}$ = - 16° (c = 1, $CHCl_3$). *Sigma-Aldrich Fine Chem.*

**4477 1,2,3,5-Tetra-O-acetyl-D-xylofuranose**
30571-56-3
$C_{13}H_{18}O_9$
Chiral intermediate. *Pfanstiehl Labs Inc.*

**4478 (+)-2,3,4,6-Tetra-O-benzoylglucopyranosyl isothiocyante**
132413-50-4

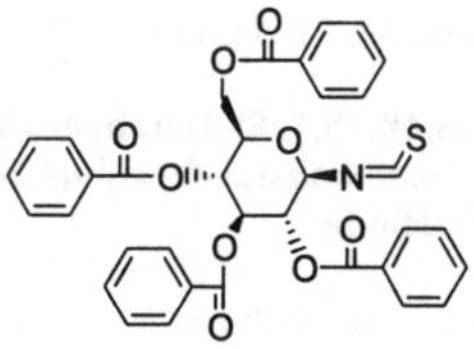

$C_{35}H_{27}NO_9S$
Chiral intermediate. mp = 152-154°; $[\alpha]^{25}$ = + 29° (c = 1, $CHCl_3$). *Sigma-Aldrich Fine Chem.*

**4479 Tetra-O-benzoyl-D-xylofuranose**
5432-87-1

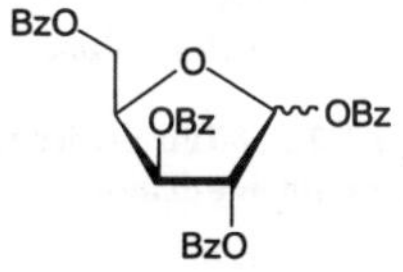

$C_{33}H_{34}O_5$
Chiral intermediate. *Pfanstiehl Labs Inc.*

**4480 2,3,4,6-Tetra-O-benzyl-D-galactopyranose**
53081-25-7

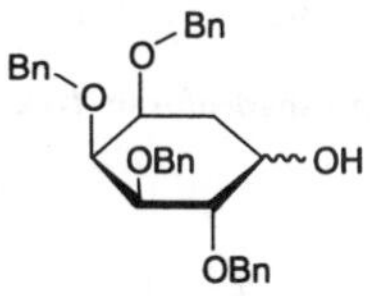

$C_{34}H_{36}O_6$
Chiral intermediate. *Pfanstiehl Labs Inc.; SECO company.*

**4481 2,3,4,6-Tetra-O-benzyl-D-glucopyranose**
6564-72-3
$C_{34}H_{36}O_6$
Chiral intermediate. mp = 151°; $[\alpha]_D^{20}$ = + 21.7° (c = 2, $CHCL_3$). *Pfanstiehl Labs Inc.; See Chem. AG.*

**4482 (+)-2,3,4,6-Tetra-O-benzylglucopyranose (mixture of anomers)**
4132-28-9

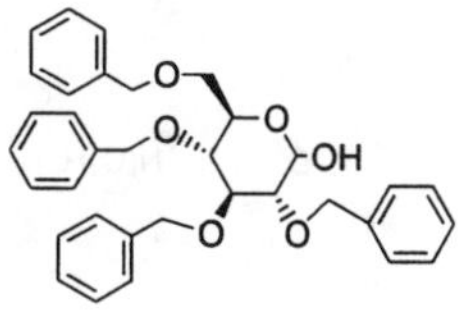

$C_{34}H_{36}O_6$
Chiral intermediate. mp = 145-149°; $[\alpha]_D^{20}$ = + 18.5° (c = 2, $CHCl_3$). *Sigma-Aldrich Fine Chem.*

**4483 2,3,4,6-Tetra-O-benzyl-D-manno-pyranoside**

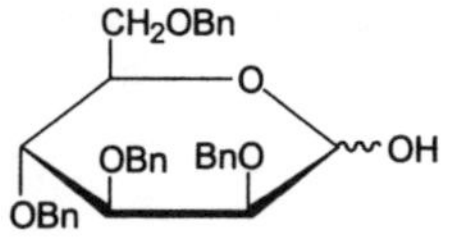

$C_{34}H_{36}O_6$
Chiral intermediate. *Pfanstiehl Labs Inc.; See Chem. AG.*

**4484 (+)-2,3,4,6-Tetra-O-pivaloylgalacto-pyranosylamine**
108342-87-6

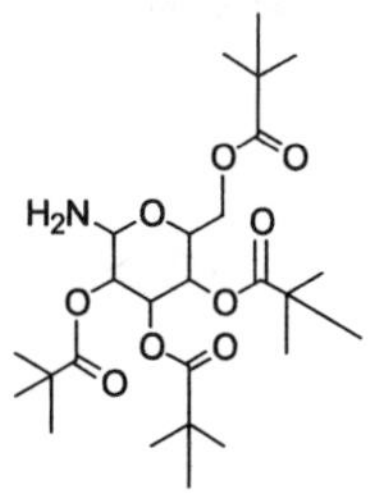

$C_{26}H_{45}NO_9$
Chiral intermediate. mp = 92-95°; $[\alpha]_D^{20}$ = + 10.4° (c = 2, $CHCl_3$). *Sigma-Aldrich Fine Chem.*

**4485 (+)-2,3,4,6-Tetra-O-benzylglucopyranosyl fluoride**

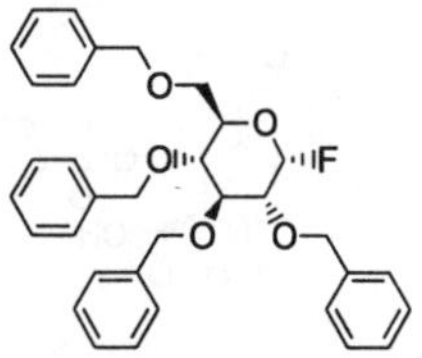

$C_{34}H_{35}FO_5$
Chiral intermediate. mp = 50-60°; $[\alpha]^{24}$ = + 1.1° (c = 1, $CHCl_3$). *Sigma-Aldrich Fine Chem.*

**4486 (-)-Tetrabutylammoniumlactate (70 wt % solution in $H_2O$)**

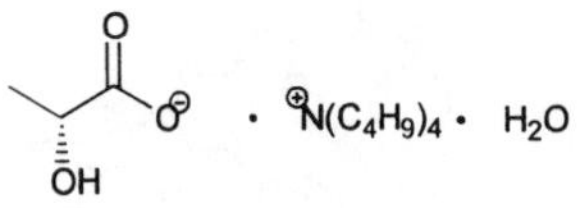

$C_{19}H_{41}NO_3$
Chiral auxiliary. d = 0.987; n = 1.438; $[\alpha]_D^{20}$ = - 4.6° (neat). *Sigma-Aldrich Fine Chem.*

**4487 (R)-Tetradecanediol**
153062-86-3

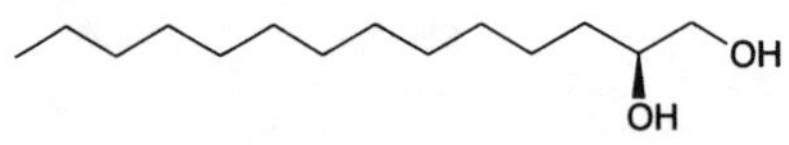

$C_{14}H_{30}O_2$
Chiral building block. *Kankyo Kagaku Center Co., Ltd.*

**4488 (S)-1,2-Tetradecanediol**
153062-87-4
$C_{14}H_{30}O_2$
Chiral building block. *Kankyo Kagaku Center Co., Ltd.*

**4489 (S)-4-Tetradecanol**

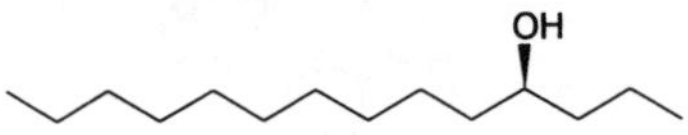

$C_{14}H_{30}O$
tetradecan-4-ol, (S)-.
(S)-4-Hydroxytetradecane. Chiral building block. *Japan Energy Corp.*

**4490 (S)-(-)-1,2,3,4-Tetrahydro-6,7-dimethoxy-3-isoquinolinecarboxylic acid p-toluenesulfonic acid salt**

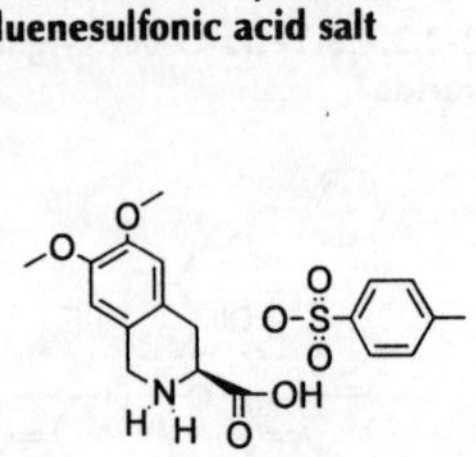

$C_{12}H_{15}NO_4$
Chiral intermediate. mp = 182-188°; $[\alpha]^{22}$ = - 63° (c = 1, 1N HCl). *Sigma-Aldrich Fine Chem.*

**4491 (-)-3aS-(3a,4,5,7a)-3a,4,5,7a-Tetrahydro-2,2-dimethyl-1,3-benzodioxole-4,5-diol**

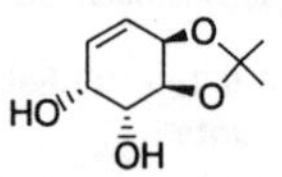

$C_9H_{14}O_4$
Chiral intermediate. mp = 68-71°; $[\alpha]_D^{20}$ = - 158° (c = 1, $CHCl_3$). *Sigma-Aldrich Fine Chem.*

**4492 (3aS,4R,7S,7aR)-Tetrahydro-2,2-dimethyl-4,7-methano-1,3-dioxolo[4.5c]pyridin-6(3aH)one**
$C_9H_{1314}O_3$
Chiral intermediate. *Celltech Chiroscience Ltd.*

**4493 (-)-3aR-(3a,5a,6a,6b)-3la,5a,6a,6b-Tetrahydro-2,2-dimethyloxirene-1,3-benzodioxole**

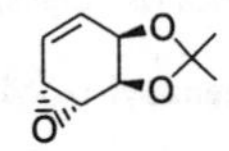

$C_9H_{12}O_3$
Chiral intermediate. $bp_{1.5}$ = 79°; d = 1.101; n = 1.476; $[\alpha]^{18}$ = - 56° (c = 1.3, $CHCl_3$). *Sigma-Aldrich Fine Chem.*

**4494 (S)-(-)-1,2,3,4-Tetrahydrodiphenyl-3-isoquinolinemethanol**

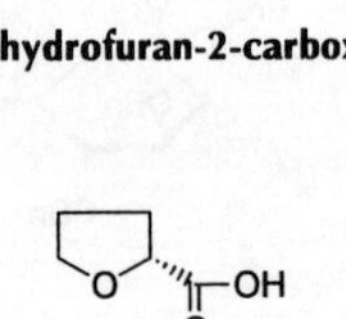

$C_{22}H_{21}NO$
Chiral intermediate. mp = 104-107°; $[\alpha]^{22}$ = - 100° (c = 1, $CH_3OH$). *Sigma-Aldrich Fine Chem.*

**4495 (R)-Tetrahydrofuran-2-carboxylic acid**
87392-05-0

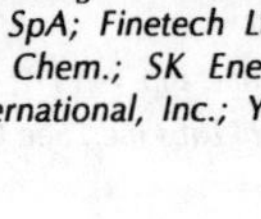

$C_5H_8O_3$
2-Tetrahydrofuroic acid. Chiral building block. $bp_{10}$ = 110-112°; d = 1.170; $[\alpha]_D^{20}$ = + 19.5 ± 0.2° (neat). *Austin Chem. Co.; Chemi SpA; Finetech Ltd.; Kiralchem Ltd.; Sigma-Aldrich Fine Chem.; SK Energy & Chem.; TCI America; Toray International, Inc.; Yamakawa Chemical Indsutry Co. Ltd.*

**4496 (S)-(-)-Tetrahydrofuran-2-carboxylic acid**

87392-07-2

$C_5H_8O_3$

(S)-(-)-2-Tetrahydrofuroic acid. Chiral building block. bp = 244-251°; $bp_{10}$ = 110-112°; d = 1.170; n = 1.46; $[\alpha]_D^{20}$ = - 3° (1, $CH_3OH$); $[\alpha]_D^{20}$ = - 19.5 ± 0.2° (neat). *Austin Chem. Co.; Finetech Ltd.; Kiralchem Ltd.; Richman Chemical Inc.; Sigma-Aldrich Fine Chem.; SK Energy & Chem.; TCI America; Vinchem; Yamakawa Chemical Indsutry Co. Ltd.*

**4497 (R)-Tetrahydrofurfuryl alcohol**

22415-59-4

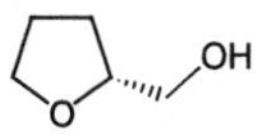

$C_5H_{10}O_2$

Furan-2-methanol, tetrahydro-, (R)-; (R)-Tetrahydro-2-furanmethanol. Chiral building block. bp = 177-178°; d = 1.054; $[\alpha]_D^{20}$ = - 2.3 ± 0.1° (neat). *Kiralchem Ltd.*

**4498 (S)-Tetrahydrofurfuryl alcohol**

57203-01-7

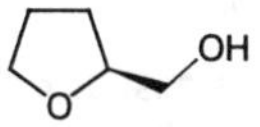

$C_5H_{10}O_2$

Furan-2-methanol, tetrahydro-, (S)-; (S)-Tetrahydro-2-furanmethanol. Chiral building block. bp = 177-178°; d = 1.054; $[\alpha]_D^{20}$ = + 2.3 ± 0.1sg (neat). *Kiralchem Ltd.*

**4499 (R)-(-)-Tetrahydrofurfurylamine**

7202-43-9

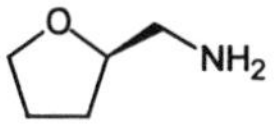

$C_5H_{11}NO$

(R)-(-)-2-(Aminomethyl)tetrahydrofuran. Chiral building block. bp = 152-153°; d = 0.9800; $n_D^{20}$ = 1.4550; $[\alpha]_D^{20}$ = - 12 ± 1° (c = 2, $CHCl_3$). *Kiralchem Ltd.; Lancaster Synthesis Ltd.; Sigma-Aldrich Fine Chem.*

**4500 (S)-(+)-Tetrahydrofurfurylamine**

7175-81-7

$C_5H_{11}NO$

(S)-(+)-2-(Aminomethyl)tetrahydrofuran. Chiral building block. bp = 152-153°; d = 0.9800; $n_D^{20}$ = 1.4550; $[\alpha]_D^{20}$ = + 12 ± 0.5° (c = 2, $CHCl_3$); $[\alpha]_D^{20}$ = + 14.3 ± 0.2° (c = 4, $CH_3OH$). *Kiralchem Ltd.; Lancaster Synthesis Ltd.; Norse Labs; Sigma-Aldrich Fine Chem.*

**4501 (-)-3,3a,8,8a-Tetrahydro-2H-indeno-1,2-doxazol-2-one**

135969-64-1

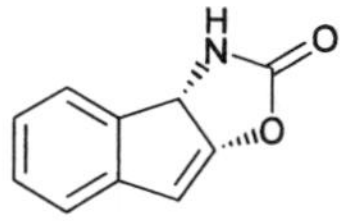

$C_{10}H_9NO_2$

Chiral intermediate. mp = 203-209°; $[\alpha]^{22}$ = - 76° (c = 0.65, EtOAc). *Sigma-Aldrich Fine Chem.*

**4502 (+)-3,3a,8,8a-Tetrahydro-2H-indeno-1,2-doxazol-2-one**

135969-65-2

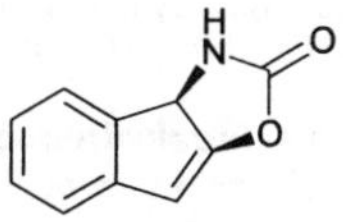

$C_{10}H_9NO_2$

Chiral intermediate. mp = 203-209°; $[\alpha]^{22}$ = + 76° (c = 0.65, EtOAc). *Sigma-Aldrich Fine Chem.*

**4503 1,2,3,4-Tetrahydroisoquinoline-(3R)-carboxylic acid**

103733-65-9

$C_{10}H_{11}NO_2$

Chiral intermediate. *Fine Organics Ltd.; Fischer Chem. AG; Great Lakes Fine Chem.*

**4504 1,2,3,4-Tetrahydroisoquinoline-(3S)-carboxylic acid**

35186-99-3

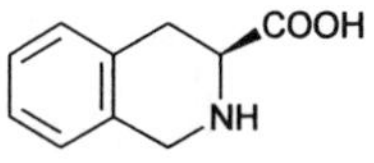

$C_{10}H_{11}NO_2$

Chiral intermediate. *Fine Organics Ltd.; Great Lakes Fine Chem.; Nippon Steel Chem. Co. Ltd.*

**4505 1,2,3,4-Tetrahydroisoquinoline-(3S)-tert-butylcarboxamide**

149057-17-0

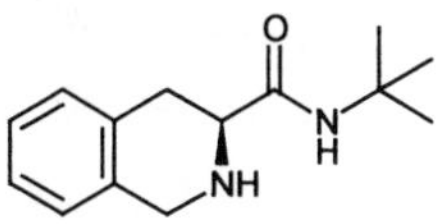

$C_{14}H_{20}N_2O$

Chiral intermediate. *Fischer Chem. AG; Great Lakes Fine Chem.*

**4506 (S)-(-)-1,2,3,4-Tetrahydro-3-isoquinoline-carboxylic acid**
74163-81-8

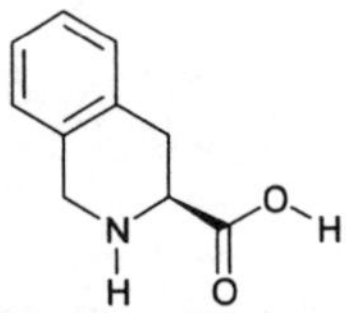

$C_{10}H_{11}NO_2$
(S)-3-Carboxy-1,2,3,4-tetrahydroisoquinoline. Chiral building block. mp = 300°; $[\alpha]_D^{20}$ = - 163 ± 3° (c = 2, 1N NaOH). *Acros Organics NV; Austin Chem. Co.; Fischer Chem. AG; Lancaster Synthesis Ltd.; See Chem. AG; Sigma-Aldrich Fine Chem.; TCI America.*

**4507 D-1,2,3,4-Tetrahydroisoquinoline-3-carboxylic acid hydrochloride**

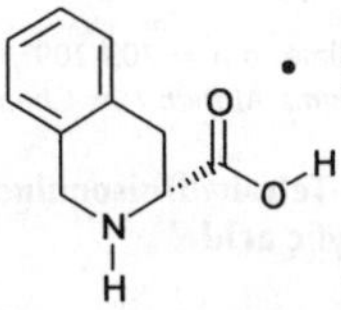

$C_{10}H_{12}ClNO_2$
Chiral intermediate. *Acros Organics NV.*

**4508 L-1,2,3,4-Tetrahydroisoquinoline-3-carboxylic acid hydrochloride**
77497-95-1

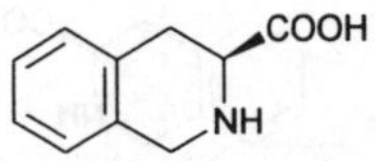

$C_{10}H_{12}ClNO_2$
Chiral intermediate. *Acros Organics NV; Fischer Chem. AG.*

**4509 (S)-(-)-1,2,3,4-Tetrahydro-2,3-isoquinoline-dicarboxylic anhydride**

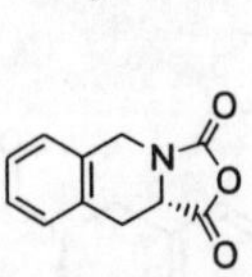

$C_{11}H_9NO_3$
Chiral intermediate. mp = 174-178°; $[\alpha]^{25}$ = - 335° (c = 0.01, $CH_2Cl_2$). *Sigma-Aldrich Fine Chem.*

**4510 (S)-(-)-1,2,3,4-Tetrahydro-3-isoquinoline-methanol**

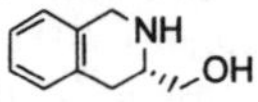

$C_{10}H_{13}NO$
Chiral intermediate. mp = 114-116°; $[\alpha]^{22}$ = - 97° (c = 1, $CH_3OH$). *Acros Organics NV; Sigma-Aldrich Fine Chem.*

**4511 [(3S)-(1,2,3,4-Tetrahydroisoquinolyl)]-N-(tert-butyl)carboxamide**
149182-72-9

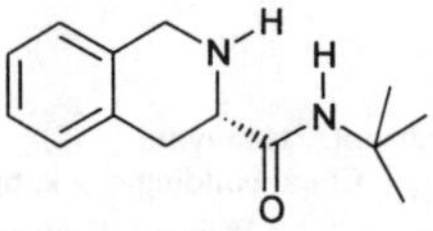

$C_{14}H_{20}N_2O$
(S)-N-tert-Butyl-1,2,3,4-tetrahydroisoquinoline-3-carboxamide. Chiral intermediate. *Acros Organics NV; Sigma-Aldrich Fine Chem.*

**4512 (3aR,4S,7R,7aS)-3a,4,7,7a-Tetrahydro-7-(methoxycarbonylamino)-2,2-dimethyl-1,3-benzodioxol-4-ol 4-acetate**

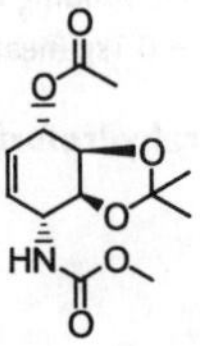

$C_{13}H_{19}NO_6$
Chiral intermediate. mp = 98-101°; $[\alpha]_D^{20}$ = + 25° (c = 1, $CHCl_3$). *Sigma-Aldrich Fine Chem.*

**4513 (+)-(4aR,10bR)-3,4,4a,10b-Tetrahydro-9-methoxy-4-propyl-2H,5H-[1]benzo-pyrano[4,3-b]-1,4-oxazine**
123748-66-3
$C_{15}H_{21}NO_3$
Chiral intermediate. *Ultrafine.*

**4514 (+)-(4aR,10bR)-3,4,4a,10b-Tetrahydro-9-methoxy-4-propyl-2H,5H-[1]benzo-pyrano[4,3-b]-1,4-oxazin-3-one**
123671-97-6
$C_{15}H_{19}NO_4$
Chiral intermediate. *Daicel Chem. Ind. Ltd; Ultrafine.*

**4515 (S)-(+)-2,3,7,7a-Tetrahydro-7a-methyl-1H-indene-1,5(6H)-dione**
17553-86-5

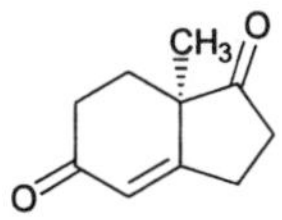

$C_{10}H_{12}O_2$
Chiral intermediate. mp = 54-59°. *Acros Organics NV.*

**4516 (S)-(+)-3,4,8,8a-Tetrahydro-8a-methyl-1,6(2H,7H)-naphthalenedione**
33878-99-8

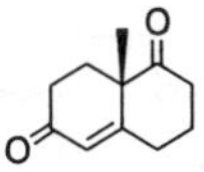

$C_{11}H_{14}O_2$
(S)-Wieland-Miescher ketone; (S)-(+)-9-Methyl-δ-5(10)-octalin-1,6-dione; (S)-Tetrahydro-8a-methyl-1,6(2H,7H)-naphthalenedione. Chiral intermediate. mp = 50-51°; $[\alpha]_D^{20}$ = + 100 ± 3° (c = 1, $C_6H_6$). *Finetech Ltd.; Lancaster Synthesis Ltd.; Sigma-Aldrich Fine Chem.*

**4517 (R)-(-)-1,2,3,4-Tetrahydro-1-naphthol**
23357-45-1

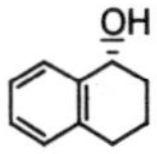

$C_{10}H_{12}O$
Chiral building block . mp = 36-38°; d = 1.09; $[\alpha]_D^{17}$ = - 32.0 ± 0.5° (c = 3,$CHCl_3$). *Kiralchem Ltd.; Sigma-Aldrich Fine Chem.*

**4518 (S)-(+)-1,2,3,4-Tetrahydro-1-naphthol**
53732-47-1

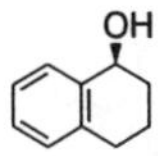

$C_{10}H_{12}O$
Chiral building block. mp = 36-38°; $bp_{1.8}$ = 92°; d = 1.09; $[\alpha]_D^{17}$ = +32.0 ± 0.5° (c = 3,$CHCl_3$). *Kiralchem Ltd.; Sigma-Aldrich Fine Chem.*

**4519 (R)-1,2,3,4-Tetrahydro-1-naphthylamine**
21966-60-9
$C_{10}H_{13}N$
(R)-1-Amino-1,2,3,4-tetrahydronaphthalene; (R)-1-Amino-tetralin. Chiral intermediate. $bp_{714}$ = 246-247°; $bp_8$ = 118-120°; $n_D^{20}$ = 1.5645. *BASF AG ; Chiragene; Lancaster Synthesis Ltd.; Sigma-Aldrich Fine Chem.*

**4520 (S)-1,2,3,4-Tetrahydro-1-naphthylamine**
21880-87-5
$C_{10}H_{13}N$
(S)-1-Amino-1,2,3,4-tetrahydronaphthalene; (S)-1-Amino-tetralin. Chiral intermediate. $bp_{714}$ = 246-247°; $bp_{7.6}$ = 118-120°; $n_D^{20}$ = 1.5645. *BASF AG ; Chiragene; Lancaster Synthesis Ltd.; Sigma-Aldrich Fine Chem.*

**4521 (5R)-3,4,5,6-Tetrahydro-5-phenyl-2H-1,4-oxazin-2-one**

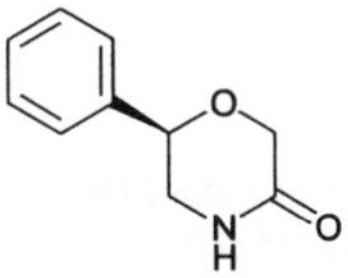

$C_{10}H_{11}NO_2$
Chiral intermediate. *Oxford Asymmetry Ltd.*

**4522 (5S)-3,4,5,6-Tetrahydro-5-phenyl-2H-1,4-oxazin-2-one**

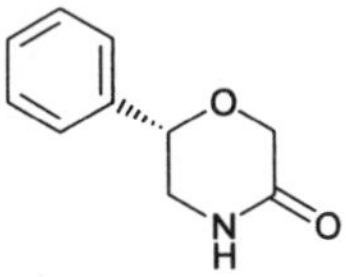

$C_{10}H_{11}NO_2$
Chiral intermediate. *Oxford Asymmetry Ltd.*

**4523 (1S,6S,7R,8R,8aR)-1,6,7,8-Tetrahydroxy-indolizidine**
79831-76-8
$C_8H_{15}NO_4$
Castanospermine. Chiral intermediate. mp = 213-217°; $[\alpha]_D^{20}$ = + 83° (c = 1, $H_2O$). *Sigma-Aldrich Fine Chem.*

**4524 (-)-3,5-O-(1,1,3,3-Tetraisopropyl-1,3-disiloxanediyl)adenosine**
69304-45-6

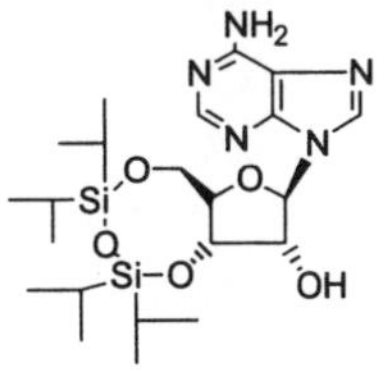

$C_{22}H_{39}N_5O_5Si_2$
Chiral intermediate. mp = 135-139°; $[\alpha]^{21}$ = - 41.6° (c = 1, $CH_3OH$). *Sigma-Aldrich Fine Chem.*

**4525 (+)-3,5-O-(1,1,3,3-Tetraisopropyl-1,3-disiloxanediyl)cytidine**

69304-42-3

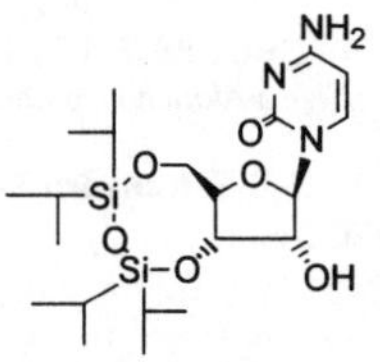

$C_{21}H_{39}N_3O_6Si_2$

Chiral intermediate. mp = 248-250°; $[\alpha]^{21}$ = + 50° (c = 1, $CH_3OH$). *Sigma-Aldrich Fine Chem.*

**4526 Tetrakis1-4-alkyl(C11-C13)phenylsulfonyl-(2R)-pyrrolidinecarboxylated rhodium(II)**

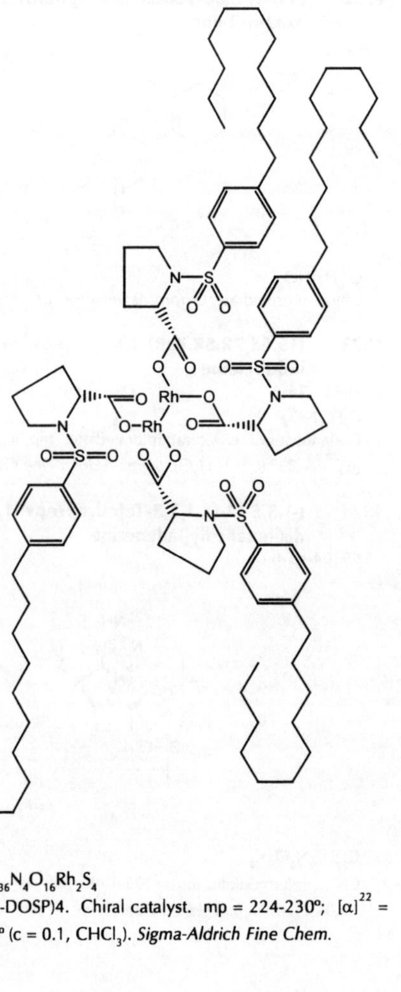

$C_{88}H_{136}N_4O_{16}Rh_2S_4$

Rh2(R-DOSP)4. Chiral catalyst. mp = 224-230°; $[\alpha]^{22}$ = + 165° (c = 0.1, $CHCl_3$). *Sigma-Aldrich Fine Chem.*

**4527 Tetrakis1-4-alkyl(C11-C13)phenylsulfonyl-(2S)-pyrrolidinecarboxylated rhodium(II)**

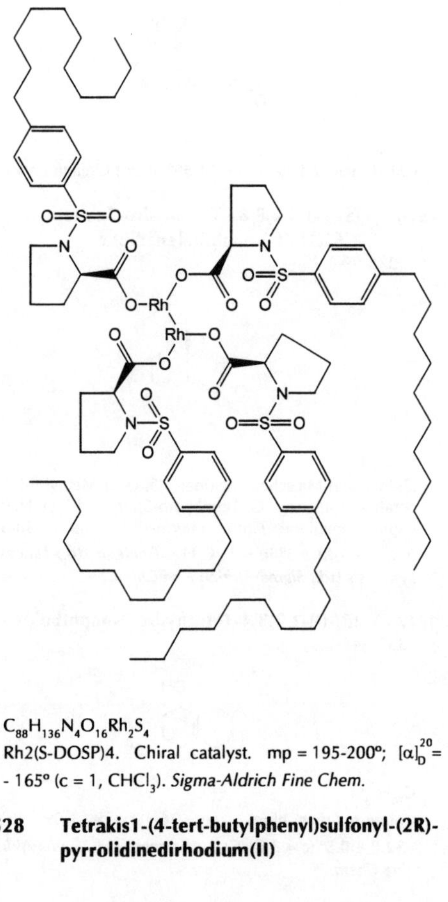

$C_{88}H_{136}N_4O_{16}Rh_2S_4$

Rh2(S-DOSP)4. Chiral catalyst. mp = 195-200°; $[\alpha]_D^{20}$= - 165° (c = 1, $CHCl_3$). *Sigma-Aldrich Fine Chem.*

**4528 Tetrakis1-(4-tert-butylphenyl)sulfonyl-(2R)-pyrrolidinedirhodium(II)**

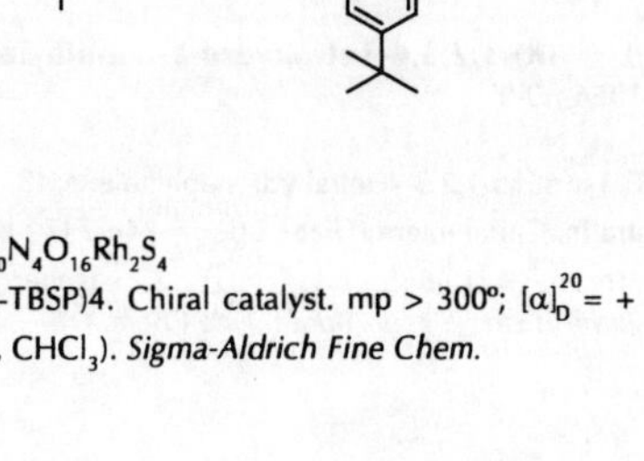

$C_{60}H_{80}N_4O_{16}Rh_2S_4$

Rh2(R-TBSP)4. Chiral catalyst. mp > 300°; $[\alpha]_D^{20}$= + 186° (c = 1, $CHCl_3$). *Sigma-Aldrich Fine Chem.*

**4529 Tetrakis1-(4-tert-butylphenyl)sulfonyl-(2S)-pyrrolidinedirhodium(II)**
154975-39-0

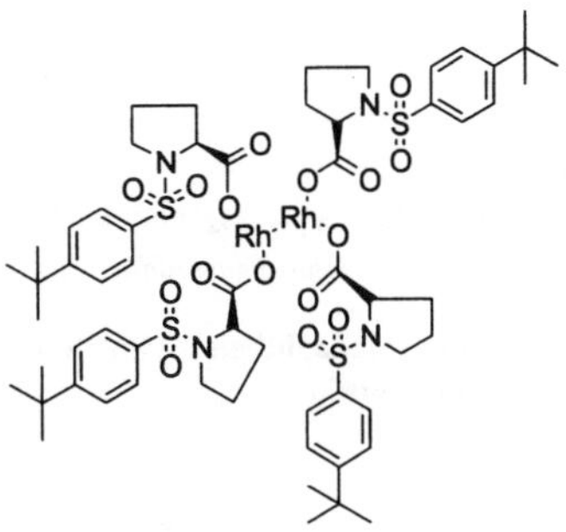

$C_{60}H_{80}N_4O_{16}Rh_2S_4$
Rh2(S-TBSP)4. Chiral catalyst. mp = 273°; $[\alpha]_D^{20}$ = - 187° (c = 0.1, $CHCl_3$). *Sigma-Aldrich Fine Chem.*

**4530 (R)-(+)-5,5'-6,6'-Tetramethyl-3,3'-di-t-butyl-1,1'-biphenyl-2,2'-diol**

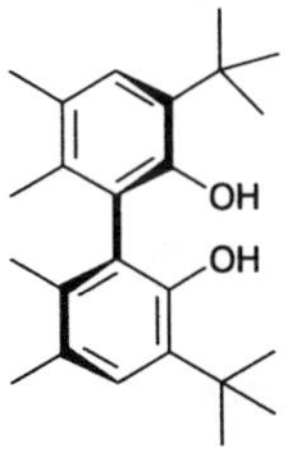

$C_{24}H_{34}O_2$
Chiral intermediate. $[\alpha]_D$ = + 64.1° (c = 0.352, THF). *Strem Chem. Inc.*

**4531 (S)-(-)-5,5'-6,6'-Tetramethyl-3,3'-di-t-butyl-1,1'-biphenyl-2,2'-diol**

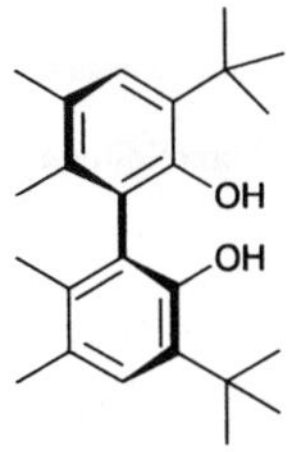

$C_{24}H_{34}O_2$
Chiral ligand. $[\alpha]_D$ = - 64.1° (c = 0.352, THF). *Strem Chem. Inc.*

**4532 4aR-trans-5-(1,5,5,8aS-Tetramethyl-2-methylenedecahydro-1-naphthalenyl)-3-R-methyl-1-penten-3-ol**
596-85-0
$C_{21}H_{36}O$
Manool. Chiral intermediate. mp = 49-52°; $[\alpha]_D^{20}$ = + 27° (c = 1, $CHCl_3$). *Sigma-Aldrich Fine Chem.*

**4533 N,N,N',N'-Tetramethyl-L-tartaramide**
26549-65-5

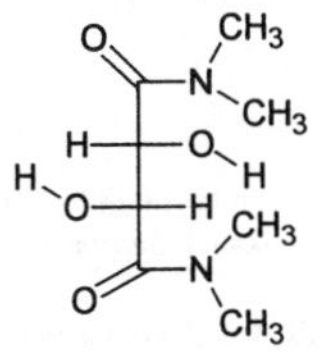

$C_8H_{16}N_2O_4$
N,N,N',N'-Tetramethyl-L-tartaric acid diamide. Chiral ligand. mp = 185-188°; $[\alpha]_D^{20}$ = + 46° (c = 3, EtOH). *Acros Organics NV; Oxford Asymmetry Ltd.; Sigma-Aldrich Fine Chem.*

**4534 N,N,N',N'-Tetramethyl-D-tartaric acid diamide**
63126-52-3

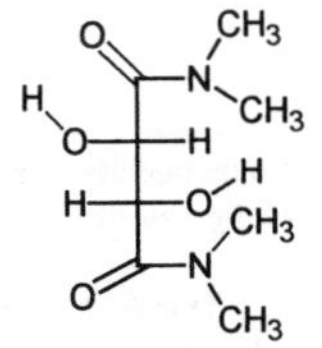

$C_8H_{16}N_2O_4$
N,N,N',N'-Tetramethyl-D-tartaramide. Chiral ligand; Chiral building block. mp = 186-188°; $[\alpha]_D^{20}$ = - 46° (c = 3, EtOH). *Acros Organics NV; Oxford Asymmetry Ltd.; Sigma-Aldrich Fine Chem.*

**4535 (S,S)-(+)-Tetrandrine**
518-34-3 9369(12)

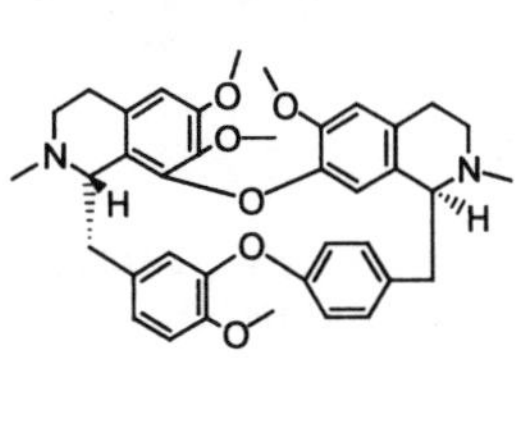

$C_{38}H_{42}N_2O_6$
Chiral intermediate. mp = 219-222°; $[\alpha]_D^{20}$ = + 285° (c = 1, $CHCl_3$). *Sigma-Aldrich Fine Chem.*

**4536 (-)-2-(2,4,5,7-Tetranitro-9-fluorenylidene-aminooxypropionic acid**
50874-31-2

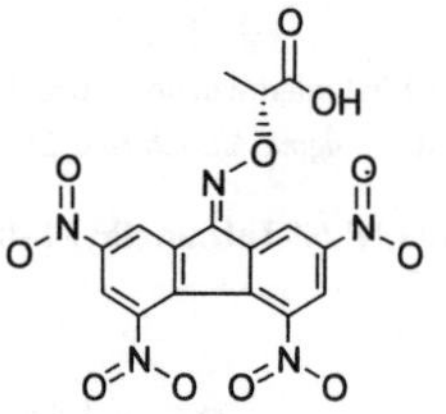

$C_{16}H_9N_5O_{11}$
(-)-TAPA. Chiral intermediate. mp = 196-199°; $[\alpha]_D^{20}$ = - 91° (c = 1.6, dioxane). *Sigma-Aldrich Fine Chem.*

**4537 (+)-2-(2,4,5,7-Tetranitro)-9-fluorenylidene-aminooxypropionic acid**
50996-73-1

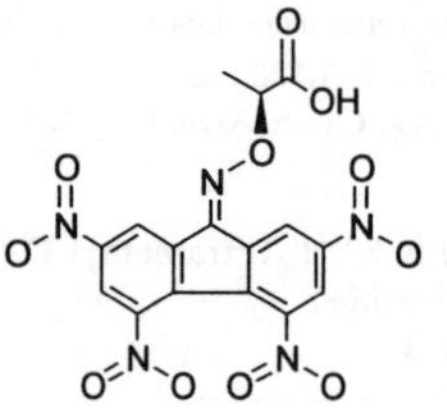

$C_{16}H_9N_5O_{11}$
(+)-TAPA. Chiral intermediate. mp = 195°; $[\alpha]^{25}$ = + 92° (c = 1.6, dioxane). *Sigma-Aldrich Fine Chem.*

**4538 (+)-2,3,4,6-Tetra-O-pivaloylgalacto-pyranosyl isothiocyanate**

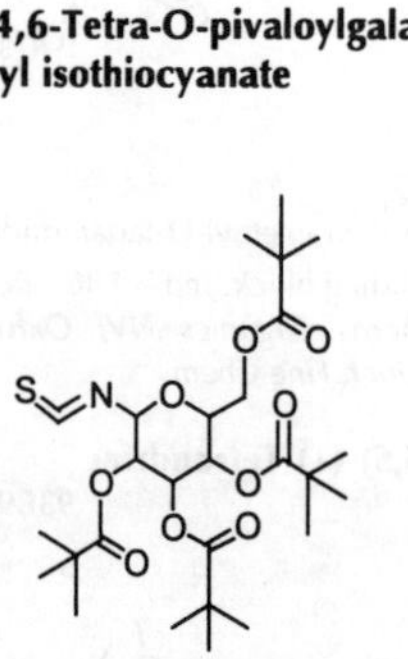

$C_{27}H_{43}NO_9S$
Chiral intermediate. mp = 142-145°; $[\alpha]_D^{20}$ = + 6.5° (c = 1, $CHCl_3$). *Sigma-Aldrich Fine Chem.*

**4539 L-Theanine**
3081-61-6 221-379-0
$C_7H_{14}N_2O_3$
Glutamine, N-ethyl-, L-.
Chiral building block. mp = 207°. *TCI America.*

**4540 (1S,4S)-2-Thia-5-azabicyclo[2.2.1]heptane hydrochloride**
125136-43-8

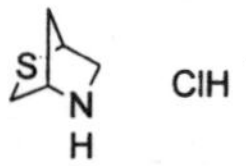

$C_5H_{10}ClNS$
Chiral intermediate. mp = 242-246°; $[\alpha]^{23}$ = + 36° (c = 1, $H_2O$). *Medinger & Sohn; Sigma-Aldrich Fine Chem.*

**4541 (-)-3-(Thianaphthen-3-yl)alanine hydrochloride**

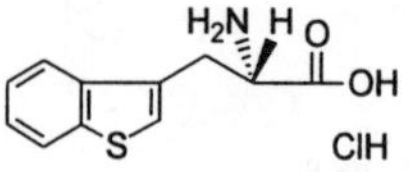

$C_{11}H_{12}ClNO_2S$
3-(Benzo[b]thiophen-3-yl)-L-alanine hydrochloride. Chiral intermediate. mp = 250°; $[\alpha]^{23}$ = - 4° (c = 7, $CH_3OH$). *Sigma-Aldrich Fine Chem.*

**4542 (+)-3-(Thianaphthen-3-yl)alanine hydrochloride**

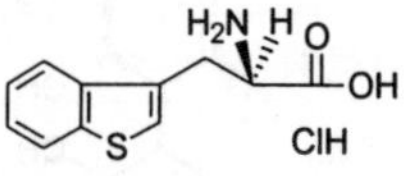

$C_{11}H_{12}ClNO_2S$
3-(Benzo[b]thiophen-3-yl)-D-alanine hydrochloride. Chiral intermediate. mp = 266°; $[\alpha]^{23}$ = + 4° (c = 7, $CH_3OH$). *Sigma-Aldrich Fine Chem.*

**4543 D-Thiaproline**
17570-08-0

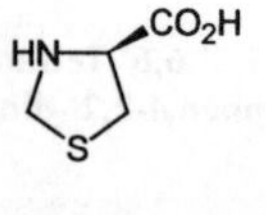

$C_4H_7NO_2S$
Chiral building block. *Omega Chem. Co. Inc.*

**4544 (R)-(-)-Thiazolidine-4-carboxylic acid**
34592-47-7 9586(12) 252-106-3

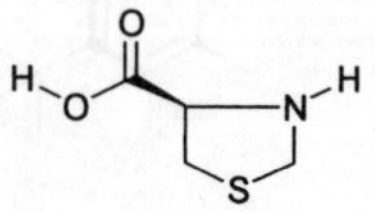

$C_4H_7NO_2S$
Thiazolidine-4-carboxylic acid, (R)-; L-Thiaproline; L-Thiazolidine-4-carboxylic acid. Chiral building block.

mp = 192-193°; $[\alpha]_D^{20}$ = - 141° (c = 1.3, $H_2O$); $[\alpha]_D^{20}$ = - 101° (c = 1, 1N HCl). *Acros Organics NV; Austin Chem. Co.; Flamma s.p.a.; Isochem; Lancaster Synthesis Ltd.; Nippon Rikagakuyakuhin; Sigma-Aldrich Fine Chem.; Sun-Orient Chemial Co. Ltd.; TCI America.*

**4545 L-2-Thiazolidinone-4-carboxylic acid**
19771-63-2 7084(12)

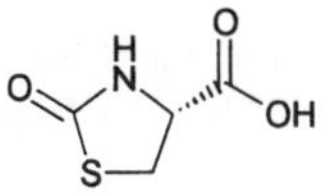

$C_4H_5NO_3S$
L-2-Oxothiazolidine-4-carboxylic acid. Chiral building block. mp = 174°; $[\alpha]^{19}$ = - 60° (c = 1, $H_2O$). *Sigma-Aldrich Fine Chem.; TCI America.*

**4546 (-)-3-(2-Thienyl)alanine**
22951-96-8

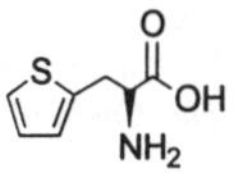

$C_7H_9NO_2S$
(S)-α-Amino-2-thiophenepropionic acid. Chiral building block. mp = 255-263°; $[\alpha]_D^{20}$ = - 29.4° (c = 1, $H_2O$). *Sigma-Aldrich Fine Chem.*

**4547 D-2-Thienylglycine**
43189-45-3 256-134-7

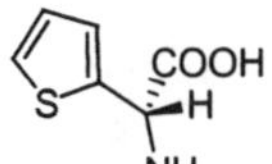

$C_6H_7NO_2S$
Thiophene-2-acetic acid, α-amino-, (S)-; Chiral building block. *Kaneka Corp.*

**4548 (-)-Thiocamphor**
53402-10-1

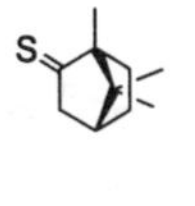

$C_{10}H_{16}S$
Resolving agent. mp = 136-138°; bp = 228-230°; $[\alpha]_D^{20}$ = - 24° (c = 3, EtOAc). *Sigma-Aldrich Fine Chem.*

**4549 (-)-Thiocholesterol**
1249-81-6 215-003-4

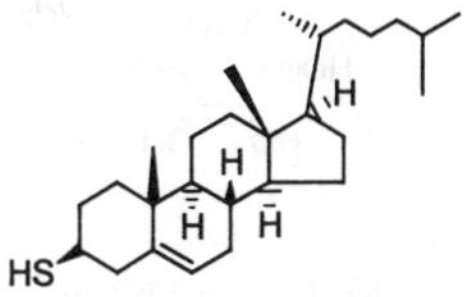

$C_{27}H_{46}S$
Pharmaceutical or derivative. mp = 97-99°; $[\alpha]_D^{20}$ = - 23° (c = 1, $CHCl_3$). *Sigma-Aldrich Fine Chem.*

**4550 L-Thiocitrulline**
156719-37-8

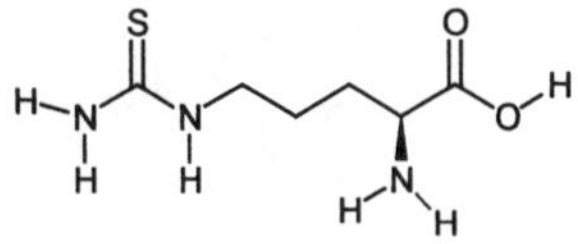

$C_6H_{13}N_3O_2S$
(3R)-cis-4,5-Dihydro-3,4-dihydroxy-2(3H)-furanone. Chiral intermediate. mp = 231-232°. *Acros Organics NV.*

**4551 1-Thio-β-D-galactopyranose sodium salt**

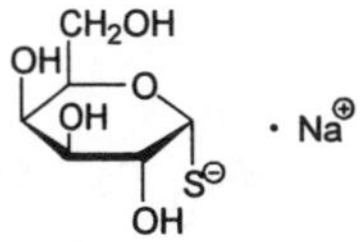

$C_6H_{11}NaO_5S$
Chiral intermediate. *See Chem. AG.*

**4552 (+)-5-Thioglucose**
20408-97-3 9470(12) 243-798-8

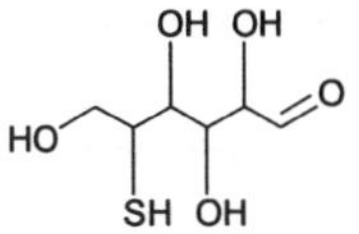

$C_6H_{12}O_5S$
Chiral building block. mp = 135-138°; $[\alpha]_D^{20}$ = + 188° (c = 1, $H_2O$, 2h). *Sigma-Aldrich Fine Chem.*

**4553 1-Thio-D-glucose sodium salt**
62778-20-5
$C_6H_{11}NaO_5S$
Glucopyranose, 1-thio-, monosodium salt, α-D-; 1-Thio-α-D-glucopyranose sodium salt; Sodium, (α-D-glucopyranosylthio)-. Listed on TSCA. Chiral intermediate. $[\alpha]$ = + 17° (c = 1.5, $H_2O$). *Acros Organics NV.*

**4554 (+)-1-Thioglucose sodium salt hydrate** 234-200-6

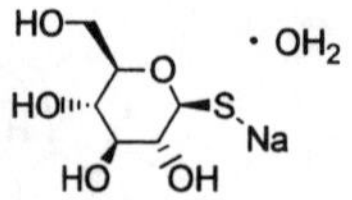

$C_6H_{11}NaO_5S$
Chiral building block. mp = 130°; $[\alpha]_D^{20}$ = + 3.5° (c = 1.5, $H_2O$). *Sigma-Aldrich Fine Chem.*

**4555 (+)-1-Thioglucose tetraacetate**
19879-84-6 243-392-0

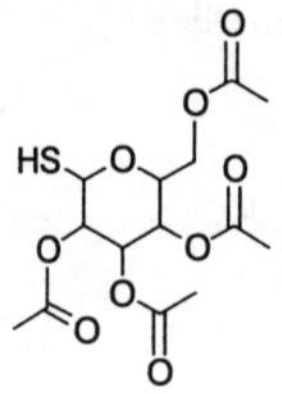

$C_{14}H_{20}O_9S$
Chiral intermediate. mp = 115-117°; $[\alpha]_D^{20}$ = + 5.8° (c = 2.2, $CHCl_3$). *Sigma-Aldrich Fine Chem.*

**4556 β-D-Thioglucose tetraacetate**
28878-90-2

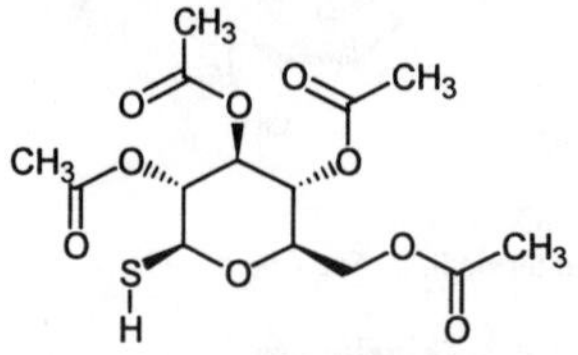

$C_{14}H_{20}O_9S$
Chiral intermediate. [α] = + 5.8° (c = 2.2, $CHCl_3$). *Acros Organics NV.*

**4557 (S)-(-)-Thiolactic acid**
57965-30-7

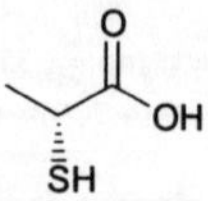

$C_3H_6O_2S$
Chiral intermediate. *Daicel Chem. Ind. Ltd.*

**4558 L-Thioleucine (S)-phenyl ester**
$C_{12}H_{17}NOS$
Chiral building block. *Sigma-Aldrich Fine Chem.*

**4559 (4R)-(-)-2-Thioxo-4-thiazolidinecarboxylic acid**
98169-56-3

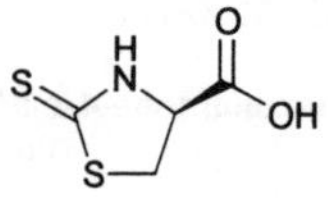

$C_4H_5NO_2S_2$
(Raphanusamic acid). Chiral intermediate. mp = 179-181°; $[\alpha]^{29}$ = - 86° (c = 2.5, 0.5N HCl). *Sigma-Aldrich Fine Chem.*

**4560 L-Threitol**
2319-57-5

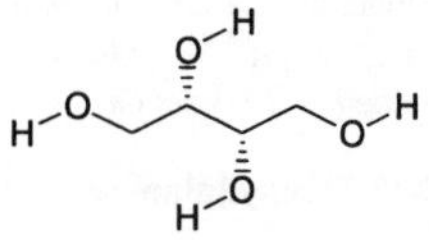

$C_4H_{10}O_4$
(2S,3S)-(-)-1,2,3,4-Butanetetrol. Chiral building block. mp = 87-88°; $[\alpha]^{19}$ = - 4° (c = 5, $H_2O$). *Acros Organics NV; Sigma-Aldrich Fine Chem.; TCI America.*

**4561 Threonic acid calcium salt**
70753-61-6
$C_4H_8O_5$
Chiral building block. mp > 300°; $[\alpha]_D^{20}$ = + 16° (c = 1, $H_2O$). *Sigma-Aldrich Fine Chem.*

**4562 D-Threonine**
632-20-2 211-171-8

$C_4H_9NO_3$
Threonine, D-; (2R,3S)-2-Amino-3-hydroxybutyric acid. Chiral building block. mp = 274°; $[\alpha]_D^{20}$ = + 29 ± 2° (c = 5, $H_2O$). *Acros Organics NV; Lancaster Synthesis Ltd.; Omega Chem. Co. Inc.; Sigma-Aldrich Fine Chem.; TCI America; Varsal Instruments, Inc.*

**4563 L-Threonine**
72-19-5 9522(12) 200-774-1

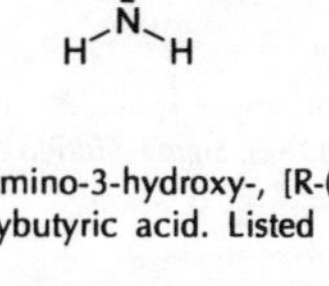

$C_4H_9NO_3$
Butanoic acid, 2-amino-3-hydroxy-, [R-(R*,S*)]-; (2S,3R)-2-Amino-3-hydroxybutyric acid. Listed on TSCA. Chiral

building block. mp = 270°; $[\alpha]_D^{20}$ = - 28 ± 1° (c = 5, $H_2O$). *Aceto Corp.; Acros Organics NV; Ajinomoto Co. Inc.; Austin Chem. Co.; Flamma s.p.a.; Kyowa Hakko Kogyo Co. Ltd.; Lancaster Synthesis Ltd.; Nippon Kayaku Co. Ltd.; Rexim/Degussa Europe; See Chem. AG; Shanghai DSL Intl. trading Co.; Sigma-Aldrich Fine Chem.; Tanabe Seiyaku Co. Ltd.; TCI America; Varsal Instruments, Inc.; Yoneyama Yakuhin Kogyo Co. Ltd.*

**4564 D-(-)-allo-Threonine**
24830-94-2 246-488-0

$C_4H_9NO_3$
Allothreonine, D-; Chiral building block. mp = 276°; $[\alpha]^{23}$ = - 8.8° (c = 2, $H_2O$). *Acros Organics NV; Sigma-Aldrich Fine Chem.*

**4565 L-(+)-allo-Threonine**
28954-12-3 249-327-2

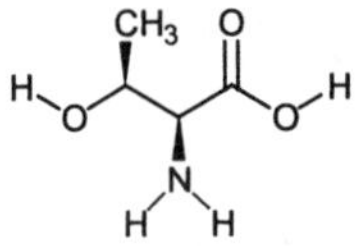

$C_4H_9NO_3$
Butanoic acid, 2-amino-3-hydroxy-, [S-(R*,R*)]-; [S-(R*,R*)]-2-Amino-3-hydroxybutanoic acid. Chiral building block. mp = 272°; $[\alpha]^{23}$ = + 9.0° (c = 2, $H_2O$). *Acros Organics NV; Sigma-Aldrich Fine Chem.; Varsal Instruments, Inc.*

**4566 L-Threonine amide hydrochloride**

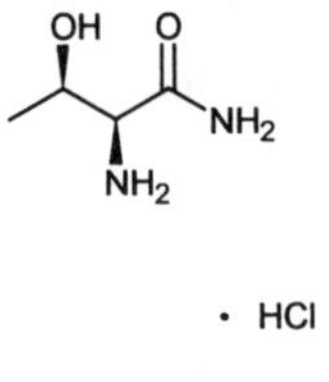

· HCl

$C_4H_{11}N_2O_2$
Chiral building block. *See Chem. AG.*

**4567 Threoninol**

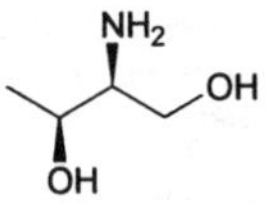

$C_4H_{11}NO_2$
Chiral building block. mp = 57.5-61.5°; $[\alpha]_D^{20}$ = + 4.3° (c = 1, $H_2O$). *Sigma-Aldrich Fine Chem.*

**4568 Thujone**
1125-12-8 9326(11) 214-405-7

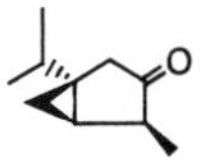

$C_{10}H_{16}O$
1-Isopropyl-4-methylbicyclo[3.1.0]hexan-3-one. Listed on TSCA. Chiral intermediate. $bp_{17}$ = 86-84°; d = 0.925; n = 1.455; $[\alpha]_D^{20}$ = - 15° (neat). *Sigma-Aldrich Fine Chem.*

**4569 Thymidine**
50-89-5 9538(12) 200-070-4

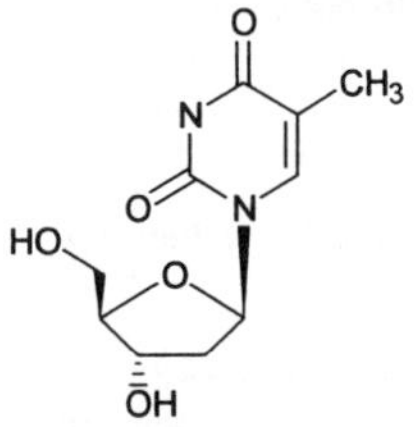

$C_{10}H_{14}N_2O_5$
Uridine, 2'-deoxy-5-methyl-; 1-(2-Deoxy-D-ribo--furanosyl)thymine(thymidine); 1-(2-Deoxy-β-D-ribofuranosyl)-5-methyluracil; 1-(2-Deoxy-β-D-erythro-pentofuranosyl)-5-methyl-2,4(1H,3H)-pyrimidinedione; 5-Methyldeoxyuridine; Deoxyribothymidine; Deoxythymidine; dT; DThyd; Thymine deoxy. Listed on TSCA. Chiral intermediate. mp = 187-189°; $[\alpha]^{16}$ = + 32.3° (c = 1, 1N NaOH); $[\alpha]^{25}$ = +18.5° (c = 1, $H_2O$). *Acros Organics NV; D & O Group; Sigma-Aldrich Fine Chem.*

**4570 Thymol-β-D-glucopyranoside**

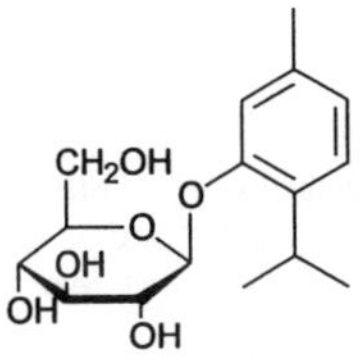

$C_{16}H_{24}O_6$
Chiral intermediate. *See Chem. AG.*

**4571 D-Thyroxine**
51-49-0 200-102-7

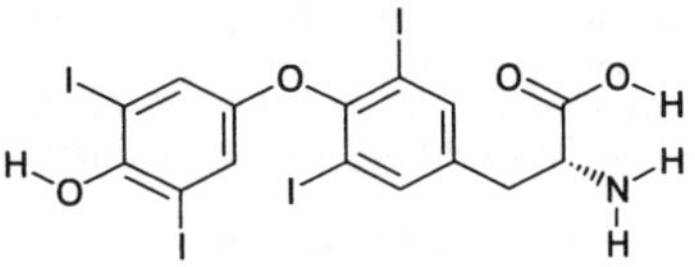

$C_{15}H_{11}I_4NO_4$
Tyrosine, O-(4-hydroxy-3,5-diiodophenyl)-3,5-diiodo-, D-; 3,3',5,5'-Tetraiodo-D-thyronine; D-4-(4-Hydroxy-3,5-

diiodophenoxy)-3,5-diiodobenzylalanine; D-T4; Dextrothyroxine. Chiral intermediate. [α] = - 19° (c = 2, 0.2N HCl - EtOH). *Acros Organics NV; TCI America.*

**4572 L-Thyroxine**
51-48-9 9555(12) 200-101-1

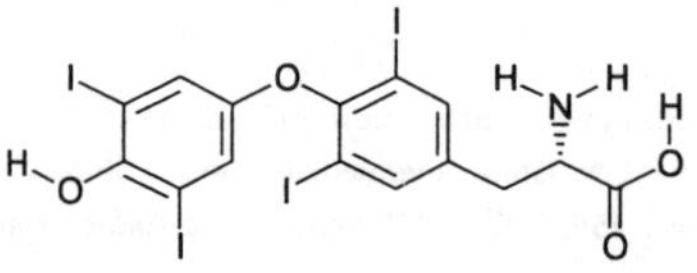

$C_{15}H_{11}I_4NO_4$
3,3',5,5'-Tetraiodo-L-thyronine. Listed on TSCA. Chiral intermediate. mp = 223°; $[\alpha]_D^{20}$ = + 21 (c = 1, 1N HCl/EtOH). *Acros Organics NV; Sigma-Aldrich Fine Chem.; TCI America.*

**4573 L-Thyroxine sodium salt**
55-03-8 200-221-4
$C_{15}H_{11}I_4NO_4$
Tyrosine, O-(4-hydroxy-3,5-diiodophenyl)-3,5-diiodo-, monosodium salt, L-; Levothyroxine sodium; L-3-[4'-(4''-Hydroxy-3'', 5''-diiodophenoxy)-3',5'-diiodophenyl] alaninesodium; Dathroid; Eltroxin; Euthyrox; Laevoxin; Letter; Levaxin; Levoroxine; Levothroid; Monosodium thyroxine; Oroxine; Sodium L-thyroxine; Sodium levothyroxine. Listed on TSCA. Chiral intermediate. *TCI America.*

**4574 Thyroxine sodium salt pentahydrate**
6106-07-6 5351(11)

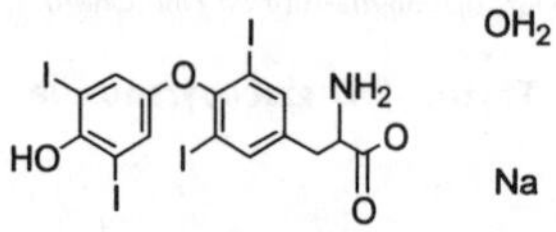

$C_{15}H_{11}I_4NO_4$
Sodium levothyroxine. Listed on TSCA. Chiral intermediate. mp = 207-210°; $[\alpha]_D^{20}$ = - 6.6° (c = 3, 1N NaOH). *Sigma-Aldrich Fine Chem.*

**4575 (S)-Timolol maleate**
26921-17-5 248-111-5
$C_{13}H_{24}N_4O_3S$
Propan-2-ol, 1-[(1,1-dimethylethyl)amino]-3-[[4-(4-morpholinyl)-1,2,5-thiadiazol-3-yl]oxy]-, (2S)-, (2Z)-2-butenedioate (1:1) (salt); (S)-1-[(1,1-Dimethylethyl)-amino]-3-[[4-(4-morpholinyl)-1,2,5-thiadiazol-3-yl]oxy]-2-propanol, (Z)-2-butenedioate (1:1) (salt); (S)-1-tert-Butylamino-3-[4-morpholino-1,2,5-thiadiazol-3-yloxy)-propan-2-ol hydrogen maleate; (S)-Timolol hydrogen maleate; Betime. Chiral intermediate. *Linnea S.A.*

**4576 (R)-(-)-p-Toluenesulfinamide**
$C_7H_9NOS$
Chiral building block. mp = 115-118°; $[\alpha]_D^{20}$ = - 99° (c = 1,$CHCl_3$). *Sigma-Aldrich Fine Chem.*

**4577 (S)-(+)-p-Toluenesulfinamide**

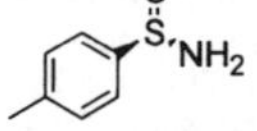

$C_7H_9NOS$
Resolving agent. mp = 118-121°; $[\alpha]^{22}$ = + 85° (c = 1, $CHCl_3$). *Sigma-Aldrich Fine Chem.*

**4578 (R)-Toluenesulfinamide benzyilidine**

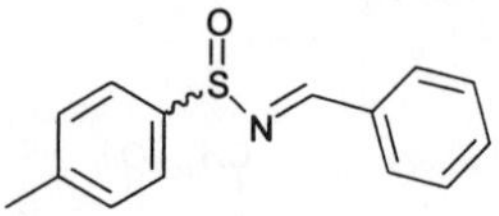

$C_{14}H_{13}NOS$
Resolving agent. *Advanced Asymmetrics, Inc.*

**4579 (S)-Toluenesulfinamide benzyilidine**

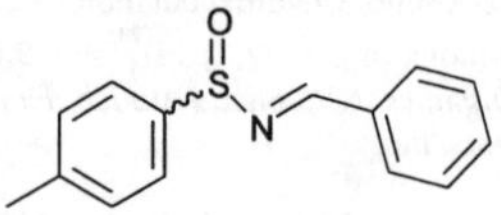

$C_{14}H_{13}NOS$
Resolving agent. *Advanced Asymmetrics, Inc.*

**4580 p-Toluenesulfonic acid (2S)-(+)-glycidyl ester**
70987-78-9

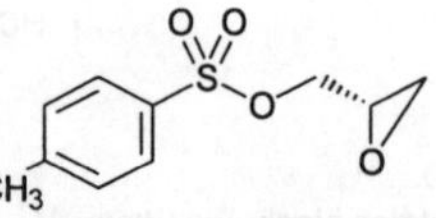

$C_{10}H_{12}O_4S$
(2S)-(+)-Glycidyl p-toluenesulfonate; (2S)-(+)-Glycidyl Tosylate. Chiral building block. mp = 48°; $[\alpha]^{19}$ = + 17° (c = 2.75, $CHCl_3$). *Acros Organics NV; Chirex Inc.; Daiso Co. Ltd.; Kaneka Corp.; Nagase & Co. Ltd.; Omega Chem. Co. Inc.; Sigma-Aldrich Fine Chem.; TCI America.*

**4581 p-Toluenesulfonic acid (S)-2-methylbutyl ester**
38261-81-3
$C_{12}H_{18}O_3S$
(S)-(+)-2-Methylbutyl p-toluenesulfonate. Chiral building block. d = 1.11. *TCI America.*

**4582 (S)-(+)-γ-Toluenesulfonylmethyl-γ-butyrolactone**
58879-34-8

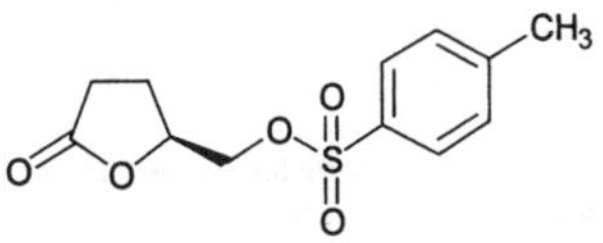

$C_{12}H_{14}O_5S$
Chiral building block. mp = 86-88°; [α] = + 45.5° (c = 1, $CHCl_3$). *Acros Organics NV; Sigma-Aldrich Fine Chem.*

**4583 L-1-4'-Tosylamino-2-phenylethylchloromethyl ketone**
402-71-1 206-954-6

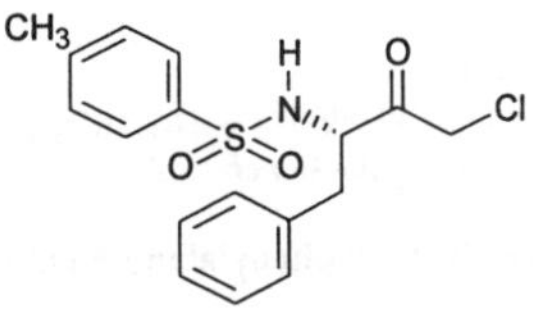

$C_{17}H_{18}ClNO_3S$
Benzenesulfonamide, N-[3-chloro-2-oxo-1-(phenylmethyl)propyl]-4-methyl-, (S)-; L-Chloromethyl-(2-phenyl-1-(p-toluenesulphonylamino)ethyl) ketone; p-Toluenesulfonamide, N-[α-(chloroacetyl)phenethyl]-, L-; Tosylphenylalanine chloromethyl ketone; TPCK. Listed on TSCA. Chiral intermediate. mp = 106-108°; [α] = - 86°. *Acros Organics NV; Sigma-Aldrich Fine Chem.*

**4584 N-α-4-Tosyl-L-arginine methyl ester hydrochloride**
1784-03-8 217-235-1

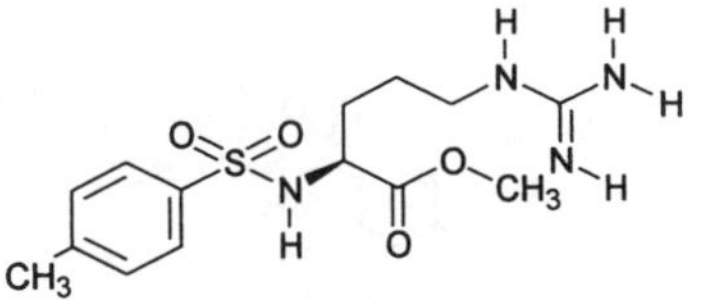

$C_{14}H_{23}ClN_4O_4S$
$N^2$-(p-tolylsulfonyl)-, methyl ester, monohydrochloride, L-; Methyl $N^2$-[(p-tolyl)sulphonyl]-L-argininate monohydrochloride; TAME hydrochloride; N-α-Toluenesulfonylarginine methyl ester hydrochloride. Chiral intermediate. mp = 145-147°; $[\alpha]_D^{20}$ = - 14° (c = 4, $H_2O$). *Acros Organics NV; Sigma-Aldrich Fine Chem.*

**4585 (1R,2R)-(-)-N-P-Tosyl-1,2-diphenylethylenediamine**

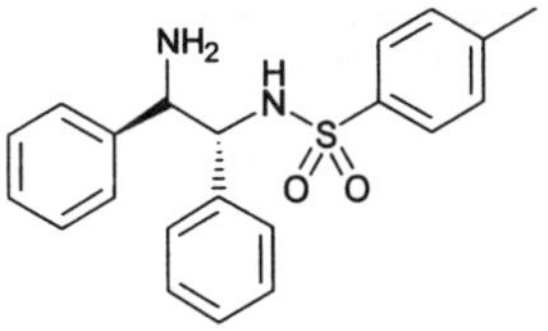

$C_{21}H_{22}N_2O_2S$
Chiral building block. mp = 128-131°; $[\alpha]_D^{20}$ = - 35° (c = 1, $CHCl_3$). *Sigma-Aldrich Fine Chem.*

**4586 (1S,2S)-(+)-N-p-Tosyl-1,2-diphenylethylenediamine**

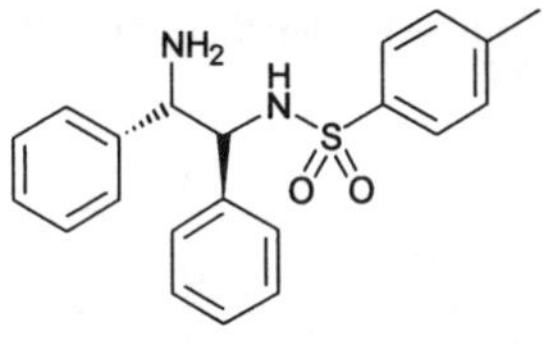

$C_{21}H_{22}N_2O_2S$
Resolving agent; Chiral building block. mp = 128-131°; $[\alpha]_D^{20}$ = + 35° (c = 1, $CHCl_3$). *Sigma-Aldrich Fine Chem.*

**4587 N-Tosyl-L-glutamic acid**
4816-80-2 225-390-1

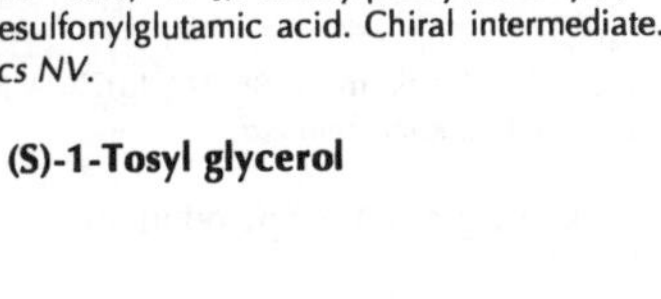

$C_{12}H_{15}NO_6S$
Glutamic acid, N-[(4-methylphenyl)sulfonyl]-, L-; p-Toluenesulfonylglutamic acid. Chiral intermediate. *Acros Organics NV.*

**4588 (S)-1-Tosyl glycerol**

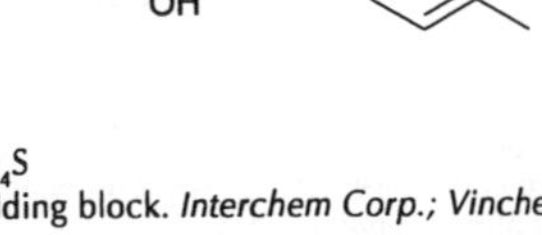

$C_{11}H_{16}NO_4S$
Chiral building block. *Interchem Corp.; Vinchem.*

**4589 L-(-)-O-Tosyllactic acid ethyl ester**
57057-80-4
$C_{12}H_{16}O_5S$
Ethyl L-(-)-O-(p-toluenesulfonyl)lacetate; L-(-)-O-(p-Toluenesulfonyl)lactic acid ethyl ester. Chiral intermediate. *TCI America.*

**4590 N-ε-1-Tosyl-L-lysine**
2130-76-9

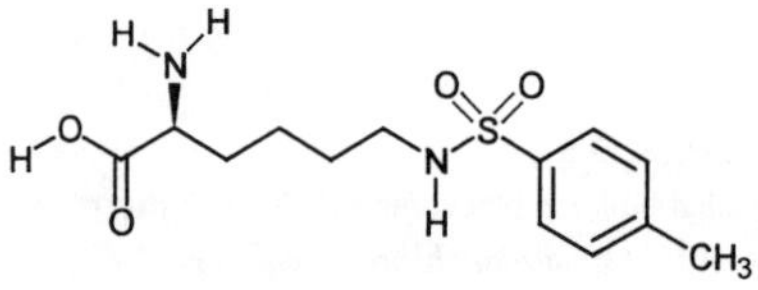

$C_{13}H_{20}N_2O_4S$
Chiral building block. *Acros Organics NV.*

**4591 (R)-(-)-γ-Tosylmethyl-γ-butyrolactone**
58879-33-7

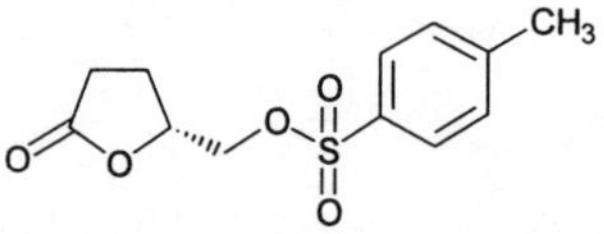

$C_{12}H_{14}O_5S$
Chiral building block. mp = 85-87°; $[\alpha]_D^{20}$ = - 36.5 ± 0.5° (c = 2, 1,2-DCE). *Acros Organics NV; Kiralchem Ltd.; Sigma-Aldrich Fine Chem.*

**4592 (S)-(+)-γ-Tosylmethyl-γ-butyrolactone**

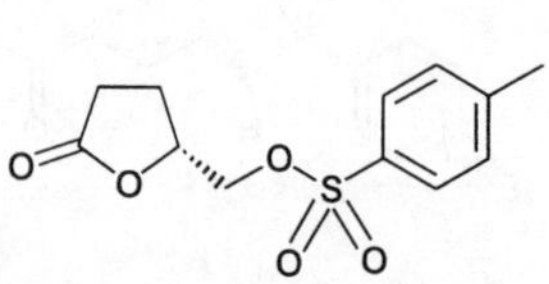

$C_{12}H_{14}O_5S$
Chiral building block. mp = 85-87°; $[\alpha]_D^{20}$ = + 36.5 ± 0.5° (c = 2, 1,2-DCE). *Kiralchem Ltd.*

**4593 (R)-Tosyloxy-1,2-epoxybutane**
213262-97-6

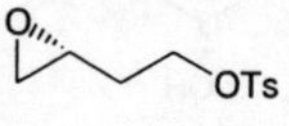

$C_{11}H_{14}O_4S$
Oxiraneethanol, 4-methylbenzenesulfonate, (R)-; Chiral building block. *Synthon Corp.*

**4594 (S)-4-Tosyloxy-1,2-epoxybutane**
91111-12-5

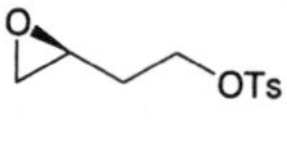

$C_{11}H_{14}O_4S$
Oxiraneethanol, 4-methylbenzenesulfonate, (S)-; Chiral intermediate. *Synthon Corp.*

**4595 (-)-N-p-Tosyl-phenylalanine**
13505-32-3

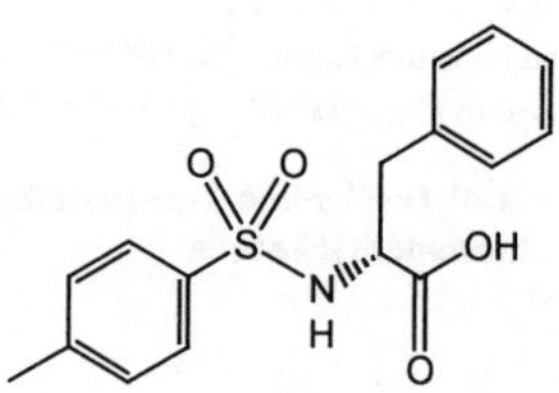

$C_{16}H_{17}NO_4S$
Chiral building block. mp = 161-163°; $[\alpha]_D^{20}$ = - 2.3° (c = 4.5, acetone). *Sigma-Aldrich Fine Chem.*

**4596 (R)-N-Tosylphenylalanine chloride**

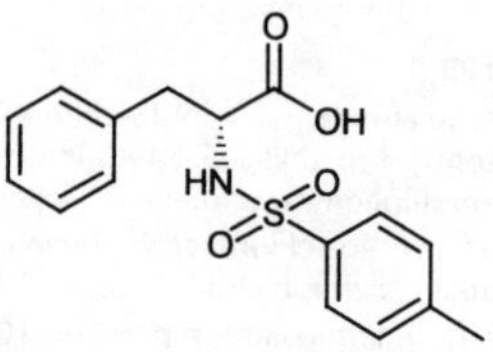

$C_{16}H_{16}ClNO_3S$
Chiral building block. *Oxford Asymmetry Ltd.*

**4597 (S)-N-Tosylphenylalanine chloride**
29739-88-6

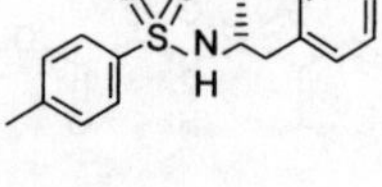

$C_{16}H_{16}ClNO_3S$
(S)-(+)-(p-Tolylsulfonylamino)hydrocinnamoyl chloride. Chiral intermediate. mp = 132°; $[\alpha]_D^{20}$ = + 11° (c = 2, $CHCl_3$). *Oxford Asymmetry Ltd.; Sigma-Aldrich Fine Chem.*

**4598 1-(p-Tosyl)-(R)-(-)-3-pyrrolidinol**
133034-00-1
$C_{11}H_{15}NO_3S$
N-(para-Tolylsulfonyl-(R)-3-pyrrolidinol. Chiral inter-

mediate. mp = 61-65°; $[\alpha]_D^{20}$ = - 26.5° (c = 1, $CH_3OH$). *Sigma-Aldrich Fine Chem.*

**4599 α,α-D-Trehalose**
99-20-7 202-739-6

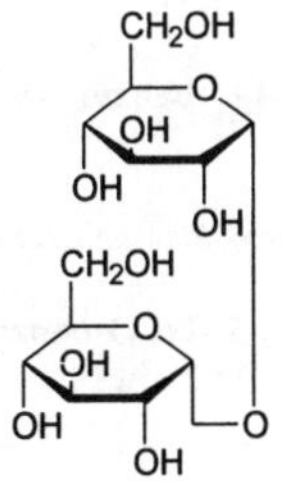

$C_{12}H_{22}O_{11}$
Glucopyranoside, α-D-glucopyranosyl, α-D-; α-D-Glucopyranosyl-α-D-glucopyranoside; Ergot sugar; Mycose; Trehaose. Listed on TSCA. Pharmaceutical or derivative. $[\alpha]_D^{20}$ = + 179.9° (c = 7, $H_2O$). *Pfanstiehl Labs Inc.; See Chem. AG; TCI America.*

**4600 (R)-1,2-4-Triacetoxybutane**
108266-50-8

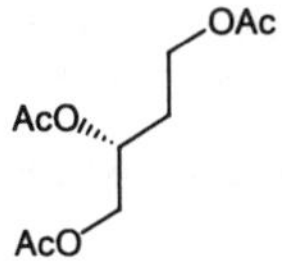

$C_{10}H_{16}O_6$
Butane-1,2,4-triol, triacetate, (R)-; Chiral intermediate. *Synthon Corp.*

**4601 (S)-1,2,4-Triacetoxybutane**
52067-45-5

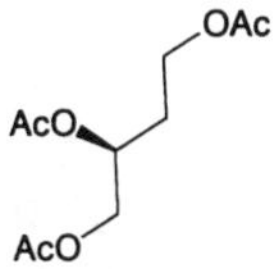

$C_{10}H_{16}O_6$
Butane-1,2,4-triol, triacetate, (S)-; Chiral intermediate. *Synthon Corp.*

**4602 (+)-Triacetylcyclodextrin**
23739-88-0

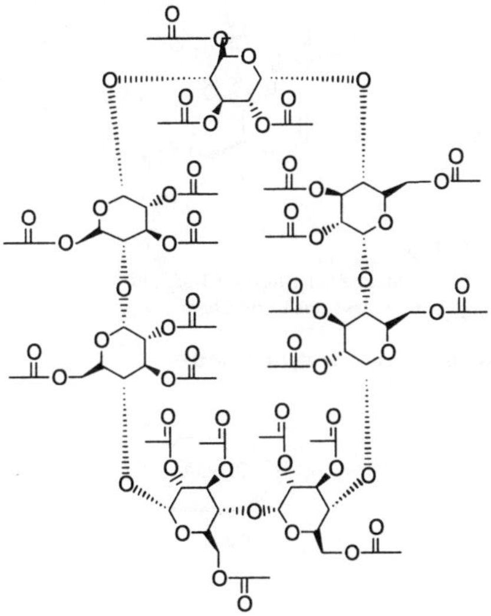

$C_{84}H_{112}O_{56}$
β-Cyclodextrin heneicosaacetate. Chiral auxiliary. mp = 204-206°; $[\alpha]^{25}$ = + 125° (c = 1, $CHCl_3$). *Sigma-Aldrich Fine Chem.*

**4603 2,3,4-Tri-O-acetyl-β-L-fucopyranosyl azide**
95581-07-0

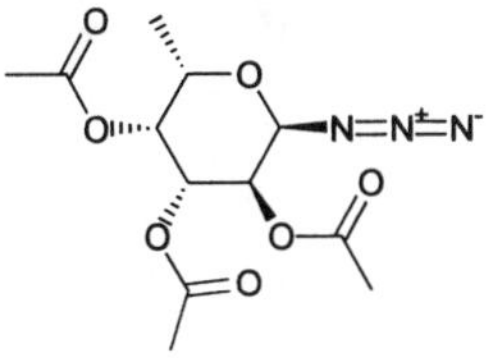

$C_{12}H_{17}N_3O_7$
Chiral intermediate. [α] = + 21° (c = 1, $CHCl_3$). *Acros Organics NV.*

**4604 (-)-Tri-O-acetylgalactal**
4098-06-0 223-859-5

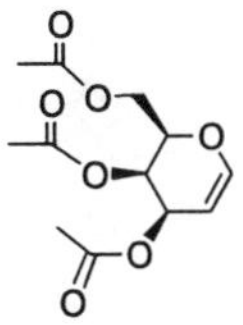

$C_{12}H_{16}O_7$
Chiral intermediate. mp = 34-38°; $[\alpha]^{22}$ = - 18° (c = 1, $CHCl_3$). *Sigma-Aldrich Fine Chem.*

**4605 Tri-O-acetyl-D-glucal**
2873-29-2 220-709-0

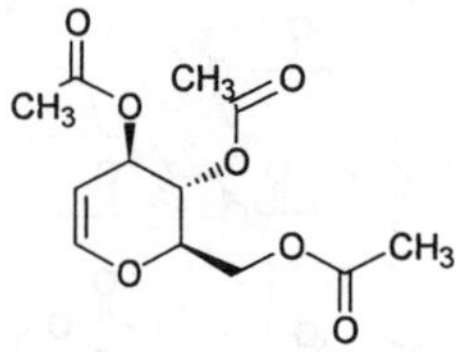

$C_{12}H_{16}O_7$
Chiral intermediate. mp = 53-55°; $[\alpha]^{25}$ = - 12° (c = 2, EtOH). *Sigma-Aldrich Fine Chem.*

**4606 Tri-O-acetyl-L-glucal**
63640-41-5

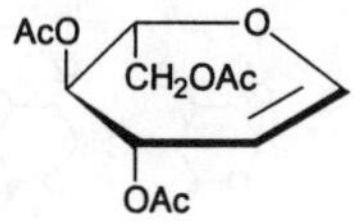

$C_{12}H_{16}O_7$
Chiral intermediate. *Pfanstiehl Labs Inc.*

**4607 2,3,4-Tri-O-acetyl-β-D-xylopyranosyl azide**
53784-33-1

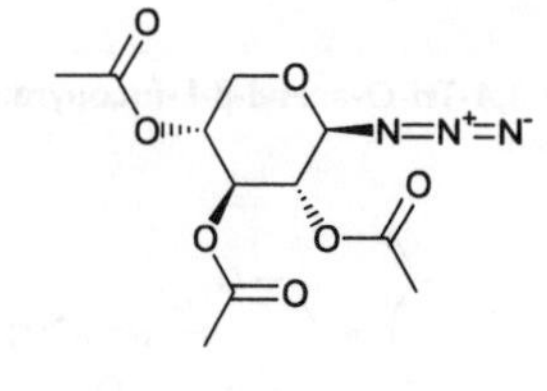

$C_{11}H_{15}N_3O_7$
Chiral intermediate. [α] = - 87.5° (c = 2, $CHCl_3$). *Acros Organics NV.*

**4608 Tri-O-benzoyl-1-O-acetyl-D-ribofuranose**
6974-32-9 230-220-4

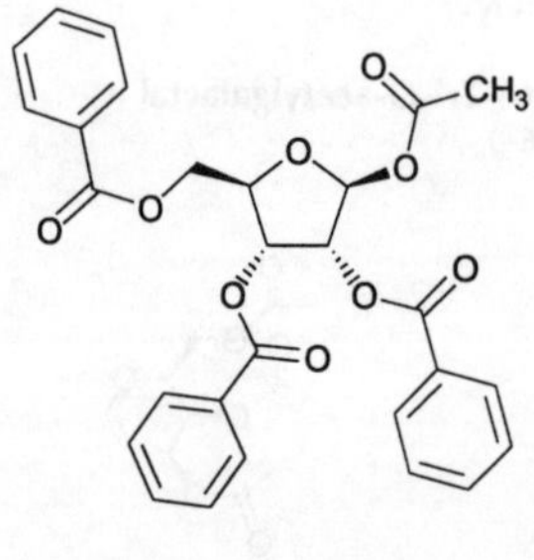

$C_{28}H_{24}O_9$
Ribofuranose, 1-acetate 2,3,5-tribenzoate, β-D-; 1-O-Acetyl-2,3,5-tri-O-benzoyl-β-D-ribofuranose; 2,3,5-Tri-O-benzoyl-β-D-ribofuranosyl acetate. Listed on TSCA. Chiral intermediate. mp = 129°; $[\alpha]_D^{20}$ = + 43.5° (c = 4, $CHCl_3$); $[\alpha]_D$ = + 24.3° (c = 1, pyridine). *Acros Organics NV; Interchem Corp.; Pfanstiehl Labs Inc.; Sigma-Aldrich Fine Chem.*

**4609 2,3,5-Tri-O-benzyl-D-arabinofuranose**
37776-25-3
$C_{26}H_{28}O_5$
Chiral intermediate. $[\alpha]_D^{20}$ = + 11.6° (c = 2, 9:1 dioxane: $H_2O$). *Pfanstiehl Labs Inc.*

**4610 2,3,5-Tri-O-benzyl-L-arabinofuranose**
89615-42-9
$C_{26}H_{28}O_5$
Chiral intermediate. *Pfanstiehl Labs Inc.*

**4611 9-(2', 3', 5'-Tri-O-benzyl-β-D-arabinofuranosyl)adenine**
3257-73-6
$C_{13}H_{31}N_5O_4$
Chiral intermediate. mp = 128°. *Pfanstiehl Labs Inc.*

**4612 2,3,4-Tri-O-benzyl-(1S)-ethylthiofucopyranoside**

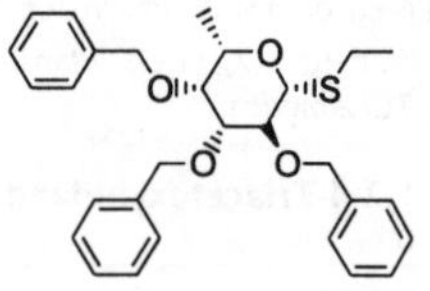

$C_{29}H_{34}O_4S$
Chiral intermediate. mp = 52-54°; $[\alpha]^{22}$ = - 15° (c = 1.5, $CH_2Cl_2$). *Sigma-Aldrich Fine Chem.*

**4613 2,3,4-Tri-O-benzyl-L-fucopyranose**
60431-34-7

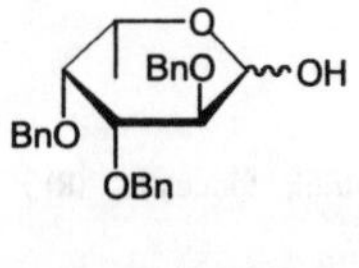

$C_{27}H_{29}O_5$
6-Deoxy-2,3,4-tri-O-benzyl-L-galactopyranose. Chiral intermediate. *Pfanstiehl Labs Inc.; See Chem. AG.*

**4614 (-)-Tri-O-benzylgalactal**
80040-79-5

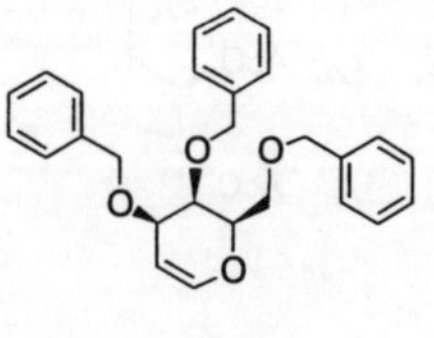

$C_{27}H_{28}O_4$
Chiral intermediate. mp = 48-54°; $[\alpha]^{22}$ = - 43° (c = 1, $CH_2Cl_2$). *Sigma-Aldrich Fine Chem.*

**4615 2,3,5-Tri-O-benzyl-1-O-p-nitrobenzoyl-D-arabinofuranose**
52522-49-3

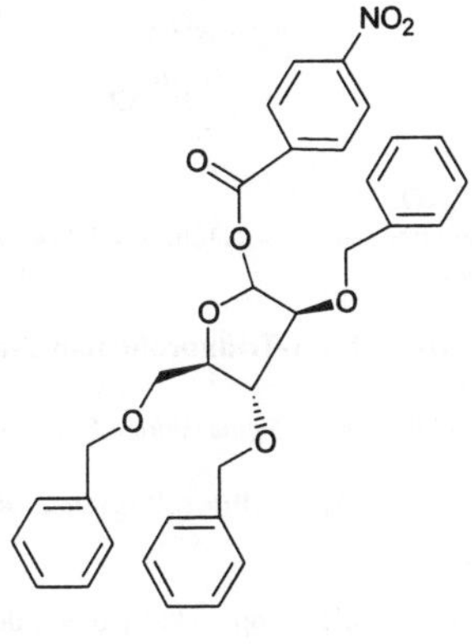

$C_{33}H_{31}NO_8$
Cinchonan-9-ol, monohydrochloride, (8α,9R)-; Chiral intermediate. mp = 205-207°; $[\alpha]^{25}$ = + 66° (c = 5.5, $CH_2Cl_2$). *Acros Organics NV; Pfanstiehl Labs Inc.; Sigma-Aldrich Fine Chem.*

**4616 2,3,5-Tri-O-benzoyl-1-O-(p-nitrobenzoyl)-D-ribofuranose**
34213-15-5
$C_{33}H_{31}NO_8$
Chiral intermediate. *Pfanstiehl Labs Inc.*

**4617 Tri-O-benzyl-L-rhamnopyranose**
59055-58-2
6-Deoxy-2,3,4-tri-O-benzyl-L-mannopyranose. Chiral intermediate. *Pfanstiehl Labs Inc.*

**4618 1,3,5-Tri-O-benzoyl-α-D-ribofuranose**
22224-41-5

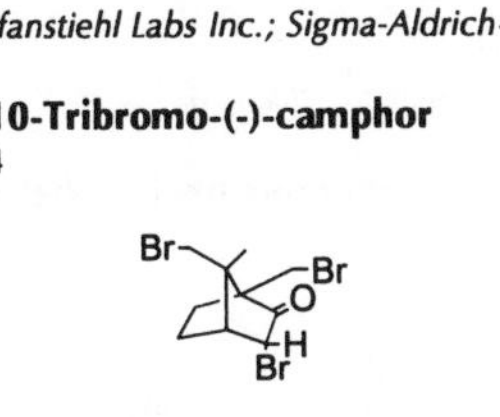

$C_{26}H_{22}O_8$
Chiral intermediate. mp = 125-129°; $[\alpha]_D^{20}$ = + 84.5° (c = 1, $CHCL_3$). *Pfanstiehl Labs Inc.; Sigma-Aldrich Fine Chem.*

**4619 3,9,10-Tribromo-(-)-camphor**
115887-80-4

$C_{10}H_{13}Br_3O$
Resolving agent. mp = 100-103°; $[\alpha]_D^{20}$ = - 62° (c = 1, $CH_3OH$). *Sigma-Aldrich Fine Chem.*

**4620 1,3:2,5:4,6-Tri-O-cyclohexylidene-D-mannitol**
70167-57-6

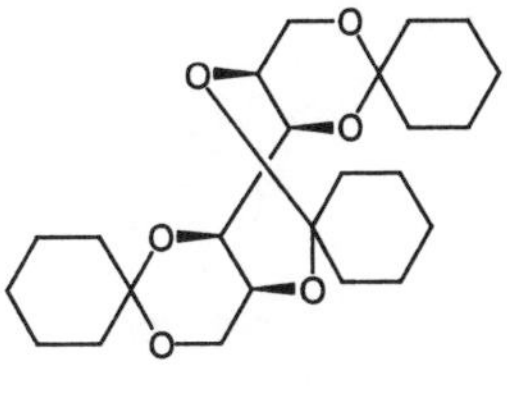

$C_{24}H_{38}O_6$
Chiral intermediate. *Acros Organics NV.*

**4621 (S)-3-Tridecanol**

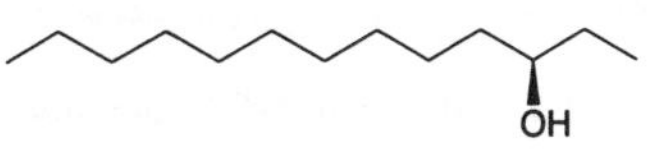

$C_{13}H_{28}O$
Tridecan-3-ol, (S)-; Chiral building block. *Japan Energy Corp.*

**4622 (S)-Triethylphosphine gold (I) 2,3,4,6-tetra-O-acetyl-1-thio-β-D-glucopyranoside**

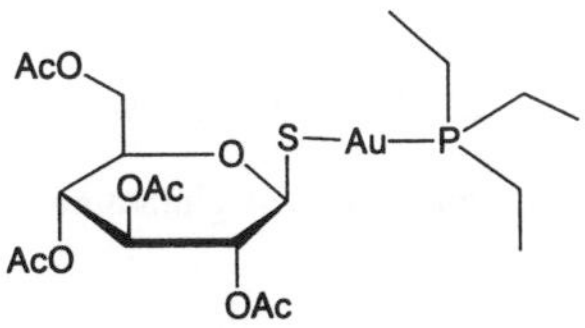

$C_{20}H_{34}AuO_9PS$
Chiral catalyst. mp = 112°. *Pfanstiehl Labs Inc.*

**4623 (3R)-(+)-3-(Trifluoroacetamido)pyrrolidine hydrochloride**
$C_6H_{10}ClF_3N_2O$
Chiral building block. *TCI America.*

**4624 (3S)-(-)-3-(Trifluoroacetamido)pyrrolidine hydrochloride**
$C_6H_{10}ClF_3N_2O$
Chiral building block. mp = 233°. *TCI America.*

**4625 (S)-(-)-2-(Trifluoroacetamido)succinic anhydride**
777-33-3

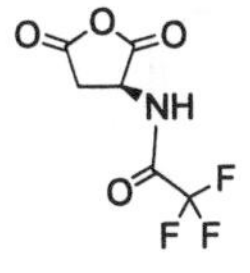

$C_6H_4F_3NO_4$
N-Trifluoroacetyl-L-aspartic acid anhydride. Chiral

intermediate. mp = 131-132°; $[\alpha]_D^{20}$ = - 28.0° (c = 1, THF). *Sigma-Aldrich Fine Chem.*

**4626 (+)-3-(Trifluoroacetyl)-D-camphor**
51800-98-7 257-429-3

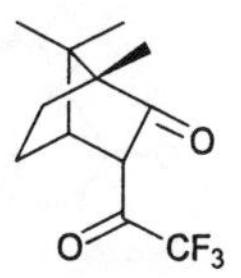

$C_{12}H_{15}F_3O_2$
Bicyclo[2.2.1]heptan-2-one, 1,7,7-trimethyl-3-(trifluoroacetyl)-, (1R)-; (1R)-1,7,7-Trimethyl-3-(trifluoroacetyl)-bicyclo[2.2.1]heptan-2-one. Chiral intermediate. $bp_{16}$ = 100-101°; d = 1.172; n = 1.451; $[\alpha]^{19}$ = + 148° (c = 2.3, $CH_2Cl_2$). *Acros Organics NV; Sigma-Aldrich Fine Chem.*

**4627 (-)-3-(Trifluoroacetyl)-L-camphor**

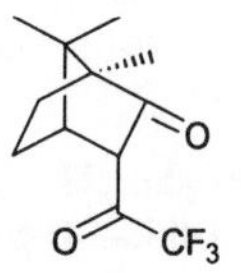

$C_{12}H_{15}F_3O_2$
Chiral intermediate. D = 1.172; n = 1.451; $[\alpha]^{19}$ = - 148.0° (c = 2.5, $CH_2Cl_2$). *Sigma-Aldrich Fine Chem.*

**4628 N-Trifluoroacetyl-L-glutamine**

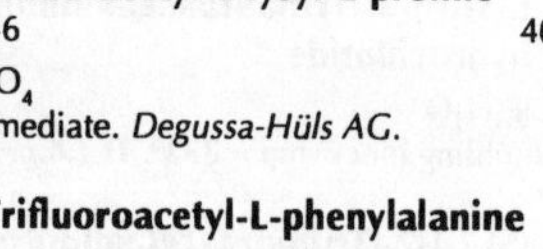

$C_7H_9F_3N_2O_4$
Chiral intermediate. *Acros Organics NV.*

**4629 ε-Trifluoroacetyl-L-lysyl-L-proline**
103300-89-6 402-920-1
$C_{13}H_{20}F_3N_3O_4$
Chiral intermediate. *Degussa-Hüls AG.*

**4630 N-Trifluoroacetyl-L-phenylalanine**

$C_{11}H_{10}F_3NO_3$
Chiral intermediate. *Acros Organics NV.*

**4631 (S)-(-)-N-(Triflouroacetyl)prolyl chloride (0.1 molar solution in $CH_2Cl_2$)**
36724-68-2

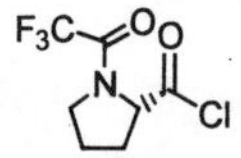

$C_7H_7ClF_3NO_2$
Chiral intermediate. d = 1.308; n = 1.424. *Sigma-Aldrich Fine Chem.*

**4632 (R)-(+)-1,1,1-Trifluoroheptan-2-ol**
$C_7H_{19}F_3O$
Chiral building block. *Sigma-Aldrich Fine Chem.*

**4633 (S)-(-)-1,1,1-Trifluoroheptan-2-ol**
130025-35-3
$C_7H_{13}F_3O$
Chiral building block. bp = 151°; d = 1.06; n = 1.38. *Sigma-Aldrich Fine Chem.*

**4634 (R)-4,4,4-Trifluoro-3-hydroxybutyric acid ethyl ester**
85571-85-3

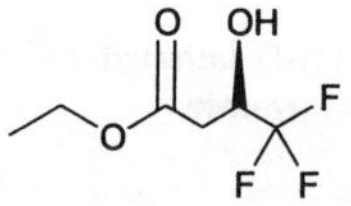

$C_6H_9F_3O_3$
Chiral building block. $bp_{21}$ = 88-90°; d = 1.259; n = 1.375; $[\alpha]^{24}$ = + 14.0° (c = 1, $CHCl_3$). *Rohner Ltd.; Sigma-Aldrich Fine Chem.*

**4635 (S)-(-)-Trifluorolactic acid**
125995-00-8
$C_3H_3F_3O_3$
(S)-(-)-3,3,3-Trifluoro-2-hydroxypropanoic acid. Chiral building block. mp = 72-76°; $[\alpha]_D^{20}$ = - 42° (c = 1, $CHCl_3$). *Sigma-Aldrich Fine Chem.*

**4636 (R)-Trifluoromethylamphetamine**

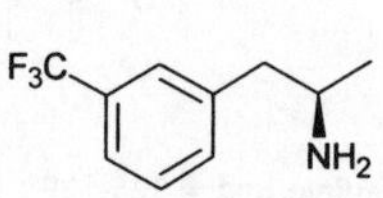

$C_{10}H_{12}F_3N$
Chiral intermediate. *Chiragene.*

**4637 (S)-3-Trifluoromethylamphetamine**

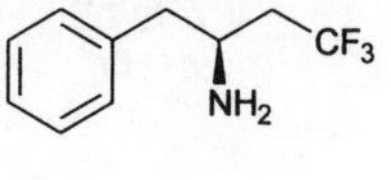

$C_{10}H_{12}F_3N$
Chiral intermediate. *Chiragene.*

**4638 2,2,2-Trifluoro-N-(S)-methylbenzyl-acetamide**
39995-51-2

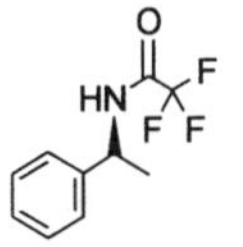

$C_{10}H_{10}F_3NO$
Chiral intermediate. mp = 93-95°; $[\alpha]_D^{20}$ = - 135° (c = 1, $CHCl_3$). *Sigma-Aldrich Fine Chem.*

**4639 (R)-(+)-1,1,1-Trifluorooctan-2-ol**
121170-45-4
$C_8H_{15}F_3O$
Chiral building block. *Sigma-Aldrich Fine Chem.*

**4640 (S)-(-)-1,1,1-Trifluorooctan-2-ol**
129443-08-9
$C_8H_{15}F_3O$
Chiral building block. *Sigma-Aldrich Fine Chem.*

**4641 (1S,2S,8R,8aR)-Trihydroxyindolizidine**
Chiral intermediate. *See Chem. AG.*

**4642 3,3',5-Trihydroxy-4'-methoxy-stilbene-3-O-β-D-glucopyranoside**
155-58-8 8340(12) 205-845-0

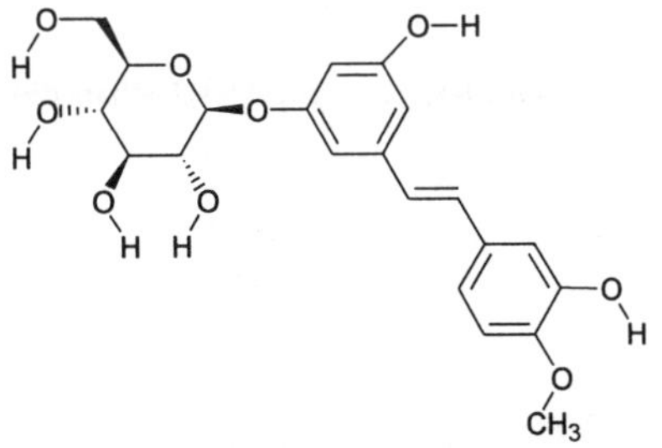

$C_{21}H_{24}O_9$
Glucopyranoside, 3-hydroxy-5-[2-(3-hydroxy-4-methoxyphenyl)ethenyl]phenyl, (E)-, β-D-; Rhapontin; 3-Hydroxy-5-[2-(3-hydroxy-4-methoxyphenyl)vinyl] phenyl-β-D-glucopyranoside; 4'-Methoxy-3,3',5-stilbenetriol-3-β-D-glucopyranoside; Ponticin. Chiral intermediate. mp = 236-240°; $[\alpha]_D^{20}$ = - 55° (c = 0.25, acetone). *Acros Organics NV; Kaden BioChem. GmbH; Sigma-Aldrich Fine Chem.*

**4643 (+)-3,3,5-Triiodo-thyronine**
6893-02-3 5535(12) 229-999-3

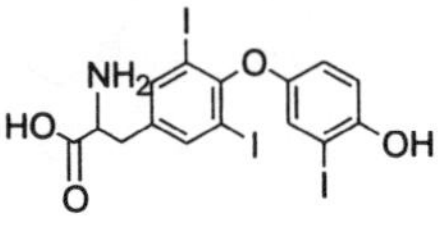

$C_{15}H_{12}I_3NO_4$
Liothyronine. Chiral intermediate. mp = 236-237°; $[\alpha]_D^{20}$ = + 23° (c = 2, 1:2 1N HCl:EtOH). *Sigma-Aldrich Fine Chem.*

**4644 (+)-3,3,5-Triiodothyronine sodium salt hydrate**
200-223-5

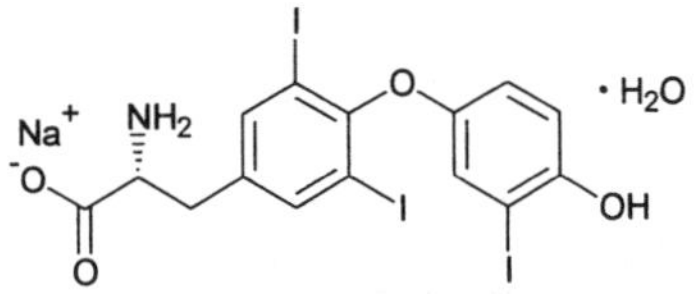

$C_{15}H_{13}I_3NO_5$
Chiral intermediate. mp = 205°; $[\alpha]^{22}$ = + 21.5° (c = 2, 1:2 1N HCl:EtOH). *Sigma-Aldrich Fine Chem.*

**4645 (+)-1,2:3,4:5,6-Tri-O-isopropylidene-mannitol**
3969-59-3

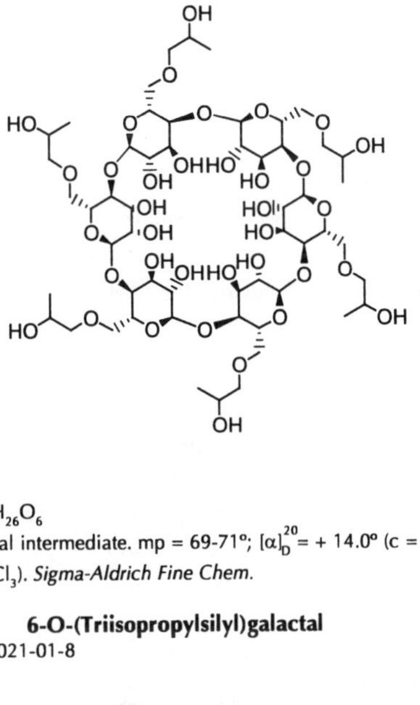

$C_{15}H_{26}O_6$
Chiral intermediate. mp = 69-71°; $[\alpha]_D^{20}$ = + 14.0° (c = 1.7, $CHCl_3$). *Sigma-Aldrich Fine Chem.*

**4646 6-O-(Triisopropylsilyl)galactal**
166021-01-8

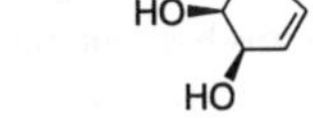

$C_{15}H_{30}O_4Si$
Chiral intermediate. n = 1.488; $[\alpha]^{22}$ = 0 ° (c = 0.9, $CHCl_3$). *Sigma-Aldrich Fine Chem.*

**4647 (-)-6-O-(Triisopropylsilyl)galactal cyclic carbonate**
149625-80-9

$C_{16}H_{28}O_5Si$
Chiral intermediate. mp = - 52°; d = 1.087; n = 1.4825; $[\alpha]^{23}$ = - 69° (c = 2.3, $CHCl_3$). *Sigma-Aldrich Fine Chem.*

**4648 (-)-6-O-(Triisopropylsilyl)glucal**
137915-37-8

$C_{15}H_{30}O_4Si$
Chiral intermediate. d = 0.726; n = 1.489; $[\alpha]^{22}$ = - 2° (c = 1, $CHCl_3$ stabilized with amylenes). *Sigma-Aldrich Fine Chem.*

**4649 1,3:2,5:4,6-Tri-O-methylene-D-mannitol**
5434-31-1 226-600-4

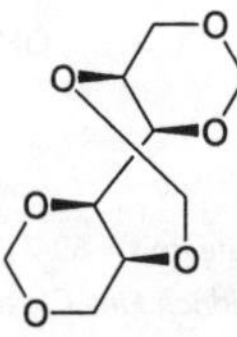

$C_9H_{14}O_6$
Mannitol, 1,3:2,5:4,6-tri-O-methylene-, D-; Chiral intermediate. mp = 230-235°; $[\alpha]^{23}$ = - 105° (c = 2, $CHCl_3$). *Acros Organics NV; Sigma-Aldrich Fine Chem.*

**4650 4β-S-trans-8,8-Trimethyl-4β,5,6,7,8,8α,9,10-octahydro-1-isopropylphenanthren-2-ol**
511-15-9
$C_{20}H_{30}O$
Totarol. Chiral intermediate. mp = 128-132°; $[\alpha]_D^{20}$ = + 41° (c = 1, $CHCl_3$). *Sigma-Aldrich Fine Chem.*

**4651 (S)-(-)-1-Trimethylsilylamino-2-methoxymethylpyrrolidine**

$C_9H_{22}N_2OSi$
Chiral intermediate. *Acros Organics NV.*

**4652 (4S,2RS)-2,5,5-Trimethylthiazolidine-4-carboxylic acid**

$C_7H_{13}NO_2S$
Chiral intermediate. *Acros Organics NV.*

**4653 (R)-(+)-1,1,2-Triphenyl-1,2-ethanediol**
95061-46-4
$C_{20}H_{18}O_2$
Chiral building block. *Advanced Asymmetrics, Inc.; Omega Chem. Co. Inc.; TCI America.*

**4654 (S)-(-)-1,1,2-Triphenyl-1,2-ethanediol**
108998-83-0

$C_{20}H_{18}O_2$
(S)-(-)-Triphenylethylene glycol. Chiral intermediate. mp = 125-127°; $[\alpha]_D^{20}$ = - 214°, (c = 1, EtOH). *Advanced Asymmetrics, Inc.; Omega Chem. Co. Inc.; Sigma-Aldrich Fine Chem.; TCI America.*

**4655 (R)-(+)-1,1,2-Triphenyl-1,2-ethanediol 2-acetate**
95061-47-5

$C_{22}H_{20}O_3$
(R)-(+)-2-Hydroxy-1,2,2-triphenylethyl acetate; (R)-

HYTRA. Chiral intermediate. mp = 237-240°; $[\alpha]_D^{20}$ = + 216 ± 5° (c = 1, pyridine). *Acros Organics NV; Advanced Asymmetrics, Inc.; Lancaster Synthesis Ltd.; Omega Chem. Co. Inc.; Oxford Asymmetry Ltd.; Sigma-Aldrich Fine Chem.*

**4656 (S)-(-)-1,1,2-Triphenyl-1,2-ethanediol 2-acetate**

95061-51-1

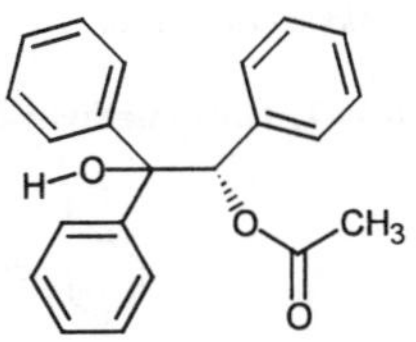

$C_{22}H_{20}O_3$
(S)-(-)-2-Hydroxy-1,2,2-triphenylethyl acetate; (S)-HYTRA. Chiral intermediate. mp = 237-240°; $[\alpha]_D^{20}$ = - 215 ± 5° (c = 1, pyridine). *Acros Organics NV; Advanced Asymmetrics, Inc.; Lancaster Synthesis Ltd.; Omega Chem. Co. Inc.; Oxford Asymmetry Ltd.*

**4657 Tris[3-(heptafluoropropylhydroxy-methylene)-D-camphorato], praseodymium derivative**

38832-94-9

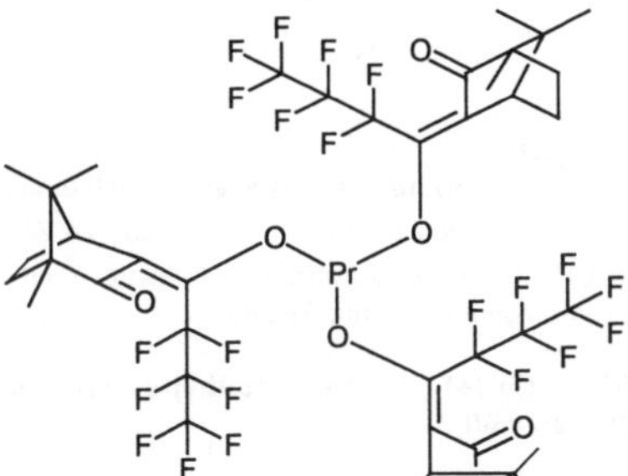

$C_{42}H_{42}F_{21}O_6Pr$
Praseodymium, tris[3-[2,2,3,3,4,4,4-heptafluoro-1-(oxo)-butyl]-1,7,7-trimethylbicyclo[2.2.1]heptan-2-onato]-; Tris-(3-heptafluorobutyryl-d-camphorato) praseodymium(III). Listed on TSCA. Chiral catalyst. mp = 187-189°; $[\alpha]^{21}$ = + 158° (c = 1, $CHCl_3$). *Acros Organics NV; Sigma-Aldrich Fine Chem.*

**4658 Tris[3-(trifluoromethylhydroxymethylene)-D-camphorato], europium(III) derivative**

34830-11-0 252-232-9

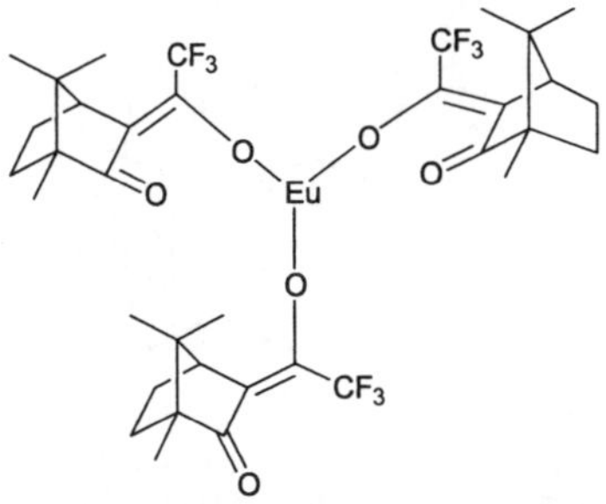

$C_{36}H_{42}EuF_9O_6$
Europium, tris[1,7,7-trimethyl-3-(trifluoroacetyl)bicyclo-[2.2.1]heptan-2-onato]-; Tris(3-trifluoroacetyl-d-camphor-ato)europium(III); tris[3-(trifluoroacetyl)bornane-2-onato-O,O']europium. Listed on TSCA. Chiral catalyst. mp = 195-198°; $[\alpha]^{22}$ = + 168.6° (c = 1, $CCl_4$). *Acros Organics NV; Sigma-Aldrich Fine Chem.*

**4659 Tris(3-(trifluoromethylhydroxymethylene)-D-camphorato), praseodymium derivative**

38053-99-5 253-762-3

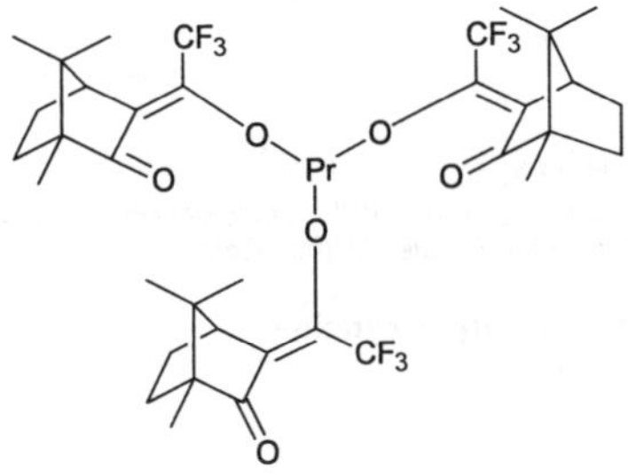

$C_{36}H_{42}F_9O_6Pr$
Praseodymium, tris[1,7,7-trimethyl-3-(trifluoroacetyl)-bicyclo[2.2.1]heptan-2-onato-O,O']-; tris[1,7,7-trimethyl-3-(trifluoroacetyl)bicyclo[2.2.1]heptan-2-onato-O,O']-praseodymium; Tris(3-trifluoroacetyl-d-camphorato)-praseodymium(III). Chiral catalyst. mp = 210-212°; [α = + 222° (c = 1.3, $CCl_4$). *Acros Organics NV; Sigma-Aldrich Fine Chem.*

**4660 Tris[3-(trifluoromethylhydroxymethylene)-D-camphorato], ytterbium(III) derivative**
38054-03-4 253-764-4

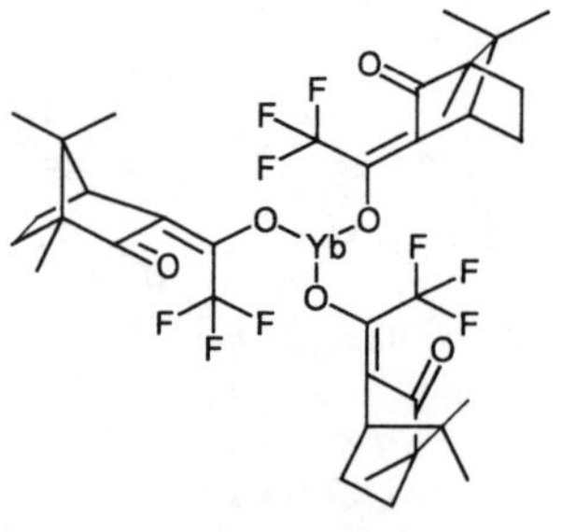

$C_{36}H_{42}F_9O_6Yb$
Ytterbium, tris[1,7,7-trimethyl-3-(trifluoroacetyl)-bicyclo[2.2.1]heptan-2-onato-O,O']-; tris[1,7,7-trimethyl-3-(trifluoroacetyl)bicyclo[2.2.1]heptan-2-onato-O,O']-ytterbium; Yb-Optishfit. Chiral catalyst. mp = 213°; [α] = + 168.4° (c = 1, $CHCl_3$). *Acros Organics NV; Sigma-Aldrich Fine Chem.*

**4661 (R)-1,2,4-Tritosyl butanetriol**

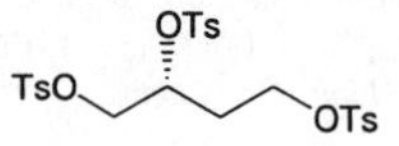

$C_{25}H_{28}O_9S_3$
Butane-1,2,4-triol, tris(4-methylbenzenesulfonate), (R)-; Chiral intermediate. *Synthon Corp.*

**4662 (S)-1,2,4-Tritosyl butanetriol**
99520-83-9

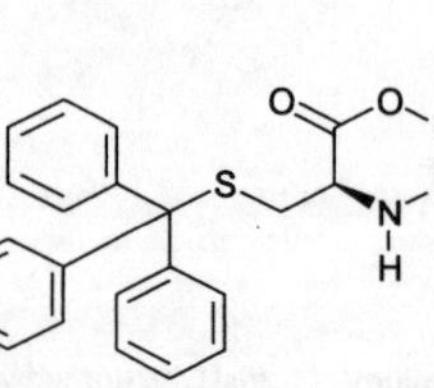

$C_{25}H_{28}O_9S_3$
Butane-1,2,4-triol, tris(4-methylbenzenesulfonate), (S)-; Chiral intermediate. *Synthon Corp.*

**4663 S-Trityl-L-cysteine**
2799-07-7

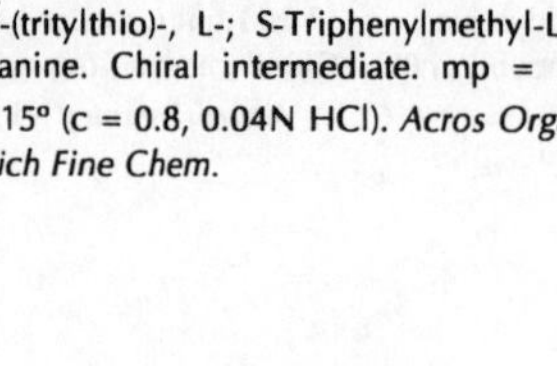

$C_{22}H_{21}NO_2S$
Alanine, 3-(tritylthio)-, L-; S-Triphenylmethyl-L-cysteine; Tritylthioalanine. Chiral intermediate. mp = 182-183°; $[\alpha]^{25}$ = + 115° (c = 0.8, 0.04N HCl). *Acros Organics NV; Sigma-Aldrich Fine Chem.*

**4664 (-)-5-O-Trityl-2-deoxy-lyxofuranosyl-thymine**
55612-11-8

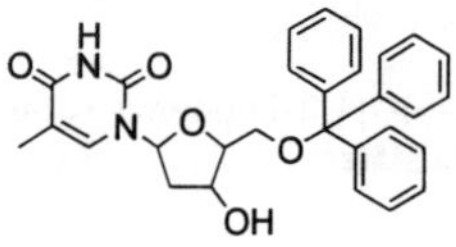

$C_{29}H_{28}N_2O_5$
Chiral intermediate. mp = 246-248°; $[\alpha]_D^{20}$ = - 39° (c = 0.5, DMF). *Sigma-Aldrich Fine Chem.*

**4665 (R)-(-)-γ-Trityloxymethyl-γ-butyrolactone**
78158-90-4

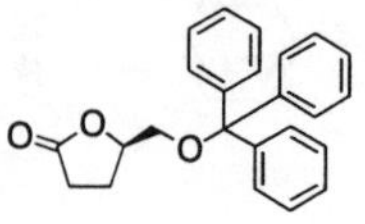

$C_{24}H_{22}O_3$
Chiral intermediate. mp = 150-152°; $[\alpha]_D^{20}$ = - 26° (c = 1, $CH_2Cl_2$). *Oxford Asymmetry Ltd.; Sigma-Aldrich Fine Chem.*

**4666 (S)-(+)-γ-Trityloxymethyl-γ-butyrolactone**
73968-62-4

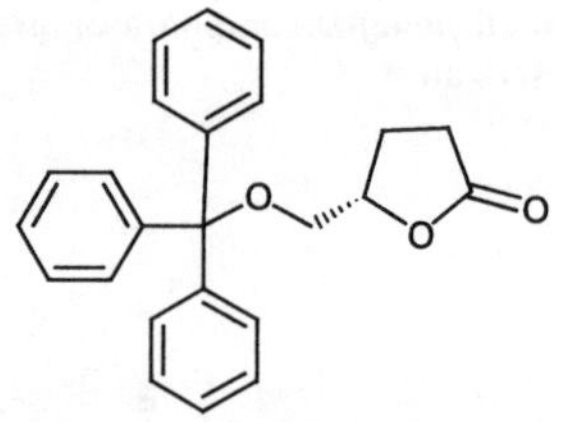

$C_{24}H_{22}O_3$
(S)-(+)-4,5-Dihydro-5-trityloxymethyl-2(3H)-furanone. Chiral intermediate. mp = 150-152°; $[\alpha]_D^{20}$ = + 26 ± 1° (c = 1, $CH_2Cl_2$). *Acros Organics NV; Lancaster Synthesis Ltd.; Oxford Asymmetry Ltd.; Sigma-Aldrich Fine Chem.*

**4667 (S)-(+)-5-Trityloxymethyl-2-pyrrolidinone**
105526-85-0

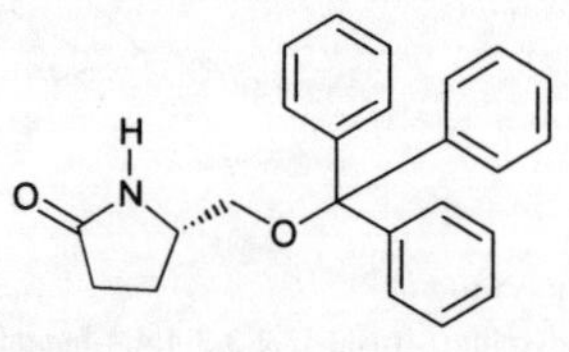

$C_{24}H_{23}NO_2$
Chiral intermediate. mp = 164-166°; $[\alpha]_D^{20}$ = + 15° (c = 1, $CHCl_3$). *Acros Organics NV; See Chem. AG; Sigma-Aldrich Fine Chem.*

**4668 N-Tritylserine methyl ester**
4465-44-5

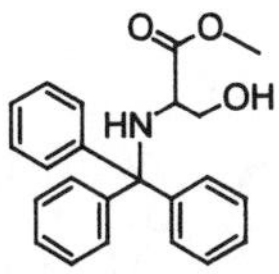

$C_{23}H_{23}NO_3$
Chiral intermediate. mp = 148-150°; $[\alpha]_D^{20}$ = + 31° (c = 1, $CH_3OH$). *Sigma-Aldrich Fine Chem.*

**4669 6-Trityl-1,2,3,4-tetra-O-acetyl-β-D-glucose**

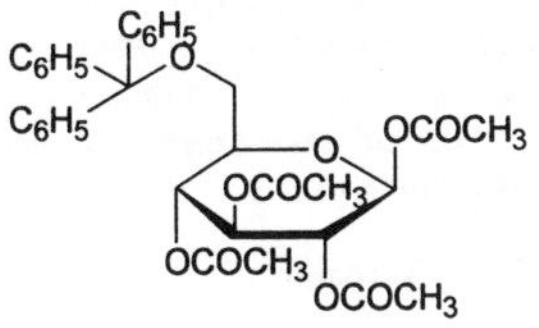

$C_{33}H_{34}O_{10}$
Chiral intermediate. *See Chem. AG.*

**4670 (R)-Trolox**

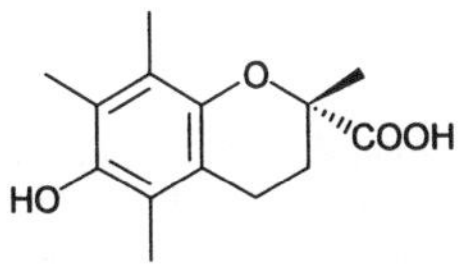

$C_{14}H_{18}O_4$
Benzo-2H-1-pyran-2-carboxylic acid, 3,4-dihydro-6-hydroxy-2,5,7,8-tetramethyl-, (R)-; (R)-6-Hydroxy-2,5,7,8-tetramethylchromane-2-carboxylic acid; (R)-3,4-Dihydro-6-hydroxy-2,5,7,8-tetramethyl-2H-1-benzopyran-2-carboxylic acid. Chiral intermediate. *Expansia UK & Ireland.*

**4671 (S)-Trolox**

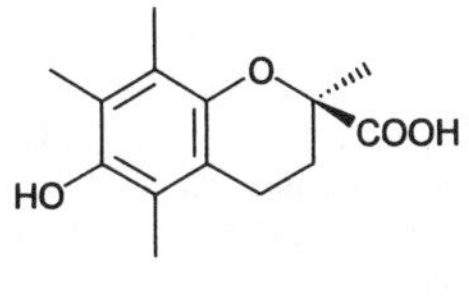

$C_{14}H_{18}O_4$
Benzo-2H-1-pyran-2-carboxylic acid, 3,4-dihydro-6-hydroxy-2,5,7,8-tetramethyl-, (S)-; (S)-6-Hydroxy-2,5,7,8-tetramethylchromane-2-carboxylic acid; (S)-3,4-Dihydro-6-hydroxy-2,5,7,8-tetramethyl-2H-1-benzopyran-2-carboxylic acid. Chiral intermediate. *Expansia UK & Ireland.*

**4672 D-Tryptophan**
153-94-6 205-819-9

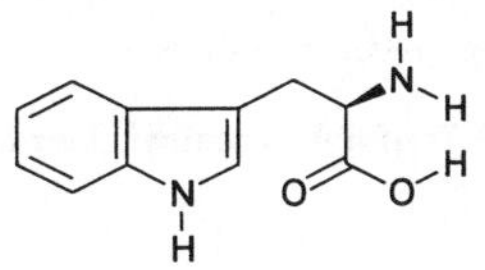

$C_{11}H_{12}N_2O_2$
D-Tryptophan; (R)-2-Amino-3-(3-indolyl)propionic acid; 3,3'-indolyl-2-alanine. Listed on TSCA. Chiral building block. mp = 282-285°; $[\alpha]_D^{20}$ = + 31.5 ± 2° (c = 1, $H_2O$). *Acros Organics NV; Ajinomoto Co. Inc.; Austin Chem. Co.; DSM Fine Chem.; Lancaster Synthesis Ltd.; Sigma-Aldrich Fine Chem.; Tanabe Seiyaku Co. Ltd.; TCI America; Yoneyama Yakuhin Kogyo Co. Ltd.*

**4673 L-Tryptophan**
73-22-3 9929(12) 200-795-6

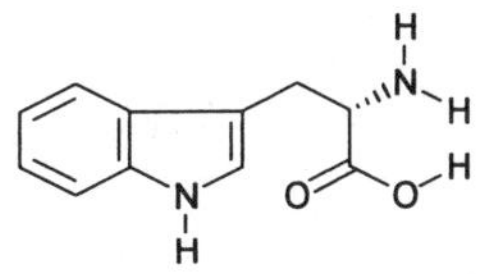

$C_{11}H_{12}N_2O_2$
Indole-3-propanoic acid, α-amino-, 1H-, (S)-; (S)-2-Amino-3-(3-indolyl)propionic acid; 1H-Indole-3-alanine, (S)-; 3-Indol-3-ylalanine; EH 121; l-β-3-Indolylalanine; L-Alanine, 3-(1H-indol-3-yl)-. Listed on TSCA. Chiral building block. mp = 281-284°; $[\alpha]_D^{20}$ = - 31.5 ± 2° (c = 1, $H_2O$). *Acros Organics NV; Ajinomoto Co. Inc.; Austin Chem. Co.; Kyowa Hakko Kogyo Co. Ltd.; Lancaster Synthesis Ltd.; Mitsui Pharm. Inc.; Nippon Kayaku Co. Ltd.; See Chem. AG; Shanghai DSL Intl. trading Co.; Sigma-Aldrich Fine Chem.; Tanabe Seiyaku Co. Ltd.; TCI America; Varsal Instruments, Inc.; Yoneyama Yakuhin Kogyo Co. Ltd.*

**4674 L-Tryptophan amide hydrochloride**

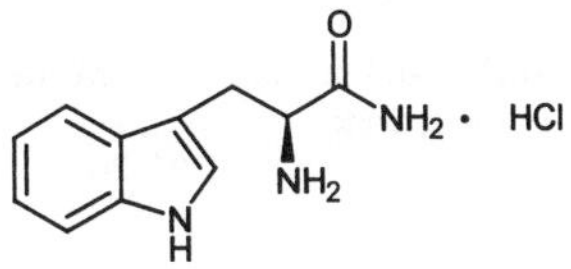

$C_{11}H_{14}ClN_3O$
Chiral building block. *See Chem. AG.*

**4675 D-Tryptophan benzyl ester**
141595-98-4
$C_{18}H_{18}N_2O_2$
Chiral building block. mp = 77-80°; $[\alpha]_D^{20}$ = - 9.9° (c = 1, EtOH). *Sigma-Aldrich Fine Chem.*

**4676 L-Tryptophan benzyl ester**
4299-69-8

$C_{19}H_{18}N_2O_2$
Chiral building block. mp = 77-80°; $[\alpha]_D^{20}$ = + 9.9° (c = 1, EtOH). *Sigma-Aldrich Fine Chem.*

**4677 D-Tryptophan methyl ester hydrochloride**
14907-17-8

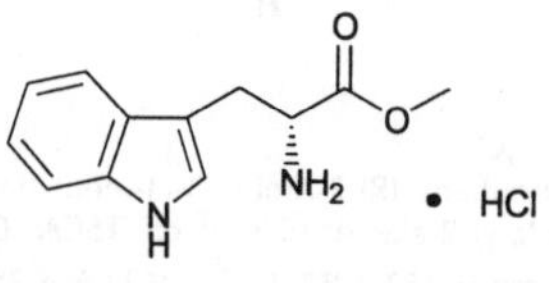

$C_{12}H_{15}ClN_2O_2$
Chiral intermediate. mp = 213-216°; $[\alpha]_D^{20}$ = - 18° (c = 5, $CH_3OH$). *Austin Chem. Co.; Norchim SA; Sigma-Aldrich.*

**4678 L-Tryptophan methyl ester hydrochloride**
7524-52-9 231-385-5

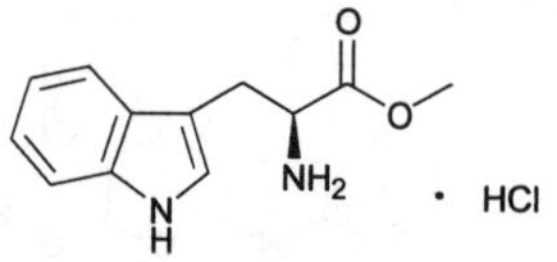

$C_{12}H_{15}ClN_2O_2$
Tryptophan, methyl ester, monohydrochloride, L-; Methyl L-tryptophanate monohydrochloride. Chiral building block. mp = 218-220°; $[\alpha]_D^{20}$ = + 18° (c = 5, $CH_3OH$). *Sigma-Aldrich Fine Chem.; TCI America.*

**4679 L-(-)-Tryptophanol**
2899-29-8

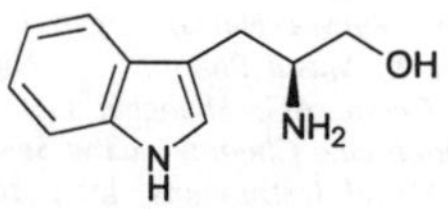

$C_{11}H_{14}N_2O$
(S)-(-)-2-Amino-3-(3-indolyl)propanol. Chiral building block. mp = 83°; $[\alpha]_D^{20}$ = - 20.5° (c = 1, $CH_3OH$). *Omega Chem. Co. Inc.; Sigma-Aldrich Fine Chem.; TCI America.*

**4680 Tu(tert-butoxycarbonyl)urarine chloride pentahydrate**
6989-98-6 9717(11) 200-356-9

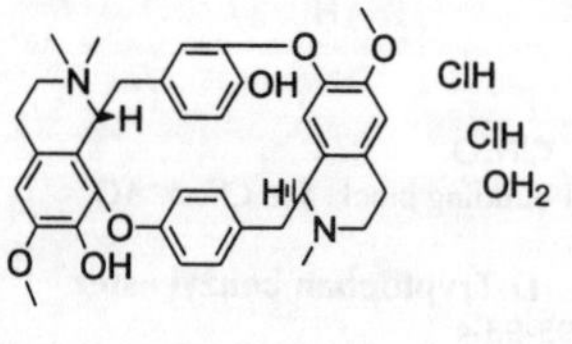

$C_{37}H_{42}Cl_2N_2O_6$
Chiral intermediate. mp = 268°; $[\alpha]^{19}$ = + 190° (c = 0.5, $H_2O$). *Sigma-Aldrich Fine Chem.*

**4681 D-(+)-Turanose**
547-25-1 9951(12) 208-918-5

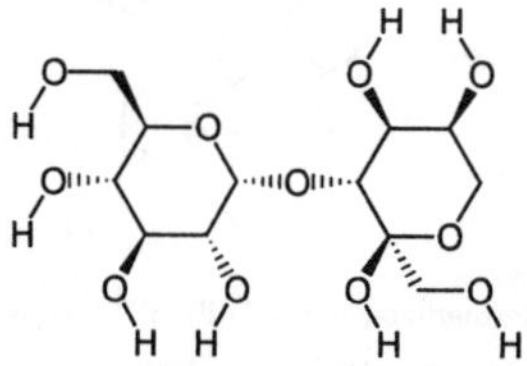

$C_{12}H_{22}O_{11}$
Fructose, 3-O-α-D-glucopyranosyl-,; 3-O-α-D-Glucopyranosyl-D-fructose. Chiral building block. mp = 170°; $[\alpha]_D^{20}$ = + 75.8° (c = 4, $H_2O$). *Acros Organics NV; Pfanstiehl Labs Inc.; See Chem. AG; Sigma-Aldrich Fine Chem.; TCI America.*

**4682 D-(+)-Tyrosine**
556-02-5 9747(11) 209-112-6

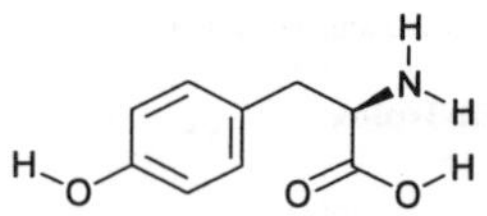

$C_9H_{11}NO_3$
Tyrosine, D-; Listed on TSCA. Chiral building block. mp > 300°; $[\alpha]_D^{20}$ = + 10.3° (c = 4, 1N HCl). *Acros Organics NV; Austin Chem. Co.; Great Lakes Fine Chem.; Rexim/Degussa Europe; Sigma-Aldrich Fine Chem.; TCI America.*

**4683 L-(-)-Tyrosine**
60-18-4 9970(12) 200-460-4

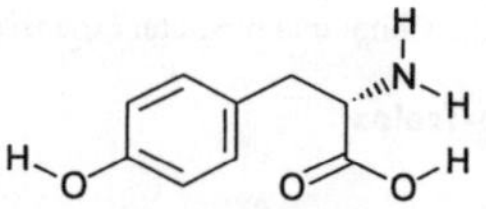

$C_9H_{11}NO_3$
Benzenepropanoic acid, α-amino-4-hydroxy-, (S)-; (S)-2-Amino-3-(4-hydroxyphenyl)propanoic acid; L-Phenylalanine, 4-hydroxy-. Listed on TSCA. Chiral building block. mp > 300°; $[\alpha]_D^{20}$ = - 10.6° (c = 4, 1N HCl). *Aceto Corp.; Acros Organics NV; Ajinomoto Co. Inc.; Austin Chem. Co.; Daesang Corp.; Freedom Chem. Diamalt GmbH; KingChem Inc.; Kyowa Hakko Kogyo Co. Ltd.; Rexim/Degussa Europe; See Chem. AG; Shanghai DSL Intl. trading Co.; Sigma-Aldrich Fine Chem.; Sochinaz SA; Sun-Orient Chemial Co. Ltd.; Tanabe Seiyaku Co. Ltd.; TCI America; Varsal Instruments, Inc.; Yoneyama Yakuhin Kogyo Co. Ltd.*

**4684 L-Tyrosine amide**

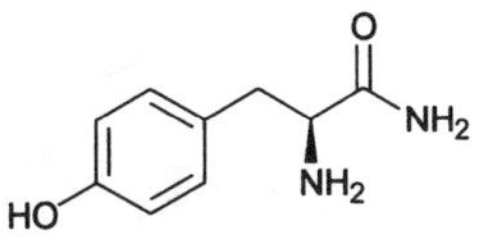

$C_9H_{12}N_2O_2$
Chiral building block. *See Chem. AG.*

**4685 L-Tyrosine disodium Salt**

$C_9H_{11}Na_2NO_3$
Chiral building block. *TCI America.*

**4686 L-Tyrosine ethyl ester**
949-67-7 213-442-6

$C_{11}H_{15}NO_3$
Tyrosine, ethyl ester, L-; Ethyl L-tyrosinate. Listed on TSCA. Chiral building block. mp = 103°. *TCI America.*

**4687 L-Tyrosine ethyl ester hydrochloride**
4089-07-0 223-820-0

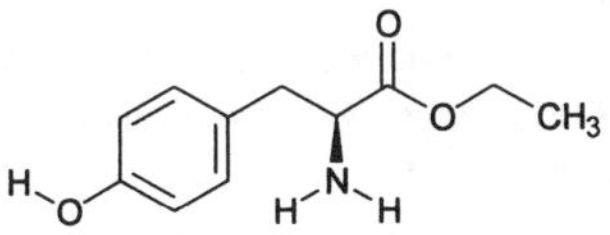

$C_{11}H_{16}ClNO_3$
Tyrosine, ethyl ester, hydrochloride, L-; Ethyl L-tyrosinate hydrochloride. Chiral building block. mp = 166°. *Acros Organics NV; Austin Chem. Co.; TCI America.*

**4688 L-Tyrosine hydrazide**
7662-51-3 231-631-1

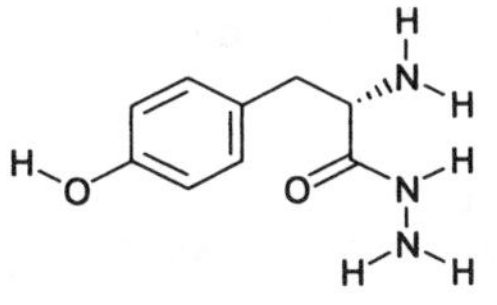

$C_9H_{13}N_3O_2$
Tyrosine, hydrazide, L-; Listed on TSCA. Chiral building block. mp = 196-198°; $[\alpha]_D^{20}$ = + 76° (c = 4.2, 1N HCl). *Acros Organics NV; See Chem. AG; Sigma-Aldrich Fine Chem.*

**4689 L-Tyrosine hydrochloride**
16870-43-2

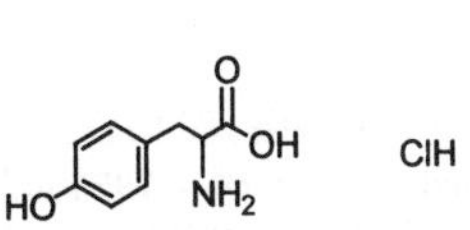

ClH

$C_9H_{12}ClNO_3$
Tyrosine, hydrochloride, L-; Chiral building block. mp = 239°; $[\alpha]_D^{20}$ = - 8° (c = 4, 1N HCl). *Sigma-Aldrich Fine Chem.; Tanabe Seiyaku Co. Ltd.*

**4690 L-Tyrosine methyl ester**
1080-06-4 214-095-3

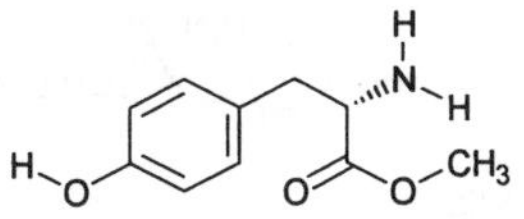

$C_{10}H_{13}NO_3$
Tyrosine, methyl ester, L-; Methyl L-tyrosinate. Chiral building block. mp = 133-135; [α] =+ 25.7° (c = 2.4, EtOH). *Acros Organics NV; Sigma-Aldrich Fine Chem.*

**4691 L-Tyrosine methyl ester hydrochloride**
3417-91-2 222-313-3

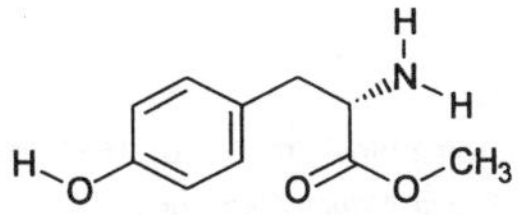

$C_{10}H_{14}ClNO_3$
Tyrosine, methyl ester, hydrochloride, L-; Methyl L-tyrosinate hydrochloride. Listed on TSCA. Chiral building block. mp = 192°; $[\alpha]^{22}$ = - 5.2° (c = 2.4, $H_2O$). *Acros Organics NV; Austin Chem. Co.; Loba Feinchemie AG; Sigma-Aldrich Fine Chem.; TCI America.*

**4692 L-Tyrosine-O-sulfate sodium salt**

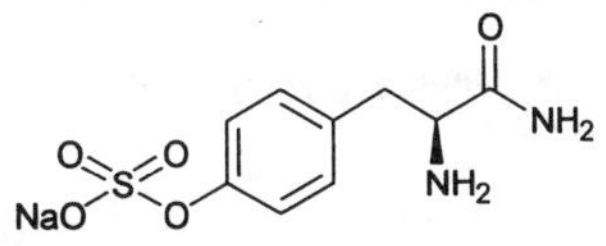

$C_9H_{11}N_2NaO_5S$
Chiral building block. *See Chem. AG.*

**4693 D-Tyrosinol**

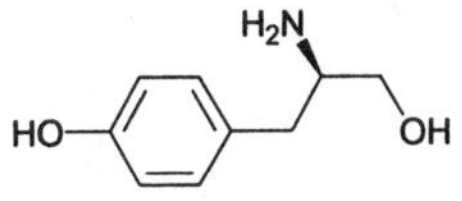

$C_9H_{13}NO_2$
Chiral building block. *Omega Chem. Co. Inc.*

**4694 L-Tyrosinol**
5034-68-4

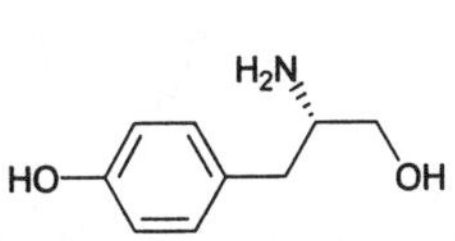

$C_9H_{13}NO_2$
Chiral building block. *Omega Chem. Co. Inc.*

**4695 (-)-Tyrosinol hydrochloride**

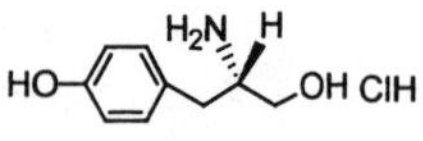

$C_9H_{14}ClNO_2$
Chiral building block. mp = 161-165°; $[\alpha]^{22}$ = - 18° (c = 1, $H_2O$). *Sigma-Aldrich Fine Chem.*

**4696 (+)-Tyrosinol hydrochloride**

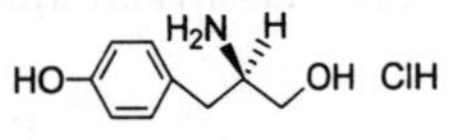

$C_9H_{14}ClNO_2$
Chiral building block. mp = 161-165°; $[\alpha]^{22}$ = + 18° (c = 1, $H_2O$). *Sigma-Aldrich Fine Chem.*

**4697 (S)-3-Undecanol**
79090-61-2

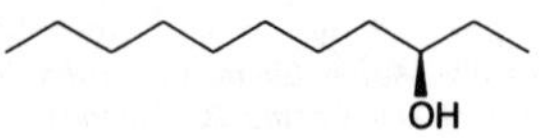

$C_{11}H_{24}O$
Undecan-3-ol, (S)-.
Chiral building block. *Japan Energy Corp.*

**4698 (R)-Undecyl glycidyl ether**
122608-92-8

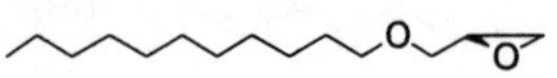

$C_{14}H_{28}O_2$
Chiral building block. $bp_4$ = 120-121°; d = 0.87; $[\alpha]_D^{20}$ = + 4.6° (neat). *Japan Energy Corp.; Sigma-Aldrich Fine Chem.*

**4699 Uridine**
58-96-8 200-407-5

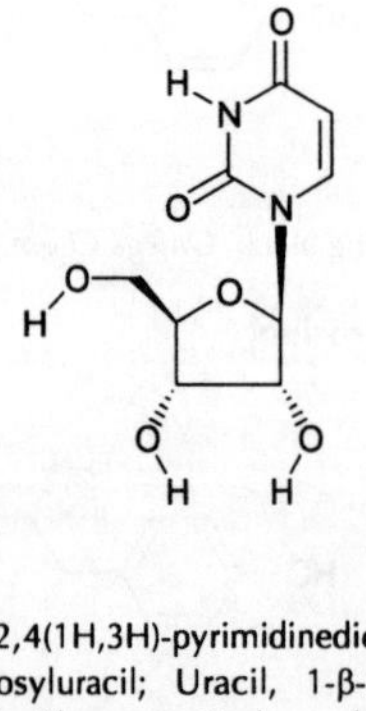

$C_9H_{12}N_2O_6$
Ribofuranoside, 2,4(1H,3H)-pyrimidinedione-1, β-D-. 1-β-D-Ribofuranosyluracil; Uracil, 1-β-D-ribofuranosyl. Listed on TSCA. Pharmaceutical or derivative. $[\alpha]$ = + 8.4° (c = 2, $H_2O$). *Acros Organics NV.*

**4700 Ursolic acid**
77-52-1 10027(11) 201-034-0

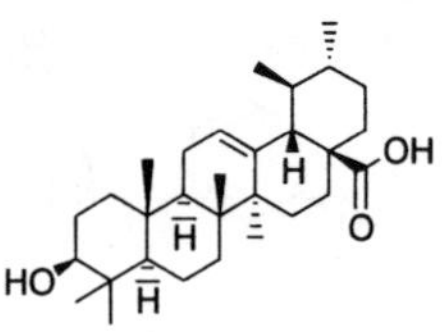

$C_{30}H_{48}O_3$
Listed on TSCA. Chiral intermediate. mp = 292°; $[\alpha]_D^{20}$ = + 67° (c = 1, alc. KOH). *Sigma-Aldrich Fine Chem.*

**4701 Usnic acid**
7562-61-0 231-456-0

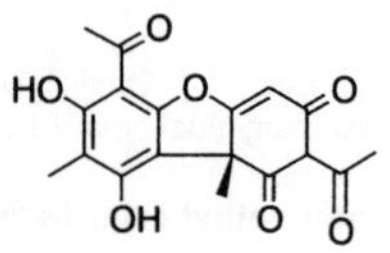

$C_{18}H_{16}O_7$
Chiral intermediate. mp = 201-203°; $[\alpha]^{25}$ = + 488° (c = 0.4, $CHCl_3$). *Kaden BioChem. GmbH; See Chem. AG; Sigma-Aldrich Fine Chem.*

**4702 Usnic acid sodium salt dihydrate**

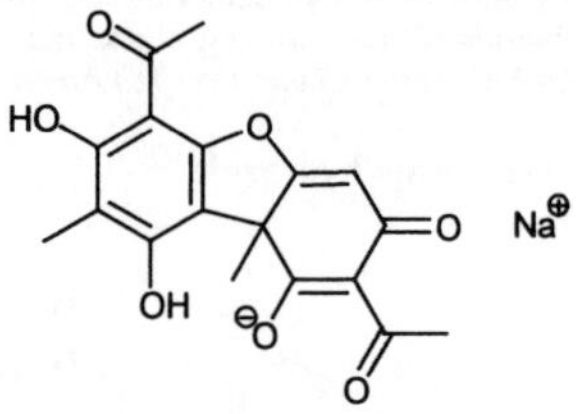

$C_{18}H_{15}NaO_7$
Chiral intermediate. *See Chem. AG.*

**4703 D-Valine**
640-68-6 211-368-9

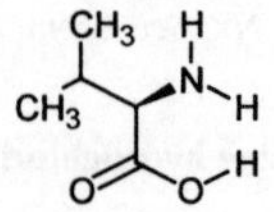

$C_5H_{11}NO_2$
Valine, D-; D-2-Aminoisovaleric acid. Listed on TSCA. Chiral building block. mp = 295°; $[\alpha]^{23}$ = - 27.35° (c = 3.4, 6N HCl). *Acros Organics NV; Austin Chem. Co.; DSM Fine Chem.; Fischer Chem. AG; Great Lakes Fine Chem.; Kaneka Corp.; Omega Chem. Co. Inc.; Recordati S.p.A.; Sigma-Aldrich Fine Chem.; Tanabe Seiyaku Co. Ltd.; TCI America; Yoneyama Yakuhin Kogyo Co. Ltd.*

**4704 L-Valine**
72-18-4 10046(11) 200-773-6

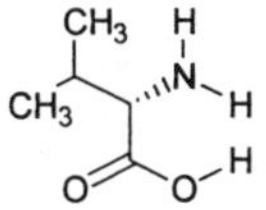

$C_5H_{11}NO_2$
Butanoic acid, 2-amino-3-methyl-, (S)-; L-2-Amino-isovaleric acid; 2-Amino-3-methylbutyric acid. Listed on TSCA. Chiral building block. mp = 295-300°; $[\alpha]_D^{20}$ = + 27.5° (c = 8, 6N HCl). *Aceto Corp.; Acros Organics NV; Ajinomoto Co. Inc.; Austin Chem. Co.; DSM Fine Chem.; Kyowa Hakko Kogyo Co. Ltd.; Rexim/Degussa Europe; See Chem. AG; Shanghai DSL Intl. trading Co.; Sigma-Aldrich Fine Chem.; Tanabe Seiyaku Co. Ltd.; TCI America; Varsal Instruments, Inc.; Yoneyama Yakuhin Kogyo Co. Ltd.*

**4705 L-Valine amide**

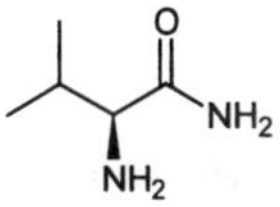

$C_5H_{12}N_2O$
Chiral building block. *Flamma s.p.a.*

**4706 L-Valine amide hydrochloride**
3014-80-0

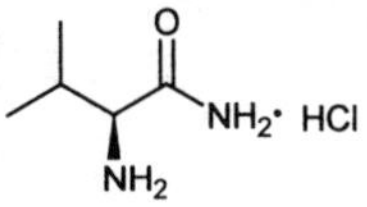

$C_5H_{13}ClN_2O$
Chiral building block. mp = 266-270°; $[\alpha]^{25}$ = + 25.8° (c = 1, $H_2O$). *See Chem. AG; Sigma-Aldrich Fine Chem.*

**4707 L-Valine ethyl ester hydrochloride**
17609-47-1 241-580-7

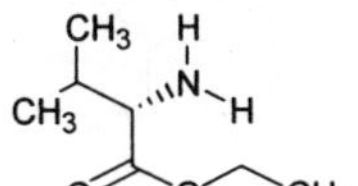

$C_7H_{16}ClNO_2$
Valine, ethyl ester, hydrochloride, L-; Ethyl L-valinate hydrochloride. Chiral building block. mp = 102-105°; $[\alpha]_D^{20}$ = + 6.7° (c = 2, $H_2O$). *Acros Organics NV; Sigma-Aldrich Fine Chem.*

**4708 L-Valine hydroxysuccinimide ester**

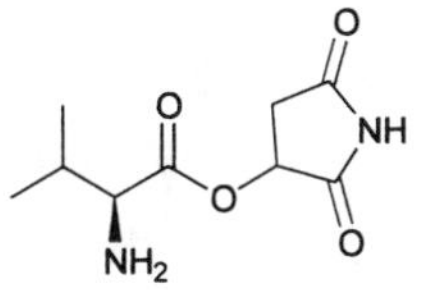

$C_9H_{14}N_2O_4$
Chiral intermediate. *Flamma s.p.a.*

**4709 D-Valine methyl ester hydrochloride**

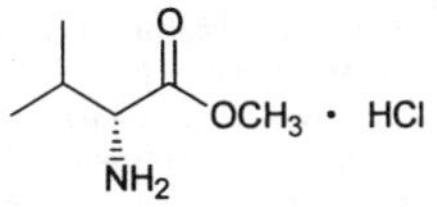

$C_6H_{14}ClNO_2$
Chiral building block. *Austin Chem. Co.*

**4710 L-Valine methyl ester hydrochloride**
6306-52-1 228-620-9

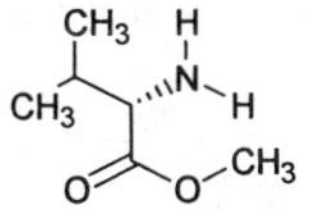

$C_6H_{13}NO_2$
Valine, methyl ester, hydrochloride, L-; L-2-Aminoisovaleric acid methyl ester hydrochloride; Methyl valinate hydrochloride. Chiral building block. mp = 171-173°; $[\alpha]^{21}$ = + 23.6° (c = 2, $CH_3OH$). *Acros Organics NV; Austin Chem. Co.; Flamma s.p.a.; Loba Feinchemie AG; Sigma-Aldrich Fine Chem.; TCI America.*

**4711 D-(-)-Valinol**
4276-09-9

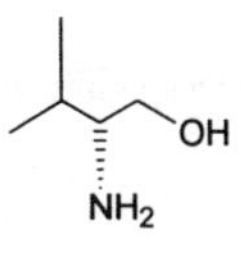

$C_5H_{13}NO$
(R)-(-)-2-Amino-3-methyl-1-butanol. Chiral building block. mp = 30-34°; bp = 189-190°; d = 0.9310; $n_D^{20}$ = 1.4550; $[\alpha]_D^{20}$ = - 11 ± 1° (c = 10, $H_2O$); $[\alpha]_D^{20}$ = - 16° (c = 10, EtOH). *Acros Organics NV; Boehringer Ingelheim Pharma KG; Fischer Chem. AG; Great Lakes Fine Chem.; Lancaster Synthesis Ltd.; Omega Chem. Co. Inc.; Oxford Asymmetry Ltd.; Sigma-Aldrich Fine Chem.; SK Energy & Chem.*

**4712 L-(+)-Valinol**
2026-48-4 217-975-5

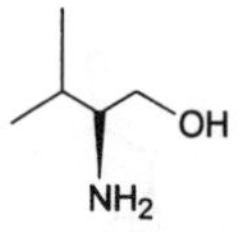

$C_5H_{13}NO$
Butan-1-ol, 2-amino-3-methyl-, (S)-; (S)-(+)-2-Amino-3-methyl-1-butanol; L-2-Amino-3-methyl-1-butanol. Chiral auxiliary; Chiral building block. mp = 30-32°; $bp_8$ = 80-81°; d = 0.9260; $n_D^{20}$ = 1.4548; $[\alpha]_D^{20}$ = + 11 ± 1° (c = 10, $H_2O$); $[\alpha]_D^{22}$ = + 14.6° (neat). *Acros Organics NV; Boehringer Ingelheim Pharma KG; Great Lakes Fine Chem.; Lancaster Synthesis Ltd.; Loba Feinchemie AG; Nippon Rikagakuyakuhin; Omega Chem. Co. Inc.; Oxford Asymmetry Ltd.; Pfanstiehl Labs Inc.; See Chem. AG; Sigma-Aldrich Fine Chem.; SK Energy & Chem.; TCI America.*

**4713 Valinomycin**
2001-95-8 217-896-6

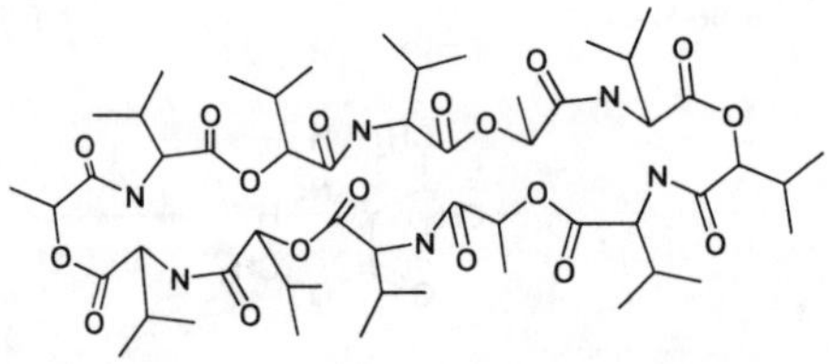

$C_{54}H_{90}N_6O_{18}$
1,7,13,19,25,31-Hexaoxa-4,10,16,22,28,34-hexaazacyclohexatriacontane-2,5,8,11,14,17,20,23,26,29,32,35-dodecone, 3,6,9,15,18,21,27,30,33-nonaisopropyl-12,24,36-trimethyl-; Cyclo(D-α-hydroxyisovaleryl-D-valyl-L-lactoyl-L-valyl-D-α-hydroxyisovaleryl-D-valyl-L-lactoyl-L-valyl-D-α-hydroxyisovaleryl-D-valyl-L-lactoyl-L-valyl); NSC 122023. Listed on TSCA. Pharmaceutical or derivative. mp = 187-189°; [α] = + 31.4° (c = 1.6, $C_6H_6$). *Acros Organics NV.*

**4714 (R)-Vanillylsuccinic acid monomethyl ester**

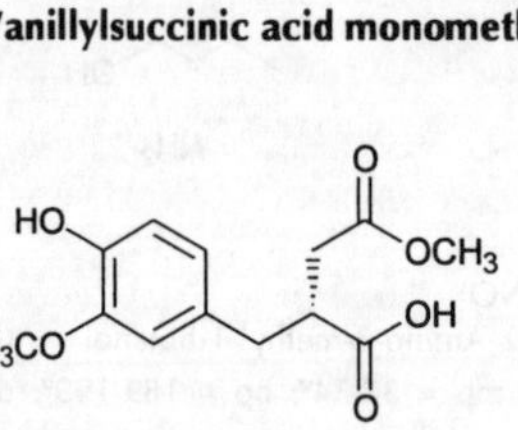

$C_{13}H_{16}O_6$
Chiral intermediate. *Mitsubishi Corp.; Toyotama Perfumery Co. Ltd.*

**4715 (S)-Vanillylsuccinic acid monomethyl ester**

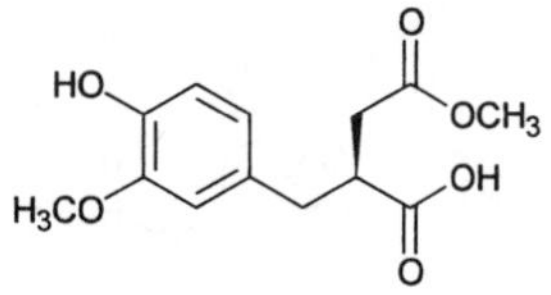

$C_{13}H_{16}O_6$
Chiral intermediate. *Mitsubishi Corp.; Toyotama Perfumery Co. Ltd.*

**4716 (R)-Veratrylsuccinic acid monomethyl ester**

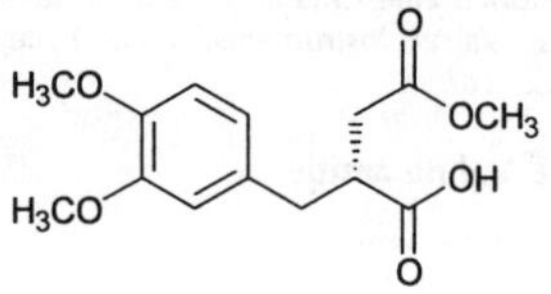

$C_{14}H_{18}O_6$
Chiral intermediate. *Mitsubishi Corp.; Toyotama Perfumery Co. Ltd.*

**4717 (S)-Veratrylsuccinic acid monomethyl ester**

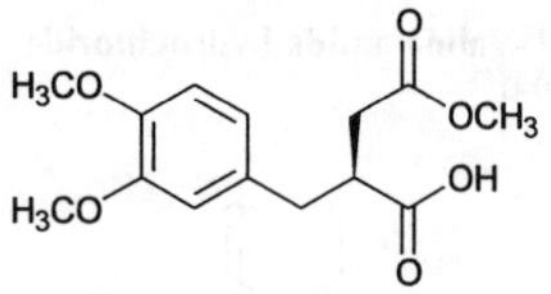

$C_{14}H_{18}O_6$
Chiral intermediate. *Mitsubishi Corp.; Toyotama Perfumery Co. Ltd.*

**4718 (S)-cis-Verbenol**
18881-04-4 242-645-2

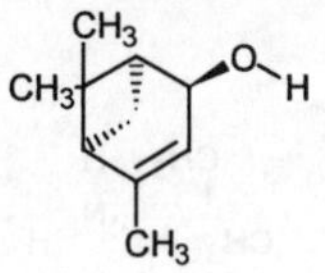

$C_{10}H_{16}O$
Bicyclo[3.1.1]hept-3-en-2-ol, 4,6,6-trimethyl-, (1S,2S,5S)-; [1S-(1α,2β,5α)]-4,6,6-Trimethylbicyclo-[3.1.1]hept-3-en-2-ol. Listed on TSCA. Chiral intermediate. mp = 62-68°. *Acros Organics NV; Sigma-Aldrich Fine Chem.*

**4719 L-(-)-Verbenone**
1196-01-6 214-807-2

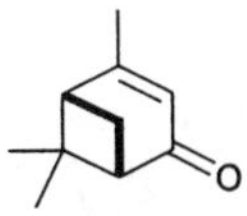

$C_{10}H_{14}O$
Verbenone, (-)-; (1S)-4,6,6-Trimethylbicyclo[3.1.1]hept-3-en-2-one; Bicyclo[3.1.1]hept-3-en-2-one, 4,6,6-trimethyl-, (1S,5S)-; Pin-2-en-4-one. Listed on TSCA. Chiral intermediate. bp = 227-228°; d = 0.974; n = 1.497; $[\alpha]^{19}$ = -193 (c = 10, EtOH); $[\alpha]_D^{20}$ = - 142° (neat). *Acros Organics NV; Sigma-Aldrich Fine Chem.*

**4720 (2R,5R)-(+)-5-Vinyl-2-quinuclidine-methanol**

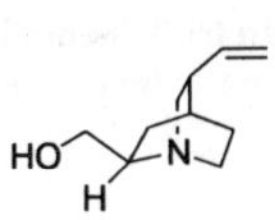

$C_{10}H_{17}NO$
Resolving agent. bp = 267°; d = 1.073; n = 1.515; $[\alpha]_D^{20}$ = + 189° (c = 1, $CH_3OH$). *Sigma-Aldrich Fine Chem.*

**4721 (2S,5R)-(+)-5-Vinyl-2-quinuclidinemethanol**

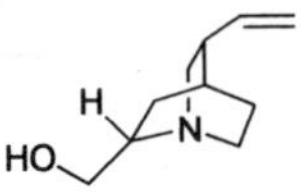

$C_{10}H_{17}NO$
Resolving agent. d = 1.073; n = 1.518; $[\alpha]^{22}$ = + 38° (c = 1, $CH_3OH$). *Sigma-Aldrich Fine Chem.*

**4722 (R)-(-)-Vinylglycine hydrochloride**

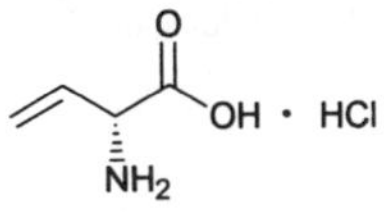

$C_4H_8ClNO_2$
Chiral building block. *See Chem. AG.*

**4723 (S)-(+)-Vinylglycine hydrochloride**

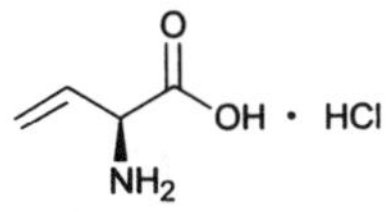

$C_4H_8ClNO_2$
Chiral building block. *See Chem. AG.*

**4724 Xanthosine dihydrate**
5968-90-1

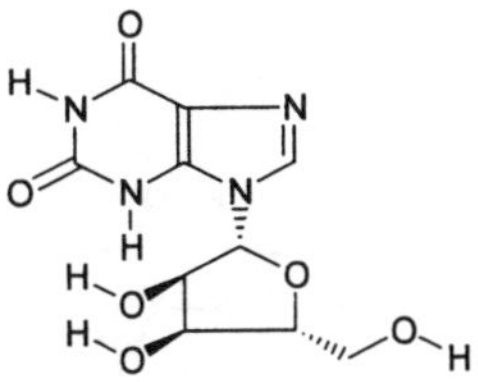

$C_{10}H_{12}N_4O_6$
Chiral intermediate. $[\alpha]^{22}$ = - 53° (c = 8, 0.3N NaOH). *Acros Organics NV.*

**4725 (S)-(+)-2-(2,6-Xylidinomethyl)pyrrolidine**
70371-56-1
$C_{13}H_{20}N_2$
(S)-(+)-N-(2-Pyrrolidinomethyl)-2,6-xylidine. Chiral intermediate. *TCI America.*

**4726 D-Xylonic acid, calcium salt hydrate**
72656-08-7

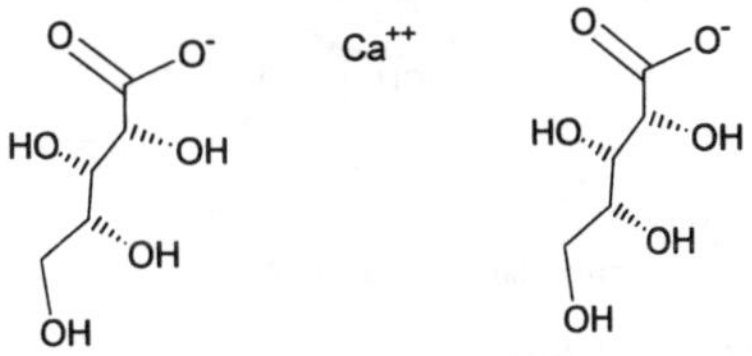

$C_{10}H_{18}CaO_{12}$
Chiral building block. *Acros Organics NV.*

**4727 D-Xylono-1,4-lactone**

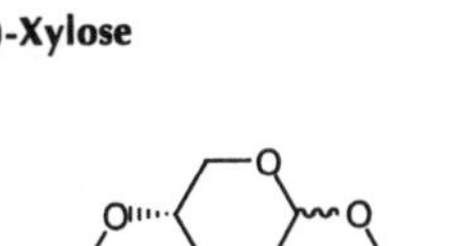

$C_5H_6O_5$
Chiral building block. $[\alpha]_D^{20}$ = + 91.6° (c = 5, $H_2O$). *Pfanstiehl Labs Inc.; See Chem. AG.*

**4728 D-(+)-Xylose**
58-86-6 200-400-7

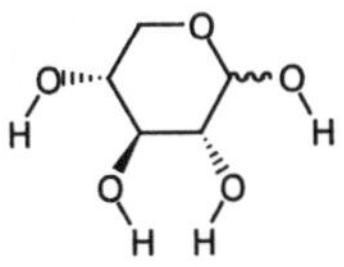

$C_5H_{10}O_5$
Xylose, D-; Wood sugar. Listed on TSCA. Chiral building block. mp = 146°; $[\alpha]_D^{20}$ = + 18.8° (c = 4, $H_2O$). *Aceto Corp.; Acros Organics NV; Austin Chem. Co.; Kaden*

*BioChem. GmbH; Pfanstiehl Labs Inc.; See Chem. AG; Sigma-Aldrich Fine Chem.; TCI America.*

**4729 L-(-)-Xylose**
609-06-3 210-174-1

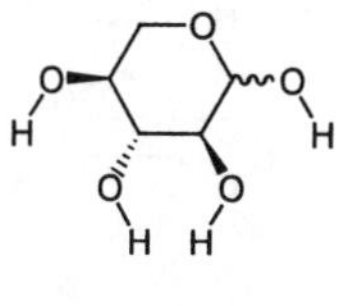

$C_5H_{10}O_5$
Xylose, L-; L-Xylopyranose. Listed on TSCA. Chiral building block. mp = 150-152°; $[\alpha]_D^{20}$ = - 18.8° (c = 4, $H_2O$). *Acros Organics NV; Kaden BioChem. GmbH; Pfanstiehl Labs Inc.; See Chem. AG; Sigma-Aldrich Fine Chem.; TCI America.*

**4730 D-Xylose tetraacetate**

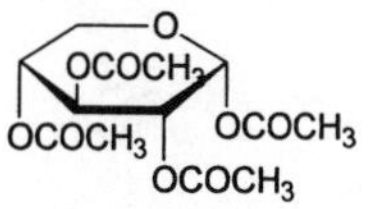

$C_{13}H_{18}O_9$
Chiral intermediate. *See Chem. AG.*

**4731 Xylulose**
527-50-4 9996(11)

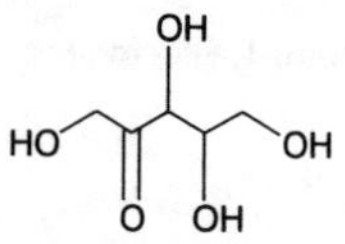

$C_5H_{10}O_5$
threo-2-Pentulose. Chiral building block. $[\alpha]_D^{20}$ = + 26° (c = 1, $H_2O$). *Sigma-Aldrich Fine Chem.*

**4732 Ytterbium tris3-(heptafluoropropyl-hydroxymethylene)-(-)-camphorate**
80464-74-0

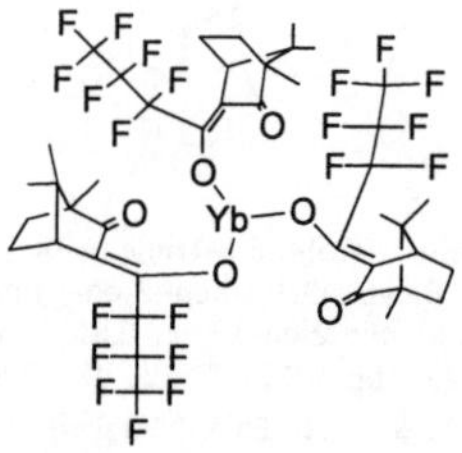

$C_{42}H_{42}F_{21}O_6YB$
(-)-Yb(hfc)3. Chiral catalyst. $[\alpha]^{18}$ = -126° (c = 1, $CH_2Cl_2$). *Sigma-Aldrich Fine Chem.*

**4733 Ytterbium tris3-(heptafluoropropyl-hydroxymethylene)-(+)-camphorate**

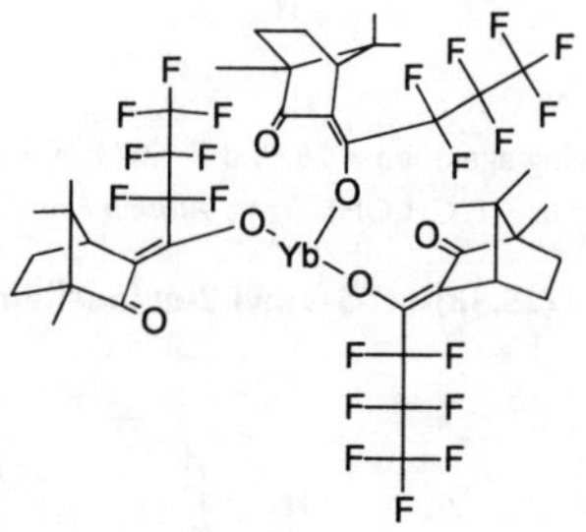

$C_{42}H_{42}F_{21}O_6YB$
(+)-Yb(hfc)3. Chiral catalyst. mp = 135-140°; $[\alpha]^{21}$ = + 126° (c = 1, $CH_2Cl_2$). *Sigma-Aldrich Fine Chem.*

**4734 Zinc α-D-glucoheptonate**
12565-63-8
$C_{14}H_{26}O_{16}Zn$
Zinc D-glycero-D-gulo-heptonate; Zinc, bis[(2x)-D-gluco-heptonato]-. Chiral building block. *Pfanstiehl Labs Inc.*

# PART III

# INDEXES

# CAS Registry Number Index

# CAS Registry Number Index

| CAS RN | Name | Record No. |
|---|---|---|
| 50-54-4 | Quinidine sulfate | 4345 |
| 50-69-1 | D-Ribose | 4375 |
| 50-70-4 | D-Sorbitol | 4412 |
| 50-81-7 | L-Ascorbic acid | 417 |
| 50-89-5 | Thymidine | 4569 |
| 50-91-9 | (+)-5-Fluorodeoxyuridine | 2855 |
| 50-99-7 | α-D-Glucose | 2944 |
| 51-31-0 | (R)-(-)-Isoproterenol | 3350 |
| 51-35-4 | trans-4-Hydroxy-L-proline | 3193 |
| 51-48-9 | L-Thyroxine | 4572 |
| 51-49-0 | D-Thyroxine | 4571 |
| 52-01-7 | (-)-Spironolactone | 4421 |
| 52-89-1 | L-Cysteine hydrochloride | 1789 |
| 52-90-4 | L-Cysteine | 1785 |
| 54-11-5 | (S)-(-)-Nicotine | 3936 |
| 54-42-2 | (+)-5-Iodo-2-deoxypuridine | 3243 |
| 54-62-6 | (+)-Aminotperin hydrate | 362 |
| 55-03-8 | L-Thyroxine sodium salt | 4573 |
| 56-41-7 | L-Alanine | 130 |
| 56-45-1 | L-Serine | 4392 |
| 56-54-2 | Quinidine | 4340 |
| 56-84-8 | L-Aspartic acid | 432 |
| 56-85-9 | L-Glutamine | 2974 |
| 56-86-0 | L-Glutamic acid | 2962 |
| 56-87-1 | L-(+)-Lysine | 3434 |
| 56-89-3 | L-(-)-Cystine | 1794 |
| 57-24-9 | l-Strychnine | 4427 |
| 57-48-7 | D-(-)-Fructose | 2887 |
| 57-50-1 | D-(+)-Saccharose | 4382 |
| 57-88-5 | (-)-Cholesterol | 1666 |
| 58-61-7 | D-Adenosine | 118 |
| 58-63-9 | D-Inosine | 3240 |
| 58-85-5 | D-(+)-Biotin | 646 |
| 58-86-6 | D-(+)-Xylose | 4728 |
| 58-96-8 | Uridine | 4699 |
| 59-14-3 | (+)-5-Bromo-2'-deoxyuridine | 810 |
| 59-23-4 | D-(+)-Galactose | 2908 |
| 59-92-7 | 3-Hydroxy-L-tyrosine | 3219 |
| 60-18-4 | L-(-)-Tyrosine | 4683 |
| 60-92-4 | Adenosine 3,5-cyclic monophosphate | 120 |
| 60-93-5 | Quinine dihydrochloride | 4352 |
| 61-76-7 | (-)-Phenylephrine hydrochloride | 4146 |
| 61-90-5 | L-Leucine | 3393 |
| 63-01-4 | 16α-Hydroxytestosterone | 3205 |
| 63-68-3 | L-Methionine | 3514 |
| 63-91-2 | L-Phenylalanine | 4108 |
| 64-86-8 | (-)-Colchicine | 1702 |
| 65-46-3 | (+)-Cytidine | 1797 |
| 65-82-7 | N-Acetyl-L-methionine | 85 |
| 67-73-2 | (+)-Fluocinolone acetonide | 2644 |
| 68-41-7 | D-(+)-Cycloserine | 1779 |
| 68-95-1 | N-Acetyl-L-proline | 98 |
| 69-65-8 | D-Mannitol | 3473 |
| 69-79-4 | 4-O-α-D-Glucopyranosyl-D-glucopyranose | 2938 |
| 70-18-8 | Reduced L-glutathione | 4368 |
| 70-26-8 | L-Ornithine | 4042 |
| 70-47-3 | L-Asparagine | 424 |
| 70-78-0 | 3-Iodo-L-tyrosine | 3252 |
| 71-00-1 | L-Histidine | 3060 |
| 72-18-4 | L-Valine | 4704 |
| 72-19-5 | L-Threonine | 4563 |
| 73-22-3 | L-Tryptophan | 4673 |
| 73-32-5 | L-Isoleucine | 3272 |
| 74-79-3 | L-(+)-Arginine | 405 |
| 77-52-1 | Ursolic acid | 4700 |
| 77-95-2 | D-(-)-Quinic acid | 4339 |
| 79-33-4 | L-Lactic acid | 3370 |
| 80-92-2 | (+)-5-β-Pregnane-3-α,20-α-diol | 4282 |
| 80-97-7 | (+)-Dihydrocholesterol | 2126 |
| 82-58-6 | D-Lysergic acid hydrate | 3431 |
| 83-48-7 | Stigmasterol | 4426 |
| 83-75-0 | Quinine ethylcarbonate | 4353 |
| 83-79-4 | (-)-Rotenone | 4379 |
| 85-31-4 | 2-Amino-6-mercaptopurine-9-D-riboside | 264 |
| 87-69-4 | L-(+)-Tartaric acid | 4439 |
| 87-79-6 | L-Sorbose | 4416 |
| 87-81-0 | D-Tagatose | 4434 |
| 87-91-2 | L-(+)-Tartaric acid diethyl ester | 4444 |
| 87-92-3 | L-(+)-Tartaric acid di-n-butyl ester | 4442 |
| 89-65-6 | D-Araboascorbic acid | 401 |
| 89-79-2 | (-)-Isopulegol | 3352 |
| 89-82-7 | (R)-(+)-Pulegone | 4326 |
| 90-39-1 | (-)-Sparteine free base | 4418 |
| 90-80-2 | D-(+)-Glucono-1,5-lactone | 2930 |
| 91-09-8 | Methyl α-D-xylopyranoside | 3872 |
| 96-82-2 | D-Lactobionic acid | 3379 |
| 97-30-3 | Methyl α-D-glucopyranoside | 3692 |
| 97-67-6 | L-(-)-Malic acid | 3460 |
| 98-09-1 | N-+le-Benzoyl-L-lysine | 487 |
| 98-79-3 | (S)-(-)-2-Pyrrolidinone-5-carboxylic acid | 4336 |
| 99-20-7 | α,α-D-Trehalose | 4599 |
| 99-48-9 | (-)-Carveol (mixture of cis and trans isomers) | 1559 |
| 107-75-5 | (-)-7-Hydroxy-3,7-dimethyloctanal | 3122 |
| 115-53-7 | Sinomenine | 4405 |
| 118-00-3 | (-)-Guanosine | 3017 |
| 118-10-5 | (+)-Cinchonine | 1686 |
| 118-65-0 | (-)-Isocaryophyllene | 3259 |
| 121-24-4 | L-(-)-Glutathione hydrate, oxidized | 2979 |
| 123-78-4 | D-erythro-Spinghosine | 4420 |
| 124-83-4 | (1R,3S)-(+)-Camphoric acid | 1360 |
| 126-07-8 | (+)-Griseofulvin | 3013 |
| 126-13-6 | D-Sucrose diacetate hexaisobutyrate | 4432 |
| 126-14-7 | D-(+)-Saccharose octaacetate | 4383 |
| 130-89-2 | Quinine monohydrochloride | 4358 |
| 130-90-5 | Quinine formate | 4354 |
| 130-95-0 | Quinine | 4347 |
| 131-48-6 | N-Acetyl-D-neuraminic acid | 92 |
| 133-05-1 | (-)-Asarinin | 416 |
| 134-03-2 | L-Ascorbic acid sodium salt | 419 |
| 134-72-5 | L-(-)-Ephedrine sulfate | 2492 |
| 137-66-6 | (+)-Ascorbic acid 6-palmitate | 418 |
| 138-09-0 | Sodium-L-malate | 4410 |
| 138-15-8 | L-(+)-Glutamic acid hydrochloride | 2969 |
| 138-52-3 | D-(-)-Salicin | 4384 |
| 138-59-0 | Shikimic acid | 4403 |
| 142-47-2 | L-Glutamic acid sodium salt | 2972 |
| 145-42-6 | Taurocholic acid, sodium salt hydrate | 4454 |
| 146-40-7 | Quinine ascorbate | 4348 |
| 147-71-7 | D-(-)-Tartaric acid | 4438 |
| 147-85-3 | L-Proline | 4288 |
| 147-94-4 | 1-β-D-Arabinofuranosyl-cytosine | 390 |
| 152-58-9 | (S)-Reichstein's substance | 4369 |
| 153-94-6 | D-Tryptophan | 4672 |
| 154-17-6 | 2-Deoxy-D-glucose | 1826 |
| 154-92-7 | N-α-Benzoyl-L-arginine | 479 |
| 155-58-8 | 3,3',5-Trihydroxy-4'-methoxy-stilbene-3-O-β-D-glucopyranoside | 4642 |
| 157-06-2 | D-(-)-Arginine | 404 |
| 299-27-4 | D-Gluconic acid, potassium salt | 2928 |
| 299-28-5 | D-Gluconic acid, calcium salt | 2926 |
| 300-39-0 | 3,5-Diiodo-L-tyrosine dihydrate | 2160 |
| 302-95-4 | (+)-Deoxycholic acid, sodium salt | 1817 |
| 303-43-5 | (-)-Cholesteryl oleate | 1674 |
| 305-84-0 | β-Alanyl-L-histidine | 147 |

# CAS Registry Number Index

# CAS Registry Number Index

# CAS Registry Number Index

# CAS Registry Number Index

# CAS Registry Number Index

# CAS Registry Number Index

# CAS Registry Number Index

# CAS Registry Number Index

# CAS Registry Number Index

# CAS Registry Number Index

# CAS Registry Number Index

# CAS Registry Number Index

# CAS Registry Number Index

# CAS Registry Number Index

# CAS Registry Number Index

# CAS Registry Number Index

# CAS Registry Number Index

# CAS Registry Number Index

# CAS Registry Number Index

# CAS Registry Number Index

# CAS Registry Number Index

# CAS Registry Number Index

# CAS Registry Number Index

# CAS Registry Number Index

# CAS Registry Number Index

# CAS Registry Number Index

# CAS Registry Number Index

# CAS Registry Number Index

# CAS Registry Number Index

# EINECS Number Index

# EINECS Number Index

# EINECS Number Index

# EINECS Number Index

# EINECS Number Index

# EINECS Number Index

# EINECS Number Index

# EINECS Number Index

# EINECS Number Index

# Name and Synonym Index

# Name and Synonym Index

# Name and Synonym Index

# Name and Synonym Index

# Name and Synonym Index

# Name and Synonym Index

# Name and Synonym Index

# Name and Synonym Index

# Name and Synonym Index

# Name and Synonym Index

# Name and Synonym Index

# Name and Synonym Index

# Name and Synonym Index

# Name and Synonym Index

# Name and Synonym Index

# Name and Synonym Index

# Name and Synonym Index

# Name and Synonym Index

# Name and Synonym Index

# Name and Synonym Index

# Name and Synonym Index

# Name and Synonym Index

# Name and Synonym Index

# Name and Synonym Index

# Name and Synonym Index

# Name and Synonym Index

# Name and Synonym Index

# Name and Synonym Index

# Name and Synonym Index

# Name and Synonym Index

# Name and Synonym Index

# Name and Synonym Index

# Name and Synonym Index

# Name and Synonym Index

# Name and Synonym Index

# Name and Synonym Index

# Name and Synonym Index

# Name and Synonym Index

# Name and Synonym Index

# Name and Synonym Index

# Name and Synonym Index

# Name and Synonym Index

# Name and Synonym Index

# Name and Synonym Index

# Name and Synonym Index

# Name and Synonym Index

# Name and Synonym Index

# Name and Synonym Index

# Name and Synonym Index

# Name and Synonym Index

# Name and Synonym Index

# Name and Synonym Index

# Name and Synonym Index

## Name and Synonym Index

# Name and Synonym Index

# Name and Synonym Index

# Name and Synonym Index

# Name and Synonym Index

# Name and Synonym Index

# Name and Synonym Index

# Name and Synonym Index

# Name and Synonym Index

# Name and Synonym Index

# Name and Synonym Index

# Name and Synonym Index

# Name and Synonym Index

# Name and Synonym Index

# Name and Synonym Index

# Name and Synonym Index

# Name and Synonym Index

# Name and Synonym Index

# Name and Synonym Index

# Name and Synonym Index

# PART IV

# MANUFACTURER AND SUPPLIER DIRECTORY

# Manufacturer and Supplier Directory

**A.H. Marks & Co. Ltd.**
Wyke
Bradford, West Yorkshire BD12 9EJ UK
44-1274-691234
(FAX) 44-1274-691176

**Aceto Corporation**
One Hollow Lane
Lake Success, NY 11042 USA
516-627-6000
(FAX) 516-627-2819

International marketer of fine, industrial and pharmaceutical chemicals; global resources include strategic relations with over 200 chemical manufacturers worldwide and a broad distribution capability; custom manufacturing, pharmaceutical intermediates, bulk pharmaceuticals, biochemicals, organic intermediates and colorants, agrochemicals and industrials.

**Acros Organics nv**
Geel, "3a, 2440"
Belgium
32-1457-5211
(FAX) 32-1459-3434

**Advanced Asymmetrics, Inc.**
P.O. Box 298
Millstadt, IL 62260-0298 USA
618-476-3920, 888-330-6324
(FAX) 618-476-7408

Specializes in the manufacture and supply of chemicals and associated products for use in chiral nonracemic synthesis; product line includes amino acids, chiral precursors, chiral catalysts, chiral auxiliaries, peptide isosteres and analogs, protecting groups; custom synthesis, to the extent of available capacity, any product useful for chiral synthesis.

**Ajinomoto Co. Inc.**
15-1, Koyobashi 1-Chome
Chuo-ku, Tokyo 104
Japan
81-3-5250-8111
(FAX) 81-3-5250-8293

**Albany Molecular Research**
21 Corporate Circle
PO Box 15098, Albany, NY 12212-5098
518-464-0279
(FAX) 518-464-0289

Contract research, small-scale manufacturing and analytical testing business providing services in medicinal chemistry, molecular modeling, organic custom synthesis, chemical development, cGMP synthesis, analytical chemistry, stability studies, scientific and technical writing services, regulatory services, auditing, and document preparation.

**Albemarle Corporation**
451 Florida Street
Baton Rouge, LA 70801-1785 USA
225-388-7402, 800-535-3030
(FAX) 225-388-7848

Core technologies include Heck chemistry alkylation/acylation, amination, halogenation, organometallics, carbonylations and inorganics; reactions include halogenation, carbonylation, carboxylation, chiral chemistry, polymerization, quaternization, and reduction

**Alfa Aesar**
30 Bond Street
Ward Hill, MA 01835-8099 USA
978-521-6300, 888-343-8025
(FAX) 978-521-6366

A Johnson Matthey Company; leading manufacturer and supplier of research chemicals, metals and materials; products include inorganic and organic research chemicals, pure elements, alloys, precious metal compounds and catalysts, rare earths, analytical products and more; perform asymmetric synthesis and chiral chemistry

**Alfa Chemicals Italiana SRL**
Viale Sarca n. 223
Milano 20126
Italy
39-0266-10-6484
(FAX) 39-0266-10-6505

**Alfa Chemicals Ltd.**
Terrrace Road South
Binfield, near Bracknell, Berks RG42 4PZ UK
44-1244-861800
(FAX) 44-1344-862010

**Altus Biologics Inc.**
625 Putnam Avenue
Cambridge, MA 02139-4807 USA
617-577-6500, 888-258-2532
(FAX) 617-577-6502

Offer biocatalysts for performing chiral resolutions of alcohols, acids, and amines and racemization-free catalytic peptide coupling reactions; also offer screening kits for identifying the best enzymes to use for a particular transformation and custom synthesis of biocatalysts

**Antibioticos**
Strada Rivoltana Km 6/7
Milan, Rodano 20090
Italy
39-02-952-35401
(FAX) 39-02-9523-7-540

Leading producer of fermented antibiotics and other pharmaceutical intermediates; chemistry includes organometallic catalysis, enzymatic catalysis, fermentation, hydrogenation, photochemistry, and basic synthetic organic reactions.

**Arran Chemical Company Ltd.**
Unit 2-3 Monksland Industrial Estate
Athlone, Co. Roscommon
Ireland
353-902-94807
(FAX) 353-902-94809

Offers a range of aroma and other specialty chemicals, pharmaceutical intermediates and custom synthesis; for asymmetric synthesis, specializes in optical resolution with resolving agents (R and S naphthylethylamine, R and S 4-bromophenethylamine, and R and S 3-methoxyphenethylamine) and stereoselective reactions.

**Artisan Industries Inc.**
73 Pond Street
Waltham, MA 02451-4594 USA
781-893-6800
(FAX) 781-647-0143

Provides technologies for very small scale to practical, commercial scale chiral resolution that includes integrated system of both simulated moving bed chromatography and solvent management; also provides contract manufacturing services under cGMP, testing and feasibility services for small demonstrations, and custom built systems.

**Arzneimittelwerk Dresden**
Meissner Strasse 35
Radebeul 01445
Germany
49-351-834-2172
(FAX) 49-351-8341899

**Asymchem**
2 Davis Dr. Suite 102
P.O. Box 12076
Research Triangle Park, NC 27709 USA
919-460-5179
(FAX) 919-319-9888

Custom synthesis producer of advanced fine organic chemical and pharmaceutical intermediates; production facilities in China; chiral products include binaphthyl derivatives and oxazolidinones.

**Austin Chemical Company, Inc.**
1565 Barclay Boulevard
Buffalo Grove, IL 60089 USA
847-520-9600
(FAX) 847-520-9160

Privately owned sales/marketing firm representing fine chemical manufacturers worldwide; serves the pharmaceutical, agricultural and specialty chemical industries, supplying key raw materials from investigational to commercial levels; some manufacturers conduct asymmetric synthesis reactions and have expertise in chiral chemistry; also provides process research and development, scale-up and manufacturing, and analytical development and quality control services.

**Avecia Life Science Molecules**
P.O. Box 42
Hexagon House, Blackley
Manchester M9 8ZS UK
44-161-740-1460
(FAX) 44-161-795-6005

Extensive experience in all standard transformations applicable to lifescience chemistry; commitment to chiral intermediates through both biotechnology and asymmetric synthesis; alcohol, amine, amide, and nitrile resolutions, microbial asymmetric ketone reductions, homochiral sulfoxides, enzyme/micro-organism screening.

**Axiva GmbH**
Industriepark Hoechst
Gebaude K 801,/Main
D-65926 Germany
49-69-30547334
(FAX) 49-69-30517206

Services include process development and optimization; product classes include anilines, aryl silanes, benzoic acids, benzyl alcohols, bisaryl compounds, bisaryl phosphanes, bromoaromatics, metallocenes, organometallics, oxothiazines, phosphinic acid derivatives, sulfonated aromatics, sulfonated ureas, and triazines.

**Bachem AG**
Haupstrasse 144
CH-4416 Bubendorf
Switzerland
41-61-931-2323
(FAX) 41-61-931-2549

Custom synthesis on an industrial scale and according to GMP guidelines; products include peptides, special amino acid derivatives, peptide mimetics, resins, and organic compounds.

**BASF Aktiengesellschaft**
CZN/E
67056 Ludwigshafen
Germany
49-621-6022666
(FAX) 49-621-6022666

Offers chemical intermediates for the pharmaceutical, agricultural, coatings, plastics, fibers, adhesives, and colorants markets, among others; custom chemical synthesis services through chemical, technological and production expertise.

**Boehringer Ingelheim Pharma KG**
Binger Strasse 173
D-55216 Ingelheim
Germany
49-6132-77-2192
(FAX) 49-6132-77-3755

Manufacturer of fine chemicals with focus on development and manufacturing of active pharmaceutical ingredients and complex, advanced intermediates under cGMP guidelines in an FDA inspected environment; strengths are enantioselective chemistry on C=X functionalities (X = O, N, C) and on heterocycles and carbohydrates.

**Borregaard Synthesis**
P.O. Box 162
N-1701 Sarpsborg Norway
47-6911-8000
(FAX) 47-6911-8970

Leading supplier of fine chemicals for both diagnostics and medicines; core technologies include multi-step organic synthesis, alkylation, aminolysis, bromination, chiral technologies, chlorination, cyanoethylation, Grignard reactions, heterogeneous catalysis, hydrobromination, hydrogenation, iodination, nitrogen heterocyclics, oxidation, and the Veilsmeir-Heck reaction.

**Calaire Chimie s.a.**
Quai dAmerique 1
B.P. 215 Calais Cedex F-62104
France
33-3-2146-2117
(FAX) 33-3-2146-2129

**Callery Chemical Company**
1420 Mars-Evans City Road
Evans City, PA 16033 USA
412-967-4141
(FAX) 412-967-4140

**Cambrex**
One Meadowlands Plaza 7th Fl.
East Rutherford, NJ 07073 USA
201-804-3060
(FAX) 201-804-3045

Leading supplier of human health, animal health, agriculture and biotechnology products worldwide to the life sciences industries and a producer of specialty chemicals through its subsidiaries in the United States and Europe; several subsidiaries perform asymmetric synthesis reactions and chiral chemistry transformations.

**Carbogen Laboratories AG**
Schachenallee 29
CH-5001 Aara
Switzerland
41-62-836-48-00
(FAX) 41-62-836-48-10

Specializes in the manufacture of experimental active pharmaceutical ingredients under cGMP protocol; asymmetric synthesis and chiral chemistry along with basic synthetic organic techniques.

**Catalytica Pharmaceuticals, Inc.**
P.O. Box 1877
Greenville, NC, 27835-1887
252-707-2488
(FAX) 252-707-2492

Fully integrated, technology based, drug development and manufacturing company; provides custom manufacturing services from process development to final drug manufacture; expertise in a wide variety of synthetic organic chemistry reactions including asymmetric catalysis, selective oxidations, selective solid acid catalysis, catalytic olefin methathesis, and many others.

**Cedarburg Laboratories, Inc.**
870 Badger Circle
Grafton, WI 53024
414-376-1467
(FAX) 414-376-1068

Custom manufacturing of complex organic molecules in a cGMP environment, process development and analytical support; reactions include asymmetric synthesis and chiral chemistry along with aldol condensations, cyanide chemistry, epoxidation, nitration, Grignard reactions, Wittig reactions, and other basic synthetic organic techniques.

**Celltech Chiroscience Ltd.**
No. 283 Cambridge Science Park
Milton Road, Cambridge CB4 4WE UK
44-1753-536632

**Central Glass Co., Ltd**
Kowa-Hitotsubashi Bldg.
Kanda-Nishikicho 3-7-1, Chiyodaku
Tokyo 101-0054
Japan
81-03-3259-7324
(FAX) 81-03-3259-7363

Plants for bulk drug production, intermediates manufacture, and bench scale processes; specializes in fluorination, chlorination triflate, and chiral Friedal-Crafts reactions.

**Chattem Chemicals**
3708 St. Elmo Ave
Chattanooga, TN 37409 USA
423-822-5000
(FAX) 423-825-0507

Custom manufactures specializes in aluminum and other metal alkoxides and their derivatives, aluminum, magnesium and calcium based antacids, wet and dry, glycine, glycine derivatives and related amino acid compounds, and purified food grade additives; specialized in reaction chemistries including addition reactions, amino acid synthesis, brominations, chelations, reductions, condensations, esterifications, selected organometallic reactions, slurry purifications and crystallizations, and spray drying.

**ChemDesign Corporation**
310 Authority Drive
Fitchburg, MA 01420 USA
508-303-4021
(FAX) 978-665-0390

A Bayer Company; custom manufacturing services for large scale enantioselective hydrogenation using ruthenium complexes of chiral ligands; temperatures up to 200 °C, pressures up to 200 bar hydrogen; also provide process development services.

**Chemi SpA**
Via Dei Lavoratori, 54
Cinisello, Balsamo 20092
Italy
39-02612-8431
(FAX) 39-02612-8960

Custom manufacturing services include contract production, process development, and production for clinical trials; chemistries include catalytic hydrogenation, epoxidation, carboxylation, high temperature reactions, glycerophospholipids, homochiral intermediates, peptides and asymmetric hydrogenation.

**Chemodynamics**
3 Crossman Road South
Sayreville, NJ 08872 USA
732-721-4700
(FAX) 732-721-6835

Services include combinatorial building blocks, custom synthesis, process development, improvement and validation, contract research, analog preparations, and multi-kilo processes.

**ChemShop BV**
Vliesvenweg 1
Weert, 6002 NM
Netherlands
31-495-460-130
(FAX) 31-495-664-233

Amino acid chemistry, chiral chemistry, classical resolutions, enzymatic resolutions, Strecker chemistry, reductions, high pressure reactions on lab scale (100 bar), hydrogenations, low temperature reactions (-180 °C), esterifications, amidations, condensations, pH controlled reactions.

**China Camphor Co. Ltd.**
P.O. Box 494
Taipei, Taiwan
886-2304-5377
(FAX) 886-2304-5375

**Chiragene, Inc.**
7 Powder Horn Drive
Warren, NJ 07059 USA
732-805-3660
(FAX) 732-805-3662

A CAMBREX Company; leading synthesizer and producer of chirally pure intermediates for the pharmaceutical industry utilizing enzymatic transformations, biocatalysis and fermentation; specializes in acylations, alkylations, ammonolysis, asymmetric synthesis, biocatalysis, biocatalyst modification, birch reductions, catalytic reductions, cyclizations, diastereomeric crystallizations, enzyme development, esterifications, Friedel-Crafts reactions, Grignard Reactions, hydrolyses, Michael Additions, and nitrosations.

**Chiral Technologies Inc.**
730 Springdale Drive
P.O. Box 564
Exton, PA 19341 USA
610-594-2100, 800-624-4725
(FAX) 610-594-2325
Custom chiral chromatographic separation services, grams to multi-hundred kilograms; also sale of proprietary chiral chromatography columns.

**Chirex, Inc.**
300 Atlantic Street, Suite 402
Stamford, CT 06901
203-351-2300
(FAX) 203-425-9996

Services include innovative process research and development, rapid response small scale manufacture, clinical trials supply and commercial manufacture with seamless technology transfer under cGMP; asymmetric technologies include hydrolytic kinetic resolution, asymmetric epoxidation, asymmetric dihydroxylation, amino acid technology, asymmetric reduction, and asymmetric ring opening.

**ChiroTech Technology Ltd.**
321 Cambridge Science Park
Milton Road, Cambridge CB4 OWG UK
44-1223-728010
(FAX) 44-1223-506701

Division of Ascot Fine Chemicals; combines biocatalysis, asymmetric chemocatalysis, crystallization technology and complex organic synthesis for custom manufacturing services; agreement with Lancaster Synthesis for offering single enantiomer building blocks; joint venture with CombiChem,Inc. for development of computationally designed libraries of single-isomer chiral compounds.

**COGNIS Corporation**
2430 N. Huachuca Dr
Tuscon, AZ 85745-1273 USA
520-629-3204
(FAX) 520-624-0912

Custom manufacturing services for most industrial market segments; reactions include alkoxylation, ammonolysis, asymmetric synthesis, chiral chemistry, chloromethylation, epichlorohydrin chemistry, ozonolysis, reduction, and other basic synthetic organic techniques.

**Conti BPC NV**
Industrepark One
Roosveld 2 B6 Landen
B-3400 Belgium
32-11-880-270
(FAX) 32-11-832-253

A CAMBREX Company; specialized in the production of bulk chemicals used in pharmaceuticals; offers generic pharmaceutical and contract manufacturing services; specializes in bulk active ingredients, optical resolution, acylation, halogenation, cyanation, dehydrogenation, Mannich reactions, reductive amination, and many other synthetic organic chemistries.

**Contract Manufacturing Services from Dow**
100 Larkin Center
1650 N. Swede Road
Midland, MI 48674 USA
517-636-6919
(FAX) 517-636-0986

Analytical services, cGMP pharmaceuticals, chiral synthesis, general multi-step synthesis, hydrogen cyanide chemistry, nitroparaffin chemistry, nucleosides, organometallic chemistry, pilot plant capabilities, peptides, route optimization and verification, and scale-up capabilties.

**CU Chemie Uetikon GmbH**
Raiffeisenstrasse 4
D-77933 Lahr
Germany
49-7821-585-238
(FAX) 49-7821-585-230

Research and development, process optimization, and production of organic fine chemicals, pharmaceutical intermediates, and active pharmaceutical ingredients; reactions include alkylation, cyanation, esterification, hydrogenation, Grignard reactions, chiral synthesis/separation, protected amino acid synthesis and reactions with butyllithium.

**D&O Group**
The Atrium
80 East Route 4
Paramus, NJ 07652 USA
201-909-0600
(FAX) 201-909-8601

The D&O Group supplies intermediates and Fine Chemicals for the Pharmaceutical and Specialty Chemical markets worldwide through a core of R&D labs, pilot plants, and manufacturers across the globe; some manufacturers represented conduct asymmetric synthesis reactions and have expertise in chiral chemistry.

**Daesang Corporation**
Daesang Bldg., 96-48
Sinsul-dong, Dongdaemun-ku
Seoul, 130-110
Korea
82-2-2220-9336-8
(FAX) 82-2-786-8421

**Daicel Chemical Ind. Ltd**
Kasumigaseki Bldg. 16F
2-5, Kasumigaseki 3-chome
Chiyoda-Ku, Tokyo 100-6077
Japan
81-3-3507-3190
(FAX) 81-3-3507-3198

Custom manufacturing services include fermentation and biotransformations such as transamination and selective asymmetric oxidation, reduction, and decarboxylation; expertise in many basic synthetic organic chemistries and peptide synthesis.

**Daiso Company, Ltd.**
1-10-8, Edobori
Nishi-ku,Osaka 550-0002
Japan
81-6-6443-5996
(FAX) 81-6-6445-5787

Offers chiral epichlorohydrins and a line of $C_3$ synthons for synthesizing chiral secondary alcohols; applying biotechnology to other substrates, expanding reduction technology to amino acid derivatives, and offering custom synthesis of advanced intermediates to broaden product offerings; also manufacture silica gels for chiral HPLC purification.

**Davos Chemical Corporation**
464 Hudson Terrace
Englewood Cliffs, NJ 07632 USA
201-569-2200
(FAX) 201-569-2201

Focus is to deliver custom intermediates, advanced pharmaceutical ingredients and process development strategies to the pharmaceutical and life science industries; specializes in unnatural amino acids (through fermentation) and synthesis of enantiomerically pure products by means of asymmetric synthesis and through the resolution of racemates (kinetic and diastereomeric salt crystallization), macrolides, peptides, steroids, heterocyclic compounds, nucleosides, and others.

**Degussa-Huls AG**
Paul-Baumann-Str. 1
D-45764 Marl
Germany
49-2365-497101
(FAX) 49-2365-4919020

Business Unit 1 provides services for the development of syntheses for complex, enantiomerically pure intermediates and bulk active substances for the pharmaceutical and agrochemcial industries; experience in scaling up production in multi-purpose facilities complying with cGMP regulations; access to sensitive raw materials.

## Manufacturer and Supplier Directory

**Delmar Chemicals**
9321 Airlee St
LaSalle, Quebec H8R 2B2
Canada
514-366-7950
(FAX) 514-366-5665

**Diagnostic Chemicals Limited**
160 Christain St
Oxford, CT 06478 USA
203-881-2020, 800-325-2436
(FAX) 203-888-1143

Provides cGMP custom manufacturing of both pharmaceutical intermediates and natural products; technical services include analytical method development and process development and optimization; reactions include chiral chemistry, epoxidation, hydrogenation, peroxidation, saponification, Grignard reactions, and other synthetic organic techniques.

**Digital Specialty Chemicals, Inc.**
P.O. Box 284
Dublin, NH 03444-0284 USA
603-563-5060
(FAX) 603-563-9288

Specializes in synthesis of phosphines, chiral phosphine ligands, other phosphorous based organic chemicals, silanes and organometallics on lab and pilot scale; reactions include chiral chemistry, photochemical reactions and Birch reductions.

**Diosynth BV**
Kloosterstraat 6
P.O. Box 20, Bhoss, 5340
Netherlands
31-412-66-20-49
(FAX) 31-412-66-25-21

Business unit of Akzo Nobel; active in industrial cell culture, fermentation, biochemical extraction and purification, and organic synthesis; core activities are based around chiral chemistry, Friedel-Craft reactions, Grignard chemistry, chlorination, and hydrogenation.

**Dipharma SpA**
Via Bissone, 5
Baranzate di Bollate, Milano 20021
Italy
39-02-382281
(FAX) 39-02-382-00198

**Diversa Corporation**
10665 Sorrento Valley Rd
San Diego, CA 92121-1609 USA
858-623-5100
(FAX) 858-626-3700

Offers recombinant enzymes in a kit format that include aminotransferases, cellulases/amylases, esterases/lipases, glycosidases, and phosphatases; kits contain stable enzymes with varying enantioselectivity and pH, temperature, and substrate specificities; kits allow screening and evaluation.

**Dr. Reddy's Group**
7-1-27, Ameerpet
Hyderabad Andhra Pradesh 500 016
India
91-40-3750984
(FAX) 91-40-3731955

Offers synthesis services across the entire production process starting from gram quantities for innovator companies up to multi-ton quantities for mature products; specialized technologies include carbohydrate and amino acid chemistry, chemical resolution of racemates, chiral synthesis, and others plus basic organic synthetic techniques.

**DSM Fine Chemicals Netherlands**
Noorderpoort 9
P.O. Box 81, Venlo 5900 AB
Netherlands
31-77-389-9555
(FAX) 31-77-389-9300

Development and production of advanced intermediates for the pharmaceutical industry and fine chemicals for agrochemicals, flavors and fragrances, dyes and pigments and other end markets; capabilities include optical resolution, asymmetric synthesis, biocatalysis, ozonolysis, vapor-phase reaction, and custom manufacturing; DSM Biologics provides contract manufacturing of biopharmaceuticals from early development to commercial supply with expertise in microbial and mammalian production technologies, including pilot to large-scale fermentation and purification.

**Dynamit Nobel GmbH**
Kalkstrasse 218
D-51377 Leverkusen
Germany
49-214-357-429
(FAX) 49-214-357-435

**Eastar Chemical Corporation**
400 Capitol Mall, Suite 900
Sacramento, CA 95814 USA
916-449-3951, 800-898-2436
(FAX) 916-686-8397

Custom manufacturer and importer of fine and specialty chemicals; reactions include asymmetric synthesis and basic synthetic organic chemistries.

**Eastman Chemical Company**
PO Box 431
Kingsport, TN 37662-5280 USA
423-229-1875
(FAX) 423-229-8133

Complex, multi-step organic chemistry for custom synthesis services to the pharmaceutical, agrochemical, and industrial fine and specialty organic chemical industries; lab scale qualification quantities through large scale commercial production.

**Elso Vegyi Industria Rt.**
POB 139
Budapest, 1443
Hungary
36-1342-4167
(FAX) 36-1342-9366

**Emmellen Biotech Pharmaceuticals Ltd.**
606-607, Gateway Plaza
Hiranandani Gardens
Powai, Mumbai 400 076
India
91-22-570-3422
(FAX) 91-22-570-3139

**EMS-Dottikon AG**
CH-5605 Dottikon,
Switzerland
41-56-616-8111
(FAX) 41-56-616-8120

**Expansia**
Route D'Avignon
B.P. 6, Aramon 30390
France
33-4-6657-0101
(FAX) 33-4-6657-0148

Manufactures fine chemical intermediates, pharmaceutical active ingredients and provides custom synthesis services; areas of special expertise include diborane chemistry, chiral chemistry, low temperature reactions, metal hydride reductions, organometallic chemistry, hydrogenation under pressure, Skraup quinoline synthesis, ethylene oxide chemistry, bromination, and synthesis of t-butyl esters of carboxylic acids.

**Farmabios S.r.l.**
Via Don Motti, 45
Gropello Cairoli (PV) 27027
Italy
39-0382-8191
(FAX) 39-0382-815886

**FermPro Manufacturing LP**
PO Box 5000,
Kingstree, SC 29556 USA
843-382-8485
(FAX) 843-382-8676

Specializes in synthesis using fermentation technology; strong experience with scale-up technology transfer and project management; expertise in dealing with a broad range of organisms and fermentation conditions; can take products from laboratory testing and process development phases to commercial production.

**Fine Organics Ltd.**
Seal Sands
Middlesbrough, Cleveland TS2 1UB UK
44-1642-546666
(FAX) 44-1642-546046

Fine chemicals manufacture for pharmaceutical, agrochemical and specialty chemical industries; organosulfur chemistry, photochemical reactions, phosgenation, thiophosgenation, chloromethylation, Grignard reactions, Friedal-Craft reactions, organolithium chemistry, oxidations and epoxidations, low temperature (-85 ºC) reactions, hydrogenations, and diborane reductions.

**FineTech, Ltd.**
Technion City
P.O. Box 3557, Haifa 31032
Israel
972-48-293880
(FAX) 972-48-343341

Chiral amines, chiral pharmaceutical intermediates, chiral amino-synthons to introduce chirality into a target molecule, β-aminotetralins and β-chromanes, and chiral resolutions for optically active arylalkylamines, arylpropanoic acids, and pro-chiral ketones.

**Finorga**
Route de Givors
BP 9, Chasse sur Rhone 38670
France
33-4-7249-1960
(FAX) 33-4-7249-1973

Enantiomeric reduction (Corey, DIP-Cl) and enantiomeric resolution; organometallic reactions, coupling with boronic acids, acetylenes, olefins, heterocyclic chemistry, and, low-temperature reactions (-90 °C).

**Fischer Chemicals AG**
Riesbachstrasse 57
CH-8034 Zurich
Switzerland
41-1-3896969
(FAX) 41-1-3896970

**Fisher Fine Chemicals**
Bishop Meadow Road
Loughborough, Leics LE11 5RG UK
44-1509-231166
(FAX) 44-1509-616015

**Flamma spa**
Via Bedeschi, 22
Chignolo D'Isola, BG 24040
Italy
39-035-499-1811
(FAX) 39-035-499-1812

Quantities ranging from lab scale to multi-ton; amino acids and related derivatives for synthesis of therapeutic peptides and peptidomimetics; provides both L- and D-amino acids; custom synthesis of these types of materials is available.

**Freedom Chemical Diamalt GmbH**
Georg-Reismuller-Strasse 32
D-80999 Munchen
Germany
49-89-81060
(FAX) 49-89-8106510

**Ganes Chemicals Inc.**
33 Industrial Park Road
Pennsville, NJ 08070-3298 USA
856-678-3601
(FAX) 856-678-4008

Custom manufacturing services to the global pharmaceutical industry including production of key intermediates and active ingredients from kg lots to commercial scale; reactions include asymmetric synthesis, chiral chemistry, acrylation, diazotization, halogenation, hydroxylation, transesterification, Hoffman degradation, Sandmeyer reaction, Williamson synthesis, and basic synthetic organic techniques.

**Genencor International**
1870-4 South Winton Rd
Rochester, NY 14618 USA
716-256-5272
(FAX) 716-244-9988

**Genzyme Pharmaceuticals**
One Kendall Square
Cambridge, MA 02139 USA
617-374-7248, 800-868-8208
(FAX) 617-252-7772

Custom manufacturing of bulk active pharmaceuticals and intermediates including peptides and phospholipids; amino acid derivatives and related products are used as the building blocks for new peptide-based pharmaceuticals; develops custom building blocks and provides route selection and process development and optimization services.

**Great Lakes Fine Chemicals**
601 East Kensington Rd
Mt. Prospect, IL 60056 USA
765-497-6100
(FAX) 765-497-6123

Chiral manufacturing at NSC Technologies; novel asymmetric synthesis, unnatural amino acids, advanced chiral intermediates, product and process research and development (chemistry and biotechnology), laboratory and pilot scale production; oxazolidinone chemistry, peptide coupling, asymmetric hydrogenation, metabolic pathway engineering, microbial expression system development, enzymatic bioconversions, resolution, biocatalysts.

**Haarmann & Reimer GmbH**
Postfach 1253
D-37601 Holzminden
Germany
49-553-1900
(FAX) 49-553-1901-649

**Hamari Chemicals**
1-4-29, Kunjima
Higashi-Yodogawa-ku
Osaka, 533-0024
Japan
81-6-6322-0191
(FAX) 81-6-6323-2115

Synthetic technology includes amino acid derivatives, asymmetric catalytic reduction, synthesis of sulfur-containing compounds, general organic reactions, and low-temperature reactions (-70 °C).

**Helsinn Chemicals SA**
Via Industria 24
6710 Biasca
Switzerland
41-91-873-0110
(FAX) 41-91-873-0111

Contract manufacturer for active ingredients and advanced intermediates, multistep organic synthesis under full cGMP, outsourcing project management and regulatory support, FDA-inspected, clean room dryers facility.

**High Force Research Ltd**
Bowburn North Industrial Estate
Bowburn, Durham DH6 5PF UK
44-191-377-9098
(FAX) 44-191-377-9099

Specializes in chemical synthesis and process R&D for pharmaceuticals and research oriented fine chemicals sectors; research chemicals for combinatorial synthesis, route selection and feasibility studies, multistep synthesis from grams to kilos, process optimization and scale-up; catalytic hydrogenation, oxidation/peroxidation, organofluorine, heterocyclic, amino acid and carbohydrate chemistry.

**Hovione SA**
Wuinta de S. Pedro, Sete Casas
Loures 2674-506
Portugal
351-21-982-9251
(FAX) 351-21-982-9259

Exclusively dedicated to the synthesis of active pharmaceutical ingredients and regulated intermediates; provides a complete solution from drug development to post-launch commercial manufacture.

**IMI (TAMI) Institute for R&D**
P.O. Box 10140
Haifa Bay 26111
Israel
972-4-8469411
(FAX) 972-4-8450078

Services include exploratory research, process research and development, and sample preparation; organic chemistry areas include biocides, chiral compounds, flame retardants, fine chemicals, organic intermediates, and synthetic peptides; chiral technologies include asymmetric synthesis, biocatalysis, chiral pool derivitization, and crystallization

**Inabata and Company, Ltd.**
8-2, Nihonbashihihoncho
2-Chome
Chuo-ku, Tokyo 103-8448
Japan
81-3-3639-6415
(FAX) 81-3-3639-6410

**Inalco S.p.A.**
Via Calabiana, 18
Milano 20139
Italy
39-02-5521-3005
(FAX) 39-02-5699-4518

Custom synthesis services; areas of expertise include pharmaceutical intermediates, complex carbohydrates, chromogenic substrates, detergents ; research products and products for biotechnologies.

**Indena**
Viale Ortles, 12
Milano 20139
Italy
39-02-574961

**Interchem Corporation**
120 Route 17 North
P.O. Box 1579
Paramus, NJ 07653-1579 USA
201-261-7333
(FAX) 201-261-7339

Acts as a manufacturer's representative for nearly 30 companies worldwide; some of the manufacturers perform asymmetric synthesis and chiral chemistry.

**Isochem**
12 Quai Henri IV
75194 Paris
France
33-1-49967280
(FAX) 33-1-49967219

Part of Group SNPE; leading player in amino acids and multistage synthesis processes; offers the pharmaceuticals industry an extremely wide range of synthesis intermediates and active ingredients; specializes in phosgenations, hydrogenation, reductions with complex hydrides, Grignard chemistry, Friedel–Craft chemistry, butyllithium metallations, protected amino acids, activation of amino acids, and peptide coupling as well as electrodialysis.

**ISP Fine Chemicals Inc.**
1979 Atlas St
Columbus, OH 43228 USA
614-876-3637, 877-477-6446
(FAX) 614-876-9532

Custom manufactures a range of penultimate intermediate and active pharmaceutical ingredients under cGMP conditions; expertise in low-temperature reactions, highly reactive organometallics, and high purity chiral molecules. Also can perform a range of basic organic synthetic techniques.

**Japan Energy Corporation**
2-10-1 Toranomon
Minato-ku
Tokyo 105-8407
Japan
81-3-5573-6632
(FAX) 81-3-5573-6814

**Kaden Biochemicals GmbH**
Porgesring 50
D-22113 Hamburg
Germany
49-40-736-045-0
(FAX) 49-40-736-045-45

**Kaneka Corporation**
3-2-4, Nakanoshima
Kita-hu
Osaka 530-8288
Japan
81-6-6226-5142
(FAX) 81-6-6226-5128

**Kankyo Kagaku Center Co., Ltd.**
5-1, Ookawa
Kanazawa-Ku, Yokahama-City, Kanagawa
Japan
81-45-701-6141
(FAX) 81-45-701-6145

**Kawaken Fine Chemicals Co. Ltd.**
Horidome Bldg., 3-3, 2-chome
Nihonbashi Horidome-cho, Chuo-ku
Tokyo 103
Japan
81-3-3663-9525
(FAX) 81-3-3661-5630

**KingChem, Inc.**
296 Kinderkamack Rd
Oradell, NJ 07649 USA
201-261-6002
(FAX) 201-262-9436

International marketing organization specializing in sourcing fine, specialty, industrial and nutritional chemicals; capabilities include process development, commercial production, and custom/toll manufacturing under cGMP conditions; some manufacturers perform asymmetric synthesis reactions and chiral chemistry.

**Kiralchem Ltd.**
17 Canalside
2-4 Orsman Road, London N1 5JQ UK
44-20-7739-2436
(FAX) 44-20-7739-2146

**Knoll AG**
P.O. Box 631
CH-4410 Liestal
Switzerland
41-61-925-0505
(FAX) 41-61-925-0323

**Kyowa Hakko Kogyo Co., Ltd.**
1-6-1 Ohtemachi
Chiyoda-ku,Toky0 100-8185
Japan
81-3-3382-0089
(FAX) 81-3-3284-1839

**Lancaster Synthesis Ltd.**
Eastgate
White Lund, Morecambe LA3 3DY UK
44-1524-36101
(FAX) 44-1524-39727

Strategic alliance with Chirotech Technology; offer process development to commercial scale production for chiral projects, with capability for multi-tonne manufacturing if required.

**Laporte Fine Chemicals**
1065 Route 22 West
Bridgewater, NJ 08807 USA
908-541-9888
(FAX) 908-541-9898

Custom synthesis activities include a number of approaches to chiral molecules; enzymatic techniques include transformation using isolated enzymes (esterases and deaminases) and classical and dynamic kinetic enzymatic resolution; chemical technologies include manipulation of available chiral molecules or use of substrates from the chiral pool, asymmetric induction using inexpensive chiral auxiliaries, resolution by direct crystallization or crystallization of diastereomeric salts, and asymmetric catalysis (hydrogenation and epoxidation).

**Linnea S.A.**
Via Cantonale
Riazzino, Locarno 6595
Switzerland
41-91-850-5050
(FAX) 41-91-850-5070

**Loba Feinchemie AG**
Fehrgasse 7
Fischamend, A-2401
Austria
43-2232-77391-0
(FAX) 43-2232-76677

Special organic intermediates including chiral amino acids, amino alcohols, and amino acid derivatives at laboratory to large pilot plant scale; highly sensitive analytical reagents and diagnostics; process development and custom synthesis including halogenation, nitration and sulfonation, organolithium and Grignard chemistry, oxidation and reduction, diazotization and Sandmeyer reaction, and others.

**Lonza Ltd**
Munchensteinerstrasse 38
CH-4002 Basel
Switzerland
41-61-316-8111
(FAX) 41-61-31-69111

Lonza is a major service company to the life-science industries with its own product line based on diketene and derivatives, HCN and pyrimidine/pyridine chemistries, and its custom manufacturing business which is based on a strong portfolio of technologies and chemistries; biotransformations and biological manufacturing.

**Macfarlan Smith Ltd.**
Wheatfield Road
Edinburgh EH11 2QA
Scotland, UK
44-131-337-2434
(FAX) 44-131-3379813

Services include separation and purification of complex mixtures from downstream process liquors and quantitative chiral separation of racemic compounds carried out via preparative simulated moving bed chromatography; facilities ideal for manufacturing trial quantities of new chemical entities and specialty intermediates; also provide custom services for extraction of natural raw materials.

**Malladi Drugs & Pharmaceuticals Ltd.**
52 Jawaharlal Nehru Road
Ekkattuthangal, Madras 600 097
India
91-44-234-8715
(FAX) 91-44-233-0819

**Maprimed S.A.**
5928 E. Garzon Av
Buenos Aires C1440AYV
Argentina
54-11-4686-1438
(FAX) 54-11-4686-5028

**Maxsyn**
6046 Youngstown-Warren Rd
Niles, OH 44446 USA
330-652-6167

**MediChem Research, Inc**
12305 S. New Ave
Lemont, IL 60439 USA
630-257-1500
(FAX) 888-650-2436

Global contract research and development company assisting customers in accelerating drug discovery research; provide chemical development, combinatorial chemistry, cGMP synthesis, and analytical services; specailizes in process development, combinatorial chemistry, natural products, carbohydrates and amino acids; licensing agreement with Regis Technologies to provide custom, optimized chiral stationary phases for the pharmaceutical industry.

**Medinger & Sohn**
Landeggerstr. 7
A-2491 Neufeld
Austria
43-2624-523420
(FAX) 43-2624-52342

**Mercian Corporation**
5-8, Kyobashi 1-Chome
Chuo-Ku, Tokyo, 104-8305
Japan
81-3-3231-3927
(FAX) 81-3-3276-0151

Custom services for microbial enzyme screening for fine chemicals, pharmaceuticals, and intermediates; contract manufacturing using microbial fermentation and scale-up fermentation (up to 100 kiloliters) under GMP conditions; contract research, strain improvement and process research.

**Merck Bulk Pharmaceutical Chemicals**
5-9 rue Anquetil
Nogent sur Marne 94736
France
33-1-4394-5161
(FAX) 33-1-4876-4352

Develop and scale up new processes for pharmaceutical drugs; expertise in cross coupling reactions, cyanation, cyclization, halogenation, hydrogenation, lithiation, low and high temperature reactions (-90 °C to 300 °C), nitration, organometallic chemistry, separation technologies, stereospecific reactions, and sulfur chemistry.

**Mitsubishi Chemical Corporation**
5-2, Marunouchi 2-chome
Chiyoda-ku, Tokyo, 100-0005
Japan
81-3-3283-6444
(FAX) 81-3-3283-6489

Expertise especially in organic synthesis and biosynthesis; develop innovative routes and optimal production processes, particularly for chiral compounds such as unnatural amino acids, sugar derivatives and other synthetic chemicals.

**Mitsubishi Rayon Co., Ltd.**
6-41, Konan-1 Chome
Minato-Ku, Tokyo 108-8606
Japan
81-3-5495-3100
(FAX) 81-3-5495-3184

Custom manufacturing services include enantioselective synthesis of optically active phenyl-substituted mandelic acids using nitrilase enzymes, optical resolution of natural and unnatural amino acid amides using amidase enzymes, optical resolution of α-substituted carboxylic acid methyl esters using esterase enzymes, and also chemical derivitization of all chiral products.

**Mitsui Chemicals, Inc.**
Kasumigaseki Bldg.
2-5 Kasumigaseki 3-chome, Chiyoda-ku
Tokyo 100-6070
Japan
81-3-3592-4060
(FAX) 81-3-3592-4211

Fine chemical producer of bulk actives/intermediates under cGMP guidelines; expertise in phosgenations, chiral synthesis, amino acid derivatives, heterocyclics, hydrogen cyanide chemistry, halogenations, and high pressure reactions.

**MultiFerm A.B.**
Box 965
Lund 220 09
Sweden
46-4618-1745
(FAX) 46-4615-9291

**Nagase & Co., Ltd.**
5-1 Nihonbashi Kobunacho
Chuo-ku, Tokyo 103
Japan
81-3-3665-3022
(FAX) 81-3-3665-3030

**Newport Synthesis Ireland Ltd.**
Baldoyle Industrial Estate
Grange Road
Baldoyle, Dublin 13
Ireland
44-132-0020
(FAX) 44-132-0026

Chiral synthesis including enzymatic reactions, reductions, and Evans Chemistry; low temperature chemistry (to -90 °C); organometallic reactions, reductions, amino alcohols, and boronic acids.

**Nippon Chemical Industrial Co., Ltd.**
2-1-15, Iwamoto-cho
Chiyoda-ku, Tokyo 101
Japan
81-3-3862-4730
(FAX) 81-3-3862-4705

**Nippon Kayaku**
Tokyo Fujimi Bldg.
11-2, Fujimi 1-Chome
Chiyoda-ku
Tokyo 102-8172
Japan
81-3-3237-5111
(FAX) 81-3-3237-5091

**Nippon Rikagakuyakuhin**
4-2-12 Nihonbashi Hon-cho
Chuo-ku, Tokyo 103
Japan
81-3-3237-5111
(FAX) 81-3-3237-5091

**Nippon Soda Co. Ltd.**
New Ohtemachi Building
2-2-1 Ohtemachi
Chiyoda-ku, Tokyo
Japan
81-3-3245-6152
(FAX) 81-3-3245-6059

**Nippon Steel Chemical Co. Ltd.**
7-21-11, Gotanda
Shinagawa-ku, Tokyo, 141-0031
Japan
81-3-5759-2741
(FAX) 81-3-5759-2777

**Nissan Chemical Industries, Ltd.**
3-7-1 Kanda-Nishiki-cho
Chiyoda-ku, Tokyo 101-0054
Japan
81-3-3296-8391, 81-3-3296-8132
(FAX) 81-3-3296-8360

Custom synthesis using chiral technologies such as chiral epoxidation, reduction, Aldol reactions, Michael reactions, optical resolution, and others.

**Nitrokemia 2000 Rt.**
P.O. Box 23
H-8184 Fuzfogyartelep
Hungary
36-88-352-011
(FAX) 36-88-351-705

**Norchim SA**
33 quai d'Amont
Saint Leu d'Esserent F-60340
France
33-3-4456-0920
(FAX) 33-3-4456-6675

Custom manufacturing services including research, development, and kilo laboratories with full analytical support; pilot plant and mulitpurpose manufacturing plants for production from kilos to several tons; reactions include asymmetric synthesis and chiral chemistry plus basic synthetic organic techniques.

**Norse Laboratories**
P.O. Box 976
Newbury Park, CA 91319 USA
805-375-3113
(FAX) 805-375-0812

**Omega Chemical Company Inc.**
8800, blvd.. de la Rive-Sud
Levis, Quebec G6V 9H1
Canada
418-837-4444, 800-661-6342
(FAX) 418-837-5196

Specializes in chiral amino acids, protected chiral amino alcohols (BOC and FMOC), chiral protected amino aldehydes, β-amino acids, amino acid analogs, chiral building blocks, and others; over 11 years of experience in custom synthesis for the pharmaceutical industry; sourcing service for bulk pharmaceutical intermediates and hard-to-find chemicals.

**Omnichem SA**
Industrial Research Park
B-1348 Louvain-la-Neuve
Belgium
32-10-483111
(FAX) 32-10-450693

Member of the Ajinomoto Group; chiral technologies include azide chemistry, catalytic hydrogenation, hydride reduction, Evans chemistry, optical resolution, and organometallic chemistry; products include amino acid derivatives, chiral building blocks for penem and prostaglandin synthesis, and intermediates for the synthesis of biphenyltetrazoles.

**Orion Corporation, Fermion**
P.O. Box 65
Espoo, 02101
Finland
358-9-4291
(FAX) 358-9-429-4328

Produces active pharmaceutical ingredients and intermediates and provides custom synthesis; have closed systems for treating hazardous substances and a unit with a Buss loop-reactor for high pressure hydrogenations; also perform alkylation, phase-transfer, resolution, and stereoselective chemistry.

**Orsynetics (Trading) Ltd.**
PO Box 21
Winchester, Hants SO22 6ZJ UK
44-1962-850733
(FAX) 44-1962-850734

**Oxford Asymmetry International plc**
151 Milton Park
Abingdon, Oxon OX14 4SD UK
44-1235-861561
(FAX) 44-1235-863139

**PCAS**
23 rue Bossuet
ZI de la Vigne aux Loups
Longjumeau, 91160 France
33-1-6909-7785
(FAX) 33-1-6448-2319

Develops and manufactures synthetic intermediates and active ingredients used in manufacturing pharmaceutical products (for human and animal use); customized development capacity and a catalog of generic molecules; expertise in key technologies of pharmaceutical synthesis (low-temperature, acrolein-based chemistry, hydrogenation, high pressure, fermentation, and others).

**Pfanstiehl Laboratories, Inc.**
1219 Glen Rock Avenue
Waukegan, IL 60085 USA
847-623-0370, 800-383-0126
(FAX) 847-623-9173

Provides a wide variety of carbohydrates and their derivatives; source of a variety of custom manufactured pharmaceutical and bulk intermediates involving complex, multi-step syntheses; custom services include process development, optimizations and scale-up.

**Pharmacia Corporation**
100 Route 206 North
Peapack, NJ 07977
908-901-8000, 888-768-5501
(FAX) 908-901-8379

**Pharmasyn, Inc.**
1840 Industrial Dr, Suite 140
Libertyville, IL 60048 USA
847-918-9018
(FAX) 847-918-1610

Specializes in the custom synthesis of fine organic compounds and biochemicals on the gram to kilogram scale; reactions include asymmetric synthesis and chiral chemistry in addition to basic organic synthetic techniques.

**Pharm-Eco Laboratories, Inc.**
128 Spring St
Lexington, MA 02421-7800 USA
781-861-9303
(FAX) 781-861-9386

Multi-disciplinary pharmaceutical chemistry outsourcing company; equipped with research and manufacturing facilities; reactions include asymmetric synthesis and chiral chemistry in addition to basic organic synthetic techniques.

**PPG - Sipsy**
One PPG Place, 36th Floor
Pittsburgh, PA 15272 USA
412-434-3696, 800-323-3487
(FAX) 412-434-2800

Pharmaceutical fine chemicals ranging from starting materials to advanced cGMP intermediates and complex APIs; chiral synthesis is a core technology with focus on asymmetric hydrogenation, Sharpless epoxidation, Corey reduction, and enzymatic resolution techniques; other specialties include reductions, organometallic chemistry, sodium amide chemistry, phosgenation, protected amino acids/esters, cyanation, and dimethyl sulfate alkylations, as well as process development.

**Profarmaco S.r.l.**
Via Cucchiari 17
Milan 20155
Italy
39-2-3310-5520
(FAX) 39-2-3310-5730

A CAMBREX company; provider of active pharmaceutical ingredients to generic prescription manufacturers and proprietary products in partnership with pharmaceutical firms; for asymmetric synthesis, specializes in optical resolution.

**Rallis India Ltd.**
Ralli House
21. D. Sukhadwala Marg
Bombay, Maharashtra 400 001
India
91-22-207-8221
(FAX) 91-22-287-5980

**Ranbaxy Laboratories Limited**
10th Floor, Devika Towers
6 Nehru Place, New Delhi 110019
India
91-11-648-2554
(FAX) 91-11-622-8150

Asymmetric synthesis, chiral separations, biocatalytic reactions (enzymes, fermentation), high-temperature reactions (250 °C), cryogenic reactions (-100 °C), high pressure reactions (100 bar), high vacuum distillations (1 mm Hg), polymer supported reactions, halogen chemistry, ozonolysis, organometallic chemistry, hazardous reactions (sodium azide, cyanides, etc.).

**RCA Reuter Chemischer Apparaebau KG**
Wippertstrasse 2
79110 Freiburg
Germany
49-761-412534
(FAX) 49-761-402549

**Recordati S.p.A.**
SS 11 Padana Superiore km 160
Cassina de'Pecchi (Milan) 20060
Italy
39-2-952581
(FAX) 39-2-95344410

**Regis Technologies, Inc.**
8210 Austin Ave
Morton Grove, IL 60053 USA
847-967-6000, 800-323-8144
(FAX) 847-967-1214

Offers preparative chiral separation technology; delivers high-purity product, resolves racemic materials into enantiopure compounds using chiral stationary phases, and isolates impurities for characterisation or toxicology studies.

**Reliable Biopharmaceutical Corporation**
1945 Walton Rd
St. Louis, MO 63114 USA
314-429-7700
(FAX) 314-429-0937

Manufacturer of nucleic acids, biochemicals, and enzymes; produces cGMP and USP biochemicals, bulk pharmaceuticals, generic drug components, proteins and enzymes; provides contract manufacturing, contract process development; specializes in deoxynucleosides, deoxynucleotides, carbohydrate derivatives, multistep organic synthesis, and viscous formulations.

**Rexim S.A. Produits Chimiques**
37-39 Avenue Marceau
Courbevoie F-92400
France
33-1-4188-1515
(FAX) 33-1-4188-1500

**Rhodia Life Science Chemicals**
201 6, Rue Georget Marrane BP 55
Venissieux, Cedex 69632
France
33-4-7278-1507
(FAX) 33-4-7278-1512

Custom manufacturing and development of active ingredients and advanced intermediates for pharmaceutical companies.

**Ricerca, LLC**
7528 Auburn Rd
PO Box 1000
Painesville, OH 44077-1000 USA
440-357-3300
(FAX) 440-354-6276

Synthetic chemistry services include de-novo synthesis, chiral synthesis, analogs, metabolites, and analytical standards as well as radiolabeled materials.

**Richman Chemical Inc.**
768 North Bethlehem Pike
Lower Gwynedd, PA 19002 USA
215-628-2946
(FAX) 215-628-4262

Chemical custom manufacturing services include cGMP manufacturing, DMF preparation, contract research and development, pilot plant production, and process development; reaction capabilities include chiral resolution/synthesis, enzymatic synthesis, alkylation, halogenation, ethoxylation, organometallic synthesis, polymerization, and other synthetic organic chemistries.

**Robinson Brothers Ltd**
Phoenix St
West Bromwich, West Midland, B70 0AH UK
44-121-553-2451
(FAX) 44-121-500-5183

Specializes in reductions including manufacturing-scale chiral reductions using homogeneous catalysts and enzymatic resolutions; also experience with high temperature and pressure reductions with heterogeneous catalysts, reductive alkylation and amination, Leukart synthesis with formaldehyde, and use of borohydrides, cyanoborohydrides, and lithium aluminum hydride.

**Rohner Ltd.**
Gempenstrasse 6
CH-4133 Pratteln
Switzerland
41-61-825-1111
(FAX) 41-61-825-1271

Subsidiary of Dynamit Nobel; new custom synthesis division is Dynamic Synthesis; core technologies include hydrogen transfer catalysis, catalytic hydrogenation, enantioselective reduction, diazotization, isolated diazonium salts, Sandmeyer reactions, Meerwein chlorosulfonations, phenylhydrazines, Grignard and Friedel–Craft reactions, cyanation reactions, cross-coupling reactions, hydrazine chemistry, indole chemistry.

**Rutgers Organics GmbH**
Sandhofer Strasse 96
68305 Mannheim
Germany
49-621-7654-349
(FAX) 49-621-7654-451

Custom chiral services: chiral pool synthesis and racemate separation via diastereomeric salts.

**Sanofi Chimie**
82, avenue Raspail
Gentilly, Cedex 94255
France
33-141-247-000
(FAX) 33-141-24-6161

**Saurefabrik Schweizerhall**
Schweizerhalle CH-4133
Switzerland
41-61-825-3111
(FAX) 41-61-821-8027

**Schweizerhall Pharma**
10 Corporate Place South
Piscataway, NJ 08854
732-981-8200, 800-243-6564
(FAX) 732-981-8282

**SEAC**
28, blvd Camelinat-BP 77
Gennevilliers, Cedex 92233
France
33-140-855-400
(FAX) 33-140-855-035

**Senn Chemicals AG**
Industriestrasse 12
P.O. Box 267
CH-8157 Dielsdorf
Switzerland
41-1-854-9054
(FAX) 41-1-854-9055

Protected amino acids, peptides, resins, carbohydrates, and biochemicals and reagents; custom synthesis reactions include derivatization and purification of amino acids and carbohydrates.

**Seres Laboratories, Inc.**
3331-B Industrial Drive
Santa Rosa, CA 95403
707-526-4526
(FAX) 707-526-4707

**Shanghai DSL International Trading Company**
Rm 2010, Shanghai Bund International Tower
No.99 Huang Pu Road, Shanghai 200080
China
86-21-6307-8833
(FAX) 86-21-6357-9799

**Shasun House**
2, Doraiswamy Rd
T. Nagar, Chennai 600 017
India
91-44-434-8930
(FAX) 91-44-434-8924

**Siegfried CMS AG**
Untere Bruehlstrasse
Zofingen CH-4800
Switzerland
41-62-647-1212
(FAX) 41-62-746-1202

Custom asymmetric synthesis services include resolution of racemates; also perform many basic conventional chemistry reactions.

**SIFA Chemicals AG**
Industriestrasse 7
Liestal 4410
Switzerland
41-61-906-9050
(FAX) 46-61-906-9044

Capacity and know-how to manufacture a wide range of fine chemicals as well as active pharmaceutical intermediates, mainly focusing on nitrates; optical resolution on pilot plant quantities; other reactions include esterification, hydrolysis, nitration, reduction, sulfur chemistry, acylation, dealkylation, elimination, dehalogenation, hydrolysis, and phase transfer processes.

**Sigma-Aldrich Fine Chemicals**
3040 Spruce St
St. Louis, MO 63103 USA
314-534-4900, 800-368-4661
(FAX) 314-652-0000

Chiral technologies and expertise in asymmetric synthesis, chiral pool elaboration, classical and enzymatic resolutions, and multi-step synthesis of complex chiral targets; expertise in the synthesis of amino alcohols, benzylic amines, binaphthyls, bis(oxazoline) ligands, chiral organoboranes, oxazolidinones, Weinreb amides, and many more.

**Sinochem Ningbo**
Flat 11-12, Hualian Building
21 Jiangxia Street, Ningbo
P.O.C. 315000
China
86-574-734-6764
(FAX) 86-574-736-6775

**Siris Limited**
L.B. Nagar
Hyderabad 500 074 A P
India
91-40-4035526
(FAX) 91-40-4035079

**SK Energy and Chemical, Inc.**
22-10 Route 208 South
Fair Lawn, NJ 07410 USA
201-796-4288
(FAX) 201-796-3291

Provides custom chemical synthesis, chemical materials synthesis under cGMP, lab and commercial scale production; available reactions include asymmetric synthesis and chiral chemistry in addition to basic organic synthetic techniques.

**Sochinaz SA**
Rte du Simplon 22
Vionnaz, Valais CH-1895
Switzerland
41-24-482-4444
(FAX) 41-24-482-4445

Primarily a quality-committed contract manufacturing company with its main activity in the production of innovative molecules with DMFs in Europe and America; active pharmaceutical ingredients, custom synthesis, specialty chemicals for foods and cosmetics, generics, and amino acids and derivatives.

**Solvias AG**
Klybeckstrasse 191
Postfach, Basel CH-4002
Switzerland
41-61-686-6161
(FAX) 41-61-686-6565

This company offers synthesis planning and optimization, feasibility studies on synthesis steps and routes, catalytic, high pressure, and fluorination reactions on mg to kg scale; also provide development of new chiral ligands for enantio- and chemoselective hydrogenation; selective catalytic oxidations for C-C and C-N coupling reactions, and separation methods for soluble metal compounds.

**SST Corporation**
635 Brighton Rd
P.O. Box 1649
Clifton, NJ 07012 USA
973-473-4300, 800-222-0921
(FAX) 973-473-4326

Custom synthesis projects from the basic research stage through commercialization; bulk drugs, pharmaceutical intermediates, and fine chemicals; reactions include asymmetric synthesis and chiral chemistry in addition to basic organic synthetic techniques.

**Strem Chemicals, Inc.**
7 Muliken Way
Dexter Industrial Park
Newburyport, MA 01950-4098 USA
978-462-3191, 800-647-8737
(FAX) 978-465-3104

Specializes in metal catalysts for organic synthesis, chiral ligands, organophosphines, metallocenes, carbonyls and derivatives, and precious metal and rare earth chemicals; can perform a variety of custom synthesis of pharmaceutical intermediates under GMP conditions; available reactions include asymmetric synthesis and chiral chemistry in addition to basic organic synthetic techniques.

**Sumitomo Chemical Co. Ltd.**
Tokyo Sumitomo Twin Building (East)
27-1, Shinkawa 2-chome
Chuo-ku
Tokyo 104-8260
Japan
81-3-5543-5500
(FAX) 81-3-5543-5901

Offers custom synthesis services for the custom-tailored manufacture of chemical compounds; utilizes organic synthesis technologies developed and perfected over many years; top priority is to understand customers' technical problems and propose optimum solutions.

**Sumitomo Seika Chemicals Co. Ltd.**
Nihonjisho Bldg. No. 1
1-13-5, Kudan Kita
Chiyoda-ku
Tokyo 102-0073
Japan

**Sun-Orient Chemical Co. Ltd.**
103, Umedo
Kawanashi-cho
Shiki-gun, Nara 636-03
Japan
81-7454-4-4121
(FAX) 81-7454-4-1070

**Synergetica International, Inc.**
P.O. Box 25824
Philadelphia, PA 19128 USA
215-487-1566
(FAX) 215-487-0238

Services to pharmaceutical, agrochemical, electronic, and specialty chemical industries; owns, contracts, and distributes fine chemicals from Chinese plants; specializes in chiral resolution and asymmetric synthesis along with heterocycle synthesis, Suzuki coupling, nitration/reduction, halogenation, alkylation on N/O, Friedel–Craft chemistry and oxidations.

**Synquest Inc.**
2225 W. Harrison St
Chicago, IL 60612 USA
312-421-1819
(FAX) 312-421-8177

Focus is on development and syntheses of complex, multi-centered organic molecules; expertise in design of multi-step organic syntheses, process development/optimization, production of intermediates and active pharmaceutical ingredients under cGMP, and drug delivery; compound types include heterocycles, prostaglandins, steroids, paclitaxel, alkaloids, vitamin D, and carbohydrates; chemistries include asymmetric syntheses; Grignard reactions, reductions, oxidations, organometallic reactions, hydrogenation, and other standard organic reactions.

**Syntai Chemicals and Pharmaceuticals**
5Fl., No. 23, Sec. 1
Mei Chuan W. Rd
W. Dist., Taichung
Taiwan
886-4-372-6181
(FAX) 886-4-372-1219

Provides array of custom synthesis technologies including asymmetric hydrogenation, chiral synthesis, Grignard chemistry, Diels-Alder chemistry, organometallic synthesis, condensation, diazonization, halogenation, esterification, Friedal-Craft acylation, oxidation/reduction, alkylation, decarboxylation, methylation, and nitration; biotechnology services including culture screening and fermentation.

**Synthetech, Inc.**
1290 Industrial Way
P.O. Box 646
Albany, OR 97321 USA
541-967-6575
(FAX) 541-967-9424

Development and large-scale production of peptide building blocks including protected amino acid derivatives, amino alcohols, specialty amino acids and small peptide fragments as intermediates for synthesis of peptides, peptidomimetics and other small molecule therapeutics; custom manufacturing and development projects for chiral fine chemicals that fit their technology base in organic synthesis and biocatalysis.

**Synthon Chiragenics Corporation**
7 Deer Park Dr
Monmouth Junction, NJ 08852 USA
732-274-0037
(FAX) 732-274-0502

Leader in carbohydrate-based chiral technology, enabling discovery, development, and manufacture of advanced pharmaceuticals; proprietary technology is based on the chiral pool approach and starts with inexpensive, readily available raw materials; available custom services include the development of custom routes, designing and producing libraries of compounds for rational drug design, key pharmacophore access, lead optimization, and a full range of cGMP production quantities.

**Taiwan Tekho Camphor Co. Ltd.**
2, Alley 17, Lane 46, Sec 1
Ti-hwa St., Taipei
Taiwan
886-2-2559-5258
(FAX) 886-2-2558-3643

**Takasago International Corporation**
3-19-22 Takanawa
Minato-ku, Tokyo
Japan
81-3-3442-1211
(FAX) 81-3-3442-1329

**Tanabe Seiyaku Co. Ltd.**
3-2-10 Dosho-machi 3-chome
Chuo-ku Osaka 541-8505
Japan
81-6-205-5555
(FAX) 81-6-205-5262

**TCI America**
9211 North Harborgate St
Portland, OR 97203 USA
503-283-1681
(FAX) 503-283-1987

Offers specialized technical assistance in developing new products; reaction capabilities include basic organic synthetic chemistries; manufactures over 13,000 different organic chemicals including many chiral compounds.

**Themis Chemicals Ltd.**
11/12, Udyognagar
S.V.Road, Goregaon (West)
Mumbai 400 062, India
91-22-875-7836
(FAX) 91-22-874-3643

**ThermoGen Inc.**
2225 West Harrison St
Chicago, IL 60612 USA
312-226-6500
(FAX) 312-226-9686

Develops and manufactures long-lived, stable enzymes for room temperature and high temperature industrial applications; uses range of new technologies to create stable proteins, to express thermophilic proteins, and to engineer enzymes by accelerated evolution to meet advanced product development needs; will screen enzyme libraries to identify appropriate candidates and offers assistance on custom biotransformation development.

**Tokyo Kasei Kogyo Co., Ltd.**
1-13-6 Nihonbashis-muromachi
Chuo-ku, Tokyo 103-0022
Japan
81-3-3278-8153
(FAX) 81-3-3278-8008

**Toray International, Inc.**
2-3-16 Nihonbashi Muromachi
Chuo-ku, Tokyo 103
Japan

81-3-3245-5849
(FAX) 81-3-3245-5198

**Torcan Chemical Ltd.**
110 Industrial Parkway North
P.O. Box 308
Aurora, ONT L4G 3H4
Canada
81-3-3245-5849
(FAX) 81-3-3245-5198

Specializes in contract research, development and manufacturing of bulk active pharmaceutical ingredients and advanced intermediates for preclinical, clinical, and small volume (few 100 kg) commercial applications; custom chiral services include preferential crystallization of diastereomeric salts, enzyme catalyzed derivatization and selective precipitation, and use of chiral reagents in synthesis.

**Toyotama Perfumery Co., Ltd.**
1-14-5, Nihonbashi Kakigara-cho
Chuo-ku, Tokyo 103
Japan
81-3-5640-5511
(FAX) 81-3-5640-5523

**Ultrafine**
Synergy House
Guildhall Close
Manchester Science Park
Manchester M15 65Y UK
44-161-226-8774
(FAX) 44-161-227-9758

Services include medicinal chemistry, custom synthesis, contract research, lead development and optimization, route selection, scale-up and process research; cGMP synthesis, combinatorial chemistry, biotransformations, analytical method devlopment and validation; reactions include asymmetric synthesis and chiral chemistry in addition to basic organic synthetic techniques.

**Uquifa S.A.**
Mallorca 262
Barcelona 08008
Spain
34-93-4879477
(FAX) 34-93-4880491

Manufacturer of bulk active pharmaceutical ingredients and key intermediates; chemical capabilities include cryogenic reactions, ozonolysis, hydrogenation, triphosgene reactions, lithium chemistry, and boron chemistry; reactions include asymmetric synthesis and chiral chemistry in addition to basic organic synthetic techniques.

**Urquima S.A.**
Dega Bahi 67
Apartado 632
Barcelona 08026
Spain
34-93-347-1000
(FAX) 34-93-456-0639

**Varsal Instruments, Inc.**
363 Ivyland Rd
Warminster, PA 18974 USA
215-957-5880
(FAX) 215-957-9111

**Vinchem**
301 Main St
PO Box 639
Chatham, NJ 07928 USA
973-635-1459
(FAX) 973-635-4841

Sales agent for manufacturers of specialty and fine chemicals, custom intermediates & bulk pharmaceuticals; some manufacturers represented conduct asymmetric synthesis reactions and have expertise in chiral chemistry.

**Yamakawa Chemical Industry Co. Ltd.**
Tanaka Building
3-1-10 Nihonbashi Muromachi
Chuo-ku, Tokyo 103-0022
Japan
81-3-3241-0191
(FAX) 81-3-3241-0170

**Yoneyama Yakuhin Kogyo Co., Ltd.**
2-3-11 Doshomachi
Osaka 541-0045
Japan
81-6-231-3555
(FAX) 81-6-223-1093

**Yuhan Corporation**
49-6, Taebang-Dong
Tongjak-Ku, Seoul
Korea
82-2-828-0191
(FAX) 82-2-828-0080

**Zambon Group SpA**
via Lillo del Duca
Bresso, Milan 10-20091
Italy
39-02-665241
(FAX) 39-02-66501492

Multinational company producing active pharmaceutical ingredients, intermediates, and finished dose forms and providing custom synthesis services; commercial scale production capacity as well as pilot plant and R&D services; specializes in chiral, mercaptan, organometallic, and nonconventional oxidation chemistry.

**Zeeland Chemicals, Inc.**
215 North Centennial St
Zeeland, MI 49464 USA
616-772-2193, 800-223-0453
(FAX) 616-772-6554

A CAMBREX Company; produces pharmaceutical intermediates and auxiliaries; specializes in alkylations, amidations, chemical resolutions, chiral synthesis, condensations, cyanoethylations, dehydrohalogenations, esterifications, high pressure reactions, hydrogenations, Kolbe carboxylations, neutralizations, process development, propoxylations, quaternizations, reductions, and reductive aminations.

**Zhejiang Chemicals Import & Export Corp.**
37 Qingchun Road
Hangzhou, Zhejiang 310009
China
86-571-704-4231
(FAX) 86-571-704-6098